[Ninth Edition]

Nester's MICROBIOLOGY

[A Human Perspective]

Denise G. Anderson
UNIVERSITY OF WASHINGTON

Sarah N. Salm
BOROUGH OF MANHATTAN COMMUNITY COLLEGE

Deborah P. Allen
JEFFERSON COLLEGE

Eugene W. Nester
UNIVERSITY OF WASHINGTON

Mc Graw Hill Education

NESTER'S MICROBIOLOGY

Published by McGraw-Hill Education, 2 Penn Plaza, New York, NY 10121.

This book is printed on acid-free paper.

1 2 3 4 5 6 7 8 9 LWI 21 20 19 18

ISBN 978-1-260-09221-9
MHID 1-260-09221-6

Cover Image: *Colored scanning electron micrograph (SEM) of Vibrio fischeri, Gram-negative, motile, bioluminescent, curved rod-shaped bacterium (prokaryote). ©Dennis Kunkel Microscopy/Science Source*

mheducation.com/highered

Robust Analytics and Reporting

- Connect Insight® generates easy-to-read reports on individual students, the class as a whole, and on specific assignments.
- The Connect Insight dashboard delivers data on performance, study behavior, and effort. Instructors can quickly identify students who struggle and can focus on material that the class has yet to master.
- Connect automatically grades assignments and quizzes, providing easy-to-read reports on individual and class performance.

©Hero Images/Getty Images

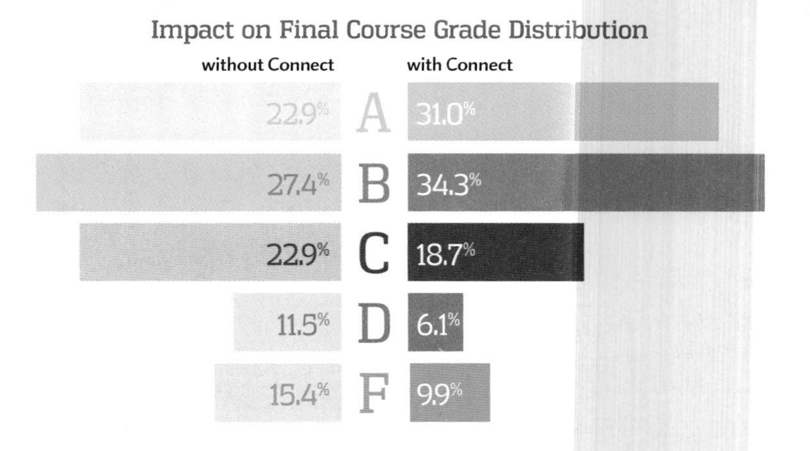

More students earn **As** and **Bs** when they use **Connect**.

Trusted Service and Support

- Connect integrates with your LMS to provide single sign-on and automatic syncing of grades. Integration with Blackboard®, D2L®, and Canvas also provides automatic syncing of the course calendar and assignment-level linking.
- Connect offers comprehensive service, support, and training throughout every phase of your implementation.
- If you're looking for some guidance on how to use Connect, or want to learn tips and tricks from super users, you can find tutorials as you work. Our Digital Faculty Consultants and Student Ambassadors offer insight into how to achieve the results you want with Connect.

McGraw-Hill Connect® is a highly reliable, easy-to-use homework and learning management solution that utilizes learning science and award-winning adaptive tools to improve student results.

Homework and Adaptive Learning

- Connect's assignments help students contextualize what they've learned through application, so they can better understand the material and think critically.
- Connect will create a personalized study path customized to individual student needs through SmartBook®.
- SmartBook helps students study more efficiently by delivering an interactive reading experience through adaptive highlighting and review.

Over **7 billion questions** have been answered, making McGraw-Hill Education products more intelligent, reliable, and precise.

Connect's Impact on Retention Rates, Pass Rates, and Average Exam Scores

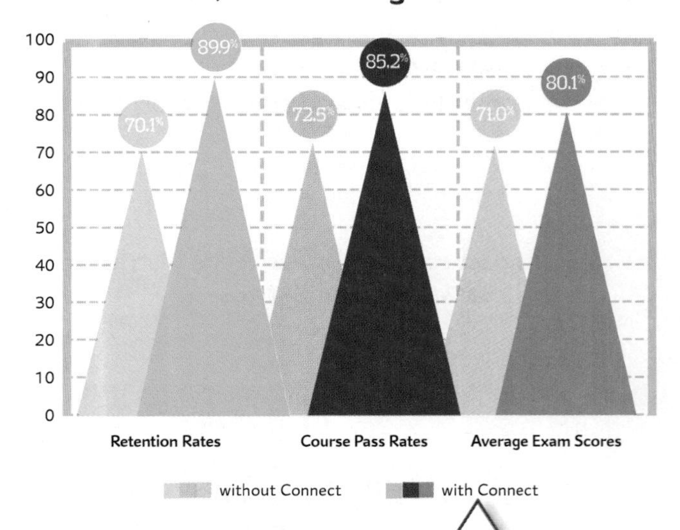

Using **Connect** improves retention rates by **19.8** percentage points, passing rates by **12.7** percentage points, and exam scores by **9.1** percentage points.

73% of instructors who use **Connect** require it; instructor satisfaction **increases** by 28% when **Connect** is required.

Quality Content and Learning Resources

- Connect content is authored by the world's best subject matter experts, and is available to your class through a simple and intuitive interface.
- The Connect eBook makes it easy for students to access their reading material on smartphones and tablets. They can study on the go and don't need internet access to use the eBook as a reference, with full functionality.
- Multimedia content such as videos, simulations, and games drive student engagement and critical thinking skills.

Brief Contents

PART I
Life and Death of Microorganisms

PART II
The Microbial World

PART III
Microorganisms and Humans

PART IV
Infectious Diseases

PART V
Applied Microbiology

About the Authors

The Nester Team:

Three Perspectives, One Vision, One Voice

The three authors of this edition—Denise Anderson, Sarah Salm, and Deborah Allen—may be a set of individuals with different insights and unique experiences, but their cooperative relationship defines the word "team." What drives them is a single shared goal: to create the most learning-friendly introductory microbiology textbook available. Each author carefully read all the chapters, looking for parts that could be tweaked for clarity. They did this with students in mind, suggesting simpler words where appropriate while maintaining the scientific rigor so important for today's healthcare professionals.

Meanwhile, Gene Nester continued to serve as "team member emeritus," keeping an eagle eye out for updates that could be incorporated into the text. His work established the text's reputation for excellence over the decades, and it lives on in this edition.

©Richard Moore

Denise Anderson

Denise Anderson is a Senior Lecturer in the Department of Microbiology at the University of Washington, where she teaches a variety of courses including general microbiology, medical bacteriology laboratory, and medical mycology/parasitology laboratory. Equipped with a diverse educational background, including undergraduate work in nutrition and graduate work in food science and in microbiology, she first discovered a passion for teaching when she taught microbiology laboratory courses as part of her graduate training. Her enthusiastic teaching style, fueled by regular doses of Seattle's famous coffee, receives high reviews from her students.

Outside academic life, Denise relaxes in the Phinney Ridge neighborhood of Seattle, where she lives with her husband, Richard Moore, and dog, Riley (neither of whom is well trained). When not planning lectures, grading papers, or writing textbook chapters, she can usually be found chatting with the neighbors, fighting the weeds in her garden, or enjoying a fermented beverage at the local pub.

©Sandy Coetzee

Sarah Salm

Sarah Salm is a Professor at the Borough of Manhattan Community College (BMCC) of the City University of New York, where she teaches microbiology, anatomy and physiology, and general biology. She earned her undergraduate and doctoral degrees at the University of the Witwatersrand in Johannesburg, South Africa. She later moved to New York, where she did postdoctoral work at the NYU School of Medicine. Her research background is diverse, and includes plant virology, prostate cancer, and bacteria in contaminated water sources.

©Mike Bohrer

Deborah Allen

Deborah Allen is a Professor at Jefferson College in Missouri, where she teaches microbiology as well as several other courses for students entering allied health careers. Her graduate work was in zoology at the University of Oklahoma and in neurobiology and behavior at Cornell University. She participated in cancer research at the University of Arkansas Medical Center before embarking on a career in publishing, working in acquisitions and development for books in the life sciences. She is now thrilled to be working on the other end of the desk with the Nester team. Away from campus, Deborah reads or listens to her favorite Eve Dallas novels, floats the rivers and listens to folk music in the Ozarks, and fully appreciates the local microbes while visiting Missouri wineries.

Courtesy Eugene Nester

Eugene Nester

Although no longer an active member of the author team, Eugene (Gene) Nester wrote the original version of the present text with Evans Roberts and Nancy Pearsall more than 30 years ago. That text, *Microbiology: Molecules, Microbes and Man,* pioneered the organ system approach to the study of infectious disease, and was developed specifically for allied health sciences.

Gene did his undergraduate work at Cornell and received his Ph.D. in microbiology from Case Western University. He then did postdoctoral work in the Department of Genetics at Stanford University with Joshua Lederberg. Following that, he joined the faculty in the Department of Microbiology at the University of Washington, where he remains active as an emeritus member. His laboratory demonstrated that *Agrobacterium* transfers DNA into plant cells—the basis for the disease crown gall—a system of gene transfer that has become a cornerstone of plant biotechnology. In recognition of his work, he was awarded the Australia Prize and the Cetus Prize in Biotechnology, and was elected to fellowship in the National Academy of Sciences, the American Academy for the Advancement of Science, the American Academy of Microbiology, and the National Academy of Sciences in India.

Presentation Tools Allow Instructors to Customize Lecture

Everything you need, in one location

Enhanced Lecture Presentations contain lecture outlines, art, photos, tables, and animations embedded where appropriate. Fully customizable, but complete and ready to use, these presentations will enable you to spend less time preparing for lecture!

Animations—More than 100 animations bring key concepts to life; available for instructors and students.

Accessible PPTs—Our lecture presentations are formatted per the latest accessibility guidelines. Alternative text, written by our textbook author team, is included for all images and static tables.

Chapter	**Animations** Animations/Videos (Animations and video files that can be easily downloaded and played without PowerPoint.)	**Labeled Images (Art, tables, and photo files as they appear in the text as zipped files)**	**Base Art Images** Art files, photos and table files without labels and leader lines.	***Accessible Lecture PPTs** (Lecture outlines tagged for screen readers.) Chapters 1-15 are available.
Chapter 01	Ch 01 Animation (10,377 KB)	Ch 01 Labeled Images (28,687 KB)	Ch 01 Baseart Images (21,510 KB)	Ch01 Accessible Lecture PPT
Chapter 02	Ch 02 Animation (35,949 KB)	Ch 02 Labeled Images (11,405 KB)	Ch 02 Baseart Images (5,932 KB)	Ch02 Accessible Lecture PPT

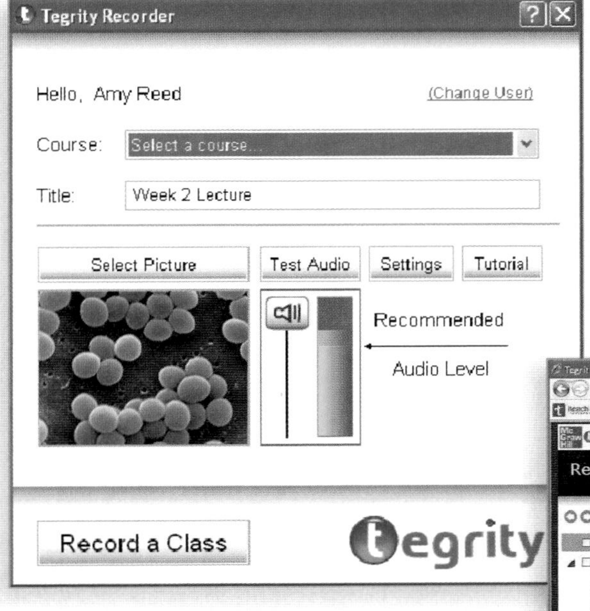

Take your course online—*easily*—with one-click Digital Lecture Capture

McGraw-Hill Tegrity® records and distributes your lecture with just a click of a button. Students can view them anytime/anywhere via computer, tablet, or mobile device. Tegrity Campus indexes as it records your slideshow presentations and anything shown on your computer so **students can use keywords to find exactly what they want to study.**

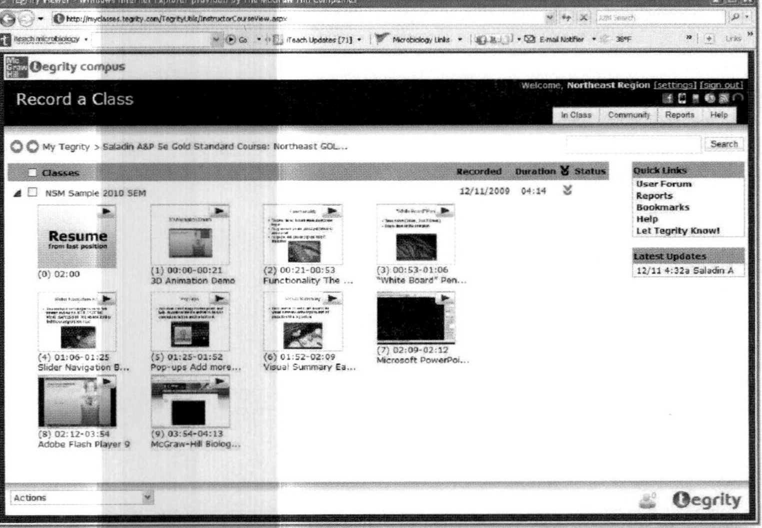

FOCUS ON UNDERSTANDING . . .

Student-Friendly Illustrations

Introduce the "big picture"

Focus figures provide an overview or highlight a key concept.

Keep the big picture in focus

A highlighted mini-version of the overview figure is often incorporated into the upper left corner of subsequent figures, helping students see how those figures fit into the big picture.

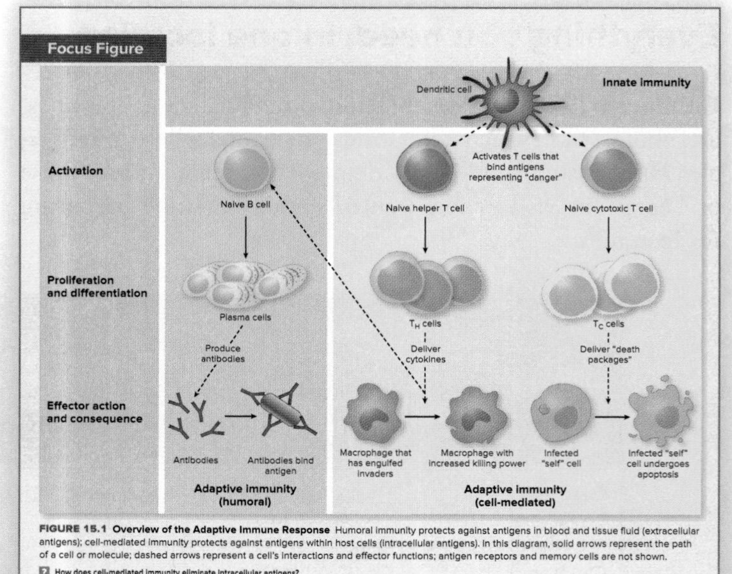

Focus Figure

Innate immunity | Dendritic cell | Activates T cells that bind antigens representing "danger"

Activation | Naive B cell | Naive helper T cell | Naive cytotoxic T cell

Proliferation and differentiation | Plasma cells | Produce antibodies | T_H cells | Deliver cytokines | T_C cells | Deliver "death packages"

Effector action and consequence | Antibodies | Antibodies bind antigen | Macrophage that has engulfed invaders | Macrophage with increased killing power | Infected "self" cell | Infected "self" cell undergoes apoptosis

Adaptive immunity (humoral) | Adaptive immunity (cell-mediated)

FIGURE 15.1 Overview of the Adaptive Immune Response Humoral immunity protects against antigens in blood and tissue fluid (extracellular antigens); cell-mediated immunity protects against antigens within host cells (intracellular antigens). In this diagram, solid arrows represent the path of a cell or molecule; dashed arrows represent a cell's interactions and effector functions; antigen receptors and memory cells are not shown.

How does cell-mediated immunity eliminate intracellular antigens?

Other descendants become **memory lymphocytes,** long-lived cells that can be activated more quickly if the antigen is encountered again. Memory cells are responsible for the effectiveness of the secondary response.

Humoral Immunity

...ell detects an extracellular antigen, that B cell ...ated, allowing it to proliferate (see figure 15.1). ...endants of activated B cells differentiate to ...ells, which are effector B cells. Plasma cells ...oteins called **antibodies.** These proteins bind ...f cells, toxins, viruses, and other antigens, ...protect the body against the effects of that ...mple, antibodies that bind diphtheria toxin ...rom the effects of the toxin (see **A Glimpse**

The structure of an antibody molecule accounts for its ability to protect against an invader. An antibody has two functional regions: the two identical arms and the single stem of the Y-shaped molecule (see figure 15.1). The ends of the arms are the parts that attach to antigens. By binding to antigens, the antibodies can neutralize their effects. For example, antibody-coated viral particles cannot attach to receptors on cells and, therefore, cannot enter the cells. Antibodies are very specific with respect to their binding, so the immune system must produce many different varieties, each with a slightly different set of "arms." Although the set of arms of different antibody molecules varies, the stem portion is functionally similar—it serves as a "red flag" that sticks out from the surface of an antibody-bound antigen. This tags the antigen for rapid elimination by macrophages or other components of the immune system. attachment of viruses

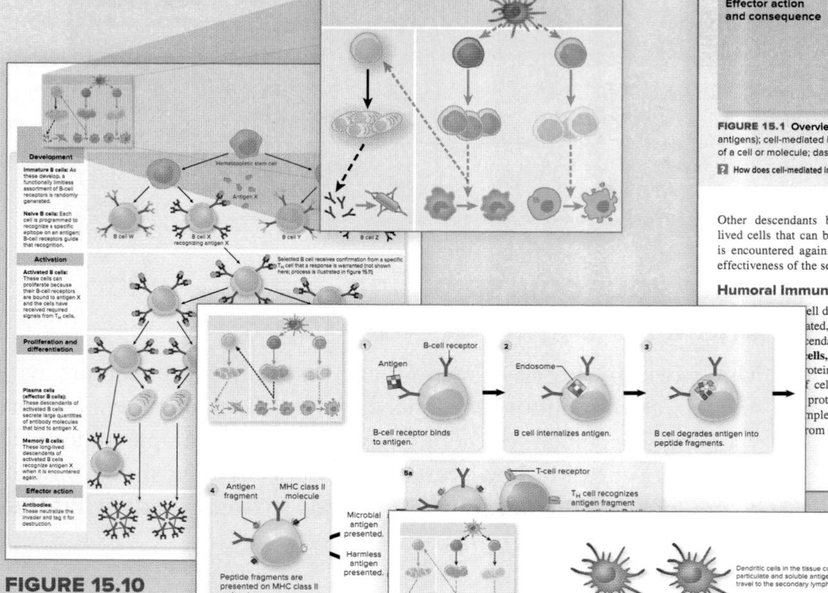

FIGURE 15.10

FIGURE 15.11

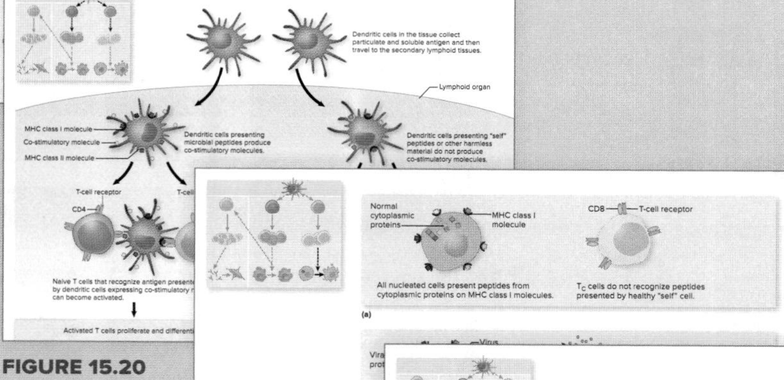

FIGURE 15.20

FIGURE 15.21

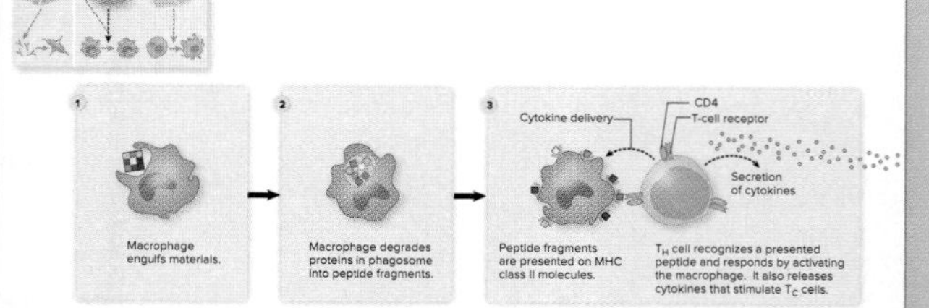

FIGURE 15.22

"Provides a logical unfolding conceptual framework that fosters better understanding."

—Jamal Bittar, University of Toldeo

Walk through the processes

Step-by-step figures direct the student using numbered icons, often with corresponding icons in the text.

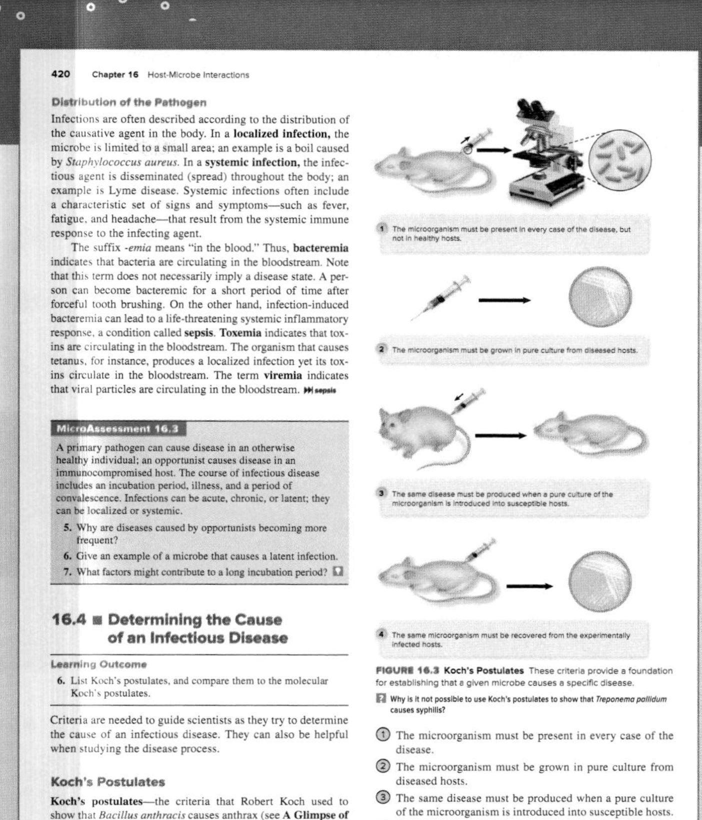

420 Chapter 16 Host-Microbe Interactions

Distribution of the Pathogen

Infections are often described according to the distribution of the causative agent in the body. In a **localized infection,** the microbe is limited to a small area; an example is a boil caused by *Staphylococcus aureus*. In a **systemic infection,** the infectious agent is disseminated (spread) throughout the body; an example is Lyme disease. Systemic infections often include a characteristic set of signs and symptoms—such as fever, fatigue, and headache—that result from the systemic immune response to the infecting agent.

The suffix *-emia* means "in the blood." Thus, **bacteremia** indicates that bacteria are circulating in the bloodstream. Note that this term does not necessarily imply a disease state. A person can become bacteremic for a short period of time after forceful tooth brushing. On the other hand, infection-induced bacteremia can lead to a life-threatening systemic inflammatory response, a condition called **sepsis**. **Toxemia** indicates that toxins are circulating in the bloodstream. The organism that causes tetanus, for instance, produces a localized infection yet its toxins circulate in the bloodstream. The term **viremia** indicates that viral particles are circulating in the bloodstream. sepsis

MicroAssessment 16.3

A primary pathogen can cause disease in an otherwise healthy individual; an opportunist causes disease in an immunocompromised host. The course of infectious disease includes an incubation period, illness, and a period of convalescence. Infections can be acute, chronic, or latent; they can be localized or systemic.

5. Why are diseases caused by opportunists becoming more frequent?
6. Give an example of a microbe that causes a latent infection.
7. What factors might contribute to a long incubation period?

16.4 ■ Determining the Cause of an Infectious Disease

Learning Outcome

6. List Koch's postulates, and compare them to the molecular Koch's postulates.

Criteria are needed to guide scientists as they try to determine the cause of an infectious disease. They can also be helpful when studying the disease process.

Koch's Postulates

Koch's postulates—the criteria that Robert Koch used to show that *Bacillus anthracis* causes anthrax (see **A Glimpse of History**)—provide a foundation for establishing that a given microbe causes a specific infectious disease (**figure 16.3**):

1 The microorganism must be present in every case of the disease, but not in healthy hosts.

2 The microorganism must be grown in pure culture from diseased hosts.

3 The same disease must be produced when a pure culture of the microorganism is introduced into susceptible hosts.

4 The same microorganism must be recovered from the experimentally infected hosts.

FIGURE 16.3 Koch's Postulates These criteria provide a foundation for establishing that a given microbe causes a specific disease.

Why is it not possible to use Koch's postulates to show that *Treponema pallidum* causes syphilis?

① The microorganism must be present in every case of the disease.
② The microorganism must be grown in pure culture from diseased hosts.
③ The same disease must be produced when a pure culture of the microorganism is introduced into susceptible hosts.
④ The microorganism must be recovered from the experimentally infected hosts.

> *"The text and illustrations are 'tight' and give each other good support."*
>
> —Richard Shipee, Vincennes University

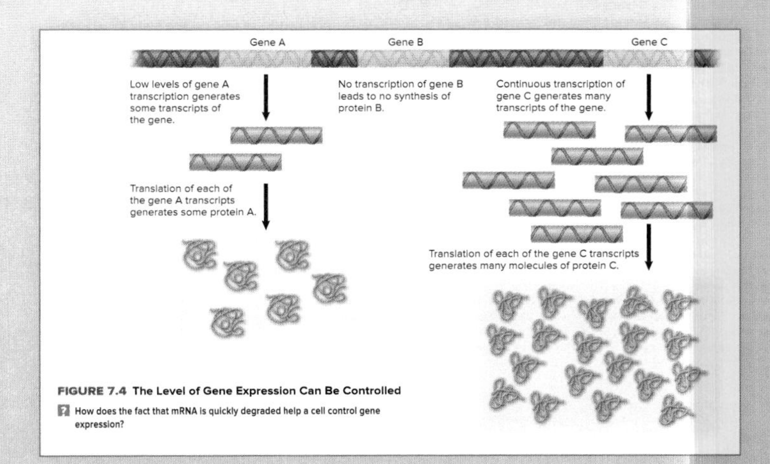

FIGURE 7.4 The Level of Gene Expression Can Be Controlled

How does the fact that mRNA is quickly degraded help a cell control gene expression?

Encourage deeper understanding

Figures have accompanying questions that encourage students to think more carefully about the concept illustrated in a figure.

Introduce the body systems

Each disease chapter includes a stunning figure that introduces the students to the anatomy of the body system.

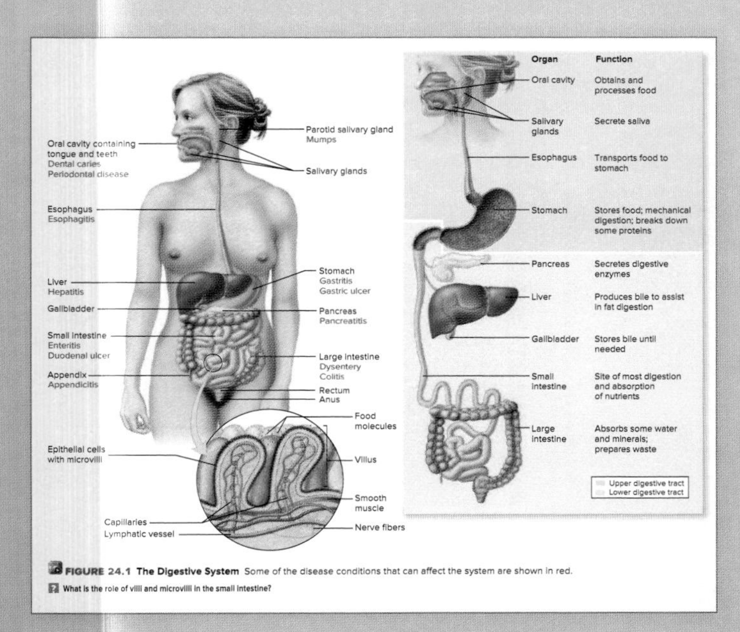

FIGURE 24.1 The Digestive System Some of the disease conditions that can affect the system are shown in red.

What is the role of villi and microvilli in the small intestine?

FOCUS ON UNDERSTANDING . . .

Student-Friendly Chapter Features

Provide the tools for understanding

Key Terms for each chapter are defined on the opening page.

Share the history

A **Glimpse of History** opens each chapter, featuring engaging stories about the men and women who pioneered the field of microbiology.

Define the expectations

Learning outcomes are found at the beginning of each numbered section, allowing organization, evaluation, and assessment of instruction.

26 Nervous System Infections

KEY TERMS

Arbovirus Arthropod-borne RNA virus, carried by vectors such as mosquitoes.

Blood-Brain Barrier Cells that work together to restrict exchange between the bloodstream and the brain.

Central Nervous System (CNS) Brain and spinal cord.

Cerebrospinal Fluid (CSF) Fluid produced in the brain that flows within and around the CNS.

Encephalitis Inflammation of the brain.

Meninges Membranes covering the brain and spinal cord.

Meningitis Inflammation of the meninges.

Peripheral Nervous System (PNS) Division of the nervous system that carries information to and from the CNS.

Transmissible Spongiform Encephalopathy (TSE) Chronic degenerative brain disease caused by prions; characterized by spongy appearance of brain tissue.

Structure of West Nile virus particles. ©*Science Picture Co/Getty Images*

A Glimpse of History

Today it is hard to appreciate the fear and loathing once attached to leprosy (*lepros*, meaning "scaly"). The Bible refers to several disfiguring skin diseases, including leprosy, and people suffering from the diseases are portrayed as filthy, outcast, or condemned by God for sin. Moses called lepers "unclean" and proclaimed they must live away from others. In the Middle Ages, lepers attended their own symbolic burial before being sent away.

Gerhard Henrik Armauer Hansen (1841–1912) was a Norwegian physician with many interests, ranging from science to religion to polar exploration. When he was 32 years old, Hansen went into medical research, and was named assistant to Dr. Daniel C. Danielson, a leading authority on leprosy. Danielson believed that leprosy was a hereditary disease of the blood and considered the idea that the disease was contagious as a "peasant superstition." Hansen, however, disproved Danielson's hypothesis in careful studies conducted over a number of years. He found a unique bacterium associated with the disease in every leprosy patient he studied. His 1873 report of the findings marked the first time that a specific bacterium was linked to a disease—almost a decade before Koch's proof of the cause of tuberculosis.

In the United States, even during the first half of the twentieth century, people diagnosed with leprosy risked having their houses burned to destroy the source of infection. Their names were changed to avoid embarrassing their family, and they were sent to a leprosarium such as the one at Carville, Louisiana, surrounded by a 12-foot fence topped with barbed wire. Sufferers were separated from spouses and children and denied the right to marry or vote. Those who attempted to escape were captured and brought back in handcuffs. The Carville leprosarium was finally closed and converted to a military-style academy for high school dropouts in 1999.

Because the word *leprosy* carries centuries of dark overtones, many people prefer to use the term *Hansen's disease*, a name that honors the discoverer of the causative bacterium.

Nervous system infections are frightening. They threaten a person's ability to move, feel, or even think. Consider poliomyelitis, which can result in a paralyzed limb or the inability to breathe without mechanical assistance. Hansen's disease (leprosy) can result in loss of fingers or toes or deformity of the face. Infections of the brain or its covering membranes can render a child deaf or intellectually disabled. Before the discovery of antibiotics, bacterial infections of the nervous system were often fatal. Fortunately, these infections are uncommon.

26.1 ■ Anatomy, Physiology, and Ecology of the Nervous System

Learning Outcomes

1. Describe how information flows through and between neurons.
2. Differentiate between the central nervous system and the peripheral nervous system.
3. Explain how bone, cerebrospinal fluid, meninges, and the blood-brain barrier protect the central nervous system.

694

MicroAssessment 3.2

Dyes are used to stain cells so they can be seen against an unstained background. The Gram stain is the most commonly used differential stain. The acid-fast stain is used to detect *Mycobacterium* species. Specific dyes and techniques are used to observe cell structures such as capsules, endospores, and flagella. Fluorescent dyes and tags can be used to observe total cells, a subset of cells, or cells that have certain proteins on their surface.

4. What are the functions of a primary stain and a counterstain?
5. Describe one error in the staining procedure that would result in a Gram-positive bacterium appearing pink.
6. What color would a Gram-negative bacterium be in an acid-fast stain?

Assess understanding

A **MicroAssessment** at the end of each numbered section summarizes the concepts and includes review questions, usually featuring one that stimulates critical thinking (indicated by a light bulb icon).

Engage the reader

MicroBytes found throughout the chapter provide small "bytes" of information, capturing the reader's attention.

MicroByte

There are more bacteria in just one person's mouth than there are people in the world!

Highlight the relevance

Focus on a Case boxes describe realistic clinical, veterinary, or environmental situations, along with questions and discussions designed to highlight the relevance of the information.

Provide perspective

Focus Your Perspective boxes show how microorganisms and their products influence our lives in many different ways.

Introduce the concepts

Focus on a Disease boxes introduce a general category of disease (pneumonia, diarrheal disease, meningitis, sexually transmitted infections), giving students a framework for understanding specific diseases.

Inspire the learner

Focus on the Future boxes describe pending challenges facing current and future microbiologists.

- **Summary** briefly reviews the key points.
- **Short Answer** questions review major chapter concepts.
- **Multiple Choice** questions allow self-testing; answers are provided in Appendix IV.
- **Application** questions provide an opportunity to use knowledge of microbiology to solve real-world problems.
- **Critical Thinking** questions encourage practice in analysis and problem solving that can be used by the student in any subject.

Review the information

End-of-chapter review encourages students to revisit the information.

Build the story

Logical chapter order helps students understand and connect the concepts.

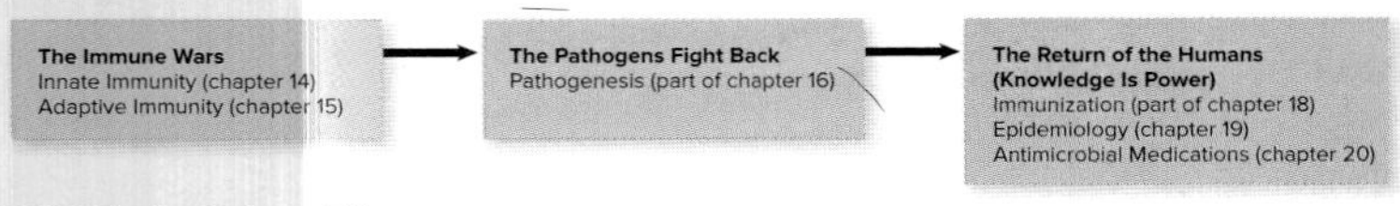

The Immune Wars
Innate Immunity (chapter 14)
Adaptive Immunity (chapter 15)

The Pathogens Fight Back
Pathogenesis (part of chapter 16)

The Return of the Humans (Knowledge Is Power)
Immunization (part of chapter 18)
Epidemiology (chapter 19)
Antimicrobial Medications (chapter 20)

FIGURE 18.1 The Host-Pathogen Trilogy

? How does immunization prevent disease?

FOCUS ON UNDERSTANDING . . .

Student-Friendly Descriptions

Include analogies

WHY? Analogies provide students a comfortable framework for making sense of difficult topics. Here's an example from chapter 14.

Innate Immunity *The innate immune system has three general components: first-line defenses, sensor systems, and innate effector actions. As a useful analogy, think of the defense systems of a high-security building or compound: The first-line defenses are the security walls surrounding the property; the sensor systems are the security cameras scattered throughout the property, monitoring the environment for signs of invasion; and the effector actions are the security teams sent to remove any invaders that have been detected, thereby eliminating the threat* ***(figure 14.1a).***

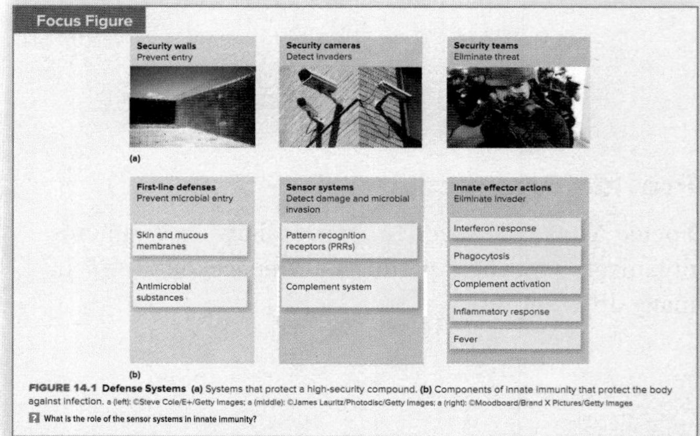

FIGURE 14.1 Defense Systems (a) Systems that protect a high-security compound. **(b)** Components of innate immunity that protect the body against infection. a (left): ©Steve Cole/E+/Getty Images; a (middle): ©James Lauritz/Photodisc/Getty Images; a (right): ©Moodboard/Brand X Pictures/Getty Images

What is the role of the sensor systems in innate immunity?

Emphasize the logic

WHY? Descriptions that emphasize the logic of processes make it easier for students to understand and retain the information. Here's an example from chapter 6.

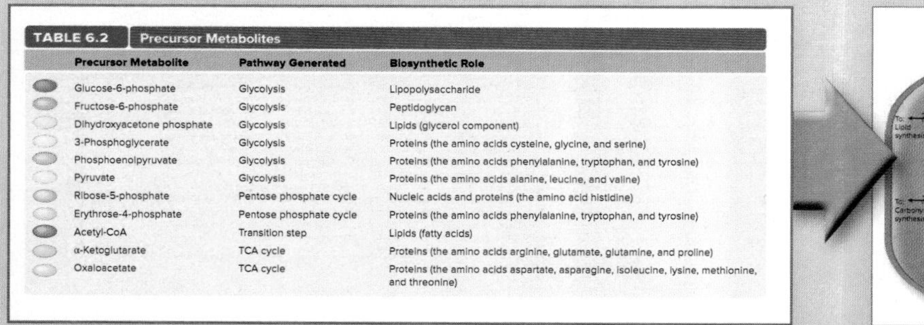

TABLE 6.2 Precursor Metabolites

Precursor Metabolite	Pathway Generated	Biosynthetic Role
Glucose-6-phosphate	Glycolysis	Lipopolysaccharide
Fructose-6-phosphate	Glycolysis	Peptidoglycan
Dihydroxyacetone phosphate	Glycolysis	Lipids (glycerol component)
3-Phosphoglycerate	Glycolysis	Proteins (the amino acids cysteine, glycine, and serine)
Phosphoenolpyruvate	Glycolysis	Proteins (the amino acids phenylalanine, tryptophan, and tyrosine)
Pyruvate	Glycolysis	Proteins (the amino acids alanine, leucine, and valine)
Ribose-5-phosphate	Pentose phosphate cycle	Nucleic acids and proteins (the amino acid histidine)
Erythrose-4-phosphate	Pentose phosphate cycle	Proteins (the amino acids phenylalanine, tryptophan, and tyrosine)
Acetyl-CoA	Transition step	Lipids (fatty acids)
α-Ketoglutarate	TCA cycle	Proteins (the amino acids arginine, glutamate, glutamine, and proline)
Oxaloacetate	TCA cycle	Proteins (the amino acids aspartate, asparagine, isoleucine, lysine, methionine, and threonine)

Introduce the players *Certain intermediates of catabolic pathways can be used in anabolic pathways. These intermediates—**precursor metabolites**—serve as carbon skeletons from which subunits of macromolecules can be made* ***(table 6.2).***

Reinforce the concept *A cell's metabolic pathways make it easy for that cell to use glucose for multiple purposes. Think of the cells as extensive biological recycling centers that routinely process millions of glucose molecules* ***(figure 6.9).*** *Molecules that remain on the central deconstruction line are oxidized completely to CO_2, releasing the maximum amount of energy. Some breakdown intermediates, however, can exit that line to be used in biosynthesis.*

Put the pieces together *Three key metabolic pathways—the **central metabolic pathways**—gradually oxidize glucose to CO_2* ***(figure 6.10).*** *The pathways are catabolic, but the precursor metabolites and reducing power they generate can also be diverted for use in biosynthesis.*

Student-Friendly Disease Presentations

Help students think like experts

Within each body system chapter, diseases are separated by major taxonomic category (bacteria, viruses, fungi, protozoa). This organization reflects a major consideration with respect to treatment options, an important consideration for students going into healthcare-related fields.

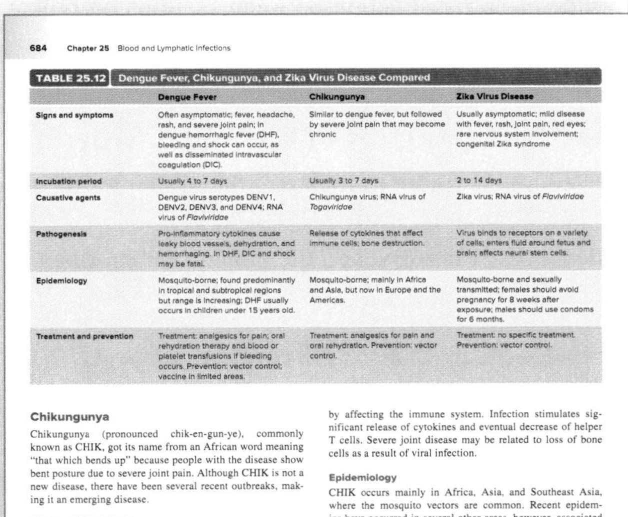

684 Chapter 25 Blood and Lymphatic Infections

TABLE 25.12 Dengue Fever, Chikungunya, and Zika Virus Disease Compared

	Dengue Fever	Chikungunya	Zika Virus Disease
Signs and symptoms	Often asymptomatic; fever, headache, rash, and severe joint pain; in dengue hemorrhagic fever (DHF), bleeding and shock can occur, as well as disseminated intravascular coagulation (DIC).	Similar to dengue fever, but followed by severe joint pain that may become chronic	Usually asymptomatic; mild disease with fever, rash, joint pain, red eyes; rare nervous system involvement; congenital Zika syndrome
Incubation period	Usually 4 to 7 days	Usually 3 to 7 days	2 to 14 days
Causative agents	Dengue virus serotypes DENV1, DENV2, DENV3, and DENV4; RNA virus of *Flaviviridae*	Chikungunya virus; RNA virus of *Togaviridae*	Zika virus; RNA virus of *Flaviviridae*
Pathogenesis	Pro-inflammatory cytokines cause leaky blood vessels, dehydration, and hemorrhaging. In DHF, DIC and shock may be fatal.	Release of cytokines that affect immune cells; bone destruction.	Virus binds to receptors on a variety of cells; enters fluid around fetus and brain; affects neural stem cells.
Epidemiology	Mosquito-borne; found predominantly in tropical and subtropical regions but range is increasing; DHF usually occurs in children under 15 years old.	Mosquito-borne; mainly in Africa and Asia, but now in Europe and the Americas.	Mosquito-borne and sexually transmitted; females should avoid pregnancy for 8 weeks after exposure; males should use condoms for 6 months.
Treatment and prevention	Treatment: analgesics for pain; oral rehydration therapy and blood or platelet transfusions if bleeding occurs. Prevention: vector control; vaccine in limited areas.	Treatment: analgesics for pain and oral rehydration. Prevention: vector control.	Treatment: no specific treatment. Prevention: vector control.

Chikungunya

Chikungunya (pronounced chik-en-gun-ye), commonly known as CHIK, got its name from an African word meaning "that which bends up" because people with the disease show bent posture due to severe joint pain. Although CHIK is not a new disease, there have been several recent outbreaks, making it an emerging disease.

Signs and Symptoms

The signs and symptoms of CHIK are similar to those of dengue and include fever that typically lasts 2 to 5 days, followed by severe joint pain (especially in the extremities—fingers, toes, wrists, and ankles) that can persist for weeks or months. Sometimes patients develop chronic joint pain that differentiates CHIK from dengue fever. Patients often develop a rash. Nonspecific symptoms include headache, conjunctivitis and photophobia, back pain, nausea, and general malaise. The disease is seldom fatal.

Causative Agent

CHIK is caused by chikungunya virus, an enveloped, single-stranded RNA virus of the *Togaviridae* family. It is transmitted by *Aedes* mosquito species, mostly *A. aegypti* and *A. albopictus*.

Pathogenesis

The disease mechanism of chikungunya is largely unknown, but it appears, like dengue fever, to cause its damaging effects by affecting the immune system. Infection stimulates significant release of cytokines and eventual decrease of helper T cells. Severe joint disease may be related to loss of bone cells as a result of viral infection.

Epidemiology

CHIK occurs mainly in Africa, Asia, and Southeast Asia, where the mosquito vectors are common. Recent epidemics have occurred in several other areas, however, associated with the inadvertent introduction and spread of the mosquito *A. albopictus*. This vector is now present in Brazil, Central America, the United States, and many European countries, where it has adapted to favorable environmental conditions. In 2013, Chikungunya virus was reported for the first time in Caribbean nations. Since then, it has spread throughout the Americas.

Treatment and Prevention

There is no specific treatment for CHIK. Analgesics and nonsteroidal anti-inflammatory drugs such as ibuprofen are used to reduce the joint pain. Fluids are given to reduce dehydration from fever. As with most mosquito-borne diseases, CHIK can be prevented by effective vector control—destroying the vector and protecting the population by use of insect repellents and insecticide-impregnated mosquito netting. A vaccine for CHIK is currently in clinical trials. The main characteristics of chikungunya are presented in table 25.12.

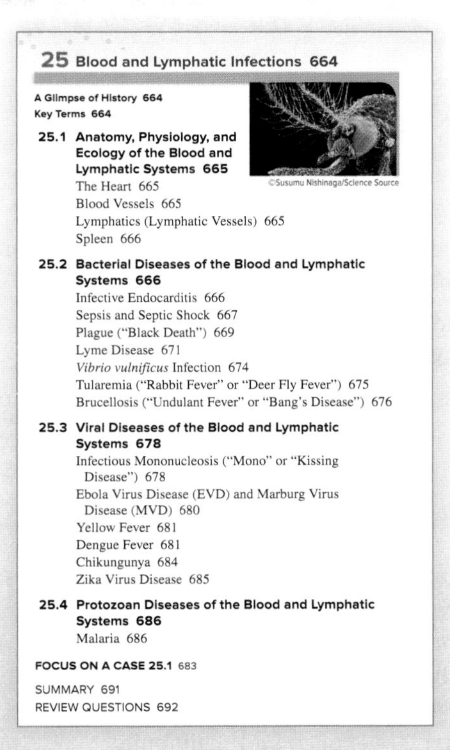

©Susumu Nishinaga/Science Source

Provide a consistent conceptual framework

Disease discussions are separated into consistent subsections, providing a conceptual framework and breaking the material into "bite-sized" pieces.

Summarize each disease's characteristics

Summary tables serve as brief reminders of the important features of each disease. Major diseases are represented with an enhanced summary table that includes an outline of the disease process keyed to a human figure, showing the entry and exit of the pathogen.

Review the diseases as a group

Each disease chapter ends with a table that summarizes the key features of the diseases discussed in that chapter.

Diseases in Review 21.1

Respiratory System Diseases

Disease	Causative Agent	Comment	Summary Table
BACTERIAL INFECTIONS OF THE UPPER RESPIRATORY TRACT			
Conjunctivitis (pink eye), otitis media (earache), sinus infection	Usually *Haemophilus influenzae* or *Streptococcus pneumoniae*	Often occur together; factors involved in the transmission are unknown.	
Streptococcal pharyngitis ("strep throat")	*Streptococcus pyogenes* (group A streptococcus)	Treated with antibiotics, partly to avoid sequelae; must be distinguished from viral pharyngitis, which cannot be treated with antibiotics.	Table 21.3
Diphtheria	*Corynebacterium diphtheriae*	Toxin-mediated disease characterized by pseudomembrane in the upper respiratory tract. Preventable by vaccination.	Table 21.4
VIRAL INFECTIONS OF THE UPPER RESPIRATORY TRACT			
Common cold	Rhinoviruses and other viruses	Runny nose, sore throat, and cough are due to the inflammatory response and cell destruction.	Table 21.5
Adenovirus pharyngitis	Adenoviruses	Similar to the common cold but with fever; spread to the lower respiratory tract can result in severe disease.	Table 21.6
BACTERIAL INFECTIONS OF THE LOWER RESPIRATORY TRACT			
Pneumococcal pneumonia	*Streptococcus pneumoniae*	Organism common in the throat of healthy people; causes disease when mucociliary escalator is impaired or with underlying conditions. Vaccine that protects against multiple strains is available.	Table 21.7
***Klebsiella* pneumonia**	*Klebsiella* species, commonly *K. pneumoniae*	Common hospital-acquired bacterium; characterized by thick, bloody, jelly-like sputum. Drug resistance is a major problem.	Table 21.7
Mycoplasmal pneumonia ("walking pneumonia")	*Mycoplasma pneumoniae*	Relatively mild pneumonia; common among college students and military recruits. Cannot be treated with medications that inhibit cell wall synthesis.	Table 21.7
Pertussis ("whooping cough")	*Bordetella pertussis*	Characterized by frequent violent coughing. Preventable by vaccination.	Table 21.8
Tuberculosis ("TB")	*Mycobacterium tuberculosis*	Most infections result in latent tuberculosis infection (LTBI), but these can reactivate to cause tuberculosis disease (TB disease). Treated using combination drug therapy, but drug resistance is an increasing problem.	Table 21.9
Legionellosis ("legionnaires' disease")	*Legionella pneumophila*	Transmitted via aerosolized water drops; smokers and those with impaired defenses are most at risk of developing disease.	Table 21.10
Inhalation anthrax	*Bacillus anthracis*	Rare zoonotic disease; may be associated with bioterrorism; high case-fatality rate.	Table 21.11
VIRAL INFECTIONS OF THE LOWER RESPIRATORY TRACT			
Influenza ("flu")	Influenza viruses	New vaccine developed yearly; viruses change seasonally due to antigenic drift; antigenic shifts cause pandemics.	Table 21.12
Respiratory syncytial virus infections	RSV	Serious disease in infants, young children, and the elderly.	Table 21.13
Hantavirus pulmonary syndrome	Hantaviruses	Acquired via inhaled dust contaminated with rodent saliva, urine, or feces. Frequently fatal.	Table 21.14
SARS and MERS	Coronaviruses	Emerging zoonotic diseases.	Table 21.15
FUNGAL INFECTIONS OF THE RESPIRATORY TRACT			
Coccidioidomycosis ("valley fever")	*Coccidioides immitis*	Environmental reservoir (soil in semi-arid desert areas); most infections are asymptomatic.	Table 21.16
Histoplasmosis ("spelunker's disease")	*Histoplasma capsulatum*	Environmental reservoir (soil enriched with bird or bat droppings); most infections are asymptomatic.	Table 21.17
***Pneumocystis* pneumonia (PCP)**	*Pneumocystis jirovecii* (formerly *carinii*)	Organism is an opportunistic fungus that causes serious lung disease in immunocompromised people, such as those with HIV/AIDS.	Table 21.18

UPDATES—Maintaining the Cutting Edge

Global Changes

- Continued "wordsmithing" to improve the clarity and readability of the descriptions
- Updated disease statistics, vaccine recommendations, treatments, and terminology
- Replaced former "Future Opportunities" boxes with "Focus on the Future" boxes
- Deleted some boxes to make room for updates and other changes

New! "Focus on . . . " Disease Boxes

These boxes cover a general category of disease, giving students a framework and the terminology for understanding the more focused coverage of individual diseases. In essence, they help students see the "forest" before learning about the "trees."

- Chapter 21—Focus on Pneumonia
- Chapter 24—Focus on Diarrheal Diseases
- Chapter 26—Focus on Meningitis
- Chapter 27—Focus on Sexually Transmitted Infections

Key Changes in Individual Chapters

Chapter 1 – Humans and the Microbial World

- Retitled and expanded the section on the normal microbiota (now The Human Microbiome)
- Added information about the National Microbiome Initiative (NMI)
- Added congenital Zika syndrome to the list of emerging diseases
- Substituted MERS for SARS in the discussion of evolution of pathogens to infect new hosts

Chapter 2 – The Molecules of Life

- Simplified the discussion of polysaccharide structures
- Updated the discussion of D-amino acids
- Added a table showing relative electronegativities of common atoms in biology (table 2.3)

Chapter 3 – Microscopy and Cell Structure

- New section on super-resolution microscopes, including an accompanying figure (figure 3.9)
- Retitled the section on atomic force microscopes (now Scanning Probe Microscopes)
- Modified the headings in the cell structure sections to provide clearer distinction between sections on prokaryotic and eukaryotic cells

Chapter 4 – Dynamics of Microbial Growth

- Created a new figure to illustrate the principles of selective and differential media (figure 4.10)
- Added a new MicroByte on the global agar shortage

Chapter 5 – Control of Microbial Growth

- Updated the information about the EPA's efforts to encourage the use of less toxic options for germicidal chemicals, and added a figure (figure 5.9) showing the new "Safer Choice" label
- Enhanced the coverage of high pressure processing (HPP)
- Updated the coverage of triclosan to include restrictions on its use in personal care products
- Updated coverage of the use of chlorine and iodine as disinfectants

Chapter 6 – Microbial Metabolism: Fueling Cell Growth

- Revised the description of vitamins

Chapter 7 – The Blueprint of Life, from DNA to Protein

- New chapter opening photo
- Modified figure 7.13 so that reading frame #1 starts with AUG
- Revised the description of capping of eukaryotic pre-mRNA

Chapter 8 – Bacterial Genetics

- Added an overview figure showing the three mechanisms of horizontal gene transfer (figure 8.18), in place of the previous summary table (was table 8.3)
- Created a new section, "Bacterial Defenses Against Invading DNA" (section 8.10), by moving and revising what was section 13.4. This change introduces restriction enzymes and CRISPR systems before they are discussed in section 9.1 (Fundamental Tools Used in Biotechnology)
- Simplified the table on DNA repair (table 8.2)
- Simplified the summary table on mobile genetic elements (now table 8.3)

Chapter 9 – Biotechnology

- Added a subsection that describes the use of CRISPR in biotechnology, including a supporting figure (figure 9.3)
- Combined what was sections 9.1–9.3 (Applications of Genetic Engineering, Techniques Used in Genetic Engineering, and Concerns Regarding Genetic Engineering)
- Retitled the subsection on next-generation sequencing methods (now High-Throughput Sequencing Methods) and expanded to include nanopore sequencing and its use in the International Space Station
- Expanded the section on PCR by adding information about RT-PCR and q-PCR
- Added a "Focus on the Future" box about Precision Medicine

Chapter 10 – Identifying and Classifying Microorganisms

- Updated figure 10.1 to include a part that illustrates the new Tree of Life based on ribosomal protein sequences
- Expanded the information about whole genome sequencing to characterize strain differences, and added information about the Genome Trakr Network

Chapter 11 – The Diversity of *Bacteria* and *Archaea*

- Added *Methanopyrus kandleri,* the current record-holder for high-temperature growth, to the section on methane-generating hyperthermophiles
- Updated the information on *Chlamydia* species to no longer state that they lack detectable peptidoglycan; also changed the question that accompanies the figure showing *Chlamydia* (now figure 11.26)

Chapter 12 – The Eukaryotic Members of the Microbial World

- Updated the description of lichens to indicate that genetic and molecular evidence suggests that they may include more than two partners
- Revised figure 12.12 to accompany an updated presentation of eukaryotic phylogeny
- Reduced the coverage of arthropod groups (some information was moved to chapters that describe arthropod-borne diseases)

Chapter 13 – Viruses, Viroids, and Prions

- Added icons that correlate steps in figure 13.5 (steps in the replication of lytic phage T4 in *E. coli*) with descriptions in the accompanying narrative
- Moved the previous section on bacterial defenses against phages to chapter 8, so that restriction enzymes and CRISPR systems are covered earlier (before chapter 9, which describes their use in biotechnology)
- Added a new rendition of the figure that illustrates animal virus replication strategies (figure 13.12) for easier understanding

Chapter 14 – The Innate Immune Response

- Moved the Focus Figure (figure 14.1) into section 14.1
- Simplified figure 14.5

Chapter 15 – The Adaptive Immune Response

- Changed the critical thinking question in MicroAssessment 15.5
- Expanded the legend for figure 15.25

Chapter 16 – Host-Microbe Interactions

- Added the definition of microbiome to the key terms
- Changed the title of section 16.2 to The Human Microbiome (was The Normal Microbiota)
- Introduced the term *dysbiosis*
- Added a subsection on damage to the host in section 16.9 (Mechanisms of Viral Pathogenesis)

Chapter 17 – Immunological Disorders

- Updated coverage of hypersensitivity reactions, particularly asthma, systemic anaphylaxis, and immune complex diseases

- Updated and increased coverage of immunotherapy to treat allergic reactions
- Added MicroByte on sublingual immunotherapy (SLIT) as an alternative to allergy shots
- Moved coverage of tuberculin skin test to the presentation of tuberculosis in section 21.4

Chapter 18 – Applications of Immune Responses

- Updated the information about polio eradication efforts by including the switch from trivalent to bivalent OPV
- Reorganized and refined the section on vaccines (section 18.2) by incorporating a new subsection titled "The Importance of Vaccines," rearranging the order of the tables, and separating the list of non-routine vaccines into a new table (table 18.5); also updated the table entries
- Added information about the monoclonal antibody recently approved for use as part of the protocol for treating inhalation anthrax
- Created a new "Focus on the Future" box that describes cancer immunotherapies, including CAR T cells

Chapter 19 – Epidemiology

- Updated the table of notifiable infectious diseases (table 19.1)
- Added a new table that lists the most common nosocomial infections (table 19.3)
- Added a subsection on visitors to the section that describes potential reservoirs for nosocomial infections
- Added new coverage of National Healthcare Safety Network (NHSN)

Chapter 20 – Antimicrobial Medications

- Expanded the section on development of antibiotics by adding information about Generating Antibiotic Incentives Now (GAIN)
- Added information about the newest glycopeptide antibiotics (dalbavancin and oritavancin) and the newest oxazolidinone (tedizolid)
- Updated the list of antiviral medications by adding NS5A inhibitors (used to treat HCV)
- Changed the "Focus on the Future" box to cover the *National Action Plan for Combating Antibiotic Resistant Bacteria*

Chapter 21 – Respiratory System Infections

- Added new box, "Focus on Pneumonia," presenting general characteristics of this important category of respiratory diseases
- Added a new section on inhalation anthrax
- Added a new section on SARS and MERS
- Rearranged section 21.2 to begin with milder diseases (pink eye, earache, sinus infections)
- Updated and focused the coverage of pneumococcal pneumonia and *Klebsiella* pneumonia
- Pertussis: revised the information about the epidemiology and the treatment and prevention; added a new photo (figure 21.16)
- Tuberculosis: increased the coverage of tuberculin skin test and added an accompanying figure (figure 21.21)

Chapter 22 – Skin Infections

- Added new section on cutaneous anthrax
- Added mention of ceftaroline as a new β-lactam antibiotic that can be used for treating MRSA
- Moved section on Lyme disease to chapter 25 (Blood and Lymphatic Infections)
- Rocky Mountain spotted fever: revised the information on the signs and symptoms and the pathogenesis
- Rubeola: revised the introduction and the information about the epidemiology and the prevention

Chapter 23 – Wound Infections

- Deleted coverage of actinomycosis (lumpy jaw)
- Removed photo showing an individual with sporotrichosis (was figure 23.14)

Chapter 24 – Digestive System Infections

- Added new box, "Focus on Diarrheal Diseases," presenting general characteristics of this important category of intestinal disease
- Periodontal disease and ANUG: revised the descriptions of causative agents
- Mumps: revised the description of epidemiology
- Cholera: added information about the new FDA-approved vaccine
- Shigellosis: updated the description of treatment
- *Clostridium difficile* infection (CDI): added information about fidaxomicin for treatment

- Norovirus: added update about the recent cultivation of the virus in the laboratory
- Hepatitis C: revised the introduction and added information about the new treatments

Chapter 25 – Blood and Lymphatic Infections

- Reorganized coverage throughout
- Revised coverage of anatomy, especially the lymphatics
- Added new section on Zika virus disease
- Dengue fever: added information about the new vaccine
- Chikungunya: standardized the organization of the section
- Added a new table that compares Dengue fever, Chikungunya, and Zika Virus disease (table 25.12)
- Added a section on Lyme disease (moved from chapter 22)
- Revised the coverage of infective endocarditis (previously SBE)

Chapter 26 – Nervous System Infections

- Added new box, "Focus on Meningitis," presenting general characteristics of this important category of nervous system diseases
- Reorganized to separate central nervous system (CNS) diseases from peripheral nervous system (PNS) diseases; superheadings added
- Pneumococcal and meningococcal meningitis: updated and focused the coverage
- Polio: updated the coverage
- Changed African sleeping sickness to African trypanosomiasis; revised the descriptions of the epidemiology and the treatment

Chapter 27 – Genitourinary Tract Infections

- Added new box, "Focus on Sexually Transmitted Infections," presenting general characteristics of this important category of genitourinary infections
- Added coverage on *Mycoplasma genitalium* infections
- Updated figure 27.21 (The Global HIV/AIDS Epidemic) to reflect changes in numbers
- Updated figure 27.25 (HIV Replication)
- Updated and modified figure 27.26 (was Deaths Due to AIDS) to include global number of people receiving ART

Chapter 28 – Microbial Ecology

- Revised the section on terrestrial habitats to emphasize the microbiome
- Changed the order of the questions in MicroAssessment 28.4

Chapter 29 – Environmental Microbiology: Treatment of Water, Wastes, and Polluted Habitats

- Deleted the "Focus on the Future" box ("Better Identification of Pathogens in Water and Wastes")

Chapter 30 – Food Microbiology

- Moved the description of starter culture
- Updated the names of the organisms used to make fermented foods
- Revised the section on making beer
- Deleted the "Focus on the Future" box ("Using Microorganisms to Nourish the World")

Acknowledgments

First and foremost, special thanks goes to Gene Nester, the leader of the team that wrote the first version of what became *Microbiology, A Human Perspective.* His efforts helped pioneer a new type of introductory microbiology textbook, designed specifically for students entering healthcare-related fields. This edition proudly builds on that original vision.

We would also like to thank the reviewers and other instructors who guided us as we developed this edition, as well as those whose input has helped the text evolve over the years. Deciding what to eliminate, what to add, and what to rearrange is always difficult, so we appreciate your suggestions.

Past students have been incredibly helpful as well. Every question helps us decide which parts of the textbook need more clarification, and every compliment lets us know when we're on the right track.

Special thanks also go to our friends, families, and colleagues for picking up the many hairs we tore out while working on the textbook. Revising a textbook is an all-consuming task—from the initial development stage to proofing the pages during production—and numerous people have acted as advisors and cheerleaders throughout. This text would not exist without the contributions of our strong group of supporters.

A list of acknowledgments is not complete without thanking our fearless leaders and friends from McGraw-Hill. Our product developer Michelle Gaseor and portfolio manager Marija Magner not only gave inspiration and sound advice, but they also laughed at our jokes and barely rolled their eyes at our idiosyncrasies. Marija Magner and Valerie Kramer helped ensure that word got out about this new edition, allowing it to find the way into your hands. It was wonderful to have Vicki Krug as our content project manager to guide us through some rocky waters on the way to publication. Additionally, we would like to thank Jonathan Miller and digital content project manager Brent dela Cruz for helping produce our digital resources that support the text and Lisa Burgess, who provided many wonderful photographs.

We hope that this text will be interesting and educational for students and helpful to instructors. Our goal is excellence, so with that in mind we would appreciate any comments and suggestions from our readers.

Denise Anderson
Sarah Salm
Deborah Allen

Reviewers for the Ninth Edition

Jason Adams, *College of Dupage*
Carollyn Boykins-Winrow, *Tidewater Community College*
Carron Bryant, *East Mississippi Community College*
Christine Clouser, *University of Minnesota*
Joyce Doan, *Bethel University*
Jose Fernandez Romero, *Borough of Manhattan Community College, CUNY*
Eric Ford, *East Mississippi Community College*
Julie Gibbs, *College of Dupage*
Dale Horeth, *Tidewater Community College*
Sara Reed Houser, *Jefferson College of Health Sciences*
Ilko Iliev, *Southern University at Shreveport*
Andrew Iverson, *William Rainey Harper College*
Kristen Joachimides, *North Seattle College*
Amine Kidane, *Columbus State*
Terrence Miller, *Central Carolina Community College*
Kari Naylor, *University of Central Arkansas*
Krista Peppers, *University of Central Arkansas*
Cynthia Ripoll, *Delgado Community College*
Rachael Romain, *Columbus State Community College*
Ben Rowley, *University of Central Arkansas*
Syed Shahabuddin, *College of Lake County*
Roger Wainwright, *University of Central Arkansas*
Kate Wilwohl, *Bethel University*
Jessica Wohlgamuth-Benedum, *Columbus State Community College*

Contents

©McGraw-Hill Education/Lisa Burgess

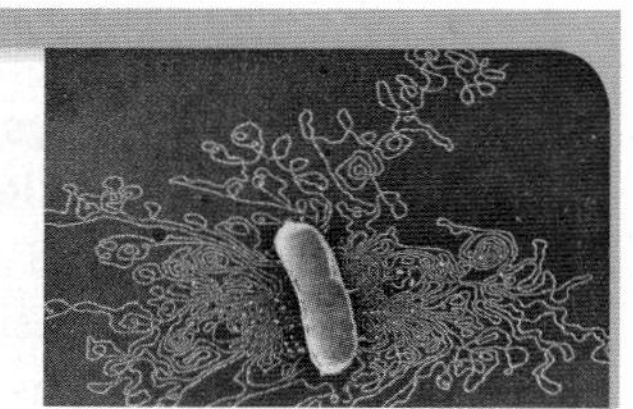
©Dr. Gopal Murti/Science Source

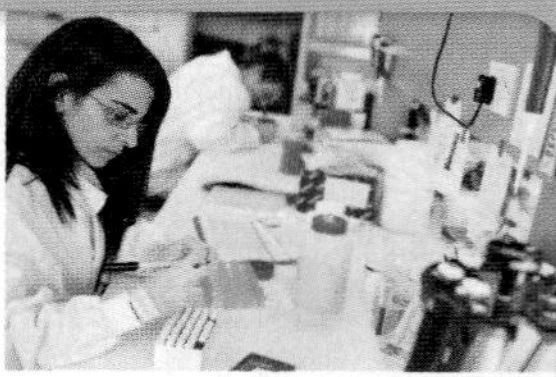
©Anthony Bradshaw/Photographer's Choice/Getty Images

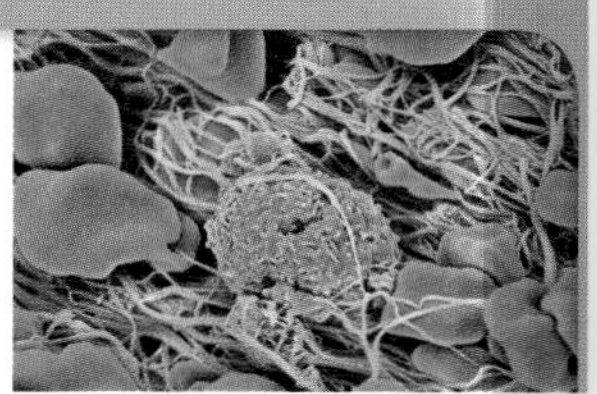
©Science Photo Library RF/Getty Images

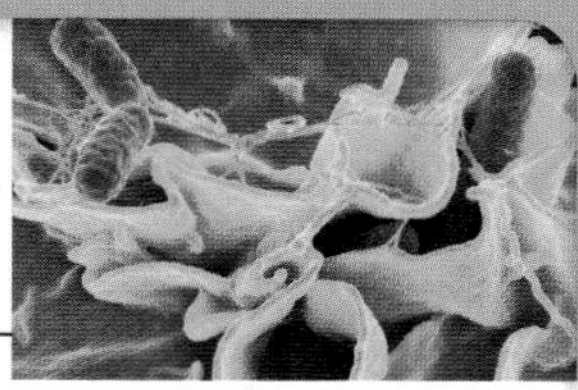
Source: NIAID, NIH, Rocky Mountain Laboratories

MICROBES, HEALTH, AND DISEASE

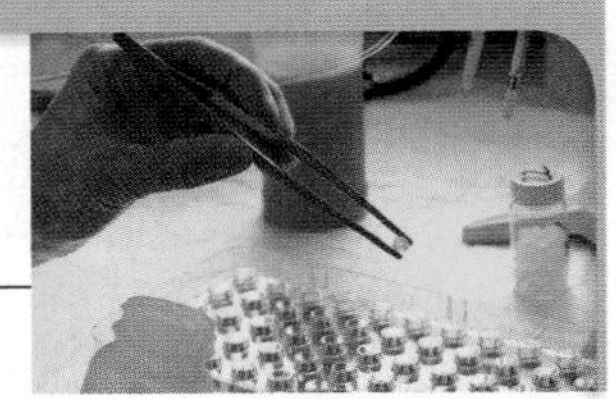
©Kevin Horan/The Image Bank/Getty Images

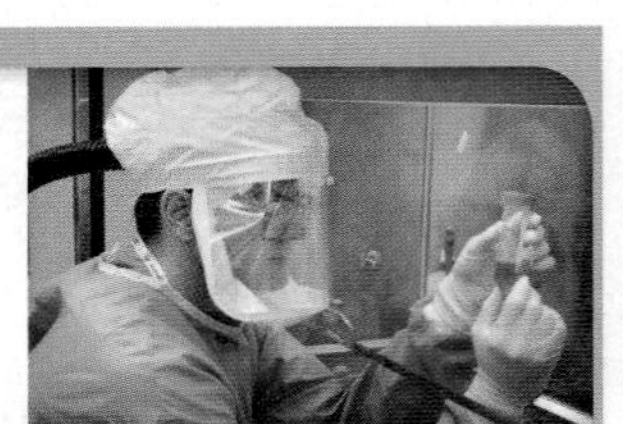
Source: James Gathany/CDC

©Science Photo Library/Getty Images

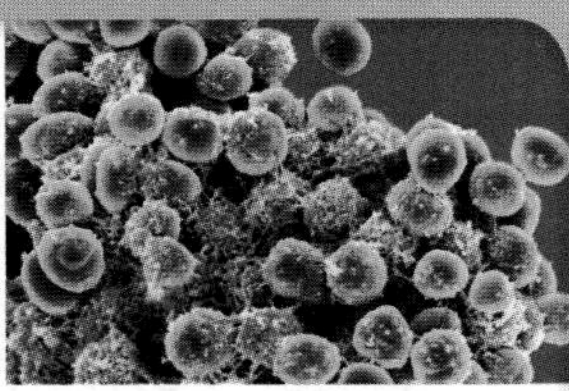
©Science Source

©Garry Watson/Science Source

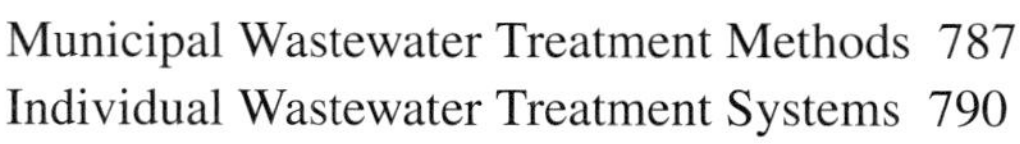

©Robert Glusic/Getty Images

©Tang Ming Tung/DigitalVision/Getty Images

1 Humans and the Microbial World

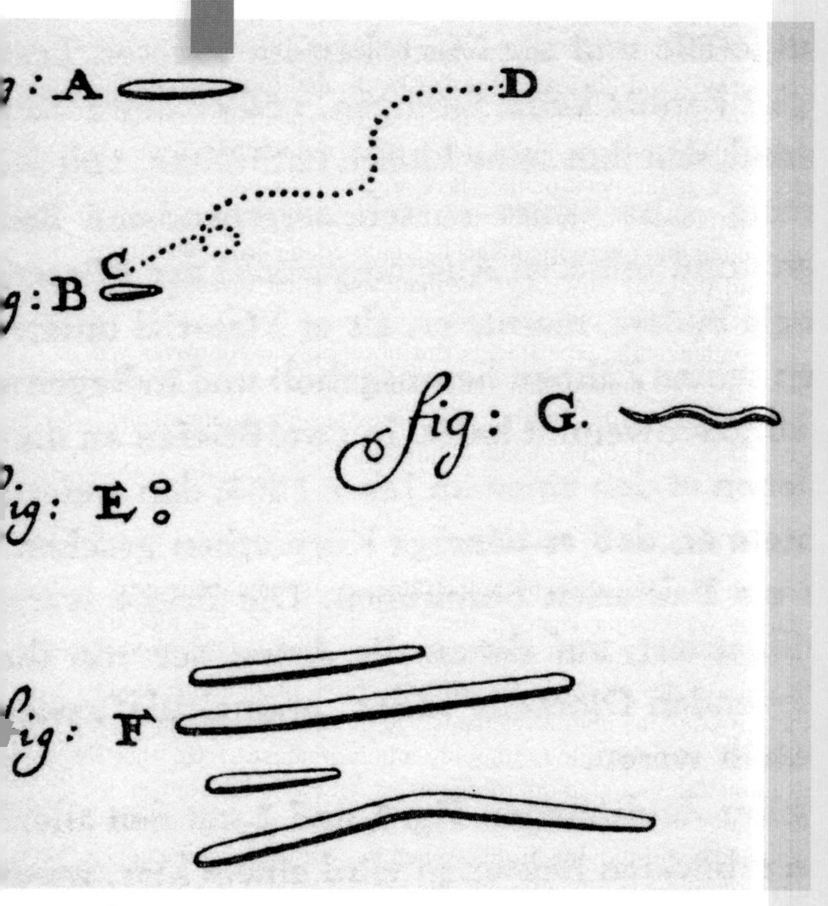

Drawings that van Leeuwenhoek made in 1683 of microorganisms he saw through his single lens microscope. He also observed organism B moving from position C to D. ©INTERFOTO/Alamy Stock Photo

KEY TERMS

Domain The highest level in biological classification. There are three domains: *Bacteria, Archaea,* and *Eukarya.*

Eukaryote Organism composed of one or more eukaryotic cells; members of the domain *Eukarya* are eukaryotes.

Eukaryotic Cell Cell type characterized by a membrane-bound nucleus.

Prion An acellular infectious agent consisting only of protein.

Prokaryote Single-celled organism consisting of a prokaryotic cell; members of the domains *Bacteria* and *Archaea* are prokaryotes.

Prokaryotic Cell Cell type characterized by the lack of a membrane-bound nucleus.

Viroid An acellular infectious agent consisting only of RNA.

Virus An acellular infectious agent consisting of nucleic acid surrounded by a protein coat.

A Glimpse of History

Microbiology as a science was born in 1674 when Antony van Leeuwenhoek, an inquisitive Dutch fabric merchant, looked at a drop of lake water through a glass lens he had carefully made. Although many people before him had used curved glass to magnify objects, Leeuwenhoek's skilled hands made a lens that uncovered a startling and amazing sight—the world of microbes. As van Leeuwenhoek wrote in a letter to the Royal Society of London, he saw

> Very many little animalcules, whereof some were roundish, while others a bit bigger consisted of an oval. On these last, I saw two little legs near the head, and two little fins at the hind most end of the body. Others were somewhat longer than an oval, and these were very slow a-moving, and few in number. These animalcules had diverse colours, some being whitish and transparent; others with green and very glittering little scales, others again were green in the middle, and before and behind white; others yet were ashed grey. And the motion of most of these animalcules in the water was so swift, and so various, upwards, downwards, and round about, that 'twas wonderful to see.

Before van Leeuwenhoek made these observations, Robert Hooke, an English microscopist, saw another kind of microorganism. In 1665, he described what he called a "microscopical mushroom." His drawing was so accurate that his specimen could later be identified as a common bread mold. Hooke also described how to make the kind of microscope that van Leeuwenhoek constructed almost 10 years later. Both men deserve equal credit for revealing the world of microbes—the organisms you are about to study.

Microbiology is the study of the microbial world—an amazing world made up of members too small to be seen without the aid of a microscope. Antony van Leeuwenhoek described this world when he observed what he called "animalcules" through his simple microscope (**figure 1.1**). What he saw were **microorganisms** (organisms too small to see with the naked eye), including bacteria, protozoa, and some fungi and algae. The microbial world also includes viruses and other infectious agents that are not considered organisms because they are not composed of cells; they are acellular. When referring to general members of the microbial world, the term **microbe** is often used.

Microorganisms are the foundation for all life on Earth. They have existed on this planet for about 3.5 billion years, and over this time, plants, animals, and modern microorganisms have evolved from them. Even today, they continue to be a driving force in the evolution of all living things. Microorganisms may be small, but as you are about to learn, our life depends on their activities.

FIGURE 1.1 Model of van Leeuwenhoek's Microscope
The original made in 1673 could magnify an object almost 300 times. The object is brought into focus with the adjusting screws. ©Tetra Images/Alamy Stock Photo

? What kinds of organisms did van Leeuwenhoek observe through his microscope?

1.1 ■ The Dispute over Spontaneous Generation

Learning Outcomes

1. Describe the key experiments of scientists who disproved spontaneous generation.
2. Explain how the successful challenge to the idea of spontaneous generation led to the Golden Age of Microbiology.
3. Describe the scientific method, using Pasteur's swan-necked flask experiment as an example.

The discovery of microorganisms in various specimens raised an interesting question: "Where did these microscopic forms originate?" Some people believed that worms and other forms of life arise from non-living material in a process referred to as **spontaneous generation.** This was challenged by an Italian biologist and physician, Francesco Redi. In 1668, he used a simple experiment to show that worms found on rotting meat originated from the eggs of flies, not from the decaying meat as supporters of spontaneous generation believed. In his experiment, Redi covered the meat with fine gauze that prevented flies from depositing their eggs. When he did this, no worms appeared.

Despite Redi's work that explained the source of worms on decaying meat, conclusive evidence that microorganisms did not arise by spontaneous generation took more than 200 years and many experiments.

Early Experiments

In 1749, John Needham, a scientist and Catholic priest, showed that flasks containing various broths (made by soaking a nutrient source such as hay or chicken in water) gave rise to microorganisms even when the flasks were boiled and sealed with a cork. At that time, brief boiling was thought to kill all organisms, so this suggested that microorganisms did indeed arise spontaneously.

In 1776, the animal physiologist and priest Lazzaro Spallanzani obtained results that contradicted Needham's experiments; no bacteria appeared in Spallanzani's broths after boiling. His experiments differed from Needham's in two significant ways: Spallanzani boiled the broths for longer periods and he sealed the flasks by melting their glass necks closed. Using these techniques, he repeatedly demonstrated that broths remained sterile (free of microorganisms). However, if the neck of the flask cracked, the broth rapidly became cloudy due to growth of the organisms. Spallanzani concluded that microorganisms had entered the broth with the air, and the corks used by Needham and other investigators did not keep them out.

Spallanzani's experiments did not stop the controversy. Some people argued that the heating process destroyed a "vital force" in the air that was necessary for spontaneous generation, and so the debate continued.

Experiments of Pasteur

One giant in science who helped disprove spontaneous generation was Louis Pasteur, the French chemist considered by many to be the father of modern microbiology. In 1861, he did a series of clever experiments. First, he demonstrated that air contains microorganisms. He did this by filtering air through a cotton plug, trapping microorganisms. He then examined the trapped microorganisms with a microscope and found that many looked identical to those described by others who had been studying broths. When Pasteur dropped the cotton plug into a sterilized broth, the broth became cloudy from the growth of these microorganisms.

Most important, Pasteur demonstrated that sterile broths in specially constructed swan-necked flasks remained sterile even when left open to air (**figure 1.2**). Microorganisms from the air settled in the bends of the flask necks, never reaching the broth. Only when the flasks were tipped would microorganisms enter the broth and grow. Pasteur's simple and elegant experiments ended the arguments that unheated air or the broths themselves contained a "vital force" necessary for spontaneous generation. They led to the theory of **biogenesis,** the production of living things from other living things (*bio* means "life"; *genesis* means "to create").

Experiments of Tyndall

Although most scientists were convinced by Pasteur's experiments, some remained skeptical because they could not reproduce his results. An English physicist, John Tyndall, finally explained the conflicting data and, in turn, showed that Pasteur was correct. Tyndall found that various types of

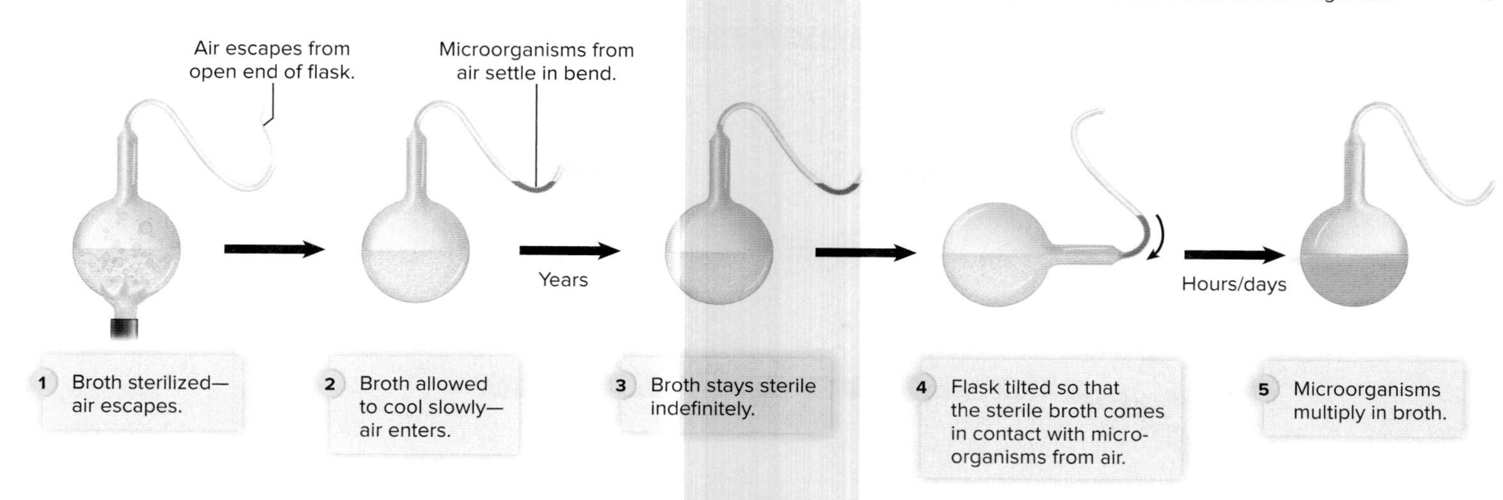

FIGURE 1.2 Pasteur's Experiment with the Swan-Necked Flask If the flask remains upright, no microbial growth occurs. If the flask is tipped, the microorganisms trapped in the neck reach the sterile broth and grow.

If the broth in Pasteur's swan-necked flasks had contained endospores, what results would have been observed?

broths required different boiling times to be sterilized. Some were sterilized by boiling for 5 minutes, whereas others, most notably broths made from hay, still contained living microorganisms even after boiling for 5 hours! Even when hay was merely present in the laboratory, broths that had previously been sterilized by boiling for 5 minutes could not be sterilized by boiling for several hours. What was going on? Tyndall finally realized that hay contained heat-resistant forms of microorganisms. When hay was brought into the laboratory, dust particles must have transferred these heat-resistant forms to the broths.

Tyndall concluded that some microorganisms exist in two forms: a cell easily killed by boiling, and one that is heat resistant. In the same year (1876), a German botanist, Ferdinand Cohn, discovered **endospores,** the heat-resistant forms of some bacteria. ⏭ endospores

The extreme heat resistance of endospores explains the differences between Pasteur's results and those of other investigators. Organisms that produce endospores are commonly found in the soil and were likely present in broths made from hay. Pasteur used only broths made with sugar or yeast extract, so his experiments probably did not have endospores. At the time, scientists did not appreciate the importance of the source of the broth, but in hindsight, the source was critical. This points out an important lesson for all scientists. In repeating an experiment, it is essential to reproduce all conditions as closely as possible. What may seem like a trivial difference may be extremely important.

The Golden Age of Microbiology

The work of Pasteur and others in disproving spontaneous generation started an era called the Golden Age of Microbiology, during which time the field of microbiology blossomed. Many important advances were made during this period, including discoveries that led to acceptance of the suggestion that microorganisms cause certain diseases, a principle now called the Germ Theory of Disease.

Figure 1.3 lists some of the important advances in microbiology made over the years in the context of other historical events. Rather than cover more history now, we will return to many of these milestones in brief stories called "A Glimpse of History" that open each chapter.

The Scientific Method

The dispute over spontaneous generation offers an excellent example of the process of science. This process, called the **scientific method,** separates science from intuition and beliefs. The scientific method involves a series of steps, including:

- **Making an observation about something and asking a question about that situation.** An example from this chapter was the observation that microorganisms were present in various examined specimens. This observation led to the question "Where did the microorganisms originate?"
- **Developing an explanation and then devising an experiment that tests the explanation.** A testable explanation of an observation is called a **hypothesis,** and experiments are done to test the hypothesis. The dispute over spontaneous generation led to two opposing hypotheses: biogenesis and spontaneous generation. Various people designed different experiments to test the hypotheses.
- **Doing the experiment, collecting the data, and drawing a conclusion.** Experiments such as the one illustrated in figure 1.2 provided data about the growth of microorganisms in previously sterile broths. In doing a

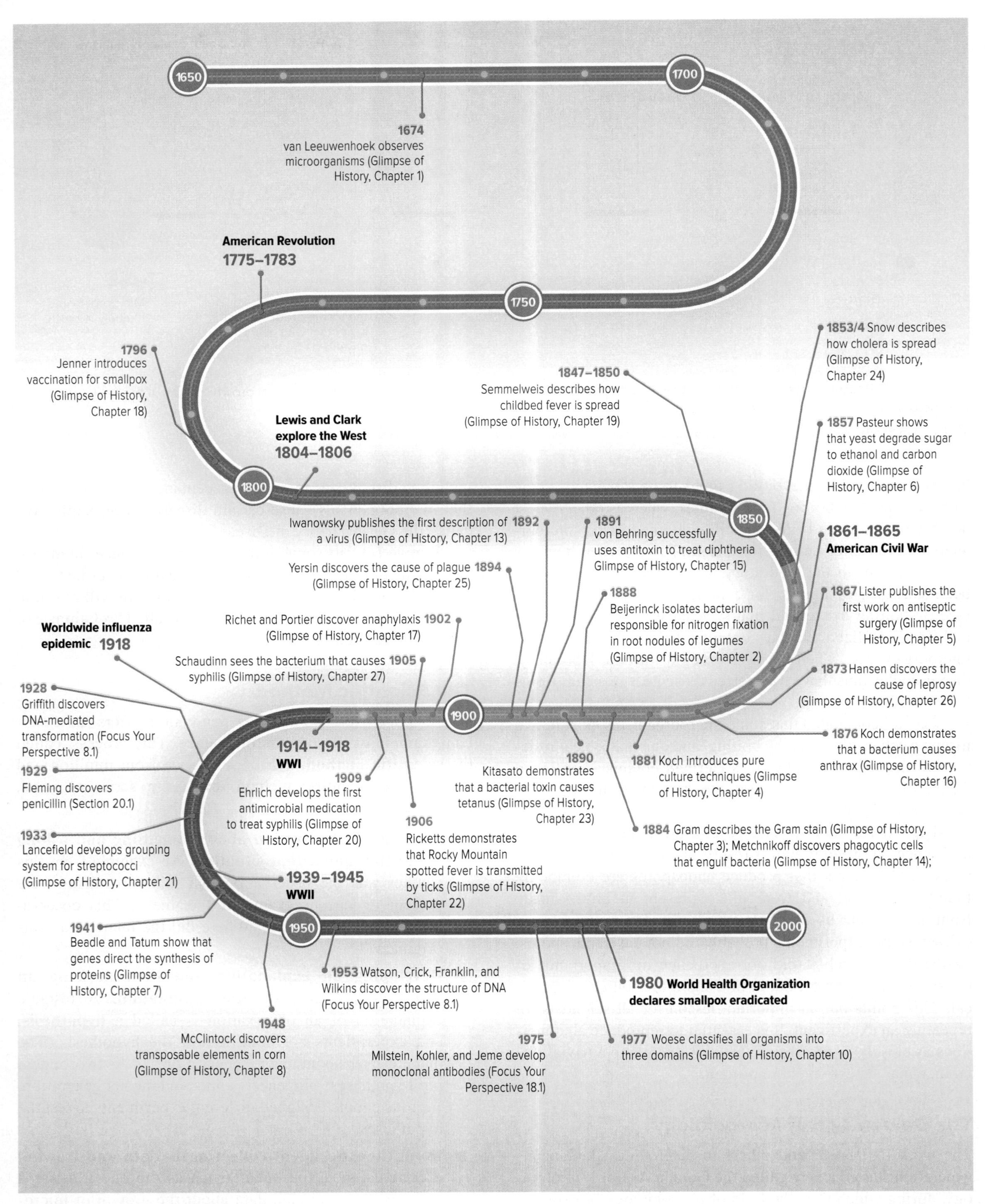

FIGURE 1.3 Historical Events in Microbiology Some major milestones in microbiology—and their timeline in relation to other historical events. The gold band indicates the Golden Age of Microbiology.

What is the Golden Age of Microbiology?

scientific experiment, a critical component is a **control.** A control helps rule out alternative explanations of the results by showing that the only feature that varied in the experiment was the characteristic being tested. Pasteur's swan-necked flask experiment was brilliantly designed because it provided the following control: After showing that the fluid in the swan-necked flasks remained sterile even when opened to air, he tipped the flasks so that bacteria could enter the fluid. By doing this, he showed that there was nothing in his original set-up that would have prevented bacteria from growing in the broth.

- **Communicating the methods, results, and conclusions.** Scientists share their work by publishing it in scientific journals. This step is particularly important because it allows other scientists to repeat the experiment to ensure the validity of the findings. Today, the respected scientific journals use a review process in which other experts in the field read communications before they are published. If deficiencies or flaws are noticed, the reviewers give suggestions for improving the experiments.

When an extensive amount of experimental evidence supports a hypothesis, that explanation may become a scientific **theory,** such as the Germ Theory of Disease. Note that the scientific meaning of the word *theory* is far different from the meaning of the word in common language, which is "a speculation or guess."

As you read the information in this textbook, continually challenge yourself by asking questions about what you have learned. If you find yourself asking a question such as "How does that happen?" try to develop a hypothesis and then devise an experiment. As you do this, consider the controls you could use. Start learning to think like a scientist!

MicroAssessment 1.1

Experiments of Pasteur and Tyndall helped disprove spontaneous generation by showing that life arises from life. Many important discoveries were made during the Golden Age of Microbiology, including ones that led to the acceptance of the Germ Theory of Disease. The scientific method uses experimental evidence, including proper controls, to support or refute hypotheses.

1. Describe Pasteur's experiment that disproved the idea that a "vital force" in air was responsible for spontaneous generation.
2. How is the meaning of the word "theory" in science different from its meaning in everyday conversation?
3. Why is it important for scientists to repeat the experiments of others?

1.2 ■ Microbiology: A Human Perspective

Learning Outcomes

4. Explain the importance of microorganisms in the health of humans and the surrounding environment.
5. List three commercial benefits of microorganisms.
6. Describe why microorganisms are useful research tools.
7. Describe the role of microbes in disease, including examples of past triumphs and remaining challenges.

Microorganisms have an enormous impact on all living things. We could not survive without them, and they also make our lives much more comfortable. At the same time, microbes can be harmful, and have killed far more people than have ever been killed in war.

The Human Microbiome

The human body carries an enormous population of microorganisms—tens of trillions of bacterial cells alone. Many sources claim that the body carries 10 times as many microbial cells as human cells, but that number is just a guess. Recent and probably more accurate estimates indicate that the ratio is likely closer to 3:1 or even 1:1. Regardless, scientists have known for years that these microorganisms, collectively referred to as the **normal microbiota** or normal flora, play an essential role in human health. For example, they prevent disease by competing with disease-causing microbes, help to degrade foods that the body otherwise could not digest, and promote the development of the immune system. In fact, studies indicate that early exposure to certain common microorganisms lessens the likelihood that an individual will develop allergies, asthma, and some other diseases. According to what is sometimes referred to as the "Old Friends" hypothesis, this early exposure helps the immune system learn to distinguish "friendly" microbes from those that can cause severe disease. In addition, animal studies suggest that the composition of the normal microbiota can affect brain chemistry and behavior, as well as the tendency to gain or lose weight.

The important role of the normal microbiota became even more obvious in recent years, thanks in part to the **Human Microbiome Project.** This coordinated set of studies, started in 2007, used DNA sequencing technologies to characterize the microbial communities that inhabit the human body. The term **microbiome** has two overlapping meanings: (1) the total genetic content of a microbial community, and (2) the microbial community itself. While the different meanings might seem confusing, they are actually quite similar because at this point the communities must be examined by studying their genetic material. The reason for this is that less than 1% of microorganisms can currently be grown in the laboratory.

So for every microbe that had been studied in the laboratory, more than 99 others could not be characterized until DNA sequencing technologies were developed.

The Human Microbiome Project has changed the way scientists view the human body and is also revealing how much more there is to discover about our microbial partners. To understand their significance, think of the earth's ecosystems (the environments and their interacting inhabitants). Over time, an interacting assortment of organisms has evolved to live in a given environment, resulting in a relatively stable community. Sudden changes can alter individual populations, often with negative consequences to the community as a whole. In turn, a disturbance in one ecosystem can affect the overall health of the planet. The human body, like a planet, is composed of various ecosystems—for example, the desert-like dry areas of the skin, and the nutrient-rich environment of the intestinal tract. An important part of these ecosystems is a population of interacting microbes. Disturbances in a microbial population can create an imbalance that may have negative consequences to that community, which, in turn, can harm a person's health. Observations such as these have led some scientists to suggest that the human body be considered a superorganism, meaning that our own cells interact with the body's normal microbiota to form a single cooperative unit.

Microorganisms in the Environment

Microorganisms are the masters of recycling, and without them we would run out of certain nutrients. For instance, humans and other animals all require oxygen gas (O_2) to breathe. However, the supply of O_2 in the atmosphere would run out if it were not continually replenished. Plants produce O_2 during photosynthesis, but so do many photosynthetic microorganisms. Another example involves nitrogen, an essential part of nucleic acids and proteins. A plentiful source of nitrogen is N_2—the most common gas in the atmosphere—yet neither plants nor animals can use it. Instead, we depend on certain microbes that convert N_2 into a form of nitrogen that other organisms can use, a process called nitrogen fixation. Without nitrogen-fixing microbes, life as we know it would not exist. ⏭ nitrogen fixation

Microorganisms are also important because they can degrade certain materials that other organisms cannot. For instance, humans and other animals cannot digest cellulose—an important component of plants. Certain microorganisms degrade cellulose, however, which is why leaves and fallen trees do not pile up in the environment. Many of the billions of microorganisms in the digestive tracts of a group of animals that includes cattle, sheep, and deer degrade cellulose; by doing so, the microorganisms help the animals digest plant material. Without cellulose-degrading microbes in their digestive tracts, these plant-eating animals would starve.

In recognition of the important role that microorganisms play in all aspects of life, the **National Microbiome Initiative (NMI)** expands the scope of microbiome research. Started in 2016, the program supports research on the microbiomes of humans as well as those of our surrounding environment.

Commercial Benefits of Microorganisms

In addition to the crucial roles microorganisms play in our very existence, they also have made life more comfortable for humans over the centuries.

Food Production

Microorganisms have been used in food production since ancient times. For example, Egyptians used yeast to make bread and beer. Virtually every population that raised milk-producing animals such as cows and goats also developed procedures to ferment milk. This allowed them to make foods such as yogurt, cheeses, and buttermilk. Today, the bacteria added to some fermented milk products are advertised as probiotics (live microorganisms that provide a health benefit), protecting against digestive disruptions. ⏭ bread, ⏭ beer, ⏭ fermented milk products, ⏭ probiotics

Biodegradation

Microorganisms play essential roles in degrading various environmental pollutants. These include materials in wastewater, as well as toxic chemicals in contaminated soil and water. Bacteria also lessen the damage from oil spills. In some cases, microorganisms are added to pollutants to hasten their decay, a process called **bioremediation.** ⏭ wastewater treatment, ⏭ bioremediation

Commercially Valuable Products from Microorganisms

Microorganisms synthesize a wide variety of different products, some of which are commercially valuable. Examples include antibiotics used in the treatment of diseases; ethanol used as a biofuel; hydrogen gas and certain oils potentially used as biofuels; amino acids used as dietary supplements; insect toxins used in insecticides; cellulose used in headphones; and polyhydroxybutyrate used in the manufacture of disposable diapers and plastics.

Biotechnology

Biotechnology—the use of microbiological and biochemical techniques to solve practical problems—depends on members of the microbial world. Information learned by studying microorganisms led to easier production of many medications, such as insulin (used to treat diabetes). In the past, insulin was isolated from pancreatic glands of cattle and pigs. Now, certain microorganisms have been genetically engineered to make human insulin. The microbe-produced insulin is easier to obtain, and patients who use it have fewer allergic reactions than occurred with the animal-derived product. Biotechnology has also led to techniques that scientists now use to genetically engineer plants to give them desirable qualities.

⏭| genetic engineering

Microbes as Research Tools

Microorganisms are wonderful model organisms to study because they have the same fundamental metabolic and genetic properties as higher life forms. All cells are composed of the same chemical elements and they synthesize their cell structures by similar mechanisms. They all duplicate their DNA, and when they degrade foods to harvest energy, they do so via the same metabolic pathways. To paraphrase a Nobel Prize–winning microbiologist, Dr. Jacques Monod—what is true of elephants is also true of bacteria, and bacteria are much easier to study! In addition, bacteria can be used to obtain results very quickly because they grow rapidly and form billions of cells per milliliter on simple inexpensive growth media. In fact, most major advances made in the last century toward understanding life have come through the study of microbes.

Microbes and Disease

Although most microbes are beneficial or not harmful, some are **pathogens,** meaning they can cause disease (a noticeable impairment in body function). The disease symptoms result from damage to the body tissues. This damage can occur either as a direct result of the pathogen's growth and products, or as a result of the body's defense mechanisms, which can inadvertently harm the host while attempting to control the pathogen.

To appreciate the effect an infectious disease can have on a population, consider that more Americans died of influenza in 1918–1919 than were killed in World Wars I and II, and the Korean, Vietnam, and Iraq wars combined. Fortunately, technological advances such as sanitation, vaccination, and antibiotic treatments have dramatically reduced the incidence of many of the most feared infectious diseases. To maintain this success, however, we must continue to develop new medications, vaccines, and disease-prevention strategies.

Past Triumphs

The Golden Age of Microbiology included an important period when scientists learned a great deal about pathogens. Between 1876 and 1918, most pathogenic bacteria were identified, and early work on viruses had begun. Once people realized that microbes could cause disease, they tried to prevent their spread. As illustrated in **figure 1.4,** the death rate due to infectious diseases has decreased dramatically over the last

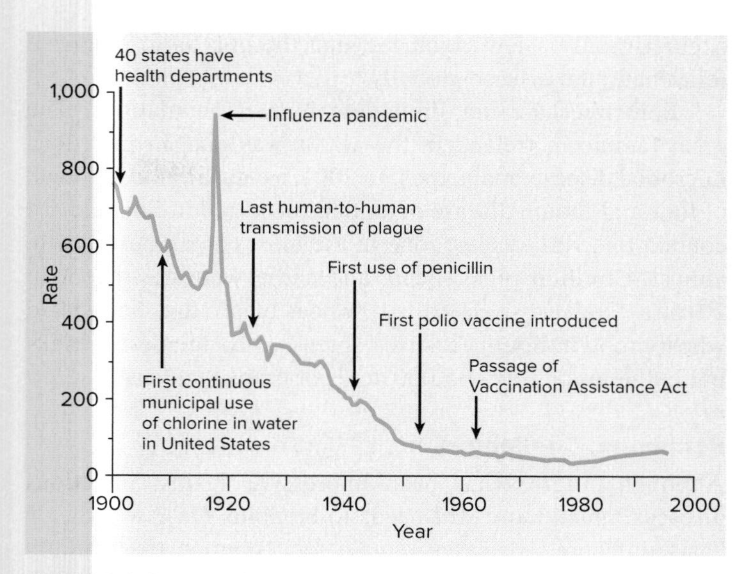

FIGURE 1.4 Trend in Death Rates Due to Infectious Diseases Crude death rate for infectious disease, United States, per 100,000 population per year.

? **Why would the creation of health departments lower the disease rate?**

100 years or so, due largely to preventing the spread of pathogens, developing vaccines to provide immunity, and using antibiotics to treat bacterial diseases when they do occur.

The viral disease smallpox was one of the most devastating diseases the world has ever known, killing about one-third of people who got it. Survivors were sometimes blinded and often left with disfiguring scars. When Europeans carried the disease to the Americas, the effect on the populations of native inhabitants who had not been exposed before was catastrophic. In recent times, an active worldwide vaccination program eliminated the disease in nature, with no cases being reported since 1977. Laboratory stocks of the smallpox virus remain, however, raising the possibility that the virus could be used in bioterrorist attacks.

Plague has been another major killer. One-third of the population of Europe, or approximately 25 million people, died of this bacterial disease in only 4 years (1347–1351). We now know that rodents can carry the bacterium, and their fleas can transmit the disease, so we take measures to control the rodent populations. We have also learned that the pneumonic form of the disease (meaning that it is in the lungs) can spread from human to human through respiratory secretions, so special precautions are taken when a patient has pneumonic plague. In addition, the discovery of antibiotics in the twentieth century made treatment possible. As a result, less than 100 people worldwide die from plague in a typical year.

Polio can cause paralysis, leading to death or disability. The disease was once relatively common, but it has been nearly eliminated because of vaccination. In fact, polio now occurs in only a few countries, and the goal is to eradicate (eliminate) the disease globally.

Epidemics are not limited to human populations. The great famine in Ireland in the 1800s was due, in part, to a microbial disease of potatoes. In 2001, a catastrophic outbreak of foot-and-mouth disease of animals occurred in England. To contain this viral disease, one of the most contagious known, almost 4 million pigs, sheep, and cattle were destroyed. In 2016, a fungal disease called "wheat blast" that devastated wheat crops in South America spread to Bangladesh, resulting in the loss of over 35,000 acres of crops that year.

Remaining Challenges

Although progress has been impressive against infectious diseases, much more still needs to be done. On a worldwide basis, infectious diseases remain too common, particularly in developing countries. Even in developed countries with sophisticated healthcare systems, infectious diseases remain a serious threat, costing lives and money.

Emerging Infectious Diseases An **emerging infectious disease (EID)** is an infectious disease that has become more common in the last 35 years. Many of these are new or newly recognized; examples include Ebola virus disease, congenital Zika syndrome, hepatitis C, Middle East respiratory syndrome (MERS), certain types of influenza, Lyme disease, acquired immunodeficiency syndrome (AIDS), hantavirus pulmonary syndrome, and mad cow disease (bovine spongiform encephalopathy) (**figure 1.5**). Others such as malaria and tuberculosis have been present for years, but have spread or become more common recently.

Some diseases arise as infectious agents evolve to infect new hosts, cause different types of damage, or become more difficult to treat because of antibiotic resistance. Genetic analysis indicates that HIV-1 (**h**uman **i**mmunodeficiency **v**irus type 1), the most common type of HIV to cause AIDS, arose from a virus that infected chimpanzees. A bacterium called *E. coli* O104:H4, which caused a severe foodborne diarrheal outbreak in Europe, appears to have gained the ability to make a specific toxin by acquiring genes from a related organism. Tuberculosis and malaria have increased in incidence in recent years, in part because the causative organisms became resistant to many of the available medications.

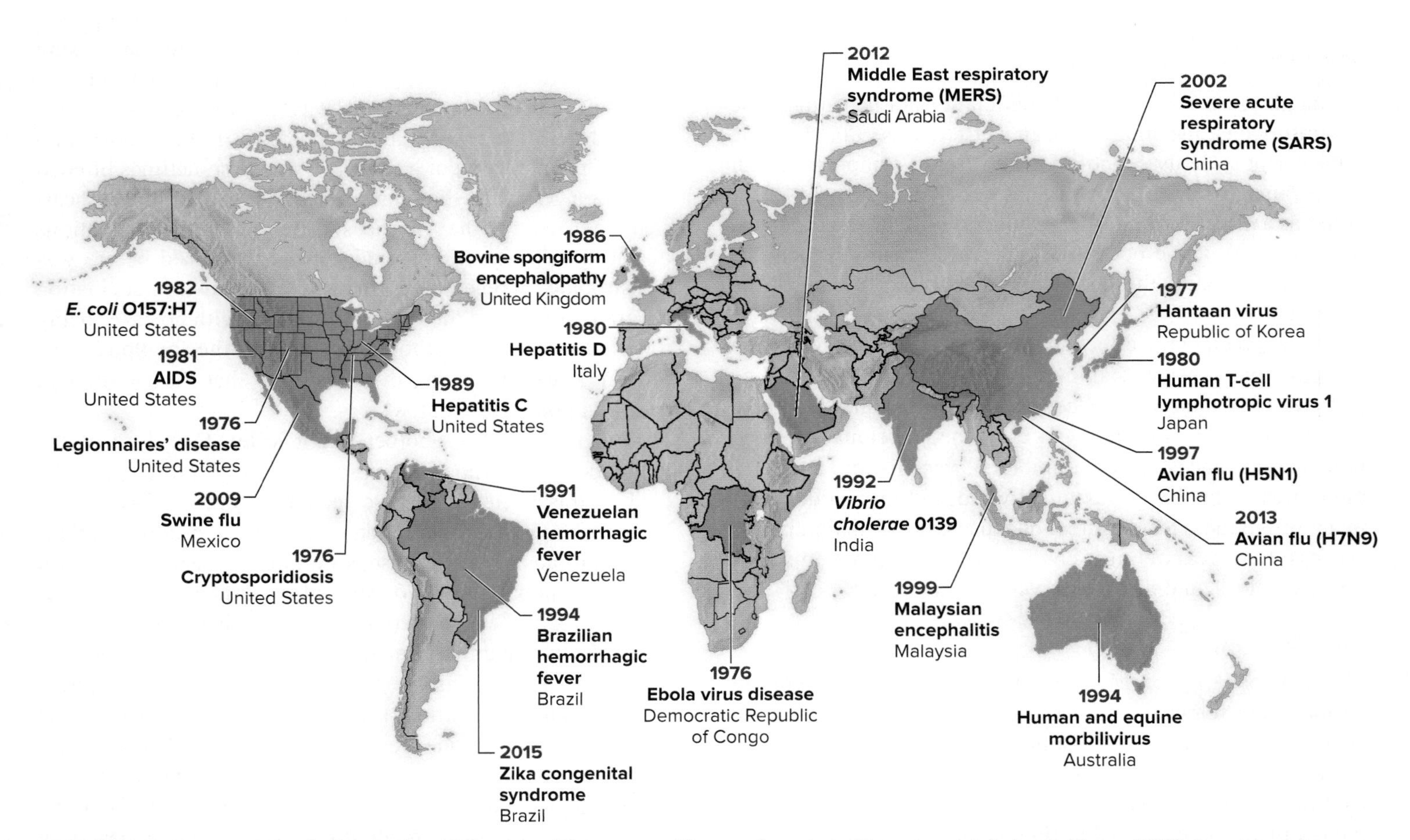

FIGURE 1.5 New and Newly Recognized Infectious Diseases or Disease Agents in Humans and Animals Since 1976 Countries where cases first appeared or were identified appear in a darker shade.

? **Why might so many of the diseases first appear or be identified in the United States and Western European countries?**

FOCUS ON A CASE 1.1

A 24-year-old woman suffered from recurrent severe episodes of an intestinal disorder called *Clostridium difficile* infection (CDI) for the past 13 months. She routinely experienced profuse watery diarrhea, abdominal pain, and fever. In addition, she was feeling tired and hopeless because she did not seem to be getting well, despite long attempts at multiple different treatments.

As with most patients who develop CDI, the woman had been taking an oral antibiotic shortly before her symptoms began—in this case, to treat a tooth infection. The antibiotic had successfully killed the bacteria that caused her tooth infection, but it also killed some members of her normal intestinal microbiota. As a result, the bacterium *Clostridium difficile*—often referred to simply as "C. diff"—thrived in her intestinal tract, growing to much higher numbers than it could before. The strain that caused her infection was able to make a toxin that damaged the lining of her intestinal tract.

When the patient first started experiencing CDI, her doctor told her to stop taking the antibiotic prescribed for her tooth infection, hoping that her CDI would resolve on its own. When that did not help, the doctor prescribed a different antibiotic that is often effective in treating CDI. The patient started feeling better, but the symptoms quickly returned when she stopped taking the medication. She also tried oral supplements containing *Lactobacillus* GG, a bacterium that sometimes appears to be effective in preventing antibiotic-associated diarrhea.

Because the patient's health was declining, doctors suggested a fecal transplant, a procedure that involves inserting feces from a healthy person into the patient's intestinal tract in order to repopulate that environment with appropriate microbes. They chose to use her sister as a fecal donor, screening both the donor and the patient to ensure that neither was infected with certain microbes, including various intestinal pathogens and HIV. Approximately ¼ cup of fresh feces was mixed with 1 quart of water and delivered to her intestinal tract via a colonoscope. Within days after the transplant, the patient began feeling better, and she soon recovered completely.

1. Why would certain oral antibiotics allow *C. difficile* to thrive in the intestinal tract?
2. Why would the doctors screen both the patient and the fecal donor for certain infectious agents?
3. Why would the doctors transplant feces rather than introducing isolated bacteria from feces to repopulate the colon?

Discussion

1. Antibiotics kill or inhibit not just pathogens, but also beneficial members of the normal microbiota, a group that protects against infection in at least two general ways. First, they quickly use nutrients that would otherwise be available to *C. difficile* and other pathogens. Also, some members of the normal microbiota make compounds that are toxic or inhibitory to other organisms. The environment of the intestinal tract is quite complex, however, so other factors might also be playing a role.
2. Physicians screen the fecal donor to decrease the likelihood that disease-causing microbes could be transferred to the patient via the procedure. The doctors screen the patient to ensure that she was not already infected with the pathogens. For example, if the patient developed symptoms of a *Salmonella* infection after the procedure, how would physicians know that she acquired the infection as a result of the procedure if they had not checked her beforehand?
3. Feces contain many types of bacteria that cannot yet be grown in the laboratory. In addition, scientists do not yet know which types of fecal bacteria protect against CDI.

Changes in society bring opportunities for infectious agents to spread, resulting in an emerging disease. More mobile populations can contribute to disease emergence as people may inadvertently carry pathogens around the globe. Diseases such as malaria, cholera, plague, and yellow fever have largely been eliminated from developed countries, but they still exist in many parts of the world. Newly infected international travelers could theoretically circle the globe, touch down in several countries, and expose many people before becoming ill themselves. Meanwhile, as suburbs of cities expand into rural areas, human populations come into closer contact with animals as well as the mosquitoes and other arthropods that normally feed on those animals. Consequently, people are exposed to pathogens they might not have encountered previously.

Infectious diseases that were under control can spread again, resulting in increased numbers of cases. Sometimes, the preventive measures become victims of their own success. For instance, decades of vaccination have nearly eliminated measles, mumps, and whooping cough in developed countries, so that most people no longer have first-hand knowledge of the dangers of the diseases. Couple this with misinformation about vaccines, and some people develop irrational fears, falsely believing that vaccines are more harmful than the diseases they prevent. When this happens, parents often refuse to vaccinate their children appropriately, leading to a situation where the diseases become more common again.

Chronic Diseases In addition to the diseases long recognized as being caused by pathogens, some illnesses once attributed to other causes may be due to microorganisms. Perhaps the best-known example is stomach ulcers, once thought to be due to stress. We now know that stomach ulcers are often caused by a bacterium *(Helicobacter pylori)* and are treatable with antibiotics. Chronic indigestion may be caused by the same bacterium. Another example is cervical cancer, which we now know is caused by human papillomavirus (HPV) infection; a vaccine against HPV prevents that cancer. Infectious microbes may play important roles in other chronic diseases as well.

Focus Figure

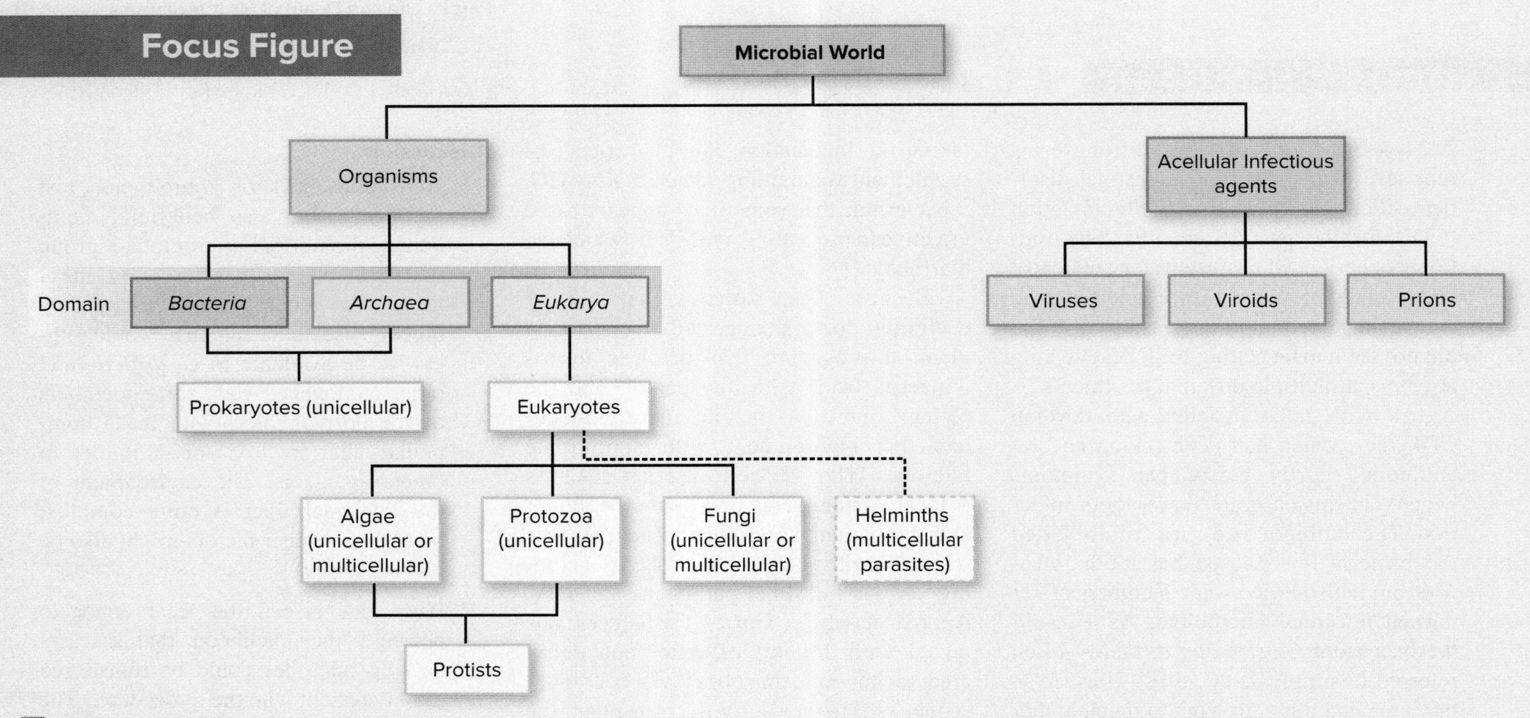

FIGURE 1.6 The Microbial World Although adult helminths (worms) can often be seen with the naked eye, some stages in the life cycle of helminths are microscopic.

Members of which two domains are prokaryotes?

MicroAssessment 1.2

Microbial activities are essential to human life as well as being commercially valuable. Microbes are important research tools. Although most microbes are beneficial or not harmful, some cause disease. Enormous progress has been made in preventing and curing infectious diseases, but some diseases are becoming more common.

4. Describe two microbial activities essential to life and three that make our lives more comfortable.
5. Describe three factors that cause certain infectious diseases to become more common.
6. Why would it seem logical, even inevitable, that at least some bacteria would attack the human body and cause disease?

1.3 ■ Members of the Microbial World

Learning Outcomes

8. Compare and contrast characteristics of members of the *Bacteria, Archaea,* and *Eukarya.*
9. Explain the features of an organism's scientific name.
10. Compare and contrast the algae, fungi, and protozoa.
11. Compare and contrast viruses, viroids, and prions.

Considering that small size is the only shared feature of all microbes, the group is tremendously diverse (**figure 1.6**). If you look at the macroscopic world around you—the plants and animals—you should be impressed by the assortment of what you see. That range of types, however, is dwarfed by the huge variety of microbes! The extent of the diversity of microbes makes sense considering that they have inhabited this planet for billions of years and have evolved to thrive in every conceivable environment—from the hydrothermal vents at the bottom of the ocean, to the icy tops of the highest mountains. Many people associate microbes with disease, but their contributions to our world go far beyond that. In fact, as section 1.2 described, we could not survive without them.

Living organisms are all composed of cells with one of two basic structures—**prokaryotic** (*pro* means "prior to" and *karyote* means "nucleus") and **eukaryotic** (*eu* means "true"). Prokaryotic cells do not have a membrane-bound nucleus. Instead, the genetic material is located in a region called the nucleoid. In contrast, the genetic material in eukaryotic cells is contained within a membrane-bound nucleus. Eukaryotic cells often have a variety of other membrane-bound organelles as well, and are typically much larger and more complex than prokaryotic cells. Organisms that consist of a prokaryotic cell are called **prokaryotes,** whereas those composed of one or more eukaryotic cells are called **eukaryotes.**

Prokaryotes fall into two very different groups—bacteria and archaea—as different from each other as they are from eukaryotes. Because of the fundamental differences between bacteria, archaea, and eukaryotes, all living organisms are now classified into three different **domains:** *Bacteria, Archaea,* and

TABLE 1.1 Characteristics of Members of the Three Domains

Characteristic	*Bacteria*	*Archaea*	*Eukarya*
Cell Type	Prokaryotic	Prokaryotic	Eukaryotic
Number of cells	Unicellular	Unicellular	Unicellular or multicellular
Membrane-bound organelles	No	No	Yes
Ribosomal RNA sequences unique to the group	Yes	Yes	Yes
Peptidoglycan in cell wall	Yes	No	No
Typical size range	0.3–2 μm	0.3–2 μm	5–50 μm

Eukarya (sometimes spelled *Eucarya*). Members of the *Bacteria* and *Archaea* are prokaryotes, whereas members of the *Eukarya* are eukaryotes. The names of the domains are italicized, with the first letter capitalized; the members of the domains are referred to as bacteria, archaea, and eukarya, respectively. **Table 1.1** compares some features of members of the three domains, but you will learn other important differences in later chapters.

The small size of microbes requires measurements not commonly used in everyday life (**figure 1.7**). Logarithms are extremely helpful in this regard, so you will find a brief discussion of them in Appendix I.

Scientific Names

When referring to microbes, we use their scientific names, which are written and pronounced in a Latin style. A pronunciation guide for these names is in Appendix II, and an audio version is available on your Connect class site. The scientific names are assigned according to the binomial (two-part name) system of nomenclature developed by Carl Linnaeus in the 1700s. The first part of the name indicates the **genus,** with the first letter always capitalized; the second part indicates the specific epithet, or **species,** and is not capitalized. Both are usually italicized or underlined—for example, *Escherichia*

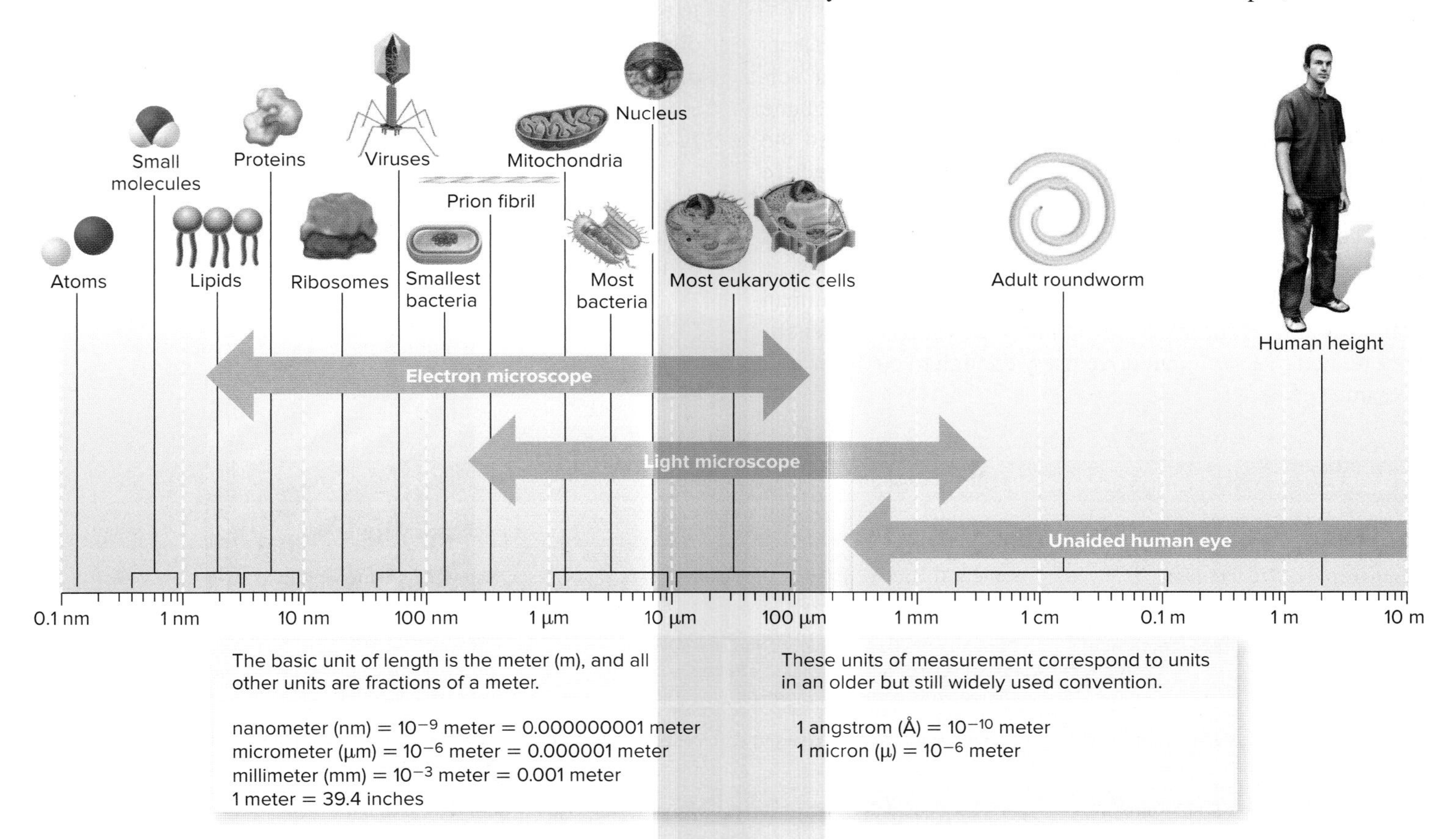

FIGURE 1.7 Sizes of Molecules, Non-Living Agents, and Organisms Note that the scale here is logarithmic (rather than linear), and each labeled increment increases by a factor of 10.

? **Why is a logarithmic scale useful when comparing sizes of members of the microbial world?**

FOCUS YOUR PERSPECTIVE 1.1

Every Rule Has an Exception

We might assume that because microorganisms have been so intensively studied over the past hundred years, no major surprises are left to be discovered. This, however, is far from the truth. In the mid-1990s, a large, peculiar-looking organism was found in the intestinal tracts of certain fish from both the Red Sea in the Middle East and the Great Barrier Reef in Australia. This organism, named *Epulopiscium fishelsoni* cannot be cultured in the laboratory. Its large size (600 μm long and 80 μm wide) makes it clearly visible without any magnification, and suggested that this organism was a eukaryote. However, it does not have a membrane-bound nucleus. A chemical analysis of the cell confirmed that it is a prokaryote and a member of the domain *Bacteria.* ⏭ ***Epulopiscium***

In 1999, a prokaryote even larger in volume was isolated from the muck of the ocean floor off the coast of Namibia in Africa. It is a spherical organism 70 times larger in volume than *E. fishelsoni.* Since it grows on sulfur compounds and contains glistening globules of sulfur, it was named *Thiomargarita namibiensis,* meaning "sulfur pearl of Namibia." ⏭ ***Thiomargarita***

In contrast to the examples of large bacteria, a unicellular alga found in the Mediterranean Sea is 1 μm in width. It is a eukaryote even though it is about the size of a typical bacterium.

How small can an organism be? A microbe discovered off the coast of Iceland is only about 400 nm (nanometers) in diameter and has one-tenth the amount of genetic information (DNA) as the common intestinal bacterium *Escherichia coli.* The tiny organism was found attached to a much larger microbe, a member of the *Archaea* growing in an ocean vent where the temperature was close to the boiling point of water. The larger organism is an *Ignicoccus* species (*igni* means "fire" and *coccus* means "sphere"). The tiny one, also a member of the *Archaea,* has been named *Nanoarchaeum equitans* (meaning "tiny archaea" and "rider"). *N. equitans* cannot be grown in the laboratory by itself, but *Ignicoccus* grows well without its *Nanoarchaeum* "rider." ⏭ **nanoarchaea**

All of these exceptions to long-standing rules point out the need to keep an open mind and not jump to conclusions! They also serve as excellent reminders that in a subject as complex as microbiology, there will almost always be exceptions!

coli. The part indicating the genus is commonly abbreviated, with the first letter capitalized—as in *E. coli.*

The origin of one or both parts of the name often reflects a characteristic of the organism or honors a particular scientist (**table 1.2**). In the case of *Escherichia coli,* the name of the genus honors Theodor Escherich, who discovered the bacterium; the species designation indicates the site where *E. coli* typically lives: the colon (large intestine). Within a given genus, there may be a number of different species. For example, the genus *Escherichia* includes other species as well, such as *E. vulneris,* which was first isolated from human wounds (*vulneris* means "of a wound"). *E. vulneris* is genetically related to *E. coli,* but not closely enough to consider it in the same species.

Members of the same species may vary from one another in minor ways, but not enough to separate the organisms into different species. In situations such as research where these differences are important, a particular microbe and its progeny may be indicated with a strain designation—for example, *E. coli* B or *E. coli* K12.

Groups of microbes are often referred to informally by names that resemble genus names but are not italicized. For instance, members of the genus *Staphylococcus* are often called staphylococci.

MicroByte

There are at least 10,000 times more bacterial species than mammalian species on Earth!

TABLE 1.2 Origin of Various Scientific Names

Name	Genus Derivation	Species Derivation
Escherichia coli **(bacterium)**	Honors Theodor Escherich, the scientist who discovered the bacterium.	Derived from the word "colon," the body site inhabited by the bacterium.
Haemophilus influenzae **(bacterium)**	Derived from *haemo* (blood) and *phil* (loving), reflecting that the bacterium requires certain components of blood for growth.	Derived from the word "influenza," the disease mistakenly thought to be caused by the bacterium; we now know that influenza is caused by a virus.
Saccharomyces cereviseae **(fungus)**	Derived from *saccharo* (sugar) and *myces* (fungus).	Derived from *cerevisia* (beer), reflecting that the fungus (a yeast) is used to make beer.
Shigella dysenteriae **(bacterium)**	Honors Kiyoshi Shiga, the scientist who discovered the bacterium.	Derived from the word "dysentery," the disease caused by the bacterium.
Staphylococcus aureus **(bacterium)**	Derived from *staphylo* (bunch of grapes) and *kokkus* (berry), reflecting the grouping and shape of the cells.	The term *aureus* (golden) indicates the common color of visible masses of the cells.

Bacteria

Bacteria (singular: bacterium) are single-celled prokaryotes. Most bacteria have specific shapes, commonly cylindrical (rod-shaped), spherical (round), or spiral. They typically have rigid cell walls that contain peptidoglycan, a compound unique to bacteria. Many of the bacteria can move using flagella (singular: flagellum), appendages that extend from the cell. ⏭ bacterial shapes, ⏭ peptidoglycan, ⏭ flagella

Bacteria typically multiply by binary fission, a process in which one cell enlarges and then divides. This forms two cells, each equivalent to the original. ⏭ binary fission

Many bacteria obtain energy from foods similar to those humans eat, but others can gain energy from seemingly unlikely sources such as hydrogen sulfide (a gas that smells like rotten eggs). Still others are photosynthetic, meaning they make cellular material using the radiant energy of sunlight.

Although most bacteria are beneficial, some cause serious diseases, and these will be a focus in the "disease chapters" of this textbook. Many of the early chapters will focus on bacteria in general, often with the aim of providing the necessary background for understanding infectious diseases.

Archaea

Like bacteria, archaea (singular: archaeon) are single-celled prokaryotes. They have similar shapes, sizes, and appearances to bacteria. In addition, they also multiply by binary fission, move primarily by means of flagella, and have rigid cell walls. Like bacteria, different groups of archaea use different energy sources; some are photosynthetic, harvesting the energy of sunlight to make cellular material. Considering how much archaea look like bacteria, scientists initially did not believe they could be so different from them. We now know, however, that there are major differences between the two groups, and the groups are only distantly related to each other. In fact, you are more closely related to plants than archaea are to bacteria!

Archaea differ from bacteria in several of their structural and functional components. For example, the archaeal cell wall does not contain peptidoglycan, whereas the bacterial wall does. Archaea also have characteristic nucleotide sequences in their ribosomal RNA (a molecule involved in protein synthesis) that differ significantly from those of bacteria. The discovery of the differences in ribosomal RNA sequences helped provide the basis for separating the two groups of prokaryotes into different domains. ⏭ ribosomal RNA

An interesting feature of many archaea is their ability to grow in extreme environments in which most other organisms cannot survive. Some, for example, can grow in salt concentrations 10 times higher than that of seawater. These organisms grow in such habitats as the Great Salt Lake and the Dead Sea. Others grow best at extremely high temperatures. One archaeon can grow at a temperature of 122°C! (100°C is the temperature at which water boils at sea level.)

TABLE 1.3 Eukaryotic Organisms Studied by Microbiologists

Organism	Characteristics
Fungi	Use organic material for energy. Size range from microscopic (yeasts) to macroscopic (molds and mushrooms).
Algae	Use sunlight for energy. Size range from microscopic (single-celled algae) to macroscopic (multicellular algae).
Protozoa	Use organic material for energy. Single-celled microscopic organisms.
Helminths	Use organic material for energy. Adult worms are typically macroscopic and often quite large, but their eggs and larval forms are microscopic.

Although the archaea that grow in extreme environments are the most intensively studied, many others are common in moderate environments. They are widely distributed in soils, the oceans, and marshes, as well as in the intestinal tracts of animals. Some archaea are part of microbial communities associated with severe periodontitis, a destructive inflammation of the gums. ⏭ periodontal disease

Eukarya

Eukarya are eukaryotes; those studied by microbiologists include fungi, algae, protozoa, and helminths (worms) (**table 1.3**). Algae and protozoa are also referred to as **protists.**

Fungi

Fungi (singular: fungus) are a diverse group of eukaryotes, ranging from single-celled yeasts that can reproduce by budding to multicellular filamentous molds (**figure 1.8**). The microscopic filaments of molds, called hyphae (singular: hypha), can branch as well as twist and turn to form a visible mat. When you see moldy foods, you are looking at the mat, sometimes along with structures that give rise to a reproductive form called conidia (also referred to as spores). The conidia easily become airborne, allowing the fungus to spread. Some fungi make macroscopic reproductive structures that we call mushrooms. ⏭ fungi

Fungi harvest energy from organic materials. To do this, they secrete enzymes that degrade the organic material, and then take in the nutrients that are released. Fungi are found in most places where organic materials, including dead plants and animals, are present.

Algae

Algae (singular: alga) are a diverse group of photosynthetic eukaryotes. Some are single-celled, whereas others are multicellular, such as seaweed (**figure 1.9**). All algae contain chloroplasts, which have chlorophyll, a green pigment. Some

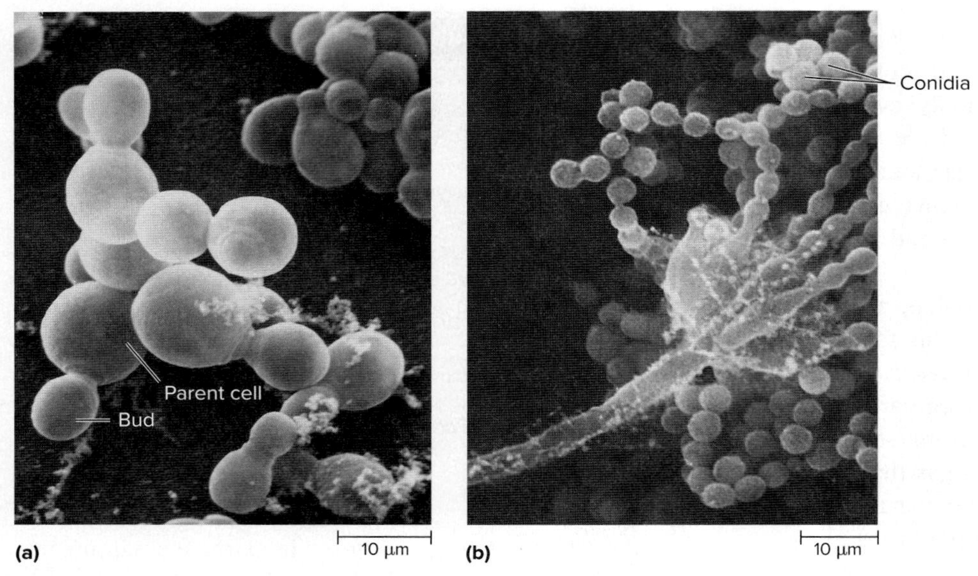

FIGURE 1.8 Fungi (a) Yeast, *Malassezia furfur.* **(b)** *Aspergillus,* a typical mold form whose reproductive structures rise above the mat of hyphae. a: Source: Janice Haney Carr/CDC; b: Source: Robert Simmons/CDC

What type of cells make up molds and yeasts?

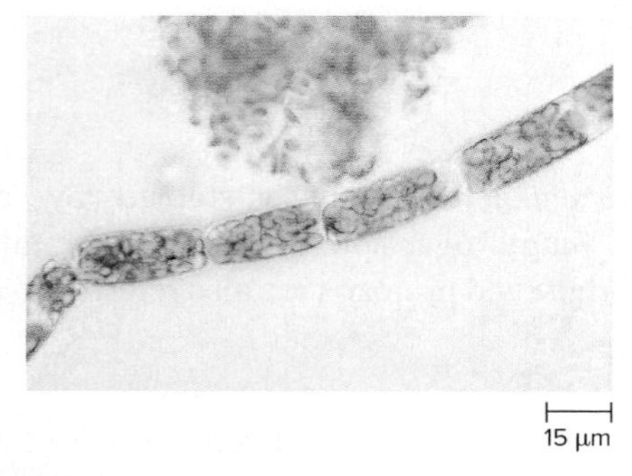

FIGURE 1.9 Alga *Ulothrix,* a filamentous green alga. ©McGraw-Hill Education/Lisa Burgess, photographer

What general features of algae distinguish them from other eukaryotic microorganisms?

also contain other pigments that give them characteristic colors. The pigments absorb the energy of light, which is used in photosynthesis. ▶▶| **algae**

Algae are usually found near the surface of either salt or fresh water or in moist terrestrial habitats. Their cell walls are rigid, but the chemical composition of the wall is quite distinct from that of bacteria and archaea. Many algae move by means of flagella, which are structurally far more complex and unrelated to flagella of prokaryotes.

Protozoa

Protozoa (singular: protozoan) are a diverse group of microscopic, single-celled eukaryotes that live in both aquatic and terrestrial environments. Although microscopic, they are very complex organisms and generally much larger than prokaryotes (**figure 1.10**). Unlike algae and fungi, protozoa do not have a rigid cell wall. Most protozoa are motile and ingest organic material as food sources. ▶▶| **protozoa**

Helminths

Parasitic **helminths** are worms that live at the expense of a host. They are an important cause of disease, particularly in developing countries. The adult worms are generally macroscopic, meaning they can be seen with the naked eye, and some of them are quite large, so technically they are not microorganisms. Microbiologists study them, however, because they cause disease and because diagnosis often involves identifying their eggs and larval forms, which are microscopic. Helminths include roundworms, tapeworms, and flukes. ▶▶| **helminths**

Acellular Infectious Agents

Viruses, viroids, and prions are acellular infectious agents, meaning that they are not composed of cells. They cannot reproduce independently and are considered non-living. By definition, an organism must be composed of one or more cells, so these acellular infectious agents are not microorganisms. Their distinguishing features are listed in **table 1.4.**

Viruses

Viruses consist of nucleic acid packaged within a protein coat (**figure 1.11**). To multiply, viruses infect living cells—referred to as **hosts**—and then use the machinery and nutrients of those

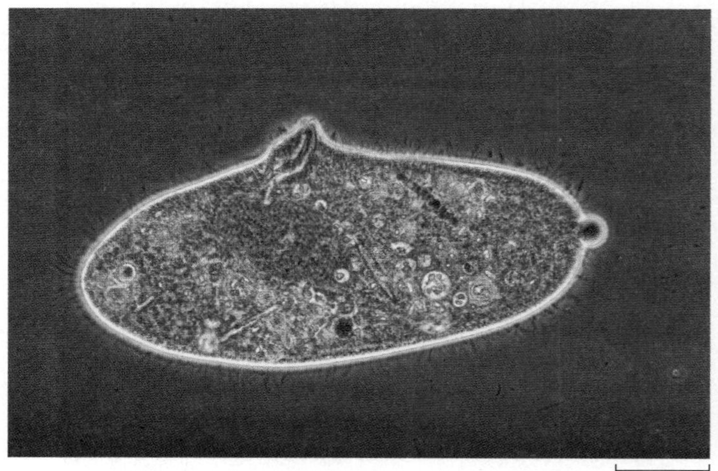

FIGURE 1.10 Protozoan A paramecium moves with the aid of hair-like appendages (called cilia) on the cell surface. ©Melba Photo Agency/Alamy Stock Photo

How do protozoa differ from both fungi and algae?

TABLE 1.4 Acellular Infectious Agents

Agent	Characteristic
Viruses	Consist of either DNA or RNA, surrounded by a protein coat. Obligate intracellular agents that use the machinery and nutrients of host cells to replicate.
Viroids	Consist only of RNA; no protein coat. Obligate intracellular agents that use the machinery and nutrients of host cells to replicate.
Prions	Consist only of protein; no DNA or RNA. Misfolded versions of normal cellular proteins that cause the normal versions to misfold.

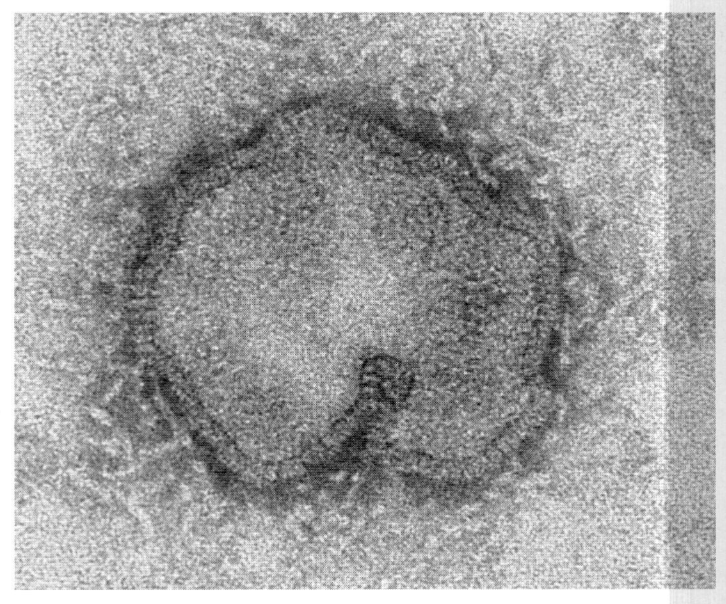

FIGURE 1.11 Virus Influenza virus, the cause of flu.
Source: Cynthia S. Goldsmith and Thomas Rowe/CDC

Why can viruses be so much smaller than cells and still replicate?

cells to replicate. Outside the hosts, however, viruses are inactive. Thus, viruses are obligate intracellular agents, meaning that they cannot replicate outside a host. ►► viruses

All forms of life, including bacteria, archaea, and eukarya, can be infected by viruses but of different types. Although viruses frequently kill the cells in which they replicate, some types can also remain within the host cell without causing obvious ill effects. As the host cells multiply, they copy the viral genetic information, passing it along to their progeny.

Viroids

Viroids are simpler than viruses, consisting of only a single, short piece of ribonucleic acid (RNA). Like viruses, they are obligate intracellular agents. Viroids cause a number of plant diseases, and some scientists speculate that they may cause diseases in humans, although no evidence for this yet exists. ►► viroids

Prions

Prions are infectious proteins that cause diseases called spongiform encephalopathies, a name that reflects the sponge-like appearance of the brain tissue (*encephalo* means "brain" and *patho* means "disease"). Perhaps the most widely recognized example is bovine spongiform encephalopathy (BSE), commonly called mad cow disease. Prions are simply misfolded versions of normal cellular proteins found in the brain. When the misfolded version comes into contact with the normal cellular protein, it causes the normal protein to also misfold. These misfolded versions bind together within the cell to form thread-like structures called fibrils (**figure 1.12**). The fibril-filled cells are not able to function and eventually die, forming spaces in the brain that lead to the characteristic sponge-like appearance. Prions are more resistant to degradation by cellular enzymes than are their normal counterparts. Prions are also resistant to the usual sterilization procedures that destroy viruses and bacteria. Some scientists speculate that Alzheimer's and Parkinson's diseases are caused by mechanisms similar to those of prion diseases. ►► prions

MicroAssessment 1.3

Microbes are given a two-part name that indicates the genus and species. Three domains of life exist: *Bacteria, Archaea,* and *Eukarya.* Members of the *Bacteria* and *Archaea* are prokaryotes, but the two groups have significant differences. Members of the *Eukarya* are eukaryotes; within this group, microbiologists study algae, fungi, protozoa, and parasitic helminths. Viruses, viroids, and prions are acellular infectious agents.

7. List two features that distinguish prokaryotes from eukaryotes.
8. Describe the chemical composition of viruses, viroids, and prions.
9. In the binomial system of classification, why is a two-part name used?

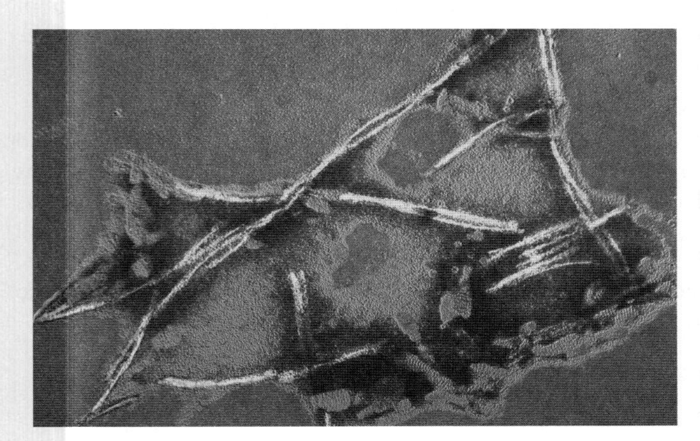

FIGURE 1.12 Prion Prion fibrils isolated from the brain of an infected cow. ©EM Unit, VLA/Science Source

How are prions different from the normal versions of the related proteins made by cells?

FOCUS ON THE FUTURE 1.1

Meet the Microbiomes!

As you study this textbook, you will probably be amazed by how much we know about the microbial world. You have already read, for example, that the human microbiome affects our well-being, and that life on this planet could not exist without microbes. But the more you learn, the more you will realize how little we actually know! Although scientists have studied microorganisms for hundreds of years, most of the advances occurred after the start of the Golden Age of Microbiology. Most studies focused on microorganisms that could be grown in the laboratory, and we now know that those examples represent less than 1% of all microbes. Complicating the matter even more is the fact that a microorganism's behavior in the laboratory can be quite different from that in a natural situation. So, yes, we know a great deal, but it really represents only the tip of the iceberg.

The depth of our understanding about microbial communities is rapidly increasing due to what could be considered the Golden Age of Microbiomes. As was already discussed, the Human Microbiome Project led to greater insight into the role of microorganisms in health and disease. It also opened up many new areas for research. For example, how many disease states are due to imbalances in our normal microbiota? Can we treat any of those by packaging certain microbes as an oral pill? Can we track the microbial profiles of an individual to predict changes in health? The National Microbiome Initiative now promises to provide additional insights, but will also lead to many new questions.

Summary

1.1 ■ The Dispute over Spontaneous Generation

The belief in **spontaneous generation** was challenged by Francesco Redi in the seventeenth century.

Early Experiments

The experiments of John Needham supported the idea of spontaneous generation while those of Lazzaro Spallazani did not.

Experiments of Pasteur

The experiments of Louis Pasteur disproved spontaneous generation and supported what is now known as the theory of **biogenesis** (figure 1.2).

Experiments of Tyndall

John Tyndall showed that some microbial forms are not killed by boiling. He and Ferdinand Cohn discovered **endospores,** the heat-resistant forms of some bacteria.

The Golden Age of Microbiology

The field of microbiology blossomed after Pasteur and others disproved spontaneous generation, leading to the Golden Age of Microbiology. Discoveries during this time led to the acceptance of the Germ Theory of Disease.

The Scientific Method

The **scientific method** includes (1) observing an occurrence and asking a question about that situation; (2) developing a **hypothesis** that explains the occurrence and devising an experiment that tests the hypothesis; (3) doing the experiment, collecting the data, and drawing conclusions; and (4) communicating the results, methods, and conclusions. A scientific **theory** is an explanation supported by a vast body of experimental evidence.

1.2 ■ Microbiology: A Human Perspective

The Human Microbiome

The **normal microbiota** is essential to human health. The **microbiome** is an interacting community of microorganisms as well as their genetic information.

Microorganisms in the Environment

Microorganisms replenish the O_2 that humans and other animals require to breathe and they convert the nitrogen gas in the air into a form that other organisms can use.

Commercial Benefits of Microorganisms

Microorganisms are used in the production of bread, wine, beer, and cheeses. Bacteria are used to degrade toxic pollutants as well as to synthesize a variety of different useful products. **Biotechnology** depends on members of the microbial world.

Microbes as Research Tools

Microorganisms are wonderful model organisms to study because they have the same fundamental metabolic and genetic properties as higher life forms. Experimental results can be obtained quickly because bacteria grow rapidly on simple, inexpensive growth media.

Microbes and Disease

Pathogens cause disease, but the death rate from infectious diseases has declined over the past 100 years or so as a result of disease-prevention efforts including vaccination (figure 1.4). More needs to be done to prevent **emerging infectious diseases,** some of which are new or newly recognized (figure 1.5). Some chronic diseases are caused by microorganisms.

1.3 ■ Members of the Microbial World

Considering that small size is the only shared feature of all microbes, the group is tremendously diverse (figure 1.6). All living organisms are classified into three **domains:** *Bacteria, Archaea,* and *Eukarya* (table 1.1). The small size of microbes requires measurements not commonly used in everyday life (figure 1.7).

Scientific Names

The first part of a scientific name indicates the **genus,** and the second part the **species;** these are written in italics or underlined (table 1.2). Members of the same species can vary, so strain designations are sometimes used.

Bacteria

Bacteria are single-celled **prokaryotes** that have peptidoglycan in their cell wall.

Archaea

Archaea are single-celled prokaryotes. Although they look like bacteria, there are significant differences. They do not contain peptidoglycan. Many archaea grow in extreme environments.

Eukarya

Eukarya are **eukaryotes** (table 1.3). **Fungi** include single-celled yeasts and multicellular molds and mushrooms; they use organic compounds as food (figure 1.8). **Algae** can be single-celled or multicellular, and use sunlight as an energy source (figure 1.9). **Protozoa** are typically motile single-celled organisms that use organic compounds as food (figure 1.10). Parasitic **helminths** are worms that live at the expense of a host.

Acellular Infectious Agents

The non-living members of the microbial world are not composed of cells (table 1.4). **Viruses** consist of nucleic acid within a protein coat (figure 1.11). **Viroids** consist of a single, short RNA molecule. **Prions** consist only of protein; apparently, they are misfolded versions of normal cellular protein (figure 1.12).

Review Questions

Short Answer

1. How did Louis Pasteur help disprove spontaneous generation?
2. Describe the scientific method.
3. Explain why life could not exist without the activities of microorganisms.
4. How is the normal microbiota important to human health?
5. List four commercially important benefits of microorganisms.
6. What characteristics of microorganisms make them important research tools?
7. List three factors that contribute to the emergence of infectious diseases.
8. In the designation *Escherichia coli* B, which part of the name indicates the genus? Which part indicates the species? Which part indicates the strain?
9. Why are viruses not microorganisms?
10. Name three non-living groups in the microbial world and describe their major properties.

Multiple Choice

1. The property of endospores that led to confusion in the experiments on spontaneous generation is their
   a) small size.
   b) ability to pass through cork stoppers.
   c) heat resistance.
   d) presence in all infusions.
   e) presence on cotton plugs.
2. The Golden Age of Microbiology was the time when
   a) microorganisms were first used to make bread.
   b) microorganisms were first used to make cheese.
   c) most pathogenic bacteria were identified.
   d) a vaccine against influenza was developed.
   e) antibiotics became available.
3. If all prokaryotes were eliminated from the planet,
   a) animals would thrive because there would be no disease.
   b) archaea would thrive because there would be no competition for nutrients.
   c) all animals would die.
   d) animals and archaea would thrive.
4. All of the following are emerging infectious diseases except
   a) smallpox.
   b) hepatitis C.
   c) Lyme disease.
   d) hantavirus pulmonary syndrome.
   e) mad cow disease.
5. All of the following are biological domains except
   a) *Bacteria.*
   b) *Archaea.*
   c) *Prokaryota.*
   d) *Eukarya.*
6. Which name is written correctly?
   a) *staphylococcus aureus*
   b) *escherichia Coli*
   c) *Staphylococcus epidermidis*
   d) <u>bacillus</u> <u>Anthracis</u>
   e) <u>Clostridium</u> <u>Botulinum</u>
7. Members of which pairing are most similar in appearance to each other?
   a) fungi and algae
   b) algae and archaea
   c) archaea and bacteria
   d) bacteria and viruses
   e) viruses and algae
8. If you wanted to increase your chances of obtaining a member of the *Archaea* (rather than a member of another domain), which would be the best site to obtain a sample?
   a) intestine of an elephant
   b) skin of an elephant
   c) a 95°C hot spring in Yellowstone
   d) a 45°C hot spring in Hawaii
   e) a raw hamburger patty
9. Viruses
   1. contain both protein and nucleic acid.
   2. infect all domains of life.
   3. can grow in the absence of living cells.
   4. are generally the same size as prokaryotes.

5. always kill the cells they infect.
    a) 1, 2
    b) 2, 3
    c) 3, 4
    d) 4, 5
    e) 1, 5

10. Antony van Leeuwenhoek could not have observed
    a) roundworms.
    b) *Escherichia coli.*
    c) yeasts.
    d) viruses.

Applications

1. The American Society for Microbiology is preparing a "Microbe-Free" banquet to emphasize the importance of microorganisms in food production. What foods could not be on the menu?
2. If you were asked to nominate one of the individuals mentioned in this chapter for the Nobel Prize, who would it be? Make a statement supporting your choice.

Critical Thinking

1. A microbiologist obtained two pure biological samples: one of a virus, and the other of a viroid. Unfortunately, the labels had been lost. The microbiologist felt she could distinguish the two by analyzing for the presence or absence of a single molecule. What molecule would she search for and why?
2. Why would archaea that grow in extreme environments be more intensively studied than those that do not?

2 The Molecules of Life

Space-filling models of water molecules. ©*McGraw-Hill Education/Lisa Burgess, photographer*

KEY TERMS

Atom The basic unit of matter.

Atomic Number The number of protons in the nucleus of an atom.

Buffer A chemical that stabilizes the pH of a solution.

Carbohydrate An organic compound composed of one or more simple sugars.

Covalent Bond A chemical bond formed when two atoms share electrons.

Hydrogen Bond The attraction between a hydrogen atom in a polar molecule and an electronegative atom in the same or another polar molecule.

Ion An atom or molecule that has gained or lost one or more electrons.

Ionic Bond A chemical bond resulting from the attraction between positively and negatively charged ions.

Isotope Form of the same chemical element that differs in its number of neutrons.

Lipid An organic molecule that is not soluble in water.

Molecule Two or more atoms held together by covalent bonds.

Nucleic Acid A macromolecule consisting of one or two nucleotide chains; DNA or RNA.

Organic Compound A compound that has a carbon atom covalently bonded to a hydrogen atom.

pH A measure of the hydrogen ion concentration or acidity of a solution on a scale of 0 to 14.

Protein A macromolecule consisting of one or more chains of amino acids.

A Glimpse of History

Farmers have understood for centuries that growing the same crop on the same piece of land year after year reduces the crop yield. Allowing a field to lie unplanted for one or more seasons lets wild plants grow, and these appear to improve the soil. During the Golden Age of Microbiology, Martinus Beijerinck, a microbiologist from the Netherlands, helped to explain the science behind what the farmers already knew. ◀◀ Golden Age of Microbiology

Beijerinck was described as a "keen observer" who was able "to fuse results of remarkable observations with a profound and extensive knowledge of biology and the underlying sciences." Fortunately for farmers, he disliked medical microbiology, preferring to study the agricultural applications of microbiology instead. Based on the work of other scientists, Beijerinck knew that certain plants improve the soil because microorganisms help them accumulate nitrogen from the air. All forms of life require nitrogen because it is an essential component of cellular material, but soils have only limited amounts. Almost 80% of Earth's atmosphere is nitrogen gas (N_2), so using nitrogen from the air might seem like an easy task, but it is far from it. N_2 is very stable, with strong chemical bonds holding the two nitrogen atoms together. Plants, animals, and most other organisms lack the ability to break those bonds, so it was unclear how microorganisms were helping plants accumulate nitrogen from the air.

Beijerinck and other scientists studied nitrogen accumulation in a group of plants called legumes. Members of this group bear seeds in pods and include peas, beans, and clover. Legume roots often have nodules (small growths that look like tumors). It had been reported that the root nodules contained microorganisms, and that these allowed the plant to accumulate nitrogen. Beijerinck made a significant breakthrough in the late 1880s by isolating a bacterium from inside a root nodule, and then showing that it was able to convert atmospheric nitrogen into a form that could be incorporated into cellular material. This process is called nitrogen fixation, and the nitrogen is said to be "fixed." Then he showed that root nodules form when the nitrogen-fixing bacterium—a member of a group now called rhizobia—is incubated with seedlings of legumes. As those plants, or even parts of the plants, die and decay, the cellular material is released, enriching the soil with nitrogen-containing nutrients.

When farmers plant crops like soybeans or allow wild clover to grow in a field, rhizobia in the nodules fix nitrogen, which can then be used by the host plant. We now know that there are many types of rhizobia, and they are specific with respect to the type of host plant with which they form a relationship. To honor Beijerinck for his achievements, a genus of nitrogen-fixing bacteria (*Beijerinckia*) is named for him.

Simply stated, chemistry is the study of matter, or the "stuff" of which the universe is composed. As you learn about microbiology, you will find that cells are masters at converting one set of chemicals to another and, in doing so, producing materials required to make new cells. This amazing ability to transform material is just one reason that microorganisms are crucial to life and that the principles described in this chapter are fundamental to information throughout the text.

2.1 ■ Elements and Atoms

Learning Outcomes

1. Describe the general structure of an atom and its isotopes.
2. Describe the importance of valence electrons.

Matter is categorized into **elements,** substances that have unique chemical properties and cannot be broken down by ordinary chemical means. There are 92 naturally occurring elements, but living organisms primarily consist of only six (**table 2.1**). As a way to remember these, think of the acronym CHONPS: carbon (the chemical symbol is C), hydrogen (H), oxygen (O), nitrogen (N), phosphorus (P), and sulfur (S). Other elements are also important in living systems, including some referred to as trace elements, reflecting the fact that they are found in very small quantities. The symbol of an element may represent more than one letter of the name, as with Ca, which represents calcium. Not all chemical symbols are derived from English, as seen with the symbol Na, which stands for *natrium* in Latin, but which we know as sodium.

Atomic Structure

An **atom** is the basic unit of all matter, and each element is composed of only one type of atom. Atoms consist of three major components:

- **Protons:** positively charged particles
- **Neutrons:** uncharged particles
- **Electrons:** negatively charged particles

Protons and neutrons together form the atomic nucleus (dense, central region of the atom), around which electrons move in a "cloud" (**figure 2.1**). Overall, the atom has no charge because the number of positive protons in the atomic nucleus is the same as the number of negative electrons in the cloud.

Each type of atom is distinguished by the number of protons in its nucleus, referred to as its **atomic number** (see table 2.1). For example, a hydrogen atom has 1 proton so its atomic number is 1; a carbon atom has 6 protons so its atomic number is 6. Each atom also has a **mass number,** the sum of the number of protons and neutrons in the nucleus of that atom. A hydrogen atom with 1 proton and no neutrons has an atomic number of 1 as well as a mass number of 1. A carbon atom with 6 protons and 6 neutrons has an atomic number of 6 and a mass number of 12. The symbols for atoms are sometimes written with the atomic number in subscript to the left, and the mass number in superscript to the left (**figure 2.2**). Note that electrons are too light to contribute to the mass of an atom.

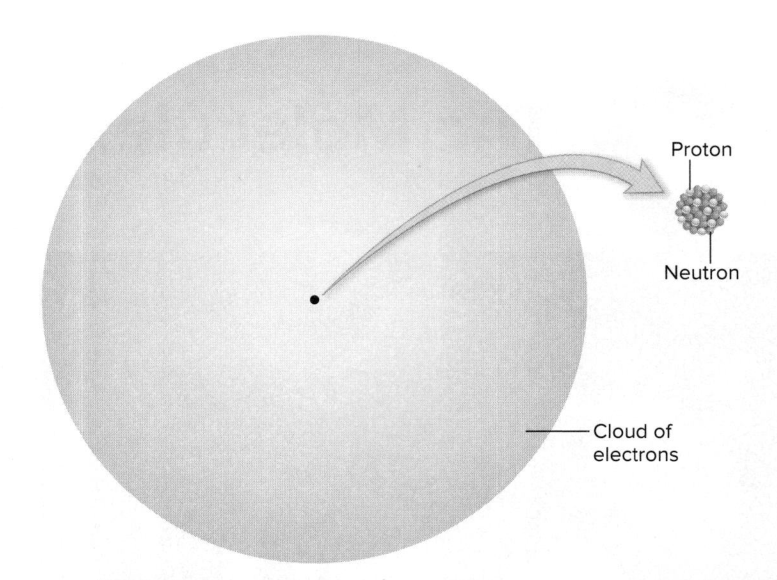

FIGURE 2.1 Atomic Structure A proton has a positive charge, a neutron has a neutral charge, and an electron has a negative charge. The electrons move around the nucleus as a cloud, arranged in shells of different energy levels.

? **How does the number of electrons in an atom compare with the number of protons?**

MicroByte

If the nucleus of an atom were a marble in the center of a football field, the electrons would occupy the entire space of the stadium.

Isotopes

All atoms of a given element have the same number of protons, but they can have different numbers of neutrons. That is, atoms of an element all have the same atomic number, but

TABLE 2.1 Characteristics of Atoms Common in Living Organisms

Atom	Symbol	Atomic Number (Number of Protons)	Mass Number (Protons + Neutrons)	Number of Covalent Bonds Formed
Hydrogen	H	1	1	1
Carbon	C	6	12	4
Nitrogen	N	7	14	3
Oxygen	O	8	16	2
Phosphorus	P	15	31	3
Sulfur	S	16	32	2

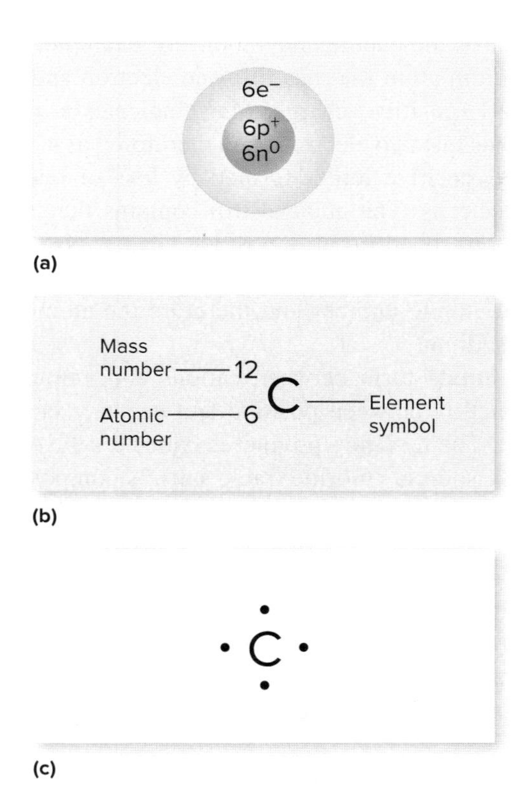

FIGURE 2.2 Depictions of a Carbon (C) Atom **(a)** The number of protons (p^+) and neutrons (n^0) in the nucleus are indicated, along with the number of electrons (e^-) in the surrounding cloud. **(b)** The mass number and atomic number are indicated on the left of the chemical symbol. **(c)** The Lewis structure highlights the number of valence electrons.

? Based on the Lewis structure of carbon, is the atom likely to enter into chemical reactions?

they can have different mass numbers. The various forms of atoms of an element are **isotopes** (*iso* means "same" and *tope* means "place"). For example, nearly 99% of naturally occurring carbon atoms have 6 neutrons, but some have 7 or 8. The fact that the atoms of an element can have different mass numbers is reflected in the term **atomic mass.** This is the average of the mass numbers of the atoms of the element, weighted according to the relative abundance of the naturally occurring isotopes. The atomic mass of carbon is 12.01.

Different isotopes behave chemically much as their more common counterparts, but their structural differences can often be detected. Because of this, researchers and clinicians use isotopes to monitor the fate of specific atoms within a population of cells. Some isotopes are unstable and emit radiation that can be detected; others are tracked using special instruments that measure differences in mass. The energy emitted from radioactive isotopes (radioisotopes) is sometimes useful in medical diagnosis. For example, to evaluate proper functioning of the human thyroid gland—which produces the iodine-containing hormone thyroxine—doctors often administer radioactive iodine. Later scans of the gland determine if the amount and distribution of iodine is normal (**figure 2.3**).

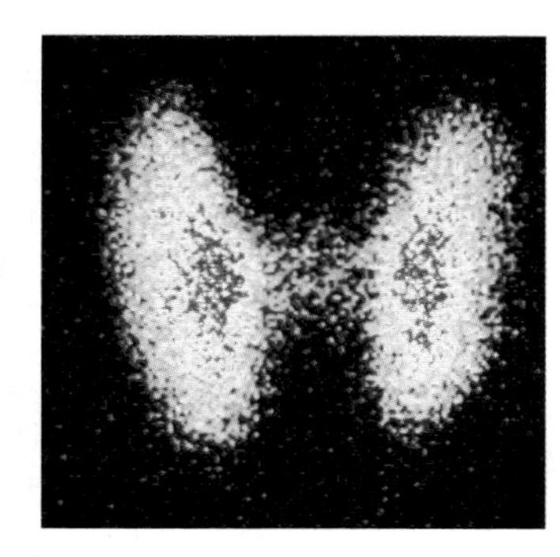

FIGURE 2.3 Radioisotopes A scan of the thyroid gland 24 hours after the patient received radioactive iodine.

? Why is radioactive iodine concentrated in the thyroid gland?

The Role of Electrons

Electrons are located in a cloud around an atom's nucleus, but at any point in time, it is impossible to determine precisely where they are (see figure 2.1). They are most likely to be in specific regions called shells, each of which is associated with a different energy level. The number of electrons that each shell can hold is limited. Electrons are attracted to the protons in the nucleus, so shells closer to the nucleus are generally filled before electrons occupy other shells. The shell closest to the nucleus holds no more than 2 electrons. These electrons are most highly attracted to the nucleus and have the lowest energy levels in that particular atom. Once that shell is filled, additional electrons occupy the next shell, which holds 8 electrons. Larger atoms have additional shells that hold even more electrons, but most atoms of biological significance follow the "octet rule," meaning they are most stable when their outer shell contains 8 electrons. An important exception is hydrogen, with a single shell; recall that the first shell has a limit of 2 electrons.

Electrons largely determine the chemical reactivity of an atom. The electrons in an atom's outer shell, called **valence electrons,** are the most important in that regard. Atoms with the maximum number of electrons in the outer shell are very stable. Atoms with an unfilled outer shell tend to react with other atoms that have unfilled outer shells. A depiction called a Lewis structure highlights the number of valence electrons in an atom. In figure 2.2c, each of the valence electrons is illustrated as a dot next to the chemical symbol.

MicroAssessment 2.1

The six most important elements in biology are carbon, hydrogen, oxygen, nitrogen, phosphorus, and sulfur. The basic unit of all matter, the atom, is composed of protons, electrons, and neutrons. The atomic number of an atom refers to the number of protons in its nucleus. The mass number of an atom refers to the number of protons plus neutrons in its nucleus. Isotopes have different numbers of neutrons. The reactivity of an atom is largely determined by the number of valence electrons in its

continued

outer shell. Atoms with an unfilled outer shell interact with other atoms in chemical reactions.

1. Why are electrons not considered in determining the mass number of an element?
2. What is the "octet rule" and its biologically important exception?
3. What do atoms of ^{14}C and ^{14}N have in common? How do they differ?

2.2 ■ Chemical Bonds and Reactions

Learning Outcomes

3. Compare and contrast ionic bonds, covalent bonds, and hydrogen bonds.
4. Explain the role of an enzyme in chemical reactions.

Atoms with an unfilled outer shell react with each other to lose, gain, or share their valence electrons to achieve a more stable state. This is the basis for chemical bond formation.

Ions and Ionic Bonds

An atom that gains or loses an electron is no longer neutral—it is an **ion** (**figure 2.4**). Atoms that gain an electron become negatively charged and are called **anions;** those that lose an electron become positively charged and are called **cations.** The type and amount of charge are indicated by a superscript to the right of the chemical symbol. For example, Na^+ indicates a sodium atom that has lost one electron and therefore carries a +1 (positive) charge; Mg^{2+} indicates a magnesium atom that has lost two electrons and therefore has a +2 charge. Note that a positive ion is formed by loss of one or more valence electrons. The nucleus still contains the same number of protons, resulting in a positive charge. Na^+ cannot be formed by gaining a proton, because gaining a proton would change the atomic number and therefore the atom would no longer be sodium.

Ionic bonds form between cations and anions because of the attraction between positive and negative charges (see figure 2.4). The resulting product is called a salt. A common type of salt, sodium chloride (table salt), is composed of Na^+ (sodium cations) and Cl^- (chloride anions) and forms a solid crystal. The structure is highly ordered because the electrical attraction between positive and negative charges brings the ions together, but the like charges repel one another and are positioned as far apart as possible. Crystals continue to grow as new ions are added. Salts such as sodium chloride dissolve in water and are called **electrolytes,** meaning that they conduct electricity.

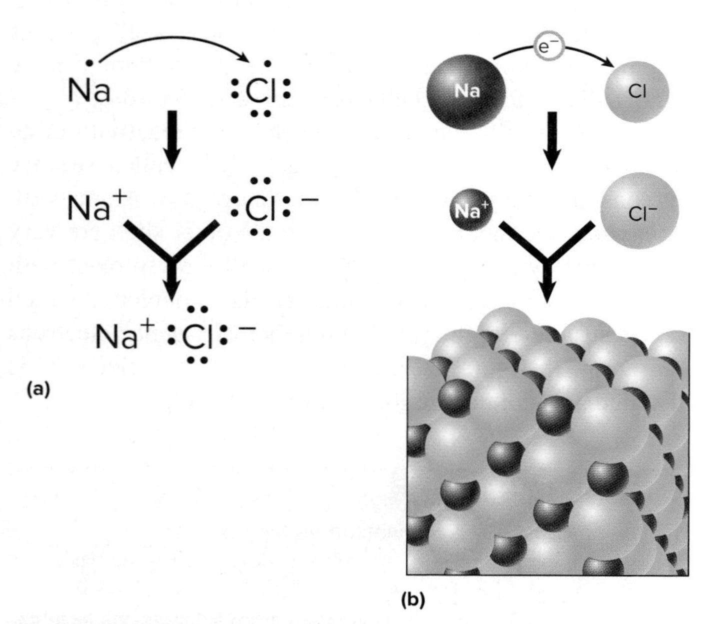

FIGURE 2.4 Ions and Ionic Bonds (a) Lewis model of sodium and chloride ions being formed, and an ionic bond between them. **(b)** Space-filling model of a salt crystal being formed by ionic bonding. Note that a cation is smaller than its neutral atom while an anion is larger.

? Which of the ions in this figure is an anion, and which is a cation?

MicroByte

Electrical charges from the heart are conducted by electrolytes and can be detected on the body surface as an electrocardiogram (ECG).

Covalent Bonds

Atoms do not always fill their valence shells by gaining or losing electrons. They may instead share pairs of valence electrons, forming **covalent bonds.** For example, a hydrogen atom (H) has 1 electron and requires 1 additional electron to fill its valence shell. If two H atoms each share their single electrons, both atoms gain stability. The single covalent bond between the two atoms is indicated by a dash, as H—H. Some atoms share more than one pair of electrons with each other, forming a double or triple covalent bond, indicated by a corresponding number of lines between the atoms. For example, an oxygen atom needs 2 electrons to fill its outer shell. If two oxygen atoms share 2 electrons with each other, then a double covalent bond is formed, represented as O=O.

Two or more atoms joined together by covalent bonds form a **molecule.** A molecule is represented by a molecular formula that indicates how many atoms of each type are present. For example, a hydrogen molecule is represented by the formula H_2. If atoms that make up a molecule are different elements, the term **compound** may be used. (Salts such as NaCl are called ionic compounds.) As an example, water is a compound that contains two hydrogen atoms and one oxygen atom; it is represented by the formula H_2O. The molecular mass of a molecule is based on the mass numbers of the component atoms. The molecular mass of most water molecules is 1 + 1 + 16, or 18.

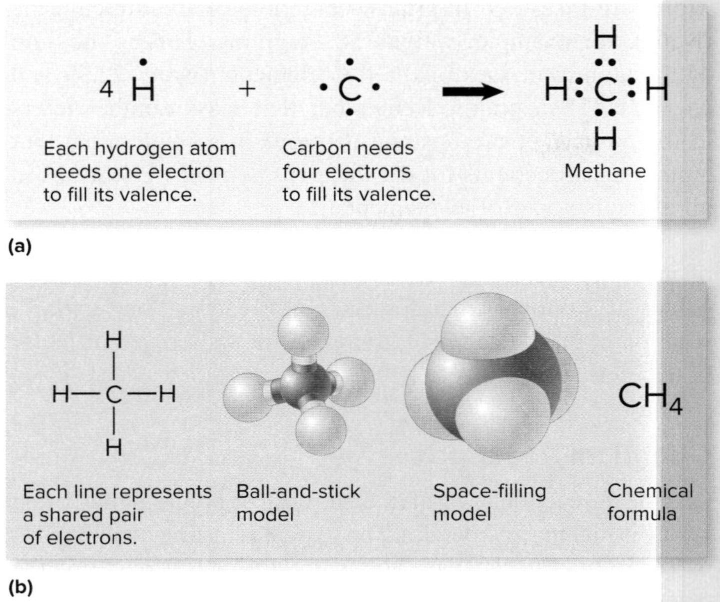

4 H + ·C· → H:C:H (with H above and below)

Each hydrogen atom needs one electron to fill its valence.

Carbon needs four electrons to fill its valence.

Methane

(a)

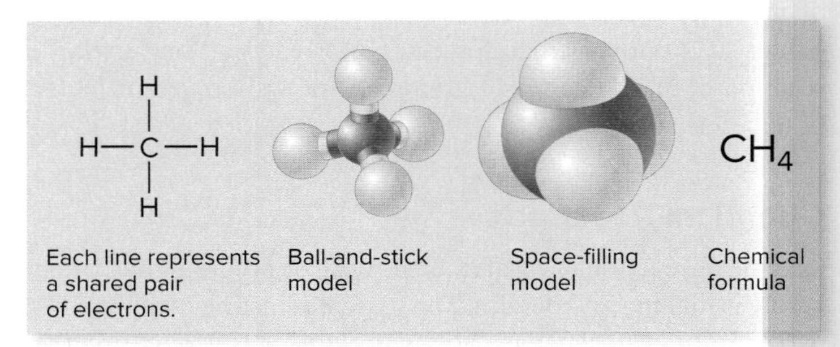

(b)

FIGURE 2.5 Covalent Bonds Covalent bonds are formed when atoms share electrons. **(a)** Methane is formed when a carbon atom fills its outer electron shell by sharing eight electrons—four belong to H atoms and four belong to the carbon atom. **(b)** Different ways of showing the methane molecule.

? **Is methane an organic molecule? Explain.**

Carbon (C) is a particularly important element in biological systems because its bonding properties provide the basis for many diverse structures. A carbon atom has four electrons in its outer shell, so it needs four more to fill it. Because of this, carbon forms four covalent bonds, typically with the main elements that make up cells: CHONPS (carbon, hydrogen, oxygen, nitrogen, phosphate, and sulfur). One C atom sharing electrons with four H atoms is methane (CH_4) (**figure 2.5**). Molecules that contain at least carbon and hydrogen are **organic compounds;** those that do not are inorganic compounds.

Electrons in a covalent bond may or may not be equally shared between the two atoms. A **non-polar covalent bond** forms when electrons are shared equally, such as when identical atoms share electrons (**table 2.2**). The same can occur between different atoms, if both have a similar attraction for electrons. Atoms with a greater attraction for electrons than others are more electronegative (**table 2.3**). If one atom in a covalent bond is significantly more electronegative than the other, then the electrons are shared unequally, resulting in a **polar covalent bond.**

The slight separation of charge resulting from a polar covalent bond is indicated by the Greek symbol delta (δ); the atom with a slight positive charge is δ^+ and the atom with the slight negative charge is δ^-. Consider the O—H bonds in water; the oxygen atom is more electronegative than the hydrogen atom (**figure 2.6**). Consequently, the oxygen atom pulls the electrons toward it, giving it a slight negative charge

TABLE 2.2 Non-Polar and Polar Covalent Bonds

Type of Covalent Bond	Atoms Involved and Charge Distribution	
Non-polar	C—C C—H H—H	C and H have similar attractions for electrons, so there is a nearly equal charge on each atom.
Polar	O—H N—H O—C N—C	The O and N atoms have a stronger attraction for electrons than do C and H, so the O and N have a slight negative charge; the C and H have a slight positive charge.

TABLE 2.3 Relative Electronegativities of Some Atoms Common in Biology*

Element	Relative Electronegativity
Hydrogen	2.1
Carbon	2.5
Nitrogen	3.0
Oxygen	3.5

*Electronegativity is the ability of an atom to attract electrons to itself in a molecule; an atom with a significantly higher electronegativity is better able to attract electrons.

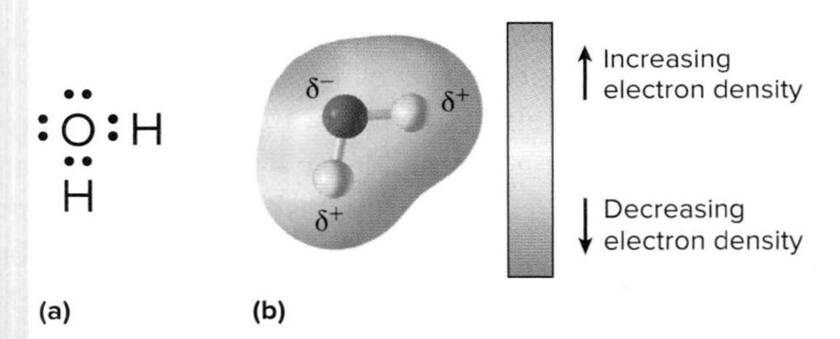

FIGURE 2.6 Polar Covalent Bonds Electrons move closer to the more electronegative atom in a compound, creating a polar molecule. **(a)** Lewis diagram of a water molecule. **(b)** Electron density model of a water molecule. The symbol δ indicates a partial charge.

? **Why is the oxygen atom in a water molecule more electron-rich than the hydrogen atoms?**

(δ^-) and leaving each of the two hydrogen atoms with a slight positive charge (δ^+). Polar covalent bonds play a key role in biological systems because they often result in formation of hydrogen bonds, discussed next.

Hydrogen Bonds

Hydrogen bonds are weak bonds formed when a hydrogen atom in a polar molecule is attracted to an electronegative atom in the same or another polar molecule (**figure 2.7;** see also table 2.2). Compounds that contain electronegative atoms, such as oxygen (O) or nitrogen (N), are common in biological systems, creating the possibility for many hydrogen bonds.

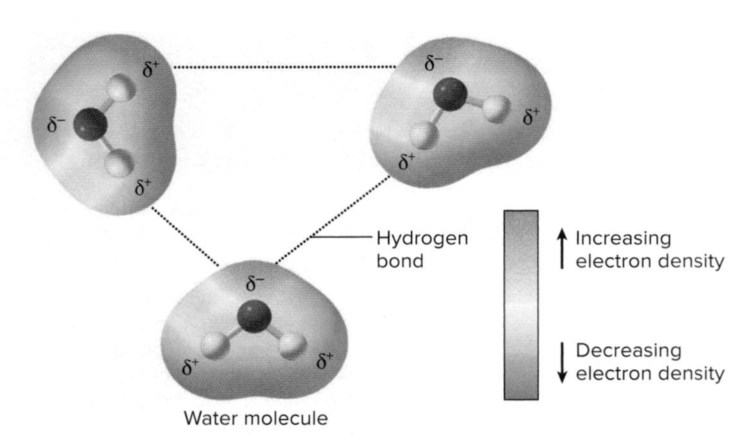

FIGURE 2.7 Hydrogen Bond Formation Hydrogen bonds form between water molecules because the electron-rich oxygen atom attracts electron-poor hydrogen atoms.

? **Explain why two identical atoms joined by a covalent bond cannot form a hydrogen bond.**

A hydrogen bond often exists for only a fraction of a second, and chemical assistance is not needed to form or break it. Hydrogen bonds between water molecules are constantly formed and broken at room temperature because the energy produced by the movement of water molecules is enough to break the bonds. Nevertheless, the abundance of the bonds gives water its unique properties. For example, the surface tension created by attraction between surface water molecules allows an insect to "walk" on a pond. Water molecules are attracted to other charged surfaces as well. This helps water travel through tiny vessels to the tops of trees. The world would look much different without the adhesive properties of water.

Although a single hydrogen bond is weak, a large number of them can hold molecules or parts of molecules together firmly. Consider the hook and loop fasteners of Velcro. A single hook-and-loop attachment does not provide much strength, but many such attachments result in a strong connection. Likewise, the two strands of a double-stranded DNA molecule are held together by many hydrogen bonds along the length of the molecule, forming a very stable interaction. However, the two strands will quickly come apart if enough energy is supplied, usually in the form of heat approaching temperatures of 100°C.

Molarity

A *mole* is a quantitative term used by chemists, much like a *dozen* is a quantitative term used by bakers. A baker may have a dozen cookies or a dozen bagels. A chemist may have a mole of glucose molecules or a mole of Na^+ ions. One **mole** is 6.022×10^{23} particles. That number is not important from a practical standpoint, but the concept is essential in chemistry—a mole of one substance has the same number of particles as a mole of any other. Scientists measure chemicals in much larger amounts than a single molecule or ion. Moles are measured in grams. One mole of sodium chloride (NaCl), for example, weighs 58.44 grams. This is the sum of the atomic masses of the two elements (Na = 22.99 and Cl = 35.45) in grams. Remember that mass number refers to the number of protons and neutrons in a single atom, but atomic mass accounts for the different number of neutrons in all of the isotopes of an element.

The **molarity** (M) of a solution is defined as the number of moles of a compound dissolved in enough water to make 1 liter of solution. Therefore, a 1 M (read as "one molar") solution of NaCl has 58.44 grams of NaCl dissolved in 1 liter of aqueous solution.

Chemical Reactions

Chemical reactions transfer electrons, a process that often involves making and breaking bonds. The starting components, or **reactants,** are changed to **products** in the course of the reaction. Depending on the nature of the reactants and products, chemical reactions can be described as synthesis reactions or decomposition reactions. Synthesis reactions combine multiple reactants to make a complex product; decomposition reactions separate a complex reactant into multiple products:

Synthesis Reaction

$$\underset{\text{(reactants)}}{A + B} \longrightarrow \underset{\text{(product)}}{AB}$$

Decomposition Reaction

$$\underset{\text{(reactant)}}{AB} \longrightarrow \underset{\text{(products)}}{A + B}$$

Covalent bonds are strong and generally do not break unless exposed to certain chemicals or large amounts of energy, generally in the form of heat. The temperatures required to break these bonds are incompatible with life, so cells use biological catalysts called **enzymes** that help break covalent bonds at lower temperatures. Enzymes may also position reactants so that they can more easily form covalent bonds. Without enzymes, reactions would proceed too slowly to maintain life. ⏭ enzymes

MicroAssessment 2.2

Ionic bonds form between positively and negatively charged ions. Covalent bonds result from sharing electrons. Hydrogen bonds form between polar molecules or portions of molecules. Chemical reactions proceed from reactants to products. Enzymes speed up chemical reactions.

4. Compare the properties of covalent, hydrogen, and ionic bonds.
5. Which type of bond requires an enzyme to break it?
6. Why are hydrogen bonds associated with polar covalent bonds, but not with non-polar covalent bonds?

2.3 ■ Water, pH, and Buffers

Learning Outcomes

5. Describe the properties of water, and explain why it is so important in biological systems.
6. Explain the concept of pH, and how the pH of a solution relates to its acidity.
7. Describe the role of buffers.

Life on Earth has always been intimately associated with water, which is a small molecule composed of two hydrogen atoms and one oxygen atom. Water (H_2O) makes up over half of the mass of a living organism and provides the medium in which countless chemical reactions fuel life processes. The properties of water that make life possible are due largely to the molecule's ability to form hydrogen bonds.

Water

The oxygen atom in a water molecule (H_2O) is quite electronegative, so it pulls electrons away from the hydrogen atoms, resulting in an asymmetrical polar molecule (see figure 2.6). Neighboring water molecules form hydrogen bonds with one another, creating a network of interacting molecules. The extent and stability of this hydrogen bonding depends on temperature. At room temperature, hydrogen bonds continually break and re-form, allowing the water molecules to move closer together as they slide past one another. At freezing temperatures, however, the molecules slow down and form the maximum number of hydrogen bonds. This places each molecule a set distance away from its neighbors, producing a lattice-like structure called ice (**figure 2.8**). Because the water molecules are farther apart in ice than in liquid water, ice is less dense than liquid water, allowing it to float. When lakes freeze in the winter, ice floats at the top, allowing organisms to survive in the denser water below.

The polar water molecule not only attracts other water molecules, it also attracts any other charged substances. This is why a crystal of table salt (NaCl) dissolves readily in water. The slight positive charge of the hydrogen atoms in the water molecules attracts the Cl^- ions in the crystal, while the slight negative charge of the oxygen atoms in the water molecules attracts the Na^+ ions (**figure 2.9**). As a result, water molecules surround the ions, pulling them away from the crystal structure and forming a solution in which water is the **solvent** and the salt ions are the **solute** (dissolved substance). In air, ionic bonds within a crystal are stable, but in water the bonds are easily broken because of the attraction to the more abundant solvent molecules.

Substances with charges can dissolve in water; they are **hydrophilic,** meaning "water-loving." Non-polar molecules are **hydrophobic,** meaning "water-fearing," and they do not dissolve in water. They can, however, be organized by the presence of water. For example, when oil drops are added to water, they simply float to the top and then often merge. This phenomenon occurs because oil molecules are pushed together as the water molecules form hydrogen bonds among themselves.

Water that contains dissolved substances freezes at a lower temperature than pure water. This is because the water molecules are attracted to the dissolved ions, so a lower temperature is required for the water molecules to take on the rigid lattice structure of ice. In nature, most water does not freeze unless the temperature drops below 0°C. Consequently, some microorganisms can multiply below 0°C, because at least some of the water remains liquid.

MicroByte

Astrobiologists who search for life elsewhere in the universe often concentrate on planets and moons where there is evidence of water.

pH of Aqueous Solutions

An important property of aqueous solutions (solutions in which water is the solvent) is their **pH,** a measure of their acidity. It is measured on a logarithmic scale from 0 to 14 in which lower numbers represent more acidic solutions (**figure 2.10**).

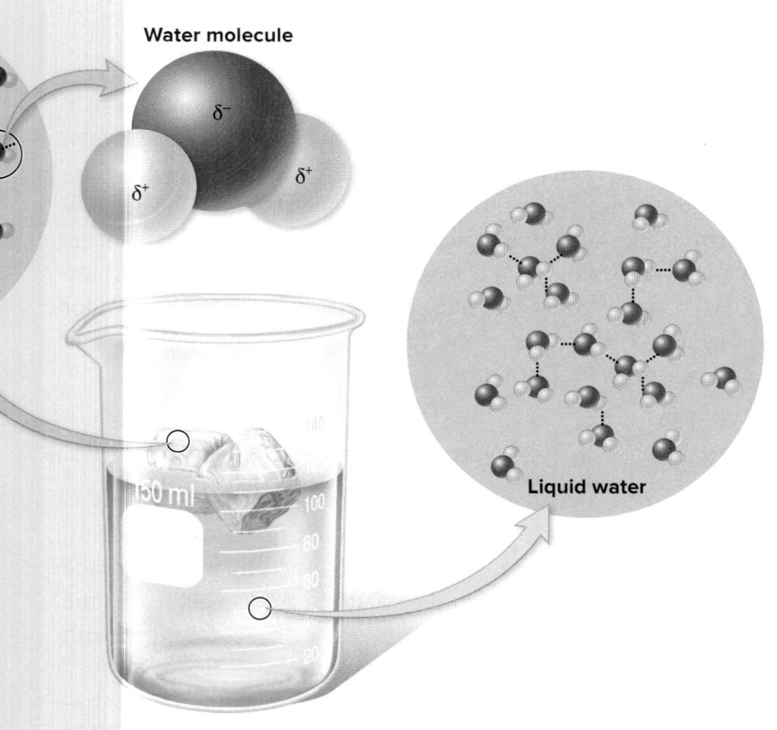

FIGURE 2.8 Water In liquid water, hydrogen bonds continuously break and re-form and the molecules can move closer together. In ice, each H_2O molecule forms hydrogen bonds to other H_2O molecules, producing a rigid crystalline structure.

? Why does ice float in water?

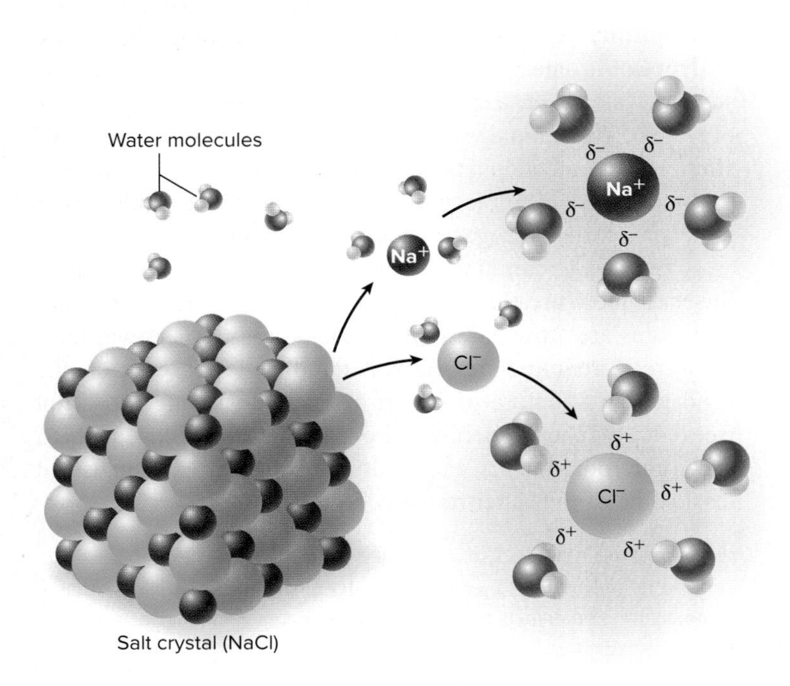

FIGURE 2.9 Salt (NaCl) Crystal Dissolving in Water In water, the Na^+ and Cl^- separate due to interaction with H_2O molecules. The Na^+ Is attracted to the slightly negatively charged O^- portion of the water molecules, and the Cl^- is attracted to the slightly positively charged H^+ portion. In the absence of water, the salt is highly structured because of ionic bonds between Na^+ and Cl^- ions.

If water were not polar, would it dissolve sodium chloride? Explain.

Water molecules break apart (dissociate) spontaneously at a low rate. When the covalent bond between the oxygen atom and one of the hydrogen atoms breaks, the strongly electronegative oxygen atom takes the electron that the weakly electronegative hydrogen atom had shared with it. This forms two ions: a proton (H^+; a hydrogen ion) and a hydroxide ion (OH^-). The two ions can then spontaneously form a new water molecule. The arrows in the following equation indicate that the reaction is reversible:

$$\underset{\text{(water)}}{H_2O} \rightleftharpoons \underset{\text{(hydrogen ion)}}{H^+} + \underset{\text{(hydroxide ion)}}{OH^-}$$

In pure water, the concentration of H^+ and OH^- ions is equal, but when substances called acids or bases are added, the balance shifts. An acid releases H^+, thereby increasing the total concentration of H^+ in the solution. A base decreases the relative concentration of H^+ ions, either by combining with H^+ or by releasing OH^- that will combine with H^+.

The pH scale ranges from 0 to 14 because the concentrations of H^+ and OH^- ions in an aqueous solution vary within these limits. Acidic solutions have a pH of less than 7, basic solutions have a pH of greater than 7, and neutral solutions have a pH of 7 (see figure 2.10). The pH, which indicates the H^+ concentration of a solution, is the negative logarithm of

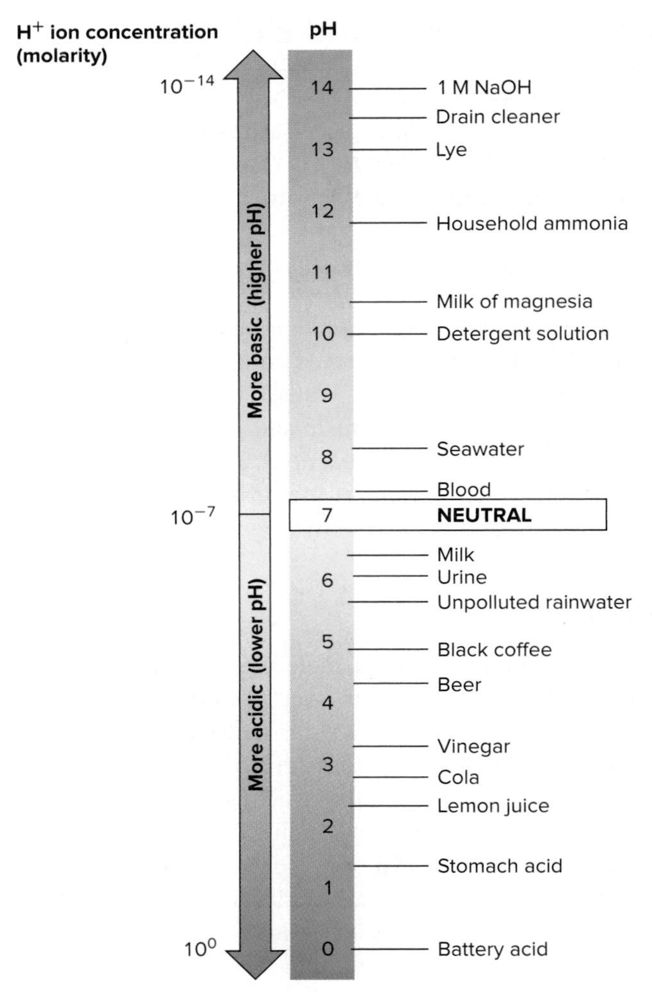

FIGURE 2.10 pH Scale The concentration of H^+ ions varies by a factor of 10 between each pH number since the scale is logarithmic.

Does the H^+ concentration increase or decrease when the pH drops from 5 to 4?

the H^+ concentration in moles per liter. Thus, pure water has a pH of 7 because the H^+ concentration (as well as the OH^- concentration) is approximately 10^{-7} moles per liter. A solution with a pH of 6 has a H^+ concentration of 10^{-6} moles per liter, and one with a pH of 5 has a H^+ concentration of 10^{-5} moles per liter. A solution with a pH of 5 is 10 times more acidic than a solution with a pH of 6. Likewise, a solution with a pH of 9 is 10 times more basic than a solution with a pH of 8.

Buffers

Products of a cell's chemical reactions are often acidic or basic, yet the interior of most cells is near neutral (pH 7). A **buffer** is a chemical that stabilizes the pH of solutions and helps to maintain a relatively constant pH. Buffers function by releasing H^+ ions to a solution when a base is added and combining with H^+ ions when an acid is added. Maintaining a specific intracellular pH is important to a cell because certain crucial molecules, such as enzymes, may change shape and lose function when the pH changes. Most bacteria can live only within a narrow pH range, usually near neutrality. Because of this, scientists also add buffers to laboratory media used to grow bacteria. ⏭| **culture media**

MicroAssessment 2.3

Weak hydrogen bonds between water molecules are responsible for the properties of water. Acidity is expressed as pH, which indicates the measure of H^+ ions in an aqueous solution. Buffers stabilize pH.

7. What characteristics of a water molecule make it polar? Give two examples of why this property is important in microbiology.
8. What is the role of a buffer in living organisms?
9. What must be added to pure water to decrease the OH^- concentration? To decrease the H^+ concentration?

2.4 ■ Organic Molecules

Learning Outcomes

8. Describe the characteristics of the different types of carbohydrates.
9. Compare and contrast the structure and function of simple lipids, compound lipids, and steroids.
10. Describe the factors that affect protein structure and function.
11. Compare and contrast the chemical compositions, structures, and major functions of DNA, RNA, and ATP.

Recall that all organic molecules contain carbon and hydrogen. Because each carbon atom forms four covalent bonds, an incredible assortment of organic molecules exists. In an organic compound with six carbon atoms, for example, those atoms may join to each other to form a linear chain, a branched chain, a ring, or any combination of those. In addition, the covalent bonds joining two carbon atoms within an organic molecule may be single, double, or even triple.

Organic molecules are often **macromolecules** (*macro* means "large"). Most macromolecules are **polymers** (*poly* means "many") formed by joining subunits or **monomers** (*mono* means "single"). Different classes of macromolecules are made up of different subunits. The four major classes of organic molecules are carbohydrates, lipids, proteins, and nucleic acids (**table 2.4**).

Cells synthesize macromolecules by covalently joining subunits together by **dehydration synthesis,** a chemical reaction that removes the equivalent of one molecule of water: a hydroxyl group (—OH) from one subunit and a hydrogen atom (—H) from an adjacent subunit (**figure 2.11**). The reverse type of reaction, called **hydrolysis,** breaks down a macromolecule into its subunits; the equivalent of H_2O is added when a covalent bond between two subunits is broken. Specific enzymes catalyze both types of chemical reactions.

The arrangement of carbon atoms in an organic compound is referred to as the carbon skeleton. Most of the atoms attached to the carbon skeleton are hydrogen atoms, but many organic molecules contain atoms of other elements as well.

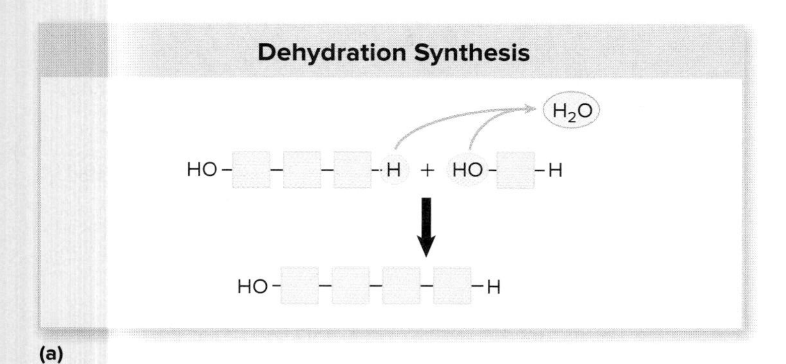

(a)

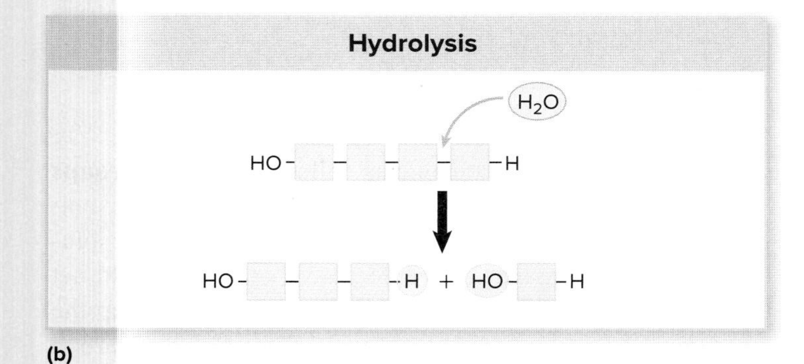

(b)

FIGURE 2.11 Synthesis and Breakdown of Macromolecules
(a) Subunits are joined together (polymerized) by removing water, a dehydration reaction. **(b)** In the reverse reaction, hydrolysis, the addition of water breaks bonds between the subunits.

? What are the four major classes of macromolecules?

TABLE 2.4 Major Classes of Organic Molecules

Name	Subunits	Major Functions
Carbohydrates	Monosaccharides	Structural component of cell walls; energy source
Lipids	Varies—subunits are not always similar	Important component of cell membranes
Proteins	Amino acids	Enzyme catalysts; structural portion of many cell components
Nucleic acids	Nucleotides	
DNA	Deoxyribonucleotides	Carrier of genetic information
RNA	Ribonucleotides	Various roles in protein synthesis; catalysis

TABLE 2.5 Biologically Important Functional Groups

Functional Group	Structure	Where Found
Aldehyde	$-C(=O)-H$	Carbohydrates
Amino	$-NH_2$	Amino acids, the subunit of protein
Carboxyl	$-C(=O)OH$	Organic acids, including amino acids and fatty acids
Hydroxyl	—OH	Carbohydrates, fatty acids, alcohol, some amino acids
Keto	$-C(=O)-$	Carbohydrates, polypeptides
Methyl	$-CH_3$	Some amino acids, attached to DNA
Phosphate	$-O-P(=O)(O^-)-O^-$	Nucleotides (subunit of nucleic acids), ATP, signaling molecules
Sulfhydryl	—S—H	Part of the amino acid cysteine

Distinctive chemical arrangements called **functional groups** contribute to the molecule's properties (**table 2.5**). For example, a carboxyl group (—COOH) may release H^+ ions in solution, and so an organic molecule that contains it could act as an acid. As you read about the subunits that make up the macromolecules described next, try to identify the carbon skeletons and the various functional groups.

Carbohydrates

Carbohydrates are a diverse group of organic compounds that includes sugars and starches. They play several important roles in living organisms:

- **Energy source.** Organisms break down carbohydrates and harvest the energy they contain. ⏭ metabolism
- **Energy storage.** Organisms can produce and store carbohydrates for later use. ⏭ storage granules
- **Source of carbon for biosynthetic products.** Many microorganisms can make all of their cell components from a single carbohydrate; that carbohydrate is glucose. ⏭ precursor metabolites
- **Component of genetic material.** The subunits of DNA and RNA contain sugars. ⏭ nucleic acids
- **Structural components of cells.** Cell walls of plants, fungi, and most bacteria contain carbohydrates. ⏭ cell wall structure

Carbohydrates contain carbon, hydrogen, and oxygen atoms in an approximate ratio of 1:2:1. The general chemical formula of carbohydrate building blocks, CH_2O, reflects this ratio. The "H_2O" in that formula is reflected in the term "carbohydrate" (meaning "hydrate of carbon").

Monosaccharides

Monosaccharides, or simple sugars, are the basic unit of a carbohydrate (*sacchar* means "sugar"). Most common monosaccharides have five or six carbon atoms, typically forming a ring (**table 2.6**). Each carbon atom of a monosaccharide is numbered using a characteristic scheme, allowing scientists to describe the positions of various functional groups attached to the molecules (**figure 2.12**).

Ribose and deoxyribose are 5-carbon sugars (pentoses) found in the nucleic acids RNA and DNA. These sugars are identical, except that deoxyribose has one less atom of oxygen than does ribose (*de* means "away from"). Ribose (a component of ribonucleic acid—RNA) has a hydroxyl group on the number 2 carbon (also called the 2 prime carbon, written 2′ carbon), whereas deoxyribose (a component of deoxyribonucleic acid—DNA) has only a hydrogen atom at that position (see figure 2.12).

Glucose, galactose, fructose, and mannose are all 6-carbon sugars (hexoses) with the molecular formula $C_6H_{12}O_6$. These are all **structural isomers,** meaning that they contain the same atoms but in different chemical arrangements, much like the words "list" and "slit" are composed of the same letters but in different arrangements (**figure 2.13**). Structural isomers result in distinct sugars with different properties. For example, glucose and mannose both have a sweet taste, but mannose has a bitter aftertaste. Glucose is an important energy source for most cells. Mannose, found on the surface of some microbes, is recognized by the body's defense system, triggering an immune response aimed at destroying invaders.

Disaccharides

Disaccharides are composed of two monosaccharides joined together by covalent bonds (see table 2.6). Two common examples are sucrose (table sugar) and lactose (milk sugar). Sucrose, which comes from sugar cane and sugar beets, is composed of the monosaccharides glucose and fructose, whereas lactose consists of glucose and galactose. Another disaccharide, maltose, composed of two glucose molecules, is a breakdown product of starch.

To form a disaccharide, two monosaccharides are joined together by a dehydration synthesis reaction between a pair of their hydroxyl groups, with the loss of the equivalent of a water molecule (**figure 2.14**). The reaction is reversible, so hydrolysis, which adds a water molecule, produces the two original monosaccharides.

TABLE 2.6 Common Monosaccharides, Disaccharides, and Polysaccharides

Name	Components	Significance
Monosaccharides		
(5-carbon)		
Ribose		Component of RNA
Deoxyribose		Component of DNA
(6-carbon)		
Glucose		Common subunit of larger carbohydrates
Galactose		Component of milk sugar (see lactose)
Fructose		Fruit sugar
Mannose		Found on the surface of some microbes
Disaccharides		
Lactose	Glucose + galactose	Milk sugar
Maltose	Glucose + glucose	Breakdown product of starch
Sucrose	Glucose + fructose	Table sugar from sugar cane and beets
Polysaccharides		
Agar	Polymer of galactose	Gelling agent in bacteriological media; extracted from the cell walls of some algae
Cellulose	Polymer of glucose; no branching	Major structural polysaccharide in plant cell walls
Chitin	Polymer of *N*-acetyl-glucosamine	Major component in fungal cell walls and exoskeleton of insects and crustaceans
Dextran	Polymer of glucose	Storage product in some bacterial cells
Glycogen	Polymer of glucose	Major storage polysaccharide in animal and bacterial cells
Starch	Polymer of glucose	Major storage product in plants

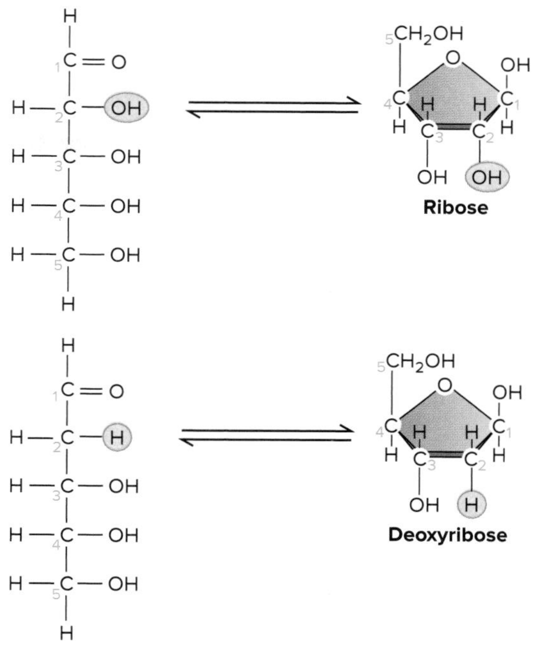

FIGURE 2.12 Ribose and Deoxyribose Note the difference at the number 2 carbon atoms. Although both linear and ring forms occur in the cell, the ring form predominates. In the diagram, the plane of the ring is perpendicular to the plane of the paper with the thick line on the ring closest to the reader.

? **What is the major chemical difference between ribose and deoxyribose?**

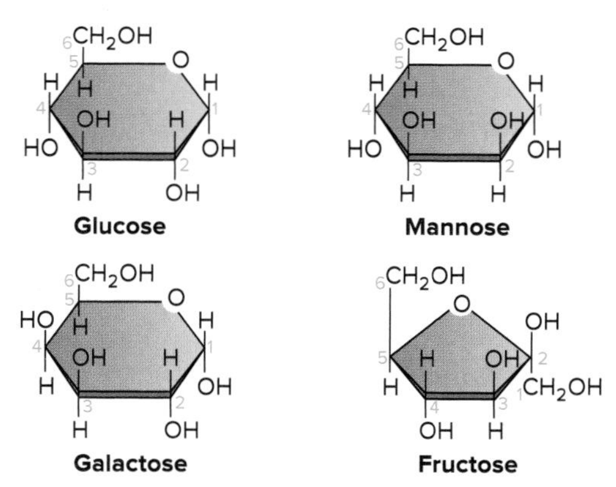

FIGURE 2.13 Common 6-Carbon Sugars These sugars are structural isomers with different properties.

? **What is a structural isomer?**

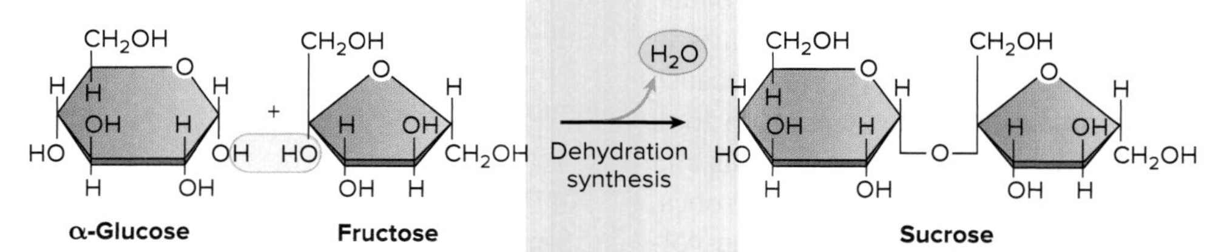

FIGURE 2.14 Formation of a Disaccharide The sucrose molecule is formed by the removal of water.

? **What type of reaction would reverse the step shown in this diagram?**

FOCUS ON A CASE 2.1

The patient was a 15-year-old girl whose mother had recently let her start dating a boy from their church. Most of the dates were to the nearby diner, where the teens ate hamburgers and drank milkshakes and then came home to watch a movie. Almost every time the patient went on a date, she started feeling bloated and crampy, had to pass gas, and later suffered from diarrhea. She was embarrassed, and afraid that she might be allergic to her new boyfriend. Her mother was afraid that her daughter had a nervous condition.

After taking a complete history, the doctor ordered a hydrogen breath test for the teen. The patient watched her diet for a few days, fasted the night before the test, and was given 25 grams (g) of lactose in water when arriving at the clinic. Several times during the next few hours, technicians measured the amount of hydrogen gas in the air she exhaled into a plastic bag. The teen entertained herself by reading magazines between tests, and as time passed, she began to feel bloated. When the doctor called the next day, she assured the mother that her daughter did not have a nervous condition. Instead, she suspected lactose intolerance and suggested that the girl replace the milkshakes she consumed on her dates with water or lemonade.

1. Why did the patient experience bloating, gas, and diarrhea?
2. How did high levels of hydrogen gas in the patient's exhaled air indicate lactose intolerance?
3. Was the patient allergic to the milkshakes?

Discussion

1. Lactose is a disaccharide found in milk, ice cream, cheese, and other dairy products. It is broken down into its constituent monosaccharides—glucose and galactose—by the enzyme lactase, which is produced in the small intestine. The monosaccharides are then absorbed into the bloodstream before reaching the large intestine. Babies, who derive much of their nutrition from milk, are generally born with the ability to produce plenty of lactase. By adulthood, however, many individuals produce much less of the enzyme and are unable to break down the amount of lactose in a small glass of milk. The disaccharide therefore enters the large intestine intact where it is fermented by intestinal bacteria, producing gases. The gases result in bloating and flatulence. The lactose also draws more water into the intestine, resulting in diarrhea.
2. The human body typically does not produce hydrogen gas (H_2) in its metabolic pathways. The hydrogen gas in the patient's exhaled air was therefore produced by bacterial metabolism. Because the patient had fasted before the test, the only nutrient source for the bacteria was the lactose ingested during the test. If the patient had had sufficient levels of lactase, her body would have broken the lactose into monosaccharides, and these would have been absorbed rather than passing through to her intestinal bacteria.
3. No, she was not suffering from an allergy because the immune system was not involved in her response. Her intolerance can be managed by simply avoiding milk products or supplementing her diet with an over-the-counter lactase supplement.

Polysaccharides

Polysaccharides are large complex carbohydrates composed of long chains of monosaccharide subunits or their derivatives (see table 2.6). Polysaccharides often contain only glucose molecules, but nevertheless are structurally diverse. Some molecules are branched. The position of the carbon atoms involved in the bonding can also differ (**figure 2.15**).

Cellulose, starch, glycogen, and dextran are all polymers of glucose. Cellulose is the principal component of plant cell walls and the most abundant organic molecule on Earth. Most organisms cannot degrade this substance because they lack the enzyme that breaks the bonds joining the subunits. Certain bacteria and fungi have that enzyme, which is why they play such an important role in recycling organic material. In contrast to the situation with cellulose, many organisms produce the enzyme that breaks the bond joining glucose subunits in starch, the energy-storage form produced by plants. Because of this, a wide variety of organisms can use starch as a food source. Glycogen is an energy-storage product of animals and some bacteria. Dextran, a storage product of some microbes, is a component of certain products used to increase the volume of blood or deliver iron to the blood of iron-deficient patients. ⏭| cellulose degradation

Chitin and agar are other important polysaccharides in microbiology. Chitin, a polymer of the glucose derivative *N*-acetylglucosamine, is in the cell walls of fungi and is a major component in the exoskeletons of insects and crustaceans. Agar, a polymer of galactose, is found in the cell walls of algae; it is extensively used as a gelling agent in media used to grow microorganisms in the laboratory. ⏭| culture media

Lipids

Lipids are a diverse group of non-polar, hydrophobic molecules. Their single common feature is that they are only slightly soluble in water. Thus, they are critically important in the structure of membranes, which function as a cell's gatekeepers. Membranes prevent cell contents from leaking out, and also keep many molecules from entering cells. Unlike other macromolecules, not all lipids are composed of similar subunits. ⏭| cytoplasmic membrane

Simple Lipids

Simple lipids contain only carbon, hydrogen, and oxygen. The most common simple lipids in nature are **triglycerides,** which are fats or oils composed of three fatty acids linked to a

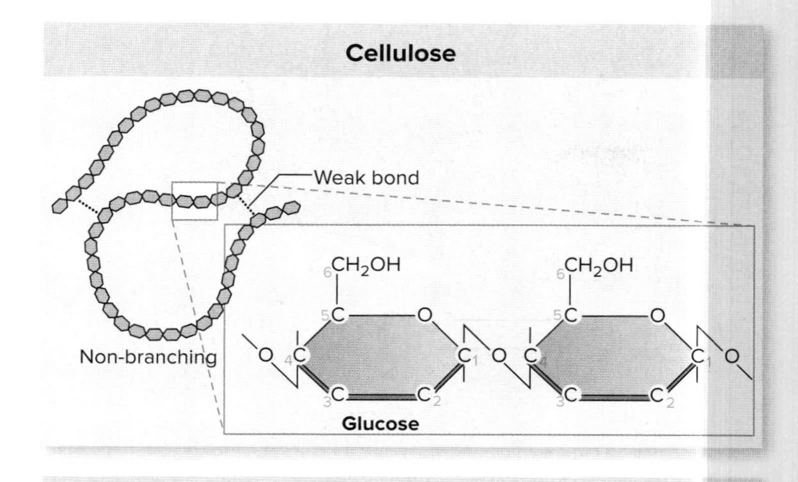

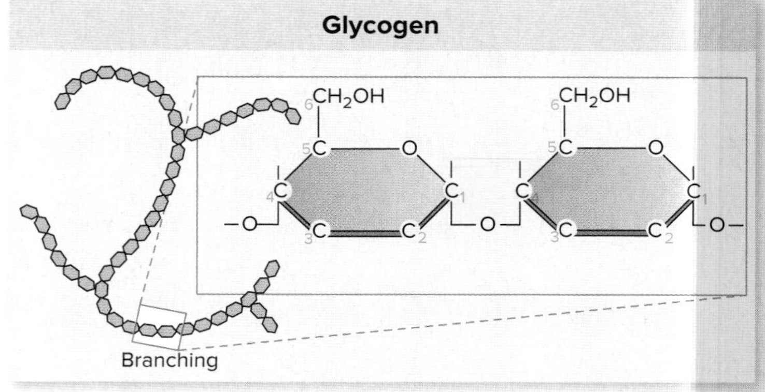

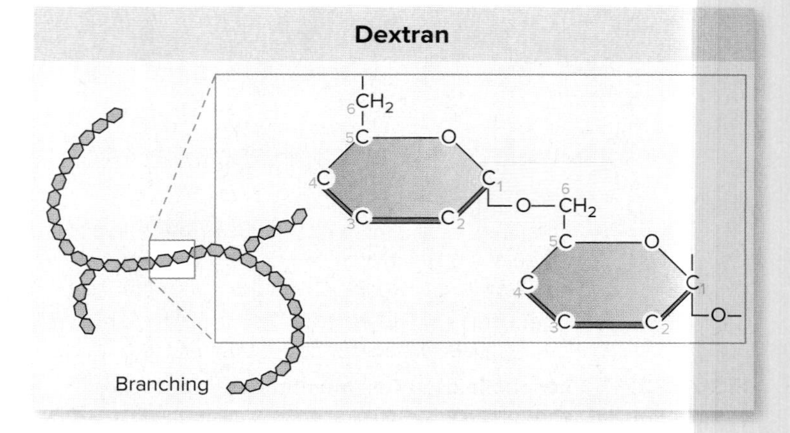

FIGURE 2.15 Three Important Polysaccharides The molecules shown have the same subunit (glucose) yet they are distinct because of differences in linkage that join the subunits, the degree of branching, and the bonds involved in branching. Weak bonding forces are also involved.

? **Where are the three polysaccharides shown above found in nature?**

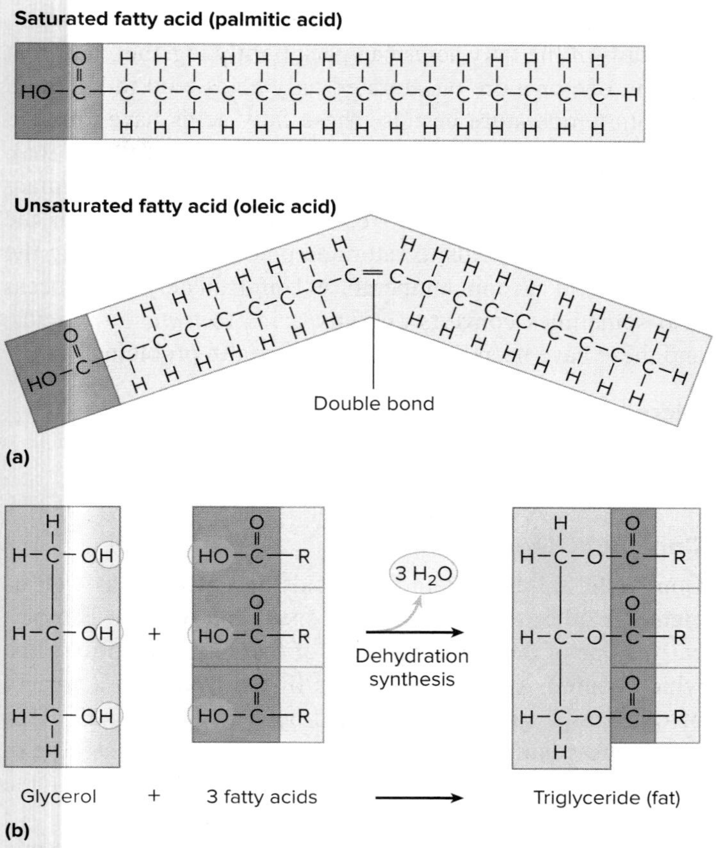

3-carbon glycerol molecule (**figure 2.16**). Fatty acids are long linear carbon skeletons with a carboxyl group (—COOH; see table 2.5) at one end. The length of the chain varies depending on the fatty acid. Glycerol is a 3-carbon molecule with a hydroxyl group (—OH) attached to each carbon. Fatty acids can join to glycerol via covalent bonds between a hydroxyl group of glycerol and the carboxyl group of the fatty acid (see figure 2.16b).

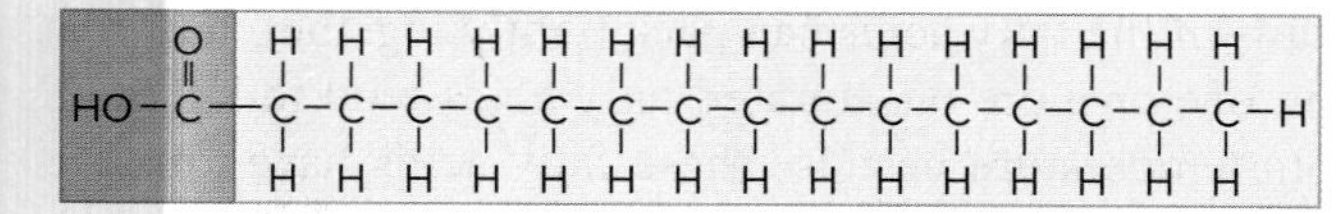

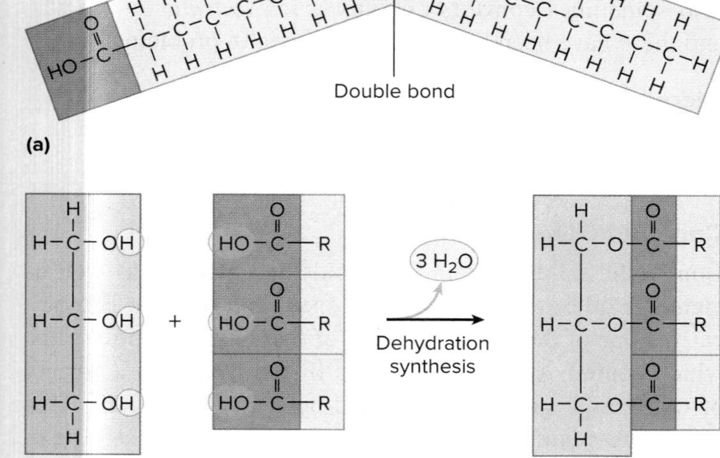

FIGURE 2.16 Fat Formation from Fatty Acids and Glycerol
(a) Most fatty acids contain an even number of carbon atoms (commonly 16 or 18) and may be saturated or unsaturated. Unsaturated fatty acids have at least one double bond between carbon atoms in the chain.
(b) Dehydration synthesis in the joining of fatty acids with glycerol; the R groups are carbon-hydrogen chains, such as those shown in (a).

? **What components of the fat in this figure make it a triglyceride?**

Although hundreds of different fatty acids exist, they can be divided into two groups based on the presence or absence of double bonds between carbon atoms. Saturated fatty acids have no double bonds between carbon atoms (see figure 2.16a). The term "saturated" means they have the maximum number of hydrogen atoms. Unsaturated fatty acids contain one or more double bonds between carbon atoms. Most naturally occurring unsaturated fatty acids are *cis*, meaning the hydrogen atoms attached to the double-bonded carbon atoms are on the same side of the bond. *Trans* fatty acids have hydrogen atoms on opposite sides of the double bond.

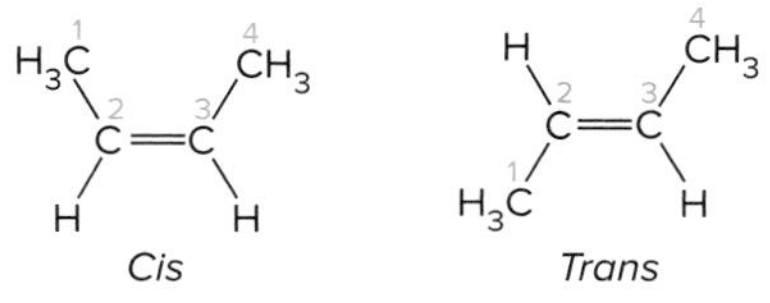

Cis *Trans*

The type of fatty acids in a triglyceride affects the melting point of the fat. Fats that contain only saturated fatty acids are typically solid at room temperature because the straight, long tails of the fatty acids can pack tightly together. Fats that contain one or more unsaturated fatty acids tend to be liquid at room temperature because these fatty acids have kinks in their long tails that prevent tight packing (see figure 2.16a). Oils are fats that are liquid at room temperature. Manufacturing processes that hydrogenate oils convert many of the unsaturated fatty acids to saturated ones, thereby making the product solid at room temperature. Unfortunately, the process also sometimes converts *cis* fatty acids to *trans* fatty acids, and these have been linked to certain health problems.

MicroByte

Diets rich in saturated fats and *trans* fats are associated with high blood cholesterol levels that may lead to clogged arteries.

Compound Lipids

Compound lipids contain fatty acids and glycerol as well as elements other than carbon, hydrogen, and oxygen. Biologically, some of the most important of these are **phospholipids,** which contain a phosphate group linked to one of a variety of other polar groups represented by R in **figure 2.17**. This phosphate-containing portion—the polar head—is soluble in water (hydrophilic). In contrast, the fatty acid portion—the non-polar "tail"—is insoluble in water (hydrophobic).

Phospholipids are an essential component of cytoplasmic membranes, the structure that separates the internal contents of a cell from the outside environment (see figure 2.17). The phospholipid molecules orient themselves in the membrane as opposing layers, forming a bilayer. The hydrophobic non-polar tails face inward, interacting with the hydrophobic tails of the phospholipid molecules in the opposing layer. The hydrophilic polar heads face outward, toward either the aqueous external environment or the aqueous cytoplasmic environment. Water-soluble substances cannot pass through the hydrophobic portion of the membrane, so cells must use special transport mechanisms to move these across the membrane. These will be discussed in chapter 3.

Other compound lipids will also be discussed in chapter 3. These include the lipoproteins (covalent associations of proteins and lipids) and lipopolysaccharides (molecules of lipid linked with polysaccharides through covalent bonds).

Steroids

Steroids are lipids with a characteristic structure consisting of four connected rings (**figure 2.18**). Their chemical structure is quite different from the structures of fats or phospholipids, but they are classified as lipids because they too are relatively insoluble in water. The hormones cortisol, estrogen, and testosterone are steroids. If a hydroxyl group (—OH) is attached to one of the rings, the steroid is a sterol. Cholesterol and ergosterol are sterols that provide rigidity to animal and fungal cell membranes, respectively.

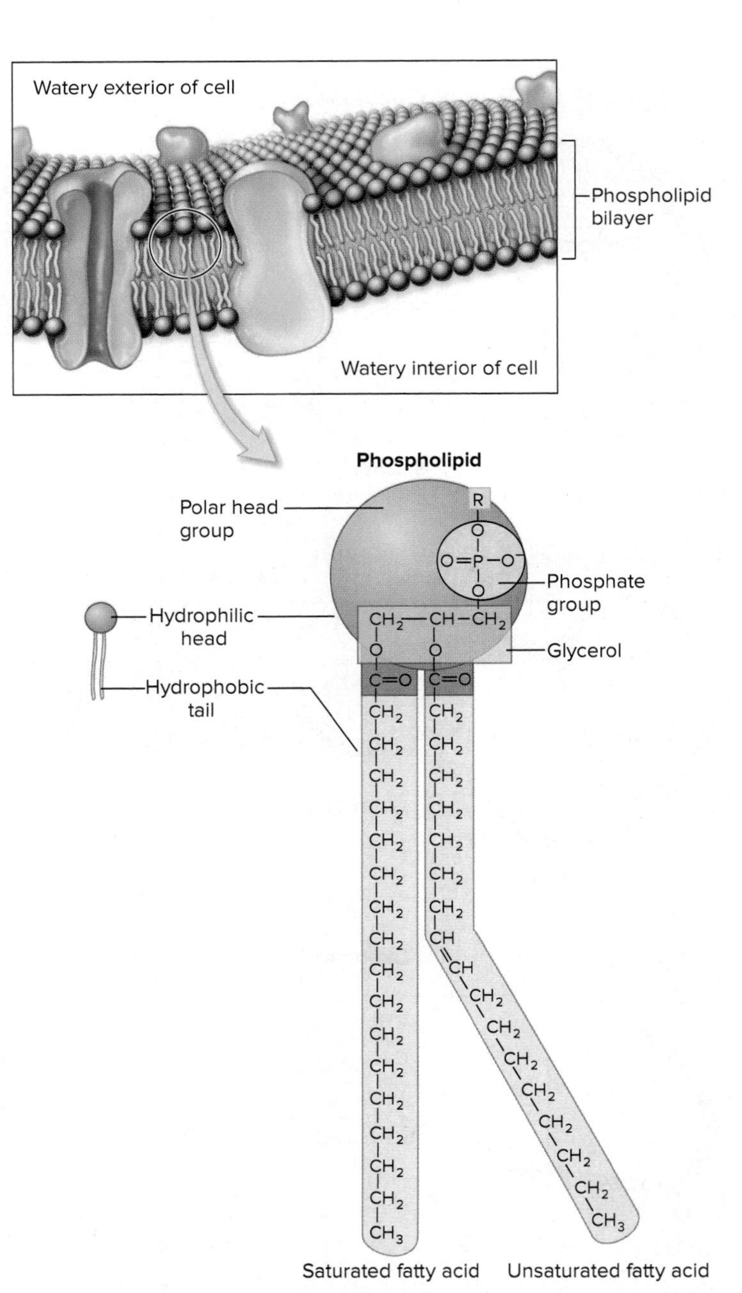

FIGURE 2.17 Phospholipids in Cell Membranes In phospholipids, two of the —OH groups of glycerol are linked to fatty acids; the third —OH group is linked to a hydrophilic head group, which contains a phosphate ion and a polar molecule (labeled R).

What features in the chemical composition of phospholipids make them ideal components of the cytoplasmic membrane?

Proteins

Even the simplest bacterial cell may contain several thousand different proteins. In spite of this great diversity, all proteins are polymer chains composed of a limited number of different subunits called **amino acids.** The chains vary in length among different proteins, and are folded into complex three-dimensional shapes based on the sequence of amino acids. Each protein has a unique shape directly related to its

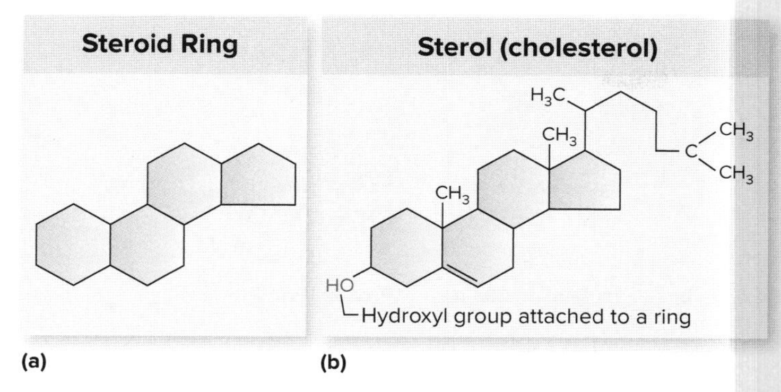

FIGURE 2.18 Steroid (a) General formula showing the four-membered ring and **(b)** the —OH group that makes the molecule a sterol. The sterol shown here is cholesterol. The carbon atoms in the ring structures and their attached hydrogen atoms are not shown.

Why are steroids classified as lipids?

function. Some of the most important functions of proteins include these:

- **Enzyme catalysis.** Enzymes that speed up chemical reactions in a cell are typically proteins. **⏭| enzymes**
- **Transport.** Transport proteins move molecules, often across a cytoplasmic membrane into or out of cells.
- **Signal reception.** Receptor proteins on the cell surface recognize conditions in the external environment.
- **Regulation.** Some proteins serve as intercellular signals or bind to DNA and regulate gene expression.
- **Motility.** Proteins are essential components of flagella and cilia, structures that move cells. **⏭| flagella**
- **Support.** Proteins make up the cytoskeleton, the structural framework of many cells. **⏭| cytoskeleton**

Amino Acids

Recall that amino acids are the subunits of proteins. There are 20 common amino acids, and these can be arranged in a nearly limitless number of combinations in a protein. All amino acids have a central carbon atom bonded to (1) a hydrogen atom, (2) a carboxyl group (—COOH), (3) an amino group (—NH_2), and (4) a side chain or R group (**figure 2.19**). The R group distinguishes the various amino acids from one another and gives each amino acid its characteristic properties. Most amino acids exist in two forms that are mirror images of one another (**Focus Your Perspective 2.1**).

Amino acids are divided into several groups based on properties of their side chains (**figure 2.20**). Non-polar amino acids are characterized by side chains that lack polar bonds; an example is the methyl group (—CH_3) of alanine. In contrast, polar amino acids have side chains that contain a polar bond; an example is the hydroxyl group (—OH) of serine. Charged amino acids carry a positive or negative charge because their side chains contain functional groups that can

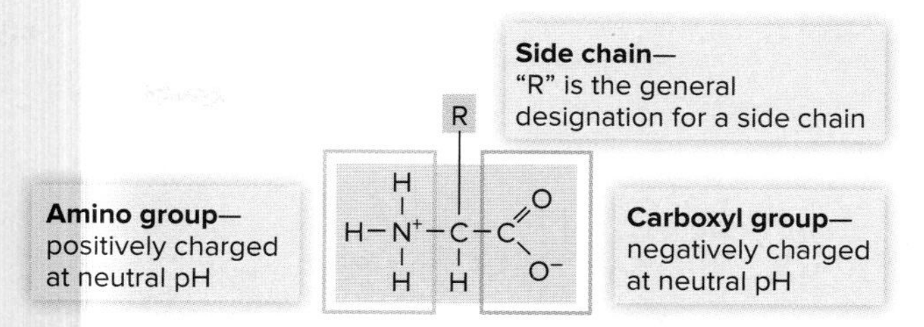

FIGURE 2.19 Generalized Amino Acid The central carbon atom in an amino acid forms four covalent bonds. Three of the following are always present: a hydrogen atom (—H), an amino group (—NH_2), and a carboxyl group (—COOH). Amino acids differ in their R groups.

Which portion of an amino acid is responsible for the unique properties of the molecule?

ionize; these include carboxyl groups (—COOH; acidic) and amino groups (—NH_2; basic). Some amino acids within those groups have other important distinctions. Within the non-polar amino acids for example, tryptophan is bulkier than the others because it contains a double-ring structure. Cysteine has a sulfur atom that can form a covalent bond with the sulfur atom of another cysteine molecule.

Peptide Bonds

Amino acids are joined together in a linear chain by covalent bonds called **peptide bonds.** Peptide bonds form by a dehydration synthesis reaction between the carboxyl group (—COOH) of one amino acid and the amino group (—NH_2) of the neighboring amino acid (**figure 2.21**). Because the R groups of amino acids are not involved in forming the peptide bonds between them, any amino acid can be joined to any other amino acid in a linear chain. This means that the 20 different amino acids can be joined together in an astounding number of combinations. Consider, for example, that you want to create beaded belts of 100 beads each and that you have unlimited numbers of 20 different types of beads to use in their construction. Arranging the beads in one long strand to make a belt, you could create 20^{100} different belts!

A short sequence of amino acids joined together by peptide bonds is called a peptide. One end of the molecule has a free amino group; this end is called the N terminal or amino terminal end. The other end of the molecule has a free carboxyl group; this end is called the C terminal or carboxyl terminal end. A longer sequence of amino acids is called a **polypeptide,** although the distinction between a peptide and a polypeptide is not precise. A **protein** is composed of one or more polypeptides folded to create a functional molecule. Although the term *protein* implies a functional entity, it is often used interchangeably with *polypeptide.*

Certain proteins have other molecules covalently bonded to the side chains of some of their amino acids. If sugar molecules are bonded, the protein is a glycoprotein; if lipids are attached, the protein is a lipoprotein. Many of these proteins are found on the surface of cells.

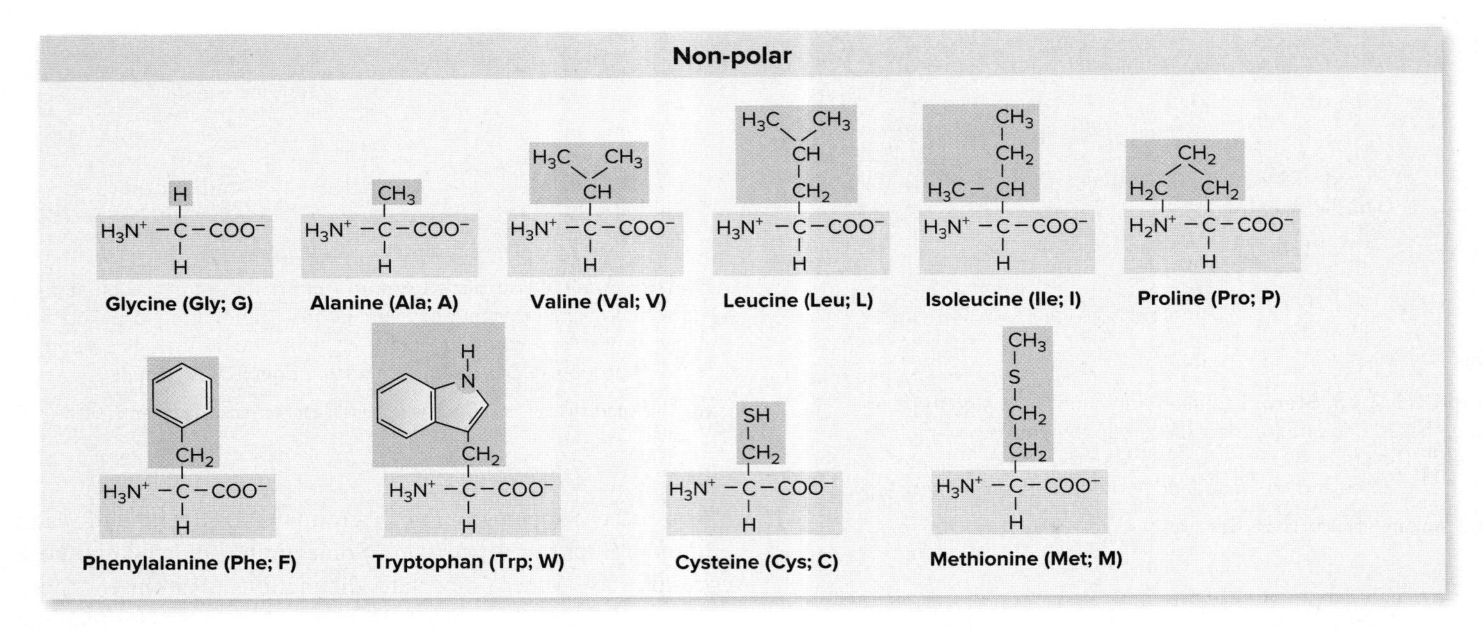

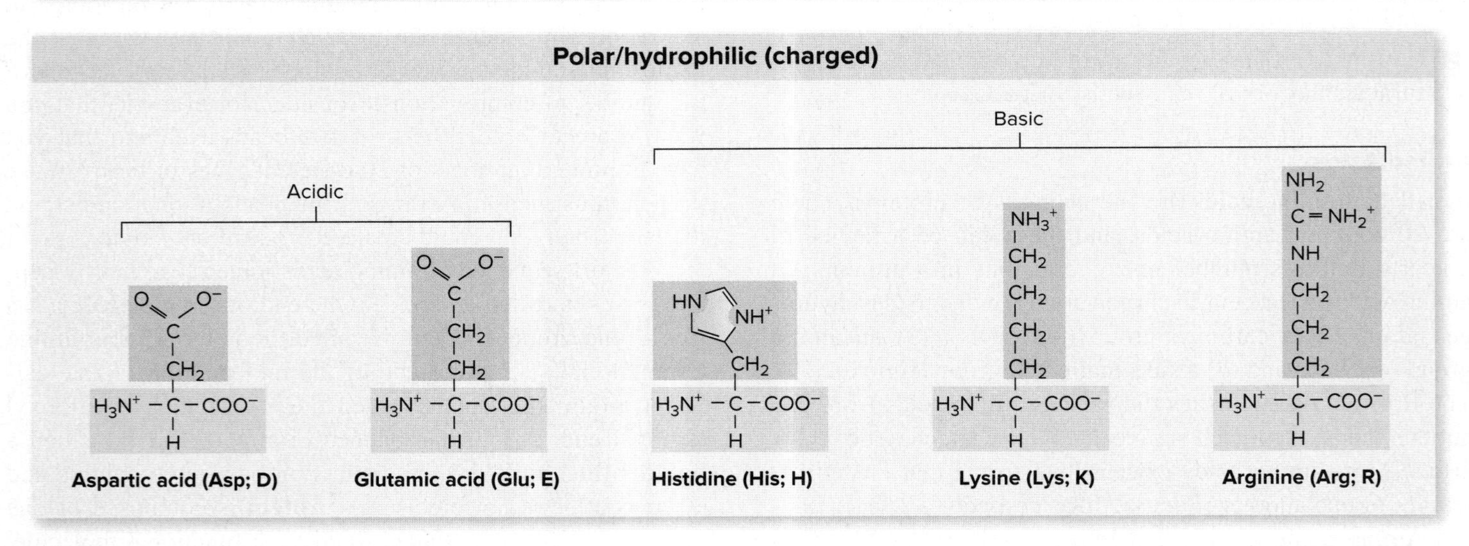

FIGURE 2.20 Common Amino Acids The groupings are based on the polarity and the overall charge of the amino acid. The basic and acidic amino acids have a net positive and a net negative charge, respectively. For simplicity, tyrosine is shown in only one group but it has both non-polar and polar characteristics. The three-letter and single-letter code names for each amino acid are given.

? What chemical groups characterize a hydrophobic amino acid? A hydrophilic amino acid?

FOCUS YOUR PERSPECTIVE 2.1

Right-Handed and Left-Handed Molecules

Although Louis Pasteur (1822–1895) is often considered the father of bacteriology, he started his scientific career as a chemist. While researching crystals for the French wine industry, he worked with tartaric and paratartaric acids, which form thick crusts within wine barrels. These two chemicals form crystals that are identical in their chemical composition, but affect polarized light very differently. When polarized light passes through tartaric acid crystals, the light rotates (twists) to the right. In contrast, paratartaric acid crystals have no effect on the light. Pasteur wanted to learn how the crystals differed.

Pasteur noticed that under a microscope, all tartaric acid crystals looked identical, whereas paratartaric acid crystals had two kinds of structures. Using tweezers, he carefully separated the two types of paratartaric acid crystals and dissolved them in separate flasks of water. He found that one solution rotated polarized light to the left and the other, to the right. Solutions containing equal numbers of the two crystals did not rotate the light at all. Pasteur concluded that paratartaric acid must be a mixture of two types of chemical structure that counteracted the light-rotating effects of each other. Both structures had the same chemical composition, but each was the mirror image of the other. We now call these **optical isomers.** Just as our right and left hands cannot be precisely superimposed upon one another (try stacking your hands on top of each other—not palm to palm—to see), the different spatial arrangements of optical isomers do not occupy space in the identical manner (**box figure 2.1**).

The interacting components of biological systems generally require a precise fit, and so, just as our right hand doesn't fit into a left-handed glove, a pair of optical isomers can have very different properties. Amino acids (except glycine) can exist in two optical isomers, designated L (left-handed) and D (right-handed). Proteins are composed of only L-amino acids. A few D-amino acids are found in the peptidoglycan component of bacterial cell walls. Some bacteria also release D-amino acids that may serve as signals in cell wall assembly. Developing insights into the roles of D-amino acids may lead to new approaches for controlling microbial growth.

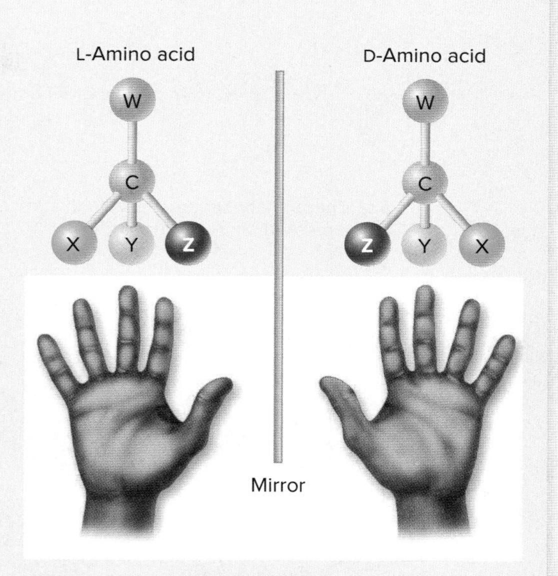

BOX FIGURE 2.1 Mirror Images of an Amino Acid The joining of a carbon atom to four different groups leads to asymmetry. Amino acids can exist in either the L or the D form, each being the mirror image of the other. The two molecules cannot be rotated in space to give two identical molecules.

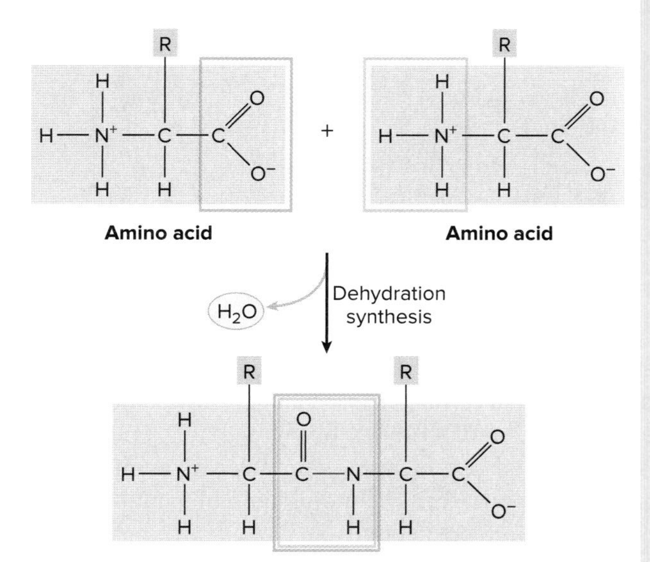

FIGURE 2.21 Peptide Bond Formation Dehydration synthesis forms the peptide bond, shown in red.

? What two chemical groups are involved in the formation of a peptide bond?

Protein Structure

Proteins have up to four levels of structure: primary, secondary, tertiary, and quaternary. The sequence and number of amino acids in the polypeptide determine its **primary structure,** which affects all other levels of protein structure (**figure 2.22**). Thus, the primary structure directs the final shape of the protein and, as a result, is responsible for its properties.

The **secondary structure** is a repeated coiling or folding in localized regions of a protein. These characteristic repeating patterns form due to weak hydrogen bonding between the carboxyl and amino groups of amino acids along the polypeptide chain (see figure 2.22b). A spiral or helical structure is an alpha (α)-helix, whereas parallel strands make up a beta (β)-pleated sheet.

Tertiary structure is the overall three-dimensional shape of a folded polypeptide (see figure 2.22c). It results largely from the interactions of R groups on amino acids. Thus, the overall shape of the protein relies on its primary structure, which is determined by which amino acids are present and

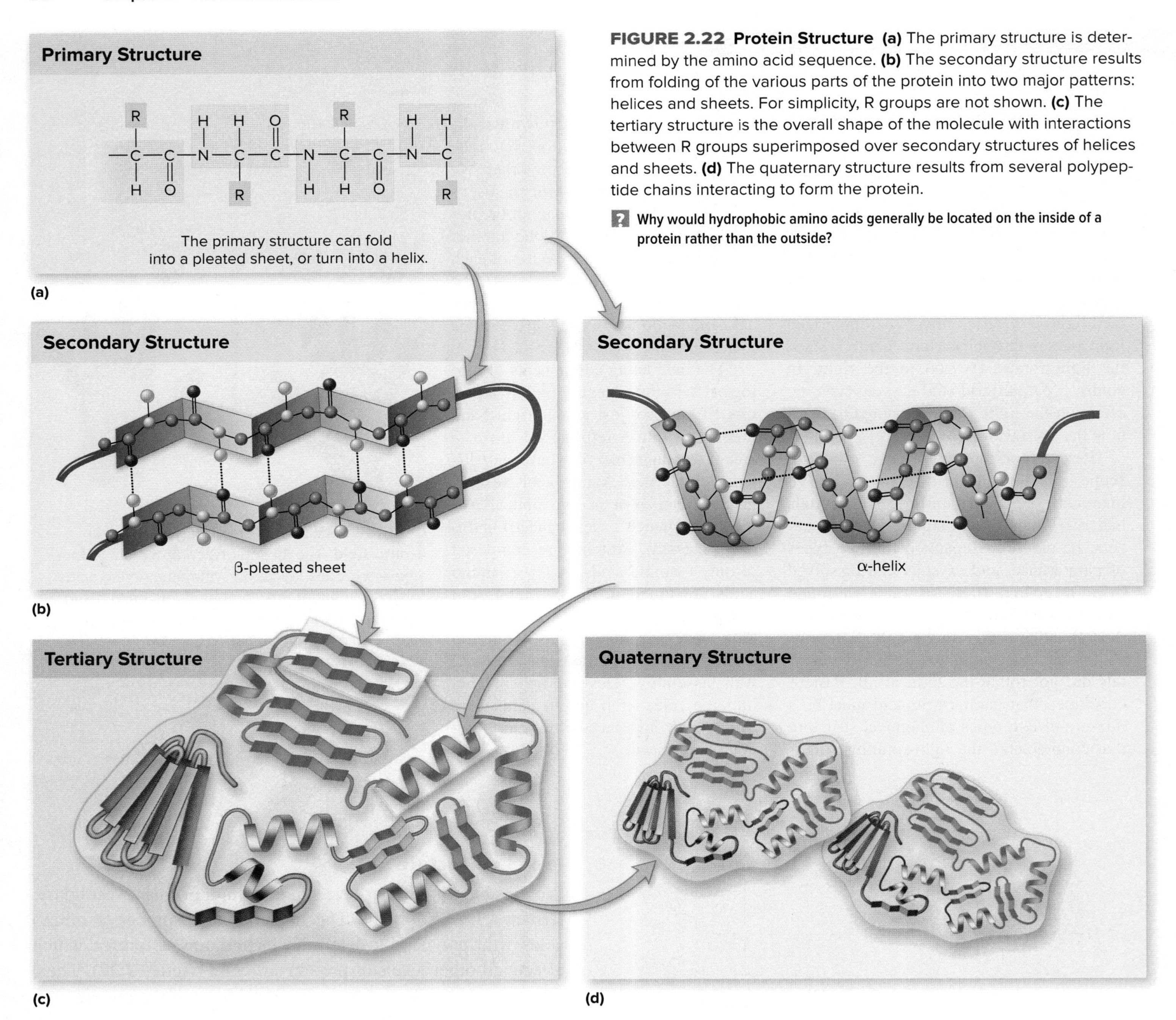

FIGURE 2.22 Protein Structure (a) The primary structure is determined by the amino acid sequence. **(b)** The secondary structure results from folding of the various parts of the protein into two major patterns: helices and sheets. For simplicity, R groups are not shown. **(c)** The tertiary structure is the overall shape of the molecule with interactions between R groups superimposed over secondary structures of helices and sheets. **(d)** The quaternary structure results from several polypeptide chains interacting to form the protein.

? Why would hydrophobic amino acids generally be located on the inside of a protein rather than the outside?

where they are in the chain. For example, large R groups occupy more space than small ones, positively charged amino acids are attracted to negatively charged ones, and a covalent bond can form between sulfur atoms in different cysteine molecules, creating a disulfide bond (S—S). Amino acids with polar side chains are hydrophilic, and are typically located on the outside of the protein molecule, where they can interact with charged polar water molecules. Amino acids with non-polar side chains are hydrophobic, so these tend to cluster inside the protein molecule, thereby avoiding water molecules. All of these interactions contribute to the complex folding patterns seen in the great diversity of proteins.

Some proteins consist of two or more polypeptide chains that are held together by many weak bonds. The specific shape that results from the association between the chains is the **quaternary structure** (see figure 2.22d). Of course, only proteins that consist of more than one polypeptide can have a quaternary structure. The individual polypeptides generally do not have biological activity.

After being synthesized, a polypeptide chain folds into its correct shape. Although many shapes are possible, typically only one functions properly. Most proteins fold spontaneously into their correct state, but protein **chaperones** are sometimes needed to help with the process.

Protein Domains

As new technologies have made it possible to study and compare the structures of a wide variety of different proteins, scientists have noticed that proteins with similar functions often have one or more substructures in common. These substructures consist of sheets and helices that fold into a stable structure independently of other parts of the molecule. A substructure associated with a particular function is referred to as a **protein domain** (**figure 2.23**). For example, a certain domain may bind DNA; another may act as a catalyst. Once a known function can be attributed to a specific domain, the role of an unknown protein with that domain can be inferred.

Large proteins sometimes have many domains and several different functions, whereas a small protein may have only one. Domains usually consist of 40–350 amino acids and are connected to each other by short lengths of polypeptides.

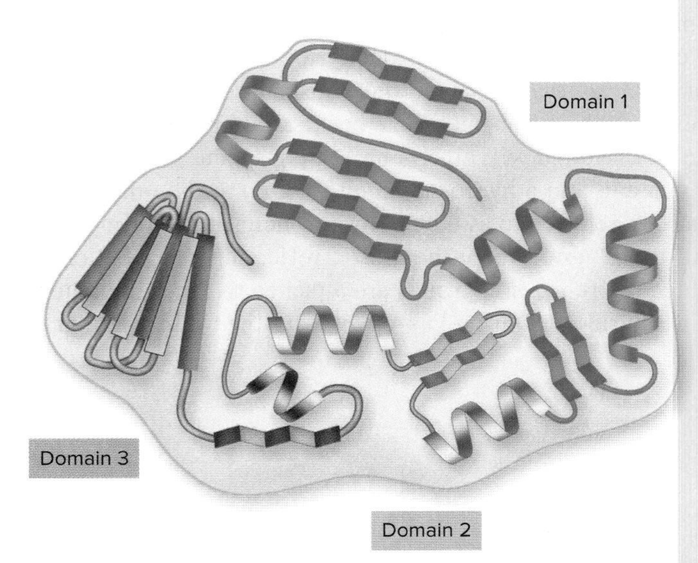

FIGURE 2.23 Domain Structure of a Protein Each domain has a different function.

? Distinguish between a polypeptide and a domain.

Protein Denaturation

High temperature, extreme pH, and certain solvents can affect the interactions between amino acids within a protein, causing it to **denature** or lose its characteristic shape (**figure 2.24**). When a protein changes shape, it may become nonfunctional. Most organisms cannot grow at very high temperatures because their enzymes denature and no longer catalyze chemical reactions. Each enzyme functions best within a specific narrow range of environmental conditions. Denaturation may be reversible, but sometimes it is not. Boiling an egg, for example, denatures the egg white protein; cooling the egg does not restore its original state.

Nucleic Acids

Nucleic acids carry genetic information. Cells decipher the encoded information and use it to link amino acid subunits in the proper order to make the various proteins.

Nucleotides

Nucleotides are the subunits of nucleic acids. Each nucleotide consists of a pentose sugar, a phosphate group, and a **nucleobase** (also called a base or a nitrogenous base). The phosphate

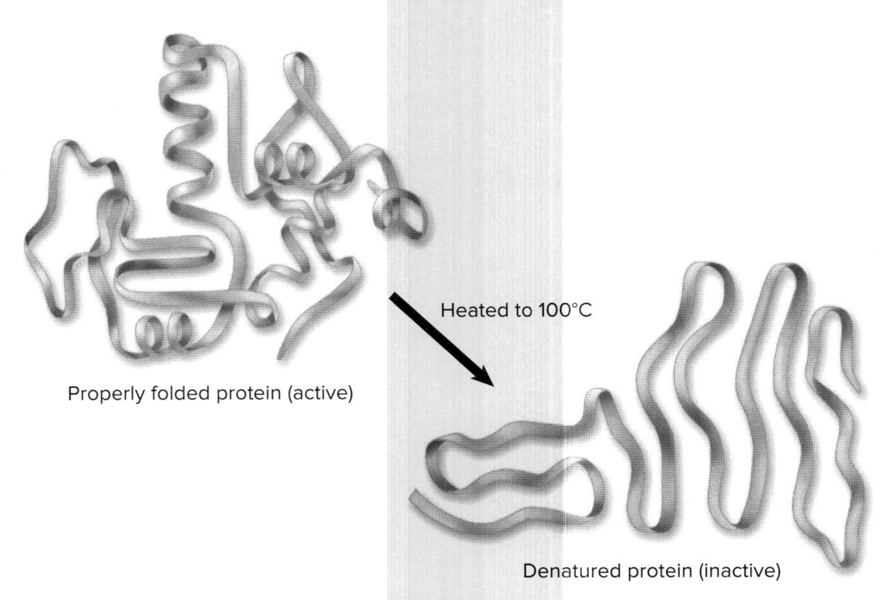

FIGURE 2.24 Denaturation of a Protein The denatured protein loses its function.

? Describe two environmental conditions that can denature a protein.

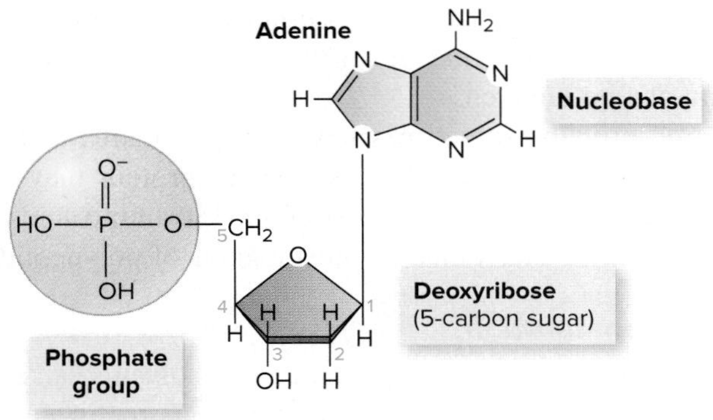

FIGURE 2.25 A Nucleotide This is one subunit of DNA. The subunit illustrated here is called adenylic acid or deoxyadenosine-5′-phosphate because the nucleobase is adenine. If the subunit lacks the phosphate group, it is called a nucleoside—in this case, deoxyadenosine.

? **What are the three components of a nucleotide?**

group is attached to the number 5 carbon of the sugar, and the nucleobase is attached to the number 1 carbon (**figure 2.25**). There are five nucleobases: two are purines, which have a double-ring structure, and three are pyrimidines, which have a single-ring structure (**figure 2.26**). The purines are adenine (A) and guanine (G); and the pyrimidines are cytosine (C), thymine (T), and uracil (U).

DNA

DNA, which stands for **deoxyribonucleic acid,** is the storage form of a cell's genetic information; the cell uses that information to make proteins. The sequence (order) of nucleotides in a cell's DNA codes for the primary structure of every protein that the cell produces. Because proteins can act as enzymes that catalyze the chemical reactions that occur in a cell, DNA determines all of the structural and functional properties of the cell. A change in the DNA sequence that results in changing even a single amino acid in a polypeptide can have a significant impact on the subsequent folding of the chain, much as changing a single letter can change the meaning of a word. For lunch, would you rather eat rice or lice?

The pentose sugar in the nucleotides of DNA is deoxyribose (which is why these nucleotides are called deoxyribonucleotides); the nucleobases are A, T, C, and G. The DNA of a typical cell is a double-stranded helical structure. Each strand is composed of a chain of deoxyribonucleotides, creating a series of alternating sugar and phosphate units that together are called the sugar-phosphate backbone (**figure 2.27**). A single DNA strand will always have a phosphate group (—PO_4) at one end and a hydroxyl group (—OH) at the other. These ends are referred to as the **5′ end** (pronounced "5 prime end") and the **3′ end** ("3 prime end"), reflecting the positions at which the functional groups are attached to the carbon atoms of deoxyribose. During DNA synthesis, the chain is elongated

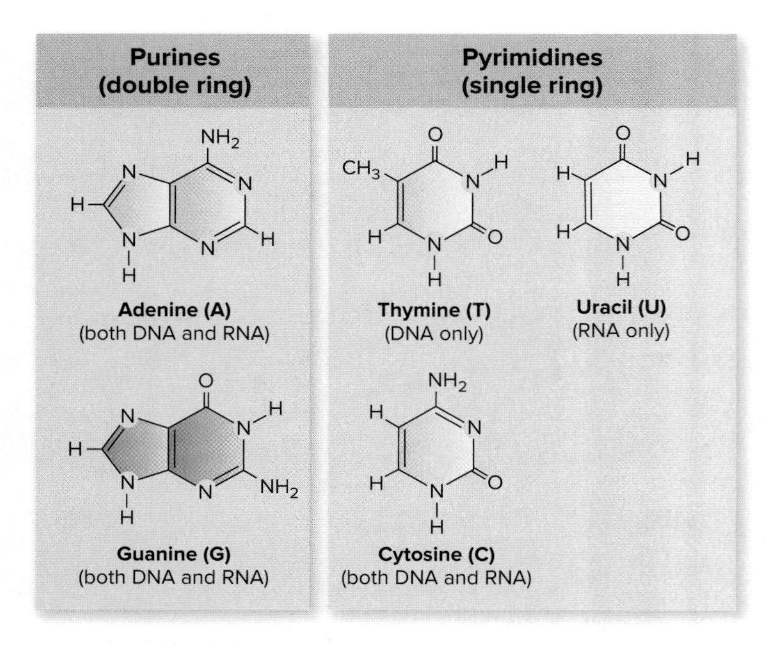

FIGURE 2.26 Purines and Pyrimidines Purines have a double-ring structure; pyrimidines have a single-ring structure.

? **Which of the nucleobases are found in DNA? In RNA?**

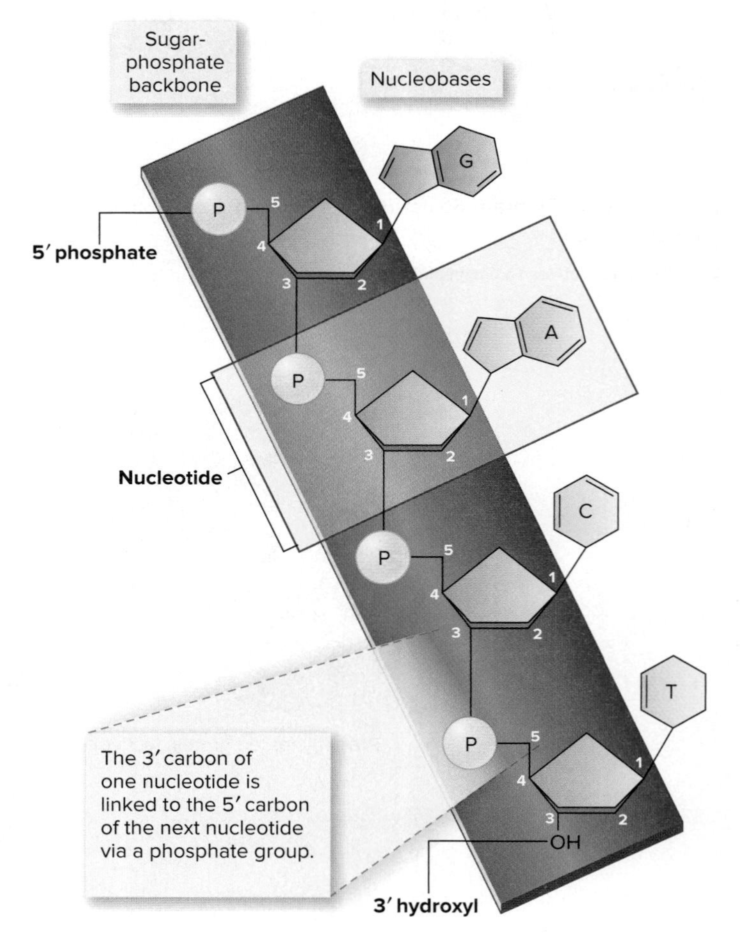

FIGURE 2.27 Single Strand of DNA Single chain of nucleotides in DNA showing the differences between the 5′ end and the 3′ end.

? **What parts of the nucleotides are joined together?**

by adding more nucleotides to the hydroxyl group at the 3′ end. This process is covered in chapter 7.

The DNA of a typical bacterium is a single double-stranded helix, arranged somewhat like a spiral staircase with two railings (**figure 2.28**). The railings represent the sugar-phosphate backbone of the molecule, and the stairs are pairs of nucleobases attached to the railings. The two strands, each about 4 million nucleotides long in a typical bacterium, are **antiparallel,** meaning they are oriented in opposite directions. One strand is oriented in the 3′ to the 5′ direction; the opposite strand is oriented in the 5′ to 3′ direction.

The two strands of DNA are also **complementary** and are held together by hydrogen bonds between the nucleobases. The term "complementary" refers to the fact that wherever an A is in one strand, a T is in the other; these opposing A-T bases are held together by two hydrogen bonds. Similarly, wherever

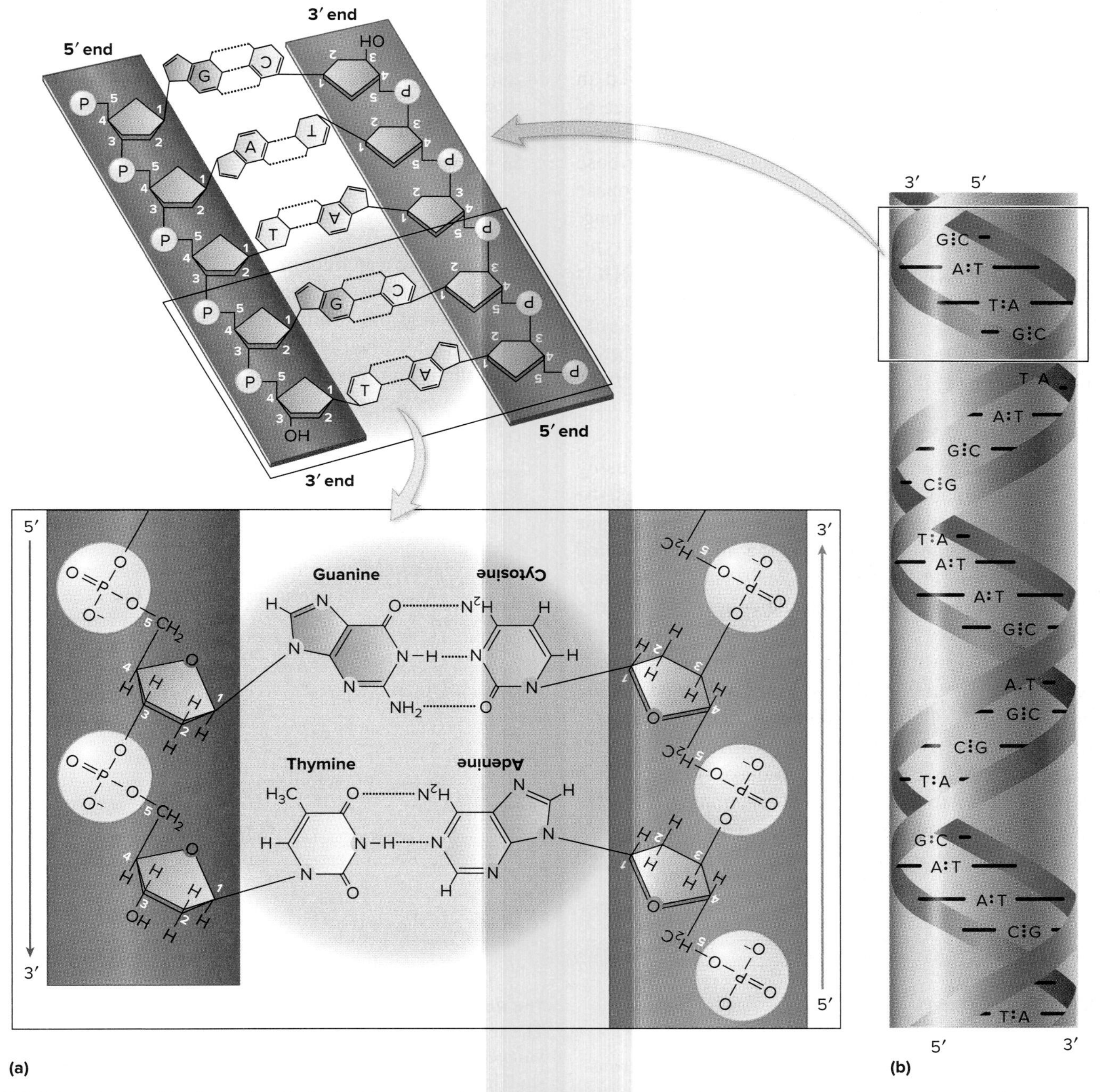

FIGURE 2.28 DNA Double-Stranded Helix (a) The sugar-phosphate backbone and the hydrogen bonding between bases. There are two hydrogen bonds between adenine and thymine and three between guanine and cytosine. **(b)** The "spiral staircase" of the sugar-phosphate backbone, with the nucleobases on the inside.

? **Which would require a higher temperature to denature—a DNA strand composed primarily of A-T base pairs or one that was the same length but composed primarily of G-C base pairs?**

a G is in one strand, a C is in the other. The G-C bases are held together by three hydrogen bonds, a slightly stronger attraction than that of an A-T pair. The double-stranded DNA molecule is generally quite stable because of the numerous hydrogen bonds that occur along its length. Over short segments of DNA, however, it is relatively easy to separate the two strands.

The characteristic bonding of A to T and G to C is called **base-pairing** and is a fundamental characteristic of DNA. Because of the rules of base-pairing, one DNA strand can be used as a template to make the complementary strand.

RNA

Several forms of **ribonucleic acid (RNA)** are involved in decoding the information in DNA to assemble a sequence of amino acids to build proteins. The basic structure of RNA is similar to that of DNA, but RNA contains the sugar ribose in place of deoxyribose. RNA also contains the nucleobase uracil in place of thymine. Further, whereas DNA is a long, double-stranded helix, RNA is considerably shorter and typically exists as a single chain of nucleotides. Although RNA is typically single-stranded, some forms contain short double-stranded regions due to hydrogen bonding between complementary nucleobases in the single strand. The types of RNA important in protein synthesis are introduced in chapter 7.

The Role of ATP

Another biologically important molecule is built using the nucleobase adenine. Adenine plus ribose produces the nucleoside called adenosine. When three phosphate groups are added to adenosine, it produces the molecule called **adenosine triphosphate,** or **ATP,** the main energy currency of cells (**figure 2.29**). To understand the concept of "energy currency," consider the currency of the United States—the dollar. You can sell a valuable item for cash, thereby converting the value of the item into dollars. You can then spend the cash, converting the dollars into another item of value. Cells do something similar. They use energy available in sunlight or in chemical compounds like sugars to make ATP, and then they "spend" the ATP to drive chemical reactions in the cell.

ATP has three phosphate groups arranged in a row, all of which are negatively charged (see figure 2.29). The charged

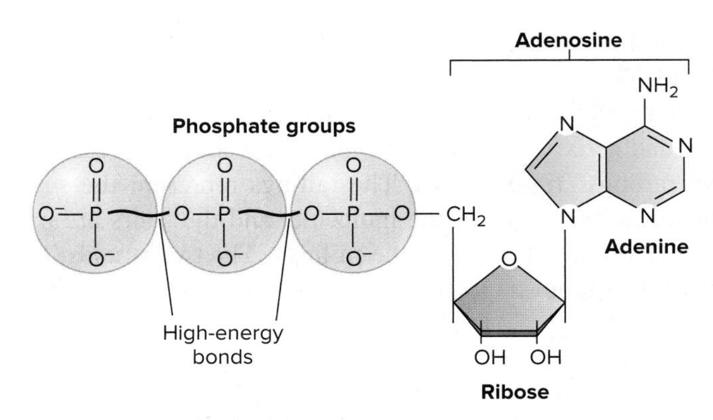

FIGURE 2.29 ATP Adenosine is a nucleoside composed of adenine and ribose. Adenosine triphosphate (ATP) serves as the energy currency of a cell. When the terminal phosphate bonds break, the energy released can be used to drive cellular reactions.

? **Why are the bonds between the phosphate groups of ATP "high energy"?**

groups repel one another, so the bonds joining them are inherently unstable. When these bonds break, they release energy that can drive a cellular process. Because of the relatively high amount of energy released, they are commonly referred to as **high-energy phosphate bonds,** indicated by the symbol ~. When the terminal phosphate bond of ATP breaks, inorganic phosphate and **adenosine diphosphate (ADP)** are formed. The processes cells use to capture the energy needed to make ATP are covered in chapter 6.

MicroAssessment 2.4

Carbohydrates serve as energy-storage molecules and structural material. Lipids are hydrophobic molecules; phospholipids are important structural components of cytoplasmic membranes. Proteins have complex structures based on the identity and sequence of amino acids. Many enzymes that catalyze chemical reactions are proteins. Nucleic acids, including DNA and RNA, are important in storing genetic information and synthesizing proteins. ATP is the main energy currency of a cell.

10. Name the subunits that are used to build a carbohydrate, a simple lipid, a protein, and a nucleic acid.

11. What level of protein structure is described as an α-helix?

12. If the DNA molecule were placed in boiling water, how would the molecule change?

Summary

2.1 Elements and Atoms

Atomic Structure

Atoms are composed of **electrons, protons,** and **neutrons** (figures 2.1, 2.2). An **element** consists of a single type of atom.

Isotopes

Isotopes are atoms of a given element that have the same number of protons, but have a different number of neutrons. Isotopes may emit energy that can be detected and tracked (figure 2.3).

The Role of Electrons

The reactivity of an atom is largely determined by the number of **valence electrons** in its outer shell. Atoms with an unfilled outer shell interact with other atoms in chemical reactions.

2.2 Chemical Bonds and Reactions

Ions and Ionic Bonds

Atoms that gain an electron become negatively charged **ions** called **anions;** those that lose an electron become positively charged, and are called **cations. Ionic bonds** form between cations and anions because of the attraction between positive and negative charges (figure 2.4).

Covalent Bonds

Covalent bonds are formed by atoms sharing electrons (figure 2.5). Molecules that contain at least carbon and hydrogen are **organic compounds;** those that do not are inorganic compounds. When atoms have an equal attraction for electrons, a **non-polar covalent bond** is formed (table 2.2). When one atom is more electronegative, a **polar covalent bond** is formed (table 2.3; figure 2.6).

Hydrogen Bonds

Hydrogen bonds are weak bonds that result from the attraction of a hydrogen atom in a polar molecule to an electronegative atom in another polar molecule (figure 2.7).

Molarity

A **mole** of one substance has the same number of particles as a mole of another substance.

Chemical Reactions

Chemical reactions proceed from **reactants** to **products. Enzymes** speed up chemical reactions.

2.3 ■ Water, pH, and Buffers

Water

Hydrogen bonding plays a very important role in the properties of water that make it important for life (figures 2.8, 2.9). Ice floats on water. Water is an effective **solvent.**

pH of Aqueous Solutions

Acidity is expressed as **pH,** which indicates the number of H^+ ions in an aqueous solution measured on a scale from 0 to 14 (figure 2.10).

Buffers

Buffers prevent a dramatic rise or fall of pH.

2.4 ■ Organic Molecules

Macromolecules are usually polymers of subunits. Different classes of macromolecules have different subunits (table 2.4). Macromolecules are made through **dehydration synthesis** and degraded by **hydrolysis** (figure 2.11). Organic molecules often contain distinct chemical arrangements called **functional groups** that contribute to the molecule's properties (table 2.5).

Carbohydrates

Monosaccharides include the 5-carbon sugars ribose and deoxyribose (figure 2.12) and the common 6-carbon sugars glucose, galactose, fructose, and mannose (figure 2.13, table 2.6). **Disaccharides** are two monosaccharides joined together by covalent bonds (figure 2.14). They include lactose, sucrose, and maltose. **Polysaccharides** are large molecules composed of chains of monosaccharide subunits or their derivatives. They include cellulose, starch, glycogen, dextran, chitin, and agar (figure 2.15).

Lipids

Lipids are a diverse group of hydrophobic, non-polar molecules. Simple lipids contain carbon, hydrogen, and oxygen. Fats consist of fatty acids linked to glycerol and may be liquid or solid at room temperature (figure 2.16). Saturated fatty acids have no double bonds, whereas unsaturated fatty acids have one or more double bonds. Compound lipids contain fatty acids and glycerol as well as elements other than carbon, hydrogen, and oxygen. **Phospholipids** are essential components of cytoplasmic membranes (figure 2.17). **Steroids** are simple lipids that have a characteristic structure consisting of four connected rings (figure 2.18). The sterol cholesterol is an example.

Proteins

Proteins are made of amino acid subunits. **Amino acids** have a central carbon atom, bonded to a carboxyl group, an amino group, and a side chain (figure 2.19). The side chain gives an amino acid its characteristic properties. Twenty major amino acids function as subunits of proteins (figure 2.20). **Peptide bonds** join the amino group of one amino acid with the carboxyl group of another (figure 2.21). The **primary structure** of a protein is its amino acid sequence (figure 2.22). The **secondary structure** is the three-dimensional shape of localized regions, characterized by helices and sheets. The **tertiary structure** is the overall three-dimensional shape, which is influenced by the R groups on amino acids. The **quaternary structure** is the shape that results from the interaction of multiple polypeptide chains. Some proteins need **chaperones** to help them fold into their functional shape. A **protein domain** is a region that folds into a stable structure independently of other parts of the molecule (figure 2.23). When bonds within a protein break, the protein changes shape and no longer functions; the protein is **denatured** (figure 2.24). Many enzymes are proteins that do not function when denatured.

Nucleic Acids

Nucleic acids carry genetic information, which is decoded to produce proteins. The nucleotide sequence of **DNA (deoxyribonucleic acid)** codes for all of the proteins that a cell produces. The nucleotides have a **nucleobase,** deoxyribose, and a phosphate group (figures 2.25, 2.26, 2.27). DNA is a double-stranded helical molecule with a sugar-phosphate backbone. The purine and pyrimidine nucleobases extend into the center of the helix (figure 2.28). The two strands of DNA are **antiparallel** and **complementary**—meaning there are G-C and A-T base pairs that are held together by hydrogen bonds between them. **RNA (ribonucleic acid)** is involved in the process that decodes the information contained in DNA. RNA is a single-stranded molecule and its nucleotides contain uracil in place of thymine and ribose in place of deoxyribose (figure 2.26). **ATP** carries energy in **high-energy phosphate bonds,** which, when broken, release the energy for use in the cell (figure 2.29).

Review Questions

Short Answer

1. Differentiate between an atom, a molecule, and a compound.
2. Why is water a good solvent?
3. Which solution is more acidic, one with a pH of 4 or one with a pH of 5? What is the concentration of H^+ ions in each?
4. What is the significance of cellulose?
5. Show how dehydration synthesis and hydrolysis reactions are related, using an example of each.
6. What is a structural isomer?

7. Name the major groups of lipids and give an example of each. What feature is common to all lipids?
8. List six functions of proteins.
9. What are the four levels of protein structure, and what is the distinguishing feature of each?
10. How do DNA and RNA differ from one another in composition and function?

Multiple Choice

1. Choose the list that goes from the lightest to the heaviest:
   a) proton, atom, molecule, electron.
   b) atom, proton, molecule, electron.
   c) electron, proton, atom, molecule.
   d) atom, electron, proton, molecule.
   e) proton, atom, electron, molecule.
2. An oxygen atom has an atomic number of 8. It typically forms
   a) no covalent bonds because it contains 8 electrons.
   b) two covalent bonds because it contains 6 electrons in its valence shell.
   c) weak hydrogen bonds with two hydrogen atoms to form a water molecule.
   d) one covalent bond with a carbon atom to form carbon dioxide.
   e) ionic bonds with another oxygen atom.
3. Dehydration synthesis is involved in the synthesis of all of the following *except*
   a) DNA.
   b) proteins.
   c) polysaccharides.
   d) lipids.
   e) monosaccharides.
4. The primary structure of a protein relates to its
   a) sequence of amino acids.
   b) length.
   c) shape.
   d) solubility.
   e) bonds between amino acids.
5. Pure water has all of the following properties *except*
   a) polarity.
   b) ability to dissolve lipids.
   c) pH of 7.
   d) covalent joining of its atoms.
   e) ability to form hydrogen bonds.
6. When the pH of a solution changes from 3 to 8, the H^+ concentration
   a) decreases as the solution becomes more basic.
   b) decreases as the solution becomes more acidic.
   c) increases as the solution becomes more acidic.
   d) increases as the solution becomes more basic.
   e) does not change as the solution becomes more basic.
7. A shortage of nitrogen (N) would make it most difficult to construct a molecule of
   a) cellulose.
   b) cholesterol.
   c) an enzyme.
   d) a triglyceride.
   e) glucose.
8. Complementarity plays a major role in the structure of
   a) proteins.
   b) lipids.
   c) polysaccharides.
   d) DNA.
   e) RNA.
9. A bilayer is associated with
   a) proteins.
   b) DNA.
   c) RNA.
   d) complex polysaccharides.
   e) phospholipids.
10. The function of a buffer is to
    a) bring the pH of a solution to neutral.
    b) speed up chemical reactions within a cell.
    c) interact with phospholipid molecules to form a cell membrane.
    d) stabilize the pH of a solution.
    e) provide energy for synthesis of ATP.

Applications

1. A group of prokaryotes known as thermophiles thrive at high temperatures that would normally destroy other organisms. Yet these thermophiles cannot survive well at the lower temperatures normally found on Earth. Propose an explanation for this observation.
2. Microorganisms use hydrogen bonds to attach to surfaces. Many of the cells lose hold of the surface because of the weak nature of these bonds. Contrast the benefits and disadvantages of using covalent bonds as a means of attaching to surfaces.

Critical Thinking

1. What properties of the carbon atom make it ideal as the key atom for so many molecules in organisms?
2. A biologist determined the amounts of several amino acids in two separate samples of pure protein. The data are shown here:

Amino Acid	Leucine	Alanine	Histidine	Cysteine	Glycine
Protein A	7%	12%	4%	2%	5%
Protein B	7%	12%	4%	2%	5%

The scientist concluded that protein A and protein B were the same protein. Do you agree with this conclusion? Justify your answer.

3. The table on the right indicates the freezing and boiling points of several molecules:

   Carbon tetrachloride and methane are non-polar molecules. How do the polarity and non-polarity of these molecules explain why the freezing and boiling points for methane and carbon tetrachloride are so much lower than those for water?

Molecule	Freezing Point (°C)	Boiling Point (°C)
Water	0	100
Carbon tetrachloride (CCl_4)	–23	77
Methane (CH_4)	–182	–164

www.mcgrawhillconnect.com

Enhance your study of this chapter with study tools and practice tests. Also ask your instructor about the resources available through Connect, including the media-rich eBook, interactive learning tools, and animations.

3 Microscopy and Cell Structure

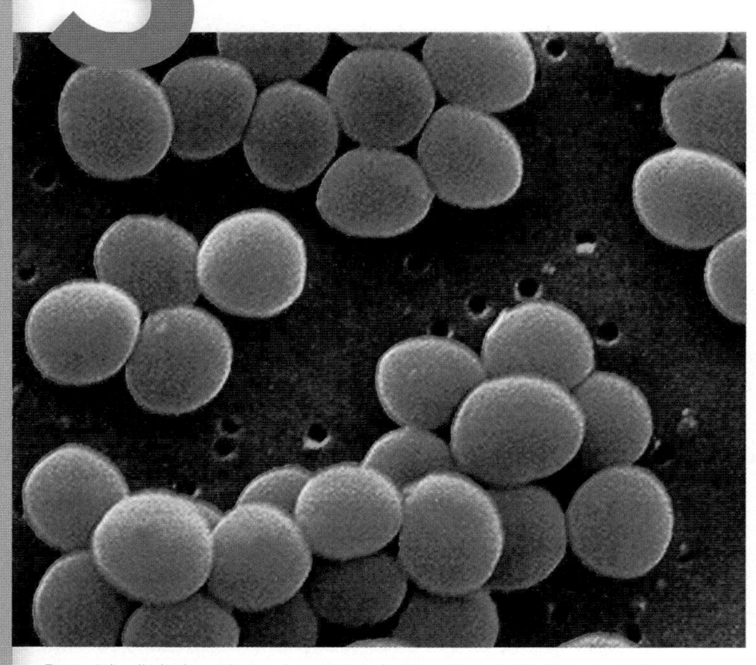

Bacterial cells (color-enhanced scanning electron micrograph). Source: *Janice Haney Carr/CDC*

KEY TERMS

Capsule A distinct, thick gelatinous material that surrounds some microorganisms.

Chemotaxis Movement of a cell toward or away from a certain chemical in the environment.

Cytoplasmic Membrane A phospholipid bilayer embedded with proteins that surrounds the cytoplasm and defines the boundary of the cell.

Endospore An extraordinarily resistant dormant cell produced by some types of bacteria.

Flagellum A type of structure used for cell movement.

Gram-Negative Bacteria Bacteria that have a cell wall characterized by a thin layer of peptidoglycan surrounded by an outer membrane; when Gram stained, these cells are pink.

Gram-Positive Bacteria Bacteria that have a cell wall characterized by a thick layer of peptidoglycan; when Gram stained, these cells are purple.

Lipopolysaccharide (LPS) Molecule that makes up the outer layer of the outer membrane of Gram-negative bacteria.

Peptidoglycan A macromolecule that provides strength to the cell wall; it is found only in bacteria.

Periplasm The gel-like material that fills the region between the cytoplasmic membrane and the outer membrane of Gram-negative bacteria.

Pili Cell surface structures that allow cells to adhere to certain material; some types are involved in a mechanism of DNA transfer.

Plasmid Extrachromosomal DNA molecule that replicates independently of the chromosome.

Ribosome Structure involved in protein synthesis.

Transport Systems Mechanisms cells use to transport nutrients and other small molecules across the cytoplasmic membrane.

A Glimpse of History

Hans Christian Joachim Gram (1853–1938) was a Danish physician working in a laboratory at the morgue of the City Hospital in Berlin, microscopically examining the lungs of patients who had died of pneumonia. He was working under the direction of Dr. Carl Friedländer, who was trying to identify the cause of pneumonia by studying patients who had died of it. Gram's task was to stain the infected lung tissue to make the bacteria easier to see under the microscope. Strangely, one of the methods he developed did not stain all bacteria equally; some types retained the first dye applied in this multistep procedure, whereas others did not. Gram's staining method revealed that two different kinds of bacteria were causing pneumonia, and these types retained the dye differently. We now recognize that this important staining method, called the Gram stain, efficiently identifies two large, distinct groups of bacteria: Gram-positive and Gram-negative. The staining outcome reflects a fundamental difference in the structure and chemistry of the cell walls of these two groups, which is why the Gram stain is a key test in the initial identification of bacterial species.

Imagine the astonishment Antony van Leeuwenhoek must have felt in the 1600s when he first observed microorganisms with his handcrafted microscopes, instruments that could magnify images approximately 300-fold. Even today, observing diverse microbes interacting in a sample of pond water can provide enormous education and entertainment.

Microscopic study of cells has revealed two fundamental types: prokaryotic and eukaryotic. The cells of all members of the domains *Bacteria* and *Archaea* are prokaryotic. In contrast, cells of all animals, plants, protozoa, fungi, and algae are eukaryotic. The similarities and differences between these two basic cell types are important from a scientific standpoint. They also have significant consequences to human health. For example, chemicals that interfere with processes unique to prokaryotic cells can be used to selectively destroy bacteria without harming humans. ⏮ prokaryotic cells, ⏮ eukaryotic cells

Prokaryotic cells are generally much smaller than most eukaryotic cells—a trait that has certain advantages as well as disadvantages. Their small size gives the cells a high surface-area-to-volume ratio, making it easier for them to take in nutrients and excrete waste products. That small size, however, also makes the cells vulnerable to a variety of threats, including predators, parasites, and competitors. To cope, prokaryotes have evolved many unique features that increase their chances of survival.

Eukaryotic cells are much more complex than prokaryotic cells. Not only are they larger, but many of their cellular processes take place within membrane-bound compartments. Eukaryotic cells are defined by the presence of one of these: the nucleus.

MICROSCOPY AND CELL MORPHOLOGY

3.1 ■ Microscopes

Learning Outcomes

1. Discuss the principles and importance of magnification, resolution, and contrast in microscopy.
2. Compare and contrast light microscopes, electron microscopes, and scanning probe microscopes.

One of the most important tools for studying microorganisms is the **light microscope,** which uses visible light and a series of lenses to magnify objects. These instruments are relatively easy to use and can magnify images approximately 1,000× (1,000-fold). They are routinely used in the laboratory to observe cell size, shape, and motility. The **electron microscope,** introduced in 1931, can magnify images in excess of 100,000×, revealing many fine details of cell structure. A major advancement came in the 1980s with the development of a scanning probe microscope, which allows scientists to produce images of individual atoms on a surface.

Principles of Light Microscopy: Bright-Field Microscopes

In light microscopy, light passes through a specimen and then through a series of magnifying lenses before entering the observer's eye. The most common type of light microscope is a **bright-field microscope,** which evenly illuminates the field of view and generates a bright background. Characteristics of this and other microscopes are summarized in **table 3.1**.

Magnification

The modern light microscope, called a **compound microscope,** has multiple magnifying lenses. The **objective lens** is a series of lenses housed in a tube immediately above the object being viewed, whereas the **ocular lens,** or eyepiece, is a lens close to the eye (**figure 3.1**). Because the objective and ocular lenses are used in combination, the total magnification is equal to the product of each lens's magnification. For example, a structure is magnified 1,000-fold when viewed through a 10× ocular lens in series with a 100× objective lens. Most compound microscopes have a selection of objective lenses that are of different powers—typically 4×, 10×, 40×, and 100×. The **condenser lens,** positioned between the light source and the specimen, does not magnify. It focuses the light on the specimen.

Resolution

The usefulness of a microscope depends primarily on its **resolving power,** which determines how much detail can be seen in the observed specimen. Resolving power is a measure of the ability to distinguish two objects that are very close together. It is defined as the minimum distance between two points at which those points can still be observed as separate objects. A high resolving power means that details of an image are clearer.

The resolving power of a microscope depends on the quality and type of lens, the wavelength of the light (shorter wavelengths give better resolution), the magnification, and how the specimen has been prepared. The maximum resolving power of the best light microscope is ordinarily 0.2 μm. This is sufficient to see the general morphology of a prokaryotic cell but too low to distinguish an object the size of most viruses. A recent and exciting advancement in light microscopy was the development of super-resolution microscopes, which use complex illumination mechanisms and merged data to construct an image with increased resolution, sometimes to 10 nm.

To obtain maximum resolution when using certain high-power objectives such as the 100× lens, immersion oil must be used to displace the air between the lens and the specimen. This prevents the refraction (bending of light rays) that occurs when light passes from glass to air (**figure 3.2**). Light rays bend when they pass from a medium of one refractive index (a measure of the relative speed of light as it passes through the medium) to another. If refraction occurs, some light rays will miss the relatively small openings of higher-power objective lenses, causing the image to look fuzzy. The immersion oil prevents refraction because it has nearly the same refractive index as glass.

Contrast

The amount of **contrast**—the difference in color intensity between an object and the background—affects how easily

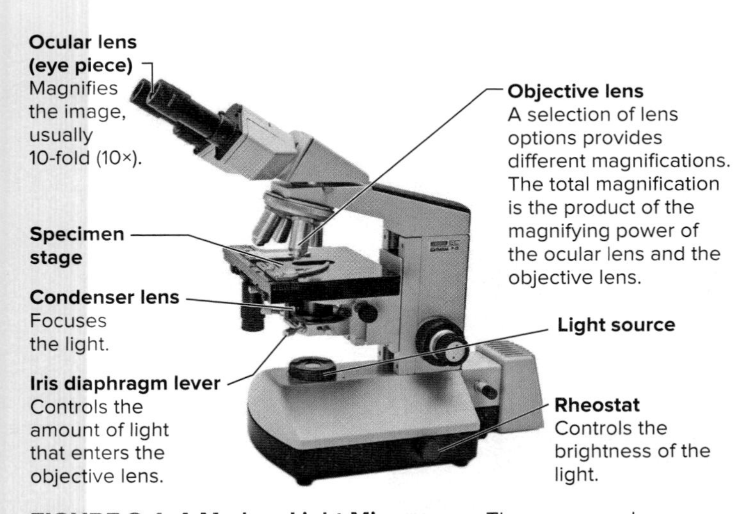

FIGURE 3.1 A Modern Light Microscope The compound microscope uses a series of magnifying lenses. ©Scenics & Science/Alamy Stock Photo

? What are the two sets of magnifying lenses called, and how do these relate to total magnification?

TABLE 3.1 A Summary of Microscopic Instruments and Their Characteristics

Instrument	Mechanism	Comment
Light Microscopes	Visible light passes through a series of lenses to produce a magnified image.	Relatively easy to use; considerably less expensive than electron and scanning probe microscopes.
Bright-field ©McGraw-Hill Education/Lisa Burgess, photographer	Illuminates the field of view evenly and generates a bright background.	Most common type of microscope.
Dark-field ©McGraw-Hill Education/Lisa Burgess, photographer	Light is directed toward the specimen at an angle.	Makes unstained cells easier to see; organisms stand out as bright objects against a dark background.
Phase-contrast ©McGraw-Hill Education/Lisa Burgess, photographer	Increases contrast by amplifying differences in refractive index.	Dense material appears darker than normal.
Differential interference contrast ©Gerd Guenther/Science Source	Two light beams pass through the specimen and then recombine.	The image of the specimen appears three-dimensional.
Fluorescence 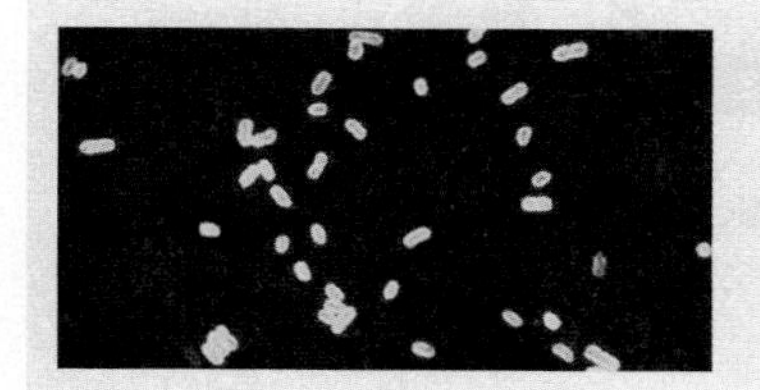 ©Evans Roberts	Projects ultraviolet light, causing fluorescent molecules in the specimen to emit longer wavelength light.	Used to observe cells stained or tagged with a fluorescent dye.

Instrument	Mechanism	Comment
Scanning laser ©A. Harrer, B. Pitts, P. Stewart/MSU-CBE	Mirrors scan a laser beam across successive regions and planes of a specimen. From that information, a computer constructs an image.	Used to obtain a three-dimensional image of a structure that has been stained with a fluorescent dye; provides detailed sectional views of intact cells.
Super-resolution ©Dr. Henrik Strah/Newcastle University	Complex illumination mechanisms and merged data are used to construct the image.	Higher resolving power than that of a conventional fluorescence microscope.
Electron Microscopes	Electron beams are used in place of visible light to produce the magnified image.	Can clearly magnify images 100,000×.
Transmission ©Lee D. Simon/Science Source	Transmits a beam of electrons through a specimen.	Complicated specimen preparation is required.
Scanning ©Dennis Kunkel Microscopy/SPL/Science Source	A beam of electrons scans back and forth over the surface of a specimen.	Used for observing surface details; produces a three-dimensional effect.
Scanning Probe Microscopes	A physical probe is used to produce detailed images of surfaces.	Produces a map showing the bumps and valleys of the sample's surface.
Atomic Force Source: Dr. Mary Ng Mah Lee, National University of Singapore/CDC	Probe moves in response to even the slightest force between it and the sample.	No special sample preparation required; produces a three-dimensional effect.

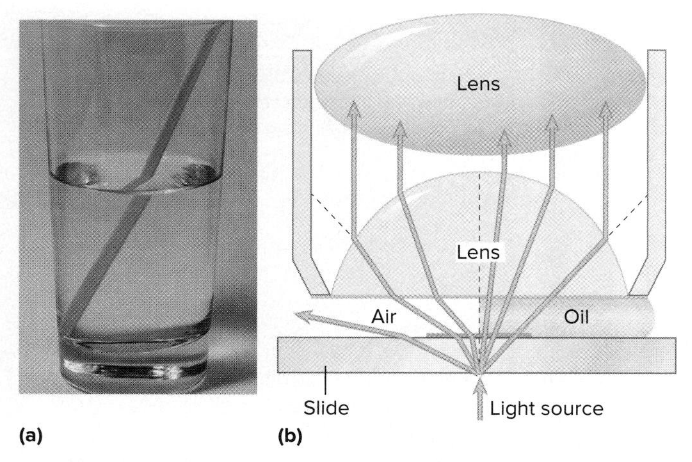

FIGURE 3.2 Refraction As light passes from one medium to another, the light rays may bend, depending on the refractive index of the two media. **(a)** The straw in water appears bent or broken because the refractive index of water is different from that of air. **(b)** Light rays bend as they pass from air to glass because of the different refractive indexes of these media. Immersion oil and glass have the same refractive index, and therefore the light rays are not bent. ©McGraw-Hill Education/Lisa Burgess, photographer

A glass rod submerged in water is easier to see than one in immersion oil. Why?

cells can be seen. As an example, colorless microorganisms are essentially transparent against a bright colorless background, so the lack of contrast makes them harder to see (**figure 3.3**). One way to overcome this difficulty is to stain the cells with one of the various dyes that will be discussed shortly. The staining process typically kills microbes, however, so it generally cannot be used to observe living cells.

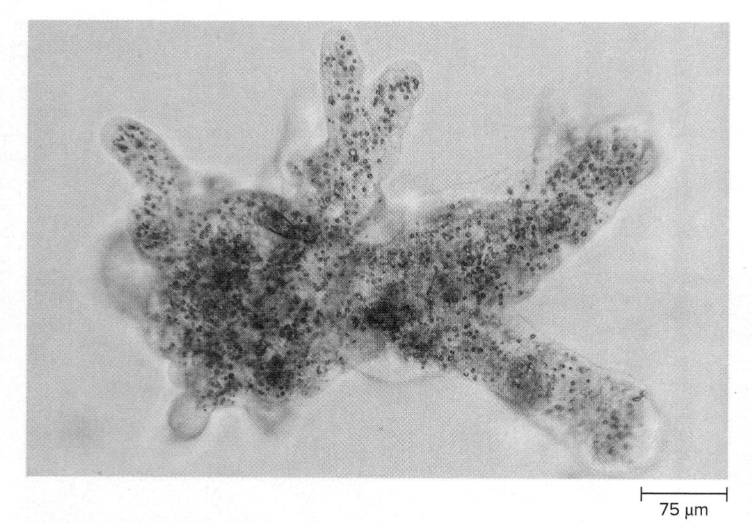

FIGURE 3.3 Bright-Field Microscopy Light illuminates the field evenly, producing a bright background. Note how difficult it is to see the transparent edges of the organism. *Amoeba proteus,* a protozoan. ©McGraw-Hill Education/Lisa Burgess, photographer

What can be done to increase the contrast before viewing microorganisms using bright-field microscopy?

Light Microscopes That Increase Contrast

Special light microscopes that increase the contrast between microorganisms and their surroundings overcome some of the difficulties of observing unstained cells. They allow scientists to easily see characteristics of living organisms such as motility. To view live organisms, the specimen is prepared as a wet mount—a drop of liquid on which a coverslip has been placed.

Dark-Field Microscopes

Cells viewed through a **dark-field microscope** stand out as bright objects against a dark background (**figure 3.4**). The microscope works in the same way that a beam of light shining into a dark room makes dust visible. For dark-field microscopy, a special mechanism directs light toward the specimen at an angle, so that only light scattered by the specimen enters the objective lens. A simple attachment called a "dark-field stop" can be placed under the condenser lens of a bright-field microscope to temporarily convert it to a dark-field microscope.

MicroByte

Dark-field microscopy is often used to see *Treponema pallidum,* which causes syphilis; the cells can be difficult to see otherwise because they are thin and do not stain well with common dyes.

Phase-Contrast Microscopes

A **phase-contrast microscope** makes cells and other dense material appear darker than normal (**figure 3.5**). It does this by amplifying the effects of the slight difference between the refractive index of dense material and that of the surrounding medium. As light passes through material, it is refracted slightly differently than when it passes through its

FIGURE 3.4 Dark-Field Microscopy Light enters at an angle, producing a dark background. *Amoeba proteus,* a protozoan. ©McGraw-Hill Education/Lisa Burgess, photographer

How does a dark-field microscope increase contrast?

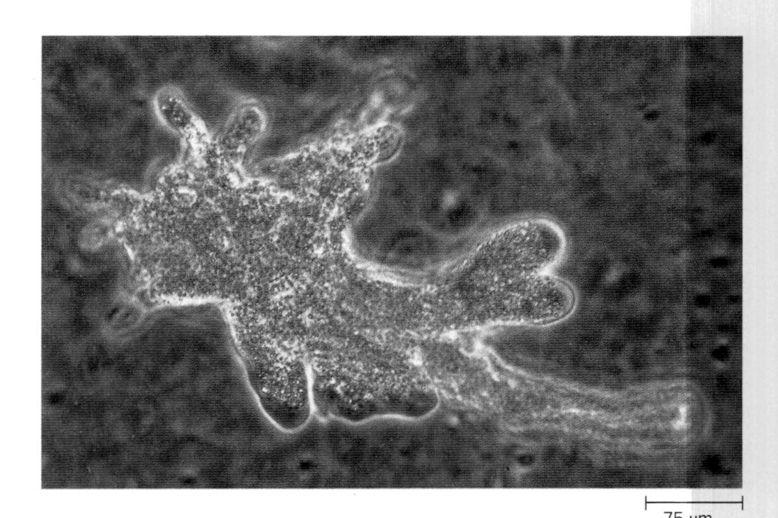

FIGURE 3.5 Phase-Contrast Microscopy Light is manipulated to darken dense material. *Amoeba proteus,* a protozoan. ©McGraw-Hill Education/Lisa Burgess, photographer

? How does a phase-contrast microscope increase contrast?

surroundings. Special optical devices increase those differences, thereby enhancing the contrast.

Differential Interference Contrast (DIC) Microscopes

A **differential interference contrast (DIC) microscope** makes the image look three-dimensional (**figure 3.6**). This microscope, like the phase-contrast microscope, depends on differences in refractive index as light passes through different materials. It has a device for separating light into two beams that pass through the specimen and then recombine. The light waves are out of phase when they recombine, thereby creating the three-dimensional appearance of the image.

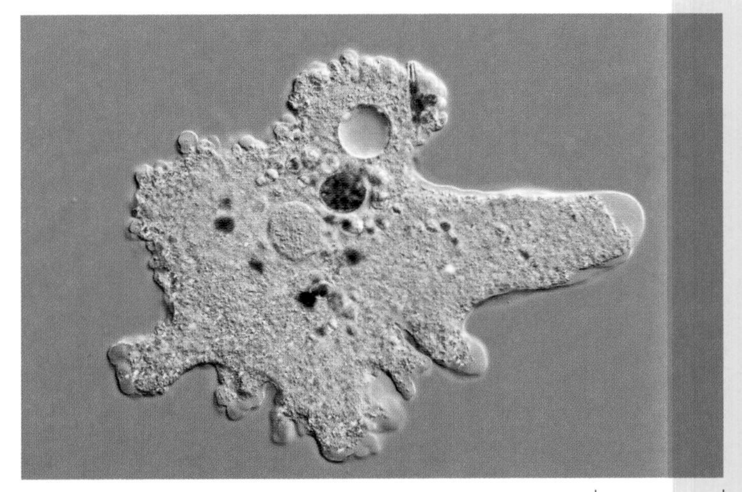

FIGURE 3.6 Differential Interference Contrast (DIC) Microscopy Light is manipulated to make the specimen appear three-dimensional. *Amoeba proteus,* a protozoan. ©Gerd Guenther/Science Source

? How does a DIC microscope increase contrast?

Light Microscopes That Detect Fluorescence

Various light microscopes that detect fluorescence are useful in certain situations.

Fluorescence Microscopes

A **fluorescence microscope** is used to observe cells or other materials that are either naturally fluorescent or stained with fluorescent dyes (**figure 3.7**). Fluorescent molecules absorb light at one wavelength (usually ultraviolet light) and then emit light of a longer wavelength. The microscope then captures only the light emitted by the fluorescent molecules, allowing fluorescent cells to stand out as bright objects against a dark background. ⏭| fluorescent dyes and tags

Most fluorescence microscopes used today are epifluorescence microscopes, meaning they project ultraviolet light onto the specimen rather than through it. Because the light does not need to travel through the specimen, cells attached to soil or other opaque particles can be observed.

Scanning Laser Microscopes (SLMs)

A **scanning laser microscope (SLM)** can be used to get detailed interior views of intact cells that have been stained with a fluorescent dye (**figure 3.8**). By using fluorescent molecules that bind only to certain compounds, the precise cellular location of those compounds can be determined. Some scanning laser microscopes can also be used to make three-dimensional images of microbial communities or other thick structures.

In **confocal microscopy,** lenses focus a laser beam to illuminate a given point on one vertical plane of a specimen. Mirrors then scan the laser beam across the specimen, illuminating successive regions and planes until the entire specimen has been scanned. Each plane corresponds to an image of one fine slice of the specimen. A computer then assembles the data and constructs a three-dimensional image, which is

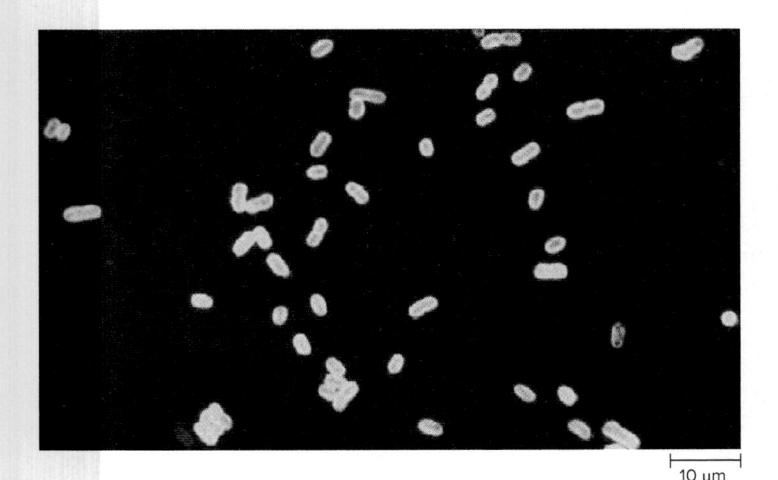

FIGURE 3.7 Fluorescence Microscopy Rod-shaped bacterial cells stained with a fluorescent molecule. ©Evans Roberts

? What is an epifluorescence microscope?

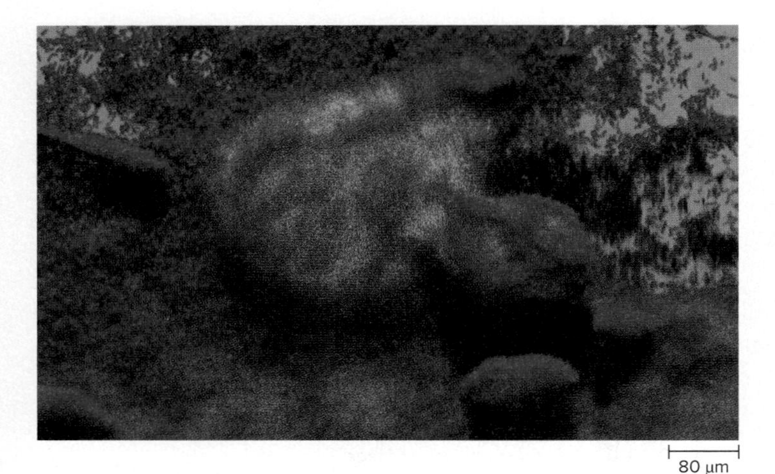

FIGURE 3.8 Scanning Laser Microscopy *Pseudomonas aeruginosa* biofilm (a polymer-encased community of microbial cells) stained with a fluorescent dye that differentiates live cells (green) and dead cells (red). The biofilm had been exposed to an antibacterial chemical, so this confocal image demonstrates that the chemical did not kill all cells in the biofilm. ©A. Harrer, B. Pitts, P. Stewart/MSU-CBE

? **How is two-photon microscopy different from confocal microscopy?**

displayed on a screen. In effect, this microscope is a miniature computerized axial tomography (CAT) scan for cells.

Two-photon microscopy (also called multiphoton microscopy) is similar to confocal microscopy, but lower-energy light is used. This light is less damaging to cells, so time-lapse images of live cells can be obtained. In addition, the light penetrates more deeply, making it possible to get interior views of relatively thick structures.

Super-Resolution Microscopes

As their name suggests, **super-resolution microscopes** achieve higher resolution than that of other light microscopes (**figure 3.9**). In recognition of this revolutionary advancement, the developers (Eric Betzig, Stefan Hell, and William Moerner) were awarded a Nobel Prize in 2014. Multiple super-resolution techniques are now available, but as an example,

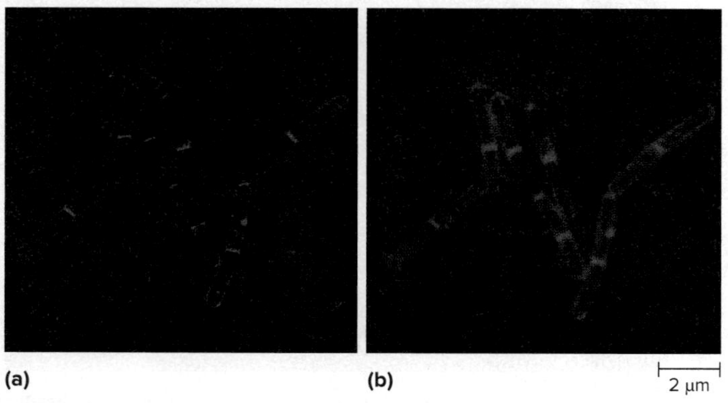

FIGURE 3.9 Super-Resolution Microscopy *Bacillus subtilis* stained with a membrane dye (Nile red). **(a)** Super-resolution image. **(b)** Conventional fluorescence image. a: ©Dr. Henrik Strah/Newcastle University; b: ©Dr. Henrik Strah/Newcastle University

? **What factors normally limit the resolving power of a light microscope?**

one uses complicated mechanisms to separately illuminate fluorescent molecules that are otherwise too close together to be seen as distinct. The data obtained are then merged to create an image with increased resolution, sometimes to 10 nm.

Electron Microscopes

Electron microscopy is in some ways comparable to light microscopy, but it can clearly magnify an object 100,000×. Rather than using glass lenses, visible light, and the eye to observe the specimen, an electron microscope uses electromagnetic lenses, electrons, and a fluorescent screen to produce the magnified image (**figure 3.10**). That image can be photographed, creating a picture called an electron photomicrograph. The image is black-and-white, but may be artificially colored to make certain components stand out or to add visual interest. ⏮ **electrons**

Electrons have a wavelength about 1,000 times shorter than visible light, so the resolving power of electron microscopes is about 1,000-fold more than light microscopes—about 0.3 nanometers (nm) or 0.3×10^{-3} µm (see figure 1.7). Consequently, considerably more detail can be observed with electron microscopy.

Electron microscopes are complex instruments because the lenses and specimen must typically be in a vacuum to avoid

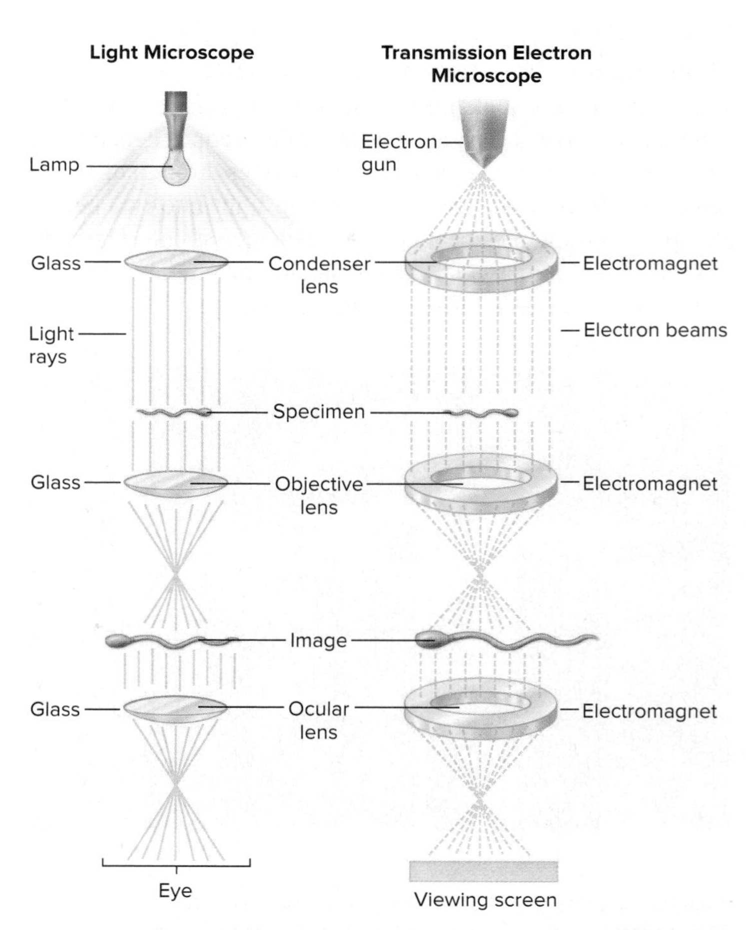

FIGURE 3.10 Principles of Light and Electron Microscopy For the sake of comparison, the light source for the light microscope has been inverted (the light is shown at the top and the ocular lens at the bottom).

? **Some electron micrographs are "color enhanced." Why would this be done?**

air molecules that would otherwise interfere with the path of electrons. The need for a vacuum means that specimen preparation is complicated and makes it impossible to observe living cells. Techniques are being developed that do not require a vacuum, making it possible to view live cells, but images obtained are not as clear as with conventional methods.

Transmission Electron Microscopes (TEMs)

A **transmission electron microscope (TEM)** is used to observe fine details of cell structure (**figure 3.11**). It works by directing a beam of electrons that either pass through the specimen or scatter (change direction), depending on the density of the region. The darker areas of the resulting image correspond to the denser portions of the specimen.

A process called thin-sectioning is used to view details of a specimen's internal structures. First, the sample is treated with a preservative and dehydrated before embedding it in plastic. Once embedded, the sample can be cut into exceptionally thin slices with a diamond or glass knife and then stained with heavy metals. Even a single bacterial cell must be cut into slices to be viewed via TEM. Unfortunately, the procedure can severely distort cells, causing artifacts (artificial structures introduced as a result of the process). Distinguishing actual cell components from artifacts that result from specimen preparation is a major concern.

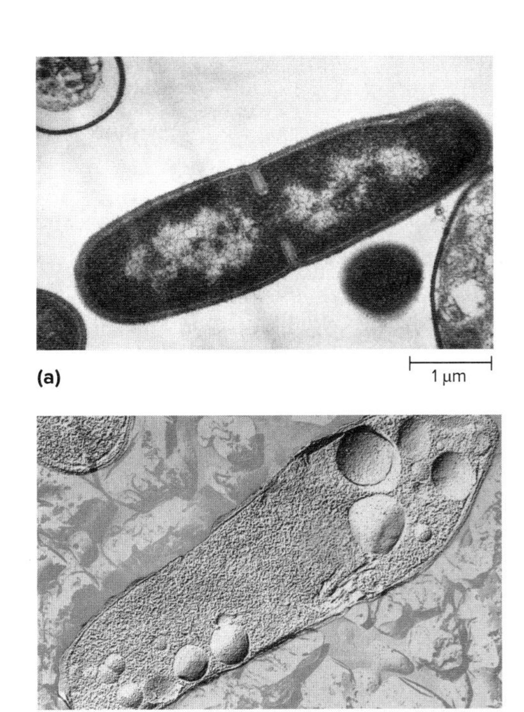

FIGURE 3.11 Transmission Electron Microscopy (TEM)
(a) *Bacillus licheniformis*, prepared by thin-sectioning; **(b)** *Pseudomonas aeruginosa*, prepared by freeze-etching. a: ©Lee D. Simon/Science Source; b: ©Dr. Tony Brain/Science Source

How is thin-sectioning different from freeze-etching?

A method called freeze-fracturing is used to observe the shape of structures within a cell. The specimen is rapidly frozen and then fractured by hitting it with a knife blade. The cells break open, usually along the middle of internal membranes. Next, the surface of the section is coated with a thin layer of carbon to create a copy of the surface. This replica is then examined in the electron microscope. A variation of freeze-fracturing is freeze-etching. In this process, the frozen surface exposed by fracturing is dried slightly under vacuum, which allows underlying regions to be exposed.

A newer method of preparing and observing specimens with an electron microscope is **cryo-electron microscopy (cryo-EM).** This method involves rapidly freezing the specimen, thereby avoiding some of the sample preparation processes that damage cells. A variation called cryo-electron tomography compiles images taken at different angles to create three-dimensional images of specimens.

Scanning Electron Microscopes (SEMs)

A **scanning electron microscope (SEM)** is used to observe surface details of cells (**figure 3.12**). A beam of electrons scans back and forth over the surface of a specimen coated with a thin film of metal. As the beam moves, electrons are released from the specimen. Those electrons are then detected, resulting in an image with a dramatic three-dimensional effect.

Scanning Probe Microscopes

Scanning probe microscopes, such as an **atomic force microscope (AFM),** use a physical probe to produce detailed

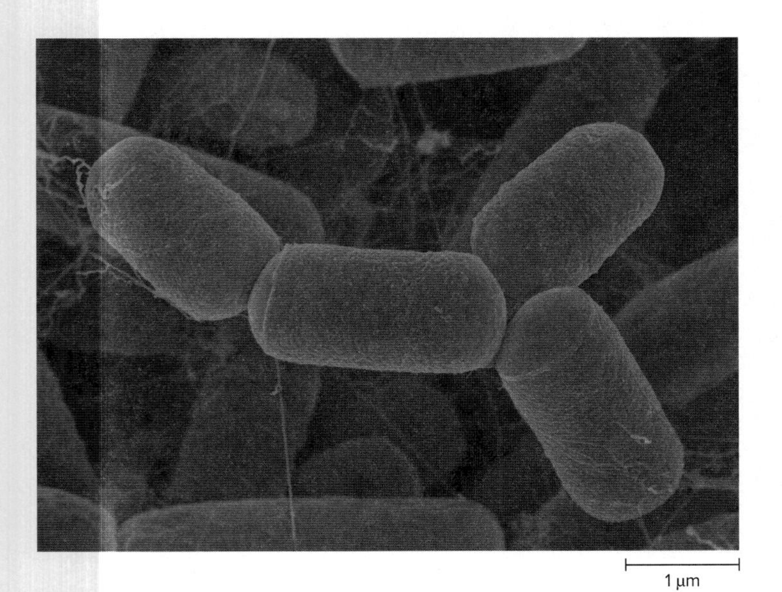

FIGURE 3.12 Scanning Electron Microscopy (SEM)
Methanobrevibacter smithii, an archaeon. ©Dennis Kunkel Microscopy/SPL/Science Source

How is scanning electron microscopy different from transmission electron microscopy?

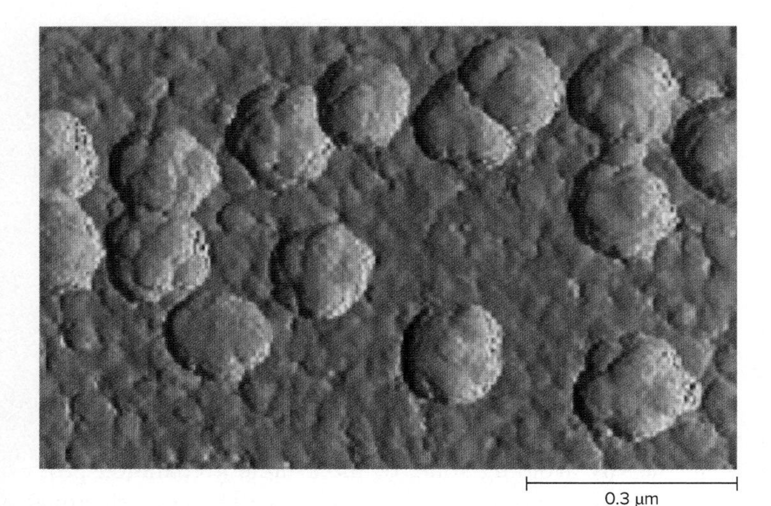

FIGURE 3.13 Scanning Probe Microscopy Atomic force micrograph. SARS virus exiting a host cell. Source: Dr. Mary Ng Mah Lee, National University of Singapore/CDC

? How does the resolving power of atomic force microscopy compare to that of electron microscopy?

images of surfaces (**figure 3.13**). An AFM has a resolving power much greater than that of an electron microscope and does not require special sample preparation. In fact, the instrument can inspect samples either in air or submerged in liquid. The mechanics of an AFM can be compared to that of a stylus mounted on the arm of a record player. A very sharp probe (stylus) moves across the sample's surface, "feeling" the bumps and valleys of the atoms. As the probe scans the sample, a laser measures its motion, and a computer produces a surface map of the sample.

MicroAssessment 3.1

The usefulness of a microscope depends on its resolving power. Light microscopes can magnify objects 1,000×. The most common type of microscope is the bright-field microscope. Dark-field, phase-contrast, and DIC microscopes increase the contrast between a microorganism and its surroundings. Fluorescence microscopes are used to observe microbes stained with a fluorescent dye. Scanning laser microscopes are used to construct three-dimensional images of thick structures that have been stained with a fluorescent dye. Super-resolution microscopes have revolutionized fluorescence microscopy by improving the resolving power. Electron microscopes can magnify an object 100,000×; they include transmission electron microscopes (TEMs) and scanning electron microscopes (SEMs). Scanning probe microscopes, such as atomic force microscopes (AFMs), produce detailed images of surfaces.

1. Why must oil be used to obtain the best resolution with a 100× lens of a light microscope?
2. What are some drawbacks of electron microscopes?
3. If an object being viewed under the phase-contrast microscope had the same refractive index as the background material, how would it appear?

3.2 ■ Preparing Specimens for Light Microscopy

Learning Outcomes

3. Describe the principles of a wet mount, a simple stain, the Gram stain, and the acid-fast stain.
4. Describe the special stains used to observe capsules, endospores, and flagella.
5. Describe the benefits of using fluorescent dyes and tags.

One of the easiest ways to microscopically examine a specimen is to use a **wet mount.** This consists of a drop of a liquid specimen placed on a microscope slide and overlaid with a coverslip, sandwiching a thin film of liquid between the slide and the coverslip. Although a wet mount makes it possible to observe living, moving microorganisms, the cells can be difficult to see because they are usually colorless and often move around quickly. To avoid this problem, the cells can be immobilized and then stained with one or more dyes.

To prepare a specimen for staining, a drop of liquid containing the organism is first placed on a microscope slide and allowed to dry (**figure 3.14**). The resulting specimen forms a film, or smear. Next, the slide is treated using heat or certain chemicals to fix (attach) the cells to the slide. Heat fixing can easily be done by passing the slide over a flame (smear side up); alternatively the slide can be placed on a slide warmer to both dry and fix the specimen. The cells are then stained, generally using one of the procedures described next and summarized in **table 3.2.**

Simple Staining

If only a single dye is used to stain a specimen, the procedure is called **simple staining.** Procedures typically use basic dyes, meaning the dyes carry a positive charge. These dyes stain cells because the positively charged dye particles are attracted to the many negatively charged cellular components. Examples of basic dyes include methylene blue, crystal violet, safranin, and malachite green.

Although acidic dyes do not stain cells, they can be used for **negative staining,** a procedure that colors the background. The cells repel the negatively charged dye, allowing the colorless cell to stand out against the background. An advantage of negative staining is that it can be done as a wet mount. This avoids the heat-fixing step, a process that can distort the shape of the cells.

Differential Staining

Differential staining is used to distinguish different groups of bacteria. The two most frequently used differential staining techniques are the Gram stain and the acid-fast stain.

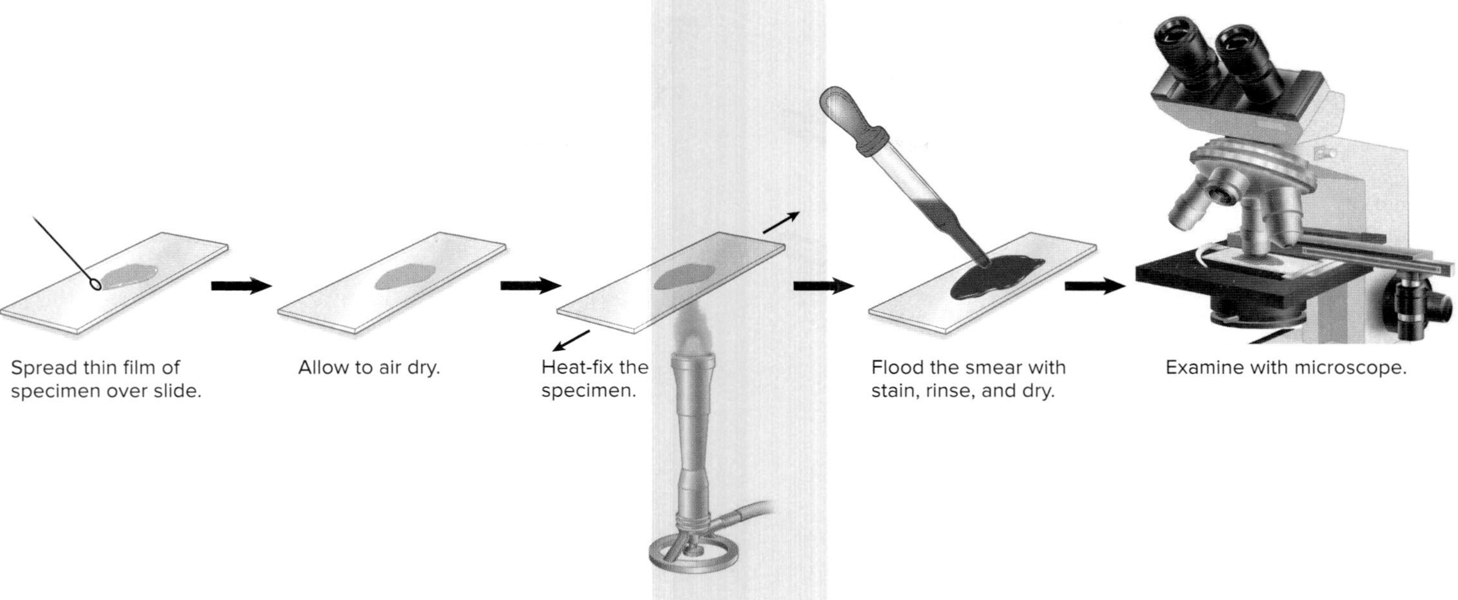

FIGURE 3.14 Staining Bacteria

What is the purpose of heating a smear before staining?

Gram Stain

The **Gram stain** is by far the most widely used procedure for staining bacteria. The basis for it was developed over a century ago by Dr. Hans Christian Gram (see **A Glimpse of History**). He showed that bacteria can be separated into two major groups: **Gram-positive bacteria** and **Gram-negative bacteria.** We now know that the difference in the staining properties of these two groups reflects a fundamental difference in the structure of their cell walls.

Gram staining involves four steps (**figure 3.15**).

1. The smear is first flooded with the **primary stain,** crystal violet in this case. The primary stain is the first dye applied in differential staining and generally stains all cells.

TABLE 3.2 A Summary of Stains and Their Characteristics

Stain	Characteristic
Simple Stains	A basic dye is used to stain cells. Easy way to increase the contrast between otherwise colorless cells and a colorless background.
Differential Stains	A multistep procedure is used to stain cells and distinguish one group of microorganisms from another.
Gram stain	Used to separate bacteria into two major groups: Gram-positive and Gram-negative. The staining characteristics of these groups reflect a fundamental difference in the chemical structure of their cell walls. This is by far the most widely used staining procedure.
Acid-fast stain	Used to detect organisms that do not easily take up stains, particularly members of the genus *Mycobacterium.*
Special Stains	A special procedure is used to stain specific cell structures.
Capsule stain	The common procedure darkens the background, so the capsule stands out as a clear area surrounding the cell.
Endospore stain	Stains endospores, a type of dormant cell that does not readily take up stains. Endospores are produced by *Bacillus* and *Clostridium* species.
Flagella stain	The staining agent adheres to and coats the otherwise thin flagella, making them visible with the light microscope.
Fluorescent Dyes and Tags	Fluorescent dyes and tags absorb ultraviolet light and then emit light of a longer wavelength.
Fluorescent dyes	Some fluorescent dyes bind to compounds found in all cells; others bind to compounds specific to only certain types of cells.
Fluorescent tags	Antibodies to which a fluorescent molecule has been attached are used to tag specific molecules.

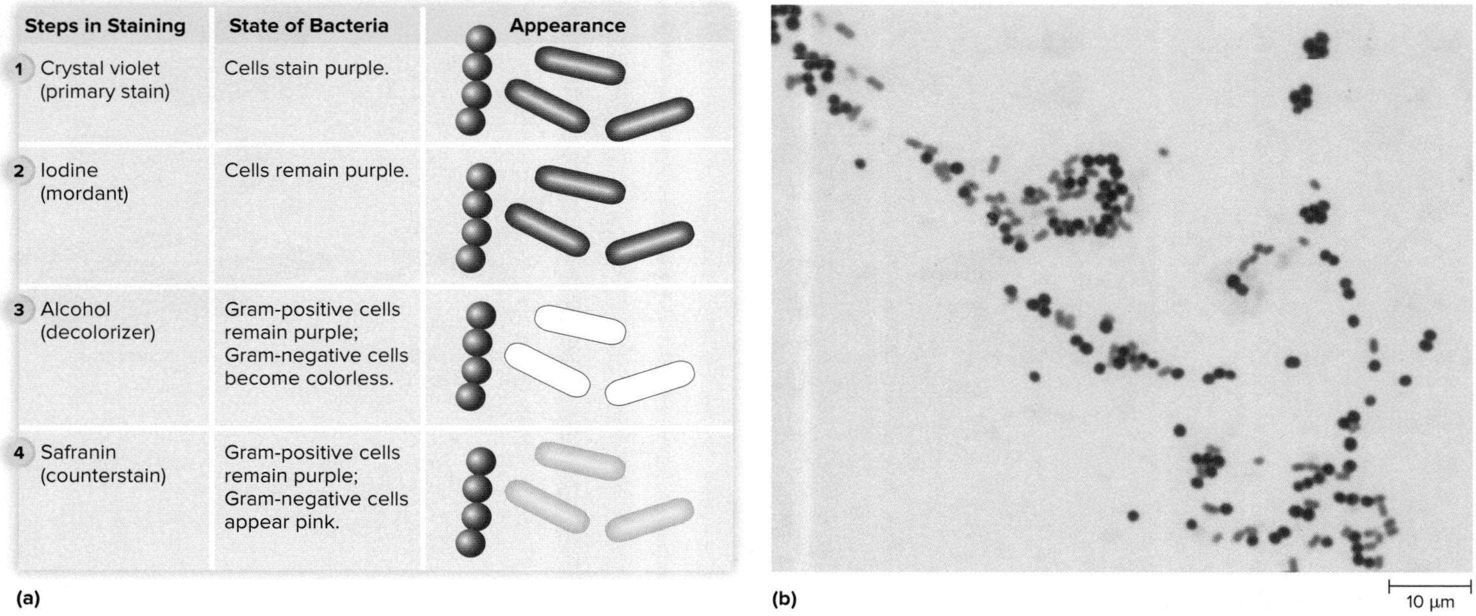

Steps in Staining	State of Bacteria	Appearance
1 Crystal violet (primary stain)	Cells stain purple.	
2 Iodine (mordant)	Cells remain purple.	
3 Alcohol (decolorizer)	Gram-positive cells remain purple; Gram-negative cells become colorless.	
4 Safranin (counterstain)	Gram-positive cells remain purple; Gram-negative cells appear pink.	

(a)

(b)

10 μm

FIGURE 3.15 Gram Stain (a) Steps in the Gram stain procedure. **(b)** Results of a Gram stain. The Gram-positive cells (purple) are *Staphylococcus aureus;* the Gram-negative cells (pink) are *Escherichia coli.* ©McGraw-Hill Education/Lisa Burgess, photographer

? **In the Gram stain procedure, which step is most time-sensitive?**

2. The smear is rinsed to remove excess dye and then flooded with a solution called Gram's iodine. The iodine is a mordant, meaning that it reacts with the dye to form a complex that is less likely to be washed from the cell.
3. The stained smear is rinsed, and then a **decolorizing agent**—usually 95% alcohol—is briefly added. This quickly removes the dye–iodine complex from the Gram-negative, but not the Gram-positive, bacteria.
4. The smear is rinsed and then a secondary stain, or **counterstain,** is applied to give a different color to the now colorless Gram-negative bacteria. For this purpose, the red dye safranin is used, which colors the cells pink. This dye enters Gram-positive cells as well, but because they are already purple, it makes little difference to their color.

To obtain reliable results, the Gram stain must be done properly. One common mistake is to decolorize a smear for the incorrect amount of time. Even Gram-positive cells can lose the crystal violet–iodine complex during prolonged decolorization. When this happens, over-decolorized Gram-positive cells will appear pink after counterstaining. In contrast, under-decolorizing results in Gram-negative cells appearing purple. Another important consideration is the age of the culture. As bacterial cells age, they lose their ability to retain the crystal violet–iodine dye complex. As a result, cells from old cultures may appear pink. Thus, Gram staining fresh cultures (less than 24 hours old) gives more reliable results.

Acid-Fast Stain

The **acid-fast stain** is a procedure used to detect a small group of organisms that do not readily take up dyes. Among these are members of the genus *Mycobacterium,* including a species that causes tuberculosis and one that causes Hansen's disease (leprosy). The cell wall of these bacteria contains high concentrations of mycolic acids—waxy fatty acids that prevent uptake of dyes such as those used in Gram staining. Therefore, harsh methods are needed to stain these organisms. Once stained, however, these same cells are very resistant to decolorization. Because mycobacteria are among the few organisms that retain the dye in this procedure, acid-fast staining can be used to presumptively identify them in clinical specimens. ⏭ **tuberculosis,** ⏭ **Hansen's disease,** ⏮ **fatty acid**

Acid-fast staining, like Gram staining, requires multiple steps. The primary stain is a red dye called carbol fuchsin, which is applied as a concentrated solution. The slide is then rinsed to remove excess dye before being flooded with acid-alcohol, a strong decolorizing agent. This agent removes the carbol fuchsin from everything but cells of a few microbial species referred to as **acid-fast.** Methylene blue is then used as a counterstain. As a result of the staining procedure, acid-fast organisms are bright reddish-pink, making them easy to distinguish from the blue non-acid-fast cells (**figure 3.16**).

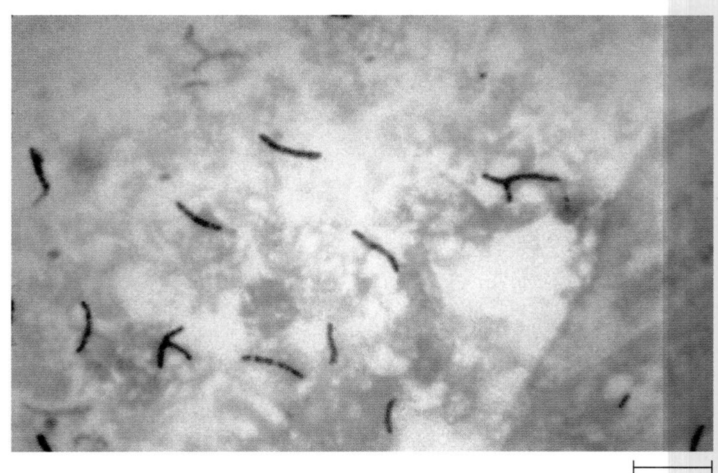

FIGURE 3.16 Acid-Fast Stain *Mycobacterium* cells keep the red primary stain, carbol fuchsin. Counterstaining with methylene blue colors the non-acid-fast cells blue. Source: Dr. George P. Kubica/CDC

What characteristic of *Mycobacterium* cells makes them acid-fast?

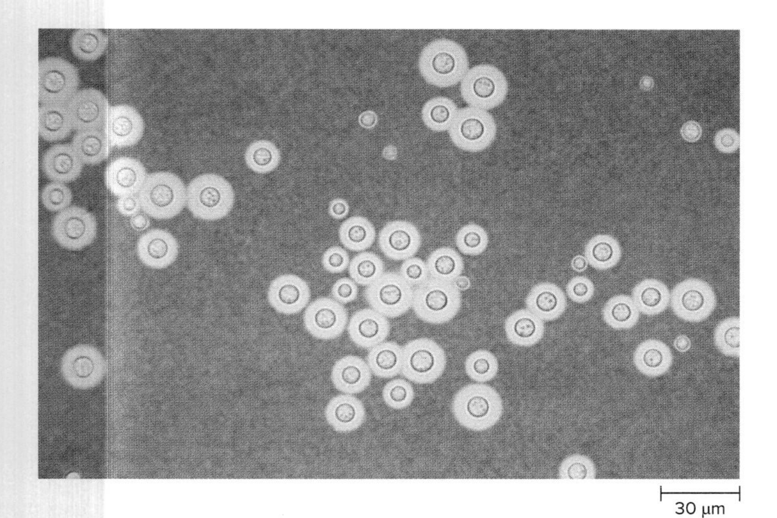

FIGURE 3.17 Capsule Stain *Cryptococcus neoformans,* an encapsulated yeast. The capsules stand out against the India ink–stained background. Source: Dr. Leanor Haley/CDC

How is the India ink capsule stain an example of a negative stain?

Special Stains to Observe Cell Structures

Special procedures can also be used to stain specific structures inside or outside the cell. The function of these structures will be discussed in more depth later in the chapter.

Capsule Stain

A capsule is a gel-like layer that surrounds certain microbes. This layer has a protective function and often increases an organism's pathogenicity. Capsules stain poorly, so **capsule stains** typically use dyes that stain the background instead, thereby making the capsule visible. In one common method, India ink is added to a suspension of cells to make a wet mount. The fine particles of ink darken the background, allowing the capsule to stand out as a clear area surrounding a cell (**figure 3.17**). ⏭ capsule, ⏮ pathogen

Endospore Stain

Members of certain genera, including *Bacillus* and *Clostridium,* form a special type of resistant, dormant cell called an endospore. These structures do not stain with the Gram stain, but they can often be seen as clear, smooth objects within stained cells. ⏭ endospore

To make endospores easier to see, an **endospore stain** is used. This multistep procedure often uses malachite green as a primary stain, with gentle heating to help the dye enter endospores. The smear is then rinsed with water, which removes the dye from everything but the endospores. After that, the smear is counterstained, most often with the red dye safranin. Using this method, the endospores will be green while all other cells will be pink (**figure 3.18**).

Flagella Stain

Flagella are appendages that provide the most common mechanism of motility for prokaryotic cells, but they are ordinarily too thin to be seen with the light microscope. The **flagella stain** uses a substance that allows the staining agent to adhere to and coat the thin flagella, making them visible using light microscopy (**figure 3.19**). ⏭ flagella

Not all species have flagella, so their presence is a distinguishing feature. In addition, the arrangement of flagella

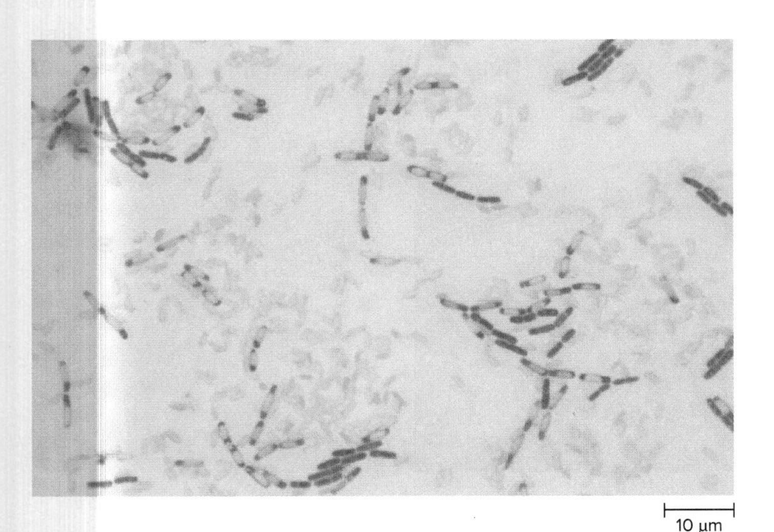

FIGURE 3.18 Endospore Stain Endospores retain the primary stain, malachite green. Counterstaining with safranin colors other cells pink. ©McGraw-Hill Education/Lisa Burgess, photographer

What color would *Staphylococcus aureus* cells be with the endospore stain shown in the photo?

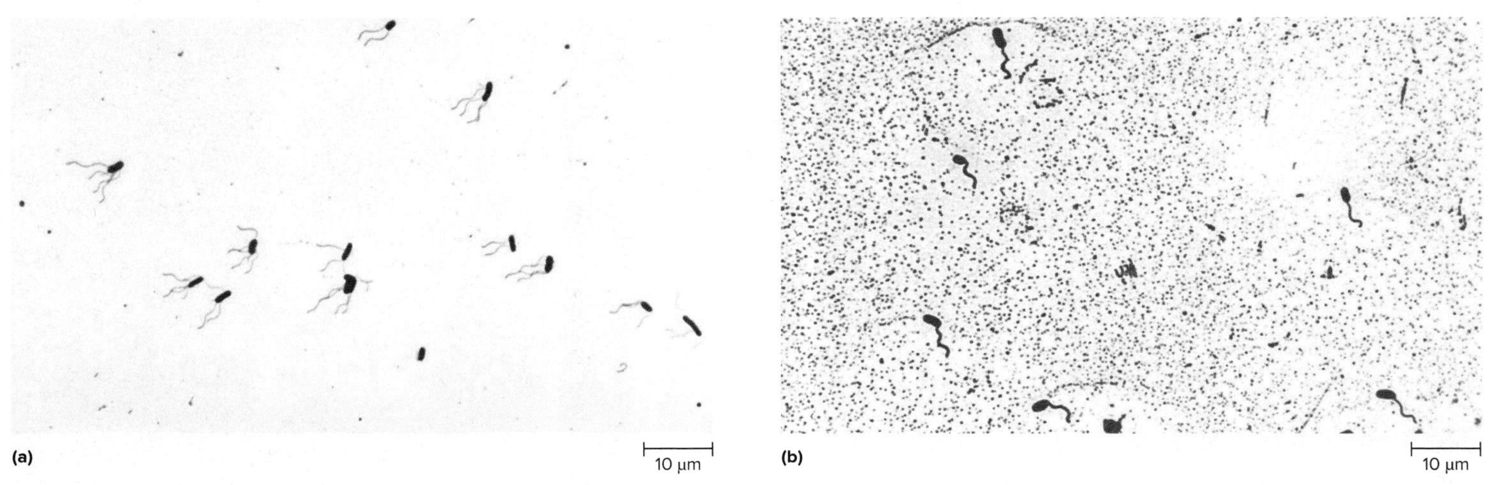

FIGURE 3.19 Flagella Stain The staining agent sticks to and coats the flagella. This increases their diameter so they can be seen with the light microscope. **(a)** *Bordetella bronchisepticum* (peritrichous flagella); **(b)** *Vibrio cholerae* (polar flagellum). Source: Dr. William A. Clark/CDC

? How can the flagella stain be helpful in identifying bacteria?

around a cell can be used to characterize an organism. For example, some species have flagella distributed around the surface of a cell—an arrangement called **peritrichous** (*peri* means "around"), whereas others have a single flagellum at one end—an arrangement called **polar** (see figure 3.19). Other arrangements also occur, including a tuft of flagella at one or both ends of the cell.

Fluorescent Dyes and Tags

Fluorescence can be used to observe total cells, a subset of cells, or cells with certain proteins on their surface, depending on the procedure (**figure 3.20**). One example is a fluorescent dye that binds to structures in all cells; it can be used for determining the total number of microbial cells in a sample. Another fluorescent dye binds to all cells but is changed by cellular processes, so it can be used to distinguish between live and dead cells (see figure 3.8). Some fluorescent dyes bind to the mycolic acids in the cell walls of *Mycobacterium* species, making the dyes useful in a staining procedure similar to the acid-fast stain. **mycolic acids**

A special technique called **immunofluorescence** is used to tag specific cell components with a fluorescent dye attached to an antibody (see figure 18.6). By tagging a protein unique to a given microbe, immunofluorescence can be used to detect and identify that organism. Antibodies, and how they are obtained, will be described in chapters 15 and 18. **antibody**

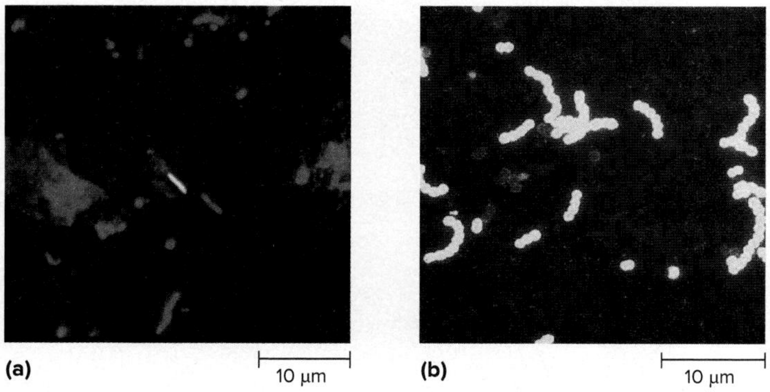

(a) 10 µm (b) 10 µm

FIGURE 3.20 Fluorescent Dyes and Tags (a) To detect a *Mycobacterium* species in a sputum sample, a dye that binds mycolic acids and fluoresces yellow is used in a modification of the acid-fast technique. In this example, acridine orange was used to stain all other organisms. **(b)** Fluorescent antibodies tag specific molecules—in this case, the antibody binds to a molecule unique to *Streptococcus pyogenes*. a: Source: CDC; b: ©Evans Roberts

? How can fluorescent dyes and tags be used to identify bacteria?

MicroAssessment 3.2

Dyes are used to stain cells so they can be seen against an unstained background. The Gram stain is the most commonly used differential stain. The acid-fast stain is used to detect *Mycobacterium* species. Specific dyes and techniques are used to observe cell structures such as capsules, endospores, and flagella. Fluorescent dyes and tags can be used to observe total cells, a subset of cells, or cells that have certain proteins on their surface.

4. What are the functions of a primary stain and a counterstain?
5. Describe one error in the staining procedure that would result in a Gram-positive bacterium appearing pink.
6. What color would a Gram-negative bacterium be in an acid-fast stain?

3.3 ■ Morphology of Prokaryotic Cells

Learning Outcomes

6. Describe the common bacterial shapes and groupings, and their significance.
7. Describe two multicellular associations of bacteria.

Prokaryotic cells come in a variety of simple shapes and often form characteristic groupings. Some aggregate, living as multicellular associations.

Shapes

Most common bacteria are one of two general shapes: spherical, called a **coccus** (plural: cocci), or cylindrical, called a **rod** (**figure 3.21**). Some cocci are slightly oval and may be flattened on one end. A rod-shaped bacterium is often called a **bacillus** (plural: bacilli), and a rod so short that it might be mistaken for a coccus is called a **coccobacillus.** The descriptive term "bacillus" should not be confused with *Bacillus,* the name of a genus. Although members of the genus *Bacillus* are rod-shaped, so are many other bacteria, including *Escherichia coli.*

Cells come in a variety of other shapes. A short, curved rod is a **vibrio** (plural: vibrios), whereas a curved rod long enough to form spirals is a **spirillum** (plural: spirilla). A long, spiral-shaped cell with a flexible cell wall and a unique mechanism of motility is a **spirochete.** Bacteria that characteristically vary in their shape are **pleomorphic** (*pleo* meaning "many" and *morphic* referring to shape).

The greatest diversity in shapes is found in low-nutrient aquatic environments. Cells there often have a large surface area, which helps them absorb nutrients more easily. For example, some aquatic bacteria have cytoplasmic extensions, giving them a star-like appearance (**figure 3.22**). Square, tile-like archaeal cells have been found in salt ponds.

Arrangements

Most prokaryotes divide by binary fission, a process in which one cell divides into two. Cells often stick to each other following division, forming characteristic arrangements (**figure 3.23**).

Cells that divide in one plane can form chains of varying length. Cocci that typically occur in pairs are routinely called **diplococci.** An important clue in the identification of *Neisseria gonorrhoeae* is its characteristic diplococcus arrangement. Other cells form long chains, a characteristic typical of some members of the genus *Streptococcus* (*strepto* means "twisted chain").

Cocci often divide in more than one plane. Those that divide in perpendicular planes form cubical packets. Members of the genus *Sarcina* form such packets. Cocci that divide in

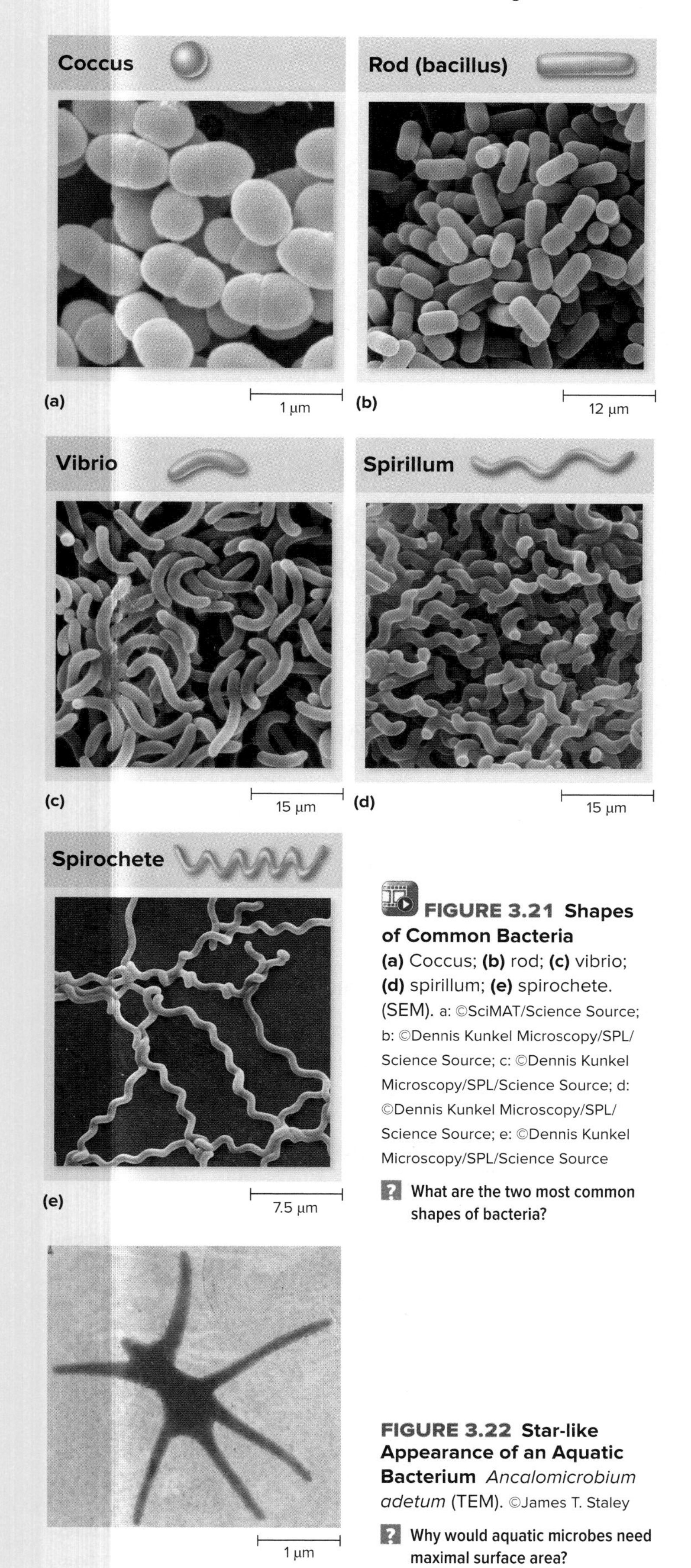

FIGURE 3.21 Shapes of Common Bacteria **(a)** Coccus; **(b)** rod; **(c)** vibrio; **(d)** spirillum; **(e)** spirochete. (SEM). a: ©SciMAT/Science Source; b: ©Dennis Kunkel Microscopy/SPL/Science Source; c: ©Dennis Kunkel Microscopy/SPL/Science Source; d: ©Dennis Kunkel Microscopy/SPL/Science Source; e: ©Dennis Kunkel Microscopy/SPL/Science Source

? What are the two most common shapes of bacteria?

FIGURE 3.22 Star-like Appearance of an Aquatic Bacterium *Ancalomicrobium adetum* (TEM). ©James T. Staley

? Why would aquatic microbes need maximal surface area?

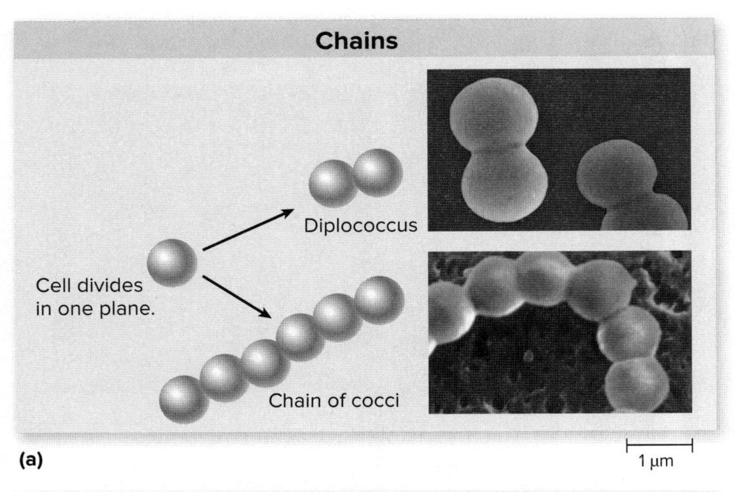

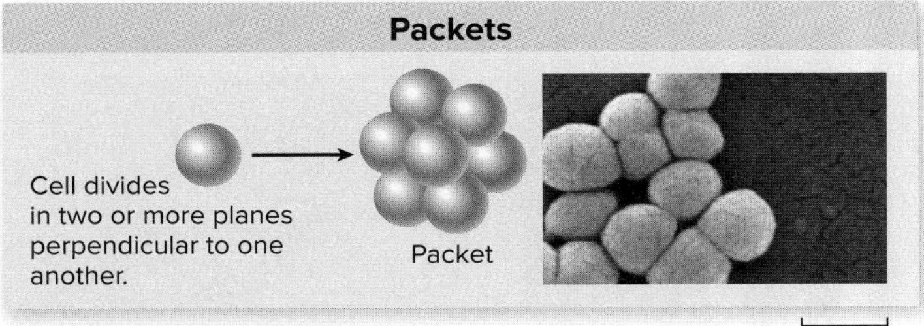

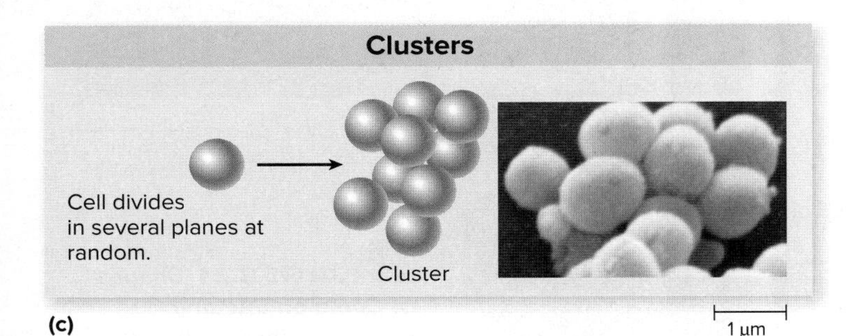

FIGURE 3.23 Common Cell Arrangements (a) Chains; **(b)** packets; **(c)** clusters. (SEM). a (top): ©Dennis Kunkel Microscopy/SPL/Science Source; a (bottom): ©BSIP SA/Alamy Stock Photo; b: Source: Betsy Crane/CDC; c: ©Eye of Science/Science Source

? How does bacterial cell division determine the characteristic cell arrangements?

several planes at random may form clusters. *Staphylococcus* species, which typically form grape-like clusters, are an example (*staphylo* means "bunch of grapes").

Multicellular Associations

Some prokaryotes characteristically live as multicellular associations. For example, members of a group of bacteria called myxobacteria glide over moist surfaces, forming swarms of cells that move as a pack. The cells release enzymes, and, as a pack, they degrade organic material, including other bacterial cells. When water or nutrients become limiting, the cells come together to form a structure called a fruiting body, which is visible to the naked eye (see figure 11.14). ⏭| **myxobacteria**

In their natural habitat, most bacteria on surfaces live in polymer-encased communities called biofilms. Details about these communities are described in chapter 4. ⏭| **biofilms**

MicroAssessment 3.3

Most common prokaryotes are cocci or rods, but other shapes include vibrios, spirilla, and spirochetes. Cells may form characteristic groupings such as chains or clusters. Some form multicellular associations.

7. What shape are *Escherichia coli* cells?
8. Describe the characteristic cell grouping of *Staphylococcus* species.
9. Based on the genus name, what is the shape and arrangement of *Streptobacillus monoliformis* cells?

PROKARYOTIC CELLS

The surface layers of a prokaryotic cell, called the **cell envelope,** consist of the cytoplasmic membrane, the cell wall, and, if present, the capsule (**figure 3.24**). Bacterial cells often have a capsule (a layer that helps protect the cell or allows it to attach to certain surfaces) but archaeal cells rarely do. Enclosed within the envelope is the **cytoplasm,** a thick substance filled with nutrients, ribosomes, and enzymes; the fluid portion of the cytoplasm is called cytosol. The **nucleoid** is the gel-like region in the cytoplasm where the cell's chromosome is found. It is not enclosed by a membrane, which distinguishes the nucleoid from the nucleus that characterizes eukaryotic cells. The cell may also have appendages that provide useful traits, including motility and the ability to adhere to certain surfaces.

Although the general structure of bacterial and archaeal cells is the same, the components have several fundamental chemical differences. Because of this, and the fact that bacteria and archaea are not closely related genetically, many scientists now feel that the two groups should not be lumped together under the general category "prokaryotic cells." Grouping them together is certainly convenient, however, particularly when trying to keep descriptions brief. With this in mind, we will cover the general characteristics of prokaryotic cell structure, using bacterial cells as a model. While doing

Focus Figure

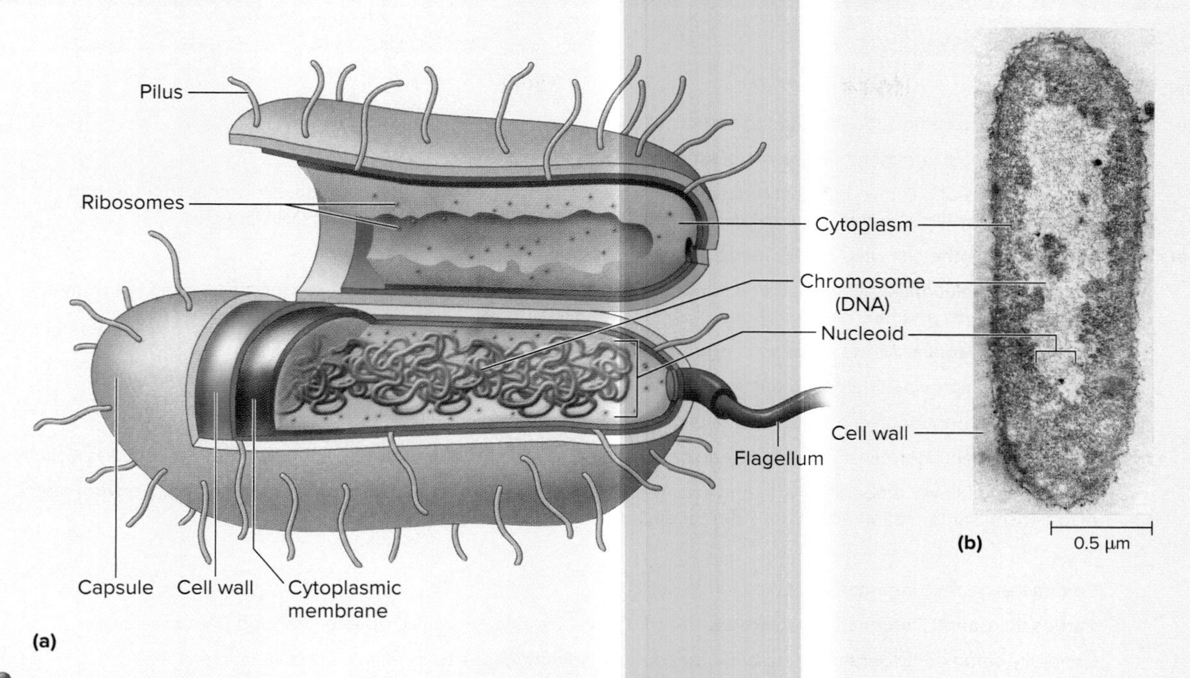

FIGURE 3.24 Typical Structures of a Prokaryotic Cell (a) Diagrammatic representation. Archaeal cells rarely have a capsule. **(b)** *E. coli* cell (TEM). ©Science Source

How does the function of the cytoplasmic membrane differ from that of the cell wall?

this, we will also mention some of the most important ways in which archaeal cells differ.

From a medical standpoint, bacterial cell components are particularly relevant because they are potential targets for antibacterial medications used to treat infectious diseases. By interfering with the function of components unique to bacteria, or at least more accessible to antibacterial medications, we can selectively kill or inhibit bacteria without harming the patient. In addition, features found in only certain bacterial groups can be used to help identify a bacterium causing an infectious disease.

We will start our discussion of prokaryotic cells by describing the cytoplasmic membrane, because it defines the boundary of the cell; it serves as the cell's gatekeeper—determining what can enter or exit the cell. After explaining the structure and function of the cytoplasmic membrane, we will turn our focus to the cell wall, which is extremely important from a medical standpoint. We will then describe the components outside of the cell wall, and, finally, the cell's internal components (**table 3.3**). Many of the cell parts are essential for growth and therefore are found in all prokaryotic cells, but others are optional. Although these optional parts are not required for growth in laboratory cultures, cells lacking them might not survive the competitive environment of the natural world.

MicroByte

Our body defenses have "alarm systems" that recognize compounds unique to bacteria, alerting our immune cells when invaders are present.

3.4 ■ The Cytoplasmic Membrane of Prokaryotic Cells

Learning Outcomes

8. Describe the structure and chemistry of the cytoplasmic membrane, focusing on how it relates to membrane permeability.
9. Describe how the cytoplasmic membrane is involved with proton motive force.
10. Compare and contrast the different types of prokaryotic transport systems: facilitated diffusion, active transport, and group translocation.
11. Explain why prokaryotic cells must secrete certain proteins.

The **cytoplasmic membrane** (or plasma membrane) is a thin, delicate structure that surrounds the cytoplasm and defines the boundary of the cell. It serves as the crucial permeability barrier between the cell and its external environment (see figure 2.17).

TABLE 3.3 Typical Prokaryotic Cell Structures

Structure	Characteristics
Structures Outside the Cell Wall	
Filamentous appendages	Composed of protein subunits that form a helical chain.
Flagella	Provide the most common mechanism of motility.
Pili	Different types of pili have different functions. The common types, often called fimbriae, allow cells to adhere to surfaces. A few types are used for twitching or gliding motility. Sex pili are involved in DNA transfer.
Capsules and slime layers	Layers outside the cell wall, usually made of polysaccharide.
Capsule	Distinct and gelatinous. Allows bacteria to adhere to specific surfaces; allows some organisms to avoid the body's defense systems and thus cause disease.
Slime layer	Diffuse and irregular. Allows bacteria to adhere to specific surfaces.
Cell Wall	Peptidoglycan provides rigidity to bacterial cell walls, preventing the cells from lysing.
Gram-positive	Thick layer of peptidoglycan that contains teichoic acids and lipoteichoic acids.
Gram-negative	Thin layer of peptidoglycan surrounded by an outer membrane. The outer layer of the outer membrane is lipopolysaccharide.
Cytoplasmic Membrane	Phospholipid bilayer embedded with proteins. Surrounds the cytoplasm, separating it from the external environment. Also transmits information about the external environment to the inside of the cell.
Internal Components	
DNA	Carries the genetic information of the cell.
Chromosome	Carries the genetic information required by a cell. Typically a single, circular, double-stranded DNA molecule.
Plasmid	Generally carries only genetic information that may be advantageous to a cell in certain situations.
Endospore	A type of dormant cell that is extraordinarily resistant to heat, desiccation, ultraviolet light, and toxic chemicals.
Cytoskeleton	Protein framework involved in cell division and control of cell shape.
Gas vesicles	Small, rigid structures that provide buoyancy to a cell.
Granules	Accumulations of high-molecular-weight polymers, synthesized from a nutrient available in relative excess.
Ribosomes	Involved in protein synthesis. Two subunits, 30S and 50S, join to form the 70S ribosome.

Structure of the Cytoplasmic Membrane

The structure of the prokaryotic cytoplasmic membrane is typical of other biological membranes: a phospholipid bilayer embedded with proteins (**figure 3.25**). The phospholipid molecules are arranged in opposing layers so that their hydrophobic tails face in, toward the other layer, and their hydrophilic heads face outward, interacting freely with aqueous solutions.
◀◀ **phospholipids**

The proteins embedded in the membrane have a variety of functions. Some act as selective gates, allowing nutrients to enter the cell and waste products to exit. Others serve as sensors of environmental conditions, providing the cell with a mechanism to monitor and adjust to its surroundings. So, while the cytoplasmic membrane functions as a permeability barrier, it also transmits information about the external environment to the inside of the cell. Membrane proteins can have other functions as well; for example, some are enzymes that catalyze essential chemical reactions. ◀◀ **protein function**

Membrane proteins are not stationary; rather, they constantly drift laterally in the phospholipid bilayer. Such movement is necessary for membrane functions. The structure and composition of the membrane, with its resulting dynamic nature, is called the **fluid mosaic model.**

Archaeal cytoplasmic membranes have the same general structure as bacterial membranes, but their phospholipid composition is distinctly different. The lipid tails of the archaeal

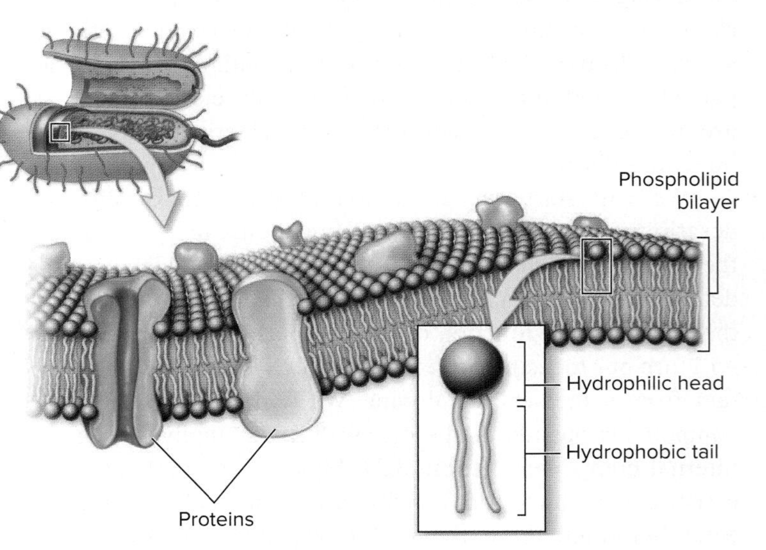

FIGURE 3.25 The Structure of the Cytoplasmic Membrane The membrane is a phospholipid bilayer embedded with proteins.

? **Which part of the membrane is hydrophobic?**

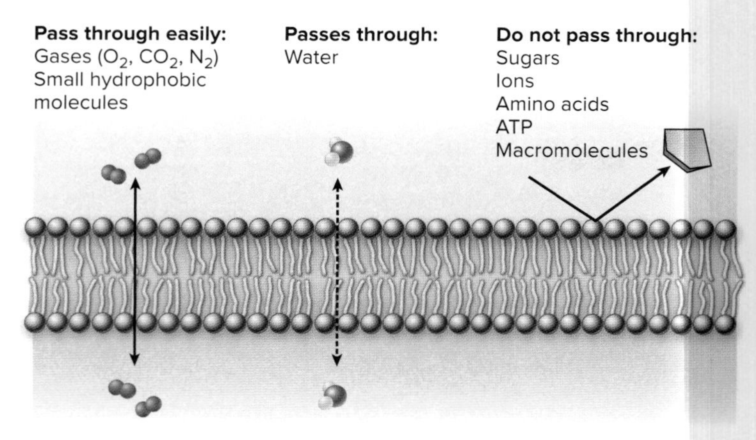

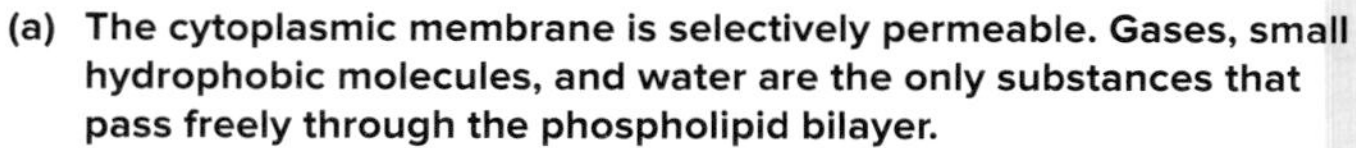
(a) The cytoplasmic membrane is selectively permeable. Gases, small hydrophobic molecules, and water are the only substances that pass freely through the phospholipid bilayer.

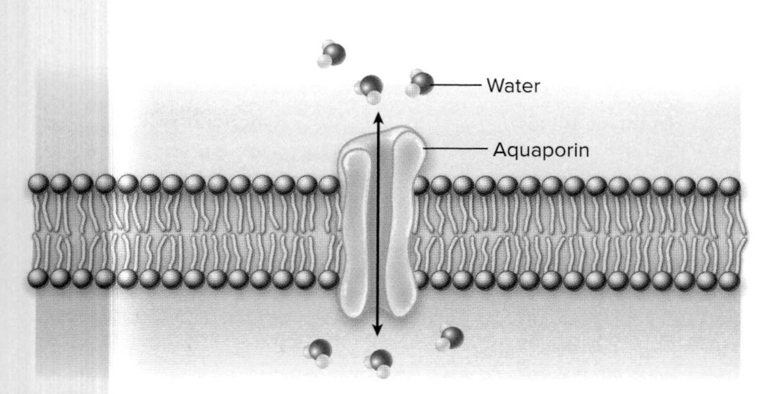

(b) Aquaporins allow water to pass through the cytoplasmic membrane more easily.

FIGURE 3.26 Permeability of the Phospholipid Bilayer (a) Selective permeability. **(b)** The role of aquaporins.

? Why are ions unable to pass through the phospholipid bilayer?

membrane lipids are connected to glycerol by a different type of chemical linkage and are not fatty acids. ⏮ glycerol, ⏮ fatty acid

MicroByte

E. coli's DNA encodes approximately 1,000 different membrane proteins, many of which are involved in transport.

Permeability of the Cytoplasmic Membrane

The cytoplasmic membrane is **selectively permeable,** meaning that only certain substances can cross it. Molecules that freely pass through the phospholipid bilayer include gases such as O_2, CO_2, and N_2, small hydrophobic compounds, and water (**figure 3.26**). Some cells facilitate water passage with **aquaporins,** pore-forming membrane proteins that specifically allow water molecules to pass through. Cells have transport systems to move certain other molecules across the membrane. ⏭ transport systems

Simple Diffusion

Molecules that freely pass through the phospholipid bilayer move in and out of the cell by **simple diffusion,** in which molecules move from a region of high concentration to one of low concentration, until equilibrium is reached. The speed and direction of movement depend on the relative concentration of molecules on each side of the membrane—the greater the difference in concentration, the higher the rate of diffusion. The molecules continue to pass through at a diminishing rate until their concentration is the same on both sides of the membrane. At equilibrium, molecules still move, but movement in one direction is balanced by movement in the other.

Osmosis

Osmosis is the diffusion of water across a selectively permeable membrane. It occurs when the concentrations of solute (dissolved molecules and ions) on two sides of a membrane are unequal. Typical of diffusion, water moves down its concentration gradient from high water concentration (low solute concentration) to low water concentration (high solute concentration). ⏮ solute

When describing osmosis, three terms are used to refer to the solutions on opposing sides of a membrane: hypotonic (*hypo* means "less"; *tonic* refers to solute), hypertonic (*hyper* means "more"), and isotonic (*iso* means "the same"). Water flows from the hypotonic solution to the hypertonic one (**figure 3.27**). No net water movement occurs between isotonic solutions.

Osmosis has important biological consequences. The cytoplasm of a cell is a concentrated solution of inorganic salts, sugars, amino acids, and various other molecules. However, the environments in which bacteria and archaea normally grow are typically very dilute (hypotonic). Water moves toward the high solute concentration, so it flows from the surrounding medium into the cell (see figure 3.27a). This inflow of water exerts tremendous osmotic pressure on the cytoplasmic membrane, much more than it typically can resist. However, the strong cell wall surrounding the membrane generally withstands such high pressure. The cytoplasmic membrane is forced against the wall but cannot balloon further. Damage to the cell wall weakens the structure, and consequently, cells may lyse (burst).

The Role of the Cytoplasmic Membrane in Energy Transformation

The cytoplasmic membrane of prokaryotic cells plays a crucial role in transforming energy—converting the energy of food

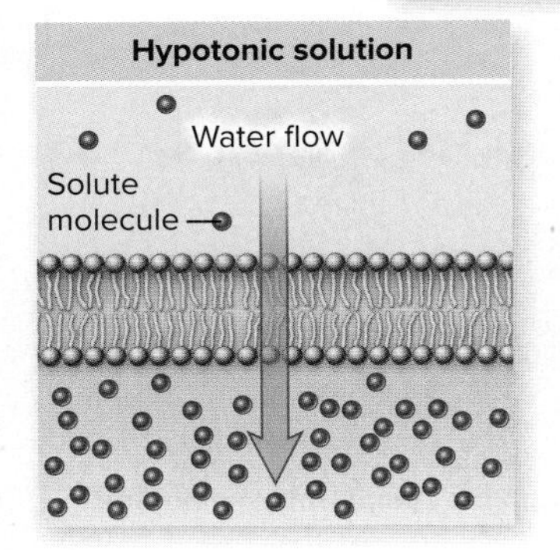

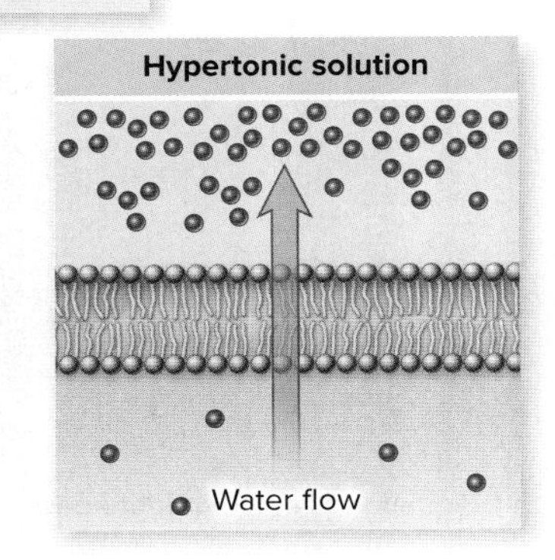

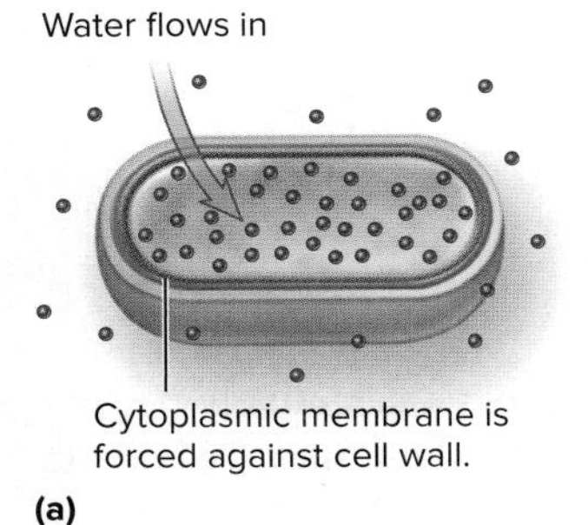

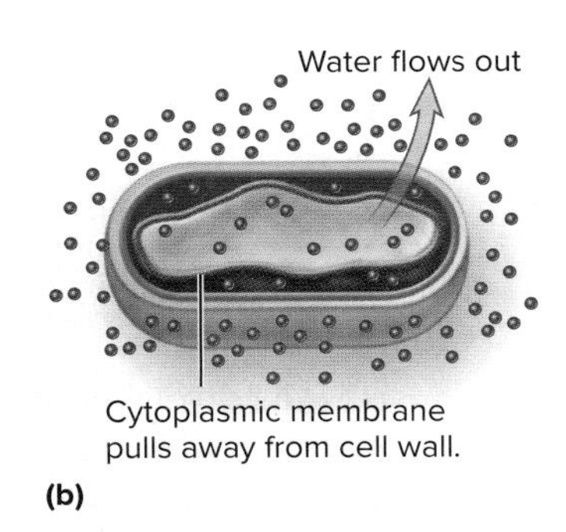

FIGURE 3.27 Osmosis (a) The effect of a hypotonic solution on a bacterial cell. **(b)** The effect of a hypertonic solution on a bacterial cell.

What might happen in part (a) if the cell wall were weakened?

or sunlight into ATP, the energy currency of a cell. This is an important difference between prokaryotic and eukaryotic cells; in eukaryotic cells, this process occurs in membrane-bound organelles, which will be discussed later in this chapter. ◀◀ ATP

As part of their energy-transforming processes, most prokaryotes have a series of protein complexes—collectively referred to as the **electron transport chain (ETC)**—embedded in their cytoplasmic membrane. Details of how the ETC works will be explained in chapter 6, but the net result is that protons are moved out of the cell. This creates an electrochemical gradient across the membrane—positively charged protons are concentrated immediately outside the membrane, whereas negatively charged hydroxide ions remain inside the cell (**figure 3.28**). The charged ions attract each other, so they stay close to the membrane. Inherent in the gradient is a form of energy called **proton motive force,** which is analogous to the energy stored in a battery. ▶▶ electron transport chain

The energy of the proton motive force is harvested by mechanisms that allow protons to move back into the cell. That energy is used to drive certain cellular processes, including ATP synthesis. It is also used to power one of the transport systems discussed next and some forms of motility.

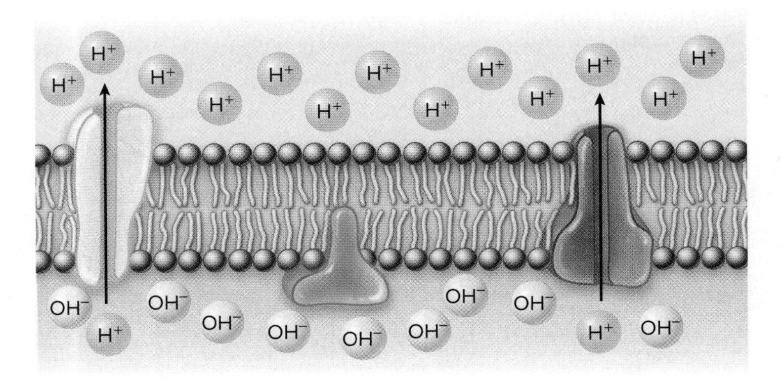

FIGURE 3.28 Proton Motive Force The electron transport chain, a series of protein complexes within the membrane, moves protons out of the cell.

Why would the protons stay close to the membrane, rather than float away?

Transport of Small Molecules Across the Cytoplasmic Membrane

Most molecules that enter or exit a cell must pass via proteins in the cytoplasmic membrane that function as selective gates. This is necessary because the membrane's phospholipid bilayer prevents nearly all substances from passing through.

Mechanisms that allow nutrients and other small molecules to enter the cell are called **transport systems.** The systems move small molecules into the cell through transport proteins, sometimes called permeases or carriers. These span the membrane, so that one end projects into the surrounding environment and the other into the cell (**figure 3.29**). The interaction between the transport protein and the molecule it carries is highly specific. Consequently, a single carrier generally transports only one type of molecule. The mechanisms of transport are summarized in **table 3.4**. Some of these are also used by cells to pump out

TABLE 3.4 Transport Mechanisms Used by Prokaryotic Cells

Transport Mechanism	Characteristics
Facilitated Diffusion	Rarely used by prokaryotes. Exploits a concentration gradient to move molecules; can only eliminate a gradient, not create one. No energy is used.
Active Transport	Energy is used to accumulate molecules against a concentration gradient.
Transporters that use proton motive force	As a proton is allowed into the cell, another substance is either brought along or expelled.
ABC transporters	ATP is used as an energy source. Binding proteins deliver a molecule to the transporter.
Group Translocation	The transported molecule is chemically altered as it passes into the cell.

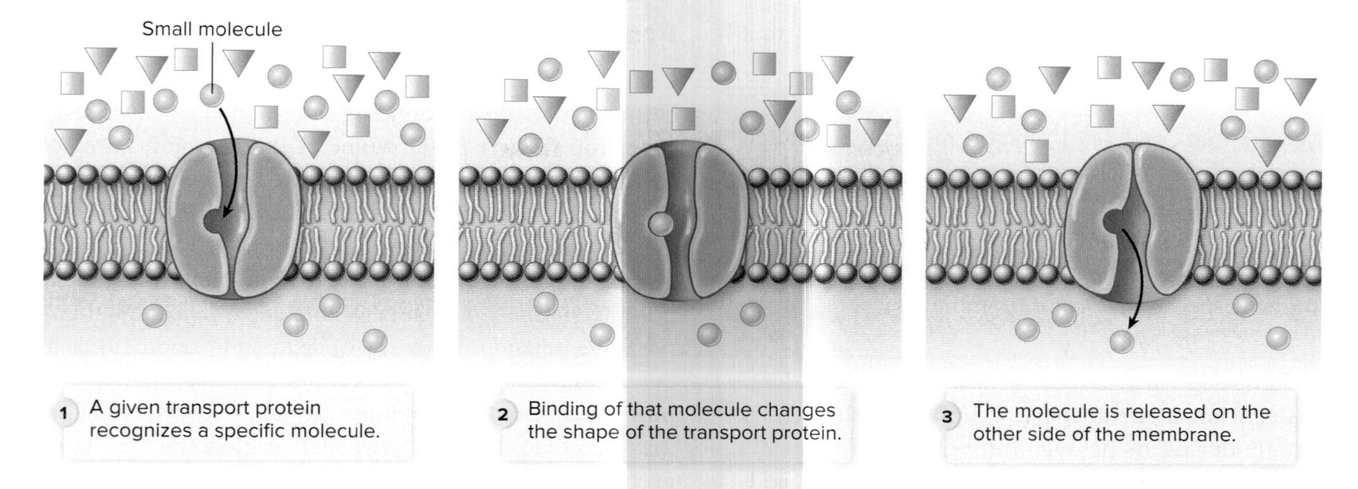

FIGURE 3.29 Transport Protein Only one type of molecule is transported across the membrane through a given transport protein.

What are some examples of small molecules that enter cells through transport proteins?

wastes and compounds such as antibiotics and disinfectants that are otherwise damaging to the cell.

Facilitated Diffusion

Facilitated diffusion is a form of passive transport, meaning that it does not require energy. It uses transport proteins to move substances that cannot diffuse through the lipid bilayer. Facilitated diffusion can only move substances down a concentration gradient (**figure 3.30a**). Molecules are transported from one side of the membrane to the other until their concentration is the same on both sides. Because prokaryotes typically grow in environments where the concentration of nutrients is lower outside the cell than inside, the cells generally do not use facilitated diffusion to take in nutrients.

Active Transport

Active transport moves compounds against a concentration gradient and, therefore, requires energy. Nearly every organic molecule that a prokaryotic cell takes in must move against a concentration gradient, so bacteria and archaea routinely use active transport.

Transport Systems That Use Proton Motive Force Many prokaryotic transport systems move small molecules and ions in or out of the cell using the energy of a proton motive force (see figure 3.30b). Transporters of this type allow a proton into the cell and simultaneously either bring along or expel another substance. Moving waste products out of the cell, on the other hand, relies on transporters that eject the compound as a proton passes in. **Efflux pumps,** which are used

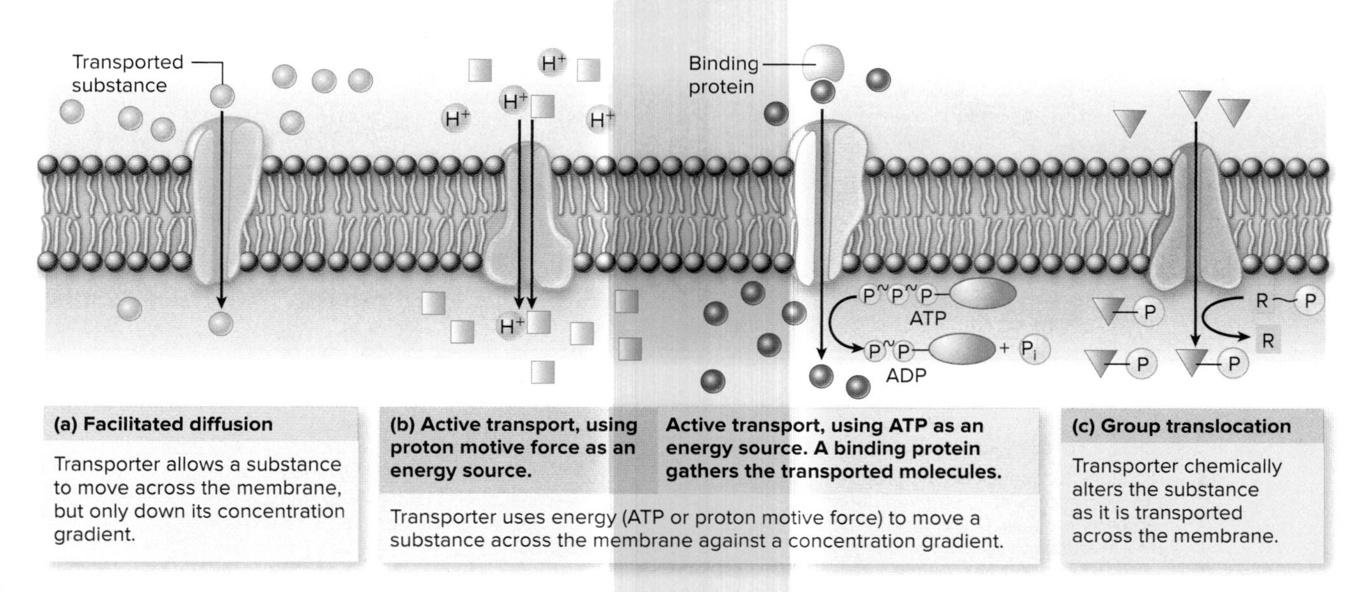

FIGURE 3.30 Types of Transport Systems **(a)** Facilitated diffusion. **(b)** Active transport. **(c)** Group translocation.

In group translocation, how would the amount of unphosphorylated R group shown in the diagram indicate the quantity of transported substance being brought into the cell?

by some bacteria to remove antimicrobial medications that have entered the cell, use this latter mechanism.

Transport Systems That Use ATP Transport mechanisms called ABC transport systems require ATP as an energy source (ABC stands for **A**TP-**B**inding **C**assette) (see figure 3.30b). These systems use specific binding proteins located immediately outside the cytoplasmic membrane to gather and deliver molecules to the respective transport complexes.

Group Translocation

Group translocation is a transport process that chemically alters a molecule during its passage through the cytoplasmic membrane (see figure 3.30c). Typically, this is done by adding a phosphate group to the molecule being transported—a chemical modification called phosphorylation. Many bacteria bring glucose into the cell using this mechanism. Although phosphorylation expends energy, that energy would be used anyway in the initial steps of glucose breakdown (the "investment phase"), a process described in chapter 6. As that chapter will describe, energy is then gained in later steps of glucose breakdown (the "payoff phase"). Although group translocation is common in bacteria, the archaea studied so far do not use this transport process.

Protein Secretion

Cells actively move certain proteins they synthesize out of the cell and into the surrounding environment—a process called **secretion** (**figure 3.31**). Some of these proteins are enzymes that function outside the cell to break down macromolecules in the external environment into their respective subunits; the smaller subunits can then be easily transported into the cell. For example, some cells secrete enzymes that break down polysaccharides into sugar molecules. The cell then transports the sugars into the cytoplasm and uses them as a nutrient source. The macromolecules are too large to transport into the cell, but the subunits are not. Other secreted proteins make up external structures such as flagella, the appendages used for motility. ⏮ polysaccharides

How does the cell machinery know which polypeptides are to be secreted? Those destined for secretion have a characteristic sequence of amino acids, typically at one end of the molecule. This sequence—a **signal sequence**—functions as a tag that directs the secretion machinery to move the preprotein (precursor protein) across the membrane. During the transport process, the signal sequence is removed and the protein ultimately folds into its functional shape. Secretion is a complex process, and a variety of different secretion systems are used by prokaryotes.

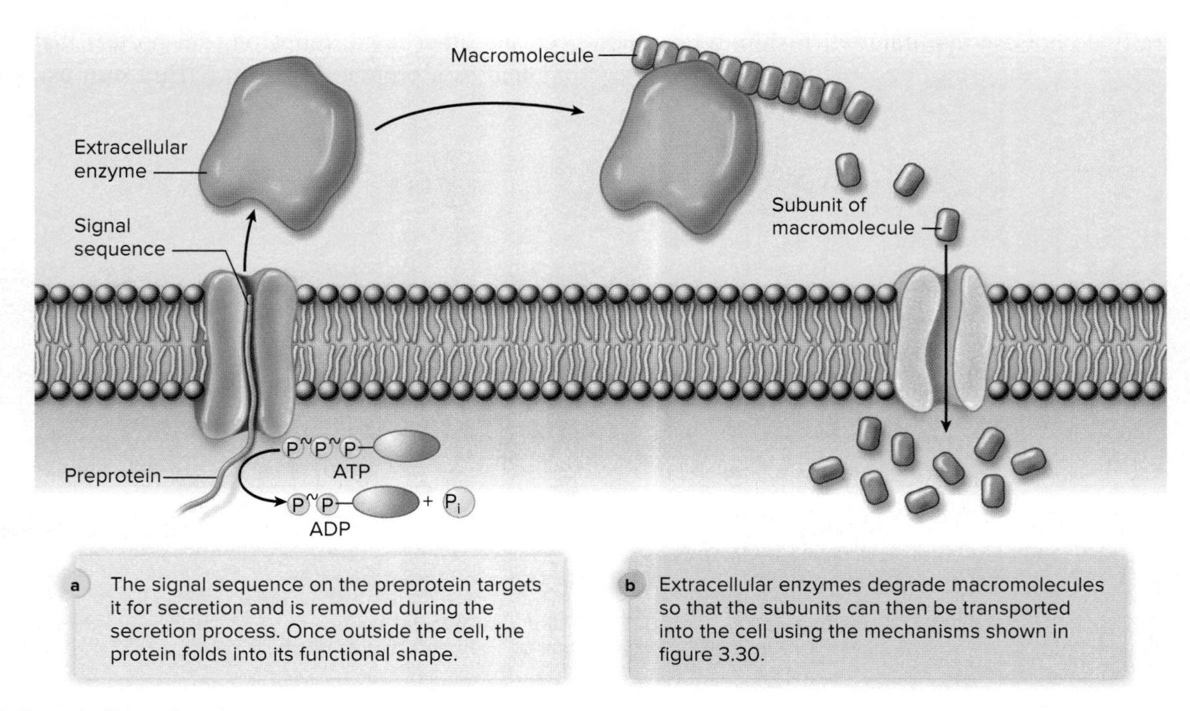

FIGURE 3.31 Protein Secretion **(a)** Generalized view of the protein secretion process. **(b)** One function of a secreted protein.

? Why would a cell secrete enzymes rather than bring intact macromolecules into the cell?

MicroAssessment 3.4

The cytoplasmic membrane is a phospholipid bilayer embedded with proteins. Molecules that pass through the bilayer move in and out by simple diffusion; osmosis is the diffusion of water across the membrane. The electron transport chain generates a proton motive force, which is used to drive several cellular processes including ATP synthesis, certain transport systems, and some types of motility. Transport systems move molecules across the membrane using facilitated diffusion, active transport, and group translocation. Proteins destined for secretion have a characteristic signal sequence.

10. Explain why the cytoplasmic membrane is described as being selectively permeable.
11. Prokaryotes rarely use facilitated diffusion. Why is this so?
12. Membrane proteins move laterally in a phospholipid bilayer but generally not up and down. Why would this be so?

3.5 ■ The Cell Wall of Prokaryotic Cells

Learning Outcomes

12. Describe the chemistry and structure of peptidoglycan.
13. Compare and contrast the structure and chemistry of the Gram-positive and Gram-negative cell walls.
14. Explain the significance of lipid A and the O antigen of LPS.
15. Explain how the cell wall affects susceptibility to penicillin and lysozyme.
16. Explain how the cell wall affects Gram staining characteristics.
17. Describe the cell walls of archaea.

The prokaryotic cell wall is a strong, somewhat rigid structure that prevents the cell from bursting. Differences in its arrangement distinguish two main groups of bacteria: Gram-positive and Gram-negative (**table 3.5**).

Peptidoglycan

The strength of the Gram-positive and Gram-negative bacterial cell walls is due to a layer of **peptidoglycan,** a material found only in bacteria (**figure 3.32**). The basic structure of peptidoglycan is an alternating series of two major subunits related to glucose: *N*-acetylmuramic acid (NAM) and *N*-acetylglucosamine (NAG). These subunits are covalently joined to one another to form a linear polymer called a glycan chain (*glyco* means "sugar"), which serves as the backbone of the peptidoglycan molecule.

Attached to each of the NAM molecules is a tetrapeptide chain (a string of four amino acids) that plays an important role in the strength of peptidoglycan. The tetrapeptide chains connect together, linking adjacent glycan chains to form a single, very large three-dimensional molecule, much like a flexible, multilayered chain-linked fence. In Gram-negative bacteria, tetrapeptides are joined directly to one another. In

TABLE 3.5 Comparison of Features of Gram-Positive and Gram-Negative Bacteria

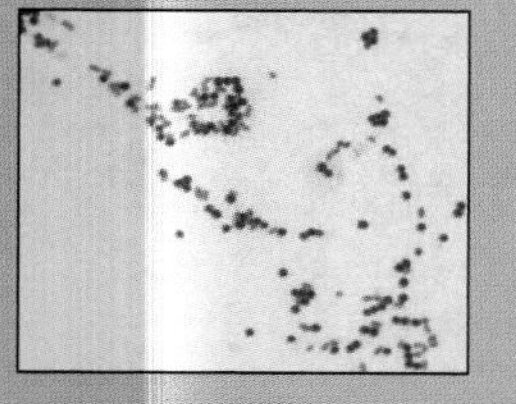

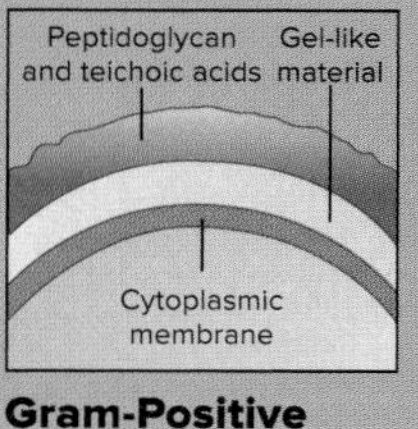

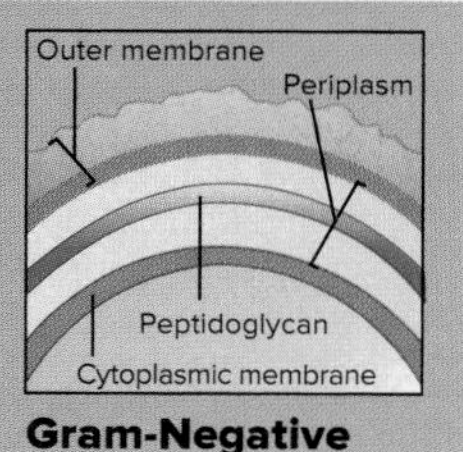

	Gram-Positive	Gram-Negative
Color of Gram-Stained Cell	Purple	Pink
Representative Genera	*Bacillus, Staphylococcus, Streptococcus*	*Escherichia, Neisseria, Pseudomonas*
Distinguishing Structures/Components		
Peptidoglycan	Thick layer	Thin layer
Teichoic acids	Present	Absent
Outer membrane	Absent	Present
Lipopolysaccharide (endotoxin)	Absent	Present
Porin proteins	Absent (unnecessary because there is no outer membrane)	Present; allow molecules to pass through outer membrane
General Characteristics		
Sensitivity to penicillin	Generally more susceptible (with notable exceptions)	Generally less susceptible (with notable exceptions)
Sensitivity to lysozyme	Yes	No

Gram-positive bacteria, they are usually linked indirectly by a peptide interbridge (a series of amino acids).

Only a few types of amino acids make up the tetrapeptide chains. One of these, diaminopimelic acid (related to lysine), is not found in any other place in nature. Some of the others are D-optical isomers, a form found in relatively few substances. ◀◀ lysine, ◀◀ D-optical isomers

The Gram-Positive Cell Wall

A relatively thick layer of peptidoglycan characterizes the Gram-positive cell wall (**figure 3.33**). As many as 30 layers of interconnected glycan chains make up the polymer. Regardless of the thickness of the layer, many small substances—including sugars and amino acids—can pass through.

In addition to peptidoglycan, the Gram-positive cell wall has **teichoic acids** (from the Greek word *teichos,* meaning "wall"). These are negatively charged chains of a common

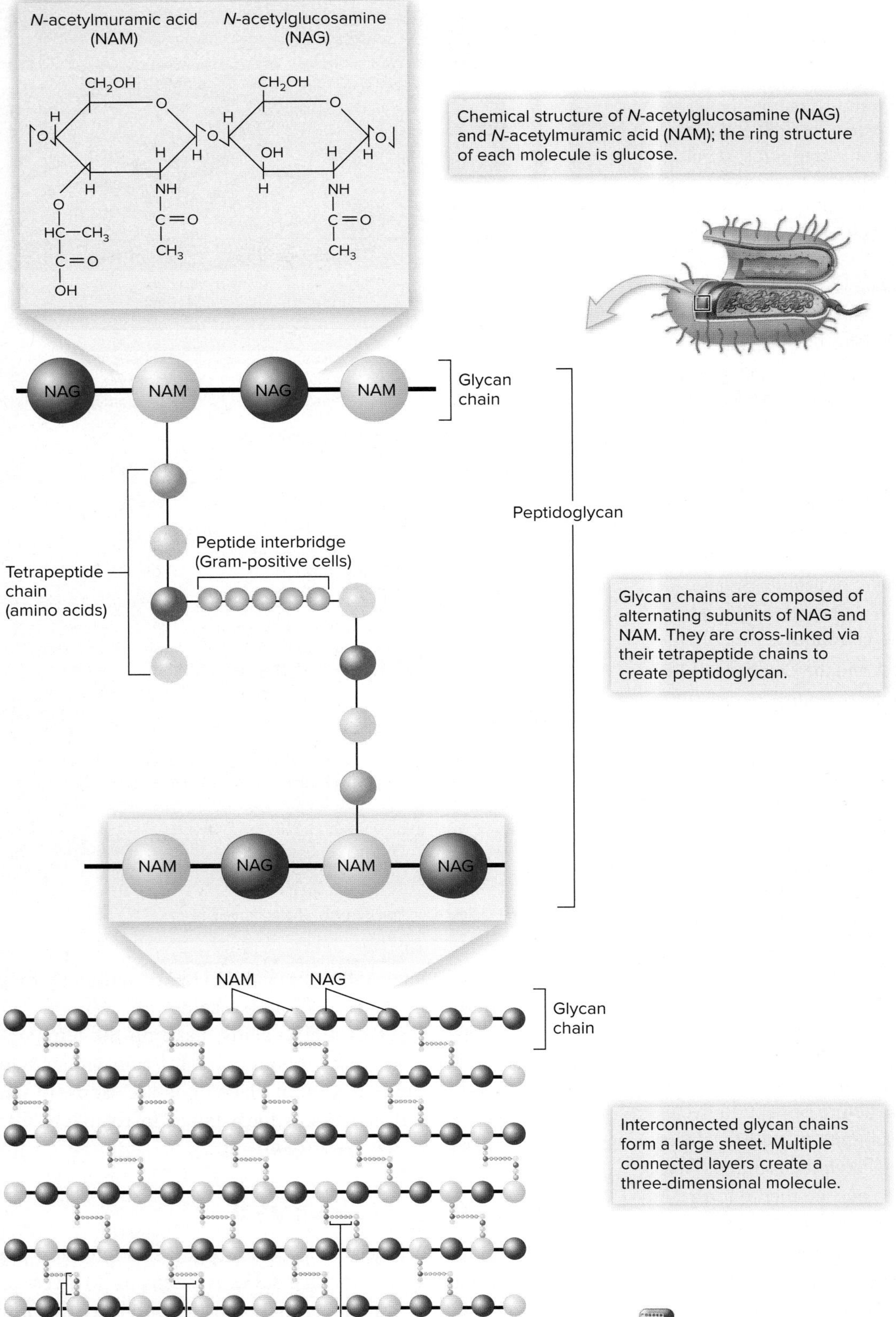

Chemical structure of *N*-acetylglucosamine (NAG) and *N*-acetylmuramic acid (NAM); the ring structure of each molecule is glucose.

Glycan chains are composed of alternating subunits of NAG and NAM. They are cross-linked via their tetrapeptide chains to create peptidoglycan.

Interconnected glycan chains form a large sheet. Multiple connected layers create a three-dimensional molecule.

FIGURE 3.32 Components and Structure of Peptidoglycan

Why might it be medically significant that peptidoglycan is found only in bacteria?

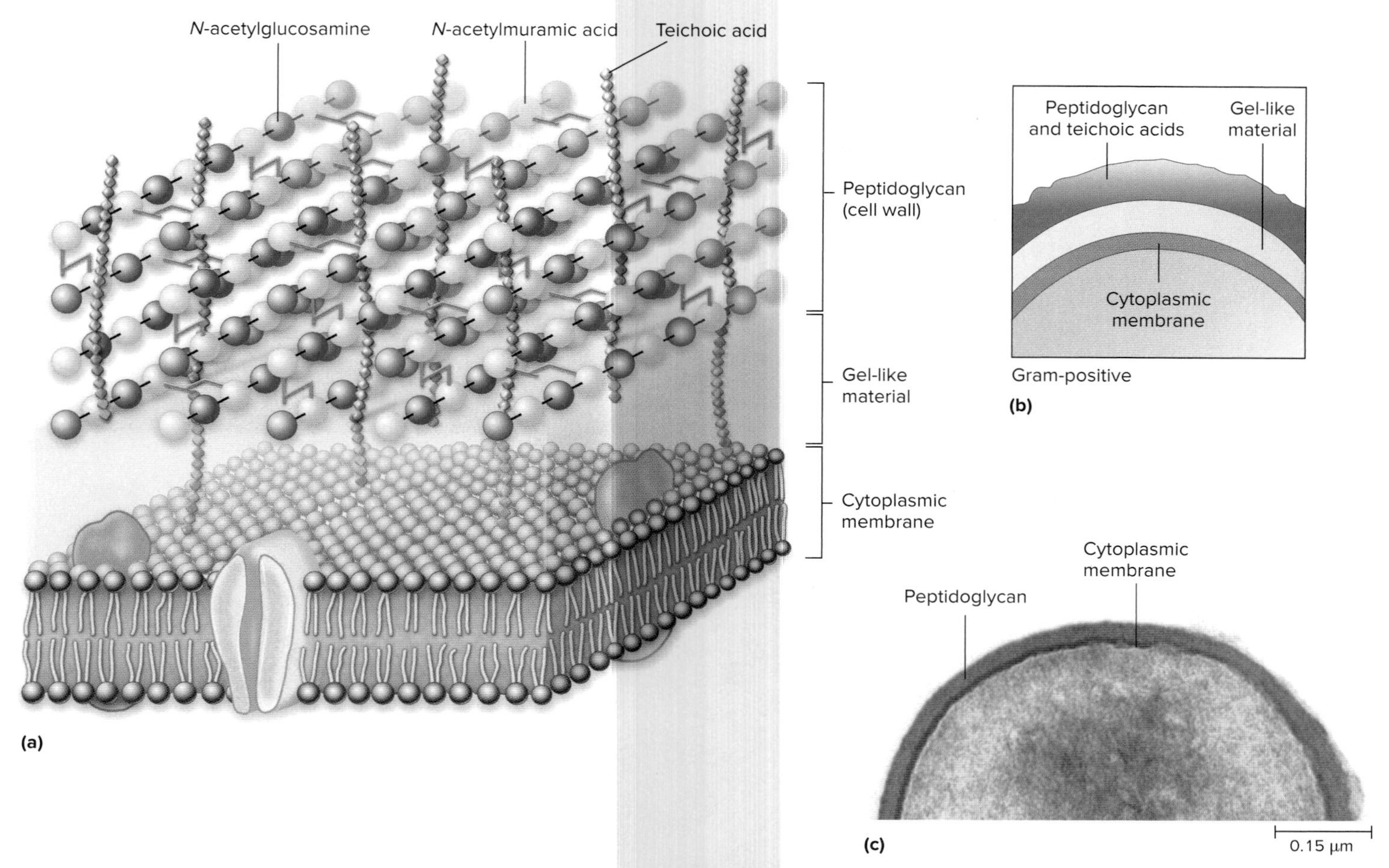

FIGURE 3.33 Gram-Positive Cell Wall (a) The Gram-positive cell wall has a relatively thick layer of peptidoglycan consisting of many sheets of interconnected glycan chains. **(b)** Simple diagram of a cross-section of the structure. **(c)** Gram-positive cell wall (TEM; *Staphylococcus aureus*). ©Egbert Hoiczyk

? What connects the glycan chains in peptidoglycan?

subunit (either ribitol-phosphate or glycerol-phosphate) to which various sugars and D-alanine are typically attached. Teichoic acids extend above the peptidoglycan layer and bind cations such as Mg^{2+}. Their function is not well understood, but they may serve as a reservoir for cations that are essential for enzyme function. They also seem to be important for cell wall construction and cell division. Some teichoic acids are covalently joined to the peptidoglycan molecule, and others are linked to the cytoplasmic membrane.

A gel-like substance is sandwiched between the cytoplasmic membrane and the peptidoglycan layer. This substance is thought to have a function similar to the periplasm associated with Gram-negative cell walls, discussed next.

The Gram-Negative Cell Wall

The Gram-negative cell wall contains only a thin layer of peptidoglycan (**figure 3.34**). Outside of that is the **outer membrane,** a unique lipid bilayer embedded with proteins. The outer membrane is joined to peptidoglycan by lipoproteins. ⏮ lipoproteins

The Outer Membrane

The outer membrane of Gram-negative bacteria is unique. Its bilayer structure is typical of other membranes, but the outside layer is made up of a molecule called **lipopolysaccharide (LPS)** rather than phospholipid. LPS is extremely important from a medical standpoint. When injected into an animal, it causes fever and other symptoms characteristic of infections by live bacteria. The symptoms arise from the body's response to LPS, a molecule the body's defense systems use as an indication that Gram-negative bacteria have invaded. If very small quantities of LPS enter the tissues, such as when a few bacterial cells contaminate a minor wound, the defense systems respond at a level that can safely eliminate the invader.

When significant amounts of the molecule spread throughout the body, however, such as when Gram-negative bacteria are growing in the bloodstream, the response can be deadly. Because of this lethal effect, LPS is called **endotoxin** (*endo* meaning "inside," although LPS is actually a component of the envelope). ⏭ endotoxin

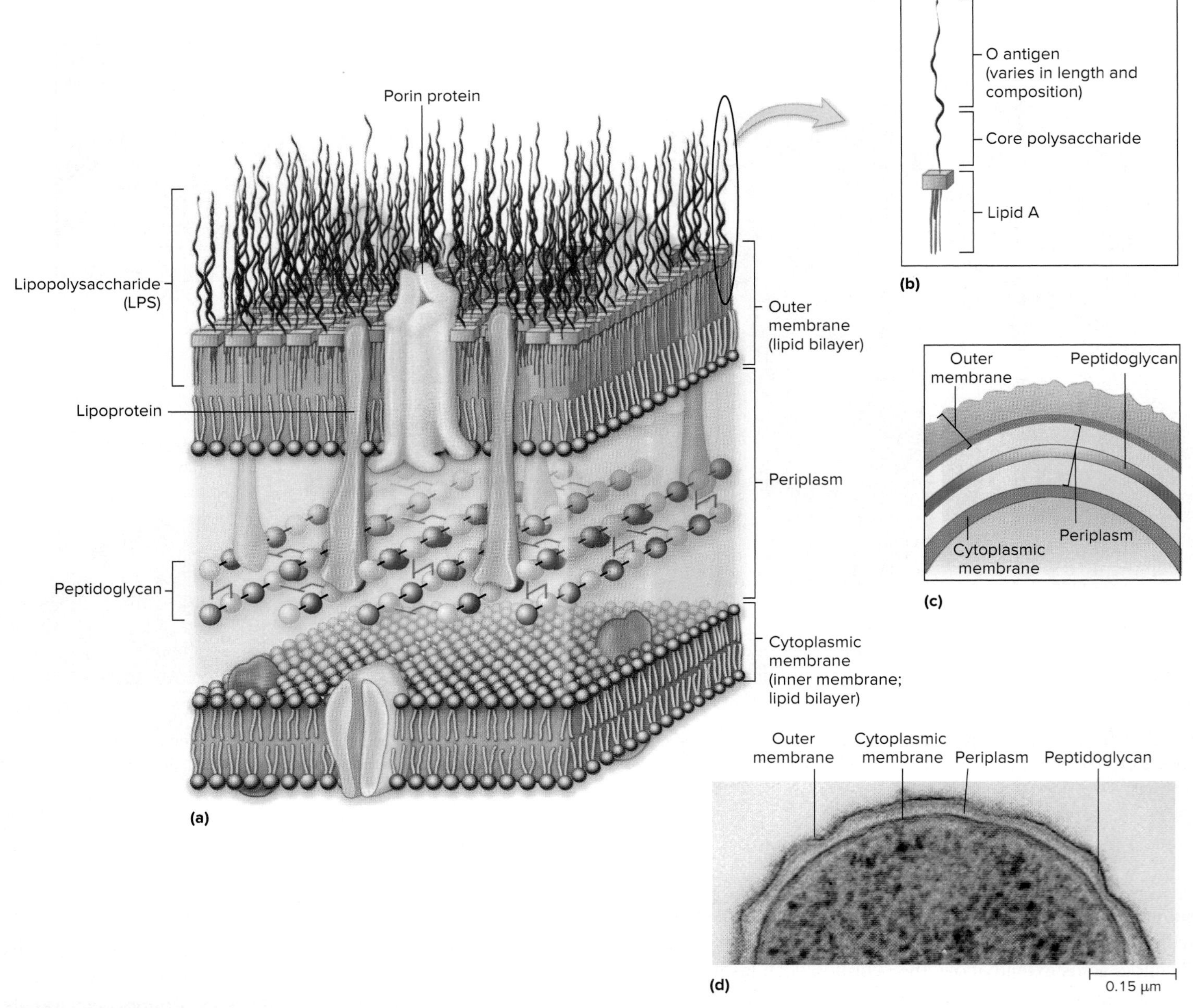

FIGURE 3.34 Gram-Negative Cell Wall (a) The Gram-negative cell wall has a thin layer of peptidoglycan made up of only one or two sheets of interconnected glycan chains. The outer membrane is a typical phospholipid bilayer, except the outer layer is lipopolysaccharide. Porins span the membrane to allow specific molecules to pass. Periplasm fills the region between the two membranes. **(b)** Structure of lipopolysaccharide. The sugars in the O antigen vary among bacterial species. **(c)** Simple diagram of a cross section of the structure. **(d)** Gram-negative cell wall (TEM; *Myxococcus xanthus*). ©Egbert Hoiczyk

? **Why is lipopolysaccharide medically significant?**

Two parts of the LPS molecule are particularly notable (see figure 3.34b):

- **Lipid A** anchors the LPS molecule in the lipid bilayer. This is the portion of the LPS molecule that the body recognizes as the sign of invading Gram-negative bacteria.
- **O antigen** is the part of LPS directed away from the membrane, at the end opposite lipid A. It is made up of a chain of sugar molecules, the number and type of which vary among different species. The differences can be used to identify certain species or strains.

MicroByte

The "O157" in *E. coli* O157:H7 refers to the characteristic O antigen.

Like the cytoplasmic membrane—which in Gram-negative bacteria is sometimes called the inner membrane—the outer membrane serves as a barrier to the passage of most molecules. It keeps out many compounds that could damage the cell, including certain antimicrobial medications. This is one reason Gram-negative bacteria are generally less sensitive to many such medications. Small molecules and ions can cross the membrane through **porins,** specialized channel-forming proteins that span the outer membrane. Some porins are specific for certain molecules; others allow many different molecules to pass.

Gram-negative bacteria have a number of unique transport mechanisms called secretion systems to move proteins across both the cytoplasmic and the outer membranes. Some of these play a crucial role in the disease process of certain pathogens, so medical microbiologists are very interested in learning more about how they function. One hope is that medications can be developed to jam them. ⏭ type III secretion systems

Periplasm

The region between the cytoplasmic membrane and the outer membrane is the periplasmic space, which is filled with a gel-like substance called **periplasm.** All exported proteins accumulate in the periplasm unless specifically moved across the outer membrane as well. For example, the enzymes that cells export to break down peptides and other molecules are in the periplasm. Similarly, the binding proteins of the ABC transport systems are found there.

Antibacterial Substances That Target Peptidoglycan

Compounds that interfere with the synthesis of peptidoglycan or alter its structural integrity weaken the molecule to a point where it can no longer prevent the cell from bursting. These substances have no effect on eukaryotic or archaeal cells because peptidoglycan is unique to bacteria.

Penicillin

Penicillin is the most thoroughly studied of a group of antibiotics that interfere with peptidoglycan synthesis. It functions by preventing the cross-linking of adjacent glycan chains (it inhibits enzymes that normally catalyze the cross-linking step). ⏭ enzyme, ⏭ penicillin

Generally, but with notable exceptions, penicillin is far more effective against Gram-positive bacteria than against Gram-negative bacteria. This is because the outer membrane of Gram-negative cells prevents the medication from reaching the peptidoglycan layer. Scientists have developed derivatives of penicillin that pass through the outer membrane, however, and these can be effective in treating infections caused by Gram-negative bacteria. ⏭ penicillin derivatives

Lysozyme

Lysozyme—an enzyme found in tears, saliva, and many other body fluids—breaks the bonds that link the alternating subunits of the glycan chains. This destroys the structural integrity of the peptidoglycan molecule. The outer membrane of Gram-negative bacteria prevents the enzyme from reaching the peptidoglycan layer, so lysozyme is generally more effective against Gram-positive bacteria. Lysozyme is sometimes used in the laboratory to remove the peptidoglycan layer from bacteria for experimental purposes.

Cell Wall Type and the Gram Stain

The type of bacterial cell wall accounts for the Gram stain reaction, but it is the inside of the cell, not the wall, that is stained by crystal violet (**figure 3.35**). Gram-positive cells retain the dye because their cell wall prevents the crystal violet–iodine complex from being washed out during decolorization. The decolorizing agent dehydrates the thick layer of peptidoglycan, and in this desiccated state the wall acts as a permeability barrier that prevents the dye from leaving the cell. In contrast, the solvent action of the decolorizing agent easily damages the outer membrane of Gram-negative bacteria. The relatively thin layer of peptidoglycan, which the decolorizing agent also damages, cannot hold back the dye complex. ⏮ Gram stain

Bacteria That Lack a Cell Wall

Some bacteria naturally lack a cell wall. *Mycoplasma* species, one of which causes a mild form of pneumonia, are flexible because they lack a rigid cell wall (**figure 3.36**). As expected, neither penicillin nor lysozyme affects these organisms. *Mycoplasma* and related bacteria can survive without a cell wall because their cytoplasmic membrane has sterols in it, making it stronger than that of most other bacteria. ⏮ sterol

Cell Walls of Archaea

As a group, archaea have a variety of cell wall types. This is probably due to the fact that they inhabit a wide range of environments, including some that are extreme. Archaea have not been studied as extensively as bacteria, however, so less is known about the structure of their walls. Archaea do not have peptidoglycan in their cell wall, but some do have a similar molecule called pseudopeptidoglycan. Many have S-layers, which are sheets of flat protein or glycoprotein subunits. These subunits self-assemble, so they may have applications in nanotechnology—a branch of science that seeks to build functional items from molecules and atoms. Some bacterial cells have S-layers as well, but the cell wall remains the primary structural component.

FOCUS ON A CASE 3.1

Small outbreaks of *Staphylococcus aureus* skin infections were reported in people who had recently been tattooed. Investigation revealed 34 similar cases in three states over a 13-month period, involving 13 unlicensed tattooists. Most of the tattooists wore gloves while creating the tattoos, but they did not follow basic infection-control procedures such as changing gloves between clients, handwashing, and disinfecting equipment. In several cases, the patient recalled seeing wounds on the tattooist's hands. In addition to the cases directly involving the tattooing procedure, an additional 10 people developed *S. aureus* infections as a result of contact with the tattoo recipients. ⏭| *Staphylococcus aureus* **skin infections**

Treating the infections was complicated by the fact that the causative agent was resistant to several types of antibiotics. *S. aureus* is a Gram-positive coccus, and when penicillin was first used medically in 1941, most strains were quite sensitive to this life-saving antibiotic. Over time, however, some strains acquired the ability to produce penicillinase, an enzyme that degrades penicillin. Because the strains can destroy the medication, they are penicillin-resistant, meaning that the medication is not an effective treatment. To counteract the problem, scientists chemically modified penicillin to create methicillin and other derivatives not destroyed by penicillinase. Unfortunately, some bacteria have now gained the ability to resist the effects of these derivatives. In the case of *S. aureus*, the resistant bacteria are referred to as MRSA (pronounced mersa), which stands for **m**ethicillin-**r**esistant *Staphylococcus aureus*. The strains in the outbreaks involving the tattoos were MRSA. ⏭| **MRSA**

Fortunately, some treatment options were still available for the patients, so in most cases, the infections cleared with simple procedures (draining the wound and/or taking certain oral antimicrobial medications). Four patients, however, were hospitalized due to a MRSA bloodstream infection that required intravenous administration of vancomycin, an antimicrobial medication typically reserved as a last resort.

1. The tattooists wore gloves, so how could they have transmitted *Staphylococcus aureus*?
2. Most *S. aureus* strains are now resistant to penicillin. Why would this be so?
3. Penicillin and its derivatives (including methicillin) all interfere with the cross-linking of adjacent glycan chains. How would this be lethal to bacterial cells?
4. MRSA strains have acquired the ability to produce a modified enzyme to make the cross-links between glycan chains. How would this allow them to be resistant to methicillin and other derivatives of penicillin?
5. Why would vancomycin be reserved only for use as a last resort?

Discussion

1. The tattooists could have contaminated their gloves in several ways. They might not have used proper procedures when putting on the gloves (specific procedures must to be used to ensure that no part of the glove is touched with bare hands except the cuff). Even if the gloves had been put on properly, the tattooists might have touched contaminated surfaces with their gloved hands. Licensed tattooists are taught to use strict safety and sanitation procedures to avoid transmitting infections.
2. When penicillin is used, it kills only the penicillin-sensitive cells. The penicillin-resistant cells can still multiply, and they can do so without competition. As penicillin is increasingly used, killing more of the sensitive cells, the resistant strains will predominate (see figure 20.13).
3. The cross-linking provides strength to the peptidoglycan molecule, just as the structure of a chain-link fence provides strength to that material. If the cross-links in peptidoglycan do not form properly, the cell wall will be too weak to prevent the cell from bursting.
4. Penicillin and its derivatives bind to the typical enzymes that form the cross-links in peptidoglycan, thereby interfering with their function. The antibiotics do not bind MRSA's modified enzyme.
5. When used to treat bloodstream or tissue infections, vancomycin must be given intravenously. Because of this, medications that can be given orally are usually a first choice. In addition, vancomycin sometimes has adverse side effects, so it should be avoided if safer alternatives are available.

Source: http://www.cdc.gov/mmwr/preview/mmwrhtml/mm5524a3.htm

MicroAssessment 3.5

Peptidoglycan is a molecule unique to bacteria that provides strength to the cell wall. The Gram-positive cell wall is composed of a relatively thick layer of peptidoglycan as well as teichoic acids. Gram-negative cell walls have a thin layer of peptidoglycan and a lipopolysaccharide-containing outer membrane. Penicillin and lysozyme interfere with the structural integrity of peptidoglycan. *Mycoplasma* species lack a cell wall. Archaea have a variety of cell wall types.

13. What is the significance of lipid A?

14. How does the action of penicillin differ from that of lysozyme?

15. Explain why penicillin kills only actively multiplying cells, whereas lysozyme kills cells in any stage of growth.

3.6 ■ Structures Outside the Cell Wall of Prokaryotic Cells

Learning Outcomes

18. Compare and contrast the structure and function of capsules and slime layers.

19. Describe the structure and arrangements of flagella, and explain how they are involved in chemotaxis.

20. Compare and contrast the structure and function of fimbriae and sex pili.

Gram-positive cell wall

Before decolorization

Thick layer of peptidoglycan

Cytoplasmic membrane

Crystal violet-iodine complex trapped in cell

Large crystal violet-iodine complexes form within the cytoplasm as a result of the first two steps of the Gram stain.

After decolorization

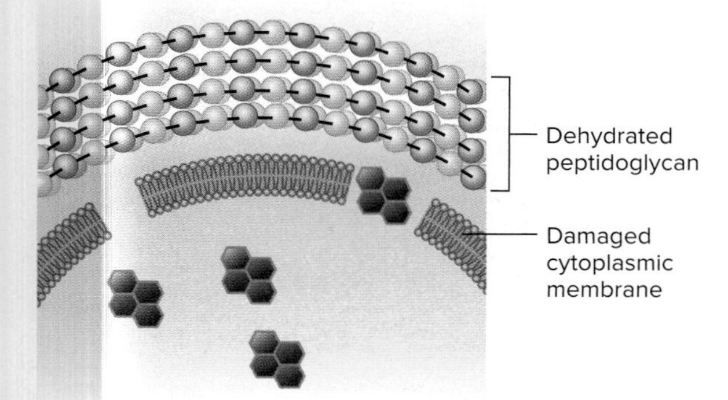

The decolorizing agent dehydrates the thick layer of peptidoglycan, causing it to become a tight mesh-like structure that prevents the crystal violet-iodine complexes from being washed out of the cell.

Gram-negative cell wall

Before decolorization

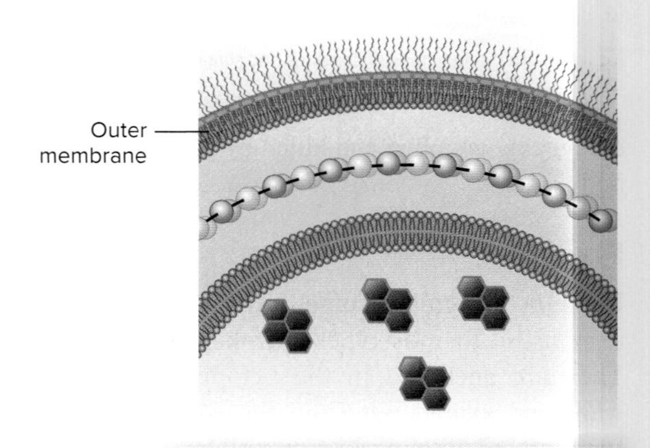

Large crystal violet-iodine complexes form within the cytoplasm as a result of the first two steps of the Gram stain.

After decolorization

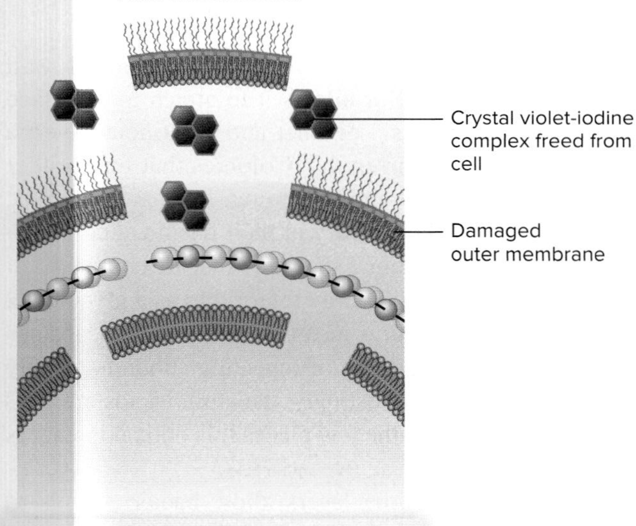

The decolorizing agent damages the outer membrane. The thin peptidoglycan layer cannot prevent the crystal violet-iodine complexes from being washed out of the cell.

FIGURE 3.35 Cell Wall Type and the Gram Stain

? The decolorizing agent damages the peptidoglycan layer. Why is this important to consider when deciding how long to decolorize a specimen as part of the Gram stain procedure?

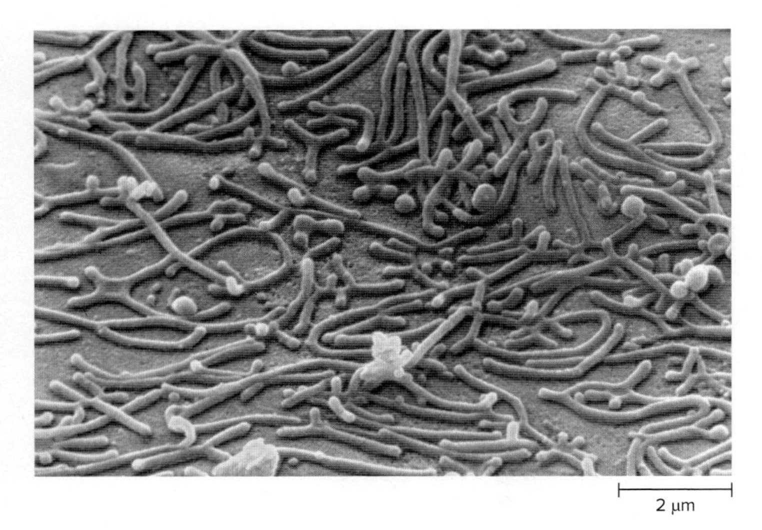

FIGURE 3.36 ***Mycoplasma pneumoniae*** These cells are flexible because they lack a cell wall (SEM). ©Don W. Fawcett/Science Source

? Would lysozyme or penicillin affect *M. pneumoniae*?

Many prokaryotes have structures outside the cell wall. These external structures are not essential to the life of microbes, but they provide a competitive advantage to cells that have them.

Capsules and Slime Layers

Bacteria often have a gel-like layer outside the cell wall that either protects the cell or allows it to attach to a surface (**figure 3.37**). A **capsule** is a distinct and gelatinous layer; a **slime layer** is diffuse and irregular. Colonies that form either of these often appear moist and glistening.

Capsules and slime layers vary in their chemical composition, depending on the microbial species. Most are composed of polysaccharides, and are commonly referred to as a **glycocalyx** (*glyco* means "sugar" and *calyx* means "shell"). A few capsules consist of polypeptides made up of repeating subunits of only one or two amino acids. Interestingly, the amino acids are generally of the D-form, one of the few places this optical isomer is found in nature. ⏮ polysaccharide, ⏮ D-amino acid

Some capsules and slime layers allow bacterial cells to adhere to specific surfaces, including teeth, rocks, and other bacteria. Once attached, the cells can grow as a biofilm, a polymer-encased community of microbes. One example is dental plaque, a biofilm on teeth. When a person ingests sucrose, *Streptococcus mutans* (a common member of the oral microbiota) can use that to make a capsule. This allows *S. mutans* to attach to other bacterial cells on the teeth. In turn, additional bacteria adhere to the sticky layer, resulting in an even greater accumulation of plaque. Acids produced by bacteria in dental plaque then damage the tooth surface. ⏭ biofilm, ⏭ dental plaque

Some capsules allow bacteria to avoid the body's defense systems that otherwise protect against infection. *Streptococcus pneumoniae,* an organism that causes bacterial pneumonia, can cause the disease only if it has a capsule. Unencapsulated cells are quickly engulfed and killed by phagocytes (a cell type of the body's defense system). ⏭ *Streptococcus pneumoniae*, ⏭ phagocytes

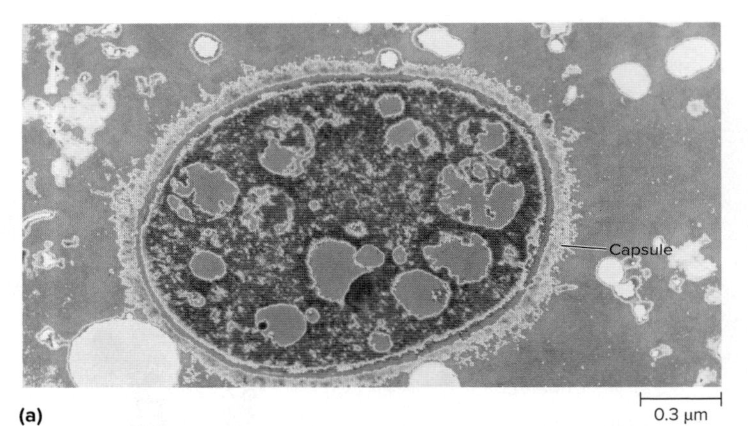

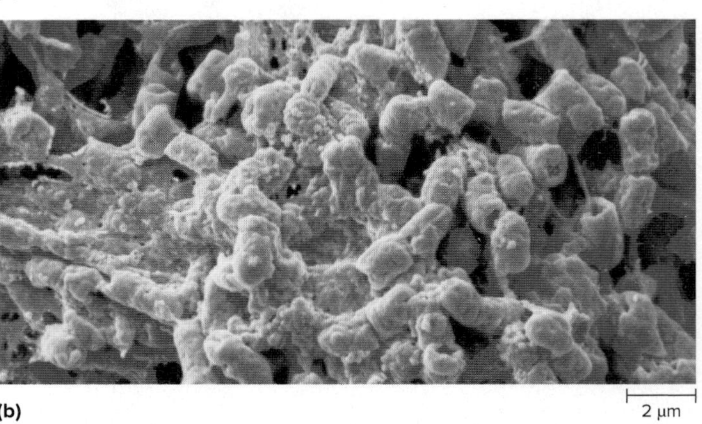

FIGURE 3.37 Capsules and Slime Layers (a) An encapsulated *Lactobacillus* species (TEM). **(b)** Masses of bacteria adhering in a layer of slime (SEM). a: ©Scimat/Science Source; b: ©Scimat/Science Source

? What are the functions of capsules and slime layers?

Flagella

Flagella (singular: **flagellum**) are long protein structures responsible for most types of prokaryotic motility (**figure 3.38**). They are anchored in the cytoplasmic membrane and cell wall, and extend out from the surface of the cell. Flagella function by spinning like propellers, pushing the cell through liquid much as a ship is driven through water. That may sound like easy work, but water has the same relative viscosity to prokaryotes that molasses has to humans!

In some cases, flagella are important in disease. For example, *Helicobacter pylori,* the bacterium that sometimes causes gastric ulcers, has powerful multiple flagella at one end of the cell. These flagella allow *H. pylori* to penetrate the thick mucous gel that coats the stomach epithelium. ⏭ *Helicobacter pylori*

MicroByte

Flagella can rotate over 100,000 rpm, propelling a cell 20 body lengths a second. This is equivalent to a 6-foot-tall man running 82 mph!

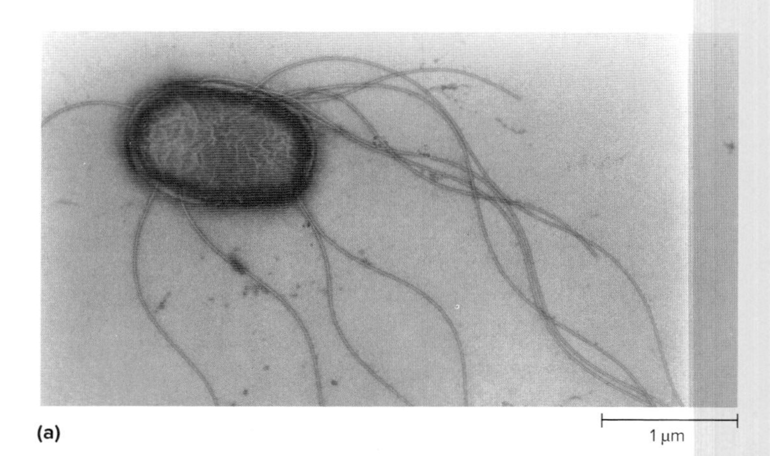

(a)

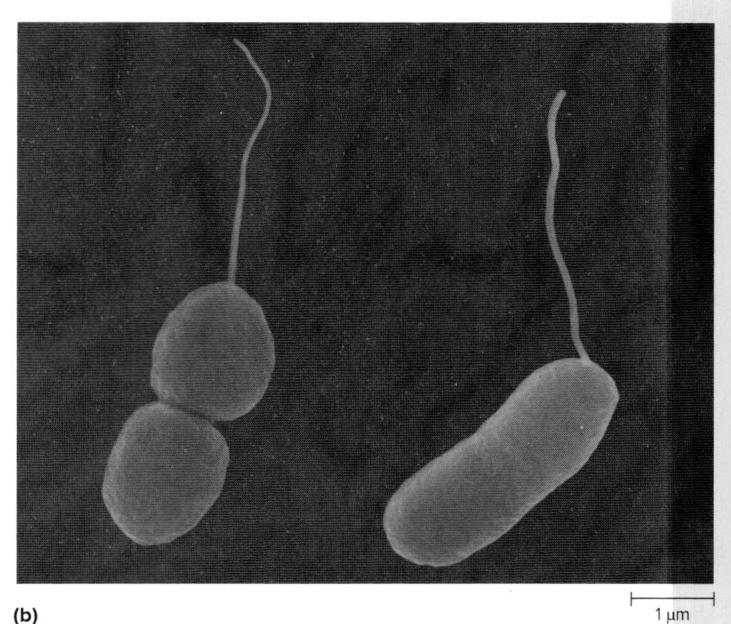

(b)

FIGURE 3.38 Flagella (a) Peritrichous flagella on *Salmonella enterica* (TEM). **(b)** Polar flagellum on *Vibrio parahaemolyticus* (SEM).
a: ©CAMR/A. Barry Dowsett/Science Source; b: ©Dennis Kunkel Microscopy/SPL/Science Source

How can flagella affect a microbe's ability to cause disease?

Structure and Arrangement of Flagella

A flagellum has three basic parts: a basal body, a hook, and a filament (**figure 3.39**). The basal body anchors the structure to the cell wall and cytoplasmic membrane. The hook is a flexible curved segment that extends out from the basal body, connecting it to the filament. The filament, which extends out into the external environment, is made up of identical subunits of a protein called flagellin. These subunits form a chain that twists into a helical structure with a hollow core.

Although the flagella of bacteria and archaea share the same general structure, they have many differences. For instance, rotation of bacterial flagella is powered by proton motive force, whereas rotation of archaeal flagella uses ATP for energy. The molecules that compose archaeal flagella and the mechanisms of their assembly are also distinct.

Recall that the numbers and arrangement of flagella can be used to characterize flagellated bacteria (see figure 3.19). As an example, *E. coli* cells have peritrichous flagella, meaning that the flagella are distributed over the entire surface. Other common bacteria have a single polar flagellum. Other arrangements include a tuft of flagella at one or both ends of the cell. ◀◀ **peritrichous flagella,** ◀◀ **polar flagellum**

Chemotaxis

Motile bacteria sense the presence of chemicals and respond by moving in a certain direction—a phenomenon called **chemotaxis.** If a compound is a nutrient, it may serve as an attractant, causing cells to move toward it. On the other hand, if the compound is toxic, it may act as a repellent, causing cells to move away.

The movement of a bacterial cell toward an attractant is not in a straight line (**figure 3.40**). When an *E. coli* cell travels, it progresses in one direction for a short time, but then stops and tumbles for a fraction of a second. As a consequence, the cell usually finds itself oriented in a completely different direction. It then moves in that direction for a short time, and tumbles again. The seemingly odd pattern of travel is due to the coordinated rotation of the flagella. When flagella rotate

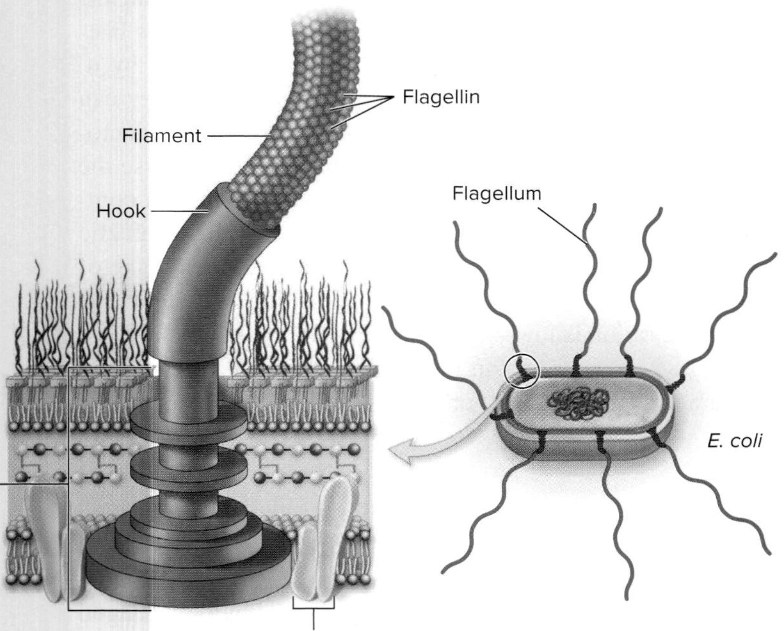

FIGURE 3.39 The Structure of a Flagellum in a Gram-Negative Bacterium The flagellum is composed of three basic parts—a filament, a hook, and a basal body.

What is the role of flagellin in the structure of a flagellum?

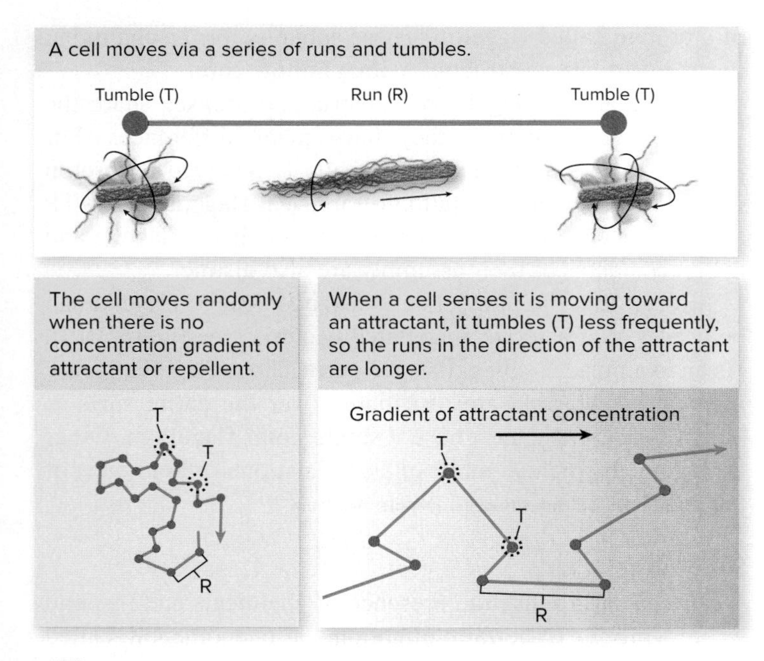

FIGURE 3.40 Chemotaxis

What mechanism causes a cell to tumble?

counterclockwise, they form a tight propelling bundle and the bacterium is pushed in a forward movement called a run. After a brief period, the direction of rotation of the flagella reverses. This change causes the cell to stop and tumble. When cells sense movement toward an attractant they tumble less frequently, so the runs in that direction are longer. Cells also tumble less frequently when they are moving away from a repellent.

In addition to reacting to chemicals, some bacteria respond to the concentration of O_2 (aerotaxis). Organisms that require O_2 for growth will move toward it, whereas bacteria that grow only in its absence tend to be repelled by it. Certain motile bacteria are magnetotactic, meaning they can react to the earth's magnetic field by the process of magnetotaxis. They contain a row of magnetic particles that cause the cells to line up in a north-to-south direction much as a compass does (**figure 3.41**). In most regions, the earth's magnetic forces have a significant vertical component, so the magnetotactic cells move downward and into sediments where the O_2 concentration is low—the environment best suited for their growth. Some bacteria can move in response to variations in temperature (thermotaxis) or light (phototaxis).

Pili

Pili (singular: **pilus**) are considerably shorter and thinner than flagella, and their function is quite different (**figure 3.42**). However, one part of their structure has a theme similar to the filament of a flagellum: a string of protein subunits arranged helically to form a long molecule with a hollow core.

Many types of pili allow cells to attach to specific surfaces; these pili are also called **fimbriae.** Strains of *E. coli* that

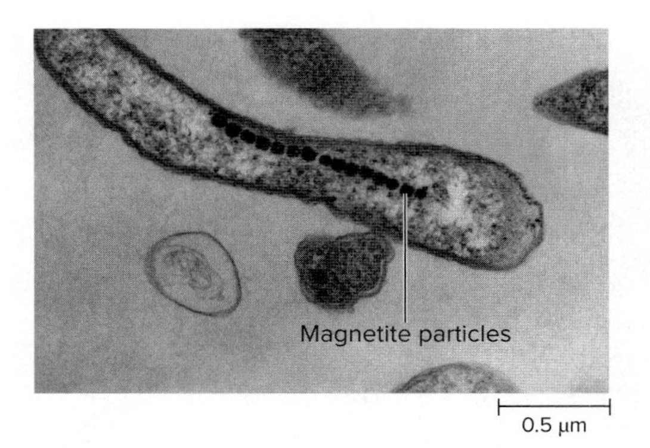

FIGURE 3.41 Magnetotactic Bacterium The chain of magnetic particles (magnetite: Fe_3O_4) within *Magnetospirillum magnetotacticum* helps align the cell along geomagnetic lines (TEM). ©Dennis Kunkel Microscopy/SPL/Science Source

How could magnetotaxis benefit a cell?

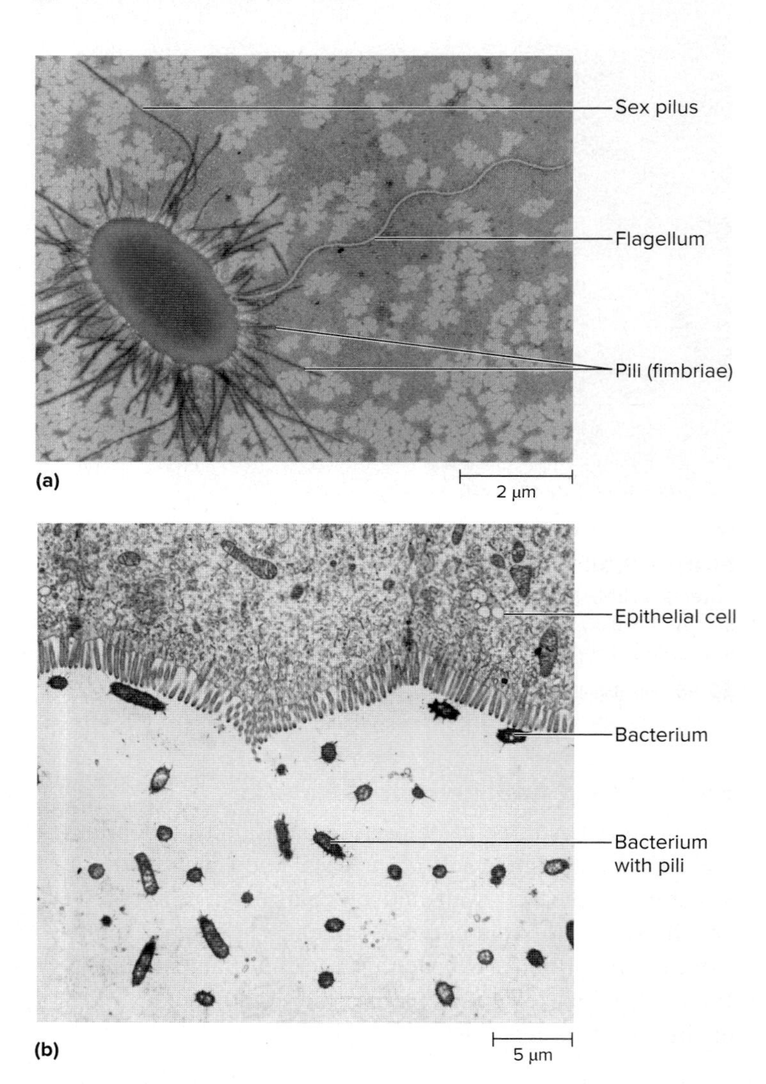

FIGURE 3.42 Pili **(a)** Pili on an *Escherichia coli* cell. The short pili (fimbriae) allow adherence, whereas the sex pilus is involved in DNA transfer (SEM). **(b)** *Escherichia coli* attaching to epithelial cells in the small intestine of a pig (TEM). a: ©Dennis Kunkel/SPL/Science Source; b: Source: Harley W. Moon/U.S. Department of Agriculture

How does the function of pili compare to that of flagella?

cause watery diarrhea have pili that allow them to attach to cells that line the small intestine. Without the ability to attach, these cells would simply move along the intestinal tract with the other intestinal contents. ⏭| enterotoxigenic *E. coli*

Some types of pili help bacterial cells move on solid media. Twitching motility (characterized by jerking movements) and certain types of smooth, gliding motility involve pili.

Another type of pilus, called a **sex pilus,** is used to join one bacterium to another for a specific type of DNA transfer. This and other mechanisms of DNA transfer will be described in chapter 8. ⏭| DNA transfer

MicroAssessment 3.6

Capsules and slime layers allow organisms to adhere to surfaces and sometimes protect bacteria from certain host defense systems. Flagella are the most common mechanism for bacterial motility. Chemotaxis is the directed movement of cells toward an attractant or away from a repellent. Pili provide a mechanism for attachment to specific surfaces.

16. How do capsules differ from slime layers?

17. *E. coli* cells have peritrichous flagella. What does this mean?

18. Explain why a sugary diet can lead to tooth decay.

3.7 ■ Internal Components of Prokaryotic Cells

Learning Outcomes

21. Describe the structure and function of the chromosome, plasmids, ribosomes, storage granules, gas vesicles, and endospores.

22. Describe the significance and processes of sporulation and germination.

Prokaryotic cells have a variety of structures within the cell. Some, such as the chromosome and ribosomes, are essential for the life of all cells, whereas others give cells certain selective advantages.

Chromosome and Plasmids

The prokaryotic **chromosome** is typically a single, circular double-stranded DNA molecule that contains all the genetic information required by a cell, as well as information that may be helpful but not required. The chromosome folds and twists to form a tightly packed mass within the cytoplasm, creating a gel-like region called the **nucleoid** (**figure 3.43**). The compact shape is due partially to nucleoid-associated proteins that bind to DNA, creating a structure that bends and folds. In addition, the DNA is twisted, or supercoiled. To understand what is meant by supercoiling, cut a rubber band and then twist one end several times before rejoining the cut ends. The twisting and coiling you see are analogous to supercoiling.

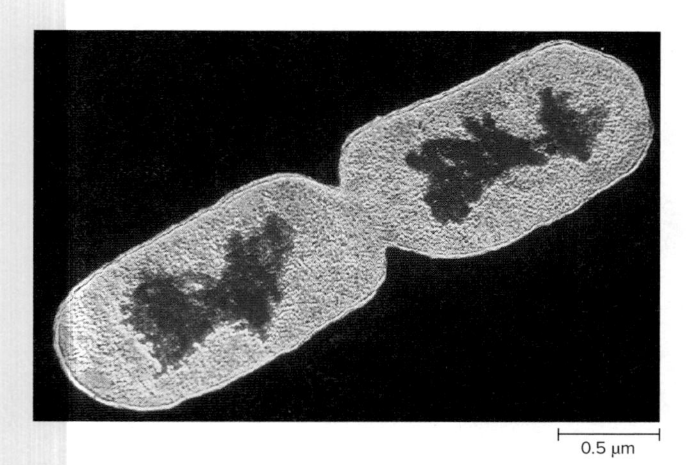

FIGURE 3.43 The Chromosome *Escherichia coli* undergoing cell division, with the DNA shown in red (TEM). ©CNRI/SPL/Science Source

? What is the gel-like region formed by the chromosome called?

Plasmids have a structure similar to the chromosome, but they are smaller and typically do not encode essential genetic information. For example, many plasmids code for the production of enzymes that destroy certain antibiotics, allowing the organism to resist the otherwise lethal effect of these medications. Because a bacterium can sometimes transfer a copy of a plasmid to another bacterial cell, this accessory genetic information can spread, which accounts in large part for the increasing frequency of antibiotic-resistant bacteria. A single cell can have more than one type of plasmid, and these can each be present in multiple copies. ⏭| plasmids

Ribosomes

Ribosomes are involved in protein synthesis, where they facilitate the joining of amino acids. Prokaryotic ribosomes are referred to as 70S, a term that reflects the relative size and density of the molecules. In that term, the "S" (for Svedberg) is a unit used to indicate how fast particles settle when spun at very high speeds in an ultracentrifuge. The faster a particle moves toward the bottom, the higher the S value and the greater the size and density. Note that S units are not strictly arithmetic; the 70S ribosome is composed of a 30S and a 50S subunit (**figure 3.44**). Prokaryotic ribosomes differ from eukaryotic ribosomes, which are 80S. The fact that they are distinct is important medically because antibiotics that interfere with the function of the bacterial 70S ribosome have no effect on the 80S molecule. Those antibiotics also have no effect on archaeal cells because archaeal ribosomes differ from their bacterial counterparts, even though they are both 70S. ⏭| role of ribosomes

Cytoskeleton

It was once thought that bacteria lacked a **cytoskeleton,** an interior protein framework that provides support and structure. Several bacterial proteins that have similarities to those of the eukaryotic cytoskeleton have now been characterized,

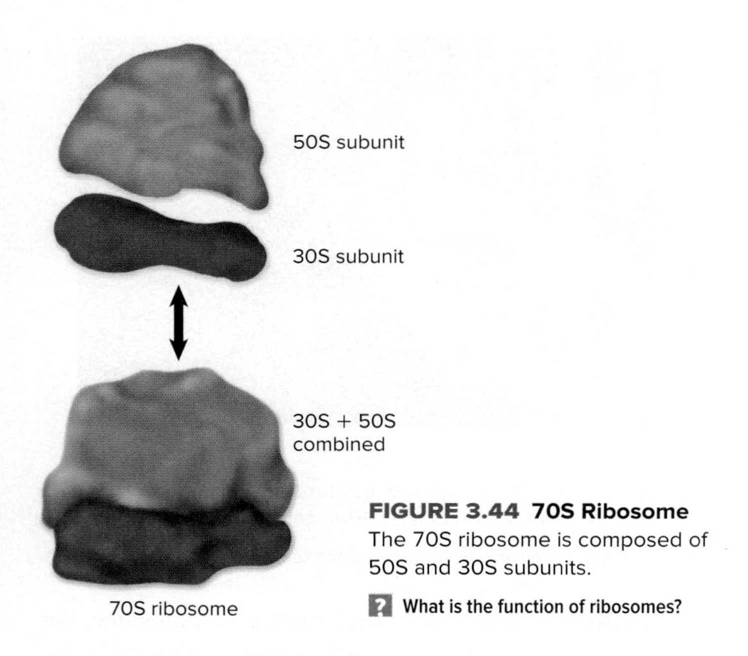

FIGURE 3.44 70S Ribosome The 70S ribosome is composed of 50S and 30S subunits.

What is the function of ribosomes?

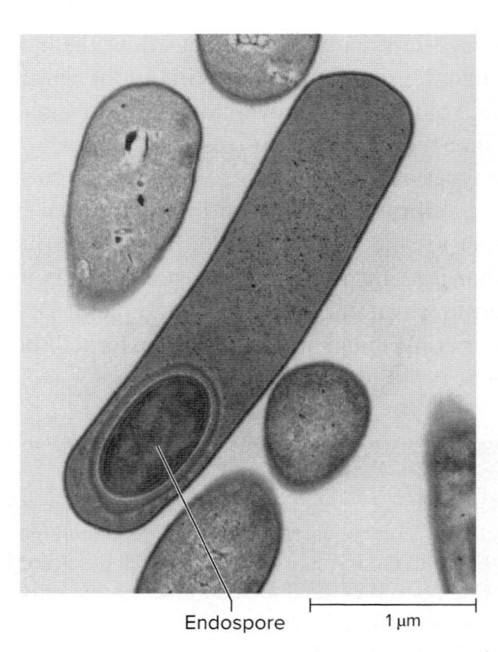

FIGURE 3.45 Endospore *Clostridium difficile* forming an endospore (TEM). ©Dr. Kari Lounatmaa/Science Source

What is the function of an endospore?

and these appear to be involved in cell division as well as in controlling cell shape.

Storage Granules

Storage granules are accumulations of high-molecular-weight polymers synthesized from a nutrient that a cell has in relative excess. If nitrogen and/or phosphorus are lacking, for example, a cell cannot multiply even if a carbon and energy source such as glucose is plentiful. Rather than waste the carbon/energy source, cells use it to produce glycogen (a glucose polymer). Later, when conditions are appropriate, cells degrade and use the glycogen. Other bacterial species store carbon and energy as poly-β-hydroxybutyrate (PHB).

Some types of granules can be easily detected by light microscopy. Volutin granules—a storage form of phosphate and sometimes energy—stain red with the dye methylene blue, whereas the surrounding cellular material stains blue. Because of this, they are often called metachromatic granules (*meta* means "change" and *chromatic* means "color").

MicroByte

Poly-β-hydroxybutyrate can be used to make biodegradable plastics. Phosphate-storing cells can be used to remove pollutants from wastewater.

Gas Vesicles

Some aquatic bacteria produce **gas vesicles:** small, rigid, protein-bound compartments that provide buoyancy to the cell. Gases, but not water, flow freely into the vesicles, thereby decreasing the density of the cell. By regulating the number of gas vesicles, a cell can float or sink to its ideal position in the water column. Bacteria that rely on sunlight for energy use gas vesicles to float closer to the surface, where light is available.

Endospores

An **endospore** is a unique type of dormant cell produced by certain bacterial species such as members of the genera *Bacillus* and *Clostridium* (**figure 3.45**). The structures may remain dormant for perhaps 100 years, or even longer, and are extraordinarily resistant to damaging conditions including heat, desiccation, toxic chemicals, and ultraviolet (UV) light. Immersion in boiling water for hours may not kill them. An endospore that survives these treatments can **germinate,** or exit the dormant stage, to become a typical multiplying cell, called a **vegetative cell.**

Because endospores can survive so long in a variety of conditions, they can be found virtually everywhere. They are common in soil, which can make its way into environments such as laboratories and hospitals and onto products such as medical devices, food, and media used to cultivate microbes. Keeping these environments and products microbe-free is very important, so special precautions must be taken to avoid or destroy endospores.

Endospores are sometimes called spores. However, this latter term is also used to refer to structures produced by unrelated microbes such as fungi. Bacterial endospores are much more resistant to environmental conditions than are other types of spores.

MicroByte

The diseases botulism, tetanus, gas gangrene, and anthrax are all caused by endospore formers.

Sporulation

Endospore formation, or **sporulation,** is a complex sequence of changes that begin when spore-forming bacteria experience limiting amounts of carbon or nitrogen. The cells sense starvation conditions, which triggers them to begin the 8-hour sporulation process.

As sporulation begins, the cell stops growing but it duplicates the DNA (**figure 3.46** ①). ② A septum forms, dividing the cell asymmetrically. ③ The larger compartment then engulfs the smaller compartment. These two compartments take on different roles in synthesizing the components that will make up the endospore. ④ As the developing endospore continues to mature, the smaller compartment develops into what is called the forespore, which will become the core of the endospore. Peptidoglycan-containing material is laid down between the two membranes that surround the forespore, forming the core wall and the cortex. Meanwhile, the mother cell makes proteins that will form the outermost layer of the endospore, the spore coat. ⑤ Ultimately, the mother cell is degraded and the endospore released.

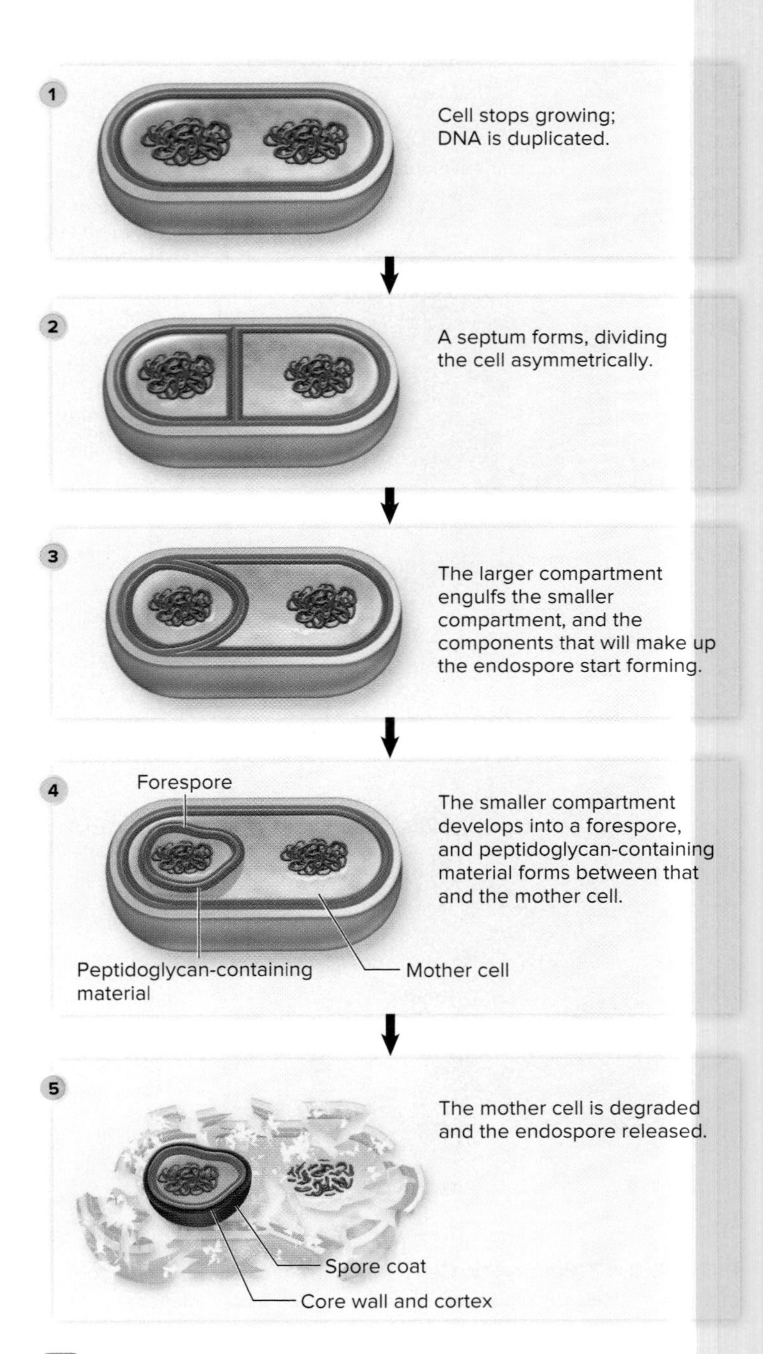

FIGURE 3.46 The Process of Sporulation

? Approximately how long does the sporulation process take?

The layers of the endospore protect it from damage. The spore coat is thought to function as a sieve, excluding molecules such as lysozyme. The cortex helps maintain the core in a dehydrated state, protecting it from the effects of heat. In addition, the core has small, acid-soluble proteins that bind to DNA, thereby protecting it from damage. The core is rich in an unusual compound (calcium dipicolinate), which seems to also play an important role in spore resistance.

Germination

Germination can be triggered by a brief exposure to heat or certain chemicals. Following such exposure, the endospore absorbs water and swells. The spore coat and cortex then crack open, and a vegetative cell grows out. As this happens, the core wall becomes the peptidoglycan layer of the vegetative cell. The process of germination is relatively fast in comparison to sporulation, taking 1–2 hours.

Endospore production and subsequent germination are not a means of cell reproduction. This is because one vegetative cell gives rise to one endospore, which then forms one vegetative cell. There is no increase in the number of cells; the process simply allows the organism to develop into a form that survives drying and other adverse conditions commonly encountered in the environment.

MicroAssessment 3.7

The prokaryotic chromosome is usually a circular, double-stranded DNA molecule that contains all of the genetic information required by a cell. Plasmids generally encode only information that is advantageous to a cell in certain conditions. Ribosomes facilitate the joining of amino acids to form a protein. The cytoskeleton is an interior framework involved in cell division as well as controlling cell shape. Storage granules are polymers synthesized from nutrients a cell has in relative excess. Gas vesicles provide buoyancy to a cell. An endospore is a highly resistant dormant stage produced by certain bacterial species.

19. Explain how glycogen granules benefit a cell.

20. Explain why endospores are an important concern for the canning industry.

21. What would be the consequence to a cell if its chromosome were damaged beyond repair, and how would the consequence differ if a plasmid were damaged instead?

EUKARYOTIC CELLS

Eukaryotic cells are generally much larger than prokaryotic cells, and their internal structures are far more complex (**figure 3.47**). One of their distinguishing characteristics is the presence of membrane-enclosed compartments, or **organelles.** The most significant of these is the nucleus, which contains the cell's genetic information (DNA). The organelles, which can occupy half the total cell volume, allow the cell to perform complex functions in physically separated regions. For example, when degradative enzymes are contained within an organelle, they can digest material without posing a threat to the integrity of the cell itself.

Each organelle contains a variety of compounds, many of which are synthesized at other locations. To deliver these to the **lumen** (interior) of another organelle, a section of the organelle buds off, forming a membrane-bound vesicle (**figure 3.48**). This carries a bit of the organelle's contents to other parts of the cell. When the vesicle encounters another organelle, the two membranes may fuse to become one unit, similar to two oil droplets merging. By doing so, the vesicle spills its contents into the organelle.

As a group, eukaryotic cells are highly variable. For example, protozoa, which are single-celled organisms, must function exclusively as self-contained units that seek and ingest food. These cells must be mobile and flexible to take in food particles, and they a lack cell wall that would otherwise provide rigidity. Animal cells also lack a cell wall, because they too must be flexible to accommodate movement. Fungi, on the other hand, are stationary and benefit from the protection provided by a strong cell wall. Fungal cell walls are composed primarily of polysaccharides including chitin (a polymer of *N*-acetylglucosamine), glucans (polymers of glucose), and mannans (polymers of mannose). Plants, which are also stationary, have cell walls composed of cellulose (a polymer of glucose). ◀◀ chitin, ◀◀ cellulose

The individual cells of a multicellular organism can be distinctly different from one another. Liver cells, for example, are obviously quite different from bone cells. Groups of cells function in cooperative associations called **tissues.** The tissue types in your body include muscle, connective, nerve, and epithelial. Various tissues work together to make up **organs,** including skin, heart, and liver. Organs and the tissues that compose them will be covered in more detail in chapters 21–27.

Full coverage of all aspects of eukaryotic cells is beyond the scope of this textbook. Instead, this section will focus on key characteristics, particularly those that directly affect the interactions of microbes with human hosts. These characteristics are summarized in **table 3.6.** The functions of prokaryotic and eukaryotic cell components are compared in **table 3.7.**

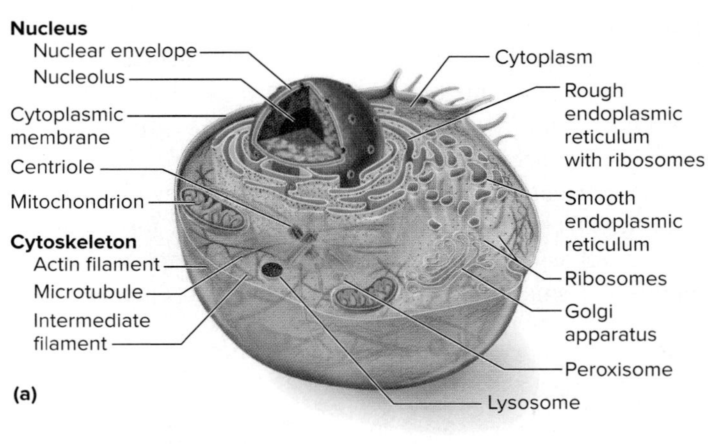

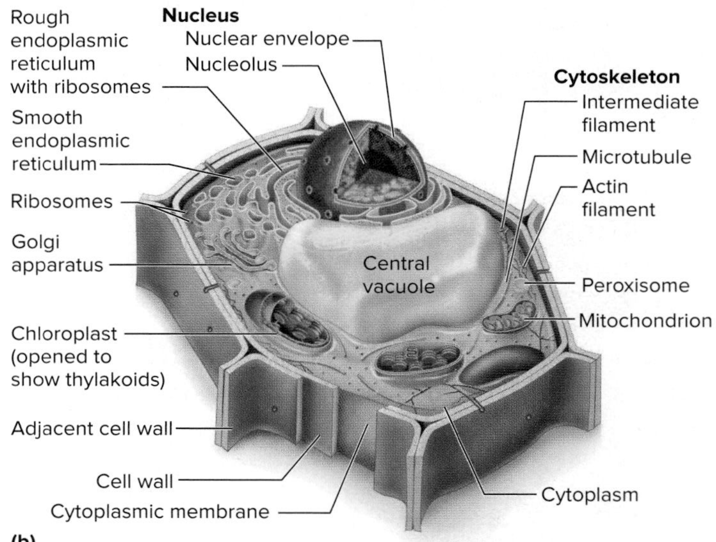

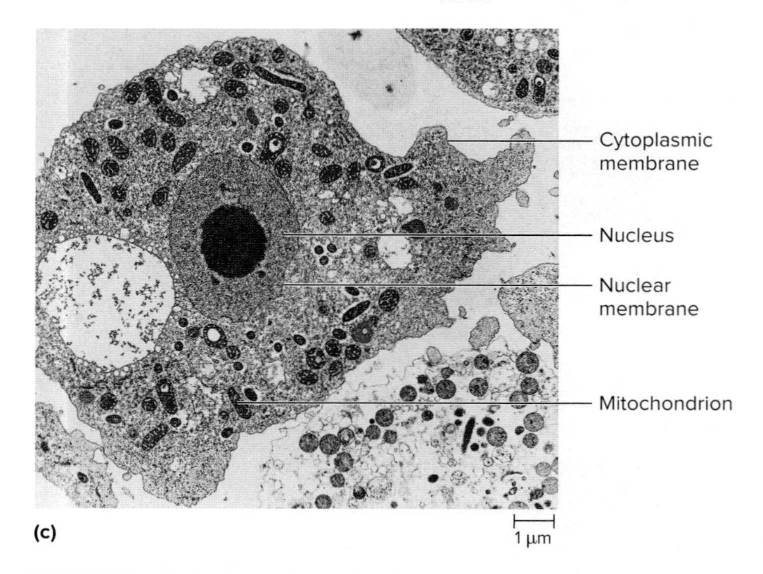

FIGURE 3.47 Eukaryotic Cells (a) Diagrammatic representation of an animal cell. **(b)** Diagrammatic representation of a plant cell. **(c)** Electron micrograph of an animal cell shows several membrane-bound structures including mitochondria and a nucleus. A prokaryotic cell is about the size of a mitochondrion (TEM). ©Dr. Thomas Fritsche

? **Which organelle contains the cell's genetic information?**

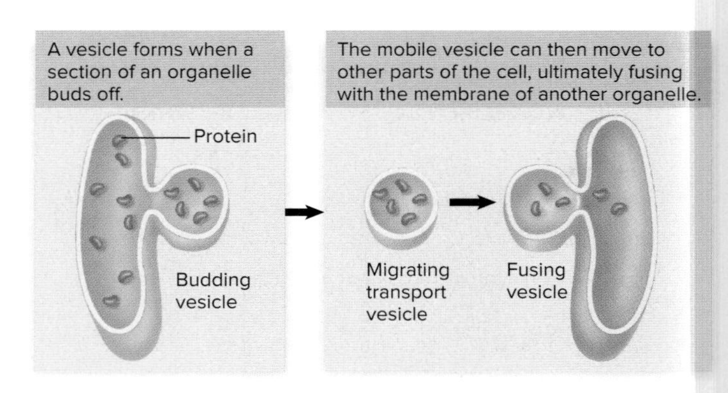

FIGURE 3.48 **Vesicle Formation and Fusion**

? The lumen is which part of an organelle?

3.8 ■ Cytoplasmic Membrane of Eukaryotic Cells

Learning Outcomes

23. Describe the structure and function of the eukaryotic cytoplasmic membrane, comparing and contrasting it with the prokaryotic counterpart.
24. Describe the mechanisms eukaryotic cells use to transfer molecules across the cytoplasmic membrane.

All eukaryotic cells have a cytoplasmic membrane, or plasma membrane, which is similar in chemical structure and function to that of prokaryotic cells.

Structure and Function of the Cytoplasmic Membrane

The eukaryotic cytoplasmic membrane is a typical phospholipid bilayer embedded with proteins. The lipid and protein composition of the layer that faces the cytoplasm, however, differs significantly from that facing the outside of the cell. The same is true for membranes that surround the organelles. The layer facing the inside of the organelle is similar to its cytoplasmic membrane counterpart that faces the outside of the cell. This lack of symmetry reflects the important role these membranes play in complex processes within the eukaryotic cell.

The membrane proteins perform a variety of functions. Some are involved in transport and others are attached to internal structures, helping to maintain cell integrity. Those that face the outside of the cell often function as **receptors.** A given receptor binds a specific molecule, which is referred to as its **ligand.** These receptor-ligand interactions are extremely important in multicellular organisms because they allow cells to communicate with each other—a process called signaling. For example, when certain cells of the human body encounter a compound unique to bacteria, they secrete specific proteins as a call for help in defending against the invader. Other cells of the body have surface receptors for these proteins, allowing them to recognize the signals, and respond accordingly. Using this cell-to-cell communication, a multicellular organism can function as a cohesive unit.

TABLE 3.6 Typical Eukaryotic Cell Structures

Structure	Characteristics
Cytoplasmic Membrane	Asymmetrical phospholipid bilayer embedded with proteins. Permeability barrier, transport, and cell-to-cell communication.
Internal Protein Structures	
Cilia	Beat in synchrony to provide movement. Composed of microtubules in a 9 + 2 arrangement.
Cytoskeleton	Dynamic filamentous network that provides structure to the cell.
Flagella	Propel or push the cell with a whip-like or thrashing motion. Composed of microtubules in a 9 + 2 arrangement.
Ribosomes	Two subunits, 60S and 40S, join to form the 80S ribosome.
Membrane-Bound Organelles	
Chloroplasts	Site of photosynthesis; the organelle harvests the energy of sunlight to generate ATP, which is then used to convert CO_2 to carbohydrates.
Endoplasmic reticulum	Site of synthesis of macromolecules destined for other organelles or the external environment.
Rough	Attached ribosomes thread proteins they are synthesizing into the lumen of the organelle.
Smooth	Site of lipid synthesis and degradation, and calcium ion storage.
Golgi apparatus	Site where macromolecules synthesized in the endoplasmic reticulum are modified before being transported in vesicles to other destinations.
Lysosome	Site of digestion of macromolecules.
Mitochondria	Harvest the energy released during the degradation of organic compounds to generate ATP.
Nucleus	Contains the genetic information (DNA).
Peroxisome	Site where oxidation of lipids and toxic chemicals occurs.

TABLE 3.7 Comparison of Typical Prokaryotic and Eukaryotic Cell Structures/Functions

	Prokaryotic	Eukaryotic
General Characteristics		
Size	Generally 0.3–2 μm in diameter.	Generally 5–50 μm in diameter.
Cell division	Chromosome replication followed by binary fission.	Chromosome replication and mitosis followed by division.
Chromosome location	Located in the nucleoid, which is not membrane-bound.	Contained within the membrane-bound nucleus.
Structures		
Cytoplasmic membrane	Relatively symmetrical with respect to the phospholipid content of the bilayers.	Highly asymmetrical; phospholipid composition of outer layer differs significantly from that of inner layer.
Cell wall	Composed of peptidoglycan *(Bacteria)*; Gram-negative bacteria have an outer membrane as well.	Absent in animal cells; composition in other cell types may include chitin, glucans, and mannans (fungi), and cellulose (plants).
Chromosome	Single, circular DNA molecule is typical.	Multiple, linear DNA molecules. DNA is wrapped around histones.
Flagella	Composed of protein subunits; attached to the cell envelope.	Made up of a 9 + 2 arrangement of microtubules; covered by an extension of the cytoplasmic membrane.
Membrane-bound organelles	Absent.	Present; includes the nucleus, mitochondria, chloroplasts (only in plant cells), endoplasmic reticulum, Golgi apparatus, lysosomes, and peroxisomes.
Nucleus	Absent; DNA resides as an irregular mass forming the nucleoid region.	Present.
Ribosomes	70S ribosomes, which are made up of 50S and 30S subunits.	80S ribosomes, which are made up of 60S and 40S subunits. Mitochondria and chloroplasts have ribosomes similar to those of bacteria.
Functions		
Degradation of extracellular substances	Enzymes are secreted that degrade macromolecules outside the cell. The resulting small molecules are transported into the cell.	Macromolecules can be brought into the cell by endocytosis. Lysosomes carry digestive enzymes.
Motility	Generally involves flagella, which are composed of protein subunits. Flagella rotate like propellers.	Involves cilia and flagella, which are made up of a 9 + 2 arrangement of microtubules. Cilia move in synchrony; flagella propel a cell with a whip-like motion or thrash back and forth to pull a cell forward.
Protein secretion	Secretion systems transport proteins across the cytoplasmic membrane.	Secreted proteins are moved to the lumen of the rough endoplasmic reticulum as they are being synthesized. From there, they are transported to the Golgi apparatus for processing and packaging. Ultimately, vesicles deliver them to the outside of the cell by the process of exocytosis.
Strength and rigidity	Peptidoglycan-containing cell wall *(Bacteria)*. Cytoskeletal components.	Cytoskeleton composed of microtubules, intermediate filaments, and microfilaments. Some have a cell wall; some have sterols in the membrane.
Transport	Primarily active transport. Group translocation *(Bacteria)*.	Facilitated diffusion and active transport. Ion channels.

The membranes of many eukaryotic cells contain sterols, which provide strength to the otherwise fluid structure. The sterol in animal cell membranes is cholesterol, whereas fungal membranes have ergosterol. This difference is important medically because some antifungal medications act by interfering with ergosterol synthesis or function. ⏭ antifungal medications

Within the cytoplasmic membrane are cholesterol-rich regions called lipid rafts. The role of these regions is still being studied, but they appear to be important in allowing the cell to detect and respond to signals in the external environment. From a microbiologist's perspective, they are also important because many viruses use these regions when they enter or exit a cell.

The cytoplasmic membrane of eukaryotes plays no role in ATP synthesis; instead, that task is performed by organelles called mitochondria. Even though the cytoplasmic membrane has no electron transport chain to generate a proton motive force, there is still an electrochemical gradient across the membrane. This is maintained by energy-requiring mechanisms that pump either sodium ions or protons to the outside of the cell. ⏭ mitochondria

Transfer of Molecules Across the Cytoplasmic Membrane

Various substances—including nutrients, signaling molecules, and waste products—must pass through the cytoplasmic membrane. Eukaryotic cells have several mechanisms to accomplish this.

Transport Proteins

The transport proteins of eukaryotic cells function as aquaporins, channels, or carriers. Aquaporins facilitate water passage. **Channels** form small pores in the membrane that allow small molecules or ions to diffuse through; the channel has a gate that can open or close in response to environmental conditions, allowing the cell to control passage. **Carriers** are analogous to the proteins in prokaryotic cells that function in facilitated diffusion and active transport. In contrast to the common situation for prokaryotes and other unicellular organisms, however, cells of multicellular organisms often take in nutrients by facilitated diffusion. This is because the organisms can often control the level of nutrients surrounding individual cells. For example, blood typically has a high enough concentration of glucose for tissue cells to bring the sugar in by facilitated diffusion. ⏮ aquaporin, ⏮ facilitated diffusion, ⏮ active transport

Endocytosis and Exocytosis

Endocytosis is the process by which a eukaryotic cell takes up material from the surrounding environment by forming invaginations (inward folds) in its cytoplasmic membrane (**figure 3.49**). It is significant because it allows cells to take in material too large to fit through transport proteins. The type of endocytosis common to most animal cells is **pinocytosis.** In this process, the small invaginations bring in liquid along with any dissolved substances. This action ultimately forms a membrane-bound compartment called an **endosome,** which then fuses with digestive organelles called lysosomes to form an **endolysosome.** Within this compartment, the enclosed material will be degraded. ⏭ lysosome

Animal cells often take up material by **receptor-mediated endocytosis.** This process allows cells to take up extracellular ligands that bind to receptors on the cell's surface. When the receptors bind their ligands, the region is internalized to form an endosome that contains the receptors along with their bound ligands. The fate of the endosomes is similar to that described for pinocytosis.

Phagocytes (a cell type of the body's defense system) and protozoa both use a specialized additional type of endocytosis called **phagocytosis,** which allows them to ingest bacteria and other relatively large debris. In phagocytosis, the cells send out arm-like extensions called **pseudopods,** which surround and enclose extracellular material, including bacteria. This action brings the material into the cell in an enclosed compartment called a **phagosome,** which ultimately fuses with lysosomes to form a **phagolysosome.** As with material in endolysosomes, the material in phagolysosomes will be degraded. ⏭ phagocytes

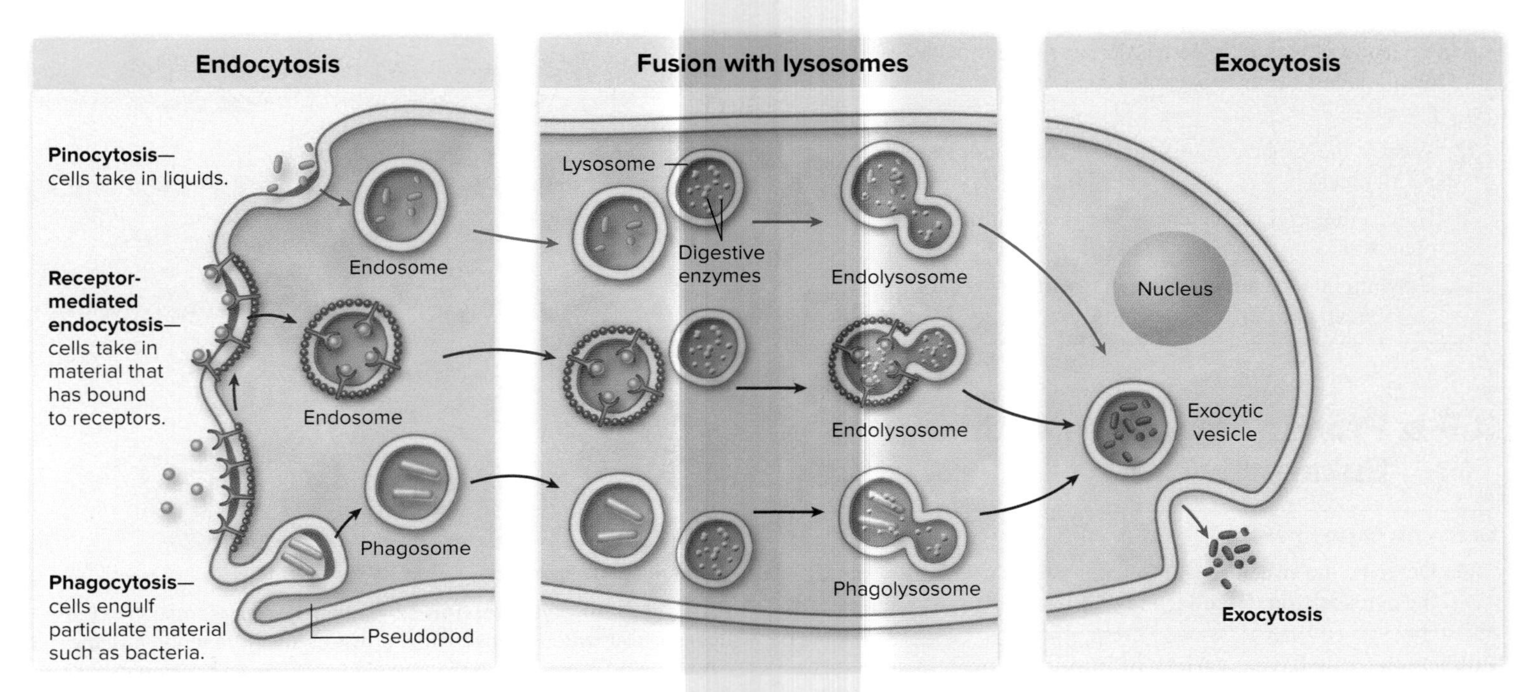

FIGURE 3.49 Endocytosis and Exocytosis These processes allow the cell to take in or remove substances too large to move through a transport protein.

? How is pinocytosis different from phagocytosis?

The process of **exocytosis** is the reverse of endocytosis. Membrane-bound exocytic vesicles inside the cell fuse with the cytoplasmic membrane and release their contents to the outside of the cell.

MicroByte

Many viruses use receptor-mediated endocytosis to enter animal cells. They bind to a specific receptor, "tricking" the cell into bringing them in.

Secretion

As in prokaryotic cells, proteins destined for secretion have a signal sequence: a characteristic amino acid sequence that functions as a tag. When ribosomes begin synthesizing a protein with such a sequence, they attach to the membrane of an organelle called the endoplasmic reticulum (ER). Then, as the polypeptide is synthesized, it is threaded through the membrane and into the ER. From there, the protein can easily be transported by vesicles to the outside of the cell. Proteins destined for various organelles also have specific amino acid tags. ▶▶| endoplasmic reticulum

MicroAssessment 3.8

The cytoplasmic membrane of eukaryotic cells is an asymmetric phospholipid bilayer embedded with proteins. Specific receptors allow cell-to-cell signaling. Transport proteins function as either channels or carriers. Pinocytosis allows cells to internalize small molecules. Protozoa and phagocytes internalize bacteria and debris by phagocytosis. Exocytosis is used to move material to the outside of the cell. Secreted proteins are threaded through the membrane of the endoplasmic reticulum as they are being made.

22. What is the medical significance of ergosterol in fungal membranes?
23. How is the origin of an endosome different from that of a phagosome?
24. How might a bacterium resist the killing effects of the environment within a phagolysosome?

3.9 ■ Protein Structures Within Eukaryotic Cells

Learning Outcomes

25. Describe the structure and function of eukaryotic ribosomes, the cytoskeleton, flagella, and cilia.

Eukaryotic cells have a variety of important protein structures within the cell. These include ribosomes, the cytoskeleton, flagella, and cilia.

Ribosomes

Ribosomes are the structures involved in protein synthesis. The eukaryotic ribosome is 80S, and is made up of a 60S and a 40S subunit. Recall that the prokaryotic ribosomes are 70S. The difference is medically relevant because certain antibacterial medications specifically interfere with the function of bacterial 70S ribosomes.

Cytoskeleton

The **cytoskeleton** forms the framework of a cell (**figure 3.50**). Its three components—(1) actin filaments (also called microfilaments), (2) microtubules, and (3) intermediate filaments—continually reconstruct to adapt to the cell's constantly changing needs.

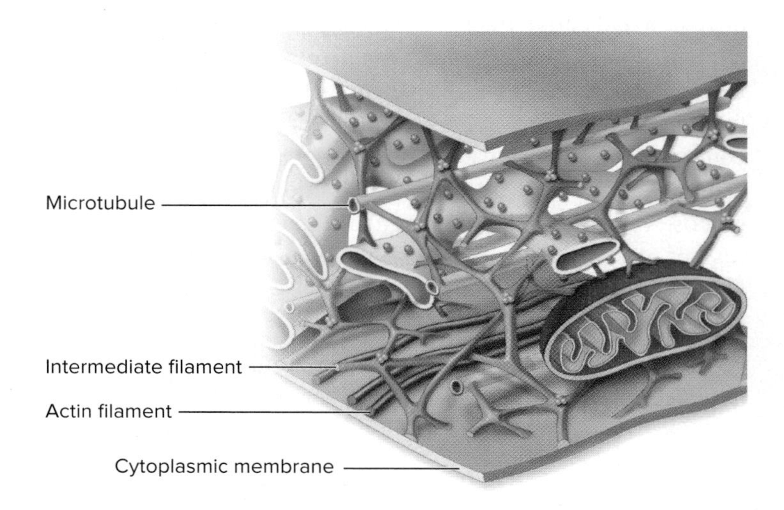

(a) Actin filament

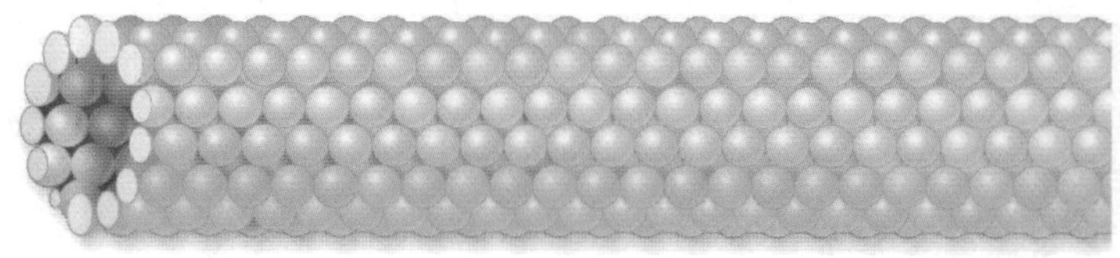

(b) Microtubule

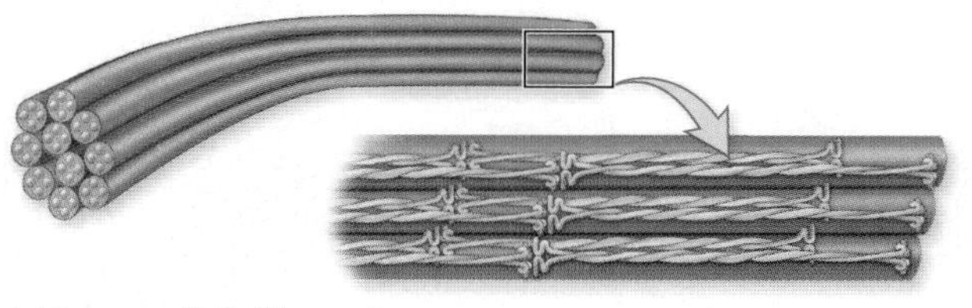

(c) Intermediate filament

FIGURE 3.50 Cytoskeleton Diagrammatic representation of the dynamic filamentous network that provides structure to the cell; the cytoskeleton is composed of three elements: **(a)** actin filaments, **(b)** microtubules, and **(c)** intermediate filaments.

What is the role of actin filaments?

FOCUS YOUR PERSPECTIVE 3.1

Pathogens Hijacking Actin

Some pathogenic bacteria have learned how to hijack the function of a eukaryotic cell's actin filaments for their own benefit. The pathogens make proteins that cause actin to polymerize or depolymerize on demand, allowing the bacteria to control movement of the eukaryotic cell's cytoplasm.

Salmonella species manipulate actin in intestinal epithelial cells (the cells that line the intestinal tract) to induce those cells to engulf them. The intestinal cells do not normally take in material using phagocytosis, but the bacteria trick them into doing so. The bacterial cells do this by injecting actin-rearranging proteins into an intestinal cell, which causes pseudopod-like membrane ruffles to form on the intestinal cell's surface (**box figure 3.1**). The ruffles extend outward and then enclose the bacteria, bringing them into the intestinal cell. Once inside an intestinal cell, the *Salmonella* can multiply, protected from many of the host defenses. ⏭| ***Salmonella* gastroenteritis**

Shigella species enter intestinal epithelial cells using a mechanism similar to that of *Salmonella* species. *Shigella* cells, however, have an additional trick. Once inside an intestinal epithelial cell, the bacteria direct their own transfer to adjacent cells. They do this by causing the host cell's actin to polymerize at one end of the bacterial cell. This forms an "actin tail" that propels the bacterium within the cell. The force of the propulsion is so great that the bacteria are often driven into the cytoplasm of neighboring cells. *Listeria monocytogenes* does the same thing (**box figure 3.2**). ⏭| **shigellosis,** ⏭| **listeriosis**

Certain strains of diarrhea-causing *E. coli,* including *E. coli* O157:H7 and a group called enteropathogenic *E. coli,* attach to intestinal epithelial cells and then inject proteins that cause a thick structure—a "pedestal"—to form underneath the attached bacterial cell (**box figure 3.3**). The pedestal is the result of rearrangement of the actin filaments just below the cell surface. The advantage the bacteria gain by causing the cell to produce pedestals is not known. ⏭| ***E. coli* gastroenteritis**

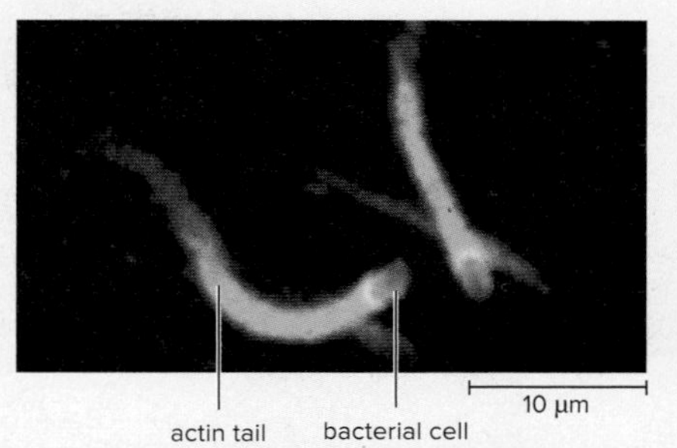

BOX FIGURE 3.2 Actin-Based Motility *Listeria* cells within host cells cause the host cell actin to polymerize at one end of the bacterial cell, propelling the bacterium with enough force to move it into the adjacent host cell.

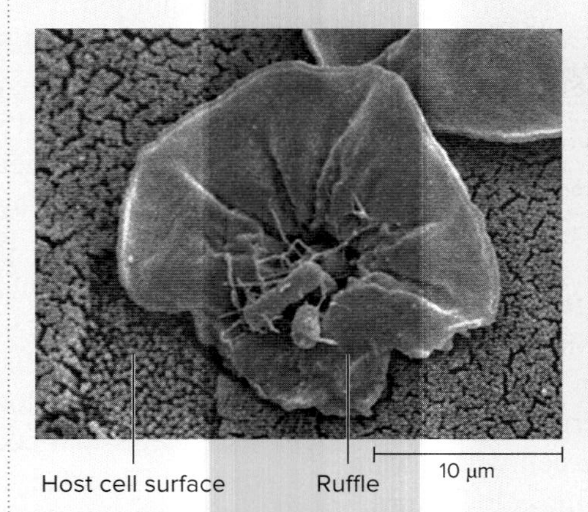

BOX FIGURE 3.1 Ruffling *Salmonella* cells enter intestinal epithelial cells by inducing rearrangement of the cells' actin, causing ruffles to form that eventually enclose the bacterial cells.

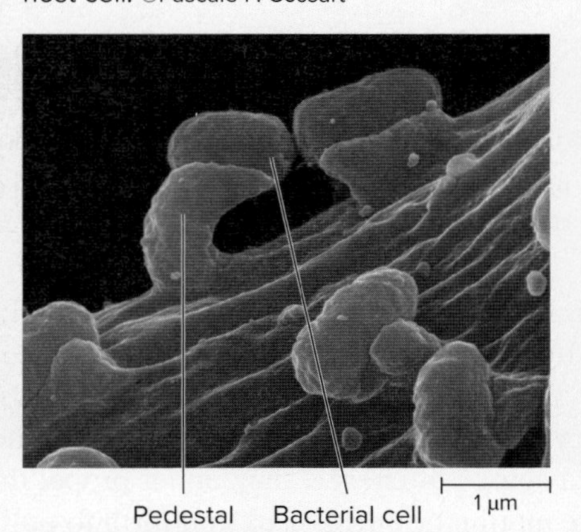

BOX FIGURE 3.3 Pedestal Formation Certain *E. coli* strains attach to intestinal epithelial cells and cause the host cell's actin to rearrange, forming a pedestal beneath the bacterial cell.

Actin filaments allow the cell cytoplasm to move. The filaments are polymers of protein subunits called actin, which rapidly assemble and subsequently disassemble to cause motion. For example, the pseudopods that form during phagocytosis develop as actin polymerizes in one part of the cell and depolymerizes in another.

Microtubules, the thickest of the cytoskeleton's components, are long, hollow structures made of protein subunits called tubulin. Microtubules form the mitotic spindles, the machinery that separates duplicated chromosomes as cells divide. Without mitotic spindles, cells could not reproduce. In addition, microtubules are the main structures that make up the cilia and flagella, the mechanisms of locomotion in some eukaryotic cells. Microtubules also function as the framework along which organelles and vesicles move within a cell. ⏭| **mitosis**

Intermediate filaments function like ropes, strengthening the cell mechanically. By providing mechanical support, they allow cells to resist physical stresses.

MicroByte

The antifungal medication griseofulvin interferes with the action of microtubules in some fungi.

Flagella and Cilia

Flagella and **cilia** are flexible structures that appear to project out of a cell yet are covered by extensions of the cytoplasmic membrane (**figure 3.51**). They are composed of long microtubules grouped in what is called a 9 + 2 arrangement: nine pairs of microtubules surrounding two individual ones. They

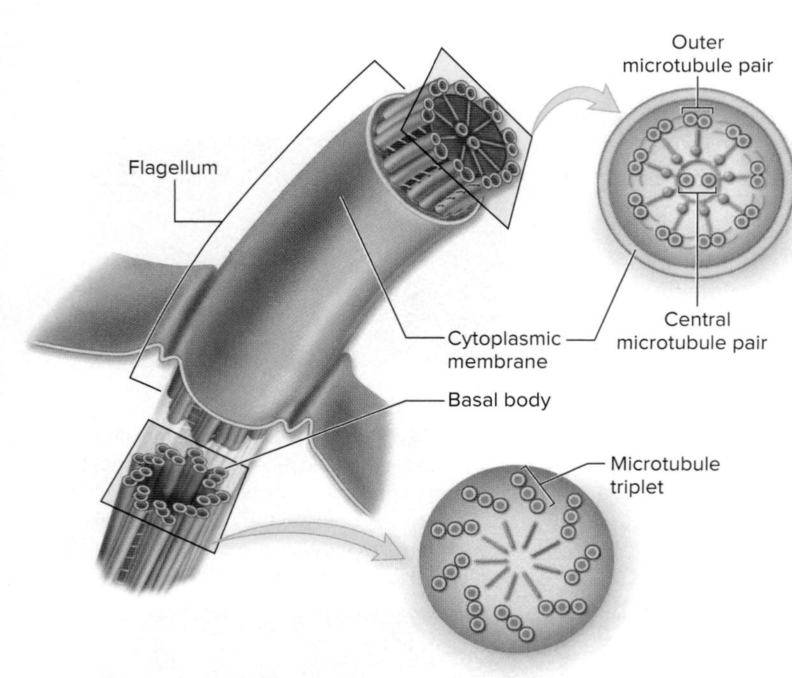

FIGURE 3.51 Flagellum A flexible structure involved in movement.

How is the structure of a eukaryotic flagellum different from its prokaryotic counterpart?

originate from basal bodies within the cell; the basal body has a slightly different arrangement of microtubules.

Although eukaryotic flagella function in motility, they are very different from their prokaryotic counterparts. Using ATP as a source of energy, they either propel the cell with a whiplike motion or thrash back and forth to pull the cell forward.

Cilia are shorter than flagella, often covering a cell and moving in synchrony (see figure 1.10). Their motion can move a cell forward in an aqueous solution, or propel surrounding material along a stationary cell. For example, epithelial cells that line the respiratory tract have cilia that beat together in a directed fashion. This moves the mucus film that covers those cells, directing it toward the mouth, where it can be swallowed. Because of this action, called the mucociliary escalator, most microbes that have been inhaled are removed before they can enter the lungs.

MicroAssessment 3.9

The 80S eukaryotic ribosome is composed of 60S and 40S subunits. The cytoskeleton is a dynamic filamentous network that provides structure to the cell; it is composed of actin filaments, microtubules, and intermediate filaments. Flagella function in motility. Cilia either propel a cell or move material along a stationary cell.

25. Explain how actin filaments are related to phagocytosis.
26. How does the action of cilia prevent microbes from entering the lungs?
27. Why would 60S and 40S ribosomal subunits make an 80S ribosome rather than a 100S ribosome?

3.10 ■ Membrane-Bound Organelles of Eukaryotic Cells

Learning Outcomes

26. Describe the function of the nucleus, mitochondria, chloroplasts, endoplasmic reticulum, Golgi apparatus, lysosomes, and peroxisomes.

Membrane-bound organelles are an important feature of eukaryotic cells that sets them apart from prokaryotic cells.

Nucleus

An important distinguishing feature of a eukaryotic cell is the **nucleus,** which contains the genetic information. The boundary of this structure is the nuclear envelope, composed of two phospholipid bilayer membranes: the inner and outer membranes (**figure 3.52**). Complex protein structures span the envelope, forming nuclear pores. These allow large molecules such as ribosomal subunits and proteins to be transported into and out of the nucleus. The nucleolus is a region where ribosomal RNAs are synthesized.

The nucleus contains multiple chromosomes, each one encoding different genetic information. Unlike the situation in most prokaryotic cells, the DNA is linear. It is wound around proteins called histones, a characteristic that adds structure and order to the long molecule.

Events that take place in the nucleus during cell division distinguish eukaryotes from prokaryotes. After DNA is replicated in eukaryotic cells, chromosomes go through a nuclear division process called mitosis, which ensures that the daughter cells receive the same number of chromosomes as the original parent. Because of mitosis, a cell that is diploid (has two copies of each chromosome) will generate two diploid daughter cells. A different process, meiosis, generates haploid daughter cells, which each have a single copy of each chromosome. In sexual reproduction, two haploid cells fuse to become one diploid cell.

Mitochondria

Mitochondria (singular: mitochondrion) function as ATP-generating powerhouses. They are complex structures bounded by two phospholipid bilayers: the outer and inner membranes (**figure 3.53**). The outer membrane is smooth, but the inner membrane is highly folded, forming invaginations called cristae. The folds increase the membrane's surface area, maximizing the ATP-generating capabilities of the organelle (the processes will be discussed in chapter 6). The mitochondrial matrix—the gel-like material enclosed by the mitochondrial inner membrane—contains DNA, ribosomes, and other molecules necessary for protein synthesis.

The mitochondrial ribosomes are similar to bacterial ribosomes. This observation, and the fact that mitochondria

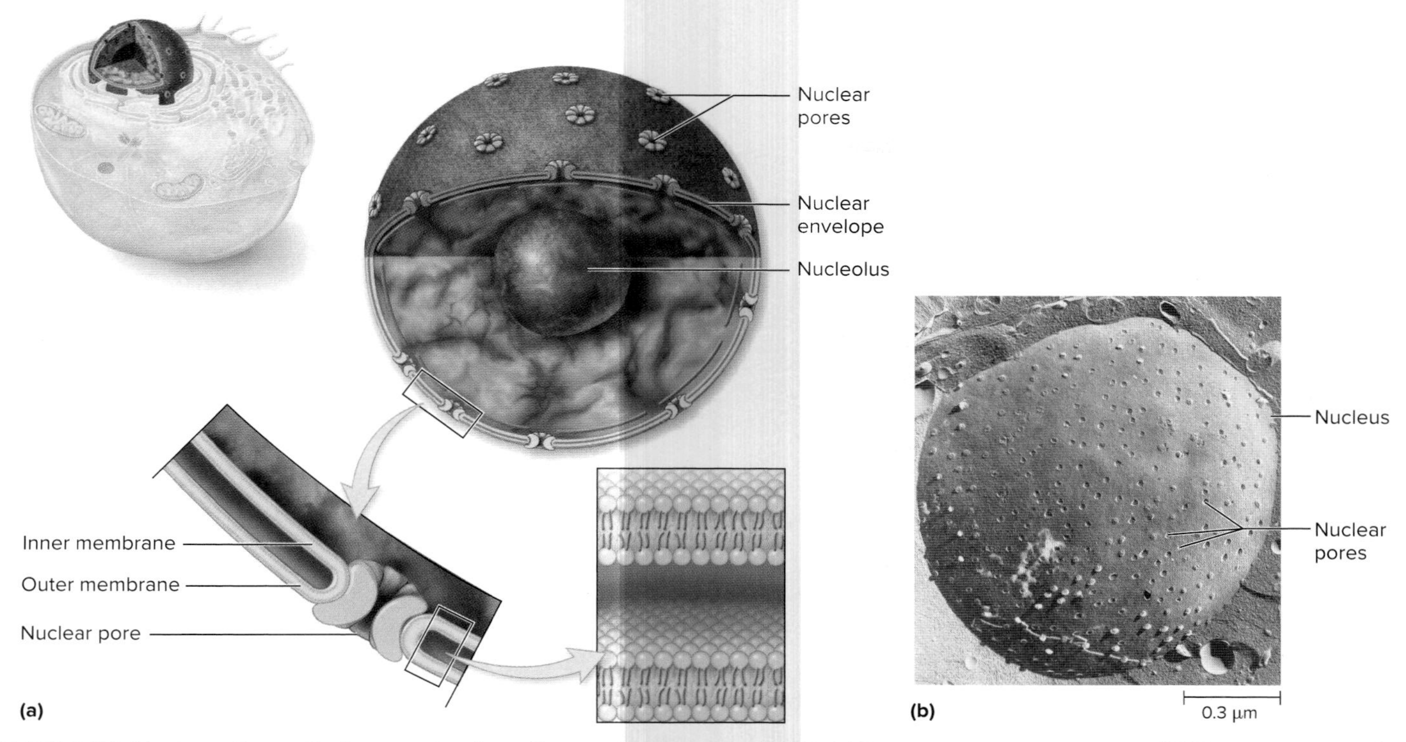

FIGURE 3.52 Nucleus Organelle that contains the cell's genetic information (DNA). **(a)** Diagrammatic representation. **(b)** Electron micrograph of a pig kidney cell by freeze-fracture technique. ©Biophoto Associates/Science Source

What is the function of nuclear pores?

divide in a fashion similar to that of bacteria, were among the first pieces of evidence that led scientists to hypothesize that mitochondria evolved from bacterial cells. Significant additional evidence has now elevated that hypothesis to a scientific theory. The **endosymbiotic theory** states that the ancestors of mitochondria as well as chloroplasts (described next) were bacteria residing within other cells in a mutually beneficial partnership. The intracellular bacterium in such a partnership is called an **endosymbiont.** As time went on, each partner became indispensable to the other, and the endosymbiont eventually lost key features such as a cell wall and the ability to replicate independently. ◀◀ scientific theory

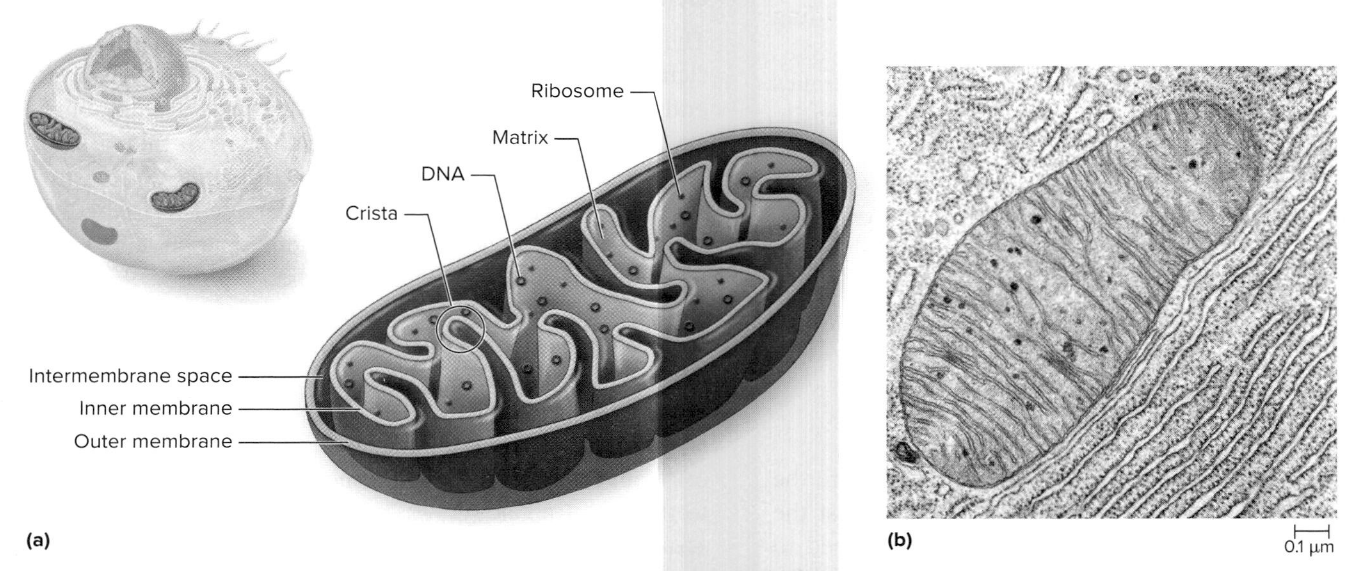

FIGURE 3.53 Mitochondria Organelles that harvest energy released during the degradation of organic compounds to synthesize ATP. **(a)** Diagrammatic representation. **(b)** Electron micrograph. ©Keith Porter/Science Source

What evidence led scientists to conclude that mitochondria evolved from bacterial cells?

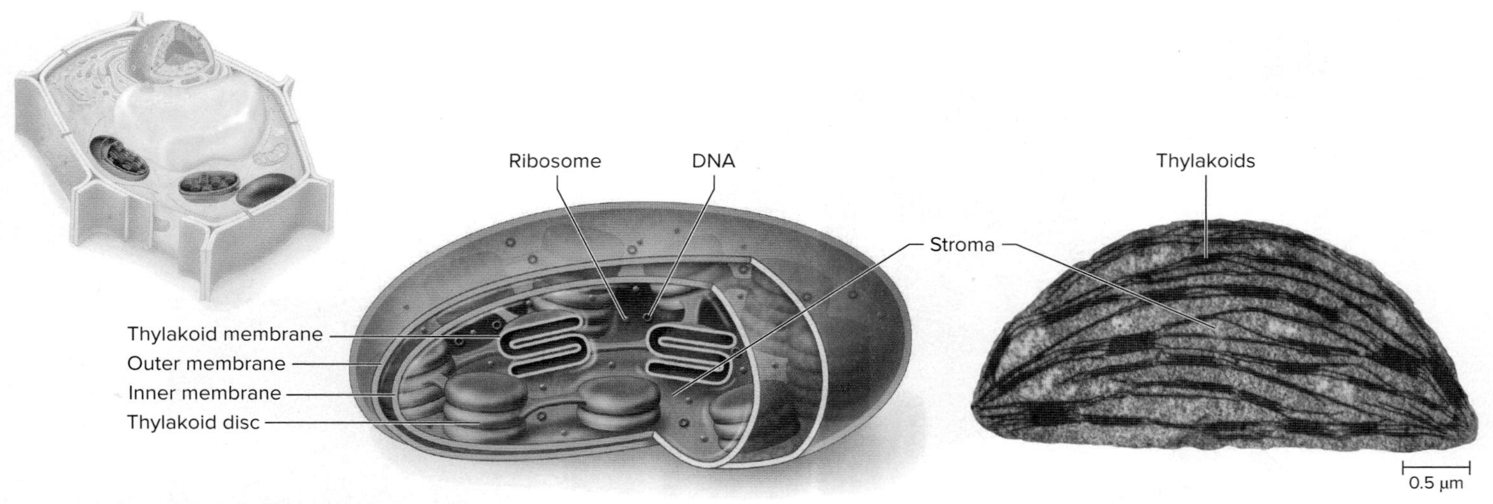

FIGURE 3.54 Chloroplasts These harvest the energy of sunlight to generate ATP, which is then used to convert CO_2 to an organic form.

Chloroplasts evolved from which group of bacteria?

Several early observations support the endosymbiotic theory. Mitochondria and chloroplasts, unlike other eukaryotic organelles, both carry some of the genetic information necessary for their function. This includes genes for certain ribosomal proteins and ribosomal RNAs that make up their ribosomes. These ribosomes contrast with the typical 80S ribosomes that characterize eukaryotic cells and, in fact, are similar to the bacterial ribosomes. In addition, the double membrane that surrounds these organelles is similar to what is seen with present-day endosymbionts, which retain their cytoplasmic membranes and live within membrane-bound compartments in their eukaryotic host cells. Like these endosymbionts, as well as other bacteria, mitochondria and chloroplasts multiply by elongating and then dividing (binary fission).

Evidence in favor of the endosymbiotic theory continues to accumulate. Modern technology allows scientists to easily determine the nucleotide sequence of DNA molecules, making it possible to compare the organelle DNA with genomes of different bacteria. This led to the discovery that some mitochondrial DNA sequences bear a striking resemblance to DNA sequences of members of a group of obligate intracellular parasites, the rickettsias. These are probably relatives of modern-day mitochondria.

Chloroplasts

Chloroplasts, found exclusively in plants and algae, are the site of photosynthesis in eukaryotic cells. They harvest the energy of sunlight to generate ATP, which is then used to convert CO_2 to sugar and starch. Like mitochondria, chloroplasts are bound by two membranes (**figure 3.54**). Within the chloroplast's stroma—the substance analogous to the mitochondrial matrix—are membrane-bound, disc-like structures called thylakoids. Chlorophyll and other pigments that capture radiant energy are embedded in the thylakoid membranes.

As with mitochondria, substantial evidence indicates that chloroplasts evolved from bacterial cells—in this case, a group of photosynthetic bacteria called cyanobacteria. Chloroplasts contain ribosomes similar to those of bacteria and DNA. In addition, they elongate and divide, and have photosynthetic mechanisms and DNA sequences similar to a group of bacteria called cyanobacteria.

Endoplasmic Reticulum (ER)

The **endoplasmic reticulum (ER)** is a complex system of flattened sheets, sacs, and tubes (**figure 3.55**). The **rough endoplasmic reticulum** has a characteristic bumpy appearance due to the many ribosomes adhering to the surface. It is the site where proteins not destined for the cytoplasm are synthesized. These include proteins targeted for secretion or transfer to an organelle's lumen. Membrane proteins such as receptors are also synthesized on the rough ER. When ribosomes begin making these proteins, they attach to the ER surface so that the polypeptides they are synthesizing can be threaded through gated pores in the membrane. This delivers the molecules to the ER's lumen, where they then fold to assume their three-dimensional shapes. Vesicles that bud off from the ER then transfer the newly synthesized proteins to the Golgi apparatus for further modification and sorting. ◀◀ polypeptide

Some regions of the ER are smooth. This **smooth endoplasmic reticulum** functions in lipid synthesis and degradation, and calcium ion storage. As with material made in

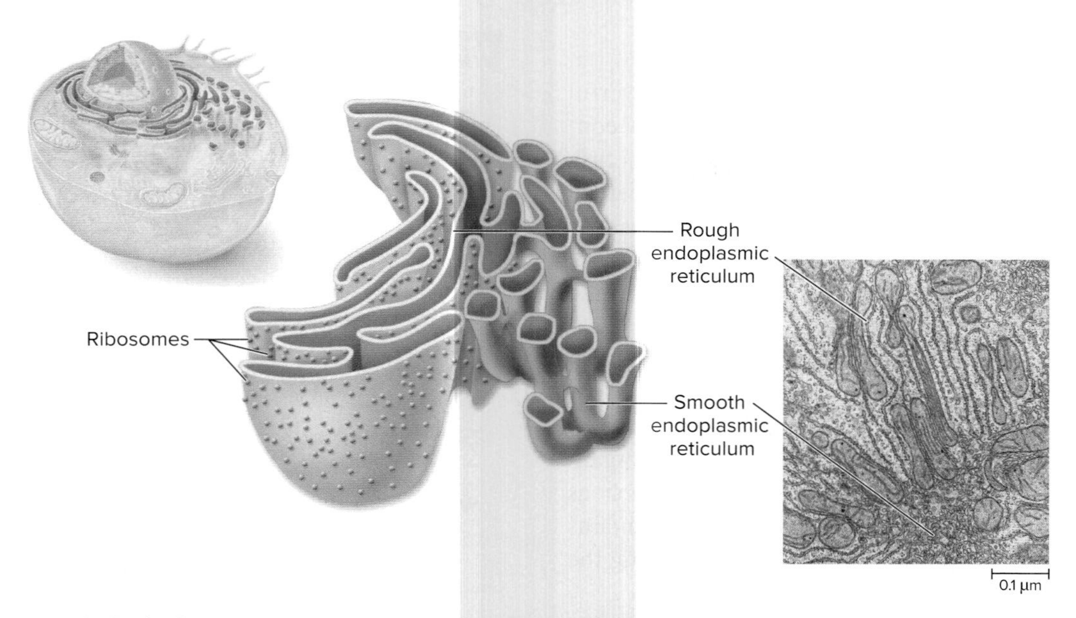

FIGURE 3.55 Endoplasmic Reticulum Site of synthesis of macromolecules destined for other organelles or the external environment.

? **What causes the bumpy appearance of the rough endoplasmic reticulum?**

the rough ER, vesicles transfer compounds from the smooth ER to the Golgi apparatus.

Golgi Apparatus

The **Golgi apparatus,** also called the Golgi complex, consists of a series of membrane-bound flattened compartments (**figure 3.56**). It is the site where macromolecules synthesized in the endoplasmic reticulum are modified before transport to other destinations. These modifications, such as the addition of carbohydrate and phosphate groups, take place in different compartments of the Golgi. Much like an assembly line, the molecules are transferred in vesicles from one compartment to another. The molecules are then sorted and delivered in vesicles to specific cellular compartments or to the outside of the cell.

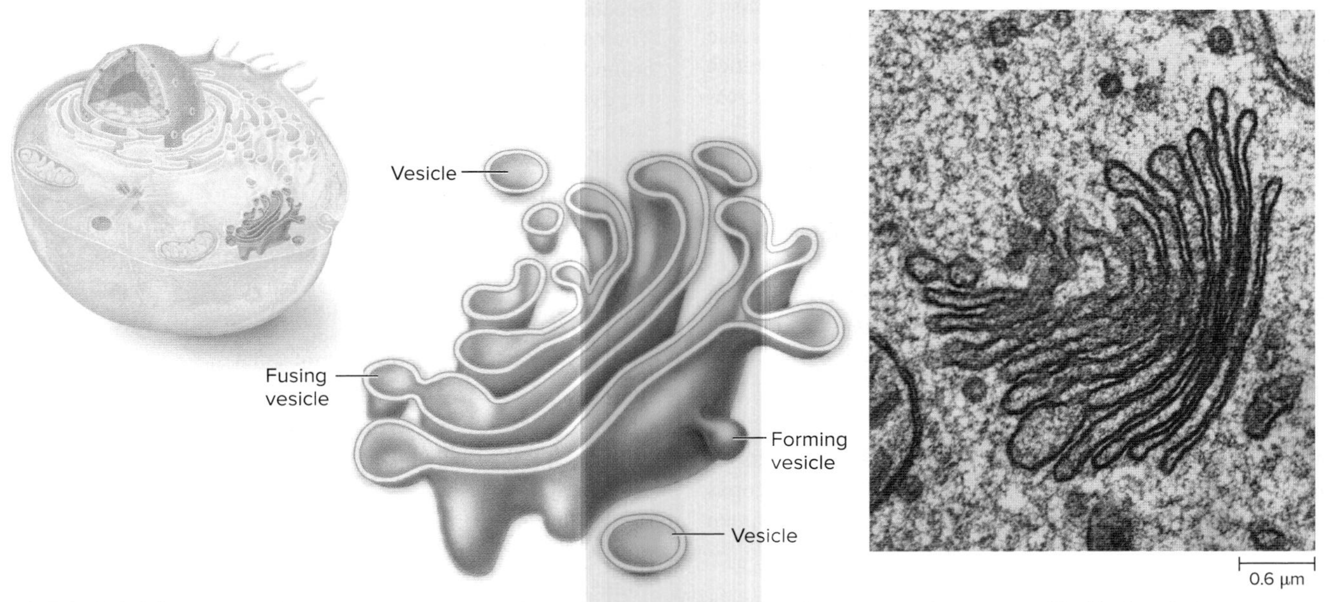

FIGURE 3.56 Golgi Apparatus Site where macromolecules synthesized in the endoplasmic reticulum are modified before being transported to other destinations.

? **How are the modified macromolecules transported from the Golgi apparatus to other sites?**

Lysosomes and Peroxisomes

Lysosomes are organelles that contain a number of powerful degradative enzymes that could destroy the cell if not contained within the organelle. Endosomes and phagosomes fuse with lysosomes, allowing materials taken up by the cell to be degraded. In a similar manner, old organelles or vesicles formed around unneeded cellular components can fuse with lysosomes. This process, autophagy, digests and recycles the cell's own contents. The scientist who described the mechanism of autophagy (Yoshinori Ohsumi) was awarded a Nobel Prize in 2016.

Peroxisomes are the organelles in which O_2 is used to help break down lipids and detoxify certain chemicals. The organelle's enzymes generate highly reactive molecules such as hydrogen peroxide and superoxide. The peroxisome contains these molecules and ultimately degrades them, protecting the cell from their toxic effects. ⏭ hydrogen peroxide, ⏭ superoxide

MicroAssessment 3.10

The nucleus, which contains the genetic information, is a distinguishing feature of a eukaryotic cell. Mitochondria are ATP-generating powerhouses. Chloroplasts are the site of photosynthesis. The rough endoplasmic reticulum is the site where proteins not destined for the cytoplasm are synthesized. The smooth ER functions in lipid synthesis and degradation, and calcium ion storage. The Golgi apparatus modifies and sorts molecules synthesized in the rough ER. Lysosomes are the organelles within which digestion takes place. In peroxisomes, O_2 is used to break down substances.

28. Describe the structure of the nucleus.

29. How does the function of the rough endoplasmic reticulum differ from that of the smooth endoplasmic reticulum?

30. Mitochondria have ribosomes similar to those of bacteria, yet antibiotics that target bacterial ribosomes are more toxic to bacteria than to humans. Why might this be so?

Summary

MICROSCOPY AND CELL MORPHOLOGY

3.1 ■ Microscopes (table 3.1)

Principles of Light Microscopy: Bright-Field Microscopes
The most commonly used type of microscope is the **bright-field microscope** (figure 3.1). The usefulness of a microscope depends largely on its **resolving power.**

Light Microscopes That Increase Contrast
Cells viewed through a **dark-field microscope** stand out against a dark background (figure 3.4). **Phase-contrast microscopes** increase differences in refraction (figure 3.5). **Differential interference contrast (DIC) microscopes** cause an image to appear three-dimensional (figure 3.6).

Light Microscopes That Detect Fluorescence
Fluorescence microscopes are used to observe cells stained with fluorescent dyes (figure 3.7). **Scanning laser microscopes (SLMs)** are used to obtain interior views of intact cells and three-dimensional images of thick structures (figure 3.8). **Super-resolution microscopes** use complex illumination mechanisms and merged data to achieve higher resolution than that of other light microscopes (figure 3.9).

Electron Microscopes
Transmission electron microscopes (TEMs) transmit electrons through a specimen (figure 3.11). **Scanning electron microscopes (SEMs)** scan a beam of electrons over the surface of a specimen (figure 3.12).

Scanning Probe Microscopes
Atomic force microscopes map surfaces on an atomic scale (figure 3.13).

3.2 ■ Preparing Specimens for Light Microscopy (table 3.2)

Simple Staining
Simple staining uses a single dye.

Differential Staining
The **Gram stain** is widely used: **Gram-positive bacteria** stain purple and **Gram-negative bacteria** stain pink (figure 3.15). The **acid-fast stain** is used to stain *Mycobacterium* species; **acid-fast** organisms stain pink and all other organisms stain blue (figure 3.16).

Special Stains to Observe Cell Structures
The **capsule stain** allows the capsule to stand out as a clear zone around a cell (figure 3.17). The **endospore stain** uses heat to help the dye enter endospores (figure 3.18). The **flagella stain** uses a substance that allows a stain to adhere to and coat the otherwise thin flagella (figure 3.19).

Fluorescent Dyes and Tags
Some fluorescent dyes bind compounds that characterize all cells; others bind compounds specific to certain cell types (figure 3.20). **Immunofluorescence** is used to tag proteins of interest with fluorescent compounds.

3.3 ■ Morphology of Prokaryotic Cells

Shapes
Most prokaryotes are **cocci** or **rods;** other common shapes are **vibrios, spirilla,** and **spirochetes** (figure 3.21).

Arrangements
Cells adhering to one another following division form characteristic arrangements such as chains, packets, and clusters (figure 3.23).

Multicellular Associations
Myxobacteria move as a pack and form fruiting bodies. Most bacteria on surfaces live as biofilms.

PROKARYOTIC CELLS (figure 3.24; table 3.3)

3.4 ■ The Cytoplasmic Membrane of Prokaryotic Cells

Structure of the Cytoplasmic Membrane (figure 3.25)

The cytoplasmic membrane is a phospholipid bilayer embedded with a variety of different proteins.

Permeability of the Cytoplasmic Membrane (figure 3.26)

The cytoplasmic membrane is **selectively permeable.** A few compounds pass through by **simple diffusion.** Some of the membrane proteins function as selective gates.

The Role of the Cytoplasmic Membrane in Energy Transformation

The **electron transport chain** generates an electrochemical gradient, a source of energy called **proton motive force** (figure 3.28).

Transport of Small Molecules Across the Cytoplasmic Membrane (figures 3.29, 3.30; table 3.4)

Facilitated diffusion moves compounds by exploiting a concentration gradient. **Active transport** moves compounds against a concentration gradient and requires energy. **Group translocation** chemically modifies a molecule during its transport.

Protein Secretion

A characteristic signal sequence targets proteins for **secretion** (figure 3.31).

3.5 ■ The Cell Wall of Prokaryotic Cells

Peptidoglycan (figure 3.32)

Peptidoglycan provides rigidity to the bacterial cell wall. It is composed of glycan strands connected via tetrapeptide chains.

The Gram-Positive Cell Wall (figure 3.33)

The Gram-positive cell wall contains a relatively thick layer of peptidoglycan. **Teichoic acids** project out of the peptidoglycan. A gel-like substance is sandwiched between the cytoplasmic membrane and the peptidoglycan layer.

The Gram-Negative Cell Wall (figure 3.34)

The Gram-negative cell wall has a relatively thin layer of peptidoglycan. An **outer membrane** contains **lipopolysaccharides**. LPS is called **endotoxin**. **Porins** permit small molecules to pass through the outer membrane. **Periplasm** fills the space between the inner and outer membranes.

Antibacterial Substances That Target Peptidoglycan

Penicillin interferes with peptidoglycan synthesis. **Lysozyme** destroys the structural integrity of peptidoglycan.

Cell Wall Type and the Gram Stain

Gram-positive cells retain the crystal violet–iodine dye complex even when subjected to decolorization, but Gram-negative cells do not (figure 3.35).

Bacteria That Lack a Cell Wall

Mycoplasma species are variable in shape and not affected by lysozyme or penicillin (figure 3.36).

Cell Walls of Archaea

Archaea have a greater variety of cell wall types than do bacteria, and they all lack peptidoglycan.

3.6 ■ Structures Outside the Cell Wall of Prokaryotic Cells

Capsules and Slime Layers

Capsules and **slime layers** allow bacteria to adhere to surfaces. Some capsules allow pathogens to avoid host defense systems (figure 3.37).

Flagella (figure 3.38)

Flagella are used for motility (figure 3.39). **Chemotaxis** is movement toward an attractant or away from a repellent (figure 3.40). Cells use phototaxis, aerotaxis, magnetotaxis, and thermotaxis to move toward light, O_2, a magnetic field, and temperature, respectively.

Pili (figure 3.42)

Many types of **pili** (**fimbriae**) allow specific attachment of cells to surfaces. **Sex pili** are involved in a form of DNA transfer.

3.7 ■ Internal Components of Prokaryotic Cells

Chromosome and Plasmids

The **chromosome** forms a region called the **nucleoid** (figure 3.43); it contains the genetic information required by a cell. **Plasmids** usually encode only genetic information that may be helpful to a cell.

Ribosomes

Ribosomes are involved in protein synthesis. The 70S bacterial ribosome is composed of a 50S and a 30S subunit (figure 3.44).

Cytoskeleton

The **cytoskeleton** is involved in cell division and regulation of shape.

Storage Granules

Storage granules are synthesized from nutrients a cell has in excess.

Gas Vesicles

Gas vesicles provide buoyancy to aquatic cells.

Endospores

Endospores are resistant to heat, desiccation, toxic chemicals, and UV light; they can **germinate** to become **vegetative cells** (figures 3.45, 3.46).

EUKARYOTIC CELLS (figure 3.47; table 3.6)

3.8 ■ Cytoplasmic Membrane of Eukaryotic Cells

Structure and Function of the Cytoplasmic Membrane

The eukaryotic cytoplasmic membrane is an asymmetrical phospholipid bilayer embedded with proteins that are involved in transport, structural integrity, and signaling.

Transfer of Molecules Across the Cytoplasmic Membrane

Gated **channels** form pores in the membrane. **Carriers** function in facilitated diffusion and active transport. **Pinocytosis** is used to take up liquids (figure 3.49). **Receptor-mediated endocytosis** is used by animal cells to take up material that binds to receptors. Protozoa and phagocytes take up material using **phagocytosis**. **Exocytosis** expels material. Proteins destined for **secretion** are made by ribosomes attached to the endoplasmic reticulum.

3.9 ■ Protein Structures Within Eukaryotic Cells

Ribosomes

The 80S ribosome is composed of 60S and 40S subunits.

Cytoskeleton

The cytoskeleton is composed of **actin filaments**, **microtubules,** and **intermediate filaments** (figure 3.50).

Flagella and Cilia

Flagella propel a cell or pull the cell forward (figure 3.51). **Cilia** move in synchrony to either propel a cell or move material along a stationary cell.

3.10 Membrane-Bound Organelles of Eukaryotic Cells

Nucleus

The **nucleus** contains the cell's genetic information (DNA) (figure 3.52).

Mitochondria

Mitochondria use the energy released during the degradation of organic compounds to generate ATP (figure 3.53). The **endosymbiotic theory** states that mitochondria descended from endosymbiotic bacteria, explaining why mitochondria have several characteristics of bacteria.

Chloroplasts

Chloroplasts capture the energy of sunlight, and then use it to synthesize ATP. This is used to convert CO_2 to an organic form (figure 3.54). As with mitochondria, chloroplasts descended from endosymbiotic bacteria.

Endoplasmic Reticulum (ER) (figure 3.55)

Proteins not destined for the cytoplasm are synthesized in the **rough endoplasmic reticulum**. The **smooth endoplasmic reticulum** is where lipids are synthesized and degraded, and calcium is stored.

Golgi Apparatus

The **Golgi apparatus** modifies and sorts molecules synthesized in the endoplasmic reticulum (figure 3.56).

Lysosomes and Peroxisomes

Lysosomes carry digestive enzymes. **Peroxisomes** are the organelles in which O_2 is used to oxidize certain substances.

Review Questions

Short Answer

1. Explain why resolving power is important in microscopy.
2. Describe the difference between a simple stain and a differential stain.
3. Describe what happens at each step in the Gram stain.
4. Compare and contrast ABC transport systems with group translocation.
5. Diagram the structure of peptidoglycan.
6. Give two reasons why the outer membrane of Gram-negative bacteria is medically significant.
7. Describe how a plasmid can help a cell.
8. How is receptor-mediated endocytosis different from phagocytosis?
9. What is the endosymbiotic theory?
10. Explain how the Golgi apparatus cooperatively functions with the endoplasmic reticulum.

Multiple Choice

1. Which of the following is most likely to be used in a typical microbiology laboratory?
   a) Bright-field microscope
   b) Confocal scanning microscope
   c) Phase-contrast microscope
   d) Scanning electron microscope
   e) Transmission electron microscope
2. When a medical technologist wants to determine if a clinical specimen contains a *Mycobacterium* species, which should be used?
   a) Acid-fast stain
   b) Capsule stain
   c) Endospore stain
   d) Gram stain
   e) Simple stain
3. *Vibrio cholerae* causes the disease cholera. Based on the name of the bacterium, what is its shape?
   a) Spherical
   b) Straight cylinder
   c) Curved rod
   d) Chains of cocci
   e) Chains of rods
4. Endotoxin is associated with
   a) Gram-positive bacteria.
   b) Gram-negative bacteria.
   c) the cytoplasmic membrane.
   d) endospores.
5. The "O157" in the name *E. coli* O157:H7 refers to the type of O antigen. From this information you know that *E. coli*
   a) has a capsule.
   b) is a rod.
   c) is a coccus.
   d) is Gram-positive.
   e) is Gram-negative.
6. Eliminating which structure is *always* deadly to cells?
   a) Flagella
   b) Capsule
   c) Cell wall
   d) Cytoplasmic membrane
   e) Fimbriae
7. If you interfered with the ability of a *Bacillus* species to form endospores, what would be the result? The bacterium would no longer be
   a) able to multiply.
   b) antibiotic resistant.
   c) Gram-positive.
   d) Gram-negative.
   e) able to withstand boiling water.

8. If a virus mimics a ligand that normally participates in receptor-mediated endocytosis, the virus might
   a) prevent pinocytosis.
   b) be taken up by the cell.
   c) damage the cytoplasmic membrane.
   d) cause the cell to form pseudopods.
   e) damage the receptor.
9. The antibiotic erythromycin prevents protein synthesis in bacterial cells. Based on this information, which of the following might be targeted by the drug?
   1) 80S ribosomes
   2) 70S ribosomes
   3) 60S ribosomal subunit
   4) 50S ribosomal subunit
   5) 40S ribosomal subunit
      a) 1, 3
      b) 1, 4
      c) 2, 3
      d) 2, 4
      e) 2, 5
10. If a eukaryotic cell were treated with a chemical that destroys tubulin, all of the following would be directly affected *except*
    a) actin.
    b) cilia.
    c) eukaryotic flagella.
    d) microtubules.
    e) More than one of these

Applications

1. You are working in a laboratory producing new antibiotics for human and veterinary use. One compound with potential value inhibits the action of bacterial ribosomes. The compound, however, was shown to inhibit the growth of animal cells in culture. What is one possible explanation for its effect on animal cells?
2. A research laboratory is investigating environmental factors that inhibit the growth of archaea. The researchers wonder if penicillin would be effective in controlling their growth. Explain the probable results of an experiment in which penicillin is added to a culture of archaea.

Critical Thinking

1. This graph shows facilitated diffusion of a compound across a cytoplasmic membrane and into a cell. As the external concentration of the compound is increased, the rate of uptake increases until it reaches a point where it slows and then begins to plateau. This is not the case with passive diffusion, where the rate of uptake continually increases as the solute concentration increases. Why does the rate of uptake slow and then eventually plateau with facilitated diffusion?

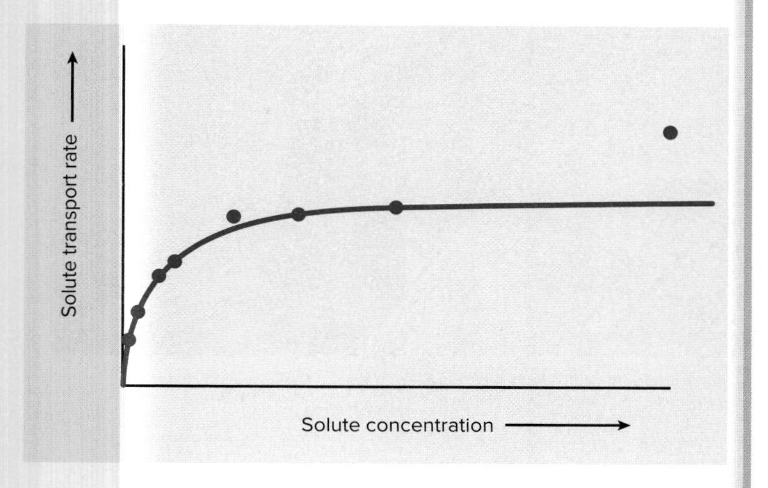

2. Most medically useful antibiotics interfere with either peptidoglycan synthesis or ribosome function. Why would the cytoplasmic membrane be a poor target for antibacterial medications?

www.mcgrawhillconnect.com

Enhance your study of this chapter with study tools and practice tests. Also ask your instructor about the resources available through Connect, including the media-rich eBook, interactive learning tools, and animations.

4 Dynamics of Microbial Growth

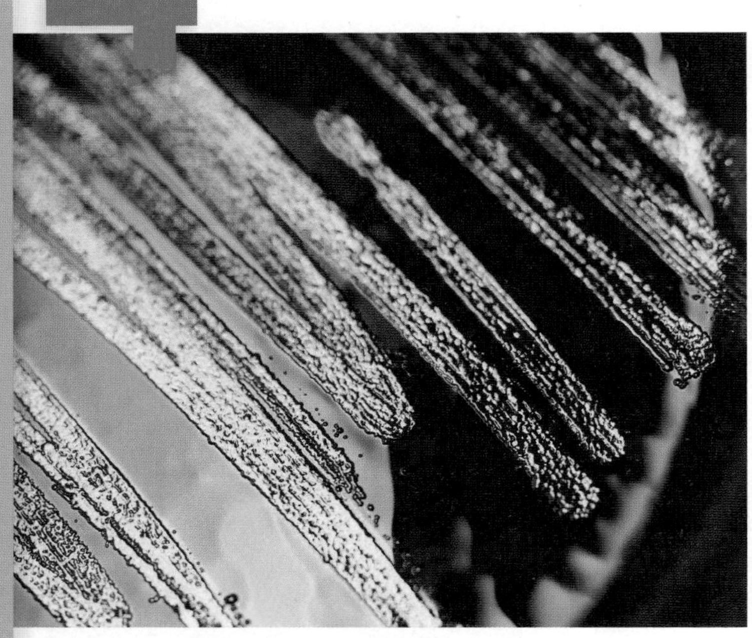

E. coli growing on eosin methylene blue (EMB) agar. ©McGraw-Hill Education/ Lisa Burgess, photographer

KEY TERMS

Biofilm Polymer-encased community of microorganisms.

Chemically Defined Medium A culture medium made of pure chemicals so that its exact chemical composition is known; generally used only for specific experiments when nutrients must be precisely controlled.

Complex Medium A culture medium that contains protein digests, extracts, or other ingredients that vary in their chemical composition.

Differential Medium A culture medium with an ingredient that certain microorganisms change in a recognizable way; used to differentiate microbes based on their metabolic traits.

Exponential (Log) Phase Stage in the growth curve during which cells divide at a constant rate; generation time is measured during this period of active multiplication.

Facultative Anaerobe Organism that grows best if O_2 is available, but can also grow without it.

Generation Time The time it takes for a population to double in number.

Obligate Aerobe Organism that requires O_2.

Obligate Anaerobe Organism that cannot multiply, and often is killed, in the presence of O_2.

Pure Culture A population descended from a single cell.

Selective Medium A culture medium with an ingredient that inhibits the growth of microbes other than the one being sought.

A Glimpse of History

The greatest contributor to methods of cultivating bacteria was Robert Koch (1843–1910), a German physician who combined a medical practice with a productive research career for which he received a Nobel Prize. Koch was primarily interested in identifying disease-causing bacteria, but to do this, he needed simple methods to isolate and grow these particular species. Koch recognized that a single bacterial cell could multiply on a solid medium in a limited area to form a distinct visible mass of descendants, so he experimented with growing bacteria on the cut surfaces of potatoes. Some species would not grow, however, because the potatoes did not contain enough nutrients, so Koch experimented with methods to solidify any liquid nutrient medium. He used gelatin initially, but there were two major problems: Gelatin melts at the temperature preferred by many medically important microorganisms, and some bacteria can digest it. In 1882, Fannie Hesse, the wife of an associate of Koch, suggested using agar. This solidifying agent was used to harden jelly at the time and proved to be the perfect answer.

Today, we take pure culture techniques for granted because of their relative ease and simplicity. Their development, however, had a major impact on microbiology. By 1900, the agents causing most of the major bacterial diseases of humans had been isolated and characterized.

Microorganisms can be found growing even in the harshest climates and most severe conditions. Environments that no unprotected human could survive, such as the ocean depths, volcanic vents, and the polar regions, have thriving microbial species. Indeed many scientists believe that if life exists on other planets, it may resemble these organisms. Each species, however, has a limited set of environmental conditions in which it can grow; even then, it will grow only if specific nutrients are available. Some microorganisms can grow at temperatures above the boiling point of water but not at room temperature. Others can grow only within an animal host, and then only in specific areas of that host.

Because of the medical significance of some microbes, as well as the nutritional and industrial use of microbial by-products, scientists must be able to grow microorganisms in culture. This is why it is important to understand the basic principles involved in microbial growth, while recognizing that much information is yet to be discovered.

4.1 ■ Principles of Microbial Growth

Learning Outcome

1. Describe binary fission and how it relates to generation time and exponential growth.

Bacteria and archaea generally multiply by the process of **binary fission,** a process in which a cell increases its size and then divides (**figure 4.1**). One cell divides into two, those two

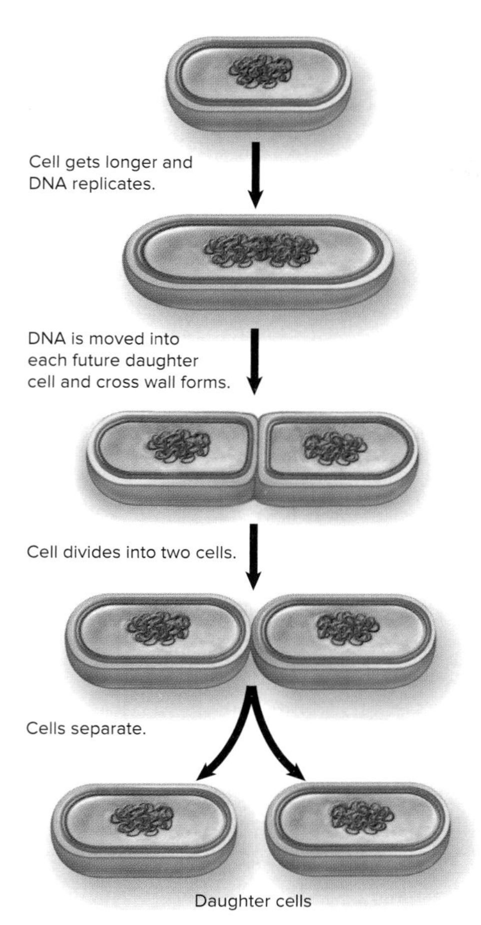

FIGURE 4.1 Binary Fission

How does the process of binary fission relate to the generation time?

divide to become four, those four become eight, and so on. In other words, the increase in cell numbers is exponential. Because it is neither practical, nor particularly meaningful, to determine the relative size of the cells in a given population, microbial growth is defined as an increase in the number of cells in a population. The time it takes for a population to double in number is the **generation time.** This varies greatly from species to species and is influenced by the conditions in which the cells are grown. For example, the generation time of a given bacterium may be 30 minutes under ideal conditions, but when the conditions are not ideal, the organism will grow much more slowly, if at all.

The exponential multiplication of bacteria has important health consequences. For instance, a mere 10 cells of a foodborne pathogen in a potato salad, sitting for 4 hours in the warm sun at a picnic, may multiply to more than 40,000 cells. Although ingesting a few cells of the pathogen might not cause disease, consuming tens of thousands could be a different matter entirely. So keep exponential multiplication in mind, and your potato salad in the cooler, the next time you go to a picnic!

To calculate how many bacterial cells will be present in a product after a certain amount of time, two factors must be known initially: the number of cells in the original population and the number of times the cells will divide during the stated period. In the potato salad example, if the pathogen has a generation time of 20 minutes, then the number of cells in the population will double every 20 minutes (**table 4.1**). This means that in a single hour, the number of cells will double three times: The initial population of 10 cells will double to become 20 cells, which will then double to become 40 cells, which will then double to become 80 cells. Over the course of 4 hours, the population will double 12 times, generating a population of 40,960 cells. For the calculation, simply multiply the number present initially $\times 2^n$, with "n" being the number of generations at that point. In other words, after 3 generations the number initially present is multiplied by $(2 \times 2 \times 2)$. So, using the potato salad example again, after 1 hour (3 generations), the number of cells is $10 \times 2^3 = 80$. After 2 hours (6 generations), the number of cells is $10 \times 2^6 = 640$. After 3 hours (9 generations) the number of cells is $10 \times 2^9 = 5{,}120$ cells. Finally, after 4 hours (12 generations), the number of cells is $10 \times 2^{12} = 40{,}960$ cells. The general equation illustrated by these examples is $N_t = N_0 \times 2^n$, with N_t being the number of cells at a given time (in minutes) and N_0 the initial number of cells.

MicroByte

Escherichia coli has a generation time of 20 minutes in ideal conditions; *Mycobacterium tuberculosis* needs at least 12 hours to double.

MicroAssessment 4.1

Most bacteria and archaea multiply by binary fission. The time required for a population to double in number is the generation time.

1. Explain why microbial growth refers to an increase in cell number rather than cell size.
2. If a bacterium has a generation time of 30 minutes, and you start with 100 cells at time 0, how many cells will you have in 30, 60, 90, and 120 minutes?

4.2 ■ Microbial Growth in Nature

Learning Outcomes

2. Describe a biofilm, and explain why biofilms are important to humans.
3. Explain why microbes that grow naturally in mixed communities sometimes cannot be grown in pure culture.

Microorganisms have historically been studied by growing them in the laboratory. Scientists now recognize, however, that the changing and complex conditions of the natural

TABLE 4.1 Example of Exponential Growth

Time in Minutes (t)	Initial Population (N_0)	Number of Generations (n)	2^n	Number of Cells in the Population (N_t)
0	10	0		10
20	10	1	2^1 (= 2)	20
40	10	2	2^2 (= 2 × 2)	40
60 (1 hour)	10	3	2^3 (= 2 × 2 × 2)	80
80	10	4	2^4 (= 2 × 2 × 2 × 2)	160
100	10	5	2^5 (= 2 × 2 × 2 × 2 × 2)	320
120 (2 hours)	10	6	2^6	640
140	10	7	2^7	1,280
160	10	8	2^8	2,560
180 (3 hours)	10	9	2^9	5,120
200	10	10	2^{10}	10,240
220	10	11	2^{11}	20,480
240 (4 hours)	10	12	2^{12}	40,960

environment, which differ greatly from the conditions in the laboratory, have significant effects on microbial growth and behavior. In fact, microbial cells are able to adjust to certain changes in their surroundings by sensing various chemicals and then responding to that input by producing materials appropriate for the situation.

Biofilms

In nature, microorganisms can live suspended in an aqueous environment, but most attach to surfaces and live in polymer-encased communities called **biofilms** (**figure 4.2**). Biofilms cause the slipperiness of rocks in a stream bed, the slimy "gunk" that coats kitchen drains, the scum that gradually accumulates in toilet bowls, and the dental plaque that forms on teeth. Biofilm formation begins when planktonic (free-floating) cells move to a surface and adhere. They then multiply and release polysaccharides, DNA, and other hydrophilic polymers to which unrelated cells may attach and grow (**figure 4.3**). The mesh-like accumulation of these polymers, referred to as extracellular polymeric substances (EPS), gives a biofilm its characteristic slimy appearance. ⏭| dental plaque

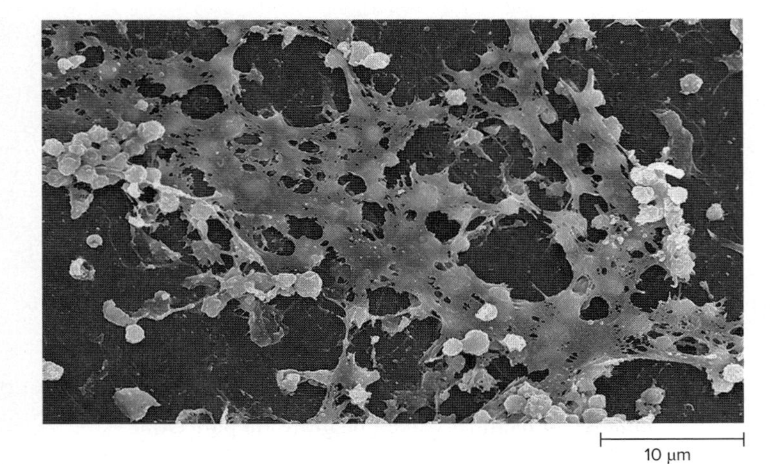

FIGURE 4.2 Biofilm on the Inside of an Indwelling Catheter Source: Rodney M. Donlan, Ph.D.; Janice Carr/CDC

? Why are infections that involve biofilms a concern?

Surprisingly, biofilms are not random mixtures of microbes in a layer of EPS, but instead have characteristic formations with channels through which nutrients and wastes pass. Cells communicate with one another by synthesizing and responding to chemical signals—an exchange important in establishing structure.

Biofilms are more than just an unsightly annoyance. Dental plaque leads to tooth decay and gum disease. Even troublesome, persistent ear infections and the complications of cystic fibrosis are due to biofilms. In fact, the majority of bacterial infections seem to involve biofilms. Treatment of these infections is difficult because microbes within the biofilm often resist the effects of antibiotics as well as the body's defenses. Biofilms are also important in industry, where their accumulations in pipes, drains, and cooling water towers can interfere with processes as well as damage equipment. The conditions within the biofilm protect the microbes, so the bacteria in a biofilm may be hundreds of times more resistant to disinfectants than are their planktonic (free-floating) counterparts. ⏭| disinfectants

Although biofilms can be damaging, they also can be beneficial. Many bioremediation efforts, which use microbes to degrade harmful chemicals, are enhanced by biofilms. So, as some industries are exploring ways to destroy biofilms,

FIGURE 4.3 Development of a Biofilm

What are extracellular polymeric substances (EPS)?

others, such as wastewater treatment facilities, are looking for ways to encourage their development. **bioremediation, wastewater treatment**

Interactions of Mixed Microbial Communities

Microorganisms in the environment regularly grow in close associations with many different species. Sometimes the interactions are cooperative, even helping the growth of certain species that otherwise could not survive. For example, organisms that cannot multiply in the presence of O_2 will grow in your mouth if neighboring microbial cells rapidly use that gas. Meanwhile, the metabolic wastes of one organism can serve as nutrients for another. Often, however, microbes in a community compete for nutrients, and some even resort to a type of biological warfare, synthesizing toxic compounds that inhibit competitors. Certain Gram-negative bacteria use a needle-like structure (called a type VI secretion system) to inject toxic compounds directly into competing bacterial cells. Understandably, the conditions that characterize these associations of mixed microbial communities are extremely difficult to reproduce in the laboratory. **microbial competition**

MicroAssessment 4.2

Biofilms have a characteristic architecture with channels through which nutrients and wastes can pass. In nature, prokaryotes often grow in associations with many different species.

3. Water bowls left out for pets sometimes develop a slimy layer if not washed regularly. What causes the slime?
4. Describe a situation in which the activities of one species benefit another.
5. Why is it more difficult to study microorganisms growing in their natural environment, as opposed to in the laboratory?

4.3 ■ Microbial Growth in Laboratory Conditions

Learning Outcomes

4. Describe how the streak-plate method is used to obtain a pure culture and how the resulting culture can be stored.
5. Describe the stages of a growth curve and compare this closed system to colony growth and continuous culture.

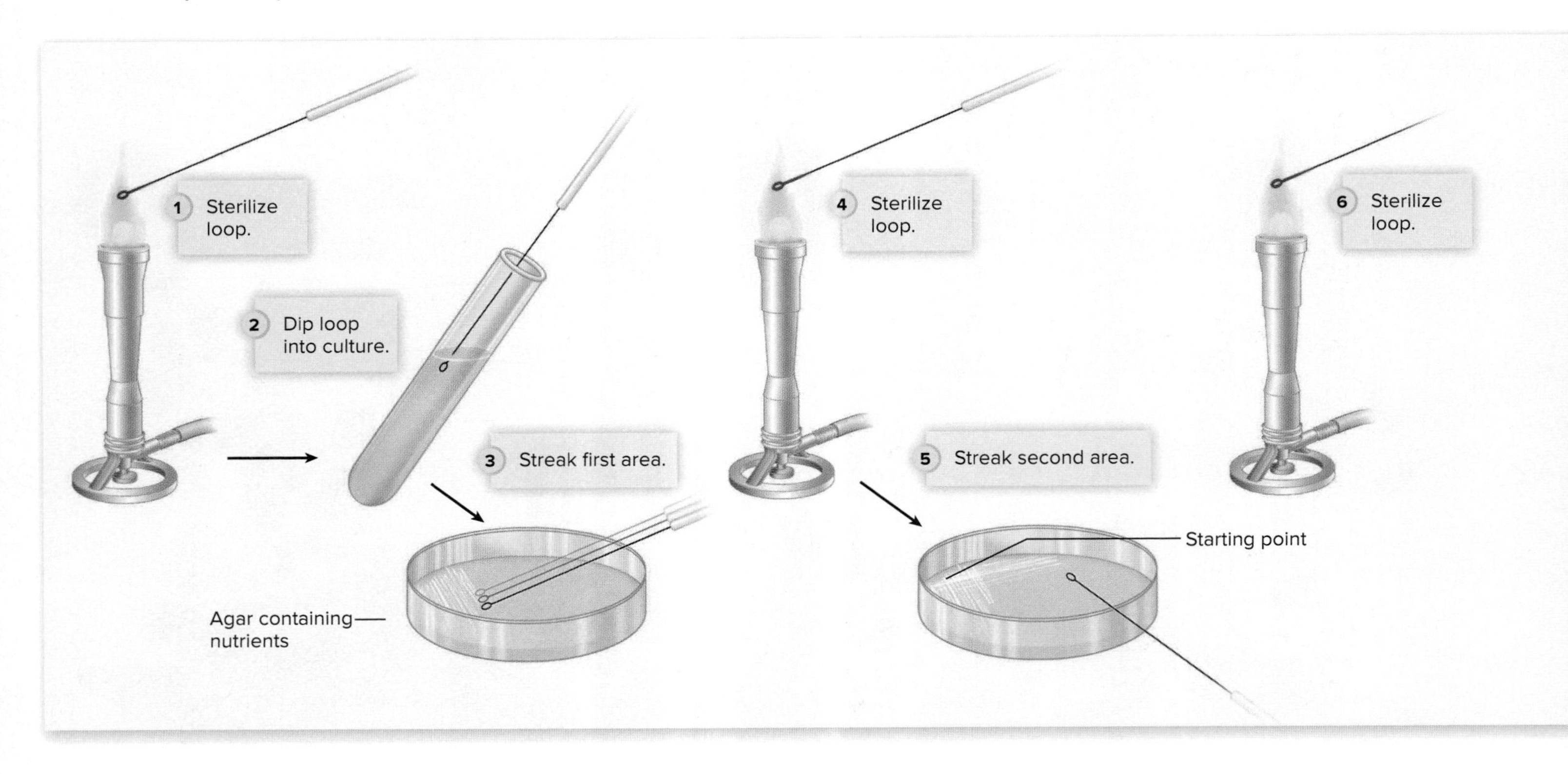

In the laboratory, bacteria and archaea are generally isolated and grown in pure culture. A **pure culture** is a population descended from a single cell and therefore contains only one species. Working with pure cultures makes it easier to identify and study the activities of a particular species. But although the results are much easier to interpret when studying pure cultures, the organisms sometimes behave differently than they do in their natural environment, as discussed earlier. Another complicating issue is that only an estimated 1% of all prokaryotes can currently be grown in culture successfully, which makes it more difficult to study the vast majority of microorganisms. Fortunately for humanity, most known medically significant bacteria can be grown in pure culture.

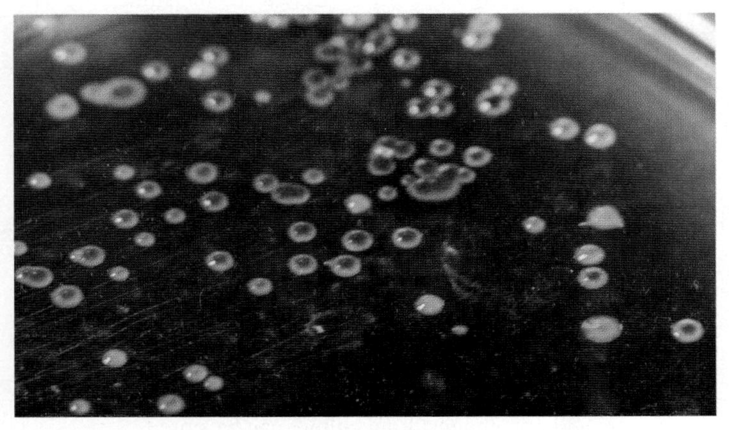

FIGURE 4.4 Colonies Growing on Agar Medium ©McGraw-Hill Education/Lisa Burgess, photographer

? **What is the purpose of agar in the medium?**

To work with a pure culture, all containers, media, and instruments must be sterile, or free of microbes, prior to use. These are then handled using **aseptic technique,** a set of procedures that minimize the chance that other organisms will be accidentally introduced. The medium the cells are grown in, or on, is called a culture medium. It consists of nutrients dissolved in water, and can be a liquid broth or a solid gel. ⏭ sterilization

Obtaining a Pure Culture

The basic requirements for obtaining a pure culture are a solid culture medium, a container to hold and maintain the medium in an aseptic condition, and a method to separate individual microbial cells. A single microbial cell, supplied with the right nutrients and conditions, will multiply on the solid medium in a limited area to form a **colony,** a distinct mass of cells (**figure 4.4**). About 1 million cells are required for a colony to be easily visible to the naked eye.

Agar, a polysaccharide extracted from seaweed, is used to solidify culture media. Unlike gelatin and other gelling agents, very few microbes can degrade agar. It is not destroyed at high temperatures and can therefore be sterilized by heating—a process that also liquefies it. Melted agar will stay liquid until cooled to a temperature below 45°C. Because of this, nutrients that would be destroyed at high temperatures can be added at lower temperatures before the agar hardens. Once solidified, an agar medium will remain so until heated above 95°C. Thus, unlike gelatin—which is liquid at 37°C—agar remains solid over the entire temperature range at which most microbes grow. ⏮ polysaccharide

FIGURE 4.5 The Streak-Plate Method The successive streaks dilute the cells. By the third set of streaks, cells should be separated enough so that isolated colonies develop after incubation.
©McGraw-Hill Education/Lisa Burgess, photographer

What is the purpose of obtaining isolated colonies?

A solid culture medium is contained in a **Petri dish,** a two-part, covered container made of glass or plastic. Although not airtight, the Petri dish does exclude airborne contaminants. A culture medium in a Petri dish is commonly referred to as a plate of that medium—for example, a nutrient agar plate or, more simply, an **agar plate.**

MicroByte

Declining harvests of the seaweed used to make agar have resulted in a worldwide agar shortage.

The Streak-Plate Method

The **streak-plate method** is the simplest and most commonly used technique for isolating microorganisms (**figure 4.5**). A sterile inoculating loop is dipped into a microbe-containing sample and then lightly drawn several times across the surface of an agar plate, creating a set of parallel streaks covering approximately one-third of the agar. The loop is then sterilized and a new series of parallel streaks is made across and at an angle to the previous ones, covering another surface section. This drags some of those cells streaked onto the first portion over to a fresh section, effectively inoculating it with a diluted sample. The loop is sterilized again, and another set of parallel streaks is made, dragging into a third area some of the organisms that had been moved into the second section. The goal is to reduce the number of cells being spread with each successive series of streaks. By the third set of streaks, cells should be separated enough so that distinct, well-isolated colonies will form.

Maintaining Stock Cultures

Once a pure culture has been obtained, it can be maintained as a stock culture, which is a culture stored for use as an inoculum in later procedures. Often, a stock culture is stored in the refrigerator as growth on the surface of an agar slant (a tube of agar that was held at an angle as it solidified). For long-term storage, stock cultures can be frozen at –70°C in a glycerol-containing solution that prevents ice crystals from forming and damaging cells. Alternatively, cells can be freeze-dried.

The Growth Curve

In the laboratory, many microorganisms are grown either on agar plates or in tubes or flasks of broth. These are considered **closed systems,** or batch cultures, because nutrients are not renewed, nor are wastes removed. As the cells grow in this type of system, the population increases in a distinct pattern of stages, and then declines. The characteristic pattern observed in a broth culture is called a **growth curve,** and is characterized by five distinct stages: lag phase, exponential or log phase, stationary phase, death phase, and phase of prolonged decline (**figure 4.6**).

Lag Phase

When microorganisms are transferred into a different medium, the cells do not immediately begin dividing. Although there is no increase in cell number during this **lag phase,** cells begin synthesizing enzymes required for growth. The length of the lag phase depends on multiple factors. For example, if cells are transferred into a medium that contains fewer nutrients, the lag phase will be longer because the cells must begin making amino acids or other components not supplied in the new medium.

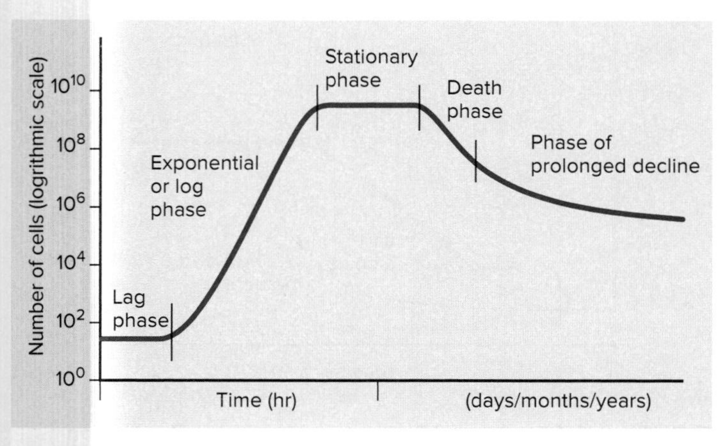

FIGURE 4.6 Growth Curve The growth curve is characterized by five distinct stages: lag phase, exponential or log phase, stationary phase, death phase, and phase of prolonged decline.

During which phase of bacterial growth is generation time measured?

Exponential Phase (Log Phase)

During the **exponential** or **log phase,** cells divide at a constant rate. Generation time is measured during this active growth phase. The phase is medically important because bacteria are most sensitive to antimicrobial medications when cells are actively multiplying.

From a commercial standpoint, the exponential phase is important because some molecules made by growing cells are valuable. For instance, amino acids can be sold as nutritional supplements, and microbial waste products such as ethanol are used as biofuels. The small molecules made by cells as they multiply are called **primary metabolites.**

In the later stages of exponential growth, nutrients gradually become depleted and waste products accumulate. Cell activities shift as this occurs. Endospore formers such as *Bacillus* and *Clostridium* species may initiate the process of sporulation. Even cells that cannot form endospores often change their activities to prepare for starvation conditions. Microbial compounds that begin accumulating at this stage are made for purposes other than growth and are called **secondary metabolites** (**figure 4.7**). Commercially, the most valuable of these are antibiotics.

Stationary Phase

Cells enter the **stationary phase** when the nutrient levels are too low to sustain growth. The total number of viable cells (cells capable of developing or multiplying) in the population remains relatively constant, but some cells are dying while others are multiplying. How can cells multiply when they have exhausted their supply of nutrients? Dead cells often burst, releasing nutrients that then fuel the growth of other cells.

During the stationary phase, the viable cells continue to synthesize secondary metabolites and maintain the altered properties they demonstrated in late log phase. The length of time cells remain in the stationary phase varies, depending on the species and environmental conditions. Some populations remain in the stationary phase for only a few hours, whereas others remain for days or longer.

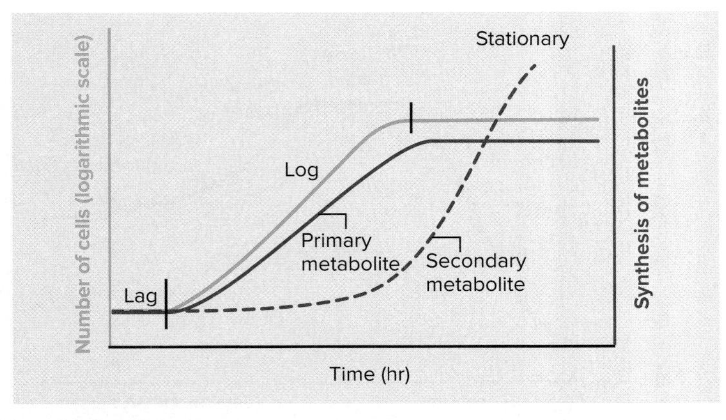

FIGURE 4.7 Primary and Secondary Metabolite Production
Primary metabolites are synthesized during the period of active multiplication. Secondary metabolites begin to be synthesized in late log phase.

? **What is the most commercially valuable secondary metabolite?**

Death Phase

The **death phase** is the period when the total number of viable cells in the population decreases as cells die off at a constant rate. Like cell growth, death is exponential, but the rate is usually much slower.

Phase of Prolonged Decline

In many cases, a fraction of the cell population survives the death phase. These cells have adapted to tolerate the worsened conditions and are able to multiply for at least a short time, using the nutrients released from the dead cells. As the conditions continue to deteriorate during this **phase of prolonged decline,** most of these survivors then die. However, the few progeny better equipped for survival can grow. This dynamic process generates successive waves of slightly modified populations, each more fit to survive than the previous ones. Thus, the statement "survival of the fittest" holds true even for closed cultures of microbes.

Colony Growth

Growth of bacteria on a solid medium is similar to that in a broth, but there are some important differences. After a lag phase, cells in a colony multiply exponentially and eventually compete with one another for available nutrients. Unlike the situation in a broth, however, the position of a single cell within a colony determines its environment. Cells multiplying on the edge of the colony face relatively little competition for O_2 and nutrients. In the center of the colony, meanwhile, the high density of cells rapidly depletes available O_2 and nutrients. Wastes such as acids accumulate, and these can be toxic. As a consequence, cells at the edge of the colony may be growing exponentially, while those in the center may be in the death phase. Cells in locations between these two extremes may be in the stationary phase.

Continuous Culture

Microbial cells can be kept in a state of continuous growth by using a **chemostat.** This device is an **open system,** meaning that nutrients can be added and waste products removed. The chemostat continually drips fresh medium into a broth culture contained in a chamber. With each drop that enters, an equivalent volume—containing cells, wastes, and spent medium—leaves through an outlet. By adjusting the nutrient content of the medium and the speed at which it enters the chamber, a constant growth rate and cell density can be maintained. This makes it possible to study a uniform population's response to different nutrient concentrations or environmental conditions. Chemostats are also used in industrial processes that harvest commercially valuable products made only during the log phase of microbial growth.

MicroAssessment 4.3

In the laboratory, the streak-plate method is used to obtain a pure culture. When grown in a closed system, a microbial population goes through distinct phases. Cells within a colony may be in any one of the phases, depending on their relative location. Chemostats are used to study uniform populations of growing cells.

6. To identify the causative agent of a given illness, a pure culture is often needed. How is a streak plate used to obtain a pure culture?
7. Explain the difference between the lag phase and the exponential phase.
8. Why would a compound that prevents bacteria from growing interfere with the function of the antibiotic penicillin?

4.4 ■ Environmental Factors That Influence Microbial Growth

Learning Outcomes

6. Describe the importance of a microorganism's requirements for temperature, O_2, pH, and water availability, and define the terms that indicate these requirements.
7. Explain the significance of reactive oxygen species, and describe the mechanisms cells use to protect against their effects.

As a group, microorganisms inhabit nearly every environment on Earth. Microbes we associate with disease and rapid food spoilage live in habitats that humans consider quite comfortable. Some microorganisms, however, live in harsh environments that would kill most other organisms. Most of the examples in this latter group, called **extremophiles** (*phile* means "loving"), are archaea.

Recognizing the major environmental factors that affect microbial growth—temperature, atmosphere, pH, and water availability—is essential for studying microbes and helps us understand their roles in the complex ecology of the planet. These factors and their associated characteristics are summarized in **table 4.2**.

TABLE 4.2 Environmental Factors That Influence Microbial Growth

Environmental Factor/ Descriptive Terms	Characteristics
Temperature	Thermostability appears to be due to protein structure.
Psychrophile	Optimum temperature between –5°C and 15°C.
Psychrotroph	Optimum temperature between 15°C and 30°C, but grows well at refrigeration temperatures.
Mesophile	Optimum temperature between 25°C and 45°C.
Thermophile	Optimum temperature between 45°C and 70°C.
Hyperthermophile	Optimum temperature of 70°C or greater.
Oxygen (O_2) Availability	Oxygen (O_2) requirement/tolerance reflects the organism's energy-harvesting mechanisms and its ability to inactivate reactive oxygen species.
Obligate aerobe	Requires O_2.
Facultative anaerobe	Grows best if O_2 is present, but can also grow without it.
Obligate anaerobe	Cannot grow in the presence of O_2.
Microaerophile	Requires small amounts of O_2, but higher concentrations are inhibitory.
Aerotolerant anaerobe (obligate fermenter)	Indifferent to O_2.
pH	Prokaryotes that live in pH extremes maintain a near-neutral internal pH by pumping protons out of or into the cell.
Neutrophile	Multiplies in the range of pH 5 to 8.
Acidophile	Grows optimally at a pH below 5.5.
Alkalophile	Grows optimally at a pH above 8.5.
Water Availability	Prokaryotes that can grow in high-solute solutions maintain the availability of water in the cell by increasing their internal solute concentration.
Halotolerant	Can grow in relatively high-salt solutions, up to approximately 10% NaCl.
Halophile	Requires high levels of sodium chloride.

Temperature Requirements

Each microbial species has a well-defined temperature range in which it grows. Within this range is the optimum growth temperature, the temperature at which the organism multiplies most rapidly. As a general rule, this optimum is close to the upper limit of the organism's temperature range.

Microorganisms are commonly divided into five groups based on their optimum growth temperatures (**figure 4.8**). Note, however, that this merely represents a convenient organization scheme. In reality, no sharp dividing line exists between each group. Furthermore, not every organism in a group can grow in the entire temperature range typical for its group. The five groups are as follows:

- **Psychrophiles** (*psychro* means "cold") have an optimum between –5°C and 15°C. These organisms grow in the cold Arctic and Antarctic regions and in lakes fed by glaciers.

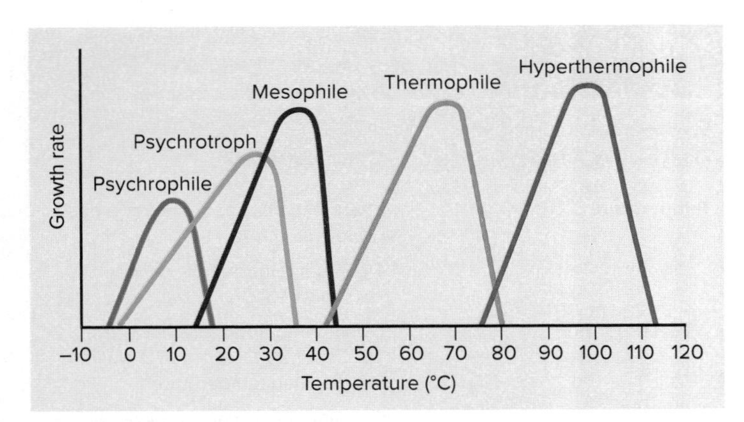

FIGURE 4.8 Temperature Requirements for Growth
Microorganisms are commonly divided into five groups based on their optimum growth temperatures. This graph shows a typical example of each group. The optimum temperature, the point at which the growth rate is highest, is near the upper limit of the range.

Most pathogens fall into which group on this chart?

- **Psychrotrophs** (*troph* means "nourishment") have an optimum between 15°C and 30°C, but grow well at lower temperatures. They are an important cause of spoilage in refrigerated foods. ⏭ **food spoilage**
- **Mesophiles** (*meso* means "middle") have an optimum between 25°C and about 45°C. *E. coli* and most other common bacteria are in this group. Pathogens, adapted to growth in the human body, typically have an optimum between 35°C and 40°C. Mesophiles that inhabit soil, a colder environment, generally have a lower optimum, close to 30°C.
- **Thermophiles** (*thermo* means "heat") have an optimum between 45°C and 70°C. These organisms commonly live in hot springs and compost heaps. ⏭ **composting**
- **Hyperthermophiles** (*hyper* means "excessive") have an optimum of 70°C or greater. These are usually archaea. One archeon, isolated from the base of a hydrothermal vent deep in the ocean, can grow at 122°C—the highest growth temperature yet recorded!

Why can some microbes withstand very high temperatures while most cannot? As a general rule, proteins from thermophiles are not denatured at high temperatures. This thermostability is due to the amino acid sequence of the protein. The amino acid sequence controls the number and position of the bonds that form within the protein, which in turn determine its three-dimensional structure. Heat-stable enzymes from thermophiles are used in high-temperature detergents. ⏮ **protein denaturation**

Temperature and Food Preservation

Refrigeration temperatures (approximately 4°C) slow spoilage because they limit the multiplication of otherwise fast-growing mesophiles. Psychrophiles and psychrotrophs can still grow at these temperatures, however, so refrigerated foods will still spoil, but more slowly. ⏭ **low-temperature storage,** ⏭ **food spoilage**

Foods and other perishable products that withstand below-freezing temperatures can be frozen for long-term storage. It is important to recognize, however, that freezing is not an effective means of destroying microbes. Recall that freezing is routinely used to preserve stock cultures.

Temperature and Disease

The temperatures of some parts of the human body are cooler than others. For example, the heart, brain, and gastrointestinal tract are near 37°C, but the temperature of the extremities is lower. For these reasons, some microbes cause disease more readily in certain parts of the body. Hansen's disease (leprosy) typically involves the coolest regions (ears, hands, feet, and fingers) because the causative bacterium *(Mycobacterium leprae)* grows best at these lower temperatures. A similar situation applies to syphilis, which is caused by a bacterium *(Treponema pallidum)* that grows best in the cooler regions of the body. Indeed, for more than 30 years the major treatment of syphilis was to induce fever by deliberately introducing the agent that causes malaria, which results in very high fevers. ⏭ **Hansen's disease,** ⏭ **syphilis,** ⏭ **fever**

Oxygen (O_2) Requirements

Like humans, some microorganisms have an absolute requirement for O_2. They grow in environments that are **aerobic,** meaning that O_2 is present. Other microorganisms thrive in environments that are **anaerobic,** meaning that little or no O_2 is present. One way of determining an organism's O_2 requirement is to grow it in a shake tube. To do this, a tube of nutrient agar is boiled, which both melts the agar and drives off the O_2. The agar is then allowed to cool to just above its solidifying temperature. Next, the test organism is added and dispersed by gentle shaking or swirling. The agar medium is then allowed to harden and the tube is incubated. Because the solidified agar slows gas diffusion, the concentration of O_2 in the tube is relatively high at the top, whereas the bottom portion is anaerobic. The cells grow in the region that has a suitable O_2 level (**table 4.3**). Based on their O_2 requirements, microorganisms can be separated into these groups:

- **Obligate aerobes** have an absolute requirement for oxygen (O_2). They use it in aerobic respiration, an energy-harvesting process. This and other ATP-generating pathways will be discussed in chapter 6. An example of an obligate aerobe is *Micrococcus luteus,* which is common in the environment. Note that obligate aerobes are often simply called aerobes. ⏭ **aerobic respiration,** ⏮ **ATP**
- **Facultative anaerobes** grow better if O_2 is present, but can also grow without it. The term "facultative" means that the organism is flexible, in this case in its requirements for O_2. Facultative anaerobes use aerobic respiration if O_2 is available, but resort to alternative types of metabolism if it is not. Growth is faster when O_2 is

TABLE 4.3 Oxygen (O_2) Requirements of Microorganisms

	Obligate aerobe	Facultative anaerobe	Obligate anaerobe	Microaerophile	Aerotolerant anaerobe
	Bacterial growth				
Growth characteristics	Grows only when O_2 is available.	Grows best when O_2 is available, but also grows without it.	Cannot grow when O_2 is present.	Grows only if small amounts of O_2 are available.	Grows equally well with or without O_2.
Use of O_2 in energy-harvesting processes	Requires O_2 for respiration.	Uses O_2 for respiration, if available.	Does not use O_2.	Requires O_2 for respiration.	Does not use O_2.
Typical mechanisms to protect against reactive oxygen species	Produces superoxide dismutase and catalase.	Produces superoxide dismutase and catalase.	Does not produce superoxide dismutase or catalase.	Produces some superoxide dismutase and catalase.	Produces superoxide dismutase but not catalase.

present because aerobic respiration produces the most ATP. *E. coli* is one of the most common facultative anaerobes in the large intestine. ⏭ fermentation, ⏭ anaerobic respiration

- **Obligate anaerobes** cannot multiply if O_2 is present; in fact, they are often killed by even brief exposure to air. Obligate anaerobes harvest energy using processes other than aerobic respiration; the details of these will be covered in chapter 6. Most inhabitants of the large intestine are obligate anaerobes, as is the bacterium that causes botulism, *Clostridium botulinum.* Note that obligate anaerobes are often simply called anaerobes.
- **Microaerophiles** require small amounts of O_2 (2% to 10%) for aerobic respiration; higher concentrations are inhibitory. An example is *Helicobacter pylori,* which causes gastric and duodenal ulcers.
- **Aerotolerant anaerobes** are indifferent to O_2. They can grow in its presence, but do not use it to harvest energy. They are also called **obligate fermenters,** because fermentation is their only metabolic option. An example is *Streptococcus pyogenes,* which causes strep throat.

MicroByte

Over half of all the cytoplasm on Earth is probably in anaerobic microbes!

Reactive Oxygen Species (ROS)

When organisms use O_2 in aerobic respiration, harmful derivatives called **reactive oxygen species (ROS)** form as by-products. Examples of these molecules include superoxide (O_2^-) and hydrogen peroxide (H_2O_2). Reactive oxygen species can damage cell components, so cells that grow aerobically must have mechanisms to protect against them. Obligate anaerobes typically do not have these mechanisms, but there are exceptions.

Virtually all organisms that grow in the presence of O_2 produce the enzyme **superoxide dismutase,** which inactivates superoxide by converting it to O_2 and hydrogen peroxide. Nearly all these organisms produce the enzyme **catalase** as well, which converts hydrogen peroxide into O_2 and water. An important exception is the aerotolerant anaerobes. The fact that they do not produce catalase is useful in the laboratory. A simple test for the enzyme can be used to distinguish two groups of medically important Gram-positive cocci that grow aerobically: *Staphylococcus* species, which are catalase positive, and *Streptococcus* species, which are catalase negative. ⏭ catalase test

pH

Each microbial species can survive within a range of pH values, and within this range is its pH optimum. Despite the pH of the external environment, however, cells maintain a constant internal pH, typically near neutral. Many prokaryotes that grow in acidic environments quickly pump out protons (H^+) that enter the cell; recall that acidic environments have relatively high concentrations of protons. Prokaryotes that grow in alkaline conditions bring in protons. ⏮ pH

Most microbes are **neutrophiles**—they live and multiply within the range of pH 5 (acidic) to pH 8 (basic), and have a pH optimum near neutral (pH 7). Food preservation methods such as pickling inhibit bacterial growth by increasing the acidity of the food.

Although most neutrophiles cannot withstand highly acidic conditions, one medically important bacterium has found a way. *Helicobacter pylori* grows in the stomach, sometimes causing ulcers. It decreases the acidity of its immediate surroundings by producing urease, an enzyme that splits urea into carbon dioxide and ammonia. The ammonia neutralizes any stomach acid surrounding the cell. ⏭ ulcers

Acidophiles grow optimally at a pH below 5.5. *Picrophilus oshimae,* a member of the *Archaea,* has an optimum pH of less than 1! This organism, which was isolated from the dry, acidic soils of a gas-emitting volcanic fissure in Japan, has an unusual cytoplasmic membrane that is unstable at a pH above 4.0.

Alkaliphiles grow optimally at a pH above 8.5. They often live in alkaline lakes and soils.

Water Availability

All microorganisms require water for growth. Even if water is present, however, it may not be available in certain environments. Dissolved substances such as salt (NaCl) and sugars, for example, interact with water molecules, making them unavailable to the cell. In many environments, particularly in certain natural habitats such as salt marshes, microorganisms are faced with this situation. If the solute concentration is higher in the medium than in the cell, water diffuses out of the cell due to osmosis. This causes the cytoplasm to dehydrate and shrink from the cell wall, a phenomenon called **plasmolysis (figure 4.9)**. ⏮ solute, ⏮ osmosis

The growth-inhibiting effect of high salt and sugar concentrations is used in food preservation. High levels of salt are added to preserve such foods as bacon, salt pork, and anchovies. Likewise, the added sugar in jams and jellies acts as a preservative. Honey naturally has a high sugar content, which is why it also has a long shelf life. ⏭ food preservation

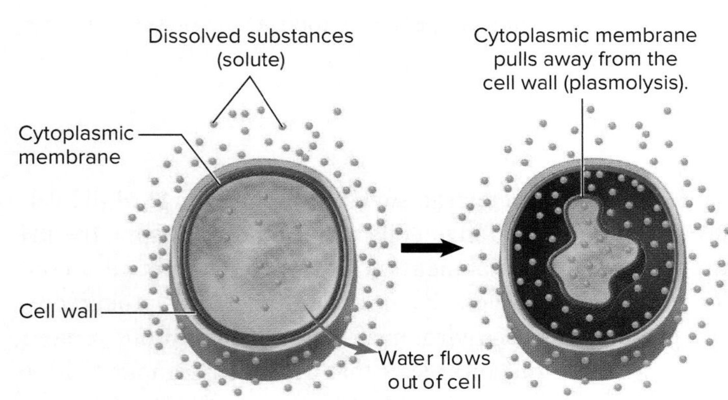

FIGURE 4.9 Effects of Solute Concentration on Prokaryotic Cells Water molecules pass freely through the cytoplasmic membrane. If the solute concentration is higher outside the cell, water moves out of the cell.

What is plasmolysis?

Although many microorganisms are inhibited by high salt concentrations, some withstand or even require them. Microbes that tolerate high salt concentrations, up to approximately 10% NaCl, are **halotolerant** (*halo* means "salt"). *Staphylococcus* species, bacteria that live on the dry, salty environment of the skin, are an example. **Halophiles** require high levels of sodium chloride. Many marine bacteria are mildly halophilic, requiring concentrations of approximately 3% sodium chloride. Certain archaea are extreme halophiles, requiring 9% sodium chloride or more. Extreme halophiles are found in environments such as the salt flats of Utah and the Dead Sea.

MicroAssessment 4.4

A microorganism can be categorized according to its optimum growth temperature as well as its O_2 requirements. Most microbes grow best at a near-neutral pH, although some prefer acidic conditions, and others grow best in alkaline conditions. Halophiles require high-salt conditions.

9. *Clostridium paradoxum* grows optimally at 55°C, pH 9.3; it will not grow in the presence of O_2. How should this bacterium be categorized with respect to its temperature, pH, and O_2 requirements?
10. What is the function of the enzyme catalase?
11. Why would hydrogen peroxide be an effective disinfectant?

4.5 ■ Nutritional Factors That Influence Microbial Growth

Learning Outcomes

8. List the required elements and give examples of common sources of required elements.
9. Explain the significance of a limiting nutrient.
10. Explain why fastidious microbes require growth factors.
11. Describe the energy and carbon sources used by photoautotrophs, chemolithoautotrophs, photoheterotrophs, and chemoorganoheterotrophs.

Growth of any microorganism depends not only on a suitable physical environment, but also on the availability of nutrients, which the cells use for biosynthesis. What sets bacteria and archaea apart from all other forms of life is their remarkable ability to use diverse sources of various elements. ⏮ elements

Required Elements

Chemical elements that make up cells are called **major elements;** these include carbon, oxygen, hydrogen, nitrogen, sulfur, phosphorus, potassium, magnesium, calcium, and iron. They are the essential components of proteins, carbohydrates, lipids, and nucleic acids (**table 4.4**).

TABLE 4.4 Representative Functions of the Major Elements

Chemical	Function
Carbon, oxygen, and hydrogen	Component of amino acids, lipids, nucleic acids, and sugars
Nitrogen	Component of amino acids and nucleic acids
Sulfur	Component of some amino acids
Phosphorus	Component of nucleic acids, membrane lipids, and ATP
Potassium, magnesium, and calcium	Required for the functioning of certain enzymes; additional functions as well
Iron	Part of certain enzymes

The source of carbon distinguishes different groups of microorganisms. **Heterotrophs** use organic carbon (*hetero* means "different" and *troph* means "nourishment"). Medically important bacteria are typically heterotrophs. **Autotrophs** (*auto* means "self") use inorganic carbon in the form of carbon dioxide (CO_2). They play a critical role in the cycling of carbon in the environment because they can convert inorganic carbon to an organic form, the process of **carbon fixation.** Without carbon fixation, the earth would quickly run out of organic carbon, which is essential to life. ⏭| carbon cycle

Nitrogen is needed to make amino acids and nucleic acids. Some microorganisms use nitrogen gas (N_2) as a nitrogen source, converting it to ammonia and then incorporating that into cellular material. This process, **nitrogen fixation,** is unique to bacteria and archaea. Like carbon fixation, it is essential to life because once the nitrogen is incorporated into cellular material such as amino acids, other organisms can easily use it. Many microbes use ammonia as a nitrogen source. Some convert nitrate to ammonia, which is then incorporated into cellular material. All of these processes that use nitrogen are important in the nitrogen cycle. ⏭| nitrogen cycle

Sulfur is a component of some amino acids. Many microbes use inorganic sulfur sources such as sulfate, but others require organic sources such as sulfur-containing amino acids.

Phosphorus is a component of nucleic acids, membrane lipids, and ATP. As with sulfur, many organisms can use inorganic phosphorus sources such as phosphate. Some, however, require organic sources, such as phosphorus-containing cell components.

Other elements, including potassium, magnesium, calcium, and iron, are required for some enzyme functions. A variety of inorganic and organic sources may be used by microbes.

Phosphorus and iron are important ecologically because they are often **limiting nutrients**—meaning they are available at the lowest concentration relative to need. To understand this concept, think about using a recipe to make chocolate chip cookies. Just as the quantity of chocolate chips available would determine the number of batches you could make (assuming the other ingredients were on hand), a limiting nutrient dictates the maximum level of microbial growth.

Trace elements are required in such small amounts that most natural environments, including water, have sufficient levels to support microbial growth. Trace elements include cobalt, zinc, copper, molybdenum, and manganese.

Growth Factors

Some microbes cannot synthesize certain organic molecules such as amino acids, vitamins, and nucleotides. Consequently, these organisms grow only if the molecules they cannot produce are available in the surrounding environment; the molecules are **growth factors.**

A microbe's growth factor requirements reflect its biosynthetic capabilities. Most *E. coli* strains, for example, can use glucose as the raw material to synthesize all of their cell components, so they do not need any growth factors. They multiply in a medium containing only glucose and several inorganic salts. In contrast, *Neisseria* species are less resourceful metabolically, and require numerous growth factors, including vitamins and amino acids. Bacteria such as *Neisseria* species are **fastidious,** meaning they have complicated nutritional requirements.

Fastidious bacteria are used to measure the quantity of vitamins in food products. To do this, a well-characterized species that requires a specific vitamin is inoculated into a medium that lacks the vitamin but is supplemented with a measured amount of the food product. The extent of growth of the bacterium is related to quantity of the vitamin in the product.

Energy Sources

Organisms harvest energy from either sunlight or chemical compounds, using processes discussed in chapter 6. **Phototrophs** obtain energy from sunlight (*photo* means "light"). They include plants, algae, and photosynthetic bacteria. **Chemotrophs** extract energy from chemical compounds (*chemo* means "chemical"). Mammalian cells, fungi, and many types of prokaryotes use organic chemicals such as sugars, amino acids, and fatty acids as energy sources. Some bacteria and archaea extract energy from inorganic chemicals such as hydrogen sulfide and hydrogen gas, an ability that distinguishes them from eukaryotes.

Nutritional Diversity

Microbiologists often group microorganisms according to the energy and carbon sources they use (**table 4.5**):

- **Photoautotrophs** use the energy of sunlight to make organic compounds from CO_2 in the atmosphere. They are primary producers, meaning they support other forms

FOCUS ON A CASE 4.1

The "patient" was Green Lake, a small lake in Seattle that sometimes had murky water. At its worst, a surface layer of greenish scum collected near the shore. In very hot weather, the area started smelling like rotting garbage—a characteristic that certainly discouraged people from enjoying the lake! The problems were due to the abundant growth of cyanobacteria, a group of photosynthetic bacteria.

Cyanobacterial blooms (accumulations of cyanobacteria), which occur most commonly in the summer months, are more than just an aesthetic annoyance. Some strains make toxins that can be deadly to animals that drink from the lake. Because of the potential danger, when a cyanobacterial bloom occurs, lakes may be closed to wading, swimming, and sailboarding. In addition, dog owners are warned to keep their pets from drinking the lake water.

To control cyanobacterial growth in Green Lake, alum (aluminum sulfate) was added to the water. Aluminum combines with phosphorus and sinks to the bottom, removing phosphorus from the water column—an action that limits cyanobacterial growth. The treatment was successful, resulting in clear water throughout the year.

1. Why do cyanobacteria blooms occur in the summer months rather than year-round?
2. What caused the greenish scum to form?
3. Why does removing phosphorus from water limit cyanobacterial growth?
4. Why did the area start smelling like rotting garbage when the weather was hot?

Discussion

1. Cyanobacteria are photosynthetic, meaning they are photoautotrophs and therefore harvest their energy from sunlight. Summer months typically have longer days with more sunshine, providing photosynthetic organisms with plenty of their energy source. In addition, the warmer temperatures allow faster growth of the bacteria.
2. The greenish scum was an accumulation of cyanobacterial cells. The bacteria float to the surface because they have gas vesicles, which they use for buoyancy. Like plants, the bacteria are green because they contain chlorophyll. In fact, extensive scientific evidence indicates that chloroplasts—the photosynthetic organelles of plants—evolved from an ancestor of modern-day cyanobacteria. ⏮ gas vesicles, ⏮ chloroplasts
3. Phosphorus is often the limiting nutrient for cyanobacteria. Carbon, which is frequently the limiting nutrient for heterotrophs, is not limiting for autotrophs because they obtain it from the CO_2 in the atmosphere. Nitrogen is often the limiting nutrient for autotrophs, but some cyanobacterial species can fix nitrogen, meaning that they can use atmospheric N_2. Considering that nitrogen-fixing cyanobacteria have readily available supplies of carbon and nitrogen, that leaves phosphorus as the most likely limiting nutrient. Without enough phosphorus, they cannot continue to synthesize their phosphorus-containing cell components, including DNA, RNA, membrane lipids, and ATP.
4. Some of the cyanobacterial cells died in the hot weather. Like any form of natural organic matter, the dead cells will decompose—a process that can generate bad odors.

of life by fixing carbon. Cyanobacteria are important primary producers that inhabit soil and aquatic environments. Many fix nitrogen as well, providing another indispensable role in the biosphere. ⏭ primary producers

- **Photoheterotrophs** use the energy of sunlight and obtain their carbon from organic compounds. Some are facultative in their nutritional capabilities. For example, some members of a group called the purple non-sulfur bacteria grow anaerobically using light as an energy source and organic compounds as a carbon source (photoheterotrophs). They can also grow aerobically in the dark using organic sources of carbon and energy (chemoheterotrophs). ⏭ purple non-sulfur bacteria
- **Chemolithoautotrophs** (*lith* means "stone"), often referred to simply as chemoautotrophs or chemolithotrophs, use inorganic compounds for energy and obtain their carbon from CO_2. These prokaryotes live in seemingly inhospitable places such as sulfur hot springs, which are rich in hydrogen sulfide, and other environments that have reduced inorganic compounds (see **Focus Your Perspective 4.1**). In regions of the ocean depths near hydrothermal vents, chemoautotrophs are the primary producers, supporting abundant life in these habitats that lack sunlight (see figure 28.12). ⏭ hydrothermal vents
- **Chemoorganoheterotrophs,** also referred to as chemoheterotrophs or chemoorganotrophs, use organic

TABLE 4.5 Energy and Carbon Sources Used by Different Groups of Microorganisms

Type	Energy Source	Carbon Source
Photoautotroph	Sunlight	CO_2
Photoheterotroph	Sunlight	Organic compounds
Chemolithoautotroph	Inorganic chemicals (H_2, NH_3, NO_2^-, Fe^{2+}, H_2S)	CO_2
Chemoorganoheterotroph	Organic compounds (sugars, amino acids, etc.)	Organic compounds

FOCUS YOUR PERSPECTIVE 4.1

Can Microorganisms Live on Only Rocks and Water?

Microorganisms have been isolated from diverse environments that previously were thought to be incapable of sustaining life. For example, members of the *Archaea* have been isolated from environments 10 times more acidic than that of lemon juice. Other archaea have been isolated from oil wells a mile below the surface of the earth at temperatures of 70°C and pressures of 160 atmospheres (at sea level, the pressure is 1 atmosphere). The discovery of these microbes suggests that thermophiles may be widespread in the earth's crust.

Perhaps the most unusual environment from which prokaryotes have been isolated is volcanic rock 1 mile below the earth's surface near the Columbia River in Washington State. What do these organisms use for food? They apparently get their energy from the H_2 produced in a reaction between groundwater and the iron-rich minerals in the rock. The groundwater also contains dissolved CO_2, which the organisms use as a source of carbon. These prokaryotes apparently exist on nothing more than rocks and water!

compounds for both energy and carbon. They are by far the most common group of microorganisms associated with humans and other animals. Individual species of chemoheterotrophs differ in the number of organic compounds they can use. For example, certain members of the genus *Pseudomonas* can obtain carbon and/or energy from more than 80 different organic compounds, including such unusual compounds as naphthalene (the ingredient associated with the smell of mothballs). At the other extreme, some organisms can degrade only a few compounds. *Bacillus fastidiosus* can use only urea and a few of its derivatives as sources of both carbon and energy.

MicroAssessment 4.5

Organisms require a source of major and trace elements. Heterotrophs use an organic carbon source, and autotrophs use CO_2. Phototrophs harvest energy from sunlight, and chemotrophs extract energy from chemicals.

12. To prevent excess phosphate from entering lakes and streams, certain laws govern the amount of phosphorus allowed in laundry and dishwasher detergents. What can happen if phosphorus levels in a lake increase?
13. How would your cells be categorized with respect to their carbon and energy sources?
14. Why would human-made materials (such as many plastics) be degraded only slowly or not at all?

4.6 ■ Cultivating Microorganisms in the Laboratory

Learning Outcomes

12. Compare and contrast complex, chemically defined, selective, and differential media.
13. Explain how aerobic, microaerophilic, and anaerobic conditions can be provided.
14. Describe the purpose of an enrichment culture.

By knowing the environmental and nutritional factors that influence the growth of specific microorganisms, it is often possible to provide appropriate conditions for their cultivation. These include a suitable growth medium and the proper atmosphere.

General Categories of Culture Media

Considering the diversity of microorganisms, it should not be surprising that hundreds of different types of media are available. Even so, some medically important organisms and most environmental ones have not yet been grown in the laboratory. **Table 4.6** summarizes the characteristics of representative examples of media.

Complex Media

A **complex medium** contains a variety of ingredients such as meat juices and digested proteins, forming what might be viewed as a tasty soup for microbes. This type of medium is easy to make and is used for routine purposes. Although a specific amount of each ingredient is in the medium, the exact chemical compositions of those substances can be highly variable. One common ingredient is peptone, a mixture of amino acids and short peptides produced by digesting any of a variety of different proteins; each batch of peptone could have a different assortment of amino acids and other substances. Extracts (the water-soluble components of a substance such as lean beef) are often included in complex media to provide vitamins and minerals. A common recipe for a complex medium—nutrient broth—consists of peptone and beef extract in distilled water. If agar is added, then nutrient agar results.

Many medically important bacteria are fastidious, requiring a medium even richer than nutrient agar. Because of this, clinical laboratories often use blood agar. As the name implies, this contains red blood cells, a source of a variety of nutrients including hemin; the medium contains other ingredients as well. A medium used for even more fastidious bacteria is chocolate agar, named for its brownish appearance

TABLE 4.6 Characteristics of Representative Media Used to Cultivate Bacteria

Medium	Characteristic
Blood agar	Complex medium used routinely in clinical labs. Differential because colonies of hemolytic organisms are surrounded by a zone of red blood cell clearing. Not selective.
Chocolate agar	Complex medium used to culture fastidious bacteria, particularly those found in clinical specimens. Not selective or differential.
Glucose-salts agar	Chemically defined medium. Used in laboratory experiments to study nutritional requirements of bacteria. Not selective or differential.
MacConkey agar	Complex medium used to isolate Gram-negative rods that typically reside in the intestine. Selective because bile salts and dyes inhibit Gram-positive organisms and Gram-negative cocci. Differential because the pH indicator turns pink-red when the sugar in the medium, lactose, is fermented.
Nutrient agar	Complex medium used for routine laboratory work. Supports the growth of a variety of nonfastidious bacteria. Not selective or differential.
Thayer-Martin agar	Complex medium used to isolate *Neisseria* species, which are fastidious. Selective because it contains antibiotics that inhibit most organisms except *Neisseria* species. Not differential.

rather than its ingredients. Chocolate agar contains lysed red blood cells and additional supplements, thereby supplying fastidious bacteria the nutrients they require.

Chemically Defined Media

A **chemically defined medium** is composed of specific amounts of pure chemicals, so its exact chemical composition is known. This type of medium is generally used only for certain research experiments when the type and quantity of nutrients must be precisely controlled. A chemically defined medium called glucose-salts broth supports the growth of *E. coli,* and contains only those chemicals listed in **table 4.7**. The cells grow more slowly in this medium than in nutrient broth because they must synthesize all of their cell components from glucose. A much longer recipe containing as many as 46 different ingredients must be used to make a chemically defined medium that supports the growth of the fastidious bacterium *Neisseria gonorrhoeae* (the cause of gonorrhea).

To maintain the pH near neutral, buffers are often added to culture media. They are especially important in defined media because some bacteria produce so much acid as a by-product of their metabolism that they inhibit their own growth. This is usually not a problem in complex media because the amino acids and other natural components provide at least some buffering action. ⏮ buffer

TABLE 4.7 Ingredients in Two Types of Media That Support the Growth of *E. coli*

Nutrient Broth (Complex Medium)	Glucose-Salts Broth (Chemically Defined Medium)
Peptone	Glucose
Beef extract	Dipotassium phosphate
Water	Monopotassium phosphate
	Magnesium sulfate
	Ammonium sulfate
	Calcium chloride
	Iron sulfate
	Water

Special Types of Culture Media

To detect or isolate a species from a mixed population, it is often necessary to make that species more prevalent or obvious. For these purposes selective and differential media are used (**figure 4.10**). They can be either complex or chemically defined, depending on the needs of the microbiologist.

Selective Media

Selective media contain an ingredient that inhibits the growth of certain species in a mixed sample, while allowing the growth of the species of interest. For example, Thayer-Martin agar is used to isolate *Neisseria gonorrhoeae* from clinical specimens. Thayer-Martin is chocolate agar to which three or more antimicrobial drugs have been added. The antimicrobials inhibit fungi, Gram-positive bacteria, and Gram-negative rods, but not most *N. gonorrhoeae* strains. Because of this, the pathogen can grow on the medium with little competition.

MacConkey agar is used to isolate Gram-negative rods from various clinical specimens such as urine. This medium has a variety of nutrients that many bacteria can use, but also includes two inhibitory components: crystal violet (a dye that inhibits Gram-positive bacteria) and bile salts (an ingredient that inhibits most non-intestinal bacteria).

Differential Media

Differential media contain a substance that certain microbes change in a recognizable way. Blood agar is differential because bacteria that produce a hemolysin (a protein that causes red blood cells to burst) destroy red blood cells in the medium around the bacterial colony, causing a zone of clearing called hemolysis (**figure 4.11**). This easily observable characteristic is important medically because some pathogens can be distinguished by their type of hemolysis. For example, the colonies of *Streptococcus pyogenes* (the cause of strep throat) produce a clear zone of hemolysis called **beta hemolysis.** This makes them stand out from *Streptococcus*

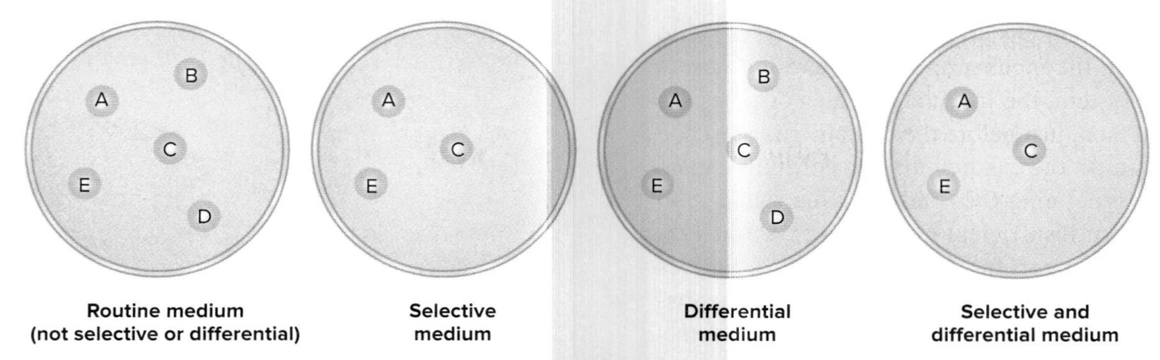

FIGURE 4.10 Selective and Differential Media A routine medium allows growth of a variety of organisms; in this example, organisms A, B, C, D, and E grow, but the colonies look similar. A selective medium contains an ingredient that inhibits the growth of certain organisms; in this example, only organisms A, C, and E can grow, but again the colonies look similar. A differential medium contains an ingredient that certain microorganisms change in a recognizable way; in this example, the colonies of organisms B and E are pink, so they look different from the rest. A selective and differential medium inhibits certain organisms, and causes colonies of some organisms to look different from others; in this example, only organisms A, C, and E can grow, and colonies of organism E are pink so they look different from colonies of organisms A and C.

? **How would you categorize nutrient agar to which penicillin had been added?**

species that reside harmlessly in the throat. Many *Streptococcus* species show **alpha hemolysis,** meaning their colonies are surrounded by a zone of greenish partial clearing; other streptococci have no effect on red blood cells. ⏭ ***Streptococcus pyogenes***

MacConkey agar is differential as well as selective (**figure 4.12**). In addition to its ingredients already mentioned, it contains lactose and a pH indicator. Bacteria that ferment the sugar produce acid, which turns the pH indicator pink-red. Therefore, colonies of lactose-fermenting bacteria are pink-red on MacConkey agar, whereas colonies of lactose-negative bacteria are colorless. ⏮ **lactose**

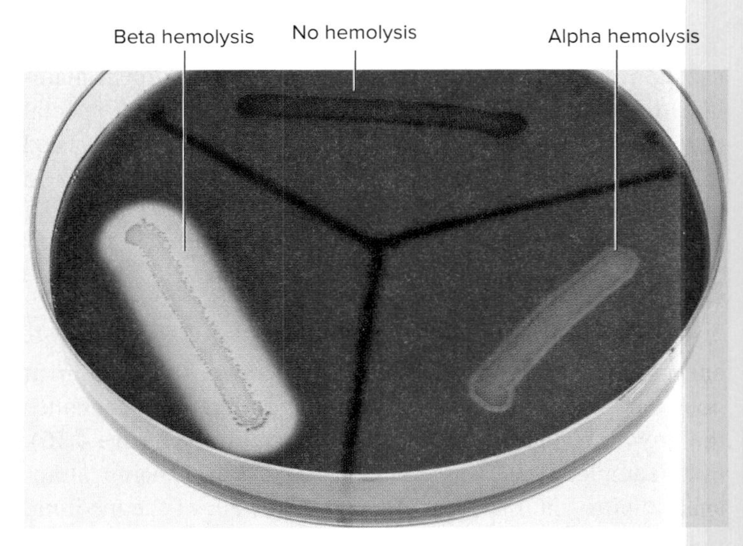

FIGURE 4.11 Blood Agar This complex medium is differential for hemolysis. A zone of colorless clearing around a colony growing on blood agar is called beta hemolysis; a zone of greenish clearing is called alpha hemolysis. ©McGraw-Hill Education/Lisa Burgess, photographer

? **Which type of hemolysis characterizes *Streptococcus pyogenes,* the bacterium that causes strep throat?**

Providing Appropriate Atmospheric Conditions

To grow microorganisms on culture media in the laboratory, appropriate atmospheric conditions must be provided.

Aerobic Conditions

When incubating most obligate aerobes and facultative anaerobes on agar media, special atmospheric conditions are not required; they can be incubated in air (approximately 20% O_2). Broth cultures of these organisms grow best when tubes or flasks containing the media are shaken, providing maximum aeration.

Many medically important bacteria, including species of *Neisseria* and *Haemophilus,* grow best in aerobic atmospheres that have additional CO_2. Some are even capnophiles, meaning

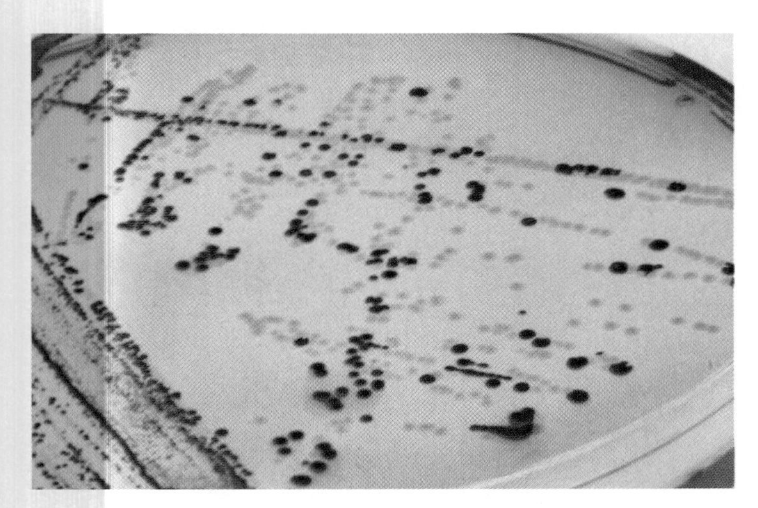

FIGURE 4.12 MacConkey Agar This complex medium is differential for lactose fermentation and selective for Gram-negative rods. ©McGraw-Hill Education/Lisa Burgess, photographer

? **What specifically causes colonies of lactose fermenters to be pink-red on MacConkey agar?**

they require increased CO_2. One of the simplest ways to build up the level of CO_2 in the incubation atmosphere is to use a candle jar. For this system, the inoculated plates or tubes are added to the jar and then, just before the container is closed, a candle within it is lit. As the candle burns, it consumes some of the O_2 in the air, generating CO_2 and H_2O; the flame soon extinguishes because of insufficient O_2. However, about 17% O_2 remains, enough to support the growth of obligate aerobes and prevent the growth of obligate anaerobes. Special incubators are also available that maintain CO_2 at prescribed levels.

Microaerophilic Conditions

Microaerophilic organisms typically require O_2 concentrations lower than those achieved in a candle jar. These microbes are often incubated in a gastight container with a special disposable packet; the packet holds chemicals that react with O_2, reducing its concentration to approximately 5–15%.

Anaerobic Conditions

Obligate anaerobes are challenging to grow because of their sensitivity to O_2-containing environments. Those that can tolerate brief exposures to O_2 are incubated in an **anaerobe container** (**figure 4.13**). This is the same type of container used to incubate microaerophiles, but the chemical composition of the disposable packet produces an anaerobic environment. An alternative method is to use a semisolid culture medium that contains a reducing agent such as sodium thioglycolate, which reacts with O_2 to make water. In many cases, an O_2-indicating dye is included as well. The medium can be heated immediately before use to remove some dissolved O_2. ⏭ reducing agent

FIGURE 4.13 Anaerobe Containers A disposable packet contains chemicals that react with O_2, thereby producing an anaerobic environment.

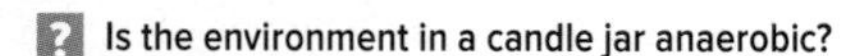

? Is the environment in a candle jar anaerobic?

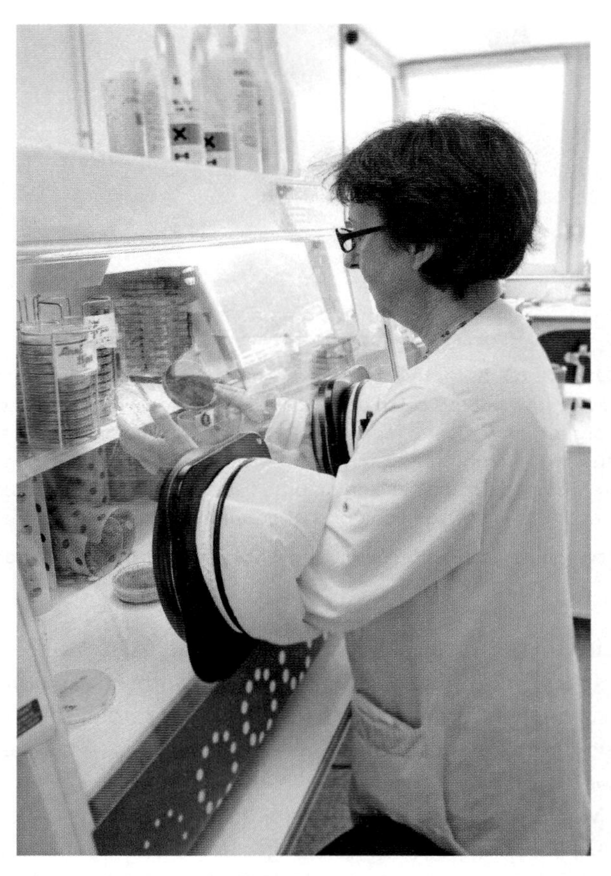

FIGURE 4.14 Anaerobic Chamber The enclosed compartment can be maintained as an anaerobic environment. A special port that can be filled with inert gas is used to add or remove items. The airtight gloves allow the researcher to handle items within the chamber.

? Why would the chamber be preferable to an anaerobe container?

A more stringent method for working with anaerobes is to use an **anaerobic chamber,** an enclosed compartment maintained as an anaerobic environment (**figure 4.14**). A special port that can be filled with an inert gas such as N_2 is used to add or remove items. Airtight gloves allow researchers to handle items within the chamber.

Enrichment Cultures

An **enrichment culture** is used to isolate an organism present as only a very small fraction of a mixed population. It does this by providing conditions that preferentially enhance the growth of that particular species in a broth (**figure 4.15**). For example, if the target microbe can grow using atmospheric nitrogen, then that element is left out of the medium. If it can use an unusual carbon source such as phenol, then that compound is added as the only carbon source. A selective agent such as bile salts may also be added. The culture is then incubated in conditions that preferentially promote the growth of the desired organism. As the microbes multiply, the relative concentration of the target organism can increase

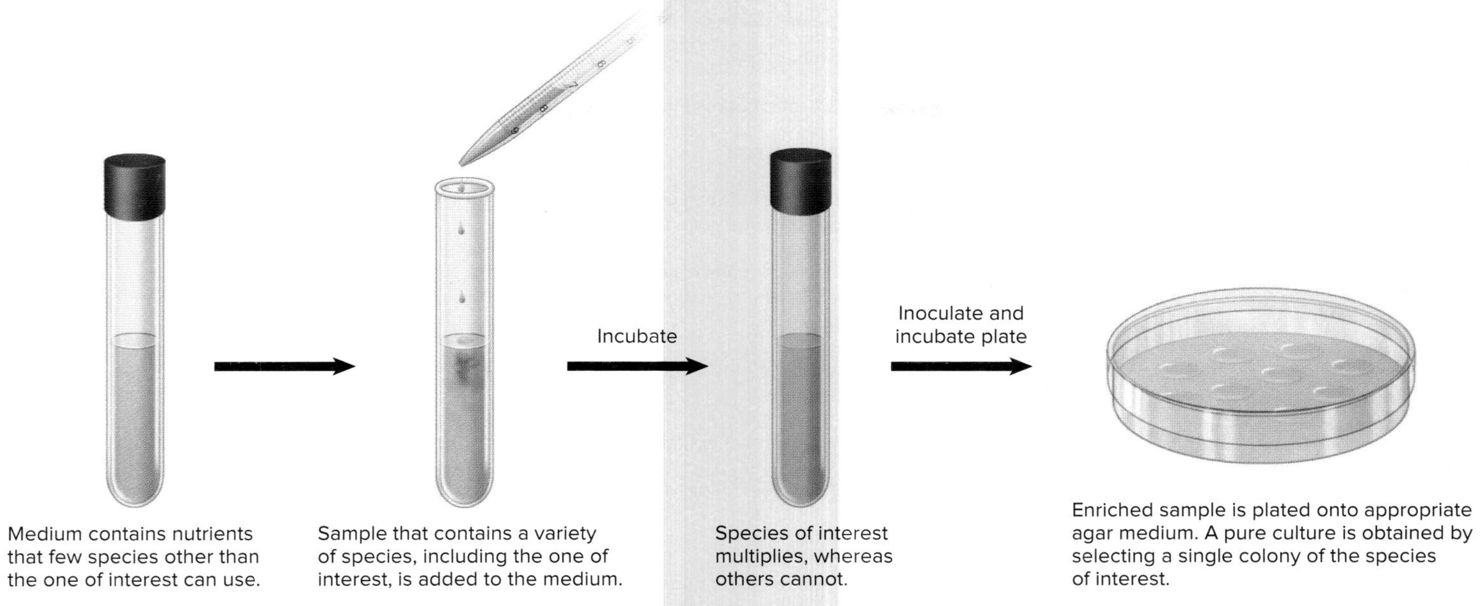

FIGURE 4.15 Enrichment Culture Medium and incubation conditions favor growth of the desired species over other microorganisms in the same sample.

? How would you enrich for an organism that can use phenol as a carbon source?

dramatically. A pure culture can then be obtained by streaking the enrichment onto an appropriate agar medium and selecting a single colony.

MicroAssessment 4.6

Culture media can be complex or chemically defined. Some media contain additional ingredients that make them selective or differential. Appropriate atmospheric conditions must be provided to grow microbes in culture. An enrichment culture increases the relative concentration of an organism growing in a broth.

15. Distinguish between complex and chemically defined media.
16. Describe two methods to create anaerobic conditions.
17. Would bacteria that cannot use lactose be able to grow on MacConkey agar?

4.7 ■ Methods to Detect and Measure Microbial Growth

Learning Outcome

15. Compare and contrast methods used for direct cell counts, viable cell counts, measuring biomass, and detecting cell products.

A variety of techniques can be used to monitor microbial growth. Characteristics of common methods are summarized in **table 4.8**.

Direct Cell Counts

Direct cell counts are particularly useful for determining the total numbers of microorganisms in a sample, including those that cannot be grown in culture. Unfortunately, the methods generally do not distinguish between living and dead cells in the specimen.

Direct Microscopic Count

One of the most rapid methods of determining the cell concentration in a suspension is the direct microscopic count (**figure 4.16**). A liquid specimen is added to a hemocytometer, a special glass slide designed specifically for counting cells. The slide has a thin chamber that holds a known volume of liquid atop a microscopic grid. The contents of the chamber can be viewed under the light microscope, so the number of cells in a given volume can be counted precisely. At least 10 million bacteria (10^7) per milliliter (mL) are usually required for enough cells to be seen in the microscope field.

Cell-Counting Instruments

A **Coulter counter** is an electronic instrument that counts cells in a suspension as they pass single file through a narrow channel (**figure 4.17**). The suspending liquid must be an electrically conducting fluid, because the machine counts the brief changes in resistance that occur when non-conducting particles such as bacteria pass.

TABLE 4.8 Methods Used to Measure Microbial Growth

Method	Characteristics and Limitations
Direct Cell Counts	Used to determine total number of cells; counts include living and dead cells.
Direct microscopic count	Rapid, but at least 10^7 cells/mL must be present to be effectively counted.
Cell-counting instruments	Coulter counters and flow cytometers count total cells in dilute solutions. Flow cytometers can also be used to count organisms to which fluorescent dyes or tags have been attached.
Viable Cell Counts	Used to determine the number of microorganisms capable of growing in a given set of conditions. Requires an incubation period of approximately 24 hours or longer. Selective and differential media can be used to determine the number of specific microbial species.
Plate count	Time-consuming but technically simple method that does not require sophisticated equipment. Generally used only if the sample has at least 10^2 cells/mL.
Membrane filtration	Used to count microorganisms in liquids that contain relatively few cells; the cells are caught on a paper-thin filter.
Most probable number	Estimates the likely concentration; it does not provide a precise count.
Measuring Biomass	Biomass can be correlated to cell number.
Turbidity	Very rapid method; used routinely. A one-time correlation with plate counts is required for turbidity to be used in determining cell number.
Total weight	Tedious and time-consuming; however, it is one of the best methods for measuring growth of filamentous microorganisms.
Detecting Cell Products	Used to detect growth, but not routinely used for quantitation. A pH indicator can detect acid production. Gases can be trapped in an inverted tube in the broth; a more sensitive method uses a fluorescent sensor that detects the slight decrease in pH that accompanies CO_2 production.

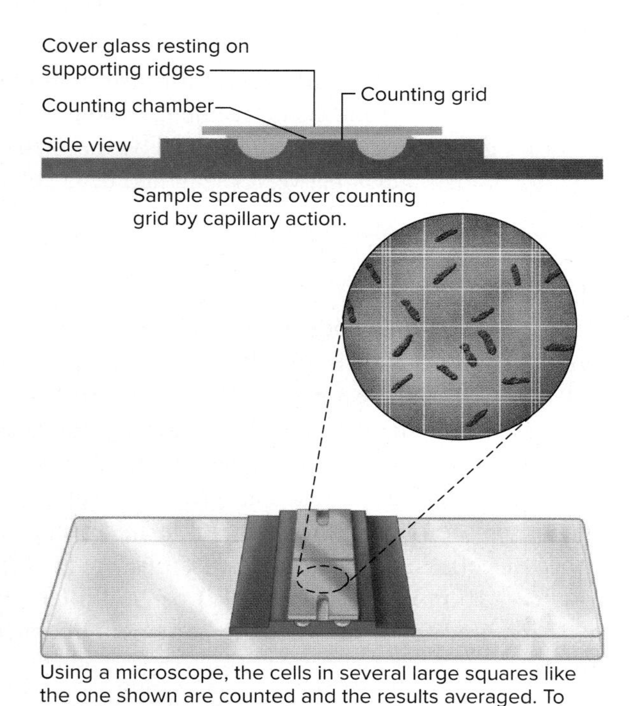

FIGURE 4.16 Direct Microscopic Count The counting chamber holds a known volume of liquid. The number of microbial cells in that volume can be counted precisely.

What is one disadvantage of doing a direct microscopic count to determine the number of cells in a suspension?

A **flow cytometer** is similar in principle to a Coulter counter except that it measures light scattered by cells as they pass a laser. The instrument can be used to count either total cells or a specific subset that has been stained with a fluorescent dye or tag. ◀◀ fluorescent dyes and tags

Viable Cell Counts

Viable cell counts determine the number of cells capable of multiplying. They are particularly valuable when working with samples such as food and water that contain too few microbes for a direct microscopic count. In addition, by using appropriate selective and differential media, these methods can be used to count the cells of a particular microbial species.

Plate Counts

Plate counts measure the number of viable cells in a sample by taking advantage of the fact that an isolated microbial cell on a nutrient agar plate will give rise to one colony. A simple count of the colonies determines how many cells were in the initial sample. Plate counts generally are done only if a sample contains more than 100 organisms/mL. Otherwise, few if any cells will be transferred to the plates. In these situations, alternative methods give more reliable results.

When counting colonies, the ideal number on a plate is between 30 and 300. Numbers outside that range are more

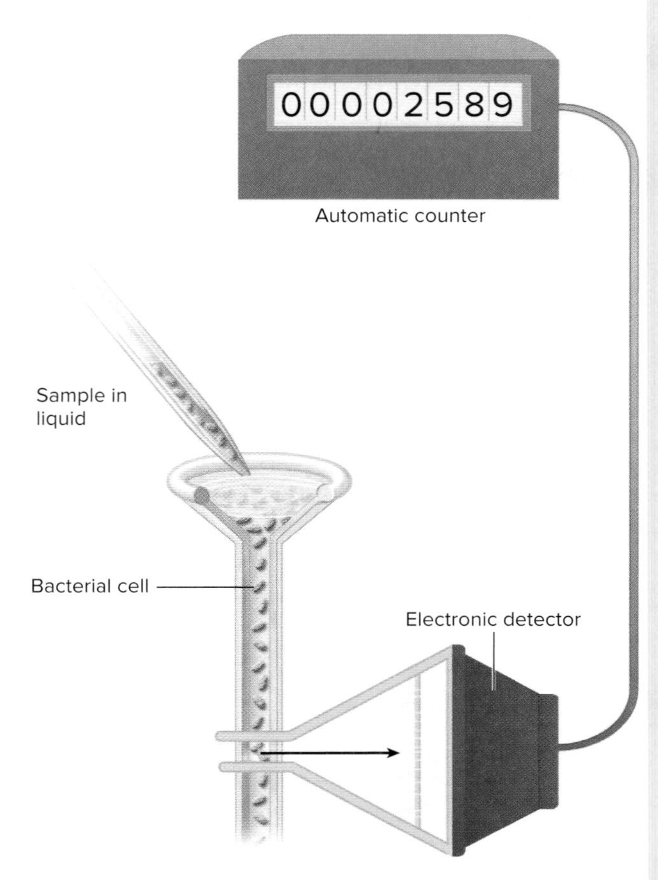

FIGURE 4.17 A Coulter Counter This instrument counts cells as they pass through a narrow channel.

? Why do the microbial cells need to be suspended in an electrically conducting fluid for this method to work?

likely to be inaccurate. Samples usually contain many more cells than that, so they generally must be diluted by a stepwise process called serial dilution (**figure 4.18**). This is done using a sterile liquid called the diluent, often physiological saline (0.85% NaCl in water). Dilutions are normally done in 10-fold increments, making the resulting math relatively simple.

Two techniques can be used to add samples to an agar plate (**figure 4.19**). In the **spread-plate method,** 0.1–0.2 mL of the diluted sample is transferred onto a plate of a solidified agar medium. It is then spread over the surface of the agar with a sterilized bent glass rod that resembles a miniature hockey stick. In the **pour-plate method,** 0.1–1.0 mL of the diluted sample is transferred to a sterile Petri dish and then a melted agar medium cooled to 50°C is added. At this temperature, agar is still liquid. The dish is then gently swirled to mix the microbial cells with the liquid agar. When the agar hardens, the individual cells become fixed in place; they form colonies when incubated. Colonies that form on the surface will be larger than those embedded in the medium.

In both methods, the plates are incubated and then the colonies are counted. The number of colonies is used to determine the concentration of the sample's **colony-forming units (CFUs)**—a measure of viable cells that accounts for the fact that microbial cells often attach to one another and then grow to form a single colony. When calculating CFUs, three things must be considered: the number of colonies, the amount the sample was diluted before being plated, and the volume plated.

In a clinical lab, a simplified version of a plate count is done on urine specimens to help diagnose urinary tract infections (UTIs). The number of CFUs in these samples is important, because small numbers are likely due to bacterial contamination of the sample during the collection process, rather than an actual infection. To determine the approximate number of CFUs per mL of urine, a calibrated loop is used to inoculate an agar plate with a tiny amount of specimen (as little as 0.001 mL). That same loop is then used to spread the sample, without flaming, to obtain isolated colonies. The labs have a set of criteria, based on the urine collection method and the patient population, for deciding how many CFUs in the specimen indicate a UTI.

Membrane Filtration

Membrane filtration is used to count microorganisms in liquid samples that contain relatively few cells, as might occur in environments such as lakes or other natural bodies of water. A known volume of the liquid is passed through a sterile membrane filter that has a pore size small enough to prevent microorganisms from passing through (**figure 4.20**). By doing this, any microorganisms in the liquid are caught on the paper-thin filter. The filter is then placed on an appropriate agar medium and incubated. The number of colonies that form on the filter indicates the number of cells in the volume filtered.

Most Probable Number (MPN)

The **most probable number (MPN) method** provides an estimate of the concentration of cells in a specimen. The procedure uses a series of dilutions to determine the point at which additional dilutions receive no cells.

To determine the MPN, three sets of three or five tubes containing a growth medium are prepared (**figure 4.21**). Each set receives a measured amount of a sample such as water, soil, or food. The amount added is determined, in part, by the expected microbial concentration in that sample. What is important is that the second set receives 10-fold less than the first, and the third set 100-fold less. In other words, each set is inoculated with an amount 10-fold less than the previous set. After incubation, the presence or absence of turbidity (cloudiness) or other indication of growth is noted; the results are then compared against an MPN table, which gives a statistical estimate of the cell concentration.

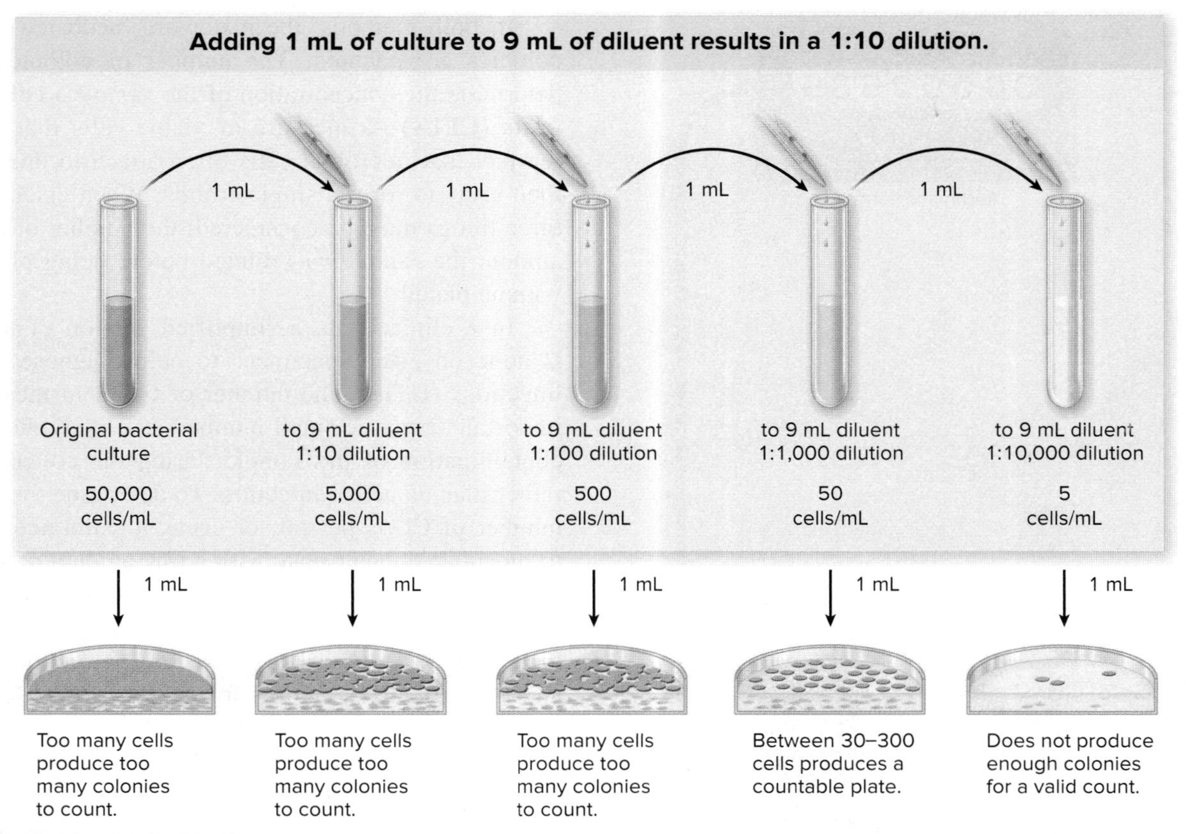

FIGURE 4.18 Making Serial Dilutions To decrease the concentration of cells in a sample, 10-fold dilutions are often used.

If the 1 mL sample had been added to 99 mL of diluent to make the first dilution (instead of 9 mL), how many cells would be in that particular dilution?

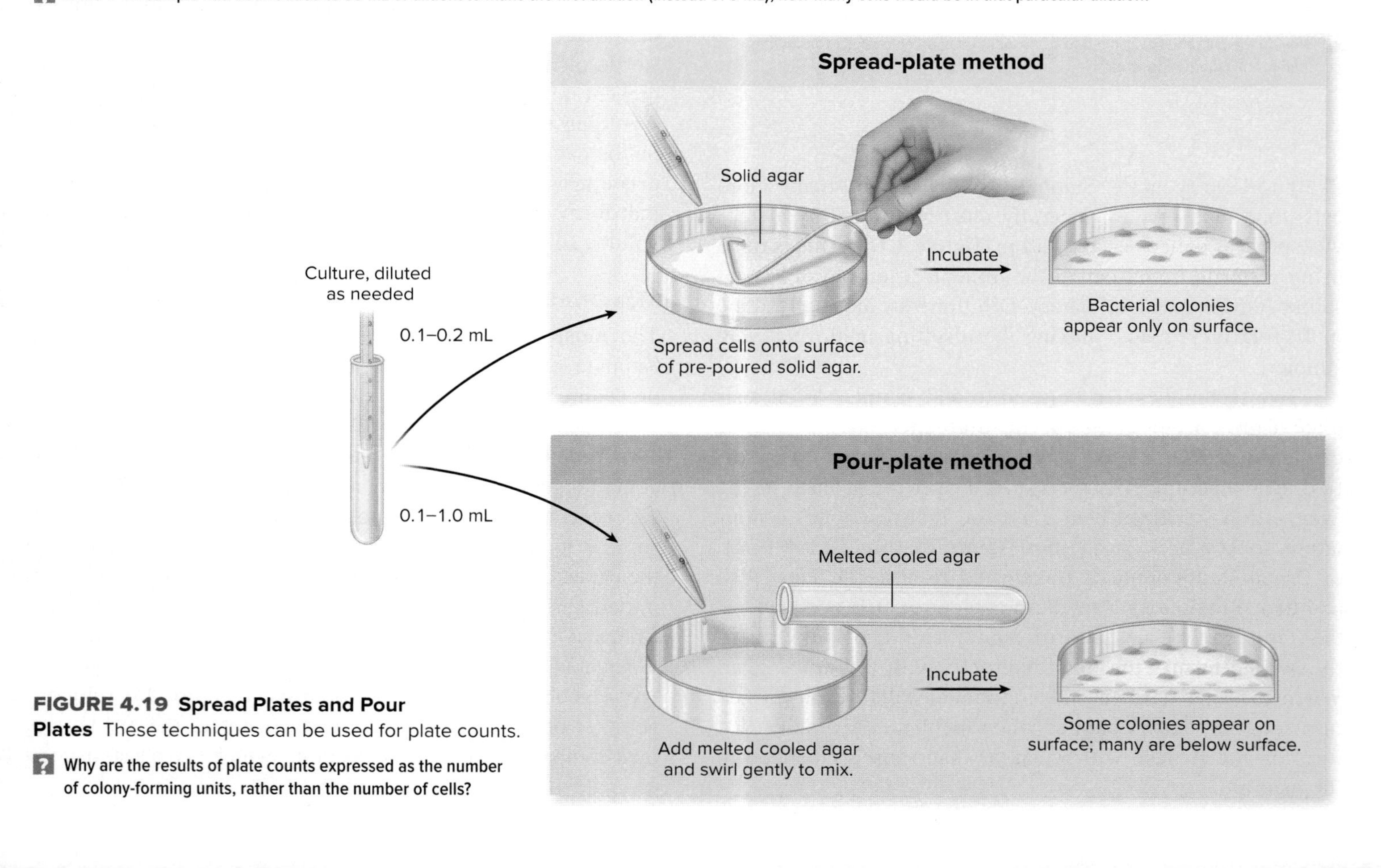

FIGURE 4.19 Spread Plates and Pour Plates These techniques can be used for plate counts.

Why are the results of plate counts expressed as the number of colony-forming units, rather than the number of cells?

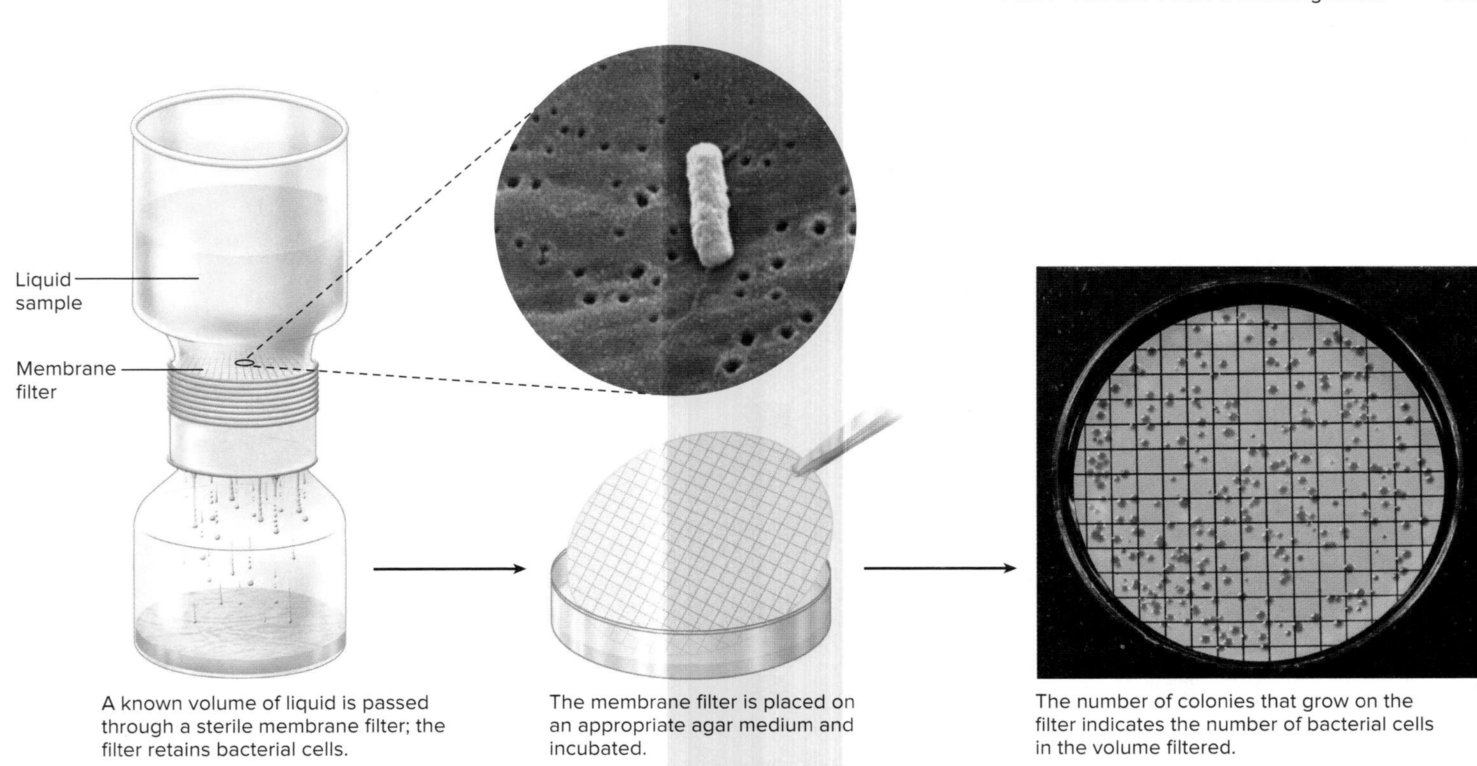

FIGURE 4.20 Membrane Filtration This technique catches the microbial cells on a membrane filter.

Source: Left: Margaret Williams, PhD; Claressa Lucas, PhD; Tatiana Travis, BS/CDC; Right: ©McGraw-Hill Education/Lisa Burgess, photographer

In what situation would membrane filtration be used rather than a plate count?

Measuring Biomass

Instead of measuring the number of cells, the cell mass can be determined.

Turbidity

The cloudiness or **turbidity** of a microbial suspension is proportional to the concentration of cells, and is measured with a spectrophotometer (**figure 4.22**). This instrument shines light through a specimen and measures the percentage that reaches a light detector. That amount—called the transmittance—is inversely proportional to the amount absorbed, which is referred to as the optical density or absorbance. In other words, the more cells present in the sample, the less light passes through to the light detector. To use turbidity to estimate cell numbers, a one-time test must be done to determine the correlation between optical density and cell concentration for the specific organism and conditions under study. Once this correlation has been done—generally using a direct microscopic count or plate count to determine cell concentration—the turbidity measurement becomes a rapid and relatively accurate assay.

One limitation of using turbidity to measure biomass is that the medium must contain a relatively high concentration of cells to be cloudy. A solution containing 1 million bacteria (10^6) per mL is still perfectly clear, and if it contains 10 times that amount, it is barely turbid. It is important to remember that although a turbid culture indicates that microbes are present, a clear solution does not guarantee their absence.

In a clinical lab, a set of reference tubes called McFarland turbidity standards are used to prepare standardized inocula for certain tests such as antimicrobial susceptibility testing. By growing a bacterial culture to the same turbidity of a given McFarland turbidity standard, a known approximate cell concentration is achieved.

Total Weight

The total weight of a culture can be used to measure growth, but the method is tedious and time-consuming. Because of this, total weight is usually used to study only filamentous organisms that do not readily separate into the individual cells necessary for a valid plate count. To measure the wet weight, cells in liquid culture are centrifuged and the liquid supernate removed. The weight of the resulting packed cell mass is proportional to the number of cells in the culture. The dry weight can be determined by heating the centrifuged cells in an oven before weighing them.

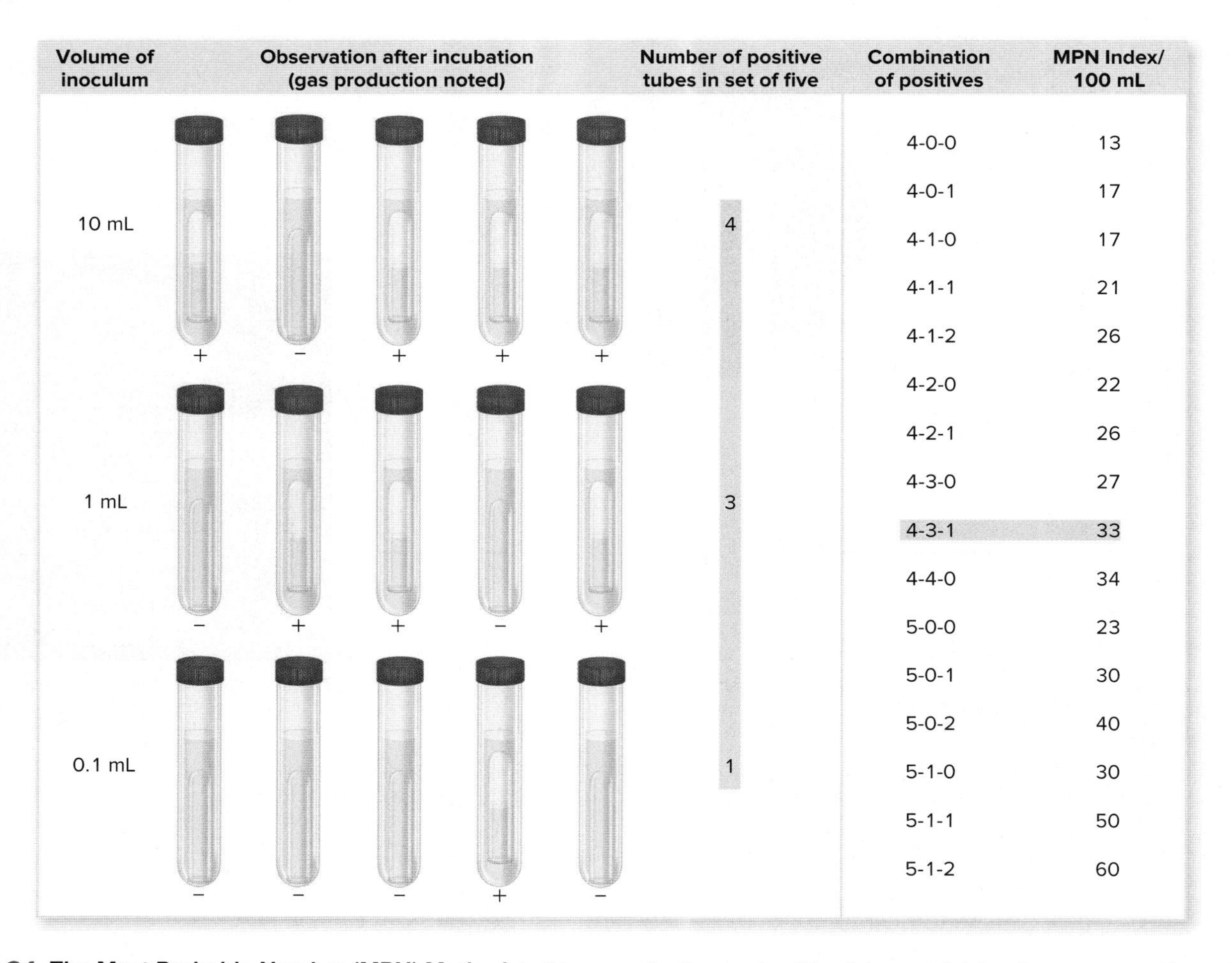

Combination of positives	MPN Index/ 100 mL
4-0-0	13
4-0-1	17
4-1-0	17
4-1-1	21
4-1-2	26
4-2-0	22
4-2-1	26
4-3-0	27
4-3-1	33
4-4-0	34
5-0-0	23
5-0-1	30
5-0-2	40
5-1-0	30
5-1-1	50
5-1-2	60

FIGURE 4.21 The Most Probable Number (MPN) Method In this example, three sets of five tubes containing the same growth medium were prepared. Each set received the indicated amount of inoculum. After incubation, the presence or absence of gas in each tube was noted. The results were then compared to an MPN table to get a statistical estimate of the concentration in the original sample of gas-producing bacteria that could grow in the medium.

If you were using a sample similar to the one illustrated, and all the tubes that received 10 mL had gas, but none that received 1.0 mL or 0.1 mL did, what would be the MPN index/100 mL?

Detecting Cell Products

Microorganisms produce a variety of acids and, sometimes, gases as a result of their metabolism. Acids can be detected by including a pH indicator in the culture medium. Several pH indicators are available, and they differ in the pH value at which their color changes. To detect gas production, inverted tubes (Durham tubes) can be used to trap gas bubbles in broth cultures. Clinical labs use more sensitive methods to detect the slight amounts of CO_2 produced by bacteria growing in patient blood samples. One method uses a fluorescent sensor to detect the slight decrease in pH that accompanies the production of CO_2.

MicroAssessment 4.7

Direct microscopic counts and cell-counting instruments generally do not distinguish between living and dead cells. Plate counts determine the number of cells capable of multiplying; membrane filtration can be used to count microorganisms in dilute liquids. The most probable number (MPN) estimates cell concentration. Turbidity of a culture can be correlated to cell number. The total weight of a culture can be correlated to the number of cells present. Microbial growth can be detected by the presence of cell products such as acids and gases.

18. When doing plate counts, why is it often necessary to dilute the sample?

19. Why is an MPN an estimate rather than an accurate number?

20. Water that accumulates in dirty bowls and dishes left in the sink often becomes cloudy over time. Besides suspended food particles, what causes the turbidity?

FOCUS ON THE FUTURE 4.1

Seeing How the Other 99% Lives

One of the biggest challenges for the future is to develop the methods to cultivate and study a wider array of bacteria and archaea. Without these microbes, humans and other animals would not be able to exist. Yet, considering their importance, we still know very little about most species, including the relative contributions of each to such fundamental processes as O_2 generation and N_2 and CO_2 fixation.

Much of our understanding of bacterial and archaeal processes comes from work with pure cultures. Yet, over 99% of prokaryotes have never been successfully grown in the laboratory. At the same time, when organisms are removed from their natural habitat, and especially when they are separated from other organisms, their environment changes drastically. Consequently, the study of pure cultures may not be the ideal for studying natural situations.

Technological advances such as flow cytometry and fluorescent labeling, along with the DNA sequencing techniques discussed in chapter 9, have made it easier to study uncultivated microorganisms. This is gradually leading to a better understanding of the diversity and the roles of microorganisms in our ecosystem. Scientists have learned a great deal since the days of Pasteur, but most of the microbial world is still a mystery.

(a) The cloudiness, or turbidity, of the liquid in the tube on the left is proportional to the concentration of cells.

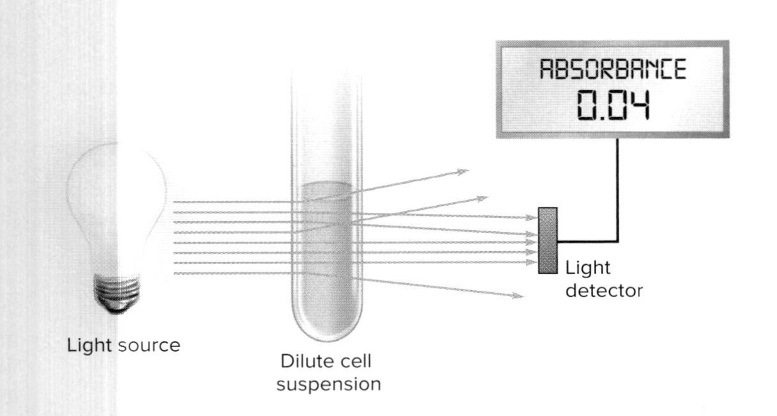

(b) A spectrophotometer is used to measure turbidity.

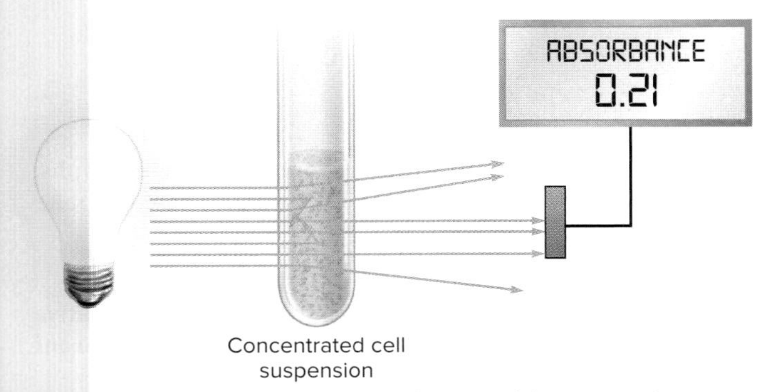

(c) The percentage of light that reaches the detector of the spectrophotometer is inversely proportional to the optical density (absorbance).

FIGURE 4.22 Measuring Turbidity with a Spectrophotometer To use turbidity to estimate cell number, a one-time test must be done to determine the correlation between optical density (absorbance) and cell concentration for the specific organism and conditions under study.

a: ©McGraw-Hill Education/Lisa Burgess, photographer; b: ©Martin Shields/Science Source

? Approximately how many bacterial cells must be in a suspension for it to be cloudy?

Summary

4.1 ■ Principles of Microbial Growth

Most bacteria and archaea multiply by **binary fission** (figure 4.1). Microbial growth is an increase in the number of cells in a population. The time required for a population to double in number is the **generation time** (table 4.1).

4.2 ■ Microbial Growth in Nature

Biofilms (figure 4.2)

Microorganisms often live in a **biofilm,** a community encased in polysaccharides and other extracellular polymeric substances.

Interactions of Mixed Microbial Communities

Microorganisms often grow in close associations containing multiple different species. The metabolic activities of one organism often affect the growth of another.

4.3 ■ Microbial Growth in Laboratory Conditions

Only an estimated 1% of prokaryotes have been grown in the laboratory.

Obtaining a Pure Culture (figure 4.5)

The **streak-plate method** is used to isolate microorganisms in order to obtain a **pure culture.**

The Growth Curve (figure 4.6)

When grown in a **closed system,** a population of prokaryotic cells goes through five phases: **lag, exponential (log), stationary, death,** and **prolonged decline.**

Colony Growth

The position of a single cell within a colony influences its environment.

Continuous Culture

Microbes can be maintained in a state of continuous growth by using a **chemostat.**

4.4 ■ Environmental Factors That Influence Microbial Growth (table 4.2)

Temperature Requirements (figure 4.8)

Organisms can be grouped as **psychrophiles, psychrotrophs, mesophiles, thermophiles,** or **hyperthermophiles** based on their optimum growth temperatures.

Oxygen (O_2) Requirements (table 4.3)

Organisms can be grouped as **obligate aerobes, facultative anaerobes, obligate anaerobes, microaerophiles,** or **aerotolerant anaerobes** based on their oxygen (O_2) requirements.

pH

Organisms can be grouped as **neutrophiles, acidophiles,** or **alkaliphiles** based on their optimum pH.

Water Availability

Halophiles are adapted to live in high-salt environments.

4.5 ■ Nutritional Factors That Influence Microbial Growth

Required Elements (table 4.4)

The **major elements** make up cell components and include carbon, nitrogen, sulfur, and phosphorus. **Trace elements** are required in very small amounts.

Growth Factors

Microorganisms that cannot synthesize cell components such as amino acids and vitamins require these as **growth factors.**

Energy Sources

Organisms harvest energy either from sunlight or from chemical compounds.

Nutritional Diversity (table 4.5)

Photoautotrophs use the energy of sunlight along with the carbon in the atmosphere to make organic compounds. **Chemolithoautotrophs** use inorganic compounds for energy and derive their carbon from CO_2. **Photoheterotrophs** use the energy of sunlight and obtain their carbon from organic compounds. **Chemoorganoheterotrophs** use organic compounds for energy and as a carbon source.

4.6 ■ Cultivating Microorganisms in the Laboratory

General Categories of Culture Media (table 4.6)

A **complex medium** contains a variety of ingredients such as peptones and extracts. A **chemically defined medium** is composed of precise mixtures of pure chemicals.

Special Types of Culture Media (figure 4.10)

A **selective medium** inhibits organisms other than the one being sought. A **differential medium** contains a substance that certain microorganisms change in a recognizable way.

Providing Appropriate Atmospheric Conditions

A candle jar provides increased CO_2, which enhances the growth of many medically important bacteria. Microaerophilic microbes are incubated in a gastight container along with a packet that generates low O_2 conditions. Anaerobes may be incubated in an **anaerobe container** or an **anaerobic chamber** (figures 4.13, 4.14).

Enrichment Cultures (figure 4.15)

An **enrichment culture** provides conditions in a broth that enhance the growth of one particular organism in a mixed population.

4.7 ■ Methods to Detect and Measure Microbial Growth (table 4.8)

Direct Cell Counts

Direct microscopic counts involve counting the number of cells viewed through a microscope (figure 4.16). Both a **Coulter counter** and a **flow cytometer** count cells as they pass through a tiny channel (figure 4.17).

Viable Cell Counts

Plate counts measure the number of viable cells by taking advantage of the fact that an isolated cell will form a single colony (figures 4.18, 4.19). **Membrane filtration** catches microbial cells in a liquid sample on a paper-thin filter; the filter is then incubated on an agar plate and the colonies counted (figure 4.20). The **most probable number (MPN) method** provides a statistical estimation of the cell concentration (figure 4.21).

Measuring Biomass

Turbidity of a culture can be correlated with the number of cells; a spectrophotometer is used to measure turbidity (figure 4.22). Wet weight and dry weight are proportional to the number of cells in a culture.

Detecting Cell Products

Products including acids and gases can indicate microbial growth.

Review Questions

Short Answer

1. Describe a detrimental and a beneficial effect of biofilms.
2. Define a *pure culture*.
3. Explain what occurs during each of the five phases of growth.
4. Explain how the environment of a colony differs from that of cells growing in a liquid broth.
5. List the five categories of optimum temperature, and describe a corresponding environment in which a representative might thrive.
6. Why would botulism be a concern with canned foods?
7. Explain why O_2-containing atmospheres kill some microbes.
8. Explain why photoautotrophs are primary producers.
9. Distinguish between a selective medium and a differential medium.
10. If the number of microorganisms in lake water were determined using both a direct microscopic count and a plate count, which method would most likely give a higher number? Why?

Multiple Choice

1. If there are 10^3 cells per mL at the middle of log phase, and the generation time of the cells is 30 minutes, how many cells will there be 2 hours later?
   a) 2×10^3
   b) 4×10^3
   c) 8×10^3
   d) 1.6×10^4
   e) 1×10^7
2. Compared with their growth in the laboratory, bacteria in nature generally grow
   a) more slowly.
   b) faster.
   c) at the same rate.
3. Cells are most sensitive to penicillin during which phase of the growth curve?
   a) Lag
   b) Exponential
   c) Stationary
   d) Death
   e) More than one of these
4. Which best describes the intestinal tract?
   a) An open system with a chemically defined medium
   b) An open system with a complex medium
   c) A closed system with a chemically defined medium
   d) A closed system with a complex medium
5. *E. coli,* a facultative anaerobe, is grown for 24 hours on the same type of solid medium, but under two different conditions: one aerobic, the other anaerobic. The size of the colonies would be
   a) the same under both conditions.
   b) larger when grown under aerobic conditions.
   c) larger when grown under anaerobic conditions.
6. The generation time of a bacterium was measured at two different temperatures. Which results would be expected of a thermophile?
   a) 20 minutes at 10°C; 220 minutes at 37°C
   b) 220 minutes at 10°C; 20 minutes at 37°C
   c) no growth at 10°C; 20 minutes at 37°C
   d) 20 minutes at 45°C; 220 minutes at 65°C
   e) 220 minutes at 37°C; 20 minutes at 65°C
7. Which of the following is *false*?
   a) *E. coli* grows faster in nutrient broth than in glucose-salts medium.
   b) Organisms require nitrogen to make amino acids.
   c) Some eukaryotes can fix N_2.
   d) An organism that grows on ham is halotolerant.
   e) Blood agar is used to detect hemolysis.
8. If the pH indicator were left out of MacConkey agar, the medium would be
   a) complex.
   b) differential.
   c) defined.
   d) defined and differential.
   e) complex and differential.
9. A soil sample is placed in liquid and the number of bacteria in the sample determined in two ways: (1) colony count and (2) direct microscopic count. How would the results compare?
   a) Methods 1 and 2 would give approximately the same results.
   b) Many more bacteria would be estimated by method 1.
   c) Many more bacteria would be estimated by method 2.
   d) Depending on the soil sample, sometimes method 1 would be higher and sometimes method 2 would be higher.
10. If the concentration of *E. coli* in a broth is between 10^4 and 10^6 cells per mL, the best way determine the precise number of living cells in the sample would be to
    a) use a counting chamber.
    b) plate out an appropriate dilution of the sample on nutrient agar.
    c) determine cell number by using a spectrophotometer.
    d) Any of these three methods would be satisfactory.
    e) None of these three methods would be satisfactory.

Applications

1. You are a microbiologist working for a pharmaceutical company and discover a new metabolite that can serve as a medication. You now must oversee its production. What are some factors you must consider if you need to grow extremely large (5,000 L) cultures of bacteria?
2. High-performance boat manufacturers know that microbes can collect on a boat, ruining its hydrodynamic properties. A boat-manufacturing facility recently hired you to help with this problem because of your microbiology background. What strategies other than routine cleaning would you pursue to come up with a long-term remedy for the problem?

Critical Thinking

1. This figure shows a growth curve plotted on a nonlogarithmic, or linear, scale. Compare this with figure 4.6. In both figures, the number of cells increases dramatically during the log or exponential phase. In this phase, the cell number increases more and more rapidly (this effect is more apparent in the accompanying figure). Why should the increase be speeding up?

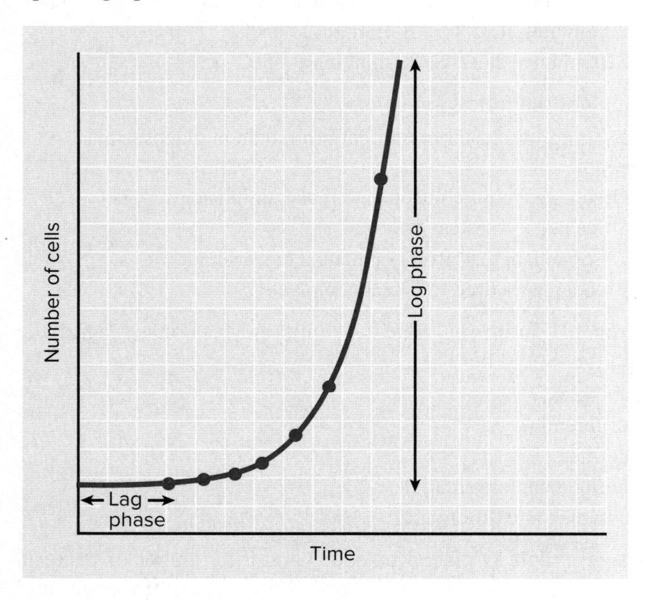

2. In question 1, how would the curve appear if the availability of nutrients were increased?

5 Control of Microbial Growth

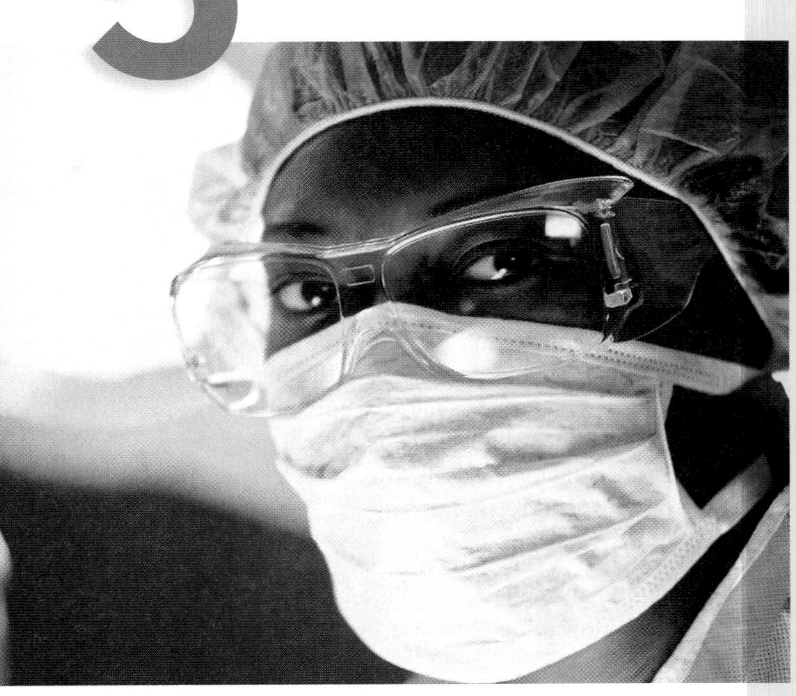

Medical settings warrant a high level of microbial control. ©Arthur Tilley/Getty Images

KEY TERMS

Antiseptic A disinfectant nontoxic enough to be used on skin.

Aseptic Technique Procedures that help prevent the accidental introduction of unwanted microbes.

Bactericide Substance that kills bacteria.

Bacteriostatic Prevents the growth of bacteria, but does not kill them.

Disinfectant A chemical that destroys many, but not all, microbes.

Pasteurization Brief heat treatment that reduces the number of spoilage organisms and destroys disease-causing microbes.

Preservation Inhibition of microbial growth to delay spoilage.

Sterile Free of all viable microbes, including endospores and viruses.

Sterilization The destruction or removal of all microbes through physical or chemical means.

A Glimpse of History

The *British Medical Journal* stated that the British physician Joseph Lister (1827–1912) "saved more lives by the introduction of his system than all the wars of the 19th century together had sacrificed." Lister revolutionized surgery by introducing methods that prevent wounds from becoming infected.

Impressed with Pasteur's work on fermentation (said to be caused by "minute organisms suspended in the air"), Lister wondered if "minute organisms" might also be responsible for the pus that formed in infected wounds. He then experimented by applying carbolic acid, a toxic compound, directly onto damaged tissues and found it prevented infections. Carbolic acid wound dressings became standard in his practice, and he took pride in the fact that his patients no longer developed gangrene. His work also provided impressive evidence for the Germ Theory of Disease, even though microorganisms specific for various diseases were not identified for another decade.

Later, Lister improved his methods even further by using surgical procedures that excluded bacteria from wounds. These procedures included sterilizing instruments before use and maintaining a clean environment in the operating room. Preventing infection was preferable to killing the bacteria after they entered wounds because the toxic effects of the disinfectant on the wound could be avoided.

The oral antiseptic Listerine was named for Joseph Lister in 1879 when it was introduced as a surgical antiseptic.

Up until the late nineteenth century, patients having even minor surgeries were at great risk of developing fatal infections due to unsanitary medical practices and hospital conditions. Physicians did not know that their hands could pass diseases from one patient to the next. Nor did they understand that airborne microscopic organisms could infect open wounds. Fortunately, today's modern hospitals use strict procedures to avoid microbial contamination, allowing most surgeries to be performed with relative safety.

Microbial growth affects more than our health. Manufacturers of a wide variety of goods recognize that microorganisms can reduce product quality. Their growth can lead to undesirable changes in the safety, appearance, taste, odor, or function of products ranging from food to lumber.

This chapter covers methods to control microbial growth on inanimate objects and some body surfaces. Most of these approaches can damage all forms of life. In contrast, antibiotics and other antimicrobial medications are specifically toxic to microbes, which is why they are so valuable in treating infectious diseases. They will be discussed in chapter 20.

5.1 ■ Approaches to Control

Learning Outcomes

1. Explain the principles of sterilization, disinfection, pasteurization, decontamination, sanitization, and preservation.
2. Compare and contrast the methods used to control microbial growth in daily life, healthcare settings, microbiology laboratories, food and food production facilities, water treatment facilities, and other industries.

The processes used to control microorganisms are either physical or chemical, or a combination of both. Physical methods include heat treatment, irradiation, filtration, and mechanical removal (washing). Chemical methods use any of a variety of antimicrobial chemicals that damage cells by denaturing proteins, altering DNA, or disrupting cell membranes. The method chosen depends on the circumstances and level of control required.

Principles of Control

Sterilization is the removal or destruction of all microorganisms and viruses on or in a product. Microbes can be removed by filtration or destroyed using heat, certain chemicals, irradiation, or high pressure. Destruction of microorganisms means they cannot be "revived" to multiply even when transferred from the sterilized product to an ideal growth medium. A **sterile** item is free of all viable microbes, including endospores. Note, however, that the term *sterile* does not consider prions. These infectious protein particles are not destroyed by standard sterilization procedures. ⏮ endospores, ⏮ prions

Disinfection is the elimination of most or all pathogens on or in a material. In practice, the term generally implies the use of antimicrobial chemicals. Unlike with sterilization, some viable microbes may remain after disinfection. **Disinfectants** are chemicals used for disinfecting inanimate objects. They are toxic to many forms of life, and therefore biocides (*bio* means "life," and *cida* means "to kill"). As they typically target harmful microbes, they are often called **germicides. Bactericides** kill bacteria, **fungicides** kill fungi, and **virucides** inactivate viruses. **Antiseptics** are antimicrobial chemicals non-toxic enough to be used on skin or other body tissue. These are routinely used to decrease bacterial numbers on skin before invasive procedures such as surgery. ⏮ pathogen

Decontamination reduces the number of pathogens to a safe level. The treatment can be as simple as thorough washing, or it may involve the use of heat or disinfectants. **Sanitization** generally implies a process that substantially reduces the microbial population to meet accepted health standards that minimize the spread of disease. Most people expect a sanitized object to look clean. This term does not indicate any specific level of control.

Preservation is the process of delaying spoilage of perishable products. Storage conditions can be adjusted to slow microbial growth. For example, food stored in a refrigerator will take longer to spoil than food left in a pantry. Alternatively, chemical preservatives can be added to a product. These are **bacteriostatic,** meaning they inhibit the growth of bacteria but do not kill them. **Pasteurization** is a brief heat treatment usually applied to food items that reduces the number of spoilage organisms and destroys pathogens without changing the characteristics of the product.

Situational Considerations

Methods used to control microbial growth vary greatly depending on the situation and level of control required (**figure 5.1**). Control measures adequate for routine circumstances of daily life might not be sufficient for situations found in hospitals, microbiology laboratories, foods and food production facilities, water treatment facilities, and other industries.

Daily Life

Washing and scrubbing with soaps and detergents is generally sufficient to control microbes in routine situations. Soap itself does not destroy many organisms. It simply aids in the mechanical removal of microbes, including most pathogens, as well as dirt, organic material, and some skin cells. Regular handwashing and bathing do not harm the beneficial normal skin microbiota, which reside more deeply on underlying layers of skin cells and in hair follicles. Other methods used to control microorganisms in daily life include cooking foods, cleaning surfaces, and refrigeration.

MicroByte

Thorough handwashing with soap and water is the most important step in stopping the spread of many infectious diseases.

Hospitals and Other Healthcare Facilities

Minimizing the numbers of microbes in healthcare settings is particularly important because of the danger of **healthcare-associated infections (HAIs).** Patients in healthcare facilities, particularly hospitals, are often more susceptible to infectious agents because of their weakened condition. In addition, patients may undergo invasive procedures such as surgeries, cutting the intact skin that would otherwise help prevent infection. Finally, pathogens are more likely to be found in healthcare settings because of the high concentration of patients with infectious disease. These patients can shed pathogens in their feces, urine, respiratory droplets, or other body secretions. ⏭ healthcare-associated infections

Healthcare facilities must be especially careful to control microorganisms in the operating rooms (**figure 5.2**). Instruments used in invasive surgical procedures must be sterile to avoid introducing even normally harmless microbes into deep body tissue where they could easily cause infection.

Prions are a relatively new concern for healthcare facilities. Fortunately, disease caused by these agents is thought to be exceedingly rare—less than 1 case per 1 million persons per year. Healthcare facilities, however, must take special precautions when handling tissue that may be contaminated with prions, because these infectious particles are very difficult to destroy.

FIGURE 5.1 Situations That Warrant Different Levels of Microbial Control **(a)** Daily home life; **(b)** foods and food production facilities; **(c)** water treatment facilities; **(d)** hospitals; **(e)** other industries. a: ©Anna Bryukhanova/Getty Images; b: ©Don Tremain/Getty Images; c: ©Marka/Alamy Stock Photo; d: ©Ryan McVay/Getty Images; e: ©Javier Larrea/Getty Images

? **Why would a hospital require a more stringent level of microbial control than daily home life?**

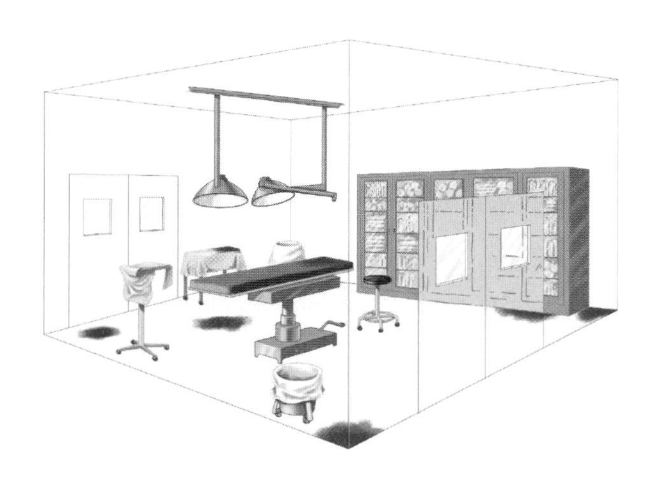

FIGURE 5.2 Contaminated Operating Room The reddish areas indicate places that were contaminated with *Pseudomonas aeruginosa* after treatment of a burn patient. Patient care rooms must be thoroughly cleaned after each use to prevent spread of infection to others.

? **Will the operating room be sterilized between patients? Explain.**

FOCUS ON A CASE 5.1

The 88-year-old patient was asleep in the emergency department with several family members sitting nearby. She had broken her wrist in a fall, but did not call for help until two days later. Doctors thought that the wrist would heal with no complications, but the patient had become dehydrated and was nauseated after taking pain medication. They wanted to admit her to the hospital, but the patient's granddaughter resisted, fearing that the woman would be exposed to dangerous infections. A doctor and nurse entered the room, passed by the germicide dispenser near the door, and approached the patient. The granddaughter stood up and said "Don't go near my grandmother until you've washed your hands!"

1. Why was the granddaughter concerned about exposure to infections in the hospital?
2. Would use of the germicide be as effective at preventing spread of infection as handwashing?
3. The doctor was wearing a necktie. Should the granddaughter be concerned about that?

Discussion

1. Healthcare facilities house a high concentration of patients with infectious disease, including some caused by microbes that have become resistant after repeated exposure to antimicrobial medications. Although hospitals use rigorous cleaning techniques to control microbial growth, vulnerable patients may still have increased risk of infection. Up to 5% of patients who enter a hospital are infected during their stay.
2. Handwashing with soap and water is the best way to avoid spread of disease. Proper handwashing physically removes microorganisms from the skin and sends them down the drain. Experts recommend singing *Happy Birthday* twice while washing so that the process lasts at least 20 seconds. When handwashing is not practical, use of an alcohol-based germicide can effectively kill most microbes on the hands.
3. Possibly. The necktie could carry microorganisms from one patient to another if it is allowed to drape over a patient during examination. Many doctors wear bow ties for this reason.

Microbiology Laboratories

Microbiology laboratories routinely work with microbial cultures and consequently must use rigorous methods to control microorganisms. To ensure that cultures remain pure, all media and instruments that contact the culture must first be sterilized to avoid contaminating the culture with environmental microbes. All materials used to grow microorganisms must again be treated before disposal to avoid contaminating workers and the environment. The use of specific methods to prevent microorganisms from contaminating an environment is called **aseptic technique.** Although all microbiology laboratory personnel must use these measures, those who work with known pathogens must be even more careful. ◀◀ aseptic technique

The Centers for Disease Control and Prevention (CDC) has established precaution guidelines for laboratories working with microorganisms. These are known as biosafety levels (BSLs) and range from BSL-1 (for work with microbes not known to cause disease in healthy people) to BSL-4 (for work with deadly pathogens for which no vaccine or specific treatment exists).

Foods and Food Production Facilities

Foods and other perishable products retain their quality longer when contaminating microbes are destroyed, removed, or inhibited. Heat treatment is the most common and reliable method used to kill microbes, but it can alter the flavor and appearance of the products. Irradiation and high pressure can be used to destroy microbes without altering the product. Chemical additives can be used to prevent microbial growth, but the risk of toxicity must always be a concern. The Food and Drug Administration (FDA) regulates many of these options to ensure safety.

Food-processing facilities need to keep surfaces relatively free of microorganisms to avoid contamination. If machinery used to grind meat is not cleaned properly, for example, it can create an environment in which bacteria multiply, eventually contaminating large quantities of product.

Water Treatment Facilities

Water treatment facilities need to ensure that drinking water is free of pathogenic microbes. Chlorine has traditionally been used to disinfect water, saving hundreds of thousands of lives by preventing the spread of waterborne illnesses such as cholera. Chlorine and other disinfectants, however, can react with naturally occurring chemicals in the water (such as organic matter) to form **disinfection by-products (DBPs).** Some of these have been linked to long-term health risks. In addition, certain pathogens, particularly *Cryptosporidium parvum,* a cause of diarrhea, can survive traditional disinfection procedures. To address these problems, water treatment regulations now require facilities to minimize the level of both DBPs and *C. parvum* in treated water. ▶▶ *Cryptosporidium parvum*

Other Industries

Many diverse industries have specialized concerns regarding microbial growth. Manufacturers of pharmaceuticals, cosmetics, deodorants, or any other product that will be ingested, injected, or applied to the skin must avoid microbial contamination that could affect the product's quality or safety.

MicroAssessment 5.1

The methods used to control microbial growth depend on the situation and the level of control required.

1. How is sterilization different from disinfection?
2. Why is it important for a microbiology laboratory technician to use aseptic technique?
3. How would a company that makes lipsticks determine if a batch had been contaminated by microorganisms during the production process?

5.2 ■ Selecting an Antimicrobial Procedure

Learning Outcome

3. Explain how the type and number of microbes, environmental conditions, risk for infection, and composition of the item influence the selection of an antimicrobial procedure.

Selecting an effective antimicrobial procedure is complicated by the fact that every procedure has disadvantages that limit its use. An ideal, multipurpose, non-toxic method simply does not exist. The ultimate choice depends on many factors, including the type and number of microbes to be controlled, environmental conditions, risk for infection, and the composition of the item to be treated.

Type of Microbes

One of the most critical considerations in selecting an antimicrobial procedure is the composition of the microbial population to be controlled. Products contaminated with microbes highly resistant to killing require a more rigorous treatment. Highly resistant microbes include the following:

- **Bacterial endospores.** The endospores of *Bacillus, Clostridium,* and related genera are the most resistant form of life typically encountered. Only extreme heat or chemical treatment ensures their complete destruction. ⏮ endospores
- **Protozoan cysts and oocysts.** Cysts and oocysts are stages in the life cycle of certain intestinal protozoan pathogens such as *Giardia lamblia* and *Cryptosporidium parvum.* These disinfectant-resistant forms appear in the feces of infected animals, including humans, and can cause diarrheal disease if ingested. Unlike endospores, protozoan cysts and oocysts are easily destroyed by boiling. ⏭ *Cryptosporidium parvum*, ⏭ *Giardia lamblia*
- ***Mycobacterium* species.** The waxy cell walls of mycobacteria make them resistant to many chemical treatments. Because of this, stronger, more toxic chemicals must be used to disinfect environments that may contain *Mycobacterium* species, including *M. tuberculosis,* the cause of tuberculosis. ⏭ tuberculosis
- ***Pseudomonas* species.** These common environmental organisms are not only resistant to some disinfectants, but in some cases actually grow in them. *Pseudomonas* species can cause serious healthcare-associated infections. ⏭ *Pseudomonas* infections
- **Non-enveloped viruses.** Viruses such as poliovirus that lack a lipid envelope are more resistant to disinfectants. Conversely, enveloped viruses, such as HIV, tend to be very sensitive to these chemicals. ⏭ non-enveloped viruses, ⏭ enveloped viruses

Number of Microorganisms

The time it takes for heat or chemicals to kill a microbial population depends in part on the number of cells present. It takes more time to kill a large population than to kill a small population, because only a fraction of organisms die during a given time interval. When using a disinfectant, it is important to apply the chemical long enough to kill the organisms. Removing most of the organisms by washing or scrubbing can minimize the time necessary to sterilize or disinfect a product.

In the commercial canning industry, the **decimal reduction time,** or **D value,** is the time required to kill 90% of a bacterial population under specific conditions (**figure 5.3**). For example, if 90% of a bacterial population of 10,000 (10^4) organisms is killed at a given temperature during the first 5 minutes, then 1,000 (10^3) organisms remain. Approximately 90% of those will be killed during the next 5 minutes, leaving 100 (10^2) organisms. After another 5 minutes, only 10 (10^1)

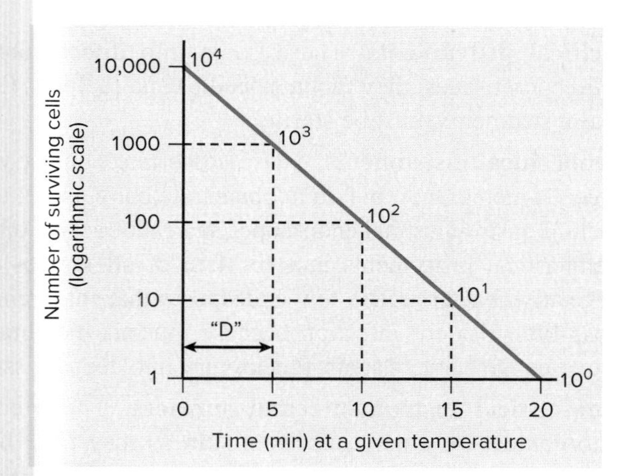

FIGURE 5.3 D Value The decimal reduction time (D value) is the time it takes to kill 90% of a microbial population under specified conditions. In this example, the D value is 5 minutes. A population of 10,000 (10^4) organisms would be reduced to only one (10^0) organism after 4 D values, or 20 minutes. A larger population requires more time for elimination under the same conditions.

How many organisms die in the interval from 10–15 minutes after application of heat?

organisms remain. After another 5 minutes, only one (10^0) organism remains. One D value reduces the number of cells by one order of magnitude. In this example, it would take 20 minutes (4 D values) to reduce a population of 10^4 cells to only one survivor.

Environmental Conditions

Dirt, grease, and body fluids such as blood can interfere with heat penetration and the action of chemical disinfectants. For this reason alone, it is important to thoroughly clean items before disinfection or sterilization.

Microorganisms in a biofilm are more resistant to chemical disinfectants than their free-living counterparts. The resistance is thought to be at least partly due to the protective nature of the accumulated extrapolymeric substances (EPS; see figure 4.3). Because of this, materials that foster the development of biofilms should be scrubbed before being treated with a chemical disinfectant.

Temperature and pH influence the effectiveness of disinfection techniques on microbial death rates. A solution of sodium hypochlorite (household bleach), for example, can kill a suspension of *M. tuberculosis* at a temperature of 55°C in half the time it would take if the suspension were held at 50°C. The hypochlorite solution is even more effective at a low pH.

Risk for Infection

To guide the selection of germicidal procedures, medical instruments are categorized according to their risk for transmitting infectious agents. Those that pose greater threats require more rigorous germicidal procedures before use. The categories are as follows:

- **Critical instruments.** These come into direct contact with body tissues; they include needles and scalpels. Critical instruments must be sterile.
- **Semicritical instruments.** These come into contact with mucous membranes, but do not penetrate body tissue; they include gastrointestinal endoscopes and endotracheal tubes. Semicritical instruments must be free of all viruses and vegetative bacteria. The few endospores that may remain pose little risk for infection because mucous membranes are effective barriers against their entry into deeper tissue.
- **Non-critical instruments and surfaces.** These come into contact only with unbroken skin so they pose little risk for infection. Countertops, stethoscopes, and blood pressure cuffs are examples of non-critical items.

Composition of the Item

Some sterilization and disinfection procedures are inappropriate for certain types of material. For example, heat treatment can damage many types of plastics and other materials. Irradiation provides an alternative to heat, but the process damages some types of plastics. Moist heat (such as boiling water) and liquid chemical disinfectants cannot be used to treat moisture-sensitive material.

MicroAssessment 5.2

The types and numbers of microorganisms initially present, environmental conditions, the potential risks associated with use of the item, and the composition of the item must all be considered when determining which sterilization or disinfection procedure to employ.

4. Describe three groups of microorganisms that are resistant to certain chemical treatments.
5. Why must critical instruments be sterile, whereas semicritical instruments need only be free of viruses and vegetative bacteria?
6. Is it safe to say that if all bacterial endospores have been killed, then all other medically important microorganisms also have been killed?

5.3 ■ Using Heat to Destroy Microorganisms and Viruses

Learning Outcomes

4. Compare and contrast the methods and outcomes of boiling, pasteurization, sterilization using pressurized steam, the commercial canning process, and dry heat.
5. Explain how heat can kill microbes.

Heat treatment is one of the most useful methods of microbial control because it is reliable, safe, relatively fast, and inexpensive, and it does not introduce potentially toxic substances into materials. Some heat-based methods sterilize the product, whereas others decrease the number of microbes. **Table 5.1** summarizes the characteristics of heat treatment and other physical methods of control.

Moist Heat

Moist heat destroys microbes by irreversibly denaturing their proteins. Examples of moist heat treatment include boiling, pasteurization, and pressurized steam.

Boiling

Boiling (100°C at sea level) easily destroys most microorganisms and viruses. Because of this, drinking water that might be contaminated during floods or other emergency situations should be boiled for at least 5 minutes. Boiling is not a method of sterilization, however, because endospores can survive the process.

TABLE 5.1 Physical Methods Used to Destroy Microorganisms and Viruses

Method	Characteristic	Use
Moist Heat	Denatures proteins. Relatively fast, reliable, safe, and inexpensive.	
Boiling	Boiling for 5 minutes destroys most microorganisms and viruses; a notable exception is endospores.	Boiling for at least 5 minutes can be used to treat drinking water.
Pasteurization	Significantly decreases the numbers of heat-sensitive microorganisms, including spoilage microbes and pathogens (except endospore formers).	Milk is pasteurized by heating it to 72°C for 15 seconds. Juices are also routinely pasteurized.
Pressurized steam (autoclaving)	Typical treatment is 121°C/15 psi for 15 minutes or longer, a process that destroys endospores.	Widely used to sterilize microbiological media, laboratory glassware, surgical instruments, and other items that steam can penetrate. The canning process renders foods commercially sterile.
Dry Heat	Destroys cell components and denatures proteins.	
Incineration	Burns cell components to ashes.	Flaming of wire inoculating loops. Also used to destroy medical wastes and contaminated animal carcasses.
Dry heat ovens	Less efficient than moist heat, requiring longer times and higher temperatures.	Laboratory glassware is sterilized by heating at 160°C to 170°C for 2 to 3 hours. Oils and dry materials are also sterilized in ovens.
Filtration	Filter retains microbes while letting the suspending fluid or air pass through small holes.	
Filtration of fluids	Various pore sizes are available; 0.2 μm is commonly used to remove bacteria.	Used for beer and wine, and to sterilize some heat-sensitive medications.
Filtration of air	HEPA filters are used to remove microbes that have a diameter of 0.3 μm or greater.	Used in biological safety cabinets, specialized hospital rooms, and airplanes. Also used in some vacuum cleaners and home air purification units.
Irradiation	Type of cell damage depends on the wavelength of the radiation.	
Ionizing radiation	Destroys DNA and possibly damages cytoplasmic membranes. Produces reactive molecules that damage other cell components. Items can be sterilized even after packaging.	Used to sterilize heat-sensitive materials including medical equipment, disposable surgical supplies, and drugs such as penicillin. Also used to destroy microbes in spices, herbs, and approved types of produce and meats.
Ultraviolet radiation	Damages DNA. Penetrates poorly.	Used to destroy microbes in the air and drinking water, and to disinfect surfaces.
High Pressure	Treatments up to 120,000 psi may denature proteins and alter the permeability of the cell. Products retain color, flavor, and nutrients.	Used to extend the shelf life of certain food products such as guacamole.

Pasteurization

Louis Pasteur developed the brief heat treatment we now call pasteurization as a way to prevent spoilage of wine without changing its flavor. Today, pasteurization is still used to destroy heat-sensitive spoilage organisms in foods and beverages, increasing the product's shelf life without significantly altering its quality. More important, pasteurization is widely used to destroy pathogens in milk and juices, protecting consumers from diseases such as brucellosis, salmonellosis, and typhoid fever. ⏭ **food spoilage**

Today, most pasteurization is by the **high-temperature–short-time (HTST) method.** With this treatment, milk is heated to 72°C and held for 15 seconds, but the parameters must be adjusted for different products. For example, ice cream is rich in fats, so it is pasteurized at 82°C for about 20 seconds.

Boxed juices and the single-serving containers of half-cream served in restaurants require no refrigeration because they are treated using the **ultra-high-temperature (UHT) method** of processing ("ultra-pasteurization"). This destroys microorganisms that can grow under normal storage conditions. The UHT process for milk requires rapidly heating the milk to 140°C, holding it at that temperature for a few seconds, and then quickly cooling it. The product is then aseptically packaged in sterile containers.

MicroByte

Clothes can be pasteurized by regulating the temperature in a washing machine.

Sterilization Using Pressurized Steam

Heat- and moisture-tolerant items such as surgical instruments, most microbiological media, and reusable glassware are sterilized using an **autoclave.** Water in a chamber in the autoclave is heated to form steam, causing the pressure in the chamber to increase (**figure 5.4**). The higher pressure, in turn, increases the temperature at which steam forms. Steam at atmospheric pressure never exceeds 100°C, but steam at an additional 15 psi (pounds per square inch) is 121°C, a temperature that kills even endospores. The pressure itself plays no direct role in the killing.

Typical conditions for sterilization are 15 psi and 121°C for 15 minutes. Longer treatment is necessary for large volumes because it takes more time for heat to completely penetrate the substance. For example, 4 L of nutrient broth in a flask takes longer to sterilize than the same volume distributed into tubes.

Other autoclave conditions can be used for specific purposes. When rapid processing is important, such as when sterile instruments must always be available in an operating room, flash sterilization with a higher temperature for a shorter time can be used. To destroy prions, autoclaving at a temperature of 132°C for 1 hour is thought to be effective.

Autoclaving is consistently effective in sterilizing most objects, if done correctly. The temperature and pressure gauges of the autoclave should both be monitored to ensure proper operating conditions. It is also critical that steam enter items to displace air. Because of this, long, thin containers should be placed on their sides, and containers should never be closed tightly.

Tape that contains a heat-sensitive indicator, which turns black at a high temperature, should be attached to items before autoclaving. This provides a visual signal that the items have been processed (**figure 5.5a**). A changed indicator, however, does not mean that the object is sterile, only that it has been heated.

Biological indicators are used to ensure that an autoclave is working properly (**figure 5.5b**). A tube containing the heat-resistant endospores of *Geobacillus stearothermophilus* can be placed near the center of an item. After that item is autoclaved, the endospores are mixed with a growth medium by manually crushing a container within the tube. If the medium changes color during incubation, microbial growth occurred, indicating an unsuccessful process.

The Commercial Canning Process

Commercial canning uses an industrial-size autoclave (called a retort). The canning process is designed to ensure that endospores of *Clostridium botulinum* are destroyed. This is critical because surviving spores can germinate in canned foods such as vegetables and meats. The resulting vegetative cells can grow in low-acid anaerobic conditions and produce botulinum toxin, one of the most potent toxins known. ⏭ botulism

In destroying endospores of *C. botulinum,* the canning process also kills all other organisms that grow under normal storage conditions. Canned foods are **commercially sterile,** meaning that the endospores of some thermophiles may survive. These are usually not a concern, however, because they grow only at temperatures well above those of normal storage.

Several factors dictate the time and temperature of the canning process. First, as discussed earlier, the higher the temperature, the shorter the time needed to kill all organisms.

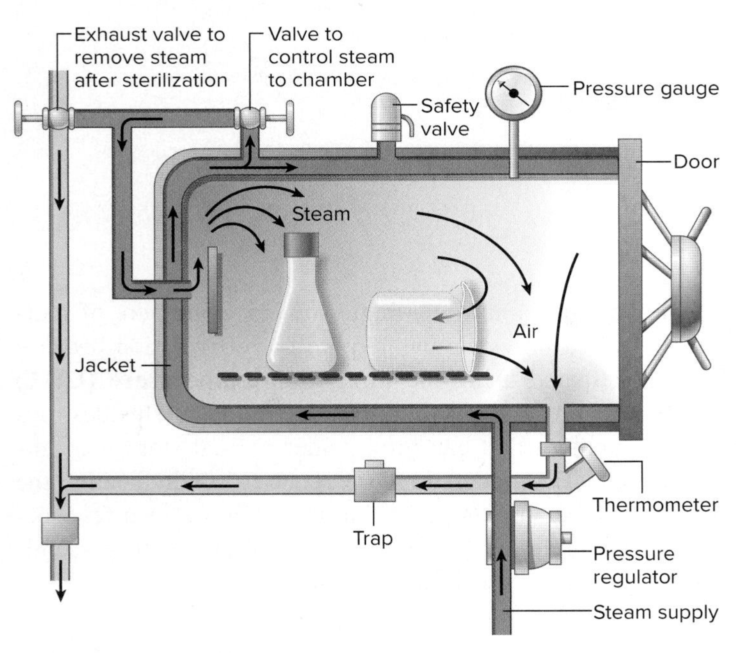

FIGURE 5.4 Autoclave Steam first travels in an enclosed layer, or jacket, surrounding the chamber. It then enters the autoclave, displacing the air downward and out through a port in the bottom of the chamber.

? Why is autoclaving more effective than boiling?

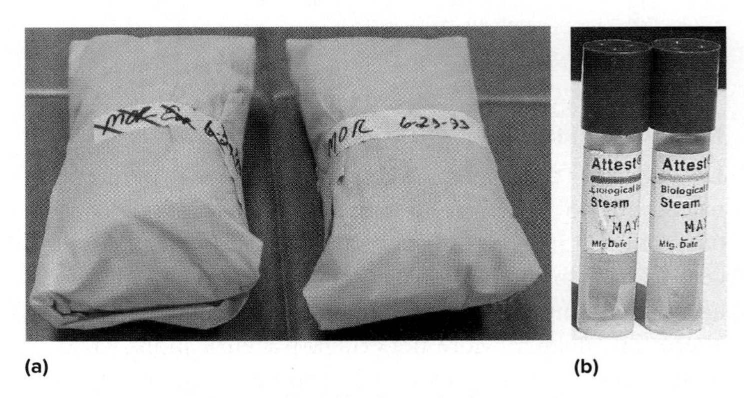

(a) **(b)**

FIGURE 5.5 Indicators Used in Autoclaving (a) Chemical indicators. The pack on the left has been autoclaved. Diagonal marks on the tape have turned black, indicating that the object was exposed to heat. **(b)** Biological indicators. Following incubation, a change of color to yellow indicates growth of the endospore-forming organism. (a&b): ©Evans Roberts

? Why would a biological indicator be better than other indicators to determine if the sterilization procedure was effective?

Second, the higher the concentration of organisms, the longer the heat treatment required to kill them all. To provide a wide safety margin, the commercial canning process is designed to reduce a population of 10^{12} *C. botulinum* endospores to only one spore. In other words, it is a 12 D process. It is virtually impossible for a food to have this many endospores.

Dry Heat

Dry heat does not penetrate microbes as well as moist heat, so it is less efficient at killing them, and therefore requires longer times and higher temperatures. For example, 200°C for 90 minutes of dry heat has the killing equivalent of 121°C for 15 minutes of moist heat.

Incineration burns the cell components to ashes. In microbiology laboratories, the wire loops continually reused to transfer bacterial cultures are sterilized by flaming—heating them in a flame until they are red hot. Alternatively, they can be heated to the same point in a benchtop incinerator designed for this purpose. Incineration is also used to destroy medical wastes and contaminated animal carcasses.

Temperatures achieved in hot air ovens destroy cell components and irreversibly denature proteins. Glass Petri dishes and glass pipettes are sterilized in ovens with non-circulating air at temperatures of 160°C to 170°C for 2 to 3 hours. Oils and dry materials such as powders are also sterilized in hot ovens. Ovens with a fan that circulates the hot air can sterilize in a shorter time because they transfer heat more efficiently.

MicroAssessment 5.3

Moist heat such as boiling water destroys most microbes. Pasteurization significantly reduces the numbers of heat-sensitive organisms, thereby slowing spoilage and killing pathogens. Autoclaves use pressurized steam to achieve high temperatures that kill microbes, including endospores. Dry heat takes longer than moist heat to kill microbes. Microorganisms are killed when cell components are destroyed or proteins are irreversibly denatured.

7. Why is it important for the commercial canning process to destroy the endospores of *Clostridium botulinum*?
8. What is the advantage of pasteurizing a food over boiling it?
9. Why does a cell die when its proteins are denatured?

5.4 ■ Using Other Physical Methods to Remove or Destroy Microbes

Learning Outcomes

6. Describe how membrane filters, depth filters, and HEPA filters are used to remove microbes.
7. Compare and contrast the use of gamma radiation, ultraviolet radiation, microwaves, and high pressure for destroying microorganisms.

Some materials cannot withstand heat treatment. For these items, other physical methods including filtration, irradiation, and high-pressure treatment can be used to destroy or remove microbes.

Filtration

Recall that membrane filtration, used to determine the number of bacteria in a liquid medium, retains bacteria while allowing the fluid to pass through. That same principle can be used to physically remove microbes from liquids or air.
◀◀ membrane filtration

Filtration of Fluids

Filtration is used extensively to remove organisms from heat-sensitive fluids such as sugar solutions, beer, and wine. Specially designed filtration units are also used by backpackers and campers to remove *Giardia* cysts and bacteria from water.

Paper-thin **membrane filters** or **microfilters** have microscopic pores that allow liquid to flow through while trapping particles too large to pass (**figure 5.6**). A vacuum is commonly used to help pull the liquid through the filter; alternatively, pressure may be applied to push the liquid. The filters are made of nitrocellulose or other polymers and come in a variety of different pore sizes, extending below the dimensions of the smallest known viruses. Pore sizes smaller than necessary should be avoided, however, because they slow the flow. Filters with a pore size of 0.2 micrometers (μm) are commonly used to remove bacteria.

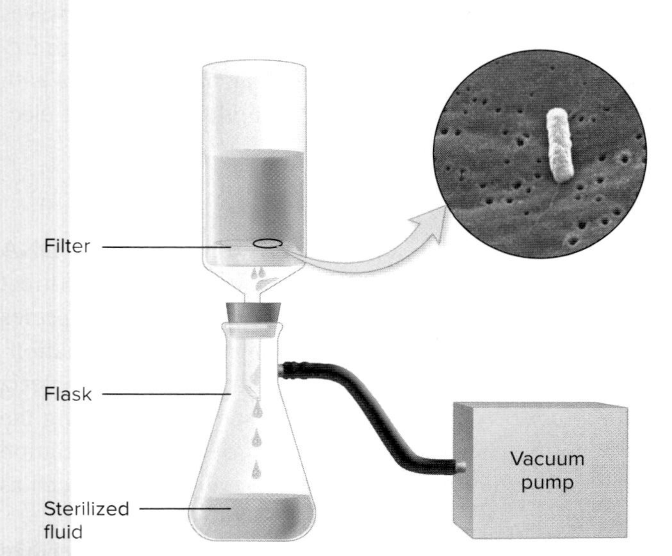

FIGURE 5.6 Sterilization of Fluids Using a Membrane Filter The liquid to be sterilized is pulled through the filter by means of a vacuum produced by a pump. Scanning electron micrograph shows a bacterial cell retained by a membrane filter. Source: Margaret Williams, PhD; Claressa Lucas, PhD; Tatiana Travis, BS/CDC

? Why would a pore size smaller than necessary be a poor choice?

Depth filters trap material within thick porous filtration material such as cellulose fibers. They have complex passages that retain microorganisms while letting the suspending fluid pass through the small holes. The diameter of the passages is often much larger than that of microbes, but the electrical charges on the walls help hold the microbial cells. ⏮ cellulose

Filtration of Air

High-efficiency particulate air (HEPA) filters remove nearly all airborne particles 0.3 μm or larger. These filters are used for keeping microorganisms out of specialized hospital rooms designed for patients extremely susceptible to infection. The filters are also used in biological safety cabinets in which laboratory personnel work with dangerous airborne pathogens such as *Mycobacterium tuberculosis.* A continuous stream of incoming and outgoing air flows through HEPA filters to hold microbes within the cabinet. The cabinets are also used to protect samples from contamination.

MicroByte

HEPA filters are used in passenger aircraft (to remove microbes from the recirculated cabin air) and in vacuum cleaners.

Irradiation

Radio waves, microwaves, visible and ultraviolet (UV) light rays, X rays, and gamma rays are all examples of **electromagnetic radiation.** This form of energy travels in waves. The amount of energy is related to the wavelength, which is the distance from crest to crest (or trough to trough) of a wave. The wavelength is inversely proportional to the frequency, the number of waves per second. Radiation with short waves, and therefore high frequency, has more energy than radiation that has long waves with low frequency. The full range of wavelengths is called the electromagnetic spectrum (**figure 5.7**).

Ionizing Radiation

Ionizing radiation harms cells directly by destroying DNA and damaging cytoplasmic membranes. It also causes indirect damage, reacting with O_2 to produce reactive oxygen species (ROS). Bacterial endospores are among the most radiation-resistant microbial forms, whereas Gram-negative bacteria such as *Salmonella* and *Pseudomonas* species are among the most susceptible. Two important forms of ionizing radiation are gamma rays (emitted from decaying radioisotopes such as cobalt-60) and X rays. ⏮ reactive oxygen species, ⏮ radioisotope

High-energy gamma rays are used extensively to sterilize heat-sensitive materials, including medical equipment, disposable surgical supplies, and drugs such as penicillin. Gamma rays break DNA and RNA strands, thereby killing microbes, but do not change the chemical composition of a product. Irradiation can generally be carried out after packaging.

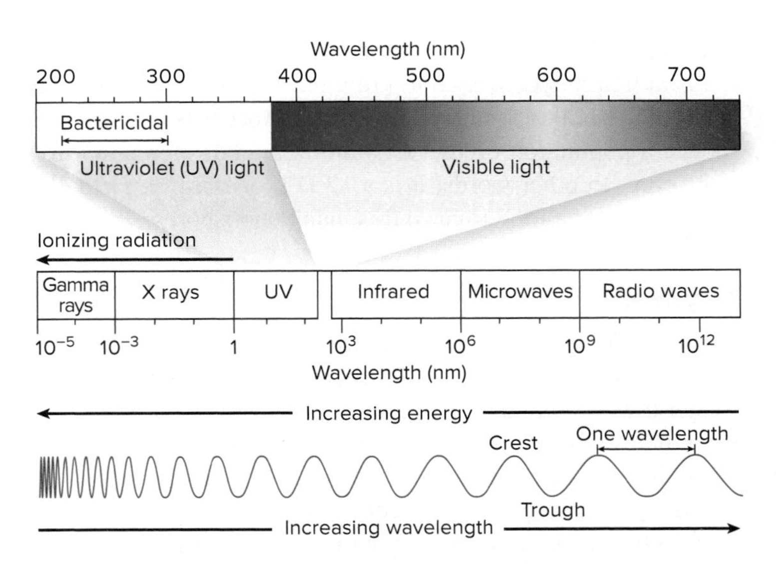

FIGURE 5.7 The Electromagnetic Spectrum Visible wavelengths include the colors of the rainbow.

Which types of radiation are ionizing?

The number of microbes destroyed by irradiating a food product depends on the dose applied. Treatments designed to sterilize food can cause unwanted flavor changes, however, which limits their usefulness. More commonly, food is irradiated as a method of eliminating pathogens and decreasing the numbers of spoilage organisms. For example, it can be used to kill pathogens such as *Salmonella* species in poultry with little or no change in taste of the product.

In the United States, irradiation has been used for many years to control microorganisms on spices and herbs. The FDA (U.S. Food and Drug Administration) has also approved irradiation of meats, fruits, vegetables, and grains. Irradiated foods sold in U.S. stores can be identified by a green radura symbol (**figure 5.8**). Many consumers refuse to accept irradiated products, even though federal and global officials

FIGURE 5.8 Radura Symbol The FDA requires display of this international symbol on foods that have been irradiated. The color and details vary among countries. ©H. S. Photos/Science Source

What types of radiation can be used to control microorganisms in food?

have endorsed the technique. Some people mistakenly believe that irradiated products are radioactive. Others think that irradiation-induced toxins or carcinogens are present in food, even though the FDA monitors the effects of radiation on both food and packaging to ensure that consumption of irradiated food is safe. Another argument raised against irradiation is that it will allow people to ignore other important food-safety practices. Irradiation, however, is intended to complement, not replace, proper food-handling procedures by producers, processors, and consumers.

Ultraviolet Radiation

Ultraviolet light in wavelengths of approximately 220–300 nm destroys microbes by damaging their DNA. Actively multiplying organisms are the most easily killed, whereas bacterial endospores are the most UV-resistant.

Ultraviolet light is used extensively to destroy microbes in the air and drinking water, and to disinfect surfaces. It penetrates poorly, however, which limits its use. Even a thin film of grease on the UV bulb or the material covering microbial cells can make it less effective. Likewise, it cannot be used to destroy microbes in solid substances or turbid liquids. Because most types of glass and plastic screen out ultraviolet radiation, UV light is most effective when used at close range against exposed microorganisms. It must be used carefully because UV rays can also damage the skin and eyes and promote the development of skin cancers.

Microwaves

Microwaves do not affect microorganisms directly, but the heat they generate can be lethal. It is important to remember, however, that microwave ovens often heat food unevenly, so even heat-sensitive cells can sometimes survive the process.

High Pressure

High-pressure processing (HPP) is used to decrease the number of microbes in some commercial food products, such as guacamole and fruit juices, without using high temperatures. The process, which typically uses pressures that may approach 120,000 psi, is thought to destroy microbes by denaturing proteins and altering cell permeability. High-pressure-treated products keep the color and flavor associated with fresh foods, as well as maintaining higher nutrient levels.

MicroAssessment 5.4

Filters can be used to remove microorganisms and viruses from liquids and air. Gamma radiation can be used to sterilize products and to decrease the number of microorganisms in foods. Ultraviolet light can be used to disinfect surfaces and air. Microwaves do not kill microbes directly but destroy them by the heat they generate. Extreme pressure can kill microorganisms.

10. What is the difference between the mechanism of a depth filter and that of a membrane filter?
11. List the advantages of using high-pressure processing for certain foods.
12. Why could sterilization by gamma irradiation be carried out even after packaging?

5.5 ■ Using Chemicals to Destroy Microorganisms and Viruses

Learning Outcomes

8. Describe the difference between sterilants, high-level disinfectants, intermediate-level disinfectants, and low-level disinfectants.
9. Describe the important factors to consider when selecting an appropriate germicidal chemical.
10. Compare and contrast the characteristics and use of alcohols, aldehydes, biguanides, ethylene oxide gas, halogens, metals, ozone, peroxygens, phenolic compounds, and quaternary ammonium compounds as germicidal chemicals.

Germicidal chemicals can be used to disinfect and, in some cases, sterilize. Most react irreversibly with proteins, DNA, cytoplasmic membranes, or viral envelopes. Their mechanisms of action, however, are often poorly understood. They are generally less reliable than heat but useful for treating large surfaces and heat-sensitive items. Some are sufficiently non-toxic to be used as antiseptics.

Effectiveness of Germicidal Chemicals

Numerous different germicidal chemicals are marketed for medical and industrial use under a variety of trade names. Frequently, they contain more than one antimicrobial chemical as well as buffers or other ingredients that can affect their antimicrobial activity. In the United States, the FDA is responsible for ensuring that chemicals used to treat medical devices work as claimed. Most other chemical disinfectants are considered pesticides and, as such, are regulated by the Environmental Protection Agency (EPA). Germicides are grouped according to their potency:

- **Sterilants.** These can destroy all microbes including endospores and viruses. They are also called sporocides. Destruction of endospores usually requires a 6- to 10-hour treatment. Sterilants are used to treat heat-sensitive critical instruments (instruments that come into contact with body tissues).
- **High-level disinfectants.** These destroy all viruses and vegetative microorganisms, but they do not reliably kill endospores. Most are simply sterilants used for short time

periods, not long enough to ensure endospore destruction. They are used to treat semicritical instruments such as gastrointestinal endoscopes.

- **Intermediate-level disinfectants.** These destroy all vegetative bacteria (including mycobacteria), as well as fungi, and most, but not all, viruses. They do not kill endospores even with prolonged exposure. This group of germicides is used to disinfect non-critical instruments such as stethoscopes.
- **Low-level disinfectants.** These destroy fungi, vegetative bacteria except mycobacteria, and enveloped viruses. They do not kill endospores, nor do they always destroy non-enveloped viruses. Intermediate-level and low-level disinfectants are also called general-purpose disinfectants. In hospitals, they are used for disinfecting furniture, floors, and walls.

To perform properly, germicides must be used strictly according to the manufacturer's instructions, especially those for dilution, temperature, and the amount of time they must be in contact with the object being treated. It is extremely important to thoroughly clean objects before they are treated to both decrease the number of microbes present and remove organic material that might interfere with the activity of the chemical.

Selecting the Appropriate Germicidal Chemical

Selecting the appropriate germicide is a complex decision. The chemical needs to kill the target microbes without damaging the material being treated, including human tissue. Ideally, the germicide is not dangerous to handle or inhale, but often the most effective options are quite toxic. Also, people generally object to bad odors, so the smell can be a factor. Some points to consider when choosing a germicide include these:

- **Toxicity.** Germicides are at least somewhat toxic to humans and the environment. Therefore, the benefit of disinfecting or sterilizing an item or surface must be weighed against the risks associated with using the chemical. For example, the risk of being exposed to pathogens in a hospital calls for using the most effective germicides, even considering the potential risks of their use. The microbiological risks associated with typical household and office situations, however, may not justify using many of those same chemicals. To encourage the use of less toxic options, the EPA allows manufacturers to place a specially designed "Safer Choice" label on products considered the least hazardous (**figure 5.9**).
- **Activity in the presence of organic matter.** Hypochlorite (bleach) and many other germicides react with organic matter, losing their effectiveness in doing so. Other chemicals such as phenolics tolerate the presence of some organic matter.
- **Compatibility with the material being treated.** Items such as electrical equipment often cannot withstand liquid chemical germicides, and so gaseous alternatives must be used. Likewise, corrosive germicides such as hypochlorite often damage some metals and rubber.
- **Residue.** Many chemical germicides leave a toxic or corrosive residue. If such a germicide is used to treat an item, the residue must be removed by washing.
- **Cost and availability.** Some germicides are less expensive and more readily available than others. Hypochlorite can easily be purchased in the form of household bleach. In contrast, ethylene oxide gas not only is expensive, but requires a special chamber, which affects its cost and practicality.
- **Storage and stability.** Some germicides are sold as concentrated stock solutions, decreasing the required storage space. The stocks are simply diluted according to the manufacturer's instructions before use. Some have a limited shelf life once prepared.
- **Environmental risk.** Germicides that retain their antimicrobial activity after use can interfere with wastewater treatment systems. The activity of those germicides must be neutralized before disposal. ⏭ wastewater treatment

FIGURE 5.9 Safer Choice Label The EPA allows manufacturers to use this label when all of the ingredients in a product have been tested to assure that they pose the least concern to health and to the environment. ©H. S. Photos/Alamy Stock Photo

? What characteristics of a product would make it safer for the environment?

Classes of Germicidal Chemicals

Germicides are represented in a number of chemical families. Each type has characteristics that make it more or less appropriate for specific uses (**table 5.2**).

Alcohols

Aqueous solutions of 60–80% ethyl or isopropyl alcohol quickly kill vegetative bacteria and fungi. They do not, however, reliably destroy bacterial endospores and some non-enveloped viruses. Alcohol denatures essential proteins such as enzymes and damages lipid membranes. Proteins are more soluble and denature more easily in alcohol mixed with water, which is why the aqueous solutions are more effective than pure alcohol.

Alcohol solutions are commonly used as antiseptics to clean skin before procedures such as injections that break intact skin. In addition, the Centers for Disease Control and Prevention (CDC) recommends that alcohol-based hand sanitizers be used routinely by healthcare personnel as a means to protect patients.

Alcohol solutions are also used as disinfectants for treating instruments and surfaces. They are relatively non-toxic and inexpensive and do not leave a residue. Unfortunately, they evaporate quickly, which limits their contact time and, consequently, their germicidal effectiveness. In addition, they can damage rubber, some plastics, and other material.

Antimicrobial chemicals such as iodine are sometimes dissolved in alcohol. These alcohol-based solutions, called tinctures, can be more effective than the corresponding aqueous solutions.

Aldehydes

The aldehydes glutaraldehyde, *ortho*-phthalaldehyde (OPA), and formaldehyde destroy microorganisms and viruses by forming chemical bonds that cross-link and inactivate

TABLE 5.2 Chemicals Used in Sterilization, Disinfection, and Preservation of Non-Food Substances

Chemical (Examples)	Characteristics	Use
Alcohols (ethanol and isopropanol)	Easy to obtain and inexpensive. Rapid evaporation limits their contact time.	Aqueous solutions of alcohol are used as antiseptics to clean skin in preparation for procedures that break intact skin, and as disinfectants for treating instruments.
Aldehydes (glutaraldehyde, *ortho*-phthalaldehyde and formaldehyde)	Capable of destroying all microbes. Irritating to the respiratory tract, skin, and eyes.	Glutaraldehyde and *ortho*-phthalaldehyde are used to sterilize medical instruments. Formalin is used in vaccine production and to preserve biological specimens.
Biguanides (chlorhexidine)	Relatively low toxicity, destroys a wide range of microbes, adheres to and persists on skin and mucous membranes.	Chlorhexidine is widely used as an antiseptic in soaps and lotions, and is often incorporated into catheters and surgical mesh.
Ethylene Oxide Gas	Easily penetrates hard-to-reach places and fabrics and does not damage moisture-sensitive material. It is toxic, is explosive, and may be carcinogenic.	Commonly used to sterilize medical devices.
Halogens (chlorine and iodine)	Chlorine solutions are inexpensive and readily available; however, organic compounds and other impurities neutralize the activity. Some forms of chlorine may react with organic compounds to form toxic chlorinated products. Iodine is more expensive than chlorine and does not reliably kill endospores.	Solutions of chlorine are widely used to disinfect inanimate objects, surfaces, drinking water, and wastewater. Tincture of iodine and iodophores can be used as disinfectants or antiseptics.
Metals (silver)	Most metal compounds are too toxic to be used medically.	Silver sulfadiazine is used in topical dressings to prevent infection of burns. Some metal compounds are used to prevent microbial growth in industrial processes.
Ozone	This unstable form of oxygen readily breaks down to an ineffective form.	Used to disinfect drinking water and wastewater.
Peroxygens (hydrogen peroxide and peracetic acid)	Readily biodegradable and less toxic than traditional alternatives. The effectiveness of hydrogen peroxide as an antiseptic is limited because the enzyme catalase breaks it down. Peracetic acid is a more potent germicide than hydrogen peroxide.	Hydrogen peroxide is used to sterilize containers for aseptically packaged juices and milk. Peracetic acid is widely used to disinfect and sterilize medical devices.
Phenolic Compounds (triclosan and hexachlorophene)	Wide range of activity, reasonable cost, remains effective in the presence of detergents and organic contaminants, leaves an active antimicrobial residue.	Triclosan is used in a variety of products. Hexachlorophene is highly effective against *Staphylococcus aureus,* but its use is limited because it can cause neurological damage.
Quaternary Ammonium Compounds (benzalkonium chloride and cetylpyridinium chloride)	Non-toxic enough to be used on food preparation surfaces. Inactivated by anionic soaps and detergents.	Widely used to disinfect inanimate objects and to preserve non-food substances.

proteins and nucleic acids. A 2% solution of alkaline glutaraldehyde is one of the most widely used liquid chemical sterilants for treating heat-sensitive medical items. Immersion in this solution for 10–12 hours destroys all microbes, including endospores and viruses. Vegetative cells can be killed with soaking times as short as 10 minutes. Glutaraldehyde is toxic, however, so treated items must be thoroughly rinsed with sterile water before use.

Ortho-phthalaldehyde is a relatively new type of disinfectant that provides an alternative to glutaraldehyde. It requires shorter processing times and is less irritating to eyes and nasal passages, but it stains proteins gray, including those of the skin.

Formaldehyde is used as a gas or as formalin, an aqueous 37% solution. It is an extremely effective germicide that kills most microbes quickly. Formalin is used to kill bacteria and inactivate viruses for use as vaccines. It can also be used to preserve biological specimens. Formalin releases irritating vapors and is a probable carcinogen, limiting its use.

Biguanides

Chlorhexidine, the most effective of a group of chemicals called biguanides, is extensively used in antiseptic products. It stays on skin and mucous membranes, is of relatively low toxicity, and destroys a wide range of microbes, including vegetative bacteria, fungi, and some enveloped viruses. Chlorhexidine is an ingredient in many products, including antiseptic skin creams, disinfectants, and mouthwashes. Chlorhexidine-impregnated catheters and implanted surgical mesh are used in medical procedures. Even tiny chips that slowly release chlorhexidine have been developed; these are inserted into periodontal pockets to treat gum disease. Adverse side effects of chlorhexidine are rare, but severe allergic reactions have been reported.

Ethylene Oxide

Ethylene oxide is an extremely useful gaseous sterilizing agent that destroys all microbes, including endospores and viruses, by reacting with proteins. As a gas, it penetrates well into fabrics, equipment, and implantable devices such as pacemakers and artificial hips. It is particularly useful for sterilizing heat- or moisture-sensitive items such as electrical equipment, pillows, and mattresses. Many disposable laboratory items, including plastic Petri dishes and pipettes, are also sterilized with ethylene oxide.

A special chamber that resembles an autoclave is used to sterilize items with ethylene oxide. This allows careful control of factors such as temperature, relative humidity, and ethylene oxide concentration, all of which influence the effectiveness of the gas. Because ethylene oxide is explosive, it is generally mixed with a non-flammable gas such as carbon dioxide. Under these carefully controlled conditions, objects can be sterilized in 3–12 hours. The toxic and potentially carcinogenic ethylene oxide must then be eliminated from the treated material using heated forced air for 8–12 hours. This is important because the chemical irritates tissues and has a persistent antimicrobial effect, which, in the case of Petri dishes and other items used for growing bacteria, is unacceptable.

Halogens

Chlorine and iodine are common disinfectants that are thought to act by reacting with proteins and other essential cell components.

Chlorine Using adequate concentrations and application times, chlorine can destroy all types of microorganisms (including survival stages such as endospores and cysts) and viruses. Chlorine-releasing compounds such as sodium hypochlorite can be used to disinfect waste liquids, swimming pool water, instruments, and surfaces. At much lower concentrations, they can be used to disinfect drinking water. Chlorine, unlike iodine, is too irritating to skin and mucous membranes to be used as an antiseptic.

Chlorine solutions are inexpensive, readily available disinfectants (**figure 5.10**). An effective solution can easily be made by diluting liquid household bleach (5.25% sodium hypochlorite) 1:100 in water, resulting in a solution of 500 ppm (parts per million) chlorine. This concentration is several hundred times the amount required to kill most pathogens, but usually necessary for fast, reliable effects. In situations when excessive organic material is present, a 1:10 dilution of bleach may be required. This is because chlorine readily reacts with organic compounds and other impurities in water, disrupting its germicidal activity. The use of high concentrations, however, should be avoided when possible, because chlorine is both corrosive and toxic. Diluted bleach deteriorates over time, so fresh solutions should be prepared regularly.

Chlorine is used at very low levels to disinfect drinking water. It is important to remember, however, that these levels are not effective against *Cryptosporidium parvum* oocysts and *Giardia lamblia* cysts, both of which may be found in clear mountain streams and other natural waters. Also, the effective amount of chlorine depends on the amount of organic material in the water.

FIGURE 5.10 Chlorine in the Form of Household Bleach ©McGraw-Hill Education/Jill Braaten, photographer

Why should bleach concentrations higher than necessary be avoided?

The presence of organic compounds is a problem because chlorine can react with some organic compounds to form trihalomethanes, disinfection by-products that are potential carcinogens. ▶▶| *Cryptosporidium parvum*, ▶▶| *Giardia lamblia*

Chlorine dioxide (ClO_2) is a strong oxidizing agent increasingly being used as a disinfectant and sterilant. Its advantage is that it does not react with organic compounds to form trihalomethanes or other toxic chlorinated products. Compressed chlorine dioxide gas, however, is explosive and liquid solutions decompose readily, so it must be generated on-site. It is used to treat drinking water, wastewater, and swimming pools.

Iodine Iodine, unlike chlorine, does not reliably kill endospores, but it may be more effective than chlorine against *Giardia lamblia* cysts. Iodine is used as a tincture, or more commonly as an iodophore, in which the iodine is linked to a carrier molecule that releases free (unbound) iodine slowly. Iodophores are not as irritating to the skin as tincture of iodine, nor are they as likely to stain. Iodophores used as disinfectants contain more free iodine (30–50 ppm) than do those used as antiseptics (1–2 ppm). Stock solutions must be strictly diluted according to the manufacturer's instructions because dilution affects the amount of free iodine available.

Surprisingly, some *Pseudomonas* species survive in concentrated stock solutions of iodophores. The reasons are unclear, but it could be due to inadequate levels of free iodine in concentrated solutions, because iodine may be released from the carrier only with dilution. *Pseudomonas* species also can form biofilms, within which cells are more resistant to chemicals. Healthcare-associated infections can result if a *Pseudomonas*-contaminated iodophore is unknowingly used to disinfect instruments. |◀◀ biofilm

MicroByte

In emergencies, drinking water can be disinfected by adding two drops of plain bleach to a liter of water; let sit for 30 minutes before ingesting.

Metal Compounds

Metal compounds kill microorganisms by combining with sulfhydryl groups (—SH) of enzymes and other proteins, thereby interfering with their function. High concentrations of most metals are too toxic to human tissue to be used medically.

Silver is one of the few metals still used as a disinfectant. Creams containing silver sulfadiazine, a combination of silver and a sulfa drug, are applied topically to prevent infection of second- and third-degree burns. Commercially available bandages with silver-containing pads can be used on minor scalds, cuts, and scrapes. For many years, doctors were required by law to add drops of another silver compound, 1% silver nitrate, into the eyes of newborns. This was to prevent ophthalmia neonatorum, an eye infection caused by *Neisseria gonorrhoeae,* acquired from infected mothers during the birth process. Drops of antibiotics have now largely replaced use of silver nitrate because they are less irritating to the eye. ▶▶| *Neisseria gonorrhoeae*

Compounds of mercury, tin, arsenic, copper, and other metals were once widely used as preservatives in industrial products and to prevent microbial growth in recirculating cooling water. Their extensive use resulted in serious pollution of natural waters, which has prompted strict controls.

Ozone

Ozone (O_3), an unstable form of oxygen, is a powerful oxidizing agent. It decomposes quickly, however, so it must be generated on-site, usually by passing air or O_2 between two electrodes. Ozone is used as an alternative to chlorine for disinfecting drinking water and wastewater.

Peroxygens

Hydrogen peroxide and peracetic acid are powerful oxidizing agents that can be used as sterilants under controlled conditions. They are readily biodegradable and, in normal concentrations of use, appear to be less toxic than the traditional alternatives (ethylene oxide and glutaraldehyde).

Hydrogen Peroxide The effectiveness of hydrogen peroxide (H_2O_2) as a germicide depends in part on whether it is used on living tissue or an inanimate object. This is because all cells that use aerobic metabolism, including tissue cells of the human body, produce catalase, the enzyme that inactivates hydrogen peroxide by breaking it down to water and O_2. Thus, when a solution of 3% hydrogen peroxide is applied to a wound, our cellular enzymes quickly break it down. When the same solution is used on an inanimate surface, however, it overwhelms the relatively low concentration of catalase produced by microbial cells. |◀◀ catalase

Hydrogen peroxide is particularly useful as a disinfectant because it leaves no residue and does not damage stainless steel, rubber, plastic, or glass. Hot solutions are commonly used in the food industry to produce commercially sterile containers for aseptically packaged juices and milk. Vapor-phase hydrogen peroxide is more effective than liquid solutions and can be used as a sterilant.

Peracetic Acid Peracetic acid is an even more potent germicide than hydrogen peroxide. A 0.2% solution of peracetic acid, or a combination of peracetic acid and hydrogen peroxide, can be used to sterilize items in less than 1 hour. It is effective in the presence of organic compounds, leaves no residue, and can be used on a wide range of materials. It has a sharp, strong odor, however, and like other oxidizing agents, it irritates the skin and eyes.

Phenolic Compounds (Phenolics)

Phenol (carbolic acid) is important historically because it was one of the earliest disinfectants (see **A Glimpse of History**), but it has an unpleasant odor and irritates the skin. Phenolics are derivatives of phenol that have greater germicidal activity. Because they are so effective, more dilute and therefore less irritating solutions can be used. Phenolics are the active ingredients in Lysol.

Phenolics destroy cytoplasmic membranes of microorganisms and denature proteins. They kill most vegetative bacteria and, in high concentrations (from 5–19%), many can kill *Mycobacterium* species. They do not, however, reliably inactivate all groups of viruses. The major advantages of phenolic compounds include their wide range of activity, reasonable cost, and ability to remain effective in the presence of detergents and organic contaminants. They also leave an active antimicrobial residue, which in some cases is desirable.

Some phenolics, such as hexachlorophene and triclosan, have been used in surgical scrubs and antiseptic skin preparations. Hexachlorophene is effective against *Staphylococcus aureus,* the leading cause of wound infections. High levels, however, have been associated with symptoms of neurotoxicity, so skin cleansers containing it are available only with a prescription. Triclosan, on the other hand, has been widely used in a variety of personal care products. That changed recently when the FDA ruled that antibacterial soap manufacturers could no longer include triclosan or 18 other chemicals in their products because they had not shown them to be safe and more effective than plain soap. The chemicals are still allowed in other personal care products, including toothpaste, as well as in antibacterial soaps sold to hospitals and restaurants. ⏭ *Staphylococcus aureus*

Quaternary Ammonium Compounds (Quats)

Quaternary ammonium compounds (quats) are cationic (positively charged) detergents non-toxic enough to be used to disinfect food preparation surfaces. Like all detergents, quats have both a charged hydrophilic region and an uncharged hydrophobic region. Because of this, they reduce the surface tension of liquids and help wash away dirt and organic material, facilitating the mechanical removal of microbes from surfaces. Unlike most common household soaps and detergents, however, which are anionic (negatively charged) and repelled by the negatively charged microbial cell surface, quats are attracted to the cell surface. They react with membranes, destroying many vegetative bacteria and enveloped viruses.

Quats are economical, effective, and widely used to disinfect clean inanimate objects and preserve non-food substances. The ingredients of many personal care products (such as shampoos and facial cleansers) include quats such as benzalkonium chloride or cetylpyridinium chloride. They also enhance the effectiveness of some other disinfectants. Cationic soaps and organic material such as gauze, however, can neutralize their activity. In addition, *Pseudomonas,* a troublesome cause of healthcare-associated infections, resists the effects of quats and can even grow in solutions preserved with them.

MicroAssessment 5.5

Germicidal chemicals can be used to disinfect and, in some cases, sterilize, but they are less reliable than heat. They are especially useful for destroying microorganisms and viruses on heat-sensitive items and large surfaces.

13. Describe the factors to consider when selecting a germicide.
14. Explain why it is essential to dilute iodophores properly.
15. Why would a heavy metal be a more serious pollutant than most organic compounds?

5.6 ■ Preservation of Perishable Products

Learning Outcome

11. Compare and contrast chemical preservatives, low-temperature storage, and reducing the available water as methods to preserve perishable products.

Preventing or slowing the growth of microorganisms extends the shelf life of products such as food, medicines, deodorants, cosmetics, and contact lens solutions. Preservatives are often added to these products to prevent or slow the growth of microbes inevitably introduced from the environment. Other common methods of slowing microbial growth include low-temperature storage such as refrigeration or freezing, and reducing available water. These methods are particularly important in preserving foods. ⏭ food spoilage, ⏭ food preservation

Chemical Preservatives

Some of the germicidal chemicals described in the previous section can be used to preserve non-food items. For example, mouthwashes often contain a quaternary ammonium compound, cosmetics may contain a mercury-containing preservative, and leather belts may be treated with one or more phenol derivatives. Food preservatives, however, must be non-toxic for safe ingestion.

Benzoic, sorbic, and propionic acids are weak organic acids sometimes added to foods such as bread, cheese, and juice to prevent microbial growth. In acidic conditions, these chemicals affect cell membrane functions. Although this action inhibits the growth of many microbes, the low pH itself prevents the growth of most bacteria. Therefore, the primary benefit of adding the preservatives is that they inhibit molds, which otherwise grow at acidic pH.

Nitrate and its reduced form, nitrite, serve a dual purpose in processed meats. From a microbiological viewpoint, their most important function is to inhibit the germination of endospores and subsequent growth of *Clostridium botulinum*. If low levels of nitrate or nitrite are not added to cured meats such as ham, bacon, bologna, and smoked fish, *C. botulinum* may grow and produce deadly botulinum toxin. At higher concentrations than required for preservation, nitrate and nitrite react with myoglobin in the meat to form a stable pigment that gives a desirable pink color associated with fresh meat. The preservatives pose a potential hazard, however, because they can be converted to nitrosamines—some of which are carcinogens—by the metabolic activities of intestinal bacteria or during cooking.

MicroByte

Swiss cheese naturally contains propionic acid, and cranberries have benzoic acid.

Low-Temperature Storage

Refrigeration inhibits the growth of many pathogens and spoilage microorganisms by slowing or stopping critical enzyme reactions. Psychrotrophic and some psychrophilic organisms, however, can grow at refrigeration temperatures. ◀◀ psychrophiles

Freezing preserves foods and other products by stopping microbial growth. The ice crystals that form kill some of the microbial cells, but those that survive can grow and spoil foods once thawed.

Reducing the Available Water

Salting and drying decrease the availability of water in food below the limits required for growth of most microbes. The high-solute environment also causes plasmolysis, which damages microbial cells (see figure 4.9). ◀◀ water availability, ◀◀ plasmolysis

Adding Sugar or Salt

Sugar and salt draw water out of cells, dehydrating them. High concentrations of these solutes are added to many foods as preservatives (**figure 5.11**). For example, fruit is made into jams and jellies by adding sugar, and fish and meats are cured by soaking them in salty water, or brine. It is important to note, however, that the food-poisoning bacterium *Staphylococcus aureus* can grow in relatively high-salt conditions. ▶▶| *Staphylococcus aureus*

FIGURE 5.11 Preserved Jams and Jellies ©Ron Bailey/E+/Getty Images

? How does a high concentration of sugar function as a preservative?

Drying Food

Historically, food was dried in the sun, but now food dehydrators are commonly used. Foods that are dried often have added salt, sugar, or chemical preservatives. For example, meat jerkies usually have added salt and sometimes sugar.

Lyophilization (freeze-drying) is widely used for preserving foods such as coffee, milk, meats, and vegetables. In the process of freeze-drying, the food is first frozen and then dried in a vacuum. When water is added to the lyophilized material, it reconstitutes. The quality of the reconstituted product is often much better than that of products dried using ordinary methods. The light weight and stability without refrigeration of freeze-dried foods make them popular with hikers.

Although drying stops microbial growth, it does not reliably kill bacteria and fungi in or on foods. For example, cases of salmonellosis have been traced to dried eggs. Eggshells and even egg yolks may be contaminated with *Salmonella* species from the gastrointestinal tract of the hen. To prevent the transmission of such pathogens, some states have laws requiring dried eggs to be pasteurized before they are sold.

MicroAssessment 5.6

Preservation techniques slow or halt the growth of microorganisms to delay spoilage and prevent foodborne illness.

16. What is the most important function of nitrate in cured meat?

17. What is the primary benefit of adding benzoic acid to foods?

18. Preservation by freezing is sometimes compared to drying. Why would this be so?

FOCUS ON THE FUTURE 5.1

Too Much of a Good Thing?

In our complex world, addressing one challenge may inadvertently lead to the creation of another. Scientists have long been pursuing less toxic alternatives to many traditional germicidal chemicals. For example, chlorhexidine is now generally used in place of hexachlorophene, and glutaraldehyde has largely replaced the more toxic formaldehyde. Meanwhile, ozone and hydrogen peroxide, which are both readily biodegradable, may eventually replace glutaraldehyde. While these less toxic alternatives are better for human health and the environment, their widespread acceptance and use may be contributing to an additional problem: the overuse and misuse of germicidal chemicals. Many products, including soaps, toothbrushes, and even clothing and toys are marketed with the claim of containing antimicrobial ingredients. Bacterial resistance to some of the chemicals included in these products already has been reported.

The issues surrounding the excessive use of antimicrobial chemicals are complicated. On one hand, there is no question that some microorganisms cause disease. Even those that are not harmful to human health can be troublesome because their metabolic end products can ruin the quality of perishable products. Based on that information, it seems wise to destroy microorganisms or inhibit their growth whenever possible. The role of microorganisms in our life, however, is not that simple. Our bodies may harbor more microbial cells than human cells, and this normal microbiota plays an important role in maintaining our health. Excessive use of antiseptics or other antimicrobials may actually predispose a person to infection by damaging the normal microbiota.

An even greater concern is that overuse of germicides will select for germicide-resistant microorganisms, a situation similar to our current problems with antibiotic resistance. By using antimicrobial chemicals indiscriminately, we may eventually make these useful tools obsolete.

The excessive use of germicides may even be contributing to the problem of antibiotic resistance. The efflux pumps that bacteria use to remove certain germicides from the cell may also eject antibiotics; this supports reconsideration of the widespread use of triclosan. ⏭ efflux pumps

Another concern is over the misguided belief that "non-toxic" or "biodegradable" chemicals cause no harm, and the common notion that "if a little is good, more is even better." For example, concentrated solutions of hydrogen peroxide, though biodegradable, can cause serious damage, even death, when used improperly. Other chemicals, such as chlorhexidine, can elicit severe allergic reactions in some people. As less toxic germicidal chemicals are developed, people must be educated on the appropriate use of these alternatives.

Summary

5.1 ■ Approaches to Control

The methods used to destroy or remove microbes can be physical, such as heat treatment, irradiation, and filtration, or chemical.

Principles of Control

A variety of terms are used to describe antimicrobial agents and processes. A **sterile** item is free of viable microbes, including endospores and viruses. **Disinfection** is the elimination of most or all pathogens on or in a material. **Antiseptics** are antimicrobial chemicals non-toxic enough to be used on body tissue.

Situational Considerations (figure 5.1)

Situations encountered in daily life, hospitals, microbiology laboratories, food production facilities, water treatment facilities, and other industries warrant different degrees of microbial control. Healthcare facilities must be especially careful to control microorganisms to prevent **healthcare-associated infections** (figure 5.2).

5.2 ■ Selecting an Antimicrobial Procedure

Type of Microbes

One of the most critical considerations in selecting a method of destroying microorganisms and viruses is the type of microbial population thought to be present on or in the product.

Number of Microorganisms

The amount of time it takes for heat or chemicals to kill a population of microorganisms is dictated in part by the number of cells initially present. Microbial death generally occurs at a constant rate. The **D value,** or **decimal reduction time,** is the time it takes to kill 90% of a population of bacteria under specific conditions (figure 5.3).

Environmental Conditions

Factors such as pH and presence of organic materials influence microbial death rates.

Risk for Infection

Medical instruments are categorized as **critical, semicritical,** and **non-critical** according to their risk of transmitting infectious agents.

Composition of the Item

Some sterilization and disinfection procedures are inappropriate for certain types of material.

5.3 ■ Using Heat to Destroy Microorganisms and Viruses

Moist Heat

Moist heat destroys microorganisms by causing irreversible coagulation of their proteins. **Pasteurization** utilizes a brief heat treatment to destroy spoilage and disease-causing organisms. Pressure cookers and **autoclaves** use pressurized steam to achieve temperatures that can kill endospores (figure 5.4, figure 5.5). The most important aspect of the commercial canning process is to ensure that endospores of *Clostridium botulinum* are destroyed.

Dry Heat

Incineration burns cell components to ashes. Temperatures achieved in hot air ovens destroy cell components and irreversibly denature proteins.

5.4 ■ Using Other Physical Methods to Remove or Destroy Microbes

Filtration (figure 5.6)

Membrane filters and **depth filters** retain microorganisms while letting the suspending fluid pass through. **High-efficiency particulate air (HEPA) filters** remove nearly all microorganisms from air.

Irradiation (figure 5.7, figure 5.8)

Ionizing radiation destroys cells by damaging cell structures and producing reactive oxygen species. Ultraviolet light damages the structure and function of nucleic acids. Microwaves kill microorganisms by generating heat.

High Pressure

High pressure is thought to destroy microorganisms by denaturing proteins and altering the permeability of the cell.

5.5 ■ Using Chemicals to Destroy Microorganisms and Viruses

Effectiveness of Germicidal Chemicals

Germicides are grouped according to their potency as **sterilants, high-level disinfectants, intermediate-level disinfectants,** or **low-level disinfectants.**

Selecting the Appropriate Germicidal Chemical

Factors that must be included in the selection of an appropriate germicidal chemical include toxicity, residue, activity in the presence of organic matter, compatibility with the material being treated, cost and availability, storage and stability, and ease of disposal.

Classes of Germicidal Chemicals (table 5.2, figure 5.9)

Solutions of 60–80% ethyl or isopropyl alcohol in water rapidly kill vegetative bacteria and fungi by coagulating enzymes and other essential proteins, and by damaging lipid membranes. Glutaraldehyde, *ortho*-phthalaldehyde, and formaldehyde destroy microorganisms and viruses by inactivating proteins and nucleic acids. Chlorhexidine is a biguanide extensively used in antiseptic products. Ethylene oxide is a gaseous sterilizing agent that destroys microbes by reacting with proteins. Sodium hypochlorite (liquid bleach) is one of the least expensive and most readily available forms of chlorine. Chlorine dioxide is used as a sterilant and disinfectant. Iodophores are iodine-releasing compounds used as antiseptics. Metals interfere with protein function. Silver-containing compounds are used to prevent wound infections. Ozone is used as an alternative to chlorine in the disinfection of drinking water and wastewater. Peroxide and peracetic acid are both strong oxidizing agents that can be used alone or in combination as sterilants. Phenolics destroy cytoplasmic membranes and denature proteins. Quaternary ammonium compounds are cationic detergents; they are non-toxic enough to be used to disinfect food preparation surfaces.

5.6 ■ Preservation of Perishable Products

Chemical Preservatives

Benzoic, sorbic, and propionic acids are sometimes added to foods to prevent microbial growth. Nitrate and nitrite are added to some foods to inhibit the germination of endospores and subsequent growth of *Clostridium botulinum.*

Low-Temperature Storage

Low temperatures above freezing inhibit microbial growth. Freezing essentially stops all microbial growth.

Reducing the Available Water

Sugar and salt draw water out of cells, preventing the growth of microorganisms (figure 5.11). Lyophilization is used for preserving food. The food is first frozen and then dried in a vacuum.

Review Questions

Short Answer

1. How is preservation different from pasteurization?
2. What is the most chemically resistant non-spore-forming bacterial pathogen?
3. Explain why it takes longer to kill a population of 10^9 cells than it does to kill a population of 10^3 cells.
4. What is the primary reason that wine is pasteurized?
5. What is the primary reason that milk is pasteurized?
6. When canning, why are low-acid foods processed at higher temperatures than high-acid foods?
7. How are heat-sensitive liquids sterilized?
8. How does microwaving a food product kill bacteria?
9. How is an iodophore different from a tincture of iodine?
10. Name two products commonly sterilized using ethylene oxide gas.

Multiple Choice

1. Unlike a disinfectant, an antiseptic
   a) sanitizes objects rather than sterilizes them.
   b) destroys all microorganisms.
   c) is non-toxic enough to be used on human skin.
   d) requires heat to be effective.
   e) can be used in food products.
2. The D value is defined as the time it takes to kill
   a) all bacteria in a population.
   b) all pathogens in a population.
   c) 99.9% of bacteria in a population.
   d) 90% of bacteria in a population.
   e) 10% of bacteria in a population.
3. Which of the following is the most resistant to destruction by chemicals and heat?
   a) Bacterial endospores
   b) Fungal spores
   c) *Mycobacterium tuberculosis*
   d) *E. coli*
   e) HIV

4. Ultraviolet light kills bacteria by
   a) generating heat.
   b) damaging DNA.
   c) inhibiting protein synthesis.
   d) damaging cell walls.
   e) damaging cytoplasmic membranes.
5. Which concentration of alcohol is the most effective germicide?
   a) 100% b) 75% c) 50% d) 25% e) 5%
6. All of the following could be used to sterilize an item *except*
   a) boiling.
   b) incineration.
   c) irradiation.
   d) sporocides.
   e) filtration.
7. All of the following are routinely used to preserve foods *except*
   a) high concentrations of sugar.
   b) high concentrations of salt.
   c) benzoic acid.
   d) freezing.
   e) ethylene oxide.
8. Aseptically boxed juices and cream containers are processed using which of the following heating methods?
   a) Commercial canning
   b) High-temperature–short-time (HTST) method
   c) Autoclaving
   d) Ultra-high-temperature (UHT) method
   e) Boiling
9. Commercial canning processes are designed to ensure destruction of which of the following?
   a) All vegetative bacteria
   b) All viruses
   c) Endospores of *Clostridium botulinum*
   d) *E. coli*
   e) *Mycobacterium tuberculosis*
10. Which of the following is *false*?
   a) A high-level disinfectant cannot be used as a sterilant.
   b. Critical items must be sterilized before use.
   c) Low numbers of endospores may remain on semicritical items.
   d) Standard sterilization procedures do not destroy prions.
   e) Quaternary ammonium compounds can be used to disinfect food preparation surfaces.

Applications

1. An agriculture extension agent is preparing pamphlets on preventing the spread of disease. In the pamphlet, he must explain the appropriate situations for using disinfectants around the house. What situations should the agent discuss?
2. As a microbiologist representing a food corporation, you have been asked to serve on a health food panel to debate the need for chemical preservatives in foods. Your role is to prepare a statement that compares the benefits of chemical preservatives and the risks. What points must you bring up that indicate the benefits of chemical preservatives?

Critical Thinking

1. This graph shows the time it takes to kill populations of the same microorganism under different conditions. What conditions would explain the differences in lines a, b, and c?

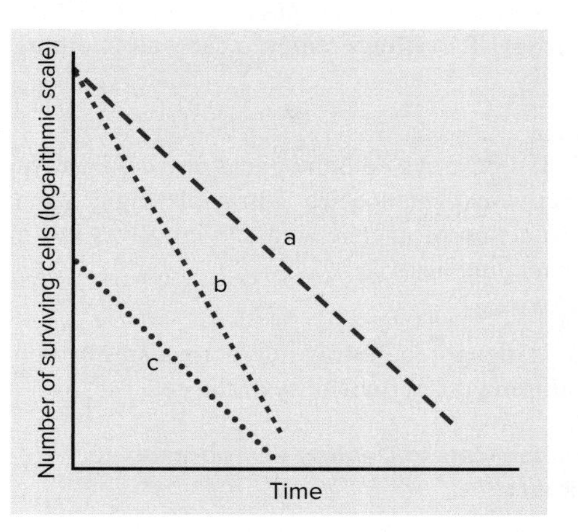

2. This diagram shows the filter paper method used to evaluate the inhibitory effect of chemical agents, heavy metals, and antibiotics on bacterial growth. A culture of a test bacterium is spread uniformly over the surface of an agar plate. Small filter paper discs containing the material to be tested are then placed on the surface of the medium. A disc that has been soaked in sterile distilled water is sometimes added as a control. After incubation, a lawn (film of growth) will cover the plate, but a clear zone will surround those discs that contain an inhibitory compound. The size of the zone reflects several factors, one of which is the effectiveness of the inhibitory agent. What are two other factors that might affect the size of the zone of inhibition? What is the purpose of the control disc? If a clear area were apparent around the control disc, how would you interpret the observation?

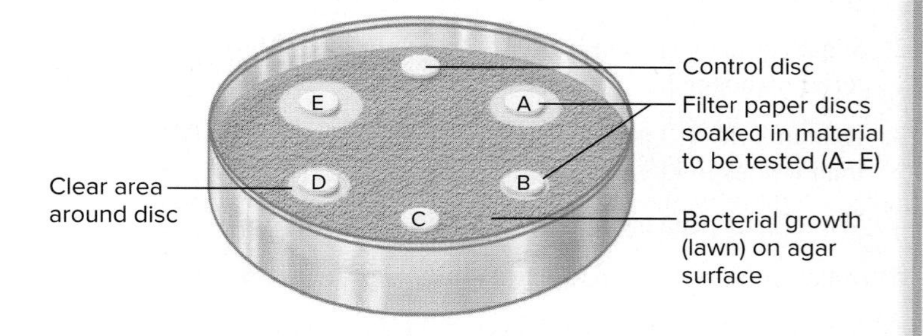

6 Microbial Metabolism: Fueling Cell Growth

Wine—a beverage produced using microbial metabolism. ©*Comstock/Getty Images*

KEY TERMS

Adenosine Triphosphate (ATP) The energy currency of cells. Hydrolysis of the bonds between its phosphate groups can be used to power endergonic (energy-consuming) reactions.

Anabolism Metabolic processes that synthesize and assemble the subunits of macromolecules, using energy of ATP; biosynthesis.

Catabolism Metabolic processes that harvest energy released during the breakdown of compounds such as glucose, using it to synthesize ATP.

Cellular Respiration Metabolic process that transfers electrons stripped from a chemical energy source to an electron transport chain, generating a proton motive force that is then used to synthesize ATP.

Electron Transport Chain (ETC) Group of membrane-embedded electron carriers that pass electrons from one to another, and, in the process, create a proton motive force.

Enzyme A molecule, usually a protein, that functions as a catalyst, speeding up a biological reaction.

Fermentation Metabolic process that stops short of oxidizing glucose or other organic compounds completely, using an organic intermediate as a terminal electron acceptor.

Oxidative Phosphorylation Synthesis of ATP using the energy of a proton motive force created by harvesting chemical energy.

Photophosphorylation Synthesis of ATP using the energy of a proton motive force created by harvesting radiant energy.

Precursor Metabolites Metabolic intermediates that can be either used to make the subunits of macromolecules or oxidized to generate ATP.

Proton Motive Force Form of energy generated as an electron transport chain moves protons across a membrane to create a chemiosmotic gradient.

Substrate-Level Phosphorylation Synthesis of ATP using the energy released in an exergonic (energy-releasing) chemical reaction during the breakdown of the energy source.

Terminal Electron Acceptor Chemical that is ultimately reduced as a consequence of fermentation or respiration.

A Glimpse of History

In the 1850s, Louis Pasteur tried to determine how alcohol develops from grape juice. Biologists had already noticed that when the juice is held in large vats, alcohol and CO_2 are produced and the number of yeast cells increases. They concluded that the multiplying cells were converting sugar in the grape juice to alcohol and CO_2. Pasteur agreed, but could not convince two very powerful and influential German chemists, Justus von Liebig and Friedrich Wöhler, who refused to believe that microorganisms caused the breakdown of sugar. Both men mocked the hypothesis, publishing pictures of yeast cells looking like miniature animals taking in grape juice through one end and releasing CO_2 and alcohol through the other.

Pasteur studied the relationship between yeast and alcohol production using a strategy commonly employed by scientists today—that is, simplifying the experimental system so that relationships can be more easily identified. First, he prepared a solution of sugar, ammonia, mineral salts, and trace elements. He then added a few yeast cells. As the yeast multiplied, the sugar level decreased and the alcohol level increased, indicating that the sugar was being converted to alcohol as the cells grew. This strongly suggested that living cells caused the chemical transformation. Liebig, however, still would not believe that the process was occurring inside microorganisms. To convince him, Pasteur tried to extract something from inside the yeast cells that would convert the sugar. He failed, like many others before him.

In 1897, Eduard Buchner, a German chemist, showed that crushed yeast cells could convert sugar to ethanol and CO_2. We now know that the cells' enzymes carried out this transformation. For these pioneering studies, Buchner was awarded a Nobel Prize in 1907.

Microbial cells, as well as all other cells, need to accomplish two fundamental tasks to grow. They must continually synthesize new parts—such as cell walls, membranes, ribosomes, and nucleic acids—that allow the cells to enlarge and eventually divide. In addition, cells need to harvest energy and convert it to a form that can power the various energy-consuming reactions, including those used to make new parts. The sum of all chemical reactions in a cell is called **metabolism.**

Microbial metabolism is important to humans for a number of reasons. For example, as scientists look for new supplies of energy, some are investigating biofuels—fuels made from renewable biological sources such as plants and organic wastes. Microbes can produce biofuels, breaking down solid materials such as corn stalks, sugar cane, and wood to make ethanol. Microbial metabolism is also important in food and beverage production; cheese-makers add *Lactococcus* and *Lactobacillus* species to milk because the metabolic wastes of these bacteria contribute to the flavor and texture of various cheeses. Bakers, brewers, vintners, and distillers use the yeast *Saccharomyces cereviseae* to make bread, beer, wine, and distilled spirits, respectively. In the laboratory, products characteristic of specific microbes can be used as identifying markers. In addition, the metabolic pathways of organisms such as *E. coli* serve as an important model for studying similar pathways in eukaryotic cells, including those of humans. Finally, microbial metabolism is medically relevant because metabolic processes unique to bacteria are potential targets for antimicrobial medications.

6.1 ■ Principles of Microbial Metabolism

Learning Outcomes

1. Compare and contrast catabolism and anabolism.
2. Describe the energy sources used by photosynthetic organisms and chemoorganoheterotrophs.
3. Describe the components of metabolic pathways (enzymes, ATP, chemical energy sources and terminal electron acceptors, and electron carriers) and the role of precursor metabolites.
4. Describe the roles of the three central metabolic pathways.
5. Distinguish between cellular respiration and fermentation.

Microbial metabolism (the sum of all chemical reactions in a cell) is easier to understand if you know the general principles and how they fit into the "big picture." With that in mind, this section will introduce you to the key concepts of metabolism and present an overview of the topic; later sections will then build on that information.

Metabolism can be separated into two components: catabolism and anabolism (**figure 6.1**). **Catabolism** is the set of chemical reactions that degrade compounds, releasing their energy. Cells capture that energy and use it to make ATP—the energy currency of the cell (see figure 2.29). **Anabolism,** or **biosynthesis,** is the set of chemical reactions that cells use to synthesize and assemble the subunits of macromolecules, using ATP for energy. As described in chapter 2, the subunits of macromolecules include amino acids, nucleotides, monosaccharides, and fatty acids. ⏮ ATP

Although catabolism and anabolism are often discussed separately, they are intimately linked. As mentioned, ATP made during catabolism is used in anabolism. In addition, some of the compounds produced during catabolism are precursor metabolites, which represent an intersection of

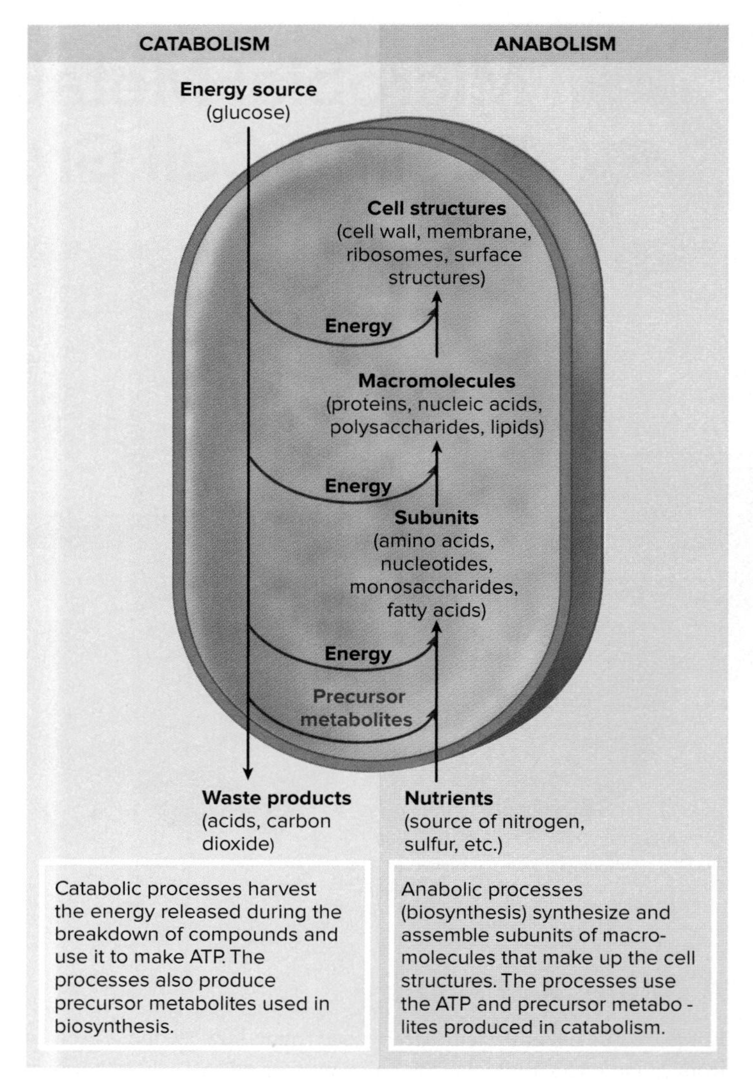

FIGURE 6.1 The Relationship Between Catabolism and Anabolism Precursor metabolites are compounds produced during catabolism that can either be further degraded or be used in anabolism to make the subunits of macromolecules.

? **Which subunits make up proteins? Which make up nucleic acids?**

catabolic and anabolic processes. As an analogy, think of what happens in a recycling center that breaks down electronic devices. Once certain components are freed from the devices, those components can be further broken down—generating raw materials such as gold that can be sold for cash—or they can be used to build refurbished devices. Precursor metabolites are chemicals that can either be further broken down to generate energy (via catabolism) or used to make certain subunits of macromolecules (via anabolism).

Energy

Energy is the capacity to do work. It can exist as potential energy (stored energy) or as kinetic energy (energy of motion) (**figure 6.2**). Potential energy can be stored in various

FIGURE 6.2 Forms of Energy Potential energy is stored energy, such as water held behind a dam. Kinetic energy is the energy of motion, such as movement of water from behind the dam.
©Farrell Grehan/Science Source

? **What is energy?**

forms, including chemical bonds, a rock on a hill, or water behind a dam.

Energy in the universe can never be created or destroyed; however, it can be changed from one form to another. In other words, although energy cannot be created, potential energy can be converted to kinetic energy and vice versa, and one form of potential energy can be converted to another. Hydroelectric dams release the potential energy of water stored behind a dam, creating the kinetic energy of moving water. This can be captured to generate an electrical current, which can then be used to charge a battery.

Metabolism involves energy transformation processes; cells take energy of one form, and convert it into another. **Photosynthetic** organisms harvest the energy of sunlight, using it to power the synthesis of organic compounds from CO_2. By doing so, they convert the kinetic energy of photons (particles that travel at the speed of light) to the potential energy of chemical bonds. **Chemoorganotrophs** obtain energy by degrading organic compounds; they then use some of that energy to make other organic compounds. In other words, they take the potential energy of certain chemical bonds and use it to create other ones. Because chemoorganotrophs depend on a constant source of organic compounds, they generally rely on the metabolic activities of photosynthetic organisms (**figure 6.3**). In some environments, however, chemolithoautotrophs play the most significant role in synthesizing organic compounds (see figure 28.12). ◀◀ chemolithoautotrophs

The amount of energy released by breaking down a compound can be explained by the concept of free energy, the energy available to do work. From a biological perspective, this is the energy released when a chemical bond is broken. In a chemical reaction, some bonds are broken and others are formed. If the starting compounds have more free energy than the products, energy is released; the reaction is **exergonic.** In contrast, if the products have more free energy than the starting compounds, the reaction requires an input of energy and is **endergonic.**

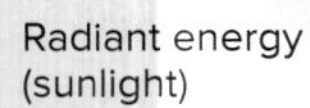

FIGURE 6.3 Most Chemoorganotrophs Depend on Photosynthetic Organisms Top: ©Robert Glusic/Getty Images; Bottom: ©Digital Vision/PunchStock

? **Why do chemoorganotrophs require the activities of photosynthetic organisms?**

The change in free energy for a given reaction is the same regardless of the number of steps involved. For example, converting glucose to CO_2 and water in a single step releases the same amount of energy as degrading it in a series of steps. Cells take advantage of this by degrading compounds step by step—a carefully controlled process that allows a cell to harvest the energy released in an exergonic reaction by having it power an endergonic one.

Components of Metabolic Pathways

The series of chemical reactions that converts a starting compound to an end product is called a **metabolic pathway** (**figure 6.4**). Pathways can be linear, branched, or cyclical. Like the flow of rivers controlled by dams, their activities can be adjusted at certain points. This allows a cell to regulate certain processes, ensuring that specific molecules are produced in precise quantities when needed. If a metabolic step is blocked, products "downstream" of that blockage will not be made.

To understand what metabolic pathways accomplish, it is important to be familiar with their essential components. These include enzymes, ATP, the chemical energy source and terminal electron acceptor, and electron carriers.

The Role of Enzymes

An **enzyme** is a molecule (usually a protein) that functions as a biological catalyst, speeding up the conversion of one substance, the **substrate,** into another, the product. A specific enzyme facilitates each step of a metabolic pathway (**figure 6.5a**). Without enzymes, energy-yielding reactions would still occur, but at rates so slow they would be insignificant. ◀◀ enzymes, ◀◀ proteins

An enzyme catalyzes a chemical reaction by lowering the **activation energy**—the energy it takes to start a reaction (see figure 6.5b). Even an exergonic reaction has an activation energy, and by lowering this barrier, an enzyme speeds the reaction. Enzymes will be described in more detail later in the chapter.

The Role of ATP

Adenosine triphosphate (ATP) is the main energy currency of cells, serving as the ready and immediate donor of free energy. The molecule is composed of ribose, adenine, and three phosphate groups arranged in a row (see figure 2.29).

Cells produce "energy currency" by using energy to add an inorganic phosphate group (P_i) to **adenosine diphosphate (ADP),** forming ATP. That energy currency can then be "spent" by removing the phosphate group, thereby releasing energy and converting the molecule back to ADP and P_i (**figure 6.6**). Cells constantly produce ATP during exergonic reactions of catabolism and then use it to power endergonic reactions of anabolism.

Chemoorganotrophs use two different processes to make ATP: substrate-level phosphorylation and oxidative phosphorylation. In **substrate-level phosphorylation,** the energy released in an exergonic reaction is used to power the addition of P_i to ADP; in **oxidative phosphorylation,** the energy

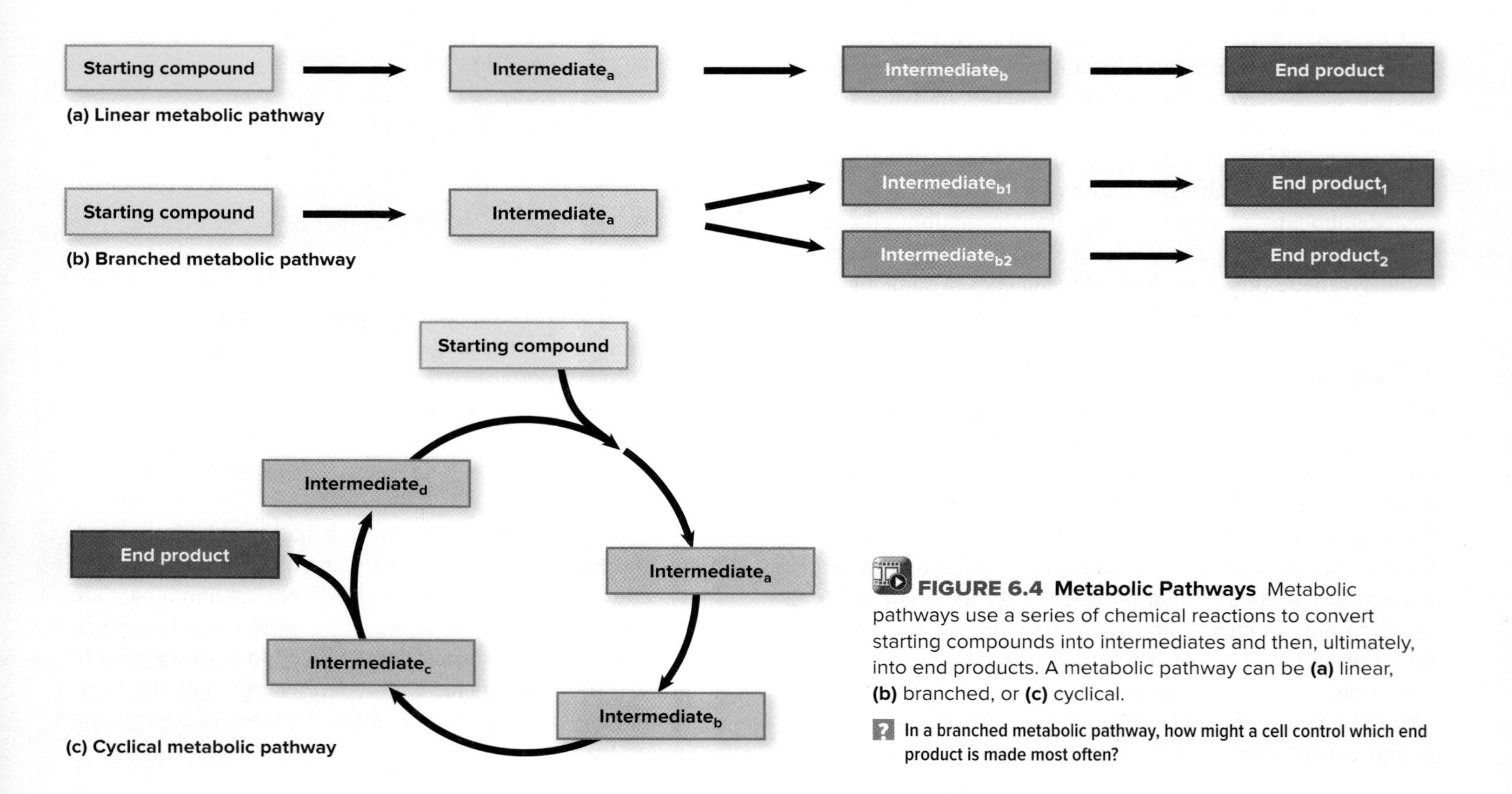

FIGURE 6.4 Metabolic Pathways Metabolic pathways use a series of chemical reactions to convert starting compounds into intermediates and then, ultimately, into end products. A metabolic pathway can be **(a)** linear, **(b)** branched, or **(c)** cyclical.

? In a branched metabolic pathway, how might a cell control which end product is made most often?

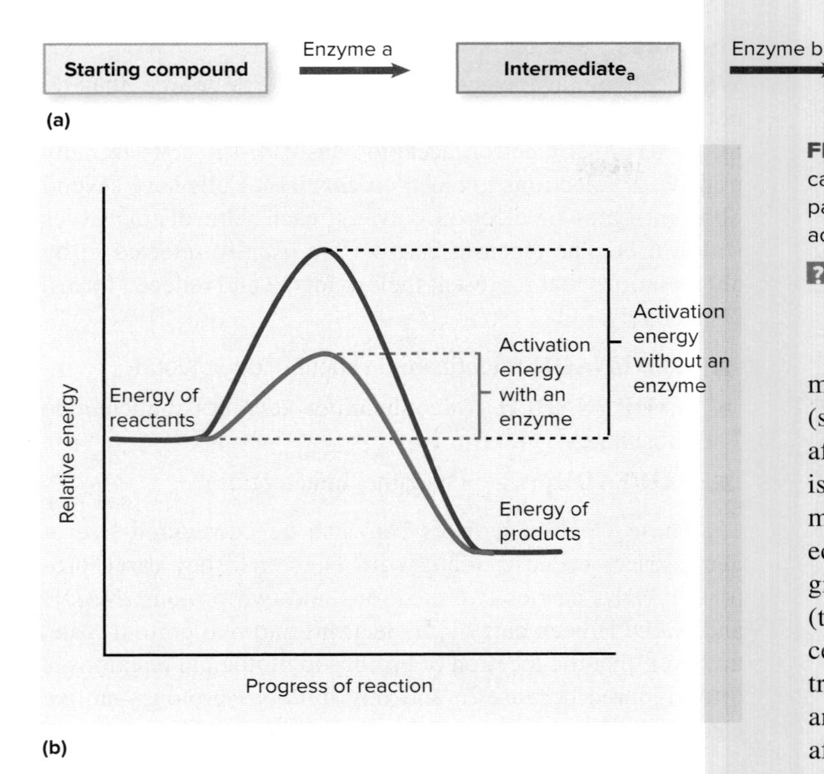

FIGURE 6.5 The Role of Enzymes Enzymes function as biological catalysts. **(a)** A specific enzyme facilitates each step of a metabolic pathway. **(b)** An enzyme catalyzes a chemical reaction by lowering the activation energy of the reaction.

? What is activation energy?

of a proton motive force drives the reaction. Recall that **proton motive force** is the form of energy that results from the electrochemical gradient established by the electron transport chain (see figure 3.28). Photosynthetic organisms can generate ATP by **photophosphorylation.** This uses the sun's radiant energy and an electron transport chain to create a proton motive force. ⏭ oxidative phosphorylation, ⏮ proton motive force, ⏭ photophosphorylation

The Role of the Chemical Energy Source and the Terminal Electron Acceptor

To understand how cells obtain energy, it is helpful to recall that certain atoms are more electronegative than others, meaning they have a greater affinity (attraction) for electrons (see table 2.3). Likewise, certain molecules have a greater affinity for electrons than do other molecules, a characteristic that relates to the electronegativities of the atoms that make up the molecules. When electrons move from a molecule that has a relatively low electron affinity (tends to give up electrons) to one that has a higher electron affinity (tends to accept electrons), energy is released. This is how cells obtain the energy used to make ATP. They remove electrons from glucose or another low electron affinity chemical, and donate them to a molecule such as O_2 that has a higher affinity. The chemical that serves as the electron donor is the **energy source,** and the one that ultimately accepts those electrons is the **terminal electron acceptor.** The greater the difference between the electron affinities of the energy source and the terminal electron acceptor, the more energy released (**figure 6.7**). ⏮ electrons, ⏮ electronegative

As a group, prokaryotes are remarkably diverse with respect to the energy sources and terminal electron acceptors they can use (see figure 6.7a and b). *E. coli* and other chemoorganotrophs use organic compounds such as glucose for an energy source, whereas chemolithotrophs use hydrogen sulfide or another inorganic chemical. Obligate aerobes use only O_2 as a terminal electron acceptor, whereas obligate anaerobes use only molecules other than O_2; facultative anaerobes use O_2 if it is available, but can also use an alternative. The metabolic diversity of bacteria and archaea is one reason they are so important ecologically. By oxidizing and reducing various chemicals, the organisms play essential roles in the cycling of biologically important elements (a topic covered in chapter 28) and in degrading compounds that could otherwise be harmful to the environment (a topic covered in chapter 29). ⏮ chemoorganotrophs, ⏮ chemolithotrophs

Cells remove electrons from the energy source through a series of **oxidation-reduction reactions,** or **redox reactions.** The substance that loses electrons is **oxidized** by the reaction; the one that gains those electrons is **reduced** (**figure 6.8**). When electrons are removed from a biological molecule, protons (H^+) often follow. The result is the removal of an electron-proton pair, or hydrogen atom. Thus, dehydrogenation (the removal of a hydrogen atom) is an oxidation. Correspondingly, hydrogenation (the addition of a hydrogen atom) is a reduction.

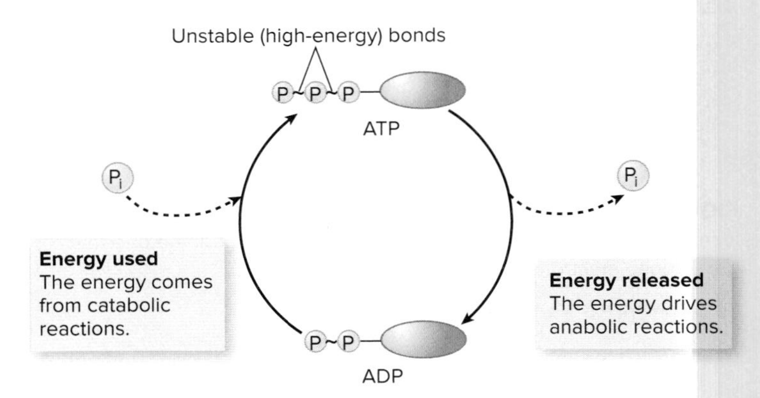

FIGURE 6.6 ATP Cells produce ATP by using energy to add inorganic phosphate (P_i) to ADP. Energy is released in the reverse reaction, which breaks the phosphate bond to convert ATP to ADP + P_i.

? What structural characteristic of ATP makes its phosphate bonds "high energy"?

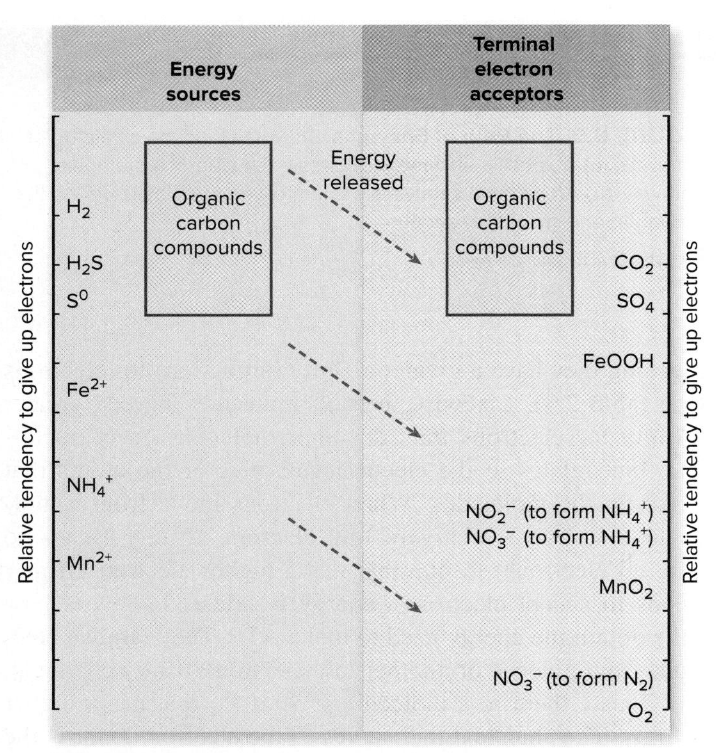

(a) Energy is released when electrons are moved from an energy source with a low affinity for electrons to a terminal electron acceptor with a higher affinity.

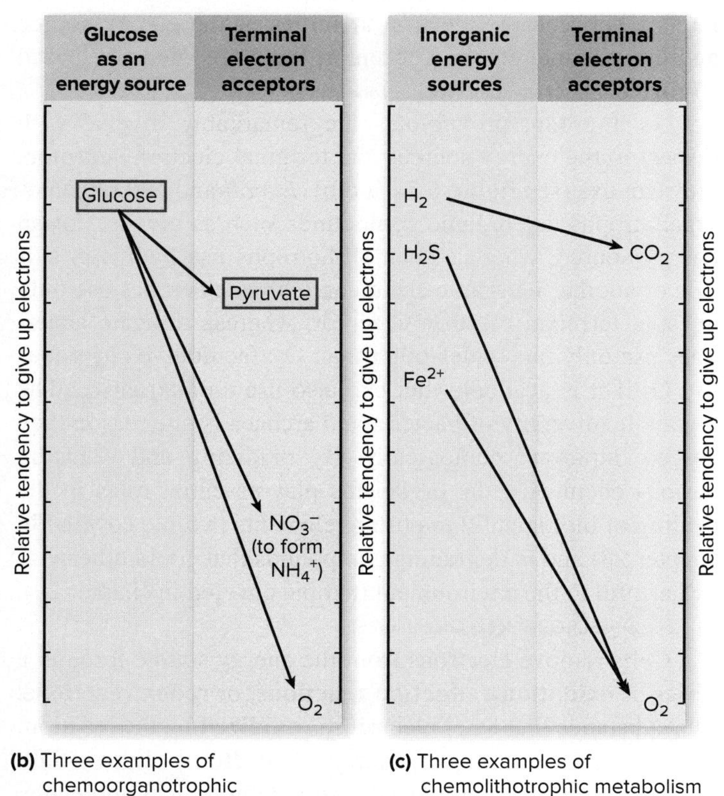

(b) Three examples of chemoorganotrophic metabolism

(c) Three examples of chemolithotrophic metabolism

FIGURE 6.7 Chemical Energy Sources and Terminal Electron Acceptors Catabolic pathways can proceed as long as the energy source has a lower electron affinity than the terminal electron acceptor. As a group, bacteria and archaea use a wide variety of energy sources and terminal electron acceptors, which is why the organisms are so important ecologically.

? **If Mn^{2+} is used as an energy source, which two molecules on the chart in (a) could serve as terminal electron acceptors?**

The Role of Electron Carriers

When cells remove electrons from an energy source, they do not remove them all at once, nor do they transfer them directly to the terminal electron acceptor. Instead, the cells initially transfer the electrons to **electron carriers.** Cells have several different types of electron carriers, each with distinct roles (**table 6.1**). The electron carriers are usually referred to by abbreviations that represent their oxidized and reduced forms, respectively:

- **NAD^+/NADH** (nicotinamide adenine dinucleotide)
- **$NADP^+$/NADPH** (nicotinamide adenine dinucleotide phosphate)
- **FAD/$FADH_2$** (flavin adenine dinucleotide)

These electron carriers can also be considered hydrogen carriers because, along with electrons, they carry protons. $FADH_2$ carries two electrons and two protons; NADH and NADPH each carry two electrons and one proton. Note, however, that the location of protons in biological reactions is often ignored because in aqueous solutions protons—unlike electrons—do not require carriers.

Reduced electron carriers represent **reducing power** because they can easily transfer their electrons to another chemical that has a higher affinity for electrons. By doing so, they raise the energy level of the recipient molecule. NADH and $FADH_2$ transfer their electrons to the electron transport chain, which then uses the energy to generate a proton motive force. In turn, this drives the synthesis of ATP in the process of oxidative phosphorylation. Ultimately the electrons are transferred to the terminal electron acceptor. The electrons

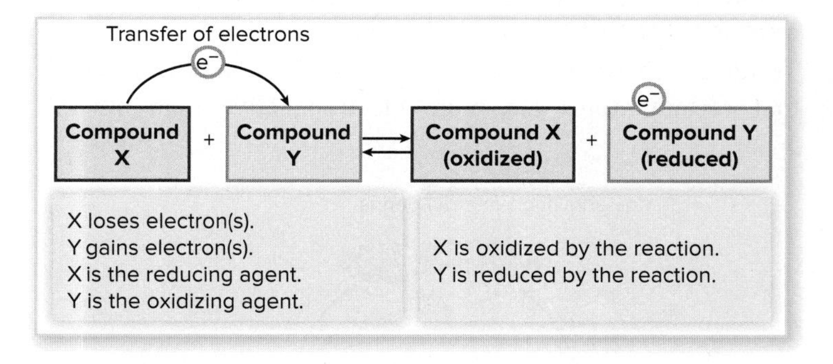

FIGURE 6.8 Oxidation-Reduction Reactions The molecule that loses one or more electrons is oxidized by the reaction; the one that gains those electrons is reduced.

? **Why is hydrogenation a reduction?**

TABLE 6.1 Electron Carriers

Carrier	Oxidized Form (Accepts Electrons)		Reduced Form (Donates Electrons)	Typical Fate of Electrons Carried
			e^- Electron, Electron carrier	
Nicotinamide adenine dinucleotide (carries 2 electrons and 1 proton)	$NAD^+ + 2\ e^- + 2\ H^+$	$\rightleftharpoons$	$NADH + H^+$	Used to generate a proton motive force that can drive ATP synthesis
Flavin adenine dinucleotide (carries 2 electrons and 2 protons; i.e., 2 hydrogen atoms)	$FAD + 2\ e^- + 2\ H^+$	$\rightleftharpoons$	$FADH_2$	Used to generate a proton motive force that can drive ATP synthesis
Nicotinamide adenine dinucleotide phosphate (carries 2 electrons and 1 proton)	$NADP^+ + 2\ e^- + 2\ H^+$	$\rightleftharpoons$	$NADPH + H^+$	Biosynthesis

carried by NADPH have an entirely different fate; they are used to reduce compounds during biosynthetic reactions. Note, however, that many microbial cells can convert reducing power in the form of NADH to NADPH.

Precursor Metabolites

Certain intermediates of catabolic pathways can be used in anabolic pathways. These intermediates—**precursor metabolites**—serve as carbon skeletons from which subunits of macromolecules can be made (**table 6.2**). For example, the precursor metabolite pyruvate can be converted to any one of three amino acids: alanine, leucine, or valine.

Recall from chapter 4 that *E. coli* can grow in glucose-salts medium, which contains only glucose and a few inorganic salts. This means that the glucose is serving two purposes in the cell: (1) the energy source, and (2) the starting point from which all cell components are made—including proteins, lipids, carbohydrates, and nucleic acids. When *E. coli* cells degrade glucose molecules, they use a series of steps that not only release energy but also form a dozen or so precursor metabolites. Other organisms use the same steps but sometimes lack the ability to convert a certain precursor metabolite into a compound needed for biosynthesis. Any essential compounds that a cell cannot synthesize must be provided from an external source. **⏮ glucose-salts medium**

A cell's metabolic pathways make it easy for that cell to use glucose for multiple purposes. Think of the cells as extensive biological recycling centers that routinely process millions of glucose molecules (**figure 6.9**). Molecules that remain on the central deconstruction line are oxidized completely to CO_2, releasing the maximum amount of energy. Some breakdown intermediates, however, can exit that line to be used in biosynthesis. The exit points are located at the steps immediately after a precursor metabolite is made. So once a precursor metabolite is made in catabolism, it can be further oxidized to release energy, or it can be used in biosynthesis.

TABLE 6.2 Precursor Metabolites

Precursor Metabolite	Pathway Generated	Biosynthetic Role
Glucose-6-phosphate	Glycolysis	Lipopolysaccharide
Fructose-6-phosphate	Glycolysis	Peptidoglycan
Dihydroxyacetone phosphate	Glycolysis	Lipids (glycerol component)
3-Phosphoglycerate	Glycolysis	Proteins (the amino acids cysteine, glycine, and serine)
Phosphoenolpyruvate	Glycolysis	Proteins (the amino acids phenylalanine, tryptophan, and tyrosine)
Pyruvate	Glycolysis	Proteins (the amino acids alanine, leucine, and valine)
Ribose-5-phosphate	Pentose phosphate cycle	Nucleic acids and proteins (the amino acid histidine)
Erythrose-4-phosphate	Pentose phosphate cycle	Proteins (the amino acids phenylalanine, tryptophan, and tyrosine)
Acetyl-CoA	Transition step	Lipids (fatty acids)
α-Ketoglutarate	TCA cycle	Proteins (the amino acids arginine, glutamate, glutamine, and proline)
Oxaloacetate	TCA cycle	Proteins (the amino acids aspartate, asparagine, isoleucine, lysine, methionine, and threonine)

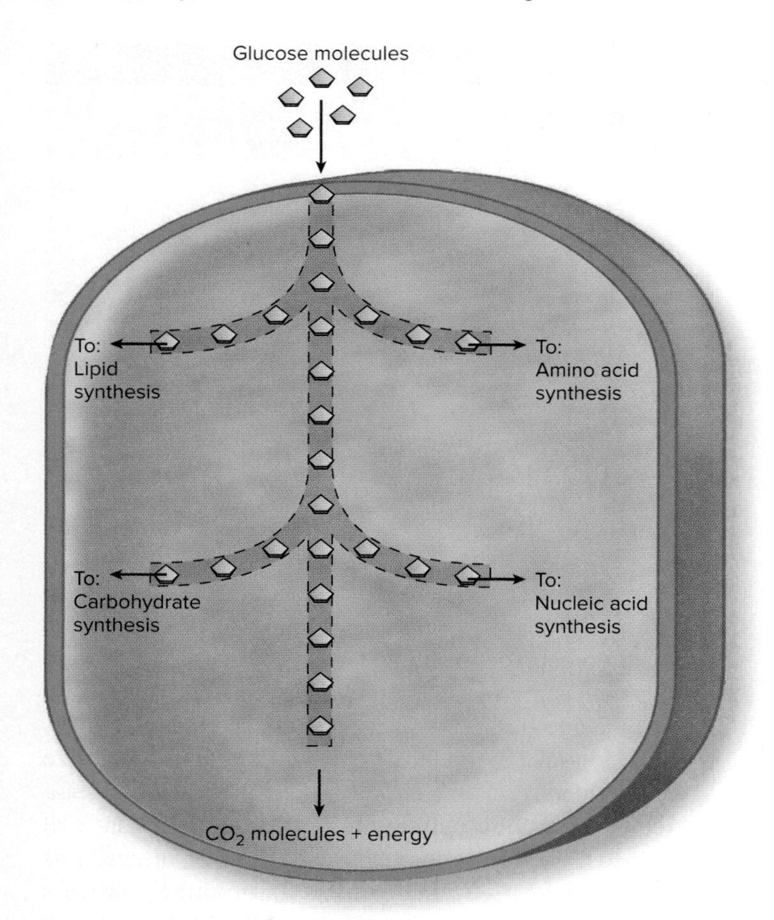

FIGURE 6.9 Cells Use Glucose for Multiple Purposes The millions of glucose molecules that continually enter a cell can have different fates. Some may be oxidized completely to release the maximum amount of energy, and others will be used in biosynthesis.

? What is the role of precursor metabolites in this process?

Overview of Catabolism

Catabolism of glucose, the preferred energy source of many cells, includes two key sets of processes:

- Oxidizing glucose molecules to generate ATP, reducing power (NADH, $FADH_2$, and NADPH), and precursor metabolites; this is accomplished in a series of reactions called the central metabolic pathways.
- Transferring the electrons carried by NADH and $FADH_2$ to the terminal electron acceptor, which is done by either cellular respiration or fermentation (recall that the electrons carried by NADPH are used in biosynthesis).

Central Metabolic Pathways

Three key metabolic pathways—the **central metabolic pathways**—gradually oxidize glucose to CO_2 (**figure 6.10**). The pathways are catabolic, but the precursor metabolites and reducing power they generate can also be diverted for use in biosynthesis. To reflect the dual role of these pathways, they are sometimes called amphibolic pathways (*amphi* meaning "both kinds"). The central metabolic pathways include the following:

- **Glycolysis.** ① This pathway splits glucose and gradually oxidizes it to form two molecules of pyruvate. Glycolysis provides the cell with a small amount of energy in the form of ATP, some reducing power, and six precursor metabolites. Some microbial cells use an alternative series of reactions (called the Entner-Doudoroff pathway) that generates a slightly different set of intermediates and end products.
- **Pentose phosphate pathway.** ② This also breaks down glucose, but its primary role in metabolism is to produce compounds used in biosynthesis, including two precursor metabolites as well as reducing power in the form of NADPH. A product of the pathway feeds into glycolysis.
- **Tricarboxylic acid (TCA) cycle.** This is also called the citric acid cycle or the Krebs cycle. (3a) Just before this cycle, a single reaction called the transition step converts the pyruvate from glycolysis into acetyl-CoA. One molecule of CO_2 is released as a result. (3b) The TCA cycle then accepts the 2-carbon acetyl group, ultimately oxidizing it to release two molecules of CO_2. The transition step and the TCA cycle together generate the most reducing power of all the central metabolic pathways. They also produce three precursor metabolites and ATP.

During the oxidation of glucose, a relatively small amount of ATP is made by substrate-level phosphorylation. The reducing power accumulated during the oxidation steps, however, can be used in cellular respiration (discussed next) to generate ATP by oxidative phosphorylation.

Cellular Respiration

④ **Cellular respiration,** also simply called **respiration,** involves transferring the electrons taken from glucose to the electron transport chain (ETC), which ultimately donates them to a terminal electron acceptor. The ETC uses the electrons to generate a proton motive force—a form of energy that can be used to make ATP by oxidative phosphorylation. In **aerobic respiration,** O_2 serves as the terminal electron acceptor. **Anaerobic respiration** is similar to aerobic respiration, but it uses a molecule other than O_2 as a terminal electron acceptor. Also, when anaerobically respiring, microbes use a modified version of the TCA cycle. Organisms that use respiration, either aerobic or anaerobic, are said to respire.

Fermentation

Cells that cannot respire are limited by their relative inability to recycle reduced electron carriers. A cell has only a limited number of carrier molecules; if electrons are not removed

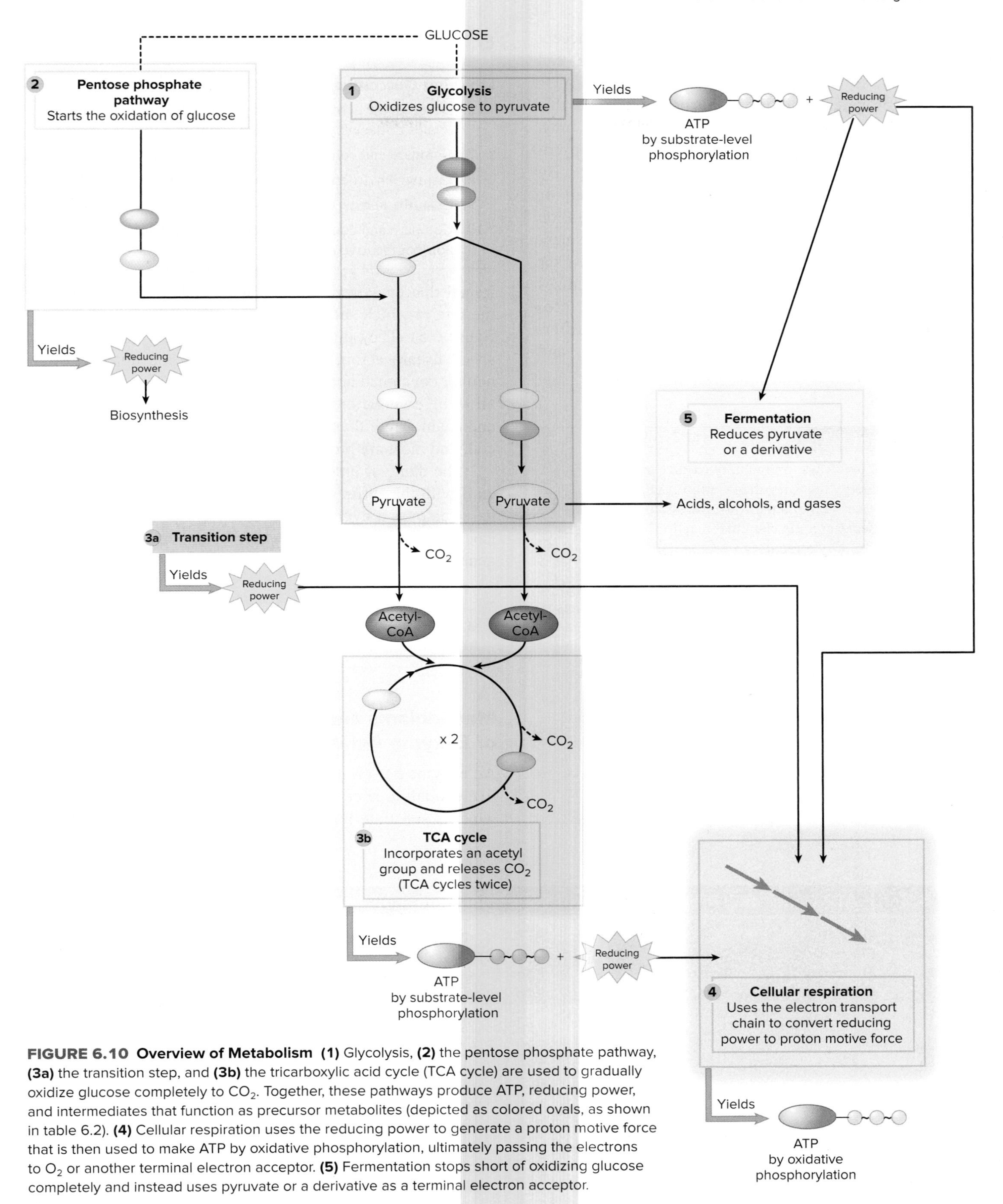

FIGURE 6.10 Overview of Metabolism **(1)** Glycolysis, **(2)** the pentose phosphate pathway, **(3a)** the transition step, and **(3b)** the tricarboxylic acid cycle (TCA cycle) are used to gradually oxidize glucose completely to CO_2. Together, these pathways produce ATP, reducing power, and intermediates that function as precursor metabolites (depicted as colored ovals, as shown in table 6.2). **(4)** Cellular respiration uses the reducing power to generate a proton motive force that is then used to make ATP by oxidative phosphorylation, ultimately passing the electrons to O_2 or another terminal electron acceptor. **(5)** Fermentation stops short of oxidizing glucose completely and instead uses pyruvate or a derivative as a terminal electron acceptor.

What is the terminal electron acceptor in aerobic respiration?

from the reduced carriers, none will be available to accept electrons. As a consequence, no more glucose molecules can be broken down, so the cell cannot continue to make ATP. **Fermentation** provides a solution to this problem. ⑤ In this process, cells break down glucose through glycolysis only, thereby generating pyruvate. The cells then use that pyruvate or a derivative of it as a terminal electron acceptor. By transferring the electrons carried by NADH to pyruvate or a derivative, NAD^+ is regenerated. Although fermentation does not involve the TCA cycle, organisms that ferment often use certain key steps of it to generate specific precursor metabolites required for biosynthesis. Fermentation oxidizes glucose only partially and, compared with respiration, produces relatively little ATP. **Table 6.3** summarizes the difference between aerobic respiration, anaerobic respiration, and fermentation.

MicroAssessment 6.1

Cells use catabolic pathways to gradually oxidize an energy source, releasing energy that they then capture. A specific enzyme catalyzes each step of a pathway. Substrate-level phosphorylation uses exergonic chemical reactions to synthesize ATP; oxidative phosphorylation uses a proton motive force to do the same. Reducing power in the form of NADH and $FADH_2$ is used to generate a proton motive force; the reducing power of NADPH is used in biosynthesis. Precursor metabolites are metabolic intermediates that can be used in biosynthesis. The central metabolic pathways generate ATP, reducing power, and precursor metabolites.

1. How does the fate of electrons carried by NADPH differ from the fate of electrons carried by NADH?
2. Why are the central metabolic pathways called amphibolic?
3. Why does fermentation supply less energy than cellular respiration?

6.2 ■ Enzymes

Learning Outcomes

6. Describe the active site of an enzyme, and explain how it relates to the enzyme-substrate complex.
7. Compare and contrast cofactors and coenzymes.
8. List two environmental factors that influence enzyme activity.
9. Describe allosteric regulation.
10. Compare and contrast competitive enzyme inhibition and non-competitive enzyme inhibition.

Recall that enzymes are biological catalysts, which increase the rate at which substrates are converted into products (see figure 6.5). They do this with extraordinary specificity and speed, usually acting on only one or a few substrates. They are neither consumed nor permanently changed during a reaction, allowing a single enzyme molecule to be rapidly used again and again. More than a thousand different enzymes exist in a cell, and most are proteins. ⏮ enzymes

The name of an enzyme usually reflects its function and ends with the suffix *-ase.* For example, isocitrate dehydrogenase removes a hydrogen atom (a proton-electron pair) from isocitrate. Some groups of enzymes are referred to by their general function—for example, proteases degrade proteins.

MicroByte

In only one second, the fastest enzymes can convert more than 10^4 substrate molecules into products.

Mechanisms and Consequences of Enzyme Action

An enzyme has on its surface an **active site,** typically a relatively small crevice (**figure 6.11**). This is the critical site to which a substrate binds by weak forces. The binding of the

TABLE 6.3 ATP-Generating Processes of Prokaryotic Chemoorganoheterotrophs

Metabolic Process	Uses an Electron Transport Chain	Terminal Electron Acceptor	ATP Generated by Substrate-Level Phosphorylation (Theoretical Maximum)	ATP Generated by Oxidative Phosphorylation (Theoretical Maximum)	Total ATP Generated (Theoretical Maximum)
Aerobic respiration	Yes	O_2	2 in glycolysis (net) 2 in the TCA cycle 4 total	34	38
Anaerobic respiration	Yes	Molecule other than O_2 such as nitrate (NO_3^-), nitrite (NO_2^-), sulfate (SO_4^{2-})	Number varies; however, the ATP yield of anaerobic respiration is less than that of aerobic respiration but more than that of fermentation.		
Fermentation	No	Organic molecule (pyruvate or a derivative)	2 in glycolysis (net) 2 total	0	2

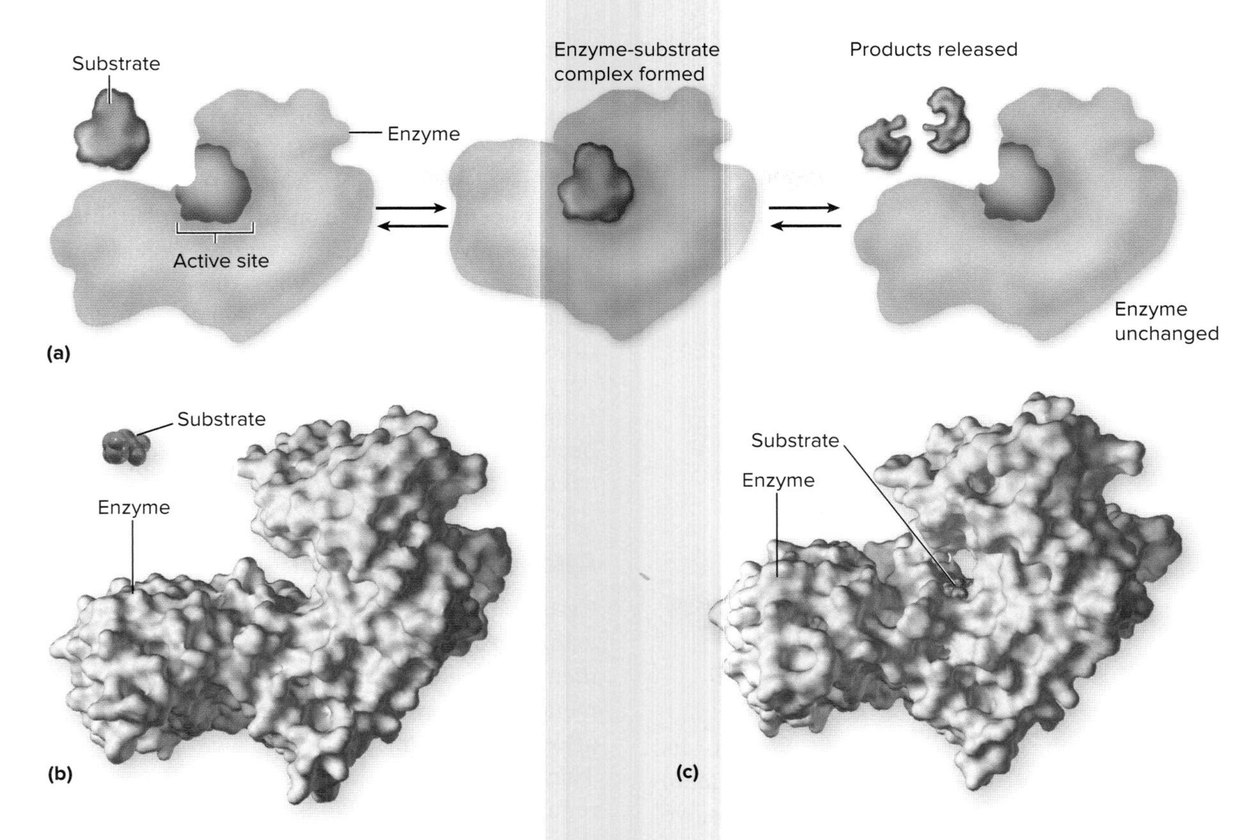

FIGURE 6.11 Mechanism of Enzyme Action **(a)** The substrate binds to the active site, forming an enzyme-substrate complex. The products are then released, leaving the enzyme unchanged and free to combine with new substrate molecules. **(b)** A model showing an enzyme and its substrate. **(c)** The binding of the substrate to the active site causes the shape of the flexible enzyme to change slightly.

? Why are enzymes so specific with respect to the reactions they catalyze?

substrate to the active site causes the shape of the flexible enzyme to change slightly. This mutual interaction, or induced fit, results in a temporary intermediate called an enzyme-substrate complex. The substrate is held within this complex in a specific orientation so that existing bonds are destabilized and new ones can easily form, lowering the activation energy of the reaction. The products are then released, leaving the enzyme unchanged and free to combine with new substrate molecules. Note that enzymes can also catalyze reactions that join two substrates to create one product. Theoretically, all enzyme-catalyzed reactions are reversible, but the free energy change of certain reactions makes them effectively non-reversible.

The interaction of an enzyme with its substrate is very specific. The substrate fits into the active site like a hand into a glove. Not only must it fit spatially, but appropriate chemical interactions such as hydrogen and ionic bonding need to occur to induce the fit. This requirement for a precise fit and interaction explains why, with few exceptions, a unique enzyme is required to catalyze each reaction in a cell. Very few molecules of any particular enzyme are needed, however, as each can be used repeatedly. ⏮ **hydrogen bonds,** ⏮ **ionic bonds**

Cofactors

Some enzymes require the assistance of an attached non-protein component called a **cofactor** (**figure 6.12**). Magnesium, zinc, copper, and other trace elements often function as cofactors. **Coenzymes** are a subset of cofactors; they are loosely attached non-protein organic compounds that help some enzymes transfer certain molecules or electrons from one compound to another. Coenzymes include the electron carriers FAD, NAD^+, and $NADP^+$, which play an important role in enzyme-catalyzed oxidation-reduction reactions. The various coenzymes function in different ways. Some remain bound to the enzyme during the transfer process, whereas others separate from the enzyme, carrying the substance being transferred along with them. The same coenzyme can assist different enzymes. Because of this, far fewer different coenzymes are required than enzymes. Like enzymes, coenzymes can be reused and, consequently, are needed only in very small quantities. ⏮ **trace elements**

Most coenzymes are derived from certain vitamins, which are organic compounds necessary for life but which the human body cannot synthesize. Without a given vitamin,

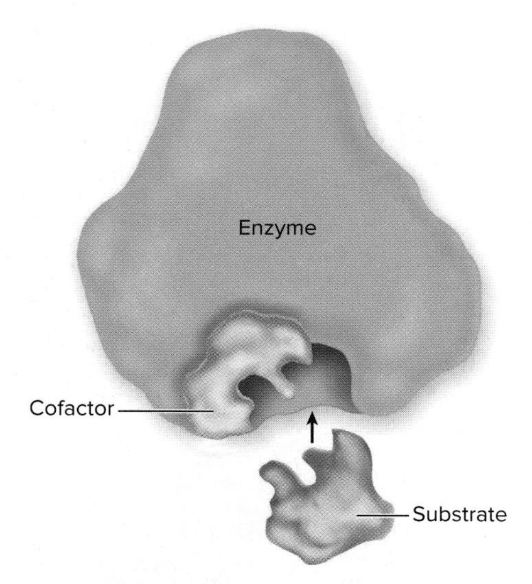

FIGURE 6.12 Some Enzymes Require the Assistance of a Cofactor Cofactors are non-protein components, either coenzymes or trace elements.

What is the function of the coenzyme FAD?

any enzyme that requires the corresponding coenzyme for its activity cannot function (**table 6.4**). Thus, a single vitamin deficiency has serious health consequences. In most cases, we must ingest vitamins as part of our diet. Many bacteria and archaea can synthesize vitamins, however, and we can absorb certain vitamins made by our intestinal microbiota.

Environmental Factors That Influence Enzyme Activity

Several environmental factors influence how well enzymes function (**figure 6.13**). Each enzyme has a narrow range of conditions—including temperature, pH, and salt concentration—within which it functions best. A 10°C rise in temperature approximately doubles the speed of enzymatic reactions, until optimal activity is reached; this explains why bacteria tend to grow more rapidly at higher temperatures. If the temperature is too high, however, proteins will denature and no longer function. Most enzymes function best at low salt concentrations and at pH values slightly above 7. Not surprisingly, most microbes grow fastest under these same conditions. Some prokaryotes, however, particularly certain members of the *Archaea,* thrive in environments where conditions are extreme. They may require high salt concentrations, very acidic conditions, or temperatures near boiling. Enzymes from these organisms can be very important commercially because they function in harsh conditions that are often typical of industrial settings. **temperature requirements, denaturation**

Allosteric Regulation

Cells can rapidly adjust the activity of certain key enzymes, using specific molecules that reversibly bind to and distort them (**figure 6.14**). This is how cells regulate the activity of metabolic pathways. The enzymes that can be controlled are **allosteric** (*allo* means "other" and *stereos* means "shape"). They have an allosteric site as well as an active site. When a regulatory molecule binds to the allosteric site, the shape of the enzyme changes. This distortion alters the relative affinity (chemical attraction) of the enzyme for its substrate. In some cases the binding of the regulatory molecule increases the affinity for the substrate; in other cases it decreases it.

Allosteric enzymes generally catalyze the first step of a pathway. If the pathway is biosynthetic, the end product generally acts as the allosteric inhibitor—a mechanism called feedback inhibition (see figure 6.14c). This allows the cell to shut down a pathway when the product begins accumulating. For example, the amino acid isoleucine is an allosteric inhibitor of the first enzyme of the pathway that converts threonine

TABLE 6.4 Some Coenzymes and Their Characteristics

Coenzyme	Vitamin from Which It Is Derived	Substance Transferred	Example of Use
Coenzyme A	Pantothenic acid (vitamin B_5)	Acyl groups	Carries the acetyl group that enters the TCA cycle
Flavin adenine dinucleotide (FAD)	Riboflavin (vitamin B_2)	Hydrogen atoms (2 electrons and 2 protons)	Carrier of reducing power
Nicotinamide adenine dinucleotide (NAD^+)	Niacin (vitamin B_3)	Hydride ions (2 electrons and 1 proton)	Carrier of reducing power
Pyridoxal phosphate	Pyridoxine (vitamin B_6)	Amino groups	Transfers amino groups in amino acid synthesis
Tetrahydrofolate	Folate/folic acid (vitamin B_9)	1-carbon molecules	1-carbon donor in nucleotide synthesis
Thiamin pyrophosphate	Thiamine (vitamin B_1)	Aldehydes	Helps remove CO_2 from pyruvate in the transition step

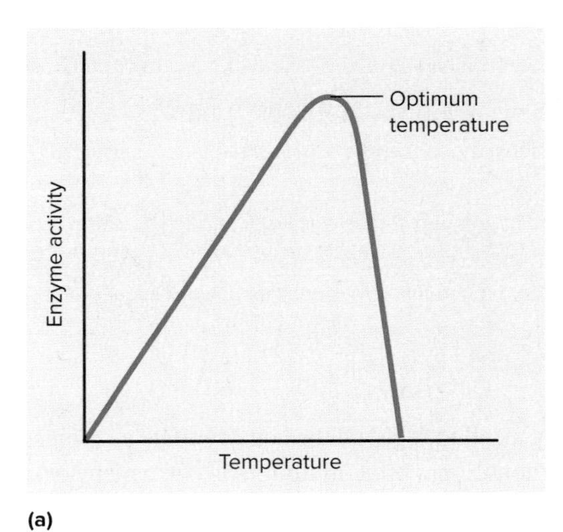

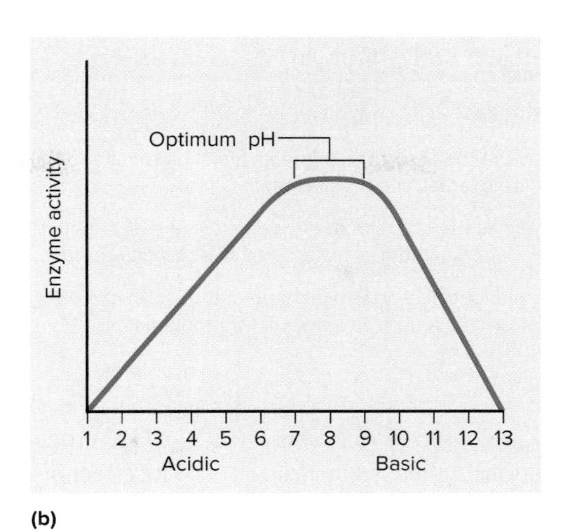

FIGURE 6.13 Environmental Factors That Influence Enzyme Activity **(a)** A rise in temperature increases the speed of enzymatic activity until the optimum temperature is reached. If the temperature gets too high, the enzyme denatures and no longer functions. **(b)** Most enzymes function best at pH values slightly above 7.

Why would an enzyme no longer function once it denatures?

to isoleucine. When the level of isoleucine is relatively high, the pathway is shut down. The binding of the isoleucine is reversible, however, so the enzyme becomes active again if isoleucine levels decrease. Cells can also control the amount of enzyme they synthesize, a topic discussed in chapter 7.

▶▶| **bacterial gene regulation**

Compounds that reflect a cell's relative energy supply often regulate allosteric enzymes of catabolic pathways. This allows cells to adjust the flow of metabolites through these pathways in response to changing energy needs. High levels of ATP inhibit certain enzymes and, as a consequence, slow down catabolic processes. In contrast, high levels of ADP warn that a cell's energy stores are low and stimulate the activity of some enzymes.

Enzyme Inhibition

Enzymes can be inhibited by a variety of compounds other than the regulatory molecules just described (**table 6.5**). These inhibitory compounds can prevent microbial growth, so they are medically and commercially valuable. The site on the enzyme to which an inhibitor binds determines whether it functions as a competitive or non-competitive inhibitor.

Competitive Inhibition

In **competitive inhibition,** the inhibitor binds to the active site of the enzyme, blocking access of the substrate to that site; in other words, the inhibitor competes with the substrate for the active site. Generally the inhibitor has a chemical structure similar to that of the normal substrate.

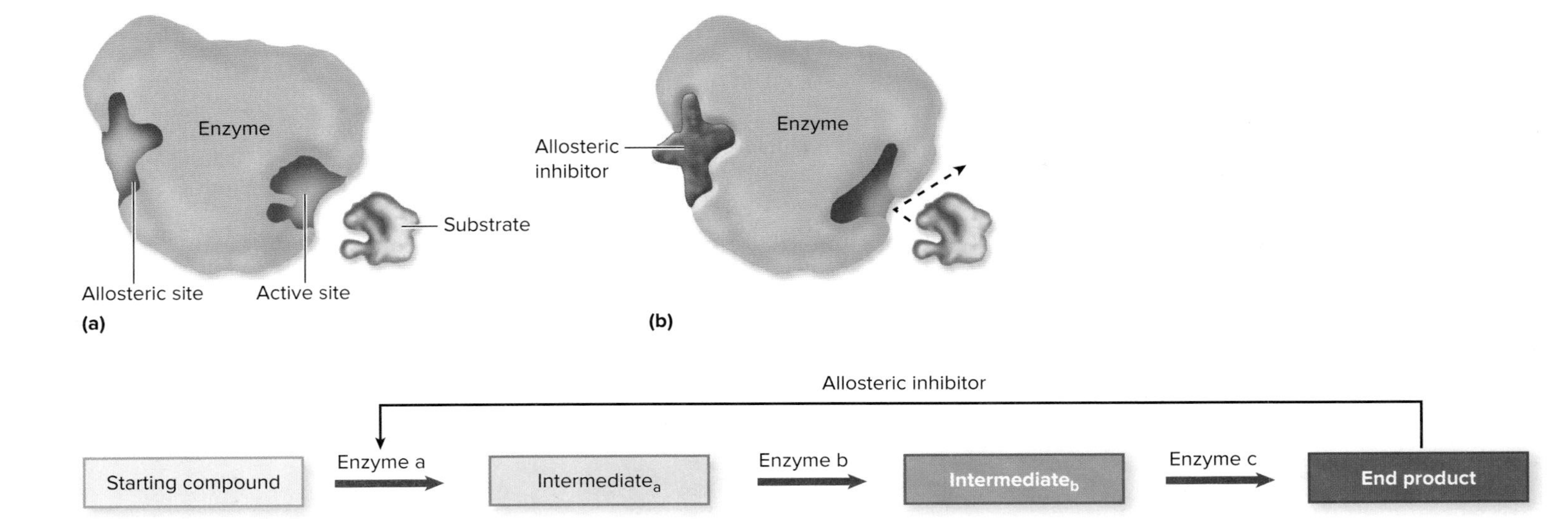

FIGURE 6.14 Regulation of Allosteric Enzymes **(a)** Allosteric enzymes have, in addition to the active site, an allosteric site. **(b)** The binding of a regulatory molecule to the allosteric site causes the shape of the enzyme to change, altering the relative affinity of the enzyme for its substrate. **(c)** The end product of a given biosynthetic pathway generally acts as an allosteric inhibitor of the first enzyme of that pathway.

Why would a cell need to regulate enzyme activity?

TABLE 6.5 Characteristics of Enzyme Inhibitors

Type	Characteristics
Competitive inhibition	Inhibitor binds to the active site of the enzyme, blocking access of the substrate to that site. Competitive inhibitors such as sulfa drugs are used as antibacterial medications.
Non-competitive inhibition (by regulatory molecules)	Inhibitor changes the shape of the enzyme, so that the substrate can no longer bind the active site. This is a reversible action that cells use to control the activity of allosteric enzymes.
Non-competitive inhibition (by enzyme poisons)	Inhibitor permanently changes the shape of the enzyme, making the enzyme non-functional. Enzyme poisons such as mercury are used in certain antimicrobial compounds.

A good example of competitive inhibition is the action of the group of antimicrobial medications called sulfa drugs (**figure 6.15**). These inhibit an enzyme in the pathway bacteria use to synthesize the vitamin folate. The drug does not affect human metabolism because humans cannot synthesize folate; it must be consumed in foods or in nutritional supplements (as folic acid). Sulfa drugs have a structure similar to *para*-aminobenzoic acid (PABA), an intermediate in the bacterial pathway for folate synthesis. Because of this, they fit into the active site of the enzyme that normally uses PABA as a substrate. By doing so, they prevent the enzyme from binding PABA. The greater the number of sulfa molecules relative to PABA molecules, the more likely it is that the active site of the enzyme will be occupied by a sulfa molecule. Once the sulfa is removed, however, the enzyme functions normally with PABA as the substrate. ⏭ sulfa drugs

Non-Competitive Inhibition

Non-competitive inhibition occurs when the inhibitor binds to a site other than the active site (see figure 6.14). The binding changes the shape of the enzyme so that the substrate can no longer bind the active site. The allosteric inhibitors discussed earlier are non-competitive inhibitors that have a reversible action and are produced by a cell to regulate the activity of its enzymes. The effect of other non-competitive inhibitors is permanent. For example, mercury oxidizes the S–H groups of the amino acid cysteine in proteins. This converts cysteine to cystine, which cannot form the important covalent disulfide bond (S–S). As a result, the enzyme shape changes, making it non-functional; the inhibitor "poisons" the enzyme. ⏮ cysteine, ⏮ disulfide bond

MicroAssessment 6.2

Enzymes speed the conversion of a substrate into a product with extraordinary specificity. They are neither consumed nor permanently changed in the reaction. The activity of some enzymes requires a cofactor. Environmental factors influence enzyme activity and, by doing so, determine how rapidly microorganisms multiply. The activity of allosteric enzymes can be regulated. A variety of different compounds and conditions affect enzyme activity.

4. Explain why sulfa drugs prevent bacterial growth without harming the human host.
5. Explain the function of a coenzyme.
6. Why is it important for a cell that allosteric inhibition be reversible?

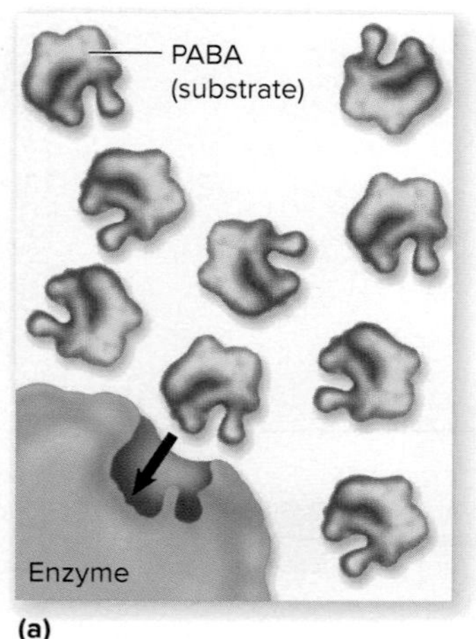

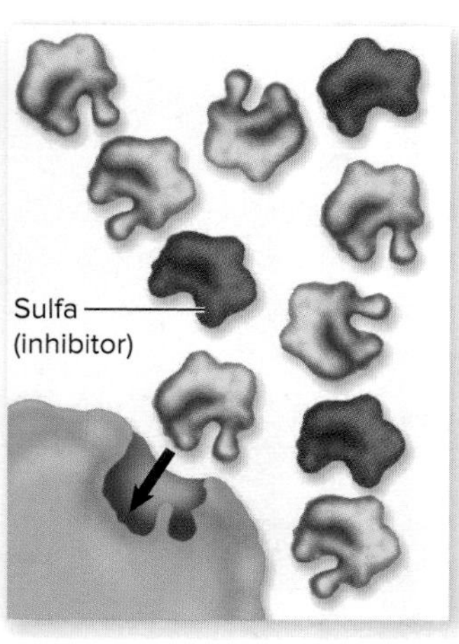

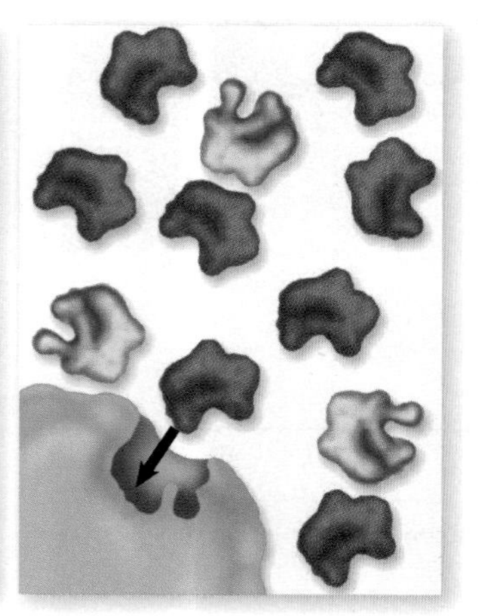
(a)

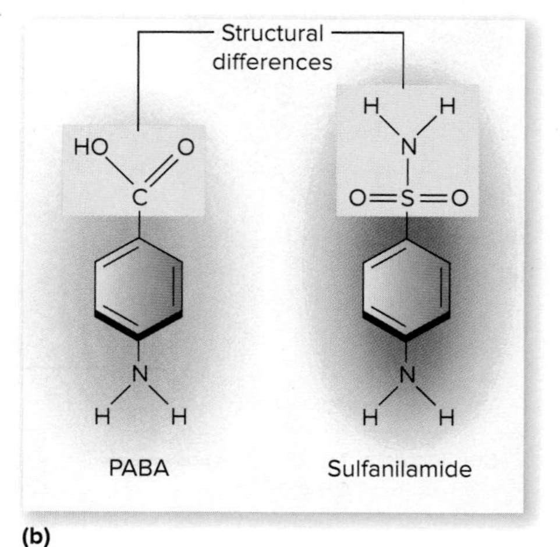

(b)

FIGURE 6.15 Competitive Inhibition of Enzymes **(a)** The inhibitor competes with the normal substrate for binding to the active site. The greater the number of inhibitor molecules relative to substrate molecules, the more likely it is that the active site of the enzyme will be occupied by an inhibitor. **(b)** A competitive inhibitor generally has a chemical structure similar to the normal substrate. Sulfanilamide, shown here, is a sulfa drug.

? **Will the illustrated enzyme function if sulfa is removed?**

6.3 ■ The Central Metabolic Pathways

Learning Outcomes

11. Diagram a simple overview that shows how the central metabolic pathways degrade glucose (include the starting and end products).
12. Compare and contrast each of the central metabolic pathways with respect to the yield of ATP, reducing power, and number of different precursor metabolites.

The three central metabolic pathways—glycolysis, the pentose phosphate pathway, and the tricarboxylic acid cycle (TCA cycle)—degrade organic molecules in a step-wise fashion, generating:

- **ATP** by substrate-level phosphorylation
- **Reducing power** in the form of NADH, $FADH_2$, and NADPH
- **Precursor metabolites** (see table 6.2)

This section describes how a molecule of glucose is broken down in the central metabolic pathways. Bear in mind, however, that many millions of molecules of glucose enter a cell, and different molecules can have different fates (see figure 6.9). For example, a cell might oxidize one glucose molecule completely to CO_2, thereby producing the maximum amount of ATP. Another glucose molecule might enter glycolysis, or perhaps the pentose phosphate pathway, only to be siphoned off as a precursor metabolite for use in biosynthesis. The step and rate at which the various intermediates are removed for biosynthesis will dramatically affect the overall energy gain of catabolism. This is generally overlooked in descriptions of the ATP-generating functions of these pathways for the sake of simplicity. However, it is important to keep in mind that because these pathways serve more than one function, the calculated energy yields are only theoretical.

The intermediates and end products of metabolic pathways are sometimes organic acids, which are weak acids. Depending on the pH of the environment, these occur as either the undissociated or the dissociated (ionized) form. Biologists often use the names of the two forms interchangeably—for example, pyruvic acid and pyruvate (an ion). Note, however, that at the near-neutral pH inside the cell, the ionized form predominates, whereas outside the cell, the acid often predominates. ⏮ion, ⏮pH

The pathways of central metabolism, along with the transition step that links glycolysis and the TCA cycle, are compared in **table 6.6.** The entire pathways with chemical formulas and enzyme names are illustrated in Appendix III.

TABLE 6.6 Comparison of the Central Metabolic Pathways

Pathway	Characteristics
Glycolysis	Glycolysis generates: • 2 ATP (net) by substrate-level phosphorylation • 2 NADH + 2 H^+ • six different precursor metabolites
Pentose phosphate cycle	The pentose phosphate cycle generates: • NADPH + H^+ (amount varies) • two different precursor metabolites
Transition step	The transition step, repeated twice to oxidize two molecules of pyruvate to acetyl-CoA, generates: • 2 NADH + 2 H^+ • one precursor metabolite
TCA cycle	The TCA cycle, repeated twice to incorporate two acetyl groups, generates: • 2 ATP by substrate-level phosphorylation (may involve conversion of GTP) • 6 NADH + 6 H^+ • 2 $FADH_2$ • two different precursor metabolites

Glycolysis

Glycolysis converts glucose to pyruvate (**figure 6.16**). For each molecule of glucose that enters the pathway, two molecules of pyruvate can be made. This generates a net gain of two molecules of ATP and two molecules of NADH. In addition, the pathway produces six different precursor molecules needed by *E. coli* (see table 6.2).

Glycolysis can be viewed as having two phases (see figure 6.16):

- **Investment** or **preparatory phase.** ① through ⑤ This uses energy because two different steps each transfer a high-energy phosphate group to the 6-carbon sugar. In eukaryotic cells, both phosphates come from ATP, as shown in figure 6.16. In bacteria, only the second phosphate comes from ATP. The first one is added as glucose is transported into the cell via group translocation; this phosphate is equivalent to one from ATP, so for simplicity, the fact that it comes from a different compound and in a different process is generally ignored. After the two phosphates are added, the 6-carbon sugar is then split to yield two 3-carbon molecules, each with a phosphate molecule. ⏮group translocation
- **Pay-off phase.** ⑥ through ⑩ This oxidizes and rearranges the 3-carbon molecules, generating 1 NADH and 2 ATP. Pyruvate is formed as a result. Note that the steps of this phase occur twice for each molecule of glucose that entered glycolysis. This is because the 6-carbon sugar was split into two 3-carbon molecules in the previous phase. In the pay-off phase, a total of 2 NADH and 4 ATP is made from 1 molecule of glucose.

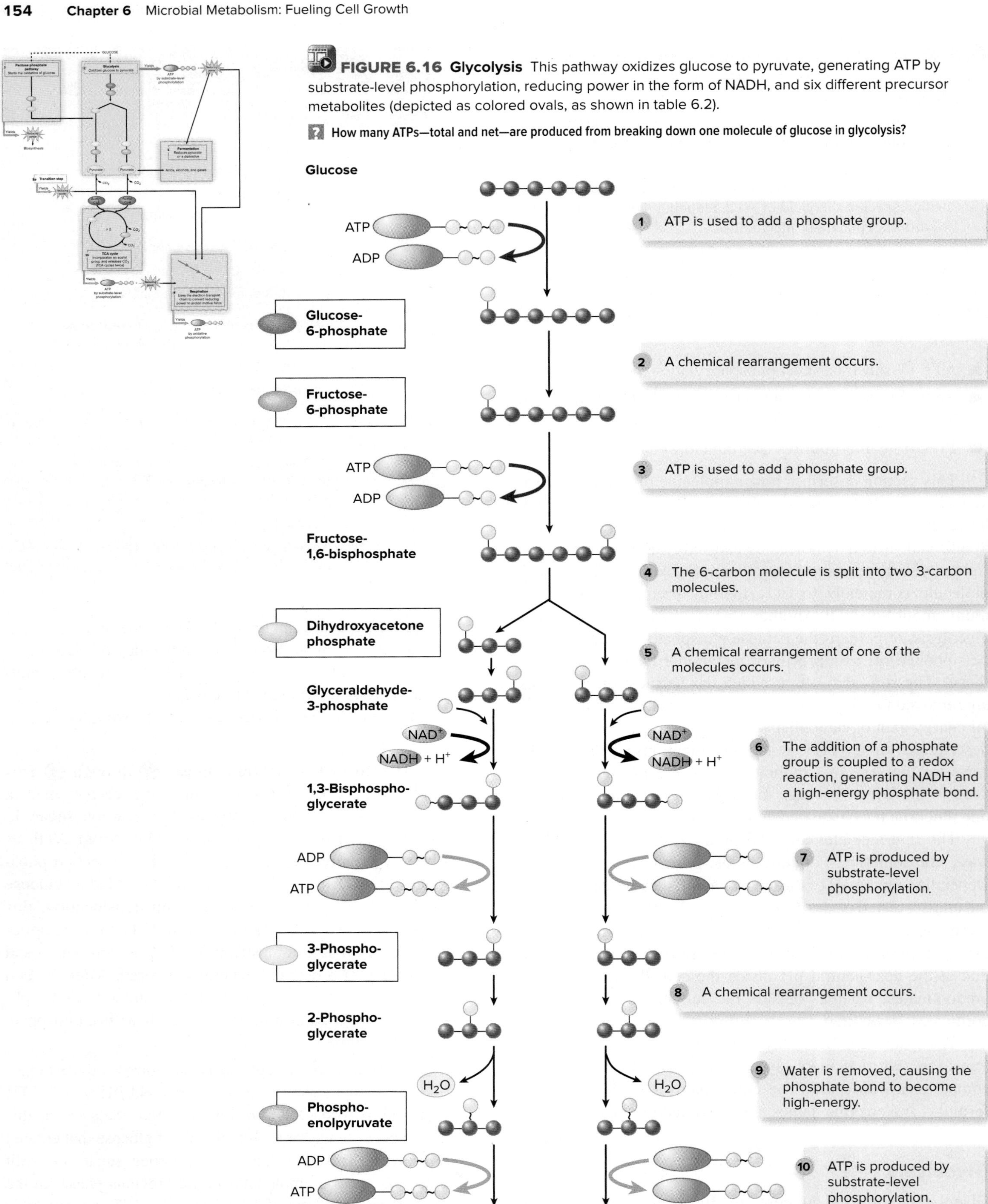

FIGURE 6.16 Glycolysis This pathway oxidizes glucose to pyruvate, generating ATP by substrate-level phosphorylation, reducing power in the form of NADH, and six different precursor metabolites (depicted as colored ovals, as shown in table 6.2).

How many ATPs—total and net—are produced from breaking down one molecule of glucose in glycolysis?

Yield of Glycolysis

For every glucose molecule degraded, the steps of glycolysis produce:

- **ATP**: 2 molecules of ATP, net gain (4 ATP molecules are made in the pay-off phase, but 2 are spent in the investment phase).
- **Reducing power**: The pay-off phase converts 2 NAD^+ to 2 NADH + 2 H^+.
- **Precursor metabolites**: Five intermediates of glycolysis as well as the end product, pyruvate, are precursor metabolites used by *E. coli.*

Pentose Phosphate Pathway

The other central metabolic pathway used by cells to break down glucose is the pentose phosphate pathway. This pathway is particularly important because of its contribution to biosynthesis. It generates reducing power in the form of NADPH, and two of its intermediates—ribose-5-phosphate and erythrose-4-phosphate—are important precursor metabolites. A product of this complex pathway is glyceraldehyde-3-phosphate (G3P), which can enter a step in glycolysis for further breakdown.

Yield of the Pentose Phosphate Pathway

The yield of the pentose phosphate pathway varies, depending upon which of several possible alternatives are taken. It can produce:

- **Reducing power:** A variable amount of reducing power in the form of NADPH.
- **Precursor metabolites:** Two intermediates of the pentose phosphate pathway are precursor metabolites.

Transition Step

The transition step, which links the previous pathways to the TCA cycle, involves several reactions catalyzed by a large multi-enzyme complex (**figure 6.17**). CO_2 is first removed from pyruvate, a step called decarboxylation. Then, a redox reaction transfers electrons to NAD^+, reducing it to NADH + H^+. Finally, the remaining 2-carbon acetyl group is joined to coenzyme A to form acetyl-CoA.

In prokaryotic cells, all the central metabolic pathways occur in the cytoplasm. In eukaryotic cells, however, the enzymes of glycolysis and the pentose phosphate pathways are located in the cytoplasm, whereas those of the transition step and the TCA cycle are within the mitochondrial matrix. Because of this, eukaryotic cells must transport pyruvate molecules into mitochondria for the transition step to occur.

⏮ **cytoplasm**, ⏮ **mitochondria**

Yield of the Transition Step

The transition step occurs twice for every molecule of glucose that enters glycolysis, oxidizing pyruvate to form acetyl-CoA. Together, these generate:

- **Reducing power:** 2 NADH + 2 H^+.
- **Precursor metabolites:** One precursor metabolite (acetyl-CoA).

Tricarboxylic Acid (TCA) Cycle

The tricarboxylic acid (TCA) cycle completes the oxidation of glucose (see figure 6.17). The cycle incorporates the acetyl groups from the transition step, ultimately releasing two molecules of CO_2. In addition to generating ATP and reducing power, the steps of the TCA cycle form two more precursor metabolites (see table 6.2).

① The TCA cycle begins when CoA transfers its acetyl group to the 4-carbon compound oxaloacetate, forming the 6-carbon compound citrate. ② Citrate is then chemically rearranged to make isocitrate. ③ This is oxidized and a molecule of CO_2 removed, producing the 5-carbon compound α-ketoglutarate. During the oxidation, NAD^+ is reduced to NADH + H^+. ④ Then, in a complex step, α-ketoglutarate is oxidized, CO_2 is removed, and CoA is added, producing the 4-carbon compound succinyl-CoA. During the process, NAD^+ is reduced to NADH + H^+. ⑤ The CoA is then removed from succinyl-CoA, and the energy released is used to produce ATP by substrate-level phosphorylation. Note that some bacteria and other cells make guanosine triphosphate (GTP) rather than ATP at this step; GTP can be converted to ATP. ⑥ Succinate is then oxidized to fumarate, as FAD is reduced to $FADH_2$. ⑦ A molecule of water is added to fumarate, producing malate. ⑧ This compound is then oxidized to oxaloacetate; note that oxaloacetate is the starting compound to which acetyl-CoA is added to initiate the cycle. During this last step, NAD^+ is reduced to NADH + H^+.

Yield of the TCA Cycle

The tricarboxylic acid cycle "turns" once for each acetyl-CoA that enters. Because two molecules of acetyl-CoA are generated for each glucose molecule that enters glycolysis, the breakdown of one molecule of glucose causes the TCA cycle to turn twice. Together, these two turns generate:

- **ATP:** 2 ATP produced in step 5.
- **Reducing power:** Redox reactions at steps 3, 4, 6, and 8 produce a total of 6 NADH + 6 H^+ and 2 $FADH_2$.
- **Precursor metabolites:** Two intermediates of the TCA cycle, formed in steps 3 and 8, are precursor metabolites.

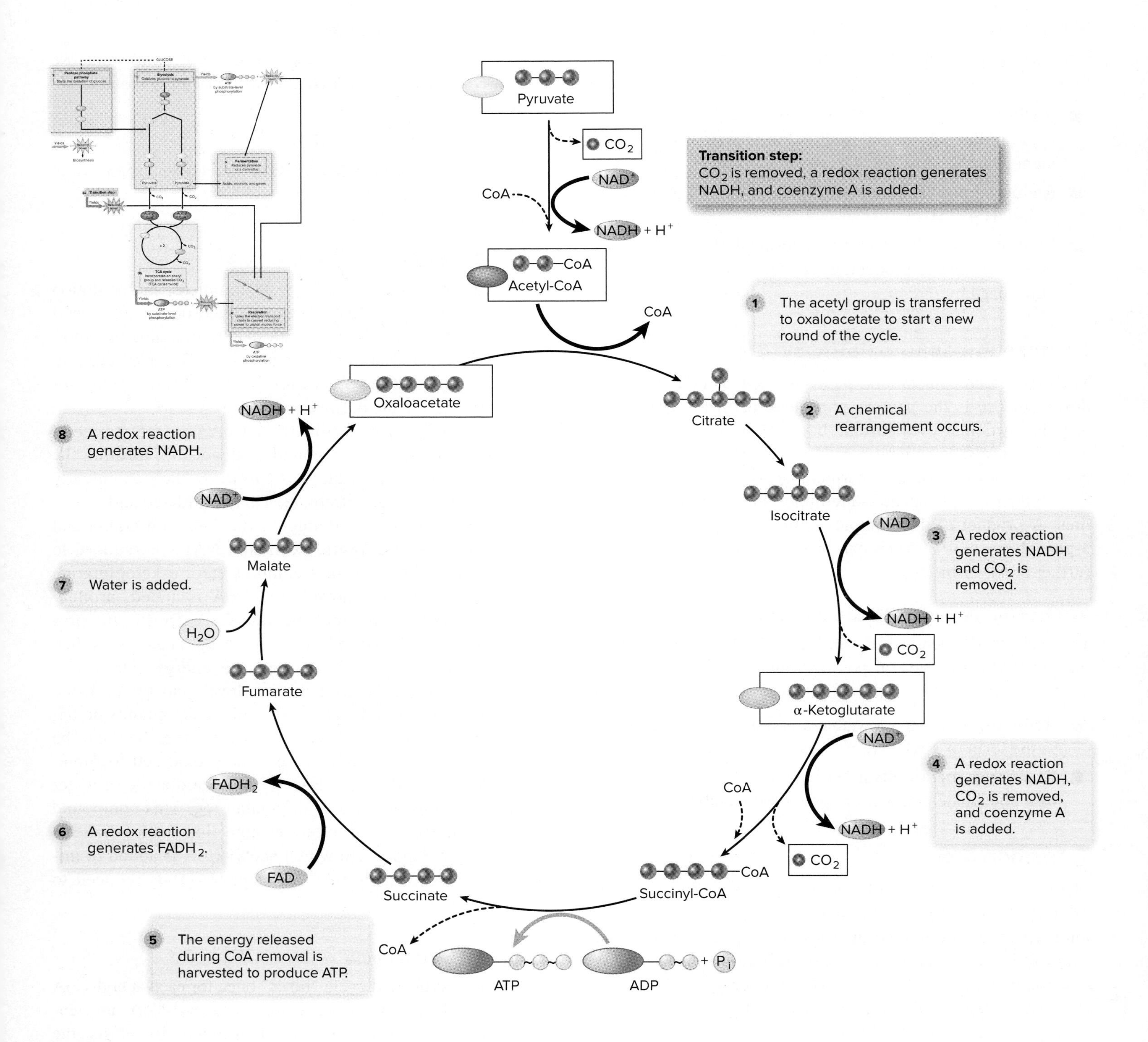

FIGURE 6.17 The Transition Step and the Tricarboxylic Acid Cycle The transition step links glycolysis and the TCA cycle, converting pyruvate to acetyl-CoA; it generates reducing power and one precursor metabolite (depicted as colored ovals, as shown in table 6.2). The TCA cycle incorporates the acetyl group of acetyl-CoA and, using a series of steps, releases CO_2; it generates ATP, reducing power in the form of both NADH and $FADH_2$, and two different precursor metabolites.

Which generates more reducing power—glycolysis or the TCA cycle?

MicroAssessment 6.3

Glycolysis oxidizes glucose to pyruvate, yielding some ATP and NADH and six different precursor metabolites. The pentose phosphate pathway also oxidizes glucose, but more importantly, it produces two different precursor metabolites and NADPH for biosynthesis. The transition step and the TCA cycle, each repeated twice, complete the oxidation of glucose, yielding some ATP, a great deal of reducing power, and three different precursor metabolites.

7. How does the "investment phase" of glycolysis effect the net yield of ATP in that pathway?
8. Which central metabolic pathway generates the most reducing power?
9. Which compound contains more free energy—glucose or oxaloacetate? On what did you base your conclusion?

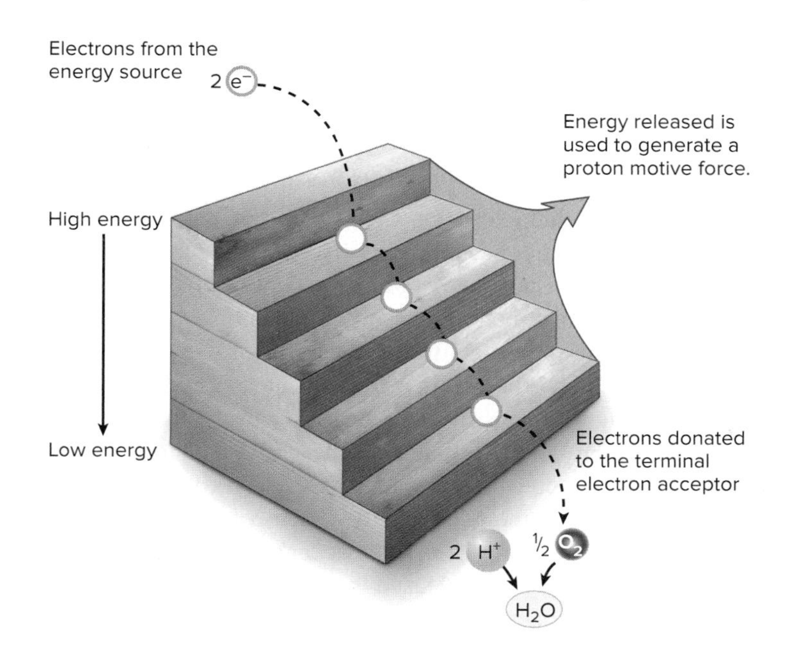

FIGURE 6.18 Electron Transport As electrons are passed along the electron transport chain, the energy released is used to establish a proton gradient.

O_2 is serving as the terminal electron acceptor in this diagram; is it being oxidized or reduced?

6.4 ■ Cellular Respiration

Learning Outcomes

13. Describe the components of the electron transport chain and how they generate a proton motive force.
14. Compare and contrast the electron transport chains of eukaryotes and prokaryotes.
15. Describe how a proton motive force is used to synthesize ATP and how the ATP yield of aerobic respiration is calculated.

Cellular respiration uses the reducing power generated in glycolysis, the transition step, and the TCA cycle to synthesize ATP. The mechanism, **oxidative phosphorylation,** involves two separate processes. First, the electron transport chain uses the reducing power of NADH and $FADH_2$ to generate a proton motive force. Then, the enzyme ATP synthase uses the energy of the proton motive force to drive the synthesis of ATP.

The association between the electron transport chain and ATP synthesis was proposed by the British scientist Peter Mitchell in 1961. His hypothesis, now called the **chemiosmotic theory,** was widely dismissed initially. Only through years of self-funded research was he able to convince others of its validity; he received a Nobel Prize in 1978.

The Electron Transport Chain (ETC)—Generating a Proton Motive Force

The **electron transport chain (ETC)** is a series of membrane-embedded electron carriers; it accepts electrons from NADH and $FADH_2$ and then passes those electrons from one carrier to the next. The transfer of electrons can be likened to a ball falling down a set of stairs; energy is released as the electrons are passed (**figure 6.18**). The energy released allows the ETC to pump protons across the membrane, generating an electrochemical gradient called a proton motive force (see figure 3.28). In prokaryotic cells, the ETC is in the cytoplasmic membrane; in eukaryotic cells, it is in the inner membrane of mitochondria. ◀◀ cytoplasmic membrane

Components of an Electron Transport Chain

Most carriers in the electron transport chain (ETC) are grouped into several large protein complexes that function as proton pumps. Others move electrons from one complex to the next. Three general types of electron carriers in the ETC are notable: quinones, cytochromes, and flavoproteins. **Quinones** are lipid-soluble organic molecules that move freely in the membrane, transferring electrons between certain protein complexes. Several types of quinones exist, one of the most common being ubiquinone (meaning "ubiquitous quinone"). Menaquinone, a quinone used in the ETC of some prokaryotes, serves as a source of vitamin K for humans and other mammals. This vitamin is required for proper blood coagulation, and mammals obtain much of their requirement by absorbing menaquinone produced by bacteria growing in the intestinal tract. **Cytochromes** are proteins that contain heme, a molecule that holds an iron atom in its center. Several different cytochromes exist, each designated with a letter, for example, cytochrome *c*. The presence of certain cytochromes can be used to distinguish certain groups of bacteria. For instance, the oxidase test—which is used in the steps to identify *Neisseria, Pseudomonas,* and *Campylobacter* species—detects the activity of cytochrome *c* oxidase (see table 10.4). **Flavoproteins** are proteins to which a flavin is attached. FAD and other flavins are synthesized from the vitamin riboflavin (see table 6.4).

General Mechanisms of Proton Pumps

An important characteristic of the electron carriers in the ETC is that some accept only hydrogen atoms (proton-electron pairs), whereas others accept only electrons. The spatial arrangement of these two types of carriers in the membrane causes protons to be moved from one side of the membrane to the other. This occurs because a hydrogen carrier receiving electrons from an electron carrier must pick up protons, which come from inside the cell (or matrix of the mitochondrion) due to the hydrogen carrier's relative location in the membrane. Conversely, when a hydrogen carrier passes electrons to a carrier that accepts electrons, but not protons, the protons are released to the outside of the cell (or intermembrane space of the mitochondrion). The net effect of these processes is that the electron transport chain pumps protons from one side of the membrane to the other, generating the concentration gradient across the membrane. Note that the gradient could not be established if energy were not released during electron transfer.

The Electron Transport Chain of Mitochondria

The electron transport chain of mitochondria has four different protein complexes, three of which function as proton pumps. In addition, two electron carriers (ubiquinone and cytochrome *c*) shuttle electrons between the complexes (**figure 6.19**):

- **Complex I** (also called NADH dehydrogenase complex). This accepts electrons from NADH, ultimately transferring them to ubiquinone (also called coenzyme Q); in the process, four protons are moved across the membrane.
- **Complex II** (also called succinate dehydrogenase complex). This accepts electrons from the TCA cycle, when

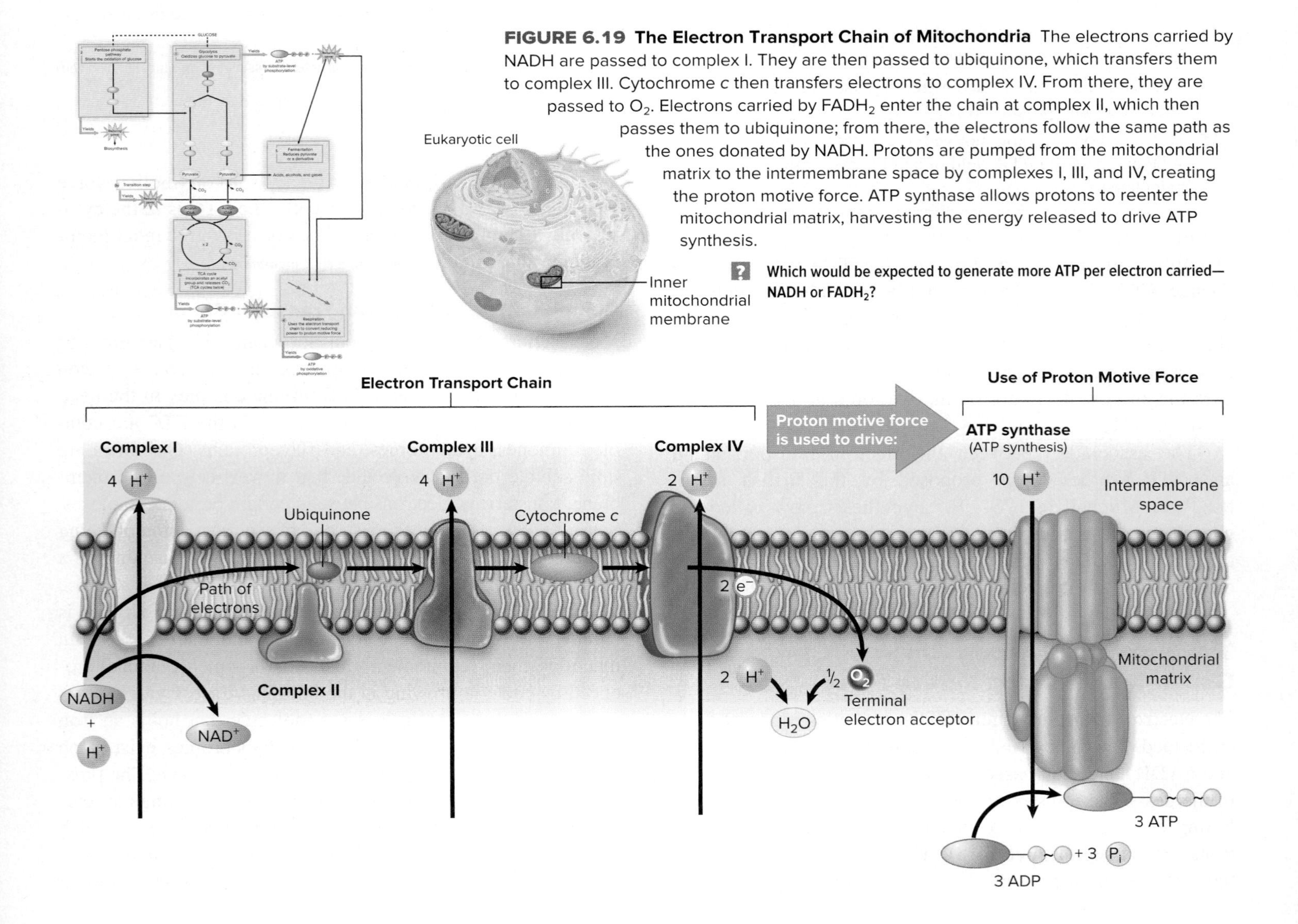

FIGURE 6.19 The Electron Transport Chain of Mitochondria The electrons carried by NADH are passed to complex I. They are then passed to ubiquinone, which transfers them to complex III. Cytochrome c then transfers electrons to complex IV. From there, they are passed to O_2. Electrons carried by $FADH_2$ enter the chain at complex II, which then passes them to ubiquinone; from there, the electrons follow the same path as the ones donated by NADH. Protons are pumped from the mitochondrial matrix to the intermembrane space by complexes I, III, and IV, creating the proton motive force. ATP synthase allows protons to reenter the mitochondrial matrix, harvesting the energy released to drive ATP synthesis.

? Which would be expected to generate more ATP per electron carried—NADH or $FADH_2$?

$FADH_2$ is formed during the oxidation of succinate (see figure 6.17, step 6). Note that the electrons carried by $FADH_2$ enter the electron transport chain "downstream" of those carried by NADH. Because of this, a pair of electrons carried by NADH results in more protons being pumped out than does a pair carried by $FADH_2$. Electrons are then transferred from complex II to ubiquinone.

- **Complex III** (also called cytochrome bc_1 complex). This accepts electrons from ubiquinone, which has carried them from either complex I or complex II. Complex III pumps four protons across the membrane before transferring the electrons to cytochrome *c*.
- **Complex IV** (also called cytochrome *c* oxidase complex). This accepts electrons from cytochrome *c* and pumps two protons across the membrane. Complex IV is a terminal oxidoreductase, meaning it transfers the electrons to the terminal electron acceptor, which, in this case, is O_2.

The Electron Transport Chains of Prokaryotes

Considering the versatility and diversity of prokaryotes, it should not be surprising that the types and arrangements of their electron transport components vary tremendously. In fact, a single species can have several alternative carriers, allowing cells to cope with ever-changing growth conditions.

The electron transport chain (ETC) of *E. coli* provides an excellent example of the versatility of some prokaryotes. This bacterium uses aerobic respiration when O_2 is available, but in the absence of O_2 it can switch to anaerobic respiration if a suitable electron acceptor such as nitrate is present. The *E. coli* ETC serves as a model for both aerobic and anaerobic respiration in bacteria.

Aerobic Respiration When growing aerobically in a glucose-containing medium, *E. coli* can use two different NADH dehydrogenases; one is a proton pump functionally equivalent to complex I of mitochondria (**figure 6.20**). *E. coli* also has a succinate dehydrogenase functionally equivalent to complex II of mitochondria. In addition to these protein complexes, *E. coli* can produce several alternatives, allowing the organism to use a variety of different energy sources, including H_2. The bacterium does not have the equivalent of complex III or cytochrome *c*. Instead, quinones move the electrons directly to one of two variations of a ubiquinol oxidase, which are functionally equivalent to complex IV of mitochondria. One variation of the ubiquinol oxidase works optimally in high O_2 conditions and pumps out four protons. The other pumps out only two protons, but it can more effectively scavenge O_2 and therefore is particularly useful in low O_2 conditions.

Anaerobic Respiration Anaerobic respiration harvests less energy than aerobic respiration because of the lower electron affinities of the terminal electron acceptors used (see figure 6.7). Some of the ETC components used during anaerobic respiration are different from those of aerobic respiration. For example, *E. coli* can synthesize a terminal oxidoreductase that uses nitrate as a terminal electron acceptor when O_2 is not available. This produces nitrite, which *E. coli* then converts to ammonia, avoiding the toxic effects of nitrite. Other bacteria can reduce nitrate even further, forming compounds such as nitrous oxide (N_2O), and nitrogen gas (N_2).

A group of obligate anaerobes called the sulfate-reducers use sulfate (SO_4^{2-}) as a terminal electron acceptor, producing hydrogen sulfide as an end product. The diversity and ecology of sulfate-reducing bacteria will be discussed in chapter 11.

⏭ sulfate-reducing bacteria

ATP Synthase—Using the Proton Motive Force to Synthesize ATP

Just as energy is required to establish a concentration gradient, energy is released as the gradient is removed or reduced. The enzyme ATP synthase uses the energy of the proton motive force to synthesize ATP. It does this by allowing protons to flow back into the bacterial cell (or matrix of the mitochondrion) in a controlled manner, simultaneously using the energy released to add a phosphate group to ADP. One molecule of ATP is formed from the entry of approximately three protons.

Theoretical ATP Yield of Oxidative Phosphorylation

By calculating the ATP yield of oxidative phosphorylation, the relative energy gains of respiration and fermentation can be compared. It is not a straightforward comparison, however, because oxidative phosphorylation has so many variables. This is particularly true for prokaryotic cells because they use the proton motive force to drive processes other than ATP synthesis. Prokaryotic cells, as a group, use different carriers in their electron transport chain and pump out a variable number of protons per pair of electrons passed.

The basis for calculating the ATP yield of oxidative phosphorylation relies on experimental studies using rat mitochondria. These studies indicate that approximately 2.5 ATP are made for each pair of electrons transferred to the electron transport chain by NADH; about 1.5 ATP are made for each pair transferred by $FADH_2$. For simplicity we will use whole numbers (3 ATP/NADH and 2 ATP/$FADH_2$) in calculations. Using these numbers, the maximum theoretical energy yield for oxidative phosphorylation in a prokaryotic cell (assuming the electron transport chain is similar to that of mitochondria) is as follows:

From glycolysis:
- 2 NADH → 6 ATP (assuming 3 for each NADH)

From the transition step:
- 2 NADH → 6 ATP (assuming 3 for each NADH)

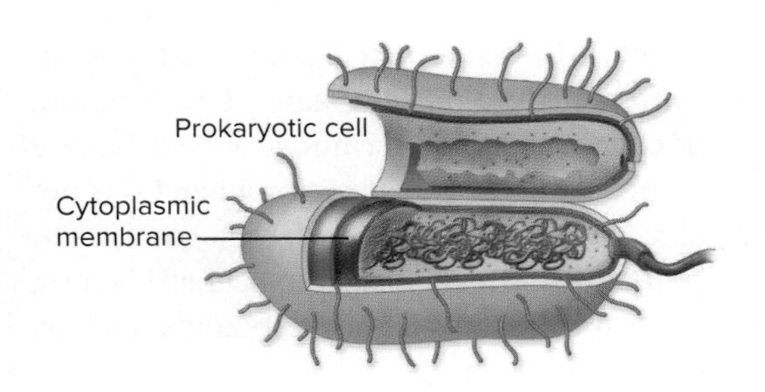

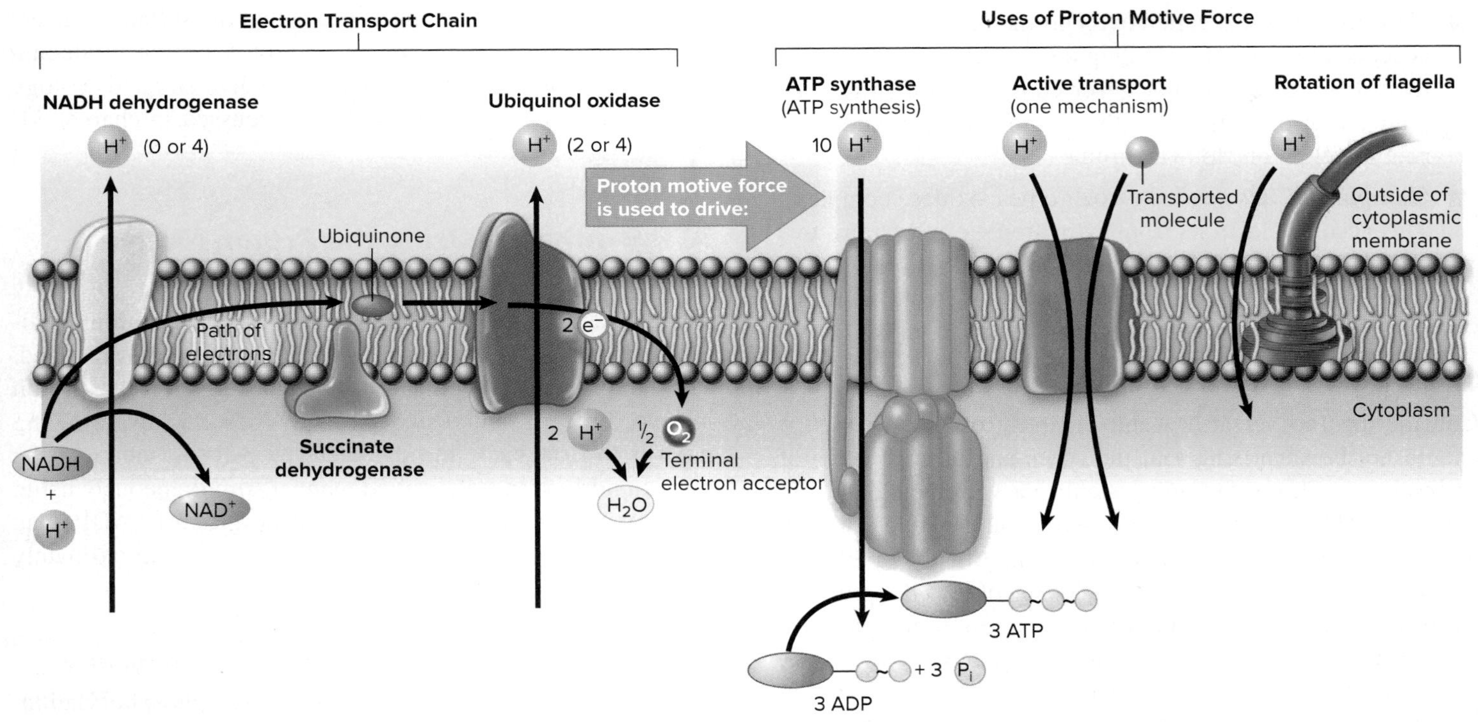

FIGURE 6.20 The Electron Transport Chain of *E. coli* Growing Aerobically in a Glucose-Containing Medium The electrons carried by NADH are passed to one of two different NADH dehydrogenases. They are then passed to ubiquinone, which transfers them to one of two ubiquinol oxidases. From there they are passed to O_2. The electrons carried by $FADH_2$ enter the chain at succinate dehydrogenase, which then transfers them to ubiquinone; from there, the electrons follow the same path as the ones donated by NADH. Protons are pumped out by one of the two NADH dehydrogenases and both ubiquinol oxidases, creating the proton motive force. ATP synthase allows protons to reenter the cell, using the energy released to drive ATP synthesis. The proton motive force is also used to drive one form of active transport and to power the rotation of flagella. *E. coli* has other components of the electron transport chain that function under different growth conditions.

? Succinate dehydrogenase is equivalent to which component of the mitochondrial electron transport chain?

From the TCA cycle:

- 6 NADH → 18 ATP (assuming 3 for each NADH)
- 2 $FADH_2$ → 4 ATP (assuming 2 for each $FADH_2$)

Total maximum ATP yield from oxidative phosphorylation = 34

The ATP gain as a result of oxidative phosphorylation will be slightly less in eukaryotic cells than in prokaryotic cells because of the fate of the reducing power (NADH) generated during glycolysis. Recall that in eukaryotic cells, glycolysis takes place in the cytoplasm, whereas the electron transport chain is located in the mitochondria. Consequently, the electrons carried by cytoplasmic NADH must be moved across the mitochondrial membrane before they can enter the electron transport chain. This uses approximately 1 ATP per NADH generated during glycolysis.

ATP Yield of Aerobic Respiration in Prokaryotes

Now that the ATP-yielding components of the central metabolic pathways have been considered, we can calculate the theoretical maximum ATP yield of aerobic respiration in prokaryotes. This yield is illustrated in **figure 6.21.**

Substrate-level phosphorylation:

- 2 ATP (from glycolysis; net gain)

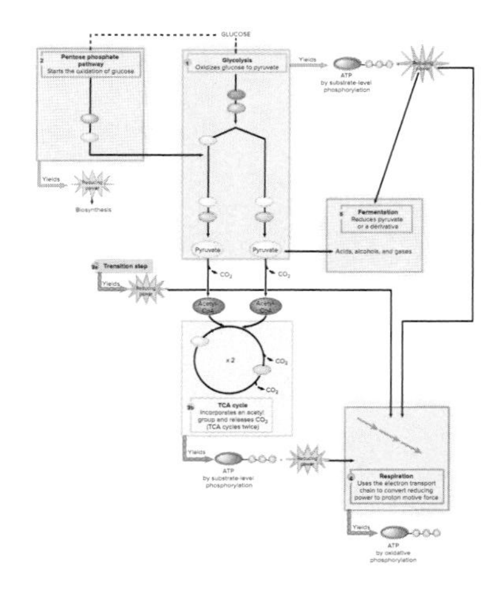

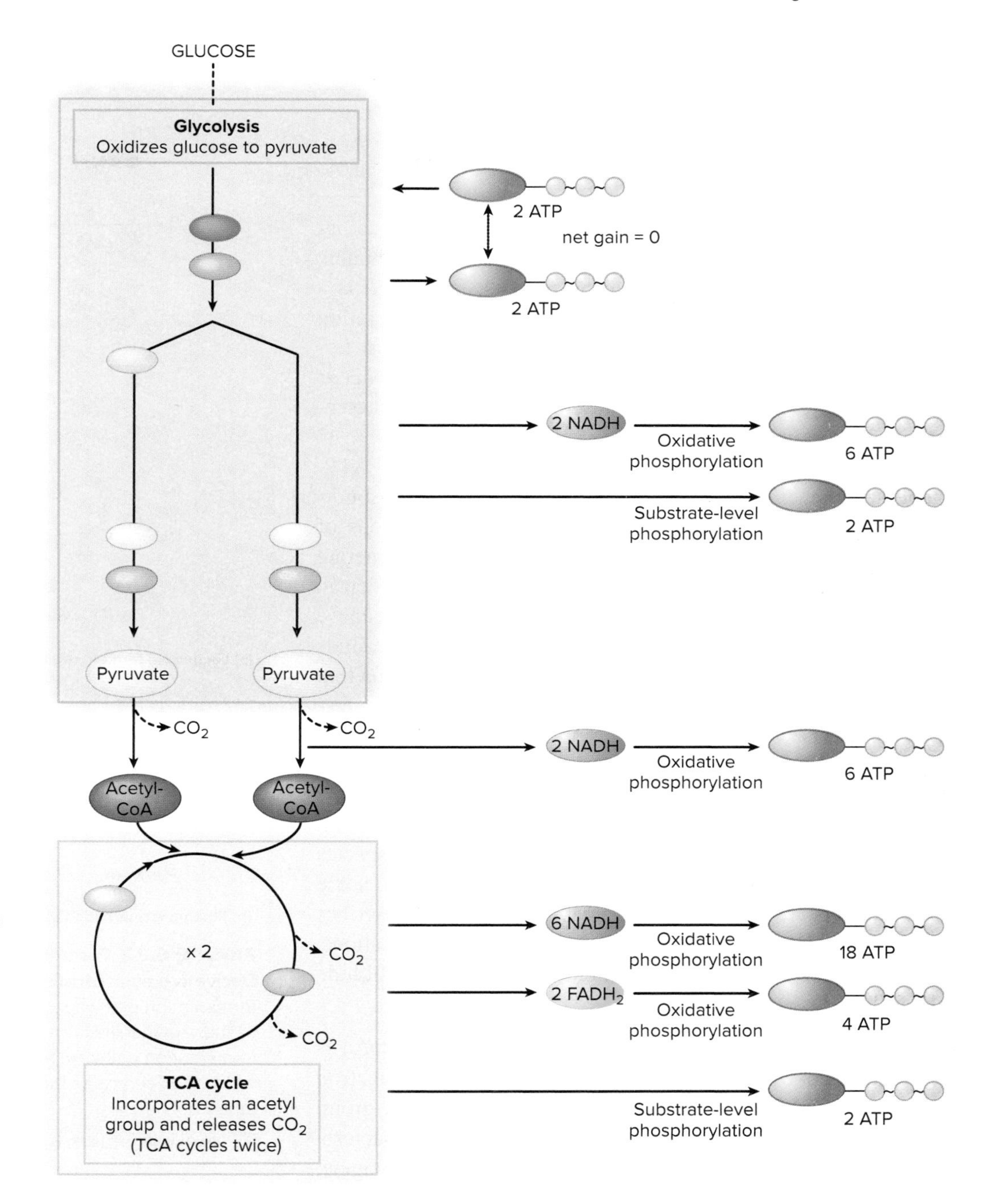

FIGURE 6.21 Maximum Theoretical Energy Yield from Aerobic Respiration in a Prokaryotic Cell This maximum energy yield calculation assumes that for every pair of electrons transferred to the electron transport chain, 3 ATP are synthesized; and for every pair of electrons donated by $FADH_2$, 2 ATP are synthesized.

Why is it difficult to calculate the actual maximum ATP yield of respiration in a prokaryotic cell?

- 2 ATP (from the TCA cycle)
- 4 total

Oxidative phosphorylation:

- 6 ATP (from the reducing power gained in glycolysis)
- 6 ATP (from the reducing power gained in the transition step)
- 22 ATP (from the reducing power gained in the TCA cycle)
- 34 total

Total ATP gain (theoretical maximum) = 38

MicroAssessment 6.4

Cellular respiration uses the NADH and $FADH_2$ generated in glycolysis, the transition step, and the TCA cycle to synthesize ATP. The electron transport chain is used to convert reducing power into a proton motive force. ATP synthase then harvests that energy to synthesize ATP. The overall process is called oxidative phosphorylation. In aerobic respiration, O_2 serves as the terminal electron acceptor; anaerobic respiration uses a molecule other than O_2.

10. In bacteria, what is the role of the molecule that serves as a source of vitamin K for humans?
11. Why is the overall ATP yield in aerobic respiration only a theoretical number?
12. Why could an oxidase also be called a reductase?

6.5 ■ Fermentation

Learning Outcomes

16. Describe the role of fermentation and the importance of the common end products.

Fermentation is used by organisms that cannot respire, either because a suitable terminal electron acceptor is not available or because they lack an electron transport chain. *E. coli* is a facultative anaerobe able to use any of three ATP-generating options: aerobic respiration, anaerobic respiration, and fermentation. In contrast, the only option for *Streptococcus pneumoniae* is fermentation because it does not have an electron transport chain.

When an organism is fermenting glucose, the only ATP-generating reactions are usually the substrate-level phosphorylations of glycolysis. The one or more additional steps of fermentation simply oxidize NADH as a means to regenerate NAD^+, using pyruvate or a derivative as a terminal electron acceptor (**figure 6.22**). This is a crucial function, because NAD^+ is needed to accept electrons in subsequent rounds of glycolysis—without it, glycolysis would stop. Note that respiring organisms do not have this problem because in respiration NAD^+ is regenerated as NADH donates its electrons to the electron transport chain.

The end products of fermentation are significant for a number of reasons (**figure 6.23**). For one thing, certain end products help identify bacterial isolates because a given species uses a characteristic fermentation pathway. In addition, some end products are commercially valuable. Chapter 30 describes how foods and beverages are produced using fermentations. Important end products of fermentation pathways include the following:

- **Lactic acid.** Lactic acid (the ionized form is lactate) is produced when pyruvate serves as the terminal electron acceptor. Lactic acid and other end products of a group of Gram-positive organisms called lactic acid bacteria contribute to the flavor and texture of cheese, yogurt, pickles, cured sausages, and other foods. Yet lactic acid also results in food spoilage, and causes tooth decay when produced by bacteria living on the teeth. ⏭ **lactic acid bacteria,** ⏭ **cheese, yogurt, and other fermented milk products,** ⏭ **pickled vegetables,** ⏭ **fermented meat products**
- **Ethanol.** Ethanol is produced in a pathway that first removes CO_2 from pyruvate, generating acetaldehyde, which then serves as the terminal electron acceptor. The end products of these reactions—which are used in making wine, beer, spirits, and bread—are ethanol and CO_2. Ethanol is also an important biofuel. *Saccharomyces* (yeast) and *Zymomonas* (bacteria) use this pathway. ⏭ **wine,** ⏭ **beer,** ⏭ **distilled spirits,** ⏭ **bread**

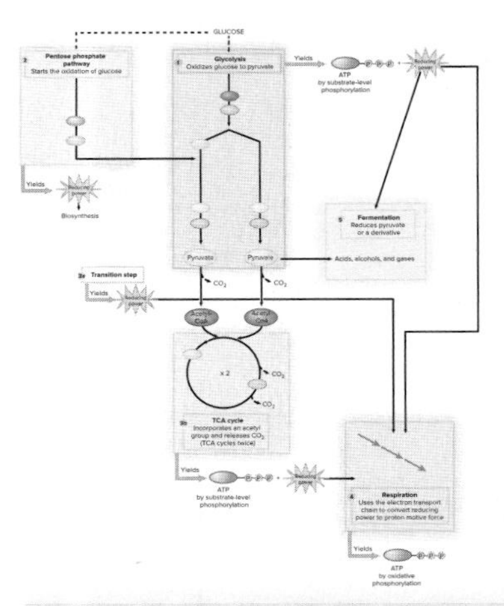

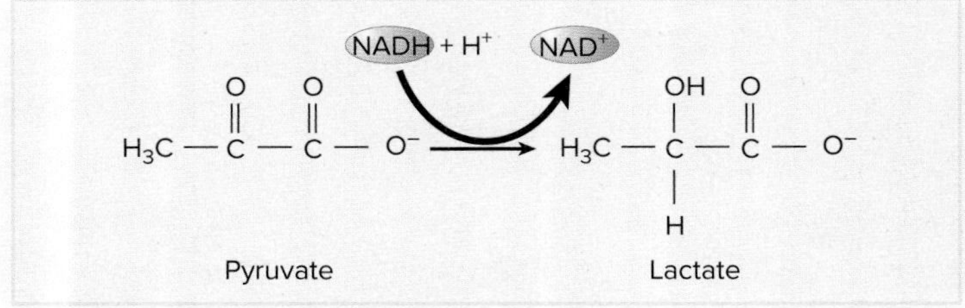

(a) Lactic acid fermentation pathway

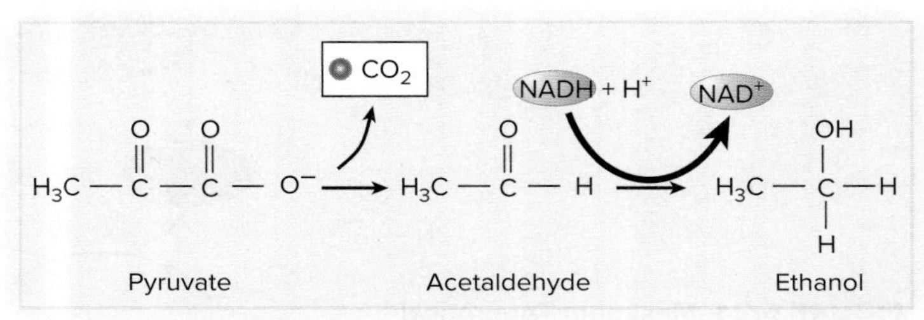

(b) Ethanol fermentation pathway

FIGURE 6.22 Fermentation Pathways Use Pyruvate or a Derivative as a Terminal Electron Acceptor (a) In the lactic acid fermentation pathway, the pyruvate generated during glycolysis serves as the terminal electron acceptor, producing lactate. **(b)** In the ethanol fermentation pathway, the pyruvate is first converted to acetaldehyde, which then serves as the terminal electron acceptor, producing ethanol.

? **Why is it important for cells to have a mechanism to oxidize NADH?**

- **Butyric acid.** Butyric acid (the ionized form is butyrate) and a variety of other end products are produced in a complex multistep pathway used by *Clostridium* species, which are obligate anaerobes. Under certain conditions, some organisms use a variation of this pathway to produce the organic solvents butanol and acetone.
- **Propionic acid.** Propionic acid (the ionized form is propionate) is generated in a multistep pathway that first adds CO_2 to pyruvate, generating a compound that then serves as a terminal electron acceptor. After this is reduced by NADH, it is further modified to form propionate. *Propionibacterium* species use this pathway, and their growth

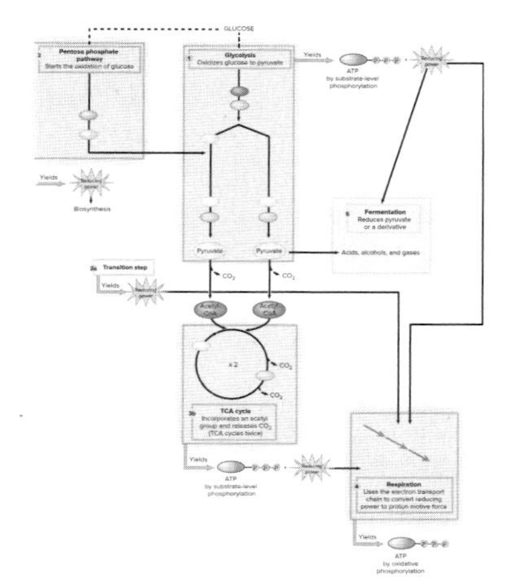

FIGURE 6.23 End Products of Fermentation Pathways

Plate, wine, beer, acetone: ©Brian Moeskau; Cheese: ©Photographer's Choice/Getty Images; Test tubes: ©McGraw-Hill Education/Auburn University Photographic Services

How can end products of fermentation help identify a bacterium?

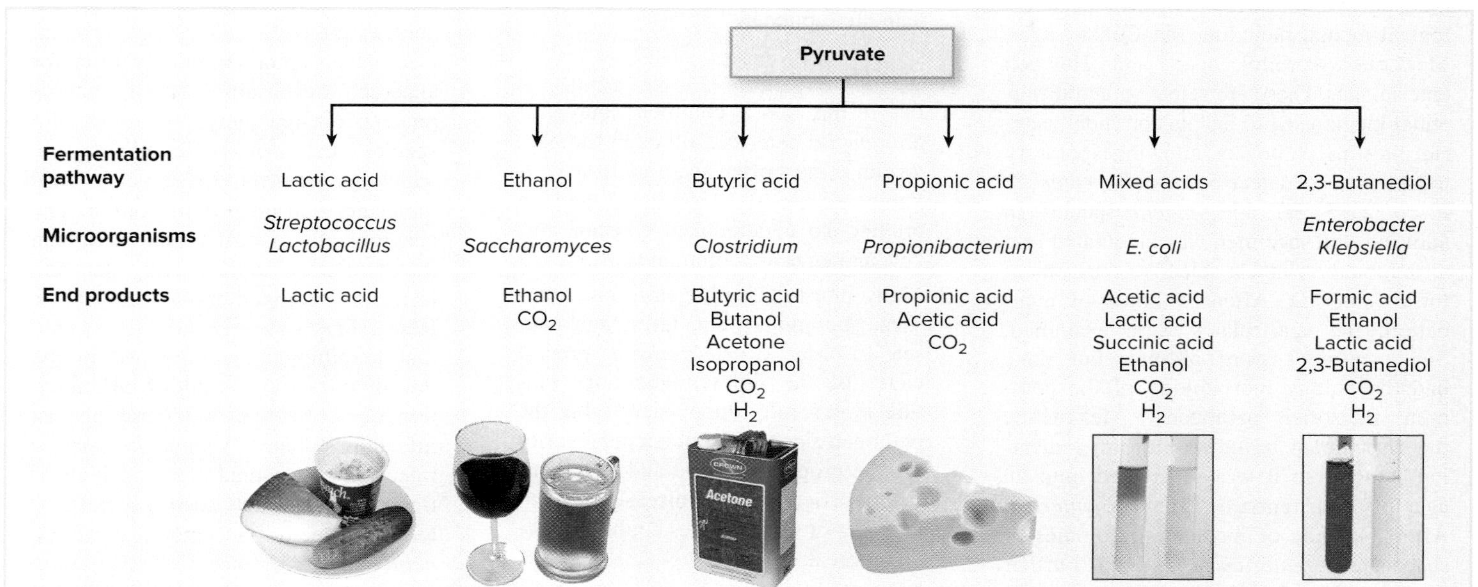

is encouraged as a part of Swiss cheese production. The CO_2 they make forms the holes, and propionic acid gives the cheese its unique flavor.

- **Mixed acids.** These are produced in a multistep branching pathway, generating a variety of different fermentation products including lactic acid, succinic acid (the ionized form is succinate), ethanol, acetic acid (the ionized form is acetate), and gases. The primary significance of this pathway is that it serves to differentiate certain members of the family *Enterobacteriaceae.* The methyl-red test detects the low pH resulting from the acidic end products, distinguishing members that use this pathway, such as *E. coli,* from those that do not, such as *Klebsiella* and *Enterobacter* (see table 10.4). ⏭ methyl-red test
- **2,3-Butanediol.** 2,3-Butanediol is produced in a multistep pathway that uses two molecules of pyruvate to generate acetoin and two molecules of CO_2. Acetoin is then used as the terminal electron acceptor. This is another pathway used to differentiate members of the family *Enterobacteriaceae;* the Voges-Proskauer test detects acetoin, distinguishing members that use this pathway (such as *Klebsiella* and *Enterobacter*) from those that do not (such as *E. coli;* see table 10.4). ⏭ Voges-Proskauer test

MicroAssessment 6.5

Fermentation stops short of the TCA cycle, using pyruvate or a derivative as a terminal electron acceptor. Many end products of fermentation are commercially valuable.

13. Why would a cell ferment rather than respire?
14. How do the methyl-red and Voges-Proskauer tests differentiate between certain members of the *Enterobacteriaceae*?
15. Fermentation is used as a means of preserving foods. Why would it slow spoilage?

FOCUS ON A CASE 6.1

A 9-year-old girl experiencing diarrhea, abdominal pain, fever, and nausea was taken to the emergency department at a local hospital. When her parents were questioned, they listed the foods she had eaten in the days before becoming ill. The doctor also asked them about animals she had touched, and the parents mentioned that she had recently been given a pet turtle. A stool sample from the child was submitted to the clinical laboratory.

In the laboratory, the stool sample was inoculated onto various types of bacteriological media, including MacConkey agar, MacConkey-sorbitol agar, and Hektoen enteric agar. These are selective and differential media used to isolate common bacterial intestinal pathogens, allowing laboratory technicians to detect *Salmonella enterica, E. coli* O157:H7, and *Shigella* species. In addition, the specimen was inoculated into tetrathionate broth, a selective enrichment for *S. enterica.* After 24 hours of incubation, the agar plates were examined. Small colonies were present, but none had the colony morphology of the common intestinal pathogens. Meanwhile, the enrichment broth was turbid, indicating growth, so it was inoculated onto an agar plate designed to isolate *S. enterica.* After 24 hours of incubation, colonies of suspected *S. enterica* were seen. Further tests confirmed the identification of *S. enterica.* ⏮ MacConkey agar, ⏮ selective media, ⏮ differential media

1. Why was it important that the child had recently been given a pet turtle?
2. Considering that the child had symptoms typical of *Salmonella* gastroenteritis, why was it necessary for the clinical lab to examine the stool specimen?
3. Tetrathionate is an oxidized form of sulfur. Why might this ingredient enrich for *S. enterica* in a broth culture?
4. Why would *S. enterica* benefit from using tetrathionate?

Discussion

1. Pet turtles are a common source of *Salmonella enterica,* although the bacterium is most often acquired from contaminated foods. It is often found on poultry, so undercooked chicken, turkey, and eggs are common sources.
2. Many intestinal pathogens cause diarrhea, abdominal pain, fever, and nausea. In order to determine the precise cause of the illness, laboratory culture is generally done. Identifying the pathogen aids the physician in prescribing the proper treatment and also helps health officials notice disease trends—a cluster of similar cases could indicate an outbreak.
3. Tetrathionate broth has been used for many years to isolate *S. enterica,* but the mechanism of how it enriches for the bacterium has only recently been determined. Unlike *E. coli* and most of its other competitors, *Salmonella* species can use tetrathionate as a terminal electron acceptor for anaerobic respiration, which gives them an advantage in a tetrathionate broth culture. In a typical broth culture, the dissolved O_2 is quickly used up by aerobically respiring bacteria. At that point, most bacteria must switch to fermentation in order to continue growing. This is because standard bacteriological media do not contain nitrate or any other terminal electron acceptor commonly used for anaerobic respiration. When tetrathionate is present, however, *Salmonella* species can anaerobically respire, which produces more ATP per glucose molecule than does fermentation. The anaerobically respiring *Salmonella* can therefore grow more quickly than the fermenting competitors.
4. Researchers have recently discovered that tetrathionate is produced in the intestinal tract as a result of inflammation, one of the body's defenses against infection. When a *Salmonella* species infects the intestinal tract, inflammation results, which then generates a terminal electron acceptor that *Salmonella* can use for anaerobic respiration. Just as in the tetrathionate broth used for enrichment, this gives the bacterium a selective advantage.

6.6 ■ Catabolism of Organic Compounds Other Than Glucose

Learning Outcomes

17. Briefly describe how polysaccharides and disaccharides, lipids, and proteins are degraded and used by a cell.

Microbes can use a variety of organic compounds other than glucose as energy sources, including polysaccharides, proteins, and lipids. To use macromolecules in the surrounding medium, a cell secretes enzymes that break the large compounds into their respective subunits (see figure 3.31). The resulting subunits are then transported into the cell and further degraded to form the appropriate precursor metabolites (**figure 6.24**). Recall that precursor metabolites can be either oxidized in one of the central metabolic pathways or used in biosynthesis. ⏮ precursor metabolites

Polysaccharides and Disaccharides

Starch and cellulose are both polymers of glucose, but different types of chemical linkages join their subunits together. The nature of this difference affects their degradation. Enzymes called amylases are made by a wide variety of organisms to digest starches. In contrast, cellulose is broken down by cellulases, which are produced by relatively few organisms. Among the organisms that produce cellulases are bacteria that reside in an organ called a rumen (found in cattle and other

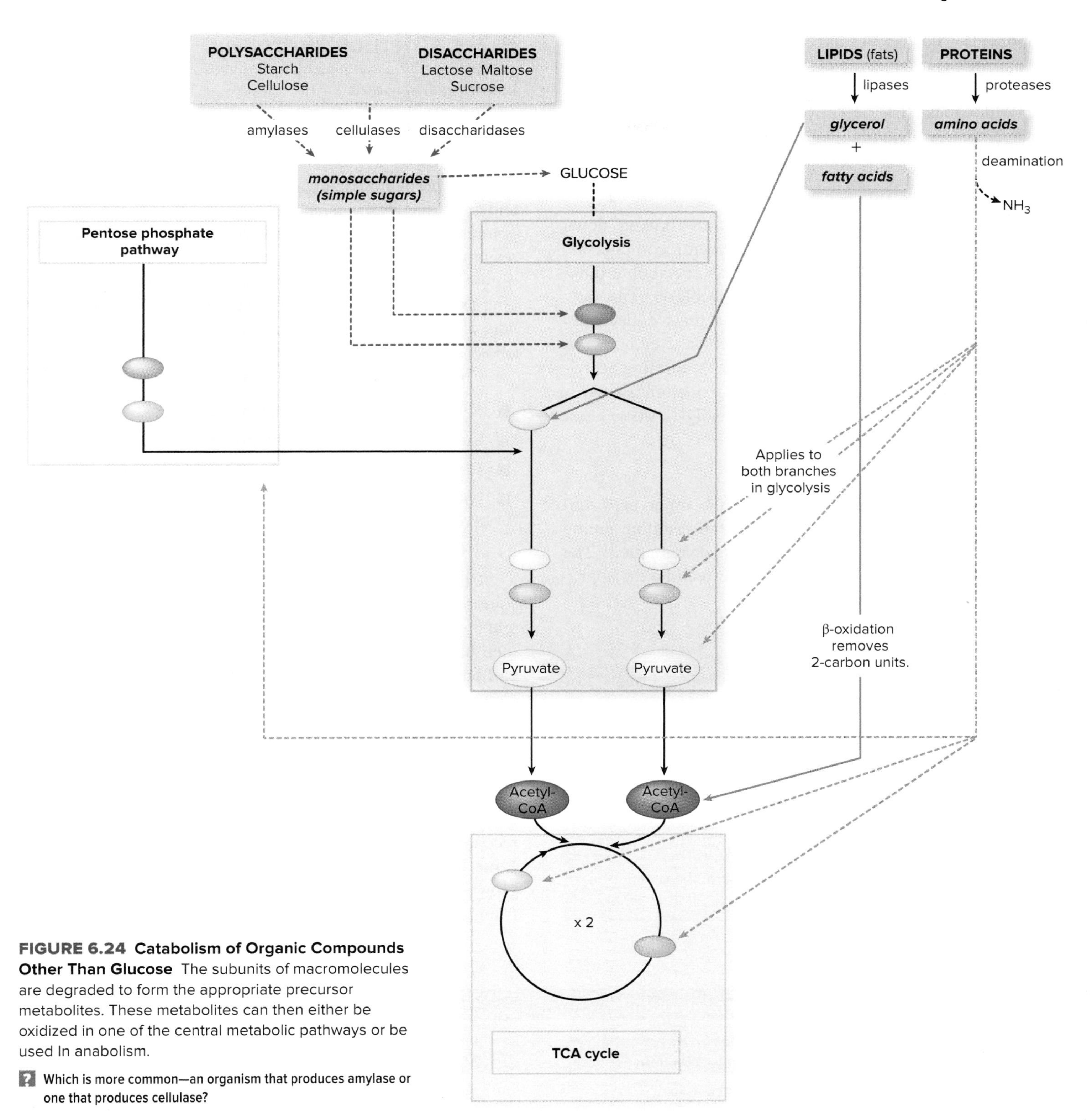

FIGURE 6.24 Catabolism of Organic Compounds Other Than Glucose The subunits of macromolecules are degraded to form the appropriate precursor metabolites. These metabolites can then either be oxidized in one of the central metabolic pathways or be used In anabolism.

? **Which is more common—an organism that produces amylase or one that produces cellulase?**

animals in the group called ruminants) and many types of fungi. Considering that cellulose is the most abundant organic compound on Earth, it is not surprising that fungi are important decomposers in terrestrial habitats. The glucose subunits released when polysaccharides are broken down can then be transported into the cell and used in glycolysis. ⏮ starch, ⏮ cellulose, ⏭ rumen

Disaccharides including lactose, maltose, and sucrose can be transported into the cell, where specific disaccharidases will then split them into their monosaccharide components. For example, the enzyme β-galactosidase breaks down lactose in the cell, forming glucose and galactose. Glucose can enter glycolysis directly, but other monosaccharides must first be converted to one of the precursor metabolites. ◂◂ disaccharides

Lipids

Fats, the most common simple lipids, are broken down by lipases into their fatty acid and glycerol components. Glycerol is then converted to the precursor metabolite dihydroxyacetone phosphate, which enters glycolysis. The fatty acids are degraded using a series of reactions collectively called β-oxidation. Each reaction transfers a 2-carbon unit from the end of the fatty acid to coenzyme A, forming acetyl-CoA, which enters the TCA cycle. Each β-oxidation is a redox reaction, generating 1 NADH + H^+ and 1 $FADH_2$. ◂◂ simple lipids

Proteins

Proteins are broken down to their respective amino acid subunits by proteases. The amino group of the resulting amino acids is removed by a reaction called a deamination. The remaining carbon skeletons are then converted into the appropriate precursor metabolites. ◂◂ protein

MicroAssessment 6.6

In order for macromolecules to be used as energy sources, they must be first broken down to their respective subunits. The subunits are then converted to the appropriate precursor metabolites, which can enter a central metabolic pathway.

16. Why do cells secrete enzymes that degrade macromolecules?

17. Explain the process used to degrade fatty acids.

18. How would cellulose-degrading bacteria in the rumen of a cow benefit the animal?

6.7 ■ Chemolithotrophs

Learning Outcomes

18. Explain how chemolithotrophs obtain energy.

Prokaryotes as a group are unique in their ability to use reduced inorganic chemicals such as hydrogen sulfide (H_2S) and ammonia (NH_3) as energy sources. Note that these compounds are the very ones produced as a result of anaerobic respiration, when inorganic molecules such as sulfate and nitrate serve as terminal electron acceptors. This is one important example of how nutrients are cycled; the waste products of one organism serve as an energy source for another. ▸▸ biogeochemical cycling and energy flow

Chemolithotrophs fall into four general groups (**table 6.7**):

- **Hydrogen bacteria** oxidize hydrogen gas.
- **Sulfur bacteria** oxidize hydrogen sulfide.
- **Iron bacteria** oxidize reduced forms of iron.
- **Nitrifying bacteria** include two groups of bacteria; one oxidizes ammonia (forming nitrite), and the other oxidizes nitrite (producing nitrate).

Chemolithotrophs extract electrons from inorganic energy sources, passing them to an electron transport chain that generates a proton motive force. The energy of this gradient is then harvested to make ATP, using the processes described earlier. As with chemoheterotrophs, the amount of energy gained depends on the energy source and the terminal electron acceptor (see figure 6.7).

Chemolithotrophs generally thrive in very specific environments where reduced inorganic compounds are found. For example, certain bacteria are found in sulfur-rich acidic environments; these organisms can be used to enhance the recovery of metals because they oxidize metal sulfides (**Focus Your Perspective 6.1**). Thermophilic chemolithotrophs that grow near hydrothermal vents of the deep ocean obtain energy from reduced inorganic compounds that spew from the vents.

TABLE 6.7 Metabolism of Chemolithotrophs

Common Name of Organism	Source of Energy	Oxidation Reaction(s) (Energy Yielding)	Important Feature(s) of Group	Common Genera in Group
Hydrogen bacteria	H_2	$H_2 + \frac{1}{2} O_2 \longrightarrow H_2O$	Can also use simple organic compounds for energy	*Hydrogenomonas*
Sulfur bacteria (non-photosynthetic)	H_2S	$H_2S + \frac{1}{2} O_2 \longrightarrow H_2O + S$ $S + 1\frac{1}{2} O_2 + H_2O \longrightarrow H_2SO_4$	Some members of this group can live at a pH of less than 1.	*Acidithiobacillus, Thiobacillus, Beggiatoa, Thiothrix*
Iron bacteria	Reduced Iron (Fe^2+)	$2\ Fe^{2+} + \frac{1}{2} O_2 + H_2O \longrightarrow 2\ Fe^{3+} + 2\ OH^-$	Iron oxide present in the sheaths of these bacteria	*Sphaerotilus, Gallionella*
Nitrifying bacteria	NH_3	$NH_3 + 1\frac{1}{2} O_2 \longrightarrow HNO_2 + H_2O$	Important in the nitrogen cycle	*Nitrosomonas*
	HNO_2	$HNO_2 + \frac{1}{2} O_2 \longrightarrow HNO_3$	Important in the nitrogen cycle	*Nitrobacter*

FOCUS YOUR PERSPECTIVE 6.1

Mining with Microbes

Microorganisms have been used to produce wine and bread for thousands of years. Only in the past several decades, however, have they been used with increasing frequency in another area: the mining industry. The mining process traditionally consists of digging crude ores (mineral-containing rocks) from the earth, crushing them, and then extracting the desired minerals. The extraction process for copper and gold frequently involves harsh conditions, such as burning the ore and then treating the remaining material with cyanide. Such activities are expensive and harmful to the environment. With the development of biomining, some of these problems are being solved.

In the process of biomining copper, the crude ore is piled outside the mine and then treated with acid. The acidic conditions encourage the growth of *Acidithiobacillus* species present naturally in the ore. These acidophilic bacteria use CO_2 as a carbon source, and they gain energy by oxidizing sulfides of iron first to sulfur and then to sulfuric acid. The sulfuric acid dissolves the insoluble copper from the ore. Similar processes are being applied to gold mining.

The current process of biomining uses microbes naturally in the ore. Many improvements should be possible. For example, oxidizing the minerals generates heat to the point that the bacteria may be killed, but it might be possible to use thermophiles to overcome this problem. In addition, many ores contain heavy metals, such as mercury, cadmium, and arsenic, which are toxic to the bacteria, but perhaps microorganisms resistant to these metals could be found. Biomining is still in its infancy.

The diversity and ecology of some chemolithotrophs will be discussed in chapter 11.

Unlike organisms that use organic molecules to fill both their energy and their carbon needs, chemolithotrophs incorporate CO_2 into an organic form. This process (carbon fixation) will be described after we cover photosynthesis (next section), because photosynthetic organisms also fix carbon.

MicroAssessment 6.7

Chemolithotrophs use reduced inorganic compounds as an energy source. They use carbon dioxide as a carbon source.

19. Describe the roles of hydrogen sulfide and carbon dioxide in chemolithoautotrophic metabolism.

20. Which energy source, Fe^{2+} or H_2S, would result in the greatest energy yield when O_2 is used as a terminal electron acceptor (*Hint:* Refer to figure 6.7)?

6.8 ■ Photosynthesis

Learning Outcomes

19. Describe the role of chlorophylls, bacteriochlorophylls, accessory pigments, reaction-center pigments, and antennae pigments in capturing radiant energy.

20. Compare and contrast the tandem photosystems of cyanobacteria and photosynthetic eukaryotes with the single photosystems of purple and green bacteria.

Plants, algae, and several groups of bacteria harvest the radiant energy of sunlight, and then use it to power the synthesis of organic compounds from CO_2. This conversion of radiant energy into chemical energy is called **photosynthesis.** The general reaction—with X indicating an atom such as oxygen or sulfur—is as follows:

$$6\ CO_2 + 12\ H_2X \xrightarrow{\text{Light Energy}} C_6H_{12}O_6 + 12\ X + 6\ H_2O$$

Photosynthetic processes are generally considered in two distinct stages. The **light reactions** (also called the light-dependent reactions) capture radiant energy and convert it to chemical energy in the form of ATP. An unrelated set of reactions (sometimes called the light-independent reactions or dark reactions) uses the ATP to synthesize organic compounds. This involves carbon fixation, which will be covered in the next section. Characteristics of various photosynthetic mechanisms are summarized in **table 6.8.**

Capturing Radiant Energy

Photosynthetic organisms are highly visible in their natural habitats because they have various colored pigments that capture the energy of light (radiant energy). The colors are due to the wavelengths reflected by the pigments—for example, pigments that absorb only blue and red light are green (see figure 5.7). Multiple pigments are involved in photosynthesis, increasing the range of wavelengths absorbed by a cell. The pigments are located in protein complexes called **photosystems** within photosynthetic membranes (**figure 6.25**). The photosystems specialize in capturing and using the energy of light.

Photosynthetic pigments include chlorophylls, bacteriochlorophylls, and accessory pigments. **Chlorophylls** are found in plants, algae, and cyanobacteria. The various types of chlorophylls are designated with a letter following the term—for example, chlorophyll *a*. **Bacteriochlorophylls** are found in anoxygenic photosynthetic bacteria ("anoxygenic" means they do not generate O_2). These pigments absorb

TABLE 6.8 Comparison of the Photosynthetic Mechanisms Used by Different Organisms

	Oxygenic Photosynthesis		Anoxygenic Photosynthesis	
	Plants, Algae	**Cyanobacteria**	**Purple Bacteria**	**Green Bacteria**
Location of the photosystem	In membranes of thylakoids, which are within the stroma of chloroplasts	In membranes of thylakoids, located within the cell	Within the cytoplasmic membrane; extensive invaginations in that membrane increase the surface area.	Primarily within the cytoplasmic membrane; chlorosomes attached to the inner surface of the membrane contain the accessory pigments.
Type of photosystem	Photosystem I and photosystem II		Similar to photosystem II	Similar to photosystem I
Primary light-harvesting pigment	Chlorophyll *a*	Chlorophyll *a*	Bacteriochlorophylls	Bacteriochlorophylls
Mechanism for generating reducing power	Non-cyclic photophosphorylation using both photosystems		Reversed electron transport	Non-cyclic use of the photosystem
Source of electrons for reducing power	H_2O	H_2O	Varies among the organisms in the group; may include H_2S, H_2, or organic compounds	
CO_2 fixation	Calvin cycle	Calvin cycle	Calvin cycle	Reversed TCA cycle
Accessory pigments	Carotenoids	Carotenoids, phycobilins	Carotenoids	Carotenoids

wavelengths not absorbed by chlorophylls, allowing the bacteria to grow in habitats where other photosynthetic organisms cannot. **Accessory pigments** increase the efficiency of light capture by absorbing wavelengths not absorbed by the other pigments. These pigments include carotenoids—found in a wide variety of photosynthetic prokaryotes and eukaryotes—and phycobilins, which are unique to cyanobacteria and red algae.

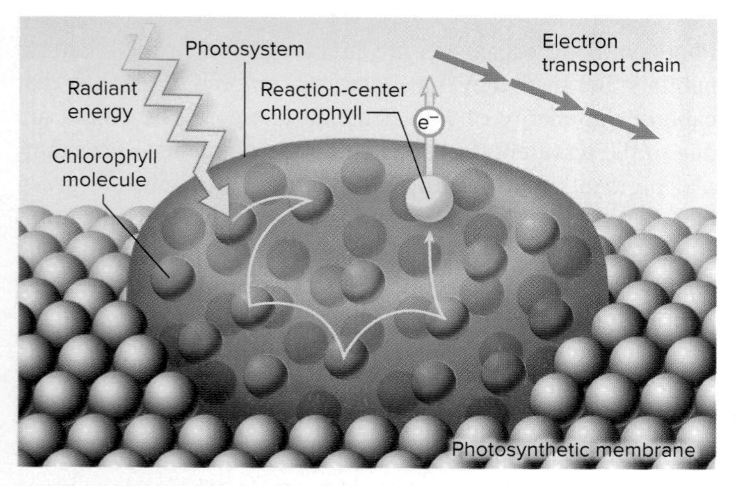

FIGURE 6.25 Photosystem Chlorophyll and other pigments capture the energy of light and then transfer it to a reaction-center pigment, which emits an electron that is then passed to an electron transport chain.

? **What is the role of an electron transport chain?**

Within the photosystems, some pigments function as reaction-center pigments and others function as antennae pigments (see figure 6.25). **Reaction-center pigments** are electron donors in the photosynthetic process. When excited by radiant energy, these emit high-energy electrons, which are then passed to an electron transport chain similar to that used in respiration. The reaction-center pigment of oxygenic photosynthetic organisms (plants, algae, and cyanobacteria) is chlorophyll *a,* whereas the anoxygenic photosynthetic organisms (purple and green bacteria) use one of the bacteriochlorophylls. **Antennae pigments** make up a complex that acts as a funnel, capturing the energy of light and then transferring it to a reaction-center pigment.

The photosystems of cyanobacteria are embedded in the membranes of structures called thylakoids located within the cells. Plants and algae also have thylakoids, in the stroma of the chloroplast (see figure 3.54). The similarity between the structure of chloroplasts and that of cyanobacteria is not surprising considering that the organelle appears to have descended from an ancestor of a cyanobacterium. ⏮ **thylakoid,** ⏮ **stroma,** ⏮ **chloroplast**

The photosystems of some types of photosynthetic bacteria are embedded in the cytoplasmic membrane. Bacteria in a group called purple bacteria have extensive invaginations in the membrane that maximize the surface area. Bacteria in a group called green bacteria have specialized structures called chlorosomes attached to the inner surface of the cytoplasmic membrane. These structures contain the accessory pigments.

Converting Radiant Energy into Chemical Energy

Photosynthetic organisms use the light reactions to accomplish two tasks:

- Capture radiant energy and use it to fuel ATP synthesis by the process of **photophosphorylation.**
- Generate the reducing power (NADPH or NADH, depending on the organism) needed to fix CO_2.

Light Reactions in Cyanobacteria and Photosynthetic Eukaryotic Cells

In cyanobacteria and chloroplasts, two distinct photosystems work together as part of the light reactions (**figure 6.26**). Energy absorbed by the photosystems raises the energy of electrons stripped from water to a high enough level that it can be used to generate a proton motive force as well as to produce reducing power. The process is oxygenic—that is, it generates O_2.

First we will consider the simplest situation, which occurs when the cell needs to synthesize ATP, but not reducing power. To accomplish this, only photosystem I is used. When radiant energy is absorbed by this photosystem, the reaction-center chlorophylls emit high-energy electrons. The electrons are then passed to an electron carrier, which transports them to a proton pump similar to complex III in the electron transport chain of mitochondria. After being used to move protons across the membrane—generating a proton motive force—the electrons are returned to photosystem I. As in oxidative phosphorylation, ATP synthase harvests the energy of the proton motive force to synthesize ATP. This overall process is called **cyclic photophosphorylation** because the electrons have followed a cyclical path: the molecule that serves as the electron donor (reaction-center chlorophyll) is also the terminal electron acceptor.

When photosynthetic cells must produce both ATP and reducing power, **non-cyclic photophosphorylation** is used. In this process, the electrons emitted by photosystem I are not passed to the proton pump but instead are used to reduce $NADP^+$ to NADPH. Although this action provides reducing power, the cell must now use another source to replenish the electrons emitted by reaction-center chlorophylls. In addition, the cell still needs to generate a proton motive force in order to synthesize ATP. Photosystem II plays an important role in this process. When photosystem II absorbs radiant energy, the reaction-center chlorophylls emit high-energy electrons that can be

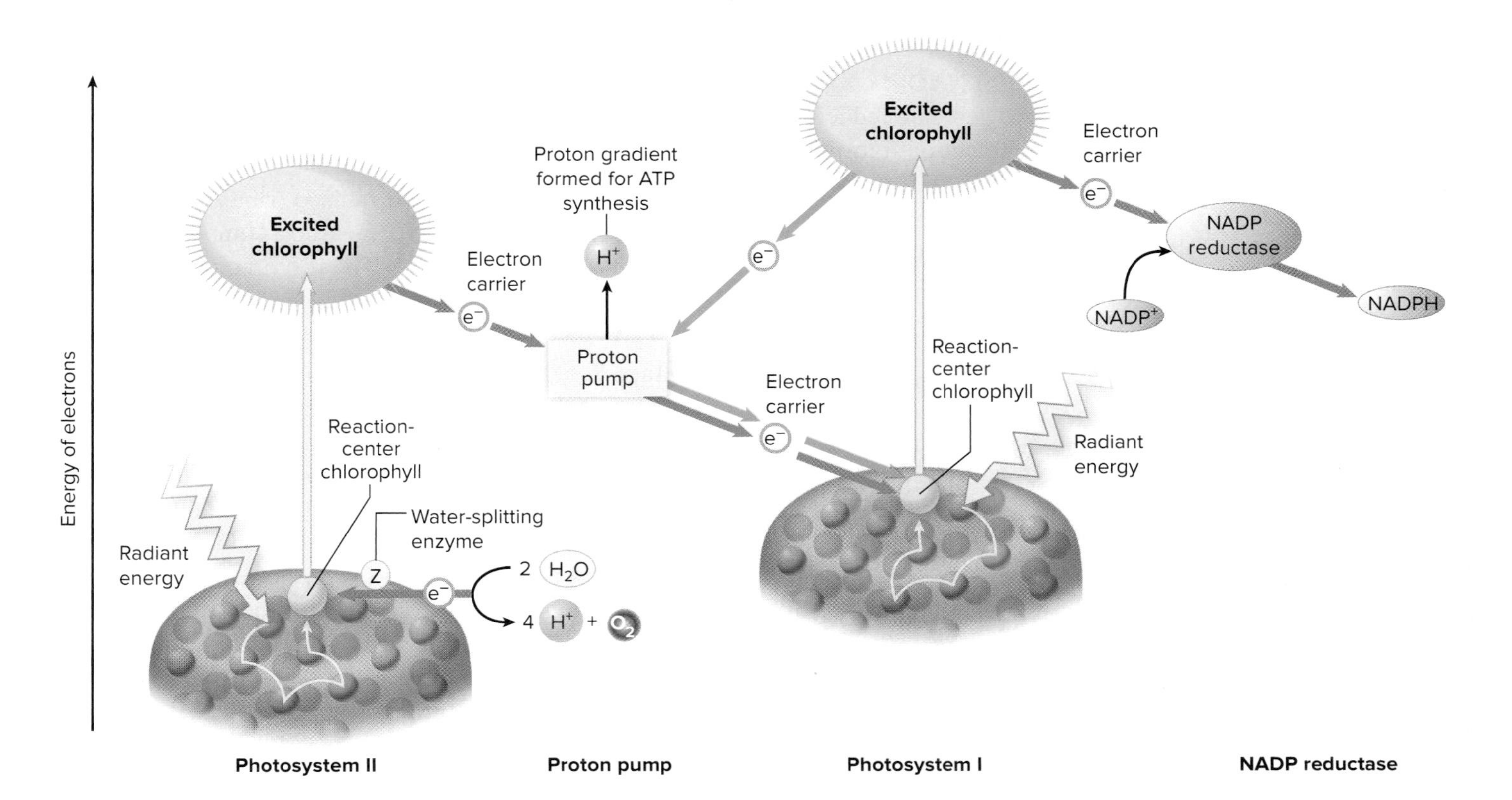

FIGURE 6.26 The Tandem Photosystems of Cyanobacteria and Chloroplasts Radiant energy captured by antennae pigments excites a reaction-center chlorophyll, causing it to emit a high-energy electron, which is then passed to an electron transport chain. In cyclic photophosphorylation, electrons emitted by photosystem I are returned to that photosystem; the path of the electrons is shown in green arrows. In non-cyclic photophosphorylation, the electrons used to replenish photosystem I are donated by radiant energy–excited photosystem II; the path of these electrons is shown in orange arrows. In turn, photosystem II replenishes its own electrons by stripping them from water, producing O_2.

When do the cells need to use non-cyclic photophosphorylation?

donated to photosystem I. First, however, the electrons are passed to the proton pump, which uses some of their energy to establish the proton motive force. The electrons emitted from photosystem II are replenished when an enzyme within that complex extracts electrons from water, donating them to the reaction-center chlorophylls. Removal of electrons from two molecules of water generates O_2. In essence, photosystem II captures the energy of light and then uses it to raise the energy level of electrons stripped from water molecules to a high enough level that they can be used to power photophosphorylation. Photosystem I then accepts those electrons, which still have some energy, and again captures the energy of light to boost the energy of the electrons to an even higher level necessary to reduce NADPH.

Light Reactions in Anoxygenic Photosynthetic Bacteria

Anoxygenic photosynthetic bacteria have only a single photosystem and cannot use water as an electron donor for reducing power. This is why they are anoxygenic (do not generate O_2). These bacteria use electron donors such as hydrogen gas (H_2), hydrogen sulfide (H_2S), and organic compounds. Two groups of anoxygenic photosynthetic bacteria are the purple bacteria and green bacteria. ⏭ purple bacteria, ⏭ green bacteria

Purple bacteria synthesize ATP using a photosystem similar to photosystem II of cyanobacteria and eukaryotes. However, this photosystem does not raise the electrons to a high enough energy level to reduce NAD^+ (or $NADP^+$), so the cells must use an alternative mechanism to generate reducing power. They use a process called reversed electron transport, using ATP to run the electron transport chain in the reverse direction, or "uphill."

Green bacteria have a photosystem similar to photosystem I. The electrons emitted from this can either generate a proton motive force or reduce NAD^+.

MicroAssessment 6.8

Photosynthetic organisms harvest radiant energy and use it to power the synthesis of organic compounds from CO_2. Various pigments are used to capture radiant energy. These pigments are arranged in complexes called photosystems. When reaction-center chlorophylls absorb the energy of light, high-energy electrons are emitted. These are then passed along an electron transport chain to generate a proton motive force, which is used to synthesize ATP. Plants and cyanobacteria use water as a source of electrons for reducing power, generating O_2. Anoxygenic photosynthetic bacteria obtain electrons from a reduced compound other than water and do not generate O_2.

21. β-Carotene is a carotenoid that mammals can use as a source of vitamin A. What is the function of carotenoids in photosynthetic organisms?
22. What is the advantage to a cell of having two photosystems that work together?
23. Energy is required to reverse the flow of the electron transport chain. Why would this be so?

6.9 ■ Carbon Fixation

Learning Outcomes

21. Describe the three stages of the Calvin cycle.

Chemolithoautotrophs and photoautotrophs incorporate carbon dioxide (CO_2) into organic compounds, the process of **carbon fixation.** In photosynthetic organisms, the steps used to do this are called the light-independent reactions. The process consumes a great deal of ATP and reducing power, which should not be surprising considering that the reverse process (oxidizing those same compounds to CO_2) releases a great deal of energy. The Calvin cycle is the most common pathway used to fix carbon, but some prokaryotes use other mechanisms. For example, the green bacteria and some members of the *Archaea* use a pathway that reverses the steps of the TCA cycle.

Calvin Cycle

The **Calvin cycle,** or Calvin-Benson cycle, named in honor of the scientists who described much of it, is a complex cycle. The easiest way to understand the outcome is to consider six "turns" of the cycle; together, these turns incorporate six molecules of CO_2 and produce the equivalent of one molecule of fructose-6-phosphate, an intermediate of glycolysis (**figure 6.27**). There are three stages of the Calvin cycle:

- **Stage 1.** ① Carbon dioxide enters the cycle when an enzyme commonly called rubisco (ribulose bisphosphate carboxylase) joins it to a 5-carbon compound, ribulose-1,5-bisphosphate (RuBP). The resulting compound spontaneously hydrolyzes to produce two molecules of a 3-carbon compound, 3-phosphoglycerate (3PG).
- **Stage 2.** ② An input of energy (ATP) and then reducing power (NADPH) converts 3PG to G3P. This molecule is also a precursor metabolite formed as an intermediate in glycolysis and can have a variety of different fates. It can be used in biosynthesis, oxidized to make other precursor compounds, or converted to a 6-carbon sugar. An important aspect of the Calvin cycle, however, is that RuBP must be regenerated from G3P for the cycle to continue. Consequently, in six turns of the cycle, a maximum of 2 G3P can be converted to a 6-carbon sugar; the rest is used to regenerate RuBP.
- **Stage 3.** ③ A series of complex reactions convert the remaining G3P to RuBP; many of the reactions in this stage are similar to ones in the pentose phosphate pathway.

Yield of the Calvin Cycle

One molecule of the 6-carbon sugar fructose can be generated for every six turns of the cycle. These six turns use 18 ATP and 12 NADPH + H^+.

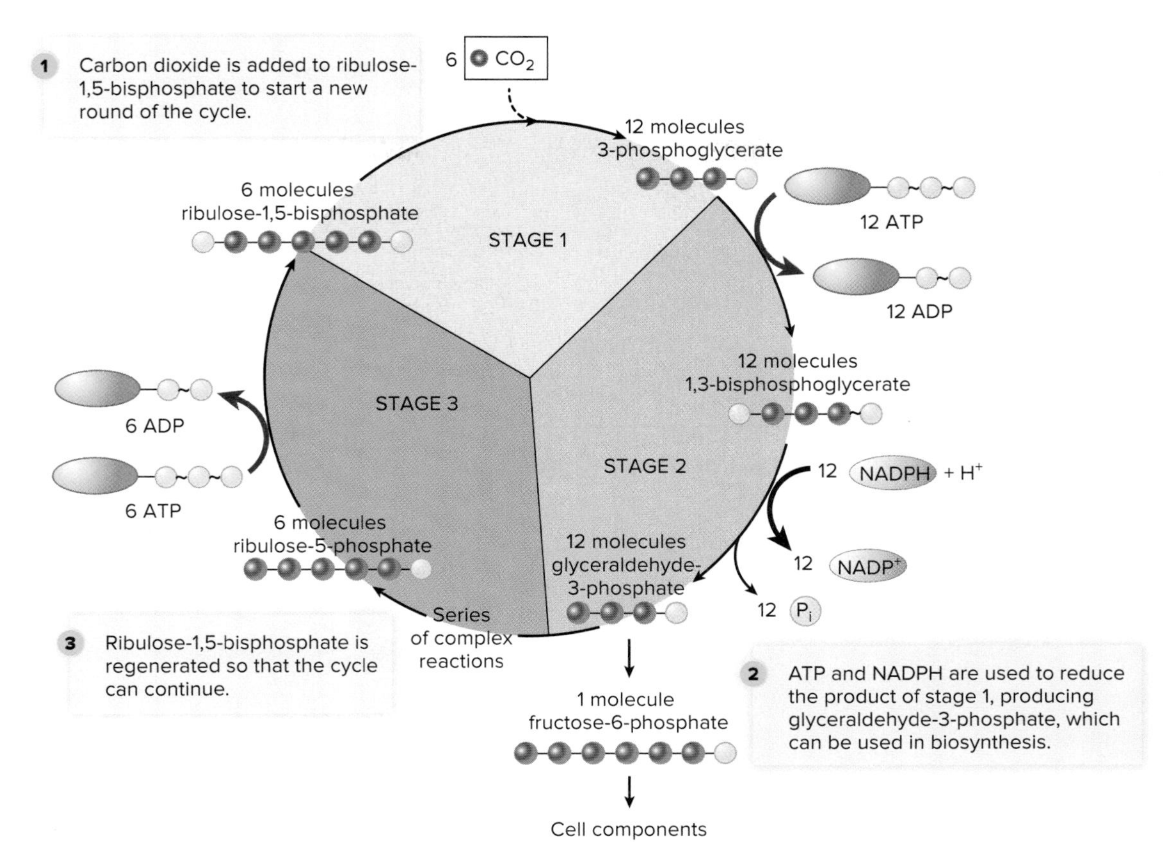

FIGURE 6.27 The Calvin Cycle The Calvin cycle has three essential stages: **(1)** Incorporation of CO_2 into an organic compound; **(2)** reduction of the resulting molecule; and **(3)** regeneration of the starting compound.

? How much ATP and NADPH must be spent to synthesize one molecule of fructose?

MicroByte

Although rubisco is unique to autotrophs, it is probably the most abundant enzyme on Earth!

MicroAssessment 6.9

The process of carbon fixation consumes a great deal of ATP and reducing power. The Calvin cycle is the most common pathway used to incorporate inorganic carbon into an organic form.

24. What is the role of rubisco?

25. What would happen if ribulose-1,5-bisphosphate (RuBP) were depleted in a cell?

6.10 ■ Anabolic Pathways—Synthesizing Subunits from Precursor Molecules

Learning Outcomes

22. Describe the synthesis of lipids, amino acids, and nucleotides.

Prokaryotes, as a group, are highly diverse with respect to the compounds they use for energy but remarkably similar in their biosynthetic processes. They synthesize the necessary subunits by starting with the same essential components: specific anabolic pathways that require ATP, reducing power in the form of NADPH, and the precursor metabolites formed in the central metabolic pathways (**figure 6.28**). Organisms lacking one or more enzymes in a given biosynthetic pathway must have the end product provided from an external source (this is why fastidious bacteria, such as lactic acid bacteria, require many different growth factors). Once the subunits are synthesized or transported into the cell, they can be assembled to make macromolecules. Various different macromolecules can then be joined to form the structures that make up the cell. ◀◀ fastidious

Lipid Synthesis

Synthesis of most lipids requires fatty acids and glycerol. To produce fatty acids, the acetyl group of acetyl-CoA (the precursor metabolite produced in the transition step) is transferred to a carrier protein. This carrier holds the developing fatty acid chain as 2-carbon units are added. When the fatty

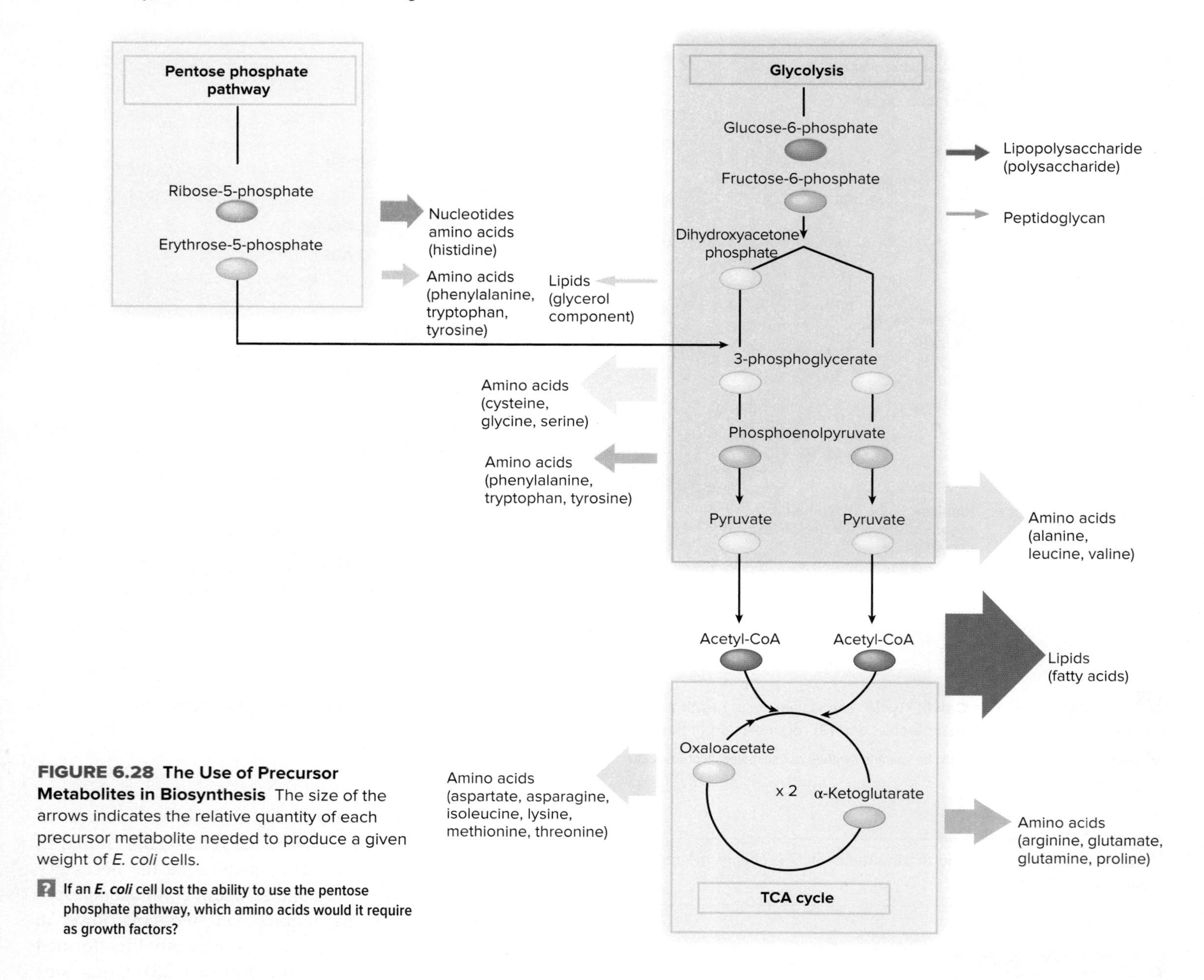

FIGURE 6.28 The Use of Precursor Metabolites in Biosynthesis The size of the arrows indicates the relative quantity of each precursor metabolite needed to produce a given weight of *E. coli* cells.

? If an *E. coli* cell lost the ability to use the pentose phosphate pathway, which amino acids would it require as growth factors?

acid reaches its required length, usually 14, 16, or 18 carbon atoms long, it is released. The glycerol component of the fat is synthesized from the precursor metabolite dihydroxyacetone phosphate, which is generated in glycolysis. ⏮ lipid

Amino Acid Synthesis

Proteins are composed of various combinations of usually 20 different amino acids. These amino acids can be grouped into structurally related families that share common pathways of biosynthesis. ⏮ amino acid

Glutamate

Amino acids are necessary for protein synthesis, but glutamate is especially important because it provides a mechanism for bacteria to incorporate nitrogen into organic material. Recall from chapter 4 that many bacteria use ammonium (NH_4^+) as their source of nitrogen; it is primarily through the synthesis of glutamate that they do this.

Glutamate is synthesized in a single-step reaction that adds ammonia to the precursor metabolite α-ketoglutarate, produced in the TCA cycle (**figure 6.29**). Once glutamate has been formed, its amino group can be transferred to other

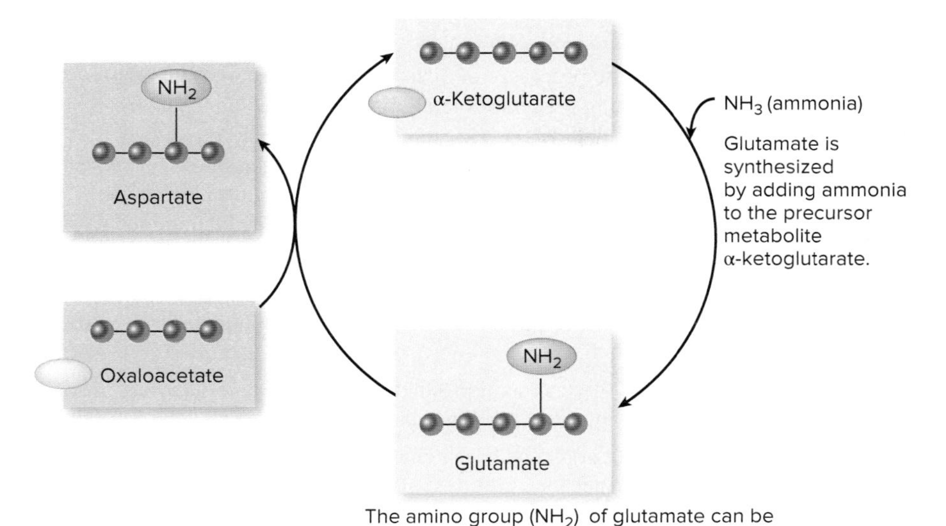

FIGURE 6.29 The Role of Glutamate in Amino Acid Synthesis Once glutamate has been synthesized from α-ketoglutarate, its amino group can be transferred to produce other amino acids.

? Alpha-ketoglutarate is produced in which central metabolic pathway?

carbon compounds to produce amino acids such as aspartate. This transfer of the amino group, a transamination, regenerates α-ketoglutarate from glutamate. The α-ketoglutarate can then be used again to incorporate more ammonia. **⏮ amino group**

Aromatic Amino Acids

Synthesis of aromatic amino acids such as tyrosine, phenylalanine, and tryptophan requires a multistep, branching pathway (**figure 6.30**). This serves as an excellent illustration of many important features of the regulation of amino acid synthesis.

The pathway begins with the joining of two precursor metabolites—phosphoenolpyruvate (3-carbon) and erythrose-4-phosphate (4-carbon)—to form a 7-carbon compound. The precursors originate in glycolysis and the pentose phosphate pathway, respectively. Then, the 7-carbon compound is modified through a series of steps until a branch point is reached. At this juncture, two options are possible. If synthesis proceeds in one direction, tryptophan is produced. In the other direction, another branch point is reached; from there, either tyrosine or phenylalanine can be made.

When an amino acid is provided to a cell, it would be a waste of carbon, energy, and reducing power for that cell to continue synthesizing it. But with branched pathways, how does the cell stop synthesizing the product of only one branch? In the pathway for aromatic amino acid biosynthesis, this occurs partly by regulating the enzymes at the branch points. Tryptophan is a feedback inhibitor of the enzyme that directs the branch to its synthesis; this sends the pathway to the steps leading to the synthesis of the other amino acids (tyrosine and phenylalanine). Likewise, these two amino acids each inhibit the first enzyme of the branch leading to their synthesis.

In addition to regulating the branch points, the three amino acids each control the first step of the pathway: the formation of the 7-carbon compound. In *E. coli,* three different enzymes can catalyze this step; each has the same active site, but they have different allosteric sites. Each aromatic amino acid acts as a feedback inhibitor for one of the enzymes. If all three amino acids are present in the environment, then very little of the 7-carbon compound will be synthesized. If only one or two of those amino acids are present, then proportionally more of the compound will be synthesized. **⏮ allosteric enzymes**

Nucleotide Synthesis

Nucleotide subunits of DNA and RNA are composed of three units: a 5-carbon sugar, a phosphate group, and a nucleobase,

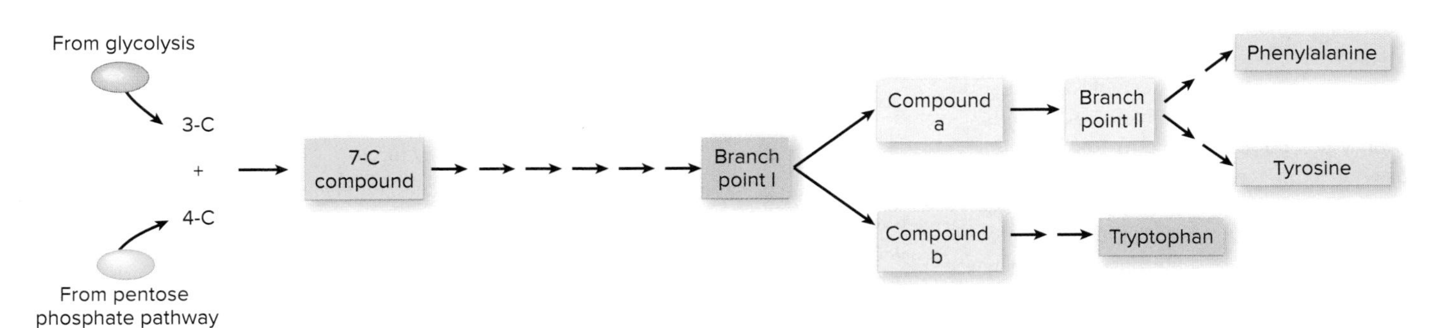

FIGURE 6.30 Synthesis of Aromatic Amino Acids A multistep branching pathway is used to synthesize aromatic amino acids. The end product of a branch inhibits the first enzyme of that branch; that end product also inhibits one of the three enzymes that catalyze the first step of the pathway.

? The synthesis of aromatic amino acids requires precursor metabolites made in which central metabolic pathways?

either a purine or a pyrimidine. They are synthesized as ribonucleotides, but these can then be converted to deoxyribonucleotides by replacing the hydroxyl group on the 2′ carbon of the sugar with a hydrogen atom. ⏮ nucleotides, ⏮ purine, ⏮ pyrimidine

The purine (double-ring structure) and pyrimidine (single-ring) nucleotides are synthesized in distinctly different manners. The starting compound of purine synthesis is ribose-5-phosphate, a precursor metabolite generated in the pentose phosphate pathway. Then, in a highly ordered sequence, atoms from the other sources are added to form the purine ring. This can then be converted to a purine nucleotide. To synthesize pyrimidine nucleotides, the pyrimidine ring is made first and then attached to ribose-5-phosphate. After one pyrimidine nucleotide is formed, the nucleobase component can be converted into any of the other pyrimidines.

MicroAssessment 6.10

Biosynthetic processes of different organisms are remarkably similar, using precursor metabolites, NADPH, and ATP to form subunits. Synthesis of the amino acid glutamate provides a mechanism for bacteria to incorporate nitrogen in the form of ammonia into organic material. Synthesis of aromatic amino acids involves branching pathways. The purine nucleotides are synthesized in a very different manner from the pyrimidine nucleotides.

26. Explain why glutamate synthesis is particularly important for a cell.

27. What three general products of the central metabolic pathways does a cell require to carry out biosynthesis?

28. With a branched biochemical pathway, why would it be important for a cell to shut down the first step as well as branching steps?

FOCUS ON THE FUTURE 6.1

Fueling the Future

Considering that microbes are masters at getting energy from seemingly unlikely sources, it should not be surprising that they are being used to help us meet our future energy needs. Bacteria, yeast, and algae play starring roles in most of the current research efforts toward developing cost-effective methods to produce **biofuels,** which are fuels made from renewable biological resources. One goal in producing and using these is to be carbon neutral, meaning that the amount of carbon released to the atmosphere when fuel is burned is the same as what goes into producing the fuel.

First-generation biofuels are made using standard technologies (such as fermentations) and readily available material (including corn and sugar cane). One example is using yeast to ferment the sugars in corn to produce bioethanol. Although this provides an alternative to fossil fuels, it uses an edible resource, contributing to food shortages and increased food prices. In addition, the farming practices used to grow corn on a large-scale basis can harm the environment, and the process for making bioethanol uses fossil fuels so the product is not carbon neutral.

In response to the problems with first-generation biofuels, newer technologies are being developed to produce what are collectively referred to as advanced biofuels. These include the following:

- **Bioethanol.** Rather than using edible foods, bioethanol produced with advanced technologies is made from cornstalks or other plant waste materials. One problem is that the main component of plant wastes is cellulose, and relatively few microbes degrade this carbohydrate. Microbes that degrade cellulose generally are more difficult to grow on a large-scale basis than yeast, so technical hurdles must be overcome for this to be a practical option.
- **Biodiesel.** Composed of fatty acid methyl esters, biodiesel has an advantage over bioethanol in that it has a higher energy content and is more compatible with current fuel storage and distribution systems. Algae and cyanobacteria—both of which are photosynthetic—capture radiant energy to make lipids that can be used to produce biodiesel. This provides a particularly attractive option because it bypasses the need for the two-stage process used to create bioethanol (plants convert sunlight to chemical energy, and then microbes ferment the plant material to produce ethanol). A problem with using photosynthetic microbes to produce biofuels, however, is that areas with enough sunlight to support their abundant growth generally lack sufficient water for the final steps of fuel production. Another option is to use a genetically engineered easy-to-grow bacterium such as *E. coli.* An *E. coli* strain has been engineered not only to produce biodiesel, but also to break down cellulose, so it can use cellulose-containing waste materials as an energy source. The efficiency is still low, however, so improvements are still necessary. Also, the engineered strain still relies on a second organism—in this case, plants—to convert radiant energy into chemical energy.
- **Biohydrogen.** Some cyanobacteria and anoxygenic phototrophs make hydrogen gas as a by-product of nitrogen fixation and grow using little more than sunlight and water. An advantage of biohydrogen is that it can be easily captured from a culture. As with the other advanced biofuels, however, there are still technical obstacles that must be overcome before the gas can be produced and stored on a large-scale basis.

Will microbes end our reliance on fossil fuels? Only time will tell, but the possibilities are exciting!

Summary

6.1 ■ Principles of Microbial Metabolism

Catabolism is the set of processes that capture and store **energy** by breaking down complex molecules. **Anabolism** includes processes that use energy to make and assemble the building blocks of a cell (figure 6.1).

Energy

Photosynthetic organisms harvest the energy of sunlight, using it to power the synthesis of organic compounds. **Chemoorganotrophs** harvest energy contained in organic compounds (figure 6.3). **Exergonic** reactions release energy; **endergonic** reactions use energy.

Components of Metabolic Pathways

A specific **enzyme** facilitates each step of a **metabolic pathway** (figure 6.5). **ATP** is the energy currency of the cell. The **energy source** is **oxidized** to release its energy (figure 6.7). The **redox** reactions **reduce** an **electron carrier** (figure 6.8). **NAD^+/NADH, $NADP^+$/NADPH,** and **FAD/$FADH_2$** are electron carriers (table 6.1).

Precursor Metabolites

Precursor metabolites are used to make the subunits of macromolecules, and they can also be oxidized to generate energy in the form of ATP (table 6.2).

Overview of Catabolism (figure 6.10)

The **central metabolic pathways** are **glycolysis,** the **pentose phosphate pathway,** and the **tricarboxylic acid cycle (TCA cycle)**. **Cellular respiration** uses the reducing power accumulated in the central metabolic pathways to generate ATP by oxidative phosphorylation. **Aerobic respiration** uses O_2 as a terminal electron acceptor; **anaerobic respiration** uses a molecule other than O_2 as a terminal electron acceptor (table 6.3). **Fermentation** uses pyruvate or a derivative as a terminal electron acceptor; this recycles the reduced electron carrier NADH.

6.2 ■ Enzymes

Enzymes function as biological catalysts; they are neither consumed nor permanently changed during a reaction.

Mechanisms and Consequences of Enzyme Action (figure 6.11)

The substrate binds to the **active site,** forming an enzyme-substrate complex that lowers the activation energy of the reaction.

Cofactors (figure 6.12; table 6.4)

Enzymes sometimes act with the assistance of **cofactors** such as **coenzymes** and trace elements.

Environmental Factors That Influence Enzyme Activity (figure 6.13)

The environmental factors most important in influencing enzyme activities are temperature, pH, and salt concentration.

Allosteric Regulation (figure 6.14)

Cells can fine-tune the activity of an **allosteric** enzyme by using a regulatory molecule that binds to the allosteric site of the enzyme.

Enzyme Inhibition

Non-competitive inhibition occurs when the inhibitor and the substrate act at different sites on the enzyme. **Competitive inhibition** occurs when the inhibitor competes with the normal substrate for the active binding site (figure 6.15).

6.3 ■ The Central Metabolic Pathways (table 6.6)

Glycolysis (figure 6.16)

Glycolysis converts one molecule of glucose into two molecules of pyruvate; the pathway produces 2 ATP, 2 NADH + H^+. Six different precursor metabolites are made.

Pentose Phosphate Pathway

The **pentose phosphate pathway** forms NADPH + 6 H^+ and two different precursor metabolites.

Transition Step (figure 6.17)

The **transition step** converts pyruvate to acetyl-CoA. Repeated twice, this produces 2 NADH + 2 H^+. The end product is a precursor metabolite.

Tricarboxylic Acid (TCA) Cycle (figure 6.17)

The **TCA cycle** completes the oxidation of glucose; two "turns" produce 6 NADH + 6 H^+, and 2 $FADH_2$, in addition to 2 ATP. Two intermediates are precursor metabolites.

6.4 ■ Cellular Respiration

The Electron Transport Chain (ETC)—Generating a Proton Motive Force

The **electron transport chain (ETC)** is a series of membrane-embedded electron carriers that move protons across the membrane. In the mitochondrial ETC, three different complexes (complexes I, III, and IV) function as proton pumps (figure 6.19). Prokaryotes vary with respect to the types and arrangements of their electron transport components (figure 6.20). Some prokaryotes can use molecules other than O_2 as terminal electron acceptors. The process of anaerobic respiration harvests less energy than aerobic respiration.

ATP Synthase—Using the Proton Motive Force to Synthesize ATP

ATP synthase permits protons to flow back across the membrane, harvesting the energy released to fuel the synthesis of ATP.

ATP Yield of Aerobic Respiration in Prokaryotes (figure 6.21)

The theoretical maximum ATP yield of aerobic respiration is 38 ATP.

6.5 ■ Fermentation

In general, the only ATP-producing reactions of fermentations are those of the glycolytic pathway; the other steps provide a mechanism for recycling NADH (figure 6.22). Some end products of fermentation are commercially valuable (figure 6.23). Certain end products help in bacterial identification.

6.6 ■ Catabolism of Organic Compounds Other Than Glucose

Specific enzymes break down macromolecules into their respective subunits (figure 6.24).

Polysaccharides and Disaccharides

Amylases digest starch, releasing glucose subunits, and are produced by many organisms. Cellulases degrade cellulose. Sugar subunits released when polysaccharides are broken down can then enter glycolysis.

Lipids

Fats are broken down by lipases, releasing glycerol and fatty acids. Glycerol is converted to the precursor metabolite dihydroxyacetone phosphate; fatty acids are degraded by β-oxidation, generating reducing power and the precursor metabolite acetyl-CoA.

Proteins

Proteins are broken down to their amino acid subunits by proteases. Deamination removes the amino group; the remaining carbon skeleton is then converted into the appropriate precursor molecule.

6.7 Chemolithotrophs

Prokaryotes, as a group, are unique in their ability to use reduced inorganic compounds such as hydrogen sulfide (H_2S) and ammonia (NH_3) as a source of energy. Chemolithotrophs are autotrophs.

6.8 Photosynthesis

The **light reactions** capture energy from light and convert it to chemical energy in the form of ATP. That energy is used for carbon fixation.

Capturing Radiant Energy

Various pigments such as **chlorophylls**, **bacteriochlorophylls**, carotenoids, and phycobilins are used to capture radiant energy. **Reaction-center pigments** function as the electron donor in the photosynthetic process; **antennae pigments** funnel radiant energy to reaction center pigments.

Converting Radiant Energy into Chemical Energy

The high-energy electrons emitted by reaction-center pigments are passed to an electron transport chain, which uses them to generate a proton motive force. The energy of a proton motive force is harvested by ATP synthase to fuel the synthesis of ATP. Photosystems I and II of cyanobacteria and chloroplasts raise the energy level of electrons stripped from water to a high enough level to be used to generate a proton motive force and produce reducing power; this process is oxygenic (figure 6.26). Purple and green bacteria use only a single photosystem; they must obtain electrons from a reduced compound other than water and therefore do not generate oxygen.

6.9 Carbon Fixation

Calvin Cycle (figure 6.27)

The most common pathway used to incorporate CO_2 into an organic form is the **Calvin cycle.**

6.10 Anabolic Pathways—Synthesizing Subunits from Precursor Molecules (figure 6.28)

Lipid Synthesis

The fatty acid components of fat are synthesized by adding 2-carbon units to an acetyl group. The glycerol component is synthesized from dihydroxyacetone phosphate.

Amino Acid Synthesis

Synthesis of glutamate from α-ketoglutarate and ammonia provides a mechanism for cells to incorporate nitrogen into organic molecules (figure 6.29). Synthesis of aromatic amino acids requires a multistep branching pathway. Allosteric enzymes regulate key steps of the pathway (figure 6.30).

Nucleotide Synthesis

Purines and pyrimidine nucleotides are made in distinctly different manners.

Review Questions

Short Answer

1. Explain the difference between catabolism and anabolism.
2. How does ATP serve as a carrier of free energy?
3. How do enzymes catalyze chemical reactions?
4. Explain how precursor molecules are involved in catabolic as well as anabolic pathways.
5. How do cells regulate enzyme activity?
6. Why do the electrons carried by $FADH_2$ result in less ATP production than those carried by NADH?
7. Name three food products produced with the aid of fermenting microorganisms.
8. In photosynthesis, what is encompassed by the term "light reactions"?
9. Unlike the cyanobacteria, the anoxygenic photosynthetic bacteria do not produce O_2. Why not?
10. What is the role of transamination in amino acid biosynthesis?

Multiple Choice

1. Which of these factors do/does not affect enzyme activity?
   a) Temperature
   b) Inhibitors
   c) Coenzymes
   d) Humidity
   e) pH
2. Which of the following statements is false? Enzymes
   a) bind to substrates.
   b) lower the energy of activation.
   c) convert coenzymes to products.
   d) speed up biochemical reactions.
   e) can be named after the kinds of reaction they catalyze.
3. Based on the name, NADH dehydrogenase is
   a) a vitamin that oxidizes NADH.
   b) a vitamin that reduces NADH.
   c) a vitamin that produces NADH.
   d) an enzyme that oxidizes NADH.
   e) an enzyme that reduces NADH.
4. What is the end product of glycolysis?
   a) Glucose
   b) Citrate
   c) Oxaloacetate
   d) α-Ketoglutarate
   e) Pyruvate
5. The central metabolic pathway(s) is/are
   a) glycolysis and the TCA cycle only.
   b) glycolysis, the TCA cycle, and the pentose phosphate pathway.
   c) glycolysis only.
   d) glycolysis and the pentose phosphate pathway only.
   e) the TCA cycle only.

6. Which of these pathways gives a cell the potential to produce the most ATP?
   a) TCA cycle
   b) Pentose phosphate pathway
   c) Lactic acid fermentation
   d) Glycolysis
7. In fermentation, the terminal electron acceptor is
   a) oxygen (O_2).
   b) hydrogen (H_2).
   c) carbon dioxide (CO_2).
   d) an organic compound.
8. In the process of oxidative phosphorylation, the energy of a proton motive force is used to generate
   a) NADH.
   b) ADP.
   c) ethanol.
   d) ATP.
   e) glucose.
9. If a bacterium loses the ability to produce $FADH_2$, which of the following cannot continue?
   a) Biosynthesis
   b) Glycolysis
   c) The pentose phosphate pathway
   d) The TCA cycle
   e) Fermentation
10. Degradation of fats as an energy source involves all of the following *except*
    a) β-oxidation. b) acetyl-CoA. c) glycerol.
    d) lipase. e) transamination.

Applications

1. A worker in a cheese-making facility argues that whey, a nutrient-rich by-product of cheese, should be dumped in a nearby pond where it could serve as fish food. Explain why this proposed action could actually kill the fish by depleting the O_2 in the pond.
2. Scientists working with DNA in vitro often store it in solutions that contain EDTA, a chelating agent that binds magnesium (Mg^{2+}). This is done to prevent enzymes called DNases from degrading the DNA. Explain why EDTA would interfere with enzyme activity.

Critical Thinking

1. A student argued that aerobic and anaerobic respiration should produce the same amount of ATP. He reasoned that they both use basically the same process; only the terminal electron acceptor is different. What is the primary error in this student's argument?
2. Chemolithotrophs near hydrothermal vents support a variety of other life-forms there. Explain how their role is analogous to that of photosynthetic organisms in terrestrial environments.

www.mcgrawhillconnect.com

Enhance your study of this chapter with study tools and practice tests. Also ask your instructor about the resources available through Connect, including the media-rich eBook, interactive learning tools, and animations.

7 The Blueprint of Life, from DNA to Protein

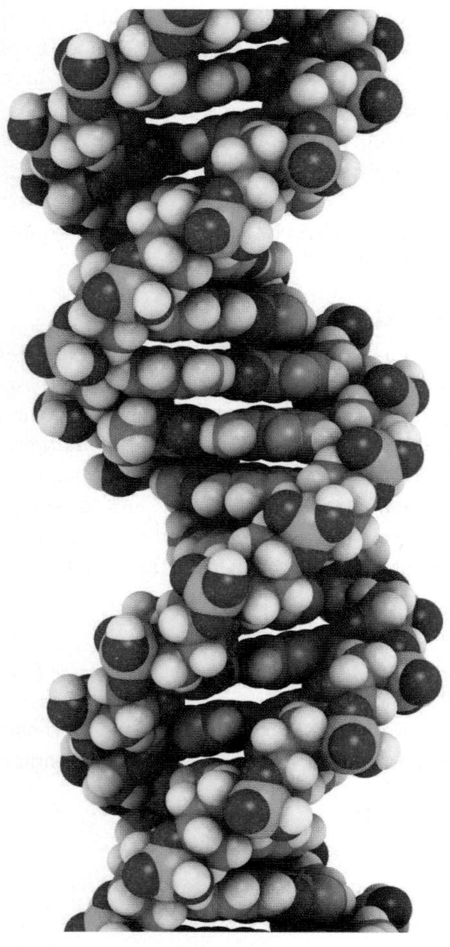

Model of DNA double helix. ©MOLEKUUL/SPL/age fotostock

KEY TERMS

Codon A series of three nucleotides that code for a specific amino acid.

DNA Polymerase Enzyme that synthesizes DNA, using an existing strand as a template to make a new complementary strand.

DNA Replication Duplication of a DNA molecule.

Gene The functional unit of the genome; it encodes a product, most often a protein.

Genome Complete set of genetic information in a cell or a virus.

Messenger RNA (mRNA) Type of RNA molecule translated during protein synthesis.

Promoter Nucleotide sequence to which RNA polymerase binds to start transcription.

Ribosomal RNA (rRNA) Type of RNA molecule present in ribosomes.

Ribosome Structure that facilitates the joining of amino acids during translation; composed of ribosomal RNA (rRNA) and protein.

RNA Polymerase Enzyme that synthesizes RNA using one strand of DNA as a template.

Transcription The process by which the information encoded in DNA is copied into RNA.

Transfer RNA (tRNA) Type of RNA molecule involved in interpreting the genetic code; each tRNA molecule carries a specific amino acid.

Translation The process by which the information carried by mRNA is used to synthesize the encoded protein.

A Glimpse of History

In 1866, the Czech-Austrian monk Gregor Mendel determined that traits are inherited as physical units, now called genes. The precise function of genes, however, was not revealed until 1941, when George Beadle and Edward Tatum published a scientific paper reporting that genes direct the production of enzymes.

Beadle and Tatum set out to discover how genes control metabolic reactions by studying common bread molds that have very simple nutritional requirements. These molds, *Neurospora* species, can grow on media containing only sucrose, inorganic salts, and the vitamin biotin. Beadle and Tatum reasoned that if they created mutant strains that required additional nutrients, they could use them to gain insights into the relationship between genes and enzymes. For example, a mutant that requires a certain amino acid probably has a defect in an enzyme required to synthesize that amino acid.

To generate mutant strains, Beadle and Tatum treated the mold cultures with X rays and then grew the resulting cells on a nutrient-rich medium that supported the growth of both the original strain and any mutants. Next, they screened thousands of the progeny to find the nutrient-requiring mutants. To identify the metabolic defect of each one, they grew them separately in various types of media containing different nutrients.

Eventually, Beadle and Tatum established that the metabolic defect was inherited as a single gene, which ultimately led to their conclusion that a single gene determines the production of one enzyme. That assumption has been modified somewhat, because we now know that some enzymes are made up of more than one polypeptide, and not all proteins are enzymes. In 1958, Beadle and Tatum were awarded a Nobel Prize, largely for these pioneering studies that ushered in the era of modern biology. ⏮ polypeptide

Consider for a moment the incredible diversity of lifeforms in our world—from the remarkable variety of microorganisms, to the plants and animals consisting of many different specialized cells. Every characteristic of each of these cells, from its shape to its function, is dictated by information within its deoxyribonucleic acid (DNA). DNA is the "blueprint," providing instructions for building an organism's components.

DNA itself is a simple structure—a linear molecule composed of only four different **nucleotides,** each containing a particular **nucleobase** (also called a nitrogenous base or simply a base): adenine (A), thymine (T), cytosine (C), or guanine (G).

Focus Figure

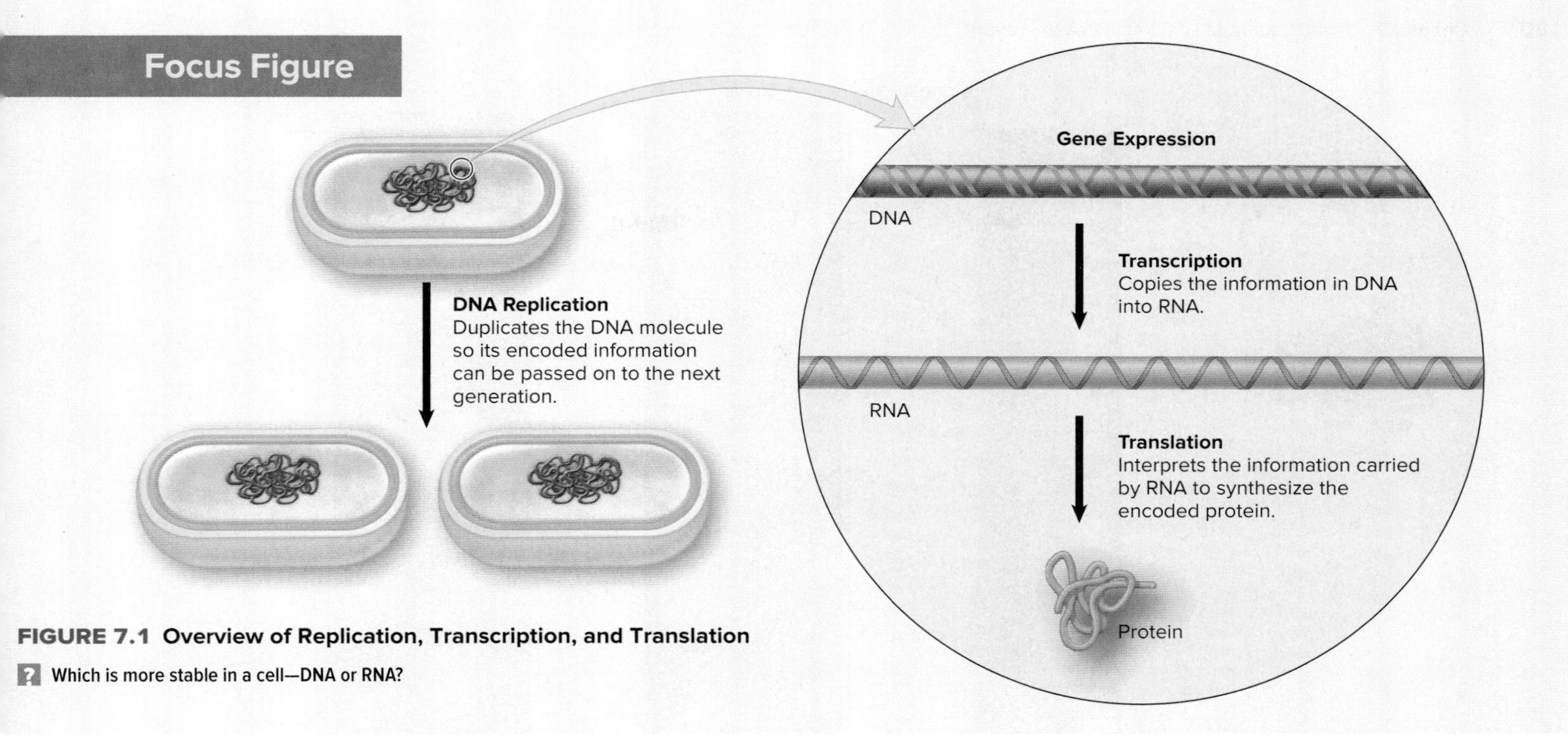

FIGURE 7.1 Overview of Replication, Transcription, and Translation

Which is more stable in a cell—DNA or RNA?

A set of three nucleotides encodes a specific amino acid; in turn, a string of amino acids makes up a protein, the structure and function of which is dictated by the order of the amino acid subunits. Some proteins serve as structural components of a cell. Others, such as enzymes, direct cellular activities including biosynthesis and energy conversion. Together, all these proteins control the cell's structure and activities, dictating the overall characteristics of that cell. ⏮ deoxyribonucleic acid (DNA), ⏮ nucleotide, ⏮ nucleobase, ⏮ amino acid, ⏮ enzymes

Although at first it might seem unlikely that the vast array of life-forms could be encoded by a molecule consisting of only four different units (the nucleotides), think about how much information can be transmitted by binary code, the language of all computers. Using only a simple series of ones and zeros, binary can code for each letter of the alphabet. String enough of these together in the right sequence and the letters become words. With longer and longer strings, the words can become complete sentences, chapters, books, or even whole libraries.

This chapter will focus on the processes bacteria use to replicate their DNA and convert the encoded information into proteins. The mechanisms used by eukaryotic cells have many similarities, but are considerably more complicated, and will be discussed only briefly. The processes in archaea are sometimes similar to those of bacteria, but often resemble those of eukaryotic cells.

7.1 ■ Overview

Learning Outcomes

1. Compare and contrast the characteristics of DNA and RNA.
2. Explain why gene regulation is important to a cell.

The complete set of genetic information of a cell is referred to as its **genome.** Technically, this includes plasmids as well as the chromosome; however, the term "genome" is often used interchangeably with chromosome. The genome of all cells is composed of DNA, but some viruses have an RNA genome. The functional unit of the genome is a **gene.** A gene encodes a product (called the gene product), most commonly a protein. The study and analysis of the nucleotide sequence of DNA is called **genomics.** ⏮ chromosome, ⏮ plasmid

All cells must accomplish two general tasks in order to multiply. First, the double-stranded DNA must be duplicated before cell division so that its encoded information can be passed on to the next generation (**figure 7.1**). This is the process of **DNA replication.** Second, the information encoded by the DNA must be decoded so that the cell can synthesize the necessary gene products. This process, **gene expression,** involves two related events: transcription and translation. **Transcription** is the process by which the information encoded in DNA is copied into a slightly different molecule: RNA. In **translation,** the information carried by the RNA is interpreted and used to synthesize the encoded protein. The flow of information from DNA → RNA → protein is often referred to as the central dogma of molecular biology.

Characteristics of DNA

DNA is usually a double-stranded, helical structure (**figure 7.2**). Each strand is composed of a chain of deoxyribonucleotide subunits, more commonly called nucleotides. Each nucleotide contains a 5-carbon sugar (deoxyribose), a phosphate group, and one of four different nucleobases (A, T, G, or C). The nucleotides are joined together by a covalent bond between

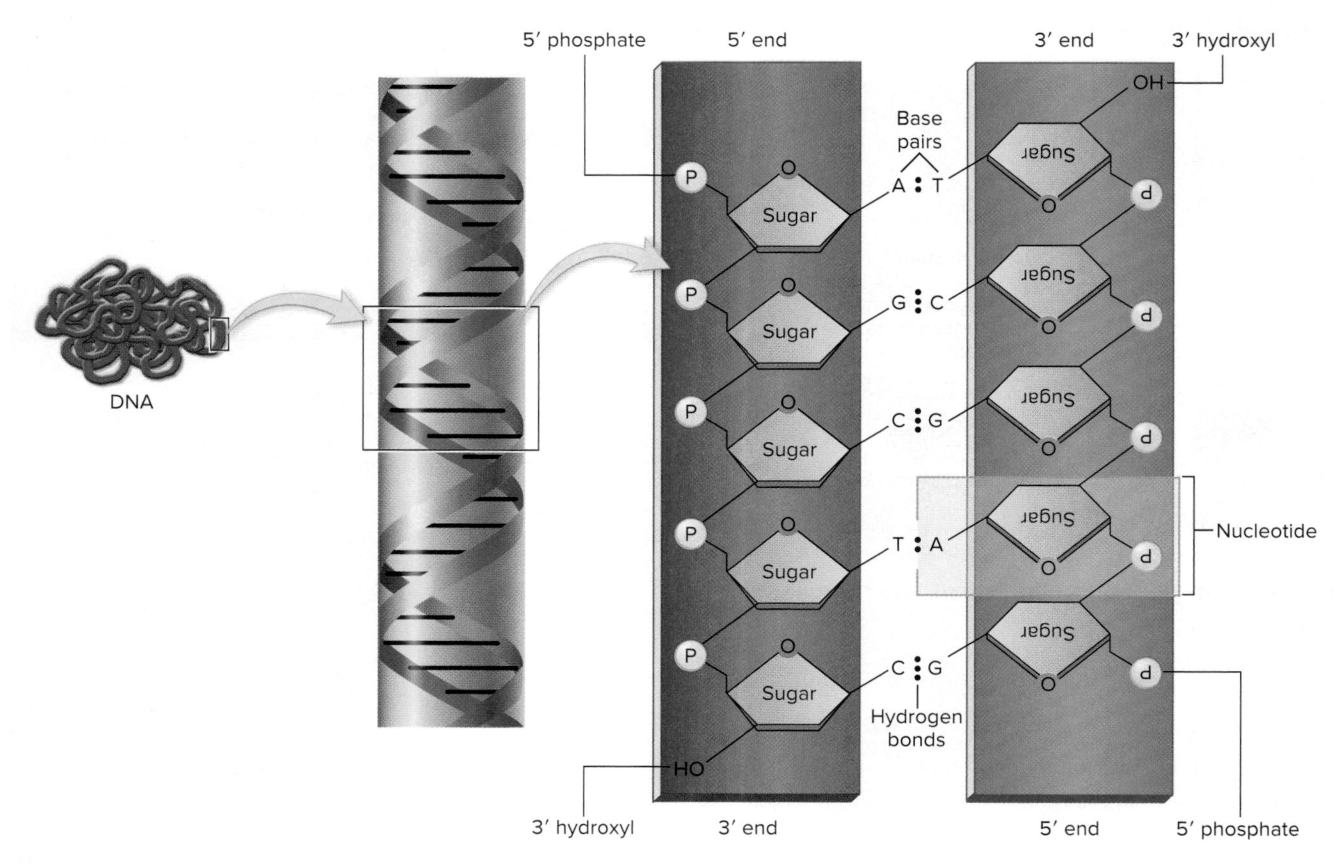

FIGURE 7.2 The Structure of DNA The two strands in the double helix are complementary. Three hydrogen bonds form between a G-C base pair and two between an A-T base pair. The strands are antiparallel; one is oriented in the 5′ to 3′ direction, and its complement is oriented in the 3′ to 5′ direction.

If a 100 base-pair double-stranded DNA fragment has 40 cytosines, how many adenines does it contain?

the 5′PO_4 (5 prime phosphate) of one nucleotide and the 3′OH (3 prime hydroxyl) of the next. The designations 5′ and 3′ refer to the numbered carbon atoms of the pentose sugar of the nucleotide (see figure 2.12). Joining the nucleotides this way creates a series of alternating sugar and phosphate units, called the sugar-phosphate backbone. Because of the chemical structure of nucleotides and how they are joined to each other, a single linear strand of DNA will always have a 5′PO_4 at one end and a 3′OH at the other. These ends are often referred to as the **5′ end** (5 prime end) and the **3′ end** (3 prime end). ◀◀ nucleotides

The two strands of DNA are **complementary** and are held together by hydrogen bonds between the nucleobases. Wherever an adenine (A) is in one strand, a thymine (T) is in the other; these opposing A-T bases are held together by two hydrogen bonds. Similarly, wherever a guanine (G) is in one strand, a cytosine (C) is in the other. These G-C bases are held together by three hydrogen bonds, a slightly stronger attraction than that of an A-T pair. The characteristic bonding of A to T and G to C is called **base-pairing** and is a fundamental characteristic of DNA. Because of the rules of base-pairing, one strand can always be used as a template for the synthesis of the opposing strand. ◀◀ hydrogen bonds

Although the two strands of DNA in the double helix are complementary, they are also **antiparallel.** That is, they are oriented in opposite directions. One strand is oriented in the 5′ to 3′ direction and its complement is oriented in the 3′ to 5′ direction.

The duplex structure of double-stranded DNA is generally quite stable because of the numerous hydrogen bonds that occur along its length. Short fragments of DNA have correspondingly fewer hydrogen bonds, so they are easily separated into single strands. Separating the two strands of DNA is called melting, or denaturing.

Characteristics of RNA

RNA, like DNA, is composed of a chain of nucleotides. It is similar to DNA in many ways, with a few important exceptions. One difference is that the sugar in the nucleotides of RNA is ribose, not deoxyribose; it has an oxygen atom that deoxyribose lacks. Another distinction is that RNA contains the nucleobase uracil in place of the thymine found in DNA. Also, RNA is usually a single-stranded linear molecule much shorter than DNA. ◀◀ ribonucleic acid (RNA)

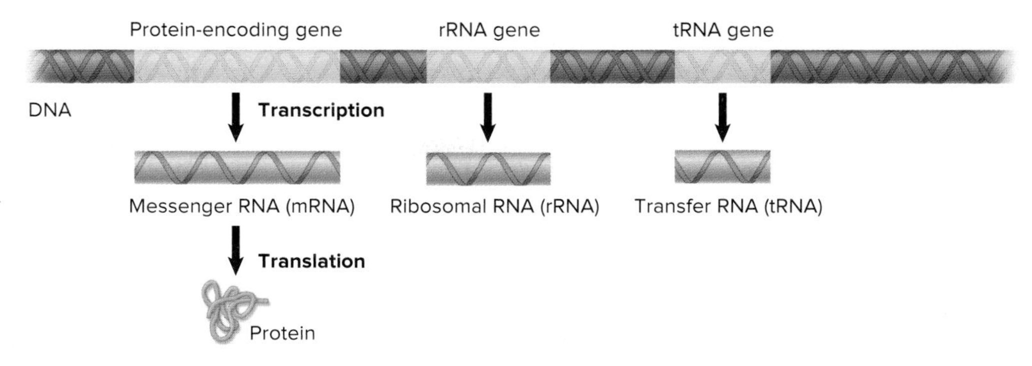

FIGURE 7.3 Three Functional Types of RNA Molecules The different functional types of RNA—messenger RNA (mRNA), ribosomal RNA (rRNA), and transfer RNA (tRNA)—are transcribed from different genes. The mRNA is translated, and the tRNA and rRNA fold into characteristic three-dimensional structures that each play a role in protein synthesis.

? Ribosomal RNA is a component of ribosomes. What are ribosomes?

RNA is synthesized using a region of one of the two strands of DNA as a template. In making the RNA molecule, or **transcript,** the base-pairing rules apply except that uracil, rather than thymine, pairs with adenine. The interaction of DNA and RNA is only temporary, however, and the transcript quickly separates from the template.

Three different functional types of RNA are required for gene expression, and these are transcribed from different sets of genes (**figure 7.3**). Most genes encode proteins and are transcribed into **messenger RNA (mRNA).** The information encrypted in mRNA is decoded according to the genetic code, which correlates each set of three nucleotides to a particular amino acid. Some genes are never translated into proteins; instead the RNAs themselves are the final products. These genes encode either **ribosomal RNA (rRNA)** or **transfer RNA (tRNA),** each of which plays a different but critical role in protein synthesis.

Regulating Gene Expression

Although a cell's DNA can encode thousands of different proteins, not all of them are needed at the same time or in equal quantities. Because of this, cells require mechanisms to regulate the expression of certain genes.

A fundamental aspect of gene regulation is the cell's ability to quickly destroy mRNA. Within minutes of being produced, transcripts are degraded by cellular enzymes. Although this might seem wasteful, it actually provides cells with an important regulatory mechanism. If transcription of a gene is turned "on," transcripts will continue to be available for translation. If it is then turned "off," the number of transcripts will rapidly decline. By simply regulating the synthesis of mRNA molecules, a cell can quickly change the levels of protein production (**figure 7.4**).

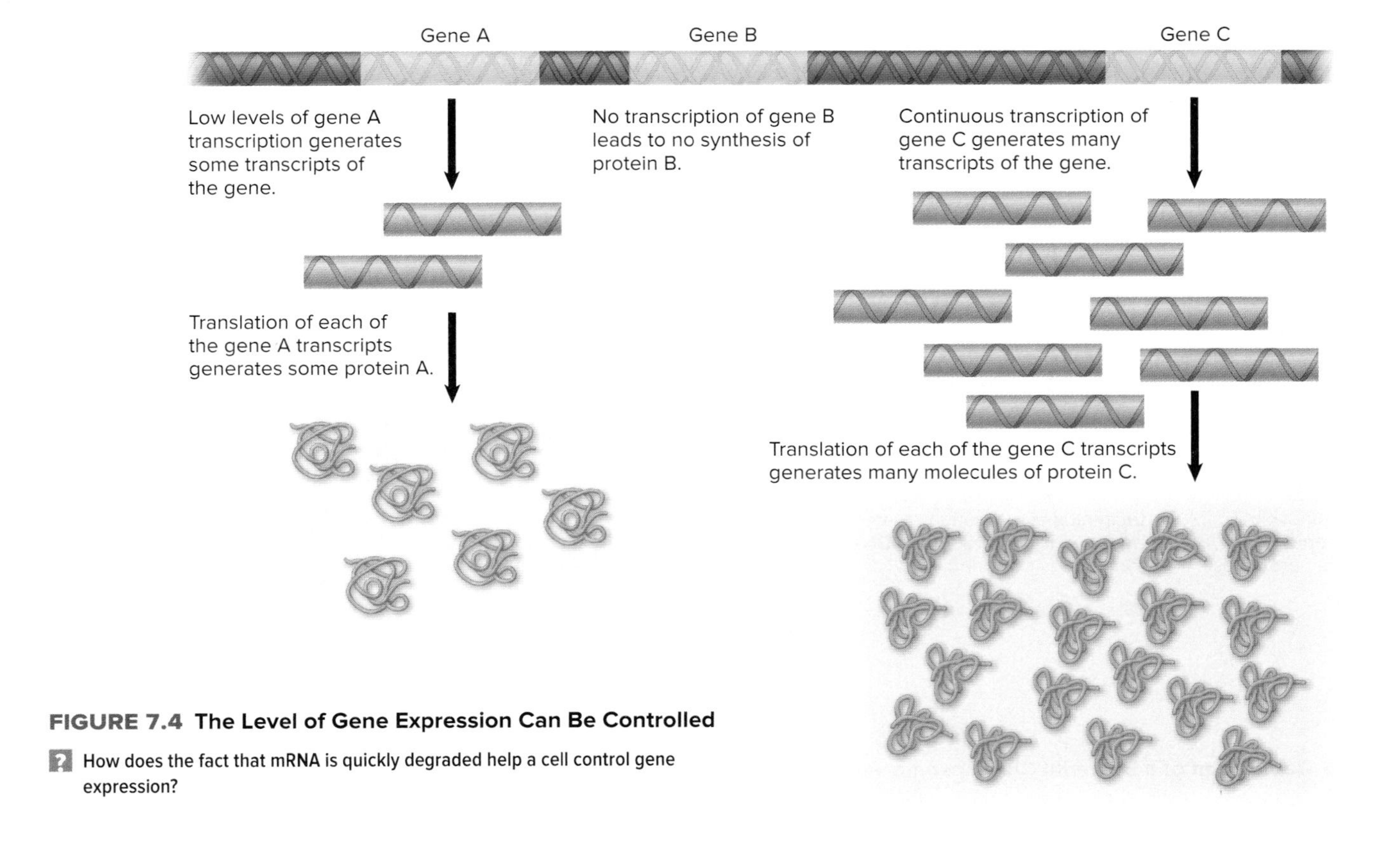

FIGURE 7.4 The Level of Gene Expression Can Be Controlled

? How does the fact that mRNA is quickly degraded help a cell control gene expression?

MicroAssessment 7.1

Replication is the process of duplicating double-stranded DNA. Transcription is the process of copying the information encoded in DNA into RNA. Translation is the process in which the information carried by mRNA is used to synthesize the encoded protein.

1. How does the 5′ end of a DNA strand differ from the 3′ end?
2. What are the base-pairing rules?
3. If the nucleotide sequence of one strand of DNA is 5′ ACGTTGCA 3′, what is the sequence of the complementary strand?

7.2 ■ DNA Replication

Learning Outcome

3. Describe the DNA replication process, including its initiation and the events that occur at the replication fork.

DNA is replicated so that each of the two cells generated during binary fission can receive one complete copy of the genome. The replication process is generally bidirectional, meaning it proceeds in both directions from a specific starting point called the **origin of replication** (**figure 7.5**). This allows a chromosome to be replicated in half the time it would take if the process were unidirectional. The progression of bidirectional replication around a circular DNA molecule creates two advancing forks where DNA synthesis is occurring. These regions, called **replication forks,** ultimately meet at a terminating site when the process is complete. ◀◀ binary fission

Each of the two DNA molecules created through replication contain one of the original strands paired with a newly synthesized strand. Because one strand of the original molecule is conserved in each molecule, replication is said to be **semiconservative.**

Initiation of DNA Replication

To initiate replication of a DNA molecule, specific proteins recognize and bind to the origin of replication (a certain DNA sequence). Bacterial chromosomes and plasmids typically contain only one of these initiating sites, and a DNA molecule that lacks the sequence will not be replicated. The proteins that bind to the bacterial origin of replication include **DNA gyrase** and **helicases,** which temporarily break and unwind the DNA helix at that site. This exposes single-stranded regions that can act

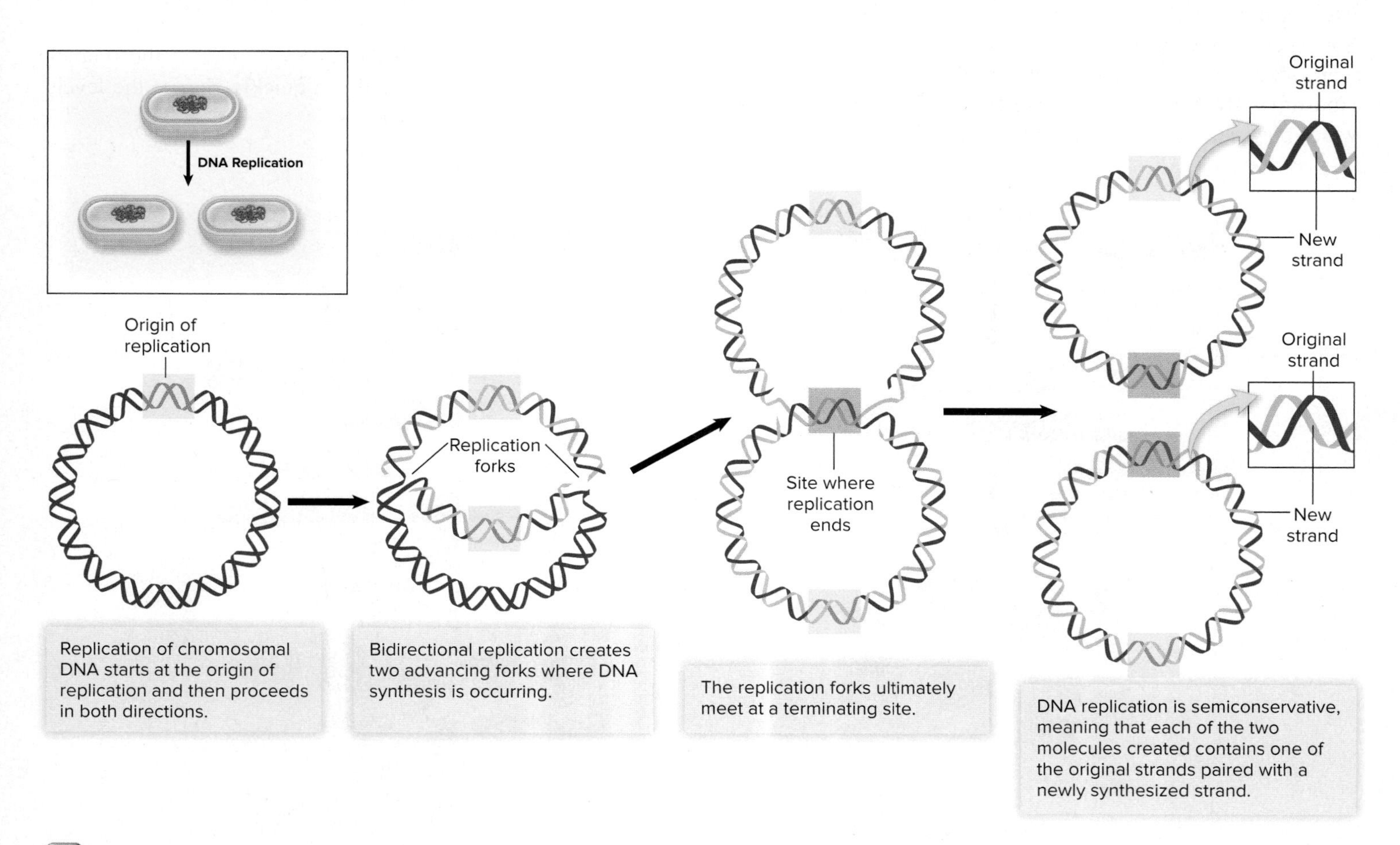

FIGURE 7.5 Replication of a Bacterial Chromosome Process of bidirectional replication.

What is a replication fork?

as templates. Enzymes called **primases** then synthesize short stretches of RNA complementary to the exposed templates. These small fragments, called **primers,** are important in the steps of replication described next. The processes used to initiate DNA replication in eukaryotic cells is similar to that in bacteria, but there are significant differences. For example, eukaryotic chromosomes typically have multiple origins of replication, and the proteins involved in the replication process are different. Likewise, the DNA replication machinery of archaea is also distinct.

MicroByte

Antibacterial medications called fluoroquinolones target the bacterial enzyme DNA gyrase.

The Process of DNA Replication

The process of DNA replication requires the coordinated action of many different enzymes and other components (**table 7.1**). Many of the proteins involved exist together in DNA-synthesizing "assembly lines" called **replisomes.**

Enzymes called **DNA polymerases** synthesize DNA in the 5′ to 3′ direction, using one parent strand as a template to make the complement (**figure 7.6**). To do this, a DNA polymerase adds nucleotides onto the 3′ end of the new strand, powering the reaction with the energy released when a high-energy phosphate bond of the incoming nucleotide is hydrolyzed. DNA polymerases add nucleotides only onto an existing nucleotide strand, so they cannot initiate synthesis. This explains why RNA primers

TABLE 7.1 Components of DNA Replication in Bacteria

Component	Comment
DNA gyrase	Enzyme that temporarily breaks the strands of DNA, relieving the tension caused by unwinding the two strands of the DNA helix.
DNA ligase	Enzyme that joins two DNA fragments together by forming a covalent bond between the sugar and phosphate residues of adjacent nucleotides.
DNA polymerases	Enzymes that synthesize DNA; they use one strand of DNA as a template to make the complementary strand. Nucleotides can be added only to the 3′ end of an existing fragment—therefore, synthesis always occurs in the 5′ to 3′ direction.
Helicases	Enzymes that unwind the DNA helix at the replication fork.
Okazaki fragment	Nucleic acid fragment produced during discontinuous synthesis of the lagging strand of DNA.
Origin of replication	Distinct region of a DNA molecule at which replication is initiated.
Primase	Enzyme that synthesizes small fragments of RNA to serve as primers for DNA synthesis.
Primer	Fragment of nucleic acid to which DNA polymerase can add nucleotides (the enzyme can add nucleotides only to an existing fragment).
Replisome	The complex of enzymes and other proteins that synthesize DNA.

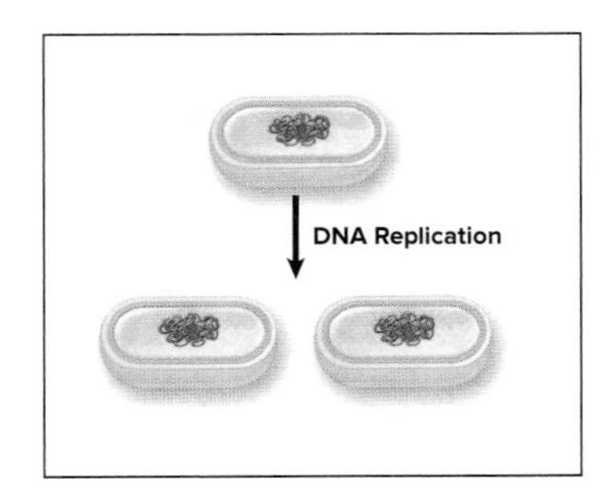

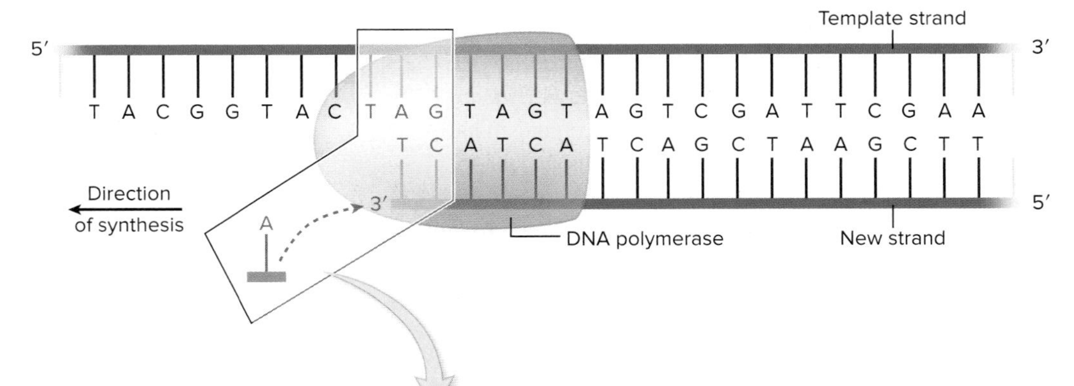

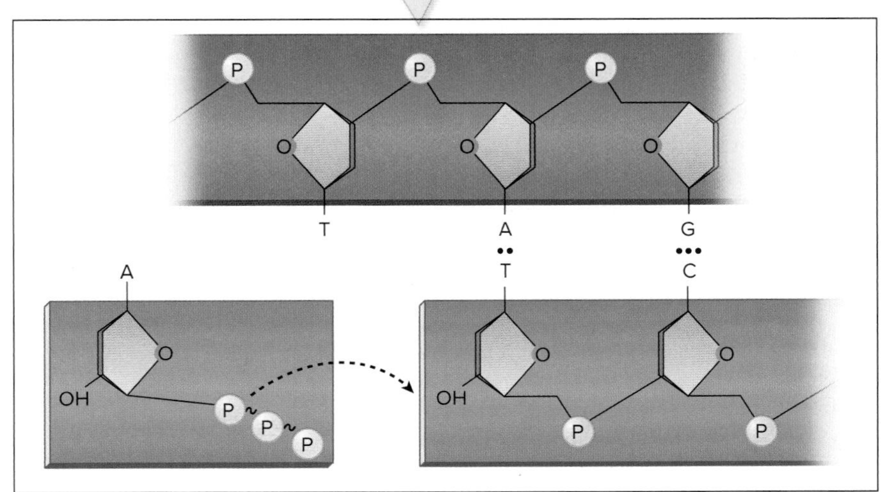

FIGURE 7.6 The Process of DNA Synthesis DNA polymerase synthesizes a new strand by adding one nucleotide at a time to the 3′ end of the elongating strand. The base-pairing rules determine the specific nucleotides that are added.

Considering that DNA is synthesized in the 5′ to 3′ direction, which direction must DNA polymerase travel along the template strand: 5′ to 3′ or 3′ to 5′?

are required at the origin of replication: They provide the DNA polymerase with a molecule to which it can add additional nucleotides. ⏮ high-energy phosphate bond

In order for replication to progress, the helicases must progressively "unzip" the DNA strands at each replication fork to reveal additional template sequences (**figure 7.7** ①). ② Synthesis of one new strand proceeds continuously as fresh template is exposed, because DNA polymerase simply adds nucleotides to the 3′ end. This strand is called the **leading strand.** ③ Synthesis of the other strand, the **lagging strand,** is more complicated. This is because DNA polymerases cannot add nucleotides to the 5′ end of a nucleotide chain, so synthesis must be reinitiated as additional template is exposed. Each time synthesis is reinitiated, another RNA primer must be made first. The result is a series of small fragments, each of which has a short stretch of RNA at its 5′ end. These fragments are called **Okazaki fragments.** ④ As DNA polymerase adds nucleotides to the 3′ end of one Okazaki fragment, it eventually reaches the 5′ end of another. A different type of DNA polymerase then removes the RNA primer nucleotides and simultaneously replaces them with deoxynucleotides. ⑤ The enzyme **DNA ligase** then seals the gaps between fragments by forming a covalent bond between the adjacent nucleotides.

When a circular bacterial chromosome is replicated, the two replication forks eventually meet at a site opposite the origin of replication. Two complete DNA molecules have been produced at this point, and these can be passed on to the two daughter cells.

It takes approximately 40 minutes for the *E. coli* chromosome to be replicated. How, then, can the organism have a generation time of only 20 minutes? This can happen because, under favorable growing conditions, a cell initiates a new round of replication before the preceding round of replication is complete. In this way, each of the two daughter cells will get one complete chromosome that has already started another round of replication. ⏮ generation time

MicroAssessment 7.2

DNA replication begins at the origin of replication and then proceeds bidirectionally, creating two replication forks. DNA polymerases synthesize DNA in the 5′ to 3′ direction, using one strand as a template to generate the complementary strand. New DNA is synthesized in continuous leading and fragmented lagging strands. Okazaki fragments formed during lagging strand synthesis are joined together by DNA ligase.

4. Why is a primer required for DNA synthesis?

5. How does synthesis of the lagging strand differ from that of the leading strand?

6. Eukaryotic chromosomes have multiple origins of replication. Why would this be the case?

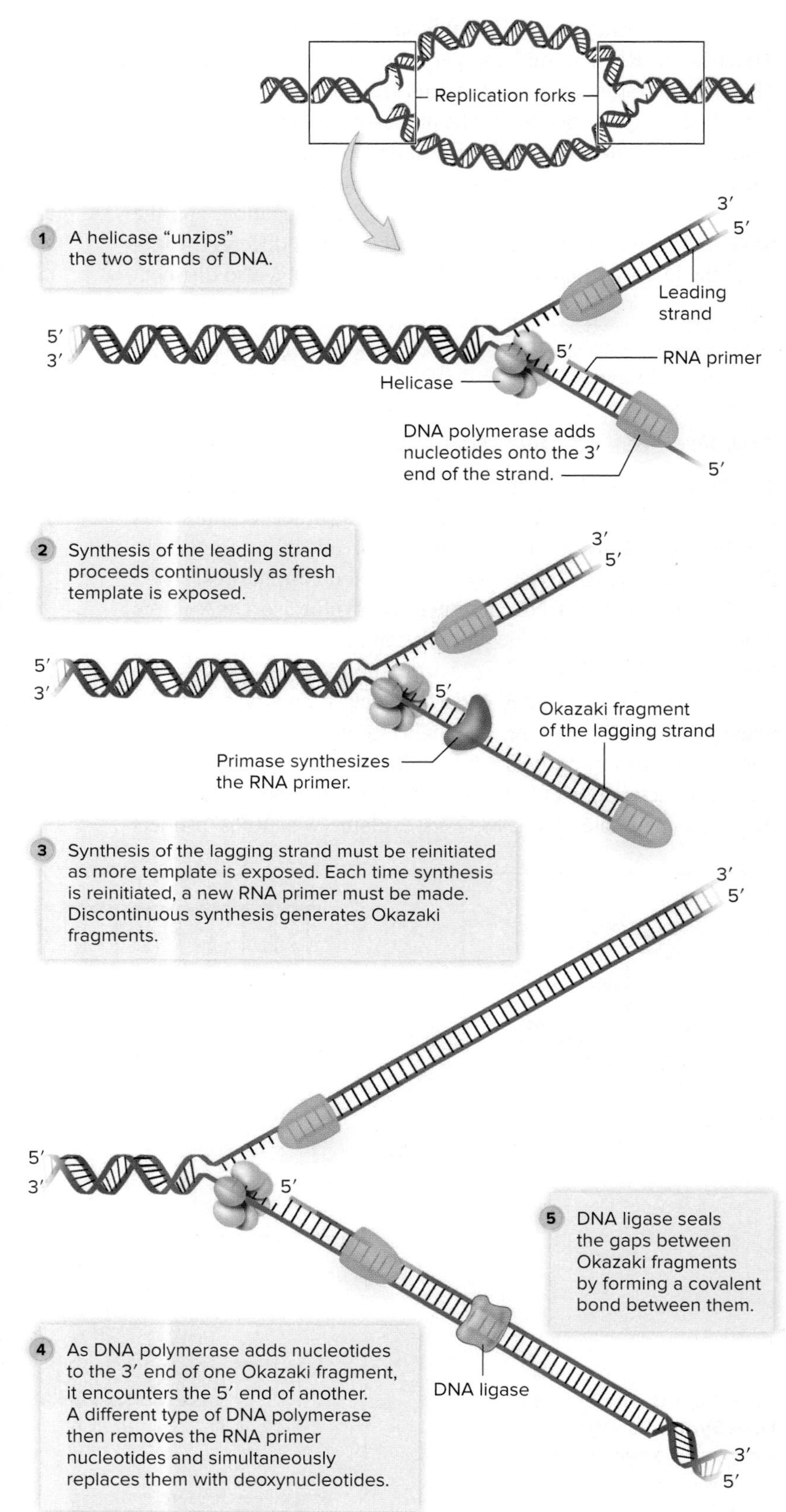

FIGURE 7.7 The Replication Fork This diagram is simplified to highlight the key differences between synthesis of the leading and lagging strands. Both strands are synthesized simultaneously.

? Synthesis of which strand requires the repeated action of DNA ligase?

7.3 ■ Gene Expression in Bacteria

Learning Outcomes

4. Describe the process of transcription, focusing on the role of RNA polymerase, sigma (σ) factors, promoters, and terminators.
5. Describe the process of translation, focusing on the role of mRNA, ribosomes, ribosome-binding sites, rRNAs, tRNAs, and codons.

Recall that gene expression involves two separate but interrelated processes: transcription and translation. Transcription is the process of synthesizing RNA from a DNA template. During translation, information encoded by an mRNA transcript is used to synthesize a protein. "Polypeptide" would be a more accurate term but the word "protein" is often used in this context for simplicity. The distinction between these two words is subtle: A polypeptide is simply a chain of amino acids, whereas a protein is a functional molecule made up of one or more polypeptides.

Transcription

In **transcription,** the enzyme **RNA polymerase** synthesizes single-stranded RNA molecules from a DNA template. Specific nucleotide sequences in the DNA direct the polymerase where to start and where to end (**figure 7.8**). The DNA sequence to which RNA polymerase can bind and initiate transcription is called a **promoter;** one that stops the process is a **terminator.** Like DNA polymerase, RNA polymerase can add nucleotides only to the 3′ end of a chain and therefore synthesizes RNA in the 5′ to 3′ direction. Unlike DNA polymerase, however, RNA polymerase can start synthesis without a primer.

The RNA sequence made during transcription is complementary and antiparallel to the DNA template (**figure 7.9**). The DNA strand that serves as the template for transcription is called the **minus (−) strand,** and its complement is called the **plus (+) strand** (**table 7.2**). Because the RNA is complementary to the (−) DNA strand, its nucleotide sequence is the same as the (+) DNA strand, except that it contains uracil rather than thymine.

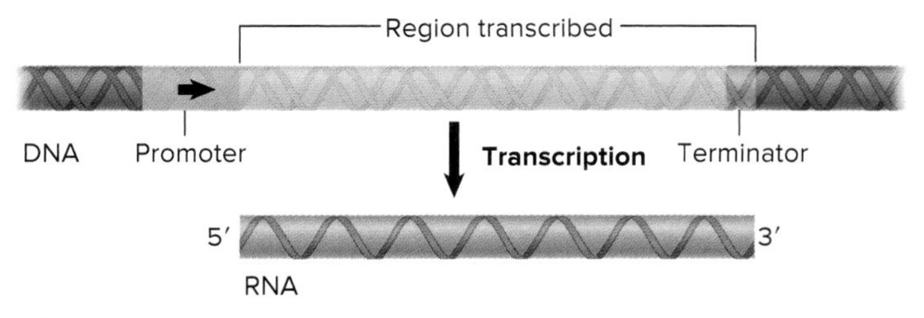

FIGURE 7.8 Nucleotide Sequences in DNA Direct Transcription The promoter is a DNA sequence to which RNA polymerase can bind in order to initiate transcription. The terminator is a sequence at which transcription stops.

? **In which direction is RNA synthesized: 5′ to 3′ or 3′ to 5′?**

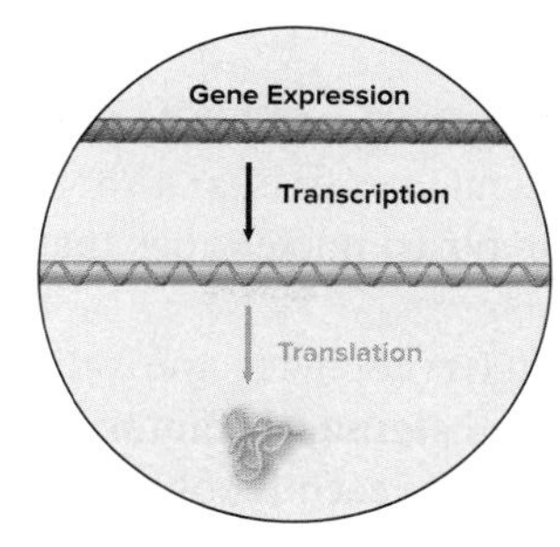

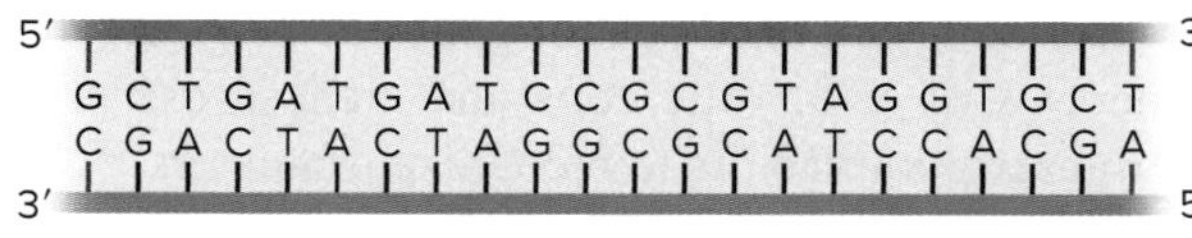

FIGURE 7.9 RNA Is Complementary and Antiparallel to the DNA Template The DNA strand that serves as a template for RNA synthesis is called the (−) strand of DNA. The complement to that is the (+) strand.

? **How does the nucleotide sequence of the (+) DNA strand differ from that of the RNA transcript?**

In prokaryotes, an mRNA molecule can carry the information for one or multiple genes. A transcript that carries one gene is called **monocistronic** (a cistron is synonymous with a gene). One that carries multiple genes is called **polycistronic.** The proteins encoded on a polycistronic message generally have related functions, allowing a cell to express related genes as one unit.

TABLE 7.2 Components of Transcription in Bacteria

Component	Comment
(−) strand of DNA	Strand of DNA that serves as the template for RNA synthesis; the resulting RNA molecule is complementary to this strand.
(+) strand of DNA	Strand of DNA complementary to the one that serves as the template for RNA synthesis; the nucleotide sequence of the RNA molecule is the same as this strand, except it has uracil rather than thymine.
Promoter	Nucleotide sequence to which RNA polymerase binds to initiate transcription.
RNA polymerase	Enzyme that synthesizes RNA using one strand of DNA as a template; synthesis always occurs in the 5′ to 3′ direction.
Sigma (σ) factor	Component of RNA polymerase that recognizes the promoter regions. A cell can have different types of σ factors that recognize different promoters, allowing the cell to transcribe specialized sets of genes as needed.
Terminator	Nucleotide sequence at which RNA synthesis stops; the RNA polymerase falls off the DNA template and releases the newly synthesized RNA.

Initiation of RNA Synthesis

Transcription is initiated when RNA polymerase binds to a promoter (**figure 7.10**). The binding denatures (melts) a short stretch of DNA, creating a region of exposed nucleotides that serves as a template for RNA synthesis.

The part of RNA polymerase that recognizes the promoter is a loosely attached subunit called **sigma (σ) factor.** A cell can produce various types of σ factors, each recognizing different promoters. By controlling which σ factors are made, cells can transcribe specialized sets of genes as needed. The RNA polymerases of eukaryotic cells and archaea use proteins called transcription factors to recognize promoters.

Promoters identify the regions of a DNA molecule that will be transcribed into RNA. In doing so, they also orient the direction of the RNA polymerase on the DNA molecule, thereby dictating which strand will be used as a template (**figure 7.11**). The direction of polymerase movement can be likened to the flow of a river. Because of this, the words "upstream" and "downstream" are used to describe relative positions of other sequences. As an example, promoters are upstream of the genes they control.

Elongation of the RNA Transcript

In the elongation phase, RNA polymerase moves along DNA, using the (−) strand as a template to synthesize a single-stranded RNA molecule (see figure 7.10). As with DNA replication, nucleotides are added only to the 3′ end; the reaction is fueled by hydrolyzing a high-energy phosphate bond of the incoming nucleotide. When RNA polymerase advances, it denatures a new stretch of DNA and allows the previous portion to renature (close). The denaturation exposes a new region of the template so elongation can continue.

Once elongation has proceeded far enough for RNA polymerase to move beyond the promoter, another molecule of the enzyme can bind, initiating a new round of transcription. Thus, a single gene can be transcribed repeatedly very quickly.

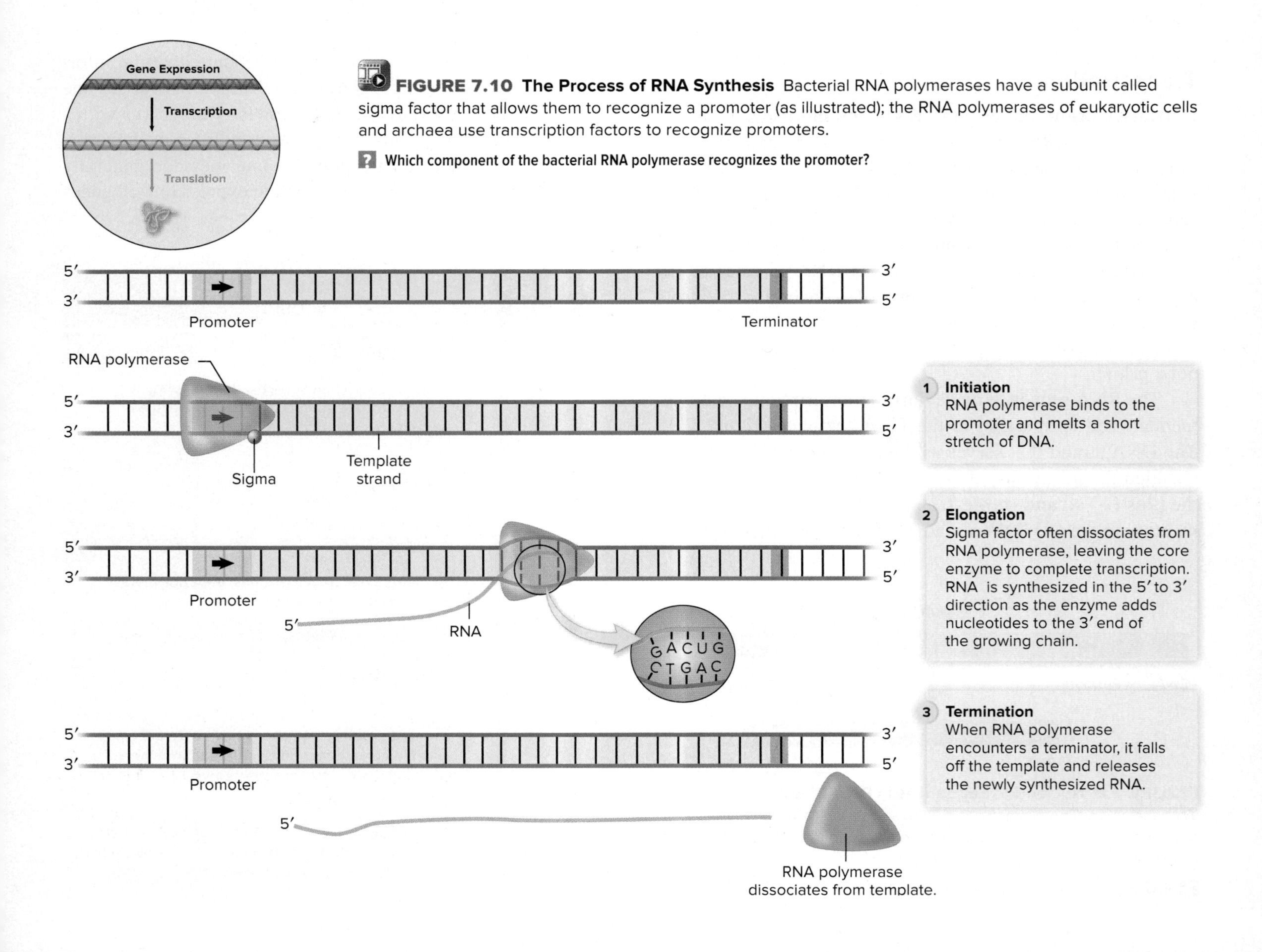

FIGURE 7.10 The Process of RNA Synthesis Bacterial RNA polymerases have a subunit called sigma factor that allows them to recognize a promoter (as illustrated); the RNA polymerases of eukaryotic cells and archaea use transcription factors to recognize promoters.

? **Which component of the bacterial RNA polymerase recognizes the promoter?**

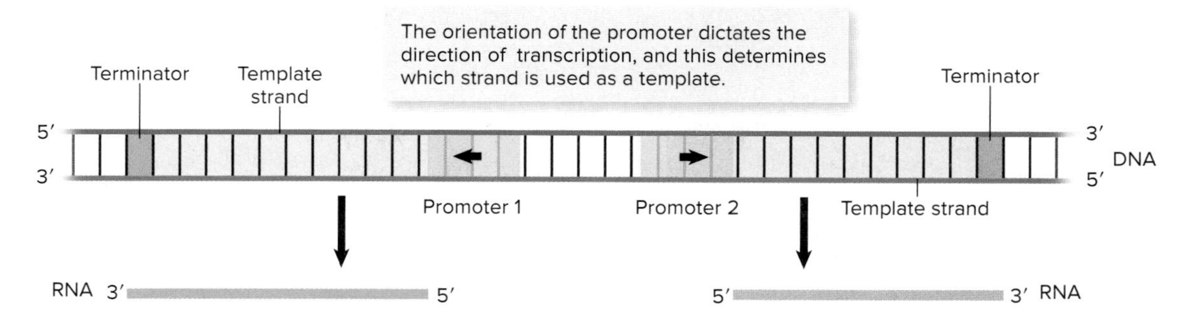

FIGURE 7.11 The Promoter Orients RNA Polymerase The orientation of RNA polymerase determines which strand will be used as the template. In this diagram, the color of the RNA molecules indicates which DNA strand was used as the template. The light blue RNA was transcribed from the red DNA strand (and is therefore analogous in sequence to the blue DNA strand), whereas the pink RNA was transcribed from the blue DNA strand (and is therefore analogous in sequence to the red DNA strand).

? The light blue RNA strand is complementary to which DNA strand in this figure? To which DNA strand is the pink RNA strand complementary?

Termination of Transcription

Just as an initiation of transcription occurs at distinct nucleotide sequences of the DNA, so does termination. When RNA polymerase encounters a sequence called a terminator, it falls off the DNA template and releases the newly synthesized RNA.

Translation

Translation is the process of decoding the information carried in the mRNA to synthesize the specified protein. The process requires three major structures—mRNA, ribosomes (which contain rRNA), and tRNAs—in addition to various other components (**table 7.3**).

The Role of mRNA

An mRNA molecule is a temporary copy of the information in DNA; it carries encoded instructions for synthesis of a specific protein, or in the case of a polycistronic message, a specific group of proteins. Recall that mRNA is composed of nucleotides, whereas proteins are composed of amino acids. Cells must decode the information in mRNA into amino acids. This is done using the **genetic code,** which correlates a series of three nucleotides—a **codon**—to one amino acid (**table 7.4**). The genetic code is practically universal, meaning that with the exception of a few minor changes, it is used by all living things.

Because a codon is a triplet of any combination of the four nucleotides, there are 64 different codons (4^3). Three are stop codons, which signal the end of translation. The remaining 61 translate to the 20 different amino acids. This means that more than one codon can code for a specific amino acid. For example, both ACA and ACG encode the amino acid threonine. Because of this redundancy, the genetic code is said to be degenerate. ⏭ **stop codon**

TABLE 7.3 Components of Translation in Bacteria

Component	Comment
Anticodon	Sequence of three nucleotides in a tRNA molecule that is complementary to a particular codon in mRNA. The anticodon allows the tRNA to recognize and bind to the appropriate codon.
mRNA	Type of RNA molecule that contains the genetic information decoded during translation.
Polyribosome (polysome)	Multiple ribosomes attached to a single mRNA molecule.
Reading frame	Grouping of a stretch of nucleotides into sequential triplets that code for amino acids; an mRNA molecule has three potential reading frames, but only one is typically used in translation.
Ribosome	Structure that facilitates the joining of amino acids during the process of translation; composed of protein and ribosomal RNA. The prokaryotic ribosome (70S) consists of a 30S and a 50S subunit.
Ribosome-binding site	Sequence of nucleotides in mRNA to which a ribosome binds; the first time the codon for methionine (AUG) appears after that site, translation generally begins.
rRNA	Type of RNA molecule present in ribosomes.
Start codon	Codon at which translation is initiated; it is typically the first AUG after a ribosome-binding site.
Stop codon	Codon that terminates translation, signaling the end of the protein; there are three stop codons.
tRNA	Type of RNA molecule involved in interpreting the genetic code; each tRNA molecule carries a specific amino acid dictated by its anticodon.

TABLE 7.4 The Genetic Code

First Letter	Second Letter: U		C		A		G		Third Letter
U	UUU	Phe Phenylalanine	UCU	Ser Serine	UAU	Tyr Tyrosine	UGU	Cys Cysteine	U
	UUC		UCC		UAC		UGC		C
	UUA	Leu Leucine	UCA		UAA	"Stop"	UGA	"Stop"	A
	UUG		UCG		UAG	"Stop"	UGG	Trp Trytophan	G
C	CUU	Leu Leucine	CCU	Pro Proline	CAU	His Histidine	CGU	Arg Arginine	U
	CUC		CCC		CAC		CGC		C
	CUA		CCA		CAA	Gln Glutamine	CGA		A
	CUG		CCG		CAG		CGG		G
A	AUU	Ile Isoleucine	ACU	Thr Threonine	AAU	Asn Asparagine	AGU	Ser Serine	U
	AUC		ACC		AAC		AGC		C
	AUA		ACA		AAA	Lys Lysine	AGA	Arg Arginine	A
	AUG	Met Methionine; "Start"	ACG		AAG		AGG		G
G	GUU	Val Valine	GCU	Ala Alanine	GAU	Asp Aspartate	GGU	Gly Glycine	U
	GUC		GCC		GAC		GGC		C
	GUA		GCA		GAA	Glu Glutamate	GGA		A
	GUG		GCG		GAG		GGG		G

Note: The genetic code correlates a codon (a series of three nucleotides) to an amino acid, as shown in the table. In some cases, the codon encodes a "stop" signal instead of an amino acid. Many amino acids are specified by more than one codon. For example, threonine is specified by four codons, which differ only in the third nucleotide (ACU, ACC, ACA, and ACG).

The nucleotide sequence of an mRNA molecule indicates where the coding region begins and ends (**figure 7.12**). The site at which it begins is particularly important because the translation "machinery" reads the mRNA in groups of three nucleotides. As a consequence, any given sequence has three possible **reading frames,** or ways in which triplets can be grouped (**figure 7.13**). If translation begins in the wrong reading frame, a very different, and generally non-functional, protein is synthesized.

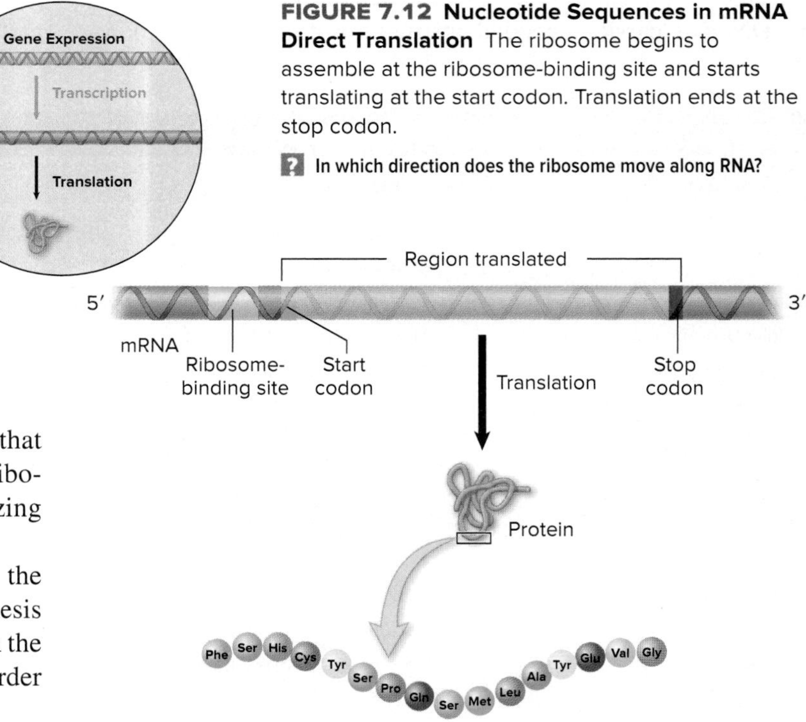

FIGURE 7.12 Nucleotide Sequences in mRNA Direct Translation The ribosome begins to assemble at the ribosome-binding site and starts translating at the start codon. Translation ends at the stop codon.

? In which direction does the ribosome move along RNA?

The Role of Ribosomes

Ribosomes serve as translation "machines": structures that string amino acids together to make a polypeptide. A ribosome does this by aligning two amino acids and catalyzing the formation of a peptide bond between them. ◀◀ ribosomes

Ribosomes also locate key punctuation sequences on the mRNA molecule, such as the points at which protein synthesis should begin. The ribosome then moves along the mRNA in the 5′ to 3′ direction, "presenting" each codon in a sequential order for decoding while maintaining the correct reading frame.

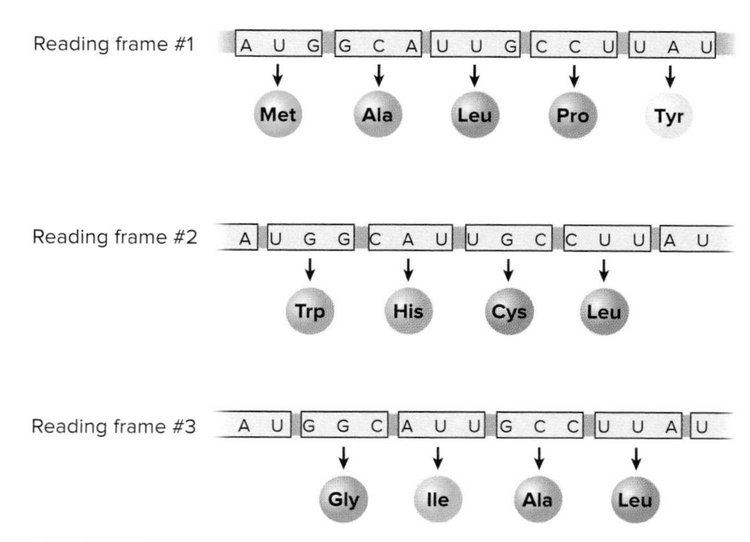

FIGURE 7.13 Reading Frames A nucleotide sequence has three potential reading frames, but only one is typically used for translation.

Why is it important that the correct reading frame is used?

Prokaryotic ribosomes are composed of a 30S subunit and a 50S subunit, each made up of protein and ribosomal RNA (rRNA) (see figures 3.44 and 10.9); the "S" stands for Svedberg unit—a measure of size. Svedberg units are not additive, which is why the 70S ribosome can have 30S and 50S subunits. ◀◀ ribosomal subunits

MicroByte

Several types of antibiotics, including tetracycline and azithromycin, interfere with the function of the bacterial 70S ribosome.

The Role of Transfer RNAs

A tRNA is an RNA molecule that carries an amino acid to be used in translation. Each tRNA is folded into a three-dimensional shape held together by hydrogen bonds, and a specific amino acid is attached at one end (**figure 7.14**). When a tRNA recognizes and base-pairs with a given codon in an mRNA molecule being translated, the ribosome transfers the amino acid carried by that tRNA onto the end of the newly forming polypeptide. The recognition between the tRNA and the mRNA occurs because each tRNA has an **anticodon:** a group of three nucleotides complementary to a codon in the mRNA. Thus, the codon sequence in an mRNA molecule determines the sequence of amino acids in the protein, according to the genetic code (see table 7.4).

Once a tRNA molecule has delivered its amino acid during translation, it can be recycled. An enzyme in the cytoplasm recognizes the tRNA and then attaches the appropriate amino acid.

Initiation of Translation

In prokaryotes, translation begins as the mRNA molecule is still being synthesized. Part of the ribosome binds to a sequence in mRNA called the **ribosome-binding site;** the first AUG

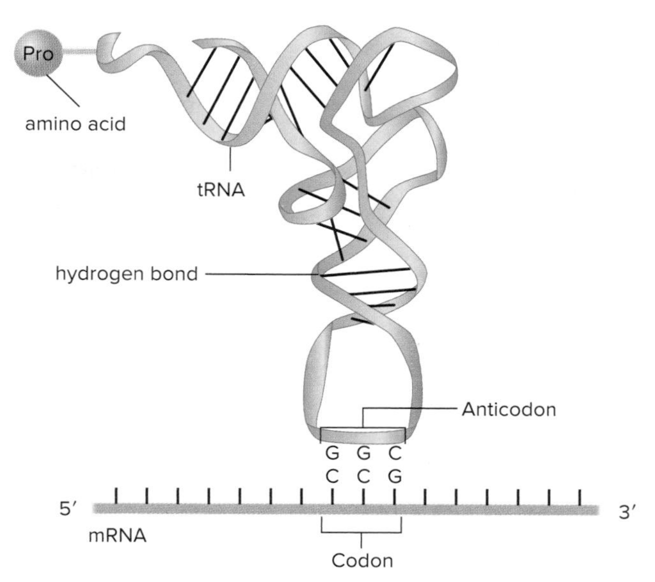

FIGURE 7.14 The Structure of Transfer RNA (tRNA) Three-dimensional illustration of tRNA. The amino acid that the tRNA carries is dictated by its anticodon. The tRNA that recognizes the mRNA codon CCG carries the amino acid proline—as specified by the genetic code.

A tRNA that has the anticodon GAG carries which amino acid?

after that site usually serves as the **start codon.** The complete ribosome assembles there, joined by an initiating tRNA that carries a chemically altered form of the amino acid methionine (*N*-formylmethionine, or f-Met). The position of the first AUG is important, as it determines the reading frame used for translation of the remainder of that protein. Note that AUG functions as a start codon only when preceded by a ribosome-binding site; at other sites, it simply encodes methionine.

Elongation of the Polypeptide Chain

The ribosome has two sites to which amino acid–carrying tRNAs can bind: the P-site and the A-site. At the start of translation, the initiating tRNA carrying the f-Met occupies the P-site (**figure 7.15**). A tRNA that recognizes the next codon on the mRNA then fills the unoccupied A-site. Once both sites are filled, an rRNA creates a peptide bond between the two amino acids carried by the tRNAs. This transfers the amino acid from the initiating tRNA to the amino acid carried by the incoming tRNA.

After the initiating tRNA has donated its amino acid to the tRNA in the A-site, the ribosome advances a distance of one codon, moving along the mRNA in a 5′ to 3′ direction. As this happens, the initiating tRNA is released through a region called the E-site. The remaining tRNA, which now carries both amino acids, occupies the P-site. The A-site is temporarily empty. A tRNA that recognizes the codon in the A-site quickly attaches there, and the process repeats.

Once translation of a gene has progressed far enough for the ribosome to clear the initiating sequences, another

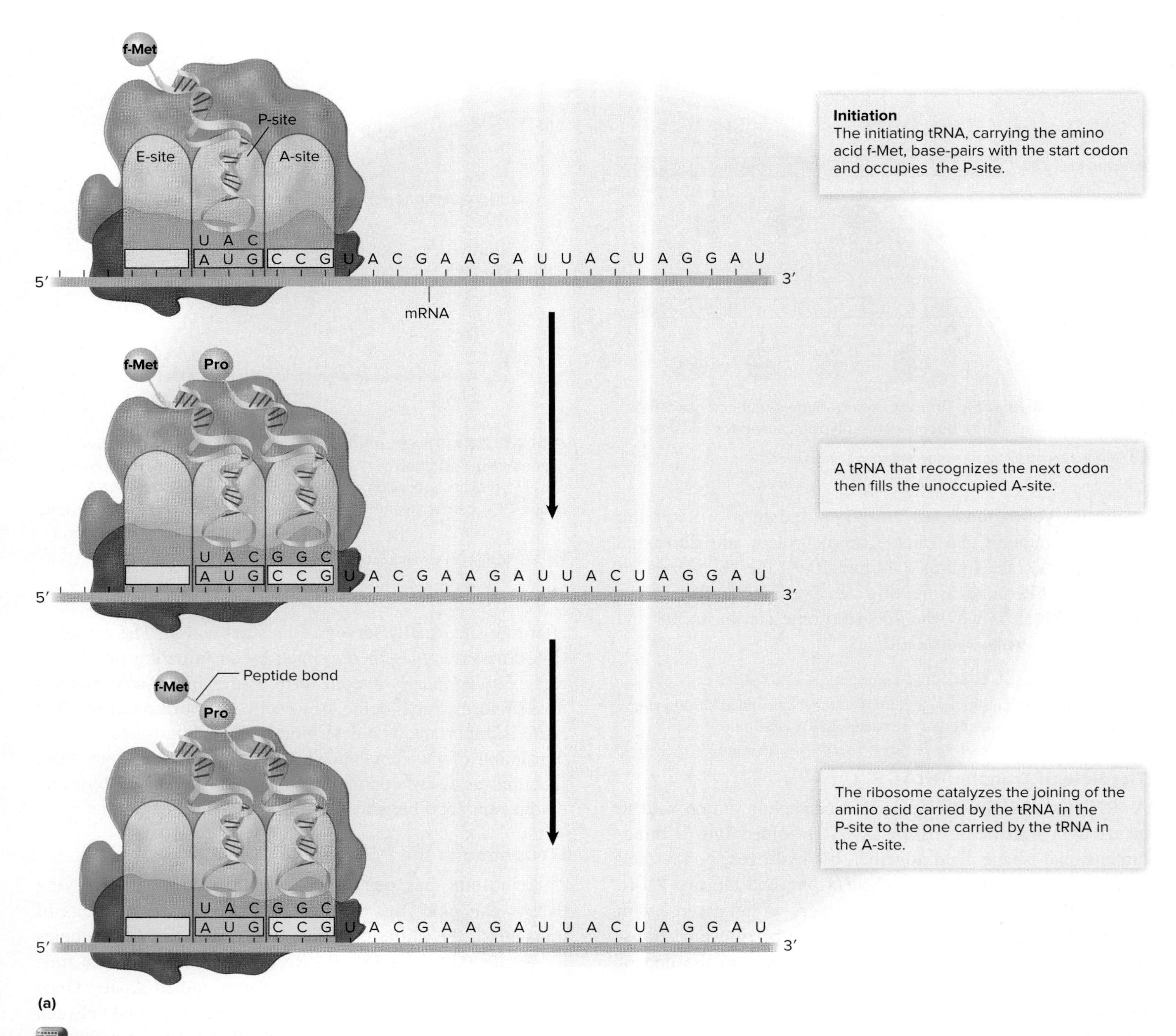

FIGURE 7.15 The Process of Translation **(a)** Initiation. **(b)** Elongation. **(c)** Termination.

What is the importance of the anticodon on the tRNA?

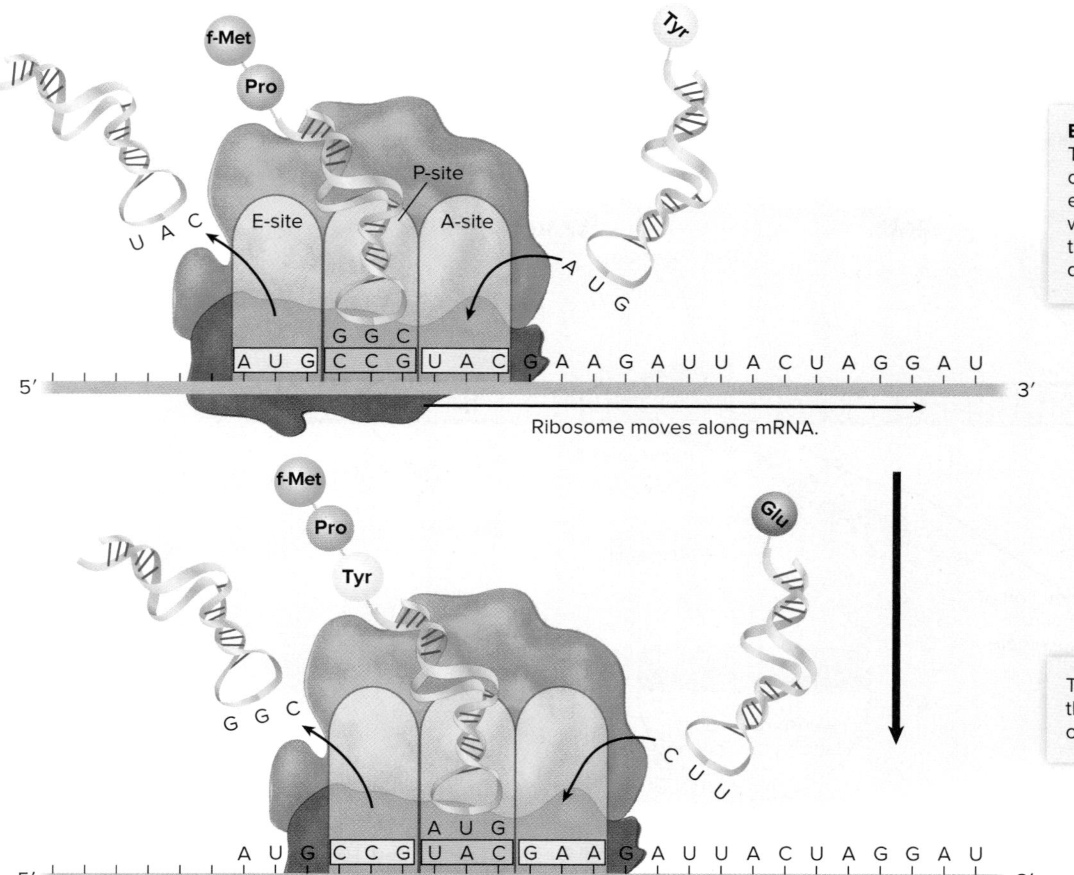

Elongation
The ribosome advances a distance of one codon. The tRNA that occupied the P-site exits through the E-site and the tRNA that was in the A-site occupies the P-site. A tRNA that recognizes the next codon quickly fills the empty A-site.

The ribosome continues advancing along the mRNA in the 5′ to 3′ direction, moving one codon at a time.

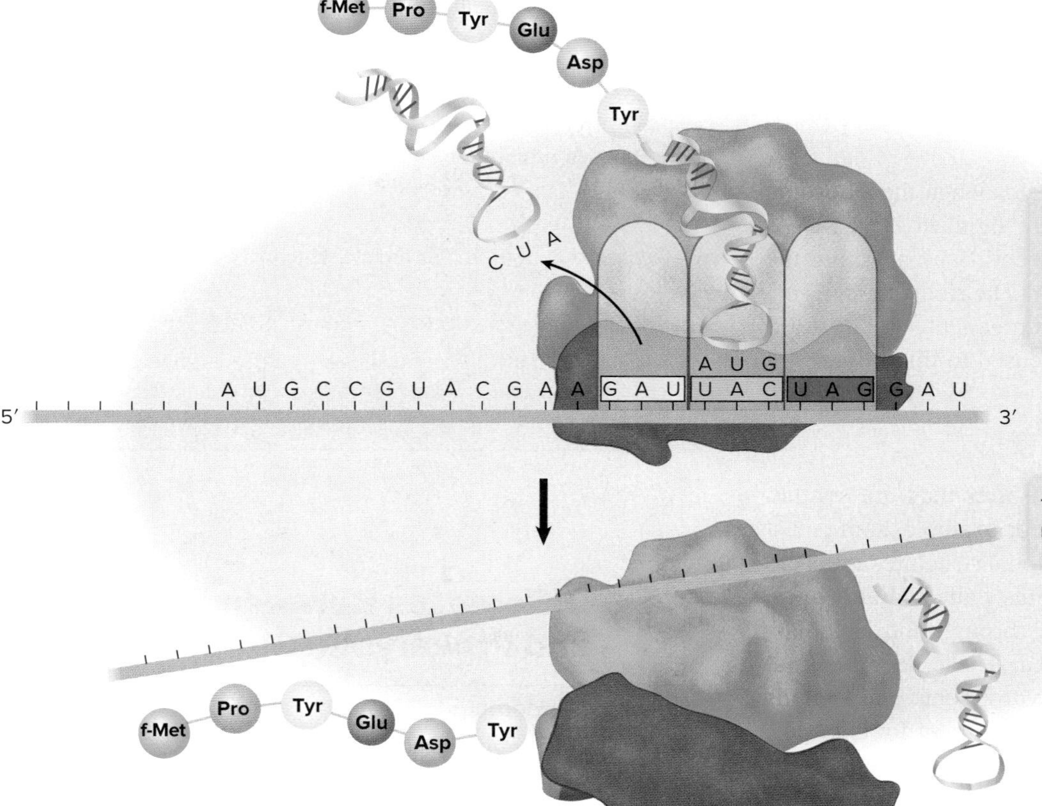

Termination
Translation continues until a stop codon is reached, signaling the end of the process. No tRNA molecules recognize a stop codon.

The components disassemble releasing the newly formed polypeptide.

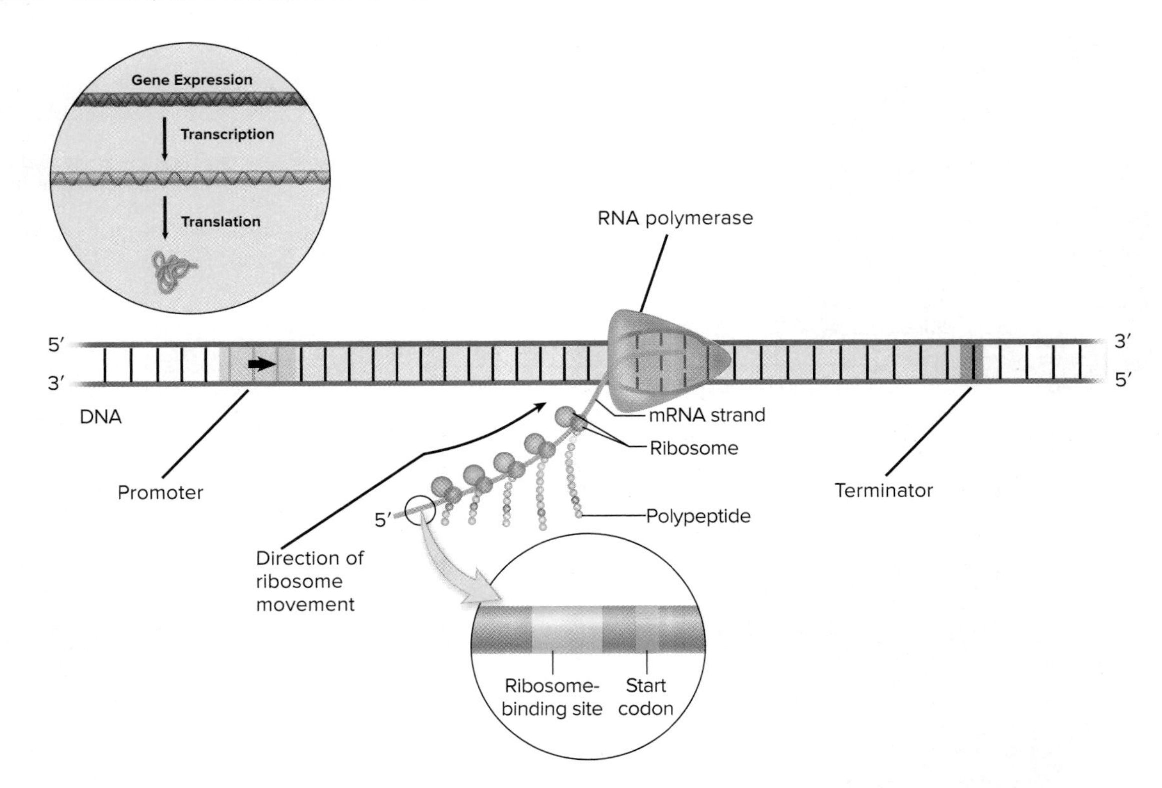

FIGURE 7.16 In Prokaryotes, Translation Begins as the mRNA Molecule Is Still Being Synthesized Ribosomes begin translating the 5′ end of mRNA before transcription is complete. More than one ribosome can be translating the same mRNA molecule.

Why is the position of the first AUG after the ribosome-binding site important?

ribosome can bind. Thus, at any one time, multiple ribosomes can be translating a single mRNA molecule (**figure 7.16**). This allows maximal protein synthesis from a single mRNA template. The assembly of multiple ribosomes attached to a single mRNA molecule is called a **polyribosome,** or a polysome.

Termination of Translation

Elongation of the polypeptide terminates when the ribosome reaches a **stop codon,** a codon not recognized by a tRNA. At this point, enzymes free the polypeptide by breaking the covalent bond that joins it to the tRNA. The ribosome falls off the mRNA, dissociating into its two component subunits (30S and 50S). The subunits can then be reused to initiate translation at other sites.

Post-Translational Modification

Polypeptides must often be modified after they are synthesized in order to become functional. For example, some must be folded into a specific three-dimensional structure, a process that requires the assistance of proteins called **chaperones.** Polypeptides destined for transport through the cytoplasmic membrane also must be modified. These have a **signal sequence,** a characteristic series of hydrophobic amino acids at their amino terminal end, which "tags" them for transport (see figure 3.31). The signal sequence must be removed by proteins in the membrane.

MicroAssessment 7.3

Gene expression involves transcription and translation. In transcription, RNA polymerase synthesizes RNA in the 5′ to 3′ direction, using one strand of DNA as a template. In translation, ribosomes synthesize proteins, using the nucleotide sequence of an mRNA molecule to determine the amino acid sequence of the encoded protein. The correct amino acid for a given codon is delivered by tRNAs. After synthesis, many proteins are modified in some way.

7. How does a promoter dictate which DNA strand is used as the template?
8. What is the role of tRNA in translation?
9. Could two mRNA molecules have different nucleotide sequences and yet code for the same protein? Explain your answer.

7.4 ■ Differences Between Eukaryotic and Prokaryotic Gene Expression

Learning Outcome

6. Describe four differences between prokaryotic and eukaryotic gene expression.

TABLE 7.5 Major Differences Between Prokaryotic and Eukaryotic Transcription and Translation

Prokaryotes	Eukaryotes
mRNA does not have a cap or a poly A tail.	Processing of the transcript (pre-mRNA) results in mRNA with a cap at the 5′ end and a poly A tail at the 3′ end.
Transcript (mRNA) does not contain introns.	Transcript (pre-mRNA) contains introns, which are removed by splicing.
Translation of mRNA begins as it is being transcribed.	mRNA is transported out of the nucleus so that it can be translated in the cytoplasm.
mRNA is often polycistronic; translation usually begins at the first AUG codon that follows a ribosome-binding site.	mRNA is monocistronic; translation begins at the first AUG.

Eukaryotes differ significantly from prokaryotes in several aspects of transcription and translation **(table 7.5)**. Eukaryotic mRNA, for example, is synthesized in a precursor form, called **pre-mRNA.** The pre-mRNA must be processed (altered) both during and after transcription to form mature mRNA (**figure 7.17**). Shortly after transcription begins, a cap (a methylated guanine derivative) is added to the 5′ end of a pre-mRNA, a process called **capping.** This cap binds specific proteins that stabilize the transcript and enhance translation. The 3′ end of the molecule is also modified, even before transcription has been terminated. This process, **polyadenylation,** cleaves the transcript at a specific sequence and then adds about 200 adenine derivatives to the new 3′ end. This creates a poly A tail, which is thought to stabilize the transcript as well as enhance translation. Another important modification is **splicing,** which removes specific segments of the transcript. Splicing is necessary because eukaryotic genes are often interrupted by non-coding sequences. These intervening sequences, **introns,** are transcribed along with the expressed regions, **exons,** and must be removed from pre-mRNA to create functional mRNA.

The mRNA in eukaryotic cells must be transported out of the nucleus before it can be translated in the cytoplasm. Thus, unlike in prokaryotes, the same mRNA molecule cannot be synthesized and translated at the same time or even in the same cellular location. The mRNA of eukaryotes is generally monocistronic, and translation of the message typically begins at the first AUG in the molecule.

The ribosomes of eukaryotes differ from those of prokaryotes. Whereas the prokaryotic ribosome is 70S, made up of 30S and 50S subunits, the eukaryotic ribosome is 80S, made up of 40S and 60S subunits. The differences in ribosome structure are medically important because certain antibiotics bind to and inactivate bacterial 70S ribosomes, but not 80S ribosomes. This explains why those antibiotics kill bacteria, usually without causing significant harm to mammalian cells.

MicroByte

Symptoms of the disease diphtheria are caused by a toxin that inactivates a protein required for 80S ribosome function.

MicroAssessment 7.4

Eukaryotic pre-mRNA must be processed, which involves capping, polyadenylation, and splicing. In eukaryotic cells, the mRNA must be transported out of the nucleus before it can be translated in the cytoplasm. Eukaryotic mRNA is monocistronic.

10. What is an intron?

11. Would a deletion of two base pairs have a greater consequence if it occurred in an intron rather than in an exon?

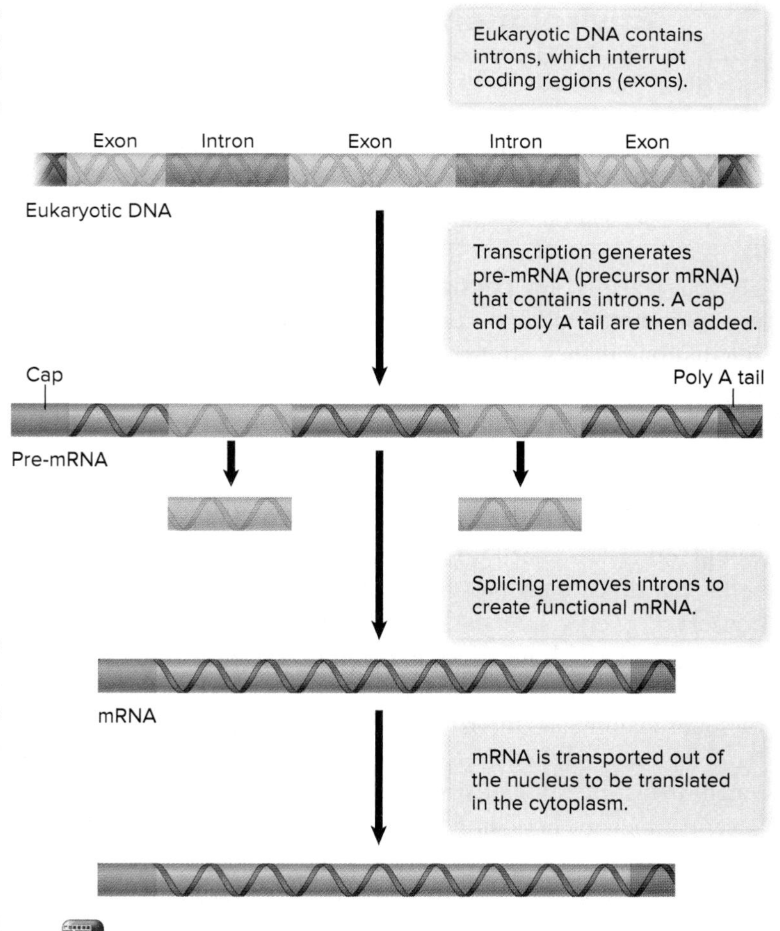

FIGURE 7.17 Processing of Eukaryotic RNA

Introns are intervening sequences; what are exons?

FOCUS YOUR PERSPECTIVE 7.1

RNA: The First Macromolecule?

The 1989 Nobel Prize in Chemistry was awarded to two Americans, Sidney Altman and Thomas Cech, who independently made the unexpected observation that RNA molecules can act as enzymes. Before their studies, only proteins were thought to have enzymatic activity.

Cech made a key observation in 1982 when he was trying to understand how introns are removed from precursor ribosomal RNA (pre-rRNA) in a eukaryotic protozoan. Because he was convinced that proteins were responsible for cutting out introns, he added all of the protein in the cells' nuclei to a suspension of pre-rRNA. As expected, the introns were cut out. For a control, he used pre-rRNA to which no nuclear proteins were added, fully expecting that nothing would happen. Much to his surprise, the introns were removed in the control as well. Based on these results, Cech could only conclude that the RNA acted on itself to cut out the introns.

The studies of Altman and his colleagues showed that RNA had catalytic properties beyond cutting out introns from pre-rRNA. Altman's group found that RNA could convert a precursor transfer RNA (pre-tRNA) molecule to its final functional state. Additional studies have shown that enzymatic reactions in which catalytic RNAs, or ribozymes, play a role are very widespread. Ribozymes have been found in the mitochondria of eukaryotic cells and shown to catalyze other reactions that resemble the polymerization of RNA.

These observations have profound implications for a long-standing question in evolutionary biology: Which came first, proteins or nucleic acids? The answer seems to be nucleic acids, specifically RNA, which acted both as a carrier of genetic information and as an enzyme. Billions of years ago, before the present universe in which DNA, RNA, and protein are found, the only macromolecule was probably RNA. Once tRNAs became available, they carried amino acids to nucleotide sequences on a strand of RNA.

7.5 ■ Sensing and Responding to Environmental Fluctuations

Learning Outcomes

7. Describe how quorum sensing and two-component regulatory systems allow cells to adapt to fluctuating environmental conditions.
8. Compare and contrast antigenic variation and phase variation.

Microorganisms are constantly faced with rapidly changing environmental conditions, and must quickly adapt to the fluctuations if they are to survive. Consider the situation of *E. coli* in the intestinal tract of mammals. In this habitat, it must cope with alternating periods of feast and famine. For a limited time after a mammal eats, the bacterial cells thrive, bathed in the mixture of amino acids, vitamins, and other nutrients. The cells actively take up these compounds they would otherwise need to synthesize. Simultaneously, the cells shut down their biosynthetic pathways, channeling the conserved energy into the rapid production of cell components such as DNA and protein. Famine, however, follows the feast. Between meals—which can be many days in the case of some mammals—the rich source of nutrients is depleted. Now the bacterial cells' biosynthetic pathways must be activated, using energy and slowing cell growth. Cells dividing several times an hour in a nutrient-rich environment might divide only once every 24 hours in a starved mammalian gut. Later, when the animal defecates, some of the *E. coli* cells are eliminated in the feces. Outside the mammalian host, the bacterial cells must cope with yet a completely different set of conditions to survive.

Signal Transduction

Signal transduction is the transmission of information from outside a cell to the inside. This allows cells to monitor and react to environmental conditions.

Quorum Sensing

Some organisms can "sense" the density of cells within their own population—a phenomenon called **quorum sensing.** This allows cells to activate genes that are useful only when expressed by a critical mass. As an example, the cooperative activities leading to biofilm formation are controlled by quorum sensing. Some pathogens use the mechanism to coordinate expression of genes involved with the infection process. ◀◀ biofilms

Quorum sensing involves a process that allows bacteria to "talk" to each other by synthesizing one or more varieties of extracellular signaling molecules. When few cells are present, the concentration of a given signaling molecule is very low. As the cells multiply in a confined area, however, the concentration of that molecule increases proportionally. Only when a signaling molecule reaches a certain level does it change the expression of specific genes (**figure 7.18**).

Some types of bacteria are able to detect and even interfere with the signaling molecules produced by other species. This allows them to "eavesdrop" on and even obstruct "conversations" of other bacteria.

Two-Component Regulatory Systems

An important mechanism that cells use to detect and react to changes in the external environment is a **two-component regulatory system** (**figure 7.19**). *E. coli* uses such a system

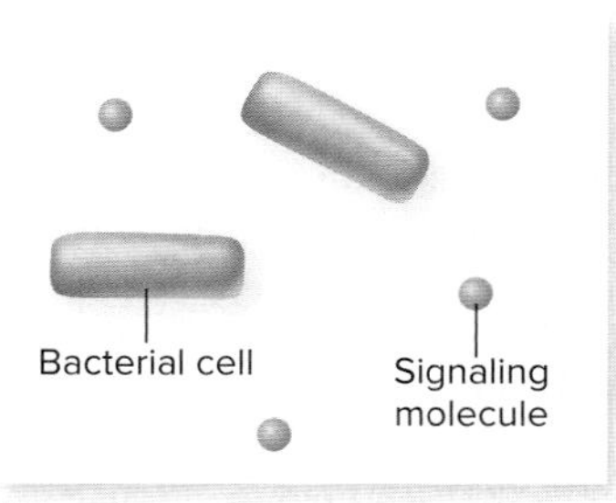

When few cells are present, the concentration of the signaling molecule is low.

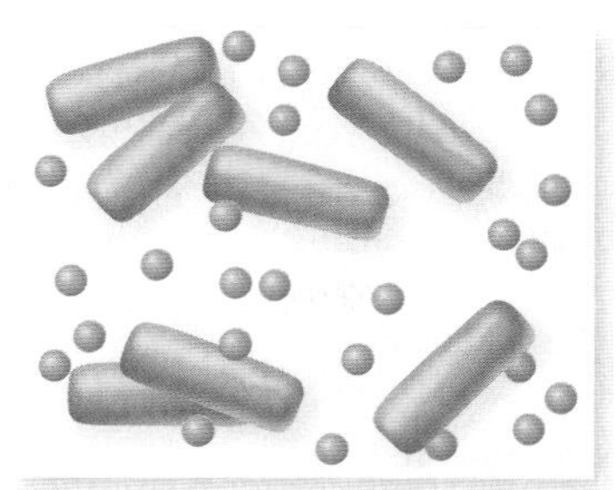

When many cells are present, the signaling molecule reaches a concentration high enough to induce the expression of certain genes.

FIGURE 7.18 Quorum Sensing

Why would it be beneficial for cells to wait until a critical population density is present before expressing certain genes?

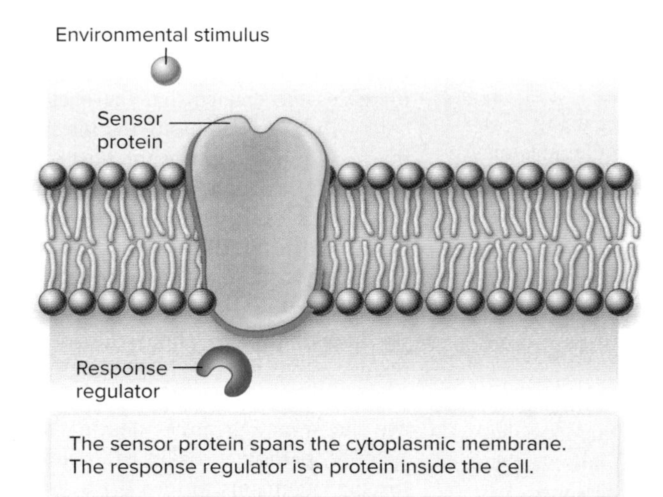

The sensor protein spans the cytoplasmic membrane. The response regulator is a protein inside the cell.

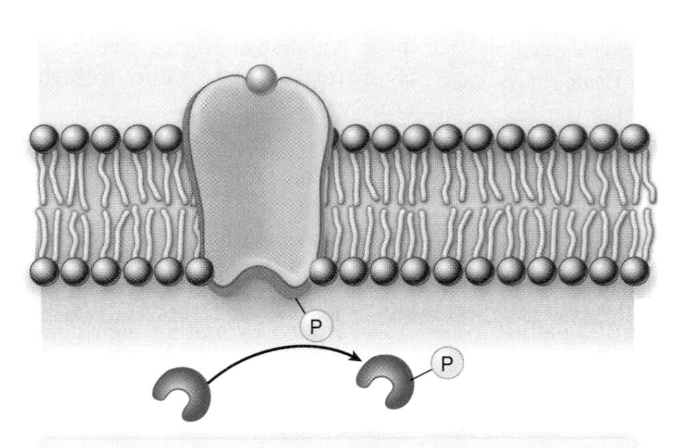

In response to a specific change in the environment, the sensor phosphorylates a region on its internal portion. The phosphate group is transferred to the response regulator, which can then turn genes on or off, depending on the system.

FIGURE 7.19 Two-Component Regulatory System

***E. coli* cells use a two-component system to sense nitrate in the environment. What would happen if they lost the ability to sense that compound?**

to control the expression of genes for its alternative types of metabolism. When nitrate is present in anaerobic conditions, the cells activate genes required to use it as the terminal electron acceptor. Some pathogens use two-component regulatory systems to sense and respond to environmental magnesium concentrations. Because the magnesium concentration within certain host cells is generally lower than that of the extracellular environment, these pathogens are able to recognize whether or not they are within a host cell. In turn, they can activate appropriate genes that help them evade host defenses. ⏮ **terminal electron acceptor**

Two-component regulatory systems consist of two different proteins: a sensor and a response regulator. The sensor spans the cytoplasmic membrane. In response to specific environmental variations, the sensor chemically modifies a region on its internal portion, usually by phosphorylating a specific amino acid. The phosphate group is then transferred to a response regulator. When phosphorylated, the response regulator can turn genes either on or off, depending on the system.

Natural Selection

Natural selection, the survival and growth of cells best adapted to live a particular environment, can also play a role in gene expression. The expression of some genes changes randomly in cells, increasing survival chances of at least a part of a population.

The role of natural selection is readily apparent in bacteria that undergo **antigenic variation,** an alteration in the characteristics of certain surface proteins. Pathogens that do this can stay one step ahead of the body's defenses by altering the very molecules our immune systems must learn to recognize. One of the most well-characterized examples is *Neisseria gonorrhoeae.* This bacterium has many different genes for pilin, the protein subunit that makes up pili, yet most of these genes are silent (not expressed). The only one expressed is in a particular chromosomal location called an expression locus. *N. gonorrhoeae* cells have a mechanism to randomly shuffle the pilin genes, moving different ones in and out of the expression locus. In a population of 10^4 cells, at least one is expressing a different type of pilin. During an infection, the body's immune system will begin to respond to the dominant pilin type, but the bacterial cells that have already "switched" to produce a different pilin type will survive and then multiply. Eventually, the immune system learns to recognize those, but by that time, another subpopulation will have switched its pilin type. ⏮ **pili,** ⏭ ***Neisseria gonorrhoeae***

Another mechanism of randomly altering gene expression is **phase variation,** the routine switching on and off of certain genes. In *E. coli,* for example, certain types of pili required for attachment to epithelial cells undergo phase variation. In an *E. coli* population adhering to an epithelial surface, some bacterial cells will spontaneously turn off the genes required for

FOCUS ON A CASE 7.1

In October 2010, about 9 months after Haiti experienced a serious earthquake, an outbreak of severe diarrhea and vomiting was reported. Patients were producing several liters of liquid stool in a day, resulting in signs of dehydration, including sunken eyes, muscle cramps, and in a few cases, convulsions and death. The illness was quickly identified as cholera—a disease caused by the bacterium *Vibrio cholerae.* Finding that organism in Haiti was a surprise, because cholera had not been reported there for at least a century. The outbreak was devastating—hundreds of thousands of people became ill with the disease, and thousands of patients died. Although the number of new cases eventually dropped significantly, cholera continues to be reported in the region.

As with most cholera outbreaks, the disease spread in Haiti through contaminated drinking water. The bacteria survive in aqueous environments by producing extracellular polymeric substances (EPS) that allow them to form a protective biofilm (see figure 4.3). Interestingly, although large numbers of live *V. cholerae* cells may be seen in contaminated water samples—some encased as a biofilm—relatively few of those cells grow in laboratory culture, suggesting that some enter a dormant stage.

After *V. cholerae*–contaminated water or food is ingested, the bacteria begin expressing different genes than they did in the external environment. They stop producing the EPS required for biofilm formation, and instead make pili that they use to attach to the epithelial cells that line the small intestine. As *V. cholerae* cells multiply in the small intestine, they make cholera toxin (CT), a toxin that causes the epithelial cells to release much more fluid than can normally be reabsorbed. The effect of the toxin results in the severe diarrhea that characterizes the disease.

1. Like many pathogens, *V. cholerae* uses quorum sensing as one of the regulatory mechanisms to control gene expression. Why would a disease-causing bacterium use quorum sensing to control expression of toxins or other molecules that allow the organism to cause disease?
2. Unlike pathogens that respond to high population density by turning on genes that allow them to cause disease, *V. cholerae* cells in a dense population turn off the genes associated with attachment and toxin production. As those genes are turned off, the cells turn on the genes required for biofilm production. How might this benefit the organism?
3. How could a better understanding of the signaling molecules involved in quorum sensing lead to new methods to control pathogens?
4. Researchers are investigating using quorum sensing signaling molecules of other bacteria to improve the isolation rate of *V. cholerae* from environmental samples. Why might these molecules induce *V. cholerae* cells to exit a dormant stage?

Discussion

1. Many bacterial pathogens do not produce the key virulence factors (the specific substances that allow the organism to cause disease) until the bacterial cell densities are high enough to overwhelm the host's defense system. The bacterial cells produce signaling molecules, and once these reach a certain level, indicating that the cell density is high, the cells produce and release the various virulence factors simultaneously. When large numbers of bacteria quickly release a toxin, for example, the host does not have time to mount an effective immune response, thereby allowing the pathogen to survive and multiply, using the affected host tissues as a source of nutrients.
2. *V. cholerae* uses quorum sensing in the opposite way from many other pathogens. At low density, the cells express genes for making attachment pili as well as cholera toxin (CT). This allows the bacteria to attach to the epithelial cells that line the small intestine, where they grow and produce CT without competition (members of the normal microbiota typically inhabit the large intestine rather than the small intestine). Once the bacteria reach a high population density, they turn off the genes for toxin and pilus production. At the same time, the cells turn on genes for proteins that help them detach from the epithelial surface as well as genes for producing extracellular polymeric substances (EPS). The detachment increases the chance that the bacterial cells will be passed along the intestinal tract, eventually leaving the host in the feces that can then contaminate water. Production of EPS allows the organisms to form biofilms, which increase their chance of survival in the external environment.
3. By knowing more about the mechanisms bacteria use to signal one another, scientists might be able to use the signaling molecules to "fool" specific pathogens. For example, by forcing organisms to turn off essential virulence genes, scientists can control the pathogenicity and survival of those organisms. A great deal of work is currently being done to achieve exactly this.
4. *Vibrio cholerae* might "eavesdrop" on the quorum-sensing communication among other bacteria as a means to assess environmental conditions. It appears that when neighboring cell populations are high, the *V. cholerae* cells respond by exiting their dormant state. This is an area of study currently being explored, and there is much more to be learned!

pili synthesis, thereby causing those cells to detach from the surface. The process is reversible, so the detached cells will later turn the genes on again, allowing the cells to colonize epithelial cells elsewhere. By altering the expression of genes such as these, at least a part of the population is ready for change.

MicroAssessment 7.5

Quorum sensing and two-component regulatory systems allow a microbial cell to respond to changing environmental conditions. The expression of some genes changes randomly, increasing the chances of survival of at least a subset of a cell population under varying environmental conditions.

12. Explain how certain bacteria "sense" the density of cells.
13. Describe antigenic variation.
14. In quorum sensing, why might a bacterium synthesize more than one type of signaling molecule?

7.6 ■ Bacterial Gene Regulation

Learning Outcomes

9. Give an example of a constitutive enzyme, an inducible enzyme, and a repressible enzyme.
10. Using the *lac* operon as a model, explain the role of inducers, repressors, and inducer exclusion.

In bacterial cells, many genes are routinely expressed, but others are regulated in response to environmental conditions. Note that scientists describe these regulated genes as capable of being turned on or off, but in reality there are no absolutes. In a population of cells, a gene that is off may still be expressed at very low levels.

A regulatory mechanism sometimes controls the transcription of only a limited number of genes, but in other cases, it controls many genes simultaneously. A set of regulated genes transcribed as a single mRNA molecule, along with the sequences that control its expression, is called an **operon.** One of the most well-characterized examples is the *lac* operon, which encodes proteins required for transporting and hydrolyzing the disaccharide lactose. Separate operons controlled by a single regulatory mechanism constitute a **regulon.** The simultaneous regulation of numerous genes is called **global control.** ⏮ disaccharide

When describing enzymes, scientists group them according to the type of regulation that controls their synthesis:

- **Constitutive.** Constitutive enzymes are synthesized constantly; the genes that encode these enzymes are always active. Constitutive enzymes usually play essential roles in the central metabolic pathways. For example, the enzymes of glycolysis are constitutive. ⏮ central metabolic pathways
- **Inducible.** Inducible enzymes are not routinely produced at significant levels; instead, their synthesis can be turned on when needed. An example is β-galactosidase, the enzyme that hydrolyzes lactose into its component monosaccharides—glucose and galactose. The genes for this enzyme are part of the *lac* operon, which is turned on only when lactose is present. This makes sense because a cell would waste precious resources if it expressed the operon when lactose was not available. Inducible enzymes are often involved in the transport and breakdown of specific energy sources.
- **Repressible.** Repressible enzymes are produced routinely, but their synthesis can be turned off when they are not required (**figure 7.20**). Repressible enzymes are generally involved in biosynthetic (anabolic) pathways, such as those that produce amino acids. Cells require a sufficient amount of a given amino acid to multiply; so, if an amino acid is not available in the environment, it needs to be produced by the cell. When the amino acid is available, however, synthesis of the enzymes used in its production would waste energy.

Mechanisms to Control Transcription

The methods a cell uses to prevent or facilitate transcription must be readily reversible, allowing cells to control the relative number of transcripts made. Two of the most common regulatory mechanisms are alternative sigma factors and DNA-binding proteins.

Alternative Sigma Factors

As described earlier, sigma factor is a loose component of RNA polymerase that functions in recognizing specific promoters. Standard sigma factors recognize promoters for genes that need to be expressed during routine growth conditions, but a cell can also produce **alternative sigma factors.** These recognize different sets of promoters, thereby controlling the expression of specific groups of genes. In the endospore former *Bacillus subtilis,* the sporulation process is controlled by a number of different alternative sigma factors. One of these controls the steps at the beginning of sporulation, whereas others then guide the stages of development in the mother cell and spore. A cell can also express anti-sigma factors, which inhibit the function of specific sigma factors. ⏮ sporulation

FIGURE 7.20 Principles of Regulation

Why would the biosynthetic enzymes of a cell be repressible rather than constitutive or inducible?

DNA-Binding Proteins

Transcription is often controlled by proteins that bind to specific DNA sequences. When a regulatory protein attaches to DNA, it can act as either a repressor, which blocks transcription, or an activator, which facilitates transcription.

Repressors A **repressor** is a regulatory protein that blocks transcription (negative regulation). It does this by binding to an operator, a specific DNA sequence located immediately downstream of a promoter. When a repressor is bound to an operator, RNA polymerase cannot progress past that DNA sequence. Repressors are allosteric proteins, however, meaning that specific molecules can attach to them and change their shape. This can change the repressor's ability to bind to operator DNA. As shown in **figure 7.21**, there are two general mechanisms by which different repressors can function:

- **Induction.** The repressor is synthesized as a form that binds to the operator, blocking transcription. When a molecule called an **inducer** attaches to the repressor, the shape of the repressor changes so that it can no longer attach to the **operator.** With the repressor unable to bind to DNA, RNA polymerase may transcribe the gene.
- **Repression.** The repressor is synthesized as a form that cannot bind to the operator. However, when a molecule termed a **corepressor** attaches to the repressor, the corepressor-repressor complex can then bind to the operator, blocking transcription.

Activators An **activator** is a regulatory protein that facilitates transcription (positive regulation). Genes controlled by an activator have an ineffective promoter preceded by an **activator-binding site.** The binding of the activator to the DNA enhances the ability of RNA polymerase to initiate transcription at that promoter. Like repressors, activators are allosteric proteins. When a molecule called an inducer binds to an activator, the shape of the activator changes so that it can then bind to the activator-binding site (**figure 7.22**). Thus, the term "inducer" applies to a molecule that turns on transcription, either by stimulating the function of an activator or by interfering with the function of a repressor.

MicroByte

In *Bacillus subtilis,* the DNA-binding protein that triggers the sporulation process controls a regulon of over 120 genes.

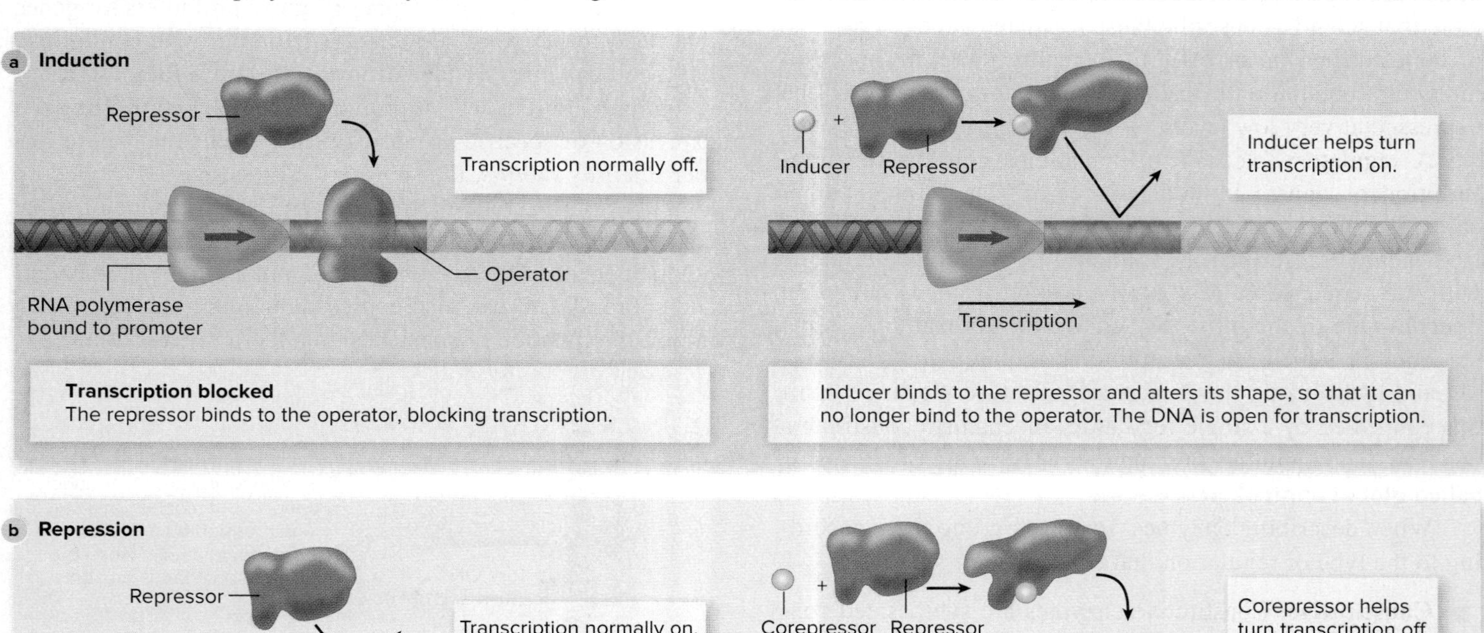

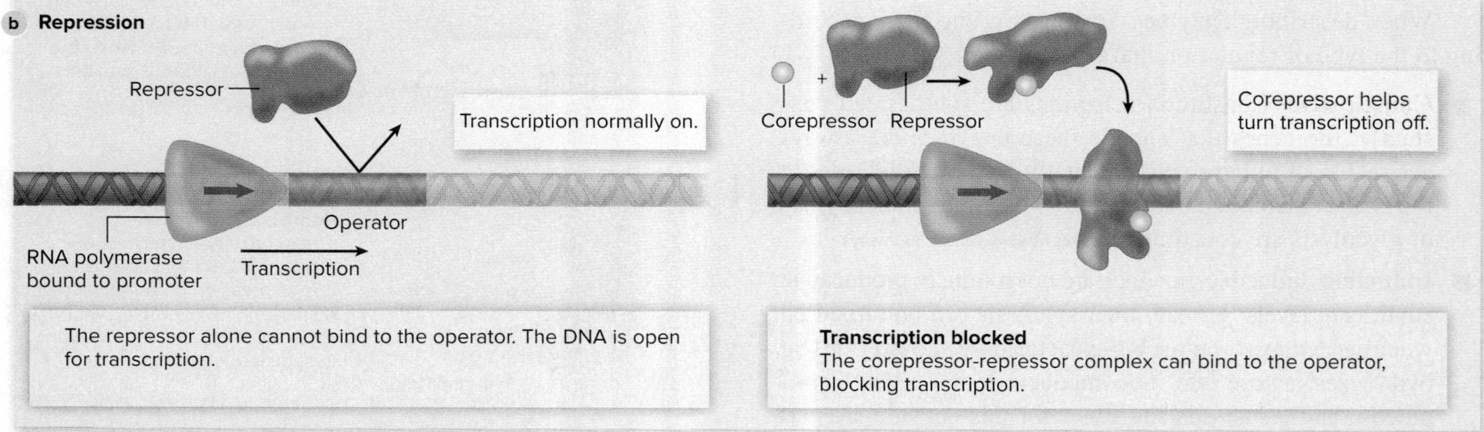

FIGURE 7.21 Transcriptional Regulation by Repressors (a) Induction. **(b)** Repression.

How is a corepressor different from an inducer? How is it similar to an inducer?

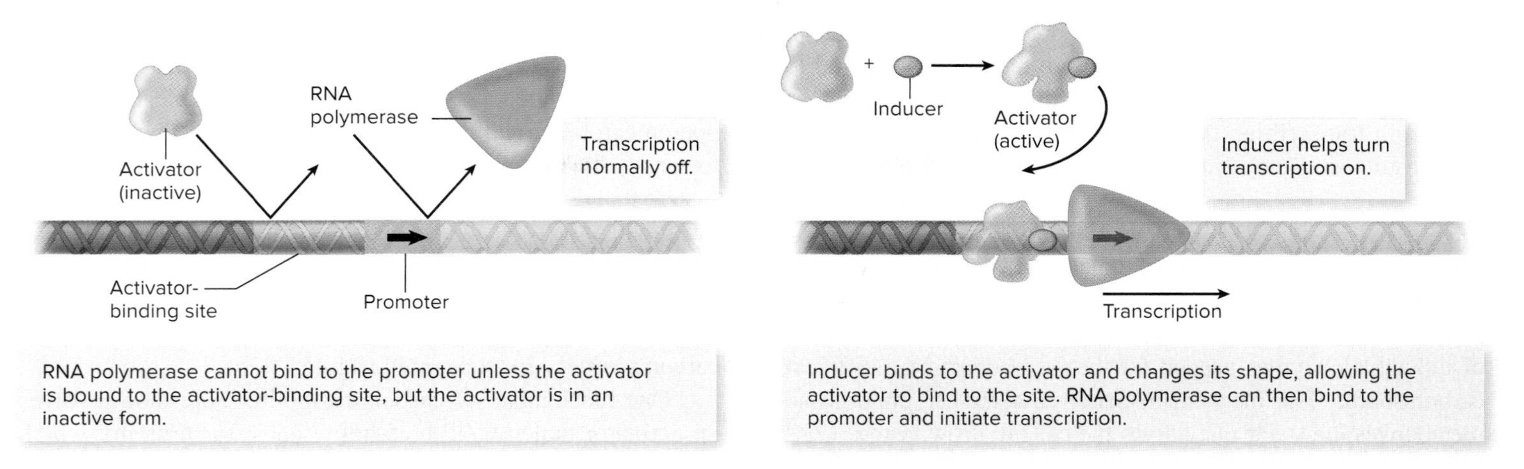

FIGURE 7.22 Transcriptional Regulation by Activators

How do activators facilitate transcription?

The *lac* Operon as a Model

Originally described in the early 1960s by François Jacob and Jacques Monod, the ***lac* operon** of *E. coli* has served as an important model for understanding the control of bacterial gene expression. This operon encodes proteins involved with the transport and degradation of lactose and is only turned on when glucose is not available but lactose is. In this situation, the cell is forced to use lactose. When glucose is available, it prevents expression of the *lac* operon genes, ensuring that cells use the most efficiently metabolized carbon source (glucose) first.

Lactose and the *lac* Operon

The *lac* operon uses a repressor that prevents transcription when lactose is not available; the repressor binds the operator, blocking RNA polymerase (**figure 7.23**). When lactose is in the cell, however, some of it is converted to allolactose, an

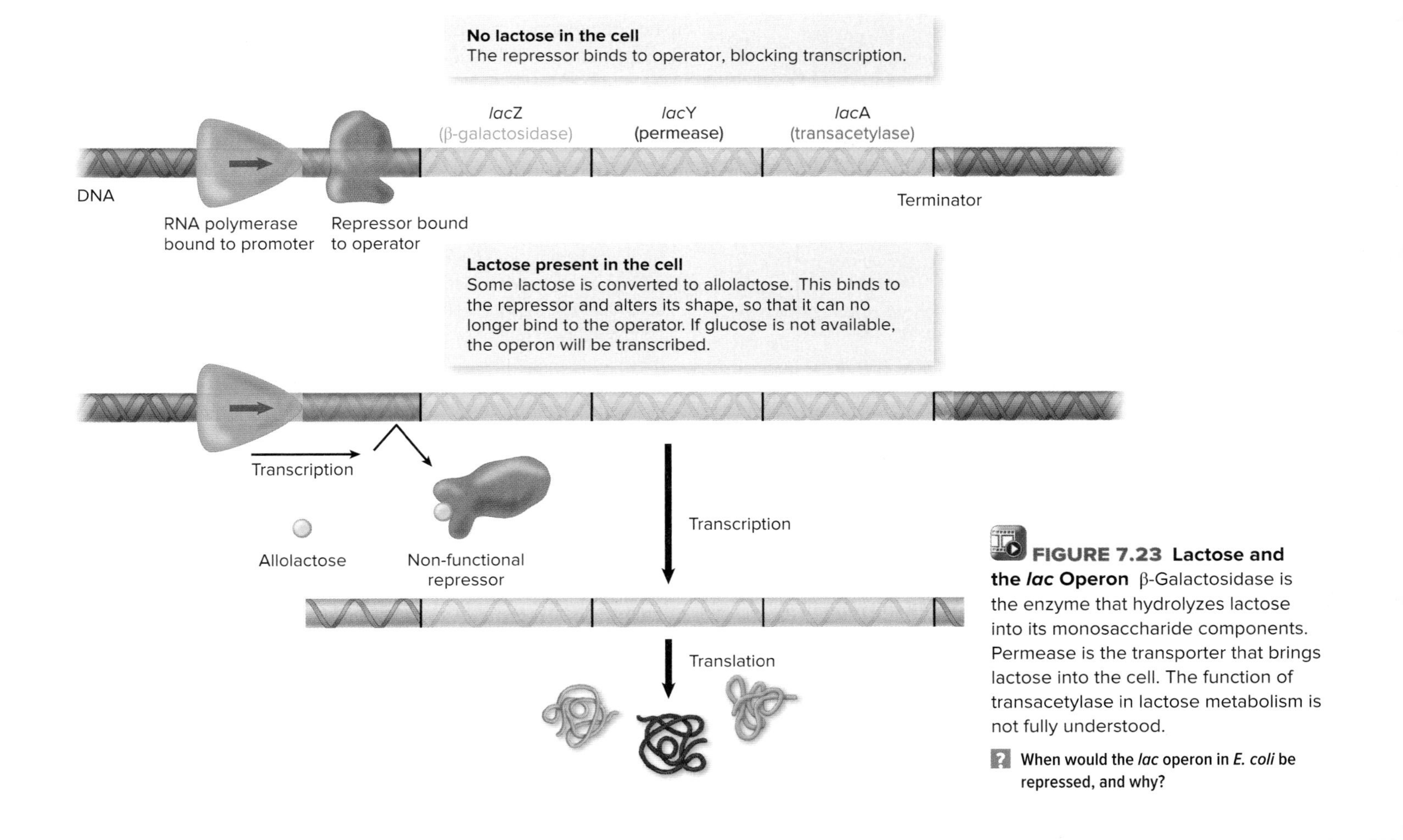

FIGURE 7.23 Lactose and the *lac* Operon β-Galactosidase is the enzyme that hydrolyzes lactose into its monosaccharide components. Permease is the transporter that brings lactose into the cell. The function of transacetylase in lactose metabolism is not fully understood.

When would the *lac* operon in *E. coli* be repressed, and why?

inducer. This compound binds the repressor and, in doing so, changes the repressor's shape so that it can no longer bind to the operator. With the operator unoccupied, RNA polymerase can begin transcribing the operon. However, this can happen only if glucose is not available in the growth medium.

Glucose and the *lac* Operon

The import and degradation of sugars by a cell are often controlled by **carbon catabolite repression (CCR),** a regulatory mechanism in which a carbon compound prevents the production of inducible enzymes needed for the metabolism of a different carbon source. This mechanism ensures that the cell imports and metabolizes the sugar that allows the cells to grow fastest. CCR can be demonstrated in *E. coli* by growing the cells in medium containing both glucose and lactose. When glucose is available, the *lac* operon is not expressed because of CCR, and the lactose is thus not used. The cells multiply initially using only glucose. Once the supply of that sugar is exhausted, growth stops for a short period as the cells synthesize enzymes needed for metabolizing lactose. Then, they begin multiplying again, this time using lactose to fuel their growth. This characteristic two-phase growth pattern is called diauxic growth (**figure 7.24**).

Carbon catabolite repression is a global control system that allows glucose to regulate expression of the *lac* operon as well as other sets of genes. Glucose does not act directly in the regulation, however. Instead, the cell's glucose transport system serves as a sensor of glucose availability. When the transport system is moving glucose molecules into the cell, catabolite repression prevents the *lac* operon from being expressed. When the transport system is idle, indicating that glucose is not available, then the *lac* operon can be turned on. How does the cell use the transporter to sense glucose levels? The transport process involves a relay system that donates a phosphate group to each incoming glucose molecule. When glucose levels are high, the unphosphorylated form of that transporter component predominates (**figure 7.25a**). Conversely, when glucose levels are low, the transporter component retains its phosphate group, indicating to the cell that other carbon sources need to be used (figure 7.25b).

One mechanism of carbon catabolite repression involves an activator called CAP (catabolite activator protein), which is required for transcription of the *lac* operon. To be functional, the activator must be bound by an inducer: an ATP derivative called cAMP (cyclic AMP). The inducer is made only when extracellular glucose levels are low, because the enzyme required for its synthesis is activated by the idle form of the glucose transporter component (see figure 7.25c).

Although a great deal of attention has been paid to the role of the activator in carbon catabolite repression, another mechanism of regulation called **inducer exclusion** might be more significant in *E. coli.* In this mechanism, when glucose is being moved into the cell, a glucose transport component binds to the lactose transporter (permease), temporarily locking it in a non-functional position. The locked permease cannot move lactose into the cell, so the *lac* operon will not be induced. Once the glucose supply diminishes, the glucose transporter becomes idle, so lactose can then be brought into the cell (see figure 7.25d). **Table 7.6** summarizes the effect of lactose and glucose levels on control of the *lac* operon.

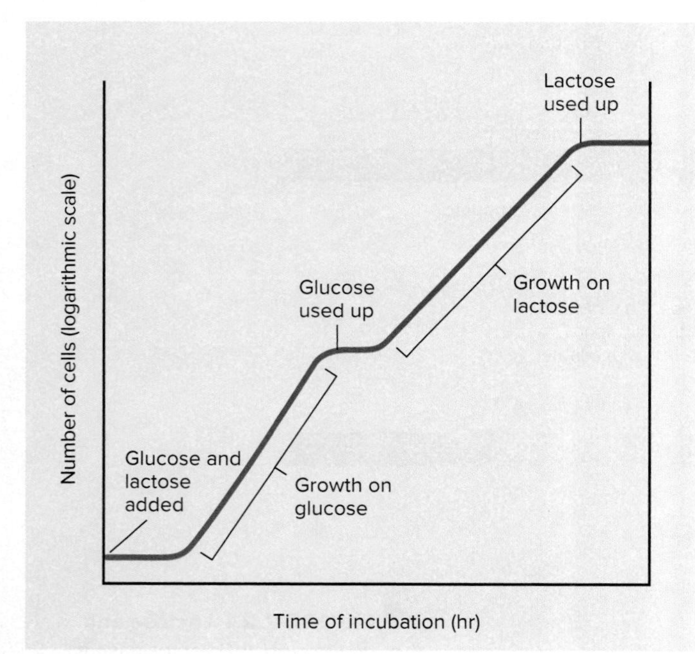

FIGURE 7.24 Diauxic Growth Curve of *E. coli* Growing in a Medium Containing Glucose and Lactose Cells preferentially use glucose. Only when the supply of glucose is used up do cells start metabolizing lactose. Note that the growth on lactose is slower than it is on glucose.

Why does bacterial growth stop temporarily when the glucose supply in the medium is used up?

MicroAssessment 7.6

Enzymes can be constitutive, inducible, or repressible. A repressor blocks transcription when it binds to an operator. An activator enhances transcription when it binds to an activator-binding site. Inducers bring about gene expression by binding either to repressors (disabling them) or to activators (allowing them to attach to the activator-binding site). The *lac* operon, a model for regulation, is controlled by a repressor, an activator, and inducer exclusion.

15. Explain the difference between a constitutive enzyme and an inducible enzyme.
16. Explain how glucose represses the *lac* operon.
17. Why would it be advantageous for a cell to control the activity of an enzyme as well as its synthesis?

7.7 ■ Eukaryotic Gene Regulation

Learning Outcome

11. Describe how RNA interference silences genes.

Considering the complexity of eukaryotic cells and the diversity of cell types found in multicellular organisms, it is not surprising that eukaryotic gene regulation is much more

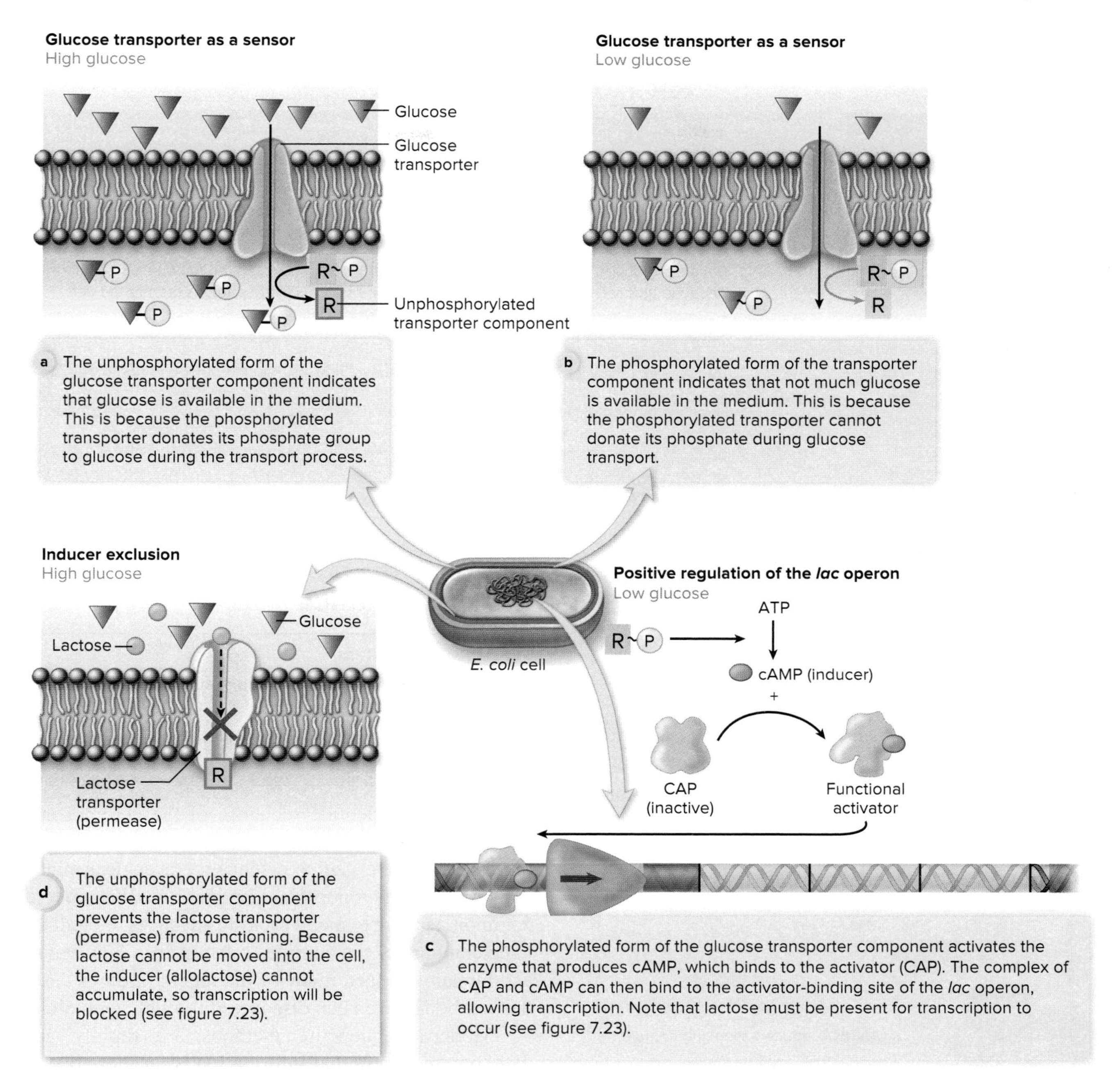

FIGURE 7.25 Glucose and the *lac* Operon

? **Why would it be advantageous for a cell to use glucose before lactose?**

complicated than that of prokaryotic organisms. Eukaryotic cells use a variety of control methods, including modifying the structure of the chromosome, regulating the initiation of transcription, and altering pre-mRNA processing and modification. We will focus only on a process called **RNA interference (RNAi),** a recent Nobel Prize-winning discovery that revolutionized views on gene regulation. Cells routinely use RNAi to destroy specific RNA transcripts, and scientists can manipulate the process to silence select genes.

In RNAi, a cell produces short single-stranded RNA pieces to locate specific RNA transcripts targeted for destruction. To function in RNAi, a short RNA strand joins a multi-protein unit called an RNA-induced silencing complex (RISC). Within a RISC, the short RNA strand serves as the probe that allows the complex to locate a specific nucleotide sequence on mRNA molecules. The short RNA strand does this by binding to complementary sequences on an mRNA molecule, tagging that transcript for destruction by enzymes in the RISC (**figure 7.26**). The components of the RISC are not destroyed in the process, so the complex is catalytic, providing a rapid and effective means of silencing genes that have already been transcribed. Two different types of RNA molecules are used in RNAi: microRNA

TABLE 7.6 Glucose and the *lac* Operon

Control of the *lac* Operon			
Glucose Level	**Lactose Level**	***lac* Operon**	**Mechanism**
High	Low/Absent	OFF	1. Because glucose is available, the activator (CAP) is not active, so the *lac* operon is not transcribed. 2. Because glucose is available, inducer exclusion prevents lactose from entering the cell. 3. Because lactose is absent, the repressor is active, thereby blocking transcription of the *lac* operon.
High	High	OFF	1. Because glucose is available, the activator (CAP) is not active, so the *lac* operon is not transcribed. 2. Because glucose is available, inducer exclusion prevents lactose from entering the cell. 3. Although lactose is present, inducer exclusion prevents it from entering the cell; because of this the repressor is active, thereby blocking transcription of the *lac* operon.
Low/Absent	Low/Absent	OFF	1. Because lactose is absent, the repressor is active, thereby blocking transcription of the *lac* operon.
Low/Absent	High	ON	1. Because glucose is not available, the activator (CAP) is active, so it facilitates transcription of the *lac* operon. 2. Because glucose is not available, inducer exclusion does not prevent lactose from entering the cell. 3. Because lactose is available in the cell, the repressor is not active, so transcription of the *lac* operon can proceed.

(miRNA) and short interfering RNA (siRNA). Although these differ in how they are produced, they are each about two dozen nucleotides in length and functionally equivalent.

MicroAssessment 7.7

RNA interference uses short strands of RNA to identify specific RNA transcripts targeted for destruction.

18. What is the role of miRNA and siRNA in regulation of gene expression?

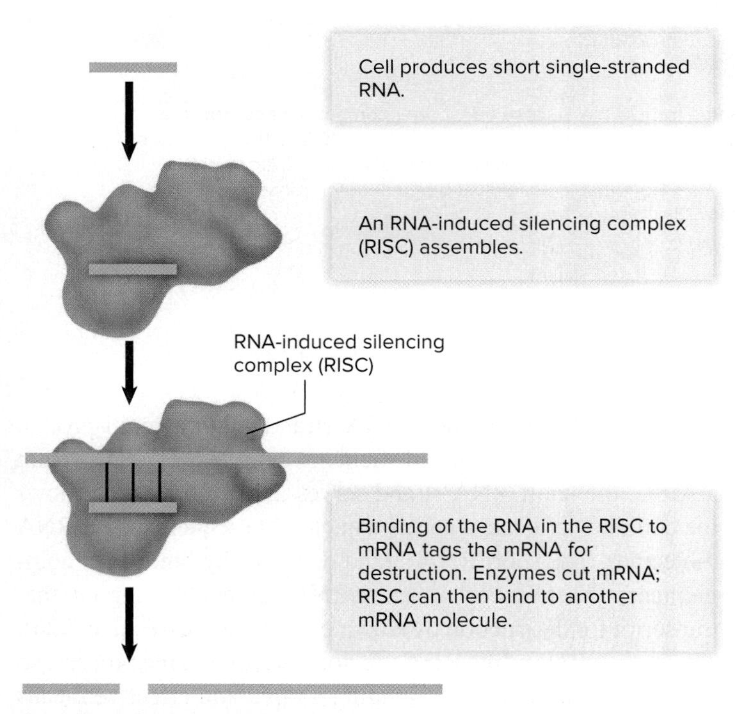

FIGURE 7.26 RNA Interference (RNAi)

? **How could RNAi be used medically?**

7.8 ■ Genomics

Learning Outcomes

12. Explain how protein-encoding regions are found when analyzing a DNA sequence.

13. Describe metagenomics and the information it can provide.

In 1995, the nucleotide sequence of the chromosome of the bacterium *Haemophilus influenzae* was published, marking the first complete genomic sequence ever determined. Since then, sequencing microbial genomes has become relatively common, leading to many exciting advances, including a better understanding of the complex relationships between microbes and humans. In fact, many of the recent findings described in this textbook have been discovered through genomics.

Modern sequencing methods are rapidly generating data, but analyzing that information is far more difficult than it might seem. Imagine trying to determine the amino acid sequence of a protein encoded by a stretch of DNA, without knowing anything about the orientation of the gene's promoter or the reading frame used for translation. Because either strand of the DNA molecule could potentially be the template strand, two entirely different RNA sequences must be considered as protein-encoding candidates. In turn, each of those two candidates has three potential reading frames, for a total of six possible reading frames to consider. Yet only one of these actually codes for the protein. Understandably, computers are an invaluable aid and are used extensively in deciphering the meaning of the raw sequence data. As a result, a new field has emerged—**bioinformatics**—which creates the computer technology to store, retrieve, and analyze nucleotide sequence data.

Analyzing a Prokaryotic DNA Sequence

When analyzing a DNA sequence, the (+) strand is used to represent the sequence of the corresponding RNA transcript. As an example, an ATG in the (+) strand of DNA indicates a possible start codon. ◀◀ plus (+) strand

Computers help locate protein-encoding regions in DNA. They search for **open reading frames (ORFs),** stretches of nucleotide sequences generally longer than 300 bp that begin with a start codon and end with a stop codon. An ORF potentially encodes a protein. Other characteristics such as an upstream sequence that can serve as a ribosome-binding site also suggest that an ORF encodes a protein.

The nucleotide sequence of an ORF can be compared with other known sequences by searching computerized databases of published sequences. Similar searches can be done using the amino acid sequence of the encoded protein. Not surprisingly, as genomes of more organisms are being sequenced, information contained in these databases is growing at a remarkable rate. If the encoded protein shows certain amino acid similarities to other characterized proteins, a presumed function can sometimes be assigned. For example, proteins that bind DNA have similar amino acid sequences in certain regions. Likewise, regulatory regions in DNA such as promoters can sometimes be identified based on similarities to known sequences.

Metagenomics

Metagenomics is the analysis of total microbial genomes in an environment. With metagenomics, researchers can study all microbes in a community, not just the relatively few that grow in culture (see figure 28.5). Imagine the insights that can be gained from this new approach: studying a person's microbiome over time to monitor changes during health and disease; comparing the microbiomes of different body sites; and even comparing microbiomes of different people around the world! Metagenomics is also being used to study microbial life in the open oceans and in soils. Analyzing these sequences gives an entirely new perspective on the extent of biodiversity and will probably lead to the discovery of new antibiotics and other medically useful compounds. ◀◀ microbiome

Although metagenomics holds a great deal of promise, the amount of new data presents tremendous challenges. New computing methods are being developed to handle the complications of analyzing such complex information.

MicroByte

Metagenomic analysis of the human microbiome indicates that the human body carries 100 × more microbial DNA than human DNA.

MicroAssessment 7.8

Sequencing methods are rapid, but analyzing the data and extracting the pertinent information is difficult.

19. What is an open reading frame?

20. Describe two things that you can learn by searching a computerized database for sequences that have similarities to a newly sequenced gene.

21. There are characteristic differences in the nucleotide sequences of the leading and lagging strands. Why might this be so?

FOCUS ON THE FUTURE 7.1

Gems in the Genomes?

From a medical standpoint, one of the most exciting challenges is to put the information obtained from sequencing microbial genomes into use. The potential gains are tremendous. For example, by studying the genomes of pathogenic microbes, scientists can learn more about specific genes that enable an organism to cause disease. Once the regulatory mechanisms of those genes are understood, scientists should be able to design drugs that turn off synthesis of key proteins, making the organism harmless.

Already, companies are searching genomic databases, a process called genome mining, to locate open reading frames (ORFs) that may encode proteins of medical value. Previously uncharacterized proteins that have sequence similarities to proteins of known therapeutic value are among the most promising. Some of these are now in clinical trials to test their efficacy, while other "gems" are probably still hidden, waiting to be discovered.

Summary

7.1 ■ Overview (figure 7.1)

Characteristics of DNA (figure 7.2)

DNA is a double helix, containing two **complementary** and **antiparallel** strands of nucleotides; a linear strand has a **5′ end** and a **3′ end.**

Characteristics of RNA

RNA, a single-stranded molecule, is transcribed from one of the two strands of DNA. Three different functional types of RNA molecules are required for gene expression: **messenger RNA (mRNA), ribosomal RNA (rRNA),** and **transfer RNA (tRNA)** (figure 7.3).

Regulating Gene Expression

Protein synthesis is generally controlled by regulating the synthesis of mRNA (figure 7.4).

7.2 DNA Replication

The bidirectional progression of replication around a circular DNA molecule creates two **replication forks** (figure 7.5). DNA replication is **semiconservative.**

Initiation of DNA Replication

DNA replication begins at the **origin of replication.**

The Process of DNA Replication

DNA polymerase synthesizes DNA in the 5′ to 3′ direction, using one strand as a **template** to generate the complementary strand (figures 7.6, 7.7).

7.3 Gene Expression in Bacteria

Transcription

During **transcription, RNA polymerase** synthesizes RNA in the 5′ to 3′ direction, producing a single-stranded RNA molecule complementary and antiparallel to the DNA template (figure 7.9). The process initiates when RNA polymerase recognizes and binds to a **promoter** (figures 7.10, 7.11). When RNA polymerase encounters a **terminator,** it falls off the DNA and releases the newly synthesized RNA.

Translation

During **translation,** the information encoded by mRNA is decoded using the **genetic code** (table 7.4). The first AUG downstream of a **ribosome-binding site** serves as a **start codon** (figure 7.12). **Ribosomes** are translation "machines." tRNAs carry specific amino acids and act as keys that interpret the genetic code (figure 7.14). The ribosome moves along mRNA in the 5′ to 3′ direction; translation terminates when the ribosome reaches a **stop codon** (figure 7.15).

7.4 Differences Between Eukaryotic and Prokaryotic Gene Expression (table 7.5)

Eukaryotic mRNA is synthesized as **pre-mRNA** that must be processed to generate a functional molecule. Processing includes **capping, polyadenylation,** and **splicing;** splicing removes **introns** (figure 7.17). In eukaryotic cells, the mRNA must be transported out of the nucleus before it can be translated in the cytoplasm.

7.5 Sensing and Responding to Environmental Fluctuations

Signal Transduction

Signal transduction allows cells to monitor and react to environmental conditions. Bacteria use **quorum sensing** to activate genes that are useful only when expressed by a critical mass (figure 7.18). **Two-component regulatory systems** use a sensor that recognizes changes outside the cell and then transmits that information to a response regulator (figure 7.19).

Natural Selection

Antigenic variation is a random alteration in the characteristics of certain surface proteins. **Phase variation** is the random switching on and off of certain genes.

7.6 Bacterial Gene Regulation

Constitutive enzymes are constantly synthesized. The synthesis of **inducible enzymes** can be turned on by certain conditions. The synthesis of **repressible enzymes** can be turned off by certain conditions (figure 7.20).

Mechanisms to Control Transcription

Repressors block transcription (figure 7.21). **Activators** enhance transcription (figure 7.22).

The *lac* Operon as a Model

The ***lac* operon** uses a repressor that prevents transcription of the genes when lactose is not available (figure 7.23). **Carbon catabolite repression (CCR)** prevents transcription of the *lac* operon when glucose is available (figure 7.25, table 7.6).

7.7 Eukaryotic Gene Regulation

Regulation in eukaryotic cells is much more complicated than that in prokaryotic cells. In **RNA interference (RNAi),** a cell synthesizes short single-stranded RNA pieces to locate specific RNA transcripts destined for destruction (figure 7.26).

7.8 Genomics

Analyzing a Prokaryotic DNA Sequence

When analyzing a DNA sequence, the (+) strand is used to represent the sequence of the corresponding RNA transcript; computers are used to search for **open reading frames (ORFs).**

Metagenomics

Metagenomics allows researchers to study all microorganisms and viruses in a community, not just the relatively few that grow in culture.

Review Questions

Short Answer

1. Explain what *semiconservative* means with respect to DNA replication.
2. What is an origin of replication?
3. Why are primers required in DNA replication but not in transcription?
4. What is polycistronic mRNA?
5. Explain why knowing the orientation of a promoter is critical when determining the amino acid sequence of an encoded protein.
6. What is the function of a sigma factor?
7. What is the fate of a protein that has a signal sequence?
8. Explain how some bacteria sense the density of cells in their own population.
9. Compare and contrast regulation by a repressor to regulation by an activator.
10. Explain why locating protein-encoding regions in a genomic sequence can be difficult.

Multiple Choice

1. All of the following are involved in transcription *except*
   a) polymerase. b) a primer. c) a promoter.
   d) a sigma factor. e) uracil.
2. All of the following are involved in DNA replication *except*
   a) a polysome. b) gyrase. c) polymerase.
   d) primase. e) a primer.
3. All of the following are directly involved in translation *except*
   a) a promoter. b) ribosomes. c) a start codon.
   d) a stop codon. e) tRNAs.
4. Using the DNA strand shown here as a template, what will be the sequence of the RNA transcript?

   promoter → 5′ GCGTTAACGTAGGC 3′
   3′ CGCAATTGCATCCG 5′

   a) 5′ GCGUUAACGUAGGC 3′
   b) 5′ CGGAUGCAAUUGCG 3′
   c) 5′ CGCAAUUGCAUCCG 3′
   d) 5′ GCCUACGUUAACGC 3′
5. A ribosome binds to the following mRNA at the site indicated by the dark box. At which codon will translation most likely begin?

   5′ ■ GCCGGAAUGCUGCUGGC

   a) GCC
   b) GGC
   c) AUG
   d) AAU
6. Which of the following statements about gene expression is *false*?
   a) More than one RNA polymerase can be transcribing a specific gene at a given time.
   b) More than one ribosome can be translating a specific transcript at a given time.
   c) Translation begins at a site called a promoter.
   d) Transcription stops at a site called a terminator.
   e) Some amino acids are coded for by more than one codon.
7. An enzyme used to synthesize the amino acid tryptophan is most likely
   a) constitutive. b) inducible.
   c) repressible. d) a and b
8. Under which of the following conditions will transcription of the *lac* operon occur?
   a) Lactose present/glucose present
   b) Lactose present/glucose absent
   c) Lactose absent/glucose present
   d) Lactose absent/glucose absent
   e) a and b
9. All of the following are characteristics of eukaryotic gene expression *except*
   a) a 5′ cap is on the mRNA.
   b) a poly A tail is on the 3′ end of mRNA.
   c) introns must be removed to create the mRNA that is translated.
   d) the mRNA is often polycistronic.
   e) translation begins at the first AUG.
10. Which of the following statements is *false*?
    a) A derivative of lactose serves as an inducer of the *lac* operon.
    b) Signal transduction provides a mechanism for a cell to sense the conditions of its external environment.
    c) Quorum sensing allows bacterial cells to sense the density of like cells.
    d) An example of a two-component regulatory system is the lactose operon, which is controlled by a repressor and an activator.
    e) An ORF is a stretch of DNA that may encode a protein.

Applications

1. A graduate student is trying to identify the gene coding for an enzyme found in a bacterial species that degrades trinitrotoluene (TNT). The student is frustrated to find that the organism does not produce the enzyme when grown in nutrient broth, making it difficult to collect the mRNA needed to help identify the gene. What could the student do to potentially increase the amount of the desired enzyme?
2. To isolate eukaryotic mRNA from cell lysates, a researcher has purchased magnetic beads that have short polymers of thymine nucleotides (T) attached. How could the beads be used to separate the mRNA from other cell components including pre-mRNA?

Critical Thinking

1. The study of protein synthesis often uses a cell-free system where cells are ground with an abrasive to release the cell contents and then filtered to remove the abrasive. These materials are added to the system, generating the indicated results:

Materials Added	Results
Radioactive amino acids	Radioactive protein produced
Radioactive amino acids *and* RNase (an RNA-digesting enzyme)	No radioactive protein produced

   What is the best interpretation of these observations?

2. In a variation of the experiment in the previous question, the following materials were added to three separate cell-free systems, generating the indicated results:

Materials Added	Results
Radioactive amino acids	Radioactive protein produced
Radioactive amino acids *and* DNase (a DNA-digesting enzyme)	Radioactive protein produced
Several hours after grinding: Radioactive amino acids *and* DNase	No radioactive protein produced

   What is the best interpretation of these observations?

8 Bacterial Genetics

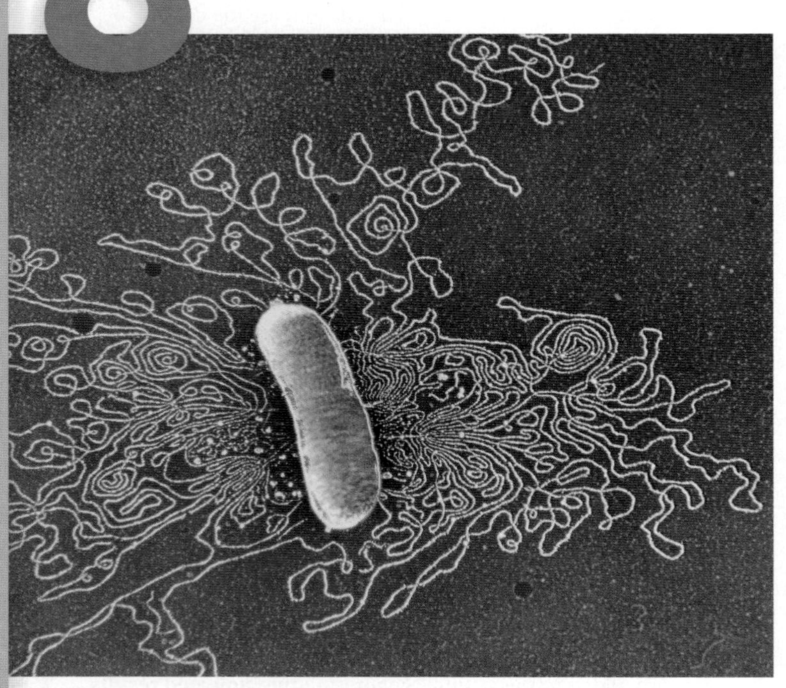

Chromosome being released from a gently lysed *E. coli* cell. ©*Dr. Gopal Murti/Science Source*

KEY TERMS

Auxotroph A microorganism that requires an organic growth factor.

Conjugation Mechanism of horizontal gene transfer in which the donor cell physically contacts the recipient cell.

DNA-Mediated Transformation Mechanism of horizontal gene transfer in which "naked" DNA is taken up by the recipient cell.

Genotype The sequence of nucleotides in an organism's DNA.

Homologous Recombination Process by which a cell replaces a stretch of DNA with a segment that has a similar nucleotide sequence.

Horizontal Gene Transfer DNA transfer from one bacterium to another by conjugation, DNA-mediated transformation, or transduction.

Mutation A change in the nucleotide sequence of a cell's DNA that is passed on to daughter cells.

Non-Homologous Recombination DNA recombination that does not require extensive nucleotide sequence similarity in the stretches that recombine.

Phenotype The observed characteristics of a cell.

Plasmid A DNA molecule that replicates independently of the chromosome.

Prototroph A microorganism that does not require any organic growth factors.

Transduction Mechanism of horizontal gene transfer in which bacterial DNA is transferred inside a phage coat.

Transposon Segment of DNA that can move from one site to another in a cell's genome.

Vertical Gene Transfer Transfer of genes from parent to offspring.

A Glimpse of History

Barbara McClintock (1902–1992) was a remarkable scientist who made several very important discoveries in genetics well before the era of large research teams and sophisticated molecular techniques. With curiosity and determination, she worked 12-hour days, 6 days a week in a small laboratory at Cold Spring Harbor on Long Island, New York.

McClintock studied color variation in corn kernels. She noticed that the kernel colors were not inherited in a predictable manner. In fact, the colors varied from one ear of corn to another. Based on extensive data, McClintock concluded that segments of DNA, now called transposons, were moving into and out of genes involved with kernel color. These moving DNA segments destroyed the function of the genes, thereby changing kernel color.

At the time that McClintock published her results in 1950, most scientists believed that chromosomal DNA was very stable and changed only through recombination. Consequently, geneticists were skeptical of her conclusions. It was not until the late 1970s that her earlier ideas began to be accepted. By that time, transposons had been discovered in many organisms, including bacteria. Although transposons were first discovered in plants, once they were found in bacteria, the field moved ahead very quickly. In 1983, at age 81, McClintock received a Nobel Prize for her discovery of transposons, popularly called "jumping genes."

In the ever-changing conditions that characterize most environments, organisms need to adapt in order to survive and multiply. If they fail to do this, competing organisms more fit to thrive in the new setting will soon predominate. This is the process of **natural selection.** Bacteria have two general means by which they routinely adjust to new circumstances: regulating gene expression (discussed in chapter 7) and genetic change, the focus of this chapter. ◀◀ bacterial gene regulation

Genetic change is the raw material for evolution, and is fundamental to all life on Earth. To study genetic change, scientists often use *E. coli* as a model system. The cells grow rapidly in small volumes of simple, inexpensive media, accumulating in very large numbers. This makes it easy to study rare events that give rise to strains differing in their genetic makeup. ◀◀ microbes as research tools

8.1 ■ Genetic Change in Bacteria

Learning Outcomes

1. Define the terms *mutation, horizontal gene transfer, genotype, phenotype, auxotroph,* and *prototroph.*

Genetic change in bacteria occurs by two mechanisms: mutation and horizontal gene transfer (**figure 8.1**). **Mutation** changes the existing nucleotide sequence of a cell's DNA, which is then passed on to the progeny (daughter cells) through **vertical gene transfer.** The modified organism and daughter cells are referred to as mutants. **Horizontal gene transfer** is the movement of DNA from one organism to another. Like mutations, the changes are then passed on to the progeny by vertical transfer.

A change in an organism's DNA alters its **genotype:** the sequence of nucleotides in the DNA. In bacterial cells, such a change can have a significant impact because bacteria are haploid, meaning that they contain only a single set of genes. There is no "backup copy" of a gene in a haploid organism. Because of this, a change in genotype often alters the organism's observable characteristics, or **phenotype.** Note, however, that the phenotype involves more than just the genetic makeup of an organism; it can also be influenced by environmental conditions. For example, colonies of *Serratia marcescens* are red when incubated at 22°C but white when incubated at 37°C. In this case, the phenotype, but not the genotype, has changed. However, if the genes responsible for pigment production are removed, the organism's phenotype and genotype both change.

Scientists often study genetic changes that alter a microorganism's nutrient requirements. For example, if a gene required for biosynthesis of the amino acid tryptophan is deleted, then the organism can multiply only if tryptophan is supplied in the growth medium. The same occurs if the gene is disrupted so that the protein it encodes no longer

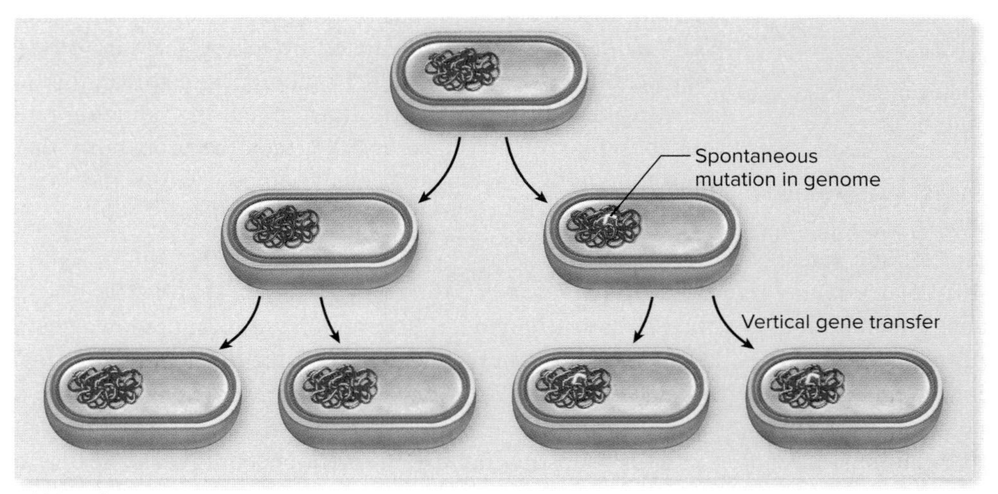

(a) Mutation

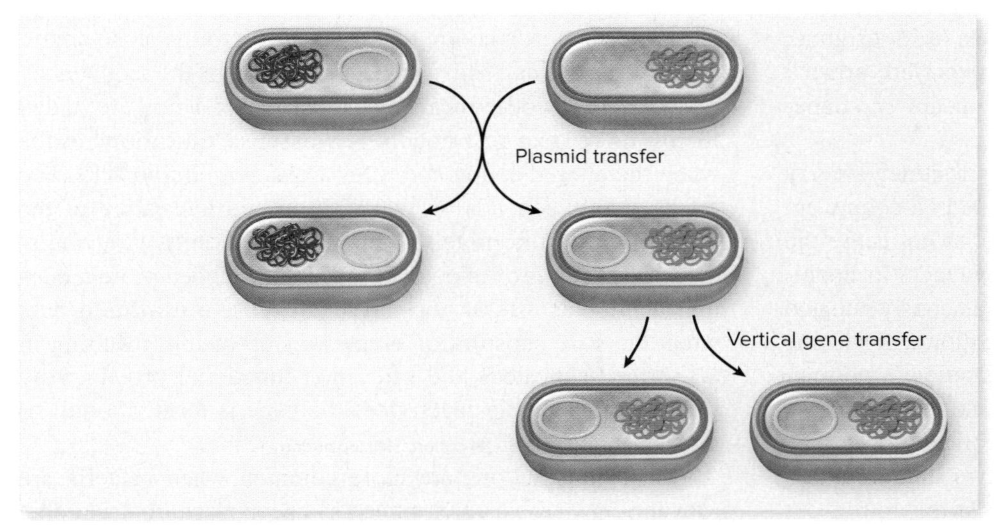

(b) Horizontal gene transfer

FIGURE 8.1 Mechanisms of Genetic Change in Bacteria (a) Mutation. **(b)** Horizontal transfer of plasmid-encoded genes; other DNA can be transferred horizontally as well.

? **Which would have a greater effect on an organism's genotype—mutation or horizontal gene transfer?**

functions properly. A mutant that requires a growth factor is an **auxotroph** (*auxo* means "increase," and *troph* means "nourishment"). This is in contrast to a **prototroph,** which does not require growth factors (*proto* means "earliest form of"). Wild-type *E. coli* is a prototroph ("wild type" refers to the typical phenotype of strains isolated from nature). By convention, a strain's characteristics are designated by three-letter abbreviations, with the first letter capitalized. For example, a strain that cannot make tryptophan is designated Trp^-. For simplicity, only required growth factors are indicated. Likewise, only if a cell is resistant to an antibiotic is it indicated; streptomycin resistance is indicated as Str^R. ⏮ growth factor

MicroAssessment 8.1

The properties of bacteria can change through either mutation or horizontal gene transfer.

1. How is mutation different from horizontal gene transfer?
2. Contrast the meaning of the terms *auxotroph* and *prototroph.*
3. Which has a more lasting effect on a cell—a change in the genotype or a change in the phenotype?

MUTATION AS A MECHANISM OF GENETIC CHANGE

8.2 ■ Spontaneous Mutations

Learning Outcomes

2. Describe three outcomes of base substitutions.
3. Describe the consequences of removing or adding nucleotides.
4. Explain how transposons cause mutations.

Spontaneous mutations are genetic changes that result from normal cell processes. They occur randomly, and genes mutate spontaneously at infrequent but characteristic rates. The mutation rate is defined as the probability that a mutation will occur in a given gene per cell division. The mutation rate of different genes usually varies between 10^{-4} and 10^{-12} per cell division. In other words, the chance that a gene will undergo a mutation when a cell replicates its DNA prior to cell division is between one in 10,000 (10^{-4}) and one in a trillion (10^{-12}). ⏭ exponents

Mutations are passed on to a cell's progeny. On rare occasions, however, a mutation will change back to its original, non-mutated state. This change is called a **reversion,** or back mutation, and like the original mutation, occurs spontaneously at low frequencies.

Because spontaneous mutations occur routinely, every large population contains mutants, so the cells in a colony are not necessarily identical. The environment does not cause the mutations but selects those cells that can grow under its conditions. For example, an organism that has spontaneously mutated to become resistant to an antimicrobial medication will become dominant in an environment where the medication is present. This happens because the antimicrobial kills the sensitive cells, allowing the resistant cells to grow without competition.

A single mutation is a rare event, so two mutations are even more unlikely. Doctors take advantage of this to prevent pathogens from developing resistance to certain antimicrobial medications. In tuberculosis treatment, for example, two or more antimicrobial medications are given simultaneously. The chance that a single cell will become resistant spontaneously to both medications is the product of the mutation rates of the two genes (calculated by taking the sum of the exponents). For example, if the mutation rate to "antibiotic X" resistance is 10^{-6} per cell division and the mutation rate to "antibiotic Y" resistance is 10^{-8}, then the probability that both mutations will spontaneously happen within the same cell is $10^{-6} \times 10^{-8}$, or 10^{-14}.

Base Substitution

Base substitution, the most common type of mutation, occurs during DNA synthesis when an incorrect nucleotide is incorporated (**figure 8.2**). If only one base pair is changed it is a **point mutation.**

Base substitution in a protein-encoding gene leads to three possible mutation outcomes: silent, missense, or nonsense (**figure 8.3**). In a **silent** or **synonymous mutation,** the change generates a codon that translates into the same amino acid as the original. This can occur because of the redundancy of the genetic code; recall that most amino acids are coded for by more than one codon. A **missense mutation** results when the altered codon codes for a different amino acid. The effect of this depends on the position and the nature of the change within the protein. In many cases, cells with a missense mutation grow slowly because the encoded protein does not function as well as normal. A **nonsense mutation** occurs when the base substitution creates a stop codon, resulting in a shorter (truncated) and often non-functional protein. Any mutation that totally inactivates the gene is termed a null or knockout mutation. ⏮ codon, ⏮ stop codon

Base substitutions are more common when bacteria are growing in aerobic environments, as opposed to anaerobic ones. This is because reactive oxygen species (ROS) such as superoxide and hydrogen peroxide are produced from O_2.

These chemicals can oxidize the nucleobase guanine, and DNA polymerase often mispairs oxidized guanine with adenine rather than with cytosine. ⏮ **reactive oxygen species,** ⏮ **nucleobase**

Deletion or Addition of Nucleotides

Deletion or addition of nucleotides during DNA replication also results in spontaneous mutations. The consequence of this depends on how many nucleotides are involved. If three nucleotide pairs are added (or deleted), this adds (or deletes) one codon. When the gene is expressed, one additional (or one less) amino acid will be in the resulting protein. The seriousness of the change's effect depends on its location in the encoded protein.

Adding or subtracting one or two nucleotide pairs causes a **frameshift mutation** (**figure 8.4**). This changes the reading frame of the corresponding mRNA molecule so that an entirely different set of codons is translated. Frequently, one of the resulting downstream codons will be a stop codon. As a consequence, a frameshift mutation likely results in a shortened, non-functional protein—a knockout mutation. ⏮ **reading frame,** ⏮ **downstream**

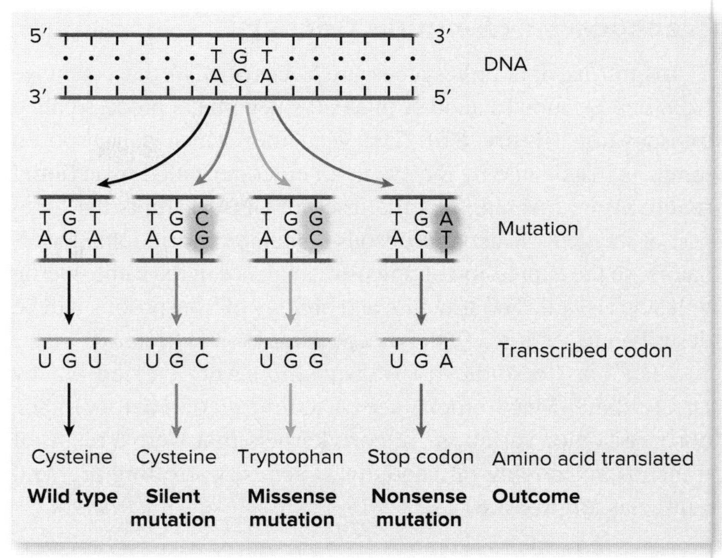

FIGURE 8.3 Potential Outcomes of Base Substitutions Outcomes include silent (also called synonymous), missense, and nonsense mutations.

Why would some missense mutations have no effect on the encoded protein but other missense mutations in the same gene result in a partially functional or even non-functional protein?

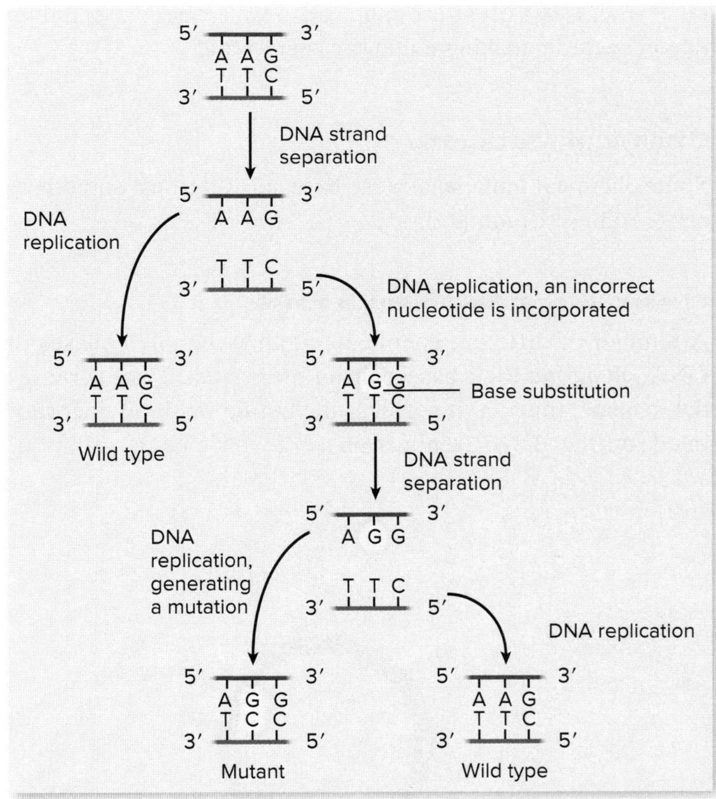

FIGURE 8.2 Base Substitution A replication error results in a mismatch between the two DNA strands. Subsequent DNA replication using the altered strand as template results in a point mutation.

What enzyme incorporated the incorrect nucleotide?

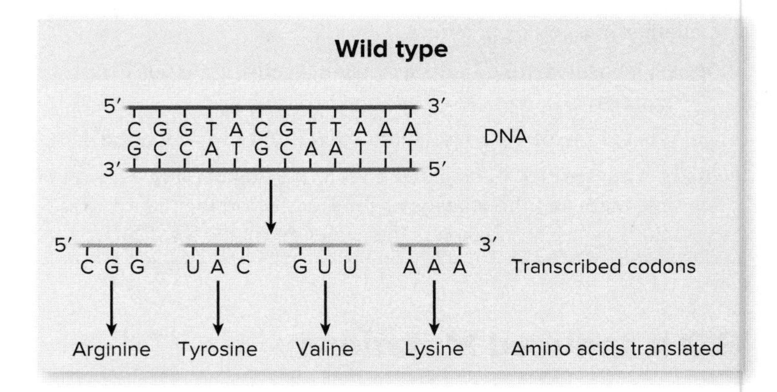

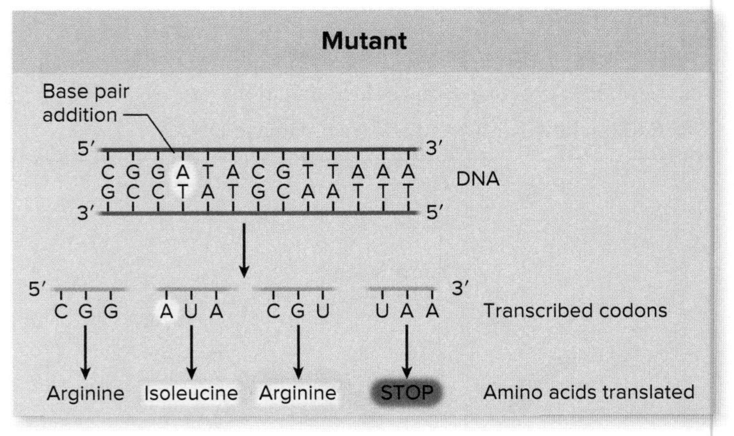

FIGURE 8.4 Frameshift Mutation as a Result of Nucleotide Addition The addition of a nucleotide pair (base pair) to the DNA results in a shift in the reading frame when the sequence is transcribed and translated. Deletion of a single nucleotide pair in the DNA would have a similar effect.

This figure shows a single base pair addition. What would happen if three base pairs were added?

Transposons (Jumping Genes)

Transposons, or jumping genes, are pieces of DNA that can move from one location to another in a cell's genome, a process called transposition (**figure 8.5**). The gene into which a transposon jumps is inactivated by the event, an outcome called **insertional inactivation,** meaning that the inserted DNA disrupts the function of the gene. Most transposons contain transcriptional terminators, so the expression of downstream genes in the same operon will stop as well. The structure and biology of transposons will be described in section 8.9. ⏮ transcriptional terminators, ⏮ operon

The classic studies of transposition were carried out by Dr. Barbara McClintock (see A Glimpse of History). She observed color variations in corn kernels that were a result of transposons moving into and out of genes controlling pigment synthesis (**figure 8.6**).

FIGURE 8.6 Transposition Detected by Changes in Seed Color Variegation in the color of corn kernels is caused by insertion of transposable elements into genes involved in pigment synthesis.

MicroAssessment 8.2

Spontaneous mutations happen during normal cell processes and can change the properties of the cell. Base substitutions that occur during DNA synthesis can lead to silent, missense, and nonsense mutations. Removing or adding nucleotides can cause a frameshift mutation. Transposons can "jump" from one location to another in a cell's genome.

4. Which is generally more serious to a cell—a missense mutation or a nonsense mutation?
5. What is the likely consequence of a frameshift mutation?
6. Is it as effective to take two antibiotics sequentially as it is to take them simultaneously, as long as the total length of time that they are both taken is the same? Explain.

8.3 ■ Induced Mutations

Learning Outcomes

5. Describe the three general groups of chemical mutagens.
6. Explain why transposons induce mutations.
7. Explain how X rays and UV light damage DNA.

Induced mutations are genetic changes that occur due to an influence outside of a cell, such as exposure to a chemical or radiation. An agent that induces the change is a **mutagen** (**table 8.1**). Geneticists—who depend on mutants to study cellular processes—often use mutagens to increase the mutation rate in bacteria, making mutants easier to find.

Chemical Mutagens

Some chemical mutagens cause base substitutions, and others cause frameshift mutations.

Chemicals That Modify Nucleobases

A number of different chemicals modify the nucleobases in DNA, changing their base-pairing properties. This increases the chance that an incorrect nucleotide will be incorporated during DNA replication. For instance a group of

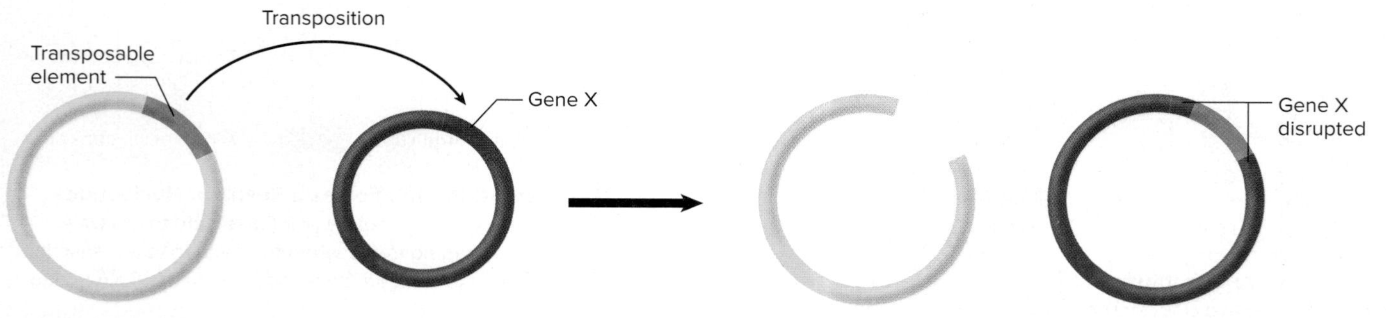

FIGURE 8.5 Transposition A transposable element has the ability to "jump" (transpose) from one piece of DNA into another.

? What effect does the transposon have on the function of gene X in this figure?

TABLE 8.1 Common Mutagens

Agent	Action	Result
Chemical Mutagens		
Chemicals that modify nucleobases	Chemical modifications change base-pairing properties of nucleobases	Nucleotide substitution
Base analogs	Base-pairing properties differ from those of nucleobases normally found in DNA	Nucleotide substitution
Intercalating agents	Insert between base pairs, pushing them apart	Addition or subtraction of nucleotides
Transposons	Randomly insert into DNA	Insertional inactivation
Radiation		
Ultraviolet (UV) light	Causes thymine dimers to form	Errors during repair process
X rays	Cause single- and double-strand breaks in DNA	Deletions

chemicals called alkylating agents adds alkyl groups (short chains of carbon atoms) onto nucleobases. An example is nitrosoguanidine, which adds a methyl group to guanine; the modified nucleobase sometimes base-pairs with thymine (**figure 8.7**).

MicroByte

Many chemical mutagens are used in cancer therapy to kill rapidly dividing cancer cells. Unfortunately, they also damage DNA in normal cells.

Base Analogs

Base analogs structurally resemble nucleobases but have different hydrogen-bonding properties. The analogs can be mistakenly used in place of the nucleobases when the cells make nucleotides, and DNA polymerase then incorporates these into DNA. When the complementary strand is synthesized, the wrong nucleotide may be incorporated opposite the base analog. For example, 5-bromouracil resembles thymine, but it often base-pairs with guanine (G) instead of with adenine (A); 2-amino purine resembles adenine but often pairs with cytosine (C) instead of with thymine (T) (**figure 8.8**). ◀◀ DNA replication

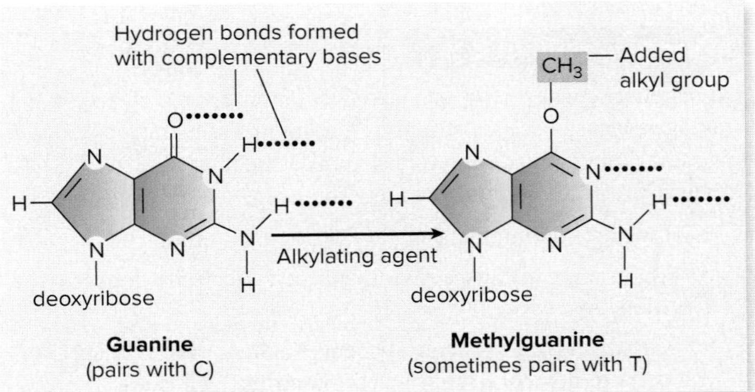

Hydrogen bonds formed with complementary bases
Added alkyl group
CH_3
Alkylating agent
deoxyribose
deoxyribose
Guanine (pairs with C)
Methylguanine (sometimes pairs with T)

(a)

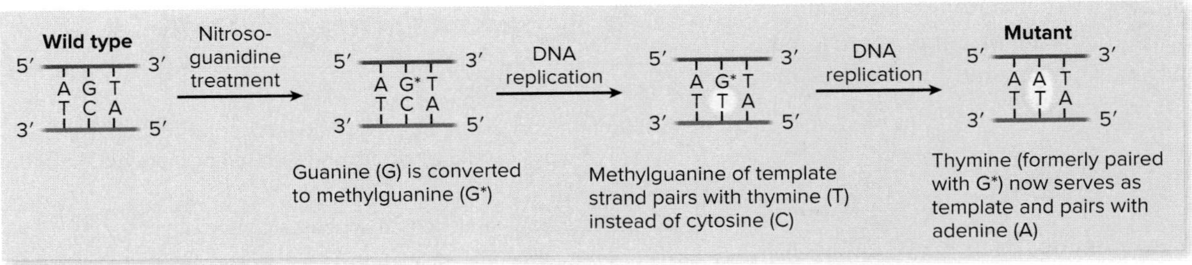

Wild type
Nitroso-guanidine treatment
DNA replication
DNA replication
Mutant
Guanine (G) is converted to methylguanine (G*)
Methylguanine of template strand pairs with thymine (T) instead of cytosine (C)
Thymine (formerly paired with G*) now serves as template and pairs with adenine (A)

(b)

FIGURE 8.7 Mutagenic Effects of the Alkylating Agent Nitrosoguanidine **(a)** Nitrosoguanidine converts guanine bases in DNA to methylguanine, which can base-pair with thymine (T) as well as with cytosine (C). **(b)** The altered base-pairing property of methylguanine can result in point mutations following DNA replication.

? What are the three possible outcomes of point mutations?

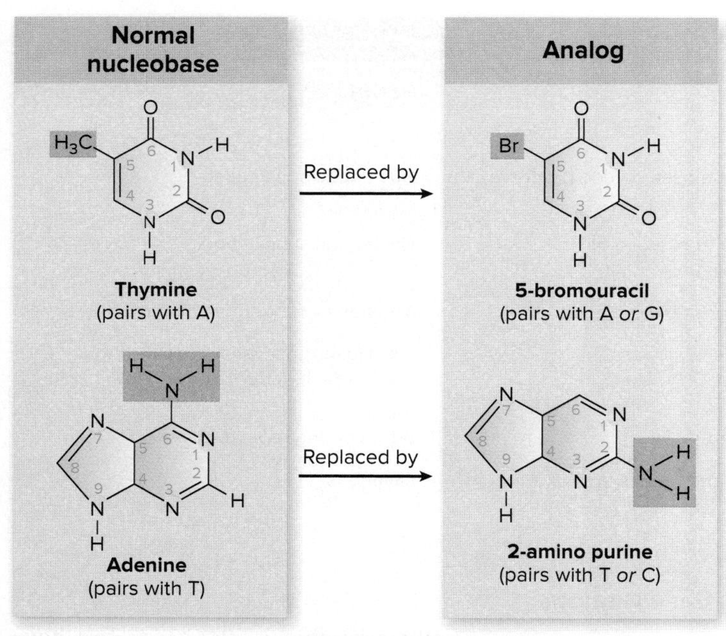

FIGURE 8.8 Two Base Analogs Used in Mutagenesis The altered base-pairing properties of the analogs can lead to point mutations by the same mechanism shown in figure 8.7b.

What base-pair substitution would 5-bromouracil generate?

Intercalating Agents

Intercalating agents increase the frequency of frameshift mutations. They do this because they are flat molecules that can insert (intercalate) between adjacent base pairs in a strand of DNA. This pushes the nucleotides apart, producing a space between bases that allows errors to be made during replication. If the intercalating agent inserts into the template strand, a base pair will be added as the new strand is synthesized. If it intercalates into the strand being synthesized, a base pair will be deleted. As in spontaneous frameshift mutants, adding or subtracting a nucleotide in DNA often results in the premature generation of a stop codon in the mRNA transcript, giving rise to a shortened protein. Chemicals used to stain DNA in the laboratory are intercalating agents.

Transposition

Transposons can be introduced intentionally into a cell in order to generate mutations. The transposon, which cannot replicate on its own because it lacks an origin of replication, inserts into the cell's genome. This generally inactivates the gene into which it inserts (see figure 8.5).

Radiation

Two kinds of radiation are commonly used as mutagens: ultraviolet (UV) light and X rays. ◀◀ radiation

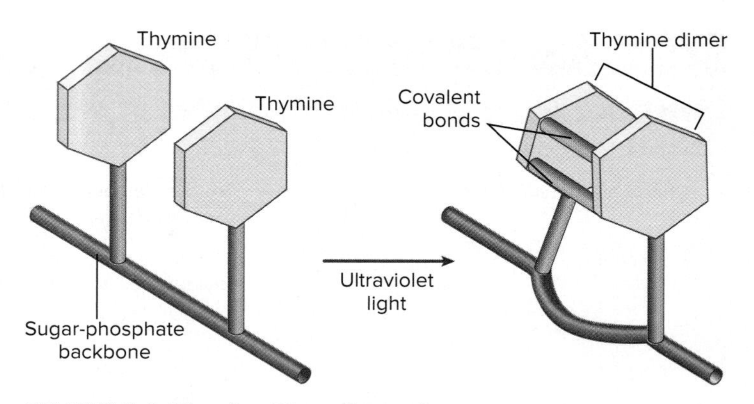

FIGURE 8.9 Thymine Dimer Formation UV light causes covalent bonds to form between adjacent thymine nucleobases on the same strand of DNA, distorting the shape of the DNA.

What effect does a thymine dimer have on DNA synthesis?

Ultraviolet Light

Exposing cells to ultraviolet light causes covalent bonds to form between adjacent thymine nucleobases on a DNA strand, producing **thymine dimers** (**figure 8.9**). Dimers distort the DNA molecule because they cannot fit properly into the double helix. Replication and transcription stop at the distortion, and as a result, the cells will die if the damage is not repaired. How then can UV light be mutagenic? Its major mutagenic action is indirect, resulting from the cell's attempt to repair the damage by SOS repair, described in the next section.

X Rays

X rays cause single- and double-strand breaks in DNA, and changes to the nucleobases. Double-strand breaks often result in deletions that are lethal to the cell.

MicroAssessment 8.3

Mutagens increase the frequency of mutations. Some chemical mutagens cause base substitutions, but intercalating agents cause frameshift mutations. Mutations from UV light are due to SOS repair; X rays cause breaks in DNA strands and alter nucleobases.

7. How does an intercalating agent cause mutations?
8. What mutagen causes thymine dimers, and why does it kill cells?
9. Why would some bacterial species be more likely than others to develop UV-induced mutations?

8.4 ■ Repair of Damaged DNA

Learning Outcomes

8. Explain how errors in nucleotide incorporation by DNA polymerase can be repaired, and the role of methylation in the process.
9. Explain how modified nucleobases can be repaired.
10. Describe three mechanisms cells use to repair thymine dimers.

The amount of spontaneous and mutagen-induced damage to DNA is enormous. This damage, if not repaired, can quickly lead to cell death and, in animals, cancer. In humans, two genes associated with breast cancer code for enzymes that repair damaged DNA. Mutations that inactivate either gene result in a high probability (about 45–80%) of developing breast cancer.

Mutations are rare because cells have multiple mechanisms to repair damaged DNA before errors are passed on to progeny (**table 8.2**). The mechanisms are not perfect, however, and certain types of DNA damage such as insertional inactivation caused by transposons cannot be repaired.

MicroByte

Every 24 hours, the genome of every cell in the human body is damaged more than 10,000 times.

Repair of Errors in Nucleotide Incorporation

DNA polymerase sometimes incorporates the wrong nucleotide as it replicates DNA. The resulting mispairing of nucleobases results in a slight distortion in the DNA helix, which can be recognized by enzymes within the cell that then repair the mistake. By quickly repairing the error before the DNA is replicated, the cell prevents the mutation. Two mechanisms for this are "proofreading" by DNA polymerase and mismatch repair.

Proofreading by DNA Polymerase

DNA polymerases are complex enzymes that not only synthesize DNA, but also check the accuracy of their actions—a process called **proofreading.** The enzymes can back up and excise (remove) a nucleotide not correctly hydrogen-bonded to the opposing nucleobase in the template strand. The DNA polymerase then inserts the correct nucleotide. Although the proofreading function of DNA polymerases is very efficient, it is not perfect.

Mismatch Repair

Mismatch repair fixes errors missed by the proofreading of DNA polymerase. A specific protein binds to the site of the mismatched nucleobase, directing an enzyme to cut the sugar-phosphate backbone of the new DNA strand. Another enzyme then degrades a short region of that DNA strand, thereby removing the misincorporated nucleotide. How does the cell know which strand is the new one? This is an important question because if the enzyme were to cut the template strand and not the new one, then the misincorporated nucleotide would remain. The key to the answer lies in methylation of the DNA nucleobases. Soon after a DNA strand is synthesized, an enzyme adds methyl groups to certain nucleobases. This takes time, however, so the new strand is still unmethylated

TABLE 8.2 Repair of Damaged DNA

	Type of Defect	Repair Mechanism	Characteristics of Repair
Spontaneous	Error in nucleotide incorporation	Proofreading by DNA polymerase	DNA polymerase removes the mispaired nucleotide and replaces it with the correct one.
		Mismatch repair	A protein binds to the site of the mismatch and cuts the unmethylated strand. A short stretch of that strand is then degraded and DNA polymerase synthesizes a replacement.
	Modified nucleobase in DNA	Action of glycosylases	A glycosylase removes the oxidized nucleobase. A short stretch of that strand is then degraded, and DNA polymerase synthesizes a replacement.
Mutagen-Induced			
Chemical	Error in nucleotide incorporation	Proofreading and mismatch repair	Same as for spontaneous mutations (see characteristics of proofreading and mismatch repair under that heading).
UV light	Thymine dimer	Photoreactivation (light repair)	The energy of light is used to break the covalent bond between the thymine nucleobases, restoring the original molecule.
		Excision repair (dark repair)	An enzyme removes a short stretch of the strand containing the thymine dimer; DNA polymerase then synthesizes a replacement.
		SOS repair	A special DNA polymerase synthesizes DNA even when the template is extensively damaged; the cell may survive but numerous mutations are introduced.

immediately after it is synthesized. Therefore, the template strand is methylated, whereas the new strand is not, allowing the repair enzyme to distinguish between the two (**figure 8.10**). After the nucleotides are removed from the new strand, the combined actions of DNA polymerase and DNA ligase then fill in that section and seal the gap. ◀◀ DNA ligase

Repair of Modified Nucleobases in DNA

Modified nucleobases can result in base substitutions if they are not repaired before the DNA is replicated. An example of a repair mechanism involves oxidized guanine. A type of enzyme called a DNA glycosylase removes that oxidized nucleobase from the sugar-phosphate backbone (**figure 8.11**). Another enzyme then recognizes that a nucleobase is missing and cuts the DNA at this site. DNA polymerase degrades a short section of this strand to remove the damage. This same enzyme synthesizes another strand with the proper nucleotides, and DNA ligase seals the gap in the single-stranded DNA.

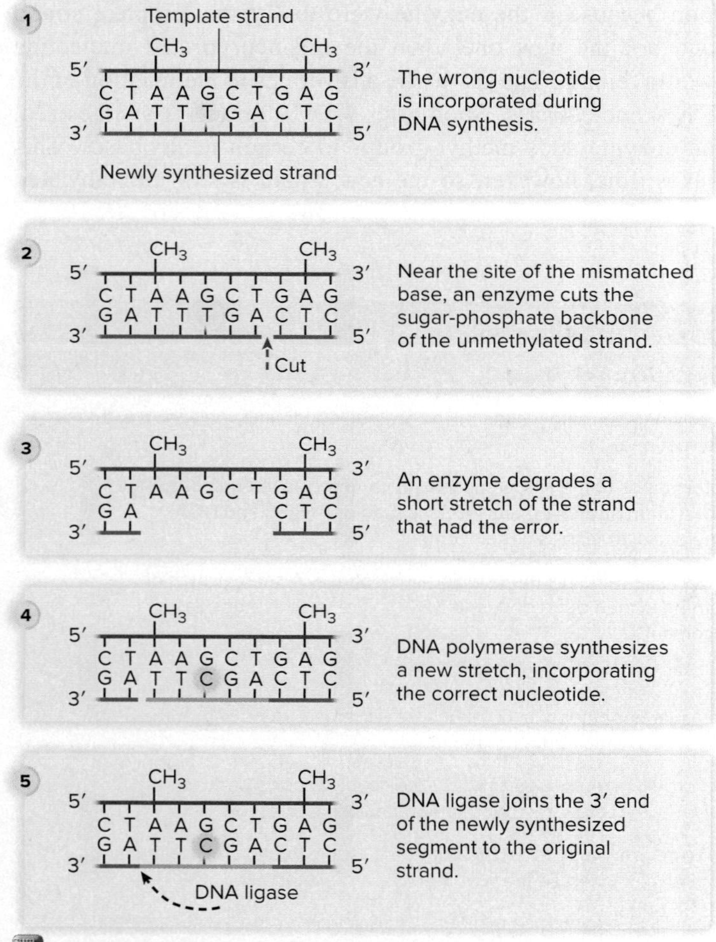

FIGURE 8.10 Mismatch Repair

What role does methylation play in mismatch repair?

MicroByte

In humans, defects in either mismatch repair or repair of modified nucleobases increase the incidence of certain cancers.

Repair of Thymine Dimers

Bacteria have several mechanisms to prevent the DNA-damaging effects of UV light, a component of sunlight. In a mechanism found only in microbes, an enzyme uses the energy of visible light to break the covalent bonds of the thymine dimer, restoring the DNA to its original state (**figure 8.12**). Because light is required for this mechanism, it is called **photoreactivation,** or light repair. In the case of a repair system called **excision repair,** or dark repair, an enzyme recognizes the major distortions in DNA that result from thymine dimer formation and then removes the DNA strand with the damaged region (figure 8.12). DNA polymerase and DNA ligase then fill in and seal the gap left by the removal of the segment.

SOS Repair

SOS repair is a last-effort attempt that bacteria use to repair extensively damaged DNA. The enzymes that carry out this repair are induced when DNA is so heavily damaged that other repair systems may not be able to correct all of the damage. DNA and RNA polymerases stall at sites of unrepaired damage, so the cells cannot replicate or transcribe their DNA. Without SOS repair, the cells would die.

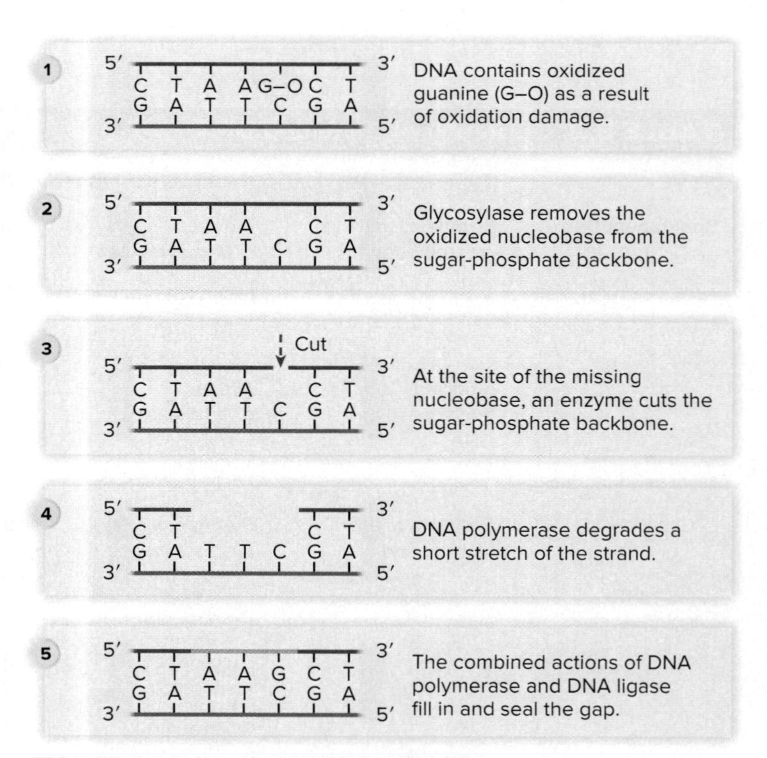

FIGURE 8.11 Repair of Oxidized Guanine

What would be the effect on cells if they did not have the glycosylase enzyme?

Damaged DNA activates expression of the several dozen genes that encode the SOS system. One component of this system is a DNA polymerase that synthesizes DNA even in extensively damaged regions. Unlike the standard DNA polymerases, however, the SOS DNA polymerase has no proofreading ability. Errors are made as a result, a process called SOS mutagenesis.

MicroAssessment 8.4

DNA polymerases have proofreading ability. Mismatch repair fixes errors missed by the proofreading mechanism; methylation distinguishes the template strand. Specific DNA glycosylases can remove modified nucleobases. Thymine dimers can be repaired through photoreactivation and excision repair; severe damage can be overcome by the SOS repair system.

10. Distinguish between photoreactivation and excision repair of thymine dimers.
11. How does UV light cause mutations?
12. To maximize the number of mutations following UV irradiation, should the irradiated cells be incubated in the light or in the dark, or does it make any difference? Explain your answer.

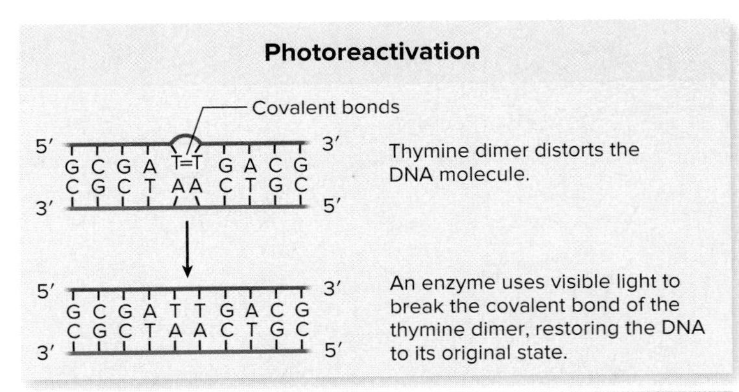

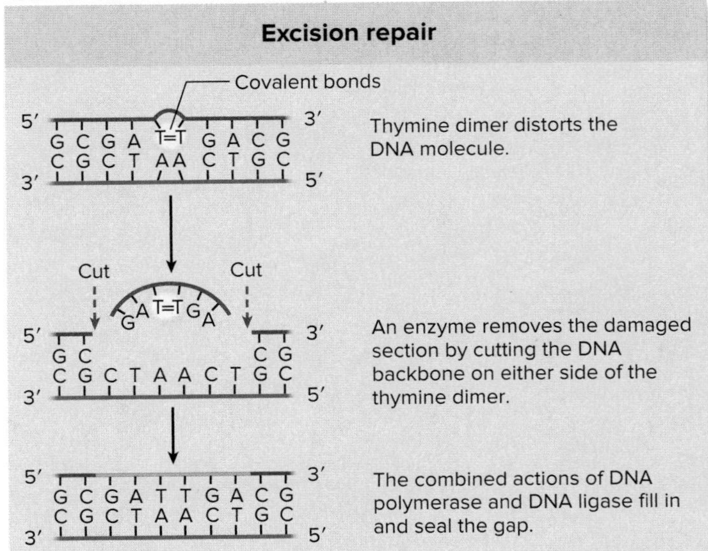

FIGURE 8.12 Repair of Thymine Dimers In photoreactivation, an enzyme uses the energy of light to break the covalent bonds. In excision repair, the section containing the dimer is replaced.

How is the mechanism of excision repair similar to that of mismatch repair?

8.5 ■ Mutant Selection

Learning Outcomes

11. Compare and contrast direct and indirect selection.
12. Describe how direct selection is used to screen for possible carcinogens.

Mutations are rare events, even when mutagens are used. This presents a challenge to a scientist who wants to isolate a desired mutant. In a culture containing several billion cells, perhaps only one cell has the mutation of interest, making it very difficult to find. Two methods—direct and indirect selection—are used to isolate such mutants.

Direct Selection

Mutants that can grow under conditions in which the parent cells cannot are usually easy to isolate by **direct selection.** In this method, cells are inoculated onto an agar medium that supports the growth of the mutant, but does not allow the parent to grow. For example, antibiotic-resistant mutants can be easily selected directly by inoculating cells onto a medium containing the antibiotic. Only the resistant cells will form colonies (**figure 8.13**).

Indirect Selection

Indirect selection is used to isolate an auxotrophic mutant from a prototrophic parent strain. This process is more difficult than direct selection because any medium that supports the growth of the mutant also allows the growth of the parent. For example, Trp^- mutants can grow only on a complex medium such as nutrient agar because this supplies the

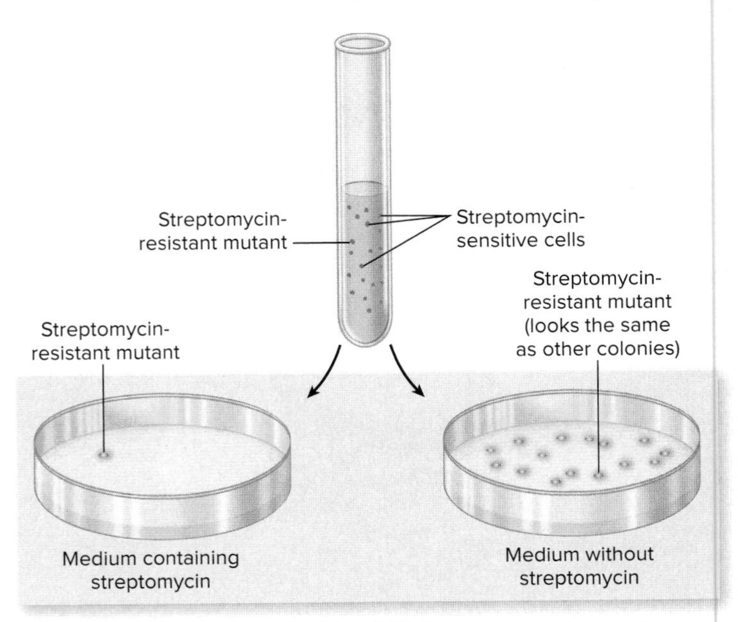

FIGURE 8.13 Direct Selection of Mutants Only cells carrying a mutation that confers resistance to streptomycin can grow on the selective medium used here.

Direct selection cannot be used to isolate auxotrophic mutants in a culture of prototrophic parent cells. Why not?

tryptophan they require, but Trp$^+$ parent cells also grow on this same medium. To overcome this problem, a technique called replica plating is used. This is sometimes preceded by penicillin enrichment, a method that increases the proportion of auxotrophs. ⏮ complex medium, ⏮ nutrient agar

Replica Plating

Replica plating is a clever method for indirect selection of auxotrophic mutants, devised by Joshua and Esther Lederberg in the early 1950s (**figure 8.14**). In this technique, the bacterial culture is first spread onto a nutrient agar plate. Mutant and non-mutant cells will grow on this medium to form colonies, creating what is referred to as the master plate. ①The master plate is then pressed onto sterile velvet, a fabric with tiny threads that stand on end like small bristles. The velvet picks up some cells of every colony. ②Next, two agar plates—one nutrient agar and one glucose-salts agar (a minimal medium lacking added nutrients)—are pressed in succession and in the same orientation onto the same velvet (see table 4.7). This transfers cells taken from the master plate to both the nutrient agar and the glucose-salts agar, creating replica (duplicate) plates. A mark on the plates is used to maintain a consistent orientation. ③The replica plates are then incubated, allowing cells to grow to form colonies; prototrophs grow on both types of media, but auxotrophs grow only on the nutrient agar. The plates are exact replicas, so colonies on the master plate that cannot grow on glucose-salts can be identified; these are auxotrophs. The particular growth factor required by a mutant can then be determined by adding nutrients individually to a glucose-salts medium and determining which one promotes growth. ⏮ glucose-salts medium, ⏮ growth factor

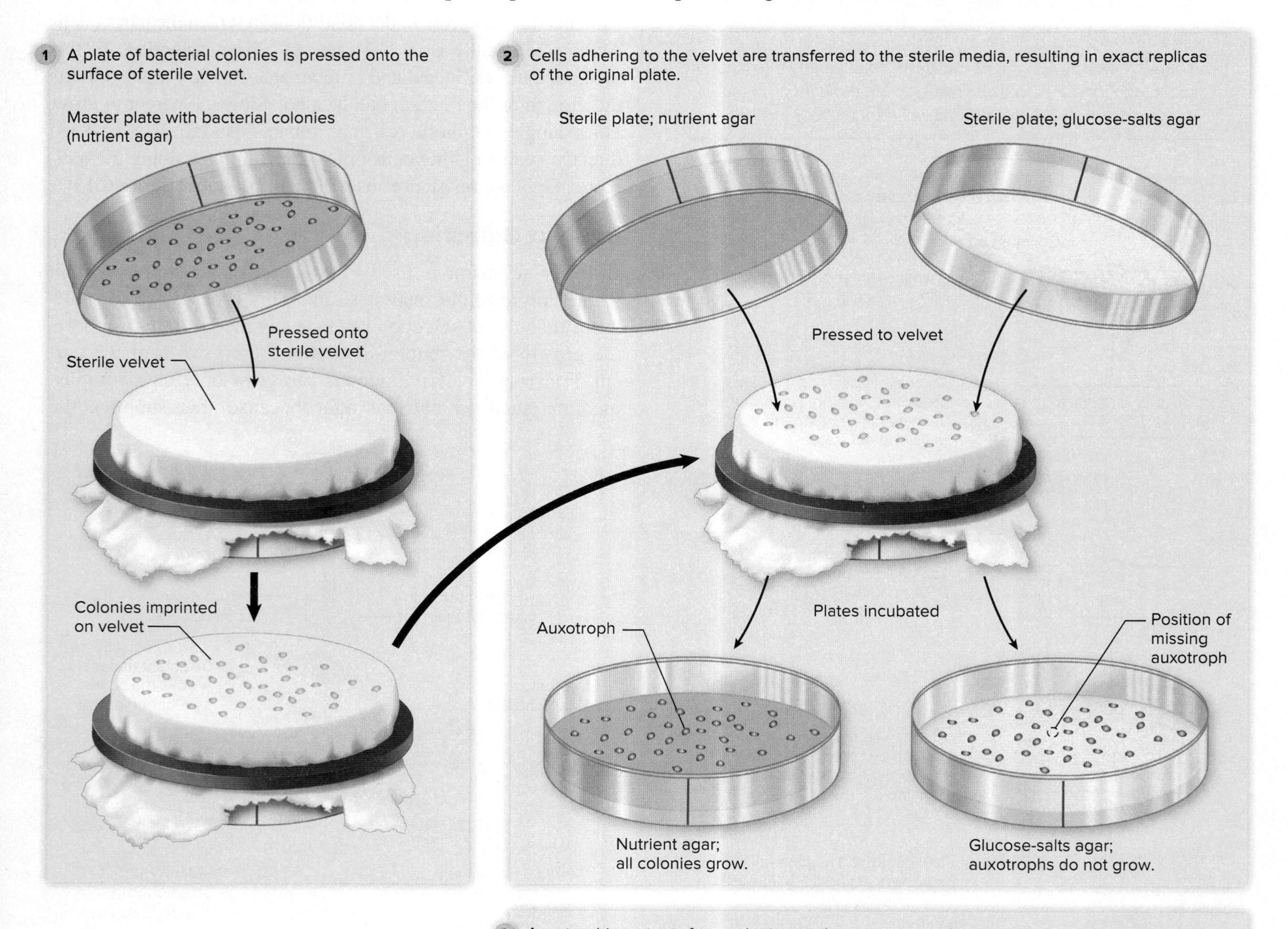

FIGURE 8.14 Indirect Selection of Mutants by Replica Plating The procedure shown was used by the Lederbergs and continues to be used today in many laboratories. Mutants are identified by comparing the growth of colonies on the two plates.

? Why go to the trouble of creating a master plate (why not simply plate the initial culture on both nutrient agar and glucose-salts agar)?

Penicillin Enrichment

Penicillin enrichment selectively kills prototrophs, and is sometimes used before replica plating to increase the proportion of auxotrophic mutants. This is helpful because even when mutagenic agents are used, the frequency of mutations in a particular gene is low, sometimes less than one in 100 million cells. When the proportion of auxotrophs is increased, these cells are easier to isolate.

Penicillin enrichment relies on the fact that penicillin kills only growing cells. The cells that have been treated with the mutagen are incubated in glucose-salts broth containing penicillin. Prototrophs can grow in this medium, so most are killed by the penicillin, whereas the auxotrophs are not killed because they cannot grow in the medium (**figure 8.15**). The enzyme penicillinase is then added to destroy the penicillin, and the cells are plated on nutrient agar to create the master plate used in replica plating. ⏮ penicillin

Screening for Possible Carcinogens

Many cancers appear to be caused by chemicals that are **carcinogens** (meaning "cancer-generating"), and most carcinogens are mutagens. Thousands of common chemicals are potentially dangerous and must be tested for carcinogenic activity. Testing in animals takes several years and it may cost $100,000 or more to test a single compound. To simplify the process, a number of quick, cheap methods have been developed to screen for possible carcinogens. All examine the mutagenic effect of the chemical in a microbiological system. The **Ames test,** devised by Bruce Ames and his colleagues in the 1960s, illustrates the concept of such tests.

The Ames test uses a histidine-requiring auxotroph of *Salmonella typhimurium* (His^-) to measure the mutagenicity of a chemical. To do the test, the chemical to be screened is added to a culture of His^- *S. typhimurium* on a glucose-salts agar plate (**figure 8.16**). This agar medium lacks histidine, so His^- auxotrophs cannot grow, whereas prototrophs can. If the test chemical is a mutagen, it will induce mutations. Some of the mutations will revert the His^- cells to the prototrophic phenotype, and these revertants will grow to form colonies on the agar. Thus, the number of colonies growing on the plate reflects the number of mutations that have occurred in the His^- culture.

As with any scientific procedure, a control must be done as part of the Ames test. The need for a control is obvious when you consider that spontaneous mutations occur (at a low rate), so at least a few revertants will likely be present even if the chemical being tested is not a mutagen. For the control, the same procedure just described is used, but without the test chemical added. After incubation, the control and test plates are compared to assess the relative amount of reversion that occurred due to the test chemical.

Animal liver extract may also be added as part of the Ames test. This is because enzymes produced by the liver modify certain chemicals, and in some cases this inadvertently transforms the chemicals into a form that is mutagenic. That is, the chemical would not normally be mutagenic, but once inside the body, the liver enzymes convert into a form that causes mutations.

If the test chemical causes mutation of the His^- cells, additional testing must be done in animals to determine if it is a carcinogen. This step is necessary because although most carcinogens are mutagens, not all mutagens are carcinogens. Thus, the Ames test is simply a valuable screening tool.

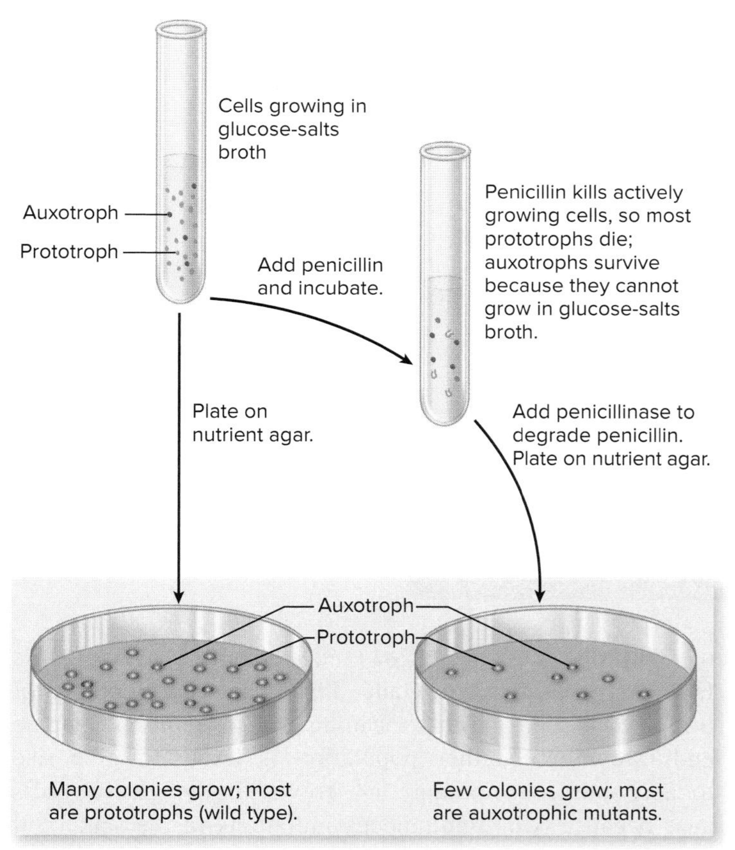

FIGURE 8.15 Penicillin Enrichment This process selectively kills prototrophs, thereby increasing the proportion of auxotrophs.

? Penicillin causes the growing cells to lyse, releasing all of their contents into the medium. How does this complicate penicillin enrichment?

MicroAssessment 8.5

Mutants can be selected using either direct or indirect techniques. Replica plating is used for indirect selection, sometimes preceded by penicillin enrichment. The Ames test is used to screen chemicals to determine which ones are possible carcinogens.

13. Distinguish between the kinds of mutants that can be isolated by direct selection from those for which indirect selection is needed.
14. In the Ames test, what information does the control provide?
15. How could you demonstrate by replica plating that the environment selects but does not mutate genes in bacteria? ?

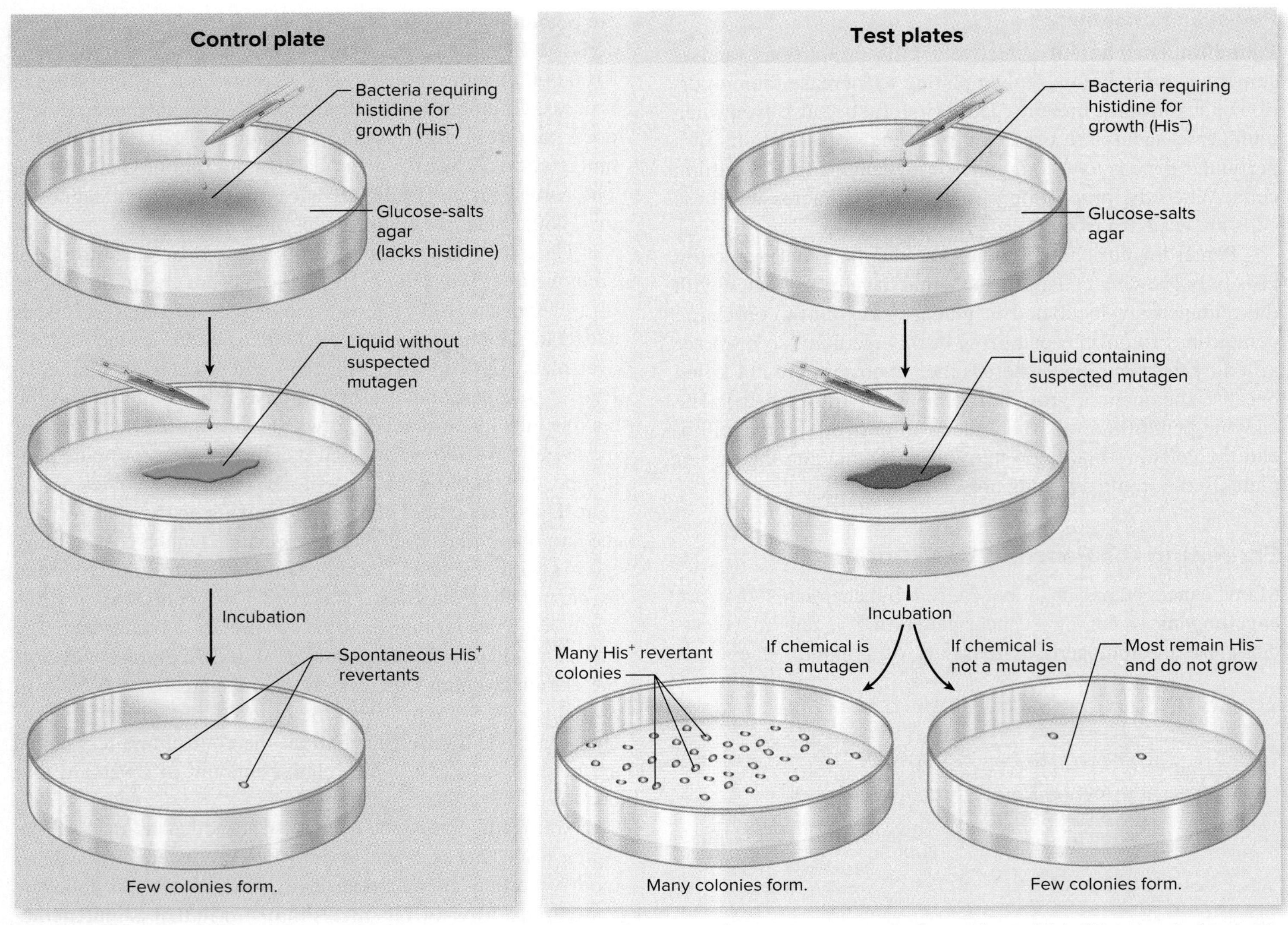

FIGURE 8.16 Ames Test to Screen for Mutagens A control plate determines the spontaneous rate of reversion to His$^+$. If the chemical tested is a mutagen, it will increase the rate of His$^+$ reversion mutations.

Why do some cells grow on the control plate in the absence of a mutagen?

HORIZONTAL GENE TRANSFER AS A MECHANISM OF GENETIC CHANGE

Microorganisms commonly acquire genes from other cells, through a process called horizontal gene transfer. To determine if the recipients have indeed acquired new characteristics, the donor and recipient cells must be genetically different, which is one reason mutants are useful. The resulting cells, **recombinants,** have properties of each of the original strains.

Figure 8.17 illustrates how horizontal gene transfer can be demonstrated. Two bacterial strains are used, neither of which can grow on a glucose-salts medium because of multiple growth factor requirements. Strain A is His$^-$ (requires histidine) and Trp$^-$ (requires tryptophan). Strain B is Leu$^-$, Thr$^-$, meaning that it requires leucine and threonine. Neither population is likely to give rise to a spontaneous mutant that grows on the glucose-salts agar because two simultaneous mutations in the same cell would be required. The strains are mixed and then spread on a glucose-salts agar plate. Colonies should form only if cells of one strain acquired genes from cells of the other strain.

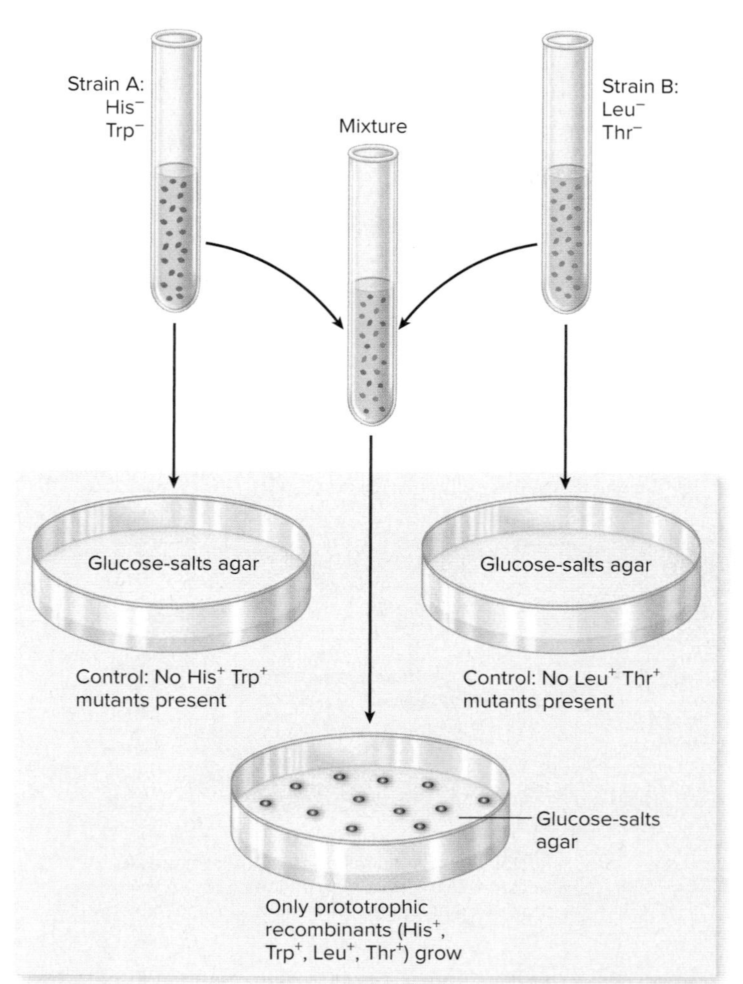

FIGURE 8.17 Experimental Demonstration of Horizontal Gene Transfer in Bacteria Recombinant colonies have genetic traits from both strains present in the mixture. Control plates demonstrate that these colonies are not a result of spontaneous mutation.

? **When demonstrating horizontal gene transfer, why is it important to use strains that each require at least two different amino acids?**

Genes can be transferred from a donor to a recipient by three different mechanisms (**figure 8.18**):

- **DNA-mediated transformation:** "Naked" DNA is taken up from the environment by a bacterial cell.
- **Transduction:** DNA is transferred from one bacterial cell to another by a bacteriophage (a virus that infects bacteria; also called a phage).
- **Conjugation:** DNA is transferred during cell-to-cell contact.

Following gene transfer, recipient cells must replicate the DNA to pass it on to daughter cells. This can happen only if the DNA is a **replicon,** meaning it has an origin of replication. Plasmids and chromosomes are replicons, but fragments of chromosomal DNA are not. If a chromosomal fragment is transferred, then it must become integrated into a replicon to be maintained in a population (**figure 8.19**). This involves a process called **homologous recombination,** which can happen only if the donor DNA is similar in nucleotide sequence to a region in the recipient cell's genome. In homologous recombination, the donor DNA becomes positioned next to the complementary region of the recipient cell's DNA. The donor DNA then replaces a homologous segment of recipient DNA, and the DNA it replaced is degraded. ⏮ **origin of replication**

8.6 ■ DNA-Mediated Transformation

Learning Outcome

13. Describe the process of DNA-mediated transformation, including the role of competent cells.

DNA-mediated transformation, commonly referred to as transformation, involves the uptake of "naked" DNA by recipient cells. Naked DNA is simply DNA that is free in the cells' surroundings; it is not contained within a cell or a virus. Naked DNA often originates from cells that have burst. As a cell bursts, its long and tightly packed chromosome breaks up into hundreds of small pieces. Another source of naked DNA is certain bacterial species that secrete small pieces of DNA, presumably as a means of promoting transformation.

The fact that the DNA is naked can be demonstrated by adding DNase (an enzyme that degrades DNA) to the donor + recipient mixture. The enzyme can destroy DNA that is free in the medium but cannot access/degrade DNA within a cell or a phage. Thus, if DNAse prevents the recipient cell from acquiring DNA, the donor's DNA must have been naked.

Competence

In order for transformation to occur, the recipient cell must be **competent**—a specific physiological state that allows the cell to take up DNA. Most competent bacteria take up DNA regardless of its source. Some species accept DNA only from closely related bacteria, recognizing it by characteristic nucleotide sequences located throughout the genome.

In the several dozen prokaryotic species that can become competent naturally, the process is tightly controlled. Some species are always competent, whereas others become so only under specific conditions, such as when the population reaches a certain density or when nutrients are in short supply. The fact that some species become competent only under precise environmental conditions highlights the remarkable ability of seemingly simple cells to sense their surroundings and adjust their behavior accordingly.

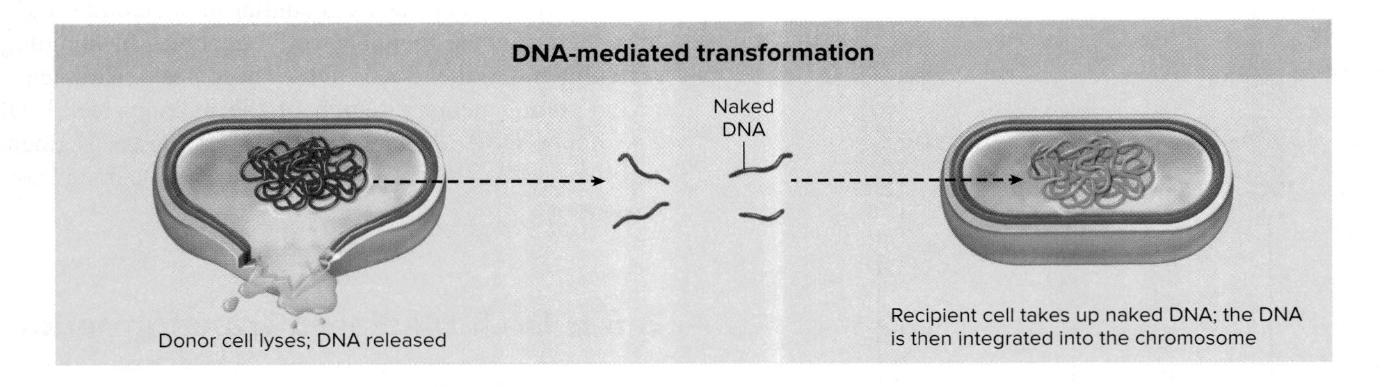

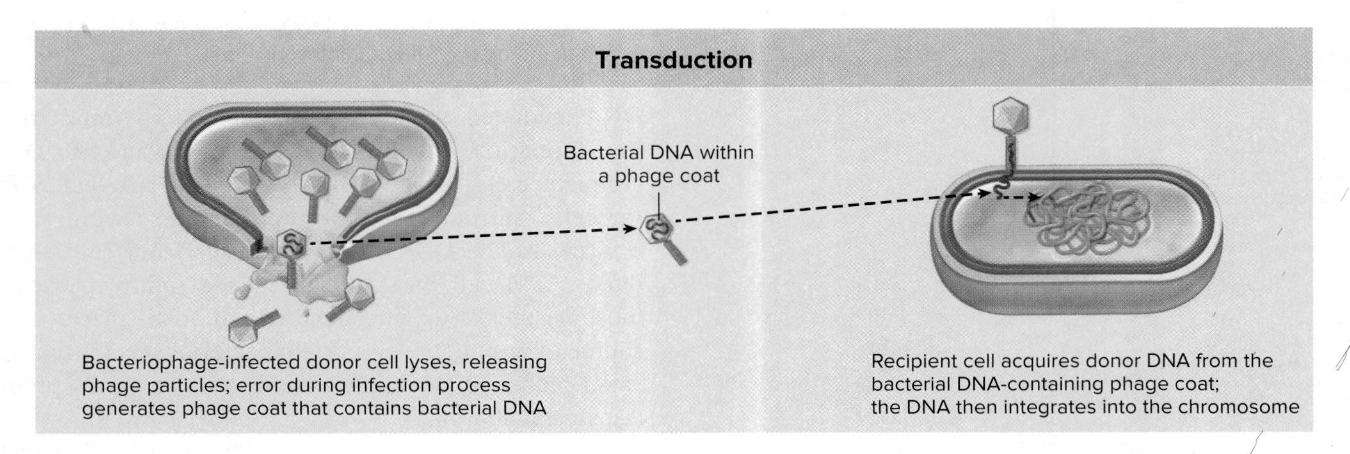

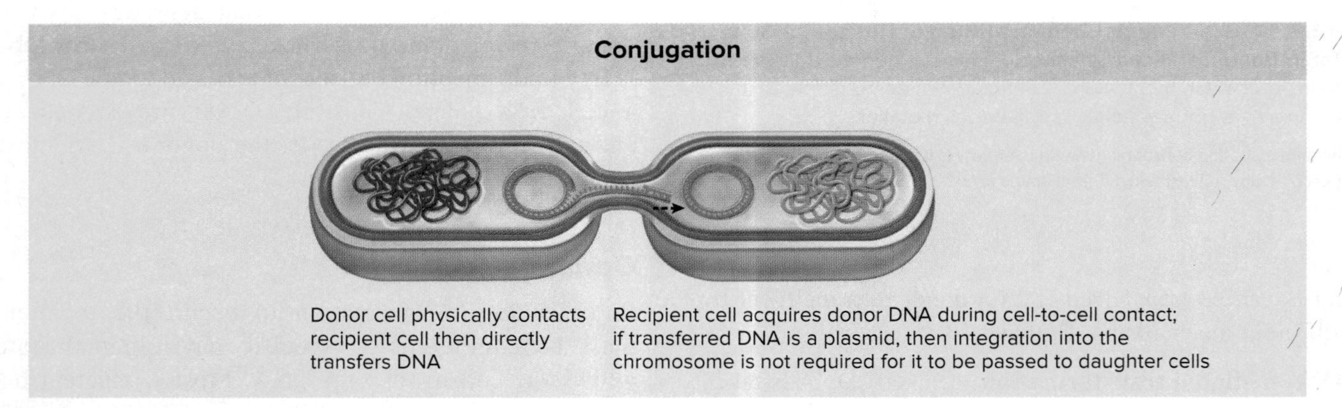

FIGURE 8.18 Mechanisms of Horizontal Gene Transfer

? DNAse (an enzyme that degrades naked DNA) would prevent which mechanism of DNA transfer?

The Process of Transformation

Double-stranded DNA molecules bind to specific receptors on the surface of competent cells (**figure 8.20** ①). ② Only one strand enters the cell, however, because nucleases at the cell surface degrade the other strand. ③ Once the donor DNA is inside the recipient cell, it integrates into the genome by homologous recombination; the strand it replaces will be degraded. ④ Because only a single strand integrates, when the chromosome is replicated and the cell divides, only one daughter cell will inherit donor DNA.

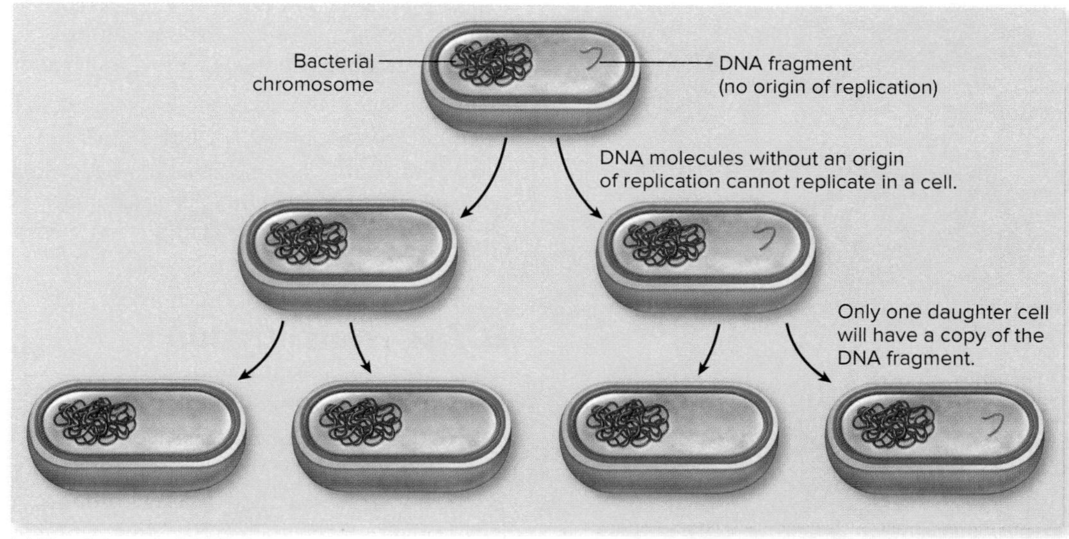

(a) Non-integrated DNA fragment

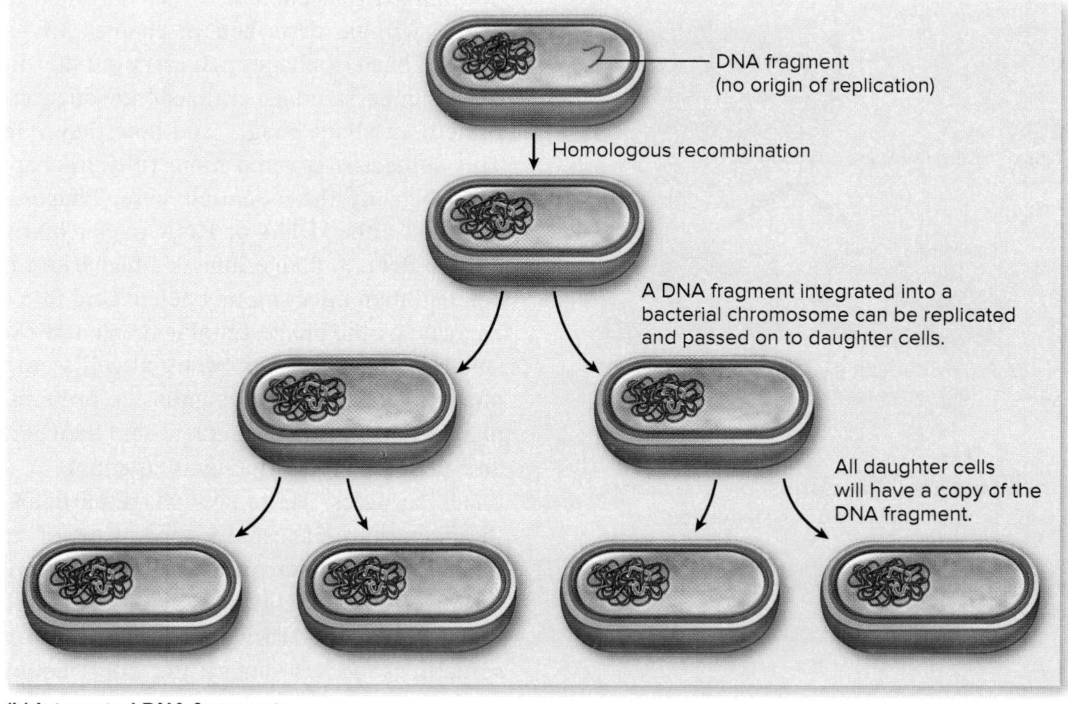

(b) Integrated DNA fragment

FIGURE 8.19 DNA Must Be Part of a Replicon to Be Maintained in a Population (a) DNA without an origin of replication will not be passed on to any additional cells during growth of the population. **(b)** If the DNA becomes integrated within a replicon of the cell, it will be inherited by all daughter cells.

? If the DNA fragment encoded penicillin resistance, how could you experimentally distinguish between (a) and (b)?

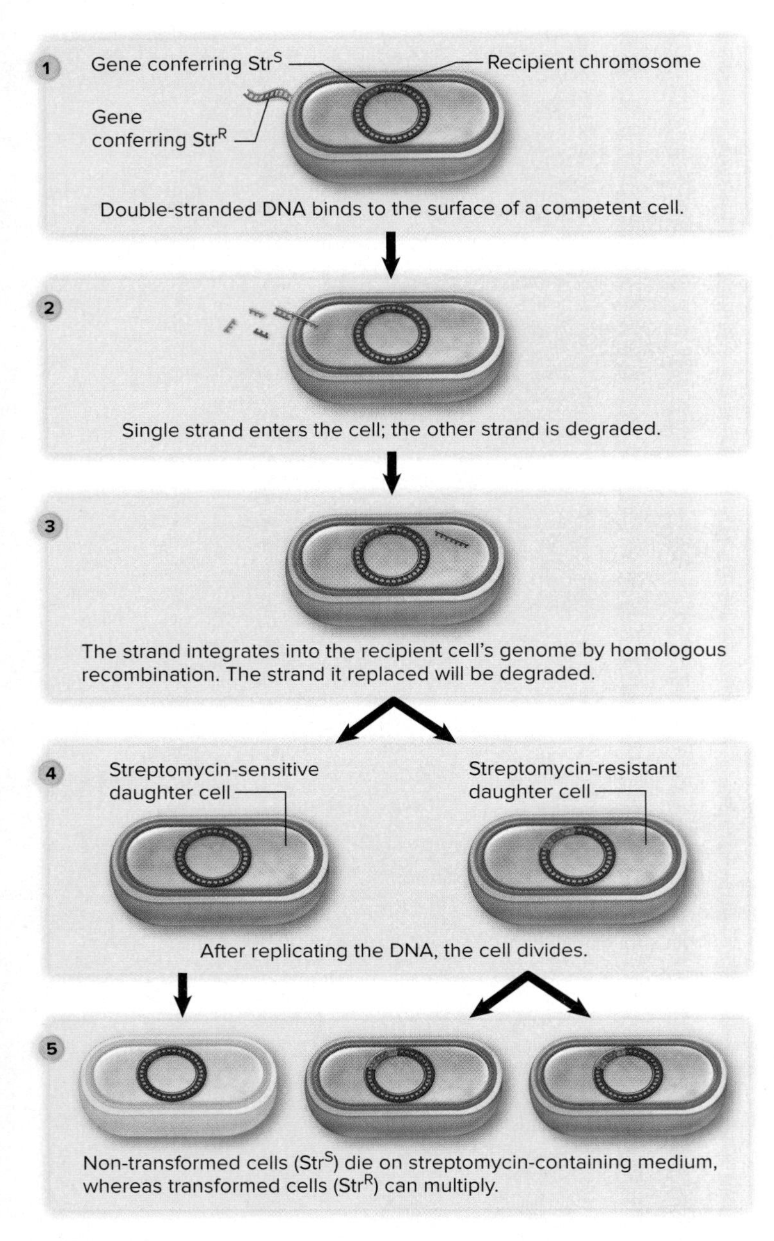

FIGURE 8.20 DNA-Mediated Transformation The donor DNA in this case contains a gene conferring resistance to streptomycin (Str^R).

How would this figure change if double-stranded DNA were incorporated into the donor cell's chromosome?

The transformed cells can be easily detected if they can multiply under selective conditions in which the non-transformed cells cannot. (5) For example, if the donor cells are Str^R and the recipient cells are Str^S, then cells transformed to Str^R will grow and form colonies on a medium that contains streptomycin. Although many other donor genes besides Str^R will be transferred and incorporated by the recipient strain, these transformants will go undetected without a mechanism to recognize them.

MicroAssessment 8.6

In DNA-mediated transformation, DNA is released from donor cells and taken up by competent recipient cells. Competent cells bind DNA and take up a single strand; that strand then integrates into the genome by homologous recombination.

16. How does DNase prevent transformation?

17. Describe two ways by which DNA can be released from cells.

18. How could DNA-mediated transformation be used to test chemicals for their mutagenic activity?

8.7 ■ Transduction

Learning Outcome

14. Describe generalized transduction.

Bacterial viruses, called **bacteriophages,** or simply **phages,** can transfer bacterial genes from a donor to a recipient by **transduction.** There are two types of transduction: generalized and specialized. In this chapter, we will describe only generalized transduction, which transfers any genes of the donor cell. Specialized transduction, which transfers only a few specific genes, will be described in chapter 13—after the infection cycle of bacteriophages that carry out this process is discussed.

To understand generalized transduction, you need to know something about phages and how they infect bacterial cells. This subject is covered more fully in chapter 13, so we will describe only the essentials here. Phages consist of genetic material, either DNA or RNA (never both), surrounded by a protein coat. A phage infects a bacterium by attaching to the cell and then injecting its nucleic acid into that cell. Enzymes encoded by the phage genome then cut the bacterial DNA into small pieces. Next, the bacterial cell's enzymes replicate the phage nucleic acid and synthesize proteins that make up the phage coat. The phage nucleic acid then enters the phage coat, and the various components assemble to produce complete phage particles. These new phage particles are released from the bacterial cell, usually as a result of host cell lysis. The phage particles then attach to other bacterial cells and begin new cycles of infection.

Generalized transduction results from a rare error that sometimes occurs during the construction of phage particles (**figure 8.21**). A fragment of bacterial DNA—produced when the phage-encoded enzyme cuts the bacterial genome—mistakenly enters the phage protein coat. This error creates what is called a **transducing particle;** it carries no phage DNA and therefore is not a phage. Like phage particles, a transducing particle will attach to another bacterial cell and inject the DNA it contains. The transducing particle, however, injects only bacterial DNA because that is all it contains. The bacterial DNA may then integrate into the recipient's chromosome by homologous recombination.

FOCUS YOUR PERSPECTIVE 8.1

The Biological Function of DNA: A Discovery Ahead of Its Time

In the 1920s, Frederick Griffith, an English bacteriologist, was studying *Streptococcus pneumoniae* ("pneumococcus"), a bacterial species that commonly causes pneumonia. These organisms are pathogenic only if they are encapsulated (make a capsule). Unencapsulated pneumococci are not pathogenic. Griffith did a four-part experiment, using mice (**box figure 8.1**). First, he injected encapsulated bacteria into mice, and the mice died. Next, he injected unencapsulated bacteria into mice, and the mice remained healthy and lived. Third, he injected heat-killed encapsulated bacteria into mice, and those mice also lived. Finally, he mixed heat-killed encapsulated (pathogenic) bacteria with live, unencapsulated (non-pathogenic) bacteria and injected them into mice. Surprisingly, the mice died. Griffith isolated live, encapsulated pneumococci from these mice.

Two years after Griffith reported these findings, another investigator, Martin H. Dawson, lysed heat-killed encapsulated pneumococci and passed the suspension of ruptured cells through a very fine filter, through which only the cytoplasmic contents of the bacteria could pass. When he mixed the filtrate (the material that had passed through the filter) with living bacteria unable to make a capsule, some bacteria began making a capsule. Moreover, the progeny of these bacteria could also make a capsule. Something in the filtrate was "transforming" the harmless unencapsulated bacteria into ones that could make a capsule.

What was this transforming principle? In 1944, after years of painstaking chemical analysis of lysates capable of transforming pneumococci, three investigators from the Rockefeller Institute, Oswald T. Avery, Colin MacLeod, and Maclyn McCarty, purified the active compound and then wrote one of the most important papers ever published in biology. In it, they reported that the transforming molecule was DNA. The significance of their discovery was not appreciated at the time, and scientists were slow to recognize its significance. None of the three investigators received a Nobel Prize, although many scientists believe that they deserved it. Their studies pointed out that DNA is a key molecule in the scheme of life and led to James Watson and Francis Crick's determination of its structure, which they published in 1953. The understanding of the structure and function of DNA revolutionized the study of biology and started the era of molecular biology.

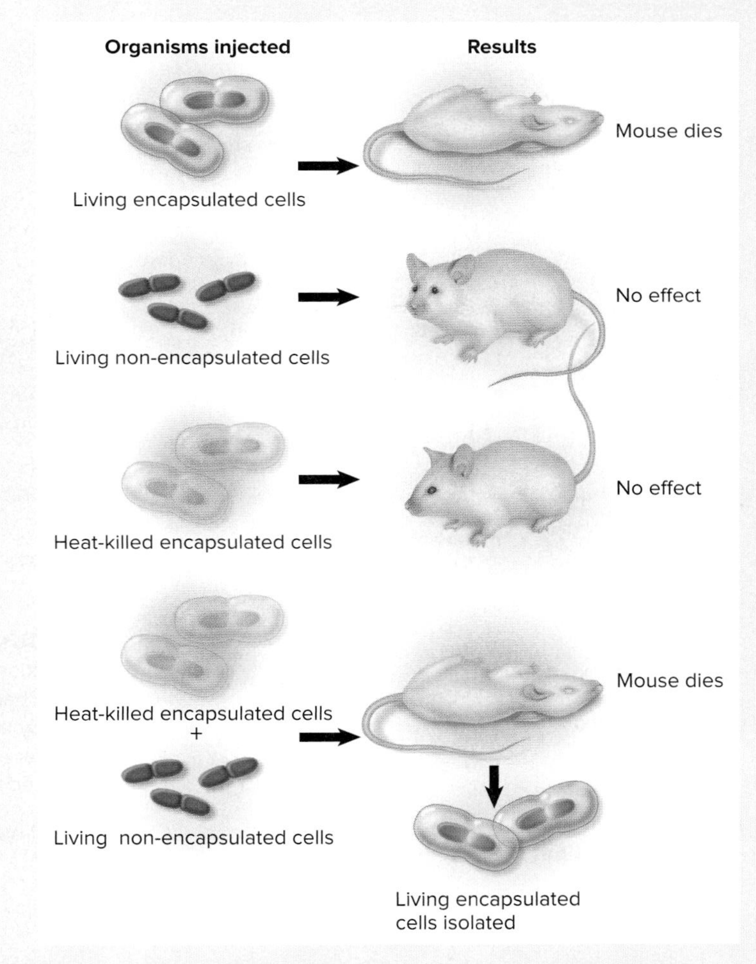

BOX FIGURE 8.1 Griffith's Demonstration of Genetic Transformation

MicroAssessment 8.7

Transduction is the transfer of bacterial DNA from one cell to another by means of a bacteriophage. It results from an error that occurs during the infection cycle of the bacteriophage.

19. What error in the phage replication cycle leads to generalized transduction?

20. What is a transducing particle?

21. Two bacterial genes are transduced simultaneously. What does this suggest about the location of the two genes relative to each other?

8.8 ■ Conjugation

Learning Outcome

15. Compare and contrast conjugation involving the following donor cells: F^+, Hfr, and F'.

Conjugation is a complex process that requires contact between donor and recipient bacterial cells. Gram-positive and Gram-negative bacteria can both transfer DNA this way, but the process is quite different in the two groups.

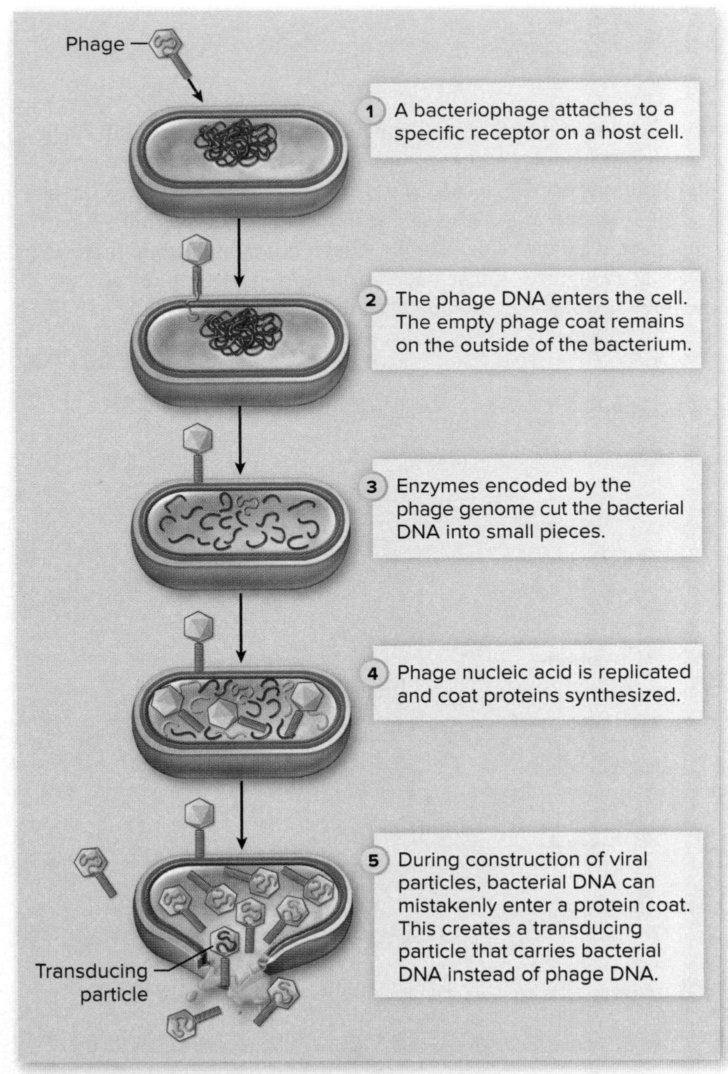

(a) Formation of a transducing particle

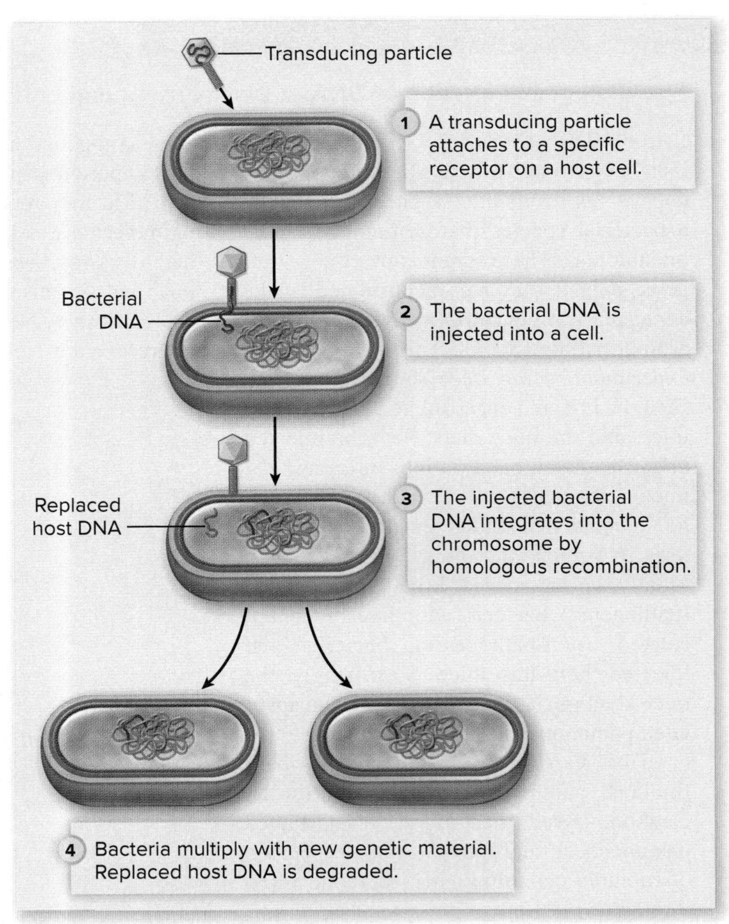

(b) The process of transduction

FIGURE 8.21 Generalized Transduction **(a)** An error during construction of phage particles produces a transducing particle, which contains bacterial DNA instead of phage DNA. **(b)** The bacterial DNA carried by the transducing particle is injected into a new host, resulting in generalized transduction. Essentially any bacterial gene can be transferred this way.

? After a phage injects its DNA into a bacterial cell, the cell begins making proteins that make up the phage coat. Why does the same thing not happen when a generalized transducing particle injects the DNA it carries?

For simplicity, however, we will consider conjugation only in the more intensely studied Gram-negative bacteria.

Plasmid Transfer

Plasmids are most frequently transferred to other cells by conjugation. These DNA molecules are replicons, so they can be replicated without integrating into the recipient's chromosome.

Conjugative plasmids direct their own transfer from donor to recipient cells. The most thoroughly studied example is the **F plasmid** (F stands for fertility) of *E. coli.* Although this plasmid does not encode any notable characteristics except those required for transfer, other conjugative plasmids encode resistance to certain antibiotics, which explains how such resistance can easily spread among a population of cells. *E. coli* cells that contain the F plasmid are designated **F$^+$,** whereas those that do not are **F$^-$.** The F plasmid encodes several proteins required for conjugation, including the **F pilus,** also referred to as the sex pilus (**figure 8.22**). ⏮ sex pilus

Plasmid transfer involves a series of steps (**figure 8.23**):

1. **Making contact.** The F pilus of the donor cell binds to a specific receptor on the cell wall of the recipient.
2. **Initiating transfer.** After contact, the F pilus retracts, pulling the two cells together. Meanwhile, a plasmid-encoded

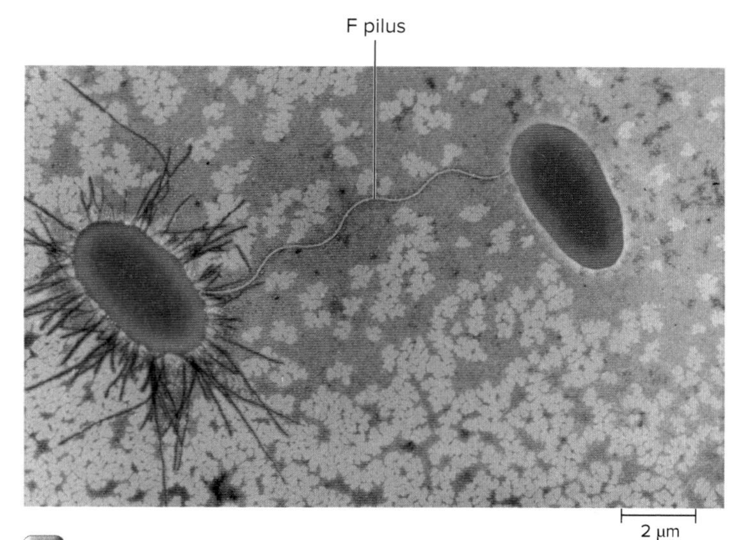

FIGURE 8.22 F Pilus Joining a Donor and Recipient Cell
©Dennis Kunkel/SPL/Science Source

What are the hair-like appendages on the cell on the left?

enzyme cuts one strand of the F plasmid at a specific nucleotide sequence, the origin of transfer.

③ **Transferring DNA.** A single strand of the F plasmid enters the F⁻ cell. Once inside the recipient cell, that strand serves as a template for synthesis of the complementary strand, generating an F plasmid. Likewise, the strand that remains in the donor serves as a template for DNA synthesis, regenerating the F plasmid. The transfer takes only a few minutes.

⑤ **Transfer complete.** Both the donor and the recipient cells are now F⁺ so they can act as donors of the F plasmid.

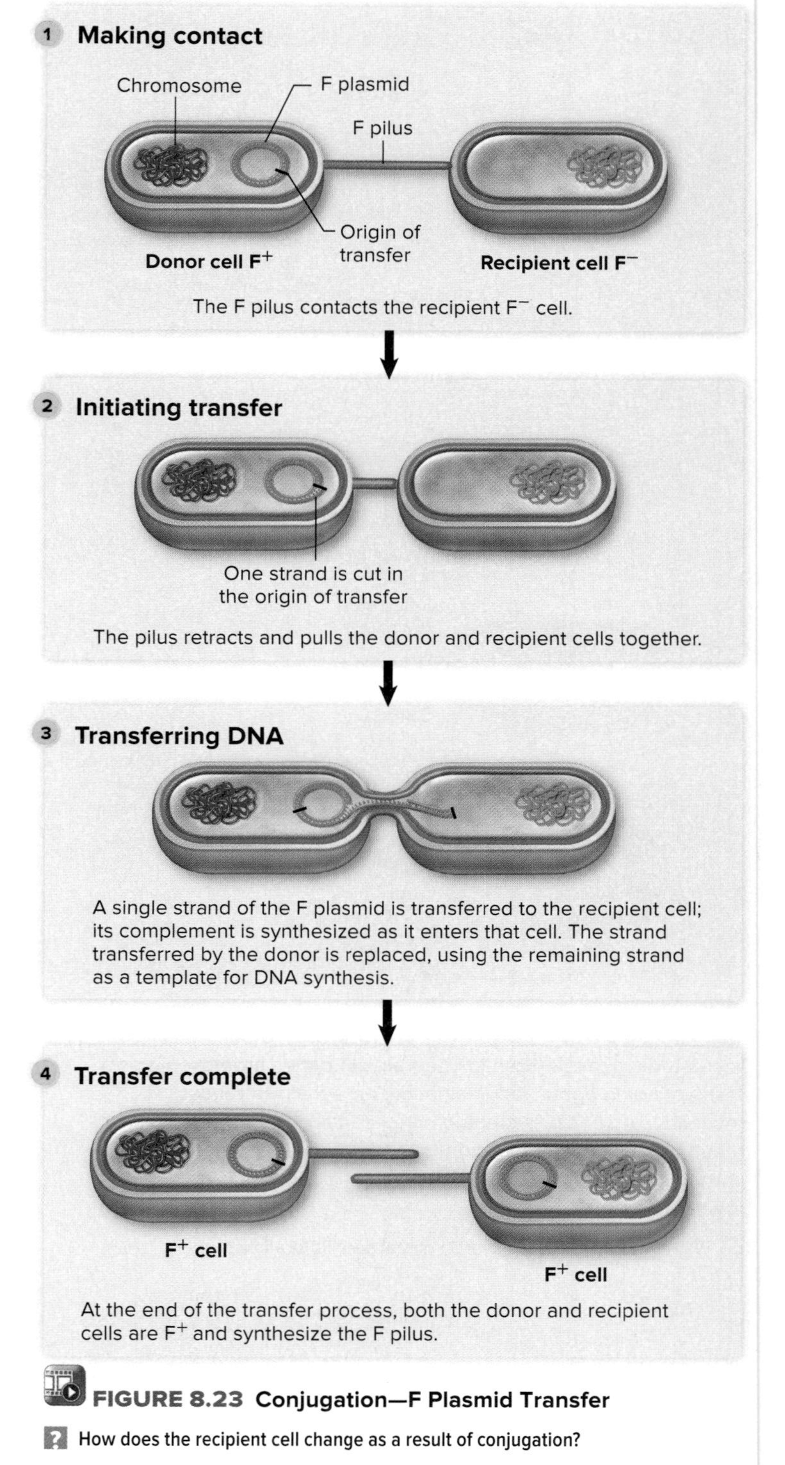

FIGURE 8.23 Conjugation—F Plasmid Transfer

How does the recipient cell change as a result of conjugation?

Chromosome Transfer

Chromosomal DNA transfer is less common than plasmid transfer and involves **Hfr cells** (meaning **h**igh **f**requency of **r**ecombination cells). These are strains in which the F plasmid has integrated into the chromosome by homologous recombination, which happens on rare occasions. As shown in **figure 8.24,** the integration of the F plasmid is reversible; the same process that generates an Hfr cell also allows the integrated plasmid to excise from the chromosome. In some cases, a mistake occurs during the excision process, creating what is called an F′ cell, the significance of which will be described in the next section.

When an Hfr cell transfers chromosomal DNA, the process involves the same general steps as transfer of the F plasmid. Like F⁺ cells, Hfr cells produce an F pilus, and the F plasmid DNA directs its transfer to the recipient cell. Because the F plasmid DNA is integrated into the chromosome, however, chromosomal DNA is also transferred, beginning with the genes on one side of the origin of transfer (**figure 8.25**). The entire chromosome is generally not transferred because it would take approximately 100 minutes for this to occur, an unlikely event because the connection between the two cells usually breaks sooner than that. Because the entire integrated

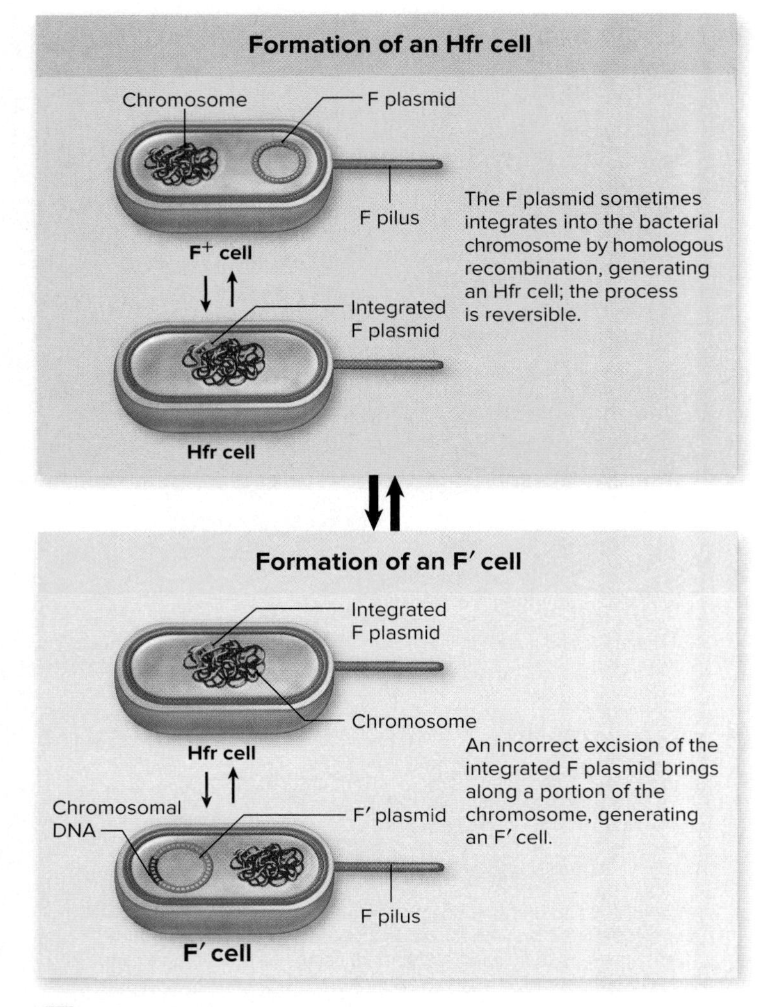

FIGURE 8.24 Formation of Hfr and F′ Cells An Hfr cell is created when the plasmid integrates into the chromosome as a result of homologous recombination; note that the process is reversible. An F′ cell is created when certain recombination events result in an incorrect excision that removes a piece of the chromosome along with the F plasmid; this process is also reversible.

? Why does the F plasmid integrate only at specific locations?

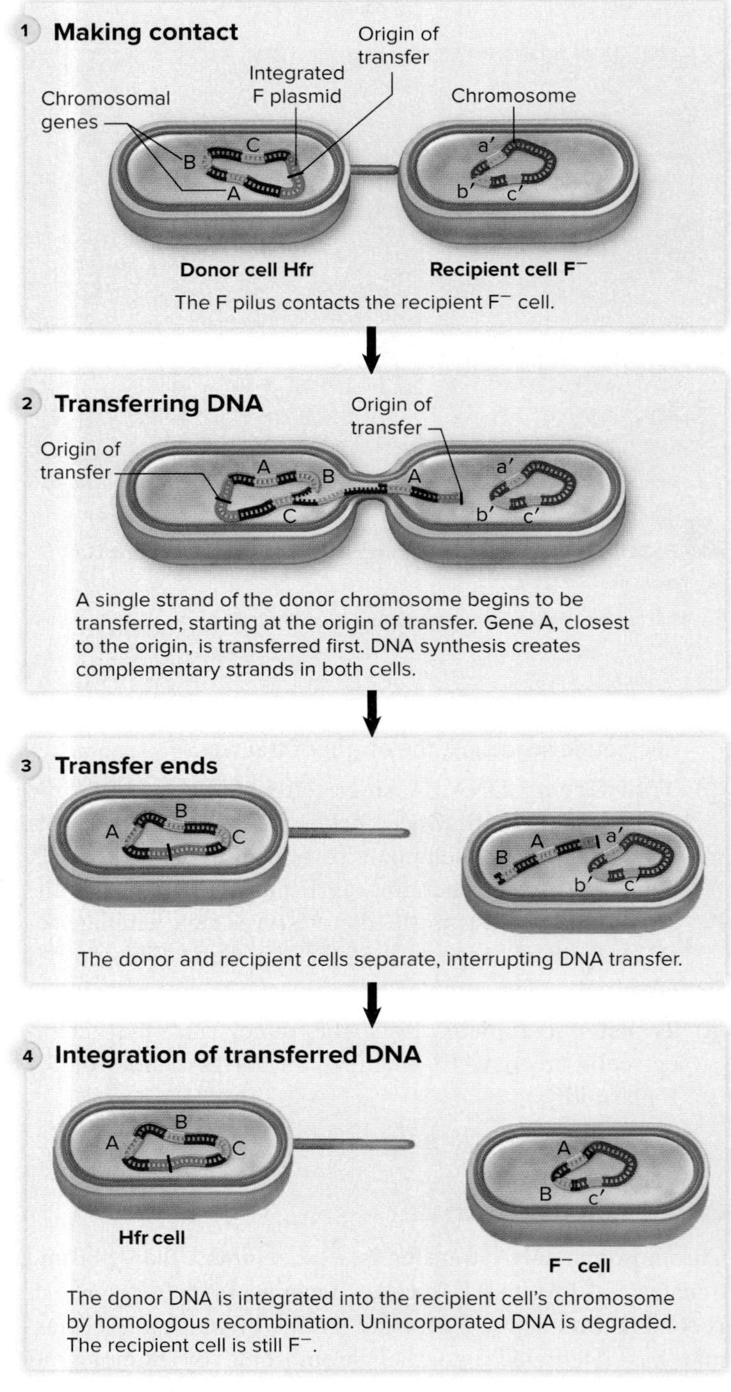

FIGURE 8.25 Conjugation—Chromosomal DNA Transfer The letters A, B, and C indicate genes in the Hfr cell; the letters a′, b′, and c′ indicate homologous but slightly different genes in the F⁻ cell.

? Why is the entire donor chromosome seldom transferred?

F plasmid is not transferred, the recipient remains F^-. The transferred chromosomal DNA is not a replicon, so it will be maintained only if it integrates into the recipient's chromosome through homologous recombination. Unincorporated DNA will be degraded.

F′ Donors

Hfr strains can revert to F^+ because the process of F plasmid integration is reversible (see figure 8.24). In some instances, however, an error occurs during excision, and a piece of the bacterial chromosome is removed along with the F plasmid DNA. This action brings a chromosomal fragment into the F plasmid, producing a plasmid called **F′** (F prime). Like the F plasmid, F′ is a replicon that is rapidly and efficiently transferred to F^- cells. In the case of F′, however, the chromosomal fragment is transferred as well.

MicroAssessment 8.8

Conjugation requires contact between donor and recipient cells. A donor cell that synthesizes an F pilus transfers the DNA to one that does not. Both plasmid and chromosomal DNA can be transferred. Following transfer, plasmids replicate, but chromosomal DNA must be integrated into a replicon to replicate.

22. The F plasmid encodes which two functions essential for conjugation?
23. Describe the outcomes of the three types of conjugation ($F^+ \times F^-$, Hfr $\times F^-$, and $F' \times F^-$).
24. Would you expect transfer of chromosomal DNA by conjugation to be more efficient if cells were plated together on solid medium (agar) or if they were mixed together in a liquid in a shaking flask? Explain.

8.9 ■ The Mobile Gene Pool

Learning Outcomes

16. Describe the concepts of a core genome and of a mobile gene pool.
17. Compare and contrast plasmids, transposons, genomic islands, and phage DNA.

Advances in genomics have uncovered surprising variation in the gene pool (the sum of all genes) of even a single species. For example, nucleotide sequence data indicate that less than half of an *E. coli* isolate's genes are shared by all strains of that species. These conserved genes make up the **core genome** of the species. The remaining ones, which vary considerably among different strains, make up the **mobile gene pool** or mobilome (**figure 8.26**). These genes can move from one DNA molecule to another, carried on various mobile genetic elements (**table 8.3**). Surprisingly, when all the non-conserved genome components of all the various *E. coli* strains are considered as a group, these sequences vastly outnumber the sequences in the core genome.

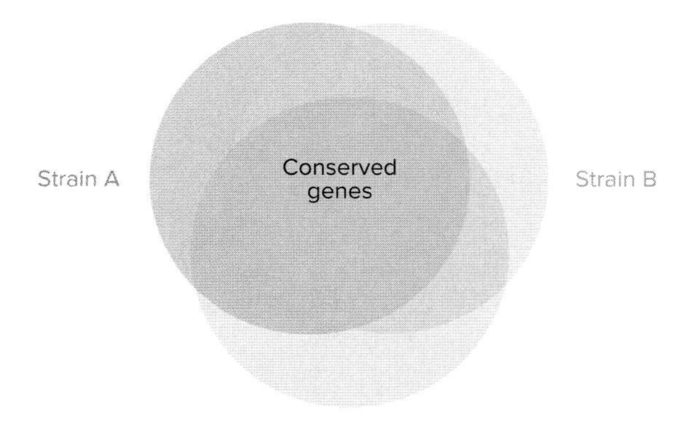

FIGURE 8.26 The Core Genome The genes that are conserved between strains make up the core genome.

What types of genes may make up the mobile gene pool?

TABLE 8.3 The Mobile Gene Pool

Mobile Genetic Element	Characteristics
Plasmid	Replicon that is independent of the chromosome and generally encodes only non-essential genetic information
Transposon	Segment of DNA that directs its own movement to another location in chromosomal or plasmid DNA
Genomic Island	Large genomic segment in a cell's genome that originated in another species
Phage DNA	Phage genome that sometimes carries additional genes

MicroByte

Almost 1,400 genes of the pathogen *E. coli* O157:H7 are not found in the laboratory strain *E. coli* K-12.

Plasmids

Plasmids are common in the microbial world and are found in many bacteria and archaea and in some eukarya, such as certain yeasts. Like chromosomes, most plasmids are circular double-stranded DNA molecules. They have an origin of replication and therefore can be replicated in a cell and passed to progeny cells. Plasmids generally do not encode information essential to the life of a cell, and therefore cells can survive their loss. They are important, however, because they provide cells with the ability to survive in a particular environment (**table 8.4**). ◀◀ plasmid

Plasmids vary with respect to their properties. Some are small and carry only a few genes; others are large and carry many genes. Low-copy-number plasmids occur in only one or a few copies per cell, whereas high-copy-number plasmids are present in many copies, perhaps 500. Most plasmids have

TABLE 8.4 Some Plasmid-Encoded Traits

Trait	Organisms in Which Trait Is Found
Antibiotic resistance	Many
Antibiotic synthesis	*Streptomyces* species
Gas vacuole production	*Halobacterium* species
Increased virulence	*Yersinia* and *Shigella* species
Insect toxin synthesis	*Bacillus thuringiensis*
Nitrogen fixation	*Rhizobium* species
Oil degradation	*Pseudomonas* species
Pilus synthesis	*E. coli, Pseudomonas* species
Toxin production	*Bacillus anthracis*
Tumor formation in plants	*Agrobacterium* species (see Focus Your Perspective 8.2)

a narrow host range, meaning they can replicate in only one species. Broad host range plasmids replicate in many different species, sometimes including both Gram-negative and Gram-positive bacteria. Some plasmids cannot be maintained in the same cell. Because of this, scientists have arranged them into different compatibility groups—members of the same compatibility group cannot be maintained in the same cell.

Many bacterial plasmids are readily transferred by conjugation. Conjugative plasmids carry all of the genetic information needed for transfer, including an origin of transfer. In contrast, mobilizable plasmids encode an origin of transfer but lack other genetic information required for transfer. However, when a conjugative plasmid is in the same cell as a mobilizable plasmid, both plasmids can be transferred. Some plasmids can transfer between unrelated species and even between Gram-positive and Gram-negative bacteria. Genes carried on one type of bacterial plasmid can even be transferred to plant cells by a process analogous to conjugation (see **Focus Your Perspective 8.2**).

FOCUS YOUR PERSPECTIVE 8.2

Bacteria Can Conjugate with Plants: A Natural Case of Genetic Engineering

The bacterium *Agrobacterium tumefaciens* uses a process analogous to conjugation to transfer certain genes into plant cells, including those of tobacco, carrots, and cedar trees. By doing so, the bacterium causes a common plant disease called crown gall, which is characterized by large tumors ("galls") on the crown of the plant (the portion above the soil line). Scientists studying crown gall tissues in the lab found that the cells multiply in the absence of added plant hormones normally required for growth. In addition, crown gall tissues synthesize large amounts of an amino acid derivative called an opine, which is not produced by the normal tissues. Although the bacterium is required to start the altered growth, it is not needed to maintain the changes to the plant cells.

How does this bacterium permanently alter plant cells? *A. tumefaciens* strains that cause crown gall tumors contain a large plasmid called the Ti (tumor-inducing) plasmid. A specific piece of that plasmid called T-DNA (transferred DNA) moves from the bacterial cell to the plant cell, where it becomes incorporated into the plant cell chromosome **(box figure 8.2)**. The T-DNA carries genes for the synthesis of the plant hormones as well as for the opine. Those genes can be expressed in plant cells but not in *A. tumefaciens,* because their promoters resemble those of plant cells rather than those of bacteria. Once the plant cells begin expressing the transferred genes, the cells produce their own source of plant hormones—explaining why they can grow in the absence of added hormones. They also begin synthesizing the opine.

◀◀ promoter

How does *A. tumefaciens* benefit by altering plant cells? This bacterium can use the opine as a source of carbon, nitrogen, and energy, whereas most other bacteria in the soil, as well as plants, cannot. In other words, *A. tumefaciens* genetically engineers the plant cells to multiply and produce food that only *A. tumefaciens* cells can use.

The *Agrobacterium*–crown gall system is important for several reasons. For one thing, it shows that DNA can be transferred from prokaryotes to eukaryotes. Many people believed that such transfer would be impossible in nature and could occur only in the laboratory. This system has also spawned an industry of plant biotechnology. The tumor formation genes in the Ti plasmid can be replaced with beneficial genes, and those genes will then be transferred and incorporated into the plant genome. Examples of genes that have been transferred into plants include those conferring resistance to viral pathogens, insects, and different herbicides. The *Agrobacterium*–crown gall system is an excellent example of how basic science research can lead to major industrial applications.

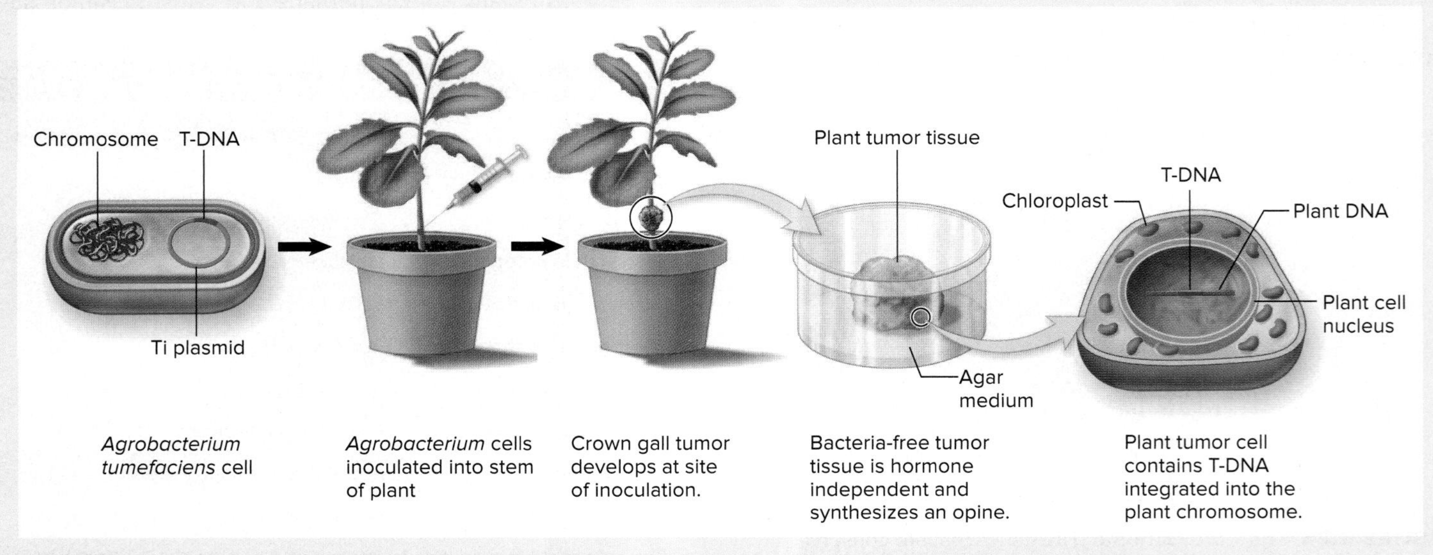

BOX FIGURE 8.2 ***Agrobacterium tumefaciens*** **Causes Crown Gall by Transferring Bacterial DNA to Plant Cells**

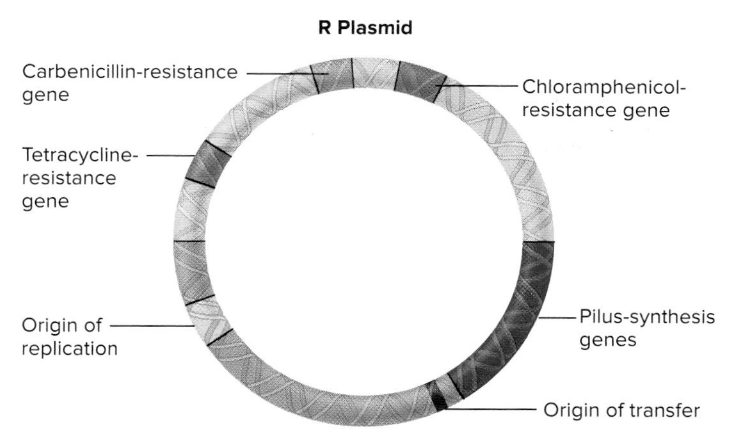

FIGURE 8.27 An R Plasmid The illustrated R plasmid carries multiple genes for antibiotic resistance as well as genes required for conjugation.

How is an R plasmid similar to an F plasmid?

Resistance or **R plasmids** are particularly important medically because they encode resistance to antimicrobial chemicals, including antibiotics (**figure 8.27**). Many of these plasmids are conjugative, carrying not only genes encoding resistance (R genes), but also genes for pilus synthesis and other properties required for conjugation. Because many R plasmids have a broad host range and carry multiple resistance genes, they can allow a wide variety of organisms to become resistant to many different antimicrobials. Members of the normal microbiota that carry R plasmids can potentially transfer them to pathogens.

Transposons

In addition to causing mutations, transposons provide a mechanism for transferring various genes. Transposons can move into other replicons in the same cell without any specificity as to where they insert.

Several types of transposons exist, varying in their structural complexity. The simplest, an **insertion sequence (IS),** encodes only transposase (the enzyme responsible for transposition) (**figure 8.28**). On each side of the gene are inverted repeats—sequences that are identical when read in the 5′ to 3′ direction. **Composite transposons** consist of one or more genes flanked by ISs. Like insertion sequences, composite transposons can move in the same replicon or from one replicon to another in the cell. Their movement is easily followed if they encode a recognizable gene product such as antibiotic resistance. They integrate into their new location through **non-homologous recombination,** a process that does not require a similar nucleotide sequence in the region of recombination. The transposon simply inserts into a stretch of DNA; it does not replace the existing sequences. If a transposon inserts into a conjugative plasmid, it can then be transferred to other cells. Any gene or group of genes flanked by ISs can move to another site, but transposons that carry genes for antibiotic resistance are particularly important medically.

Genomic Islands

Genomic islands are large DNA segments in a cell's genome that originated in other species. This conclusion is based on the fact that their nucleotide composition is quite different from the rest of the cell's genome. In general, each bacterial species has a characteristic proportion of G-C (guanine-cytosine) base pairs, so a large segment of DNA that has a very different G-C ratio suggests the segment originated from a foreign source and was transferred to the cell through horizontal gene transfer. G + C content

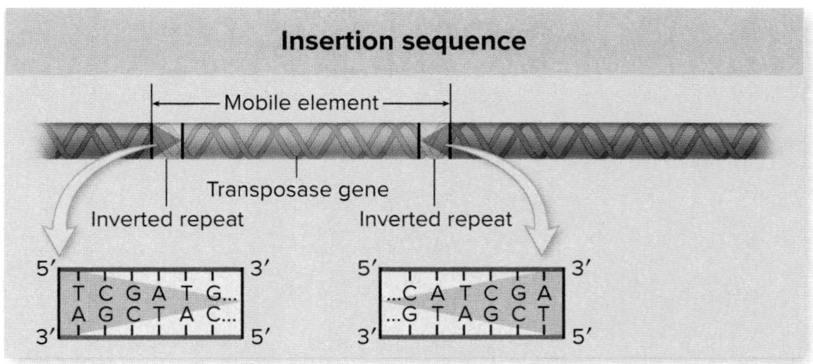

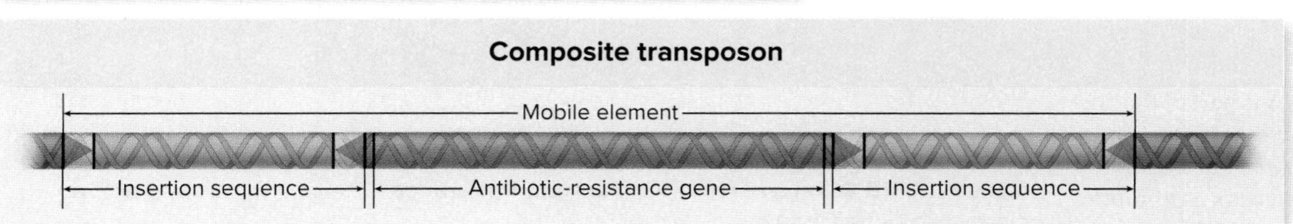

FIGURE 8.28 Transposons The borders of insertion sequences are defined by inverted repeats 15–20 nucleotides in length. The first six nucleotides are shown here in expanded view to demonstrate their inverted orientation. Composite transposons consist of two IS elements and the DNA between them, all of which move as a single unit.

Why are some transposons medically important?

The characteristics encoded by genomic islands include use of specific energy sources, acid tolerance, development of symbiosis, and ability to cause disease. Genomic islands that encode the latter are called **pathogenicity islands.**

Phage DNA

As you will learn in chapter 13, certain types of phages can insert their DNA into the host cell chromosome. When this occurs, the phage DNA becomes part of the host cell's genome, which will be replicated and passed on to progeny cells.

FOCUS ON A CASE 8.1

The patient was a 40-year-old woman with multiple health problems, including diabetes, chronic foot ulcers that repeatedly became infected, and chronic renal failure that required regular dialysis. Because of the recurring infections in her toes, multiple surgeries had been done to amputate those toes.

While hospitalized after one of the surgeries, the patient developed bacteremia (bacteria in the blood) as well as an abscess as a complication of a procedure used in dialysis. The causative agent in both infected sites was found to be methicillin-resistant *Staphylococcus aureus* (MRSA). Relatively few options exist for treating MRSA infections, so in these cases doctors often use the antibiotic vancomycin as a medication of "last resort." Past medical history indicated that the patient's foot ulcers had been treated with multiple courses of antibiotics, including vancomycin, over a span of a year before the surgery.

The patient's bacteremia was successfully treated with vancomycin and another antimicrobial medication, but she then developed an infection of the dialysis catheter exit-site. Vancomycin-resistant *S. aureus* (VRSA) and vancomycin-resistant *Enterococcus faecalis* (VRE) were isolated from the catheter tip, and then from two of her foot ulcers.

1. Based on the name *E. faecalis,* where would you think the bacterium is usually found?
2. How might the MRSA strain have become vancomycin-resistant?

Discussion

1. *Enterococcus faecalis* is part of the normal intestinal microbiota (*entero* means "intestine," and *faecal* indicates that it is associated with feces). Antibiotic-resistant strains are relatively common, particularly in hospitals and other environments where the medications are frequently used.
2. The only obvious difference between the two *S. aureus* strains (MRSA and VRSA) was that the latter had a gene encoding resistance to vancomycin inserted into a plasmid present in both strains. Further studies showed that the vancomycin resistance gene of VRSA was part of a transposon that was identical to one in the VRE isolated from the same patient.

In both VRSA and VRE, the transposon was integrated into a plasmid, but the plasmids in the two organisms were different. It appears that VRE transferred its transposon-containing plasmid to the vancomycin-sensitive *S. aureus* by conjugation (**box figure 8.3**). This entering plasmid was then destroyed by enzymes in *S. aureus,* but before that happened the transposon jumped to the plasmid already in the *S. aureus* cell.

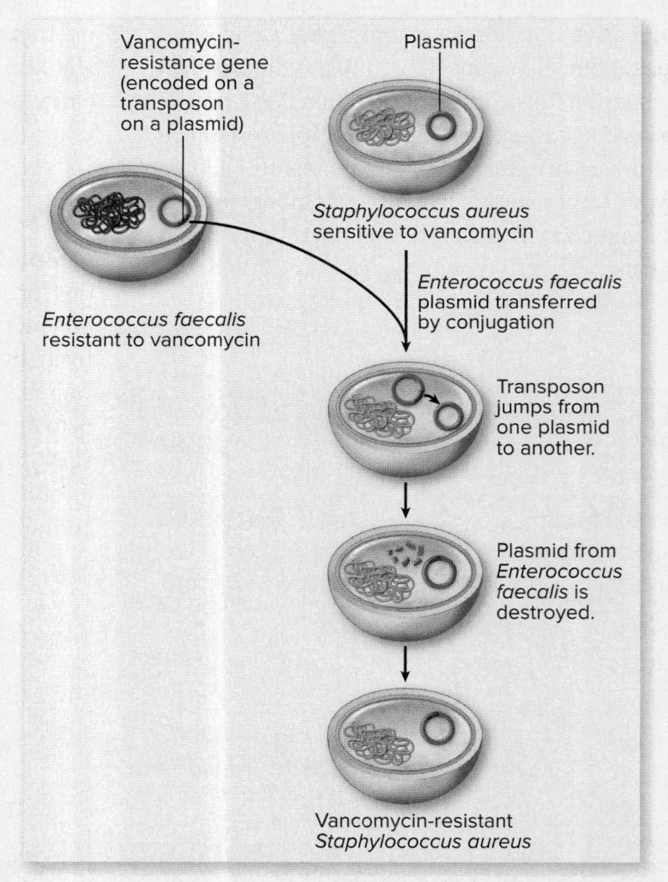

BOX FIGURE 8.3 Transfer of Vancomycin Resistance. The transfer involved both a plasmid and a transposon.

What role did the transposon play in the transfer of vancomycin resistance?

Source: MMWR: 51(26):565–567 ["*Staphylococcus aureus* Resistant to Vancomycin—United States, 2002" *Morbidity and Mortality Weekly Report*], July 5, 2002.

MicroAssessment 8.9

Plasmids vary in size, copy number, host range, compatibility to coexist with other plasmids, and ability to be transferred to other cells. R plasmids are important medically because they encode resistance to various antimicrobial medications. Transposons can move from one location to another in the same replicon or to other replicons. Genomic islands are large DNA segments thought to have originated in other species. Phage DNA can integrate into bacterial chromosomes.

25. What functions must a plasmid encode to be self-transmissible?

26. What characteristic of a genomic island suggests that it originated in another species?

27. Considering that *Staphylococcus epidermidis* does not typically cause disease in a healthy person, why would it be significant if it carried an R plasmid?

8.10 ■ Bacterial Defenses Against Invading DNA

Learning Outcome

18. Explain how restriction modification systems and CRISPR systems recognize and destroy invading DNA.

In some cases, foreign DNA that enters a bacterial cell is destroyed before it can be replicated or integrated into the cell's genome. This is because many bacteria are able to recognize and destroy invading DNA, using systems that likely evolved as defenses against phages. As you will read in chapter 9, the defense mechanisms are also important in biotechnology because they allow scientists to cut DNA at precise nucleotide sequences. Once a cut has been made, the nucleotide sequence at that site can be manipulated.

Restriction-Modification Systems

Restriction-modification systems were discovered by scientists studying why certain bacterial strains are relatively resistant to phage infection. Their research showed that those strains degrade foreign DNA through the combined action of two types of enzymes: a restriction enzyme and a modification enzyme. The restriction enzyme recognizes a specific short nucleotide sequence within the cell and then cuts the DNA molecule at that sequence. The modification enzyme protects that cell's own DNA from the action of the restriction enzyme by adding methyl groups to the nucleobases recognized by the restriction enzyme. Because restriction enzymes cannot degrade methylated DNA, a restriction enzyme will destroy incoming foreign DNA but not the host DNA. Occasionally, the modification enzyme will methylate the incoming foreign DNA before the restriction enzyme has acted so that the invading DNA will not be degraded (**figure 8.29**). Different bacteria have different versions of restriction enzymes and modification enzymes, so hundreds of varieties exist, each recognizing different sequences in DNA.

⏭ restriction enzymes

CRISPR Systems

CRISPR systems were discovered when scientists recognized that certain bacterial genomes include very small pieces of phage DNA. Presumably, the bacteria have survived certain phage infections and retained small segments of that invader's DNA, incorporating them into the bacterial genome. There, the segments are used to recognize and destroy that specific invading DNA in the future, providing the cell with a form of adaptive immunity.

Several versions of CRISPR systems have been characterized, but they all function in the same general manner. The first time a phage genome or other invading DNA enters a cell, a complex of specific Cas proteins (named for **C**RISPR-**a**ssociated **s**equences) cuts that DNA into short fragments (**figure 8.30**①). Some of those DNA fragments are then inserted into a chromosomal region called a CRISPR array (named for the characteristic **c**lusters of **r**egularly **i**nterspersed **s**hort **p**alindromic **r**epeats); an integrated segment of captured DNA is called a spacer. In the cases of phage infections, the spacer DNA provides a historical record of past infections. It allows the surviving cell and its descendants to recognize specific phage DNA if it is encountered again.

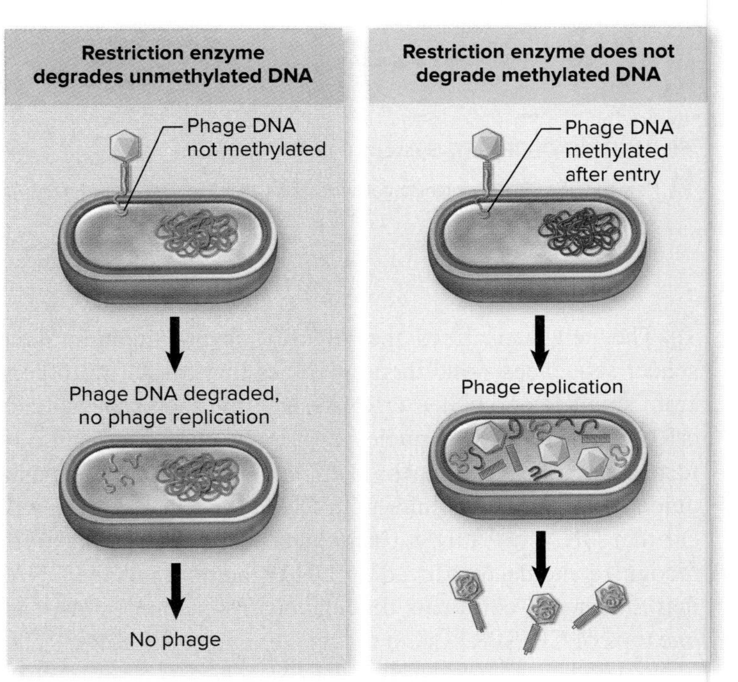

FIGURE 8.29 Restriction-Modification System

How is the DNA modified by this system?

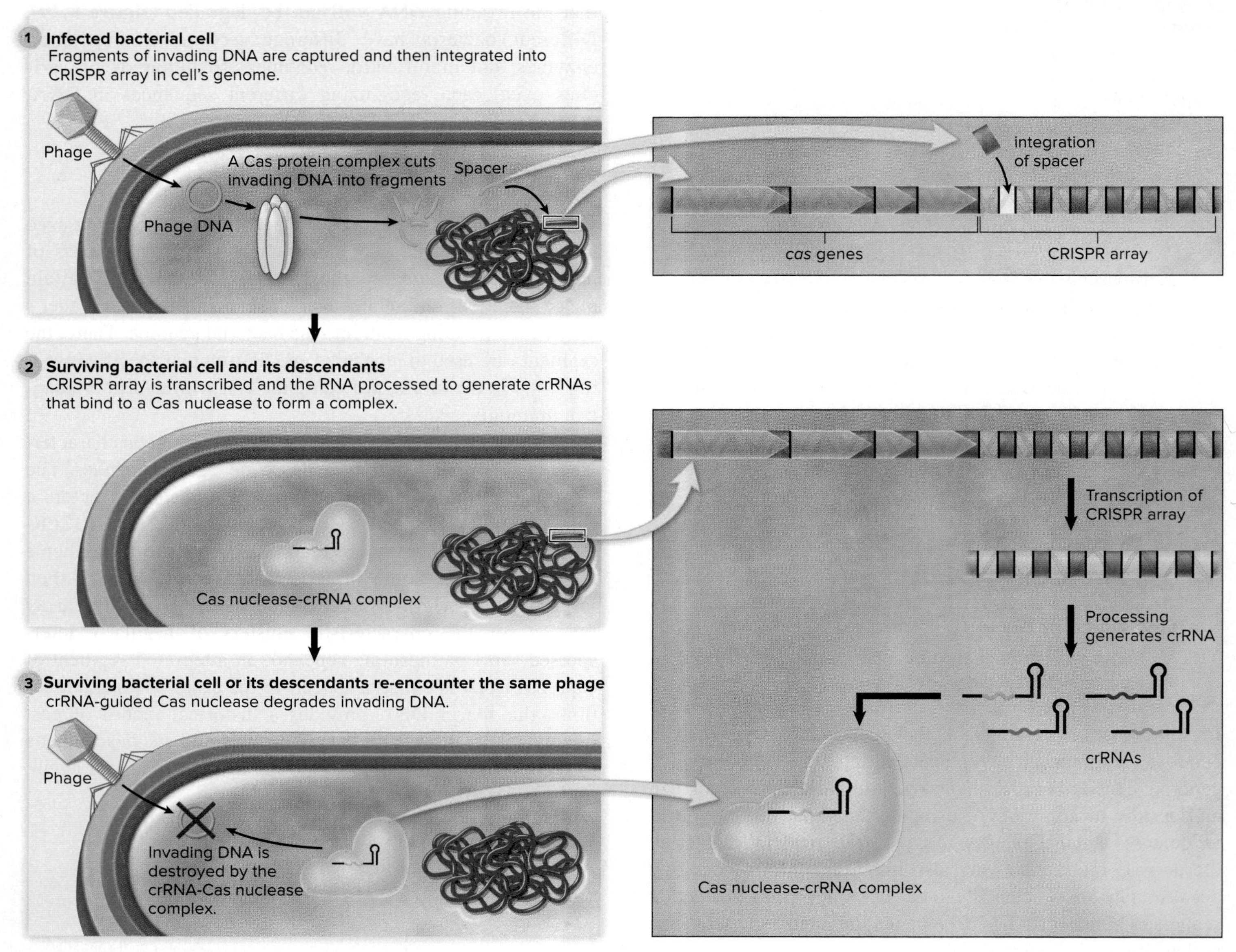

FIGURE 8.30 CRISPR System

How does the CRISPR system target nucleic acid of an invading phage for destruction?

② The cell transcribes the CRISPR array—including the spacer sequences—and then processes that transcript to generate small RNAs called crRNAs, one for each spacer. Each crRNA binds to a Cas nuclease (a DNA-cutting enzyme) to form a complex. Within the complex, crRNA functions as a guide; if crRNA base-pairs with DNA, the Cas nuclease will cut that DNA. ③ Thus, a Cas nuclease-crRNA complex will recognize and destroy invading DNA (including phage DNA) that a bacterial cell or its descendants re-encounter. At least one type of CRISPR system recognizes and cuts foreign RNA.

MicroAssessment 8.10

Bacteria use restriction modification systems and CRISPR systems to defend against invading DNA.

28. How do modification enzymes protect host cell DNA from restriction enzymes?
29. How does a bacterial cell acquire a historical record of phage infections?
30. How is the mechanism of a CRISPR system similar to the mechanism of RNA interference?

Summary

8.1 ■ Genetic Change in Bacteria

The **genotype** of bacteria can change through either **mutation** or **horizontal gene transfer** (figure 8.1). Bacteria are haploid, so any changes in DNA can easily alter the **phenotype.** A mutant that requires a growth factor is an **auxotroph;** in contrast, a **prototroph** does not require growth factors.

MUTATION AS A MECHANISM OF GENETIC CHANGE

8.2 ■ Spontaneous Mutations

Spontaneous mutations occur as a result of normal cell processes. They are stable but occasionally revert back to the non-mutant form. The chance that two given spontaneous mutations will occur within the same cell is the product of the individual mutation rates.

Base Substitution

Base substitutions occur during DNA synthesis (figure 8.2). They result in silent (synonymous), missense, and nonsense mutations (figure 8.3).

Deletion or Addition of Nucleotides

Deleting or adding one or two nucleotides causes a **frameshift mutation,** changing the reading frame of the encoded protein (figure 8.4). This often results in a shortened non-functional protein.

Transposons (Jumping Genes)

Transposons can move from one location to another in a cell's genome. The gene into which the transposon jumps is insertionally inactivated by the event (figure 8.5).

8.3 ■ Induced Mutations

Induced mutations are caused by **mutagens** (table 8.1).

Chemical Mutagens

Some chemicals modify nucleobases, altering their hydrogen-bonding properties (figure 8.7). **Base analogs** can be mistakenly incorporated in place of the usual nucleobases, and they have different hydrogen-bonding properties (figure 8.8). **Intercalating agents** insert into the double helix and push nucleotides apart, resulting in frameshift mutations.

Transposition

Transposons can be introduced intentionally into a cell in order to inactivate genes.

Radiation

Exposure to ultraviolet light results in **thymine dimer** formation (figure 8.9). The repair mechanism can cause mutations. X rays cause single- and double-strand breaks.

8.4 ■ Repair of Damaged DNA (table 8.2)

Repair of Errors in Nucleotide Incorporation

DNA polymerases have a **proofreading** function. **Mismatch repair** removes a portion of the strand that has a misincorporated nucleotide. A new DNA strand is then synthesized (figure 8.10).

Repair of Modified Nucleobases in DNA

Specific DNA glycosylases remove modified nucleobases in DNA (figure 8.11).

Repair of Thymine Dimers

In **photoreactivation,** an enzyme uses the energy of light to break the bonds of the thymine dimer (figure 8.12). In **excision repair,** the damaged single-stranded segment is removed and replaced.

SOS Repair

SOS repair is a last-effort repair mechanism that uses a special DNA polymerase that has no proofreading ability but can bypass the damaged DNA. Consequently, the newly synthesized DNA has many mutations.

8.5 ■ Mutant Selection

Direct Selection

Direct selection is used to obtain mutants that grow under conditions in which the parent cell cannot (figure 8.13).

Indirect Selection

Indirect selection uses **replica plating** to isolate an auxotroph from a prototrophic parent strain (figure 8.14). **Penicillin enrichment** increases the proportion of auxotrophic mutants in a culture (figure 8.15).

Screening for Possible Carcinogens

The **Ames test** is used to determine if a chemical is a mutagen and therefore a possible **carcinogen** (figure 8.16).

HORIZONTAL GENE TRANSFER AS A MECHANISM OF GENETIC CHANGE (figure 8.18)

For newly acquired DNA to replicate in a cell, it must either be a replicon or integrate into the cell's genome (figure 8.19).

8.6 ■ DNA-Mediated Transformation

DNA-mediated transformation transfers "naked" DNA.

Competence

A cell must be **competent** to take up DNA, and only certain species naturally become competent.

The Process of Transformation

Short strands of double-stranded DNA bind to cells, but only one strand enters (figure 8.20).

8.7 ■ Transduction

Transduction is the transfer of bacterial DNA by a **bacteriophage.** There are two types: **generalized transduction** and **specialized transduction** (figure 8.21).

8.8 ■ Conjugation

Conjugation requires cell-to-cell contact.

Plasmid Transfer

F^+ cells synthesize an **F pilus,** encoded on an **F plasmid;** the F plasmid is transferred from an F^+ to an F^- cell (figure 8.23).

Chromosome Transfer

Hfr cells have the F plasmid integrated into the chromosome (figure 8.24). When the F plasmid is transferred, chromosomal DNA moves into a recipient cell along with it (figure 8.25).

F′ Donors

An **F′** donor carries a modified F plasmid that contains a piece of chromosomal DNA.

8.9 ■ The Mobile Gene Pool (table 8.3)

Conserved genes make up the **core genome** of a species; the remaining ones make up the **mobile gene pool** (figure 8.26).

Plasmids

Plasmids are replicons that typically encode only non-essential information (table 8.4); many can be transferred by conjugation. **R plasmids** code for antibiotic resistance (figure 8.27).

Transposons

Transposons provide a mechanism for transferring genes. An **insertion sequence (IS)** encodes only transposase (figure 8.28). A **composite transposon** has one or more genes flanked by insertion sequences.

Genomic Islands

Genomic islands are large DNA segments in a cell's genome that originated in other species.

Phage DNA

DNA of certain phages can integrate into a bacterial chromosome.

8.10 ■ Bacterial Defenses Against Invading DNA

Restriction-Modification Systems

Restriction-modification systems allow bacterial cells to quickly distinguish and destroy invading DNA. Restriction enzymes recognize and cut specific DNA sequences; modification enzymes protect the hosts' DNA from the action of the restriction enzymes by adding methyl groups to certain nucleobases (figure 8.29).

CRISPR Systems

CRISPR systems allow bacterial cells to destroy invading DNA that has been encountered previously. The cell transcribes a chromosomal region called a CRISPR array that contains fragments captured from invading DNA; processing of the transcript generates RNA fragments that bind to a Cas nuclease, allowing it to recognize and destroy that DNA (figure 8.30).

Review Questions

Short Answer

1. How is an auxotroph different from a prototroph?
2. Why is deleting one nucleotide generally more detrimental than deleting three?
3. What type of mutation in an operon is most likely to affect the synthesis of more than one protein?
4. What is meant by "proofreading" with respect to DNA polymerase?
5. Why would a cell use SOS repair, considering that it introduces mutations?
6. Why is replica plating used to isolate an auxotrophic mutant from a prototrophic parent?
7. What is transduction?
8. How is an F^+ strain different from an Hfr strain?
9. Name four mobile genetic elements.
10. Why are R plasmids important?

Multiple Choice

1. UV light exposure forms
   a) covalent bonds between the two strands of DNA.
   b) hydrogen bonds between the two strands of DNA.
   c) covalent bonds between adjacent thymine nucleobases on the same strand of DNA.
   d) covalent bonds between adjacent guanine and cytosine nucleobases on the same strand of DNA.
   e) hydrogen bonds between adjacent guanine and cytosine nucleobases on the same strand of DNA.
2. If cells were exposed to UV light, the highest frequency of mutations would be obtained if the cells were immediately
   a) placed in the dark.
   b) exposed to visible light.
   c) shaken vigorously.
   d) incubated at a temperature below their optimum for growth.
3. Penicillin enrichment of mutants works on the principle that the antibiotic
   a) kills only Gram-positive cells.
   b) primarily kills cells during the lag phase of growth.
   c) kills most Gram-negative cells.
   d) kills only growing cells.
   e) inhibits formation of the lipopolysaccharide layer.
4. Mechanisms that repair errors in nucleotide incorporation are
   1) mismatch repair.
   2) proofreading by DNA polymerase.
   3) light repair.
   4) SOS repair.
   5) excision repair.
      a) 1, 2
      b) 2, 3
      c) 3, 4
      d) 4, 5
      e) 1, 5
5. You are trying to isolate a mutant of wild-type *E. coli* that requires histidine for growth. This can best be done using
   1) direct selection.
   2) replica plating.
   3) penicillin enrichment.
   4) the Ames test.
   5) reversion.
      a) 1, 2
      b) 2, 3
      c) 3, 4
      d) 4, 5
      e) 1, 5
6. All plasmids
   1) carry genes for antimicrobial resistance.
   2) encode mechanisms for transmission to other bacteria.
   3) occur in multiple copies in the cells.

4) code for non-essential functions.
5) are replicons.
   a) 1, 2
   b) 2, 3
   c) 3, 4
   d) 4, 5
   e) 1, 5

7. Adding DNase to a mixture of donor and recipient cells will prevent gene transfer via
   a) DNA-mediated transformation.
   b) conjugation.
   c) generalized transduction.
8. An F pilus or its equivalent is essential for
   1) DNA-mediated transformation.
   2) chromosome transfer by conjugation.
   3) plasmid transfer by conjugation.
   4) generalized transduction.
   5) cell movement.
      a) 1, 2
      b) 2, 3
      c) 3, 4
      d) 4, 5
      e) 1, 5
9. A plasmid that can replicate in *E. coli* and *Pseudomonas* is most likely a/an
   a) broad host range plasmid.
   b) self-transmissible plasmid.
   c) high-copy-number plasmid.
   d) essential plasmid.
   e) low-copy-number plasmid.
10. The frequency of transfer of an F′ molecule by conjugation is closest to the frequency of transfer of
    a) chromosomal genes by conjugation.
    b) an F plasmid by conjugation.
    c) an F plasmid by transformation.
    d) an F plasmid by transduction.
    e) an R plasmid by DNA transformation.

Applications

1. Some bacteria may have higher mutation rates than others following exposure to UV light. Discuss a reason why this might be the case. What experiments could you do to determine whether this is a likely possibility?
2. A pharmaceutical researcher is disturbed to discover that the major ingredient of a new drug formulation causes frameshift mutations in bacteria. What other information would the researcher want before looking for a substitute chemical?

Critical Thinking

1. You have the choice of different kinds of mutants for use in the Ames test to determine the frequency of reversion by suspected carcinogens. You can choose a deletion, a point mutation, or a frameshift mutation. Would it make any difference which one you chose? Explain.
2. You have isolated a strain of *E. coli* that is resistant to penicillin, streptomycin, chloramphenicol, and tetracycline. You also observe that when you mix this strain with cells of *E. coli* that are sensitive to the four antibiotics, they become resistant to streptomycin, penicillin, and chloramphenicol but remain sensitive to tetracycline. Explain what is going on.

9 Biotechnology

Scientists working at a bench in a DNA laboratory. ©*Anthony Bradshaw/ Photographer's Choice/Getty Images*

KEY TERMS

Colony Blotting Technique used to determine which colonies on an agar plate contain a given nucleotide sequence.

DNA Cloning Procedure in which a fragment of DNA is inserted into a vector and then transferred into another cell, where it then replicates.

DNA Gel Electrophoresis A procedure used to separate DNA fragments according to their size.

DNA Microarray A probe-based technique used to study gene expression patterns.

DNA Probe Single-stranded piece of DNA, tagged with an identifiable marker, that is used to detect a complementary sequence.

DNA Sequencing Process of determining the nucleotide sequence of a DNA molecule.

Fluorescence In Situ Hybridization (FISH) Technique that uses a fluorescent probe to detect a given nucleotide sequence within intact cells on a microscope slide.

Genetic Engineering Deliberately altering an organism's genetic information using in vitro techniques.

Polymerase Chain Reaction (PCR) In vitro technique used to repeatedly duplicate (amplify) a specific region of a DNA molecule, increasing the number of copies exponentially.

Recombinant DNA Molecule DNA molecule created by joining DNA fragments from two different sources.

Restriction Enzyme Type of enzyme that recognizes a specific nucleotide sequence and then cuts the DNA within or near that site.

Vector DNA molecule, often a plasmid, that functions as a carrier of cloned DNA.

A Glimpse of History

In 1976, Argentine newspapers reported a violent shootout between soldiers and the occupants of a house in suburban Buenos Aires, leaving the five extremists inside dead. Conspicuously absent from those reports were the identities of the "extremists": a young couple and their three children, ages 6 years, 5 years, and 6 months. Over the next 7 years, similar scenarios recurred as the military junta that ruled Argentina killed thousands of citizens it perceived as threats. This "Dirty War," as it came to be known, finally ended in 1983 with the collapse of the military junta and the election of a democratic government. The new leaders opened previously sealed records that confirmed what many had already suspected—more than 200 children had survived the bloodshed and had in fact been kidnapped and placed with families that supported the junta.

Dr. Mary-Claire King was at the University of California at Berkeley when she was enlisted to help in the effort to return the children to the surviving members of their biological families. Dr. King and others recognized that DNA technology could be used for this important humanitarian cause. Blood and tissue samples from one individual can be distinguished from those of another by comparing certain DNA sequences. These same principles can also be used to show that a particular child is the progeny of a given set of parents. Because a person has two copies of each chromosome—one inherited from each parent—half of a child's DNA will represent maternal sequences and the other half will represent paternal sequences. The case of the Argentine children was complicated, however, because most of the parents were dead or missing. Often, the only surviving relatives were aunts and grandmothers, and it is difficult to use chromosomal DNA to show genetic relatedness between a child and such relatives. Dr. King decided to investigate mitochondrial DNA (mtDNA). This organelle DNA, unlike chromosomal DNA, is inherited only from the mother. A child will have the same nucleotide sequence of mtDNA as his or her siblings, the mother and her siblings, and the maternal grandmother.

By comparing the nucleotide sequences of mtDNA in different individuals, Dr. King was able to locate key positions that varied extensively among unrelated people, but were similar in maternal relatives. Dr. King's technique, developed out of a desire to help reunite families victimized by war, has now found many uses. Today her lab, now at the University of Washington, still uses molecular biology techniques for humanitarian efforts, identifying the remains of victims of massacres around the world.

Biotechnology uses microbiological and biochemical techniques to solve practical problems and make useful products. In the past, this often meant labor-intensive searches for naturally occurring mutants that had desirable characteristics. Today, various DNA-based technologies make it possible to genetically alter organisms to give them more useful traits. In fact, biotechnology is now nearly synonymous with **genetic engineering,** the process of deliberately altering an organism's genetic information using in vitro techniques.

Table 9.1 summarizes the applications of the DNA-based technologies described in this chapter. The information and innovations these have generated affect society in numerous ways—from agricultural practices and medical diagnoses to evidence used in courtrooms.

9.1 ■ Fundamental Tools Used in Biotechnology

Learning Outcome

1. Describe the role of restriction enzymes, gel electrophoresis, and CRISPR in biotechnology.

To appreciate the applications of biotechnology, it helps to understand some of the basic components of a molecular biologist's "tool kit." As we describe these, remember that diagrams typically focus on only one or a few DNA molecules to illustrate what is happening at a molecular level. In reality, scientists are often working with millions of molecules.

Restriction Enzymes

Restriction enzymes are a powerful tool in biotechnology because they allow scientists to easily cut DNA in vitro, in a predictable and controlled manner. Recall that these enzymes are produced by bacteria as a mechanism to destroy invading DNA, particularly phage genomes (see figure 8.29).

A restriction enzyme cuts each strand of DNA within or near a specific sequence called a **recognition sequence** (**figure 9.1a**). This action digests the DNA to generate **restriction fragments.** Researchers use restriction enzymes not only to cut DNA, but also to create **recombinant DNA molecules,** which are molecules made by joining DNA from two different sources. The restriction enzymes commonly used produce a staggered cut in the recognition sequence, resulting in ends with short overhangs of usually four nucleotides (figure 9.1b). The overhangs are called "sticky" ends, or cohesive ends, because they will **anneal** (form base pairs) with one another. Any two complementary cohesive ends can anneal, even those from two different organisms. The enzyme **DNA ligase** is used to form a covalent bond between adjacent nucleotides, joining the two molecules. Thus, if restriction enzymes are viewed as scissors that cut DNA into fragments, then DNA ligase is the glue that pastes the fragments together.

⏮ restriction enzymes, ⏮ DNA ligase

Each restriction enzyme recognizes a specific 4- to 6-base-pair nucleotide sequence (**table 9.2**). The sequences are typically palindromes, meaning they are the same on both strands when read in the 5′ to 3′ direction. The name of a particular restriction enzyme represents the bacterium from which it was first isolated. The first letter is the first letter of the genus name, and the next two are the first letters of the species name. Other numbers or letters indicate the strain and order of discovery. For example, EcoRI is from *E. coli* strain RY13.

DNA Gel Electrophoresis

DNA gel electrophoresis separates DNA fragments by size. This allows researchers to determine the

TABLE 9.1 Applications of DNA-Based Technologies

Technology	Application
Genetic engineering	Genetically engineered microorganisms are used to produce medically and commercially valuable proteins, to produce specific DNA sequences, and as a tool for studying gene function and regulation.
DNA sequencing	Once the nucleotide sequence of a DNA segment has been determined, the information can be used to decipher the amino acid sequence of the encoded proteins. The nucleotide sequence can also be compared to that from other organisms, which gives insights into genetic relatedness.
Polymerase chain reaction (PCR)	The presence of a specific segment of DNA can be detected, and the size determined, in only a matter of hours. This can be used in diagnosis of an infectious disease if DNA specific to a pathogen can be amplified (repeatedly duplicated). PCR is also used to type ("fingerprint") DNA for forensic evidence. Variations of PCR are used to study gene expression.
Probe technologies	Colony blots are used to detect colonies that contain a specific DNA sequence; fluorescence in situ hybridization (FISH) is used to identify cells directly in a specimen; DNA microarrays are used to study gene expression.

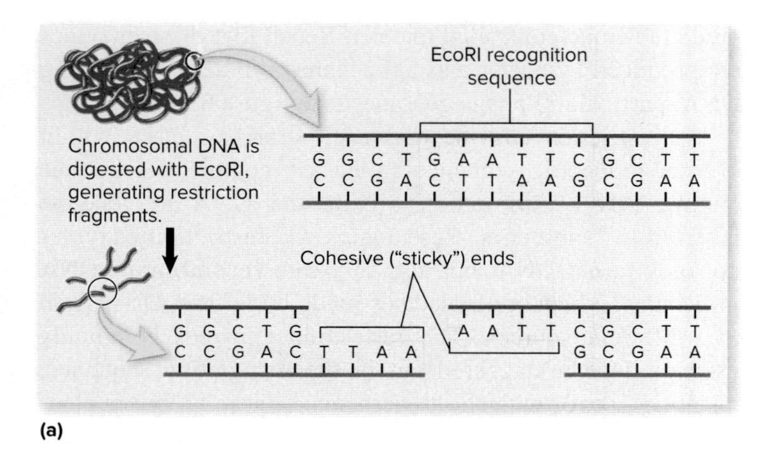

(a)

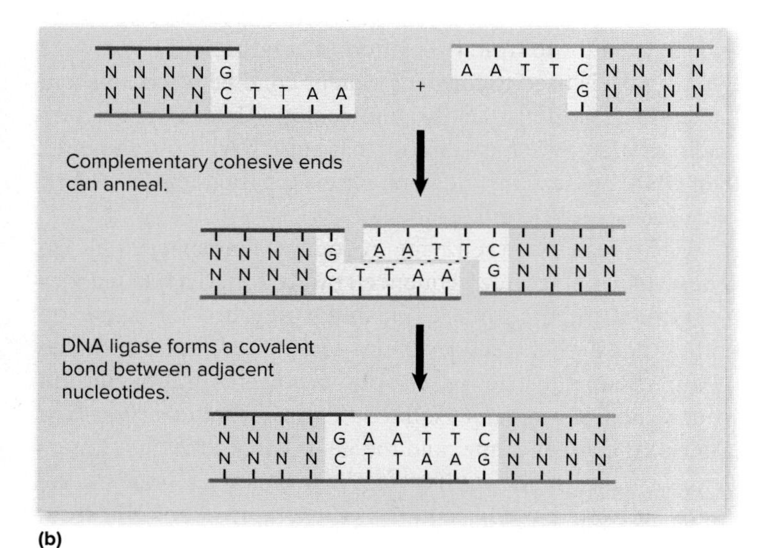

(b)

FIGURE 9.1 Action of Restriction Enzymes (a) In vitro digestion of DNA with a restriction enzyme generates restriction fragments. **(b)** Fragments that have complementary cohesive ends can anneal, regardless of their original source (N = nucleotide, meaning that any of the four nucleobases could be in that position as long as base-pairing rules are followed).

? How do restriction enzymes make it easier for scientists to create recombinant DNA molecules?

fragment sizes generated as a result of digesting a DNA sample with restriction enzymes, or to isolate certain DNA fragments.

The DNA samples are added to wells positioned near one end of a slab of a gelatin-like substance—either agarose (a highly purified form of agar) or polyacrylamide—that has been placed in an apparatus containing a buffer (an electrically conductive solution) (**figure 9.2**). A size standard (a mixture of DNA fragments of known sizes) is added to one or more unfilled wells; this serves as a basis for comparison, allowing the researcher to determine the sizes of the various DNA fragments in the samples. An electric current is then

TABLE 9.2 Examples of Common Restriction Enzymes

Enzyme	Microbial Source	Recognition Sequence (Arrows Indicate Cleavage Sites)
AluI	*Arthrobacter luteus*	↓ 5′ A G C T 3′ 3′ T C G A 5′ ↑
BamHI	*Bacillus amyloliquefaciens* H	↓ 5′ G G A T C C 3′ 3′ C C T A G G 5′ ↑
EcoRI	*Escherichia coli* RY13	↓ 5′ G A A T T C 3′ 3′ C T T A A G 5′ ↑

run through the gel. DNA is negatively charged, so the fragments move toward the positively charged electrode, which is positioned at the end of the slab opposite the wells. Short fragments of DNA move easily through the gel matrix, and so migrate quickly through the gel. Larger fragments of DNA move more slowly through the matrix, so do not migrate as far in the gel in a given period of time and remain closer to the wells. The DNA is not visible in the gel unless it is stained. To do this, the gel can be immersed in a solution containing one of several commercially available dyes that intercalate into the DNA helix and fluoresce under UV light. Each visible band on the gel represents millions of molecules of a specific-sized fragment of DNA. The basic procedure of gel electrophoresis can also be used to separate other macromolecules, specifically RNA and proteins, according to their size. ⏮ intercalating agents

CRISPR/Cas9

Scientists searching for ways to modify genomes in vivo have a new tool, thanks to the bacterial **CRISPR systems;** recall that bacterial cells use these to recognize and destroy invading DNA. What makes the CRISPR systems so useful in the biotechnologist's tool kit is that a Cas nuclease (a DNA-cutting enzyme) called Cas9 has been genetically modified so that it can be used to locate and alter very specific sites in a DNA molecule. The modified Cas9 and another required component can be introduced into a cell as a complete complex, or the genes for the components can be moved in. ⏮ CRISPR systems

One modified version of Cas9 can be used for **gene editing,** a process that introduces targeted changes into the existing

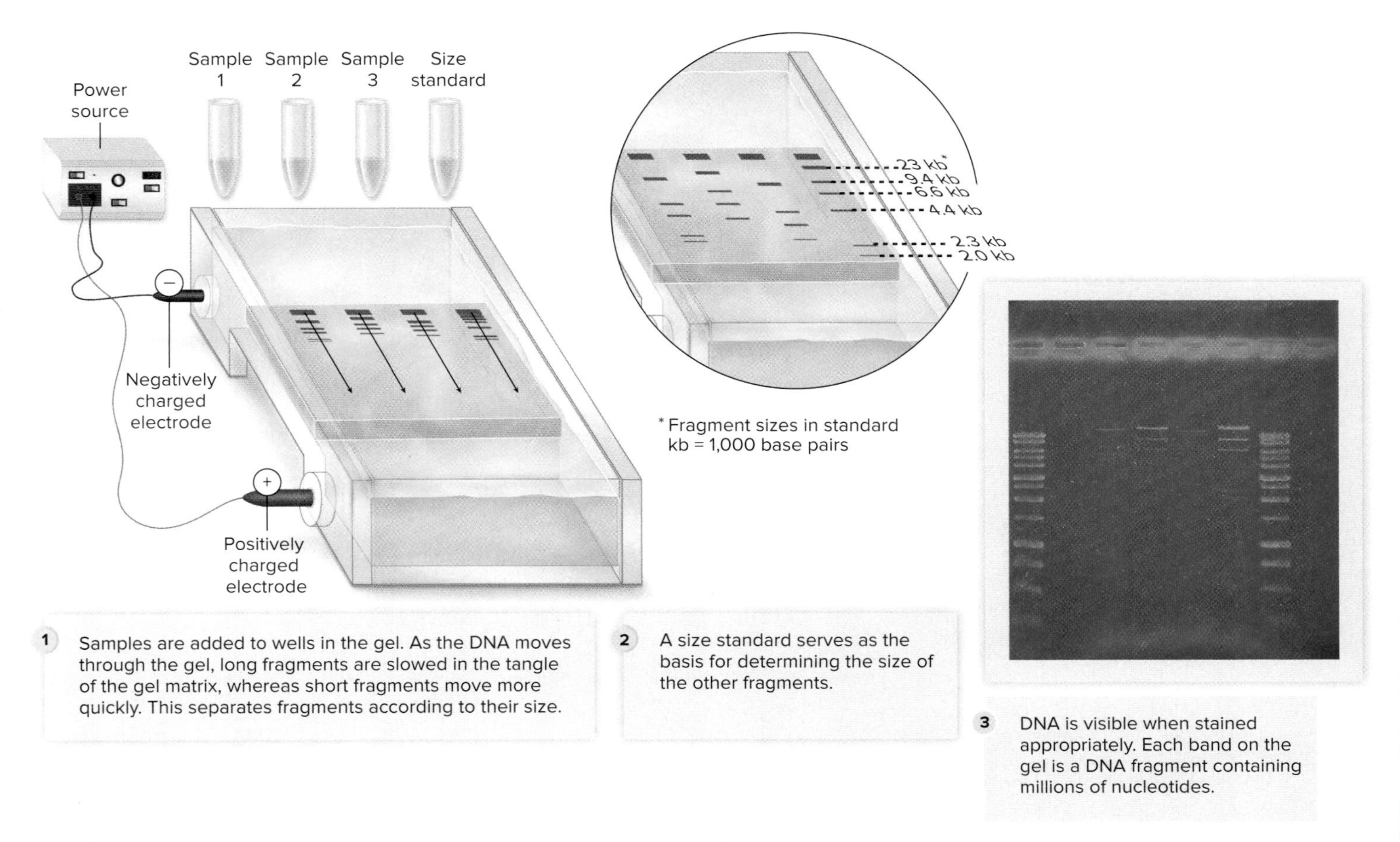

FIGURE 9.2 DNA Gel Electrophoresis Gel electrophoresis separates DNA fragments according to size. DNA in the gel is visible when stained with a dye that fluoresces under UV light.

How does gel electrophoresis separate DNA fragments according to size?

nucleotide sequence of a cell's genome (**figure 9.3a**). Like the typical Cas9 nuclease, the modified form uses a piece of single-stranded RNA as a guide to recognize a particular DNA sequence within a cell. The nuclease then functions as molecular scissors, making a double-stranded cut in that DNA. Scientists can control the precise location at which the cut occurs simply by synthesizing a specific RNA molecule to act as the guide RNA. Once the cut is made, the nucleotide sequence at that site can be changed, using processes that manipulate the cell's normal DNA repair mechanisms.

Modified forms of the Cas9 nuclease that do not cut DNA are used in studies of gene function. These inactivated forms are called "dead" Cas9, or dCas9, to indicate that they lack the nuclease function. Like the other Cas9 systems, however, they use a guide RNA to locate and bind to a specific DNA sequence within a cell. One version of dCas9 binds DNA and physically blocks RNA polymerase from transcribing a gene, thus turning the gene off (see figure 9.3b). Other versions have specific molecules attached, allowing those molecules to be delivered to certain chromosomal locations (see figure 9.3c). For example, dCas9 that carries an activator can turn a gene on when it binds DNA upstream of a promoter. Another can add methyl groups to DNA, allowing researchers to study the effects of DNA modification. A dCas9 has also been engineered to carry a fluorescent marker, allowing researchers to observe the location of the gene within a cell.

MicroAssessment 9.1

Restriction enzymes recognize specific nucleotide sequences, and then cut the DNA at or near those sequences, generating restriction fragments. Gel electrophoresis is used to separate the DNA fragments according to their size. CRISPR is used to modify genomes and study gene function.

1. What is the importance of sticky ends in genetic engineering?
2. How does gel electrophoresis separate different-sized DNA fragments?
3. What should a restriction enzyme isolated from *Staphylococcus aureus* strain 3A be called?

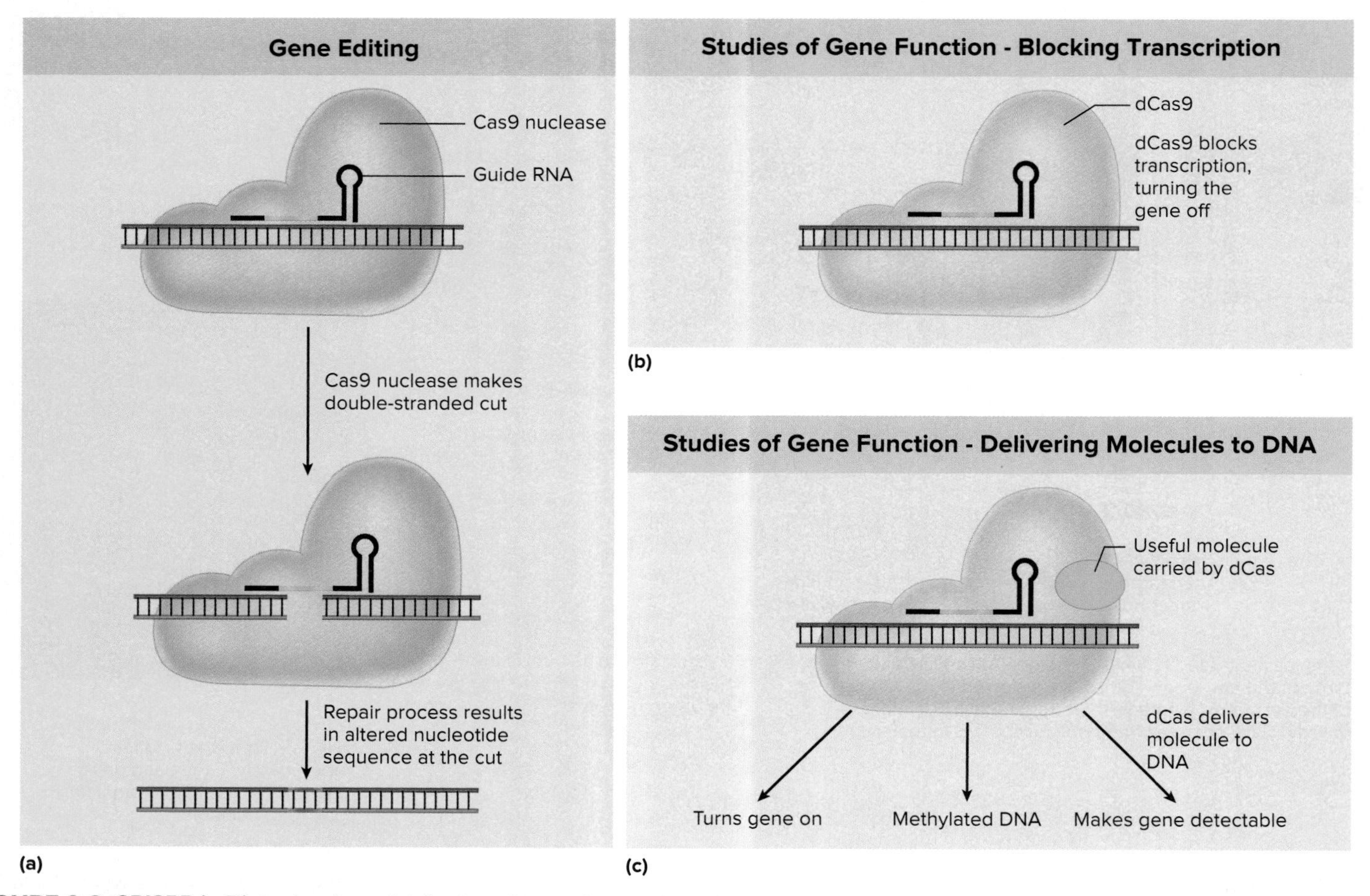

FIGURE 9.3 CRISPR in Biotechnology (a) Cas9 nuclease, along with guide RNA, can be used to change a specific DNA sequence within a cell. **(b)** dCas9 can turn off a gene's expression by blocking transcription. **(c)** Versions of dCas9 that carry certain molecules can be used to deliver those molecules to specific chromosomal locations.

? How is dCas9 different from Cas9?

9.2 ■ Genetic Engineering

Learning Outcomes

2. Describe the applications of genetic engineering.
3. Outline the steps used to genetically engineer bacteria.
4. Describe some concerns regarding DNA technologies.

Genetic engineering brought biotechnology into a new era by allowing microorganisms to be altered for medical, industrial, and research uses (**table 9.3**). Plants and animals can now be genetically engineered as well, creating what are commonly referred to as genetically modified organisms (GMOs). Technically, however, a GMO is any organism that has been genetically engineered, including bacteria and yeasts.

Genetically Engineered Bacteria

Genetically engineered bacteria have a variety of uses, including protein production, DNA production, and serving as a tool for research. In fact, much of the information described in this textbook was revealed through research using genetically engineered bacteria.

Although CRISPR/Cas9 technologies can now be used to genetically engineer bacteria, most genetic engineering up to this point has relied on **DNA cloning.** This process involves isolating DNA from one organism, using a restriction enzyme to cut that DNA, and then transferring a fragment of that DNA into a different organism (**figure 9.4**). As part of the process, the transferred DNA must replicate in the recipient in order to be passed to progeny, thereby generating a cell population that carries copies of the DNA fragment (see figure 8.19). Most DNA fragments, however, will not contain an origin of replication and therefore will not replicate independently in a cell. In addition, they generally have no similarity to chromosomal sequences in the cell, so they will not integrate into the chromosome by homologous recombination. To generate a cloned nucleotide sequence that will be replicated in a cell, the DNA fragment can be inserted into a plasmid or other independently replicating DNA molecule to form a recombinant molecule. The DNA molecule used as a carrier of the

TABLE 9.3 Some Applications of Genetic Engineering

Example	Use
PROTEIN PRODUCTION	
Pharmaceutical Proteins	
Alpha interferon	Treating cancer and viral infections
Beta interferon	Treating multiple sclerosis
Deoxyribonuclease	Treating cystic fibrosis
Erythropoietin	Treating some types of anemia
Factor VIII	Treating hemophilia
Gamma interferon	Treating cancer
Glucocerebrosidase	Treating Gaucher disease
Growth hormone	Treating dwarfism
Insulin	Treating diabetes
Platelet-derived growth factor	Treating foot ulcers in diabetics
Streptokinase	Dissolving blood clots
Tissue plasminogen activator	Dissolving blood clots
Vaccines	
Foot-and-mouth disease	Preventing foot-and-mouth disease in animals
Hepatitis B	Preventing hepatitis
HPV	Preventing cervical cancer
Other Proteins	
Bovine somatotropin	Increasing milk production in cows
Chymosin	Cheese-making
Restriction enzymes	Cutting DNA into fragments
DNA PRODUCTION	
DNA for study	Determining nucleotide sequences; obtaining DNA probes
RESEARCHING GENE FUNCTION AND REGULATION	
Creating gene fusions	Studying the conditions that affect gene activity
TRANSGENIC PLANTS	
Herbicide-resistant plants	Modified plants are not killed by biodegradable herbicides used to kill weeds
Pest-resistant plants	Modified plants resist certain insects
Plants with improved nutritional value	Modified rice produces vitamin A and iron

cloned DNA is called a **vector** and is commonly a plasmid that has been genetically modified. DNA that has been incorporated into the vector is called an insert. ⏮ **origin of replication,** ⏮ **homologous recombination**

If a gene coding for a valuable protein is inserted into a high-copy-number vector, a bacterium that carries the recombinant molecule can potentially make large amounts of that

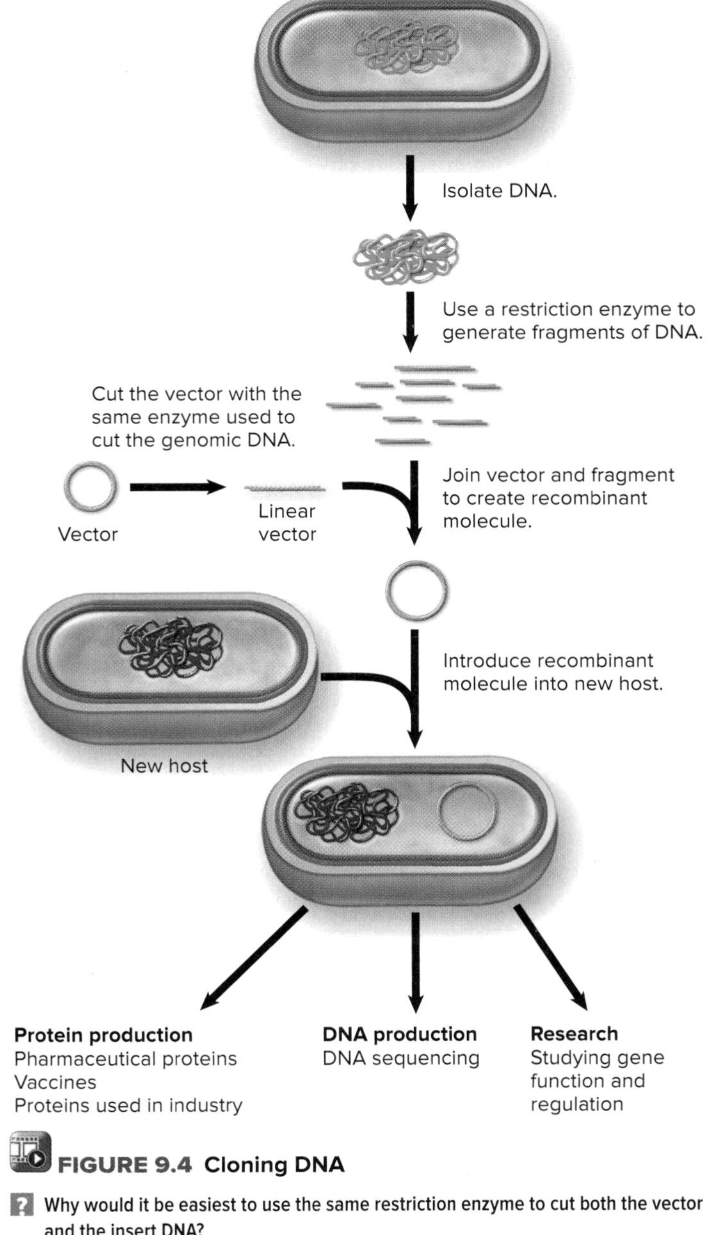

FIGURE 9.4 Cloning DNA

Why would it be easiest to use the same restriction enzyme to cut both the vector and the insert DNA?

protein. This is because each gene copy can be transcribed and translated. ⏮ **high-copy-number plasmid**

Protein Production

A number of different pharmaceutical proteins are now produced by genetically engineered microorganisms. In the past, these proteins were extracted from live animal or cadaver tissues, which made them expensive and limited in supply. Human insulin, used in treating diabetes, was one of the first important pharmaceutical proteins to be produced through genetic engineering. The original commercial product was extracted from pancreatic glands of cattle and pigs and sometimes caused allergic reactions in people receiving it. Once the gene for human insulin was cloned into bacteria, microbes became the major source of insulin. The product is safer and more economical than the version extracted from animal tissues.

One of the most widely used proteins made by genetically engineered organisms is chymosin (rennin), an enzyme used in cheese production. It causes milk to coagulate and produces desirable changes in the cheeses as they ripen. Traditionally, chymosin was obtained from the stomachs of calves, but genetically engineered bacteria are now the main source of the enzyme. The microbial product is less expensive and more reliably available.

Genetically engineered microorganisms can be used in vaccine production. Vaccines protect against disease by exposing a person's immune system to killed or weakened forms of the pathogen, or to parts of the pathogen. Although vaccines are generally composed of whole microbes, only specific proteins are necessary to immunize (induce protection) against the disease. The genes coding for these proteins can be cloned into yeast or bacteria, allowing the organisms to produce large amounts of the protein, which can then be purified. This type of vaccine is currently used to prevent hepatitis B and cervical cancer in humans, and foot-and-mouth disease of domestic animals. ⏭ vaccines

MicroByte

The insulin yield from a 2,000-liter culture of *E. coli* cells with a cloned insulin gene is the same as 1,600 pounds of pancreatic glands!

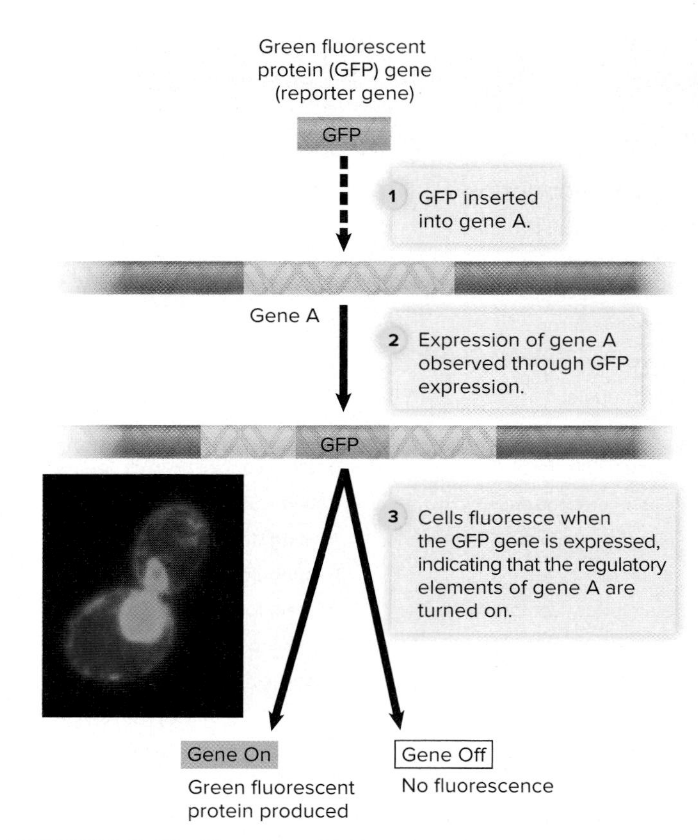

FIGURE 9.5 The Function of a Reporter Gene The regulatory elements are not shown here. Courtesy of Pamela Silver and Jason Karana, Dana Farber Cancer Institute

? **Why is the green fluorescent protein useful as a reporter gene?**

DNA Production

In many cases, researchers are interested in obtaining readily available supplies of certain DNA fragments. By cloning a segment of DNA into a well-characterized bacterium such as *E. coli,* that DNA can then be used for study and further manipulation. For example, a human cell contains an estimated 25,000 genes, whereas *E. coli* contains only 4,500 genes; thus, a human gene cloned into the bacterium on a high-copy-number vector represents a much higher percentage of the total DNA in the recipient cell than it represented in the original cell. This makes it easier to isolate the DNA as well as the gene product.

Random samples of DNA from any environment can be cloned into *E. coli* (a procedure called shotgun cloning) and then the nucleotide sequence determined. By doing this, the genomic characteristics of some of the 99% of bacteria that have not been grown in culture are now being studied. Shotgun cloning is the first step in metagenomics, the study of the total genomes in a sample. ⏮ metagenomics

Research

The function and regulation of genes can be studied with experiments involving gene cloning. For example, gene regulation can be studied by creating a gene fusion—joining the gene of interest and a reporter gene (**figure 9.5**). The reporter gene encodes a product that is easy to see, such as green fluorescent protein (GFP). It is expressed only when the gene of interest to which it is joined is activated. This makes it relatively simple to detect gene expression and, in turn, determine the conditions that affect gene activity.

Genetically Engineered Eukaryotes

Multicellular organisms can also be genetically engineered. A plant or animal that carries a cloned gene is **transgenic.** CRISPR/Cas9 technologies can now be used to genetically engineer eukaryotic organisms, but most genetically modified organisms (GMOs) currently in use have been developed through older methods.

Transgenic plants are of particular interest to microbiologists because their development started with basic research studying *Agrobacterium tumefaciens;* the Ti plasmid of this bacterium has now been genetically manipulated so it can be used as a vector to deliver desirable genes to plant cells (see Focus Your Perspective 8.2). Using the Ti plasmid as a vector, corn, cotton, and potatoes have been engineered to produce a biological insecticide called Bt toxin, which is naturally produced by the bacterium *Bacillus thuringiensis*. Unlike many chemically synthesized insecticides, Bt toxin is toxic only to insects and their larvae.

Soybeans, cotton, and corn have been engineered to resist the effects of the herbicide glyphosate (Roundup). This allows growers to apply this biodegradable herbicide, which kills weeds and other non-engineered plants, in place of alternative herbicides that remain in the environment longer. The herbicide can be applied through the growing season, so the soil can be tilled less frequently, preventing erosion.

Plants with improved nutritional value are also being developed. Genes that code for the synthesis of β-carotene, a precursor of vitamin A, have been introduced into rice. The rice was also engineered to provide more dietary iron. The diet of much of the world's population is deficient in these essential nutrients, so advances such as these could profoundly affect global health.

Techniques Used to Clone DNA

An approach frequently used to clone a specific gene into a bacterium such as *E. coli* is to make a **DNA library,** a collection of clones that together contain the entire genome (**figure 9.6**). This is made by using restriction enzymes to cut the DNA of the organism being studied and then cloning the entire set of restriction fragments into a population of *E. coli* cells. Although each cell in the resulting population contains only one fragment of the genome, the entire genome is represented in the population as a whole. Once a DNA library has been prepared, colony blots (which will be described later) can be used to determine which cells contain the gene of interest.

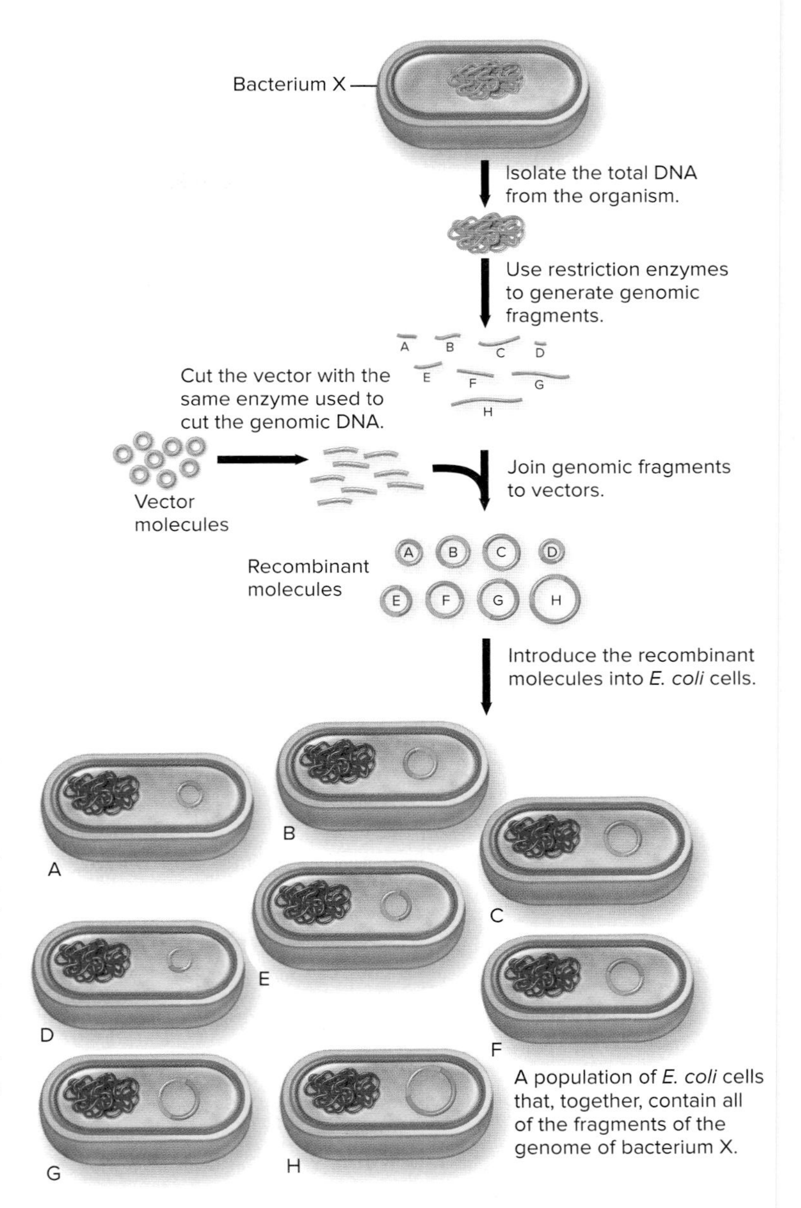

FIGURE 9.6 A DNA Library Each cell contains one fragment of a given genome.

What enzyme is used to join the genomic fragments to vector molecules?

Obtaining DNA

The first step of DNA cloning is to obtain DNA from the bacterium of interest. This is done by adding a detergent to cells in a broth culture to lyse them. As the cells burst, the relatively fragile DNA is inevitably sheared into many pieces of varying lengths.

When cloning eukaryotic genes into bacteria for protein production, a researcher must first obtain a copy of DNA that lacks introns. To do this, the researcher isolates mature mRNA from the appropriate tissue; recall that introns have been removed from these molecules. Then, a strand of DNA complementary to the mRNA is synthesized in vitro using reverse transcriptase, an enzyme from retroviruses. Reverse transcriptase copies an RNA template to make a single-stranded piece of DNA. That strand of DNA is then used as a template for synthesis of its complement, creating double-stranded DNA. The resulting copy of DNA, or **cDNA,** encodes the same protein as the original DNA but lacks introns (**figure 9.7**). ◂◂ introns, ▸▸ reverse transcriptase

Generating a Recombinant DNA Molecule

As described earlier, restriction enzymes and DNA ligase are used to create recombinant molecules. The **vector,** usually a modified plasmid or bacteriophage, has an origin of replication and functions as a carrier of the cloned DNA (**figure 9.8**). The vector must have at least one restriction enzyme recognition site. This allows the circular vector to be cut, so that the DNA of interest (called an insert) can be joined to it. Many vectors have been engineered to contain a multiple-cloning site, which is a short sequence that has the recognition sequences of several different restriction enzymes. The value of a multiple-cloning site is its versatility—it can be

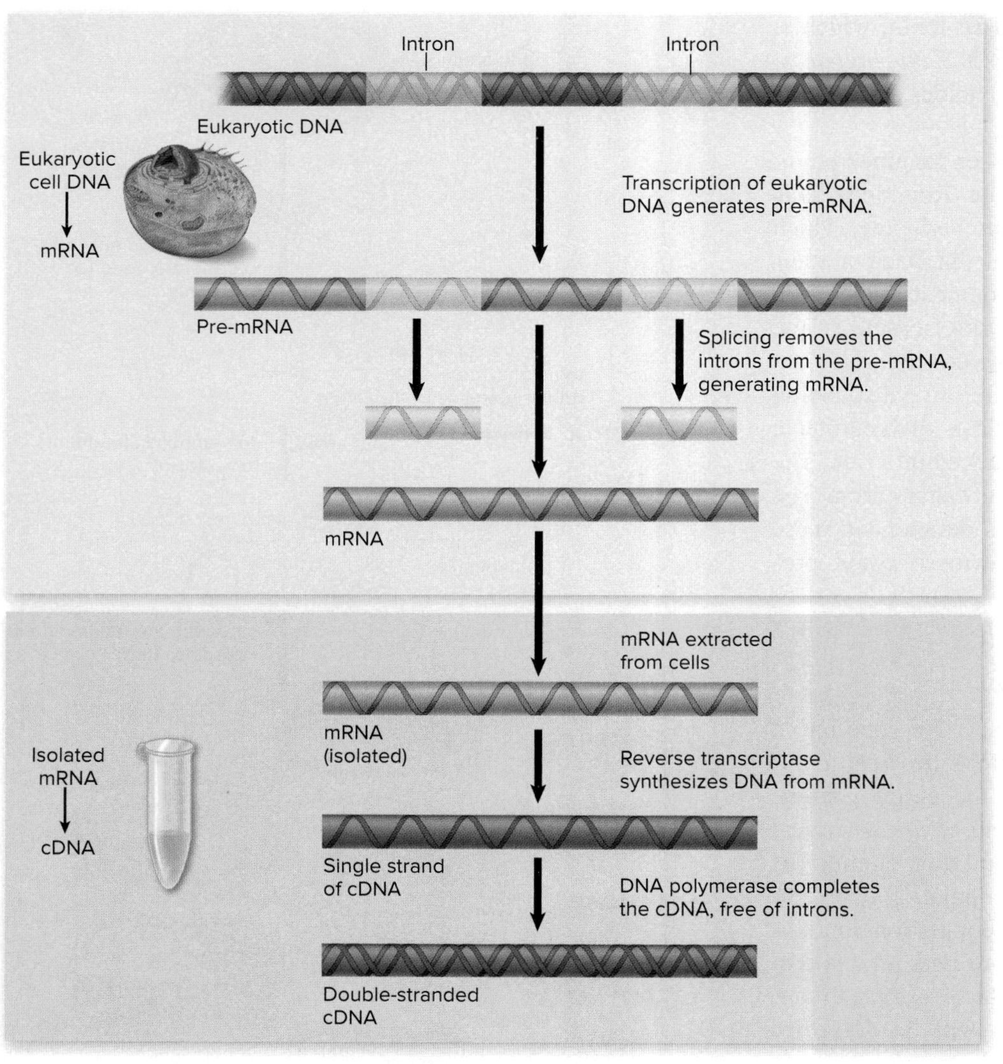

FIGURE 9.7 Making cDNA from Eukaryotic mRNA The cDNA encodes the same protein as the original DNA but lacks introns.

What enzyme is used to make cDNA from pre-mRNA?

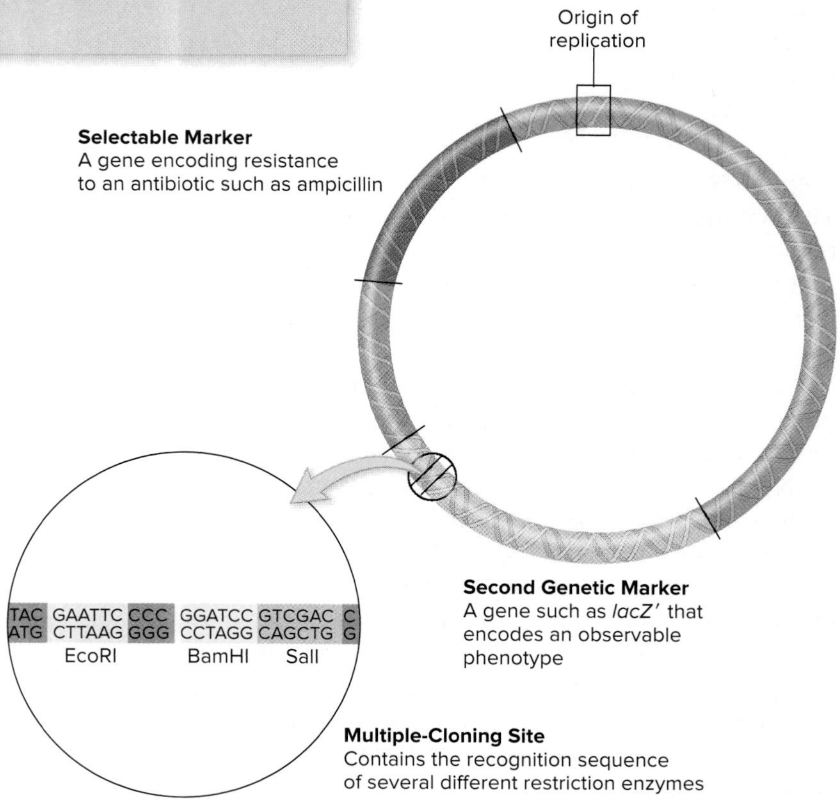

Origin of replication

Selectable Marker
A gene encoding resistance to an antibiotic such as ampicillin

TAC GAATTC CCC GGATCC GTCGAC C
ATG CTTAAG GGG CCTAGG CAGCTG G
EcoRI BamHI SalI

Second Genetic Marker
A gene such as *lacZ'* that encodes an observable phenotype

Multiple-Cloning Site
Contains the recognition sequence of several different restriction enzymes

FIGURE 9.8 Typical Properties of an Ideal Vector Most vectors have an origin of replication, a selectable marker, and a multiple-cloning site into which the DNA of interest can be inserted. The multiple-cloning site is usually situated within a second genetic marker and allows the researcher to identify cells containing a recombinant vector.

When an insert has been successfully joined with a vector, will the selectable marker still be functional? Will the second genetic marker still be functional?

used for cloning an insert obtained by digesting the DNA with any of a number of different restriction enzymes.

Vectors typically have a **selectable marker,** which is usually a gene that encodes resistance to an antibiotic such as ampicillin (see figure 9.8). The selectable marker allows the researcher to eliminate any cells that have not taken up molecules containing vector sequences, which is important because even under ideal conditions, most cells in a population do not take up DNA. Cells that have taken up molecules that contain vector sequences are able to grow on a selective medium that contains the antibiotic. ⏭ ampicillin

Most vectors have a second genetic marker that will be made non-functional if an insert is in the multiple-cloning site (see figure 9.8). This disruption of the marker's function, called **insertional inactivation,** allows researchers to distinguish colonies of cells that contain recombinant plasmids from colonies of cells containing a vector lacking an insert. Being able to distinguish the two is important because both can grow on the selective medium and will likely be present. The vector lacking an insert is an unwanted product that forms when the two ends of the cut vector anneal (stick) to each other, regenerating the circular vector.

Various vectors are available for cloning eukaryotic DNA into bacterial cells, and the choice depends largely on the purpose of the procedure. If the goal is to produce the encoded protein, then a vector designed to optimize transcription and translation of the insert DNA is used. To create a DNA library of a human or other eukaryotic genome, a vector that can carry a large insert is generally used.

Obtaining Cells That Contain Recombinant DNA Molecules

A common method of introducing DNA into a bacterial host is DNA-mediated transformation. *E. coli* cells are not naturally competent, however, and must be treated with certain chemicals to induce them to take up DNA. An alternative technique is to introduce the DNA by electroporation, a procedure that creates temporary pores in the cytoplasmic membrane by exposing the cells to an electric current. The bacteria are then grown on a medium that both selects for cells containing vector sequences and differentiates those carrying recombinant plasmids. ⏮ DNA-mediated transformation

A good illustration of the value of an appropriate vector in obtaining cells that have a recombinant plasmid is provided by a vector called pUC18 (**figure 9.9**). This vector's selectable marker is a gene encoding ampicillin resistance. Thus, cells that carry intact vector or a recombinant molecule will grow on media containing ampicillin. The second genetic marker is a gene called *lacZ′*, the product of which helps cleave a colorless chemical, X-gal, to form a blue compound. Cells that carry intact vector (circular vector without an insert) have a functional *lacZ′* gene and therefore form blue colonies when grown on X-gal, whereas those that contain a recombinant molecule form white colonies. When transformed *E. coli* cells are grown on a medium containing both ampicillin and X-gal, cells that form white colonies (rather than blue) should contain a recombinant molecule; these are then further characterized to determine if they carry the gene of interest.

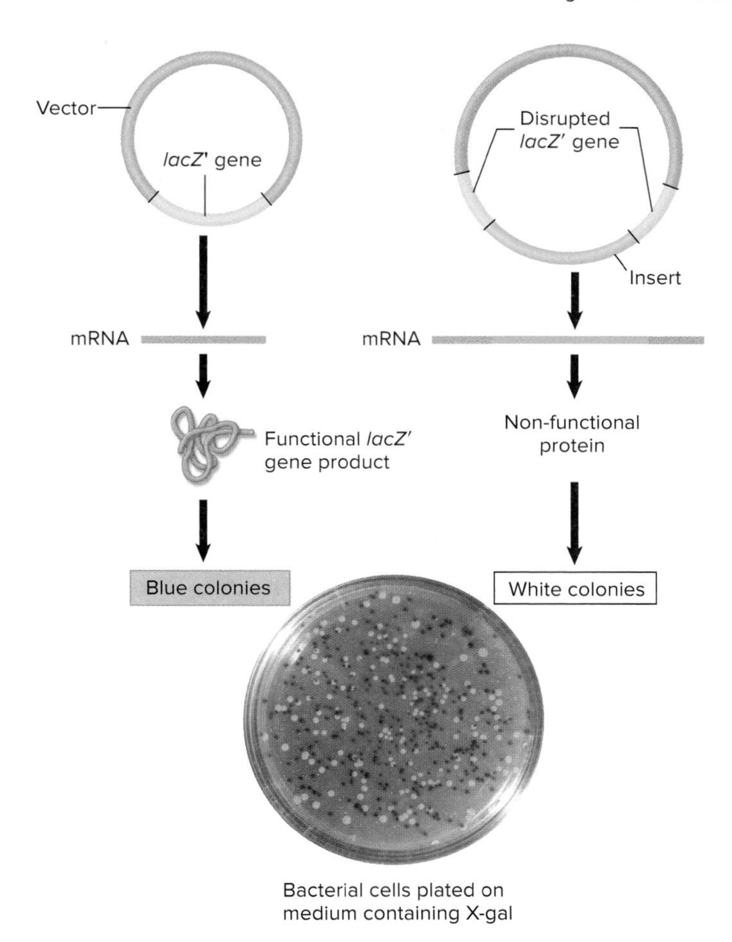

FIGURE 9.9 The Function of the *lacZ′* Gene in a Vector
The *lacZ′* gene is used to differentiate cells that contain recombinant plasmid from those that contain vector alone.

? What color colonies will cells that contain a recombinant plasmid form?

Concerns Regarding Genetic Engineering

Any new technology should be tested to ensure it is safe and effective. When recombinant DNA technologies first made gene cloning possible over four decades ago, there were many concerns about their use and possible abuse. Even the scientists who developed the technologies were concerned about potential dangers. In response, the National Institutes of Health (NIH) formed the Recombinant DNA Advisory Committee (RAC) to develop guidelines for conducting research involving recombinant DNA techniques and gene cloning. Today, we are enjoying the benefits of many of those technologies, as can be seen by the list of commercially available products in table 9.3. It should be noted, however, that although the technologies can be used to make life-saving products, there is no guarantee they will not be used for malicious purposes. Today, the idea that new infectious agents are being created for the purpose of bioterrorism is a disturbing possibility.

The development of genetically modified organisms (GMOs) has also raised concerns—some logical and others not. For example, some people have expressed fear over the fact that GM foods "contain DNA." Considering that DNA is consumed routinely in every plant and animal we eat, this is obviously an irrational concern. Others worry that unanticipated allergens could be introduced into food products, posing a health threat. To address this issue, the FDA has implemented strict guidelines, such as requiring producers to show that GM products intended for human consumption do not cause unexpected allergic reactions. Despite precautions, GM corn not approved for human consumption has been detected in products such as tortilla chips, and this causes continued concern about the effectiveness of regulatory control. Another concern about GM products is their possible unintended effects on the environment. Some laboratory studies have shown that pollen from plants genetically modified to produce Bt toxin can inadvertently kill monarch butterflies; other studies, however, have refuted the evidence. In addition, there are indications that herbicide-resistance genes can be transferred to weeds, decreasing the usefulness of the herbicide. As with any new technology, the impact of GMOs will need to be carefully scrutinized to avoid negative consequences. The new capabilities that come with CRISPR/Cas9 technologies bring additional concerns because the genetic changes may be more difficult to detect. It is important that ongoing discussions about these complex issues continue as the technologies advance.

MicroAssessment 9.2

Genetically engineered organisms can be used to produce medically and commercially important proteins, vaccines, and DNA for study; they are also used as research tools. Plants can be genetically engineered to be pest resistant, to be herbicide resistant, or to produce certain vitamins. To clone a eukaryotic gene into bacteria, introns can be removed from the DNA to create cDNA. Most vectors used in cloning have an origin of replication, a selectable marker, a multiple-cloning site, and a second genetic marker. Concerns about genetic engineering include possible introduction of allergens into food products and unintended effects on the environment.

4. Name a disease that is treated with a protein produced by a genetically engineered microorganism.
5. Explain the role of ampicillin and the *lacZ′* gene in cloning.
6. What would happen if a bacterial cell took up a fragment of foreign DNA that was not inserted into a vector?

9.3 ■ DNA Sequencing

Learning Outcomes

5. Describe two applications of DNA sequencing.
6. Describe the automated dideoxy chain termination method of DNA sequencing, and compare its use to that of high-throughput methods.

DNA sequencing is the process of determining the order of the nucleotides in a DNA molecule. The **Human Genome Project,** the completed undertaking to sequence the human genome, resulted in highly automated and efficient techniques. These allowed scientists to more readily determine the genomic sequence of other organisms, including both prokaryotes and eukaryotes, fueling the rapidly growing field of genomics. The resulting explosion of data gave rise to a new field— bioinformatics—that uses computer-based methods to organize and analyze the information. ◀◀ genomics, ◀◀ bioinformatics

Once the DNA sequence of a genome has been determined, the amino acid sequence of the encoded proteins can be established. This makes it possible to compare characteristics of various proteins in different organisms. Non-coding sequences and mobile genetic elements can be compared as well. DNA sequences can also be used to determine the evolutionary relatedness of organisms, a topic discussed in chapter 10. ◀◀ mobile genetic elements

Sequencing technologies have become so efficient that they are now being used in large projects such as the National Microbiome Initiative (NMI). Recall that the NMI expands the scope of microbiome research beyond that of the Human Microbiome Project (HMP) by also studying microbiomes of our surrounding environment. ◀◀ National Microbiome Initiative, ◀◀ Human Microbiome Project

MicroByte

The Genetic Information Nondiscrimination Act (GINA) protects Americans against discrimination in healthcare coverage and employment based on their genome.

DNA Sequencing Methods

A wide variety of in vitro DNA sequencing techniques are available today—ranging from the relatively simple dideoxy chain termination method to highly automated rapid methods collectively referred to as high-throughput sequencing.

Dideoxy Chain Termination Method

The **dideoxy chain termination method**—also referred to as Sanger sequencing in honor of the Nobel Prize–winning scientist who developed it—was the standard sequencing method for decades. High-throughput sequencing has now replaced it for large-scale projects, but the method is still commonly used on a smaller scale. It requires the following components:

- **Template DNA:** Single-stranded DNA from which complementary copies are synthesized. ⏮ template
- **DNA polymerase:** The enzyme that synthesizes DNA. ⏮ DNA polymerase
- **Primer:** Short DNA molecule that anneals to the complementary sequence in the single-stranded template. The primer allows the researcher to choose where synthesis starts on the template (**figure 9.10**). Recall that DNA polymerase can add nucleotides only to an existing fragment of DNA, thereby extending the length of the fragment. ⏮ primer
- **Deoxynucleotides:** The four deoxynucleotides used in DNA synthesis—dATP, dGTP, dCTP, and dTTP.
- **Dideoxynucleotides:** Derivatives of deoxynucleotides that differ only in that they lack the 3′OH group.

If only the first four components were in the reaction, full-length molecules complementary to the template DNA would always be synthesized. The dideoxynucleotides, however, function as **chain terminators.** They lack the chemical group to which subsequent nucleotides would be added. Therefore, when a dideoxynucleotide is added to a growing strand of DNA, no additional nucleotides can be incorporated. Elongation of that strand stops (**figure 9.11**).

In a modern dideoxy sequencing reaction, DNA polymerase, template DNA, primer, the four deoxynucleotides, and a very small amount of the four dideoxynucleotides are mixed together (**figure 9.12** ①). The different dideoxynucleotides each carry a distinct fluorescent marker. When the reaction is incubated at an appropriate temperature, the primer molecules anneal to template molecules and DNA synthesis begins. ② Each nucleotide chain is elongated until a chain terminator—a dideoxynucleotide—is incorporated. Termination happens infrequently, however, because of the small amount of ddNTPs relative to dNTPs. The significant aspect is that the color of the fluorescent marker on the chain terminator can be used to determine which nucleotide was incorporated at the terminating position. ③ After the sequencing reaction, the sample is heated, which denatures the DNA. Gel electrophoresis is then used to separate the DNA fragments and determine their relative size. The conditions (pH, temperature, and gel concentration) keep the DNA single-stranded and allow fragments that differ in length by only one nucleotide to be separated. A laser is used to detect the colors of fluorescent bands, recording their intensity as a peak. The order of the colored peaks reflects the nucleotide sequence of the DNA (**figure 9.13**).

High-Throughput Sequencing Methods

The term high-throughput sequencing refers to several highly automated methods that generate huge amounts of DNA

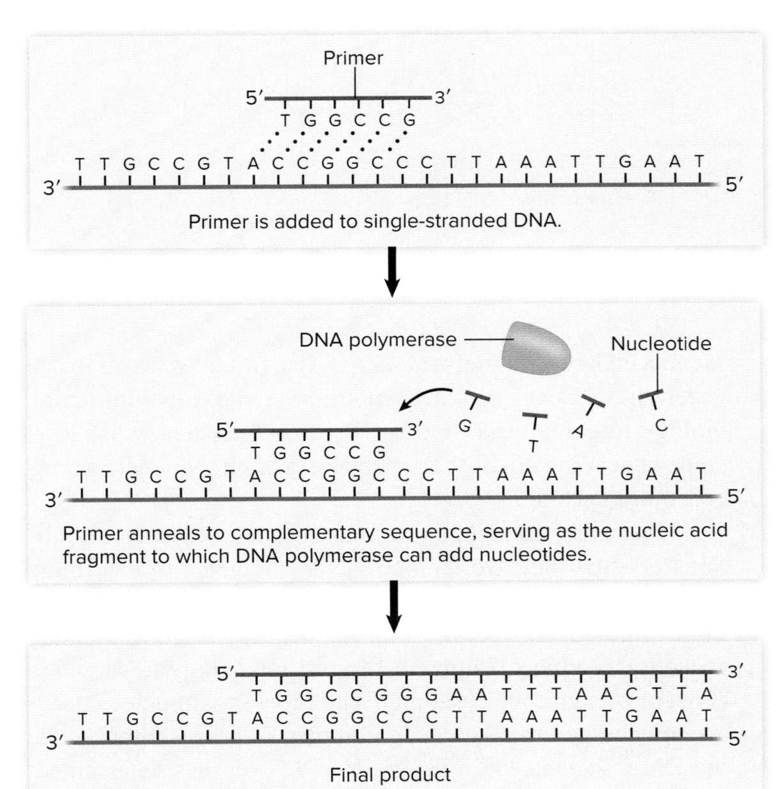

FIGURE 9.10 Primers Through appropriate primer selection, a researcher can choose the site where in vitro DNA synthesis will start.

? Why are primers required in DNA sequencing reactions?

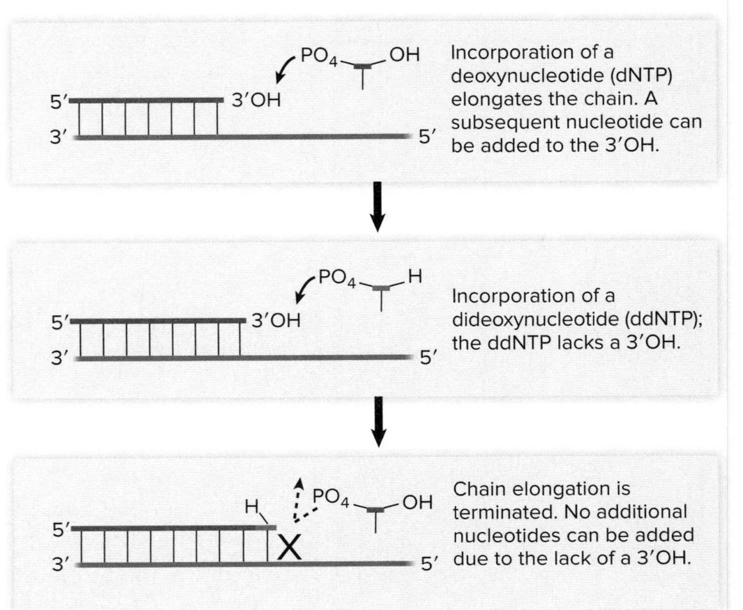

FIGURE 9.11 Chain Termination by a Dideoxynucleotide Once a dideoxynucleotide is incorporated into a growing strand, additional nucleotides cannot be added.

? How is a dideoxynucleotide different from a deoxynucleotide?

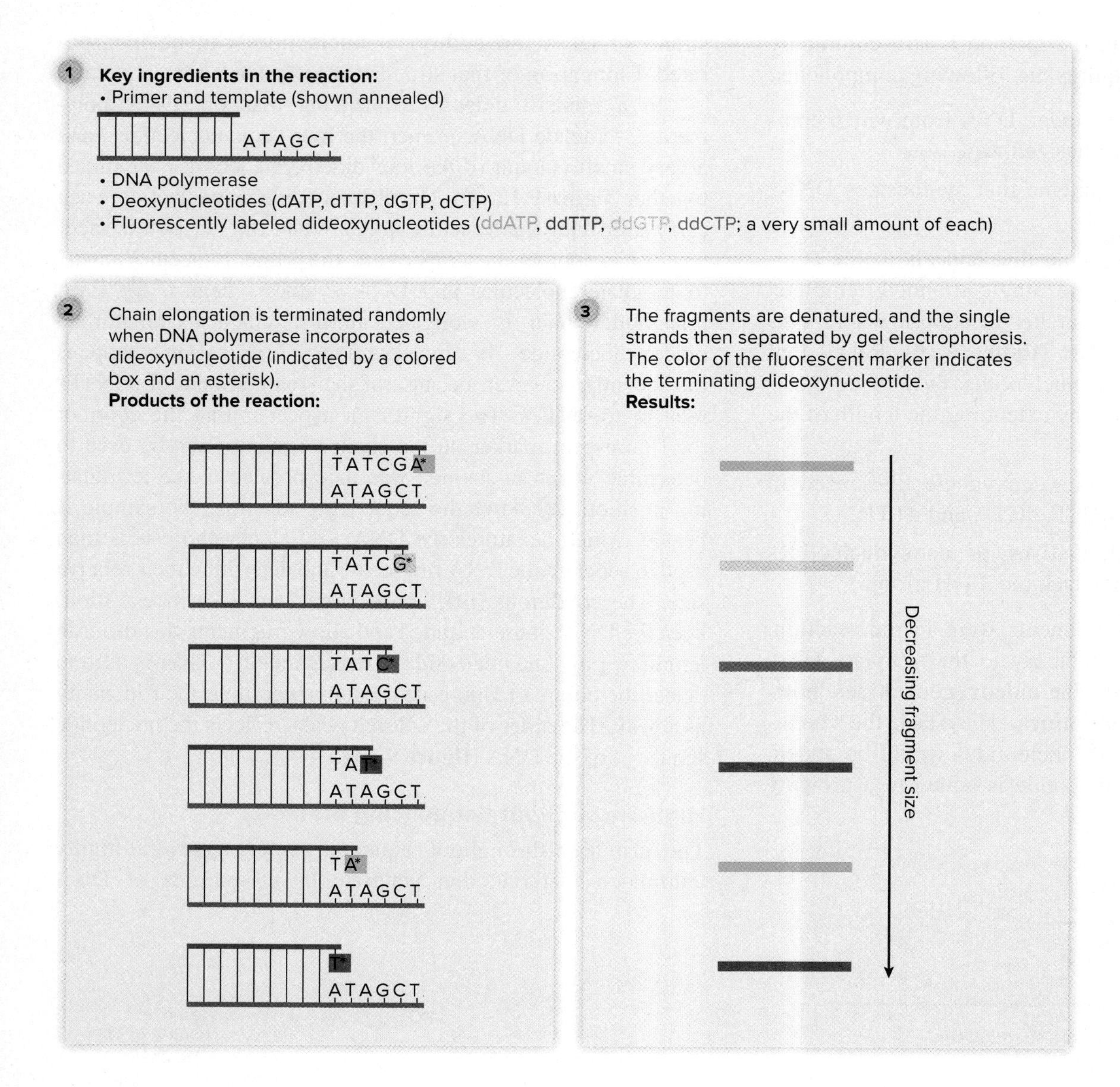

FIGURE 9.12 Dideoxy Chain Termination Method of DNA Sequencing This figure illustrates the principles of the automated method, which uses dideoxynucleotides that carry a fluorescent label.

? **What would happen if the deoxynucleotides were left out of the sequencing reaction?**

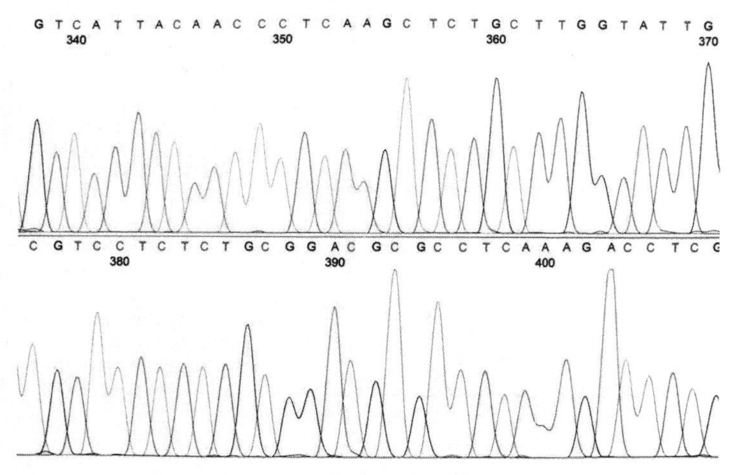

FIGURE 9.13 Results of Automated DNA Sequencing The order of the colored peaks reflects the nucleotide sequence of the DNA. ©Alan John Lander Phillips/E+/Getty Images

? **What allows the peaks to be different colors?**

sequence data relatively quickly. Also referred to as next-generation ("next-gen") sequencing, they are revolutionizing biology research because their speed and efficiency has lowered sequencing costs, thereby making large sequencing projects more feasible.

Most high-throughput sequencing methods analyze millions of small but overlapping DNA fragments to determine their sequences. Computers are then used to align and merge those data to create one long sequence—a process called sequence assembly. **Figure 9.14** uses an analogy to show the concept of sequence assembly. The process is much easier if a reference genome has been completed. For example, once the DNA sequence of one strain of *E. coli* was determined, it became easier to assemble the sequences of other strains because the likely order of the fragments was known.

A relatively new high-throughput method is nanopore sequencing, which determines the nucleotide sequence of

Data	Alignment	Result
reading th keep read ing the textbook	reading th keep read ing the textbook	keep reading the textbook

FIGURE 9.14 Analogy for DNA Sequence Assembly Here, letters and spaces are being aligned and merged to form a phrase—an analogy for the process used in DNA sequence assembly.

Why is it useful to have a reference sequence when assembling DNA sequencing data?

long fragments of DNA. This method passes a single-stranded DNA fragment in an electrically conductive solution through a microscopic pore in a membrane. As the different nucleotides in the strand move through the pore, they temporarily block the electrical current to varying degrees. The recorded pattern of disruptions reflects the nucleotide sequence of the fragment. A nanopore sequencing device was recently used to sequence microbial DNA aboard the International Space Station.

Errors are common in high-throughput sequencing, so each genome must be analyzed multiple times to increase the reliability of the results. For example, if the identity of a particular nucleotide is determined 10 times, and one time it is shown to be an "A," but the other nine times it is a "T," then it most likely is a T.

MicroAssessment 9.3

Efficient DNA sequencing methods have fueled the rapidly growing field of genomics. The dideoxy chain termination method was once the most widely used sequencing technique, but high-throughput sequencing is now used for large-scale projects.

7. How is the National Microbiome Initiative different from the Human Microbiome Project?
8. How does a ddNTP terminate DNA synthesis?
9. What would happen in the sequencing reaction if the relative concentration of a dideoxynucleotide were increased?

9.4 ■ Polymerase Chain Reaction (PCR)

Learning Outcomes

7. Describe how PCR can be used to diagnose diseases.
8. Explain how PCR can be used to exponentially amplify a select region of DNA.

The **polymerase chain reaction (PCR)** makes it possible to create more than a billion copies of a given region of DNA—often referred to as the target DNA or target sequence—in a matter of hours. Those copies, the **PCR product,** are generated in sufficient concentration to be visible when DNA fragments in the sample are separated by gel electrophoresis, stained with a fluorescent compound, and illuminated with UV light.

One important application of PCR is disease diagnosis. For example, if a person is suspected of having the sexually transmitted disease gonorrhea, PCR can be used to detect the causative agent *(Neisseria gonorrhoeae)* in a specimen from the infected site. This is done by treating the specimen to release the DNA in cells and then using PCR to amplify (increase the amount of) a selected sequence unique to *Neisseria gonorrhoeae.* Successful amplification of that target indicates that the pathogen is present (**figure 9.15**).

Variations of Conventional PCR

Since PCR was introduced, numerous variations have been developed, each with specific applications. Two of the most widely used are RT-PCR and qPCR.

RT-PCR (reverse-transcription PCR) is used to detect certain mRNA sequences in a sample. It is based on conventional PCR, but first the enzyme reverse transcriptase is used to synthesize cDNA from the mRNA template. The cDNA then serves as the template for amplification. ⏭ **reverse transcriptase**

qPCR (quantitative PCR), also called real-time PCR, is used to not only amplify the target sequence but also monitor the PCR product's accumulation. Unlike conventional PCR,

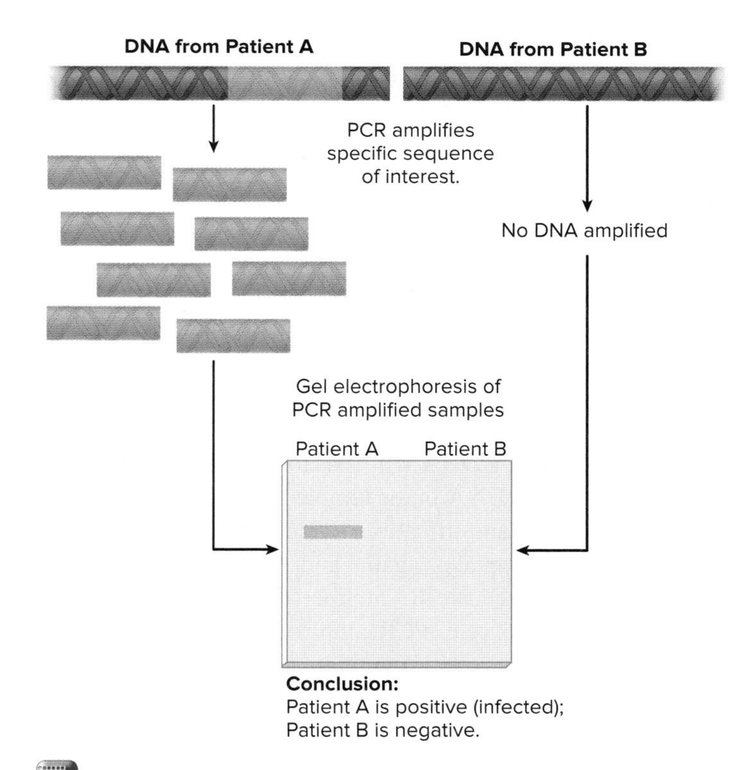

FIGURE 9.15 PCR Amplifies a Selected Sequence

How can PCR be used to diagnose an infectious disease?

qPCR includes a fluorescent marker that can be used to detect the PCR product. The amount of fluorescence increases as the PCR product accumulates, thereby allowing the researcher to (1) track the amplification in "real time," rather than using gel electrophoresis at the end; and to (2) determine the relative amount of target DNA initially in the sample, which is reflected by the amount of fluorescence generated.

The methods of RT-PCR and qPCR can be combined, allowing researchers to study changes in gene expression. Reverse transcription is used to create cDNA copies of mRNA isolated from cells grown under a given set of conditions, and then the relative amounts of certain mRNA molecules are determined using qPCR techniques.

PCR Methods

The polymerase chain reaction starts with a double-stranded DNA molecule that serves as a template from which more than a billion copies of the target sequence can be produced. The process involves DNA synthesis reactions and the following key ingredients:

- **Double-stranded DNA.** This contains the target sequence and serves as a template for DNA synthesis.
- ***Taq* polymerase.** This heat-stable DNA polymerase is from the thermophile *Thermus aquaticus.*
- **Primers.** These short DNA molecules, chosen by the researcher, determine which portion of the template DNA is amplified.
- **Deoxynucleotides.** The four deoxynucleotides used in DNA synthesis—dATP, dGTP, dCTP, and dTTP—are required to make copies of the target.

The Three-Step Amplification Cycle

PCR uses a repeating cycle consisting of three steps (**figure 9.16**). ① In the first step, the sample is heated to near-boiling (about 95°C) in order to denature the DNA. ② In the second step, the temperature is lowered (to about 50°C); within seconds, the primers anneal to their complementary sequences on the denatured target DNA. ③ In the third step, the temperature is raised to the optimal temperature of *Taq* DNA polymerase (about 70°C), allowing DNA synthesis to occur. After one three-step cycle, the target DNA in the template is duplicated.

In subsequent amplification cycles, the original DNA strands serve as templates again, and so do the newly synthesized strands. Because the number of template molecules increases with each cycle, PCR amplifies the DNA exponentially. After a single cycle of the three-step reaction, there will be two double-stranded DNA molecules for every original; after the next cycle, there will be four; after the next there will be eight; and so on.

A critical factor in PCR is *Taq* polymerase, the heat-stable DNA polymerase of *Thermus aquaticus.* This polymerase, unlike the DNA polymerase of *E. coli,* is not destroyed at the high temperature used to denature the DNA in the first step of each amplification cycle. If a heat-stable polymerase were not used, fresh polymerase would need to be added after that step in each cycle of the reaction. The discovery and characterization of *T. aquaticus* through basic research was key

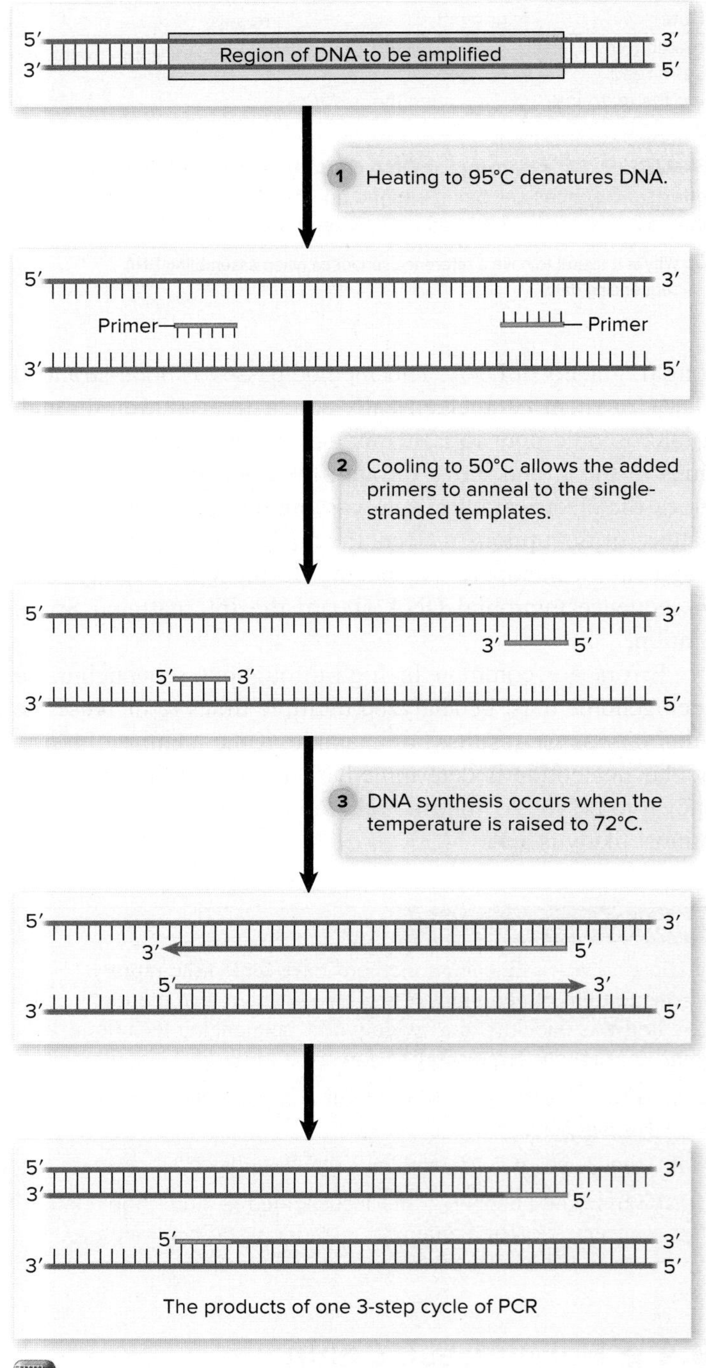

FIGURE 9.16 Steps of a Single Cycle of PCR

? **Considering that PCR can be used to make over a billion copies of target DNA in only a matter of hours, approximately how many molecules of primers must be included in the reaction?**

to developing this widely used and commercially valuable method. ◀◀ **thermophile**

Generating the PCR Product

Although the preceding description explains how PCR makes copies of DNA containing a target sequence, it does not show how fragments containing only that target sequence are generated. Recall that these amplified fragments—the PCR product—allow researchers to detect target DNA in a sample.

To understand how the PCR product is made from a long piece of DNA, you must visualize at least three cycles of PCR while also considering the exact sites to which the primers anneal to the template (**figure 9.17**). ① In the first cycle, two new strands are synthesized. Note, however, that their 5′ ends are primer DNA; the strands are shorter than the original templates but longer than the target DNA. These mid-length strands will be made whenever the original full-length molecule is used as a template, which is once per three-step cycle.

In the next cycle, the full-length molecules will again be used as templates, repeating the process just described. ② More importantly, the mid-length strands created during the first cycle will be used as templates for DNA synthesis. As before, the primers will anneal to these strands and then nucleotides will be added to the 3′ end. Elongation, however, will stop at the 5′ end of the template molecule, because DNA synthesis requires a template. Recall that the 5′ end of the template is primer DNA. Thus, whenever a mid-length strand is used as a template, a short strand that includes only the target sequence is generated. The 5′ and 3′ ends of this strand are determined by the sites to which the primers initially annealed.

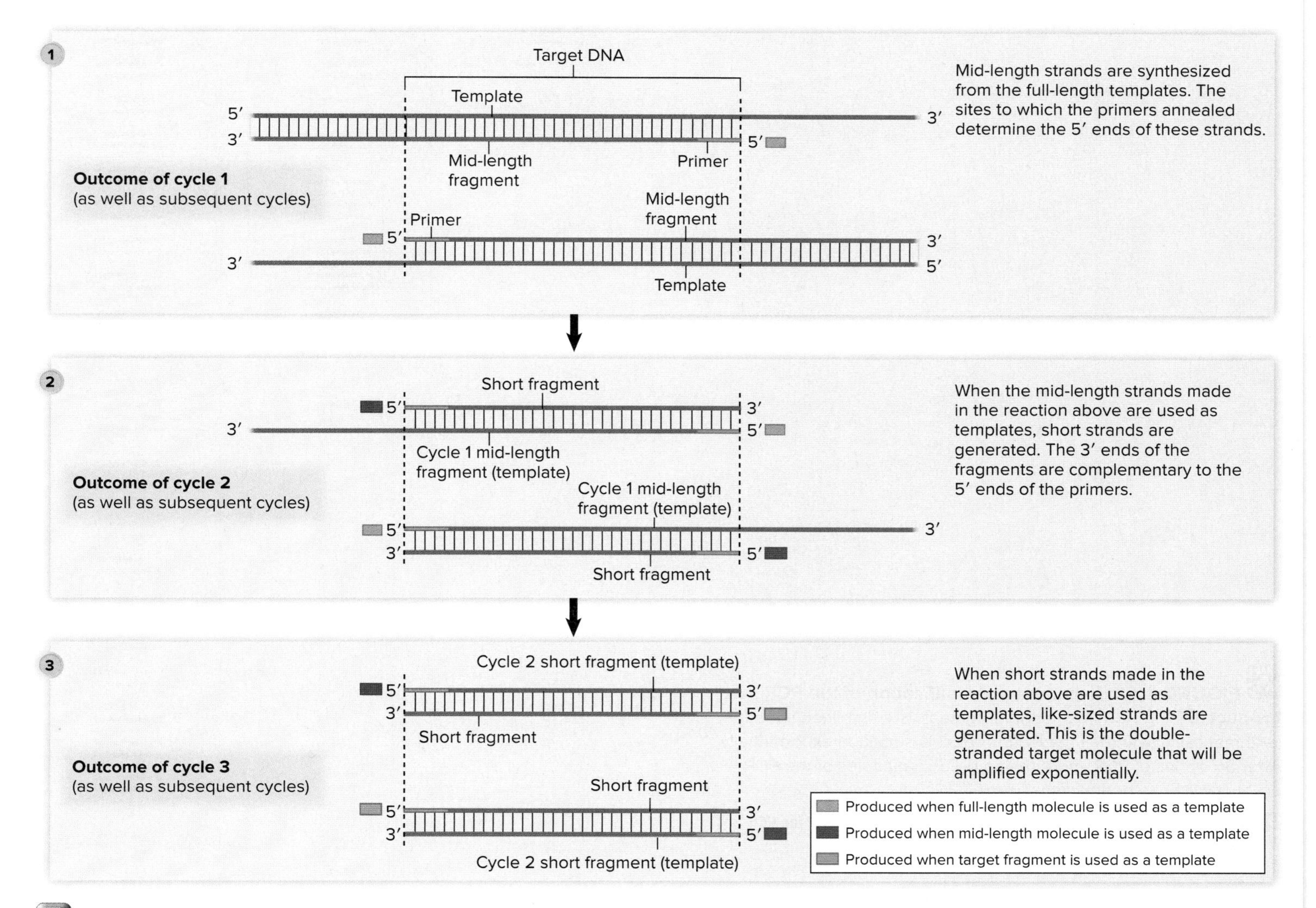

FIGURE 9.17 Generating the First Copies of the PCR Product The positions to which the primers anneal to the template dictate the size and sequence of the fragment amplified exponentially.

? **If two mid-length strands are present at the end of cycle 1, how many will be present at the end of cycle 3?**

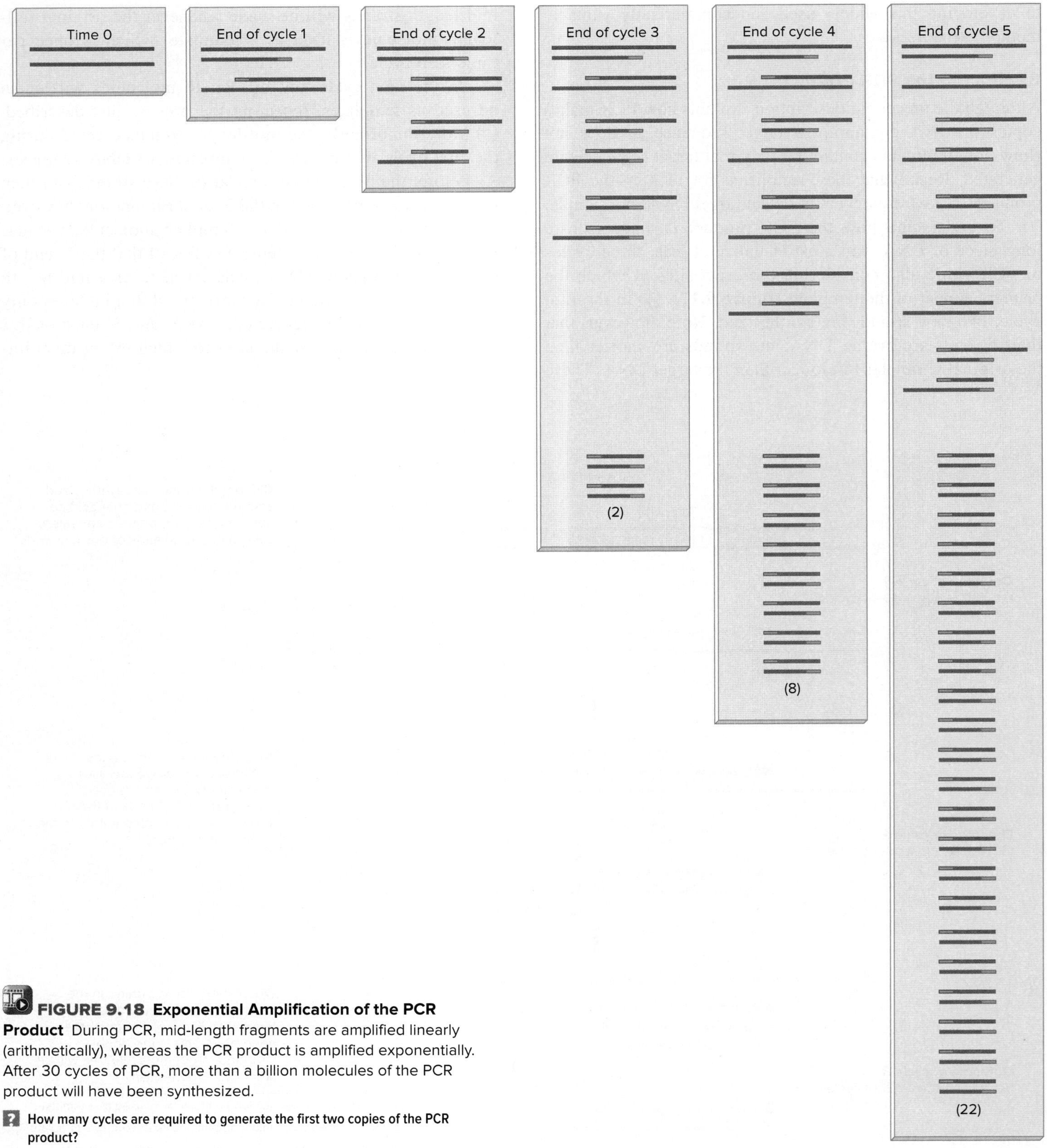

Time 0

End of cycle 1

End of cycle 2

End of cycle 3

(2)

End of cycle 4

(8)

End of cycle 5

(22)

FIGURE 9.18 Exponential Amplification of the PCR Product During PCR, mid-length fragments are amplified linearly (arithmetically), whereas the PCR product is amplified exponentially. After 30 cycles of PCR, more than a billion molecules of the PCR product will have been synthesized.

How many cycles are required to generate the first two copies of the PCR product?

In the third round of replication, the full-length and the mid-length strands again will be used as templates, repeating the processes just described. (3) The short strands generated in the preceding round will also be used as templates, however, generating the double-stranded PCR product. Continuing to follow the events in further rounds of replication will reveal that this fragment is exponentially amplified (**figure 9.18**).

FOCUS ON A CASE 9.1

After serving more than 10 years on a rape charge, a wrongfully convicted young man was released from prison when a new PCR-based technique for DNA typing (also called DNA profiling or DNA fingerprinting) indicated that he was not guilty. It showed that DNA from the semen taken from the rape victim did not match that of the accused man. Indeed, a DNA database showed a match with another man currently in prison for an unrelated rape charge.

The PCR-based method of DNA typing amplifies certain chromosomal regions that contain short tandem repeats (STRs). STRs are short nucleotide sequences (typically two to six base pairs) that repeat consecutively a variable number of times, usually within an intron or other untranslated region. For example, the sequence AATG repeats consecutively 5 to 14 times on chromosome 2 in an intron within the thyroid peroxidase gene. In one individual, there may be 9 of these STRs in one copy of that chromosome and 7 in the other, whereas another individual may have 11 and 5. This variation (called a polymorphism) makes the repeats a useful genetic marker for distinguishing individuals. The FBI now catalogs PCR-based DNA profiles from unsolved crimes and convicted violent offenders, making it easier to track or link the crimes of serial offenders. The database, called CODIS (**Com**bined **DNA** **I**ndex **S**ystem), uses the amplification pattern of 20 different STR loci (chromosomal locations). It is nearly impossible for two people to share the same number of repeats at all these loci, unless they are identical twins.

1. Why would a PCR-based method for DNA typing be valuable in forensics?
2. How is PCR-based typing used to detect STR polymorphisms?
3. How can the method analyze 20 different loci simultaneously?

Discussion

1. Results can be obtained in less than 5 hours from a sample as small as a drop of body fluid the size of a pinhead.
2. Primers that bind regions on each side of the STRs are used to determine the number of repeats (**box figure 9.1**). A fragment that contains 9 repeats, for example, will be longer than one that contains only 7. The PCR-amplified fragments can be quickly separated using a rapid type of gel electrophoresis called capillary electrophoresis. A size standard is included so that the sizes of the PCR-amplified fragments can be determined.
3. Commercially available kits contain fluorescently labeled primers that allow simultaneous amplification and subsequent recognition of each of the 20 loci. A laser detects the color of each amplified fragment as it moves out of the capillary gel, and computer analysis generates a pattern of peaks that reflect the STR profile of the DNA sample.

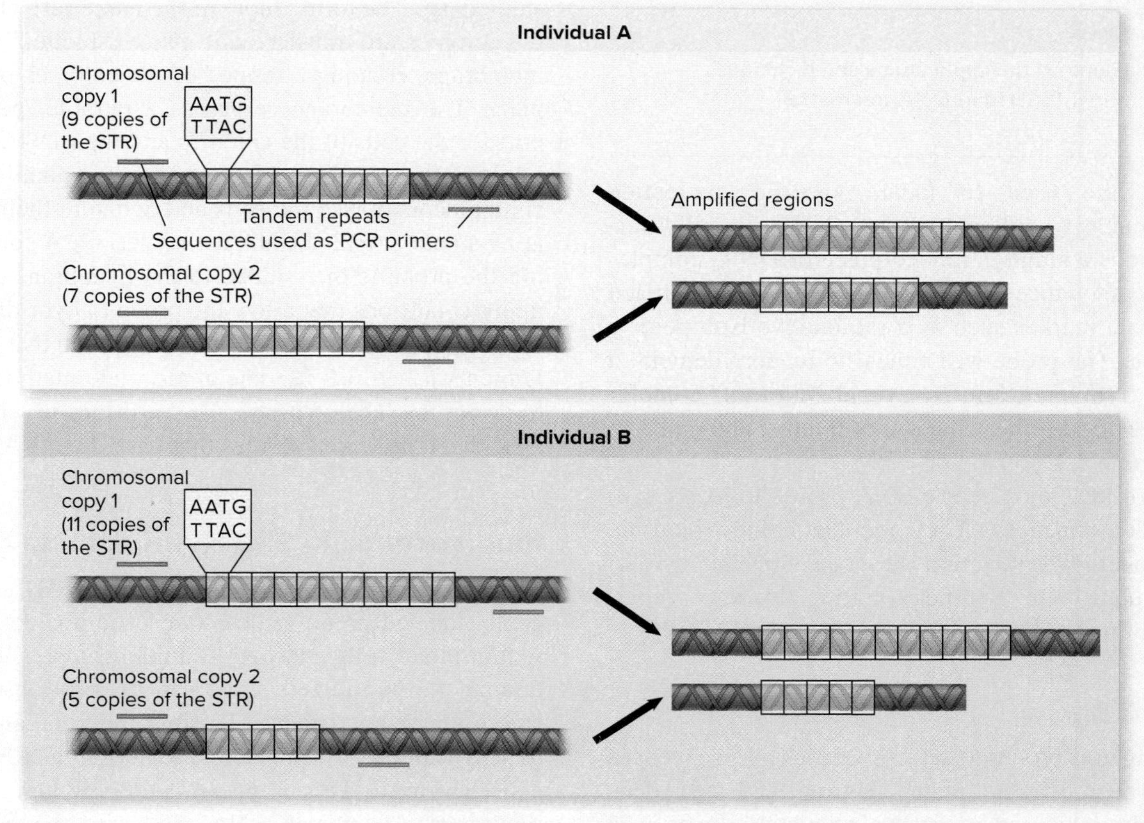

BOX FIGURE 9.1 Using PCR as Part of DNA Typing PCR is used to amplify certain chromosomal regions containing short tandem repeats (STRs). The number of copies of a given STR varies among people, resulting in corresponding differences in the length of the amplified fragments. Typically, at least 20 different STR locations are analyzed.

Selecting Primer Pairs

The nucleotide sequences of the two primers are critical because the primers dictate which part of the DNA is amplified. Each primer must be complementary to one end of the target DNA, so that DNA synthesis will extend across that stretch of DNA (see figure 9.16). If a researcher wants to amplify a DNA sequence that encodes a specific protein, he or she must first determine the nucleotide sequences at the ends of the gene. That information can be used to obtain the appropriate pair of primers.

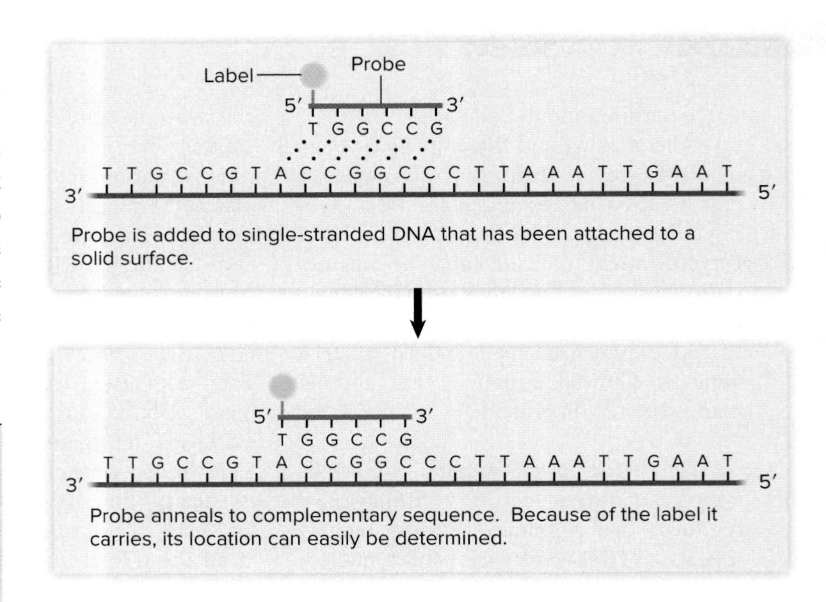

FIGURE 9.19 DNA Probes These single-stranded pieces of DNA tagged (labeled) with a detectable marker are used to detect specific nucleotide sequences in DNA or RNA samples that have been attached to a solid surface.

Why is it important that the probe be labeled?

MicroAssessment 9.4

The polymerase chain reaction (PCR) is used to rapidly increase the amount of a specific segment of DNA in a sample.

10. What is the role of the enzyme obtained from *Thermus aquaticus* in PCR?

11. Explain why it is important to use a polymerase from a thermophile in the PCR reaction.

12. Sequencing reactions can be done using PCR. In this case, would two primers be necessary? Explain.

9.5 ■ Probe Technologies

Learning Outcomes

9. Compare and contrast the applications and techniques of colony blotting, FISH, and DNA microarray technologies.

DNA probes are used to locate specific nucleotide sequences in nucleic acid samples attached to a solid surface. The probe is a single-stranded piece of DNA, complementary to the sequence of interest, that has been labeled with a detectable marker such as a radioactive isotope or a fluorescent dye. The probe will anneal to its complement, a process called **hybridization.** By hybridizing to its complement, the probe "finds" the sequence of interest and makes it detectable (**figure 9.19**).

A variety of technologies use DNA probes to locate specific nucleotide sequences. They include colony blotting, fluorescence in situ hybridization (FISH), and DNA microarrays. The technique called Southern blotting also uses probes, but its applications have been largely replaced by PCR.

Colony Blotting

Colony blotting uses probes to detect specific DNA sequences in colonies grown on agar plates (**figure 9.20** ①). This method is commonly used to determine which clones in a DNA library or other collection contain a sequence being studied. ② The term "blot" in the name reflects the fact that the colonies are transferred in place ("blotted") onto a nylon membrane, creating a replica of the colonies on the original plate. The membrane serves as a durable, permanent support for the cells of the colonies and their DNA. ③ After the transfer, the membrane is soaked in an alkaline solution to simultaneously lyse the cells and denature their DNA, generating single-stranded DNA molecules. ④ A solution containing the probe is then added to the membrane and incubated under conditions that allow the probe to hybridize to complementary sequences on the filter. Any probe that has not bound is then washed off. ⑤ The appropriate method is then used to detect the labeled probe. The positions to which the probe hybridized indicate colonies that have the DNA of interest.

Fluorescence In Situ Hybridization (FISH)

Fluorescence in situ hybridization (FISH) uses a fluorescently labeled probe to detect specific nucleotide sequences within intact cells attached to a microscope slide. Cells containing the hybridized probe can then be observed using a fluorescence microscope. To study microorganisms, a probe that hybridizes to sequences on ribosomal RNA (rRNA) is generally used. This is because multiplying cells can have thousands of copies of rRNA, increasing the technique's sensitivity. Other characteristics of rRNA that make it useful

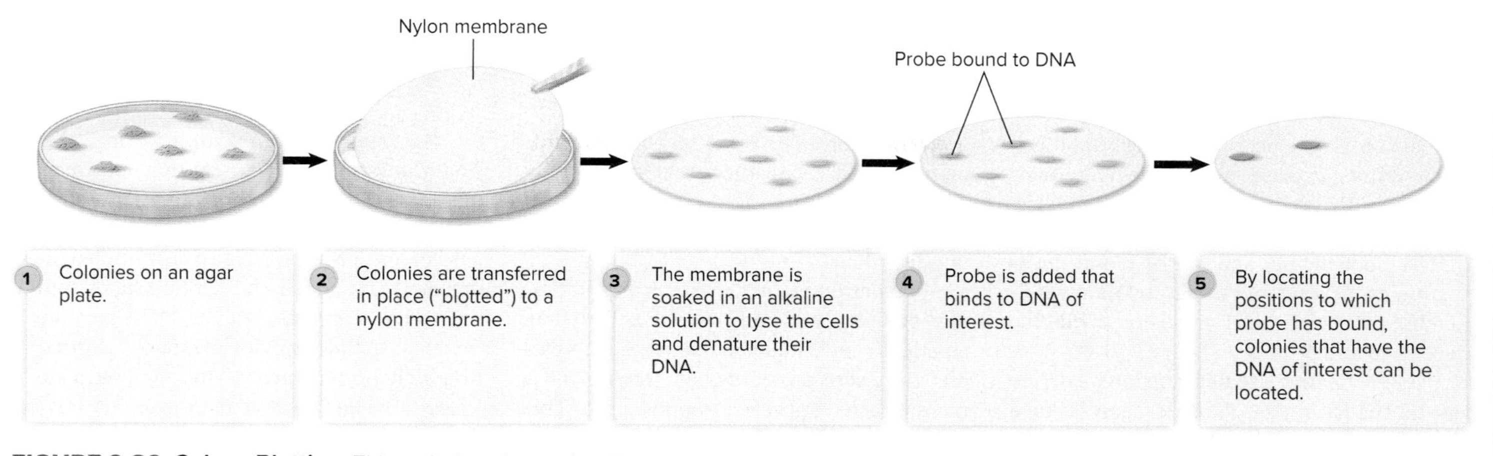

FIGURE 9.20 Colony Blotting This technique is used to determine which colonies on an agar plate contain a given DNA sequence.

What is the purpose of soaking the membrane in an alkaline solution?

for identifying microorganisms are described in chapter 10. ◀◀ fluorescence microscope

FISH is revolutionizing microbial ecology research and holds great promise in clinical laboratories. It provides a way to rapidly identify microorganisms directly in a specimen, bypassing the need to grow them in culture. FISH can be used to detect either a group of related organisms or a specific species, depending on the nucleotide sequence of the probe. For example, FISH can be used to determine the relative proportion of two different groups of microorganisms in a specimen by using two separate probes—one specific for each group—labeled with fluorescent markers of different colors (**figure 9.21**). It can also be used in clinical studies. For example, FISH is used to identify cells of the bacterium that causes tuberculosis in a sputum specimen. ▶▶ microbial ecology, ▶▶ tuberculosis

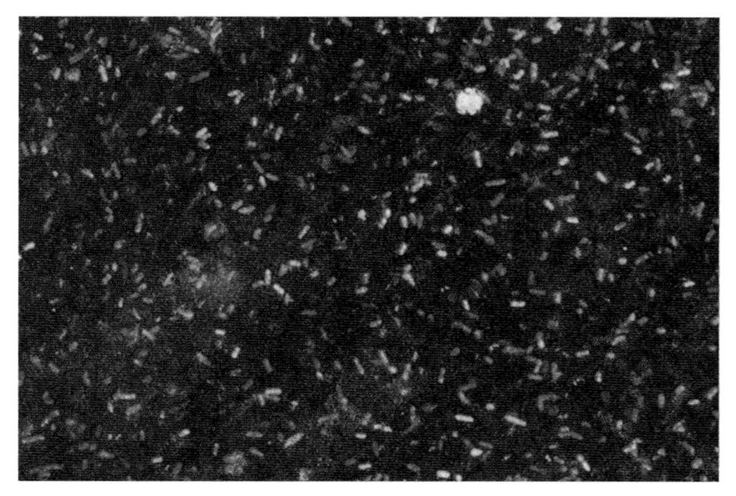

FIGURE 9.21 Fluorescence In Situ Hybridization (FISH) Different probes, each labeled with a unique colored fluorescent marker, were used to distinguish various types of bacteria in an earthworm gut. ©Seana Davidson

Why does FISH typically require a probe that hybridizes to rRNA?

To analyze a sample using FISH, the sample must first be treated with chemicals to preserve the shape of the cells, inactivate enzymes that might otherwise degrade the nucleic acid, and make the cells more permeable so that the labeled probe molecules can easily enter. Once the specimen has been prepared, it is put on a glass slide, bathed with a solution containing the labeled probe, and incubated under conditions that allow hybridization to occur. Unbound probe is then washed off. Finally, the specimen is viewed using a fluorescence microscope.

DNA Microarrays

DNA microarrays (also called gene chips or biochips) are primarily used to study gene expression in organisms whose genomes have been sequenced. The microarray is a glass slide or other small solid support carrying an arrangement of tens or hundreds of thousands of DNA spots, each of which contains many copies of a specific oligonucleotide (a short DNA sequence); the oligonucleotide functions like a probe. Each spot has a different oligonucleotide, so the microarray carries as many different probes as there are spots. To study gene expression, mRNA is isolated from an organism and used to make fluorescently labeled, single-stranded cDNA. These cDNA fragments are then added to the microarray and allowed to hybridize. Any fragments that are complementary to the oligonucleotides will anneal; meanwhile, those that are not complementary will not anneal and will be removed during a subsequent washing step. Thus, the oligonucleotides to which cDNAs hybridize are nucleotide sequences of genes that the organism was expressing. The locations of the labeled cDNA molecules are identified and

FOCUS ON THE FUTURE 9.1

Precision Medicine

Precision medicine is a medical approach that tailors disease prevention and treatment to the individual, often using the patient's genetic makeup as a basis. As DNA sequencing and probe-based technologies become increasingly more efficient, the benefits of precision medicine are expanding.

One effective use of precision medicine is cancer treatment. Genetic analysis of a patient's cancer cells can sometimes reveal the type of abnormality that led to the uncontrolled cell growth. In turn, this information allows selection of therapy specific for that defect. For example, certain breast cancers that arise due to over-expression of a protein called HER2 (human epidermal growth factor receptor 2) respond to treatment with a medication that specifically blocks the protein. Such a treatment is preferable to one that simply kills all rapidly dividing cells.

Prevention is also an extremely important goal of precision medicine. By using a patient's individual characteristics to predict risk, care providers can recommend a more targeted approach to prevention. This application of precision medicine is still in its infancy, however, because although some genetic risk factors are known, there is still a great deal to discover.

In recognition of the importance of individualized prevention and treatment approaches, the White House established the *Precision Medicine Initiative* in 2015. The aim is to develop new targeted treatments and also learn more about individual factors that correlate to disease risk. As part of the project, at least a million volunteers from across the United States will provide researchers with a DNA sample for analysis, along with personal information including medical history and intimate lifestyle details such as data from activity-tracking devices.

In the future, your medical records might include a complete genetic profile, allowing physicians to prescribe preventions and treatments designed specifically for your circumstances!

analyzed using a computerized scanner. If the experiment is done using cDNA from cultures grown under different sets of conditions—each cDNA preparation labeled with a different fluorescent marker—variations in gene expression can be revealed (**figure 9.22**). ◀◀ cDNA

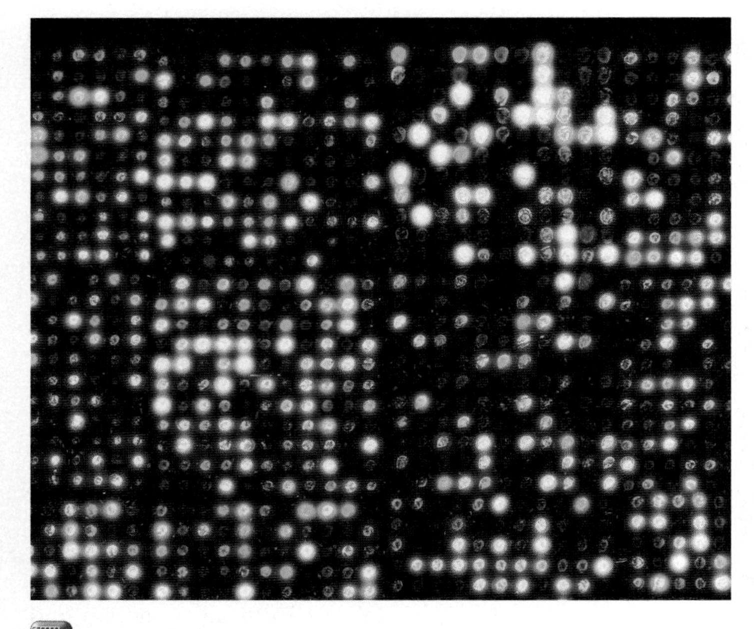

FIGURE 9.22 A DNA Microarray This is an example of an array used to study gene expression. Two different cDNA samples (one labeled with a red fluorescent marker and the other with a green fluorescent marker) were simultaneously hybridized to fragments in the microarray. Red dots indicate the positions to which one sample hybridized, and green dots indicate the positions to which the other hybridized. Yellow dots indicate that both samples hybridized.

? **How can DNA microarrays be used to determine which genes of a bacterial pathogen are expressed both inside and outside a host cell?**

MicroByte

Using DNA microarrays, researchers discovered that pathogens express some genes only when in certain locations within the human body.

MicroAssessment 9.5

Colony blotting is used to identify colonies that contain a given DNA sequence. Fluorescence in situ hybridization is used to observe individual cells that contain a given nucleotide sequence. DNA microarrays are primarily used to study gene expression.

13. What role does colony blotting play in cloning?

14. Why is a probe that binds to rRNA used in FISH?

15. In FISH, what would happen if unbound probe molecules were not washed off?

Summary

9.1 Fundamental Tools Used in Biotechnology

Restriction Enzymes (figure 9.1)

Restriction enzymes cut DNA into fragments. Cohesive (sticky) ends will **anneal** to one another, making it possible to join DNA from two different organisms.

DNA Gel Electrophoresis (figure 9.2)

DNA gel electrophoresis separates DNA fragments according to their size.

CRISPR/Cas9 (figure 9.3)

Scientists use a genetically modified Cas9 nuclease of a **CRISPR system** to modify genomes and to study gene function.

9.2 Genetic Engineering (table 9.3)

Genetically Engineered Bacteria

Bacteria can be engineered to produce pharmaceutical proteins, vaccines, and other proteins more efficiently (figure 9.4). When a segment of DNA is cloned into *E. coli,* it is readily available for study and further manipulation.

Genetically Engineered Eukaryotes

Transgenic plants have been engineered to resist pests and herbicides, have improved nutritional value, and function as edible vaccines.

Techniques Used to Clone DNA

A **DNA library** is a collection of clones that together contain the entire genome of an organism (figure 9.6). To isolate DNA, cells are lysed by adding a detergent. To obtain eukaryotic DNA without introns, reverse transcriptase is used to make a **cDNA** from an mRNA template (figure 9.7). DNA ligase is used to join the **vector** (frequently a plasmid) and the insert (figures 9.8, 9.9). The recombinant molecule is introduced into the new host, usually *E. coli,* using transformation or electroporation. The transformed cells are grown on medium that both selects for cells containing vector sequences and differentiates those that carry recombinant molecules.

Concerns Regarding Genetic Engineering

Genetic engineering can be used to make life-saving products, but there is no guarantee that it will not be used for malicious purposes. Genetically modified organisms hold many promises, but concerns exist about the inadvertent introduction of allergens into a food product and adverse effects on the environment.

9.3 DNA Sequencing

DNA sequencing is the process of determining the order of nucleotides in a DNA molecule. The sequence can be used to decipher the amino acid sequence of the encoded proteins. DNA sequences can also be used to determine the evolutionary relatedness of organisms.

DNA Sequencing Methods

A key component of the dideoxy sequencing reaction is a dideoxynucleotide, a nucleotide that lacks the 3′OH and therefore functions as a chain terminator (figure 9.11). The sizes of fragments in a sequencing reaction indicate the positions of the terminating nucleotide (figures 9.12, 9.13). High-throughput sequencing methods allow for faster, cheaper sequencing (figure 9.14).

9.4 Polymerase Chain Reaction (PCR)

PCR is used to exponentially amplify target DNA (figure 9.15). Over a billion copies of the target can be generated in a matter of hours. Researchers can then detect that **PCR product.**

Variations of Conventional PCR

RT-PCR (reverse transcription PCR) is used to detect a given mRNA sequence in a sample. **qPCR (quantitative PCR)** (also called real-time PCR) allows researchers to track the PCR product accumulation during the course of a PCR reaction.

PCR Methods

PCR amplifies target DNA by repeatedly denaturing double-stranded DNA, annealing specific primers to their complementary sequences, and synthesizing DNA (figures 9.16, 9.17, 9.18). The primers dictate which part of the DNA is amplified.

9.5 Probe Technologies

DNA probes are used to locate specific nucleotide sequences (figure 9.19).

Colony Blotting

Colony blotting uses a probe to identify colonies that contain a DNA sequence of interest (figure 9.20).

Fluorescence In Situ Hybridization (FISH)

Fluorescence in situ hybridization (FISH) uses a fluorescently labeled probe to detect specific nucleotide sequences within intact cells attached to a microscope slide (figure 9.21).

DNA Microarrays

DNA microarrays contain tens or hundreds of thousands of oligonucleotides that each function in a manner analogous to a probe (figure 9.22).

Review Questions

Short Answer

1. Why are restriction enzymes useful in biotechnology?
2. In the CRISPR/Cas9 system, how is the Cas9 used for gene editing different from the one used to study gene function?
3. Describe three general uses of genetically engineered bacteria.
4. Describe three uses of transgenic plants.
5. What is a DNA library?
6. What is cDNA? Why is it used when cloning eukaryotic genes?
7. How many different temperatures are used in each cycle of the polymerase chain reaction?
8. How does PCR eventually generate a discrete-sized fragment from a much longer piece of DNA?
9. What is the function of a DNA probe?
10. How does a DNA microarray function as a set of probes?

Multiple Choice

1. What is the function of a vector?
   a) Destroys cells that do not contain cloned DNA
   b) Allows cells to take up foreign DNA
   c) Carries cloned DNA, allowing it to replicate in cells
   d) Encodes herbicide resistance
   e) Encodes Bt toxin
2. The Ti plasmid of *Agrobacterium tumefaciens* is used to genetically engineer which of the following cell types?
   a) Animals
   b) Bacteria
   c) Plants
   d) Yeast
   e) All of these
3. Which of the following can be used to generate a DNA library?
   a) PCR
   b) Sequencing
   c) Colony blotting
   d) Microarrays
   e) Cloning
4. An ideal vector has all of the following *except*
   a) an origin of replication.
   b) a gene encoding a restriction enzyme.
   c) a gene encoding resistance to an antibiotic.
   d) a multiple-cloning site.
   e) the *lacZ′* gene.
5. Which of the following describes the function of the *lacZ′* gene in a cloning vector?
   a) Means of selecting for cells that contain vector sequences
   b) Means of distinguishing cells that have taken up recombinant molecules
   c) Site required for the vector to replicate
   e) Mechanism by which cells take up the DNA
   d) Gene for a critical nutrient required by transformed cells
6. Which is used for cloning eukaryotic genes but not prokaryotic genes?
   a) Restriction enzymes
   b) DNA ligase
   c) Reverse transcriptase
   d) Vector
   e) Selectable marker
7. Lack of which of the following distinguishes a dideoxynucleotide from its deoxynucleotide counterpart?
   a) $5'PO_4$
   b) $3'OH$
   c) $5'OH$
   d) $3'PO_4$
   e) c and d
8. In a sequencing reaction, the dATP was left out of the tube. What would be the result of this error?
   a) No synthesis would occur.
   b) Synthesis would never continue past the first A.
   c) Synthesis would not stop until the end of the template.
   d) Synthesis would terminate randomly, regardless of the nucleotide incorporated.
   e) The error would have no effect.
9. The polymerase chain reaction uses *Taq* polymerase rather than a DNA polymerase from *E. coli,* because *Taq* polymerase
   a) introduces fewer errors during DNA synthesis.
   b) is heat-stable.
   c) can initiate DNA synthesis at a wider variety of sequences.
   d) can denature a double-stranded DNA template.
   e) is easier to obtain.
10. The polymerase chain reaction generates a fragment of a distinct size even when an intact chromosome is used as a template. What determines the boundaries of the amplified fragment?
    a) The concentration of one particular deoxynucleotide in the reaction
    b) The duration of the elongation step in each cycle
    c) The position of a termination sequence, which causes the *Taq* polymerase to fall off the template
    d) The sites to which the primers anneal
    e) The temperature of the elongation step in each cycle

Applications

1. Two students in a microbiology class are arguing about the origins of biotechnology. One student argued that biotechnology started with the advent of genetic engineering. The other student disagreed, saying that biotechnology was as old as ancient civilization. What was the rationale for the argument by the second student?
2. A student wants to clone gene X. On both sides of the gene are the recognition sequences for AluI and BamHI (look at table 9.2). Which enzyme would be easier to use for the cloning experiment and why?

Critical Thinking

1. Discuss some potential issues regarding gene editing to correct genetic defects in humans.
2. An effective DNA probe can sometimes be developed based on the amino acid sequence of the protein encoded by the gene. A student argued that this is too time-consuming since the complete amino acid sequence must be determined in order to create the probe. Does the student have a valid argument? Why or why not?

10 Identifying and Classifying Microorganisms

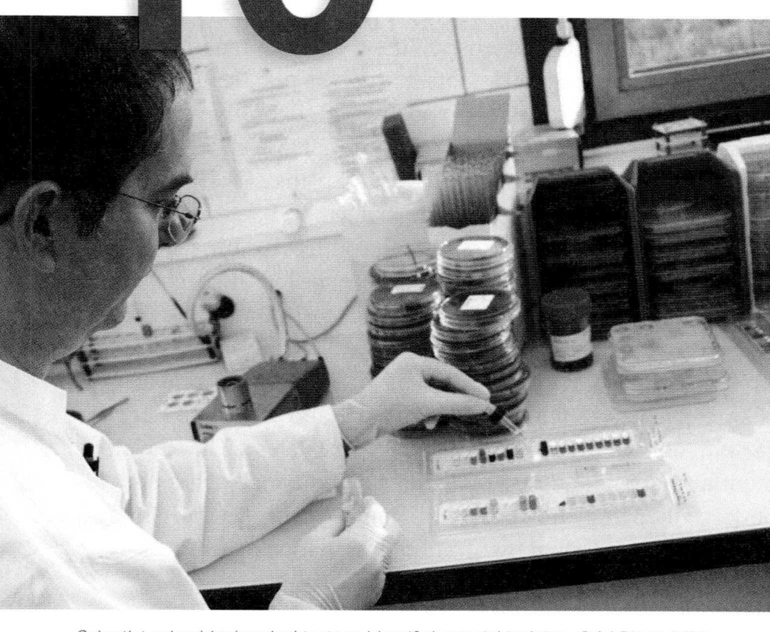

Scientist using biochemical tests to identify bacterial isolates. ©AJ Photo/HOP American/SPL/Science Source

KEY TERMS

Classification The process of arranging organisms into similar or related groups (taxa), primarily to make it easier to identify them for study.

Domain A collection of similar kingdoms; there are three domains: *Bacteria, Archaea,* and *Eukarya.*

Genus A collection of related species.

Identification The process of characterizing an isolate in order to determine the group (taxon) to which it belongs.

Nomenclature The system of assigning names to organisms.

Phylogeny Evolutionary relatedness of organisms.

Species A group of closely related strains; the basic unit of taxonomy.

Strain A pure culture isolate; subgroup within a species.

Taxonomy The science of characterizing organisms in order to arrange them into hierarchical groups (taxa); involves three interrelated areas: identification, classification, and nomenclature.

A Glimpse of History

In the early 1870s, the German botanist Ferdinand Cohn wrote several papers on bacterial classification, in which he grouped microorganisms according to shape—spherical, short rods, elongated rods, and spirals. That scheme was not adequate, however, because there were too many different kinds of bacteria with similar shapes.

The second major attempt at bacterial classification was initiated by the Danish scientist Sigurd Orla-Jensen. In 1908, he proposed that bacteria be classified according to their physiological properties. The idea was promising, but there were definite drawbacks. Organisms that were classified as similar based on certain physiological traits would appear unrelated when comparing other traits.

A quarter of a century later, two Dutch microbiologists, Albert Kluyver and C. B. van Niel, proposed classification systems based on evolutionary relationships. They faced a serious problem, however, because they could not distinguish "resemblance" from "relatedness." The fact that two prokaryotes look or act alike does not mean they are genetically related.

In 1970, Roger Stanier, a microbiologist at the University of California, Berkeley, pointed out that relationships could be determined by comparing either physical traits, such as proteins and cell walls, or nucleotide sequences. At that time, most microbiologists, including Stanier, assumed that all prokaryotes are basically similar. However, chemical analysis of various prokaryotic structures—including the cell wall, cytoplasmic membrane, and ribosomal RNA—showed that this was not the case.

In the late 1970s, Carl Woese and his colleagues at the University of Illinois determined the nucleotide sequence of ribosomal RNA from a wide variety of organisms. Based on the data, they recognized that prokaryotes could be divided into two major groups that differ from each other as much as they differ from eukaryotic cells. This led to a revolutionary system of classification that separates prokaryotes into two domains: *Archaea* and *Bacteria.* All eukaryotic organisms, including humans, are in a third domain: *Eukarya.*

Information that is logically organized is easier to use. Newspapers, for instance, do not scatter various subjects throughout the paper; instead, the information is grouped by general topic, such as local news, sports, and entertainment. A large library would be extremely difficult to use if the locations of the many books were not organized by subject matter. Likewise, scientists have sorted living organisms into different groups, to better show the relationships among the species.

Take a moment to think about how you would group bacteria if you were to create a classification system. Would you group them according to shape? Or would it make more sense to consider their motility? Perhaps you would group them according to their medical significance. But then, how would you classify two apparently identical organisms that differed in their pathogenicity? This chapter describes some of the methods currently used by microbiologists to classify microorganisms.

10.1 ■ Principles of Taxonomy

Learning Outcome

1. Describe how microorganisms are identified, classified, and assigned names.

Taxonomy is the science of characterizing and naming organisms in order to arrange them into hierarchical groups (taxa). Organisms with similar properties are grouped together and separated from ones that are different. Taxonomy can be viewed as three separate but interrelated areas:

- **Identification**: The process of characterizing an isolate (a population of cells descended from a single cell) to determine the group (taxon) to which it belongs.
- **Classification**: The process of arranging organisms into similar or related groups, primarily to make it easier to identify them for study.
- **Nomenclature**: The system of assigning names to organisms.

Strategies Used to Identify Microorganisms

In practical terms, identifying the genus and species of a microorganism may be more important than understanding its genetic relationship to other microbes. For example, a food manufacturer is most interested in detecting microbial contaminants that can spoil a food product. In a clinical laboratory, identifying a pathogen quickly is crucial so that a patient can be treated appropriately.

To identify microorganisms, many different procedures may be used, including microscopic examination, culture characteristics, biochemical tests, and nucleic acid analysis. In a clinical laboratory, the patient's disease symptoms play an important role as well. For example, pneumonia in an otherwise healthy adult is typically caused by *Streptococcus pneumoniae,* a bacterium easily differentiated from others using a few specific tests. In contrast, diagnosing the cause of a wound infection is often more difficult, because several different microorganisms could be involved. Often, however, it is necessary only to detect organisms known to cause a particular disease, rather than to conclusively identify each and every one. For instance, a fecal specimen from a patient with diarrhea and fever would generally be tested only for the presence of organisms that cause those symptoms. ⏭ *Streptococcus pneumoniae*

Strategies Used to Classify Microorganisms

Understanding the evolutionary relatedness, or **phylogeny**, of microorganisms is important in creating a classification scheme that reflects their evolution and biology. Such a scheme is more useful than one that simply groups organisms by random characteristics, because it is less prone to the bias of human perceptions. It also makes it easier to classify newly recognized species and allows scientists to make predictions, such as the likelihood that genes will be acquired from a given organism.

Unfortunately, determining evolutionary relatedness among microorganisms is more difficult than doing so for plants and animals. Not only do microorganisms have few differences in size and shape, they do not undergo sexual reproduction. In higher organisms such as plants and animals, the basic taxonomic unit, a **species**, is generally considered to be a group of morphologically similar organisms capable of interbreeding to produce fertile offspring. Obviously, it is not possible to apply these same criteria to bacteria, archaea, and other microorganisms that reproduce by binary fission rather than sexual reproduction. Because of this, classification of these organisms is challenging, and other criteria must be used to define a species.

Historically, taxonomists have relied on phenotypic characteristics—such as cell wall type and ability to degrade certain compounds—to classify prokaryotes. The development and application of molecular techniques such as nucleotide sequencing, however, are finally making it possible to determine the phylogeny of microorganisms. ⏮ phenotype

Taxonomic Hierarchies

Taxonomic classification categories are arranged in a hierarchical order, with the species being the basic unit. The species designation gives a formal taxonomic status to a group of related isolates, or **strains**, which, in turn, allows for their identification. Without classification, scientists and others would not be able to communicate efficiently about organisms. Taxonomic categories include:

- **Species**: A group of closely related strains. Note that members of a species are not all identical; individual strains may differ from one another. The difficulty for the taxonomist is to decide how different two isolates must be in order to be classified as separate species rather than strains of the same species. In certain cases, a species may be divided into two or more subspecies. This is particularly true for the commercially valuable organisms used to make fermented foods. ⏭ microorganisms in food and beverage production
- **Genus**: A collection of similar species.
- **Family**: A collection of similar genera. In prokaryotic nomenclature, the name of the family ends in the suffix *-aceae.*
- **Order**: A collection of similar families. In prokaryotic nomenclature, the name of the order ends in the suffix *-ales.*
- **Class**: A collection of similar orders.
- **Phylum** or **Division**: A collection of similar classes.

TABLE 10.1 Taxonomic Ranks of the Bacterium *Escherichia coli*

Formal Rank	Example
Domain	*Bacteria*
Phylum	*Proteobacteria*
Class	*Gammaproteobacteria*
Order	*Enterobacteriales*
Family	*Enterobacteriaceae*
Genus	*Escherichia*
Species	*coli*

- **Kingdom**: A collection of similar phyla or divisions.
- **Domain**: A collection of similar kingdoms. The domain reflects the characteristics of the cells that make up the organism (see **A Glimpse of History**).

Note, however, that microbiologists often group microorganisms into informal categories based on one or more distinctive characteristics, rather than using the higher taxonomic ranks such as order, class, and phylum. Examples of informal groupings of bacteria include the lactic acid bacteria, the anoxygenic phototrophs, the endospore formers, and the sulfate reducers. Organisms within these groupings share similar phenotypic and physiological characteristics, but may not be genetically related. ⏭ **lactic acid bacteria,** ⏭ **anoxygenic phototrophs,** ⏭ **sulfate reducers**

Table 10.1 shows how a particular bacterial species is classified. The rank of kingdoms is not shown because that level of classification in *Bacteria* and *Archaea* is still in a state of flux.

Classification Systems

Classification systems change over the years as new information is discovered. There is no "official" classification system, and as new ones are introduced, others become outdated. The classification scheme currently favored by most microbiologists is the three-domain system, which designates all organisms as belonging to one of the three domains: *Bacteria, Archaea,* and *Eukarya.* The system is based on comparison of nucleotide sequences in ribosomal RNA (rRNA) from a wide variety of organisms. The rRNA data are consistent with other differences between bacteria and archaea, including the chemical compositions of their cytoplasmic membranes and cell walls (**table 10.2**). ⏮ **bacteria,** ⏮ **archaea,** ⏮ **eukarya**

Figure 10.1a shows a tree of life (a diagram that illustrates the relatedness of organisms) based on rRNA sequence data that support the three-domain system. Figure 10.1b shows a new and expanded version of the tree that was developed by comparing the amino acid sequences of ribosomal proteins. Because this new version includes many bacteria that have not yet been grown in pure culture, it highlights the extraordinary diversity of the microbial world, particularly within the *Bacteria.* As metagenomics allows additional novel organisms to be characterized, the tree is expected to change even more. ⏮ **metagenomics**

Before the three-domain system was introduced, the most widely accepted classification scheme was the five-kingdom system, proposed by R. H. Whittaker in 1969. The five kingdoms in this system are Plantae, Animalia, Fungi, Protista (mostly single-celled eukaryotes), and Prokaryotae. Although the five-kingdom system recognizes the obvious morphological differences between plants and animals, it does not reflect the genetic insights of the ribosomal data, which indicate that plants and animals are more closely related to each other than archaea are to bacteria.

Bergey's Manual of Systematic Bacteriology Microbiologists generally rely on the classifications listed in the reference text *Bergey's Manual of Systematic Bacteriology.* All known bacterial and archaeal species are described there. If the properties of a recently isolated prokaryotic organism do not agree with any description in *Bergey's Manual,* then that organism likely represents a new discovery. The newest edition of this comprehensive manual was published in five volumes and classifies bacteria and archaea according to the most recent information on their phylogeny. In addition to containing descriptions of organisms, all volumes include information on the ecology, methods of enrichment, culture, and isolation of the organisms as well as methods for their maintenance and

TABLE 10.2 A Comparison of Properties Typical of Members of the Three Domains

Cell Feature	*Archaea*	*Bacteria*	*Eukarya*
Cytoplasmic membrane lipids	Hydrocarbons (not fatty acids) linked to glycerol by ether linkage	Fatty acids linked to glycerol by ester linkage	Fatty acids linked to glycerol by ester linkage
Membrane-bound nucleus	No	No	Yes
Peptidoglycan cell wall	No	Yes	No
Presence of introns	Sometimes	No	Yes
Ribosomes	70S	70S	80S

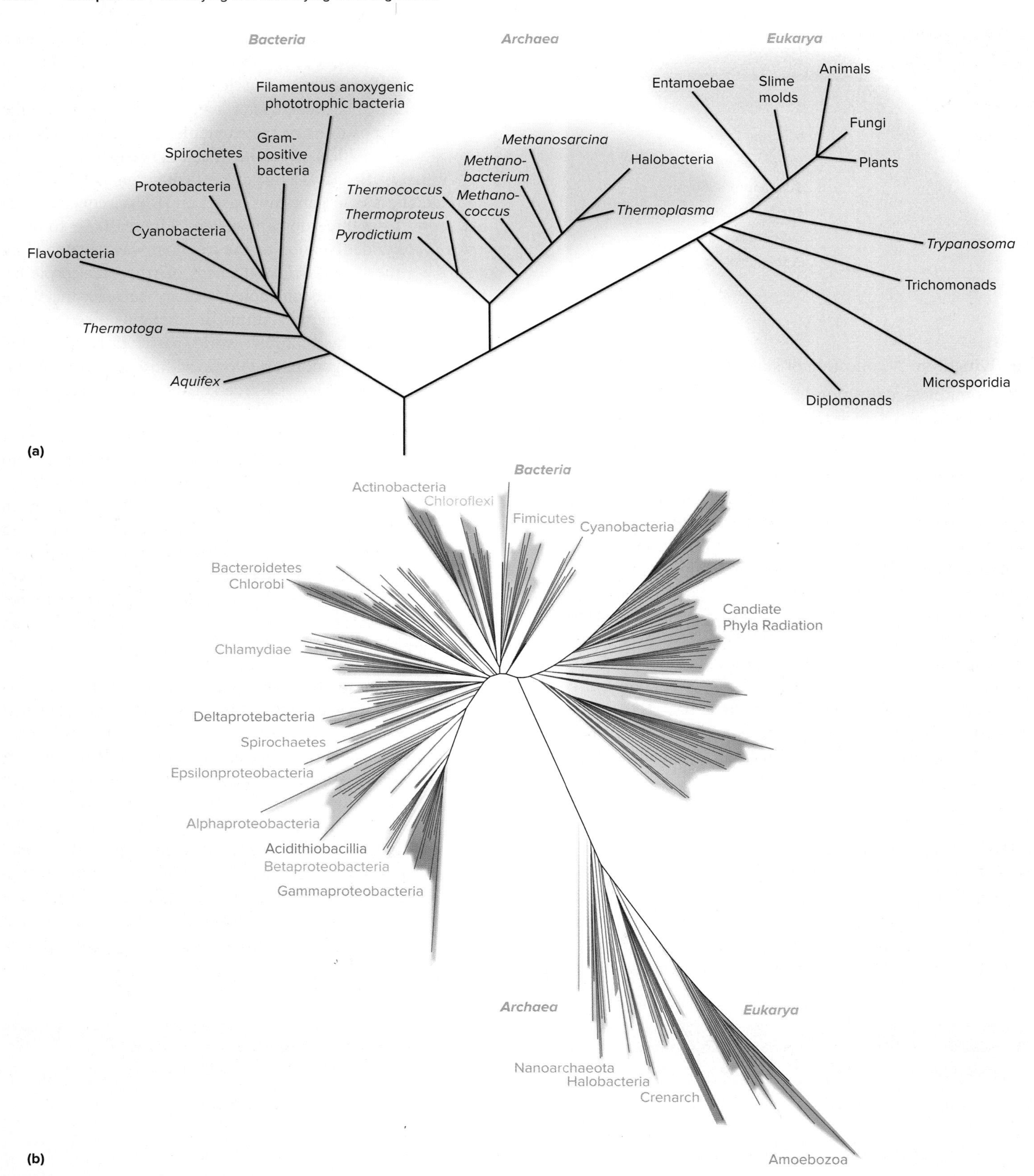

FIGURE 10.1 Tree of Life (a) Based on ribosomal RNA data. **(b)** Based on ribosomal protein sequence data, including information from many organisms that have not been grown in pure culture. The colors of the branches are arbitrary. For simplicity, only some of the names of the various organisms have been included here. The group of bacteria referred to as Candidate Phyla Radiation (CPR) have small genomes and appear to lack crucial genes required to multiply independently of a host cell. Source for (b): Hug et al., "A New View of the Tree of Life." *Nature Microbiology* 1, March, 2016.

? Why would members of the Candidate Phyla Radiation be difficult to grow in pure culture?

preservation. However, the heart of the work is a description of all characterized bacteria and archaea and their groupings.

Nomenclature

Bacteria and archaea are given names according to a set of internationally recognized rules: the *International Code of Nomenclature of Bacteria.* The names may originate from any language, but must include a Latin suffix. In some cases, the name reflects the organism's habitat or other characteristic, but it often honors a researcher.

Just as classification is always in a state of flux, so is the assignment of names. Although revising names may increase scientific accuracy, it often leads to confusion, particularly when names of medically important organisms are changed. To ease the transition after a name change, the former name may be included in parentheses. For example, *Lactococcus lactis,* a bacterium that was once included in the genus *Streptococcus,* is sometimes indicated as *Lactococcus (Streptococcus) lactis.*

MicroAssessment 10.1

Taxonomy consists of three interrelated areas: identification, classification, and nomenclature. In clinical laboratories, identifying the genus and species of an isolate is more important than understanding its evolutionary relationship to other organisms.

1. Why might it be easier to determine the cause of pneumonia than to determine the cause of a wound infection?
2. What is *Bergey's Manual*?
3. Some biologists have been reluctant to accept the three-domain system. Why might this be the case?

10.2 ■ Identification Methods Based on Phenotype

Learning Outcome

2. Describe how phenotypic characteristics—including microscopic morphology, culture characteristics, metabolic capabilities, serology, and protein profile—can be used to identify microorganisms.

Phenotypic characteristics can be used to help identify microorganisms, without the need for sophisticated equipment. These and other identification methods are summarized in **table 10.3.**

Microscopic Morphology

An important initial step in identifying a microorganism is to determine its size, shape, and staining characteristics. Microscopic examination gives information very quickly and is sometimes enough to make a presumptive identification.

Size and Shape

The size and shape of a microorganism can easily be determined microscopically. Based only on the size and shape, the organism can be identified as a prokaryote, fungus, or protozoan. In a clinical lab, this can sometimes be enough to diagnose certain eukaryotic infections. For example, a wet mount of stool is examined for the eggs of parasites when certain roundworms are suspected (**figure 10.2**). ⏮ wet mount

Gram Stain

The Gram stain distinguishes between Gram-positive and Gram-negative bacteria (see figure 3.15). This relatively rapid

TABLE 10.3 Methods Used to Identify Microorganisms

Method	Comments
Phenotypic Characteristics	Most of these methods do not require sophisticated equipment and can easily be done anywhere in the world.
Microscopic morphology	Size, shape, and staining characteristics can give suggestive information as to the identity of the organism. Further testing, however, is needed to confirm the identification.
Culture characteristics	Colony morphology can give initial clues to the identity of an organism.
Metabolic capabilities	A set of biochemical tests can be used to identify a microorganism.
Serological testing	Proteins and polysaccharides that make up a microorganism are sometimes characteristic enough to be considered identifying markers. These can be detected using specific antibodies.
Protein profile	MALDI-TOF MS separates and sorts an organism's proteins by mass, generating a profile that provides a fast way to identify an organism grown in culture.
Genotypic Characteristics	These methods are increasingly being used to identify microorganisms. Even an organism that occurs in very low numbers in a mixed culture can be identified.
Detecting specific nucleotide sequences	Nucleic acid probes and nucleic acid amplification tests can be used to identify a microorganism grown in culture. In some cases, the method is sensitive enough to detect the organism directly in a specimen.
Sequencing rRNA genes	This requires amplifying and then sequencing rRNA genes, but it can be used to identify organisms that have not yet been grown in culture.

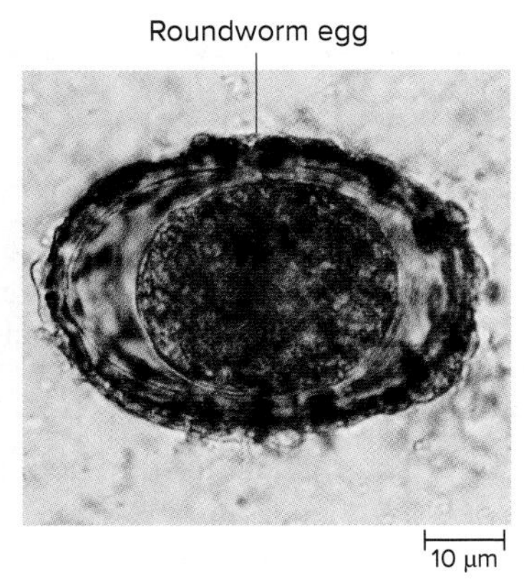

FIGURE 10.2 Wet Mount of a Clinical Specimen Roundworm *(Ascaris)* egg in a stool specimen. ©BSIP/UIG via Getty Images

Roundworms belong to which domain?

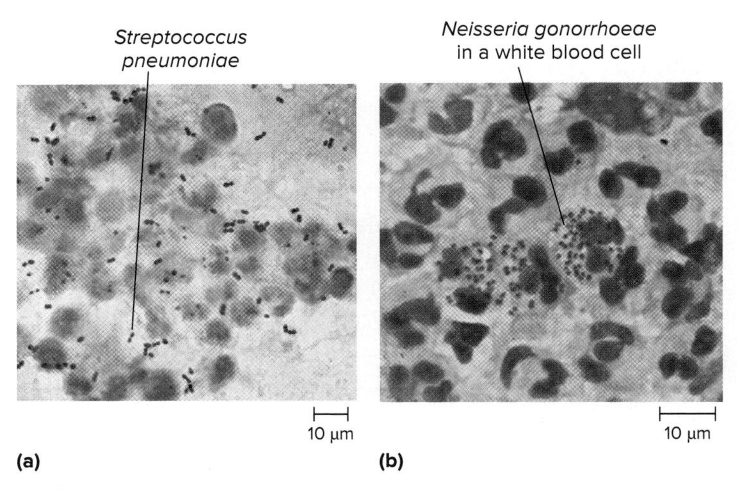

FIGURE 10.3 Gram Stains of Clinical Specimens (a) Sputum showing Gram-positive *Streptococcus pneumoniae* and **(b)** male urethral discharge showing Gram-negative *Neisseria gonorrhoeae* inside white blood cells. a: ©Eye of Science/Science Source; b: Source: Bill Schwartz/CDC

What disease does the patient in (a) have? What disease does the patient in (b) have?

test narrows the list of possible identities of an organism, an essential step in the identification process. **⏮ Gram stain**

In a clinical lab, the Gram stain of a specimen is generally not sensitive or specific enough to diagnose the cause of most infections, but it is still an extremely useful tool. The technician can see the Gram reaction, the shape and arrangement of the bacterial cells, and whether the organisms appear to be growing as a pure culture or with other bacteria and/or cells of the host. However, most medically important bacteria cannot be identified by Gram stain alone. For example, *Streptococcus pyogenes,* the bacterium that causes strep throat, cannot be distinguished microscopically from the streptococci that are part of the normal throat microbiota. A Gram stain of a stool specimen cannot distinguish *Salmonella* from *E. coli.* These organisms generally must be isolated in pure culture and tested by other means for accurate identification. **⏭ strep throat**

In certain cases, the Gram stain result gives enough information to start appropriate antimicrobial treatment. For example, a Gram stain of sputum showing numerous white blood cells and Gram-positive diplococci is highly suggestive of *Streptococcus pneumoniae,* a bacterium that causes pneumonia (**figure 10.3a**). In some other cases, the Gram stain result is enough for diagnosis. For instance, the presence of Gram-negative diplococci clustered in white blood cells in a sample of a urethral discharge from a man may be considered diagnostic for gonorrhea, the sexually transmitted infection caused by *Neisseria gonorrhoeae* (figure 10.3b). This diagnosis can be made because *N. gonorrhoeae* is the only Gram-negative diplococcus found within white blood cells in the male urethra. **⏭ pneumonia, ⏭ gonorrhea**

Special Stains

Certain microorganisms have unique characteristics that can be detected with special staining procedures. As an example, members of the genus *Mycobacterium* are some of the few acid-fast microorganisms (see figure 3.16). If a patient has symptoms of tuberculosis, then an acid-fast stain will be done on a sputum sample to help identify *Mycobacterium tuberculosis.* **⏮ acid-fast stain**

Culture Characteristics

Colony morphology can give initial clues to the identity of the organism. For example, colonies of the bacterium *Serratia marcescens* are often red when incubated at 22°C due to the production of a pigment. The bacterium *Pseudomonas aeruginosa* often produces a soluble greenish pigment, which discolors the growth medium (see figure 11.11). In addition, cultures of *P. aeruginosa* have a distinct fruity odor. **⏮ growth of colonies**

In a clinical lab, where rapid but accurate diagnosis is essential, specimens are inoculated onto differential media as a preliminary step in the identification process. A specimen taken by swabbing the throat of a patient complaining of a sore throat is inoculated onto blood agar. This makes it possible to detect the characteristic β-hemolytic colonies of *Streptococcus pyogenes,* the bacterium that causes strep throat (see figure 4.11). Urine collected from a patient suspected of having a urinary tract infection (UTI) is plated onto MacConkey agar. *E. coli,* the most common cause of UTIs, can grow on that medium, where it ferments lactose to form pink colonies (see figure 4.12). **⏮ differential media, ⏮ blood agar, ⏮ MacConkey agar**

Metabolic Capabilities

Colony morphology can give clues as to the identity of a microorganism, but other tests are generally necessary for a more reliable identification. The metabolic capabilities of a microorganism—such as the types of sugars it ferments or the end products it makes—are commonly used in identification procedures, particularly for bacteria. Using metabolic capabilities to identify bacteria is somewhat analogous to identifying people based on characteristics such as hair color, food preferences, and hobbies. The traits are simply easy-to-detect features, and some are more useful in certain situations than others. For example, when distinguishing among several non-fermenting bacteria, it would not be helpful to compare the types of sugars fermented; other tests must be used instead.

Biochemical Tests

A variety of biochemical tests can be used to determine an isolate's metabolic capabilities (**table 10.4**). Some of the tests are quite rapid, because they are designed to detect enzymes already present at sufficient levels in a bacterial colony. Most, however, require an incubation period of at least 18 hours.

One of the easiest and fastest biochemical tests is an assay for the enzyme catalase (**figure 10.4a**). Recall that many organisms produce catalase to protect against hydrogen peroxide (see table 4.3). To detect catalase, a small portion of a colony is transferred to a microscope slide or the inside of a Petri dish, and then a drop of hydrogen peroxide (H_2O_2) is added. If catalase is present, it immediately breaks down the hydrogen peroxide to form O_2 and water; the O_2 can be observed as bubbles in the reagent. Most bacteria that grow in the presence of O_2 are catalase-positive. Important exceptions are the lactic acid bacteria, which include members of the genus *Streptococcus.* Thus, if β-hemolytic colonies grow from a throat culture but testing reveals they are all catalase-positive, then *Streptococcus pyogenes* has been ruled out.
⏮ **catalase,** ⏭ ***Streptococcus pyogenes***

Most biochemical tests rely on a chemical indicator that changes color when a compound is degraded. The test for the ability of an organism to ferment a given sugar involves adding the organism to a broth growth medium containing the sugar and a pH indicator. If the organism ferments the sugar, acid is produced, which lowers the pH, resulting in a color change; an inverted tube traps any gas produced (see figure 10.4b). A medium designed to detect urease (an enzyme that degrades

TABLE 10.4 Characteristics of Some Important Biochemical Tests

Biochemical Test	Principle of the Test	Positive Reaction
Catalase	Rapid test that detects the activity of the catalase, an enzyme that breaks down hydrogen peroxide to form O_2 and water.	The reagent bubbles.
Citrate	Determines whether or not citrate can be used as a sole carbon source.	Growth, usually accompanied by the color change of a pH indicator.
Gelatinase	Detects enzymatic breakdown of gelatin to polypeptides.	The solid gelatin is converted to liquid.
Hydrogen sulfide production	Detects H_2S released as sulfur-containing amino acids are degraded.	A black precipitate forms due to the reaction of H_2S with iron salts in the medium.
Indole	Detects the enzymatic removal of the amino group from tryptophan.	The product, indole, reacts with an added chemical reagent, turning the reagent a deep red color.
Lysine decarboxylase	Detects the enzymatic removal of the carboxyl group from lysine.	The medium becomes more alkaline, causing a pH indicator to change color.
Methyl red	Detects mixed acids, the characteristic end products of a particular fermentation pathway. ⏮ **mixed acids**	The medium becomes acidic (pH below 4.5); a red color develops upon the addition of a pH indicator.
Oxidase	Rapid test that detects the activity of cytochrome c oxidase, a component of the electron transport chain of specific organisms. ⏮ **cytochrome c**	A dark color develops after a specific reagent is added.
Phenylalanine deaminase	Detects the enzymatic removal of the amino group from phenylalanine.	The product of the reaction, phenylpyruvic acid, reacts with ferric chloride to give the medium a green color.
Sugar fermentation	Detects the acidity resulting from fermentation of a specific sugar incorporated into the medium; also detects gas production.	The medium becomes acidic, causing a pH indicator to change color. An inverted tube traps any gas that is made.
Urease	Detects the enzymatic degradation of urea to carbon dioxide and ammonia.	The medium becomes alkaline, causing a pH indicator to change color.
Voges-Proskauer	Detects acetoin, an intermediate of the fermentation pathway that leads to 2,3-butanediol production. ⏮ **2,3-butanediol**	A red color develops after chemicals that detect acetoin are added.

(a)

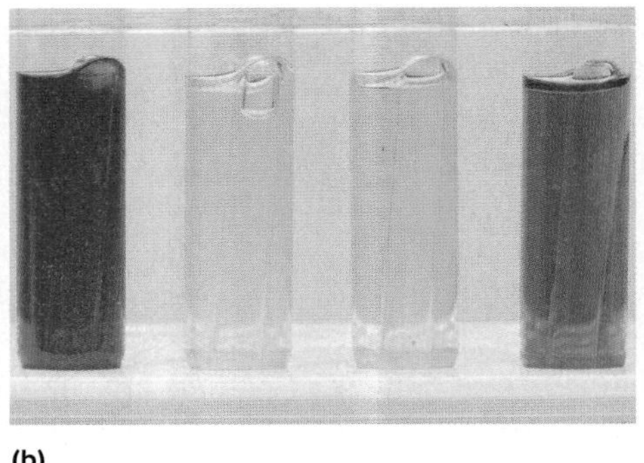

(b)

(c)

FIGURE 10.4 Biochemical Tests Results, left to right: **(a)** Catalase positive and negative, **(b)** Sugar fermentation negative, positive (acid and gas), positive (acid only), and uninoculated control. **(c)** Urease positive and uninoculated control. a: ©Denise Anderson; b and c: ©McGraw-Hill Education/Lisa Burgess, photographer

? What specifically causes the color to change in the sugar fermentation and urease tests?

urea to produce carbon dioxide and ammonia) contains urea and a pH indicator (see figure 10.4c).

The basic strategy for identifying bacteria based on biochemical tests relies on a **dichotomous key**, a series of alternative choices that lead to the identification of an organism (**figure 10.5**). In the case of bacterial identification, the choices are results of laboratory tests such as those listed in table 10.4. Because each test often requires an incubation period, however, it would be too time-consuming to proceed one step at a time. In addition, relying on a single biochemical test at each step could lead to misidentification. For example, if a strain that normally gives a positive result for a certain test lost the ability to produce a key enzyme, it would instead have a negative result. Therefore, several different biochemical tests are inoculated at the same time in order to identify the organism faster and more conclusively.

In certain cases, biochemical testing can be done without culturing the organism. *Helicobacter pylori,* the cause of most stomach ulcers, can be detected using the breath test, which assays for the presence of urease. The patient drinks a solution containing urea labeled with an isotope of carbon. If *H. pylori* is present, its urease breaks down the urea, releasing labeled CO_2, which escapes through the airway. Several hours after drinking the solution, the patient exhales into a balloon, and that air is then tested for labeled CO_2. This test is less invasive and, consequently, much cheaper and faster than the stomach biopsy that would be needed to culture the organism.

⏭ *Helicobacter pylori*, ⏮ isotope

Commercial Variations of Traditional Biochemical Tests

Several less labor-intensive commercial variations of traditional biochemical tests are available. The API system uses a strip holding a series of tiny cups that contain dried media (**figure 10.6**). A liquid suspension of the test bacterium is added to each compartment, inoculating as well as rehydrating the media. The media are similar to those used in traditional tests, giving rise to comparable color changes. After

FIGURE 10.5 Dichotomous Key This shows an example of steps that can be used to distinguish some of the common causes of urinary tract Infections. Additional tests may be done to confirm the identity of the pathogen.

? When identifying organisms, why would a lab technician inoculate most biochemical tests at the same time, rather than waiting for one result before starting the next test?

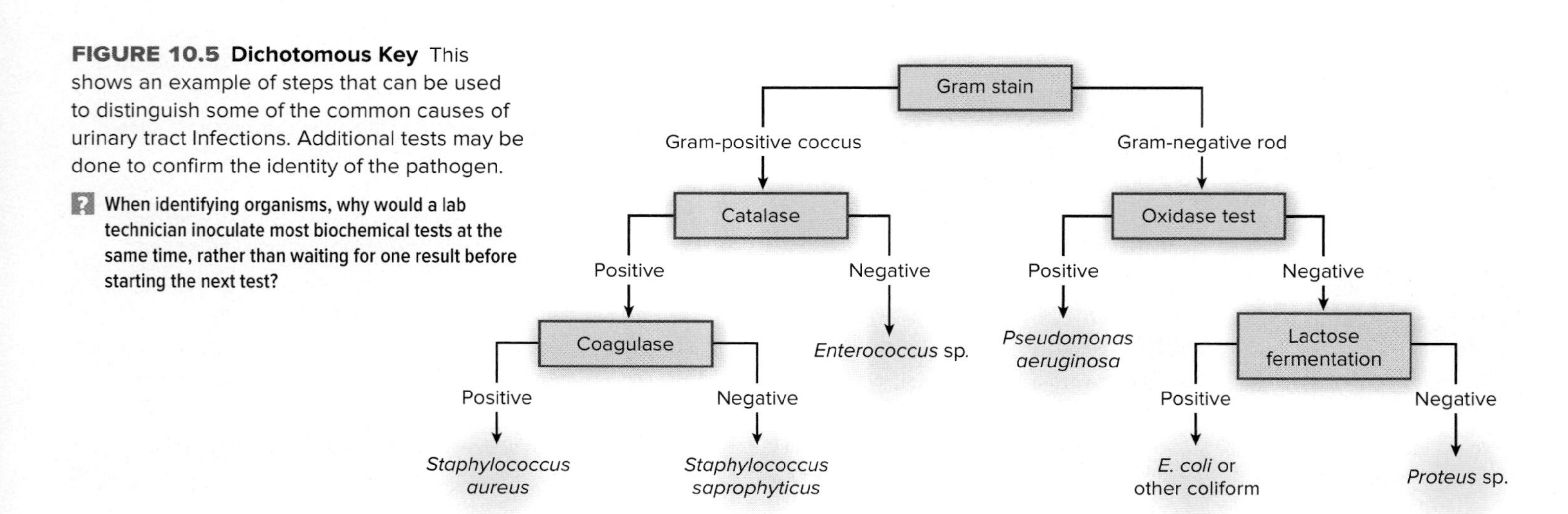

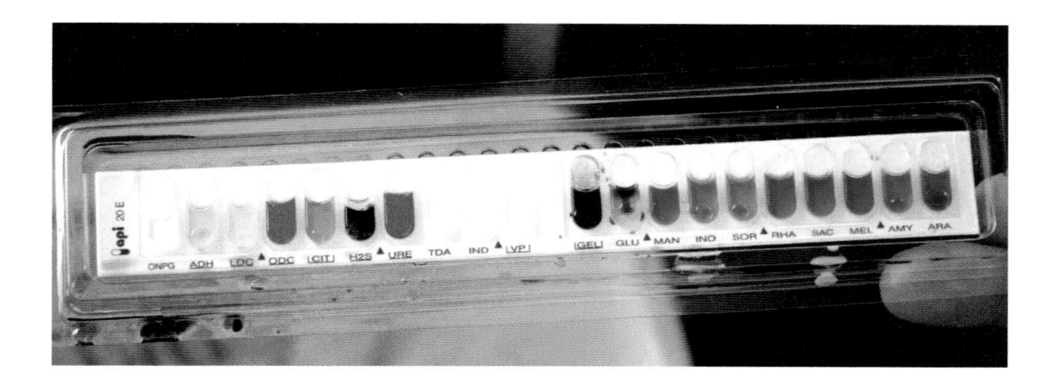

FIGURE 10.6 API Test Strip Each small cup contains a dried medium similar in formulation to the traditional tests. ©BSIP/Science Source

? **What advantage do commercial variations of biochemical tests have over traditional methods?**

a 16-hour incubation of the inoculated test strip, the results are determined by inspection. The pattern of results is converted to a numerical score, which is then entered into a computer to identify the organism. A system by Biolog uses a microtiter plate, a small tray containing nearly one hundred wells, to assay simultaneously an organism's ability to use a wide variety of carbon sources. Modifications of these plates allow researchers to characterize the metabolic capabilities of microbial communities, such as those in soil, water, or wastewater.

Highly automated systems also are available. The VITEK 2 system uses a miniature card that contains multiple wells with different types of dried media. After a relatively short incubation period (at least several hours, but sometimes longer), a computer reads the growth pattern in the wells.

Serological Testing

Serological testing uses antibodies to detect specific proteins and polysaccharides, a topic covered in chapter 18. A microbial cell's proteins and polysaccharides—particularly those that make up surface structures including the cell wall, capsule, flagella, and pili—are sometimes characteristic enough to be used as identifying markers. For example, certain species of *Streptococcus* contain a unique carbohydrate as part of their cell wall, and antibodies can be used to detect this molecule. Some of the serological methods, such as those used to confirm the identity of *S. pyogenes,* are quite simple and rapid. ▶▶| antibodies

Protein Profile

A relatively new technology that determines a microorganism's protein profile is revolutionizing microbial identification because of its speed—microorganisms in a colony can often be identified in less than 15 minutes. The technology, called **MALDI-TOF MS** (**m**atrix-**a**ssisted **l**aser **d**esorption **i**onization **t**ime **o**f **f**light **m**ass **s**pectrometry), is particularly important in clinical labs where rapid identification of bacterial isolates is crucial for patient care.

MALDI-TOF is a type of mass spectrometry, which is used to determine the chemical composition of a sample by measuring the masses of the various components. The process is not only faster than other rapid identification methods, it also requires less technical skill. First, a portion of a colony, along with a solution called a matrix, is spotted onto a sample plate (**figure 10.7** ①). ② That plate is then set into an apparatus that exposes the sample to a laser beam, a treatment that both converts molecules to a gaseous state (desorption) and ionizes them. The procedure is referred to as "soft ionization" because it leaves the large molecules relatively intact, rather than breaking them into small pieces. The ions then travel through what is called a flight tube. Because small ions travel faster than larger ones, this separates and sorts the molecules by mass ("time of flight"). A detector records the ions as they arrive, generating a pattern of peaks called a mass spectrum. ③ This is essentially a "fingerprint" or profile of the proteins and other macromolecules in the cell, particularly ribosomal proteins. Computer software then uses that profile to identify a given microbial colony by comparing its mass spectrum to a database of results from known organisms. |◀◀ ion

MicroAssessment 10.2

The size, shape, and staining characteristics of a microorganism give important clues to its identity. Culture characteristics provide additional information, but conclusive identification often relies on multiple biochemical tests. The proteins and polysaccharides that make up a bacterium are sometimes unique enough to be considered identifying markers. MALDI-TOF mass spectrometry characterizes an organism's proteins, and the profile generated can be used as an identifying trait.

4. How does MacConkey agar help identify the cause of a urinary tract infection?

5. Describe two methods to test for the enzyme urease.

6. A sample must contain many microbial cells for any to be seen by microscopic examination. Why?

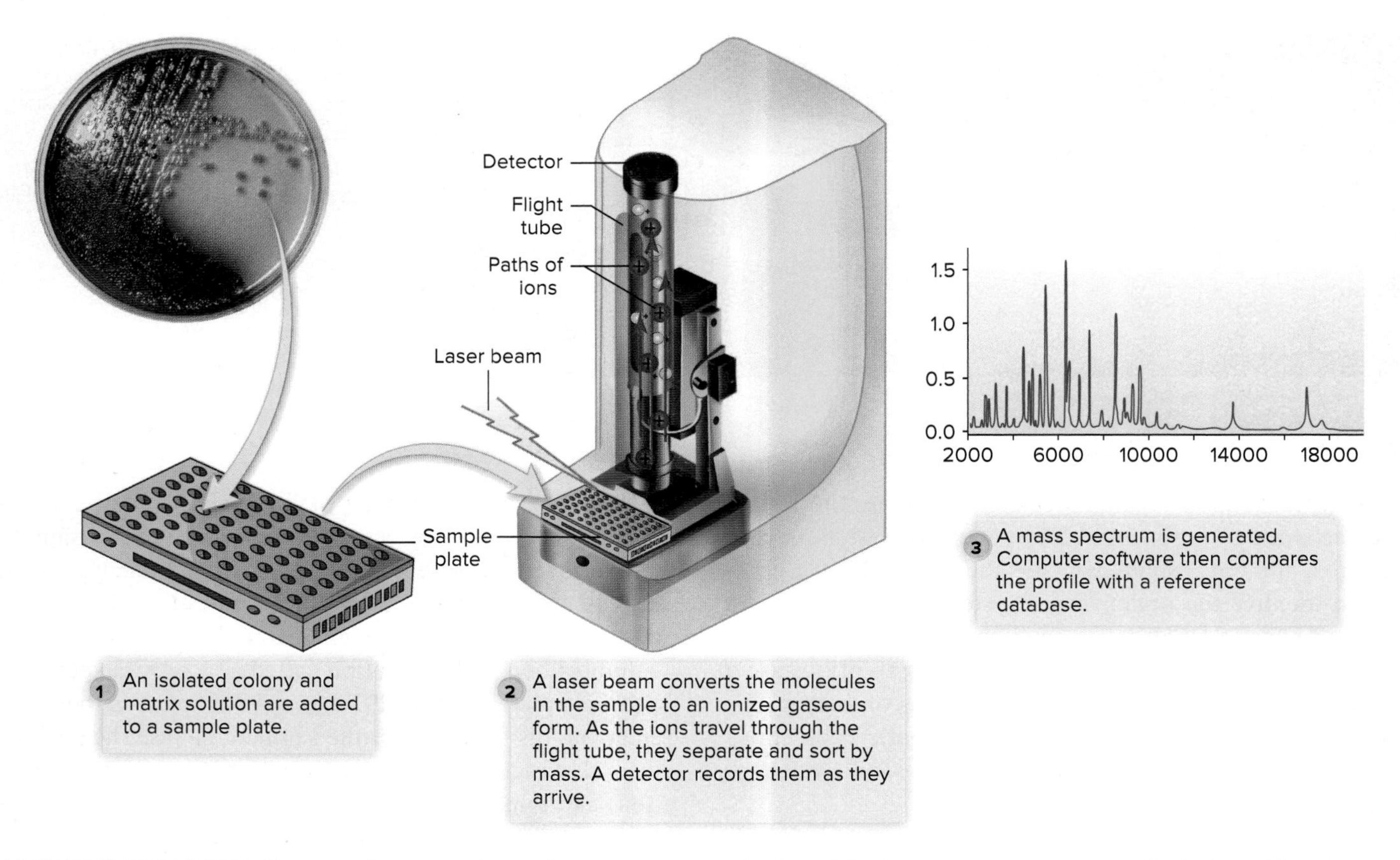

FIGURE 10.7 MALDI-TOF The procedure separates and sorts a microorganism's proteins by mass to generate a profile that provides a fast way to identify a colony. ©McGraw-Hill Education/Lisa Burgess, photographer

In a clinical lab, what advantage does MALDI-TOF have over traditional biochemical testing?

10.3 ■ Identification Methods Based on Genotype

Learning Outcome

3. Describe how nucleic acid probes, NAATs, and sequencing 16S rRNA genes can be used to identify microorganisms.

Many of the technologies discussed in chapter 9 can be used to identify a microorganism based on its genotype. Some of these methods even make it possible to identify organisms that cannot yet be grown in culture.

Detecting Specific Nucleotide Sequences

Nucleic acid probes and nucleic acid amplification tests (NAATs) can both be used to detect nucleotide sequences unique to a given species or related group. A significant limitation, however, is that each probe or amplification detects only a single possibility. If the organism in question could be one of five different species or related groups, then five distinguishable probes or amplifications would be needed. ◀◀ DNA probe

Nucleic Acid Probes

A nucleic acid probe can locate a nucleotide sequence that characterizes a particular species or group (**figure 10.8**). The probe is a single-stranded piece of nucleic acid, usually DNA, labeled with a detectable tag such as a radioisotope or a fluorescent dye. It is complementary to the sequence of interest.

Most methods that use nucleic acid probes to detect DNA sequences rely on a step that increases the amount of DNA in the sample. This can be done by inoculating the specimen onto an agar medium so that each microbial cell multiplies, forming a colony. Alternatively, a preliminary in vitro DNA amplification step such as PCR can be done. ◀◀ polymerase chain reaction (PCR)

Fluorescence in situ hybridization (FISH) often uses probes that bind 16S ribosomal RNA (rRNA). Because of this, an amplification step is not needed since numerous copies of rRNA are naturally present in multiplying cells. Various different probes that bind rRNA are available, each specific for a given **signature sequence.** A signature sequence is a nucleotide sequence in rRNA that characterizes either a certain species or a group of related organisms (see figure 9.21). ◀◀ fluorescence in situ hybridization (FISH)

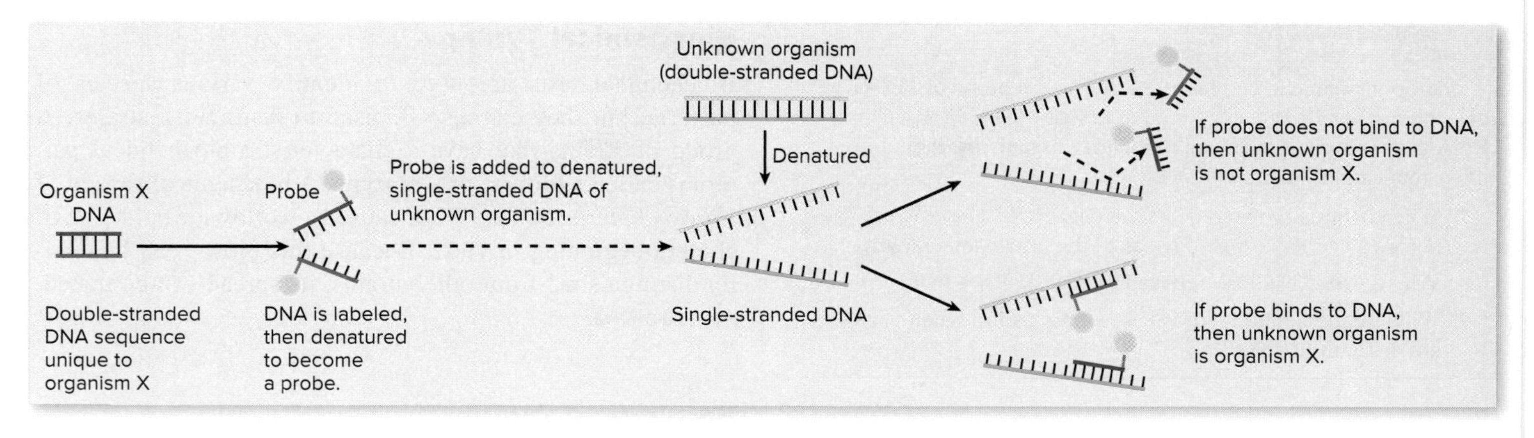

FIGURE 10.8 Nucleic Acid Probes Detect Specific DNA Sequences The probe, a single-stranded piece of nucleic acid labeled with a detectable marker, is used to locate a unique nucleotide sequence that identifies a particular microbial species.

What type of label is used for fluorescence in situ hybridization (FISH) probes?

Nucleic Acid Amplification Tests (NAATs)

Several methods, referred to as **nucleic acid amplification tests (NAATs),** can be used to increase the number of copies of specific DNA sequences that serve as identifying markers. This allows researchers to detect the sequences in samples such as body fluids, soil, food, and water. These methods can be used to detect organisms present in extremely small numbers as well as those that cannot yet be grown in culture. In most cases, the amplified fragment is visible as a distinct band when the DNA is separated on an agarose gel, stained, and illuminated with UV light. Alternatively, a DNA probe can be used to detect the amplified DNA. ⏮ **DNA gel electrophoresis**

One of the most common NAATs is the polymerase chain reaction (PCR; see figure 9.15). To use PCR to detect a microbe of interest, a sample is treated to release and denature the DNA. Specific primers and other ingredients are then added (see figure 9.16). After 30 cycles of PCR, the target DNA will have been amplified approximately a billion-fold (see figure 9.18).

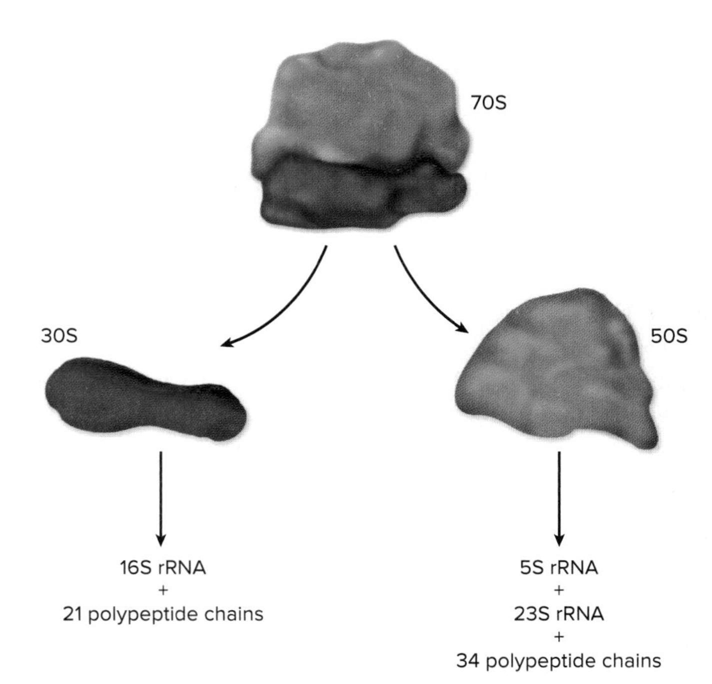

FIGURE 10.9 Ribosomal Components The 70S ribosome of prokaryotes is composed of three types of rRNA (5S, 16S, and 23S) and an assortment of different ribosomal proteins.

Which type of ribosomal RNA is most often used in taxonomy?

Sequencing Ribosomal RNA Genes

The nucleotide sequence of ribosomal RNA molecules **(rRNAs),** or the DNA that encodes them **(rDNAs),** can be used to identify microbes (**figure 10.9**). rRNAs are useful in microbial classification and identification because their sequences are relatively stable; the ribosome would not function with too many mutations. Of the different prokaryotic rRNAs (5S, 16S, and 23S), the 16S molecule is the most useful in taxonomy because of its moderate size (approximately 1,500 nucleotides). 16S RNA and its eukaryotic counterpart, 18S RNA, are called small subunit (SS or SSU) rRNAs because they are part of the small subunit of the ribosomes. Once the nucleotide sequence of an unknown organism's SSU rRNA has been determined, a researcher can compare it with sequences of known organisms by searching extensive computerized databases.

Using rDNA to Identify Uncultivated Organisms

The vast majority of microbes cannot yet be grown in culture. However, the DNA from these organisms can be amplified, cloned, and sequenced, making it possible to detect and identify them. The bacterium that causes Whipple disease, a rare intestinal illness, was identified this way. The organism was given the name *Tropheryma whipplei,* and a specific probe was then developed to detect it in intestinal tissue, well before it could be grown in culture.

MicroAssessment 10.3

A microorganism can be identified by using a probe or NAAT to detect a nucleotide sequence unique to that organism. Ribosomal RNA genes can be sequenced to identify organisms, including those that cannot be grown in culture.

7. When using a probe to detect an organism, why is it necessary to have some idea as to the organism's identity?
8. Why is 16S RNA also referred to as SSU RNA?
9. Why are molecular methods especially useful when bacteria are difficult to grow?

10.4 ■ Characterizing Strain Differences

Learning Outcome

4. Describe five distinct methods to distinguish different strains.

The ability to distinguish different strains of a given species is useful in some situations, particularly when tracking the source of foodborne diseases. In 2016, for example, 33 reported cases of severe diarrheal disease across 23 states in the United States involved a specific strain of *Salmonella enterica*. Most patients reported eating a powdered meal replacement prior to their illness, leading to a recall of the product. Linking the 33 cases among the many thousands of severe diarrheal cases that occur nationwide each year would be impossible without methods to distinguish different strains.

Characterizing strain differences is not limited to investigations of foodborne illness. It also plays an important role in forensic investigations of bioterrorism and other biocrimes, and in diagnosing certain diseases. The methods used to characterize different strains are summarized in **table 10.5**.

Biochemical Typing

Biochemical tests are used to identify various species of bacteria, but they can also be used to distinguish strains. A group of strains that have a characteristic biochemical pattern is called a **biovar**, or a **biotype**. A biochemical variant of *Vibrio cholerae* called El Tor caused a worldwide epidemic of cholera beginning in 1961. Because this biovar can be readily distinguished from other strains, its spread can be traced.

⏭ *Vibrio cholerae*

Serological Typing

Proteins and carbohydrates that vary among strains can be used as distinguishing markers. For example, different strains of *E. coli* and related bacteria can be distinguished by the antigenic type of their flagella, capsules, and lipopolysaccharide molecules (**figure 10.10**). The "O157:H7" designation of *E. coli* O157:H7 refers to the antigenic type of its lipopolysaccharide (the O antigen) and flagella (the H antigen). A group of strains that have cell surface antigens different from other strains is called a **serovar**, or a **serotype**. ⏭ antigen, ⏮ lipopolysaccharide

Molecular Typing

Subtle differences in DNA sequences can be used to distinguish among phenotypically identical strains. One method of doing this is to compare the patterns of fragment sizes produced when the same restriction enzyme is used to digest DNA from each isolate; the different patterns are called **restriction fragment length polymorphisms (RFLPs)**.

Gel electrophoresis is used to separate the fragments so they can be observed. Isolates of the same species that have different RFLPs are considered different strains, whereas those that have identical RFLPs may or may not be the same strain (**figure 10.11**). ⏮ restriction enzymes

TABLE 10.5 Summary of Methods Used to Characterize Different Strains

Method	Comment
Biochemical typing	Biochemical tests are most commonly used to identify various species of bacteria, but in some cases they can be used to distinguish different strains. A group of strains that have a characteristic biochemical pattern is called a biovar or a biotype.
Serological typing	Proteins and carbohydrates that vary among strains can be used to differentiate strains. A group of strains that have a characteristic serological type is called a serovar or a serotype.
Molecular typing	Gel electrophoresis can be used to detect restriction fragment length polymorphisms (RFLPs). Whole-genome sequencing (WGS) is increasingly being used to detect differences.
Phage typing	Strains of a given species sometimes differ in their susceptibility to various types of bacteriophages.
Antibiograms	Antibiotic susceptibility patterns can be used to characterize strains.

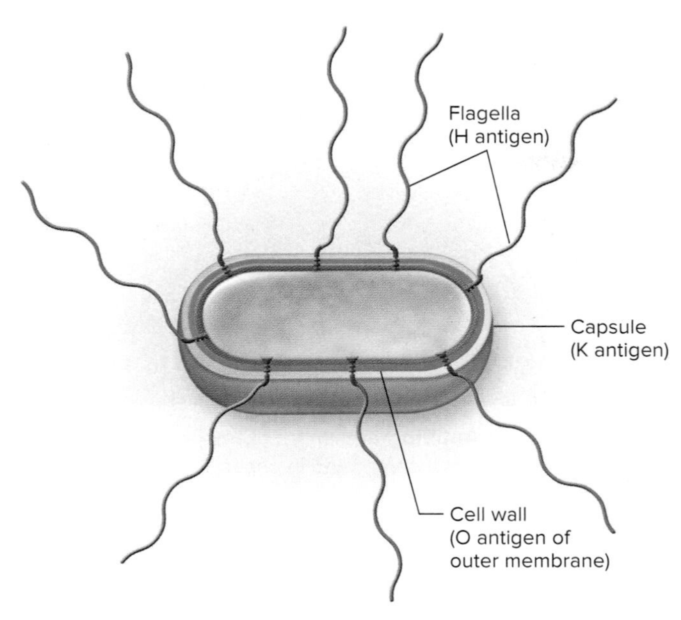

FIGURE 10.10 Serotypes The cell structures used to distinguish different strains of members of the family *Enterobacteriaceae* are shown.

What structures are reflected in the "O157:H7" of *E. coli* O157:H7?

The CDC established a program called **PulseNet** to make it easier for public health and food regulatory agencies to use RFLP data to track foodborne disease outbreaks. Laboratories from around the country can submit RFLP patterns of foodborne pathogens to a computer database and quickly receive information about other isolates showing the same pattern. Using this database, multistate foodborne disease outbreaks can more readily be recognized and traced. This is how the diarrheal cases that led to the powdered meal replacement product recall were found to be related. ⏭ CDC

Advances in DNA sequencing methods are making it practical to sequence and compare the DNA of different isolates. Whole genome sequencing (WGS) has been used to trace outbreaks caused by strains so closely related that RFLP patterns cannot reliably distinguish them. The FDA recently began organizing what is called **Genome Trakr,** a network of laboratories that collect WGS data for use in tracking outbreaks caused by various foodborne pathogens.

Phage Typing

Strains of a given species may differ in their susceptibility to bacteriophages. **Bacteriophages,** or phages, are viruses that infect bacteria, often lysing them; they will be described in more detail in chapter 13. The susceptibility of an organism to a particular type of phage can be easily determined. First, a culture of the test organism is inoculated into melted, cooled nutrient agar and poured onto the surface of an agar plate, creating a uniform layer of cells. Drops of different types of bacteriophage are then placed on the surface of the

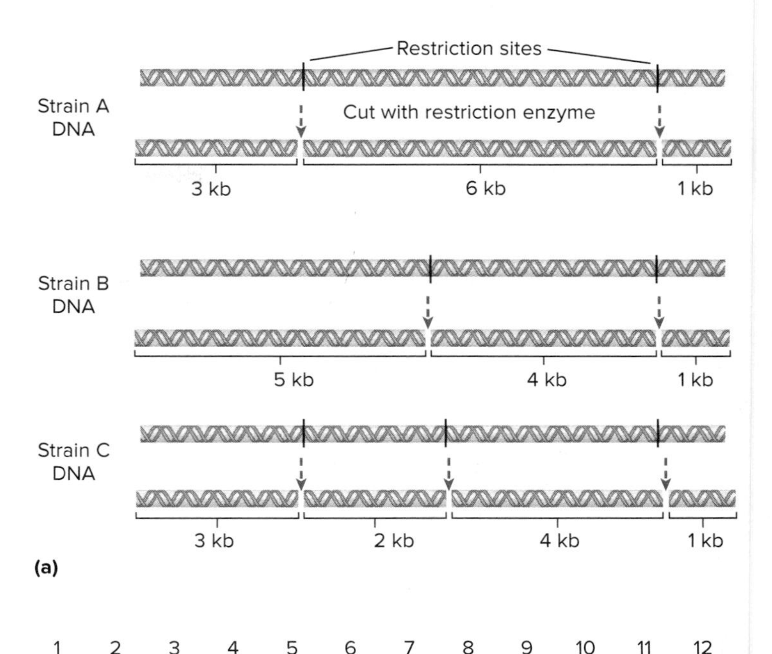

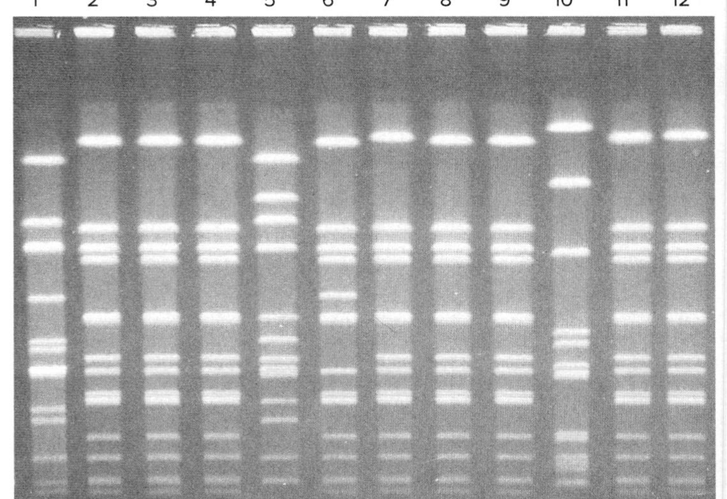

FIGURE 10.11 Restriction Fragment Length Polymorphisms (RFLPs) **(a)** Different strains of a species may have subtle variations in nucleotide sequences that give rise to a slightly different assortment of restriction fragment sizes (1 kb = 1,000 base pairs). **(b)** In the method shown, genomic DNA is digested with a restriction enzyme that cuts infrequently; the resulting fragments are separated by gel electrophoresis and then stained with ethidium bromide. The results in lanes 2, 3, 4, 7, 8, 9, 11, and 12 indicate that those isolates likely originated from the same source. ©Patricia Moor, Gadi Research Center, University of Canberra

How does a database of RFLP patterns help scientists detect multistate foodborne disease outbreaks?

agar. During incubation, the bacteria multiply, forming a visible layer of turbidity. If the bacterial strain is susceptible to a specific type of phage that lyses its host, a clear area will form at the spot where the drop was added. The patterns

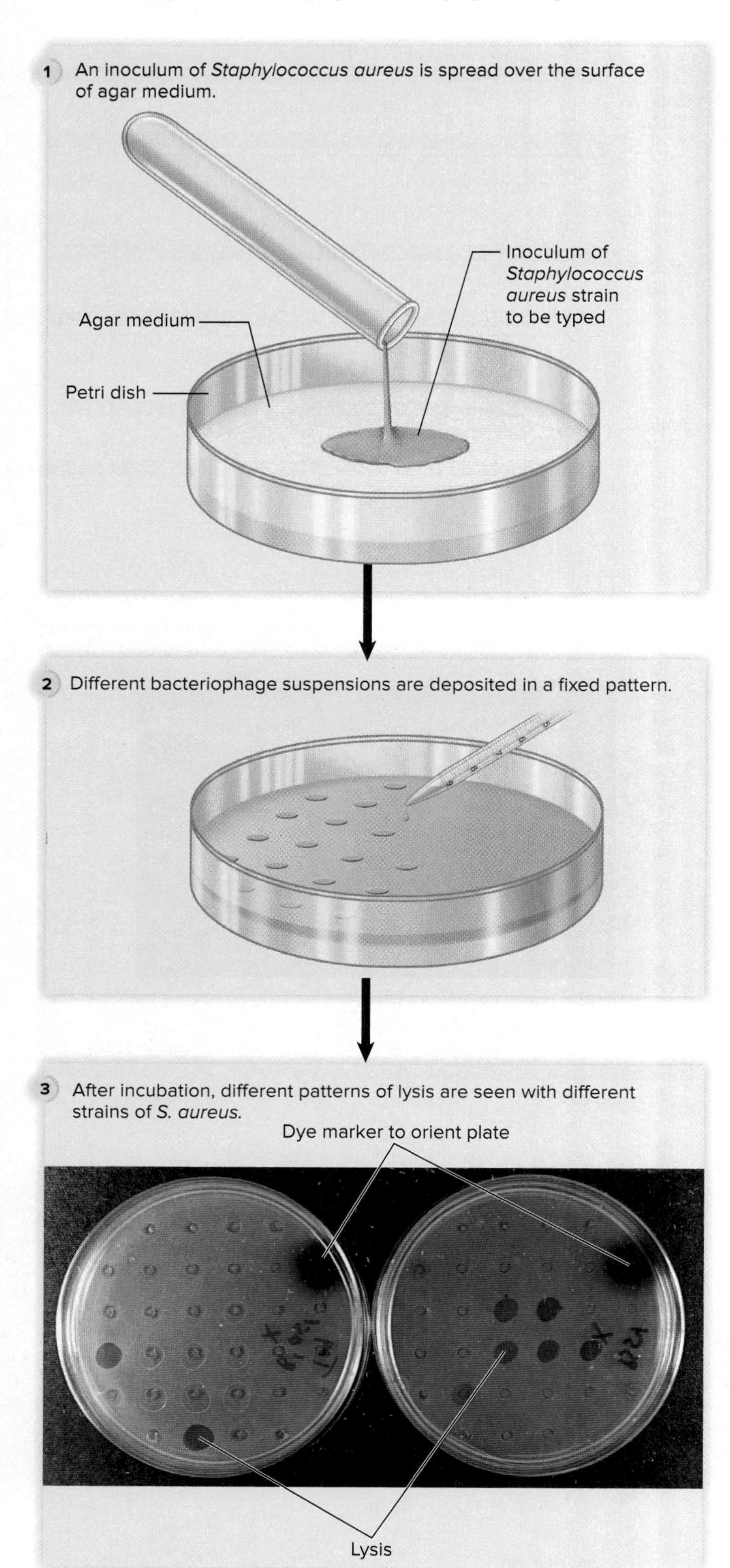

FIGURE 10.12 Phage Typing ©Evans Roberts

Why would a lab technician do phage typing rather than molecular typing?

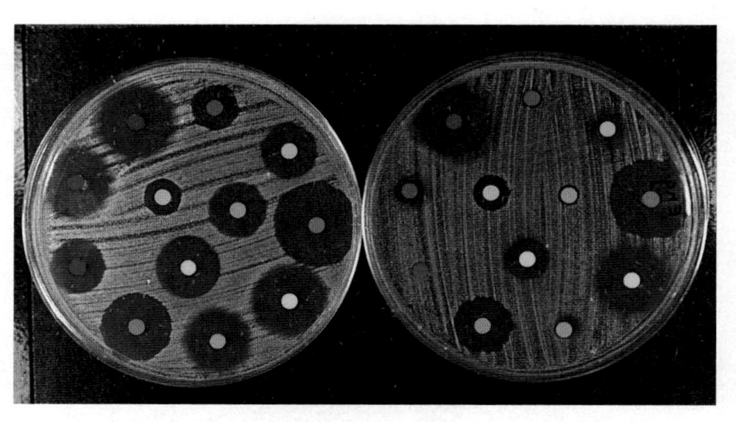

FIGURE 10.13 An Antibiogram In this example, 12 different antimicrobial medications incorporated in paper discs have been placed on two plates containing different cultures of *Staphylococcus aureus*. Clear areas are where bacterial growth has been inhibited. ©Evans Roberts

How can you tell that these *S. aureus* isolates are different strains?

of clearing indicate the susceptibility of the test organism to different phages (**figure 10.12**). Bacteriophage typing has now largely been replaced by molecular methods that detect genomic differences, but it is still a useful tool, particularly for laboratories that lack equipment to do molecular typing. ⏭| bacteriophage

Antibiograms

Antimicrobial susceptibility patterns, or **antibiograms**, can distinguish different strains. As with phage typing, this method has largely been replaced by molecular techniques. To determine the antibiogram, a culture is uniformly inoculated onto the surface of nutrient agar. Paper discs containing different antibiotics or other antimicrobial medications are then placed on the agar. After incubation, clear areas (zones of inhibition) will be visible around discs of antimicrobials that inhibit or kill the organism (**figure 10.13**). ⏭| antibiotic

MicroAssessment 10.4

Strains of a given species may differ in phenotypic characteristics such as biochemical capabilities, protein and polysaccharide components, susceptibility to bacteriophages, and sensitivity to various antimicrobial medications. Molecular techniques can be used to detect genomic differences between strains that are phenotypically identical.

10. Explain the difference between a biotype and a serotype.
11. Describe the significance of RFLPs in distinguishing between strains.

FOCUS ON A CASE 10.1

Wisconsin state health authorities alerted the CDC about an *E. coli* O157:H7 outbreak. Soon thereafter, officials from Oregon and New Mexico reported a similar outbreak. Within days, the CDC determined that the strains from Wisconsin matched those of the other two states, and that the patients involved recalled eating pre-packaged fresh spinach. The CDC then issued a press release advising people not to eat bagged fresh spinach, and a company that produces several brands of bagged spinach announced a voluntary recall of all fresh spinach products. In the end, the implicated strain was found to have caused 205 cases of illness in 26 states, resulting in three deaths. ▶▶| *E. coli* O157:H7

1. How could the clinical laboratory separate and distinguish *E. coli* O157:H7 from the many other *E. coli* strains that normally inhabit the human intestine?
2. How could the CDC determine that the strains from the three states originated from the same source?

Discussion

1. To identify *E. coli* O157:H7 in a stool specimen, the sample is inoculated onto a special agar medium designed to distinguish it from strains that typically inhabit the large intestine. One such medium is sorbitol-MacConkey, a modified version of MacConkey agar that contains the carbohydrate sorbitol in place of lactose. On this medium, most *E. coli* O157:H7 isolates are colorless because they do not ferment sorbitol. In contrast, common strains of *E. coli* ferment sorbitol, giving rise to pink colonies. Serological testing is then used to determine if the colorless *E. coli* colonies are serotype O157; strains that test positive are then generally tested to confirm they are serotype H7.
2. To determine if two isolates of *E. coli* O157:H7 originated from the same source, DNA is extracted and purified from each isolate and is then digested with restriction enzymes. Gel electrophoresis is generally used to compare the resulting restriction fragment length polymorphism (RFLP) patterns of the isolates (see figure 10.11). Those that have identical patterns are presumed to have originated from the same source. Patients from whom those isolates originated can then be questioned to determine where they likely contacted the disease-causing organism. After that, culture methods are used to try to isolate the organism from the suspected food source. If those attempts are successful, the RFLP patterns of the resulting isolates are then compared with those of the related cases.

10.5 ■ Classifying Microorganisms

Learning Outcome

5. Describe how sequence analysis of ribosomal components, DNA hybridization, and G + C content are used to classify microorganisms.

Phylogeny (evolutionary relatedness) of prokaryotes was difficult to determine before molecular methods such as DNA sequencing techniques were developed, so classification schemes historically grouped these organisms by shape, staining characteristics, metabolic capabilities, and other phenotypic traits. Although this was convenient, phenotypic differences can be due to only a few gene products, and a single mutation resulting in a non-functional enzyme can change an organism's capabilities. In addition, phenotypically similar organisms may be only distantly related; conversely, those that appear dissimilar may be closely related.

Molecular techniques such as DNA sequencing revolutionized classification by providing insight into the amount of time that has passed since two organisms diverged from a common ancestor. Random mutations cause an organism's nucleotide sequences to change over time, so the more time that has passed since two organisms diverged, the greater the differences in their sequences. By comparing DNA sequences of different organisms, scientists can construct a **phylogenetic tree,** a diagram that depicts the evolutionary heritage of organisms (**figure 10.14**).

Table 10.6 summarizes some of the methods used to determine the relatedness of microorganisms for the purpose of classification.

Sequence Analysis of Ribosomal Components

The sequences of ribosomal components (ribosomal RNA and ribosomal proteins) are considered reliable indicators of evolutionary relatedness because ribosomes are present in all organisms and perform crucial and functionally constant tasks. Only a limited number of mutations in the genes that encode the components can happen without affecting the survival of an organism. Because of this, some of the sequences will still be similar even in groups of organisms that diverged long ago. In addition, the ribosomal genes are not commonly horizontally transferred—an event that would complicate an analysis of evolutionary relatedness. |◀◀ horizontal gene transfer

Ribosomal RNA (rRNA)

Nucleotide sequence analysis of **rRNAs** (ribosomal RNAs) revolutionized microbial classification by giving rise to the

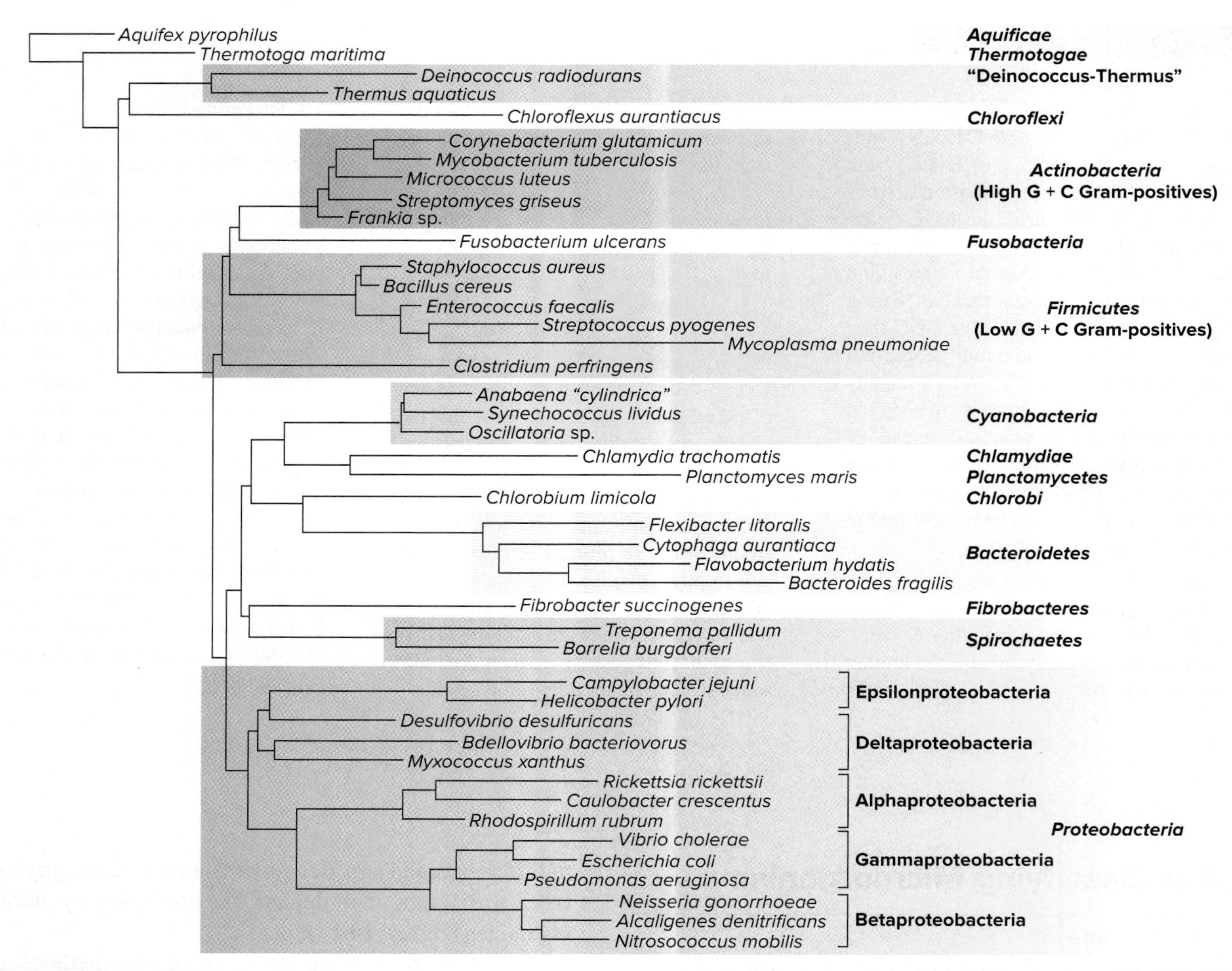

FIGURE 10.14 Phylogenetic Tree of the Bacteria Each branch represents the evolutionary distance between two species.

? Based on this tree, which are more closely related—*Deinococcus radiodurans* and *Thermus aquaticus* or *Bacillus cereus* and *Clostridium perfringens*?

three-domain system. With respect to classification, small ribosomal subunit rRNAs (SSU rRNAs) are the most useful because of their moderate size (see figure 10.9). Recall that in bacteria and archaea, the SSU rRNA is 16S; in eukaryotes it is 18S. Today, **rDNAs** (the genes that encode rRNAs) are usually studied because of the ease of DNA sequencing methods.

Even microorganisms that cannot yet be grown in culture can be tentatively classified by SSU rDNA sequence analysis.

TABLE 10.6 Methods Used to Determine the Relatedness of Different Microorganisms

Method	Comment
Genotypic Characteristics	Differences in DNA sequences can be used to determine the point in time at which two organisms diverged from a common ancestor.
Sequence analysis of ribosomal components	Certain regions of the SSU rDNA (16S in bacteria and archaea and 18S in eukaryotes) can be used to determine distant relatedness of diverse organisms; other regions can be used to determine more recent divergence. Amino acid sequences of ribosomal proteins can also be compared.
DNA hybridization	The extent of nucleotide sequence similarity between two isolates can be determined by measuring how completely single strands of their DNA hybridize to one another.
G + C content	Determining the G + C content offers a crude comparison of genomes. Organisms with identical G + C contents can be entirely unrelated, however.
Phenotypic Characteristics	Traditionally, relatedness of different bacteria has been decided by comparing phenotypes. These characteristics, however, do not necessarily reflect the evolutionary relatedness of organisms.

DNA can be extracted from environmental samples such as soil and water, and the SSU rDNA then amplified using PCR, cloned, and sequenced. The sequences can then be compared with databases containing SSU rDNA sequences of known organisms.

Although SSU rDNA sequence analysis has been very helpful in determining the phylogeny of distantly related organisms, it is often unreliable for distinguishing closely related species. This is because closely related but genetically distinct microorganisms can have identical SSU rDNA sequences. In these cases, DNA hybridization is a better tool to assess relatedness.

Ribosomal Proteins

With the widespread availability of whole and partial genome sequences, scientists can now use the deduced amino acid sequences of various ribosomal proteins as a basis for determining relatedness. An advantage of this approach is that it accounts for novel organisms not detected by standard techniques used to amplify SSU rDNA. In addition, each organism has an assortment of different ribosomal proteins (see figure 10.9), so more gene products can be compared. The greatly expanded tree of life shown early in the chapter is based on comparisons of the amino acid sequences of 16 different ribosomal proteins from thousands of different species (see figure 10.1b). As more microbial genome sequences are determined, the tree is likely to become even more complex.

DNA Hybridization

The extent of nucleotide sequence similarity between two organisms can be determined by measuring how completely single strands of their DNA hybridize to each other. Just as the complementary strands of DNA from one organism will anneal (base-pair), so will homologous DNA of a different organism. The extent of hybridization reflects the degree of sequence similarity. Two bacterial strains that show at least 70% similarity are generally considered to be members of the same species. Surprisingly, DNA hybridization studies have shown that members of the genera *Shigella* and *Escherichia,* which are quite different based on biochemical tests, should actually be grouped in the same species. Note that human and chimpanzee DNA have approximately 99% similarity by DNA hybridization studies. Therefore, by the criteria used to classify bacteria, humans and chimpanzees would be members of the same species! ⏮ DNA hybridization

G + C Content

One way to roughly compare the genomes of different bacteria is to determine their **G + C content** (also called the GC content), which is the percentage of G-C base pairs in an organism's DNA. If the GC content of two organisms differs by more than a small percentage, they cannot be closely related. A similarity of base compositions, however, does not necessarily mean that the organisms are related, because the nucleotide sequences and genome sizes could differ greatly.

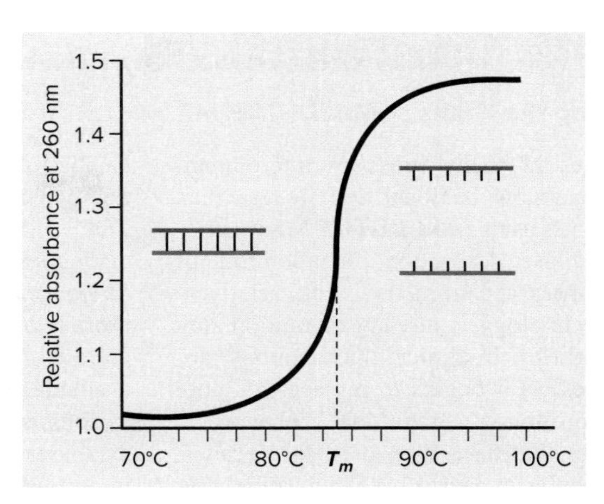

FIGURE 10.15 A DNA Melting Curve The absorbance (relative absorbance at 260 nm) rapidly increases as double-stranded DNA denatures (melts).

What characteristic of a DNA sequence does the T_m reflect?

The GC content is often measured by determining the temperature at which the double-stranded DNA denatures, or melts. DNA that has a high GC content melts at a higher temperature because G-C base pairs are held together by three hydrogen bonds, whereas A-T pairs are held together by only two hydrogen bonds. The temperature at which double-stranded DNA melts can be determined by measuring the absorbance of UV light by a solution of DNA as it is heated. The absorbance rapidly increases as the DNA denatures (**figure 10.15**).

Phenotypic Methods

Classification schemes that group microorganisms by phenotype have largely been replaced by methods that rely on DNA sequence data. Some taxonomists, however, believe that classification should be based on more than just genotypic traits. Regardless of the classification scheme used, phenotypic methods are still important because they provide a foundation for microbial identification.

MicroAssessment 10.5

Analyzing and comparing the sequences of ribosomal RNA components has revolutionized the classification of microorganisms. Other characteristics used to determine relatedness include DNA hybridization, GC content, and phenotypic properties.

12. Explain why ribosomal components are useful for determining phylogeny.
13. Explain the role of DNA hybridization in classification.
14. Why would it be easier to sequence rDNA than to sequence rRNA?

FOCUS ON THE FUTURE 10.1

Pushing the Limits of MALDI-TOF MS

Colonies of many types of microorganisms can now be identified in less than 15 minutes using **MALDI-TOF MS** (matrix-assisted laser desorption ionization time of flight mass spectrometry). This relatively new technology is already revolutionizing the workflow in clinical microbiology laboratories, as it begins to replace the more time-consuming traditional phenotypic methods that have been used for decades. The current impact of MALDI-TOF, however, may be only the tip of the iceberg. ⏮ MALDI-TOF MS

One of the limitations of MALDI-TOF MS relates to the quality of the reference databases. To identify an unknown organism, the protein profile of that organism must be matched to the profile of a known organism in a preexisting database. If a database does not contain information for certain microorganisms, this limits the usefulness of the technology. Thus, the more MALDI-TOF MS is used, the better the databases and the more reliable the technology will become. For example, MALDI-TOF MS is currently unreliable for distinguishing viridans streptococci from *Streptococcus pneumoniae* because protein profiles of the organisms are so similar. As more profiles from different strains become available, it may be easier to distinguish the various isolates. In addition, the reference databases currently focus on medically important microorganisms, but as the protein profiles of a wider range of microorganisms become available, MALDI-TOF MS could also prove to be an invaluable tool for studying other groups of microbes as well. ⏭ viridans streptococci, ⏭ *Streptococcus pneumoniae*

MALDI-TOF MS also shows promise for identifying microorganisms directly in clinical specimens such as urine and blood. Currently, the procedure requires colonies of pure cultures, which means that an organism cannot be identified until it has formed colonies on an agar plate—a step that requires a relatively long incubation period. In some cases, it should be possible to bypass the need for culture, and instead identify pathogens directly in clinical specimens. Urine and blood are prime candidates for this, because infections in those sites are typically monomicrobial, meaning that only one type of microbe is causing the infection.

In addition to using MALDI-TOF MS to identify microorganisms, the technology may be reliable for rapidly detecting resistance to antimicrobial medications. The resistance is often due to a microbe-produced enzyme that destroys the medication; in some cases, that enzyme-induced change can be detected using MALDI-TOF MS. Traditional methods of determining a microorganism's antimicrobial resistance typically take at least one day. Faster results allow appropriate treatments to start sooner.

Summary

10.1 ■ Principles of Taxonomy

Taxonomy consists of three interrelated areas: **identification**, **classification**, and **nomenclature**.

Strategies Used to Identify Microorganisms

To characterize and identify microorganisms, a wide assortment of technologies is used, including microscopic examination, cultural characteristics, biochemical tests, and nucleic acid analysis.

Strategies Used to Classify Microorganisms

Taxonomic classification categories are arranged in a hierarchical order, with the species being the basic unit. Taxonomic categories include **species**, **genus**, **order**, **class**, **phylum** (or **division**), **kingdom**, and **domain**. Individual **strains** within a species vary in minor properties (table 10.1).

Nomenclature

Microorganisms are assigned names governed by official rules.

10.2 ■ Identification Methods Based on Phenotype (table 10.3)

Microscopic Morphology

The size, shape, and staining characteristics of a microorganism yield important clues as to its identity (figures 10.2, 10.3).

Culture Characteristics

Selective and differential media used in the isolation process can provide information that helps identify an organism.

Metabolic Capabilities

Most biochemical tests rely on a pH indicator or chemical reaction that shows a color change when a compound is degraded. The basic strategy for identification using biochemical tests relies on the use of a **dichotomous key** (figures 10.4, 10.5; table 10.4).

Serological Testing

Proteins and polysaccharides that make up a prokaryote's surface are sometimes characteristic enough to be identifying markers.

Protein Profile

MALDI-TOF MS (matrix assisted laser desorption ionization time of flight mass spectrometry) generates a profile of a colony's proteins and macromolecules, which can be used to identify the organism (figure 10.7).

10.3 ■ Identification Methods Based on Genotype

Detecting Specific Nucleotide Sequences

A probe complementary to a sequence unique to a given microbe is used to detect that organism (figure 10.8). **Nucleic acid amplification tests (NAATs)** such as PCR increase the number of copies of a specific nucleotide sequence, thereby determining if a given organism is present.

Sequencing Ribosomal RNA Genes

The nucleotide sequence of **ribosomal RNA (rRNA)** can be used to identify prokaryotes. Newer techniques simply sequence **rDNA.** Organisms that cannot yet be cultivated can be identified by amplifying, cloning, and then sequencing specific regions of rDNA.

10.4 Characterizing Strain Differences (table 10.5)

Biochemical Typing

A group of strains that has a characteristic biochemical variation is called a **biovar**, or a **biotype**.

Serological Typing

A group of strains that differs serologically from other strains is called a **serovar,** or a **serotype** (figure 10.10).

Molecular Typing

Two isolates that have different **restriction fragment length polymorphisms (RFLPs)** are considered different strains (figure 10.10). Whole-genome sequencing (WGS) can now be done to compare isolates.

Phage Typing

The susceptibility to various types of bacteriophages can be used to demonstrate strain differences (figure 10.12).

Antibiograms

Antimicrobial susceptibility patterns can be used to distinguish strains (figure 10.13).

10.5 Classifying Microorganisms (table 10.6)

DNA sequences can be used to construct a **phylogenetic tree** (figure 10.14).

Sequence Analysis of Ribosomal Components

Analyzing and comparing the nucleotide sequences of **rRNA** and, more recently, **rDNA** has revolutionized the classification of organisms. The amino acid sequences of ribosomal proteins can also be compared.

DNA Hybridization

The extent of nucleotide similarity between two organisms can be determined by measuring how completely single strands of their DNA will anneal to each other.

G + C Content

The **G + C content** can be measured by determining the temperature at which double-stranded DNA melts (figure 10.15).

Phenotypic Methods

Classification schemes that group microorganisms by phenotype have largely been replaced by methods that rely on DNA sequence data.

Review Questions

Short Answer

1. Name and describe the three areas of taxonomy.
2. Explain the basis for the three-domain systems of classification.
3. Describe how a dichotomous key is used when identifying bacteria.
4. Why is MALDI-TOF MS particularly important in clinical microbiology?
5. Describe the difference between using a probe and using PCR to detect a specific sequence.
6. Explain how signature sequences are used in bacterial identification.
7. Describe the function of PulseNet.
8. What is a phylogenetic tree?
9. List two advantages of using sequence analysis of ribosomal components in classification.
10. Describe how the GC content of DNA can be measured.

Multiple Choice

1. Which of the following is the newest taxonomic unit?
   a) Strain
   b) Family
   c) Order
   d) Species
   e) Domain
2. If an acid-fast bacterium is detected in a clinical sample, then the organism could be
   a) *Cryptococcus neoformans.*
   b) *Mycobacterium tuberculosis.*
   c) *Neisseria gonorrhoeae.*
   d) *Streptococcus pneumoniae.*
   e) *Streptococcus pyogenes.*
3. The "breath test" for *Helicobacter pylori* infection detects the presence of which of the following?
   a) Antigens
   b) Catalase
   c) Hemolysis
   d) Lactose fermentation
   e) Urease
4. The "O157:H7" of *E. coli* O157:H7 refers to the
   a) biotype.
   b) serotype.
   c) phage type.
   d) antibiogram.
5. When hydrogen peroxide is placed on a colony of an unknown bacterium, bubbles form. Based on this information, you can conclude that the bacterium
   a) is *Staphylococcus epidermidis.*
   b) is a lactic acid bacterium.
   c) is beta-hemolytic.
   d) is catalase-positive.
   e) can cause strep throat.

6. Which of the following is best to use for determining the evolutionary relatedness of organisms?
   a) Ability to form endospores
   b) 16S ribosomal RNA sequence
   c) Sugar degradation
   d) Motility
7. If the GC content of two organisms is 70%, the
   a) organisms are definitely related.
   b) organisms are definitely not related.
   c) AT content is 30%.
   d) organisms likely have extensive DNA homology.
   e) organisms likely have many characteristics in common.
8. If two microbial isolates have similar but different 16S rRNA sequences, they are probably
   a) both motile.
   b) both pathogens.
   c) both cocci.
   d) members of the same domain.
   e) the same strain.
9. The sequence of which ribosomal genes are most commonly used for establishing phylogenetic relatedness?
   a) 5S
   b) 16S
   c) 23S
   d) All of these are commonly used.
10. Which of the following is false?
   a) *Tropheryma whipplei* could be identified before it had been grown in culture.
   b) The GC content of DNA can be measured by determining the temperature at which double-stranded DNA melts.
   c) Sequence differences between organisms can be used to assess their relatedness.
   d) Based on DNA homology studies, members of the genus *Shigella* should be in the same species as *Escherichia coli.*
   e) Gel electrophoresis is used to determine the serotype of an organism.

Applications

1. Microbiologists debate the use of biochemical similarities and cell features as a way of determining the taxonomic relationships among prokaryotes. Explain why some microbiologists believe these similarities and differences are a powerful taxonomic indicator, whereas others think they are not very useful for that purpose.
2. A researcher interested in investigating the genetic relationship of mitochondria to bacteria must decide on the best method to study this. What advice would you give the researcher?

Critical Thinking

1. In figure 10.15, how would the curve appear if the GC content of the DNA sample were increased? How would the curve appear if the AT content were increased?
2. When DNA probes are used to detect specific sequence similarities in bacterial DNA, the probe is heated and the two strands of DNA are separated. Why must the probe DNA be heated?

11 The Diversity of *Bacteria* and *Archaea*

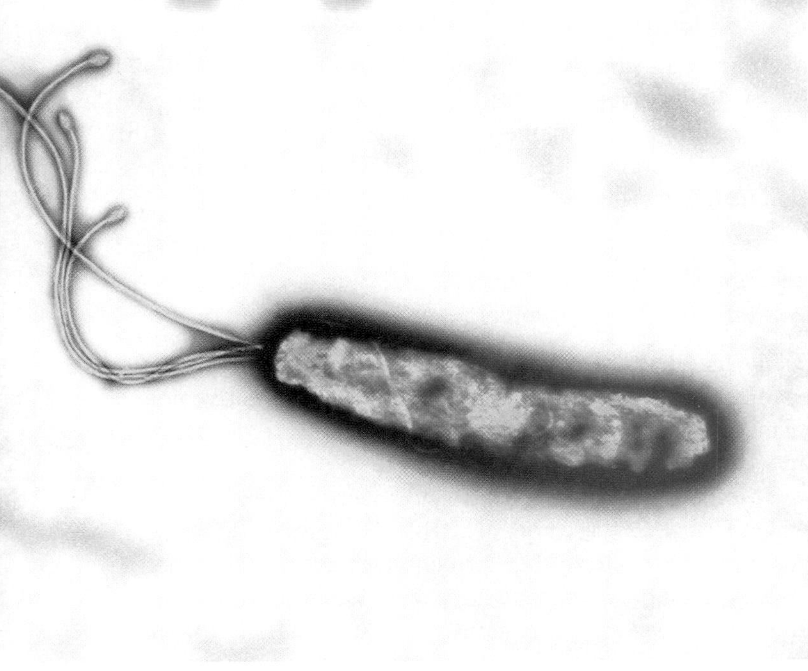

Helicobacter pylori (color-enhanced transmission electron micrograph). ©*Science Photo Library RF/Getty Images*

KEY TERMS

Anoxygenic Phototrophs Photosynthetic organisms that do not produce O_2.

Chemolithotrophs Organisms that harvest energy by oxidizing inorganic chemicals.

Chemoorganotrophs Organisms that harvest energy by oxidizing organic chemicals.

Cyanobacteria Gram-negative oxygenic phototrophs; genetically related to chloroplasts.

Lactic Acid Bacteria Gram-positive bacteria that generate lactic acid as a major end product of their fermentative metabolism.

Methanogens Archaea that obtain energy by oxidizing hydrogen gas, using CO_2 as a terminal electron acceptor, thereby generating methane.

Myxobacteria Gram-negative bacteria that form complex multicellular structures called fruiting bodies.

Nitrifiers Gram-negative bacteria that obtain energy by oxidizing inorganic nitrogen compounds such as ammonium or nitrite.

Oxygenic Phototrophs Photosynthetic organisms that produce O_2.

Prosthecate Bacteria Gram-negative bacteria that have extensions projecting from the cells, thereby increasing their surface area.

Spirochetes Long, helical bacteria that have flexible cell walls and endoflagella.

Sulfur-Oxidizing Bacteria Gram-negative bacteria that obtain energy by oxidizing elemental sulfur and reduced sulfur compounds, generating sulfuric acid.

A Glimpse of History

Cornelius B. van Niel (1897–1985) earned his Ph.D. from the Technological University in Delft, Holland, the home of an approach to microbiology now commonly referred to as "the Delft School." The outstanding program there was chaired in succession by two well-known microbiologists: Martinus Beijerinck and Albert Kluyver.

As Kluyver's student, van Niel was influenced by his mentor's belief that biochemical processes were fundamentally the same in all cells and that microorganisms could be important research tools, serving as a model to study biochemical process. Thirty years later, Kluyver and van Niel presented lectures that would be published in the book *The Microbe's Contribution to Biology*.

Shortly after completing his dissertation in 1928, van Niel accepted a position at the Hopkins Marine Station in California. There, he continued work he started under Kluyver's direction on the photosynthetic activities of the brightly colored purple bacteria. He demonstrated that these microbes require light for growth, yet, unlike plants and algae, do not produce O_2. He also showed that the purple bacteria oxidize hydrogen sulfide as they fix CO_2. Furthermore, van Niel noted that the photosynthetic reactions in all photosynthetic organisms are remarkably similar, except the purple bacteria use hydrogen sulfide in place of water and produce oxidized sulfur compounds instead of O_2. This finding raised the possibility that O_2 generated by plants and algae did not come from carbon dioxide, as was believed at the time, but rather from water.

In addition to his scientific contributions, van Niel was an outstanding teacher. During the summers at Hopkins Marine Station, he taught a bacteriology course, inspiring many microbiologists with his enthusiasm for the diversity of microorganisms and their importance in nature. His sharp memory and knowledge of the literature, along with his appreciation for the remarkable abilities of microbes, allowed him to convey the wonders of the microbial world to his students.

Scientists are only beginning to understand the vast diversity of microbial life. Although well over a million species of prokaryotes are thought to exist, only a fraction of them have been described and classified. Traditional culture and isolation techniques have not supported the growth and subsequent study of the vast majority. This situation is changing as new molecular techniques make it easier to discover and characterize previously unrecognized species. The sheer volume of information uncovered by modern technologies, however, can be daunting for scientists and students alike.

This chapter covers a variety of prokaryotes, focusing primarily on their extraordinary diversity rather than the phylogenetic relationships discussed in chapter 10. Note, however, that no single chapter could describe all known prokaryotes and, consequently, only a relatively small selection is presented.

METABOLIC DIVERSITY

As a group, prokaryotes use an impressive range of mechanisms to harvest energy in order to produce ATP. This section will highlight the metabolic diversity by describing select prokaryotes. **Table 11.1** summarizes characteristics of the archaea and bacteria covered in this section.

11.1 ■ Anaerobic Chemotrophs

Learning Outcomes

1. Compare and contrast the characteristics and habitats of methanogens, sulfur- and sulfate-reducing bacteria, *Clostridium* species, lactic acid bacteria, and *Propionibacterium* species.

For approximately the first 1.5 billion years that prokaryotes inhabited Earth, the atmosphere was **anoxic,** meaning it lacked O_2. In that anaerobic environment, some early **chemotrophs** (organisms that harvest energy by oxidizing chemicals) probably used pathways of anaerobic respiration, using terminal electron acceptors such as carbon dioxide or elemental sulfur, which were plentiful in the environment. Others may have used fermentation, passing the electrons to an organic molecule such as pyruvate. ⏮ chemotrophs, ⏮ anaerobic respiration, ⏮ terminal electron acceptor, ⏮ fermentation

Today, anaerobic habitats are still common. Mud and tightly packed soil limit the diffusion of gases, and any O_2 that penetrates is quickly converted to water by aerobically respiring organisms. This creates anaerobic conditions just below the surface. Aquatic environments may also become anaerobic if they contain nutrients that promote the rapid growth of O_2-consuming microbes. This is evident in polluted lakes, where fish may die because of a lack of dissolved O_2. The human body also provides many anaerobic environments, such as in the intestinal tract. Even the skin and the oral cavity, which are routinely exposed to O_2, have anaerobic microenvironments. These are created via the localized depletion of O_2 by aerobes.

MicroByte

Approximately 99% of the prokaryotes that inhabit the intestinal tract are obligate anaerobes.

Anaerobic Chemolithotrophs

Chemolithotrophs obtain energy by oxidizing inorganic chemicals such as hydrogen gas (H_2) to obtain energy. Those growing anaerobically obviously cannot use O_2 as a terminal electron acceptor and instead must use an alternative such as carbon dioxide or sulfur. Relatively few anaerobic chemolithotrophs have been discovered, and most are members of the domain *Archaea.* Some bacterial examples that inhabit aquatic environments will be discussed later. ⏮ chemolithotrophs

Methanogens

Methanogens are a group of archaea that generate ATP by oxidizing hydrogen gas, using CO_2 as a terminal electron acceptor. Their name is derived from the fact that this process generates methane (CH_4), a colorless, odorless, flammable gas:

$$\underset{\text{(energy source)}}{4\,H_2} + \underset{\text{(terminal electron acceptor)}}{CO_2} \longrightarrow CH_4 + 2\,H_2O$$

Many methanogens can also use alternative energy sources such as formate, methanol, or acetate. Representative genera of methanogens include *Methanobrevibacter* (see figure 3.12) and *Methanosarcina* (**figure 11.1**).

Methanogens are found in anaerobic environments where H_2 and CO_2 are both available. Because these gases are generated by bacteria that ferment organic material, methanogens often grow in association with these microbes. Methanogens, however, are generally not found in environments containing high levels of sulfate, nitrate, or other inorganic electron acceptors. This is because microorganisms that use these electron acceptors to oxidize H_2 have a competitive advantage; the use of CO_2 as an electron acceptor releases comparatively little energy (see figure 6.7). Environments

FIGURE 11.1 Methanogen *Methanosarcina mazei* (SEM). ©Eye of Science/Science Source

? What roles do hydrogen gas and carbon dioxide play in the metabolism of methanogens?

TABLE 11.1 Metabolic Diversity of Prokaryotes

Group/Genera	Characteristics	Phylum
Anaerobic Chemolithotrophs		
Methanogens—*Methanospirillum, Methanosarcina*	Members of the *Archaea* that oxidize hydrogen gas, using CO_2 as a terminal electron acceptor to generate methane. ⏭ **methanogens**	*Euryarchaeota*
Anaerobic Chemoorganotrophs—Anaerobic Respiration		
Sulfur- and sulfate-reducing bacteria—*Desulfovibrio*	Use sulfate as a terminal electron acceptor, generating hydrogen sulfide. Found in anaerobic muds rich in organic material. Gram-negative. ⏭ **sulfur- and sulfate-reducing bacteria**	*Proteobacteria*
Anaerobic Chemoorganotrophs—Fermentation		
Clostridium	Endospore-forming obligate anaerobes. Inhabitants of soil. Gram-positive. ⏭ **the genus *Clostridium***	*Firmicutes*
Lactic acid bacteria—*Streptococcus, Enterococcus, Lactococcus, Lactobacillus, Leuconostoc*	Produce lactic acid as the major end product of their fermentative metabolism. Aerotolerant anaerobes. Several genera are used by the food industry. Gram-positive. ⏭ **lactic acid bacteria**	*Firmicutes*
Propionibacterium	Obligate anaerobes that produce propionic acid as their main fermentation end product. Used in the production of Swiss cheese. Gram-positive. ⏭ **the genus *Propionibacterium***	*Actinobacteria*
Anoxygenic Phototrophs		
Purple sulfur bacteria—*Chromatium, Thiospirillum, Thiodictyon*	Form colored masses in sulfur-rich aquatic habitats and use sulfur compounds as a source of electrons when making reducing power. Gram-negative. ⏭ **purple sulfur bacteria**	*Proteobacteria*
Purple non-sulfur bacteria—*Rhodobacter, Rhodopseudomonas*	Grow in aquatic habitats, preferentially using organic compounds as a source of electrons for reducing power. Many are metabolically versatile. Gram-negative. ⏭ **purple non-sulfur bacteria**	*Proteobacteria*
Green sulfur bacteria—*Chlorobium, Pelodictyon*	Found in habitats similar to those preferred by the purple sulfur bacteria. Gram-negative. ⏭ **green sulfur bacteria**	*Chlorobi*
Filamentous anoxygenic phototrophic bacteria—*Chloroflexus*	Characterized by their filamentous growth. Gram-negative. ⏭ **filamentous anoxygenic phototrophic bacteria**	*Chloroflexi*
Others—*Heliobacterium*	Have not been studied extensively. ⏭ **other anoxygenic phototrophs**	*Firmicutes*
Oxygenic Phototrophs—Cyanobacteria		
Anabaena, Synechococcus	Important primary producers. Some fix N_2. Gram-negative. ⏭ **cyanobacteria**	*Cyanobacteria*
Aerobic Chemolithotrophs		
Filamentous sulfur oxidizers—*Beggiatoa, Thiothrix*	Oxidize sulfur compounds as energy sources. Found in sulfur springs and sewage-polluted waters. Gram-negative. ⏭ **filamentous sulfur oxidizers**	*Proteobacteria*
Unicellular sulfur oxidizers—*Thiobacillus, Acidithiobacillus*	Oxidize sulfur compounds as energy sources. Some species produce enough acid to lower the pH to 1.0. Gram-negative. ⏭ **unicellular sulfur oxidizers**	*Proteobacteria*
Nitrifiers—*Nitrosomonas, Nitrosococcus, Nitrobacter, Nitrococcus*	Oxidize ammonia or nitrite as energy sources. This converts certain fertilizers to a form easily leached from soils, and depletes O_2 in waters polluted with ammonia-containing wastes. Genera that oxidize nitrite prevent the toxic buildup of nitrite. Gram-negative. ⏭ **nitrifiers**	*Proteobacteria*
Hydrogen-oxidizing bacteria—*Aquifex, Hydrogenobacter*	Thermophilic bacteria that oxidize hydrogen gas as an energy source. One of the earliest bacterial forms to exist on earth. ⏭ **hydrogen-oxidizing bacteria**	*Aquifacae*
Aerobic Chemoorganotrophs—Obligate Aerobes		
Micrococcus	Widely distributed; common laboratory contaminants. Gram-positive. ⏭ **the genus *Micrococcus***	*Actinobacteria*
Mycobacterium	Waxy cell wall resists staining; acid-fast. ⏭ **the genus *Mycobacterium***	*Actinobacteria*
Pseudomonas	Common environmental bacteria that, as a group, can degrade a wide variety of compounds. Gram-negative. ⏭ **the genus *Pseudomonas***	*Proteobacteria*
Thermus	*Thermus aquaticus* is the source of *Taq* polymerase (used in PCR). Stains Gram-negative. ⏭ **the genera *Thermus* and *Deinococcus***	*Deinococcus-Thermus*
Deinococcus	Resistant to the damaging effects of gamma radiation. Stains Gram-positive. ⏭ **the genera *Thermus* and *Deinococcus***	*Deinococcus-Thermus*
Aerobic Chemoorganotrophs—Facultative Anaerobes		
Corynebacterium	Widespread in nature. Gram-positive. ⏭ **the genus *Corynebacterium***	*Actinobacteria*
The *Enterobacteriaceae*—*Escherichia, Enterobacter, Klebsiella, Proteus, Salmonella, Shigella, Yersinia*	Most reside in the intestinal tract. Those that ferment lactose are coliforms; their presence in water serves as an indicator of fecal pollution. Gram-negative. ⏭ **the family *Enterobacteriaceae***	*Proteobacteria*
Vibrio	Typically found in marine environments because most species require at least low levels of Na^+ for growth. Gram-negative. ⏭ **the genus *Vibrio***	*Proteobacteria*

where methanogens are commonly found include swamps, marine sediments, rice paddies, and the digestive tracts of humans and other animals. The methane produced can be seen as bubbles rising in swamp waters and is part of the 10 cubic feet of gas discharged from a cow's digestive system each day. As a by-product of wastewater treatment plants, methane gas can be collected and used for heating and generating electricity.

Studying methanogens is challenging because they are very sensitive to O_2. Special techniques, including anaerobe chambers, are used to culture them. ⏮ culturing anaerobes

Anaerobic Chemoorganotrophs—Anaerobic Respiration

Chemoorganotrophs oxidize organic compounds such as glucose to obtain energy. Those that grow anaerobically often use sulfur or sulfate as a terminal electron acceptor. ⏮ chemoorganotrophs

Sulfur- and Sulfate-Reducing Bacteria

When sulfur compounds are used as terminal electron acceptors, they become reduced to form hydrogen sulfide, the compound responsible for the rotten-egg smell of many anaerobic environments. An example of this reaction is:

organic compounds (energy source) + S (terminal electron acceptor) ⟶ $CO_2 + H_2S$

In addition to its unpleasant odor, H_2S is a problem to industry because it reacts with metals, corroding pipes and other structures. Ecologically, however, prokaryotes that reduce sulfur compounds are an essential component of the sulfur cycle. ⏭ sulfur cycle

Sulfate- and sulfur-reducing bacteria generally live in mud that has organic material and oxidized sulfur compounds. The H_2S they produce causes mud and water to turn black when it reacts with iron molecules. At least a dozen genera are recognized in this group, the most extensively studied of which are species of *Desulfovibrio*. These are Gram-negative curved rods.

Some representatives of the *Archaea* also use sulfur compounds as terminal electron acceptors, but the characterized examples generally do not inhabit the same environments as their bacterial counterparts. Although most of the sulfur-reducing bacteria are either mesophiles or thermophiles, the characterized sulfur-reducing archaea are hyperthermophiles, inhabiting extreme environments such as hydrothermal vents. They will be discussed later in the chapter. ⏮ thermophiles, ⏮ hyperthermophiles

Anaerobic Chemoorganotrophs—Fermentation

Many types of anaerobic bacteria obtain energy by fermentation, producing ATP only by substrate-level phosphorylation. There are many variations of fermentation, using different organic energy sources and producing characteristic end products, but one example is:

glucose (energy source) ⟶ pyruvate (terminal electron acceptor) ⟶ lactic acid

The Genus *Clostridium*

Members of the genus *Clostridium* are Gram-positive rods that can form endospores (see figure 3.45). They are common soil inhabitants, and the vegetative cells live in the anaerobic microenvironments created when aerobic organisms consume available O_2. Clostridial endospores can tolerate O_2 and survive for long periods by withstanding levels of heat, drying, chemicals, and irradiation that would kill vegetative bacteria. When conditions become favorable, these endospores germinate, and the resulting vegetative bacteria multiply. Vegetative cells that develop from endospores are responsible for a variety of diseases, including tetanus (caused by *C. tetani*), gas gangrene (caused by *C. perfringens*), botulism (caused by *C. botulinum*), and antibiotic-associated diarrheal disease (caused by *C. difficile*). Some species of *Clostridium* are normal inhabitants of the intestinal tract of humans and other animals. ⏮ endospores, ⏭ tetanus, ⏭ gas gangrene, ⏭ botulism, ⏭ *Clostridium difficile* infection

As a group, *Clostridium* species ferment a wide variety of compounds, including sugars and cellulose. Some of the end products are commercially valuable—for example, *C. acetobutylicum* produces acetone and butanol. Some species can ferment amino acids by an unusual process that oxidizes one amino acid, using another as a terminal electron acceptor. This generates a variety of bad-smelling end products associated with rotting flesh.

Lactic Acid Bacteria

Gram-positive bacteria that produce lactic acid as a major end product of their fermentative metabolism make up a group called the **lactic acid bacteria.** This includes members of the genera *Streptococcus, Enterococcus, Lactococcus, Lactobacillus,* and *Leuconostoc*. Most can grow in aerobic environments, but they typically only carry out fermentation. They can be easily distinguished from other bacteria that grow in the presence of O_2 because they lack the enzyme catalase (see figure 10.4a). ⏮ obligate fermenters, ⏮ catalase

Streptococcus species are cocci that typically grow in chains of varying lengths (**figure 11.2**). They inhabit the oral

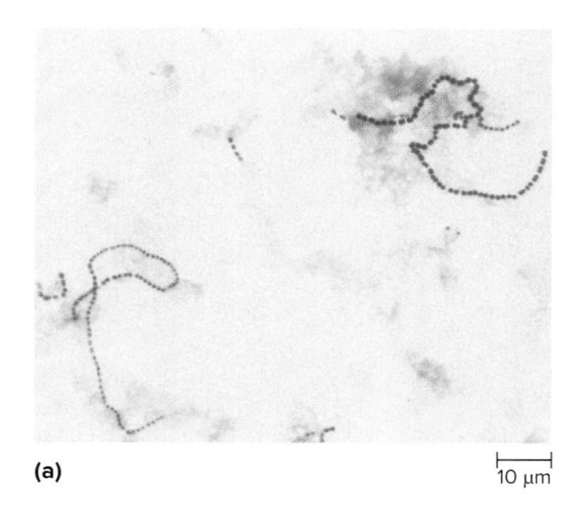

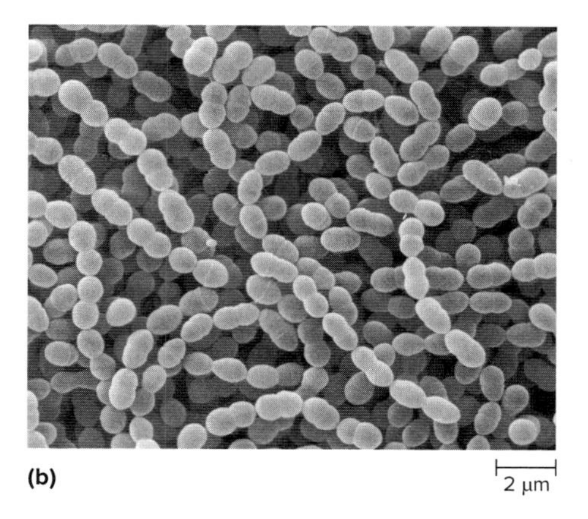

FIGURE 11.2 ***Streptococcus*** **Species** **(a)** Gram stain. **(b)** Scanning electron micrograph. a: ©McGraw-Hill Education/Lisa Burgess; b: ©Custom Medical Stock Photo/Getty Images

What is the major metabolic end product of *Streptococcus* species?

cavity, generally as part of the normal microbiota. *S. thermophilus* is important to the food industry because it is used to make yogurt. Some streptococci, however, are pathogens. One of the most important is *S. pyogenes* (group A strep), which causes pharyngitis (strep throat) and other diseases. Unlike the streptococci that typically inhabit the throat, *S. pyogenes* is β-hemolytic, an important characteristic used to distinguish it from most members of the normal microbiota (see figure 4.11). ◀◀ hemolysis

Species of *Lactococcus* and *Enterococcus* were at one time included in the genus *Streptococcus*. The genus *Lactococcus* now includes species used to make cheeses. *Enterococcus* species, commonly referred to as enterococci, typically inhabit the intestinal tract of humans and other animals.

Members of the genus *Lactobacillus* are rod-shaped bacteria. They are common members of the microbiota in the mouth and the healthy vagina during childbearing years. In the vagina, they break down glycogen that has been deposited in the vaginal lining in response to estrogen (the female sex hormone). The resulting low pH helps prevent vaginal infections. Lactobacilli are also often present in decomposing plant material, milk, and other dairy products. Like some other lactic acid bacteria, they are important in the production of fermented foods (**figure 11.3**).

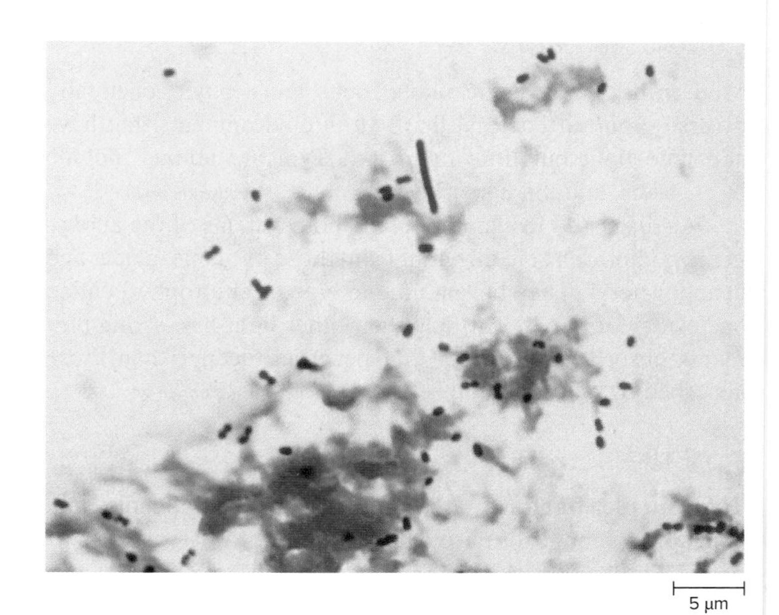

FIGURE 11.3 Lactic Acid Bacteria in Yogurt This Gram stain shows the Gram-positive rod *Lactobacillus delbrueckii* subsp. *bulgaricus* and the Gram-positive coccus *Streptococcus thermophilus*. ©McGraw-Hill Education/Lisa Burgess

Lactobacillus **species are common members of the normal microbiota of which human body sites?**

The Genus *Propionibacterium*

Propionibacterium species are Gram-positive pleomorphic (irregular-shaped) rods that produce propionic acid as their main fermentation end product. They can also ferment lactic acid, and so can extract energy from a waste product of other bacteria.

Propionibacterium species are valuable to the dairy industry because their fermentation end products are important in Swiss cheese production. The propionic acid gives the typical flavor of the cheese, and CO_2, also a product of the fermentation, creates the characteristic holes. Skin microbiota that were included in the genus *Propionibacterium* have been reclassified into a new genus, *Cutibacterium*.

MicroAssessment 11.1

Methanogens are archaea that oxidize hydrogen gas, using CO_2 as a terminal electron acceptor, to generate methane. The sulfur- and sulfate-reducing bacteria oxidize organic compounds, with sulfur or sulfate serving as a terminal electron acceptor, to generate hydrogen sulfide. *Clostridium* species, the lactic acid bacteria, and *Propionibacterium* species oxidize organic compounds, with an organic compound serving as a terminal electron acceptor.

1. What metabolic process creates the rotten-egg smell characteristic of many anaerobic environments?
2. Describe two beneficial contributions of the lactic acid bacteria.
3. Relatively little is known about many obligate anaerobes. Why would this be so?

11.2 ■ Anoxygenic Phototrophs

Learning Outcomes

2. Compare and contrast the characteristics of the purple bacteria and the green bacteria.

The earliest photosynthesizing organisms were likely **anoxygenic phototrophs.** Rather than using water as a source of electrons when making reducing power for biosynthesis, these organisms use hydrogen sulfide or organic compounds, and therefore do not generate O_2. For example:

$$\underset{\text{(carbon source)}}{6\,CO_2} + \underset{\text{(electron source)}}{12\,H_2S} \longrightarrow C_6H_{12}O_6 + 12\,S + 6\,H_2O$$

Modern-day anoxygenic phototrophs are a phylogenetically diverse group of bacteria that live in environments that have adequate light but little or no O_2. Typical habitats include bogs, lakes, and the upper layer of muds. ⏮ reducing power

As discussed in chapter 6, the photosystems of the anoxygenic phototrophs are different from those of plants, algae, and cyanobacteria. They have a unique type of chlorophyll called bacteriochlorophyll. This and their other light-harvesting pigments absorb wavelengths that penetrate deeper than those absorbed by chlorophyll *a*. ⏮ photosystems, ⏮ chlorophyll *a*

Purple Bacteria

The **purple bacteria** are Gram-negative organisms that appear red, orange, or purple due to their light-harvesting pigments. Unlike other anoxygenic phototrophs, the components of their photosynthetic apparatus are all contained within the cytoplasmic membrane. Folds in this membrane increase the surface area available for the photosynthetic processes.

Purple Sulfur Bacteria

Purple sulfur bacteria can sometimes be seen growing as colored masses in sulfur-rich aquatic habitats (**figure 11.4**). The cells are relatively large, sometimes larger than 5 µm in diameter, and some are motile by flagella. They may also have gas vesicles, allowing them to move up or down to their preferred level in the water column. Most store sulfur in intracellular granules. ⏮ gas vesicles

The purple sulfur bacteria preferentially use hydrogen sulfide to generate reducing power, although some can use other inorganic molecules (such as H_2) or organic compounds (such as pyruvate). Many are strict anaerobes and phototrophs, but some can grow in the absence of light aerobically, oxidizing reduced inorganic or organic compounds as a source of energy. Representative genera include *Chromatium, Thiospirillum,* and *Thiodictyon.*

Purple Non-Sulfur Bacteria

The purple non-sulfur bacteria are found in a wide variety of aquatic habitats, including moist soils, bogs, and rice paddies.

FIGURE 11.4 Purple Sulfur Bacteria Photograph of purple bacteria growing in a bog. ©Johnny Madsen/Alamy Stock Photo

? What causes the purple color of the bacterial masses?

One important characteristic that distinguishes them from the purple sulfur bacteria is that they preferentially use a variety of organic molecules rather than hydrogen sulfide as a source of electrons for reducing power. In addition, they lack gas vesicles, and if they store sulfur, the granules form outside the cell.

Purple non-sulfur bacteria are remarkably versatile metabolically. Not only do they grow as phototrophs using organic molecules as a source of electrons, but many can use a metabolism similar to the purple sulfur bacteria, using hydrogen gas or hydrogen sulfide as an electron source. In addition, most can grow aerobically in the absence of light using chemotrophic metabolism. Representative genera of purple non-sulfur bacteria include *Rhodobacter* and *Rhodopseudomonas.*

Green Bacteria

The **green bacteria** are Gram-negative organisms that are typically green or brownish in color.

Green Sulfur Bacteria

Green sulfur bacteria are found in habitats similar to those preferred by the purple sulfur bacteria. Like the purple sulfur bacteria, they use hydrogen sulfide as a source of electrons for reducing power and they form sulfur granules. The granules, however, form outside of the cell. The accessory pigments of the green sulfur bacteria are located in structures called chlorosomes. The bacteria lack flagella, but many have gas vesicles. All are strict anaerobes, and none can use a chemotrophic metabolism. Representative genera include *Chlorobium* and *Pelodictyon.* ⏮ accessory pigments

Filamentous Anoxygenic Phototrophic Bacteria

Filamentous anoxygenic phototrophic bacteria form multicellular arrangements and exhibit gliding motility. The most thoroughly studied of this group are members of the genus *Chloroflexus,* particularly the thermophilic strains that grow in hot springs. Many of the filamentous anoxygenic phototrophs have chlorosomes, which initially led scientists to believe they were related to the green sulfur bacteria. Their 16S rDNA sequences indicate otherwise. As a group, filamentous anoxygenic phototrophs are diverse metabolically. Some preferentially use organic compounds to generate reducing power and can also grow in the dark aerobically using chemotrophic metabolism.

Other Anoxygenic Phototrophs

Although the green and purple bacteria have been studied most extensively, other types of anoxygenic phototrophs exist. Among these are members of the genus *Heliobacterium,* Gram-positive endospore-forming rods related to members of the genus *Clostridium.*

MicroAssessment 11.2

Anoxygenic phototrophs harvest the energy of sunlight, but do not generate O_2. The purple sulfur bacteria and the green sulfur bacteria use hydrogen sulfide as a source of electrons to generate reducing power; the purple non-sulfur bacteria and many of the filamentous anoxygenic phototrophs preferentially use organic compounds.

4. Describe a structural characteristic that distinguishes the purple sulfur bacteria from the green sulfur bacteria.
5. What is the function of gas vesicles?
6. How do anoxygenic phototrophs benefit by having light-harvesting pigments that absorb wavelengths that penetrate deeper than those absorbed by chlorophyll *a*?

11.3 ■ Oxygenic Phototrophs

Learning Outcomes

3. Describe the characteristics of cyanobacteria, including how nitrogen-fixing species protect their nitrogenase enzyme from O_2.

Nearly 3 billion years ago, the Earth's atmosphere began changing as O_2 was gradually introduced to the previously anoxic environment. This was probably due to the evolution of the cyanobacteria, thought to be the earliest **oxygenic phototrophs.** These photosynthetic organisms use water as a source of electrons for reducing power, generating O_2:

$$\underset{\text{(carbon source)}}{6\,CO_2} + \underset{\text{(electron source)}}{6\,H_2O} \longrightarrow C_6H_{12}O_6 + 6\,O_2$$

Cyanobacteria still play an essential role in the biosphere. As **primary producers,** they harvest the energy of sunlight, using it to convert CO_2 into organic compounds. They were initially thought to be algae and were called blue-green algae until electron microscopy revealed their prokaryotic structure. ⏭ primary producers

Cyanobacteria

Cyanobacteria are a diverse group of photosynthetic Gram-negative bacteria that inhabit a wide range of environments, including freshwater and marine habitats, soils, and the surfaces of rocks. Many are able to convert nitrogen gas (N_2) to ammonia, which can then be incorporated into cell material. This process, called **nitrogen fixation,** is an exclusive ability of prokaryotes. ⏭ nitrogen fixation

General Characteristics of Cyanobacteria

Cyanobacteria are morphologically diverse. Some are unicellular, with typical prokaryotic shapes such as cocci, rods, and spirals. Others form filamentous multicellular associations called trichomes that may or may not be enclosed within a sheath (a tube that holds and surrounds a chain of cells) (**figure 11.5**). Motile trichomes glide as a unit. Cyanobacteria that live in aquatic environments often have gas vesicles, allowing them to move vertically within the water column. When large numbers of cyanobacteria accumulate

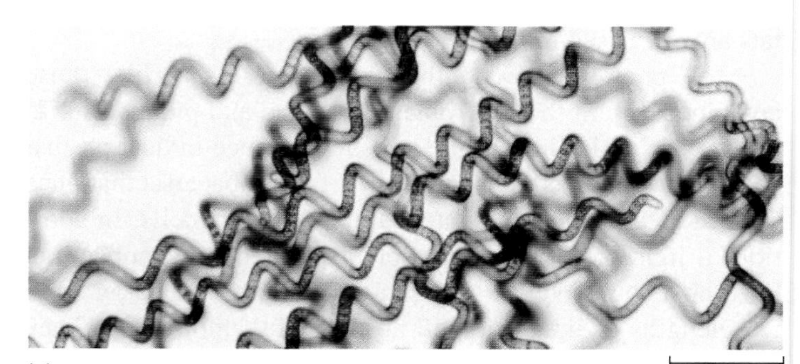

FIGURE 11.5 Cyanobacteria (a) The spiral trichome of *Spirulina* species. **(b)** Differential interference contrast photomicrograph of a species of *Oscillatoria*. Note the arrangement of the individual cells in the trichome. a: ©Sinclair Stammers/Science Source; b: ©M. I. Walker/Science Source

? What is a trichome?

FIGURE 11.6 Cyanobacterial Bloom Excessive growth of cyanobacteria in an aquatic environment leads to buoyant masses of cells rising to the surface. ©Wayne Carmichael (Wright State University), Mark Schneegurt (Wichita State University), and Cyanosite (www-cyanosite.bio.purdue.edu)

? **What allows cyanobacterial cells to rise to the surface?**

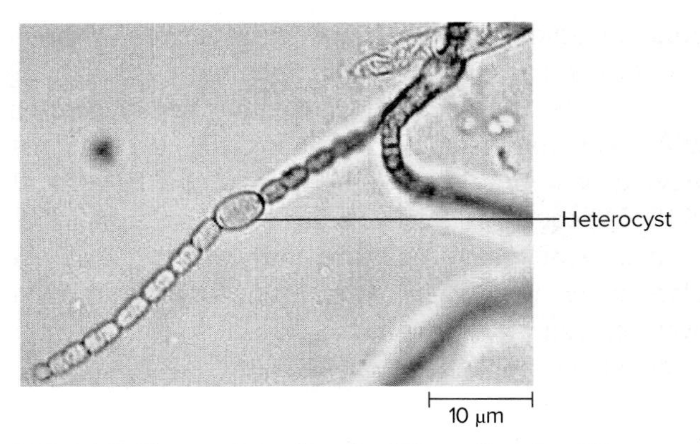

FIGURE 11.7 Heterocyst of an *Anabaena* Species Nitrogen fixation occurs within these specialized cells. ©Roger Burks (University of California at Riverside), Mark Schneegurt (Wichita State University), and Cyanosite (www-cyanosite.bio.purdue.edu)

? **Why is it important for heterocysts to not have a functional photosystem II?**

in stagnant lakes or other freshwater habitats, it is called a bloom (**figure 11.6**). In the bright, hot conditions of summer, the buoyant cells lyse and decay, creating a bad-smelling scum. The ecological effects of these blooms on aquatic habitats are discussed in chapter 28. ▶▶| aquatic habitats

The photosystems of the cyanobacteria are like those contained within the chloroplasts of algae and plants. This is not surprising in light of the genetic evidence indicating that chloroplasts evolved from a species of cyanobacteria that once resided as an endosymbiont within eukaryotic cells. In addition to light-harvesting chlorophyll pigments, cyanobacteria have phycobiliproteins. These pigments absorb energy from wavelengths of light not well absorbed by chlorophyll. They contribute to the blue-green, or sometimes reddish, color of the cyanobacteria. |◀◀ chloroplast, |◀◀ photosystem

Nitrogen-Fixing Cyanobacteria

Nitrogen-fixing cyanobacteria are very important ecologically. They can incorporate both N_2 and CO_2 into organic material, so they generate a form of these nutrients that can then be used by other organisms. Thus, their activities can ultimately support the growth of a wide range of organisms in environments that would otherwise lack usable nitrogen and carbon. Also, like all cyanobacteria, they help limit atmospheric CO_2 buildup by using the gas as a carbon source.

Nitrogenase, the enzyme complex that catalyzes nitrogen fixation, is destroyed by O_2; therefore, nitrogen-fixing cyanobacteria must protect the enzyme from the O_2 they generate. Species of *Anabaena,* which are filamentous, isolate nitrogenase by restricting the process of nitrogen fixation to specialized thick-walled cells called **heterocysts** (**figure 11.7**). Heterocysts lack photosystem II and consequently do not generate O_2. The heterocysts of some species form at very regular intervals within the filament, reflecting the ability of cells within a trichome to communicate. One species of *Anabaena, A. azollae,* forms an intimate relationship with the water fern *Azolla.* The bacterium grows and fixes nitrogen within the protected environment of a special sac in the fern, providing *Azolla* with a source of available nitrogen. *Synechococcus* species fix nitrogen only in the dark. Consequently, nitrogen fixation and photosynthesis occur at different times of the day. |◀◀ photosystem II

Other Notable Characteristics of Cyanobacteria

Cyanobacteria have various other notable characteristics—some beneficial, others damaging. Filamentous cyanobacteria are responsible for maintaining the structure and productivity of soils in cold desert areas such as the Colorado Plateau. Their sheaths persist in soil, creating a sticky fibrous network that prevents erosion. In addition, these bacteria provide an important source of nitrogen and organic carbon in otherwise nutrient-poor soils. On the negative side, some cyanobacteria produce geosmin, a chemical that has a distinctive "earthy" odor and makes drinking water taste odd. Some aquatic species such as *Microcystis aeruginosa* produce toxins that can be deadly to an animal when consumed.

MicroAssessment 11.3

The photosystems of cyanobacteria generate O_2 and are similar to those of algae and plants. Many cyanobacteria can fix nitrogen.

7. What is the function of a heterocyst?

8. How do cyanobacteria prevent erosion in cold desert regions?

9. How could heavily fertilized lawns foster the development of cyanobacterial blooms? **?**

11.4 ■ Aerobic Chemolithotrophs

Learning Outcomes

4. Compare and contrast the characteristics of sulfur-oxidizing bacteria, nitrifiers, and hydrogen-oxidizing bacteria.

Aerobic chemolithotrophs obtain energy by oxidizing reduced inorganic chemicals, using O_2 as a terminal electron acceptor.

Sulfur-Oxidizing Bacteria

The **sulfur-oxidizing bacteria** are Gram-negative rods or spirals, which sometimes grow in filaments. They obtain energy by oxidizing elemental sulfur and reduced sulfur compounds, including hydrogen sulfide and thiosulfate. O_2 serves as a terminal electron acceptor, generating sulfuric acid. An example of this reaction is:

$$\underset{\text{(energy source)}}{S} + \underset{\text{(terminal electron acceptor)}}{1\tfrac{1}{2}O_2} + H_2O \longrightarrow H_2SO_4$$

These bacteria are important in the sulfur cycle. ⏭| sulfur cycle

Filamentous Sulfur Oxidizers

Beggiatoa and *Thiothrix* species are filamentous sulfur oxidizers that live in sulfur springs, in sewage-polluted waters, and on the surface of marine and freshwater sediments. They store sulfur, depositing it as intracellular granules, but differ in the nature of their filamentous growth (**figure 11.8a**). The filaments of *Beggiatoa* species move by gliding motility, a mechanism that does not require flagella. The filaments may bend and twist to form a cluster of filaments wrapped together. In contrast, the filaments of *Thiothrix* species are immobile; they fasten at one end to rocks or other solid surfaces. Often they attach to other cells, forming characteristic rosette arrangements of filaments (figure 11.8b). Progeny cells detach from the ends of these filaments and use gliding motility to move to new locations, where they form additional filaments. Overgrowth of these filamentous organisms in wastewater at treatment facilities causes a problem called bulking. Because the masses of filamentous organisms do not settle easily, bulking interferes with the process that separates the solid and liquid portions of the waste. ⏭| wastewater treatment

Unicellular Sulfur Oxidizers

Acidithiobacillus species are found in both terrestrial and aquatic habitats, where their ability to oxidize metal sulfides can be used to recover metals (see Focus Your Perspective 6.1). The bacteria oxidize insoluble metal sulfides such as gold sulfide, producing sulfuric acid. This lowers the pH, which converts the metal to a soluble form. *Acidithiobacillus* species can also be used to prevent acid rain, a problem that occurs when sulfur-containing coals and oils are burned. To remove the sulfur from the fuels, the bacteria are allowed to oxidize it to sulfate, a form that can then be extracted.

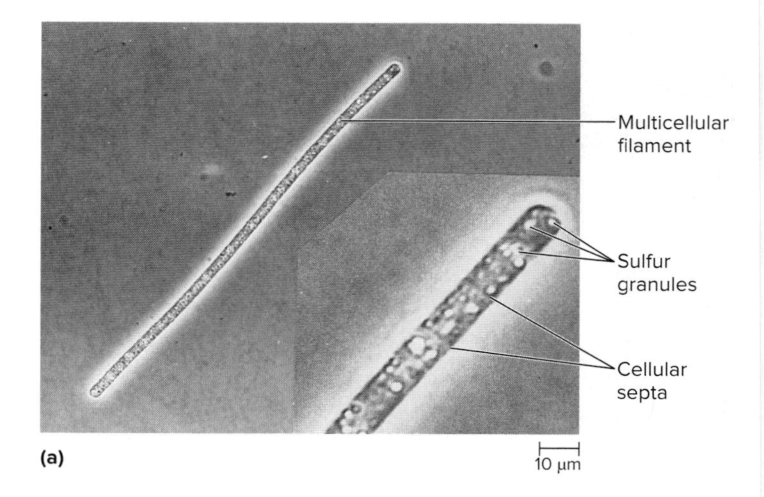

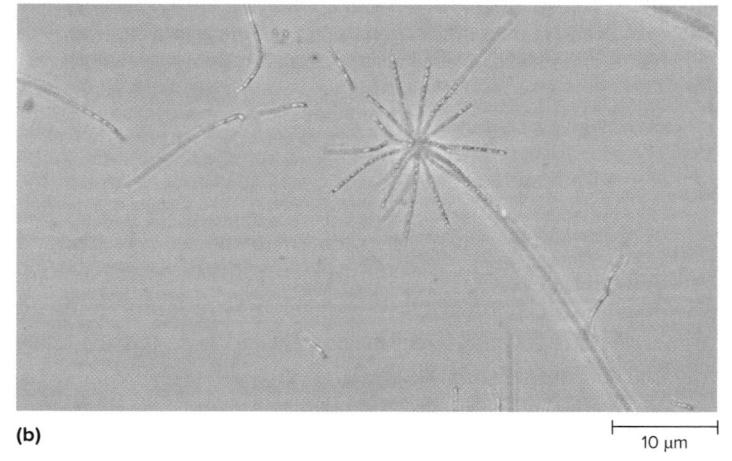

FIGURE 11.8 Filamentous Sulfur-Oxidizing Bacteria Phase-contrast photomicrographs. **(a)** Multicellular filament of a *Beggiatoa* species; the septa separate the cells. **(b)** Multicellular filaments of a *Thiothrix* species, forming a rosette arrangement. ©James T. Staley

? What is the role of sulfur in the metabolism of sulfur-oxidizing bacteria?

Acidithiobacillus species can also cause severe environmental problems. For example, the strip mining of coal exposes metal sulfides, which the bacteria can then oxidize to produce sulfuric acid; some species produce enough acid to lower the pH to 1.0. The resulting runoff can acidify nearby streams, killing trees, fish, and other wildlife (**figure 11.9**). The runoff may also contain toxic metals made soluble by the bacteria.

Nitrifiers

Nitrifiers are a diverse group of Gram-negative bacteria that obtain energy by oxidizing inorganic nitrogen compounds such as ammonium or nitrite. These bacteria are a concern to farmers who fertilize their crops with compounds containing ammonium, a form of nitrogen retained by soils because

FIGURE 11.9 Acid Drainage from a Mine Sulfur-oxidizing bacteria oxidize exposed metal sulfides, generating sulfuric acid. The yellow-red color is due to insoluble iron oxides. ©Ronald Karpilo/Alamy Stock Photo

How can the metabolic activities that result in this acid drainage be used in a commercially valuable manner?

its positive charge causes it to adhere to negatively charged soil particles. The potency and longevity of the fertilizer are affected by nitrifying bacteria converting the ammonium to nitrate. Although plants use the nitrate more easily, it is rapidly washed out of soils.

Nitrifying bacteria are also an important consideration when disposing of wastes that have a high ammonium concentration. As nitrifying bacteria oxidize nitrogen compounds, they use O_2, so waters polluted with nitrogen-containing wastes can quickly become hypoxic (low in dissolved O_2).

The nitrifiers include two metabolically distinct groups that typically grow in close association. Together, they can oxidize ammonium to form nitrate. The ammonia oxidizers, which include the genera *Nitrosomonas* and *Nitrosococcus,* convert ammonium to nitrite in the following reaction:

$$\underset{\text{(energy source)}}{NH_4^+} + \underset{\text{(terminal electron acceptor)}}{1\tfrac{1}{2}O_2} \longrightarrow NO_2^- + H_2O + 2\,H^+$$

The nitrite oxidizers, which include the genera *Nitrobacter* and *Nitrococcus,* then convert nitrite to nitrate as follows:

$$\underset{\text{(energy source)}}{NO_2^-} + \underset{\text{(terminal electron acceptor)}}{\tfrac{1}{2}O_2} \longrightarrow NO_3^-$$

The latter group is particularly important in preventing the buildup of nitrite in soils, which is toxic and can contaminate ground water. The oxidation of ammonium to nitrate (nitrification) is an important part of the nitrogen cycle. ▶▶| nitrogen cycle

Hydrogen-Oxidizing Bacteria

Members of the Gram-negative genera *Aquifex* and *Hydrogenobacter* are among the few hydrogen-oxidizing bacteria that are obligate chemolithotrophs. An example of the reaction in their metabolism is:

$$\underset{\text{(energy source)}}{H_2} + \underset{\text{(terminal electron acceptor)}}{\tfrac{1}{2}O_2} \longrightarrow H_2O$$

These related organisms are thermophilic and typically live in hot springs. Some *Aquifex* species have a maximum growth temperature of 95°C, the highest of any bacteria. Based on 16S rRNA studies, hydrogen-oxidizing bacteria were one of the earliest bacterial forms to exist on Earth. The fact that they require O_2 seems contradictory to their evolutionary position, but in fact, the low amount they require might have been available early on in certain niches due to photochemical processes that split water.

MicroAssessment 11.4

Sulfur oxidizers use sulfur compounds as energy sources, generating sulfuric acid. The nitrifiers oxidize nitrogen compounds such as ammonium or nitrite. Hydrogen-oxidizing bacteria oxidize H_2.

10. Describe two beneficial roles of sulfur oxidizers.
11. Why would farmers be concerned about nitrifying bacteria?
12. Why would sulfur-oxidizing bacteria store sulfur?

11.5 ■ Aerobic Chemoorganotrophs

Learning Outcomes

5. Describe the representative obligate aerobes and facultative anaerobes.
6. Describe the family *Enterobacteriaceae,* and explain what distinguishes coliforms from other members of this family.

Aerobic chemoorganotrophs oxidize organic compounds to obtain energy, using O_2 as a terminal electron acceptor:

$$\underset{\text{(energy source)}}{\text{organic compounds}} + \underset{\text{(terminal electron acceptor)}}{O_2} \longrightarrow CO_2 + H_2O$$

They include a wide assortment of bacteria, ranging from some that inhabit very specific environments to others that are ubiquitous. This section will profile only representative genera found in a variety of different environments. Later sections will describe examples that thrive in specific habitats.

Obligate Aerobes

Obligate aerobes obtain energy using respiration exclusively; none of them can ferment.

The Genus *Micrococcus*

Members of the genus *Micrococcus* are Gram-positive cocci found in soil and on dust particles, inanimate objects, and skin. Because they are often airborne, they can easily contaminate bacteriological media. There, they typically form pigmented colonies, a characteristic that helps identify them. The colonies of *M. luteus,* for example, are generally yellow (**figure 11.10**). They tolerate dry conditions and can grow in salty environments such as 7.5% NaCl.

The Genus *Mycobacterium*

Mycobacterium species are widespread in nature and include harmless saprophytes, which live on dead and decaying matter, as well as pathogens. They stain poorly because of waxy lipids (mycolic acids) in their unusual cell wall, but special procedures can be used to increase the penetration of certain dyes. Once stained, the cells resist destaining, even with acidic decolorizing solutions. Because of this, *Mycobacterium* species are called **acid-fast,** and the acid-fast staining procedure is an important step in identifying them (see figure 3.16). *Nocardia* species, a related group of bacteria common in soil, are also acid-fast. ⏮ **acid-fast**

Mycobacterium species are generally pleomorphic rods; they often occur in chains that sometimes branch, or bunch together to form cord-like groups. Several species are notable for their effect on human health, including *M. tuberculosis,* which causes tuberculosis, and *M. leprae,* which causes Hansen's disease (leprosy). *Mycobacterium* species are more resistant to disinfectants than most other vegetative bacteria. In addition, they are resistant to many of the most common antimicrobial medications.

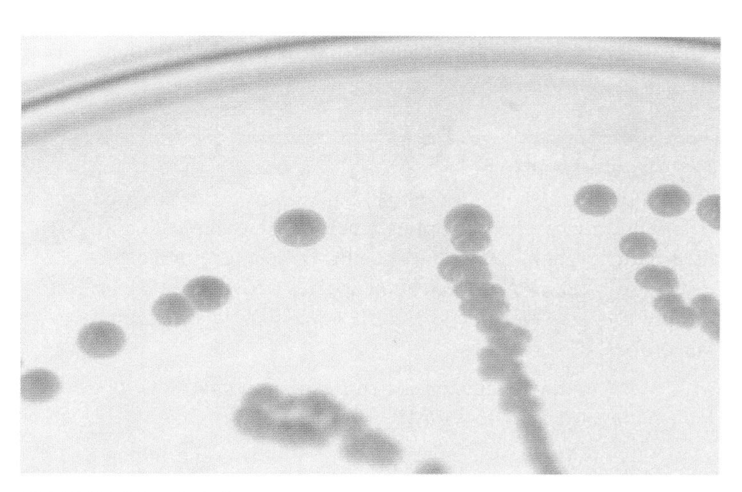

FIGURE 11.10 ***Micrococcus luteus* Colonies** ©McGraw-Hill Education/Lisa Burgess

? Why is *Micrococcus luteus a* common contaminate on bacteriological media?

The Genus *Pseudomonas*

Pseudomonas species are Gram-negative rods that have polar flagella and often produce pigments (**figure 11.11**). Although most are strict aerobes, some can grow anaerobically if nitrate is available as a terminal electron acceptor. They do not ferment and are oxidase-positive, characteristics that help distinguish them from members of the family *Enterobacteriaceae,* including *E. coli.* ⏮ **oxidase test**

As a group, *Pseudomonas* species have extremely diverse biochemical capabilities. Some can metabolize more than 80 different substrates, including unusual sugars, amino acids, and compounds containing aromatic rings. Because of this, *Pseudomonas* species play an important role in the degradation of many synthetic and natural compounds that resist breakdown by most other microorganisms. The ability to carry out some of these degradations is encoded on plasmids.

Pseudomonas species are widespread, typically inhabiting soil and water. Although most are harmless, some cause disease in plants and animals. Medically, the most significant species is *P. aeruginosa.* It is a common opportunistic pathogen, meaning that it primarily infects people who have underlying medical conditions. Unfortunately, it can grow in nutrient-poor environments, such as water used in respirators, and is resistant to many disinfectants and antimicrobial medications. Because of this, hospitals must be very careful to prevent it from infecting patients. ⏭ ***Pseudomonas aeruginosa,*** ⏭ **opportunistic pathogen**

The Genera *Thermus* and *Deinococcus*

Thermus and *Deinococcus* are related genera that have scientifically and commercially important characteristics. *Thermus* species are thermophilic, as their name implies, and this trait is valuable because of their heat-stable enzymes. The bacteria have an unusual cell wall, and stain Gram-negative.

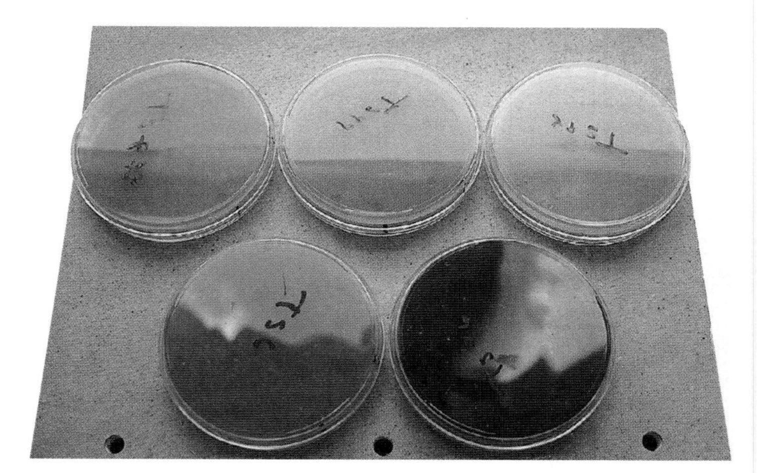

FIGURE 11.11 Pigments of *Pseudomonas* Species Cultures of different strains of *Pseudomonas aeruginosa*. Note the different colors of the water-soluble pigments. ©Evans Roberts

? Why is the fact that *Pseudomonas* species can grow in nutrient-poor environments medically important?

Deinococcus species' unusual cell wall has multiple layers, and they stain Gram-positive. They are unique in their extraordinary resistance to the damaging effects of gamma radiation. *D. radiodurans* can survive a radiation dose several thousand times that lethal to a human being. The dose literally shatters the organism's genome into many fragments, yet enzymes in the cells can repair the extensive damage. Scientists anticipate that through genetic engineering, *Deinococcus* species may eventually help clean up soil and water that have been contaminated by radioactive wastes.

MicroByte

The DNA polymerase of *T. aquaticus* (*Taq* polymerase) plays a key role the polymerase chain reaction (PCR).

Facultative Anaerobes

Facultative anaerobes preferentially use aerobic respiration if O_2 is available. As an alternative, however, they can ferment.

The Genus *Corynebacterium*

Members of the genus *Corynebacterium* are widespread in nature. They are Gram-positive pleomorphic rods, often club-shaped (*koryne* is Greek for "club") and arranged to form V shapes or palisades (side-to-side stacks) (**figure 11.12**). Bacteria that exhibit this characteristic morphology are referred to as **coryneforms** or **diphtheroids.** *Corynebacterium* species are generally facultative anaerobes, but some are strict aerobes. Many *Corynebacterium* species live harmlessly in the throat, but toxin-producing strains of *C. diphtheriae* can cause the disease diphtheria. ⏭ diphtheria

The Family *Enterobacteriaceae*

Members of the family *Enterobacteriaceae,* often referred to as **enterics** or enterobacteria, are Gram-negative rods. Their name reflects the fact that most are found in the intestinal tract of humans and other animals (in Greek, *enteron* means "intestine"), although some thrive in rich soil. Enterics that are part of the normal intestinal microbiota include *Enterobacter, Klebsiella,* and *Proteus* species as well as most strains of *E. coli.* Those that cause diarrheal disease include *Shigella* species, *Salmonella enterica,* and some strains of *E. coli.* Life-threatening systemic diseases include typhoid fever, caused by *Salmonella enterica* serotype Typhi, and both the bubonic and pneumonic forms of plague, caused by *Yersinia pestis.* ⏭ diarrheal disease, ⏭ typhoid fever, ⏭ plague

Members of the *Enterobacteriaceae* are facultative anaerobes that ferment glucose and, if motile, generally have peritrichous flagella. The family includes over 40 recognized genera that can be distinguished using biochemical tests. Within a given species, many different strains have been described. These are often distinguished using serological tests that detect differences in cell walls, flagella, and capsules (see figure 10.10). ⏮ peritrichous flagella

Enteric bacteria that characteristically ferment lactose are included in a group called **coliforms.** This is an informal grouping of certain common intestinal inhabitants such as *E. coli* that are easy to detect in food and water; for years, regulatory agencies have considered them to be an indicator of fecal pollution. Their presence indicates a possible health risk because fecal-borne pathogens might also be present. ⏭ coliform

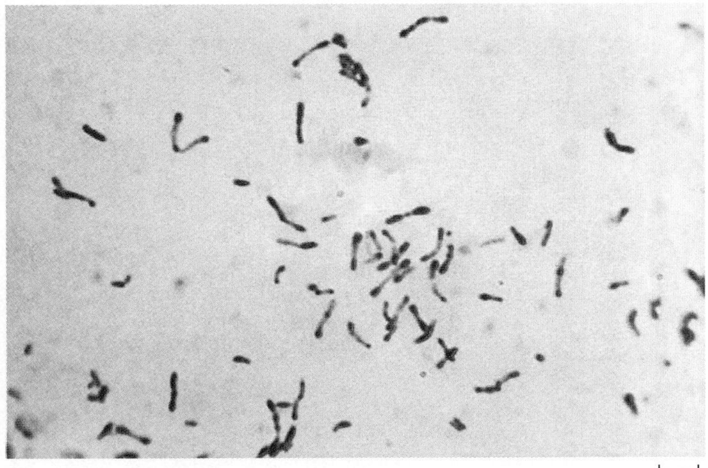

FIGURE 11.12 ***Corynebacterium*** The Gram-positive pleomorphic rods are often arranged to form V shapes or palisades.
Source: Dr. P. B. Smith/CDC

? How does the name of this genus reflect the shape of the bacteria?

The Genus *Vibrio*

Members of the genus *Vibrio* are typically found in marine environments because most species require at least low levels of Na^+ for growth. They are Gram-negative straight or slightly curved rods and are facultative anaerobes. Pathogens include *V. cholerae,* which causes the severe gastrointestinal disease cholera, *V. parahaemolyticus,* which causes a milder gastrointestinal disease, and *V. vulnificus,* which causes a systemic illness, particularly in patients who have an underlying illness such as liver disease. Some *Vibrio* species are bioluminescent, and these will be described in a later section. ⏭ cholera, ⏭ bioluminescent bacteria

MicroAssessment 11.5

Micrococcus, Mycobacterium, Pseudomonas, Thermus, and *Deinococcus* species are obligate aerobes that harvest energy by degrading organic compounds, using O_2 as a terminal electron acceptor. Most *Corynebacterium* species and all members of the family *Enterobacteriaceae* and the genus *Vibrio* are facultative anaerobes.

13. What unique characteristic makes members of the genus *Deinococcus* noteworthy?
14. What is the significance of finding coliforms in drinking water?
15. Why would it be an advantage for a *Pseudomonas* species to encode enzymes for degrading certain compounds on a plasmid rather than on the chromosome?

ECOPHYSIOLOGICAL DIVERSITY

As a group, prokaryotes show remarkable diversity in their physiological adaptations to a wide range of habitats. Bacteria and archaea have evolved to live in virtually all environments—from the hydrothermal vents of deep oceans to the frozen expanses of Antarctica.

This section will highlight the physiological mechanisms bacteria and archaea use to live in terrestrial and aquatic environments; the study of these adaptations is called **ecophysiology.** The section will also describe some examples of bacteria that use animals as habitats. **Table 11.2** summarizes characteristics of the bacteria covered in this section.

11.6 ■ Thriving in Terrestrial Environments

Learning Outcomes

7. Describe the bacterial groups that form resting stages.
8. Compare and contrast *Agrobacterium* species and rhizobia.

Microorganisms that live in soil must tolerate a variety of conditions. Daily and seasonally, soil can routinely alternate between wet and dry as well as warm and cold. Nutrient availability can also cycle from plentiful to scarce. To thrive in this

TABLE 11.2 Ecophysiological Diversity

Group/Genera	Characteristics	Phylum
Thriving in Terrestrial Environments		
Endospore formers—*Bacillus, Clostridium*	*Bacillus* species include both obligate aerobes and facultative anaerobes; *Clostridium* species are obligate anaerobes. Gram-positive. ⏭ **endospore formers**	*Firmicutes*
Azotobacter	Form cysts. Notable for their ability to fix nitrogen in aerobic conditions. Gram-negative. ⏭ **the genus *Azotobacter***	*Proteobacteria*
Myxobacteria—*Chondromyces, Myxococcus, Stigmatella*	Cells come together to form a fruiting body, within which the cells differentiate to become dormant microcysts. Gram-negative. ⏭ **myxobacteria**	*Proteobacteria*
Streptomyces	Resemble fungi in their pattern of growth; produce antibiotics. Gram-positive. ⏭ **the genus *Streptomyces***	*Actinobacteria*
Agrobacterium	Cause plant tumors. Scientists use their Ti plasmid to move genes into plant cells. Gram-negative. ⏭ **the genus *Agrobacterium***	*Proteobacteria*
Rhizobia—*Rhizobium, Sinorhizobium, Bradyrhizobium, Mesorhizobium, Azorhizobium*	Fix nitrogen; form a symbiotic relationship with legumes. Gram-negative. ⏭ **rhizobia**	*Proteobacteria*
Thriving in Aquatic Environments		
Sheathed bacteria—*Sphaerotilus, Leptothrix*	Form chains of cells enclosed within a protective sheath. Swarmer cells move to new locations. Gram-negative. ⏭ **sheathed bacteria**	*Proteobacteria*
Prosthecate bacteria—*Caulobacter, Hyphomicrobium*	Appendages increase their surface area. Gram-negative. ⏭ **prosthecate bacteria**	*Proteobacteria*
Bdellovibrio	Predators of other bacteria. Gram-negative. ⏭ **the genus *Bdellovibrio***	*Proteobacteria*
Bioluminescent bacteria—*Aliivibrio fischeri, Vibrio harveyi*, and *Photobacterium phosphoreum*	Some form symbiotic relationships with specific types of squid and fish. Gram-negative. ⏭ **bioluminescent bacteria**	*Proteobacteria*
Legionella	Often reside within protozoa. Gram-negative. ⏭ **the genus *Legionella***	*Proteobacteria*
Epulopiscium	Very large cigar-shaped bacteria that multiply by releasing several daughter cells; each cell has thousands of copies of the genome. Gram-positive. ⏭ **the genus *Epulopiscium***	*Firmicutes*
Free-living spirochetes—*Spirochaeta, Leptospira* (some species)	Long spiral-shaped bacteria that move by means of endoflagella. Gram-negative. ⏭ **spirochetes**	*Spirochaetes*
Magnetospirillum	Magnetic crystals allow them to move in water and sediments. Gram-negative. ⏭ **magnetotactic bacteria**	*Proteobacteria*
Spirillum	Spiral-shaped, microaerophilic bacteria. Gram-negative. ⏭ **the genus *Spirillum***	*Proteobacteria*
Thioploca, Thiomargarita	Use novel mechanisms to compensate for the fact that their energy source (reduced sulfur compounds) and terminal electron acceptor (nitrate) do not coexist. ⏭ **the genera *Thioploca* and *Thiomargarita***	*Proteobacteria*
Animals as Habitats		See table 11.3.

ever-changing environment, microbes have evolved mechanisms to cope with unfavorable conditions and to use plants as sources of nutrients.

Bacteria That Form a Resting Stage

Several genera that live in soil can form a resting stage that allows them to survive the dry periods that occur in many soils. Of the various types of dormant cells, endospores are by far the most resistant to environmental extremes.

Endospore Formers

Bacillus and *Clostridium* species are the most common Gram-positive rod-shaped bacteria that form endospores; the position of the spore in the cell can help in identification (**figure 11.13**).

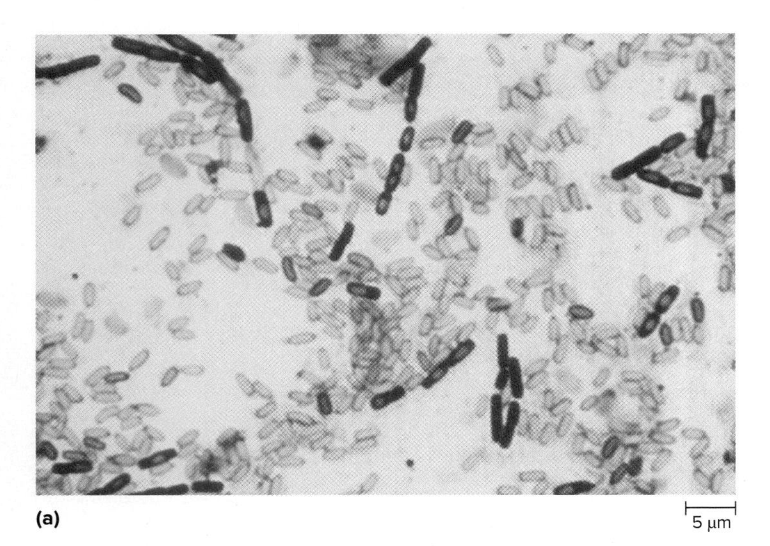

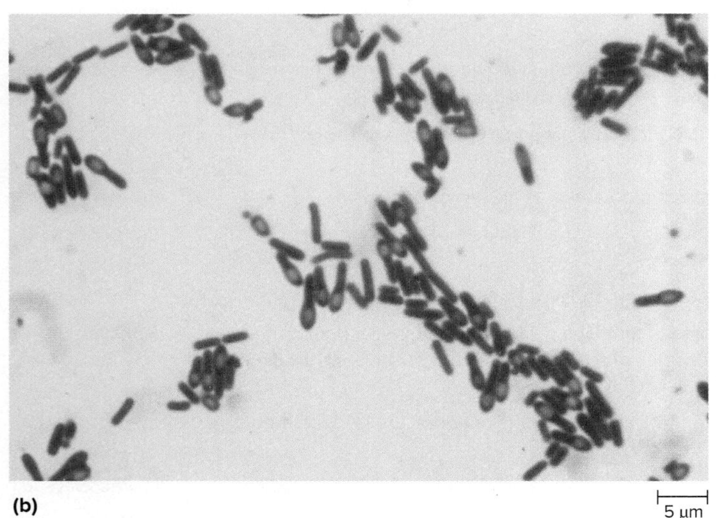

FIGURE 11.13 Endospore Formers (a) Endospores forming in the mid-portion of cells *(Bacillus megaterium)*. **(b)** Endospores forming at the ends of cells *(Clostridium tetani)*. ©McGraw-Hill Education/Lisa Burgess

Members of which genus are endospore-forming obligate anaerobes?

Clostridium species, which are obligate anaerobes, were discussed earlier. *Bacillus* species include both obligate aerobes and facultative anaerobes, and some are medically important. *B. anthracis* causes the disease anthrax, which can be acquired from contacting its endospores in soil or in animal hides or wool. Unfortunately, the spores have also been used as an agent of domestic bioterrorism. **◀◀ endospores, ▶▶ anthrax**

The Genus *Azotobacter*

Azotobacter species are Gram-negative pleomorphic, rod-shaped bacteria that live in soil. They can form a type of resting cell called a cyst that can withstand drying and ultraviolet radiation but not high heat.

Azotobacter species are also notable for their ability to fix nitrogen in aerobic conditions; recall that the enzyme nitrogenase is inactivated by O_2. Apparently, the extremely high respiratory rate of *Azotobacter* species consumes O_2 so rapidly that a low O_2 environment is maintained inside the cell. In addition, a protein in the cell binds nitrogenase, thereby protecting it from O_2 damage.

Myxobacteria

The **myxobacteria** are a group of aerobic Gram-negative rods that have a unique developmental cycle as well as a resting stage. When conditions are favorable, cells secrete a slime layer that other cells then follow, creating a swarm of cells. But then, when nutrient levels are low, the behavior of the group changes. The cells begin to come together, and then pile up to form a multicellular structure called a **fruiting body,** which is often brightly colored (**figure 11.14**). In some species, the fruiting body is quite complex, consisting of a mass of cells elevated and supported by a stalk made of a hardened slime. The cells within the fruiting body differentiate to become spherical, dormant forms called microcysts. These are much more resistant to heat, drying, and radiation

FIGURE 11.14 Myxobacterial Fruiting Body
©Mary F. Lampe

How do fruiting bodies help myxobacteria survive unfavorable conditions?

than are the vegetative cells of myxobacteria, but are much less resistant than bacterial endospores.

Myxobacteria are important in nature as degraders of complex organic substances; they can digest bacteria and certain algae and fungi. Scientifically, these bacteria serve as an important model for studying developmental biology. Included in the myxobacteria are the genera *Chondromyces, Myxococcus,* and *Stigmatella.*

The Genus *Streptomyces*

The genus *Streptomyces* includes more than 500 species of aerobic Gram-positive bacteria that resemble fungi in their pattern of growth. Like the fungi, they form a mycelium (a visible mass of branching filaments), and this gives *Streptomyces* colonies a chalky surface (**figure 11.15a**). The filaments that make up the mycelium are called hyphae. At the tips of these, chains of characteristic dormant spores called conidia develop (figure 11.15b). Conidia are resistant to drying and are easily spread in air currents. Note that even though this pattern of growth resembles fungi, which are eukaryotes, *Streptomyces* species are much smaller and are prokaryotes.

Streptomyces species produce a variety of extracellular enzymes that allow them to degrade a wide range of organic compounds. They are also responsible for the characteristic "earthy" odor of soil; like the cyanobacteria, they produce geosmin. One species of *Streptomyces, S. somaliensis,* can cause an infection of subcutaneous tissue called an actinomycetoma. ⏮ geosmin

Streptomyces species naturally produce a wide array of medically useful antibiotics, including streptomycin, tetracycline, and erythromycin. The role these compounds play in the life cycle of *Streptomyces* is not entirely understood, but at the low levels produced in soils, they appear to be involved in cell signaling.

Bacteria That Associate with Plants

Members of two related genera use very different means to obtain nutrients from plants. *Agrobacterium* species are plant pathogens that cause tumor-like growths, whereas *Rhizobium* species form a mutually beneficial relationship with certain types of plants.

The Genus *Agrobacterium*

Agrobacterium species are Gram-negative rod-shaped bacteria that have an unusual mechanism of gaining a competitive advantage in soil. They cause plant tumors, the outcome of their ability to genetically alter plants for their own benefit (**figure 11.16**). They do this by attaching to wounded plant tissue, and then transferring a part of a plasmid to a plant cell; in *A. tumefaciens* the plasmid is called the **Ti plasmid** (for "tumor-inducing"). The transferred DNA encodes the ability to synthesize plant growth hormones, causing uncontrolled growth of the plant tissue and resulting in a tumor. The transferred DNA also encodes enzymes that direct the synthesis of an opine, an unusual amino acid derivative; *Agrobacterium* can then use this compound as a nutrient source (see Focus Your Perspective 8.2).

MicroByte

Scientists have modified the Ti plasmid, turning it into a commercially valuable tool used to genetically engineer plant cells.

Rhizobia

Rhizobia are a group of Gram-negative rod-shaped bacteria that often fix nitrogen and form intimate relationships with legumes (plants that bear seeds in pods). This group of bacteria includes members of the genera *Rhizobium, Sinorhizobium, Bradyrhizobium, Mesorhizobium,* and *Azorhizobium.*

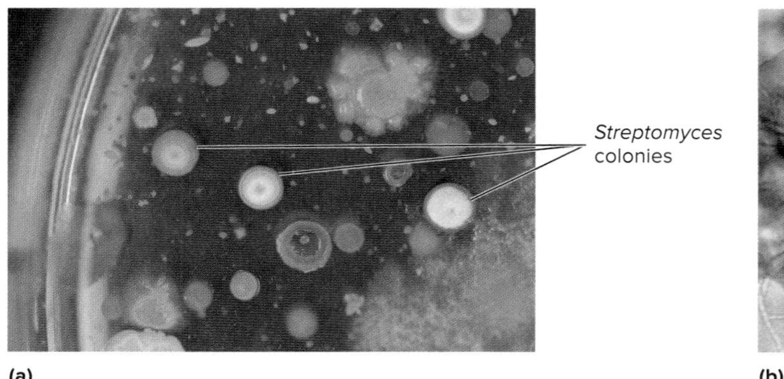

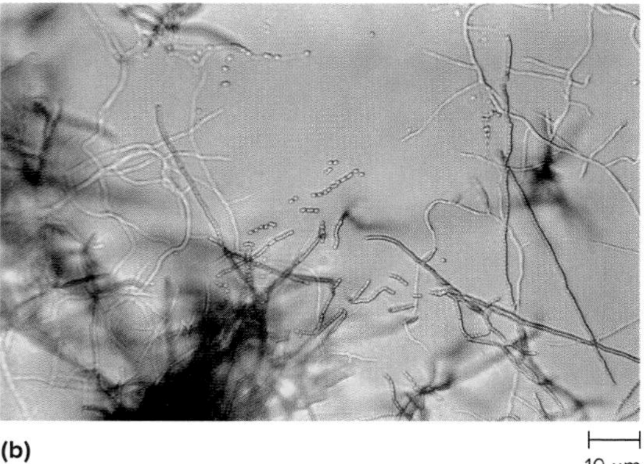

FIGURE 11.15 ***Streptomyces*** **(a)** Colonies, which typically have a chalky appearance and may be white or colored. **(b)** A photomicrograph showing the spherical conidia at the ends of the filamentous hyphae. a: ©McGraw-Hill Education/Lisa Burgess; b: Source: Dr. David Berd/CDC

? Why are *Streptomyces* species extremely important medically?

FIGURE 11.16 Plant Tumor Caused by *Agrobacterium tumefaciens* ©Custom Life Science Images/Alamy Stock Photo

What is the role of the Ti plasmid in plant tumor formation?

The bacteria live within cells in nodules formed on the roots of the plants (**figure 11.17**). The plants synthesize the protein leghemoglobin, which binds and controls the levels of O_2 (see figure 28.13). Within the resulting microaerobic environment of nodules, the bacteria are able to fix nitrogen. Rhizobia residing within plant cells are examples of endosymbionts, organisms that provide a benefit to the cells in which they reside. **symbiotic nitrogen-fixers**

MicroAssessment 11.6

Bacillus and *Clostridium* species make endospores, the most resistant type of dormant cell known. *Azotobacter* species, myxobacteria, and *Streptomyces* species all produce dormant cells that tolerate some unfavorable conditions but are less resistant than endospores. *Agrobacterium* species and rhizobia obtain nutrients from plants, but the former are plant pathogens and the latter benefit the plant.

16. Why are myxobacteria important in nature?

17. How does *Agrobacterium* benefit from inducing a plant tumor?

18. If you wanted to determine the number of endospores in a sample of soil, what could you do before plating it?

11.7 ■ Thriving in Aquatic Environments

Learning Outcomes

9. Describe examples of how bacteria that thrive in aquatic environments obtain and store nutrients.

Most aquatic environments lack a steady supply of nutrients. The bacteria that live in these habitats have evolved various mechanisms for obtaining and storing nutrients more efficiently.

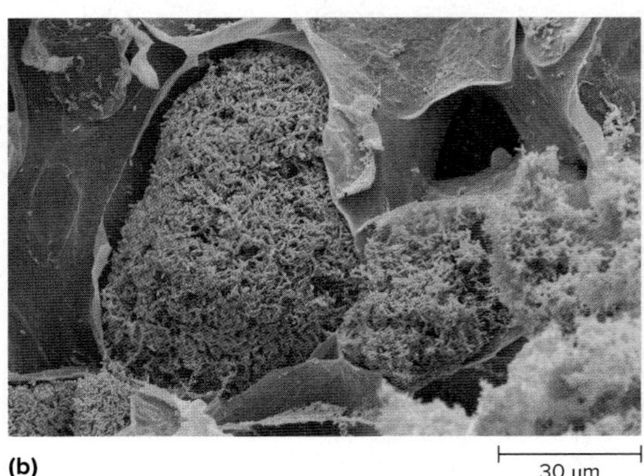

FIGURE 11.17 Symbiotic Relationship Between Rhizobia and Certain Plants (a) Root nodules. **(b)** Rhizobial cells within a nodule (color-enhanced SEM). a: ©McGraw-Hill Education/Lisa Burgess; b: ©Andrew Syred/Science Source

How do rhizobia benefit plants?

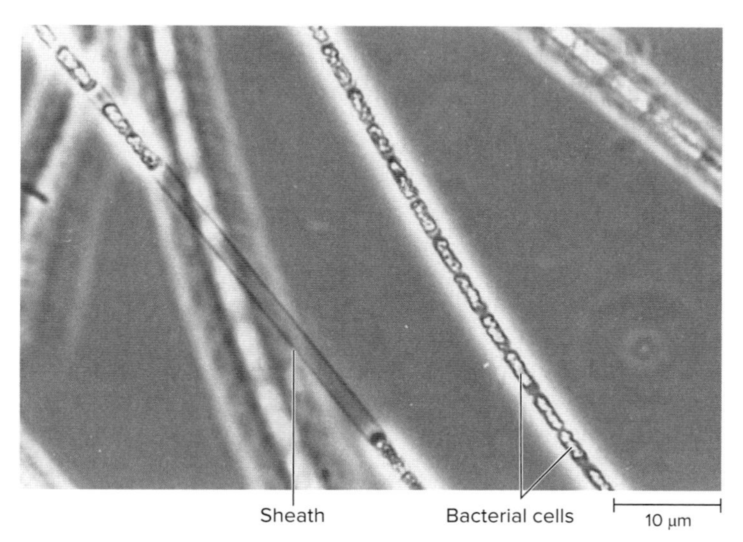

FIGURE 11.18 Sheathed Bacteria Phase-contrast photomicrograph of a *Sphaerotilus* species. ©James T. Staley

What is the role of the sheath?

Sheathed Bacteria

Sheathed bacteria form chains of cells encased within a tube, or sheath (**figure 11.18**). The sheath plays a protective role, helping the bacteria attach to solid objects in favorable habitats while sheltering them from attack by predators. Masses of filamentous sheaths can often be seen streaming from rocks in flowing water polluted by nutrient-rich wastes. They often interfere with sewage treatment and other industrial processes by clogging pipes. Sheathed bacteria include species of *Sphaerotilus* and *Leptothrix,* which are Gram-negative rods.

Sheathed bacteria spread by forming motile cells called swarmer cells that exit through the unattached end of the sheath. These then move to a new solid surface, where they attach. If enough nutrients are present, they can multiply and form a new sheath, which gets longer as the chain of cells grows.

Prosthecate Bacteria

The **prosthecate bacteria** are a diverse group of Gram-negative bacteria that have projections called prosthecae, which are extensions of the cytoplasm and cell wall. These extensions provide increased surface area to facilitate absorption of nutrients. Some prosthecae allow the organisms to attach to solid surfaces.

The Genus *Caulobacter*

Because of their remarkable life cycle, *Caulobacter* species serve as a model for research on cellular differentiation. Entirely different events occur in an orderly fashion at opposite ends of the cell.

Caulobacter cells have a single polar prostheca, commonly called a stalk (**figure 11.19**). At the tip of the stalk is an adhesive holdfast, a structure that provides a mechanism for attachment. To multiply, the cell elongates and divides by binary fission, producing a motile swarmer cell at the end opposite the stalk. This swarmer cell has a flagellum, located at the pole opposite the site of division. The swarmer cell detaches and moves to a new location, where it adheres via a holdfast near the base of its flagellum. It then loses its flagellum, replacing it with a stalk. Only then can the daughter cell replicate its DNA and repeat the process. In favorable conditions, a single cell divides and produces daughter cells many times.

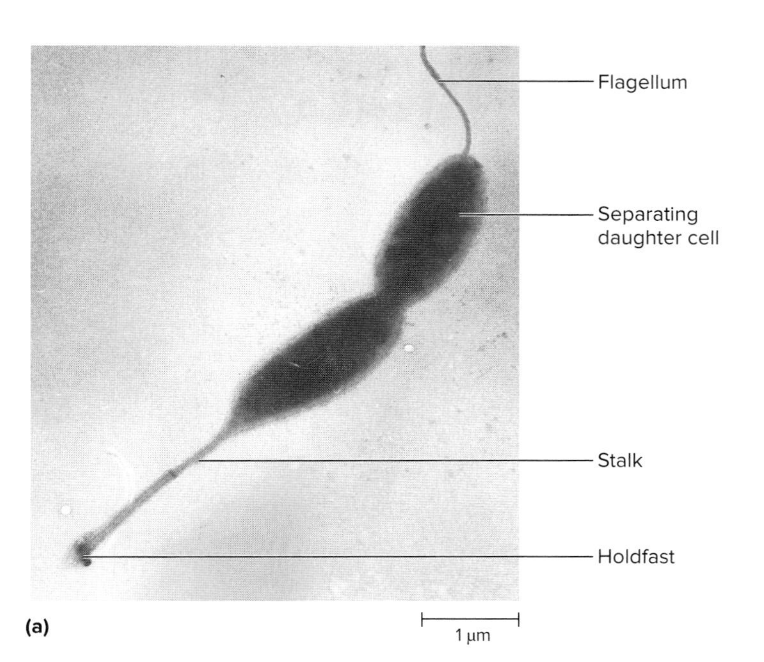

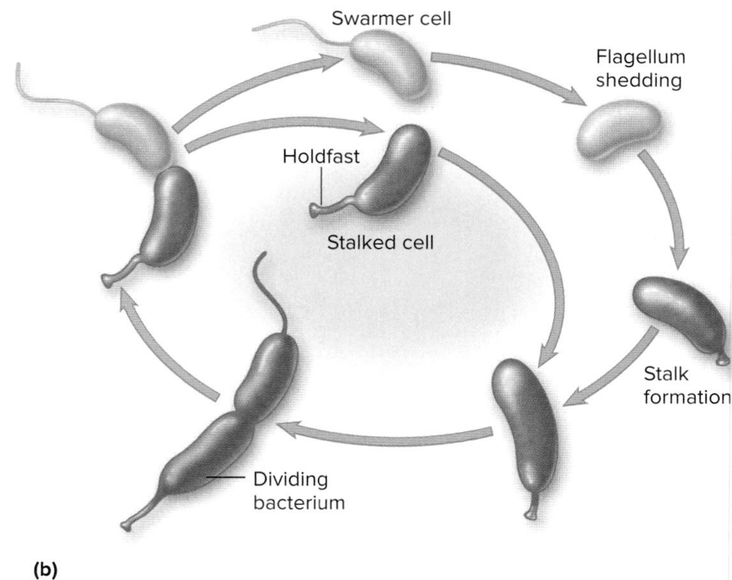

FIGURE 11.19 *Caulobacter* **(a)** Transmission electron micrograph. **(b)** Life cycle. ©James T. Staley

What characteristic of *Caulobacter* species makes them important research models?

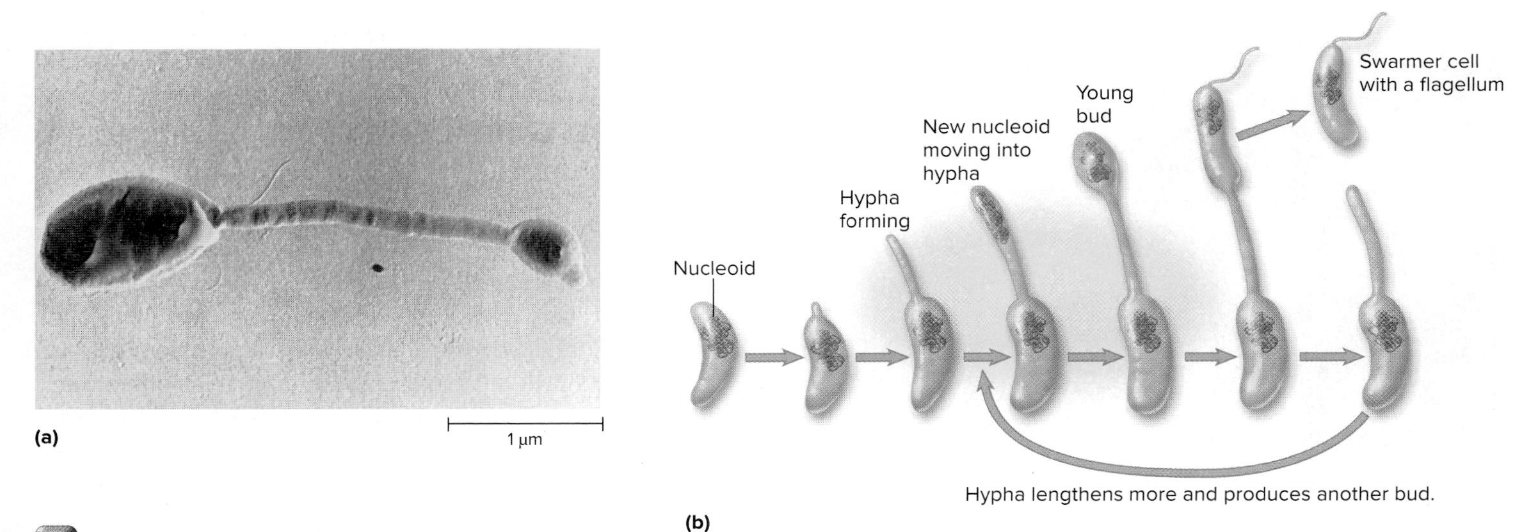

FIGURE 11.20 ***Hyphomicrobium*** **(a)** Electron micrograph. Note the bud forming at the tip of the polar prostheca. **(b)** Life cycle. ©James T. Staley

? What will happen to the swarmer cell?

The Genus *Hyphomicrobium*

Hyphomicrobium species are in many ways similar to *Caulobacter* species, except they have a distinct method of reproduction. The single polar prostheca of the parent cell enlarges at the tip to form a bud (**figure 11.20**). This continues enlarging and also develops a flagellum, eventually giving rise to a motile daughter cell. The daughter cell (swarmer cell) then detaches and moves to a new location, eventually losing its flagellum and forming a polar prostheca at the opposite end to repeat the cycle. As with *Caulobacter* species, a single cell can repeatedly produce daughter cells.

Bacteria That Derive Nutrients from Other Organisms

Some bacteria obtain nutrients directly from other organisms. Examples include *Bdellovibrio* species, bioluminescent bacteria, *Epulopiscium* species, and *Legionella* species.

The Genus *Bdellovibrio*

Bdellovibrio species (*bdello,* from the Greek word for "leech") are highly motile Gram-negative curved rods that prey on *E. coli* and other Gram-negative bacteria (**figure 11.21**). When a

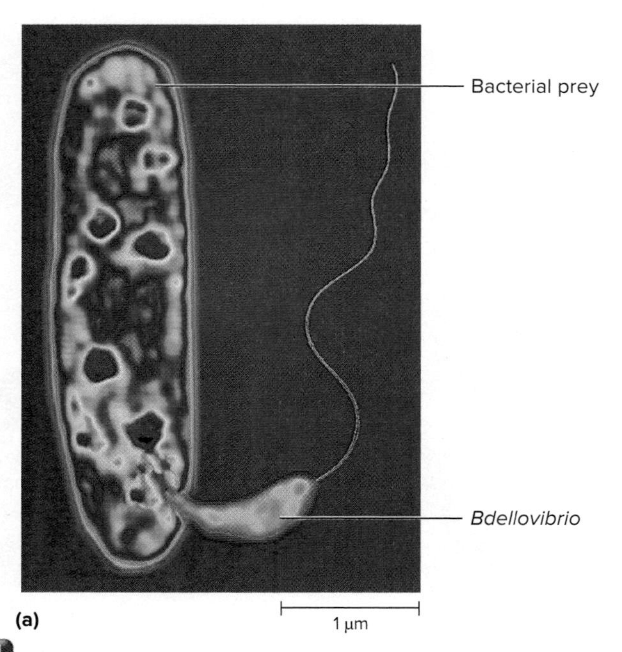

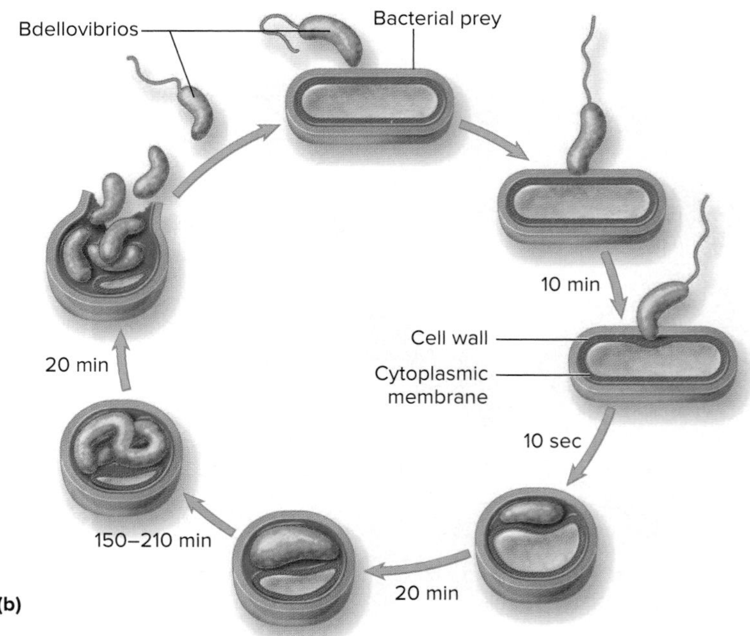

FIGURE 11.21 ***Bdellovibrio*** **(a)** Color-enhanced transmission micrograph of a *Bdellovibrio* cell attacking its prey. **(b)** Life cycle of *Bdellovibrio*. Note that the diagram exaggerates the size of the space in which *Bdellovibrio* multiplies. ©Alfred Pasieka/Photolibrary/Getty Images

? How does a *Bdellovibrio* cell penetrate the prey?

Bdellovibrio cell attacks, it strikes its prey with such force that it propels the prey a short distance. The parasite then attaches to its host and rotates with a spinning motion. At the same time, it makes digestive enzymes that break down lipids and peptidoglycan, eventually forming a hole in the cell wall of the prey. This allows the parasitic bacterium to penetrate the peptidoglycan, lodging in the periplasm. There, over a period of several hours, *Bdellovibrio* degrades and utilizes the prey's cellular contents. It derives energy by aerobically oxidizing amino acids and acetate. The parasite increases in length, ultimately dividing to form several motile daughter cells. When the host cell lyses, the *Bdellovibrio* progeny are released to find new hosts, repeating the cycle. ⏮ periplasm

Bioluminescent Bacteria

Certain bacteria are **bioluminescent,** meaning they emit light (**figure 11.22**). This ability plays an important role in the mutually beneficial relationship between some of these bacteria and specific types of fish and squid. For example, certain types of squid have a specialized organ within their ink sac that is colonized by the bioluminescent bacterium *Aliivibrio (Vibrio) fischeri.* The light produced by the bacterial cells in the organ is thought to serve as a type of camouflage, masking the squid's contrast against the light from above and any shadow it might otherwise cast. Meanwhile, the squid provides nutrients to the bacteria in the organ. Another example is the flashlight fish, which carries bioluminescent bacteria in a light organ below its eye. By opening and closing a lid that covers the light organ, the fish can control the amount of light released, a tactic believed to confuse predators and prey.

Luminescence is catalyzed by the enzyme luciferase. Studies revealed that the genes encoding it are expressed only when the density of the bacterial population reaches a critical point. This phenomenon of quorum sensing is now recognized as an important mechanism by which a variety of different bacteria regulate the expression of certain genes. ⏮ quorum sensing

In addition to *Aliivibrio fischeri,* other examples of bioluminescent bacteria include *Vibrio harveyi* and *Photobacterium phosphoreum.* All of these are Gram-negative, straight or curved rods and are facultative anaerobes. Bioluminescent bacteria typically inhabit marine environments.

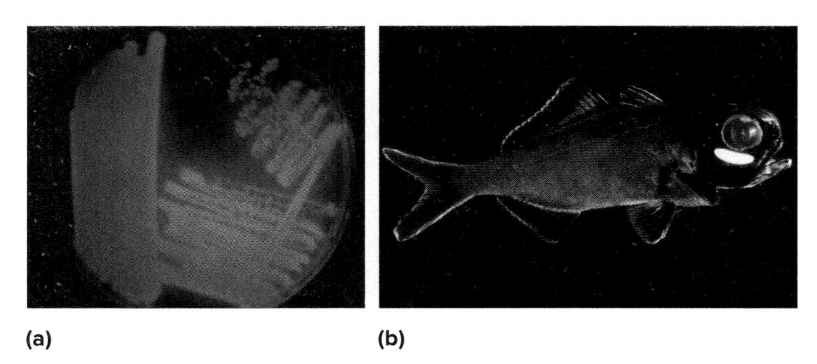

(a) (b)

FIGURE 11.22 Bioluminescent Bacteria (a) Plate culture of bioluminescent bacteria. **(b)** Flashlight fish; under the eye is a light organ colonized with bioluminescent bacteria.

? What role does quorum sensing play in bioluminescence?

The Genus *Legionella*

Legionella species are common in aquatic environments, where they often live within protozoa. They have even been isolated from water in air conditioners and produce misters. They are Gram-negative obligate aerobes that use amino acids, but not carbohydrates, as a source of carbon and energy. *Legionella pneumophila* can cause legionellosis, a respiratory disease, when inhaled in aerosolized droplets. ⏭ legionellosis

The Genus *Epulopiscium*

Epulopiscium species are Gram-positive cigar-shaped bacteria that live in the intestinal tract of surgeonfish. They are much larger than most prokaryotes (600 μm × 80 μm), and each cell has thousands of copies of the chromosome scattered throughout the cell. This means that the cell has thousands of copies of each gene, allowing the necessary proteins to be synthesized even in the far reaches of the large cell.

Epulopiscium species have an unusual life cycle. Rather than undergoing typical binary fission, they get very large, finally lysing to release up to seven daughter cells. They have not yet been grown in culture.

Bacteria That Move by Unusual Mechanisms

Some bacteria have unique mechanisms of motility that allow them to easily move to desirable locations. These organisms include the spirochetes and the magnetotactic bacteria.

Spirochetes

The **spirochetes** (Greek *spira* for "coil" and *chaete* for "hair") are a group of Gram-negative bacteria with a spiral shape and a unique motility mechanism that allows them to move through thick, viscous environments such as mud. Distinguishing characteristics include their spiral shape, flexible cell wall, and motility by means of **endoflagella** (also called axial filaments). Unlike typical flagella, endoflagella are contained within the periplasm. Either a single endoflagellum or a tuft originates at each end of the cell, and the structures extend toward each other, overlapping in the mid-region of the cell. Rotation of the endoflagella within the limited area between the cytoplasmic and outer membranes causes the cell to move like a corkscrew, sometimes bending and twisting. Many spirochetes are very slender and can be seen only by using special methods such as dark-field microscopy (**figure 11.23**). Many are also difficult or impossible to grow in culture. ⏮ periplasm

Spirochetes include free-living species that inhabit aquatic environments, as well as ones that reside on or in animals. *Spirochaeta* species are anaerobes or facultative

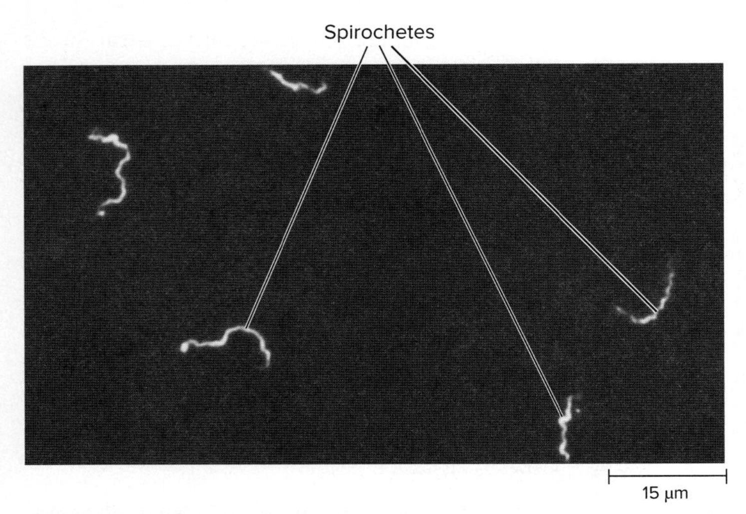

FIGURE 11.23 Spirochetes Dark-field photomicrograph of spirochetes. Source: C. W. Hubbard/CDC

Why are many spirochetes difficult to see with bright-field microscopy?

anaerobes that live in muds and anaerobic waters. *Leptospira* species are aerobic; some are free-living in aquatic environments, whereas others grow within animals. *L. interrogans* causes the disease leptospirosis, which can be transmitted in the urine of infected animals. Spirochetes adapted to live in body fluids of humans and other animals will be discussed later. ⏭ leptospirosis

Magnetotactic Bacteria

Magnetotactic bacteria such as *Magnetospirillum (Aquaspirillum) magnetotacticum* contain a string of magnetic crystals that align cells with the Earth's magnetism (see figure 3.41). This allows them to move up or down in the water or sediments. It is thought that this unique type of movement allows the cells to locate the microaerophilic habitats they require. *Magnetospirillum* species are Gram-negative, spiral-shaped organisms. ⏮ magnetotaxis

Bacteria That Form Storage Granules

A number of aquatic bacteria form granules that store nutrients. Recall that anoxygenic phototrophs often store sulfur granules, which can later be used as a source of electrons for reducing power. Some bacteria store phosphate, and others store compounds that can be used to generate ATP.

The Genus *Spirillum*

Spirillum species are Gram-negative spiral-shaped, microaerophilic bacteria. *Spirillum volutans* forms volutin granules, which are storage forms of phosphate. These are sometimes called metachromatic granules to reflect their characteristic staining with the dye methylene blue. The cells of *S. volutans* are typically large, over 20 µm in length. In wet mounts, *Spirillum* species may be seen moving to a narrow zone near the edge of the coverslip, where O_2 is available in the optimum amount. ⏮ volutin

The Genera *Thioploca* and *Thiomargarita*

Thioploca and *Thiomargarita* species live in marine environments and store both sulfur (their energy source) and nitrate (their terminal electron acceptor). This allows the bacteria to compensate for the fact that their energy source and terminal electron acceptor are not usually found in the same location. Reduced sulfur compounds are often plentiful in anaerobic marine sediments, whereas the waters above have nitrate. By storing both sulfur and nitrate, the cells always have a supply of each.

Thioploca cells "commute" between the sulfur-rich sediments and the nitrate-rich waters, gathering and storing their energy source and terminal electron acceptor as needed. They do this by forming long sheaths within which the cells move between the two environments.

The cells of the huge bacterium *Thiomargarita namibiensis* ("sulfur pearl of Namibia") are a pearly white color due to globules of sulfur in their cytoplasm (see Focus Your Perspective 1.1). Each cell contains a large nitrate storage vacuole that takes up about 98% of the cell volume. These organisms, which can reach a diameter of 0.75 mm, are not motile and instead rely on storms or other disturbances to bring them into contact with nitrate-rich waters.

MicroByte

Thiomargarita namibiensis can store a 3-month supply of both its energy source (sulfur) and its terminal electron acceptor (nitrate).

MicroAssessment 11.7

Sheathed bacteria attach to solid objects in favorable locations. Prosthecate bacteria produce extensions that maximize the absorptive surface area. *Bdellovibrio* species, bioluminescent bacteria, and *Legionella* species use nutrients from other organisms. Spirochetes and magnetotactic bacteria move by unusual mechanisms. Some organisms form storage granules.

19. How do squid benefit from having a light organ colonized by bioluminescent bacteria?

20. What is the habitat of *Legionella* species?

21. The genomes of free-living spirochetes are larger than those of ones that live within an animal host. Why would this be so?

FOCUS ON A CASE 11.1

The "patient" was an enormous region of Lake Erie. An inches-thick green scum covered the surface during the summer months, and fish and other aquatic animals in the area were dying. Microscopic analysis revealed that the green scum was a bloom of cyanobacteria, a group of photosynthetic bacteria. The aquatic animals were dying because the waters had insufficient dissolved O_2, causing the animals to suffocate. In turn, this created a large "dead zone"—a region that lacks animal life.

The source of the problem was high levels of phosphorus in the lake. When excess phosphorus is available, cyanobacteria may grow to high numbers, an event referred to as a bloom. There are many potential sources of this type of pollution, ranging from industries to individuals. For example, manure and chemical fertilizers used by farms and homeowners to promote plant growth often contain phosphate, a source of phosphorus for both plants and other organisms, including cyanobacteria. Fertilizers applied to soils can be washed into the lake by rain. Another source of phosphate is wastewater treatment plants; detergents as well as biological materials can contain phosphate, and most wastewater treatment processes do not remove this substance. Liquid from septic systems also contains phosphate.

The obvious solution to the problem of excess cyanobacterial growth is to identify and control the most significant levels of phosphorus pollution. Although that might seem simple, it requires that communities and individuals recognize and accept responsibility for a widespread problem, and then work together to develop a solution.

1. Would "dead zones" lack all life, or only animal life?
2. Why would sunlight in combination with phosphorus-containing pollution promote the growth of cyanobacteria?
3. Why would cyanobacterial growth lead to decreased levels of O_2 in the water?
4. How could microorganisms be used to remove phosphate in wastewater?
5. What health risks could cyanobacterial blooms pose for residents of cities that draw their drinking water from the lake?

Discussion

1. Dead zones still have living organisms, just not those that depend on O_2 for respiration. For example, fermenting and anaerobically respiring microorganisms can survive in environments that have low or no O_2.
2. Cyanobacteria are photosynthetic, so they harvest energy from the radiant energy of the sun and use CO_2 as a carbon source. In the summer months, there is usually plenty of sunshine, so blooms are more common during summer. Many cyanobacterial species can fix nitrogen, converting N_2 in the atmosphere into forms that cells can incorporate into organic material. Considering that nitrogen-fixing cyanobacteria have virtually unlimited supplies of carbon and nitrogen, as well as plenty of their energy source in the summer, phosphorus is often the limiting nutrient. ⏮ **limiting nutrient**
3. Some of the cyanobacteria eventually burst, providing organic materials that can be used as an energy source by various chemoheterotrophic microorganisms in the water. When these organisms aerobically respire, they use O_2 dissolved in the water as a terminal electron acceptor, reducing it to H_2O. This lowers the concentration of dissolved O_2, eventually to the point that not enough is available to support the requirements of fish and other aquatic animals.
4. *Spirillum volutans* and some other bacteria make volutin granules, which are storage forms of phosphate. These bacteria could be added to wastewater, where they would take up and accumulate the phosphate. The bacterial cells could then be removed, thereby removing the phosphate. Chemical methods for removing phosphate are also available, and these are increasingly being used as an additional step in wastewater treatment.
5. Some types of cyanobacteria make toxins. For a few days in August, 2014, over 400,000 residents of Toledo, Ohio, were told not to use their tap water because it contained unsafe toxin levels due to a cyanobacterial bloom in Lake Erie.

11.8 ■ Animals as Habitats

Learning Outcomes

10. Compare and contrast the examples of bacteria that use animals as habitats.

The bodies of animals, including humans, provide a wide variety of ecological habitats for microbes—from dry, O_2-rich surfaces to moist, anaerobic folds and depressions. **Table 11.3** lists the medically important bacteria covered in this and other sections of the chapter.

Bacteria That Inhabit the Skin

The skin is typically dry and salty, providing an environment inhospitable to many microorganisms. Members of the genus *Staphylococcus,* however, thrive under these conditions. The propionic acid bacteria, which were discussed earlier, inhabit anaerobic microenvironments of the skin. ⏭ **anatomy, physiology, and ecology of the skin**

The Genus *Staphylococcus*

Staphylococcus species are Gram-positive cocci that are facultative anaerobes. Most, such as *S. epidermidis,* reside

TABLE 11.3 Medically Important Bacteria

Organism	Medical Significance	Phylum
Gram-Negative Rods		
Bacteroides species	Obligate anaerobes that commonly inhabit the mouth, intestinal tract, and genital tract. Cause abscesses and bloodstream infections. ⏭ **the genus *Bacteroides***	*Bacteroidetes*
Bordetella pertussis	Causes pertussis (whooping cough). ⏭ **the genus *Bordetella***	*Proteobacteria*
Enterobacteriaceae	Most live in the intestinal tract. ⏮ **the family *Enterobacteriaceae***	*Proteobacteria*
Enterobacter species	Normal microbiota of the intestinal tract.	
Escherichia coli	Normal microbiota of the intestinal tract. Some strains cause urinary tract infections; some strains cause specific types of intestinal disease; some cause meningitis in newborns.	
Klebsiella pneumoniae	Normal microbiota of the intestinal tract. Causes pneumonia.	
Proteus species	Normal microbiota of the intestinal tract. Cause urinary tract infections.	
Salmonella enterica serotype Enteritidis	Causes gastroenteritis. Grows in the intestinal tract of infected animals; acquired by consuming contaminated food.	
Salmonella enterica serotype Typhi	Causes typhoid fever. Grows in the intestinal tract of infected humans; transmitted in feces.	
Shigella species	Cause dysentery. Grow in the intestinal tract of infected humans; transmitted in feces.	
Yersinia pestis	Causes bubonic plague, which is transmitted by fleas, and pneumonic plague, which is transmitted in respiratory droplets of infected individuals.	
Haemophilus influenzae	Causes ear infections, respiratory infections, and meningitis in children. ⏭ **the genus *Haemophilus***	*Proteobacteria*
Haemophilus ducreyi	Causes chancroid, a sexually transmitted disease. ⏭ **the genus *Haemophilus***	*Proteobacteria*
Legionella pneumophila	Causes legionellosis, a respiratory disease. Grows within protozoa; acquired by inhaling contaminated water droplets. ⏮ **the genus *Legionella***	*Proteobacteria*
Pseudomonas aeruginosa	Causes burn, urinary tract, and bloodstream infections. Common in the environment. Grows in nutrient-poor aqueous solutions. Resistant to many disinfectants and antimicrobial medications. ⏮ **the genus *Pseudomonas***	*Proteobacteria*
Gram-Negative Rods—Obligate Intracellular Parasites		
Chlamydophila (Chlamydia) pneumoniae	Causes atypical pneumonia, or "walking pneumonia." Acquired from an infected person. ⏭ **the genera *Chlamydia* and *Chlamydophila***	*Chlamydiae*
Chlamydophila (Chlamydia) psittaci	Causes psittacosis, a form of pneumonia. Transmitted by birds. ⏭ **the genera *Chlamydia* and *Chlamydophila***	*Chlamydiae*
Chlamydia trachomatis	Causes a sexually transmitted disease that mimics the symptoms of gonorrhea. Also causes conjunctivitis in newborns and trachoma (a serious eye infection). ⏭ **the genera *Chlamydia* and *Chlamydophila***	*Chlamydiae*
Coxiella burnetii	Causes Q fever. Acquired by inhaling organisms shed by infected animals. ⏭ **the genus *Coxiella***	*Proteobacteria*
Ehrlichia chaffeensis	Causes human ehrlichiosis. Transmitted by ticks. ⏭ **the genera *Rickettsia*, *Orientia*, and *Ehrlichia***	*Proteobacteria*
Orientia tsutsugamushi	Causes scrub typhus. Transmitted by mites. ⏭ **the genera *Rickettsia*, *Orientia*, and *Ehrlichia***	*Proteobacteria*
Rickettsia prowazekii	Causes epidemic typhus. Transmitted by lice. ⏭ **the genera *Rickettsia*, *Orientia*, and *Ehrlichia***	*Proteobacteria*
Rickettsia rickettsii	Causes Rocky Mountain spotted fever. Transmitted by ticks. ⏭ **the genera *Rickettsia*, *Orientia*, and *Ehrlichia***	*Proteobacteria*
Wolbachia pipientis	Resides within the filarial worms that cause river blindness and elephantiasis. ⏭ **the genus *Wolbachia***	*Proteobacteria*
Gram-Negative Curved Rods		
Campylobacter jejuni	Causes gastroenteritis. Grows in the intestinal tract of infected animals; acquired by consuming contaminated food. ⏭ **the genera *Campylobacter* and *Helicobacter***	*Proteobacteria*
Helicobacter pylori	Causes peptic ulcers. Neutralizes stomach acid by producing urease. ⏭ **the genera *Campylobacter* and *Helicobacter***	*Proteobacteria*
Vibrio cholerae	Causes cholera, a severe diarrheal disease; acquired by drinking contaminated water. ⏮ **the genus *Vibrio***	*Proteobacteria*
Vibrio parahaemolyticus	Causes gastroenteritis. Acquired by consuming contaminated seafood. ⏮ **the genus *Vibrio***	*Proteobacteria*
Vibrio vulnificus	Causes a systemic disease, particularly in people who have liver failure or other underlying complications. ⏮ **the genus *Vibrio***	*Proteobacteria*
Gram-Negative Cocci		
Neisseria meningitidis	Causes meningitis. ⏭ **the genus *Neisseria***	*Proteobacteria*
Neisseria gonorrhoeae	Causes gonorrhea, a sexually transmitted infection. ⏭ **the genus *Neisseria***	*Proteobacteria*

TABLE 11.3 Medically Important Bacteria (*continued*)

Organism	Medical Significance	Phylum
Gram-Positive Rods		
Bacillus anthracis	Causes anthrax. Acquired by inhaling endospores in soil, animal hides, and wool. Also acquired by touching or ingesting the endospores. Bioterrorism agent. ⏮ **endospore formers**	*Firmicutes*
Bifidobacterium species	Predominant member of the intestinal tract in breast-fed infants. Thought to play a protective role in the intestinal tract by excluding pathogens. ⏮ **the genus *Bifidobacterium***	*Actinobacteria*
Clostridium botulinum	Causes botulism. Foodborne disease results from ingesting toxin-contaminated foods, typically canned foods that have been improperly processed. ⏮ **the genus *Clostridium***	*Firmicutes*
Clostridium perfringens	Causes gas gangrene. Acquired when soil-borne endospores contaminate a wound. ⏮ **the genus *Clostridium***	*Firmicutes*
Clostridium tetani	Causes tetanus. Acquired when soil-borne endospores are inoculated into deep tissue. ⏮ **the genus *Clostridium***	*Firmicutes*
Clostridium difficile	Causes *Clostridium difficile* infection (CDI), which is associated with antibiotic use and can result in severe diarrhea. ⏮ **the genus *Clostridium***	*Firmicutes*
Corynebacterium diphtheriae	Toxin-producing strains cause diphtheria. ⏮ **the genus *Corynebacterium***	*Actinobacteria*
Gram-Positive Cocci		
Enterococcus species	Normal microbiota of the intestinal tract. Cause urinary tract infections. ⏮ **lactic acid bacteria**	*Firmicutes*
Micrococcus species	Found on skin as well as in a variety of other environments; often contaminate bacteriological media. ⏮ **the genus *Micrococcus***	*Actinobacteria*
Staphylococcus aureus	Leading cause of wound infections. Causes boils, carbuncles, food poisoning, and staphylococcal toxic shock syndrome. ⏭ **the genus *Staphylococcus***	*Firmicutes*
Staphylococcus epidermidis	Normal microbiota of the skin. ⏭ **the genus *Staphylococcus***	*Firmicutes*
Staphylococcus saprophyticus	Causes urinary tract infections. ⏭ **the genus *Staphylococcus***	*Firmicutes*
Streptococcus pneumoniae	Causes pneumonia and meningitis. ⏮ **lactic acid bacteria**	*Firmicutes*
Streptococcus pyogenes	Causes pharyngitis (strep throat), rheumatic fever, wound infections, glomerulonephritis, and streptococcal toxic shock. ⏮ **lactic acid bacteria**	*Firmicutes*
Acid-Fast Rods		
Mycobacterium tuberculosis	Causes tuberculosis. ⏮ **the genus *Mycobacterium***	*Actinobacteria*
Mycobacterium leprae	Causes Hansen's disease (leprosy); peripheral nerve invasion is characteristic. ⏮ **the genus *Mycobacterium***	*Actinobacteria*
Spirochetes	Characterized by spiral shape, flexible cell wall, and endoflagella. ⏮ **spirochetes**	
Treponema pallidum	Causes syphilis, a sexually transmitted disease. The organism has never been grown in culture. ⏭ **the genera *Treponema* and *Borrelia***	*Spirochaetes*
Borrelia burgdorferi	Causes Lyme disease, a tick-borne disease. ⏭ **the genera *Treponema* and *Borrelia***	*Spirochaetes*
Borrelia recurrentis and *B. hermsii*	Causes relapsing fever. Transmitted by arthropods. ⏭ **the genera *Treponema* and *Borrelia***	*Spirochaetes*
Leptospira interrogans	Causes leptospirosis, a waterborne disease. Excreted in urine of infected animals. ⏮ **spirochetes**	*Spirochaetes*
Cell Wall-less		
Mycoplasma pneumoniae	Causes atypical pneumonia ("walking pneumonia"). Not susceptible to penicillin because it lacks a cell wall. ⏭ **the genus *Mycoplasma***	*Tenericutes*

harmlessly as part of the normal microbiota of the skin. Like other bacteria that aerobically respire, *Staphylococcus* species are catalase-positive. This distinguishes them from *Streptococcus, Enterococcus,* and *Lactococcus* species, which are also Gram-positive cocci but lack the enzyme catalase. Several species of *Staphylococcus* are notable for their medical significance. *Staphylococcus aureus* causes a variety of diseases, including skin and wound infections, as well as food poisoning. *Staphylococcus saprophyticus* causes urinary tract infections.

Bacteria That Inhabit Mucous Membranes

Mucous membranes of the respiratory, genitourinary, and intestinal tracts provide a habitat for numerous kinds of bacteria, many of which have already been discussed. For example, *Streptococcus* and *Corynebacterium* species reside in the respiratory tract, *Lactobacillus* species inhabit the vagina, and *Clostridium* species and members of the family *Enterobacteriaceae* thrive in the intestinal tract. Some of the other genera are discussed next.

The Genus *Bacteroides*

Bacteroides species are small, strictly anaerobic, Gram-negative rods and coccobacilli. They inhabit the mouth, intestinal tract, and genital tract of humans and other animals. *Bacteroides* species, which make up about a third of the bacteria in human feces, play an important role in digestion. In addition, they are often responsible for abscesses and bloodstream infections that follow appendicitis and abdominal surgery. Many are killed by brief exposure to O_2, so they are difficult to study.

The Genus *Bifidobacterium*

Bifidobacterium species are Gram-positive, irregular, rod-shaped anaerobes that reside primarily in the intestinal tract of humans and other animals. They are the most common members of the intestinal microbiota of breast-fed infants and are thought to provide a protective function by excluding disease-causing bacteria. Formula-fed infants are also colonized with members of this genus, but generally the concentrations are lower.

The Genus *Bordetella*

Bordetella species are small, Gram-negative coccobacilli that only grow aerobically. Special media must be used for their cultivation because they are nutritionally fastidious. The most significant species medically is *Bordetella pertussis,* which causes whooping cough, a highly contagious infectious disease of humans. *Bordetella bronchiseptica* causes a respiratory infection in dogs, an illness commonly called "kennel cough." ⏮ fastidious

The Genera *Campylobacter* and *Helicobacter*

Members of the genera *Campylobacter* and *Helicobacter* are curved Gram-negative rods. As microaerophiles, they require specific atmospheric conditions to grow in culture. *Campylobacter jejuni* causes diarrheal disease in humans. It typically lives in the intestinal tract of domestic animals, particularly poultry. *Helicobacter pylori* inhabits the stomach, where it can cause peptic ulcers (lesions in the stomach and duodenum); it has also been linked to stomach cancer. An important factor in its ability to survive in the stomach is its production of the enzyme urease. This breaks down urea to produce ammonia, which neutralizes the acid in the cell's immediate surroundings. ⏭ peptic ulcers

The Genus *Haemophilus*

Haemophilus species are Gram-negative coccobacilli that, as their name reflects, are "blood loving." They require hematin and/or NAD, which are found in blood. Many species are common microbiota of the respiratory tract. *H. influenzae* causes ear infections, respiratory infections, and meningitis, primarily in children. *Haemophilus ducreyi* causes the sexually transmitted disease chancroid. ⏭ ear infections, ⏭ meningitis, ⏭ chancroid

The Genus *Mycoplasma*

Members of the genus *Mycoplasma* lack a cell wall, but most have sterols in their membrane to provide strength and rigidity. They are among the smallest forms of life, and their genomes are thought to be about the minimum size for encoding the essential functions for a free-living organism.

Mycoplasmas are a particular concern in laboratories studying viruses and eukaryotic cell physiology. This is because the small bacteria can pass through the filters used to sterilize tissue culture media, so they easily contaminate the media and thereby compromise experimental results. Mycoplasmas can grow to very high numbers in media without causing cloudiness, so contamination might not be noticed. When grown on agar media, *Mycoplasma* cells in the center of a colony grow into the medium, producing a dense central area that gives the characteristically small colony a "fried egg" appearance (**figure 11.24**). ⏭ tissue culture

Medically, one of the most significant mycoplasmas is *M. pneumoniae,* which, as its name implies, causes a form of pneumonia. This type of pneumonia, often called "walking pneumonia," cannot be treated with penicillin or other antibiotics that interfere with peptidoglycan synthesis, because these organisms lack a cell wall.

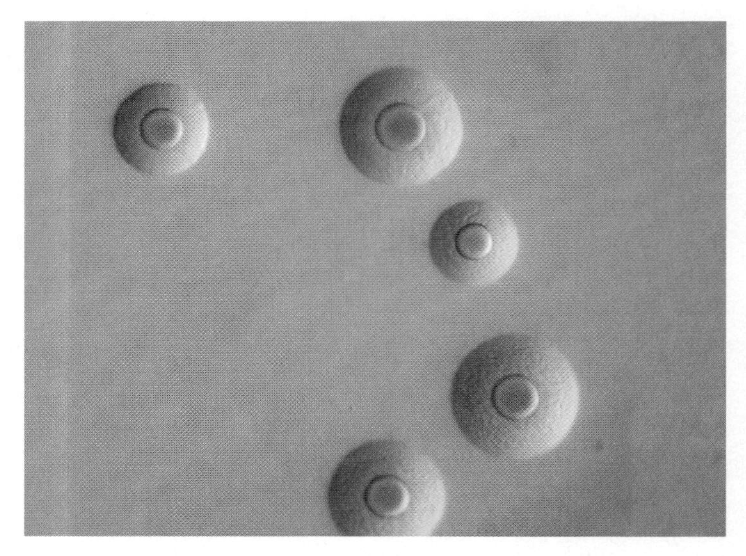

FIGURE 11.24 ***Mycoplasma* Colonies** The characteristically small colonies have a dense center, resulting in a "fried egg" appearance. ©Christine Citti

? Why are members of this genus not affected by penicillin?

MicroByte

The genome of *Mycoplasma genitalium* is only 5.8×10^5 base pairs, approximately one-eighth the size of the *E. coli* genome.

The Genus *Neisseria*

Neisseria species are Gram-negative bacteria, typically kidney-bean-shaped cocci in pairs. They are common microbiota of animals including humans, growing on mucous membranes. *Neisseria* species are typically aerobes, but some can grow anaerobically if a suitable terminal electron acceptor such as nitrite is present. Those noted for their medical significance include *N. gonorrhoeae,* which causes the sexually transmitted disease gonorrhea, and *N. meningitidis,* which causes meningitis; both are nutritionally fastidious. ◀◀ fastidious, ▶▶| gonorrhea, ▶▶| meningitis

The Genera *Treponema* and *Borrelia*

Members of the genera *Treponema* and *Borrelia* are spirochetes that typically inhabit body fluids and mucous membranes of humans and other animals. Recall that spirochetes are characterized by their corkscrew shape and endoflagella. Although they have a Gram-negative cell wall, they are often too thin to be seen using standard microscopy and staining methods.

Treponema species are obligate anaerobes or microaerophiles that often inhabit the mouth and genital tract. The species that causes syphilis, *T. pallidum,* is difficult to study because it has never been grown in culture. Analysis of its genome sequence indicates that the organism is a microaerophile with a metabolism highly dependent on its host. It lacks critical enzymes of the TCA cycle and a variety of other metabolic pathways. ▶▶| syphilis

Three *Borrelia* species are pathogens, transmitted by arthropods such as ticks and lice. *B. recurrentis* and *B. hermsii* both cause relapsing fever; *B. burgdorferi* causes Lyme disease. An unusual feature of *Borrelia* species is their genome—a linear chromosome and many linear and circular plasmids. ▶▶| Lyme disease

Obligate Intracellular Parasites

Obligate intracellular parasites cannot reproduce outside a host cell. By living within host cells, the parasites are supplied with a source of compounds they would otherwise need to synthesize for themselves. As a result, most intracellular parasites have lost the ability to make substances needed for extracellular growth. Bacterial examples of this group of organisms include members of the genera *Rickettsia, Orientia, Ehrlichia, Coxiella, Chlamydia, Chlamydophila,* and *Wolbachia,* which are all tiny Gram-negative rods or coccobacilli.

The Genera *Rickettsia, Orientia,* and *Ehrlichia*

Species of *Rickettsia, Orientia,* and *Ehrlichia* are responsible for several serious human diseases spread by blood-sucking arthropods such as ticks and lice. *Rickettsia rickettsii* causes Rocky Mountain spotted fever, *R. prowazekii* causes epidemic typhus, *O. tsutsugamushi* causes scrub typhus, and *E. chaffeensis* causes human ehrlichiosis.

The Genus *Coxiella*

The only characterized species of *Coxiella, C. burnetii,* is an obligate intracellular bacterium that survives well outside the host cell. During its intracellular growth, *C. burnetii* forms spore-like structures called small-cell variants (SCVs) that later allow it to survive in the outside environment. The structures, however, lack the extreme resistance to heat and disinfectants characteristic of endospores (**figure 11.25**). *Coxiella burnetii* causes Q fever of humans, a disease most often acquired by inhaling bacteria shed from infected animals. This is particularly a problem when animals give birth because high numbers of the bacteria can be found in the placenta of infected animals.

The Genera *Chlamydia* and *Chlamydophila*

Chlamydia and *Chlamydophila* species are transmitted directly from person to person and have a unique growth cycle (**figure 11.26**). Inside the host cell, they initially exist as non-infectious reticulate bodies, which reproduce by binary fission. Later in the infection, the bacteria differentiate into smaller, dense-appearing elementary bodies, which are the infectious form released when the host cell ruptures.

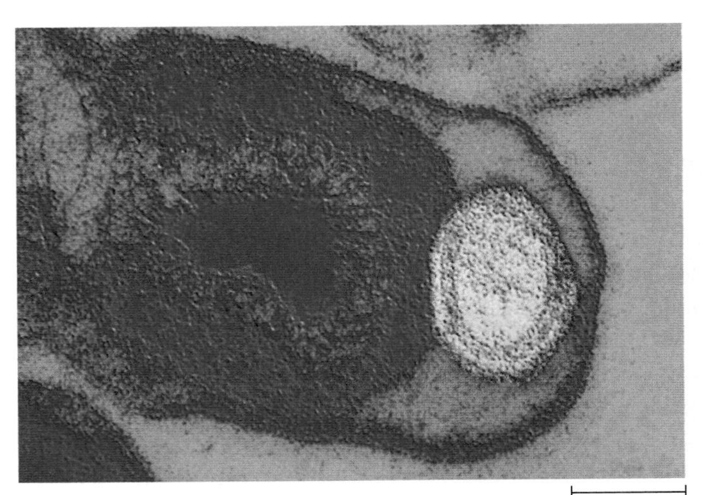

FIGURE 11.25 ***Coxiella*** Color-enhanced TEM of *C. burnetii*. The oval orange-colored object is the spore-like structure. ©Alfred Pasieka/Science Photo Library/Science Source

? What disease does *C. burnetii* cause?

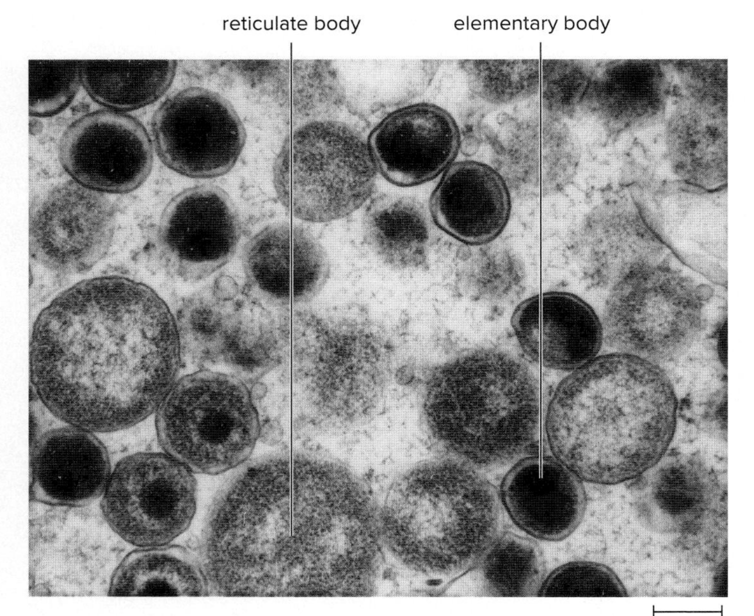

FIGURE 11.26 ***Chlamydia*** **Growing in Tissue Cell Culture** Reticulate bodies and elementary bodies are shown (TEM). ©David M. Phillips/Science Source

Which form of *Chlamydia*—the reticulate body or elementary body—is infectious?

Chlamydia trachomatis causes eye infections and a sexually transmitted infection that mimics gonorrhea; *Chlamydophila pneumoniae* causes atypical pneumonia; and *Chlamydophila psittaci* causes psittacosis, a form of pneumonia. ***Chlamydia trachomatis***

The Genus *Wolbachia*

The only known species of *Wolbachia, W. pipientis,* infects arthropods (including insects, spiders, and mites) and parasitic worms. It is primarily transmitted maternally, via the eggs of infected females to their offspring.

In arthropods, *W. pipientis* uses unique strategies to increase the overall population of infected females, including killing male embryos, allowing infected females to reproduce asexually, and causing infected males to gain female traits. In addition, the parasite destroys embryos resulting from the mating of an infected male with either an uninfected female or a female infected with a different strain.

Wolbachia pipientis does not infect mammals, but its medical importance was recognized with the discovery that it resides within filarial worms that cause the diseases river blindness and elephantiasis. The chronic and debilitating inflammation associated with these diseases appears to result from the immune response directed against the bacterial cells in the invading worms. This discovery paves the way for new treatments, because eliminating the bacteria not only lessens the symptoms but also kills the filarial worms.

MicroAssessment 11.8

Staphylococcus species thrive in the dry, salty conditions of the skin. *Bacteroides* and *Bifidobacterium* species live in the gastrointestinal tract; *Campylobacter* and *Helicobacter* species can cause disease when they reside there. *Neisseria* species, mycoplasmas, and spirochetes inhabit other mucous membranes. Obligate intracellular parasites—including *Rickettsia, Orientia, Ehrlichia, Coxiella, Chlamydia, Chlamydophila,* and *Wolbachia* species—are unable to reproduce outside a host cell.

22. How does *Helicobacter pylori* withstand stomach acidity?

23. What characteristic of *Mycoplasma* species separates them from other bacteria?

24. Why would breast feeding affect the composition of a baby's intestinal microbiota?

11.9 ■ Archaea That Thrive in Extreme Conditions

Learning Outcomes

11. Compare and contrast the characteristics and habitats of the extreme halophiles and the extreme thermophiles.

Characterized members of the *Archaea* typically thrive in extreme environments (**table 11.4**). These include conditions of high heat, acidity, alkalinity, and salinity. An exception is the methanogens, discussed earlier in the chapter; they inhabit anaerobic niches shared with members of the *Bacteria.* In addition to the characterized archaea, many others have been detected in a variety of non-extreme environments using molecular techniques.

Extreme Halophiles

The extreme halophiles are found in high numbers in salty environments such as salt lakes, soda lakes, and brines used for curing fish. Most grow well in saturated salt solutions (32% NaCl), and they require a minimum of about 9% NaCl. Because they produce pigments, their growth can be seen as red patches on salted fish and pink blooms in concentrated saltwater ponds (**figure 11.27**).

Extreme halophiles are aerobic or facultatively anaerobic chemoheterotrophs, but some also can obtain additional energy from light. These organisms have the light-sensitive pigment bacteriorhodopsin, which absorbs energy from sunlight and uses it to pump protons from the cell. This creates a proton gradient that can be used to drive flagella or synthesize ATP.

Extreme halophiles come in a variety of shapes, including rods, cocci, discs, and triangles. They include genera such as *Halobacterium, Halorubrum, Natronobacterium,* and *Natronococcus;* members of these latter two genera are extremely alkaliphilic as well as halophilic.

TABLE 11.4 Archaea

Group/Genera	Characteristics	Phylum
Methanogens—*Methanospirillum, Methanosarcina*	Generate methane when they oxidize hydrogen gas as an energy source, using CO_2 as a terminal electron acceptor.	*Euryarchaeota*
Extreme halophiles—*Halobacterium, Halorubrum, Natronobacterium, Natronococcus*	Found in salt lakes, soda lakes, and brines. Most grow well in saturated salt solutions.	*Euryarchaeota*
Extreme thermophiles—*Methanothermus, Methanopyrus, Pyrodictium, Pyrolobus, Sulfolobus, Thermophilus, Picrophilus, Nanoarchaeum*	Found near hydrothermal vents and in hot springs; some grow at temperatures above 100°C. Includes examples of methane-generating, sulfur-reducing, and sulfur-oxidizing archaea, as well as extreme acidophiles.	*Crenarchaeota, Euryarchaeota,* and *Nanoarchaeota*

Extreme Thermophiles

The extreme thermophiles (hyperthermophiles) are found near volcanic vents and fissures that release sulfurous gases and other hot vapors. Because these regions are thought to closely mimic early Earth's environment, scientists are interested in studying the prokaryotes that live there. Others are found in hydrothermal vents in the deep sea and hot springs.

⏮ hyperthermophiles

Methane-Generating Hyperthermophiles

In contrast to the mesophilic methanogens discussed earlier, some methanogens are extreme thermophiles. For example, *Methanothermus* species, which can grow in temperatures as high as 97°C, grow optimally at approximately 84°C. *Methanopyrus kandleri* can grow at 122°C— the current record for highest recorded growth temperature.

Sulfur-Reducing Hyperthermophiles

The sulfur-reducing hyperthermophiles are obligate anaerobes that use sulfur as a terminal electron acceptor, generating H_2S. They harvest energy by oxidizing organic compounds and/or H_2. These archaea can be isolated from hot sulfur-containing environments such as sulfur hot springs and hydrothermal vents. They include some of the most thermophilic organisms known, a few even growing above 100°C. One example, *Pyrolobus fumarii,* was isolated from a "black smoker" 3,650 m (about 12,000 feet) deep in the Atlantic Ocean. It grows between 90°C and 113°C. Another hydrothermal vent isolate, *Pyrodictium occultum,* has an optimum temperature of about 105°C and cannot grow below 82°C. Its disc-shaped cells are connected by hollow tubes, forming a web-like network (**figure 11.28**). A hydrothermal vent isolate called "strain 121" (also referred to as *Geogemma barossii*) grows at 121°C, which was previously the highest recorded growth temperature.

Nanoarchaea

The discovery of an archaeum so unique that it represents an entirely new phylum, *Nanoarcheota* ("dwarf archaea"), was made possible by the earlier discovery of a new genus of sulfur-reducing hyperthermophiles, *Ignicoccus* ("the fire sphere"). *Nanoarchaeum equitans* ("rider") grows as 400 nm spheres attached to the surface of—presumably parasitizing—an *Ignicoccus* species.

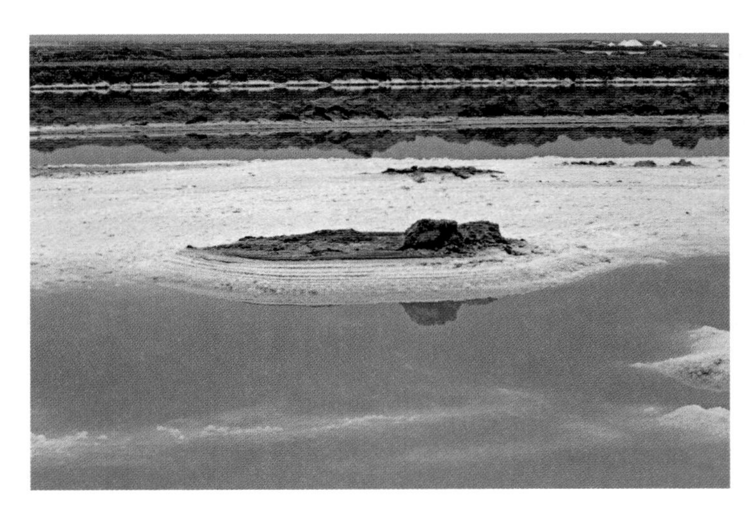

FIGURE 11.27 Solar Evaporation Pond ©Armands Pharyos/Alamy Stock Photo

? What causes the pink color in the pond?

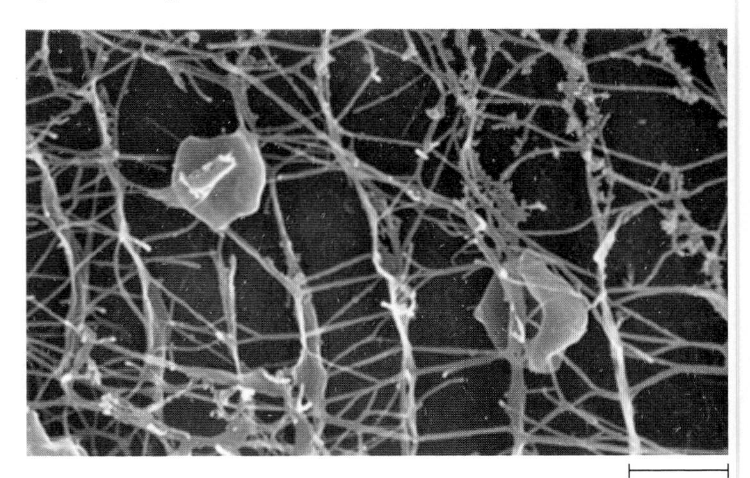

FIGURE 11.28 *Pyrodictium* The disc-shaped cells are connected by hollow tubes (SEM). ©Dr. Karl O. Stetter/University of Regensburg

? How would members of this genus be grouped with respect to their temperature preference?

FIGURE 11.29 Typical Habitat of *Sulfolobus* Sulfur hot spring in Yellowstone National Park. ©blanscape/Getty Images

Why must *Solfolobus* species be able to tolerate acidic conditions?

Sulfur Oxidizers

Sulfolobus species are found at the surface of acidic sulfur-containing hot springs such as many of those found in Yellowstone National Park (**figure 11.29**). They are obligate aerobes that oxidize sulfur compounds, using O_2 as a terminal electron acceptor to generate sulfuric acid. In addition, they are thermoacidophilic, only growing above 50°C and at a pH between 1 and 6.

Thermophilic Extreme Acidophiles

Members of two genera, *Thermoplasma* and *Picrophilus,* are notable for growing in extremely acidic, hot environments. *Thermoplasma* species grow optimally at pH 2; in fact, *T. acidophilum* lyses at neutral pH. It was originally isolated from a coal waste pile. *Picrophilus* species tolerate conditions even more acidic, growing optimally at a pH below 1. Two species isolated in Japan inhabited acidic areas in regions that spew sulfurous gases.

MicroAssessment 11.9

Many characterized archaea inhabit extreme environments. These include conditions of high salinity, heat, acidity, and alkalinity.

25. What is the habitat of *Nanoarchaeum equitans*?

26. At which relative depth in a sulfur hot spring would a sulfur oxidizer likely be found?

27. What characteristic of the methanogens makes it logical to discuss them with the *Bacteria* rather than with the *Archaea* described in this section?

FOCUS ON THE FUTURE 11.1

Astrobiology: Searching for Life Beyond Earth

If life as we know it exists elsewhere in the universe, it will likely be microbial. The task, then, is to figure out how to find and detect such extraterrestrial microorganisms.

Considering that we still know relatively little about the microbial life on our own planet, coupled with the extreme difficulty of obtaining or testing extraterrestrial samples, this is a daunting challenge with many as yet unanswered questions. What is the most likely source of life beyond Earth? What is the best way to preserve specimens for study on Earth? What will be the culture requirements to grow such organisms? Astrobiology, the study of life in the universe, is a relatively new field that brings together scientists from a wide range of disciplines, including microbiology, geology, astronomy, biology, and chemistry. The goal is to determine the origin, evolution, distribution, and destiny of life in the universe. Astrobiologists are also developing lightweight, dependable, and meaningful testing devices to be used in future space missions.

To prepare for researching life on other planets and their moons, microbiologists have turned to some of the most extreme environments here on Earth. These include glaciers and ice shelves, hot springs, deserts, volcanoes, deep ocean hydrothermal vents, and subterranean features such as caves. Because select microorganisms can survive in these environs, which are analogous to conditions expected beyond Earth, they are good testing grounds for the technology to be used on future missions.

Summary

METABOLIC DIVERSITY (table 11.1)

11.1 Anaerobic Chemotrophs

Anaerobic Chemolithotrophs

The **methanogens** are a group of archaea that harvest energy by oxidizing H_2, using CO_2 as a terminal electron acceptor (figure 11.1).

Anaerobic Chemoorganotrophs—Anaerobic Respiration

Desulfovibrio species reduce sulfur compounds to form hydrogen sulfide.

Anaerobic Chemoorganotrophs—Fermentation

Clostridium species form endospores. The **lactic acid bacteria** produce lactic acid as their primary fermentation end product (figures 11.2, 11.3). *Propionibacterium* species produce propionic acid as their primary fermentation end product.

11.2 ■ Anoxygenic Phototrophs

Purple Bacteria

The **purple bacteria** appear red, orange, or purple; the components of their photosynthetic apparatus are all within the cytoplasmic membrane.

Green Bacteria

The **green bacteria** are typically green or brownish. Their accessory pigments are often located in chlorosomes.

Other Anoxygenic Phototrophs

Other anoxygenic phototrophs have been discovered, including some that form endospores.

11.3 ■ Oxygenic Phototrophs

Cyanobacteria (figures 11.5, 11.6)

Genetic evidence indicates that chloroplasts evolved from a species of **cyanobacteria.** Nitrogen-fixing cyanobacteria are important ecologically, because they provide an available source of both carbon and nitrogen (figure 11.7). Filamentous species maintain the structure and productivity of some soils. Some species of cyanobacteria produce toxins that can be deadly to animals that ingest contaminated water.

11.4 ■ Aerobic Chemolithotrophs

Sulfur-Oxidizing Bacteria

The filamentous sulfur oxidizers *Beggiatoa* and *Thiothrix* live in sulfur springs, in sewage-polluted waters, and on the surface of marine and freshwater sediments (figure 11.8). *Acidithiobacillus* species are found in both terrestrial and aquatic habitats (figure 11.9).

Nitrifiers

Ammonia oxidizers convert ammonium to nitrite; they include *Nitrosomonas* and *Nitrosococcus.* Nitrite oxidizers convert nitrite to nitrate; they include *Nitrobacter* and *Nitrococcus.*

Hydrogen-Oxidizing Bacteria

Aquifex and *Hydrogenobacter* species are thought to be among the earliest bacterial forms to exist on Earth.

11.5 ■ Aerobic Chemoorganotrophs

Obligate Aerobes

Micrococcus species are found in soil and on dust particles, inanimate objects, and skin (figure 11.10). *Mycobacterium* species are widespread in nature. They are **acid-fast.** *Pseudomonas* species are widespread in nature and have extremely diverse metabolic capabilities (figure 11.11). *Thermus aquaticus* is the source of *Taq* polymerase, a component of PCR. *Deinococcus radiodurans* can survive radiation exposure several thousand times that lethal to a human being.

Facultative Anaerobes

Corynebacterium species are widespread in nature (figure 11.12). Members of the family *Enterobacteriaceae* typically inhabit the intestinal tract of animals, although some reside in rich soil. **Coliforms** are used as indicators of fecal pollution. *Vibrio* species typically live in marine environments.

ECOPHYSIOLOGICAL DIVERSITY (table 11.2)

11.6 ■ Thriving in Terrestrial Environments

Bacteria That Form a Resting Stage

Bacillus and *Clostridium* species form endospores, the most resistant dormant form known (figure 11.13). *Azotobacter* species form cysts. Cells of **myxobacteria** come together to form a fruiting body, within which the cells become dormant microcysts (figure 11.14). *Streptomyces* species form chains of conidia at the end of hyphae. Many species naturally produce antibiotics (figure 11.15).

Bacteria That Associate with Plants

Agrobacterium species cause plant tumors (figure 11.16). They transfer a portion of the **Ti plasmid** to plant cells, genetically engineering the plant cells to produce opines and plant growth hormones. **Rhizobia** reside as endosymbionts in nodules on the roots of legumes, fixing nitrogen (figure 11.17).

11.7 ■ Thriving in Aquatic Environments

Sheathed Bacteria

Sheathed bacteria attach to solid objects in favorable habitats; the sheath shelters them from attack by predators (figure 11.18).

Prosthecate Bacteria

Caulobacter species have a single polar prostheca called a stalk; at the tip of the stalk is a holdfast. The cells divide by binary fission (figure 11.19). *Hyphomicrobium* species divide by forming a bud at the tip of their single polar prostheca (figure 11.20).

Bacteria That Derive Nutrients from Other Organisms

Bdellovibrio species prey on other bacteria (figure 11.21). Certain species of bioluminescent bacteria form symbiotic relationships with specific types of squid and fish (figure 11.22). *Epulopiscium* species live within the intestinal tract of surgeonfish. *Legionella* species often reside within protozoa; they can cause respiratory disease when inhaled.

Bacteria That Move by Unusual Mechanisms

Spirochetes move by means of endoflagella (figure 11.23). Magnetotactic bacteria contain a string of magnetic crystals that allow them to move up or down in water or sediments to the microaerophilic niches they require.

Bacteria That Form Storage Granules

Spirillum volutans form volutin granules, a storage form of phosphate. *Thioploca* species "commute" to nitrate-rich waters. *Thiomargarita namibiensis,* the largest bacterium known, stores sulfur and has a nitrate-containing vacuole.

11.8 ■ Animals as Habitats (table 11.3)

Bacteria That Inhabit the Skin

Staphylococcus species are facultative anaerobes.

Bacteria That Inhabit Mucous Membranes

Bacteroides species inhabit the mouth, intestinal tract, and genital tract of humans and other animals. *Bifidobacterium* species reside in the intestinal tract of animals, including humans, particularly breast-fed infants. *Campylobacter* and *Helicobacter* species are microaerophilic. *Haemophilus* species require compounds found in blood. *Neisseria* species are nutritionally fastidious aerobes that grow in the oral cavity and genital tract. *Mycoplasma* species lack a cell wall; they often have sterols in their membrane that provide strength and rigidity (figure 11.24). *Treponema* and *Borrelia* species are spirochetes that typically inhabit mucous membranes and body fluids of humans and other animals.

Obligate Intracellular Parasites

Species of *Rickettsia, Orientia,* and *Ehrlichia* are spread when a flea or tick transfers bacteria during a blood meal. *Coxiella burnetii* survives well outside the host due to spore-like structures (figure 11.25). *Chlamydia* and *Chlamydophila* species are transmitted directly from person to person (figure 11.26). *Wolbachia pipientis* alters the reproductive biology of infected arthropods; although it does not infect mammals, it lives within the filarial worms that cause river blindness and elephantiasis.

11.9 Archaea That Thrive in Extreme Conditions

Extreme Halophiles

Extreme halophiles are found in salt lakes, soda lakes, and brines used for curing fish (figure 11.27).

Extreme Thermophiles

Methanothermus and *Methanopyrus* species are hyperthermophiles that make methane. Sulfur-reducing hyperthermophiles are obligate anaerobes that use sulfur as a terminal electron acceptor. Nanoarchaea grow as spheres attached to an *Ignicoccus* species. Sulfur-oxidizing hyperthermophiles use O_2 as a terminal electron acceptor, generating sulfuric acid. Thermophilic extreme acidophiles have an optimum pH of 2 or below.

Review Questions

Short Answer

1. What kind of bacteria might compose the subsurface scum of polluted ponds?
2. What kind of bacterium might be responsible for plugging the pipes in a sewage treatment facility?
3. Give three examples of energy sources used by chemolithotrophs.
4. Name two genera of endospore-forming bacteria. How do they differ?
5. How is the life cycle of *Epulopiscium* species unusual?
6. What unique motility structure characterizes the spirochetes?
7. In what way does the metabolism of *Streptococcus* species differ from that of *Staphylococcus* species?
8. How have species of *Streptomyces* contributed to the treatment of infectious diseases?
9. What characteristics of *Azotobacter* species protect their nitrogenase enzyme from inactivation by O_2?
10. Compare and contrast the relationships of *Agrobacterium* and *Rhizobium* species with plants.

Multiple Choice

1. A catalase-negative colony growing on a plate that was incubated aerobically could be which of these genera?
   a) *Bacillus*
   b) *Escherichia*
   c) *Micrococcus*
   d) *Staphylococcus*
   e) *Streptococcus*
2. All of the following genera are spirochetes *except*
   a) *Borrelia.*
   b) *Caulobacter.*
   c) *Leptospira.*
   d) *Spirochaeta.*
   e) *Treponema.*
3. If you examined the acidic runoff from a coal mine, which of the following would you most likely find growing there?
   a) *Clostridium*
   b) *Escherichia*
   c) Lactic acid bacteria
   d) *Thermus*
   e) *Acidithiobacillus*
4. The dormant forms of which of the following genera are the most resistant to environmental extremes?
   1) *Azotobacter*
   2) *Bacillus*
   3) *Clostridium*
   4) Myxobacteria
   5) *Streptomyces*
   a) 1, 2
   b) 2, 3
   c) 3, 4
   d) 4, 5
   e) 1, 5
5. If you read that coliforms had been found in a lake, the report could have been referring to which of the following genera?
   a) *Bacteroides*
   b) *Bifidobacterium*
   c) *Clostridium*
   d) *Escherichia*
   e) *Streptococcus*
6. Which of the following genera preys on other bacteria?
   a) *Bdellovibrio*
   b) *Caulobacter*
   c) *Hyphomicrobium*
   d) *Photobacterium*
   e) *Sphaerotilus*
7. All of the following genera are obligate intracellular parasites *except*
   a) *Chlamydia.*
   b) *Coxiella.*
   c) *Ehrlichia.*
   d) *Mycoplasma.*
   e) *Rickettsia.*
8. Which of the following genera are known to fix nitrogen?
   1) *Anabaena*
   2) *Azotobacter*
   3) *Deinococcus*

4) *Mycoplasma*
5) *Rhizobium*
   a) 1, 3, 4
   b) 1, 2, 5
   c) 2, 3, 5
   d) 2, 4, 5
   e) 3, 4, 5

9. Which of the following archaea would most likely be found coexisting with bacteria?
   a) *Nanoarchaeum*
   b) *Halobacterium*
   c) *Methanococcus*
   d) *Picrophilus*
   e) *Sulfolobus*
10. *Thermoplasma* and *Picrophilus* grow best in which of the following extreme conditions?
    a) Low pH
    b) High salt
    c) High temperature
    d) a and c
    e) b and c

Applications

1. A student argues that it makes no sense to be concerned about coliforms in drinking water because they are harmless members of our normal microbiota. Explain why regulatory agencies are concerned about coliforms.
2. A friend who has lakefront property and cherishes her lush green lawn complains of the green foul-smelling scum on the lake each summer. Explain how her lawn might be contributing to the problem.

Critical Thinking

1. Soil often goes through periods of extreme dryness and extreme wetness. What characteristics of *Clostridium* species make them well suited for these conditions?
2. Some organisms use sulfur as an electron donor (a source of energy), whereas others use sulfur as an electron acceptor. How can this be if there must be a difference between the electron affinity of electron donors and acceptors for an organism to obtain energy?

12 The Eukaryotic Members of the Microbial World

Red blood cells containing a protozoan that causes malaria. ©*Biophoto Associates/ Science Source*

KEY TERMS

Algae Eukaryotic photosynthetic organisms with relatively simple reproductive structures and no organized vascular system.

Arthropod Animal with an external skeleton and jointed appendages such as an insect or an arachnid; may act as a vector in transmitting disease

Definitive Host Organism in which a parasite undergoes sexual reproduction or matures to its adult form.

Fungus Heterotrophic eukaryotic organism with a chitinous cell wall.

Helminth A worm; parasitic helminths often have complex life cycles.

Intermediate Host Organism in which a parasite undergoes asexual reproduction or is found only in its immature form.

Mycosis Disease caused by fungal infection.

Phytoplankton Microscopic free-floating photosynthetic organisms.

Protists Usually single-celled eukaryotes that are not fungi, plants, or animals.

Protozoa Heterotrophic protists.

Saprophyte Organism that takes in nutrients from dead and decaying matter.

Yeasts Unicellular fungi without motile sex cells.

A Glimpse of History

Sometimes, a single event can change the course of history. A water mold, *Phytophthora infestans,* which causes a disease called late blight in potatoes, contributed to just such an event—the Irish Potato Famine. From 1845 to 1847, the potato crop in Ireland was ruined by the disease, and the Irish faced starvation. Estimates vary, but it is likely that 1.5 million people died during the famine, while more than 1 million moved to other countries—primarily the United States and Canada. The population of Ireland fell by about 25% during this period.

Two hundred years earlier, Spaniards had brought potatoes to Europe from South America. The potato became the main source of nutrition for many Irish people. It was easy to grow and provided a nearly complete food source that could be stored for months. Irish dependence on this single food left them open to a major disaster.

Phytophthora infestans affects every part of the potato plant, including the leaves, stems, and tubers (potatoes). The potato turns to a black, mushy mess. Because the tuber is underground, an infection is often not noticed until it has destroyed the entire plant. Once a field is contaminated, ridding it of the infectious agent is very difficult. The infection is transmitted by spores, which can persist for years in the soil and are spread by wind as well as by those who handle the potatoes.

Plant breeders and scientists still battle potato blight. In 2009, the potato genome was sequenced, and scientists are working to locate and promote expression of resistance genes in crop potatoes. That same year the genome of *Phytophthora infestans* was also sequenced, revealing long expanses of repetitive genes coding for enzymes that attack the plants. These provide sources of variation that can make it difficult for a plant to develop resistance to the pathogen. The battle continues as late blight results in global losses of up to $10 billion per year, causing devastation in developing countries where many still rely on potatoes as their main source of nutrition.

All eukaryotic organisms are in the domain *Eukarya,* a diverse group ranging from microscopic members to plants and animals. Microscopic eukaryotes are found in various groups, including fungi, algae, protozoa, slime molds, and water molds. In this chapter, we also include multicellular worms and certain arthropods. Although usually visible to the naked eye, many worms and arthropods cause or transmit human disease. Moreover, they are often carried or transmitted in forms that are indeed microscopic. ◀◀ **eukaryotes,** ◀◀ **domains**

Eukaryotic organisms have traditionally been classified based on anatomical characteristics. Such classification has always been problematic because it does not necessarily reflect evolutionary relationships. Modern DNA sequencing makes it possible to examine these organisms at the molecular level, but developing an accurate classification scheme is still challenging. Because of this, some microscopic eukaryotes are discussed in informal groups that do not reflect their evolutionary relatedness. For example, the term **algae** is used to collectively refer to simple autotrophic (photosynthetic) eukaryotes, **fungi** are heterotrophic organisms with chitin in their cell wall, and **protozoa** are unicellular heterotrophic organisms that are not fungi. **Protists** make up a diverse group of usually single-celled eukaryotes that are not fungi, plants, or animals. ⏮ photosynthesis

Eukaryotes are characterized by a membrane-bound nucleus (*eukaryote* means "true nucleus"). This organelle contains their genetic information, packaged in structures called chromosomes. Because their DNA is located in the nucleus and the ribosomes are in the cytoplasm, the processes of transcription and translation cannot occur simultaneously, as they do in prokaryotic cells. Eukaryotes also have lysosomes, Golgi, and other membrane-bound organelles that allow compartmentalization of cell processes. Energy transformation in eukaryotes primarily occurs in membrane-rich mitochondria and chloroplasts where electron transport chains operate; in prokaryotes, these are associated with the cytoplasmic membrane. Many eukaryotes do not have a cell wall, and when they do, it lacks peptidoglycan, the molecule that characterizes the bacterial cell wall. Finally, most eukaryotic cells have a well-developed cytoskeleton that is important in movement and maintenance of structure (see figure 3.50). ⏮ organelles, ⏮ transcription, ⏮ translation

The mode of reproduction in eukaryotes is fundamentally different from that of prokaryotes. Recall that prokaryotes are typically haploid, meaning they carry only one copy of the genome. They reproduce by binary fission (see figure 4.1). Eukaryotes may be haploid (carrying one copy of the genome) or diploid (carrying two copies of the genome). They often progress through a complex life cycle involving both forms.

Eukaryotes may reproduce asexually by a type of nuclear division called **mitosis** in which each daughter cell receives the same number of chromosomes as the parent cell. When a haploid cell reproduces by mitosis, two haploid cells are produced. Likewise, when a diploid cell reproduces by mitosis, two diploid cells are produced. Eukaryotic cells may also reproduce using a type of nuclear division called **meiosis,** in which diploid cells give rise to haploid cells. These haploid cells can develop into haploid organisms, or they can be used as haploid sex cells called **gametes** that are involved in sexual reproduction (**figure 12.1**). Gametes are often flagellated and highly mobile. Fusion of two gametes forms a diploid cell, allowing recombination of genetic material. This sexual process increases genetic variation that can be used as the raw material for natural selection, the basis of evolution.

MicroByte

Eukaryotic pathogens can be difficult to target with medication because their cell components are often the same as those of humans.

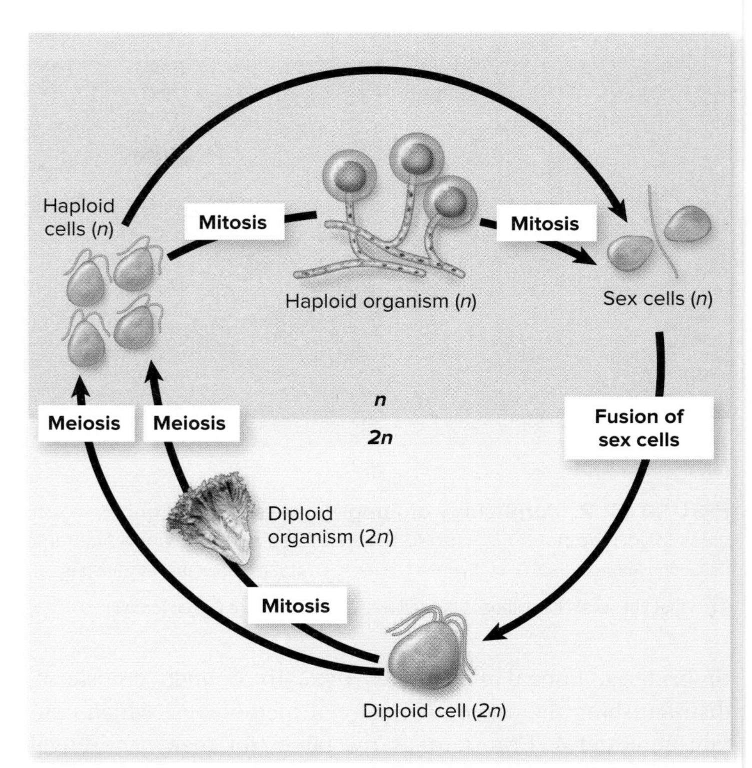

FIGURE 12.1 Cell Division Meiosis gives rise to haploid cells that can grow into a haploid organism by mitosis or can be used as sex cells (gametes) in sexual reproduction. Fusion of gametes forms a diploid cell that can grow into a diploid organism by mitosis or can undergo meiosis to form haploid cells.

? **In humans, can a gamete be diploid?**

12.1 ■ Fungi

Learning Outcomes

1. Describe the structure, habitat, and reproductive strategies of the various types of fungi.
2. Explain the important symbiotic relationships of fungi.
3. Name and describe the roles of economically and medically important fungi.

The study of fungi (singular: **fungus**) is known as **mycology.** Fungi include molds, single-celled yeasts, and the familiar mushroom. These terms are unrelated to the classification of fungi, but instead refer to their morphological forms (**figure 12.2**):

- **Yeasts** are single-celled fungi.
- **Molds** are filamentous fungi.
- **Mushrooms** are simply the reproductive structures of certain fungi, similar to a peach on a peach tree. Like peaches, some large mushrooms are edible.

Fungi are characterized by a cell wall that contains chitin—the same molecule found in the exoskeleton of insects. It is somewhat stronger than the cellulose-based cell wall of plants, and chemically distinct from the peptidoglycan cell wall

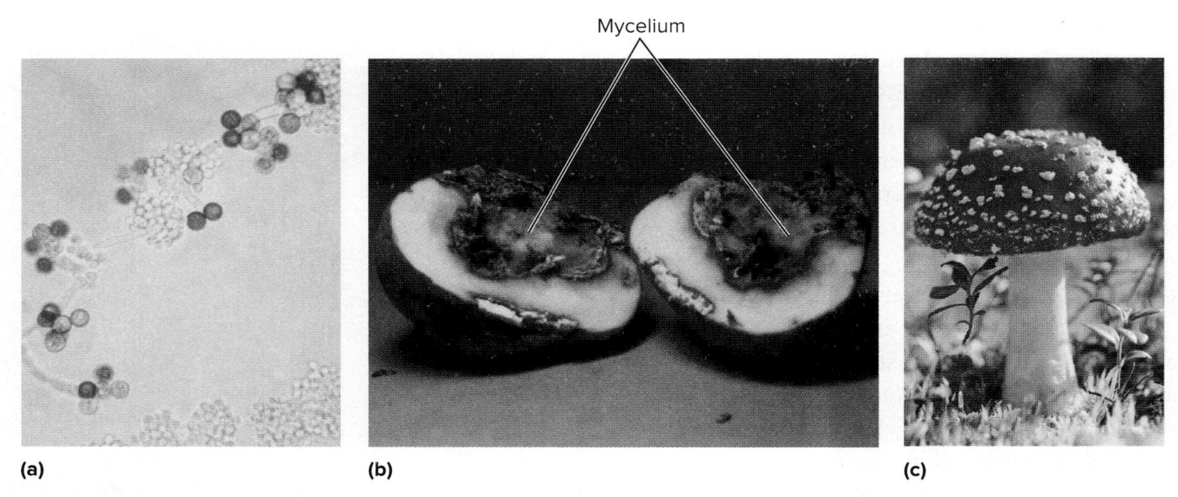

FIGURE 12.2 Morphology of Fungi **(a)** Microscopic *Candida albicans;* the blue circles are asexual reproductive cells. **(b)** The cottony white mass inside the potato is a mold, characterized by intertwined filaments (mycelium). **(c)** *Amanita muscaria,* a highly poisonous mushroom. a: Source: Dr. Godon Roberstad/CDC; b: ©Martha Nester; c: ©Jorgen Bausager/Getty Images

What chemical characterizes the cell wall in each of these fungal forms?

of bacteria. Fungal membranes typically contain ergosterol, distinguishing them from animal cell membranes, which contain cholesterol. Ergosterol is the target for many antifungal medications.

To obtain nutrients, fungi secrete enzymes into the environment to break large molecules into smaller ones that they can absorb. Fungi, along with bacteria, are the main decomposers of organic compounds. They can degrade both cellulose and lignin, the main components of wood. This decomposition releases carbon dioxide into the atmosphere and nitrogen compounds into the soil. These compounds are then taken up by plants and again converted into organic compounds. Without this recycling of organic material, the earth would quickly be overrun with organic waste. ⏭ biogeochemical cycles

In their key role as decomposers, fungi are **saprophytic,** meaning that they use nutrients from dead or decaying organic matter. Some fungi, however, can absorb nutrients from living tissue, acting as parasites (meaning they live at the expense of their host). Relatively few fungi infect humans—although athlete's foot and vaginal yeast infections are not uncommon—but many plant species suffer devastating fungal infections. More often, fungi partner with other organisms in mutually beneficial relationships. For example, fungi and algae form lichens that can grow on surfaces where neither can survive alone. ⏭ parasitism, ⏭ mutualism

Types of Fungi

Classification of fungi is in a state of flux and will continue to be so until the genomes of many more species have been sequenced. About 100,000 species of fungi have been described, and over 100 fungal genomes have been sequenced, but gene studies indicate that there are perhaps 1.5 million species in nature. We will describe only four major fungal groups: ascomycetes (Ascomycota), basidiomycetes (Basidiomycota), zygomycetes (Zygomycota), and chytrids (Chytridiomycota) (**table 12.1**). The real diversity within fungi is much more complex.

The ascomycetes (sac fungi) are a diverse group with about 75% of all known fungi. They include *Penicillium,* the source of the earliest recognized antibiotic: penicillin. Some ascomycetes are pathogens such as those that cause Dutch elm disease. Others are highly prized for the flavor of their reproductive structures (morels and truffles). Many ascomycetes form lichens in association with a photosynthetic partner.

The basidiomycetes (club fungi) are commonly recognized by their reproductive structures—mushrooms—that are often collected or grown for food. The group also includes plant parasites like rusts and smuts, which earn their common names because of their appearance on the infected plant. Significant losses in wheat, rye, and corn crops have been caused by these fungi.

The zygomycetes include the common black bread mold, *Rhizopus,* and other organisms that often spoil fruits and vegetables. The black growth that you see on bread is composed of reproductive structures called sporangia that support and house hundreds of asexual reproductive cells called sporangiospores (**figure 12.3**).

The chytrids usually live in water, but some live in the guts of mammalian herbivores, where they help to digest plant material. Some are parasitic. For example, *Batrachochytrium dendrobatidis* can infect the skin of frogs and is believed to be responsible for the catastrophic decline in frog populations over the past two decades. Chytrids are the only type of fungus with motile forms; the reproductive cells have flagella.

Structure of Fungi

Most fungi are molds—multicellular organisms composed of thread-like filaments called **hyphae** (singular: hypha)

TABLE 12.1 Characteristics of Major Groups of Fungi

Group and Representative Member(s)	Appearance	Usual Habitat	Some Distinguishing Characteristics	Asexual Reproduction	Sexual Reproduction
Ascomycetes *Neurospora, Saccharomyces cerevisiae* (baker's yeast), *Penicillium, Aspergillus*		Terrestrial, on fruit and other organic materials	Unicellular and multicellular; multicellular forms have septate hyphae	Budding is common; conidia are produced	Involves the formation of an ascus (sac) on specialized hyphae
Basidiomycetes *Agaricus campestris* (meadow mushroom), *Cryptococcus neoformans*		Terrestrial	Some unicellular; multicellular forms have septate hyphae; group includes mushrooms, smuts, rusts	Commonly absent	Produce basidiospores that are borne on club-shaped structures
Zygomycetes *Rhizopus stolonifer* (black bread mold)		Terrestrial	Multicellular; hyphae generally lack septa	Sporangiospores develop in sporangia on the tips of aerial hyphae	Sexual spores known as zygospores can remain dormant in adverse environments
Chytridiomycetes *Batrachochytrium dendrobatidis*		Aquatic, guts of herbivores, parasitic	Most are unicellular but some are multicellular; motile reproductive cells	Motile zoospores	Flagellated gametes in male and female

(figure 12.4). A visible mass of hyphae is a **mycelium** (plural: mycelia). Fungi are generally not motile, so they cannot move toward a food source. Instead, the tips of the hyphae respond to a nutrient source by growing very quickly in that direction. Most fungal species have septate hyphae, meaning that cross walls (septa) separate the cells; pores in the septa allow cytoplasm to flow from cell to cell. In some species, septa are absent because the cells undergo mitosis without accompanying cell division, resulting in a filament with many nuclei.

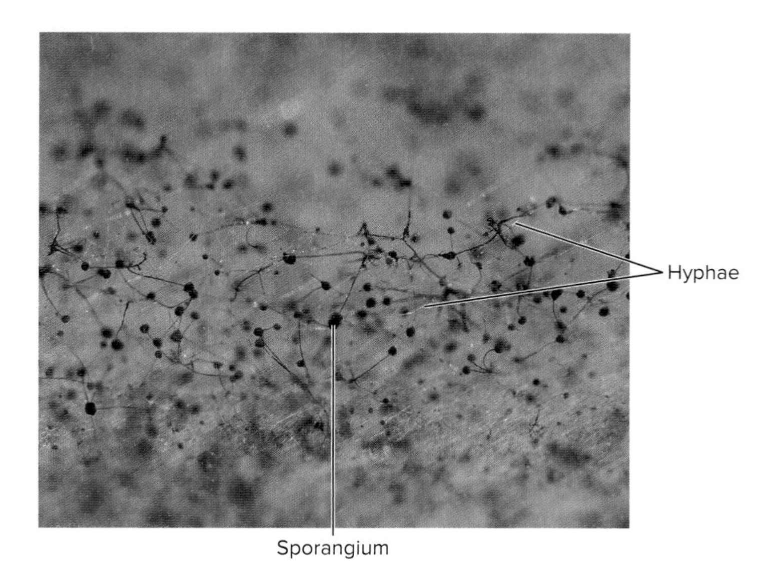

FIGURE 12.3 Black Bread Mold *Rhizopus stolonifer,* showing thread-like hyphae (singular: hypha) and black sporangia housing asexual reproductive cells called sporangiospores. ©McGraw-Hill Education/Lisa Burgess

? Why would this mold be considered a saprophyte?

The high surface-to-volume ratio of hyphae makes them well adapted to absorb nutrients. The enzymes that they release not only break down nutrient molecules, but also repel the growth of other hyphae. As a result, hyphae spread throughout the food source, ensuring that each hypha will have maximum access to nutrients. When you see mold growing on a food, you mainly see hyphae that give rise to spores (a generic term for the reproductive cells). Most of the hyphae are within the food, where they act in nutrient absorption.

Fungi are most successful in moist environments. Mildew and mold become problems, for example, in a damp basement. Chronic exposure to fungal components can lead to allergies or other health problems.

Fungi that parasitize plants have specialized hyphae called **haustoria** that protrude into host cells to gain nutrients. Haustoria do not penetrate the host cell's cytoplasmic membrane, but are surrounded by it, much like a hand in a mitten. Saprophytic fungi sometimes have specialized hyphae called rhizoids, which anchor them to the substrate. Fungal rhizoids do not function in exchange of materials.

Dimorphic fungi can grow as single yeast cells or multicellular hyphae, depending on the environmental conditions. Some of these fungi cause systemic disease in humans. For example, *Histoplasma capsulatum* grows in the soil as a mold, and its spores easily become airborne. When they are inhaled into the warm, moist environment of the lungs, they develop into the yeast form, which can cause disease. ⏭ histoplasmosis

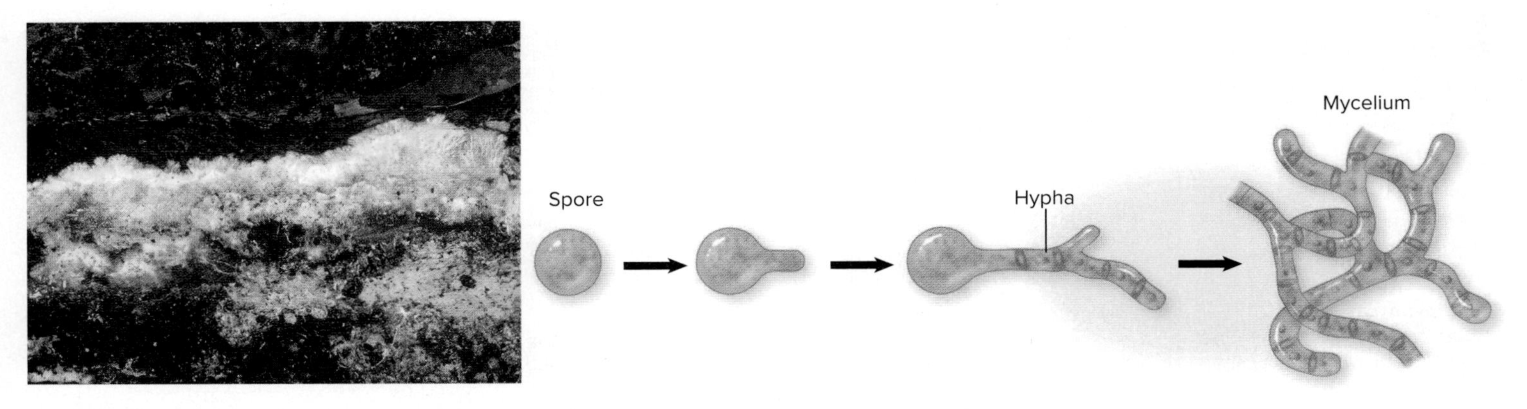

FIGURE 12.4 Formation of Hyphae and Mycelium The white mass is called the mycelium. The mycelium forms when fungal spores (reproductive structures) germinate and then elongate to form filaments called hyphae that intertwine. ©McGraw-Hill Education/Steven P. Lynch

? Where do fungi digest their food?

MicroByte

The largest organism on Earth, called the humongous fungus, has a mycelium that spans over 2,200 acres in Oregon.

Fungal Habitats

Fungi are mainly terrestrial organisms, but they are found in nearly every habitat on Earth, including the thermal pools at Yellowstone National Park, volcanic craters, and bodies of water with high salt content, such as the Great Salt Lake and the Dead Sea. Some fungal species occur only on a particular strain of one genus of plants. Others are widespread because they are extremely versatile in what they can degrade and use as a source of carbon and energy. As a group, fungi can degrade leather, cork, hair, wax, ink, jet fuel, carpet, drywall, and even some synthetic plastics.

Some fungal species can grow in concentrations of salts, sugars, or acids that inhibit the growth of most bacteria. Because of this, fungi are often responsible for spoiling pickles, jams, and other foods. They also grow better than bacteria on acidic fruits and vegetables. Different fungi can grow from a pH as low as 2.2 to one as high as 9.6, and they usually grow well at a pH of 5.0 or lower. ⏭ food spoilage

Most fungi prefer an environment from 20°C to 35°C, but they can easily survive at refrigerator temperatures and below. Some are resistant to pasteurization, and others can grow at temperatures below the freezing point of water. ⏮ pasteurization

Most fungi are aerobic, but some of the yeasts are facultative anaerobes, producing ethanol by fermentation. Some obligate anaerobes live in the rumen (stomach compartment) of cows, where they are important in the digestion of the plant material. ⏮ oxygen (O_2) requirements, ⏭ rumen

Symbiotic Relationships of Fungi

Lichens are symbiotic life forms that result from the association of fungus with a photosynthetic organism such as an alga or a cyanobacterium. Recent molecular and genetic study of these classic symbionts has revealed that lichens are not necessarily composed of only two partners. Rather, the association may include yeasts and non-photosynthetic bacteria in addition to the already recognized fungal and photosynthetic members. Further study of this complex interaction will certainly improve our understanding of the biology of lichens, and may also provide valuable information about symbiotic relationships.

Lichens are found in terrestrial habitats all over the world. The fungal partner in the lichen absorbs water and minerals for the symbiotic pair, in addition to providing protection. The photosynthetic member supplies the fungus with organic nutrients (**figure 12.5**). Lichens can grow in extreme ecosystems where neither partner could survive alone. For example, they grow in sub-Arctic tundra where they are the primary diet of reindeer. Lichens are often the first life to appear on bare rock, where they can begin the process of soil formation. In spite of their hardiness in certain environments, lichens are often a good indicator of air quality. Polluted air is lethal to lichens because they absorb—but are unable to excrete—its common components, including toxic metals, sulfur dioxide, and ozone. Lichens are not common in industrial cities.

Some fungi grow in a mutually beneficial association with plant roots, forming **mycorrhizas** (**figure 12.6**). The high surface area of fungal hyphae increases the plant's ability to absorb water and minerals. The fungus also supplies the plant with nitrogen and phosphorus from the breakdown of organic material in the soil. The plant, in return, supplies the fungus with organic compounds. It is estimated that 80% of vascular plants have some type of mycorrhizal association. Plants with mycorrhizas grow better than those without. Most orchids cannot germinate without mycorrhizas that help provide nutrients for the young plant. ⏭ mycorrhizas

Certain insects also depend on symbiotic relationships with fungi. For example, leaf-cutter ants farm their own fungal gardens (**figure 12.7**). The ants cannot eat tropical vegetation because the leaves are often poisonous. Instead, the ants

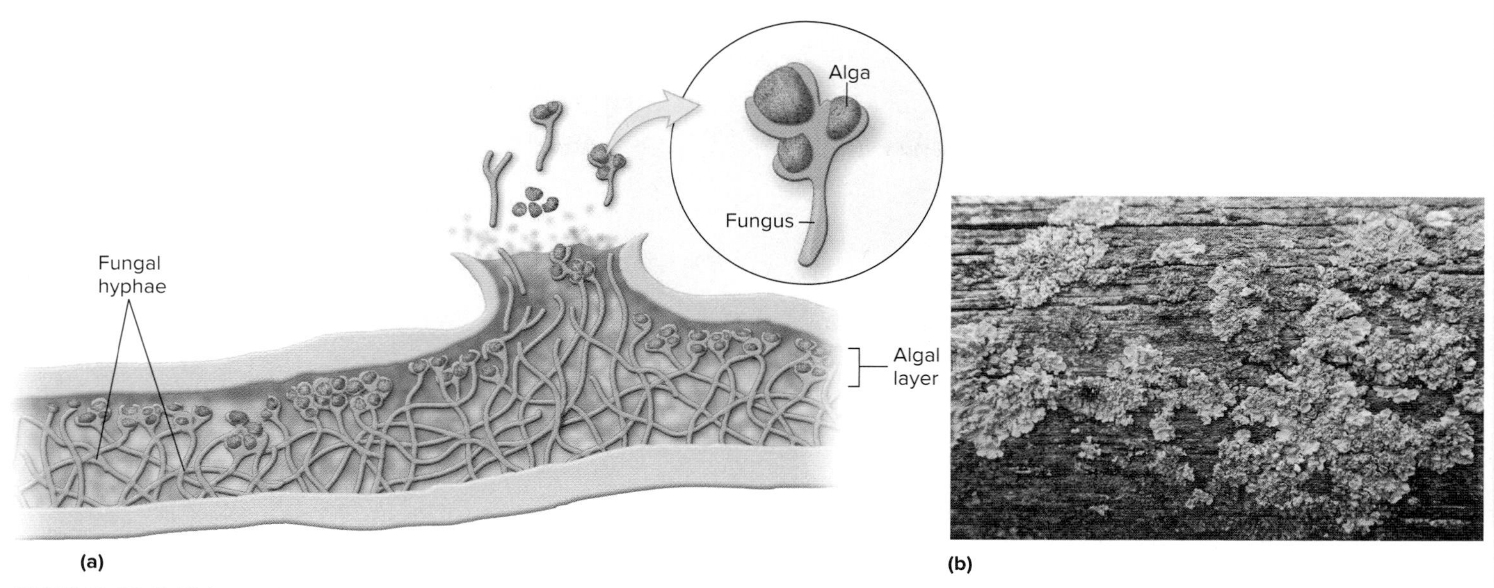

FIGURE 12.5 Lichens **(a)** A lichen includes photosynthetic cells intertwined within the hyphae of a fungal partner. **(b)** Lichens on a fallen tree. b: ©Núria Talavera/Getty Images

What does the fungus contribute to the lichen?

chop the plant leaves into bits and add a mycelium. The fungi grow, secreting enzymes that digest the plant material, and eventually produce reproductive structures that the ant then uses as its food source.

Reproduction in Fungi

The reproductive forms of fungi are very important in identification. In medical microbiology, the most important of these are the cells that develop asexually, because they can be seen in pure cultures grown in the laboratory. However, the sexual forms play an important role in fungal classification and some are commercially valuable. Mushrooms, for example, are structures that contain the sexual reproductive forms of basidiomycetes (**figure 12.8**). The sexual forms of some fungi have never been seen, which made it difficult to classify these organisms before DNA sequencing methods were developed. Further complicating matters, many fungi have been inadvertently "discovered" twice—once based on the sexual reproductive forms and once based on the asexual forms—so they are known by two different names!

The asexual reproductive cells of molds come in a variety of sizes and shapes. These cells are typically called **conidia** (or in zygomycetes, sporangiospores). However, the term *spore* is often used generically to refer to reproductive cells that are formed either sexually or asexually. The conidia are

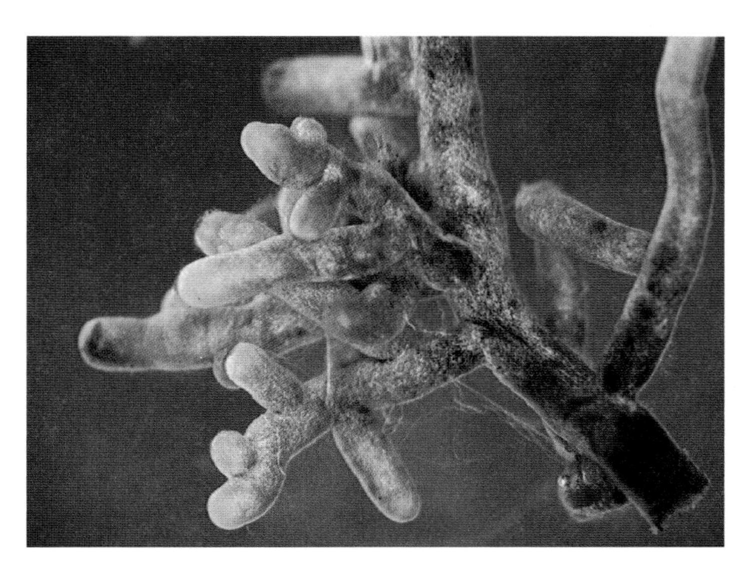

FIGURE 12.6 Mycorrhizas Fungi (seen here in white) form intimate relationships with the roots of most green plants. They supply the plant with nitrogen and phosphorus and increase the plant's ability to absorb water. ©Ellen Larsson

Are mycorrhizas considered parasites? Explain.

FIGURE 12.7 Leaf-Cutter Ants These ants carry food for fungi, whose reproductive structures then serve as a food source for the ants. ©Glenn Oakley/Getty Images

Do leaf-cutter ants eat leaves? Explain.

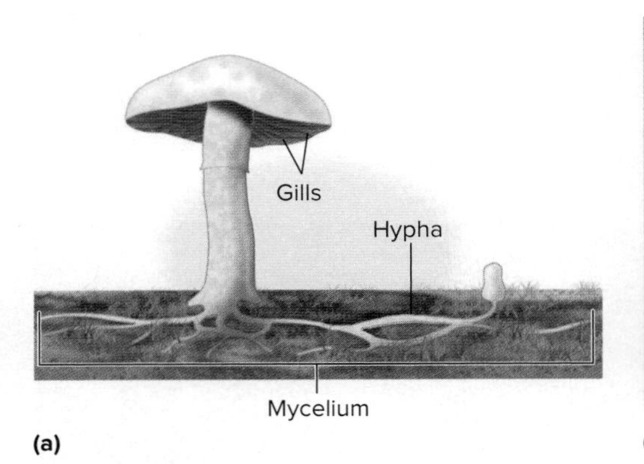

(a)

(b)

FIGURE 12.8 Mushrooms (a) The extensive underground mycelium gives rise to fruiting bodies called mushrooms, composed of dense hyphae protecting the spores produced under the gills. **(b)** Fruiting bodies. b: ©Paulina Lenting-Smulders/ Getty Images

What group of fungi produces mushrooms?

small and numerous and easily carried by wind or water. They are somewhat resistant to drying and other unfavorable conditions and can persist for years until conditions improve. When fungal conidia germinate, they begin to grow hyphae in the direction of a food source. As the food source is used up, conidia formation typically increases. Molds may also reproduce asexually by fragmentation. A piece of the parent breaks off and grows as a separate organism. Hyphae may grow from a piece of mycelium placed on a suitable nutrient source, as in the farming technique of leaf-cutter ants.

Yeast cells reproduce asexually by budding. In this process, the nucleus divides by mitosis and one nucleus migrates into a smaller daughter cell, or bud. The daughter cell pinches off from a larger parent cell, which eventually dies after producing a number of buds (**figure 12.9**).

Most fungal cells are not motile; sexual reproduction results when hyphae from two different mating types grow toward one another and fuse. Because the two mating types have no obvious differences, mycologists refer to them as "+" and "–." Often after this fusion, the cells of the fungus will house both haploid nuclei, forming a distinct fungal structure called a dikaryon (meaning two nuclei). Eventually, the nuclei fuse and then undergo meiosis, forming haploid spores. Note that fungal spores are quite different from bacterial endospores, which are much more resistant to environmental conditions and are not a means of reproduction. **⏮ endospore**

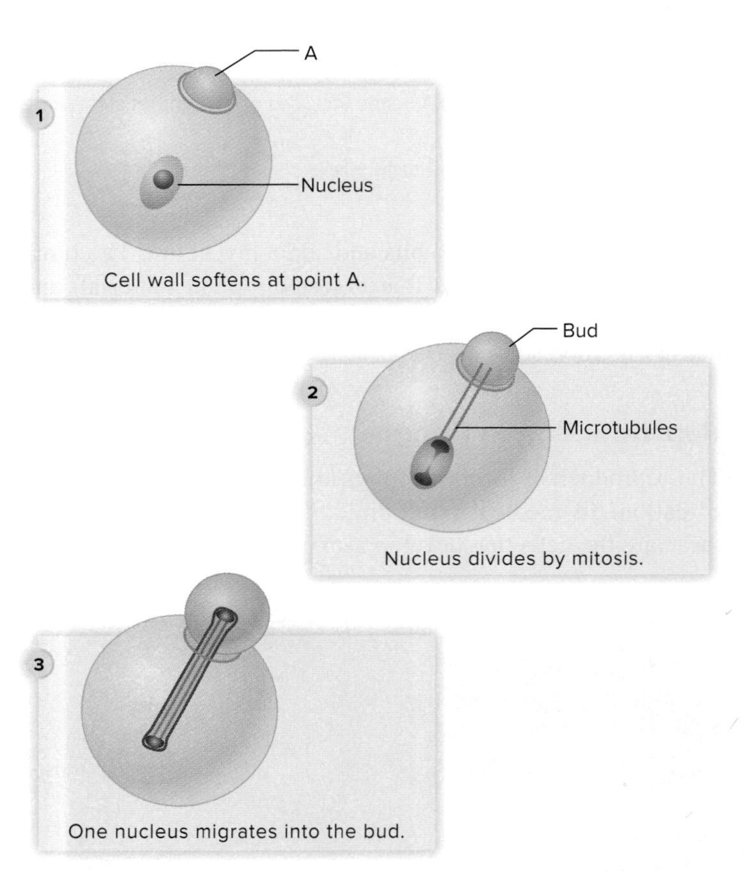

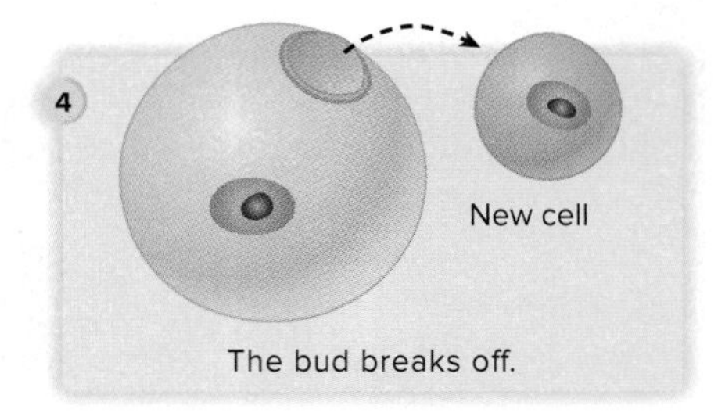

FIGURE 12.9 Budding in Yeast The nucleus divides by mitosis, and one nucleus moves into the bud. A cell wall is laid down beneath the bud. When the bud breaks off, it forms a new cell.

What information in the figure indicates that the process generates a bud that has the same number of chromosomes as the original cell?

Economic Importance of Fungi

Many fungi are important commercially, as some synthesize important antimicrobial medicines like penicillin. More recently, yeasts have been genetically engineered to produce a variety of important molecules, such as human insulin and a vaccine against hepatitis B. The first eukaryotic genome sequenced was that of the yeast *Saccharomyces cerevisiae,* a model organism used in a variety of genetic and biochemical studies. Fungi are useful tools for studying complex eukaryotic events—such as cancer and aging—within a simple cell. **⏭ penicillin, ⏮ genetic engineering**

Saccharomyces cerevisiae, also known as brewer's yeast or baker's yeast, has long been used in the production of wine, beer, and bread. Other fungal species are useful in making the large variety of cheeses produced throughout the world. Ironically, fungi are also among the greatest spoilers of food products; tons of food are thrown away each year because fungal growth has made it inedible. ⏭| *Saccharomyces*

Fungi cause a number of plant diseases that cost billions of dollars from losses due to crop damage or expenditures on preventive measures. Grain crops are particularly vulnerable to fungal infection. Dutch elm disease, caused by the fungus *Ophiostoma ulmi* (*Ulmus* is a genus of elm), dramatically changed the landscape of many U.S. cities when all of the elm trees lining streets and surrounding public buildings were killed. The cost of removing these diseased and dying trees was significant. More recently, a fungal disease called coffee rust devastated coffee crops in Central America, causing significant losses, from 2012 to 2014.

Medical Importance of Fungi

Because relatively few fungal species infect humans, and many fungi produce important antimicrobial medications, their impact on human health is probably a net positive. Still, some fungal diseases are relatively common, such as athlete's foot and jock itch. Life-threatening fungal infections such as cryptococcal meningitis are rare in otherwise healthy people. In immunocompromised patients, however, fungal diseases are much more common and devastating. ⏭| *Cryptococcus*

Fungi cause human illnesses in three general ways:

- **Hypersensitivity reaction.** The person develops an allergic or asthmatic reaction to the fungal components.
- **Infection.** The fungus grows on or in the human body, causing fungal disease or **mycosis.**
- **Intoxication.** The fungus produces a toxin that a person ingests.

Fungal spores or conidia are found everywhere on Earth and up to altitudes of more than 7 miles. The air we breathe may contain more than 10,000 of these cells per cubic meter. People with allergies often monitor published pollen and mold counts and avoid unnecessary outdoor activity when levels are high. Fungal spores can also trigger asthma attacks. ⏭| asthma

Superficial (cutaneous) mycoses affect the hair, skin, or nails (see figure 22.20). They are caused by skin-invading molds called dermatophytes. Systemic mycoses affect tissues deep within the body and are usually caused by inhalation of spores. The names of mycoses often reflect the causative agent (**table 12.2**). ⏭| dermatophytes

Some fungi produce toxins. **Aflatoxins,** produced by *Aspergillus* species, are the most studied of these toxins and are considered carcinogenic. The U.S. Food and Drug Administration (FDA) monitors levels of aflatoxins in foods such as grains and peanuts. The rye mold *Claviceps purpurea,* also known as **ergot,** produces a toxin with hallucinogenic properties. Strange

TABLE 12.2 Some Medically Important Fungal Diseases

Disease	Causative Agent
Candidial skin infection	*Candida albicans*
Coccidioidomycosis	*Coccidioides immitis*
Cryptococcal meningoencephalitis	*Cryptococcus neoformans* or *C. gattii*
Histoplasmosis	*Histoplasma capsulatum*
Pneumocystis pneumonia	*Pneumocystis jirovecii*
Sporotrichosis	*Sporothrix schenckii*
Tinea versicolor	*Malassezia furfur*
Vulvovaginal candidiasis	*Candida albicans*

behavior in people eating contaminated rye may have led to the accusations of witchcraft in Salem, Massachusetts, in the late 1600s. The active chemical has been purified from *C. purpurea* to yield the drug ergotamine, which decreases blood flow. It is used to control uterine bleeding and relieve migraine headaches. Some fungi, such as *Amanita* species, produce powerful toxins that can cause fatal liver damage (see figure 12.2c).

MicroAssessment 12.1

Fungi have chitin in their cell walls and absorb nutrients after secreting digestive enzymes onto a food source. As saprophytes, they are important recyclers of carbon and other elements. Fungi can also form important relationships with other organisms, such as those seen in lichens or mycorrhizas. Fungi may be single-celled yeasts or multicellular molds that grow by extending tube-like hyphae into a food source. They may reproduce sexually by fusion of hyphae of different mating types, or asexually. Fungi cause significant plant diseases and food spoilage, but also are a food source and are used in food production. Relatively few fungi cause human disease.

1. How can fungi cause disease in humans?
2. Explain how the symbiotic association between fungi and green plants benefits each partner.
3. Why does conidia formation increase when food supplies of the fungus are diminishing?

12.2 ■ Algae

Learning Outcomes

4. Explain how structure and habitat differ among various types of algae.
5. Describe how algae can affect human health.

The term **algae** refers to simple photosynthetic eukaryotes. Algae differ from other eukaryotic photosynthetic organisms such as land plants because they lack an organized vascular system and they have relatively simple reproductive structures. Most algae are aquatic; they may be microscopic or macroscopic. Algae do not directly infect humans, but some produce toxins that may be concentrated in other animals that humans use for food.

Algae contain pigments that absorb radiant energy. That energy is used in the process of photosynthesis, converting CO_2 and H_2O to organic material and O_2. Algae are essential for maintaining life in aquatic environments because other organisms depend upon the organic molecules for food. ◀◀ photosynthesis

Types of Algae

Algae are a diverse group of protists that share some fundamental characteristics, but are not necessarily related (see figure 12.12). The various types of algae are often characterized by the pigments they contain. All algae contain chlorophyll *a,* a pigment also found in green plants and cyanobacteria. Different types of algae also contain other pigments that absorb different wavelengths of light and give them their characteristic appearance. These pigments extend the range of light waves that can be used for photosynthesis. Some of the general characteristics of algal groups are summarized in **table 12.3**.

Structure of Algae

Algae can be either unicellular or multicellular. All algae, however, contain chloroplasts with photosynthetic pigments and usually have rigid cell walls mostly composed of cellulose. The cell walls of some multicellular species contain large amounts of other compounds, a few of which are commercially valuable. For example, the cell walls of red algae contain agar used to solidify growth media, and those of brown algae contain alginic acid used to provide the consistency in products like ice cream and cosmetics. ◀◀ chloroplasts, ◀◀ agar

Microscopic Algae

Microscopic algae can be single-celled organisms floating free or propelled by flagella, or they can grow in long multicellular chains or filaments. The unicellular algae—including diatoms, as well as some green algae, dinoflagellates and euglenids, and a few red algae—are well adapted to an aquatic environment. As single cells, they have relatively large, absorptive surfaces, thus effectively using the dilute nutrients available. Some microscopic algae such as *Volvox* form colonies of 500–60,000 biflagellated cells, which can be visible to the naked eye (**figure 12.10**).

Diatoms are abundant in aquatic environments and are distinguished by silicon dioxide incorporated into their cell walls. When these organisms die, they sink to the bottom of the ocean, where their remains accumulate because the cell wall does not readily decompose. Deposits of diatoms are mined for a substance known as **diatomaceous earth,** used for filtering systems, abrasives in polishes, and many other purposes. Diatoms that sank to the ocean floor millions of years ago have become a major source of crude oil and natural gas.

Macroscopic Algae

Macroscopic algae include multicellular brown algae, green algae, and red algae (**figure 12.11**). Brown algae include many species of kelp that form "forests" under the water, providing food and a habitat for other marine organisms. Large multicellular algae possess a structure called a holdfast, which looks like a root system, but it is not used to obtain water and nutrients for the organism. It functions simply to anchor the organism to a firm substrate. The stalk (known as the stipe) of a multicellular alga usually has leaflike structures or blades attached to it. The blades are the main site of photosynthesis, although some also bear reproductive structures. Many large algae have gas-containing bladders that keep the blades floating on the surface of the water to maximize exposure to sunlight.

Algal Habitats

Algae are found in both fresh and salt water, as well as in moist soil. Because the oceans cover more than 70% of the earth's surface, aquatic algae are major producers of molecular oxygen as well as important users of carbon dioxide. Algae must live at water

TABLE 12.3 Characteristics of Major Groups of Algae

Group	Usual Habitat	Principal Pigments (in addition to chlorophyll *a*)	Storage Product	Cell Wall
Brown algae	Salt water	Xanthophylls, especially fucoxanthin	Starch-like carbohydrates; mannitol; fats	Cellulose and pectin; alginic acid
Diatoms, golden brown algae	Fresh water; salt water; soil; higher plants	Carotenes	Starch-like carbohydrates	Pectin, often containing silica or calcium
Dinoflagellates	Mostly salt water but common in fresh water	Carotenes; xanthophylls	Starch; oils	Cellulose and pectin
Euglenids	Fresh water	Chlorophyll *b;* carotenes; xanthophylls	Fats; starch-like carbohydrates	Lacking
Green algae	Fresh water; salt water; soil; tree bark; lichens	Chlorophyll *b;* carotenes; xanthophylls	Starch	Cellulose and pectin
Red algae, corallines	Mostly salt water, several genera in fresh water	Phycobilins; carotenes; xanthophylls	Starch-like carbohydrates	Cellulose and pectin; agar; carrageenan

FIGURE 12.10 Volvox A colony of cells forms a hollow sphere. Daughter colonies develop within the spheres. ©Stephen Durr

? What pigments give the green color to *Volvox?*

depths that allow penetration of light. Algae with different pigments can occupy different depths and thus avoid competition.

Small organisms that float or drift near the surface of aquatic environments are called plankton (*planktos* means "drifting"). Unicellular algae make up a significant part of the **phytoplankton** (*phyton* means "plant"), the free-floating, photosynthetic organisms found in plankton. Phytoplankton forms the base of aquatic food chains, with all other organisms dependent upon them. For example, microscopic heterotrophs in the **zooplankton** (*zoo* means "animal") graze on phytoplankton, and then both become food for other organisms, including filter-feeding whales, some of the world's largest mammals. ⏭| food chain

MicroByte

More O_2 is produced by algae in the oceans than by all forests on Earth combined.

Algal Reproduction

Some algae, especially multicellular filamentous species, reproduce asexually by fragmentation. Many algae alternate between a haploid generation and a diploid generation. Sometimes, as is the case with *Ulva* (sea lettuce), the generations look physically similar and can be distinguished only by microscopic examination. In other cases, the two forms look quite different.

Medical Importance of Algae

Although algae do not directly cause disease in humans, some do so indirectly. Disease most often occurs during algal blooms, when the numbers of phytoplankton increase dramatically due to changes in water conditions. An upwelling of water due to warmer temperatures often brings more nutrients such as phosphorus and nitrogen from the bottom to the surface. When organisms encounter warmer waters and additional nutrients, they multiply rapidly. The runoff of fertilizers along waterways and coastlines, or pollution from untreated sewage, may also cause algae to proliferate. Algal blooms of dinoflagellates are commonly known as **red tides.**

Dinoflagellates of *Gonyaulax* species produce neurotoxins such as saxitoxin and gonyautoxin, some of the most potent non-protein poisons known. Shellfish such as clams, mussels, scallops, and oysters feed on these dinoflagellates without apparent harm and, in the process, accumulate the neurotoxin in their tissues. When humans eat the shellfish, they suffer symptoms of paralytic shellfish poisoning, including

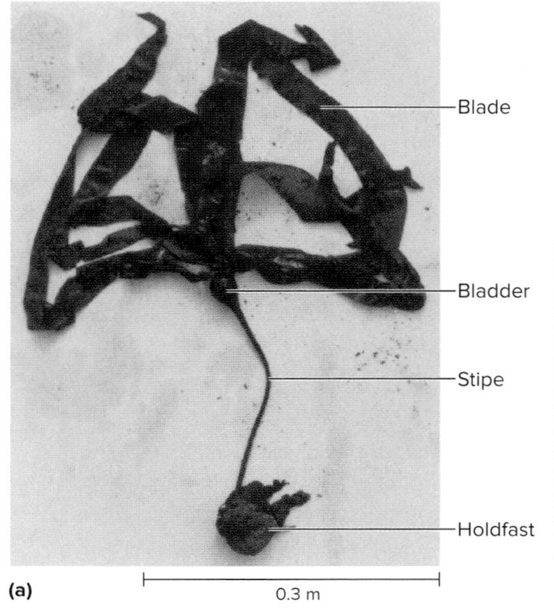

FIGURE 12.11 Types of Algae (a) This young brown alga, *Nereocystis luetkeana* (bladder kelp), has a large bladder filled with gas. The blades are the most active sites for photosynthesis. The holdfast anchors the kelp to rocks or other surfaces. In a single season, kelp can grow to lengths of 5–15 m. **(b)** Sea lettuce, a green alga. **(c)** A type of red alga. a: ©Martha Nester; b: ©Steven P. Lynch; c: ©Dan Ippolito

? What is the function of the gas bladder in kelp?

numbness, dizziness, muscle weakness, and impaired respiration. Death can result from respiratory failure. Cooking the shellfish does not destroy these toxins. *Gonyaulax* species are found in both the North Atlantic and the North Pacific and have seriously affected the shellfish industry on both coasts.

MicroAssessment 12.2

Algae are a diverse group of aquatic eukaryotic organisms that contain chlorophyll *a* and carry out photosynthesis. Algae form the base of aquatic food chains and produce much of our atmospheric O_2. Algae do not cause disease directly, but they can produce toxins that are harmful when ingested by humans.

4. What primary characteristics distinguish algae from other organisms?
5. What harmful effects can algae have on humans?
6. Why do algae have a greater variety of photosynthetic pigments than land plants?

12.3 ■ Protozoa

Learning Outcomes

6. Characterize the phylogeny of protozoa.
7. Describe the structures, habitats, and modes of reproduction in protozoa.
8. Describe the medical importance of protozoa.

DNA sequencing studies have confirmed that the organisms classically considered **protozoa** ("animal-like" unicellular organisms) form a diverse group in which most members bear little relationship to the others. Protozoa are most easily defined by what they are not. In this chapter, protozoa are unicellular heterotrophic organisms that are not fungi, slime molds, or water molds.

Types of Protozoa

Protozoa are **protists,** along with the algae, slime molds, and water molds. As with algae, protozoa are not a unified group. Instead, they include members of a number of different taxonomic groups (**figure 12.12**). Protozoa have traditionally been grouped based on their mode of locomotion. Examination of ultrastructure and DNA sequences coding for rRNA, however, show that organisms may not be closely related just because they all use flagella, cilia, or pseudopodia for locomotion. Protozoa are extremely diverse, but we will concentrate on members of this group that cause human disease.

Apicomplexans are parasites with a structure called an apical complex at one end that helps them to penetrate the cell membrane of host cells. Many have complex life cycles that alternate between sexual and asexual forms. This group includes *Plasmodium* species that cause malaria, making them one of the most significant causes of infectious disease in the world. It also includes the human pathogens *Toxoplasma gondii, Cryptosporidium parvum,* and *Cyclospora cayetanensis.* ⏭ malaria, ⏭ toxoplasmosis, ⏭ cryptosporidiosis, ⏭ cyclosporiasis

Diplomonads and **parabasalids** are groups of flagellated protists that lack mitochondria. Diplomonads typically have two nuclei and live within hosts where conditions are anaerobic, or in stagnant water where O_2 levels are low. The group includes *Giardia lamblia,* a common cause of diarrhea in campers who drink water directly from streams or lakes (see figure 24.21). Parabasalids live within a host organism. Some, for example, live in the guts of termites where they digest cellulose for their hosts. Others, such as *Trichomonas vaginalis,* may cause disease. Parabasalids have a unique structure called a hydrogenosome, which, in the absence of true mitochondria, produces some ATP while generating molecular hydrogen (H_2). Both groups reproduce asexually; a few parabasalids also reproduce sexually. ⏭ giardiasis, ⏮ trichomoniasis

Kinetoplastids have at least one flagellum and are characterized by a distinctive mass of DNA in their large single mitochondrion. The structure of this mitochondrial DNA is quite unusual; it is a series of thousands of interlocking rings. Because the complex structure is unique to the kinetoplastids, it may be a useful target for medications to treat diseases caused by this group of eukaryotic pathogens. Examples of kinetoplastids include: *Trypanosoma brucei,* which is transmitted by tsetse flies and causes African sleeping sickness (trypanosomiasis; see figure 26.14); *T. cruzi,* which is transmitted by "kissing" bugs and causes Chagas' disease; and *Leishmania* species, which are transmitted by sandflies and cause leishmaniasis. ⏭ African trypanosomiasis

Loboseans and **heteroloboseans** have an amoeboid (flexible) body form, but they are only distantly related to one another. Loboseans change shape as they move by extending and retracting pseudopodia and engulfing food particles by phagocytosis. *Entamoeba histolytica* infects humans, causing diarrhea ranging from mild asymptomatic disease to severe dysentery. Heteroloboseans exist in an amoeboid form during part of their life cycle, but also form flagellated cells. *Naegleria fowleri* can swim through the water in its flagellated form but assumes its amoeboid form upon entering the human body, where it literally eats the brain of its host (**figure 12.13**). ⏭ *Entamoeba histolytica,* ⏭ *Naegleria fowleri*

MicroByte

Certain antibiotics may target the apical complex—a structure derived from prokaryotes through endosymbiosis.

Structure of Protozoa

Protozoa lack the rigid cellulose cell wall found in algae or the chitinous cell wall found in fungi. Members of a group of amoeboid protists called foraminifera, however, secrete a hard shell of calcium carbonate. Even though protozoa typically lack a cell wall, they often have a specific shape determined by the material beneath the cytoplasmic membrane. Many protozoa have specialized structures for movement, such as cilia, flagella, or pseudopodia. As described in chapter 3, eukaryotic flagella and cilia are distinctly different from prokaryotic flagella (see figures 3.39 and 3.51).

Protozoan Habitats

The majority of protozoa are free-living aquatic organisms. They are essential as decomposers in many ecosystems. Some species, however, are parasitic, living on or in other host organisms. In marine environments, protozoa make up part of the zooplankton. On land, protozoa are abundant in soil as well as in or on plants and animals. Other protozoan habitats include the guts of termites, roaches, and ruminants such as cattle.

Protozoa are an important part of the food chain. They help maintain ecological balance by devouring large numbers of bacteria and algae and, in turn, serve as food for larger species. A single paramecium (a type of protozoan) can ingest as many as 5 million bacteria in one day. ⏭| **food chain**

Protozoan Reproduction

Some protozoa have complex life cycles involving more than one habitat or host. Some protozoan species can exist as a **trophozoite** (vegetative or feeding form) or as a **cyst** (resting form). Environmental conditions—such as the lack of nutrients, moisture, or O_2; low temperature; or the presence of toxic chemicals—may trigger the development of a protective cyst. Some protozoa, such as *Cryptosporidium* and *Entamoeba,* develop a protective cell wall during the cyst stage of their life cycle. This shields the organism as it is transferred from host to host and also helps it withstand stomach acid as it enters a new host. Stomach acid may assist in removal of the cell wall so that the trophozoite can emerge in time to infect the host's intestines. Many parasitic protozoa are transmitted to new hosts during their cyst stage. ⏭| **cryptosporidiosis,** ⏭| **amebiasis**

Both asexual and sexual reproduction are common in protozoa and may alternate during the life cycle of some organisms. Binary fission takes place in many groups of protozoa.

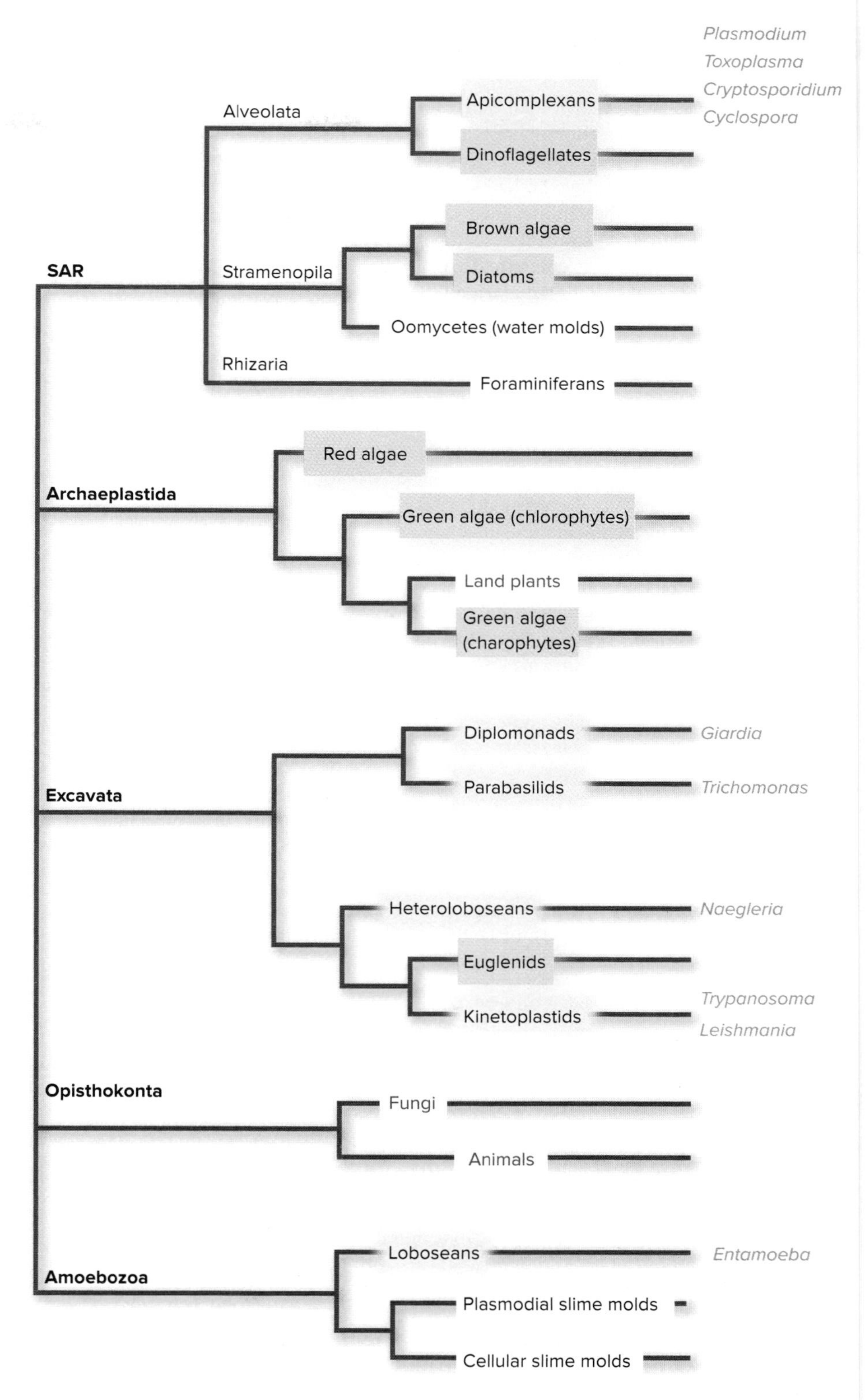

FIGURE 12.12 A Phylogeny of the Eukaryotes Based on Ribosomal RNA Comparisons Algae (highlighted in green) and protozoa (highlighted in blue) are found in a number of classification groups. SAR is an acronym for the taxonomic groups it includes. Significant disease-causing protozoan genera are listed on the right. Non-protists are shown in red. (Not all taxonomic groups of eukarya are shown.)

? What do the diplomonads and the parabasalids have in common?

FOCUS ON A CASE 12.1

The patient was Dudley, a Golden Retriever mix brought to the veterinarian's office by his owner, Sam. "The Dude" was usually playful, but for several days he had simply sprawled in front of the sliding glass door and watched ground squirrels scamper across the patio. Sam was concerned when Dudley did not want to eat, and alarmed by Dudley's recent explosive diarrhea.

The vet asked if Dudley had done anything unusual in the past few weeks. Sam explained that they had gone camping in the state park a couple of weeks earlier. Further questioning revealed that while hiking, Dudley had lapped water directly from a clear stream. Sam groaned as the vet nodded his head and asked the technician to gather supplies to take a stool sample from Dudley. The vet suspected that "The Dude" might be suffering from giardiasis caused by ingesting cysts of *Giardia lamblia,* a protozoan sometimes found in clear streams.

1. What types of conditions might be revealed from a stool sample?
2. What happens to ingested cysts of *Giardia lamblia*?
3. How can a stool sample confirm the vet's suspicions that Dudley was infected with *Giardia*?
4. The veterinarian wrote a prescription for metronidazole (Flagyl) for Dudley, a medication used to treat infections caused by anaerobic organisms. Why would this medication be a common choice for treating giardiasis?

Discussion

1. A stool sample can confirm digestive system infections or the presence of parasites. Abnormal absorption of nutrients or the presence of blood may reveal other conditions that affect the digestive tract and related organs like the liver and pancreas.
2. Ingested *Giardia lamblia* cysts are surrounded by a tough outer wall that protects them from the acidity of stomach secretions. Cysts pass through the stomach and develop into the active form of the parasite (trophozoites) in the small intestine. These attach to the intestinal wall and multiply, often causing diarrhea and cramping. After detaching from the host, the trophozoites slowly develop once again into the cyst form as they pass through the intestine. The cysts contained in feces are highly infectious and, if ingested with contaminated water or food, begin the cycle in a new host.
3. If Dudley has diarrhea due to rapid movement of waste materials through the digestive tract, microscopic examination of a fecal smear may reveal the trophozoite form of the parasite. If waste materials are moving slowly enough to form solid stools, then the trophozoites will have time to develop into cysts before exiting the body. A fecal flotation test can reveal the presence of cysts. Feces are mixed with a test solution that is denser than the cysts. The cysts float to the top of the mixture where they can be skimmed off and examined under the microscope. Shedding of the parasite is often intermittent, so examination of a stool sample may not reveal its presence. If chronic infection is suspected, presence of the parasite may be confirmed by a test called ELISA that detects specific antibodies (see figure 18.8).
4. *Giardia lamblia* is a diplomonad that lacks mitochondria and therefore does not rely upon aerobic cellular respiration. Unlike most other eukaryotes, *Giardia* produces an enzyme similar to one made by anaerobic bacteria. The enzyme results in production of a chemical called ferrodoxin that can transfer an electron to metronadizole. This activates the medication, which then binds to and damages DNA, thus killing the parasite.

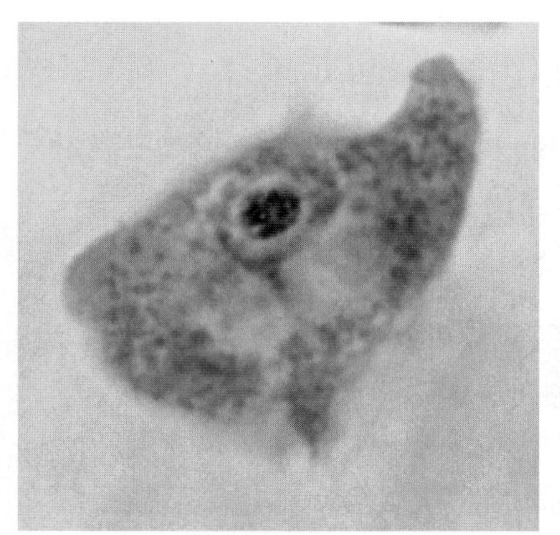
(a)

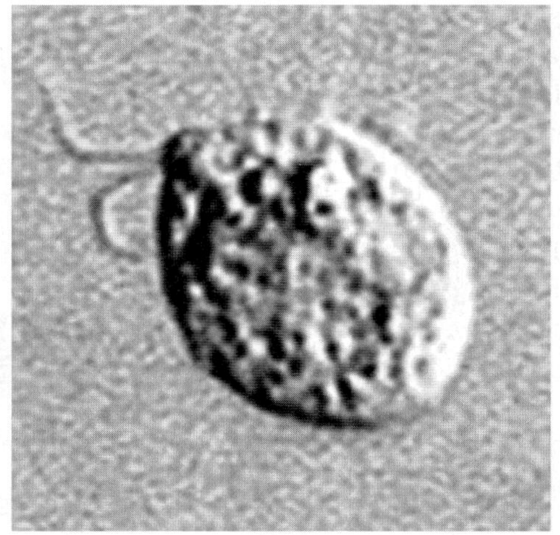
(b)

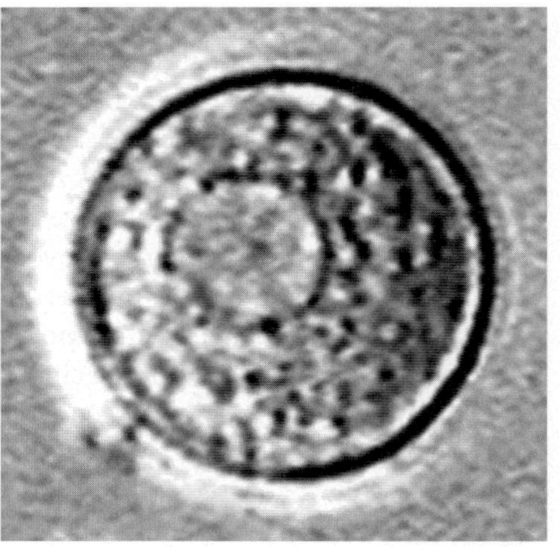
(c)

FIGURE 12.13 ***Naegleria fowleri*** **(a)** In human tissues, the organism exists in the form of an amoeba (10–11 µm at its widest diameter). **(b)** After a few minutes in water, the flagellate form appears. **(c)** Under adverse conditions, a dormant resistant stage (cyst) is formed. a: Source: Dr. George Healy/CDC; b: Source: CDC; c: Source: CDC

? What is the advantage of flagella to a protozoan in water?

TABLE 12.4 Protozoa of Medical Importance

rRNA Classification	Genus of Disease-Causing Protozoan	Disease Caused by Protozoan
Apicomplexan	*Plasmodium*	Malaria
	Toxoplasma	Toxoplasmosis
	Cryptosporidium	Cryptosporidiosis
	Cyclospora	Cyclosporiasis
Diplomonad	*Giardia*	Giardiasis
Heterolobosean	*Naegleria*	Primary amebic meningoencephalitis
Kinetoplastid	*Trypanosoma*	African sleeping sickness
Lobosean	*Entamoeba*	Amebiasis
Parabasalid	*Trichomonas*	Trichomoniasis

Some protozoa divide by multiple fissions, or **schizogony,** in which the nucleus divides a number of times and then the cell produces many single-celled organisms. Multiple fission of the asexual forms of *Plasmodium* in humans results in the release of large numbers of parasites into the host's circulation at regular intervals, producing the cyclic symptoms of malaria. ⏭ malaria

Medical Importance of Protozoa

The majority of protozoa do not cause disease, but those that do have significant impact on world health. Although the mortality rate for malaria fell by 60% in populations at risk between 2000 and 2015, malaria remains one of the greatest killers of humans through the ages. In 2015, about 214 million individuals suffered from malaria, and an estimated 438,000 deaths resulted, most of them in sub-Saharan Africa. The chapter-opening photo shows the characteristic ring form of a *Plasmodium* parasite within red blood cells.

Amebiasis (amebic dysentery) affects nearly 50 million people worldwide each year, claiming up to 100,000 lives. *Cryptosporidium* and *Giardia* are also among the leading causes of diarrhea worldwide. Trypanosomes that cause sleeping sickness (trypanosomiasis) have made some regions of Africa uninhabitable for centuries. **Table 12.4** lists some of the most significant disease-causing protozoa.

MicroAssessment 12.3

Protozoa are a diverse group of single-celled, non-photosynthetic organisms. They occupy a variety of habitats and are a very important part of food chains. A number of significant human diseases, such as malaria, are caused by protozoa.

7. What is the function of the apical complex seen in apicomplexans?
8. What is the role of a cyst in protozoan reproduction?
9. Why might a parasitic protozoan lack mitochondria?

12.4 ■ Slime Molds and Water Molds

Learning Outcome

9. Compare and contrast the two types of slime molds with one another and with water molds.

The slime molds and the water molds are protists that were once considered types of fungi. Although they may look and act like fungi, at a cellular and molecular level, they are completely unrelated to them (see figure 12.12). The shared characteristics of fungi and water molds, in particular, are good examples of **convergent evolution**—a process that occurs when organisms develop similar characteristics independently as they adapt to similar environments.

Slime Molds

Slime molds are terrestrial organisms composed of amoeboid cells that live on soil, leaf litter, and the surfaces of decaying vegetation, where they ingest organic matter by phagocytosis. They are important links in the terrestrial food chain because they ingest microorganisms and, in turn, serve as food for larger predators.

There are two types of slime molds—cellular and plasmodial:

- **Cellular slime molds** have a vegetative form composed of single, amoeba-like cells. When they run out of food, the single cells aggregate into a mass of cells called a slug. Some single cells then form a fruiting body, while others differentiate into spores. These look very much like fungal fruiting bodies and spores (**figure 12.14a**). The model eukaryotic organism, *Dictyostelium discoideum,* is a cellular slime mold important for the study of cell aggregation and multicellular development.
- **Plasmodial slime molds** are large multinucleated "super-amoebae" that may easily reach 0.5 m in diameter. They are widespread and readily visible in their natural environment due to their large size and often their bright color (figure 12.14b). Following germination of haploid spores, the cells fuse to form a diploid cell in which the nucleus divides repeatedly, forming a multinucleated stage called a **plasmodium.** The plasmodium oozes over the surface of decaying wood and leaves, ingesting organic debris and microorganisms. When food or water is in short supply, the plasmodium is stimulated to form spore-bearing fruiting bodies, and the process begins again.

MicroByte

Japanese researchers have shown that a slime mold takes the shortest path between two nutrients.

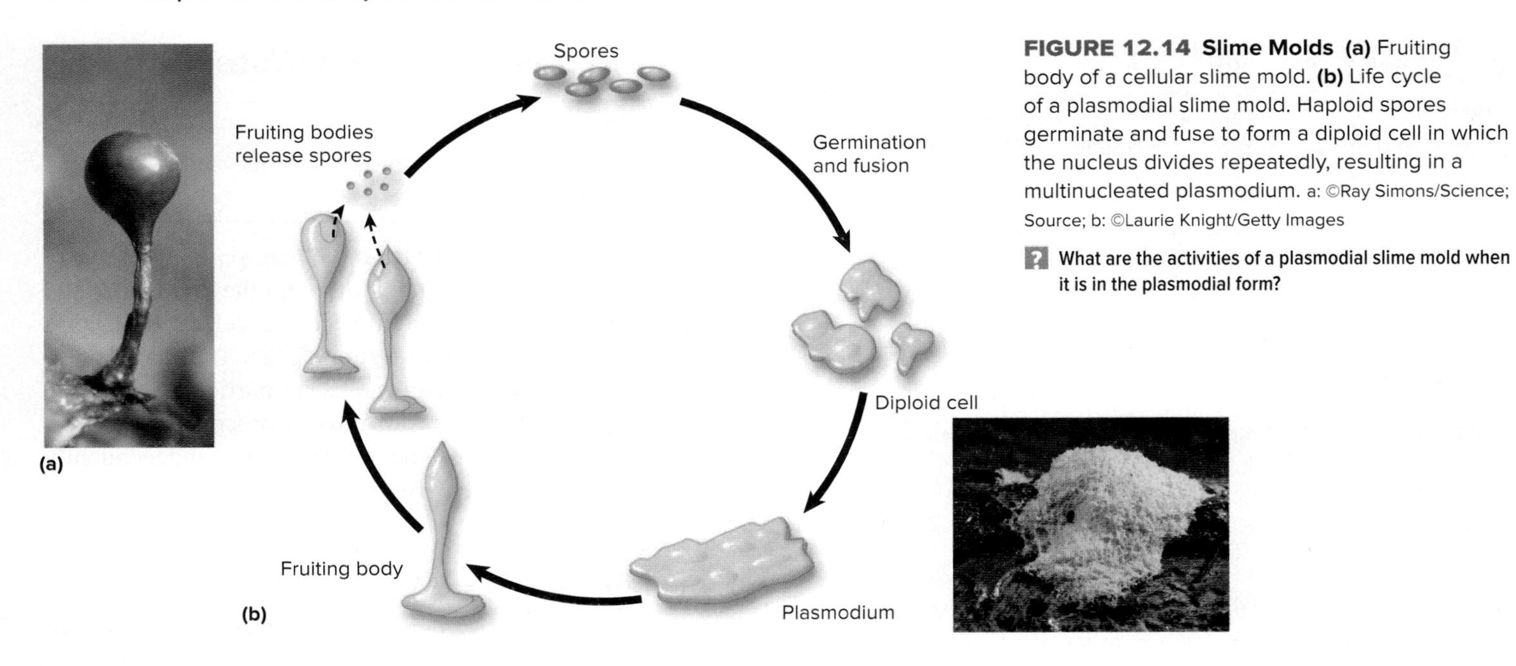

FIGURE 12.14 Slime Molds (a) Fruiting body of a cellular slime mold. **(b)** Life cycle of a plasmodial slime mold. Haploid spores germinate and fuse to form a diploid cell in which the nucleus divides repeatedly, resulting in a multinucleated plasmodium. a: ©Ray Simons/Science Source; b: ©Laurie Knight/Getty Images

What are the activities of a plasmodial slime mold when it is in the plasmodial form?

Water Molds

The **water molds,** or oomycetes, form masses of white threads on decaying material (**figure 12.15**). Like fungi, they secrete digestive enzymes onto a substrate and absorb small molecules for nutrients. The cytoplasm in their filaments is continuous with many nuclei. However, water molds have cellulose in their cell walls rather than chitin. They lack chloroplasts and have flagellated reproductive cells.

Oomycetes cause some serious diseases of food crops, such as late blight of potato and downy mildew of grapes. The late blight of potato was a factor in the potato famine in Ireland in the 1840s that sent waves of immigrants to the United States (see A Glimpse of History).

FIGURE 12.15 Water Mold The white "threads" secrete digestive enzymes used to break down organic compounds. ©Noble Proctor/Science Source

How is a water mold like a fungus; how is it different from one?

MicroAssessment 12.4

Cellular slime molds may exist as single cells or may form an aggregation called a slug. Plasmodial slime molds form a multinucleated "super-amoeba." Both engulf food by phagocytosis and form fruiting bodies similar to those of fungi. Water molds secrete enzymes and absorb organic nutrients; they form masses of white threads on organic matter.

10. What do an amoeba and a slime mold have in common?

11. How are slime molds similar to, and different from, fungi?

12. What environmental conditions led to the convergent evolution of fungi and water molds?

12.5 ■ Multicellular Parasites: Helminths

Learning Outcomes

10. Explain the roles of a definitive host, an intermediate host, and a dead-end host in the life cycles of helminths.

11. Explain how disease-causing helminths can be transmitted to humans.

12. Compare and contrast the structures of a roundworm, a tapeworm, and a fluke; describe one disease caused by each of these.

Helminths are worms, a type of animal. Some are parasitic and can invade human tissues—an action that causes disease or robs the body of nutrients. Although not microorganisms, parasitic helminths are often identified using microscopic and immunological techniques familiar to microbiologists.

TABLE 12.5 Nematodes, Cestodes, and Trematodes

Infectious Agents	Disease	Disease Characteristics
Nematodes (roundworms)		
Ascaria *(Ascaris lumbricoides)*	Ascariasis	Abdominal pain, vomiting, intestinal blockage
Filaria (*Wuchereria bancrofti* and *Brugia malayi*)	Filariasis	Fever, swelling of lymphatic structures, genitals, and extremities
Hookworm (*Necator americanus* and *Ancylostoma duodenale*)	Hookworm disease	Anemia, weakness, fatigue, physical and intellectual disability in children
Pinworm *(Enterobius vermicularis)*	Enterobiasis	Anal itching, restlessness, irritability, nervousness, poor sleep
Threadworm *(Strongyloides stercoralis)*	Strongyloidiasis	Skin rash at site of penetration, cough, abdominal pain, weight loss
Trichinella *(Trichinella spiralis)*	Trichinellosis	Fever, swelling of upper eyelids, muscle soreness
Whipworm *(Trichuris trichiura)*	Trichuriasis	Abdominal pain, bloody stools, weight loss
Cestodes (tapeworms)		
Beef tapeworm *(Taenia saginata)*	Tapeworm disease	Few or no symptoms
Fish tapeworm *(Diphyllobothrium latum)*	Tapeworm disease	Few or no symptoms, sometimes anemia
Pork tapeworm *(Taenia solium)*	Tapeworm disease	Few or no symptoms
	Cysticercosis	Variable symptoms depending on location and number of eggs that form larval cysts (cysticerci) in the body
Trematodes (flukes)		
Blood fluke (*Schistosoma* species)	Schistosomiasis	Liver damage, malnutrition, weakness, and accumulation of fluid in the abdominal cavity
Cercaria that infect birds and other animals	Swimmer's itch	Inflammation of the skin, itching

The microscopic eggs (ova) of each helminth species are distinctive and can be identified by size, shape, and other features. The helminths are divided into two groups: the **roundworms** (nematodes) and the **flatworms.** The flatworms are further divided into the tapeworms (cestodes) and the flukes (trematodes). Multicellular parasites have been largely controlled in industrialized nations, but they still cause suffering in and the death of many millions each year in developing parts of the world. **Table 12.5** lists the major diseases caused by helminths.

Life Cycles and Transmission of Helminths

Some helminths have complex life cycles involving one or more **intermediate hosts** that house a sexually immature stage of the parasite and are necessary for its development. Snails serve as an intermediate host for the fluke *Schistosoma mansoni,* a cause of schistosomiasis. Sexual reproduction of the parasite takes place in humans, its **definitive host.** A human may also become an accidental or **dead-end host** if infected by a parasite that normally completes its life cycle in another host. For example, swimmers are sometimes infected with the larvae of flukes that typically complete their life cycles in fish or water birds. These flukes cannot complete their life cycle in humans, but when they burrow under the skin, they cause a local inflammation called "swimmer's itch."

Helminths enter the body in a number of ways. Hookworm larvae (immature forms), for example, live in the soil and can burrow through human skin. They multiply in the human digestive tract and are eliminated with feces. When sanitation is poor and people are bare-footed, the parasite is easily transmitted. Hookworms can be found in about 740 million individuals, mostly in tropical and subtropical regions.

Some helminths are inadvertently eaten with food, as when the larva of the nematode *Trichinella spiralis* is ingested in the flesh of animals. Eating undercooked pork is the most common cause of trichinellosis. More often, helminth eggs are ingested on the surface of contaminated foods. Children with pinworms *(Enterobius vermicularis),* for example, may pick up eggs by touching their anus and transmit them to a surface. A food handler who does not use proper handwashing may then inadvertently transfer the eggs from that surface to the food.

Some helminths are transmitted through insect bites. One example is *Wuchereria bancrofti,* a type of filarial (thin or thread-like) nematode that lives outside the intestine and is transmitted by mosquitoes. The adult worms live in the lymphatic vessels, resulting in blockage of the lymphatic

FOCUS YOUR PERSPECTIVE 12.1

What Causes River Blindness?

Female black flies, sometimes called buffalo gnats, swarm around their hosts and bite repeatedly to obtain blood needed for development of their eggs. They require rapidly flowing water for development of their larvae and so are most often found near rivers. These flies are associated with a disease called river blindness that affects at least 18 million individuals, 99% of whom live in Africa. Of those, the World Health Organization estimates that about 270,000 are blind as a result of their infection and up to another 1 million people have impaired vision. River blindness is the second leading cause of infectious blindness. (The leading cause is trachoma caused by the bacterium *Chlamydia trachomatis.*) But are the black flies the cause of this disease?

When biting, the flies may transmit larvae of a filarial nematode, *Onchocerca volvulus,* to a human host, leading to river blindness, also called onchocerciasis. Once in a human host, the larvae reside in nodules and mature to adulthood in about a year, at which time adult females produce millions of microfilariae. These can migrate through the skin where they can be picked up by another bite of a black fly to continue the life cycle. When infections are heavy, the microfilariae can also be found in the blood and in the eye. So are the *Onchocerca* larvae the real cause of the disease?

As microfilariae move throughout the bodies of their hosts, they carry a bacterial population of *Wolbachia pipientis* that is necessary for fertility and viability of the worms. When microfilariae die, they release the *Wolbachia.* The bacteria cause an inflammatory response that can damage sensitive host tissues. When this happens in the eye, the result is vision impairment and blindness. Ultimately, it is the bacteria carried by *Onchocerca* larvae that lead to the symptoms of river blindness. ◀◀ *Wolbachia*

Efforts to control river blindness have focused on eliminating the black fly vector and providing a medication called ivermectin that targets the worm. Ivermectin reduces the number of microfilariae for a few months, but does not destroy the adults, so repeated medication is needed. An alternative approach is to target the *Wolbachia* bacteria with common antimicrobial medications such as doxycycline or tetracycline. This not only reduces effects of the disease, but it sterilizes the worms so that they can no longer produce microfilariae, thus disrupting the life cycle.

drainage. The buildup of fluid can cause massive swelling in various parts of the body, a condition called elephantiasas (**figure 12.16**). Another example is *Onchocerca volvulus*—a nematode that is spread by flies and can cause river blindness (see **Focus Your Perspective 12.1**).

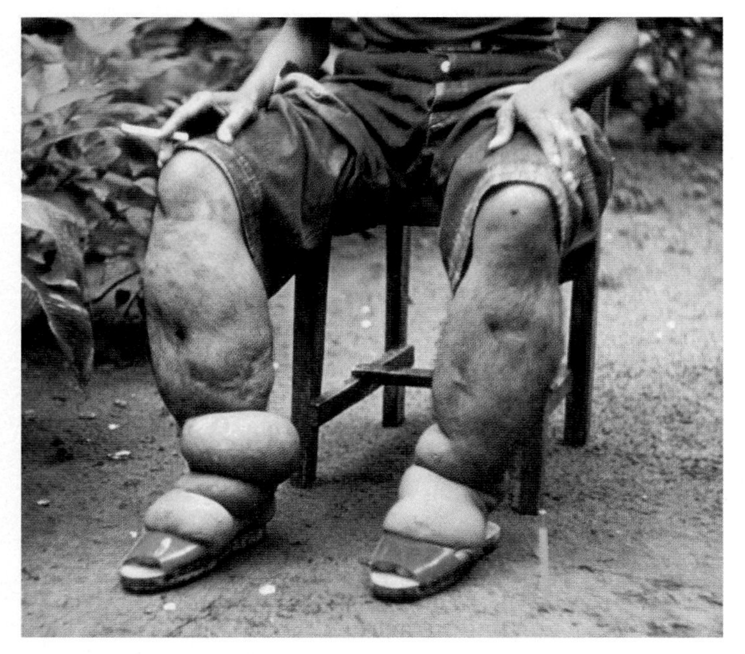

FIGURE 12.16 Elephantiasis Buildup of fluid has occurred as a result of adult worms living in the lymphatic vessels leading from the limbs. Source: CDC

? What type of worm causes this condition?

Roundworms (Nematodes)

A **roundworm,** or **nematode,** has a cylindrical, tapered body with a digestive tract that extends from the mouth to the anus. The nematode *Caenorhabditis elegans* is a model eukaryotic organism that has been the subject of numerous studies in genetics and development because it matures quickly, its genome has been sequenced, and all of its 959 body cells can be identified. Many nematodes are free-living in soil and water. Others are parasites and produce serious disease.

Ascariasis, caused by *Ascaris lumbricoides,* is the most common human disease caused by roundworms. Females are larger than males and may exceed 30 cm long while producing more than 200,000 eggs per day that are eliminated in the feces. Ingested eggs hatch in the digestive tract, releasing immature worms that burrow into the bloodstream (**figure 12.17**). After they reach the lungs, they can be coughed up and swallowed. When the immature worms again reach the intestine, they grow and begin to produce eggs that can again be released with the feces. Although they do not feed on human tissue, they do rob the body of nutrients by feeding on material that passes through the digestive tract. They may cause choking and pulmonary symptoms when they enter the respiratory tract.

MicroByte

Over 1 billion people worldwide are believed to carry the roundworm *Ascaris lumbricoides.*

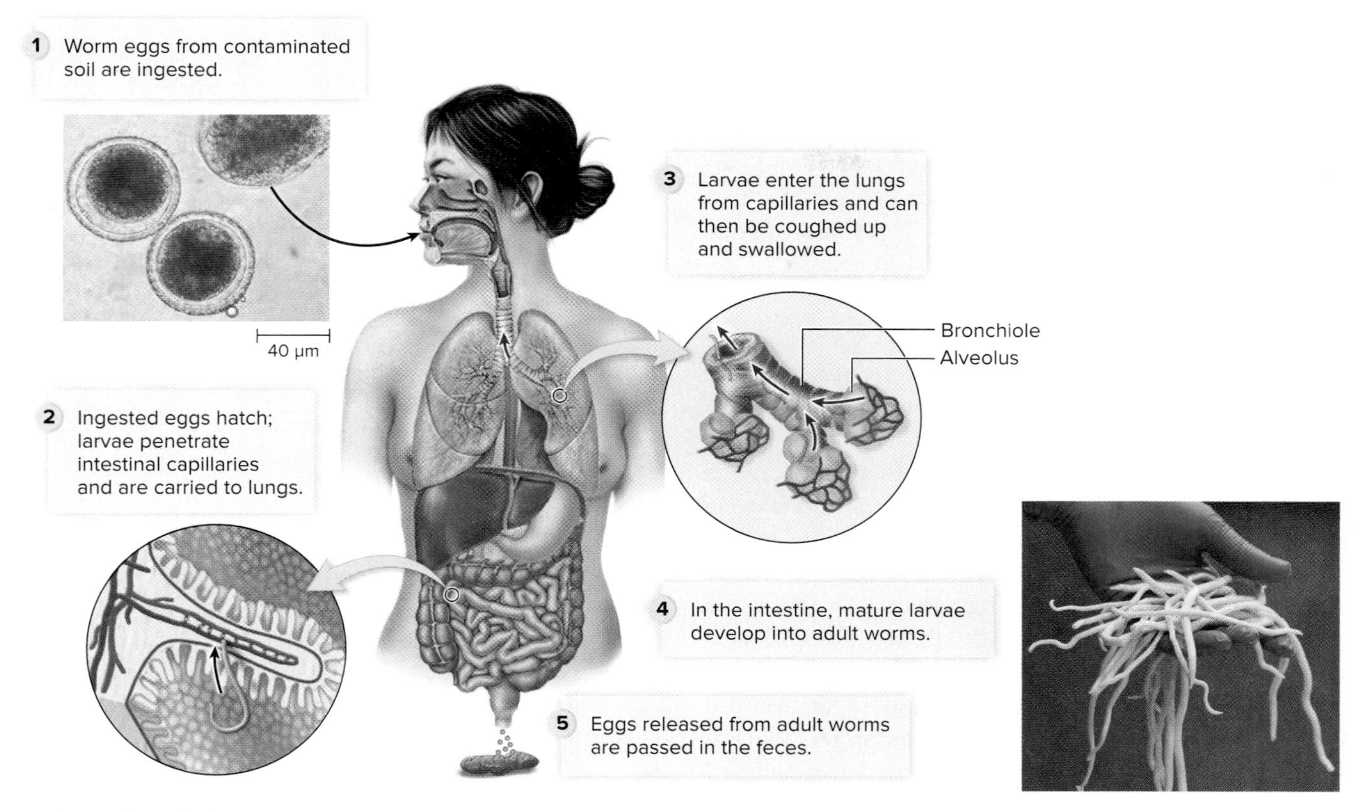

FIGURE 12.17 Ascariasis Life cycle of *Ascaris lumbricoides,* the largest roundworm infecting the human intestine, averaging about 30 cm. Larvae hatching in the intestine migrate through the lungs and back to the intestine before maturing to adulthood. The thick-walled ova are nearly spherical. (eggs): ©McGraw-Hill Education/Lisa Burgess; (worms): ©McGraw-Hill Education/Lisa Burgess

How does *Ascaris* get from the lungs to the intestines?

Tapeworms (Cestodes)

Tapeworms, or **cestodes,** have flat, ribbon-shaped bodies. They have no digestive system, and do not feed directly on the tissues of their host. Rather, adult tapeworms attach within the intestines of the definitive host and absorb predigested nutrients through their body. Some types have reached over 15 meters in length while living for over a decade in the intestine of a host. The **scolex** (head end) of the worm often has suckers and hooks for attachment (**figure 12.18**). Attached to the scolex is the neck and then a number of segments called **proglottids** that contain both male and female reproductive structures. The proglottids farthest from the scolex contain fertilized eggs. As the worm grows, these segments break off and are eliminated in the feces along with tapeworm eggs. When a suitable intermediate host ingests the eggs, the eggs hatch, releasing a larval form that penetrates the intestinal wall and migrates into tissues. These larval forms are infectious when consumed by a definitive host.

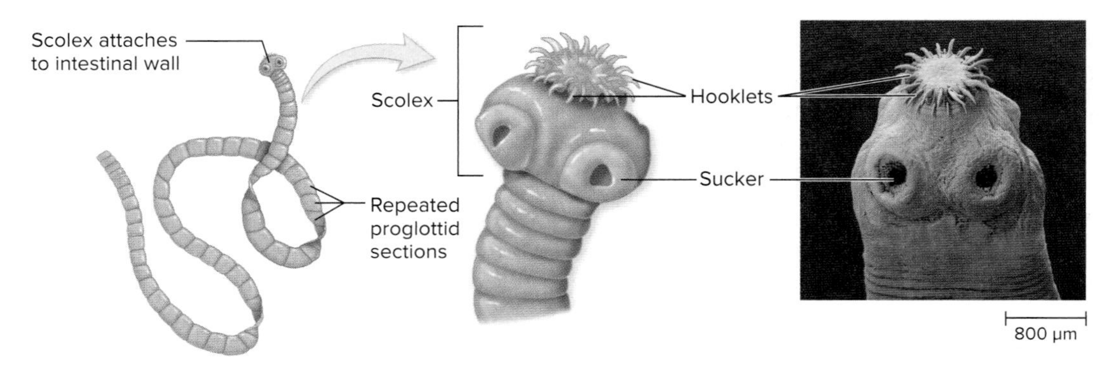

FIGURE 12.18 Tapeworm The scolex holds the tapeworm to the intestinal surface. Proglottids contain reproductive structures. They are shed in the feces as the tapeworm elongates by adding new segments. ©Eye of Science/Science Source

Why can a tapeworm live without a digestive system?

The most common tapeworms of humans have intermediate hosts of cattle, pigs, and fish. Humans (the definitive host) become infected when they eat raw or undercooked meat containing the larval forms. Unfortunately, humans can also serve as the accidental intermediate host of the pork tapeworm *(Taenia solium).* If someone inadvertently ingests eggs of this tapeworm, the eggs can hatch, releasing larvae. These may migrate to the brain, resulting in serious neurological symptoms.

Flukes (Trematodes)

Flukes, or **trematodes,** are flat leaf-shaped worms with two suckers that the adult worms use to both attach to and move along a surface of their definitive host. The worms have a mouth, but no anus, so nutrients enter and wastes exit from the same opening. The parasites typically have a complex life cycle that involves at least two hosts: a snail host in which the organism asexually reproduces, and a mammalian or other vertebrate host in which the organism sexually reproduces. There are two general categories of flukes: tissue flukes and blood flukes. The tissue flukes are hermaphroditic (have both sex organs in the same worm), whereas the blood flukes have separate sexes.

Although many flukes are medically important, schistosomes (blood flukes) are particularly significant because of the devastating disease they cause—schistosomiasis (also called bilharzia). Spread of the disease requires a suitable snail host, along with sewage-contaminated fresh water. Snails infected with a schistosome release thousands of a larval form of the parasite called cercaria. These swim in search of an appropriate definitive host, and when they encounter a human, they burrow through the skin, and enter the circulatory system. There, the worms mature, and then the male and female worms find each other and mate. The mating pair moves to a specific region of the body, where the female deposits her fertilized ova in tiny veins. A strong inflammatory response develops, similar to what happens when a foreign object such as a sliver lodges in the skin or other body site, sometimes pushing the ova through the walls of the vein. In the case of schistosomes that deposit their ova in the veins near the intestine, this pushes the eggs into the intestinal tract, where they are then eliminated with feces. For the schistosome that deposits its ova in the veins near the bladder, the eggs are pushed into the bladder, where they are eliminated with urine. In either case, if viable ova reach fresh water, they hatch to release ciliated larvae. These larval forms then infect specific snail species, completing the life cycle.

The most severe outcomes of schistosomiasis are due to eggs not released into the intestinal tract or the bladder. As the inflammatory response to the ova continues, normal tissues are destroyed and replaced with scar tissue. A particularly damaging situation occurs if the blood circulatory system moves ova deposited near the intestinal tract to the liver. There, the inflammatory response gradually destroys liver tissue, leading to life-threatening liver damage.

MicroAssessment 12.5

Helminths, including the roundworms, tapeworms, and flukes, cause serious diseases in humans. They may enter a human host through ingestion with food or water, by an insect bite, or by piercing the skin. Many helminths have a complex life cycle with more than one host.

13. What are the major differences among nematodes, cestodes, and trematodes?
14. Differentiate between a definitive host and an intermediate host.
15. Why do so many helminth diseases occur in the tropics?

12.6 ■ Arthropods

Learning Outcomes

13. Describe two ways in which arthropods are related to disease in humans.
14. Give an example of a disease transmitted by each of the following: mosquitoes, flies, fleas, lice, and ticks.

Arthropods are animals that include the insects (such as flies, mosquitoes, lice, and fleas) and the arachnids (such as ticks and mites). Their main role in disease is to serve as **vectors** that can transmit microorganisms and viruses to humans. An arthropod may act as a mechanical vector that simply transfers a pathogen from one surface to another, or it may be a biological vector that plays an essential role in the life cycle of the pathogen. For example, species of *Plasmodium* that cause malaria multiply within an *Anopheles* mosquito during their life cycle, and species of trypanosomes that cause African sleeping sickness multiply within the tsetse fly (*Glossina* species). ⏭ vector, ⏭ malaria, ⏭ African sleeping sickness

When an arthropod vector feeds on an infected host, it may pick up pathogens and then transfer them to humans in a later bite. Some arthropods bite only one type of host. Certain mosquitoes that carry *Plasmodium* bite only humans; typically, only female mosquitoes take a blood meal. However, fleas that carry *Yersinia pestis,* the bacterium responsible for plague, bite both humans and small mammals such as rats.

The incidence of vector-borne diseases can be decreased by controlling the vector or the infected hosts. The risk of mosquito-borne encephalitis, for example, can be minimized by eliminating standing water and using insect repellent. These do not act on the virus that causes encephalitis, but they do reduce the incidence of vector transmission. Plague in the United States has been controlled mostly by eliminating rat populations that may infect their fleas with the bacterium *Yersinia pestis.* Cases of plague today occur from occasional transmission by fleas on wild rodents such as rock squirrels

TABLE 12.6 Some Arthropods That Transmit Infectious Agents

Arthropod	Infectious Agent	Disease and Characteristic Features
Insects		
Black fly (*Simulium* species)	*Onchocerca vulvulus*	Onchocerciasis (river blindness)—rash, itching, visual impairment
Flea (*Xenopsylla cheopis*)	*Yersinia pestis*	Plague—fever, headache, confusion, enlarged lymph nodes, skin hemorrhage
Louse (*Pediculus humanus*)	*Rickettsia prowazekii*	Typhus—fever, hemorrhage, rash, confusion
Mosquito (*Aedes* species)	Flavivirus	Yellow fever—fever, vomiting, jaundice, bleeding
Mosquito (*Aedes* species)	Flavivirus	Dengue fever—high fever; headache; joint, muscle, and bone pain
Mosquito (*Anopheles* species)	*Plasmodium* species	Malaria—chills, bouts of recurring fever
Mosquito (*Culex* species)	Togavirus	Viral encephalitis—fever, nausea, convulsions, coma
Sand fly (*Phlebotomus* species)	*Leishmania*	Leishmaniasis—ulcers, nosebleeds, diarrhea, fever, cough
Tsetse fly (*Glossina* species)	Trypanosomes	African sleeping sickness—sleepiness, headache, coma
Arachnids		
Tick (*Dermacentor* species)	*Rickettsia rickettsii*	Rocky Mountain spotted fever—fever, hemorrhagic rash, confusion
Tick (*Ixodes* species)	*Borrelia burgdorferi*	Lyme disease—fever, rash, joint pain, nervous system impairment

or prairie dogs. Examples of some important arthropods, the agents they transmit, and the resulting diseases are shown in **table 12.6**. ⏭| plague

Some arthropods cause disease even when they do not act as a vector. For example, the pubic louse, *Phthirus pubis,* is commonly transmitted during sexual intercourse. It is not a vector of infectious agents, but it can cause an unpleasant itch associated with "crabs." Similarly, dust mites do not transmit infectious disease, but inhalation of the mites and their waste products can sometimes trigger asthma. The larvae of some mites are called "chiggers" and may cause intense itching where they attach and feed on fluids within skin cells. Scabies**,** a disease caused by another mite, is easily transmitted by personal contact. Allergic reactions to female mites that have burrowed into the outer layers of skin are largely responsible for the itchy rash of scabies.

MicroAssessment 12.6

Arthropods such as flies, mosquitoes, fleas, lice, and ticks act as vectors for the spread of disease. Some participate in the life cycle of an infectious agent and transmit disease through saliva when biting or burrowing. Infestations of mites and lice may cause itching.

16. How can arthropods spread disease in humans?

17. What causes the itching of scabies?

18. Why are diseases transmitted by insect vectors more common in the summer than in the winter?

Summary

Algae, fungi, and **protozoa** are not precise classification groups when the rRNA sequences of these organisms are considered. Cell structure in eukaryotic organisms is different from that seen in prokaryotes. Eukaryotes have a membrane-bound nucleus. Reproduction may be asexual using **mitosis** or sexual using **meiosis,** which forms **gametes** (figure 12.1).

12.1 ■ Fungi

Yeast, mold, and **mushroom** are common terms that indicate morphological forms of fungi (figure 12.2). Fungi have chitinous cell walls and are often **saprophytes,** secreting enzymes onto a surface and absorbing nutrients.

Types of Fungi (table 12.1)

Classification of fungi is in flux. Zygomycetes (figure 12.3), ascomycetes, basidiomycetes, and chytridiomycetes are distinctive types of fungi.

Structure of Fungi

Fungal filaments are called **hyphae,** and a tangled mass of hyphae is called a **mycelium** (figure 12.4). **Dimorphic fungi** can grow either as single cells (yeast) or as mycelia.

Fungal Habitats

Fungi occupy just about every ecological habitat and can spoil a large variety of food materials because they can grow in high concentrations of sugar, salt, and acid. Fungi can be found in moist environments, at a wide range of temperatures, and at pH from 2.2 to 9.6. Fungi can degrade most organic materials.

Symbiotic Relationships of Fungi

Lichens result from an association of fungus with a photosynthetic organism such as an alga or a cyanobacterium (figure 12.5). **Mycorrhizas** are an intimate association of a fungus and the roots of a plant (figure 12.6). Leaf-cutter ants grow gardens of fungus for food (figure 12.7).

Reproduction in Fungi

Asexual reproduction may occur by a variety of methods (figures 12.8, 12.9). Sexual reproduction may involve fusion of hyphae from different mating types.

Economic Importance of Fungi

The yeast *Saccharomyces cerevisiae* is used in the production of beer, wine, and bread. *Penicillium* and other fungi synthesize antibiotics. Fungi spoil many food products and cause diseases of plants such as Dutch elm disease and coffee rust. Fungi have been useful tools in genetic and biochemical studies.

Medical Importance of Fungi (table 12.2)

Relatively few fungi cause human disease, but fungi cause devastating diseases in plants. Fungi may produce hypersensitivity reactions, or infections called **mycoses** (such as candidiasis). Fungi also can produce toxins that make humans ill. These include **aflatoxin, ergot,** and those in poisonous mushrooms.

12.2 ■ Algae

Algae are a diverse group of photosynthetic organisms.

Types of Algae (table 12.3)

Types of algae differ in their major photosynthetic pigments, but all contain chlorophyll *a*.

Structure of Algae

Algae may be microscopic (figure 12.10) or macroscopic (figure 12.11). Their cell walls are made of cellulose and other commercially important materials such as agar and alginic acid.

Algal Habitats

Algae are found in fresh and salt water as well as in soil. Unicellular algae make up a significant part of the **phytoplankton.**

Algal Reproduction

Algae reproduce asexually as well as sexually.

Medical Importance of Algae

Algae do not directly cause disease, but produce toxins during algal blooms that are ingested by fish and shellfish. Consumption of these shellfish may result in paralytic shellfish poisoning; cooking does not destroy the toxins.

12.3 ■ Protozoa

Protozoa are a diverse group of microscopic, unicellular organisms that lack chlorophyll.

Types of Protozoa (figure 12.12)

In classification schemes based on rRNA, protozoa are not a single group of organisms. **Apicomplexans** *(Plasmodium, Toxoplasma, Cryptosporidium, Cyclospora)* are parasites with an apical complex that helps them to penetrate host cells. **Diplomonads** *(Giardia)* and **parabasalids** *(Trichomonas)* have no mitochondria. **Kinetoplastids** *(Trypanosoma, Leishmania)* have distinct mitochondrial DNA. **Loboseans** *(Entamoeba)* and **heteroloboseans** *(Naegleria)* move by pseudopodia at some stage in their lives (figure 12.13).

Structure of Protozoa

Protozoa lack a cell wall, but most maintain a definite shape using the material lying just beneath the cytoplasmic membrane.

Protozoan Habitats

Most protozoa are free-living and are found in marine and fresh water as well as terrestrial environments. They are important decomposers in many ecosystems and are a key part of the food chain.

Protozoan Reproduction

Life cycles are often complex and include more than one habitat or host. In some cases, a vegetative **trophozoite** can develop into a resting **cyst.** Reproduction is often by binary fission; some reproduce by multiple fissions or **schizogony.**

Medical Importance of Protozoa (table 12.4)

Protozoa cause diseases such as malaria, African sleeping sickness (trypanosomiasis), and amebiasis.

12.4 ■ Slime Molds and Water Molds

Slime molds and water molds were once considered types of fungi.

Slime Molds (figure 12.14)

Cellular slime molds exist as amoeba-like single cells, but when food supplies run low, they aggregate into a slug in which some cells differentiate into spores. **Plasmodial slime molds** form a multinucleated **plasmodium** that oozes over a surface, ingesting organic material. When food runs short, they form fruiting bodies that bear spores.

Water Molds

Oomycetes, or **water molds,** can cause serious plant diseases (figure 12.15). Water molds and fungi are an example of **convergent evolution.**

12.5 ■ Multicellular Parasites: Helminths (table 12.5)

Helminths are worms, a type of animal. Some are parasitic.

Life Cycles and Transmission of Helminths

Helminths can be transmitted by burrowing through the skin, being ingested, or being transmitted through insect bites (figure 12.16). Some helminths have complex life cycles with asexual stages occurring in one or more **intermediate hosts** and the sexual or adult stage occurring in the **definitive host.** A human being may be a **dead-end host** in which the organism cannot complete its life cycle.

Roundworms (Nematodes)

Most **nematodes** or roundworms are free-living, but they may cause serious disease such as hookworm disease and ascariasis (figure 12.17).

Tapeworms (Cestodes)

Cestodes are tapeworms with segmented bodies and hooks to attach to the wall of the intestine (figure 12.18). Most tapeworm infections occur in persons who eat uncooked or undercooked meats.

Flukes (Trematodes)

Trematodes, or flukes, often have complicated life cycles that require more than one host. The larvae of *Schistosoma* species can penetrate the skin of persons wading in infected waters and cause serious disease.

12.6 ■ Arthropods

Arthropods can act as **vectors** for disease (table 12.6). Some arthropods can infest the body and cause itching and a rash.

Review Questions

Short Answer

1. What are the major differences between a prokaryotic cell and a eukaryotic cell?
2. What are the differences among a yeast, a mold, and a mushroom?
3. Why are few lichens found in industrial areas?
4. In what ways are fungi economically important?
5. What is a mycosis? Give an example.
6. What characteristics do all algae have in common?
7. Compare and contrast the organisms that cause malaria and African sleeping sickness and their transmission.
8. Name a disease for which humans are an intermediate host and another for which humans are a definitive host. Give an example of a disease in which humans are a dead-end host.
9. Describe the life cycle of *Schistosoma* species.
10. Explain how an insect might act as a mechanical vector for one disease and a biological vector for another.

Multiple Choice

1. Members of this group have chitinous cell walls.
   a) Algae
   b) Protozoa
   c) Fungi
   d) Helminths
2. Members of this group are photosynthetic.
   a) Algae
   b) Protozoa
   c) Fungi
   d) Helminths
3. Which of the following statements regarding protozoa is *false*?
   a) Some protozoa lack mitochondria.
   b) Some protozoa contain chlorophyll.
   c) Some protozoa are parasitic.
   d) Malaria is caused by a protozoan.
4. Which of the following statements about fungi is *false*?
   a) Fungi may cause disease when growing in or on the human body.
   b) Some fungi produce toxins that can kill humans.
   c) Fungal spores may cause allergic responses in humans.
   d) Systemic mycoses are common in otherwise healthy adults.
5. Which of the following is mismatched?
   a) *Plasmodium*—malaria
   b) Trypanosomes—dysentery
   c) Dinoflagellates—paralytic shellfish poisoning
   d) Nematode—trichinellosis
6. Which of the following is mismatched?
   a) Trematode—fluke
   b) Mosquito—malaria
   c) Baker's yeast—algae
   d) Apicomplexan—protozoa
7. The current phylogeny of eukaryotes is based upon
   a) rRNA sequence comparisons.
   b) possession of photosynthetic pigments.
   c) mode of reproduction.
   d) locomotor structures.
8. All algae have
   a) chlorophyll *a*.
   b) cell walls that contain agar.
   c) holdfasts.
   d) red tides.
9. Which of the following statements regarding protists is *false*?
   a) They include both autotrophic and heterotrophic organisms.
   b) They include both microscopic and macroscopic organisms.
   c) They are an intermediate host in schistosomiasis.
   d) They include algae and protozoa.
10. Which of the following statements regarding tapeworms is *false*?
    a) They absorb nutrients from the host through their body wall.
    b) They complete their life cycles in a single host.
    c) They are hermaphroditic.
    d) They cannot be transmitted from human to human.

Applications

1. A molecular biologist working for a government-run fishery in Vietnam is interested in controlling the dinoflagellate *Pfisteria* in fish farms. *Pfisteria* produces toxins that stun the fish and then causes the skin to slough off, allowing the parasites to dine on the tissues of the fish. He needs to develop a treatment that kills *Pfisteria* without harming the fish or the beneficial green algae that serve as food for the young fish. What strategy should the biologist consider for developing a selective treatment?
2. Paper recycling companies refuse to collect paper products that are contaminated with food or have been sitting wet for a day. A college sorority member who is running a recycling program on campus wishes to know the reason for this. What reason did the chemist who works for the recycling company probably give her for this policy?

Critical Thinking

1. If you discovered a new type of nucleated cell in a lake near your home, how would you determine whether the cell was from a fungus, an alga, a protozoan, or a water mold?
2. Fungi are known for growing and reproducing in a wide range of environmental extremes in temperature, pH, and osmotic pressure. What does this tolerance for extremes indicate about fungal enzymes?

13 Viruses, Viroids, and Prions

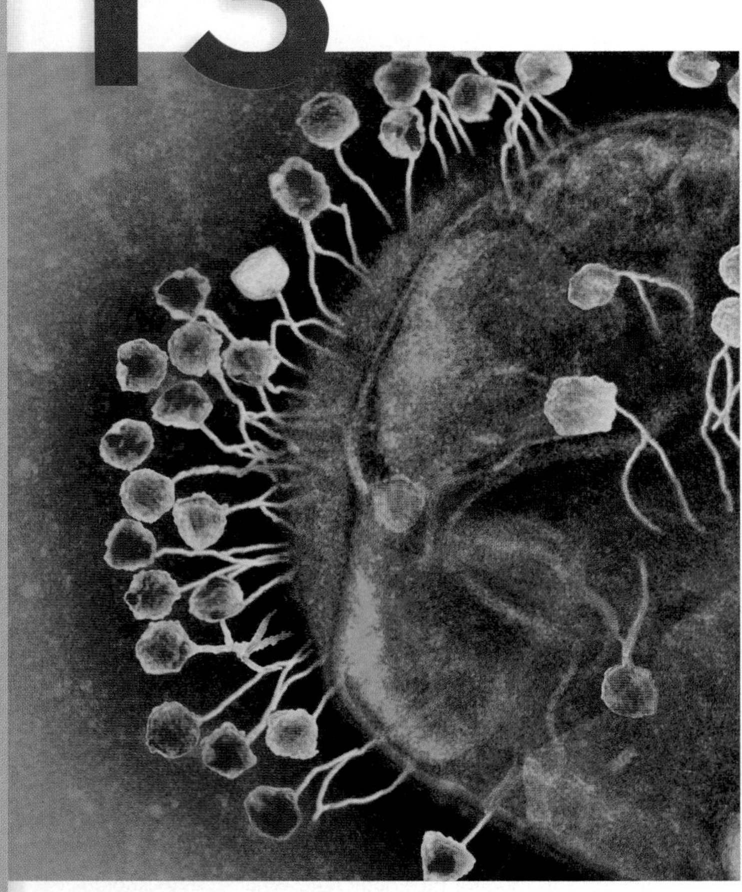

Phage particles attached to a bacterium (color-enhanced scanning electron micrograph). ©*AMI Images/Science Source*

KEY TERMS

Bacteriophage A virus that infects bacteria; often shortened to *phage.*

Latent Infection Viral infection in which the viral genome is present but not active, so new viral particles are not being produced.

Lysogen A bacterium that carries phage DNA (a prophage) integrated into its genome.

Lysogenic Conversion A change in the properties of a bacterium, conferred by a prophage.

Lytic Infection Viral infection of a host cell with a subsequent production of more viral particles and lysis of the cell.

Prion An infectious protein that causes a neurodegenerative disease.

Productive Infection Viral infection in which more viral particles are produced.

Viroid An infectious agent of plants that consists only of RNA.

Virion A complete virus in its inert non-replicating form; also referred to as a viral particle.

A Glimpse of History

During the late nineteenth century, many bacteria, fungi, and protozoa were identified as infectious agents. Most of these organisms could be easily seen using a light microscope, and could be grown in the laboratory. In the 1890s, however, Dimitri Iwanowsky and Martinus Beijerinck found a mosaic disease of tobacco plants that was caused by an unusual agent. The agent was too small to be seen with the light microscope. Furthermore, it passed through filters that retained most known bacteria, and could be grown only in media that contained living cells. Beijerinck called the agent a "filterable virus." About 10 years later, Frederick W. Twort and Félix d'Herelle discovered a filterable virus that destroys bacteria. *Virus* means "poison," a term once applied to all infectious agents. With time, the adjective *filterable* was dropped and now we use only the word *virus* to describe these agents.

Viruses have many features more characteristic of complex chemicals than of cells. For example, tobacco mosaic virus (TMV) can be precipitated from a suspension with ethyl alcohol and still remain infective. A similar treatment destroys the infectivity of bacteria. In 1935, Wendell Stanley crystallized TMV. Its physical and chemical properties obviously differed from those of cells, which cannot be crystallized. Surprisingly, the crystallized TMV could still cause disease.

In the simplest of terms, viruses can be viewed as genetic information—either DNA or RNA—contained within a protective protein coat. They are inert particles, which means that they are incapable of metabolism, replication, or motility. When a viral genome enters a host cell, however, it can hijack that cell's replication machinery, inducing the cell to produce more viral particles. In essence, viruses straddle the definition of life. Outside a host cell they are inert, but inside a cell they direct activities that have a profound effect on that cell. So although they are infectious agents, they are not organisms.

Viruses can be broadly grouped into two general types based on the category of cells they infect. Some infect prokaryotic cells, and others infect eukaryotic cells. Although both groups are viruses, those that infect bacteria are also referred to as **bacteriophages,** or simply **phages** (*phage* means "to eat").

The fact that viruses are obligate intracellular parasites makes them very difficult to study. Unlike most bacteria and eukaryotic cells, which can be grown in pure culture, viruses require live organisms as hosts. In addition, viruses are too small to be seen with a light microscope and can be visualized only with an electron microscope.

The first section of this chapter describes the general characteristics of viruses. The primary focus of the remaining

FOCUS YOUR PERSPECTIVE 13.1

Microbe Mimicker

Like bacteria, viruses vary tremendously in size and complexity. This is becoming increasingly apparent as viruses from a variety of animal and bacterial hosts are studied in greater detail. One of the most unusual viruses recently characterized came from *Acanthamoeba polyphaga,* a free-living amoeba. The virus was first thought to be a Gram-positive bacterium because of its staining characteristics and hairy appearance, with many fibrils sticking out from its capsid (see figure 13.1). Its name, mimivirus, reflects this original misconception (**mi**micking **mi**crobe). Mimivirus has a diameter of approximately 800 nm. This is twice the size of a small bacterium, and these virus particles can be observed using a light microscope. Its genome size is enormous for a virus—1.2 million base pairs, which is large enough to encode almost 1,000 proteins. Many of the encoded proteins have never been found in viruses before, and include enzymes of nucleic acid synthesis, DNA repair, translation, and polysaccharide biosynthesis. However, the mimi virion does not contain ribosomes and it cannot replicate without a host cell. This virus has been placed in a taxonomic group of other large DNA viruses (the *Megaviridae*), members of which infect distinctly different organisms, including vertebrates and algae.

After the discovery of mimiviruses, a mimivirus-infected amoeba was found infected with an additional virus, a parasite of the mimivirus. Its discoverers named the new virus Sputnik and called it a virophage. They named the infected virus a mamavirus (**box figure 13.1**). Sputnik can replicate in the amoeba only when mimivirus has infected the same cell. When this happens, fewer mimivirus particles are produced and they display unusual morphologies. Thus, Sputnik behaves as a true parasite. Interestingly, Sputnik appears to contain genes from other viruses, raising an intriguing question: Is horizontal gene transfer between viruses a possibility? ◀◀ horizontal gene transfer

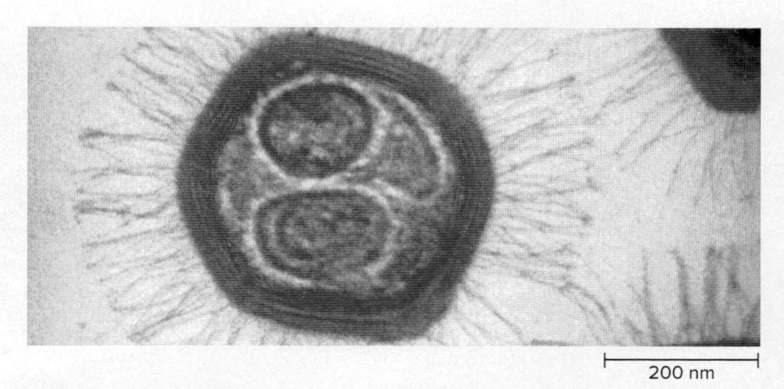

BOX FIGURE 13.1 Mamavirus (Large Mimivirus) with Sputnik Virophages ©Dr. Raoult/Science Source

The mimivirus and virophages are not the last unusual viruses discovered. In 2013, two new amoeba-infecting megaviruses were discovered and named Pandoraviruses. These viruses have genomes approximately twice the size of mamaviruses. Other virus families likely remain undiscovered. Their discovery and characterization will not only expand our knowledge of viruses but also challenge our definition of what is living and non-living—and perhaps provide an answer to the question, "Where did viruses originate?"

sections is on bacteriophages and animal viruses. Animal viruses are obviously important because of the diseases they cause, but why learn about bacteriophages? We study them because they are easy to cultivate in the laboratory, and they serve as an important model to understand the molecular biology of animal viruses and their relationships with their hosts. What you learn about them will help you understand similar relationships between the medically important viruses and the cells they infect. Bacteriophages are also important because they serve as a vehicle for horizontal gene transfer in bacteria, which was described in chapter 8. In addition, they are important because they kill bacteria, thereby reducing bacterial populations in nature. This is significant ecologically, but also has medical applications. In recent years, the Food and Drug Administration (FDA) has approved the use of species-specific preparations of bacteriophages for control of bacteria during food production and storage. These preparations have been used to prevent the growth of food-contaminating pathogens such as *Listeria monocytogenes, Salmonella enterica,* and *E. coli* O157:H7 on various food products. The use of bacteriophages as an alternative to antibiotics in treating bacterial infections has also been explored. ◀◀ horizontal gene transfer

MicroByte

There are an estimated 10^{31} bacteriophages on Earth; they are the most numerous of all biological entities.

13.1 ■ General Characteristics of Viruses

Learning Outcomes

1. Describe the general features of viral architecture.
2. Describe how viruses are classified and named.

Most viruses are extremely small (**figure 13.1**). In fact, they are approximately 100- to 1,000-fold smaller than the cells they infect. The smallest viruses are about 10 nm in diameter, and contain very little nucleic acid, perhaps as few as 10 genes. The largest known viruses are about 800 nm, the size of the smallest bacterial cells. Indeed, one very large virus is so big it was first identified as a bacterium (see **Focus Your Perspective 13.1**).

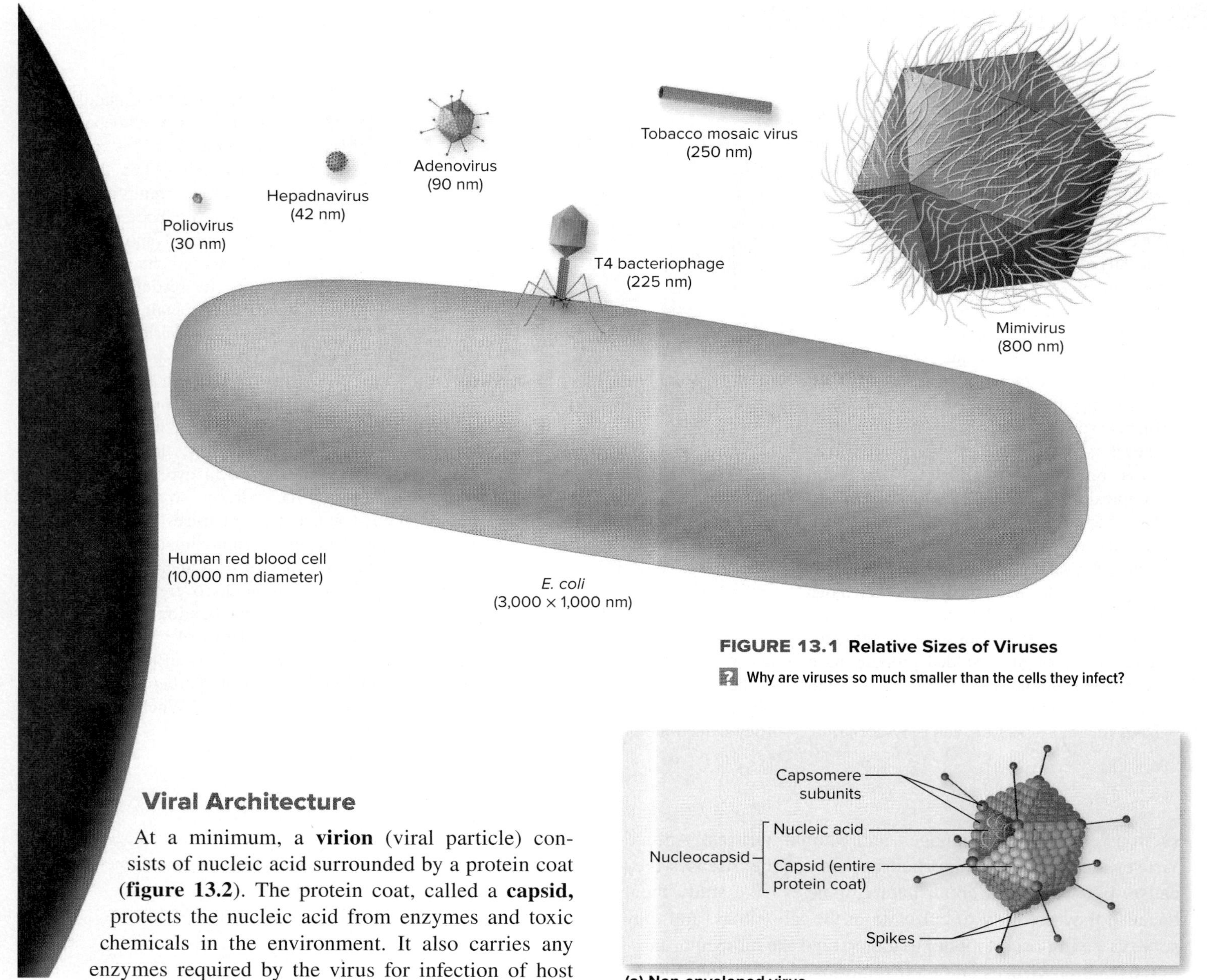

FIGURE 13.1 Relative Sizes of Viruses

? Why are viruses so much smaller than the cells they infect?

Viral Architecture

At a minimum, a **virion** (viral particle) consists of nucleic acid surrounded by a protein coat (**figure 13.2**). The protein coat, called a **capsid,** protects the nucleic acid from enzymes and toxic chemicals in the environment. It also carries any enzymes required by the virus for infection of host cells. The capsid together with the nucleic acid it encloses is called the **nucleocapsid.** Given the fact that viral genomes are relatively small in size and therefore can encode only a limited number of different proteins, it is not surprising that capsids are simple in chemical structure. They are composed of identical protein subunits, called capsomeres, arranged in a precise manner to form the capsid.

Some viruses have an envelope—a lipid bilayer outside the capsid (see figure 13.2b). These **enveloped viruses** obtain that bilayer from the host cell. Sandwiched between the nucleocapsid and the envelope is the matrix protein, which is unique to enveloped viruses. Viruses that do not have an envelope are called **non-enveloped** or **naked viruses.** Nearly all phages are non-enveloped. In general, enveloped viruses are more susceptible to disinfectants because these chemicals damage the envelope, making the viruses non-infectious.

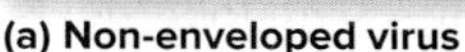

(a) Non-enveloped virus

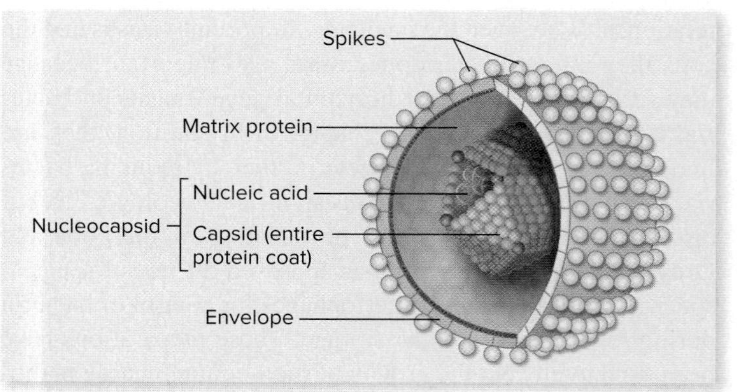

(b) Enveloped virus

FIGURE 13.2 Two Different Types of Viral Architecture

? Why would a disinfectant that destroys the envelope (thereby removing the spikes from the particle) make the remaining particle non-infectious?

Viruses contain only a single type of nucleic acid—either RNA or DNA—but never both. This provides a useful method for classifying viruses, which are frequently referred to as either RNA or DNA viruses. The genome may be linear or circular, either double-stranded or single-stranded.

Viruses have specific protein components that allow the virion to attach to specific receptor sites on host cells. Phages, for example, have tail fibers that attach to host cells, and many animal viruses have protein structures called **spikes** that stick out from either the lipid bilayer of enveloped viruses or the capsid of non-enveloped viruses (see figures 13.1 and 13.2).

A virus generally is one of three different shapes: icosahedral, helical, or complex (**figure 13.3**). Icosahedral viruses appear spherical when viewed with the electron microscope, but their surface is actually 20 flat triangles arranged in a manner somewhat similar to that of a soccer ball. Helical viruses appear cylindrical when viewed with the electron microscope. Their capsomeres are arranged in a helix, somewhat similar to a spiral staircase. Some helical viruses are short and rigid, whereas others are long and filamentous. Complex viruses have more complicated structures. Phages are the most common examples of this, many

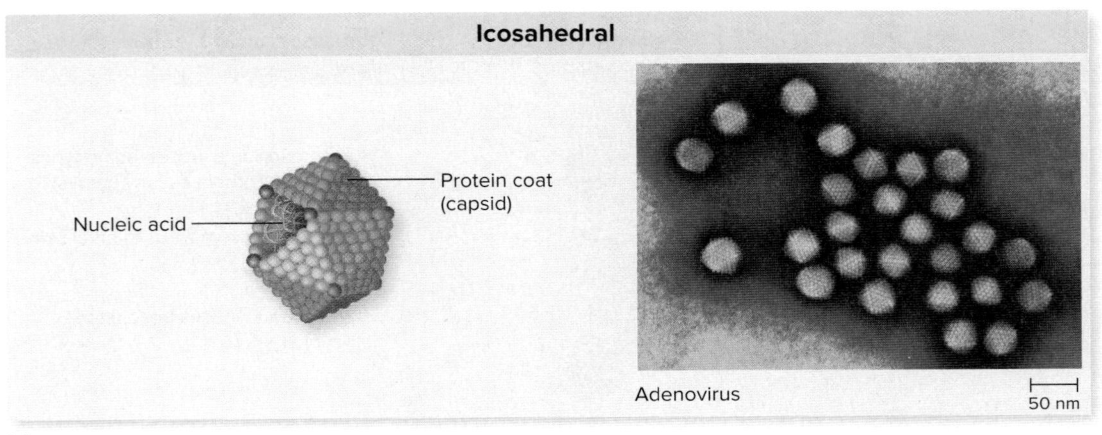

(a)

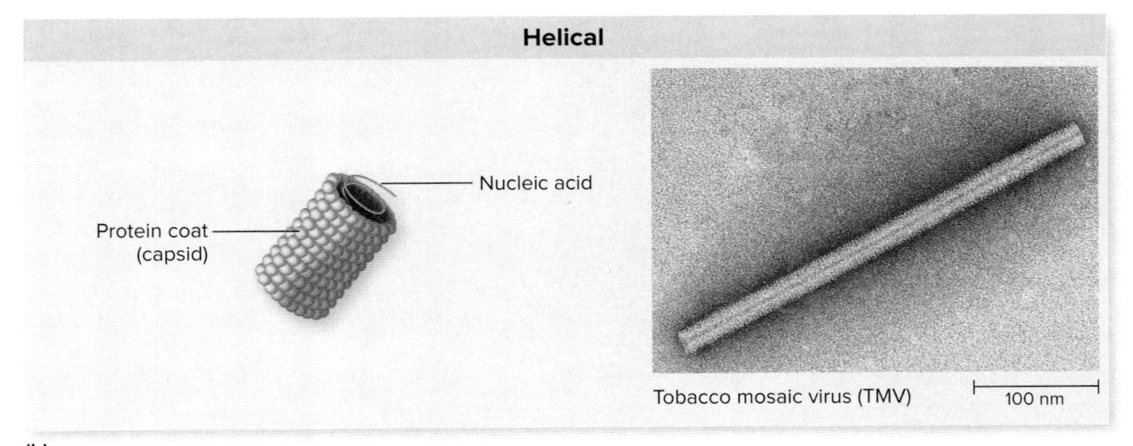

(b)

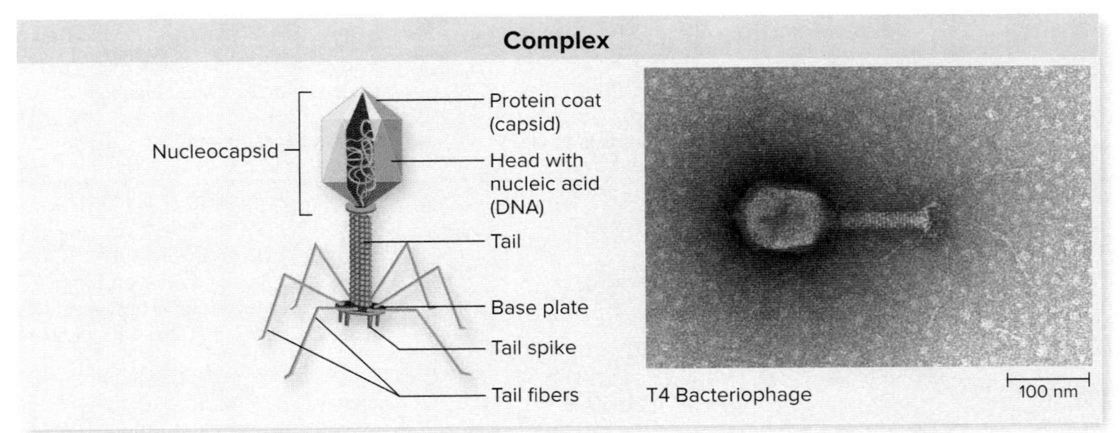

(c)

FIGURE 13.3 Common Shapes of Viruses a: ©Eye of Science/Science Source; b: ©Science Source; c: ©Biophoto Associates/ Science Source

What determines the shape of the virus?

having an icosahedral nucleocapsid, referred to as the head, with a long helical protein component, the tail.

Viral Taxonomy

Although viruses are not living organisms, they are biological entities that are classified to provide easy identification and study (**table 13.1**). The International Committee on Taxonomy of Viruses (ICTV) keeps an online database and publishes a report describing the key features, classification, and nomenclature of recognized viruses. The ICTV 2015 report describes over 6,000 viruses belonging to 3,704 species, 609 genera, 27 subfamilies, 111 families, and 7 orders. Key characteristics used in the current classification scheme are the genome structure (type of nucleic acid and strandedness) and the hosts they infect (bacteria, archaea, animals, plants). A variety of other characteristics including viral shape and disease symptoms are also considered.

TABLE 13.1 Classification of Some Human Viruses

Nucleic Acid	Outer Covering	Family	Drawing of Virion (Not to Scale)	Representative Member (and Disease Caused)
DNA Viruses				
Double-stranded DNA	Non-enveloped	*Papillomaviridae*		Human papillomaviruses (some types cause warts; others cause cancers)
		Polyomaviridae		Merkel cell polyomavirus (Merkel skin cancer)
		Adenoviridae		Human adenoviruses (respiratory infections)
	Enveloped	*Herpesviridae*		Herpes zoster virus (chickenpox); herpes simplex viruses (cold sores, genital herpes)
		Poxviridae		Smallpox virus (smallpox)
Single-stranded DNA	Non-enveloped	*Parvoviridae*		Human parvovirus B19 (fifth disease)
RNA Viruses				
Double-stranded	Non-enveloped	*Reoviridae*		Human rotaviruses (diarrheal disease)
Single-stranded (plus strand)	Non-enveloped	*Picornaviridae*		Polioviruses (poliomyelitis); rhinovirus (colds); hepatitis A virus (hepatitis A)
		Caliciviridae		Norovirus (gastroenteritis)
	Enveloped	*Togaviridae*		Rubella virus (rubella); Chikungunya virus (Chikungunya fever)
	Enveloped	*Flaviviridae*		Yellow fever virus (yellow fever); dengue virus (dengue fever); hepatitis C virus (hepatitis C); Zika virus (Zika virus disease)
		Coronaviridae		Severe acute respiratory syndrome coronavirus (SARS), Middle East respiratory syndrome coronavirus (MERS)

TABLE 13.1 Classification of Some Human Viruses (*Continued*)

Nucleic Acid	Outer Covering	Family	Drawing of Virion (Not to Scale)	Representative Member (and Disease Caused)
Single-stranded (minus strand)	Enveloped	*Rhabdoviridae*		Rabies virus (rabies)
		Filoviridae		Ebola virus (Ebola virus disease)
		Paramyxoviridae		Mumps virus (mumps); measles virus (measles)
		Orthomyxoviridae		Influenza virus (influenza)
		Bunyaviridae		Hantavirus (hantavirus pulmonary syndrome)
		Arenaviridae		Lassa virus (Lassa fever)
Reverse Transcribing Viruses				
DNA	Enveloped	*Hepadnaviridae*		Hepatitis B virus (hepatitis B)
RNA	Enveloped	*Retroviridae*		Human immunodeficiency virus (AIDS)

Our discussion of classification will focus on the viruses that infect animals. The names of virus families are derived from a variety of sources, but they all end in the suffix *-viridae* and are italicized. Their names follow no consistent pattern. In some cases, the name indicates the appearance of the viruses in the family—for example, *Coronaviridae* coming from *corona,* which means "crown." In other cases, the virus family is named for the geographic area from which a member was first isolated. *Bunyaviridae* is derived from Bunyamwera, in Uganda, Africa. Each family contains numerous genera whose names end in *-virus,* making it a single word—for example, *Enterovirus.* The species name is often the name of the disease the virus causes—for example, *poliovirus* causes poliomyelitis. Formal scientific names are italicized, making them easy to recognize.

In contrast to bacterial nomenclature, in which an organism is referred to by its genus and species name, viruses are commonly referred to only by their species name or by an informal name, neither of which is capitalized—for example, rhabdovirus. In addition, informal terms are often used to refer to groups of animal viruses that are not taxonomically related but share critical characteristics such as the primary route of transmission (**table 13.2**). Viruses transmitted via the fecal-oral route, for instance, are referred to as enteric viruses (enteric means relating to the intestines), and viruses transmitted through the respiratory route are called respiratory viruses. Zoonotic viruses cause zoonoses, which are diseases transmitted from an animal to a human. One group of viruses, the **arboviruses** (meaning **ar**thropod **bo**rne), is so named because they are spread by arthropods such as mosquitoes, ticks, and sandflies. The arthropods are biological vectors; when an infected arthropod takes a blood meal from an animal, it transmits the virus. In many cases, these viruses can infect widely different species. The same arthropod might bite birds, reptiles, and mammals and transfer viruses among those different groups. Arboviruses cause important diseases such as West Nile encephalitis, La Crosse encephalitis, yellow fever, and dengue fever.

TABLE 13.2 Grouping of Human Viruses Based on Route of Transmission

Virus Group	Mechanism of Transmission	Common Viruses Transmitted
Enteric	Fecal-oral route	Enteroviruses (polio); noroviruses; rotaviruses (diarrhea)
Respiratory	Respiratory or salivary route	Influenza; measles; rhinoviruses (colds)
Sexually transmitted	Sexual contact	Herpes simplex virus type 2 (genital herpes); HIV
Zoonotic	Vector (such as arthropods)	West Nile encephalitis; Zika virus disease, dengue fever
	Animal to human directly	Rabies; cowpox

MicroAssessment 13.1

Viruses consist of nucleic acid surrounded by a protein coat and replicate only inside living cells. Viruses are typically icosahedral, helical, or complex in shape. Almost all phages are non-enveloped (naked), whereas animal viruses are either non-enveloped or enveloped. Viruses are classified based primarily on the characteristics of their genome, such as type of nucleic acid and strandedness. Viruses are often grouped by their route of transmission.

1. Where does an enveloped virus get its envelope?
2. List three ways in which viruses can be transmitted from one organism to another.
3. Most enteric viruses are non-enveloped. Why would this be so?

13.2 ■ Bacteriophages

Learning Outcomes

3. Compare and contrast lytic, temperate, and filamentous phage infections.
4. Describe two consequences of lysogeny.

An enormous variety of different phages exist, each with a characteristic shape, size, genome structure, and replication strategy. The general strategies for phage replication are similar in concept to those of viruses that infect animal cells, and they result in two possible outcomes: (1) a **productive infection,** in which new viral particles are produced, and (2) a **latent state,** in which the viral genome remains silent within the cell but is replicated along with the host cell genome (**figure 13.4**). For a productive infection, some types of viruses kill their host but others do not.

Lytic Phage Infections: T4 Phage as a Model

By definition, **lytic** or **virulent phages** exit the host at the end of the infection cycle by lysing the cell. Because these viral infections result in the formation of new viral particles, they are productive infections. A well-studied lytic phage is T4, a double-stranded DNA phage. During its infection cycle, the phage takes over ("hijacks") a bacterial cell, directing that cell to synthesize new phage particles. The infection cycle of T4, which is similar to that of other lytic phages, can be viewed as a five-step process: attachment, genome entry, synthesis, assembly, and release (**figure 13.5**). The entire process from entry of the phage nucleic acid to the exit of the phage takes about 30 minutes.

Attachment ①

Phage particles collide with their host cells by chance. On contact, a phage attaches to a receptor on the host cell surface or to an appendage such as a pilus. In the case of T4, the receptor is on the bacterial cell wall. The receptors used by phages normally perform important functions for the cell—the phages merely exploit the molecules for their own use. Cells that lack the receptor used by a particular phage are resistant (not susceptible) to infection by that specific phage.

Genome Entry ②

Following attachment, a bacteriophage injects its genome into the cell. T4 does this by degrading a small part of the bacterial cell wall, using an enzyme located in the tip of its tail. This enzyme, T4 lysozyme, is functionally similar to the lysozyme found in tears and eggs, and degrades peptidoglycan. The tail contracts so that the phage particle appears to "squat" (sit) on the surface of the cell. This action injects the phage DNA through the host's cell wall and membrane, and into the cell. The capsid remains on the outside of the cell. The separation of the nucleic acid from its protein coat before replication is a feature of all viruses. ◀◀ lysozyme

Synthesis of Phage Proteins and Genome ③

Within minutes after the T4 DNA is injected into a host cell, some of the genes on that viral genome are transcribed and translated by the infected cell. The phage-encoded proteins are made in a specific time sequence to control the course of infection. Early on, the phage directs the bacterial cell to synthesize proteins (early proteins) needed to initiate phage replication. For example, a nuclease that degrades the host cell's DNA is made. The phage genome is then synthesized. Toward the end of the infection cycle, the phage directs the bacterial cell to synthesize late proteins. These are structural proteins that make up the phage, including those that make up the capsid and the tail. ◀◀ promoters

Focus Figure

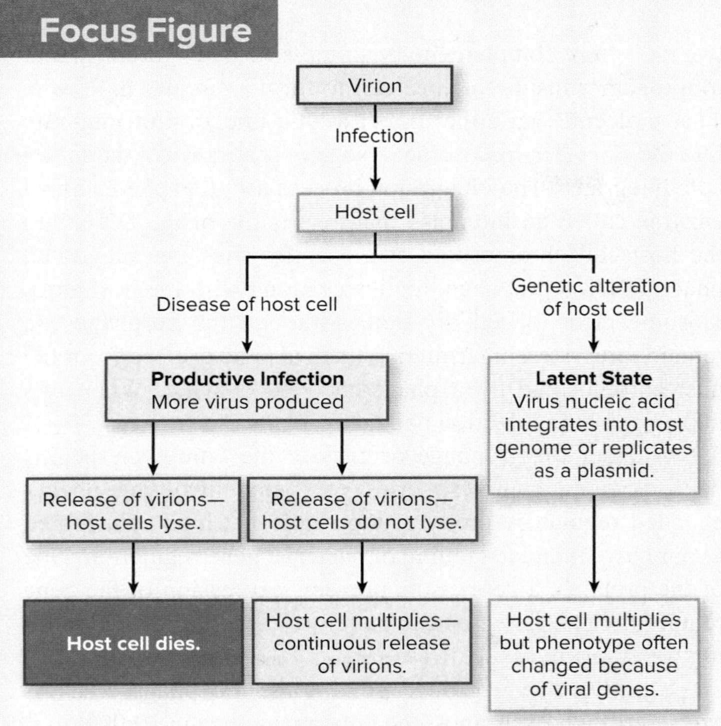

FIGURE 13.4 Major Types of Relationships Between Viruses and the Cells They Infect

? Which type of interaction is least harmful to the host?

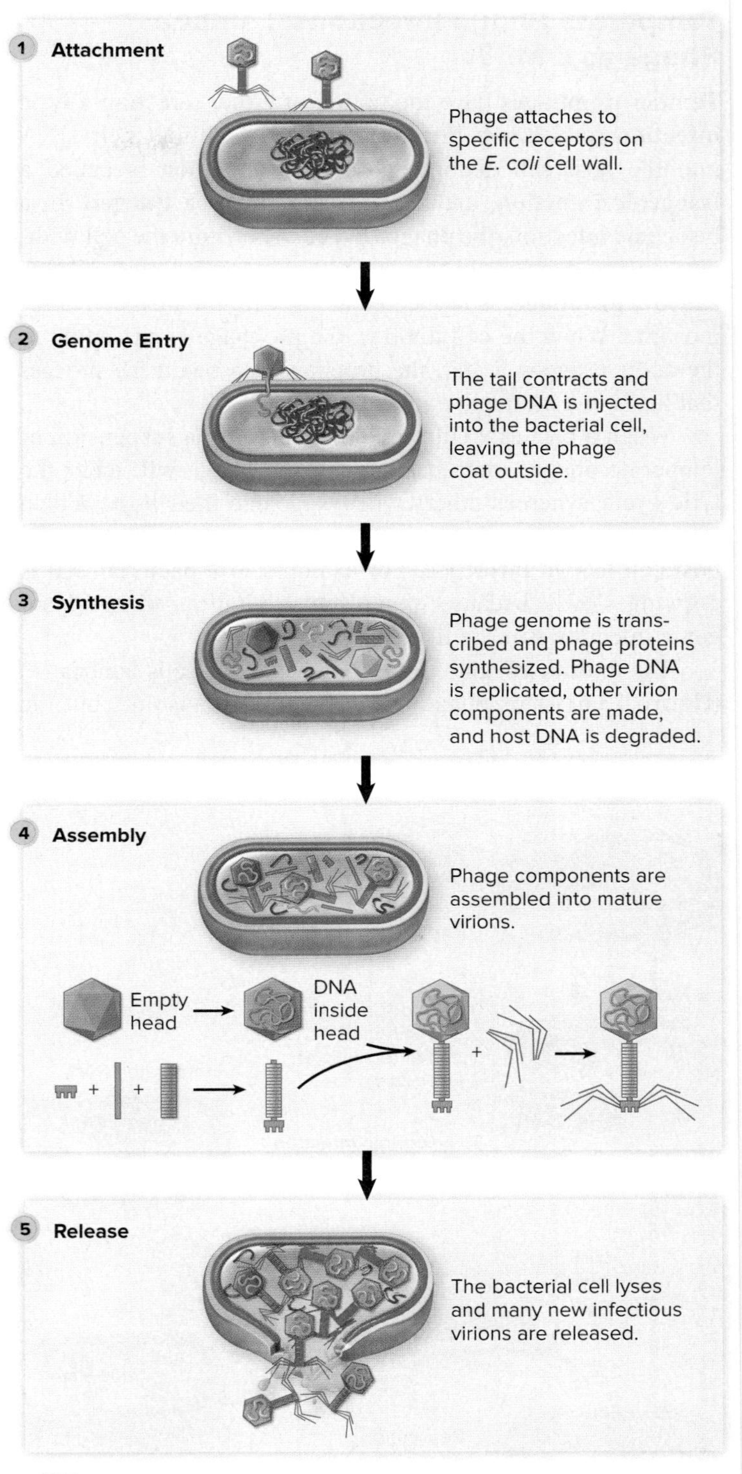

FIGURE 13.5 Steps in the Replication of the Lytic Phage T4 in *E. coli*

? If one phage particle infects one bacterial cell, how many particles are inside the cell 5 minutes after infection? Explain.

Assembly (Maturation) ④

Once multiple copies of the phage genome and the various structural components of the phage are made, they assemble to produce new phage particles. This is a complex, multistep process. Once the phage head is formed, DNA is packed into it; the tail is then attached, followed by the addition of the tail spikes. Some of these components self-assemble, meaning that the proteins join together spontaneously to form a specific structure. In other steps, certain phage proteins serve as scaffolds on which various protein components associate. The scaffolds themselves do not become a part of the final structure, much as scaffolding required to build a house does not become part of the structure.

MicroByte

A phage's nucleic acid is so tightly packed inside the capsid that the internal pressure is 10 times higher than the pressure in a champagne bottle!

Release ⑤

Late in infection, the phage-encoded enzyme lysozyme is produced. This enzyme digests the host cell wall from within, causing the cell to lyse, thereby releasing phage. In the case of the T4 phage, the **burst size**—the number of phage particles released—is about 200. These phage particles then infect any susceptible cells in the environment, and the process of phage replication is repeated.

Temperate Phage Infections: Lambda Phage as a Model

Temperate phages have the option of either directing a **lytic infection** (productive infection) or incorporating their DNA into the host cell genome. The latter situation is called a **lysogenic infection,** and the infected cell is a **lysogen.** In a lysogenic infection, the phage DNA exists within the cell without causing damage. In this state, the integrated phage DNA, called a **prophage,** is replicated along with the host cell chromosome. When the cell divides, the prophage is passed on to the cell's progeny. Later, the prophage can begin the process that leads to a productive infection.

When a bacterial culture is infected with a suspension of temperate phage, some particles of the phage will enter the lytic cycle, whereas others will lysogenize their host. Which cycle occurs is largely random, but the metabolic state of the host cell has an influence. For example, if a bacterial cell is growing slowly because of nutrient limitation, then a lysogenic infection is more likely to occur.

The most thoroughly studied temperate phage is lambda (λ) (**figure 13.6**). This phage has a linear chromosome, but the two ends have complementary single-stranded overhangs that join together inside the host cell to form a circular molecule. That molecule can either direct a lytic infection or integrate into the *E. coli* chromosome; lysogeny occurs when the molecule integrates. The integration process uses a phage-encoded enzyme called an integrase that inserts the phage DNA into the host cell chromosome at a specific site. The integrated phage DNA, a prophage, replicates along with the host chromosome prior to cell division. Although the prophage can remain integrated indefinitely, it can also be excised from the host chromosome by a phage-encoded enzyme. When this happens, a lytic infection begins.

Whether the prophage persists or the lytic cycle begins depends on a complex series of events involving phage-encoded regulatory proteins, the study of which contributed to our current understanding of bacterial gene regulation. One of the proteins, a **repressor,** prevents expression of the gene required for excision, and is therefore essential for maintaining the lysogenic state. ⏮ bacterial gene regulation, ⏮ repressor

Under ordinary growth conditions, the phage DNA is excised from the chromosome only about once in 10,000 divisions of the lysogen. If, however, a lysogenic culture is treated with a DNA-damaging agent such as ultraviolet light, the SOS repair system comes into play. This system activates a protease that destroys the repressor protein responsible for maintaining the integration of the prophage. As a consequence, the prophage is excised from the chromosome—allowing the phage to enter the lytic cycle—and a productive infection results. This process, **phage induction,** lets the phage escape from a damaged host. ⏮ ultraviolet light, ⏮ SOS repair, ⏮ protease

Consequences of Lysogeny

Although a lysogen is morphologically identical to an uninfected cell, lysogeny has consequences. These include immunity to superinfection (infection by the same type of phage) and lysogenic conversion.

Lysogens are protected against infection by the same phage because the repressor that maintains the prophage in the integrated state will also bind to the operator on incoming phage DNA.

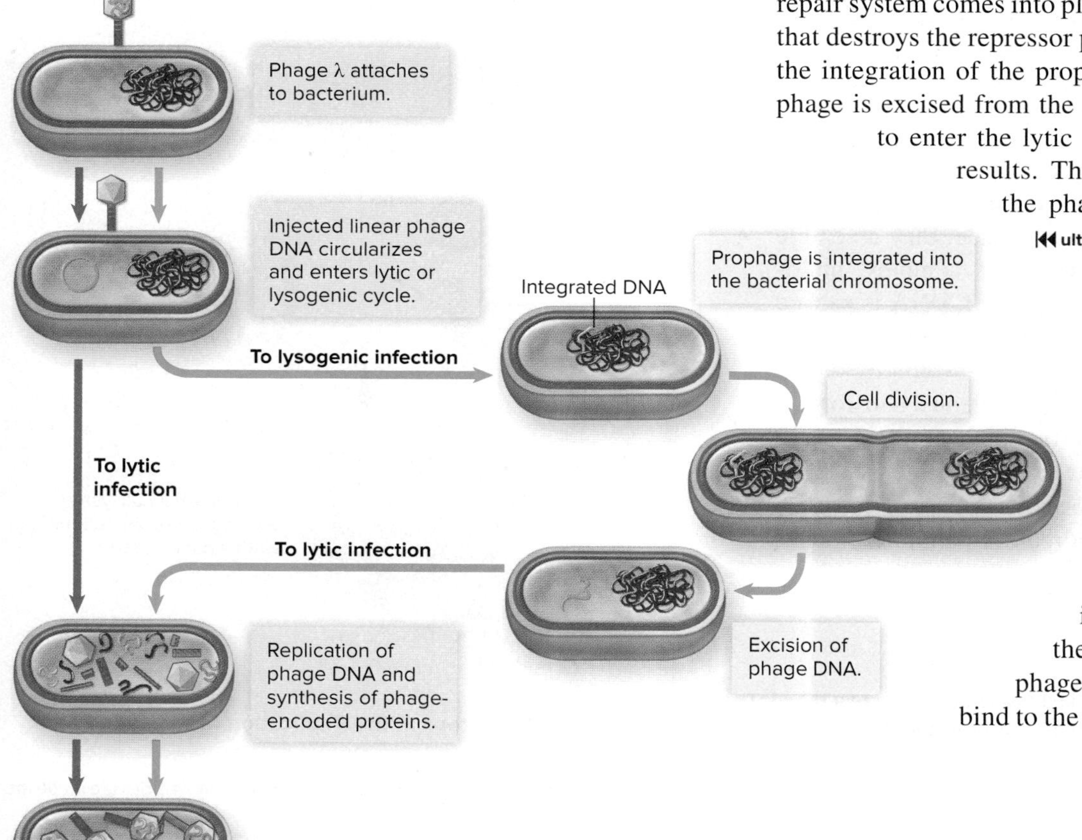

FIGURE 13.6 Steps in the Replication of the Temperate Phage λ in *E. coli*

When is it advantageous for a temperate phage to excise its DNA from the host chromosome?

TABLE 13.3 Some Properties Encoded by Prophage (Lysogenic Conversion)

Microorganism	Medical Importance	Property Encoded by Phage
Clostridium botulinum	Causes botulism	Botulinum toxin
Corynebacterium diphtheriae	Causes diphtheria	Diphtheria toxin
Escherichia coli O157:H7	Causes hemolytic uremic syndrome	Shiga toxin
Salmonella enterica	Causes food poisoning	Modification of lipopolysaccharide of outer membrane
Streptococcus pyogenes	Causes scarlet fever	Streptococcal pyrogenic exotoxins (SPEs)
Vibrio cholerae	Causes cholera	Cholera toxin

The operator is a regulatory region that controls the expression of the genes that direct a lytic infection. Thus, if another λ phage injects its DNA into a λ lysogen, that DNA will not be expressed. This phenomenon is called immunity to superinfection. Note that this type of immunity is different from that of the human immune system, which is discussed in chapters 14 and 15. ⏮ operator

Lysogenic conversion is a change in the phenotype of a lysogen as a consequence of the specific prophage it carries. For example, only strains of *Corynebacterium diphtheriae* that are lysogenic for a certain phage synthesize the toxin that causes diphtheria. If a toxin is encoded exclusively by phage genes, then only bacterial strains that carry the prophage will synthesize the toxin. Some examples of lysogenic conversion are given in **table 13.3.** ⏭ diphtheria

Filamentous Phage Infections: M13 Phage as a Model

Filamentous phages are single-stranded DNA phages that look like long fibers. They cause productive infections, but the process does not kill the host cells. Infected cells, however, grow more slowly than uninfected cells.

M13 is a filamentous phage that initiates infection by attaching to a protein on the F pilus of *E. coli* (**figure 13.7**). Its single-stranded DNA genome then enters the cytoplasm of the bacterial cell, where a host cell DNA polymerase synthesizes the complementary strand. This double-stranded DNA is referred to as the replicative form (RF). One strand of the RF is then used as a template to make mRNA as well as multiple copies of the phage's single-stranded genome (**figure 13.8**).

M13 particles are assembled during a process called extrusion, which the virus uses to exit the cell. In this process, phage coat protein molecules are inserted into the host cell's cytoplasmic membrane. At the same time, other phage-encoded proteins form pores that span the cytoplasmic and outer membrane. Then, as phage DNA is secreted through the pores, the coat protein molecules cover the single-stranded DNA to form the nucleocapsids.

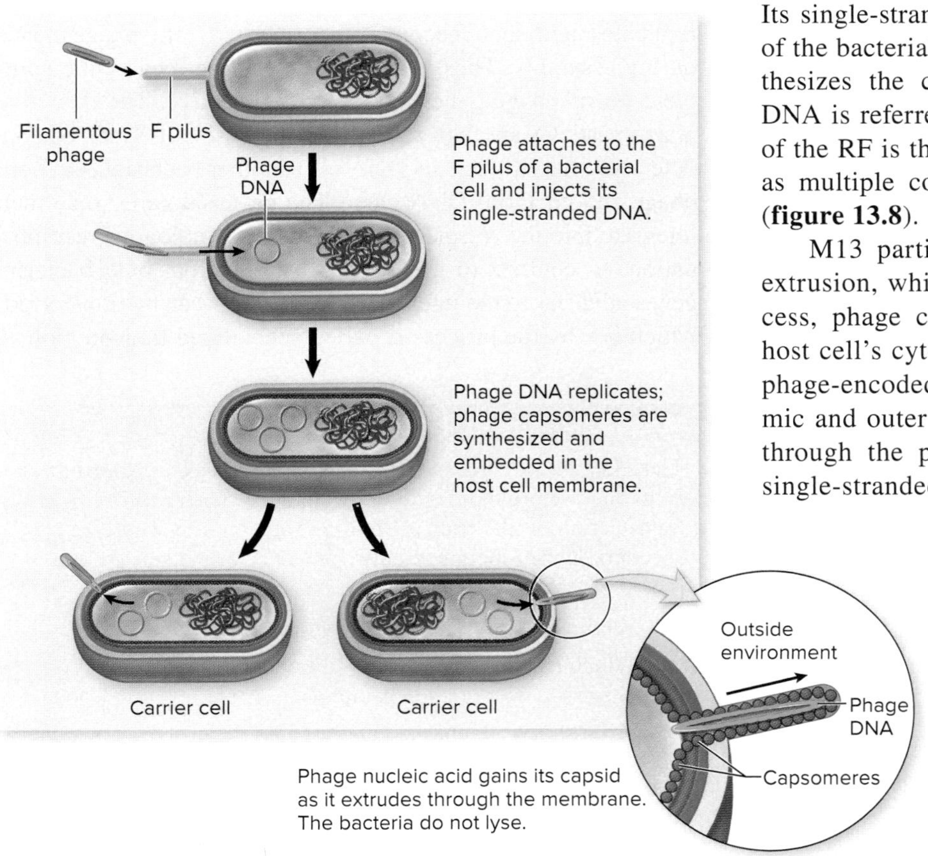

Filamentous phage
F pilus
Phage DNA
Phage attaches to the F pilus of a bacterial cell and injects its single-stranded DNA.
Phage DNA replicates; phage capsomeres are synthesized and embedded in the host cell membrane.
Carrier cell
Carrier cell
Outside environment
Phage DNA
Capsomeres
Phage nucleic acid gains its capsid as it extrudes through the membrane. The bacteria do not lyse.

FIGURE 13.7 Replication of a Filamentous Phage

? Why might infected cells grow more slowly than uninfected cells?

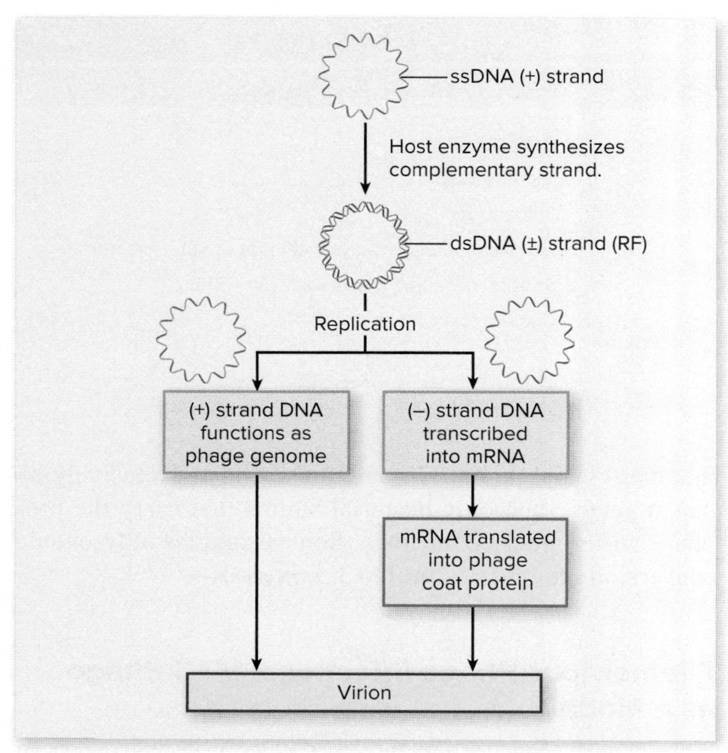

FIGURE 13.8 Macromolecule Synthesis in Filamentous Phage Replication

Which host enzyme synthesizes the complement to the ssDNA to make the RF?

MicroAssessment 13.2

Lytic phages lyse their host cells, whereas temperate phages either lyse their host or integrate their DNA into the host cell's genome. Prophage DNA often codes for gene products that confer new properties on the host. Filamentous single-stranded DNA phages are extruded from the host cell without killing the cell.

4. How can a productive phage infection not kill a host cell?
5. Give an example of a virulent phage and of a temperate phage.
6. Describe how the replication cycle of T4 phage is different from that of lambda.

13.3 ■ The Roles of Bacteriophages in Horizontal Gene Transfer

Learning Outcome

5. Compare and contrast generalized and specialized transduction.

Errors during the phage replication process play an important role in horizontal gene transfer. As briefly discussed in chapter 8, phages can transfer DNA from one bacterial cell (the donor) to another (the recipient) in the process called transduction. There are two types of transduction—generalized and specialized—and these are due to two different types of errors. ◂◂ horizontal gene transfer

Generalized Transduction

Generalized transduction results from a packaging error during phage assembly. Some phages degrade the bacterial chromosome into many fragments during lytic infection. Any of these short bacterial DNA fragments can be mistakenly packaged into the phage head during assembly (see figure 8.21). Phage heads that contain only bacterial genes instead of phage genes cannot direct a phage replication cycle. Because of this, they are called **transducing particles** rather than phage particles. ◂◂ generalized transduction

Following the release of the transducing particle from the phage-infected host, the particle binds to another bacterial cell and injects its DNA. That bacterial DNA may then integrate into the recipient cell by homologous recombination. Any gene of the donor cell can be transferred this way, which is why this mechanism is called generalized transduction. ◂◂ homologous recombination

Specialized Transduction

Specialized transduction results from an excision mistake made by a temperate phage during its transition from a lysogenic to a lytic cycle. Following induction, the phage DNA is usually excised precisely from the bacterial chromosome. On rare occasions, however, a short piece of bacterial DNA on each side of the phage DNA is taken, and a piece of phage DNA remains in the bacterial chromosome (**figure 13.9**). The excised DNA—containing both bacterial and phage genes—replicates and then becomes incorporated into phage heads during assembly. These phage particles do not carry the complete set of phage genes, so they are defective. The defective phage particles are then released as the host cells lyse. When a defective phage injects its DNA into another bacterial cell, both phage and bacterial DNA enter. The bacterial genes may then integrate into the recipient's genome via homologous recombination. In contrast to generalized transduction, only bacterial genes adjacent to the integrated phage DNA can be transferred, which is why the process is called specialized transduction.

MicroAssessment 13.3

Transduction is a process by which bacterial DNA is transferred from one bacterium to another by a phage. Generalized transduction results from a DNA packaging error, whereas specialized transduction results from an error in excision of a prophage.

7. With respect to transduction, what is a defective phage?
8. Which is more likely to be a specialized transducing phage—a lytic or a temperate phage?
9. Most temperate phages integrate into the host chromosome, whereas some replicate as plasmids. Which kind of relationship do you think would be more likely to maintain the phage in the host cell? Why?

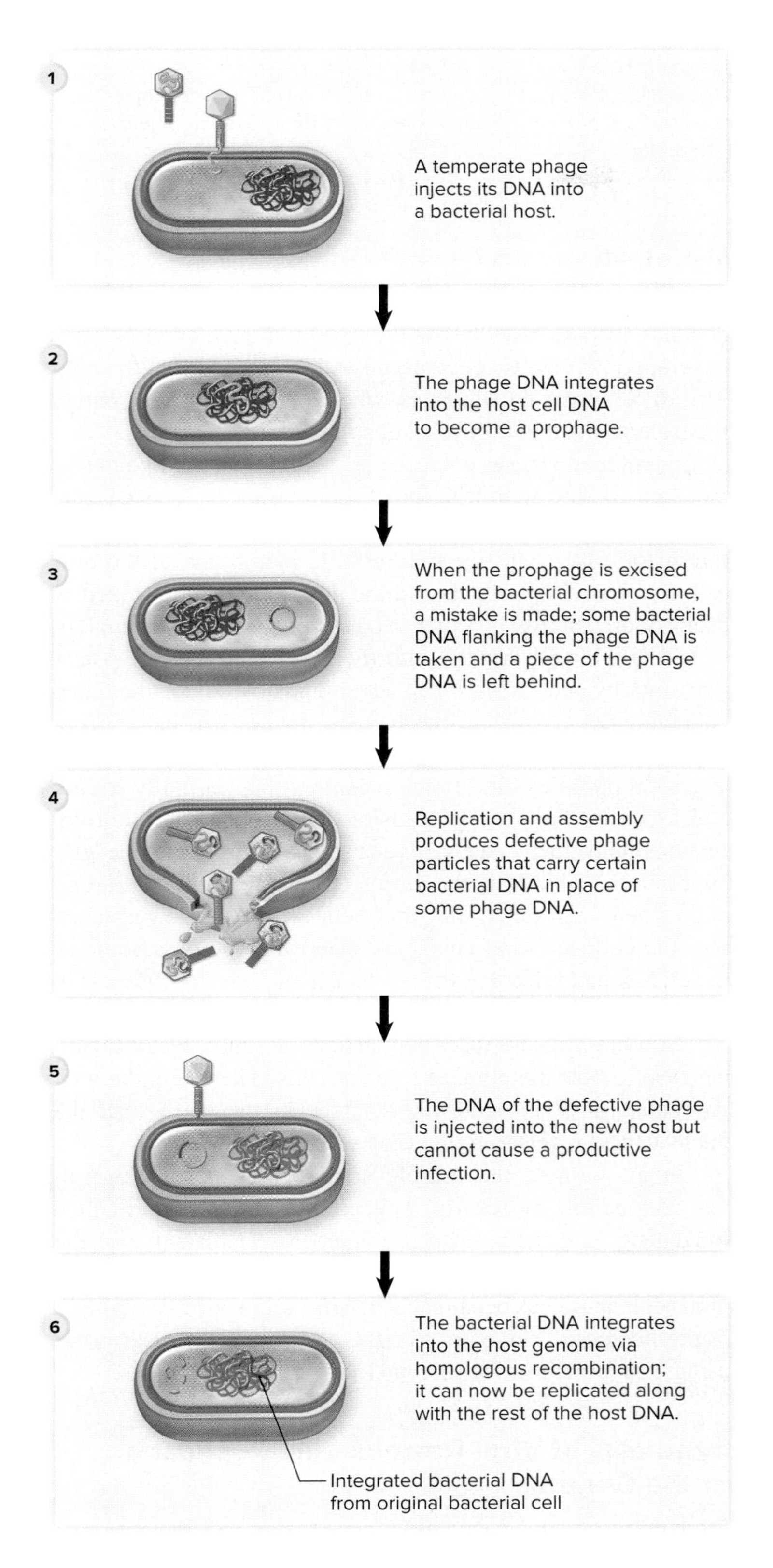

FIGURE 13.9 Specialized Transduction

? What mistake in the temperate phage replication cycle leads to specialized transduction?

13.4 ■ Methods Used to Study Bacteriophages

Learning Outcome

6. Describe how a plaque assay is used to detect and estimate the numbers of phage particles.

Viruses can multiply only inside living actively metabolizing cells, so studying viruses in the laboratory requires cultivating appropriate host cells. It is easier to grow bacterial cells than to grow animal cells, which is one reason that the study of bacteriophages advanced much faster than investigations on animal viruses.

Plaque assays are routinely used to determine the concentration of phage particles in samples such as sewage, seawater, and soil. In this type of assay, a melted, cooled soft agar medium (contains less solidifying agent) is inoculated with both a bacterial host and the phage-containing specimen. That mixture is then poured over the surface of a nutrient agar plate, creating a top layer over a base plate. The bacteria present in the top layer multiply rapidly, producing a dense lawn (turbid layer) of bacterial growth. Phage particles in the specimen infect susceptible bacteria and lyse them. The lysed bacteria release progeny phage that diffuse through the soft agar layer to neighboring bacteria, which they then infect. Lysis of these cells leads to the formation of **plaques,** which are circular zones of clearing in the bacterial lawn (**figure 13.10**). Each plaque represents a plaque-forming unit (PFU) initiated by a single phage particle infecting a cell.

In many cases, different dilutions of the phage suspension are used when doing a plaque assay. This is to ensure that one plate will have a statistically valid number of plaques to be accurately counted. The number is used to determine the **titer,** which is the concentration of infectious phage particles in the original phage suspension.

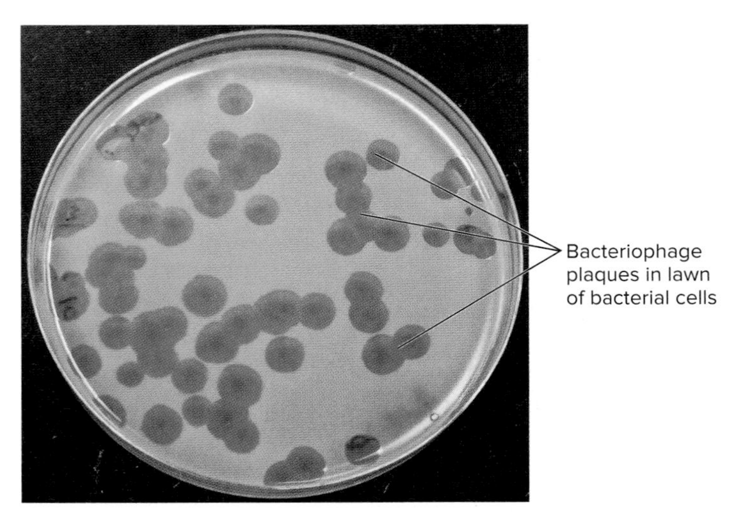

FIGURE 13.10 Phage Plaques ©McGraw-Hill Education/Lisa Burgess

? Why are plaques limited in size?

MicroAssessment 13.4

Bacteriophages require bacterial host cells. Plaque assays are used to quantitate phage particles.

10. What are plaque-forming units?

11. Is it important to have fewer phages than bacterial host cells when doing a quantitative plaque assay? Explain.

13.5 ■ Animal Virus Replication

Learning Outcomes

7. Describe the steps of a generalized infection cycle of animal viruses.

8. Compare and contrast the replication strategies of DNA viruses, RNA viruses, and reverse-transcribing viruses.

Understanding the infection cycle of animal viruses is particularly important from a medical standpoint because virus replication often depends on virally encoded enzymes, which are potential targets of antiviral drugs. By interfering with the activities of the enzymes required for viral replication, antiviral medications can slow the progression of a viral infection, often giving host defense systems enough time to eliminate the virus before symptoms appear.

As we discuss the infection cycles of animal viruses, it is helpful to recall the cycles described for bacteriophages, because they have features in common. A generalized infection cycle of animal viruses can be viewed as a five-step process: attachment, genome entry, synthesis, assembly, and release.

Attachment

The process of attachment (adsorption) is basically the same in all virus-cell interactions. Animal viruses have attachment proteins (spikes) on their surfaces (see figure 13.2). The receptors to which these proteins bind are usually glycoproteins on the host cell cytoplasmic membrane, and often more than one receptor is required for effective attachment. HIV, for example, must bind to two different molecules before it can enter the cell. As with the receptors used by phages, the normal function of these receptors is completely unrelated to their role in virus attachment. ⏮ glycoproteins, ⏭ HIV replication cycle

Because a virion must bind to specific receptors, a particular virus may be able to infect only a single or a limited number of cell types and tissues (called tissue tropism). Likewise, most viruses can infect only a single species (the viral host range). This explains the resistance that some animals have to certain diseases. For example, dogs do not contract measles from humans, nor do humans contract distemper from cats. Some viruses, such as the rabies virus, however, can infect unrelated animals such as horses and humans, with serious consequences in both. ⏭ measles, ⏭ rabies

MicroByte

A cell can have tens to hundreds of thousands of receptor copies on its surface, providing many opportunities for a virion to encounter a receptor.

Penetration and Uncoating

The mechanism an animal virus uses to enter its host cell depends in part on whether the virion is enveloped or non-enveloped. In all cases, the entire virion is taken into the cell. This differs from most phages, where only nucleic acid enters the bacterial cell and the capsid stays outside.

Enveloped viruses enter the host cell by one of two mechanisms: fusion with the host membrane or endocytosis (**figure 13.11**). In the case of fusion, the lipid envelope of the virion fuses with the cytoplasmic membrane of the host cell after the virion attaches to the host cell receptor, much as drops of oil can fuse together. As a result of fusion, the nucleocapsid is released directly into the cell's cytoplasm. Viruses that enter by endocytosis take advantage of receptor-mediated endocytosis, a normal mechanism by which cells bring certain extracellular material into the cell (see figure 3.49). The viral particles bind to the receptors that normally trigger and facilitate the process, causing the cell to take them up. For example, influenza viruses use their HA envelope glycoproteins to bind to the host cell membrane, facilitating endocytosis and entry into that cell. After a virion is taken into the cell, the viral envelope fuses with the membrane of the endosome, releasing the nucleocapsid into the cytoplasm.
⏮ receptor-mediated endocytosis

Non-enveloped viruses, which have no lipid envelope, cannot fuse to host membranes to enter cells. Therefore, the virions enter only via endocytosis. Once in the endosome, the nucleocapsid is released into the cytoplasm.

In all viruses, the nucleic acid separates from its protein coat before the start of replication. This process is called **uncoating,** and may occur simultaneously with entry of the virion into the cell, or after the final intracellular destination for viral replication has been reached. Most DNA viruses multiply in the nucleus. They enter the nucleus through nuclear pores using viral proteins that have nuclear localization signals.

Synthesis of Viral Proteins and Replication of the Genome

Production of viral particles in an infected cell requires two distinct but interrelated events: (1) expression of viral genes to produce structural and catalytic proteins, such as capsid proteins and any enzymes required for replication; and (2) synthesis of multiple copies of the viral genome. The viral proteins are sometimes synthesized as a polyprotein that is subsequently cleaved by viral proteases into individual proteins. The viral proteases that cleave the polyproteins can be

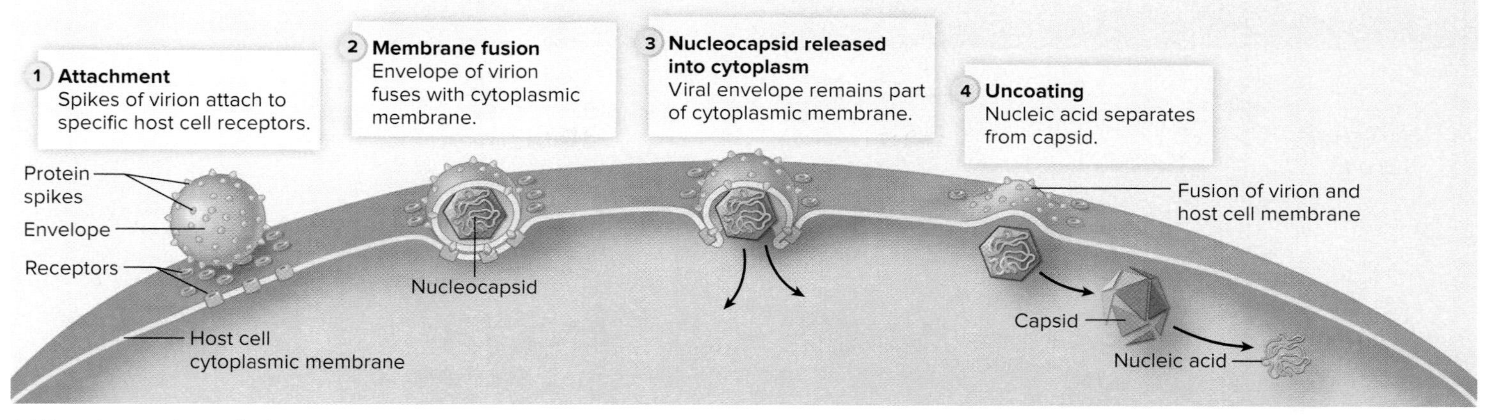

(a) Entry by membrane fusion

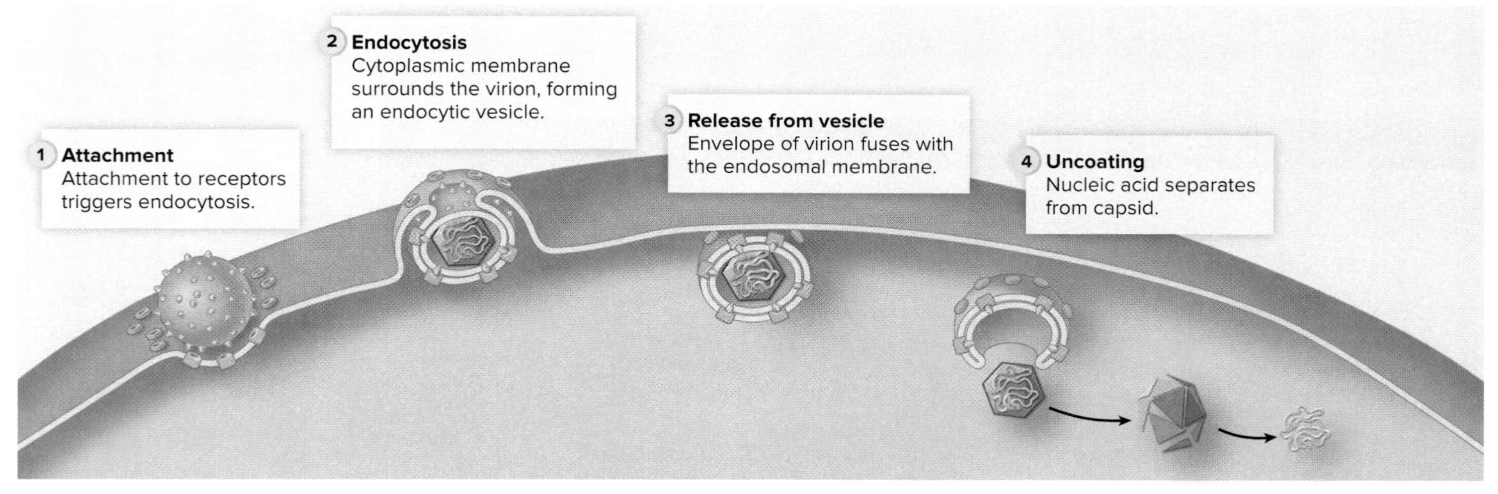

(b) Entry by endocytosis

FIGURE 13.11 Entry of Enveloped Animal Viruses into Host Cells **(a)** Entry following membrane fusion. **(b)** Entry by endocytosis.

Could phages enter bacteria by the process of membrane fusion? Explain.

useful targets for antiviral medications. For example, several medications used to treat HIV infection inhibit the HIV protease. Likewise, a new medication for treating hepatitis C inhibits the HCV protease.

The replication strategy of viruses can be divided into three general categories: those used by (1) DNA viruses, (2) RNA viruses, and (3) reverse transcribing viruses (**figure 13.12**). As you will see in the following discussion, the type of genome has a significant influence on the viral replication strategy, a primary reason that virus classification takes genome structure into account (see table 13.1).

Replication of DNA Viruses

Most DNA viruses replicate in the nucleus of the host cell and use host cell components (such as ribosomes and enzymes) for DNA synthesis and gene expression. These viruses often encode their own DNA polymerase, which allows them to replicate even if the host cell is not actively duplicating its own chromosome.

Replication of double-stranded DNA viruses is fairly simple because it follows the central dogma of molecular biology, described in chapter 7. Like cells, the genome of double-stranded DNA viruses has two complementary strands. Recall that complementary DNA strands can be referred to as (+) strands and (−) strands, designations that indicate which strand is used as the template for RNA synthesis (see figure 7.9). As shown in figure 13.12a (top panel), the genome of a double-stranded DNA virus can be referred to as (+/−) to indicate that it has both a (+) strand and a (−) strand. These (+/−) genomes can be transcribed to produce mRNA (+RNA). The mRNA is then translated to make proteins. In addition, the double-stranded DNA genome serves as a template for DNA replication. ◀◀ central dogma of molecular biology

Replication of single-stranded DNA viruses is quite similar to that of double-stranded DNA viruses, except that a complement to the single-stranded DNA molecule must be synthesized to generate a double-stranded (+/−) DNA molecule. Once that has occurred, the genes can be expressed to produce the

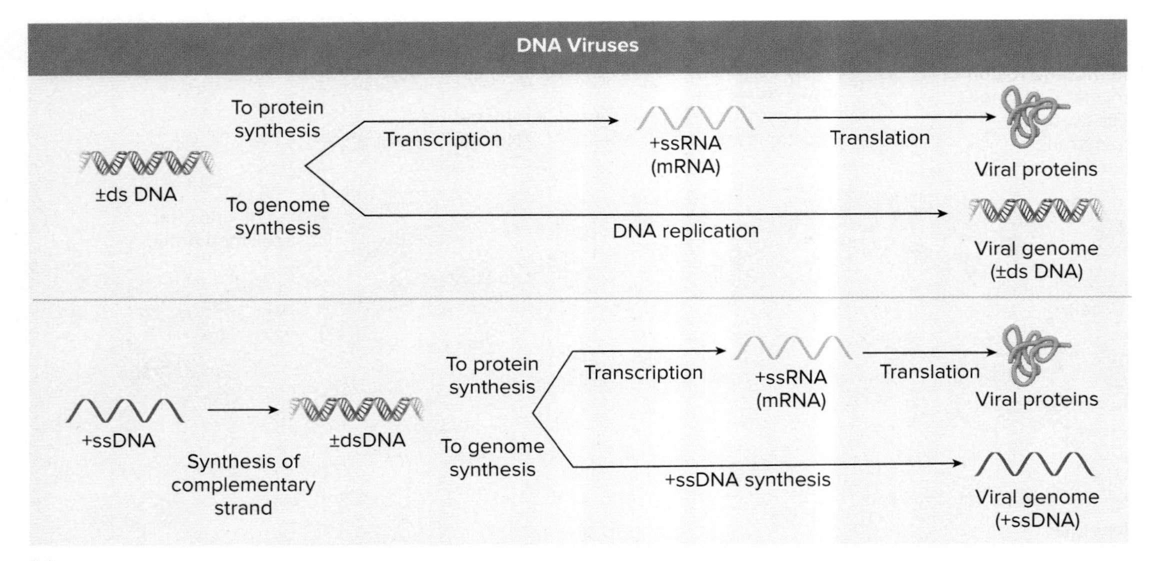

(a)

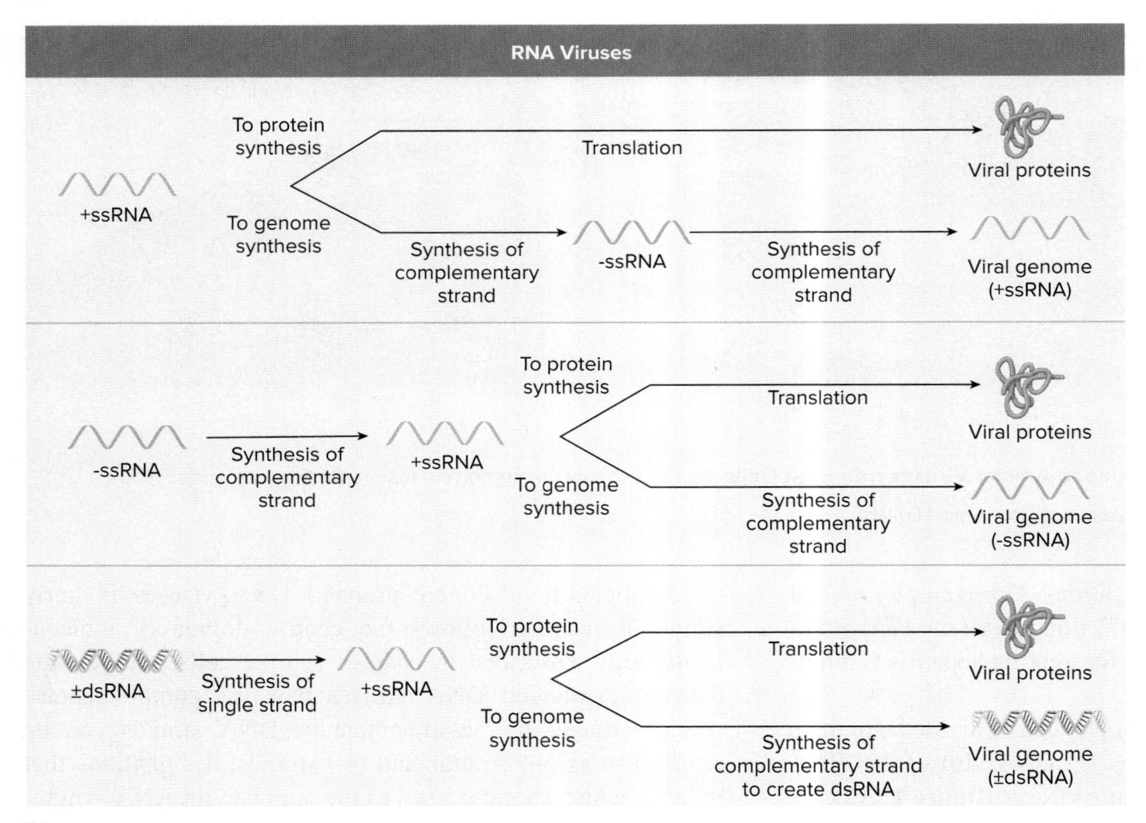

(b)

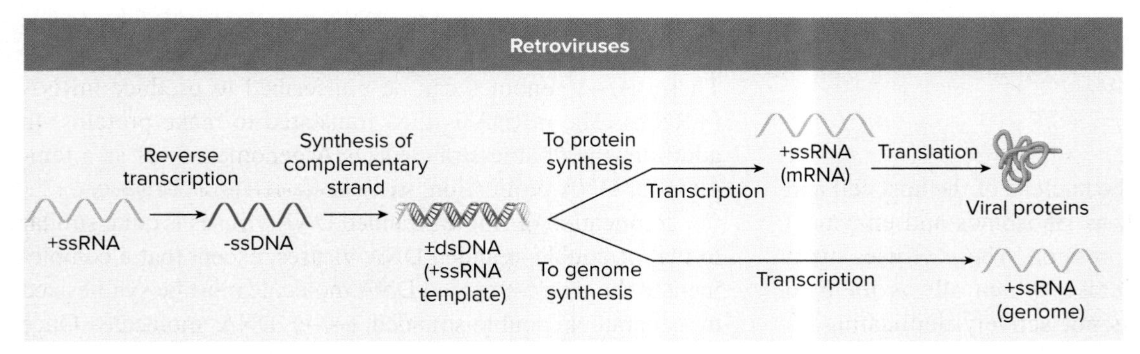

(c)

FIGURE 13.12 Viral Replication Strategies

? **Why do RNA viruses need to encode a polymerase in order to replicate their genome, while DNA viruses do not?**

encoded proteins (see figure 13.12a, bottom panel). Meanwhile, the newly synthesized DNA strand—which is complementary to the single-stranded DNA genome—also acts as a template for producing more single-stranded DNA genome copies.

Replication of RNA viruses

Most RNA viruses are single-stranded, and they replicate in the cytoplasm. Their replication always requires a virally encoded RNA polymerase, often called a **replicase.** The replicase is an RNA-dependent RNA polymerase, meaning that it uses an RNA template to synthesize a new strand of RNA. Compare this with the RNA polymerase involved in transcription—that enzyme synthesizes a strand of RNA from a DNA template and is thus a DNA-dependent RNA polymerase.

The replication strategy for RNA viruses depends on the type of viral genome. Of the single-stranded RNA viruses, some types have a (+) RNA strand genome, which also serves as mRNA; others have a (−) RNA strand genome, meaning that the genome is the complement to mRNA. Double-stranded RNA viruses have a genome that consists of both a (+) RNA strand and a (−) RNA strand.

In the case of RNA viruses that have a single strand (+) RNA genome, the replication strategy is relatively simple because the genome functions as mRNA (see figure 13.12b, top panel). The viral RNA can immediately bind to host cell ribosomes and be translated to make proteins. One of the proteins encoded is a viral replicase, and once this is made, it can be used to synthesize more copies of the viral genome. To do this, it first makes multiple complementary (−) RNA strands using the original (+) RNA strand as a template. These (−) RNA strands then act as templates to produce more (+) RNA strands. The newly synthesized (+) RNA strands can either be translated to make more viral proteins, or be packaged as genomes into new virions being formed.

Replication of a single-stranded (−) RNA virus is more complicated because a (−) RNA strand cannot be translated directly. Instead, it must first be copied into a (+) RNA strand (see figure 13.12b, middle panel). This is done by a replicase carried into the host cell by the virus. Once the (+) RNA strand has been produced, it can be translated to make viral proteins, and can also be used as a template for synthesizing new (−) RNA strands. One of the proteins made during translation is replicase, and as the new viral particles are assembled, a molecule of the replicase is packaged along with a (−) RNA molecule into the capsid.

Double-stranded RNA viruses, which are relatively uncommon, must also carry their own replicase because the host cell machinery is unable to translate double-stranded RNA. The replicase immediately uses the (−) RNA strand of the double-stranded RNA molecule as a template to make (+) strand RNA. This molecule is then translated to make more replicase, and the infection cycle can continue (see figure 13.12b, bottom panel).

Replicases lack proofreading ability, and therefore make more mistakes than DNA polymerases, generating mutations during replication. These mutations lead to antigenic variation and allow some RNA viruses to adapt to selective pressures. Influenza viruses, for example, exhibit a type of antigenic variation called **antigenic drift.** This occurs as mutations accumulate in genes encoding key viral surface proteins that are recognized by the immune system. Because of such changes, a person whose immune response protected him or her against influenza virus one year may not be protected against the variant that circulates the next year. ⏭ antigenic variation, ⏭ antigenic drift

Some RNA viruses have segmented genomes, meaning that their genome consists of more than one piece of RNA. If two different strains of such a virus infect the same host cell, the new viral particles that are synthesized in that cell can have a mix of segments from each of the two original virus strains—an outcome called reassortment. For example, consider a virus strain that has RNA segments designated "a" through "e" and a different strain that has segments "A" through "E." All new viral particles being generated must have a total of five segments, but as these can originate from either of the parent strains, new viruses can have different combinations of capitalized and lowercase letters—aBCDE, abcDE, AbcDE, and so on. Influenza virus is an example of a virus with a segmented genome that can undergo reassortment. When a new subtype of a virus is formed by reassortment between different strains of a virus or even between different viruses, the phenomenon is known as **antigenic shift.** ⏭ antigenic shift

Replication of Reverse-Transcribing Viruses

Reverse-transcribing viruses encode the enzyme **reverse transcriptase,** an RNA-dependent DNA polymerase, which synthesizes DNA from an RNA template. Recall that in regular transcription, an RNA molecule is synthesized from a DNA template. In reverse transcription, a DNA molecule is synthesized from an RNA template.

Retroviruses, which include the human immunodeficiency virus (HIV), have a (+) strand RNA genome, and carry reverse transcriptase within the virion. After entering the host cell, the reverse transcriptase uses the RNA genome as a template to make one strand of DNA. The complement to that DNA strand is then synthesized to make double-stranded DNA, which integrates into the host cell chromosome (see figure 13.12c). Once integrated, the viral DNA may remain in a latent state (similar to what occurs with phage λ), or it may be transcribed into RNA that is translated to synthesize viral proteins needed for production of new virions. Either way, the integration is permanent, so the viral DNA copy cannot be eliminated from the cell. ⏭ replication of HIV

Replication of hepatitis B virus (HBV) includes a step that uses reverse transcriptase, but the process is very different

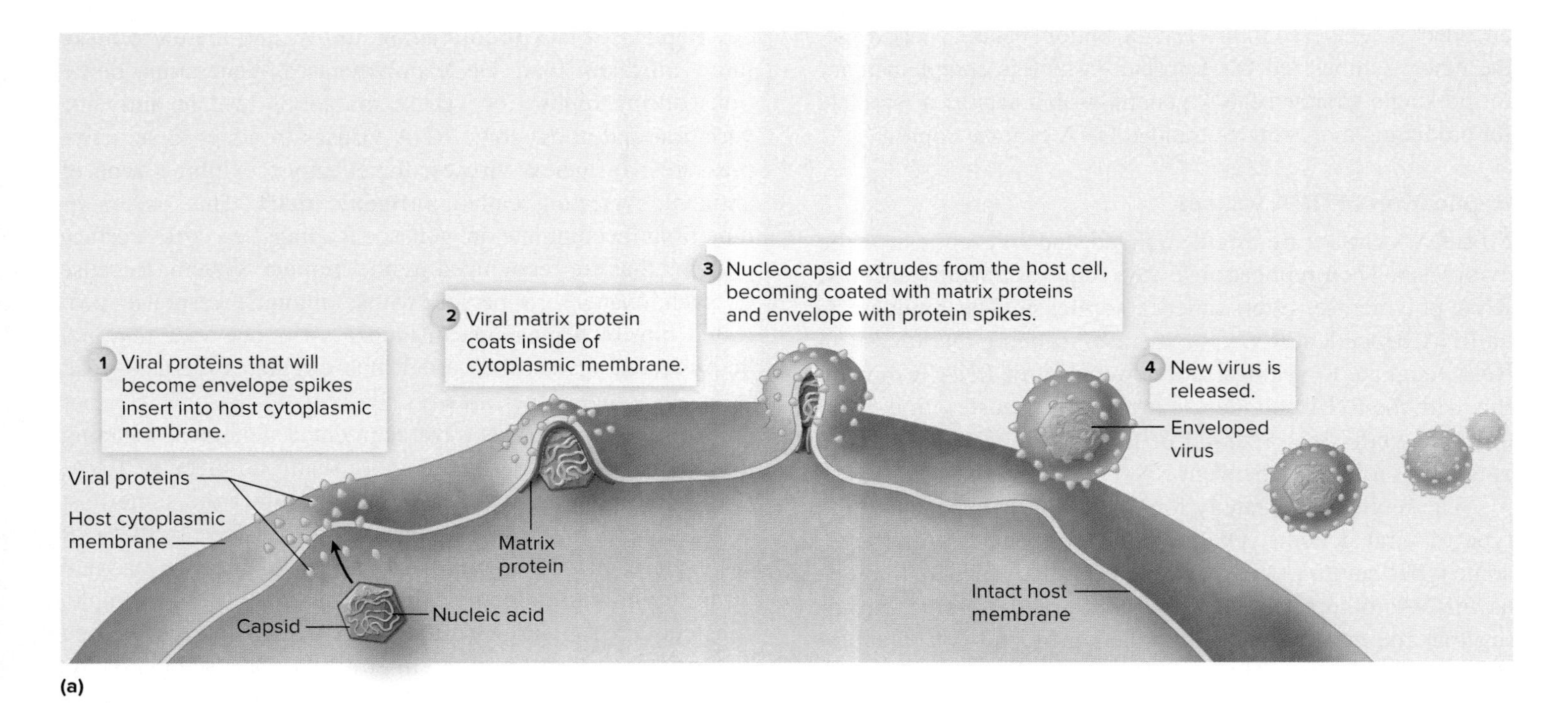

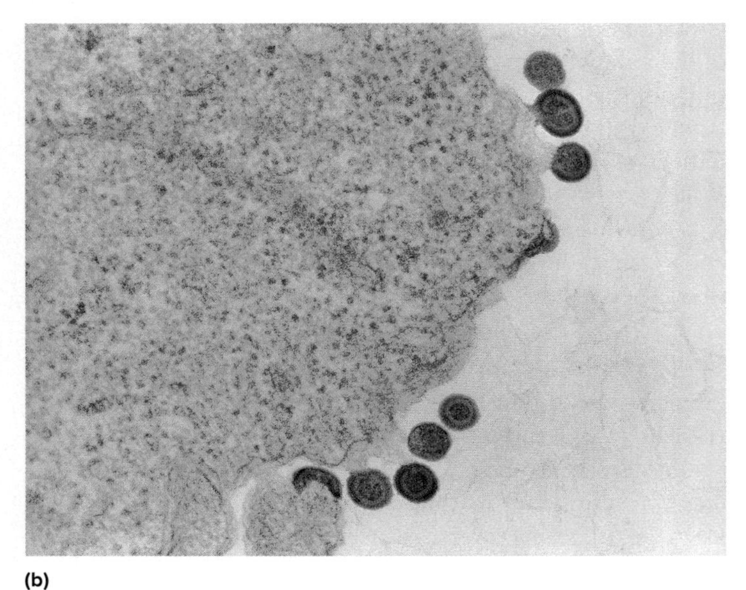

FIGURE 13.13 Mechanisms for Releasing Enveloped Virions **(a)** Process of budding. **(b)** Virions budding from the surface of a human cell (color-enhanced TEM). b: ©Dennis Kunkel Microscopy/SPL/Science Source

What component of the virion is gained in the process of budding?

from other reverse-transcribing viruses. For the sake of simplicity, we will not describe replication of HBV here; instead, it is included in a separate section about the virus in a later chapter (see figure 24.19). ⏭ replication of hepatitis B virus

Assembly and Maturation

Viral assembly and maturation are important steps in the replication of viruses. They involve bringing together newly formed viral nucleic acid with capsid proteins and packaging them to form the nucleocapsid. Even though viruses are structurally diverse, the assembly process is similar in all viruses. The process is a spontaneous self-assembly that occurs when an appropriate amount of viral nucleic acid and capsid proteins have accumulated in the host cell. Despite the fact that the assembly is spontaneous, it occurs in an ordered and stepwise manner.

The site of assembly and maturation differs according to the virus. Non-enveloped viruses mature fully in the host cell cytoplasm. In the case of enveloped viruses, some maturation steps occur as the virion leaves the host cell. The virus deposits protein into the host cytoplasmic membrane and the assembled nucleocapsid then binds to these regions of the membrane before exiting the host cell. Some DNA viruses, such as herpesviruses, assemble their nucleocapsids in the nucleus of the host cell. The site of virion assembly and the type of virus affect the way in which viral particles are released from the host cell (covered next).

Release

Just as the entry mechanism into a host cell depends on whether the virion is enveloped or non-enveloped, so does its release. Most enveloped viruses are released by **budding,** a process whereby the virus acquires its envelope (**figure 13.13**). Before budding occurs, virally encoded protein spikes insert into specific regions of the host cell's membrane. Matrix protein accumulates on the inside surface of those same regions. Assembled nucleocapsids are then extruded from the cell at these regions, becoming covered with a layer of matrix protein and lipid envelope in the process. Not all enveloped viruses have cytoplasmic membrane-derived envelopes, however. Some obtain their envelope from the membrane of an organelle such as the Golgi apparatus or the rough endoplasmic reticulum. They do this by budding into the organelle. From there, they are transported in vesicles to the outside of the cell. Budding may not destroy the cell because the membranes can be repaired after the viral particles exit. ⏮ **rough endoplasmic reticulum,** ⏮ **Golgi apparatus**

Non-enveloped viruses are released when the host cell dies. How this occurs, however, is sometimes quite different from how bacteriophages kill their hosts. Many viruses trigger a normal cellular process called **apoptosis,** or programmed cell death, prior to the release of the viral particles. The immune response of an animal, which is directed toward eliminating the virus, can also lead to the same process. The virions released from the dead cells may then invade any healthy cells in the area. ⏭ **apoptosis**

To be maintained in nature, infectious virions must leave one animal host and be transmitted to another. Viral particles may be shed in feces, urine, genital secretions, blood, or mucus and saliva released from the respiratory tract during coughing or sneezing. From there, the virus enters the next host to begin another round of infection.

MicroAssessment 13.5

Animal viruses and bacteriophages share similar features in their infection cycle. Because many animal viruses are enveloped and phages are not, differences exist in their entry and exit. The genomes of animal viruses are diverse, influencing viral replication strategies.

12. Why are virally encoded enzymes medically important?

13. How do enveloped viruses exit a cell?

14. Why can viruses not replicate independently of living cells?

13.6 ■ Categories of Animal Virus Infections

Learning Outcome

9. Compare and contrast acute infections and the two types of persistent infections caused by animal viruses.

Animal virus infections can be divided into two major categories: acute and persistent (**figure 13.14**). **Acute infections** are characterized by the sudden onset of symptoms of a relatively short duration. In contrast, **persistent infections** can continue with or without symptoms for years, or even for the life span of the host.

Although we often categorize viral infections as either acute or persistent, they do not always fall neatly into a single category. For instance, when a person is first infected with HIV, the virus replicates to high levels, causing acute symptoms, including fever, fatigue, swollen lymph nodes, and headache. The immune system soon eliminates most virions, and the symptoms subside. However, the DNA copy of the viral genome integrates into the host cell chromosome, resulting in a persistent infection with subsequent symptoms associated with acquired immunodeficiency syndrome (AIDS) developing after many years. Thus, HIV infection has features of both acute and persistent infections.

Acute Infections

Acute infections result in a burst of virions being released from infected host cells. Although the virus-infected cells often die, the host may survive. This is because the immune system of the animal host may gradually eliminate the virus over a period of days to months. Examples of diseases due to acute viral infections include influenza, mumps, and poliomyelitis. Symptoms of the diseases result from localized or widespread tissue damage following cell death as well as damage caused by the immune response itself. ⏭ **influenza,** ⏭ **mumps,** ⏭ **polio**

Persistent Infections

Persistent infections remain for years, or even the lifetime of the host—sometimes without any symptoms. In laboratory studies, two general types of persistent infections can be identified: chronic and latent (**table 13.4**). These categories may overlap in an actual infection, however, with different cells experiencing different types of infection.

Chronic infections are characterized by the continuous low-level production of viral particles. In some cases, the infected cell survives and slowly releases viral particles. In other cases, the infected cell lyses, but only a small proportion of cells is infected at any given time; the result is that a low number of viral particles are continuously released. From a practical standpoint, the important aspect of either type of persistent infection is the continuous production of infectious viral particles, often in the absence of disease symptoms. Consequently, a person can transmit the virus to others even in the absence of symptoms. For example, some people infected with hepatitis B virus develop a chronic infection. They become carriers of the virus, able to pass it to other people through blood and body fluids. ⏭ **hepatitis B virus**

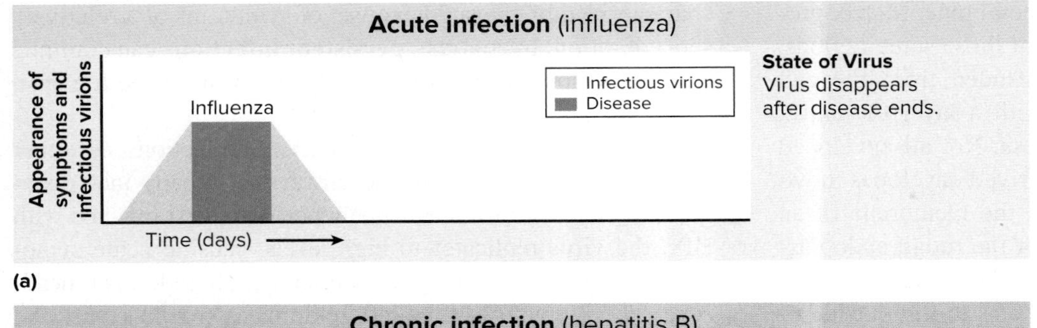

(a)

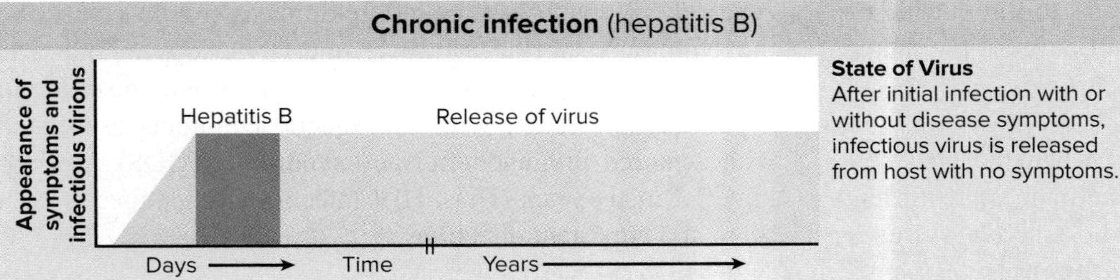

(b)

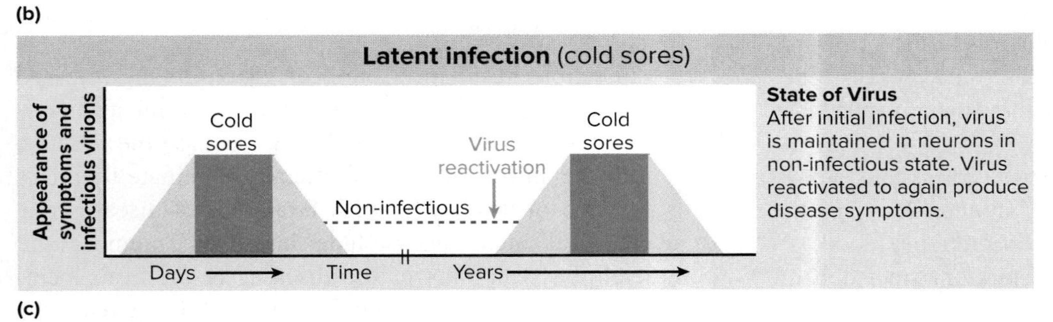

(c)

FIGURE 13.14 Types of Infection by Animal Viruses (a) Acute infection. **(b)** Persistent infection—chronic. **(c)** Persistent infection—latent.

? To show the outcome of infection by the virus that causes chickenpox, which part of this figure (a, b, or c) would you relabel, and what changes would you make to that part?

In **latent infections** the viral genome remains silent within a host cell, yet can reactivate to cause a productive infection. Some viruses do not integrate into the host cell chromosome; rather, they replicate independently of the host genome, much like a plasmid. The silent viral genome is called a **provirus.** The fact that a provirus cannot be eliminated from the body means that the disease can recur even after an extended period without symptoms. Consider the disease shingles. A person can get shingles only if he or she has previously had chickenpox—a disease caused by varicella zoster virus (VZV). Following the acute infection that caused chickenpox, the VZV genome goes dormant in nerve ganglia. Under certain circumstances, the virus can reactivate—newly formed VZV particles move down peripheral nerves and spread locally, causing the painful skin lesions characteristic of shingles. Latent infection also explains why cold sores, which are usually caused by herpes simplex type 1 virus (HSV-1), can recur. HSV-1 causes an acute infection in mucosal epithelial cells, leading to the typical symptoms of cold sores. From there, the virus can spread to sensory nerve cells where it remains latent. Later, the latent virus can reactivate to cause another episode of cold sores (**figure 13.15**). What reactivates the virus is not clear, but certain physiological or immunological changes in the host are often involved.

⏭ **herpes simplex type 1 virus**

TABLE 13.4 Examples of Persistent Infections

Virus	Type	Cells Involved	Disease
Cytomegalovirus (CMV; *Herpesviridae* family)	Latent	Salivary glands, kidney epithelium, leukocytes	CMV pneumonia, eye infections, congenital CMV infection
Epstein-Barr virus	Latent	B cells, which are involved in antibody production	Mononucleosis, Burkitt's lymphoma
Hepatitis B virus	Chronic	Hepatocytes (liver cells)	Hepatitis, cirrhosis, hepatocellular carcinoma
Hepatitis C virus	Chronic	Hepatocytes (liver cells)	Hepatitis, cirrhosis, hepatocellular carcinoma
Herpes simplex virus type 1	Latent	Neurons of sensory ganglia	Primary oral herpes and recurrent herpes simplex (cold sores)
Herpes simplex virus type 2	Latent	Neurons of sensory ganglia	Genital herpes and recurrent genital herpes
Human immunodeficiency virus (HIV)	Mixed	Chronic: Activated helper T cells, macrophages	AIDS
		Latent: Memory helper T cells	
Varicella zoster (*Herpesviridae* family)	Latent	Satellite cells of sensory ganglia	Chickenpox and shingles

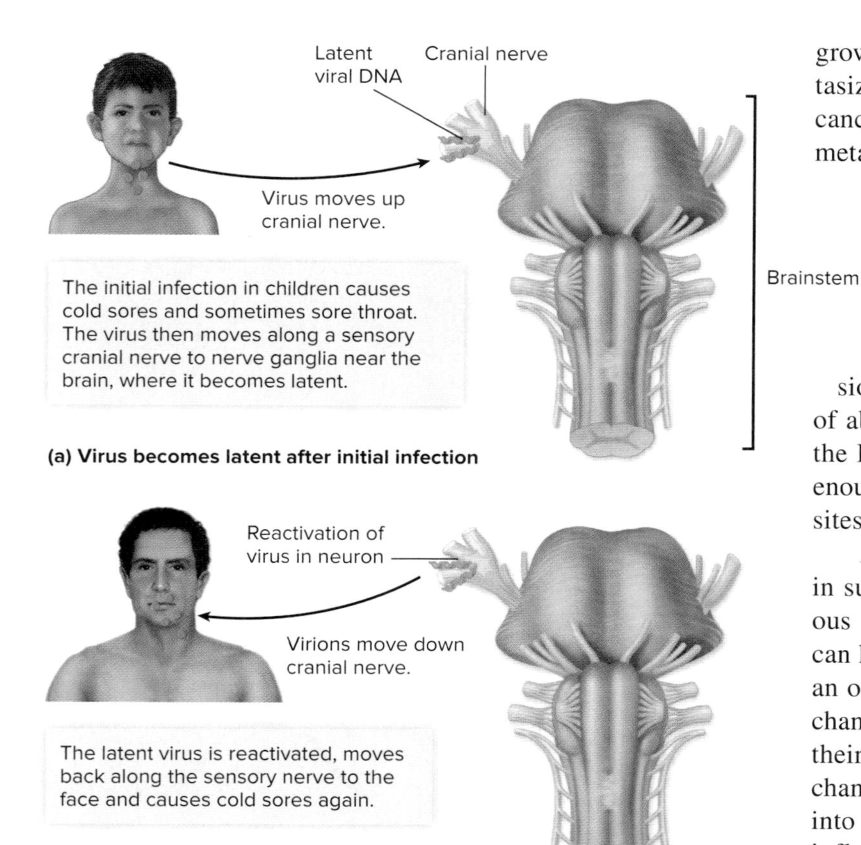

(a) Virus becomes latent after initial infection

(b) Activation of latent virus

FIGURE 13.15 Infection Cycle of Herpes Simplex Virus, HSV-1

? What is a provirus?

MicroAssessment 13.6

Many viruses cause acute infections in which viruses multiply and spread rapidly in the host. Some viruses cause persistent infections; these long-term infections can be chronic or latent.

15. Distinguish between acute and persistent viral infections at the cellular level.

16. How are latent viral infections different from chronic infections?

17. Could the same type of virus cause both an acute and a persistent infection? Explain.

13.7 ■ Viruses and Human Tumors

Learning Outcome

10. Describe the roles of proto-oncogenes and tumor suppressor genes in controlling cell growth, and discuss how some viruses can circumvent this control.

A tumor is an abnormal growth of tissue resulting from a malfunction in the normally highly regulated process of cell growth. Some tumors are benign, meaning they do not metastasize (spread) or invade nearby normal tissue. Others are cancerous or malignant, meaning they have the potential to metastasize.

Control of cell growth involves genes that stimulate cell growth, called proto-oncogenes, and others that inhibit cell growth, termed tumor suppressor genes. Expression of these genes is coordinated to regulate growth and cell division. Mutations that either increase the expression of proto-oncogenes or decrease the expression of tumor suppressor genes are the most common cause of abnormal and/or uncontrolled growth. A single change in the DNA sequence of these regulatory genes is probably not enough to cause a tumor; rather, multiple changes at different sites are required.

An **oncogene** is a proto-oncogene that has been changed in such a way that it promotes uncontrolled growth. Numerous events, including spontaneous and induced mutations, can lead to conversion of a host's normal proto-oncogene into an oncogene, leading to tumor formation. In some cases, the change is associated with a viral infection. Viruses that insert their genome into the host cell's chromosome may cause changes at the insertion site that convert a proto-oncogene into an oncogene. DNA damage resulting from long-term inflammation associated with chronic infections can also cause the conversion. In addition, certain viruses carry an oncogene, so the infection interferes with the host cell's own growth-control mechanisms.

Viruses that lead to cancer formation are called oncogenic viruses. Most virus-induced tumors are caused by certain DNA viruses (**table 13.5**). The majority of tumors are not caused by oncogenic viruses, however, but by mutations in host genes that regulate cell growth.

At least 15 human papillomaviruses (HPVs) are associated with the development of cancers. Although the mechanism by which the viruses cause cancer is not entirely clear, several HPV-encoded proteins appear to interfere with the function of an important tumor suppressor gene product. A vaccine that protects against the HPV serotypes most highly associated with cervical cancers and some other cancers is available. Other viruses are implicated in cancer and are listed in table 13.5. In all of these cases, only a small percentage of infected individuals develop the cancer, indicating that other important factors must be involved.

◀◀ papillomaviruses, ▶▶| HPV vaccines

The various effects that animal viruses can have on their host cells are shown in **figure 13.16.**

MicroByte

In 2008, a new virus called Merkel cell polyomavirus was described; it causes a rare and aggressive form of skin cancer in some people.

TABLE 13.5 Viruses Associated with Cancers in Humans

Virus	Type of Nucleic Acid	Kind of Tumor
Epstein-Barr	DNA	Burkitt's lymphoma; nasopharyngeal carcinoma; B-cell lymphoma
Hepatitis B	DNA	Hepatocellular carcinoma
Human herpesvirus type 8	DNA	Kaposi's sarcoma
Human papillomaviruses (HPVs)	DNA	Different kinds of tumors, caused by different HPV types
Hepatitis C	RNA	Hepatocellular carcinoma
HTLV-1	RNA (retrovirus)	Adult T-cell leukemia (rare)

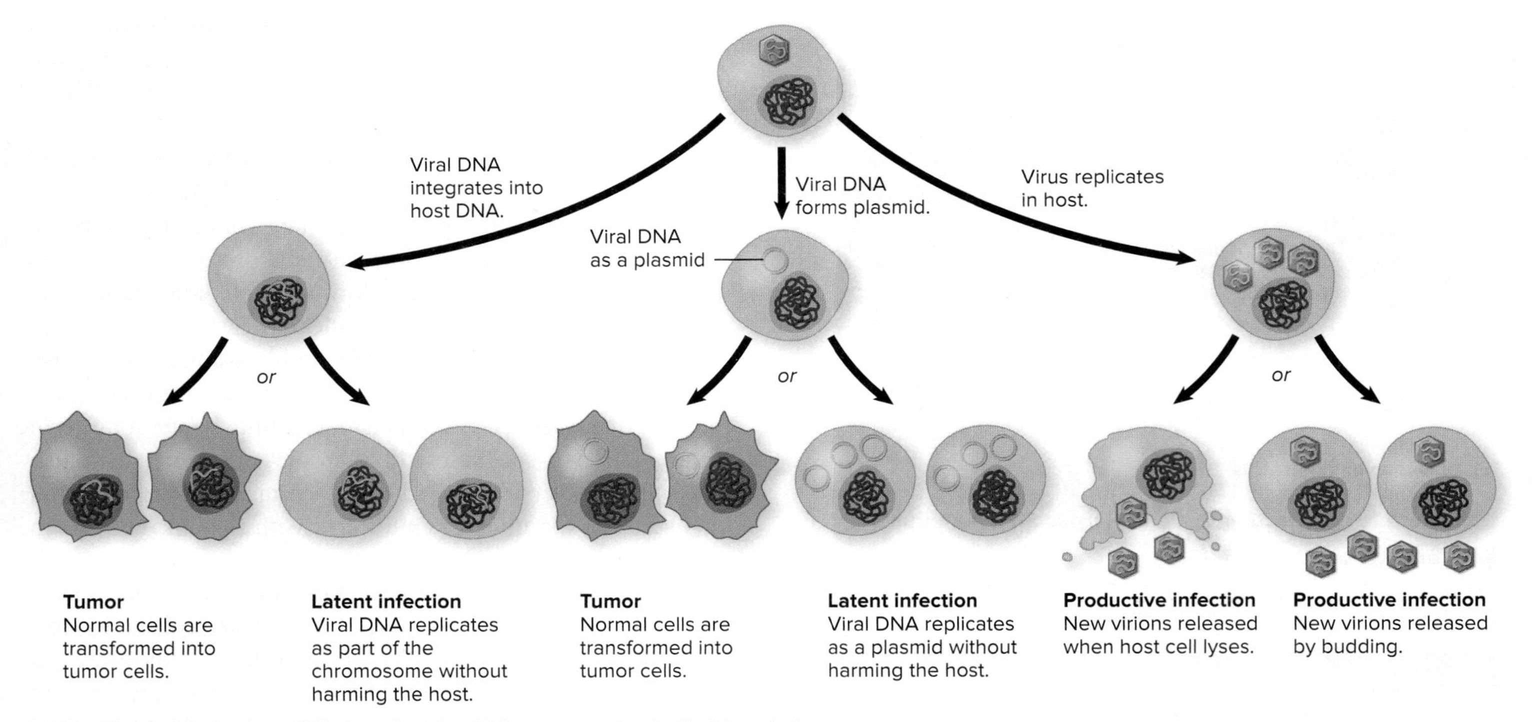

FIGURE 13.16 Various Effects of Animal Viruses on the Cells They Infect

? What relationship is the most destructive to the infected cell?

MicroAssessment 13.7

The majority of tumors are not caused by viruses; they are caused by mutations in certain host genes that regulate cell growth. The most common viral causes of tumors are certain DNA viruses.

18. What is the function of a proto-oncogene?

19. Explain why a vaccine can prevent cervical cancer.

20. Why would it be advantageous to a virus to interfere with the function of proto-oncogenes or tumor suppressor genes?

13.8 ■ Cultivating and Quantitating Animal Viruses

Learning Outcomes

11. Describe how animal cells are cultivated in the laboratory.

12. Describe the methods used to quantitate animal viruses.

To study viruses, they must be grown in an appropriate host in the laboratory. This is much more difficult to do with animal viruses than with phages.

Cultivating Animal Viruses

Initially, the only way to study animal viruses was to inoculate live animals with a suspension of viral particles. Later, embryonated (fertilized) chicken eggs were used to grow viruses, which made research much easier. Although improved techniques have largely replaced these methods, they are still used to cultivate certain viruses. For example, part of the process for making most flu vaccines involves growing influenza viruses in embryonated chicken eggs. ⏭ vaccine, ⏭ influenza virus

Today, **cell culture,** or **tissue culture,** is commonly used to cultivate most animal viruses. Animal cells—grown in a liquid medium contained in special screw-capped flasks—are

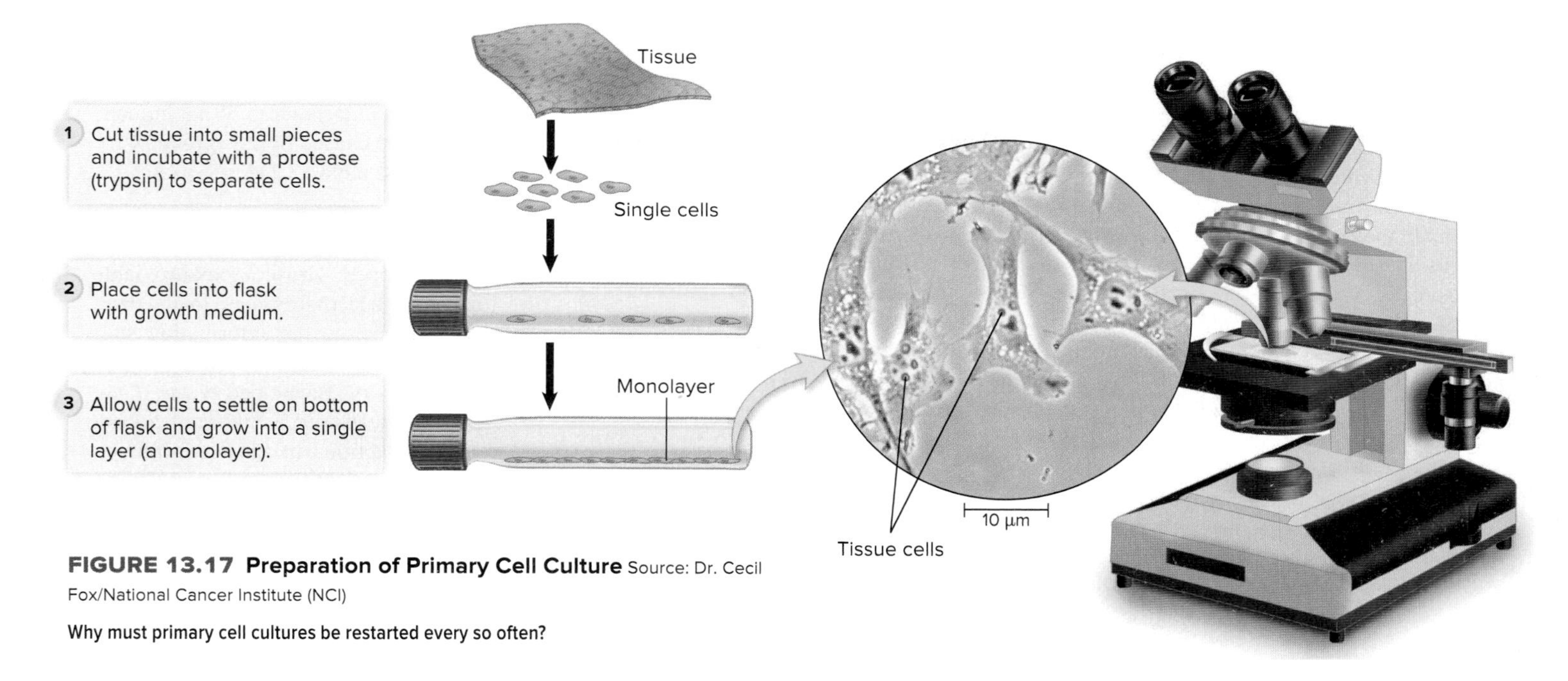

FIGURE 13.17 Preparation of Primary Cell Culture Source: Dr. Cecil Fox/National Cancer Institute (NCI)

Why must primary cell cultures be restarted every so often?

used as host cells for the virus culture. Much like bacteria, animal cells provided with the proper nutrients can divide repeatedly. They grow much more slowly than most bacteria, however, dividing no faster than once every 24 hours (*E. coli* can divide every 20 minutes). Animal cell types that make up solid tissues typically grow as a monolayer—a single sheet of cells adhering to the bottom of the flask. White blood cells grow as single cells in suspension. Although cell culture is generally easier than growing viruses in living animals, not all viruses can be grown in this way.

Obtaining and Maintaining Cell Cultures

One way to obtain animal cells for culture is to remove tissue from an animal, and then process it to get individual cells. These cells can then be grown in a flask with a liquid nutrient medium. Cells acquired in this way form primary cultures (**figure 13.17**). A problem with this approach is that normal cells can divide only a limited number of times, and so new primary cultures must regularly be made. To avoid this problem, tumor cells are often used in cell culture. These cells multiply indefinitely in vitro, resulting in what is called an established cell line.

MicroByte

The commonly used HeLa cell line was derived from cervical cancer cells taken from a patient named **He**nrietta **La**cks, who eventually died from the cancer.

Effects of Viral Replication on Cell Cultures

Many viruses can be detected by their effect on cells in cell cultures. A virus propagated in cell culture often causes distinct morphological alterations in infected cells, called a **cytopathic effect** (**figure 13.18**). The host cells, for example,

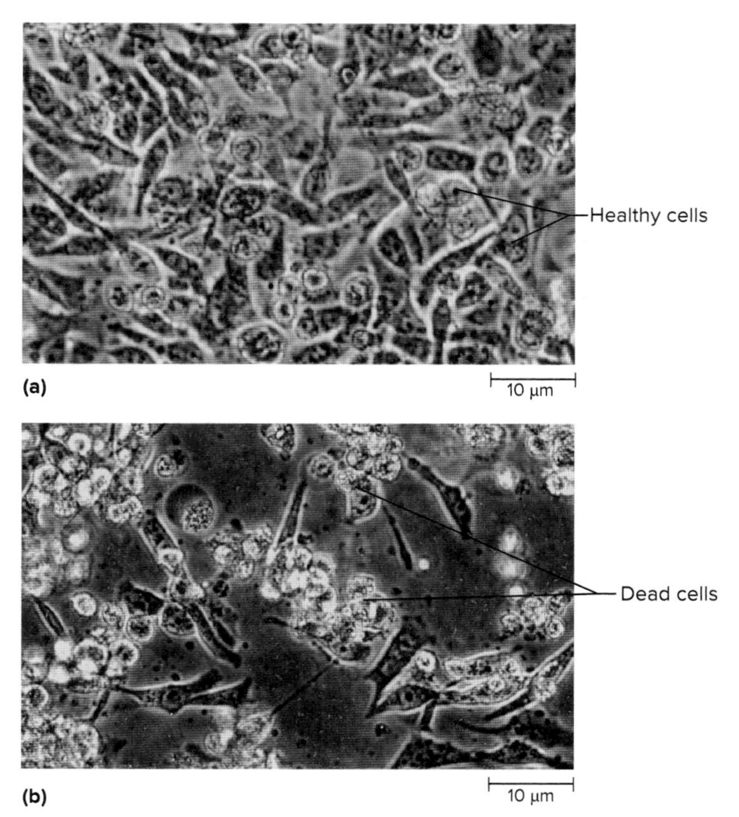

FIGURE 13.18 Cytopathic Effects of Virus Infection on Cells in Culture **(a)** Healthy HeLa cells **(b)** HeLa cells infected with a bunyavirus. a: Modified from "The Genome Sequence of Lone Star Virus, a Highly Divergent Bunyavirus Found in the *Amblyomma americanum* Tick," PLOS ONE, April 2013, Volume 8, Issue 4, e62083, with permission from University of California, San Francisco, and the Centers for Disease Control and Prevention; b: Modified from "The Genome Sequence of Lone Star Virus, a Highly Divergent Bunyavirus Found in the *Amblyomma americanum* Tick," PLOS ONE, April 2013, Volume 8, Issue 4, e62083, with permission from University of California, San Francisco, and the Centers for Disease Control and Prevention

FOCUS ON A CASE 13.1

In August 2007, a foot-and-mouth disease (FMD) outbreak was confirmed in cattle on a farm in Surrey, in the United Kingdom. Although the number of infected animals was small, the news was significant, particularly to livestock farmers, because the disease spreads easily and affects cattle, goats, sheep, and pigs. Infected animals develop high fever, followed by blisters in the mouth and on the feet. The animals sometimes die. Control measures are aimed at preventing the spread of the disease and include destroying at-risk animals and restricting trade—actions that obviously create a hardship for the affected farms. In a previous outbreak in 2001, the tourist industry was also devastated because travel in the British countryside, home to many small farms, was restricted.

The virus that causes the disease is called FMDV (foot-and-mouth disease virus), and the strain implicated in this outbreak was identified as O_1 BFS67-like, a strain that had not circulated in animals in the United Kingdom since 1967. Of particular note was the fact that the same strain was being used for research and vaccine development by laboratories located at a site very near the Surrey farm where the FMD outbreak was identified. A later investigation concluded that wastewater from the laboratories was the probable source of the virus; a drainage pipe had leaked and contaminated the surrounding soil. From there, vehicles, workers, or wild animals likely carried the virus to nearby farms. In the end, all susceptible livestock from the affected farms were destroyed, which successfully stopped the outbreak.

1. What does the proposed mechanism of virus spread from the laboratory to the farm animals suggest about the structure of FMDV (enveloped versus non-enveloped)? Explain.
2. Developing a long-lasting vaccine against FMDV is difficult because the virus has a relatively high mutation rate; a vaccine against one strain might not protect against another one that develops later. What does the high mutation rate suggest about the viral genome structure? Explain.
3. FMDV is a picornavirus. What human diseases are caused by this group of viruses (see table 13.1)?

Discussion

1. Non-enveloped viruses remain viable longer in the environment than enveloped viruses, so the information suggests that the virus is non-enveloped (which it is). Enveloped viruses are generally destroyed easily, and do not remain infectious in the environment; they would not still be infective after a trip through wastewater and into soil, and then to the farms.
2. RNA viruses typically have a higher mutation rate than DNA viruses, so the information suggests that FMDV has an RNA genome (which it does). RNA virus must encode replicase, an enzyme that uses RNA as a template to synthesize RNA. Replicases lack proofreading, so viruses that have RNA genomes often have a high mutation rate.
3. Poliomyelitis, colds, and hepatitis A.

may change shape, detach from the surface, or lyse. Infected cells may fuse into a giant multinuclear cell (syncytium), a mechanism of viral spread. Several viruses, such as HIV and measles, cause this cytopathic effect.

Certain viruses cause an infected cell to form a distinct region called an **inclusion body,** the site of viral replication. The position of an inclusion body in a cell depends on the type of virus. Because cytopathic effects are often characteristic of a particular virus, they are useful to scientists studying and identifying viruses.

Quantitating Animal Viruses

One of the most precise methods for determining the concentration of animal viruses in a sample is the plaque assay. This is similar in principle to the method described for quantitating bacteriophages, but in the case of animal viruses, a monolayer of cultured tissue cells serves as the host. Again, clear zones surrounded by uninfected cells are counted to determine the viral titer.

If a sample contains a high enough concentration of viruses to be seen with an electron microscope, direct counts can be used to determine the number of viral particles in a suspension (**figure 13.19**).

A viral titer can be estimated using a **quantal assay.** In this method, several dilutions of the virus preparation

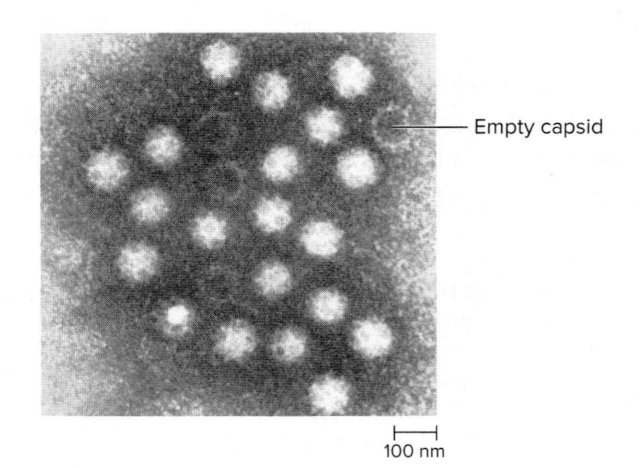

FIGURE 13.19 Electron Micrograph of Calicivirus Note that empty capsids are also easy to see. Source: National Exposure Research Laboratory/EPA

? Is a virion with an empty capsid infectious?

are administered to a number of animals, cells, or chick embryos, depending on the host specificity of the virus. The titer of the virus (the endpoint) is the dilution at which 50% of the inoculated hosts are infected or killed. This can be reported as either the **ID_{50}** (infective dose), or the **LD_{50}** (lethal dose).

Certain viruses cause red blood cells to agglutinate (clump). This phenomenon, called **hemagglutination,** occurs when individual viral particles attach to surface molecules of multiple red blood cells simultaneously, connecting the cells to form an aggregate (**figure 13.20**). Hemagglutination is visible only at high concentrations of viruses, and can be used to determine only the relative concentration of viral particles. Hemagglutination is measured by mixing serial dilutions of the viral suspension with a standard amount of red blood cells. The highest dilution showing maximum agglutination is the titer of the virus. One group of animal viruses that can agglutinate red blood cells is the orthomyxoviruses, of which the influenza virus is a member.

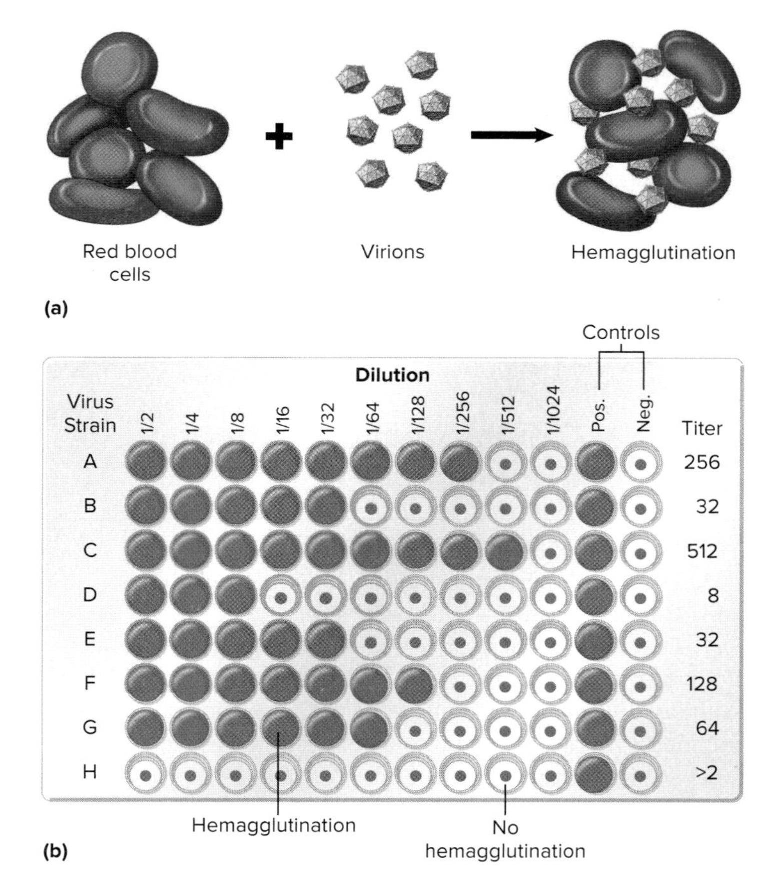

FIGURE 13.20 Hemagglutination (a) Diagram showing virions combining with red blood cells, resulting in hemagglutination. **(b)** Assay of viral titer. Red blood cells normally form a pellet when they sink to the bottom of the wells, whereas agglutinated ones do not.

? What is the titer of virus strain F?

MicroAssessment 13.8

Various hosts are required to grow different viruses. Viruses that lyse their host cells can be assayed by counting plaques. Other methods include direct counts, quantal assays, and hemagglutination.

21. In tissue culture, why is it advantageous to use tumor cells rather than normal cells?

22. Discuss two methods used to quantitate animal viruses.

23. How might syncytia formation of infected cells benefit viruses?

13.9 ■ Plant Viruses

Learning Outcomes

13. Compare and contrast the mechanisms by which plant and animal viruses enter host cells.

14. Discuss the ways that plant viruses can be transmitted to their hosts.

Viral diseases of plants are economically important, particularly when they affect crop plants such as corn, wheat, rice, soybeans, and sugar beets. Over half of the crop yields can be lost when a serious plant virus spreads.

Viral infection of plants can be recognized through various visible signs, including yellowing of foliage with irregular lines appearing on the leaves and fruits (**figure 13.21**). Individual cells or specialized organs of the plant may die, and tumors may appear. Usually, infected plants are stunted in their growth, although in a few cases growth is stimulated, leading to deformed structures. Plants generally do not recover from viral infections, because unlike animals, plants are not capable of developing specific immunity to rid themselves of invading viruses. In severely infected plants, virions may accumulate in enormous quantities. For example, as much as 10% of the dry weight of a tobacco mosaic virus-infected plant may consist of virus. The virus, which causes a serious disease of tobacco, retains its infectivity for at least 50 years, which explains why it is usually difficult to eliminate the virions from a contaminated area. Other plant viruses are also extraordinarily stable in the environment.

In a few cases, plants have been purposely maintained in a virus-infected state. The most well-known example involves tulips, in which a virus transmitted through the bulbs can cause a desirable color variegation of the flowers (**figure 13.22**). The infecting virus was transmitted through bulbs for a long time before the cause of the variegation was even suspected.

In contrast to phages and animal viruses, when plant viruses infect a cell they do not attach to specific receptors. Instead, they enter through wound sites in the cell wall, which

(a)

(b)

FIGURE 13.21 Signs of Viral Diseases of Plants (a) Typical mosaic pattern and ring lesions on a tobacco leaf resulting from infection by tobacco mosaic virus. **(b)** Wheat plant showing yellowing caused by wheat streak mosaic virus. a: ©Nigel Cattlin/Alamy Stock Photo; b: ©Nigel Cattlin/Science Source

How do most plant viruses enter plants?

FIGURE 13.22 Tulip with Symptoms Resulting from a Viral Infection ©Dr. Jeremy Burgess/Science Source

What other symptoms can plant viruses cause?

is otherwise very tough and rigid. Infection in the plant can then spread from cell to cell through channels (the plasmodesmata) that interconnect cells.

Plant viruses can be transmitted through soil contaminated by prior growth of infected plants, and by growers themselves. A small percentage of the known plant viruses are transmitted through contaminated seeds, tubers, or pollen. Viruses can also spread through grafting of healthy plant tissue onto diseased plants. Tobacco mosaic virus is transmitted to healthy seedlings on the hands of workers who have been in contact with the virus from infected plants. The most important transmitters of plant viruses are probably insects; thus, insect control is a critical tool for preventing the spread of plant viruses.

MicroByte

Tobacco in cigarettes, chewing tobacco, and cigars may carry the tobacco mosaic virus. Smokers who work with plants should wash their hands thoroughly before handling them.

MicroAssessment 13.9

Plant viruses cause many plant diseases and are of major economic importance. They invade through wound sites in the cell wall and are spread through soil, insects, and growers.

24. How does a plant virus penetrate the tough outer coat of the plant cell?
25. How are plant viruses transmitted?
26. Why is it especially important for plant viruses that they remain stable outside the plant?

13.10 ■ Other Infectious Agents: Viroids and Prions

Learning Outcomes

15. Describe the chemical structure of viroids and prions.
16. Compare hosts of viroids and prions.
17. Describe the process by which prions accumulate in tissues.

Although viruses are composed of only nucleic acid surrounded by a protective protein coat, viroids and prions have an even simpler structure.

Viroids

Viroids consist solely of a small single-stranded RNA molecule that forms a closed ring. They are about one-tenth the size of the smallest infectious viral RNA genome known. All known viroids infect only plants, where they cause serious diseases, including potato spindle tuber, chrysanthemum stunt, citrus exocortis, cucumber pale fruit, hopstunt, and cadang-cadang. Like plant viruses, they enter plants through wound sites rather than binding to specific receptors. A great deal is known about the structure of viroid RNA, but many questions remain. How do viroids replicate? How do they cause disease? How did they originate? The answers to these questions will provide insights into new and fascinating features of these unusual members of the microbial world.

Prions

Prions are composed only of protein, which is reflected in the name (derived from **pro**teinaceous **in**fectious agent). They have no nucleic acid. These agents have been linked to a number of slow, always fatal, human diseases including Creutzfeldt-Jakob disease and kuru, as well as to animal diseases such as scrapie (sheep and goats), mad cow disease or bovine spongiform encephalopathy (cattle), and chronic wasting disease (deer and elk) (**table 13.6**). In all these diseases, prion proteins accumulate in neural tissue. For unknown reasons, neurons die and brain function deteriorates as the tissues develop characteristic holes (**figure 13.23**). The characteristic sponge-like appearance of the brain tissues gave rise to the general term **transmissible spongiform encephalopathies,** which refers to all prion diseases. ⏭ prions

Considering that prions lack nucleic acid, the mechanism by which they accumulate in tissue and cause disease has long been an intriguing question. An answer began to emerge with the discovery of a normal neuronal protein that is identical in amino acid sequence to a prion protein. The infectious protein is simply a misfolded version of the normal protein, but that altered shape has significant consequences. For one thing, it makes the infectious protein resistant to degradation by the host cell proteases that normally destroy older proteins as new ones are synthesized. The misfolding also causes the molecules to become insoluble, causing them to aggregate in cells. The infectious prion proteins are now referred to as **PrP^{SC}** (for **pr**ion **p**rotein, **sc**rapie) and the normal counterparts as **PrP^{C}** (for **pr**ion **p**rotein, **c**ellular). A significant factor in PrP^{SC} accumulation is the outcome of that protein's interaction with PrP^{C}. It appears that the interaction, perhaps along with unknown factors, changes the folding of PrP^{C} molecules, converting them to PrP^{SC} (**figure 13.24**).

In most cases, a prion disease is transmitted only to members of the same species. However, the barrier to prion transmission between species also depends on the strain of prion. It is now clear that the prion that caused mad cow disease in England also killed more than 170 people by causing a disease very similar to Creutzfeldt-Jakob disease; to reflect the

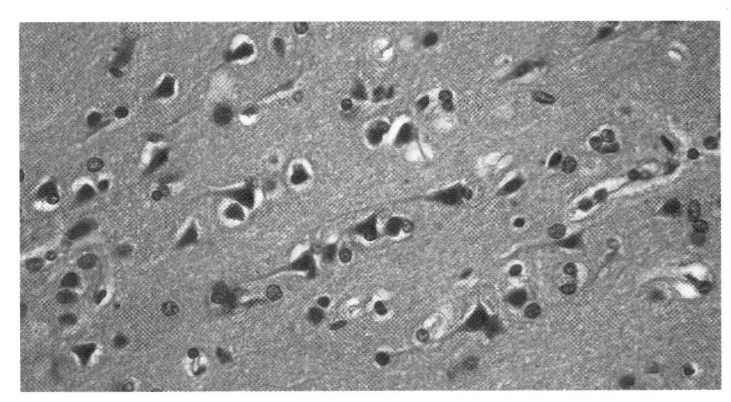

(a)

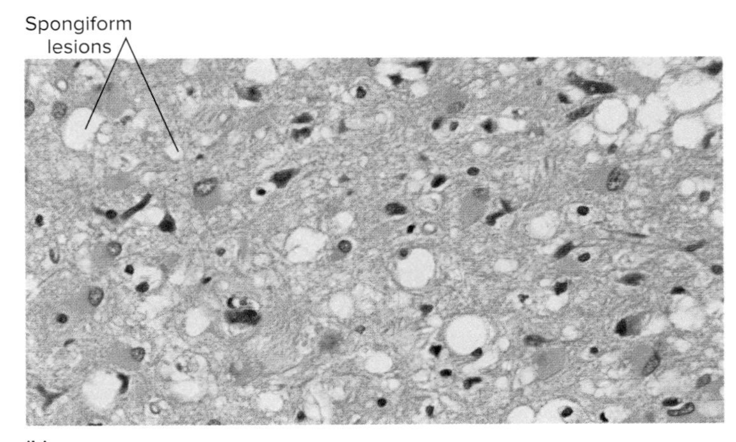

(b)

FIGURE 13.23 Appearance of a Brain with Spongiform Encephalopathy (a) Normal brain section. **(b)** Brain section of patient with spongiform encephalopathy. a: ©M. Abbey/Science Source; b: Source: Teresa Hammett/CDC

? From these photos, why has this disease been given the name it has?

TABLE 13.6 Prion Diseases

Disease	Host
Scrapie	Sheep and goats
Bovine spongiform encephalopathy (mad cow disease)	Cattle
Chronic wasting disease	Deer and elk
Transmissible mink encephalopathy	Ranched mink
Exotic ungulate encephalopathy	Antelope in South Africa
Feline spongiform encephalopathy	Cats
Kuru	Humans (transmitted by cannibalism)
Variant Creutzfeldt-Jakob disease	Humans (transmitted by consumption of prion-contaminated beef)
Creutzfeldt-Jakob disease	Humans (inherited)
Gerstmann-Sträussler-Scheinker syndrome	Humans (inherited)
Fatal familial insomnia	Humans (inherited)

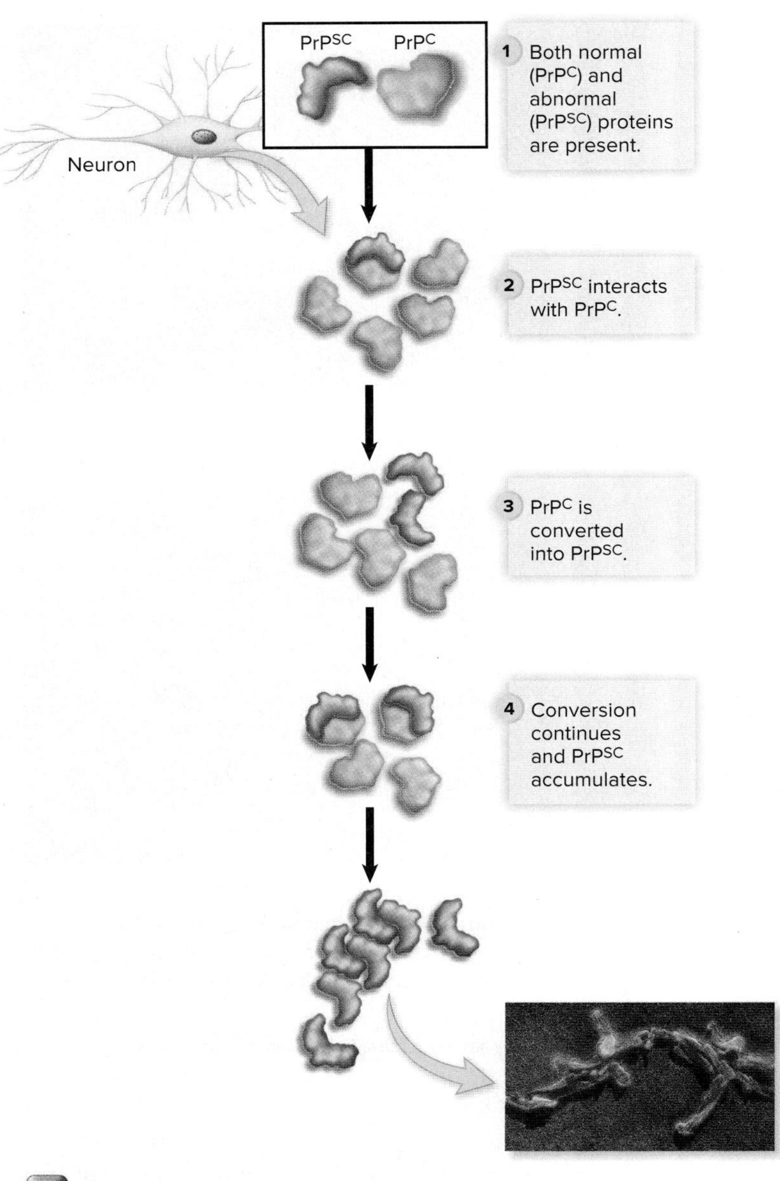

FIGURE 13.24 Proposed Mechanism by Which Prions Propagate ©Dr. Raoult/Science Source

Why would PrP^{SC} accumulate when PrP^{C} does not?

similarity, the disease was called variant Creutzfeldt-Jakob disease. Presumably these people ate infected beef. So far, no human deaths have been attributed to eating sheep infected with the scrapie agent or deer and elk infected with the prion causing chronic wasting disease (CWD).

A good example of how a prion disease spreads in the United States is CWD. Deer at a research facility in Colorado were the first animals to show signs and symptoms of CWD in 1967. Within 40 years, the disease spread from Colorado to wild deer populations through their natural migration and as a result of transport of captive farmed animals located in more than a dozen states and parts of Canada. The disease is spread through direct animal-to-animal (nose-to-nose) contact and as a result of indirect exposure to prions in the environment. Feed, water sources, and soil may be contaminated with prions shed in saliva, urine and feces, or decomposing carcasses. Prions are unusually resistant to heat and chemical treatments, making them difficult, if not impossible, to eliminate from the environment. Additional information about the role of prions in human disease is covered in chapter 26.

MicroAssessment 13.10

Two infectious agents that are structurally simpler than viruses are viroids and prions. Viroids contain only single-stranded RNA and no protein; prions contain protein and no nucleic acid.

27. Distinguish between a viroid and a prion in terms of structure and hosts.

28. Discuss how prion proteins accumulate in nervous tissue.

29. Must all prion diseases result from eating infected food? Explain.

Summary

13.1 General Characteristics of Viruses

Most viruses are approximately 100- to 1,000-fold smaller than the cells they infect (figure 13.1).

Viral Architecture

At a minimum, a **virion** (viral particle) consists of nucleic acid surrounded by a **capsid.** Capsids are composed of capsomeres. Some viruses have an envelope surrounding the **nucleocapsid;** other viruses are **non-enveloped** or **naked** (figure 13.2). Viruses contain either RNA or DNA, but never both. The shape of a virus is generally icosahedral, helical, or complex (figure 13.3).

Viral Taxonomy

Viruses are classified primarily on the basis of their genome structure and hosts they infect (table 13.1). The names of virus families end in *-viridae.* Viruses are also given informal names and are sometimes grouped based on their routes of transmission (table 13.2).

13.2 Bacteriophages

Lytic Phage Infections: T4 Phage as a Model

Lytic or **virulent phages** exit the host at the end of the cycle by lysing the host, resulting in a **productive infection.** The infection proceeds through five steps: attachment, genome entry, synthesis of phage proteins and genome, assembly (maturation), and release (figure 13.5).

Temperate Phage Infections: Lambda Phage as a Model

Temperate phages have the option of either directing a productive infection or initiating a **lysogenic infection** (figure 13.6); the infected cell is a **lysogen.** A **repressor** maintains the **prophage** in an integrated state, but the prophage can be excised to initiate a lytic infection. Lysogens are immune to superinfection. **Lysogenic conversion** occurs if a prophage carries genes that change the phenotype of the host cell (table 13.3).

Filamentous Phage Infections: M13 Phage as a Model

Filamentous phages cause productive infections, but the viral particles are continually extruded from the host in the assembly process and the cells are not killed (figure 13.7).

13.3 The Roles of Bacteriophages in Horizontal Gene Transfer

Generalized Transduction

Generalized transduction results from a phage packaging error during assembly. Generalized transducing particles can transfer any gene of a donor cell.

Specialized Transduction

Specialized transduction results from an excision mistake made by a temperate phage during its transition from a lysogenic to a lytic cycle (figure 13.9). Only genes located near the site at which the temperate phage integrates are transduced.

13.4 Methods Used to Study Bacteriophages

Plaque assays are used to quantitate the phage particles in samples (figure 13.10). Each plaque represents a single phage particle infecting a cell.

13.5 Animal Virus Replication

The generalized infection cycle of animal viruses can be viewed as a five-step process.

Attachment

Attachment proteins or spikes on the viral particle attach to specific receptors on the cell surface.

Penetration and Uncoating

In the case of animal viruses, the entire virion enters the cell. Enveloped viruses either fuse with the host membrane or are taken in by receptor-mediated endocytosis (figure 13.11). Non-enveloped virions enter by receptor-mediated endocytosis. **Uncoating** releases the nucleic acid from the protein coat.

Synthesis of Viral Proteins and Replication of the Genome (figure 13.12)

DNA viruses generally replicate in the nucleus and use the host cell machinery for DNA synthesis as well as gene expression, although they often encode their own DNA polymerase. Replication of single-stranded DNA viruses is similar to that of double-stranded DNA viruses, but the complementary strand must be synthesized first. RNA viruses usually replicate in the cytoplasm. Replication requires a virally encoded **replicase** to synthesize the complementary RNA strand. This enzyme lacks proofreading ability and makes more mistakes in replication than DNA polymerase does. Reverse-transcribing viruses encode **reverse transcriptase,** which synthesizes DNA from an RNA template. As with replicases, these enzymes are error-prone.

Assembly and Maturation

Capsids are formed, and then the genome and any necessary proteins are packaged within it. The process may take place in the cytoplasm, in the nucleus, or in a variety of organelles.

Release

Enveloped virions most often exit by **budding** (figure 13.13). Non-enveloped virions are released when the host cell dies.

13.6 Categories of Animal Virus Infections (figure 13.14)

Acute infections are characterized by sudden onset of symptoms of relatively short duration. **Persistent infections** can continue for the lifetime of the host, with or without symptoms.

Acute Infections

On a cellular level, acute infections can be compared to productive lytic infections by bacteriophages, but even though the cells often die, the host may survive because of the immune response.

Persistent Infections

Chronic infections are characterized by the continuous production of low levels of viral particles; **latent infections** are analogous to lysogeny in bacteriophages.

13.7 Viruses and Human Tumors

Some viruses carry **oncogenes** that interfere with the ability of the cell to control growth. Most virus-induced tumors are caused by certain DNA viruses (table 13.5). A vaccine against human papillomaviruses (HPVs) prevents many cervical cancers.

13.8 Cultivating and Quantitating Animal Viruses

Cultivating Animal Viruses

Cell culture or **tissue culture** is commonly used to cultivate most viruses (figure 13.17). This can be done using **primary cultures** or established cell lines. Many viruses can be detected by their effect on cells in culture, called a **cytopathic effect** (figure 13.18). Certain viruses cause an infected cell to form an **inclusion body.**

Quantitating Animal Viruses

The plaque assay is one of the most precise methods for determining the concentration of animal viruses in a sample. In some cases, virions can be counted with an electron microscope (figure 13.19). **Quantal assays** estimate the titer by determining the $\mathbf{ID_{50}}$ or $\mathbf{LD_{50}}$. The concentration of viruses able to cause **hemagglutination** can be measured by determining the highest dilution that clumps red blood cells (figure 13.20).

13.9 Plant Viruses

Viral infection of plants can be recognized by outward signs such as yellowing foliage, stunted growth, and tumor formation; plants usually do not recover from the infection (figure 13.21). Virions do not bind to receptor sites on plant cells but enter through wound sites. Insects are probably the most important transmitter of the viruses.

13.10 Other Infectious Agents: Viroids and Prions

Viroids

Viroids are plant pathogens that consist of small, circular, single-stranded RNA molecules.

Prions

Prions are composed solely of protein, and cause a number of **transmissible spongiform encephalopathies** (table 13.6). Prions accumulate by converting PrP^C to PrP^{SC}, proteins that are less susceptible to proteases and form aggregates (figure 13.24).

Review Questions

Short Answer

1. Why are non-enveloped viruses generally more resistant to disinfectants than are enveloped viruses?
2. How is the replication cycle of lambda phage different from that of T4?
3. What is lysogenic conversion?
4. How is specialized transduction different from generalized transduction?
5. What is the difference between acute and persistent infections?
6. Why must (−) strand but not (+) strand RNA viruses bring their own replicase into a cell?
7. Why are RNA viruses and retroviruses more error-prone in their replication than DNA viruses?
8. What is the role of a prophage in persistent infections?
9. How do oncogenes differ from proto-oncogenes?
10. Describe how prions propagate.

Multiple Choice

1. Capsids are composed of
   a) DNA.
   b) RNA.
   c) protein.
   d) lipids.
   e) polysaccharides.
2. The tail fibers on phages are associated with
   a) attachment.
   b) penetration.
   c) transcription of phage DNA.
   d) assembly of a virus.
   e) lysis of the host.
3. Classification of viruses is based on all of the following *except*
   a) type of nucleic acid.
   b) shape of virus.
   c) size of virus.
   d) host infected.
   e) strandedness of nucleic acid.
4. Temperate phages can do all of the following *except*
   a) lyse their host cells.
   b) change the properties of their hosts.
   c) integrate their DNA into the host DNA.
   d) bud from their host cells.
   e) become prophages.
5. All phages must be able to
   1) inject their nucleic acid into a host cell.
   2) kill the host cell.
   3) multiply in the absence of living bacteria.
   4) lyse the host cell.
   5) have their nucleic acid replicate in the host cell.
   a) 1, 2 b) 2, 3 c) 3, 4
   d) 4, 5 e) 1, 5
6. Filamentous phages
   a) infect animal and bacterial cells.
   b) cause their host cells to grow more quickly.
   c) are extruded from the host cell.
   d) undergo assembly in the cytoplasm.
   e) degrade the host cells' DNA.
7. Influenza vaccines must be changed yearly because the amino acid sequence of the viral proteins changes gradually over time. Based on this information, which is the most logical conclusion? The influenza virus
   a) is enveloped.
   b) is non-enveloped.
   c) has a DNA genome.
   d) has an RNA genome.
   e) causes a persistent infection.
8. Acute infections of animals
   1) are a result of productive infection.
   2) generally lead to long-lasting immunity.
   3) result from integration of viral nucleic acid into the host.
   4) are usually followed by chronic infections.
   5) often lead to tumor formation.
   a) 1, 2
   b) 2, 3
   c) 3, 4
   d) 4, 5
   e) 1, 5
9. Determining viral titers of both phage and animal viruses frequently involves
   a) plaque formation.
   b) quantal assays.
   c) hemagglutination.
   d) determining the ID_{50}.
   e) counting of virions by microscopy.
10. Prions
    a) contain only nucleic acid without a protein coat.
    b) replicate like HIV.
    c) integrate their nucleic acid into the host genome.
    d) cause diseases of humans.
    e) cause diseases of plants.

Applications

1. A public health physician isolated large numbers of phages from rivers used as a source of drinking water in western Africa. The physician is very concerned that humans might become ill from drinking this water, although she knows that phages specifically attack bacteria. Why is she concerned?
2. Researchers debate the evolutionary value to the virus of its ability to cause disease. Many argue that viruses accidentally cause disease and only in animals that are not the natural host. They state that this strategy may eventually prove fatal to the

virus's future in that host. It is reasoned that the animals will eventually develop immune mechanisms to combat the virus and prevent its spread. Another group of researchers supports the view that disease is a way to enhance the survival of the virus. What position would you take, and what arguments would you give to support your view?

Critical Thinking

1. Viruses that infect bacterial cells do not infect human cells, and viruses that infect human cells do not infect bacterial cells. Explain why this should be the case.
2. Why is it virtually impossible to eradicate (eliminate) a disease caused by a zoonotic virus?

14 The Innate Immune Response

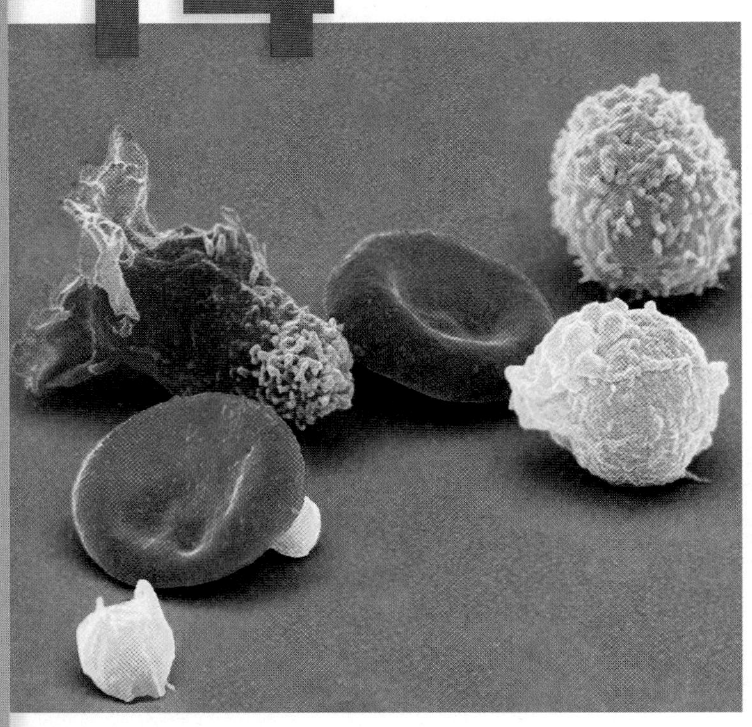

Human blood cells (color-enhanced scanning electron micrograph). ©*Science Photo Library RF/Getty Images*

KEY TERMS

Apoptosis Programmed death of "self" cells that does not cause inflammation.

Complement System Series of proteins in blood and tissue fluids that can be activated to help destroy and remove invading microbes.

Cytokines Proteins that function as chemical messengers, allowing cells involved in host defenses to communicate.

Inflammatory Response Coordinated innate response with the purpose of containing a site of damage, localizing the response, eliminating the invader, and restoring tissue function.

Innate Immunity Host defenses involving anatomical barriers, sensor systems that recognize patterns associated with microbes or tissue damage, phagocytic cells, the inflammatory response, and fever.

Macrophage Type of phagocytic cell that resides in tissues and has multiple roles, including scavenging debris and producing pro-inflammatory cytokines.

Membrane Attack Complexes (MACs) Complement system components assembled to form pores in membranes of invading cells.

Neutrophil Major type of phagocytic cell in blood; neutrophils quickly move to infected tissues, where they destroy invading microbes.

Opsonization Coating of an object with molecules for which phagocytes have receptors, making it easier for phagocytosis to occur.

Pattern Recognition Receptors (PRRs) Proteins on or in cells that recognize specific compounds unique to microbes or tissue damage.

Phagocyte Cell type that specializes in engulfing and digesting microbes and cell debris (a process called phagocytosis).

A Glimpse of History

As soon as microorganisms were shown to cause disease, scientists worked to explain how the body defends itself against their invasion. Ilya Metchnikoff, a Russian-born scientist, hypothesized that specialized cells in the body destroy invading organisms. His ideas arose while he was studying the larval form of starfish. As he looked at the larvae under the microscope, he could see amoeba-like cells within the bodies. He described his observations:

> . . . I was observing the activity of the motile cells of a transparent larva, when a new thought suddenly dawned on me. It occurred to me that similar cells must function to protect the organism against harmful intruders. . . . I thought that if my guess was correct a splinter introduced into the larva of a starfish should soon be surrounded by motile cells much as can be observed in a man with a splinter in his finger. No sooner said than done. In the small garden of our home . . . I took several rose thorns that I immediately introduced under the skin of some beautiful starfish larvae which were as transparent as water. Very nervous, I did not sleep during the night, as I was waiting for the results of my experiment. The next morning, very early, I found with joy that it had been successful.

Metchnikoff reasoned that certain cells in animals are able to ingest and destroy foreign material. He called these cells phagocytes ("cells that eat") and proposed they were primarily responsible for the body's ability to destroy invading microbes. He then studied the process by watching phagocytes ingest and destroy invading yeast cells in transparent water fleas. In 1884, Metchnikoff published a paper supporting his belief that phagocytic cells were primarily responsible for destroying disease-causing organisms. He spent the rest of his life studying this process and other biological phenomena. Metchnikoff was awarded a Nobel Prize in 1908 for his studies of immunity.

From a microorganism's standpoint, the tissues and fluids of the human body are much like a culture flask filled with a warm nutrient-rich solution, but guarded by armies of cells. Those armies generally keep the interior of the body—including blood, muscles, and bones—sterile. If this were not the case, microbes would simply degrade our tissues, just as they readily break down the bodies of dead animals. When the body's defense systems work to eliminate an invader, however, they must not cause excessive damage to the body's own tissues. Likewise, they must not wage war with the normal microbiota that lives on the body's surfaces, because the resulting constant conflict would damage the tissues. The system must maintain a delicate balancing act,

Focus Figure

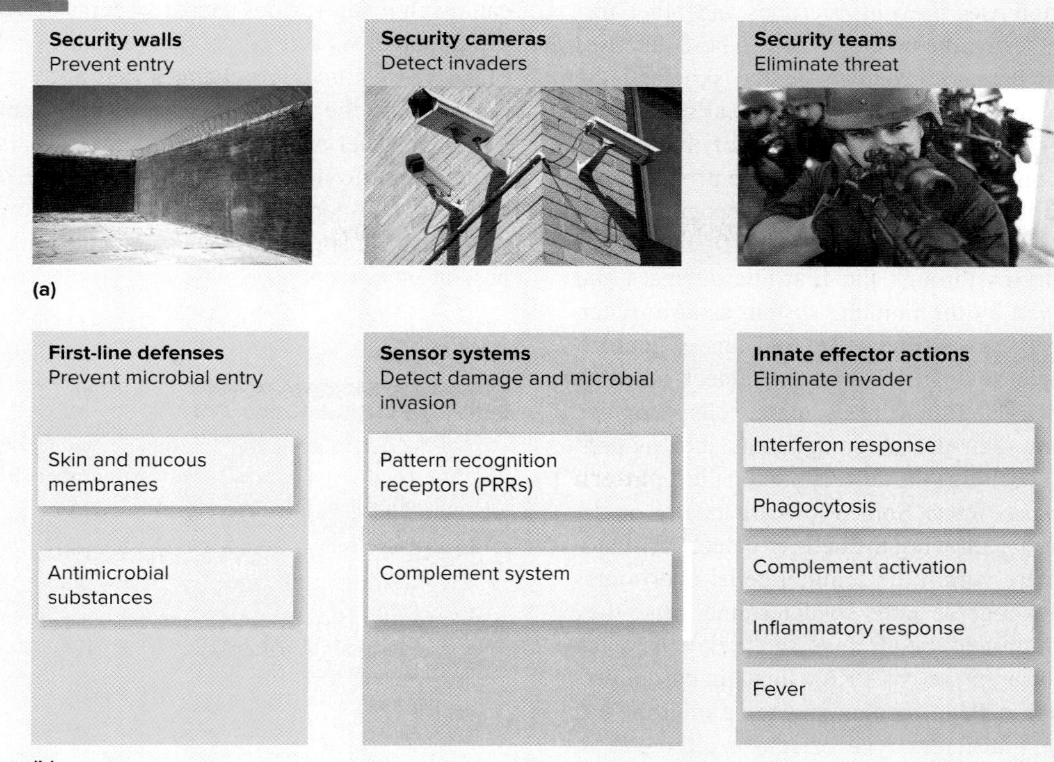

FIGURE 14.1 Defense Systems (a) Systems that protect a high-security compound. **(b)** Components of innate immunity that protect the body against infection. a (left): ©Steve Cole/E+/Getty Images; a (middle): ©James Lauritz/Photodisc/Getty Images; a (right): ©Moodboard/Brand X Pictures/Getty Images

? What is the role of the sensor systems in innate immunity?

destroying pathogens while maintaining relatively stable conditions within the human body—a state called homeostasis (*homeo* means "similar" and *stasis* means "standing still"). Failure to do that can be deadly.

Like other multicellular organisms, the human body has several mechanisms of defense. **Innate immunity** is the routine protection present at birth; it is germ-line encoded, meaning that it is passed from one generation to the next, and includes anatomical barriers as well as certain cell types and chemicals. In addition to innate immunity, vertebrates have evolved a more specialized defense system, providing protection called **adaptive immunity.** This develops throughout life as a result of exposure to microbes or certain other types of foreign material, and substantially increases the host's ability to defend itself. The substance that causes an immune response is called an **antigen.** Each time the body is exposed to an antigen, the adaptive defense system first "learns" and then "remembers" the most effective response to that specific antigen; it then reacts accordingly if the antigen is encountered again. An important action of the adaptive immune response is the production of proteins called **antibodies.** These bind specifically to antigens, thereby targeting them for destruction or removal by other host defenses. The adaptive immune response can also destroy the body's own cells—referred to as **host cells** or **"self" cells**—that are infected with a virus or other invader.

To simplify the description of the immune system, it is helpful to consider it as a series of individual parts. This chapter will focus almost exclusively on innate immunity. Remember, however, that although the various parts are discussed separately, their actions are connected and coordinated. In fact, as you will see in chapter 15, certain components of the innate defenses educate the adaptive defenses, helping them recognize that a particular antigen represents a microbial invader.

14.1 ■ Overview of the Innate Immune Defenses

Learning Outcome

1. Outline the fundamental components of the innate defenses.

The innate immune system has three general components: first-line defenses, sensor systems, and innate effector actions. As a useful analogy, think of the defense systems of a high-security building or compound: The first-line defenses are the security walls surrounding the property; the sensor systems are the security cameras scattered throughout the property, monitoring the environment for signs of invasion; and the effector actions are the security teams sent to remove any invaders that have been detected, thereby eliminating the threat (**figure 14.1a**).

The **first-line defenses** prevent microbes and other foreign material from entering the body's tissues. These defenses include the physical barriers provided by the skin and the mucous membranes, along with the antimicrobial substances that bathe them (see figure 14.1b). Members of the normal microbiota residing on the surfaces also provide protection.

Sensor systems allow the immune system to recognize when the first-line defenses have been breached (see figure 14.1b). Any microbe that passes through the first-line defenses and into tissue is perceived by the immune system as an invader. Certain host cells serve as **sentinel cells** (lookouts or guards), positioned at strategic sites in the body to detect invading microbes in blood or tissue fluids. The sentinel cells recognize microbes by detecting their unique components such as peptidoglycan, using a special group of receptors called **pattern recognition receptors (PRRs).** Some PRRs are located on the surface of sentinel cells, allowing the cells to detect surrounding invaders; others are within the sentinel cells' endosomes or phagosomes, allowing the cells to determine what they have engulfed. In addition to the PRRs of sentinel cells, many cell types have a different set of PRRs in their cytoplasm. These PRRs allow the cells to recognize when a microbe has invaded them. A very different type of sensor circulates in blood and tissue fluids. This sensor is a set of proteins collectively called the **complement system,** and was named because it can "complement" (act in combination with) the adaptive immune defenses. The complement system proteins circulate in an inactive form, but they become activated in response to certain stimuli, setting off a chain of events that results in removal and destruction of invading microbes. ◀◀ endosome, ◀◀ phagosome

Innate effector actions eliminate invaders (see figure 14.1b). When one or more of the sensor systems detects an invading microbe, various effector mechanisms may be called into action. Because the sensors typically recognize patterns associated with certain groups of microbes, the effector actions can be tailored to defend against those groups. For example, when a host cell recognizes that it is infected by a virus, that cell produces an **interferon (IFN),** a type of protein that warns nearby cells about the virus. Those neighboring cells react by preparing to shut down their biosynthetic activities if they too become infected. By doing so, the cells can deprive the virus of a mechanism to replicate. In response to sensor signals that indicate a bacterial infection or tissue damage, a group of cells called **phagocytes** are recruited to the site of invasion or damage. These cells specialize in engulfing and digesting microbes and cell debris, a process called **phagocytosis.** Some types of phagocytes play multiple roles, not only destroying invaders, but also serving as sentinel cells. Certain activated complement system proteins recruit phagocytic cells; others bind to foreign material—an action that makes it easier for phagocytes to engulf the material. A coordinated response called the **inflammatory response** can result when various sensor systems detect infection or tissue damage. As a result of this response, cells that line local blood vessels undergo changes that allow fluids to leak out into tissues; the fluids contain complement system components and other proteins. Phagocytic cells also move from the bloodstream to the tissues. Another innate response is **fever,** a higher-than-normal body temperature. Fever interferes with the growth of some pathogens and can enhance the effectiveness of other responses.

MicroAssessment 14.1

First-line defenses prevent microbes from entering tissues. Sensor systems recognize invading microbes. Innate effector actions eliminate the invader.

1. List two sensor systems of the innate defenses.
2. Describe three innate effector actions.
3. In addition to peptidoglycan, which molecules unique to bacteria might pattern recognition receptors recognize?

14.2 ■ First-Line Defenses

Learning Outcome

2. Describe the first-line defenses, including the physical barriers, antimicrobial substances, and normal microbiota.

The body's borders are the first line of defense against invading microbes (**figure 14.2**). Some of these borders are thought of as being "inside" the body, but they directly contact the external environment. For example, the digestive tract, which begins at the mouth and ends at the anus, is simply a hollow tube that runs through the body, allowing intestinal cells to absorb nutrients from food that passes through (see figure 24.1); the respiratory tract is an open system that allows O_2 and CO_2 to be exchanged (see figure 21.1).

In this section, we will describe the general physical and chemical aspects of the anatomical barriers, as well as the protective contributions of the normal microbiota. These are described in more detail in the chapters dealing with each body system.

Physical Barriers

All exposed surfaces of the body are lined with epithelial cells (**figure 14.3**). These cells are tightly packed together and rest on a thin layer of fibrous material, the basement membrane.

Skin

The skin, an obvious visible barrier, is the most difficult for microbes to penetrate. It is composed of two main layers:

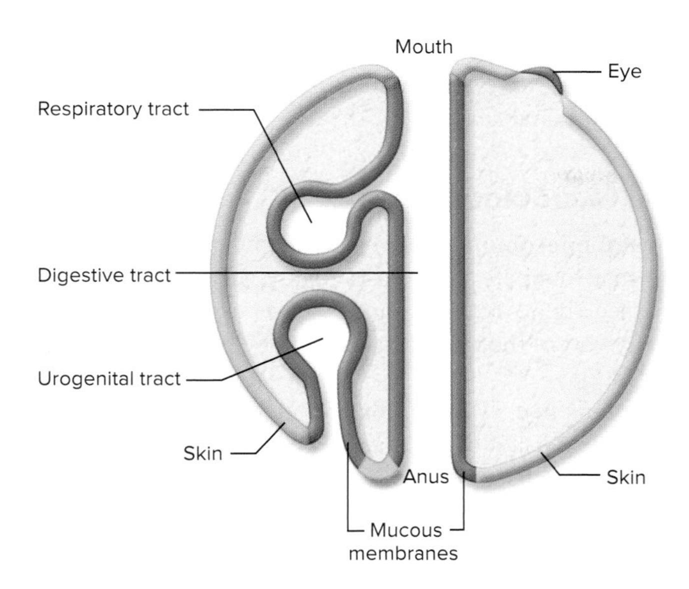

FIGURE 14.2 The Body's Borders These borders separate the interior of the body from the surrounding environment; they are the initial obstacles microorganisms must overcome to invade tissues. The skin is shown in tan, and mucous membranes in pink.

Why are the contents of the digestive tract considered to be in contact with the external environment?

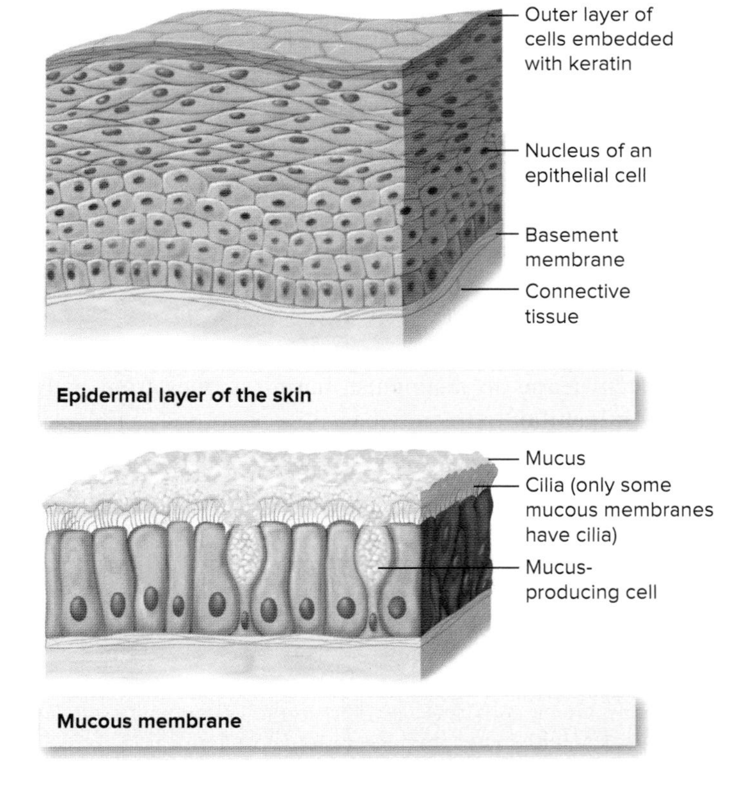

FIGURE 14.3 Epithelial Barriers Cells of these barriers are tightly packed together and rest on a layer of thin fibrous material, the basement membrane.

What is the purpose of the cilia on the respiratory epithelium?

the dermis and the epidermis (see figure 22.1). The dermis contains tightly woven fibrous connective tissue, making it extremely tough and durable (the dermis of cattle is used to make leather). The epidermis is composed of many layers of epithelial cells that become progressively flattened toward the surface. The outermost sheets are made up of dead cells filled with a water-repelling protein called keratin; the result is that the skin is a dry environment. The cells continually flake off, taking with them any microbes that might be adhering.

▶▶| anatomy, physiology, and ecology of the skin

Mucous Membranes

Mucous membranes line the digestive tract, respiratory tract, and genitourinary tract. They are constantly bathed with mucus or other secretions that help wash microbes from the surface. Most mucous membranes have mechanisms that move microbes toward areas where they can be eliminated. For example, peristalsis—the contractions of the intestinal tract—propels food and liquid toward the anus and also helps remove microbes. The respiratory tract is lined with ciliated cells; the hair-like cilia constantly beat in an upward motion, moving materials away from the lungs to the throat, where they can then be swallowed. This movement out of the respiratory tract is referred to as the mucociliary escalator. As with skin cells, there is a constant turnover of mucosal epithelial cells, and cells that are shed take attached microbes with them. **|◀◀ cilia**

MicroByte

A person sheds approximately 1 billion intestinal cells per hour.

Antimicrobial Substances

Skin and mucous membranes are protected by a variety of substances that inhibit or kill microorganisms (**figure 14.4**). For example, the salty residue that accumulates on skin as perspiration (sweat) evaporates inhibits all but salt-tolerant microbes.

Lysozyme, an enzyme that degrades peptidoglycan, is in tears, saliva, and mucus. It is also found within the body, in phagocytic cells, blood, and the fluid that bathes tissues. **|◀◀ lysozyme**

Peroxidases are part of systems that form antimicrobial compounds, using hydrogen peroxide (H_2O_2) in the process. Microorganisms that produce the enzyme catalase are less susceptible to the lethal effects of peroxidase systems because they can potentially convert hydrogen peroxide to water and O_2 before peroxidases have a chance to use it. Peroxidase systems are found in saliva, milk, body tissues, and phagocytes. **|◀◀ catalase**

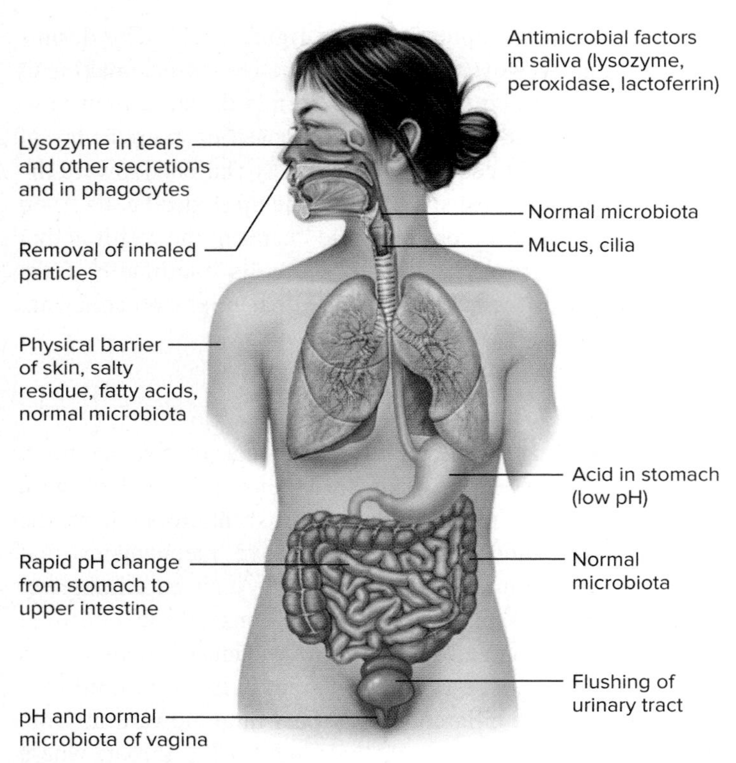

FIGURE 14.4 Antimicrobial Substances and the Normal Microbiota These play important roles in protecting the body's borders.

How is lysozyme antibacterial?

Lactoferrin is an iron-binding protein in saliva, mucus, milk, and some types of phagocytes. A similar compound, **transferrin,** is in blood and tissue fluids. By binding to iron, these proteins make it unavailable to microorganisms. Recall that iron is one of the major elements required by organisms, so withholding it prevents microbial growth. Some microorganisms can capture iron from the host, however, counteracting this defense. ⏮ major elements

Antimicrobial peptides (AMPs) are short chains of amino acids (usually 15–20 amino acids long) that have antimicrobial activity and are produced by a wide range of organisms. The role of these peptides in host defenses is not completely understood, but a group of AMPs called **defensins** are known to be very important in protecting epithelial borders. Defensins are positively charged AMPs that insert into microbial membranes, forming pores that damage cells. Certain epithelial cells produce and release defensins, preventing invasion of those body surfaces. Expression of defensins increases when microbial invasion is detected, helping the body to eliminate the infection. Defensins are also produced by phagocytes, which use them to destroy the microorganisms they have ingested. In addition to directly killing invading microorganisms, defensins and other AMPs also promote and direct various immune responses. Vitamin D plays a role in regulation of expression of some AMPs, which might explain why people who have a vitamin D deficiency are more susceptible to certain diseases.

Normal Microbiota (Flora)

The **normal microbiota (flora)** is the population of microorganisms that routinely grow on the body surfaces of healthy humans (see figure 16.1). Although these organisms are not technically part of the immune system, they provide considerable protection.

One protective effect of the normal microbiota is the competitive exclusion of pathogens. For example, the normal microbiota prevents pathogens from adhering to host cells by covering binding sites that might otherwise be used for attachment. The population also consumes available nutrients that could otherwise support the growth of less desirable organisms.

Some members of the normal microbiota produce compounds toxic to other bacteria. In the hair follicles of the skin, for instance, *Cutibacterium* species degrade lipids, releasing fatty acids that inhibit the growth of many pathogens. In the gastrointestinal tract, some strains of *E. coli* synthesize colicins, a group of proteins toxic to certain bacteria. *Lactobacillus* species growing in the vagina produce lactic acid as a fermentation end product, resulting in an acidic pH that inhibits the growth of some pathogens.

Disruption of the normal microbiota, which occurs when antibiotics are used, can predispose a person to various infections. Examples include antibiotic-associated diarrhea and pseudomembranous colitis, caused by the growth of toxin-producing strains of *Clostridium difficile* in the intestine, and vulvovaginitis, caused by excessive growth of *Candida albicans* in the vagina. ⏭ *Clostridium difficile* infection, ⏭ vulvovaginal candidiasis

The normal microbiota is also essential to the development of the immune system. As certain microbes are encountered, the system learns to distinguish harmless ones from pathogens. An inability to tolerate harmless microbes can result in chronic inflammatory conditions such as Crohn's disease.

MicroAssessment 14.2

Physical barriers that prevent microbes from entering the body include skin and mucous membranes. Antimicrobial substances, including lysozyme, peroxidase enzymes, lactoferrin, and defensins, are on body surfaces. The normal microbiota excludes pathogens and promotes immune system development.

4. How does peristalsis protect against intestinal infection?
5. What is the role of lactoferrin and transferrin?
6. How would damage to the ciliated cells of the respiratory tract predispose a person to infection?

14.3 ■ The Cells of the Immune System

Learning Outcome

3. Describe the characteristics and roles of granulocytes, mononuclear phagocytes, dendritic cells, and lymphocytes.

The cells of the immune system can move from one part of the body to another, traveling through the body's circulatory systems like vehicles on an extensive interstate highway system. They are always found in normal blood, but their numbers usually increase during infections, recruited from reserves of immature cells in the bone marrow. Some of the cell types are found primarily in the blood, but others leave the blood circulatory system and take up residence in various tissues. Certain cell types play dual functions, having crucial roles in both innate and adaptive immunity.

The formation and development of blood cells is called **hematopoiesis** (Greek for "blood" and "to make"). All blood cells, including those important in the body's defenses, originate from the same cell type, the **hematopoietic stem cell,** found in the bone marrow (**figure 14.5**). As with other types of stem cells, hematopoietic cells are capable of long-term self-renewal, meaning they can divide repeatedly. Hematopoietic stem cells are induced to develop into the various types of blood cells by a group of proteins called colony-stimulating factors (CSFs).

The general categories of blood cells and their derivatives include red blood cells, platelets, and white blood cells. Red blood cells, or **erythrocytes,** carry O_2 in the blood. Platelets, which are actually fragments arising from large cells called megakaryocytes, are important for blood clotting. White blood cells, or **leukocytes,** are important in all host defenses. Leukocytes can be divided into three broad groups: granulocytes, mononuclear phagocytes, and lymphocytes (**table 14.1**). Dendritic cells are a type of mononuclear phagocyte, but we will discuss them separately because of their distinct function.

Granulocytes

Granulocytes contain cytoplasmic granules filled with various compounds important for the cells' protective functions. The three types of granulocytes—neutrophils, basophils, and eosinophils—are named based on the staining properties of their granules.

Neutrophils efficiently engulf and destroy bacteria and other material. Their granules, which stain poorly, contain many enzymes and antimicrobial substances that help destroy the engulfed materials. Neutrophils are the most numerous

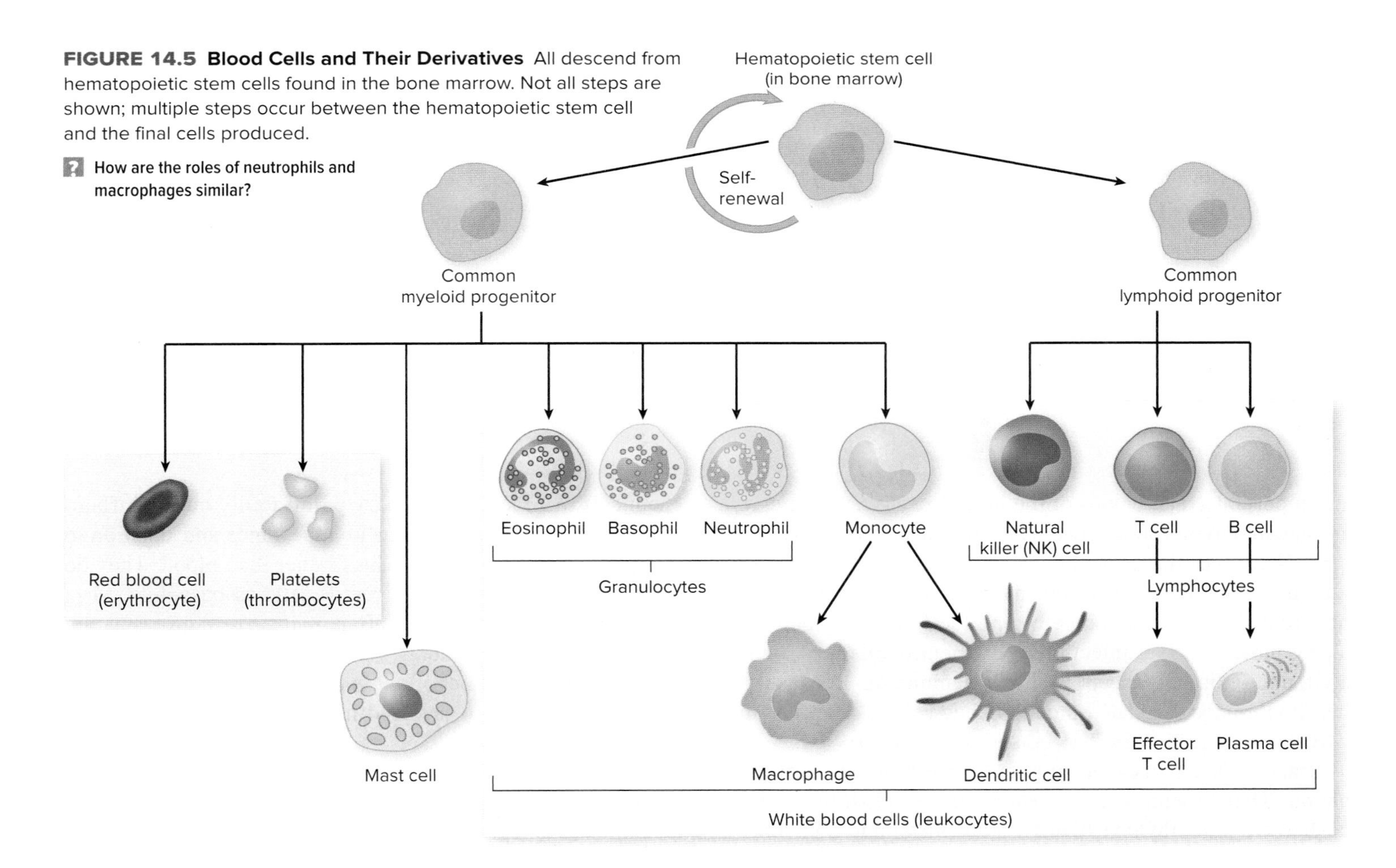

FIGURE 14.5 Blood Cells and Their Derivatives All descend from hematopoietic stem cells found in the bone marrow. Not all steps are shown; multiple steps occur between the hematopoietic stem cell and the final cells produced.

? How are the roles of neutrophils and macrophages similar?

TABLE 14.1 Leukocytes and Their Derivatives

Cell Type (Percentage of Blood Leukocytes)	Major Function and Other Characteristics
Granulocytes	
Neutrophils (polymorphonuclear neutrophilic leukocytes or PMNs, often called polys; 55–65%)	Phagocytosis; they also release substances that trap and destroy microbial invaders. Most abundant leukocyte in blood.
Eosinophils (2–4%)	Release chemicals that destroy eukaryotic parasites. Found mainly in tissues below the mucous membranes.
Basophils (0–1%), mast cells	Release histamine and other inflammation-inducing chemicals. Basophils are found in blood, whereas mast cells are present in most tissues.
Mononuclear Phagocytes	
Monocytes (3–8%)	Phagocytosis. Found in blood; they differentiate into either macrophages or dendritic cells when they migrate into tissues.
Macrophages	Phagocytosis; an important type of sentinel cell. Found in tissues; sometimes known by different names based on the tissue in which they are found.
Dendritic cells	Collect antigens from the tissues and then bring them to lymphocytes that gather in the secondary lymphoid organs (e.g., lymph nodes, spleen, appendix, tonsils); an important type of sentinel cell.
Lymphocytes (25–35%)	
B and T cells	Participate in the adaptive responses. Found in lymphoid organs (e.g., lymph nodes, spleen, appendix, tonsils, thymus, bone marrow); also in blood.
Innate lymphoid cells (NK cells are an example)	Various subsets have different roles and different locations.

and important granulocytes of the innate responses. They are also called polymorphonuclear neutrophilic leukocytes, polys, or PMNs, names that reflect the appearance of multiple lobes of their single nucleus. They normally account for over half of the circulating white blood cells, and their numbers increase during most bacterial infections. Few neutrophils are generally found in tissues, except during inflammation. Because of their importance in innate immunity, they will be described in more detail in section 14.7. ⏭ characteristics of neutrophils

Basophils are involved in allergic reactions and inflammation. Their granules, which stain dark purplish-blue with the basic dye methylene blue, contain histamine and other chemicals that increase capillary permeability during inflammation. **Mast cells** are similar in appearance and function to basophils but are found in tissues rather than blood. They do not come from the same precursor cells as basophils. Mast cells are important in the inflammatory response and are responsible for many allergic reactions.

Eosinophils are important in ridding the body of parasitic worms. They are also involved in allergic reactions, causing some of the symptoms associated with allergies, but reducing others. The granules of eosinophils, which stain red with the acidic dye eosin, contain antimicrobial substances and also histaminase, an enzyme that breaks down histamine.

Mononuclear Phagocytes

Mononuclear phagocytes make up the mononuclear phagocyte system (MPS) (**figure 14.6**). This grouping includes **monocytes**—which circulate in the blood—and the cell types that develop from them as they leave the bloodstream and migrate into tissues. **Macrophages** are a differentiated form of monocytes, meaning they have gained specialized properties. They are an important type of sentinel cell, present in nearly all tissues, and are particularly abundant in the liver, spleen, lymph nodes, lungs, and peritoneal (abdominal) cavity. When residing in tissues, they are sometimes given different names based on their location (see figure 14.6). Macrophages will be discussed in more detail in section 14.7. ⏭ characteristics of macrophages

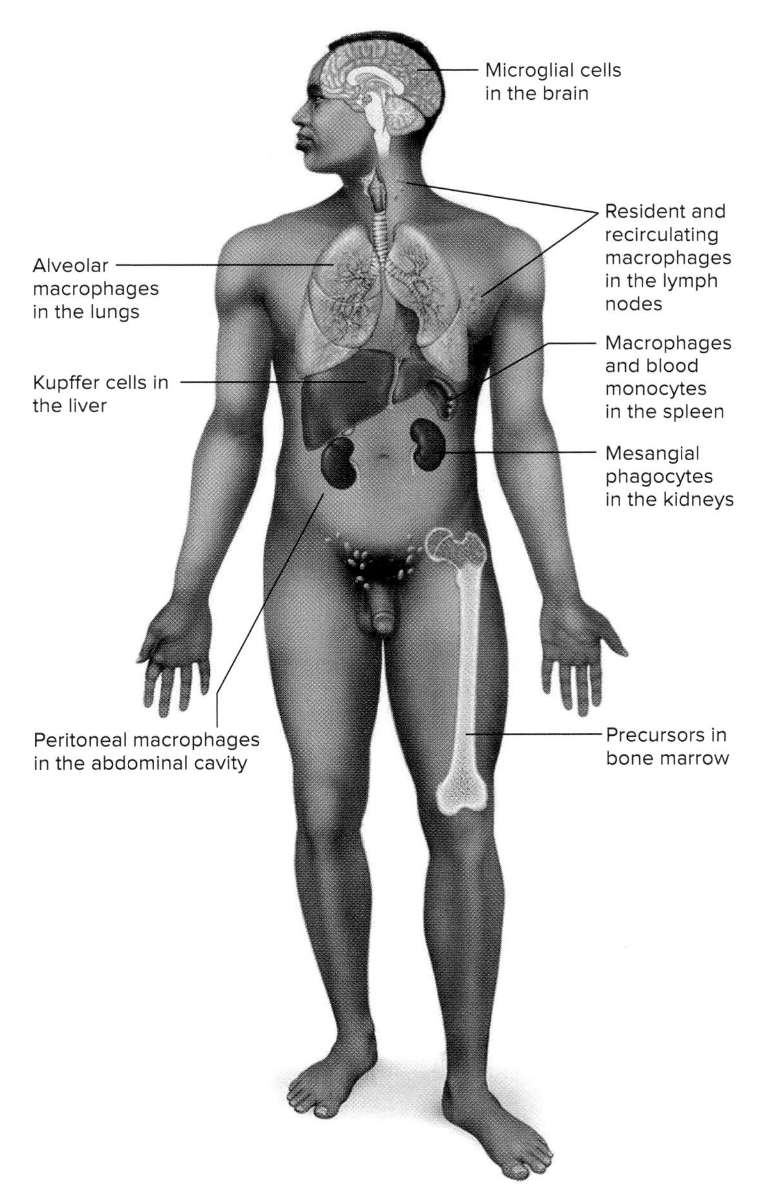

FIGURE 14.6 Mononuclear Phagocyte System Cells in this system sometimes have special names depending on their location—for example, Kupffer cells (in the liver) and alveolar macrophages (in the lung).

? Macrophages develop from which type of blood cell?

Dendritic Cells

Dendritic cells are sentinel cells that function as "scouts." They engulf material in the tissues and then bring it to the cells of the adaptive immune system for "inspection." Most dendritic cells develop from monocytes, but some develop from other cell types. Details regarding the interactions of dendritic cells with the cells of the adaptive immune response will be discussed in chapter 15.

Lymphocytes

Specific groups of **lymphocytes** are responsible for adaptive immunity—the focus of chapter 15. In contrast to the generic pattern recognition that characterizes the innate defenses, cells of the two major groups of lymphocytes, **B cells** and **T cells,** are remarkably specific in their recognition of antigen. These cell types generally reside in lymph nodes and other lymphatic tissues. A group of lymphocytes called **innate lymphoid cells (ILCs)** differ from B and T cells in that they lack specificity in their mechanism of antigen recognition. Several subsets of ILCs have been identified quite recently; they are common near mucous membranes, and appear to have multiple roles that can promote a balanced inflammatory response. One type of ILC, called a **natural killer (NK) cell,** has been recognized for quite some time. As its name implies, it kills certain types of cells—a role that will be described in chapter 15. ⏭ lymphatic tissues

MicroAssessment 14.3

Granulocytes include neutrophils, basophils, and eosinophils. Mononuclear phagocytes include monocytes and macrophages. Dendritic cells function as scouts for the adaptive immune system. Lymphocytes are responsible for adaptive immunity.

7. Which type of granulocyte is the most abundant?
8. How are most dendritic cells related to macrophages?
9. Why can bone marrow transplants be used to replace defective lymphocytes?

14.4 ■ Cell Communication

Learning Outcome

4. Describe the characteristics and roles of surface receptors, cytokines, and adhesion molecules in innate immunity.

Immune cells must communicate with each other in order to mount a coordinated response to microbial invasion. They do this through surface receptors, cytokines, and adhesion molecules.

Surface Receptors

Surface receptors can be viewed as the "eyes" and "ears" of a cell. They are proteins that generally span the cytoplasmic membrane, connecting the outside of the cell with the inside, allowing the cell to sense and respond to external signals. Each receptor is specific with respect to the compound or compounds it will bind; a molecule that can bind to a given receptor is called a **ligand** for that receptor. When a ligand binds to its surface receptor, the internal part of the receptor is modified. This change triggers some type of response by the cell, such as chemotaxis. Cells can alter the types and numbers of surface molecules they make, allowing them to respond to signals relevant to their immediate situation. ⏮ chemotaxis

Cytokines

Cytokines can be viewed as the "voices" of a cell. A cytokine produced by one cell diffuses to another and binds to the appropriate cytokine receptor of that cell. Binding of a cytokine to its receptor induces a change in the cell such as growth, differentiation, movement, or cell death. Cytokines act at extremely low concentrations, having local, regional, or systemic effects. Cytokines include:

- **Chemokines.** These are important in chemotaxis of immune cells. Certain types of cells have receptors for chemokines, allowing the cells to sense the location where they are needed, such as an area of inflammation.
- **Colony-stimulating factors (CSFs).** These are important in the multiplication and differentiation of leukocytes (see figure 14.5). When more leukocytes are needed during an immune response, a variety of different colony-stimulating factors direct immature cells into the appropriate maturation pathways.
- **Interferons (IFNs).** These were discovered because of their antiviral effects, but they have several roles in the host defenses. They are important in a number of regulatory mechanisms, stimulating the responses of some cells and inhibiting others. In section 14.5, we will focus on the effector action of IFNs during viral infection. ⏭ the interferon response
- **Interleukins (ILs).** These are produced by leukocytes and have diverse, often overlapping, functions. As a group, they are important in both innate and adaptive immunity.
- **Tumor necrosis factor (TNF).** This was discovered because of its role in killing tumor cells, a characteristic reflected by the name, but it has multiple roles. It helps initiate the inflammatory response and triggers one process of "cell suicide," a programmed cell death called apoptosis. ⏭ apoptosis

Groups of cytokines often act together or in sequence to generate a response. For example, certain cytokines referred to as **pro-inflammatory cytokines** (TNF, IL-1, IL-6, and others) contribute to inflammation. Others are involved in promoting antibody responses (IL-4 and others). A different group stimulates certain types of T cells (IL-2, an IFN, and others). The source and effects of various cytokines are listed in **table 14.2.**

MicroByte

HIV takes advantage of two chemokine receptors, CCR5 and CXCR4, using them as attachment sites for infection.

Adhesion Molecules

Adhesion molecules on the surface of cells allow those cells to "grab" other cells. For example, when phagocytic cells in the blood are needed in tissues, the endothelial cells that line the blood vessels synthesize adhesion molecules that bind to passing phagocytic cells. This slows the rapidly moving phagocytes, allowing them to then leave the bloodstream. Cells also use adhesion molecules to attach to other cells so that one cell can deliver cytokines or other molecules directly to another cell.

TABLE 14.2 Some Important Cytokines

Cytokine	Source	Effect
Chemokines	Various cells	Chemotaxis
Colony-Stimulating Factors (CSFs)	Various cells	Stimulate growth and differentiation of different kinds of leukocytes
Interferons	Various cells	Regulate immune responses; antiviral
Interleukins (ILs)		
IL-1	Macrophages, epithelial cells	T-cell activation; macrophage activation; induces fever
IL-2	T cells	T-cell proliferation
IL-4	T cells, mast cells	Promotes antibody responses
IL-6	T lymphocytes, macrophages	T- and B-cell growth; inflammatory response; fever
Tumor Necrosis Factor (TNF)	Macrophages, T cells, NK cells	Promotes inflammation; cytotoxic for some tumor cells; regulates certain immune functions

MicroAssessment 14.4

Surface receptors allow a cell to detect molecules present outside that cell. Cytokines provide cells with a mechanism of communication. Adhesion molecules allow one cell to adhere to another.

10. What is a ligand?

11. How do cytokines function?

12. How could colony-stimulating factors be used as a therapy?

14.5 ■ Pattern Recognition Receptors (PRRs)

Learning Outcomes

5. Describe the significance of pattern recognition receptors (PRRs) in the immune response.
6. Compare and contrast the various PRRs that monitor a cell's surroundings, the material ingested by the cell, and the cell's cytoplasm.
7. Explain how the interferon response prevents viral replication.

Pattern recognition receptors (PRRs) are sensors that allow the body's cells to "see" signs of microbial invasion. If a cell detects microbial invasion, it produces cytokines to alert other components of the defense system. Although PRRs were discovered relatively recently and much is still being learned about them, they have stimulated tremendous interest in innate immunity. In fact, their discovery changed the way that immunologists view the immune system. In the past, the topics covered in this chapter were considered "non-specific," but we now know that the PRRs of innate immunity help the body's cells to recognize the general category of an infectious agent, which plays an important role in shaping the immune response to that agent.

Many PRRs detect components of certain groups of microbes—for example, cell wall–associated compounds (peptidoglycan, teichoic acid, lipolysaccharide, and lipoproteins), flagellin subunits, and microbial nucleic acid. These compounds are called **microbe-associated molecular patterns (MAMPs),** reflecting the fact that they originate from microbes. They are also called PAMPS (for pathogen-associated molecular patterns), but they are not exclusive to pathogens, which is why many microbiologists prefer the term MAMPs. Some PRRs recognize **damage-associated molecular patterns (DAMPs):** molecules that indicate host cell damage.

The various PRRs and the cell types that express them allow immune responses to be tailored to the category of pathogen and the situation. For example, if a macrophage's PRRs detect bacterial products, then that cell produces pro-inflammatory cytokines, leading to an inflammatory response. If a dendritic cell's PRRs detect bacterial products, then that cell relays that information to lymphocytes, allowing those cells to mount an appropriate response. If a virally infected cell's PRRs detect viral nucleic acid, then that cell produces an interferon. The interferon alerts immune cells to the presence of a virus, and also promotes an antiviral response in nearby tissue cells.

The outcome of certain diseases is influenced by PRR-generated signals from sentinel cells and infected cells. In some cases the signals induce a protective response, but in other cases the response can be excessive and therefore damaging. In addition, people who have mutations in the genes encoding some PRRs are more likely to develop certain inflammatory diseases or autoimmune diseases. Because of this, scientists are working to learn more about the mechanisms and outcomes of pattern recognition, in hopes that a better understanding will lead to new treatment options for a wide range of disease states—from illnesses caused by pathogens, to those caused by immune system dysfunction.

PRRs are located in three distinct locations on or in cells: (1) on the cell surface, (2) in endosomes and phagosomes, and (3) free in the cytoplasm (**figure 14.7**). Because of these locations, PRRs provide cells with information about not only which microbes are present, but also whether the microbes are inside or outside a host cell. The signals from PRRs in different locations can complement each other, provoking a stronger response when an invader is detected.

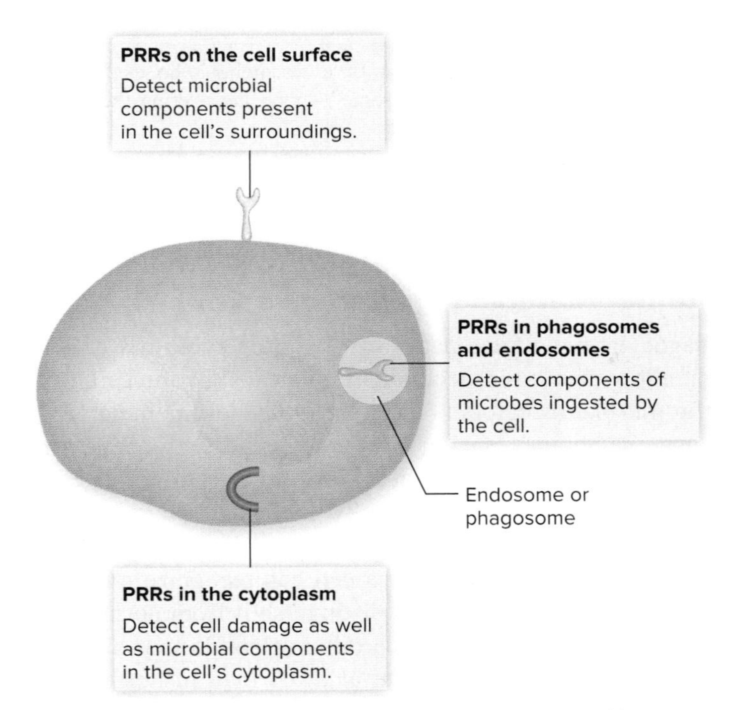

FIGURE 14.7 Locations and Roles of Pattern Recognition Receptors (PRRs)

Why would a cell produce cytokines when a PRR detects microbial components?

Pattern Recognition Receptors (PRRs) that Monitor a Cell's Surroundings

Sentinel cells such as phagocytes and cells that line blood vessels and other sterile body sites have pattern recognition receptors (PRRs) on their cell surface, anchored in the cytoplasmic membrane. These allow the cells to detect invaders in the surrounding environment. The most well-characterized of these PRRs are the **toll-like receptors (TLRs).** A number of different TLRs have been described (at least 10 in humans); some are on the cell surface, and others (discussed in the next section) are in endosomes and phagosomes. Each TLR recognizes a distinct compound or group of compounds associated with microbes (**figure 14.8**). TLRs anchored in the cytoplasmic membrane generally detect components of the outermost layers of microbial cells, including lipopolysaccharide (LPS), lipoproteins, and flagellin.

Other PRRs that monitor the cell's surroundings have also been discovered, such as membrane-anchored versions of a group called C-type lectin receptors (CLRs). These receptors bind to certain carbohydrate molecules often found on the surface of microorganisms. Dendritic cells have TLRs as well as CLRs, so they are able to gather a great deal of information about invaders they encounter. They then pass that information on to lymphocytes, thereby helping to shape that response.

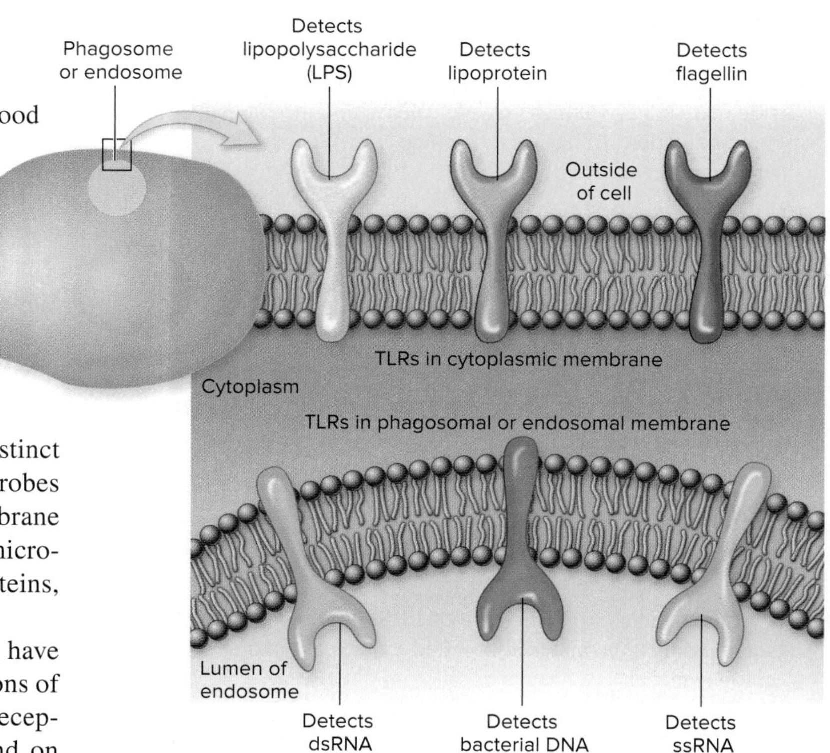

FIGURE 14.8 Toll-Like Receptors (TLRs) These pattern recognition receptors are anchored in membranes of sentinel cells, allowing these cells to "see" microbial compounds that originated outside the cell. Not all of the compounds recognized by TLRs are shown in this figure.

? From the standpoint of defending the human body, why would it be beneficial for the cells that line the blood vessels to have TLRs on their surface?

MicroByte

"Toll-like" reflects the reaction of a researcher who exclaimed "toll"—German slang for "awesome"—upon discovering the gene for a similar receptor in a fruit fly.

Pattern Recognition Receptors (PRRs) That Monitor Material Ingested by a Cell

Phagocytic cells have pattern recognition receptors (PRRs) that allow them to inspect material ingested by the cell. Specific toll-like receptors (TLRs) are in phagosomal and endosomal membranes, facing the lumen of the organelle (the inside of the organelle, which is exposed to the organelle's contents). These TLRs typically recognize characteristics of nucleic acids that indicate a microbial origin. Although it might seem surprising that a cell can recognize microbial nucleic acid, several features distinguish it from normal host nucleic acid. For one thing, certain nucleotide sequences are much more common in bacterial DNA than in normal host cell DNA, and these can be recognized by a TLR. In addition, the genome of RNA viruses is often double-stranded during the viral replication cycle. Even DNA viruses may generate long dsRNA because both strands of DNA are sometimes used as templates for transcription, leading to production of complementary RNA molecules. Long pieces of cellular RNA are typically not double-stranded because only one DNA strand in a gene is used as a template for mRNA synthesis.

⏮ endosome, ⏮ phagosome

Certain TLRs in endosomes or phagosomes are not fully functional until the vesicle fuses with lysosomes, exposing the contents of the vesicle to low pH and digestive enzymes (see figure 3.49). Because of this requirement, the TLRs are less likely to contact host cell nucleic acid, thereby avoiding an inappropriate response.

Pattern Recognition Receptors (PRRs) That Monitor a Cell's Cytoplasm

Most host cells have at least one category of pattern recognition receptors (PRRs) in their cytoplasm. These PRRs are

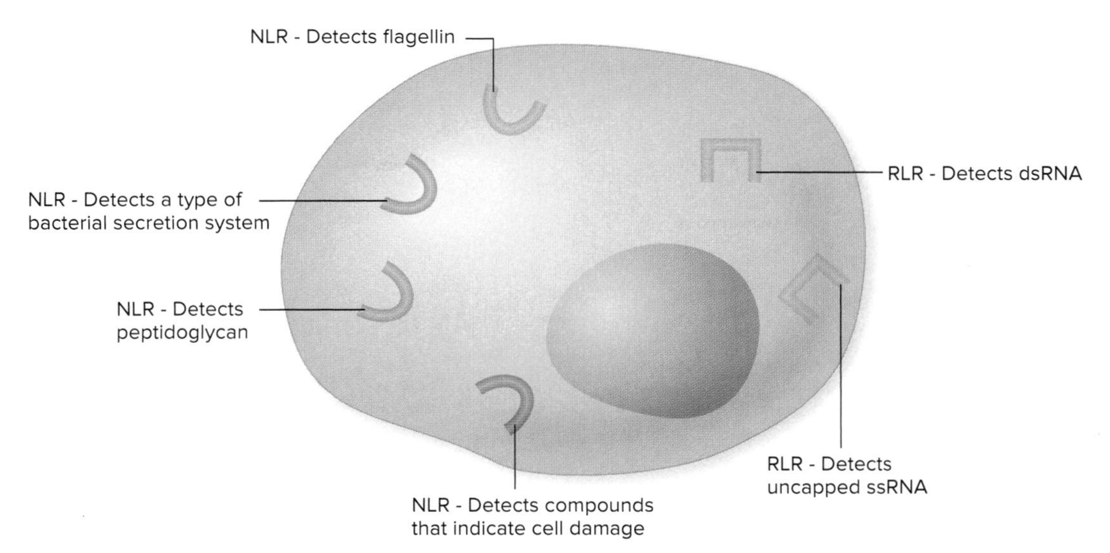

FIGURE 14.9 NOD-Like Receptors (NLRs) and RIG-Like Receptors (RLRs) These pattern recognition receptors are found within cells and detect either microbial components or signs of cell damage. Not all NLRs and RLRs are shown in the illustration.

With respect to the source of microbial compounds detected, how do NLRs and RLRs differ from TLRs?

not nearly as well-characterized as the membrane-associated PRRs, but it is clear that they allow cells to monitor their own cytoplasmic contents for signs of invasion. Examples of cytoplasmic PRRs include RIG-like receptors (RLRs) and NOD-like receptors (NLRs) (**figure 14.9**).

RIG-like receptors (RLRs) are cytoplasmic proteins that detect viral RNA, and are found in most cell types. Because RLRs are so widespread, they represent a very important early-warning system for viral infections—any virally infected cell can alert neighboring cells that a virus is present. RLRs can distinguish viral RNA from normal cellular RNA because at least two characteristics differ. As mentioned earlier, viral RNA is often double-stranded. In addition, viral RNA often lacks a cap; recall that a process called capping normally modifies the 5′ end of cellular RNA after transcription. ⏮ capping

NOD-like receptors (NLRs) are cytoplasmic proteins that detect either microbial components or signs of cell damage. They are found in a variety of cell types, but are particularly important in macrophages and dendritic cells. At least 23 NLRs have been described, but details about the roles of most of them are still being uncovered. The microbial products detected by the most well-characterized NLRs include peptidoglycan, flagellin, and components of a secretion system that some pathogenic bacteria use to inject various molecules into host cells. When certain NLRs in macrophages and dendritic cells detect invasion, they combine with other proteins in the cytoplasm to form a structure called an **inflammasome.** This structure triggers inflammation by activating a potent pro-inflammatory cytokine. ⏭ secretion system

An Outcome of Cytoplasmic Pattern Recognition: The Interferon Response

When a cell's cytoplasmic pattern recognition receptors (PRRs) detect viral RNA, the cell responds by synthesizing and secreting a type of interferon (IFN) that induces nearby cells to develop an antiviral state (**figure 14.10**). Interferon molecules attach to specific receptors on neighboring cells, causing the cells to express what can be viewed as inactive "suicide enzymes." For convenience, we will refer to these collectively as inactive antiviral proteins (iAVPs). These iAVPs can be activated by viral dsRNA. Once activated, the antiviral proteins (AVPs) degrade mRNA and stop protein synthesis, leading to apoptosis of that cell. A key feature of this response is that the iAVPs are activated by long dsRNA, which is typically found only in virally infected cells. Thus, when cells bind interferon, only the infected ones will be sacrificed. Their uninfected counterparts remain functional but are prepared to undergo apoptosis should they become infected. ⏭ apoptosis

MicroAssessment 14.5

Sentinel cells use pattern recognition receptors (PRRs) to detect microbial components in the surroundings and in ingested material. Many cells use PRRs to determine if they are infected. Viral RNA triggers an interferon response.

13. Give three examples of MAMPs.

14. If a cell produces antiviral proteins (AVPs), what happens to that cell when those proteins encounter long dsRNA?

15. Why would the discovery of TLRs alter the view that innate immunity is non-specific?

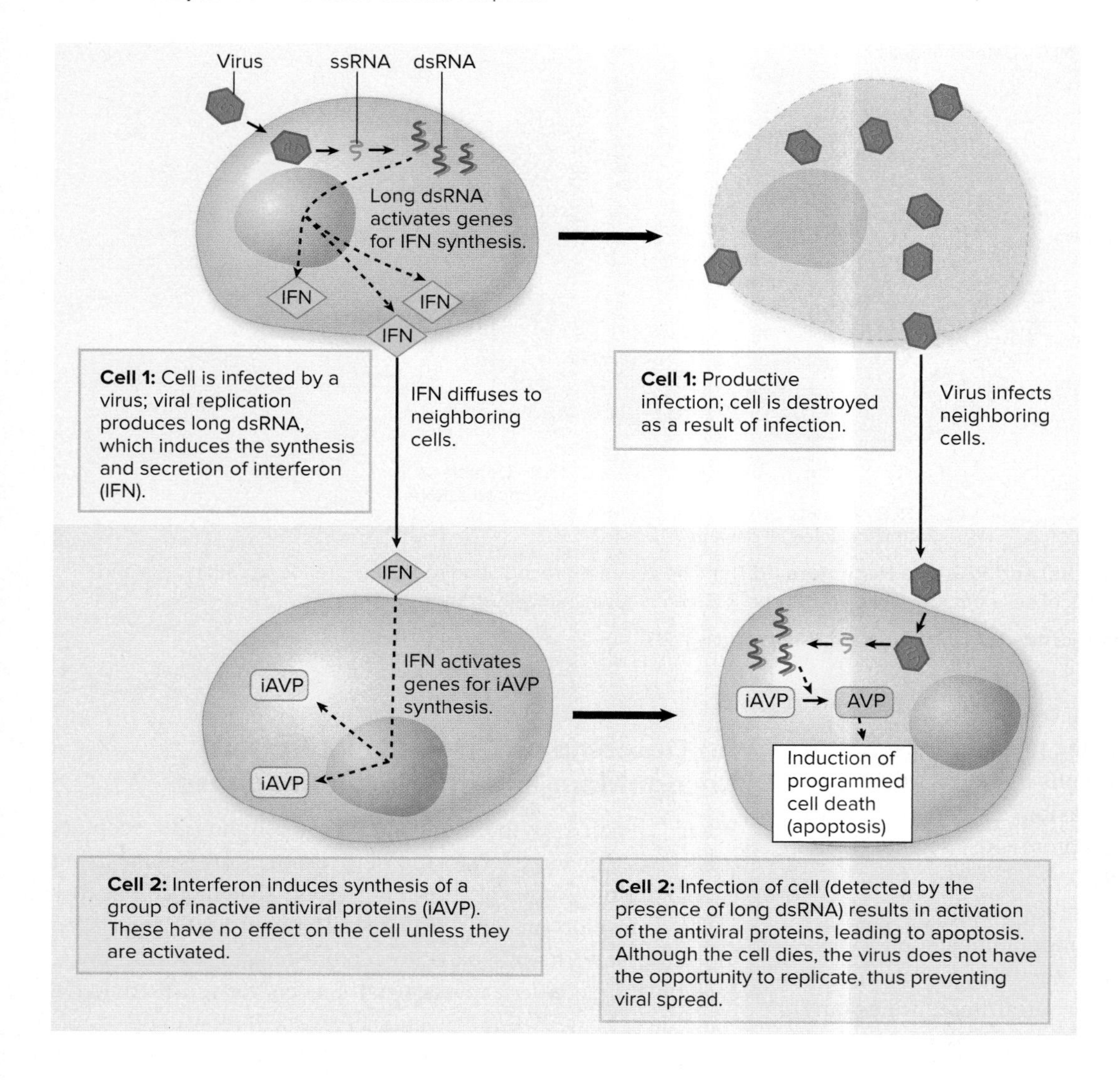

FIGURE 14.10 Antiviral Effects of Interferon

Why would it be beneficial to a host for a virally infected cell to undergo apoptosis?

14.6 ■ The Complement System

Learning Outcome

8. Describe the three pathways that lead to complement system activation and the three outcomes of activation.

The **complement system,** often simply referred to as complement, is a series of proteins that circulate in the blood and the fluid that bathes the tissues (**figure 14.11**). The proteins routinely circulate in an inactive form, but certain signals that indicate the presence of microbial invaders start a reaction cascade that rapidly activates the system. The activated complement proteins have specialized functions that help remove and destroy the invader.

The name of the complement system is derived from the observation that the system "complements" the function of antibodies, a component of adaptive immunity. Each of the major complement system proteins has been given a number along with the letter C (for complement). The nine major proteins, C1 through C9, were numbered in the order of their discovery and not the order in which they react. When a complement protein is split into two fragments, those fragments are distinguished by adding a lowercase letter to each name. For example, C3 is split into C3a and C3b.

Complement System Activation

The complement system can be activated by three different pathways that converge when a complex called C3 convertase is formed (see figure 14.11). C3 convertase then splits C3, leading to additional steps of the activation cascade.

Alternative Pathway

The name of the **alternative pathway** may seem to imply that the pathway is "second choice," but it actually reflects the fact that the pathway was not discovered first. The pathway is quickly and easily triggered, providing vital early warning that an invader is present. The alternative pathway is triggered when C3b binds to foreign cell surfaces. The binding of C3b allows other complement proteins to then attach, eventually forming the C3 convertase. What might seem confusing is the fact that C3b is a product of complement activation, yet it also triggers the alternative pathway. How can it be both a product and a trigger? This can occur because C3 is somewhat unstable, and spontaneously splits to C3a and C3b at a low rate even when the complement system has not been activated. The C3a and C3b formed this way are rapidly inactivated by regulatory proteins, but some C3b is always present to trigger the alternative pathway when needed.

Lectin Pathway

Activation of the complement system via the **lectin pathway** involves pattern recognition molecules called mannose-binding lectins (MBLs). These bind to certain arrangements of multiple mannose molecules; mannose is a type of carbohydrate commonly found on the surface of some microbial cells, particularly bacteria and fungi. Once an MBL attaches to a surface,

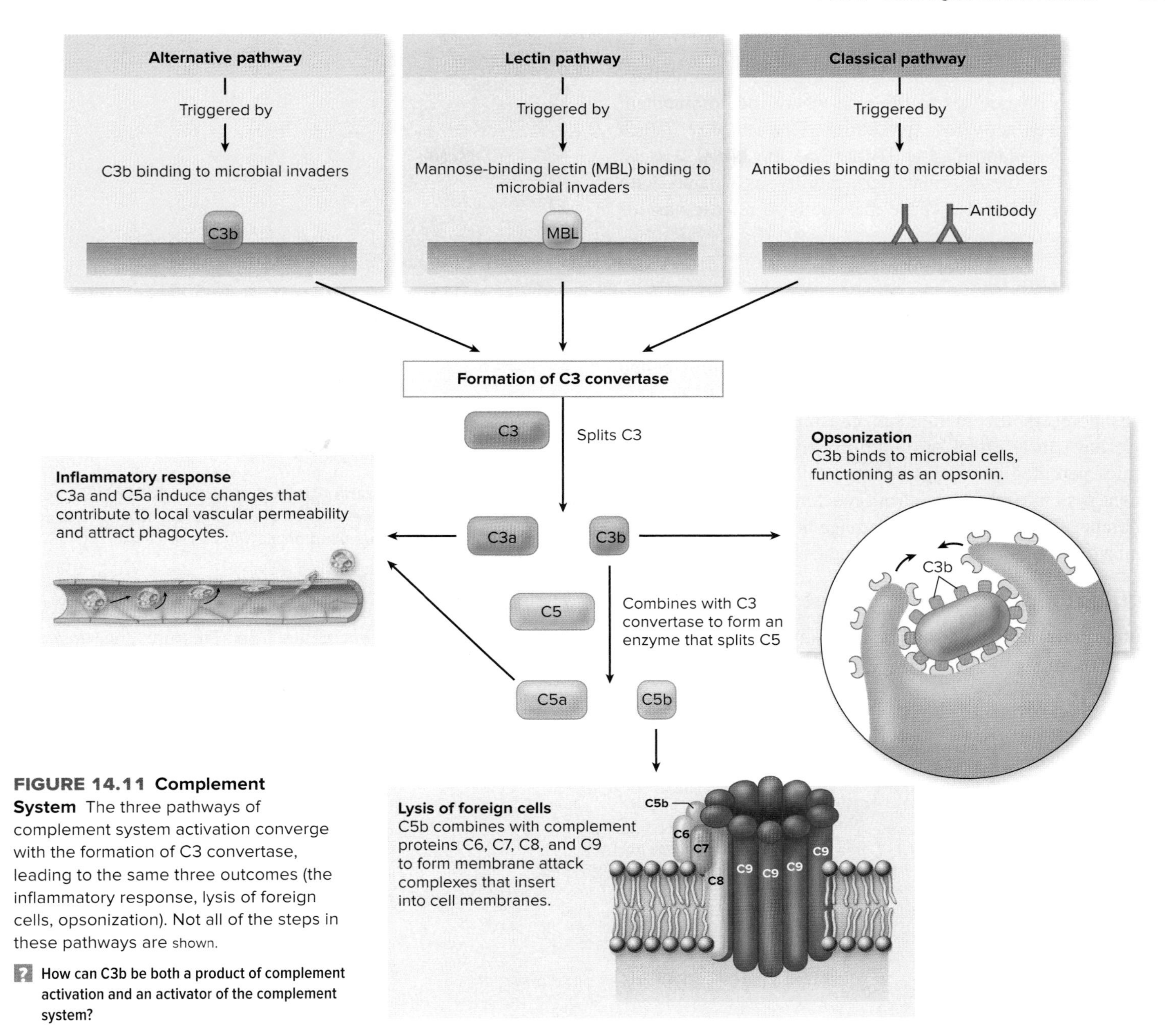

FIGURE 14.11 Complement System The three pathways of complement system activation converge with the formation of C3 convertase, leading to the same three outcomes (the inflammatory response, lysis of foreign cells, opsonization). Not all of the steps in these pathways are shown.

? How can C3b be both a product of complement activation and an activator of the complement system?

it can interact with other complement system components to form a C3 convertase. ⏮ **mannose**

Classical Pathway

Complement system activation by the **classical pathway** requires antibodies. When antibodies bind to an antigen (forming an antigen-antibody complex, also called an immune complex), they interact with the same complement system component involved with the lectin pathway to form a C3 convertase.

Effector Functions of the Complement System

Activation of the complement system eventually leads to three major protective outcomes: opsonization, an inflammatory response, and lysis of foreign cells (see figure 14.11).

Opsonization

The C3b concentration increases substantially when the complement system is activated, and these molecules bind to bacterial cells or other foreign particles. This has two effects: (1) continued complement activation via the alternative pathway, and (2) **opsonization.** Material that has been opsonized (meaning "prepared for eating") is easier for phagocytes to bind to and engulf. This is because phagocytes have receptors that attach specifically to molecules referred to as **opsonins** (in this case, C3b).

MicroByte

Opsonization is like coating a microbe with one side of a strip of Velcro, and a phagocyte with the other, allowing phagocytes to easily attach to microbes.

Inflammatory Response

The complement component C5a is a potent chemoattractant, drawing phagocytes to the area where the complement system has been activated. In addition, C3a and C5a induce changes in the endothelial cells that line the blood vessels, contributing to the vascular permeability associated with inflammation. They also cause mast cells to release various pro-inflammatory cytokines.

Lysis of Foreign Cells

Complexes of complement system proteins (C5b, C6, C7, C8, and multiple C9 molecules) spontaneously assemble in cell membranes, forming doughnut-shaped structures called **membrane attack complexes (MACs)** (**figure 14.12**). This creates pores in the membrane, causing the cells to lyse. MACs have little effect on Gram-positive bacteria because the thick peptidoglycan layer of these cells prevents the complement system components from reaching their cytoplasmic membranes. In contrast, MACs damage both the outer and the cytoplasmic membranes of Gram-negative bacteria.

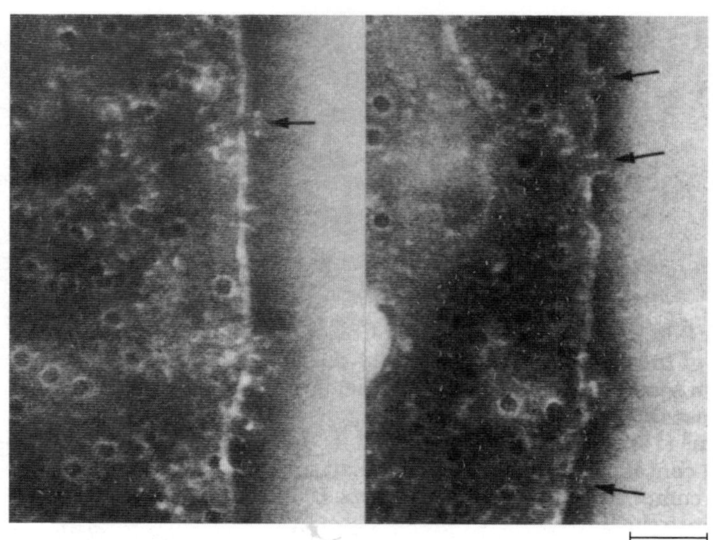

FIGURE 14.12 Membrane Attack Complexes (MACs) Electron micrographs of MACs in membranes of two different cells (arrows point to side views; dark dots are head-on views). ©Sucharit Bhakdi

? How do MACs cause cells to lyse?

Regulation of the Complement System

A number of different control mechanisms prevent host cells from activating the complement system and also protect them from the effector functions of the complement proteins. For example, molecules in host cell membranes bind regulatory proteins that quickly inactivate C3b. This prevents host cell surfaces from triggering the alternative pathway of complement regulation (**figure 14.13**). It also prevents host cells from

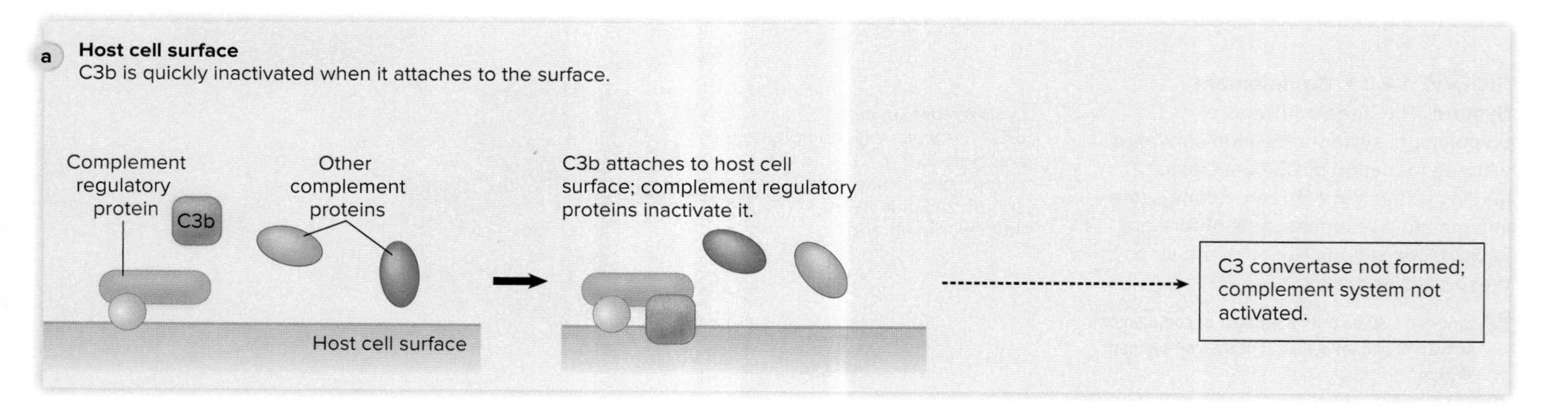

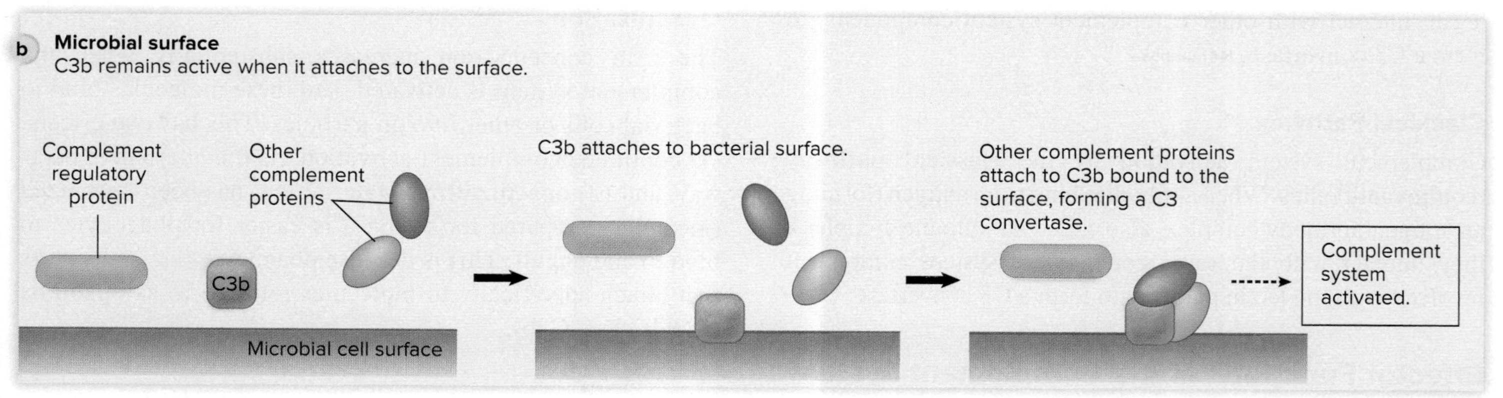

FIGURE 14.13 Complement Regulatory Proteins Inactivate C3b **(a)** Molecules in host cell membranes bind regulatory proteins that quickly inactivate C3b. **(b)** Most microbial cell surfaces do not bind the regulatory proteins, so they trigger the alternative pathway of complement activation.

? Some pathogens attract complement regulatory proteins to their surfaces. How would this help the pathogens avoid destruction?

being opsonized. Most microbial cell surfaces do not bind the regulatory proteins, but as we will discuss in chapter 16, some pathogens have mechanisms to hijack the host's protective mechanisms.

MicroAssessment 14.6

The complement system can be activated by three different pathways that each lead to opsonization, an inflammatory response, and lysis of foreign cells.

16. Describe the outcome of opsonization.

17. The body's own cells do not trigger the alternative pathway of complement system activation. Explain why this is so.

18. Some pathogens produce C5a peptidase, an enzyme that destroys C5a. How would this benefit the pathogen?

14.7 ■ Phagocytosis

Learning Outcomes

9. Outline the steps of phagocytosis.

10. Compare and contrast the roles of macrophages and neutrophils.

Phagocytes routinely engulf and digest material, including invading microbes. In routine situations, such as when microbes enter through a minor skin wound, resident macrophages in the tissues destroy the relatively few invaders that enter. If the microbes are not rapidly cleared, macrophages produce cytokines to recruit additional phagocytes—particularly neutrophils—for extra help.

The Process of Phagocytosis

Phagocytosis involves a series of steps (**figure 14.14**). These are particularly important medically, because most pathogens have evolved the ability to evade one or more of them, a topic explored in chapter 16.

Chemotaxis ①

Phagocytic cells are recruited to the site of infection or tissue damage by chemicals that act as chemoattractants. These include products of microorganisms, phospholipids released by injured host cells, chemokines, and the complement system component C5a.

Recognition and Attachment ②

Phagocytic cells use various receptors to bind invading microbes either directly or indirectly. For example, direct binding occurs when a phagocyte's receptors bind mannose. Indirect binding happens when a particle has first been opsonized. Opsonins are extracellular proteins that tag particles

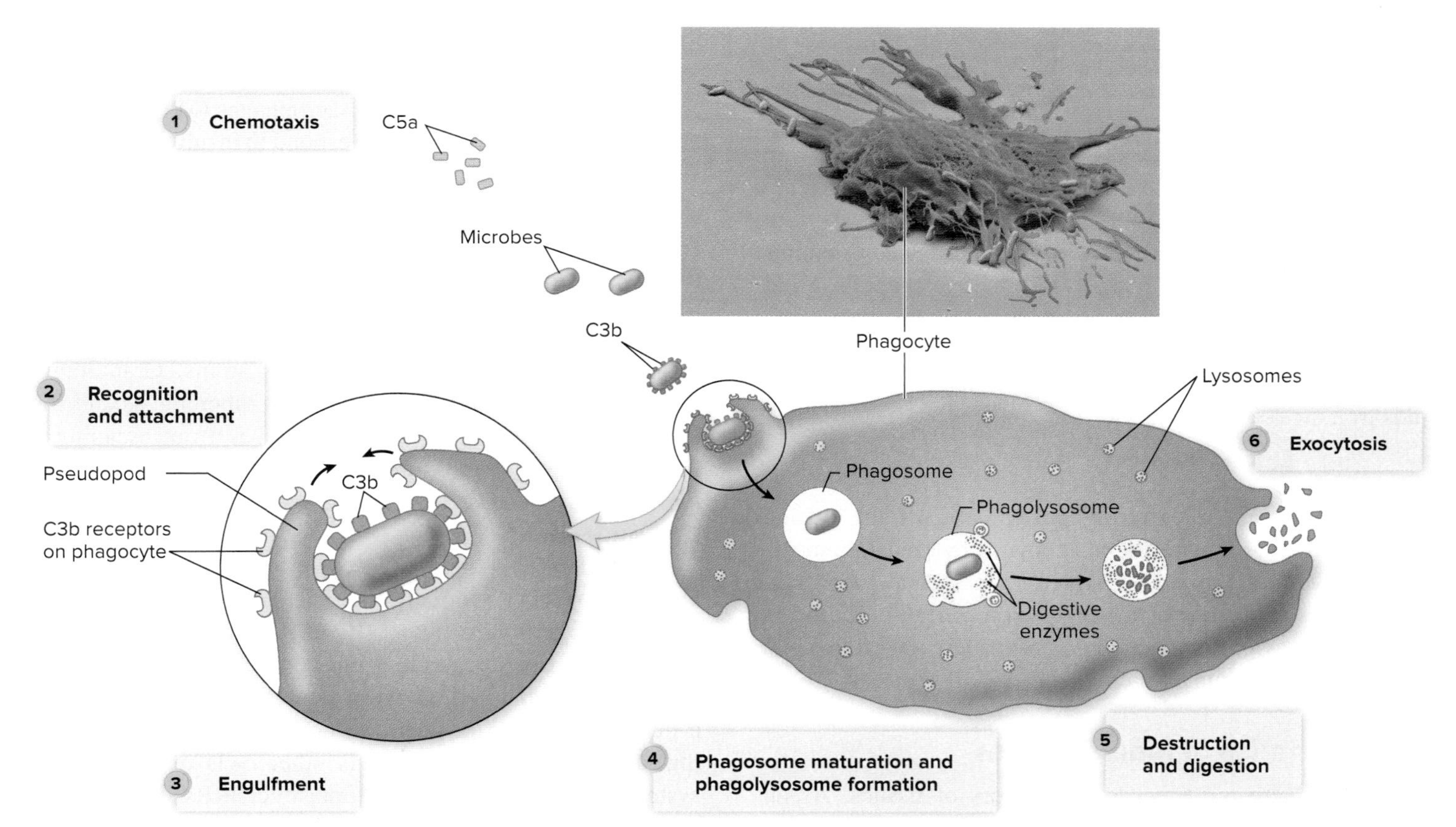

FIGURE 14.14 Phagocytosis This diagram shows a microbe that has been opsonized by the complement protein C3b; certain classes of antibodies can also function as opsonins. ©Meckes/Ottawa/SPL/Science Source

What would happen if a bacterium could prevent the phagosome from fusing with lysosomes?

for phagocytosis and include the complement component C3b and certain classes of antibody molecules. Phagocytes have specific receptors for opsonins, so opsonized material is easier for the phagocyte to attach to and engulf.

Engulfment ③

Once the phagocyte has attached to a particle, it sends out pseudopods that surround and engulf the material. This action brings the material into the cell, enclosed in a **phagosome.** If a phagocyte encounters something too large to engulf, it releases its toxic contents as a means of destroying it. ⏮ pseudopod, ⏮ phagosome

Phagosome Maturation and Phagolysosome Formation ④

Initially, a phagosome has no antimicrobial capabilities, but it matures to develop these; for example, the pH becomes progressively more acidic. The maturation stages are highly regulated, and depend on the type of material ingested. If a phagocyte's toll-like receptors (TLRs) indicate that a phagosome contains microbial components, for instance, then that phagosome will have a different fate than if the contents are only host cell material. Eventually, the phagosome fuses with enzyme-filled lysosomes, forming a **phagolysosome.** ⏮ lysosome

Destruction and Digestion ⑤

A number of factors within the phagolysosome work together to destroy an engulfed invader. O_2 consumption increases dramatically—a phenomenon called respiratory burst—allowing an enzyme to produce reactive oxygen species (ROS), which are toxic. Another enzyme makes nitric oxide, which reacts with ROS to produce additional toxic compounds. Special pumps move protons into the phagolysosome, further lowering the pH. The various enzymes contributed by the lysosomes degrade peptidoglycan and other components. Antimicrobial peptides damage membranes of the invader, and lactoferrin binds iron. ⏮ reactive oxygen species

Exocytosis ⑥

The phagolysosome releases undigested debris to the outside of the cell by fusing with the phagocyte's cytoplasmic membrane. As you will learn in chapter 15, a special process in macrophages puts some of the ingested material on the cell's surface as a way of displaying bits of invaders to certain cells of the adaptive immune system. ⏮ exocytosis

Characteristics of Macrophages

Macrophages are the everyday protectors of tissues. As an analogy, they would be the "beat cops" that protect city streets. Macrophages routinely phagocytize dead cells and debris, but are ready to destroy invaders and call in reinforcements when needed. They are always present in tissues, where they either slowly wander or remain stationary. These phagocytic cells play an essential role in every major tissue in the body.

Macrophages live for weeks to months, or even longer. They maintain their killing power by continually regenerating their lysosomes. As macrophages die, circulating monocytes—which can differentiate into macrophages—leave the blood and migrate to the tissues to replace them. Monocyte migration increases in response to invasion and tissue damage.

Macrophages can develop into **activated macrophages** if surrounding cells produce certain cytokines or other chemicals. For example, certain pro-inflammatory cytokines activate macrophages, giving them greater killing power and encouraging a continued inflammatory response; these "killer macrophages" are sometimes referred to as M1 macrophages. Other chemicals cause macrophages to produce cytokines that lessen the inflammatory response, as well as promote wound healing and tissue repair; these "healer macrophages" are sometimes referred to as M2 macrophages. Note, however, that the differentiation of M1 versus M2 macrophages is not fixed; a macrophage can change its role in response to environmental cues, so presenting the types as two polarized alternatives is not entirely accurate. Nevertheless, the relative activities of macrophages seem to drive certain disease states. In type 2 diabetes, for example, an overabundance of inflammation-promoting (M1) macrophages in the adipose tissue decreases the sensitivity of various tissues to insulin. In contrast, anti-inflammatory macrophages (M2) promote an environment where tissue cells respond to insulin. ⏭ macrophage activation

If activated macrophages fail to destroy microbes, the phagocytes can fuse together to form **giant cells.** Macrophages, giant cells, and T cells form concentrated groups called **granulomas** that wall off and retain organisms or other material that cannot be destroyed; again, this is an example of the cooperation between defense systems. Granulomas, which are part of the disease process in tuberculosis and several other illnesses, prevent the microbes from escaping to infect other cells (see figure 21.20). Unfortunately, they also harm the host because they interfere with normal tissue function. ⏭ tuberculosis

In addition to their phagocytic activities, macrophages are very important sentinel cells. They are well-equipped with pattern recognition receptors (PRRs), allowing them to "see" microbes in the surrounding environment, in material they have ingested, and in their cytoplasm. If a macrophage detects a microbe that has invaded the body, it produces cytokines that alert and stimulate various other cells of the immune system. If a macrophage's cytoplasmic PRRs detect microbial components, indicating that the phagocytic cell itself is infected, an inflammasome forms, an event that triggers a strong inflammatory response. ⏮ inflammasome

Characteristics of Neutrophils

Neutrophils can be compared to a SWAT team—quick to move into an area of trouble and ready to eliminate the invaders. These phagocytic cells play a crucial role during the early stages of inflammation, being the first cell type recruited to the site of damage from the bloodstream. They have more killing power than macrophages. The cost for their effectiveness, however, is a relatively short life span of only a few days in the tissues; once they have used their granules, they die. Fortunately, many more neutrophils are in reserve.

Neutrophils not only kill microbes through phagocytosis, they can also release the contents of their granules. In essence, they can behave as mobile grenades, exploding in an area of infection to release destructive enzymes and antimicrobial peptides. They also release their DNA to form neutrophil extracellular traps (NETs). The DNA strands in the NET ensnare microbes, allowing the granule contents that accumulate within the NET to destroy them.

MicroByte

For every neutrophil in the circulatory system, about 100 more are waiting in the bone marrow, ready to be mobilized when needed.

MicroAssessment 14.7

The process of phagocytosis includes chemotaxis, recognition and attachment, engulfment, phagosome maturation and phagolysosome formation, destruction and digestion, and exocytosis. Macrophages are long-lived and always present in tissues; they can be activated to have more power. Neutrophils are highly active, short-lived phagocytic cells that must be recruited to the site of damage.

19. How does a phagolysosome differ from a phagosome?

20. Tuberculosis is characterized by granulomas called tubercles. What is a granuloma?

21. What could a microorganism do to avoid engulfment by a phagocyte?

14.8 ■ The Inflammatory Response

Learning Outcomes

11. Describe the inflammatory process, focusing on the factors that initiate the response and the outcomes of inflammation.

12. Compare and contrast apoptosis and pyroptosis.

When microbes are introduced into normally sterile body sites, or when tissues are damaged, **inflammation** occurs. The purpose of this is to contain a site of damage, localize the response, eliminate the invader, and restore tissue function. Everyone has experienced the signs of inflammation; in fact, the Roman physician Celsus described the four cardinal signs of it about two thousand years ago: swelling, redness, heat, and pain. A fifth sign, loss of function, is sometimes present.

Factors That Trigger an Inflammatory Response

A number of factors trigger an inflammatory response, but they often involve the pattern recognition receptors (PRRs) described in section 14.5. Recall that these receptors detect microbe-associated molecular patterns (MAMPs) and damage-associated molecular patterns (DAMPs). The triggers of inflammation cause host cells to release **inflammatory mediators,** a collective term for various pro-inflammatory cytokines and chemicals such as histamine. Inducers of the inflammatory process include:

- **Microbes.** When PRRs of sentinel cells such as macrophages detect MAMPs, the cells produce inflammatory mediators. One of these, tumor necrosis factor (TNF), induces the liver to synthesize acute-phase proteins, a group of proteins that facilitate phagocytosis and complement activation. Meanwhile, microbial surfaces trigger complement activation, also leading to an inflammatory response.
- **Tissue damage.** The sensors of DAMPs are not well understood, but seem to involve NOD-like receptors (NLRs). When cells detect DAMPs, they respond by releasing inflammatory mediators. If blood vessels are injured, two enzymatic cascades are activated. One is the coagulation cascade, which results in blood clotting, and the other produces several molecules that increase blood vessel permeability.

MicroByte

Aspirin and other non-steroidal anti-inflammatory drugs (NSAIDs) interfere with the production of prostaglandins, a group of inflammatory mediators.

The Inflammatory Process

The inflammatory process involves a sequence of events that result in dilation of small blood vessels, leakage of fluids from those vessels, and the migration of leukocytes out of the bloodstream and into the tissues (**figure 14.15**).

The diameter of local blood vessels increases due to the action of inflammatory mediators. This results in greater blood flow to the area, causing the heat and redness associated with inflammation. It also slows the blood flow in the capillaries. Because of the dilation, normally tight junctions between endothelial cells are disrupted, allowing more fluid to leak from the vessels and into the tissue. This fluid

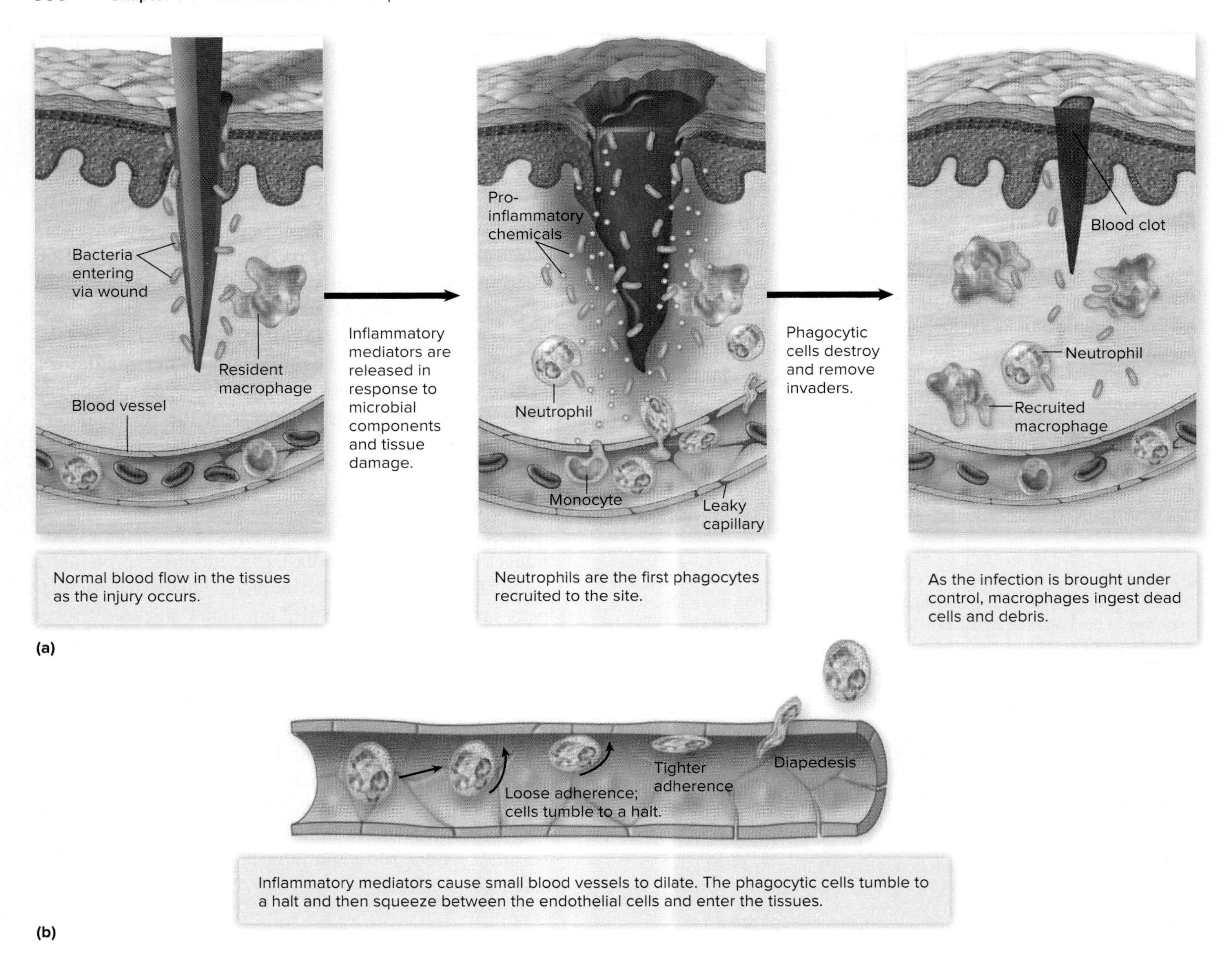

FIGURE 14.15 The Inflammatory Response **(a)** The process of inflammation. **(b)** Phagocytes leave the blood vessels and move to the site of infection.

? **Which type of phagocyte is the first to be recruited to a site of inflammation?**

contains various substances, such as transferrin, complement system proteins, and antibodies, that help counteract invading microbes. The increase of fluids in the tissues causes the swelling and pain associated with inflammation. Pain also results from the direct effect of certain chemicals on sensory nerve endings.

Some of the pro-inflammatory cytokines cause endothelial cells of the local blood vessels to produce adhesion molecules that loosely "grab" phagocytes. The phagocytes normally flow rapidly through the vessels but slowly tumble to a halt as the adhesion molecules attach to them. The phagocytic cells themselves begin producing a different type of adhesion molecule that strengthens the attachment. Then, in response to various chemoattractants, the phagocytes leave the blood vessels and move into the surrounding tissues. They do this by squeezing between the cells of the dilated vessel, a process called diapedesis. Neutrophils are the first to arrive at the site of infection, and they actively phagocytize foreign material. Monocytes (which mature into macrophages at the site of infection) and lymphocytes arrive later. Clotting factors in the fluid that leaks into the tissues initiate clotting reactions in the surrounding area, walling off the site of infection. This helps prevent bleeding and stops the spread of invading microbes. As the inflammatory process continues, large quantities of dead neutrophils accumulate. Those dead cells, along with tissue debris, make up pus. A large

FOCUS ON A CASE 14.1

A 9-year-old boy with cystic fibrosis—a genetic disease that causes a number of problems, including the buildup of thick sticky mucus in the lungs—complained of increasing fatigue, shortness of breath, and worsening cough. When his mother took him to the doctor, she mentioned that his cough was productive, meaning that it contained sputum (pronounced *spew-tum*). She was particularly concerned that the sputum was a blue-green color. His doctor immediately suspected a lung infection by *Pseudomonas aeruginosa*—a common complication of cystic fibrosis. A sputum sample was collected and sent to the clinical laboratory.

In the clinical laboratory, the sample was plated onto MacConkey agar and blood agar and incubated. Mucoid colonies surrounded by a bluish-green color grew on both types of agar media. The colonies on MacConkey had no pink coloration, so the medical technologist concluded that the cells did not ferment lactose. She noted the blue-green color on the agar plates and in the sputum, knowing that *P. aeruginosa* makes several pigmented compounds that give rise to colors ranging from yellow to blue. One of the pigments functions as a siderophore, which is a molecule that binds iron. Another is important for biofilm formation. Further testing showed that the bacterium was an oxidase-positive, Gram-negative rod, consistent with the physician's initial suspicions.

The patient was treated with antibiotics, with only limited success. Like most cystic fibrosis patients, he developed a chronic lung infection that continued to require repeated treatment.

1. What role did cystic fibrosis play in the disease process?
2. What is the significance of the mucoid phenotype of the colonies?
3. How would the siderophore (the iron-binding compound) benefit the bacterium?
4. Why would the boy's lung infection make his pre-existing respiratory problems even worse?

Discussion

1. Cystic fibrosis patients often have an accumulation of thick mucus in their lungs, which interferes with the mucociliary escalator and other first-line defenses. With a compromised (weakened) mucociliary escalator, microbes that are inhaled are not easily removed. In addition, the accumulated mucus serves as a nutrient source for bacteria.
2. The mucoid colonies suggest that the bacterium produces an extracellular material that forms a capsule or a slime layer. This material, also referred to as extracellular polymeric substances (EPS), allows *Pseudomonas aeruginosa* cells to form biofilms. The biofilm protects the bacterial cells from various components of the immune system, including antimicrobial peptides and phagocytes. Bacteria growing within a biofilm are much more difficult for the immune system to destroy. ⏮ EPS
3. Siderophores help the bacterium obtain iron from the host. Recall that the body produces iron-binding proteins, including lactoferrin and transferrin; this prevents microbes from using the host's iron and thereby limits their growth. Microorganisms that make siderophores essentially engage in a "tug-of-war" with the body over iron. This tug-of-war is especially important for *P. aeruginosa* because iron levels influence biofilm formation. When iron is limiting, *P. aeruginosa* cells are motile and do not initiate biofilm formation.
4. In response to a bacterial infection in the lungs, an inflammatory response develops. The capillaries in the area become leaky, allowing fluids to enter the tissues. Those fluids cover the respiratory surfaces, thereby interfering with gas exchange. In addition, inflammation recruits neutrophils to the area. Some of the neutrophils will die, releasing the destructive enzymes in their granules as a result.

amount of pus in a confined region constitutes an abscess (see figure 22.2). ⏭ abscess

The extent of inflammation varies depending on the nature of the injury, but the response is localized, begins immediately upon injury, and increases rapidly. A short-term inflammatory response is called **acute inflammation** and is characterized by an abundance of neutrophils. As the infection is brought under control, resolution of inflammation begins. Neutrophils stop entering the area, and macrophages clean up the damage by ingesting dead cells and debris. As the area heals, new capillaries grow, destroyed tissues are replaced, and scar tissue forms.

If the body's defenses cannot limit the infection, **chronic inflammation** occurs. This is a long-term inflammatory process that can last for years. In chronic inflammation, macrophages and giant cells accumulate, and granulomas form. ⏮ giant cells, ⏮ granulomas

Damaging Effects of Inflammation

The inflammatory process can be compared to a sprinkler system that prevents fire from spreading in a building. Although the process usually limits damage and restores function, the response itself can cause significant harm. One undesirable consequence is that some enzymes and toxic products contained within phagocytic cells are inevitably released, damaging tissues.

If inflammation is limited, such as in a response to a cut finger, the damage caused by the process is normally minimal. If the process occurs in a delicate system, however, such

FOCUS YOUR PERSPECTIVE 14.1

For *Schistosoma*, the Inflammatory Response Delivers

Schistosoma species, the parasitic flatworms that cause the disease schistosomiasis (also called snail fever or bilharzia), use the immune response to assist them in completing one portion of their complex life cycle. ⏮ schistosomiasis

A person can become infected with schistosomes by wading or swimming in water that contains a larval form of the parasite called cercariae, which are released from infected snails. Cercariae penetrate skin by burrowing through with the aid of digestive enzymes. They then move into the bloodstream, where they mature into adult worms that can live for over 25 years. Adult worms mask themselves from the immune system by coating themselves with various blood proteins, an ability that provides them with a primitive stealth "cloaking device."

Schistosoma species have separate sexes, and the male and female worms manage to locate each other in the bloodstream. The male's body has a deep longitudinal groove in which he clasps his female partner to live in life-long embrace (schistosoma means "split-body," referring to the long slit). To reproduce, the worms migrate to the tiny veins of either the intestines or the bladder (depending on the schistosome species), where the female lays hundreds of ova per day. In contrast to the adult worms—which effectively hide from the immune system—the eggs provoke a strong inflammatory response. This pushes the eggs to the closest body surface, in a manner similar to what is experienced as a sliver in the skin works its way out. In the case of species that deposit ova in veins near the intestine, the eggs are pushed out into the intestinal tract, where they are eliminated in feces. Ova of species that deposit the eggs near the bladder are pushed into the bladder, where they are eliminated in urine. If untreated sewage that contains schistosome eggs reaches fresh water, the ova can hatch, releasing a ciliated larval stage that infects and multiplies asexually in a specific freshwater snail host. The infected snail then releases large numbers of cercariae, which can infect a human host to complete the parasite's life cycle.

The symptoms of schistomiasis are due to the many ova that are not expelled. If these ova are swept to the liver by the bloodstream, the resulting inflammatory process and granuloma formation gradually destroy liver cells. The cells are replaced with scar tissue, causing the liver to malfunction. In turn, this results in a fluid buildup in the abdominal cavity, as well as malnutrition. Chronic schistosomiasis can also damage the lungs and bladder, and occasionally, the central nervous system.

Despite their complex life cycle, *Schistosoma* species are highly successful parasites. Not only are they adept at avoiding certain immune responses that would otherwise lead to their destruction, they have learned to exploit the inflammatory response for their own spread.

as the membranes that surround the brain and spinal cord, the consequences can be severe, even life-threatening. As you learn more about infectious diseases, you will notice that many of the most severe effects of infection result from the inflammatory response.

Cell Death and the Inflammatory Response

In addition to traumatic cell death (necrosis) that results from tissue damage, host cells can self-destruct. This capability allows the host to eliminate any cells no longer needed, and it also serves as a mechanism for sacrificing "self" cells that might otherwise spread an infection. One type of programmed cell death avoids an inflammatory response, whereas another type promotes one.

Apoptosis (*apo* means "off"; *ptosis* means "falling") is a programmed cell death that does not trigger an inflammatory response. During apoptosis, the dying cells undergo certain changes. For example, the shape of the cell changes, enzymes cut the DNA, and pieces of the cell bud off, effectively shrinking the cell. Some changes appear to serve as a signal to macrophages that the remains of the cell are to be engulfed without the events associated with inflammation.

If the pattern recognition receptors in a macrophage's cytoplasm are triggered, that cell might initiate **pyroptosis** (*pyro* means "fire"). Unlike apoptosis, this programmed self-destruction triggers an inflammatory response. Thus, pyroptosis sacrifices infected macrophages so that they cannot play host to an invader, and it also recruits various components of the immune system to the region.

MicroAssessment 14.8

The inflammatory response is initiated when microbes invade or tissues are damaged. The outcome is dilation of small blood vessels, leakage of fluids from those vessels, and migration of leukocytes out of the bloodstream and into the tissue. Inflammation helps contain an infection, but the response itself can be damaging. Apoptosis destroys "self" cells without initiating inflammation; pyroptosis triggers an inflammatory response.

22. Describe two general events that can initiate inflammation.
23. Describe two changes in cells undergoing apoptosis.
24. Infection of the fallopian tubes can lead to infertility. Why would this be so?

14.9 ■ Fever

Learning Outcome

13. Describe the induction and outcomes of fever.

Fever is an important host defense mechanism and a strong indication of infectious disease, particularly a bacterial one.

Human body temperature is normally kept around 37°C by a temperature-regulation center in the brain. During an infection, the regulating center "sets" the body's thermostat at a higher level. An oral temperature above 37.8°C is regarded as fever.

A higher temperature setting results when macrophages release pro-inflammatory cytokines in response to microbial products. The cytokines act as messages carried in the bloodstream to the brain, where the temperature-regulating center raises the body temperature in response. The rise in temperature prevents microbes with lower optimum temperatures from growing, giving the immune system time to eliminate the microbes before they cause too much harm. A moderate fever has also been shown to enhance several protective processes, including the inflammatory response, phagocytic activity, multiplication of lymphocytes, release of substances that attract neutrophils, and production of interferons and antibodies. Release of leukocytes into the blood from the bone marrow also increases. These effects are likely due to increased rates of enzymatic reactions. ◀◀ **temperature requirements**

Fever-inducing cytokines and other substances are **pyrogens** (*pyro* means "fire" or "heat" and *gen* means "to generate"). Pyrogenic cytokines are endogenous pyrogens, meaning the body makes them, whereas microbial products such as lipopolysaccharide (LPS) are exogenous pyrogens, meaning they are introduced from external sources.

MicroAssessment 14.9

Fever results when macrophages release pro-inflammatory cytokines; this occurs when macrophages detect microbial products.

25. What is a pyrogen?

26. How does fever inhibit the growth of pathogens?

27. Syphilis was once treated by infecting the patient with the parasite that causes malaria, a disease characterized by repeated cycles of fever, shaking, and chills. Why would this treatment control syphilis?

Summary

14.1 Overview of the Innate Immune Defenses

First-line defenses prevent entry into the body, sensor systems detect invasion, and effector mechanisms destroy and remove invaders (figure 14.1).

14.2 First-Line Defenses (figure 14.2)

Physical Barriers

The skin is composed of two main layers—the dermis and the epidermis. Mucous membranes are constantly bathed with mucus and other secretions that help wash microbes from the surfaces (figure 14.3).

Antimicrobial Substances

Lysozyme, peroxidases, lactoferrin, and **antimicrobial peptides** inhibit or kill microorganisms (figure 14.4).

Normal Microbiota (Flora)

Members of the **normal microbiota** competitively exclude pathogens and stimulate the host defenses.

14.3 The Cells of the Immune System (figure 14.5, table 14.1)

Granulocytes

Granulocytes include **neutrophils, basophils,** and **eosinophils.**

Mononuclear Phagocytes

Monocytes circulate in blood; **macrophages** are in tissues (figure 14.6).

Dendritic Cells

Dendritic cells develop from monocytes; some have other origins.

Lymphocytes

Lymphocytes, which include **B cells, T cells,** and **innate lymphoid cells (ILCs),** are involved in adaptive immunity.

14.4 Cell Communication

Surface Receptors

Surface receptors bind **ligands,** allowing the cell to detect substances.

Cytokines (table 14.2)

Cytokines include **chemokines, colony-stimulating factors (CSFs), interferons (IFNs), interleukins (ILs),** and **tumor necrosis factor (TNF).**

Adhesion Molecules

Adhesion molecules allow cells to adhere to other cells.

14.5 Pattern Recognition Receptors (PRRs) (figure 14.7)

Pattern recognition receptors (PRRs) are sensors that allow the body's cells to "see" signs of microbial invasion. Many PRRs detect **microbe-associated molecular patterns (MAMPs),** and some detect **damage-associated molecular patterns (DAMPs).** The outcome of certain diseases is influenced by PRR-generated signals from sentinel cells and infected cells.

Pattern Recognition Receptors (PRRs) That Monitor a Cell's Surroundings

Some **toll-like receptors (TLRs)** are anchored in the cytoplasmic membranes of sentinel cells such as phagocytes and cells that line blood vessels (figure 14.8). These TLRs detect certain microbial surface components. Membrane-anchored C-type lectin receptors (CLRs) detect certain carbohydrates molecules found on the surface of some microbial cells.

Pattern Recognition Receptors (PRRs) That Monitor Material Ingested by a Cell

Some toll-like receptors (TLRs) are anchored in endosomal and phagosomal membranes. These TLRs typically recognize characteristics of microbial nucleic acid.

Pattern Recognition Receptors (PRRs) That Monitor a Cell's Cytoplasm

RIG-like receptors (RLRs) detect viral RNA in a cell's cytoplasm. **NOD-like receptors (NLRs)** detect microbial components or signs of damage in a cell's cytoplasm (figure 14.9). Certain NLRs in macrophages and dendritic cells allow formation of an **inflammasome.**

An Outcome of Cytoplasmic Pattern Recognition: The Interferon Response

Virally infected cells respond by making interferons, causing nearby cells to prepare to undergo apoptosis if they become infected with a virus (figure 14.10).

14.6 The Complement System

Complement System Activation

The **complement system** detects microbial cells and antibodies bound to antigens, and is activated in response (figure 14.11).

Effector Functions of the Complement System

The major protective outcomes of complement system activation include **opsonization,** an inflammatory response, and lysis of foreign cells (figure 14.12).

Regulation of the Complement System

Complement regulatory proteins prevent host cell surfaces from activating the complement system via the alternative pathway (figure 14.13).

14.7 Phagocytosis

The Process of Phagocytosis (figure 14.14)

The steps of **phagocytosis** include chemotaxis, recognition and attachment, engulfment, **phagosome** maturation and **phagolysosome** formation, destruction and digestion, and exocytosis.

Characteristics of Macrophages

Macrophages are always present in tissues to some extent but can call in reinforcements when needed. A macrophage can become an **activated macrophage.** Macrophages, **giant cells,** and T cells form **granulomas** that wall off and retain material that cannot be destroyed. Macrophages are important sentinel cells.

Characteristics of Neutrophils

Neutrophils are the first cell type recruited from the bloodstream to the site of damage.

14.8 The Inflammatory Response

Swelling, redness, heat, and pain are the signs of **inflammation,** the body's attempt to contain a site of damage, localize the response, eliminate the invader, and restore tissue function.

Factors That Trigger an Inflammatory Response

Inflammation is initiated when microbes are detected by pattern recognition receptors (PRRs) or the complement system, or when tissue damage occurs.

The Inflammatory Process

The inflammatory process results in dilation of small blood vessels, leakage of fluids from those vessels, and movement of leukocytes from the bloodstream into the tissues (figure 14.15). **Acute inflammation** is characterized by an abundance of neutrophils; **chronic inflammation** is characterized by macrophage and giant cell accumulation, and granuloma formation.

Damaging Effects of Inflammation

The inflammatory response can be damaging to the host, and in some cases this is life-threatening.

Cell Death and the Inflammatory Response

Apoptosis is a mechanism of eliminating "self" cells without triggering an inflammatory response; **pyroptosis** triggers an inflammatory response.

14.9 Fever

Fever results when macrophages release certain pro-inflammatory cytokines. It inhibits the growth of many pathogens and increases the rate of various body defenses.

Review Questions

Short Answer

1. Describe how the skin protects against infection.
2. What factors in saliva aid in protection against microbes?
3. Why is iron availability important in body defenses?
4. Name two categories of cytokines and give their effects.
5. What is the function of pattern recognition receptors?
6. Contrast the pathways of complement activation.
7. How do complement proteins cause foreign cell lysis?
8. How do phagocytes enter tissues during an inflammatory response?
9. How is acute inflammation different from chronic inflammation?
10. Describe the function of apoptosis.

Multiple Choice

1. Lysozyme does which of the following?
   a) Disrupts cell membranes
   b) Hydrolyzes peptidoglycan
   c) Waterproofs skin
   d) Propels gastrointestinal contents
   e) Propels the cilia of the respiratory tract
2. The hematopoietic stem cells in the bone marrow can develop into which of the following cell types?
   1) Red blood cell
   2) T cell
   3) B cell
   4) Monocyte
   5) Macrophage
   a) 2, 3
   b) 2, 4
   c) 2, 3, 4, 5
   d) 1, 4, 5
   e) 1, 2, 3, 4, 5
3. All of the following refer to the same type of cell *except*
   a) macrophage.
   b) neutrophil.
   c) poly.
   d) PMN.

4. Considering the role of pattern recognition receptors (PRRs), which of the following are they the least likely to detect?
   a) Peptidoglycan
   b) Glycolysis enzymes
   c) Lipopolysaccharide
   d) Flagellin
   e) Certain nucleotide sequences
5. The direct/immediate action of interferon on a cell is to
   a) interfere with the replication of the virus.
   b) prevent the virus from entering the cell.
   c) stimulate synthesis of inactive "suicide enzymes."
   d) stimulate the immune response.
   e) stop the cell from dividing.
6. A pathogen that can avoid the complement component C3b would directly protect itself from
   a) opsonization.
   b) triggering inflammation.
   c) lysis.
   d) inducing interferon.
   e) antibodies.
7. Which of the following statements about phagocytosis is *false*?
   a) Phagocytes move toward an area of infection by chemotaxis.
   b) Digestion of invaders occurs within a phagolysosome.
   c) Phagocytes have receptors that recognize C3b bound to bacteria.
   d) Phagocytes have receptors that recognize antibodies bound to bacteria.
   e) Macrophages die after phagocytizing bacteria, but neutrophils regenerate their lysosomes and survive.
8. If you are analyzing the cell types in a granuloma, which of the following are you least likely to find?
   a) Neutrophils
   b) Macrophages
   c) Giant cells
   d) T cells
9. All of the following trigger an inflammatory response *except*
   a) engagement of PRRs.
   b) complement system activation.
   c) interferon induction of antiviral protein synthesis.
   d) tissue damage.
10. Which of the following statements about inflammation is *false*?
   a) Vasodilation results in leakage of blood components.
   b) The process can damage host tissue.
   c) Neutrophils are the first to migrate to a site of inflammation.
   d) Apoptosis induces inflammation.
   e) The signs of inflammation are redness, swelling, heat, and pain.

Applications

1. Patients who have recently had a bone marrow transplant are extremely susceptible to infection. Why would this be so?
2. A cattle farmer sees a sore on the leg of one of his cows. The farmer feels the sore and notices that the area just around the sore is warm to the touch. A veterinarian examines the wound and explains that the warmth may be due to inflammation. The farmer wants an explanation of the difference between the localized warmth and fever. What would be the vet's explanation to the farmer?

Critical Thinking

1. Why would it benefit the body to have an adaptive immune system in addition to an innate immune system?
2. Some bacterial cells avoid the killing effects of activated complement proteins. How might they do this?

15 The Adaptive Immune Response

Blood clot, with erythrocytes, fibrin filaments, and a lymphocyte (color-enhanced scanning electron micrograph). ©*Science Photo Library RF/Getty Images*

KEY TERMS

Adaptive Immunity Protection provided by immune responses that improve due to exposure to antigens; involves B cells and T cells.

Antibody Y-shaped protein that binds antigen.

Antigen Molecule that reacts specifically with either an antibody or an antigen receptor on a lymphocyte.

Antigen-Presenting Cells (APCs) Cells such as B cells, macrophages, and dendritic cells that can present exogenous antigens to naive or memory T cells, activating them.

B Cell Type of lymphocyte programmed to make antibodies.

Cell-Mediated Immunity (CMI) Immunity involving a T-cell response.

Clonal Selection Process in which a lymphocyte's antigen receptor binds to an antigen, allowing the lymphocyte to multiply.

Cytotoxic T Cell Type of lymphocyte programmed to destroy infected or cancerous "self" cells.

Dendritic Cell Cell type responsible for activating naive T cells.

Effector Lymphocyte Differentiated descendant of an activated lymphocyte; it has properties that allow it to help eliminate antigen.

Helper T Cell Type of lymphocyte programmed to activate B cells and macrophages, and assist other parts of the adaptive immune response.

Humoral Immunity Immunity involving B cells and an antibody response.

Lymphocytes A group of white blood cells (leukocytes) involved in adaptive immunity; B cells and T cells are examples.

Major Histocompatibility Complex (MHC) Molecules Host cell surface proteins that present antigen to T cells.

Memory Lymphocytes Long-lived descendants of activated lymphocytes that can quickly respond when a specific antigen is encountered again.

Plasma Cell Effector form of a B cell; it functions as an antibody-secreting factory.

T_C Cell Effector form of a cytotoxic T cell; it induces apoptosis in infected or cancerous "self" cells.

T_H Cell Effector form of a helper T cell; it activates B cells and macrophages, and releases cytokines that stimulate other cells of the immune system.

A Glimpse of History

Near the end of the nineteenth century, diphtheria was a terrifying disease that killed many infants and small children. The first symptom was a sore throat, often followed by the development of a gray membrane that could come loose and block the airway. Frederich Loeffler, working in Robert Koch's laboratory in Berlin, found club-shaped bacteria growing in the throats of people with the disease but not elsewhere in their bodies. He hypothesized that the organisms were making a poison that spread through the bloodstream. In Paris, at the Pasteur Institute, Emile Roux and Alexandre Yersin followed up by growing the bacteria and extracting the poison, or toxin, from culture fluids. When the toxin was injected into guinea pigs, it generally killed them. ⏭ **Alexandre Yersin**

Back in Berlin, Emil von Behring injected diphtheria toxin into guinea pigs that had recovered from lab-induced diphtheria. These animals did not become ill, suggesting that something in their blood protected them; von Behring called it *antitoxin.* To test this idea, he mixed toxin with serum (the liquid portion of clotted blood) from a guinea pig that had recovered from diphtheria, and injected it into one that had not had the disease. The animal remained well. In further experiments, he cured animals with diphtheria by giving them antitoxin.

The effectiveness of antitoxin was put to the test in late 1891, when a diphtheria epidemic broke out in Berlin. On Christmas night, antitoxin was first given to an infected child, who then recovered from the dreaded disease. The substances with antitoxin properties were then given the name *antibodies,* and materials that induced antibody production were called *antigens.*

Emil von Behring received the first Nobel Prize in Medicine in 1901 for his work on antibody therapy. It took many more decades of investigation to reveal the biochemical nature of antibodies. In 1972, Rodney Porter and Gerald Edelman were awarded a Nobel Prize for their part in determining the structure of antibodies.

In contrast to the innate immune response, which is always ready to respond to patterns that signify invasion or damage, the adaptive immune response improves as a result of exposure to microbial invaders or other foreign material. When such material is first encountered, specific **lymphocytes**—the main participants in adaptive immunity—recognize the material and then proliferate (multiply) in response. This allows the most effective lymphocytes to eliminate a specific invader when it is encountered. The protection provided by this response is called **adaptive immunity.** On first exposure to a given microbe or other antigen (a general term for a molecule to which an immune system reacts), adaptive immunity takes a week or more to build. During this delay, the host depends on innate immunity for protection, which may not be sufficient to prevent disease. In some cases, a person may not survive long enough for the adaptive immune response to reach an effective level.

An important characteristic of the adaptive immune response is molecular specificity, meaning that the recognition of the antigen is precise. For example, if a person has the viral disease measles, a specific adaptive immune response develops, which eliminates the measles virus but does not protect against another disease such as mumps.

The characteristic of adaptive immunity that prevents diseases from recurring is immunological memory, which produces a stronger response to re-exposure. People who survive diseases such as measles, mumps, or diphtheria generally never develop the same disease again. This is because the first exposure allowed the adaptive immune system to "learn" the best response. Today, vaccination prevents many diseases by exposing a person's immune system to relatively harmless forms of a pathogen or its products. The vaccine causes a specific adaptive immune response to develop, so that if the vaccine recipient is then exposed to the actual pathogen, the memory response protects against that agent. It is true that a person can get certain diseases multiple times, but that is generally due to the pathogen's ability to avoid recognition by the host defenses, a topic discussed in chapter 16. ⏭ vaccination

Another important aspect of adaptive immunity is **tolerance,** the ability to ignore any given molecule. Most significantly, the immune system can distinguish normal host cells from invading microbes, an ability referred to as self versus non-self recognition. A more accurate description might be "healthy self" versus "dangerous," the latter including invading microbes as well as cancerous or other "corrupt" self cells. Regardless, the ability to develop tolerance is crucial because without it the immune system would routinely turn against the body's own cells, attacking them just as it does invading microbes. It would also regularly attack harmless substances such as pollen. Tolerance is not a fail-safe system, however, which explains the occurrence of autoimmune diseases and hypersensitivity reactions such as allergies and gluten intolerance. ⏭ autoimmune disease

The adaptive immune system is extraordinarily complex, involving a network of cells, cytokines, and other compounds. The need for such complexity is logical given the task of the system, but it makes the system difficult to explain and understand! Although it might seem that any description should start at the beginning and then continue in a linear progression, that approach can be confusing to someone new to the subject. As an analogy, imagine explaining how to build a car to someone who has never even seen one; the person would have no idea why certain parts were being put together. With that in mind, we have chosen to start this chapter with a general overview of the adaptive immune response, focusing on the essential characteristics of the process and how the response eliminates invading microbes.

15.1 ■ Overview of the Adaptive Immune Response

Learning Outcome

1. Compare and contrast the general characteristics of humoral immunity and cell-mediated immunity.

The first adaptive response to a particular antigen is called the **primary response.** As a result of this initial encounter, the adaptive immune system "remembers" the mechanism that was effective against that specific antigen. Consequently, if the same antigen is encountered later in life, a stronger antigen-specific adaptive immune response occurs, called the **secondary response.** ⏮ antigen

The adaptive immune response uses two basic strategies for eliminating foreign material: humoral immunity and cell-mediated immunity. As illustrated in **figure 15.1,** different cell types are responsible for the outcomes of these (the figure is an overview of the information in this section, so you should refer to it frequently while reading the descriptions). **Humoral immunity** (*humor* means "fluids") eliminates microbial invaders as well as their toxins from the bloodstream and tissue fluids; these invaders are considered extracellular antigens because they are not within a host cell. Humoral immunity involves **B cells (B lymphocytes),** a cell type that develops in the bone marrow in mammals. **Cell-mediated immunity (CMI),** or cellular immunity, deals with invaders residing within a host cell. These invaders are considered intracellular antigens and include viruses and bacteria that are replicating within a host cell. The cell-mediated response also attacks cancerous cells. Cell-mediated immunity involves **T cells (T lymphocytes),** a name that reflects the fact that they mature in the thymus.

Humoral and cell-mediated immunity are both powerful and, if misdirected, can damage the body's own tissues. Because of this, the adaptive immune response is carefully regulated; a **naive lymphocyte** (meaning one that has not encountered antigen previously) cannot react until it receives specific signals to confirm that the antigen is a microbe or other harmful substance. Once an individual B cell or T cell receives those signals, it can become activated (see figure 15.1). An **activated lymphocyte** can proliferate (multiply), so that for every single naive lymphocyte that becomes activated, millions of progeny are formed. Some of those descendants differentiate (develop distinctive characteristics) to become **effector lymphocytes**—short-lived cells that express specific traits that help eliminate invaders.

Focus Figure

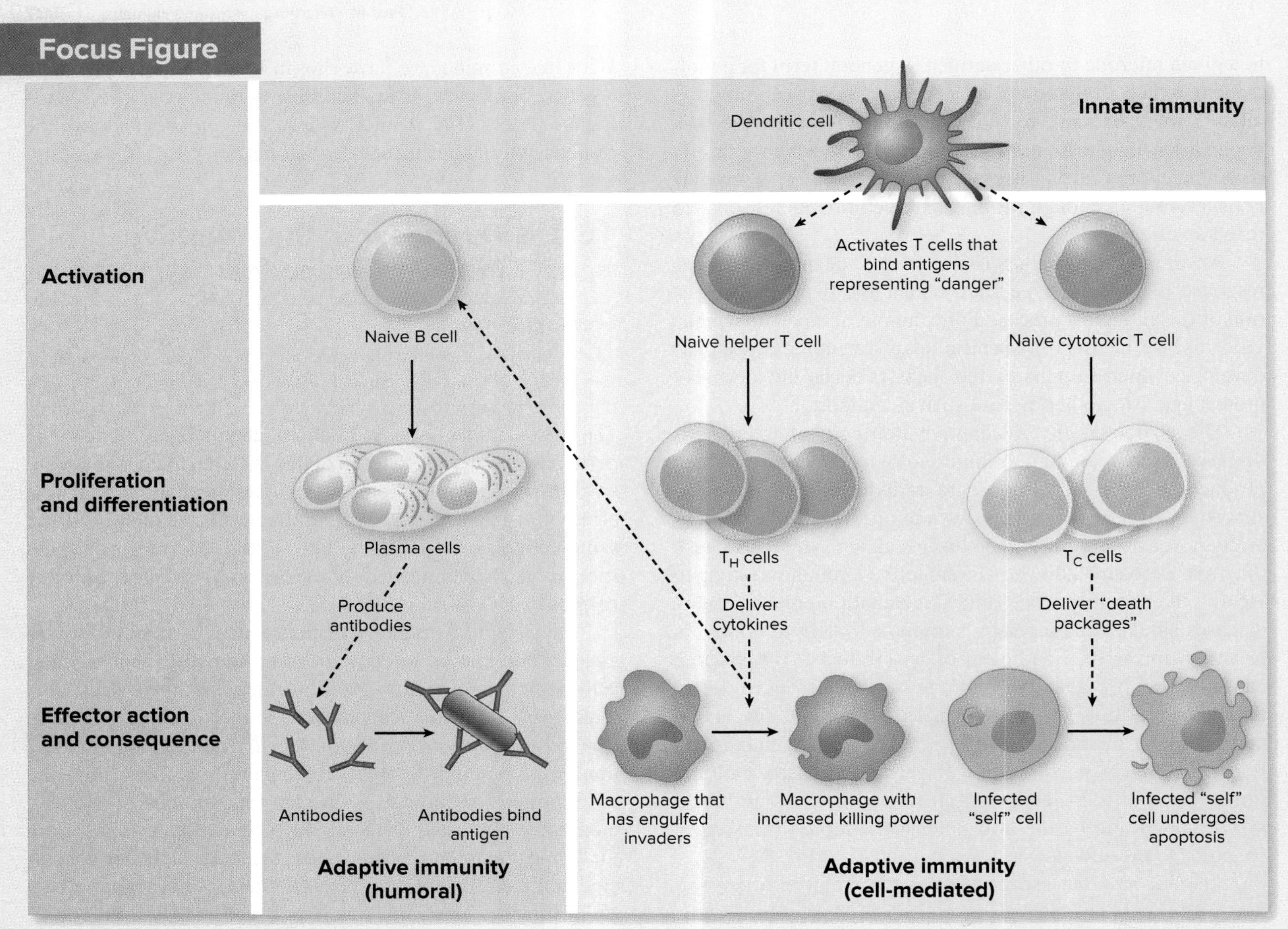

FIGURE 15.1 Overview of the Adaptive Immune Response Humoral immunity protects against antigens in blood and tissue fluid (extracellular antigens); cell-mediated immunity protects against antigens within host cells (intracellular antigens). In this diagram, solid arrows represent the path of a cell or molecule; dashed arrows represent a cell's interactions and effector functions; antigen receptors and memory cells are not shown.

How does cell-mediated immunity eliminate intracellular antigens?

Other descendants become **memory lymphocytes,** long-lived cells that can be activated more quickly if the antigen is encountered again. Memory cells are responsible for the effectiveness of the secondary response.

Humoral Immunity

When a naive B cell detects an extracellular antigen, that B cell may become activated, allowing it to proliferate (see figure 15.1). Some of the descendants of activated B cells differentiate to become **plasma cells,** which are effector B cells. Plasma cells make Y-shaped proteins called **antibodies.** These proteins bind to the surfaces of cells, toxins, viruses, and other antigens, and in doing so, protect the body against the effects of that antigen. For example, antibodies that bind diphtheria toxin protect patients from the effects of the toxin (see **A Glimpse of History**).

The structure of an antibody molecule accounts for its ability to protect against an invader. An antibody has two functional regions: the two identical arms and the single stem of the Y-shaped molecule (see figure 15.1). The ends of the arms are the parts that attach to antigens. By binding to antigens, the antibodies can neutralize their effects. For example, antibody-coated viral particles cannot attach to receptors on cells and, therefore, cannot enter the cells. Antibodies are very specific with respect to their binding, so the immune system must produce many different varieties, each with a slightly different set of "arms." Although the set of arms of different antibody molecules varies, the stem portion is functionally similar—it serves as a "red flag" that sticks out from the surface of an antibody-bound antigen. This tags the antigen for rapid elimination by macrophages or other components of the immune system. ⏮ attachment of viruses

How does a naive B cell know when to respond? The B-cell surface has numerous copies of a **B-cell receptor (BCR),** which is a membrane-bound version of the antibody that the B cell is genetically programmed to make (**figure 15.2**). When a naive B cell encounters an antigen that its BCR binds, the interaction signals the cell that it should respond. Before the B cell can begin multiplying, however, it requires a second signal, usually from a specific type of T cell (see figure 15.1). The second signal confirms that the antigen is indeed something that needs to be eliminated. The two signals can be likened to the actions needed to move a parked car forward—before the car will move, the brake must be released and the gas pedal depressed. In the case of a B cell, the B-cell receptor must bind antigen (which "releases the brake"), and the T lymphocyte must confirm that a response is warranted (which "depresses the gas pedal"). Once the B cell receives confirmation from the T cell, that specific B cell then becomes activated and begins multiplying. Many of the resulting clones (cell copies) eventually differentiate to become plasma cells that produce antibody; some instead become memory B cells that respond quickly to a later exposure to the same antigen.

Cell-Mediated Immunity

Two subsets of T cells help eliminate antigens: cytotoxic T cells and helper T cells (see figure 15.1). **Cytotoxic T cells** are responsible for destroying host cells that contain viruses or are otherwise "corrupt." **Helper T cells** are responsible for directing and assisting the various responses of humoral and cell-mediated immunity. These two T-cell groups look the same but differ in their types of surface proteins called CD markers, which allows scientists to distinguish them.

Each T cell has numerous copies of a surface molecule called a **T-cell receptor (TCR),** which binds a specific antigen (see figure 15.2). The TCR is functionally similar to the B-cell receptor; however, a TCR does not recognize antigen that is free in fluids. Instead, the antigen must be presented by one of the body's own cells.

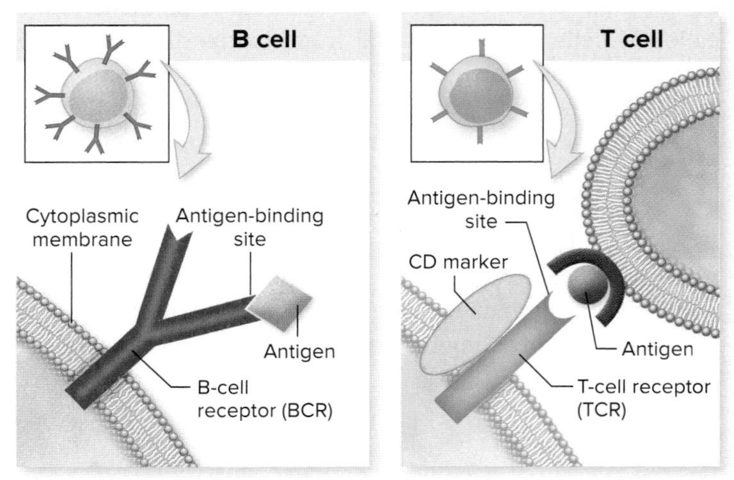

FIGURE 15.2 Antigen Receptors on Lymphocytes B cells and T cells have surface receptors that allow them to recognize specific antigens. The CD marker identifies the type of T cell.

? How is a B-cell receptor similar to an antibody?

To activate a naive T cell, a dendritic cell presents the antigen to the cell along with stimulatory signals; recall that dendritic cells are a part of innate immunity (see figure 15.1). Recognition of the presented antigen by the T cell serves as the first signal ("releasing the brake"); the stimulatory signals of the presenting dendritic cell provide the second signal ("depressing the gas pedal"), confirming that the antigen is from an invader. ⏮ dendritic cell

After a naive cytotoxic T cell is activated, it proliferates, and then some of its descendants differentiate to become an effector form called a **T_C cell,** also referred to as a CTL (for cytotoxic T lymphocyte). When a T_C cell encounters an infected "self" cell, it instructs that cell to undergo apoptosis. Sacrificing the infected cell prevents the intracellular invader from multiplying. Some descendants of activated cytotoxic cells become memory cytotoxic cells. ⏮ apoptosis

After a naive helper T cell is activated, it proliferates, and then some of its descendants differentiate to become an effector form called a **T_H cell.** These cells produce specific cytokines that activate B cells and macrophages. They also produce cytokines that direct and support other cells of the immune system, including other T cells. Some descendants of activated helper T cells become memory helper T cells.

A third T-cell subset, **regulatory T cells** (formerly T suppressor cells), was described relatively recently and is currently the focus of a great deal of research. Regulatory T cells are similar to the other T cells in that they have TCRs, but their role is entirely different. Instead of fostering an immune response, they help prevent one. This is important because it stops the immune system from overreacting and responding to harmless substances. Regulatory T cells (T_{reg}), often referred to as "T regs," are involved in some of the disorders described in chapter 17. The topics covered in this chapter primarily involve cytotoxic and helper T cells.

MicroAssessment 15.1

B cells are programmed to make antibodies that help eliminate extracellular antigens. Cytotoxic T cells are programmed to destroy infected "self" cells as a means of eliminating intracellular antigens. Helper T cells are programmed to orchestrate the adaptive immune response. Recognition of antigen by a naive B cell or T cell serves as a signal to that cell that it needs to respond; in order to become activated, however, the cell requires a confirming second signal.

1. How is the cell-mediated response different from the humoral response?
2. Why is it important that B cells and T cells become activated before they can begin multiplying in response to an antigen?
3. How would you expect a T_C cell to respond if it encountered a T_H cell that was infected with a virus?

15.2 ■ Anatomy of the Lymphatic System

Learning Outcome

2. Compare and contrast the roles of lymphatic vessels, the various secondary lymphoid organs, and primary lymphoid organs.

The **lymphatic system** is a collection of tissues and organs that bring populations of B cells and T cells into contact with antigens (**figure 15.3**). This is important because each lymphocyte recognizes only one or a few different antigens. Before the immune system can develop an effective response, the appropriate lymphocyte must encounter the given antigen. ◀◀ tissues, ◀◀ organs

Lymphatic Vessels

Flow within the lymphatic system occurs via the **lymphatic vessels,** or lymphatics. These vessels carry **lymph,** a colorless fluid derived from tissue fluid (see figure 25.1).

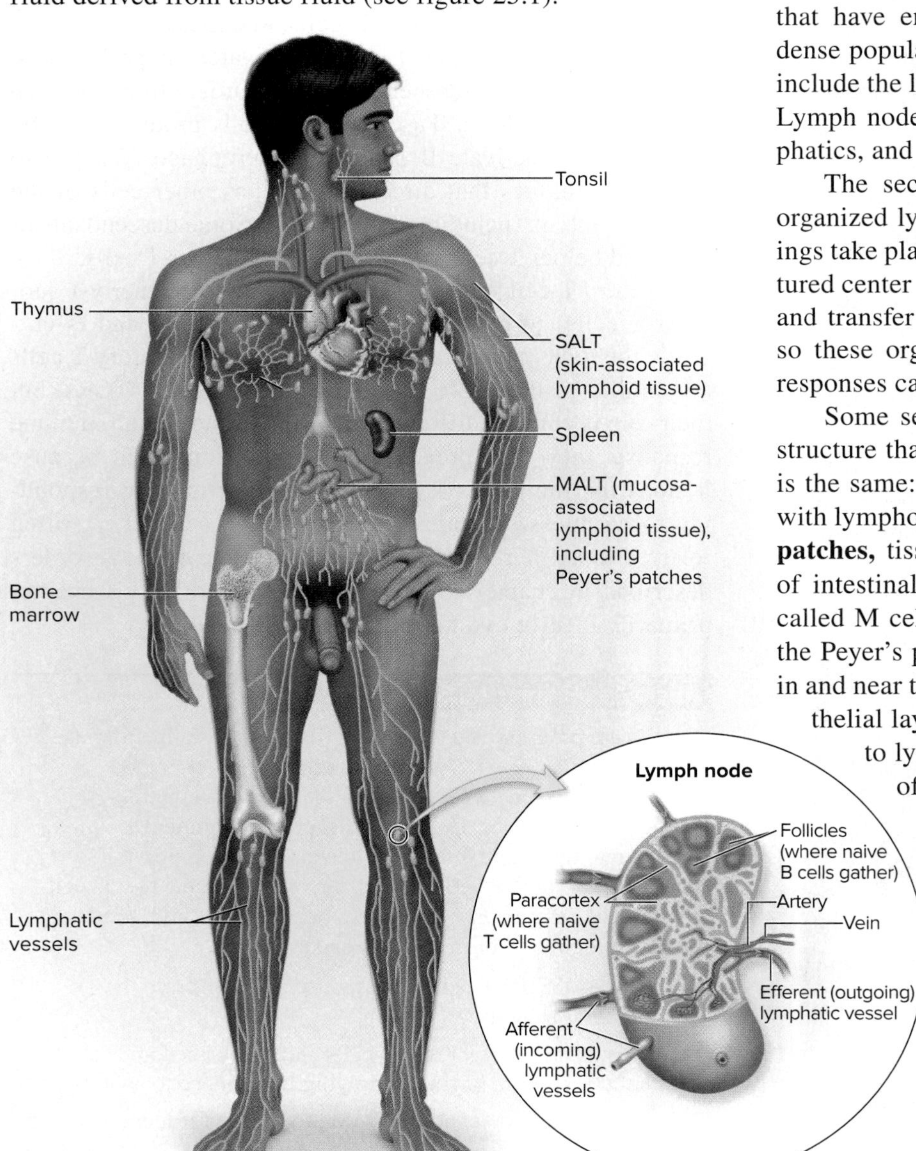

FIGURE 15.3 Anatomy of the Lymphatic System Lymph flows through a system of lymphatic vessels, passing through lymph nodes and lymphoid tissues.

? What is the function of the lymph nodes?

Lymph forms as a result of the body's circulatory system (**figure 15.4**). As blood enters capillaries, some of the liquid is forced out to join the tissue fluid (the extracellular fluid that bathes the tissues). Most of the liquid will re-enter the capillaries, but some is left in the tissues. Excess tissue fluid enters lymphatic vessels, carrying various antigens and other material along with it, to become lymph. As the lymph is pushed through the lymphatic vessels, it passes through lymph nodes, a type of secondary lymphoid organ (discussed next). It then empties back into the blood circulatory system at a large vein behind the left collarbone. Note that the inflammatory response causes more fluid to enter the tissues at the site of inflammation; this causes a corresponding increase in the antigen-containing fluids that enter lymphatic vessels.

Secondary Lymphoid Organs

Secondary lymphoid organs are the sites where antigens that have entered the body are brought into contact with dense populations of lymphocytes. Examples of these organs include the lymph nodes, spleen, and tonsils (see figure 15.3). Lymph nodes, for example, capture materials from the lymphatics, and the spleen collects materials from blood.

The secondary lymphoid organs are like busy, highly organized lymphoid coffee shops where many cellular meetings take place. The anatomy of these organs provides a structured center for various cells of the immune system to interact and transfer cytokines. No other places in the body do this, so these organs are the only sites where adaptive immune responses can be initiated. ◀◀ cytokines

Some secondary lymphoid organs are less organized in structure than the lymph nodes and spleen, but their purpose is the same: to capture antigens and bring them into contact with lymphocytes. Among the most important are the **Peyer's patches,** tissues in the intestinal walls that inspect samples of intestinal contents. Specialized intestinal epithelial cells called M cells transfer material from the intestinal lumen to the Peyer's patches (**figure 15.5**). In addition, dendritic cells in and near the Peyer's patches can also reach through the epithelial layer and grab material in the intestine to present it to lymphocytes. Peyer's patches are part of a network of lymphoid tissues called **mucosa-associated lymphoid tissue (MALT).** This network plays a crucial role in **mucosal immunity,** the immune response that prevents microbes from invading the body via the mucous membranes.

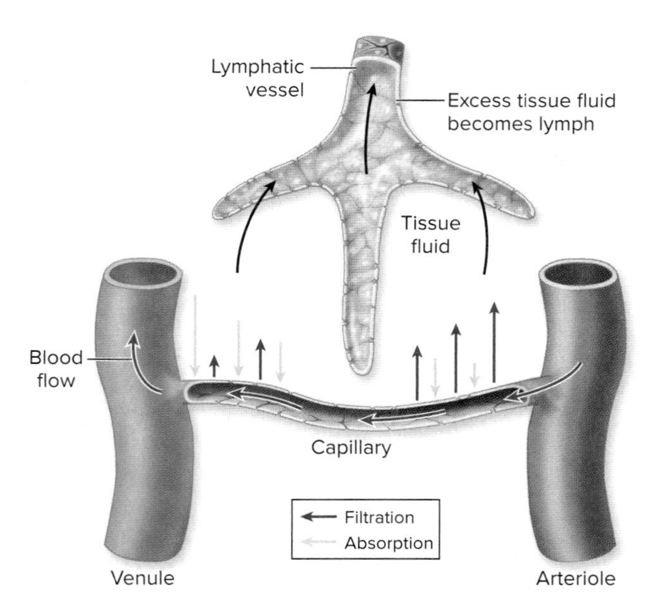

FIGURE 15.4 Formation of Lymph As blood enters the capillaries, the pressure forces some fluid out of the vessels and into the tissues; most of that fluid then re-enters the capillaries, but some enters the lymphatic vessels, becoming lymph.

? Where do antigens that enter a lymphatic vessel contact lymphocytes?

Lymphoid tissues under the skin are called **skin-associated lymphoid tissue (SALT).** ⏮ lumen

Primary Lymphoid Organs

Primary lymphoid organs include the bone marrow and the thymus. The bone marrow is where hematopoietic stem cells reside; recall that these give rise to all blood cells, including lymphocytes (see figure 14.5). B cells and T cells all originate in the bone marrow but only B cells mature there; immature T cells migrate to the thymus and mature there. Once mature, the lymphocytes gather in the secondary lymphoid organs just described, waiting to encounter antigens. ⏮ hematopoietic stem cells

MicroAssessment 15.2

Lymphatic vessels carry antigen-containing fluid collected from tissues to the lymph nodes. Secondary lymphoid organs are where antigens that have entered the body are brought into contact with dense populations of lymphocytes. Primary lymphoid organs are where T and B cells develop.

4. What is lymph?
5. What are Peyer's patches?
6. How would mucosal immunity prevent diarrheal disease?

15.3 ■ The Nature of Antigens

Learning Outcome

3. Define the terms *antigen, T-dependent antigen, T-independent antigen, immunogenic,* and *epitope.*

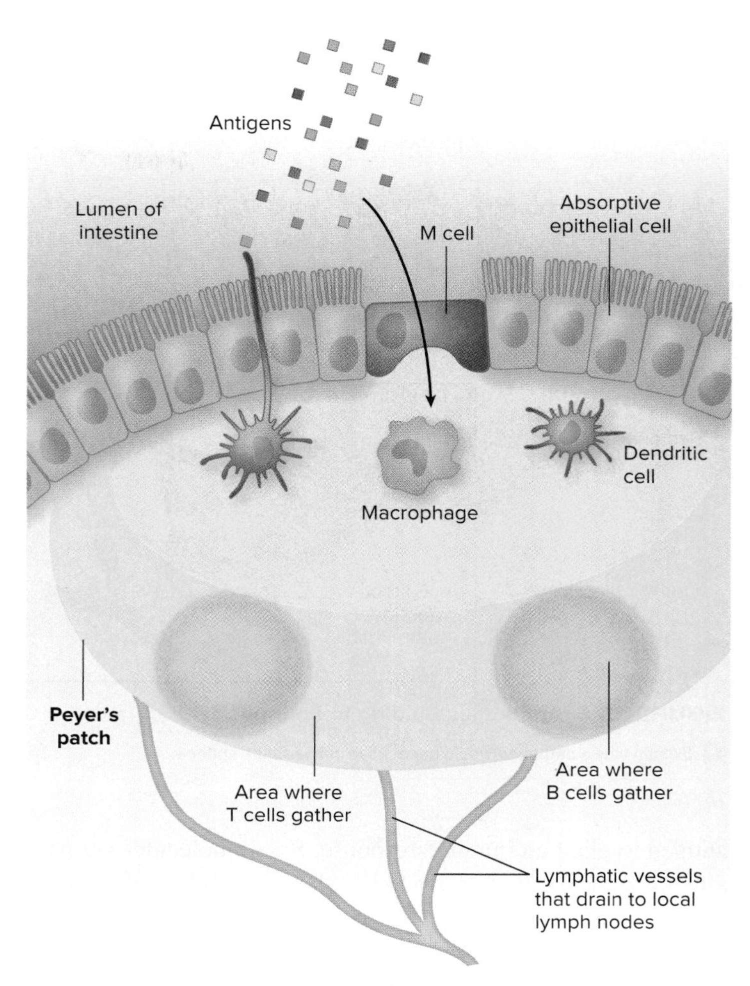

FIGURE 15.5 Peyer's Patch M cells transfer samples of intestinal contents to lymphocytes that reside in Peyer's patches. Dendritic cells can reach through the epithelial layer and grab material in the intestine to present it to lymphocytes.

? What is mucosal immunity?

The term **antigen** was first used to refer to compounds that induce antibody production; it is derived from the descriptive expression **anti**body **gen**erator. Today, the term is used more broadly to describe any molecule that reacts specifically with an antibody, a B-cell receptor, or a T-cell receptor; it does not necessarily imply that the molecule can induce an immune response.

Antigens include an enormous variety of materials, from invading microbes and their products to plant pollens, but they fall into two general categories. Most antigens are **T-dependent antigens,** meaning that the responding B cell requires a signal from a T_H cell as part of the activation process. T-dependent antigens characteristically have a protein component. In contrast, **T-independent antigens** can activate B cells without T_H cell help. They include lipopolysaccharide (LPS) and molecules with identical repeating subunits, such as some carbohydrates.

Various antigens differ in their effectiveness in stimulating an immune response. Proteins generally induce a strong response, whereas lipids and nucleic acids often do not. The term **immunogenic** is used to describe the relative ability of an

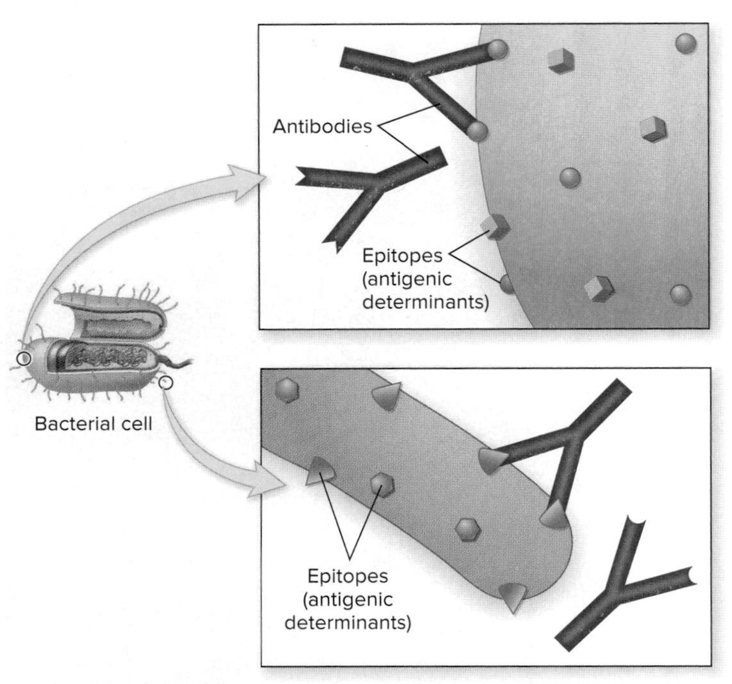

FIGURE 15.6 Antibodies Binding to Epitopes

Explain why a pathogen would never have only a single epitope.

antigen to elicit an immune response. Small molecules are usually not immunogenic, meaning that they do not elicit a response.

Although antigens are generally large molecules, the adaptive immune system recognizes distinct regions of the molecule known as **epitopes,** or antigenic determinants (**figure 15.6**). Some epitopes are stretches of 10 or so amino acids, whereas others are three-dimensional shapes such as a region that sticks out in a molecule. A bacterial cell usually has a diverse assortment of macromolecules on its surface, each with a number of distinct epitopes, so the entire cell has an enormous number of different epitopes.

MicroAssessment 15.3

The adaptive immune response is directed against epitopes on antigens.

7. How is an epitope different from an antigen?
8. Would a denatured antigen be expected to have the same epitopes as its native (undenatured) counterpart?

15.4 ■ The Nature of Antibodies

Learning Outcomes

4. Diagram an antibody, labeling the various functional regions.
5. Describe six protective outcomes of antibody-antigen binding.
6. Compare and contrast the five classes of immunoglobulins.

The structure and characteristics of antibody molecules are crucial for their functional properties. Knowing about these features will help you understand their essential roles in the host defenses.

Structure and Properties of Antibodies

Antibodies, also called immunoglobulins, are Y-shaped proteins that have two general parts—the arms and the stem (**figure 15.7a**). The two identical arms, called the **Fab regions**, have ends that bind antigen. The stem is the **Fc region.** These names were assigned following early studies that showed that enzymatic digestion of antibodies yielded two types of fragments: fragments that were antigen-binding (Fab) and fragments that could be crystallized (Fc).

All antibodies have the same basic Y-shaped structure, called an antibody monomer. It consists of two copies of a high-molecular-weight polypeptide chain, called the **heavy chain,** and two copies of a lower-molecular-weight polypeptide chain, called the **light chain** (see figure 15.7b). The amino acids in the chains fold into characteristic domains, referred to as immunoglobulin domains; the light chains have two domains each, and most heavy chains have four. Each light chain is linked to a heavy chain by a disulfide bond. The fork of the Y is a flexible stretch called the hinge region, and one or more disulfide bonds link the two heavy chains there. ⏮ protein domain, ⏮ disulfide bond

When the amino acid sequences of antibody molecules that bind to different epitopes are compared, tremendous variation can be seen in the portion referred to as the variable region. The other portion is known as the constant region.

Variable Region

The **variable region** is the portion at the ends of the Fab regions; it accounts for the antigen-binding specificity of the antibody (see figure 15.7b). Part of this region is the antigen-binding site, the portion that attaches to a specific epitope. The fit needs to be very precise, because the interaction depends on numerous non-covalent bonds to keep the molecules together. Nevertheless, the antigen-antibody interaction is reversible, and the molecules can separate, leaving both antigen and antibody unchanged.

Constant Region

The **constant region** includes the entire Fc region, as well as part of the two Fab regions (see figure 15.7b). The consistent nature of this region allows other components of the immune system to recognize the otherwise diverse antibody molecules.

There are five general types of constant regions, and these correspond to the major classes (also called isotypes) of immunoglobulin (Ig) molecules: IgM, IgG, IgA, IgD, and IgE. Each class has distinct functions and properties that will be described later, after we consider some general properties of antibodies.

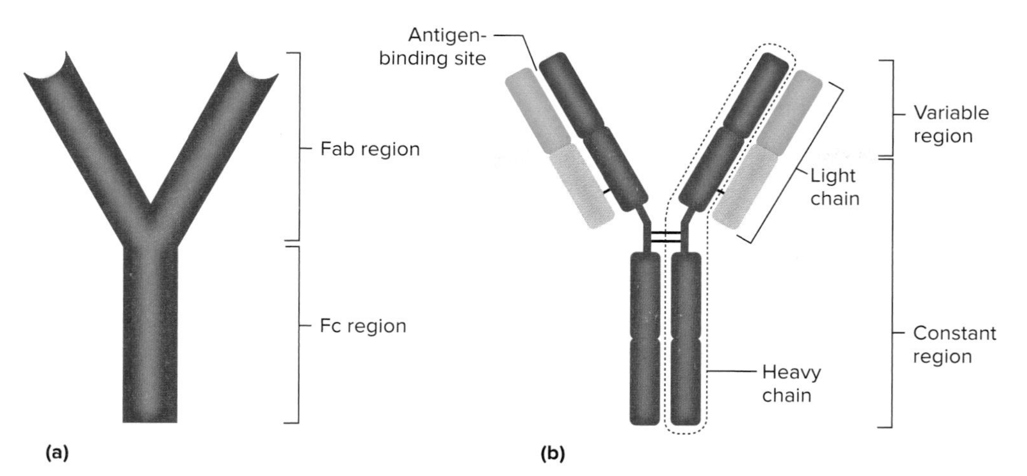

FIGURE 15.7 Structure of an Antibody Molecule **(a)** Simplified Y-shaped structure; the arms of the Y make up the Fab regions, and the stem is the Fc region. **(b)** The molecule is made up of two identical heavy chains and two identical light chains. Disulfide bonds link the chains. The amino acid sequence of the variable region accounts for the antigen-binding specificity; the amino acid sequence of the constant region determines the class of the molecule.

? **How does the function of the constant region differ from that of the variable region?**

Protective Outcomes of Antibody-Antigen Binding

The protective outcomes of antibody-antigen binding (**figure 15.8**) depend partly on the antibody class, and include:

- **Neutralization.** Toxins and viruses must bind specific molecules on a cell surface before they can damage that cell. A toxin or virus coated with antibodies cannot attach to cells and is said to be neutralized.
- **Opsonization.** Phagocytic cells have receptors for the Fc region of IgG molecules, making it easier for the phagocyte to engulf antibody-coated antigens. Recall from chapter 14 that the complement protein C3b opsonizes antigens; IgG molecules have a similar effect. ⏮ opsonization by C3b
- **Complement system activation.** Antigen-antibody complexes (commonly called immune complexes) can trigger the classical pathway of complement system activation. When multiple molecules of certain antibody classes are

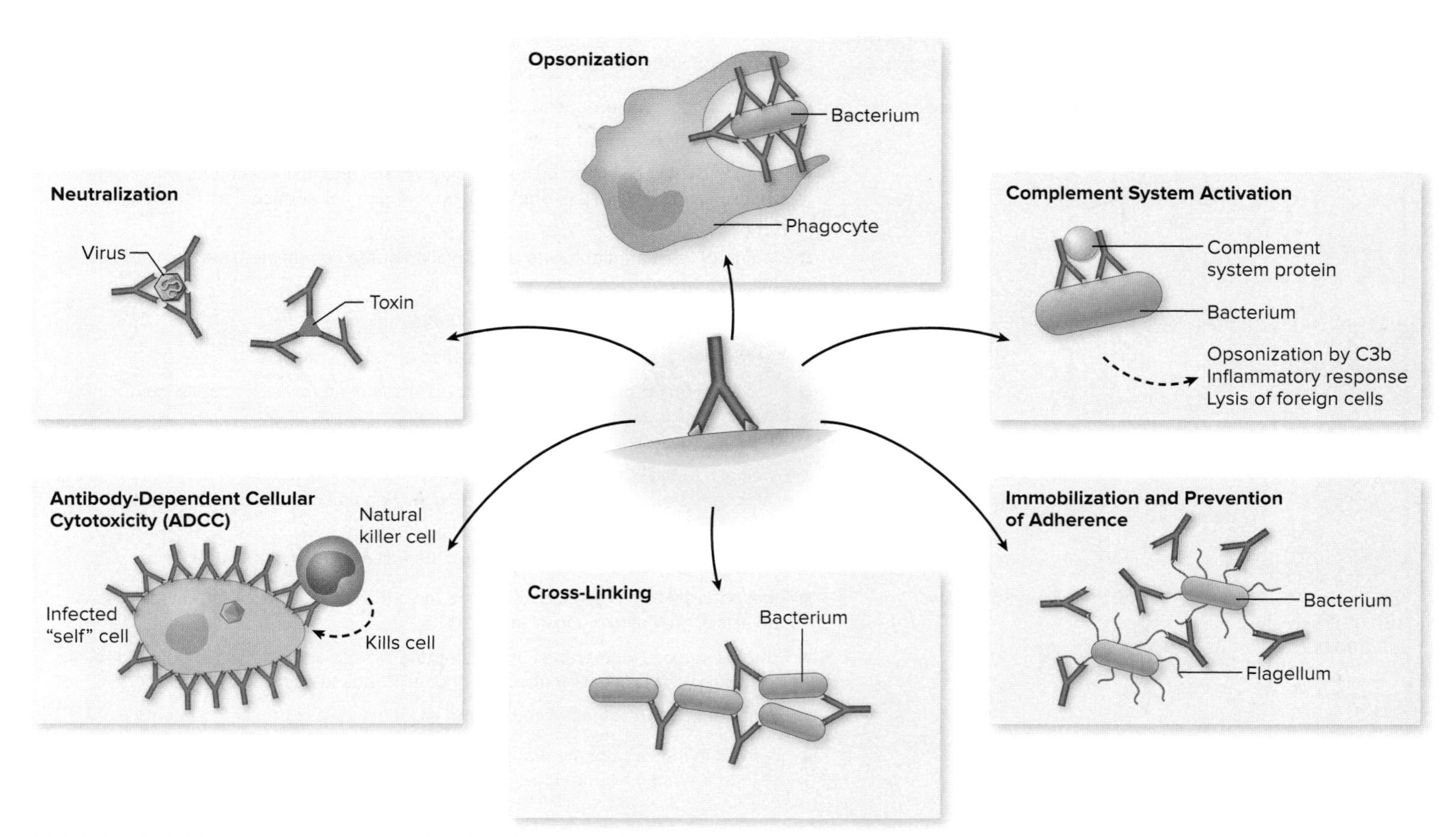

FIGURE 15.8 Protective Outcomes of Antibody-Antigen Binding

? Which pathway of complement activation is depicted in this figure?

bound to a cell surface, a specific complement system protein attaches to side-by-side Fc regions. This activates the cascade of reactions that lead to production of the opsonin C3b, initiation of an inflammatory response, and formation of membrane attack complexes. ⏮ complement system

- **Immobilization and prevention of adherence.** Binding of antibodies to flagella interferes with a microbe's ability to move; binding to pili prevents a bacterium from attaching to surfaces. Movement and attachment are often necessary for a pathogen to infect a host, so antibodies that bind to flagella or pili can prevent infection.
- **Cross-linking.** The ends of the two arms of an antibody can bind separate but identical antigen molecules, linking them. The overall effect is that large antigen-antibody complexes form, creating big "mouthfuls" of antigens for phagocytic cells to engulf.
- **Antibody-dependent cellular cytotoxicity (ADCC).** When multiple IgG molecules bind to a virally infected cell or a tumor cell, that cell becomes a target for destruction by **natural killer (NK) cells.** The NK cell attaches to the Fc regions of IgG and, once attached, kills the target cell by delivering destructive compounds directly to it. ⏭ natural killer cells

Immunoglobulin Classes

Antibody molecules in all five major immunoglobulin classes have the same basic monomeric structure, but each class has a different constant portion of the heavy chain. Some of the immunoglobulins are made up of multiple copies of the basic monomeric structure. Characteristics of the various immunoglobulin classes are summarized in **table 15.1.**

IgM

IgM is the first class produced during the primary response to an antigen. It is the principal class produced in response to some T-independent antigens. IgM accounts for 5% to 13% of the circulating antibodies.

IgM is a pentamer. It does not typically cross from the bloodstream into tissues, so its main role is to control bloodstream infections. The five monomeric subunits give IgM a total of 10 antigen-binding sites, so it cross-links antigens very effectively. It is the most efficient class in triggering the classical pathway of complement system activation.

TABLE 15.1 Characteristics of the Main Classes of Antibodies

Class and Molecular Weight (Daltons)	Structure	Percentage of Total Serum Immunoglobulin (Half-Life in Serum)	Properties and Functions
IgM 970,000		5–13% (10 days)	■ First antibody class produced during the primary response. ■ Principal class produced in response to T-independent antigens. ■ Provides direct protection by neutralizing viruses and toxins, immobilizing motile organisms, preventing microbes from adhering to cell surfaces, and cross-linking antigens. ■ Binding of IgM to antigen leads to activation of the complement system (classical pathway).
IgG 146,000		80–85% (21 days)	■ Most abundant class in the blood and tissue fluids. ■ Provides longest-term protection because of its long half-life. ■ Provides direct protection by neutralizing viruses and toxins, immobilizing motile organisms, preventing microbes from adhering to cell surfaces, and cross-linking antigens. ■ Binding of IgG to antigen facilitates phagocytosis, leads to activation of the complement system (classical pathway), and allows antibody-dependent cellular cytotoxicity. ■ Transported across the placenta, providing protection to a developing fetus; long half-life extends the protection through the first several months after birth.
IgA monomer 160,000; secretory IgA 390,000		10–13% (6 days)	■ Most abundant class produced, but the majority of it is secreted into mucus, tears, and saliva, providing mucosal immunity. ■ Protects mucous membranes by neutralizing viruses and toxins, immobilizing motile organisms, and preventing attachment of microbes to cell surfaces. ■ Component of breast milk; protects the intestinal tract of breast-fed infants.
IgD 184,000		<1% (3 days)	■ Involved in the development and maturation of the antibody response. Its functions in blood are not well understood.
IgE 188,000		<0.01% (2 days)	■ Binds via the Fc region to mast cells and basophils. This bound IgE allows those cells to detect parasites and other antigens and respond by releasing their granule contents. ■ Involved in many allergic reactions.

IgG

IgG is the most abundant serum immunoglobulin (serum is the liquid portion of clotted blood), accounting for about 80% to 85% of the total serum antibodies. It circulates in the blood but exits the vessels to enter the tissues as well.

IgG provides the longest-term protection of any antibody class; its half-life is 21 days, meaning that a given number of IgG molecules will be reduced by about 50% after 21 days. In addition, IgG is generally the first and most abundant circulating antibody class produced during the secondary response. The basis for this will be discussed in section 15.6. IgG provides protection by neutralization, opsonization, complement activation, immobilization and prevention of adherence, cross-linking, and ADCC (see figure 15.8).

An important characteristic of IgG is that it is transported across the placenta into the fetus's bloodstream, so it protects the developing fetus against infections. Women who are not already immune to a given pathogen lack IgG against that microbe, so they are warned to take extra precautions during pregnancy to avoid pathogens that can infect and damage a fetus. For example, pregnant women are advised not to eat raw meat or change a cat's litter box. This is to avoid a primary infection by *Toxoplasma gondii,* a parasite found in raw meat as well as the feces of infected cats. ⏭ *Toxoplasma gondii*

Maternal IgG protects not only the developing fetus, but also the newborn. The maternal antibodies present at birth gradually degrade over a period of about 6 months, but during this time the infant begins producing protective antibodies (**figure 15.9**).

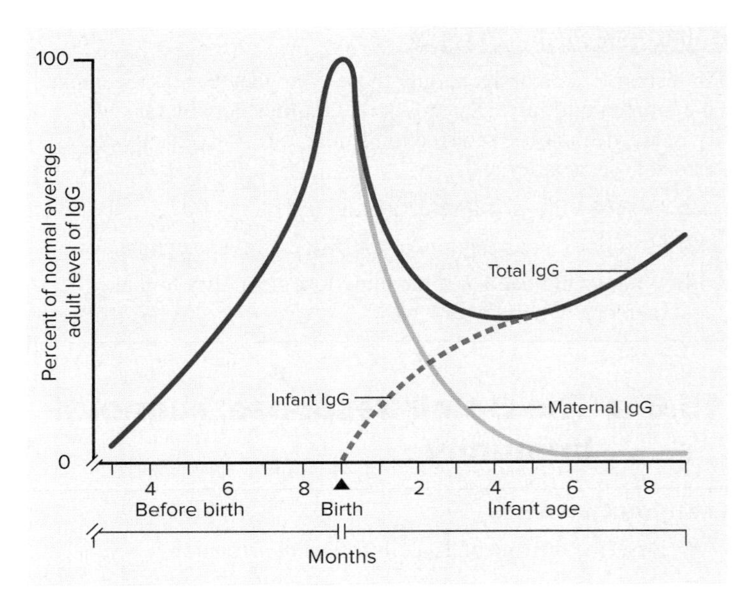

FIGURE 15.9 Immunoglobulin G Levels in the Fetus and Infant As the fetus develops, maternal IgG is transported across the placenta to provide protection. After birth, the infant begins producing antibodies.

? Why does the level of maternal antibodies decrease over time?

IgG is also in colostrum, the first breast milk produced after giving birth. The newborn's intestinal tract absorbs this antibody.

IgA

Most IgA is the secreted form, a dimer called **secretory IgA (sIgA).** This form is important in mucosal immunity and is found on the mucous membranes that line the gastrointestinal, genitourinary, and respiratory tracts. It is also in secretions such as saliva, tears, and breast milk. Secretory IgA in breast milk protects breast-fed infants against intestinal pathogens. ⏮ mucosal immunity

Protection by secretory IgA is primarily due to the direct effects of its binding. These include neutralizing toxins and viruses and interfering with the attachment of microbes to host cells.

IgA is produced mainly by the plasma cells that reside in the mucosa-associated lymphoid tissues (MALT). Recall that plasma cells are the antibody-secreting form of B cells. As IgA is transported across the mucosa (mucous membranes), a polypeptide called the secretory component is added. This attaches the antibody to the layer of mucus that coats the mucosal surface and protects it from destruction by enzymes there. Although IgA is the most abundant immunoglobulin class produced, relatively little is in the serum; only about 10% to 13% of antibodies in the serum are IgA, and these are the monomeric form. ⏮ MALT

IgD

IgD accounts for less than 1% of all serum immunoglobulins. It is involved with the development and maturation of the antibody response, but its functions in serum have not been clearly defined.

IgE

IgE is barely detectable in serum, because most is tightly bound via the Fc region to basophils and mast cells, rather than being free in the circulation. The bound IgE molecules allow these cells to detect and respond to antigens. For example, when antigen binds to two adjacent IgE molecules carried by a mast cell, the cell releases histamine and other inflammatory mediators. IgE-mediated responses seem to be important in eliminating parasites, particularly parasitic worms (helminths). ⏮ inflammatory mediators, ⏮ helminths

Unfortunately for allergy sufferers, basophils and mast cells also release their chemicals when IgE binds to normally harmless materials such as foods, dusts, and pollens, leading to immediate reactions such as coughing, sneezing, and tissue swelling. In some cases these allergic, or hypersensitivity, reactions can be life-threatening. ⏭ hypersensitivities

MicroAssessment 15.4

Antibody-antigen binding results in neutralization, immobilization and prevention of adherence, cross-linking, opsonization, complement activation, and antibody-dependent cytotoxicity. Immunoglobulin classes include IgM, IgG, IgA, IgD, and IgE.

9. Why is IgM particularly effective at cross-linking antigens?
10. Which maternal antibody classes protect a breast-fed newborn?
11. In opsonization with IgG, why is it important that phagocytes recognize the antibodies only after they bind antigen?

15.5 ■ Clonal Selection and Expansion of Lymphocytes

Learning Outcome

7. Outline the process of clonal selection and expansion.

Early on, immunologists recognized that the immune system can make a seemingly unlimited range of antibody specificities. The clonal selection theory describes how this occurs (**figure 15.10**). Each B cell is programmed to make only a single specificity of antibody, so each B cell can recognize only one epitope. When antigen is introduced, only the B cells capable of making the antibodies that bind to the antigen can begin multiplying. This generates a population of clones—copies of the specific B cells capable of making the appropriate antibodies.

Clonal selection is a critical theme in the adaptive immune response, applying to both B cells and T cells. As lymphocytes mature in the primary lymphoid organs, a population of cells able to recognize a functionally limitless variety of antigens is generated; each individual cell, however, recognizes and responds to only one epitope. Thus, if a person's immune system can make antibodies to billions of different epitopes, that person must have billions of different B cells, each interacting with a single specific epitope.

Each lymphocyte residing in the secondary lymphoid organs is waiting for the "antigen of its dreams"—an antigen that has an epitope to which that particular lymphocyte is programmed to respond. When an antigen enters a lymphoid organ, only those rare lymphocytes that recognize it can respond; the specificity of the antigen receptor they carry on their surface (B-cell receptor or T-cell receptor) determines this recognition. Lymphocytes that do not recognize the antigen remain inactive. Recall that in most cases, lymphocytes require additional signals—confirmation from another cell type—in order to multiply. This helps prevent the immune system from mounting a response against "self" molecules by mistake.

Some progeny of the lymphocytes that encountered their "dream antigen" leave the secondary lymphoid organs and migrate to the tissues, where they continue responding for as long as the antigen is present. Once the antigen has been eliminated, these cells will undergo apoptosis. ◀◀ apoptosis

The activities of individual lymphocytes change as they encounter antigen. As a means of clarifying discussions of lymphocyte characteristics, descriptive terms are sometimes used:

- **Immature lymphocytes.** These have not fully developed their antigen-specific receptors.
- **Naive lymphocytes.** These have antigen receptors, but have not yet encountered the antigen to which they are programmed to respond.
- **Activated lymphocytes.** These are able to proliferate; they recognize that a specific antigen is present because their antigen receptor has attached to it, and they have received the second signal confirming that the antigen requires a response.
- **Effector lymphocytes.** These are descendants of activated lymphocytes, armed with the ability to produce specific cytokines or other protective substances. Plasma cells are effector B cells, T_C cells are effector cytotoxic T cells, and T_H cells are effector helper T cells.
- **Memory lymphocytes.** These are long-lived descendants of activated lymphocytes; they can quickly become activated when an antigen is encountered again. Memory lymphocytes are responsible for the speed and effectiveness of the secondary response.

MicroByte

The body is thought to have about 1 billion B cells, and only one or a few will recognize a given epitope the first time it is encountered.

MicroAssessment 15.5

In response to an antigen, only those lymphocytes that recognize the antigen multiply. Depending on their developmental stage, lymphocytes may be referred to as immature, naive, activated, effector, or memory cells.

12. Describe the clonal selection theory.
13. How does a naive lymphocyte differ from an activated one?
14. What would happen if the body lost the ability to make memory lymphocytes?

15.6 ■ The B-Cell Response: Humoral Immunity

Learning Outcomes

8. Describe the role of T_H cells in B-cell activation.
9. Compare and contrast the primary and the secondary B-cell responses.
10. Compare and contrast the response to T-dependent antigens and T-independent antigens.

FIGURE 15.10 Clonal Selection and Expansion During the Antibody Response

What is meant by clonal selection?

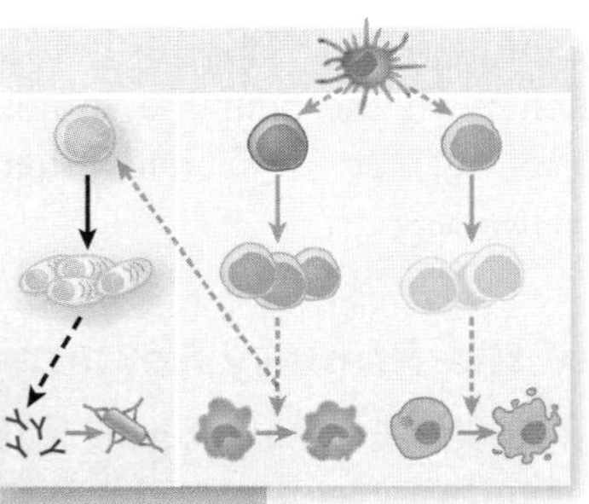

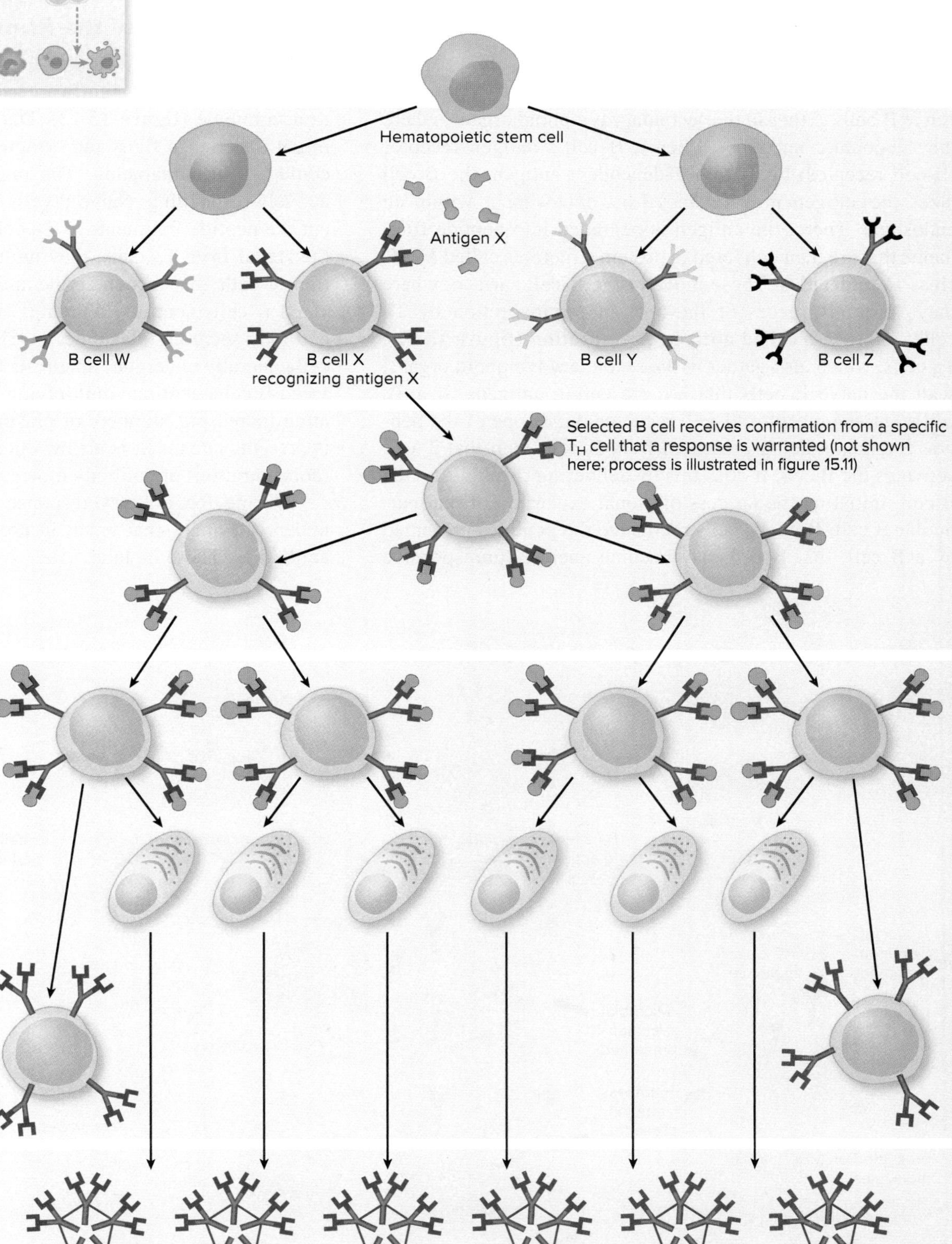

This section will focus mainly on the type of B-cell response that requires a second signal from T_H cells, because this is the most common kind; the antigens involved are called **T-dependent antigens** and are typically proteins. The B-cell response to a different group of antigens (T-independent antigens) will be described at the end of this section.

B-Cell Activation

Naive B cells gather in the secondary lymphoid organs, where they encounter antigens. When a B cell's antigen receptor (B-cell receptor) binds to a T-dependent antigen, the B cell takes the antigen in by endocytosis, enclosing it within an endosome. There, the antigen is degraded into peptide fragments that are then delivered to protein structures called **MHC class II molecules.** These move to the B-cell surface, where they "present" pieces of the antigen for inspection by T_H cells—a process called **antigen presentation** (**figure 15.11**). T_H cells, which also gather in the secondary lymphoid organs, scan the naive B cells that are presenting antigens. If a T_H cell's antigen receptor (T-cell receptor) binds one of the peptide fragments being presented by a B cell, then that T cell activates the B cell. It does this by delivering cytokines to the B cell, initiating the process of clonal expansion of that particular B cell. If no T_H cells recognize the peptides presented by a B cell, that B cell may become anergic (unresponsive to future exposure to the antigen). This results in tolerance to that antigen, a mechanism the adaptive immune system uses to avoid responses against "self" and other harmless antigens. ⏮ endocytosis, ⏮ tolerance

Characteristics of the Primary Response

In the first (primary) exposure to an antigen, it takes about 10 to 14 days for a significant concentration of antibodies to accumulate (**figure 15.12**). During this delay, the person might experience signs and symptoms of an infection, which could be life-threatening. The immune system, however, is actively responding. Naive B cells that bind the antigen present the peptide fragments to T_H cells. Once those B cells get the signal from T_H cells, they multiply, generating a population of cells that recognize the antigen. As some of the activated B cells continue dividing, others differentiate to form antibody-secreting plasma cells (**figure 15.13**). Each plasma cell generally undergoes apoptosis after several days, but activated B cells continue multiplying and differentiating, generating increasing numbers of plasma cells as long as antigen is present. The result is a slow but steady increase in the titer (concentration) of antibody molecules.

During the primary response, the proliferating B cells undergo changes that result in production of more effective antibodies. These include:

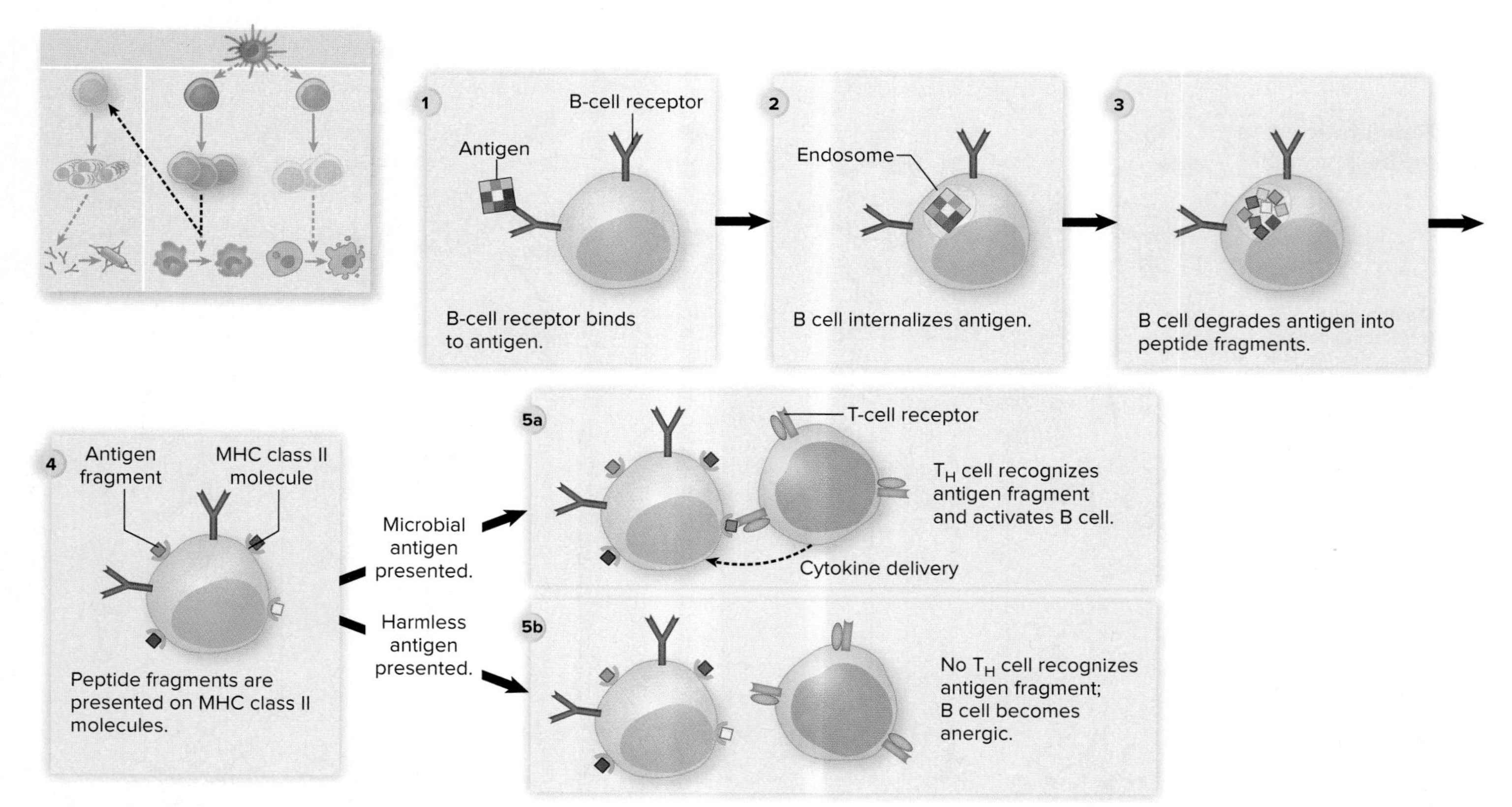

FIGURE 15.11 B-Cell Activation The B cell processes the antigen and presents it to T_H cells. If a T_H cell recognizes the antigen, it activates the B cell, allowing it to undergo clonal expansion.

? **Why is it beneficial for a B cell to become anergic if no T_H recognizes an antigen fragment that the B cell presents?**

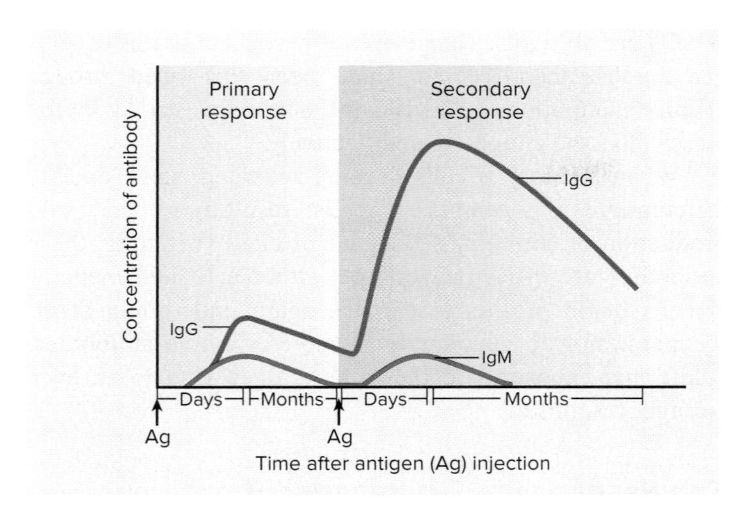

FIGURE 15.12 The Primary and Secondary Responses to Antigen The first exposure to antigen elicits relatively low amounts of IgM, followed by IgG. The second exposure, which characterizes the memory of the adaptive immune system, elicits rapid production of relatively large quantities of IgG.

? **How would this graph be different if it illustrated antibody levels on a mucosal surface?**

- **Affinity maturation**. This is a form of natural selection among proliferating B cells (**figure 15.14**). As activated B cells multiply, spontaneous mutations commonly occur in certain regions of the antibody genes. Some of these result in slight changes in the antigen-binding site of the antibody (and therefore the B-cell receptor). B cells that bind antigen for the longest duration are most likely to proliferate, and these give rise to plasma cells that secrete antibodies that bind antigen more effectively.
- **Class switching**. All B cells are initially programmed to differentiate into plasma cells that secrete IgM. Cytokines produced by T_H cells, however, induce some activated B cells to switch that genetic program, causing them to differentiate into plasma cells that secrete other antibody classes. B cells in the lymph nodes most commonly switch to IgG production (**figure 15.15**). B cells in the mucosa-associated lymphoid tissue (MALT) generally switch to IgA production, providing mucosal immunity.

After class switching, some B cells become memory B cells. These persist in the body for years and accumulate in high enough numbers to give a fast secondary response if the same antigen is encountered again later.

The antibody response begins to decrease as the antibodies clear the antigen. As fewer molecules of antigen remain to stimulate the lymphocytes, the activated lymphocytes undergo apoptosis. Memory B cells, however, are long-lived even in the absence of antigen.

MicroByte

A plasma cell secretes thousands of identical antibody molecules per second.

Characteristics of the Secondary Response

The secondary response is much faster and more effective than the primary response. In fact, repeat invaders are generally eliminated before they cause noticeable harm. This is why a person who has recovered from a particular disease typically has long-lasting immunity to the microbe that caused it. Vaccination takes advantage of this naturally occurring phenomenon. ⏭ vaccination

Memory B cells are responsible for the efficiency of the secondary response. For one thing, there are more cells—memory B cells as well as memory helper T cells—that can respond to a specific antigen. In addition, the memory

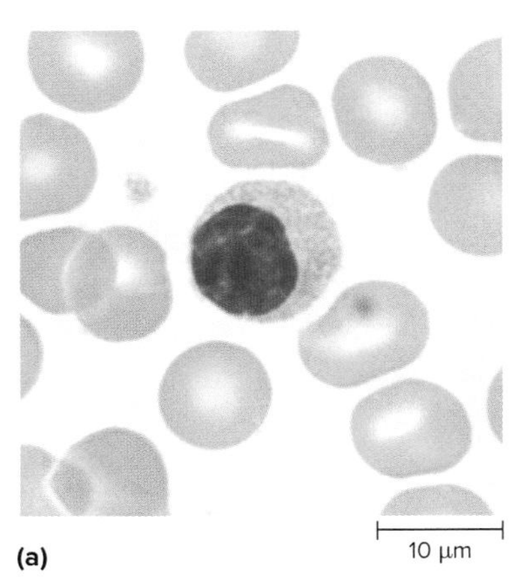

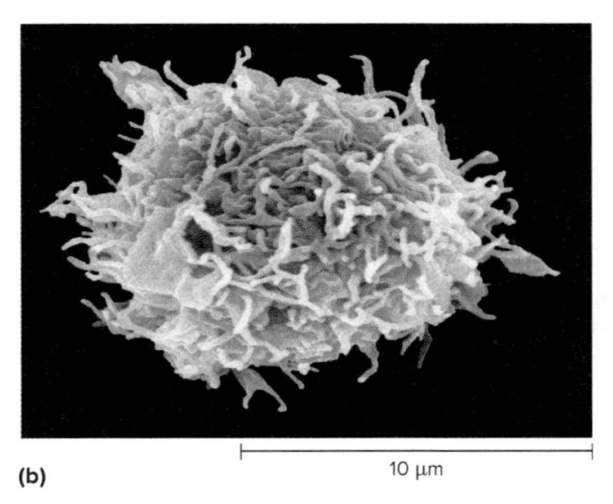

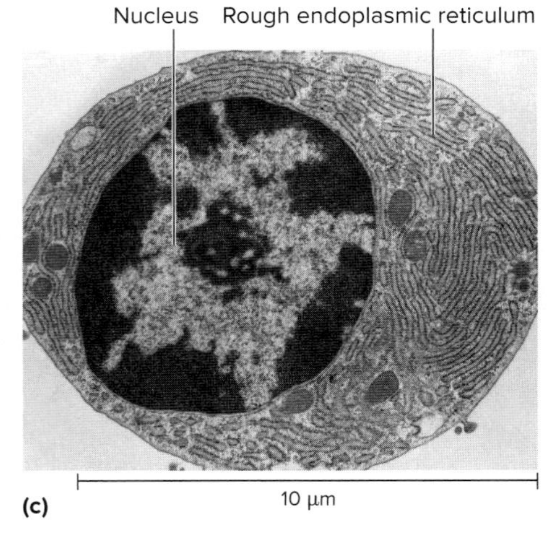

FIGURE 15.13 Lymphocytes and Plasma Cells (a) Light micrograph of a T cell. The morphology is the same as that of a B cell. **(b)** Scanning electron micrograph of a B cell. **(c)** Plasma cell, an effector B cell, which secretes antibody molecules. Note the large amount of rough endoplasmic reticulum, the site of protein synthesis.

? **Would a single plasma cell produce antibody molecules of the same specificity or of different specificities? Explain your answer.**

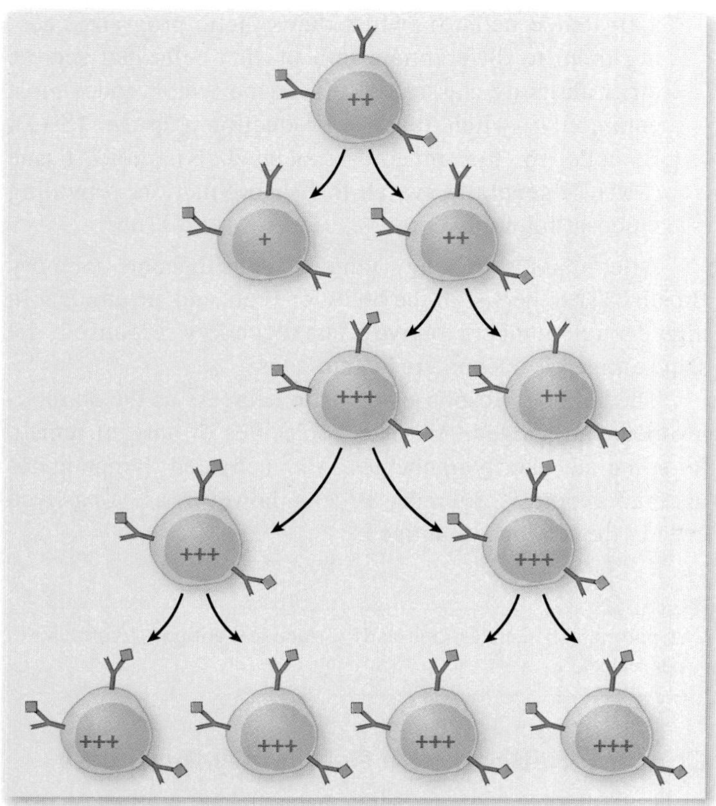

FIGURE 15.14 Affinity Maturation B cells that bind antigen for the longest duration are the most likely to proliferate. The plus signs indicate the relative quality of binding of the B-cell receptor (and therefore antibody) to the antigen; those in green indicate the most "fit" to continue proliferating.

What accounts for the change in a B cell's ability to bind antigen?

B cells are able to scavenge even low concentrations of antigen because their receptors have been fine-tuned through affinity maturation. Likewise, the antibodies coded for by these cells bind antigen more effectively.

When memory B cells become activated, some quickly differentiate to form plasma cells, resulting in the rapid production of antibodies. Because of class switching, these antibodies are often IgG or IgA. Other activated memory B cells begin proliferating, once again undergoing affinity maturation to generate even more effective antibodies. Additional exposures to the same antigen lead to an even stronger response.

The Response to T-Independent Antigens

T-independent antigens can activate B cells without the aid of T_H cells. Relatively few antigens are T-independent, but they can be very important medically.

Polysaccharides and other molecules that have many identical evenly spaced epitopes are one type of T-independent antigen. Because of the arrangement of epitopes, clusters of

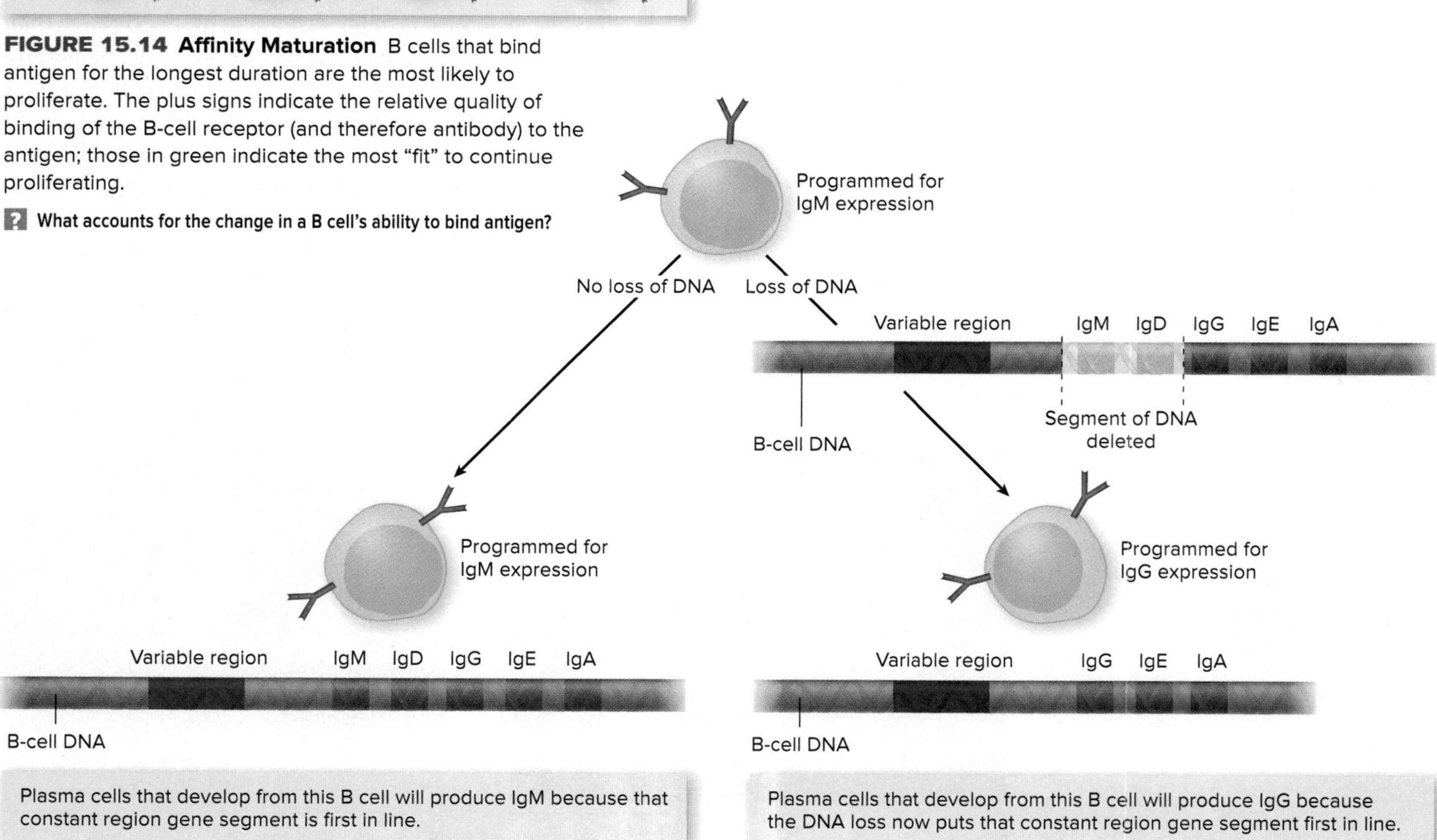

FIGURE 15.15 Class Switching Naive B cells are programmed to produce IgM antibodies. Activated B cells undergo class switching. Class switching does not alter the variable region and therefore has no effect on the antibody specificity. The plasma cells descended from B cells in lymph nodes most commonly produce IgG after class switching.

How would this illustration change if it showed class switching by B cells residing in Peyer's patches?

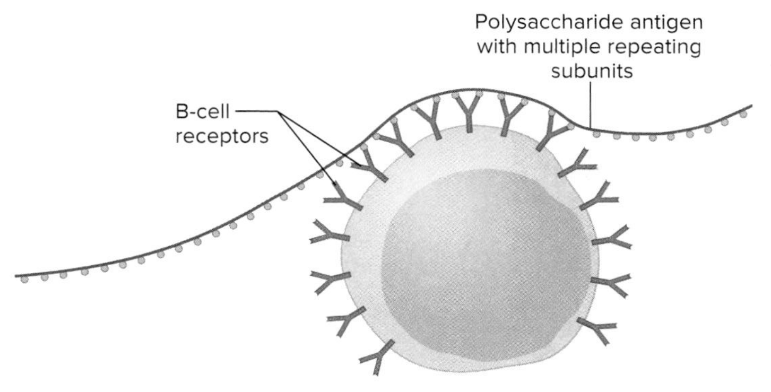

FIGURE 15.16 T-Independent Antigens Antigens such as some polysaccharides have multiple repeating epitopes. Because of the arrangement of epitopes, clusters of B-cell receptors bind to the antigen simultaneously, leading to B-cell activation without the involvement of T_H cells.

Why is the response to T-independent antigens important medically?

B-cell receptors bind the antigen simultaneously, an event that leads to activation of the B cell without the involvement of helper T cells (**figure 15.16**). These T-independent antigens are particularly significant because they are not very immunogenic in young children. This is why children less than 2 years of age are more susceptible to pathogens such as *Streptococcus pneumoniae* and *Haemophilus influenzae,* which coat themselves in polysaccharide capsules. Antibodies that bind the capsules would be protective if the child's immune system could make them. Vaccines made from purified capsules are available, but, likewise, they do not stimulate an immune response in young children. Fortunately, newer vaccines designed to induce a T-dependent response have been developed.

Lipopolysaccharide (LPS), a component of the outer membrane of Gram-negative bacteria, is another type of T-independent antigen. The constant presence of antibodies against LPS is thought to provide an early defense against invading Gram-negative bacteria.

MicroAssessment 15.6

The binding of a T-dependent antigen to a B cell's receptor is an initial signal to that B cell that an invader is present. Before that B cell can become activated, however, it must get a second signal in the form of cytokines delivered by a T_H cell that recognizes the same antigen. Activated B cells proliferate, ultimately becoming either plasma cells that secrete antibody molecules or long-lived memory cells. Affinity maturation and class switching occur in the primary response; these allow a swift and more effective secondary response. B cells that bind to T-independent antigens can become activated without the aid of T_H cells.

15. Describe the significance of class switching.
16. How does the ability to bind antigen increase as B cells multiply?
17. Why should B cells residing in the mucosa-associated lymphoid tissues produce IgA?

15.7 ■ The T-Cell Response: Cell-Mediated Immunity

Learning Outcomes

11. Describe the importance of T-cell receptors and CD markers.
12. Describe the role of dendritic cells in T-cell activation.
13. Compare and contrast T_H and T_C cells with respect to antigen recognition and the response to antigen.

The role of T cells is very different from that of B cells (see figure 15.1). For one thing, T cells never produce antibodies. Instead, effector T cells directly interact with other cells—target cells—to cause distinct changes in those cells (**table 15.2**).

General Characteristics of T Cells

Like B cells, T cells have multiple copies of a receptor on their surface that recognizes a specific epitope. The **T-cell receptor (TCR)** has two polypeptide chains (a set of either alpha and beta or gamma and delta), each with a variable and constant region (**figure 15.17**). From a structural standpoint,

TABLE 15.2 Characteristics of T Cells

T Cell Type/ CD Marker	Antigen Recognition	Effector Form	Potential Target Cells	Effector Function	Source of Antigen Recognized by Effector Cell
Cytotoxic/CD8	Peptides presented on MHC class I molecules	T_C cell	All nucleated cells	Induces target cell to undergo apoptosis	Endogenous (produced within the target cell)
Helper/CD4	Peptides presented on MHC class II molecules	T_H cell	B cells, macrophages	Activates target cell	Exogenous (produced outside of the target cell)

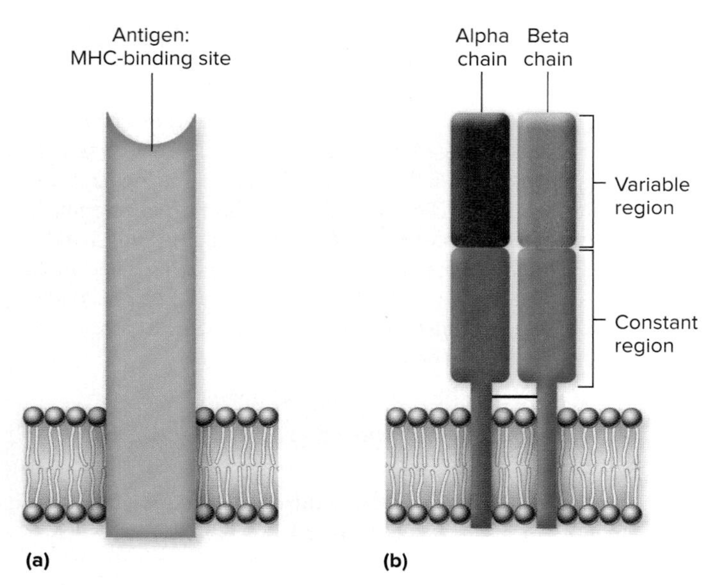

FIGURE 15.17 Structure of a T-Cell Receptor (a) Simplified structure, showing the antigen:MHC-binding site. **(b)** The molecule is made of two different chains, each with a constant region and a variable region, linked by a disulfide bond.

A T-cell receptor is similar to which part of a B-cell receptor?

the T-cell receptor can be compared to one "arm" of the B-cell receptor.

Unlike the B-cell receptor, the T-cell receptor does not interact with free antigen. Instead, the antigen must be "presented" by another host cell, as described in section 15.6 (see figure 15.11). The host cell does this by partly degrading (processing) the antigen and then displaying (presenting) individual peptides of the antigen's proteins. The peptides are cradled in the groove of proteins called **major histocompatibility complex (MHC) molecules,** which are on the surface of the presenting cell.

Two types of MHC molecules present antigen: **MHC class I** and **MHC class II** (**figure 15.18**). Both are shaped somewhat like an elongated bun and hold the peptide lengthwise, like a bun holds a hot dog. When a T cell recognizes an antigen, the cell is actually recognizing both the peptide and the MHC molecule simultaneously. In other words, the T-cell receptor recognizes the "whole sandwich"—the peptide:MHC complex.

MHC class I molecules present endogenous antigens (antigens made within the cell). MHC class II molecules present exogenous antigens (antigens taken up by a cell). All nucleated cells produce MHC class I molecules, but only specialized cell types (dendritic cells, B cells, and macrophages)—collectively referred to as **antigen-presenting cells (APCs)**—make MHC class II molecules (see **Focus Your Perspective 15.1**).

Recall from the introductory section that two functionally distinct T-cell populations are involved in eliminating antigen: cytotoxic T cells and helper T cells. These differ in their roles, and also in how they recognize antigen. Cytotoxic T cells recognize antigen presented on MHC class I molecules, whereas helper T cells recognize antigen presented on MHC class II molecules. Because of these recognition characteristics, effector cytotoxic T (T_C) cells respond to endogenous antigens, whereas effector helper T (T_H) cells respond to exogenous antigens (**figure 15.19**).

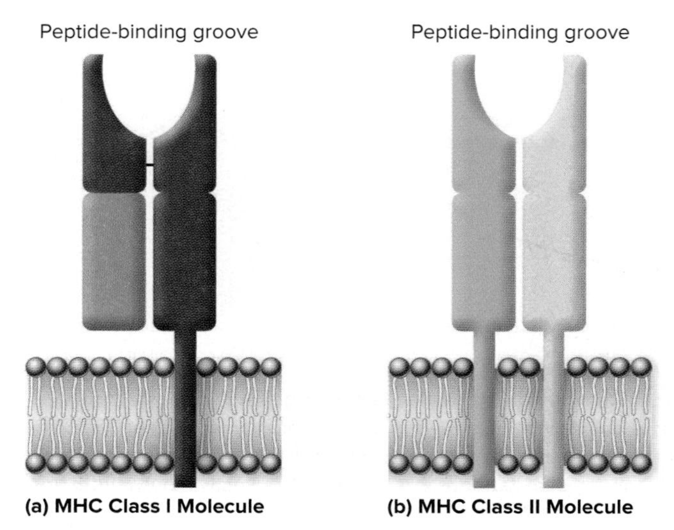

FIGURE 15.18 MHC Molecules (a) MHC class I molecule; cytoplasmic proteins (endogenous antigens) are presented in the groove of these molecules. **(b)** MHC class II molecule; proteins taken in by the cell (exogenous antigens) are presented in the groove of these molecules.

Which cell type recognizes peptides presented in MHC class I molecules? Which cell type recognizes peptides presented in MHC class II molecules?

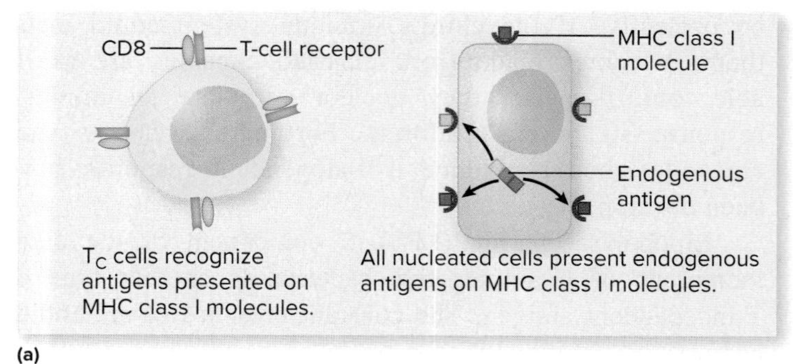

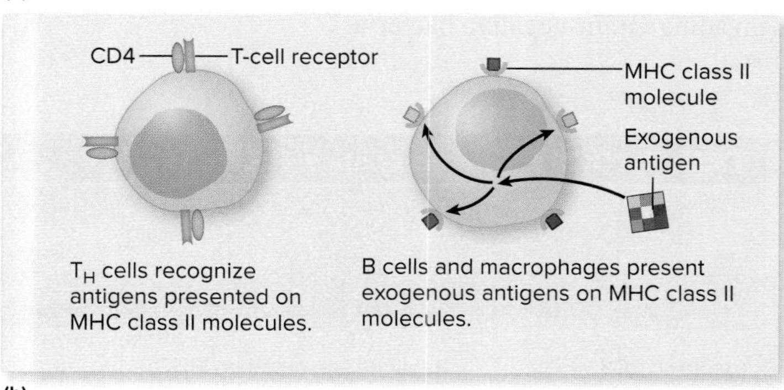

FIGURE 15.19 Antigen Recognition by Effector T Cells (a) Cytotoxic T (T_C) cell. **(b)** Helper T (T_H) cell.

What is the fate of a cell that presents antigen recognized by a T_C cell? What is the fate of a cell that presents antigen recognized by a T_H cell?

FOCUS YOUR PERSPECTIVE 15.1

What Flavors Are Your Major Histocompatibility Complex Molecules?

The major histocompatibility molecules were discovered over half a century ago, long before scientists knew about their essential role in adaptive immunity. During World War II, bombing raids caused serious burns in many people, stimulating research into skin transplants to replace burned tissue. That research quickly expanded to include transplants of a variety of other tissues and organs. Unfortunately, the transplanted material was generally rejected by the recipient's immune system because certain molecules on the donor cells differed from those on the recipient's cells; the donor tissue was perceived as an "invader" by the recipient's immune system. To overcome this problem, researchers tried to more closely match donor and recipient tissues, relying on newly developed tissue-typing tests. The typing tests looked for leukocyte surface molecules called **human leukocyte antigens (HLAs),** which serve as markers for tissue compatibility. Later, researchers determined that HLAs were encoded by a cluster of genes, now called the major histocompatibility complex. Unfortunately, the terminology can be confusing because the molecules that transplant biologists refer to as HLAs are called MHC molecules by immunologists.

It is highly unlikely that two random individuals will have identical MHC molecules. This is because the genes encoding them are polygenic, meaning they are encoded by more than one locus (position on the chromosome), and each locus is highly polymorphic (multiple forms). As an analogy, if MHC molecules were candy, each cell would be covered with pieces of chocolate, taffy, and lollipop. The type of chocolate could be dark, white, milk, or a number of various flavors; the taffy could be peppermint, raspberry, or cinnamon, and so on. As you might imagine, the number of different possible combinations is enormous.

There are three loci of MHC class I genes, designated HLA-A, HLA-B, and HLA-C (**box figure 15.1**). There are at least 890 alleles (forms) for HLA-A, 1,400 for HLA-B, and 620 for HLA-C. In addition, the loci are all co-dominantly expressed. In other words, the set of the three MHC class I genes you inherited from your mother and the set you inherited from your father are both expressed. Putting this all together, your cells most likely express six different varieties of MHC class I molecules—two of the over 890 known HLA-A possibilities, two of the over 1,400 HLA-B possibilities, and two of the over 620 HLA-C possibilities. As you can imagine, the chance that anyone you know will have those same MHC class I molecules is extremely low, unless you have an identical twin.

Why is there so much diversity in MHC molecules? The answer lies in the complex demands of antigen presentation. MHC class I molecules bind peptides that are only 8 to 10 amino acids in length; MHC class II molecules bind peptides that are only 13 to 25 amino acids in length. Somehow, within that limitation, the MHC molecules must bind as many different peptides as possible in order to ensure that a representative selection from the proteins within a cell can be presented to T cells. The ability to bind a wide variety of peptides is particularly important considering how readily microbes evolve in response to selective pressure. For example, if a single alteration in a viral protein prevented all MHC molecules from presenting peptides from that protein, then a virus with that mutation could overwhelm the defenses. The ability to bind different peptides is not enough, however, because, ideally, a given peptide should be presented in several slightly different orientations so that distinct parts of the three-dimensional structures can be inspected by T-cell receptors. No single variety of MHC molecule can accomplish all of these aims, which explains the need for their diversity.

The variety of MHC molecules that a person has on his or her cells affects that individual's adaptive response to certain antigens. This is not surprising because MHC molecules differ in the array of peptides they can bind and in the manner in which those peptides are held in the molecule. Thus, characteristics of the MHC molecules affect what the T cells actually "see." In fact, the severity of certain diseases has been shown to correlate with the MHC type of the infected individual. For example, rheumatic fever, which can occur as a consequence of *Streptococcus pyogenes* infection, develops more frequently in individuals with certain MHC types. The most serious outcomes of schistosomiasis have also been shown to correlate with certain MHC types. Epidemics of life-threatening diseases such as plague and smallpox have dramatically altered the relative proportion of MHC types in certain populations, killing those whose MHC types ineffectively present peptides from the causative agent. ⏭ rheumatic fever, ⏮ schistosomiasis

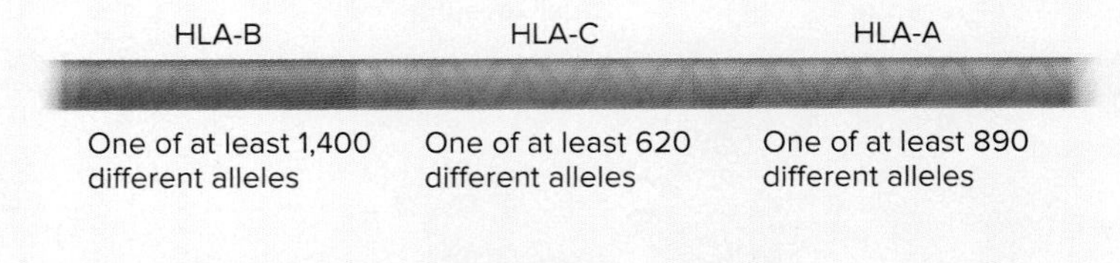

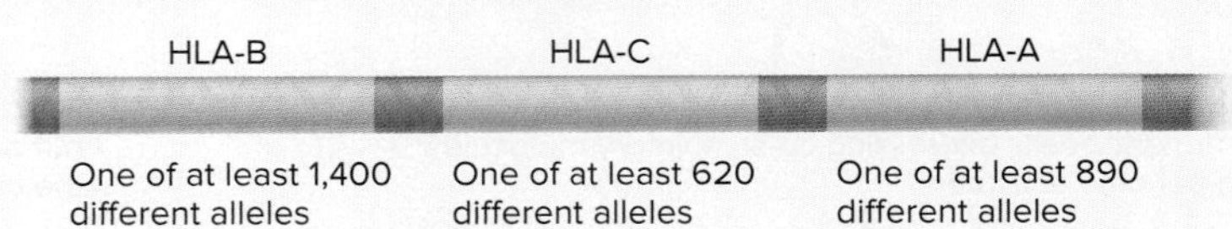

BOX FIGURE 15.1 MHC Polymorphisms The order of the MHC class I genes on the chromosome is B, C, and A.

Cytotoxic and helper T cells are identical microscopically, so scientists distinguish them based on the presence of surface proteins called **cluster of differentiation (CD) markers.** Most cytotoxic T cells have the CD8 marker and are frequently referred to as CD8 T cells; most helper T cells carry the CD4 marker and are often called CD4 T cells. Note that CD4 is also a receptor for HIV, which explains why the virus infects helper T cells.

Activation of T Cells

Dendritic cells, the scouts of innate immunity, play a crucial role in T-cell activation. Immature dendritic cells reside in peripheral tissues, particularly under the skin and the mucosa, gathering various materials from those areas. The cells use both phagocytosis and pinocytosis to take up particulate and soluble material that could contain foreign proteins. Dendritic cells located just below the mucosal barriers are even able to send tentacle-like extensions between the epithelial cells of the barriers. Using this action, the dendritic cells gather material from the respiratory tract and the lumen of the intestine. The dendritic cells have TLRs (toll-like receptors) and other pattern recognition receptors that allow them to recognize pathogens. If invading microbes are detected, the dendritic cells take up even more material, and then enter lymphatic vessels, which transport them to secondary lymphoid organs where they will encounter naive T cells. Dendritic cells near the end of their life span do this as well. ⏮ dendritic cells, ⏮ pattern recognition receptors

En route to the secondary lymphoid organs, the dendritic cells mature into a form that presents antigen to naive T cells (**figure 15.20**). Dendritic cells that detected invading microbes produce surface proteins called **co-stimulatory molecules.** These molecules function as "emergency lights" that interact with the T cell, communicating that the material being presented indicates "danger." Naive T cells that

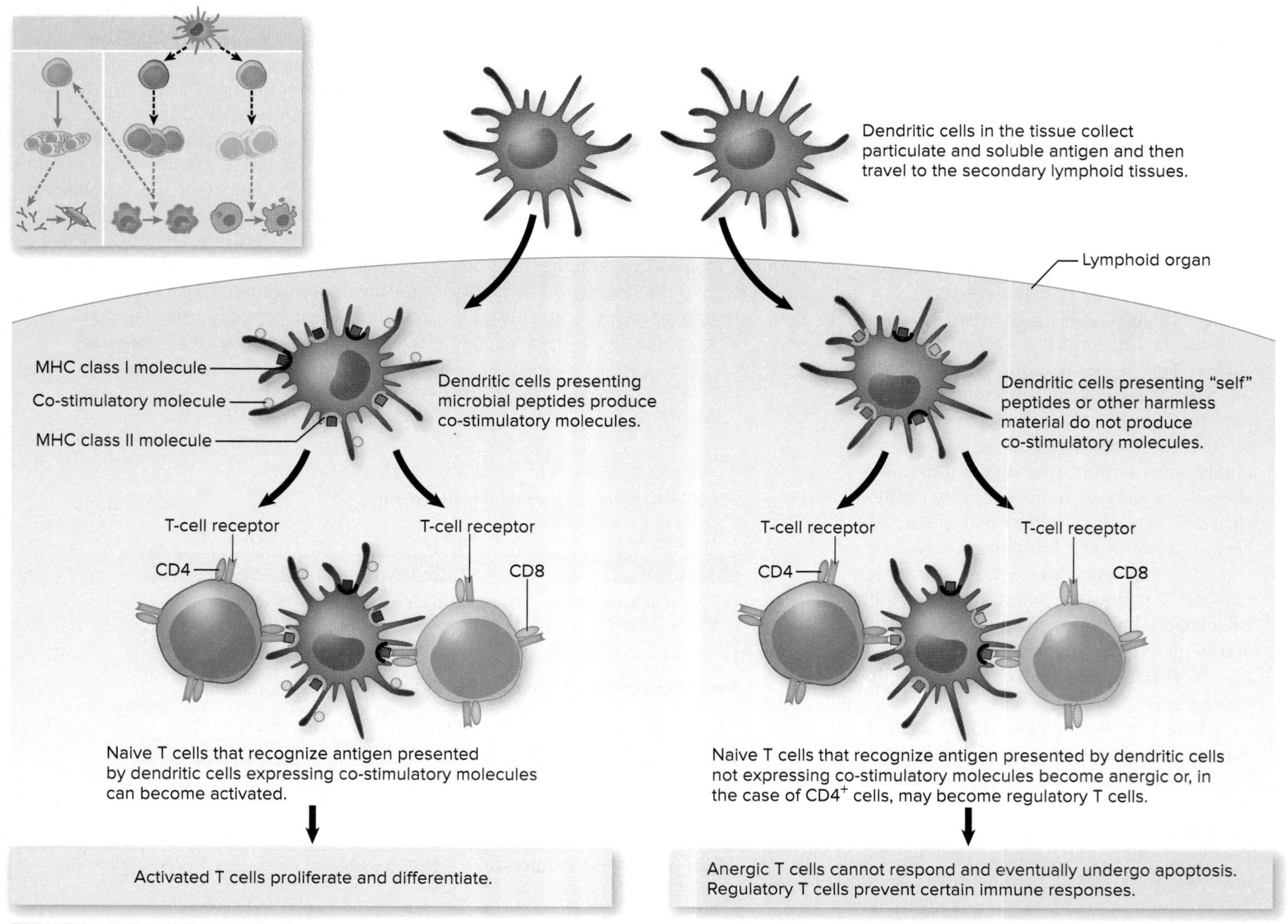

FIGURE 15.20 T-Cell Activation

? What causes a dendritic cell to produce co-stimulatory molecules?

recognize antigen presented by dendritic cells displaying co-stimulatory molecules can become activated. In contrast, naive T cells that recognize antigen presented by a dendritic cell not displaying co-stimulatory molecules become anergic (unresponsive) and eventually undergo apoptosis. Recall that inducing anergy is one way the adaptive immune response eliminates lymphocytes that recognize "self" proteins, and is an important mechanism of self-tolerance. Alternatively, instead of becoming anergic, some of the responding $CD4^+$ T cells will become regulatory T cells (T_{reg} cells). The role of T_{reg} cells is to prevent certain immune responses, so this outcome provides another mechanism for developing tolerance.

Dendritic cells are able to "cross present" antigens, meaning they can present peptides on both types of MHC molecules—class I and class II—regardless of the origin. This allows them to present antigen to, and therefore activate, cytotoxic T cells as well as helper T cells.

Once a T cell is activated, it undergoes clonal selection and expansion as described previously, eventually forming effector cells and memory cells. These can leave the secondary lymphoid organs and circulate in the bloodstream. They can also enter tissues, particularly at sites of infection.

The effector helper T cells can take on different roles, depending on the signals given by the dendritic cell. This results in at least three subsets of T_H cells—T_H1, T_H2, and T_H17—that produce different ranges of cytokines and therefore promote different types of immune responses. In general, T_H1 cells promote a response effective against intracellular invaders, T_H2 cells promote a response effective against parasitic worms (helminths), and T_H17 cells promote a response effective against extracellular invaders. For simplicity, however, we will first describe the general functions of T_H cells as a group, rather than focus on the individual subsets. The important thing to remember at this point is that dendritic cells are able to guide the responses of the helper T cells, tailoring the response based on the microbe-associated molecular patterns (MAMPs) that the dendritic cell detected as it gathered the antigen. ⏮ MAMP

Macrophages and B cells also present exogenous antigens on MHC class II molecules and can produce co-stimulatory molecules. Based on this information, it seems they should also be able to activate naive T cells, but studies suggest they do not contact naive T cells in vivo. They do encounter memory T cells, however, so they probably activate these cells during the secondary response.

MicroByte

Of the approximately 10^{10} T cells in the body, only a few will recognize a given epitope the first time it is encountered.

Effector Functions of T_C (CD8) Cells

T_C cells induce apoptosis in infected "self" cells. They also destroy cancerous "self" cells. How do T_C cells distinguish infected or cancerous cells from their normal counterparts? The answer lies in the significance of antigen presentation on MHC class I molecules (**figure 15.21**). All nucleated cells routinely degrade a portion of the proteins they produce (endogenous proteins). They then load the resulting peptides into the groove of MHC class I molecules and deliver them to the cell surface for inspection by circulating T_C cells (see figure 15.19a). If a "self" cell is producing only normal proteins, then the peptides being presented should not be recognized by any circulating T_C cells. If the "self" cell is infected with a replicating virus or microorganism, however, then peptides from the invading microbe will be presented on MHC class I molecules, and those will be recognized by circulating T_C cells. This makes the presenting cell a target for the lethal effector functions of T_C cells. The same thing can occur if the "self" cell is producing abnormal proteins that characterize cancerous cells.

When a T_C cell encounters a cell presenting a peptide it recognizes, the T_C cell binds to that cell, and then delivers a "death package" to it. The T_C cell does this by releasing perforin, a molecule that forms pores in the target cell membrane, along with several proteases. The pores made by perforin allow the proteases to enter the target cell, where they cause reactions that induce that cell to undergo apoptosis. In addition, a specific molecule on the T_C cell can engage a "death receptor" on the target cell, also initiating apoptosis. Killing the target cell by inducing apoptosis rather than lysis minimizes the number of intracellular microbes that might spill into the surrounding area and infect other cells. Most microbes remain in the cell remnants until they are ingested by macrophages. The T_C cell survives and can go on to kill other targets.

In addition to inducing apoptosis in the target cell, the T_C cell will also produce various cytokines that strengthen the "security system." One cytokine increases antigen processing and presentation in nearby cells, making it easier for T_C cells to find other infected cells. Another cytokine activates local macrophages whose TLRs have been triggered. Note that a more efficient mechanism of macrophage activation involves an effector function of T_H cells (discussed next).

Effector Functions of T_H (CD4) Cells

T_H cells orchestrate the immune response. They activate B cells and macrophages and direct the activities of B cells, macrophages, and T cells.

T_H cells recognize antigen presented on MHC class II molecules. Recall that these are found only on antigen-presenting cells (APCs) such as B cells and macrophages, which gather, process, and present exogenous antigens (see figure 15.19b). When a T_H cell recognizes a peptide presented by a B cell or a macrophage, it delivers cytokines that activate the cell. Various cytokines are also released, depending on the subset of T_H cell.

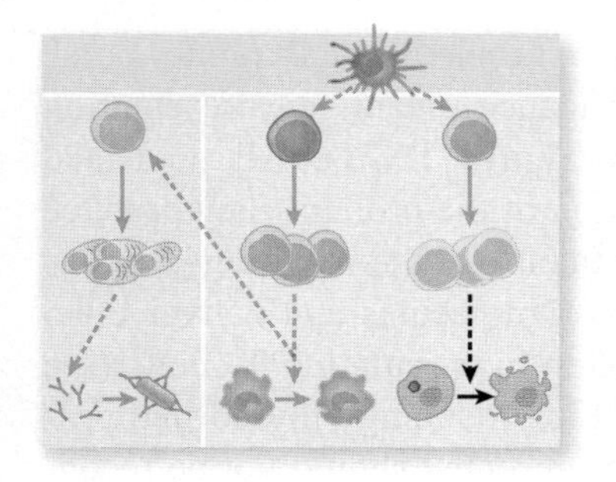

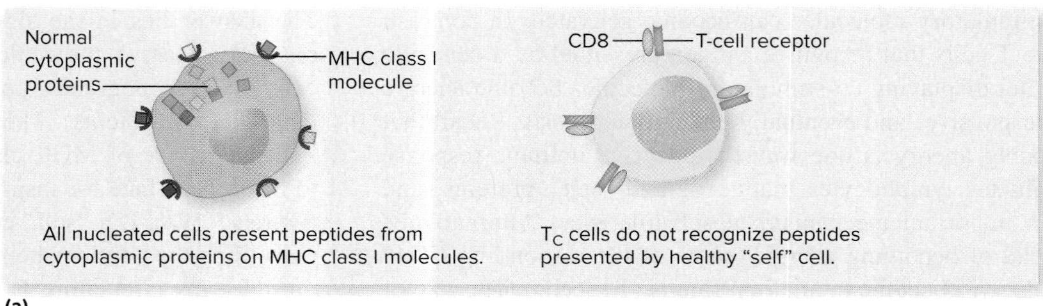

(a)

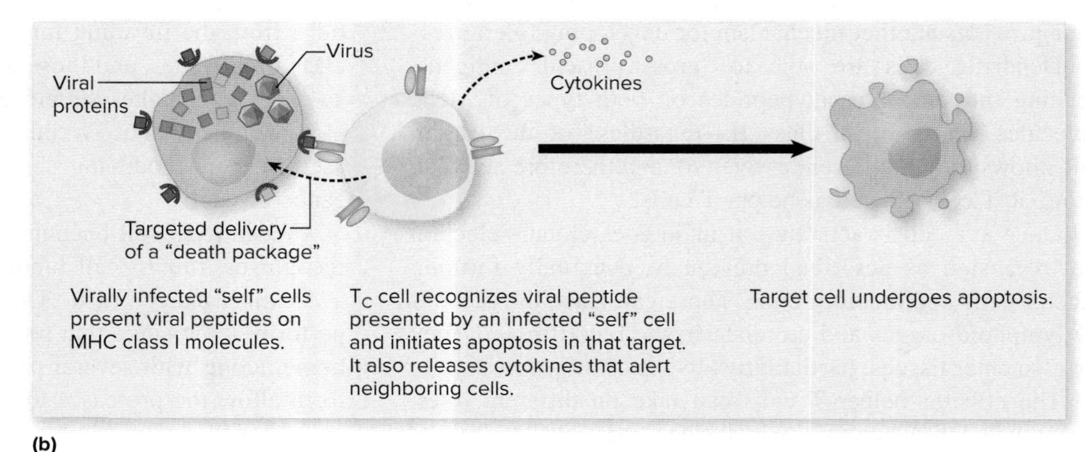

(b)

FIGURE 15.21 Functions of T_C Cells **(a)** T_C cells ignore healthy "self" cells. **(b)** T_C cells induce apoptosis in virally infected "self™" cells.

Could a T_C cell induce apoptosis in a "self" cell that lost the ability to produce MHC class I molecules?

The Role of T_H Cells in B-Cell Activation

B-cell activation was described earlier (see figure 15.11). This section reviews that process, but includes more details specific to the role of T_H cells.

When a naive B cell binds antigen via its B-cell receptor, the cell takes the antigen in by endocytosis. Proteins within the endosome are then degraded to produce short peptides that can be loaded into the groove of MHC class II molecules. If a T_H cell encounters a B cell presenting a peptide it recognizes, it delivers cytokines to that cell. These activate the B cell, allowing it to proliferate and undergo class switching. The cytokines also drive the formation of memory B cells.

A T_H cell does not need to recognize the same epitope as the B cell it activates. This is because the B cell presents many peptides from the antigen, and the T-cell receptor could bind any of those. In fact, a responding B cell probably recognizes an epitope on a pathogen's surface, whereas a T_H cell could very well recognize a peptide from within the pathogen.

Understanding the role of antigen processing and presentation led to a vaccine that now prevents infection by what was once the most common cause of meningitis in children: *Haemophilus influenzae.* Recall that young children are particularly susceptible to meningitis caused by this organism because it produces a polysaccharide capsule, an example of a T-independent antigen to which this age group responds poorly. A polysaccharide antigen can be converted to a T-dependent antigen by covalently attaching it to a large protein molecule, making a **conjugate vaccine.** The polysaccharide component of the vaccine binds to a B-cell receptor and the entire molecule is taken in. The protein component will then be processed and presented to T_H cells. Although the B cell recognizes the polysaccharide component of the vaccine, the T cells recognize peptides from the protein component. This recognition leads to B-cell activation and subsequent production of antibodies that bind the capsule. Conjugate vaccines against other pathogens have also been developed. ▶▶| conjugate vaccines

Antigen processing and presentation also explains how some people develop allergies to penicillin. This medication is a **hapten,** a small molecule that elicits an immune response only when it combines with a large carrier compound. In the body, penicillin can react with proteins, forming a penicillin-protein conjugate. If IgE antibodies are formed, their binding to penicillin can result in allergic reactions, ruling out the further use of the antimicrobial medication in these individuals.
▶▶| allergy, |◀◀ penicillin

The Role of T_H Cells in Macrophage Activation

As discussed in chapter 14, macrophages routinely engulf and degrade invading microbes, clearing most organisms even before an adaptive response is mounted. If this alone is not enough to control the invader, macrophages can be activated by T_H cells, allowing the phagocytes to produce more potent destructive mechanisms. |◀◀ macrophages

FOCUS ON A CASE 15.1

A 34-year-old man who had recently returned from a safari in Africa experienced what he assumed was severe jet lag, feeling dizziness, a headache, and body aches. His condition gradually worsened over several days as he developed a fever and became nauseated. At his family's urging, he made an appointment to see his primary care physician. The doctor questioned him about his trip to Africa, concerned that he had acquired his illness there. The patient explained that he had visited a travel clinic before he left, where he was vaccinated against several diseases, including polio, typhoid fever, meningococcal meningitis, yellow fever, and hepatitis A. He was also given information about measures to prevent malaria. There is currently no vaccine against malaria, in part because the mosquito-borne protozoan parasite has a complex life cycle that includes multiple antigenically distinct stages. The man was prescribed an anti-malarial medication as a prophylaxis (preventive measure), but he stopped taking it after only a few days. He did, however, follow the clinic's advice to use an insect repellent because mosquitoes transmit the disease, but he ignored their suggestion of using a product that contains DEET (*N, N*-diethyl-*meta*-toluamide). Instead, he chose a natural plant-based product that was not very effective.

Based on the patient's symptoms and travel history, the physician suspected malaria. This disease is caused by several different species of *Plasmodium,* a type of parasite that infects red blood cells (RBCs). The physician took a blood sample from the patient, which he immediately sent to the clinical lab for testing. When the laboratory technician did a microscopic examination of the specimen, she saw many *Plasmodium*-infected RBCs. She also saw banana-shaped gametocytes (the form that is infectious to mosquitoes)—a morphology that identifies the parasite as *P. falciparum.* She quickly reported the results to the physician, recognizing that malaria caused by this species can be a medical emergency.

The patient was admitted to the hospital, where he was immediately started on an intravenous antimalarial medication. His condition continued to deteriorate, however, and he began experiencing episodes of chills and fever, and was often drenched with sweat. Within a matter of days, he developed neurological symptoms, including confusion, anxiety, and seizures. He soon slipped into a coma and died.

Autopsy results indicated that the patient had developed cerebral malaria, a result of *P. falciparum*–infected RBCs clogging the capillaries in the brain. This deadly complication sometimes occurs with *P. falciparum* infection when infected RBCs stick to the capillary walls in the brain. The RBCs do this because a protein made by *P. falciparum* inserts into the RBC cytoplasmic membrane and then binds to the surface of cells that line the capillaries.

1. What is the advantage to the parasite of having several antigenic forms in its life cycle?
2. Why would it be difficult to develop a vaccine against malaria?
3. What is the advantage to the parasite of multiplying within red blood cells?
4. What is the advantage to the parasite of having the RBCs stick to the capillaries rather than circulate?
5. Why might the patient have died even though he was being treated?

Discussion

1. The multiple antigenic forms of the parasite make it more difficult for the immune response to eliminate the organism. By the time an effective adaptive response has developed against one stage in the parasite's life cycle, the organism has progressed to the next form. Over a period of weeks, however, there is an effective immune response against the form that infects red blood cells, thereby limiting disease progression.
2. A vaccine has been difficult to develop for several reasons. If the vaccine protects against only a single stage of the parasite, and even one *Plasmodium* cell in that stage escapes immune detection, then the parasite can continue its life cycle in the infected person. Developing a vaccine against multiple stages of the parasite is possible, but considerably more difficult. Another hurdle is the fact that the complex life cycle of the parasite makes studying the organism difficult. For example, one stage replicates in liver cells, which have been difficult to grow in culture until recently. Scientists are still discovering the mechanisms that the parasite uses to avoid the immune system, but every advancement brings researchers closer to an effective vaccine.
3. By hiding within one of the body's own cells, the parasite avoids humoral immunity. Multiplying within a red blood cell provides an additional advantage: RBCs do not make MHC molecules, so the infected RBCs are not a target of cytotoxic T cells. ⏮ **humoral immunity**
4. By sticking to the capillaries, the red blood cells do not circulate in the bloodstream. Circulating blood passes through the spleen, a secondary lymphoid organ. In the spleen, old and damaged RBCs are removed. The parasite has multiple genes for the protein that attaches to the walls of the capillaries, allowing the organism to produce dozens of antigenically distinct varieties; as the immune response begins to make antibodies that recognize one version of the protein, some of the parasites will have switched to make a different antigenic type.
5. The damage had probably already started before treatment was begun. Once the RBCs clog the capillaries, the obstruction deprives brain cells of O_2. An inflammatory response develops in the damaged tissues, which makes the problem even worse. It is also possible that the parasite was resistant to the medications being delivered; drug resistance in *Plasmodium* species is an increasing problem.

The steps of macrophage activation are very similar to those described for B-cell activation. When macrophages engulf material, they enclose it within a membrane-bound phagosome (**figure 15.22**). The proteins within the phagosome are then degraded, and the resulting peptides presented on MHC class II molecules. If a T_H cell recognizes one of the peptides, it delivers cytokines that activate the macrophage. ⏮ phagosome

When a macrophage is activated, it gets larger, the plasma membrane becomes ruffled and irregular, and the cell increases its metabolism so that the lysosomes—which contain antimicrobial substances—increase in number. The activated macrophage also begins producing nitric oxide, a potent antimicrobial chemical, along with various compounds that can be released to destroy extracellular microorganisms.

If the response is still not sufficient to control the infection, activated macrophages fuse together, forming **giant cells.** These, along with other macrophages and T cells, can form granulomas that wall off the offending agent, preventing infectious microbes from escaping to infect other cells. Activated macrophages are important in the immune response against the bacterium that causes tuberculosis *(Mycobacterium tuberculosis)* and other microorganisms that can survive within regular macrophages. ⏮ giant cell, ⏮ granuloma

Subsets of T_H Cells

The various subsets of effector helper T cells (T_H1, T_H2, T_H17) produce different groups of cytokines, thereby directing the immune system toward an appropriate response for a given antigen. For example, the cytokines produced by T_H1 cells activate macrophages and stimulate T_C cells, thereby promoting a response against intracellular pathogens; T_H2 cells direct a response against multicellular pathogens by recruiting eosinophils and basophils; and T_H17 cells recruit neutrophils, thereby directing a response against extracellular pathogens. The outcome of some conditions, such as Hansen's disease (leprosy), appears to correlate with the type of helper T-cell response.

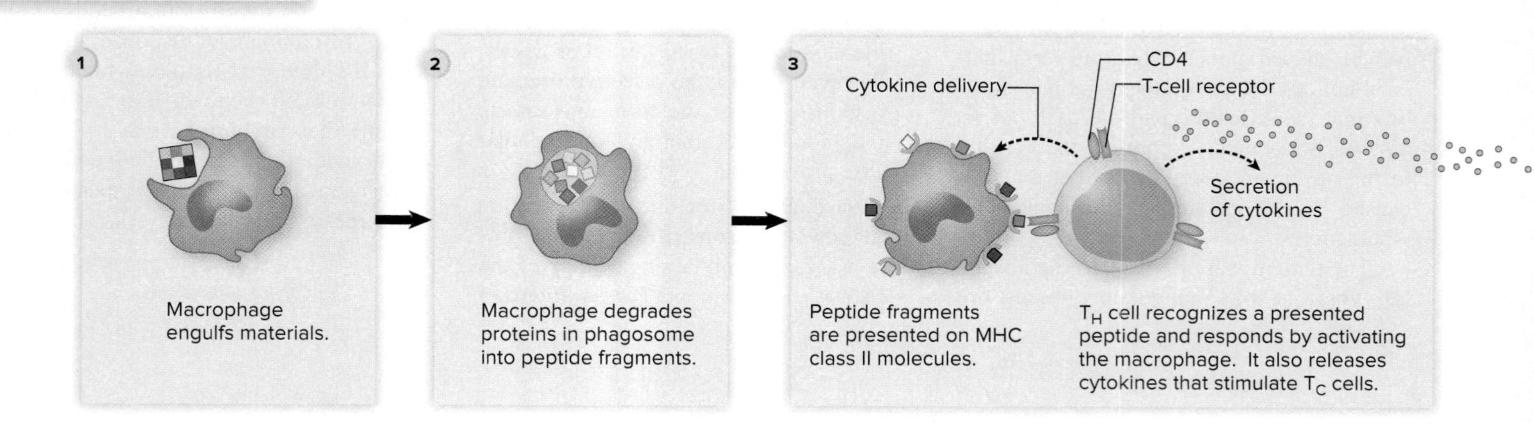

FIGURE 15.22 The Role of T_H Cells in Macrophage Activation

How does macrophage activation help prevent disease?

MicroAssessment 15.7

Dendritic cells expressing co-stimulatory molecules activate T cells that recognize the presented antigen. T_C (CD8) cells recognize antigen presented on MHC class I molecules; they induce apoptosis in target cells and produce cytokines that increase the level of surveillance. T_H (CD4) cells recognize antigen presented on MHC class II molecules (found on B cells and macrophages); they activate the target cells and secrete various cytokines that orchestrate the immune response.

18. Name three types of antigen-presenting cells.

19. If an effector CD8 cell recognizes antigen presented on an MHC class I molecule, how should it respond?

20. Why would a person who has AIDS be more susceptible to the bacterium that causes tuberculosis?

15.8 ■ Natural Killer (NK) Cells

Learning Outcome

14. Describe two distinct protective roles of NK cells.

Natural killer (NK) cells are innate lymphoid cells (ILCs), a group of lymphocytes lacking the antigen-specific receptors that characterize B cells and T cells. The activities of NK cells assist the adaptive immune responses. ⏮ innate lymphoid cells

NK cells induce apoptosis in antibody-bound "self" cells. This process, antibody-dependent cellular cytotoxicity (ADCC), allows NK cells to destroy host cells that have viral or other foreign proteins inserted into their membrane (see figure 15.8). NK cells can do this because they have Fc receptors for IgG molecules on their surface; recall that Fc receptors bind the "red flag" portion of antibody molecules. The NK cell attaches to the antibodies and then delivers perforin- and protease-containing granules directly to the antibody-bound cell, initiating apoptosis. ⏮ ADCC

NK cells also recognize and destroy stressed host cells that do not have MHC class I molecules on their surface (**figure 15.23**). This is important because some viruses have evolved mechanisms to avoid the action of cytotoxic T cells by interfering with the process of antigen presentation; cells infected with such a virus will essentially lack MHC class I molecules and thus cannot be a target of cytotoxic T cells. The NK cells recognize the absence of MHC class I molecules on those cells, along with the presence of certain molecules that indicate the cells are under stress, and induce the infected cells to undergo apoptosis.

Recent evidence indicates that NK cells are more than killing machines. For example, they produce cytokines that help regulate and direct certain immune responses. Unfortunately, studying these actions is difficult because there are different subsets of NK cells, and the activities of the various subsets are influenced by cues in their local environment.

MicroAssessment 15.8

Natural killer (NK) cells can kill antibody-bound cells by antibody-dependent cellular cytotoxicity (ADCC). NK cells also kill stressed cells that lack MHC class I molecules on their surface.

21. What mechanism do NK cells use to kill "self" cells?

22. What can cause a "self" cell to not display MHC class I molecules?

23. Why might a virus that interferes with the production of the host's MHC class I molecules encode a non-functional version?

15.9 ■ Lymphocyte Development

Learning Outcomes

15. Describe the roles of gene rearrangement, imprecise joining, and combinatorial associations in the generation of diversity.

16. Describe positive and negative selection of lymphocytes.

As descendants of hematopoietic stem cells develop into B cells and T cells, they acquire their ability to recognize distinct epitopes. B cells undergo the developmental stages in the bone marrow; T cells go through the maturation processes described in this section in the thymus. ⏮ hematopoietic stem cells

The events involved in the adaptive immune response, from the maturation of lymphocytes to the development of their effector functions, are summarized in **figure 15.24.**

Generation of Diversity

The mechanisms lymphocytes use to produce a seemingly limitless assortment of antibodies and antigen-specific receptors were first discovered in studies using B cells. Because the processes in B cells are very similar to those for T cells, we will use B cells as a general model to describe the generation of diversity with respect to antigen recognition.

Each B cell responds to only one epitope, yet the population of B cells within the body appears able to respond to more than 100 million different epitopes. Based on this number and the information presented in chapter 7, it might seem logical to assume that the human genome has over 100 million different antibody genes, each encoding specificity for a single epitope. This is impossible, however, because the human genome has only 3 billion nucleotides and contains about 25,000 genes.

The question of how such tremendous diversity in antibodies could be generated puzzled immunologists until Dr. Susumu Tonegawa solved the mystery. For this work, he was awarded a Nobel Prize in 1987.

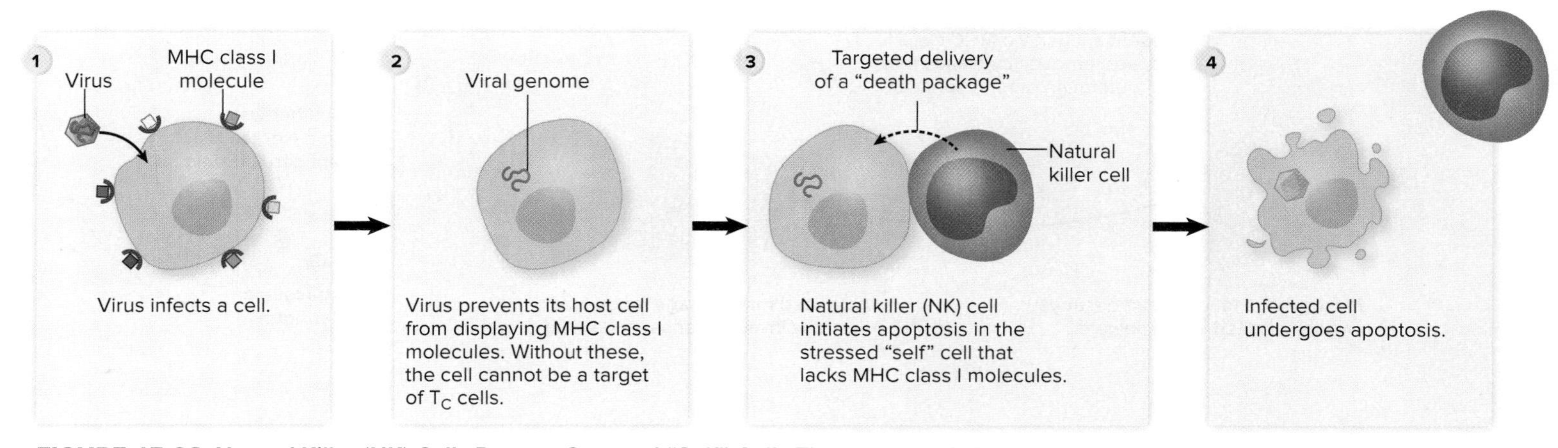

FIGURE 15.23 Natural Killer (NK) Cells Destroy Stressed "Self" Cells That Lack MHC Class I Molecules

Why would a virus that can interfere with a host cell's production of MHC class I molecules be at an advantage?

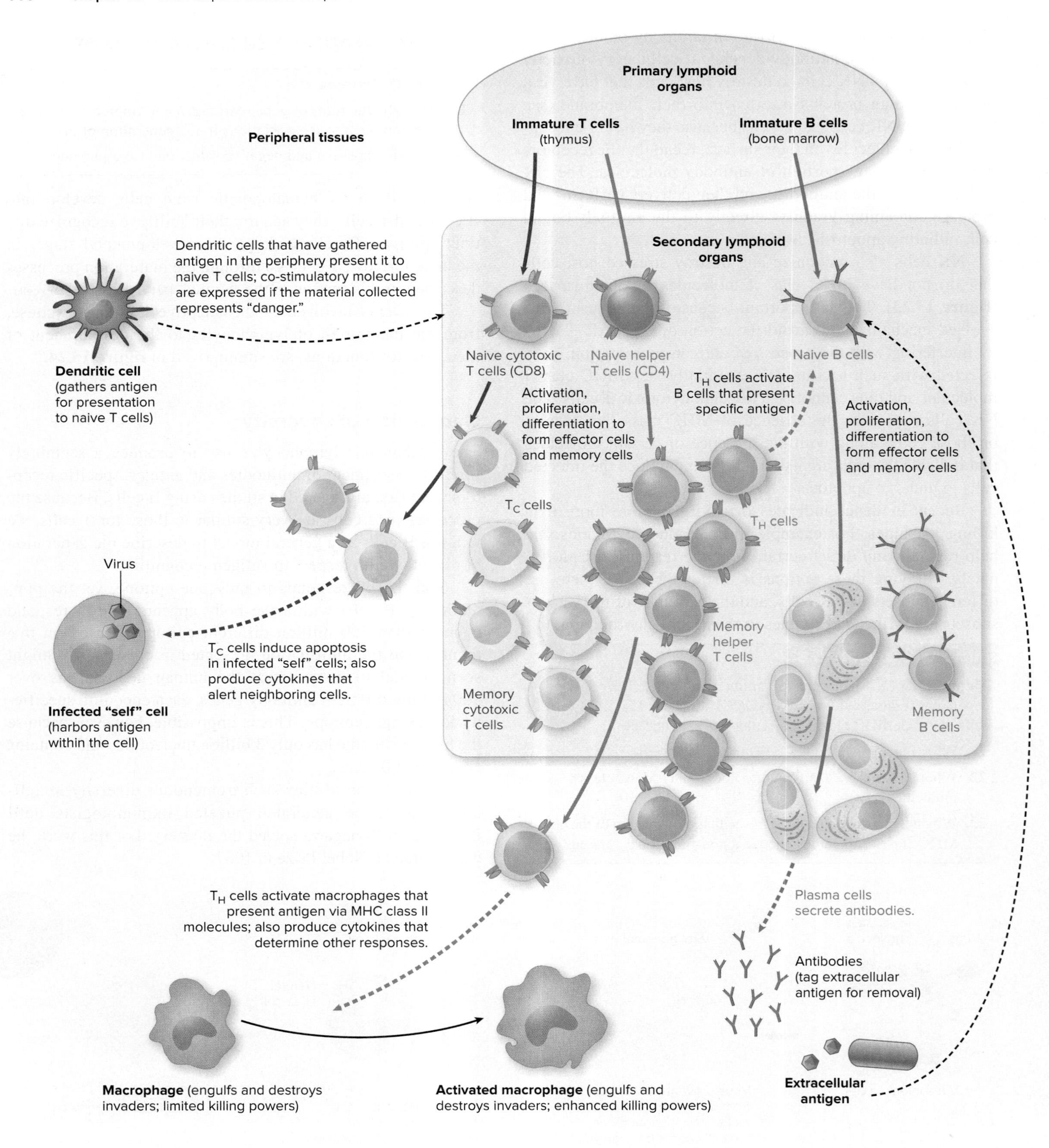

FIGURE 15.24 Summary of the Adaptive Immune Response

? How would the adaptive immune response be affected if memory cells could not be produced?

Gene Rearrangement

A primary mechanism for generating a wide variety of different antibodies using a limited-size region of DNA relies on a strategy similar to that of a practical and well-dressed traveler living out of a small suitcase. By mixing and matching different shirts, pants, and shoes, the traveler can create a wide variety of unique outfits from a limited number of components. Likewise, a B cell expresses three gene segments, one each from DNA regions called V (variable), D (diversity), and J (joining), to form a nearly unique variable region of the heavy chain of an antibody (**figure 15.25**). Recall that a version of this molecule also serves as the B-cell receptor.

A human hematopoietic stem cell has about 40 different V segments, 25 different D segments, and 6 different J segments in the DNA that encodes the variable region of the heavy chain. As a B cell develops, however, two large regions of DNA are permanently removed, thereby joining distinct V, D, and J regions. The joined segments encode the heavy chain of the antibody that the mature B cell is programmed to make. Thus, one B cell could express the combination V3, D1, and J2 to produce its heavy chain, whereas another B cell might use V19, D25, and J6; each combination would result in a unique antibody specificity.

Just as DNA segments rearrange in the heavy chain genes during B-cell development, specific segments move in the light chain genes as well.

Imprecise Joining

As the DNA segments are joined during gene rearrangement, nucleotides are often deleted or added between the sections. This imprecise joining changes the reading frame of the encoded polypeptide so that two B cells that have the same V, D, and J segments for their heavy chain could potentially give rise to antibodies with very different specificities. Likewise, the segments that encode the light chain often join imprecisely.

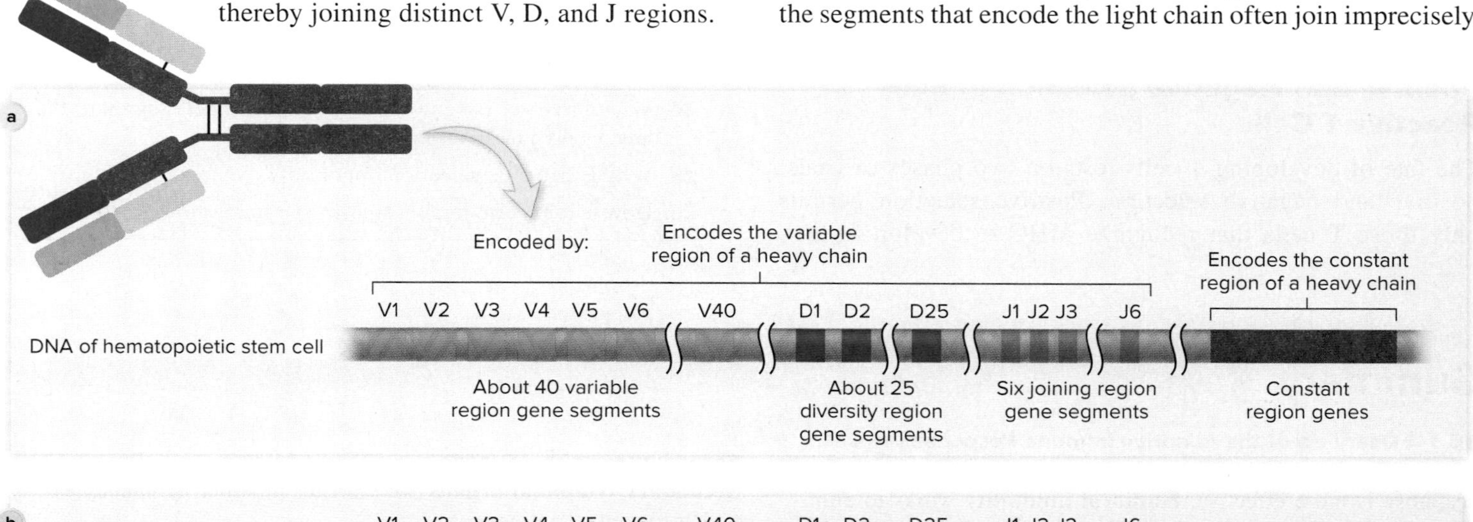

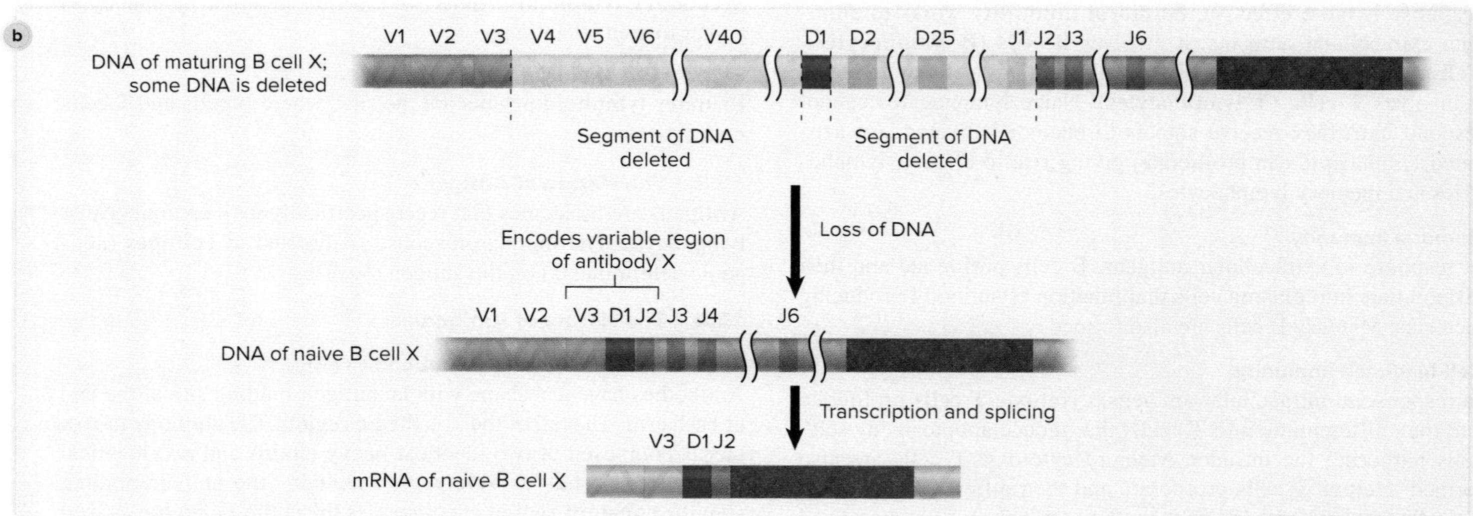

FIGURE 15.25 Antibody Diversity (a) The variable region of the heavy chain of an antibody molecule is encoded by one each of three gene segments: V (variable), D (diversity), and J (joining). **(b)** As a hematopoietic stem cell differentiates to become a maturing B cell, some of the DNA in the antibody-encoding region is lost, resulting in the joining of a V segment, a D segment, and a J segment. Together, the joined segments encode the variable region of the heavy chain of the antibody that that particular B cell is programmed to make. This diagram shows only the gene segments for the heavy chain and is not drawn to scale.

? **Which three features of this process in developing B cells contribute to antibody diversity?**

Combinatorial Associations

Combinatorial association refers to the specific groupings of light chains and heavy chains that make up the antibody molecule. Both types of polypeptides independently acquire diversity through gene rearrangement and imprecise joining. Additional diversity is then introduced when these two polypeptides join, because the combination of the two chains creates the antigen-binding site (see figure 15.7b).

Negative Selection of Self-Reactive B Cells

Once a maturing B cell has developed its antigen receptor (B-cell receptor), it undergoes a process called negative selection, which eliminates any B cell that binds "self." Most developing B cells fail negative selection and, as a consequence, are induced to undergo apoptosis. If these cells are not eliminated, then the immune system may attack "self" substances by mistake, resulting in autoimmunity. ⏭| autoimmunity

Positive and Negative Selection of Self-Reactive T Cells

The fate of developing T cells rests on two phases of trials: positive and negative selection. Positive selection permits only those T cells that recognize MHC to develop further. Recall that the T-cell receptor, unlike the B-cell receptor, recognizes a peptide:MHC complex. The T-cell receptor, therefore, must show at least some recognition of an MHC molecule regardless of the peptide it is carrying. T cells that show insufficient recognition fail positive selection and, as a consequence, are eliminated. Each T cell that passes positive selection is also subjected to negative selection. T cells that recognize "self" peptides presented on MHC molecules are eliminated. Positive and negative selection processes are so strict that over 95% of developing T cells undergo apoptosis in the thymus.

MicroAssessment 15.9

Lymphocytes generate diversity of antigen specificity through rearrangement of gene segments, imprecise joining of those segments, and combinatorial associations of heavy chains and light chains. Negative selection eliminates B cells and T cells that recognize normal "self" molecules. Positive selection permits only those T cells that recognize the MHC molecules to develop further.

24. What three gene segments encode the variable region of the heavy chain of an antibody molecule?

25. Why is negative selection important?

26. How is imprecise joining similar to a frameshift mutation?

Summary

15.1 Overview of the Adaptive Immune Response (figure 15.1)

As a result of the **primary response** to an antigen, the **secondary response** is more effective. **Humoral immunity** works to eliminate extracellular antigens; it involves **B cells (B lymphocytes)**. **Cell-mediated immunity (CMI)** deals with intracellular antigens; it involves **T cells (T lymphocytes).** Naive lymphocytes cannot respond until they receive signals to become activated. An activated lymphocyte can proliferate, giving rise to effector lymphocytes and memory lymphocytes.

Humoral Immunity

In response to extracellular antigens, B cells proliferate and then differentiate into **plasma cells** that function as antibody-producing factories. Memory B cells are also formed.

Cell-Mediated Immunity

In response to intracellular antigens, **cytotoxic T cells** proliferate and then differentiate into **T_C cells** that induce apoptosis in "self" cells harboring the intruder. Memory cytotoxic T cells are also formed. **Helper T cells** proliferate and then differentiate to form **T_H cells** that help orchestrate the various responses of humoral and cell-mediated immunity. Memory helper T cells are also formed.

15.2 Anatomy of the Lymphatic System (figure 15.3)

Lymphatic Vessels

Lymph, which contains antigens that have entered tissues, flows in the **lymphatic vessels** to the lymph nodes (figure 15.4).

Secondary Lymphoid Organs

Secondary lymphoid organs are the sites at which lymphocytes gather to contact antigens.

Primary Lymphoid Organs

Primary lymphoid organs are the sites where B cells and T cells mature.

15.3 The Nature of Antigens

Antigens are molecules that react specifically with an antibody or lymphocyte. The immune response is directed to **epitopes** (antigenic determinants) on the antigen (figure 15.6).

15.4 The Nature of Antibodies

Structure and Properties of Antibodies (figure 15.7)

Antibodies have a Y shape with an antigen-binding site at the end of each arm. The tail of the Y is the **Fc region.** The antibody monomer is composed of two identical **heavy chains** and two identical **light chains.** The **variable region** contains the antigen-binding site; the **constant region** encompasses the entire Fc region as well as part of the Fab regions.

Protective Outcomes of Antibody-Antigen Binding (figure 15.8)

Antibody-antigen binding results in neutralization, **opsonization,** complement activation, immobilization and prevention of adherence, cross-linking, and **antibody-dependent cellular cytotoxicity (ADCC).**

Immunoglobulin Classes (table 15.1)
The five major antibody classes—IgM, IgG, IgA, IgD, and IgE—each have distinct functions.

15.5 ■ Clonal Selection and Expansion of Lymphocytes

When an antigen enters a secondary lymphoid organ, only the lymphocytes that specifically recognize that antigen via their antigen receptor will respond (figure 15.10). Lymphocytes may be immature, naive, activated, effector, or memory cells.

15.6 ■ The B-Cell Response: Humoral Immunity

Most antigens are **T-dependent antigens,** meaning the B cells that recognize them require help from T_H cells.

B-Cell Activation
B cells present peptides from T-dependent antigens to T_H cells for inspection. If a T_H cell recognizes a peptide, it delivers cytokines to the B cell, initiating the process of clonal expansion, which ultimately gives rise to plasma cells that produce antibodies (figure 15.11).

Characteristics of the Primary Response
In the primary response, the expanding B-cell population undergoes **affinity maturation.** Under the direction of T_H cells, **class switching** and memory cell formation also occur (figures 15.14, 15.15).

Characteristics of the Secondary Response
Memory cells are responsible for the swift and effective secondary response, eliminating invaders before they cause noticeable harm (figure 15.12).

The Response to T-Independent Antigens
T-independent antigens include polysaccharides that have multiple identical evenly spaced epitopes, and LPS (figure 15.16).

15.7 ■ The T-Cell Response: Cell-Mediated Immunity

General Characteristics of T Cells (table 15.2, figure 15.19)
Cytotoxic (CD8) T cells recognize antigen presented on **major histocompatibility complex (MHC) class I molecules.** Helper (CD4) T cells recognize antigen presented on **MHC class II molecules.**

Activation of T Cells (figure 15.20)
Dendritic cells sample material in tissues and then travel to secondary lymphoid organs to present antigens to naive T cells. The dendritic cells that detect molecules associated with danger produce **co-stimulatory molecules** and are able to activate both subsets of T cells.

Effector Functions of T_C (CD8) Cells
T_C cells induce apoptosis in cells that present peptides they recognize on MHC class I molecules; they also produce cytokines that increase the level of surveillance (figure 15.21). All nucleated cells present peptides from endogenous proteins in the groove of MHC class I molecules.

Effector Functions of T_H (CD4) Cells (figures 15.11, 15.22)
T_H cells activate cells that present peptides they recognize on MHC class II molecules; various cytokines are released, depending on the subset of the responding T_H cell. Macrophages and B cells present peptides from exogenous proteins in the groove of MHC class II molecules. Subsets of T_H cells direct the immune system to an appropriate response for a given antigen.

15.8 ■ Natural Killer (NK) Cells

NK cells mediate antibody-dependent cellular cytotoxicity (ADCC). NK cells also induce apoptosis in host cells that are not bearing MHC class I molecules on their surface (figure 15.23).

15.9 ■ Lymphocyte Development

Generation of Diversity
Mechanisms used to generate the diversity of antigen specificity in lymphocytes include rearrangement of gene segments, imprecise joining of those segments, and combinatorial associations of heavy and light chains (figure 15.25).

Negative Selection of Self-Reactive B Cells
Negative selection occurs as B cells develop in the bone marrow; if the B-cell receptor of a developing B cell binds material in the bone marrow, that B cell undergoes apoptosis.

Positive and Negative Selection of Self-Reactive T Cells
Positive selection permits only those T cells that show moderate recognition of the MHC molecules to develop further. Negative selection also occurs.

Review Questions

Short Answer

1. What is a secondary lymphoid organ?
2. Diagram an IgG molecule and label (a) the Fc region and (b) the areas that combine with antigen.
3. Describe the protective outcomes of antibody-antigen binding.
4. Which antibody class is the first produced during a primary response?
5. Which antibody class neutralizes viruses in the intestinal tract?
6. Describe clonal selection and expansion in the immune response.
7. How do T-independent antigens differ from T-dependent antigens?
8. What are antigen-presenting cells (APCs)?
9. Describe the role of dendritic cells in T-cell activation.
10. How does the role of natural killer cells differ from that of cytotoxic T cells?

Multiple Choice

1. The variable regions of antibodies are located in the
   1) Fc region.
   2) Fab region.
   3) light chain.
   4) heavy chain.
   5) light chain *and* heavy chain.
      a) 1, 3
      b) 1, 5
      c) 2, 3
      d) 2, 4
      e) 2, 5

2. Which of the following statements about antibodies is *false*?
   a) If you removed the Fc portion, antibodies would no longer be capable of opsonization.
   b) If you removed the Fc portion, antibodies would no longer be capable of activating the complement system.
   c) If you removed an Fab portion, an antibody would no longer be capable of cross-linking antigen.
   d) If IgG were a pentamer, it would bind antigens more efficiently.
   e) If IgE had a longer half-life, it would protect newborn infants.
3. Which class of antibody can cross the placenta?
   a) IgA
   b) IgD
   c) IgE
   d) IgG
   e) IgM
4. A person who has been vaccinated against a disease should have primarily which of these types of serum antibodies against that agent 2 years later?
   a) IgA
   b) IgD
   c) IgE
   d) IgG
   e) IgM
5. Which of the following statements about B cells/antibody production is *false*?
   a) A B cell initially has the potential to make more than one class of antibody.
   b) In response to antigen, all B cells located close to the antigen begin dividing.
   c) Each B cell is programmed to make a single specificity of antibody.
   d) The B-cell receptor allows B cells to detect antigen.
   e) The cell type that makes and secretes antibody is called a plasma cell.
6. Which term describes the loss of specific heavy chain genes?
   a) Affinity maturation
   b) Apoptosis
   c) Clonal selection
   d) Class switching
7. Which of the following specifically refers to an effector lymphocyte?
   a) B cell
   b) CD8 T cell
   c) CD4 T cell
   d) Plasma cell
8. Which markers are found on all nucleated cells?
   a) MHC class I molecules
   b) MHC class II molecules
   c) CD4
   d) CD8
9. Which of the following are examples of an antigen-presenting cell (APC)?
   1) Macrophage
   2) Neutrophil
   3) B cell
   4) T cell
   5) Plasma cell
   a) 1, 2
   b) 1, 3
   c) 2, 4
   d) 3, 5
   e) 1, 2, 3
10. Which is an appropriate response if an antigen is presented on MHC class II molecules?
    a) An effector CD8 cell kills the presenting cell.
    b) An effector CD4 cell kills the presenting cell.
    c) An effector CD8 cell activates the presenting cell.
    d) An effector CD4 cell activates the presenting cell.

Applications

1. Many dairy operations keep cow's milk for sale and use formula and feed to raise any calves. One farmer noticed that calves raised on the formula and feed needed to be treated for diarrhea more frequently than calves left with their mothers to nurse. He had some tests run on the diets and discovered no differences in the calories or nutritional content. The farmer called a veterinarian and asked him to explain the observations. What was the vet's response?
2. What kinds of diseases would be expected to occur as a result of lack of T or B lymphocytes?

Critical Thinking

1. The development of primary and secondary immune responses to an antigen differ significantly. The primary response may take a week or more to develop fully and establish memory. The secondary response is rapid and relies on the activation of clones of memory cells. Would it not be better if clones of reactive cells were maintained regardless of prior exposure? In this way, the body could always respond rapidly to *any* antigen exposure. Would there be any disadvantages to this approach? Why or why not?
2. Early investigators proposed two hypotheses to explain the specificity of antibodies. The clonal selection hypothesis states that each lymphocyte can produce only one specificity of antibody. When an antigen binds to that B-cell receptor, the lymphocyte is selected to give rise to a clone of plasma cells producing the antibody. The template hypothesis states that any antigen can interact with any lymphocyte and act as a template, causing newly forming antibodies to be specific for that antigen. In one experiment to test these hypotheses, an animal was immunized with two different antigens. After several days, lymphocytes were removed from the animal and individual cells placed in separate small containers. Then, the original two antigens were placed in the containers with each cell. What result would support the clonal selection hypothesis? The template hypothesis?

16 Host-Microbe Interactions

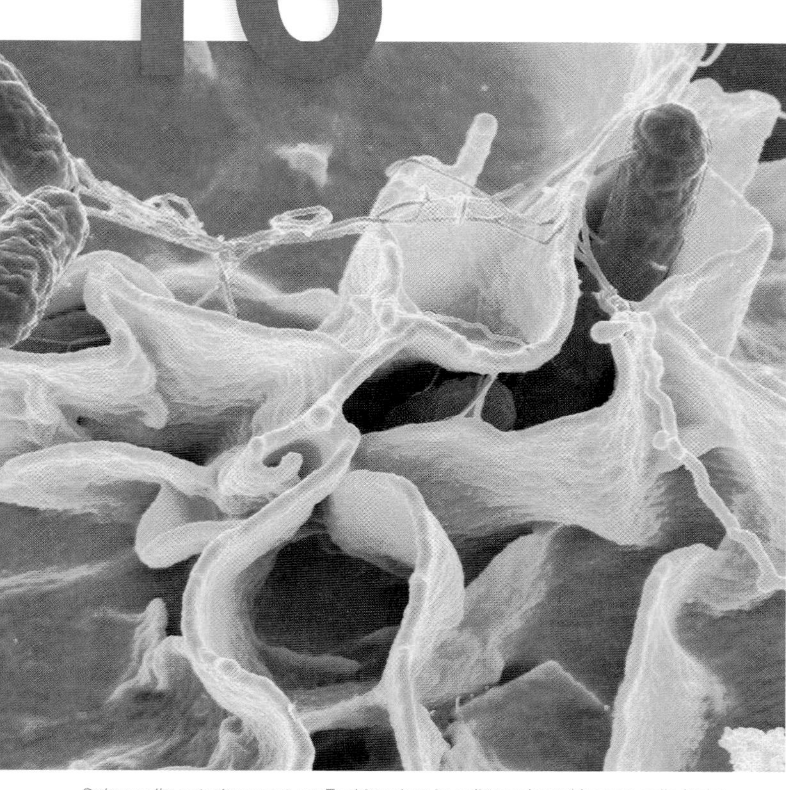

Salmonella enterica serotype Typhimurium invading cultured human cells (color-enhanced SEM). *Source: NIAID, NIH, Rocky Mountain Laboratories*

KEY TERMS

Acute Infection An infection characterized by symptoms that develop fairly quickly and last a relatively short time.

Chronic Infection An infection that generally develops slowly and lasts for months or years.

Colonization Establishment and growth of a microorganism on a surface.

Endotoxin The lipopolysaccharide (LPS) component of the outer membrane of Gram-negative bacteria; lipid A is responsible for the toxic properties of LPS.

Exotoxin A toxic protein produced by a microorganism; often simply referred to as a *toxin.*

Immunocompromised Having a weakness or defect in the innate or adaptive defenses.

Infection Colonization by a pathogen on or within the body.

Infectious Disease An infection that prevents the body from functioning normally.

Latent Infection Infection in which the infectious agent is present but not causing symptoms.

Microbiome The total genetic information of a community of microorganisms in a given environment; also the community itself, in which case the term is often used synonymously with normal microbiota.

Normal Microbiota The group of microorganisms that routinely colonize the body of a healthy individual.

Opportunistic Pathogen A microbe that causes disease only when introduced into an unusual location or into an immunocompromised host.

Primary Pathogen A microbe able to cause disease in an otherwise healthy individual.

Virulence Factors Traits of a microbe that promote pathogenicity.

A Glimpse of History

Throughout history, people have tried to understand the spread of diseases. In ancient times, people thought that illnesses were divine punishment for their sins. Gradually it became obvious, however, that at least some diseases could be transmitted by contact with people who had the disease. By the Middle Ages, many people accepted this and fled cities to escape diseases.

With van Leeuwenhoek's discovery of microorganisms in the late seventeenth century, people began to suspect that microorganisms might cause disease, but the techniques of the times could not prove this. It was not until 1876 that Robert Koch offered convincing evidence of what is now known as the germ theory of disease. He showed that *Bacillus anthracis* causes anthrax, an often fatal disease of humans, sheep, and other animals. With his microscope, Koch observed *B. anthracis* cells in the blood and spleen of dead sheep. He then inoculated mice with the infected sheep blood, and recovered *B. anthracis* from the blood of those mice. In addition, he grew the bacteria in pure culture and showed that they caused anthrax when injected into healthy mice. From these experiments and later work with *Mycobacterium tuberculosis,* Koch formalized what is now known as Koch's postulates—a group of criteria for establishing the cause of an infectious disease.

Every day we come into contact with an enormous number and variety of microorganisms. Some enter our respiratory system as we breathe; others are ingested with each bite of food or sip of drink; and still more adhere to our skin whenever we touch an object or surface. It is important to recognize, however, that the vast majority of these microbes cause no harm. Some may colonize the body surfaces, taking up residence with the variety of other harmless microbes that live there; others are shed with dead epithelial cells. Most of those swallowed are either killed in the stomach or eliminated in feces.

Relatively few microbes cause noticeable damage to the human body; those that can are called **pathogens.** They have distinct characteristics that allow them to avoid at least some of the body's defenses. Research into how these microbes evade our innate and adaptive defenses is leading to many fascinating discoveries. The knowledge gained will hopefully make it possible to develop therapies targeted at specific pathogens.

This chapter will explore some of the ways in which microbes colonize the human host, living either as members of the normal microbiota or causing disease. It will also describe how pathogens evade or overcome the immune responses and damage the host.

MICROBES, HEALTH, AND DISEASE

Many people think of microorganisms as "germs" that should routinely be killed or avoided. Most microbes are harmless, however, and many are beneficial. It is a delicate balancing act, though, because even microbes that are typically harmless can cause disease if the opportunity arises. Weaknesses or defects in the innate or adaptive defenses can leave people vulnerable to invasion; such individuals are said to be **immunocompromised.** Factors that can cause a person to become immunocompromised include malnutrition, cancer, AIDS or other diseases, surgery, wounds, genetic defects, alcohol or drug abuse, and immunosuppressive therapy that accompanies procedures such as organ transplants.

16.1 ■ The Anatomical Barriers as Ecosystems

Learning Outcome

1. Compare and contrast mutualism, commensalism, and parasitism.

The skin and mucous membranes are barriers against invading microorganisms, but they are also part of a complex ecosystem—an interacting biological community along with the environment that shapes it. The intimate relationships between the microorganisms and the human body are an example of **symbiosis,** meaning "living together."

Microorganisms can have a variety of symbiotic relationships with one another and with the human host. These relationships may take on different characteristics depending on the closeness of the association and the relative advantages to each partner. Symbiotic associations can take several forms, and these may change, depending on the state of the host and the traits of the microbes:

- **Mutualism** is an association in which both partners benefit. Members of the normal microbiota contribute to our health in many ways. As just one example, some bacteria in the large intestine synthesize vitamin K and certain B vitamins that are then available for our body to absorb. The microorganisms benefit as well, supplied with warmth and a variety of different energy sources.
- **Commensalism** is an association in which one partner benefits but the other remains unharmed. Many microbes living on the skin are neither harmful nor helpful to the human host, but they obtain food and other necessities from the host.
- **Parasitism** is an association in which one organism, the parasite, benefits at the expense of the other. All pathogens are parasites, but medical microbiologists often reserve the word *parasite* for eukaryotic pathogens such as certain protozoa and helminths (worms).

MicroAssessment 16.1

The symbiotic relationship between a host and a microbe can be described as mutualism, commensalism, or parasitism, depending on the relative benefit to each partner.

1. How is mutualism different from commensalism?

16.2 ■ The Human Microbiome

Learning Outcomes

2. Describe how the composition of a person's microbiome can change over time.
3. Describe the importance of the microbiome in protecting against infection, promoting oral tolerance, aiding digestion, and producing substances beneficial to human health.

The **normal microbiota** is the group of microorganisms routinely found growing on the body of healthy individuals (**figure 16.1**). The community is also called the **microbiome** (the term is also used to refer to the total genetic information of

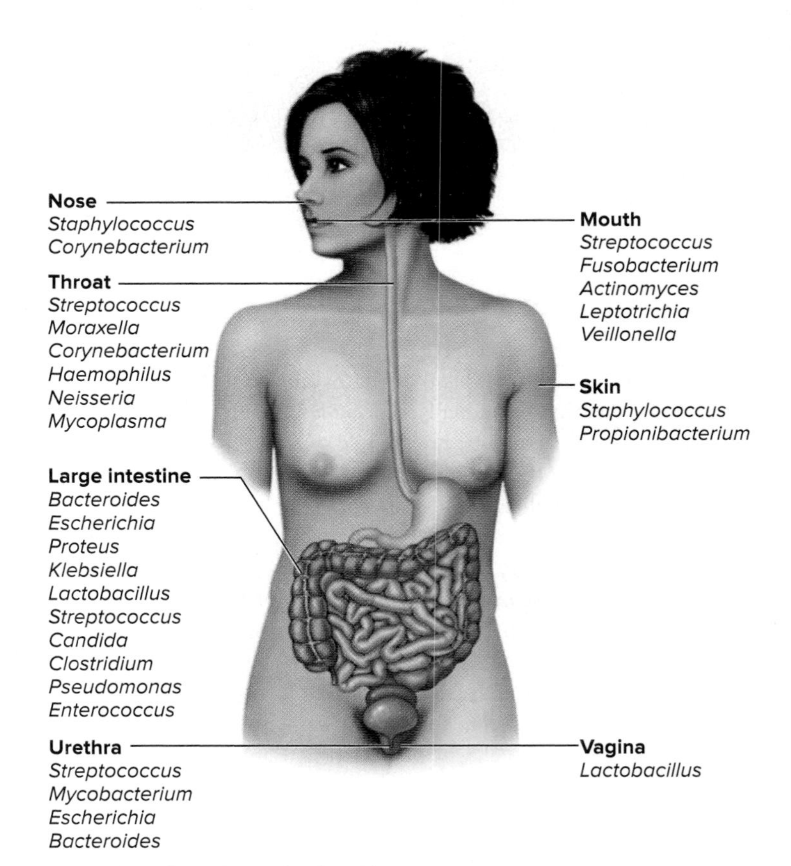

FIGURE 16.1 Normal Microbiota This shows only some of the common genera; many others may also be present.

What can happen if members of the normal microbiota in the intestinal tract are killed or their growth is suppressed?

the community). Microbes that typically inhabit body sites for extended periods are resident microbiota, whereas temporary occupants are transient microbiota.

Considering how important the microbiome is to human health, relatively little is known about its members. As described in chapter 1, that is quickly changing, as several large-scale research efforts are studying this diverse population. The studies typically use **metagenomics,** the sequence analysis of the DNA extracted directly from a given environment, which allows scientists to investigate all microbes in a sample, including those that have not yet been grown in culture. ⏮ metagenomics

MicroByte

There are more bacteria in just one person's mouth than there are people in the world!

Composition of the Microbiome

Babies begin acquiring their microbiota at birth, when they first encounter a wide variety of microorganisms. During passage through the mother's birth canal, the baby is exposed to lactobacilli and an assortment of other microbes that take up residence in its digestive tract and on the skin. A baby delivered by cesarean section is not exposed to the genital fluids, so microbes from the mother's skin and the environment establish themselves as residents on the newborn instead. Preterm infants, whether delivered vaginally or by cesarean section, have a different microbiome from that of full-term infants.

Breast feeding also influences the composition of a baby's microbiome. Recent studies show that breast milk contains a remarkable variety of microorganisms as well as certain carbohydrates that appear to specifically nourish a healthy microbiome in the infant intestine. Babies who are breast-fed have higher concentrations of *Bifidobacterium* species and lactic acid bacteria in their intestinal tract. ⏮ the genus *Bifidobacterium*, ⏮ lactic acid bacteria

The composition of a baby's microbiome changes over time, as microbes encountered in various foods, on pets, on other humans, and in the environment establish themselves. Starting at 1 year of age and continuing over a period of about 2 years, the makeup of a child's intestinal microbiome slowly begins to resemble that of an adult.

An average adult carries over 100 trillion microorganisms, and there is considerable diversity in the composition of the microbiome, both among different individuals and in one person over time. Comparisons of microbial populations, however, reveal certain consistencies. For example, the two most common phyla represented in the adult intestinal tract are the *Firmicutes* (a phylum that includes *Clostridium* and *Bacillus* species) and *Bacteroidetes* (a phylum that includes *Bacteroides* species). The variety of species within these phyla, however, can vary significantly. The makeup of the community represents a balance of many forces that can alter the microbial population's quantity and composition. Changes occur in response to physiological variations within the host (such as hormonal changes), and as a direct result of the activities of the human host (such as consuming food). An intriguing example of the dynamic nature of the microbiome was the discovery that the composition of the intestinal microbiota of obese people differs from that of lean people. Obese people have more members of the *Firmicutes,* whereas thin individuals typically have more members of the *Bacteroidetes.* As obese people lose weight, their intestinal microbiome changes to resemble that of typically lean people.

Researchers are currently attempting to determine if certain microbiome compositions result in a given disease state. For example, there seems to be a link between intestinal **dysbiosis** (an imbalance in the microbiome) and inflammatory bowel disease. Although some relationships have been observed, it is too early to tell if these simply represent correlation rather than showing cause and effect.

Beneficial Roles of the Human Microbiome

Scientists have long recognized that the normal human microbiome protects against infection, but recent studies show that its importance goes far beyond that. Members of this community also help with digestion and produce beneficial substances that are important to human health. In fact, some researchers are suggesting that the intestinal microbiome be considered an organ of the human body.

Protecting Against Infection

As discussed in chapter 14, the normal microbiota excludes pathogens by (1) covering binding sites that might otherwise be used for attachment, (2) consuming available nutrients, and (3) producing compounds toxic to other bacteria. When members of the normal microbiota are killed or their growth is suppressed, as can happen during antibiotic treatment, pathogens may colonize and cause disease. For instance, certain antibiotics inhibit the *Lactobacillus* species that normally predominate in the vagina of mature females. These bacteria normally suppress the growth of the yeast *Candida albicans,* and without their protective action, the yeast cells can multiply to high numbers, resulting in vulvovaginal candidiasis. Oral antibiotics can also inhibit members of the normal intestinal microbiota, allowing the overgrowth of toxin-producing strains of *Clostridium difficile* that cause antibiotic-associated diarrhea and colitis. ⏭ vulvovaginal candidiasis, ⏭ *Clostridium difficile* infections

Another crucial role of the microbiome is to stimulate the adaptive immune system. For example, when small numbers of bacteria that normally live on the skin enter tissues through scrapes and cuts, the body develops an immune response against them. Surface proteins of pathogens are often similar to those of members of the normal microbiota, so antibodies against skin bacteria will bind to any entering pathogens as well. The importance of the microbiome in stimulating the adaptive immune system can be shown in mice reared in a microbe-free environment. These "germ-free" animals have greatly underdeveloped mucosa-associated lymphoid tissue (MALT). ⏮ MALT

Promoting Oral Tolerance

The microbiome appears to play an important role in the development of oral tolerance by the immune system. In a complex series of events, our defenses learn to lessen the immune response to the many microbes that routinely inhabit the gut, as well as to foods that pass through. Recent studies into the actions of regulatory T cells indicate that early and consistent exposure to certain microbes in the gut stimulates these T cells, thereby preventing the immune system from overreacting to harmless microbes and substances. This idea is the basis of the **hygiene hypothesis,** which proposes that insufficient exposure to microbes can lead to allergies and autoimmune diseases. It is a fine balance, however, because contact with certain pathogens can be deadly. ⏮ tolerance

Aiding Digestion

The human intestinal microbiome allows the body to extract more energy from foods. One reason for this is that the human genome encodes relatively few enzymes that degrade complex carbohydrates. Instead, the body relies on microorganisms to break down most types of dietary fiber. In the anaerobic environment of the intestinal tract, the microbes ferment the products, making short-chain fatty acids that the body then absorbs. Of the various microbes that inhabit the intestinal tract, *Bacteriodes* species appear to make the greatest variety of carbohydrate-degrading enzymes.

Producing Substances Important for Human Health

As mentioned earlier, bacteria in the intestinal tract produce vitamin K and certain B vitamins that can be used by the host. In addition, bacterial fermentation of dietary fiber produces butyrate, an important energy source for the epithelial cells that line the large intestine.

MicroAssessment 16.2

The microbiome protects against potentially harmful organisms, stimulates the immune system, promotes oral tolerance, aids digestion, and produces substances important for human health.

2. What is an advantage of using metagenomics over using culture methods when studying the normal microbiota?
3. What factor favors overgrowth of *Clostridium difficile* in the intestine?
4. Some research suggests that babies delivered by cesarean section and not breast-fed are more prone to developing allergies. What could explain this effect?

16.3 ■ Principles of Infectious Disease

Learning Outcomes

4. Define the terms *primary pathogen, opportunistic pathogen,* and *virulence.*
5. Describe the characteristics of infectious diseases, including the course of disease, duration of symptoms, and distribution of the pathogen.

The term **colonization** refers to the establishment and growth of a microbe in a particular environment. If the microbe has a parasitic relationship with the host, then the term **infection** can be used. That is, a member of the normal microbiota is said to have colonized the host, but a pathogen is described as having either colonized or infected the host. Infection does not always lead to illness. It can be **subclinical,** meaning that symptoms either do not appear or are mild enough to go unnoticed.

An infection that results in disease (a condition that prevents the body from functioning normally) is called an **infectious disease.** Diseases are characterized by symptoms and signs; **symptoms** are the subjective effects of the disease experienced by the patient, such as pain and nausea, whereas **signs** are the objective evidence, such as rash, pus formation, and swelling.

Effects of one disease may leave a person predisposed to developing another. For example, a respiratory illness that damages the mucociliary escalator makes a person more likely to develop pneumonia. The initial infection is a **primary infection;** an additional infection that occurs as a result of the primary infection is a **secondary infection.** ⏮ mucociliary escalator

Pathogenicity

A **primary pathogen,** or more simply, a pathogen, is a microorganism or virus that causes disease in otherwise healthy individuals. Diseases such as malaria, measles, influenza, strep throat, plague, tetanus, and tuberculosis are caused by primary pathogens.

An **opportunistic pathogen,** or opportunist, causes disease only when the body's innate or adaptive defenses are compromised, or when introduced into an unusual location. Opportunists can be members of the normal microbiota or they can be from the environment. For instance, *Pseudomonas* species are common environmental bacteria that routinely come into contact with healthy individuals without harmful effect, yet they can cause fatal infections in individuals who have the genetic disease cystic fibrosis and also in burn patients (see figure 23.6). Ironically, as our healthcare systems improve, extending the life span of patients through surgery and immunosuppressive medications, diseases caused by opportunists are becoming more common. Also, many organisms not previously known to cause disease have now been shown to do so in severely immunocompromised patients.

The term **virulence** refers to the degree of pathogenicity of an organism. An organism described as highly virulent is more likely to cause disease, particularly severe disease, than might otherwise be expected. *Streptococcus pyogenes* causes strep throat, for example, but certain strains are particularly virulent, causing diseases such as necrotizing fasciitis ("flesh-eating disease"). **Virulence factors** are the traits of a microorganism that specifically allow it to cause disease. The genes encoding these traits can sometimes be transferred horizontally. ⏭ necrotizing fasciitis, ⏮ horizontal gene transfer

Characteristics of Infectious Disease

Infectious diseases that spread from one host to another are called **communicable** or **contagious diseases.** Some contagious diseases, such as colds and measles, are easily transmitted. The ease with which a contagious disease spreads partly reflects the **infectious dose**—the number of microbes necessary to establish an infection. For example, the intestinal disease shigellosis is quite contagious in humans because only 10 to 100 cells of a *Shigella* species need be ingested to establish an infection; in contrast, salmonellosis does not spread as easily because as many as 10^6 cells of *Salmonella enterica* serotype Enteritidis must be ingested to cause illness. The difference in these infectious doses reflects, in part, the pathogen's ability to survive the acidic conditions encountered as the cells pass through the stomach. Generally, the infectious dose is expressed as the ID_{50}, an experimentally derived figure that indicates the number of microbial cells administered that resulted in disease in 50% of the test population. ⏭ *Shigella*, ⏭ *Salmonella enterica* serotype Enteritidis

Progression of Infectious Disease

The progression of an infectious disease includes several stages (**figure 16.2**). The time between introduction of a microbe to a susceptible host and the onset of signs and symptoms is the **incubation period.** This varies considerably, from only a few days (for the common cold), to several weeks (for hepatitis A), to many months (for rabies), and even years (for Hansen's disease). The length of the incubation period depends on a variety of factors, including the growth rate of the pathogen, the host's condition, and the number of infectious cells or virions encountered. ⏭ Hansen's disease

A phase of **illness** follows the incubation period. During this period, a person will experience the signs and symptoms of the disease. In some cases, onset of illness is preceded by a **prodromal phase**—a period of early, vague symptoms such as malaise (a general feeling of illness) and headache. After the illness subsides, there is a period of **convalescence,** the stage of recuperation and recovery from the disease. Even though there is no indication of infection during the incubation and convalescent periods, many infectious agents can still be spread during these stages. Some individuals, called **carriers,** harbor an infectious agent for months or years and continue to spread the pathogen, even though they appear healthy. ⏭ carriers

Following recovery from infection, the person's immune system normally has produced enough protective antibodies and memory lymphocytes to prevent reinfection with the same microbe. The same is true after vaccination.

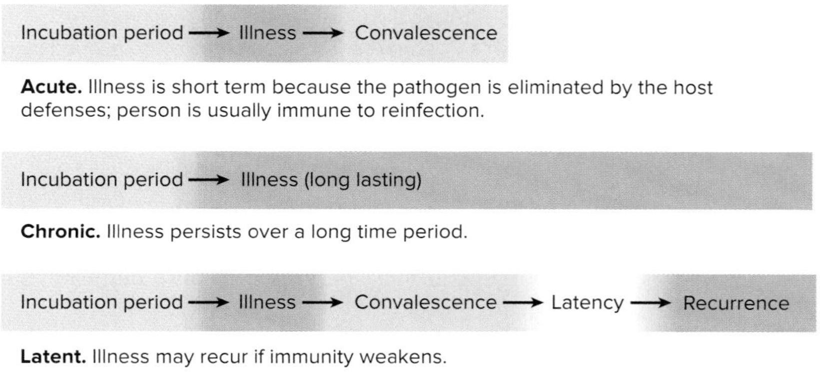

Incubation period ⟶ Illness ⟶ Convalescence

Acute. Illness is short term because the pathogen is eliminated by the host defenses; person is usually immune to reinfection.

Incubation period ⟶ Illness (long lasting)

Chronic. Illness persists over a long time period.

Incubation period ⟶ Illness ⟶ Convalescence ⟶ Latency ⟶ Recurrence

Latent. Illness may recur if immunity weakens.

FIGURE 16.2 The Progression of Infectious Diseases Infections can be acute, chronic, or latent.

? Some diseases include a prodromal phase. Where would this phase fit in the figure?

Duration of Symptoms

Infections and the associated diseases are often described according to the timing and duration of the symptoms (see figure 16.2):

- **Acute infections** are characterized by symptoms that develop quickly but last only a short time; an example is strep throat.
- **Chronic infections** develop slowly and last for months or years; an example is tuberculosis.
- **Latent infections** are never completely eliminated; the microbe continues to exist in host tissues, often within host cells, without causing any symptoms. If there is a decrease in immunity sometime later, the latent infection may reactivate and become symptomatic. Note that the symptomatic phase of the disease may be either acute or chronic. For example, the infection caused by the varicella-zoster virus results in the characteristic symptoms of chickenpox, an acute illness. That illness is stopped by an effective immune response, leaving the host immune to reinfection. The virus, however, is not completely eliminated. It hides in sensory nerves, held in check by the immune system. Later in life, infectious viral particles may be produced again, causing the skin disease shingles (herpes zoster). In tuberculosis, the mycobacteria are often initially confined within a small area by host defenses, causing no symptoms; much later, the bacteria may begin multiplying again, resulting in a chronic illness. Other diseases in which the causative agent becomes latent include cold sores and genital herpes. ⏭ chickenpox, ⏭ tuberculosis

Distribution of the Pathogen

Infections are often described according to the distribution of the causative agent in the body. In a **localized infection,** the microbe is limited to a small area; an example is a boil caused by *Staphylococcus aureus*. In a **systemic infection,** the infectious agent is disseminated (spread) throughout the body; an example is Lyme disease. Systemic infections often include a characteristic set of signs and symptoms—such as fever, fatigue, and headache—that result from the systemic immune response to the infecting agent.

The suffix *-emia* means "in the blood." Thus, **bacteremia** indicates that bacteria are circulating in the bloodstream. Note that this term does not necessarily imply a disease state. A person can become bacteremic for a short period of time after forceful tooth brushing. On the other hand, infection-induced bacteremia can lead to a life-threatening systemic inflammatory response, a condition called **sepsis**. **Toxemia** indicates that toxins are circulating in the bloodstream. The organism that causes tetanus, for instance, produces a localized infection yet its toxins circulate in the bloodstream. The term **viremia** indicates that viral particles are circulating in the bloodstream. ⏭ sepsis

MicroAssessment 16.3

A primary pathogen can cause disease in an otherwise healthy individual; an opportunist causes disease in an immunocompromised host. The course of infectious disease includes an incubation period, illness, and a period of convalescence. Infections can be acute, chronic, or latent; they can be localized or systemic.

5. Why are diseases caused by opportunists becoming more frequent?
6. Give an example of a microbe that causes a latent infection.
7. What factors might contribute to a long incubation period?

16.4 ■ Determining the Cause of an Infectious Disease

Learning Outcome

6. List Koch's postulates, and compare them to the molecular Koch's postulates.

Criteria are needed to guide scientists as they try to determine the cause of an infectious disease. They can also be helpful when studying the disease process.

Koch's Postulates

Koch's postulates—the criteria that Robert Koch used to show that *Bacillus anthracis* causes anthrax (see **A Glimpse of History**)—provide a foundation for establishing that a given microbe causes a specific infectious disease (**figure 16.3**):

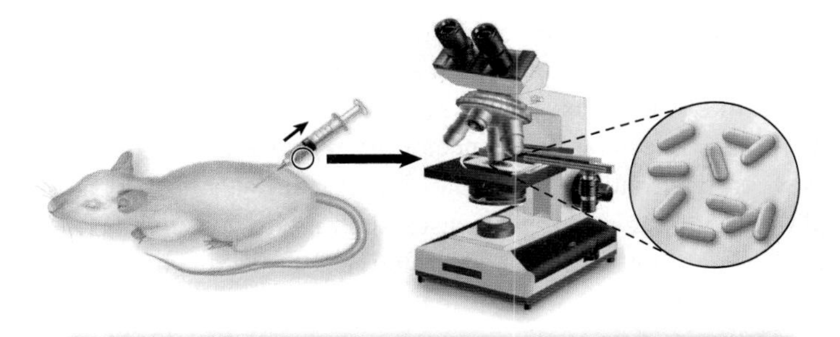

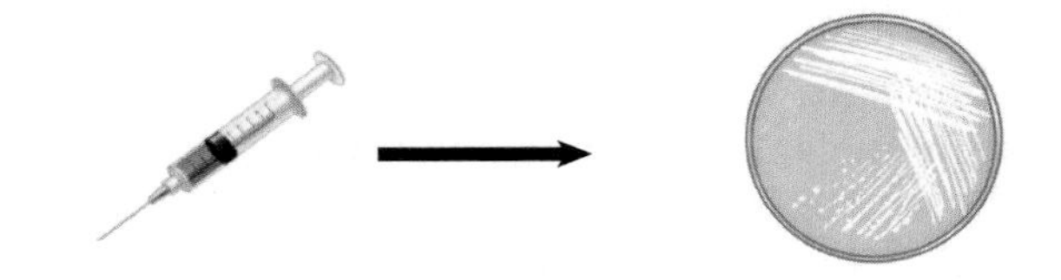

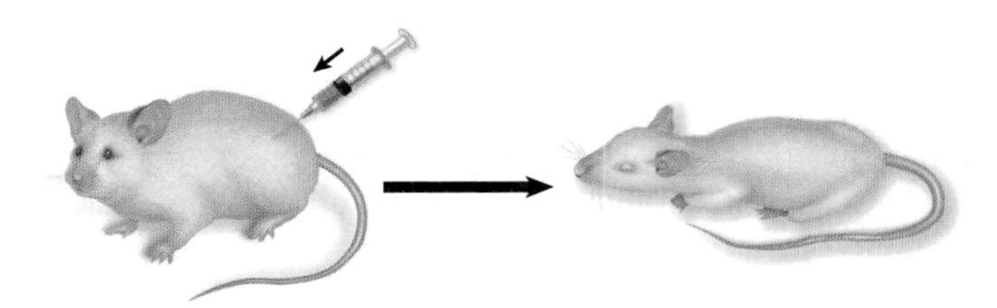

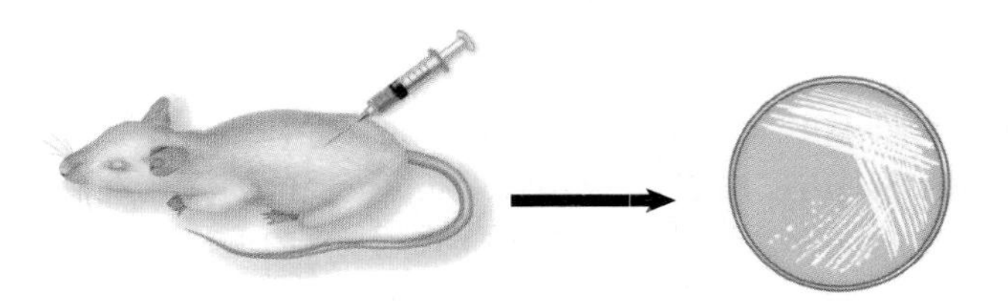

FIGURE 16.3 Koch's Postulates These criteria provide a foundation for establishing that a given microbe causes a specific disease.

Why is it not possible to use Koch's postulates to show that *Treponema pallidum* causes syphilis?

① The microorganism must be present in every case of the disease.

② The microorganism must be grown in pure culture from diseased hosts.

③ The same disease must be produced when a pure culture of the microorganism is introduced into susceptible hosts.

④ The microorganism must be recovered from the experimentally infected hosts.

When Koch studied anthrax, he grew *B. anthracis* from all cases examined; he introduced pure cultures of the organisms into healthy susceptible mice, causing them to develop anthrax. Finally, he recovered the organism from the experimentally infected mice.

It is important to note that there are many situations in which Koch's postulates cannot be carried out. For example, the second postulate cannot be fulfilled for organisms that cannot be grown in laboratory medium, such as *Treponema pallidum* (causes syphilis). In other cases, the third postulate does not always hold true. There are many examples, including cholera and polio, in which some infected people do not have signs or symptoms of disease. In addition, some diseases are polymicrobial, meaning that multiple species act together to cause the illness; an example is chronic periodontal disease. Also, suitable experimental animal hosts are not available for some diseases and it would not be ethical to test the postulates on humans because of safety concerns. Nevertheless, despite the limitations of the postulates, they have provided scientists with a logical framework for determining the causes of infectious diseases. ⏭ syphilis, ⏭ cholera, ⏭ polio, ⏭ periodontal disease

Molecular Koch's Postulates

Molecular Koch's postulates are similar in principle to Koch's postulates, but they rely on molecular techniques to study a microbe's virulence factors. They are particularly relevant in the study of pathogens such as *E. coli* and *Streptococcus pyogenes,* which can cause several different diseases depending on the virulence factors of a given strain. Molecular Koch's postulates are as follows:

1. The virulence factor gene or its product should be found in pathogenic strains of the microorganism.
2. Mutating the virulence gene to disrupt its function should reduce the virulence of the pathogen.
3. Reversion of the mutated virulence gene or replacement with a wild-type version should restore virulence to the strain.

As with the traditional Koch's postulates, it is not always possible to apply all of these criteria, but they provide an approach to studying how infectious agents cause disease.

MicroAssessment 16.4

Koch's postulates can be used to establish that a given microbe causes a specific infectious disease. Molecular Koch's postulates are used to identify the virulence factors responsible for disease.

8. How were Koch's postulates used to prove the cause of anthrax?
9. Why can Koch's postulates not be used to identify the causes of diseases due to polymicrobial infections?

MECHANISMS OF PATHOGENESIS

From a microbe's perspective, the interior of the human body is a rich source of nutrients guarded by the innate and adaptive defenses. The ability to get past these defenses and cause damage is what distinguishes pathogens from other microbes. Understanding how they do this helps illustrate why only certain microbes can cause disease in a healthy host. Pathogenic mechanisms generally follow one of several patterns:

- **Production of toxins that are then ingested.** The microbe does not grow on or in the host, so this is not an infection but rather a foodborne intoxication, a form of food poisoning. The only relevant virulence determinant in this situation is toxin production. Relatively few bacteria cause foodborne intoxication; these include *Clostridium botulinum* (causes botulism) and toxin-producing strains of *Staphylococcus aureus* (cause staphylococcal food poisoning). ⏭ botulism, ⏭ *Staphylococcus aureus* foodborne intoxication
- **Colonization of mucous membranes of the host, followed by toxin production.** The microbe adheres to a mucous membrane such as the lining of the intestinal or upper respiratory tracts and multiplies to high numbers. There, it produces a toxin that interferes with cell function. Examples of bacteria that do this include *Vibrio cholerae* (causes cholera), *E. coli* O157:H7 (causes bloody diarrhea), and *Corynebacterium diphtheriae* (causes diphtheria). ⏭ cholera, ⏭ *E. coli* O157:H7 diarrhea, ⏭ diphtheria
- **Invasion of host tissues.** The microbe penetrates the first-line defenses and then multiplies within the tissues. Organisms that do this generally have mechanisms to avoid destruction by macrophages; some also have mechanisms to avoid antibodies. There are numerous examples of bacteria that invade, including *Mycobacterium tuberculosis* (causes tuberculosis), *Yersinia pestis* (causes plague), and *Salmonella enterica* (most strains cause gastroenteritis and one causes typhoid fever). ⏭ *Mycobacterium tuberculosis,* ⏭ *Yersinia pestis,* ⏭ *Salmonella enterica*
- **Invasion of host tissues, followed by toxin production.** These microbes are similar to those in the previous category, but in addition to invading, they also make toxins. Examples include *Shigella dysenteriae* (causes diarrhea) and *Clostridium tetani* (causes tetanus). ⏭ *Shigella dysenteriae,* ⏭ *Clostridium tetani*

A successful pathogen needs to overcome the host defenses only long enough to multiply and then exit the host. In fact, a pathogen that completely overwhelms the host defenses is actually at a disadvantage because it will likely kill the host. If the host dies, the pathogen loses an exclusive source of nutrients and perhaps the opportunity to be transmitted.

Pathogens and their hosts generally evolve over time to a state of **balanced pathogenicity.** The pathogen becomes less virulent while the host becomes less susceptible. This was demonstrated when the myxoma virus was intentionally introduced into Australia in the early 1950s to kill the rapidly increasing rabbit population. As expected, the rabbit population dropped dramatically after the virus was introduced. Eventually, however, the numbers of rabbits again began rising. Viruses isolated from these rabbits were shown to be less virulent than the original strain, and the rabbits were more resistant to the original virus strain.

The next sections will describe how pathogens adhere to and colonize host tissue, avoid innate defenses, avoid adaptive defenses, and cause the damage associated with disease. We will focus on mechanisms of bacterial pathogenesis because these are by far the most thoroughly characterized; later in the chapter, we will discuss pathogenesis of viruses and eukaryotic organisms. As we describe various virulence factors, recognize that their roles are not mutually exclusive—a single structure can serve more than one purpose. Also note that one microbe can have more than one virulence factor, and various strains of the same species can have different virulence factors.

16.5 ■ Establishing Infection

Learning Outcomes

7. Describe the requirements for adherence and colonization.
8. Explain the role of type III secretion systems in infection.

To cause disease, most pathogens must first adhere to a body surface and then multiply. In some cases, they deliver molecules to epithelial cells, causing changes in those cells.

Adherence

The first-line defenses are very effective in sweeping microbes away, so pathogens must adhere to host cells to initiate infection. Microbes that attach to cells, however, do not necessarily cause disease. For example, members of the normal microbiota often adhere to epithelial cells with no ill effect whatsoever. Other factors such as toxin production or invasion generally must come into play before disease results. ⏮ first-line defenses

Bacteria use **adhesins** to attach to host cells (**figure 16.4**). These are often located at the tips of pili (pili used for attachment are often called fimbriae; see figure 3.42). Adhesins can also be a component of other surface structures such as capsules or various cell wall proteins (see figure 3.37). ⏮ pili, ⏮ capsule

The molecule to which an adhesin attaches is called the **receptor.** Note that receptors have distinct roles for the host cells; the microbes merely exploit the molecules for their own use. For example, the normal role of the receptor used by *Neisseria gonorrhoeae* is to help protect host cells from damage by the complement system. Receptors are typically glycoproteins or glycolipids, and the adhesin binds to the sugar portion. ⏮ complement system, ⏮ glycoprotein

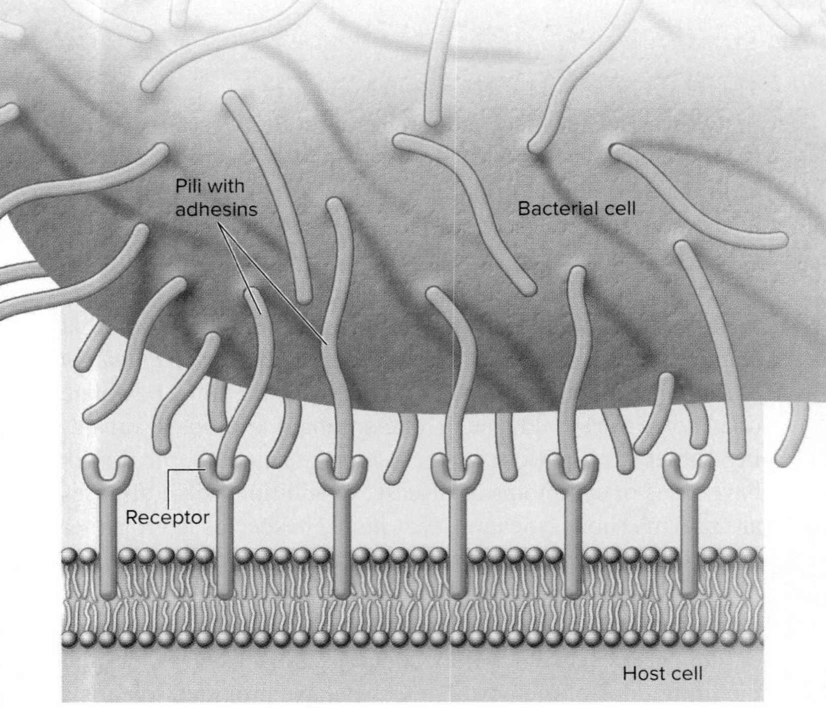

FIGURE 16.4 Adhesins In this illustration, adhesins at the tips of pili (fimbriae) attach to receptors on the host cell surface.

? **What would happen if a pathogen lost the ability to produce adhesins?**

Adhesin-receptor binding is highly specific, dictating the type of cells to which the bacterium can attach. For instance, the adhesin of common *E. coli* strains allows them to adhere to cells that line the large intestine, where the strains multiply as part of the normal microbiota. Pathogenic *E. coli* strains have additional adhesins, broadening the range of tissues to which they can attach. Strains that cause urinary tract infections have pili that attach to the bladder, and strains that cause watery diarrhea have pili that adhere to cells of the small intestine.

Colonization

A microorganism must multiply in order to colonize the host. In many cases, pathogens grow in biofilms. ⏮ biofilms

To colonize a mucosal surface, the pathogen must deal with the host's defenses that protect those surfaces. Recall, for example, that the body uses lactoferrin and transferrin to bind iron, thereby limiting the growth of microbes. Some pathogens respond by producing their own iron-binding molecules, called **siderophores;** others can use the iron bound to the host proteins. ⏮ lactoferrin and transferrin

Secretory IgA also protects mucosal surfaces. Pathogens, however, have evolved mechanisms to avoid those antibodies. Mechanisms include rapid turnover of pili (to shed any bound antibody), antigenic variation, and **IgA proteases** (enzymes that cleave IgA antibodies). ⏮ IgA, ⏮ antigenic variation

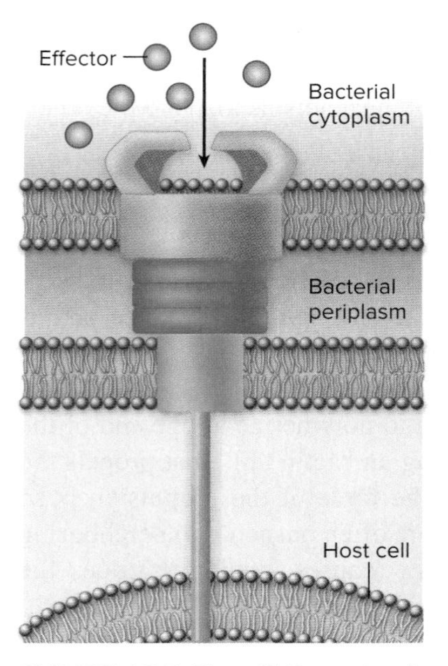

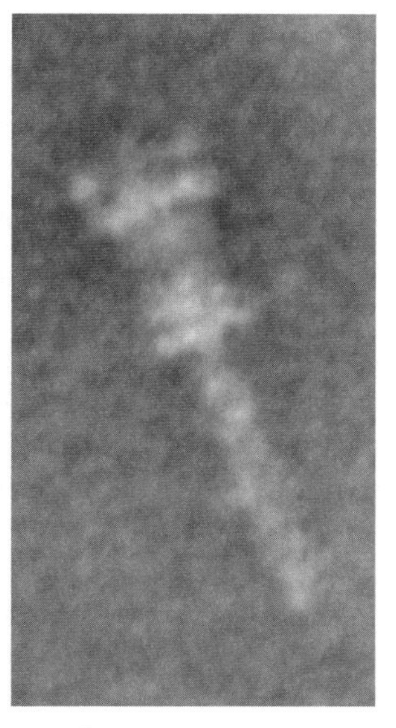

FIGURE 16.5 Type III Secretion Systems Gram-negative bacteria use type III secretion systems to deliver certain molecules directly to host cells, inducing changes in those cells. Peptidoglycan is not shown in this figure. ©Chihiro Sasakawa, University of Tokyo

What bacterial structure do type III secretion systems resemble?

If the body site has normal microbiota, the new arrival must compete for space and nutrients. It must also tolerate any toxic products such as fatty acids produced by the competitors.

Delivering Effector Proteins to Host Cells

Some Gram-negative pathogens deliver proteins directly into host cells using secretion systems. A **type III secretion system,** or injectisome, is a syringe-like structure that injects proteins into eukaryotic cells (**figure 16.5**). The injected proteins, referred to as effector proteins, induce changes such as altering the cell's cytoskeleton structure. Some effector proteins direct the host cell to engulf the bacterial cell, a process discussed in the next section. Several types of secretion systems have been discovered, and some can inject molecules other than proteins. ⏮ **secretion,** ⏮ **cytoskeleton**

MicroAssessment 16.5

Pathogens use adhesins, often on pili, to bind to a body surface. To colonize a surface, the pathogen must often compete with the normal microbiota, prevent binding of secretory IgA, and obtain iron. Some bacteria deliver effector proteins to epithelial cells, inducing a specific change in those cells.

10. What are siderophores?

11. What is a type III secretion system?

12. From a microbe's perspective, why is it a good strategy to adhere to a receptor that plays a critical function for a host cell?

16.6 ■ Invasion—Breaching the Anatomical Barriers

Learning Outcome

9. Describe the mechanisms pathogens use to penetrate the skin and mucous membranes.

Some bacterial pathogens cause disease while remaining on the mucosal surfaces, but many others penetrate the anatomical barriers. By crossing the epithelial barrier, invading microbes can multiply in the nutrient-rich tissues without competition.

Penetrating the Skin

Skin is the most difficult anatomical barrier for microbes to penetrate. Bacterial pathogens that invade via this route rely on skin-damaging injury. As examples, *Staphylococcus aureus* enters tissues via a cut or other wound and *Yersinia pestis* is injected by infected fleas. ⏭ **plague**

Penetrating Mucous Membranes

Mucous membranes are the entry points for most pathogens, but the invasive processes are complex and difficult to study. It appears, however, that there are at least two mechanisms used for invasion: directed uptake by cells and exploiting antigen-sampling processes.

Directed Uptake by Cells

Some pathogens induce non-phagocytic cells to engulf them. The pathogen first attaches to a cell, then triggers the process of endocytosis. ⏮ **endocytosis**

Gram-negative bacteria often inject effector proteins that induce engulfment by host cells. *Salmonella* species, for example, use a type III secretion system to deliver specific proteins to intestinal epithelial cells. These cause actin molecules in the host cell cytoplasm to rearrange, resulting in characteristic **membrane ruffling** on the cell's surface (**figure 16.6**). The ruffles enclose the bacterial cells, bringing them into the intestinal cell. ⏮ **actin**

Exploiting Antigen-Sampling Processes

As described in chapter 15, mucosa-associated lymphoid tissue (MALT) samples material from the mucosal surface. Some pathogens use this process to cross the membranes. ⏮ **MALT**

Several pathogens use M cells to cross the intestinal barrier. Recall that M cells transport material from the lumen of the intestine to the Peyer's patches (see figure 15.5). Most microbes delivered this way are destroyed by the macrophages in the Peyer's patches, but certain pathogens have mechanisms to avoid this fate. When *Shigella* cells are transferred to the macrophages, for instance, the bacteria survive, and eventually induce the phagocyte to undergo apoptosis (**figure 16.7**). The freed bacterial cells then bind to the base of the mucosal epithelial cells and cause these non-phagocytic cells to engulf them, using a mechanism similar to that of *Salmonella*. ⏮ **M cell,** ⏮ **Peyer's patches**

FIGURE 16.6 Ruffling *Salmonella enterica* serotype Typhimurium inducing ruffles on an M cell (a specialized intestinal epithelial cell), leading to uptake of the bacterial cells. ©Mark A. Jepson, Universty of Bristol

How do *Salmonella* cells induce ruffling?

Some pathogens invade by means of alveolar macrophages, which engulf material that enters the lungs. *Mycobacterium tuberculosis* cells produce surface proteins that direct macrophages to take up the bacteria. Although this might seem to be a disadvantage to the bacterial cells, it actually allows them to avoid a process that could otherwise lead to macrophage activation. *Mycobacterium* cells survive within macrophages that have not been activated.

MicroAssessment 16.6

Skin is the most difficult barrier for microbes to penetrate. Some pathogens cross the mucous membrane barrier by inducing mucosal epithelial cells to engulf them or by taking advantage of antigen-sampling processes.

13. How do *Shigella* species enter intestinal epithelial cells?
14. Why do *Mycobacterium tuberculosis* cells direct their own engulfment by macrophages?
15. Why would it be difficult to study invasion of mucous membranes?

16.7 ■ Avoiding the Host Defenses

Learning Outcome

10. Describe mechanisms that bacteria use to avoid destruction by phagocytes, killing by complement system proteins, and recognition by antibodies.

Inside the body, invading microorganisms soon encounter the innate and adaptive immune defenses. Pathogens as a group have evolved a wide variety of mechanisms to avoid the otherwise lethal effects of these defenses.

Hiding Within a Host Cell

Some pathogens enter host cells, where they hide from complement proteins, phagocytes, and antibodies. Once a *Shigella* cell is within an intestinal epithelial cell, it directs its own transfer to adjacent cells (see figure 16.7). It does this by causing the host cell actin to polymerize at one end of the bacterial cell, thereby forming an "actin tail" that propels the bacterium within the cell. The force of the propulsion is so great that the bacterial cells are often pushed into neighboring cells. *Listeria monocytogenes* (causes meningitis) does the same thing (**figure 16.8**). ◀◀ actin, ▶▶ listeriosis

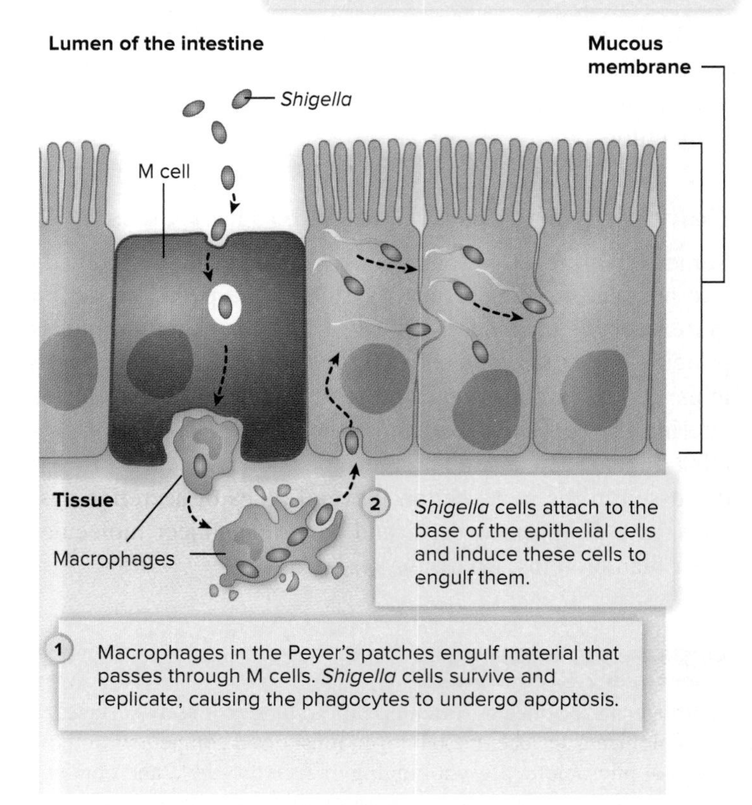

FIGURE 16.7 Antigen-Sampling Processes Provide a Mechanism for Invasion *Shigella* species use M cells to move across the intestinal epithelial barrier. Once the bacterial cells are on the other side, macrophages ingest them, but the bacteria are able to escape and then infect other cells.

What is the normal function of M cells?

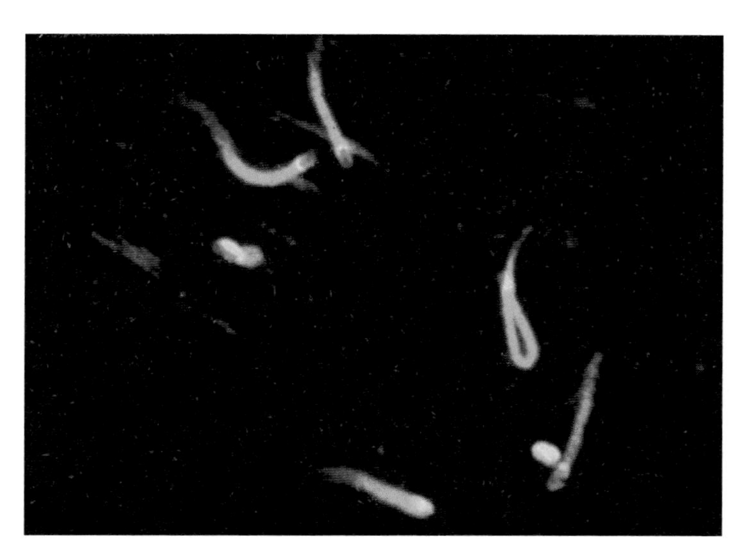

FIGURE 16.8 Actin Tail of Intracellular *Listeria monocytogenes*
Rapid polymerization of host cell actin (green) at one end of the bacterial cell (orange) propels the bacterium within the cell. ©Pascale F. Cossart

How does an actin tail benefit a bacterial cell?

Avoiding Destruction by Phagocytes

Phagocytes destroy microbes, using a process that involves multiple steps: chemotaxis, recognition and attachment, engulfment, and fusion of the phagosome with lysosomes (see figure 14.14). Pathogens have evolved several mechanisms to avoid destruction (**figure 16.9**).

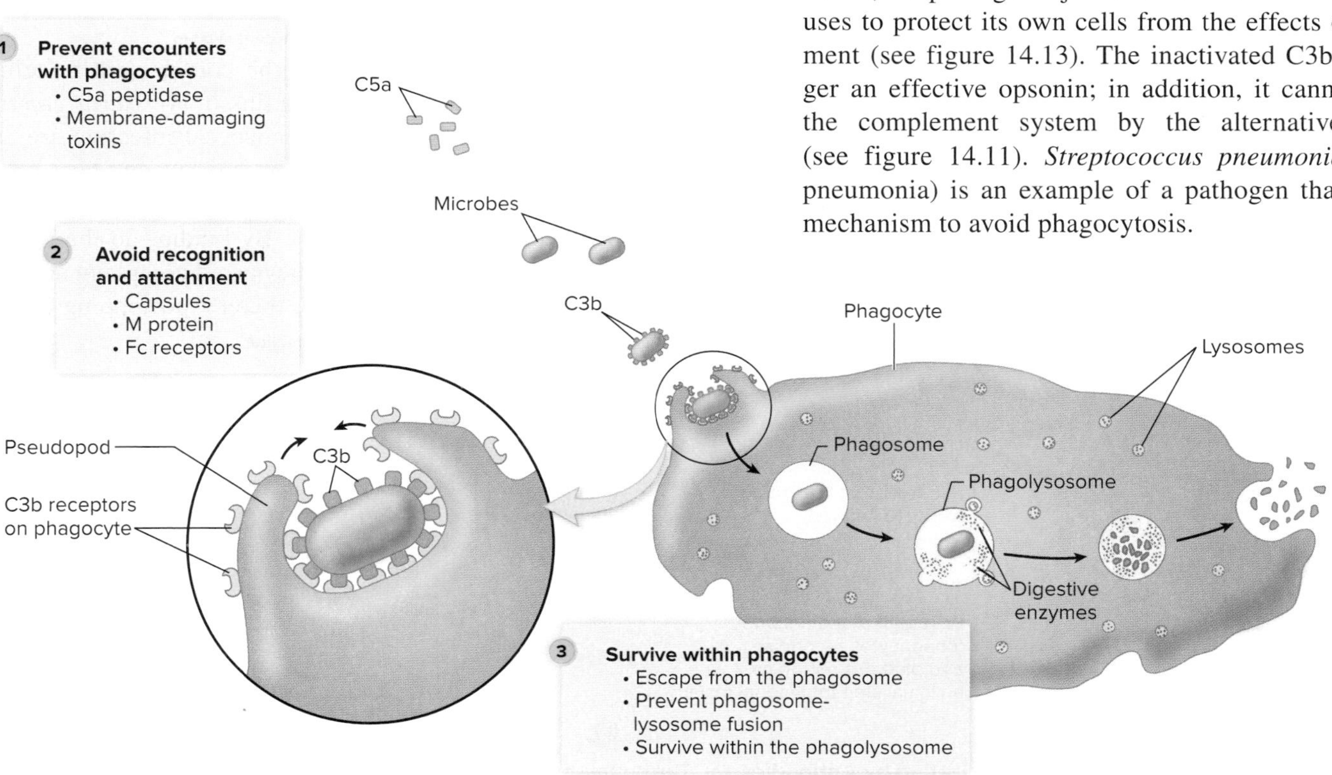

FIGURE 16.9 Avoiding Destruction by Phagocytes

Mycobacterium tuberculosis recruits phagocytes to the site of infection. Based on this information, which of the three methods shown here does the bacterium likely use to avoid being destroyed?

① Preventing Encounters with Phagocytes

Some pathogens prevent phagocytosis by avoiding macrophages and neutrophils altogether. The mechanisms include:

- **C5a peptidase.** This enzyme degrades the complement system component C5a, a chemoattractant that recruits phagocytic cells. *Streptococcus pyogenes* (causes strep throat) makes C5a peptidase. ⏮ C5a, ⏭ *Streptococcus pyogenes*
- **Membrane-damaging toxins.** These kill phagocytes and other cells, often by forming pores in their membranes. The leaky cells swell and then lyse. *S. pyogenes* makes a membrane-damaging toxin called streptolysin O. ⏭ membrane-damaging toxins

② Avoiding Recognition and Attachment

Some pathogens avoid being recognized by phagocytes. Recall that phagocytes recognize and attach to foreign material more efficiently if opsonins such as C3b or antibodies coat it. Mechanisms that bacteria use to avoid opsonization include: ⏮ opsonins

- **Capsules.** These have long been recognized for their ability to prevent phagocytosis. Scientists now know that capsules often do this by interfering with opsonization. In some cases, the capsule binds the host's complement regulatory proteins that inactivate C3b—in other words, the pathogen hijacks the mechanism that the body uses to protect its own cells from the effects of complement (see figure 14.13). The inactivated C3b is no longer an effective opsonin; in addition, it cannot activate the complement system by the alternative pathway (see figure 14.11). *Streptococcus pneumoniae* (causes pneumonia) is an example of a pathogen that uses this mechanism to avoid phagocytosis.

- **M protein.** This component of the cell wall of *Streptococcus pyogenes* functions in a manner similar to that described for capsules. It binds a complement regulatory protein that inactivates C3b, thereby preventing it from being an effective opsonin. Also, that inactivated C3b cannot activate the complement system by the alternative pathway.
- **Fc receptors.** These proteins bind the Fc region of antibodies, interfering with their function as opsonins (**figure 16.10**). Recall that antibodies have two parts: the Fab region, which binds specifically to antigens, and the Fc region, which functions as a "red flag" (see figure 15.7). In a normal situation, when antibody molecules are bound to a microbial cell, the Fc region sticks out, serving as a marker to the immune system that the cell is an invader. Bacterial cells that have Fc receptors on their surface bind the Fc portion of random antibody molecules, orienting the antibodies so that the Fab region sticks out. The phagocytic cell has no mechanism for recognizing the Fab regions, so this masks the bacterial cell from phagocytes. *Staphylococcus aureus* and *Streptococcus pyogenes* make Fc receptors (protein A and protein G, respectively). ◀◀ Fc and Fab regions

③ Surviving Within Phagocytes

Some bacteria make no attempt to avoid engulfment by phagocytes, instead using it as an opportunity. It allows them to hide from antibodies, control some aspects of the immune response, and be transported to other locations in the body. Mechanisms used to survive within phagocytes include:

- **Escape from the phagosome.** Some pathogens escape from the phagosome before it fuses with lysosomes. The bacteria then multiply within the cytoplasm of the phagocyte, protected from other host defenses. *Listeria monocytogenes* produces a molecule that forms pores in the phagosomal membrane, allowing the bacterial cells to escape. *Shigella* species lyse the phagosome before it fuses with lysosomes.
- **Preventing phagosome-lysosome fusion.** Bacteria that prevent phagosome-lysosome fusion avoid the otherwise inevitable exposure to the destructive components of lysosomes. *Salmonella* species can sense that they have been ingested by a macrophage, and then respond by producing a protein that blocks the fusion process.
- **Surviving within the phagolysosome.** Relatively few microbes can survive the destructive environment within the phagolysosome. *Coxiella burnetii* (causes Q fever), however, is able to withstand the conditions. Once the organism has been ingested by a macrophage, it delays fusion of the phagosome with the lysosome, allowing additional time for the microbe to equip itself for growth within the phagolysosome.

Avoiding Killing by Complement System Proteins

As described in chapter 14, activation of the complement system leads to three primary outcomes: lysis of foreign cells by membrane attack complexes (MACs), opsonization, and inflammation (see figure 14.11). The latter two outcomes are associated with phagocytosis, and the mechanisms to avoid destruction by phagocytes were just described, so here we will focus on avoiding MACs. Gram-negative bacteria are susceptible to MACs because their outer and cytoplasmic membranes serve as targets; the thick layer of peptidoglycan of Gram-positive bacteria prevents the MACs from reaching the cytoplasmic membrane. ◀◀ complement system

Bacteria that avoid killing by the complement proteins are said to be **serum resistant;** strains of *Neisseria gonorrhoeae* that cause disseminated gonococcal infection are an example. These strains hijack the mechanism that host cells use to prevent their own surfaces from activating the complement system (see figure 14.13). By binding to the host's complement regulatory proteins, they avoid complement activation by the alternative pathway, thereby postponing MAC formation. ▶▶ disseminated gonococcal infection

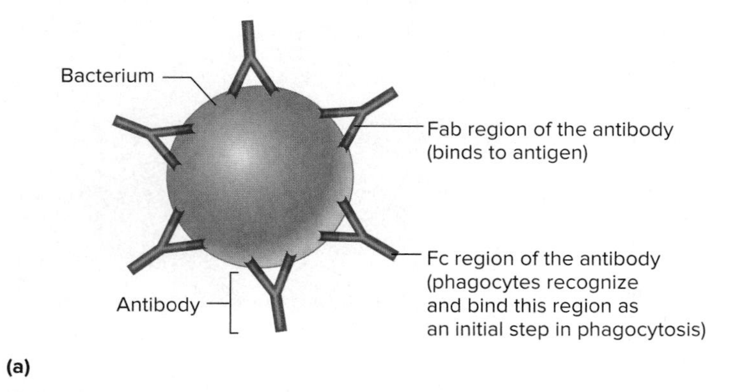

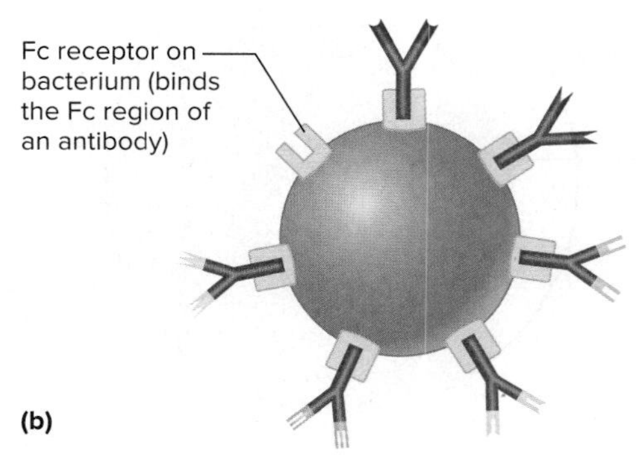

FIGURE 16.10 Fc Receptors Prevent Opsonization by Antibodies (a) The normal orientation of antibody molecules on the surface of a bacterium; note that the Fc region of the antibody sticks out from the bacterial cell, making it available for a phagocyte to recognize and bind. **(b)** The effect of Fc receptors on a bacterial cell's surface; the receptors bind the Fc portion of antibodies, regardless of their specificity.

? Are antibodies that bind to Fc receptors specific for the bacterium that has the receptors on its surface?

Avoiding Recognition by Antibodies

Pathogens that survive the innate defenses soon encounter an additional obstacle, the adaptive defenses. For bacteria, the most important of these are antibodies. Mechanisms for avoiding them include:

- **IgA protease.** This enzyme cleaves IgA, the class of antibody found in mucus and other secretions. *Neisseria gonorrhoeae* and a variety of other pathogens produce IgA protease. This enzyme may also have other roles.
- **Antigenic variation.** Some pathogens routinely alter the structure of their surface antigens. This allows them to stay ahead of antibody production by altering the very molecules that antibodies would otherwise recognize. *Neisseria gonorrhoeae* is able to vary the antigenic structure of its pili; antibodies produced by the infected host in response to one variation of the pili cannot bind effectively to another. ⏮ antigenic variation
- **Mimicking host molecules.** Pathogens sometimes produce and cover themselves with molecules similar to those normally found in the host. This molecular mimicry takes advantage of the fact that the immune system typically does not mount an attack against "self" molecules. Certain strains of *Streptococcus pyogenes* have a capsule composed of hyaluronic acid, a polysaccharide found in human tissues.

MicroAssessment 16.7

Mechanisms bacteria use to avoid destruction by phagocytes include preventing encounters with phagocytes, avoiding recognition and attachment, and surviving within the phagocyte. Serum-resistant bacteria avoid the killing effects of complement system proteins. Mechanisms for avoiding antibodies include IgA protease, antigenic variation, and mimicking host molecules.

16. Describe how Fc receptors help a microbe avoid phagocytosis.
17. Describe three mechanisms pathogens may use to survive within phagocytic cells.
18. Encapsulated organisms can be phagocytized once antibodies against the capsule have been produced. Why would this be so?

16.8 ■ Damage to the Host

Learning Outcomes

11. Describe the difference between exotoxins and endotoxins.
12. Compare and contrast neurotoxins, enterotoxins, and cytotoxins, giving two examples of each.
13. Explain how inflammation and antibodies can cause damage.

Damage due to infection can be the result of direct effects of the pathogen, such as toxins produced, or indirect effects, such as the immune response. In many cases, the damage helps the organism exit the host, allowing it to spread to others. For example, *Vibrio cholerae* (causes cholera) induces watery diarrhea—up to 20 liters of microbe-containing fluid in one day! In areas of the world without adequate sewage treatment, this can lead to contaminated water supplies and widespread outbreaks. *Bordetella pertussis* (causes whooping cough) causes severe bursts of coughing, propelling the respiratory pathogens into the air.

Exotoxins

A number of Gram-positive and Gram-negative pathogens make **exotoxins**—proteins that have very specific damaging effects (**table 16.1**). Exotoxins are often a major cause of damage to an infected host.

Exotoxins are often secreted by the bacterium but they can also leak into the surrounding fluid following lysis of the bacterial cell. In most cases, the pathogen must colonize a body surface or tissue to produce enough toxin to cause damage. With foodborne intoxication, however, the bacterial cells multiply in a food product where they produce toxin that is then consumed. In the case of botulism, ingestion of even tiny amounts of botulinum toxin is sufficient to cause paralysis. Like most other exotoxins, botulinum toxin can be destroyed by heating. ⏭ botulism

Exotoxins can act locally, or they may be carried in the bloodstream throughout the body, causing systemic effects. *Corynebacterium diphtheriae* (causes diphtheria) grows and releases its exotoxin in the throat. There, the toxin destroys local cells, leading to the accumulation of dead host cells, pus, and blood. This forms a "pseudomembrane" in the throat that sometimes dislodges and blocks the airway. The toxin can be absorbed and carried to the heart, nervous system, and other organs, causing additional damage. ⏭ diphtheria

Because exotoxins are proteins, the immune system can generally produce neutralizing antibodies against them (see figure 15.8). Unfortunately, many exotoxins are so powerful that fatal damage occurs before an adequate immune response is mounted. This is why vaccination is so important. It prevents otherwise common and often fatal diseases such as tetanus and diphtheria (see table 18.3). The vaccines against tetanus and diphtheria are **toxoids,** which are inactivated toxins. The toxoid induces production of specific antibodies, and these will immediately bind the toxin if it is encountered later. A vaccine against botulinum toxin is also available, but it is not part of routine vaccination because the risks of developing the disease are extremely low if sensible food preparation procedures are followed. If a person develops symptoms of a toxin-mediated disease, he or she can be treated with **antitoxin,** a suspension of neutralizing antibodies. ⏭ vaccines, ⏭ antitoxin

Many exotoxins can be grouped into functional categories according to the tissues they affect (see table 16.1). **Neurotoxins** damage the nervous system, causing symptoms

TABLE 16.1 Exotoxins Produced by Various Primary Pathogens

Example	Name of Disease; Name of Toxin	Characteristics of the Disease	Mechanism
A-B TOXINS—Composed of two subunits, A and B. The A subunit is the toxic, or active, part; the B subunit binds to the target cell.			
Neurotoxins			
Clostridium botulinum	Botulism; botulinum toxin	Flaccid paralysis	Blocks transmission of nerve signals to the muscles by preventing the release of acetylcholine.
Clostridium tetani	Tetanus; tetanospasmin	Spastic paralysis	Blocks the action of inhibitory neurons by preventing the release of neurotransmitters.
Enterotoxins			
Enterotoxigenic *E. coli* (ETEC)	Traveler's diarrhea; heat-labile enterotoxin (cholera-like toxin)	Severe watery diarrhea	Modifies a regulatory protein in intestinal cells, causing those cells to continuously secrete electrolytes and water.
Vibrio cholerae	Cholera; cholera toxin	Severe watery diarrhea	Modifies a regulatory protein in intestinal cells, causing those cells to continuously secrete electrolytes and water.
Cytotoxins			
Bacillus anthracis	Anthrax; edema factor, lethal factor	Inhaled form—septic shock; cutaneous form—skin lesions	Edema factor modifies a regulatory protein in cells, causing accumulation of fluids in the tissues. Lethal factor inactivates proteins involved in cell signaling functions.
Bordetella pertussis	Pertussis (whooping cough); pertussis toxin	Sudden bouts of violent coughing	Modifies a regulatory protein in respiratory cells, causing accumulation of respiratory secretions and mucus. Other factors also contribute to the symptoms.
Corynebacterium diphtheriae	Diphtheria; diphtheria toxin	Pseudomembrane in the throat; heart, nervous system, kidney damage	Inhibits protein synthesis by inactivating an elongation factor of eukaryotic cells. Kills local cells (in the throat) and is carried in the bloodstream to various organs.
Shiga toxin–producing *E. coli* (STEC)	Bloody diarrhea, hemolytic uremic syndrome; Shiga toxin	Diarrhea that may be bloody; kidney damage	Inactivates the 60S subunit of eukaryotic ribosomes, stopping protein synthesis.
Shigella dysenteriae	Dysentery, hemolytic uremic syndrome; Shiga toxin	Diarrhea that contains blood, pus, and mucus; kidney damage	Inactivates the 60S subunit of eukaryotic ribosomes, stopping protein synthesis.
MEMBRANE-DAMAGING TOXINS (cytotoxins)—Disrupt plasma membranes, causing leakiness that results in cell lysis.			
Clostridium perfringens	Gas gangrene; α-toxin	Extensive tissue damage	Removes the polar head group on the phospholipids in the membrane, damaging membrane structure.
Staphylococcus aureus	Wound and other infections; leukocidin	Accumulation of pus	Inserts into membranes, forming pores that allow fluids to enter the cells.
Streptococcus pyogenes	Pharyngitis and other infections; streptolysin O	Accumulation of pus	Inserts into membranes, forming pores that allow fluids to enter the cells.
SUPERANTIGENS—Override the specificity of the T-cell response.			
Staphylococcus aureus (certain strains)	Foodborne intoxication; staphylococcal enterotoxins	Nausea and vomiting	Not well understood with respect to how the ingested toxins lead to the characteristic symptoms of foodborne intoxication.
Staphylococcus aureus (certain strains)	Staphylococcal toxic shock; toxic shock syndrome toxin (TSST)	Fever, vomiting, diarrhea, muscle aches, rash, low blood pressure	Systemic toxic effects due to the resulting massive release of cytokines.
Streptococcus pyogenes (certain strains)	Streptococcal toxic shock; streptococcal pyrogenic exotoxins (SPE)	Fever, vomiting, diarrhea, muscle aches, rash, low blood pressure	Systemic toxic effects due to the resulting massive release of cytokines.
OTHER TOXIC PROTEINS			
Staphylococcus aureus	Scalded-skin syndrome; exfoliatin	Separation of the outer layer of skin	Destroys material that holds the layers of skin together.
Various organisms	Various diseases; proteases, lipases, and other hydrolases	Tissue damage	Degrades proteins, lipids, and other compounds that make up tissues.

such as paralysis. **Enterotoxins** cause symptoms associated with intestinal disturbance, such as diarrhea and vomiting. **Cytotoxins** damage a variety of different cell types, either by interfering with essential cellular mechanisms or by lysing cells. Some exotoxins do not fall into any of these groups, instead causing symptoms associated with an overly strong—and therefore damaging—immune response. Most exotoxins fall into three general categories that reflect their structure and general mechanism of action: A-B toxins, membrane-damaging toxins, and superantigens.

MicroByte

Botox is a dilute suspension of botulinum toxin (see Focus Your Perspective 30.1).

A-B Toxins

A-B toxins consist of two parts: the A subunit is the toxic (active) portion and the B subunit binds to a specific surface molecule on cells (**figure 16.11**). In other words, the A subunit, usually an enzyme, is responsible for the effects of the toxin on a cell, whereas the B subunit dictates the type of cell to which the toxin is delivered.

The structure of A-B toxins offers novel approaches for the development of vaccines and therapies. For example, a fusion protein that contains diphtheria toxin is used to treat a type of cancer called cutaneous T-cell lymphoma. By fusing the toxin to a cytokine that binds T cells, the toxin is delivered to those cells, including cancerous ones. In some countries, the B subunit of cholera toxin is used as a component of an orally administered vaccine against cholera. Antibodies that bind the B subunit prevent cholera toxin from attaching to intestinal cells, thus protecting the vaccine recipient. Researchers are now experimenting with joining medically useful compounds to B subunits, allowing medications to be delivered specifically to the cell type targeted by the B subunit.

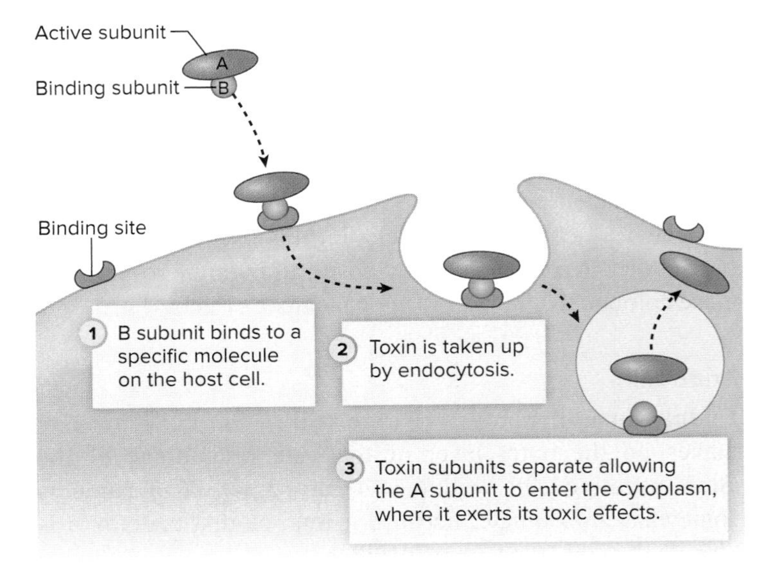

FIGURE 16.11 The Action of an A-B Toxin The structure and mechanisms of uptake of different A-B toxins may vary slightly.

? Would an antibody response against the B subunit protect against the effects of the toxin? Why or why not?

Membrane-Damaging Toxins

Membrane-damaging toxins are cytotoxins that disrupt eukaryotic cytoplasmic membranes, causing the cell to lyse. Many lyse red blood cells, causing hemolysis that can be observed when the organisms are grown on blood agar; these toxins are often referred to as hemolysins. ◀◀ hemolysis, ◀◀ blood agar

Pore-forming toxins insert into the phospholipid bilayer of membranes, forming channels that allow uncontrolled passage of fluids. One pore-forming toxin is streptolysin O, the compound responsible for the characteristic β-hemolysis of *Streptococcus pyogenes* grown anaerobically on blood agar (see figure 4.11). Recall that streptolysin O also helps *S. pyogenes* avoid phagocytosis.

Phospholipases hydrolyze phospholipids in the cytoplasmic membrane. The α-toxin of *Clostridium perfringens* (causes gas gangrene) is an example. ◀◀ phospholipid, ▶▶ gas gangrene

Superantigens

Superantigens are exotoxins that stimulate an abnormally high number of T_H cells (effector helper T cells), causing a massive release of cytokines (a "cytokine storm"). This leads to fever, nausea, vomiting, and diarrhea. The effects of a cytokine storm can be life-threatening, leading to organ failure and circulatory collapse. Examples of superantigens include toxic shock syndrome toxin (TSST) as well as several other toxins produced by *Staphylococcus aureus* and *Streptococcus pyogenes*. ◀◀ helper T cells

Superantigens function by overriding the normal specificity of helper T cells' antigen recognition. They do this by binding simultaneously to the outer portion of the major histocompatibility (MHC) class II molecule on antigen-presenting cells and the T-cell receptor (**figure 16.12**). The T-cell machinery interprets the binding to mean that the T-cell receptor recognizes the antigen presented on the MHC molecule, when it probably does not. An antigen usually stimulates about one in 10,000 helper T cells, whereas superantigens stimulate as many as one in five. Many T cells undergo apoptosis following the stimulation, thereby suppressing the immune response. Superantigens are also suspected of contributing to autoimmune diseases. By overriding the normal control mechanisms of adaptive immunity, superantigens may promote proliferation of T cells that respond to healthy "self." ◀◀ MHC class II molecules, ◀◀ apoptosis

The exotoxins produced by *Staphylococcus aureus* strains that cause foodborne intoxication are superantigens. Although they result in nausea and vomiting and are therefore referred to as enterotoxins, their structure and action are very different from the enterotoxins of *Vibrio cholerae* and pathogenic *E. coli* strains. The mechanism by which they induce vomiting

FOCUS ON A CASE 16.1

The patient was a 14-year-old boy who had been coughing frequently for about 2 weeks. He initially developed what his parents assumed was the common cold, with symptoms of a runny nose, malaise, low fever, and mild cough. His cough gradually worsened and was particularly severe at night. Some of his coughing episodes were so intense that they caused him to vomit.

The physician considered the possibility that the boy had pertussis (whooping cough), a disease caused by the bacterium *Bordetella pertussis*. The boy had been vaccinated against the disease as a child, but the physician knew that recent outbreaks suggest that the vaccine used since the 1990s might not provide long-lasting immunity. The physician swabbed the boy's nasopharynx (where the nasal passage connects with the throat) and sent the sample to the clinical lab for culture and PCR testing. In talking to the boy's parents about the next steps, she told them to watch him for signs such as a spike in fever, which could indicate a secondary infection.

Culture and PCR results came back positive for *B. pertussis,* and the boy was treated with an antibacterial medication. His cough persisted for several more weeks, but he gradually improved and made a full recovery.

1. Why would it be beneficial to *Bordetella pertussis* to cause coughing?
2. The bacterium colonizes the ciliated cells of the upper respiratory tract. How might it avoid being swept away by the mucociliary escalator?
3. The bacteria do not invade the epithelial cells, yet they damage those cells. What might cause the damage?
4. Why would the physician be concerned about a secondary infection?

Discussion

1. Coughing propels the bacterial cells into the air, where another person can inhale them. In this way, the bacterium can move to the next susceptible host. Because whooping cough is easily transmitted, it spreads quickly among unvaccinated populations, so contacts of people who have the disease should be notified. Signs and symptoms of pertussis are often mild in adults, mimicking the common cold, so these people can unknowingly serve as a source of *B. pertussis* to others. Infants cannot receive their first whooping cough vaccination until age 2 months, so they are at particular risk for developing the disease. This is a concern because the disease can be quite severe, and even deadly, in infants.
2. The bacterium produces multiple adhesins that specifically allow it to colonize ciliated cells. The attachment structures include a protein called filamentous hemagglutinin (FHA), named for its ability to agglutinate red blood cells, and fimbriae.
3. *Bordetella pertussis* produces several toxins that damage cells. One of them, tracheal cytotoxin (TCT), is toxic to ciliated epithelial cells. TCT is a fragment of peptidoglycan that *B. pertussis* specifically releases during growth. Another is pertussis toxin (PTX), an A-B toxin. This toxin modifies a regulatory protein in respiratory cells, causing accumulation of respiratory secretions and mucus. The combination of increased fluid production and decreased ciliary action results in the characteristic cough of pertussis, because only the cough reflex remains for clearing respiratory secretions. The classical "pertussis cough" includes a rapid series of coughs—a reflex for clearing fluid from the airways—followed by a whooping sound as the patient gasps for air.
4. Ciliated respiratory cells are an important part of the first-line defenses. When the ciliated cells are damaged, the mucociliary escalator does not function normally, therefore making it more likely that other pathogens will colonize the respiratory tract.

is poorly understood. Unlike most other exotoxins, the enterotoxins produced by *Staphylococcus aureus* are heat-stable. Even thorough cooking of foods contaminated with these toxins will not prevent illness. ⏭| *Staphylococcus aureus* food poisoning

Other Toxic Proteins

Various proteins that are not A-B toxins, membrane-damaging toxins, or superantigens can have damaging effects. An important example is the toxin produced by strains of *Staphylococcus aureus* that cause scalded skin syndrome. This toxin, **exfoliatin,** destroys material that binds together the layers of skin, causing the outer layer to peel (see figure 22.4). The bacteria might be growing in a small, localized lesion but the toxin spreads systemically.

Enzymes can also be damaging. Various hydrolytic enzymes, including proteases, lipases, and collagenases, break down tissue components. In addition to destroying tissues, some of these enzymes help the bacteria spread.

Endotoxin and Other Bacterial Cell Wall Components

The host defenses are primed to respond to various bacterial cell wall components, including lipopolysaccharide and peptidoglycan. A strong and widespread immune response to these compounds, however, can have toxic effects.

Endotoxin

Endotoxin is lipopolysaccharide (LPS), the molecule that makes up the outer layer of the outer membrane of the Gram-negative cell wall (see figure 3.34). The name is somewhat unfortunate, because it implies that endotoxin is "inside the cell," and, conversely, that exotoxins are "outside the cell." This is misleading, because endotoxin is an integral part of the outer membrane, whereas exotoxins are proteins that may or may not be secreted by the bacterial cell. Unlike most exotoxins, endotoxin cannot be converted to an

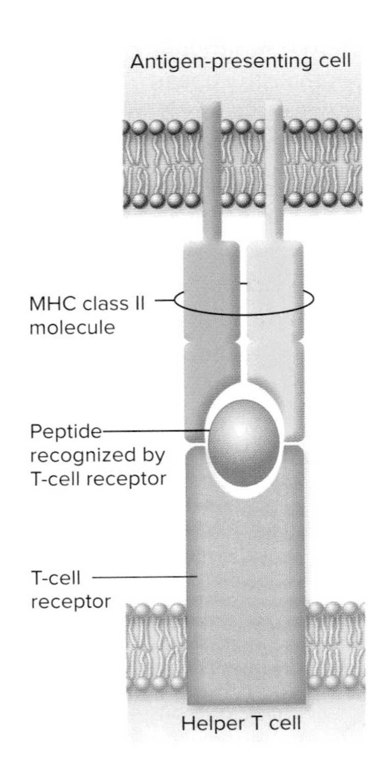

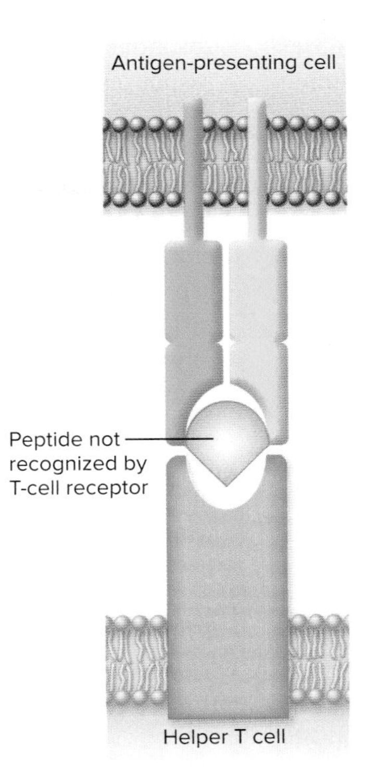

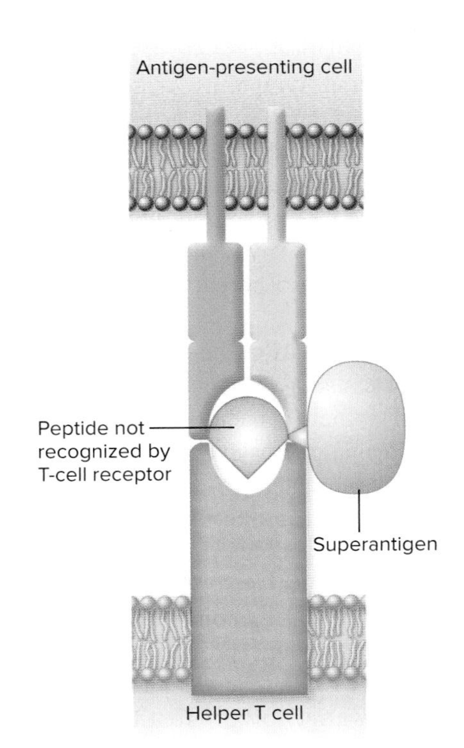

a Helper T cell that recognizes peptide releases cytokines in response.

b Helper T cell that does not recognize peptide is not activated and does not release cytokines

c Superantigen binds to the outer portion of the MHC class II molecule and the T cell receptor, causing the T cell to respond as though it recognizes the peptide even when it does not.

FIGURE 16.12 Superantigens (a) Normal situation: antigen recognition. **(b)** Normal situation: lack of antigen recognition. **(c)** Effect of a superantigen.

? What specifically causes the toxic effects of superantigens?

effective toxoid for immunization. **Table 16.2** summarizes some of the other differences between exotoxins and endotoxin. ◀◀ **lipopolysaccharide**

Recall from chapter 3 that the lipopolysaccharide molecule contains lipid A. This component triggers an inflammatory response and is responsible for the damaging effects of LPS. When lipid A is present in a localized region, the response helps clear an infection. It is a different situation entirely, however, when the infection is systemic, as in a bloodstream infection. Imagine a state in which inflammation occurs throughout the body—extensive leakage of fluids from permeable blood vessels and widespread activation

TABLE 16.2 Comparison of Exotoxins and Endotoxin

Property	Exotoxins	Endotoxin
Bacterial source	Gram-positive and Gram-negative species	Gram-negative species only
Location in the bacterium	Synthesized in the cytoplasm; may or may not be secreted	Component of the outer membrane of the Gram-negative cell wall
Chemical nature	Protein	Lipopolysaccharide (the lipid A component)
Ability to form a toxoid	Generally, yes	No
Heat stability	Generally inactivated by heat	Heat-stable
Mechanism	A distinct toxic mechanism for each	Innate immune response; a systemic response can lead to fever, a dramatic drop in blood pressure, and disseminated intravascular coagulation
Toxicity	Generally very potent; some are among the most potent toxins known	Small amounts in a localized area lead to an appropriate immune response that helps clear an infection, but systemic distribution can be deadly

of the coagulation cascade. This is the response that causes the signs and symptoms of sepsis. Such an overwhelming systemic response causes a dramatic drop in blood pressure, disseminated intravascular coagulation (DIC), and fever (see figure 25.3). When this happens, it is called **septic shock** (shock is a life-threatening condition resulting from insufficient blood flow); when the response is triggered by endotoxin, it may also be called endotoxic shock. ⏭ sepsis, ⏭ DIC, ⏭ septic shock

Lipid A is part of the Gram-negative outer membrane and does not cause a response unless it is released. This occurs primarily when a bacterium lyses, which can happen as a result of phagocytosis, activation of the complement system (due to membrane attack complexes), and treatment with certain types of antibiotics. Once released from a cell, the LPS molecules can activate the innate and adaptive defenses by a variety of mechanisms. Monocytes, macrophages, and other cells have toll-like receptors (TLRs) that detect LPS, inducing the cells to produce pro-inflammatory cytokines. LPS also functions as a T-independent antigen; at high concentrations it activates a variety of different B cells, regardless of the specificity of their B-cell receptor. ⏮ TLRs, ⏮ T-independent antigens

Endotoxin is heat-stable; it is not destroyed by autoclaving. Thus, even though autoclaved material is sterile, it may still contain endotoxin. Disastrous results including death have resulted from administering intravenous (IV) fluids contaminated with endotoxin. To verify that such fluids are not contaminated, a very sensitive test known as the Limulus amoebocyte lysate (LAL) assay is done. This uses proteins extracted from blood of the horseshoe crab *(Limulus polyphemus)* that form a gel-like clot when exposed to endotoxin; as little as 10 to 20 picograms (1 picogram = 10^{-12} grams) of endotoxin per milliliter can be detected using the LAL. Horseshoe crabs are one of this planet's more unique and ancient life-forms. The essential role they play in this test has led to an increased awareness of the importance of their habitat. A non-lethal system of capture, blood sampling, and release has been developed.

Other Bacterial Cell Wall Components

Peptidoglycan and other bacterial cell wall components can cause symptoms similar to those that characterize the response to endotoxin. The systemic response can lead to sepsis and septic shock.

Damaging Effects of the Immune Response

Although the immune response eliminates invading microbes, it can inadvertently damage host tissues as well. The reactions to endotoxin and other cell wall components are examples of damaging effects of the immune response, but are typically considered a toxic effect of the bacterium because the reactions can be immediate and overwhelming. The damaging responses discussed next take longer to appear.

Damage Associated with Inflammation

The inflammatory response itself can destroy tissue because phagocytic cells recruited to the area release some of the enzymes and toxic products they contain. The life-threatening aspects of bacterial meningitis, for example, are due to the inflammatory response itself. Complications of certain sexually transmitted infections are also due to the damage associated with inflammation. For example, if *Neisseria gonorrhoeae* or *Chlamydia trachomatis* infections involve the fallopian tubes, the inflammatory response can lead to scarring that obstructs the tubes, either preventing fertilization or predisposing a woman to an ectopic pregnancy (meaning the fertilized egg implants outside the uterus).

Damage Associated with Adaptive Immunity

The adaptive immune response can also lead to damaging effects. Mechanisms include:

- **Immune complexes.** When antibodies bind to antigens, the complexes can settle in the kidneys and joints, where they activate the complement system, causing destructive inflammation. An example is acute glomerulonephritis, a complication that can follow skin and throat infections caused by *Streptococcus pyogenes;* the immune complexes that form as a result of the infection trigger a response that damages kidney structures called glomeruli. ⏭ acute glomerulonephritis
- **Cross-reactive antibodies.** Certain antibodies produced in response to an infection bind to the body's own tissues, promoting an autoimmune response. Acute rheumatic fever—a complication that can follow strep throat—is thought to be the result of antibodies against *S. pyogenes* binding to normal tissue proteins. This occurs most frequently in people with certain MHC types (see Focus Your Perspective 15.1). ⏭ acute rheumatic fever

MicroAssessment 16.8

Damage can be due to exotoxins, including A-B toxins, superantigens, membrane-damaging toxins, and other toxic proteins. Endotoxin and other cell wall components in the bloodstream can result in septic shock. The inflammatory response can cause tissue damage that leads to scarring. Immune complexes can trigger damaging inflammation in the kidneys and joints; cross-reactive antibodies can lead to an autoimmune response.

19. Cholera toxin is an A-B toxin. What does that tell you about its structure?
20. How is an enterotoxin different from endotoxin?
21. Home-canned foods should be boiled before consumption to prevent botulism. Considering that boiling does not destroy endospores, why would it prevent the disease?

16.9 ■ Mechanisms of Viral Pathogenesis

Learning Outcomes

14. Describe how viruses bind to host cells, and how they spread to other cells.
15. Describe how viruses avoid immune responses and ultimately damage cells.

To infect a host, a virus must enter an appropriate cell, use the cell's machinery for replication, keep the host from recognizing and destroying the infected cell, and then move to new cells or hosts. Damage to the host can occur as a direct result of infection or from the immune response.

Binding to Host Cells and Invasion

As discussed in chapter 13, viruses attach to target cells via specific receptors. Only cells that have the receptor can be infected, so this influences the host range and tissue specificity of a particular virus. For example, HIV's receptor is CD4, a molecule found on helper T cells and macrophages. Reoviruses bind to receptors on M cells in the Peyer's patches of the intestinal tract. Once attached, viruses then enter the cells using either receptor-mediated endocytosis or fusion with the cytoplasmic membrane (see figure 13.11). ⏮ CD4

Virions released from a cell can either infect neighboring cells or spread to other tissues via the bloodstream or lymphatic system. Polioviruses, for example, initially infect cells in the throat and intestinal tract. Upon release from these cells, the virus enters the bloodstream and may then spread to infect motor nerve cells of the brain and spinal cord, causing paralysis. ⏭ polio

Avoiding Immune Responses

Viruses must avoid the host defenses that detect and eliminate invaders. These include interferon, apoptosis, and antibodies. ⏮ interferon, ⏮ apoptosis, ⏮ antibodies

Avoiding the Antiviral Effects of Interferons

Early in viral infection, interferons play an important role in limiting viral spread (see figure 14.10). They cause cells to produce enzymes that, when activated, prevent viral replication. To avoid this, some viruses encode proteins that shut down expression of host genes. Others avoid interferon's effects by interfering with activation of the enzymes.

Regulating Host Cell Death

Many viruses regulate host cell death. Some kill the host cell once it has produced virions, and others prevent the host cell from dying prematurely.

Recall that the interferon response induces apoptosis in virally infected cells. Some viruses prevent this from happening, thereby giving the virus more time to replicate. They do this by controlling a protein called p53 that regulates apoptosis in cells. When this protein is inhibited, tumors sometimes develop. Papillomaviruses, a group of viruses that cause certain types of cancer and various types of warts, interfere with the normal function of p53.

To avoid being detected by the cell-mediated immune response, many types of viruses interfere with antigen presentation by MHC class I molecules. Herpesviruses, for example, block the movement of the molecules to the surface of the infected cell, so that T_C cells (effector cytotoxic T cells) cannot inspect the proteins being made. The immune system, however, is prepared for such a strategy. Natural killer (NK) cells recognize stressed cells that lack MHC class I molecules and destroy them (see figure 15.23). Perhaps not surprisingly, some viruses have methods to trick NK cells. Cytomegalovirus (CMV), which causes disease in immunocompromised people, encodes production of "fake" MHC class I molecules. These decoy versions are displayed on the surface of the host cell, tricking the immune system into believing that all is well in the cell (**figure 16.13**). ⏮ antigen presentation by MHC class I, ⏮ natural killer cells

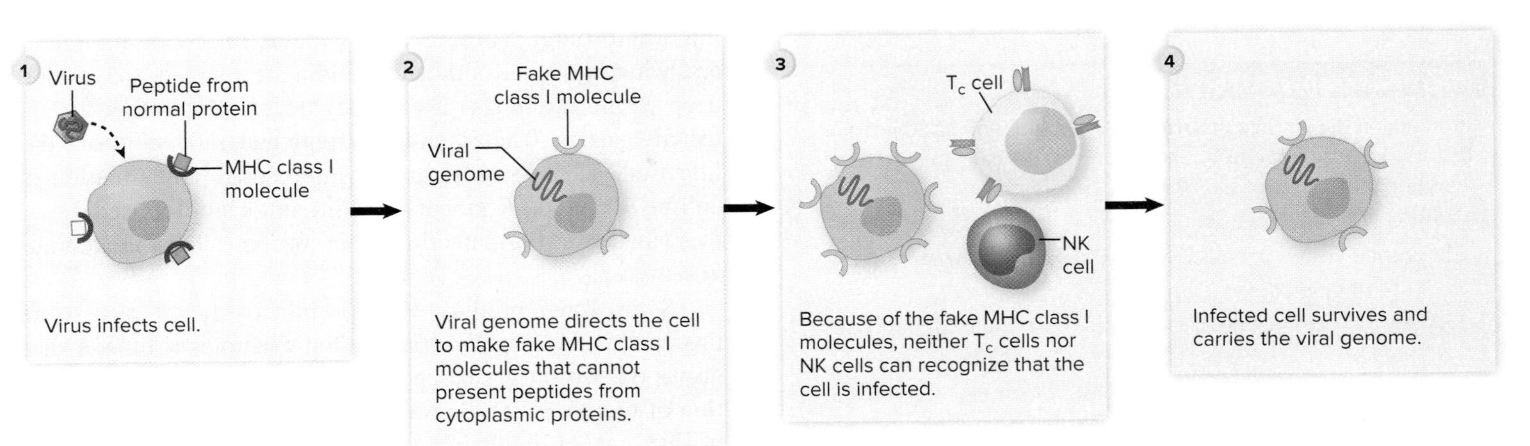

FIGURE 16.13 The Role of MHC Class I Decoys in Viral Pathogenesis

? What is the role of MHC class I molecules in adaptive immunity?

Avoiding Antibodies

Antibodies generally control the spread of viruses by neutralizing extracellular virions (see figure 15.8). To avoid antibodies, some viruses move directly from one cell to its immediate neighbors. Other viruses remain intracellular by forcing cellular neighbors to fuse, forming a large multinucleated cell called a **syncytium.** HIV induces syncytia formation. ⏮ syncytium

The surface antigens of some viruses change rapidly, outpacing the body's capacity to produce effective neutralizing antibodies. This rapid evolution occurs because the replicases of RNA viruses and the reverse transcriptase of HIV have no proofreading ability, so mutations occur fairly often. Thus, as the viruses replicate, they give rise to pools of genetically altered virions. An immune response against essential proteins that cannot tolerate extensive mutations may eventually clear the viruses. This does not seem to be the case for HIV, however. An HIV-infected person produces many antibodies against HIV, but they are not effective in clearing the virus. ⏮ replicase, ⏮ reverse transcriptase

Some viruses actually take advantage of antibodies, using them to help enter macrophages. Opsonized viral particles are engulfed by phagocytes, which they then infect. Dengue hemorrhagic fever, a severe illness caused by dengue viruses, is associated with such antibody-dependent enhancement of infection. ⏭ dengue hemorrhagic fever

Damage to the Host

As viruses replicate, they often damage host cells. In addition, the immune response to the infection can also be damaging. Many viruses cause what are referred to as **flu-like symptoms**—malaise, fever, and body aches—which are a result of the body's response to the infection. Some viruses elicit an overproduction of cytokines—referred to as a **cytokine storm**—resulting in an exaggerated immune response that can be deadly.

MicroAssessment 16.9

Proteins on the surface of virus particles function as adherence factors. Viruses, as a group, have several mechanisms for avoiding the effects of interferon, regulating host cell death, and avoiding antibodies.

22. From a virus's perspective, why would it be beneficial to prevent apoptosis?
23. How do cytomegaloviruses avoid the cellular immune response?
24. Why would various unrelated viral infections often include a similar set of symptoms (fever, headache, fatigue, and runny nose)?

16.10 ■ Mechanisms of Eukaryotic Pathogenesis

Learning Outcomes

16. Describe the mechanisms of pathogenesis of dermatophytes, *Candida albicans,* and the dimorphic fungi.
17. Compare and contrast the mechanisms of pathogenesis of protozoa and helminths.

Pathogenesis of eukaryotic cells, including fungi and protozoa, involves the same basic scheme as that of bacterial pathogens: colonization, evasion of host defenses, and damage to the host. The mechanisms, however, are often not as well understood.

Fungi

Most fungi, such as yeasts and molds, are saprophytes, meaning they acquire nutrients from dead and dying material; disease-causing fungi are generally opportunists, although notable exceptions exist.

Members of a group of fungi referred to as dermatophytes cause superficial infections of hair, skin, and nails, but do not invade deeper tissues. These fungi have keratinases, enzymes that break down the keratin in superficial tissues, allowing the fungi to use this protein as a source of nutrition. Dermatophytes cause diseases such as ringworm and athlete's foot.

Fungi in the normal microbiota, especially the yeast *Candida albicans,* can cause disease in immunocompromised hosts. Factors that can lead to excessive growth of *C. albicans* include AIDS, uncontrolled diabetes, severe burns, and inhibition of normal microbiota due to hormonal influences or antibiotic treatment. *C. albicans* generally infects the mucous membranes, causing thrush (an infection of the throat and mouth) or vulvovaginitis.

Cryptococcus species make a very large capsule. This capsule, like those produced by certain bacteria, interferes with phagocytosis.

Most of the serious fungal infections are caused by dimorphic fungi. These occur as molds in the environment, and when the small airborne conidia are inhaled, they lodge deep within the lungs. There, they develop into other forms, usually yeasts. The immune system generally controls the infection, so most cases are asymptomatic. The situation is entirely different if the person is immunocompromised, however, because the infections can become life-threatening. ⏮ dimorphic fungi, ⏮ conidia

Some fungi produce toxins, collectively referred to as mycotoxins. *Aspergillus flavus,* for example, a fungus that grows on certain grains and nuts, produces aflatoxin. Ingestion of this toxin can damage the liver and increases the risk of liver cancer. Fungal conidia and other products can cause hypersensitivities in some people. ⏭ hypersensitivities

Protozoa and Helminths

Most pathogenic protozoa and helminths either live within the intestinal tract or enter the body's tissues via the bite of an arthropod. *Schistosoma* species, however, can enter the skin directly (see Focus Your Perspective 14.1).

Like bacteria and viruses, eukaryotic parasites attach to host cells via specific receptors. *Plasmodium vivax,* one of the two most common causes of malaria, attaches to the Duffy blood group antigen on red blood cells. Most people of West African ancestry lack this antigen and are therefore resistant to infection by this species. *Giardia lamblia* uses a disc that functions as a suction cup to attach to the intestinal surface; it also has an adhesin associated with the disc that helps the initial attachment.

Protozoa and helminths use a variety of mechanisms to avoid antibodies. Some hide within cells, thus avoiding exposure to antibodies as well as certain other defenses. Malarial parasites, for example, produce enzymes that allow them to penetrate red blood cells. This cell type does not present antigen to T_C cells, so the parasite escapes antibodies as well as the cell-mediated immune defenses. *Plasmodium falciparum* makes proteins that insert into the infected red blood cells (RBCs). These proteins cause the RBCs to stick to the lining of the capillaries, thereby preventing the infected RBCs from circulating through the spleen, where they would likely be destroyed. *Leishmania* species survive and multiply within macrophages when phagocytized. Parasites such as the African trypanosomes, the cause of sleeping sickness, prevent antibodies from recognizing them by routinely varying their surface antigens through antigenic variation. *Schistosoma* species coat themselves with host proteins, thereby disguising themselves. Some parasitic worms appear to suppress immune responses in general.

The extent and type of damage caused by parasites vary tremendously. In some cases, the parasites compete for nutrients in the intestinal tract, contributing to malnutrition of the host. Helminths may accumulate in high enough numbers or grow long enough to block the intestines or other organs. Some parasites produce enzymes that digest host tissue, causing direct damage. In other cases, damage is due to the immune response; examples include the high fevers that characterize malaria, and the granulomatous response to *Schistosoma* eggs.

⏮ granulomas

MicroAssessment 16.10

Pathogenic mechanisms of fungi, protozoa, and helminths involve the same basic scheme as described for bacteria: colonization, evasion of host defenses, and damage to the host.

25. What is the importance of keratinase?

26. How do the African trypanosomes avoid the effects of antibodies?

27. Why would relatively few people of West African ancestry have the Duffy blood group antigen?

FOCUS ON THE FUTURE 16.1

The Potential of Probiotics

Considering the important protective roles the normal microbiota play in human health, it is not surprising that administering live beneficial microbes, referred to as **probiotics,** has been suggested as therapy for diarrheal and other diseases. In vitro studies certainly support their use, indicating that some strains of bacteria interfere with the growth or toxin production of certain pathogens. Animal studies using probiotics have also shown promising results. For example, chicks fed a mixture of *Lactobacillus* species that normally inhabit poultry intestines were less likely to be subsequently colonized by the pathogen *Salmonella enterica.* But which probiotics will be successful in treating humans?

Advertisers often make claims about the benefits of probiotics, but many questions are still unanswered. For example, will a microbe that has beneficial effects in vitro be stable in a food or supplement so that an adequate dose can be consumed? And even if it is stable in the package, will it survive passage through the stomach so that viable cells enter the intestinal tract? If it does access the intestines, can it colonize there, or will it simply pass through?

Beneficial effects of probiotics observed in humans so far are species and strain specific. One species may be protective, whereas another closely related organism is not. For example, several small studies indicate that consumption of *Lactobacillus rhamnosus* GG can shorten the duration of certain diarrheal illnesses. In contrast, studies that used a mixture containing *L. bulgaricus* and *L. acidophilus* generally showed no helpful effect. Because of this, products that list ingredients such as "live active culture" or "lactic acid bacteria" do not necessarily contain a beneficial strain.

Complicating matters is that probiotics are considered "dietary supplements," so there is little regulatory control over health claims. Early studies such as those using *L. rhamnosus* GG certainly highlight their potential, but larger and well-controlled scientific studies of probiotics are needed to provide more data as to their effectiveness in curing and preventing various medical conditions.

Summary

MICROBES, HEALTH, AND DISEASE

16.1 ■ The Anatomical Barriers as Ecosystems

In **mutualism,** both partners benefit; in **commensalism,** one partner benefits while the other is unaffected; and in **parasitism,** the parasite benefits at the expense of the host.

16.2 ■ The Human Microbiome (figure 16.1)

Composition of the Microbiome

Babies begin acquiring their microbiome during delivery and feeding. The microbial community of adults shows considerable diversity. *Firmicutes* and *Bacterioidetes* are commonly represented phyla in the intestinal tract of adults.

Beneficial Roles of the Human Microbiome

The human microbiome protects against infection by excluding pathogens, priming the adaptive immune system to react against pathogens, helping the immune system develop oral tolerance to harmless substances, aiding digestion, and producing certain vitamins and other substances important for health.

16.3 ■ Principles of Infectious Disease

Infectious diseases have characteristic **signs** and **symptoms.** A **secondary infection** can occur as the result of a **primary infection.**

Pathogenicity

A **primary pathogen** causes disease in otherwise healthy individuals; an **opportunistic pathogen** causes disease only when the body's innate or adaptive defenses are compromised. **Virulence** refers to the degree of pathogenicity of an organism.

Characteristics of Infectious Disease

Communicable or **contagious diseases** spread from one host to another; ease of spread partly reflects the **infectious dose.** Stages of infectious disease include the **incubation period, illness,** and **convalescence;** during the illness a person experiences signs and symptoms of the disease (figure 16.2). **Infections** can be described as **acute, chronic,** or **latent,** depending on the timing and duration of symptoms. Infections can be **localized** or **systemic.**

16.4 ■ Determining the Cause of an Infectious Disease

Koch's Postulates

Koch's postulates are used to determine the cause of infectious disease (figure 16.3).

Molecular Koch's Postulates

Molecular Koch's postulates are used to identify virulence factors that contribute to disease.

MECHANISMS OF PATHOGENESIS

16.5 ■ Establishing Infection

Adherence

Bacteria use **adhesins** to bind to host cells (figure 16.4).

Colonization

Rapid turnover of pili, antigenic variation, and **IgA proteases** allow bacteria to avoid the effects of secretory IgA. **Siderophores** enable microbes to scavenge iron.

Delivering Effector Proteins to Host Cells

Type III secretion systems of Gram-negative bacteria allow them to deliver proteins directly to host cells (figure 16.5).

16.6 ■ Invasion—Breaching the Anatomical Barriers

Penetrating the Skin

Some pathogens enter tissues via a bite or other skin-damaging injury.

Penetrating Mucous Membranes

Some pathogens induce mucosal epithelial cells to engulf bacterial cells; others exploit antigen-sampling processes (figures 16.6, 16.7).

16.7 ■ Avoiding the Host Defenses

Hiding Within a Host Cell

Some bacteria can evade the innate defenses, as well as certain adaptive defenses, by entering host cells.

Avoiding Destruction by Phagocytes (figure 16.9)

Mechanisms to prevent encounters with phagocytes include C5a peptidase and membrane-damaging toxins. Mechanisms to avoid recognition and attachment by phagocytes include **capsules, M protein,** and **Fc receptors** (figure 16.10). Mechanisms to survive within the phagocyte include escape from the phagosome, preventing phagosome-lysosome fusion, and surviving within the phagolysosome.

Avoiding Killing by Complement System Proteins

Some **serum resistant** bacteria postpone the formation of membrane attack complexes by interfering with activation of the complement system via the alternative pathway.

Avoiding Recognition by Antibodies

Mechanisms to avoid antibodies include **IgA protease, antigenic variation,** and mimicking "self."

16.8 ■ Damage to the Host

Exotoxins (table 16.1)

Exotoxins are proteins that have very specific damaging effects; they may act locally or cause dramatic systemic effects. Many can be grouped into categories such as **neurotoxins, enterotoxins,** or **cytotoxins.** The toxic activity of **A-B toxins** is mediated by the A subunit; the binding to specific cells is mediated by the B subunit (figure 16.11). **Membrane-damaging toxins** disrupt cell membranes either by forming pores or by removing the polar head group on phospholipids in the membrane. **Superantigens** override the specificity of the T-cell response, causing systemic effects due to the massive release of cytokines (figure 16.12).

Endotoxin and Other Bacterial Cell Wall Components

The symptoms associated with **endotoxin** are due to a vigorous host response. Lipid A of lipopolysaccharide is responsible for its toxic properties. Peptidoglycan and certain other components induce various cells to produce pro-inflammatory cytokines.

Damaging Effects of the Immune Response

The release of enzymes and toxic products from phagocytic cells can damage tissues. **Immune complexes** can cause kidney and joint damage; cross-reactive antibodies can result in an autoimmune response.

16.9 ■ Mechanisms of Viral Pathogenesis

Binding to Host Cells and Invasion

Viruses attach to specific receptors on the target cell.

Avoiding Immune Responses

Some viruses can avoid the effects of interferon; some can regulate apoptosis of the host cell (figure 16.13). To avoid antibodies, some viruses transfer directly from cell to cell; the surface antigens of some viruses change quickly, outpacing the production of antibodies.

Damage to the Host

Viruses often damage host cells directly. The immune response to many viruses causes **flu-like symptoms;** some viruses elicit a potentially deadly **cytokine storm.**

16.10 ■ Mechanisms of Eukaryotic Pathogenesis

Fungi

Saprophytes are generally opportunists; dermatophytes cause superficial infections of skin, hair, and nails. The most serious fungal infections are caused by dimorphic fungi.

Protozoa and Helminths

Eukaryotic parasites attach to host cells via specific receptors. They use a variety of mechanisms to avoid antibodies; the extent and type of damage they cause vary tremendously.

Review Questions

Short Answer

1. Describe three types of symbiotic relationships.
2. Describe two situations that can lead to changes in the composition of a person's microbiome.
3. How are acute, chronic, and latent infections different from one another?
4. Why are Koch's postulates not sufficient to establish the cause of all infectious diseases?
5. Describe the four general mechanisms by which microorganisms cause disease.
6. Describe two mechanisms that bacteria use to invade via mucous membranes.
7. Explain how a capsule can allow an organism to be serum resistant and avoid phagocytosis.
8. Give an example of a neurotoxin, an enterotoxin, and a cytotoxin.
9. Describe two mechanisms a virus might use to prevent the induction of apoptosis in an infected cell.
10. How do *Schistosoma* species avoid antibodies?

Multiple Choice

1. Opportunistic pathogens are *least* likely to affect which of the following groups?
   a) AIDS patients
   b) Cancer patients
   c) College students
   d) Drug addicts
   e) Transplant recipients
2. Capsules and M protein are thought to interfere with which of the following?
   a) Opsonization by complement proteins
   b) Opsonization by antibodies
   c) Recognition by T cells
   d) Recognition by B cells
   e) Phagosome-lysosome fusion
3. The C5a peptidase enzyme of *Streptococcus pyogenes* breaks down C5a, resulting in
   a) lysis of the *Streptococcus* cells.
   b) lack of opsonization of *Streptococcus* cells.
   c) killing of phagocytes.
   d) decreased accumulation of phagocytes.
   e) inhibition of membrane attack complexes.
4. All of the following are known mechanisms of avoiding the effects of antibodies *except*
   a) antigenic variation.
   b) mimicking "self."
   c) synthesis of an Fc receptor.
   d) synthesis of IgG protease.
   e) remaining intracellular.
5. Which of the following statements about diphtheria toxin is *false*?
   a) It is an example of an endotoxin.
   b) It is produced by a species of *Corynebacterium.*
   c) It inhibits protein synthesis.
   d) It can cause local damage to the throat.
   e) It can cause systemic damage (that is, to organs such as the heart).
6. Which of the following statements about botulism is *true*?
   a) It is caused by *Bacillus botulinum,* an obligate aerobe.
   b) The toxin is heat-resistant, withstanding temperatures of 100°C.
   c) The organism that causes botulism can cause disease without encountering the immune response.
   d) Vaccinations are routinely given to prevent botulism.
   e) Symptoms of botulism include uncontrolled contraction of muscles.
7. Superantigens
   a) are exceptionally large antigen molecules.
   b) cause a very large antibody response.
   c) elicit a response from a large number of T cells.
   d) attach non-specifically to B-cell receptors.
   e) assist in a protective immune response.
8. Which of the following statements about endotoxin is *true*?
   a) It is an example of an A-B toxin.
   b) It is a component Gram-positive bacteria.
   c) It can be converted to a toxoid.
   d) It is heat-stable.
   e) It causes T cells to release cytokines.

9. The tissue damage caused by *Neisseria gonorrhoeae* is primarily due to
   a) cross-reactive antibodies.
   b) exotoxins.
   c) hydrolytic enzymes.
   d) the inflammatory response.
   e) all of these.
10. Which of the following statements about viruses is *false*?
   a) They may colonize the skin.
   b) They may enter host cells by endocytosis.
   c) They may enter host cells by fusion of the viral envelope with the cell membrane.
   d) They may induce apoptosis in infected host cells.
   e) They may suppress expression of MHC class I molecules on host cells.

Applications

1. A group of smokers suffering from *Staphylococcus aureus* infections are suing the cigarette companies. They claim that the disease was aggravated by smoking. The group is citing studies indicating that phagocytes are inhibited in their action by compounds in cigarette smoke. A statement prepared by their lawyers states that the *S. aureus* would not have caused such a severe disease if the phagocytes were functioning properly. During the proceedings, a microbiologist was called in as a professional witness for the court. What were her conclusions about the validity of the claim?
2. A microbiologist wrote a grant proposal to identify molecules that jam type III secretion systems. Her principal rationale for the research was that interfering with secretion would disrupt a pathogen's ability to cause disease without harming the normal microbiota. Is this a reasonable proposal? Why or why not?

Critical Thinking

1. A student argued that no distinction should be made between commensalism and parasitism. Even in commensalism, the microorganisms are gaining some benefit (such as nutrients) from the host, and this represents a loss to the host. In this sense, the host is being damaged. Does the student have a valid argument? Why or why not?
2. A microbiologist argued that there is no such thing as "normal" microbiota of the human body, since the microbiome is dynamic and is constantly changing, depending on diet and external environment. What would be an argument against this microbiologist's view?

17 Immunological Disorders

Asthmatic individual using an inhaler. ©*Science Photo Library/Getty Images*

KEY TERMS

Allergy A damaging immune response to a usually harmless environmental antigen (allergen).

Allograft Transplanted tissue or organ from a non-identical donor of the same species.

Autoimmune Disease Damaging reaction of the immune system against "self" antigens.

Delayed-Type Hypersensitivity Exaggerated immune response of antigen-specific T cells.

Desensitization Allergy treatment that induces IgG production by gradual exposure to small amounts of allergen; IgG competes with IgE.

Hypersensitivity Exaggerated immune response that damages tissue.

Immune Complex Combination of antibody and soluble antigen capable of triggering the classical pathway of complement system activation.

Immunodeficiency A condition in which the immune system provides an inadequate response, leaving an individual vulnerable to infection.

Systemic Anaphylaxis Immediate hypersensitivity reaction caused by IgE attached to circulating basophils.

Tolerance Decreased reactivity of the immune system to a specific antigen.

A Glimpse of History

Pasteur is widely quoted as saying, "Chance favors the prepared mind." This was the case when Charles Richet (1850–1935) discovered hypersensitivities, sometimes called allergies. Richet was a French physiologist who performed early experiments on toxins and the immune responses to them. He and his colleague, Paul Portier, were cruising on Prince Albert of Monaco's yacht when they hypothesized in 1901 that the Portuguese man-of-war jellyfish must produce a toxin that causes the swollen painful reactions to its stings. Prince Albert encouraged them to study the reactions, so they made an extract of the jellyfish tentacles and showed that it could indeed be toxic.

Upon returning to France, Richet and Portier continued their work by studying the effects of the toxin from a common sea anemone. When they tested it on dogs, some survived the first exposure to the potent toxin, but when given a small second dose at least 3 weeks later, the dogs died within minutes. Richet and Portier had worked with toxins earlier so they had prepared minds. They recognized that the dogs' reactions were probably caused by an immune response. Unlike protective immune responses they had seen before, however, this one was destructive. They called the dangerous response anaphylaxis, indicating it was the opposite of *phylaxis,* the Greek term for protection. Richet received a Nobel Prize in 1913 for his work on anaphylaxis.

As you learned in chapters 14 and 15, the host defenses involve complex interactions between many types of cells and chemical messengers. When those interactions are coordinated and appropriate, innate immunity and adaptive immunity provide excellent protection against microbial invaders. In some cases, however, the immune system fails to operate appropriately, resulting in disease. This chapter covers three general categories of immune system disorders: exaggerated immune responses that damage normal host tissue (hypersensitivities), misdirected immune responses that attack normal host tissues (autoimmune diseases), and lack of adequate immune responses that leave an individual vulnerable to infection (immunodeficiencies).

17.1 ■ Hypersensitivities

Learning Outcomes

1. Compare and contrast the immunological reactions involved in types I through IV hypersensitivities.
2. Describe two examples of each type of hypersensitivity.

The immune system does a superb job protecting the body from infection, but these same protective mechanisms can be harmful if uncontrolled, as Richet and Portier showed when they discovered anaphylaxis (see **A Glimpse of History**). Just as a building's sprinkler system that protects against fire can cause significant water damage if set off by accident, inappropriate immune responses can also become destructive. An exaggerated immune response that injures tissue is called **hypersensitivity** and can be categorized into one of four groups according to the mechanisms and timing of the response (**table 17.1**).

TABLE 17.1 Some Characteristics of the Major Types of Hypersensitivities

Characteristic	Type I Hypersensitivity: Immediate IgE-Mediated	Type II Hypersensitivity: Cytotoxic	Type III Hypersensitivity: Immune Complex–Mediated	Type IV Hypersensitivity: Delayed-Type Cell-Mediated
Effector	B cells	B cells	B cells	T cells
Type of antigen	Soluble	On cell surface	Soluble	Soluble or on cell surface
Type of antibody	IgE	IgG, IgM	IgG, IgM	None
Other immune cells involved	Basophils, mast cells	NK cells	Phagocytes	Dendritic cells
Mediators	Histamine, leukotrienes, prostaglandins	Complement, ADCC	Complement, neutrophil enzymes, cytokines	Cytokines
Time of reaction after challenge with antigen	Immediate, up to 30 minutes	Hours to days	Hours to days	Peaks at 48 to 72 hours
Skin reaction	Wheal and flare	None	Arthus reaction	Induration, inflammation
Examples	Hives, hay fever, asthma, anaphylactic shock	Transfusion reaction, hemolytic disease of the newborn	Serum sickness, disseminated intravascular coagulation	Contact dermatitis, tissue transplant rejection

Type I Hypersensitivities: Immediate IgE-Mediated

A **type I hypersensitivity** involves IgE and is commonly referred to as an allergic reaction, or **allergy.** This is a rapid exaggerated immune response to an **allergen**—an antigen that is usually a harmless environmental substance. Although any type of hypersensitivity can technically be called an allergy, we will follow the common practice of using the term to refer only to type I hypersensitivities. ⏮ IgE

Allergic reactions occur within 24 hours of exposure, often within minutes. A variety of common substances can trigger the reactions, including:

- Inhaled substances such as pollen, pet dander, or mold
- Injected substances such as insect venom or certain medications such as penicillin
- Ingested substances such as peanuts or seafood

Allergic reactions occur only in individuals who are sensitized by prior exposure to a specific allergen. **Sensitization** begins when the first contact with the allergen induces an antibody response (**figure 17.1** ①). Helper T cells activate naive B cells. These are initially programmed to produce IgM molecules, but as activated B cells multiply, they undergo class switching (see figure 15.15). ② In people who are prone to allergies, activated B cells often switch such that the plasma cells that descend from them as well as their memory cell derivatives make IgE. ③ The Fc portion of this antibody class binds to Fc receptors on either mast cells or basophils, positioning the IgE molecules so that their antigen-binding sites are available to interact with a specific allergen. ⏮ class switching, ⏮ basophil, ⏮ mast cell

Once a person is sensitized to a given allergen, the next exposure to that allergen will cause an allergic response. Mast cells and basophils contain granules holding powerful inflammatory mediators such as histamine, leukotrienes, and prostaglandins. ④ When adjacent captured IgE molecules on a mast cell or basophil bind to an antigen, the IgE molecules are cross-linked. ⑤ This triggers the cell to undergo **degranulation,** a process that releases its inflammatory mediators. ⑥ These mediators can have a variety of effects, including fluid accumulation due to leaky dilated blood vessels, contraction of smooth muscles, itching and pain due to effects on nerve endings, and increased mucus production. Note that the degranulation response of mast cells and basophils is beneficial and effective against parasitic worms that are too large to be engulfed by phagocytosis. Unfortunately, when IgE binds to allergens rather than to an invading parasite, the result is like a sprinkler system set off by accident. ⏮ inflammatory mediators

Food allergies affect about one out of every five children, but they often go away before a child starts school. The tendency to develop allergic reactions is inherited, but the specific allergen to which a person reacts is related to environmental exposures rather than to genetic traits. Note that a food allergy is not the same as a food intolerance. Lactose intolerance, for example, may produce discomfort, but it does not involve the immune system (see Focus on a Case 2.1).

Localized Allergic Reactions

Localized allergic reactions often occur in the skin or in the respiratory tract. Outcomes include hives, hay fever, and asthma, depending on the site of the reaction.

Hives (urticaria) is an allergic skin condition characterized by the formation of a wheal and flare. The wheal is an

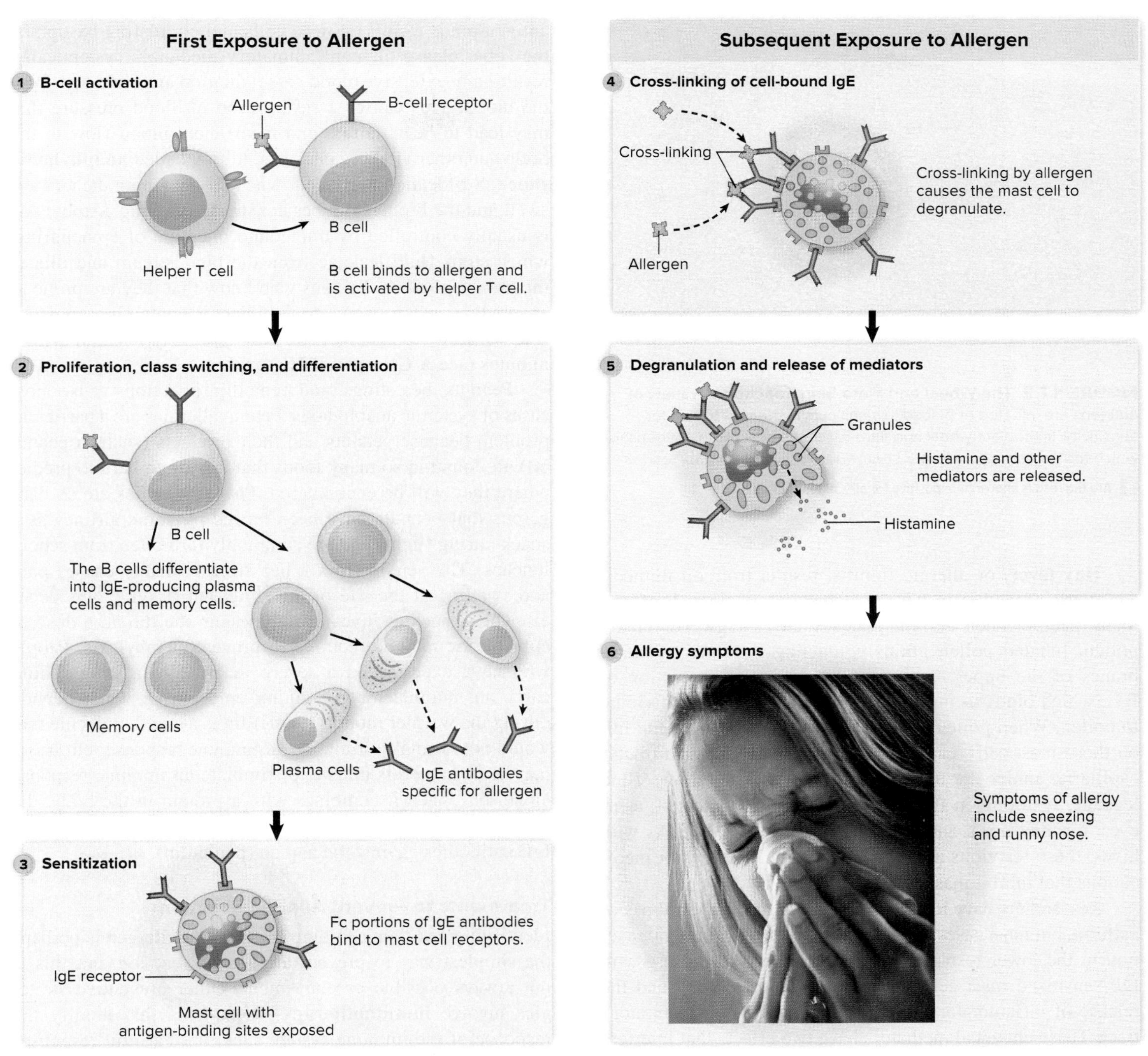

FIGURE 17.1 Type I Hypersensitivity: Immediate IgE-Mediated First exposure to an allergen induces an IgE antibody response, leading to sensitization. With subsequent exposures to the allergen, cross-linking of IgE molecules on mast cells results in release of histamine and other substances that can cause itching, swelling, and pain, and conditions such as asthma, hay fever, hives, and anaphylactic shock. ©Stockbyte/Getty Images

? Why are antihistamines sometimes used to combat allergy symptoms?

itchy swelling generally resembling a mosquito bite, surrounded by redness, the flare. The wheal and flare reaction is seen also in positive skin tests for allergens (**figure 17.2**). Hives may occur when a person with a food allergy ingests that food. Allergens are absorbed from the intestinal tract into the bloodstream. When they reach the skin, the allergens bind to IgE carried by mast cells. The mast cells degranulate, releasing inflammatory mediators, including histamine, which causes swelling as fluid leaks from dilated capillaries. Extensive swelling along with smooth muscle contraction in the throat and larynx may obstruct breathing. Antihistamine medications, such as diphenhydramine (Benadryl), are often used to relieve symptoms of allergy by blocking binding sites for histamine.

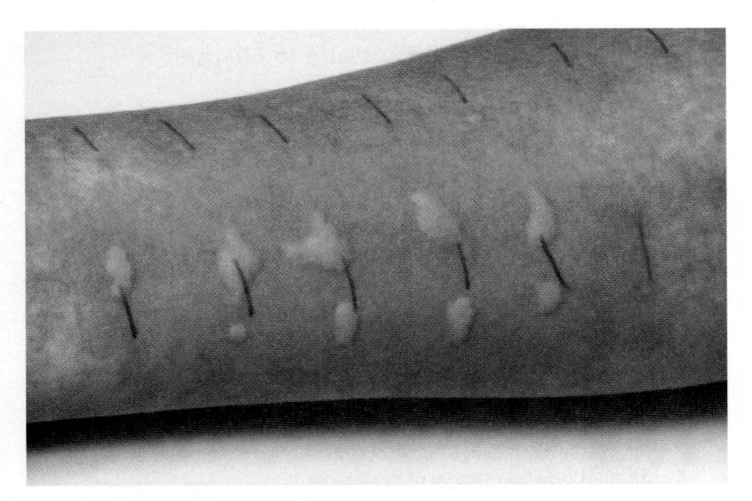

FIGURE 17.2 The Wheal and Flare Skin Reaction A variety of antigens are injected or placed in small cuts in the skin to test for sensitivity. Immediate wheal and flare reactions occur with antigens to which the person is sensitive. ©Southern Illinois University/Science Source

Are the results shown here positive for all antigens tested?

Hay fever, or allergic rhinitis, results from an immediate allergic reaction in the upper respiratory tract. It affects many people when certain plants such as ragweed release pollen. Inhaled pollen grains contacting the mucous membranes of the upper respiratory tract trigger production of IgE, which binds to mast cells and sensitizes that individual to pollen. When pollen is contacted again, it binds to the IgE on those mast cells, causing release of histamine that dilates capillaries under the mucous membranes. This allows fluid leakage, resulting in the familiar signs of hay fever—teary eyes, a runny nose, and sneezing (see figure 17.1). As with hives, these reactions are blocked by antihistamines or medications that inhibit mast cell degranulation.

Researchers have identified a number of different types of **asthma,** but most cases result from an immediate allergic reaction in the lower respiratory tract. An allergen reacting with IgE-sensitized mast cells causes their degranulation and the release of inflammatory mediators into the lower respiratory tract. These chemical mediators have two effects that interfere with breathing: immediate spasms of smooth muscle tissue lining the bronchial tubes and increased mucus production. At the onset of an asthma attack, individuals may use an inhaler to self-deliver medication that will relax the constricted muscles and help open the airways. Other inhalers are routinely used for the prevention of asthma attacks. These contain corticosteroids that reduce inflammation, and possibly a long-acting bronchodilator as well (see the chapter-opening photo).

Systemic Anaphylaxis

Systemic anaphylaxis is a rare but serious form of IgE-mediated allergy that can occur when an antigen enters the bloodstream and spreads throughout the body. When the antigen binds to IgE on mast cells and circulating basophils, the cells release their inflammatory mediators systemically, resulting in extensive blood vessel dilation and fluid loss from the blood. This causes a severe drop in blood pressure that may lead to heart failure and insufficient blood flow to the brain and other vital organs—a condition called **anaphylactic shock.** Suffocation may occur when tissues lining the airways swell and the bronchial tubes constrict. Systemic anaphylaxis is usually controlled by immediate injection of epinephrine, which stops fluid leakage from the bloodstream and dilates the bronchial tubes. Persons who know that they are prone to anaphylaxis often carry an autoinjector containing epinephrine. Otherwise, anaphylactic reactions may be fatal within minutes (see **A Glimpse of History**).

Peanuts, bee stings, and penicillin injections cause most cases of systemic anaphylaxis. Peanut allergies are a particular problem because peanuts and their products (such as peanut oil) are found in so many foods that it is often hard to predict where they will be encountered. Peanut allergies are so dangerous that peanuts have been barred by some airlines as a snack during flights and are commonly forbidden from school lunches. The venom from a bee sting normally causes pain and swelling at the site of the sting, but if symptoms occur elsewhere, such as hives or tightness in the throat, a dose of epinephrine may be required to prevent anaphylaxis. People who have experienced a severe reaction to a sting should carry an autoinjector containing epinephrine when outside during the warmer months. Penicillin is a small molecule that would not normally stimulate an immune response, but it is a hapten. A hapten is unable to stimulate an immune response alone, but when it combines with a protein in the body, the hapten-protein complex can trigger widespread production of IgE antibodies against the hapten (penicillin). ⏮ hapten

Treatments to Prevent Allergic Reactions

Identifying and then avoiding the relevant allergen is perhaps the simplest way to prevent an allergic reaction, but this is not always possible or convenient. Other preventive strategies involve **immunotherapy**—treatments that modify the response of the immune system such that harmful responses are decreased or helpful responses are increased. For example, a common treatment for allergy is **desensitization** (hyposensitization), a procedure that causes the immune system to produce IgG against the allergen. The IgG antibodies probably protect the patient by binding to the offending antigen, thus coating it and facilitating its removal before it can attach to bound IgE on mast cells or basophils (**figure 17.3**). Desensitization therapy involves "allergy shots," or injecting the person with the allergen in extremely dilute, but gradually increasing, concentrations over a period of several months. The antigen must be diluted enough to avoid an anaphylactic reaction. As the concentration in the injections increases during the course of treatment, the person becomes less and

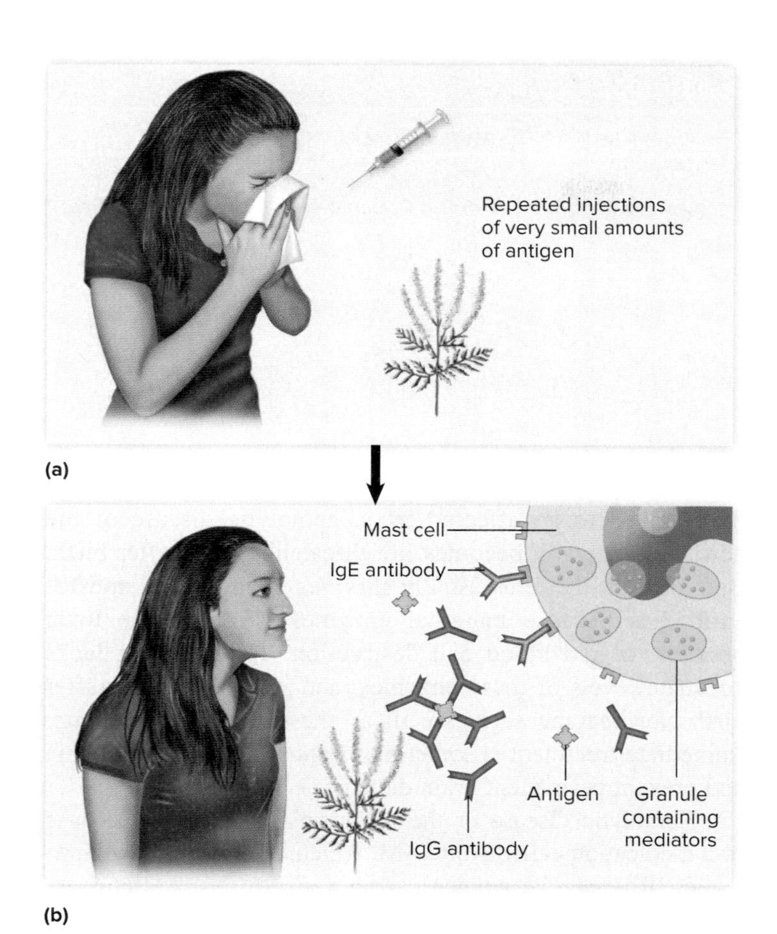

FIGURE 17.3 Immunotherapy for IgE Allergies (a) Repeated injections of very small amounts of antigen are given over several months. **(b)** This regimen leads to the formation of specific IgG antibodies. The IgG reacts with antigen before it can bind to IgE, and therefore it blocks the IgE reaction.

? Why are only very small amounts of antigen injected?

less sensitive and may even lose the hypersensitivity entirely. Activated regulatory T cells may also play a role through release of cytokines that suppress the IgE response. Unfortunately, desensitization does not always work, and it may require continued therapeutic exposure over several years to maintain effectiveness. Also, desensitization does not work well for food allergies. ◀◀ IgG, ◀◀ regulatory T cells

The type of immunotherapy used in fighting severe asthma that does not respond well to inhaled corticosteroids and long-term bronchodilators involves recombinant humanized monoclonal antibodies (rhuMabs). These are animal-derived antibodies that have been modified so that the human immune system is more likely to tolerate them. Omalizumab (Xolair), for example, contains a genetically engineered IgG molecule that binds specifically to the Fc portion of IgE molecules, thereby blocking the site that would otherwise attach to mast cells and basophils. Other rhuMabs that target specific interleukins associated with inflammatory eosinophil activity are being developed. Often used in addition to traditional treatments, rhuMabs are reserved for the most severe asthma cases because they are expensive and, on rare occasions, they may cause anaphylactic shock. ▶▶ rhuMab

MicroByte

Sublingual immunotherapy (SLIT) may in some cases replace allergy shots with a tablet containing allergen placed under the tongue.

Type II Hypersensitivities: Cytotoxic

In type II hypersensitivity reactions, antibodies bind to molecules on the surface of a host cell and trigger its destruction by the complement system or by antibody-dependent cellular cytotoxicity (ADCC). Because the responses destroy cells, type II reactions are called cytotoxic hypersensitivities. Recall that antibodies bound to an antigen trigger the classical pathway of complement activation. One outcome of activation is cell lysis due to the assembly of membrane-attack complexes (MACs) in cell membranes. In ADCC, the Fc regions of antibodies that are bound to a cell mark the cell for destruction. Natural killer (NK) cells bind to the exposed Fc regions and then deliver chemicals that destroy the marked cell. Transfusion reactions, hemolytic disease of the newborn, and some autoimmune diseases involve type II reactions. ◀◀ complement system, ◀◀ antibody-dependent cellular cytotoxicity (ADCC)

Transfusion Reactions

Erythrocytes (red blood cells) have various surface antigens that can differ from one person to another. Human ABO blood types, for example, are determined by the A or B antigens present on red blood cells. People who have type A antigens are blood type A. At the same time, human plasma contains antibodies to the A or B antigens that are not present on the red blood cells. Individuals with blood type A carry anti-B antibodies. Similarly, those who have type B antigens are blood type B and have anti-A antibodies. People with blood type O lack A and B antigens, but have antibodies to both. People with blood type AB have both antigens, and have antibodies to neither.

Anti-A and anti-B antibodies are called natural antibodies because they occur without obvious pre-exposure. They probably arise due to multiple exposures to bacteria, dust, and food that have components similar to blood group antigens. Anti-A and anti-B antibodies are not present at birth but generally appear within 6 months. They are mostly of the class IgM, and therefore cannot cross the placenta. ◀◀ IgM

When a person receives a transfusion of red blood cells with different antigens from his or her own, a **transfusion reaction** occurs when natural antibodies bind to those cells, resulting in agglutination (clumping). The transfused cells are rapidly destroyed by membrane attack complexes or NK cells. Release of hemoglobin into the bloodstream and the presence of damaged membranes can block blood vessels and initiate

TABLE 17.2 Antigens and Antibodies in Human ABO Blood Groups

			Incidence of Blood Type in United States			
Blood Type	Antigen on Erythrocyte	Antibody in Plasma	Among Asians	Among Blacks	Among Caucasians	Among Hispanics
A	A	Anti-B	28%	26%	40%	31%
B	B	Anti-A	25%	19%	11%	10%
AB	A and B	Neither anti-A nor anti-B	7%	4%	4%	2%
O	Neither	Both anti-A and anti-B	40%	51%	45%	57%

clotting reactions. This may damage the kidneys, produce fever, or cause respiratory and digestive problems. Because of the potentially life-threatening consequences of giving the wrong blood to a patient, donor and recipient blood types are carefully cross-matched in clinical settings. The cells and serum of the donor are mixed with the cells and serum of the recipient. Any clumping indicates an incompatibility between the two blood types, either in the ABO group (**table 17.2**) or in another of the many antigens on the surface of the red blood cell. That blood would not be used in the transfusion.

Hemolytic Disease of the Newborn

Another important antigen on red blood cells is the Rh (rhesus) antigen, also called the D antigen. People with the Rh antigen are Rh-positive; those without it are Rh-negative. In contrast to the ABO system, Rh-negative individuals do not have natural antibodies to the Rh antigen. These antibodies develop only when an Rh-negative individual is exposed to Rh-positive cells. Just as donor and recipient bloods are matched for the ABO system, they are also matched for Rh system antigens. If Rh-positive blood is given to an Rh-negative recipient, then that person will develop antibodies to the Rh-positive red blood cells. If the person is exposed to Rh-positive blood again, the immune system will destroy the transfused cells. More commonly, an Rh-negative person is exposed to Rh-positive cells with pregnancy.

An Rh-negative woman who carries an Rh-positive baby is likely to develop antibodies to the Rh antigen when she is exposed to the baby's blood. This happens when the placenta ruptures during childbirth. She may also be exposed during pregnancy or after induced or spontaneous abortion. Antibodies produced by the woman will not affect her red blood cells because her cells lack the Rh antigen. They will not affect her first baby because IgM, which is produced upon first exposure, cannot cross the placenta. If the woman carries a second Rh-positive baby, however, her anti-Rh IgG antibodies can cross the placenta and damage the red blood cells of her developing fetus, causing **hemolytic disease of the newborn** (**figure 17.4**). This is also called erythroblastosis fetalis because the affected fetus responds by producing immature red blood cells called erythroblasts.

Although transfusions of Rh-negative blood are sometimes given to an affected fetus, hemolytic disease of the newborn generally becomes life-threatening only after birth. Before birth, the fetus usually survives the attack by anti-Rh antibodies because maternal enzymes eliminate the toxic products of red blood cell destruction. The newborn lacks adequate levels of these enzymes, and within 36 hours after birth can become seriously ill as these toxins accumulate. Immediate treatment is sometimes required to correct anemia and prevent permanent brain damage.

Hemolytic disease of the newborn is rare today because of a medication called RhoGAM, which contains anti-Rh antibodies. Rh-negative women carrying an Rh-positive fetus are injected with RhoGAM during pregnancy and again shortly after delivery (see figure 17.4). The anti-Rh antibodies bind to any Rh-positive erythrocytes that may have entered the mother's circulation from the baby. This prevents these red blood cells from stimulating a primary immune response in the mother with the development of memory cells. Unfortunately, injecting anti-Rh antibody is not effective if the Rh-negative mother has already formed memory cells. ⏮ primary immune response

Type III Hypersensitivities: Immune Complex–Mediated

Type III hypersensitivities involve the formation of **immune complexes** consisting of antibodies bound to soluble antigen. The exposed Fc regions of the antibodies bind to Fc receptors on phagocytes, which engulf and destroy the complexes, thus removing them from the bloodstream. When there is more antigen than antibody, however, the complexes may not be quickly destroyed but remain in circulation or at their sites of formation in tissue. Circulating immune complexes can lodge in blood vessel walls or in areas such as kidneys where the capillaries are small and densely packed. They also can activate the complement system, causing inflammation and damage to the tissue in which they are lodged (**figure 17.5**). ⏮ immune complexes, ⏮ Fc region

Immune complex disease may arise when the two antigen-binding sites of antibody molecules bind to separate antigen molecules, causing the molecules to clump. Immune

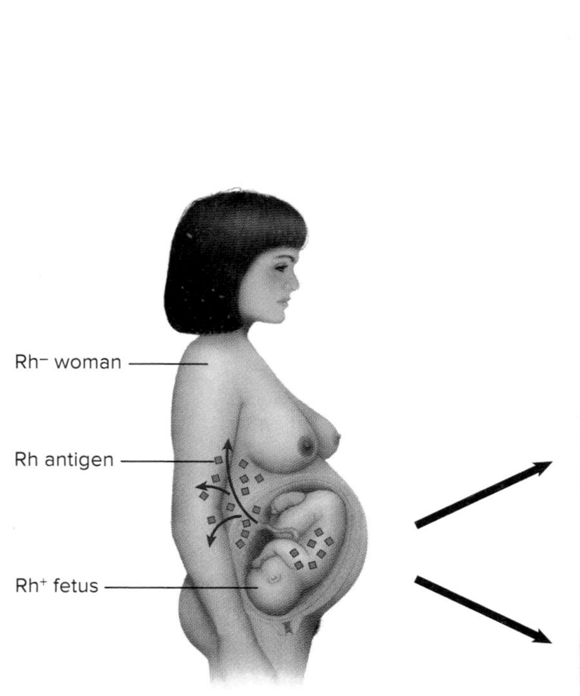

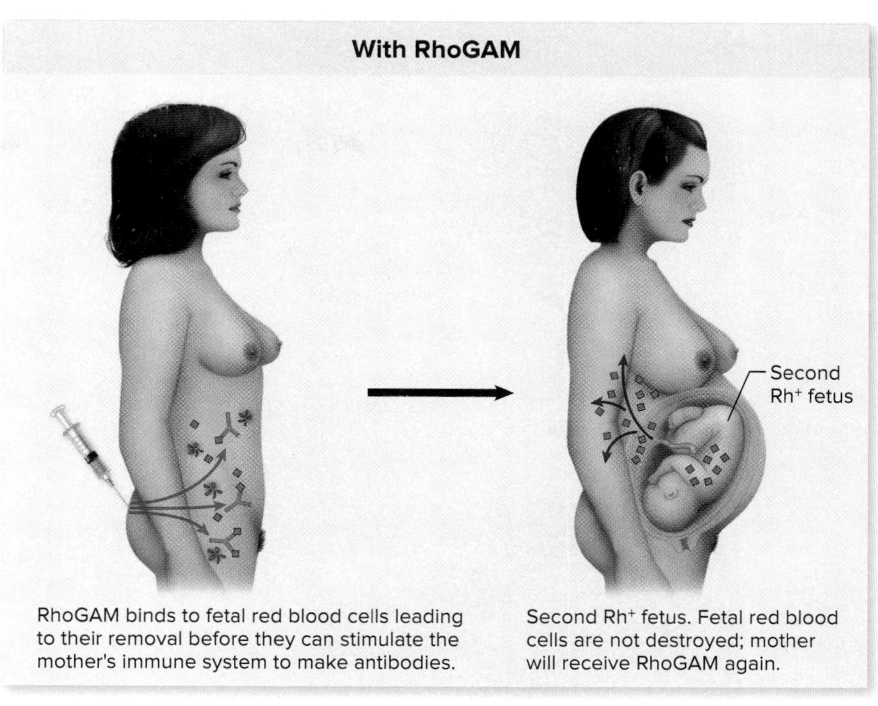

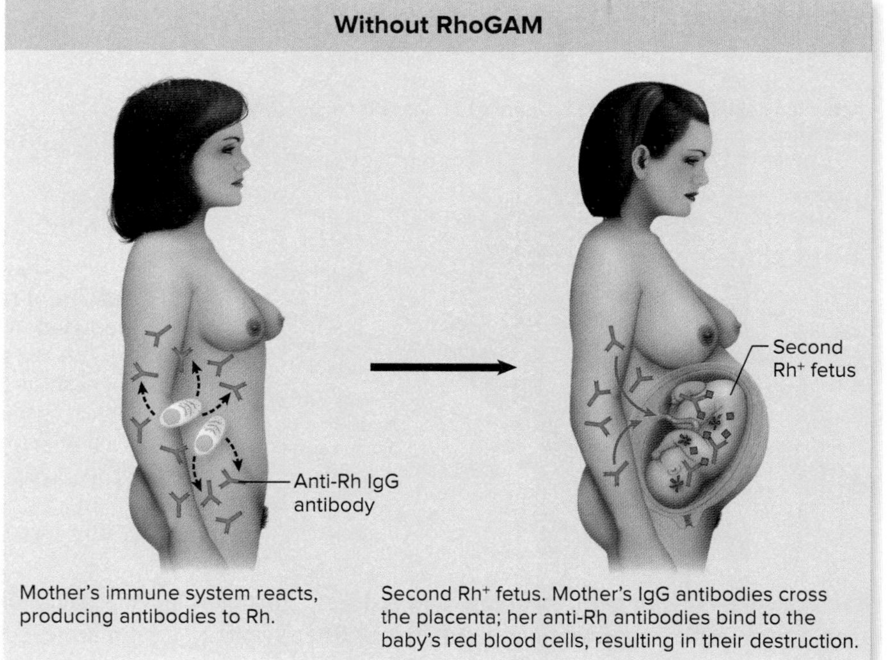

FIGURE 17.4 Hemolytic Disease of the Newborn RhoGAM contains antibodies to Rh antigens and removes them from the mother's blood before her immune system can respond to them.

? **Would a mother need RhoGAM again with a third Rh$^+$ fetus?**

complexes are responsible for the rashes, joint pains, and other symptoms seen in a number of diseases, such as systemic lupus erythematosus, infective endocarditis, early rubella infection, and malaria. When deposited in the kidneys, immune complexes cause glomerulonephritis. This dangerous inflammation is associated with certain infections and autoimmune diseases. Immune complexes can also trigger a devastating condition, disseminated intravascular coagulation (DIC), in which clots form in small blood vessels, leading to failure of vital organs. ⏭ glomerulonephritis, ⏭ disseminated intravascular coagulation

The **Arthus reaction** is a localized immune complex reaction. If antigen is injected into a previously immunized person who already has high levels of circulating specific antibody, immune complexes can form at the site of injection where antigen levels are high. The result is a reaction

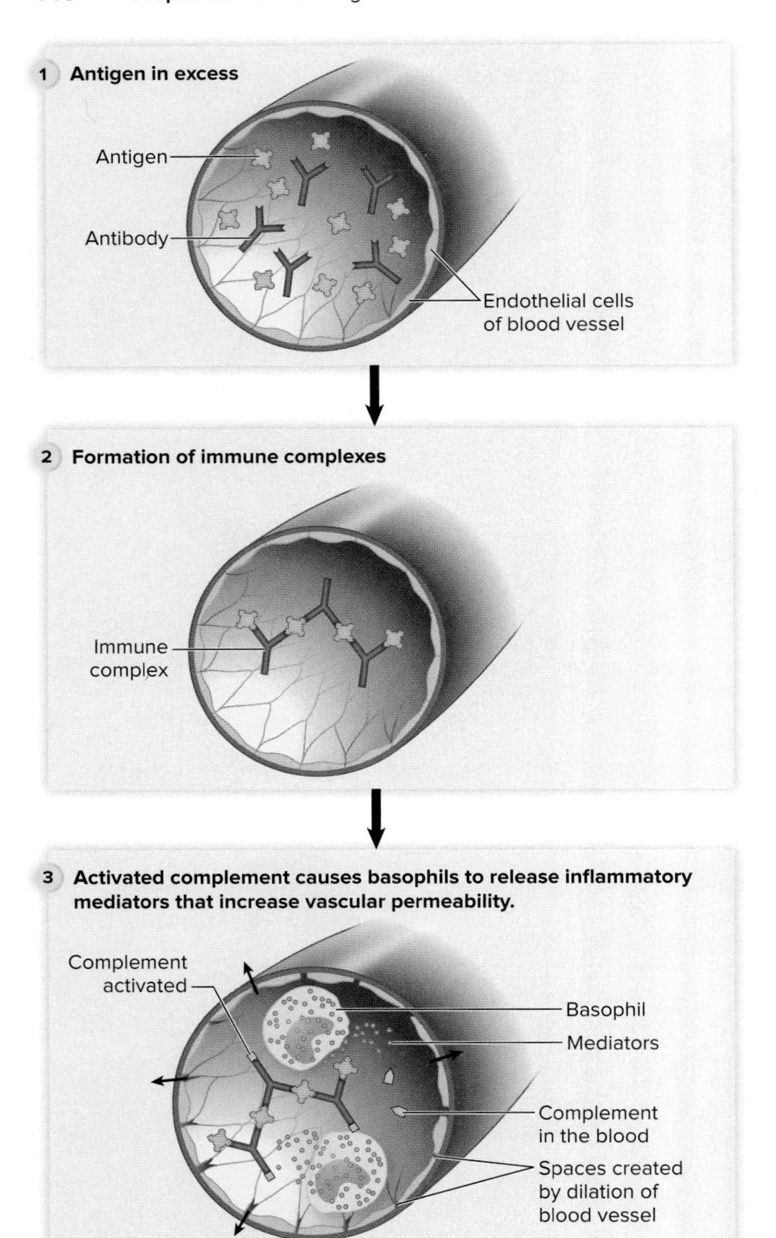

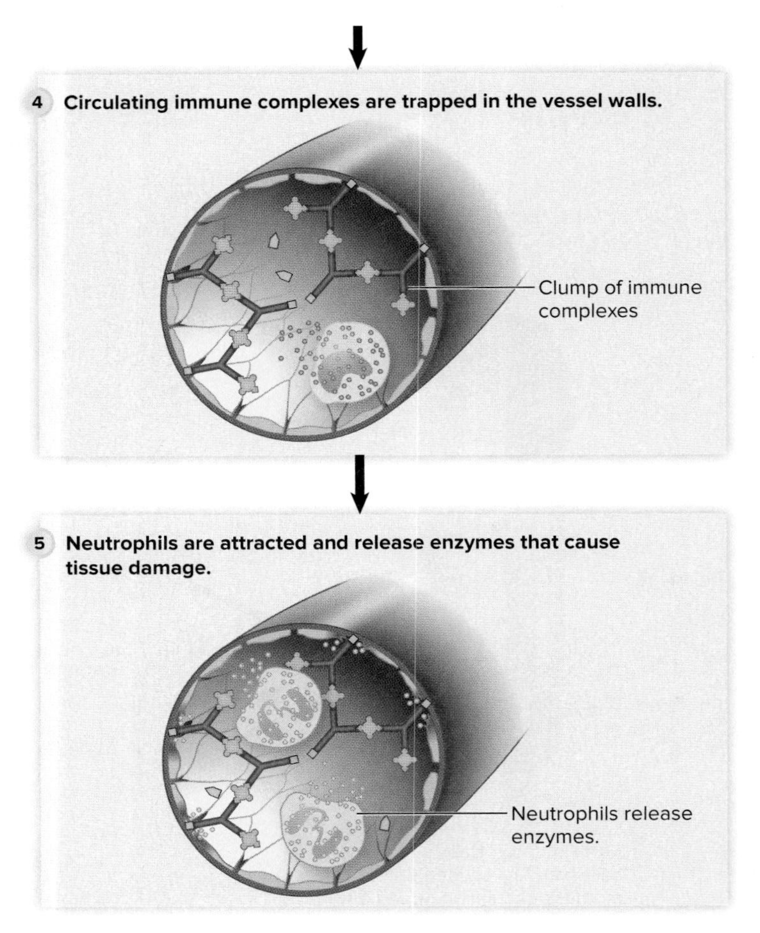

FIGURE 17.5 Type III Hypersensitivity: Immune Complex–Mediated When antigen is in slight excess **(1),** immune complexes form **(2)** and activate complement, resulting in increased vascular permeability **(3).** Immune complexes are trapped in the blood vessels **(4)** and complement initiates inflammation and attracts neutrophils that release enzymes, damaging the tissue **(5).**

? **Why do immune complexes form only when an antigen is soluble?**

characterized by severe pain, swelling, and redness. This can happen, for example, when tetanus-diphtheria booster vaccine is given to a person too frequently. The immune complexes form in the skin tissues outside the blood vessels, where they activate complement, resulting in attraction of neutrophils. The release of neutrophil enzymes contributes to a local inflammatory response that peaks in 6 to 12 hours.

Serum sickness is a systemic immune complex disease. It can occur following injection of antibodies from a non-human animal into a person to provide passive immunity that is immediate but short-term. For example, horse serum that contains antibodies against botulinum toxin protects against the disease botulism. Unfortunately, in addition to containing the protective antibodies, non-human animal serum contains a variety of antigens to which the recipient may mount an immune response. When antibodies are produced against antigens in the serum, immune complexes form. After 7 to 10 days, enough immune complexes form to cause signs such as fever, inflammation of blood vessels, arthritis, and kidney damage. People with serum sickness usually recover as the antigens of the animal serum are cleared from the body. Because of the potential to cause serum sickness, non-human animal serum is replaced with hyperimmune human globulin whenever possible. The serum sickness form of hypersensitivity is rare now, but can occur following treatment of heart attack patients with the bacterial enzyme streptokinase to dissolve clots, or after injection of horse-derived anti-venom to treat snakebites. ⏭ passive immunity, ⏭ hyperimmune globulin

Type IV Hypersensitivities: Delayed-Type Cell-Mediated

Delayed-type hypersensitivity reactions (type IV hypersensitivities) are due to antigen-specific T-cell responses and can occur almost anywhere in the body. Whereas immediate hypersensitivities involve antibodies and occur within minutes, these cell-mediated reactions peak 2 to 3 days following antigen exposure. ⏮ cell-mediated immunity

Recall from chapter 15 that cell-mediated immunity plays a central role in combating intracellular microbial infections. Effector cytotoxic T cells are particularly important because they destroy the infected host cells. Although this prevents the infection from spreading, it also damages tissues. With chronic infections, delayed-type hypersensitivity causes extensive host cell destruction and progressive loss of tissue function. These reactions can also be seen in some autoimmune diseases and in rejection of transplanted tissue.

Contact Hypersensitivities

Contact hypersensitivity, also called **allergic contact dermatitis,** is due to the response of effector T cells to small molecules that penetrate intact skin and act as sensitizing agents. The mechanisms of tissue damage are similar to those of other types of delayed-type hypersensitivities, but damage to skin cells causes an irritating rash and sometimes blisters.

Familiar substances like the nickel of metal jewelry, the chromium salts in certain leather products, components of some cosmetics, or oily substances on the surface of poison ivy and poison oak leaves are common sensitizing agents in contact dermatitis (**figure 17.6**). All of these materials shed small chemicals that act as haptens and combine chemically

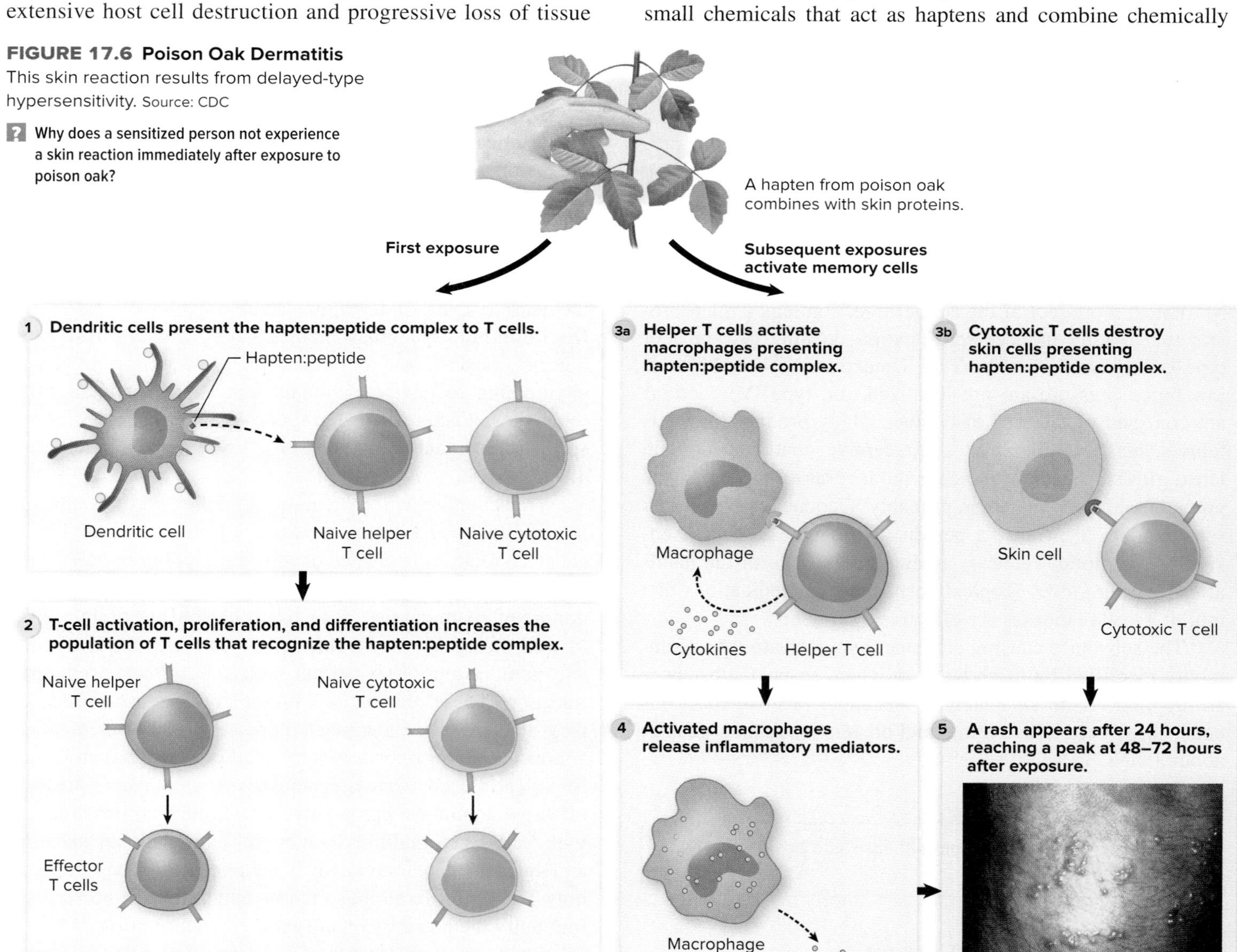

FIGURE 17.6 Poison Oak Dermatitis
This skin reaction results from delayed-type hypersensitivity. Source: CDC

? Why does a sensitized person not experience a skin reaction immediately after exposure to poison oak?

FOCUS YOUR PERSPECTIVE 17.1

The Fetus as an Allograft

One allograft that does not face the usual threat of rejection is a very special one—the mammalian fetus. Half of the fetal antigens are of paternal origin and are likely to differ from those of the mother. In spite of immunological differences, the fetus develops in the uterus and is not rejected. The mechanisms for survival of the fetal allograft have long been the subject of research, but they are not yet fully understood.

Do paternal antigens from the fetus reach the mother's immune system where they might cause a response? Clearly, they do. Mothers make antibodies to paternal antigens, such as the Rh antigen on red blood cells. Furthermore, small numbers of fetal cells are found in the maternal circulation during pregnancy. The placenta, however, prevents most fetal cells from entering the mother and most maternal T cells from reaching the fetus. The outer layer of the placenta acts as a buffer zone between the fetus and the mother. Cells of this layer do not express MHC class I or II molecules and are not subject to T-cell attack. They also avoid destruction by natural killer cells. The uterus of a pregnant female is an immunologically privileged site, meaning that tolerance to antigens is favored over destruction. Allografts are usually safe from rejection in an immunologically privileged (isolated) site. The mechanisms of immunological privilege in the maternal uterus are complex, involving immunosuppressive cytokines and other chemicals, a relative lack of responsive dendritic cells in crucial regions, and an increase in the number of T_{reg} cells (the subset of T cells that promote tolerance). It is well recognized that maternal immune responses are somewhat suppressed during pregnancy, and scientists are getting closer to understanding how it happens.

with proteins in the skin. Dendritic cells take up and process these hapten:protein complexes and present the resulting hapten:peptides to T cells in nearby lymph nodes. Activated T cells form memory cells, sensitizing the individual to the substance. Once the person has been sensitized, a second exposure causes a damaging cell-mediated response. ◀◀ hapten

Latex, a product of the rubber tree, contains a plant protein that readily induces sensitization, leading to type I or type IV hypersensitivities. Type I reactions such as hives or low blood pressure are not localized, but type IV reactions are confined to the area of contact. Many products, such as fabrics, elastics, toys, and contraceptive condoms, contain latex; gloves worn by healthcare and laboratory workers, food preparers, and many others probably account for most latex sensitization. Typically, a person with a type IV hypersensitivity will notice redness, itching, and a rash on the hands after wearing gloves. Topical cortisone-like medications that inhibit T-cell responses are effective treatments.

The substance causing a contact hypersensitivity is commonly identified by patch tests, in which suspect substances are applied to the skin under a bandage. Positive reactions consisting of redness, itching, and blisters reach their peak in about 3 days.

MicroByte

Approximately half of the people with latex allergies are allergic to bananas as well.

Rejection of Transplanted Tissues

A special case of delayed-type cell-mediated hypersensitivity occurs with the rejection of transplanted organs or tissues. Most medical transplants involve **allografts** in which the tissues of the donor and those of the recipient are both human, but not genetically identical. Antigenic differences, particularly in the major histocompatibility complex (MHC) molecules, lead to rejection of the graft (see Focus Your Perspective 15.1). Larger differences result in more vigorous rejection. Autografts (tissue transplanted from elsewhere on the recipient's body) and isografts (grafts donated by an identical twin) cause the least danger of rejection. Xenografts involving tissues from non-human animals induce strong immune responses, but they may become more common as scientists use genetic engineering to develop animals that express human MHC molecules. **Focus Your Perspective 17.1** discusses a very special transplantation situation: the fetus as an allograft. ◀◀ MHC molecules

Transplant rejection involves a complex set of events, primarily involving effector cytotoxic T cells and natural killer cells. To reduce the chance of rejection, typing of MHC molecules in tissues and ABO antigens in blood confirms that donor and recipient tissues match as closely as possible. Even in well-matched tissue transplants, however, recipients must use immunosuppressive drugs indefinitely to prevent graft rejection. These medications minimize tissue rejections, but they also make the recipient more susceptible to infections and cancer. Cyclosporin A (produced by a fungus) and tacrolimus (produced from a species of *Streptomyces*) are both effective immunosuppressants. These medications interfere with cellular signaling, thereby inhibiting clonal selection and expansion of activated T lymphocytes. They suppress only T-cell proliferation, so the patient still has some level of immunity mediated by other types of immune cells.

Bone marrow transplants have saved the lives of many thousands of people. Bone marrow contains hematopoietic stem cells that can replace the recipient's diseased cells. Generally, the patient's own bone marrow and immune system cells are intentionally destroyed by radiation and chemotherapy before transplantation of the new bone marrow. Donor bone marrow is infused through the subclavian vein under the left collarbone (clavicle). The hematopoietic stem cells circulate until they reach the recipient's bone marrow, where they leave the bloodstream and begin multiplying. Close monitoring is required for months after a transplant because the recipient is prone to infection and bleeding while the grafted marrow restores the supply of red and white blood cells and platelets. Recovery may be complicated by a graft-versus-host reaction in which the graft-derived immune system recognizes the patient's tissues as an invader and attacks them. ⏮ **hematopoietic stem cells**

MicroAssessment 17.1

Hypersensitivity reactions are exaggerated immune reactions that cause tissue damage. Type I hypersensitivity reactions, commonly called allergies, occur immediately after exposure to antigen. Type II hypersensitivity reactions cause lysis of cells. Type III hypersensitivities arise when immune complexes persist in the tissues and activate the complement system, triggering an inflammatory response. Type IV delayed-type hypersensitivity depends on the action of sensitized T cells. Examples are allergic contact dermatitis and transplant rejection.

1. Describe the process of desensitization that reduces the IgE response to an allergen.
2. Describe the events that result in a rash after exposure to poison ivy.
3. Why is the recipient of a kidney transplant unlikely to suffer from a graft-versus-host reaction?

17.2 ■ Autoimmune Disease

Learning Outcomes

3. Define *autoimmunity,* and explain some possible causes and treatments.
4. Give five examples of autoimmune diseases and the mechanism of tissue injury in each.

Autoimmune disease results when the immune system mistakenly attacks the body's own normal tissues. In the process of self-tolerance, the immune system routinely eliminates or silences any lymphocytes that recognize **autoantigens,** which are the "self" antigens found in cells and on body tissues. In autoimmune disease, self-tolerance is faulty. Autoantigens may stimulate a humoral immune response, resulting in the production of **autoantibodies**—antibodies against self-antigens. Or they may activate T cells, resulting in cell damage and inflammation in the body's tissues. Or they may do both.

Self-tolerance occurs by at least two mechanisms: (1) negative selection of developing lymphocytes in the thymus results in apoptosis (programmed death) of cells that recognize autoantigens, and (2) naive lymphocytes that recognize antigen but do not receive the necessary second signals become unresponsive or, in the case of helper T cells, may develop into regulatory T (T_{reg}) cells. T_{reg} cells inhibit responses to self-antigens. ⏮ **negative selection of lymphocytes,** ⏮ **regulatory T cells**

The causes of autoimmune disease are not clear and involve multiple factors. They often are seen in relatives who share genes for major histocompatibility (MHC) molecules, pointing to a genetic factor. Environmental factors, including infection, may also play a role. For example, a pathogen may cover itself with molecules that resemble those on host tissues in order to evade the immune system. An immune response against that pathogen may inappropriately attack healthy tissues as well, causing inflammation and damage even in the absence of the pathogen. In other cases, the action or control of T_{reg} cells may be deficient. ⏮ **MHC,** ⏮ **regulatory T cells**

The Range of Autoimmune Diseases

Autoimmune diseases can be organ-specific or systemic, depending on whether the autoantigens are localized to an organ or found throughout the body:

- **Type 1 diabetes mellitus.** Once termed insulin-dependent or juvenile diabetes, this is an organ-specific autoimmune disease caused by the destruction of pancreatic β cells by cytotoxic T cells. Pancreatic β cells produce insulin, a hormone that allows cells to take up glucose from the bloodstream. Lack of insulin increases blood glucose levels, drawing water from the cells and leading to symptoms of increased thirst and urination. Because the cells do not take in glucose that would be used as their energy source, symptoms also include extreme hunger, fatigue, and weight loss. Persons with diabetes are vulnerable to high blood pressure, stroke, blindness, kidney disease, and conditions that lead to limb amputation. Insulin injections are used to treat the disease.
- **Graves' disease.** This is an organ-specific autoimmune disease that affects the thyroid gland. In most cases, antibodies are directed at receptors for thyroid-stimulating hormone (TSH) on the thyroid gland. Attachment of the antibody to the receptor activates the receptor inappropriately, leading to increased thyroid hormone production and enlargement of the gland. An enlarged thyroid is sometimes evident as a goiter (**figure 17.7**). Metabolic rate increases in affected individuals, resulting in signs and symptoms that include weight loss, fatigue, irritability, heat intolerance, rapid heartbeat, and bulging eyes.
- **Rheumatoid arthritis.** This is a systemic disease in which cell-mediated mechanisms and antibodies target connective tissues, most often within joints (**figure 17.8**). T cells infiltrate the joints and, when stimulated by specific antigens, they release cytokines that cause inflammation.

Antibodies to tissue in the joints form immune complexes characteristic of type III hypersensitivity that further damage the joint. This crippling inflammatory condition is one of the most common autoimmune diseases, and unlike arthritis that arises from wear and tear of joints, it may occur at any age. Rheumatoid arthritis is most common in women ages 30 to 50. ⏮ cytokines, ⏮ immune complexes

- **Systemic lupus erythematosus (SLE).** As the name implies, this is a systemic disease characterized by a variety of autoantibodies and immune complexes. It affects mainly women. Autoantibodies against molecules found in the nuclei of cells enable SLE to affect many different body systems. Symptoms vary considerably, but typically include rashes, along with joint pain and swelling. Severe cases involve damage to the kidneys or other organs. Treatment may address specific symptoms, suppress immunity, or target inflammatory mediators.
- **Myasthenia gravis.** The name means "grave muscle weakness" and reflects the fact that this systemic disease involves a disruption in nerve transmission to muscles. Antibodies bind to acetylcholine receptors at the neuromuscular junction, thereby blocking nerve impulses that normally cause muscle contraction. Moreover, binding of the antibodies may activate the complement system, which then damages the receptors. Drugs that inhibit the enzyme cholinesterase (the enzyme that degrades acetylcholine) allow acetylcholine to accumulate at the neuromuscular junction and may lessen the effects of the disease.

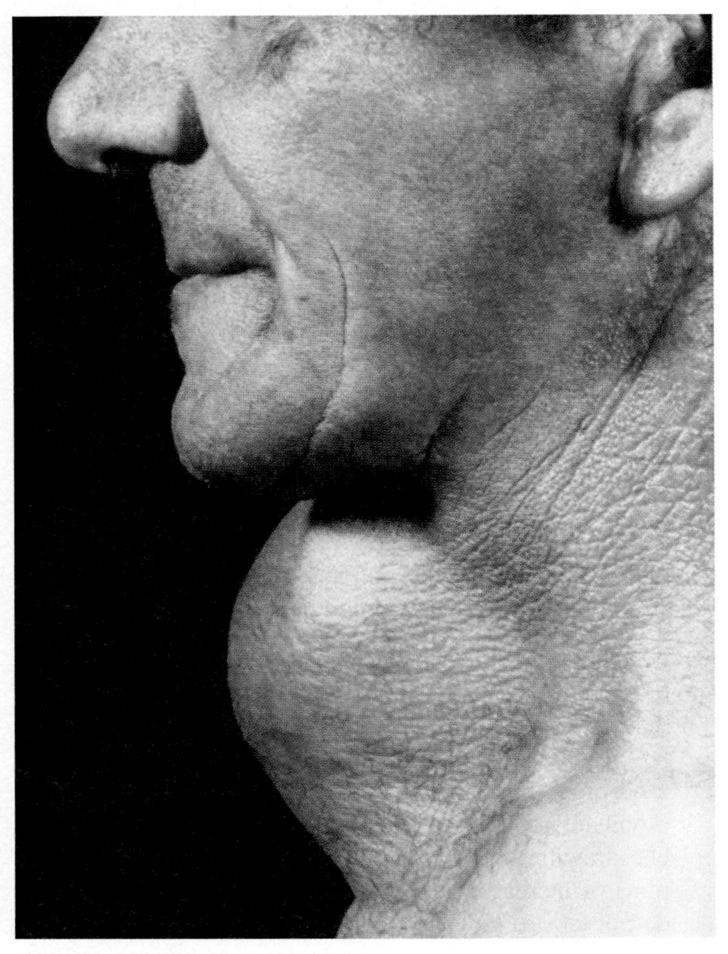

FIGURE 17.7 Goiter A goiter is an enlarged thyroid gland that may result from overstimulation of the gland. ©Biophoto Associates/Science Source

? **What receptors in the thyroid gland are stimulated in Graves' disease?**

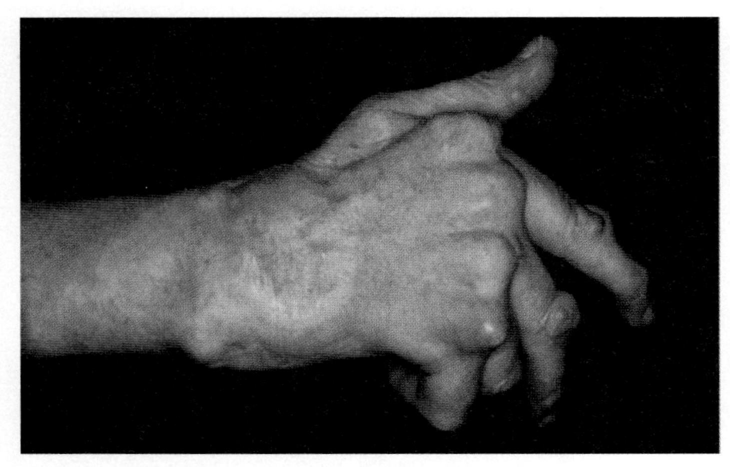

FIGURE 17.8 Rheumatoid Arthritis Autoimmune reactions cause chronic inflammation and destruction of the joints. ©Princess Margaret Rose Orthopaedic Hospital/Science Source

? **How do antibodies contribute to the damage to the joints in this hand?**

MicroByte

Babies born to mothers with myasthenia gravis experience temporary muscle weakness because of IgG autoantibodies that crossed the placenta.

Characteristics of these autoimmune diseases are summarized in **table 17.3.**

TABLE 17.3 Characteristics of Some Autoimmune Diseases

Disease	Organ Specificity	Major Mechanism of Tissue Damage
Type 1 diabetes mellitus	Pancreas	T-cell destruction of pancreatic β cells
Graves' disease	Thyroid	Autoantibodies bind thyroid-stimulating hormone receptors, overstimulating the thyroid
Rheumatoid arthritis	Systemic, especially joints	Lymphocyte destruction of joint tissues
Systemic lupus erythematosus	Systemic	Autoantibodies to DNA and other nuclear components form immune complexes in small blood vessels
Myasthenia gravis	Muscle	Autoantibodies bind to acetylcholine receptors on muscle, preventing muscle contraction

FOCUS ON A CASE 17.1

A 28-year-old white female reported to her physician that she had no energy and was tired all the time. She had a slight fever and could not shake a cough that was disturbing her sleep at night. Her joints ached, and her hands and feet were often swollen. She was taking over-the-counter ibuprofen for the pain. The patient has two older sisters. One suffers from Graves' disease, as did her mother, and the other has rheumatoid arthritis.

The patient's total white blood cell count was not elevated. A differential white blood cell count revealed 18% lymphocytes (the normal range is 25–33%). Clinical tests further showed that the patient's red blood cell count was 3.5 million/mm^3, which is below the normal range of 4–6 million/mm^3. The levels of complement proteins in her blood were below normal, but she had significant protein levels in her urine. A positive antinuclear antibody (ANA) test indicated the presence of autoimmune disease.

After reviewing the patient's complete history as well as her physical exam and test results, the physician diagnosed her as having systemic lupus erythematosus (SLE). The patient was given a prescription for prednisone and told to stop taking ibuprofen. The doctor also ordered a flu shot and a vaccination against pneumococcal pneumonia.

1. What evidence led the physician to diagnose SLE?
2. How might SLE have caused the fatigue in this patient?
3. What is a likely cause of the abnormal protein levels in the urine and blood of this patient?
4. Why did the physician give his patient a flu shot and a pneumococcal vaccination?

Discussion

1. Systemic lupus erythematosus (SLE) is a widespread disorder of connective tissue that may affect many different organs. It is difficult to diagnose because it mimics a number of other conditions. The antinuclear antibody (ANA) test is not specific for lupus, but a positive test is seen in 97% of those with the disorder. When combined with physical symptoms such as joint pain, and a family history of autoimmune disorders, such as rheumatoid arthritis and Graves' disease, the results are more reliable.
2. The patient was anemic, possibly due to a type II hypersensitivity reaction targeting red blood cells for destruction by autoantibodies.
3. Elevated protein levels in the urine indicate that the kidneys are not functioning properly. This could indicate a type III hypersensitivity reaction caused by autoantibodies forming immune complexes that lodge in the glomerular capillaries and damage the kidneys. Low levels of complement protein in the blood are consistent with systemic inflammation caused by SLE.
4. The differential white blood cell count indicates that the immune system could also be targeting lymphocytes for destruction. This would make the patient more vulnerable to infections. Moreover, prednisone inhibits inflammation and proliferation of T lymphocytes, thus further suppressing immunity. The patient's cough indicated that she may already have some fluid in her lungs. The physician was minimizing the patient's risk for common respiratory infections.

Treatment of Autoimmune Diseases

Autoimmune diseases are usually treated by limiting the immune response. This can be done with anti-inflammatory or immunosuppressant medications. Replacement therapy may be necessary to restore damaged tissues or tissue products, such as when patients with type 1 diabetes receive insulin. Because the risk of infection and cancer increases when the immune system is suppressed, other options offer more promising approaches to the treatment of autoimmune disease.

One strategy being explored to treat autoimmune diseases is to induce **tolerance,** a process that decreases the reactivity of the immune system to a specific antigen. A promising approach is to introduce the antigen by the oral route, so that the immune system "learns" to tolerate it, just as it does the many antigens ingested in food. This antigen-specific immunotherapy causes fewer side effects than drugs that generally suppress the immune system. Trials designed to induce oral tolerance in humans have successfully relieved the symptoms of several autoimmune diseases, but there is much still to learn about the immunological mechanisms, antigen preparations, doses, and duration of treatment. ◀◀ immunological tolerance

Scientists have tried curing type 1 diabetes by replacing the tissues destroyed by immune cells. Transplanting the pancreas or the insulin-producing cells of that organ has been successful in many cases, but dangerous immunosuppressive agents must be given to prevent rejection. In fact, the transplants are typically only attempted for patients with advanced diabetes who also require a kidney transplant, so therefore must take the immunosuppressive drugs anyway. Research efforts are directed toward developing better methods of transplantation, such as injecting pancreas cells harvested from cadavers into the vein that carries blood to the liver. The cells remain in the liver and produce insulin, often eliminating or decreasing the need for injected insulin. However, the cells usually die in months or a few years, and the side effects of immunosuppressants are often severe. Research on manipulating stem cells to produce insulin-secreting β cells holds greater promise for curing type 1 diabetes.

MicroAssessment 17.2

Autoimmune disease can result when the immune system attacks autoantigens. Damage from autoimmune reactions may be caused by antibody action or by cell-mediated immune functions, or both. Some autoimmune diseases are organ-specific, but others are systemic. Typical treatment involves drugs that suppress immune responses, but these have dangerous side effects.

4. What might make an autoimmune reaction general rather than tissue-specific?
5. What are the negative side effects of taking immunosuppressive drugs?
6. How could viruses or bacteria potentially trigger autoimmune disease?

TABLE 17.4 Immunodeficiency Diseases

Disease	Immune Component Involved
Primary Immunodeficiencies	
Chronic granulomatous disease (CGD)	Phagocytes (defect)
Congenital agammaglobulinemia	B cells (deficiency)
DiGeorge syndrome	T cells (deficiency)
Hereditary angioedema	Complement regulator (deficiency)
Leukocyte adhesion deficiency	Phagocytes (deficiency)
Selective IgA deficiency	B cells making IgA (deficiency)
Severe combined immunodeficiency (SCID)	Bone marrow stem cells (defect)
Secondary Immunodeficiencies	
Acquired immunodeficiency syndrome (AIDS)	Helper T cells (destroyed during HIV infection)
Monoclonal gammopathy	B cells (multiply out of control)

17.3 ■ Immunodeficiency Disorders

Learning Outcomes

5. Compare and contrast primary and secondary immunodeficiencies.
6. Explain how immunodeficiency can lead to multiple and unusual infections.

In contrast to hypersensitivities and autoimmune diseases that arise from an exaggerated or inappropriate immune response, **immunodeficiencies** develop when the body cannot launch or sustain an immune response. Primary immunodeficiency disorders are present from birth, resulting from a genetic defect or developmental abnormality. Secondary immunodeficiencies occur after birth, perhaps due to infection, malignancy, or other stresses on the immune system. People with either type of immunodeficiency tend to have repeated infections; the types of infections often depend on which part of the immune system is affected. For example, a B-cell deficiency will result in less antibody production and more infections by bacteria that travel in the bloodstream. Some important immunodeficiency diseases are listed in **table 17.4.**

Primary Immunodeficiencies

Primary immunodeficiencies, which are present from birth, are generally rare. They may affect B cells, T cells, natural killer (NK) cells, phagocytes, or complement components. Scientists know many of the gene defects that cause primary immunological disorders, and work is underway to correct them. Deficiencies and defects can occur at any point in the complex steps that lead to an effective immune response. A few categories of primary immunodeficiencies include the following:

- **Antibody deficiencies.** One of the most common primary immunodeficiencies is selective IgA deficiency, in which very little or no IgA is produced. People with this disorder may appear healthy, but they often develop repeated infections of the respiratory, gastrointestinal, and genitourinary tracts, where secretory IgA normally protects against colonization by pathogens. Another antibody deficiency is agammaglobulinemia, a disease in which few or no antibodies are produced.
- **Lymphocyte deficiencies.** A disease called severe combined immunodeficiency (SCID) results when hematopoietic stem cells in the bone marrow fail to produce functional lymphocytes. Children with SCID generally die of infectious disease at an early age unless they are successfully treated with a bone marrow transplant to populate the bone marrow with healthy cells. A variety of gene defects can cause SCID. One is a mutation in a gene coding for an enzyme important in the proliferation of B and T cells. Scientists have achieved some success against this type of SCID using gene therapy. In this process, the patient's own hematopoietic stem cells are modified in vitro, using a retrovirus to insert a gene that encodes the missing enzyme. The modified cells are introduced into the patient. The process is not perfect, but results are promising, and the possibility of treating other severe disorders with gene therapy is exciting.

  Another disease involving lymphocyte deficiency is DiGeorge syndrome, which results when the thymus fails to develop. T cells do not differentiate and are absent in circulation. Affected individuals have other developmental defects as well, such as heart and blood vessel abnormalities, and the inability to maintain proper levels of calcium and phosphorus. Those with severe thymic deficiency may be treated with a transplant of thymus tissue.

As expected from a lack of T cells, affected people are susceptible to infections by eukaryotic pathogens, such as *Pneumocystis jirovecii* and other fungi, as well as viruses and obligate intracellular bacteria. ⏭ *Pneumocystis jirovecii*

- **Defects in phagocytic cells.** Chronic granulomatous disease (CGD) is caused by a defect that results in failure of phagocytes to function properly. Surprisingly, these patients can mount an immune response to most types of infections, but they are unable to respond effectively to a few common pathogens such as *Staphylococcus aureus* or molds of the genus *Aspergillus.* Patients also produce an abundance of non-functional phagocytes, forming granulomas. Treatments for CGD are aimed at reducing the potential for infection by giving the patient antibiotics, and boosting the immune response by giving interferons. In leukocyte adhesion deficiencies, white blood cells fail to leave the circulation to concentrate at sites of infection. ⏮ reactive oxygen species
- **Defects in complement system components.** People with deficiencies in the early components of the complement system (such as C1 and C2) may develop immune complex diseases, because these components normally help clear immune complexes from the circulation. People who lack late components (C5, C6, C7, C8) have recurrent *Neisseria* infections, because membrane attack complexes typically destroy these bacteria. People who lack certain important regulatory proteins of the complement system experience uncontrolled complement activation. This causes fluid accumulation and potentially fatal tissue swelling, a condition called hereditary angioedema. ⏮ complement system

MicroByte

SCID is sometimes called "bubble boy disease" after a patient who survived for 12 years in a sterile inflated plastic bubble.

Secondary Immunodeficiencies

Secondary, or acquired, immunodeficiency diseases develop after birth. For example, cancers involving the lymphatic system often decrease effective antibody-mediated immunity. Multiple myeloma is a cancer arising from a single plasma cell that proliferates out of control and produces large quantities of immunoglobulin. This overproduction of a single kind of molecule (monoclonal gammopathy) results in the use of resources to produce immunoglobulin of a single specificity at the expense of others needed to fight infection. The result is an overall immunodeficiency.

Some viral infections will deplete certain cells of the immune system, resulting in secondary immunodeficiency. The measles virus, for example, replicates in various cells of the immune system, killing many of them and leaving the body temporarily more vulnerable to other infections. Syphilis, leprosy, and malaria affect the T-cell population and also macrophage function, causing defects in cell-mediated immunity.

One of the most serious and widespread secondary immunodeficiencies is AIDS (acquired immunodeficiency syndrome), caused by human immunodeficiency virus (HIV). Helper T cells are destroyed in AIDS, leaving the affected person highly susceptible to opportunistic infections. Some typical opportunistic infections that characterize the onset of AIDS are listed in **table 17.5.** ⏮ opportunistic infection

TABLE 17.5 Opportunistic Infections Characteristic of AIDS Patients

Infection	Causative Agent
Bacterial Infections	
Disseminated *Mycobacterium avium* complex (MAC) infection	*Mycobacterium avium* complex
Tuberculosis	*Mycobacterium tuberculosis*
Fungal Infections	
Candidiasis (thrush)	*Candida albicans*
Coccidioidomycosis	*Coccidioides immitis*
Cryptococcosis	*Cryptococcus* species
Histoplasmosis	*Histoplasma capsulatum*
Pneumocystis pneumonia	*Pneumocystis jirovecii*
Protozoan Infections	
Cryptosporidiosis	*Cryptosporidium parvum*
Toxoplasmosis	*Toxoplasma gondii*
Viral Infections	
Cytomegalovirus disease	Cytomegalovirus

MicroAssessment 17.3

Primary immunodeficiencies may be caused by genetic or developmental defects in components of the immune response. Secondary, or acquired, immunodeficiencies result from infection, malignancies, or environmental influences.

7. Compare and contrast severe combined immunodeficiency disease (SCID) and chronic granulomatous disease (CGD).
8. How can a person with multiple myeloma be immunodeficient?
9. Why are people with B-cell deficiencies more prone to bacterial infections, whereas people with T-cell deficiencies are more prone to viral infections?

Summary

17.1 Hypersensitivities (table 17.1)

An exaggerated immune response that injures tissue is called **hypersensitivity.**

Type I Hypersensitivities: Immediate IgE-Mediated

Type I hypersensitivity reactions occur when IgE attached to mast cells or basophils reacts with specific antigen, resulting in the release of powerful mediators of the allergic reaction (figure 17.1). Localized allergic reactions include **hives** (urticaria), **hay fever** (allergic rhinitis), and **asthma** (figure 17.2). **Systemic anaphylaxis** is a rare but serious reaction that can lead to shock and death. **Desensitization** is a form of **immunotherapy** that is often effective in decreasing the type I hypersensitivity state (figure 17.3). Newer treatments involving rhuMabs that alter the immune response are being used to fight asthma.

Type II Hypersensitivities: Cytotoxic

Type II hypersensitivity reactions, or cytotoxic reactions, occur when antibodies bound to the surface of a host cell trigger its destruction by complement proteins or antibody-dependent cellular cytotoxicity (ADCC). The ABO blood group antigens have been the major cause of **transfusion reactions** (table 17.2). The rhesus blood group antigens (Rh) are usually responsible for hemolytic disease of the newborn (figure 17.4). Injected anti-Rh antibody (RhoGAM) helps prevent Rh sensitization of Rh-negative mothers.

Type III Hypersensitivities: Immune Complex–Mediated

Type III hypersensitivity reactions are mediated by small antigen-antibody complexes that activate complement, attract neutrophils, and contribute to inflammation. The **immune complexes** are often deposited in small blood vessels, where they cause inflammatory disease such as glomerulonephritis in the kidney (figure 17.5).

Type IV Hypersensitivities: Delayed-Type Cell-Mediated

Delayed-type hypersensitivity reactions depend on the actions of sensitized T lymphocytes. Contact hypersensitivity, or **allergic contact dermatitis,** occurs frequently in response to substances such as poison ivy, nickel in jewelry, and chromium salts in leather products (figure 17.6). Transplantation rejection of **allografts** is caused largely by type IV cellular reactions.

17.2 Autoimmune Disease

Responses against autoantigens can lead to **autoimmune diseases.**

The Range of Autoimmune Diseases

Autoimmune diseases can be organ-specific (table 17.3, figure 17.7) or widespread (figure 17.8). Some autoimmune diseases are caused by antibodies that target body components, and others by cell-mediated reactions or a combination of antibodies and immune cells.

Treatment of Autoimmune Diseases

Autoimmune diseases are usually treated with drugs that suppress the immune and/or inflammatory responses.

17.3 Immunodeficiency Disorders (table 17.4)

Immunodeficiencies can be primary genetic or developmental defects in any components of the immune response, or they can be secondary (acquired).

Primary Immunodeficiencies

B-cell immunodeficiencies result in diseases involving a lack of antibody production, such as agammaglobulinemias and selective IgA deficiency. T-cell deficiencies result in diseases such as DiGeorge syndrome. Lack of both T- and B-cell functions results in combined immunodeficiencies, which are generally severe. Defective phagocytes are found in chronic granulomatous disease.

Secondary Immunodeficiencies

Acquired immunodeficiencies can result from malnutrition, immunosuppressive agents, infectious diseases (such as AIDS), and malignancies such as multiple myeloma. AIDS or other immunosuppressive states can increase the incidence of opportunistic infections (table 17.5).

Review Questions

Short Answer

1. How does treatment for hay fever differ from treatment for asthma?
2. List two physical responses of systemic anaphylaxis that can lead to rapid death.
3. What are the major differences between an IgE-mediated skin reaction, such as hives, and a delayed-type hypersensitivity reaction, such as rejection of transplanted tissue?
4. Compare and contrast the Arthus reaction and serum sickness.
5. What causes hemolytic disease of the newborn?
6. Compare and contrast the autoimmune processes causing myasthenia gravis and Graves' disease.
7. Give an example of an organ-specific autoimmune disease and one that is widespread, involving a variety of tissues and organs.
8. Explain why a patient who received a successful lung transplant might subsequently die from pneumonia.
9. Why does an IgA deficiency predispose a person to diarrheal diseases?
10. What is the difference between a primary immunodeficiency and a secondary immunodeficiency? Give an example of each.

Multiple Choice

1. An IgE-mediated allergic reaction
   a) reaches a peak within minutes after exposure to antigen.
   b) occurs only in response to polysaccharide antigens.
   c) requires complement activation.
   d) requires considerable macrophage participation.
   e) is characterized by induration.

2. Which of the following statements is true of the ABO blood group system in humans?
   a) A antigen is present on type O red blood cells.
   b) People with blood group A have A antigens on their red blood cells and natural anti-A antibodies in their plasma.
   c) Natural anti-A and anti-B antibodies are of the class IgG.
   d) People with blood group O do not have natural antibodies against A and B antigens.
   e) In blood transfusions, incompatibilities cause complement-mediated lysis of red blood cells.
3. Repeated injections of very small amounts of antigen are an effective therapy for combating allergies because
   a) IgG production is stimulated; the IgG outcompetes IgE for the antigen.
   b) IgG binds to mast cells before IgE can do so.
   c) repeated exposure will cause production of IgE to decrease.
   d) antigen binds to mast cells and prevents their degranulation.
   e) repeated exposure to antigen prevents the production of histamine.
4. Which of the following could be the result of a type III hypersensitivity reaction?
   a) DiGeorge syndrome
   b) glomerulonephritis
   c) diabetes type 1
   d) hemolytic disease of the newborn
   e) anaphylactic shock
5. Delayed-type hypersensitivity reactions in the skin
   a) are characterized by a wheal and flare reaction.
   b) peak at 4 to 6 hours after exposure to antigen.
   c) require complement activation.
   d) show induration because of the influx of cells.
   e) depend on activities of the Fc portion of antibodies.
6. Organ transplants, such as of kidneys,
   a) are experimental at present.
   b) can be successful only if there are exact matches between donor and recipient.
   c) survive best if irradiated before the transplant.
   d) survive best if B cells are suppressed in the recipient.
   e) may be rejected by a complex process in which cellular mechanisms predominate.
7. All of the following are true of autoimmune disease *except*
   a) some seem to have a genetic component.
   b) induction of tolerance may alleviate symptoms.
   c) damage to organs occurs due to long-term exaggerated production of IgE.
   d) disease may result from reaction to viral antigens that are similar to autoantigens.
   e) some are organ-specific and some are widespread in the body.
8. All of the following approaches are used to treat autoimmune diseases *except*
   a) immunosuppressant drugs.
   b) induction of tolerance.
   c) antibiotics.
   d) anti-inflammatory medications.
   e) replacement therapy, as with insulin in diabetes.
9. If the thymus is removed from a 2-year-old child, you might expect which of the following to occur?
   a) Increased success of an organ transplant and increased incidence of bacterial infection
   b) Increased success of an organ transplant and decreased incidence of bacterial infection
   c) Decreased success of an organ transplant and increased incidence of viral infection
   d) Decreased incidence of cancer and decreased incidence of viral infection
   e) Decreased success of an organ transplant and increased incidence of cancer
10. In which one of the following combinations of hypersensitivity reactions are B cells involved?
    a) Type I, but not types II, III, and IV hypersensitivity
    b) Type IV, but not types I, II, and III hypersensitivity
    c) Types I, II, and III, but not type IV hypersensitivity
    d) Types II, III, and IV, but not type I hypersensitivity
    e) B cells are involved in all types of hypersensitivity.

Applications

1. Jack and Jill were badly burned in an accident at the well and both were taken to the burn unit of the local hospital. The burns covered only a small area of skin so grafts were prepared for both patients from the skin of Jack's thigh. Jack's graft was successful and his burn healed completely. Jill, however, rejected the grafted skin. Explain the immune responses of both patients to these grafts. What treatments could have helped Jill to avoid rejection of her graft?
2. Horse serum containing specific antibody to snake venom has been a successful approach to treating snakebite in humans. How do you think this antivenom could be generated? What are some advantages of using horses instead of humans to produce the antibody? Why might it be unsafe to administer the antivenom more than once?

Critical Thinking

1. Hypersensitivity reactions, by definition, lead to tissue damage. Can they also be beneficial? Explain.
2. Why might malnutrition result in immunodeficiencies?

18 Applications of Immune Responses

An immunoassay. ©*Kevin Horan/The Image Bank/Getty Images*

KEY TERMS

Active Immunity Immunity that results from an immune response upon exposure to an antigen.

Antiserum A preparation of serum that contains protective antibodies.

Attenuated Vaccine Vaccine composed of a weakened form of the pathogen that is generally unable to cause disease.

Enzyme-Linked Immunosorbent Assay (ELISA) Technique that uses enzyme-labeled antibodies to detect given antigens or antibodies.

Fluorescent Antibody (FA) Test Technique that uses fluorescently labeled antibodies to detect specific antigens on cells attached to a microscope slide.

Immunoassay A test that takes advantage of the specificity of antigen-antibody interactions by using known antibodies or antigens to detect or quantify given antigens or antibodies.

Inactivated Vaccine Vaccine composed of killed bacterial cells, inactivated viral particles, fractions of the pathogen, or inactivated toxin.

Passive Immunity Immunity that results when antibodies are transferred to an individual.

Vaccine A preparation of a pathogen or its products used to induce active immunity.

Western Blotting Procedure that uses labeled antibodies to detect specific antigens in a mixture of proteins separated according to their molecular weight.

A Glimpse of History

Long before people knew that microbes cause disease, they recognized that individuals who recovered from a disease such as smallpox rarely got it a second time. Old Chinese writings dating from the Sung dynasty (960–1280) describe a procedure known as variolation, in which small amounts of the powdered scabs from smallpox lesions were inhaled or placed into a scratch made in the skin. The resulting disease was usually mild, and the person was then immune to smallpox. Occasionally, however, severe disease developed, often resulting in death. In addition, the person became contagious, so the disease could spread.

Although variolation was practiced in China and the Mideast a thousand years ago, it was not widely used in Europe until after 1719. At that time, Lady Mary Wortley Montagu, wife of the British ambassador to Turkey, had their children immunized against smallpox in this way. Variolation then became popular in Europe. Because of the dangers, however, and the fact that the procedure was expensive, many people remained unprotected.

As an apprentice physician, Edward Jenner noted that milkmaids who had recovered from cowpox (a disease of cows that caused few or no symptoms in humans) rarely got smallpox. Then, in 1796, long before viruses had been discovered, he conducted a classic experiment in which he deliberately transferred material from a cowpox lesion on the hand of a milkmaid, Sarah Nelmes, to a scratch he made on the arm of a young boy named James Phipps. Six weeks later, Jenner inoculated Phipps with pus from a smallpox victim, but the boy did not develop smallpox. Phipps had been made immune to smallpox when he was inoculated with pus from the cowpox lesion.

Using the less dangerous cowpox material in place of the scabs from smallpox cases, Jenner and others worked to spread the practice of variolation. Later, Pasteur used the word vaccination (from the Latin vacca for "cow") to describe any type of protective inoculation. By the twentieth century, most of the industrialized world was generally free of smallpox as the result of routine vaccination.

In 1967, the World Health Organization (WHO) started a program of intensive smallpox vaccination. Because there were no animal hosts and no non-immune humans to whom it could be spread, the disease died out. The last naturally contracted case occurred in Somalia, Africa, in 1977, and 2 years later WHO declared the world free of smallpox. Nevertheless, a few laboratories around the world still have the virus. In this age of bioterrorism concerns, some see smallpox as a major threat should the deadly virus ever be released into the largely unprotected populations of the world. Because of this, large supplies of the vaccine have been stockpiled in the United States to be used in case of an emergency.

Chapters 14 and 15 discussed the innate and adaptive defense systems, describing antibodies and lymphocytes. This chapter will consider how **immunization,** the process of inducing immunity, can be used to protect against disease. In fact, immunization has probably had the greatest impact on human health of any medical procedure, and it is just one example of how knowledge is power with respect to fighting disease (**figure 18.1**). We will also explore some useful applications of immunological reactions in diagnostic tests.

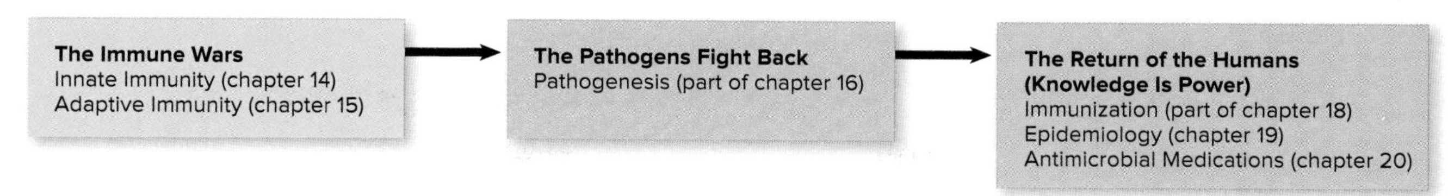

FIGURE 18.1 The Host-Pathogen Trilogy

How does immunization prevent disease?

IMMUNIZATION

18.1 ■ Principles of Immunization

Learning Outcome

1. Compare and contrast naturally acquired active immunity, artificially acquired active immunity, naturally acquired passive immunity, and artificially acquired passive immunity.

Naturally acquired immunity is the gain of adaptive immunity through normal events, such as exposure to an infectious agent. Immunization mimics those same events, protecting against disease by inducing **artificially acquired immunity** (**figure 18.2**). The protection provided by immunization can be either active or passive.

Active Immunity

Active immunity is the result of an immune response in a person who has been exposed to antigen. As a result of the exposure, specific B and T cells are activated and they then multiply, giving the person lasting protection due to immunological memory. Active immunity can develop either naturally from an actual infection or artificially from vaccination. ⏮ immunological memory

Passive Immunity

Passive immunity occurs naturally during pregnancy; the mother's IgG antibodies cross the placenta and protect the fetus. These antibodies remain active in the newborn during the first few months of life, when his or her own immune responses are still developing. This is why a number of infectious diseases typically do not develop until a baby is 3 to 6 months of age, after the maternal antibodies have been degraded. Passive immunity also occurs as a result of breast feeding; the secretory IgA in breast milk protects the digestive tract of the child. Note that passive immunity provides no memory; once the transferred antibodies are degraded, the protection is lost.

Artificially acquired passive immunity involves injecting a person with antibodies produced by other people or animals. This can be used to (1) prevent disease immediately before or after likely exposure to a pathogen, (2) limit the duration of certain diseases, and (3) block the action of microbial toxins. A preparation of serum (the fluid portion of blood that remains after blood clots) containing the protective antibodies is referred to as **antiserum.** One that protects against a toxin is called an **antitoxin.**

Two kinds of antisera (or antitoxins) are used. **Hyperimmune globulin**—prepared from the sera of donors with high amounts of antibodies to certain disease agents—is used to prevent or treat specific diseases. Examples of hyperimmune globulin include tetanus immune globulin (TIG), rabies immune globulin (RIG), and hepatitis B immune globulin (HBIG). **Immune globulin,** the IgG fraction of blood plasma pooled from many donors, has a variety of antibodies due to typical infections and vaccines experienced by the donors. It is used to protect unvaccinated people who have been recently exposed to certain diseases, including measles and hepatitis A. It is also administered to immunosuppressed people who have low levels of antibodies. ⏮ incubation period

MicroAssessment 18.1

Active immunity occurs naturally in response to infections, and artificially in response to vaccination. Passive immunity occurs naturally from maternal antibodies transferred during pregnancy and breast feeding, and artificially through administration of immune globulin or hyperimmune globulin.

1. How is naturally acquired active immunity different from artificially acquired active immunity?
2. What is antitoxin?
3. If an unvaccinated person suffers a deep puncture wound that puts him or her at risk for tetanus, why would TIG be advised rather than simply vaccinating the individual?

18.2 ■ Vaccines and Immunization Procedures

Learning Outcomes

2. Describe the role of vaccines in providing herd immunity.
3. Compare and contrast the characteristics of attenuated and inactivated vaccines.

Focus Figure

Active Immunity	Passive Immunity

Natural Active Immunity

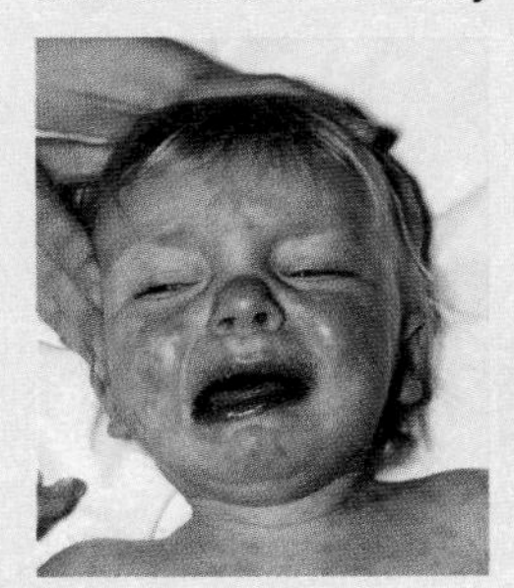

Immunity that results from an immune response in an individual after exposure to an infectious agent.

Artificial Active Immunity

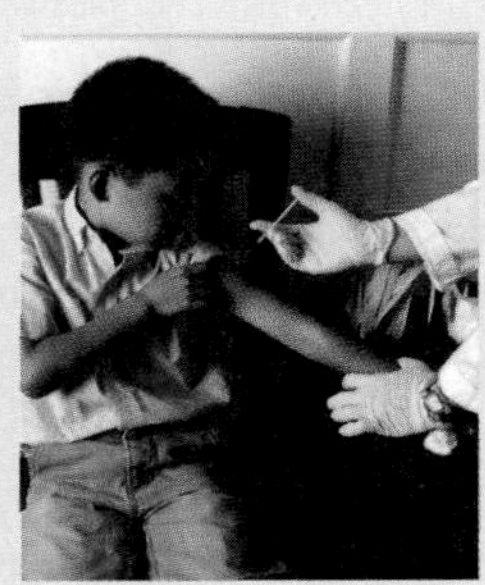

Immunity that results from an immune response in an individual after vaccination.

Natural Passive Immunity

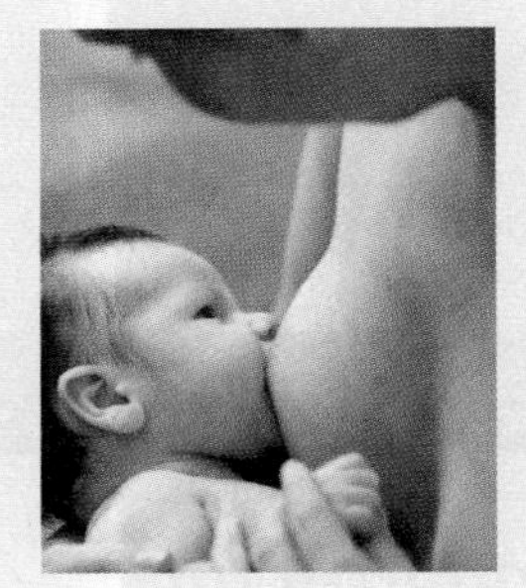

Immunity that results when antibodies from a woman are transferred to her developing fetus during pregnancy or to an infant during breast feeding.

Artificial Passive Immunity

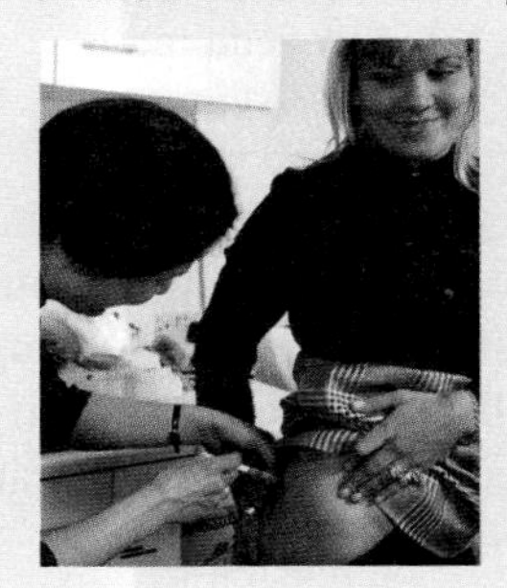

Immunity that results when antibodies contained in the serum of other people or animals are injected into an individual.

FIGURE 18.2 Acquired Immunity Acquired immunity can be natural or artificial, active or passive. (top left): ©SPL/Science Source; (top right): ©Jose Luis Pelaez Inc/Blend Images LLC; (bottom left): ©McGraw-Hill Education/Jill Braaten; (bottom right): ©SPL/Science Source

Why does active immunity last longer than passive immunity?

4. Explain the importance of vaccinations.
5. Compare and contrast the inactivated polio vaccine (IPV) and the oral polio vaccine (OPV).

A **vaccine** is a preparation of a pathogen or its products used to induce active immunity. Vaccines not only protect an individual against disease, they can also prevent diseases from spreading in a population. This is because **herd immunity** develops when a critical portion of a population is immune to a disease, either through natural immunity or vaccination. The infectious agent is unable to spread because there are not enough susceptible hosts. Herd immunity is responsible for dramatic declines in childhood diseases, both in the United States and in developing countries. Unfortunately, some of these diseases reappear and spread as a direct consequence of parents' failure to have their children vaccinated.

Effective vaccines should be safe, with few side effects, while giving lasting protection against the illness. They should induce protective antibodies or immune cells, or both, as appropriate for controlling the infection. Of course, vaccines ideally should be inexpensive, stable with a long shelf life, and easy to administer. ◀◀ cell-mediated immunity

Vaccines fall into two general categories—attenuated and inactivated—based on whether or not the immunizing agent can replicate (attenuated agents can replicate whereas inactivated agents cannot). Each type has characteristic advantages and disadvantages (**table 18.1**).

Attenuated Vaccines

An **attenuated vaccine** is a weakened form of the pathogen that generally cannot cause disease. The attenuated strain replicates in the vaccine recipient, causing an infection with undetectable or mild disease that typically results in long-lasting immunity. Because infection with the attenuated strain mimics that of a wild-type strain, it induces the appropriate type of immune response. For instance, attenuated vaccines given orally induce mucosal immunity (a secretory IgA

TABLE 18.1 A Comparison of Characteristics of Attenuated and Inactivated Vaccines

Characteristic	Attenuated Vaccines	Inactivated Vaccines
Types	Attenuated viruses, attenuated bacteria	Inactivated whole agents, toxoids, subunits, VLPs (virus-like particles), polysaccharides, conjugates
Route of administration	Injection, oral, or nasal	Injection
Need for adjuvant	No	Yes
Antibody response (memory)	IgG; secretory IgA if administered orally or nasally	IgG
Cell-mediated immune response	Good	Poor
Relative duration of protection	Longer	Shorter
Number of doses	Usually one or two	Multiple
Risk of mutation to virulence	Very low	Absent
Risk to immunocompromised recipient	Can be significant	Absent
Stability in warm temperatures	Poor	Good

response), protecting against pathogens that infect via the gastrointestinal tract. Some attenuated vaccines are able to stimulate cytotoxic T cells, inducing cell-mediated immunity. ◀◀ mucosal immunity

Attenuated strains are often produced by growing a microbe under conditions that cause mutations to accumulate, making the microbe less pathogenic. Viruses of humans can sometimes be attenuated by growing them in cells of a different animal species; the mutations that allow the viruses to multiply in the other animal cells often cause them to grow poorly in human cells. Genetic manipulation is now also being used to produce strains of pathogens with low virulence. Specific genes are mutated and used to replace wild-type genes. The inserted mutant genes are engineered so that the microbes cannot revert to the wild type.

Attenuated vaccines have several advantages compared to their inactivated counterparts. For one thing, only one or two doses of an attenuated agent are often enough to induce relatively long-lasting immunity. This is because the microbe multiplies in the body, causing the immune system to be exposed to the antigen for a longer period and in greater amounts than with inactivated agents. In addition, the vaccine strain has the added potential of being spread from an individual being immunized to other non-immune people, thereby immunizing the contacts of the vaccine recipient.

The disadvantages of attenuated agents are that they sometimes cause disease in immunosuppressed people and that they can occasionally mutate to become pathogenic again. Attenuated vaccines are generally not advised for pregnant women because of the possibility that the vaccine strain may cross the placenta and damage the developing fetus. Another disadvantage of attenuated vaccines, especially in developing countries where they are desperately needed, is that refrigeration is usually required to keep them active.

Attenuated vaccines currently in widespread use include those against measles, mumps, rubella, chickenpox, and rotavirus gastroenteritis. The adenovirus vaccine given to military recruits is non-attenuated replicating virus, but the dose of the respiratory pathogen is taken orally, a route that does not result in disease.

Inactivated Vaccines

An **inactivated vaccine** is unable to replicate, but retains the immunogenicity of the pathogen or toxin. The advantage of inactivated vaccines is that they cannot cause infections or revert to pathogenic forms. Because they do not replicate, however, there is no amplification of the dose in vivo, so the magnitude of the immune response is limited. To compensate for the relatively low effective dose, regular booster doses are usually needed to induce sufficient immunity to be protective. Some inactivated vaccines include the whole infectious agent, and others include only fractions of the agent. Examples currently in use include:

- **Inactivated whole agent vaccines.** These contain killed microorganisms or inactivated viruses. The vaccines are made by treating the pathogen with formalin or another chemical that does not significantly change the surface epitopes. The treatment leaves the agent immunogenic even though it cannot reproduce. Vaccines in this category include those against influenza, rabies, and hepatitis A. ◀◀ formalin
- **Toxoid vaccines.** These are inactivated toxins used to protect against diseases caused by bacterial toxins. They are prepared by treating the toxins to destroy the toxic part of the molecules while retaining the antigenic epitopes. Diphtheria and tetanus vaccines are toxoids.
- **Subunit vaccines.** These consist of key protein antigens or antigenic fragments of a pathogen. Obviously, they can be developed only after research has revealed which of the microbe's components are most important in triggering a protective immune response. Their advantage is that cell parts that may cause undesirable side effects are not included. The vaccine currently used to prevent whooping

cough (pertussis) is a subunit vaccine, referred to as the acellular pertussis (aP) vaccine. It does not cause the side effects that sometimes occurred with the killed whole-cell vaccine used previously. Subunit vaccines produced using genetically engineered microorganisms are called recombinant subunit vaccines. An example is the vaccine against the hepatitis B virus; it is made using yeast cells engineered to produce part of the viral protein coat.

- **VLP (virus-like particle) vaccines.** These contain empty capsids. Laboratory microorganisms are genetically engineered to produce the major capsid proteins of a virus, which then self-assemble. The human papillomavirus (HPV) vaccine is a VLP vaccine. ⏮ capsid
- **Polysaccharide vaccines.** These contain the polysaccharides that make up the capsules of certain organisms. They are not effective in young children because polysaccharides are T-independent antigens; recall that these antigens generally elicit a poor response in this age group. The pneumococcal vaccine given to adults is a polysaccharide vaccine. ⏮ T-independent antigens
- **Conjugate vaccines.** These are polysaccharides linked to proteins, a modification that converts the polysaccharides into T-dependent antigens. The first conjugate vaccine developed was against *Haemophilus influenzae* type b (Hib) and has nearly eliminated Hib meningitis in children. The conjugate vaccine developed against certain *Streptococcus pneumoniae* strains promises to do the same for a variety of infections caused by those strains. ⏭ *Haemophilus influenzae* type b, ⏭ *Streptococcus pneumoniae*

In addition to the types of inactivated vaccines currently used, research is underway to develop novel versions. One of the most promising is **nucleic acid–based vaccines,** which are segments of naked nucleic acid—either DNA or mRNA—from an infectious agent. Studies have shown that when the nucleic acid is injected into tissue, some cells express the genes for a short period of time, producing the encoded microbial antigens. These then induce an immune response.

Many inactivated vaccines contain an **adjuvant,** a substance that enhances the immune response to antigens (*adjuvare* means "to help"). These are necessary additives because purified antigens such as toxoids and subunit vaccines are often poorly immunogenic by themselves because they lack the "danger" signals—the MAMPs (microbe-associated molecular patterns) that characterize invading microbes. Recall that the MAMPs cause dendritic cells to produce co-stimulatory molecules, allowing them to activate helper T cells, which, in turn, activate B cells (see figure 15.20). Adjuvants are thought to function by providing the danger signals to dendritic cells. Some adjuvants appear to adsorb the antigen, releasing it at a slow but constant rate to the tissues and surrounding blood vessels. Unfortunately, many effective adjuvants trigger an intense inflammatory response, making them unsuitable for use in vaccines for humans. Alum (aluminum hydroxide and aluminum phosphate) is the most common adjuvant used, but others, including one that uses a derivative of lipid A, have recently been developed. ⏮ MAMP, ⏮ dendritic cells, ⏮ lipid A

The Importance of Vaccines

Before vaccines were available to prevent common diseases, thousands of people died or were permanently disabled from these diseases (**table 18.2**). Unfortunately, many people still become ill or even die from vaccine-preventable diseases. One reason is that some parents refuse to have their children vaccinated, fearing that vaccination might be harmful. In situations such as this, vaccines have become victims of their own success. They have been so effective at preventing diseases that many people no longer realize how serious these diseases can be. Reports of adverse effects of vaccination have led some people to falsely believe that the risk of vaccination is greater than the risk of diseases.

There will always be at least some risk associated with almost any medical procedure, but there is no question that the benefits of routine vaccinations greatly outweigh the very slight risks. Data show that a child with measles has a 1:2,000 chance of developing serious brain inflammation, compared with a 1:1,000,000 chance from the measles vaccine. Between 1989 and 1991, measles immunization rates dropped 10% and an outbreak of 55,000 cases occurred, with 120 deaths. Once the vaccination rates increased again, measles outbreaks became rare. The suggestion that certain vaccines are associated with autism again threatened the acceptance of immunization. It is important to note, however, that numerous scientific studies show no evidence of a link between the two.

Table 18.3 lists bacterial diseases prevented by routine vaccines; **table 18.4** lists viral diseases prevented by

TABLE 18.2 The Effectiveness of Universal Immunization in the United States

Disease	Cases per Year Before Immunization	Decrease After Immunization
Diphtheria	175,885 (1920–1922)	Nearly 100%
Haemophilus influenzae type b invasive disease	20,000 (estimated)	98.8%
Measles	503,282 (1958–1962)	Nearly 100%
Mumps	152,209 (1968)	99.2%
Pertussis (whooping cough)	147,271 (1922–1925)	77.6%*
Poliomyelitis	16,316 (1951–1954)	100%
Rubella (German measles)	50,230 (1966–1969)	Nearly 100%
Smallpox	48,164 (1900–1904)	100%
Tetanus	1,314 (1922–1926)	98%

*Reflects the recent increase in number of cases.

TABLE 18.3 Bacterial Diseases Prevented by Routine Vaccines

Disease	Vaccine Type (Vaccine Acronym)	Persons Who Should Receive the Vaccine
Diphtheria	Toxoid (the "D" in the DTaP, and the "d" in Td and Tdap; the lowercase letter indicates a lower dose)	DTaP: infants, toddlers, and preschoolers Td: adolescents and adults; every 10 years as a booster (Tdap should be used once instead) ⏭ **diphtheria**
Haemophilus influenzae type b disease	Conjugate (Hib)	Infants and toddlers; older children and adults with certain medical conditions that put them at increased risk ⏭ ***Haemophilus influenzae* type b disease**
Meningococcal disease	Conjugate against serogroups ACWY (MenACWY) Polysaccharide against serogroups ACWY (MPSV4) Recombinant subunit against serogroup B (MenB)	MenACWY: adolescents age 11 or 12, with booster at age 16; anyone 2 months or older who is at increased risk due to certain medical conditions or potential exposures (MPSV4 is the preferred alternative for some older adults) MenB: people 10 years or older who are at increased risk due to certain medical conditions or potential exposure ⏭ **meningococcal disease**
Pertussis (whooping cough)	Subunit (the "aP" in DTaP and the "ap" in Tdap; the lowercase letter indicates a lower dose)	DTaP: infants, toddlers, and preschoolers Tdap: adolescents, pregnant women, and anyone who did not receive it as an adolescent ⏭ **pertussis (whooping cough)**
Pneumococcal disease	Conjugate (PCV13) Polysaccharide (PPSV23)	PCV13: infants and toddlers; adults age 65 and over; people with certain health conditions PPSV23: adults age 65 and over; people with certain health conditions ⏭ **pneumococcal disease**
Tetanus	Toxoid (the "T" in DTaP, Td, and Tdap)	DTaP: infants, toddlers, and preschoolers Td: adolescents and adults; every 10 years as a booster (Tdap should be used once instead) ⏭ **tetanus**

TABLE 18.4 Viral Diseases Prevented by Routine Vaccines

Disease	Vaccine Type (Vaccine Acronym)	Persons Who Should Receive the Vaccine
Hepatitis A	Inactivated virus (HepA)	Toddlers; older children and adolescents who have not received the vaccine; adults who are at increased risk due to potential exposure ⏭ **hepatitis A**
Hepatitis B	Recombinant subunit (HepB)	Newborns, infants, and toddlers; older children and adolescents who have not received the vaccine; adults who are at increased risk due to certain medical conditions or potential exposure ⏭ **hepatitis B**
Human papillomavirus (HPV)–related diseases	VLPs of 9 serotypes (9vHPV)	Adolescents age 11 or 12 (but may be given ages 9 through 26) ⏭ **human papillomavirus (HPV)–related diseases**
Influenza	Inactivated virus (IIV is produced in chicken eggs; ccIV is from cell-culture) Recombinant subunit (RIV) Attenuated virus (LAIV), taken as a nasal mist; not always recommended due to lack of effectiveness	People 6 months of age or older (vaccine type used depends on patient's age, allergies, or other characteristics; given yearly because the antigens of the virus change frequently) ⏭ **influenza**
Measles	Attenuated virus (an "M" in MMR and MMRV)	Toddlers and preschoolers; older children who have not received the vaccine; adults without evidence of immunity ⏭ **measles**
Mumps	Attenuated virus (an "M" in MMR and MMRV)	Toddlers and preschoolers; older children who have not received the vaccine; adults without evidence of immunity ⏭ **mumps**
Polio	Inactivated virus (IPV); an attenuated virus (OPV) is used in some parts of the world	Infants, toddlers, and preschoolers; older children who have not received the vaccine; adults at increased risk due to potential exposure ⏭ **polio**
Rotavirus gastroenteritis	Attenuated virus (RV)	Infants ⏭ **rotavirus gastroenteritis**
Rubella (German measles)	Attenuated virus (an "R" in MMR and MMRV)	Toddlers and preschoolers; older children who have not received the vaccine; adults without evidence of immunity ⏭ **rubella (German measles)**
Shingles	Attenuated virus (HZV)	Adults age 60 and over ⏭ **shingles**
Varicella-zoster (chickenpox)	Attenuated virus (the "V" in the MMRV vaccine; also VAR)	Toddlers and preschoolers; school-age children and adults without evidence of immunity ⏭ **varicella-zoster (chickenpox)**

routine vaccines. Some routine vaccines are administered in combination, thereby minimizing the number of injections required. For example, MMRV is a combination vaccine used to immunize against measles, mumps, rubella, and varicella. DTaP-IPV-Hib is a combination vaccine used to immunize against diphtheria, tetanus, pertussis, polio, and *Haemophilus influenzae* type B disease.

The U.S. Centers for Disease Control and Prevention (CDC) regularly publishes recommended immunization schedules for children, adolescents, and adults. These are updated regularly as vaccines are developed and modified. Because of the complexity of the schedules and the frequency at which they have been updated recently, it is important to know how to access the most current schedule. They are available at the following CDC website: http://www.cdc.gov/vaccines/schedules/. Vaccines listed in the CDC's recommended immunization schedule are generally covered by the National Vaccine Injury Compensation Program. This no-fault alternative for resolving vaccine injury claims was established to stabilize the U.S. vaccine market. A tax on every vaccine dose purchased funds the program.

Table 18.5 lists diseases that can be prevented by non-routine vaccines. The relatively few people at increased risk for these diseases generally work in certain occupations or travel to parts of the world where the diseases are known to exist.

TABLE 18.5 Diseases Prevented by Non-Routine Vaccines

Disease	Vaccine Type	Persons Who Should Receive the Vaccine
Adenovirus respiratory infection	Two serotypes of active virus (taken orally)	Military recruits entering basic training and certain other military personnel at higher risk for infection
Anthrax	Subunit	Adults in occupations that put them at risk of exposure, such as some military personnel
Cholera	Attenuated bacterium (taken orally)	People age 18–64 traveling to endemic areas
Japanese encephalitis	Inactivated virus	Certain people traveling to Asia where the disease occurs
Rabies	Inactivated virus	People at high risk for exposure, such as veterinarians and other animal handlers; people who have recently been exposed to the virus
Smallpox	Vaccinia virus	Used only in response to biological warfare threat
Tuberculosis	Attenuated bacterium (BCG strain)	Used only in special circumstances in the United States; widely used in other countries
Typhoid fever	Two forms—attenuated bacterium (taken orally) and polysaccharide	People traveling to parts of the world where typhoid fever is common; people living with a typhoid carrier; laboratory personnel who work with *Salmonella* Typhi
Yellow fever	Attenuated virus	Certain people traveling to parts of the world where the disease is known to exist

An Example of Vaccination Strategy—The Campaign to Eliminate Poliomyelitis

Vaccines against poliomyelitis provide an excellent illustration of the complexity of vaccination strategies. The virus that causes this disease enters the body orally, infects the cells that line the throat and intestinal tract, and then invades the bloodstream. From there, it can invade nerve cells and cause the disease poliomyelitis (see figure 26.8). There are three serotypes of poliovirus, any of which can cause the disease. The Salk vaccine, developed in the mid-1950s, consists of inactivated virus particles of all three serotypes. This vaccine, now called the **inactivated polio vaccine (IPV),** successfully lowered the rate of the disease dramatically but had the disadvantage of requiring a series of injections for maximum protection. In 1961, the Sabin vaccine became available. This vaccine, now called the **oral polio vaccine (OPV)** consists of attenuated strains that replicate in cells that line the throat and intestinal tract. This vaccine has the advantage of cheaper oral administration. ⏭| poliomyelitis

OPV and IPV both cause the immune system to produce antibodies that protect against viral invasion of the central nervous system and consequent poliomyelitis. OPV, however, has a distinct advantage over IPV in that it induces better mucosal immunity (secretory IgA response). Thus, the antibodies in the throat and intestinal tract can neutralize wild poliovirus before it infects cells, thereby preventing the virus from replicating and then being transmitted in feces. Because of this, OPV provides better herd immunity. A disadvantage of OPV is that the attenuated viruses can mutate to become virulent. Approximately one case of poliomyelitis arises for every 2.4 million doses of OPV administered. An obvious way to avoid vaccine-related poliomyelitis is to abandon OPV in favor of IPV. As is often the case, however, the situation is not as simple as it might seem. If only IPV is used, wild poliovirus can still replicate in the vaccine-recipient's throat and intestinal lining and then be transmitted to others in feces, thereby spreading in a population. Therefore, polio eradication (complete elimination) depends on OPV because it prevents spread of wild poliovirus.

A campaign to eliminate polio using OPV was so successful that by 1980 the United States was free of wild-type poliovirus (see figure 26.9). By 1991, the virus had been eliminated from the Western Hemisphere. Because of the continued risk of vaccine-associated poliomyelitis, a vaccine strategy that

FOCUS ON A CASE 18.1

Patient A, an unvaccinated university student, had recently traveled to Europe. He became worried when he developed a fever and swelling on half his face, so he visited the campus clinic. The staff there diagnosed a bacterial infection and prescribed antibiotics. A week later, Patient A returned to the clinic, this time reporting pain in his testicles. At that point, the staff suspected mumps, a viral disease that often affects the parotid glands (the largest salivary glands) and sometimes other sites, including the testicles. There is no effective treatment for mumps, but the staff referred the patient for serological testing. He did not follow through.

A month later, the roommate of Patient A visited the clinic, also complaining of pain, fever, and swelling on half his face. Although he had been vaccinated, mumps was suspected, and serological testing was done. Those results were inconclusive: The mumps IgM test was negative, but the mumps IgG test was positive. A short time later, three other students visited the clinic, experiencing the same symptoms, and again, mumps was suspected. Their diagnosis was confirmed using polymerase chain reaction (PCR).

Health officials were concerned because the mumps cases appeared to be linked, so an investigation was done to determine the extent of the outbreak. A total of 29 cases were found, approximately half of which were confirmed by PCR; one was confirmed by finding IgM antibodies specific for the mumps virus. Seven of the ill individuals either were unvaccinated or had not received the recommended number of MMR (measles, mumps, and rubella) doses.

All students at the university were then encouraged to get an additional dose of the MMR vaccine, and vaccine clinics were set up around the campus.

1. Why was it significant that Patient A had not been vaccinated and had traveled to Europe?
2. What is the significance of finding IgM antibodies?
3. Why might the roommate's serology test results have been negative for mumps IgM and positive for IgG?
4. What might explain the fact that some students became ill with mumps even though they had been adequately vaccinated?

Discussion

1. Without vaccination against mumps, Patient A was susceptible to the disease. He was unlikely to contract the disease in the United States, because of herd immunity—widespread vaccination has made most individuals immune to the disease, and the lack of susceptible hosts prevents the virus from spreading. Unfortunately, unfounded concerns over vaccination have decreased the vaccination rate in some countries, leading to outbreaks of vaccine-preventable diseases, including mumps. Patient A had traveled to a country that was experiencing an outbreak, and he likely contracted the disease there, bringing the infecting virus back with him when he returned to the United States.
2. The presence of pathogen-specific IgM antibodies indicates a current or very recent infection, because IgM has a relatively short half-life. In contrast, IgG has a longer half-life, so those antibodies remain circulating longer. In general, high levels of IgM antibodies indicate a current infection, whereas high levels of IgG antibodies indicate a current infection, past infection, or vaccination.
3. The roommate had been vaccinated, so he likely had at least some circulating mumps-specific IgG antibodies as a result. In addition, the vaccine would have induced production of mumps-specific memory cells, which would have quickly made more IgG upon exposure to the virus. Some IgM was also likely made upon infection, but the levels must not have been high enough to be detectable at the time he was tested. Although the patient had circulating IgG molecules, the quantities were not sufficient to protect him from the disease.
4. Mumps cases in vaccinated people could be due to waning immunity. Although mumps vaccination is part of the childhood vaccination schedule, the resulting immunity lessens over time. In addition, some vaccinated individuals never received the recommended number of doses. Finally, some vaccine recipients do not mount a strong enough response to completely protect them from the disease.

Source: "Mumps Outbreak on a University Campus - California, 2011" Morbidity and Mortality Weekly Report (MMWR) December 7, 2012 / 61(48); 986-989. Centers for Disease Control and Prevention http://www.cdc.gov/mmwr/preview/mmwrhtml/mm6148a2.htm?s_cid=mm6148a2_e

attempted to capture the best of both vaccines was adopted in the United States. Children first received doses of IPV, protecting them from the disease. Following these doses, OPV was given, providing mucosal protection while also boosting immunity. The routine use of OPV was then discontinued altogether in the United States.

The original goal of global eradication of polio by 2000 was not achieved, but substantial progress has been made and efforts continue. For these eradication programs, OPV must be used in regions where wild-type poliovirus might still be present because that vaccine prevents transmission of the virus. In 2015, the type 2 polio strain was declared globally eradicated. Because of that, a new OPV that only contains serotypes 1 and 3 has now replaced the previous version that contained all three serotypes. Chapter 26 has more information about polio and the progress toward its elimination.

MicroAssessment 18.2

An attenuated vaccine is a weakened form of the pathogen. An inactivated vaccine is unable to replicate but retains the immunogenicity of the pathogen or toxin; examples include killed microorganisms, inactivated viruses, and fractions of the agents, including toxoids. Routine childhood immunizations have prevented millions of cases of disease and many deaths during the past decades.

4. What is the difference between an attenuated and an inactivated vaccine?
5. Childhood diseases such as measles and mumps are rare now, so why is it important for children to be immunized against them?
6. In 2004, Israel switched from using OPV to using IPV. Then, after wild-type poliovirus was found to be circulating in the population in 2013, a supplementary OPV program was begun for children who had received only IPV. What would be the rationale for an OPV supplementary program?

IMMUNOLOGICAL TESTING

Immunological testing takes advantage of the specificity of antibody-antigen interactions, using it for diagnosis. One of the earliest yet still relevant examples is the tuberculin skin test (also called the PPD test or the Mantoux test), which can be used to detect infection by the bacterium *Mycobacterium tuberculosis.* A person who is infected typically develops a strong cell-mediated response to the bacterium and its products. When a purified protein derivative (PPD) from the organism is injected into the upper layers of skin of that person, an area of firm swelling surrounded by redness usually develops at the injection site (see figure 21.21). In contrast, people who are not infected typically show little, if any, response. An exception is people who have received the BCG vaccine against tuberculosis (see table 18.5); some of these individuals may show a positive tuberculin skin test. ◀◀ tuberculin skin test

18.3 ■ Principles of Immunoassays

Learning Outcomes

6. Describe how known antibodies can be used to identify an unknown antigen, and vice versa.
7. Explain how the antibody titer is determined.
8. Compare and contrast polyclonal and monoclonal antibodies.

An **immunoassay** is an in vitro test that that takes advantage of the specificity of antigen-antibody interactions to detect or quantify given antigens or antibodies in a sample. As an example, one way to diagnose the sexually transmitted infection syphilis is to detect the presence of the causative bacterium, *Treponema palladium,* in fluid from a genital lesion on the patient. To do this, antibodies specific to the bacterium are added to the specimen; if the antibodies attach to an organism in the specimen, then the organism is *T. pallidum* (**figure 18.3**). In the reverse situation, the fluid from the patient can be tested for antibodies that bind specifically to *T. pallidum.* If antibodies are present, then the patient's immune system must have responded to the microbe at some point, indicating either previous or current infection.

To determine if a person has certain specific antibodies in her or his blood, either the serum or the plasma is tested. **Serum** is the fluid portion that remains after blood clots; **plasma** is the fluid portion of blood treated with an anticoagulant to prevent clotting. Because serum is so often used as a source of antibodies, the study of in vitro antibody-antigen interactions is referred to as **serology,** and immunoassays are also referred to as serological tests. Most frequently serological testing implies examining a patient's blood for specific antibodies to diagnose a disease.

A person who has not been exposed to a given pathogen typically lacks specific antibodies against that microbe and is referred to as seronegative. Once infected, that person will begin producing detectable levels of specific antibodies about a week or two later, becoming seropositive. This change from seronegative to seropositive is referred to as **seroconversion.** As the infection progresses, increasing amounts of specific antibodies are produced. A rise in the concentration (titer) of specific antibodies is characteristic of an active infection. In contrast, low but steady levels of specific antibodies indicate a previous infection or vaccination.

Quantifying Antigen-Antibody Reactions

The concentration of antibody molecules in a specimen such as serum is usually determined by making serial dilutions, similar to what is done to count bacterial cells (see figure 4.18). A series of 2-fold or 10-fold dilutions are used to dilute the specimen, and then antigen is added to each dilution. The **titer** (concentration) is expressed as the reciprocal of the last dilution that gives a detectable antigen-antibody reaction. Thus, if a positive reaction is observed in the dilution 1:256 but not in 1:512, then the antibody titer is 256.

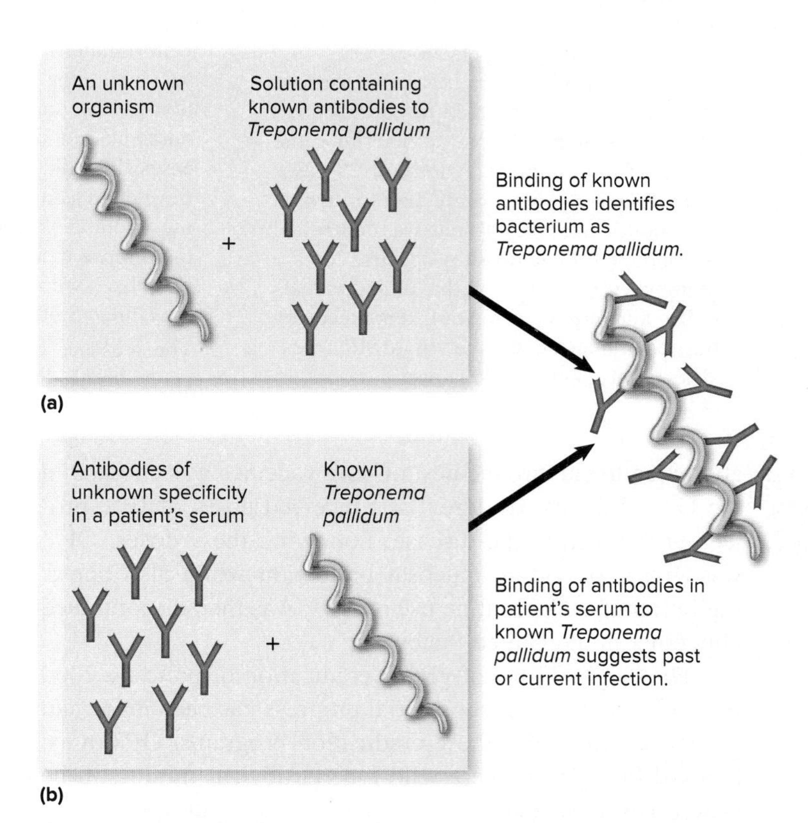

FIGURE 18.3 Principles of Immunoassays These assays can be used to **(a)** detect and therefore identify unknown bacteria (or other antigens); **(b)** detect specific antibodies.

? What three occurrences could account for a person having antibodies to a specific infectious agent?

Immunoassays can be done in test tubes, but this requires many tubes and large amounts of reagents. Therefore, the assays are usually done using microtiter plates, which are plastic plates that have many tiny wells (**figure 18.4**). The volumes used in each well are a mere fraction of those needed for even a small test tube, so tests can be done on very small samples. Special equipment can be used to mix the reagents and read the results.

Obtaining Known Antibodies

Laboratory animals are used to obtain known antibodies that bind to a given infectious agent. The choice of polyclonal antibodies versus monoclonal antibodies (discussed next) depends largely on the intended use of the product.

Polyclonal Antibodies

Polyclonal antibodies bind to various epitopes on an antigen. To obtain them, animals such as rabbits or goats are immunized with either the whole agent or part of the agent, and their serum is collected to harvest the antibodies. Multiple naive B cells of the animal will have responded to the immunization, giving rise to a mixture of antibodies that together bind a variety of epitopes on the antigen. The more complex the antigen, the greater the number of different epitopes recognized by the polyclonal antibodies. For instance, injecting whole bacteria results in a wider range of antibody specificities than injecting purified toxin. A variety of polyclonal antibody preparations are commercially available from laboratory supply companies.

One problem with polyclonal antibodies is that some may bind to closely related organisms, resulting in false-positive reactions. As an example, *Shigella* species have outer membrane proteins in common with *E. coli,* so an animal immunized with whole *Shigella* cells would produce some antibodies that also bind *E. coli* cells. If those antibodies were used in a diagnostic test for *Shigella,* a specimen containing *E. coli* but not *Shigella* would give a false-positive result.

FIGURE 18.4 Microtiter Plate Immunoassays can be done in the wells of these small plates. ©McGraw-Hill Education/Lisa Burgess

? **What is the advantage of doing immunoassays in a microtiter plate?**

Certain serological tests discussed in this chapter use **anti-human IgG antibodies,** which bind to the constant region of any human IgG molecules. Anti-human IgG antibodies are obtained from animals that have been immunized with IgG from human serum. The human antibodies are antigenic to the animal and, because of this, the animal produces antibodies against the human IgG antibodies. Anti-human IgG antibodies are commercially available, as are antibodies that bind to the other immunoglobulin classes.

Monoclonal Antibodies

Monoclonal antibodies (MAbs, Mabs, or mAbs) recognize only a single epitope. They are difficult and expensive to develop, so a given specificity is generally available only if it has commercial value due to widespread use.

MAbs are obtained through a complicated process that involves taking B cells from an immunized animal, and then fusing those short-lived B cells with other cells that will divide repeatedly in culture (see **Focus Your Perspective 18.1**). Because each B cell is programmed to make antibody molecules against only a single epitope, antibody preparations produced by the descendants (clones) of a single B cell are all identical.

MAbs are often used in immunoassays, but they can also be used therapeutically. Medications composed of MAbs have names ending in "-mab." Obiltoxaximab, a monoclonal antibody that neutralizes anthrax toxin, has recently been approved for use as part of the protocol for treating inhalation anthrax. Some monoclonal antibodies are tagged with a drug or toxic substance and then used to deliver that tag to a specific cell type in vivo. These tagged MAbs—called antibody-drug conjugates (ADCs)—are being developed to treat certain types of cancers (see **Focus on the Future 18.1**).

The monoclonal antibodies used therapeutically are often "humanized," a process that uses recombinant DNA techniques to replace most of the animal-derived antibody molecule with human equivalents. This creates what is called a **rhuMAb** (**r**ecombinant **hu**manized **m**onoclonal **a**nti**b**ody). In humans, rhuMAbs have a longer half-life than standard monoclonal antibodies because the human immune system is less likely to destroy them. Medications composed of rhuMAbs have names ending in "-zumab." Examples of rhuMabs used to treat certain types of cancer are described in Focus on the Future 18.1.

MicroByte

Antibody-drug conjugates behave as "smart bombs," delivering toxins only to cells that have a given antigen on their surface and, in doing so, destroying only specific types of cells.

FOCUS YOUR PERSPECTIVE 18.1

Monoclonal Antibodies

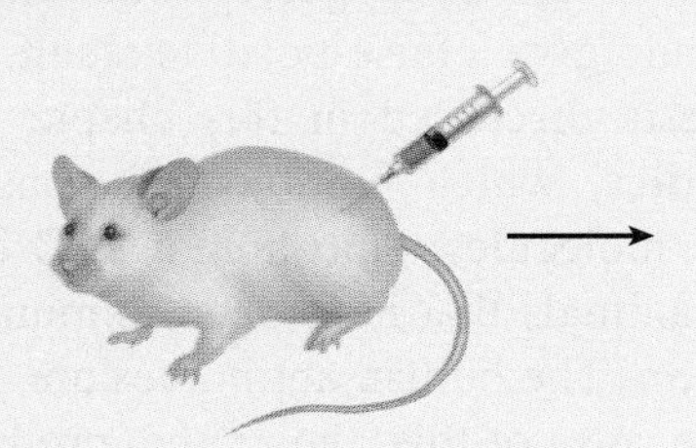

Immunize a mouse with antigen X to activate and induce proliferation of specific B cells.

B cells from spleen. These are capable of making anti-X antibodies, but die after several generations.

Myeloma cells. These abnormal plasma cells grow indefinitely, cannot make antibodies, and have a mutation that makes them susceptible to the drug aminopterin.

B cells die.

Myeloma cells die.

Mix the two cell types along with a chemical that induces their fusion, and then incubate in a medium that contains aminopterin. The B cells and myelomas die, but hybridomas proliferate.

Hybridoma cells. These are fusions of B cells and myeloma cells.

Select single hybridoma cell that recognizes desired anti-X epitope and maintain it in culture.

Harvest antibodies made by the hybridoma cells.

Monoclonal antibodies. These all have the same constant and variable regions, and therefore recognize the same epitope and have the same functional characteristics.

BOX FIGURE 18.1 Production of Monoclonal Antibodies

When an animal is injected with an antigen, its immune system responds by making antibodies directed against the antigen's different epitopes. Even though there is a single antigen, a variety of different B cells respond, resulting in the production of polyclonal antibodies. Unfortunately, this makes it difficult to standardize experimental results, because the antibody composition is different each time the antiserum is made.

In 1975, Georges Köhler and Cesár Milstein overcame the problem of variable antiserum preparations by developing a technique to make monoclonal antibodies. These are antibodies produced by a single B clone, so all molecules in a preparation will have the same constant and variable regions and, thus, the same functional characteristics and specificity. With such consistency, tests can be standardized more easily and with greater reliability.

To make monoclonal antibodies, a laboratory animal is immunized with the agent being studied, and that animal's B lymphocytes are then isolated (**box figure 18.1**). These are then fused with myeloma cells, which are malignant (cancerous) plasma cells. Unlike normal plasma cells, these myeloma cells can divide repeatedly in culture and do not make antibodies. In addition, they have lost the capacity to produce a critical enzyme, so they cannot grow in a medium that contains the drug aminopterin. When the B cells and the myeloma cells are mixed and grown in a medium that contains aminopterin, only the fusion products, called hybridomas, can proliferate. A hybridoma cell retains critical traits from the fused cells: The B cell supplies the genes for the specific antibody production, and the myeloma cell supplies the cellular machinery for producing the antibodies and multiplying indefinitely.

In the laboratory, monoclonal antibodies are the basis of a number of diagnostic tests. For example, monoclonal antibodies against a hormone can detect pregnancy only 10 days after conception. Specific monoclonal antibodies are used for rapid diagnosis of a wide variety of infectious diseases and are now being used in treatments as well. Köhler and Milstein won the Nobel Prize in 1984 for their work.

MicroAssessment 18.3

Immunoassays are used to detect or quantify given antigens or antibodies in a sample. The sample may be serially diluted to determine the antibody titer. Antibodies used in immunoassays can be either polyclonal or monoclonal.

7. What is the significance of a rise in titer of specific antibodies in serum samples taken at different times?
8. How are polyclonal antibodies different from monoclonal antibodies?
9. Would antibodies produced by a patient in response to infection be monoclonal or polyclonal?

18.4 ■ Common Types of Immunoassays

Learning Outcomes

9. Explain how labeled antibodies are used in direct and indirect tests.
10. Compare and contrast fluorescent antibody tests, ELISAs, and Western blots.
11. Describe how the fluorescence-activated cell sorter is used in immunoassays.
12. Compare and contrast precipitation reactions and agglutination reactions.

Antibodies are small molecules, so they cannot be seen using conventional microscopy. Because of this limitation, many clever assays have been developed that allow scientists to monitor antigen-antibody interactions. The assays fall into two general categories:

1. **Immunoassays that use labeled antibodies.** When antibodies are labeled with a detectable marker such as an enzyme, fluorescent dye, or radioactive isotope, their location can be tracked. Scientists can then determine if the antibodies are attached to an antigen. An advantage of tests that use labeled antibodies is that they are relatively sensitive, meaning they are more likely than other methods to detect the antigen in question, if it is present. Various labeled antibodies are extensively used in research and clinical laboratories, so many classes and specificities are commercially available.
2. **Immunoassays that involve visible antigen-antibody aggregates.** Many tests in this category were extensively used in the past, but have now largely been replaced by newer more sensitive tests that use labeled antibodies. However, immunoassays that involve antigen-antibody aggregates are relatively inexpensive and technically simple, so some are still extremely useful in certain situations.

Immunoassays That Use Labeled Antibodies

Diagnostic tests that use labeled antibodies are routine in clinical laboratories, so understanding the basic principles is important. We will start by describing those principles, and then turn our focus to specific tests and their applications.

Basic Principles

Labeled antibodies can be used to detect a given bacterium or other antigen as a way of identifying that antigen. Alternatively, they can be used to detect antibodies that bind to a given antigen as a way of confirming that a patient's immune system has mounted a response against that antigen. The tests can be either direct or indirect.

Direct immunoassays are typically used to identify an unknown antigen (for example, a certain pathogen) in a clinical specimen. The general procedure is illustrated in **figure 18.5a.** In this example, the unknown antigen is suspected to be "antigen X." The specimen containing the unknown antigen is first attached to a solid surface, and then labeled antibodies that will bind only antigen X are added. A washing step removes all unbound antibodies. The bound antibodies are then located by testing for the label. If the label is detected, then the specimen contained antigen X; conversely, if the label is not detected, then the specimen did not contain antigen X.

Indirect immunoassays are used to detect antibodies of a given specificity in a patient's serum. The tests are called indirect because they require a labeled secondary antibody to detect the unlabeled first (primary) antibody. The general procedure is illustrated in figure 18.5**b.** In this example, the patient's serum is being tested for IgG antibodies against "antigen X." To detect these, known antigen X must be attached to a solid surface. The patient's serum (containing antibodies) is added to this; any anti-X antibodies in the serum will bind the antigen on the solid surface. A washing step then removes any unbound antibodies. The next steps are aimed at detecting any bound IgG antibodies (primary antibodies). To do this, labeled anti-human IgG antibodies are used; these secondary antibodies bind to any human IgG molecules, regardless of their source or specificity. A washing step then removes any unbound molecules. In the final step, a test is done to detect the label. If it can be detected, then the labeled secondary antibodies are present. That, in turn, indicates that the primary antibodies are present. ◀◀ anti-human IgG antibodies

Fluorescent Antibody (FA) Test

The **fluorescent antibody (FA) test** uses fluorescence microscopy to locate fluorescently labeled antibodies bound to antigens fixed to a microscope slide (**figure 18.6**). The fact that the labeled microbes are observed by microscopy is an advantage over the other methods because the size and shape of the antigens can be seen (**table 18.6**). A disadvantage is that the procedure is relatively time-consuming, because each specimen must be individually examined using a fluorescence microscope. ◀◀ fluorescence microscope

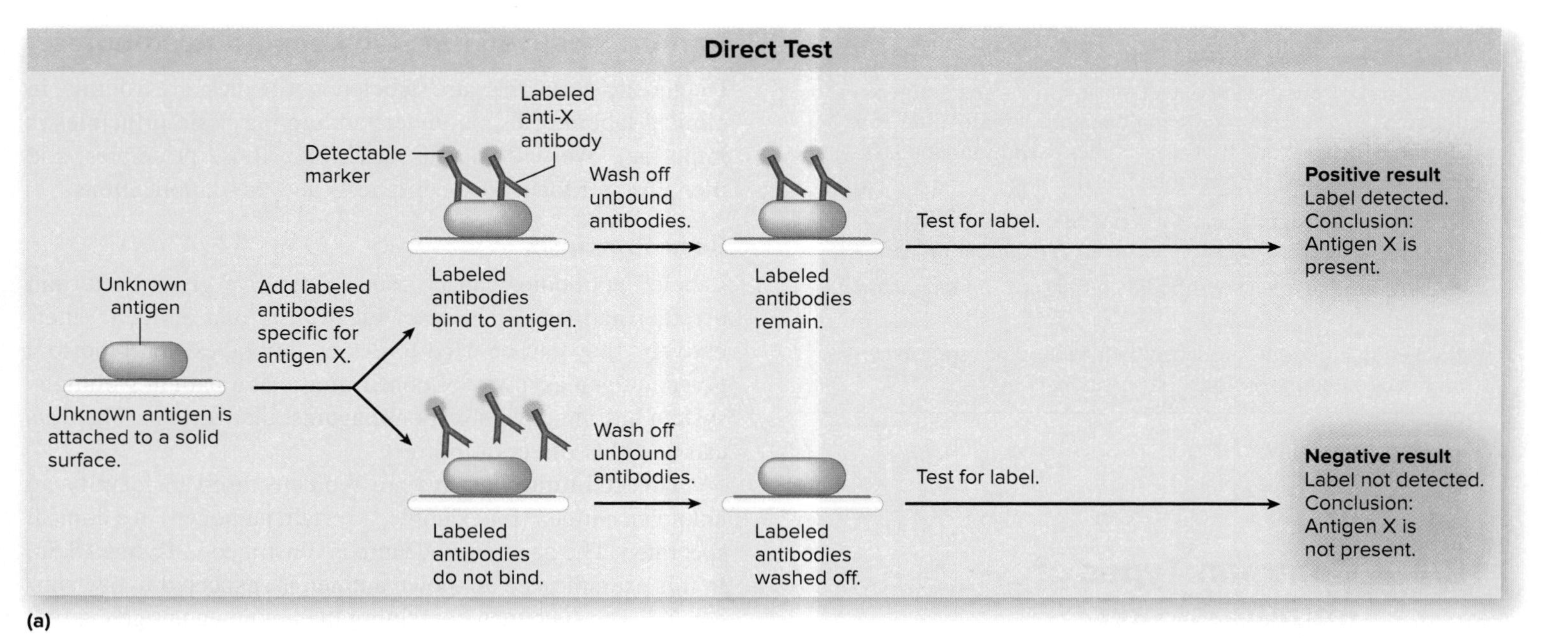

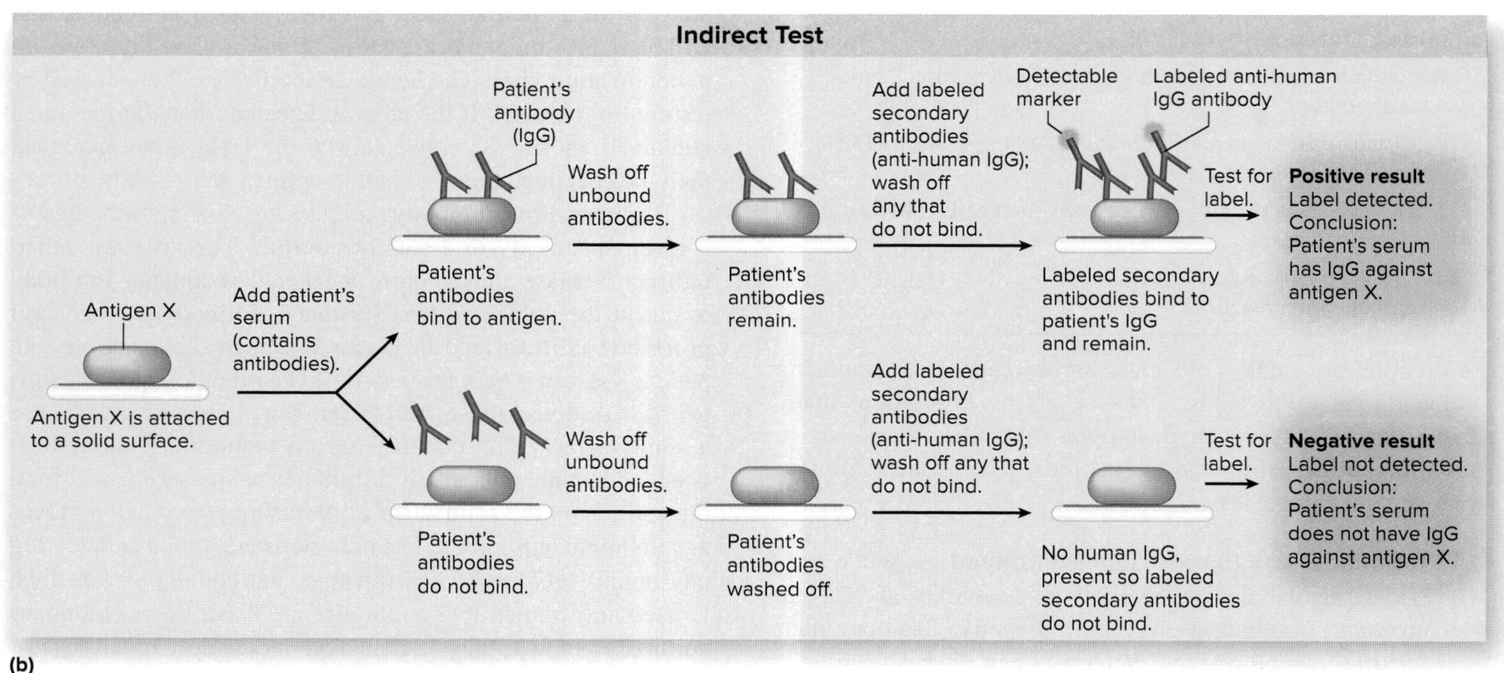

FIGURE 18.5 Basic Principles of Tests That Use Labeled Antibodies to Detect Antigen-Antibody Interactions **(a)** Positive and negative results of a direct test. **(b)** Positive and negative results of an indirect test.

? In (b), why would it be more efficient to use labeled anti-human IgG rather than label the patient's antibodies?

The direct FA is used to detect certain microorganisms, including the bacterium that causes syphilis. It can also be used to diagnose a viral infection, but in those cases the antigen-containing tissues are observed. For example, FA is used to detect rabies virus in brain tissue. Infected tissues will fluoresce, but individual viral particles cannot be seen.

The indirect FA test is used to detect antibodies against certain pathogens. For example, it can be used to determine if a patient has been infected with *Treponema pallidum,* the bacterium that causes syphilis. Several different fluorescent dyes, including fluorescein (fluoresces green) and rhodamine (fluoresces red), can be used to label the antibodies. By using various fluorescent dyes, different antigens in the same preparation can be located. ⏮ fluorescent dyes and tags

Enzyme-Linked Immunosorbent Assay (ELISA)

As the name implies, **enzyme-linked immunosorbent assays (ELISAs)** use enzyme-labeled antibodies—in other words, antibodies with an enzyme attached to them. The assays are also simply referred to as enzyme immunoassays (EIAs).

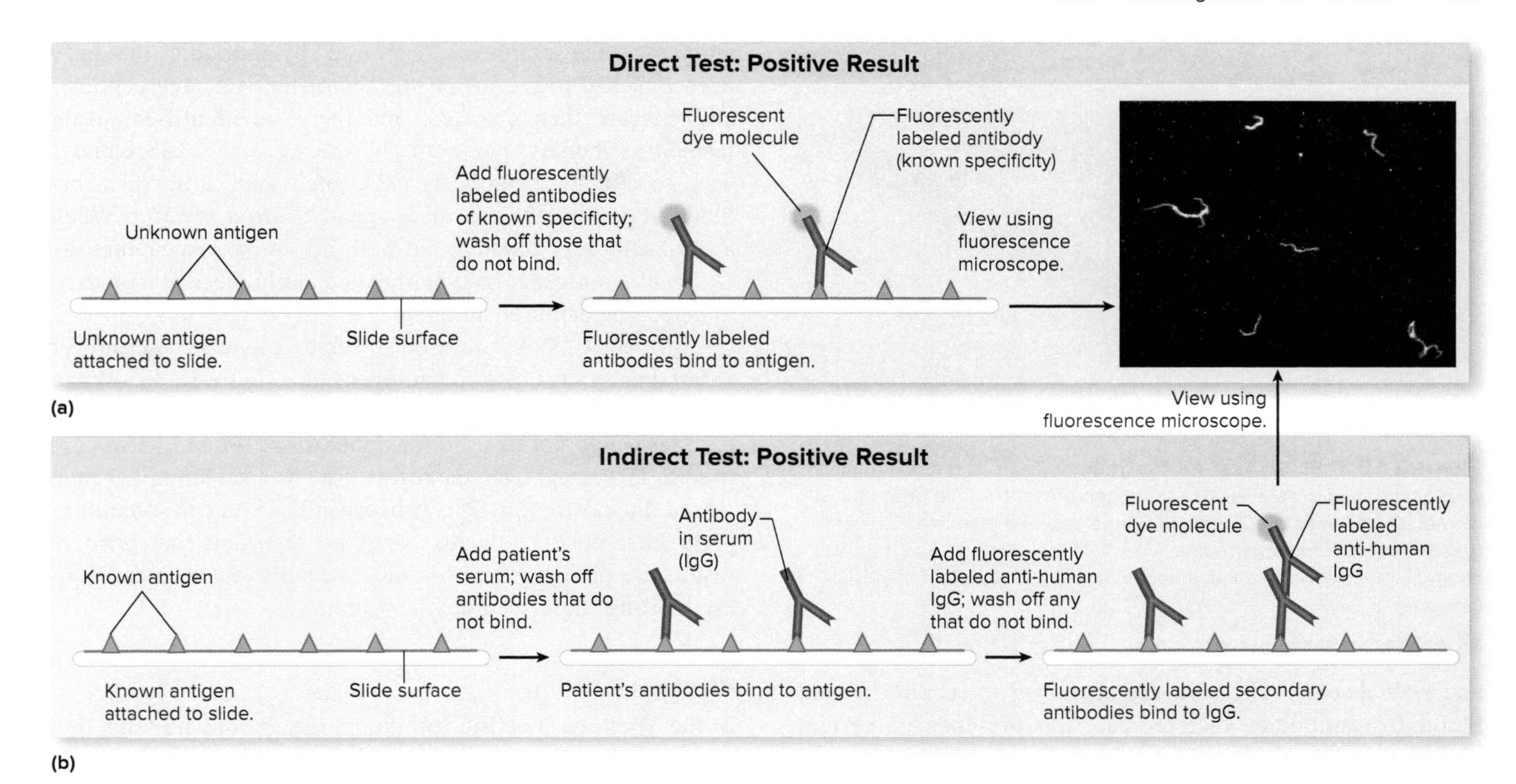

FIGURE 18.6 Fluorescent Antibody (FA) Test (a) Positive direct FA test. **(b)** Positive indirect FA test. Source: CDC

In this diagram, why is anti-human IgG used in the indirect test but not in the direct test?

An enzyme commonly used to label antibodies is peroxidase from the horseradish plant. To detect the enzyme, a colorimetric assay is used to measure the enzymatic conversion of a colorless substrate into a colored product. The assays have advantages over other common immunoassays in that they are easy to do and often require minimal technical skills (see table 18.6). In fact, they provide the basis for several simple strip tests, including the rapid group A strep tests done in doctor's clinics and over-the-counter home pregnancy tests (**figure 18.7**). Clinical labs often do ELISAs as part of the diagnoses of a variety of different diseases, including Lyme disease, West Nile fever, giardiasis, measles, hepatitis B and C,

TABLE 18.6 Comparison of Immunoassays That Use Labeled Antibodies

Method	Characteristics	Examples of Use
Fluorescent antibody (FA) tests	Fluorescence microscopy is used to locate fluorescently labeled antibodies bound to antigens. The size and shape of the antibody-bound antigens can often be determined, but each specimen needs to be examined individually.	Used in clinical labs to diagnose certain diseases including syphilis and rabies.
Enzyme-linked immunosorbent assay (ELISA)	A colorimetric assay is used to locate enzyme-labeled antibodies bound to antigens. When done in microtiter plates, large numbers of specimens can be screened all at once, but in some cases false positives can be a problem. Commercially available options are easy to perform and widely used.	Used in over-the-counter pregnancy tests; in physicians' offices to diagnose strep throat; in clinical labs as part of the diagnosis of a wide range of diseases, including Lyme disease, West Nile fever, hepatitis B and C, and HIV disease. Also used to screen donated blood for antibodies that suggest that the blood may contain HIV, hepatitis B virus, or hepatitis C virus.
Western blot	Various methods are used to detect labeled antibodies bound to proteins that have been separated by size. The results are generally more reliable than with the ELISA because they provide more information, but fewer samples can be tested at a time.	Used in clinical labs to confirm certain positive ELISA tests, such as for HIV and Lyme disease.
Fluorescence-activated cell sorter	Separates and determines the relative concentrations and certain characteristics of fluorescent-antibody-labeled cells. Widely used to diagnose and monitor the progression of diseases involving white blood cells.	Used in clinical labs to track progression of HIV disease, to characterize different types and stages of lymphomas and leukemias, and to diagnose immunodeficiency diseases.

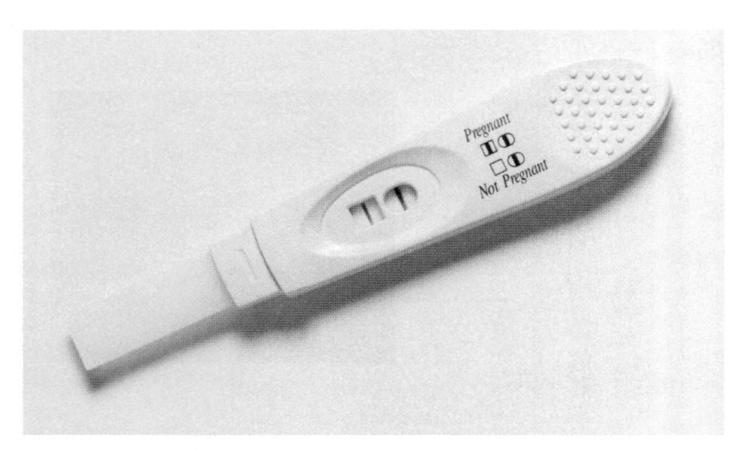

FIGURE 18.7 ELISA Test for Pregnancy The test detects human chorionic gonadotropin (HCG), an antigen present only in pregnant women. A urine sample is applied on the left. Two pink lines indicate reaction of HCG with antibodies, a positive test. A single line indicates absence of HCG in the urine, a negative test. ©Raimund Koch/The Image Bank/Getty Images

? What captures the antigen on the strip?

and HIV disease. The tests are also used to screen donated blood for antibodies that indicate the presence of certain viruses, including HIV, hepatitis B virus, and hepatitis C virus. The tests are often done in microtiter plates, allowing multiple samples to be tested all at once.

Direct ELISAs are often done using what is called a "sandwich method" (**figure 18.8a**). With this, specific antigens in the sample are "captured" by antibodies that have been attached to the inside surface of the well; the captured antigens are then detected, and therefore identified, using labeled antibodies. For example, the wells of ELISA plates used to diagnose giardiasis are coated with antibodies that bind antigens of the causative agent, *Giardia lamblia.* When a stool sample is added to the well, the antibodies capture the *G. lamblia* antigens. Enzyme-labeled antibodies are then used to detect the captured antigens.

Indirect ELISAs are used to screen blood and serum for antibodies against certain pathogens (see figure 18.8b). When donated blood is tested for antibodies against HIV, a positive product would not be used for transfusion. When ELISAs are used in HIV diagnosis, the test is used as a screen rather than a final diagnostic test. This is because ELISAs can sometimes yield false-positive results, so it is important that positive results are confirmed with a more reliable test such as Western blotting (described next).

Western Blotting

In the **Western blotting** technique, the various proteins that make up an antigen are separated by size before reacting them with antibodies. This makes it possible to determine exactly which proteins the antibodies are recognizing, an essential part of accurate HIV testing. A drawback of the procedure is that relatively few samples can be tested at a time (see table 18.6).

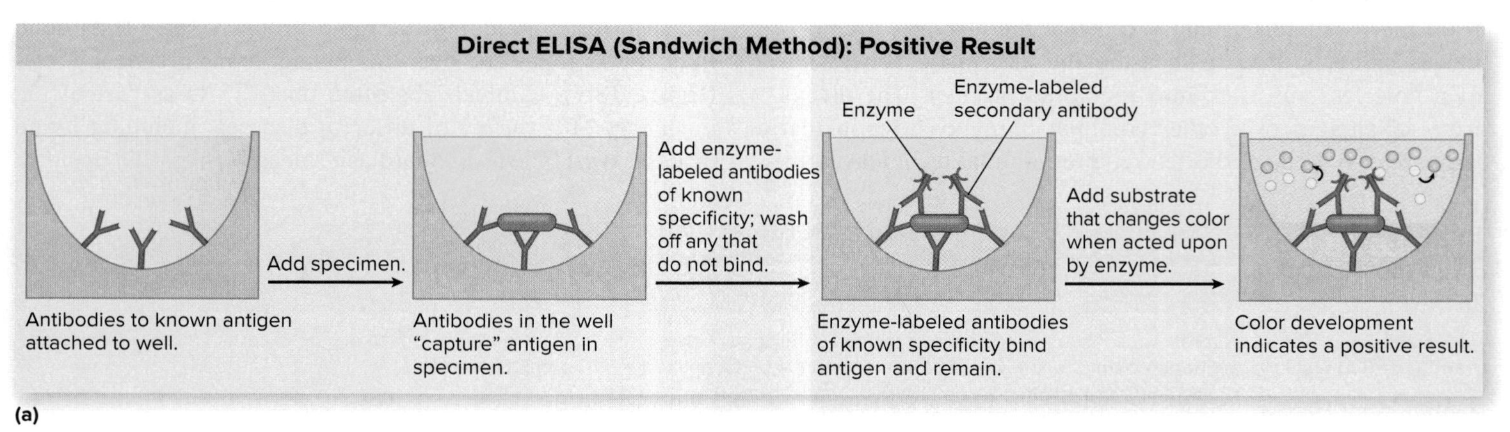

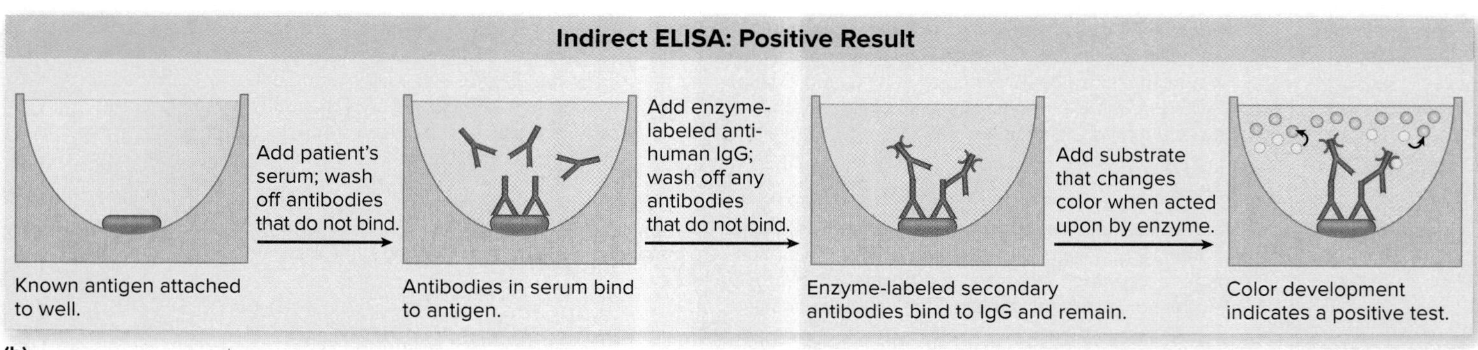

FIGURE 18.8 Enzyme-Linked Immunosorbent Assay (ELISA) **(a)** Positive direct ELISA. **(b)** Positive indirect ELISA.

? In the first panel of part (a), what is the purpose of the antibodies attached to the well?

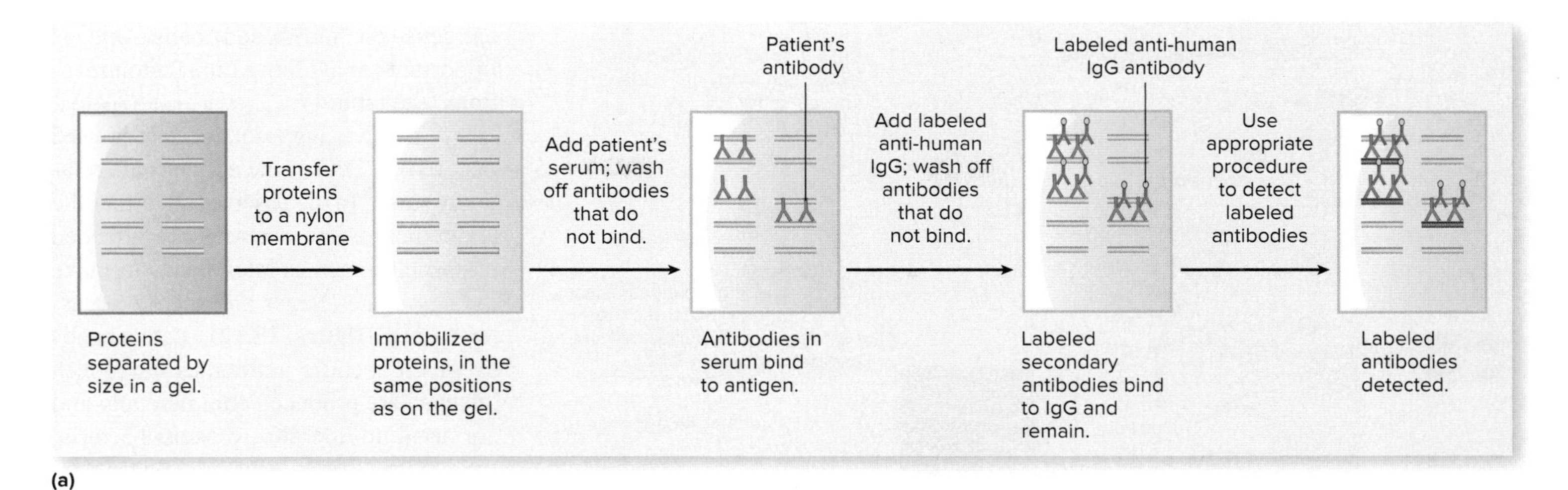

(a)

(b)

FIGURE 18.9 Western Blotting **(a)** Procedure. **(b)** Results of HIV testing. Each vertical strip is from a different patient. The dark bands represent proteins recognized by the patients' antibodies, and the numbers to the right indicate protein sizes. b: ©Hank Morgan/Science Source

Why are the results of the Western blot generally more reliable than those of the ELISA?

The basic steps of the Western blotting procedure are illustrated in **figure 18.9a.** First, a special type of gel electrophoresis called SDS-PAGE is used to separate the proteins. To do this, samples are loaded onto a polyacrylamide gel matrix, and an electrical current is then run through the gel. Smaller proteins move faster through the gel matrix than the larger ones, so the proteins separate on the basis of size. The separated proteins in the gel are then transferred ("blotted") to a nylon membrane to immobilize them in the same positions they were in the gel, creating what is referred to as a blot (a membrane containing the immobilized proteins). The steps after that are very similar to those of an ELISA. To determine if a patient's serum has antibodies specific for any of the proteins in the sample, some of that serum is added to the blot, after which unbound antibodies are washed off. Enzyme- or radioactively labeled anti-human IgG antibodies are then added, which attach to any serum antibodies that have attached to the proteins. Unbound labeled antibodies are then washed off. Finally, the label is detected, generating a bar code–like pattern that reflects which proteins on the blot were recognized by the patient's antibodies (figure 18.9b). The detection method depends on the type of molecule used to label the antibodies.

Fluorescence-Activated Cell Sorter (FACS)

A **fluorescence-activated cell sorter (FACS)** can be used for a variety of purposes, including diagnosing and monitoring the progression of diseases involving white blood cells. For example, it can be used to characterize different types and stages of lymphomas and leukemias as well as to diagnose immunodeficiency diseases. In addition, it can be used to track the progression of HIV disease by determining the serum levels of CD4 T cells. ⏮ CD markers

A FACS is a specialized version of a flow cytometer that sorts and counts cells labeled with fluorescent antibodies (**figure 18.10**). Perhaps more important, it also measures the amount of fluorescence, along with cell size and density, allowing researchers to compare the concentrations and characteristics of the labeled cells or other particles. This is why it is so useful for monitoring differences in various types of white blood cells (see table 18.6). ⏮ flow cytometry

Immunoassays That Involve Visible Antigen-Antibody Aggregates

As described in chapter 15, antibodies can cross-link antigens, thereby creating large "mouthfuls" for phagocytic cells (see figure 15.8). These clumped antigen-antibody complexes can be observed in agglutination and precipitation reactions.

Agglutination Reactions

Agglutination reactions take advantage of the visible clumping that occurs when antibodies cross-link relatively large particles such as cells. The reactions are used in red blood cell typing, and to identify certain types of microorganisms.

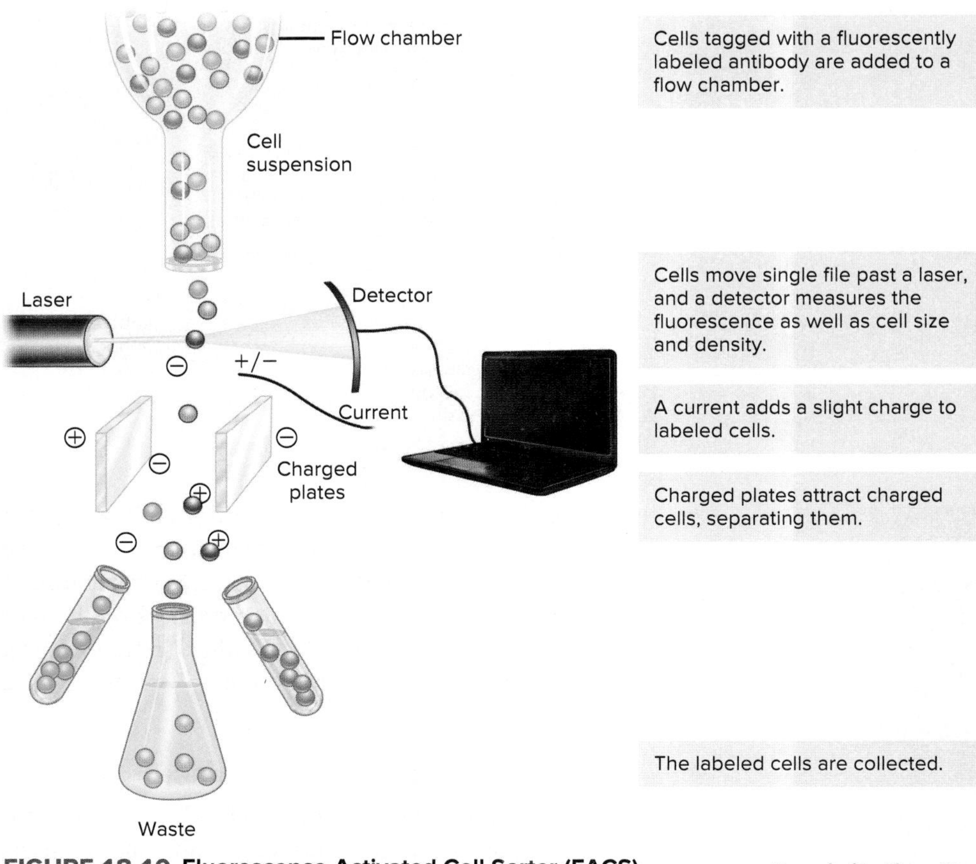

FIGURE 18.10 Fluorescence-Activated Cell Sorter (FACS)

In a direct agglutination test (DAT), an antibody suspension is mixed with the large antigen molecules (**figure 18.11a**). If the antibodies bind to the antigens, visible clumping will occur—a positive test. The agglutination of red blood cells by antibody binding or other means is referred to as hemagglutination and is used in blood typing (figure 18.11b). Recall that blood typing involves detecting certain carbohydrate antigens on red blood cells, and is important for avoiding transfusion reactions (see table 17.2). ◀◀ transfusion reactions

A passive agglutination test is used instead of a DAT if the antigens are relatively small. In the passive test, either the antibodies or the antigens are attached to particles such as latex beads to make the aggregates larger and therefore easier to see (**figure 18.12**). Latex beads to which specific antibodies have been attached are produced commercially and are used to identify various bacteria, fungi, viruses, and parasites. They are also used to test for hormones, drugs, and other substances in body fluids. The beads are mixed with a drop of a body fluid or suspended microbial culture. If the specific antigen is present, visible clumps form. Latex agglutination tests are commonly used to identify certain *Streptococcus* species (figure 18.12b). In the reverse type of reaction, antigens are attached to latex beads, making it possible to detect specific antibodies.

Precipitation Reactions

When antibodies bind to soluble antigens—that is, molecules that are dissolved rather than suspended particles—large lattice-like complexes may form that precipitate out of solution (**figure 18.13**). This is the basis of **precipitation reactions.** The complexes can take several hours to form and develop only at certain relative concentrations of antibody and antigen molecules. If there is a great excess of either, the complexes do not form, and consequently, no precipitate will be seen.

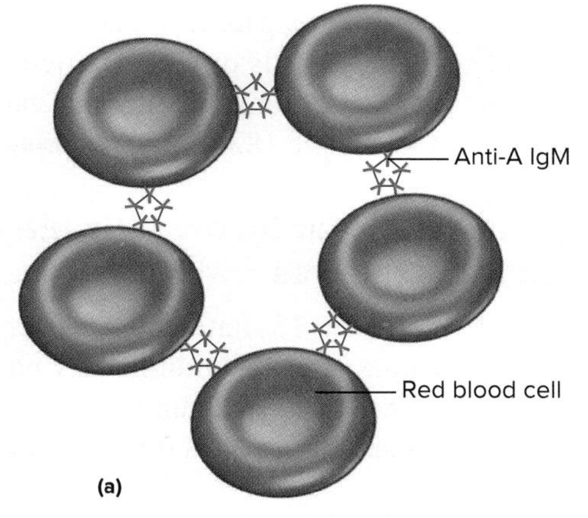

(a)

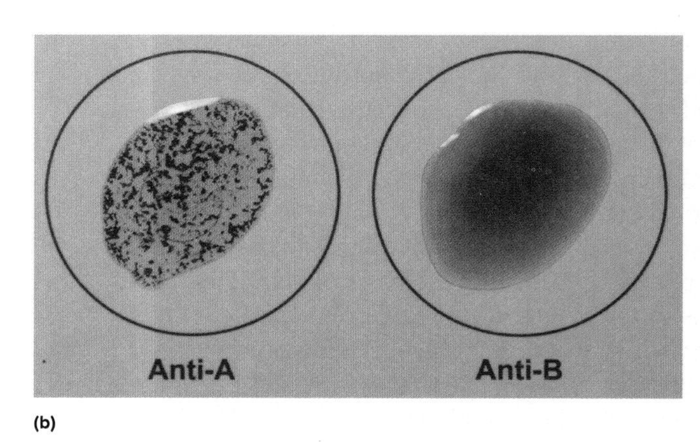

(b)

FIGURE 18.11 Direct Agglutination In this example, red blood cells (RBCs) are being tested for ABO blood groups by mixing them with separate antibody suspensions specific for either A or B antigens. **(a)** If antibodies bind an antigen on the RBCs, the resulting cross-linking will cause agglutination. **(b)** The RBCs agglutinated when mixed with anti-A antibodies, but did not agglutinate when mixed with anti-B antibodies, indicating that the blood group is type A. b: ©McGraw-Hill Education/Lisa Burgess

What term is used to describe the agglutination of red blood cells?

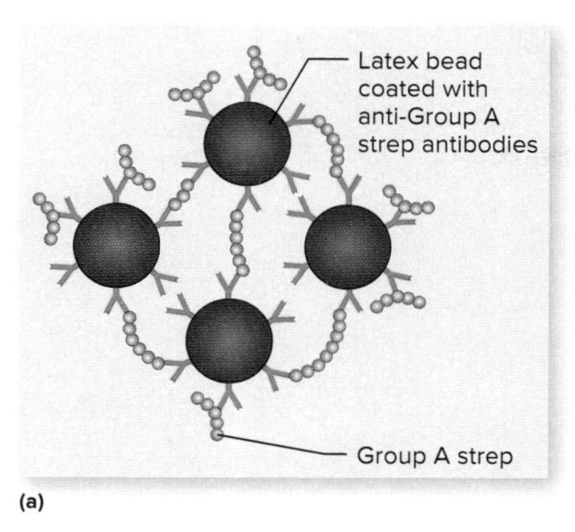

(a)

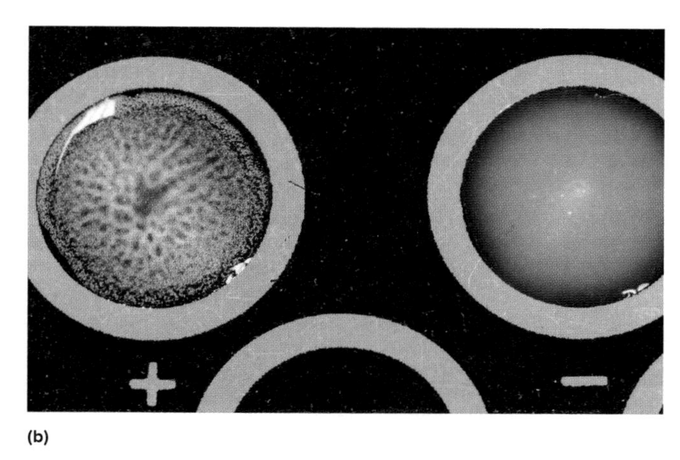

(b)

FIGURE 18.12 Passive Agglutination In this example, antibody-coated latex beads are used to identify *Streptococcus pyogenes.* **(a)** The latex beads are coated with antibodies that bind specifically to cell wall antigens of the bacteria. **(b)** Visible clumping (shown on the left) confirms that the organism is *S. pyogenes.* Negative test results are shown on the right. b: ©McGraw-Hill Education/Lisa Burgess

Why coat latex beads with antibodies rather than simply adding the antibodies to a bacterial suspension?

The easiest way to get the proper concentrations is to place the antigen and antibody suspensions near each other in a gel and let the molecules diffuse toward each other—a process called immunodiffusion. A precipitate will form in a distinct region called the zone of optimal proportions.

Precipitation reactions are not commonly used in diagnosis today because they have largely been replaced by methods that rely on labeled antibodies. However, the principle of the reactions is nicely demonstrated by the Ouchterlony technique, which can be done in a Petri dish (**figure 18.14**). Antigen and antibody solutions are placed into separate wells cut in the gel contained in the dish. The two solutions will gradually diffuse outward, meeting between the wells. If the antibody molecules recognize the antigen, a line of precipitation will form at the zone of optimal proportions. Because there are often multiple antigens present in the sample, as well as different specificities of antibodies in the serum, more than one line can form, each in its area of optimal proportions. The Ouchterlony test can be used to detect autoantibodies associated with certain connective tissue disorders. ◀◀ autoantibodies

MicroAssessment 18.4

Antibodies labeled with a detectable marker can be used to identify an antigen (direct test) or to detect a patient's antibodies to a known antigen (indirect test). Examples of methods that use labeled antibodies include ELISA, Western blotting, fluorescence antibody tests, and fluorescence-activated cell sorters. Agglutination and precipitation reactions both depend on formation of visible antigen-antibody complexes.

10. Why is the ELISA used so commonly relative to the Western blot?

11. How is a direct agglutination test different from a passive agglutination test?

12. Why is a false positive more significant in HIV testing of patients than in screening donated blood for transfusions?

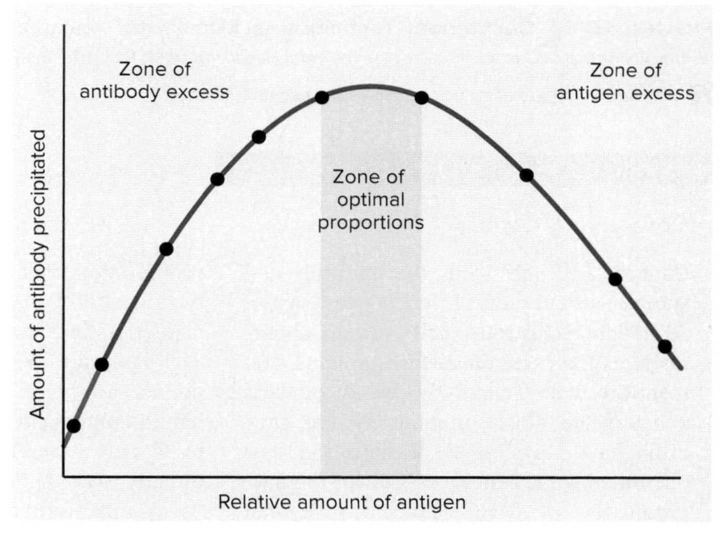

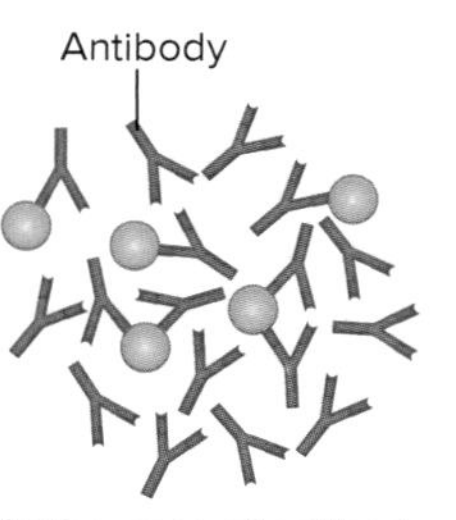

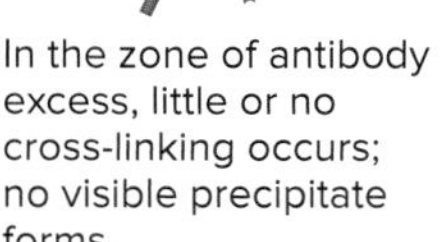

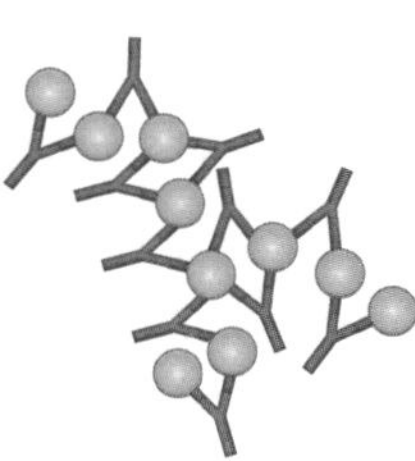

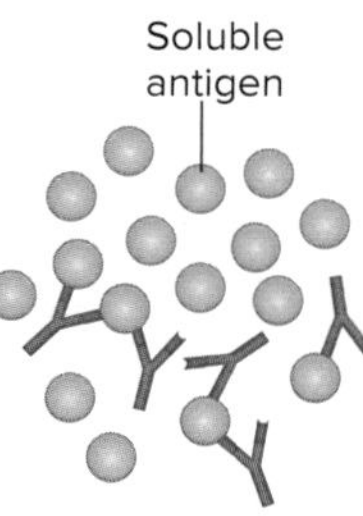

In the zone of antibody excess, little or no cross-linking occurs; no visible precipitate forms.

In the zone of optimal proportions, extensive cross-linking occurs; a visible precipitate forms.

In the zone of antigen excess, little or no cross-linking occurs; no visible precipitate forms.

FIGURE 18.13 Antigen-Antibody Precipitation Reactions The maximum amount of precipitate forms in the zone of optimal proportion.

It takes fewer molecules of IgM than of IgG to cause precipitation. Why would this be so?

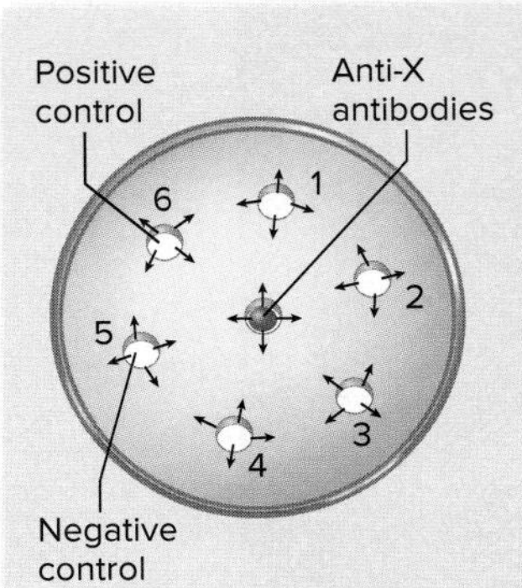

Antibody and antigen solutions are placed into separate wells cut into the gel. The antigens and antibodies diffuse toward each other. In this example, antibodies that bind antigen X (called anti-X antibodies) were added to the well in the center of the gel. Samples that contain unknown antigens are added to wells 1 through 4. A negative control (contains no antigen X) was added to well 5, and a positive control (contains known antigen X) was added to well 6.

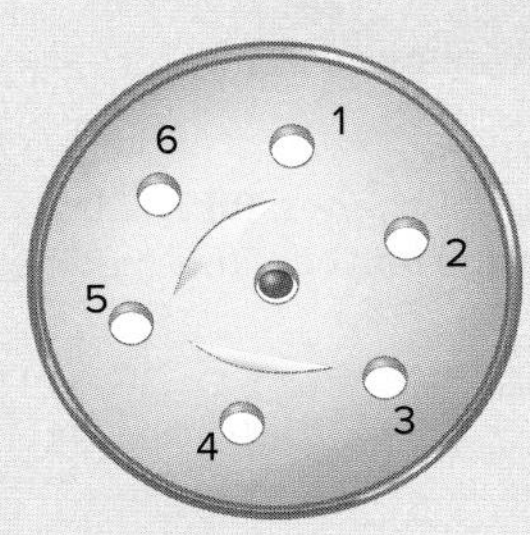

When antibody molecules that recognize the antigen meet at the zone of optimal proportions, antigen-antibody complexes precipitate out of solution, forming a visible line. In this example, a line has formed between the center well and well 6 (the positive control). A line has also formed between the center well and the sample in well 4, indicating that the sample contains antigen X. The other samples do not contain detectable amounts of antigen X.

(a)

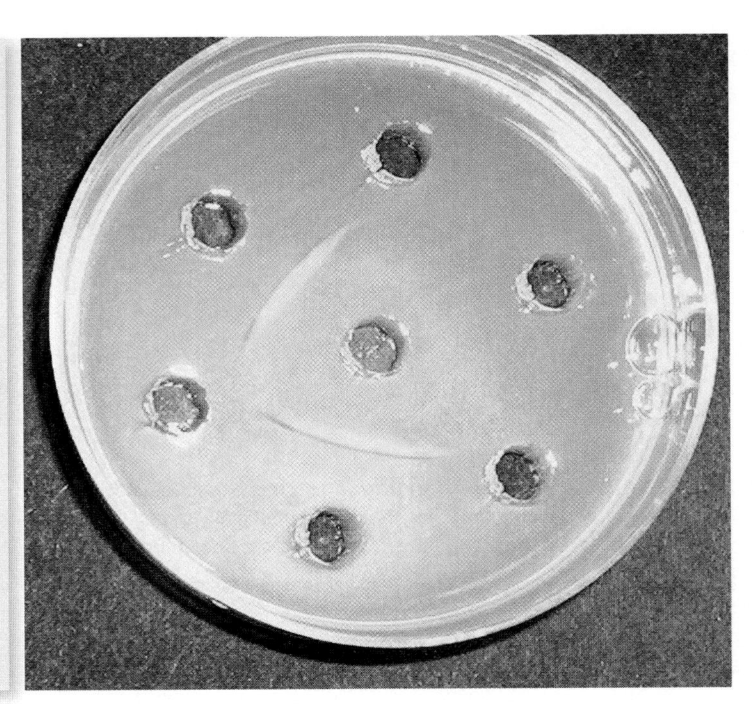

(b)

FIGURE 18.14 Ouchterlony Technique (a) Method. **(b)** Photograph of results. Source: Reproduced by permission from Schadler, D. L. 2003. Antigen-antibody testing: A visual simulation or virtual reality. *The Plant Health Instructor.* DOI: 10.1094/PHI-K-2003-0224-01

What is the purpose of including positive and negative controls?

FOCUS ON THE FUTURE 18.1

Conquering Cancer

Cancers are "self" cells that multiply and spread without control, leading to disease. Although cancerous cells often abnormally express certain surface proteins, and therefore can potentially be recognized and targeted by the immune system, they often have ways to evade detection. For one thing, some cancer cells down-regulate production of MHC molecules so that less antigen is presented. In addition, the immune system has mechanisms to avoid responses against "self" cells, thereby interfering with an effective response. Scientists are developing ways to manipulate the immune system so that it more effectively attacks cancerous cells—an approach called immunotherapy. Components of cancer immunotherapies currently in use include cytokines, monoclonal antibodies, and genetically modified T cells.

Recall that cytokines direct certain immune cell activities. Aldesleukin, a medication used to treat certain types of kidney cancer and melanoma, is the cytokine interleukin 2 (IL-2). It binds to and stimulates proliferation of T cells, an important cell type in fighting cancer. Denileukin diftitox is interleukin 2 attached to diphtheria toxin. This is used to treat cutaneous T-cell lymphoma, a type of T-cell cancer. The interleukin component of the medication binds to T cells, thereby delivering the toxin directly to them.

Another form of cancer immunotherapy involves monoclonal antibodies, some of which have additional molecules attached. Examples and mechanisms of monoclonal antibodies in cancer immunotherapy include:

- **Naked antibodies.** These carry no additional molecules. In certain cases, by binding to a cancer cell, the antibody molecules mark the cell for destruction via antibody-dependent cellular cytotoxicity (ADCC). An example is rituximab, which is used to treat some types of B-cell cancers. It binds to a B-cell surface protein called CD20, thereby marking B cells for destruction. Other naked antibodies function as "checkpoint inhibitors," meaning that they interfere with processes that normally help prevent T cells from attacking "self" cells. Pembrolizumab binds to a programmed death receptor (PD-1) on T cells that certain cancer cells otherwise take advantage of to avoid being attacked. Some naked antibodies interfere with a function required for cancer cell proliferation. Trastuzumab, which is used to treat breast cancers that result from overexpression of the HER2 (human epidermal growth factor receptor 2), binds to the receptor, thereby preventing the cell from receiving the signal that otherwise stimulates growth of the cell.
- **Antibody-drug conjugates.** These have been constructed to deliver a toxin or other molecule to a cancerous cell. An example is ado-trastuzumab emtansine; the trastuzumab component

continued

binds HER2 and thereby delivers emtansine, a cytotoxin that enters the cell and interferes with the function of tubulin.

- **Bispecific antibodies.** These have been constructed to have two different antigen binding sites. An example is blinatumomab, which is used to treat some types of acute lymphocytic leukemia. One of the antigen binding sites attaches to CD3, a protein found on T cells; the other binds to CD19, a surface protein found on some cancerous B cells involved in the leukemia. Thus, the bispecific antibodies connect the two cell types, thereby increasing the chance that the T cells will destroy the cancerous B cells. Blinatumomab is also referred to as bispecific T-cell engager (BiTE).

A new entry into the immunotherapy arsenal is chimeric antigen receptor (CAR) T cells. These are a patient's own T cells that have been genetically engineered to express a modified T-cell receptor: the extracellular part of the modified receptor is composed of one "arm" of an antibody, and the intracellular part is the signaling domain of a T-cell receptor. Thus, a CAR T cell has been engineered to recognize and respond to a specific epitope, without the need for antigen presentation or co-stimulation. Once the CAR T cells have been created, they are put back into the patient, where they will then seek out and destroy any host cells that have the given epitope on their surface. This type of therapy shows tremendous promise in treating certain types of blood cancers.

Summary

IMMUNIZATION

18.1 ■ Principles of Immunization (figure 18.2)

Active Immunity

Active immunity occurs naturally in response to infections or other natural exposure to antigens, and artificially in response to vaccination.

Passive Immunity

Passive immunity occurs naturally during pregnancy and through breast feeding, and artificially by transfer of preformed antibodies, as in **immune globulin** and **hyperimmune globulin.**

18.2 ■ Vaccines and Immunization Procedures (table 18.1)

A **vaccine** is a preparation of a pathogen or its products used to induce active immunity. It protects an individual against disease, and can also provide **herd immunity**.

Attenuated Vaccines

An **attenuated vaccine** is a weakened form of the pathogen that can replicate but is generally unable to cause disease.

Inactivated Vaccines

Inactivated vaccines are unable to replicate but they retain the immunogenicity of the infectious agent or toxin. They include inactivated whole agents, toxoids, subunits, VLPs, polysaccharides, and conjugates. **Adjuvants** increase the intensity of the immune response to the antigen in a vaccine.

The Importance of Vaccines (tables 18.3, 18.4, 18.5)

Routine childhood immunizations have prevented millions of cases of disease and many deaths (table 18.2).

An Example of Vaccination Strategy—The Campaign to Eliminate Poliomyelitis

Vaccines against poliomyelitis provide an excellent illustration of the complexity of vaccine strategies.

IMMUNOLOGICAL TESTING

18.3 ■ Principles of Immunoassays

A seronegative individual becomes seropositive after initial infection with an agent; this **seroconversion** usually takes about a week to 10 days. To determine if a patient has antibodies in the blood against a specific infectious agent, the patient's **serum** or **plasma** is tested.

Quantifying Antigen-Antibody Reactions

The concentration of antibody molecules in a specimen is usually determined by making serial dilutions; the last dilution that gives a detectable antigen-antibody reaction reflects the **titer.**

Obtaining Known Antibodies

Polyclonal antibodies recognize multiple epitopes, whereas **monoclonal antibodies** recognize only a single epitope. **Anti-human IgG antibodies** are used to detect IgG molecules in a patient specimen. Recombinant DNA techniques have been used to produce humanized monoclonal antibodies.

18.4 ■ Common Types of Immunoassays (table 18.6)

Immunoassays That Use Labeled Antibodies

Direct immunoassays are typically used to identify unknown antigens (figure 18.5a). **Indirect immunoassays** are typically used to detect antibodies of a given specificity in a patient's serum (figure 18.5b). The **fluorescent antibody (FA) test** relies on fluorescence microscopy to locate fluorescently labeled antibodies bound to antigens fixed to a microscope slide (figure 18.6). The **enzyme-linked immunosorbent assay (ELISA)** uses antibodies labeled with a detectable enzyme (figure 18.7, 18.8). In the **Western blot** technique, the various proteins that make up an antigen are separated by size before reacting them with antibodies (figure 18.9). The **fluorescence-activated cell sorter (FACS)** can be used to count and separate antigens labeled with fluorescent antibodies, as well as to determine concentrations and characteristics of the labeled particles (figure 18.10).

Immunoassays That Involve Visible Antigen-Antibody Aggregates

Antibodies bound to particulate antigens cause obvious aggregates to form. Examples of **agglutination reactions** include direct agglutination tests and passive agglutination tests (figures 18.11 and 18.12). Antibodies bound to soluble antigens may form complexes that precipitate out of solution (figure 18.13); an example of a test that involves a **precipitation reaction** is the Ouchterlony technique (figure 18.14).

Review Questions

Short Answer

1. How is immune globulin different from hyperimmune globulin?
2. Describe two advantages of an attenuated vaccine over an inactivated one.
3. Describe two advantages of an inactivated vaccine over an attenuated one.
4. What is herd immunity?
5. Describe how both active and passive immunization can be used to combat tetanus.
6. Why are humanized monoclonal antibodies better for therapy than the original versions?
7. What is the purpose of anti-human IgG antibodies in immunological testing?
8. What is an advantage of the fluorescent antibody test over an ELISA?
9. Is blood typing an example of a precipitation reaction or an agglutination reaction?
10. In a precipitation reaction, what is meant by "zone of optimal proportions"?

Multiple Choice

1. Which is an example of immunization that elicits active immunity?
   a) Giving antibodies against diphtheria
   b) Immune globulin injections to prevent hepatitis
   c) Inactivated polio vaccine
   d) Rabies immune globulin
   e) Tetanus immune globulin
2. Breast feeding provides which of the following to an infant?
   a) Artificial active immunity
   b) Artificial passive immunity
   c) Natural active immunity
   d) Natural passive immunity
3. Vaccines ideally should be all of the following *except*
   a) effective in protecting against the disease.
   b) inexpensive.
   c) stable.
   d) living.
   e) easily administered.
4. Severely immunosuppressed people should not receive the measles vaccine. Based on this information, the vaccine is
   a) an inactivated whole agent.
   b) a toxoid.
   c) a subunit vaccine.
   d) a genetically engineered vaccine against hepatitis B.
   e) an attenuated vaccine.
5. All of the following are attenuated vaccines *except*
   a) chickenpox.
   b) mumps.
   c) rubella.
   d) pertussis.
   e) measles.
6. *Haemophilus influenzae* is an encapsulated bacterium that causes infant meningitis. Considering this, which would be best to protect infants against this organism?
   a) Conjugate vaccine
   b) Polysaccharide vaccine
   c) Attenuated vaccine
   d) Subunit vaccine
   e) Toxoid
7. In quantifying antibodies in a patient's serum,
   a) total protein in the serum is measured.
   b) the antibody is usually measured in grams per mL.
   c) the serum is serially diluted.
   d) both antigen and antibody are diluted.
   e) the titer refers to the amount of antigen added.
8. Which of the following about immunological testing is *false*?
   a) Polyclonal antibody preparations recognize multiple epitopes.
   b) Monoclonal antibodies recognize a single epitope.
   c) Serum and plasma can both be tested for antibodies.
   d) The direct ELISA uses anti-human IgG antibodies.
   e) A rise in specific antibody titer indicates an active infection.
9. All of the following are matching pairs *except*
   a) ELISA—radioactive label.
   b) fluorescence-activated cell sorter—flow cytometry.
   c) fluorescent antibody test—microscopy.
   d) Western blot—gel electrophoresis.
10. Which of the following would be most useful for screening thousands of specimens for antibodies that indicate a certain disease?
    a) Western blot
    b) Fluorescent antibody
    c) ELISA
    d) All of the above
    e) None of the above

Applications

1. A new parent asks you which vaccines the CDC recommends for a 2-month-old infant. Based on the latest CDC's recommended immunization schedules found online, what is your answer?
2. In tests to determine if a patient has measles, the laboratory looks for IgM. Why would finding IgM be more significant than finding IgG?

Critical Thinking

1. An ELISA test is used to screen patient specimens for HIV. A positive ELISA test is confirmed by a Western blot test. Why not the other way around, with the ELISA second?
2. *Staphylococcus aureus* makes a protein called protein A, which binds to the Fc region of antibody molecules from a wide variety of species. How could protein A be exploited in immunoassays?

19 Epidemiology

CDC microbiologist studies influenza virus. *Source: James Gathany/CDC*

KEY TERMS

Attack Rate The proportion of susceptible persons developing illness in a population exposed to an infectious agent.

Case-Fatality Rate The proportion of persons diagnosed with a specific disease who die from that disease.

Communicable Disease An infectious disease that can be transmitted from one host to another.

Endemic A disease or other occurrence that is constantly present in a population.

Epidemic A disease or other occurrence that has a much higher incidence than usual.

Herd Immunity Protection of an entire population based upon a critical concentration of immune hosts that prevents the spread of an infectious agent.

Incidence The number of new cases of a disease in a population at risk during a specified time period.

Morbidity Illness; most often expressed as the rate of illness in a defined population.

Mortality Death; most often expressed as the rate of death in a defined population.

Outbreak A group of cases occurring during a brief time interval and affecting a specific population.

Pandemic An epidemic spread over several continents.

Prevalence The total number of cases of a disease in a given population at any time or for a specific period.

Reservoir of Infection The natural habitat of a pathogen; sum of the potential sources of an infectious agent.

A Glimpse of History

Puerperal fever, a bacterial infection of the uterus following childbirth, rose to epidemic proportions when more women chose to deliver their babies in hospitals. By the mid-1800s in the hospitals of Vienna, the medical hub of the world at that time, about one of every eight women died of puerperal fever following childbirth.

In 1841, Ignaz Semmelweis, a Hungarian, traveled to Vienna to study medicine. After medical school, he worked for Professor Johann Klein, who, with a staff of medical students, ran one section of a maternity hospital. Midwives and midwifery students served a second section. Semmelweis noticed that the incidence of puerperal fever in Dr. Klein's section was as high as 18%, four times that in the section served by the midwives. During this period, a friend of Semmelweis incurred a scalpel wound while doing an autopsy and died of symptoms very similar to those of puerperal fever. Semmelweiss reasoned that the "poison" that killed his friend probably also contaminated the hands of the medical students who did autopsies. Perhaps these students were transferring the "poison" from the cadavers to the women in childbirth. Midwives, after all, did not perform autopsies. This was before Pasteur and Koch established the germ theory of disease, so Semmelweis had no way of knowing that the "poison" being transferred was probably *Streptococcus pyogenes,* a common cause of many infections, including puerperal fever.

To test his hypothesis that physicians and students were transferring "poison" to patients, Semmelweis had them wash their hands with a strong disinfectant before attending patients. The incidence of puerperal fever dropped to one-third its previous level. Instead of appreciating these findings, Semmelweis's colleagues refused to admit responsibility for the deaths of so many patients. His work was so fiercely attacked that he was forced to leave Vienna and return to his native Hungary. Although his disinfection techniques achieved a remarkable reduction in the number of deaths from puerperal fever there as well, Semmelweis became increasingly outspoken and bitter. He was finally confined to a mental institution, where he died one month later of a generalized infection similar to the kind that had killed his friend and the many women who had contracted puerperal fever following childbirth.

Epidemiology is the study of the distribution and causes of disease in populations. Epidemiologists—the "disease detectives"—collect and compile data about sources of disease and associated risk factors. Using that information, they design strategies to prevent or predict the spread of disease. They approach a disease outbreak much as a criminal detective describes the scene of a crime using expertise in diverse disciplines including ecology, microbiology, sociology, statistics, and psychology. Many of our daily habits, from handwashing to waste disposal, are based on the work of epidemiologists.

19.1 ■ Basic Concepts of Epidemiology

Learning Outcome

1. Explain why epidemiologists are most concerned with the rate of disease rather than the number of cases.

Diseases that can be transmitted from one host to another, such as measles, colds, and influenza, are **contagious,** or **communicable, diseases.** Transmission may be direct or indirect. For example, the virus that causes cold sores may be transmitted directly through a kiss, or indirectly by drinking from a glass just used by a symptomatic individual. **Non-communicable diseases** do not spread from one host to another. Microorganisms that cause these diseases most often arise from an individual's normal microbiota or from the environment. For example, legionellosis is caused by the bacterium *Legionella pneumophila,* which can be contracted from warm natural waters or from the water systems of buildings. Unlike most other respiratory infections, it typically does not spread from person to person.

Epidemiologists generally are less concerned with the number of cases of a disease than with the rate of infection in a population. For example, to have 100 people in a city of 1,000,000 develop genital herpes in a given time period is not as alarming as to have 100 people in a town of 5,000 develop the same disease over the same time period. The much higher rate or proportion of the population infected is of greater concern to the epidemiologist.

The **attack rate** describes the number of susceptible people who become ill in a population after exposure to an infectious agent. For example, if 100 people at a party eat chicken contaminated with *Salmonella,* and 10 people develop symptoms of salmonellosis, then the attack rate is 10%. The attack rate reflects many factors, including the infectious dose of the organism and the general health of the host population. Communicable diseases may have a significant secondary attack rate—the proportion of people in contact with infected individuals who then go on to develop the disease.

The **incidence** of a disease is the number of new cases in a specific time period in a given population; the **prevalence** of a disease is the total number of cases at any time or for a specific period in a given population. Usually these are expressed as a rate, the number of cases per 100,000 people. Incidence provides a useful measure of the risk that an individual will contract the disease. Prevalence reflects the overall impact of a disease on society because it includes both old and new cases, taking the duration of the disease into account.

Morbidity refers to illness. Most often it is expressed as a rate—the incidence of disease in a defined population. Contagious diseases such as influenza often have a high morbidity rate because each infected individual may transmit the infection to several others. **Mortality** refers to death. Again, it is also generally expressed as a rate—the number of people in a defined population who die during a given period. **Case-fatality rate** is the proportion of persons diagnosed with a specific disease who die from that disease. Diseases such as Ebola virus disease (EVD) are feared because of their very high case-fatality rate.

Endemic diseases are constantly present in a given population. For example, the common cold is endemic in the United States; malaria is not. When cases occur only from time to time, they are **sporadic.** An unusually large number of cases in a population constitutes an **epidemic.** Epidemics may be caused by diseases not normally present in a population (such as when cholera was reintroduced to the Western Hemisphere), or by fluctuations in the incidence of endemic diseases such as influenza and pneumonia (**figure 19.1**). An **outbreak** describes a group of cases occurring during a brief time interval and affecting a specific population; an outbreak may signal the onset of an epidemic. When an epidemic spreads over several continents, as AIDS has, it is called a **pandemic.**

MicroByte

About 1,500 malaria cases are reported in the United States each year, mostly following travel to countries where the disease is endemic.

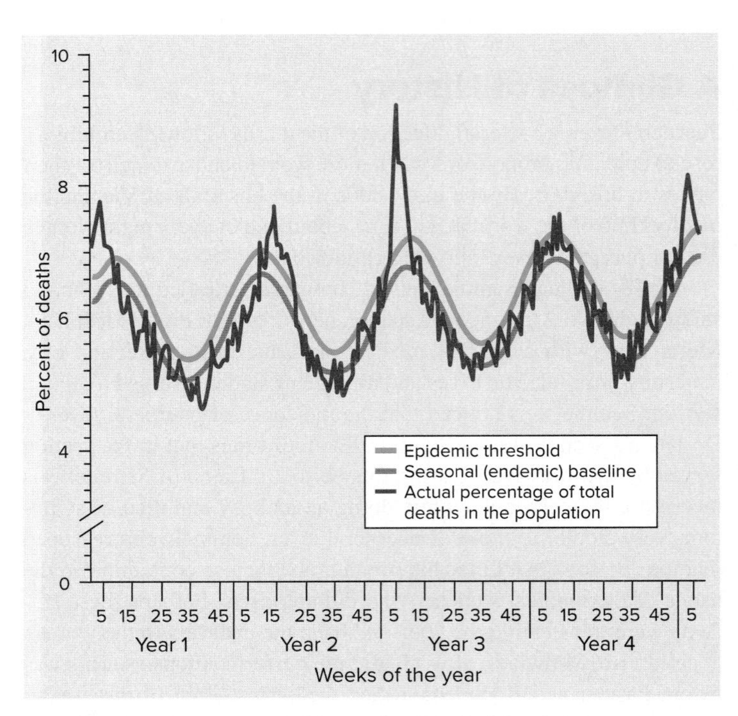

FIGURE 19.1 Endemic Disease Can Be Epidemic Example of yearly fluctuation of pneumonia and influenza deaths (expressed as a percentage of all deaths).

? Based on this graph, during what season(s) of the year is a pneumonia epidemic most likely?

MicroAssessment 19.1

Communicable diseases can be transmitted from one host to another. Rates of disease within a population are a concern of epidemiologists who study disease patterns. An unusually large number of cases within a population constitutes an epidemic.

1. Name two typical sources of non-communicable diseases.
2. Explain the difference between incidence of a disease and prevalence of the disease.
3. Why might a disease be endemic in one region, but not in another?

19.2 ■ Chain of Infection

Learning Outcome

2. Describe the steps involved in the chain of infection.

The spread of an infectious disease follows a series of steps often termed the chain of infection. First, there must be a source, or reservoir, of an infectious agent. If the reservoir is an infected host, the agent must leave that host through a portal of exit, be transmitted to a new host, and colonize the new host or enter that host through a portal of entry (**figure 19.2**). Knowing the chain of infection for a given disease allows researchers and public health workers to determine where links in the chain can be broken, thereby stopping or slowing the spread of the disease.

Reservoirs of Infection

The natural habitat of a pathogen, the **reservoir of infection,** may be on or in an animal (including humans) or in an environment such as soil or water (**figure 19.3**). The reservoir of infection affects the extent and distribution of a disease. Once the reservoir is identified, susceptible people can be prevented from coming into contact with the disease source.

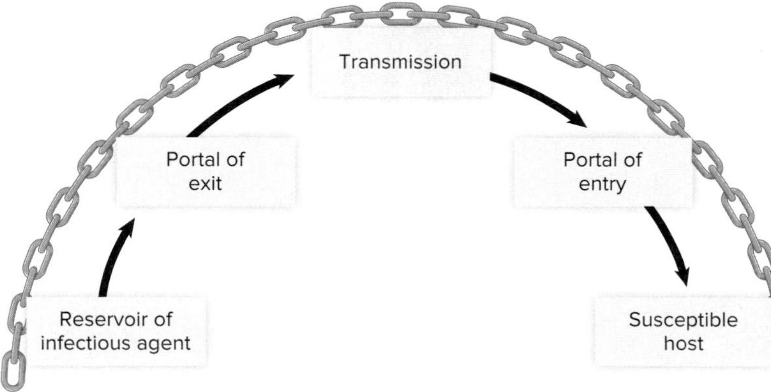

FIGURE 19.2 Chain of Infection If any link in this chain is broken, disease transmission is slowed or stopped.

How could disease transmission be stopped at the portal of exit?

FIGURE 19.3 Reservoirs of Infection (top): ©Amanda Clement/Getty Images; (bottom): ©Rudmer Zwerver/Shutterstock

How might you be a reservoir of infection?

Human Reservoirs

Infected humans are a significant reservoir of most communicable diseases. In some cases, humans are the only reservoir. In other cases, the pathogen can also exist in non-human animals and, occasionally, in the environment as well. When infected humans are the only reservoir, the disease can be easier to control because it is easier to set up prevention and control programs in humans than in wild animals. The eradication of smallpox is an excellent example. The combined effects of widespread vaccination programs (which resulted in fewer susceptible people) and the isolation of those who became infected eliminated the smallpox virus because it no longer had a reservoir in which to multiply.

Symptomatic Infections People with symptomatic illnesses are an obvious source of infectious agents. Ideally they understand the importance of taking precautions to avoid transmitting their illness to others. Resting at home while ill both helps the body recover and protects others from exposure to the disease-causing agent. Even conscientious people, however, can unintentionally be a source of infection. For example, people who are in the incubation period of mumps shed virus before symptoms appear. ◀◀ incubation period, ▶▶ mumps

Asymptomatic Carriers A person can harbor a pathogen with no ill effects, acting as a **carrier** of the disease agent. These people may shed the organism intermittently or constantly for months, years, or even a lifetime.

Some carriers have an asymptomatic infection; their immune system is actively responding to the invading microorganism, but they have no obvious clinical symptoms. Because these people often have no reason to consider themselves a reservoir, they move freely about, spreading the pathogen. People with asymptomatic infections are a significant complicating factor in the control of sexually transmitted infections (STIs) such as gonorrhea. Up to 60% of women infected with *Neisseria gonorrhoeae* have no symptoms and may unknowingly transmit the organism to their sexual partners. In contrast, infected men are more often symptomatic and seek medical treatment. ⏭ *Neisseria gonorrhoeae*

Some potentially pathogenic microbes can colonize the skin or mucosal surfaces, establishing themselves as part of a person's microbiota. For instance, many people carry *Staphylococcus aureus* as a part of their nasal or skin microbiota. Carriers of *S. aureus* may never have any illness or disease as a result of the organism, but they remain a potential source of infection to themselves and others. Unfortunately, ridding a colonized carrier of the infectious organism is often difficult, even with the use of antimicrobial medications. ⏭ *Staphylococcus aureus*

Non-Human Animal Reservoirs

Non-human animal reservoirs are the source of many pathogens. These can be very difficult to control, particularly in wild animal populations. Epidemics of plague no longer occur in the United States, in part because populations of wild rats, rock squirrels, and prairie dogs are controlled. These animals and their fleas are the natural sources of *Yersinia pestis,* the bacterium that causes plague. Raccoons, skunks, and bats are reservoirs of the rabies virus. Rodents, particularly the deer mouse, are the reservoir for hantavirus. Poultry are a reservoir of gastrointestinal pathogens such as species of *Campylobacter* and *Salmonella.* ⏭ *Yersinia pestis*

Diseases such as plague and rabies that can be transmitted to humans but exist primarily in other animals are called zoonotic diseases, or **zoonoses.** Zoonotic diseases are often more severe in humans than in the typical animal host because the infection in humans is accidental; there has been no evolution toward the balanced pathogenicity that normally exists between host and parasite. ⏮ balanced pathogenicity

Environmental Reservoirs

Some pathogens have environmental reservoirs. *Clostridium botulinum,* which causes botulism, and *Clostridium tetani,* which causes tetanus, are both widespread in soils. Pathogens that have environmental reservoirs are difficult or impossible to eliminate. ⏭ *Clostridium botulinum,* ⏭ *Clostridium tetani*

Portals of Exit

A pathogen must leave its reservoir to be transmitted to a susceptible host. If the reservoir is an animal, the body orifice or surface from which a microbe is shed is called the **portal of exit** (**figure 19.4**). Microorganisms that inhabit the intestinal tract are shed in the feces. *Vibrio cholerae,* an intestinal pathogen that causes massive volumes of watery diarrhea, can contaminate drinking water and food. Pathogens such as *Mycobacterium tuberculosis* and various respiratory viruses exit the body in droplets of saliva and mucus when people talk, laugh, sing, sneeze, or cough. Organisms that inhabit the skin are constantly shed on skin cells. Even as you read this text you are shedding skin cells, some of which may have *Staphylococcus aureus* on their surface. Genital pathogens such as *Neisseria gonorrhoeae* can be carried in urethral and vaginal secretions. ⏭ *Vibrio cholerae,* ⏭ *Mycobacterium tuberculosis*

Disease Transmission

An infectious agent must somehow be transmitted from its reservoir to the next host. **Vertical transmission** is the transfer of an infectious agent from a pregnant female to her fetus, or from a mother to her infant during childbirth or breast feeding. Prenatal care includes interventions to prevent vertical transmission of pathogens. **Horizontal transmission** refers to transfer by any other method, including person to person or environment to person (**figure 19.5**).

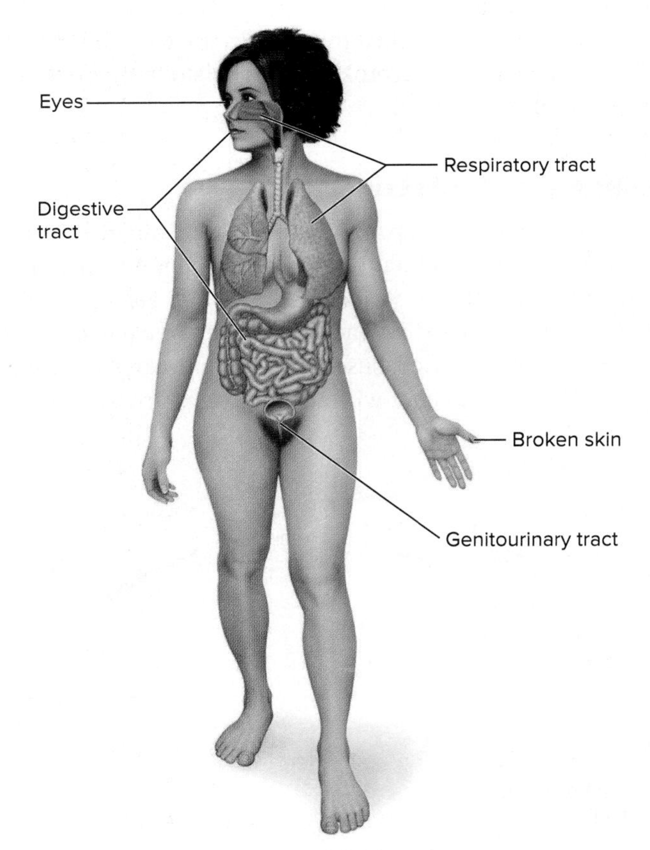

FIGURE 19.4 Portals of Exit or Entry A portal of exit is the route by which an infectious agent leaves the host. Some pathogens cause disease only when they enter a host through a specific portal of entry.

? **Name two portals of exit used by organisms that infect the digestive tract.**

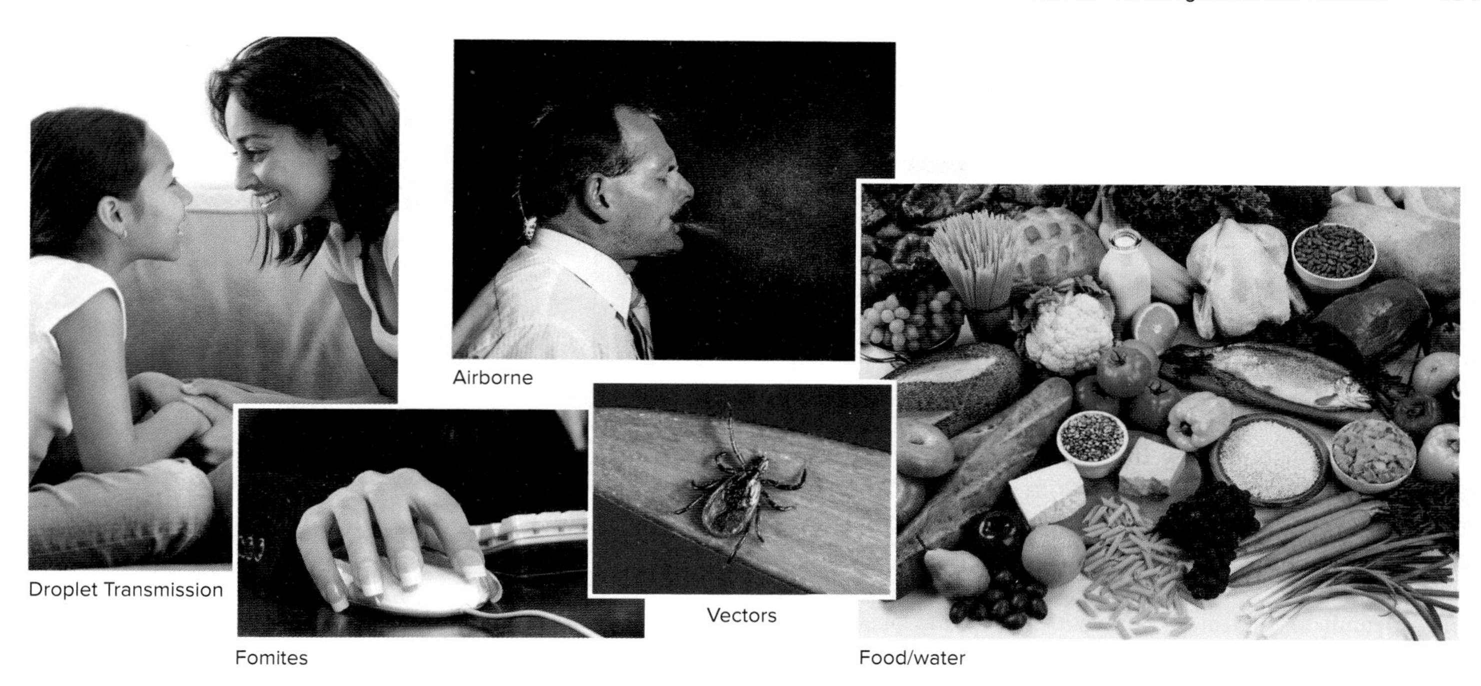

FIGURE 19.5 Horizontal Transmission of Pathogens (talking): ©Fuse/Getty Images; (computer mouse): ©Ingram Publishing; (sneeze): ©Kent Wood/Science Source; (tick): Source: CDC; (foods): ©Mitch Hrdlicka/Getty Images

? List three potential fomites that you contacted today.

Direct Transmission

Direct transmission of a pathogen from one host to another typically involves immediate transfer of the infectious agent to an appropriate portal of entry. This could occur via direct physical contact or projection of respiratory droplets onto mucous membranes.

Direct Contact Touching is **direct contact.** It can be as simple as a handshake or as intimate as sexual intercourse. Organisms with a low infectious dose—meaning that very few microbial cells are needed to initiate an infection—are often transmitted by direct contact. An example is *Shigella* species, which are easily transmitted by direct contact, particularly in day-care settings where young children play together. Handwashing, a simple routine that physically removes microbes, is important in preventing diseases that spread by direct contact. Even washing in plain water reduces the numbers of potential pathogens on the hands, which decreases the possibility of transferring or ingesting enough cells to establish an infection. Routine handwashing is considered to be the single most important measure for preventing the spread of infectious disease. When handwashing is not possible, use of alcohol gels may provide protection, but they are not effective against all pathogens.

Pathogens that cannot survive for extended periods in the environment must generally be transmitted through direct contact. *Treponema pallidum,* which causes syphilis, and *Neisseria gonorrhoeae,* which causes gonorrhea, both die quickly when exposed to a relatively cold, dry environment. Because of this, intimate sexual contact or other activity involving direct mucous-membrane-to-mucous-membrane contact is generally required to transmit them. ⏭ *Treponema pallidum,* ⏭ *Neisseria gonorrhoeae*

Droplet Spread When people talk, laugh, sing, sneeze, or cough, they continually discharge microorganisms in liquid droplets. **Droplet transmission** can spread respiratory disease if someone nearby inhales pathogen-laden respiratory droplets. This is a particularly important source of contamination in densely populated buildings such as schools and military barracks. Large droplets generally fall to the ground no farther than a meter (approximately 3 feet) from release. Desks or beds in such locations should be spaced more than 1.5 m apart to minimize the transfer of infectious agents. The spread of respiratory diseases is minimized if people cover their mouths (preferably not with their bare hands) when they cough or sneeze.

Indirect Transmission

Indirect transmission of a pathogen from one host to another, or from the surrounding environment to a new host, occurs through the air or via contaminated food, water, or inanimate objects. Other organisms such as mosquitoes, flies, or ticks can also transmit pathogens from one host to another.

Airborne Respiratory diseases are often transmitted through the air when small particles are inhaled into lungs, where any pathogens they carry can potentially cause disease. When larger particles (larger than 10 μm) are inhaled, they are usually trapped in the mucus that lines the nose and throat before they can enter the lungs. The mucociliary escalator then moves them toward the throat where they are eventually swallowed.

People continually discharge respiratory microorganisms in liquid droplets during normal activities such as talking. Large droplets quickly fall, but smaller droplets remain in the air as fluid evaporates from them. These airborne particles, called **droplet nuclei,** are composed of microbes attached to a thin coat of the dried material. Droplet nuclei can remain suspended indefinitely in the presence of even slight air currents. Other airborne particles, including dead skin cells, household dust, and soil disturbed by the wind, may also carry respiratory pathogens. Crowded conditions increase the risk of transmitting communicable respiratory diseases such as influenza or the common cold.

Understandably, airborne transmission of pathogens is very difficult to control. To prevent the buildup of airborne pathogens, modern public buildings have ventilation systems that constantly change the air. The air pressure in hospital microbiology laboratories can be lowered so that air flows in from the corridors, preventing microorganisms and viruses from being swept out of the lab to other parts of the building. Air in specialized hospital rooms, jetliners, and some laboratories is circulated through high-efficiency particulate air (HEPA) filters to remove airborne organisms that may be present. ⏮ **HEPA filters**

Vehicle-borne Pathogens can be transmitted via vehicles including **fomites** (inanimate objects), foods, and water.

Fomites, including clothing, keyboards, cell phones, doorknobs, and drinking glasses, can become contaminated when someone touches them. For example, when a carrier of *Staphylococcus aureus* touches a skin lesion or a colonized nostril, the bacterial cells now on that person's fingers can be transferred to a fomite. Another person then handling that object can acquire the microbes. Handwashing is an important control measure for preventing this type of indirect transmission.

Foods can become contaminated in a number of ways. Animal products such as meat and eggs may carry pathogens that originated from the animal's intestinal tract. This is the case with poultry contaminated with species of *Salmonella* or *Campylobacter* and hamburger contaminated with *E. coli* O157:H7. Pathogens can also be unintentionally added during food preparation. *Staphylococcus aureus* carriers who do not wash their hands prior to preparing food can easily contaminate the food. **Cross-contamination** results when pathogens from one food are transferred to another. A cutting board used first to chop raw chicken and then to cut salad ingredients can transfer *Salmonella* cells from the chicken onto the salad (**figure 19.6**). Because many foods are a rich nutrient source, microorganisms can multiply to high numbers if the

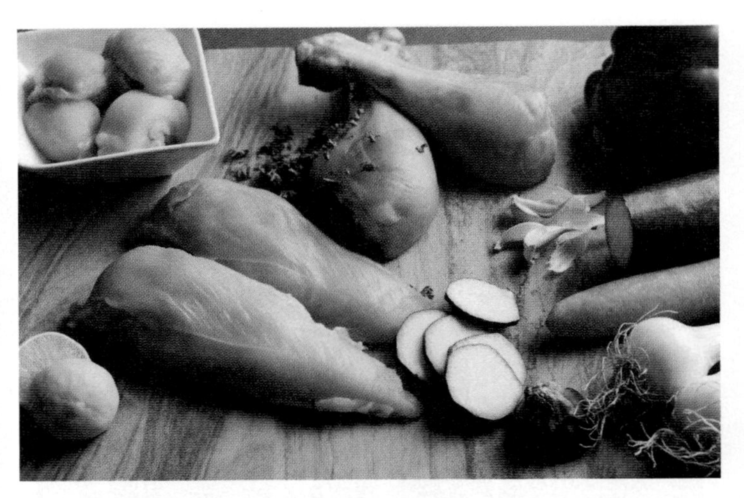

FIGURE 19.6 Cross-Contamination Vegetables exposed to juices from raw chicken may be a source of foodborne infection. ©Glow Cuisine/Getty Images

? What is the source of *Salmonella* species in raw chicken?

contaminated food is not refrigerated. Sound food-handling methods, including sanitary preparation as well as thorough cooking and proper storage, can prevent foodborne diseases. ⏭ **foodborne illness**

Waterborne disease outbreaks can involve large numbers of people because municipal water systems distribute water to widespread areas. The 1993 waterborne outbreak of cryptosporidiosis from *Cryptosporidium parvum,* an intestinal parasite, in Milwaukee, Wisconsin, was estimated to have affected more than 400,000 people. Prevention of waterborne diseases requires disinfection and filtration of drinking water and proper disposal and treatment of sewage. ⏭ ***Cryptosporidium parvum,*** ⏭ **drinking water treatment,** ⏭ **sewage treatment**

Vector-borne A **vector** is any living organism that can carry a disease-causing microbe, but most commonly the term is used to refer to arthropods such as mosquitoes, flies, fleas, lice, and ticks. A vector can carry a pathogen externally or internally.

Flies that land on feces can pick up intestinal pathogens such as *Escherichia coli* O157:H7 or *Shigella* species on their legs. If the fly then lands on food and transfers the microorganisms, it serves as a **mechanical vector,** carrying the microbe on its body from one place to another (**figure 19.7a**).

Diseases such as plague, malaria, and Lyme disease are transmitted via arthropods that harbor the pathogen internally. The vector often injects the infectious agent while taking a blood meal; for example, infected fleas transmit *Yersinia pestis* when biting. In the case of malaria, caused by protozoa in the genus *Plasmodium,* the mosquito not only transmits the parasite but also plays an essential role in its reproductive life cycle. Such vectors are called **biological vectors** (figure 19.7b). The malarial parasite multiplies to high numbers within this vector and also progresses to a developmental stage that is infective for humans. ⏭ ***Plasmodium***

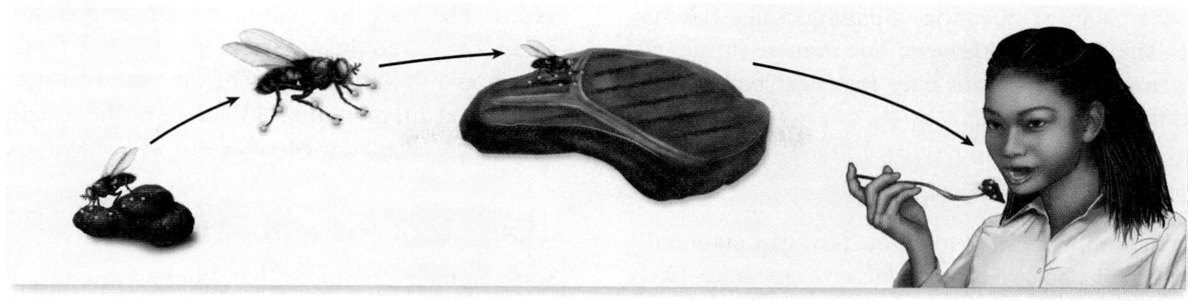

(a) Mechanical vector

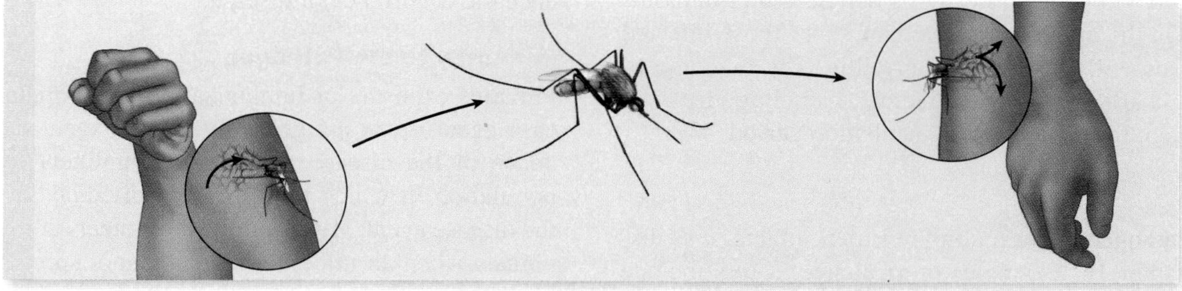

(b) Biological vector

FIGURE 19.7 Vector Transmission **(a)** A mechanical vector moves microbes from one place to another. **(b)** A biological vector participates in the life cycle of the pathogen and provides a place for it to multiply.

Name one disease spread by a mechanical vector and one disease involving a biological vector.

Prevention of vector-borne disease relies largely on control of arthropods. Malaria, once endemic in the continental United States, was successfully eliminated from the nation through a combination of mosquito control and prompt treatment of infected patients. Unfortunately, worldwide eradication efforts that initially showed great promise ultimately failed, in part due to the decreased vigilance that accompanied the dramatic but short-lived decline of the disease.

MicroByte

On average, two new mosquito-borne diseases that infect humans are discovered every year.

Portals of Entry

The last step in the chain of infection is colonization of a new host. In some cases, the pathogen grows on the outer surface of the new host, but often it must gain access to the host through a **portal of entry** (see figure 19.4). Respiratory pathogens released into the air during a cough generally cause disease only when someone inhales them. For these pathogens, the nose is the typical portal of entry. Intestinal pathogens such as *Vibrio cholerae* or *Shigella* species must usually be ingested. For *Shigella* species, the infectious dose is about 10 to 100 cells, a number easily transferred by casual contact. If cells of a *Shigella* species were transferred by touch, however, they would cause disease only if they were then transported to the mouth. In this case, the mouth is the portal of entry. When fecal organisms are inadvertently ingested, the transfer is called **fecal-oral transmission.** ⏭| *Shigella*

Many organisms can cause disease if they enter one body site, but are harmless if they enter another. For instance, *Enterococcus faecalis* may cause a bladder infection if it enters the urinary tract, but it is harmless when it is ingested and then colonizes the large intestine, where it often resides as part of the normal microbiota.

MicroAssessment 19.2

Spread of a disease can be prevented by breaking the chain of infection. The reservoir of a disease agent can be infected people, other animals, or the environment. To spread, infectious microbes must exit one reservoir, be transmitted to a susceptible host, and then enter that host. Handwashing and vector control can prevent many diseases; airborne transmission of pathogens is difficult to control.

4. How can an insect act as both a mechanical vector and a biological vector?
5. How does handwashing reduce the spread of disease?
6. Considering that circulating blood is not normally released from the body, describe how blood-borne microbes might exit.

19.3 ■ Factors That Influence the Epidemiology of Disease

Learning Outcome

3. Explain how characteristics of a pathogen, a host, or the environment can influence the epidemiology of a disease.

To complete the chain of infection, a pathogen needs a susceptible host. The spread of disease, however, is influenced by various factors involving not only the host, but also the pathogen and the environment.

Characteristics of the Pathogen

An infectious agent transmitted to a new host can potentially cause disease in that host. Some pathogens are more likely to cause disease than others. For example, nearly 100% of people infected with the measles virus will develop symptoms of measles. On the other hand, less than 1% of those infected with poliovirus will develop poliomyelitis. The outcome of transmission is affected by many factors, including virulence of the pathogen, the dose, and the incubation period.

Virulence

Successful pathogens have multiple virulence factors, as discussed in chapter 16. Examples of virulence factors include mechanisms for attaching to host surfaces, avoiding recognition or destruction by the immune system, and damaging the host. The array of virulence factors expressed by an organism affects the type, severity, and ease of transmission of disease. ⏮ virulence factors

The Dose

The probability of infection and disease is generally lower when an individual is exposed to smaller numbers of a pathogen. This is because a certain minimum number of cells of the pathogen are required to produce enough damage to cause disease. If 30 bacterial cells enter the body, for example, but 1,000,000 cells are required to produce symptoms, then it will take some time for the bacterial population to increase to that number. As the bacterial cells are multiplying, the host defenses are working to eliminate them, and may successfully do so before symptoms appear. This is why small doses often result in asymptomatic infections.

There are few if any infections for which immunity is guaranteed. An unusually large exposure to a pathogen, such as can occur in a laboratory accident, may cause disease in a person who is ordinarily immune to that microbe. Therefore, even immunized persons should take precautions to minimize exposure to infectious agents. This is especially important for medical workers taking care of patients with infectious diseases.

The Incubation Period

The extent of the spread of an infectious agent is influenced by its incubation period. Diseases with long incubation periods can spread far and wide before the first symptomatic cases appear. An excellent example was a typhoid fever outbreak that originated at a ski resort in Switzerland in 1963. As many as 10,000 people were exposed to drinking water containing small numbers of the bacterium *Salmonella enterica* serotype Typhi. The long incubation period of typhoid fever, 10 to 14 days, allowed the organism to be spread by the skiers, who flew home to various parts of the world before they became ill. As a result, more than 430 cases of the disease appeared in at least six countries. ⏭ typhoid fever, ⏮ incubation period

Characteristics of the Host

Some populations more than others are likely to be affected by a given pathogen. Many population characteristics influence the occurrence of disease.

Immunity to the Pathogen

Previous exposure or immunization of a population to a disease agent or an antigenically related agent decreases incidence of the disease. A disease is unlikely to spread in a population in which 90% of the individuals are immune to the disease agent. **Herd immunity** protects non-immune individuals when an infectious agent cannot spread in a population because most potential hosts are immune. Unfortunately, some infectious agents can undergo antigenic variation and thereby overcome herd immunity. ⏮ antigenic variation

General Health

Malnutrition, overcrowding, and fatigue increase people's susceptibility to infectious diseases, enhancing the spread. Infectious diseases are more of a problem in developing areas of the world where individuals are crowded together without proper food or sanitation. Factors that promote good general health result in increased resistance to diseases such as tuberculosis. When infection does occur in a healthy individual, it is more likely to be asymptomatic or result in mild disease.

Age

The very young and the elderly are generally more susceptible to infectious agents. In the case of young children, the immune system is not fully developed and, consequently, they are vulnerable to certain diseases. For example, young children are particularly susceptible to meningitis caused by *Haemophilus influenzae.* The introduction of Hib vaccine dramatically decreased the incidence of meningitis in this age group. The elderly are more prone to disease because immunity wanes over time. Influenza outbreaks in nursing homes often have high case-fatality rates. Older adults are also less likely to update their immunizations, making them more susceptible to diseases such as tetanus (**figure 19.8**). To prevent tetanus, a booster vaccine is needed every 10 years. ⏮ Hib vaccine, ⏭ tetanus

Gender

Gender may influence disease distribution. For example, women are more likely to develop urinary tract infections because their urethra (the tube that connects the urinary bladder to the external environment) is relatively short. Microbes can ascend the urethra into the bladder.

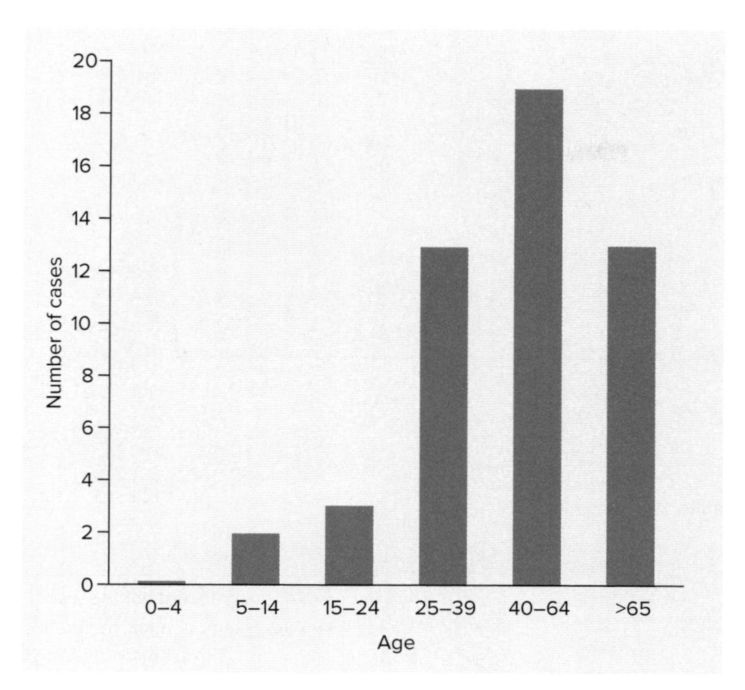

FIGURE 19.8 New Cases of Tetanus by Age Group

Why do you think individuals from 40–64 contract tetanus more often than younger people?

Behavioral Practices

Behavioral practices can significantly influence the rate and type of disease transmission. An infant who is breast-fed is less likely to have infectious diarrhea because of the protective antibodies in the mother's milk. Groups who eat traditional dishes made from raw freshwater fish are more likely to acquire the tapeworm *Diphyllobothrium latum,* a parasite normally killed by cooking.

Genetic Background

Natural immunity can vary with genetic background, but it is usually difficult to determine the relative importance of genetic, behavioral, and environmental factors. In a few instances, however, the genetic basis for resistance to infectious disease is known. For example, many people of black African ancestry are not susceptible to malaria caused by *Plasmodium vivax* because they lack a specific red blood cell receptor used by the infecting organism. Some populations of Northern European ancestry are less susceptible to HIV infection because they lack a certain receptor on their white blood cells.

Characteristics of the Environment

Environmental factors can influence the epidemiology of disease. Factors in the physical environment, including temperature, water and nutrient supply, radiation, and the availability of light and O_2, determine which organisms can exist and reproduce in a given environment. In addition to physical factors, the environment includes other organisms, including humans. Overcrowding in human populations is a major risk factor for infectious disease as transmission is often easier under crowded conditions, and poor sanitation creates opportunities for microbial spread and growth.

Unless transmission is by direct contact, the infectious agent must be able to survive in the environment between hosts. Some organisms, for example, form endospores that allow them to withstand damaging environmental conditions. Endospores may survive for decades and are resistant to extremes in temperature, radiation, lack of water, and other harsh conditions. A powerful environmental influence on an infectious agent is exposure to antibiotics that kills many susceptible organisms, allowing resistant organisms to multiply with less competition. ⏮ **endospores**

MicroByte

People who take antacids or drugs that decrease production of stomach acid have increased risk of gastrointestinal infections.

MicroAssessment 19.3

The outcome of disease transmission is affected by the virulence, dose, and incubation period of the infecting agent; various characteristics of the host population; and environmental conditions around the pathogen and the host.

7. Explain how a low dose of an infectious agent can result in an asymptomatic infection.
8. Why are diseases with long incubation periods more likely to result in an epidemic?
9. Why do influenza outbreaks in nursing homes typically have higher case-fatality rates than influenza outbreaks in a college dormitory?

19.4 ■ Epidemiological Studies

Learning Outcomes

4. Compare and contrast descriptive studies, analytical studies, and experimental studies.
5. Describe how a common-source epidemic can be distinguished from a propagated epidemic.

British physician John Snow illustrated the power of a well-designed epidemiological study over 150 years ago. Long before the relationship between microbes and disease was accepted, he documented that the cholera epidemics occurring in England from 1849 to 1854 were associated with contaminated water supplies. He did this by carefully comparing the conditions of households that were affected by cholera to those that were not, eventually determining that the primary difference was their water supply. Preventing access to the contaminated water broke the chain of infection. ⏭ **cholera**

Descriptive Studies

When a disease outbreak occurs, epidemiologists conduct a **descriptive study** by collecting data about the time, the place, and the individuals affected during the occurrence. That information, often collected from case reports, is used to compile a list of possible risk factors involved in the spread of disease.

The Person

Determining the profile of those who become ill is critical to defining the population at risk. Variables such as age, gender, ethnicity, occupation, personal habits, previous illnesses, socioeconomic class, and marital status may all yield clues about risk factors for developing the disease. For example, in the Swiss ski resort epidemic of typhoid fever mentioned earlier, cases occurred only in tourists because the local people rarely drank water, preferring wine instead.

The Place

The location of disease occurrence helps pinpoint the exact source. It may also give clues about potential reservoirs, vectors, or geographic boundaries that might affect disease transmission. For example, malaria can be transmitted only in regions that have an appropriate mosquito vector. If a specific location cannot be identified, it may take longer to control an outbreak. This can occur, for example, when contaminated commercial products are released over a wide area.

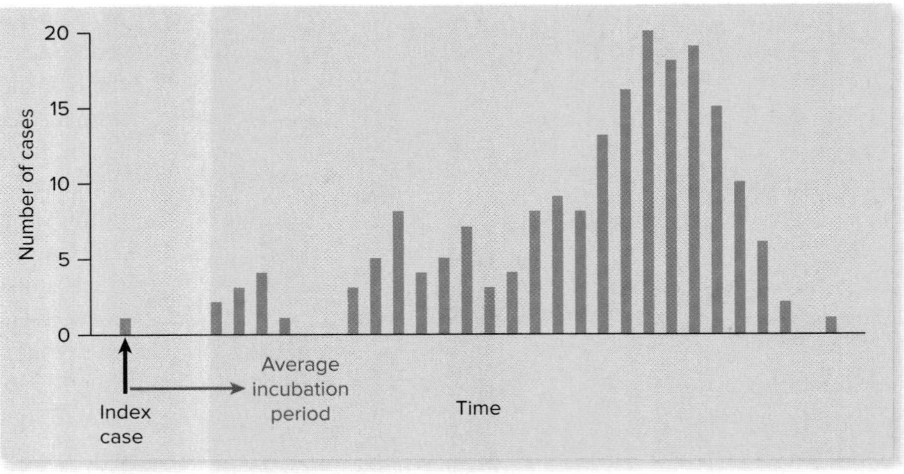

(a) Propagated Epidemic

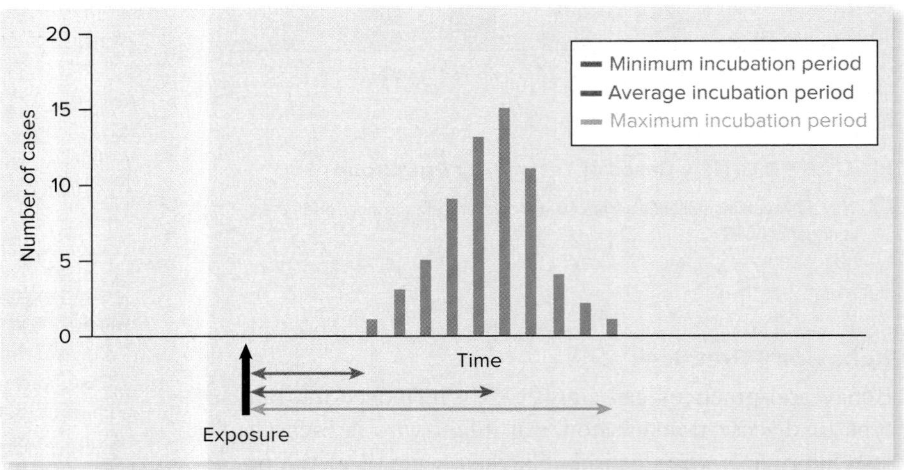

(b) Common-Source Epidemic

FIGURE 19.9 Comparison of Propagated Versus Common-Source Epidemics (a) The time between the index case and subsequent cases reflects the average incubation period of the disease. As the disease is propagated, this interval is blurred. **(b)** Cases in a common-source epidemic appear only within the incubation period.

? Why does the number of cases increase over several incubation periods in a propagated epidemic?

The Time

The timing of the appearance of cases during an outbreak yields important clues about the nature of the disease. To study this, epidemiologists make a graph that illustrates the time of onset of disease in each case. The graph, called an "epi" curve (for "epidemic curve"), helps to determine how the disease is being spread.

A **propagated epidemic** occurs when a disease is contagious, with one person transmitting it to several others, who then transmit it to several more, and so on. In this situation, the number of ill people rises gradually, with higher peaks in the number of cases over time (**figure 19.9a**). The first case in such an outbreak is called the **index case.** The average time between symptom onset in the index case and symptom onset of the next cases reflects the average incubation period of the disease. As the disease continues to spread, the gaps between new cases disappear.

A **common-source epidemic** occurs when all of the cases result from exposure to a single source of the infectious agent. The classic example of this type of epidemic is illness that results when people eat contaminated food at a picnic or a party. The resulting epi curve will show a rapid rise in the number of people who become ill (figure 19.9b). Not all common-source epidemics involve a single exposure to the infectious agent. In some cases, contact is continuous over a long period of time or is intermittent, such as with a faulty drinking water disinfection system. Also, some epidemics may start with a single source, but then those infected individuals transmit the disease to their contacts, leading to a propagated epidemic.

The season in which the epidemic occurs can also be significant. Respiratory diseases including influenza, respiratory syncytial virus infections, and the common cold are more easily transmitted in crowded indoor conditions during the winter. Conversely, vector-borne and foodborne diseases are often transmitted in warm weather when people are more likely to be exposed to mosquitoes and ticks, or to eat picnic food that has not been stored properly (**figure 19.10**).

Analytical Studies

Analytical studies are designed to determine which of the potential risk factors identified by the descriptive studies are actually relevant in the spread of the disease. They include three general types: case-control studies, cross-sectional studies, and cohort studies.

Case-Control Studies

A **case-control study** starts with the disease and attempts to determine the chain of events leading to it (**figure 19.11**). The investigator works backward, trying to identify what put the cases (people who developed the disease) at risk for the disease. To do this, the activities of the cases are compared with those of controls (similar people who remained healthy). If an activity is common among the cases, but not the controls, that activity may have been a factor in the development of the disease. An important consideration in the design of case-control studies is the selection of controls. The controls must be matched to cases with respect to variables such as age, gender, and socioeconomic status. The aim is to ensure that all controls had equal probability of coming in contact with the disease agent and had the same potential to become cases.

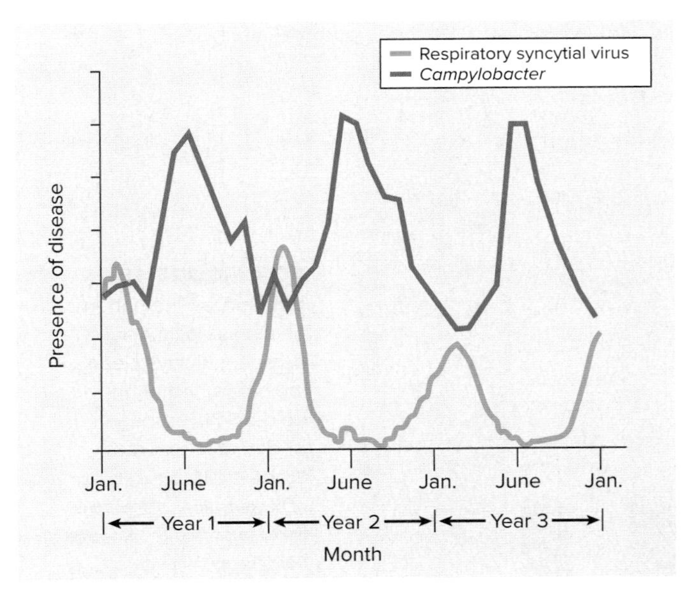

FIGURE 19.10 Seasonal Occurrence of Respiratory and Gastrointestinal Infections

Why does the level of foodborne disease increase in the summer?

Cross-Sectional Studies

A **cross-sectional study** collects data about a population at a given point in time. It does not attempt to establish the cause of a disease, but instead provides a snapshot of the population's characteristics—such as behaviors, exposure, and disease—that might suggest risk factors for the disease (see figure 19.11).

Cohort Studies

Cohort studies look ahead to see if previously identified risk factors actually predict a tendency to develop the disease (see figure 19.11). Cohort groups (sets of people whose exposure to the risk factor is known) are selected and then followed over time. The incidence of disease in those who were exposed to the risk factor and those who were not is compared. By following cohort groups, epidemiologists attempt to determine if the suspected cause does indeed correlate with the expected effect. This type of study is less likely to be influenced by poor memory than studies following a disease outbreak. It is generally more time-consuming and expensive, however, particularly when examining a disease with a long incubation period.

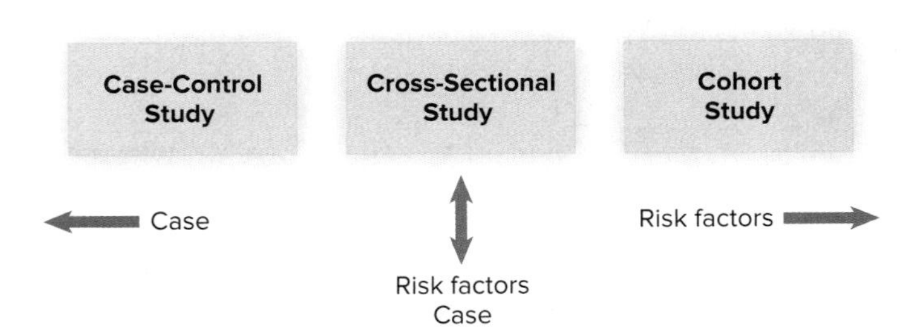

FIGURE 19.11 Analytical Studies When groups of disease cases are compared, epidemiologists can test their predictions about the nature of a disease. Case-control studies consider past events that may have led up to the onset of disease. Cross-sectional studies gather information about possible risk factors and the occurrence of disease at one point in time. Cohort studies most often follow groups with specific risk factors into the future to see if disease develops.

In a case-control study, how do the cases differ from the controls?

FOCUS ON A CASE 19.1

A 32-year-old woman came to the emergency department complaining of fatigue, nausea, vomiting, and abdominal pain. The nurse recorded a fever of 100°F and noticed a yellowish skin color. Suspecting that it could be a type of hepatitis, a contagious viral disease that affects the liver, the physician ordered several tests, including one for liver enzyme activity and one that detects antibodies against hepatitis A virus (HAV). Results of the tests confirmed a diagnosis of hepatitis A, a disease that is spread through fecal-oral transmission. As required by law, the physician reported the case to the state health officials.

The source of hepatitis A can be difficult to track because of the long incubation period—usually 3 to 5 weeks. After interviewing the patient, health officials determined that she was one of 162 people who were diagnosed with hepatitis A during a 4-month period. The source of infection was determined to be a mixture of frozen fruits marketed as an organic antioxidant blend. The epi curve is shown in **box figure 19.1.**

1. Why is reporting the case to public health officials required by law?
2. How could public health officials determine the source of the infection?
3. The epi curve for this outbreak looks different from the common-source and propagated epi curves shown in figure 19.9. How can you explain this?
4. Nearly all of the cases were in western states, with only 4 of the 162 cases in non-western states. How can the occurrence of cases in non-western states be explained?

Discussion

1. Collection of disease data can detect outbreaks early and possibly prevent their spread by identifying the source. Further, analysis of data can be used to develop programs to control future outbreaks. If the source is a commercial product, as it was in this case, a recall can be issued to eliminate the primary source of infection.
2. Health officials define a case and then gather and examine information from those individuals who meet the defined case criteria. Officials would gather descriptive information about when and where cases occurred and who became ill. This information would allow development of hypotheses about the source of the outbreak. Additional questioning would be directed at testing those hypotheses. Hepatitis A is usually spread by consuming contaminated food or water. In this instance, officials determined which foods these individuals had consumed and where they got it. The outbreak was tracked to pomegranate seeds shipped from Turkey and included in the antioxidant product that was distributed to a specific chain of markets in the western United States.
3. An epi curve may provide significant information about a pattern of disease occurrence, but sometimes it is ambiguous. The antioxidant product that was the source of this outbreak was consumed over several weeks before the outbreak was noticed. Because of the long and variable incubation period of hepatitis A, early cases were spread over a relatively long period compared with a typical common-source outbreak with an abrupt start. Also, hepatitis A is contagious and can be spread to secondary contacts. However, the long incubation period can minimize propagation of the infection because those with known exposure can be vaccinated or receive preventive treatment. In this case the cycles of infection relative to incubation period were blurred.
4. People from these states had likely traveled to or through the western states experiencing the outbreak. Alternatively, infected individuals from western states could have transmitted the infection to others while traveling outside the outbreak area.

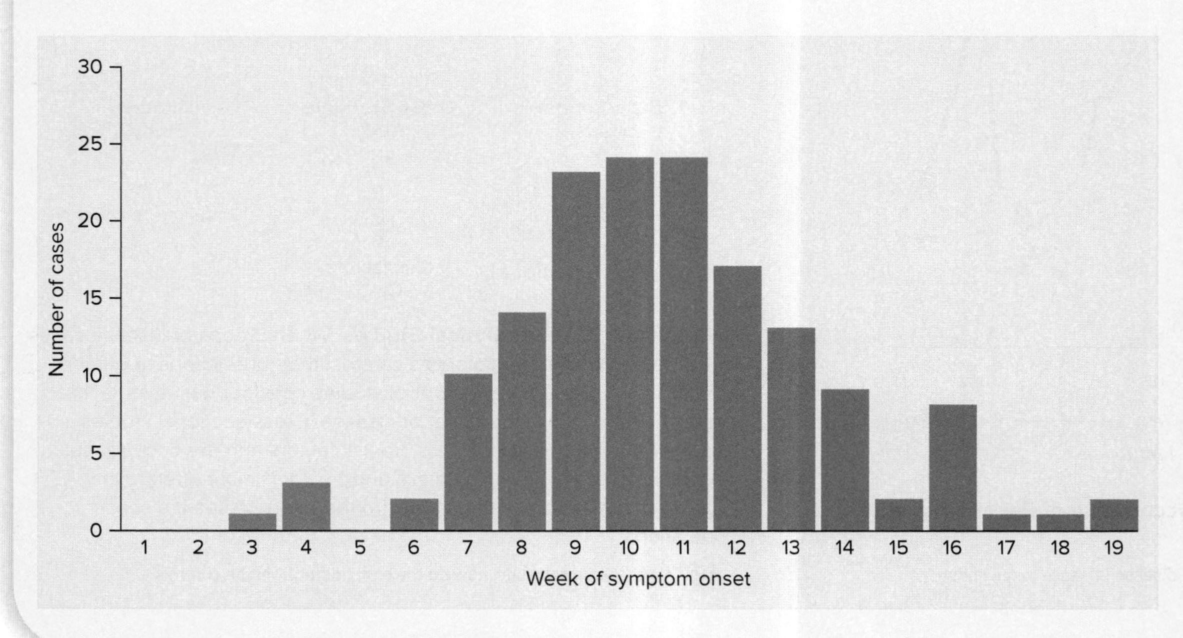

BOX FIGURE 19.1. Epi Curve The pattern of disease occurrence does not always clearly indicate whether an epidemic is propagated or common-source.
Source: "Multistate outbreak of hepatitis A virus infections linked to pomegranate seeds from Turkey (Final Update)," Centers for Disease Control and Prevention, October 23, 2013. http://www.cdc.gov/hepatitis/outbreaks/2013/a1b-03-31/epi.html

Experimental Studies

An **experimental study** is sometimes used to judge the cause-and-effect relationship between the risk factors and the development of disease, but most frequently it is done to assess the value of a particular intervention or treatment (such as antimicrobial drug therapy). The effectiveness of the treatment is compared with one of known value or with a **placebo.** A placebo is a mock drug—it looks and tastes like the experimental drug but has no medicinal value. To assess the value of the experimental drug, a group of patients is divided into two subgroups; one group is given the treatment and the other receives an alternative or a placebo. To avoid bias, the study should ideally be **double-blind,** where neither the researchers nor the patients know who is receiving the experimental treatment. Ethical issues sometimes make it necessary to use animals rather than human patients in experimental studies.

MicroAssessment 19.4

Descriptive epidemiological studies attempt to identify the potential risk factors that lead to disease. Analytical studies try to determine which suspected factors are actually relevant to disease development. Experimental studies are generally used to evaluate the effectiveness of a treatment or an intervention in preventing disease.

10. On what three factors can a descriptive study focus?
11. How is the timing of a propagated epidemic related to the incubation period of the pathogen?
12. Why is it important to include a placebo in a scientific study to assess the effectiveness of a drug?

19.5 ■ Infectious Disease Surveillance

Learning Outcomes

6. Compare and contrast the roles of the Centers for Disease Control and Prevention and state public health departments.
7. Describe the activities of the World Health Organization.
8. Describe the conditions that may allow eradication of a disease.

Infectious disease surveillance, nationally and worldwide, is one of the most important aspects of disease prevention. It involves both recognizing and reporting disease cases to public health authorities. This information can alert officials to changes in incidence or prevalence of a disease, or it may provide clues as to the cause of disease. In some cases, cooperative surveillance efforts coupled with global immunization programs and isolation of cases can result in eradication of disease, as was the case with smallpox.

National Disease Surveillance Network

Infectious disease control depends on a network of agencies across the country to monitor disease development. It is partly because of this network that infectious diseases do not claim more lives in the United States.

Centers for Disease Control and Prevention

The **Centers for Disease Control and Prevention (CDC)** in Atlanta, Georgia, is part of the U.S. Department of Health and Human Services. It provides support for infectious disease laboratories in the United States and abroad and collects data on diseases that impact public health. Each week, the CDC publishes the *Morbidity and Mortality Weekly Report (MMWR),* which summarizes the status of a number of diseases. The *MMWR* is available online (http://www.cdc.gov/mmwr/), making it readily accessible to anyone in the world.

The number of new cases of over 50 notifiable diseases is reported to the CDC by individual states (**table 19.1**). The list of diseases considered notifiable is determined through collaborative efforts of the CDC and state health departments. Typically the diseases are of relatively high incidence or pose potential danger to public health. The data collected by the CDC are published in the *MMWR* along with historical numbers to reflect any trends. Potentially significant case reports, such as the 1981 report of a cluster of opportunistic infections in young homosexual men that marked the start of the AIDS epidemic, are also included in the *MMWR.* This publication is an invaluable aid to physicians, public health agencies, teachers, students, and others concerned about infectious disease or public health. In fact, many of the epidemiological charts and stories in this textbook are from the *MMWR.*

The CDC also conducts research relating to infectious diseases and can dispatch teams worldwide to assist with identifying and controlling epidemics. As the Ebola epidemic of 2014 showed, an outbreak anywhere in the world is a potential threat. The CDC also provides refresher courses for laboratory and infection control personnel.

MicroByte

Chlamydial infection is the most commonly reported sexually transmitted infection in the United States, with over 1.5 million reported cases in 2015.

Public Health Departments

Each state has a state epidemiologist (or equivalent) who oversees a network of public health laboratories involved in infection surveillance and control as well as other health-related activities. Individual states have the authority to mandate which diseases must be reported by physicians to the state. The prompt response of health authorities in Washington State that helped stop an outbreak of *Escherichia coli* O157:H7 in 1993 caused by contaminated hamburger

TABLE 19.1 Notifiable Infectious Diseases, 2016

Individual states and territories require physicians to report cases of these notifiable diseases. In turn, the number of cases is reported to the CDC, where they are collated and published in the *MMWR*.

- Anthrax
- Arboviral neuroinvasive and non-neuroinvasive diseases
- Babesiosis
- Botulism
- Brucellosis
- Campylobacteriosis
- Chancroid
- *Chlamydia trachomatis* infection
- Cholera
- Coccidioidomycosis (valley fever)
- Cryptosporidiosis
- Cyclosporiasis
- Dengue virus infections
- Diphtheria
- Ehrlichiosis/anaplasmosis
- Giardiasis
- Gonorrhea
- *Haemophilus influenzae,* invasive disease
- Hansen's disease (leprosy)
- Hantavirus infection (non-HPS)
- Hantavirus pulmonary syndrome (HPS)
- Hemolytic uremic syndrome, post-diarrheal
- Hepatitis, viral (acute and chronic)
- HIV infection
- Influenza-associated pediatric mortality
- Invasive pneumococcal disease
- Legionellosis (Legionnaires' disease)
- Leptospirosis
- Listeriosis
- Lyme disease
- Malaria
- Measles (rubeola)
- Meningococcal disease
- Mumps
- Novel influenza A infections
- Pertussis (whooping cough)
- Plague
- Poliomyelitis, paralytic
- Poliovirus infection, nonparalytic
- Psittacosis
- Q fever
- Rabies, animal and human
- Rubella (German measles)
- Rubella, congenital syndrome
- Salmonellosis
- Severe acute respiratory syndrome–associated coronavirus (SARS-CoV) disease
- Shiga toxin–producing *Escherichia coli* (STEC)
- Shigellosis
- Smallpox (variola)
- Spotted fever rickettsiosis
- Streptococcal toxic shock syndrome
- Syphilis
- Syphilis, congenital
- Tetanus
- Toxic shock syndrome (other than streptococcal)
- Trichinellosis (trichinosis)
- Tuberculosis
- Tularemia
- Typhoid fever
- Vancomycin-intermediate *Staphylococcus aureus* (VISA) and vancomycin-resistant *Staphylococcus aureus* (VRSA) infection
- Varicella (chickenpox)
- Vibriosis
- Viral hemorrhagic fever (VHF)
- Yellow fever
- Zika virus disease and Zika virus, congenital infection

patties was partly because Washington then was one of the few states with surveillance and reporting measures for that organism. The epidemic had actually started in other states but had gone unrecognized. Likewise, the fungus *Cryptococcus gattii* appeared in Washington State in 2007, causing lung infections and meningitis in healthy people. Cases now must be reported in Washington and Oregon and they are closely watched by health authorities in other states. ⏭| ***Cryptococcus gattii***

Other Components of the Public Health Network

The public health network also includes public schools, which report absentee rates, and hospital laboratories, which report on the isolation of pathogens with epidemiological significance. In conjunction with these local activities, the news media alert the general public to the presence of infectious disease.

Worldwide Disease Surveillance

The **World Health Organization (WHO),** an agency of the United Nations with 193 member states, is devoted to achieving the highest possible level of health around the globe. It has four main functions: (1) provide worldwide guidance in the field of health; (2) set global standards for health; (3) cooperatively strengthen national health programs; and (4) develop and transfer appropriate health technology. To accomplish its goals, the WHO provides education and technical assistance to member countries.

The WHO distributes information through a series of periodicals and books as well as through its website (www.who.int). For example, the *Weekly Epidemiological Record* reports timely information about epidemics of public health importance, particularly those of global concern.

Reduction and Eradication of Disease

Humans have been enormously successful in eliminating or reducing the occurrence of certain diseases through efforts in improved sanitation, reservoir and vector control, vaccination, and antibiotic treatment (see figure 1.4). In the United States, many diseases that were once common and claimed many lives are now relatively rare. Successful vaccination programs have led to dramatic decreases in the number of deaths caused by *Haemophilus influenzae, Corynebacterium diphtheriae, Clostridium tetani, Bordetella pertussis,* and others. Meanwhile, recognizing and controlling the source of diseases such as malaria, plague, and cholera have been effective in limiting their spread.

One human disease—smallpox—has been globally eradicated, eliminating the natural occurrence of a disease that had a 25% case-fatality rate and disfigured many who survived (**figure 19.12**). Because humans were the only reservoir for the smallpox virus, programs of extensive immunization along with isolation of cases were able to eliminate the reservoir and therefore eradicate the disease. The WHO hopes to eradicate polio and measles soon. Dracunculiasis, caused by a helminth (worm), may be the next disease to be globally eradicated. There were nearly 3.5 million cases in 1986, but in 2016 there were only 25. Political and social upheaval, complacency, and lack of financial support can result in a resurgence of diseases unless the pathogens are completely eliminated. ⏭ polio, ⏭ measles

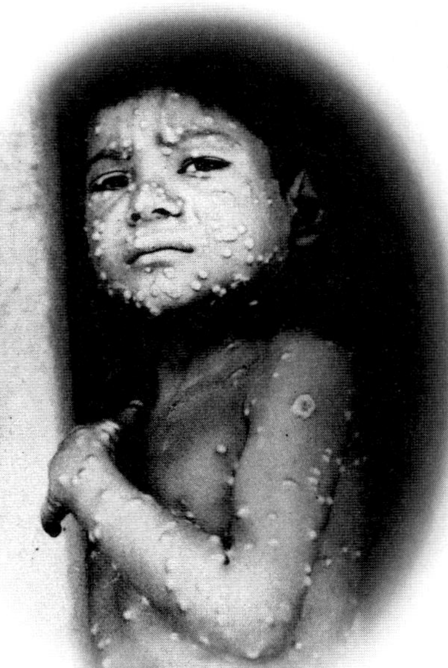

FIGURE 19.12 Case of Smallpox, a Disease That Has Now Been Eradicated ©World History Archive/Newscom

? What characteristics make a disease a good candidate for eradication?

MicroAssessment 19.5

Across the United States, a network of agencies, including the Centers for Disease Control and Prevention (CDC) and state and local public health departments, monitors disease development. The World Health Organization (WHO) is devoted to achieving the highest possible level of health for people around the globe. Humans have been successful in reducing or eliminating certain diseases, but must remain vigilant to prevent their resurgence.

13. What is the *MMWR*?

14. Explain why smallpox was successfully eradicated, but rabies probably never will be.

15. Explain why we have relatively accurate data on the number of cases of measles that occur in the United States but not on the number of cases of the common cold.

19.6 ■ Emerging Infectious Diseases

Learning Outcomes

9. Explain how microbial evolution can lead to emergence of disease.

10. Describe how human behavior can contribute to the emergence and reemergence of disease.

11. Explain how climate can influence disease emergence.

Scientific advances made over the past several decades led to speculation that the war against infectious diseases, particularly those caused by bacteria, had been won. Microorganisms, however, have occupied this planet far longer than humans, evolving to reside in every habitat with the potential for life, including the human body. It is no surprise that new and previously unrecognized pathogens are emerging, and that some of those we had previously controlled are now making a comeback.

Emerging infectious diseases are those that are novel or have recently increased in incidence. New or newly recognized diseases include Middle East respiratory syndrome (MERS), Ebola virus disease, and Zika virus disease. Familiar diseases such as malaria and tuberculosis are increasing in incidence after years of decline (see figure 1.5). Microbes are adept at taking advantage of new opportunities to thrive. Some factors that contribute to the emergence of diseases include:

- **Microbial evolution.** The emergence of some diseases follows the natural evolution of microbes. For example, a relatively new serotype of *Vibrio cholerae,* designated O139, gained the ability to produce a protective capsule. Immunity against the earlier strain does not guarantee immunity to the newer one. In the case of avian influenza, scientists are concerned that the virus will evolve to spread from person to person. Resistance to antimicrobial drugs is contributing to the reemergence of many diseases, including malaria. ⏭ avian flu, ⏭ malaria
- **Complacency and public health efforts.** As an infectious disease is controlled and therefore of less concern, complacency can develop, paving the way for resurgence of the disease. The early success of the plan to eliminate tuberculosis in the United States by the year 2000 caused news reports, education, and research money to be diverted to more common diseases. At the same time, certain social welfare programs were cut. As a result, poor health and living conditions put more people at risk of developing active tuberculosis and the disease reemerged. Fortunately, new public health measures have brought the disease back under control in the United States. ⏭ tuberculosis
- **Changes in human society.** Day-care centers, where diapered infants mingle, unmindful of sanitation and hygiene, are a relatively new component of American society. For obvious reasons, these centers can be hotbeds of contagious diseases. This is particularly true for the intestinal pathogens *Giardia* and *Shigella,* which have a low infectious dose, because infants often explore through taste and touch and are thus likely to ingest fecal organisms (**figure 19.13**). Moreover, many young children have not yet acquired immunity to communicable diseases such as

FIGURE 19.13 Children in a Day-Care Center ©Loic Venance/AFP/Getty Images

Why are *Shigella* and *Giardia* species readily transmitted in day-care centers?

colds and diarrhea that are readily transmitted among this susceptible population. ⏭ *Giardia*, ⏭ *Shigella*

- **Advances in technology.** Technology may make life easier, but can inadvertently create new habitats for microorganisms. For example, the introduction of contact lenses gave microorganisms the opportunity to grow in a new location—the lenses and storage solutions of people who did not use proper disinfection techniques. In turn, this resulted in new types of eye infections.
- **Population expansion.** As populations increase, people often move into areas where they are more likely to come into contact with reservoirs of disease. When new homes are built along the borders of forests, for example, humans are exposed to more deer that carry ticks, the host for *Borrelia burgdorferi,* the cause of Lyme disease. ⏭ Lyme disease

MicroByte

About 75% of emerging infectious diseases affecting humans are of animal origin.

- **Development.** Dams, which provide important sources of power necessary for economic development, have inadvertently extended the range of certain diseases. The life cycle of the parasite that causes the disease schistosomiasis involves an aquatic snail. Construction of dams such as the Aswan Dam on the Nile River has increased the habitat for the snail, thus extending the range of the parasite. ⏮ schistosomiasis
- **Mass production, widespread distribution, and importation of food.** Foodborne illness has always existed, but the ease with which we can now transport items worldwide from a single production plant can create new problems. Widespread distribution of foods contaminated with pathogens can result in a similarly broad outbreak of disease. Contaminated cantaloupes shipped from a single grower in 2011 led to a multi-state epidemic of listeriosis that killed 33 people.
- **War and civil unrest.** War and civil unrest can disrupt the infrastructure on which disease prevention relies. Refugee camps that crowd people into substandard living quarters lacking toilet facilities and safe drinking water are hotbeds of infectious diseases such as cholera and dysentery. Unfortunately, war also disrupts disease-eradication efforts.
- **Climate changes.** Warm temperatures favor the reproduction and survival of some arthropods, which can serve as vectors for diseases such as malaria, West Nile encephalitis, and Zika virus disease. The heavy rainfall and flooding caused by the effects of El Niño may be related to surges of cholera cases in Africa.

MicroAssessment 19.6

Microbes are adept at responding to change. Evolution of the pathogen, changes in human behavior, or climate change can provide new opportunities for pathogens to thrive.

16. Why might a person immunized against cholera contract the disease anyway?
17. How would you expect a trend toward warmer climates to affect the spread of vector-borne disease?
18. What political and societal factors might lead to a decrease in childhood immunizations?

19.7 ■ Healthcare-Associated Infections

Learning Outcomes

12. Identify the most common types of healthcare-associated infections and their causes.
13. Describe four reservoirs of infectious agents in healthcare settings and three mechanisms by which the agents can be transferred to patients.
14. Describe the roles of Infection Control Committees and the CDC in preventing nosocomial infections.

Healthcare-associated infections (HAIs) are infections that individuals acquire while receiving treatment in a healthcare setting such as a hospital, therapy center, urgent care facility, or long-term care residence. They are among the top 10 causes of death in the United States. Perhaps this should not be surprising considering that hospitals, in particular, are densely populated communities of unusually susceptible people where the most antimicrobial resistant and virulent pathogens can potentially circulate. Some of these are listed in **table 19.2.**

Hospital-acquired infections, or **nosocomial infections** (derived from the Greek word for hospital), affect up to 5% of patients. Many organisms that cause nosocomial infections seldom cause disease in healthy people, but a sick or immunocompromised individual undergoing surgery or being treated with invasive medical procedures is highly vulnerable to infection. Nosocomial infections often result from medical devices such as catheters or ventilators that breach the first-line barriers of normal host defenses. Common types of nosocomial infections are listed in **table 19.3.**

Reservoirs of Infectious Agents in Healthcare Settings

The organisms that cause healthcare-associated infections can originate from a number of different sources, including other patients, the healthcare environment, healthcare workers, visitors, and the patient's own microbiota.

Other Patients

Other patients in a healthcare setting can be a source of pathogenic microbes. Patients who have an infectious disease, perhaps the reason for their hospitalization, may discharge pathogens into the environment via skin cells, respiratory droplets, and other body secretions and excretions. Thorough cleaning and the use of disinfectants minimize the spread of these pathogens.

Healthcare Environment

Some Gram-negative rods, particularly the common opportunistic pathogen *Pseudomonas aeruginosa,* can thrive in healthcare environments such as sinks, ventilators, and toilets. Not only is *P. aeruginosa* resistant to many disinfectants and antimicrobial medications, it requires few nutrients, enabling it to multiply in environments containing little other than water. Many nosocomial infections have been traced to soaps, disinfectants, and other aqueous solutions that were contaminated with the organism. ⏮ ***Pseudomonas aeruginosa***

Healthcare Workers

Outbreaks of healthcare-associated infections are sometimes traced to infected personnel. Clearly, those who report to work with even a mild case of influenza can expose patients to an infectious agent that may have serious or fatal consequences to those in poor health. A more troublesome source of infection is a healthcare worker who is an asymptomatic carrier of a pathogen such as *Staphylococcus aureus.* These personnel may not recognize that they pose a risk to patients until they are implicated in an outbreak. Carriers who are members of a surgical

TABLE 19.2 Common Causes of Healthcare-Associated Infections (HAIs)

Infectious Agent	Comments
Acinetobacter baumannii	Environmental bacterium found on skin of healthy people; causes a variety of HAIs; most strains are multi-drug resistant
Candida species	Yeasts; part of the normal microbiota; common cause of bloodstream infections; some are resistant to antifungal medications ⏭ ***Candida albicans***
Clostridium difficile	Toxin-producing strains can cause diarrhea and colitis in people taking antibiotics ⏭ ***Clostridium difficile***
Enterobacteriaceae species *(E. coli, Klebsiella pneumoniae)*	Part of the normal intestinal microbiota; cause a variety of HAIs; some are not easily treated because they are resistant to last-resort antibiotics ⏮ ***Enterobacteriaceae***
Enterococcus species	Part of the normal intestinal microbiota; cause a variety of HAIs; some have developed antimicrobial resistance ⏮ **enterococci**
Pseudomonas species	Grow in moist, nutrient-poor environments such as the humidifier of a mechanical ventilator; cause a variety of HAIs, including infections of burn wounds; some are multi-drug resistant ⏮ ***Pseudomonas***
Staphylococcus aureus	Many people are carriers; transmissible on fomites; often resistant to a variety of antimicrobial drugs; causes a variety of HAIs ⏭ ***Staphylococcus aureus***

TABLE 19.3 Typical Nosocomial Infections

Type	Comments
Urinary tract infections (UTIs)	Infections of the bladder or kidneys; most hospital-acquired UTIs are associated with catheter use
Bloodstream infections	Infectious agents in the blood can lead to sepsis, particularly in sick or immunocompromised individuals; hospital cases are often associated with central lines (catheters placed near the heart) that may allow infectious agents to enter the bloodstream directly.
Surgical site infections	Barriers such as skin and mucous membranes are breached, increasing vulnerability to infection; normal microbiota from the skin or gastrointestinal (GI) tract are common causes.
Pneumonia	Lung infection; in hospitals, many cases are associated with ventilator use; most common cause of death from nosocomial infection in critically ill patients
Gastrointestinal infections	Most hospital-acquired GI infections are due to *Clostridium difficile;* related to antibiotic use ⏭ ***Clostridium difficile***
Other infections	Infection sites including skin, eyes, uterus, burns, and others

team pose a particular threat because transfer of a pathogen directly into a surgical site can result in a systemic infection.

Visitors

Patients can be exposed to infectious agents by visitors to their hospital rooms. To minimize this risk, the number of visitors may be limited, especially if visitors show signs of respiratory infection that could be easily spread by droplets. Like healthcare personnel, visitors should wash their hands before and after touching a patient. Healthcare facilities may ban live plants or animals in some patient rooms.

Patient Microbiota

Nearly any invasive procedure can transmit organisms of a patient's normal microbiota to otherwise sterile body sites. When intravenous fluids are administered, for example, *Staphylococcus epidermidis,* a member of the skin microbiota, can potentially gain access to the bloodstream. The immune system can usually eliminate small numbers of these organisms, but the underlying illness of many hospitalized patients compromises their immunity and they can develop bacteremia. Similarly, some patients can inadvertently inhale their normal oral microbiota, resulting in healthcare-associated pneumonia. Severely immunocompromised patients, such as people who have undergone cancer chemotherapy or are on immunosuppressive drugs, are prone to activation of latent infections that their immune system could previously control. ⏮ bacteremia, ⏮ latent infection

Transmission of Infectious Agents in Healthcare Settings

Transmission of HAI-causing infectious agents may be direct through contact with others, including healthcare personnel. It may also be indirect, through contact with medical devices or other items in the healthcare environment, or through the air.

Direct Transmission—Healthcare Personnel

Healthcare personnel must be extremely vigilant to avoid transmitting infectious disease agents, particularly from patient to patient. What Ignaz Semmelweis found to be true in the 1800s is equally true today—handwashing between contact with individual patients helps prevent the spread of disease (see **A Glimpse of History**). Healthcare personnel should routinely wash or disinfect their hands after touching one patient before going to the next. Patient rooms typically have alcohol gel dispensers for use upon entering, but soap and water is more effective in preventing the transmission of some agents, such as *Clostridium difficile*. Recall that this organism produces endospores that are not destroyed by alcohol, but can be removed by washing. Healthcare professionals should perform a more thorough hand scrubbing (lasting 10 minutes with a strong disinfectant) before participating in surgery, or when working in a nursery or an intensive care or isolation unit. They are further instructed to wear gloves when they have contact with blood, mucous membranes, broken skin, or body fluids.

Indirect Transmission—Medical Devices

Diagnostic and therapeutic procedures can potentially transmit infectious agents to patients. This is particularly true in intensive care units (ICUs), where patients generally have indwelling catheters used to deliver intravenous fluids or monitor the patient's condition (**figure 19.14**). Urinary tract infections, one of the most common types of nosocomial infection, can result if a urinary catheter inadvertently introduces microbes into the normally sterile urinary bladder. Microbes can enter the bloodstream if the tip of an intravenous (IV) catheter is colonized with normal skin microbiota or if IV lines or the fluids they carry are contaminated. Even normally benign microbes can cause life-threatening bloodstream infections. Mechanical ventilators that assist a patient's breathing by pumping air directly into the trachea can potentially deliver microbes. If inadequately sterilized instruments are used in an invasive procedure such as surgery or a biopsy, they can transmit infectious agents.

Indirect Transmission—Airborne

Most hospitals are designed to minimize the airborne spread of microbes. Airflow to operating rooms is usually regulated so that it is supplied under slight pressure, thereby preventing

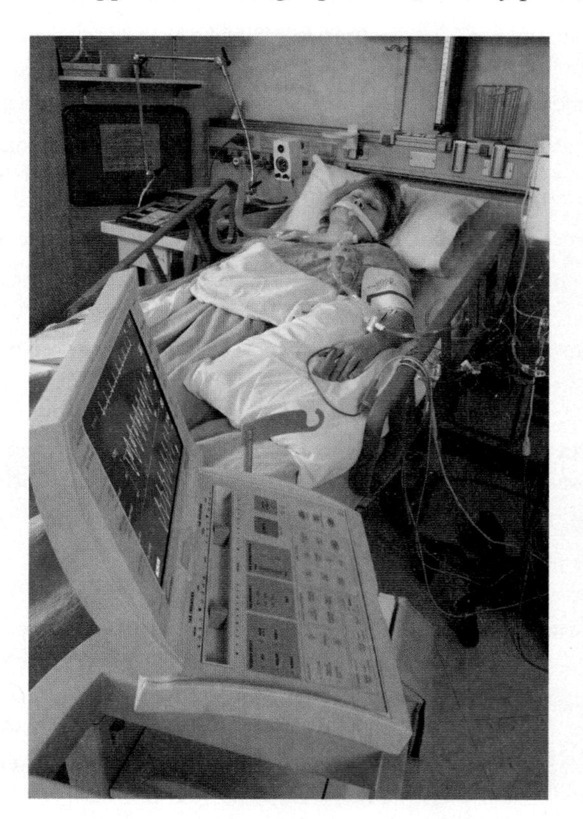

FIGURE 19.14 Patient in an Intensive Care Unit ©David Joel/Photographer's Choice RF/Getty Images

? To what sources of nosocomial infections is this patient exposed?

FOCUS YOUR PERSPECTIVE 19.1

Standard Precautions—Protecting Patients and Healthcare Personnel

One of the biggest challenges for a hospital has always been to prevent spread of disease within that confined setting. Practices called Standard Precautions are now used in the care of all patients as a means to prevent infection of both patients and healthcare personnel. A set of supplementary measures called Transmission-Based Precautions are used in addition to the Standard Precautions if a patient is, or might be, infected with a highly transmissible or epidemiologically important pathogen. The Transmission-Based Precautions are separated into three sets—Airborne Precautions, Droplet Precautions, and Contact Precautions—that are used singly or in combination as appropriate.

The Standard Precautions can be summarized as:

- **Hand hygiene.** During the delivery of healthcare, avoid unnecessary touching of surfaces in close proximity to the patient. When hands are visibly dirty or contaminated, wash them with soap and water; if non-antimicrobial soap is used, decontaminate hands with an alcohol-based hand rub. If hands are not visibly soiled, decontaminate hands before direct contact with patients; after contact with blood, body fluids, secretions, excretions, or contaminated items; immediately after removing gloves; and between patient contacts. If contact with spores is likely to have occurred, wash hands with soap and water.
- **Personal protective equipment (PPE).** Gloves are worn when touching blood, body fluids, secretions, excretions, and contaminated items; and for touching mucous membranes and non-intact skin. A gown is worn during procedures and patient-care activities when contact of clothing/exposed skin with blood/body fluids, secretions, and excretions is anticipated. Mask, goggles, or a face shield is worn during procedures and patient-care activities likely to generate splashes or sprays of blood, body fluids, or secretions.
- **Respiratory hygiene/cough etiquette.** Instruct symptomatic persons to cover mouth/nose when sneezing/coughing; use tissues and dispose in a no-touch receptacle; observe hand hygiene after soiling of hands with respiratory secretions; wear surgical mask if tolerated or maintain spatial separation of more than 3 feet if possible.
- **Patient placement.** Prioritize for single-patient room if patient is at increased risk of transmission, is likely to contaminate the environment, does not maintain appropriate hygiene, or is at increased risk of acquiring infection or developing adverse outcome following infection.
- **Patient-care equipment and instruments/devices.** If the equipment is soiled, handle in a manner that prevents transfer of microorganisms to others and to the environment; wear gloves if visibly contaminated; perform hand hygiene.
- **Care of the environment.** Develop procedures for routine care, cleaning, and disinfection of environmental surfaces, especially frequently touched surfaces in patient-care areas.
- **Textiles and laundry.** Handle in a manner that prevents transfer of microorganisms to others and to the environment.
- **Safe injection practices.** Use aseptic technique to avoid contamination of sterile injection equipment. Specific precautions describe how medications and IV solutions are stored and administered.
- **Infection control practices for special lumbar puncture procedures.** Wear a surgical mask when placing a catheter or injection material into the spinal canal or subdural space.
- **Worker safety.** Adhere to federal and state requirements for protection of healthcare personnel from exposure to bloodborne pathogens.

From Siegel, J.D., Rhinehart, E., Jackson, M., Chiarello, L., and the Healthcare Infection Control Practices Advisory Committee, 2007 Guidelines for Isolation Precautions: Preventing Transmission of Infectious Agents in Healthcare Settings, June 2007. http://www.cdc.gov/hicpac/pdf/isolation/isolation2007.pdf

contaminated air in the corridors from flowing into the room. Floors are washed with a damp mop or floor washer to avoid sweeping microbes into the air. High-efficiency particulate air (HEPA) filters, which remove most airborne particles, are used to exclude airborne microbes from rooms in which extremely susceptible patients reside. ⏮ HEPA filters

Preventing Healthcare-Associated Infections

Healthcare-associated infections (HAIs) pose significant risk to patients, increase the time people spend in a hospital, and add billions to the price of healthcare every year. Most of these infections, however, are preventable. The key steps in preventing HAIs are to first detect their occurrence and then establish policies to prevent them. To this end, nearly every hospital has an Infection Control Committee composed of various professionals, including nurses, physicians, dietitians, housekeeping staff, epidemiologists, and microbiology laboratory personnel. Hospitals may employ an infection control practitioner (ICP) to perform active surveillance of the types and numbers of infections that arise in the hospital. The Infection Control Committee, along with the ICP, drafts and implements preventive policies following the guidelines suggested by the Standard Precautions and the Transmission-Based Precautions (see **Focus Your Perspective 19.1**). The set of precautions appropriate for a given patient are often posted outside the patient's door to inform staff and visitors of the policies. In addition, the CDC assists the efforts of healthcare institutions by issuing progress reports that summarize national and statewide trends in HAIs.

The CDC also takes an active role in preventing HAIs, with the overall goal of eliminating them. The Healthcare Infection

FOCUS ON THE FUTURE 19.1

Maintaining Vigilance Against Bioterrorism

Today, an unfortunate challenge in epidemiology is to maintain vigilance against **bioterrorism**—the deliberate release of infectious agents or their toxins as a means to cause harm. Microbes pose a threat as agents of bioterrorism. Hopefully, future attacks will never occur, but it is crucial to be prepared for the possibility. Prompt recognition of such an event, followed by rapid and appropriate isolation and treatment procedures, can help to minimize the consequences. The CDC, in cooperation with the Association for Professionals in Infection Control and Epidemiology (APIC), has prepared a bioterrorism readiness plan to be used as a template by healthcare facilities. Many of the recommendations are based on the Standard Precautions already used by hospitals to prevent the spread of infectious agents (see Focus Your Perspective 19.1).

The CDC separates bioterrorism agents into three categories based on the ease of spread and severity of disease. **Category A agents** pose the highest risk because they are easily spread or transmitted from person to person and result in high mortality.

These agents include:

- ***Bacillus anthracis.*** Endospores of this bacterium are easily found in nature and are very stable. The most severe outcome, inhalation anthrax, results when an individual breathes in the airborne spores. It can lead to a rapidly fatal systemic illness. Cutaneous anthrax, which occurs when the organism enters the skin, manifests as a blister that develops into a skin ulcer with a black center. Although this usually heals without treatment, it can also progress to a fatal bloodstream infection. Gastrointestinal anthrax results from consuming contaminated food, leading to vomiting of blood and severe diarrhea; it is not common but has a high case-fatality rate. Anthrax can be prevented by vaccination, but that option is not widely available. Preventive treatment with antimicrobial medications is possible for those who might have been exposed, but this requires prompt recognition of exposure. Fortunately, person-to-person transmission of the agent is not likely.
- ***Clostridium botulinum.*** Botulism is caused naturally by the ingestion of botulinum toxin, produced by *Clostridium botulinum.* Any mucous membrane can absorb the toxin, so aerosolized toxin could be used as a weapon. Botulism can be prevented by vaccination, but that option is not widely available. An antitoxin is also available in limited supplies. Botulism is not contagious.
- ***Yersinia pestis.*** Pneumonic plague, caused by inhalation of *Yersinia pestis,* is the most likely form of plague to result from a biological weapon. Although no effective vaccine is available, post-exposure preventive treatment with antimicrobial medications is possible. Special isolation precautions must be used for patients who have pneumonic plague because the disease is easily transmitted by respiratory droplets.
- **Smallpox virus.** Although a vaccine is available to prevent infection with this virus, routine immunization was stopped over 40 years ago because the natural disease has been eradicated. As is the case with nearly all infections caused by viruses, effective drug therapy is not available. Special isolation precautions must be used for smallpox patients because the virus can be acquired through droplet, airborne, or contact transmission.
- ***Francisella tularensis.*** This bacterium, naturally found in animals such as rodents and rabbits, causes the disease tularemia. Inhalation of fewer than 50 cells results in severe pneumonia, which is incapacitating but would probably have a lower case-fatality rate than inhalation anthrax or pneumonic plague. A vaccine is not available, but post-exposure preventive treatment with antimicrobial medications is possible. Fortunately, person-to-person transmission of the agent is not likely.
- **Viruses that cause hemorrhagic fevers.** These include various viruses such as those that cause Ebola virus disease and Marburg virus disease. Symptoms vary depending on the virus, but severe cases show signs of bleeding from many sites. There are no vaccines against these viruses, and generally no treatment. Some of these viruses can be transmitted from person to person, so patient isolation in these cases is important.

Category B agents pose moderate risk because they are relatively easy to spread and cause moderate morbidity. **Category C agents** are emerging pathogens that could be engineered for easy dissemination.

Control Practices Advisory Committee (HICPAC) provides advice to hospitals and recommends guidelines for surveillance, prevention, and control of healthcare-associated infections. HICPAC issues revised recommendations in response to emerging threats such as Ebola virus disease (EDV) in 2014 and Zika virus in 2016. Established in 2005, the National Healthcare Safety Network (NHSN) tracks data, identifies problem areas, assesses progress in prevention of HAIs, and publishes updates. In 2014, based on data from over 17,000 hospitals and other healthcare institutions, reports showed decreases in almost every type of HAI, with some instances of decreases over 50%. We face great challenges, particularly with the emergence of antimicrobial resistance, but with knowledge, determination, and vigilance, hospitals are becoming safer places to be sick.

MicroAssessment 19.7

Invasive treatments and resistant microorganisms contribute to the incidence of healthcare-associated infections. These infections may originate from other patients, the healthcare environment, healthcare workers, or the patient's own normal microbiota. Diagnostic and therapeutic procedures can potentially transmit infectious agents. The most important steps in preventing healthcare-associated infections are to first detect their occurrence and then establish policies to prevent their development.

19. Explain why an IV catheter poses a risk to a patient.

20. Describe two ways in which infectious agents can be transmitted to a patient.

21. Explain why the rate of nosocomial infections is often relatively high in emergency room settings.

Summary

19.1 ■ Basic Concepts of Epidemiology

Epidemiologists study the frequency and distribution of disease in order to identify its cause, source, and route of transmission. They focus on the rate of disease. Diseases that are constantly present in a population are **endemic;** an unusually large number of cases in a population constitutes an **epidemic** (figure 19.1).

19.2 ■ Chain of Infection

Spread of a disease can be prevented by breaking the chain of infection (figure 19.2).

Reservoirs of Infection (figure 19.3)

Preventing susceptible people from coming in contact with a **reservoir of infection** can prevent infectious disease. People who have asymptomatic infections or are colonized with a pathogen are **carriers** of the infectious agent. **Zoonoses** such as plague and rabies can be transmitted to humans but exist primarily in other animals. Pathogens with environmental reservoirs are probably impossible to eliminate.

Portals of Exit (figure 19.4)

Pathogens may be shed in feces, in respiratory droplets, on skin cells, in genital secretions, and in urine.

Disease Transmission (figure 19.5)

Vertical transmission occurs between mother and infant. **Horizontal transmission** includes direct transmission and indirect transmission. **Direct transmission** includes direct contact, which occurs when one person physically touches another, and droplet spread of respiratory pathogens. **Indirect transmission** may be airborne, vehicle-borne, or vector-borne. Airborne transmission of pathogens is the most difficult to control. Vehicles of transmission include **fomites,** food (figure 19.6), and water. Handwashing is a key control measure in preventing diseases that are spread through direct contact, fomites, or contaminated food. Foodborne pathogens can originate from the animal reservoir or from contamination during food preparation. Waterborne pathogens often originate from sewage contamination. A **mechanical vector** (figure 19.7) carries a microbe on its body from one place to another. Pathogens can multiply to high numbers in a **biological vector.** Prevention of vector-borne disease relies on vector control.

Portals of Entry

Many organisms can cause disease if they enter one body site, but are harmless if they enter another.

19.3 ■ Factors That Influence the Epidemiology of Disease

The spread of disease is influenced by various factors involving the pathogen, the host, and the environment.

Characteristics of the Pathogen

Successful pathogens have multiple virulence factors. The probability of infection and disease is generally lower if an individual is exposed to small numbers of pathogens. Diseases with a long incubation period can spread before the first cases appear.

Characteristics of the Host

A disease is unlikely to spread very widely in a population showing **herd immunity** in which most people are immune to the disease agent. Malnutrition, overcrowding, and fatigue increase the susceptibility of people to infectious diseases. The very young and the elderly are generally more susceptible to infectious agents (figure 19.8). Natural immunity can vary with genetic background, but it is difficult to determine the relative importance of genetic, cultural, and environmental factors.

Characteristics of the Environment

The environment includes physical factors and other organisms. Unless transmitted by direct contact, a pathogen must survive in the environment between hosts.

19.4 ■ Epidemiological Studies

Descriptive Studies

Descriptive studies are used to identify potential risk factors that correlate with the development of disease. Determining the time that the illness occurred helps distinguish a **common-source epidemic** from a **propagated epidemic** (figure 19.9). Some epidemics are seasonal (figure 19.10).

Analytical Studies (figure 19.11)

Analytical studies are designed to determine which risk factors are actually relevant to disease development. A **case-control study** compares the past activities of cases with controls to determine the cause of the epidemic. **Cross-sectional studies** survey a range of people at a defined point in time. **Cohort studies** compare groups to determine if the identified risk factors predict a tendency to develop disease.

Experimental Studies

Experimental studies are generally used to evaluate the effectiveness of a treatment or intervention in preventing disease.

19.5 ■ Infectious Disease Surveillance

National Disease Surveillance Network

The **Centers for Disease Control and Prevention (CDC)** collects data on diseases of public health importance and summarizes their status in the *Morbidity and Mortality Weekly Report (MMWR);* other activities of the CDC include research, assistance in controlling epidemics, and support for infectious disease laboratories. State public health departments are involved in infection surveillance and control (table 19.1).

Worldwide Disease Surveillance

The **World Health Organization (WHO)** is devoted to achieving the highest possible level of health for people around the globe.

Reduction and Eradication of Disease

Smallpox has been eradicated (figure 19.12). The WHO hopes to soon eliminate polio, measles, and dracunculiasis.

19.6 ■ Emerging Infectious Diseases

Emerging infectious diseases are new or newly recognized, or are reemerging after years of decline. Factors that contribute to the emergence and reemergence of diseases include microbial evolution, the breakdown of public health infrastructure, changes in human behavior, advances in technology, population expansion, economic development, mass distribution and importation of food, war, and climate changes (figure 19.13).

19.7 ■ Healthcare-Associated Infections (table 19.2; table 19.3)

Healthcare-associated infections (HAIs) are acquired by individuals in a healthcare setting. **Nosocomial infections** are acquired in a hospital.

Reservoirs of Infectious Agents in Healthcare Settings
The organisms that cause healthcare-associated infections may originate from other patients, the healthcare environment, healthcare workers, or the patient's own microbiota.

Transmission of Infectious Agents in Healthcare Settings
Healthcare-associated infections can result from medical devices that breach the first-line barriers of the normal host defense (figure 19.14). Healthcare personnel should routinely wash or disinfect their hands after touching one patient before going to the next.

Preventing Healthcare-Associated Infections
The most important steps in preventing healthcare-associated infections are to first recognize their occurrence and then establish policies to prevent both their development and their spread.

Review Questions

Short Answer

1. Compare the impact on a society of an endemic debilitating disease with high incidence to the impact on society of a similar disease with high prevalence.
2. What is the epidemiological significance of people who have asymptomatic infections?
3. Explain why zoonotic diseases are often severe in humans.
4. List the main portals of exit from the human body.
5. Name the most important control measure for preventing person-to-person transmission of a disease.
6. Describe the factors within a population that may make it more susceptible to infectious disease.
7. Draw representative epi curves (time versus number of people ill) depicting a propagated and a common-source epidemic.
8. Differentiate among a case-control study, a cross-sectional study, and a cohort study.
9. What information is available in the *Weekly Epidemiological Record*?
10. How does the assignment of individuals to groups in a cohort study differ from assignment of individuals to groups in an experimental study?
11. Describe the factors that contribute to the emergence or reemergence of disease.
12. What are the main reservoirs of nosocomial infections?

Multiple Choice

1. Which of the following is an example of a fomite?
   a) Table
   b) Flea
   c) *Staphylococcus aureus* carrier
   d) Water
   e) Air
2. Which of the following would be the easiest to eradicate?
   a) A pathogen that is common in wild animals but sometimes infects humans
   b) A disease that occurs exclusively in humans, always resulting in obvious symptoms
   c) A mild disease of humans that often results in no obvious symptoms
   d) A pathogen found in marine sediments
   e) A pathogen that readily infects both wild animals and humans
3. Which of the following methods of disease transmission is the most difficult to control?
   a) Airborne
   b) Foodborne
   c) Waterborne
   d) Vector-borne
   e) Direct person to person
4. Which of the following statements is *false*?
   a) A botulism epidemic that results from improperly canned green beans is an example of a common-source outbreak.
   b) Droplet nuclei fall quickly to the ground.
   c) Congenital syphilis is an example of a disease acquired through vertical transmission.
   d) Plague is endemic in the rock squirrel population in parts of the United States.
   e) The first case in an outbreak is called the index case.
5. Which of the following statements is *false*?
   a) A disease with a long incubation period might spread extensively before an epidemic is recognized.
   b) A person exposed to a low dose of a pathogen might not develop disease.
   c) The young and the aged are more likely to develop certain diseases.
   d) Malnourished populations are more likely to develop certain diseases.
   e) Herd immunity occurs when a population does not engage in a given behavior, such as eating raw fish, that would otherwise increase their risk of disease.
6. The purpose of an analytical study is to
   a) identify the person, place, and time of an outbreak.
   b) identify risk factors that result in high frequencies of disease.
   c) assess the effectiveness of preventive measures.
   d) determine the effectiveness of a placebo.
   e) None of the above
7. If you and your family all develop infectious diarrhea, the most likely portal of entry for the pathogen was the
   a) large intestine.
   b) mouth.
   c) skin.
   d) respiratory tract.
   e) nose.
8. All of the following are thought to contribute to the emergence of disease *except*
   a) advances in technology.
   b) breakdown of public health infrastructure.
   c) construction of dams.
   d) mass distribution and importation of food.
   e) widespread vaccination programs.

9. Which of the following common causes of healthcare-associated infections is an environmental organism that grows readily in nutrient-poor solutions?
   a) *Enterococcus*
   b) *Escherichia coli*
   c) *Pseudomonas aeruginosa*
   d) *Staphylococcus aureus*
10. An endemic disease
    a) is typically transmitted by asymptomatic carriers.
    b) is more prevalent in the winter months than in the summer months.
    c) requires transmission by a vector.
    d) has been eradicated.
    e) is always present at some level in a given population.

Applications

1. A news station reported about a potentially fatal epidemic disease occurring in a small African village. An epidemiologist from the CDC was interviewed to discuss the disease and was very distressed that it was not being contained. Why did the epidemiologist feel the disease was a concern for people in North America?
2. An international team was gathered to discuss how funding should be spent to eliminate human infectious disease. There is only enough funding to eliminate one disease. How would the scientists go about choosing the next disease to be eliminated from the planet?

Critical Thinking

1. A student disagreed with the presentation of the examples in figure 19.9. She claimed that the number of cases from a common-source outbreak could remain high over a much longer period of time in some cases and not decrease to zero. Is the student's claim reasonable? Why or why not?
2. As shown in the graphs that follow, the proportion of deaths due to cancer and heart disease in the United States today is much higher than it was in 1900. Explain the factors that most likely contribute to this difference.

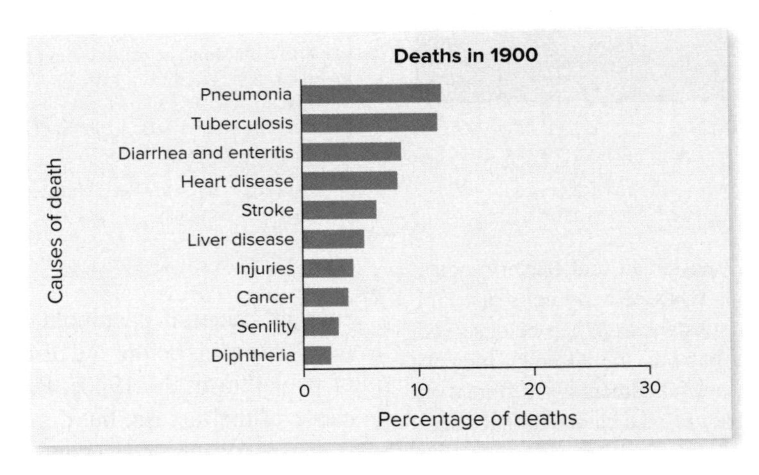

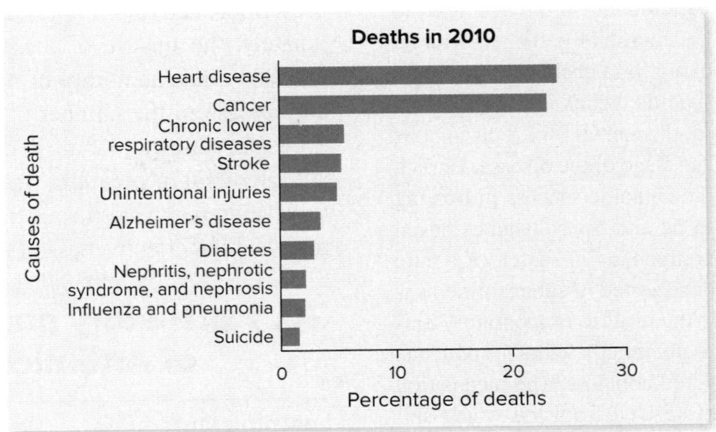

20 Antimicrobial Medications

Antibiotic susceptibility testing. ©*Science Source*

KEY TERMS

Acquired Resistance Resistance that develops due to genetic changes, including mutation and horizontal gene transfer.

Antibiotic A compound naturally produced by molds or bacteria that inhibits the growth of or kills other microorganisms.

Antimicrobial Medication An antibiotic or other chemical that acts by inhibiting or killing microorganisms and is used to treat an infectious disease; also called an antimicrobial drug.

Antiviral Medication A chemical that acts by interfering with the infection cycle of a virus and is used to treat a viral infection; also called an antiviral drug.

Bactericidal Describes a chemical that kills bacteria.

Bacteriostatic Describes a chemical that inhibits the growth of bacteria.

Broad-Spectrum Antimicrobial An antimicrobial medication that is effective against a wide range of microorganisms, often including both Gram-positive and Gram-negative bacteria.

Chemotherapeutic Agent A chemical used to treat disease.

Intrinsic (Innate) Resistance Resistance due to inherent characteristics of the organism.

Narrow-Spectrum Antimicrobial An antimicrobial medication that is effective against a limited range of microorganisms.

R Plasmid A plasmid that encodes resistance to one or more antimicrobial medications.

A Glimpse of History

Paul Ehrlich (1854–1915), a German physician and bacteriologist, became intrigued with the way various types of body cells differ in their ability to take up dyes and other substances. When he observed that certain dyes stain bacterial cells but not animal cells, indicating that the two cell types are somehow fundamentally different, it occurred to him that it might be possible to find a chemical that selectively harms bacteria without affecting human cells.

Ehrlich began searching for a "magic bullet," a term he used to describe a medication that would kill a microbial pathogen without harming the human host. He began looking for a chemical that would cure the sexually transmitted disease syphilis, which is caused by the spirochete *Treponema pallidum*. Much of the mental illness during this time resulted from tertiary syphilis, a late stage of the disease. Ehrlich knew that an arsenic compound had shown some success in treating a protozoan disease of animals, and so he and his colleagues began testing hundreds of different arsenic compounds in search of a cure for syphilis. In 1909, the 606th compound tested, arsphenamine, was found to be highly effective in treating the disease in laboratory animals. Although the compound itself was potentially lethal for patients, it did cure infections previously considered hopeless. The medication was given the name Salvarsan, a term derived from the words salvation and arsenic. Ehrlich's discovery proved that some chemicals could indeed selectively kill microbes.

Think back to the last time you were prescribed an **antimicrobial medication,** a chemical used to treat an infectious disease. Could you have recovered without the medication? The prognosis for people with common diseases such as bacterial pneumonia and severe staphylococcal infections was grim before the discovery and widespread availability of penicillin in the 1940s. Physicians were able to identify the cause of the disease, but the only treatment option was usually bed rest. Today, however, antimicrobial medications are routinely prescribed, and this simple cure is often taken for granted. Unfortunately, the misuse of these life-saving medications, coupled with the amazing ability of microorganisms to adapt, has led to an increase in the number of resistant organisms. Some people even speculate that we are in danger of seeing an end to the era of antimicrobial medications. In response, scientists are scrambling to develop new varieties of the medications while also trying to maintain the effectiveness of ones we already have.

20.1 ■ History and Development of Antimicrobial Medications

Learning Outcomes

1. Describe the discovery of antimicrobial medications, including antibiotics.
2. Explain how new antimicrobial medications are developed.

To appreciate the important role of antimicrobial medications in modern life, it helps to understand the history and development of these life-saving remedies.

Discovery of Antimicrobial Medications

The development of Salvarsan by Paul Ehrlich was the first documented example of a chemical used successfully as an antimicrobial medication (see **A Glimpse of History**). The next breakthrough came almost 25 years later, when the German chemist Gerhard Domagk discovered that a red dye called Prontosil could be used to treat streptococcal infections in animals. Surprisingly, Prontosil had no effect on streptococci growing in test tubes. It was later discovered that enzymes in the blood of the animal split the Prontosil molecule, producing a smaller molecule called sulfanilamide; this breakdown product acted against the infecting streptococci. Thus, the discovery of sulfanilamide, the first of a group of chemicals now called sulfa drugs, was based on luck as well as scientific effort. If Prontosil had been screened only against bacteria in test tubes and not given to infected animals, its effectiveness might never have been discovered.

Salvarsan and Prontosil are **chemotherapeutic agents,** meaning chemicals used to treat disease. Because they are used to treat microbial infections, they can also be called antimicrobial medications, antimicrobial drugs, or, more simply, antimicrobials.

Discovery of Antibiotics

In 1928, Alexander Fleming, a British scientist, was working with cultures of *Staphylococcus aureus* when he noticed that colonies growing near a contaminating mold looked as if they were dissolving (**figure 20.1**). Recognizing that the mold might be secreting a substance that killed bacteria, he proceeded to study it more carefully. He identified the mold as a species of *Penicillium* and found it was indeed producing a bacteria-killing substance; he called this penicillin. Even though Fleming was unable to purify penicillin, he showed that it was remarkably effective in killing many different bacterial species and did not cause adverse effects when injected into rabbits and mice. Fleming recognized the potential medical significance of his discovery, but eventually stopped studying it after he became discouraged with his inability to purify the compound.

About 10 years after Fleming's discovery of penicillin, two other scientists in Britain, Ernst Chain and Howard Florey, successfully purified the compound. In 1941, the chemical was tested for the first time on a police officer with a life-threatening *Staphylococcus aureus* infection. The patient improved so dramatically that within 24 hours his illness seemed under control. Unfortunately, the supply of purified penicillin ran out, and the man eventually died of the infection. Later, with greater supplies of penicillin, two deathly ill patients were successfully cured.

The need for effective antimicrobial medications to treat soldiers wounded in World War II caused British and American scientists to help each other determine the chemical structure of penicillin and develop the means for its large-scale production. Several different penicillins were found in the *Penicillium* cultures, and were designated alphabetically. Penicillin G (or benzyl penicillin) was found to be the most suitable for treating infections. This was the first of what we now call **antibiotics**—antimicrobial medications naturally produced by microorganisms.

Soon after the discovery of penicillin, Selman Waksman isolated a bacterium from soil, *Streptomyces griseus,* that produced an antibiotic he called streptomycin. The realization that bacteria as well as molds could produce antibiotics prompted researchers to begin screening hundreds of thousands of different microbial strains for antibiotic production. Even today, pharmaceutical companies examine soil samples from around the world in hopes of finding microbes that produce previously undiscovered antibiotics.

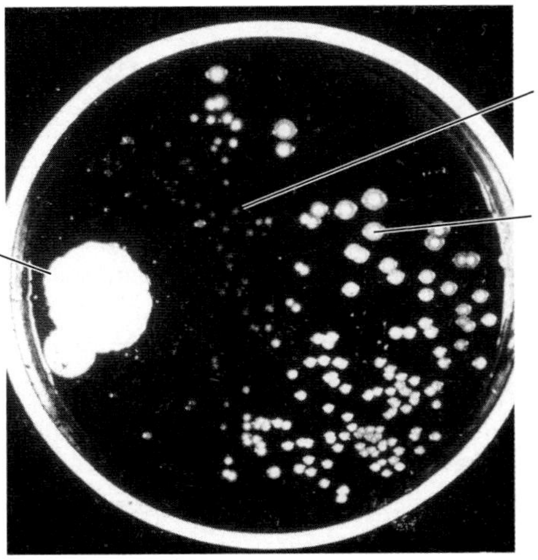

FIGURE 20.1 Mold Affecting the Growth of *Staphylococcus aureus* Alexander Fleming's photograph of his contaminated plate that led to the discovery of penicillin. ©Biophoto Associates/Science Source

? **Why are only the colonies growing close to the mold inhibited?**

Development of New Antimicrobial Medications

Most antibiotics come from microorganisms that normally live in the soil, including species of *Streptomyces* and *Bacillus* (bacteria), and *Penicillium* and *Cephalosporium* (fungi). To commercially produce an antibiotic, a carefully selected strain of the appropriate species is grown in a huge vat of broth medium. The antibiotic is then extracted from the medium and purified.

In the 1960s, scientists began altering the chemical structure of certain antibiotics to give the medications new properties. For example, penicillin G, which is active mainly against Gram-positive bacteria, was altered to create ampicillin, a version that kills a variety of Gram-negative species as well. Another change to penicillin created methicillin, which is less susceptible to

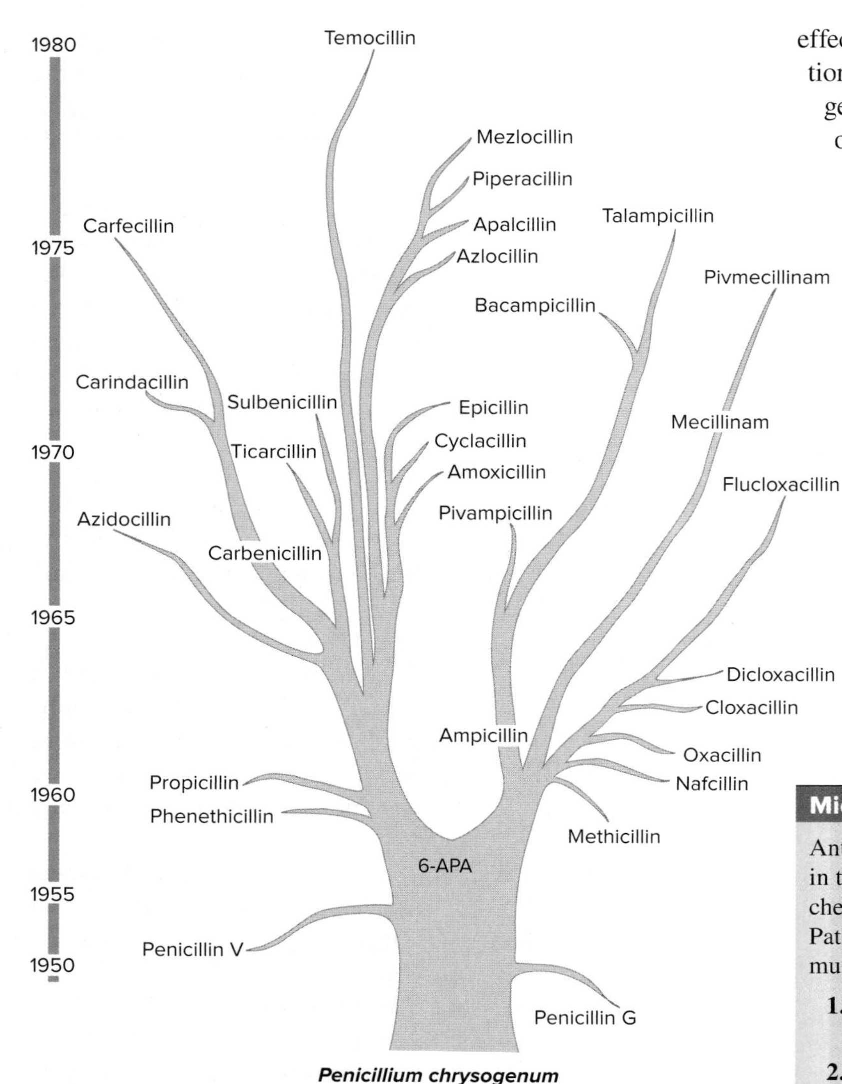

FIGURE 20.2 Family Tree of Penicillins All of the derivatives contain 6-aminopenicillanic acid (6-APA), the core portion of penicillin G.

Why is it necessary to develop new generations of antimicrobial medications?

enzymes used by some bacteria to destroy penicillin. Today a variety of penicillin-like medications exist, making up what is referred to as the family of penicillins, or simply penicillins (**figure 20.2**). Other unrelated antibiotics have also been altered to give them new characteristics. These chemically modified compounds are called semisynthetic. In some cases, the entire substance can be synthesized in the laboratory. By convention, these partially or totally synthetic chemicals are still called antibiotics because microorganisms can produce the core structure naturally.

Microorganisms continually evolve, allowing pathogens to develop resistance to the available antimicrobial medications. As a result, scientists must constantly work to create new versions. This task has grown increasingly difficult, however, because the obvious options have already been discovered. The problem is compounded by the fact that the process of developing new antimicrobial medications is financially risky—multiple stages of clinical trials are required to show that a product is safe and effective before it can receive U.S. Food and Drug Administration (FDA) approval. Even if the effort is successful, pathogens will likely develop resistance, which lessens the return on investment. In addition, new antimicrobial medications are sometimes reserved as a last resort for treating certain severe infections in order to decrease the chance that organisms will become resistant to the new medication and then spread. Because of these challenges, everyone must cooperate to protect the medications we currently have and to promote research to develop new options.

In response to the increasing problem of antibiotic resistance, the U.S. government enacted a new law—Generating Antibiotics Incentives Now (GAIN)—as a way to encourage companies to develop antimicrobial medications that target certain pathogens. Under GAIN, if an antimicrobial medication under development receives the designation Qualified Infectious Disease Product (QIDP), a high-priority process is used for its review. The QIDP designation also adds potential commercial value because any QIDP-designated medication that receives FDA approval can be exclusively marketed for an additional 5 years.

MicroAssessment 20.1

Antimicrobials are chemotherapeutic agents that are effective in treating microbial infections. Antibiotics are antimicrobial chemicals naturally produced by particular microorganisms. Pathogens develop resistance to antimicrobials, so new options must continue to be developed.

1. How is the microbe that makes penicillin different from the one that makes streptomycin?
2. Define and contrast the terms *chemotherapeutic agent, antimicrobial medication,* and *antibiotic.*
3. How might *Streptomyces griseus* cells protect themselves from the effects of streptomycin?

20.2 ■ Characteristics of Antimicrobial Medications

Learning Outcome

3. Describe selective toxicity; antimicrobial action; spectrum of activity; tissue distribution/metabolism/excretion; effects of combinations; adverse effects; and resistance to antimicrobials.

Many different antimicrobial medications are available, each with characteristics that make it more or less suitable for a given clinical situation. Hundreds of tons and many millions of dollars' worth of the chemicals are now produced each year.

Selective Toxicity

Antimicrobial medications exhibit **selective toxicity,** which means that the substance causes greater harm to microbes than to the human host. They do this by interfering with essential structures or biochemical processes that are common or accessible in microbes but not in human cells.

Although the ideal antimicrobial medication is non-toxic to humans, most can be harmful at high concentrations. In other words, selective toxicity is a relative term. The toxicity of a given medication is expressed as the **therapeutic index,** which is the lowest dose toxic to the patient divided by the dose typically used for therapy. Antimicrobials that have a high therapeutic index are less toxic, often because the medication acts against an essential biochemical process of microorganisms that does not exist in human cells. For example, penicillin G, which interferes with bacterial cell wall synthesis, has a very high therapeutic index. A related term is **therapeutic window,** which is the range between the dose used therapeutically and the toxic dose; a medication that has a high therapeutic index has a wide therapeutic window.

When an antimicrobial that has a low therapeutic index is used, the concentration in the patient's blood must be carefully monitored to make sure it does not reach a toxic level. Medications too toxic for systemic use can sometimes be used for topical applications (meaning applied to a body surface), such as first-aid antibiotic skin ointments.

Antimicrobial Action

Some antimicrobial medications kill microbes, whereas others only inhibit their growth. Both actions are medically important.

Bacteriostatic chemicals inhibit bacterial growth. A patient taking a bacteriostatic medication must rely on his or her body's defense systems to kill or eliminate the pathogen after its growth has been stopped. Sulfa drugs, for example, are frequently prescribed for treating urinary tract infections. They prevent bacteria in the bladder from growing, so that urination can more effectively eliminate them.

Bactericidal chemicals kill bacteria. These are particularly useful when the host defenses cannot be relied on to eliminate pathogens. Bactericidal medications are sometimes only inhibitory, depending on the drug concentrations and the stage of bacterial growth.

Spectrum of Activity

Antimicrobial medications vary with respect to the range of microorganisms they kill or inhibit. Some affect a narrow range of microorganisms, such as only Gram-positive bacteria, whereas others affect a wide range, generally including both Gram-positive and Gram-negative organisms.

Broad-spectrum antimicrobials affect a wide range of bacteria. These medications are important for treating acute life-threatening diseases when immediate antimicrobial treatment is essential and there is no time to culture and identify the pathogen. The disadvantage of broad-spectrum antimicrobials is that by affecting a wide range of microbes they also disrupt the microbiome. This beneficial community plays an important role in excluding pathogens so disrupting it puts the patient at risk of developing other infections. ⏮ microbiome

Narrow-spectrum antimicrobials affect a limited range of bacteria, so they are less disruptive to the microbiome. Their use, however, depends on knowing the antimicrobial susceptibility of the pathogen. A patient may be started on a broad-spectrum antimicrobial and then switched to a narrow-spectrum one later, once the relevant information has been determined.

Effects of Antimicrobial Combinations

Combinations of antimicrobials are sometimes used to treat infections, but these must be chosen carefully because some options counteract the effects of others. Bacteriostatic antimicrobials that prevent cell division, for example, interfere with the action of bactericidal medications that kill only actively dividing cells. Counteracting combinations such as this are antagonistic. In contrast, combinations in which the activity of one medication enhances the activity of the other are synergistic. Combinations that are neither synergistic nor antagonistic are additive.

Tissue Distribution, Metabolism, and Excretion of the Medication

Antimicrobials differ not only in their action and activity, but also in how they are distributed in tissues, metabolized, and excreted by the body. Only some medications cross from the blood into the cerebrospinal fluid, an important factor in treating meningitis. Medications that are unstable at low pH are destroyed by stomach acid when swallowed, so these options are typically given by intravenous or intramuscular injection. ⏭ meningitis

Another important characteristic of an antimicrobial medication is its rate of elimination, expressed as the half-life. The half-life of a medication is the time it takes for the serum concentration of that chemical to decrease by 50%. This dictates the frequency of doses required to maintain an effective level in the body. Penicillin V, which has a very short half-life, needs to be taken four times a day, whereas azithromycin, which has a half-life of over 24 hours, is taken only once a day or less. Patients who have kidney or liver dysfunction often excrete or metabolize medications more slowly, and so the dosages must be adjusted accordingly to avoid toxic levels.

Adverse Effects

As with any medication, several concerns and dangers are associated with antimicrobials. It is important to remember, however, that the medications have saved countless lives when properly prescribed and used.

Allergic Reactions

Some people develop allergies to antimicrobials. An allergy to penicillin or related medication usually results in a fever or rash but can abruptly cause life-threatening systemic anaphylaxis. For this reason, people who have allergic reactions to a given antimicrobial must alert their physicians and pharmacists so an alternative can be prescribed. They should also wear a bracelet or necklace that records that information in case of emergency. ⏮ systemic anaphylaxis

Toxic Effects

Several antimicrobials are toxic at high concentrations or occasionally cause adverse reactions. Aminoglycosides such as streptomycin can damage kidneys, impair the sense of balance, and even cause irreversible deafness. Patients taking these medications must be closely monitored because of the low therapeutic index. Some antimicrobials have such severe potential side effects or are so toxic that they are used only for life-threatening conditions when no other options are available. In rare cases, for example, chloramphenicol causes the potentially lethal condition aplastic anemia, in which the body is unable to make white and red blood cells. Polymyxin E (colistin) was once considered too toxic to be used systemically, but as bacteria have become resistant to other antibiotics, it is sometimes the only remaining choice. In addition, pregnant women should avoid certain antimicrobial medications because of the potential adverse effects on the developing fetus.

Dysbiosis

Taking an antimicrobial can lead to **dysbiosis,** an imbalance in the microbiome. Scientists are still studying the effects of this, but one obvious outcome is that pathogens normally unable to compete may multiply to high numbers. For example, patients who take certain broad-spectrum antimicrobials orally sometimes develop diarrheal disease caused by *Clostridium difficile.* This bacterium generally cannot establish itself in the intestine due to competition from other bacteria. When the microbiome is disrupted, however, *C. difficile* can sometimes grow to high numbers and cause serious intestinal damage, resulting in symptoms that range from mild diarrhea to life-threatening colitis. ⏮ *Clostridium difficile* infection (CDI)

Resistance to Antimicrobials

Certain bacteria are inherently resistant to the effects of some antimicrobial medications, a trait called **intrinsic (innate) resistance.** As an example, *Mycoplasma* species lack a cell wall, so they are resistant to penicillin and any other antimicrobial that interferes with peptidoglycan synthesis. Many Gram-negative bacteria are intrinsically resistant to certain medications because the lipid bilayer of their outer membrane prevents the molecules from entering (see figure 3.34). ⏮ the genus *Mycoplasma,* ⏮ Gram-negative cell wall

In contrast to intrinsic resistance, **acquired resistance** refers to the development of resistance in a previously sensitive organism. This occurs through spontaneous mutation or horizontal gene transfer (see figure 8.1). Acquired resistance is a significant and ongoing problem—as pharmaceutical companies create a greater variety of antimicrobial medications, microorganisms continue to evolve, developing mechanisms to avoid the drugs' effects. This very important topic will be discussed in section 20.5. ⏭ resistance to antimicrobial medications

MicroAssessment 20.2

When choosing an antimicrobial to prescribe, a variety of factors must be considered, including the therapeutic index, antimicrobial action, spectrum of activity, effects of combinations, tissue distribution, half-life, adverse effects, and resistance of the microbe.

4. Which would be safer to use: an antimicrobial that has a low therapeutic index or one that has a high therapeutic index? Why?
5. In what clinical situation is it most appropriate to use a broad-spectrum antimicrobial?
6. Why would antimicrobials that have toxic effects be used at all?

20.3 ■ Mechanisms of Action of Antibacterial Medications

Learning Outcomes

4. Describe the β-lactam antibiotics and other antimicrobials that inhibit cell wall synthesis.
5. Describe the antimicrobial medications that interfere with the following: (1) protein synthesis, (2) nucleic acid synthesis, (3) metabolic pathways, (4) cell membrane integrity.
6. Describe the antibacterial medications used to treat *Mycobacterium tuberculosis* infections.

This section describes the mechanisms of action of various classes of antibacterial medications, highlighting their bacterial targets (**figure 20.3**). A group of medications called β-lactam antibiotics will be covered in the greatest detail, because their features serve as excellent examples of some important general concepts.

Inhibit Cell Wall Synthesis

Bacterial cell walls are unique in that they contain peptidoglycan (see figures 3.33 and 3.34). Recall that this molecule is composed of glycan chains that are cross-linked via peptide bridges between NAM molecules of adjacent chains (see figure 3.32). Antimicrobial medications that interfere with

Focus Figure

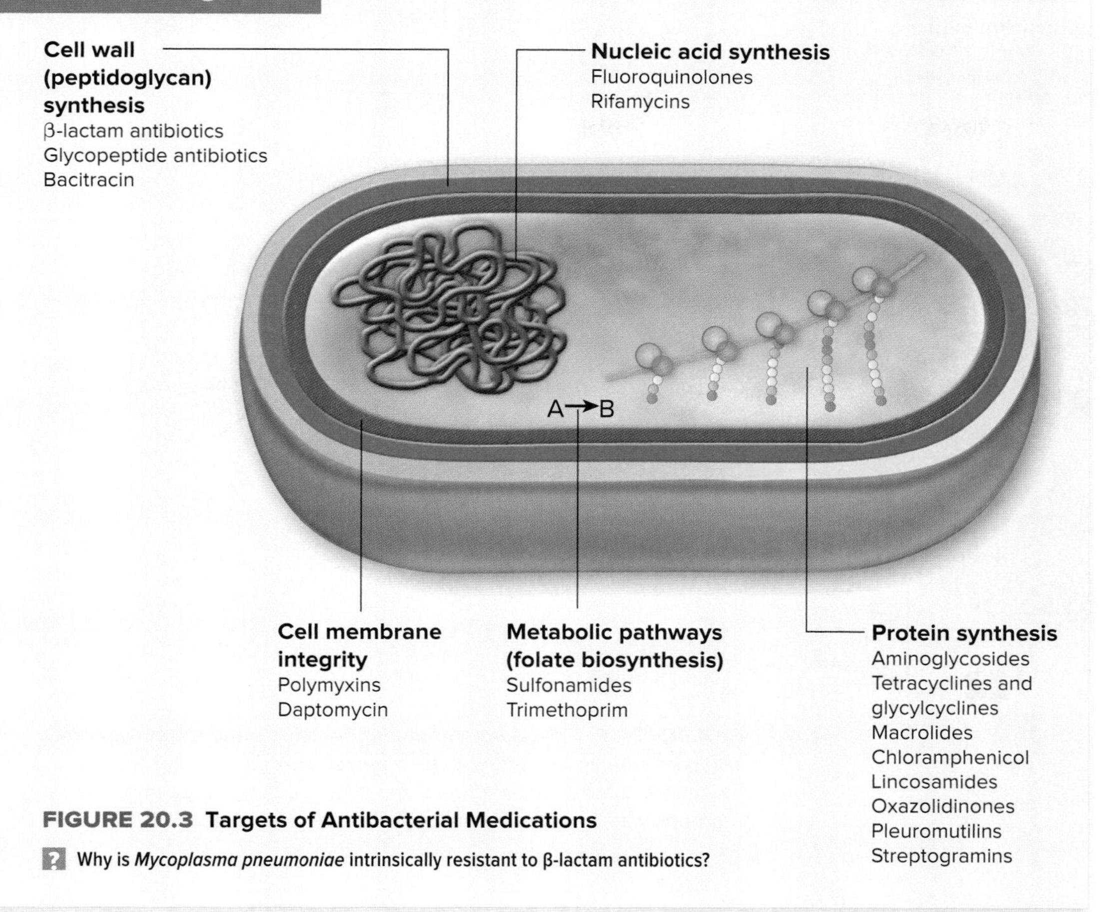

FIGURE 20.3 Targets of Antibacterial Medications

? Why is *Mycoplasma pneumoniae* intrinsically resistant to β-lactam antibiotics?

peptidoglycan synthesis include β-lactam antibiotics, glycopeptide antibiotics, and bacitracin (**figure 20.4**).

β-Lactam Antibiotics

Penicillins, cephalosporins, carbapenems, and monobactams are **β-lactam antibiotics,** meaning that they have a shared chemical structure called a β-lactam ring (**figure 20.5**). They all have a high therapeutic index, so they are often the preferred option for treating bacterial infections.

All β-lactam antibiotics interfere with peptidoglycan synthesis by preventing cross-links from forming between adjacent glycan chains. Specifically, they competitively inhibit a group of enzymes that catalyze the formation of peptide bridges between NAM molecules on the chains. The enzymes are commonly called **penicillin-binding proteins (PBPs),** reflecting the fact that they bind penicillin and were initially discovered during experiments to study the effects of the medication. By interfering with the cross-linking function of the PBPs, a β-lactam antibiotic weakens the cell wall to the point where the cell may burst. The β-lactam antibiotics are typically bactericidal only against growing bacteria, because these cells continuously synthesize peptidoglycan.

◀◀ competitive inhibition

The different β-lactam antibiotics vary in their spectrum of activity. One reason for this is the cell wall structure of the bacteria. The peptidoglycan of Gram-positive organisms is exposed to the outside environment, so the β-lactam antibiotics can directly contact the enzymes that synthesize the molecule. In contrast, the outer membrane of Gram-negative bacteria prevents some types of the medications from reaching their target. Another factor is the type of PBPs. The PBPs of Gram-positive bacteria differ somewhat from those of Gram-negative bacteria, and the PBPs of obligate anaerobes differ from those of aerobes. The various PBPs have different affinities (amounts of attraction) for the β-lactam antibiotics. Differences in affinity can even exist among related organisms.

Some bacteria resist the effects of certain β-lactam drugs by synthesizing a **β-lactamase,** an enzyme that breaks the β-lactam ring, destroying the activity of the antibiotic. Just as there are many β-lactam antibiotics, there are various β-lactamases, and these differ in the range of medications they destroy. Penicillinase is a β-lactamase that inactivates only members of the penicillin family. In contrast, extended-spectrum β-lactamases (ESBLs) inactivate a wide variety of β-lactam antibiotics, including penicillins, cephalosporins, and monobactams. Carbapenemases are β-lactamases that inactivate the greatest variety of β-lactam antibiotics; some inactivate all classes, including carbapenems, penicillins, cephalosporins, and monobactams. As a whole, Gram-negative bacteria produce a much wider variety of β-lactamases than Gram-positive organisms can.

Penicillins All members of the penicillin family share a common basic structure (see figure 20.5a). This structure's side chain has been chemically modified to create penicillin derivatives, each with unique characteristics (**figure 20.6**). The derivatives can be loosely grouped into several categories:

- **Natural penicillins.** These are the original penicillins produced naturally by the mold *Penicillium chrysogenum.* Natural penicillins are narrow-spectrum antibiotics, effective against Gram-positive and a few Gram-negative bacteria. Penicillin V is more stable in acid and, therefore,

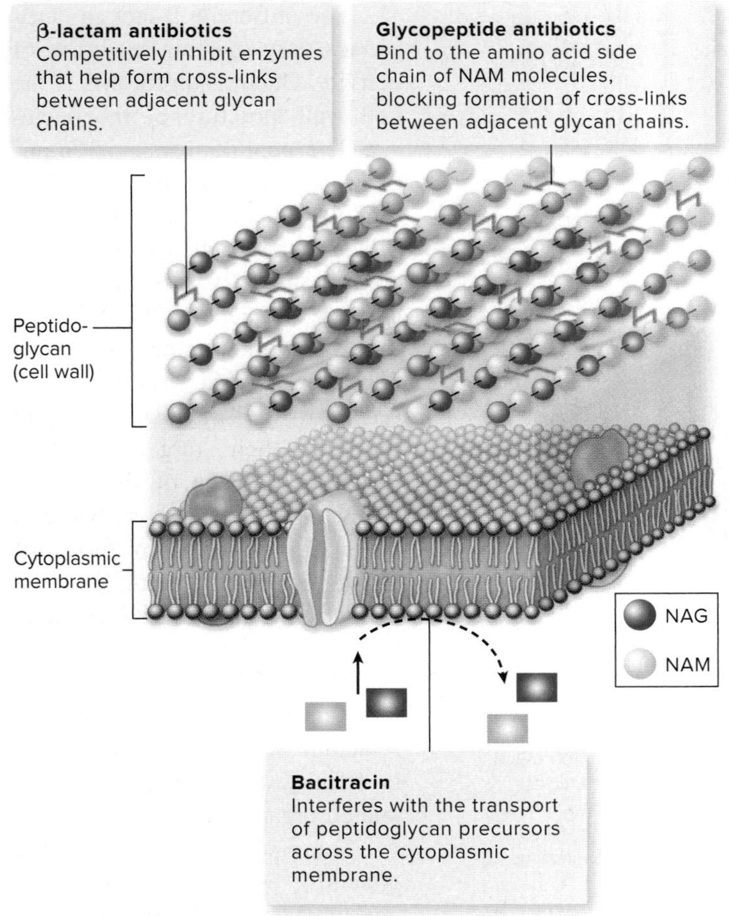

FIGURE 20.4 Antibacterial Medications That Interfere with Cell Wall Synthesis

? What is the function of penicillin-binding proteins?

Penicillin

R — C(=O) — NH — CH — CH — S — C(CH₃)₂ — CH — COOH; C(=O) — N (β-lactam ring)

(a)

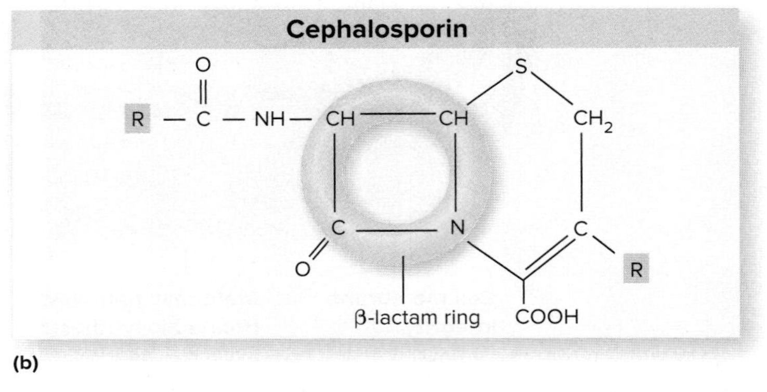

(b)

FIGURE 20.5 The β-Lactam Ring of Penicillins and Cephalosporins The core chemical structure of **(a)** a penicillin; **(b)** a cephalosporin. In the diagram, the β-lactam rings are indicated by an orange ring. The R groups vary among different penicillins and cephalosporins.

? Why is it not surprising that penicillins and cephalosporins have a high therapeutic index?

better absorbed than penicillin G when taken orally. Bacteria that produce penicillinase are resistant to the natural penicillins.

- **Penicillinase-resistant penicillins.** Scientists developed these in response to the problem of penicillinase-producing *Staphylococcus aureus* strains. Penicillinase-resistant penicillins include methicillin and dicloxacillin. Unfortunately, some *S. aureus* strains acquired the ability to make an altered PBP (PBP2a) to which most β-lactam antibiotics do not bind as well. These strains are called MRSA (methicillin-resistant *S. aureus*).
- **Broad-spectrum penicillins.** These are active against not only penicillin-sensitive Gram-positive bacteria, but also many Gram-negative bacteria. Unfortunately, they can be inactivated by many β-lactamases. Broad-spectrum penicillins include ampicillin and amoxicillin.
- **Extended-spectrum penicillins.** These have greater activity against most of the *Enterobacteriaceae* as well as *Pseudomonas aeruginosa.* This is important because these Gram-negative bacteria are common causes of healthcare-associated infections and are often resistant to many other antimicrobial medications. Extended-spectrum penicillins, however, have less activity against Gram-positive bacteria. Like the broad-spectrum penicillins, many β-lactamases destroy them. Examples of extended-spectrum penicillins include ticarcillin and piperacillin.
- **Penicillins + β-lactamase inhibitor.** This is a combination of agents. The β-lactamase inhibitor interferes with the activity of some types of β-lactamases, thereby protecting the penicillin against enzymatic destruction. An example is Augmentin, a combination of amoxicillin and clavulanic acid.

Cephalosporins The chemical structure of these antibiotics protects them from destruction by certain β-lactamases. Some cephalosporins are not very effective against Gram-positive bacteria, however, because the antibiotics have a low affinity for their PBPs.

The cephalosporins have been chemically modified, giving rise to several different groups, referred to as first-, second-, third-, and fourth-generation versions. These include cephalexin and cefazolin (first-generation), cefaclor and cefprozil (second-generation), cefixime and ceftriaxone (third-generation), and cefepime (fourth-generation). The later generations are generally more effective against Gram-negative bacteria and less susceptible to destruction by some β-lactamases.

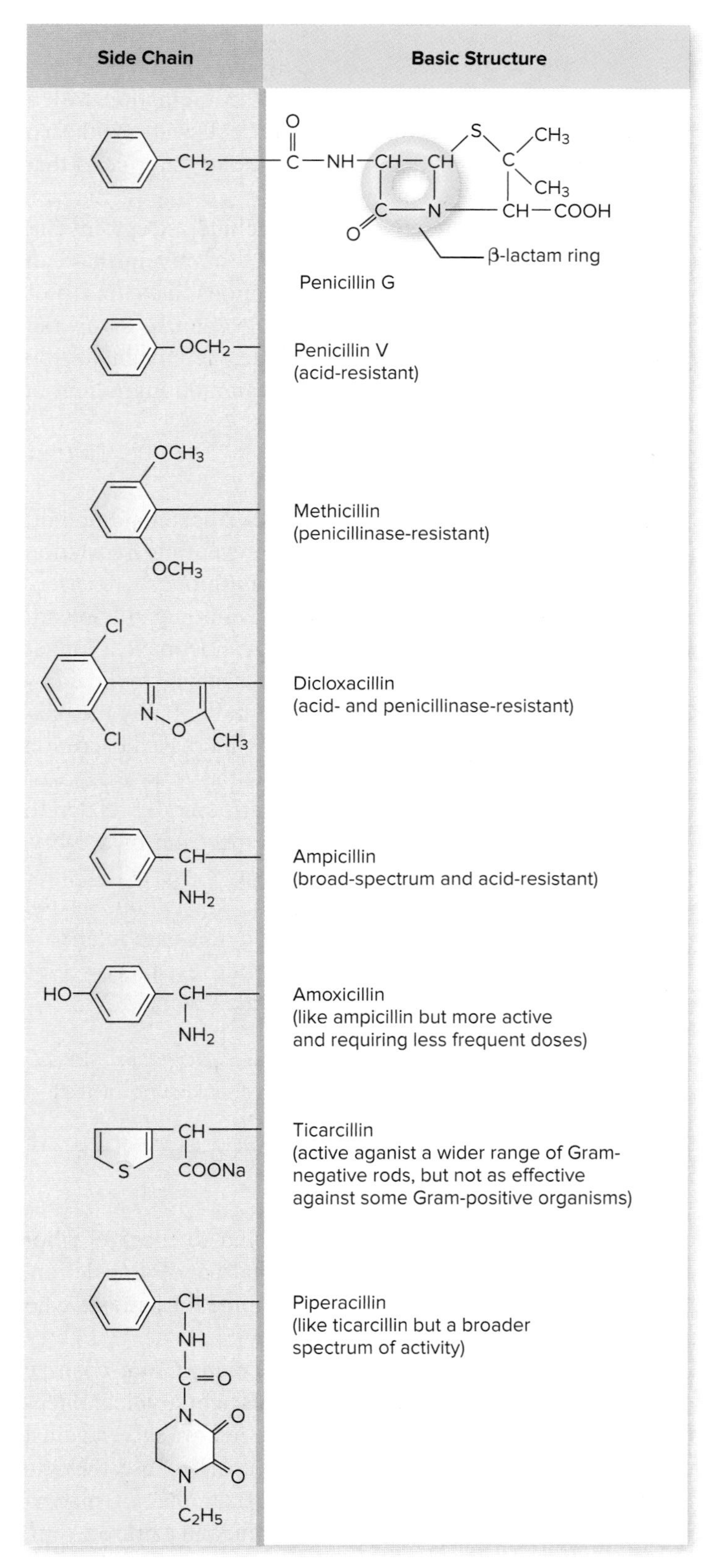

FIGURE 20.6 Chemical Structures and Properties of Representative Members of the Penicillin Family The entire structure of penicillin G and the side chains of other penicillins are shown.

? Could penicillin be used to treat MRSA (methicillin-resistant *Staphylococcus aureus*) infections? Explain.

Two newer cephalosporins, sometimes referred to as fifth-generation cephalosporins, have the additional advantage of being active against MRSA. The only one currently approved in the United States is ceftaroline, although the other recently received QIDP status. In addition, certain cephalosporins are now available in combination with a β-lactamase inhibitor. A combination called Zerbaxa was the first antimicrobial medication approved under the GAIN act. ⏮ QIDP, ⏮ GAIN

Carbapenems These are effective against a wide range of Gram-negative and Gram-positive bacteria. Carbapenems are not inactivated by the extended-spectrum β-lactamases (ESBLs) produced by certain Gram-negative bacteria, so they are usually reserved as a last resort for treating severe diseases caused by ESBL-producing organisms. Carbapenems include imipenem, ertapenem, meropenem, and doripenem.

Bacteria that produce a carbapenemase are resistant to all carbapenems and often all other β-lactam antibiotics as well. These bacteria also usually have mechanisms to resist other antimicrobials, meaning that they are resistant to all, or nearly all, conventional antimicrobial medications. In some cases, the only treatment option is an antibiotic normally considered too toxic to use systemically.

Monobactam The only monobactam used medically, aztreonam, is primarily effective against members of the family *Enterobacteriaceae.* Bacteria that produce an extended-spectrum β-lactamase (ESBL) or certain carbapenemases are resistant to aztreonam.

Glycopeptide Antibiotics

Glycopeptide antibiotics interfere with peptidoglycan synthesis by binding to the amino side chain of NAM molecules, blocking the formation of cross-links between adjacent glycan chains. This weakens the cell wall, causing cell lysis. These antibiotics are effective against only Gram-positive bacteria because the outer membrane of Gram-negative bacteria prevents them from reaching their target. They have side effects unrelated to their activity against peptidoglycan, so they have a relatively low therapeutic index. Glycopeptides are usually reserved for treating serious infections caused by Gram-positive bacteria resistant to β-lactam antibiotics.

In the United States, the most widely used glycopeptide is vancomycin. Acquired resistance is typically due to a change in the peptide side chain of the NAM molecule that prevents the antibiotic from binding. Vancomycin is poorly absorbed from the intestinal tract, so it must be administered intravenously except when used to treat intestinal infections. Newer glycopeptide antibiotics include telavancin and the recently approved dalbavancin and oritavancin. They share vancomycin's effect on peptidoglycan synthesis, but in addition they insert into the bacterial membrane, causing the cell to become leaky.

Bacitracin

Bacitracin inhibits cell wall biosynthesis by interfering with the transport of peptidoglycan precursors across the cytoplasmic membrane. Its toxicity typically limits its use to topical applications (used only on the surface of the skin); however, it is a common ingredient in over-the-counter (non-prescription) first-aid skin ointments.

Inhibit Protein Synthesis

Several types of antibacterial medications inhibit prokaryotic protein synthesis (**figure 20.7**). Although all cells synthesize proteins, the structure of the bacterial 70S ribosome—composed of a 30S and a 50S subunit—is different enough from the eukaryotic 80S ribosome to make it a suitable target for selective toxicity. The mitochondria of eukaryotic cells have ribosomes similar to bacterial ribosomes, however, which may partially account for the toxicity of some of these drugs. ⏮ ribosome structure

Aminoglycosides

Aminoglycosides are bactericidal antibiotics that irreversibly bind to the 30S ribosomal subunit, causing it to distort and malfunction. This blocks the initiation of translation and causes misreading of mRNA by ribosomes that have already passed the initiation step. Unfortunately, the medications can cause severe side effects including hearing loss and dizziness (vertigo) as a result of damage to sensory components of the inner ear. They can also cause kidney damage. Consequently, aminoglycosides are typically used only when less toxic alternatives are not available.

Aminoglycosides are generally not effective against anaerobes, enterococci, and streptococci because they enter bacterial cells by a process that requires respiratory metabolism. To extend their spectrum of activity, the aminoglycosides are sometimes used in a synergistic combination with a penicillin. The penicillin interferes with cell wall synthesis, which, in turn, allows the aminoglycoside to enter cells that would otherwise be resistant.

Examples of aminoglycosides include streptomycin, gentamicin, and tobramycin. A form of tobramycin that can be inhaled makes treatment of lung infections in cystic fibrosis patients caused by *Pseudomonas aeruginosa* safer and more effective. Another aminoglycoside, neomycin, is too toxic for systemic use; however, it is a common ingredient in over-the-counter first-aid skin ointments.

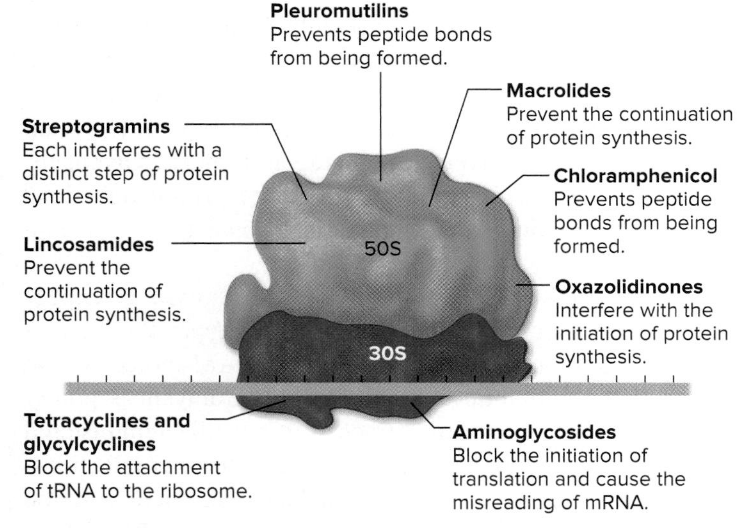

FIGURE 20.7 Antibacterial Medications That Inhibit Bacterial Protein Synthesis These medications bind to the bacterial ribosome.

? **Some medications that inhibit bacterial protein synthesis have a low therapeutic index. Why would this be so?**

Tetracyclines and Glycylcyclines

Tetracyclines reversibly bind to the 30S ribosomal subunit, blocking the attachment of tRNA and preventing translation from continuing. These bacteriostatic antibiotics are effective against certain Gram-positive and Gram-negative bacteria. Some tetracyclines, including doxycycline, have a longer half-life, allowing less-frequent doses. Resistance to the tetracyclines is due to a decrease in their accumulation by the bacterial cell—by either decreased uptake or increased excretion (efflux)—or to structural alteration of the target. ⏮ efflux pumps

Glycylcyclines are functionally and structurally related to the tetracyclines but they have a wider spectrum of activity. In addition, they are effective against many bacteria that have acquired resistance to the tetracyclines. Tigecycline is the only example currently approved. Glycylcylines are relatively new, so acquired resistance is still rare, but resistance seems to be due to increased efflux (elimination).

MicroByte

Tetracyclines and glycylcyclines can cause discoloration in teeth when used by young children.

Macrolides

Most macrolide antibiotics reversibly bind to the 50S ribosomal subunit and prevent the continuation of translation. They often serve as the medication of choice for patients who are allergic to penicillins.

Macrolides are bacteriostatic against many Gram-positive bacteria as well as the most common causes of atypical pneumonia ("walking pneumonia"). They are not effective against members of the family *Enterobacteriaceae* because they do not pass through the outer membrane. Examples of macrolides include erythromycin, clarithromycin, and azithromycin. Both clarithromycin and azithromycin have a longer half-life than erythromycin, so that they can be taken less frequently. Resistance can occur through modification of the ribosomal RNA target, production of an enzyme that chemically modifies the medication, or alterations that result in decreased uptake by the bacterial cell. ⏭ walking pneumonia

Chloramphenicol

Chloramphenicol binds to the 50S ribosomal subunit, preventing peptide bonds from being formed and, consequently, blocking translation. The action of the antibiotic is bacteriostatic.

Chloramphenicol is active against a wide range of bacteria, but it is generally only used as a last resort for life-threatening infections in order to avoid a rare but lethal side effect. This complication, aplastic anemia, can occur in response to even a small amount of chloramphenicol and is characterized by the inability of the body to form white and red blood cells.

Lincosamides

Lincosamides are bacteriostatic antibiotics that inhibit a variety of Gram-negative and Gram-positive bacteria by binding to the 50S ribosomal subunit, preventing the continuation of translation. They are particularly useful for treating infections resulting from intestinal perforation because they inhibit *Bacteroides fragilis,* a member of the normal intestinal microbiota that is frequently resistant to other antimicrobials. Unfortunately, the risk of developing *Clostridium difficile* infection (CDI) is greater for people taking lincosamides than some other antimicrobials because most *C. difficile* strains are resistant to the lincosamides. The most commonly used lincosamide is clindamycin. Resistance is often due to enzymatic modification of the ribosomal target.

Oxazolidinones

Oxazolidinones are synthetic antimicrobials that bind to the 50S ribosomal subunit, interfering with the initiation of translation. They are bacteriostatic against a variety of Gram-positive bacteria, and are particularly useful for treating infections caused by bacteria that are resistant to β-lactam antibiotics and vancomycin. Two oxazolidinones are approved by the FDA: linezolid and tedizolid. The latter, which was the second antimicrobial medication approved under the GAIN act, is the more potent of the two.

Pleuromutilins

Pleuromutilins are antibiotics that have been used to treat animals for many years, but one derivative has now been approved for use in humans. It can only be used topically, but other derivatives that can be used systemically are in clinical trials. Pleuromutilins bind to the 50S ribosomal subunit, preventing peptide bonds from being formed during translation. The antibiotics are active against many types of Gram-positive bacteria.

Streptogramins

Streptogramins are antibiotics that bind to the 50S ribosomal subunit, inhibiting translation. A combination of two—quinupristin and dalfopristin—acts as a synergistic duo by binding to two different sites on the 50S subunit; individually, each drug is bacteriostatic but together they are bactericidal. Quinupristin-dalfopristin is effective against a variety of Gram-positive bacteria, but its use is generally reserved for treating infections caused by strains that are resistant to other antimicrobials. Mechanisms of resistance to streptogramins include increased efflux and enzymatic alteration of the target.

Inhibit Nucleic Acid Synthesis

Enzymes required for nucleic acid synthesis are the targets of some groups of antimicrobial medications.

Fluoroquinolones

Fluoroquinolones are synthetic compounds that inhibit one or more of a group of enzymes called topoisomerases, which maintain the supercoiling of DNA within the bacterial cell. One type of topoisomerase, DNA gyrase, breaks and rejoins strands to relieve the strain caused by the localized unwinding of DNA during replication and transcription. Consequently, inhibition of this enzyme prevents these essential cell processes. ⏮ **supercoiled DNA**

The fluoroquinolones are bactericidal against a wide variety of bacteria, including both Gram-positive and Gram-negative organisms. Examples of fluoroquinolones include ciprofloxacin, levofloxacin, and moxifloxacin. Acquired resistance is most commonly due to an alteration in the DNA gyrase target.

Rifamycins

Rifamycins are antibiotics that block bacterial RNA polymerase from initiating transcription. Rifampin, the most widely used rifamycin, is bactericidal against many Gram-positive and some Gram-negative bacteria as well as members of the genus *Mycobacterium.*

Rifampin is primarily used to treat tuberculosis and Hansen's disease (leprosy) and to prevent meningitis in people who have been exposed to *Neisseria meningitidis.* In some patients, a reddish-orange pigment appears in urine and tears. Bacteria quickly develop resistance to the antibiotic, due to a mutation in the gene that encodes RNA polymerase.

Fidaxomicin

Fidaxomicin is a relatively new bactericidal antibiotic that interferes with transcription by binding to RNA polymerase. It passes through the intestinal tract without being absorbed and is primarily effective against *Clostridium* species, making it particularly useful for treating *C. difficile* infection (CDI). ⏭ ***Clostridium difficile* infection**

Metronidazole

Metronidazole (Flagyl) is a synthetic compound that interferes with DNA synthesis and function, but only in anaerobic microorganisms. The selective toxicity is due to the fact that anaerobic metabolism is required to convert the medication to its active form. The active form then binds DNA, interfering with synthesis and causing damaging breaks. Metronidazole is used to treat bacterial vaginosis and *Clostridium difficile* infection (CDI). ⏭ **bacterial vaginosis**

Interfere with Metabolic Pathways

Relatively few antibacterial medications interfere with metabolic pathways. Among the most useful are the folate inhibitors—sulfonamides and trimethoprim. These synthetic compounds each inhibit different steps in the pathway that leads initially to the synthesis of folate and ultimately to the synthesis of a coenzyme required for nucleotide biosynthesis (**figure 20.8**). The combination of a sulfonamide and trimethoprim has a synergistic effect so both are included in a medication called **co-trimoxazole.** Animal cells lack the enzymes in the folate synthesis portion of the pathway, which is why folate is a dietary requirement. ⏮ coenzyme

Sulfonamides

Sulfonamides and related compounds, collectively referred to as **sulfa drugs,** inhibit the growth of many Gram-positive and Gram-negative bacteria. They are structurally similar to *para*-aminobenzoic acid (PABA), a substrate in the pathway for folate biosynthesis. Because of this similarity, the enzyme that normally binds PABA binds sulfa drugs instead, an example of competitive inhibition (see figure 6.15). Human cells do not have this enzyme, providing the basis for the selective toxicity of the sulfonamides. Resistance to the sulfonamides is often due to the acquisition of a plasmid that encodes an enzyme the medication does not bind to as well. ⏮ competitive inhibition

Trimethoprim

Trimethoprim inhibits the bacterial enzyme that catalyzes a metabolic step following the one inhibited by sulfonamides. Fortunately, the medication has little effect on the enzyme's counterpart in human cells. The most common mechanism of resistance is a plasmid-encoded alternative enzyme that the medication does not bind to as well. Unfortunately, the genes encoding resistance to trimethoprim and to sulfonamide are often carried on the same plasmid.

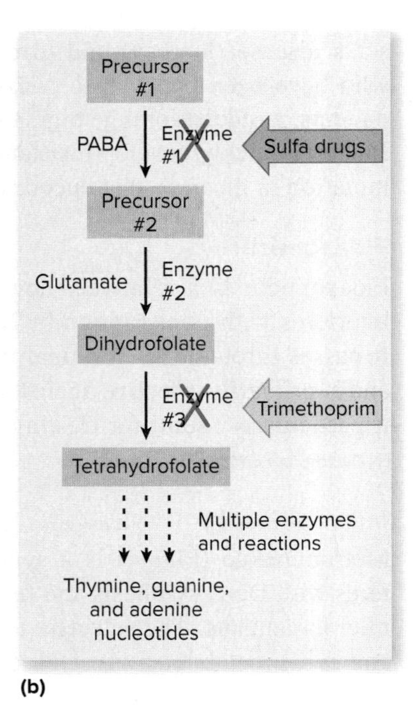

FIGURE 20.8 Inhibitors of the Folate Pathway (a) The chemical structure of PABA and a sulfa drug (sulfanilamide). **(b)** The sulfa drugs and trimethoprim interfere with different steps of the pathway.

? Sulfa drugs have a high therapeutic index. Why would this be so?

Interfere with Cell Membrane Integrity

A few antibiotics damage bacterial membranes. They cause the cells to leak, leading to cell death.

Daptomycin inserts into bacterial cytoplasmic membranes, and is used to treat certain infections caused by Gram-positive bacteria resistant to other medications. It is not effective against Gram-negative bacteria because it cannot pass through the outer membrane.

Polymyxins bind to the membranes of Gram-negative cells. Unfortunately, these antibiotics also bind to eukaryotic cells, though to a lesser extent, which generally limits their use. Polymyxin B is a common ingredient in first-aid skin ointments. Despite their toxicity, polymyxin B and polymyxin E (colistin) are sometimes used as a last resort to treat certain life-threatening bacterial infections that do not respond to other antimicrobials.

As mentioned previously, the two newest glycopeptide antibiotics—albavancin and oritavancin—disrupt cell membranes in addition to interfering with peptidoglycan synthesis. ⏮ glycopeptide antibiotics

Effective Against *Mycobacterium tuberculosis*

Relatively few antimicrobials are effective against *Mycobacterium tuberculosis.* This is due to several factors, including the organism's waxy cell wall (which prevents the entry of many drugs) and slow growth. A group of medications serve as the **first-line drugs** for treating tuberculosis (TB), meaning they are preferred because they are the most effective as well as the least toxic. Four of these are usually given in combination for the first 8 weeks of treatment, and then a combination of two are given for the next 18 weeks. This combination therapy decreases the chance that resistant mutants will develop; if some cells in the infecting population spontaneously develop resistance to one medication, the other one will eliminate them. The **second-line drugs** are used for strains resistant to the first-line drugs; however, they either are less effective or have greater risk of toxicity. ⏭ *Mycobacterium tuberculosis*

The core first-line drugs for treating TB have different targets. Isoniazid (INH) inhibits the synthesis of mycolic acids, a primary component of the cell wall. Ethambutol (EMB) inhibits enzymes required for synthesis of other mycobacterial cell wall components. Recent studies indicate that pyrazinamide (PZA) interferes with a process that mycobacterial cells use to restart stalled ribosomes. Another first-line drug, rifampin (RIF), was discussed earlier in this section. ⏮ rifampin

Characteristics of the antimicrobial medications described in this section are summarized in **table 20.1.**

TABLE 20.1 Characteristics of Antibacterial Medications

Target/Class/Example(s)	Comments/Characteristics
Cell Wall Synthesis	
β-lactam antibiotics	Bactericidal against a variety of bacteria; interfere with peptidoglycan synthesis by inhibiting enzymes called penicillin-binding proteins (PBPs) that help form cross-links between adjacent glycan chains.
Penicillins Penicillin G, methicillin, dicloxacillin, ampicillin, amoxicillin, ticarcillin, piperacillin	Groups include natural penicillins, penicillinase-resistant penicillins, broad-spectrum penicillins, and extended-spectrum penicillins; these differ in spectrum of activity and susceptibility to β-lactamases.
Cephalosporins Cephalexin, cefazolin, cefaclor, cefprozil, cefixime, ceftriaxone, cefepime, ceftaroline	The later generations are usually more effective against Gram-negative bacteria and less susceptible to destruction by certain β-lactamases.
Carbapenems Imipenem, meropenem, ertapenem, and doripenem	Broad spectrum of activity; not destroyed by most β-lactamases including extended-spectrum β-lactamases, but susceptible to carbapenemases.
Monobactams Aztreonam	Primarily active against members of the family *Enterobacteriaceae*. Not inactivated by most β-lactamases but susceptible to extended-spectrum β-lactamases and some carbapenemases.
Glycopeptide antibiotics Vancomycin, televancin, dalbavancin, and oritavancin	Bactericidal against Gram-positive bacteria; interfere with peptidoglycan synthesis by binding to the amino acid side chain of NAM molecules, blocking formation of cross-links between adjacent glycan chains. Dalbavancin and oritavancin also disrupt membrane integrity. Used to treat serious infections caused by Gram-positive bacteria that are resistant to most other options.
Bacitracin	Bactericidal against Gram-positive bacteria; interferes with the transport of peptidoglycan precursors across the cytoplasmic membrane. Common ingredient in non-prescription antibiotic ointments.
Protein Synthesis	
Aminoglycosides Streptomycin, gentamicin, tobramycin, neomycin	Bactericidal against aerobic and facultative bacteria; bind to the 30S ribosomal subunit, blocking the initiation of translation and causing the misreading of mRNA. Toxicity limits the use. Neomycin is commonly used in non-prescription topical antibiotic ointments.
Tetracyclines and glycylcyclines Tetracyline and doxycycline (tetracyclines), tigecycline (a glycylcycline)	Bacteriostatic against some Gram-positive and Gram-negative bacteria; bind to the 30S ribosomal subunit, blocking the attachment of tRNA.
Macrolides Erythromycin, clarithromycin, azithromycin	Bacteriostatic against many Gram-positive bacteria as well as the most common causes of atypical pneumonia; bind to the 50S ribosomal subunit, preventing the continuation of protein synthesis.
Chloramphenicol	Bacteriostatic and broad-spectrum; binds to the 50S ribosomal subunit, preventing peptide bonds from being formed. Generally used only as a last resort for life-threatening infections.
Lincosamides Lincomycin, clindamycin	Bacteriostatic against a variety of Gram-positive and Gram-negative bacteria. Bind to the 50S ribosomal subunit, preventing the continuation of protein synthesis.
Oxazolidinones Linezolid, tedizolid	Bacteriostatic against a variety of Gram-positive bacteria. Useful for treating infections caused by bacteria that are resistant to other options. Bind to the 50S ribosomal subunit, interfering with the initiation of protein synthesis.
Pleuromutilins Retapamulin	Bacteriostatic against a variety of Gram-positive bacteria. Bind to the 50S ribosomal subunit, preventing peptide bonds from being formed. The only derivative currently approved for humans is for topical use only.
Streptogramins Quinupristin, dalfopristin	Used as a synergistic bactericidal combination; the two drugs bind to two different sites on the 50S ribosomal subunit. Effective against a variety of Gram-positive bacteria, but generally reserved for strains that are resistant to other antimicrobials.
Nucleic Acid Synthesis	
Fluoroquinolones Ciprofloxacin, levofloxacin, moxifloxacin	Bactericidal against a wide variety of Gram-positive and Gram-negative bacteria; inhibit topoisomerases.
Rifamycins Rifampin	Bactericidal against Gram-positive and some Gram-negative bacteria. Bind RNA polymerase, blocking the initiation of RNA synthesis.
Fidaxomicin	Bactericidal, particularly against *Clostridium* species; interferes with RNA polymerase activity.
Metronidazole	Bactericidal against anaerobes. Activated by anaerobic metabolism and then binds DNA, interfering with synthesis and causing damaging breaks.

(continued)

TABLE 20.1 Characteristics of Antibacterial Medications (*Continued*)

Target/Class/Example(s)	Comments/Characteristics
Folate Biosynthesis	
Sulfonamides	Bacteriostatic against a variety of Gram-positive and Gram-negative bacteria. Structurally similar to *para*-aminobenzoic acid (PABA) and therefore inhibit the enzyme for which PABA is a substrate.
Trimethoprim	Often used with a sulfonamide for a synergistic effect, a combination called co-trimoxazole. Inhibits the enzyme that catalyzes a step following the one inhibited by the sulfonamides.
Cell Membrane Integrity	
Daptomycin	Bactericidal against Gram-positive bacteria by damaging the cytoplasmic membrane.
Polymyxins Polymyxin B, polymyxin E (colistin)	Bactericidal against Gram-negative bacteria by damaging cell membranes. Toxicity limits their use.
Mycobacterium tuberculosis	
Ethambutol	Inhibits the synthesis of a component of the mycobacterial cell wall.
Isoniazid	Inhibits synthesis of mycolic acids, a major component of the mycobacterial cell wall.
Pyrazinamide	Interferes with a process that mycobacterial cells use to restart stalled ribosomes.

MicroAssessment 20.3

The targets of antimicrobial medications include biosynthetic pathways for peptidoglycan, protein, nucleic acid, and folate and the integrity of membranes. Medications used to treat tuberculosis often interfere with processes unique to *Mycobacterium tuberculosis.*

7. Why are β-lactam antibiotics bactericidal only to growing bacteria?
8. What is the target of the macrolides?
9. Considering that all β-lactam antibiotics have the same target, why do they vary in their spectrum of activity?

20.4 ■ Antimicrobial Susceptibility Testing

Learning Outcomes

7. Describe the Kirby-Bauer disc diffusion test.
8. Describe how the minimum inhibitory concentration (MIC) and the minimum bactericidal concentration (MBC) are determined.

Susceptibility of a pathogen to a specific antimicrobial medication is often unpredictable. In these cases, laboratory tests are used to determine the susceptibility of the organism to various antimicrobials. Once the results are known, the most appropriate medication can be chosen. In our discussion, we will focus on methods used to determine the susceptibility of bacteria; note that similar procedures are used for yeasts.

Conventional Disc Diffusion Method

The **Kirby-Bauer disc diffusion test** is a relatively simple method routinely used to determine the susceptibility of a given bacterial strain to a variety of antimicrobial medications. A standard inoculum of the strain is first uniformly spread on the surface of an agar plate. Then 12 or so discs, each containing a known amount of a different antimicrobial, are placed on the surface of the medium (**figure 20.9**). During incubation, the various antimicrobials diffuse outward,

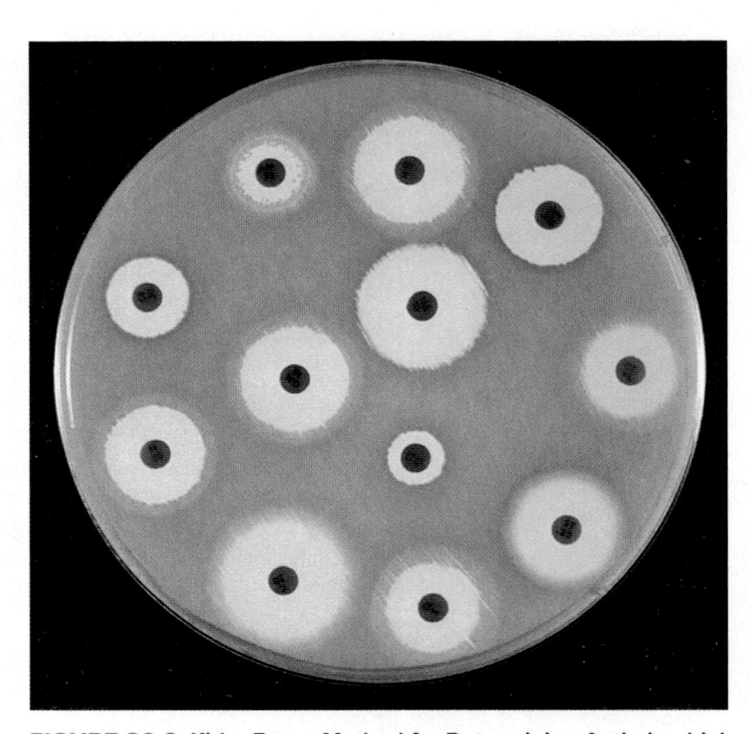

FIGURE 20.9 Kirby-Bauer Method for Determining Antimicrobial Susceptibility The size of the zone of inhibition surrounding the disc reflects, in part, the sensitivity of the bacterial strain to the medication.
Source: Gilda L. Jones/CDC

When using the Kirby-Bauer test to determine antimicrobial susceptibility, a chart must be consulted. Why not simply choose the option that gives the largest zone size?

forming a concentration gradient around each disc. Meanwhile the bacterial cells multiply, eventually forming a film of growth on the plate, except in regions around the discs where the bacteria were killed or their growth was inhibited. This clear area in which no visible growth occurs is called a **zone of inhibition.**

The size of the zone of inhibition around an antimicrobial disc reflects, in part, the degree of susceptibility of the organism to the medication. The zone size is also influenced by characteristics of the chemical, including its molecular weight and stability, as well as the amount in the disc.

Special charts have been prepared correlating the diameter of the zone of inhibition to susceptibility of bacteria to the medication. Based on the size of the zone and information in the chart, a bacterium can be described as susceptible, intermediate, or resistant to a particular antimicrobial. The procedures discussed next can be used to gain more precise information about an organism's susceptibility.

Minimum Inhibitory and Minimum Bactericidal Concentrations (MIC and MBC)

If the Kirby-Bauer test gives unusual results, or when a company is developing a new antimicrobial medication, then a broth dilution test to determine the **minimum inhibitory concentration** (MIC) may be necessary. The MIC is the lowest concentration of a specific antimicrobial medication needed to prevent the visible growth of a given bacterial strain in vitro. It is determined by growing the test strain in broth cultures containing different concentrations of the antimicrobial (**figure 20.10**). To do this, serial dilutions are used to generate decreasing concentrations of the medication in tubes

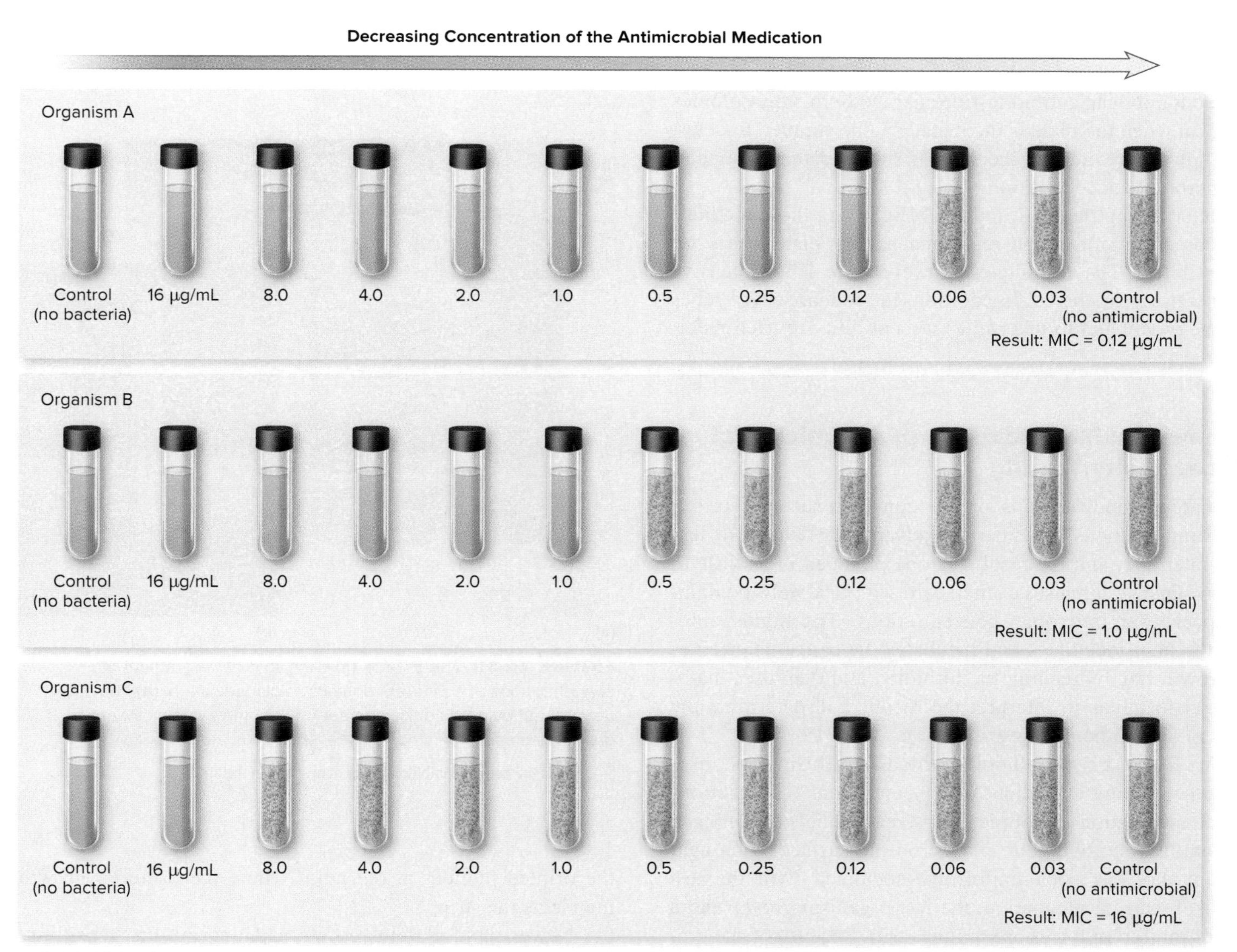

FIGURE 20.10 Determining the Minimum Inhibitory Concentration (MIC) of an Antimicrobial Medication The lowest concentration of the antimicrobial that prevents growth of the culture is the MIC.

? Considering that organism C does not grow in the tube that contains 16 μg/mL of the medication, why might the bacterium still be considered resistant to the medication?

containing a suitable growth medium. Then, a standard inoculum of bacterial cells is added to each tube. The tubes are incubated for at least 16 hours and then examined for turbidity (cloudiness), which indicates growth. The lowest concentration of the medication that prevents growth of the microorganism is the MIC.

From a medical standpoint, a microbial strain is considered susceptible to a given antimicrobial only if the organism is inhibited by concentrations used clinically. For example, an organism with an MIC of 16 µg/mL would be considered resistant to an antimicrobial if the highest level that could be achieved in vivo were less than that. Microbes that have an MIC between susceptible (treatable) and resistant (untreatable) are called intermediate.

The **minimum bactericidal concentration (MBC)** is the lowest concentration of a specific antimicrobial medication that kills 99.9% of cells of a given bacterial strain in vitro. The MBC is determined by finding out how many live organisms remain in tubes from the MIC test that showed no growth. A small sample from each of those tubes is transferred to a plate containing an antibiotic-free agar medium. Any colonies that form when the plate is incubated can be counted to determine how many cells in the sample survived that particular antimicrobial concentration.

Determining the MIC and the MBC using these methods gives precise information regarding an organism's in vitro susceptibility. The techniques, however, are labor-intensive and therefore expensive. In addition, individual sets of tubes must be inoculated to determine susceptibility to each medication tested.

Commercial Modifications of Antimicrobial Susceptibility Testing

Commercial modifications of the conventional susceptibility testing methods offer certain advantages. They are less labor-intensive, and the results can be obtained in as little as 4 hours. One system uses a small card with tiny wells containing specific antimicrobial concentrations. The highly automated system inoculates and incubates the cards, determines the growth rate by reading the turbidity, and then uses mathematical formulas to interpret the results and determine the MICs in 6 to 15 hours (**figure 20.11**).

The E test, a modification of the disc diffusion test, uses a strip containing a gradient of concentrations of an antimicrobial medication. Multiple strips, each one containing a different antimicrobial, are placed on the surface of an agar medium that has been uniformly inoculated with the test organism. During incubation, the test organism grows, and a zone of inhibition forms around the strip. Because of the gradient of antimicrobial concentrations, the zone of inhibition is shaped somewhat like a teardrop that intersects the strip at some point (**figure 20.12**). The MIC is determined by reading the printed number at the point where the bacterial growth intersects the strip.

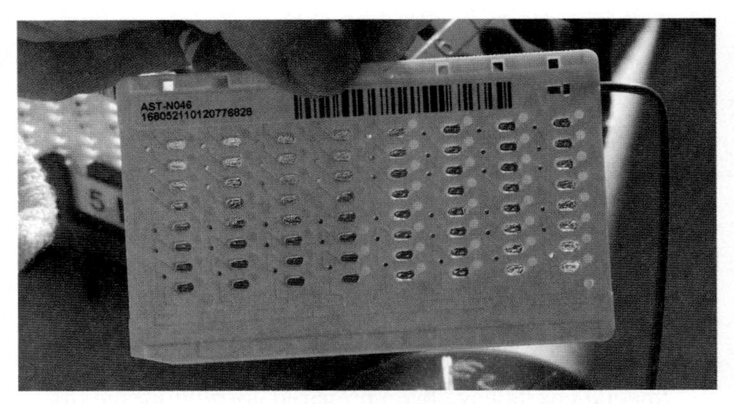

FIGURE 20.11 Automated Tests Used to Determine Antimicrobial Susceptibility The tiny wells in the card contain specific concentrations of an antimicrobial. An automated system inoculates and incubates the cards, determines the growth rate by reading turbidity, and uses mathematical formulas to interpret the results and derive the MICs. ©BSIP/UIG Via Getty Images

What are two advantages of automated tests used to determine antimicrobial susceptibility?

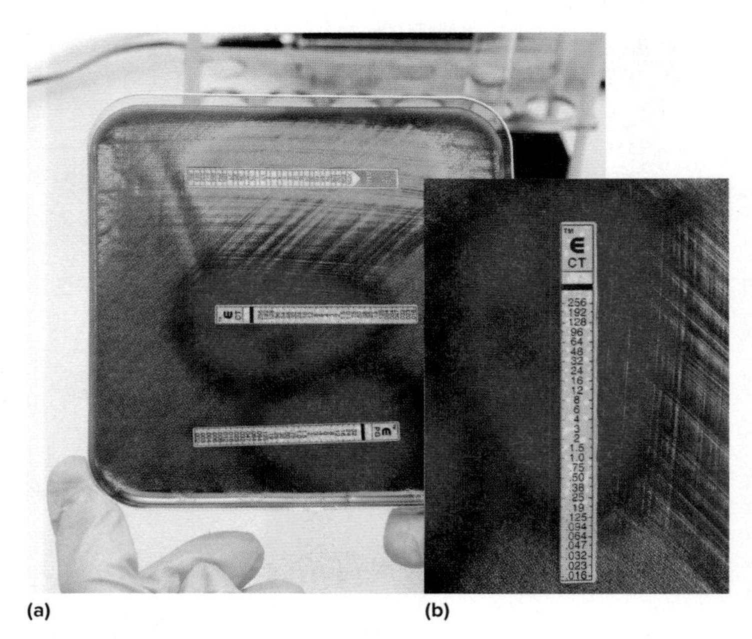

(a) (b)

FIGURE 20.12 The E Test (a) Each strip has a gradient of concentrations of a different antimicrobial medication. **(b)** The MIC is determined by reading the number on the strip at the point at which growth intersects the strip. a: ©BSIP/Newscom; b: ©BSIP/Newscom

Why is the zone of inhibition larger at one end of the strip?

Newer rapid systems do not examine in vitro sensitivity and instead detect genes encoding antibiotic resistance. Some of these systems use microarrays to detect the genes; others use PCR. ⏮ microarray, ⏮ PCR

FOCUS YOUR PERSPECTIVE 20.1

Measuring the Concentration of an Antimicrobial Medication in Blood or Other Body Fluids

Sometimes it is necessary to determine the concentration of an antimicrobial medication in a patient's blood or other body fluid. For example, patients who are being administered an aminoglycoside must often be carefully monitored to ensure that the concentration of the antimicrobial in their blood does not reach an unsafe level, particularly if they have kidney or liver dysfunction that interferes with normal elimination. Likewise, newly developed medications must be tested to determine achievable levels in the blood, urine, or other body fluids.

A diffusion bioassay is used to measure the concentration of an antimicrobial in a fluid specimen. The test relies on the same principle as the Kirby-Bauer test, except in this case it is the concentration of medication being determined, not the sensitivity of the organism.

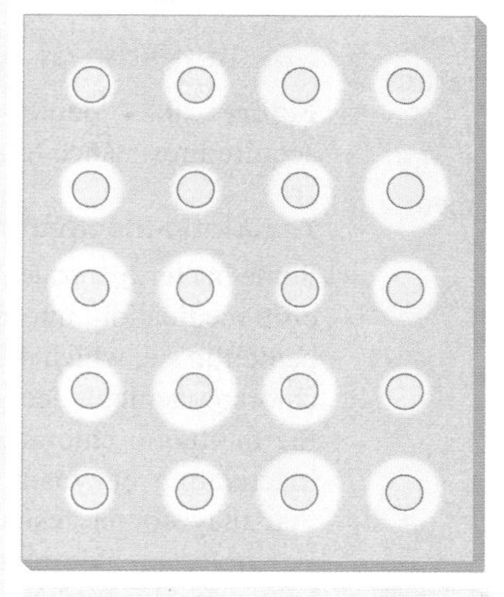

Standards and patient's serum are added to agar that has been seeded with a susceptible strain of bacteria.

(a)

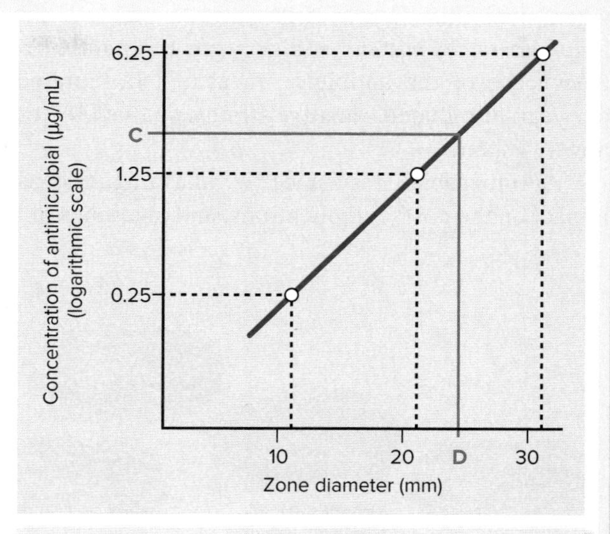

A standard curve that correlates the zone diameter with the antimicrobial concentration is constructed. The antimicrobial concentration in the serum can be determined using the line relating zone size to concentration.

(b)

BOX FIGURE 20.1 Diffusion Bioassay

To do a diffusion bioassay, a culture of a stock organism highly susceptible to the medication in question is added to melted cooled agar. The mixture is then poured into an agar plate and allowed to solidify. This results in a solid medium uniformly inoculated with the sensitive organism. Cylindrical holes are then punched out of the agar, creating wells. Standards (known concentrations of the medication) are then added to some of the wells, while others are filled with the body fluid being tested. As the bacteria grow during overnight incubation, zones of inhibition develop around the wells. The size of a zone corresponds to the concentration of medication in the well—the higher the concentration, the larger the zone of inhibition (**box figure 20.1**). The zone sizes around the standards are measured, and from this a standard curve is constructed by plotting the zone sizes against the corresponding antimicrobial concentrations. A line relating zone size to concentration is obtained, from which the concentration of the antimicrobial in the body fluid can be read.

MicroAssessment 20.4

Disc diffusion tests can determine whether an organism is susceptible, intermediate, or resistant to a variety of different antimicrobials. The MIC and the MBC are quantitative measures of a bacterial strain's susceptibility to an antimicrobial medication. Commercial tests for determining antimicrobial sensitivity are less labor-intensive and often more rapid.

10. List two factors other than a strain's sensitivity that influence the size of the zone of inhibition around an antimicrobial disc.

11. Explain the difference between the MIC and the MBC.

12. Why would it be important for the Kirby-Bauer disc diffusion test to use a standard concentration of the bacterial strain being tested?

20.5 ■ Resistance to Antimicrobial Medications

Learning Outcomes

9. Describe four general mechanisms of antimicrobial resistance.

10. Describe how antimicrobial resistance can be acquired.

11. List five examples of emerging antimicrobial resistance.

12. Describe how the emergence and spread of antimicrobial resistance can be slowed.

After sulfa drugs and penicillin were introduced, people hoped that such medications would eliminate most bacterial diseases. We now realize, however, that resistance limits the usefulness of all known antimicrobials.

As antimicrobial medications are increasingly used and misused, resistant organisms have a selective advantage over their sensitive counterparts (**figure 20.13**). For example, when penicillin G was first introduced, less than 3% of *Staphylococcus aureus* strains were resistant to its effects. Heavy use of the antibiotic, measured in hundreds of tons per year, eliminated sensitive strains, so that 90% or more are now resistant.

Antimicrobial resistance is alarming because of the impact on the cost, complications, and outcomes of treatment. Dealing with the problem requires an understanding of the mechanisms and spread of resistance. This section focuses on bacterial resistance, but the principles apply to other microbes as well.

FIGURE 20.13 The Selective Advantage of Resistance to Antimicrobial Medications When antimicrobials are used, organisms that are resistant (R) to their effects have a selective advantage over their sensitive (S) counterparts.

How does overuse of antibiotics contribute to increasing numbers of antibiotic-resistant bacteria?

Mechanisms of Acquired Resistance

Figure 20.14 depicts the most common mechanisms of acquired resistance to antimicrobial medications.

Antibiotic-Inactivating Enzymes

Some bacteria produce enzymes that chemically modify a specific medication, interfering with its function. One example is penicillinase, which destroys penicillin. Another is the enzyme chloramphenicol acetyltransferase, which chemically alters the antibiotic chloramphenicol, making it ineffective. Some inactivating enzymes have an extended spectrum, meaning that they confer resistance to a wide variety of antibiotics. Examples include extended-spectrum β-lactamases and carbapenemases. Both of these groups inactivate many different β-lactam antibiotics.

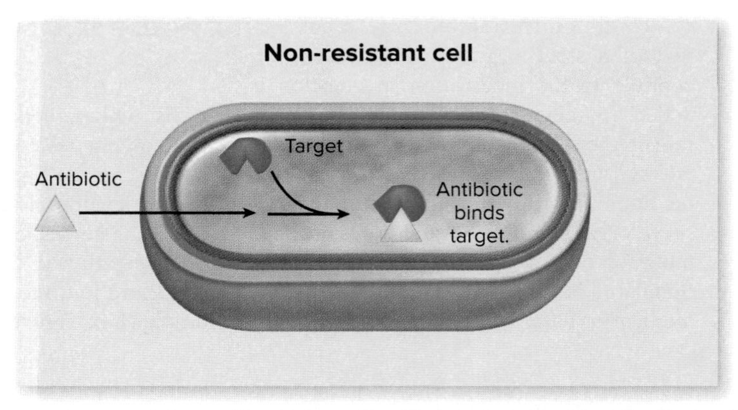

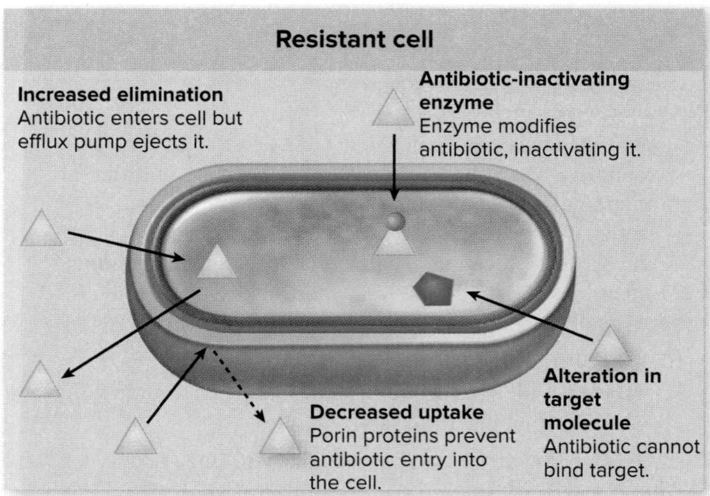

FIGURE 20.14 Common Mechanisms of Acquired Resistance to Antibiotics and Other Antimicrobial Medications

As a group, bacteria use which three methods to avoid the effects of β-lactam antibiotics?

Alteration in the Target Molecule

Antimicrobial medications generally act by binding to specific target molecules in a microorganism, interfering with the target's function. Minor structural changes in the target can prevent the medication from binding, thereby protecting the organism from its effects. For instance, modifications in the penicillin-binding proteins (PBPs) prevent β-lactam antibiotics from binding to them. Similarly, a change in ribosomal RNA, the target for the macrolides, prevents those medications from interfering with ribosome function. In some cases, an alteration in one target allows a bacterium to become resistant to multiple classes of antibiotics simultaneously. For example, some bacteria have acquired the ability to produce an enzyme that adds a methyl group to an rRNA molecule of the 50S ribosome. By doing this, the bacteria prevent binding of macrolides, lincosamides, and streptogramins to the ribosome, and are therefore resistant to all these antibiotics.

Decreased Uptake of the Medication

Recall that porin proteins in the outer membrane of Gram-negative bacteria selectively permit small molecules to pass through that membrane and enter the cell's periplasm; from there the molecules may cross the cytoplasmic membrane to enter the cytoplasm. Changes in the porin proteins can therefore prevent certain antimicrobials from entering the cell's periplasm or cytoplasm. By stopping entry of an antimicrobial, an organism avoids its effects. ⏮ **porins**

Increased Elimination of the Medication

Bacteria use **efflux pumps** to transport antimicrobials and other damaging compounds out of a cell. When a cell makes more of these pumps, the compound is ejected faster. In addition, structural changes in the pumps can influence the range of compounds that can be pumped out. Resistance that develops by this mechanism sometimes allows an organism to become resistant to several different antimicrobials simultaneously. ⏮ **efflux pumps**

Acquisition of Resistance

Antimicrobial resistance can be acquired by either spontaneous mutation, which alters existing genes, or gaining new genes (see figure 8.1).

Spontaneous Mutation

As cells replicate, spontaneous mutations happen at a relatively low rate. Those few mutations that occur, however, can have a significant effect on the resistance of a bacterial population to an antimicrobial medication. ⏮ **spontaneous mutation**

Acquired resistance to streptomycin (an aminoglycoside) is a good example of the consequence of spontaneous mutation. A single base-pair change in the gene encoding a ribosomal protein alters the target enough to make the cell streptomycin-resistant. When a streptomycin-sensitive strain is grown in streptomycin-free medium to a population of 10^9 cells, at least one cell in the population probably has that particular mutation. If streptomycin is then added to the medium, only that cell and its descendants will still be able to replicate, giving rise to a streptomycin-resistant population.

Medications for which a single point mutation causes resistance are sometimes used in combination with one or more other drugs, an approach called **combination therapy.** If any cell spontaneously develops resistance to one medication, another one will still kill it. Combination therapy is effective because the chance that a cell will simultaneously develop mutational resistance to multiple antimicrobial medications is extremely low.

When an antimicrobial medication has several different targets or can bind to multiple sites on a single target, resistance due to spontaneous mutation is less likely to occur. This is because multiple mutations would be required to prevent binding of the antimicrobial to its target. If a sensitive strain is grown to a population of 10^9 cells, a few of the cells might have a mutation that allows them to be slightly less sensitive to the medication, but it is unlikely that a single cell could have accumulated the precise combination of mutations necessary to be resistant. Thus, appropriate levels of the medication for a long enough period of time would still be effective against all the cells in that particular population. If doses are skipped, however, the medication level may drop to sub-inhibitory levels. In this environment, the least sensitive cells may multiply, which increases the chance that they can accumulate the needed mutations to be resistant. Likewise, if the treatment is stopped too soon, the least sensitive cells may not have been eliminated, and when these cells multiply, additional mutations may make them even less sensitive. Progressive misuse of the medications can eventually lead to resistant strains.

Gene Transfer

Genes encoding resistance to antimicrobial medications can spread to different strains, species, and even genera, most commonly through conjugative transfer of **R plasmids.** These plasmids often carry several different resistance genes, each one encoding resistance to a specific antimicrobial. Thus, when an organism acquires an R plasmid, it may become resistant to several different medications simultaneously. ⏮ **R plasmid,** ⏮ **conjugation**

The origin of some resistance genes carried on R plasmids was probably a bacterium that developed resistance through spontaneous mutation. Others may have come from the microbe that naturally produces the antibiotic. A gene coding for an enzyme that chemically modifies an aminoglycoside likely came from the *Streptomyces* species that produces the drug.

Examples of Emerging Resistance

Many organisms are developing resistance to antimicrobials. The situation has become so worrisome that the CDC recently published a document, *Antibiotic Resistance Threats in the United States,* which lists the most serious threats, categorizing them by level: urgent, serious, concerning (**table 20.2**).

TABLE 20.2 Threats of Resistance to Antimicrobial Medications

Microorganism	Comments
CDC's Threat Level: Urgent	
Clostridium difficile	*C. difficile* causes antibiotic-associated intestinal infections that can sometimes be life-threatening; the organism is considered an urgent threat because *Clostridium difficile* infection (CDI) is related to the use of antimicrobials.
Carbapenem-resistant *Enterobacteriaceae* (CRE)	Members of the family *Enterobacteriaceae* are a common cause of healthcare-associated infections (HAIs). Carbapenems are often a last resort for treating *Enterobacteriaceae* that are resistant to other options, so CRE infections are very difficult or impossible to treat effectively.
Drug-resistant *Neisseria gonorrhoeae*	*N. gonorrhoeae* causes gonorrhea, a sexually transmitted infection. Multi-drug resistance is so widespread that combination therapy using ceftriaxone (a cephalosporin) and azithromycin (a macrolide) is now recommended.
CDC's Threat Level: Serious	
Multi-drug-resistant *Acinetobacter*	*Acinetobacter* species are a common cause of healthcare-associated infections (HAIs); many strains are resistant to multiple antimicrobial medications.
Drug-resistant *Campylobacter*	*Campylobacter* species are a common cause of foodborne diarrhea; resistance to ciprofloxacin and azithromycin is increasing.
Fluconazole-resistant *Candida* (a fungus)	*Candida* species (fungi) are a common cause of healthcare-associated infections (HAIs); relatively few antifungal medications are available, so acquired resistance to even one type is significant.
Extended-spectrum β-lactamase (ESBL) producing *Enterobacteriaceae*	Members of the family *Enterobacteriaceae* are a common cause of healthcare-associated infections (HAIs). ESBL-producing *Enterobacteriaceae* are resistant to all the β-lactam antibiotics, leaving carbapenem as the last treatment option.
Vancomycin-resistant *Enterococcus* (VRE)	*Enterococcus* species are a common cause of healthcare-associated infections (HAIs), including urinary tract and bloodstream infections. Enterococci are innately resistant to many antimicrobials, and VRE strains are resistant to nearly all antimicrobial medications.
Multi-drug-resistant *Pseudomonas aeruginosa*	*P. aeruginosa* is a common cause of healthcare-associated infections (HAIs); it is also a common cause of pneumonia in cystic fibrosis patients. *P. aeruginosa* is innately resistant to many antimicrobials, so multi-drug resistance makes a difficult situation even more challenging.
Drug-resistant non-typhoidal *Salmonella*	Non-typhoidal *Salmonella* strains are a common cause of foodborne diarrhea. Uncomplicated infections are not usually treated with antimicrobial medications, but infections that spread to the bloodstream can be life-threatening and should be treated.
Drug-resistant *Salmonella* Typhi	*Salmonella* Typhi causes typhoid fever, a life-threatening systemic disease. Most illnesses are travel-related, and strains are becoming increasingly resistant to certain treatment options.
Drug-resistant *Shigella*	*Shigella* species cause diarrheal diseases that spread easily in conditions of poor sanitation. Widespread antimicrobial resistance makes treatment more difficult.
Methicillin-resistant *Staphylococcus aureus* (MRSA)	*S. aureus* is a common cause of healthcare-associated (HA) and community-acquired (CA) skin and wound infections. MRSA is resistant to nearly all β-lactam antibiotics; HA-MRSA strains are generally resistant to many other antimicrobial medications as well, so severe infections are typically treated with vancomycin; CA-MRSA strains are usually susceptible to some other options.
Drug-resistant *Streptococcus pneumoniae*	*S. pneumoniae* causes pneumonia and meningitis, as well as bloodstream, ear, and sinus infections. Some strains are resistant to a variety of antibiotics, including penicillins and macrolides.
Drug-resistant tuberculosis	*M. tuberculosis* causes tuberculosis. Treatment has always been a long and complicated process because the organism is slow-growing and innately resistant to many antimicrobials. Treatment of infections caused by drug-resistant strains takes longer, is more expensive, and carries more health risks. Drug-resistant strains are an increasing problem worldwide, but are relatively rare in the United States.
CDC's Threat Level: Concerning	
Vancomycin-resistant *Staphylococcus aureus* (VRSA)	S. *aureus* causes hospital-acquired (HA) and community-acquired (CA) infections. VRSA strains are resistant to nearly all antimicrobial medications.
Erythromycin-resistant Group A streptococcus	Group A streptococcus causes a variety of infections, including pharyngitis, impetigo, and tissue infections. A main concern regarding antibiotic resistance is invasive infections (infections of otherwise sterile body sites).
Clindamycin-resistant Group B streptococcus	Group B streptococcus is an important cause of neonatal meningitis; pregnant women who are colonized with the organism are treated to prevent infection of the newborn.

These categories reflect multiple factors, including the public health significance of the resistant strains as well as the likelihood that the organisms will continue to spread. Some of the problems associated with the increasing antimicrobial resistance are highlighted by the following examples.

Enterococci

Enterococci are part of the normal intestinal microbiota and a common cause of healthcare-associated infections. They are intrinsically less susceptible to many common antimicrobials. For example, their penicillin-binding proteins have low affinity for certain β-lactam antibiotics. In addition, many enterococci have R plasmids. Some strains, called **vancomycin-resistant enterococci (VRE),** are even resistant to vancomycin. Recall that this antibiotic is usually reserved as a last resort for treating serious infections caused by Gram-positive bacteria resistant to all β-lactam antibiotics. Because vancomycin resistance in VRE strains is encoded on a plasmid, the resistance is transferable to other organisms.

Enterobacteriaceae

Members of the family *Enterobacteriaceae* are intrinsically resistant to many antimicrobials because the outer membrane of these Gram-negative bacteria prevents the medications from entering cells. The situation became more complicated when some enterics developed the ability to produce a β-lactamase, allowing the strains to resist the effects of ampicillin and other penicillins. Some strains then developed the ability to produce extended-spectrum β-lactamases (ESBLs), making them resistant to most cephalosporins and monobactams, as well as to penicillins. More recently, strains referred to as **carbapenem-resistant *Enterobacteriaceae* (CRE)** have been discovered. Many of these produce an enzyme that inactivates carbapenems as well as all other β-lactam antibiotics. The strains have additional mechanisms to resist all or nearly all other routine medications, which is why colistin—a medication normally considered too toxic to be used systemically—is now sometimes administered. The recent discovery of plasmid-encoded resistance to colistin has caused understandable concern that a CRE strain will acquire resistance to that medication as well.

Mycobacterium tuberculosis

Tuberculosis (TB) treatment has always been a long and complicated process, requiring a combination of two or more different antimicrobial medications taken for a period of 6 months or more. Unfortunately, *Mycobacterium tuberculosis* can easily become resistant to the first-line anti-TB drugs (the preferred medications) through spontaneous mutation. Large numbers of bacterial cells are found in an active infection, so it is likely that at least one cell has developed spontaneous resistance to a drug, which is why combination therapy is required. The length of treatment is due to the very slow growth of *M. tuberculosis.* ⏭ *Mycobacterium tuberculosis*

Many tuberculosis patients make the mistake of skipping doses or stopping treatment too soon. As a consequence, strains of *M. tuberculosis* develop resistance to the first-line drugs. This results in even longer, more expensive treatments that are also less effective.

Tuberculosis that is resistant to treatment by two of the first-line anti-TB medications—isoniazid and rifampin—is called **multi-drug-resistant tuberculosis (MDR-TB).** To prevent the emergence of resistant *M. tuberculosis* strains, some cities are using directly observed therapy (DOT); with DOT, healthcare workers watch patients swallow the prescribed pills to ensure treatment compliance. **Extensively drug-resistant tuberculosis (XDR-TB)** is an even greater concern. This is defined as tuberculosis resistant to treatment with isoniazid and rifampin as well as three or more of the second-line anti-TB medications. ⏭ directly observed therapy

Neisseria gonorrhoeae

Fifty years ago, gonorrhea, the sexually transmitted infection caused by *Neisseria gonorrhoeae,* was very easy to treat because the bacterium was quite susceptible to penicillin. Then, some strains developed resistance to penicillin through mutation, and others acquired a plasmid that encoded production of penicillinase. As these strains spread, the infection could no longer reliably be treated with penicillin. Several other treatment options were available, including tetracyclines, macrolides and fluoroquinolones, but as these antimicrobial medications were increasingly used, some *N. gonorrhoeae* strains gradually developed resistance to them as well. Today, only certain cephalosporins are reliably effective against most strains, but strains resistant to those have now been isolated. Because of the concerns over resistance, treatment of gonorrhea now relies on combination therapy: a single intramuscular dose of ceftriaxone (a third-generation cephalosporin) along with an oral dose of azithromycin (a macrolide). ⏭ gonorrhea

Staphylococcus aureus

Staphylococcus aureus, another common cause of healthcare-associated infections, is becoming increasingly resistant to antimicrobials. Although nearly all strains were susceptible to penicillin when it was first introduced 70 years ago, most are now resistant due to their acquisition of a gene encoding penicillinase. Until recently, infections by these strains could be treated with methicillin or other penicillinase-resistant penicillins. New strains have emerged, however, that not only produce penicillinase, but also make a penicillin-binding protein called PBP2a, which has a low affinity for most β-lactam antibiotics. These strains, called **methicillin-resistant *Staphylococcus aureus* (MRSA),** are resistant to methicillin as well as to nearly all other β-lactam antibiotics. Ceftaroline, a new cephalosporin, is an important exception because it binds PBP2a. ⏭ *Staphylococcus aureus*

FOCUS ON A CASE 20.1

A 36-year-old woman came to the emergency department complaining of severe shortness of breath, cough, and chest pain. She said she had been feeling ill for 3 months and had lost 20 pounds as a result. When questioned about her travel history, she said she had immigrated to the United States from India when she was 10 and had since taken several trips back there to visit family members. A chest X ray was done, and lesions in one of her lungs were consistent with pulmonary tuberculosis, a lung infection caused by *Mycobacterium tuberculosis*. ▶▶| tuberculosis

The patient's cough was not productive (meaning it was dry), so physicians sampled material from her lower respiratory tract using bronchoalveolar lavage (BAL). This involves delivering sterile saline into the lungs via a long thin tube and then collecting the fluid. The BAL fluid was then sent to the clinical laboratory for examination.

In the clinical laboratory, a technician prepared a smear of the BAL fluid on a microscope slide, in preparation for doing an acid-fast stain. He also inoculated the fluid onto media that support the growth of *M. tuberculosis*. After doing the acid-fast stain, the technician carefully examined the stained specimen microscopically. He could not find any acid-fast bacilli (AFB), so he reported the result as negative. The technician then did a nucleic acid amplification test (NAAT) designed to detect *M. tuberculosis* DNA; that result was positive. |◀◀ acid-fast stain, |◀◀ NAAT

The patient was prescribed the standard four-drug treatment for tuberculosis disease: isoniazid (INH), rifampin (RIF), pyrazinamide (PZA), and ethambutol (EMB). After 3 weeks, she reported feeling slightly better, but was still coughing. A month later, the culture results were positive for *M. tuberculosis*. Antimicrobial susceptibility testing was done, and the infecting strain was found to be resistant to INH and RIF. Two other medications were then substituted, and the local public health office was contacted to provide directly observed therapy (DOT) for the remainder of her treatment.

1. Why was the patient's travel history important?
2. Why are *Mycobacterium tuberculosis* infections more difficult to treat than most other bacterial infections?
3. Why was a combination of antimicrobial medications prescribed?
4. Why did it take so long for culture results to be available?
5. What is the significance of the fact that the strain was resistant to isoniazid and rifampin?
6. Why was directly observed therapy used?

Discussion

1. Although tuberculosis is no longer endemic in the United States, it is still common in many parts of the world, including India. The patient could have contracted the infection as a child in India or during her trips to visit family members.
2. *Mycobacterium tuberculosis* has a very waxy cell wall, which prevents many antibacterial medications from penetrating. In addition, it is extremely slow growing, and most antimicrobial medications are most effective against rapidly multiplying bacteria.
3. Combinations of antimicrobials are used to prevent the development of resistant mutants. Consider a situation where a single point mutation in a gene allows a bacterium to become resistant to a given antimicrobial, and the chance of that mutation occurring is one in a million (10^{-6}). If a susceptible bacterial cell is grown to a population of a billion cells, then about 1,000 cells will have spontaneously developed resistance to that medication. Meanwhile, if another medication is prescribed, but the rate of spontaneous mutations to resistance for that one is 10^{-7}, then 100 cells will have developed resistance to that option. However, the chance that a single cell will spontaneously develop resistance to both medications simultaneously is the product of the two rates, meaning 10^{-13}.
4. *Mycobacterium tuberculosis* grows very slowly, with a generation time of over 16 hours, so it takes weeks for visible colonies to form. This is why detection methods that do not require culture are so important for early diagnosis.
5. Isoniazid and rifampin are first-line anti-tuberculosis drugs, the preferred options for treatment. When *M. tuberculosis* is resistant to the first-line drugs, then the patient must be treated with second-line drugs, and these are not as effective and are also more toxic and more expensive. Tuberculosis caused by strains resistant to isoniazid and rifampin is called MDR-TB (multi-drug-resistant tuberculosis).
6. Directly observed therapy (DOT) ensures that patients comply with the prescribed treatment, thereby preventing additional resistance from developing. In addition, patients who comply with their treatment become non-infectious sooner, so there is less opportunity for the infectious agent to be transmitted. From a public health standpoint, this is extremely important because of the increased costs and less favorable outcomes associated with treating MDR-TB.

There are two categories of MRSA strains—healthcare-associated (HA-MRSA) and community acquired (CA-MRSA). HA-MRSA strains are generally resistant to a wide range of antimicrobial medications, so severe infections are often treated with vancomycin. A few hospitals, however, have reported isolates that are no longer susceptible to normal levels of vancomycin. So far, strict hospital guidelines designed to immediately halt the spread of these **vancomycin-intermediate *S. aureus* (VISA)** and **vancomycin-resistant *S. aureus* (VRSA)** strains have been successful. Fortunately, most CA-MRSA strains are currently susceptible to certain medications other than β-lactam antibiotics.

Streptococcus pneumoniae

Until recently, *Streptococcus pneumoniae,* the leading cause of pneumonia in adults, has remained very sensitive to penicillin.

Some isolates, however, are now resistant to the antibiotic. This acquired resistance is due not to the production of a β-lactamase, but to changes in the chromosomal genes coding for the targets of penicillin—the penicillin-binding proteins. The modified targets have lower affinities for the medication. The nucleotide changes do not appear to have come from point mutations as one might expect; instead, they are due to the acquisition of chromosomal DNA from other species of *Streptococcus.* As you may recall from earlier reading, *S. pneumoniae* can acquire DNA through DNA-mediated transformation. In addition to penicillin resistance, some *S. pneumoniae* strains have developed resistance to other antibiotics, including macrolides. ⏭ *Streptococcus pneumoniae,* ⏮ *DNA-mediated transformation*

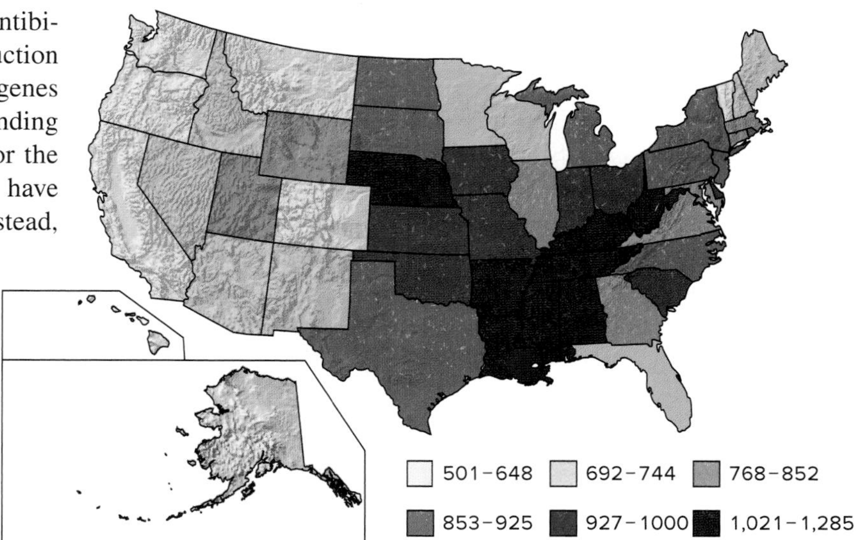

FIGURE 20.15 Rate of Antimicrobial Prescriptions (per 1,000 Persons) in the United States, by State, 2014

What suggestions might an antimicrobial stewardship program give to decrease high rates of antimicrobial prescriptions?

Preventing Resistance

To reverse the alarming trend of increasing antimicrobial resistance, everyone must cooperate. On an individual level, physicians as well as the general public must take more responsibility for the appropriate use of these life-saving medications. On a global scale, countries around the world need to make important policy decisions about what is, and what is not, an appropriate use of these medications.

The Responsibilities of Physicians and Other Healthcare Workers

Physicians and other healthcare workers need to increase their efforts to identify the cause of a given infection and, only if appropriate, prescribe suitable antimicrobials. Recent studies show that the rate at which physicians prescribe antibiotics varies widely across the United States, perhaps reflecting inappropriate use of these life-saving medications in certain regions (**figure 20.15**). In response to the problem of increasing antimicrobial resistance, hospitals and healthcare systems are implementing antimicrobial stewardship programs to ensure appropriate treatments. Although these efforts may be more expensive in the short term, they will ultimately save both lives and money.

The Responsibilities of Patients

Patients need to carefully follow the instructions that accompany their prescriptions, even if those instructions seem inconvenient. When a patient skips a scheduled dose, the blood level of that medication may not remain high enough to inhibit the growth of the least-sensitive members of the microbial population. If these less-sensitive organisms then have a chance to grow, they will give rise to a population that is not as sensitive as the original. Likewise, failure to complete the prescribed course of treatment may not kill the least-sensitive organisms, allowing their subsequent multiplication. Misusing antimicrobials by skipping doses or failing to complete the prescribed treatment increases the likelihood that resistant mutants will develop.

The Importance of an Educated Public

A greater effort must be made to educate people about the role and limitations of antimicrobial medications. First and foremost, people need to understand that antibiotics are not effective against viruses. Taking antibiotics will not cure the common cold or any other viral illness. A few antiviral medications are available, but they are effective against only certain viruses. Unfortunately, far too many people mistakenly believe that antibiotics are effective against viruses, and often seek prescriptions to "cure" viral infections. This misuse only selects for antibiotic-resistant bacteria in the normal microbiota. Even though these organisms typically do not cause disease, they can serve as a reservoir for R plasmids, eventually transferring their resistance genes to an infecting pathogen.

Global Impacts of the Use of Antimicrobial Medications

The overuse of antibiotics and other antimicrobial medications is a worldwide concern. Countries may vary in their laws and customs, but antimicrobial resistance recognizes no political boundaries. An organism that develops resistance in one country can quickly be transported globally.

In many parts of the world, particularly in developing countries, antimicrobial medications are available on a nonprescription basis. Because of the consequences of inappropriate use, restricting or eliminating the over-the-counter availability of these medications should be considered.

Another worldwide concern is the use of antimicrobials in animal feeds. Low levels of these substances in feeds enhance the growth of animals, a seemingly attractive option. This use, like any other, however, selects for drug-resistant organisms, which has caused many scientists to question its ultimate wisdom. In fact, infections caused by drug-resistant *Salmonella* strains have been linked to animals whose feed was supplemented with those medications. In response to these concerns, the FDA developed and implemented a plan that has now phased out the growth-enhancing use of medically important antimicrobials in animals raised as food in the United States. In addition, there is growing pressure worldwide to ban the use of antimicrobials in animal feeds.

MicroAssessment 20.5

Mutations and transfer of genetic information allow microorganisms to become resistant to antimicrobial medications. Antimicrobial resistance affects the outcomes of medical treatment. Slowing the emergence and spread of resistant microbes involves the cooperation of healthcare personnel, educators, and the general public.

13. Explain how using a combination of two antimicrobial medications helps prevent the development of spontaneously resistant mutants.
14. Explain the significance of a member of the normal microbiota that harbors an R plasmid.
15. A student argued that "spontaneous mutation" meant that an antimicrobial could cause mutations. Is the student correct? Why or why not?

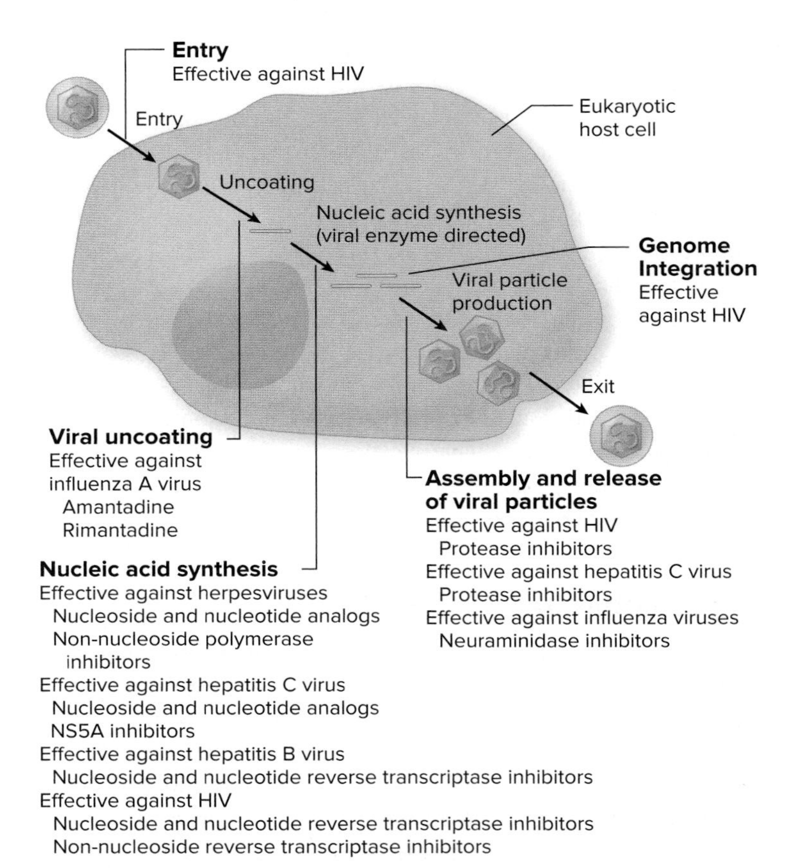

FIGURE 20.16 Targets of Antiviral Medications

Why are there relatively few options for antiviral medications?

20.6 ■ Mechanisms of Action of Antiviral Medications

Learning Outcome

13. Describe the antiviral medications that interfere with the following: (1) viral uncoating, (2) nucleic acid synthesis, (3) genome integration, (4) assembly and release of viral particles.

Viruses rely almost exclusively on the host cell's metabolic machinery for their replication, making it difficult to find a target for selective toxicity. They have no cell wall, ribosomes, or any other structure targeted by antibiotics. Because of this, viruses are completely unaffected by antibiotics. Many encode their own polymerases, however, and these, along with a few other viral proteins, are potential targets of **antiviral medications** (**figure 20.16**). These medications, also called antiviral drugs or antivirals, are generally effective against only a specific type of virus. Perhaps not surprisingly, the greatest variety of antivirals at this point are directed at HIV (human imunodeficiency virus). The names of anti-HIV medications are typically shortened to a three-letter abbreviation or acronym. ⏭ HIV

An important limitation of antivirals is that they are only effective against replicating viruses. Viruses such as herpesviruses and HIV can remain latent in cells, so the medications do not cure these infections; they simply shorten the the duration of the active infection. Latent virus can still reactivate, causing symptoms to recur. ⏮ latent viral infections

For viruses such as HIV and HCV (hepatitis C virus) that evolve rapidly to develop resistance to single medications, combination therapy is typically used. Manufacturers are now developing fixed-dose combinations that make it easier for patients to adhere to prescribed treatment protocols. ⏭ HCV, ⏮ combination therapy

Prevent Viral Entry

Some medications effective against HIV are entry inhibitors, meaning they prevent the virus from entering host cells. Enfuvirtide (ENF), a medication referred to as a fusion inhibitor, does this by binding to an HIV protein that promotes fusion of the viral envelope with the cell membrane. Maraviroc (MCV), referred to as a CCR5 antagonist, blocks the HIV co-receptor CCR5. ⏭ HIV attachment and entry

Interfere with Viral Uncoating

Although not currently used due to widespread resistance, two medications—amantadine and rimantadine—interfere with uncoating of the influenza A virus. Recall that uncoating is the process by which the nucleic acid of a viral particle is released from the protein coat. ⏭ influenza A virus

Interfere with Nucleic Acid Synthesis

Many of the most effective antiviral medications target virally encoded enzymes used during replication of viral nucleic acid. These antivirals are generally limited to treating herpesviruses, HBV (hepatitis B virus), HCV (hepatitis C virus), or HIV infections.

Nucleoside and Nucleotide Analogs

Several antiviral medications are nucleoside or nucleotide analogs—chemicals structurally similar to the building blocks of DNA and RNA. A nucleoside analog can be phosphorylated in vivo by a virally encoded or normal cellular enzyme to form a nucleotide analog that interferes with nucleic acid synthesis. For example, many nucleotide analogs act as chain terminators during DNA synthesis; when the analog is incorporated into a growing nucleotide chain, additional nucleotides cannot be added to that chain. The basis for the selective toxicity of most analogs is the fact that error-prone virally encoded polymerases are more likely than the host cell enzymes to incorporate them. Therefore, more damage is done to the rapidly replicating viral genome than to the host cell genome. ⏮ nucleotide

Acyclovir, famciclovir, and valacyclovir are nucleoside analogs used to speed the healing of the characteristic lesions associated with shingles, chickenpox, cold sores, and genital herpes—diseases all caused by herpesviruses. The medications cause little harm to uninfected cells because normal cellular enzymes are not likely to convert the chemicals into nucleotide analogs. Instead, a virally encoded enzyme found only in infected cells converts the medications into their active form.

Sofosbuvir is a relatively new medication that has revolutionized the treatment of hepatitis C because it is so effective when used in combination with at least one other anti-HCV medication. This nucleotide analog interferes specifically with the HCV's replicase (an RNA-dependent RNA polymerase).

Recall that HIV and HBV are both reverse-transcribing viruses, meaning that a step in their replication cycle requires the virally encoded enzyme reverse transcriptase. Nucleoside and nucleotide analogs that interfere with the activity of reverse transcriptase are referred to as NRTIs (nucleoside/nucleotide reverse transcriptase inhibitors). In the case of HIV treatment, two NRTIs are often used in combination with at least one other antiretroviral medication; examples of NRTIs used to treat HIV infection include abacavir (ABC), emtricitabine (FTC), lamivudine (3TC), tenofovir alafenamide (TAF), and zidovudine (AZT or ZDV). NRTIs used to treat chronic hepatitis B (caused by HBV) include tenofovir and lamivudine. ⏮ replication of reverse-transcribing viruses

Certain nucleoside analogs are used only for severe infections because of their significant side effects. An example is ganciclovir, which is used to treat life- or sight-threatening cytomegalovirus (CMV) infections in immunocompromised patients.

Non-Nucleoside Polymerase Inhibitors

Non-nucleoside polymerase inhibitors are compounds that inhibit the activity of viral polymerases by binding to a site other than the nucleotide-binding site. A relatively new example, dasabuvir, is a component of a fixed-dose combination to treat HCV. Another example, foscarnet, is used to treat severe infections caused by CMV and HSV strains that are resistant to other medications.

Non-Nucleoside Reverse Transcriptase Inhibitors (NNRTIs)

Non-nucleoside reverse transcriptase inhibitors (NNRTIs) inhibit the activity of reverse transcriptase by binding to a site other than the nucleotide-binding site of the enzyme. Examples of NNRTIs used to treat HIV infections include efavirenz (EFV), etravirine (ETR), nevirapine (NVP), and rilpivirine (RPV).

NS5A Inhibitors

NS5A inhibitors offer a relatively new option for treating HCV infections. As the name implies, they inhibit NS5A, an HCV-encoded protein required for replication of the viral genome. Examples include elbasvir, ombitasvir, and velpatasvir.

Prevent Genome Integration

Integrase strand transfer inhibitors (INSTIs), also simply called integrase inhibitors, interfere with the action of the HIV-encoded enzyme integrase, thereby preventing the virus from inserting the DNA copy of its genome into that of the host cell. Examples of INSTIs include dolutegravir (DTG), elvitegravir (EVG), and raltegravir (RAL). ⏭ HIV integrase

Prevent Assembly and Release of Viral Particles

Virally encoded enzymes required for the assembly and release of viral particles are the targets of medications used to treat certain viral infections.

Protease Inhibitors

Protease inhibitors (PIs) are chemicals that inhibit virally encoded proteases. Recall that some viruses transcribe and translate certain groups of genes as a single unit to form a polyprotein. Those viruses then use virally encoded proteases to cleave the viral polyproteins in order to release the individual proteins.

Viral proteases are virus-specific, so each type of protease inhibitor is virus-specific as well. Protease inhibitors used to treat HIV infection include atazanavir (ATV), darunavir (DRV), and ritonavir (RTV). Those used to treat HCV infections include grazoprevir, paritaprevir, and simeprevir. ⏭ HIV protease, ⏭ HCV protease

Neuraminidase Inhibitors

Neuraminidase inhibitors inhibit neuraminidase, an enzyme encoded by influenza viruses. This enzyme is important for the release of viral particles from infected cells. Three neuraminidase inhibitors are currently available: oseltamivir (taken orally), zanamivir (inhaled), and peramivir (given by injection). All shorten the infection when used within 2 days of the onset of symptoms.

Characteristics of antiviral medications described in this section are summarized in **table 20.3.**

TABLE 20.3 Characteristics of Antiviral Medications

Target/Class/Example(s)	Comments/Characteristics
Viral Entry	
Fusion Inhibitors	Prevent viral particles from fusing with host cell.
Enfuvirtide (ENF), maraviroc (MVC)	Used to treat HIV infections.
Viral Uncoating	
Uncoating inhibitors	Prevent release of viral nucleic acid from the protein coat.
Amantadine and rimantadine	Used in the past to reduce severity and duration of influenza A infections.
Nucleic Acid Synthesis	
Nucleoside and nucleotide analogs	Incorporated by error-prone virally encoded polymerases, resulting in defective viral nucleic acid.
Acyclovir, famciclovir, valacyclovir	Used to treat shingles, chickenpox, cold sores, and genital herpes—diseases all caused by herpesviruses.
Sofosbuvir	Used to treat HCV infections.
Nucleoside and nucleotide reverse transcriptase inhibitors (NRTIs): abacavir (ABC), emtricitabine (FTC), lamivudine (3TC), tenofovir alafenamide (TAF), and zidovudine (AZT, ZDV)	Used to treat HIV infections; some are used to treat chronic hepatitis B.
Ganciclovir	Used to treat life- or sight- threatening cytomegalovirus infections.
Non-nucleoside polymerase inhibitors	Inhibit the activity of viral polymerases by binding to a site other than the nucleotide-binding site.
Dasabuvir	Used in a fixed-dose combination to treat HCV infections.
Foscarnet	Used to treat severe infections caused by CMV and HSV strains that are resistant to other medications.
Non-nucleoside reverse transcriptase inhibitors (NNRTIs)	Inhibit the activity of reverse transcriptase by binding to a site other than the nucleotide-binding site.
Efavirenz (EFV), etravirine (ETR), nevirapine (NVP), and rilpivirine (RPV)	Used to treat HIV infections.
NS5A inhibitors	Inhibit NS5A, an HCV-encoded protein required for replication of the viral genome.
Elbasvir, ombitasvir, velpatasvir	Used to treat HCV infections.
Genome Integration	
Integrase strand transfer inhibitors	Interfere with the HIV-encoded enzyme integrase
Dolutegravir (DTG), elvitegravir (EVG), and raltegravir (RAL)	Used to treat HIV infections.
Assembly and Release of Viral Particles	
Protease inhibitors	Inhibit virally encoded proteases, which cleave viral polyproteins to release individual proteins.
Atazanavir (ATV), darunavir (DRV), ritonavir (RTV)	Used to treat HIV infections.
Grazoprevir, paritaprevir, simeprevir	Used to treat hepatitis C.
Neuraminidase inhibitors	Inhibit an influenza virus–encoded enzyme required for release of viral particles from infected cells
Oseltamivir, zanamivir, peramivir	Used to treat influenza.

MicroAssessment 20.6

Viral replication generally uses host cell machinery; because of this, there are few targets for selectively toxic antiviral medications. Available antivirals are virus-specific; targets include viral entry, viral uncoating, nucleic acid synthesis, integrase, and the assembly and release of viral particles.

16. Explain why acyclovir has fewer side effects than do other nucleoside analogs.
17. How do protease inhibitors interfere with the production of infectious viral particles?
18. Why are nucleoside analogs active only against replicating viruses?

20.7 ■ Mechanisms of Action of Antifungal Medications

Learning Outcome

14. Describe the antifungal medications that interfere with the following: (1) fungal cytoplasmic membrane synthesis and function, (2) cell wall synthesis, (3) cell division, (4) nucleic acid synthesis, (5) protein synthesis.

Eukaryotic pathogens such as fungi more closely resemble human cells than do bacteria, which is why there are so few targets for antifungal medications (**figure 20.17**).

Interfere with Cytoplasmic Membrane Synthesis and Function

The target of most antifungal medications is ergosterol, a sterol found in the cytoplasmic membrane (plasma membrane) of fungal but not human cells. ⏮ sterol

Azoles

The azoles, a class of synthetic compounds, include two families of antifungal compounds: the imidazoles and the newer, less toxic triazoles. Both families inhibit ergosterol synthesis, resulting in defective fungal membranes that leak cytoplasmic contents. Four triazoles—fluconazole, voriconazole, posaconazole, and isavuconazole—are extremely important for treating systemic fungal infections. Each has a slightly different spectrum of activity. Another triazole, eficonazole, was recently approved as a topical agent for treating nail infections. Imidazoles, including miconazole and clotrimazole, are commonly used in non-prescription creams, ointments, and suppositories to treat vaginal yeast infections. They are also used on skin to treat dermatophyte infections. ⏭ dermatophyte

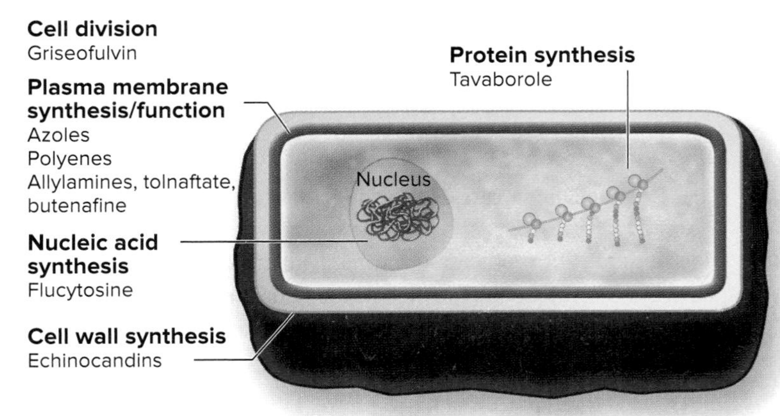

FIGURE 20.17 Targets of Antifungal Medications

What compound in the fungal cytoplasmic membrane (plasma membrane) is the target of most antifungal medications?

Polyenes

The polyenes, a group of antibiotics produced by certain species of *Streptomyces,* bind to ergosterol. This disrupts the fungal membrane, killing the cell by allowing its cytoplasmic contents to leak out. Unfortunately, the polyenes are quite toxic to humans, which limits their use. The side effects of the polyene amphotericin B are so severe that the drug is usually reserved for treating life-threatening infections for which no other option is available. Newer lipid-based emulsions are less toxic but more expensive. Current formulations of another polyene, nystatin, are too toxic to be used systemically, but because the medication is not absorbed, it is used in skin creams, lozenges, and oral suspensions to treat *Candida albicans* infections.

Allylamines, Tolnaftate, Butenafine

The allylamines inhibit an enzyme in the pathway of ergosterol synthesis; tolnaftate and butenafine appear to do the same. They can be applied to the skin to treat dermatophyte infections, but terbinafine (an allyamine) can also be taken orally.

Interfere with Cell Wall Synthesis

Fungal cell walls contain some components not produced by animal cells. Medications in a family called echinocandins interfere with synthesis of the fungal cell wall component β-1,3 glucan, causing the cells to burst. The current examples—caspofungin, micafungin, and anidulafungin—are particularly useful in treating systemic *Candida* infections. Caspofungin can also be used for treating invasive aspergillosis that does not respond to other medications.

Interfere with Cell Division

The target of one antifungal medication, griseofulvin, is cell division. Griseofulvin interferes with the action of tubulin, the protein that polymerizes to form microtubules. Because tubulin is a part of all eukaryotic cells, the selective toxicity of the medication may be due to its greater uptake by fungal cells. When taken orally for months, it is absorbed and concentrated in the dead keratinized layers of the skin. Fungi that then invade keratin-containing structures such as skin and

TABLE 20.4 Characteristics of Antifungal Medications

Target/Class/Example(s)	Comments
Cytoplasmic Membrane	
Azoles	Interfere with ergosterol synthesis, leading to defective fungal membranes.
Imidazoles: miconazole, clotrimazole	Used in non-prescription creams, ointments, and suppositories to treat vaginal yeast infections; also used on skin to treat dermatophyte infections.
Triazoles: fluconazole, voriconazole, posaconazole, and isavuconazole, eficonazole	Most are extremely important for treating a variety of systemic fungal infections; eficonazole is only used topically.
Polyenes	Bind to ergosterol, causing the fungal membrane to leak. Very toxic
Amphotericin B, nystatin	Amphotericin is used as a last resort for treating life-threatening infections; nystatin is used topically.
Allylamines, tolnaftate, butenafine	Inhibit an enzyme in the pathway of ergosterol synthesis.
Naftifine, terbinafine (allylamines); tolnaftate; butenafine	Used topically to treat dermatophyte infections; terbinafine can be taken orally.
Cell Wall Synthesis	
Echinocandins	Interfere with β-1,3 glucan synthesis.
Caspofungin, micafungin, and anidulafungin	Used to treat *Candida* infections; some are used to treat invasive aspergillosis that resists other medications.
Cell Division	
Griseofulvin	Used to treat skin and nail infections. Active only against fungi that invade keratinized cells.
Nucleic Acid Synthesis	
Flucytosine	Used to treat systemic yeast infections; inhibits an enzyme required for nucleic acid synthesis.
Protein Synthesis	
Tavaborole	Used topically to treat nail infections.

nails take up the medication, which prevents them from multiplying. Griseofulvin is active only against fungi that invade keratinized cells and is used to treat skin and nail infections. ◂◂ tubulin, ◂◂ keratin

Interfere with Nucleic Acid Synthesis

Nucleic acid synthesis is a common feature of all eukaryotic cells, so chemicals that interfere with the process would generally be too toxic for medical use. However, the medication flucytosine (a synthetic derivative of the nucleobase cytosine) is taken up by yeast cells and then converted by yeast enzymes to a form that inhibits an enzyme required for nucleic acid synthesis. Unfortunately, resistant mutants are common, and therefore, flucytosine is used mostly in combination with amphotericin B or an alternative. Side effects can be significant, so it is used only for serious yeast infections.

Interfere with Protein Synthesis

A new antifungal medication, tavaborole, inhibits protein synthesis by interfering with an enzyme that "charges" a tRNA molecule by attaching the correct amino acid. It is used topically to treat nail infections (onychomycosis).

The characteristics of various antifungal medications are summarized in **table 20.4.**

MicroAssessment 20.7

Because fungi are eukaryotic cells, there are relatively few targets for selectively toxic antifungal medications. Most antifungal medications interfere with ergosterol function or synthesis. Other targets include cell wall synthesis, cell division, and nucleic acid synthesis.

19. Why is amphotericin B used only for treating life-threatening infections?

20. Why is flucytosine generally used only in combination with other drugs?

20.8 ■ Mechanisms of Action of Antiprotozoan and Antihelminthic Medications

Learning Outcome

15. Describe the targets of most antiparasitic medications.

Most antiparasitic medications probably interfere with biosynthetic pathways of protozoan parasites or the neuromuscular function of worms. Unfortunately, compared with antibacterials, antifungals, and antivirals, relatively little research and development go into antiparasitic medications, because most parasitic diseases are concentrated in the poorer areas of the world where people simply cannot afford to spend money on expensive medications.

Some of the most important antiparasitic medications and their characteristics are summarized in **table 20.5.**

TABLE 20.5 Characteristics of Some Antiprotozoan and Antihelminthic Medications

Causative Agent/Medication	Comments
Intestinal protozoa	
Iodoquinol	Used as a follow-up treatment to eliminate the cysts of the protozoan that causes intestinal amebiasis; mechanism unknown.
Metronidazole, tinidazole	Used to treat amebiasis, giardiasis, and trichomoniasis; activated by the metabolism of anaerobic organisms, forming reactive compounds that alter DNA.
Nitazoxanide	Used to treat cryptosporidiosis and giardiasis; interferes with anaerobic energy metabolism.
Parmomycin	Used as a follow-up treatment to eliminate the cysts of the protozoan that causes intestinal amebiasis; mechanism against cysts unclear.
Plasmodium* (Malaria) and *Toxoplasma	
Artemether-lumefantrine	A type of ACT (artemisinin-based combination therapy) used to treat malaria, particularly that acquired in regions where chloroquine resistance occurs; the combination therapy is intended to prevent development of additional resistance. Artemether, an artemisinin derivative, is thought to be activated by heme digestion products released by the parasite, forming a reactive molecule that then damages parasite components. Lumefantrine action is thought to be associated with heme.
Atovaquone-proguanil	A synergistic combination used to treat malaria, particularly that acquired in regions where chloroquine resistance occurs. Atovaquone interferes with mitochondrial electron transport, and proguanil disrupts folate synthesis.
Chloroquine	Used to treat malaria; thought to interfere with the parasite's ability to detoxify heme, a substance that accumulates as the parasite grows in red blood cells.
Mefloquine	Used to treat malaria, particularly that acquired in regions where chloroquine resistance occurs; thought to interfere with the parasite's ability to detoxify heme.
Primaquine	Used as an additional treatment component for the relapsing forms of malaria; it kills the dormant liver stage, but the mechanism is not understood.
Pyrimethamine, sulfonamide	A synergistic combination used to treat toxoplasmosis and malaria; interferes with folate metabolism.
Trypanosomes and *Leishmania*	
Amphotericin B	Used to treat some types of leishmaniasis.
Benznidazole	Used to treat acute Chagas' disease; enzymes in the parasite convert it into a reactive form that damages certain cell components.
Eflornithine	Used to treat some types/stages of African trypanosomiasis; inhibits the enzyme ornithine decarboxylase.
Melarsoprol	Used to treat advanced stages of some types/stages of African trypanosomiasis but treatment can be lethal; inactivates enzymes by binding sulfhydryl groups.
Miltefosine	Used to treat some types of leishmaniasis; mechanism of action likely involves interaction with lipids and interference with mitochondrial function.
Nifurtimox	Used to treat acute Chagas' disease and, in combination with other medications, some types of African trypanosomiasis; thought to form reactive compounds that damage certain cell components.
Parmomycin	Used to treat some types of leishmaniasis; mechanism of action against the parasite unclear.
Pentamidine	Used to treat infections caused by some types/stages of African trypanosomiasis; mechanism is poorly understood, but appears to involve inhibition of nucleic acid synthesis.
Sodium stibogluconate	Used to treat some types of leishmaniasis; complex mechanism of action that involves sulfhydryl groups of cell components.
Suramin	Used to treat infections caused by some types/stages of African trypanosomiasis; mechanism is poorly understood, but appears to involve inhibition of certain enzymes.
Intestinal and Tissue Helminths	
Albendazole, mebendazole	Albendazole is used to treat a variety of helminth infections, including tissue and intestinal infections; mebendazole is poorly absorbed, so it is only used to treat intestinal helminths; both bind to tubulin of helminths, resulting in immobilization of the worm.
Diethylcarbamazine	Used to treat infections caused by certain filarial nematodes; results in damage to the outer surface of the microfilaria, making it easier for the immune system to eliminate them.

(continued)

TABLE 20.5 Characteristics of Some Antiprotozoan and Antihelminthic Medications (*Continued*)

Causative Agent/Medication	Comments
Ivermectin	Used to treat infections caused by *Strongyloides* and certain tissue nematodes; interferes with nerve transmission in the worm, causing flaccid (limp) paralysis.
Praziquantel	Effective in eliminating a variety of trematodes (flukes) and cestodes (tapeworms); results in sustained contractions of the worm.
Pyrantel pamoate	Used to eliminate certain intestinal nematodes (roundworms); interferes with neuromuscular activity of worms, causing paralysis.

FOCUS ON THE FUTURE 20.1

Combating Antibiotic Resistance

Considering the difficulty of developing new antibacterial medications, a critically important focus must be to prevent antibiotic resistance. The U.S. government now has a defined role in this effort, as outlined in the *National Action Plan for Combating Antibiotic-Resistant Bacteria,* released in 2015. This 5-year plan seeks to significantly reduce the number of infections caused by bacteria listed as urgent and serious threats by the CDC (see table 20.2).

The *National Action Plan* has five general goals that can be summarized as follows:

- Prevent the emergence and spread of antibiotic-resistant bacteria.
- Strengthen surveillance efforts that pertain to antibiotic-resistant bacteria.
- Develop rapid tests to detect and characterize resistant bacteria.
- Increase research that focuses on developing new antibiotics, other therapeutics, and vaccines.
- Improve international collaborations that aim to combat antibiotic resistance.

In addition, the *National Action Plan* includes specific objectives for each goal, as well as a list of 1-, 3-, and 5-year milestones to be met. The CDC will have a significant role in implementing parts of the plan, but other public agencies, healthcare providers, agricultural industry leaders, veterinarians, and manufacturers will be involved as well.

The problems due to infections caused by antibiotic-resistant bacteria—from costs to medical complications—are significant. Hopefully, the increased national focus on antibiotic resistance will lead to long-term solutions.

Summary

20.1 ■ History and Development of Antimicrobial Medications

Discovery of Antimicrobial Medications

Salvarsan, developed by Paul Ehrlich, was the first documented example of an **antimicrobial medication.**

Discovery of Antibiotics

Alexander Fleming discovered that a species of the fungus *Penicillium* produces an antimicrobial chemical he called penicillin, the first example of an **antibiotic** (figure 20.1).

Development of New Antimicrobial Medications

Most modern antibiotics come from species of *Streptomyces* and *Bacillus* (bacteria) and *Penicillium* and *Cephalosporium* (fungi). Antimicrobial medications can be chemically modified to give them new properties (figure 20.2).

20.2 ■ Characteristics of Antimicrobial Medications

Selective Toxicity

Medically useful antimicrobials have **selective toxicity**; the relative toxicity of a medication is expressed as the **therapeutic index.**

Antimicrobial Action

Bacteriostatic chemicals inhibit the growth of bacteria; chemicals that kill bacteria are **bactericidal.**

Spectrum of Activity

Broad-spectrum antimicrobials affect a wide range of bacteria; those that affect a narrow range are called **narrow-spectrum.**

Effects of Antimicrobial Combinations

Combinations of antimicrobial medications can be synergistic, antagonistic, or additive.

Tissue Distribution, Metabolism, and Excretion of the Medication

Some antimicrobials cross the blood-brain barrier into the cerebrospinal fluid; these can be used to treat meningitis. Medications that are unstable in acid must be administered through injection. The half-life of a medication affects how frequently it must be taken.

Adverse Effects

Some people develop allergies to certain antimicrobials. Some antimicrobials can have potentially damaging side effects such as kidney damage. By disrupting the microbiome, antimicrobials can cause **dysbiosis.**

Resistance to Antimicrobials

Certain types of bacteria have **intrinsic (innate) resistance** to a particular antimicrobial medication. **Acquired resistance** is due to spontaneous mutation or the acquisition of new genetic information.

20.3 ■ Mechanisms of Action of Antibacterial Medications
(figure 20.3, table 20.1)

Inhibit Cell Wall Synthesis (figure 20.4)

The **β-lactam antibiotics,** which include penicillins, cephalosporins, carbapenems, and monobactams, inhibit enzymes that help form cross-links between adjacent glycan chains, ultimately leading to cell lysis. Other medications that interfere with cell wall synthesis include glycopeptide antibiotics and bacitracin.

Inhibit Protein Synthesis (figure 20.7)

Antibiotics that inhibit protein synthesis by binding to the bacterial 70S ribosome include aminoglycosides, tetracyclines, macrolides, chloramphenicol, lincosamides, oxazolidinones, pleuromutilins, and streptogramins.

Inhibit Nucleic Acid Synthesis

Antimicrobials that inhibit nucleic acid synthesis include fluoroquinolones, rifamycins, fidaxomicin, and metronidazole.

Interfere with Metabolic Pathways (figure 20.8)

Sulfa drugs and trimethoprim block different enzymes in a metabolic pathway required for nucleotide biosynthesis. The two medications are often used together as a synergistic combination called **co-trimoxazole.**

Interfere with Cell Membrane Integrity

Daptomycin and polymixins damage bacterial membranes, as do two of the new glycopeptide antibiotics.

Effective Against *Mycobacterium tuberculosis*

First-line drugs used to treat tuberculosis include isoniazid, ethambutol, and pyrazinamide. **Second-line drugs** are used for strains resistant to the first-line drugs.

20.4 ■ Antimicrobial Susceptibility Testing

Conventional Disc Diffusion Method (figure 20.9)

The **Kirby-Bauer disc diffusion test** determines the susceptibility of a bacterial strain to a battery of antimicrobial medications.

Minimum Inhibitory and Minimum Bactericidal Concentrations (MIC and MBC) (figure 20.10)

The **minimum inhibitory concentration (MIC)** is the lowest concentration of a specific antimicrobial needed to prevent the growth of a bacterial strain in vitro. The **minimum bactericidal concentration (MBC)** is the lowest concentration of a specific antimicrobial that kills 99.9% of cells of a given strain of bacterial in vitro.

Commercial Modifications of Antimicrobial Susceptibility Testing

Automated methods can determine antimicrobial susceptibility in as little as 4 hours (figure 20.11).

20.5 ■ Resistance to Antimicrobial Medications (figure 20.13)

Mechanisms of Acquired Resistance (figure 20.14)

Enzymes that chemically modify an antimicrobial destroy the activity of the medication. Structural changes in the target can prevent the medication from binding. Altered porin proteins prevent drugs from entering cells. **Efflux pumps** actively pump drugs out of cells.

Acquisition of Resistance

Resistance can be acquired through spontaneous mutation or horizontal gene transfer. The most common mechanism of transfer of antibiotic resistance genes is through the conjugative transfer of **R plasmids.**

Examples of Emerging Resistance (table 20.2)

Strains of many organisms, including enterococci, *Enterobacteriaceae, Mycobacterium tuberculosis, Neisseria gonorrhoeae, Staphylococcus aureus,* and *Streptococcus pneumoniae* are becoming increasingly resistant to antimicrobial medications.

Preventing Resistance

Physicians should prescribe antimicrobial medications only when appropriate. Patients need to carefully follow prescribed instructions when taking antimicrobials. The public must be educated about the appropriateness and limitations of antimicrobial therapy.

20.6 ■ Mechanisms of Action of Antiviral Medications
(figure 20.16, table 20.3)

Prevent Viral Entry

Entry inhibitors are used to treat HIV infections; they include a fusion inhibitor and a CCR5 antagonist.

Interfere with Viral Uncoating

Amantadine and rimantadine block the uncoating of influenza A virus after it enters a cell.

Interfere with Nucleic Acid Synthesis

Many of the most effective antiviral medications target virally encoded enzymes used during replication of viral nucleic acid; these include nucleoside and nucleotide analogs, non-nucleoside polymerase inhibitors, non-nucleoside reverse transcriptase inhibitors, and NS5A inhibitors. Certain options are available to treat herpesviruses, HCV, or HIV infections.

Prevent Genome Integration

Integrase strand transfer inhibitors (INSTIs) interfere with the function of HIV integrase.

Prevent Assembly and Release of Viral Particles

Protease inhibitors bind to and inhibit virally encoded proteases; some options are used to treat HIV, and others are used to treat HCV. Neuraminidase inhibitors interfere with the release of influenza virus particles from a cell.

20.7 ■ Mechanisms of Action of Antifungal Medications
(figure 20.17, table 20.4)

Interfere with Cytoplasmic Membrane Synthesis and Function

Azoles interfere with ergosterol synthesis. Polyenes disrupt fungal cell membranes by binding to ergosterol. Allylamines, tolnaftate and butenafine all inhibit an enzyme in the pathway of ergosterol synthesis.

Interfere with Cell Wall Synthesis

Echinocandins interfere with the synthesis of β-1,3 glucan.

Interfere with Cell Division

Griseofulvin is concentrated in keratinized skin cells, where it inhibits fungal cell division.

Interfere with Nucleic Acid Synthesis

Flucytosine is taken up by yeast cells and converted by yeast enzymes to an active form.

Interfere with Protein Synthesis

Tavaborole inhibits protein synthesis by interfering with an enzyme that "charges" a tRNA molecule by attaching the correct amino acid.

20.8 ■ Mechanisms of Action of Antiprotozoan and Antihelminthic Medications (table 20.5)

Most antiparasitic medications are thought to interfere with biosynthetic pathways of protozoan parasites or the neuromuscular function of worms.

Review Questions

Short Answer

1. Describe the difference between the terms *antibiotic* and *antimicrobial.*
2. Define *therapeutic index* and explain its importance.
3. Explain how penicillin-binding proteins (PBPs) are involved in a bacterium's susceptibility to β-lactam antibiotics.
4. Name three classes of antimicrobial medications that target bacterial ribosomes.
5. Explain the roles of the first-line drugs versus the second-line drugs in the treatment of tuberculosis.
6. Compare and contrast the method for determining the minimum inhibitory concentration (MIC) with the Kirby-Bauer disc diffusion test.
7. Name three targets that can be altered sufficiently via spontaneous mutation to result in resistance to an antimicrobial medication.
8. What is MRSA? Why is it significant?
9. Why is it difficult to develop antiviral medications?
10. Explain the difference between the mechanism of action of an azole and that of a polyene.

Multiple Choice

1. Which of the following targets would you expect to be the most selective with respect to toxicity?
   a) Cytoplasmic membrane function
   b) DNA synthesis
   c) Glycolysis
   d) Peptidoglycan synthesis
   e) 70S ribosome
2. Penicillin has been modified to make derivatives that differ in all of the following *except*
   a) spectrum of activity.
   b) resistance to β-lactamases.
   c) potential for allergic reactions.
   d) a and c.
3. Which of the following is the target of β-lactam antibiotics?
   a) Peptidoglycan synthesis
   b) DNA synthesis
   c) RNA synthesis
   d) Protein synthesis
   e) Folic acid synthesis
4. Which of the following statements is *false*?
   a) A bacteriostatic chemical stops the growth of a microorganism.
   b) The lower the therapeutic index, the less toxic the medication.
   c) Broad-spectrum antibiotics are associated with the development of *Clostridium difficile* infections.
   d) Azithromycin has a longer half-life than does penicillin V.
   e) Chloramphenicol can cause a life-threatening type of anemia.
5. All of the following interfere with the function of the bacterial ribosome *except*
   a) fluoroquinolones. b) lincosamides. c) macrolides.
   d) streptogramins. e) tetracyclines.
6. The target of the sulfonamides is
   a) cytoplasmic membrane proteins.
   b) folate synthesis.
   c) gyrase.
   d) peptidoglycan biosynthesis.
   e) RNA polymerase.
7. Routine antimicrobial therapy to treat tuberculosis involves taking
   a) one medication for 10 days.
   b) two or more medications for 10 days.
   c) one medication for at least 6 months.
   d) two or more medications for at least 6 months.
   e) five medications for 2 years.
8. *Staphylococcus aureus* strains referred to as HA-MRSA are sensitive to
   a) methicillin. b) penicillin. c) most cephalosporins.
   d) vancomycin. e) none of the above
9. Acyclovir is a
   a) nucleoside analog.
   b) non-nucleoside polymerase inhibitor.
   c) protease inhibitor.
   d) none of the above
10. The antifungal medication griseofulvin is used to treat
    a) vaginal infections. b) systemic infections.
    c) nail infections. d) eye infections.

Applications

1. A physician was treating one young woman and one elderly patient for urinary tract infections caused by the same type of bacterium. Although the patients had similar body dimensions and weight, the physician gave a smaller dose of the antibacterial medication to the older patient. What was the physician's rationale for this decision?
2. Many scientists have criticized the use of low-dosage antibiotics and other antimicrobial agents to enhance the growth of cattle and chickens. Why would they be against this practice? Why would producers be reluctant to stop it?

Critical Thinking

1. Sulfonamides are not as effective in the presence of high-nutrient material such as pus or dead tissue. Why might this be the case?
2. Figure 20.12 shows the E-test procedure for determining an MIC value. How would the zone of inhibition appear if the medication concentrations in the strip were decreased slightly?

21 Respiratory System Infections

View of the throat in an individual with streptococcal pharyngitis. *Source: Heinz F. Eichenwald, MD/CDC*

KEY TERMS

Antigenic Drift Minor changes that occur naturally in influenza virus antigens as a result of mutation.

Antigenic Shift Major changes in the antigenic composition of influenza viruses that result from reassortment of viral RNA during infection of the same host cell by different viral strains.

Directly Observed Therapy Short-Course (DOTS) Method used to ensure that tuberculosis (TB) patients comply with their treatment; the healthcare worker watches while the patient takes each dose of medication.

Extensively Drug-Resistant Tuberculosis (XDR-TB) Tuberculosis caused by strains of *Mycobacterium tuberculosis* resistant to the first-line anti-TB medications isoniazid and rifampicin, and at least three of the second-line anti-TB medications.

Mucociliary Escalator Moving layer of mucus propelled by cilia lining the respiratory tract that traps bacteria and other particles and carries them toward the throat.

Multi-Drug-Resistant Tuberculosis (MDR-TB) Tuberculosis caused by strains of *Mycobacterium tuberculosis* resistant to isoniazid and rifampicin—two of the first-line anti-TB medications.

Otitis Media Inflammation of the middle ear.

Pharyngitis Inflammation of the throat.

Pneumonia Inflammation of the lungs accompanied by filling of the air sacs with fluids such as pus and blood.

Sputum Thick fluid containing mucus, pus, and other material coughed up from the lungs.

Tubercle Granuloma formed in tuberculosis; granulomas are collections of lymphocytes and macrophages found in a chronic inflammatory response, an attempt by the body to wall off and contain persistent organisms and antigens.

A Glimpse of History

Rebecca Lancefield (1895–1981) was a prominent microbiologist, famous for demonstrating that streptococci can be classified according to their cell wall carbohydrates. This system, known as the "Lancefield grouping," is still used today.

Lancefield attended Wellesley College in 1912, and then went to Columbia University, where she obtained her master's degree. She discontinued her studies during World War I, but earned her Ph.D. from Columbia in 1925 while maintaining a position at Rockefeller Institute of Medical Research (now Rockefeller University), where her work focused on the streptococci. At the time, streptococcal classification was largely based on the type of hemolysis displayed when organisms were grown on blood agar. β-hemolytic strep colonies produce a colorless clearing of red blood cells (β-hemolysis), whereas α-hemolytic strep colonies produce a greenish partial clearing of the red blood cells (α-hemolysis). A turning point in her research came when she recognized that streptococci have important antigens on their surface, some of which correlate with the disease-causing potential of a strain. She noticed that one particular antigen—a cell wall carbohydrate—could be used to separate the various streptococci into different groups. Most strains of β-hemolytic streptococci isolated from human infections have the same "A" cell wall carbohydrate; others have "B" carbohydrate; those from cattle, horses, and guinea pigs have "C" carbohydrate; human normal microbiota streptococci have "D" carbohydrate; and so on. The "Lancefield grouping" was a much better predictor of pathogenic potential than hemolysis on blood agar. Lancefield also discovered M protein on the surface of *Streptococcus pyogenes,* a group A streptococcus (GAS). M protein shows great variation among strains of *S. pyogenes* and plays a significant role in its virulence.

Rebecca Lancefield was the first female president of the American Association of Immunologists and in 1970 was elected to the prestigious National Academy of Sciences.

Respiratory infections include an enormous variety of illnesses ranging from the trivial to the fatal. They can be divided into infections of the upper respiratory tract (in the head and neck) and infections of the lower respiratory tract (in the chest). Common upper respiratory infections such as colds or sinus infections are uncomfortable, but they are not life-threatening. Lower respiratory infections such as pneumonia and tuberculosis, however, are often serious, and may be fatal.

21.1 ■ Anatomy, Physiology, and Ecology of the Respiratory System

Learning Outcomes

1. Outline the structures of the upper and lower respiratory tracts and their functions.
2. Describe the components of the mucociliary escalator and explain its importance.

The major function of the respiratory system is gas exchange. Each breath brings in O_2, which replenishes the supply in the blood, and then releases CO_2, the waste product of cellular metabolism. In addition, movement of air over the vocal cords makes sound, which allows us to speak, and sensors in the nose detect odors. Although the mouth can also be used for breathing, it is considered part of the digestive tract, so it is covered in chapter 24.

The respiratory system consists of the upper respiratory tract and the lower respiratory tract. We also include the eyes and the ears in our discussion. Even though these are not part of the respiratory tract, they are associated with ducts that open onto mucous membranes lining the nasal cavity and throat. Due to that connection, they can serve as portals of entry to the respiratory tract. ⏮ portals of entry

The Upper Respiratory Tract

The upper respiratory tract includes the nose and nasal cavity, pharynx (throat), and epiglottis (**figure 21.1**). Mucous membranes line the respiratory tract. These are coated with mucus, a slimy glycoprotein material that traps airborne dust and other particles, including microorganisms. Mucus is produced by specialized cells called **goblet cells** that are scattered among the cells of the membrane. Their name reflects their shape, which is narrow at the base and wide at the surface, like a wine glass or goblet. ⏮ mucous membrane

Ciliated epithelium lines most mucous membranes of the respiratory system. Cells of this lining have cilia—tiny, hair-like projections—along their exposed free border. The cilia beat synchronously, continually propelling mucus away from the lungs. The mucus, along with any trapped microbes, is then swallowed and digested. This mechanism, the **mucociliary escalator,** is an important defense that helps prevent microorganisms that have been inhaled from reaching the lungs. Smoking, alcohol or narcotic abuse, and viral infections all impair ciliary movement and increase the chance of infection. ⏮ cilia

The tonsils are lymphoid organs, located so that they come into contact with microbes entering the upper respiratory tract. They are important in the immune response but can also be the sites of infection, resulting in tonsillitis. Inflamed tonsils at the opening of the throat are visible in the chapter opening photo. Tonsils in the upper throat are called adenoids.

Some of the bacterial genera that inhabit the upper respiratory system are listed in **table 21.1.** Although generally harmless, they are opportunists that can cause disease when host defenses are impaired. ⏮ opportunists

The Nose and Nasal Cavity

Air enters the respiratory system at the nostrils and flows into the nasal cavity. When cold air enters the nose, the blood flow immediately increases due to nervous reflexes. This warms the air in the nose to near body temperature while passage over mucous membranes saturates it with water vapor.

The nasal entrance usually contains diphtheroids and staphylococci. Up to 30% of healthy people consistently carry *Staphylococcus aureus* in their noses; an even higher percentage of hospital personnel are likely to carry this important hospital- or community-acquired pathogen. Farther inside the nasal passages, the normal microbiota is similar to that of the throat. Infection of the nasal passages, usually by viruses, causes inflammation or rhinitis, resulting in the familiar "runny nose." ⏮ diphtheroids

TABLE 21.1 Normal Microbiota of the Upper Respiratory System

Genus	Characteristics	Comments
Corynebacterium	Pleomorphic, often club-shaped Gram-positive rods; non-motile	Aerobic or facultatively anaerobic. Those that look similar to *C. diphtheriae*, the cause of diphtheria, are included in a group called diphtheroids.
Fusobacterium, Porphyromonas, Prevotella	Gram-negative rods	Obligate anaerobes. Most often found in the space between teeth and gums.
Haemophilus	Small, Gram-negative rods	Facultative anaerobes. Commonly include the potential pathogen *H. influenzae*.
Moraxella	Gram-negative diplococci and diplobacilli	Aerobic. Some microscopically resemble pathogenic *Neisseria* species such as *N. meningitidis*.
Staphylococcus	Gram-positive cocci in clusters	Facultative anaerobes. Potential pathogen *Staphylococcus aureus* commonly inhabits the nostrils.
Streptococcus	Gram-positive cocci in chains	Aerotolerant (obligate fermenters). Viridans streptococci, which are α-hemolytic (greenish partial hemolysis), and non-hemolytic streptococci are common. β-hemolytic (clear hemolysis) streptococci and the potential pathogen *S. pneumoniae* may also be present.

FIGURE 21.1 Anatomy and Infections of the Respiratory System **(a)** Lateral view of upper respiratory tract. **(b)** Frontal view of the upper and lower respiratory tracts. Some of the diseases that can affect the system are shown in red.

Why would inflammation in the lungs be life-threatening?

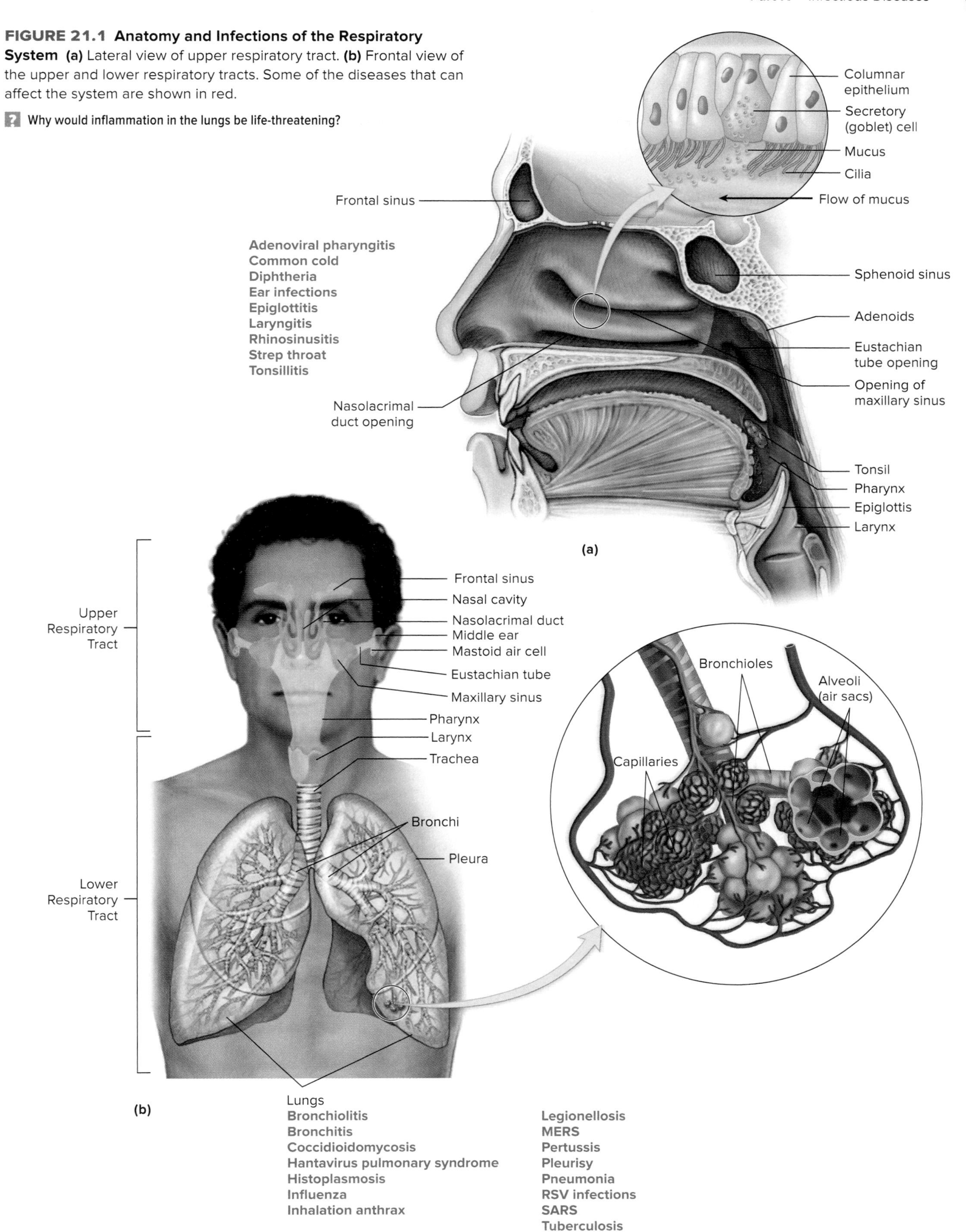

The skull contains air-filled cavities called sinuses that are lined with ciliated epithelium continuous with that of the nasal cavity. Infection of the sinuses is called rhinosinusitis, or sinusitis, and may cause a "sinus headache" or postnasal "drip."

Pharynx (Throat) and Epiglottis

From the nose, air moves to the throat (pharynx). Inflammation of the throat, **pharyngitis,** is commonly the result of viral infection. The throat is shared by the respiratory system and the digestive system. During swallowing, a small flap called the epiglottis covers the glottis, the opening that allows air to pass between the upper and lower respiratory tracts. This prevents swallowed material from entering the lungs. Inflammation of the epiglottis, called epiglottitis, can be life-threatening because the swollen flap can block the glottis and close the airway.

Streptococci, including viridans streptococci (a group of α-hemolytic streptococci; *viridis* means "green") and non-hemolytic species, are common members of the normal microbiota of the throat. A variety of other bacteria are also found there, including *Moraxella* species, diphtheroids, and various obligate anaerobes. Some common opportunistic pathogens also colonize the throat, including *Streptococcus pneumoniae, Haemophilus influenzae,* and *Neisseria meningitidis.* ⏮ α-hemolysis

MicroByte

Cilia beat at a rate of about 1,000 times a minute, propelling a film of mucus along the mucociliary escalator.

Eyes

The surface of the eyes and the lining of the eyelids are covered by mucous membranes called conjunctiva. Infection of the conjunctiva is called **conjunctivitis.** Surprisingly, even though the eyes are constantly exposed to large numbers of microorganisms, the conjunctiva of healthy people usually have very few bacteria. This is because the eye is bathed with tears rich in lysozyme and secretory IgA, and is cleaned by the blinking reflex, which wipes the eye surface like a car windshield wiper cleans a windshield. From the conjunctiva, microbes may be swept into the tear duct and then into the nasal cavity. ⏮ lysozyme

Ears

The ear has three parts: external, middle, and inner ear. The external ear includes the outer portion of the ear and the auditory canal, which is protected from microbes by cerumen (ear wax). The middle ear, which is separated from the auditory canal by the eardrum, is sterile. It is connected by the Eustachian tubes to the upper portion of the pharynx. These tubes equalize the pressure in the middle ear and drain normal mucous secretions. They are also a route through which microbes can enter the middle ear, leading to an infection called **otitis media.** Enlargement of the adenoids can contribute to middle ear infections by interfering with normal drainage from the Eustachian tubes. The fluid-filled inner ear is also microbe-free. Occasionally, however, viruses or bacteria enter the inner ear, often following an upper respiratory tract infection. Ear infections may extend to nearby structures such as the mastoid air cells in the skull, causing mastoiditis.

The Lower Respiratory Tract

The lower respiratory tract includes the larynx (voice box), trachea (windpipe), bronchi, and lungs (figure 21.1b). Inflammation of the larynx is called laryngitis and results in hoarseness. The trachea is a continuation of the larynx and branches into two bronchi. Inflammation of the bronchi is called bronchitis, commonly the result of viral infection or chronic coughing due to smoking. The bronchi branch repeatedly, becoming bronchioles, the site of an important viral infection called bronchiolitis. The smallest bronchioles end in the alveoli, which are tiny, thin-walled air sacs where gas exchange occurs. Inflammation of the lungs is called pneumonitis. Pneumonitis that causes the alveoli to fill with pus and fluid is called **pneumonia.** Lung tissues have many macrophages that readily move into the alveoli and airways to engulf infectious agents, helping to prevent pneumonia. The lungs are surrounded by two membranes called pleura; one adheres to the lung and the other to the chest wall and diaphragm. The pleura normally slide over each other as the lung expands and contracts. Inflammation of the pleura is called pleurisy (pleuritis), characterized by severe chest pain. The action of the mucociliary escalator and the antimicrobial action of alveolar secretions help to keep microorganisms out of the lungs.

MicroAssessment 21.1

The respiratory system provides a warm, moist environment for microorganisms. It is protected by tonsils and adenoids, and by the mucociliary escalator. The upper respiratory tract contains highly diverse microbiota. Some members of the normal microbiota are opportunistic pathogens. The mucociliary escalator helps prevent microorganisms that have been inhaled from reaching the lungs.

1. What is the normal function of the tonsils and adenoids?
2. Describe the normal microbiota of the upper respiratory tract.
3. How would paralysis of the cilia affect the occurrence of pneumonia?

UPPER RESPIRATORY TRACT INFECTIONS

21.2 ■ Bacterial Infections of the Upper Respiratory System

Learning Outcomes

3. List the parts of the upper respiratory system commonly infected by *Streptococcus pneumoniae* and *Haemophilus influenzae.*
4. Compare the distinctive characteristics of strep throat and diphtheria.

A number of bacterial species can infect the upper respiratory tract. Some, such as *Haemophilus influenzae,* can cause sore throats, but the infections generally do not require treatment because the bacteria are quickly eliminated by the immune system. Others, such as *Streptococcus pyogenes,* cause infections that require treatment because they can cause serious complications.

Pink Eye, Earache, and Sinus Infections

Bacterial infections leading to **conjunctivitis** ("pink eye"; inflammation of the eye surface), **otitis media** (inflammation of the middle ear), and sinusitis (inflammation of the sinuses) are very common, often occur together, and frequently have the same causative agent. Pink eye is easily spread and is therefore a concern when it occurs in a day-care facility or school. Otitis media is especially common in children, and is responsible for millions of doctor visits per year in the United States. Sinusitis is common in both adults and children.

Signs and Symptoms

The signs and symptoms of acute bacterial conjunctivitis include increased tears, redness of the conjunctiva, swollen eyelids, sensitivity to bright light, and large amounts of pus (**figure 21.2**). Acute bacterial conjunctivitis differs from viral conjunctivitis, in which eyelid swelling and pus are usually minimal.

Otitis media causes a severe earache—in a typical case, a young child wakes from sleep screaming with pain. The intense pain often causes vomiting. Children may also show ear tugging and difficulty hearing. Fever may be mild or absent.

In sinusitis, facial pain and a pressure sensation occur in the region of the involved sinus. Headache and severe malaise also occur. A thick green nasal discharge that may contain pus and blood sometimes develops as well.

Causative Agents

Pink eye, earache, and sinus infections are often caused by two common bacterial pathogens: (1) *Haemophilus influenzae,* a tiny Gram-negative rod; and (2) *Streptococcus pneumoniae,* the Gram-positive encapsulated diplococcus known as pneumococcus. Strains that infect the conjunctiva have adhesins that allow them to attach firmly to the epithelium. ⏭ *Haemophilus influenzae,* ⏭ *Streptococcus pneumoniae,* ⏮ adhesins

Conjunctivitis can also be caused by a variety of other organisms, including *Moraxella lacunata,* enterobacteria, and *Neisseria gonorrhoeae.* Although not common, some eye infections are caused by environmental microbes that contaminate eye medications and contact lens solutions. When these infections occur they are usually serious and can lead to permanent eye damage. People who wear contact lenses should follow the instructions for using cleansing solutions exactly; cloudy or outdated solutions and eye medications should be thrown away.

Otitis media and sinusitis can also be caused by *Mycoplasma pneumoniae, Streptococcus pyogenes, Moraxella catarrhalis,* and *Staphylococcus aureus.* About one-third of the cases are caused by respiratory viruses, explaining why some infections do not respond to antibiotics, which have no effect on viruses.

Pathogenesis

Few details are known about the pathogenesis of acute bacterial conjunctivitis. Organisms are probably inoculated directly onto the conjunctiva from airborne respiratory droplets or transferred from contaminated hands. They resist destruction by lysozyme. Attachment may be aided by degradation of mucin, a protective component of surface mucus. Following attachment, the bacteria release various tissue-damaging enzymes, sometimes combined with toxins, that further harm eye tissue. ⏮ lysozyme

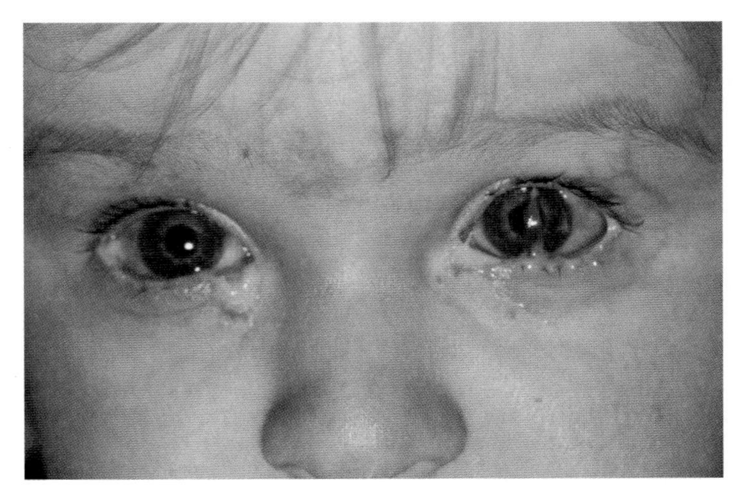

FIGURE 21.2 Bacterial Conjunctivitis Redness, swollen eyelids, and pus are characteristic of bacterial conjunctivitis.

? **What symptom shown here is usually minimal in viral conjunctivitis?**

Otitis media and sinusitis are usually preceded by infections of the nasal cavity and upper pharynx (see figure 21.1). Infection damages the ciliated cells, resulting in inflammation and swelling. When the Eustachian tube cannot move secretions from the middle ear, fluid and pus collect behind the eardrum. This leads to a buildup of pressure, causing the ear to ache. The eardrum may perforate (burst), discharging blood or pus from the ear and giving immediate relief from pain. With treatment, holes in the eardrum usually heal quickly. In some people, the causative organisms create a biofilm, leading to chronic infections that may be difficult to treat. Fluid behind the eardrum may impair hearing ability, causing a delay in speech development in some young children. Both middle ear and sinus infections sometimes spread to the membranes covering the brain, causing meningitis. ⏭ meningitis

Epidemiology

The epidemiological factors involved in the appearance and spread of the eye, ear, and sinus infections caused by *H. influenzae* and *S. pneumoniae* are largely unknown; many people carry the bacteria without ill effects. The virulence of the bacteria, crowding of potential hosts, and presence of respiratory viruses are probably all important factors in these epidemics.

A preceding or simultaneous viral illness is common in otitis media and sinusitis; the virus probably damages the mucociliary mechanism that would normally protect against bacterial infection. Otitis media is rare in the first month of life, but it becomes very common in early childhood. Children who use pacifiers after the age of 2 have an increased risk of developing otitis media. Conditions that cause inflammation of the nasal mucosa—including viral infections, nasal allergies, and exposure to air pollution or cigarette smoke—play a role in some cases. Older children develop immunity to *H. influenzae,* and the bacterium rarely causes otitis media in children beyond age 5. Sinusitis tends to affect adults and older children in whom the sinuses are more fully developed.

Treatment and Prevention

Bacterial conjunctivitis is effectively treated with eyedrops or ointments containing an antimicrobial medication to which the infecting strain is sensitive. An antibiotic such as amoxicillin is generally effective against both otitis media and sinusitis. Alternative medications are available where antibiotic-resistant strains of *H. influenzae* and *S. pneumoniae* are common. Unfortunately, if the causative agent forms a biofilm, the disease often recurs. In general, antibiotics have been used indiscriminately in treating otitis media. However, when properly used, they decrease the risk of complications such as meningitis. Decongestants and antihistamines generally are ineffective and can be harmful because they reduce the immune response.

General preventive measures for conjunctivitis include handwashing, and protecting the eyes from contamination by avoiding rubbing or touching them, particularly with shared towels. Bacterial conjunctivitis is highly contagious, so people suspected of having it should be kept home from school or day-care settings for diagnosis and start of treatment. Otitis media during the "flu" season can be substantially decreased by giving influenza vaccine to infants in day-care facilities. Surgical removal of enlarged adenoids improves drainage from the Eustachian tubes and can help prevent recurrences in certain patients. In those with chronically malfunctioning Eustachian tubes and hearing loss, tiny plastic ventilation tubes are often inserted through the eardrums so that pressure can equalize **(figure 21.3).** There are no proven preventive measures for sinusitis.

Streptococcal Pharyngitis ("Strep Throat")

Sore throat is one of the most common reasons that people in the United States seek medical care. Many of these visits are due to reasonable concerns about streptococcal pharyngitis, commonly known as strep throat. Streptococcal infections may lead to dangerous post-streptococcal sequelae. These are complications that develop after the initial infection and will be discussed in the next section.

Signs and Symptoms

Strep throat is characterized by a sore throat, difficulty swallowing, and fever, which develop after an incubation period of 2 to 5 days. The throat is red, with patches of pus and scattered tiny

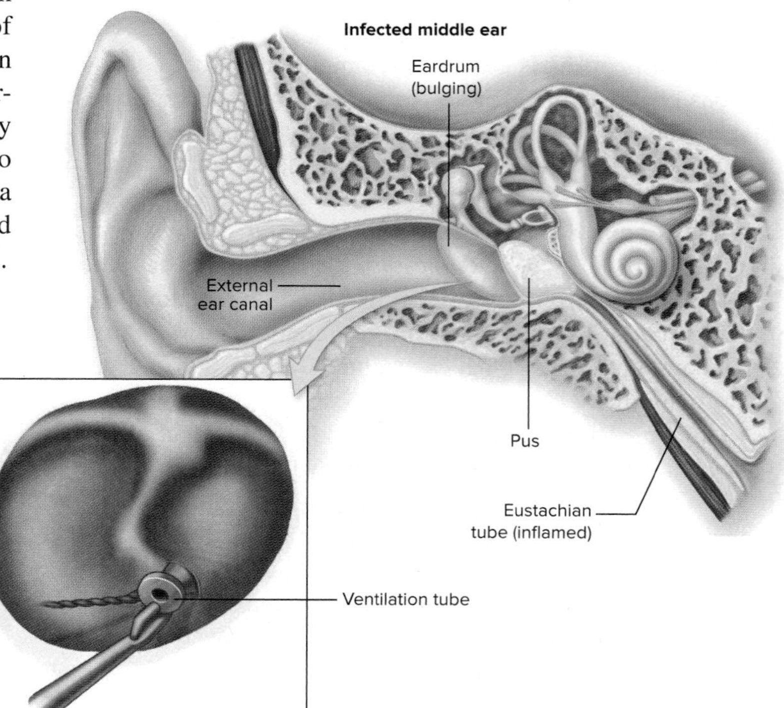

FIGURE 21.3 Otitis Media Inset shows a ventilation tube placed in the eardrum to equalize middle ear pressure in individuals with chronically malfunctioning Eustachian tubes.

Why is otitis media painful?

hemorrhages. The lymph nodes in the neck are enlarged and tender. Abdominal pain or headache may occur in older children and young adults. Patients do not usually have a cough, weepy eyes, or runny nose. Most patients with strep throat recover spontaneously after about a week. In fact, many infected people have only mild symptoms or no symptoms at all.

Causative Agent

Strep throat is caused by *Streptococcus pyogenes,* a Gram-positive coccus that grows in chains (**figure 21.4**). The organism can be differentiated from other streptococci that normally inhabit the throat by its colony morphology on blood agar— *S. pyogenes* colonies are surrounded by a characteristic clear zone of β-hemolysis (**figure 21.5**). In contrast, most species of *Streptococcus* that are typically part of the normal throat microbiota are either α-hemolytic, producing a zone of greenish partial clearing around colonies on blood agar, or non-hemolytic. A few other streptococci and some other bacteria are also β-hemolytic, so further tests are needed to identify *S. pyogenes.* ◀◀ hemolysis, ◀◀ blood agar

Streptococcus pyogenes is commonly referred to as group A streptococcus (GAS), reflecting its Lancefield grouping (see **A Glimpse of History** in this chapter). This grouping system uses antibodies to distinguish the cell wall carbohydrates in streptococcal species. *S. pyogenes* is characterized by the "A" carbohydrate in its cell wall. Antibodies that bind to the group A carbohydrate are the basis for many of the rapid diagnostic tests done on throat specimens in a physician's office.

Different *S. pyogenes* strains within GAS are distinguished by variations of a surface antigen called **M protein,** an important virulence factor. Strains with certain types of M protein are the cause of strep throat, whereas others with different M protein are more likely to cause skin infections.

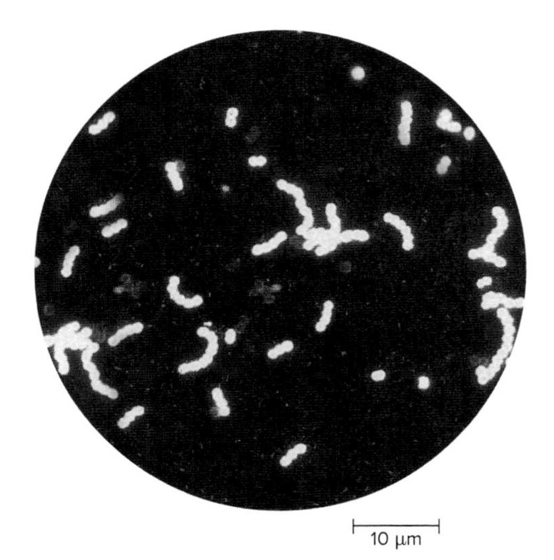

FIGURE 21.4 ***Streptococcus pyogenes*** Chain formation by *S. pyogenes*, revealed by fluorescence microscopy. ©Evans Roberts

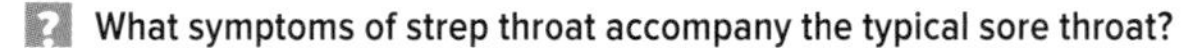
What symptoms of strep throat accompany the typical sore throat?

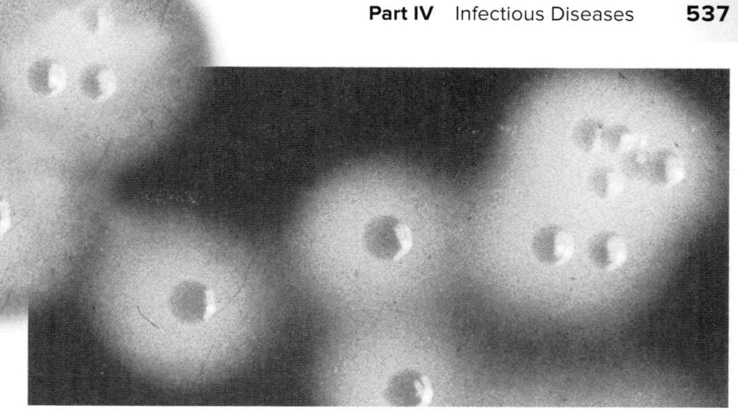

FIGURE 21.5 ***Streptococcus pyogenes*** **Colonies on Blood Agar** The colonies are surrounded by a wide clear zone of β-hemolysis. ©Evans Roberts

How can α-hemolysis be differentiated from β-hemolysis?

Pathogenesis

Streptococcus pyogenes has many virulence factors that enhance its ability to cause disease (**table 21.2**). Some are structural components of the cell wall that allow the bacterium to avoid host defenses (**figure 21.6**). Others are destructive enzymes and toxins released by the bacterial cell that damage or kill host cells. ◀◀ virulence factor

Proteins in the cell wall of *S. pyogenes* allow the bacteria to attach to host cells. M protein is an important adhesin involved in attachment—host antibodies that bind to it prevent infection. Unfortunately, there are more than 80 antigenic types of M protein among the many strains of *S. pyogenes,* and antibodies to one type do not prevent infection by a strain that has a different type. Another protein, protein F, allows *S. pyogenes* to attach to cells of the throat by adhering to fibrin, a protein found on epithelial cells. Once *S. pyogenes* colonizes

TABLE 21.2 Virulence Factors of *Streptococcus pyogenes*

Product	Effect
C5a peptidase	Inhibits recruitment of phagocytes by destroying complement component C5a
Hyaluronic acid capsule	Inhibits phagocytosis
M protein	Interferes with phagocytosis by causing inactivation of complement component C3b, an opsonin; involved in attachment to host cells
Protein F	Responsible for attachment to host cells
Protein G	Binds to Fc portion of antibodies, thereby interfering with opsonization
Streptococcal pyrogenic exotoxins (SPEs)	Superantigens responsible for scarlet fever, toxic shock, "flesh-eating" fasciitis
Streptolysins O and S	Lyse leukocytes and erythrocytes
Tissue-degrading enzymes	Enhance spread of bacteria by breaking down DNA, proteins, blood clots, tissue, hyaluronic acid

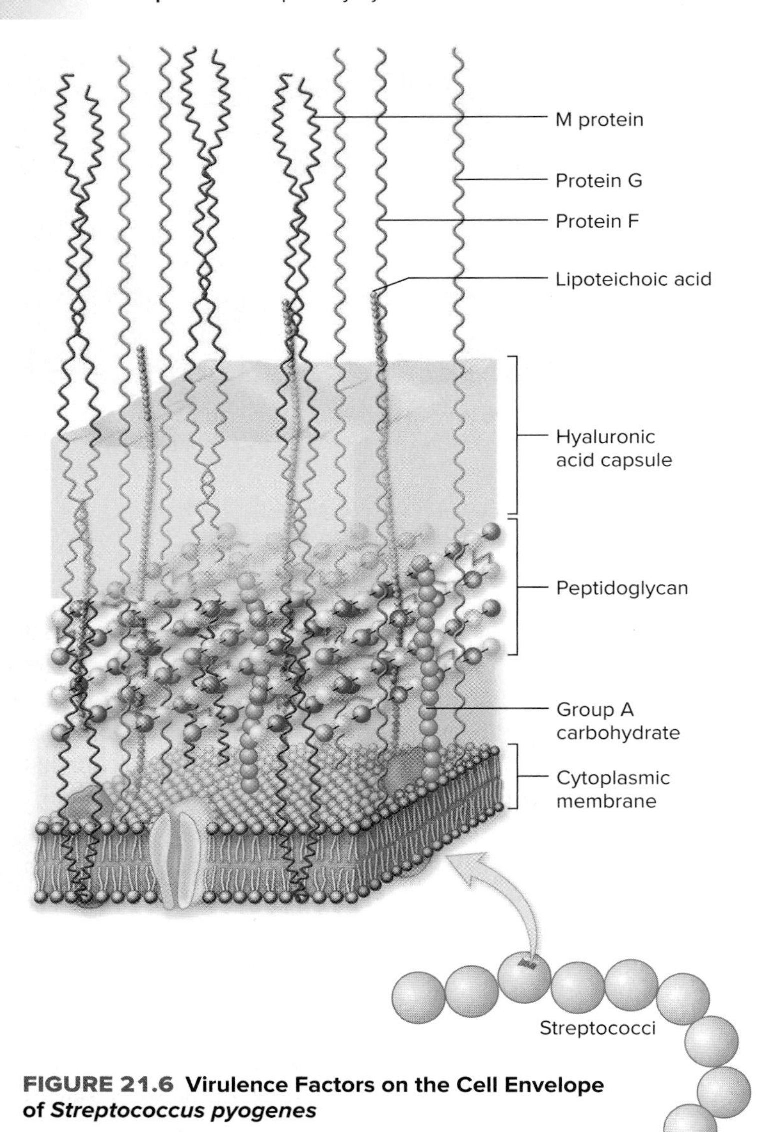

FIGURE 21.6 Virulence Factors on the Cell Envelope of *Streptococcus pyogenes*

? What is the role of M protein in pathogenesis?

host tissues, it produces enzymes such as DNase, hyaluronidase, and proteases that break down intracellular connections and allow the organism to spread rapidly to other cells. This spread is assisted by streptokinase, which causes the breakdown of blood clots.

In addition to factors that promote attachment and colonization of a host, *S. pyogenes* also has mechanisms for avoiding the host immune system. The cells have a hyaluronic acid capsule that prevents phagocytosis. Hyaluronic acid is a normal component of human tissue, so the capsule acts as a cloaking disguise, making it difficult for the host immune system to detect the pathogen. M protein, in addition to aiding attachment to the host cell, also interferes with phagocytosis. It prevents complement component C3b (an opsonin that would usually increase phagocytosis) from being deposited on the bacterial cell wall. Another cell wall protein—protein G—is an Fc receptor that binds the Fc portion of IgG, preventing opsonization by antibodies (see figure 16.10). The organism further inhibits phagocytosis by releasing C5a peptidase, an enzyme that destroys complement component C5a (normally responsible for attracting phagocytes to the site of a bacterial infection). It also produces streptolysins O and S—enzymes that destroy blood cells by making holes in their cell membranes. Leukocyte (white blood cell) destruction by the streptolysins inhibits the immune response; destruction of erythrocytes (red blood cells) causes the β-hemolysis exhibited by *S. pyogenes* (see figure 21.5). ◀◀ complement system, ◀◀ C5a peptidase, ◀◀ membrane-damaging toxins, ◀◀ opsonins

A few strains of *S. pyogenes* produce **streptococcal pyrogenic exotoxins (SPEs),** a family of exotoxins that cause severe streptococcal diseases characterized by high fever (*pyro* means "fire"). SPEs are encoded by bacteriophages that have incorporated their DNA into the bacterial chromosome, an example of lysogenic conversion. These toxins are superantigens, causing massive activation of T cells. The resulting uncontrolled release of cytokines (called a cytokine storm) is probably responsible for the seriousness of these infections. A person with strep throat caused by an SPE-producing strain of *S. pyogenes* may develop scarlet fever, characterized by high fever, roughening of the skin, and a pink-red rash. Although the toxin that causes these symptoms is an SPE, it is traditionally referred to as erythrogenic toxin (*erythro* means "red"). The rash of scarlet fever is found on the head, neck, chest, and thighs, but usually not around the mouth. The toxin also causes the tongue to look like a ripe strawberry—red and spotted. Some SPE-producing strains of *S. pyogenes* have additional virulence factors that allow them to cause severe invasive diseases such as streptococcal toxic shock syndrome, and "flesh-eating disease," or necrotizing fasciitis. ◀◀ lysogenic conversion, ◀◀ superantigens, ▶▶ necrotizing fasciitis

MicroByte

Streptococcal streptokinase is used as a clot-dissolving medication in treating some heart attacks and pulmonary embolisms.

Epidemiology

Streptococcus pyogenes naturally infects only humans. Strains that cause strep throat spread easily by respiratory droplets generated by shouting, coughing, and sneezing. Peak incidence of strep throat occurs in winter or spring and is highest in grade school children. Epidemics have also originated from food contaminated with *S. pyogenes.*

Nasal carriers of *S. pyogenes* are more likely than pharyngeal carriers to spread the organisms. Anal carriers are not common, but have been implicated in healthcare-associated infections. People may be asymptomatic carriers of *S. pyogenes* for weeks if they are not treated. Some people become long-term carriers; in these cases, the infecting strain usually becomes deficient in M protein and is not a threat to the carrier or to others. ◀◀ healthcare-associated infections

Treatment and Prevention

People with fever and sore throat should be seen by a physician so that a throat swab can be taken for a rapid strep test or rapid antigen detection test (RADT) and throat culture. Confirmed strep throat is usually treated with an oral penicillin, which eliminates the organism in most cases. Treatment given as late as 9 days after the onset prevents certain post-streptococcal sequelae (covered next).

Adequate ventilation and avoiding crowds help to control the spread of streptococcal infections. No vaccine is available, although some candidates are in clinical trials. **Table 21.3** summarizes important information about strep throat.

Post-Streptococcal Sequelae

Post-streptococcal sequelae—complications that can develop after strep throat or other streptococcal infections—are thought to be a result of immune responses to *Streptococcus pyogenes*. The specific complications that develop are related to the target of the immune responses, including the heart, kidneys, joints, brain, or skin. Sequelae are uncommon in developed countries because of quick identification and treatment of *S. pyogenes* infections. Two of the most common are discussed here.

Acute Rheumatic Fever

Acute **rheumatic fever** usually begins about 3 weeks after recovery from strep throat. Signs and symptoms include fever, joint pains, chest pain, rash, and nodules under the skin. Uncontrollable body movements (chorea, also called St. Vitus' dance) can occur and, when present, are major criteria for diagnosis. Carditis (inflammation of heart tissue), the most serious complication, develops in about 30–50% of patients. This can lead to chronic rheumatic heart disease in which one or more of the heart valves are damaged, causing them to leak and resulting in heart failure later in life. The damaged valves are also prone to infection, usually by bacteria from the normal skin or mouth microbiota, resulting in a heart valve infection called infective endocarditis. ⏭| infective endocarditis

Rheumatic fever develops only in people who are genetically predisposed to it; some MHC class II alleles are involved in susceptibility (see Focus Your Perspective 15.1). The pathogenesis of rheumatic fever is not well understood but it is thought to be an autoimmune response involving antibodies to *S. pyogenes* that cross-react with host tissue antigens similar to the pathogen antigens. In chronic rheumatic heart disease, for example, antibodies to streptococcal M protein may also recognize and bind to myosin in the heart. The heart tissue is then targeted for attack by the host's own immune

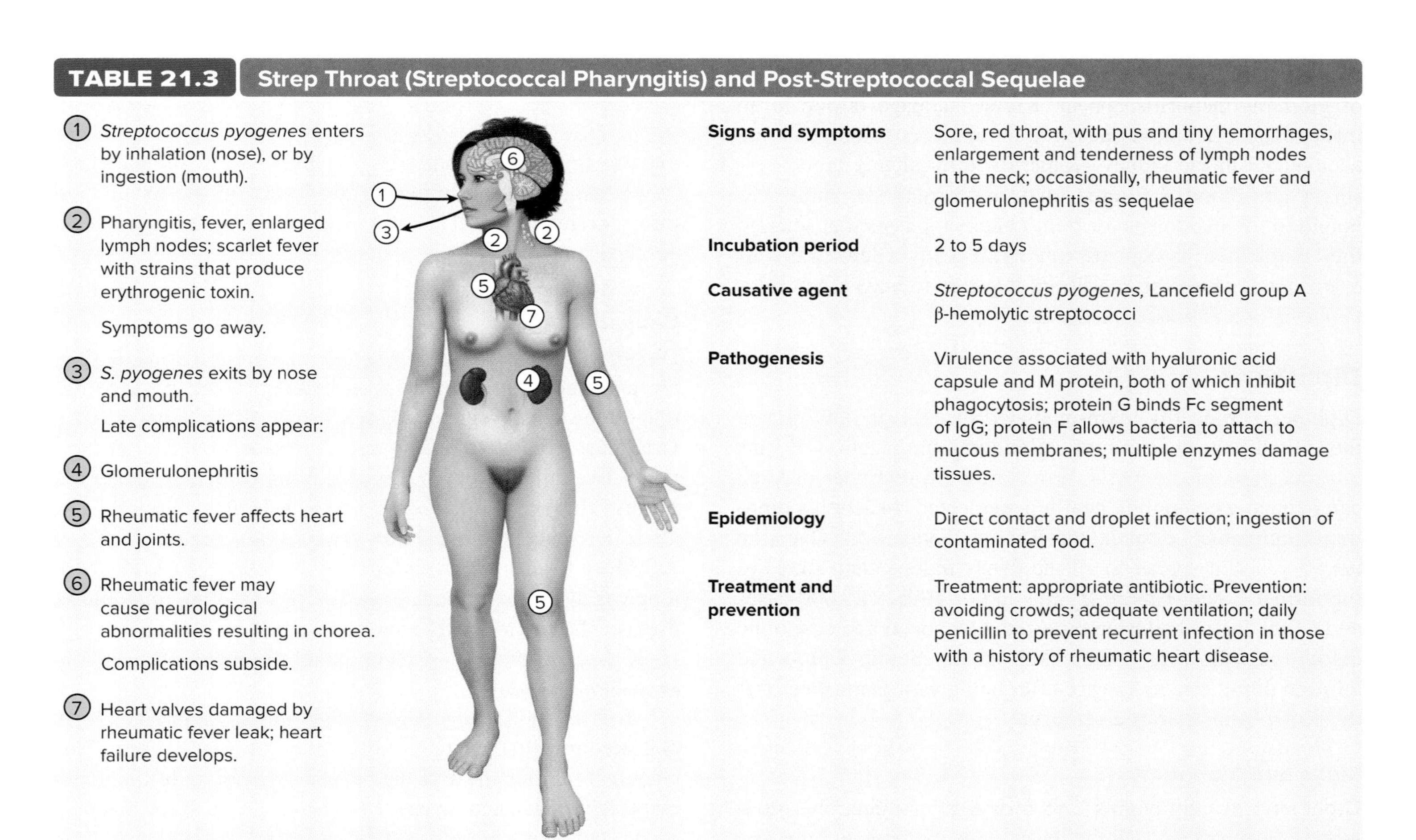

Signs and symptoms	Sore, red throat, with pus and tiny hemorrhages, enlargement and tenderness of lymph nodes in the neck; occasionally, rheumatic fever and glomerulonephritis as sequelae
Incubation period	2 to 5 days
Causative agent	*Streptococcus pyogenes,* Lancefield group A β-hemolytic streptococci
Pathogenesis	Virulence associated with hyaluronic acid capsule and M protein, both of which inhibit phagocytosis; protein G binds Fc segment of IgG; protein F allows bacteria to attach to mucous membranes; multiple enzymes damage tissues.
Epidemiology	Direct contact and droplet infection; ingestion of contaminated food.
Treatment and prevention	Treatment: appropriate antibiotic. Prevention: avoiding crowds; adequate ventilation; daily penicillin to prevent recurrent infection in those with a history of rheumatic heart disease.

system. As a result, effector helper T (T_H) cells release pro-inflammatory cytokines, causing inflammation that leads to permanent damage to the local tissues, particularly the heart valves. **◀◀ MHC, ◀◀ inflammation, ◀◀ helper T cells**

Incidence of rheumatic fever has declined in the United States. This is probably a result of quick treatment of strep throat with antibiotics, and decreased prevalence of the strains associated with the disease. Overall, the risk of developing acute rheumatic fever after severe untreated strep throat is 3% or less, and those with untreated mild pharyngitis have an even lower risk. Nonetheless, 10 to 20 million new cases of rheumatic fever occur globally each year, mostly in developing countries. Signs and symptoms usually decrease with rest and anti-inflammatory medicines such as aspirin. People with cardiac damage due to acute rheumatic fever take penicillin daily for years to prevent recurrence of the disease.

Acute Post-Streptococcal Glomerulonephritis

Acute post-streptococcal glomerulonephritis occasionally follows strep throat but is more often an aftermath of a strep skin infection. It begins 7–21 days after the initial infection, usually about 10 days after strep throat. Signs and symptoms include fever, fluid retention, high blood pressure, and blood and protein in the urine (making the urine look like tea or cola). There are no streptococci in the urine or the diseased kidney tissues, and the body's immune response has generally eliminated *S. pyogenes* from the throat before the symptoms of glomerulonephritis appear. Kidney damage is due to an inflammatory reaction caused by streptococcal antigens that accumulate in the kidney glomeruli (tufts of tiny blood vessels where blood is filtered for urine formation); antibodies bound to these antigens form immune complexes that activate the complement system (**figure 21.7**). Only a few strains of *S. pyogenes* cause the condition; it is rare to have glomerulonephritis twice. **◀◀ immune complexes**

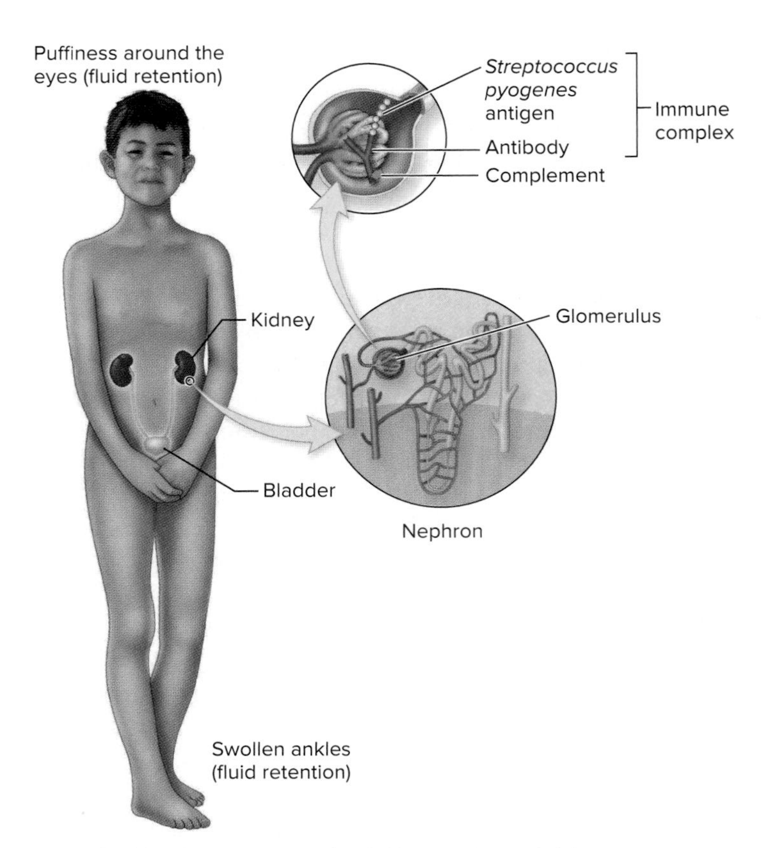

FIGURE 21.7 Pathogenesis of Glomerulonephritis Immune complexes are deposited in the kidney glomeruli, causing an inflammatory response.

? Why might blood appear in the urine during glomerulonephritis?

Diphtheria

Diphtheria, a deadly toxin-mediated disease, is now rare in the United States because of childhood immunization. Events in other parts of the world, however, are a reminder of what can happen when public health is neglected. In 1990, a diphtheria epidemic began in the Russian Federation. Over the next 5 years, it spread to all the newly independent states of the former Soviet Union. By the end of 1995, 125,000 cases and 4,000 deaths had been reported. The social and economic disruption following the breakup of the Soviet Union had allowed diphtheria to emerge after being well controlled over the previous quarter of a century.

Signs and Symptoms

Diphtheria usually begins 2 to 6 days after infection with a mild sore throat and slight fever, with extreme fatigue and malaise (general discomfort). The neck swells dramatically. A whitish-gray **pseudomembrane** forms on the tonsils and throat or in the nasal cavity. Heart and kidney failure and paralysis may occur later.

Causative Agent

Diphtheria is caused by *Corynebacterium diphtheriae,* a pleomorphic, non-motile, non-spore-forming, Gram-positive rod that often stains irregularly. One reason for this is that the bacteria have storage granules (metachromatic granules) at their poles that stain darkly. The organisms are club-shaped (*koryne* means "club") and often occur side by side in "palisades" (like a wooden fence) (**figure 21.8**). **◀◀ metachromatic granules**

Most strains of *C. diphtheriae* release diphtheria toxin, a powerful exotoxin that causes the serious symptoms of the disease. The gene for this toxin is carried by a specific lysogenic bacteriophage, an example of lysogenic conversion. **◀◀ lysogenic conversion**

To isolate *C. diphtheriae* from throat material, a special medium that contains potassium-tellurite is used. This chemical inhibits most of the normal microbiota of the upper respiratory tract and causes *C. diphtheriae* to form black or brown colonies on the medium, aiding in identification. The

FOCUS ON A CASE 21.1

The patient was a college student who complained to her roommate about a sore throat. Although the student had developed a case of strep throat just 2 years previously, she was not too worried because she did not feel particularly ill and she did not plan to go to the campus clinic. Her roommate, a microbiology student, argued that a visit to the clinic was important, noting that her instructor had said that strep throat should be treated with antibiotics. Eventually, the ill student agreed to visit the clinic.

At the clinic, the student was given a rapid strep test (RST)—the doctor swabbed the back of the student's throat and used a kit to immediately determine whether *Streptococcus pyogenes,* the causative agent of strep throat, was present. The test was negative. The doctor told the student that he would send the throat swab away for culture and that he would have a definitive answer for her soon. Meanwhile, he sent her home with a prescription for penicillin, but told her not to fill it until she heard back from him. A day later, the clinic doctor called the student and told her that she did indeed have strep throat and to begin taking the penicillin immediately.

1. Considering that the student had suffered from strep throat just 2 years previously, why would she develop it again?
2. Why would the RST be negative when the student did in fact have strep throat?
3. Why was the student given penicillin to treat her infection?
4. Why is it important to treat strep throat promptly?
5. Why are there no vaccines against *Streptococcus pyogenes*?

Discussion

1. There are many antigenically distinct strains of *Streptococcus pyogenes,* and immunity to one strain does not protect against another. From an infection standpoint, the most important antigen of *S. pyogenes* is M protein, a major virulence factor that allows the bacterial cells to avoid phagocytosis. Antibodies to M protein prevent opsonization and protect against infection, but there are more than 80 different varieties of the protein. M protein variation means that a person can get strep throat more than once—in this case, the student's current strep throat was likely due to a *S. pyogenes* strain that had an M protein type different from the strain that had infected her previously.
2. The rapid strep test (RST) is very specific, but is not as sensitive as culture. Specificity indicates the reliability of a positive test; the high specificity of RST means that a positive result is likely a true positive—*S. pyogenes* is present in the specimen. Sensitivity indicates the reliability of a negative test; the low sensitivity of RST means that a negative result could be a false negative—*S. pyogenes* could be present in the specimen, but the test was simply unable to detect it. That is why laboratory culture, which has both high specificity and high sensitivity, is used as a backup for diagnosis. *S. pyogenes* is easily recognizable on blood agar by the zone of β-hemolysis surrounding the colonies on this medium. Other bacteria are also β-hemolytic, so catalase tests and group A–specific latex agglutination are done to confirm the identification.
3. The student was given penicillin because *S. pyogenes* is still sensitive to this antibiotic. Penicillin is generally the medication of choice for treating strep throat because it is inexpensive, safe, and effective. Had the student been allergic to penicillin, the doctor would have likely prescribed azithromycin, also effective in treating strep throat.
4. Although strep throat itself is not a very serious disease, it can lead to post-streptococcal sequelae or other complications. Treating with antibiotics may lessen the chance of these complications. The sequelae, including rheumatic fever and glomerulonephritis, are due to the immune response against *S. pyogenes* and occur after the infection has resolved. Rheumatic fever is an autoimmune reaction that results from molecular mimicry of *S. pyogenes*—the epitopes of some variations of M protein are similar to epitopes on self proteins. When a person is infected with one of these strains, the immune response to the strain may result in the production of antibodies that also attach to self cells, targeting them for destruction. Glomerulonephritis results from an inflammatory reaction in the glomeruli of the kidneys. In the case of post-streptococcal glomerulonephritis, an accumulation of immune complexes consisting of antibodies bound to streptococcal antigens is thought to trigger the inflammation. Another concern regarding strep throat is that a few strains produce streptococcal pyogenic exotoxins (SPEs), and these can cause scarlet fever and certain severe invasive diseases.
5. Developing a vaccine against *S. pyogenes* has not been possible because of the great variation in the M protein of these organisms. A vaccine against one M type might not protect against another. An additional complication is the risk of autoimmunity, such as that seen in rheumatic fever. An appropriate vaccine would have to protect against many different strains of *S. pyogenes* without causing an autoimmune response.

organism can also be grown on Loeffler's medium, which enhances the formation of the metachromatic granules.

Pathogenesis

Corynebacterium diphtheriae has little invasive ability, and rarely enters the blood or tissues. The disease is caused by the diphtheria exotoxin released by the bacteria growing in the throat. The pseudomembrane that forms is made up of dead epithelial cells (killed by the exotoxin) and clotted blood, along with fibrin and the leukocytes that accumulate during inflammation. This membrane may come loose and obstruct the airways, causing the patient to suffocate. The toxin may

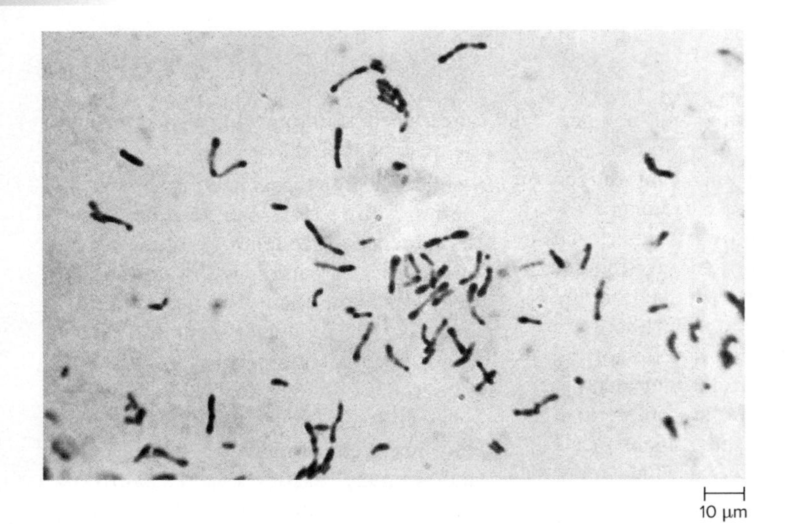

FIGURE 21.8 ***Corynebacterium diphtheriae*** Source: CDC

? How do strains of *C. diphtheriae* acquire the gene for toxin production?

be absorbed into the bloodstream, allowing it to access and damage heart, nerve, and kidney tissues.

Diphtheria toxin is an A-B toxin (**figure 21.9**). The B subunit attaches to specific receptors on a host cell membrane, and the entire toxin molecule is taken into the cell by endocytosis. Cells lacking these receptors do not take up the toxin and are unaffected by it, explaining why some tissues are not affected in diphtheria, while others are severely damaged. Once the toxin is inside the cell, the A subunit separates from the B subunit. It then becomes a functional enzyme that inactivates elongation factor 2 (EF-2), which is required for movement of the eukaryotic ribosome on mRNA. This stops protein synthesis, and the cell dies. The toxin is extremely potent because the A subunit, an enzyme, is not used up in the reaction, so a single molecule can inactivate nearly all of the cell's EF-2. ⏮ A-B toxins, ⏮ endocytosis

Epidemiology

Humans are the primary reservoir for *Corynebacterium diphtheriae.* The organisms are typically spread by air from infected people. They are then acquired either by inhalation or from fomites. *C. diphtheriae* can also cause cutaneous diphtheria; chronic skin ulcers that develop may become a source of infection to others. ⏮ fomites

Treatment and Prevention

Diphtheria is treated by injecting the patient with antiserum against diphtheria toxin. This is done immediately once the disease is suspected. Because the toxin is so potent, delaying treatment to wait for culture results can be fatal. The bacteria are sensitive to various antibiotics, including erythromycin and penicillin, but such treatment has no effect on toxin that has already been absorbed. Even with treatment, about 10% of diphtheria patients die. ⏮ antiserum

Because diphtheria results from toxin production rather than microbial invasion, it can be effectively prevented by immunization with toxoid. The toxoid, prepared by formalin treatment of diphtheria toxin, causes the body to produce antibodies that specifically neutralize the toxin. The childhood vaccination DTaP consists of diphtheria and tetanus toxoids along with components of *Bordetella pertussis,* the bacterium that causes pertussis (whooping cough). Unfortunately, these immunizations are often neglected, and serious epidemics of diphtheria have occurred periodically. Since the 1980s, there has been an active campaign in most of the United States to ensure that children who are entering school are immunized against diphtheria. As a result, the incidence of the disease has been reduced to only a few cases a year. Immunity decreases after childhood, however, so booster injections must be given every 10 years to maintain protection. **Table 21.4** summarizes the main features of diphtheria. ⏮ toxoid

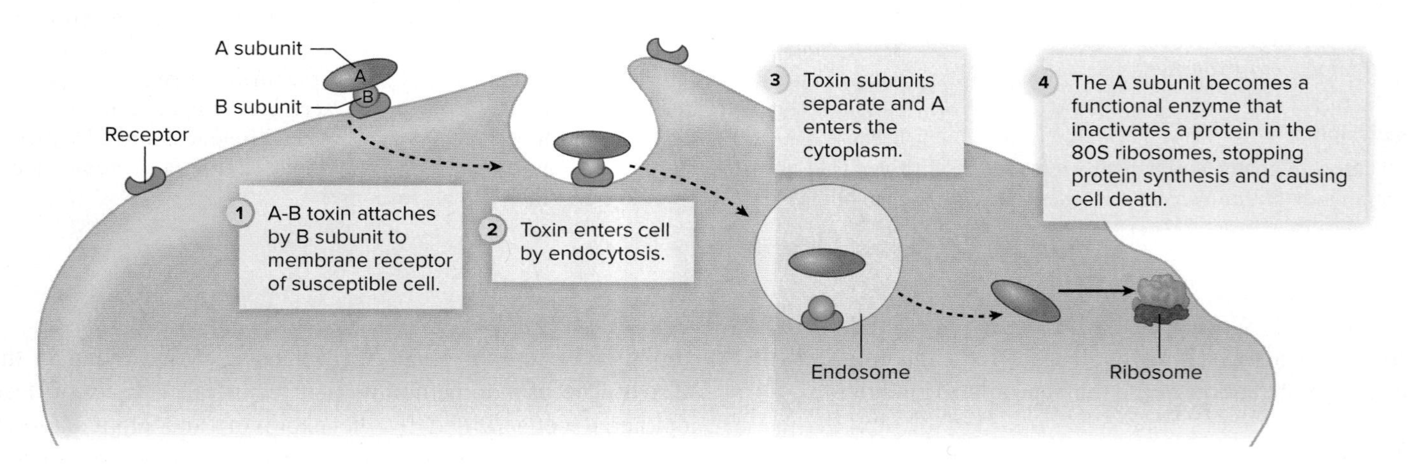

FIGURE 21.9 Mode of Action of Diphtheria Toxin

? Why is diphtheria toxin so potent?

TABLE 21.4 Diphtheria

1. *Corynebacterium diphtheriae* enters by inhalation.
2. Infection established in throat.
3. Toxin released locally, pseudomembrane forms.
4. Circulating toxin may cause paralysis, or damage heart muscle, kidneys, nerves.
5. Membrane may come loose and obstruct breathing.
6. Exit from body by respiratory secretions.

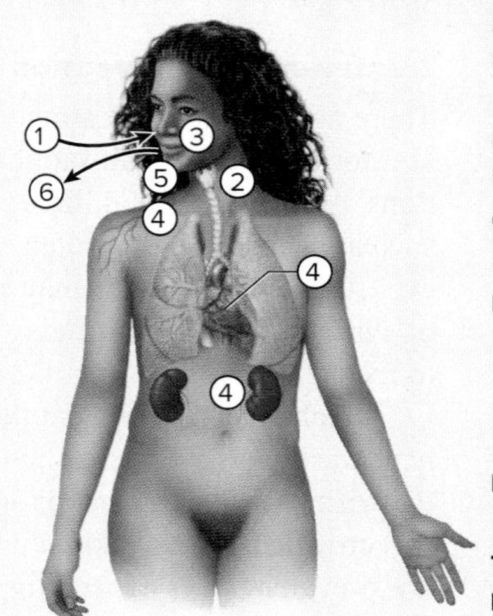

Signs and symptoms	Sore throat, fever, fatigue, and malaise; pseudomembrane forms on tonsils and throat or in nose; paralysis, heart and kidney failure
Incubation period	2 to 6 days
Causative agent	*Corynebacterium diphtheriae,* an A-B toxin–producing, non-spore-forming Gram-positive rod
Pathogenesis	Infection in upper respiratory tract; exotoxin is released and absorbed by bloodstream; toxin kills cells by interfering with protein synthesis; affects cells that have receptors for the toxin—mainly heart, kidney, and nerve tissue.
Epidemiology	Inhalation of infectious droplets; indirect contact with fomites.
Treatment and prevention	Treatment: antitoxin; appropriate antibiotic to prevent transmission. Prevention: immunization of infants and children with toxoid; boosters for adults.

MicroAssessment 21.2

Conjunctivitis, otitis media, and sinusitis are common infections that often occur together and are caused by the same pathogens. *Streptococcus pyogenes,* a group A streptococcus with many virulence factors, causes a sore throat commonly known as strep throat. Untreated *S. pyogenes* infections can sometimes cause serious diseases (sequelae) in other parts of the body long after the initial infections have resolved. In diphtheria, toxin is absorbed into the bloodstream and circulates throughout the body, selectively damaging certain tissues such as heart, kidneys, and nerves.

4. Name two post-streptococcal sequelae.
5. How does diphtheria toxin kill cells?
6. How would adequate ventilation help to prevent the spread of streptococcal infections?

21.3 ■ Viral Infections of the Upper Respiratory System

Learning Outcomes

5. List the strategies helpful in avoiding common colds.
6. Give the distinctive characteristics of adenoviral respiratory tract infections.

The average person in the United States gets two to five viral upper respiratory infections each year. Although hundreds of different viruses cause these infections, they produce similar symptoms and generally resolve without treatment. However, they impair respiratory tract defenses and allow for more serious secondary bacterial infections.

The Common Cold

The common cold is the most frequent infectious disease in humans, and accounts for more than half of upper respiratory tract infections every year. Colds are the leading cause of absences from school and result in people missing 150 million workdays per year in the United States.

MicroByte

The average adult gets between two and four common colds a year, whereas children can get as many as eight colds in this time.

Signs and Symptoms

Colds begin 1 to 2 days after infection with malaise, followed by a runny nose, sneezing, coughing, a mildly sore throat, and hoarseness. The nasal secretions are initially profuse and watery, but may thicken and become cloudy as the cold progresses. There is no fever unless secondary bacterial infection occurs. Symptoms typically last about a week, but a mild cough may continue a bit longer.

Causative Agents

Viruses that cause the common cold are often simply called "cold viruses." Between 30% and 50% of colds are caused by the 100 or more types of human rhinoviruses (*rhino* means "nose," as in rhinoceros, or "horny nose") (**figure 21.10**). These are members of the family *Picornaviridae* (*pico* means "small" and *rna* is ribonucleic acid; thus, "small RNA viruses"), a group of non-enveloped viruses with a single-stranded RNA genome. Rhinoviruses can usually be grown in cell cultures under conditions that mimic the upper respiratory tract (33°C and a slightly acid pH). They are inactivated if the pH drops below 5.3, and therefore are usually destroyed

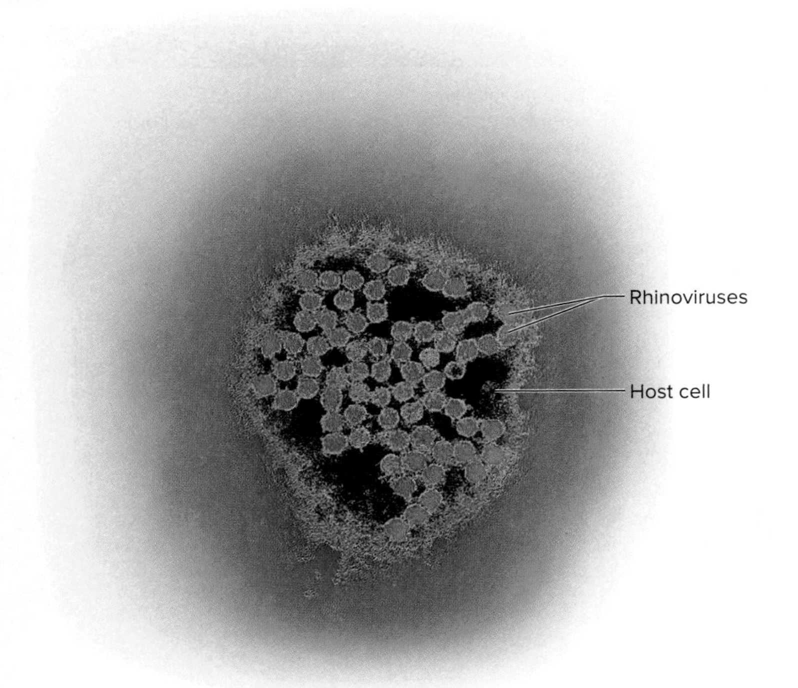

FIGURE 21.10 Rhinovirus-Infected Cell Color-enhanced TEM. ©A.B. Dowsette/SPL/Science Source

What kinds of cells do rhinoviruses infect?

in the stomach. Many other viruses, such as various adenoviruses and coronaviruses, and some bacterial species can also produce signs and symptoms of the common cold.

Pathogenesis

Rhinoviruses infect epithelial cells that line the upper respiratory tract. Ciliary motion in the infected cells stops and the cells may die and slough off. The damage causes the release of pro-inflammatory cytokines and stimulates nervous reflexes, resulting in increased nasal secretions, tissue swelling that partially or completely obstructs the airways, and sneezing. The infection is stopped by immune responses, but it can spread into the ears, the sinuses, or even the lower respiratory tract before this occurs. Rhinoviruses may cause life-threatening pneumonia in immunocompromised people.

Epidemiology

Humans are the only source of cold viruses, which are spread when airborne virus-containing droplets are inhaled or when secretions from infected people are accidentally rubbed into the eyes or nose by contaminated hands. Young children transmit cold and other respiratory viruses very easily because they are often careless with their respiratory secretions. Infected people have high concentrations of virus in their nasal secretions and on their hands during the first 2 or 3 days of a cold and are most likely to transmit the virus. Virus levels are often undetectable after 4 or 5 days, but low levels can be present for 2 weeks. Although only a few virions are sufficient to infect the nasal epithelium, colds are actually not highly contagious if one takes reasonable preventive measures such as handwashing.

Treatment and Prevention

There are no proven treatments for the common cold. Viruses are not affected by antibiotics or other antibacterial medications. Analgesics (painkillers) and antipyretics (fever-reducers) such as aspirin and ibuprofen can help reduce symptoms. However, these medications may prolong symptoms and duration of virus shedding, and delay antibody production that leads to recovery.

Mechanisms to prevent the spread of rhinoviruses include handwashing (even in plain water) to physically remove the viruses, keeping hands away from the face, and avoiding crowded places like subways when respiratory diseases are prevalent. It is especially important to avoid people with colds during the first few days of their symptoms, when they are shedding high numbers of viral particles. It has not been possible to develop a vaccine because such a large number of immunologically different viruses cause colds. However, all known rhinovirus strain genomes have been sequenced, revealing new possibilities for vaccines. **Table 21.5** summarizes some important features of the common cold.

Adenovirus Respiratory Tract Infections

Adenoviruses are widespread and can cause a variety of different types of infections, depending on the viral serotype. For example, some cause a sore throat, whereas others cause eye infections. They are representative of the many viruses that cause upper respiratory tract infections characterized by fever.

Signs and Symptoms

Adenoviruses that infect the upper respiratory tract generally cause a runny nose, but unlike the common cold, fever is typically present. Signs and symptoms appear 5 to 10 days after infection. The throat is usually sore, with regions of gray-white pus on the pharynx and tonsils; these characteristics can be confused with those of strep throat. The lymph nodes of the neck become large and tender and a mild cough is

TABLE 21.5	The Common Cold
Signs and symptoms	Malaise, runny nose, sneezing, cough, sore throat, hoarseness
Incubation period	1 to 2 days
Causative agent	Mainly rhinoviruses—more than 100 types; many other viruses; some bacteria
Pathogenesis	Viruses attach to respiratory epithelium; ciliary action stops, and cells slough; secretion of mucus increases, and inflammatory reaction occurs.
Epidemiology	Inhalation of infectious droplets; transfer of infectious mucus to nose or eye by contaminated fingers.
Treatment and prevention	Treatment: no treatment except for control of symptoms. Prevention: handwashing, avoiding people with colds, and keeping hands away from face.

common. In some epidemics, conjunctivitis or diarrhea may occur. Patients sometimes develop a severe cough with chest pain, symptoms that can be confused with those of pneumonia. Recovery usually takes about 1 to 3 weeks.

Causative Agent

More than 50 antigenic types of adenoviruses infect humans. The viruses are non-enveloped, with double-stranded DNA. Like many other non-enveloped viruses, they can remain infectious in the environment for long periods of time and are resistant to destruction by detergents and alcohol solutions. They are easily inactivated, however, by heat (56°C), adequate levels of chlorine, and various other disinfectants.

Pathogenesis

Adenoviruses attach to and infect epithelial cells. Once inside the cell, the genome is transported to the host cell nucleus, where the virus multiplies. Adenoviruses have many mechanisms for avoiding host defenses, including delaying apoptosis, blocking interferon function, and interfering with antigen presentation by MHC class I molecules. Once replication is complete, a virally encoded "death protein" is produced that causes host cell lysis. In severe infections, extensive cell destruction and inflammation occur. Different serotypes of adenoviruses affect different tissues. ⏮ apoptosis, ⏮ interferon, ⏮ MHC class I molecule

Epidemiology

Humans are the only reservoir of adenoviruses. Because these viruses can persist in the environment, healthcare personnel must take precautions to prevent transferring them from one patient to another on medical instruments. Adenoviruses commonly infect schoolchildren, usually resulting in sporadic cases, but occasionally causing outbreaks in winter and spring. Summer epidemics can occur when viruses are transmitted in inadequately chlorinated swimming pools. Adenoviral disease is spread by respiratory droplets so it can be a problem in military recruits and other people living together in crowded conditions. Asymptomatic infections are common, which fosters epidemic spread. The viruses are shed from the respiratory tract during the acute illness and continue to be eliminated in the feces for months thereafter.

TABLE 21.6 Adenovirus Respiratory Tract Infections

Signs and symptoms	Fever, sore throat, cough, swollen lymph nodes of neck, pus on tonsils and throat
Incubation period	5 to 10 days
Causative agent	Adenoviruses—more than 50 serotypes—non-enveloped, double-stranded DNA genome
Pathogenesis	Virus multiplies in host cells; cell destruction and inflammation occur; different serotypes produce different symptoms.
Epidemiology	Inhalation of infectious droplets; possible spread from gastrointestinal tract.
Treatment and prevention	Treatment: no treatment except for control of symptoms. Prevention: handwashing, avoiding crowds.

Treatment and Prevention

As with colds, there is no specific treatment for adenovirus disease, and most patients recover on their own. Secondary bacterial infections may occur, however, and these require treatment with an antibacterial medication.

An orally administered attenuated vaccine against two serotypes most likely to cause severe disease is available, but is currently given only to military recruits. **Table 21.6** summarizes some important features of adenoviral infections.

MicroAssessment 21.3

The common cold may be caused by many different viruses, particularly rhinoviruses. Transmission requires close personal contact during which respiratory droplets from an infected person or respiratory secretions on contaminated hands are transferred to the nasal epithelium. Symptoms typically last for about a week before the disease is stopped by immune responses. Adenovirus infections may resemble colds, but fever is present.

7. Why are there no vaccines for the common cold?
8. How is an adenovirus infection treated?
9. Contrary to your grandmother's warnings, going outside in cold weather without a coat is not likely to cause a cold. Why not?

LOWER RESPIRATORY TRACT INFECTIONS

21.4 ■ Bacterial Infections of the Lower Respiratory System

Learning Outcomes

7. Compare the distinctive features of pneumococcal, *Klebsiella,* and mycoplasmal pneumonia.
8. Outline the pathogenesis and treatment of pertussis, tuberculosis, legionellosis, and inhalation anthrax.

Bacterial infections of the lower respiratory tract are less common than those of the upper respiratory tract, but they are generally much more serious. An earache or a sore throat is unlikely to be life-threatening, but the same causative agents can cause serious illnesses in the lungs. Pneumonias top the list of fatal community-acquired infections in the United States and are common healthcare-associated infections.

FOCUS ON PNEUMONIA

Pneumonia is a disease of the lower respiratory tract in which the alveoli (air sacs) of the lungs fill with fluids such as pus and blood. It typically results from an inflammatory response to microbial infection of the lungs, and is the leading cause of death due to infectious disease in the United States.

Signs and Symptoms

The signs and symptoms of pneumonia generally include cough, chills, shortness of breath, fever, and chest pain. In severe cases, the patient may develop cyanosis (bluish skin color) due to poor blood oxygenation. Pneumonia ranges from mild to life-threatening, depending largely on the causative agent but also on any underlying health problems of the patient. Some pathogens cause what is referred to as atypical pneumonia or "walking pneumonia," a term that reflects the mild symptoms. Pneumonias are often accompanied by a productive cough, meaning that a pus- and mucus-containing fluid called **sputum** comes up from the lungs.

To diagnose pneumonia, a physician uses a stethoscope to listen for a characteristic crackling or bubbling sound that occurs in the lungs as air passes by fluid in the alveoli. A chest X ray will likely be done to determine which parts of the lung are infected; areas of infection usually appear as white shadows. The patient may also be asked to give a sputum sample, which can be examined microscopically and inoculated onto appropriate laboratory media as part of the process to identify a bacterial or fungal cause of pneumonia.

Pathogenesis

Microbes that cause pneumonia are often opportunists that cause disease only when the mucociliary escalator or other defenses of the respiratory tract are not functioning optimally. Conditions that interfere with the function of the mucociliary escalator include alcohol and narcotic use as well as viral respiratory infections such as influenza. That is why patients who have influenza or other illnesses that damage the respiratory tract or interfere with the patient's ability to cough are more susceptible to developing pneumonia. Organisms that cause pneumonia frequently make a capsule. Because of the capsule, alveolar macrophages that normally destroy invading microbes in the lungs cannot effectively eliminate the pathogen initially. Once opsonizing antibodies are produced during a B-cell response, however, phagocytes can remove the microbes. ⏮ mucociliary escalator, ⏮ capsules, ⏮ opsonization by antibodies

The damaging effects of pneumonia are largely a result of the inflammatory response to the causative agent. As the capillaries become leaky during inflammation, fluids collect in the alveoli and interfere with O_2 and CO_2 exchange. In addition, phagocytes and other leukocytes are recruited to the site of infection and mucus production increases. Accumulating leukocytes and mucus create a thick substance that may clog the alveoli, a condition called consolidation. Consolidation is most common in severe bacterial pneumonia. The inflammatory response seen in severe pneumonia often affects nerve endings in the pleura, causing pain.

Epidemiology

Pneumonias are often categorized as either community-acquired, meaning that they develop in members of the general public, or healthcare-associated, meaning that they develop in hospitalized patients or other people within the healthcare system. Some types of community-acquired pneumonia (CAP) are contagious. Most, however, originate from the patient's own upper respiratory microbiota. These organisms may gain access to the lungs when a person inadvertently inhales his or her own throat secretions. As with CAPs, healthcare-associated pneumonias (HAPs) often occur when the patient inadvertently inhales his or her own upper respiratory microbiota. Patients at particular risk are those on mechanical ventilators used to help breathing, because the ventilator tube allows a portal for microorganisms to enter the lower airways.

Treatment and Prevention

Bacterial and fungal pneumonias are treated with antimicrobial medications, chosen according to the susceptibility of the causative agent. Unfortunately, bacteria that cause healthcare-associated pneumonias are often multi-drug-resistant. In general, there are no effective treatments for viral pneumonias.

Vaccines are available to prevent pneumococcal disease but not the other types of pneumonia. Prevention therefore involves staying out of crowded situations and avoiding other respiratory illnesses.

Pneumococcal Pneumonia

Pneumococci are an important cause of community-acquired pneumonia (CAP), accounting for about 60% of the adult pneumonia patients requiring hospitalization.

Signs and Symptoms

Typical signs and symptoms of pneumococcal pneumonia start after an incubation of 1 to 3 days. They are usually preceded by 1 to 2 days of runny nose and upper respiratory congestion, after which the person experiences sudden fever and shaking chills. The severe chest pain is aggravated by each breath or cough, causing shallow, rapid breathing. Blood from the lungs makes the sputum pinkish or rust-colored; poor blood oxygenation leads to cyanosis (bluish skin color). People who survive without treatment sweat heavily as temperature falls to normal after 7 to 10 days.

Causative Agent

Pneumococcal pneumonia is caused by *Streptococcus pneumoniae,* a Gram-positive diplococcus known as pneumococcus. The cells, which are often elongated with a tapered end, are referred to as lancet-shaped (**figure 21.11**). (A lancet is a surgical instrument with a pointed end.) The most striking characteristic of *S. pneumoniae* is its thick polysaccharide capsule, which is responsible for the organism's virulence. There are over 90 different serotypes of *S. pneumoniae,* each

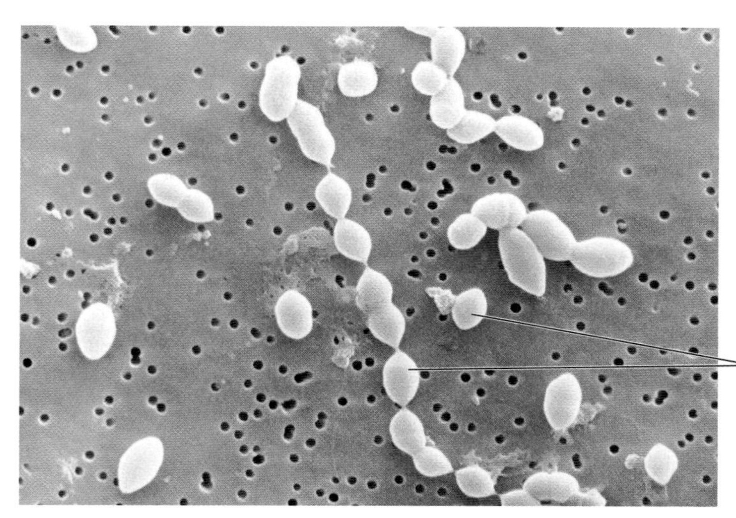

FIGURE 21.11 ***Streptococcus pneumoniae*** Scanning electron micrograph shows lancet-shaped cells trapped by filtering. ©BSIP/Science Source

? **With a Gram stain, would *Streptococcus pneumoniae* appear purple or pink?**

with different capsular antigens. Certain serotypes are more commonly associated with pneumonia and other invasive diseases that infect normally sterile body sites. Strains of the organism that lack a capsule do not cause invasive disease.

Pathogenesis

Encapsulated pneumococci have a number of virulence factors that contribute to pathogenicity. The bacteria are resistant to phagocytosis because their capsule interferes with the action of the complement system component C3b, an important opsonin. Pneumococcal surface protein (PspA) also interferes with the action of C3b. The bacteria produce pneumolysin, a membrane-damaging toxin that destroys ciliated epithelium. As with most severe pneumonias, the inflammatory response leads to an accumulation of fluid and phagocytic cells in the lung alveoli, causing breathing difficulty. This fluid can be seen as abnormal shadows on chest X rays of patients (**figure 21.12**). Sputum coughed from the lungs increases in amount and contains pus, blood, and many pneumococci. ⏮ complement C3b, ⏮ opsonin

Pneumococci may enter the bloodstream from the inflamed lungs, causing three complications that are often fatal: sepsis (a bloodstream infection); endocarditis (an infection of the heart valves); and meningitis (an infection of the membranes covering the brain and spinal cord). People who do not develop complications usually produce enough specific anti-capsular antibodies within about a week to allow phagocytosis and destruction of the pneumococci, resulting in complete recovery. ⏭ sepsis, ⏭ meningitis, ⏭ endocarditis

Epidemiology

Up to 30% of healthy people carry encapsulated pneumococci in their throat. These bacteria seldom reach the lungs because the mucociliary escalator effectively removes them. When this defense mechanism is impaired, the risk of pneumococcal pneumonia rises dramatically. There is also an increased risk of the disease in people over 50, or in those with underlying heart or lung disease, diabetes, or cancer.

Treatment and Prevention

Most pneumococcal infections can be cured with penicillin if given early in the illness. However, strains of pneumococci resistant to one or more antibiotics are becoming increasingly common. ⏮ *S. pneumoniae* antibiotic resistance

Children under the age of 2 years, adults over age 65, and people with health conditions that put them at high risk for pneumococcal pneumonia should receive the PCV13 vaccine (**p**neumococcal **c**onjugate **v**accine, **13** serotypes). The conjugate vaccine contains capsular polysaccharides from 13 pneumococcal serotypes attached to bacterial proteins (thereby creating T-dependent antigens). This is important in children younger than age 2 because polysaccharides are T-independent antigens, and children in this age group do not produce an effective immune response against them. Another vaccine containing capsule polysaccharides from the 23 most common pneumococcal serotypes is recommended for all adults over age 65, as well as some high-risk individuals age 2 and older. This vaccine, called PPSV23 (**p**neumococcal **p**olysaccharide **v**accine, **23** serotypes), should be given at least 1 year after receiving PCV13. **Table 21.7** describes some features of pneumococcal pneumonia. ⏮ conjugate vaccine, ⏮ T-independent antigens

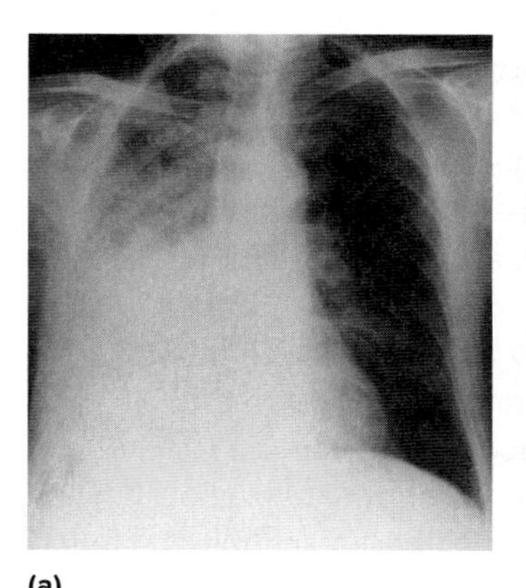
(a)

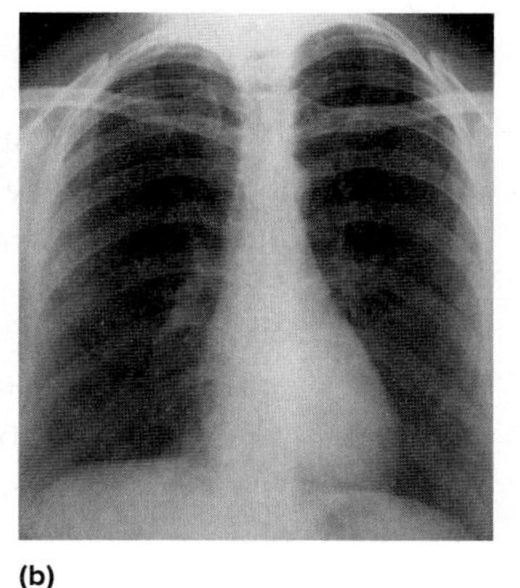
(b)

FIGURE 21.12 Chest X Ray Appearance in Pneumococcal Pneumonia **(a)** Pneumonia. The right lung (left side of figure) appears white because the alveoli are filled with fluid. **(b)** Normal X ray after recovery. a: ©Biophoto Associates/Science Source; b: ©Dick Luria/Science Source

? **Why does the sputum of a pneumonia patient become pinkish or rust-colored?**

TABLE 21.7 Pneumococcal, *Klebsiella*, and Mycoplasmal Pneumonias Compared

	Pneumococcal Pneumonia	*Klebsiella* Pneumonia	Mycoplasmal Pneumonia
Signs and symptoms	Cough, sudden chills and fever, shortness of breath, chest pain, cyanosis, rust-colored sputum from blood	Cough, repeated chills, fever, shortness of breath, chest pain, cyanosis, bloody jelly-like sputum	Gradual onset of dry cough, fever, fatigue, headache, and muscle aches
Incubation period	1 to 3 days	1 to 3 days	2 to 3 weeks
Causative agent	*Streptococcus pneumoniae* (pneumococcus); encapsulated strains	*Klebsiella pneumoniae*, encapsulated enterobacterium	*Mycoplasma pneumoniae;* lacks cell wall
Pathogenesis	Inhalation of encapsulated pneumococci; colonization of alveoli triggers an inflammatory response; fluid and inflammatory cells fill the alveoli.	Inhalation of colonized mucus droplets from the throat. Destruction of lung tissue and abscess formation common; infection spreads via blood to other body tissues.	Inhalation of infected droplets. Bacterial cells attach to receptors on respiratory epithelium; inhibition of ciliary motion and destruction of cells.
Epidemiology	High carrier rates for *S. pneumoniae*. Risk of pneumonia increases with conditions such as alcoholism, narcotic use, and viral infections that impair the mucociliary escalator. Other risk factors are chronic heart or lung disease, diabetes, and cancer.	*Klebsiella* species and other Gram-negative rods are common causes of fatal healthcare-associated pneumonias. Often resistant to antibiotics.	Mild infections are common; infected people spread the disease.
Treatment and prevention	Treatment: antibiotics. Prevention: PCV13: conjugate vaccine against 13 serotypes; PPSV23: polysaccharide vaccine against 23 serotypes.	Treatment: a combination of antibiotics; resistance is a problem. Prevention: no vaccine available.	Treatment: antibiotics, excluding cell wall synthesis inhibitors. Prevention: no vaccine available; avoid crowding in schools and military facilities.

Klebsiella Pneumonia

Enterobacteria such as *Klebsiella* species and other Gram-negative rods can cause pneumonia, especially if host defenses are impaired. They are common hospital-acquired pathogens and cause most of the deaths from healthcare-associated infections. ◀◀ **enterobacteria,** ◀◀ **healthcare-associated infections**

Signs and Symptoms

The general signs and symptoms of *Klebsiella* pneumonia—cough, chills, shortness of breath, fever, chest pain, and cyanosis—are the same as those of pneumococcal pneumonia and also appear after an incubation of 1 to 3 days. *Klebsiella* pneumonia patients typically have repeated chills, however, and they produce thick, bloody, jelly-like sputum.

Causative Agent

Klebsiella pneumoniae is a Gram-negative rod with a large capsule that produces big, noticeably mucoid colonies when grown on agar (**figure 21.13**). It is commonly part of the normal microbiota of the gastrointestinal tract and may be found in the mouth or the throat.

Pathogenesis

Klebsiella pneumoniae is contracted through secretions transmitted by person-to-person contact, or from medical equipment such as ventilators. Organisms first colonize the throat and gain access to the lung via inhaled air or mucus. Specific adhesins aid colonization. The capsule is an essential virulence factor, probably interfering with the action of complement system component C3b. In addition, *Klebsiella* produces a specific siderophore with a very high affinity for iron, allowing it to "steal" the essential element from host cells. Recent studies indicate that iron depletion causes a type of cellular stress that induces inflammation and enhances spread of the bacterium. *Klebsiella* pneumonia results in

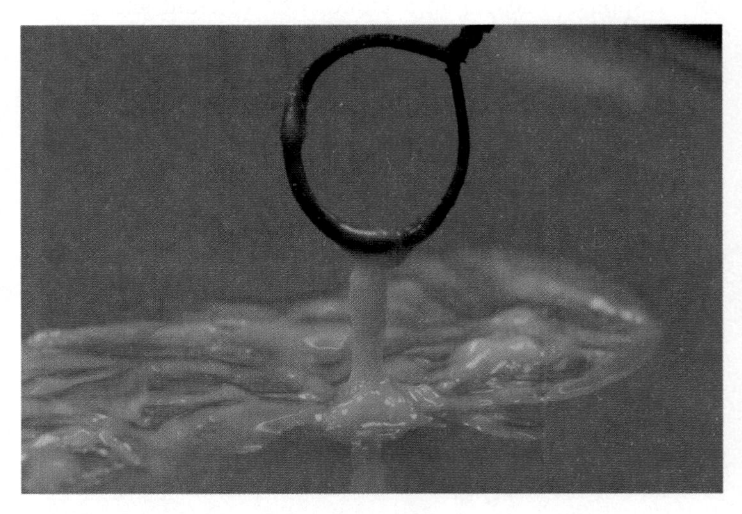

FIGURE 21.13 ***Klebsiella pneumoniae*** Growth on an agar plate. Notice the mucoid nature of the bacterial growth. ©McGraw-Hill Education/Lisa Burgess

? **Why is *Klebsiella* pneumonia often fatal?**

more tissue damage than pneumococcal pneumonia causes, as well as in rapid formation of lung abscesses. Therefore, even with effective antibacterial medication, the lung can be permanently damaged and the patient may die. Infection often spreads to the bloodstream, causing abscesses in other tissues such as the liver and the brain, and also causing septic shock. ⏮ complement C3b, ⏭ abscess, ⏭ septic shock, ⏮ siderophore

Epidemiology

Klebsiella species are widespread in nature. In humans, they are commonly part of the normal microbiota. Typically, people who contract *Klebsiella* pneumonia are very old, are very young, or have a compromised immune system (such as alcoholics, or those in a hospital or other institutional setting). The strains that circulate in hospitals and nursing homes are frequently resistant to antimicrobial medications and are increasingly multi-drug-resistant.

Treatment and Prevention

Klebsiella pneumonia is treated with antibiotics. Drug sensitivity tests must be carried out to determine which medications should be used. In seriously ill patients, immediate combination antibiotic therapy is given. Unfortunately, some of the medications used in these combinations, such as aminoglycosides, have severe side effects. Surgery may be required to drain abscesses.

Treatment of *Klebsiella* infections is becoming very challenging because the strains are developing increasing antibiotic resistance. *Klebsiella* species commonly produce a plasmid-encoded β-lactamase, an enzyme that makes the bacteria resistant to certain β-lactam antibiotics (primarily the penicillins). Many strains also produce an extended-spectrum β-lactamase (ESBL), which makes them resistant to many of the cephalosporins as well. Some strains produce versatile β-lactamases called carbapenemases, which makes them resistant to carbapenems and other β-lactam drugs. These strains, referred to as CRKP (carbapenem-resistant *K. pneumoniae*), are included in the CRE (carbapenem-resistant *Enterobacteriaceae*), which the CDC lists as an urgent threat with respect to antimicrobial resistance (see table 20.2). Unfortunately, carbapenems are usually given as a last resort, so very few effective treatment choices remain for CRKP infections. The case-fatality rate for CRKP pneumonia, even with treatment, is as high as 50%, and patients tend to die more quickly than other pneumonia patients. ⏮ plasmid, ⏮ β-lactamase

There are no specific preventive measures such as vaccination for *Klebsiella* pneumonia. To prevent spread between patients, healthcare workers must follow infection control measures such as wearing gloves and gowns when in a room with a *Klebsiella* patient, and washing hands. Disinfecting the environment, using sterile respiratory equipment, and using antimicrobial medications only when necessary help control the organisms and their development of resistance in hospitals. See table 21.7 for a description of the main features of this disease.

Mycoplasmal Pneumonia ("Walking Pneumonia")

Mycoplasmal pneumonia is the most common kind of pneumonia in children and young adults. The risk is greatest in crowded conditions that might occur in a college dormitory or military barracks. The disease is generally mild (as reflected by its popular name "walking pneumonia") and seldom requires hospitalization. Because it does not show the severe signs and symptoms of most other pneumonias, it is also called atypical pneumonia.

Signs and Symptoms

The onset of mycoplasmal pneumonia is typically gradual, as long as 2 to 3 weeks after infection. The first symptoms are sore throat, chills, fever, headache, muscle pain, and fatigue. After several days, a dry cough begins, but mucoid sputum may be produced later. Some people also develop otitis media.

Causative Agent

Mycoplasmal pneumonia is caused by *Mycoplasma pneumoniae,* a small bacterium with no cell wall (see figure 3.36). It grows slowly and is aerobic. Like other mycoplasmas, *M. pneumoniae* colonies on agar look like fried eggs (**figure 21.14**). ⏮ mycoplasmas

Pathogenesis

Only a few inhaled *M. pneumoniae* cells are required to start an infection. Cells attach by means of adhesion proteins to

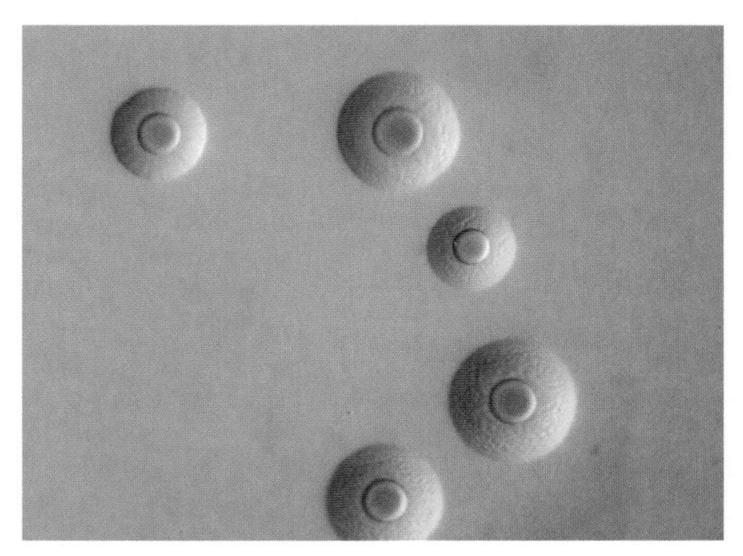

FIGURE 21.14 ***Mycoplasma pneumoniae*** **Colonies** The tiny colonies each have a dense center, giving them a "fried egg" appearance. ©Christine Citti

? Why is penicillin ineffective in treating mycoplasmal pneumonia?

specific receptors on the respiratory epithelium. There, they interfere with ciliary action and cause the ciliated cells to slough off. Damage to the mucociliary escalator makes the patient more vulnerable to secondary bacterial infections. An inflammatory response characterized by accumulation of lymphocytes and phagocytes causes the walls of the bronchial tubes and alveoli to thicken.

Epidemiology

Mycoplasma pneumoniae is spread by aerosolized droplets of respiratory secretions. The organisms are shed in these secretions for a long time period, ranging from about 1 week before symptoms begin to many weeks afterward, thereby increasing the likelihood of transmission. Mycoplasmal pneumonia accounts for about one-fifth of bacterial pneumonias, and has a peak incidence in young people. Immunity after recovery is not permanent, and repeat infections have occurred within 5 years.

Treatment and Prevention

Mycoplasma pneumoniae lacks a cell wall, so antibiotics that act against bacterial cell wall synthesis are not effective. Macrolides shorten the illness if given early.

No practical preventive measures exist for mycoplasmal pneumonia, except avoiding crowding in schools and military facilities. See table 21.7 for a description of the main features of this disease.

Pertussis ("Whooping Cough")

Whooping cough is the common name for pertussis (*per* means "intensive"; *tussis* means "cough"). This vaccine-preventable disease is still endemic in many countries, including developed nations such as the United States. Worldwide, the disease causes up to half a million deaths yearly.

Signs and Symptoms

Pertussis has an incubation period of 1 to 2 weeks and has three stages:

- **Catarrhal stage** (meaning inflammation of the mucous membranes). This typically lasts 1 to 2 weeks, with signs and symptoms that resemble an upper respiratory tract infection—runny nose, sneezing, low fever, and mild cough.
- **Paroxysmal stage** (meaning repeated sudden attacks). This lasts for about 2 to 4 weeks, or longer, and includes frequent bursts of violent, uncontrollable coughing. The cough is dry, but is severe enough to burst small blood vessels in the eyes. The tongue protrudes and the neck veins stand out. The coughing spasm is followed by forceful attempts to inhale, causing the characteristic "whoop" of this disease. Vomiting and seizures can occur during this stage and the patient may also become cyanotic (blue from lack of O_2).
- **Convalescent stage** (meaning recovery). During this stage the person is no longer contagious. The coughing attacks gradually become less frequent, and the person slowly recovers over a period of several weeks.

Causative Agent

Whooping cough is caused by *Bordetella pertussis,* a tiny, encapsulated, strictly aerobic, Gram-negative rod. The cells are sensitive to drying and sunlight, and die quickly outside the host.

Pathogenesis

When *Bordetella pertussis* is inhaled, it attaches specifically to ciliated cells of the respiratory epithelium. Attachment is aided by two required colonization factors: (1) filamentous hemagglutinin (FHA), a pilus that extends from the bacterial surface; and (2) fimbriae. The areas colonized by *B. pertussis* include the upper throat, trachea, bronchi, and bronchioles. The organisms grow in dense masses on the epithelial surface, but generally do not invade the cells. Instead, they release three toxins that play critical roles in the disease process.

Pertussis toxin (PT) is an A-B exotoxin (**figure 21.15**). The B subunits attach to receptors on the host cell surface, allowing the A subunit to move through the cytoplasmic

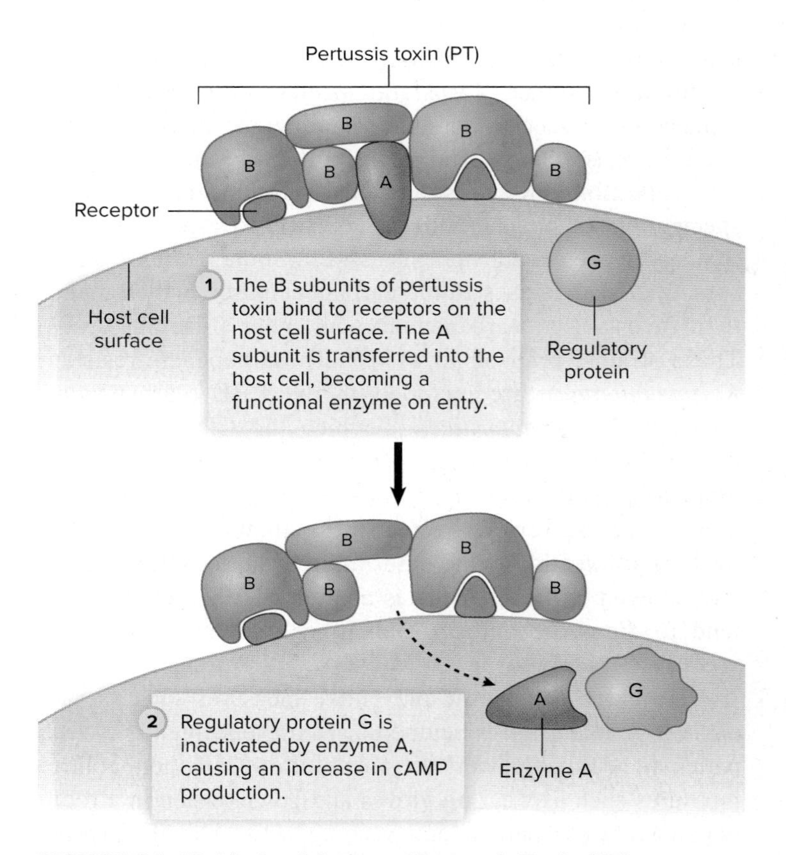

FIGURE 21.15 Mode of Action of Pertussis Toxin (PT)

What is the result of increased cAMP production?

membrane. As it does so, it activates a membrane-bound regulatory protein that controls production of cyclic adenosine monophosphate (cAMP), leading to increased production of this molecule. High levels of cAMP interfere with cell signaling pathways, resulting in significant increase in mucus output, decreased killing ability of phagocytes, massive release of lymphocytes into the bloodstream, ineffectiveness of natural killer cells, and low blood sugar. ⏮ A-B toxin, ⏭ cAMP

Adenylate cyclase toxin (ACT) is both a membrane-damaging toxin and an enzyme. This toxin reduces phagocytosis by causing lysis of accumulating leukocytes. Inside the cell, it also catalyzes the reaction that converts ATP to cAMP.

Tracheal cytotoxin (TCT) is a fragment of peptidoglycan that *B. pertussis* releases during growth, and it causes host cells to release a fever-inducing cytokine (interleukin-1; IL-1). It is also toxic to ciliated epithelial cells, causing them to die and slough off, resulting in a rapid decline in ciliary action. ⏮ interleukin-1

Increased mucus production along with decreased ciliary action results in the severe cough that characterizes pertussis because only the cough reflex remains for clearing secretions from the lungs. Some bronchioles are completely obstructed by mucus, resulting in small areas of collapsed lung. Spasms or partial mucus plugs in other bronchioles let air enter but not escape. Pneumonia due to *B. pertussis* or, more commonly, secondary bacterial infection may cause death.

Epidemiology

Pertussis is highly contagious, spread via respiratory secretions suspended in air. Patients are most infectious during the catarrhal stage, when large numbers of organisms are present in mucus secretions. Pertussis is classically a disease of infants, and most fatal cases occur in infants under 1 year of age. A milder form of the disease occurs in older children and adults, and may be mistaken for a persistent cold, asthma, or bronchitis. These infected people may unknowingly serve as a source of infection for high-risk populations (such as infants that are not yet vaccinated), potentially leading to outbreaks of severe disease. Despite the availability of a successful vaccine, the reported incidence of pertussis is steadily increasing. Possible factors contributing to this trend include better surveillance and diagnostic techniques, suboptimal vaccination (especially in developing nations), decreasing immunity in vaccinated people, and development of *B. pertussis* strains for which the current vaccine is not effective.

Treatment and Prevention

Macrolides such as erythromycin taken during the catarrhal stage usually eliminate *B. pertussis* from the respiratory secretions and limit its spread. Antibiotics are ineffective once the paroxysmal stage is reached. Pertussis in infants can be life-threatening; severe cases often require intensive supportive therapy **(figure 21.16).**

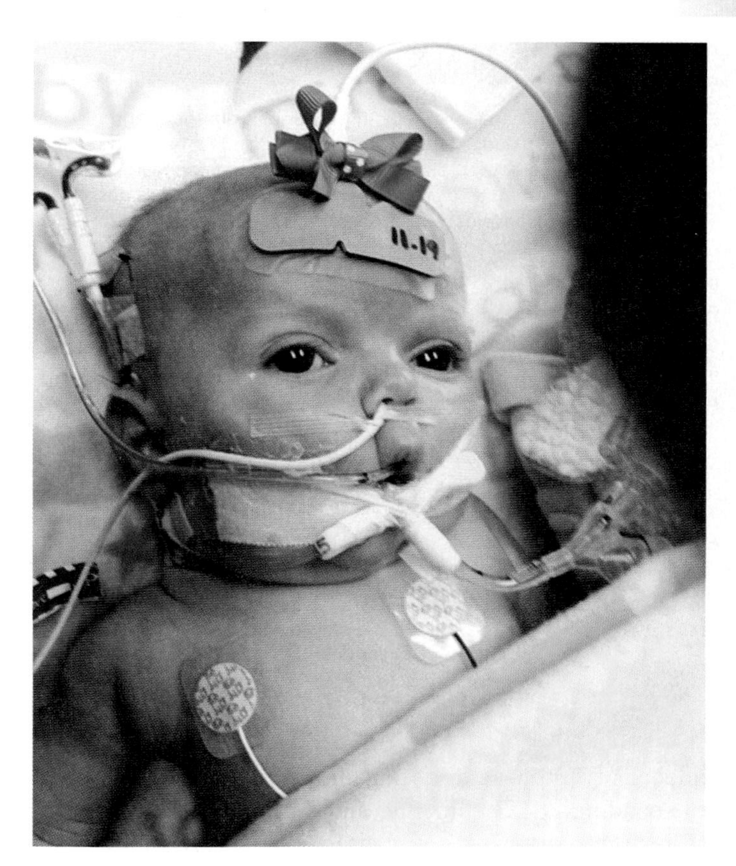

FIGURE 21.16 Infant with Severe Pertussis This baby's blood is being oxygenated outside the body while the lungs are impaired. Source: CDC

? What is the effect of *B. pertussis* on ciliated respiratory epithelial cells?

Pertussis is effectively prevented with a vaccine. The original vaccine, composed of whole *B. pertussis* cells, produced a dramatic decrease in pertussis cases, but sometimes caused severe side effects. It has been replaced with a newer acellular (subunit) pertussis vaccine (aP), that includes only part of the bacterium instead of whole cells. It is given in combination with diphtheria and tetanus toxoids—a grouping referred to as the DTaP vaccine—at 2, 4, 6, and 15 to 18 months, and again at 4 years old. Evidence indicates that immunity induced by the acellular vaccine decreases over time. For this reason, adolescents should receive a booster with a lower dose of pertussis antigen, which is given in combination with a tetanus and diphtheria booster (a vaccine referred to as Tdap). That same booster is given to women in their third trimester of pregnancy as a means to provide passive immunity to the newborn. The main features of pertussis are shown in **table 21.8.** ⏮ subunit vaccine

Tuberculosis ("TB")

Tuberculosis (TB) was once a very common disease, but the number of cases gradually declined in developed countries as living standards improved. In 1985, however, the

TABLE 21.8 Pertussis ("Whooping Cough")

Signs and symptoms	Characterized by three stages: catarrhal stage includes a runny nose, cough, and fever; paroxysmal stage consists of spasms of violent coughing, sometimes leading to vomiting and convulsions; convalescent stage is indicated by less frequent coughing as the person recovers.
Incubation period	1 to 2 weeks
Causative agent	*Bordetella pertussis,* a tiny Gram-negative rod
Pathogenesis	Colonization of the surfaces of the upper respiratory tract; ciliary action slowed; toxins released by *B. pertussis* cause death of epithelial cells and increased cAMP; fever, excessive mucus output, and a rise in the number of lymphocytes result.
Epidemiology	Inhalation of infected droplets; older children and adults have mild symptoms.
Treatment and prevention	Treatment: certain antibiotics given before coughing spasms start. Prevention: acellular vaccine (DTaP) for immunization of infants and children; Tdap booster for adolescents and pregnant women.

incidence of TB began to rise again, the trend associated with the expanding AIDS epidemic and increasing prevalence of drug-resistant strains of the causative agent (**figure 21.17**). In response, the CDC developed a plan to increase efforts in identifying and treating cases among high-risk groups, particularly people living in poverty, people with AIDS, prisoners, and immigrants from countries with high rates of TB. By 1993, the incidence began to decrease; by 2013, incidence had leveled off at about 3.0 cases per 100,000 population. ⏭ AIDS

MicroByte

One-third of the global population is believed to be infected with *Mycobacterium tuberculosis.*

Signs and Symptoms

The initial infection with *Mycobacterium tuberculosis* typically results in an asymptomatic lung infection. The immune response generally controls this primary infection but is not able to eliminate it entirely. The person is left healthy but with a latent infection, known as **latent tuberculosis infection (LTBI).** Later in life, the person may develop **tuberculosis disease (TB disease),** also called active tuberculosis. TB disease is a chronic illness characterized by slight fever, progressive weight loss, night sweating, and persistent cough, often producing blood-streaked sputum. Some people, especially children or those with compromised immune systems, may develop TB disease upon initial infection. *M. tuberculosis* mainly infects the lung, but if the bacterial cells get into the bloodstream, they may spread to cause disease in other tissues, including kidney, bone, joints, and the central nervous system. This disseminated disease is called miliary TB because the lesions caused by the bacteria resemble millet seeds (grain) in the tissues. ⏮ latent infection

Causative Agent

Tuberculosis is caused by *Mycobacterium tuberculosis,* commonly called the tubercle bacillus. The organism is a slender, acid-fast, rod-shaped bacterium (**figure 21.18**). It is a strict aerobe that grows very slowly, with a generation time of over 16 hours, making rapid diagnosis of TB difficult. The organism has an unusual cell wall that contains a large amount of complex glycolipids called mycolic acids. These make it unusually resistant to drying, disinfectants, and strong acids and alkali, although it is easily killed by pasteurization. The mycolic acids are also responsible for its acid-fast staining. A group of related mycobacteria, called *Mycobacterium avium* complex (MAC)—consisting of several strains of two closely related species (*M. avium* and *M. intracellulare*)—can also cause opportunistic disease in certain people. ⏮ acid-fast staining

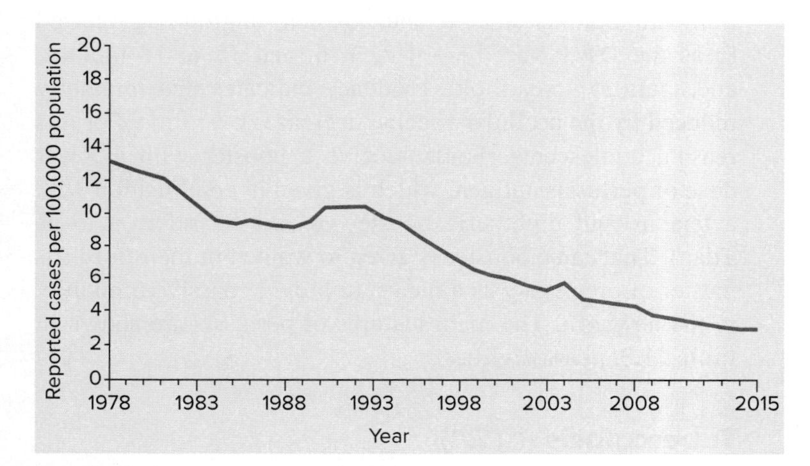

FIGURE 21.17 Incidence of TB Disease, United States, 1978–2015

What caused the rise in TB disease incidence between 1988 and 1993?

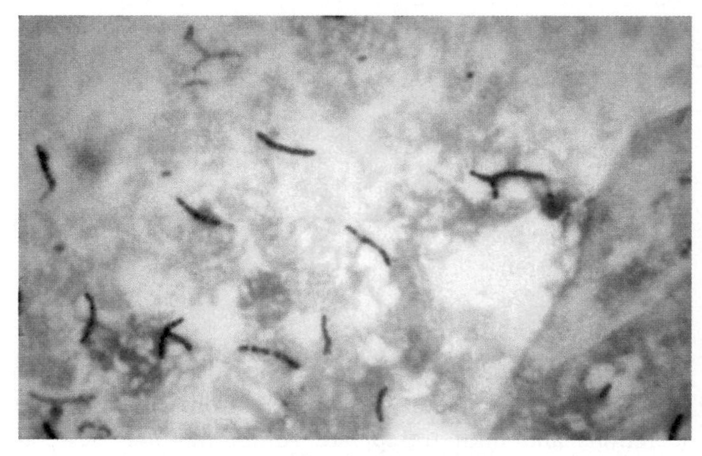

FIGURE 21.18 *Mycobacterium tuberculosis* The cells appear reddish with acid-fast stain. Source: Dr. George P. Kubica/CDC

Why is *M. tuberculosis* so resistant to drying?

Pathogenesis

When airborne *M. tuberculosis* cells are inhaled, they may enter the lungs. Alveolar macrophages quickly engulf the bacteria but are unable to destroy them because the mycolic acids in the bacterial cell wall prevent fusion of the phagosome with lysosomes. The bacteria leave the phagosome and multiply within the cytoplasm of the macrophages (**figure 21.19** ①).

The multiplying bacilli cause an inflammatory response, recruiting more macrophages to the site, thereby providing *M. tuberculosis* with additional host cells in which to multiply. Some of the macrophages fuse together to form giant multinucleated cells. Others are induced by the bacteria to accumulate large numbers of oil droplets, becoming foamy macrophages. The lipids in foamy macrophages are thought to help the bacteria survive within the cells. ② Lymphocytes collect around the macrophages, walling off the infected area from the surrounding tissue. This localized collection of inflammatory cells—a granuloma—is the body's characteristic response to microorganisms and other foreign substances that resist destruction and removal by phagocytosis. The granulomas of tuberculosis are called **tubercles.** ③ Within the tubercle, effector helper T cells release cytokines that activate macrophages to destroy the bacteria infecting them. At this time, a fibrous layer forms around the macrophages, keeping the lymphocytes outside the tubercle. This tissue calcifies and can be seen on X rays as Ghon foci. If the adjacent lymph nodes are involved, the focus is called a Ghon complex. Some of the mycobacteria in the calcifying tubercles survive, but they are prevented from multiplying by conditions in the tubercle, including low pH and low available O_2. The bacteria remain in this state for many years, causing a latent TB infection (LTBI). People with LTBI are asymptomatic and noninfectious. In many cases, the infection resolves. ◀◀ granuloma

TB disease results if the inflammatory response cannot contain or destroy the mycobacteria. This can occur during primary infection but can also happen in a person with LTBI—the infection reactivates (reactivation TB) if the person's immunity becomes impaired by stress, advanced age, or disease such as AIDS. ④ Within the tubercle, macrophages containing mycobacteria die, releasing bacteria, enzymes, and cytokines. An area of necrosis (death) is formed in the

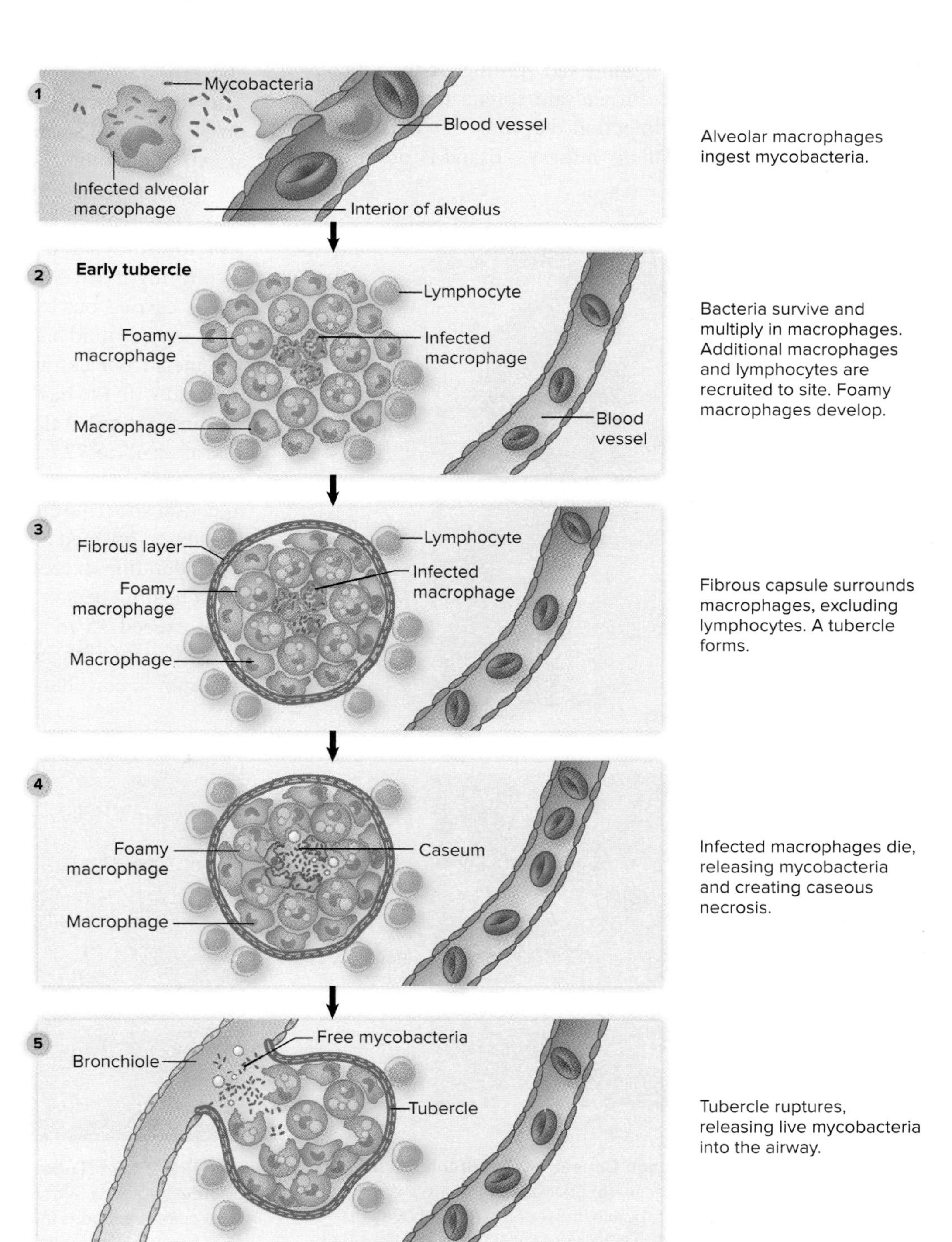

FIGURE 21.19 Pathogenesis of Tuberculosis

? Why is prolonged treatment necessary for tuberculosis?

center of the tubercle; this has the texture of soft cheese and is referred to as caseous necrosis. It is thought that foamy macrophages play an important role in necrosis formation and that the caseum contains lipids from these cells. ⑤ The tubercle then ruptures, releasing the bacteria and dead material into the airways. This causes a large lung defect called a tuberculous cavity that spreads the bacteria to other parts of the lung (**figure 21.20**). Lung cavities characteristically persist, slowly enlarging for months or years and shedding bacterial cells into the bronchi. The organisms can then be transmitted to other people by coughing and spitting. If the tubercle bacilli enter the bloodstream and are spread throughout the body, they cause foci of infection that look like small white granules in the tissues. This is miliary TB and is often fatal, despite treatment.

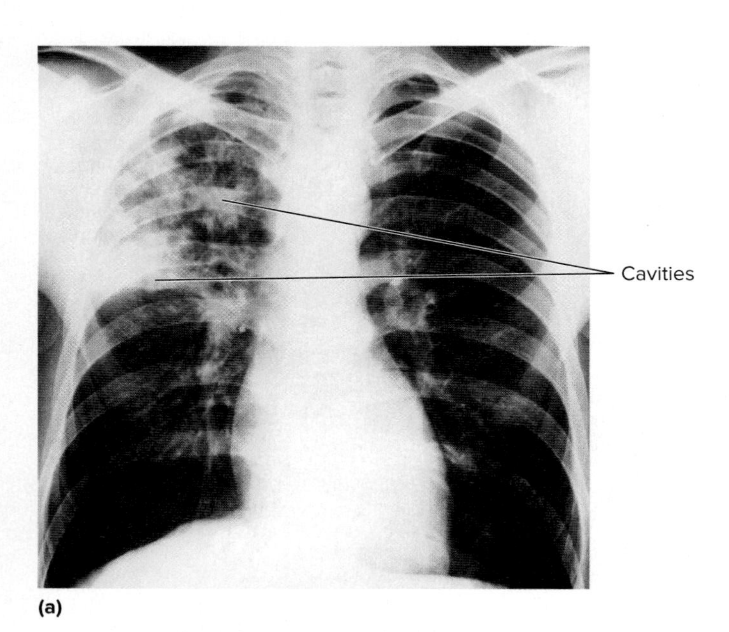

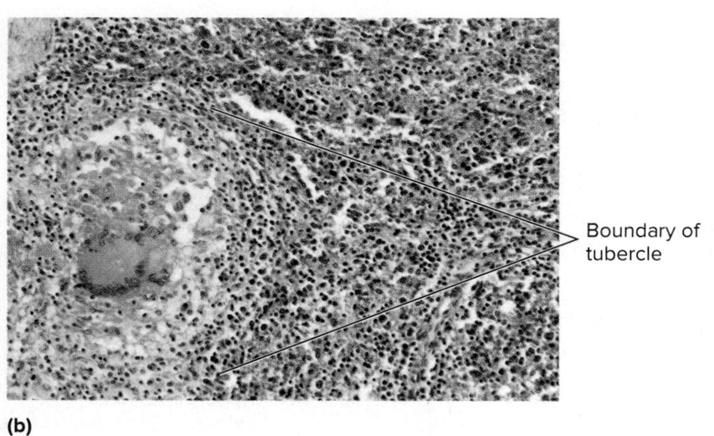

FIGURE 21.20 Lung Damage Caused by Tuberculosis (a) Chest X ray of a person with TB disease. **(b)** Stained lung tissue showing tubercle. In the center of the tubercle, most of the nuclei have disappeared because the cells are dead and the tissue has begun to liquefy. a: ©Walter Dawn/Science Source; b: ©Steve Gschmeissner/Science Source

? What is a tubercle?

Epidemiology

An estimated 13 million Americans have LTBI, but only 5–10% of latent infections will reactivate to cause active tuberculosis. Foreign-born U.S. residents have much higher incidence of LTBI than those born in the United States. Transmission of *M. tuberculosis* occurs almost entirely by the respiratory route; 10 or fewer inhaled cells are enough to cause infection if they reach the alveoli. Factors important in transmission include the frequency of coughing, the adequacy of ventilation (transmission is unlikely to occur outdoors), and the degree of crowding. Immunodeficiency increases activation of *M. tuberculosis* in those with LTBI, a significant problem in AIDS patients. TB disease in a person with HIV is an AIDS-defining condition.

The tuberculin skin test (TST), also known as the Mantoux (pronounced man-too) test, is an extremely important tool for studying the epidemiology of the disease and for identifying those who are infected with *M. tuberculosis.* The test is carried out by injecting into the skin a small amount of a sterile fluid called purified protein derivative (PPD), obtained from cultures of *M. tuberculosis.* People who are infected with the bacterium develop redness and a firm swelling (induration) at the injection site that reaches a peak intensity after 48 to 72 hours (**figure 21.21**). This reaction is due to the accumulation of macrophages and T lymphocytes at the injection site, caused by a delayed-type hypersensitivity reaction to the injected antigens. The test is considered positive if the induration exceeds a certain diameter within three days. For a healthy person with no risk factors for TB, the diameter must exceed 15 mm to be considered a positive test. In an AIDS patient with suppressed immunity, however, a diameter of 5 mm is considered positive. A positive reaction does not

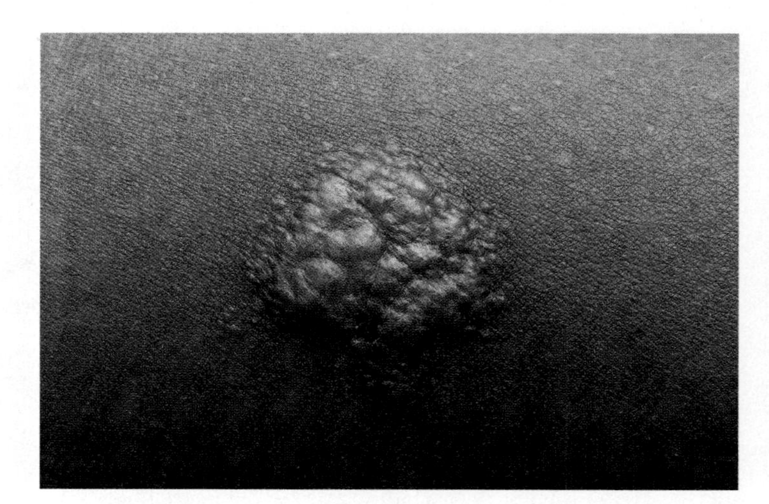

FIGURE 21.21 Tuberculin Skin Test This positive test is caused by a delayed-type hypersensitivity reaction to *Mycobacterium tuberculosis* antigens injected into the skin. ©BrazilPhotos.com/Alamy Stock Photo

? Explain why a tuberculin skin test on an AIDS patient with tuberculosis may show a negative result.

necessarily mean that the person has TB disease, only that the person may have been infected by *M. tuberculosis* at some time in the past; they could have either LTBI or TB disease. ⏮ delayed-type hypersensitivity

Besides the tuberculin skin test, blood tests are available for diagnosing *M. tuberculosis* infection. Interferon gamma–release assays (IGRAs) detect interferon gamma produced in response to *M. tuberculosis.* IGRAs generally give results similar to TST but more quickly and with greater specificity. This is the preferred type of test for people who have received a TB vaccine called the BCG vaccine (next section), or who cannot return for a TST follow up. ⏮ interferon

A newer assay called the Xpert MTB/RIF is used to detect the DNA of *M. tuberculosis* and closely related species, as well as genes for resistance to the antibiotic rifampicin, a first-line medication used to treat TB disease. Nucleic acid amplification methods are fast, sensitive, and accurate, and are used on clinical specimens. Continued research is already producing additional rapid diagnostic assays. ⏮ nucleic acid amplification tests

Treatment and Prevention

Combination therapy in which multiple antimicrobial medications are given at the same time is typically used in the treatment of tuberculosis. This is done because people with TB disease have high numbers of *M. tuberculosis* cells, so there is a strong likelihood that some of the cells will have acquired resistance to a given drug by spontaneous mutation. A cell is much less likely to develop resistance to multiple drugs given simultaneously. Treatment must be continued for months to clear the infection because of the long generation time of *M. tuberculosis* and its resistance to destruction by body defenses. ⏮ combination therapy, ⏮ antibiotic resistance

The most effective and least toxic first-line medications for treating TB disease include rifampin (RIF), isoniazid (INH), pyrazinamide (PZA), and ethambutol (EMB), which are bactericidal against actively growing organisms. Some kill metabolically inactive intracellular organisms as well. In the initial treatment phase of TB disease, all four first-line drugs are given for 8 weeks. After that, INH and RIF are given in combination for another 18 weeks. ⏮ antibacterial medications effective against *Mycobacterium tuberculosis*

Strains resistant to one or more first-line medications often evolve when people fail to comply with the complex treatment regime. What commonly happens is that once a person feels better and the symptoms of TB disease disappear, he or she becomes careless about continuing to take the prescribed medications. To combat this problem, **DOTS (directly observed therapy short-course)** may be used; healthcare workers who supply the medications watch the patients swallow the prescribed tablets. DOTS implementation has been enormously successful in improving national TB control programs and in helping control TB globally. The global TB mortality rate decreased by 47% between 1990 and 2015. In 2006, the World Health Organization (WHO) devised the Global Plan to Stop TB strategy, which builds on previous DOTS successes and addresses the current key challenges in TB control.

Despite the success of DOTS, the problem of drug-resistant *M. tuberculosis* strains has nevertheless reached alarming proportions in some areas. During the 1990s, **multi-drug-resistant TB (MDR-TB)** became an increasing problem. MDR-TB is caused by *M. tuberculosis* strains resistant to rifampin and isoniazid—the two most effective first-line drugs available—and therefore must be treated with less effective, more toxic (causing liver damage), and more expensive second-line medications. By the end of the decade, **extensively drug-resistant TB (XDR-TB)** caused by strains that resist first-line as well as many of the second-line drugs appeared. These antibiotic-resistant *M. tuberculosis* strains now threaten tuberculosis control efforts around the world. Fortunately, promising new anti-tuberculosis medications are being developed. In 2012, the FDA approved a new medication called bedaquiline for treating MDR-TB. The approval was "fast-tracked," a system used by the FDA to approve medications for life-threatening diseases that have new treatment options. ⏮ MDR-TB, ⏮ XDR-TB

In the United States, TB prevention involves identifying unsuspected cases using skin or blood tests and lung X rays. People with active disease are then treated, thereby interrupting the spread of the causative agent. People who have LTBI are also treated, reducing the risk of developing TB disease later in life. The preferred treatment for LTBI is isoniazid for 9 months.

The CDC collects and analyzes a significant amount of TB data in an effort to track and control the disease. Individual TB cases are tracked, for example, to measure the performance of preventive and control measures, including evidence of resistance. Incoming information is shared about suspected TB outbreaks so that they can be managed more efficiently.

Prevention and control of TB are global challenges and many countries lack the resources to track and treat both TB disease and LTBI. Vaccination against TB has been widely used in many countries. The vaccine used, BCG (Bacille Calmette-Guérin), is a live attenuated vaccine derived from *M. bovis,* a cattle-infecting species that has little virulence in humans. Although this vaccine prevents childhood TB disease, it appears ineffective in preventing LTBI, which can later reactivate. Use of the vaccine is discouraged in the United States, because people who receive it usually develop a positive tuberculin skin test. By causing a positive test, this vaccination eliminates an important way of diagnosing TB early in the disease when it can most easily be treated. BCG is also not safe to use in severely immunocompromised patients. Several new genetically engineered vaccines are being developed, many of which are currently being tested in TB-endemic regions. The main features of tuberculosis are shown in **table 21.9.**

TABLE 21.9 Tuberculosis ("TB")

1. Airborne *Mycobacterium tuberculosis* cells are inhaled and lodge in the lungs.
2. The bacteria are phagocytized by lung macrophages and multiply within them, protected by lipid-containing cell walls and other mechanisms.
3. Infected macrophages are carried to various parts of the body such as the kidneys, brain, lungs, and lymph nodes; release of *M. tuberculosis* occurs.
4. Delayed-type hypersensitivity develops; wherever *M. tuberculosis* has lodged, an intense inflammatory reaction develops.
5. The bacteria are surrounded by macrophages and lymphocytes; growth of the bacteria ceases.
6. Intense inflammatory reaction and release of enzymes can cause caseous necrosis and cavity formation.
7. With uncontrolled or reactivated infection, *M. tuberculosis* exits the body through the mouth with coughing.

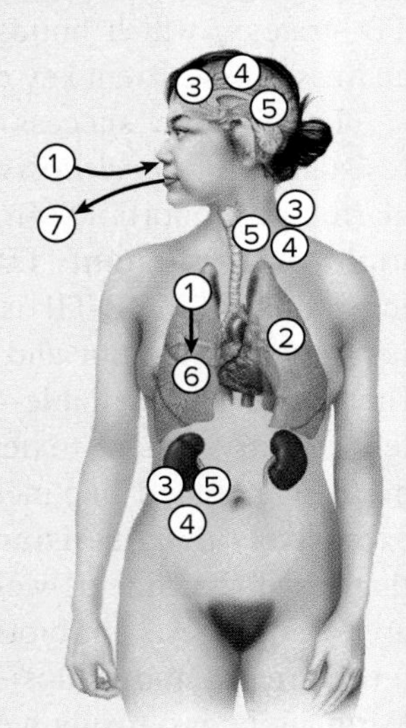

Signs and symptoms	Chronic fever, weight loss, cough, sputum production
Incubation period	2 to 10 weeks
Causative agent	*Mycobacterium tuberculosis;* unusual cell wall with high lipid content
Pathogenesis	Colonization of alveoli causes inflammatory response; ingestion by macrophages follows; organisms survive in macrophages and are carried to lymph nodes, lungs, and other body tissues; infecting bacteria multiply; granulomas form.
Epidemiology	Inhalation of airborne organisms; latent infections can reactivate.
Treatment and prevention	Treatment: lengthy process requiring four and then two anti-TB medications; DOTS. Prevention: tuberculin (Mantoux) skin test, interferon-gamma release assays (IGRAs), and rapid tests for detection of infection allows early treatment of active TB; treatment of latent infections.

Legionellosis ("Legionnaires' Disease")

Legionellosis (also called Legionnaires' disease) was unknown until 1976, when a number of people attending an American Legion convention in Philadelphia developed a mysterious pneumonia that was fatal in many cases. Months of scientific investigation eventually paid off when the cause was discovered to be a previously unknown bacterium commonly present in the natural environment.

Signs and Symptoms

Legionellosis typically begins after an incubation of 2 to 10 days with headache, muscle aches, high fever, confusion, and shaking chills. A dry cough develops that later produces small amounts of sputum, sometimes containing blood. Shortness of breath is common. Pleurisy can also occur. About 25% of the cases also have some digestive tract symptoms such as diarrhea, abdominal pain, and vomiting. Recovery is slow, and weakness and fatigue last for weeks.

Causative Agent

Legionellosis is caused by *Legionella pneumophila,* a fastidious Gram-negative rod that requires a special medium for laboratory culture. The organism stains poorly with conventional dyes, but it can be detected using immunofluorescence (**figure 21.22**). *L. pneumophila* is a facultative intracellular parasite and survives well in freshwater protozoa such as *Acanthamoeba.* These protozoa form cysts during adverse environmental conditions, allowing the bacteria within them to survive. *Legionella* also persists in biofilms—if the biofilm is disturbed, huge numbers of *Legionella* are released into the water.

Pathogenesis

Legionella pneumophila is acquired by inhaling aerosolized water contaminated with the organism. Healthy people are quite resistant to infection, but smokers and those with impaired host defenses are susceptible.

The bacteria lodge in and near the alveoli of the lung, where they promote their uptake by alveolar macrophages.

FIGURE 21.22 ***Legionella pneumophila*** **Stained with Fluorescent Antibody** The bacterium does not stain well with common methods used to examine tissue or sputum. Source: Dr. William Cherry/CDC

How do the organism's staining characteristics and fastidious nature help explain why it was unrecognized as a cause of disease for so long?

One of their surface proteins, macrophage invasion potentiator (Mip), aids entry into the macrophages. The bacterial cells also bind complement component C3b, an opsonin. Once ingested by the macrophages, the bacteria survive by preventing phagosome-lysosome fusion. They also manipulate other events within the phagocyte, creating an environment in which they can multiply. The host cell eventually dies, releasing bacterial cells that can then infect other tissues. Necrosis (tissue death) of alveolar cells and an inflammatory response result, causing multiple small abscesses, pneumonia, and pleurisy. Bacteremia is often present. Fatal respiratory failure (meaning the lungs can no longer adequately oxygenate the blood or expel CO_2) occurs in about 15% of hospitalized cases.

Epidemiology

Legionella pneumophila is widespread in warm natural waters containing other microorganisms such as protozoa, in which the bacteria live and multiply. The organism also survives well in the water systems of buildings, particularly in hot water systems, where chlorine levels are generally low. Increased chlorine levels sometimes fail to decontaminate water systems, probably because the *L. pneumophila* cells are protected inside protozoa. Legionellosis cases have originated from contaminated aerosols from large central air-conditioning systems, nebulizers, and sprays to freshen produce, and from showers and water faucets. Direct person-to-person spread, however, does not occur.

Treatment and Prevention

Legionellosis is typically treated with a macrolide or a fluoroquinolone. *L. pneumophila* produces a β-lactamase, which makes it resistant to many penicillins and some cephalosporins. Also, because the bacteria multiply inside the alveolar macrophages, medication must be able to accumulate within these cells to be effective, which β-lactam antibiotics do poorly. Some patients require O_2 therapy. ⏮ β-lactamase

Most efforts at control of legionellosis have focused on designing equipment to minimize the risk of infectious aerosols and on disinfecting procedures. The main features of legionellosis are given in **table 21.10.**

Inhalation Anthrax

Anthrax is primarily a disease of domestic livestock and other herbivores, but it also occurs in humans. It is of particular significance in the United States because of the 2001 domestic bioterrorism attack linked to mail containing purified *B. anthracis* spores. Eleven people developed inhalation anthrax as a result of the attack, and five of those died. In addition, 11 people developed cutaneous anthrax. The causative agent, *Bacillus anthracis,* is now listed as a Category A bioterrorism agent. ⏮ bioterrorism, ⏭ cutaneous anthrax

Signs and Symptoms

Inhalation anthrax typically begins with flu-like symptoms—discomfort, mild fever, and a nonproductive cough. These typically begin within a week after exposure, but may take up to 2 months to appear. Fever then increases within a few days, and shortness of breath and chest pain develop. Respiratory distress may lead to cyanosis as tissues are deprived of O_2. In about half the cases, anthrax meningitis develops, indicated by severe headache and evidence of blood when the cerebrospinal fluid is examined.

TABLE 21.10	Legionellosis ("Legionnaires' Disease")
Signs and symptoms	Muscle aches, headache, fever, cough, shortness of breath, chest and abdominal pain, diarrhea, vomiting
Incubation period	2 to 10 days
Causative agent	*Legionella pneumophila*, a Gram-negative bacterium that stains poorly
Pathogenesis	Organism multiplies within phagocytes; released with death of the cell; necrosis of cells lining the alveoli; inflammation and formation of multiple small abscesses.
Epidemiology	Originates mainly from warm water contaminated with protozoa, such as found in air-conditioning systems.
Treatment and prevention	Treatment: antibiotics. Prevention: avoiding contaminated water aerosols; regular cleaning and disinfection of humidifying devices.

Causative Agent

Bacillus anthracis is an endospore-forming, Gram-positive, non-hemolytic, non-motile, rod-shaped bacterium (**figure 21.23**). The spores can remain in the environment indefinitely. Vegetative cells produce a capsule that protects them from phagocytosis, but unlike most capsules, it is composed of an amino acid polymer (poly-D-glutamic acid) rather than a polysaccharide. ⏮ endospore

Pathogenesis

Once inhaled deep into the lungs, *B. anthracis* endospores are taken up by alveolar macrophages and carried to lymph nodes in the chest, where they germinate. In some cases, they may remain in the lung for up to 60 days before germinating. In general, clumps of endospores do not cause disease when inhaled, because particles larger than 5 μm are trapped and removed by the mucociliary escalator or other defenses of the upper respiratory tract before reaching the lungs.

Vegetative *B. anthracis* cells make three proteins—protective antigen (PA), edema factor (EF), and lethal factor (LF)— that function together as anthrax toxin to kill phagocytes and damage surrounding tissues. First, PA binds to host cells, forming a pore that allows EF or LF to enter the cells, where they disrupt critical cell functions. This results in pulmonary edema, and triggers a series of events that cause death of the cell.

The toxin-producing encapsulated *B. anthracis* cells multiply and spread into the bloodstream, overwhelming the body's defense mechanisms and damaging tissues. Without treatment, the inhalation anthrax case-fatality rate is near 95%; early treatment reduces it to about 45%.

Epidemiology

Anthrax is a zoonotic disease that can be transmitted to humans who work with animals or animal products, particularly in countries where *Bacillus anthracis* is prevalent. People who work in slaughterhouses or leather tanneries, for example, are most likely to be affected. Although the disease occurs naturally in the United States, it is extremely rare, so bioterrorism is suspected whenever a case occurs. Fortunately, anthrax is not transmitted from person to person.

Treatment and Prevention

Inhalation anthrax should be treated as soon as possible after infection, because the toxins released by vegetative bacteria can cause death within a few days. The antimicrobial medications doxycycline or ciprofloxacin are given along with a monoclonal antibody that binds to protective antigen (PA). The antibody-bound PA cannot attach to host cells, so LF and EF can no longer enter cells to cause further damage. Treatment may also include mechanical ventilation and drainage of fluid from the lungs.

If someone is suspected of having been exposed to *B. anthracis* endospores, as was the case with many postal workers in the 2001 attack mentioned earlier, then prophylactic (preventive) antimicrobial treatment is given for 60 days. That duration is important because inhaled endospores may remain in the lungs for some time before germinating. If administered before symptoms arise, antibiotic treatment generally prevents development of the disease.

An important control mechanism is to reduce infection in livestock, particularly through vaccination programs using an attenuated vaccine. That vaccine is not appropriate for humans because of the risks. An acellular (subunit) anthrax vaccine is available for people who are at an increased risk of contact with anthrax endospores, such as members of the U.S. military, people who handle certain animals or animal products, and some laboratory workers. The main features of inhalation anthrax are shown in **table 21.11.** ⏮ attenuated vaccine, ⏮ subunit vaccine

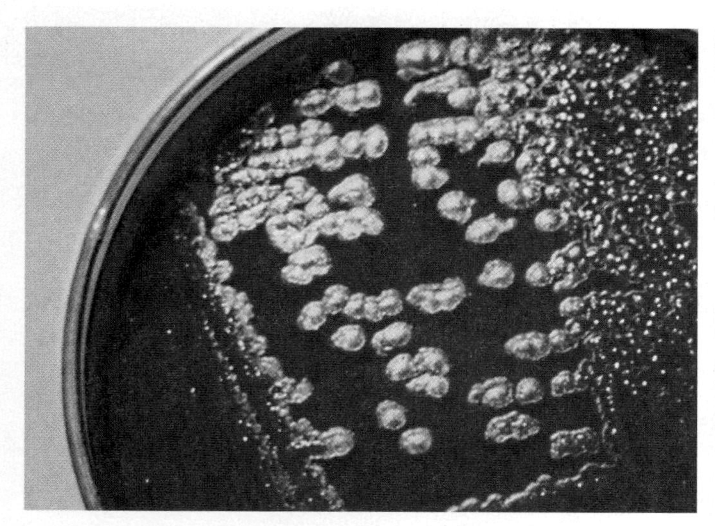

FIGURE 21.23 ***Bacillus anthracis*** Colonies grown on laboratory media have a rough surface, often with comma-shaped extensions. ©Scott Camazine/Science Source

? **What color would these organisms appear to be if stained with a Gram stain?**

TABLE 21.11	Inhalation Anthrax
Signs and symptoms	Fever, nonproductive cough; progressing to shortness of breath, chest pain; meningitis in about 50% of cases
Incubation period	3–60 days
Causative agent	*Bacillus anthracis,* an endospore-forming Gram-positive rod
Pathogenesis	Endospores engulfed by macrophages in lung germinate and produce toxins that kill macrophages, cause pulmonary edema, and damage respiratory cells.
Epidemiology	Rare zoonotic disease that may be transmitted by animals or animal products; no human transmission; possible agent of bioterrorism.
Treatment and prevention	Treatment: antimicrobial medications; monoclonal antibody that binds to PA. Prevention: vaccinate livestock; acellular vaccine given to at-risk groups and military.

MicroAssessment 21.4

Pneumococcal pneumonia is typically community-acquired in adults. Pneumonia due to *Klebsiella* species is mainly hospital-acquired and leads the causes of death from healthcare-associated infections. Symptoms of mycoplasmal pneumonia are usually relatively mild. Whooping cough (pertussis) is mainly a threat to infants but is commonly spread by adults; childhood immunization against the disease protects them, but immunity often does not persist to adulthood. Tuberculosis is a chronic disease; most infections become latent, posing the risk of reactivation throughout life. *Legionella pneumophila,* the bacterium that causes legionellosis, originates from water containing other microorganisms, where it can grow within protozoa. Inhalation anthrax is rare in the United States, but has been related to bioterrorism.

10. What structural feature of the *Streptococcus pneumoniae* cell is responsible for its virulence?

11. Outline the pathogenesis of tuberculosis.

12. Why is pertussis toxin not eliminated by the mucociliary escalator?

21.5 ■ Viral Infections of the Lower Respiratory System

Learning Outcomes

9. Describe antigenic drift and antigenic shift, and discuss how they affect the epidemiology of influenza.
10. Compare the distinctive characteristics of respiratory syncytial virus infection, hantavirus pulmonary syndrome, SARS, and MERS.

DNA viruses such as the adenoviruses sometimes cause serious pneumonias, but RNA viruses are of greater overall importance because of the large number of people they infect and their potential for serious outcomes. The following section covers some diseases caused by RNA viruses.

Influenza ("Flu")

Influenza is a good example of the constantly changing interaction between people and infectious agents. Antigenic changes in the influenza viruses are responsible for serious annual epidemics of the disease—almost 20% of the world population gets infected with flu virus every year. The three major influenza virus types (A, B, and C) differ in their protein coats. Type A, considered here, causes the most serious disease, and its epidemics are widespread. Type B outbreaks occur each year, but they are less extensive and the disease is not as severe. Type C strains cause only mild disease. Influenza viruses do not cause "stomach flu."

Signs and Symptoms

After a short incubation period averaging 2 days, influenza typically begins with headache and muscle aches, fever, sore throat, and fatigue, peaking in 6 to 12 hours. A dry cough develops and worsens over a few days. These acute symptoms usually go away within a week, leaving the patient with a lingering cough, fatigue, and generalized weakness for additional days or weeks.

Causative Agent

Influenza A virus is an enveloped single-stranded RNA virus belonging to the family *Orthomyxoviridae.* The genome is segmented into eight pieces, each of which contains one or two genes (**figure 21.24**). Embedded in the envelope are two kinds of glycoprotein spikes—hemagglutinin antigen (HA) and neuraminidase antigen (NA)—which have a role in viral pathogenesis. The HA spikes allow the virus to recognize and attach to specific receptors on ciliated host epithelial cells, initiating infection. HA spikes cause red blood cells to stick together (hemagglutination), a useful characteristic for virus identification. NA is an enzyme that plays a critical role in the release of newly formed virions from host cells. As new virions are made, they bud out of the host cell but remain bound to surface receptors in the host membrane. NA destroys these receptors, allowing the virions to leave the infected host cell, and aiding the spread of the virus to uninfected host cells.

There are different subtypes of influenza A viruses, characterized by antigenically distinct HA and NA spikes. The subtypes are given numbers according to these variations—H1, H2, N1, N2, and so on. For example, the "avian flu" epidemic of 1997 was caused by influenza virus H5N1, whereas the "swine (pig) flu" epidemic of 2009 was caused by

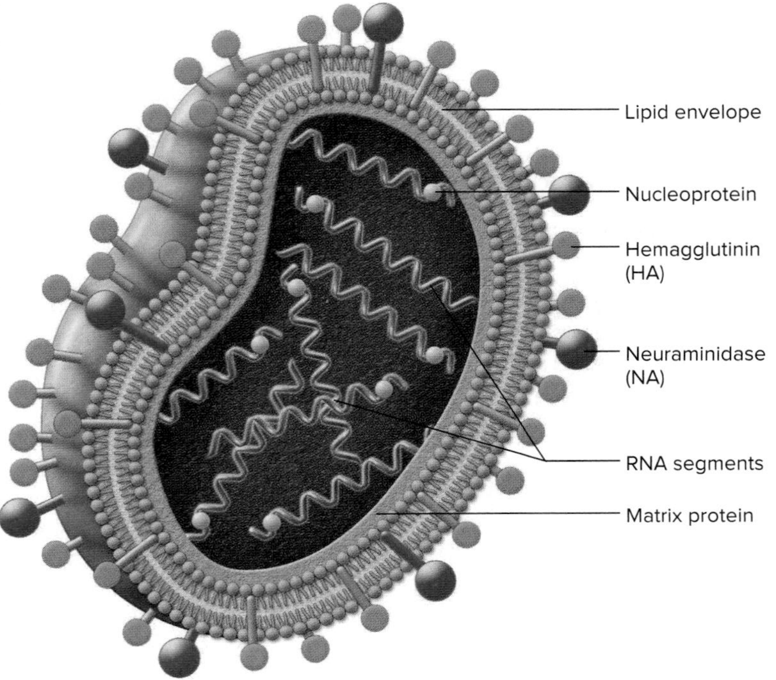

FIGURE 21.24 Structure of Influenza Virus

What are the functions of the HA and NA spikes?

influenza virus H1N1. There are 18 HA and 11 NA subtypes, but only H1, H2 and H3, and N1 and N2 typically spread from person to person.

Pathogenesis

People acquire influenza virus by inhaling aerosolized respiratory secretions from someone who has the disease or from fomites. The virions attach by their HA spikes to specific receptors on ciliated respiratory epithelial cells and enter the cell by endocytosis. New viral parts are quickly synthesized, and regions of the host cell membrane become embedded with virally encoded HA and NA glycoproteins. Within 6 hours, mature virions bud from the host cell, acquiring host cell–derived membrane containing these HA and NA spikes as they do so. The virus spreads rapidly to nearby cells, including mucus-secreting cells and cells of the alveoli. The common flu signs and symptoms are caused by tissue damage as well as pro-inflammatory cytokines produced by virus-infected cells. Infected epithelial cells die and slough off, thus destroying the mucociliary escalator. The damage to this important first line of defense makes the person susceptible to secondary respiratory infections. The immune response quickly controls the influenza virus infection in most cases, although complete recovery of the respiratory epithelium may take 2 months or more.

Epidemiology

The case-fatality rate for influenza is typically very low, but many people fall ill during an epidemic so the total number of deaths is high. Although influenza virus infection alone can kill otherwise healthy people, most deaths are due to bacterial secondary infections such as pneumonia. People are predisposed to these infections because of the virally induced damage to the respiratory epithelium. In fact, the bacterium *Haemophilus influenzae* got its name because it was often found in the lungs of people who died after having influenza, leading to the incorrect conclusion that it caused the disease.
⏮ secondary infection

Influenza epidemics occur every year. Pandemics occur periodically over the years, marked by rapid spread of the viruses around the globe and higher-than-normal morbidity. Several factors are involved in the spread of influenza viruses, but major attention has focused on their antigenic changeability. Two types of variation occur— antigenic drift and antigenic shift:

- **Antigenic drift.** This is caused by minor mutations in the genes that code for the HA and NA antigens and is responsible for the yearly occurrence of influenza outbreaks, called **seasonal influenza.** The mutations happen during normal viral replication and often cause a change in only a single amino acid in the HA or NA spikes. They occur frequently, however, and are enough to make immunity developed to virus strains of previous years less effective. Strains that arise because of antigenic drift are named to indicate the year and location where they were isolated. For example, A/Texas/77 (H3N2) is an influenza A strain isolated in 1977 in Texas, whereas A/Bangkok/79 (H3N2) was isolated in Bangkok in 1979. The Bangkok/79 strain has minor HA spike mutations that distinguish it from the Texas/77 strain. The antibodies produced by people who have recovered from A/Texas/77 (H3N2) are only partially effective against the A/Bangkok/79 (H3N2). Thus, the newer Bangkok strain might be able to cause a minor epidemic in populations previously exposed to the Texas strain.
- **Antigenic shift.** This is an uncommon but more dramatic change that occurs as a result of viral genome reassortment and is the cause of **pandemic influenza.** Recall that the influenza virus genome is segmented, meaning that viral proteins are encoded on eight different RNA segments rather than being encoded on one long molecule. When two different influenza viruses infect a cell at the same time, the progeny produced can have RNA segments from either of the viruses. From an infectious disease standpoint, this is particularly problematic when a genome segment from a virus strain that normally does not infect humans is acquired by a virus strain that does. For example, if a pig is simultaneously infected with two virus strains—one that infects only pigs and birds and another that infects pigs and humans—the viruses that emerge will still have eight genome segments, but each segment may originate from either of the initial infecting strains. This can result in a strain that infects humans with novel HA and/or NA antigens for which human populations have no immunity **(figure 21.25).** In 2009, a swine flu virus called H1N1 emerged. This virus was found to have genes from avian, swine, and human flu viruses. The pandemic that resulted from the H1N1 virus ended in 2010, but not before an estimated quarter of a million people died, mostly in Africa and Southeast Asia.

MicroByte

An H1N1 flu virus caused the 1918 "Spanish" flu pandemic that killed 50–100 million people around the globe.

Ecological studies show that all the known influenza A virus types exist in aquatic birds, generally causing chronic intestinal infections. Bird influenza viruses (avian flu) readily infect domestic fowl, and from them can potentially infect other domestic animals and humans. In 1997, a "bird flu" epidemic in Hong Kong involved an H5N1 virus from chickens that spread to humans, sometimes causing fatal infections. Different H5N1 strains have been categorized as low pathogenic avian influenza (LPAI) and highly pathogenic avian

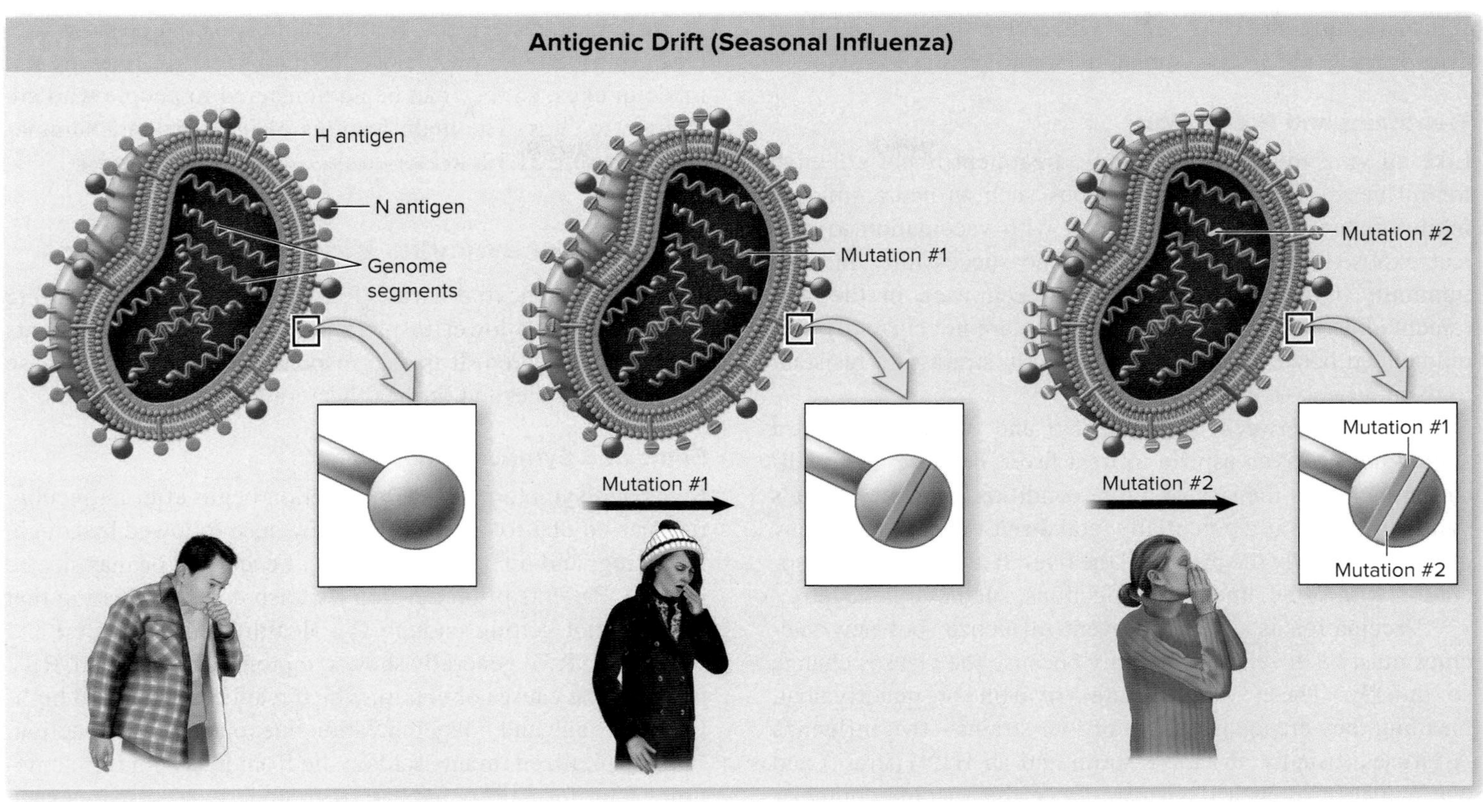

(a)

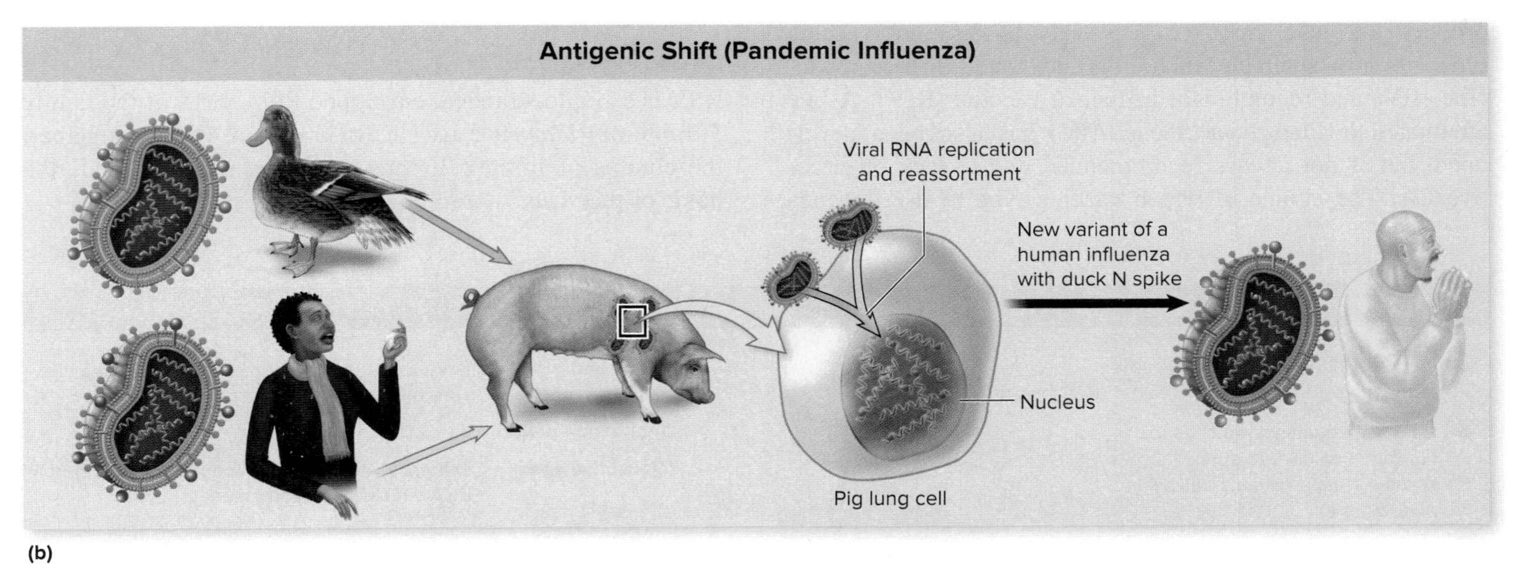

(b)

FIGURE 21.25 Influenza Virus: Antigenic Drift and Antigenic Shift With drift, repeated mutations cause a gradual change in the HA and/or NA spikes, so that antibodies against the original virus become progressively less effective. With shift, there is a sudden major change in the spikes because the virus acquires a new genome segment.

Why can antigenic shifts cause pandemics?

influenza (HPAI), depending on whether they cause fatal disease or not. Fortunately, the HPAI strain that started the Hong Kong outbreak did not spread easily from person to person, and so the number of human cases remained relatively low. In 2013, a novel bird flu (H7N9) arose in China. Although it mostly affects birds, over 1,200 cases have been reported in humans, with a 30–40% case-fatality rate. In the first 2 months of 2017 alone, China reported over 300 cases of H7N9 influenza. Fortunately, these cases have mostly been linked to contact with infected birds and not to human transmission. HPAI strains have been detected in birds in the United States, but as of late 2017, no human cases had been reported. A constant concern is that an HPAI strain will evolve through antigenic drift or antigenic shift to spread person to person.

Treatment and Prevention

Like all viral infections, antibiotic treatment is not effective for influenza. Antiviral medications such as neuraminidase inhibitors are generally used along with vaccination to protect exposed people until the vaccine successfully induces immunity. Other medications have been used in the past (amantadine and rimantadine), but they are not currently recommended because many circulating flu strains are resistant to them. ⏮ neuraminidase inhibitors

Children between the ages of 5 and 15 with influenza should not be given aspirin to treat fever, because this medication increases the risk of a rare condition known as **Reye's syndrome.** This is a potentially fatal disease that affects many organs, especially the brain and the liver. It also occurs in association with some other viral infections, such as chickenpox.

Vaccination is used to prevent influenza, but new vaccines must be developed annually because the viruses change so quickly. These vaccines are trivalent or quadrivalent, meaning they are against three or four strains—two influenza A viruses (usually an H3N2 strain and an H1N1 strain) and one or two influenza B viruses. Once the vaccine composition is chosen, it takes 6 to 9 months for manufacturers to produce adequate amounts. A variety of different vaccine types are now available, including inactivated influenza vaccine (IIV) and recombinant influenza vaccine (RIV). A live attenuated influenza vaccine (LAIV) has also been developed, but is not always recommended due to lack of effectiveness. The choice of which vaccine type to use depends on patient characteristics. For example, a new high-dose vaccine is available for people over 65, and some vaccines are not made in eggs, so they can be administered to people who are allergic to eggs. The main features of influenza are summarized in **table 21.12.** ⏮ inactivated vaccine, ⏮ attenuated vaccine

Respiratory Syncytial Virus (RSV) Infections

Respiratory syncytial virus (RSV) infection is the leading cause of serious lower respiratory tract infections in infants and young children. It is also responsible for serious disease in elderly people, and for healthcare-associated epidemics.

Signs and Symptoms

Signs and symptoms of RSV infection begin after an incubation period of 1 to 4 days with runny nose followed by cough, wheezing, and difficulty breathing. Fever may or may not be present. Patients often develop a bluish color, indicating that they are not getting enough O_2. Healthy older children and adults with RSV generally show symptoms of a bad cold. RSV is one of the causes of **croup,** which manifests as a loud high-pitched cough and noisy inhalation due to airway obstruction.

Hospitalized infants seldom die from RSV, but it is sometimes fatal for elderly patients with underlying diseases such as heart and lung disease, cancer, and immunodeficiency.

Causative Agent

RSV is a single-stranded, enveloped RNA virus of the family *Paramyxoviridae.* It causes cells in culture to fuse together; the clumps of fused cells are known as **syncytia,** thus the name of the virus.

TABLE 21.12 Influenza ("Flu")

1. Influenza virus is inhaled and carried to the lungs.
2. Viral hemagglutinin attaches to receptors on ciliated epithelial cells and the virus enters the cell by endocytosis.
3. Host cell synthesis is diverted to producing virions.
4. Newly formed virions bud from infected cells; they are released by viral neuraminidase and infect ciliated epithelium, mucus-secreting, and alveolar cells.
5. Infected cells ultimately die and slough off; recovery of the mucociliary escalator may take weeks.
6. Secondary bacterial infection of the lungs, ears, and sinuses is common.
7. The virus exits with coughing.

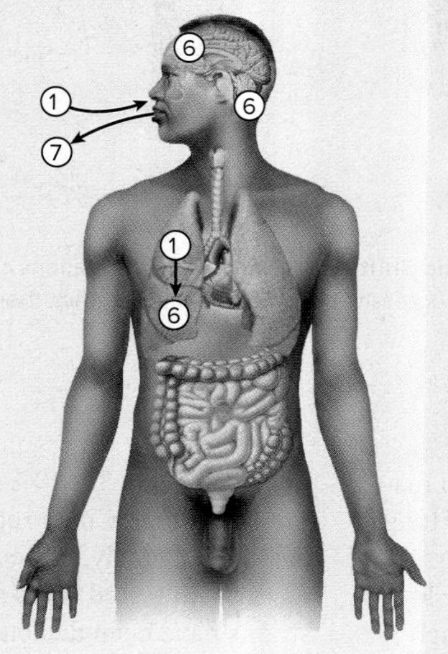

Signs and symptoms	Fever, muscle aches, lack of energy, headache, sore throat, nasal congestion, cough
Incubation period	1 to 2 days
Causative agent	Influenza virus, an enveloped single-stranded RNA virus with a segmented genome
Pathogenesis	Infection of respiratory epithelium; cells destroyed and virus released to infect other cells. Secondary bacterial infection results from damaged mucociliary escalator.
Epidemiology	Antigenic drift and antigenic shift prevent immunity.
Treatment and prevention	Treatment: antiviral medications somewhat effective for treatment when given early in the disease. Prevention: vaccines developed and given annually.

Pathogenesis

The virus enters the body by inhalation and infects the respiratory tract epithelium, causing cells to die and slough off. Bronchiolitis is a common feature of the disease; the inflamed bronchioles become partially plugged by sloughed cells, mucus, and fluid that has oozed from the walls of the bronchi. The initial obstruction causes wheezing when air rushes through the narrowed passageways, sometimes causing the condition to be confused with asthma. The obstruction often acts like a one-way valve, allowing air to enter the lungs, but not leave. In many cases the inflammatory process extends into the alveoli, causing pneumonia. There is a high risk of secondary infection because of the damaged mucociliary escalator. ⏮ mucociliary escalator

Epidemiology

RSV outbreaks are common from late fall to late spring, peaking in mid-winter. Recovery from infection produces only weak and short-lived immunity, so that infections can recur throughout life. Healthy children and adults usually have mild illness and readily spread the virus to others.

Treatment and Prevention

There are no effective antiviral medications for treating RSV.

Preventing healthcare-associated RSV illness requires strict isolation techniques. People with underlying illnesses and infants at high risk can be protected from the disease by giving them passive immunity via monthly injections of immune globulin or a recombinant humanized monoclonal antibody (RhuMab) called palivizumab. No vaccines are available. The main features of RSV infections are summarized in **table 21.13.** ⏮ passive immunity, ⏮ RhuMab

TABLE 21.13	Respiratory Syncytial Virus (RSV) Infections
Signs and symptoms	Runny nose, cough, fever, wheezing, difficulty breathing, bluish color
Incubation period	1 to 4 days
Causative agent	Respiratory syncytial virus (RSV), a paramyxovirus that produces syncytia
Pathogenesis	Sloughing of respiratory epithelium and inflammatory response plug bronchioles, cause bronchiolitis; pneumonia from bronchiolar or alveolar inflammation; secondary infection.
Epidemiology	Yearly epidemics during the cool months; readily spread by otherwise healthy older children and adults with mild symptoms; no lasting immunity.
Treatment and prevention	Treatment: no satisfactory antiviral treatment. Prevention: no vaccine. For high-risk individuals, passive immunization with immune globulin or a monoclonal antibody.

Hantavirus Pulmonary Syndrome

In the spring of 1993, a newly emerging disease made a dramatic appearance in the "Four Corners" region of the American Southwest, an area where the states of Arizona, Colorado, New Mexico, and Utah come together. It was a small outbreak, but quite alarming because most victims were vigorous young adults who developed influenza-like symptoms, and then died within days. Scientists from the CDC rushed to join local epidemiologists and health officials investigating the outbreak. Their studies quickly established that the disease was associated with exposure to mice. They soon learned that the patients had been infected with a hantavirus closely related to one that plagued American troops during the 1950s Korean War.

Signs and Symptoms

After an incubation period ranging from 3 days to 6 weeks, hantavirus pulmonary syndrome begins with fever, muscle aches (especially in the lower back), nausea, vomiting, and diarrhea. Unproductive cough and increasingly severe shortness of breath appear within a few days, often followed by shock and death.

Causative Agents

Hantavirus pulmonary syndrome is caused by a hantavirus called the Sin Nombre virus (SNV)—sin nombre means "no name" in Spanish. Hantaviruses are enveloped viruses of the family *Bunyaviridae.* Their genome consists of three segments of single-stranded RNA. In nature, these viruses primarily infect rodents, causing lifetime infections, without any apparent harm to the animals. Each type of hantavirus generally infects a particular rodent species.

Pathogenesis

The virus enters the body by inhalation of airborne dust contaminated with the urine, feces, or saliva of infected rodents. It then enters the circulation and is carried throughout the body, infecting the cells that line capillaries. Massive amounts of the viral antigen appear in lung capillaries. The inflammatory response to the viral antigen causes the vessels to leak large amounts of fluid into the lungs, suffocating the patient and causing the blood pressure to fall. Shock and death occur in about 30% of the cases. Fortunately, despite the large amount of viral antigen in the lung capillaries, few mature infectious virions enter the air passages of the lung, so person-to-person transmission is rare.

Epidemiology

Hantavirus pulmonary syndrome is a zoonosis. It is considered an emerging disease because of its recent discovery and apparent increase in frequency but has probably existed for centuries. Since the description of the syndrome, cases have been identified from Canada to Argentina, including several hundred from the United States (**figure 21.26**). Most of the cases have occurred west of the Mississippi River and were

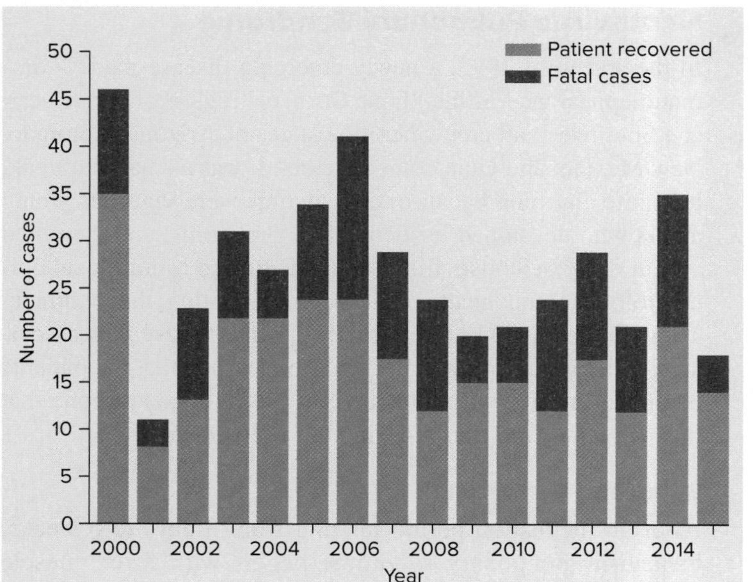

FIGURE 21.26 Total Hantavirus Pulmonary Syndrome Cases, United States, 2000–2015

? Which animals most commonly carry hantaviruses?

TABLE 21.14	Hantavirus Pulmonary Syndrome
Signs and symptoms	Fever, muscle aches, vomiting, diarrhea, cough, shortness of breath, shock
Incubation period	3 days to 6 weeks
Causative agent	Sin Nombre and related hantaviruses; enveloped viruses with segmented single-stranded RNA
Pathogenesis	Viral antigen localizes in capillary walls in the lungs; inflammation.
Epidemiology	Zoonotic disease; epidemics associated with increases in mouse populations near housing; generally no person-to-person spread.
Treatment and prevention	Treatment: no antiviral treatment. Prevention: avoiding contact with rodents; sealing access to houses, food supplies; good ventilation; avoiding dust; using disinfectants in cleaning rodent-contaminated areas.

due to a type of hantavirus carried by deer mice, but hantavirus carried by other rodents can cause disease as well. Epidemics have been associated with increases in mouse populations in poor communities with substandard housing. Thirty percent or more of the mice can become carriers of the disease. Complex ecological factors such as numbers of foxes, owls, snakes, and other predators and weather (which affect food supply), all play a role in mouse population levels. The emergence of hantavirus pulmonary syndrome is a convincing example of how environmental change can result in infectious human disease. ◂◂ zoonoses, ◂◂ emerging diseases

Treatment and Prevention

There is no proven antiviral treatment for hantavirus pulmonary syndrome, a disease that is often fatal.

Prevention of the syndrome is based on minimizing exposure to rodents and dusts contaminated by their urine, saliva, and feces. Rodent populations should be controlled by keeping foods in containers and making buildings as mouse-proof as possible. When cleaning an area where rodents are found, maximal ventilation should be ensured, and the area should be mopped with a disinfectant solution rather than being swept with brooms and vacuum cleaners that can stir up dust. Lethal traps and poisons may be necessary to decrease the rodent population in the area. The main features of hantavirus pulmonary syndrome are summarized in **table 21.14.**

SARS and MERS

Several coronaviruses commonly infect humans, typically causing mild upper respiratory infections. In 2003, however, a novel coronavirus was identified as the cause of a new disease that became known as SARS (severe acute respiratory syndrome). A brief SARS pandemic involved at least 27 countries, resulting in over 8,000 cases and almost 800 deaths. Fortunately, international infection control efforts limited the spread of the disease, and no new cases have been reported in the world since 2004. Still, it serves as an excellent reminder that a disease can emerge in one part of the world and rapidly spread across the globe. In 2012, another novel coronavirus was identified as the cause of a disease that became known as Middle East respiratory syndrome (MERS). By 2016, the epidemic had spread to more than 25 countries, resulting in over 1,800 cases and 600 deaths.

Signs and Symptoms

SARS and MERS usually begin 2 to 7 days after infection, with typical flu-like signs and symptoms of fever, malaise, muscle aches, a non-productive cough, and shortness of breath. Many patients, particularly those who are elderly or have other medical conditions, develop pneumonia and acute respiratory distress, often leading to death.

Causative Agents

SARS coronavirus (SARS-CoV) and MERS coronavirus (MERS-CoV) are members of the *Coronaviridae* family, a group of enveloped, single-stranded RNA viruses characterized by spikes on their surface. The name coronavirus reflects the crown-like appearance of the spikes (**figure 21.27**). Most coronaviruses infect only a single animal species, but SARS-CoV and MERS-CoV infect humans as well as other animals.

Pathogenesis

Relatively little is known about the mechanisms of pathogenesis of SARS-CoV and MERS-CoV compared to most other viruses that cause human diseases, partly because they are so

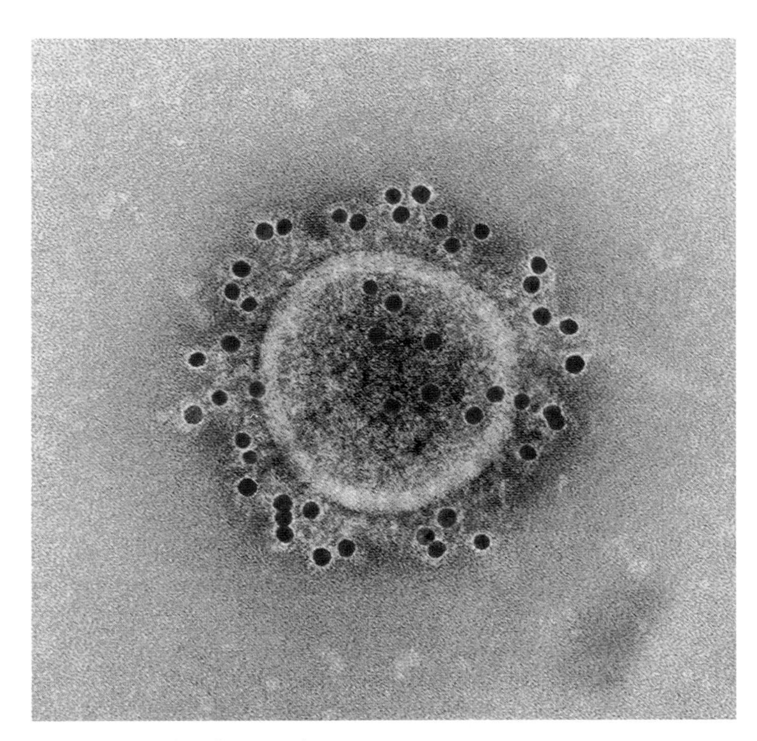

FIGURE 21.27 Coronavirus Source: NIAID

Besides humans, what animals are believed to host SARS-CoV or MERS-CoV?

newly recognized. They infect cells in the lower respiratory tract, and replicate their genome within virus-induced vesicles, a feature that allows them to remain hidden from the pattern recognition receptors that cells normally use to detect double-stranded RNA. Ultimately, however, the infection directly damages cells and elicits a strong immune response that causes further lung damage. It appears that a vigorous cytokine response (a "cytokine storm") is largely responsible for the most severe effects of the disease. ◀◀ pattern recognition receptor

Epidemiology

SARS-CoV and MERS-CoV are zoonotic viruses that likely originated in bats, which are known to harbor several different coronaviruses. Genetic and epidemiological evidence suggests that a bat coronavirus similar to SARS-CoV was transmitted to a civet (a cat-like animal) or some other animal sold in live markets in China. That animal then transmitted the virus to humans. In the case of MERS, evidence indicates that a bat coronavirus was transmitted to a dromedary camel (a camel with one hump), where it spread and then infected humans.

SARS-CoV and MERS-CoV are likely transmitted in respiratory droplets during close person-to-person contact. Some infected people appear to be "super-spreaders," meaning that they are more infectious than others. Both diseases spread in healthcare settings, but MERS primarily affects patients rather than staff, suggesting that underlying health problems might be risk factors.

TABLE 21.15 SARS and MERS

Signs and symptoms	Fever, muscle aches, cough; eventual respiratory distress or pneumonia
Incubation period	2–7 days
Causative agent	Coronaviruses: SARS-CoV and MERS-CoV
Pathogenesis	Alveolar damage, inflammation, fluid accumulation; cytokine storm may damage tissue.
Epidemiology	Zoonotic viruses; spread by respiratory droplets; seen in healthcare facilities.
Treatment and prevention	Treatment: supportive therapy. Prevention: no vaccine.

Treatment and Prevention

There are no specific treatments for SARS or MERS, but experimental medications are often given in severe cases. In addition, patients may receive supportive therapies, such as a mechanical ventilator to aid in breathing.

Typical prevention measures such as handwashing, frequently disinfecting surfaces, and avoiding close contact with sick people are recommended to prevent infection. No vaccines are available to protect against coronavirus diseases. Based on preliminary efforts to develop a SARS-CoV vaccine, however, research is underway to develop vaccines against MERS-CoV. Strategies developed to combat these two novel coronaviruses may help to protect against any additional coronavirus diseases that may emerge. **Table 21.15** presents some important features of SARS and MERS.

MicroAssessment 21.5

The speed with which influenza travels around the world and the rapid evolution of the virus due to antigenic drift and shift make the disease an extremely serious threat. Most deaths from influenza are caused by secondary bacterial infections. Respiratory syncytial virus is the leading cause of serious respiratory disease in infants and young children. Hantavirus pulmonary syndrome is often fatal. It is contracted from inhalation of dust contaminated by urine, feces, or saliva from mice infected with certain hantaviruses. SARS and MERS are caused by novel coronaviruses.

13. Why are there so many deaths from influenza when it is generally a mild disease?
14. What is the source of the virus that causes hantavirus pulmonary syndrome?
15. Why might you expect an influenza epidemic to be more severe following an antigenic shift in the virus rather than after antigenic drift?

21.6 ■ Fungal Infections of the Lung

Learning Outcomes

11. Describe the pathogenesis of histoplasmosis.
12. Outline the epidemiology of coccidioidomycosis and *Pneumocystis* pneumonia.

Serious lung diseases caused by fungi are quite unusual in healthy, immunocompetent people. Symptomatic and asymptomatic infections that subside without treatment, however, are common.

Coccidioidomycosis ("Valley Fever")

In the United States, coccidioidomycosis occurs mainly in Arizona, California, Nevada, New Mexico, Utah, and West Texas. People who are exposed to dust and soil, such as farm workers, are most likely to become infected, but only 40% develop symptoms.

Signs and Symptoms

"Flu-like" signs and symptoms such as fever, cough, chest pain, and loss of appetite and weight are common manifestations of coccidioidomycosis. Other signs and symptoms include joint pain, rash on upper trunk, and painful nodules on the extremities. The majority of people with the disease recover spontaneously within a month. A small percentage of patients, however, develop chronic disease.

Causative Agent

Coccidioidomycosis is caused by *Coccidioides immitis,* a dimorphic fungus that grows in soil as a mold. Its hyphae give rise to numerous barrel-shaped, highly infectious structures called arthroconidia (**figure 21.28a**). These become airborne and can be inhaled. In infected tissues, the arthroconidia develop into thick-walled spherules that may contain several hundred small cells called endospores, not to be confused with bacterial endospores (figure 21.28b). ⏮ dimorphic fungi

(a)

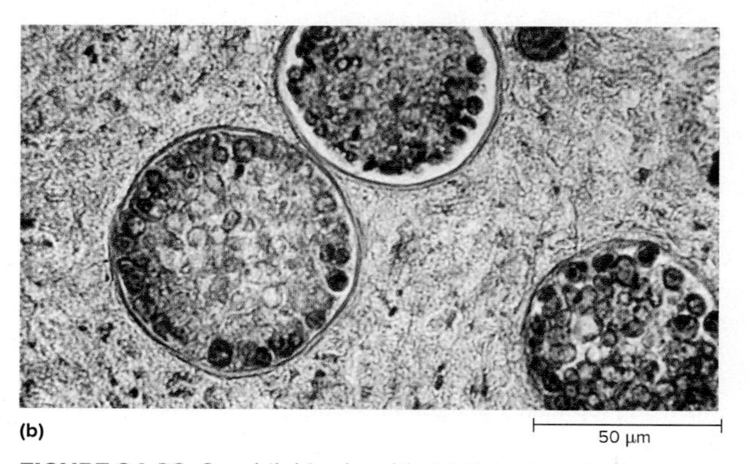

(b)

FIGURE 21.28 ***Coccidioides immitis*** **(a)** Mold-phase hyphae giving rise to barrel-shaped arthroconidia. **(b)** Spherules in tissues.
a: ©Michael Abbey/Science Source; b: Source: Dr. Lucille K. Georg/CDC

Which form of this fungus is found in host tissues?

Pathogenesis

Arthroconidia enter the lung with inhaled air and develop into spherules. Endospores are produced within the spherules and are released when the spherules eventually rupture. Each endospore then develops into a new endospore-containing spherule, and the process is repeated. Each time this cycle occurs, an inflammatory response is provoked. Tissue injury that results from hypersensitivity to fungal antigens causes nodules and joint pain.

The organisms are usually eliminated by body defenses, but caseous necrosis (dead tissue with cheese-like consistency) occurs in a few people, resulting in lung cavities similar to those seen in tuberculosis. Occasionally, organisms are carried throughout the body by the bloodstream and infect the skin, mucous membranes, brain, and other organs. This disseminated form of the disease occurs more often in people with AIDS or other immunodeficiencies and is fatal without treatment. ⏮ caseous necrosis

Epidemiology

Coccidioides immitis grows only in semi-arid desert areas of the Western Hemisphere. In these areas, infections occur during the hot, dry, dusty seasons when airborne arthroconidia are easily dispersed from the soil. Dust stirred up by earthquakes can result in epidemics. Rainfall encourages growth of the fungus, which then produces increased numbers of arthroconidia when dry conditions return. People can contract coccidioidomycosis by simply traveling through the endemic area. Infectious arthroconidia have unknowingly been transported to other areas, but the organism apparently is unable to thrive in moist climates.

Treatment and Prevention

Medications approved for treatment of serious cases of coccidioidomycosis include fluconazole or itraconazole, which are given for long periods of time. Even with treatment, disseminated disease can reactivate months or years later.

Preventive measures include avoiding dust in the endemic areas, and watering and planting vegetation to aid in dust control. **Table 21.16** describes the main features of coccidioidomycosis.

TABLE 21.16	**Coccidioidomycosis ("Valley Fever")**
Signs and symptoms	Fever, cough, chest pain, loss of appetite and weight; less frequently, painful nodules on extremities, pain in joints; skin, mucous membranes, brain, and internal organs sometimes involved
Incubation period	2 days to 3 weeks
Causative agent	*Coccidioides immitis,* a dimorphic fungus
Pathogenesis	Inhaled arthroconidia develop into spherules that mature and rupture to release endospores, each of which then develops into another spherule; inflammatory response damages tissue; hypersensitivity to fungal antigens causes painful nodules and joint pain.
Epidemiology	Occurs only in certain semi-arid regions of the Western Hemisphere.
Treatment and prevention	Treatment: antifungal medications. Prevention: dust control methods such as grass planting and watering.

Histoplasmosis ("Spelunker's Disease")

Histoplasmosis, like coccidioidomycosis, is usually benign but occasionally mimics tuberculosis. Rare, serious forms of the disease suggest that the patient has an underlying immunodeficiency such as AIDS. The distribution is more widespread than that of coccidioidomycosis and is associated with different soil types and climate.

Signs and Symptoms

Most infections are asymptomatic. Fever, cough, and chest pain are the most common symptoms, sometimes with shortness of breath. Mouth sores may develop, especially in children.

Causative Agent

Histoplasmosis is caused by the dimorphic fungus *Histoplasma capsulatum,* although the name is misleading because the fungus does not have a capsule. This organism grows in soils contaminated by bat or bird droppings, but it is not pathogenic for these animals. In pus or tissue from people with active disease, *H. capsulatum* is a tiny oval yeast that grows within macrophages (**figure 21.29a**). The mold form of the organism characteristically produces two kinds of conidia: macroconidia, which often have numerous projecting knobs (figure 21.29b), and tiny microconidia. ⏮ dimorphic fungus

Pathogenesis

Histoplasma capsulatum microconidia inhaled into the lungs are taken up by resident macrophages. The fungus then develops into the yeast form, which multiplies within the phagocytes. Granulomas develop, closely resembling those seen in tuberculosis, sometimes even showing caseous necrosis.

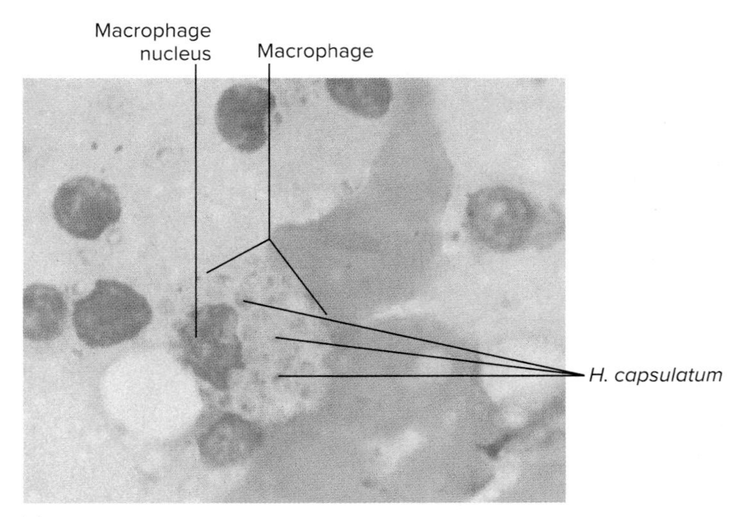

(a)

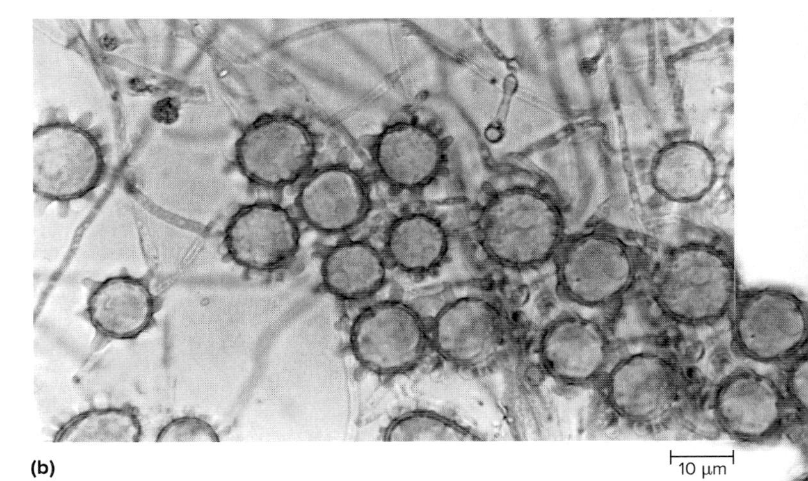

(b)

FIGURE 21.29 ***Histoplasma capsulatum*** **(a)** Yeast-phase organisms In the cytoplasm of a macrophage. **(b)** Mold phase, showing macroconidia. a: ©Evans Roberts; b: ©Evans Roberts

? Is this fungus encapsulated?

Eventually, the lesions are replaced with scar tissue, and many calcify, becoming visible on X rays. In rare cases, particularly in those who are immunodeficient, the disease spreads throughout the body.

Epidemiology

The distribution of histoplasmosis is quite different from that of coccidioidomycosis. **Figure 21.30** shows the distribution of histoplasmosis in the United States, but the disease also occurs in tropical and temperate zones scattered around the world. Cave explorers (spelunkers) are at risk for histoplasmosis because many caves contain soil contaminated with bat droppings. Most cases in the United States have occurred in the Mississippi and Ohio River drainage areas and in South Atlantic states. Skin tests reveal that millions of people living in these areas have been infected.

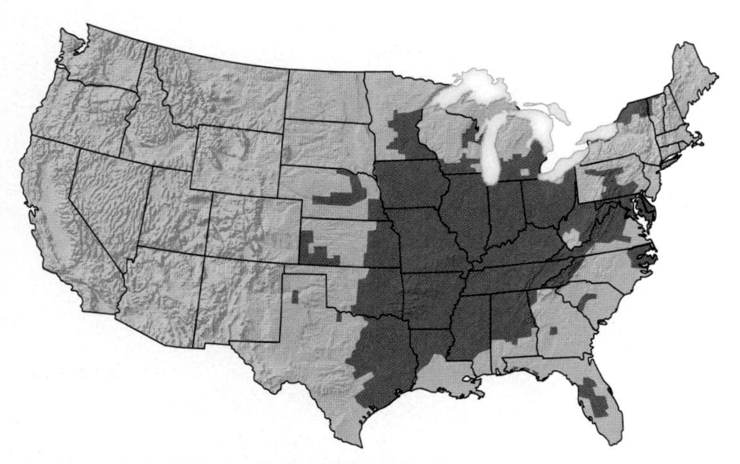

FIGURE 21.30 Geographic Distribution of *Histoplasma capsulatum* in the United States

Why is histoplasmosis also called "spelunker's disease"?

Treatment and Prevention

Treatment of histoplasmosis is similar to that of coccidioidomycosis. Itraconazole is used for treating mild to moderate disease. In severe cases, amphotericin B is given until the patient improves. Then it is replaced with itraconazole, which is less toxic.

No proven preventive measures are known other than to avoid areas where soil is heavily enriched with bat and bird droppings, especially if the soil has been left undisturbed for a long period. Some researchers have recommended placing several inches of clay soil over soils containing large quantities of old droppings. The main features of histoplasmosis are described in **table 21.17.**

Pneumocystis Pneumonia (PCP)

Pneumocystis pneumonia (PCP), a severe, infectious lung disease, was recognized just after World War II in Europe when it killed malnourished, premature infants in hospitals. Subsequently, occasional cases were identified among immunodeficient patients. With the start of the AIDS epidemic, the incidence of PCP increased rapidly, and it is still a common opportunistic infection in AIDS patients who are not receiving preventive care.

TABLE 21.17	Histoplasmosis ("Spelunker's Disease")
Signs and symptoms	Mild respiratory symptoms; less frequently, fever, chest pain, cough, chronic mouth sores
Incubation period	5 to 8 days
Causative agent	*Histoplasma capsulatum,* a dimorphic fungus
Pathogenesis	Microconidia inhaled, change to yeast phase, multiply in macrophages; granulomas form; disease spreads throughout the body in people with AIDS or other immunodeficiencies.
Epidemiology	The fungus grows in soil contaminated by bird or bat droppings, especially in Ohio and Mississippi River valleys, and in the southeast United States. Found in many other countries. Spelunkers are at risk.
Treatment and prevention	Treatment: antifungal medications. Prevention: avoiding soils contaminated with bird or bat droppings.

Signs and Symptoms

Many people are infected with the fungus that causes PCP, but the infection is latent because it is kept under control by the immune system. In immunocompromised people, however, the organism causes disease. The signs and symptoms of *Pneumocystis* pneumonia typically begin slowly after an incubation period of 4 to 6 weeks, with gradually increasing shortness of breath and rapid breathing. Fever is usually slight or absent, and only about half of the patients have a cough, which is non-productive. As the disease progresses, a dusky coloration of the skin and mucous membranes appears and gradually gets worse—this is caused by poor oxygenation of the blood that can become fatal.

Causative Agent

Pneumocystis pneumonia is caused by *Pneumocystis jirovecii,* a tiny yeast-like fungus (**figure 21.31**). The organism was formerly classified as *P. carinii* and it is still sometimes known by that name (explaining the acronym PCP for this type of pneumonia). *P. jirovecii* differs from many fungi in the chemical makeup of its cell wall, which makes it resistant to medications often used against fungal pathogens. The organism forms cysts that have a characteristic appearance, helping in its identification. It has not reliably been cultivated in vitro.

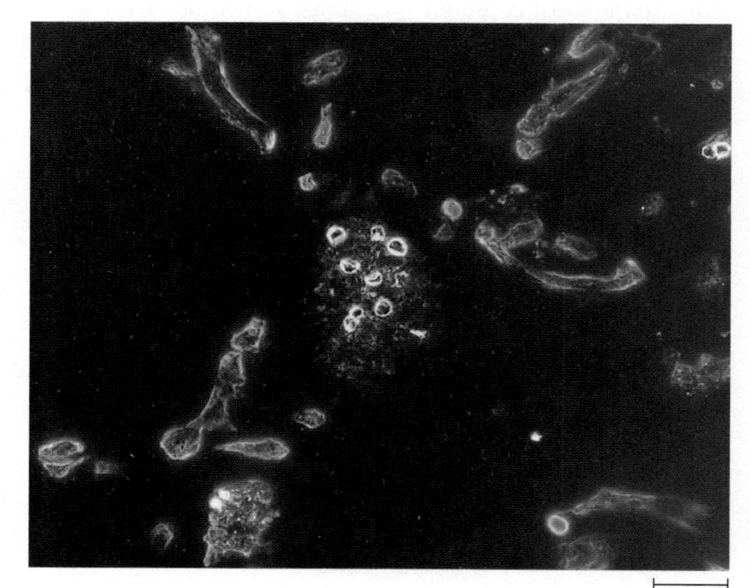

FIGURE 21.31 Fluorescent Antibody Stain of *Pneumocystis jirovecii* The yellow circles are *P. jirovecii* cysts. ©Cecil H. Fox/Science Source

Why is *Pneumocystis* pneumonia known by the acronym PCP?

Pathogenesis

P. jirovecii cells are easily inhaled into lung tissue. In experimental infections, these attach to the alveolar walls, and the alveoli fill with fluid, macrophages, and masses of *P. jirovecii* cells in various stages of development. Later, the alveolar walls become thickened and scarred, preventing the free passage of O_2.

Epidemiology

Pneumocystis jirovecii is distributed worldwide. Most children are infected with the organism at an early age although the infection is asymptomatic. The source and transmission of human infections are unknown. Most cases of PCP occur in people with immunodeficiency, but it is uncertain whether their disease is caused by reactivation of a latent infection, or new infection from inhalation of cells. Epidemics among hospitalized malnourished infants and elderly nursing home residents suggest airborne spread, and *P. jirovecii* has been detected in indoor and outdoor air by using the polymerase chain reaction (PCR). ⏮ PCR

Treatment and Prevention

PCP is most often treated with co-trimoxazole (TMP-SMX). Alternative medications are given to people who cannot tolerate TMP-SMX because of its side effects—mainly rash, nausea, and fever. For unknown reasons, people with HIV disease are more likely to develop these side effects than others. After treatment for PCP, individuals with HIV disease must receive lower prophylactic doses of their medication indefinitely, or until they have a sustained rise in $CD4^+$ T-cell count to above 200 cells/μL (a $CD4^+$ cell count of 200 cells/μL or lower indicates AIDS). ⏮ co-trimoxazole

To prevent PCP, HIV-infected people are typically started on medication as soon as they become immunodeficient, or until characteristic opportunistic diseases appear. The main features of *Pneumocystis* pneumonia are presented in **table 21.18.**

TABLE 21.18 *Pneumocystis* Pneumonia (PCP)

Signs and symptoms	Gradual onset, shortness of breath, rapid breathing, non-productive cough, slight or absent fever, and dusky color of skin and mucous membranes
Incubation period	4 to 8 weeks
Causative agent	*Pneumocystis jirovecii* (previously known as *P. carinii*)—a fungus
Pathogenesis	Inhaled fungal cells attach to alveolar walls, causing alveoli to fill with fluid, macrophages, and fungal cells. Alveolar walls thicken, impairing O_2 exchange.
Epidemiology	Most humans become infected in early childhood—source of human infections is unknown. Epidemics can arise in immunodeficient groups.
Treatment and prevention	Treatment: antimicrobial medications. Prevention: prophylactic doses to those with AIDS of the same medications used for treatment.

The key features of the diseases covered in this chapter are highlighted in the **Diseases in Review 21.1** table.

MicroAssessment 21.6

Coccidioidomycosis and histoplasmosis are caused by fungi that live in the soil. The body responds to these infections in a way that mimics tuberculosis. *Pneumocystis* pneumonia (PCP) affects immunocompromised people.

16. Why should an immunodeficient person avoid traveling through hot, dry, dusty areas of the southwestern United States?

17. Why might cave exploration increase the risk of histoplasmosis?

18. Several students staying in a hotel next to a bulldozing operation developed histoplasmosis. How might the bulldozing explain the outbreak?

Diseases in Review 21.1

Respiratory System Diseases

Disease	Causative Agent	Comment	Summary Table
BACTERIAL INFECTIONS OF THE UPPER RESPIRATORY TRACT			
Conjunctivitis (pink eye), otitis media (earache), sinus infection	Usually *Haemophilus influenzae* or *Streptococcus pneumoniae*	Often occur together; factors involved in the transmission are unknown.	
Streptococcal pharyngitis ("strep throat")	*Streptococcus pyogenes* (group A streptococcus)	Treated with antibiotics, partly to avoid sequelae; must be distinguished from viral pharyngitis, which cannot be treated with antibiotics.	Table 21.3
Diphtheria	*Corynebacterium diphtheriae*	Toxin-mediated disease characterized by pseudomembrane in the upper respiratory tract. Preventable by vaccination.	Table 21.4
VIRAL INFECTIONS OF THE UPPER RESPIRATORY TRACT			
Common cold	Rhinoviruses and other viruses	Runny nose, sore throat, and cough are due to the inflammatory response and cell destruction.	Table 21.5
Adenovirus pharyngitis	Adenoviruses	Similar to the common cold but with fever; spread to the lower respiratory tract can result in severe disease.	Table 21.6
BACTERIAL INFECTIONS OF THE LOWER RESPIRATORY TRACT			
Pneumococcal pneumonia	*Streptococcus pneumoniae*	Organism common in the throat of healthy people; causes disease when mucociliary escalator is impaired or with underlying conditions. Vaccine that protects against multiple strains is available.	Table 21.7
***Klebsiella* pneumonia**	*Klebsiella* species, commonly *K. pneumoniae*	Common hospital-acquired bacterium; characterized by thick, bloody, jelly-like sputum. Drug resistance is a major problem.	Table 21.7
Mycoplasmal pneumonia ("walking pneumonia")	*Mycoplasma pneumoniae*	Relatively mild pneumonia; common among college students and military recruits. Cannot be treated with medications that inhibit cell wall synthesis.	Table 21.7
Pertussis ("whooping cough")	*Bordetella pertussis*	Characterized by frequent violent coughing. Preventable by vaccination.	Table 21.8
Tuberculosis ("TB")	*Mycobacterium tuberculosis*	Most infections result in latent tuberculosis infection (LTBI), but these can reactivate to cause tuberculosis disease (TB disease). Treated using combination drug therapy, but drug resistance is an increasing problem.	Table 21.9
Legionellosis ("legionnaires' disease")	*Legionella pneumophila*	Transmitted via aerosolized water drops; smokers and those with impaired defenses are most at risk of developing disease.	Table 21.10
Inhalation anthrax	*Bacillus anthracis*	Rare zoonotic disease; may be associated with bioterrorism; high case-fatality rate.	Table 21.11
VIRAL INFECTIONS OF THE LOWER RESPIRATORY TRACT			
Influenza ("flu")	Influenza viruses	New vaccine developed yearly; viruses change seasonally due to antigenic drift; antigenic shifts cause pandemics.	Table 21.12
Respiratory syncytial virus infections	RSV	Serious disease in infants, young children, and the elderly.	Table 21.13
Hantavirus pulmonary syndrome	Hantaviruses	Acquired via inhaled dust contaminated with rodent saliva, urine, or feces. Frequently fatal.	Table 21.14
SARS and MERS	Coronaviruses	Emerging zoonotic diseases.	Table 21.15
FUNGAL INFECTIONS OF THE RESPIRATORY TRACT			
Coccidioidomycosis ("valley fever")	*Coccidioides immitis*	Environmental reservoir (soil in semi-arid desert areas); most infections are asymptomatic.	Table 21.16
Histoplasmosis ("spelunker's disease")	*Histoplasma capsulatum*	Environmental reservoir (soil enriched with bird or bat droppings); most infections are asymptomatic.	Table 21.17
***Pneumocystis* pneumonia (PCP)**	*Pneumocystis jirovecii* (formerly *carinii*)	Organism is an opportunistic fungus that causes serious lung disease in immunocompromised people, such as those with HIV/AIDS.	Table 21.18

Summary

21.1 ■ Anatomy, Physiology, and Ecology of the Respiratory System

The Upper Respiratory Tract

The upper respiratory tract includes the nose and nasal cavity, pharynx (throat), and epiglottis (figure 21.1). Ciliated cells line much of the respiratory tract and remove microorganisms by constantly propelling mucus out of the respiratory system. A wide variety of microorganisms colonize parts of the tract (table 21.1).

The Lower Respiratory Tract

The lower respiratory tract includes the larynx, trachea, bronchi, and lungs. Pleural membranes surround the lungs. Viruses and microorganisms are normally absent from the lower respiratory tract.

UPPER RESPIRATORY TRACT INFECTIONS

21.2 ■ Bacterial Infections of the Upper Respiratory System

Pink Eye, Earache, and Sinus Infections

Conjunctivitis (pink eye) is usually caused by *Haemophilus influenzae* or *Streptococcus pneumoniae* (pneumococcus) (figure 21.2). **Otitis media** and sinusitis often develop when infection spreads from the upper portion of the pharynx (figure 21.3).

Streptococcal Pharyngitis ("Strep Throat") (tables 21.2, 21.3; figure21.6)

Streptococcus pyogenes is a β-hemolytic Gram-positive coccus that causes strep throat (figures 21.4, 21.5). Strains that produce **SPEs (streptococcal pyrogenic exotoxins)** can cause scarlet fever. Other diseases are associated with SPEs as well.

Post-Streptococcal Sequelae

Post-streptococcal sequelae, including rheumatic fever and glomerulonephritis, may follow strep throat and are due to the immune response and molecular mimicry (figure 21.7).

Diphtheria (table 21.4)

Diphtheria, caused by *Corynebacterium diphtheriae,* is a toxin-mediated disease that can be prevented by immunization (figures 21.8, 21.9).

21.3 ■ Viral Infections of the Upper Respiratory System

The Common Cold (table 21.5)

The common cold can be caused by many different viruses, rhinoviruses being the most common (figure 21.10).

Adenovirus Respiratory Tract Infections (table 21.6)

Adenoviruses cause illnesses that can resemble a common cold or strep throat, with symptoms varying from mild to severe.

LOWER RESPIRATORY TRACT INFECTIONS

Focus on Pneumonia

In **pneumonia** the alveoli of the lungs fill with fluids. Signs and symptoms generally include cough, chills, shortness of breath, fever, and chest pain. A cough often brings up **sputum** from the lungs. Pneumonia is often caused by opportunists that cause disease when the mucociliary escalator is not functioning optimally. Vaccines are available to prevent pneumococcal disease but not the other types of pneumonia.

21.4 ■ Bacterial Infections of the Lower Respiratory System

Pneumococcal Pneumonia (table 21.7; figures 21.11, 21.12)

Streptococcus pneumoniae is a very common cause of community-acquired pneumonia.

***Klebsiella* Pneumonia** (table 21.7)

Klebsiella pneumonia is representative of many healthcare-associated pneumonias that cause permanent damage to the lung (figure 21.13). Serious complications such as lung abscesses and bloodstream infection are more common than with many other bacterial pneumonias.

Mycoplasmal Pneumonia ("Walking Pneumonia") (table 21.7, figure 21.14)

Mycoplasmal pneumonia is often called walking pneumonia; serious complications are rare. Penicillins and cephalosporins are not useful in treatment because the causative agent, *M. pneumoniae,* lacks a cell wall.

Pertussis ("Whooping Cough") (table 21.8; figures 21.15, 21.16)

Whooping cough is characterized by violent spasms of coughing and gasping. Childhood immunization against the causative agent, *Bordetella pertussis,* prevents the disease. Mild disease is common in adults.

Tuberculosis ("TB") (table 21.9; figures 21.17–21.21)

Tuberculosis, caused by the acid-fast rod *Mycobacterium tuberculosis,* is generally slowly progressive or heals and remains latent, presenting the risk of later reactivation. A group of related mycobacteria, including *M. avium* and *M. intracellulare,* form a complex (MAC) that causes opportunistic disease in those with AIDS or other immune-compromising conditions.

Legionellosis ("Legionnaires' Disease") (table 21.10, figure 21.22)

Legionellosis occurs when there is a high infecting dose of the causative microorganisms or an underlying lung disease. The cause, *Legionella pneumophila,* is a rod-shaped bacterium common in the environment.

Inhalation Anthrax (table 21.11, figure 21.23)

Inhalation anthrax is an animal disease that rarely occurs in humans, but it has been associated with bioterrorism and is a Category A agent. A high percentage of cases lead to anthrax meningitis.

21.5 ■ Viral Infections of the Lower Respiratory System

Influenza ("Flu") (table 21.12; figures 21.24, 21.25)

Widespread epidemics are characteristic of influenza A viruses. **Antigenic drift** is responsible for **seasonal influenza**; **antigenic shift** can cause **pandemic influenza.** Deaths are usually but not always caused by secondary infection.

Respiratory Syncytial Virus (RSV) Infections (table 21.13)

RSV is the leading cause of serious respiratory disease in infants and young children.

Hantavirus Pulmonary Syndrome (table 21.14, figure 21.26)

Hantavirus pulmonary syndrome is an often fatal disease contracted when airborne dust contaminated by urine from mice infected with a hantavirus is inhaled.

SARS and MERS (table 21.15, figure 21.27)
Novel coronaviruses emerged in 2003 to cause severe acute respiratory syndrome (SARS) and in 2012 to cause Middle East respiratory syndrome (MERS). International infection control efforts limited the spread of these emerging diseases.

21.6 Fungal Infections of the Lung

Coccidioidomycosis ("Valley Fever") (table 21.16, figure 21.28)
Coccidioidomycosis occurs in hot, dry areas of the Western Hemisphere and is initiated by airborne arthroconidia of the dimorphic soil fungus *Coccidioides immitis.*

Histoplasmosis ("Spelunker's Disease") (table 21.17, figure 21.29)
Histoplasmosis is similar to coccidioidomycosis but occurs in tropical and temperate zones around the world (figure 21.30). The causative fungus, *Histoplasma capsulatum,* is dimorphic and is found in soils contaminated by bat or bird droppings.

***Pneumocystis* Pneumonia (PCP)** (table 21.18, figure 21.31)
Pneumocystis pneumonia occurs as an opportunistic infection in immunocompromised people, including those with HIV/AIDS.

Review Questions

Short Answer

1. How does contamination of the eye lead to upper respiratory infection?
2. After you recover from strep throat, can you get it again? Explain.
3. Where is the gene for diphtheria toxin production?
4. Why has no vaccine been developed for the common cold?
5. How do alcoholism and cigarette smoking predispose a person to pneumonia?
6. Give a mechanism by which *Klebsiella pneumoniae* can become antibiotic-resistant.
7. Why does the incidence of whooping cough rise promptly when pertussis immunizations are stopped?
8. Why is combination therapy used to treat TB disease?
9. Why did it take so long to discover the cause of legionellosis?
10. Why is it important to get a new influenza vaccination every year?

Multiple Choice

1. The following are all complications of streptococcal pharyngitis *except*
   a) glomerulonephritis.
   b) scarlet fever.
   c) damaged heart valves.
   d) acute rheumatic fever.
   e) Reye's syndrome.
2. All of the following are true of diphtheria *except*
   a) a membrane that forms in the throat can cause suffocation.
   b) a toxin is produced that interferes with ribosome function.
   c) the causative organism typically invades the bloodstream.
   d) immunization with a toxoid prevents the disease.
   e) it may form chronic skin ulcers.
3. Adenoviral infections and the common cold are both
   a) caused by picornaviruses.
   b) often associated with fever.
   c) associated with severe sore throat.
   d) lower respiratory infections.
   e) avoided by handwashing.
4. All are true of mycoplasmal pneumonia *except*
   a) it is a mycosis.
   b) it usually does not require hospitalization.
   c) penicillin is ineffective for treatment.
   d) it is the leading cause of bacterial pneumonia in college students.
   e) the infectious dose of the causative organism is low.
5. All of the following are true of legionellosis *except*
   a) the causative organism can grow inside protozoa.
   b) it spreads readily from person to person.
   c) it is more likely to occur in long-term cigarette smokers than in nonsmokers.
   d) it is often associated with diarrhea or other intestinal symptoms.
   e) it can be contracted from household water supplies.
6. Which of the following infectious agents is most likely to cause a pandemic?
   a) Influenza A virus
   b) *Streptococcus pyogenes*
   c) *Histoplasma capsulatum*
   d) Sin Nombre virus
   e) SARS-CoV
7. Respiratory syncytial virus
   a) is a leading cause of serious lower respiratory tract infections in infants.
   b) is an enveloped DNA virus of the adenovirus family.
   c) attaches to host cell membranes by means of neuraminidase.
   d) poses no threat to elderly people.
   e) mainly causes disease in the summer months.
8. In the United States, hantaviruses
   a) are limited to southwestern states.
   b) are carried only by deer mice.
   c) infect human beings with a fatality rate of about 30%.
   d) were first identified in the early 1970s.
   e) are contracted mainly in bat caves.
9. All of the following are true of coccidioidomycosis *except*
   a) it is contracted by inhaling arthroconidia.
   b) it is caused by a dimorphic fungus.
   c) endospores are produced within a spherule.
   d) it is more common in Maryland than in California.
   e) it is often associated with painful nodules on the legs.
10. The disease histoplasmosis
    a) is caused by an encapsulated bacterium.
    b) is contracted by inhaling arthroconidia.

c) occurs mostly in hot, dry, and dusty areas of the American Southwest.
d) is a threat to AIDS patients living in areas bordering the Mississippi River.
e) is commonly fatal for pigeons and bats.

Applications

1. A physician is advising the family on the condition of a diphtheria patient. How would the physician explain why the disease affects some tissues and not others?
2. How should a physician respond to a mother who asks if her daughter can get pneumococcal pneumonia again?

Critical Thinking

1. If all transmission of *Mycobacterium tuberculosis* from one person to another were stopped, how long would it take for the world to be rid of the disease?
2. Medications that prevent and treat influenza by binding to neuraminidase on the viral surface act against all the kinds of influenza viruses that infect humans. What does this imply about the nature of the interaction between the medications and the neuraminidase molecules?

22 Skin Infections

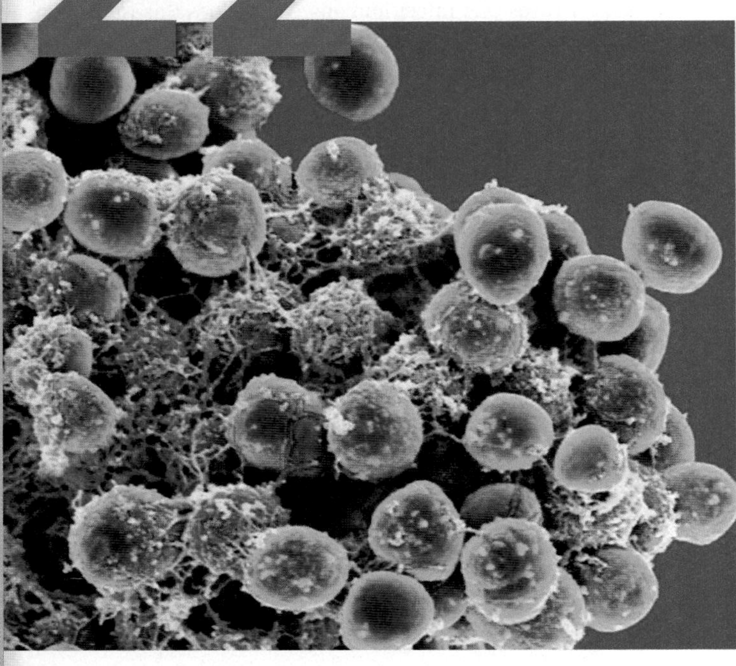

Staphylococcus epidermidis cells (color-enhanced scanning electron micrograph).

KEY TERMS

Abscess A localized collection of pus within a tissue.

Carbuncle A cluster of boils; painful infection of the skin and subcutaneous tissues.

Exfoliatin A bacterial toxin that causes peeling of the outer epidermis.

Folliculitis Inflammation of hair follicles.

Furuncle A boil; a localized skin infection that penetrates into subcutaneous tissue.

Impetigo A superficial skin disease characterized by thin-walled vesicles, oozing blisters, and yellow crusts.

Pyoderma Any skin disease characterized by production of pus.

A Glimpse of History

Howard T. Ricketts was born in Ohio in 1871. He studied medicine in Chicago, and then specialized in pathology, the study of the nature of disease and its causes. In 1902, he was appointed to the faculty of the University of Chicago, where his research interests turned to Rocky Mountain spotted fever (RMSF).

Patients with RMSF have a dramatic rash and often die. The disease was poorly understood, but Ricketts noticed tiny rod-shaped bacteria in the blood of people and laboratory animals with the disease. Because he could transmit RMSF by injecting laboratory animals with blood from an infected person, Ricketts was sure these tiny invaders caused the disease.

Scientists suspected that the disease was contracted from tick bites, and Ricketts showed that certain species of ticks could transmit the disease from one animal to another. The infected ticks remained healthy, and, surprisingly, their eggs frequently contained large numbers of the same bacteria found in the RMSF patients' blood. When these eggs were fertilized, they developed into infected ticks. Ricketts was never able to cultivate these bacteria for further studies so he declined to give them a scientific name and traveled to Mexico to study a very similar disease—louse-borne typhus. Unfortunately, he contracted typhus and died at the age of 39. Stanislaus Prowazek, a Czech scientist who was also studying typhus, met the same fate at almost the same age.

The martyrdom of the two young scientists struggling to understand infectious diseases is memorialized in the name of the louse-borne typhus agent, *Rickettsia prowazekii.* Both the genus and the species names of the RMSF agent, *Rickettsia rickettsii,* recognize Howard Ricketts. We now know that these bacteria are obligate intracellular parasites. Because most culture media lack live cells, the bacteria did not grow in the culture media used by Ricketts and Prowazek.

Much of the human body's contact with the outside world occurs at the skin surface. This tough, flexible outer covering provides a remarkable barrier to invasion and infection. Its exposed state, however, leaves it vulnerable to a variety of injuries. Cuts, punctures, burns, chemical injury, and insect or tick bites can break this barrier and provide a way for pathogens not only to infect the skin itself, but to access underlying tissues. Alternatively, skin infections may occur when microorganisms or viruses that enter the body from another site (such as the respiratory or gastrointestinal systems) are carried by the bloodstream to the skin.

22.1 ■ Anatomy, Physiology, and Ecology of the Skin

Learning Outcomes

1. Describe the function of skin in health and disease.
2. Explain the role of normal skin microbiota in health and disease.

The skin is far more than a passive wrapping for the body. It is a barrier that prevents the entry of microbes, regulates body temperature, and restricts the loss of fluid from body tissues. In addition, it has many sensory receptors that provide the central nervous system with information about the environment. The skin also plays an essential role in the function of the immune system. If the skin barrier breaks, resident macrophages and dendritic cells produce cytokines that trigger an immune response. Collections of lymphocytes found in skin-associated lymphoid tissues (SALT) are ready to proliferate when needed. ⏮ cytokines, ⏮ skin-associated lymphoid tissue

The skin is composed of two layers: the **epidermis** and the **dermis** (**figure 22.1**). The subcutaneous layer, or hypodermis, lies beneath the dermis and supports the skin:

- **Epidermis.** This is the surface layer, made up of multiple layers of flat epithelial cells. The outermost cells are dead and filled with keratin, a tough, water-resistant protein also found in hair and nails. These skin cells regularly flake off, exposing their replacements immediately below. Microbes living on the outermost cells are shed with the flaking skin cells. This process is a valuable defense against infection. Cells at the boundary between the dermis and the epidermis are actively dividing. These newly formed cells migrate to the skin's surface, become flattened, and die as keratin fills their cytoplasm.
- **Dermis.** The dermis is directly below the epidermis. This layer is composed of connective tissue and contains many tiny nerves, glands, blood vessels, and lymphatic vessels. The dermis provides the epidermis with nutrients and removes wastes from it.
- **Subcutaneous tissue** (*cutaneous* relates to the skin). Below the dermis is supportive tissue containing blood vessels, fat, and other types of cells.

The outermost layers of skin are bathed in secretions produced by glands in the dermis. Sweat glands produce sweat, which is delivered via fine tubules to the surface of the skin. There, the sweat evaporates to leave a salty residue that inhibits many microbes. Sebaceous glands produce the oily secretion called **sebum.** The glands open into hair follicles, so that sebum flows up through the follicles and out over the skin surface. Sebum keeps the hair and skin soft, flexible, and water-repellent.

In addition to being a physical barrier, skin is a distinct ecological habitat. Compared with the moist, warm conditions in the respiratory tract, most areas of the skin are cool and dry. Members of the normal skin microbiota are uniquely suited to thrive in this environment. They use the substances in sebum and sweat as nutrients to fuel their growth, and in doing so, inhibit other microbes. For example, resident bacteria degrade lipids in sebum, producing fatty acids that are toxic to many bacteria. In fact, the skin surface is an unfriendly habitat for most pathogens, being too dry, salty, acidic, and toxic for their survival. Those that tolerate the conditions are often shed with the dead skin cells. ⏮ normal microbiota

Although members of the normal microbiota play an essential protective role on the skin, they can also be a problem. Some are opportunistic pathogens, causing disease in people with impaired body defenses. In addition, metabolic

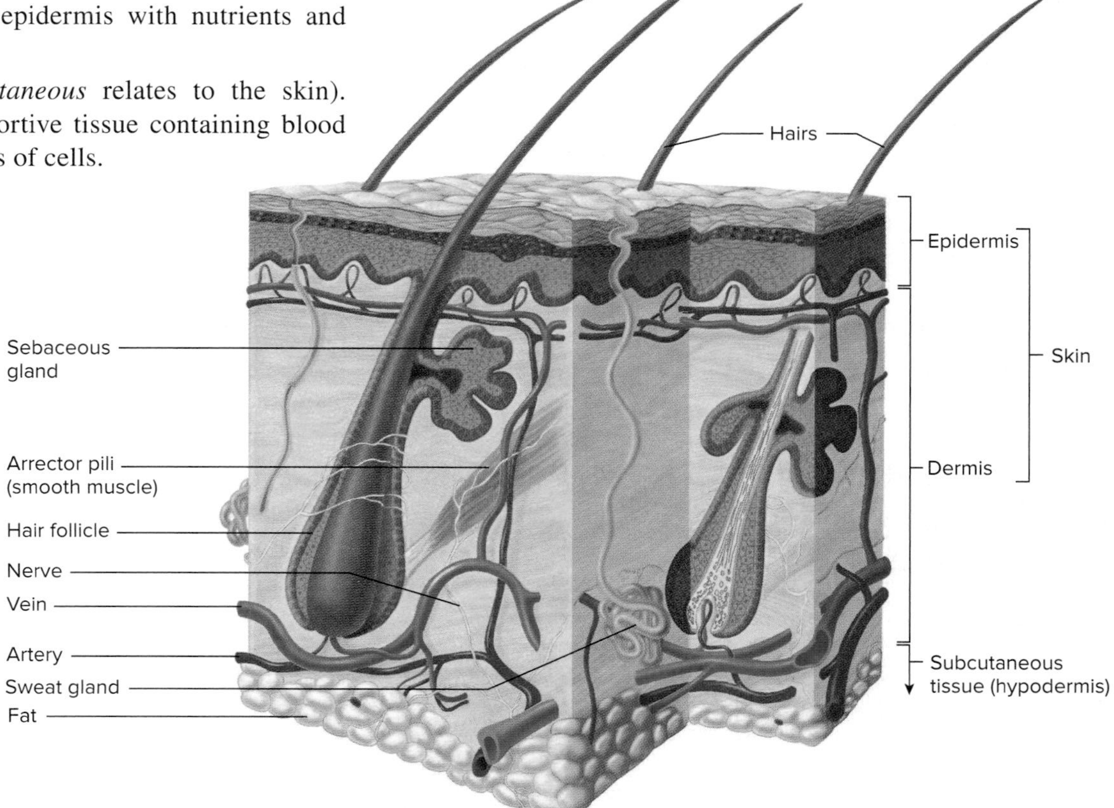

FIGURE 22.1 Microscopic Anatomy of the Skin

? How can anaerobic bacteria grow on human skin?

products of normal microbiota are responsible for body odor. Sweat is odorless when first secreted, but bacteria degrade some of its components, producing foul-smelling compounds. Most antiperspirants prevent body odor by decreasing the number and metabolic activity of bacteria at the site of application. ◀◀ opportunistic pathogens

Different regions of the skin can be compared to unique neighborhoods, made up of distinct numbers and types of inhabitants. Depending on the body location and amount of moisture, the number of bacteria on the skin surface may range from only about 1,000 organisms per square centimeter (on the back) to more than 10 million (in the groin and the armpit, where there is plenty of moisture). Most of the microbial skin inhabitants can be categorized in three groups: diphtheroids, staphylococci, and fungi (**table 22.1**).

In oily regions like the forehead, upper chest, and back, diphtheroids are common. The name of this informal grouping refers to the fact that these Gram-positive bacteria physically resemble *Corynebacterium diphtheriae,* the pathogen that causes diphtheria. Among the most common diphtheroids on the skin are *Cutibacterium* (formerly *Propionibacterium*) species. These obligate anaerobes grow within hair follicles, where O_2 is limited. ◀◀ *Corynebacterium,* ◀◀ diphtheroids, ◀◀ anaerobe

Staphylococci such as *Staphylococcus epidermidis* are salt-tolerant Gram-positive cocci that grow well on the dry environment of the skin surface. Staphylococci use available nutrients on the skin, preventing pathogen colonization. They also produce antimicrobial substances active against other Gram-positive bacteria, helping to maintain a balance among the microbial inhabitants of the skin ecosystem. ◀◀ *Staphylococcus*

Malassezia species are tiny lipid-dependent yeasts (fungi) present on human skin from late childhood onward. These yeasts can be grown on laboratory media containing fatty substances such as olive oil.

MicroByte

The average person sheds about 40,000 skin cells daily, or 1,500 during a typical microbiology lecture.

TABLE 22.1 Principal Members of the Normal Skin Microbiota

Name	Characteristics
Diphtheroids	Variably shaped, non-motile, Gram-positive rods of the *Cutibacterium (Proprionibacterium)* and *Corynebacterium* genera
Staphylococci	Gram-positive cocci arranged in packets or clusters; facultatively anaerobic
Fungi	Small yeasts of the genus *Malassezia* that require oily substances for growth are common

MicroAssessment 22.1

The skin resists colonization by most microbial pathogens, and provides a physical barrier to infection. Members of the normal skin microbiota help protect the skin from colonization by pathogens. Resident microbes are responsible for body odor and can cause disease when the host's defenses are impaired.

1. Describe four characteristics of skin that help it resist infection.
2. Name and describe three groups of microorganisms generally present on normal skin.
3. Would you expect a woman living in the tropics or one living in the desert to have larger numbers of bacteria living on the surface of her skin?

22.2 ■ Bacterial Diseases of the Skin

Learning Outcomes

3. Compare and contrast acne and hair follicle infections.
4. Describe the characteristics of staphylococcal scalded skin syndrome and impetigo.
5. Compare and contrast the pathology and epidemiology of Rocky Mountain spotted fever with those of cutaneous anthrax.

Few bacterial species invade intact skin directly. However, strands of hair provide a route of invasion by extending past the epidermis to hair follicles that penetrate into the dermis. Acne and hair follicle infections both begin with the hair shaft and are two of the most common infections that occur due to direct skin invasion.

Acne Vulgaris

Acne in its most common form begins at puberty in association with a rise in sex hormones. Although the incidence of acne drops dramatically after puberty, some people have acne well into their 30s and 40s.

Signs and Symptoms

Acne is characterized by enlarged sebaceous glands and increased secretion of sebum. The hair follicle epithelium thickens and sloughs off in clumps, gradually blocking the flow of sebum to the skin surface. Continued sebum production by the infected gland can force a plug of material to the surface, where it is visible as a blackhead. With complete obstruction, the follicle fills with sebum, causing the epidermis to bulge outward. This produces a whitish lesion called a whitehead.

Causative Agent

The species typically associated with acne is *Cutibacterium (Propionibacterium) acnes.* The cells are Gram-positive rods that grow anaerobically in and around hair follicles. They produce lipases that break down the oily sebum in the sebaceous glands and use the resulting fatty acids and glycerol as a food source. Sebum accumulates when a gland is blocked, providing more food for *C. acnes.* These bacteria then multiply to enormous numbers in the trapped sebum.

Pathogenesis

The metabolic products of the dividing bacteria cause an inflammatory response, attracting neutrophils whose enzymes damage the wall of the enlarged follicle. This can cause the follicle to burst, releasing its contents into the surrounding tissue. The result is an **abscess**—a collection of pus surrounded by inflamed tissue. The pus is made up of living and dead neutrophils, bacteria, and tissue debris. Most abscesses will eventually heal, but may leave a scar. ⏮ **inflammatory response,** ⏮ **neutrophils**

Epidemiology

Most people have *C. acnes* on their skin throughout their lives. The increased incidence of acne during puberty is probably due to excess sebum production caused by increased hormone levels, particularly testosterone. Infants may also have acne caused by maternal hormones stimulating sebum production.

Treatment and Prevention

Acne is usually a relatively mild infection that will resolve without treatment. The length of infection can be limited by medications such as antibiotics and benzoyl peroxide that inhibit the growth of *C. acnes.* Other medications such as isotretinoin act by reducing sebum production. Isotretinoin is generally prescribed for only the most serious cases of acne because it has potentially serious side effects. Squeezing acne lesions is ill-advised, because it may cause the inflamed follicles to rupture, leading to more acne scars. It can also introduce bacteria into the bloodstream, which could spread the infection to multiple locations throughout the body.

Hair Follicle Infections

Hair follicle infections are generally mild and commonly clear up without treatment. In some instances, however, they progress into severe or even life-threatening disease.

Signs and Symptoms

There are three outcomes of hair follicle infections, listed by increasing severity: folliculitis, furuncles, and carbuncles. **Folliculitis** is inflammation of hair follicles, causing small red bumps, or pimples. The hair can be pulled from its follicle, releasing a small amount of pus. The infection often goes away without further treatment. If the infection extends from the follicle to adjacent tissues, it will cause a localized redness, swelling, severe tenderness, and pain. This lesion is called a **furuncle** or boil. Pus may drain from the boil along with a plug of inflammatory cells and dead tissue. A furuncle may worsen to form a **carbuncle,** a large area of redness, swelling, and pain with several sites of draining pus. Carbuncles usually develop in areas of the body where the skin is thick, such as the back of the neck. Fever is often present, along with other signs of a serious infection. ⏮ **fever**

Causative Agent

Most furuncles and carbuncles, as well as many cases of folliculitis, are caused by *Staphylococcus aureus.* The name derives from the Greek root *staphyle* or "a bunch of grapes," referring to the arrangement of the bacteria as seen in stained smears. The species name, *aureus,* or "golden," refers to the typical creamy color of the *S. aureus* colonies. This bacterium is an extremely important potential pathogen and is mentioned frequently throughout this text as the cause of a number of medical conditions (**table 22.2**).

Staphylococcus aureus can be distinguished from most other staphylococci because it produces **coagulase** and **clumping factor.** Coagulase causes a large clot to form in plasma inoculated with *S. aureus* during several hours of incubation. In contrast, clumping factor on the cell surface causes rapid clotting when plasma is mixed into a concentrated suspension of *S. aureus.* Clumping factor is easier to test for in the lab and is often called "slide coagulase." Both coagulase and clumping factor are important virulence factors for *S. aureus.* Other staphylococcal species, such as *Staphylococcus epidermidis,* cause disease infrequently and lack the genes for coagulase and clumping factor. ⏮ **virulence factor**

Pathogenesis

Infection begins when *Staphylococcus aureus* attaches to the cells of a hair follicle, multiplies, and spreads to involve the sebaceous glands. The infection induces an inflammatory response with swelling and redness, followed by attraction and accumulation of neutrophils. If the infection continues, the follicle becomes a plug of inflammatory cells and dead tissue overlying a small abscess (**figure 22.2**). The infectious

TABLE 22.2 Some Diseases Often Caused by *Staphylococcus aureus*

Disease	Comments
Food poisoning	Intoxication from consuming exotoxin in food ⏭ **food poisoning**
Hair follicle infections	Folliculitis, carbuncles, furuncles ⏮ **hair follicle infections**
Impetigo	Superficial skin disease with pus production ⏭ **impetigo**
Infective endocarditis	Infection of heart valves or inner lining of heart ⏭ **infective endocarditis**
Staphylococcal scalded skin syndrome	Toxin-mediated skin disease ⏭ **staphylococcal scalded skin syndrome**
Toxic shock syndrome	Superantigens causing low blood pressure and organ failure ⏭ **staphylococcal toxic shock syndrome**
Wound infections	Colonization of wounds; possible systemic complications ⏭ **wound infections**

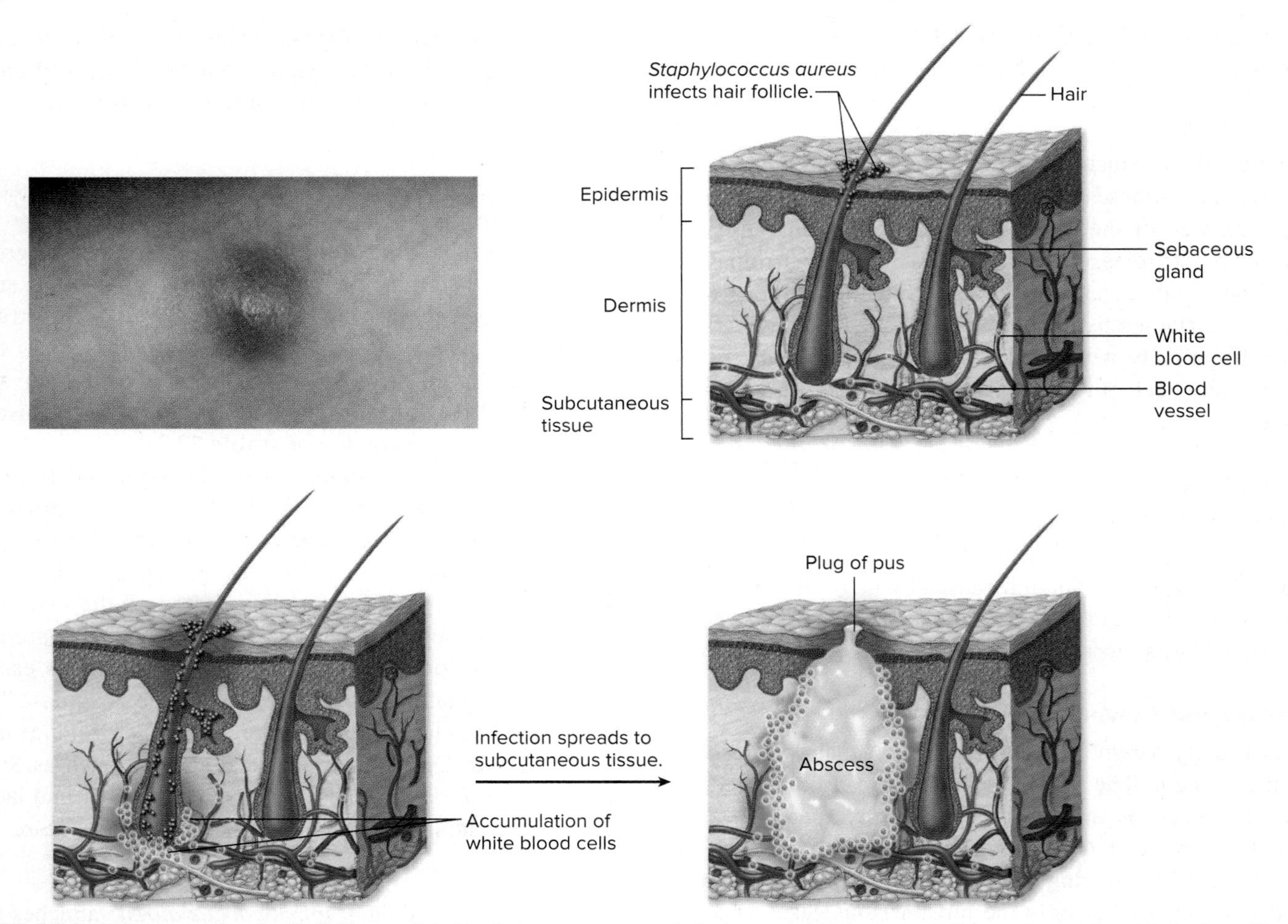

FIGURE 22.2 Pathogenesis of a Boil (Furuncle) *Staphylococcus aureus* infects a hair follicle through its opening on the skin surface. The infection produces a plug of necrotic material, a small abscess in the dermis, and finally, a larger abscess in the subcutaneous tissue. ©Dr P. Marazzi/Science Source

? **Why is it a bad idea to squeeze acne lesions?**

process sometimes spreads deeper, reaching the subcutaneous tissue, where a large abscess forms. This subcutaneous abscess is responsible for the painful localized swelling of a boil. Without effective treatment, pressure within the abscess may increase, causing it to expand to other hair follicles and form a carbuncle. If organisms enter the bloodstream, the infection can spread to other parts of the body, such as the heart, bones, or brain.

S. aureus strains can have many different virulence factors, although not all pathogenic strains make the same factors (**table 22.3**). Nearly all strains have **protein A,** a cell wall component that interferes with phagocytosis by binding to the Fc portion of antibodies and preventing opsonization (see figure 16.10). Many also synthesize a polysaccharide capsule that inhibits phagocytosis (**figure 22.3**).

Coagulase and clumping factor allow *S. aureus* to colonize soft tissues while avoiding the immune response. Coagulase activates a blood protein called prothrombin to form thrombin. Thrombin then converts fibrinogen to fibrin—the fibrous protein found in blood clots. Although most coagulase is secreted, some is tightly bound to the bacterial cells and coats their surface with fibrin when they come into contact with blood. This helps disguise the bacteria, hiding them from the immune system. Fibrin-coated staphylococci resist phagocytosis. Clumping factor on the surface of *S. aureus* cells is a virulence factor because it attaches to fibrinogen and fibrin present in damaged tissues, and helps the bacteria colonize wound surfaces. Because plastic devices such as intravenous catheters and artificial heart valves quickly become coated with fibrinogen once in a patient, they also become targets for colonization.

S. aureus secretes many enzymes that can damage host tissues. Hyaluronidase degrades hyaluronic acid, a component of host tissue that helps hold the cells together. Proteases degrade various host proteins including collagen, the white fibrous protein found in skin, tendons, and connective tissue. Lipases degrade lipids, forming fatty acids and glycerol that provide a food source for *S. aureus* when it colonizes oily hair follicles.

TABLE 22.3 Properties of *Staphylococcus aureus* Implicated in Its Virulence

Product	Effect
Capsule	Inhibits phagocytosis
Clumping factor	Attaches the bacterium to fibrin, fibrinogen, and plastic devices
Coagulase	May slow progress of leukocytes into infected area by producing clots in the surrounding capillaries
Enterotoxins	Superantigens cause food poisoning if ingested, cause toxic shock if systemic
Exfoliatin	Destroys material that binds outer layers of the epidermis together, causing scalded skin syndrome
Fibronectin-binding protein	Attaches bacterium to acellular tissue substances, endothelium, epithelium, clots, indwelling plastic devices
Hyaluronidase	Breaks down hyaluronic acid component of tissue, thereby allowing infection to spread
Leukocidin	Kills neutrophils or causes them to release their enzymes
Lipase	Breaks down fats by hydrolyzing the bond between glycerol and fatty acids
Proteases	Degrade collagen and other tissue proteins
Protein A	Binds to Fc portion of antibody, thereby interfering with opsonization that otherwise facilitates phagocytosis
Toxic shock syndrome toxin	Causes rash, diarrhea, and shock
α-Toxin	Makes holes in host cell membranes

Most strains of *S. aureus* also produce one or more toxins that damage or kill host cells. The membrane-damaging α-toxin kills host cells by making holes in their membranes. Leukocidins kill white blood cells. A relatively small percentage of *S. aureus* strains produce one or more additional disease-specific toxins. One of these, exfoliatin, causes staphylococcal scalded skin syndrome, the next disease described in this chapter. ⏮ **membrane-damaging toxin**

Epidemiology

Staphylococcus aureus can be found in the nostrils of almost everyone at one time or another. Up to 30% of healthy adults

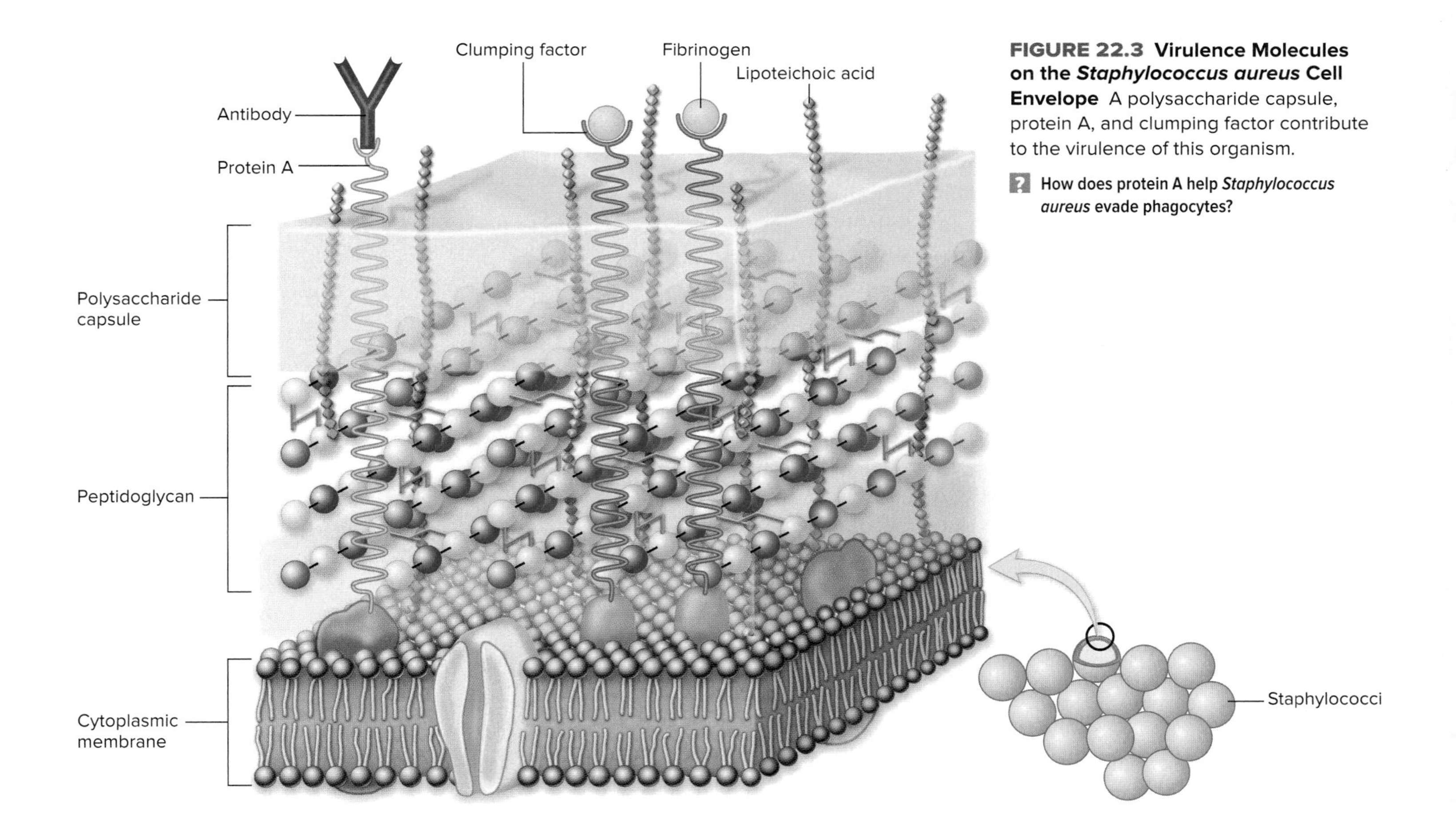

FIGURE 22.3 Virulence Molecules on the *Staphylococcus aureus* Cell Envelope A polysaccharide capsule, protein A, and clumping factor contribute to the virulence of this organism.

? **How does protein A help *Staphylococcus aureus* evade phagocytes?**

have continually positive nasal cultures for a year or more, whereas over 60% will be colonized at some time during a given year. The organisms are then spread to other parts of the body and to the environment by the hands. Although the nostrils seem to be the preferred habitat of *S. aureus,* moist areas of skin are also frequently colonized. People with boils and other staphylococcal infections shed large numbers of *S. aureus* and should not work with food, or near patients with surgical wounds or chronic illnesses. Staphylococci survive well in the environment, so they are easily transferred from one host to another via fomites. ◀◀ **fomites**

Because *S. aureus* is so common, tracing the source of a staphylococal outbreak requires precise identification of the strain involved. This can be done using techniques such as molecular typing, phage typing, and antibiogram determination (see section 10.4). ◀◀ **characterizing strain differences**

Treatment and Prevention

Treatment of boils and carbuncles may require minor surgery to drain the pus from the lesion. Afterward, patients are usually given oral antibiotics.

Treating *S. aureus* infections can be complicated because many strains are resistant to multiple antibacterial medications. When penicillin was first used to treat *S. aureus* infections, nearly all strains were susceptible to the medication. However, strains that produce a penicillinase (a type of β-lactamase) spread quickly, so that most isolates are now resistant to it. Strains producing PBP2a, a modified version of a penicillin-binding protein, also appeared. These strains, referred to as MRSA (methicillin-resistant *Staphylococcus aureus*), are resistant to nearly all β-lactam antibiotics (the exception is ceftaroline, a new cephalosporin). Many MRSA strains are resistant to a number of other medications as well. Before 1997, even the most resistant examples were reliably treated with vancomycin. Since then, however, the first vancomycin-intermediate *S. aureus* (VISA) and vancomycin-resistant *S. aureus* (VRSA) strains were identified. Several medications active against VRSA have been marketed, including linezolid, daptomycin, and tigecycline. ◀◀ **β-lactamase,** ◀◀ **β-lactam antibiotics,** ◀◀ **penicillin-binding proteins,** ◀◀ **VISA and VRSA**

When MRSA strains first appeared, most could be traced to hospitals and clinics. These strains are referred to as HA-MRSA (hospital-acquired MRSA). More recently, however, completely different strains have become widespread among healthy carriers in the community. These new strains are called CA-MRSA (community-acquired MRSA). In the United States, all common CA-MRSA isolates appear to be progeny of a strain type called USA-300. In general, they are resistant only to β-lactam antibiotics (except ceftaroline) and macrolides. They are more virulent than HA-MRSA and have a group of genes that codes for a leukocyte-destroying leukocidin. There is some debate as to whether this leukocidin contributes to higher morbidity and mortality or is simply a trait shared by most strains of CA-MRSA.

In an effort to control the spread of MRSA, many hospitals screen patients at admission. The patients who carry a MRSA strain are isolated and given appropriate antibacterial treatment to prevent the bacterium from spreading to other people. Some hospitals also screen patients for MRSA when they are discharged from the hospital to make sure they do not take a MRSA strain home with them. ◀◀ **carriers**

Preventing staphylococcal skin disease is very difficult because so many individuals carry the organism. Applying an antibacterial cream to the nostrils and using soaps containing an antibacterial agent such as hexachlorophene to wash the skin reduce the bacterial burden and may eliminate the carrier state. Because *S. aureus* is most commonly transmitted by the hands, handwashing and regular use of hand sanitizers limit the spread of these bacteria.

MicroByte

People colonized with *S. aureus* may have as many as 10^8 cells of the bacterium per nostril.

Staphylococcal Scalded Skin Syndrome

Staphylococcal scalded skin syndrome (SSSS) is a potentially fatal, toxin-mediated disease that occurs mainly in infants.

Signs and Symptoms

As the name suggests, staphylococcal scalded skin syndrome causes a patient's skin to look as though it has been burned with hot liquid or steam, so that the outer layer blisters and peels (**figure 22.4**). It begins after a variable incubation period with a generalized redness of the skin, sometimes affecting the

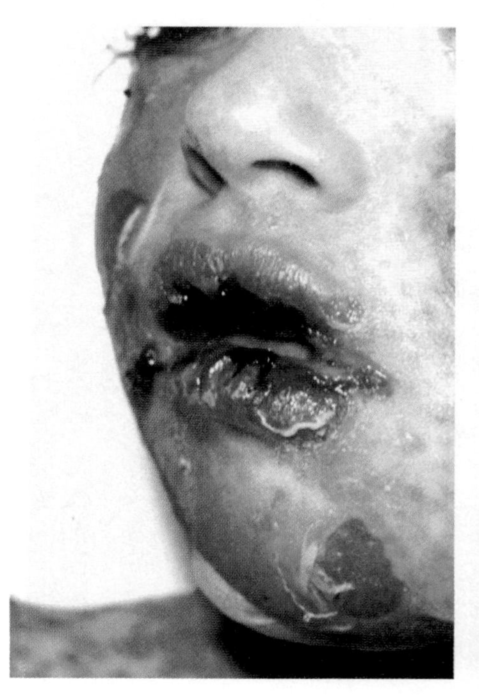

FIGURE 22.4 Staphylococcal Scalded Skin Syndrome (SSSS) A toxin called exfoliatin, produced by certain strains of *Staphylococcus aureus,* causes the outer layer of skin to blister and peel.

? **Why is SSSS sometimes fatal?**

TABLE 22.4 Staphylococcal Scalded Skin Syndrome

Signs and symptoms	Malaise, irritability, fever, sensitive red rash with sandpaper texture, large blisters, peeling of skin
Incubation period	Variable, usually days
Causative agent	Strains of *Staphylococcus aureus* that produce exfoliatin toxin
Pathogenesis	Exfoliatin is carried by the bloodstream to the epidermis, where it causes the outer layer to blister and peel; loss of body fluid and secondary infections contribute to mortality.
Epidemiology	Person-to-person transmission; mainly in infants.
Treatment and prevention	Treatment: antibiotics; removal of dead tissue. Prevention: isolation of the patient to prevent secondary infection and to limit the spread of the disease to others.

entire body. Other early signs and symptoms include malaise—a vague feeling of discomfort and uneasiness—as well as irritability and fever. The nose, mouth, and genitalia may be painful for one or more days before the typical features of the disease become apparent. Within 48 hours after the redness appears, the skin becomes wrinkled, and large blisters filled with clear fluid develop. The skin is tender to the touch, peels easily, and feels like sandpaper.

Causative Agent

Staphylococcal scalded skin syndrome is caused by *Staphylococcus aureus* strains that produce a toxin called **exfoliatin.** Only about 5% of *S. aureus* strains produce this toxin, which destroys material that binds together the outer layers of epidermis. At least two kinds of exfoliatins exist. One is coded for by a plasmid gene, and the other by a chromosomal gene. ◀◀ plasmids

Pathogenesis

Exfoliatin causes the epidermis to split just below the dead keratinized outer layer. Even though the bacteria generally colonize a relatively small area of the skin, the exfoliatin that they produce is carried throughout the body in the bloodstream, so the localized infection results in widespread skin damage. Due to skin loss, there is significant loss of body fluid and danger of secondary infection. ◀◀ secondary infection

Epidemiology

Staphylococcal scalded skin syndrome can affect anyone, but occurs most frequently in newborn infants. Elderly and immunocompromised individuals are also at increased risk. Transmission is generally by direct contact from person to person. The disease usually appears in isolated cases, but small epidemics in nurseries sometimes occur.

Treatment and Prevention

Initial treatment for staphylococcal scalded skin syndrome includes an antibiotic such as a penicillinase-resistant penicillin. As with severe burns, dead skin is removed to help prevent secondary infection. Although the disease can be fatal, quick treatment usually leads to full recovery. There are no preventive measures except to place patients suspected of having SSSS in protective isolation. This limits the spread of the pathogen to others and helps prevent secondary infections in the isolated patient. **Table 22.4** describes the main features of this disease.

Impetigo

Impetigo is the most common type of **pyoderma,** a superficial skin disease characterized by pus production (**figure 22.5**). Pyodermas can result from infection of an insect bite, burn, scrape, or other wound. Sometimes, the injury is so slight it is not even noticeable.

Signs and Symptoms

Signs and symptoms of impetigo result from inflammation in patches of epidermis just beneath the dead, keratinized outer layer. Thin-walled blisters develop 2 to 5 days after infection. These then break and are replaced by oozing, yellowish crusts of drying plasma. Usually, patients experience little fever or pain, but lymph nodes near the involved areas often enlarge. ◀◀ plasma

Causative Agent

Many cases of impetigo, even epidemics, are due to *Streptococcus pyogenes.* These Gram-positive, chain-forming organisms are group A streptococci, bacteria characterized

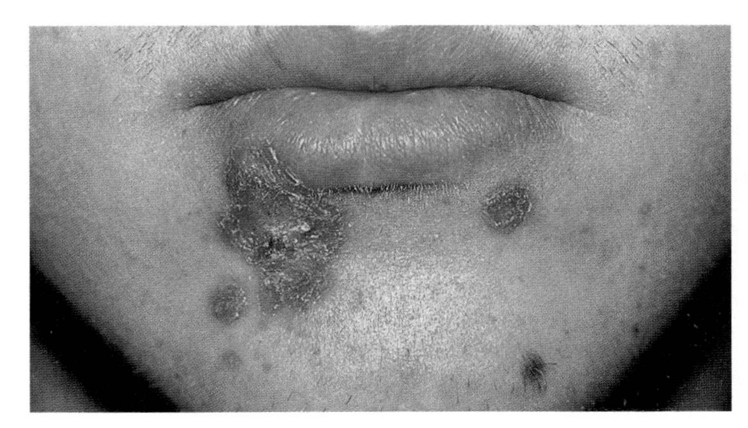

FIGURE 22.5 Impetigo This type of pyoderma is often caused by *Streptococcus pyogenes.* ©SPL/Science Source

? How is impetigo spread in a population?

TABLE 22.5 ***Streptococcus pyogenes* Versus *Staphylococcus aureus***

	Streptococcus pyogenes	*Staphylococcus aureus*
Morphology	Chains of Gram-positive cocci	Clusters of Gram-positive cocci
Growth characteristics	Beta-hemolytic colonies; catalase-negative, coagulase-negative, obligate fermenter	Golden, beta-hemolytic colonies, catalase-positive, coagulase-positive, facultative anaerobe
Virulence factors	Cell wall contains group A polysaccharide, an Fc receptor (protein G), and M protein. Bacterium produces hemolysins (streptolysins O and S), streptokinase, DNase, hyaluronidase, and others	Cell wall contains an Fc receptor (protein A). Bacterium produces, hemolysins, leukocidin, hyaluronidase, nuclease, protease, and others
Diseases	Impetigo, strep throat, wound infections, scarlet fever, puerperal fever, toxic shock, and necrotizing fasciitis. Complications: glomerulonephritis and rheumatic fever	Boils, staphylococcal scalded skin syndrome, wound infections, abscesses, bone infections, impetigo, food poisoning, and staphylococcal toxic shock syndrome

by the presence of a polysaccharide called group A carbohydrate in their cell walls. Like *Staphylococcus aureus, S. pyogenes* causes a variety of different diseases and is mentioned in numerous places throughout this text. The most detailed description is in chapter 21. Impetigo may also be caused by *S. aureus.* **Table 22.5** compares *S. aureus* and *S. pyogenes.* ⏮ *Streptococcus pyogenes*

Pathogenesis

Many strains of *S. pyogenes* exist, some of which can colonize the skin. Minor injuries introduce the bacteria into deeper layers of epidermis, leading to infection. Scratching an infected area may spread the bacteria to other areas of the skin.

The surface components of *S. pyogenes,* including a hyaluronic acid capsule and a cell wall component called M protein, interfere with phagocytosis of the organism (see figure 21.6). A number of extracellular products contribute to the virulence of *S. pyogenes,* similar to those described for *S. aureus.* These include enzymes such as proteases, nucleases, and hyaluronidase. None of them appear to be essential for infection, however, because antibodies against them fail to protect experimental animals. A more detailed description of these virulence factors is in chapter 21. ⏮ virulence factors of *S. pyogenes,* ⏮ M protein

Even though impetigo is limited to the epidermis, streptococcal products are absorbed into the circulation. The immune response to these proteins is thought to cause post-streptococcal sequelae, particularly acute glomerulonephritis. ⏮ post-streptococcal sequelae, ⏮ glomerulonephritis

Epidemiology

Impetigo is most common among poor young children living in hot, humid areas. Person-to-person contact spreads the disease, as do insects, and fomites such as toys and towels. ⏮ fomites

Treatment and Prevention

So far, *S. pyogenes* strains remain susceptible to penicillin. For patients allergic to penicillin, erythromycin can be substituted.

People with impetigo should avoid contact with others to prevent spreading the bacteria. The transmission of impetigo can be limited by keeping skin clean and avoiding sharing personal items such as towels and washrags. Antiseptics and wound cleansing also decrease the chance of infection. **Table 22.6** summarizes the main features of impetigo.

Rocky Mountain Spotted Fever

Rocky Mountain spotted fever (RMSF) was first recognized in the Rocky Mountain area of the United States—thus its name. The disease is caused by an obligate intracellular bacterium that is transmitted by certain species of hard-shelled ticks.

TABLE 22.6 **Impetigo**

Signs and symptoms	Blisters that break, releasing plasma and pus; formation of yellowish crusts; lymph node enlargement
Incubation period	2 to 5 days
Causative organisms	*Streptococcus pyogenes, Staphylococcus aureus*
Pathogenesis	Organisms entering the skin through minor breaks; certain strains of *S. pyogenes* that cause impetigo can also cause glomerulonephritis.
Epidemiology	Spread by direct contact, insects, and fomites.
Treatment and prevention	Treatment: appropriate antibiotic. Prevention: keeping clean; treating skin injuries.

Signs and Symptoms

Rocky Mountain spotted fever begins suddenly with a headache, pains in the muscles and joints, and fever. In most cases, within a few days of infection, a rash characterized by flat pink spots first appears on the wrists and ankles, and then spreads to the palms and soles of the feet (which is not typical of most other diseases characterized by a rash), as well as up the arms and legs to the rest of the body. The spots may develop into small bumps (**figure 22.6**). In about 50% of cases, these may be followed by purplish spots called **petechiae** that form as small blood vessels hemorrhage under the skin. Damage to the heart, kidneys, brain, and other body tissues can cause a drop in blood pressure, shock, and death unless treatment is given promptly.

Causative Agent

RMSF is caused by *Rickettsia rickettsii* (**figure 22.7**). The organisms are tiny, Gram-negative, non-motile coccobacilli. They are obligate intracellular parasites, and their dependence on host cells makes the bacteria very difficult to grow in culture (see A Glimpse of History). *R. rickettsii* can sometimes be identified early in an infection by microscopic examination of skin lesion biopsies (pieces of tissue surgically removed from the lesion site). Also, their DNA can be amplified by the polymerase chain reaction (PCR) and identified with a probe.

⏮ PCR, ⏮ DNA probe

Pathogenesis

Rocky Mountain spotted fever is transmitted by the bite of a tick infected with *R. rickettsii.* The tick attaches and feeds, but the process is usually painless and goes unnoticed. The pathogen typically is transmitted to the host after 4–10 hours. The bacteria are released into capillary blood with the tick saliva and induce endothelial cells lining the host's small blood vessels to engulf them. Once inside the endothelial cells, rickettsias leave the phagosome and multiply. Like several other pathogens, they cause host cell actin to polymerize, forming an "actin tail" that propels the bacterial cells into adjacent host cells (see Focus Your Perspective 3.1). Eventually, the host cell membrane is so damaged by this process that the cell ruptures, releasing bacteria to travel through the bloodstream, infecting more cells.

Infection can cause inflammation in the blood vessels (vasculitis) that leads to clotting and small areas of necrosis (tissue death) due to lack of O_2. This process causes a hemorrhagic skin rash but may also cause tissue damage in the brain, heart, kidneys, and other organs. In severe cases, endotoxin (lipopolysaccharide) released into the bloodstream from the rickettsial cell walls results in systemic inflammation. This leads to life-threatening shock (due to insufficient blood flow) as well as clotting throughout the body, an organ-damaging condition called disseminated intravascular coagulation (DIC).

⏮ lipopolysaccharide, ⏭ disseminated intravascular coagulation

Epidemiology

Rocky Mountain spotted fever occurs in a sporadic distribution throughout the Americas. Despite the name of the disease, the highest incidence in the United States has generally been in the south Atlantic and south-central states (**figure 22.8**).

RMSF is a zoonosis, maintained in nature in various species of ticks and mammals. Rickettsias generally do not

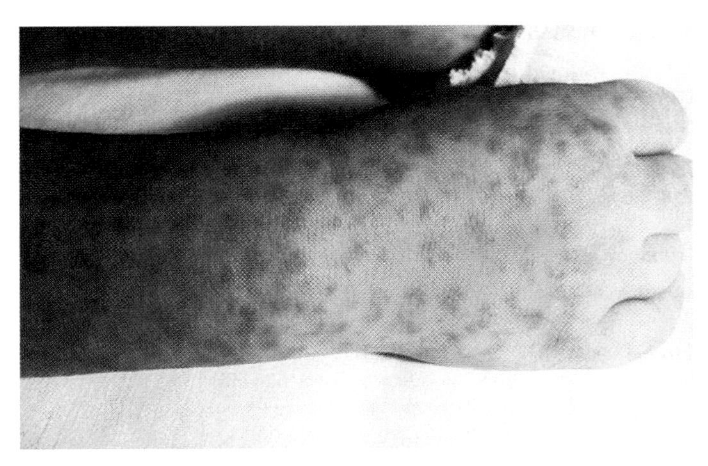

FIGURE 22.6 Rash Caused by Rocky Mountain Spotted Fever (RMSF) Early in the disease, most people have a rash of flat pink non-itchy spots. The rash typically begins on the arms and legs, and spreads centrally. Source: CDC

? How might the rash change over the course of the disease?

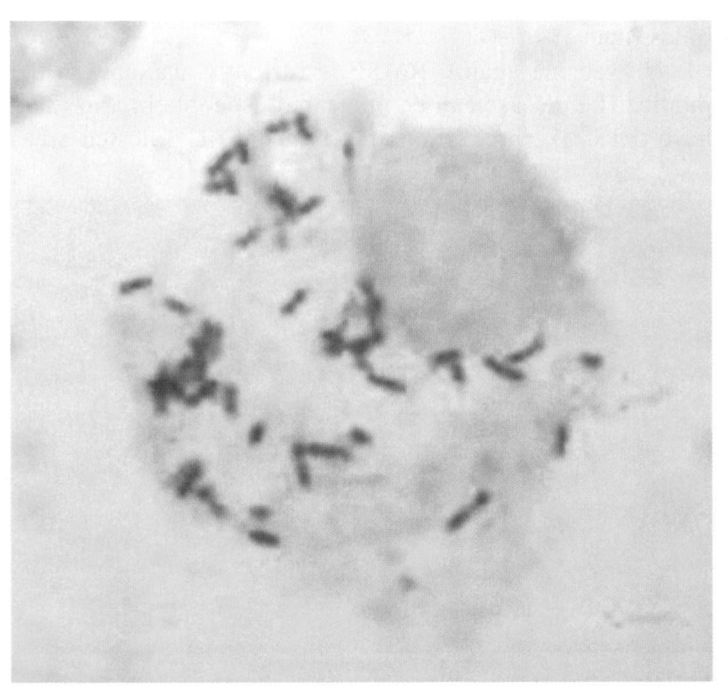

FIGURE 22.7 The Obligate Intracellular Bacterium *Rickettsia rickettsii* Source: CDC

? Do you think *Rickettsia rickettsii* can grow on blood agar? Explain.

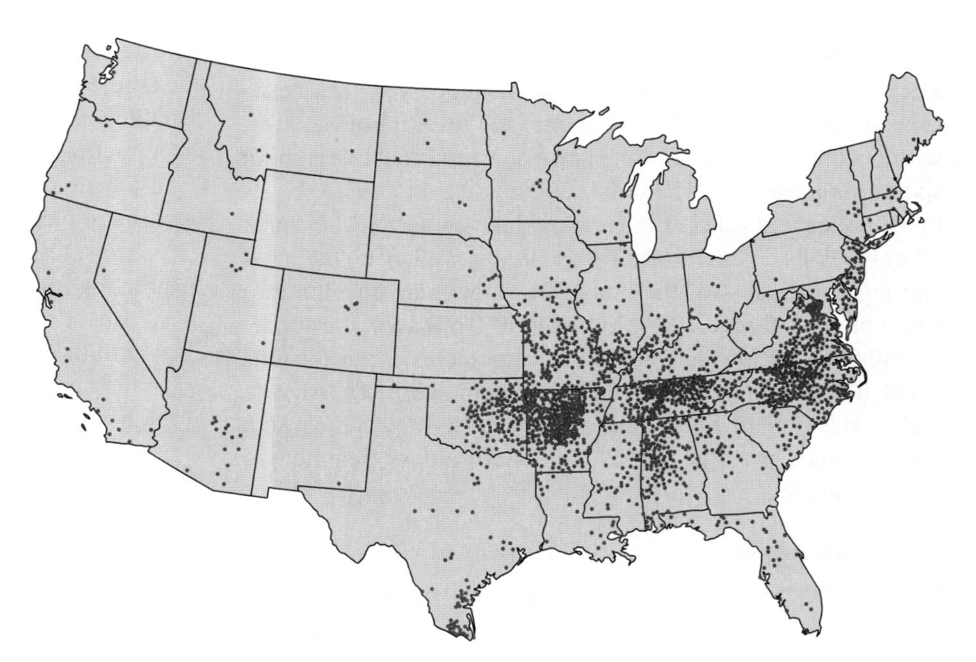

FIGURE 22.8 Rocky Mountain Spotted Fever Reported cases in the United States, 2014; each dot represents one case reported in the person's county of residence. Source: "Tickborne Diseases of the United States: Rocky Mountain Spotted Fever," Centers for Disease Control and Prevention. www.cdc.gov/ticks/tickbornediseases/rmsf.html CDC

Why do most cases of RMSF occur in the summer months?

cause disease in their natural hosts—the host and pathogen have co-evolved to reach a state of balanced pathogenicity. Because they are accidental hosts, humans often develop severe disease. ◀◀ balanced pathogenicity

Several species of ticks transmit RMSF to humans. The main vector in the western United States is the wood tick, *Dermacentor andersoni* (**figure 22.9**), while in the East it is the dog tick, *D. variabilis*. Once infected, ticks remain so for life, transmitting *R. rickettsii* from one generation to the next through their eggs. Ticks are most active from April to September. Not surprisingly, most cases of RMSF occur during this time of year. ◀◀ zoonoses

Treatment and Prevention

The antibiotic doxycycline is very effective in treating Rocky Mountain spotted fever if given early in the disease, before irreversible damage to vital organs has occurred. Without treatment, the overall case-fatality rate from the disease is about 20%, but it can be much higher in elderly patients. With early diagnosis and treatment, the case-fatality rate is reduced to less than 5%.

No vaccine against RMSF is currently available to the public. The disease can be prevented if people in areas that have ticks take precautions to (1) avoid tick-infested areas when possible; (2) wear protective clothing; (3) use tick repellents such as DEET (*N,N*-diethyl-*meta*-toluamide); (4) carefully inspect their bodies (especially the scalp, armpits, and groin) for ticks several times daily; and (5) remove attached ticks carefully to avoid crushing them and thereby contaminating the bite wound with their infected tissue fluids. Ticks can be removed using blunt tweezers to grasp the mouthparts and then gently pulling away from the skin. The site of the bite should be treated with an antiseptic. Touching the tick with a hot object, gasoline, or whiskey is ineffective. The main features of RMSF are summarized in **table 22.7.**

FIGURE 22.9 *Dermacentor andersoni*, the Wood Tick The wood tick is the main vector of RMSF in the western United States. Source: Dr. Christopher Paddock/CDC

What can be done to limit exposure to these ticks?

Cutaneous Anthrax

Anthrax is found in animal populations around the world, but is rare in humans. It may, however, occur when endospores

TABLE 22.7 Rocky Mountain Spotted Fever (RMSF)

Signs and symptoms	Headache, pains in muscles and joints, and fever, followed by a hemorrhagic rash that begins on the extremities
Incubation period	4 to 8 days
Causative organism	*Rickettsia rickettsii*, an obligate intracellular bacterium
Pathogenesis	Organisms multiply at site of tick bite, invade the bloodstream, and then infect endothelial cells; blood vessel involvement and systemic inflammatory response damage tissues.
Epidemiology	A zoonosis transmitted by bite of infected tick, usually *Dermacentor* species.
Treatment and prevention	Treatment: appropriate antibiotics. Prevention: avoiding tick-infested areas; using tick repellent; removing ticks within 4 hours.

of *Bacillus anthracis* enter the body by inhalation, ingestion, or contact with the skin. Cutaneous anthrax, in which spores are acquired through a skin lesion, is the most common type in humans, but inhalation anthrax has a much higher case-fatality rate. ⏮ inhalation anthrax

Signs and Symptoms

Within a few days after skin exposure to *B. anthracis* endospores, a small red bump (papule) that resembles an insect bite appears at the site of infection. Over a few days, the bump will develop vesicles or blisters and eventually form an ulcer (open sore) surrounded by swelling. A characteristic black eschar (dead tissue resembling a flat scab) will cover the center of the area and may persist for several weeks (**figure 22.10**). The lesions are painless and seldom result in scarring.

Causative Agent

Bacillus anthracis is an endospore-forming, Gram-positive non-motile rod-shaped or slightly curved bacterium with blunt ends. Its species name is derived from the Greek word for coal because it forms black "coal-like" scabs on infected skin (see figure 22.10). The infectious form of the organism is an endospore that can remain in the environment indefinitely. In the host, the endospores germinate to form vegetative cells capable of producing exotoxins.

Pathogenesis

Virulence of *Bacillus anthracis* is largely determined by the capsule that protects the organism from phagocytosis and the exotoxin complex that the vegetative bacteria produce. These are described in chapter 21 in the discussion of inhalation anthrax. All types of anthrax have the potential to become systemic, but this is much less likely with cutaneous anthrax.

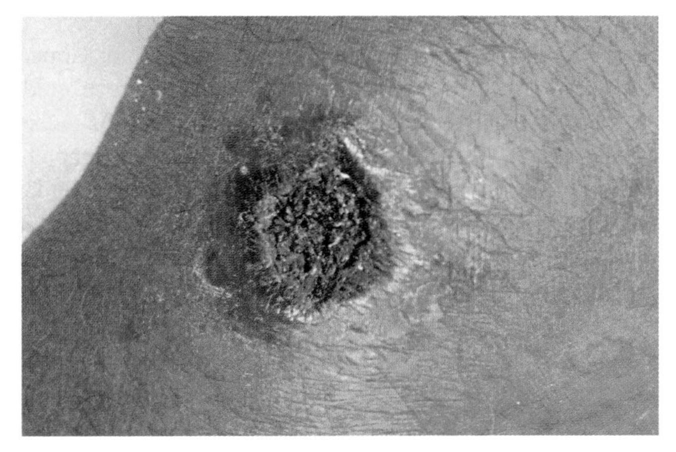

FIGURE 22.10 Anthrax Lesion This lesion on the forearm shows the distinctive tissue erosion and black eschar associated with *Bacillus anthracis* infection. Source: James H. Steele/CDC

? Will this lesion leave behind a scar?

MicroByte

Injection anthrax in heroin users looks like cutaneous anthrax, but can spread quickly throughout the body.

Epidemiology

Cutaneous anthrax accounts for over 90% of human anthrax cases. It is usually seen in people who work with animals or animal products such as farmers, veterinarians, leather tanners, or those who work with wool. Cases are usually mild and require no treatment. A small percentage of cases become systemic, most often leading to death. Routine person-to-person transmission of cutaneous anthrax does not occur.

Treatment and Prevention

Cutaneous anthrax can be treated with 7–14 days of ciprofloxacin (or other fluoroquinolone), doxycycline, or penicillin. Antimicrobial medications have little effect on healing of the cutaneous lesion, but may prevent the organism from spreading systemically. Typical cases require no hospitalization. As discussed with inhalation anthrax, the vaccine is given only to people at greatest risk of exposure. The main features of cutaneous anthrax are summarized in **table 22.8.**

TABLE 22.8	Cutaneous Anthrax
Signs and symptoms	Small red bump develops into painless black eschar
Incubation period	Usually 1–7 days
Causative agent	*Bacillus anthracis,* an endospore-forming, Gram-positive rod
Pathogenesis	Exotoxins cause cell death.
Epidemiology	Seen in people who work with animals or animal products; no person-to-person transmission.
Treatment and prevention	Treatment: antimicrobial medications. Prevention: vaccine for at-risk populations.

MicroAssessment 22.2

Folliculitis, furuncles, and carbuncles are usually caused by *Staphylococcus aureus.* Some strains of *S. aureus* produce exfoliatin and can cause staphylococcal scalded skin syndrome. Impetigo is a pyoderma often caused by *Streptococcus pyogenes* or *S. aureus.* Rocky Mountain spotted fever is caused by *Rickettsia rickettsii* and is transmitted by ticks. Cutaneous anthrax is rare in humans.

4. List four extracellular products of *Staphylococcus aureus* that contribute to its virulence.
5. Describe the pattern of skin rash used to diagnose Rocky Mountain spotted fever.
6. Patients with staphylococcal scalded skin syndrome often do not have *Staphylococcus aureus* growing in the affected areas. Why do you think this is the case?

22.3 ■ Viral Diseases of the Skin

Learning Outcomes

6. Compare and contrast chickenpox, measles, and rubella.
7. Compare and contrast fifth disease and roseola.
8. Describe the process that leads to warts.

Several different viruses initially infect the upper respiratory tract, but are carried in the blood to the skin, where they cause distinctive skin rashes. These diseases are usually diagnosed based on the appearance of the rash and other clinical findings. When the disease is not typical, however, immunological testing can be used to identify antibodies against each virus. Other viruses enter through skin lesions. ◀◀ immunological testing

Varicella (Chickenpox)

Chickenpox is the popular name for varicella, a rash that was common in childhood before the varicella vaccine was introduced in 1995. The causative agent is a member of the *Herpesviridae* family. All herpesviruses produce latent infections that can reactivate long after recovery from the initial illness. ◀◀ latent infections

Signs and Symptoms

Most cases of childhood chickenpox are mild. Early signs and symptoms include fever, headache, and malaise. A characteristic rash develops 10 to 21 days after infection, beginning with small red spots called macules that progress to little bumps called papules. These become small blisters called vesicles or pustules that then crust over. When the rash is present, lesions are seen at various stages of development from macule to crusted vesicle (**figure 22.11**). Lesions can occur anywhere on the body, but usually first appear on the head, chest, and back. They also may occur on mucous membranes such as those lining the mouth and eyelids. Patients may have only a few lesions or as many as a few hundred. The lesions are pruritic (itchy), and scratching them can lead to serious, even fatal, secondary infection by *Streptococcus pyogenes* or *Staphylococcus aureus*.

The signs and symptoms of chickenpox tend to be more severe in older children and adults. About 20% of adults develop pneumonia. This may lead to rapid breathing, cough, shortness of breath, and a dusky skin color from lack of O_2. The pneumonia subsides with the rash, but respiratory signs and symptoms often persist for weeks. Chickenpox is a threat to immunocompromised patients of any age—the virus damages the lungs, heart, liver, kidneys, and brain, resulting in death in about 20% of the cases. ◀◀ pneumonia

A person who has been infected with the chickenpox virus is at risk of later developing **shingles,** a disease that results from reactivation of the latent virus. This disease can occur at any age but becomes increasingly common later in life. It begins with pain or tingling in the skin region supplied by a sensory nerve that houses the latent virus. After a few days to 2 weeks, a rash characteristic of chickenpox appears, but unlike with chickenpox, the rash is usually restricted to the area supplied by the branches of the involved sensory nerve (**figure 22.12**). These nerves most often serve the chest or abdomen, but shingles can appear on a limb or on the face. Facial shingles can pose risk to the eyes. Unlike the chickenpox rash, the shingles rash is very painful. Although the lesions generally subside within a week, the pain may persist for months, or longer. In people with AIDS or other immunodeficiency, the rash may spread over the entire body. Shingles does not spread from person to person, because it is caused by a reactivation of latent virus.

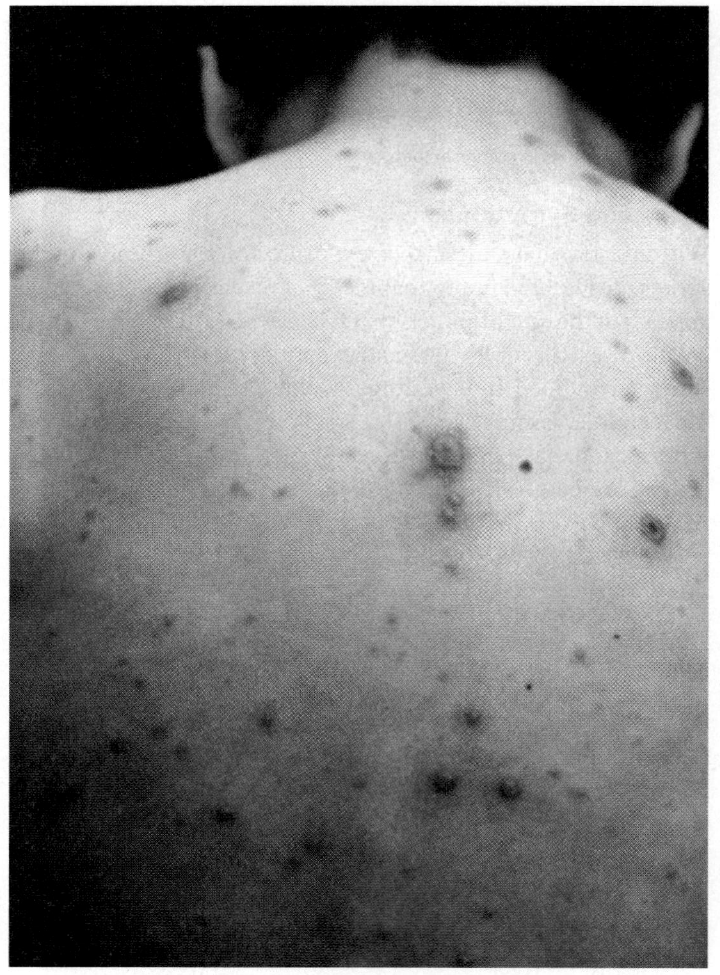

FIGURE 22.11 A Child with Chickenpox (Varicella) Characteristically, lesions in various stages of development—macules, papules, vesicles, and pustules—are present. Source: CDC

? Is chickenpox caused by a poxvirus?

Causative Agent

Chickenpox and shingles (herpes zoster) are caused by varicella-zoster virus (VZV), a member of the *Herpesviridae*

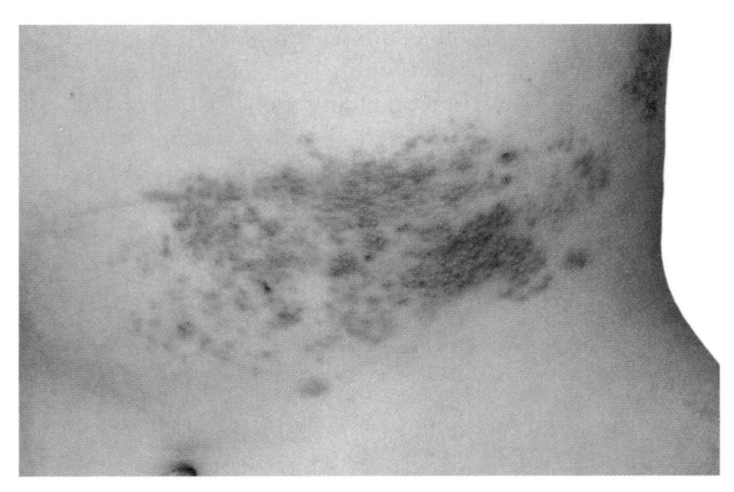

FIGURE 22.12 Shingles The painful rash typically is limited to the surface region of the body associated with a sensory nerve housing the latent virus. ©franciscodiazpagador/Getty Images

Can you catch shingles from a person with shingles?

family. It is an enveloped, double-stranded DNA virus, identical to other herpesviruses in appearance (**figure 22.13**).

Pathogenesis

Varicella-zoster virus enters the body through the respiratory tract, establishes an infection, replicates in regional lymph nodes, and travels to the skin via the bloodstream. After the skin is infected, the virus spreads locally and the characteristic skin lesions appear.

Stained preparations of infected cells show **intranuclear inclusion bodies** (pink bodies in the nucleus) where the virus reproduces. Some infected cells fuse together, forming multinucleated giant cells. The infected cells swell and eventually lyse. When an area of skin infection involves a sensory nerve ending, the virus enters the nerve. From there, it travels to ganglia (singular: ganglion), which are collections of nerve cell bodies, located near the spinal cord. Conditions inside the ganglia prevent full expression of the VZV genome. However, viral DNA is present and fully capable of coding for mature infectious virus. The mechanism of viral replication suppression within the nerve cell is not known but is probably under the control of immune cells.

Shingles is most likely to occur when cell-mediated immunity declines. With a weakened cellular immune system, the virus can reactivate in an infected nerve cell. The virus replicates in the nucleus of the nerve cell, and then moves via the nerve cell back to the skin, where it can spread locally, causing skin lesions. Unlike the situation with chickenpox, however, the replicating virus originates from a single infected nerve cell rather than from the bloodstream, which is why the rash of shingles occurs in only a limited region, as opposed to the widespread rash of chickenpox. With the appearance of the skin lesions, a prompt, intense secondary (memory) response of adaptive immunity begins. As immune cells respond, the signs and symptoms of shingles disappear. Sometimes, however, the inflammatory response leads to scars and chronic pain. ⏮ secondary response, ⏮ cell-mediated immunity

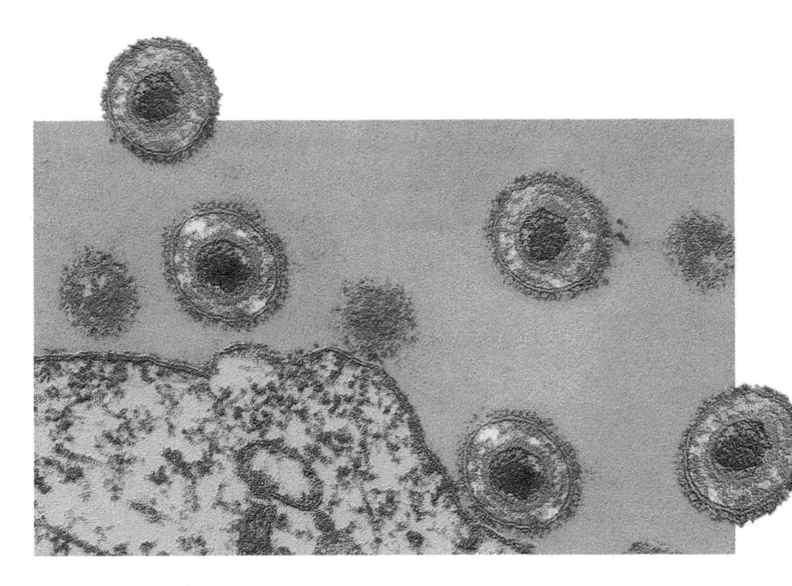

FIGURE 22.13 Varicella-Zoster Virus Colored-enhanced TEM of the virus that causes chickenpox and shingles. ©Oliver Mecks/Science Source

How does this virus cause both shingles and chickenpox?

Epidemiology

Chickenpox is a notifiable disease, requiring that confirmed cases are reported to the CDC. However, many cases are so mild that they go unreported. In the early 1990s, before a vaccine was available, there were approximately 4 million cases of chickenpox each year in the United States, resulting in 10,000 hospitalizations and approximately 100 to 150 deaths. Within 10 years after the release of the vaccine, chickenpox caused fewer than half a million cases and 8 deaths each year.

VZV is highly contagious and can be transmitted by both respiratory secretions and skin lesions. As with many diseases transmitted by the respiratory route, most cases occur in the winter and spring months. Humans are the only reservoir. The incubation period averages about 2 weeks, with a range of 10 to 21 days. People are infectious from 1 to 2 days before the rash appears until all the lesions have crusted (usually 4 days after the onset). ⏮ reservoir

Chickenpox is a major threat to babies. It has a case-fatality rate as high as 30% in newborn infants. Babies born to mothers who had chickenpox early in pregnancy occasionally develop congenital varicella syndrome. These babies may be born with underdeveloped head and limbs, and cataracts.

The ability of VZV to form a latent infection allows it to persist within a population. A person who had chickenpox in the past can develop shingles, creating a source of the virus that can be transmitted to susceptible individuals. Those newly infected people will then develop chickenpox. Thus, epidemics of chickenpox can reappear from cases of shingles if enough susceptible individuals are present.

Treatment and Prevention

The antiviral medications acyclovir and famciclovir, among others, are helpful in preventing and treating VZV infections. Certain over-the-counter medications may be given to treat fever, but aspirin should not be given to children between the ages of 5 and 15 because it increases the risk of Reye's syndrome. ◀◀ Reye's syndrome

An attenuated chickenpox vaccine has been used in the United States since 1995. All healthy children and adults without a history of chickenpox are advised to be vaccinated against the disease. The first dose for children is given when they are between 12 and 15 months old, and the second dose is given when they are between 4 and 6 years old. The doses are typically administered as a single-agent vaccine, but a combination vaccine called MMRV that also immunizes against measles, mumps, and rubella (German measles) can be used. HIV-infected children and adults should receive the chickenpox vaccine if their immune system has not been compromised. It should not be given to people with immunodeficiencies although healthy, non-immune contacts of such people should be vaccinated. ◀◀ attenuated vaccines

By preventing chickenpox, the vaccine reduces the chance of developing shingles. A vaccine is available to prevent shingles in individuals 60 years of age or older. This vaccine is composed of the same attenuated virus used in the chickenpox vaccine, but in a much higher dose. It is given in a single dose and reduces the risk of developing shingles by 50%.

People with impaired immunity are at risk of severe disseminated VZV infections. These individuals include newborns, people with AIDS or organ transplants, or cancer patients undergoing chemotherapy. Immunocompromised individuals can be partially protected from severe disease by being passively immunized with a hyperimmune globulin with high concentrations of antibody to VZV. The main features of chickenpox are summarized in **table 22.9.** ◀◀ passive immunity, ◀◀ hyperimmune globulin

Rubeola (Measles)

Measles (sometimes called "hard measles" or "red measles") is a common name for rubeola. The dramatic reduction in measles cases as a result of immunizing children with an attenuated vaccine against the disease is a great success story. In 2000, measles was no longer considered endemic in the Western Hemisphere. The number of cases in the United States, however, has increased in recent years, due in large part to some parents' failure to vaccinate their children. The number of measles outbreaks more than doubled in the years 2009–2014 compared to the years 2000–2008.

Signs and Symptoms

Measles begins after an incubation period of 10 to 12 days with fever, runny nose, cough, and swollen, red, weepy eyes. Diarrhea may occur. Within a few days, a fine red rash

TABLE 22.9 Varicella (Chickenpox)

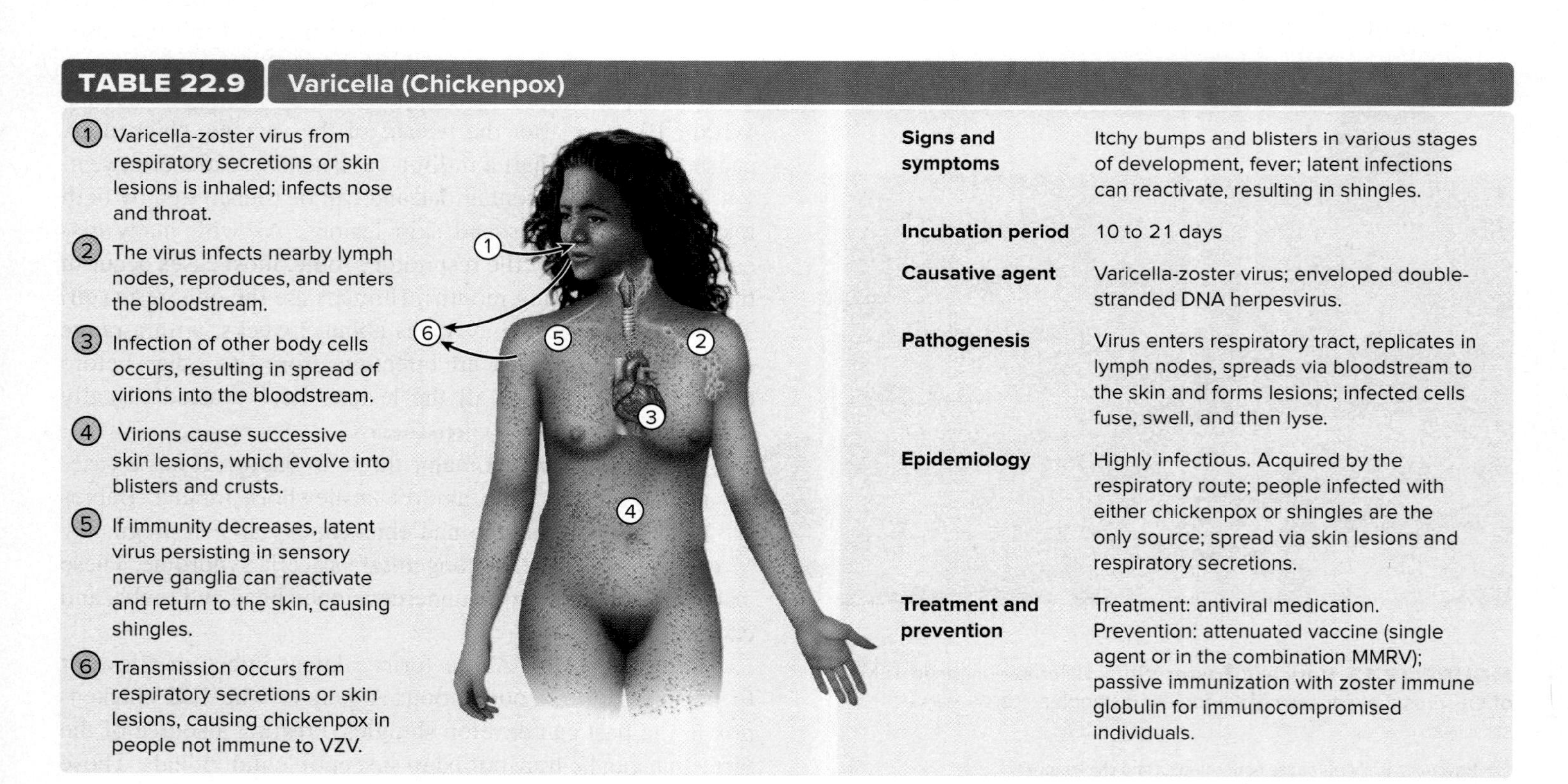

Signs and symptoms	Itchy bumps and blisters in various stages of development, fever; latent infections can reactivate, resulting in shingles.
Incubation period	10 to 21 days
Causative agent	Varicella-zoster virus; enveloped double-stranded DNA herpesvirus.
Pathogenesis	Virus enters respiratory tract, replicates in lymph nodes, spreads via bloodstream to the skin and forms lesions; infected cells fuse, swell, and then lyse.
Epidemiology	Highly infectious. Acquired by the respiratory route; people infected with either chickenpox or shingles are the only source; spread via skin lesions and respiratory secretions.
Treatment and prevention	Treatment: antiviral medication. Prevention: attenuated vaccine (single agent or in the combination MMRV); passive immunization with zoster immune globulin for immunocompromised individuals.

FOCUS YOUR PERSPECTIVE 22.1

The Ghost of Smallpox: An Evil Shade

Historically, smallpox epidemics have been devastating to the Americas. In the 1500s, smallpox virus introduced into Central and South America infected and killed many thousands of people and may have contributed to the downfall of the Inca and Aztec nations. An epidemic that swept the Massachusetts coast in the 1600s killed so many Native Americans that in some communities there were not enough survivors to bury the dead. Even in the 1700s, during the Revolutionary War, a smallpox epidemic raged through the American colonies. General George Washington suspected that the virus had been deliberately introduced by the British. So many of his men were ill after his defeat at Quebec in 1777 that he ordered the mass variolation of remaining troops. ⏮ smallpox, ⏮ variolation

The last case of smallpox occurred in 1978, the result of a laboratory accident. Why, then, are we still concerned about this disease? The answer is that the smallpox virus still exists, locked in high-security laboratories in the United States and Russia, and perhaps held in secret locations by countries or individuals that could use it to harm others. A number of factors are notable regarding the possibility that the smallpox virus might be used as an agent of bioterror.

Factors That Might Encourage Use of Smallpox Virus for Bioterrorism

- It spreads easily from person to person, mainly through close contact with respiratory secretions, but also by airborne virus from the respiratory tract, skin lesions, and contaminated bedding or other objects.
- It can be highly lethal, with case-fatality rates generally above 25%. After the virus establishes infection of the respiratory system, it enters the lymphatics and the bloodstream, causing lesions of the skin and throughout the body.
- The virus is relatively stable, probably remaining infective for hours in the air of a building; viable smallpox has been demonstrated in dried crusts from skin lesions after storage for 10 years at room temperature in ordinary envelopes.
- Large numbers of people are highly susceptible to the virus. Routine vaccination against smallpox was discontinued in the United States several decades ago.
- The relatively large genome of the smallpox virus probably allows genetic modifications that could enhance its virulence.

Factors That Discourage Use of Smallpox Virus for Bioterrorism

- Culturing the smallpox virus is dangerous and requires advanced knowledge and laboratory facilities.
- A highly effective vaccine is available. A protective antibody response occurs rapidly and prevents disease for 10 years or more. Smallpox is prevented even when the vaccine is administered up to 4 days after exposure to the virus.
- Infected people do not spread the disease during the long incubation period, generally 12 to 14 days; they become infectious only with the onset of fever.
- The disease can usually be diagnosed rapidly, by the characteristic appearance of skin lesions that predominate on the face and hands, and by laboratory examination of material from skin lesions.
- There is already widespread experience on how to watch for and contain the disease.

In a simulated attack on an American city in the summer of 2001, only 24 primary cases of smallpox increased in 2 months to 3 million, with 1 million deaths. This study probably greatly exaggerated the risk. Nevertheless, even though unlikely, the potential danger from a smallpox introduction has caused the United States to begin preparations for this possibility, including markedly expanding its stockpile of smallpox vaccine.

appears on the face and spreads to the neck and trunk and then to the extremities (**figure 22.14**). The rash gradually changes to a brownish color. Unless complications occur, signs and symptoms generally disappear in about a week. Unfortunately, many cases are complicated by secondary infections caused mainly by *Staphylococcus aureus, Streptococcus pneumoniae, Streptococcus pyogenes,* and *Haemophilus influenzae*. These bacteria easily invade the body because the measles virus damages the respiratory mucous membranes. Secondary infections most commonly cause earaches and bacterial pneumonia.

In a small number of cases, the measles virus causes viral pneumonia, with rapid breathing, shortness of breath, and dusky skin color from lack of adequate O_2 exchange in the lungs. Encephalitis (inflammatory disease of the brain) is another serious complication, marked by fever, headache, confusion, and seizures. This complication sometimes results in permanent brain damage, with mental disability, deafness, and epilepsy.

Very rarely, measles is followed 2 to 10 years later by a disease called subacute sclerosing panencephalitis (SSPE). This disease is marked by slow, progressive brain degeneration, generally resulting in death within 2 years. Defective measles virus particles can be detected in the brains of these patients along with high levels of measles antibodies in their blood. Immunization programs in the United States have decreased SSPE, but there is concern that the recent increase in measles cases may lead to more SSPE in the future.

Measles that occurs during pregnancy increases the risk of miscarriage, premature labor, and low-birth-weight babies. Birth defects, however, generally do not occur.

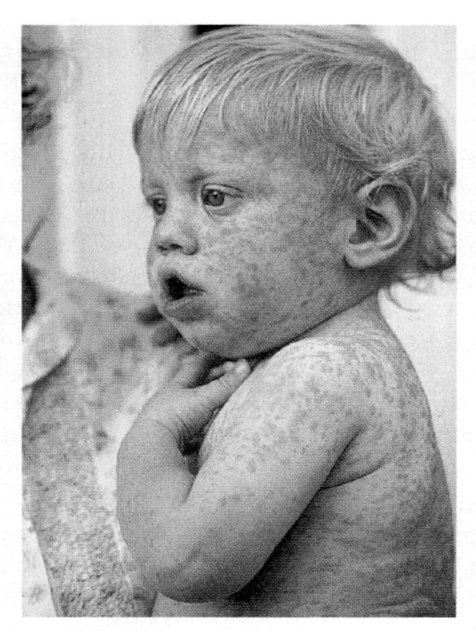

FIGURE 22.14 A Child with Measles (Rubeola) The rash is usually accompanied by fever, runny nose, and a bad cough. ©Lowell Georgia/Science Source

Why is it important to continue immunizing children in the United States against measles, even though it is a rare disease in the Western Hemisphere?

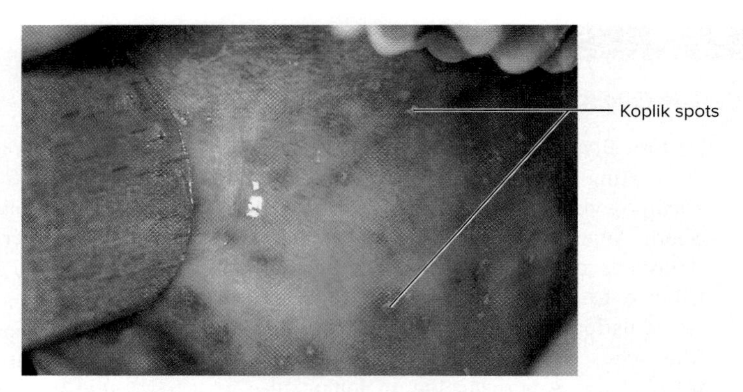

FIGURE 22.15 Koplik Spots, Characteristic of Measles (Rubeola) These spots resemble grains of salt on a red base. ©Dr. P. Marazzi/SPL/Science Source

Can someone have measles without developing Koplik spots?

Causative Agent

Measles is caused by rubeola virus, an enveloped, single-stranded RNA virus of the *Paramyxoviridae* family. Two biologically important proteins are found on the viral envelope: hemagglutinin and fusion protein. The virus uses hemagglutinin to attach to host cells and the fusion protein to fuse the viral envelope with the host cell's cytoplasmic membrane for entry into the cell. The fusion protein also causes adjacent infected host cells to join together, producing multinucleated giant cells.

Pathogenesis

Rubeola virus is acquired by the respiratory route. It replicates in the upper respiratory epithelium, spreads to lymphatic tissues, and eventually spreads to all parts of the body. The rash is caused by a cell-mediated immune response against viruses multiplying in the skin.

Mucous membrane involvement is responsible for an important early diagnostic sign, **Koplik spots** (**figure 22.15**). These look like grains of salt on red and rough oral mucosa, resembling red sandpaper, and are usually best seen in the back of the mouth, opposite the molars. The spots are temporary and may go unnoticed as they can disappear within a day of developing.

The measles virus temporarily suppresses cell-mediated immunity, making the person more susceptible to other diseases. For example, cold sores often reappear and latent tuberculosis infection reactivates in some individuals. It also damages the respiratory mucous membranes, thereby increasing the susceptibility of measles patients to secondary bacterial infections, especially infection of the middle ear and lung. Damage to the intestinal epithelium may explain the diarrhea that sometimes occurs in measles and contributes to high measles death rates in poor countries. In the United States, deaths from measles occur in about one to two of every 1,000 cases, mainly from pneumonia and encephalitis. ⏭ cold sore, ⏮ tuberculosis

Epidemiology

Humans are the only natural host of rubeola virus. Before vaccination became widespread in the 1960s, probably less than 1% of the global population escaped infection with this highly contagious virus. At one time, measles caused approximately 2.6 million deaths worldwide each year. It is still one of the leading causes of death in children globally, and in 2014, measles killed nearly 115,000 people, most of them under 5 years old. Continued global use of the measles vaccine has resulted in a progressive decline in the number of cases, but failure to vaccinate against measles in recent years has led to a general global increase in incidence.

Endemic measles was eliminated in the United States in 2000, but the incidence of measles in the United States is on the rise. This is largely caused by importation of the disease by travelers from countries where it is still endemic. Once introduced, the disease can spread, primarily among nonimmune populations including (1) children too young to be vaccinated; (2) children and adults inadequately vaccinated; and (3) unvaccinated people. Measles is a notifiable disease and must be reported to the CDC. In 2014, 667 cases of measles were reported in the United States, the majority of which occurred in unvaccinated people.

MicroByte

In February 2015, 125 cases of measles were related to a California theme park that draws many international visitors.

TABLE 22.10 Rubeola (Measles)

1. Rubeola virus spread by air or direct contact infects the upper respiratory tract, then the lymph nodes in the region.
2. Virus enters the bloodstream and is carried to all parts of the body, including the brain, lungs, and skin.
3. Skin cells infected with the rubeola virus are attacked by cytotoxic T cells, causing a generalized rash.
4. Virus replicating in the lungs can cause pneumonia; the brain can also be infected.
5. In rare cases, virus persisting in the brain causes subacute sclerosing panencephalitis, months or years after the acute infection.
6. Secondary infection of the ears and lungs is common.
7. Transmission is by respiratory secretions.

Signs and symptoms	Rash, fever, weepy eyes, cough, nasal discharge, and sometimes diarrhea.
Incubation period	10 to 12 days
Causative agent	Rubeola virus, an enveloped single-stranded RNA paramyxovirus
Pathogenesis	Virus multiplies in respiratory tract; spreads to lymphatic tissues, then to all parts of body, notably skin, lungs, and brain; damage to respiratory tract epithelium leads to secondary infection of lungs (pneumonia) and ears.
Epidemiology	Acquired by respiratory route; highly contagious; humans only source.
Treatment and prevention	Treatment: no antiviral treatment. Prevention: attenuated vaccine (MMR or MMRV).

Treatment and Prevention

No antiviral treatment is approved for measles, but an attenuated vaccine can prevent the disease. The measles vaccine is given together with mumps and rubella vaccines (MMR), and sometimes with varicella vaccine as well (MMRV). The first injection of vaccine is given near an infant's first birthday. Since 1989, a second injection of vaccine is given when children enter elementary school. The two-dose regimen has resulted in the immunity of at least 99% of the recipients. In an epidemic, vaccine is given to babies as young as 6 months, who are then reimmunized before their second birthday. Students entering high school or college are often required to get a second dose of vaccine if they have not had one earlier. Those at high risk of acquiring measles, such as medical personnel, should be immunized unless they definitely have had the disease or have laboratory proof of immunity. Some features of measles are summarized in **table 22.10.**

Rubella (German Measles)

German measles and "three-day measles" are common names for rubella. The term *German measles* arose because the disease was first described in Germany. In contrast to chickenpox and measles, rubella is typically a mild, often unrecognized disease that is difficult to diagnose. It is of concern because infection of pregnant women can have tragic consequences for the fetus.

Signs and Symptoms

Characteristic signs and symptoms of German measles develop 2 to 3 weeks after infection and typically last only a few days. These include slight fever, mild cold symptoms, and enlarged lymph nodes behind the ears and on the back of the neck. After about a day, a faint pink rash appears over the face, chest, and abdomen (**figure 22.16**). Unlike measles, there are

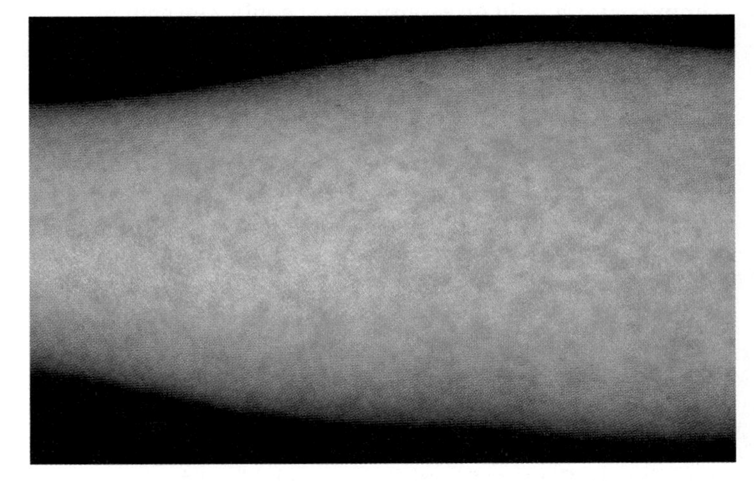

FIGURE 22.16 Adult with German Measles (Rubella) Symptoms are often very mild, but the effects on a fetus can be devastating.

? Is congenital rubella syndrome more likely to occur if a pregnant woman is infected early or late in the pregnancy?

FOCUS ON A CASE 22.1

A 20-year-old man was immunized against measles as a requirement for starting college. He had received his first dose of measles vaccine at approximately 1 year of age. Past medical history revealed that he had previously contracted the human immunodeficiency virus (HIV). Laboratory tests showed that he had a very low CD4+ lymphocyte count, indicating a severely damaged immune system.

About a month after his pre-college immunization, he developed *Pneumocystis* pneumonia, a lung infection characteristic of AIDS, was hospitalized, had a good response to treatment, and was discharged. Ten months later, he was again hospitalized for symptoms of a severe lung infection. He had no rash. Multiple laboratory tests to determine the cause of his infection were negative. Finally, a lung biopsy was performed and revealed "giant cells"—very large cells with multiple nuclei. These results were highly suggestive of measles pneumonia, and measles virus subsequently was recovered from cell cultures of the biopsy material. Other studies showed it to be the vaccine strain of measles virus. The patient received intravenous immune globulin and improved. Subsequently, however, his condition deteriorated, and he died of presumed complications of AIDS.

1. Is measles immunization a good idea for people with immunodeficiency?
2. Is it surprising that the vaccine virus was still present in this patient 11 months after vaccination? Explain.
3. Despite the severe infection, there was no rash. Why?

Discussion

1. Measles is often disastrous for people with AIDS or other immunodeficiencies. HIV patients should be immunized as soon as possible in their illness, before the immune system becomes so weakened it cannot respond effectively to the vaccine. Also, as this and other cases have shown, the vaccine virus can itself cause disease when immunodeficiency is severe. With the worldwide effort to eliminate measles, the risk of exposure to the wild-type measles virus, as opposed to the laboratory-derived vaccine virus, is declining, but outbreaks in colleges and other institutions still occur. A severely immunodeficient individual can be passively immunized against measles with immune globulin if exposed to the wild-type virus.
2. Measles is a persistent viral infection, meaning that following infection the virus can remain in the body for months or years in a slowly replicating form. It has been suggested but not proved that this explains the typical lifelong immunity conferred by measles. In immunocompromised people, persistent infection leads to damage to the brain, lungs, and liver, and possibly to the intestine.
3. Following acute infection, the measles virus floods the bloodstream and is carried to various tissues of the body, including the skin. The rash that characterizes measles is caused by T cells attacking measles virus antigen lodged in the skin capillaries. In the absence of functional T cells, the rash does not occur.

no diagnostic Koplik spots. Adults commonly develop painful joints, with pain generally lasting 3 weeks or less. Up to 50% of infected individuals may show no symptoms at all. Although it is a mild disease in most people, German measles poses a significant threat to developing fetuses.

Causative Agent

German measles is caused by the rubella virus, a member of the *Togaviridae* family. The virus is an enveloped, single-stranded RNA virus that can easily be grown in cell cultures. Proteins on the viral surface cause in vitro hemagglutination, which can be inhibited by specific antibody. This allows serological identification of the virus.

Pathogenesis

The rubella virus enters the body via the respiratory tract. It multiplies in the throat and in local lymph nodes, causing a sustained viremia as it enters the bloodstream. The virus travels to various body tissues, including the skin and the joints. Humoral and cell-mediated immunity develop against the virus, and the resulting immune complexes probably cause the rash and joint symptoms. ◀◀ viremia, ◀◀ immune complexes

If a woman is infected early in her pregnancy, virus particles in the bloodstream can cross the placenta and infect the fetus. Infections within the first 6 weeks of pregnancy will damage most fetuses, commonly causing deafness. This is less likely to happen later in pregnancy. Virtually all types of fetal cells are susceptible to infection; some cells are killed, whereas others develop a persistent infection in which cell division is impaired and chromosomes are damaged. The result is a characteristic pattern of fetal abnormalities referred to as the **congenital rubella syndrome (CRS).** The abnormalities include cataracts and other eye defects, brain damage, deafness, heart defects, and low birth weight despite normal gestation. Babies may be stillborn. Those that live continue to excrete rubella virus in throat secretions and urine for many months. Even infants who are apparently healthy can infect others with the excreted virus.

Epidemiology

Humans are the only natural host for rubella virus. Typically, infected people can transmit the virus from 1 week before the rash appears until 1 week afterward. The disease is highly contagious, although less so than measles; it is estimated that before widespread use of the vaccine began in 1969, only 10% to 15% of people reached adulthood without being

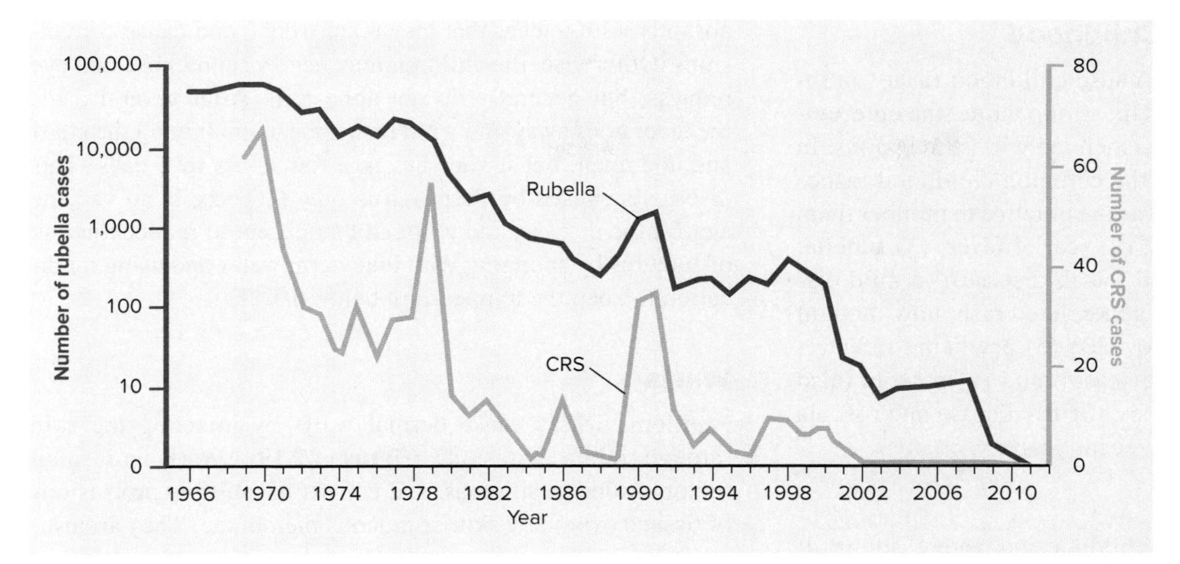

FIGURE 22.17 Reported Cases of German Measles (Rubella) and Congenital Rubella Syndrome (CRS), United States, 1966–2011. The spikes that occurred in 1978 and 1991 resulted from failure to vaccinate in certain populations. The number of rubella cases is now less than 10 per year.

? **Why are there fewer cases of CRS than of German measles at every time tested?**

infected. Major epidemics arose periodically. One epidemic in 1964 resulted in about 30,000 cases of congenital rubella syndrome. This caused a "rubella bulge" of hearing-impaired children that affected the educational system in the United States for decades. Since 2004, rubella has not been considered endemic in the United States. Nevertheless, as with measles, rubella may still be spread among unvaccinated populations when travelers enter the region from countries where the disease is still endemic. Cases of rubella or CRS must be reported to the CDC.

Treatment and Prevention

No specific antiviral therapy for rubella is available. The disease can be prevented by an attenuated rubella virus given to babies at 12–16 months old with a second dose at age 4–6 years (as part of the MMR or MMRV vaccine). The vaccine produces long-lasting immunity in nearly everyone who gets it. Women who are planning to have children are counseled to ensure that they are immune to rubella. If they are not, they are given the vaccine, unless they are already pregnant. The vaccine contains attenuated virus and is not given to pregnant women for fear it might result in congenital defects. As an added precaution, women are advised not to become pregnant for 28 days after receiving the vaccine. ◀◀ MMRV vaccine

Use of the vaccine has markedly reduced the incidence of rubella in the United States to generally less than 10 cases per year (**figure 22.17**). Some features of German measles are summarized in **table 22.11.**

TABLE 22.11 Rubella (German Measles)

1. Airborne rubella virus infects nose and throat.
2. Virus enters lymph nodes in the region.
3. Rubella virus multiples and enters the bloodstream.
4. Circulating virus reacts with antibodies, resulting in immune complexes.
5. Immune complexes lodge in the skin, causing a mild rash, and in the joints, causing pain.
6. In pregnant women, rubella virus crosses the placenta, infecting the fetus, resulting in congenital rubella syndrome.
7. Transmission to others is by respiratory secretions.

Signs and symptoms	Mild fever and cold symptoms, rash beginning on forehead and face, enlarged lymph nodes behind the ears; joint pain in adults
Incubation period	14 to 21 days
Causative agent	Rubella virus, an enveloped RNA virus
Pathogenesis	Following replication in the upper respiratory tract, virus spreads to all parts of the body and crosses the placenta; surviving fetuses often develop abnormally and excrete the virus for months after birth.
Epidemiology	Virus typically present in nose and throat from 1 week before rash to 1 week after; infection occurs via the respiratory route; humans are the only source.
Treatment and prevention	Treatment: no specific antiviral treatment. Prevention: attenuated vaccine (MMR or MMRV).

Other Viral Rashes of Childhood

The kinds of viruses that can cause childhood rashes probably number in the hundreds. One group alone, the enteroviruses, has about 50 members associated with skin lesions. In the early 1900s the causes of the common childhood rashes were largely unknown, and it was the practice to number them 1 to 6 as follows: (1) rubeola; (2) scarlet fever; (3) rubella; (4) Duke's disease (also called fourth disease)—a mild disease with fever and bright red generalized rash, now thought to have been due to an enterovirus; (5) erythema infectiosum (commonly called fifth disease); and (6) roseola (also called sixth disease). The causes of fifth disease and roseola have been established only in recent years.

Fifth Disease

Fifth disease occurs in both children and young adults. It begins with mild fever, runny nose, and headache. During this time the disease is spread by respiratory droplets. A diffuse redness appears on the cheeks, making the face appear as if it has been slapped (**figure 22.18**). The rash commonly spreads to other parts of the body, especially the extremities, and it may come and go for 2 weeks or more before recovery. As the centers of the red blotches begin to clear, the rash appears "lacy." Adults with fifth disease often have joint pains. The disease is caused by parvovirus B-19, a non-enveloped, single-stranded DNA virus that preferentially infects certain bone marrow cells, leading to a temporary reduction in red blood cell production. This is a major threat to people with sickle cell anemia or other anemias because these individuals are already deficient in red blood cells. It may also cause anemia in a developing fetus; up to 5% of women infected with the virus during pregnancy suffer spontaneous abortion.

Roseola

Roseola is a common but mild disease in children 6 months to 3 years old. It makes parents very anxious because it begins abruptly with a fever that may reach 105°F and cause convulsions. Otherwise, the children may have a runny nose and eye redness, but generally do not appear ill. After several days, the fever goes away and a red rash appears mainly on the chest and abdomen, but it vanishes in a few hours to 2 days. This disease is caused by herpesvirus type 6. There is no vaccine against the disease, and no treatment except to reduce the risk of seizures by sponging with lukewarm water and using medication to keep the temperature below 102°F.

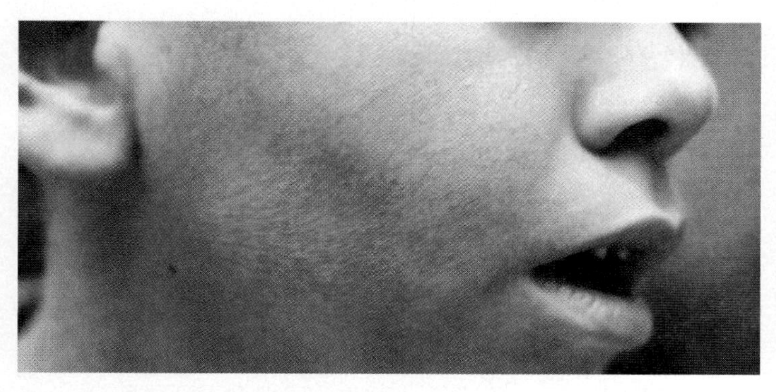

FIGURE 22.18 Fifth Disease "Slapped cheek" appearance of the rash on the face. Source: CDC

? Why is this disease a major threat to individuals with sickle cell anemia?

Warts

Papillomaviruses cause dermal warts by infecting the skin through minor abrasions (**figure 22.19**). Warts are small tumors called papillomas, and consist of multiple protrusions of tissue covered by skin or mucous membrane. They are usually benign (non-cancerous). About half of the time, dermal warts (on the skin) disappear within 2 years without treatment. Papillomaviruses are non-enveloped, double-stranded DNA viruses. More than 100 different papillomaviruses infect humans. They are difficult to study because they grow poorly in cell cultures or experimental animals. Warts of other animals are generally not infectious for humans.

Papillomaviruses can remain infectious on inanimate objects such as wrestling mats, towels, and shower floors, and infection can be acquired from such contaminated objects. The viruses infect the deeper cells of the epidermis and reproduce in the nuclei. Some of the infected cells

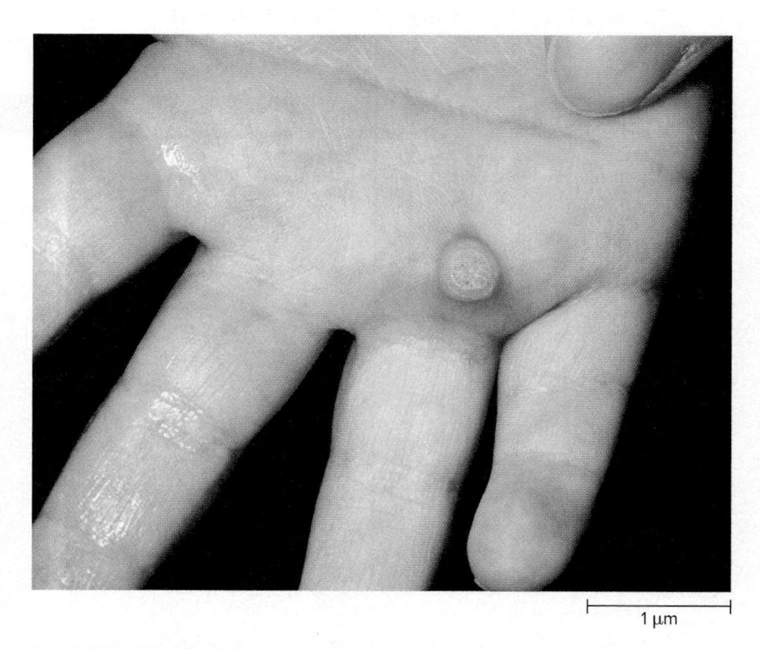

FIGURE 22.19 Dermal wart. Warts are contagious but benign skin growths. ©Science Photo Library/Science Source

? Which group of viruses causes warts?

grow abnormally, forming the wart. The incubation period ranges from 2 to 18 months. Infectious virus is present in the wart and can contaminate fingers or objects that pick or rub the lesions.

Like other tumors, warts can be treated effectively only by killing or removing all of the abnormal cells. This can usually be done by freezing the wart with liquid nitrogen, by cauterization (burning the tissue usually with an electrically heated needle), or by surgically removing the wart. Virus generally remains in the adjacent normal-appearing skin, however, and may cause additional warts.

Warts that grow on the soles of the feet are called plantar warts (often mistakenly called planter's warts; *plantar* is a term that refers to the sole of the foot). These warts are very difficult to treat because the pressure of standing on them causes them to grow wide and deep.

MicroAssessment 22.3

The virus that causes varicella (chickenpox) becomes latent and can reactivate to cause shingles. Measles (rubeola) makes a person more susceptible to other infections. Rubella (German measles) during early pregnancy can cause devastating fetal damage. Other common childhood rashes include fifth disease and roseola. Dermal warts are benign tumors that often resolve without treatment.

7. What important diagnostic sign is often present in the mouth of measles (rubeola) patients?
8. Does exposure to a person with shingles transmit shingles or chickenpox?
9. Why is it a good idea to immunize both boys and girls against rubella?

22.4 ■ Fungal Diseases of the Skin

Learning Outcomes

9. Describe the characteristics of superficial cutaneous mycoses, including the role of dermatophytes.
10. Describe the condition tinea versicolor and its causative agent, *Malassezia furfur.*

Diseases caused by fungi are called mycoses. Several fungi are responsible for mild to serious infections of the skin. The condition of the host's defenses against infection determines the severity of most fungal infections.

Superficial Cutaneous Mycoses

A group of molds called **dermatophytes** can invade hair, nails, and the keratinized portion of the skin. The resulting mycoses have common names such as jock itch, athlete's foot, and ringworm, as well as Latinized names that describe their location: tinea capitis (scalp), tinea barbae (beard), tinea axillaris (armpit), tinea corporis (body), tinea cruris (groin), and tinea pedis (feet), to list a few. Tinea simply means "worm," which probably reflects early incorrect ideas about the cause.

Signs and Symptoms

Most people colonized by dermatophytes have no signs or symptoms at all. Others complain of itching, a bad odor, or a rash. In ringworm, a rash occurs at the site of the infection and consists of a scaly area surrounded by redness at the outer margin, producing irregular rings or a lacy pattern on the skin. On the scalp, patchy areas of hair loss can occur, with a fine stubble of short hair left behind. Infected nails become thickened and brittle and may separate from the nailbed. Sometimes, a rash consisting of fine papules and vesicles develops distant from the infected area. This rash is referred to as a dermatophytid, or "id" reaction, a result of allergy to products of the infecting fungus.

Causative Agents

Dermatophytes are a group of skin-invading molds including members of the genera *Epidermophyton, Microsporum,* and *Trichophyton* (**figure 22.20**). They can be grown on media especially designed for molds and are usually identified by their colonial and microscopic appearance. In some cases their nutritional requirements and biochemical tests are used for identification as well. Identification is not always necessary, however, as treatment for all dermatophyte infections is the same.

Pathogenesis

The normal skin is generally resistant to invasion by dermatophytes. Some species, however, are relatively virulent and can even cause epidemics, especially among children. In moist areas of the skin, dermatophytes can invade keratin-containing cells and structures. They produce an enzyme called keratinase that breaks down the protein, allowing the fungi to use it as a nutrient. Dermatophytes can invade the epidermis down to the level of the keratin-producing cells. Hair is invaded at the follicle, which is relatively moist.

Fungal products diffuse into the dermis and provoke an immune reaction, which probably explains why adults tend to be more resistant to infection than children. Children are more likely to have hypersensitivities, like asthma and **eczema,** a weepy skin rash, with formation of crusts, usually due to an allergy.

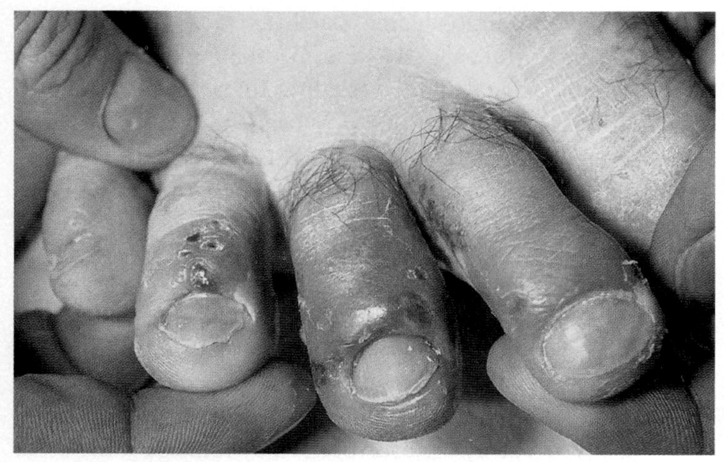

(a)

(b) 20 µm

FIGURE 22.20 Dermatophytosis **(a)** Tinea pedis, usually caused by species of *Trichophyton.* **(b)** Large boat-shaped conidia of *Microsporum gypseum,* a cause of scalp ringworm in children.

What is the common name for tinea pedis?

Epidemiology

Patient age, virulence of the infecting strain of mold, and moisture availability are important factors in determining the course of infection. Common causes of excessive moisture include obesity where folds of skin lie together, tight clothing, and plastic or rubber footwear. Potentially pathogenic molds may be present in soil and on pets such as young cats and dogs. Fungi acquired from these sources tend to cause more noticeable signs and symptoms in humans.

Treatment and Prevention

Numerous prescription and over-the-counter medications can be used to treat superficial skin infections. However, nail infections are often much more difficult to cure. Oral medications may be required for months, and sometimes the nail must be surgically removed.

Keeping skin and nails clean and dry effectively prevents most dermatophyte infections. Powders, open shoes, changing of socks, and applying rubbing alcohol after bathing may help prevent toenail infections.

Other Fungal Diseases

Although *Malassezia furfur* is generally harmless and commonly found on the skin, in some people, it causes skin conditions such as a scaly face rash, dandruff, or tinea versicolor (**figure 22.21**). The latter is a common skin disease characterized by patchy scaliness and increased pigment in light-skinned people, or a decrease in pigment in dark-skinned people. Scrapings of the affected skin show large numbers of *M. furfur,* both in its yeast form and as short filaments called hyphae. Unknown host factors are important in these diseases because most people carry the organism on their skin without any disease. AIDS patients often have a severe rash with pus-filled pimples caused by *Malassezia* yeasts, and the organisms may even infect internal organs in patients receiving lipid-containing intravenous feedings. ⏮ yeasts, ⏮ hyphae

The yeast *Candida albicans* may live harmlessly among the normal microbiota of the skin, but in some people it invades the deep layers of the skin and subcutaneous tissues (**figure 22.22**). In many people with candidal skin infections, the cause for the invasion cannot be determined.

The key features of the diseases covered in this chapter are highlighted in the **Diseases in Review 22.1** table that follows.

MicroAssessment 22.4

The fungi that cause skin mycoses can commonly colonize skin without causing signs or symptoms. Molds called dermatophytes can invade the hair, nails, and skin. The best protection against fungal skin infections is to maintain normal skin dryness.

10. What is a mycosis?

11. Name several of the superficial cutaneous mycoses caused by dermatophytes.

12. Why do you think it is so much more difficult to treat nail infections than to treat other superficial dermatophytoses?

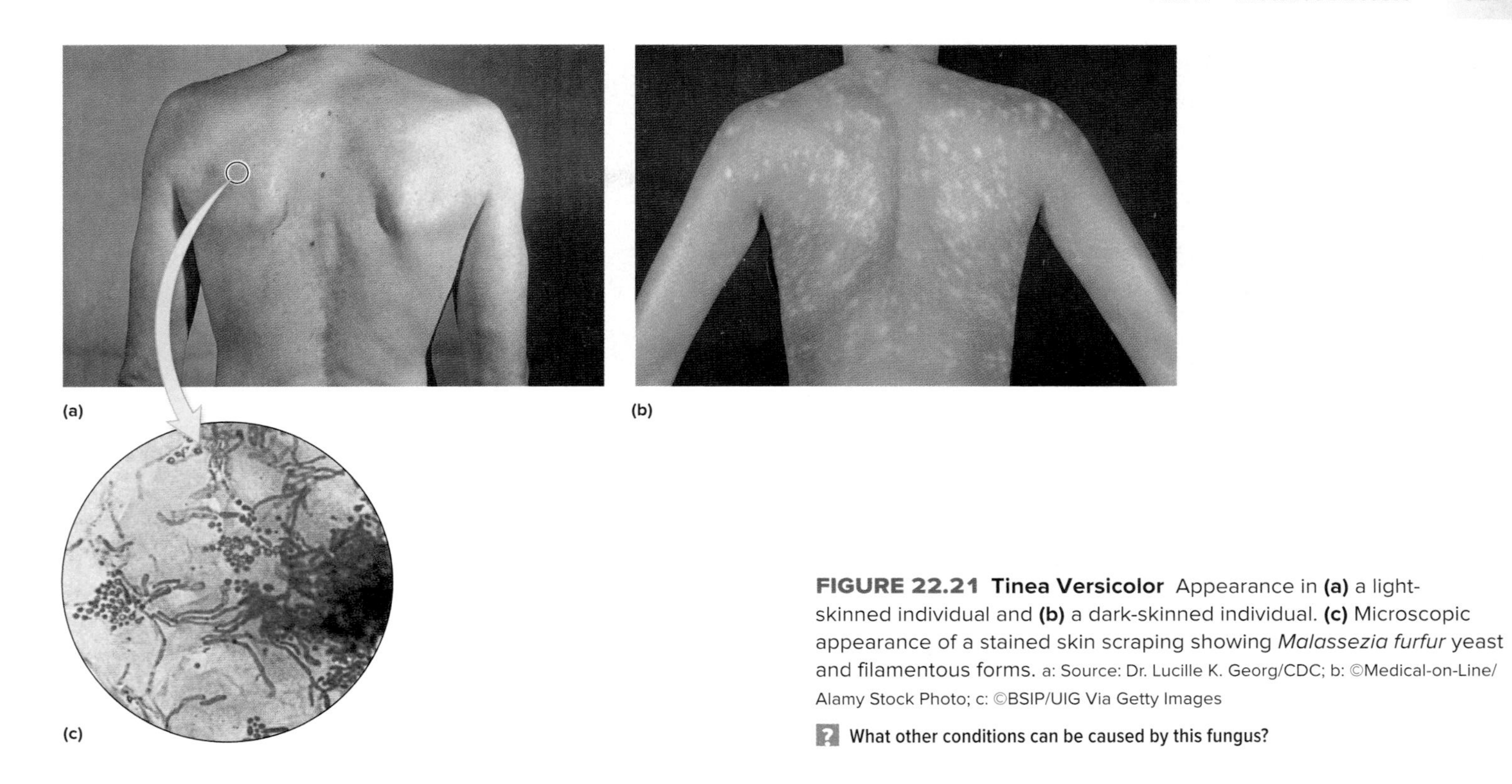

FIGURE 22.21 Tinea Versicolor Appearance in **(a)** a light-skinned individual and **(b)** a dark-skinned individual. **(c)** Microscopic appearance of a stained skin scraping showing *Malassezia furfur* yeast and filamentous forms. a: Source: Dr. Lucille K. Georg/CDC; b: ©Medical-on-Line/Alamy Stock Photo; c: ©BSIP/UIG Via Getty Images

? **What other conditions can be caused by this fungus?**

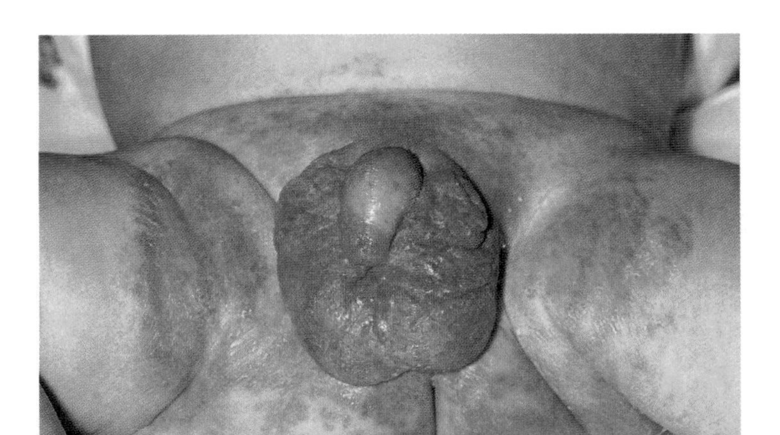

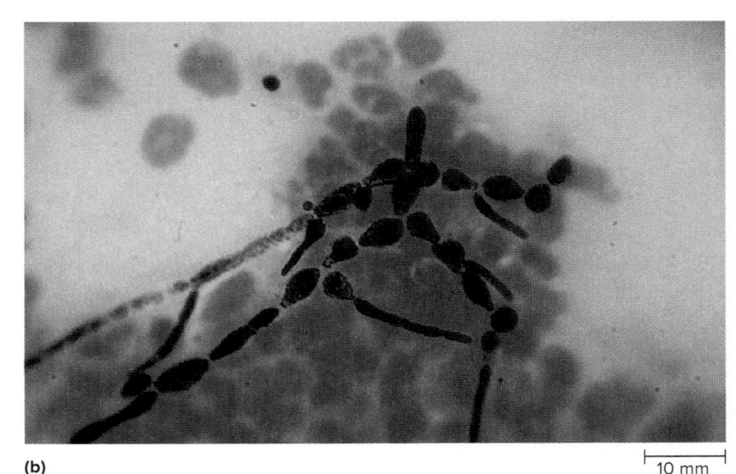

FIGURE 22.22 ***Candida albicans.*** **(a)** Causing a diaper rash. **(b)** Gram stain of pus showing *C. albicans* yeast forms and filamentous forms called pseudohyphae. a: ©Dr. P. Marazzi/SPL/Science Source; b: ©Evans Roberts

? **Would an antibiotic ointment containing penicillin be effective against this candidal infection?**

Diseases in Review 22.1

Common Bacterial, Viral, and Fungal Skin Diseases

Disease	Causative Agent	Comment	Summary Table
BACTERIAL SKIN DISEASES			
Acne vulgaris	*Cutibacterium* (formerly *Propionibacterium*) *acnes* commonly associated	Most common during puberty, probably due to excess sebum secretion in response to increased hormone levels.	
Hair follicle infections	*Staphylococcus aureus*	Causative agent commonly colonizes the nostrils and moist skin areas; skin infections include folliculitis, furuncles, and carbuncles. Organism is often resistant to multiple antibiotics.	
Staphylococcal scalded skin syndrome	Exfoliatin-producing strains of *S. aureus*	Characterized by peeling of the outer layer of skin; occurs in infants, as well as the elderly and immunocompromised.	Table 22.4
Impetigo	Usually *Streptococcus pyogenes;* sometimes *Staphylococcus aureus*	Characterized by blisters, which break and are replaced by oozing yellow crusts; some people later develop glomerulonephritis.	Table 22.6
Rocky Mountain spotted fever (RMSF)	*Rickettsia rickettsii*	Spread by ticks; characterized by rash that spreads and then becomes hemorrhagic.	Table 22.7
Cutaneous anthrax	*Bacillus anthracis*	Zoonotic disease that is rare in humans; characterized by black eschar.	Table 22.8
VIRAL SKIN DISEASES			
Varicella (chickenpox)	Varicella-zoster virus (VZV)	Virus enters via the respiratory tract; infection characterized by itchy skin lesions; VZV becomes latent and can later reactivate to cause shingles. Preventable by vaccination.	Table 22.9
Rubeola (measles)	Rubeola virus	Virus enters via the respiratory tract; characterized by respiratory symptoms and a spreading rash; Koplik spots occur in mouth. Preventable by vaccination.	Table 22.10
Rubella (German measles)	Rubella virus	Virus enters via the respiratory tract, causes mild respiratory symptoms, joint pain, and fine rash; can damage developing fetus (congenital rubella syndrome). Preventable by vaccination.	Table 22.11
Warts (dermal warts)	Papillomaviruses	Warts are benign skin tumors that can be removed by freezing, burning, surgery, or topical medication.	
FUNGAL SKIN DISEASES			
Superficial cutaneous mycoses	Usually *Epidermophyton, Microsporum,* or *Trichophyton* species	Fungi invade keratinized skin, causing conditions that are commonly known as athlete's foot, jock itch, and ringworm.	
Other fungal diseases	*Malassezia furfur, Candida albicans*	Both organisms are usually harmless on the skin but *M. furfur* sometimes causes skin conditions such as scaly face rash, dandruff, or tinea versicolor. *C. albicans* sometimes invades deeper layers and subcutaneous tissues.	

Summary

22.1 ■ Anatomy, Physiology, and Ecology of the Skin

The skin prevents the entry of microbes, regulates body temperature, restricts the loss of fluid from body tissues, and plays an essential role in the function of the immune system. It is composed of the **epidermis** and the **dermis** (figure 22.1). Common members of the skin microbiota include diphtheroids, staphylococci, and fungi (table 22.1).

22.2 ■ Bacterial Diseases of the Skin

Acne Vulgaris

Acne is characterized by enlarged sebaceous glands. **Sebum** accumulation within these glands allows *Cutibacterium* (formerly *Propionibacterium*) *acnes* to grow to high numbers. Their metabolic products cause an inflammatory response.

Hair Follicle Infections

Folliculitis, **furuncles,** and **carbuncles** are caused by *Staphylococcus aureus,* which is coagulase-positive and often resistant to penicillin and other antibiotics (figures 22.2 and 22.3; table 22.3). There are many different strains that vary in virulence. A carbuncle is more serious because the infection is more likely to be carried to the heart, brain, or bones.

Staphylococcal Scalded Skin Syndrome (table 22.4)

Staphylococcal scalded skin syndrome results from **exfoliatin** produced by certain strains of *Staphylococcus aureus* (figure 22.4).

Impetigo

Impetigo is a superficial skin disease caused by *Streptococcus pyogenes* or *Staphylococcus aureus* (figure 22.5; table 22.5, table 22.6).

Rocky Mountain Spotted Fever (table 22.7; figure 22.6)

Rocky Mountain spotted fever, caused by the obligate intracellular bacterium *Rickettsia rickettsii* (figure 22.7), is an often fatal disease transmitted to humans by the bite of an infected tick (figure 22.8, figure 22.9).

Cutaneous Anthrax (table 22.8)

Bacillus anthracis infects the skin causing a small red bump that develops into an eschar resembling a flat black scab (figure 22.10). Cutaneous anthrax is rare in humans; person-to-person transmission does not occur.

22.3 ■ Viral Diseases of the Skin

Varicella (Chickenpox) (table 22.9)

Chickenpox, once a common disease of childhood, is caused by the varicella-zoster virus (figure 22.11). **Shingles** can occur months or years after chickenpox and is due to reactivation of the virus (figure 22.12). Shingles cases can be sources of chickenpox epidemics.

Rubeola (Measles) (table 22.10)

Rubeola (measles) is a potentially dangerous viral disease that can lead to serious secondary bacterial infections, and fatal lung or brain damage (figures 22.14, 22.15). Measles can be controlled by vaccinating with an attenuated vaccine.

Rubella (German Measles) (figure 22.16, table 22.11)

German measles (rubella), if contracted by a woman early in pregnancy, often results in birth defects, making up the **congenital rubella syndrome.** Immunization with an attenuated virus protects against this disease (figure 22.17).

Other Viral Rashes of Childhood

Numerous childhood rashes are caused by viruses (figure 22.18). Examples include fifth disease (erythema infectiosum) and roseola (sixth disease).

Warts

Warts are skin tumors caused by a number of papillomaviruses (figure 22.19). They are usually benign and may disappear without treatment. Warts on the soles of the feet are called plantar warts.

22.4 ■ Fungal Diseases of the Skin

Superficial Cutaneous Mycoses

Dermatophytes cause athlete's foot, ringworm, and invasions of the hair and nails (figure 22.20).

Other Fungal Diseases

Malassezia species can cause tinea versicolor and dandruff, as well as serious skin disease in AIDS patients (figure 22.21). *Candida albicans* may live harmlessly among the normal microbiota, but it can invade deeper layers of the skin and subcutaneous tissues (figure 22.22).

Review Questions

Short Answer

1. What is the difference between a furuncle and a carbuncle?
2. Why do only certain strains of *Staphylococcus aureus* cause scalded skin syndrome?
3. How is impetigo spread?
4. How does the fact that Rocky Mountain spotted fever is a zoonosis relate to the relative severity of the disease symptoms?
5. Who is most at risk for contracting cutaneous anthrax?
6. Describe the progression of the rash of varicella.
7. What is the relationship between chickenpox (varicella) and shingles (herpes zoster)?
8. Why are many cases of measles complicated by secondary infections?
9. What is the significance of rubella viremia during pregnancy?
10. How does a person contract warts?

Multiple Choice

1. Which of the following conditions is important in the ecology of the skin?
   a) Temperature
   b) Salt concentration
   c) Lipids
   d) pH
   e) All of the above

2. *Staphylococcus aureus* can be responsible for all of the following conditions *except*
   a) impetigo.
   b) food poisoning.
   c) toxic shock syndrome.
   d) scalded skin syndrome.
   e) athlete's foot.
3. The main effect of staphylococcal protein A is to
   a) interfere with phagocytosis.
   b) enhance the attachment of the Fc portion of antibody to phagocytes.
   c) coagulate plasma.
   d) kill white blood cells.
   e) degrade collagen.
4. Which of the following contributes to the virulence of *Streptococcus pyogenes*?
   a) Protease
   b) Hyaluronidase
   c) DNase
   d) All of the above
   e) None of the above
5. Which of the following statements is true of streptococcal impetigo?
   a) It is caused by a Gram-negative rod.
   b) It cannot be transmitted from one person to another.
   c) Pathogenic streptococci all produce coagulase.
   d) All of the above.
   e) None of the above.
6. All of the following are true of Rocky Mountain spotted fever *except*
   a) the disease is most prevalent in the western United States.
   b) it is caused by an obligate intracellular bacterium.
   c) it is a zoonosis transmitted to humans by ticks.
   d) those with the disease characteristically develop a hemorrhagic rash.
   e) antibiotic therapy is usually curative if given early in the disease.
7. If a woman begins to develop shingles, which of the following can you conclude?
   a) She was recently exposed to someone who had chickenpox.
   b) She was recently exposed to someone who had shingles.
   c) She had chickenpox in the past.
   d) She had shingles in the past.
   e) The rash will likely spread over her entire trunk and extremities.
8. Which of the following statements is more likely to be true of measles (rubeola) than of German measles (rubella)?
   a) Koplik spots are present.
   b) It causes birth defects.
   c) It causes only a mild illness.
   d) Humans are the only natural host.
   e) Attenuated virus vaccine is available for prevention.
9. All of the following must be cultivated in cell cultures instead of cell-free media *except*
   a) *Rickettsia rickettsii.*
   b) rubella virus.
   c) varicella-zoster virus.
   d) *Candida albicans.*
   e) rubeola virus.
10. All of the following might contribute to development of ringworm or other superficial cutaneous mycoses *except*
   a) obesity.
   b) playing with kittens.
   c) rubber boots.
   d) using skin powder.
   e) dermatophyte virulence.

Applications

1. A school administrator in a small Iowa community prohibited a child with chickenpox from attending school. He said this was the first case of chickenpox in the school in 6 years and he did not want to have an outbreak. Several parents argued to the school board that an outbreak would benefit the school in the long term. Discuss the pros and cons of allowing this child to attend school.
2. A public health official was asked to speak about immunization during a civic group luncheon. One parent asked if rubella was still a problem. In answering, the official cautioned women planning to have another child to have their present children immunized against rubella. Why did the official suggest this?

Critical Thinking

1. In the past, teenagers with acne were told to stop eating chocolate and French fries and to vigorously scrub their skin. Explain how you think these recommendations would affect the teenagers' acne.
2. Why might it be more difficult to eliminate a disease like Rocky Mountain spotted fever from the earth than to eliminate rubeola or rubella?

23 Wound Infections

Infected leg wound. ©Garry Watson/Science Source

KEY TERMS

Abscess A localized collection of pus within a tissue.

Fasciitis Inflammation of the fascia, which are bands of fibrous tissue that underlie the skin and surround muscle and body organs. When fasciitis leads to death of tissue, it is called necrotizing fasciitis.

Granulation Tissue New tissue formed during healing of an injury.

Pus Yellowish fluid composed of proteins, living and dead leukocytes, and tissue debris.

Pyogenic Pus-producing.

Superantigens Molecules that bind to and stimulate helper T cells, resulting in overproduction of cytokines and sometimes fatal shock.

Synergistic Infection An infection in which two or more species of pathogens act together to produce an effect greater than the sum of effects if each pathogen were acting alone.

A Glimpse of History

Endospores of *Clostridium tetani,* the bacterium that causes tetanus (lockjaw), are found in soil and dust—virtually everywhere. Tetanus used to be common before its cause and pathogenesis were understood, and it often ended in an agonizingly painful death. Dr. Shibasaburo Kitasato (1853–1931), studying the disease in Robert Koch's laboratory in Germany, was the first to discover that *C. tetani* is an obligate anaerobe. This critical information helped him develop a method to grow the bacterium in pure culture, an essential step toward characterizing a pathogen and learning how it causes disease.

Kitasato showed that laboratory animals injected with *C. tetani* developed tetanus. He was puzzled, however, by a surprising finding: Although the animals died of generalized disease, there were no *C. tetani* cells anywhere in the animal other than the injection site. By doing experiments in which he injected the tails of mice and then removed the inoculated tissue at hourly intervals, he showed that the animals developed tetanus only if the bacteria remained in the animals for more than an hour. He also showed that the organisms stayed at the site of inoculation; at no time were they found in the rest of the body. Kitasato reasoned that something other than bacterial invasion was causing the disease.

While Kitasato was working with tetanus, another scientist, Emil von Behring, was busy investigating how *Corynebacterium diphtheriae* caused the disease diphtheria. Together, Kitasato and von Behring showed that toxins produced by the bacteria caused both diseases. The concept that a bacterial toxin could cause disease was an extremely important advance in the understanding and control of infectious diseases.

Most people occasionally suffer wounds that cause breaks in the skin or the mucous membranes. Microorganisms almost always enter the wound, originating from either the damaging object or the environment. Infection may result, depending on several factors, including (1) virulence of the microbes; (2) number of microbial cells in the wound; (3) status of the host's immune system; and (4) the type of wound, especially whether the tissues are crushed or the wound contains foreign matter (such as dirt or bits of cloth). Wounds that contain foreign material usually become infected and do not heal until the matter is removed. This is because the foreign substance creates surfaces for biofilm development and provides places for microorganisms to multiply out of the reach of phagocytes and other immune defenses. It may also reduce available O_2, allowing the growth of anaerobic pathogens and interfering with phagocyte function. Clean wounds often heal without treatment despite microbial colonization, but sometimes even a minor wound can result in a severe, or possibly fatal, infection.

23.1 ■ Anatomy, Physiology, and Ecology of Wounds

Learning Outcomes

1. Name three tissue components exposed by wounds to which pathogens specifically attach.
2. Describe the beneficial and harmful aspects of abscess formation.

Wounds vary in their characteristics, severity, and associated risks. The general categories of wounds include:

- **Incisions:** Produced by a knife or other sharp object.
- **Punctures:** Result from penetration by a small sharp object, such as a needle or a nail.
- **Lacerations:** Occur when the tissue is torn.
- **Contusions:** Produced by a blow that crushes tissue.
- **Abrasions:** Occur when the epidermis is scraped off.
- **Gunshot wounds:** Caused by bullets or other projectiles.
- **Burns:** Caused by heat (thermal burns), electricity, chemicals, radiation, or friction.

Wounds expose tissue components usually protected by skin or mucous membranes, uncovering surfaces to which pathogens can attach and then colonize. These components include collagen, fibronectin, fibrinogen, and fibrin. **Collagen** is a fibrous material, the main supportive protein of skin, tendons, scars, and other body structures. **Fibronectin** is a fibrous glycoprotein found both as a circulating form and in tissue, where it binds cells and other tissue substances together. **Fibrinogen** is a blood protein; when a wound occurs, this protein is converted to **fibrin,** which forms clots in the damaged vessels. The clots stop the flow of blood as the first step in the wound repair process.

Wound healing begins as new capillaries form in the area and connective tissue cells (fibroblasts) multiply. This produces a red, translucent (clear) fibrous material called **granulation tissue.** In clean wounds, granulation tissue fills the space created by the wound. This tissue gradually shrinks and is converted to collagen, a component of scar tissue that is eventually covered by skin or mucous membrane (**figure 23.1**). When a wound results in extensive tissue damage or severe infection, a medical procedure called **debridement** is used to remove dead, damaged, and infected tissue. Debridement helps wounds heal faster.

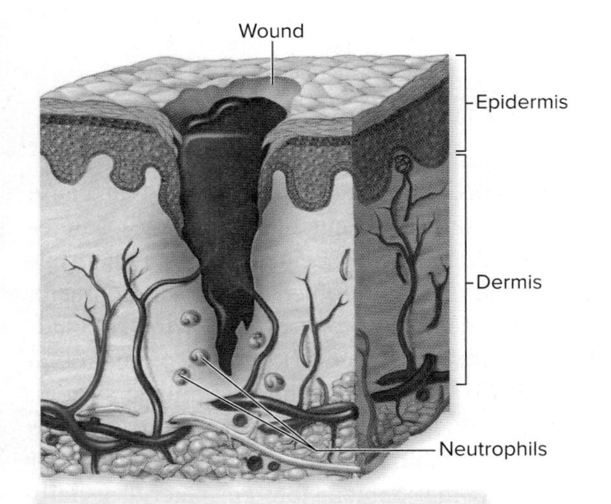

1 Cut blood vessels bleed into the wound.

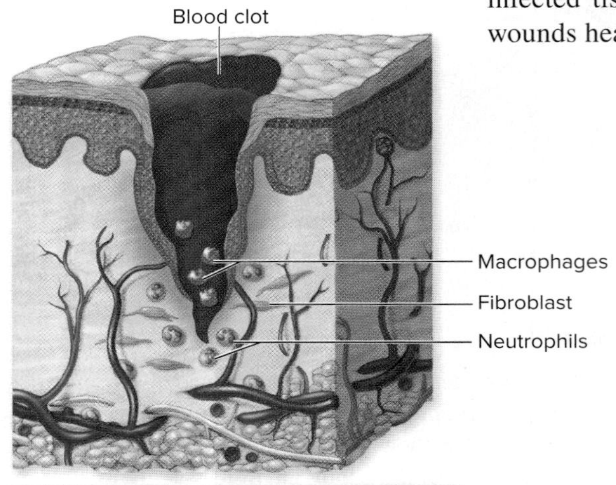

2 Blood clot forms in wound, and phagocytes destroy microbes.

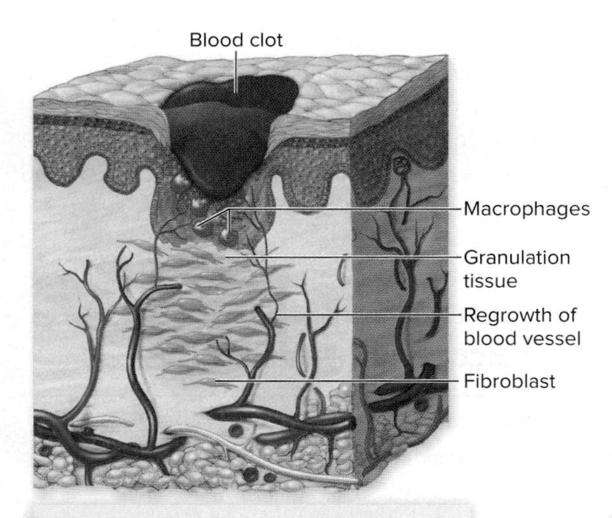

3 Wound fills with granulation tissue and blood vessels regrow.

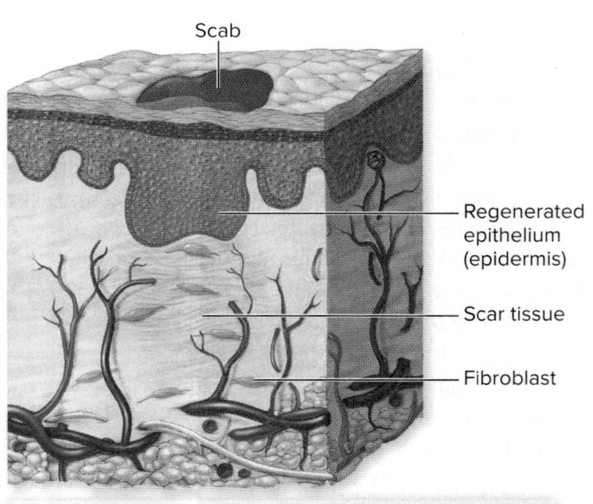

4 Scar tissue forms. Collagen contracts and epithelium regenerates.

FIGURE 23.1 The Process of Wound Repair

? What is the function of granulation tissue?

MicroByte

In maggot debridement therapy (MDT), live, sterile maggots (fly larvae) are placed on a wound to remove dead tissue and aid in wound healing.

Wound Abscesses

An **abscess** (**figure 23.2**) is a localized collection of **pus** surrounded by inflamed tissue. The pus—a thick yellowish fluid—contains living and dead leukocytes, tissue debris, and proteins. Abscesses form as a result of the body's immune defenses and usually indicate an infection. Although an abscess helps localize the infection and prevents its spread, it also indicates a potentially serious situation. If microbial cells escape the abscess, they can enter the blood or the lymph, leading to infection in other parts of the body.

The very nature of abscesses makes them difficult to treat. They have no blood vessels, because the developing pus pocket destroys them or pushes them aside, and adjacent blood vessels are often blocked by clots. This lack of blood circulation makes it difficult for antimicrobial medications to reach the infected site. Even if the medications do enter the site, the chemical nature of pus interferes with the action of some antibiotics. In addition, antimicrobial medications are often ineffective against microorganisms in the abscess because the microbes stop multiplying (and most antimicrobial medications work against actively dividing cells). Abscesses usually must therefore burst to a body surface or be drained surgically to be effectively treated. ⏮ antimicrobial medications

Anaerobic Wounds

An important feature of many wounds is that they are relatively anaerobic, which allows the growth of obligate anaerobes such as *Clostridium tetani*. The anaerobic conditions

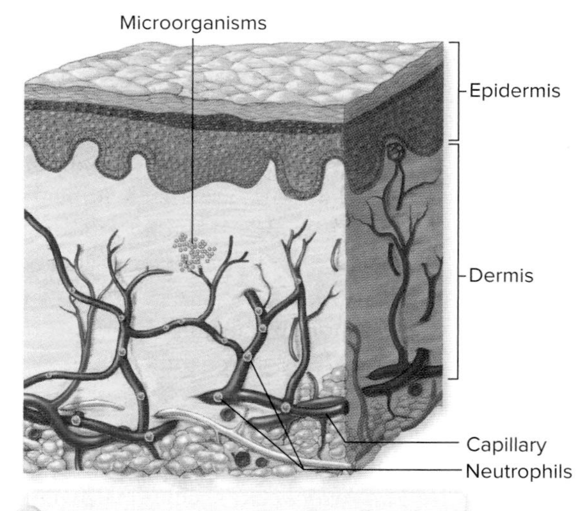

1 Microorganisms enter the tissue from a wound or from the bloodstream.

2 Blood vessels dilate, and leukocytes migrate to the area of the developing infection.

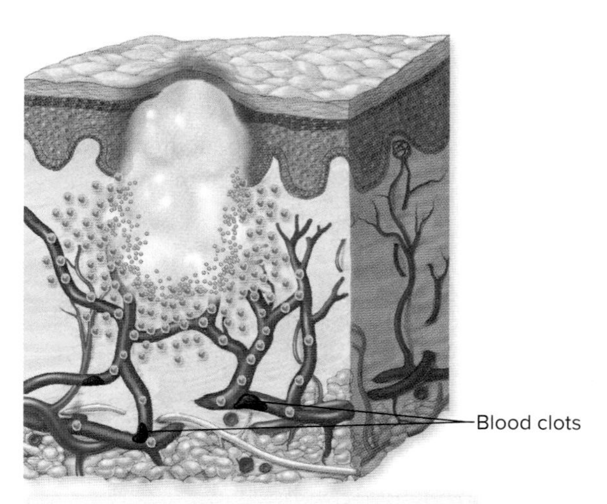

3 Pus forms and an abscess develops; clotting occurs in adjacent blood vessels.

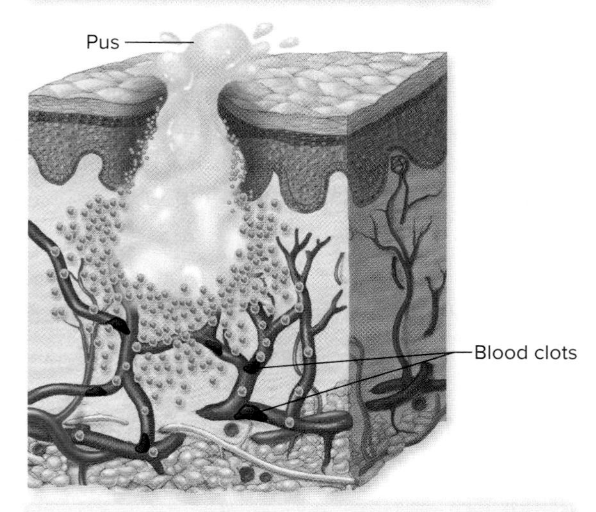

4 Buildup of pressure causes the abscess to expand in the direction of least resistance; if it reaches a body surface, it may rupture and discharge its contents.

FIGURE 23.2 Abscess Formation

? What is the composition of pus?

can result from numerous factors, including extensive tissue damage, small but deep penetration associated with puncture wounds, and contamination with dirt. Growth of multiple types of organisms in a wound (polymicrobial infection) can also create anaerobic conditions if aerobic or facultative microorganisms use up the available O_2 during cellular respiration.

MicroAssessment 23.1

Wounds can be classified as incisions, punctures, lacerations, contusions, abrasions, gunshots, or burns. Wounds expose tissue components to which pathogens can attach. Granulation tissue fills a healing wound. Abscess formation provides a way of isolating infections and preventing their spread. Anaerobic conditions in wounds are created by the presence of dead tissue and foreign material.

1. Name and describe two substances in wounds to which pathogens attach.
2. Give two reasons an abscess might not respond to antibiotic treatment.
3. Why is it important that the granulation tissue shrinks after it is formed?

23.2 ■ Common Bacterial Infections of Wounds

Learning Outcomes

3. Give distinctive characteristics of wound infections caused by *Staphylococcus aureus, Streptococcus pyogenes,* and *Pseudomonas aeruginosa.*
4. Discuss how *Staphylococcus epidermidis* is able to form biofilms on medical devices such as artificial joints.

If a wound becomes infected, several serious consequences are possible. These include (1) delayed healing, (2) formation of abscesses, and (3) spread of the bacteria or their toxins to other areas of the body. Infected surgical wounds often split open, as swelling causes the stitches to pull through tissues weakened by the infection. The infection can spread to devices such as an artificial hip or knee, which may then

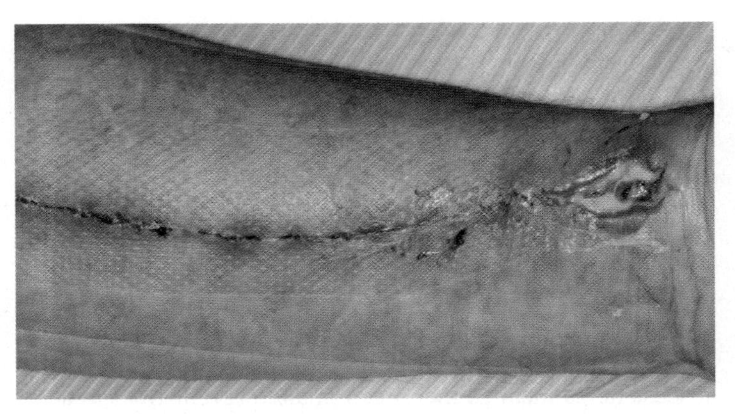

FIGURE 23.3 Surgical Wound Infection ©Dr. P. Marazzi/Science Source

Why do infected surgical wounds sometimes split open?

have to be removed before the infection can be eliminated. **Table 23.1** summarizes the characteristics of the leading causes of wound infections.

Staphylococcal Wound Infections

Staphylococcus species, common inhabitants of the nostrils and the skin, are the leading causes of wound infections (**figure 23.3**). Of the 30 or more recognized *Staphylococcus* species, only two cause most human wound infections: *Staphylococcus aureus* and *Staphylococcus epidermidis.* The more important of these is *S. aureus,* which was overed extensively in chapter 22. ◀◀ the genus *Staphylococcus*, ◀◀ bacterial skin diseases

Signs and Symptoms

Staphylococcus species are **pyogenic,** meaning that they cause the production of pus (*pyo* means "pus" and *genic* means "generating"). Staph infections are usually characterized by an inflammatory reaction, with swelling, redness, and pain. Fever occurs if the infected area is large or if the infection has spread to the blood or the lymph.

Toxic shock syndrome can occur if the wound is infected with a toxin-producing *S. aureus* strain. Signs and symptoms of this include high fever, muscle aches, a life-threatening

TABLE 23.1 Leading Causes of Wound Infections

Causative Organism	Characteristics	Consequences
Staphylococcus aureus	Gram-positive cocci in clusters, coagulase-positive	Delayed healing; abscess formation; extension into tissues, artificial devices, or bloodstream; some strains can cause toxic shock syndrome
Streptococcus pyogenes	Gram-positive cocci in chains; Lancefield group A	Same as above, except some strains can cause "flesh-eating" necrotizing fasciitis
Pseudomonas aeruginosa	Gram-negative rods; pigments produced result in a green color	Delayed healing; abscess formation; extension into tissues, artificial devices, or bloodstream; septic shock

drop in blood pressure, and shock. Sometimes the infected person will also have a rash and diarrhea. ⏭ **staphylococcal toxic shock syndrome**

Causative Agents

Staphylococcus aureus and *S. epidermidis* are Gram-positive cocci that grow in clusters (**figure 23.4**). They are facultative anaerobes and are quite hardy, having evolved to thrive on skin, which is dry and salty. They survive well in the environment and are easily transferred from person to person. ⏮ **facultative anaerobe**

The coagulase test is used to distinguish *S. aureus* from other staphylococci. *S. aureus* is often referred to as "coag-positive staph" because it makes the protein coagulase, whereas the other staphylococcal species are collectively referred to as "coag-negative staph." Of the coag-negative staphylococci, *S. epidermidis* is the most common cause of healthcare-associated infections, including those of surgical wounds. However, *S. aureus* causes serious wound infections much more commonly than *S. epidermidis.* ⏮ **coagulase,** ⏮ **healthcare-associated infections**

Pathogenesis

Staphylococcus aureus *S. aureus* produces multiple virulence factors that act together in the disease process (see table 22.3). These virulence factors are covered extensively in chapter 22 (skin infections), but it is important to recognize that they play a critical role in wound infections as well. For example, clumping factor and other proteins allow the cells to attach to clots and tissue components, an initial step in colonization. Lipases, proteases, and hyaluronidases together cause tissue damage. Capsules, coagulase, and protein A protect the cells from the complement system, phagocytes, and antibodies. Immunity to *S. aureus* infection is generally weak or nonexistent, probably because the organism evades the immune defenses so effectively. ⏮ ***S. aureus*** **pathogenesis,** ⏮ **complement system**

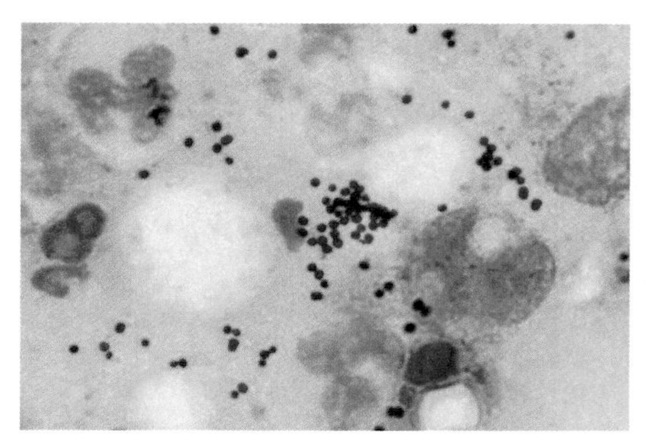

FIGURE 23.4 ***Staphylococcus aureus*** Gram stain of pus.
©Lisa Burgess-Scimeca

? What does the name *Staphylococcus* indicate about the typical growth arrangement of the bacterial cells?

Staphylococcus aureus infections sometimes cause systemic complications. Bacteria growing in a wound can spread to other parts of the body, leading to abscesses in the heart, the bones, or other tissues. Some *S. aureus* strains produce **superantigens** that can enter the circulation and activate many helper T cells. The T cells then release large amounts of cytokines (a cytokine storm), resulting in an overreaction that leads to toxic shock syndrome. ⏮ **superantigens,** ⏮ **cytokines**

Staphylococcus epidermidis *S. epidermidis* is not particularly virulent and cannot invade healthy tissues. However, the bacterium often causes minor abscesses around sutures. It can also form biofilms on medical devices, including indwelling catheters and artificial joints. Biofilm formation begins when the organism adheres to and then colonizes these devices by binding to fibronectin, the blood protein that quickly coats surgical implants in the body. The bacteria then produce a slime layer, or glycocalyx—a critical step in biofilm formation. Biofilms are a serious problem for several reasons. Diffusion of antibacterial medications into them is slow and inefficient. Even if the medication does enter the biofilm, it may be ineffective because bacteria within the biofilm are often metabolically inactive. In addition, bacteria in biofilms can come loose, and are then carried by the bloodstream to the heart and other tissues. In people with a compromised immune system, this can result in multiple tissue abscesses, which require surgical treatment, or in infective endocarditis. In healthy people, wound infections by *S. epidermidis* are usually cleared by host defenses without additional treatment. ⏮ **glycocalyx,** ⏮ **biofilm,** ⏭ **infective endocarditis**

MicroByte

It takes more than 100,000 *S. aureus* cells injected into skin to cause an abscess, whereas only 100 injected into a suture site to do the same.

Epidemiology

Some people are *S. aureus* carriers, putting them at greater risk for surgical wound infections caused by this species. Other factors that increase a person's risk include advanced age, poor general health, immunosuppression, an infection at another body site, and a prolonged stay in the hospital before surgery. Additional information about the epidemiology of *S. aureus* is discussed in chapter 22. ⏮ ***S. aureus*** **epidemiology**

S. epidermidis is found on the skin and mucous membranes of most people, residing as part of their normal microbiota. It is an opportunist that can cause disease in people who are immunocompromised. ⏮ **opportunist**

Treatment and Prevention

Treating staphylococcal infections can often be difficult because many strains are resistant to antibiotics. **Methicillin-resistant *S. aureus* (MRSA)** is a serious problem in wound infections. This organism is resistant to nearly all β-lactam antibiotics (an exception is ceftaroline, a new cephalosporin), and is therefore difficult to treat. MRSA may be healthcare-associated (HA-MRSA) or community-acquired (CA-MRSA). Both HA-MRSA and CA-MRSA are resistant to multiple antibiotics, although CA-MRSA is more susceptible than HA-MRSA and can usually be treated successfully with sulfa drugs, tetracyclines, or clindamycin. HA-MRSA is sometimes resistant even to these and is often susceptible only to a few medications, including vancomycin and ceftaroline. **Vancomycin-intermediate *S. aureus* (VISA)** and **vancomycin-resistant *S. aureus* (VRSA)** have emerged, making treatment of these infections extremely difficult. New medications, including linezolid, daptomycin, and tigecycline, are still effective in treating them. ⏮ antibacterial resistance, ⏮ MRSA, ⏮ VISA and VRSA, ⏮ HA-MRSA, ⏮ CA-MRSA

To reduce the chance of infection, including staphylococcal infection, wounds should be thoroughly cleaned, removing any dirt or dead tissue. Clean, deep wounds and surgical wounds should be quickly closed by sutures to help avoid infection. Surgical wound infection rates can be greatly reduced if the patient is given an effective anti-staphylococcal medication immediately before and after surgery. For unknown reasons, the infection rate is actually increased if the medication is given more than a few hours before or after surgery. **Table 23.2** summarizes the main features of staphylococcal wound infections.

Group A Streptococcal "Flesh-Eating Disease"

Streptococcus pyogenes is another common cause of wound infections. *S. pyogenes* infections are generally easy to treat because all known strains of the organism are still susceptible to penicillin. Occasionally, however, the infections can progress rapidly, even leading to death despite antimicrobial treatment. These more severe infections are called "invasive" because they spread into tissues and organs, causing pneumonia, meningitis, puerperal fever, **fasciitis**—inflammation of the fascia that surround muscles and body organs—and streptococcal toxic shock. This section will focus on **necrotizing fasciitis** ("flesh-eating disease"), a rare but serious complication of *S. pyogenes* infection (**figure 23.5**). ⏮ *Streptococcus pyogenes*

Signs and Symptoms

Signs and symptoms of necrotizing fasciitis are very serious. In fulminant disease (the term means "sudden and extreme"), the symptoms appear and progress within a few hours. In other cases (acute and subacute disease), the symptoms appear and progress over several days. Severe pain develops at the site of the wound, which sometimes can be so minor that no break in the skin is even seen. Within a short time, swelling occurs, and the injured person develops fever and confusion. Patients may also suffer from fatigue and vomiting. The overlying skin becomes stretched and discolored because of the swelling. Unless treatment is started quickly, shock and death usually follow in a short time.

Causative Agent

Streptococcus pyogenes is a β-hemolytic, Gram-positive, chain-forming, aerotolerant anaerobe, with Lancefield group A cell wall carbohydrate and a hyaluronic acid capsule. Strains of *S. pyogenes* that cause invasive disease are more virulent because they produce various enzymes and toxins that cause severe tissue damage (covered next). ⏮ aerotolerant anaerobe, ⏮ β-hemolysis, ⏮ pyrogens

Pathogenesis

The pathogenesis of *Streptococcus pyogenes* has been covered in detail in chapter 21. The bacterium produces F protein that helps it colonize the wound (see table 21.2). It also produces a variety of enzymes that destroy the subcutaneous fatty tissue

TABLE 23.2 Staphylococcal Wound Infections

Signs and symptoms	Redness, swelling, and pain at wound site; pus formation; sometimes fever; occasionally shock
Incubation period	Variable
Causative agent	*Staphylococcus aureus* and *S. epidermidis*
Pathogenesis	*S. aureus* produces many virulence factors that cause tissue damage and destruction; *S. epidermidis* is less virulent but may form biofilms.
Epidemiology	*S. aureus* carriers at greater risk for wound infection; predisposing factors also include advanced age, general poor health, immunosuppression, and prolonged hospital stay. *S. epidermidis* is opportunistic and affects the immunocompromised.
Treatment and prevention	Treatment: antibiotics; resistance a problem with *S. aureus*. Prevention: cleaning wounds, pre- and post-operative wound care.

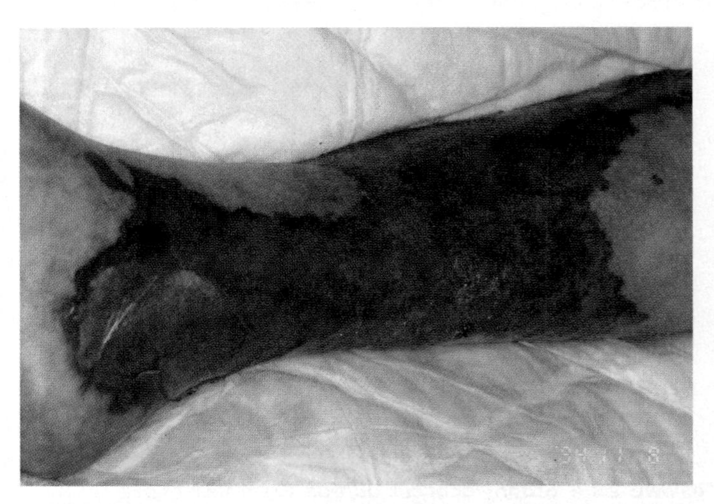

FIGURE 23.5 Individual with *Streptococcus pyogenes* "Flesh-Eating Disease" (Necrotizing Fasciitis) ©DermNet New Zealand

? Why does necrotizing fasciitis require immediate surgery?

and fascia. These enzymes include streptokinases (which break down blood clots), hyaluronidases (which break down connections between cells), deoxyribonucleases (which break down DNA), and streptolysins (which break down red blood cells and leukocytes). In some cases, these enzymes may also damage muscle tissue.

Strains of *S. pyogenes* that cause necrotizing fasciitis also produce a variety of streptococcal pyrogenic exotoxins (SPEs). SPE A is a superantigen that causes helper T cells to release large amounts of cytokines, leading to toxic shock. SPE B is a protease that causes tissue death and breakdown, leading to fluid accumulation in the area and intense swelling. The organisms multiply in the dead tissue, using the breakdown products as nutrients.

As the bacteria grow, they shed M protein, a surface protein that attaches to fibrinogen. The M protein–fibrinogen complexes bind to neutrophils, causing them to release strong inflammatory molecules that increase the vascular permeability in the host. The blood vessels leak fluid, resulting in a life-threatening drop in blood pressure and shock. M protein also helps the organism avoid phagocytosis by promoting the inactivation of complement component C3b.

Epidemiology

Cases of necrotizing fasciitis in the United States are rare and sporadic. Of the deaths caused by invasive *S. pyogenes* infections in the United States yearly, very few are due to necrotizing fasciitis. The risk of developing necrotizing fasciitis and other invasive *S. pyogenes* infections is increased with predisposing conditions that include diabetes, cancer, alcoholism, AIDS, recent surgery, abortion, childbirth, chickenpox, and injected-drug abuse. Invasive infections seldom occur in healthy individuals with minor injuries.

Treatment and Prevention

The toxins produced by *S. pyogenes* spread with such speed that immediate surgery is often essential to reduce the pressure of the swollen tissue and to remove dead tissue. Amputation is sometimes necessary to quickly remove the source of toxins. A combination of broad-spectrum antibiotics is given intravenously. The patient also receives supportive care, including fluid replacement, cardiac monitoring, and nutritional support.

There are no proven preventive measures for *S. pyogenes* fasciitis, although M protein vaccines are being tested. Generating such vaccines is challenging, but several candidates have proved promising in animal studies. **Table 23.3** summarizes the main features of necrotizing fasciitis.

TABLE 23.3 Necrotizing Fasciitis

Signs and symptoms	Severe pain and swelling at wound site; skin discoloration; fever, confusion, and shock
Incubation period	Varies; several hours in fulminant disease to several days in less severe cases
Causative agent	*Streptococcus pyogenes,* Lancefield group A β-hemolytic streptococci
Pathogenesis	Organism attaches and colonizes using protein F, then destroys muscle and organ fascia by releasing destructive enzymes and exotoxins; strong inflammatory response caused by M protein.
Epidemiology	Very rare; most cases occur in people with predisposing conditions such as diabetes, cancer, AIDS, and alcoholism, among others.
Treatment and prevention	Treatment: aggressive treatment essential with removal of affected tissue (debridement), or sometimes amputation, and intravenous antibiotics. Supportive care also given. Prevention: no proven preventive measures.

Pseudomonas aeruginosa Infections

Pseudomonas aeruginosa, an opportunistic pathogen, is a significant cause of healthcare-associated infections. In hospitals, *P. aeruginosa* is an important cause of lung infections and a common cause of wound infections, especially of thermal burns. Burns have large exposed areas of dead tissue that are not protected by normal body defenses and are therefore highly susceptible to bacterial infection. Almost any opportunistic pathogen can infect burns, but *P. aeruginosa* is among the most common and most difficult to treat. ◀◀ *Pseudomonas*

P. aeruginosa also occasionally causes community-acquired infections. Such infections include skin rashes and external ear canal infections from contaminated swimming pools and hot tubs, and eye infections from contaminated contact lens solutions. Other infections caused by this organism include those of bones (resulting from puncture wounds), heart valves (in injection-drug abusers), and soft tissues (from procedures such as ear piercing). *P. aeruginosa* also frequently forms biofilms in the lungs of people with the inherited disease cystic fibrosis, causing serious infection.

Signs and Symptoms

Signs and symptoms of *Pseudomonas aeruginosa* infection are serious and include chills, fever, skin lesions, and shock, which are caused by bloodstream invasion by this organism. A very noticeable symptom of *P. aeruginosa* infection of burns and other wounds is pus of a characteristic green color, caused by water-soluble pigments produced by the organisms (**figure 23.6a;** see also figure 11.11). These pigments, which are virulence factors, include yellow pyoverdin and blue pyocyanin, which together are green.

Causative Agent

Pseudomonas aeruginosa is a motile Gram-negative rod with a single polar flagellum (figure 23.6c). It is found in a wide variety of environments such as soil and water, where it grows easily and fast. The bacterium is classified as an aerobe.

FIGURE 23.6 ***Pseudomonas aeruginosa*** **(a)** Infected skin graft. **(b)** Culture showing green discoloration of an agar medium. **(c)** Stained cells with their single polar flagellum. a: ©Matt Meadows/Science Source; b: ©McGraw-Hill Education/Lisa Burgess c: ©McGraw-Hill Education/Lisa Burgess

? **What causes the green discoloration of the wound and the culture medium?**

However, it also respires anaerobically in the absence of O_2 if nitrate is present—an important factor in the organism's ability to form biofilms. ◀◀ anaerobic respiration

Pathogenesis

Pseudomonas aeruginosa infection of burns and other wounds expands tissue damage, delays healing, and increases the risk of septic shock. The organism has numerous virulence factors. It uses its polar flagellum and pili for attachment and colonization of host tissues. Once established, it produces an exotoxin (exotoxin A) that stops host cell protein synthesis. It also uses a type III secretion system (TTSS) to deliver into the host cell various effector proteins that change the activities of that cell. Different strains produce different effector proteins, but most strains that infect burns produce one called ExoS (exoenzyme S). This protein prevents phagocytosis of the pathogen and causes apoptosis of host cells. The organism also produces membrane-damaging toxins, including phospholipase C, which destroys lecithin, an important lipid component of cell membranes. Destruction of lecithin leads to cytoplasmic membrane disruption and cell death. Finally, *P. aeruginosa* produces enzymes such as proteases and elastases that destroy collagen and elastin, which help the organism spread in the tissues. ◀◀ type III secretion systems

The pigments produced by *P. aeruginosa* also contribute to its virulence. Pyocyanin plays a role in biofilm formation, allowing the organism to avoid host defenses as well as antimicrobial medications. Pyoverdin is a siderophore that the bacterium uses to acquire iron, which is often a limiting nutrient in the host. ◀◀ biofilms, ◀◀ siderophores

Epidemiology

Pseudomonas aeruginosa is widespread in nature. The organism can grow in most places where there is moisture, including soaps, ointments, eyedrops, contact lens solutions, cosmetics, disinfectants, swimming pools, hot tubs, and even distilled water. Because *P. aeruginosa* can be introduced into hospitals on flowers, potted plants, and fruit baskets, visitors are not allowed to take these items into hospital burn wards or intensive care units. The organism can also be found on many kinds of hospital equipment, on the soles of shoes, and in illegal injectable drugs, all of which have been sources of serious infections.

Treatment and Prevention

Pseudomonas aeruginosa infections are treated with antimicrobial medications. Established infections, however, are very difficult to treat because *P. aeruginosa* is resistant to a wide range of antibiotics. The situation is made even worse by the fact that the organism forms biofilms, and bacteria growing within biofilms are even more resistant. Only a few antibiotics are effective against this pathogen, including fluoroquinolones, gentamicin, and imipenem—and even these medications may not be effective against all strains. Antibiotic sensitivity tests are done to determine the most appropriate medication for treatment. For systemic infections, antibacterial medications must usually be given intravenously in high doses.

P. aeruginosa infections can be prevented by eliminating possible sources of the bacterium and by prompt care of wounds. Removing dead tissue from burn wounds, followed by application of an antibacterial cream, also helps prevention of infection by this organism. **Table 23.4** summarizes the main features of *Pseudomonas* infections.

TABLE 23.4 *Pseudomonas* **Infections**

Signs and symptoms	Chills, fever, skin lesions, and shock; green-colored pus
Incubation period	Variable, depending on site of infection
Causative agent	*Pseudomonas aeruginosa;* a Gram-negative motile rod that can respire both aerobically and anaerobically
Pathogenesis	Uses flagellum and pili for attachment and colonization. Secretes exotoxin A and injects effector proteins that alter cell activities; produces toxins that disrupt host cytoplasmic membranes; pigments contribute to pathogenicity.
Epidemiology	Widespread organism carried on vegetation; difficult to avoid.
Treatment and prevention	Treatment: antibiotics, although resistance and biofilm formation are problematic. Prevention: eliminating bacterial source; proper wound care, including removal of dead tissue.

MicroAssessment 23.2

Staphylococcus aureus is the most important cause of wound infections because it is commonly carried by humans, transfers easily from one person to another, and has multiple virulence factors. *Staphylococcus epidermidis* forms biofilms on foreign materials, protecting the bacteria from body defenses and antibacterial medications. "Flesh-eating" strains of *Streptococcus pyogenes* are uncommon but can cause life-threatening disease. *Pseudomonas aeruginosa,* a pigment-producing organism, is widespread in the environment and is a major cause of healthcare-associated infections.

4. Why are antibacterial medications not effective for treating necrotizing fasciitis?
5. Why do wound infections caused by *Pseudomonas aeruginosa* sometimes have green pus?
6. Why is it not surprising that staphylococci are the most common cause of wound infections?

23.3 ■ Diseases Due to Anaerobic Bacterial Wound Infections

Learning Outcomes

5. Describe the conditions that lead to the development of anaerobic wound infections.
6. Discuss why it is difficult to treat wounds infected with toxin-producing bacteria.

Wounds are often anaerobic and may be colonized by certain strictly anaerobic species of bacteria. This section describes two distinctive diseases that result from anaerobic bacterial wound infections: tetanus and gas gangrene.

Tetanus ("Lockjaw")

Tetanus is often fatal. Fortunately, it is uncommon in economically developed countries. Exposure to the causative organism cannot be avoided because it produces endospores that are widespread in dust and dirt, frequently contaminating clothing, skin, and wounds. Even a minor wound in a nonimmunized person can result in tetanus if the wound provides conditions that allow germination of the spores. ⏮ **endospores**

Signs and Symptoms

Tetanus is characterized by continuous, painful, and uncontrollable cramp-like muscle spasms that are usually generalized but may be limited to one area of the body (**figure 23.7**). The spasms often begin with the jaw muscles, giving the disease the common name "lockjaw." The early symptoms usually appear about a week after infection and include restlessness, irritability, difficulty swallowing, spasms of the jaw muscles, and sometimes seizures. As more muscles go into sustained contraction (called tetany), breathing becomes difficult, abnormal heart rhythms may occur, and in some cases

FIGURE 23.7 Infant with Neonatal Tetanus Source: CDC

How does a newborn infant contract tetanus?

bones can fracture. After a period of severe pain, the infected person often dies of pneumonia or from lung damage caused by regurgitation of stomach contents into the lung. Surviving patients remain in the hospital for long periods of time, which increases their risk of healthcare-associated infections. ⏮ healthcare-associated infections

Newborns can also develop a form of tetanus called neonatal tetanus. About 4 to 14 days after birth, the baby becomes irritable and develops muscle rigidity with spasms (see figure 23.7). As the muscles of the face and mouth are affected, the infant can no longer feed. Finally the baby stops breathing and dies.

Causative Agent

Tetanus is caused by *Clostridium tetani,* an anaerobic, spore-forming, Gram-positive, rod-shaped bacterium (**figure 23.8**). The bacterium has two characteristic features: (1) a spherical endospore that forms at the end of the cell, and (2) swarming growth that quickly spreads over the surface of solid media. Diagnosis of tetanus is typically by signs and symptoms. However, identification of *C. tetani* depends on detecting its toxin that plays a key role in pathogenesis (discussed next). ⏮ *Clostridium*

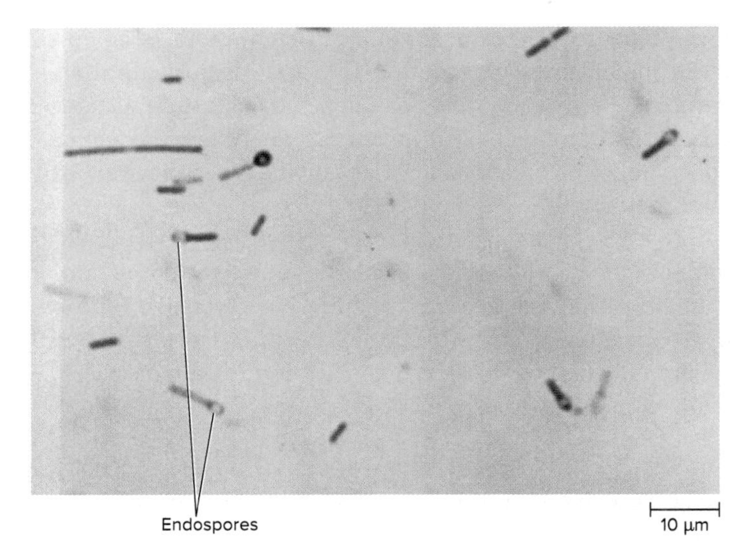

FIGURE 23.8 ***Clostridium tetani*** Terminal endospores are characteristic of this species. Source: Dr. Holdeman/CDC

? How is the position of the endospores shown here different from endospores of other clostridial species?

Pathogenesis

Clostridium tetani is not invasive, and colonization is generally localized to a wound. Its pathological effects are caused by **tetanospasmin,** a toxin that causes muscles to contract without control (a condition known as spastic paralysis). Tetanospasmin is an A-B exotoxin and is released from multiplying vegetative cells. The B subunit attaches to receptors on motor neurons, which then take up the A portion by endocytosis. The toxin is carried to the neuron cell body in the spinal cord. There, the motor neuron contacts other neurons that normally control its action by means of chemicals called neurotransmitters. In normal situations, some of these other neurons stimulate the motor nerve cell, causing muscle contraction, while others make the motor neuron resistant to stimulation, inhibiting muscle contraction. Tetanospasmin prevents the release of the neurotransmitter from inhibitory neurons, resulting in sustained muscle contraction and the characteristic muscle spasms (**figure 23.9**). ⏮ A-B toxins

The toxin generally spreads across the spinal cord to the side opposite the wound and then downward. This initially causes muscle spasms on only one side of the wound, but as toxin levels increase, the spasms spread to the muscles on the opposite side, and then downward. Tetanospasmin released from the infected wound often enters the bloodstream, which

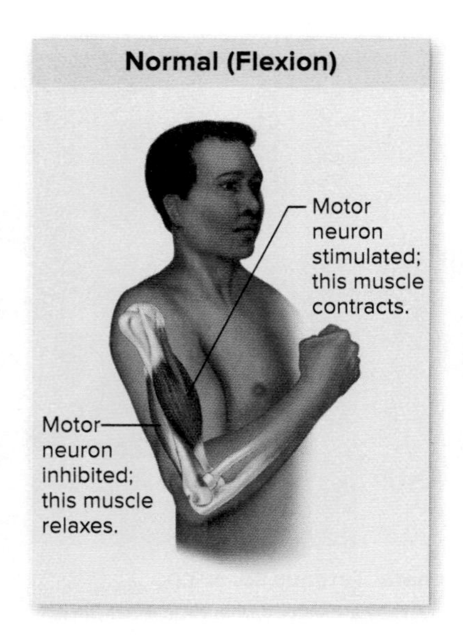

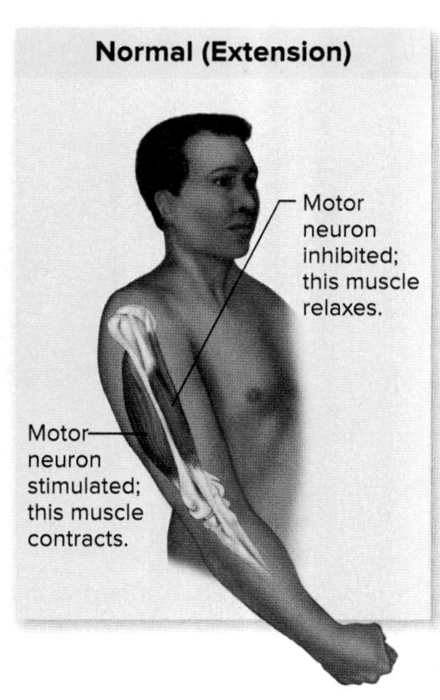

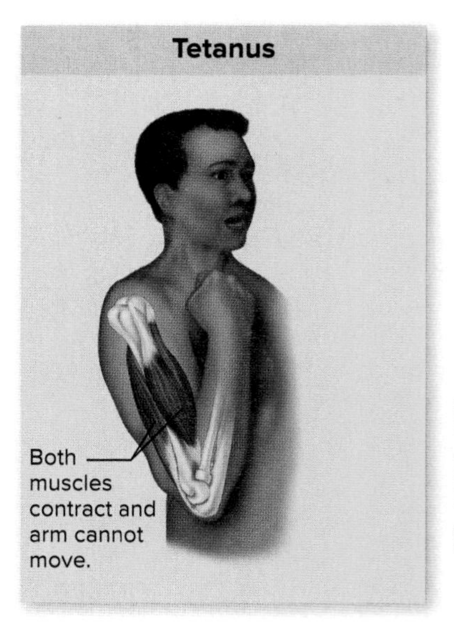

FIGURE 23.9 Tetanus and Inhibitory Neuron Function

? How does tetanospasmin cause muscle contraction?

carries it to the central nervous system. In these cases, inhibitory neurons of the brain are first affected, and the muscles of the jaw are among the first to develop spasms.

Epidemiology

Clostridium tetani spores are found not only in dirt and dust, but also in the gastrointestinal tract of humans and other animals that have eaten foods contaminated with its spores. Therefore, fecal contamination is a potential source of infection. Many cases of tetanus result from anaerobic puncture wounds, which occur in stepping on a nail or other sharp object, body piercing, tattooing, animal bites, splinters, injected-drug abuse, and insect stings. Cases can also occur following surgery. Surface abrasions and burns can result in tetanus if the wound is anaerobic because of dirt or dead tissue. Neonatal tetanus may occur if *C. tetani* spores are unintentionally introduced into the umbilical cord wound, either when the cord is cut with unsterilized instruments or when it is covered with an unsterile dressing.

People who get tetanus have inadequate immunity, meaning they never received a tetanus vaccine or did not receive a scheduled booster shot. In the case of neonatal tetanus, the baby was born to a woman who had inadequate immunity; recall that maternal antibodies remain active in the newborn, providing passive immunity for the first several months of life. Perhaps not surprisingly, tetanus occurs most frequently in less developed countries, because vaccination is not always available, and access to adequate healthcare is often lacking. ⏮ passive immunity

MicroByte

World Health Organization education initiatives in developing countries have reduced the cases of neonatal tetanus by almost 95% in the last 20 years.

Treatment and Prevention

The wound is thoroughly cleaned of all dead tissue and foreign material that could cause anaerobic conditions. If tetanus develops, it is treated by injecting the affected person with human tetanus immune globulin (TIG), a preparation of antibodies to tetanospasmin. The antibodies bind to toxin molecules not yet attached to nerve cells, thereby neutralizing their effects and providing passive immunity. TIG, however, cannot neutralize tetanospasmin that is already attached to nerve tissue. Because of this, the patient is given muscle relaxants and supportive care, including being placed on a ventilator if needed. Over time, the affected nerves repair themselves. An antibacterial medication such as metronidazole is given to kill any actively multiplying clostridia, preventing the production of more tetanospasmin. The person is also given tetanus vaccine. Neonatal tetanus is treated with antibiotics, antitoxin, and supportive care, but treatment is mostly ineffective. ⏮ TIG

Tetanus can easily be avoided by vaccination with tetanus toxoid, which is inactivated tetanospasmin. Immunization with tetanus toxoid in combination with diphtheria toxoid and acellular pertussis vaccine, called DTaP, is recommended in children under the age of 7. DTaP is given at 2, 4, 6, and 18 months of age, with a booster dose when children enter school. Older children and adults should get additional booster doses at 10-year intervals to maintain adequate immunity. For this, Tdap or Td is used (the lowercase letters indicate a lower dose of diphtheria and pertussis antigens). People who have recovered from tetanus are not immune to the disease and must be immunized. People who have been injured or burned or are planning surgery may need a booster shot of the tetanus vaccine. **Table 23.5** summarizes prevention of tetanus by wound management. Neonatal tetanus can be prevented by immunizing a woman either before or during her pregnancy, as well as using clean equipment and sterile practices during delivery. **Table 23.6** describes the main features of tetanus. ⏮ toxoid

TABLE 23.5 Tetanus Prevention in the Management of Wounds

Immunization History	CLEAN MINOR WOUNDS		ALL OTHER WOUNDS	
	Toxoid	TIG	Toxoid	TIG
Unknown, or fewer than three injections of toxoid	Yes	No	Yes	Yes
Fully vaccinated (three or more injections of toxoid)				
■ 5 years or less since last dose	No	No	No	No
■ 5 to 10 years since last dose	No	No	Yes	No
■ More than 10 years since last dose	Yes	No	Yes	No

Clostridial Myonecrosis ("Gas Gangrene")

Before antibiotics were discovered, clostridial myonecrosis (commonly called "gas gangrene") killed many soldiers wounded in wars. Endospores of *Clostridium perfringens,* the bacterium that can cause this disease, are found in soil and dust everywhere, and frequently can be isolated from wounds. However, contamination with *C. perfringens* rarely leads to gas gangrene, because like *C. tetani,* the organism grows only in anaerobic conditions. Gas gangrene occurs mainly in neglected wounds with fragments of bone, foreign material, and serious tissue damage. It also occurs occasionally in surgical sites, especially in people with underlying diseases.

Signs and Symptoms

Gas gangrene signs and symptoms appear 1 to 5 days after infection and are dramatic. They include severe pain

TABLE 23.6 Tetanus ("Lockjaw")

1. *Clostridium tetani* spores from dust or dirt enter a wound.
2. In anaerobic wounds, the spores germinate, and vegetative bacteria release an exotoxin called tetanospasmin.
3. Tetanospasmin is carried to the central nervous system by motor nerve axons or by the bloodstream.
4. The toxin prevents any inhibitory neurons it reaches from functioning.
5. Other neurons that cause muscle contraction act unopposed, causing muscle spasms.
6. The result is a sustained, painful cramp-like muscle spasm.

Symptoms	Restlessness, irritability, difficulty swallowing; muscle pain and spasm in jaw, abdomen, back, or entire body.
Incubation period	3 days to 3 weeks; average 8 days
Causative agent	*Clostridium tetani,* an anaerobic, spore-forming, Gram-positive rod
Pathogenesis	Tetanus results from tetanospasmin, an A-B exotoxin produced by the bacterium. The toxin is carried to the brain and spinal cord by motor nerve axons or circulating blood; toxin acts against nerve cells that normally inhibit muscle contraction. Other nerves that normally cause muscle contraction then act unopposed, causing muscle spasms.
Epidemiology	Organisms common in soil; spores contaminate wounds, germinate in those having anaerobic conditions, particularly dirty or puncture wounds.
Treatment and prevention	Treatment: antibiotics and tetanus immune globulin (TIG); supportive care also given. Prevention: immunization of infants and children with toxoid; boosters for adults.

that quickly increases in the infected wound, followed by swelling in the area, and a thin, bloody or brownish fluid that leaks from the wound. This fluid may look frothy because of gas bubbles released by the organism. The overlying skin becomes stretched tight and mottled with black (**figure 23.10**). Although seriously ill, the patient remains quite alert until late in the illness when, near death, he or she becomes delirious and goes into a coma.

Causative Agent

Several species of *Clostridium* can cause gas gangrene when they invade injured muscle, but by far the most common is *C. perfringens,* an encapsulated Gram-positive anaerobic toxin-producing rod. Although *C. perfringens* is an endospore-forming organism, it usually does not produce spores in wounds or cultures.

Pathogenesis

Two main factors lead to the development of gas gangrene: (1) the presence of dirt and dead tissue in the wound, and (2) long delays before the wound is treated. *Clostridium perfringens* is unable to infect healthy tissue but grows easily in poorly oxygenated and dead tissues that provide it with a source of nutrients. It releases α-toxin, an enzyme that destroys lecithin in host cell membranes, leading to cell lysis. The toxin diffuses from the area of infection, killing leukocytes and tissue cells. Several enzymes produced by the pathogen, including

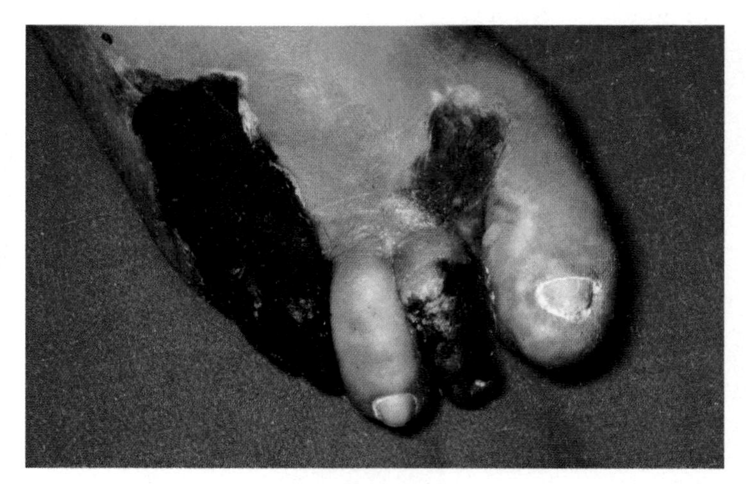

FIGURE 23.10 Foot of Person with Gas Gangrene (Clostridial Myonecrosis) ©Central Manchester University Hospitals NHS Foundation Trust, UK/Science Source

? How is gas gangrene treated?

collagenase and hyaluronidase, break down the host tissues. As the bacterial cells multiply, they ferment the tissue breakdown products, producing large amounts of hydrogen and carbon dioxide. These gases accumulate in the tissue and cause a rise in pressure, which aids in spread of the infection. Without quick surgical treatment, massive amounts of α-toxin diffuse into the bloodstream and destroy red blood cells, tissue capillaries, and other structures throughout the body. This systemic involvement leads to wide-scale cytokine release, resulting in shock and organ failure. It is not reversed by treatment with antibody to α-toxin.

Epidemiology

Clostridium perfringens is widespread in soil and is common in the intestinal tract of many animals and humans. It is sometimes found in the vagina of healthy women. Besides neglected trauma wounds and a few surgical wounds, gas gangrene of the uterus is fairly common after self-induced abortions, and occasionally occurs after miscarriages and childbirth. In other cases, diseases such as arteriosclerosis or diabetes—which lead to poor oxygenation of tissue from restricted blood flow—are predisposing factors. Cancer patients also have increased susceptibility to gas gangrene.

Treatment and Prevention

Treatment of gas gangrene includes prompt debridement of all dead and infected tissues, and may require amputations. High doses of antibiotics, often a combination of penicillin and clindamycin, are given to help stop bacterial growth and toxin production. However, the medications may not diffuse well into large areas of dead tissue and do not inactivate toxin. Hyperbaric oxygen treatment is sometimes used; the patient is placed in a special chamber and breathes pure O_2 under increased pressure. The pressure treatment inhibits growth of the clostridia, and so stops the release of toxin. The high levels of O_2 also improve oxygenation of injured tissues, lessening anaerobic conditions. The use of hyperbaric treatment is controversial as it is not always successful and may lead to other complications.

There is no available vaccine for gas gangrene. The disease can be prevented by prompt cleaning and debridement of dead tissue from wounds. **Table 23.7** describes the main features of this disease.

TABLE 23.7 Clostridial Myonecrosis ("Gas Gangrene")

1. *Clostridium perfringens* spores enter a wound that has two essential characteristics: dead tissue and anaerobic conditions.
2. The spores germinate, and the vegetative bacteria multiply in dead tissue, producing α-toxin.
3. α-Toxin diffuses into normal tissue and kills it. The infection spreads into the dead tissues as bacterial enzymes break down host cells.
4. Swelling occurs; gas produced by fermentation aids rapid progress of the infection.
5. Massive amounts of α-toxin are produced and diffuse into the bloodstream, destroying blood cells and other cells throughout the body. Systemic response to the damage leads to shock and organ failure.

Signs and Symptoms	Severe pain; gas and fluid seep from wound, blackening of overlying skin; shock and death commonly follow.
Incubation period	Usually 1 to 5 days
Causative agent	Usually *Clostridium perfringens;* other clostridia less frequently.
Pathogenesis	Organism grows in dead and poorly oxygenated tissue and releases α-toxin; toxin kills leukocytes and normal tissue cells by degrading lecithin in their cell membranes; systemic response to the damage leads to shock and organ failure.
Epidemiology	Trauma wounds; dirt contamination of wounds, tissue death, impaired circulation to tissue as in people with diabetes or arteriosclerosis; self-induced abortions.
Treatment and prevention	Treatment: surgical removal of dirt and dead tissues of primary importance; hyperbaric O_2 of possible value; antibiotics to kill vegetative *C. perfringens.* Prevention: prompt cleaning and debridement of wounds; no vaccine available.

FOCUS ON A CASE 23.1

A 63-year-old woman, healthy except for mild diabetes, had her gallbladder removed. The surgery went well, but within 72 hours the surgical incision became swollen and pale. Within hours, the swollen area widened and developed a bluish discoloration. The surgeon suspected gas gangrene. Antibiotic treatment was started, and the patient was rushed back to the operating room, where the entire swollen area, including the repaired operative incision, was surgically removed. After that, the wound healed normally, although she required a skin graft to close the large wound. High numbers of *Clostridium perfringens* grew from the wound culture.

Six days later, a 58-year-old woman had surgery in the same operating room for a malignant tumor of the colon. The surgery was performed without difficulty, but 48 hours later she developed rapidly advancing swelling and bluish discoloration of her surgical wound. As with the first case, gangrene was suspected and she was treated with antibiotics and surgical removal of the affected tissue. She also needed a skin graft. Her wound culture also showed *C. perfringens.*

Because the surgery department had never had any of its patients develop surgical wound infections with *C. perfringens* before this, the hospital epidemiologist was asked to do an investigation. Among the findings of the investigation:

- Cultures of surfaces in the operating room grew large numbers of *C. perfringens.*
- Unknown to the medical staff, a worker had recently serviced a fan in the ventilation system of the operating room, and for a time air was allowed to flow into the operating room, rather than out of it.
- A road outside the hospital was being repaired using heavy machinery, creating clouds of dust.

As a result of these findings, the operating room and its ventilating system were cleaned and upgraded. No further cases of surgical wound gas gangrene developed.

1. Was the surgeon's diagnosis correct?
2. Many other patients had surgery in the same operating room. Why did only these two patients develop wound gas gangrene?
3. What could be done to help identify the source of the patients' infections?

Discussion

1. *Clostridium perfringens* is commonly cultivated from wounds without any evidence of infection. However, in the cases presented here, there was not only a heavy growth of the organism but also a clinical picture that suggested gas gangrene. The surgeon's diagnosis was correct.
2. In both cases, there was an underlying condition that increased the two patients' susceptibility to infection: cancer in one, and diabetes in the other. Moreover, both had a recognized source for the organism. Cultures of as many as 20% of diseased gallbladders are positive for *Clostridium perfringens,* and the organism is commonly found in large numbers in the human intestine—a potential source in the case involving removal of the colon tumor.
3. The surgeon thought that the infecting organism might have come from the patients themselves because such strains are often much more virulent than strains that live and sporulate in the soil. On the other hand, the gross contamination of the operating room, as revealed by the cultures of its surfaces, could indicate a very large infecting dose at the operative site, possibly compensating for lesser virulence. In addition, no further cases occurred after cleaning the operating room and fixing the ventilation system. Unfortunately, in this case, no cultures were done from the removed gallbladder or colon tumor, nor were the strains isolated from the wounds and the environment compared. Comparing the antibiotic susceptibility, toxin production, and other characteristics of the different isolates could have helped identify the source of the infections.

MicroAssessment 23.3

Tetanus ("lockjaw") is caused by an anaerobic, spore-forming bacterium with little invasive ability. The organisms release an exotoxin called tetanospasmin that is carried to the nervous system, causing uncontrolled muscle spasms and possibly death. Gas gangrene (clostridial myonecrosis), caused by another anaerobic spore-former, usually starts in neglected wounds containing dead tissue and foreign material, and then spreads into normal muscle tissue.

7. Why do many tetanus patients fail to show immediate improvement when given tetanus antitoxin?
8. What factors lead to the development of gas gangrene?
9. Do babies need to be immunized against tetanus if their mother had been immunized against the disease? Explain.

23.4 ■ Bacterial Infections of Bite Wounds

Learning Outcomes

7. Explain why human mouth microbiota can cause serious bite wound infections.
8. Describe two zoonotic wound infections.

Each year, more than 3 million people in the United States are bitten by animals. The biggest concern about these bites is the viral disease, rabies. Human bites are also common. The infection risk following a bite depends partly on the type of injury (crushing, lacerated, or puncture), and partly on the kinds of pathogens introduced into the bite wound. ⏭ rabies

Human Bites

Wounds caused by human teeth (such as bites that break the skin, hitting the teeth of another person, or being injured with objects that have been in a person's mouth) are common and can result in very serious infections. These infections are caused by normal mouth microbiota. Occasionally, diseases such as syphilis, hepatitis B, and hepatitis C are transmitted this way, because the causative agents of these diseases are found in the blood or saliva.

Signs and Symptoms

The wound may appear minor at first but within a few hours becomes painful and swells massively (**figure 23.11**). Foul-smelling pus often leaks out of it.

Causative Agents

The most frequent causes of infection in human bite wounds are members of the normal mouth microbiota, including streptococci, fusiforms, spirochetes, and *Bacteroides* species, often in association with *Staphylococcus aureus.*

Pathogenesis

Bite wounds are usually infected with both aerobic and anaerobic species. Crushed tissue and the presence of facultative anaerobes, which reduce available O_2, create conditions that allow anaerobic bacteria to establish infection at the site. Although most members of the mouth microbiota are harmless alone, together they produce large numbers of toxins and enzymes that damage the host tissues and cells involved in the immune response. These include leukocidin, collagenase, hyaluronidase, ribonuclease, various proteases, neuraminidase, and enzymes that destroy antibodies and proteins of the complement system. The result of all these factors is a **synergistic infection,** meaning that the sum effect of all the organisms acting together is greater than their individual effects. Irreversible destruction of tissues such as tendons and permanent loss of function can be the result. ⏮ facultative anaerobes

Epidemiology

Most serious human bite infections occur because of fights, or during forcible restraint, as found in law enforcement and in mental institutions. The risk is greatly increased when the biting individual has poor mouth care and extensive dental disease. Bites by young children are usually not serious, mostly because children seldom break the skin or cause crushing wounds when they bite.

Treatment and Prevention

Bite wounds should be treated immediately to avoid serious infections. This involves opening the infected area with a scalpel, washing the wound thoroughly with sterile fluid, and removing dirt, foreign matter (such as broken teeth), and dead tissue. Antibacterial medications should also be given. Usually more than one medication is used because many bacteria that cause bite wound infections are resistant to various drugs such as penicillin; a medication effective against anaerobes should be included. If bite wounds are not treated quickly, serious complications may occur, including infection of muscles and tendons, which may require surgery.

Prevention of bite wound infections involves avoiding situations that lead to biting and hitting. Prompt cleaning of any bite wound followed by application of an antiseptic can help prevent infection. The main features of human bite wound infections are presented in **table 23.8.**

Pasteurella multocida Bite Wound Infections

Infections caused by bacteria that live in the mouth of an animal are very common. Surprisingly, bite infections from a number of kinds of animals are generally caused by a single bacterial species, *Pasteurella multocida.*

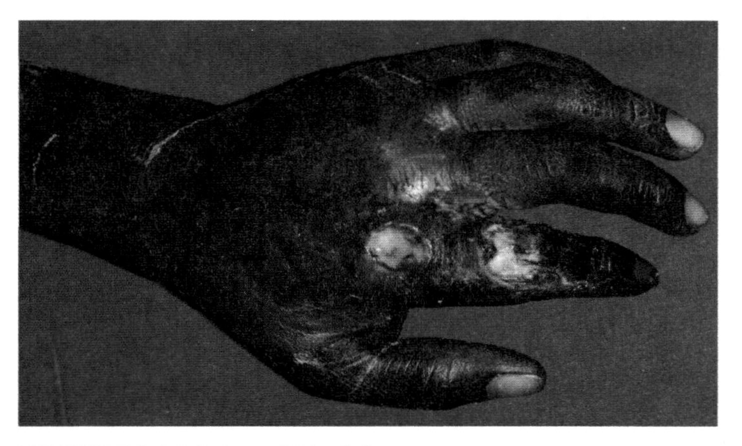

FIGURE 23.11 Infected Bite Injury Hand with infected bite injury to the index finger. ©Dr. M.A. Ansary/Science Source

? Why do human bite wounds often contain both aerobic and anaerobic bacteria?

TABLE 23.8 Human Bite Wound Infections

Signs and symptoms	Rapid onset, pain, massive swelling, drainage of foul-smelling pus
Incubation period	Usually 6 to 24 hours
Causative agent	Mixed mouth microbiota: anaerobic streptococci, fusiforms, spirochetes, anaerobic Gram-negative rods; sometimes *Staphylococcus aureus*
Pathogenesis	Various mouth bacteria act synergistically to destroy tissue.
Epidemiology	Fighting; forcible restraint; poor mouth care and extensive dental disease.
Treatment and prevention	Treatment: antibiotics and surgery. Prevention: no proven preventive measures except to avoid fights: prompt cleaning of wound and application of antiseptic is advised.

FOCUS YOUR PERSPECTIVE 23.1

Infection Caused by a Human "Bite"

A 26-year-old man injured a finger of his right hand during a bar fight when he punched someone in the mouth. The man did not seek medical help until more than 36 hours after the fight. At that time, his entire hand was very swollen, red, and tender, and the swelling was spreading to his arm. The surgeon cut open the infected tissues, allowing pus to drain. He removed the damaged tissue and washed the wound with sterile fluid. Smears and cultures of the infected material showed aerobic and anaerobic bacteria characteristic of mouth normal microbiota, including species of *Bacteroides* and *Streptococcus.* The patient was given antibiotics to fight infection, but the wound did not heal well and continued to drain pus. Several weeks later, X rays revealed that infection had spread to the bone at the base of the finger. To cure the infection, the finger had to be amputated.

Signs and Symptoms

Early signs of infection in wounds caused by animal bites develop within 24 hours and include spreading redness, tissue tenderness, and swelling, followed by pus discharge. Abscesses commonly form. Without quick treatment, the infection can lead to bloodstream invasion or permanent loss of function.

Causative Agent

Pasteurella multocida is a Gram-negative, facultatively anaerobic encapsulated coccobacillus. The organisms are classified into five serogroups based on the antigenic type of their capsule.

Pathogenesis

The details of *P. multocida* pathogenesis are not yet known. Some strains produce a toxin that kills host cells by affecting protein synthesis and cell cycle regulation. The organisms have an antiphagocytic capsule, but when antibodies bind to them, phagocytes are able to ingest and kill the opsonized bacteria. If untreated, *P. multocida* can cause bacteremia, leading to endocarditis or meningitis.

MicroByte

Using *Pasteurella multocida,* Pasteur demonstrated that attenuated organisms could be used as a vaccine.

Epidemiology

Humans—and many other healthy animals such as cats, dogs, and monkeys—carry *P. multocida* as normal oral and upper respiratory microbiota. Both diseased and healthy animals are reservoirs for human infections. Cats are more likely to carry *P. multocida* than dogs, and so cat bites more frequently cause the infection. *P. multocida* also causes diseases in a number of other animal species, sometimes causing epidemics of fatal pneumonia and bloodstream infection in them.

Treatment and Prevention

Unlike many Gram-negative pathogens, *P. multocida* is susceptible to penicillin, so a medication that is a combination of a penicillin derivative (amoxicillin) and a β-lactamase inhibitor (clavulanate) is used. The clavulanate is helpful because many bite wounds often also contain strains of β-lactamase-producing *Staphylococcus aureus.* This and other antibacterial medications are effective if given early in the infection, so treatment is started even before the diagnosis has been confirmed by culturing the bacterium.

No vaccines for *P. multocida* are available for use in humans. Immediate cleaning of bite wounds and quick medical attention usually prevent the development of serious infection. **Table 23.9** gives the main features of *Pasteurella multocida* bite wound infections.

Bartonellosis ("Cat Scratch Disease")

Bartonellosis, also known as "cat scratch disease," is a bacterial disease that can be acquired after being bitten, licked, or scratched by an infected cat.

Signs and Symptoms

Cat scratch disease begins within a week of a scratch or bite with the appearance of a pus-filled pimple at the site of injury. Painful enlargement of the lymph nodes of the area develops in 1 to 7 weeks. Patients may develop fever, and in some cases

TABLE 23.9 *Pasteurella multocida* **Bite Wound Infections**

Signs and symptoms	Spreading redness, tenderness, swelling, discharge of pus.
Incubation period	24 hours or less
Causative agent	*Pasteurella multocida,* a Gram-negative, facultatively anaerobic, encapsulated coccobacillus
Pathogenesis	*P. multocida* introduced by bite, resists phagocytes because encapsulated; may produce a cytotoxin that kills host cells.
Epidemiology	Carried by many animals in their mouth or upper respiratory tract.
Treatment and prevention	Treatment: antibiotics, effective if given promptly. Prevention: no vaccine available for humans; prompt wound care.

the lymph nodes become pus-filled and soft. The disease generally disappears without treatment in 2 to 4 months. A small number of patients develop irritation of an eye with local lymph node enlargement, epileptic seizures and coma due to encephalitis, or acute or chronic fever associated with bloodstream or heart valve infection.

Complications can occasionally arise from cat scratch disease, mostly in people with underlying issues such as AIDS. These include peliosis hepatis, in which blood-filled cysts form in the liver, and bacillary angiomatosis, in which nodules composed of proliferating blood vessels develop in the skin and other parts of the body.

Causative Agent

Cat scratch disease is caused by *Bartonella henselae,* a curved, Gram-negative rod.

Pathogenesis

The virulence factors of *B. henselae* and the process by which it causes disease are not yet known. The organisms can enter the body by a cat bite or scratch, or when cat saliva contaminates a mucous membrane. The bacteria are carried to the lymph nodes, and the disease is resolved by the immune system in most cases. Spreading by the bloodstream occurs in some people, however, causing serious complications.

Epidemiology

Cat scratch disease occurs worldwide, often in children and young adults. In the United States, it is the most common cause of chronic localized lymph node enlargement in young children. The disease is a zoonosis, transmitted to humans mainly by bites or scratches from cats. Person-to-person spread does not occur. Asymptomatic bacteremia is common in cats, however, and transmission from cat to cat occurs by cat fleas. Fleas may also transmit the disease to people who handle infected cats.

Treatment and Prevention

Any cat scratch or bite should be promptly cleaned with soap and water and then treated with an antiseptic. If signs of infection develop, prompt medical evaluation is needed. Severe *B. henselae* infections can usually be treated with antibiotics such as ampicillin, but some strains are resistant to this medication.

There are no proven preventive measures for cat scratch disease, other than to avoid handling stray cats, especially those with fleas. The main features of cat scratch disease are presented in **table 23.10.**

TABLE 23.10	Bartonellosis ("Cat Scratch Disease")
Signs and symptoms	Pimple appears at the bite or scratch site, followed by local lymph node enlargement; nodes may soften and drain pus; prolonged fever and convulsions indicate spread to other body parts. Complications (peliosis hepatis or bacillary angiomatosis) may occur, mostly in people with AIDS.
Incubation period	Usually less than 1 week
Causative agent	*Bartonella henselae,* a Gram-negative rod
Pathogenesis	Bacteria enter with cat bite or scratch and reach lymph nodes, where the disease is usually stopped.
Epidemiology	A zoonosis, spread among cats by fleas. No person-to-person spread.
Treatment and prevention	Treatment: antibiotics. Prevention: promptly washing wounds, applying antiseptic; avoiding rough play with cats and keeping them free of fleas.

Streptobacillary Rat Bite Fever

Rat bites are fairly common among poor people in large cities and among workers who handle laboratory rats.

Signs and Symptoms

The bite wound usually heals quickly without any problems. Two to 10 days later, however, the person develops chills, fever, head and muscle aches, and vomiting. The fever comes and goes. A rash usually appears after a few days, followed by joint pain.

Causative Agent

The cause of streptobacillary rat bite fever is *Streptobacillus moniliformis,* a facultative anaerobic, Gram-negative rod that varies in shape (pleomorphic). The organism is unique in that it spontaneously develops L-forms. These are variants that lack a cell wall, first identified at the famous Lister Institute (hence, the *L* in *L-form*).

Pathogenesis

Streptobacillus moniliformis enters the body through a bite or scratch, and sometimes by ingestion. There is typically little enlargement of the local lymph nodes, and the bacteria quickly enter the bloodstream and spread throughout the body. The majority of people recover without treatment in about 2 weeks, but some people develop serious complications such as brain abscesses or infection of the heart valves. These cases may be fatal.

Epidemiology

Many healthy laboratory rats, mice, and other rodents carry *S. moniliformis* in their nose and throat. Laboratory and pet store workers are at increased risk of contracting the disease. Rat bites and scratches are the usual source of human infections. Epidemics of the disease have occurred when people

TABLE 23.11	Streptobacillary Rat Bite Fever
Signs and symptoms	Chills, fever, muscle aches, headache, and vomiting; later, rash and pain in one or more of the large joints
Incubation period	Usually 2 to 10 days (range, 1 to 22) after a rat bite
Causative agent	*Streptobacillus moniliformis*, a highly pleomorphic, Gram-negative bacterium
Pathogenesis	*S. moniliformis* invades bloodstream, spreads throughout body. Most patients recover without treatment but occasionally life-threatening complications (brain abscesses, endocarditis) occur.
Epidemiology	Transmitted by rats or other animals that prey on them; may also be spread by food or drink contaminated with rat feces.
Treatment and prevention	Treatment: antibiotics. Prevention: controlling wild rats and mice; care in handling laboratory animals.

have eaten or drunk anything contaminated with *S. moniliformis* from rodent feces. Cases of rat bite fever have also been associated with exposure to animals that eat rodents, including cats and dogs.

Treatment and Prevention

Penicillin, given intravenously, is used to treat cases of rat bite fever. Prevention of rat bite fever includes control of wild rat populations and care in handling laboratory rats and their feces. The main features of streptobacillary rat bite fever are presented in **table 23.11.**

MicroAssessment 23.4

Human bite infections can be dangerous because certain members of the mouth microbiota, which have little invasive ability when growing alone, can invade and destroy tissue when growing synergistically. *Pasteurella multocida* can infect bite wounds caused by a number of different animals, especially cats. Cat bites and scratches can also transmit *Bartonella henselae,* the cause of cat scratch disease, which is characterized typically by local lymph node enlargement. Streptobacillary rat bite fever, caused by *Streptobacillus moniliformis,* is acquired from bites of rats and mice, and from animals that eat them. It is characterized by fever that comes and goes and a rash.

10. What Gram-negative organism commonly infects wounds caused by bites of any of a variety of different animals?
11. What is the most common cause of chronic localized lymph node enlargement in young children?
12. Why are members of the normal mouth microbiota a cause of serious infection in human bites?

23.5 ■ Fungal Wound Infections

Learning Outcome

9. Give the distinctive features of sporotrichosis.

Fungal infections of wounds are more serious than dermal mycoses. Typically, they are caused by soil fungi that enter through injuries that allow them to invade subcutaneous tissue. Sporotrichosis is a common fungal infection that occurs worldwide. It is typically a mild disease but may become life-threatening under certain circumstances.

Sporotrichosis ("Rose Gardener's Disease")

Sporotrichosis, also known as "rose gardener's disease," occurs around the world and is mostly associated with activities that lead to puncture wounds from plant material. Although many cases are sporadic, the disease can occur in groups of people working in the same occupation, such as farmers, gardeners, and agricultural workers. In addition to vegetation sources, sporotrichosis can be transmitted by infected animals, especially cats. The disease is classified as an emerging zoonosis.

Signs and Symptoms

The site of infection in most cases is a hand or arm, but the body, legs, and face can also be involved. Typically, a chronic ulcer forms at the wound site, followed by a slowly progressing series of ulcerating nodules that develop sequentially toward the center of the body. Lymph nodes in the region of the wound enlarge, but patients generally do not become ill. If they have AIDS or other immunodeficiency, however, the disease can spread throughout the body, infecting joints and the central nervous system, and becoming life-threatening.

Causative Agent

Sporotrichosis is caused by the dimorphic fungus *Sporothrix schenckii,* which lives in soil and on vegetation (**figure 23.12**).

⏮ dimorphic fungus

Pathogenesis

Sporothrix schenckii conidia enter the body through an injury caused by plant material. After an incubation period that usually ranges from 1 to 3 weeks but can be longer, the multiplying fungi cause a small pimple to form at the site of the injury. This slowly enlarges and ulcerates, producing a painless, red lesion that bleeds easily. There is little or no pus unless the ulcer becomes secondarily infected with bacteria. After a week or longer, the process repeats itself—progression of the disease usually follows the flow of a lymphatic vessel. In healthy people, the process does not continue beyond the lymph node. ⏮ lymphatic vessel

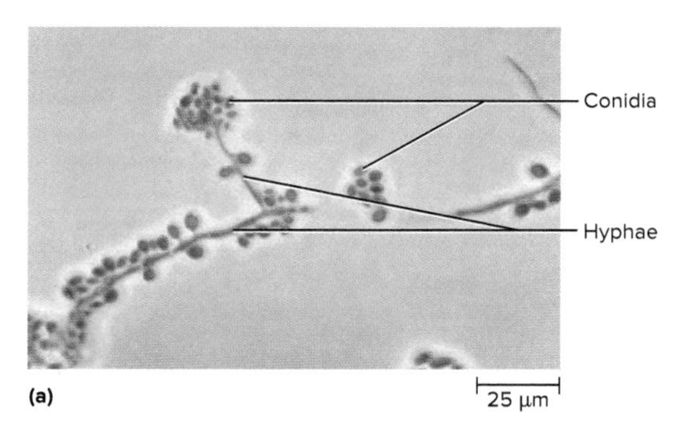

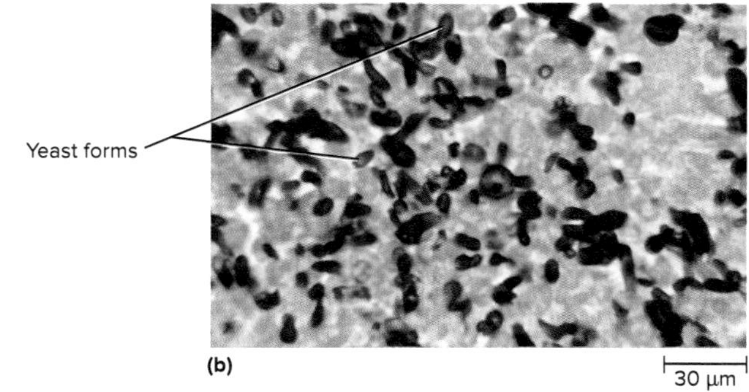

FIGURE 23.12 ***Sporothrix schenckii*** **(a)** Mold form. **(b)** Yeast form, as seen in infected tissue from a laboratory mouse. a: Source: Dr. Libero Ajello/CDC; b: Source: Dr. Libero Ajello/CDC

? ***Sporothrix schenckii*** **is dimorphic; what does this mean?**

Epidemiology

Sporotrichosis is distributed worldwide, mostly in the warmer regions but extending into temperate climates. It is an occupational disease of farmers, carpenters, gardeners, greenhouse workers, and others who work with plant materials. People who handle infected animals are also at risk. Risk factors for the disease besides occupation include diabetes, immunosuppression, and alcoholism. Individuals with chronic lung disease can contract *S. schenckii* lung infections from inhaling dust from hay or cattle feed. Deaths from sporotrichosis are rare.

Treatment and Prevention

Itraconazole is typically used to treat sporotrichosis, with the addition of amphotericin B in certain severe cases. Oral potassium iodide (KI) was commonly used in the past, and is still important in resource-poor countries, but it can cause uncomfortable side effects. Without treatment, the disease can become chronic.

Sporotrichosis can be prevented by wearing protective gloves and a long-sleeved shirt when working with soil and vegetation. Veterinarians should avoid scratches from infected animals, especially cats. **Table 23.12** presents some of the main features of sporotrichosis.

The key features of the diseases covered in this chapter are highlighted in the **Diseases in Review 23.1** table that follows.

TABLE 23.12 Sporotrichosis ("Rose Gardener's Disease")

Signs and symptoms	Painless, ulcerating nodules appearing in sequence in a linear pattern
Incubation period	Usually 1 to 3 weeks
Causative agent	*Sporothrix schenckii,* a dimorphic fungus
Pathogenesis	Fungal cells multiply at site of introduction, causing a small nodule that ulcerates. Fungi carried by lymph flow repeat the process along the lymphatic vessel. May spread beneath the skin irrespective of lymphatic vessels.
Epidemiology	Distributed worldwide in tropical and temperate climates. Veterinarians and people in occupations that require contact with sharp plant materials are at particular risk.
Treatment and prevention	Treatment: antifungal medications. Prevention: protective gloves and clothing when handling plant material, avoiding scratches by infected animals.

MicroAssessment 23.5

Rose gardener's disease (sporotrichosis), caused by the dimorphic fungus *Sporothrix schenckii,* is widely distributed around the world, affecting mainly those associated with occupations that expose them to splinters and sharp vegetation. Unlike most invasive fungal infections, sporotrichosis is usually easy to treat.

13. List some of the risk factors associated with contracting rose gardener's disease.

14. What is unique about the treatment of sporotrichosis?

15. Why is sporotrichosis often misdiagnosed?

Diseases in Review 23.1

Wound Infections

Disease	Causative Agent	Comment	Summary Table
BACTERIAL INFECTIONS			
Staphylococcal infections	*Staphylococcus aureus* and *S. epidermidis*	*S. aureus* (described in chapter 22) is the most common cause of wound infections, but *S. epidermidis,* part of the normal skin microbiota, can form biofilms on medical devices.	Table 23.2
Group A streptococcal "flesh-eating disease"	SPE-producing strains of *Streptococcus pyogenes*	SPEs and enzymes of *S. pyogenes* (described in chapter 21) cause inappropriate T-cell activation, tissue breakdown, and toxic shock.	Table 23.3
***Pseudomonas aeruginosa* infections**	*Pseudomonas aeruginosa*	Widespread environmental organism that commonly infects burned tissues; produces pigments that result in a green color; resistant to a wide variety of antimicrobial medications.	Table 23.4
DISEASES DUE TO ANAEROBIC BACTERIAL WOUND INFECTIONS			
Tetanus ("lockjaw")	*Clostridium tetani*	Endospore-forming anaerobe grows in wound-damaged tissues; tetanus toxin blocks the action of inhibitory neurons, leading to continuous painful muscle contractions; typically fatal without treatment. Preventable by vaccination (DTaP and boosters).	Tables 23.5 and 23.6
Clostridial myonecrosis ("gas gangrene")	Usually *Clostridium perfringens*	Endospore-forming anaerobe that grows in necrotic and poorly oxygenated tissue; gases produced accumulate in tissues; typically fatal without treatment.	Table 23.7
BACTERIAL INFECTIONS OF BITE WOUNDS			
Human bites	Mixed aerobic and anaerobic species, often members of the normal mouth microbiota	Crushing nature of wound causes tissue damage and anaerobic conditions; synergistic infection adds to that tissue damage.	Table 23.8
***Pasteurella multocida* bite wound infections**	*Pasteurella multocida*	Characterized by abscess at the site of the wound; causative agent is a common member of the normal mouth microbiota of animals; results most often from cat bites.	Table 23.9
Bartonellosis ("cat scratch disease")	*Bartonella henselae*	Most common cause of lymph node enlargement at one site in children; acquired via a cat bite or scratch.	Table 23.10
Streptobacillary rat bite fever	*Streptobacillus moniliformis*	Characterized by relapsing fevers, rash, and joint pain. Acquired via a rat bite or scratch, or ingestion of food or water contaminated with rodent feces.	Table 23.11
FUNGAL WOUND INFECTIONS			
***Sporotrichosis* ("rose gardener's disease")**	*Sporothrix schenckii*	Characterized by painless ulcerating nodules that develop sequentially along a lymphatic vessel; dimorphic fungus lives on soil and vegetation and enters the body through an injury.	Table 23.12

Summary

23.1 ■ Anatomy, Physiology, and Ecology of Wounds

Wounds expose tissue components to which pathogens specifically attach. Wounds heal by forming **granulation tissue,** which contracts to minimize scar tissue (figure 23.1).

Wound Abscesses

An **abscess,** composed of **pus,** localizes an infection within tissue, preventing its spread (figure 23.2).

Anaerobic Wounds

Anaerobic conditions are likely to occur in wounds containing dead tissue or foreign material, and in wounds with limited exposure to the air.

23.2 ■ Common Bacterial Infections of Wounds (table 23.1)

Wound infections may lead to delayed healing, abscess formation, spread of bacteria or toxins throughout the body, opening of surgical wounds, and creation of biofilms on artificial devices.

Staphylococcal Wound Infections (figures 23.3, 23.4)

Staphylococcus aureus and *S. epidermidis* are the most common wound-infecting staphylococcal species. *Staphylococcus aureus* has many virulence factors (table 23.2). *Staphylococcus epidermidis* is less virulent but can form biofilms on catheters and other devices.

Group A Streptococcal "Flesh-Eating Disease" (figure 23.5)

Necrotizing fasciitis is caused by strains of *Streptococcus pyogenes* that produce a variety of toxins, including SPE B, a protease thought to be responsible for the tissue destruction.

***Pseudomonas aeruginosa* Infections** (figure 23.6)

Pseudomonas aeruginosa is a motile, aerobic, opportunistic Gram-negative rod. It is widespread in the environment, and a cause of both healthcare-associated and community-acquired infections. Infections are characterized by production of green pus.

23.3 ■ Diseases Due to Anaerobic Bacterial Wound Infections

Tetanus ("Lockjaw") (tables 23.5, 23.6)

Tetanus is characterized by sustained, painful muscle spasms (figure 23.7), caused by **tetanospasmin.** This exotoxin is produced by *Clostridium tetani,* a non-invasive, anaerobic, Gram-positive rod (figure 23.8). The toxin prevents release of neurotransmitter by inhibitory neurons, causing uncontrolled muscle contraction (figure 23.9). The disease is prevented by vaccination with toxoid, followed by regular booster injections.

Clostridial Myonecrosis ("Gas Gangrene") (table 23.7)

Gas gangrene is usually caused by the toxin-producing anaerobe *Clostridium perfringens* and is characterized by discharge of a thin, brown, bubbly fluid and dark discoloration of the overlying skin (figure 23.10); α-toxin causes tissue necrosis, and gas is produced from bacterial fermentation of nutrients in the dead tissue. It is treated by debridement of dead tissue and prevented through proper wound care. There is no vaccine.

23.4 ■ Bacterial Infections of Bite Wounds

The kind of bite wound infection depends on the types of infectious agents in the mouth of the biting animal, and the nature of the wound—whether punctured, crushed, or torn.

Human Bites (table 23.8)

Human bites can result in very serious infections, with pain, massive swelling, and foul-smelling pus (figure 23.11). Infections are usually caused by synergistic activity among normal mouth microbiota introduced into the wound, including anaerobic streptococci, fusiforms, spirochetes, *Bacteroides* sp., and *Staphylococcus aureus.* They are treated with antibiotics and prevented by prompt cleansing and use of an antiseptic.

***Pasteurella multocida* Bite Wound Infections** (table 23.9)

Pasteurella multocida, a Gram-negative rod, can infect bite wounds inflicted by a number of animal species. *P. multocida* causes disease in animals, but many animals are asymptomatic carriers.

Bartonellosis ("Cat Scratch Disease") (table 23.10)

Cat scratch disease is caused by *Bartonella henselae,* and begins with a pimple at the site of a bite or scratch, followed by enlargement of local lymph nodes, which often become pus-filled. Patients (often children) usually recover without treatment.

Streptobacillary Rat Bite Fever (table 23.11)

Streptobacillary rat bite fever is characterized by recurring fevers, head and muscle aches, and vomiting. A rash and joint pains often develop. It is usually caused by *Streptobacillus moniliformis,* a pleomorphic, Gram-negative rod.

23.5 ■ Fungal Wound Infections

Fungal wound infections are usually more serious than dermal mycoses. They are typically caused by soil fungi.

Sporotrichosis ("Rose Gardener's Disease") (table 23.12)

Sporotrichosis ("rose gardener's disease") is a chronic fungal disease mainly of people who work with soil or vegetation. It is characterized by painless, ulcerating nodules that develop along the path of a lymphatic vessel. It is caused by the dimorphic fungus *Sporothrix schenckii* (figure 23.12).

Review Questions

Short Answer

1. What property of *Staphylococcus epidermidis* helps it to colonize plastic materials used in medical procedures?
2. What is the relationship between the superantigens of *S. aureus* and the organism's production of toxic shock?
3. Name two underlying conditions that predispose a person to *Streptococcus pyogenes* flesh-eating disease.
4. Give two sources of *Pseudomonas aeruginosa.*
5. Outline the pathogenesis of tetanus.
6. Explain why *C. tetani* can be cultivated from wounds in the absence of tetanus.

7. What characteristics of bite wounds lead to anaerobic infections?
8. What is the causative agent of cat scratch disease? Why is it a threat to patients with AIDS?
9. What is a synergistic infection? How might one be acquired?
10. Why is sporotrichosis sometimes called rose gardener's disease?

Multiple Choice

1. Which of the following about *Staphylococcus aureus* is *false*?
   a) It is generally coagulase-positive.
   b) Its infectious dose is increased in the presence of foreign material.
   c) Some strains infecting wounds can cause toxic shock.
   d) Nasal carriers have an increased risk of surgical wound infection.
   e) It is pyogenic.
2. Which of these statements about *Streptococcus pyogenes* is *false*?
   a) It is a Gram-positive coccus occurring in chains.
   b) Some strains that infect wounds can cause toxic shock.
   c) Some strains that infect wounds can cause necrotizing fasciitis.
   d) It can cause puerperal fever.
   e) A vaccine is available for preventing *S. pyogenes* infections.
3. Choose the one *false* statement about *Pseudomonas aeruginosa*.
   a) It is widespread in nature.
   b) Some strains can grow in distilled water.
   c) It is a Gram-positive rod.
   d) It produces pigments that together are green.
   e) Under certain circumstances, it can grow anaerobically.
4. Which of these statements about tetanus is *true*?
   a) It can start from a bee sting.
   b) Immunization is carried out using tiny doses of killed *C. tetani*.
   c) Those who recover from the disease are immune for life.
   d) Tetanus immune globulin does not prevent the disease.
   e) It is easy to avoid exposure to spores of the causative organism.
5. Choose the one *true* statement about gas gangrene.
   a) There are few or no leukocytes in the wound drainage.
   b) It is best to rely on antibacterial medications and avoid disfiguring surgery.
   c) A toxoid is generally used to protect against the disease.
   d) Only one antitoxin is used for treating all cases of the disease.
   e) It is easy to avoid spores of the causative agent.
6. Which of the following statements about human bite wound infections is *false*?
   a) They are caused by synergistic bacterial infections.
   b) The causative agents are normal mouth microbiota.
   c) They are seldom serious and often resolve without treatment.
   d) They may sometimes be contaminated with debris.
   e) The causative agents include both aerobes and anaerobes.
7. Which of the following statements about *Pasteurella multocida* is *false*?
   a) Infections generally respond to a penicillin.
   b) It can cause epidemics of fatal disease in domestic animals.
   c) It is commonly found in the mouths of biting animals, including humans.
   d) A vaccine is used to prevent *P. multocida* disease in people.
   e) Cat bites are more likely than dog bites to result in *P. multocida* infections.
8. Which of these statements about cat scratch disease is *false*?
   a) It is a common cause of chronic lymph node enlargement in children.
   b) It is a serious threat to people with AIDS.
   c) Cat scratches are the only mode of transmission to humans.
   d) It does not spread person to person.
   e) It can affect the brain or heart valves in a small percentage of cases.
9. The following statements about *Streptobacillus moniliformis* are all true *except*
   a) it can be transmitted by food contaminated with rodent feces.
   b) pet store workers are at increased risk of infection.
   c) it can be transmitted by the bites of animals other than rats.
   d) human infection is characterized by irregular fevers, rash, and joint pain.
   e) it is a Gram-positive spore-forming rod.
10. Which statement concerning sporotrichosis is *false*?
    a) It is characterized by ulcerating lesions along the path of a lymphatic vessel.
    b) Person-to-person transmission is common.
    c) It can occur in epidemics.
    d) It can persist for years if not treated.
    e) The causative organism is a dimorphic fungus.

Applications

1. Clinicians become concerned when the laboratory reports that organisms capable of digesting collagen and fibronectin are present in a wound culture. What is the basis of their concern?
2. An army field nurse working at a mobile surgical hospital asks this question of all the ambulance drivers: "Was the soldier wounded while in a field with cows?" Why does the nurse ask this question?

Critical Thinking

1. In what way would the incidence of tetanus at various ages in a developing country differ from age incidence in developed countries?
2. Could colonization of a wound by a non-invasive bacterium cause disease? Explain your answer.

24 Digestive System Infections

Yellow color of eye and skin as seen in jaundice. ©*Garry Watson/Science Source*

KEY TERMS

Cariogenic Causes dental caries (tooth decay).

Cirrhosis Scarring of the liver that interferes with normal liver function.

Dysbiosis An imbalance in the normal microbiota; may be caused by taking antimicrobial medications.

Dysentery A serious form of diarrhea characterized by blood, pus, and mucus in the feces.

Gastroenteritis Acute inflammation of the stomach and intestines; the syndrome of nausea, vomiting, diarrhea, and abdominal pain.

Gingivitis Inflammation of the gums.

Hemolytic Uremic Syndrome (HUS) Serious condition characterized by red blood cell breakdown and kidney failure.

Hepatitis Inflammation of the liver.

Microvilli Tiny extensions from the surfaces of cells such as those lining the intestinal villi; they increase surface area of the mucosa.

Oral Rehydration Therapy (ORT) A treatment used to replace fluid and electrolytes lost due to diarrheal disease.

A Glimpse of History

Cholera is a very old disease, thought to have originated in the Far East thousands of years ago. Sanskrit writings indicate it existed in India many centuries before Christianity. With the increased shipping of goods and the mobility of people during the nineteenth century, cholera spread from Asia to Europe and then to North America. The disease caused major epidemics in the nineteenth century.

John Snow (1813–1858), a London physician, demonstrated that cholera was transmitted by contaminated water. He observed that almost all people who contracted the disease during an outbreak in 1854 got their water from a well on Broad Street, whereas neighbors who got their water elsewhere were unaffected. Even though the germ theory of disease had not yet been described, Snow was able to convince local authorities to remove the suspected pump handle so that people were forced to get their water elsewhere. The number of new cases decreased. Because the outbreak had already begun subsiding before the handle was removed, Snow's explanation that cholera was a waterborne disease was not accepted by most doctors and government officials. By 1866, however, it was obvious that cholera occurred in areas where water had been contaminated with sewage, just as Snow had proposed. Public health agencies then played a major role in preventing epidemic cholera. In 1883 Robert Koch, who provided convincing evidence for the germ theory of disease, isolated *Vibrio cholerae,* the bacterium that causes cholera.

In October 2010, less than a year after Haiti was devastated by a major earthquake, a strain of toxigenic *Vibrio cholerae* resulted in an outbreak of cholera that was quickly tracked by local health officials, supported by the CDC. Within a month, over 16,000 Haitians had been hospitalized and almost 1,000 cholera deaths were reported. A number of public health groups worked to set up treatment and prevention facilities, educate the population about how to prevent infection, and train hundreds of healthcare workers across the country to deal with the growing number of cholera cases. Although the epidemic has slowed, new cases are still occurring, and natural disasters such as Hurricane Matthew in October 2016 only increase those numbers. Since 2010, over 800,000 cases of cholera have been reported in Haiti, with over 9,400 deaths. Education about treatment and prevention of the disease continues, but a long-term solution to the problem will require significant improvements to the country's water treatment and sanitation facilities. In 2016, up to 50% of the residents of Haiti still lacked reliable access to clean water.

24.1 ■ Anatomy, Physiology, and Ecology of the Digestive System

Learning Outcomes

1. Describe the functions of the main components of the upper and lower digestive tract.
2. Identify the functions of the liver and other accessory organs.
3. Describe the significance of the normal intestinal microbiota.

The main purpose of the digestive system is to convert the food we eat into a form that the body's cells can use as a source of energy and raw materials for growth. It does this by first physically breaking down food into small particles, then chemically breaking down those particles even further, and finally absorbing the available nutrients. The waste material that remains is eliminated as feces.

The digestive system includes two general components: the digestive tract and the accessory organs (**figure 24.1**). The digestive tract is a hollow tube that starts at the mouth and ends at the anus. When referring only to the stomach and the intestines, the term gastrointestinal tract is often used. The accessory organs, which include the salivary glands, liver, and pancreas, support the process of digestion by producing vital enzymes and other substances that help break down food.

Like the respiratory system and the skin, the digestive tract is one of the body's major boundaries with the environment.

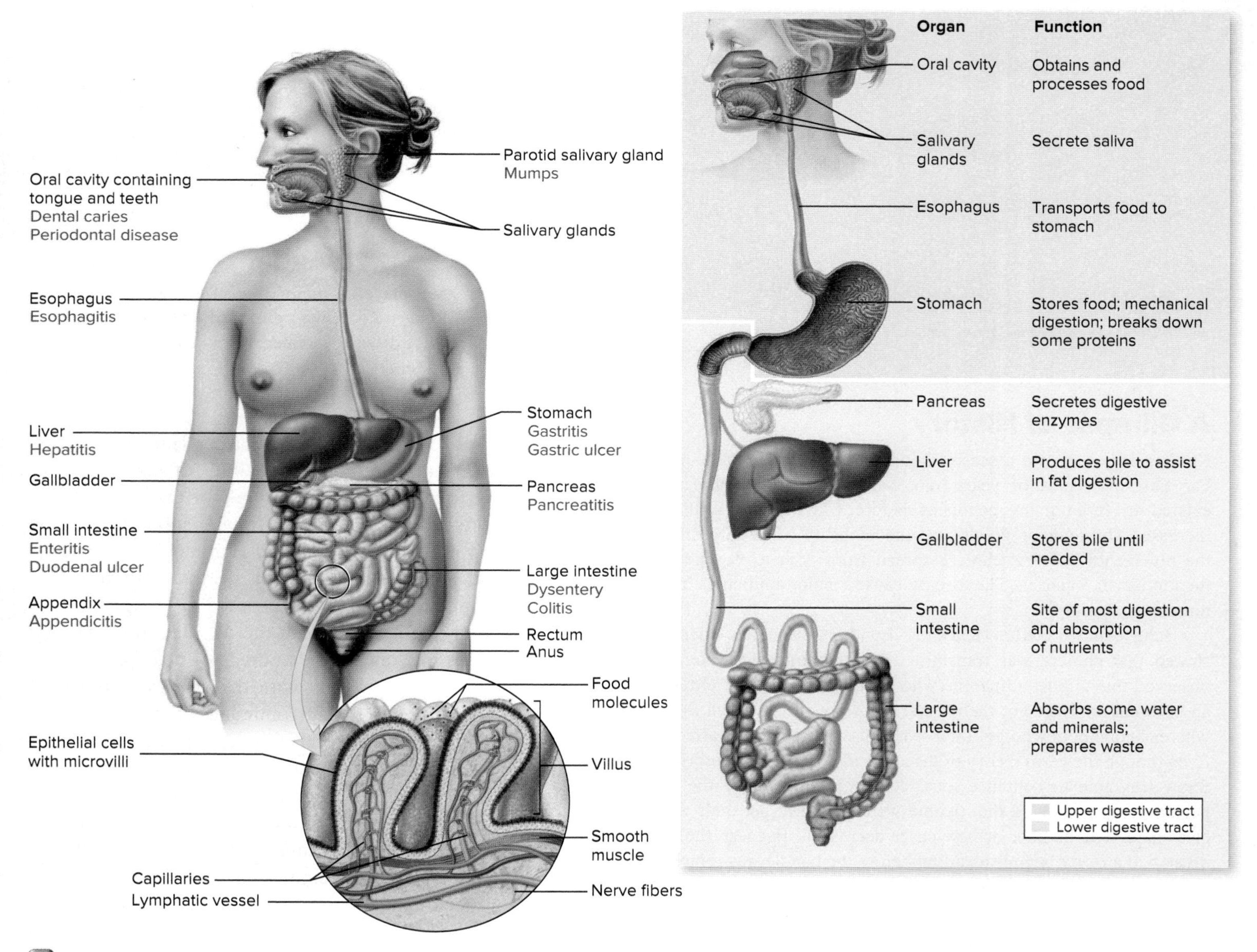

FIGURE 24.1 The Digestive System Some of the disease conditions that can affect the system are shown in red.

What is the role of villi and microvilli in the small intestine?

It is distinct, however, in that the foods being digested by the host provide a ready source of carbon and energy for microbial growth. Every mouthful of food you ingest feeds not only yourself but the microbial population in your digestive tract as well.

Most microbes that inhabit the digestive tract live in harmony with the host, but the balance in this complex ecosystem is delicate. In many regions, a mucous membrane only a single cell layer thick separates the microbial population from underlying tissues, so damage to this layer allows resident microbes to penetrate deeper tissue. Ingested pathogens often have mechanisms to breach intact barriers.

The Upper Digestive System

The upper digestive system includes the mouth, salivary glands, esophagus, and stomach.

The Mouth and Salivary Glands

The mouth is mainly a grinding apparatus that begins the physical and chemical digestion of food. The act of chewing allows the teeth to grind the food into smaller pieces, while the saliva moistens the food and provides amylase, an enzyme that helps start the breakdown of starches.

The outer portion of teeth is made up of a hard protective substance called enamel (**figure 24.2**). Proteinaceous material from the saliva adheres to the enamel, creating a thin film or pellicle. Various types of bacteria can attach to receptors on the pellicle, colonizing the tooth surface to create a biofilm called **dental plaque.** Over time, mineral salts deposit in plaque, creating a crusty substance called dental calculus or tartar. Routine brushing and flossing can remove plaque, but tartar removal requires professional cleaning.

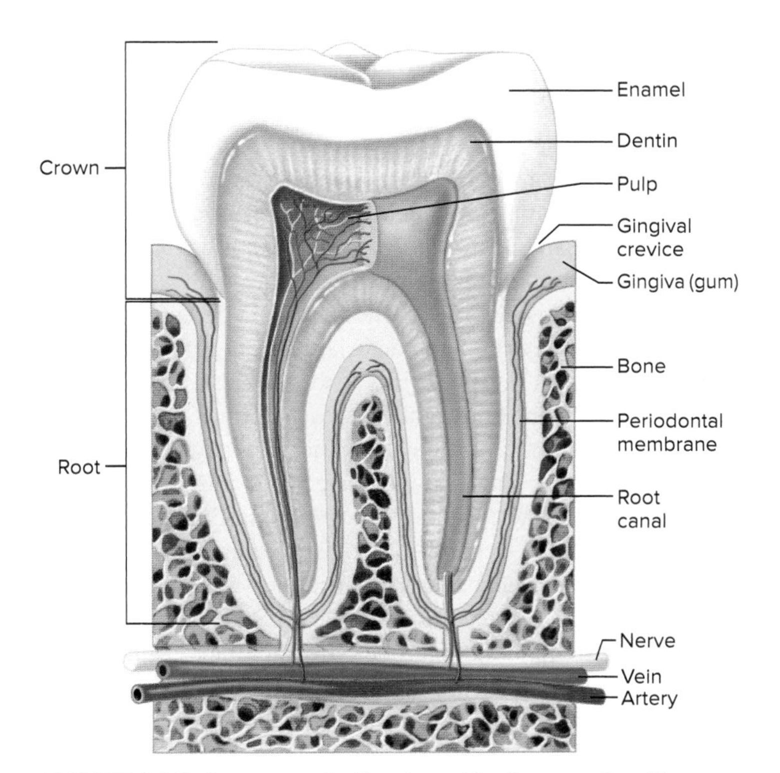

FIGURE 24.2 Structure of a Tooth and Its Surrounding Tissues

? What is the difference between dental plaque and tartar?

When the tooth enamel deteriorates or is otherwise damaged, microorganisms can enter the substance of the tooth. Tooth decay, or **dental caries,** can usually be repaired with dental fillings. If bacteria reach the root canal, which is a region of the tooth filled with pulp (tissue in the tooth that contains blood vessels and nerves), either the tooth must be removed or the pulp must be replaced with an inert substance. This replacement procedure—like the region that is filled—is called a root canal.

Another frequent site of microbial accumulation and infection is the gingival crevice, the space between the gum and a tooth. In response to accumulated plaque in that region, the gums become inflamed, a condition called **gingivitis.** The irritated gums may recede from the tooth root, allowing bacteria to reach that region as well, which can ultimately result in tooth loss.

Saliva is produced by various salivary glands that empty into the mouth. Normally, 1,500 mL of saliva are secreted each day, equivalent to 75% of a 2-liter bottle of soda. The largest of these glands, the parotid glands, are notable because the mumps virus infects them. In addition to saliva's function in digestion, it plays an important role in protecting the oral cavity against microbes. Saliva is rich in secretory IgA (the antibodies that provide mucosal immunity), and it contains antibacterial compounds such as lysozyme and lactoferrin. People with poor saliva production, such as those undergoing radiation therapy for head and neck cancer, are subject to severe tooth decay. ⏮ lactoferrin, ⏮ secretory IgA, ⏮ lysozyme

Microorganisms that colonize the mouth bind specifically to molecules on host tissues or to attached microbes, allowing them to resist the flushing effect of salivary flow and the scrubbing action of food and the tongue. They are also able to withstand the antibacterial effects of saliva. Although constant shedding of the superficial layers of cells of the mucous membranes limits the numbers of microbes in the mouth, over 600 types of bacteria can be found there. Members of the genus *Streptococcus*—which are obligate fermenters—are among the most common. In addition to *Streptococcus* species, obligate anaerobes such as *Porphyromonas* species are also surprisingly common in the mouth. The anaerobes thrive in complex communities, protected because aerobic species consume available O_2. The foul-smelling end products of anaerobic metabolism are associated with halitosis, also known as bad breath.

MicroByte

There can be up to 100 billion bacteria per gram of plaque.

The Esophagus

The esophagus is a collapsible muscular tube about 10 inches long, connecting the mouth to the stomach. Food is pushed to the stomach by peristalsis, the rhythmic muscular contractions of the digestive tract. The esophagus has a relatively sparse microbial population, consisting mostly of bacteria from the mouth and upper respiratory tract. One reason for the relative lack of microbes is the secretory IgA–containing mucus and saliva that bathe the lining of the tube. Microbes trapped in those secretions are propelled by peristalsis through the esophagus, along with food and liquids. The esophagus rarely becomes infected except in individuals with AIDS or other immunodeficiencies. ⏭ AIDS

The Stomach

The stomach is an expandable sac-like structure with a muscular wall. Its primary function is to break down and store food particles as they await controlled entry into the small intestine. The gastric juices are highly acidic, which denatures the proteins in food particles. The acidic environment also activates pepsinogen produced in the stomach to form pepsin, a protein-splitting enzyme. The cells that line the stomach protect themselves from the acid and the enzymes by secreting a thick alkaline mucus. Most bacterial cells cannot survive the hostile environment, so a normal empty stomach has few microorganisms.

The Lower Digestive System

The lower digestive system includes the small and large intestines, as well as the pancreas and the liver.

The Small Intestine

Digestive fluids from the pancreas and the liver are added to the stomach contents as they enter the first part of the small intestine (the duodenum). The alkaline fluid from the pancreas neutralizes stomach acid and contains digestive enzymes. The liver contributes **bile,** an emulsifying agent that helps to physically break down fat globules. Bile salts help the intestine absorb oils, fats, and fat-soluble vitamins. Many types of bacteria are killed by bile, but those adapted to live in the intestinal tract resist its bactericidal effects. This has practical applications in the laboratory, because bile salts are used in selective media designed to isolate intestinal bacteria. ⏮ bactericidal, ⏮ selective media

Additional enzymes are produced by intestinal cells. Some are secreted into the intestinal lumen (the region inside the tube), but many are permanently attached to the cell membranes. Microbial infections that damage these cells interfere with digestion.

If the intestinal tract were a simple pipe, it would have a surface area smaller than that of a bath towel. Instead, although only about 6 meters in length, the small intestine has an enormous surface area, approximately 250 square meters, or the size of a tennis court. The inside surface of the small intestine (sometimes referred to as the brush border) has many tiny, finger-like projections called **villi.** Each of these is lined with cells that have cytoplasmic projections called **microvilli** (see figure 24.1). Bundles of actin filaments make up the structural core of these projections. At the base of the villi are pits called crypts. These are small glands that continuously secrete large amounts of enzyme-containing fluids and mucus into the lumen. Certain cells in the crypts give rise to new intestinal epithelial cells. Intestinal epithelial cells are completely replaced every 9 days—one of the fastest turnover rates in the body. ⏮ actin

A major role of the small intestine is nutrient and fluid absorption. Cells that line the villi take in amino acids and monosaccharides, bringing in sodium ions (Na^+) simultaneously. Fatty acids, vitamins, and minerals such as iron are absorbed as well. In addition, the small intestine absorbs fluids. Taking into account the volume of digestive juices, as well as saliva and the fluids ingested as food and drink, this amounts to approximately 9 liters per day! Understandably, disruption in this fluid balance can result in diarrhea.

Few microbes reside in the upper small intestine because the flushing action of rapidly passing digestive juices limits their ability to colonize the surface. As the involuntary muscle action of peristalsis propels the intestinal contents toward the large intestine, the movement slows and the bacterial population increases. The immune system monitors this population with numerous dendritic cells sampling the environment and M cells delivering small amounts of the lumen's contents to Peyer's patches. ⏮ dendritic cells, ⏮ M cells, ⏮ Peyer's patches

The Large Intestine

The main function of the large intestine is to absorb water as well as vitamins produced by the resident microbiota. Because the small intestine absorbs so much water, only 300 to 1,000 mL of fluid normally reach the large intestine per day, and most of this is absorbed. The semisolid feces, composed of indigestible material and bacteria, remain. Infection of the large intestine can interfere with absorption and stimulate the painful contractions known as "stomach cramps."

Microbes flourish in the large intestine, supported by the abundance of nutrients in undigested and indigestible food material. In fact, bacteria make up about one-third of the fecal weight, reaching concentrations of approximately 10^{11} cells per gram! Anaerobic bacteria, particularly members of the genus *Bacteroides,* make up about 99% of the microbial population. The remaining microbes are facultative anaerobes, primarily *Escherichia coli* and other members of the *Enterobacteriaceae.* ⏮ *Bacteroides,* ⏮ *Enterobacteriaceae*

As part of the Human Microbiome Project, scientists studied the diverse community of microorganisms that inhabit the human intestines. The intestinal microbiota can degrade a wide variety of foods. Some high-fiber foods, like beans and broccoli, contain substances indigestible by the gastric and intestinal juices but readily degraded by intestinal organisms. Microbial breakdown of these materials often produces large amounts of gas, causing abdominal discomfort and intestinal gas (flatus). Bacterial enzymes can also convert various substances in food to carcinogens and therefore may be involved in the development of intestinal cancers. ⏮ **Human Microbiome Project**

The intestinal microbes are important to human health because they synthesize a number of useful vitamins, including niacin, thiamine, riboflavin, pyridoxine, vitamin B_{12}, folic acid, pantothenic acid, biotin, and vitamin K. They are also essential for normal development of mucosal immunity. At the same time, many members of the intestinal microbiota are opportunistic pathogens that can cause disease if they gain access to other body sites such as the urinary tract.

The normal microbiota helps prevent pathogens from colonizing the large intestine. Antibiotic treatment, especially with broad-spectrum medications, disrupts the normal microbiota, creating an imbalance called **dysbiosis** that may lead to mild to severe **antibiotic-associated diarrhea.** In some cases, this is caused by toxin-producing strains of *Clostridium difficile,* which readily colonize the intestine of people whose normal intestinal microbiota has been reduced by antimicrobial chemotherapy. Suppression of intestinal microbiota with antibacterial medications can also increase susceptibility to other pathogens such as *Salmonella enterica.* ⏭ ***Clostridium difficile* infection, ⏭ *Salmonella enterica***

The Pancreas

The pancreas is located behind the stomach. Some pancreatic cells produce hormones, and others make digestive enzymes. About 2 liters of pancreatic digestive juices discharge directly into the upper portion of the small intestine each day. These fluids, as well as digestive juices of the small intestine itself, are alkaline and neutralize the stomach acid as it passes into the small intestine.

The Liver

The liver is a large organ located in the upper right portion of the abdomen. One of its roles is to produce bile, which is then concentrated and stored in a sac-like structure called the gallbladder. The bile then flows through a system of tubes into the upper small intestine. Severe liver disease or obstruction of the bile ducts can cause **jaundice,** a yellow color of the skin and eyes caused by buildup of a bile component (bilirubin) in the blood (see chapter-opening photo).

The liver also inactivates toxic substances that enter the bloodstream, generally from the digestive tract. For example, ammonia produced by intestinal bacteria and then absorbed into the bloodstream could reach poisonous levels if not detoxified by the liver. These same processes can also chemically alter and remove many medications. Therefore, if the liver is damaged, as it might be by a viral infection, lower medication doses might be needed to prevent an accidental overdose.

MicroAssessment 24.1

The digestive tract is a complex ecosystem and a major route for pathogens to enter the body. Bacteria colonizing the tooth surface can create a biofilm called plaque. People with poor saliva production risk severe tooth decay. Infections of the esophagus are so unusual that their occurrence suggests immunodeficiency. The stomach stores food; acidic gastric fluid destroys most microbes before they reach the intestines. The small intestine carries out the majority of the digestion and absorption of nutrients. Undigested material becomes feces in the large intestine. Disruption of the normal microbiota through antibiotic treatment can result in antibiotic-associated diarrhea.

1. What causes halitosis (bad breath)?
2. What makes the surface area of the small intestine so large, when it is only about 6 meters long?
3. Two patients recently had spinal surgery, but one is jaundiced. Which one would likely need a lower dose of acetaminophen?

UPPER DIGESTIVE SYSTEM INFECTIONS

24.2 ■ Bacterial Diseases of the Upper Digestive System

Learning Outcomes

4. Compare and contrast dental caries, periodontal disease, and acute necrotizing ulcerative gingivitis.
5. Describe *Helicobacter pylori* gastritis and how it relates to gastric ulcers and stomach cancer.

Bacterial diseases of the upper digestive system often involve the teeth and gums (**figure 24.3**). Some of these local infections—which often go unnoticed for years—may have consequences for the rest of the body as well. For example, some studies suggest that chronic gum infections contribute to arterial disease, rheumatoid arthritis, and premature births. In addition, members of the normal oral microbiota can enter the bloodstream during dental procedures and cause infective endocarditis. Another site of upper digestive infections is the stomach. ⏭ **infective endocarditis**

Dental Caries (Tooth Decay)

Dental caries (tooth decay) is the most common chronic disease and the main reason for tooth loss. In the United States, about 60% of all teenagers have tooth decay, making it four times more common than asthma in this group.

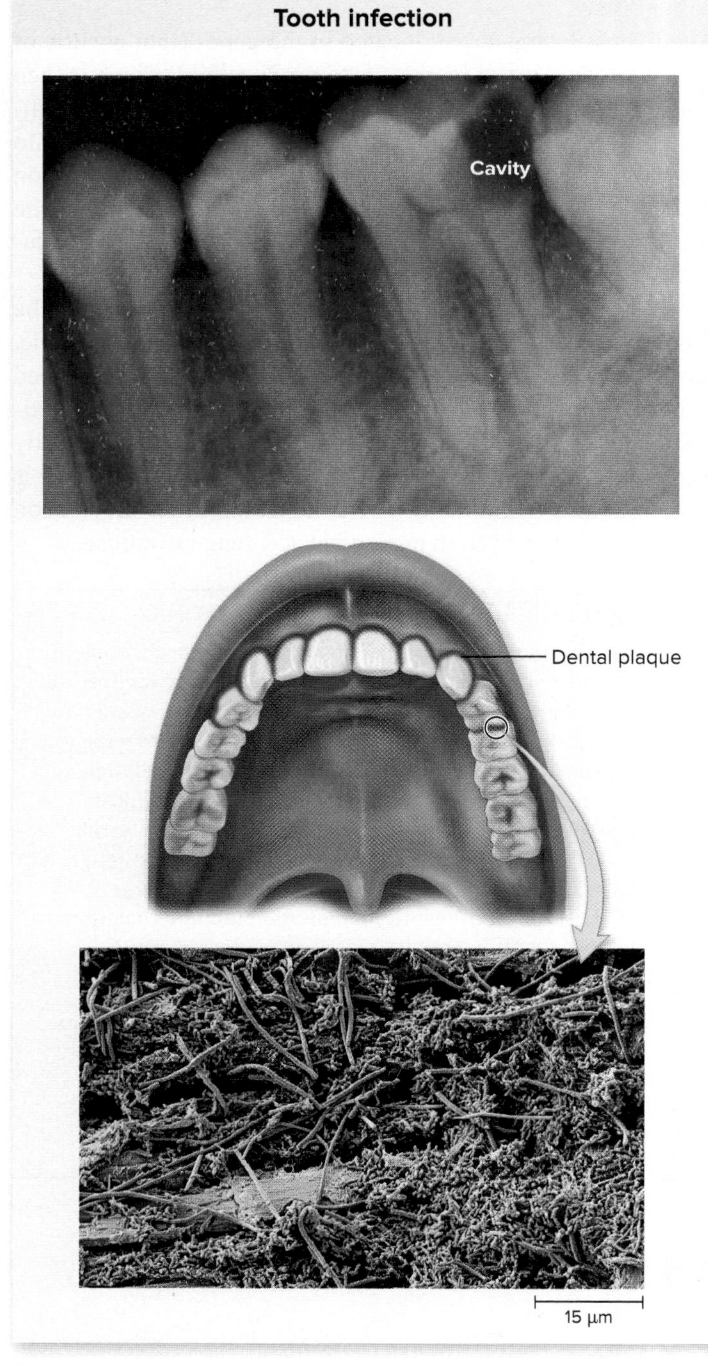

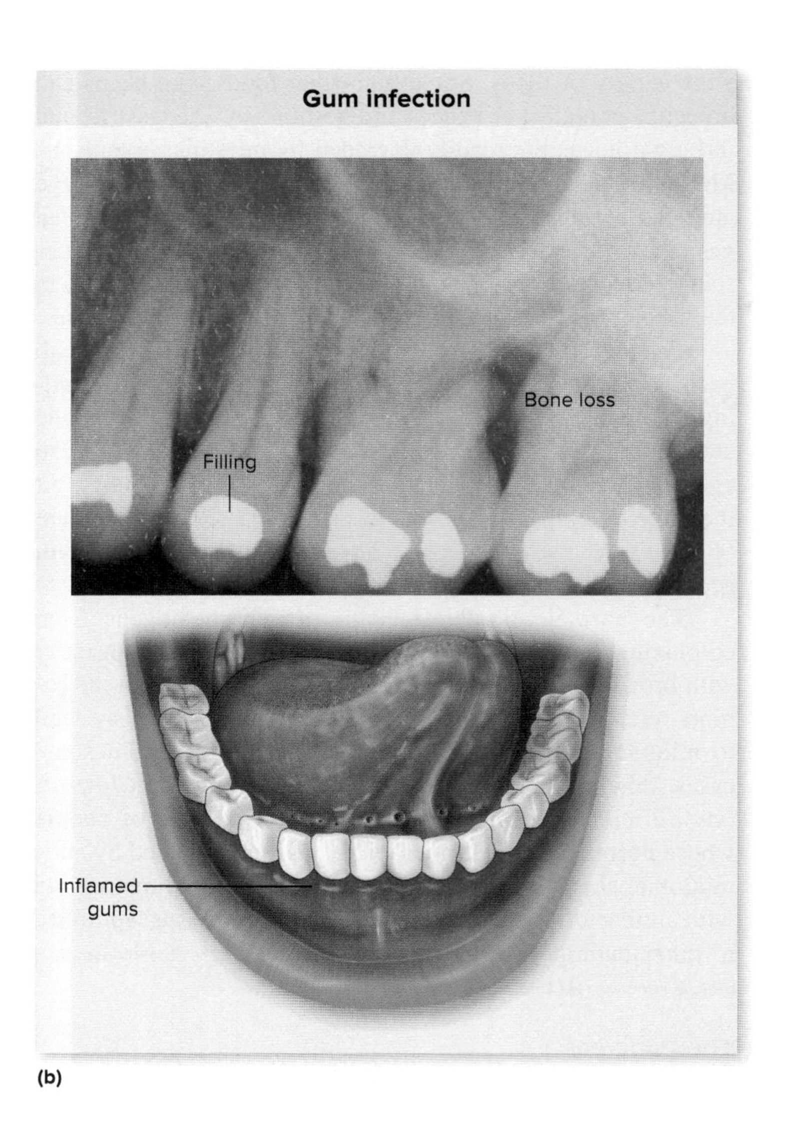

FIGURE 24.3 Infections of the Teeth and Gums (a) Note the different types of bacteria in the community forming dental plaque (shown in purple). **(b)** If plaque hardens between teeth and gingiva, it can increase the likelihood of periodontal disease, which may result in bone loss and loosening or loss of teeth. a (top): ©Dr. Larry Hoffman; a (bottom): ©Steve Gschmeissner/Science Source; b: ©Dr. Larry Hoffman

How does brushing away plaque decrease the chance of developing cavities?

Signs and Symptoms

Dental caries can become very advanced in the months before signs and symptoms develop. Sometimes there is noticeable discoloration, roughness, or defect, and sometimes a tooth can break during chewing. However, the severe, throbbing pain of a toothache is often the first symptom.

Causative Agent

Streptococcus mutans and related species are **cariogenic** (meaning "caries generating"). These Gram-positive cocci live only on teeth and therefore do not colonize the mouth in the absence of teeth. Unlike many other bacteria, they thrive in acidic environments (below pH 5), which allows them to survive the conditions that result from the lactic acid they produce during fermentation. They convert sucrose (table sugar) into extracellular insoluble polysaccharides called glucans. Glucans are essential for the development of dental caries on smooth tooth surfaces. ◀◀ polysaccharides, ◀◀ fermentation

Pathogenesis

Formation of dental caries is a complex process. First, oral streptococci adhere to the pellicle (the thin film of proteinaceous material that adheres to tooth enamel) to create the biofilm called dental plaque, but this alone does not lead to caries (see figure 24.3). If dietary sucrose is present, extracellular enzymes produced by *S. mutans* split the sucrose into its two monosaccharide components: glucose and fructose. The glucose is polymerized to make glucans, and the fructose is fermented, producing lactic acid. The glucans help create an even thicker biofilm by binding a mixed population of microbes together and to the tooth, making the plaque impenetrable to saliva. ⏮ **dental plaque**

Plaque that contains cariogenic microbes can act as a tiny acid-soaked sponge closely applied to the tooth. In fact, when sugar enters the mouth, the pH of the plaque drops from its normal value of about 7 to below 5 within minutes (**figure 24.4**). This 100-fold increase in acidity begins dissolving the enamel of the teeth. After food leaves the mouth, the pH of the plaque rises slowly to neutrality. The delay is due to continued acid production from fermentation of polysaccharides held in storage granules within cariogenic bacteria. ⏮ **storage granules**

Both *S. mutans* and a sucrose-rich diet are required to produce dental caries on smooth surfaces of the teeth. In deep fissures or pits, plaque can accumulate in the absence of *S. mutans*, and tooth decay can occur if lactic acid–producing bacteria and fermentable substances are present.

Epidemiology

Dental caries is worldwide in distribution, but the incidence varies significantly depending mainly on intake of dietary sucrose and access to preventive dental care. Genetics also plays an important role because some people inherit resistance to the disease. The incidence of dental caries peaks during teen years and falls off with age. This is probably because the pits and fissures on the tooth—ideal sites for bacterial growth—wear down with time.

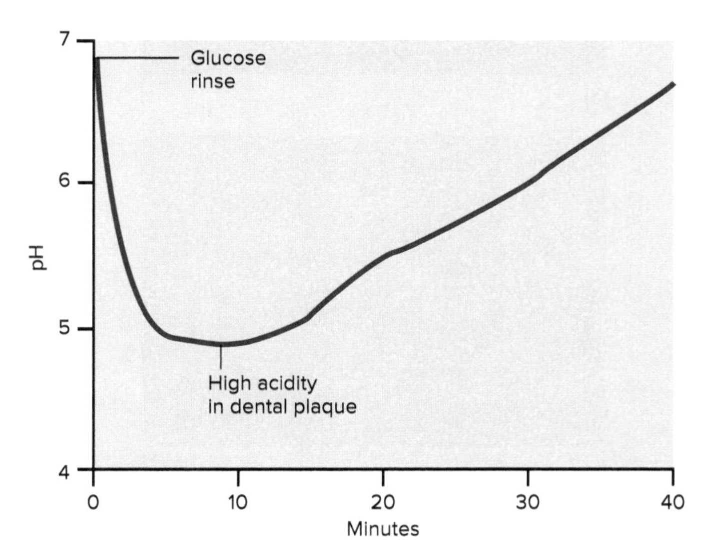

FIGURE 24.4 Acidity of Dental Plaque Plaque becomes more acidic after the mouth is rinsed with a glucose solution. Tooth enamel begins to dissolve at about pH 5.5.

? **Would you expect this dip in pH if *Streptococcus mutans* metabolized sugar via aerobic respiration rather than via fermentation?**

Treatment and Prevention

Treatment of dental caries requires drilling out the cavity, filling the defect with material such as composite resin (tooth-colored) or dental amalgam (silver-colored), and restoring the contour of the tooth.

The most important method for controlling dental caries is restricting dietary sucrose and other refined carbohydrates. This reduces *Streptococcus mutans* colonization as well as acid production by cariogenic plaque. However, the quantity of sugar in the diet is not the most important factor leading to dental caries—the frequency of eating sugary foods and the length of time the food stays on the teeth are more critical. Chewing sugar-free gum reduces dental caries, probably because it increases the flow of saliva.

Trace amounts of fluoride prevent dental caries by making tooth enamel harder and more resistant to dissolving in acid. In the United States, more than half the population is supplied with fluoridated public drinking water, resulting in a 60% reduction in dental caries. This saves billions of dollars in dental costs each year. In areas where fluoridated drinking water is not available, fluoride tablets or solutions can be used. To have optimum effect, children should begin receiving fluoride before their permanent teeth erupt. Fluoride applied to tooth surfaces in the form of mouthwashes, gels, or toothpaste is generally less effective.

Toothbrushing and flossing disrupt and remove plaque, but toothbrush bristles cannot remove plaque from the deep narrow pits and fissures normally present in children's teeth, which is where most childhood tooth decay starts. Caries can be prevented by using a sealant—a kind of epoxy glue that seals the fissures, kills the bacteria in plaque, and prevents bacterial recolonization. Older people have lower incidence of fissure-related caries, but are prone to receding gums that expose root surfaces, which become sites for dental caries.

> **MicroByte**
> Eliminating sucrose-containing sweets reduces dental caries by 90%.

Periodontal Disease

Periodontal disease often results from plaque accumulation near the gum margin (periodontal means "around the tooth"). Bacterial products in plaque trigger an inflammatory response in the gums and the local tissues, leading to at least some degree of tissue damage. Periodontal disease includes **gingivitis** (swelling and redness of the gums) and **chronic periodontitis,** a destructive response that progressively damages the structures that support the teeth (see figure 24.3). Chronic periodontitis is an important cause of tooth loss from middle age onward.

Signs and Symptoms

Gingivitis is marked by gums that are tender and bleed easily. This can occur within days of plaque accumulation and typically goes away once plaque is removed.

Chronic periodontitis is a long-term response to plaque and is characterized by bad breath, red shiny gums that bleed easily, and bone loss that leads to loosening of the teeth. The base of the teeth becomes discolored—ranging from yellowish to black—and the gums recede. Exposed roots of the teeth are susceptible to dental caries.

Causative Agent

The plaques associated with periodontal disease are typically communities of Gram-negative anaerobes, making them quite different from plaque associated with dental caries. Moreover, a recent study that examined 16S rDNA showed that the plaques associated with gingivitis differ from those that characterize chronic periodontitis. Determining the causative agents and their roles is complicated, however, because the infections are polymicrobial, meaning that multiple species of interacting bacteria are involved. Hundreds of different species have been identified in the plaques, most of which have not been cultivated. Genera that various studies have implicated in chronic periodontitis include *Porphyromonas, Tannerella, Prevotella, Treponema, Selemonas,* and *Actinobacillus.*

As plaque (and tartar) accumulates at the gum margin—especially in hard-to-clean areas between the teeth—it gradually extends into the gingival crevice. Bacterial products trigger an inflammatory response, leading to the symptoms of gingivitis. If the level of plaque remains small, the host response limits the process. ◀◀ tartar

When the inflammatory response does not control the microbial population, chronic periodontitis leads to tissue destruction. The process appears to involve many factors, including host characteristics and the composition of the microbial population. Tissue-degrading enzymes released by the plaque microbes weaken the gingival tissue and cause the gingival crevice to widen and deepen. This allows the plaque to spread farther toward the root of the tooth. The pattern recognition receptors of surrounding cells detect the lipopolysaccharide of the Gram-negative outer membrane (endotoxin), causing the release of pro-inflammatory cytokines. Some of these induce such a strong inflammatory response that nearby tissues are damaged. With progressive tissue damage, the membrane that attaches the tooth root to the bone weakens, and the bone gradually softens. The tooth becomes loose and may fall out. ◀◀ pattern recognition receptors

Epidemiology

Periodontal disease such as gingivitis can begin in childhood. After age 65, almost 90% of individuals have some degree of chronic periodontitis. Smokers and those with underlying defects in acquired or innate immunity (including those with AIDS) often have severe periodontitis that leads to tooth loss.

Treatment and Prevention

Periodontal disease can be treated in its early stages by cleaning out the inflamed gingival crevice and removing plaque and tartar. In advanced cases, minor surgery is usually required to expose and clean the roots of the teeth. These physical treatment methods may be used along with antibiotic therapy in some patients. Careful flossing and toothbrushing is thought to reduce the incidence of periodontal disease, especially when combined with twice-yearly polishing and removal of tartar at a dental office.

Acute Necrotizing Ulcerative Gingivitis

Acute necrotizing ulcerative gingivitis (ANUG) is a severe, acute condition distinct from other forms of periodontitis (**figure 24.5**). The disease was first called trench mouth because it was common among soldiers living in trenches during World War I, unable to care for their teeth and gums. Since 2005, dentists have noticed a dramatic increase in ANUG cases associated with methamphetamine use, coining the term "meth mouth" as a form of ANUG. Meth mouth is further complicated by rampant dental caries.

Signs and Symptoms

ANUG begins abruptly and is characterized by bleeding painful gums, abscessed and broken teeth, and extremely bad breath.

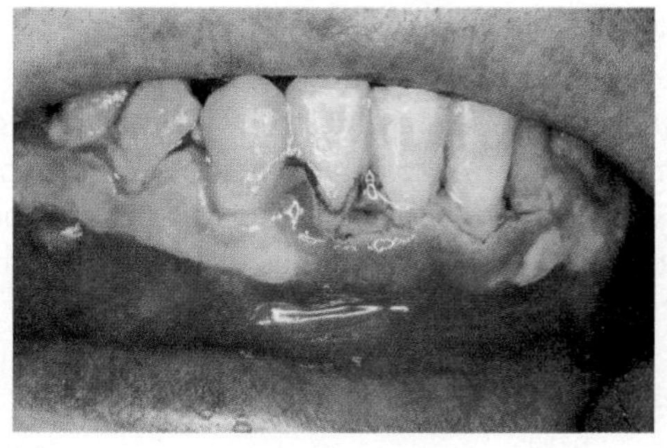

(a)

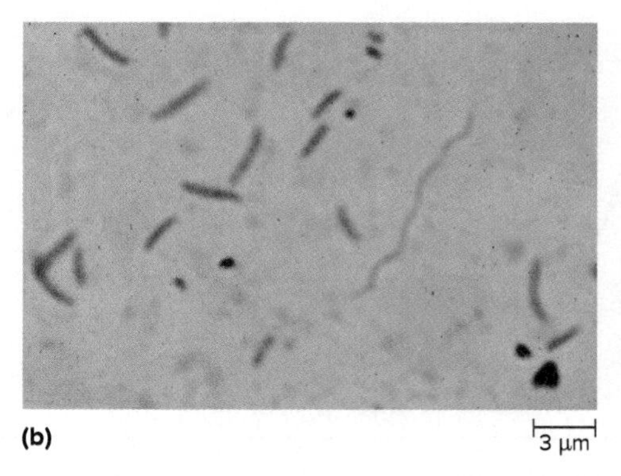

(b)

FIGURE 24.5 Acute Necrotizing Ulcerative Gingivitis (ANUG)
(a) Red, swollen gums with loss of tissue, especially between the teeth. **(b)** Gram stain of material from the gingival crevice showing a spirochete and rod-shaped bacteria. a: ©Biophoto Associates/Science Source; b: ©Evans Roberts

? **What aspects of methamphetamine abuse increase the risk of developing ANUG?**

Causative Agent

ANUG is associated with heavy growth of anaerobes at the gum line. As with periodontal disease, the infections are polymicrobial. Bacteria that appear to play a role include spirochetes, fusiforms (rods with tapered ends), and *Prevotella intermedia.*

Pathogenesis

The spirochetes and the other anaerobes associated with ANUG are presumed to act together to destroy tissues in the oral cavity, but the precise mechanisms are unknown. Plaque is always present, but its bacterial composition shows much larger numbers of spirochetes and other anaerobes than are present in chronic periodontal disease. The spirochetes invade the tissue, causing necrosis and ulceration, mainly of the gums between the teeth. Inflammation may lead to bone loss around the teeth.

Methamphetamine use often lowers saliva production, leading to a dry mouth and increased risk of biofilm formation, tooth decay, and ANUG. Tooth grinding, also associated with methamphetamine use, increases the risk of fracturing teeth weakened by severe decay and gingivitis.

Epidemiology

ANUG can occur at any age in association with poor oral hygiene, particularly when combined with poor nutrition, high sugar diet, chronic stress, and immunodeficiency. Methamphetamine use appears to intensify the problems.

Treatment and Prevention

Most bacteria at the gingival crevice are anaerobes, so local hydrogen peroxide treatment rapidly relieves ANUG symptoms. Patients with systemic symptoms will benefit from antibiotic treatment as well. Long-term cures include removing plaque and tartar. Prevention begins with daily brushing and flossing, and twice-yearly professional cleaning. The main features of teeth and gum infections are presented in **table 24.1.** ⏮ **hydrogen peroxide**

Helicobacter pylori Gastritis

Helicobacter pylori is present in many people, in whom it may cause gastritis (inflammation of the stomach), peptic ulcers (lesions in the lining of the stomach or duodenum), or even cancer—or it may produce no disease at all. The association of *Helicobacter pylori* with ulcers was demonstrated in the 1980s when Barry Marshall, one of the scientists who discovered the bacterium, intentionally drank a culture of it and developed the disease.

Signs and Symptoms

Most *Helicobacter pylori* infections are asymptomatic. Gastritis can result in belching, bloating, and sometimes vomiting. Development of peptic ulcers is marked by localized abdominal pain, tenderness, and bleeding. Chronic gastritis can lead to stomach cancer.

Causative Agent

Helicobacter pylori is a short, curved, Gram-negative, microaerophilic bacterium. It has multiple polar flagella that are unusual because they are covered by sheaths. ⏮ ***Helicobacter***

TABLE 24.1 Important Infections of the Teeth and Gums

	Dental Caries	Periodontal Diseases	Acute Necrotizing Ulcerative Gingivitis
Signs and symptoms	None until advanced disease; then roughness, discoloration, broken tooth, throbbing pain	Gingivitis characterized by tender, bleeding gums. Chronic periodontitis characterized by bad breath; red, shiny gums that bleed easily; loose teeth; and exposed bases of teeth	Pain, bleeding gums, abscessed and broken teeth, bone loss, and extremely bad breath
Incubation period	Months before cavity is detectable	Gingivitis—days; periodontitis—months or years	Undetermined
Causative agent	Cariogenic bacteria in dental plaque, particularly *Streptococcus mutans*	Anaerobic microbes in dental plaque	Probably a spirochete of the genus *Treponema* acting with other anaerobes
Pathogenesis	Bacteria in plaque produce acid from dietary sugars; slowly dissolves the tooth enamel; sucrose is critical for cariogenic plaque formation.	Plaque formed at the gum margins causes the inflammatory response associated with gingivitis. If plaque extends into the gingival crevices, the strong inflammatory response damages tissues that support the teeth, causing chronic periodontitis.	The spirochetes and certain other anaerobes act synergistically to destroy tissues. The spirochetes invade tissue, causing necrosis and ulceration.
Epidemiology	Worldwide distribution, incidence depending on dietary sucrose, natural or supplemental fluoride. More prevalent in the young.	Gingivitis can begin in childhood; periodontitis primarily affects older people, particularly smokers and immunodeficient individuals.	Associated with poor oral hygiene, malnutrition, immunodeficiency, or methamphetamine use. All ages are susceptible.
Treatment and prevention	Treatment: mechanical removal of plaque. Prevention: sealing pits and fissures in children's teeth, restriction of dietary sucrose, supplemental fluoride.	Treatment: surgical treatment in severe cases of periodontitis to clean tooth roots. Prevention: avoid buildup of plaque.	Treatment: local treatment with possible antibiotic therapy, followed by removal of plaque and tartar. Prevention: avoid buildup of plaque.

Pathogenesis

Helicobacter pylori cells survive the acidic environment of the stomach by (1) producing urease, an enzyme that converts urea to ammonia, thereby creating an alkaline microenvironment, and (2) burrowing within the layer of mucus that coats the stomach lining (**figure 24.6**). Urea is normally found in gastric juices because it is released as proteins are degraded. *H. pylori* cells use their flagella to move through the mucus, following the pH gradient from the acidic gastric lumen to the nearly neutral underlying epithelial cells. The bacterial cells then attach to the mucus-secreting epithelium or multiply adjacent to it. To colonize the stomach, *H. pylori* must be able to both move away from the acidic lumen and produce ammonia from urea.

H. pylori has several characteristics that allow it to manipulate the host's immune response and cause chronic disease. Its lipopolysaccharide (LPS) and flagella have modifications that prevent them from being recognized by certain pattern recognition receptors of the innate immune system, a trait that helps the cells limit the inflammatory response to infection. The bacterium also produces proteins that damage host cells or alter their activities. One of these, VacA (**vac**uolating cytotoxin), is a membrane-damaging toxin that promotes the flow of urea into the stomach, induces apoptosis in epithelial cells, and interferes with the function of T cells. Strains associated with increased risk of cancer produce CagA (**c**ytotoxin-**a**ssociated **g**ene), which is delivered directly to host cells via a secretion system. Its effects include altering the host cell cytoskeleton, interfering with host cell signaling, and promoting inflammation. ◀◀ pattern recognition receptors, ◀◀ membrane-damaging toxins

As the local epithelial cells are damaged due to the combined effects of the *H. pylori* products, stomach acid, and the chronic inflammation, mucus production decreases. The thinning of the protective mucous layer and host cell damage probably account for the development of peptic ulcers.

H. pylori infections persist for years, often for life. The outcome of the infection is quite variable. Only about one in six infected persons develops ulcers. A small percentage of chronically infected people develops stomach cancer, but more than 90% of those with stomach cancer are infected.

Epidemiology

Overall, about one in five adults in the United States is infected with *H. pylori,* but the incidence increases with age, reaching almost 80% for those over age 75. Infections tend to cluster in families, and rates are highest in low socioeconomic groups. Transmission of *H. pylori* probably occurs by the fecal-oral route, or possibly by the oral-oral route. The bacteria have also been found in well water.

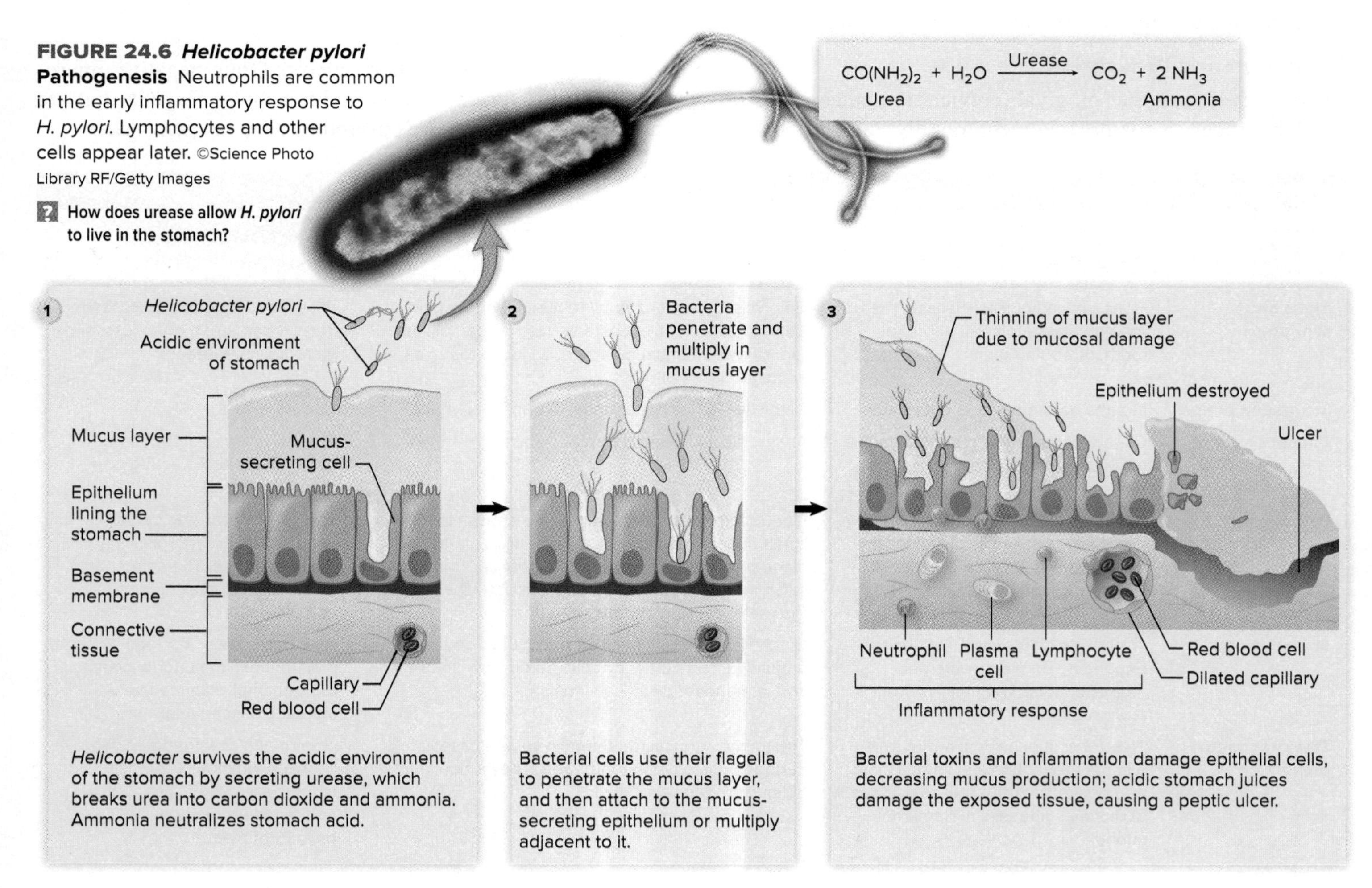

FIGURE 24.6 *Helicobacter pylori* Pathogenesis Neutrophils are common in the early inflammatory response to *H. pylori.* Lymphocytes and other cells appear later. ©Science Photo Library RF/Getty Images

? How does urease allow *H. pylori* to live in the stomach?

FOCUS ON A CASE 24.1

The patient was a 35-year-old man who consulted his physician because of upper abdominal pain. The pain was described as a steady burning or gnawing sensation, like a severe hunger pain. Usually it came on $1\frac{1}{2}$ to 3 hours after eating, sometimes waking him from sleep. Generally, it was relieved in a few minutes by food or antacid medicines.

On examination, the patient appeared well, without evidence of weight loss. The only notable finding was tenderness slightly to the right of the midline in the upper part of the abdomen. A test of the patient's feces was positive for blood. The remaining laboratory tests were normal.

Endoscopy, a procedure that uses a long, flexible fiber-optic device passed through the mouth, showed a patchy redness of parts of the stomach lining. A biopsy specimen was taken. The endoscopy tube was then passed through the stomach and into the duodenum. About 2 cm into the duodenum, there was a red lesion 8 mm in diameter that lacked a mucous membrane and appeared to be "punched out." A blood clot was also present. After the endoscopy, a portion of the biopsy specimen was placed on urea-containing medium. Within a few minutes, the medium began to change color, indicating a developing alkaline pH.

1. What is the diagnosis for this patient?
2. What would you expect microscopic examination and culture of the gastric mucosa biopsy to show?
3. How do flagella and the enzyme urease function as bacterial virulence factors in the development of peptic ulcers?
4. It took a long time for doctors to accept that this condition was caused by infection. Why?

Discussion

1. This patient had a duodenal ulcer. The ulcer had penetrated deeply beyond the mucosa, involving small blood vessels and causing bleeding. This was apparent from the clot seen at endoscopy and the positive test for blood in the feces.
2. Microscopic examination of the biopsy showed curved bacteria, confirmed by culture to be *Helicobacter pylori*.
3. Flagella propel the bacterial cells through the mucous layer, moving them from the acidic environment of the lumen (approximately pH 2) to the epithelial surface, where the pH is nearly neutral (about pH 7.4). Non-motile strains do not cause disease because they are removed with mucus turnover. Urease breaks down urea in the stomach to produce ammonia, forming an alkaline environment around the cell that neutralizes the acidity of gastric fluid. This protects the organism until it can burrow below the mucous layer. Strains that are urease-deficient do not cause disease unless they are injected directly into the mucous layer.
4. Claude Bernard, a scientist of Pasteur's time, put it this way: "It is that which we do know which is the greatest hindrance to our learning that which we do not know." In 1983, when Barry Marshall proclaimed before an international gathering of infectious disease experts that a bacterium caused stomach and duodenal ulcers, everyone "knew" it could not be true because they thought no organism could survive stomach acidity and enzymes. Indeed, almost everyone already "knew" the cause of ulcers to be psychosomatic. There is much still to be learned about the cause of ulcers, however, and Bernard's statement remains relevant.

TABLE 24.2 *Helicobacter pylori* **Gastritis**

Signs and symptoms	Infections are often asymptomatic, but peptic ulcers can cause abdominal pain, tenderness, and bleeding.
Incubation period	Usually undetermined
Causative agent	*Helicobacter pylori*, a curved, Gram-negative, microaerophilic bacterium, with multiple polar flagella covered by sheaths
Pathogenesis	*H. pylori* cells survive stomach acidity by producing urease and burrowing within the stomach's mucous coating. Bacterial products and inflammation damage the mucosal layer, which can lead to peptic ulcers. Cancer rarely develops, but most with stomach cancer are infected with *H. pylori*.
Epidemiology	Probably fecal-oral transmission. Incidence increases with age.
Treatment and prevention	Treatment: a combination of antibiotics and a medication that inhibits stomach acid production. Prevention: no proven preventive.

Treatment and Prevention

Helicobacter pylori infections can be treated with a combination of antibiotics and a medication that inhibits stomach acid production. This usually results in clearing of the gastritis and healing of any ulcers. There are no proven preventive measures. The main features of *H. pylori* gastritis are shown in **table 24.2.**

MicroAssessment 24.2

The pathogenesis of tooth decay depends on both dietary sucrose and acid-forming bacteria in biofilms on teeth. Periodontal disease is an inflammatory disease associated with plaque at the gum margin and can lead to loosening of teeth. ANUG destroys gingival tissue. Chronic infection by *Helicobacter pylori* is a key factor in the development of peptic ulcers and stomach cancer.

4. Why are new cavities less common, but loss of teeth more common, in people over age 65?
5. What is the relationship between *H. pylori* and stomach cancer?
6. Your mother probably told you that candy causes cavities in your teeth. Is her claim true? Explain your answer.

24.3 ■ Viral Diseases of the Upper Digestive System

Learning Outcome

6. Compare and contrast herpes simplex (cold sores) and mumps.

In this section, we focus on two viral diseases that have dramatic signs and symptoms involving the upper digestive system. Herpes simplex (cold sores) is characterized by painful oral ulcers. Mumps causes enlarged and painful parotid glands.

Some viral diseases involve the upper digestive system but produce more dramatic symptoms elsewhere in the body. For example, measles causes an obvious skin rash and respiratory symptoms, as well as Koplik spots in the mouth. Chickenpox causes a skin rash and oral blisters. Infectious mononucleosis causes impressively enlarged lymph nodes and spleen and may give rise to oral ulcers and bleeding gums. ⏮ chickenpox, ⏭ infectious mononucleosis, ⏮ measles

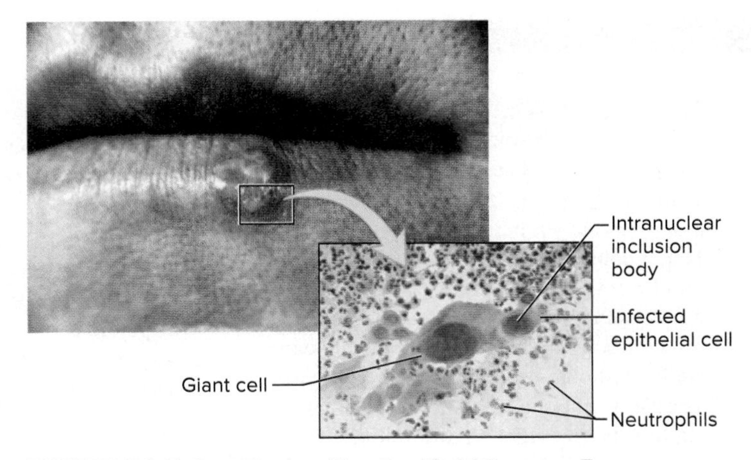

FIGURE 24.7 Oral Herpes Simplex (Cold Sores or Fever Blisters) The photomicrograph shows stained material from a herpes simplex lesion, including a multinucleated giant cell and intranuclear inclusion bodies, the sites of viral replication. Source: CDC; (inset): ©Frederick C. Skvara, M.D.

? **How do the signs and symptoms of recurrent herpes simplex differ from those of the primary infection?**

Oral Herpes Simplex (Cold Sores)

Oral herpes simplex is a common disease with many symptoms and several names. Herpes simplex labialis indicates the location of recurrent lesions on the lips (labia). The lesions are also referred to as cold sores or fever blisters. They typically begin in the mouth and throat. Involvement of the esophagus suggests AIDS or other immunodeficiency. Although the disease is usually insignificant, it can have tragic consequences in newborn infants or people with immunodeficiency.

Signs and Symptoms

The initial signs and symptoms appear after an incubation of 2 to 20 days and include fever and small blisters in the mouth. The blisters break within a day or two, producing superficial ulcers that can be so painful they make it difficult to eat or drink. Although the lesions heal without treatment within about 10 days, the virus persists as a latent infection; the affected person may suffer recurrent cold sores (**figure 24.7**). The signs and symptoms of recurrences are usually less severe than those of initial infection. They include tingling, itching, and burning on the lips, followed by blisters and painful ulcerations, which usually heal within 7 to 10 days.

Causative Agent

Cold sores are caused by herpes simplex viruses (HSVs), enveloped viruses containing double-stranded, linear DNA. There are two types of the virus: HSV-1 and HSV-2. Most oral infections are due to HSV-1; HSV-2 usually causes genital infections. Both types persist throughout life as latent infections that reactivate periodically. ⏭ genital herpes, ⏮ latent infection

Pathogenesis

HSV multiplies in the epithelium of the mouth or throat, and destroys the cells. Some epithelial cells fuse together, producing large, multinucleated giant cells. The nucleus of an infected cell typically contains a deeply staining area called an intranuclear inclusion body—the site of viral replication (see figure 24.7). The characteristic blisters contain many infectious virions. Although an immune response develops and quickly limits the infection, some virions enter the sensory nerves in the area.

Viral DNA persists in nerve cells in a latent form—non-infectious and non-replicating. This latent virus can occasionally reactivate, at which point it is carried by the nerves to the skin or the mucous membranes, where it then multiplies to produce the recurrent disease. Stresses that can precipitate recurrences include menstruation, sunburn, and any illness associated with fever.

Epidemiology

The initial infection with HSV typically occurs during childhood. The virus is widespread and infects up to 90% of some U.S. populations, usually resulting in mild symptoms. Approximately one in three Americans suffers from cold sores. The virus is transmitted primarily by close physical contact, but transmission by fomites such as plastic or cloth is also possible. The greatest risk of infection is from contact with lesions or saliva from patients within a few days of disease onset, when virions are most abundant. Even the saliva of asymptomatic people can be infectious, posing a risk to dentists and other healthcare workers. ⏮ fomite

HSV can infect almost any body tissue. For example, herpetic whitlow (a painful finger infection) is not uncommon among nurses. Wrestlers can develop infections at almost any skin site because saliva containing HSV can contaminate wrestling mats and get rubbed into abrasions. Blindness can result if the virus is rubbed into the eye. Although uncommon, HSV is the most frequently identified cause of sporadic viral encephalitis, a serious brain disease.

Treatment and Prevention

Medications such as acyclovir target the DNA polymerase of HSV and are useful for treating severe cases and for preventing disabling recurrences. They do not affect the latent virus, however, and therefore cannot cure the infection. Sunlight exposure can trigger disease recurrence, so sunscreens are sometimes a helpful preventive. Some of the main features of oral herpes simplex are shown in **table 24.3.** ◀◀ acyclovir

Mumps

Mumps is an acute viral illness that affects glands such as the parotid salivary glands. Formerly common in the United States, the disease is now relatively rare because of routine childhood immunization. Outbreaks do occur, however, mostly due to fading immunity in college students and other young adults. In 2010, about 3,000 high-school-aged students in New York City reported mumps. In 2014–2016, a number of outbreaks involving several hundred people were reported, largely involving college campuses.

Signs and Symptoms

The onset of mumps begins 15 to 21 days after infection and is marked by fever, loss of appetite, and headache. These symptoms are typically followed by painful swelling of one or both parotid glands (**figure 24.8**). Spasm of the underlying muscle makes it difficult to chew or talk, perhaps giving rise to the name of the disease (*mump* means "to mumble" or "whisper"). Symptoms usually fade in about a week.

Painful parotid swelling is characteristic of mumps, but symptoms can arise elsewhere in the body with or without parotid swelling. For example, headache and stiff neck indicate that the virus is causing meningitis—infection of the coverings of the brain. Pregnant women with mumps often miscarry. Rare but serious consequences of mumps, such as death from brain infection, are most likely to occur in older people. Sudden-onset deafness due to asymptomatic infection has also been reported. ▶▶ meningitis

Mumps symptoms are generally more severe in people past the onset of puberty. About 25% of cases in post-pubertal males are complicated by orchitis—a rapid, intensely painful swelling of one or both testicles to three to four times their normal size. Atrophy (shrinkage) of the involved testicles commonly develops after recovery from the illness, and in rare cases causes infertility. In post-pubertal females, ovarian involvement occurs in about one of 20 cases and is characterized by pelvic pain.

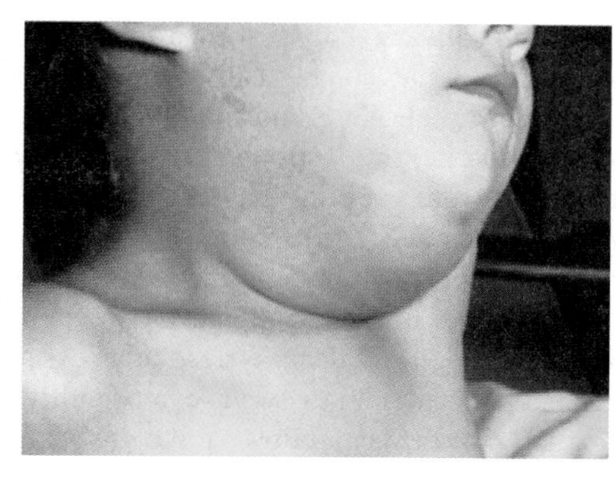

FIGURE 24.8 A Child with Mumps The swelling below the earlobe is due to salivary gland infection. Source: CDC

? **When might mumps result in infertility?**

TABLE 24.3 Oral Herpes Simplex (Cold Sores)

Signs and symptoms	Initial infection: fever, ulcerations of the mouth and throat. Recurrences: itching, tingling, or pain usually on the lip, then blisters that break, leaving a painful sore.
Incubation period	2 to 20 days
Causative agent	Herpes simplex virus (HSV), usually type 1
Pathogenesis	Virus multiplies in the epithelium and destroys the cells. An immune response limits the infection, but HSV DNA persists in sensory nerves. This DNA becomes the source of infectious virions that are carried to the skin or mucous membranes, usually of the lip, causing recurrent sores.
Epidemiology	Widespread distribution; transmitted by close physical contact. Blisters contain large numbers of infectious virions. Saliva of asymptomatic people can be infectious.
Treatment and prevention	Treatment: medications that target HSV DNA polymerase can shorten the duration of the symptoms or prevent recurrences. Prevention: sunscreen can prevent recurrences due to ultraviolet exposure.

Causative Agent

Mumps virus is an enveloped single-stranded RNA virus. It is a member of the *Paramyxoviridae* family, a group that includes the rubeola and respiratory syncytial viruses. Only one serotype of the mumps virus is known.

Pathogenesis

The mumps virus enters the body when virus-containing droplets of saliva are inhaled. It reproduces first in the upper respiratory tract, then spreads throughout the body in the bloodstream. Symptoms begin only after tissues such as the parotid glands, meninges, pancreas, ovaries, or testicles are infected. This is why the incubation period is relatively long.

In the parotid salivary glands, the virus multiplies in the epithelium of ducts that convey saliva to the mouth. This destroys these cells, releasing enormous quantities of virus into the saliva and eliciting a strong inflammatory response, which causes the severe swelling and pain characteristic of the disease. A similar sequence of events occurs in the testicles, where the virus infects the tubules that convey the sperm. The swelling and pressure often impair the blood supply, which can lead to hemorrhage and death of testicular tissue. Kidney tubules are also infected, and the virus can be cultivated from the urine for 10 or more days following the onset of illness.

Epidemiology

Humans are the only natural host of the mumps virus. It is often spread by individuals who have asymptomatic infections, but secrete the virus in their saliva and continue to mingle with other people. In symptomatic patients, the virus can be present in saliva from almost a week before symptoms appear to 2 weeks afterward. Peak infectivity however, is from 1 to 2 days before parotid swelling until the gland begins to return to normal size. A person who has recovered from mumps appears to have lifelong immunity.

Treatment and Prevention

There is no effective treatment for mumps. However, an effective attenuated vaccine has been available in the United States since 1967. It is part of the measles, mumps, rubella, and varicella vaccine (MMRV), as well as the measles, mumps, and rubella vaccine (MMR). **Figure 24.9** shows how the incidence of mumps has generally declined, although it increased in the 1980s because of cuts to funding for vaccinations. Mumps is a good candidate for eradication because it infects only humans, latent infections do not occur, and there is only one serotype. The main features of mumps are presented in **table 24.4.** ⏮ eradication

TABLE 24.4 Mumps

Signs and symptoms	Fever, headache, loss of appetite, followed by painful swelling of parotid gland(s). Meningitis can occur. Painful enlargement of the testicles; pelvic pain in women.
Incubation period	Generally 15 to 21 days
Causative agent	Mumps virus, a single-stranded RNA virus
Pathogenesis	The virus initially replicates in the upper respiratory tract, then spreads throughout the body in the bloodstream. In the salivary glands, the virus multiplies in the cells that line the ducts and kills them. Inflammatory responses cause swelling and pain.
Epidemiology	Humans are the only source of the virus. Infections are often asymptomatic, so the disease can be spread unknowingly.
Treatment and prevention	Treatment: no antiviral therapy is available. Prevention: an effective attenuated vaccine is available.

MicroAssessment 24.3

Herpes simplex (cold sores) is characterized by acute disease followed by lifelong latency and the possibility of recurrences. Infectious virus is often present in saliva in the absence of symptoms. Mumps virus infections characteristically cause enlargement of the parotid glands, but they can involve the brain, testicles, and ovaries, and cause miscarriages.

7. In what type of cell does HSV-1 persist?
8. Why is mumps virus a good candidate for worldwide eradication?
9. Why would medication fail to cure HSV infections even though it prevents recurrent cold sores?

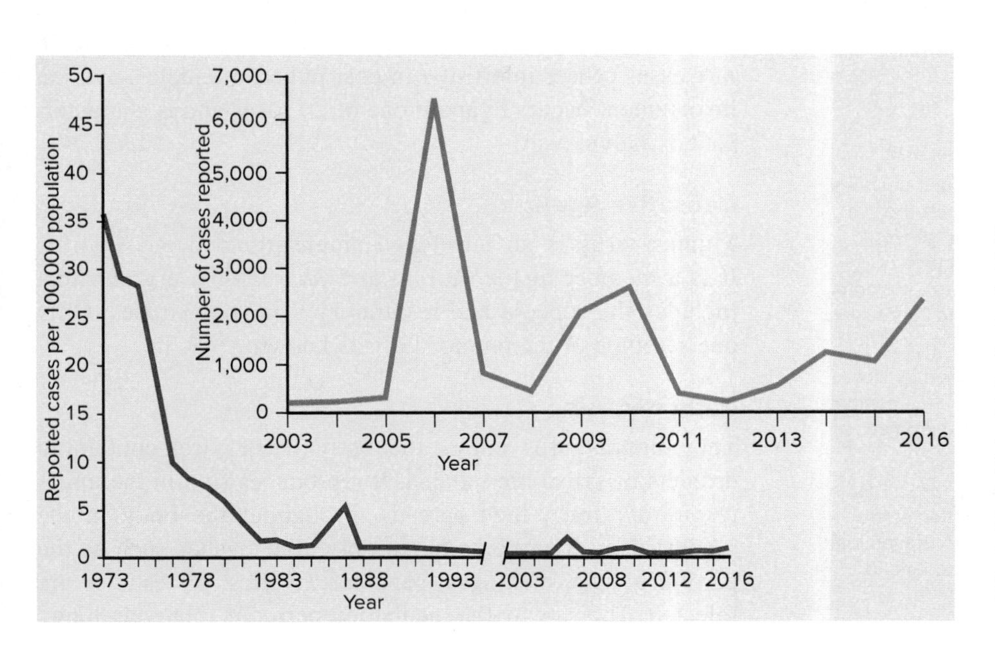

FIGURE 24.9 Reported Cases of Mumps, United States, 1973–October 2016 Mumps vaccine was licensed in 1967. Two doses of MMR vaccine were recommended in 1989 following an outbreak of over 20,000 cases in 1986–1987. An epidemic among college students resulted in more than 6,000 cases of mumps in 2006. Another outbreak of about 3,000 cases occurred in 2009–2010, prompting one New York county to begin offering a third dose of MMR vaccine. Sporadic outbreaks continue to occur, often associated with college campuses.

What events or factors might contribute to mumps outbreaks among college students?

LOWER DIGESTIVE SYSTEM INFECTIONS

24.4 ■ Bacterial Diseases of the Lower Digestive System

Learning Outcomes

7. Describe the general characteristics of diarrheal diseases.
8. Compare and contrast cholera, shigellosis, the various types of *E. coli* gastroenteritis, *Salmonella* gastroenteritis, typhoid and paratyphoid fevers, campylobacteriosis, and *Clostridium difficile* infection.

Diarrheal illness is a common result of bacterial infection of the intestinal tract. Bacteria can also use the intestines as an entry to the rest of the body, thereby causing other types of illness. This section focuses on intestinal infections, and chapter 30 gives examples of foodborne intoxications that can cause diarrhea. Tapeworm and roundworm infections of the intestines are discussed in chapter 12. ⏭ **foodborne intoxication,** ⏮ **helminth infections**

FOCUS ON DIARRHEAL DISEASES

Some physicians refer to diarrheal disease as **gastroenteritis** (*gastro-*, "stomach," *entero-*, "intestine," *-itis,* "inflammation"), whereas others use the term "stomach flu" ("stomach flu" has no relationship to "the flu," or influenza).

Signs and Symptoms

The signs and symptoms of gastroenteritis include diarrhea, loss of appetite, nausea and vomiting, and sometimes fever. The incubation period is typically a day or two. Diarrhea can be abundant and watery when infection involves the small intestine. Large intestine invasion typically results in small amounts of diarrhea that contains mucus, pus, and sometimes blood. **Dysentery** is characterized by blood and pus in the feces.

Pathogenesis

The pathogenesis of diarrheal disease varies according to the causative agent. In the case of bacteria, many of the responsible virulence genes are on plasmids, phages, or other mobile genetic elements, which can be transferred from one species or strain to another by horizontal gene transfer. Because of this, even strains within the same species can differ in the way they cause disease.

The infectious dose of an intestinal pathogen (the number of cells required to establish infection) is related to its ability to survive exposure to stomach acid. Acid-tolerant organisms are more likely to survive passage through the stomach; surviving cells then multiply in the intestines to high enough numbers to cause disease symptoms. In contrast, acid-sensitive organisms are often killed in the stomach; only when millions of cells are consumed will enough survive to reach the intestines, where they multiply and cause disease. People with low gastric acidity are more susceptible to intestinal infections, even by organisms that normally have a high infectious dose. ⏮ **infectious dose**

Attachment of the pathogen to the intestinal surface is typically needed for infection, and the site involved influences the outcome of the disease. Infections of the small intestine generally result in a watery diarrhea because the fluid exchange that normally occurs there is disrupted. The loss of water and electrolytes (salts) draws fluid from the bloodstream, resulting in dehydration. In severe cases, the reduced blood volume may not be sufficient to keep vital organs working properly, a potentially fatal situation. In contrast, infection of the large intestine often results in damage that elicits a strong inflammatory response and results in dysentery.

Common mechanisms of pathogenesis seen in intestinal disease include alterations in intestinal epithelial cells, cell invasion, and exotoxin production. Toxins involved in intestinal infections fall into two groups: **enterotoxins,** which cause water and electrolytes to flow from intestinal cells; and **cytotoxins**, which cause cell death. Some types of cytotoxins produced in the intestine can be absorbed into the bloodstream, resulting in systemic effects. ⏮ **exotoxin**

Epidemiology

Hundreds of thousands of children around the world die of diarrheal illnesses each year. Most are infants, but no age group is spared. Fatalities are less common in the United States, but millions of diarrhea cases still occur each year.

Diarrheal diseases are typically transmitted by the fecal-oral route, commonly by ingestion of food or water that has been contaminated with animal or human feces. Sexual practices that lead to oral-anal contact can also transmit intestinal pathogens. Organisms with a low infectious dose—perhaps 100 cells—are additionally transmitted through casual person-to-person contact. Pathogens that have a low infectious dose are a particular problem in day-care centers and other places where hygiene standards might be difficult to maintain.

Sewage treatment, handwashing, and chlorinating drinking water are important measures to control diarrheal diseases. In the United States, surveillance using PulseNet—a DNA subtyping resource—helps track illness caused by specific intestinal pathogens. This helps public health agencies detect foodborne outbreaks so that intervention strategies can be implemented. ⏮ **PulseNet**

Treatment and Prevention

Oral rehydration therapy (ORT)—giving appropriate fluids by mouth—can be used to counteract the loss of fluid and electrolytes due to diarrhea. If the patient drinks plain water, however, the intestinal tract cannot absorb enough to keep pace with the amount lost. Fortunately, researchers discovered that glucose increases the absorptive capacity of the intestine. This breakthrough led to the development of what is called oral rehydration salts (ORS) solution, a highly effective lifesaver in severe

(Continued)

FOCUS ON DIARRHEAL DISEASES (continued)

diarrhea, regardless of cause. ORS is a mixture of glucose and various salts (sodium chloride, potassium chloride, and trisodium citrate) that is commercially available as pre-measured packets to be dissolved in clean water. The World Health Organization has also developed a list of recommended home fluids (RHF) that can prevent dehydration. If oral rehydration cannot be tolerated—for example, if the patient is vomiting—then intravenous hydration may be required.

Antimicrobial medications are not helpful in most intestinal infections, and often prolong the illness because they suppress the normal microbiota. These drugs can be lifesaving, however, if microorganisms invade beyond the intestine. For children in developing countries, zinc supplements decrease the length and severity of diarrheal disease. No treatments are available for viral diarrhea.

Only a few vaccines are available to prevent diarrheal diseases, and these are pathogen-specific.

Cholera

Cholera causes potentially fatal diarrhea. Seven cholera pandemics have occurred since the early 1800s, with the last one beginning in 1961 in Indonesia and spreading to South Asia, the Middle East, and parts of Europe and Africa. South America had remained cholera-free for 100 years until January 1991, when the disease abruptly appeared in Peru. The source of the bacterium was likely a freighter that discharged bilgewater into a Lima harbor. Lima's water supply was not chlorinated and quickly became contaminated. The disease then spread rapidly, so that in 2 years more than 700,000 cases and 6,323 deaths had been reported in South and Central America. In 2011, the cholera epidemic in Haiti (see chapter introduction) led to the highest number of recorded cases in a country in a single year. That figure was surpassed during the first seven months of 2017 in Yemen, where civil war disrupted sanitation, water treatment, and access to health care. ◀◀ pandemic disease, ◀◀ cholera epidemic in Haiti

Signs and Symptoms

Cholera is a classic example of severe watery diarrheal disease. Signs and symptoms develop after a relatively short incubation period of about 12 to 48 hours. Huge volumes of diarrhea are abruptly produced—up to 20 liters a day—leading to severe dehydration that can result in shock, multiple organ failure, and death. The diarrheal fluid is described as "rice water stool" because of its clear and watery appearance. Vomiting also occurs in most people at the onset of the disease, and many people suffer muscle cramps caused by loss of fluid and electrolytes.

Causative Agent

The causative agent of cholera is *Vibrio cholerae,* a curved, Gram-negative rod with a single flagellum. There are several different serotypes, grouped according to differences in their O antigen. The two serotypes that account for current cholera epidemics are O1 (which is pandemic) and O139 (which is largely limited to Asia). *V. cholerae* is halotolerant and can grow in alkaline conditions, two characteristics used to design appropriate selective media. ◀◀ O antigen, ◀◀ selective media

Pathogenesis

Vibrio cholerae cells are killed by acid, so large numbers must be ingested before enough survive passage through the stomach to establish infection. Once in the small intestine, surviving bacteria adhere to the small intestinal epithelium using pili and other surface proteins. The bacteria multiply on the epithelial cells but do not visibly damage them. However, the bacteria produce cholera toxin, an enterotoxin (**figure 24.10**). This toxin activates ion transport channels in the epithelial cell membrane, causing chloride and other electrolytes to exit the cell. Water follows the electrolytes, resulting in an outpouring of fluid and salts into the intestinal lumen. Although the toxin does not affect the large intestine, the volume of fluid is too much to be absorbed, causing diarrhea that results in severe dehydration.

Cholera toxin is an A-B toxin encoded by a bacteriophage that infects *V. cholerae,* an example of lysogenic conversion. The B (binding) portion attaches irreversibly to specific receptors on the microvilli of the epithelial cells, allowing the A (active) portion to enter the cells. It causes severe diarrhea by activating a G protein, which normally functions as a sophisticated on/off switch that controls activities inside the cell in response to external signals. Cholera toxin chemically modifies the G protein, locking it in the "on" position (see figure 24.10). This, in turn, results in nonstop activity of an enzyme called adenylate cyclase, which converts ATP to cAMP. The net effect is high levels of cAMP at the cell membrane, a condition that causes the cell to continually secrete chloride. The normal turnover of intestinal cells eventually removes the toxin. ◀◀ lysogenic conversion, ◀◀ A-B toxins

Epidemiology

Fecally contaminated water is the most common source of cholera infection, although foods such as contaminated crab, oysters, and vegetables have also been implicated in outbreaks. A person with cholera can discharge a million or more *V. cholerae* cells in each milliliter of feces. Although cholera is relatively common worldwide, effective water and wastewater treatment can eliminate its spread. Most U.S. cases involve international travel.

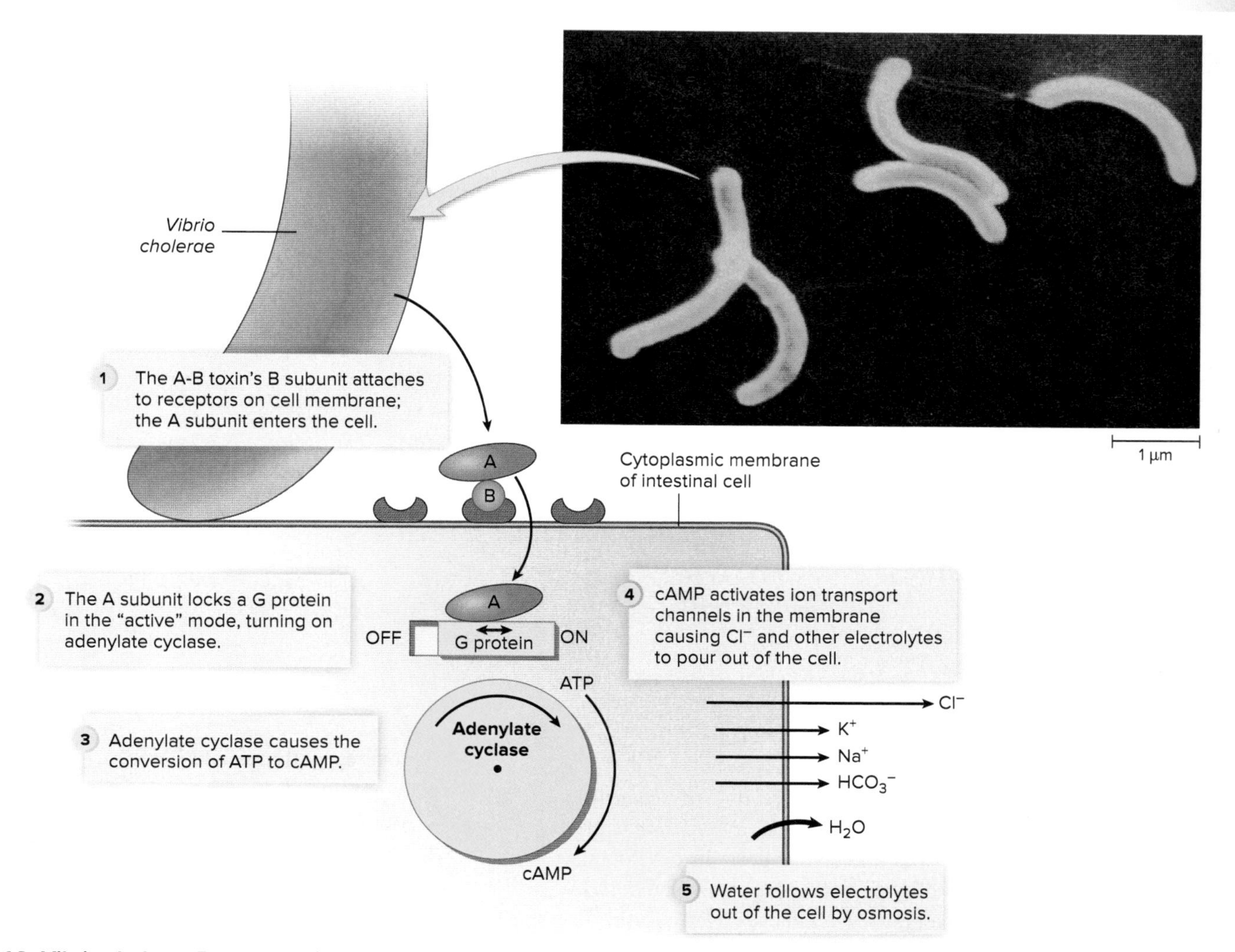

FIGURE 24.10 ***Vibrio cholerae*** **Pathogenesis** *Vibrio cholerae* attaches to the small intestine using pili and begins to produce an A-B toxin. ©London School of Hygiene & Tropical Medicine/Science Source

How would vaccination against the B subunit of cholera toxin help prevent disease?

Treatment and Prevention

Treatment of cholera depends primarily on the rapid replacement of fluid and electrolytes before irreversible damage to vital organs can occur. If patients are vomiting or are too weak to drink oral rehydration solutions, intravenous lines are started. The prompt administration of intravenous or oral rehydration therapy decreases the mortality of cholera from over 30% to less than 1%. In severe cases, antibiotics may be administered in addition to rehydration therapy.

Cholera control depends largely on adequate sanitation and safe, clean water supplies. Travelers to areas where cholera is endemic are advised to cook food immediately before eating it. Crabs should be cooked for no less than 10 minutes. No fruit should be eaten unless peeled personally by the traveler, and ice should be avoided unless known to be made from boiled water. Cholera vaccines have been available since the early 1990s in many countries other than the United States. In 2016, the FDA approved a live attenuated oral vaccine against cholera caused by serogroup O1 that is intended for use by U.S. travelers between 18 and 64 years of age who are visiting areas endemic for cholera. The main features of cholera are summarized in **table 24.5.**

Shigellosis

Shigellosis is found all over the world, most commonly in areas lacking adequate sewage treatment. Reported cases in the United States currently average about 20,000 per year, but the true prevalence is much higher, because diarrhea often goes unreported. Globally, shigellosis is estimated to cause about 500,000 deaths per year.

Signs and Symptoms

Shigellosis has an incubation period of 1 to 3 days. It classically involves dysentery, but some *Shigella* species cause watery diarrhea. Other signs and symptoms include headache, vomiting, fever, stiff neck, convulsions, and joint pain. The disease is often fatal for infants in developing countries.

▶▶| dysentery

TABLE 24.5 Cholera

1. *Vibrio cholerae* enters the mouth with fecally contaminated food or drink.
2. The bacteria attach to epithelial cells of the small intestine.
3. Cholera toxin enters the cells and causes them to continuously secrete chloride ions. Other electrolytes and water follow.
4. The outpouring of water and electrolytes into the intestinal lumen causes watery diarrhea.
5. Fluid loss causes severe dehydration, resulting in shock and death unless the fluid can be replaced.
6. The bacteria exit the body with feces.

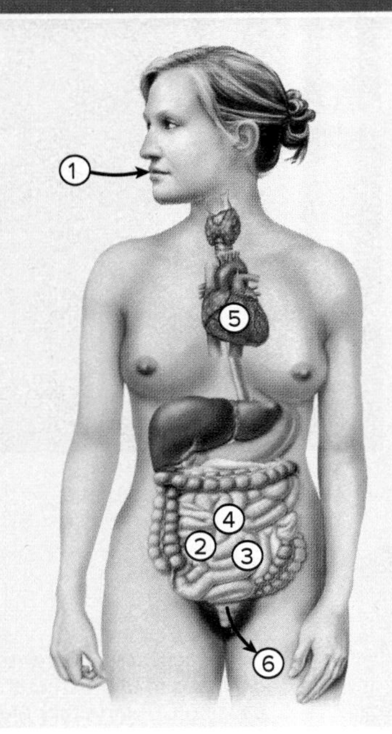

Signs and symptoms	Abrupt onset of massive diarrhea, vomiting, muscle cramps
Incubation period	Short, generally 12 to 48 hours
Causative agent	*Vibrio cholerae*, an alkali- and salt-tolerant, curved Gram-negative rod
Pathogenesis	Cholera toxin causes chloride, other electrolytes, and water to pour out of the intestinal epithelium and into the lumen; leads to dehydration and shock.
Epidemiology	Ingestion of fecally contaminated food or water; sometimes associated with marine crustaceans.
Treatment and prevention	Treatment: replacement of fluid and electrolytes using intravenous or oral rehydration therapy. Prevention: purifying water and handwashing; vaccines available.

Causative Agent

Shigellosis is caused by the four species of *Shigella—S. dysenteriae, S. flexneri, S. boydii,* and *S. sonnei.* Of these, *S. dysenteriae* is the most virulent, and *S. sonnei* the least. Shigellosis in developing countries is typically caused by *S. dysenteriae* and *S. flexneri.* In the United States, however, *S. sonnei* causes over two-thirds of the cases. Like other members of the family *Enterobacteriaceae, Shigella* species are Gram-negative rods.

Pathogenesis

Shigella species invade intestinal epithelial cells, causing a strong inflammatory response. To initiate invasion, the bacteria take advantage of the antigen sampling function of M cells, which normally transfer microbes to macrophages in Peyer's patches (**figure 24.11**). Once *Shigella* cells enter macrophages, they escape from the phagosome and multiply in the cytoplasm. These "hidden" bacteria are released near the bases of epithelial cells when the infected macrophages die. ⏮ M cells

Shigella cells attach to specific receptors on the intestinal epithelial cells and induce those cells to take them in (see figure 24.11). Once inside, the *Shigella* cells escape into the cytoplasm of epithelial cells, where they multiply. Although non-motile, the bacterial cells produce a protein that polymerizes host cell actin to form an "actin tail" that propels the bacterium within the cell, sometimes with enough force to move it into a neighboring cell (see Focus Your Perspective 3.1). The overall result of invasion and spread is the death and sloughing of patches of epithelium. The bare areas become intensely inflamed, covered with pus and blood, which accounts for the signs and symptoms of dysentery.

Some strains of *Shigella dysenteriae* produce a potent cytotoxin known as Shiga toxin, a chromosomally encoded A-B toxin. The toxin enters the bloodstream and then the B portion binds to endothelial cells that line the small blood vessels, particularly in the kidneys. This allows the A subunit to then enter the cells, where it interacts with ribosomes, halting protein synthesis, which leads to cell death. Shiga toxin is responsible for **hemolytic uremic syndrome (HUS),** an often fatal condition that can follow *Shigella dysenteriae* infection. In HUS, red blood cells lyse in the tiny blood vessels, resulting in anemia and kidney failure. Symptoms of HUS sometimes include paralysis or other signs of nervous system injury. Shiga toxin is also produced by strains of *Escherichia coli* that cause HUS.

Epidemiology

Shigellosis is almost exclusively a disease of humans, transmitted by the fecal-oral route. Unlike *V. cholera, Shigella* cells are not easily killed by stomach acid, so the infectious dose is very small. Shigellosis spreads readily in overcrowded populations with poor sanitation, such as in refugee camps and day-care centers. People who engage in anal intercourse are also prone to contracting the disease. Fecally contaminated food and water have caused numerous shigellosis outbreaks.

Treatment and Prevention

Shigellosis generally clears on its own, but antimicrobial medications are useful in severe cases because they shorten

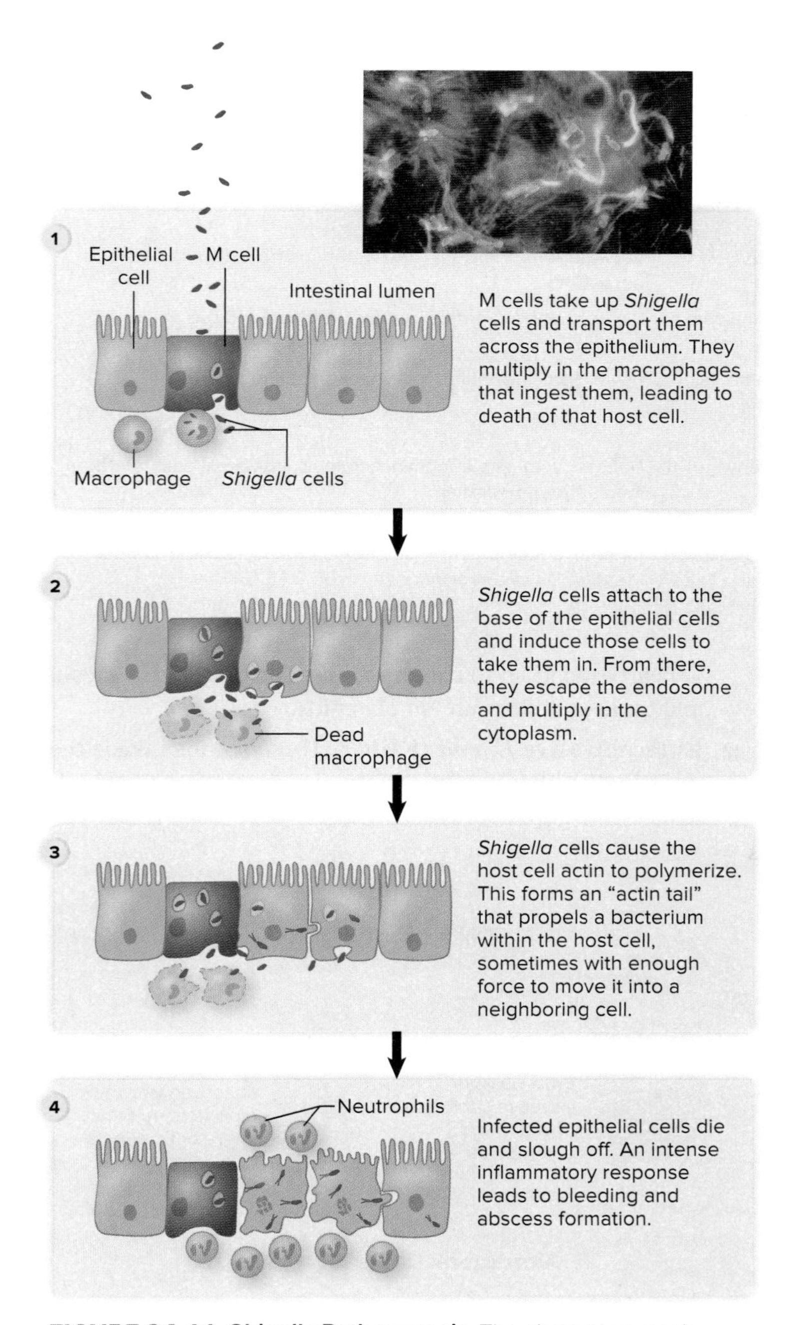

FIGURE 24.11 ***Shigella* Pathogenesis** The photomicrograph shows the actin tails (green) that form on intracellular *Shigella* cells (orange) and rapidly push these non-motile bacteria from cell to cell. Dr. Philippe J. Sansonetti, Professeur Institut Pasteur

? **Why does shigellosis cause dysentery, whereas cholera causes watery diarrhea?**

the duration of symptoms and the time during which the pathogens are discharged in the feces. Unfortunately, many strains have R plasmids that encode resistance to a variety of medications. ⏮ R plasmids

The spread of *Shigella* is controlled by sanitary measures that reduce fecal contamination of food and water supplies. Two specimens of feces, collected at least 48 hours after stopping antimicrobial medicines, must be negative for *Shigella* before a person is allowed to return to a day-care center or food-handling job. Shigellosis can be tracked through PulseNet, making it easier for public health agencies to detect outbreaks. **Table 24.6** describes the main features of shigellosis. ⏮ PulseNet

TABLE 24.6 Shigellosis

Signs and symptoms	Fever, dysentery, vomiting, headache, stiff neck, convulsions, and joint pain
Incubation period	1 to 3 days
Causative agent	Four species of *Shigella,* Gram-negative, members of the *Enterobacteriaceae*
Pathogenesis	Bacteria cross the intestinal mucous membrane via M cells; induce uptake by epithelial cells; "actin tails" allow bacteria to spread to neighboring cells, leading to death and sloughing of epithelium. Some strains make Shiga toxin.
Epidemiology	Fecal-oral transmission; low infectious dose; humans generally the only source.
Treatment and prevention	Treatment: antimicrobial medications shorten duration of symptoms and pathogen excretion; many strains have R plasmids. Prevention: spread is controlled by sanitary measures.

MicroByte

Ingestion of as few as 10 cells of *Shigella* can result in shigellosis.

Escherichia coli Gastroenteritis

Escherichia coli strains are almost universal residents of the intestines of humans and a number of other animals. Although most are harmless, certain strains produce specific virulence factors that allow them to cause intestinal disease. Other strains—with different virulence factors—cause urinary tract infections, septicemia, and meningitis. ⏮ virulence factors

Signs and Symptoms

The incubation periods, signs, symptoms, and severity of *E. coli* gastroenteritis depend on the infecting strain. Some strains cause watery diarrhea and others cause dysentery. One group can cause hemolytic uremic syndrome (HUS), marked by anemia due to lysis of red blood cells and kidney failure. ⏭ dysentery, ⏭ HUS

Causative Agent

Escherichia coli is a Gram-negative rod, closely related to *Shigella* species. Most strains ferment lactose, an easily observable trait that distinguishes them from *Shigella* species.

Pathogenesis

Escherichia coli strains that cause intestinal disease can be grouped into six pathovars (pathogenic varieties), based on their array of virulence factors (**table 24.7**).

TABLE 24.7 Characteristics of Diarrhea-Causing *Escherichia coli*

Designation	Characteristic Features	Clinical Picture
Diffusely adhering *E. coli* (DAEC)	Grows as a diffuse layer in a thick mucus-associated biofilm on the intestinal epithelium; produces toxins	Diarrhea, particularly in children
Enteroaggregative *E. coli* (EAEC)	Grows in brick-like aggregations in a thick mucus-associated biofilm on the intestinal epithelium; produces toxins	Variable symptoms; nausea, watery diarrhea (both acute and persistent)
Enteroinvasive *E. coli* (EIEC)	Invades the intestinal epithelium, causing a disease very similar to shigellosis	Fever, cramps, diarrhea containing blood and pus
Enteropathogenic *E. coli* (EPEC)	Colonizes the small intestine; induces changes in actin filaments, causing the microvilli to be replaced by pedestals under the bacterial cells (A/E lesions)	Fever, vomiting, watery diarrhea containing mucus
Enterotoxigenic *E. coli* (ETEC)	Colonizes the small intestine and produces toxins, one nearly identical to cholera toxin	Nausea, vomiting, abdominal cramps, massive watery diarrhea leading to dehydration
Shiga toxin–producing *E. coli* (STEC)	Same as EPEC, except it produces Shiga toxin and colonizes the large intestine rather than the small intestine	Fever, abdominal cramps, bloody diarrhea without pus; some patients develop hemolytic uremic syndrome; most strains identified in outbreaks are serotype O157:H7

The characteristics of *E. coli* strains that are capable of producing gastroenteritis are described below:

- **Shiga toxin–producing *E. coli* (STEC).** These strains, also referred to as EHEC or enterohemorrhagic *E. coli,* produce Shiga toxins, a family of functionally identical toxins that includes the toxin of *Shigella dysenteriae.* In STEC strains, the genes for Shiga toxins are encoded by various related prophages, an example of lysogenic conversion. Some STEC strains produce multiple types of Shiga toxins, and even make other toxins as well (**figure 24.12**). Most of the strains identified in outbreaks belong to a single serotype, O157:H7, but there are important exceptions. STEC strains typically colonize the large intestine, where they inject effector proteins that cause **attaching and effacing (A/E) lesions.** The intestinal damage results in diarrhea, which becomes bloody (hemorrhagic diarrhea) due to the action of Shiga toxins on the local blood vessels. About 5% to 10% of people infected with STEC develop hemolytic uremic syndrome (HUS). The STEC strain that caused the 2011 outbreak in Europe (serotype O104:H4) is unusual because it does not produce A/E lesions and has characteristics of the EAEC pathovar, which will be discussed shortly. ▶▶| Shiga toxin, |◀◀ lysogenic conversion
- **Enterotoxigenic *E. coli* (ETEC).** These strains make pili that allow them to attach to and colonize the small intestine. In addition, they secrete enterotoxins, one of which is nearly identical to cholera toxin. The genes for adhesin and toxin synthesis are on plasmids.
- **Enteroinvasive *E. coli* (EIEC).** These strains invade the intestinal epithelium, causing a disease similar to shigellosis (see figure 24.12).

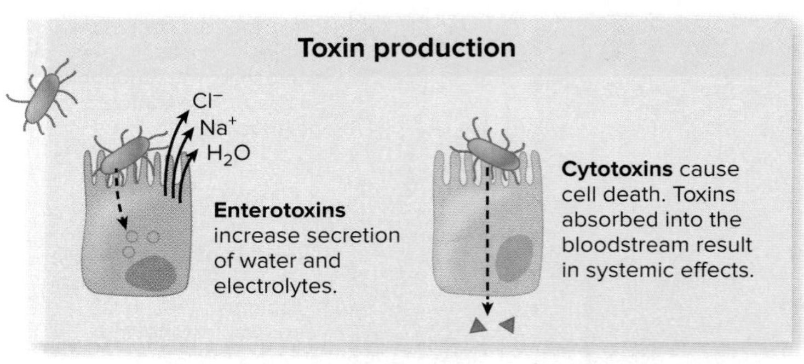

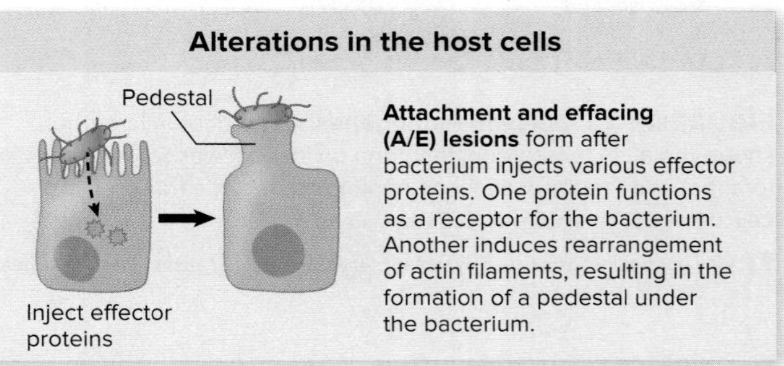

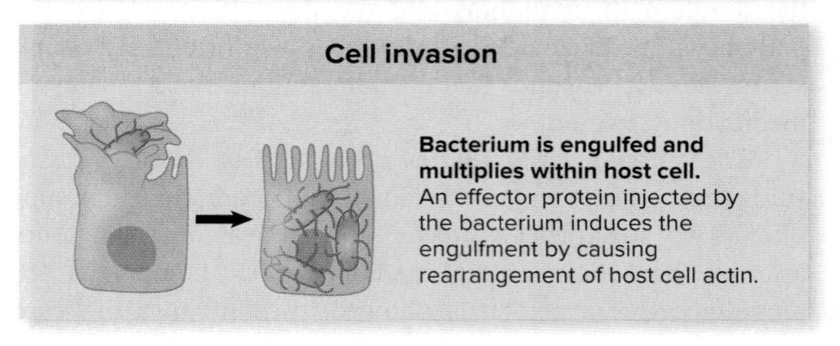

FIGURE 24.12 Common Mechanisms of Pathogenesis of Intestinal Pathogens Together, various strains of *E. coli* show each of these mechanisms. Invasion or cell damage elicits a strong inflammatory response that also contributes to the severity of disease.

? **Why is adhesion generally a prerequisite for pathogenicity?**

- **Enteropathogenic *E. coli* (EPEC).** These strains produce pili that allow them to colonize the small intestine, where they inject effector proteins that cause A/E lesions.
- **Enteroaggregative *E. coli* (EAEC).** These strains produce pili that allow them to adhere to the intestinal epithelium. There, they grow in characteristic aggregations ("brick-like") in a thick mucus-associated biofilm. In addition, they produce enterotoxins and cytotoxins, damaging the intestinal cells and evoking an inflammatory response.
- **Diffusely adhering *E. coli* (DAEC).** These strains are similar to EAEC, but rather than forming aggregations, they grow as a diffuse layer.

Epidemiology

Just as the symptoms and pathogenesis of *E. coli* gastroenteritis depend on the infecting strain, so does the epidemiology.

STEC strains can be foodborne, and epidemics have involved contaminated ground beef, unpasteurized milk, apple juice, bean sprouts, and green leafy vegetables. The initial source of infection is often untreated cow manure, reflecting the fact that cattle are an important reservoir. The infectious dose of STEC strains is typically very low, so in addition to foodborne transmission, the bacteria are easily spread by direct contact. ⏮ reservoir of infection

ETEC strains commonly cause diarrhea in infants in developing countries as well as travelers visiting those regions. Their infectious dose is relatively high, so they typically do not spread by direct contact. ETEC are species-specific, and those that infect animals are responsible for significant mortality in young livestock.

EIEC strains primarily cause disease in young children in developing countries. When compared with some *E. coli* strains, they are not easily spread by direct contact, indicating that the infectious dose is not very low. Humans appear to be the only source of infection.

EPEC strains are an important cause of chronic diarrhea in infants. The infection is uncommon in breast-fed infants, probably because of protective antibodies in breast milk. Outbreaks have occurred in hospital nurseries, so the infectious dose for infants is thought to be very low. A variety of animals have been shown to harbor EPEC, but their importance as a source of infection is not known.

EAEC strains cause diarrhea in children, travelers, and AIDS patients. A variety of animals harbor EAEC strains, but it is not clear that the animal strains cause human disease.

DAEC strains have caused diarrhea outbreaks in children, but relatively little is known about their epidemiology.

Treatment and Prevention

Treatment of *E. coli* gastroenteritis varies according to the symptoms and the infecting strain, but as with any disease, fluid lost from vomiting and diarrhea should be replaced. Antibacterial medications are typically not used because most cases are self-limiting. Patients with STEC infections who were treated with antibiotics showed an overall worse outcome.

Measures to prevent *E. coli* gastroenteritis include handwashing, pasteurization of drinks, and thorough cooking of food. PulseNet helps track infections caused by *E. coli* O157:H7 and other STEC strains. Traveler's diarrhea can usually be prevented with bismuth preparations (such as Pepto-Bismol). Unfortunately, the widespread use of antibiotics to prevent diarrhea has promoted development of resistant strains. Some features of *E. coli* gastroenteritis are summarized in **table 24.8.**

Salmonella Gastroenteritis

Salmonella gastroenteritis is caused by numerous serotypes of *Salmonella enterica* and can be acquired from many animal sources. An estimated 1.2 million cases occur in the United States each year, with 400 deaths. Large outbreaks are usually due to commercially distributed foods contaminated by animal feces.

Signs and Symptoms

Signs and symptoms of *Salmonella* gastroenteritis include diarrhea (sometimes bloody), abdominal cramps, nausea, vomiting, headache, and fever. The disease is often short-lived and mild, but that varies depending on the virulence of the infecting strain and the number of cells ingested. Similarly, the incubation period varies from 6 hours to 3 days.

Causative Agent

Salmonella enterica is a Gram-negative rod and, like *Shigella* and *E. coli,* a member of the *Enterobacteriaceae.* The various *Salmonella* strains are subdivided into more than 2,400 serotypes based on differences in their cell wall (O), flagellar (H), and capsular (K) antigens (see figure 10.10). Each serotype was once considered a separate species and given a distinct name. However, DNA sequence data indicate there are only two species: *S. enterica* and *S. bongori,* the latter only rarely isolated from humans. ⏮ serotypes

The serotype of a *Salmonella* strain is significant from both an epidemiological and a disease standpoint. For instance, certain serotypes cause enteric fever (discussed next) instead of gastroenteritis. Because of the significance, the serotype is often included with the name—for example, *Salmonella enterica* serotype Dublin. For convenience, the name is often shortened, as in *Salmonella* Dublin. Note that the serotype is not italicized and the first letter is capitalized, which distinguishes it from a species name. *Salmonella* serotype Typhimurium and *Salmonella* serotype Enteritidis are those most commonly isolated in the United States.

TABLE 24.8 *Escherichia coli* **Gastroenteritis**

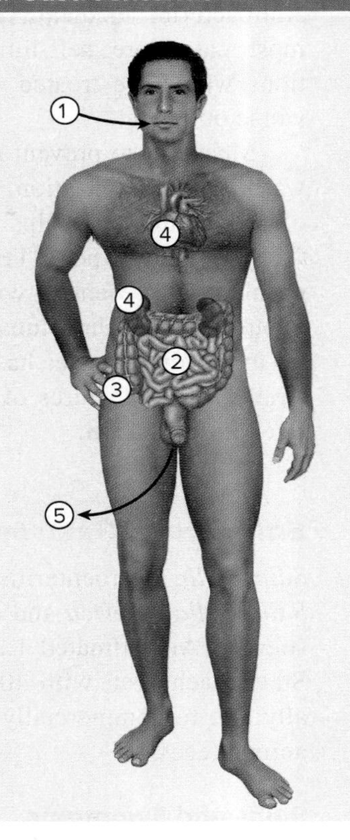

(1) Pathogenic strain of *E. coli* enters by the fecal-oral route, either directly from an infected person or with contaminated food or drink.

(2) Most strains colonize the small intestine and produce watery diarrhea.

(3) Others invade the large intestine and cause dysentery.

(4) Some strains produce Shiga toxin, which is absorbed by the bloodstream and causes hemolytic-uremic syndrome.

(5) The bacteria exit the body with feces.

Signs and symptoms	Watery diarrhea or dysentery
Incubation period	2 hours to 6 days
Causative agent	*Escherichia coli,* certain strains only
Pathogenesis	Various mechanisms; attachment to small intestinal cells allows colonization; some strains produce one or more enterotoxins; some strains invade large intestinal epithelium; others alter actin polymerization to cause attachment and effacing lesions, and may produce Shiga toxin.
Epidemiology	Common in travelers; can be foodborne or waterborne; fecal-oral route transmission; some strains have an animal source.
Treatment and prevention	Treatment: replacement of fluids and electrolytes. Prevention: handwashing; pasteurization of drinks, thorough cooking of meats. Bismuth compounds help prevent traveler's diarrhea.

Pathogenesis

Most *Salmonella* serotypes are sensitive to acid, so millions of cells must generally be ingested for enough to survive passage through the stomach to colonize the intestines. Upon reaching the distal small intestine (the region farthest from the stomach), some of the bacterial cells attach to specific receptors on the surface of the epithelial cells. Contact activates a type III secretion system that transfers bacterial effector proteins into the epithelial cell. Within minutes, the epithelial cell takes in the bacterial cells by endocytosis. The bacteria multiply within a phagosome and are discharged from the base of the cell by exocytosis. Some bacterial cells escape the phagosome and multiply in the cytoplasm. Macrophages and neutrophils take up any bacteria that are released, but the macrophages are often destroyed as a result. The infection, however, remains localized. The inflammatory response increases epithelial cell fluid secretion, causing diarrhea. ⏮ type III secretion system

Not all of the infecting *Salmonella* cells invade the epithelial cells. Those that do not can benefit from the inflammatory response caused by the invading bacteria. The oxidizing environment created by neutrophils recruited during inflammation generates tetrathionate—a compound that *Salmonella* can use as a terminal electron acceptor for anaerobic respiration. Most normal intestinal microbiota cannot use this compound and are therefore restricted to fermenting in the anaerobic environment. The ability to anaerobically respire gives *Salmonella* cells a competitive advantage in the densely populated intestinal tract; recall that respiration generates more metabolic energy than fermentation (see table 6.3). ⏮ anaerobic respiration

Epidemiology

Most cases of *Salmonella* gastroenteritis originate from a nonhuman animal source; the bacteria live in the intestinal tract of a number of different types of domestic animals. The bacteria sometimes survive for months in soil and water, so they can be spread in untreated manure. Poultry often carry *S. enterica,* and eggs can be contaminated as well. Other contaminated products—including tomatoes, brewer's yeast, alfalfa sprouts, protein supplements, raw eggs, dry milk, and even a red dye used to diagnose intestinal disease—have started outbreaks. Children are commonly infected by colonized but seemingly healthy pets—particularly lizards, snakes, and turtles—that shed the bacteria in their feces. The disease has been on the rise, largely due to mass production and distribution of foods.

Treatment and Prevention

Most people with *Salmonella* gastroenteritis recover without antimicrobial treatment, which is fortunate because *S. enterica* strains have shown increasing plasmid-mediated resistance to antimicrobial medications. Many isolates of *Salmonella* serotype Typhimurium are resistant to five or more medications. Antibiotic-resistant strains are often associated with more severe illness. Resistance is likely due to the widespread use of subtherapeutic levels of antibiotics in animal feeds, which is why several countries have now banned this practice. Antibiotics are not advised for treating humans except in cases involving tissue or bloodstream invasion.

Control of *Salmonella* gastroenteritis depends on sanitary handling of animal carcasses, pasteurizing or irradiating animal products, and testing products for contamination. Surveillance and tracing contaminated sources using PulseNet are also important. Adequate cooking kills *Salmonella* cells, but the center of cooked food does not always reach the 160° Fahrenheit recommended to kill foodborne bacteria. **Table 24.9** presents the characteristics of *Salmonella* gastroenteritis.

Typhoid and Paratyphoid Fevers

Typhoid fever and paratyphoid fever are **enteric fevers**—systemic diseases that originate in the intestine. They are caused by specific serotypes of *Salmonella enterica* and spread from person to person through fecal-oral transmission. Although rare in the United States, millions of cases occur in developing countries.

Signs and Symptoms

Following an incubation period of 1 to 4 weeks, patients develop a gradually increasing fever over several days, severe headache, constipation, and abdominal pain. In severe cases, intestinal rupture, internal bleeding, shock, and death may follow.

Causative Agent

Typhoid fever is caused by *Salmonella* serotype Typhi, whereas paratyphoid fever is caused by *Salmonella* serotype Paratyphi. These are systemic diseases, so cases are confirmed by blood culture rather than stool culture.

TABLE 24.9 ***Salmonella* Gastroenteritis**

Signs and symptoms	Diarrhea, vomiting, headache, abdominal pain, and fever
Incubation period	Usually 6 to 72 hours
Causative agent	*Salmonella enterica*, Gram-negative, member of the *Enterobacteriaceae*
Pathogenesis	Induce uptake by epithelial cells in the distal small intestine; bacteria multiply in the phagosome and then are discharged at the base of the cell; inflammatory response increases fluid secretion
Epidemiology	Ingestion of food contaminated by animal feces, especially poultry
Treatment and prevention	Treatment: antimicrobial medication is generally not advised. Prevention: relies on adequate cooking and proper handling of food.

Pathogenesis

Bacteria that cause enteric fever colonize the intestines, cross the mucous membrane via M cells, and then multiply within macrophages. The bacteria are then carried in the bloodstream to locations throughout the body. The systemic infection causes fever, abscesses, sepsis (a systemic inflammatory response due to bloodstream infection), and shock, often with little or no diarrhea. The Peyer's patches are sometimes destroyed, leading to rupture of the intestine, hemorrhage, and death. Surprisingly little is known about the bacterial mechanisms that lead to enteric fever, but researchers have recently identified a toxin (typhoid toxin) that is produced by the bacterial cells only when they are within a host cell. The role of this toxin in the disease process is still being studied.
⏭ sepsis, ⏮ Peyer's patches

Epidemiology

Humans are the only known host for *Salmonella* serotypes Typhi and Paratyphi, so the bacteria are spread from person to person. Often, this occurs via contaminated food or water.

Some patients surviving typhoid or paratyphoid fever remain colonized with the causative agents. The bacteria reside in the gallbladder, where they multiply free of competition because most other bacteria are killed or inhibited by concentrated bile. Carriers can shed high numbers of the bacteria for years. Mary Mallon—"Typhoid Mary"—a young Irish cook living in New York State in the early 1900s, was a notorious *Salmonella* serotype Typhi carrier. She was responsible for at least 53 cases of typhoid fever over a 15-year period. In her day, about 350,000 cases occurred in the United States each year. The incidence is now less than 400 cases per year, and most of these are related to international travel.

Treatment and Prevention

Antimicrobial medications are used to treat enteric fever, but some *Salmonella* serotype Typhi strains are multi-drug-resistant, so susceptibility testing must be done. Most strains in the United States are susceptible to fluoroquinolones. Surgical removal of the gallbladder and months of antibiotic therapy are often necessary to rid *Salmonella* serotype Typhi carriers of their infection.

An attenuated live oral vaccine for preventing typhoid fever is about 50% to 75% effective. An equally effective injectable vaccine composed of *Salmonella* serotype Typhi capsular polysaccharide is also available. There is no vaccine against *Salmonella* serotype Paratyphi. **Table 24.10** summarizes some of the features of enteric fever. ⏮ attenuated vaccine

Campylobacteriosis

Campylobacter jejuni was first isolated from a diarrheal stool in 1972, but it then took 5 years to develop a suitable culture protocol. Once this was widely used, *C. jejuni* infection was found to be a common cause of diarrhea.

TABLE 24.10	Typhoid and Paratyphoid Fevers
Signs and symptoms	Fever, severe headache, constipation, and abdominal pain; sometimes followed by intestinal rupture, internal bleeding, shock, and death
Incubation period	1 to 4 weeks
Causative agent	*Salmonella* serotype Typhi (typhoid fever) and *Salmonella* serotype Paratyphi (paratyphoid fever), Gram-negative, members of the *Enterobacteriaceae*
Pathogenesis	Cross the intestinal epithelium via M cells; multiply within macrophages and carried in the bloodstream throughout the body. The Peyer's patches are sometimes destroyed, leading to rupture of the intestine and hemorrhage.
Epidemiology	Humans are the only source of infection. Person-to-person fecal-oral transmission.
Treatment and prevention	Treatment: antimicrobial medications. Prevention: attenuated and inactivated vaccines for typhoid fever; no vaccine for paratyphoid fever.

Signs and Symptoms

The incubation period can vary from 1 to 11 days, but symptoms usually appear in 2–5 days. These include fever, vomiting, diarrhea, and abdominal cramps. Dysentery occurs in about half the cases.

Causative Agent

Campylobacter jejuni is a spirally curved, Gram-negative rod (**figure 24.13**). It can be cultivated from feces under microaerophilic conditions using a selective medium to suppress the growth of other intestinal organisms. ◀◀ selective media

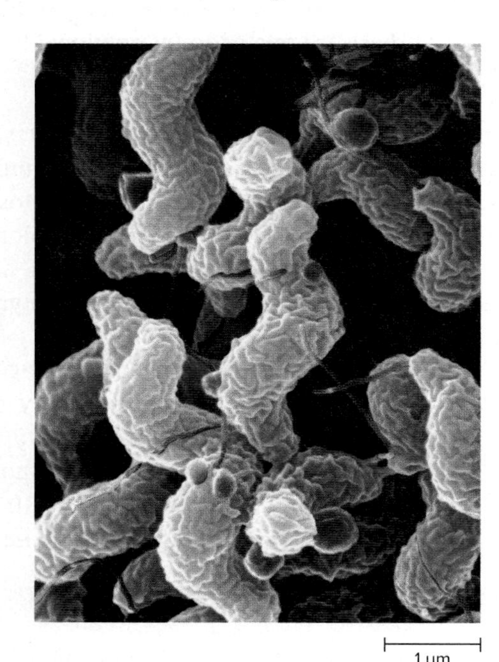

FIGURE 24.13 ***Campylobacter jejuni*** Color-enhanced scanning electron micrograph. ©Science Source

? **Why was the incidence of this common disease largely unknown for decades?**

Pathogenesis

Once ingested, *Campylobacter jejuni* passes through the stomach and penetrates the epithelial cells of the small and large intestines. The bacteria multiply within and beneath the epithelium and cause a localized inflammatory reaction. Penetration into the bloodstream is uncommon.

A mysterious consequence of *C. jejuni* infection, Guillain-Barré syndrome, occurs in about 0.1% of cases. Up to 40% of all Guillain-Barré cases are preceded by campylobacteriosis, and autoimmunity is likely involved. The syndrome begins within about 10 days of the onset of diarrhea, with tingling of the feet followed by progressive paralysis of the legs, the arms, and the rest of the body. Most patients require hospitalization, but recover completely; about 5% of patients die despite treatment.

Epidemiology

Campylobacteriosis is a leading bacterial diarrheal illness in the United States, with an estimated 1.3 million cases each year. Infection does not typically result in death, but still about 100 people die from it each year, mostly among the elderly and those with AIDS or other immunodeficiencies. Numerous foodborne and waterborne outbreaks of *C. jejuni* have been reported, involving as many as 3,000 people. Most cases, however, are sporadic.

C. jejuni lives in the intestines of a variety of domestic animals, including pets and migratory birds. Poultry is a common source of infection, and up to 90% of raw poultry products contain the organism. The infectious dose of *C. jejuni* is as low as 500 organisms, so one drop of juice from raw chicken meat can easily result in infection. Unpasteurized milk and nonchlorinated surface water have also started epidemics. Despite a low infectious dose, person-to-person spread of *C. jejuni* is rare.

Treatment and Prevention

Campylobacter jejuni gastroenteritis is typically self-limiting and leaves the patient immune to further infection. Azithromycin or fluoroquinolones are used to treat severe cases.

Campylobacteriosis can be prevented by cooking and handling raw poultry properly to avoid cross-contamination. Chicken should be cooked until no longer pink (160° Fahrenheit). Pet owners should wash their hands after contact with animal feces, and outdoor play areas should be kept free of bird droppings. Chlorinating drinking water and pasteurizing beverages are also important control measures. PulseNet helps track *Campylobacter* isolates. The main features of campylobacteriosis are listed in **table 24.11.** ◀◀ cross-contamination

Clostridium difficile Infection (CDI)

Clostridium difficile has been recognized as a cause of antibiotic-associated diarrhea for many years, but the severity of the symptoms and the number of outbreaks have been increasing since the early 2000s. The disease is particularly a

TABLE 24.11 Campylobacteriosis

Signs and symptoms	Diarrhea, fever, abdominal cramps, vomiting, bloody stools
Incubation period	Usually 2 to 5 days
Causative agent	*Campylobacter jejuni,* a curved Gram-negative, microaerophilic rod
Pathogenesis	Low infectious dose. The bacteria multiply within and beneath the epithelial cells, causing an inflammatory response. Bloodstream invasion is uncommon. Complicated by Guillain-Barré syndrome on rare occasions.
Epidemiology	Large foodborne and waterborne outbreaks originate from poultry and other animals; person-to-person spread rare
Treatment and prevention	Treatment: antibiotics in severe cases. Prevention: relies on adequate cooking and proper handling of food, particularly poultry. Water chlorination and pasteurization of beverages are effective control measures.

problem in healthcare settings, where *C. difficile* is the most common cause of diarrhea. **healthcare-associated infections (HAIs)**

Signs and Symptoms

The severity of the signs and symptoms due to *Clostridium difficile* infection (CDI) range widely. Some patients have only mild diarrhea, often accompanied by fever and abdominal pain and usually beginning less than a week after infection. In other cases, the disease progresses to colitis (inflammation of the colon), sometimes with patchy areas on the colon called pseudomembranes, a characteristic of **pseudomembranous colitis** (**figure 24.14**). In severe cases, CDI is life-threatening.

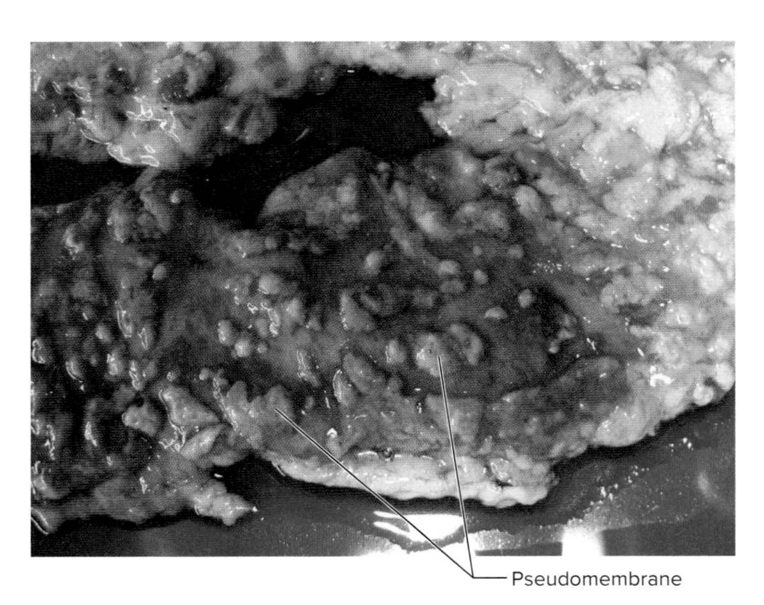

FIGURE 24.14 Pseudomembranous Colitis Pseudomembranes are composed of dead epithelium, inflammatory cells, and clotted blood.
©McGraw-Hill Education

? How does antibiotic therapy predispose a person to this condition?

Causative Agent

Clostridium difficile (commonly referred to as "C. diff") is a Gram-positive, rod-shaped, endospore-forming obligate anaerobe. The endospores are highly resistant to common disinfectants and environmental conditions, making spread of the disease difficult to control. When the spores are ingested, they develop into vegetative bacterial cells that can sometimes colonize the large intestine.

Many strains of *C. difficile* exist, but these can be separated into several large groups that differ in their pathogenicity. Those that cause CDI produce one or more characteristic toxins, and tests that detect the toxins or toxin-encoding genes are often used to diagnose the disease. A group of hypervirulent strains have been increasing in prevalence. Unlike most other strains of *C. difficile,* these are resistant to fluoroquinolones, and frequent use of the medications may have given the strains a selective advantage. Because of the severity of the disease, and its relationship to antibiotic use, the CDC considers *Clostridium difficile* to be an urgent health threat (see table 20.2).

Pathogenesis

When antibiotics have disrupted the normal microbiota in the large intestine, creating an imbalanced condition called dysbiosis, *C. difficile* can proliferate. Pathogenic *C. difficile* strains release toxins that result in a variety of signs and symptoms, including diarrhea. At least two toxins (toxin A and toxin B) appear to cause symptoms in *Clostridium difficile* infection (CDI). Both toxins are A-B toxins that disrupt regulation of host cell actin polymerization and various other signaling pathways, causing lethal effects to the intestinal epithelium. The toxins also cause macrophages to release proinflammatory cytokines, inducing a strong inflammatory response. In some cases, the net result is formation of pseudomembranes composed of dead epithelium, inflammatory cells, and clotted blood on the inner wall of the intestine. A third toxin, sometimes called binary toxin because it is composed of separate proteins, is produced by the hypervirulent strains. Like toxins A and B, this toxin also interferes with actin polymerization. **dysbiosis**

Epidemiology

Clostridium difficile infection (CDI) primarily occurs in hospitalized patients on antibiotic therapy, probably because the organism can only grow to high numbers when the normal intestinal microbiota has been disrupted. Some patients carry it as a minor component of their normal microbiota, but most acquire the microorganism while in the hospital. Infectious endospores are shed in the feces and can be transmitted through the environment or by contact with healthcare workers or other patients. An increasing number of cases of CDI are community acquired; many of these patients have an underlying chronic intestinal condition such as Crohn's disease.

Treatment and Prevention

When practical, the antibiotics that predisposed the patient to *Clostridium difficile* infection (CDI) are stopped. Simply discontinuing these often causes symptoms to disappear. If this is not an option or it fails, a different antibiotic may be used to control the primary infection while also killing *C. difficile.* Vancomycin and metronidazole (given orally) are common alternatives. Fidaxomicin, a narrow-spectrum antibiotic with less effect on the normal intestinal microbiota, was approved in 2011 to treat CDI.

For recurrent or severe cases of CDI that do not resolve with antibiotic therapy, one promising treatment involves replacing the intestinal microbial community. A procedure called fecal transplantation, or fecal microbiota transplant (FMT), introduces feces from a healthy donor into the colon of the patient via an enema, colonoscope, or nasogastric tube (see Focus on a Case 1.1). By doing this, a healthy, normal microbiota is added to the disrupted system, creating a balanced population that can compete with *C. difficile.* Success rates for fecal transplants have been encouraging for treating difficult cases of CDI. Trials in which the fecal sample was delivered in capsule form has a success rate of about 90%; this method has not yet been approved by the FDA, but may be available for use under strict guidelines.

Measures to prevent CDI include minimizing the use of antibiotics, and avoiding transmission of *C. difficile* by handwashing, wearing gloves, and keeping surfaces disinfected. Characteristics of *Clostridium difficile* infection are summarized in **table 24.12.**

TABLE 24.12 *Clostridium difficile* Infection (CDI)

Signs and symptoms	Variable; mild diarrhea to pseudomembranous colitis, a potentially fatal inflammation of the colon
Incubation period	Usually less than a week
Causative agent	*Clostridium difficile,* a Gram-positive, rod-shaped, endospore-forming anaerobe
Pathogenesis	Toxins disrupt host cell actin and cell signaling, causing lethal effects to the intestinal epithelium.
Epidemiology	Primarily occurs in patients on antibiotic therapy.
Treatment and prevention	Treatment: when practical, stop antimicrobial medications; otherwise, an antimicrobial medication that targets *C. difficile.* Fecal transplant to restore normal intestinal microbiota. Prevention: includes handwashing and disinfection of surfaces.

MicroAssessment 24.4

Shigella, Salmonella, various *E. coli* strains, *Vibrio cholerae,* and *Campylobacter jejuni* account for most intestinal bacterial infections. Pathogenic mechanisms of these bacteria include attachment, enterotoxin or cytotoxin production, cell invasion, and destruction of microvilli. Dehydration can be treated with oral rehydration therapy. Typhoid and paratyphoid fevers are systemic, life-threatening diseases. *Clostridium difficile* causes antibiotic-associated diarrhea.

10. Explain how *Vibrio cholerae* causes cholera.

11. How does the epidemiology of *Salmonella* gastroenteritis differ from that of typhoid fever?

12. How would you devise a selective medium for *Vibrio cholerae*?

24.5 ■ Viral Diseases of the Lower Digestive System—Intestinal Tract

Learning Outcome

9. Compare and contrast the diseases caused by rotaviruses and noroviruses.

Viral infections of the lower digestive system are common in all age groups, resulting in millions of cases of gastroenteritis each year in the United States alone.

Rotavirus Gastroenteritis

Most cases of viral gastroenteritis in infants and children around the world are caused by rotaviruses. Before a vaccine became available in 2006, rotavirus infections in the United States resulted in half a million emergency room or clinic visits, and over 55,000 hospital admissions each year. Worldwide, more than 200,000 children still die from rotavirus infection each year.

Signs and Symptoms

After an incubation period of about 24 to 48 hours, rotaviral gastroenteritis begins abruptly with vomiting and slight fever, followed in a short time by profuse, watery diarrhea. Signs and symptoms are generally gone in about a week, but fatal dehydration can occur if fluids are not replaced.

Causative Agent

Rotaviruses are non-enveloped viruses with a double-walled capsid, and a double-stranded, 11-segment RNA genome (**figure 24.15**). The viruses represent a major subgroup of the family *Reoviridae.*

Pathogenesis

Rotaviruses mainly infect the epithelial cells that line the upper part of the small intestine. Infection causes epithelial cell death and decreased production of digestive enzymes. The damaged lining fails to absorb fluids, leading to watery diarrhea. In addition, one of the viral proteins appears to

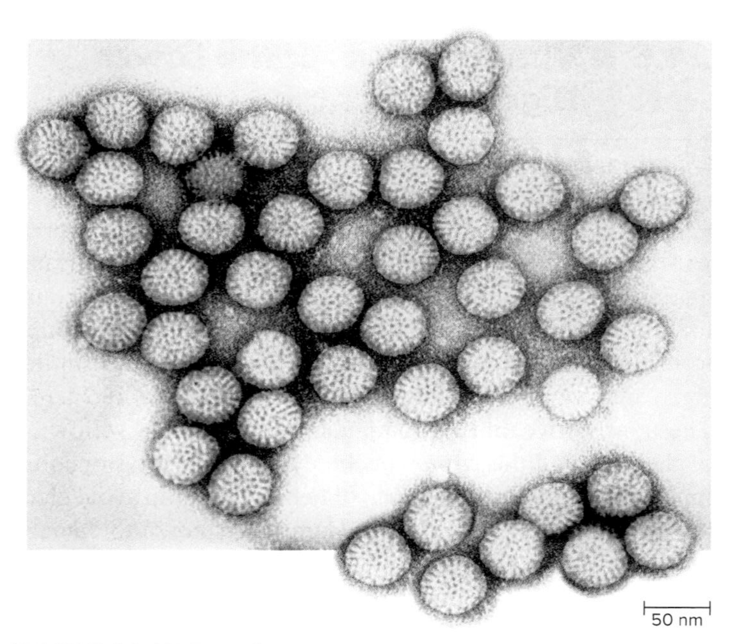

FIGURE 24.15 Rotavirus Color-enhanced electron micrograph shows a distinctive rim of radiating capsomere units that serve as spikes (*rota* means wheel). Source: Dr. Erskine Palmer & Byron Skinner/CDC

What age group is most commonly infected with rotavirus?

function as an enterotoxin, causing fluid secretion in a manner somewhat similar to that of cholera toxin.

Epidemiology

Rotaviruses are transmitted by the fecal-oral route. Childhood epidemics generally occur in winter in temperate climates, probably because children are often indoors in groups where the viruses can spread easily. In regions where vaccination is not common, most children are infected before age 5. The infection results in some immunity, so the second and third rotaviral infections cause much milder diarrhea than the first. Rotaviruses also cause about 25% of traveler's diarrhea cases.

Rotaviruses infect a wide variety of young wild and domestic animals in addition to humans. Rotavirus strains that infect other species typically do not infect humans, but reassortment of genetic segments has been shown to occur when a host contains two different strains of the virus.

Treatment and Prevention

Although there is no direct treatment of rotavirus infection, some infants and small children are hospitalized and given intravenous fluids to prevent dehydration.

Handwashing, disinfectant use, and other sanitary measures help limit the spread of rotaviruses. Attenuated vaccines administered to infants have resulted in a substantial decline in this disease.

Norovirus Gastroenteritis

Noroviruses are the most common cause of viral gastroenteritis in the United States, responsible for an estimated 21 million cases annually. They were originally called "Norwalk viruses," from Norwalk, Ohio, the place where they were first implicated in an epidemic of gastroenteritis. The viruses are designated a Category B bioterrorism agent because they spread easily, with the potential of causing large, demoralizing outbreaks. ◀◀ bioterrorism agents

Signs and Symptoms

Norovirus gastroenteritis usually causes abrupt onset of nausea, vomiting, and watery diarrhea. The incubation period is generally 12 to 48 hours. Vomiting is typically most severe in older children and adults, and generally resolves within the first day or two. Other symptoms take several days to subside. Dehydration may occur.

Causative Agent

Noroviruses are non-enveloped, single-stranded RNA viruses (**figure 24.16**). These viruses represent a group of gastroenteritis-producing viruses within the *Caliciviridae*. Since the mid-2000s, new strains have emerged every 2 to 3 years, sometimes—but not always—causing a significant increase in the number of gastroenteritis cases.

Pathogenesis

Noroviruses infect the epithelium of the upper small intestine, causing epithelial cell death and decreased production of digestive enzymes. The epithelium generally recovers fully within about 2 weeks. For reasons that are not understood, immunity to noroviruses is only short term, lasting just several months. Because of this, individuals can have repeated infections. The virus has only recently been cultivated in

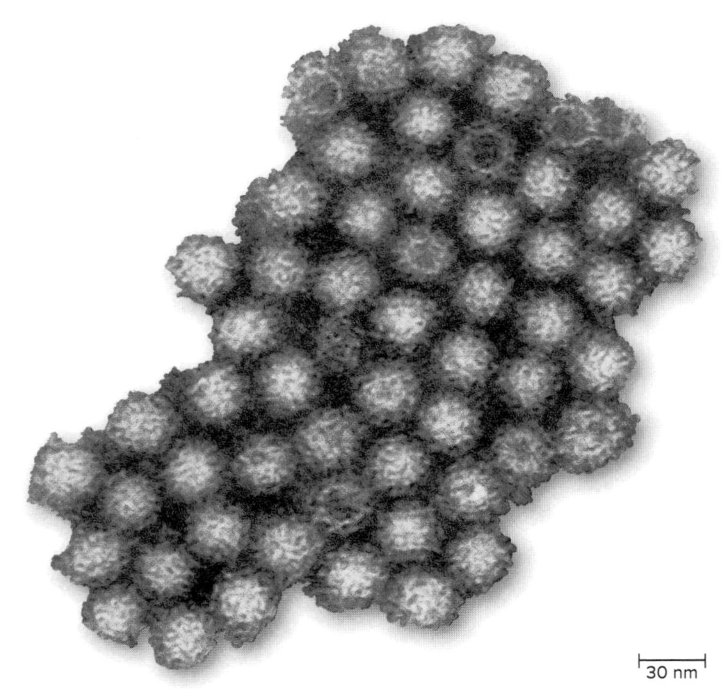

FIGURE 24.16 Norovirus Color-enhanced electron micrograph. Source: Charles D. Humphrey/CDC

Why are norovirus epidemics common in crowded settings such as cruises and college dormitories?

TABLE 24.13 Rotavirus and Norovirus Gastroenteritis Compared

Causative Agent	Rotavirus	Norovirus
Virus characteristics	Double-walled capsid; double-stranded segmented RNA genome	Single-stranded RNA genome
Incubation period	24 to 48 hours	12 to 48 hours
Signs and symptoms	Vomiting, abdominal cramps, diarrhea, lasting 3 to 8 days	Vomiting, abdominal cramps, diarrhea, lasting 1 to 3 days
Epidemiology	Most children infected before age 5; milder disease in older children	Low infectious dose; limited immunity; epidemics common
Prevention	Attenuated vaccine	No vaccine

the laboratory. Now that it has, studying the mechanisms of pathogenesis will be easier, which may lead to new methods of treatment and prevention.

Epidemiology

Transmission of noroviruses is primarily by the fecal-oral route. Vomit also contains infectious viral particles so transmission through aerosols and contaminated surfaces can also occur. The disease is highly contagious due to the low infectious dose of the virus—less than 20 viral particles. The virions are relatively resistant to destruction and are stable in the environment.

Norovirus epidemics are common on cruise ships and in college dormitories. Their easy spread also makes them a concern for healthcare facilities. More than 50% of foodborne disease outbreaks are due to norovirus, so in 2009 the CDC started CaliciNet, a national surveillance network to trace norovirus strains.

Treatment and Prevention

There are no proven anti-noroviral medications. There is no vaccine, and natural immunity to the viruses is short-lived. Thorough handwashing using soap and water is a vital preventive measure. Disinfectants and other sanitary measures also minimize viral transmission. Infected food workers should be restricted from preparing food for 72 hours after their symptoms subside. **Table 24.13** compares noroviruses and rotaviruses.

MicroAssessment 24.5

Rotaviruses are the leading cause of viral gastroenteritis in infants and children, and they also are a common cause of traveler's diarrhea. Noroviruses are the most common cause of viral gastroenteritis. They are highly infectious and easily spread.

13. Why are rotavirus infections common in young children but not in adults?

14. Why is handwashing an important means to control the spread of norovirus?

15. Why might it be more difficult to develop an effective vaccine against noroviruses than against rotaviruses?

24.6 ■ Viral Diseases of the Lower Digestive System—Liver

Learning Outcome

10. Compare and contrast hepatitis A, B, and C.

At least five different viruses can cause **hepatitis,** an inflammation of the liver, but three types—A, B, and C—account for most cases (**figure 24.17**). The viruses are unrelated to one another, but all damage the liver and cause similar signs and symptoms during an acute infection (**table 24.14**). The most noticeable sign is jaundice, which is a yellowing of the skin and the whites of the eyes (see chapter opening image). Patients with any form of hepatitis should avoid alcohol, acetaminophen, and other chemicals known to damage the liver.

Hepatitis A

Hepatitis A, formerly called infectious hepatitis, is found around the world. Introduction of an effective vaccine in 1995 significantly lowered the incidence of the disease.

Signs and Symptoms

Hepatitis A is an acute illness with no known chronic form or carrier state. Older children and adults with the disease usually develop jaundice, fever, fatigue, clay-colored feces, and vomiting after an incubation period of about 1 month. Most young children (younger than 6 years old) and many older children (ages 6 to 14) are asymptomatic. About one in five infected adults requires hospitalization. Patients generally recover within 2 months, but some take up to 6 months.

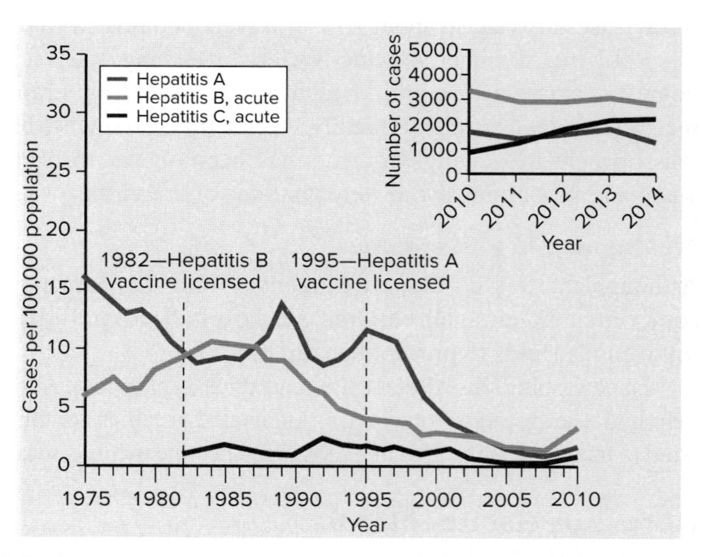

FIGURE 24.17 Incidence of Viral Hepatitis in the United States, 1975–2014

What organ is affected by hepatitis?

TABLE 24.14 Viral Hepatitis

	Hepatitis A	Hepatitis B	Hepatitis C
Causative agent	Non-enveloped, single-stranded RNA picornavirus, HAV	Enveloped, double-stranded DNA hepadnavirus, HBV	Enveloped, single-stranded RNA flavivirus, HCV
Transmission	Fecal-oral	Blood, semen	Blood, possibly semen
Incubation period	3 to 5 weeks (range, 2 to 7 weeks)	10 to 15 weeks (range, 6 to 23 weeks)	6 to 7 weeks (range, 2 to 24 weeks)
Prevention	Inactivated vaccine; immune globulin	Subunit vaccine; hepatitis B immune globulin (HBIG)	No vaccine
Comments	Usually mild symptoms, but often prolonged; full recovery; no chronic carriers; combined hepatitis A and B vaccine available	Acute symptoms often more severe than in hepatitis A; chronic disease can lead to cirrhosis and cancer; chronic carriers; can cross the placenta; combined hepatitis A and B vaccine available	Usually few or no symptoms; progressive liver damage can lead to cirrhosis and cancer; chronic carriers; highly effective treatments are newly available
	Hepatitis D	**Hepatitis E**	
Causative agent	Defective single-stranded RNA virus, HDV	Non-enveloped, single-stranded RNA calicivirus, HEV	
Transmission	Blood, semen	Fecal-oral	
Incubation period	2 to 12 weeks	2 to 6 weeks	
Prevention	No vaccine	No vaccine	
Comments	Prior or concurrent HBV infection necessary; can cause worsening of hepatitis B; can cross the placenta	Similar to hepatitis A, except severe disease in pregnant women	

Causative Agent

Hepatitis A is caused by the hepatitis A virus (HAV), a non-enveloped single-stranded RNA virus of the family *Picornaviridae.* There is one serotype of HAV.

Pathogenesis

Following ingestion, the virus reaches the liver by an unknown route. The liver is the main site of replication and the only tissue known to be damaged by the infection. The virus is released into the bile and eliminated with the feces.

Epidemiology

Hepatitis A virus spreads by the fecal-oral route, principally via fecally contaminated hands, food, or water. Many outbreaks have been traced to restaurants where infected food handlers failed to wash their hands. Raw shellfish are also a frequent source of infection because they concentrate the virus from fecally polluted seawater. Groups at high risk of contracting the disease include children in day-care centers, residents in nursing homes, international travelers, and individuals having sexual contact with an infected person. Because of the long incubation period, HAV can spread widely through a population before being detected (see Focus on a Case 19.1). Infected infants and children can shed the virus in their feces for several months.

Treatment and Prevention

No antiviral treatment for hepatitis A is available, but post-exposure prophylaxis (PEP) can prevent the disease from developing if given within 2 weeks of exposure to the virus. Two PEP options now exist for hepatitis A: passive immunization with immune globulin and induction of active immunity with a vaccine. Passive immunization gives immediate but short-term protection. Vaccination provides long-term protection, but is only useful in patients who will develop an adequate immune response to the vaccine. ⏮ **immune globulin**

An effective vaccine, composed of inactivated HAV, is available and is recommended for all children between 1 and 2 years of age; for older children and adolescents who have not received the vaccine; and for high-risk groups, such as people traveling to areas of high incidence or people in occupations that put them at high risk of exposure. Older people who have not been vaccinated are still at risk, but since the introduction of vaccination, the number of reported cases of hepatitis A has dropped to historic lows. ⏮ **inactivated vaccine**

Hepatitis B

Hepatitis B, formerly known as serum hepatitis, can cause acute or chronic infection. Unlike HAV, which is passed through the fecal-oral route, the hepatitis B virus (HBV) is transmitted through contact with body fluids.

Signs and Symptoms

Signs and symptoms of acute hepatitis B are similar to those of other forms of hepatitis, ranging from asymptomatic to severe. The incubation period varies considerably—from about 2 to 5 months—depending on the dose received. The acute disease is rarely fatal, and the virus is usually cleared within weeks to months of initial symptoms. It can become chronic, particularly in infants and children. One in five people with a chronic infection develop **cirrhosis** (scarring of the liver), liver failure, liver cancer, or other chronic liver disease.

Causative Agent

Hepatitis B is caused by hepatitis B virus (HBV), an enveloped virus that has a mostly double-stranded DNA genome (a portion is single-stranded). Unlike most enveloped viruses, HBV is remarkably resistant to environmental conditions—the virions can still be infectious after a week outside the body. HBV is a member of the family *Hepadnavirus* (*hepa-*, "liver," and *-dna-*, "DNA").

Three components of HBV are notable because they serve as useful markers of infection:

- **Hepatitis B surface antigen (HBsAg).** This envelope protein is produced during viral replication in amounts far in excess of that needed for virion production. The antigen appears in the bloodstream days or weeks after infection, often long before signs of liver damage are evident. Antibodies to HBsAg confer immunity to HBV.
- **Hepatitis B core antigen (HBcAg).** IgM antibodies against this antigen indicate active viral replication.
- **Hepatitis B e antigen (HBeAg).** High levels of this soluble component of the viral core correlate with increased risk of liver damage and increased risk of spreading the disease.

The complete hepatitis B virion is also referred to as a Dane particle (**figure 24.18**). This term provides a contrast to another common viral product of infection: viral envelopes with HBsAg, but lacking DNA. These empty envelopes can be 1,000 times more common in the bloodstream than Dane particles, which is why HBsAg is a useful marker of infection.

Pathogenesis

Following its entry into the body, HBV is carried to the liver by the bloodstream. Surface antigen (HBsAg) allows the virus to attach to and enter host cells.

The next steps of HBV replication are complex, but they help explain why certain antiviral medications can decrease disease symptoms. The replication process begins when the viral genome is transported to the host cell nucleus, and the single-stranded gap filled in (**figure 24.19**). At this point, the genome

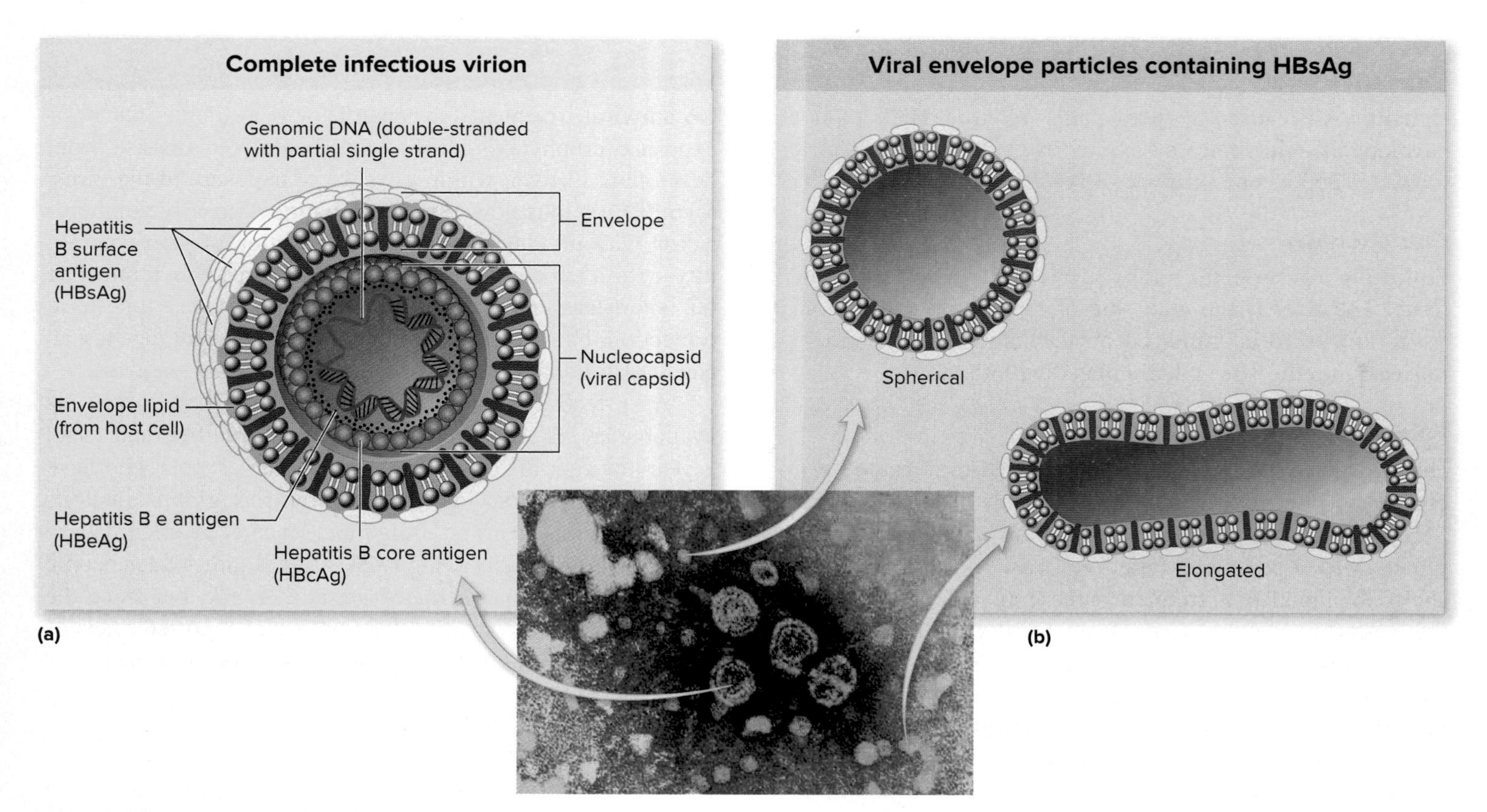

FIGURE 24.18 Hepatitis B Virus Components Found in the Blood of Infected Individuals **(a)** Complete infectious virion also known as a Dane particle. **(b)** Smaller spherical and elongated envelope particles lacking DNA. Source: Dr. Erskine Palmer/CDC

? **What does circulating HBeAg indicate about individuals with chronic HBV infection?**

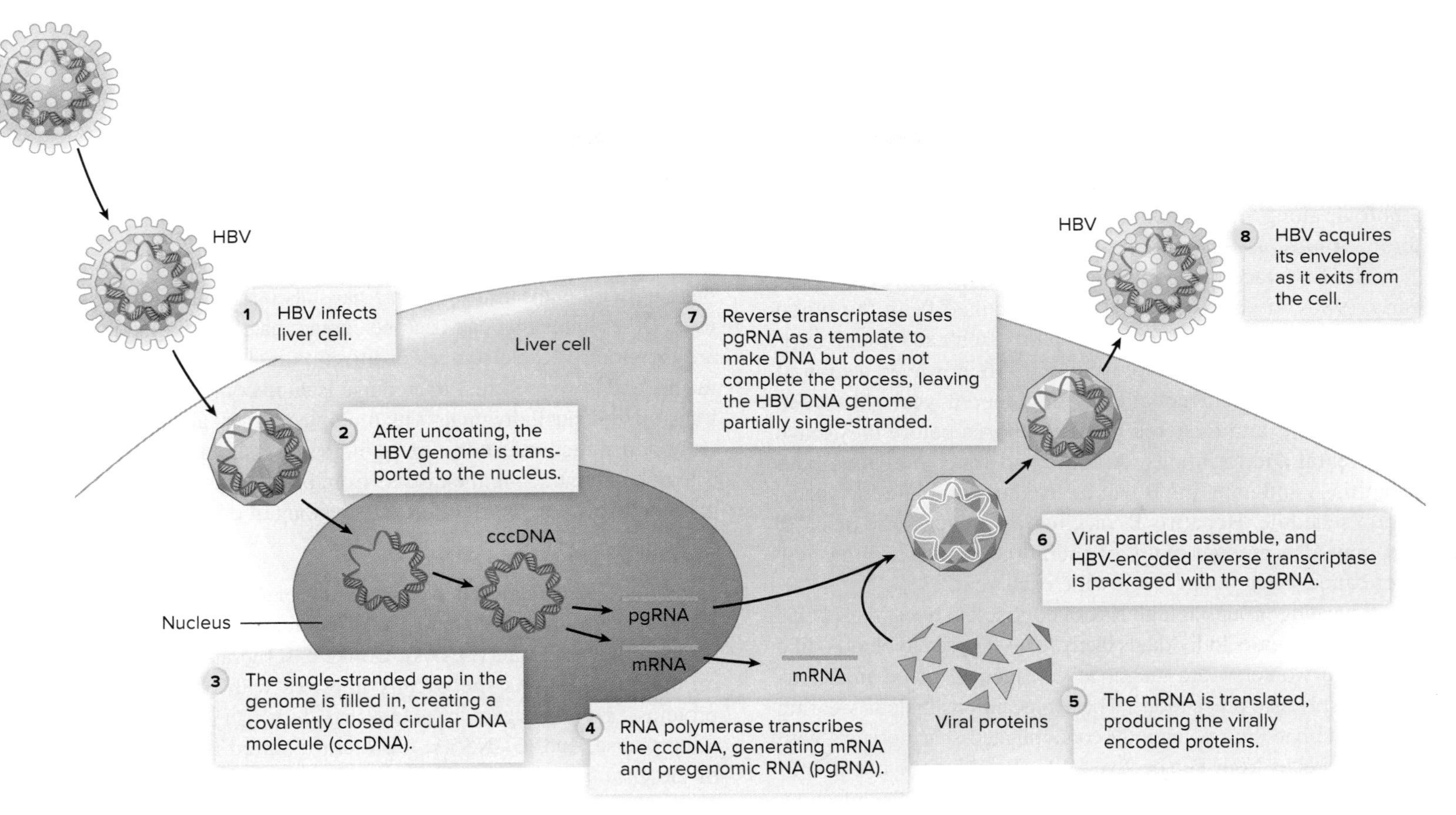

FIGURE 24.19 Replication of Hepatitis B Virus

? With respect to the replication strategy, what unusual feature does HBV have in common with HIV?

is a double-stranded covalently closed circular DNA molecule (cccDNA). A host cell RNA polymerase uses the cccDNA as a template to produce mRNA molecules that are then translated to make the various viral proteins. The significant and unusual aspect of HBV is that an RNA molecule is produced to serve as a template for DNA synthesis; in other words, HBV is a reverse-transcribing virus. The RNA molecule—referred to as pregenomic RNA (pgRNA)—is packaged into the nucleocapsid along with an HBV-encoded **reverse transcriptase.** This enzyme uses the pgRNA as a template to make one strand of DNA; it then uses that DNA molecule as a template to synthesize the complementary strand. The process is not finished inside the virion, however, leaving the HBV genome only partly double-stranded. ⏮ reverse transcriptase

Liver damage from HBV infection is likely due to the cell-mediated immune response, as effector cytotoxic T cells attack infected liver cells. The liver cell destruction leads to cirrhosis. The role of the virus in the development of liver cancer is not well understood, but in most cases of the cancer, the viral DNA has integrated into the host cell genome.

Epidemiology

Hepatitis B can be transmitted in body fluids, such as saliva, blood, blood products, and semen. Activities that mix these fluids from two individuals are considered risk factors—examples include sharing needles, toothbrushes, razors, or towels, and using unsterile tattooing or ear-piercing instruments. Unprotected sex is a particularly risky behavior, with nearly half of new hepatitis B cases in the United States acquired through sexual intercourse.

When HBV infection becomes chronic, the virus continues replicating—circulating in the blood for many years—even if the patient is asymptomatic. These infected individuals are extremely important in the spread of hepatitis B because they are often unaware of their infection. The failure to clear the infection is age related. More than 90% of infants who contract the virus from an infected mother at or shortly after birth develop a chronic infection. In contrast, 25% to 50% of children infected between the ages of 1 and 5, and 6% to 10% of individuals infected as older children or adults, will develop a chronic infection.

A progressive rise in hepatitis B cases was reported in the United States from 1965 to the mid-1980s. Since that time, the HBV vaccine has lowered the incidence of the disease. In recent years, the CDC estimates 19,000 new acute HBV infections have occurred annually, and about 1 million people in the United States are chronically infected.

Treatment and Prevention

No curative antiviral treatment currently exists for hepatitis B, but post-exposure prophylaxis (PEP) with hepatitis B immune globulin (HBIG) and vaccine can offer protection. Also, some chronically infected patients show remarkable improvement when given certain reverse transcriptase inhibitors such as tenofovir along with injections of genetically engineered interferon. ⏮ reverse transcriptase inhibitors, ⏮ interferon, ⏮ hyperimmune globulin

The first vaccine against hepatitis B was approved in the early 1980s. It consisted of HBsAg obtained from the blood of chronic carriers; 1 mL of their blood often had enough antigen to immunize eight people! Since 1986, however, a subunit vaccine produced in genetically engineered yeast has been available. It is administered to all newborns before they leave the hospital and to infants. A combined vaccine against both hepatitis A and hepatitis B can be used to complete the vaccination schedule. Hepatitis B vaccination prevents the disease, and may also prevent many of the 500,000 to 1 million new liver cancers that occur worldwide each year.

Educating groups at high risk of contracting hepatitis B helps prevent the disease. Individuals likely to be exposed to blood—like healthcare workers—are taught to consider all blood infectious and use universal precautions. These precautions include wearing gloves and handling potentially contaminated sharp objects with care (see Focus Your Perspective 19.1). Teaching the importance and proper use of condoms also helps limit spread of the infection.

Hepatitis C

Hepatitis C is the most common chronic blood-borne infection in the United States. Some estimate that more than 3 million Americans are infected with HCV.

Signs and Symptoms

The symptoms of hepatitis C are similar to those of hepatitis A and B except they are generally milder. The incubation period averages about 6 weeks (range, 2 weeks to 6 months). About 65% of infected individuals have no symptoms of acute infection; only about 25% have jaundice. In many cases, the infections become chronic. Before effective treatments were available, cirrhosis and liver cancer developed in 10% to 20% of patients.

Causative Agent

HCV is an enveloped, single-stranded RNA virus of the family *Flaviviridae,* first cultivated in vitro in 2005. There is considerable genetic variability; types and subtypes differ in pathogenicity as well as in response to treatment.

Pathogenesis

HCV infection generally occurs from exposure to contaminated blood. Within the host cell, the viral genome is translated as a single polyprotein that is then cleaved by HCV protease into multiple proteins. Although most people lack symptoms with acute infection, more than 80% develop chronic infections. The virus infects the liver and triggers inflammatory and immune responses. After that, the disease process starts and stops—at times the liver seems to return to normal, then, weeks or months later, shows marked inflammation.

Epidemiology

Although HCV is transmitted in blood, the mechanism of exposure is not always obvious. Approximately 50% of infections in the United States are due to sharing of syringes. Tattoos and body piercing with unclean instruments have also transmitted the disease. Other items that can become contaminated with blood and are therefore possible sources of infection include toothbrushes, razors, and towels. Sexual intercourse is an uncommon means of transmission, but individuals with multiple partners and other sexually transmitted infections are at increased risk for the disease. The risk of contracting the disease from blood transfusion is now extremely low because of effective screening of donated blood.

Treatment and Prevention

Hepatitis C treatment has been revolutionized by the development of a group of new anti-HCV medications referred to as direct acting antivirals (DAAs). Unlike past therapies that relied on interferon, DAAs specifically interfere with HCV-encoded proteins. HCV targets of DAAs include: (1) polymerase, (2) protease, and (3) NS5A, a protein required for replication of the viral genome. Specific combinations of DAAs are taken orally for treatment, some of which are available in fixed-dose tablets. For example, one formulation includes sofosbuvir (a nucleotide analog) along with velpatasvir (an NS5A inhibitor); another includes elbasvir (an NS5A inhibitor) and grazoprevir (a protease inhibitor). Different treatment combinations are related to differences in HCV genotype. In early studies, the new treatments resulted in greater than 97% cure rate.

No vaccine is available for preventing hepatitis C. However, physicians often recommend vaccination against hepatitis A and B to help prevent a combination of infections that might severely damage the liver.

MicroAssessment 24.6

Hepatitis viruses are a diverse, unrelated group that cause liver inflammation. Most cases of hepatitis are caused by hepatitis viruses A, B, or C. Hepatitis A is transmitted by the fecal-oral route. Hepatitis B, transmitted by exposure to body fluids, may be acute or chronic. A combination hepatitis A and hepatitis B vaccine is available. Hepatitis C is transmitted by blood and occasionally by sexual intercourse; up to 80% of infections become chronic. Chronic hepatitis often results in cirrhosis and liver cancer.

16. What behaviors might transmit hepatitis C?

17. At what stage in the replication of hepatitis B does a reverse transcriptase inhibitor act?

18. Why is it theoretically possible to immunize multiple people from the blood of one hepatitis B patient?

24.7 ■ Protozoan Diseases of the Lower Digestive System

Learning Outcome

11. Compare and contrast giardiasis, cryptosporidiosis, cyclosporiasis, and amebiasis.

Protozoa—single-celled eukaryotes—are important causes of human intestinal disease. Intestinal protozoan pathogens are all transmitted by the fecal-oral route. ⏮ protozoa

Giardiasis

Giardiasis is one of the most commonly identified waterborne illness in the United States. It can be contracted from clear mountain streams, chlorinated city water that has not been filtered, and person-to-person contact. The disease has worldwide distribution and is responsible for many cases of traveler's diarrhea.

Signs and Symptoms

In epidemics of giardiasis, about two-thirds of exposed individuals develop symptoms. The incubation period is generally 6 to 20 days. Symptoms can range from mild (indigestion, "gas," and nausea) to severe (vomiting, explosive diarrhea, abdominal cramps, fatigue, and weight loss). The symptoms usually end without treatment in 1 to 4 weeks, but some cases become chronic. Both symptomatic and asymptomatic persons can become long-term carriers, unknowingly excreting infectious cysts with their feces.

Causative Agent

Giardia lamblia is a flagellated protozoan shaped like a pear cut lengthwise. It has two side-by-side nuclei that resemble eyes and an adhesive disc on its undersurface that together give the organism a distinctive appearance (**figure 24.20**). It can exist in two forms: a growing, feeding **trophozoite** and a dormant **cyst.** The cysts have thick walls composed of a tough, flexible, chitin-like polysaccharide that protects the organism from harsh environmental conditions (see Focus on a Case 12.1). ⏮ chitin

Because *G. lamblia* lacks mitochondria, it was thought to have evolved before eukaryotes acquired these organelles. More recent evidence appears to refute this idea because the protozoan contains atypical energy-metabolizing structures called mitosomes that likely evolved from mitochondria.

Pathogenesis

The cysts of *G. lamblia* are infectious because, unlike the trophozoites, they are resistant to stomach acid. From each cyst that reaches the upper part of the small intestine, two trophozoites emerge. Some of these attach to the epithelium by their adhesive disc; others use their flagella to move freely in the intestinal mucus. Some may even migrate up the bile duct to the gallbladder and cause cramping or jaundice.

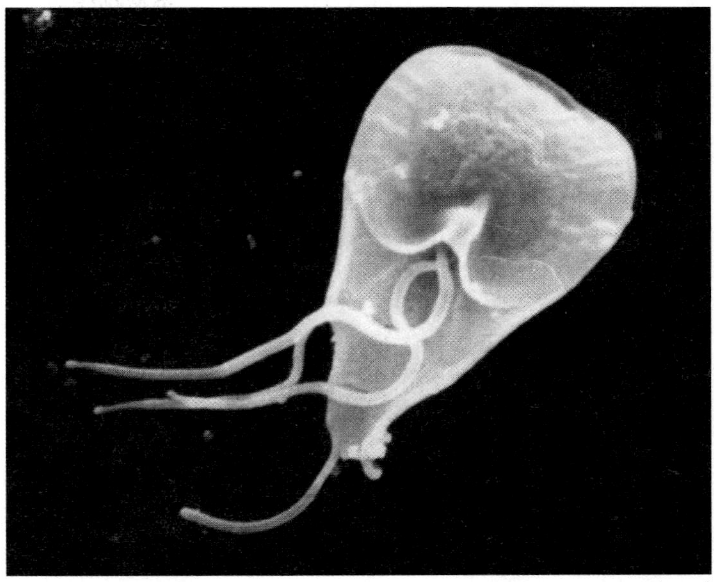

FIGURE 24.20 ***Giardia lamblia*** Scanning electron micrograph (SEM). Source: Janice Haney Carr/CDC

? Does this image show a cyst or a trophozoite?

The trophozoites interfere with the intestine's ability to absorb nutrients and secrete digestive enzymes. The result is bulky feces containing fat, excessive intestinal gas from bacterial digestion of unabsorbed nutrients, and malnutrition. Some intestinal impairment may be the result of the immune response to the parasite.

When trophozoites detach from the epithelium, they are carried by intestinal contents toward the large intestine. If the transit time is long enough, they develop into cysts. Thus, a person who has formed stools is more likely to eliminate cysts in the feces, whereas a person with diarrhea is apt to eliminate trophozoites (**figure 24.21**).

Epidemiology

Giardia lamblia is transmitted by the fecal-oral route and spreads easily because only about 10 cysts are required to establish infection. Water contaminated with human feces is a common source of infection, but feces from animals such as beavers, raccoons, dogs, and cats can also be implicated. The cysts are infectious and can remain viable in cold water for more than 2 months. Hikers who drink from streams, even in remote areas thought to contain safe water, are at risk of contracting giardiasis.

Although waterborne outbreaks are the most common, person-to-person contact can also transmit the disease. This is especially likely in day-care centers where workers' hands become contaminated while changing diapers. Sexual

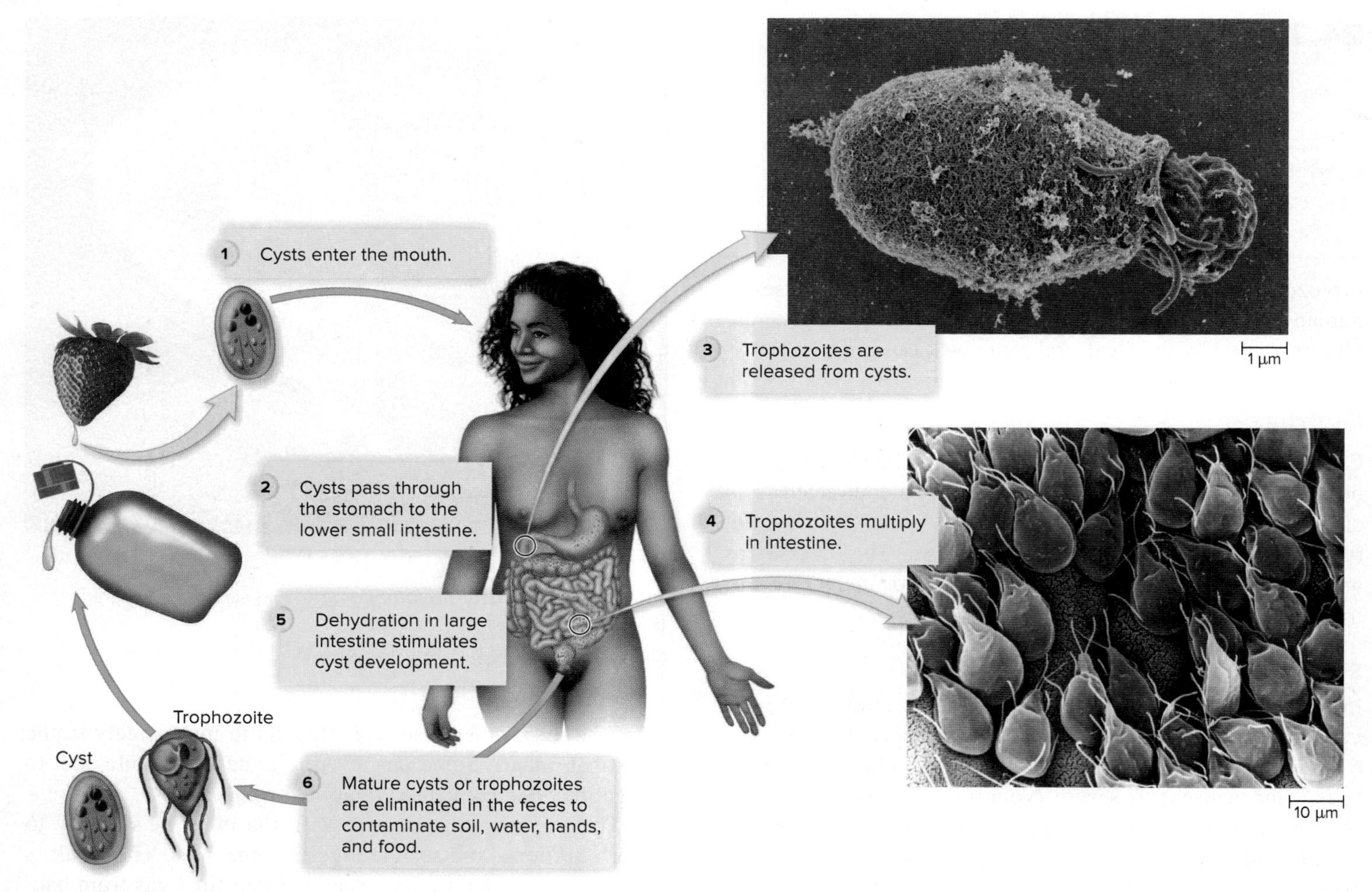

FIGURE 24.21 Life Cycle of *Giardia lamblia* (3): Dr. Stan Erlandsen/CDC; (4): Dr. Stan Erlandsen/CDC

? Why is an infected person with diarrhea less likely to transmit giardiasis than one who has formed stools?

practices that lead to oral-anal contact can also transmit giardiasis. Transmission by fecally contaminated food has also been reported. Good personal hygiene, especially handwashing, decreases the chance of spreading the infection.

MicroByte

A single human stool can contain 300 million *G. lamblia* cysts.

Treatment and Prevention

Several medicines can be used to treat giardiasis, including tinidazole, metronidazole (Flagyl), and the newest option—nitazoxanide.

The level of chlorine used to treat municipal water supplies does not destroy *Giardia* cysts, so the water is generally filtered to remove them. For hikers, the best way to make drinking water safe from giardiasis is to boil it for 1 minute or use a portable filter with a pore size of 1 µm or smaller. Other methods, including adding commercial water-purifying tablets, tincture of iodine, or household bleach are time-consuming and much less reliable. As in all chemical microbial control procedures, time and temperature are important. Only an hour may be necessary to treat warm water, but many hours are required to kill cysts in cold water. **Table 24.15** describes the main features of giardiasis.

Cryptosporidiosis ("Crypto")

Cryptosporidiosis, commonly called "crypto," is another leading cause of waterborne disease in the United States. It was first identified as a threat to humans when the AIDS epidemic struck in the early 1980s, but the disease is a health risk not only to those with immunodeficiency but also to the public at large. A 1993 waterborne outbreak in Milwaukee affected more than 400,000 people!

Signs and Symptoms

Signs and symptoms of cryptosporidiosis include fever, loss of appetite, nausea, abdominal cramps, and profuse watery diarrhea. These begin after an incubation period of 4 to 12 days. The symptoms generally last 10 to 14 days, but in people with immunodeficiency diseases, they can last for months and be life-threatening.

TABLE 24.15 Giardiasis

Signs and symptoms	Mild: indigestion, intestinal gas, nausea. Severe: vomiting, diarrhea, abdominal cramps, weight loss
Incubation period	6 to 20 days
Causative agent	*Giardia lamblia*, a flagellated pear-shaped protozoan with two nuclei
Pathogenesis	Ingested cysts survive stomach passage; trophozoites emerge from cysts in the small intestine, where some attach to epithelium; mucosal function is impaired by adherent protozoa and immune response.
Epidemiology	Ingestion of fecally contaminated water; low infectious dose; person-to-person spread.
Treatment and prevention	Treatment: several antimicrobial medication options. Prevention: boiling, filtering, or disinfecting drinking water.

Causative Agent

Cryptosporidiosis is caused by *Cryptosporidium parvum,* an apicomplexan. The organism multiplies intracellularly in the small intestinal epithelium. Unlike other Apicomplexa, its entire life cycle occurs in a single host. The oocyst stage is an acid-fast sphere that contains four banana-shaped sporozoites (**figure 24.22**). ⏮ **Apicomplexa,** ⏮ **acid-fast**

Pathogenesis

The digestive fluids of the small intestine release sporozoites from ingested oocysts. The sporozoites then invade the epithelial cells of the small intestine, altering the epithelium and intestinal villi and causing inflammation. Water and electrolyte secretion increases, and nutrient absorption decreases. Cell-mediated immunity is important in controlling the infection.

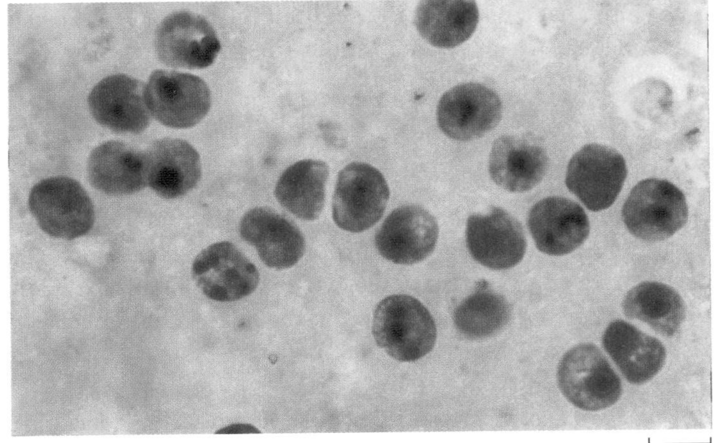

FIGURE 24.22 Oocysts of *Cryptosporidium parvum* Feces contain oocysts. Person-to-person and waterborne transmission are common. ©Michael Abbey/Science Source

? Why is person-to-person transmission more likely with *Cryptosporidium* infection than with *Cyclospora cayetanensis*?

Epidemiology

Oocysts of *Cryptosporidium parvum* passed in feces are immediately infectious. The infectious dose is as few as 10, so person-to-person spread readily occurs with poor sanitation. The organism is responsible for many cases of traveler's diarrhea.

Infected individuals often have prolonged diarrhea and can continue to eliminate infectious oocysts for 2 weeks or more after the diarrhea ceases. These oocysts can survive for up to 6 months in food and water, and are even more resistant to chlorine than *Giardia* cysts are. They are too small to be removed from drinking water by some filtration methods.

C. parvum is particularly difficult to control because it has a wide host range, infecting domestic animals such as dogs, pigs, and cattle. Feces from these animals, as well as from humans, can contaminate food and drinking water. Epidemics have arisen from drinking water, swimming pools, a water slide, a zoo fountain, day-care centers, unpasteurized apple juice, and other food and drink.

Treatment and Prevention

Nitazoxinide may be prescribed for individuals with healthy immune systems; the effectiveness of the medication for immunocompromised patients has not been established. Most otherwise healthy people, however, will recover without treatment.

Effective measures for preventing the disease include sanitary disposal of human and animal feces and treating water using filtration, ultraviolet radiation, or ozone. Pasteurizing liquids for consumption is also an important control measure. All food handlers should wash their hands with soap and water regularly, and those with diarrhea should not handle food until symptom-free. Immunodeficient individuals are advised to avoid contact with animals, to boil or filter drinking water using a 1 μm or smaller pore size filter, and to avoid recreational water activities. The main features of cryptosporidiosis are shown in **table 24.16.**

Cyclosporiasis

Cyclosporiasis first came to medical attention in the late 1980s, with widely scattered epidemics of severe diarrhea. For several years, the causative organism was known only as a "cyanobacterium-like body." Later, the bodies were shown to be the oocysts of a protozoan, *Cyclospora cayetanensis.* ⏮ **cyanobacteria**

Signs and Symptoms

Cyclosporiasis begins after an incubation period of about 1 week, with fatigue, loss of appetite, slight fever, vomiting, and watery diarrhea, followed by weight loss. The diarrhea usually subsides in 3 to 4 days, but relapses occur for up to 4 weeks.

TABLE 24.16 Cryptosporidiosis

Signs and symptoms	Fever, loss of appetite, nausea, abdominal cramps, watery diarrhea
Incubation period	4 to 12 days
Causative agent	*Cryptosporidium parvum,* an apicomplexan. Its life cycle takes place entirely within the epithelial cells of the small intestine.
Pathogenesis	Ingested oocysts release sporozoites in the intestine. The sporozoites invade epithelial cells, causing an inflammatory response. Secretion of water and electrolytes increases; absorption of nutrients decreases.
Epidemiology	Oocysts of *C. parvum* are immediately infectious. The infectious dose is low, so person-to-person spread occurs easily. Infected individuals can discharge the oocysts for weeks after symptoms subside. The oocysts resist chlorination, and pass through many municipal water filtration systems. Wide host range.
Treatment and prevention	Treatment: antimicrobial medication. Prevention: pasteurization of beverages, boiling or filtering drinking water. Sanitary disposal of human and animal feces.

Causative Agent

Cyclospora cayetanensis is an apicomplexan, and its oocysts are similar to those of *Cryptosporidium parvum* but larger. However, *C. cayetanensis* oocysts are not yet infectious when passed in the feces. With favorable conditions outside the body, sporocysts containing sporozoites develop within the infectious oocyst.

Pathogenesis

Little is known about the pathogenesis of *Cyclospora cayetanensis* because there is no good animal model for the disease. No hosts other than humans are known. Analysis of biopsies from infected patients confirms that sexual and asexual forms of the protozoan are both present in the intestinal epithelium.

Epidemiology

The oocysts of *Cyclospora cayetanensis* are immature and not infectious when eliminated in the stool, so person-to-person spread does not occur. In temperate regions, most infections happen in spring and summer months and among travelers to tropical areas. This is likely because warm, moist conditions favor maturation of the oocysts. Fresh produce (raspberries) has been implicated in a number of epidemics. In most instances, the produce was imported from a tropical region, but the source of the contaminating organisms is unknown. *C. cayetanensis* has been found in natural waters, but the source could not be determined.

TABLE 24.17 Cyclosporiasis

Signs and symptoms	Fatigue, loss of appetite, vomiting, watery diarrhea, and weight loss. Symptoms improve in 3 to 4 days, but relapses can occur for up to a month.
Incubation period	Usually about 1 week (range, 1 to 12 days)
Causative agent	*Cyclospora cayetanensis,* an apicomplexan
Pathogenesis	Little is known. Biopsies show sexual and asexual stages in the intestinal epithelium.
Epidemiology	The oocysts of *C. cayetanensis* are not infectious when discharged in the feces, and so person-to-person spread does not occur. Travelers to tropical countries are at risk of infection. Produce, especially raspberries imported from tropical Central America, has been implicated in most North American outbreaks.
Treatment and prevention	Treatment: antimicrobial medication. Prevention: use of boiled or filtered drinking water is advised in the tropics. Thorough washing of imported fruits and vegetables.

Treatment and Prevention

Co-trimoxazole (trimethoprim plus sulfamethoxazole) effectively treats most cases of cyclosporiasis.

No specific preventive measures are available. People should boil or filter drinking water and thoroughly wash produce such as berries and leafy vegetables during an outbreak. The main features of cyclosporiasis are presented in **table 24.17.**

Amebiasis

Amebiasis is caused by *Entamoeba histolytica,* a parasite that is believed to infect up to 50 million people worldwide, mostly in developing countries. Although it usually results in asymptomatic or mild disease, life-threatening disease can occur in areas where virulent strains spread easily in crowded, unsanitary living conditions.

Signs and Symptoms

Patients with *Entamoeba histolytica* infections are commonly asymptomatic. Some patients suffer from chronic, mild diarrhea, however, lasting months or years. Acute dysentery occurs in the most severe cases, which can be fatal. The incubation period varies from a few days to a few months.

Causative Agent

Entamoeba histolytica forms cysts that have a chitin-containing wall. Mature cysts—the infectious form for the next host—have four nuclei (**figure 24.23**).

Pathogenesis

Ingested cysts of *E. histolytica* survive passage through the stomach. The cysts then excyst, releasing trophozoites. Upon

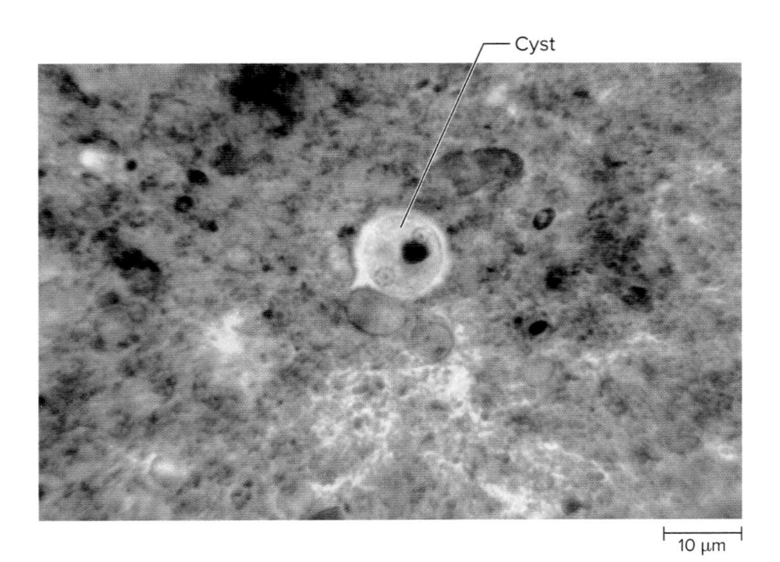

FIGURE 24.23 **Cyst of *Entamoeba histolytica*** Source: CDC

Why is the organism transmitted in the cyst form and not as a trophozoite?

reaching the large intestine, these trophozoites begin feeding on mucus and intestinal bacteria. The irritating effect of the trophozoites on the cells causes intestinal cramps and diarrhea. Many strains produce a cytotoxic enzyme that kills intestinal epithelium on contact. The organisms may also invade the intestinal lining, causing ulcerations and bloody diarrhea, a condition referred to as **amebic dysentery.** Sometimes the invading amoebae penetrate deeper, entering blood vessels that carry them to the liver or other organs. This leads to the development of amebic abscesses in those organs, a result of tissue damage caused as the organisms multiply.

Epidemiology

Entamoeba histolytica has a low infectious dose, spreads by the fecal-oral route, and is distributed worldwide. The disease is more common in tropical areas where sanitation is poor. In the United States, cases occur mainly in poverty-stricken areas, and among migrant farm workers and men who have sex with men. Humans are the only important reservoir.

TABLE 24.18 Amebiasis

Signs and symptoms	Diarrhea, abdominal pain, blood in feces
Incubation period	2 days to several months
Causative agent	*Entamoeba histolytica*
Pathogenesis	Ingested cysts excyst, releasing trophozoites; these feed on mucus and bacteria in the large intestine; cytotoxic enzyme kills intestinal cells. The trophozoites may penetrate the intestinal wall and are sometimes carried to the liver and other organs, resulting in abscesses.
Epidemiology	Ingestion of fecally contaminated food or water; disease associated with poverty, migrant workers, and men who have sex with men.
Treatment and prevention	Treatment: antimicrobial medication. Prevention: good sanitation and personal hygiene.

Treatment and Prevention

Medications such as metronidazole and paromomycin are available for treatment. Prevention of amebiasis depends on sanitary measures and avoiding fecal contamination of foods and drinking water. **Table 24.18** gives the main features of amebiasis.

The key features of the diseases covered in this chapter are highlighted in the **Diseases in Review 24.1** table.

MicroAssessment 24.7

Giardiasis and cryptosporidiosis are common waterborne diseases that can also be transmitted person to person. Cyclosporiasis is a similar disease, but is not transmissible person to person. Amebiasis is caused by an amoeba that ulcerates the large intestinal epithelium, resulting in dysentery.

19. What causes the excessive intestinal gas that characterizes giardiasis?

20. Why is cryptosporidiosis difficult to control?

21. How could the fact that *Cryptosporidium parvum* is acid-fast be used in diagnosis?

Diseases in Review 24.1

Digestive System Diseases

Disease	Causative Agent	Comment	Summary Table
BACTERIAL INFECTIONS OF THE UPPER DIGESTIVE SYSTEM			
Dental caries (tooth decay)	*Streptococcus mutans*	Cariogenic organisms adhere to teeth and produce acids that damage tooth surfaces; incidence correlates with sugar consumption and lack of preventive dental care.	Table 24.1
Periodontal disease	Certain groups of Gram-negative anaerobes	Gingivitis is the result of the inflammatory response to plaque and tartar at the gum line; chronic periodontitis damages the structures that support the teeth, leading to tooth loss.	Table 24.1
Acute necrotizing ulcerative gingivitis (ANUG)	*Treponema* sp. with other anaerobes	Characterized by painful, bleeding gums, abscessed and broken teeth, and extremely bad breath; associated with poor oral hygiene, particularly in combination with other stresses.	Table 24.1
***Helicobacter pylori* gastritis**	*Helicobacter pylori*	Causes peptic ulcers; associated with gastric cancers; *H. pylori* withstands stomach acid by producing urease.	Table 24.2
VIRAL INFECTIONS OF THE UPPER DIGESTIVE SYSTEM			
Oral herpes simplex (cold sores)	Herpes simplex virus (usually type 1)	Initial infection typically occurs during childhood; latent virus reactivates, causing recurrent cold sores.	Table 24.3
Mumps	Mumps virus	Characterized by painful swelling of the parotid glands; virus enters via the respiratory tract and then spreads to glands; can cause infertility in males; preventable by vaccination.	Table 24.4
BACTERIAL INFECTIONS OF THE LOWER DIGESTIVE SYSTEM			
Cholera	*Vibrio cholerae*	Classic example of watery diarrhea; *V. cholerae* colonizes small intestine and produces cholera toxin, causing secretion of water and electrolytes; rehydration is critical.	Table 24.5
Shigellosis	*Shigella* species	Classic example of bacterial dysentery; *Shigella* species invade cells of large intestine; low infectious dose; some species produce Shiga toxin.	Table 24.6
***Escherichia coli* gastroenteritis**	Certain strains of *E. coli*	Various symptoms depending on the pathovar; STEC strains cause bloody diarrhea and HUS; symptoms of ETEC are similar to those of cholera; symptoms of EIEC are similar to those of shigellosis.	Table 24.7, Table 24.8
***Salmonella* gastroenteritis**	*Salmonella enterica* (not serotypes Typhi or Paratyphi)	Foodborne, particularly poultry	Table 24.9
Typhoid and paratyphoid fevers	*Salmonella enterica* serotypes Typhi and Paratyphi	*Salmonella* serotypes Typhi and Paratyphi cross intestinal mucosa and invade bloodstream, causing life-threatening enteric fever; vaccine for typhoid fever is available.	Table 24.10
Campylobacteriosis	*Campylobacter jejuni*	Foodborne, particularly poultry	Table 24.11
***Clostridium difficile* infection (CDI)**	*Clostridium difficile*	Typically occurs only in patients on antibiotic therapy; most common cause of diarrhea in healthcare settings.	Table 24.12
VIRAL INFECTIONS OF THE LOWER DIGESTIVE SYSTEM—INTESTINAL TRACT			
Rotavirus gastroenteritis	Rotaviruses	Most common cause of viral gastroenteritis in infants and children in United States; preventable by vaccination.	Table 24.13
Norovirus gastroenteritis	Noroviruses	Most common cause of viral gastroenteritis in the United States; highly infectious; epidemics are common on cruise ships and college dormitories.	Table 24.13
VIRAL INFECTIONS OF THE LOWER DIGESTIVE SYSTEM—LIVER			
Hepatitis A	Hepatitis A virus (HAV)	Spreads via fecal-oral route; usually mild symptoms; no chronic carriers; preventable by vaccination.	Table 24.14
Hepatitis B	Hepatitis B virus (HBV)	Spreads in blood and semen; reverse transcribing virus; chronic infections can result in cirrhosis and liver cancer; chronic carriers; preventable by vaccination.	Table 24.14
Hepatitis C	Hepatitis C virus (HCV)	Spreads in blood, possibly semen; chronic infections can result in cirrhosis and liver cancer; chronic carriers.	Table 24.14
PROTOZOAN INFECTIONS OF THE LOWER DIGESTIVE SYSTEM			
Giardiasis	*Giardia lamblia*	Common waterborne illness in the United States; low infectious dose; cysts are chlorine-resistant.	Table 24.15
Cryptosporidiosis	*Cryptosporidium parvum*	Wide host range; oocysts are chlorine-resistant; low infectious dose.	Table 24.16
Cyclosporiasis	*Cyclospora cayetanensis*	No person-to-person spread because oocysts must mature in the environment to become infectious; imported produce has been implicated in outbreaks.	Table 24.17
Amebiasis	*Entamoeba histolytica*	Most common in tropical regions where sanitation is poor; low infectious dose; amebic dysentery results from intestinal damage.	Table 24.18

Summary

24.1 ■ Anatomy, Physiology, and Ecology of the Digestive System

The digestive system encompasses the digestive tract and accessory organs (figure 24.1).

The Upper Digestive System

The upper digestive system includes the mouth and salivary glands, the esophagus, and the stomach. Bacteria grow on teeth, attached to the pellicle that adheres to the enamel (figure 24.2), forming a biofilm called **dental plaque.** Peristalsis in the esophagus propels microbes to the stomach. The acidity and enzymes in the stomach destroy most bacterial cells.

The Lower Digestive System

The lower digestive system includes the small and large intestines, as well as the pancreas and the liver. As the stomach contents enter the small intestine, digestive fluids from the pancreas and the liver are mixed in. **Villi** and **microvilli** increase the surface area for absorption in the small intestine. The large intestine absorbs remaining nutrients and more water; feces are eliminated. Microbes make up about one-third of the weight of feces. The liver produces **bile,** which is stored in the gallbladder.

UPPER DIGESTIVE SYSTEM INFECTIONS

24.2 ■ Bacterial Diseases of the Upper Digestive System

Dental Caries (Tooth Decay) (table 24.1)

Plaque that contains *Streptococcus mutans* or other cariogenic microbes can act as a tiny acid-soaked sponge applied closely to the tooth (figures 24.3, 24.4). Control of dental caries depends mainly on restricting dietary sucrose, supplying fluoride, flossing, and brushing teeth.

Periodontal Disease (table 24.1)

Periodontal disease is caused by an inflammatory response to the plaque bacteria at the gum line; it is an important cause of tooth loss in older people (figure 24.3).

Acute Necrotizing Ulcerative Gingivitis (table 24.1)

Acute necrotizing ulcerative gingivitis (ANUG) can occur at any age in association with poor mouth care (figure 24.5).

Helicobacter pylori Gastritis (table 24.2)

Helicobacter pylori predisposes the stomach and the uppermost part of the small intestine to peptic ulcers (figures 24.6). Treatment with antimicrobial medications can cure the infection and prevent peptic ulcer recurrence.

24.3 ■ Viral Diseases of the Upper Digestive System

Oral Herpes Simplex (Cold Sores) (table 24.3; figure 24.7)

Herpes simplex is caused by an enveloped DNA virus. HSV persists as a latent infection inside sensory nerves; active disease occurs when the body is stressed.

Mumps (table 24.4; figure 24.8)

Mumps is caused by an enveloped RNA virus that infects the parotid glands and a variety of other body tissues. Mumps virus generally causes more severe disease in persons beyond the age of puberty; it can be prevented using an attenuated vaccine (figure 24.9).

LOWER DIGESTIVE SYSTEM INFECTIONS

Focus on Diarrheal Diseases

Bacterial infections of the small intestine typically cause watery diarrhea, whereas invasion of the large intestine causes **dysentery.** The bacteria are transmitted via the fecal-oral route, often through contaminated foods and water. Intestinal pathogens that have a low infectious dose can be transmitted by direct contact as well. Dehydration can be treated with **oral rehydration therapy (ORT),** using oral rehydration salts (ORS).

24.4 ■ Bacterial Diseases of the Lower Digestive System

Cholera (table 24.5)

Cholera is a severe form of watery diarrhea caused by a toxin of *Vibrio cholerae* that acts on the small intestinal epithelium (figure 24.10).

Shigellosis (table 24.6)

Shigella species invade the epithelium of the large intestine, causing dysentery (figure 24.11). *Shigella dysenteriae* produces Shiga toxin, which causes **hemolytic uremic syndrome (HUS).**

Escherichia coli Gastroenteritis (tables 24.7, 24.8; figure 24.12)

Escherichia coli strains that cause intestinal disease can be grouped into six pathovars: STEC (or EHEC), ETEC, EIEC, EPEC, EAEC, and DAEC. STEC strains cause hemolytic uremic syndrome.

Salmonella Gastroenteritis (table 24.9)

Most cases of *Salmonella* gastroenteritis originate from animals. The organisms are often foodborne, commonly in poultry and eggs.

Typhoid and Paratyphoid Fevers (table 24.10)

Typhoid fever is caused by *Salmonella* serotype Typhi, and paratyphoid fever is caused by *Salmonella* serotype Paratyphi. Both of these infect only humans. The diseases are characterized by high fever, headache, and abdominal pain. Untreated, they can be fatal. Vaccination helps prevent typhoid fever.

Campylobacteriosis (table 24.11)

Campylobacter jejuni is the most common bacterial cause of diarrhea in the United States; it usually originates from domestic animals (figure 24.13).

Clostridium difficile Infection (CDI) (table 24.12)

Clostridium difficile causes diarrhea and **pseudomembranous colitis,** primarily in patients on antibiotic therapy (figure 24.14).

24.5 ■ Viral Diseases of the Lower Digestive System—Intestinal Tract

Rotavirus Gastroenteritis (table 24.13)

Rotaviral gastroenteritis is the main diarrheal illness of infants and young children but also can involve adults, as in traveler's diarrhea. Rotaviruses are segmented RNA viruses (figure 24.15).

Norovirus Gastroenteritis (table 24.13)

Norovirus is the most common cause of viral gastroenteritis in the United States. Noroviruses are RNA viruses (figure 24.16).

24.6 ■ Viral Diseases of the Lower Digestive System—Liver (figure 24.17)

Hepatitis A (table 24.14)
Hepatitis A virus (HAV) is a picornavirus spread by fecal contamination of hands, food, or water. Infection is often asymptomatic in young children; disease in older children and adults is prolonged.

Hepatitis B (table 24.14)
Hepatitis B virus (HBV) is a hepadnavirus spread by blood, blood products, and semen (figures 24.18, 24.19). Chronic infection can lead to **cirrhosis** and liver cancer.

Hepatitis C (table 24.14)
Hepatitis C virus (HCV) is a flavivirus transmitted mainly by blood. Acute infection is often asymptomatic, and many infections become chronic, leading to cirrhosis and liver cancer.

24.7 ■ Protozoan Diseases of the Lower Digestive System

Giardiasis (table 24.15)
Transmission of *Giardia lamblia* is usually via drinking water contaminated by feces. It is a common cause of traveler's diarrhea (figures 24.20, 24.21). The **cysts** survive in chlorinated water but can be removed by filtration.

Cryptosporidiosis ("Crypto") (table 24.16)
Cryptosporidium parvum oocysts are infectious, resist chlorination, and are too small to be removed by many filters (figure 24.22). It is a cause of many waterborne and foodborne epidemics, and traveler's diarrhea.

Cyclosporiasis (table 24.17)
Cyclospora cayetanensis is transmitted via fecally contaminated water or produce such as berries; it causes traveler's diarrhea. Oocysts are not infectious when passed in feces, so there is no person-to-person spread; no hosts other than humans are known.

Amebiasis (table 24.18)
Entamoeba histolytica is an important cause of dysentery; infection can spread to the liver and other organs (figure 24.23). Cysts are infectious.

Review Questions

Short Answer

1. Describe two characteristics of *Streptococcus mutans* that contribute to its ability to cause dental caries.
2. Describe the process of periodontal disease.
3. How does *Helicobacter pylori* cause stomach ulcers?
4. Besides the ducts of the salivary glands, what other locations in the body are sometimes infected by the mumps virus?
5. What characterizes the solutions used for oral rehydration therapy?
6. How do *Shigella* cells move from one host cell to another even though they are non-motile?
7. Name four different pathogenic groups of *Escherichia coli*.
8. What predisposes someone to a *Clostridium difficile* infection?
9. Name two kinds of hepatitis that can be prevented by vaccines.
10. Contrast the cause and epidemiology of giardiasis and amebiasis.

Multiple Choice

1. Which of the following about intestinal bacteria is *false*?
   a) They produce vitamins.
   b) They can produce carcinogens.
   c) They are mostly aerobes.
   d) They produce gas from indigestible substances in foods.
   e) They include potential pathogens.
2. All of the following attributes of *Streptococcus mutans* are important in tooth decay *except*
   a) it produces endotoxin, which triggers an inflammatory response.
   b) it can grow at pH below 5.
   c) it produces lactic acid.
   d) it synthesizes glucan.
   e) it stores fermentable polysaccharide.
3. *Helicobacter pylori* has all of the following characteristics *except*
   a) it is a helical bacterium with sheathed flagella.
   b) it has not been cultivated in vitro.
   c) it produces a powerful urease.
   d) it causes long-term infections, lasting for years.
   e) it can cause stomach ulcers.
4. *Vibrio cholerae* pathogenesis involves all of the following *except*
   a) attachment to the small intestinal epithelium.
   b) production of cholera toxin.
   c) lysogenic conversion.
   d) acid resistance.
   e) watery diarrhea.
5. Which of the following statements concerning *Salmonella enterica* serotype Typhi is *false*?
   a) It is commonly acquired from domestic animals.
   b) It can colonize the gallbladder.
   c) It is highly resistant to being killed by bile.
   d) It can destroy Peyer's patches.
   e) It causes typhoid fever.
6. Which statement about rotavirus gastroenteritis is *false*?
   a) A vaccine is available to prevent the disease.
   b) On a worldwide basis, most of the deaths are due to dehydration.
   c) Most cases of the disease occur in infants and children.
   d) The causative agent infects mainly the stomach.
   e) The disease is transmitted by the fecal-oral route.
7. Which of the following statements about noroviruses is *false*?
   a) They are the most common cause of viral gastroenteritis in the United States.
   b) They have a low infectious dose.
   c) They generally cause vomiting lasting 1 to 2 weeks.
   d) Immunity does not last long.
   e) They are a Category B bioterrorism agent.

8. Which of the following statements about hepatitis is *false*?
   a) Both RNA and DNA viruses can cause hepatitis.
   b) Some kinds of hepatitis can be prevented by vaccines.
   c) HCV infections are often associated with injected-drug abuse.
   d) Lifelong carriers of hepatitis A are common.
   e) Hepatitis A spreads by the fecal-oral route.
9. Which of the following statements about hepatitis B virus is *false*?
   a) Replication involves reverse transcriptase.
   b) Infected persons may have large numbers of non-infectious viral particles circulating in their bloodstream.
   c) In the United States, infection rates have been steadily increasing over the last few years.
   d) Asymptomatic infections can last for years.
   e) Infection can result in cirrhosis.
10. Choose the most accurate statement about cryptosporidiosis.
   a) Waterborne transmission is unlikely.
   b) The host range of the causative agent is narrow.
   c) It is prevented by chlorination of drinking water.
   d) Person-to-person spread does not occur.
   e) The life cycle of the causative agent occurs within small intestinal epithelial cells.

Applications

1. One reason given by Peruvian officials for not chlorinating their water supply is that chlorine can react with substances in water or in the intestine to produce carcinogens. How do you assess the relative risks of chlorinating or not chlorinating drinking water?
2. A medical scientist is designing a research program to determine the effectiveness of hepatitis B vaccine in preventing liver cell cancer. Because liver cell cancer probably has multiple causes, how would you measure the success of an anticancer vaccination program?

Critical Thinking

1. Why does the lack of a brown color in feces indicate hepatitis?
2. Mutant strains of *Helicobacter pylori* that lack the ability to produce urease fail to cause infection when they are swallowed. Infection occurs, however, if a tube is used to introduce them directly into the layer of mucus that overlies the stomach epithelium. What does this imply about the role of urease in the bacterium's pathogenicity?

25 Blood and Lymphatic Infections

Color-enhanced SEM of the head of a female mosquito. ©*Susumu Nishinaga/ Science Source*

KEY TERMS

Bubo An enlarged, tender lymph node characteristic of plague and some sexually transmitted infections.

Disseminated Intravascular Coagulation (DIC) Condition in which clots form in small blood vessels throughout the body, causing organ failure due to lack of O_2 and bleeding due to shortage of clotting proteins.

Infective Endocarditis Inflammation of the heart valves or lining of the heart chambers resulting from infection.

Lymphangitis Inflammation of lymphatic vessels.

Paroxysm A sudden recurrence or worsening of symptoms as seen in the cycle of chills, fever, and sweats of malaria.

Petechiae Small purple spots on the skin and mucous membranes caused by hemorrhage from small blood vessels.

Sepsis Acute illness caused by an inflammatory response that results when pathogens or their products circulate in the bloodstream.

Septic Shock A range of effects that result from a systemic inflammatory response to a bloodstream infection or circulating endotoxin; effects include fever, drop in blood pressure, and DIC.

A Glimpse of History

Alexandre Emile John Yersin (1863–1943) is one of the most interesting, though relatively unknown, contributors to the understanding of infectious diseases. He developed an interest in science when, as a young boy, he found a microscope and dissecting instruments that had belonged to his dead father. A local physician befriended Yersin and influenced him to study medicine. Once he had become a physician, Yersin volunteered at the Pasteur Institute in Paris, where he was hired by Èmile Roux, a coworker of Louis Pasteur. Yersin became well recognized for his work at the Institute, but was bored with research and did not want to practice medicine because he felt it was wrong for physicians to make money from other people's sickness.

Yersin left the Pasteur Institute and was employed as a physician on a ship sailing from France to Vietnam. He became enchanted with Vietnam and its people, and made it his home for the rest of his life, even establishing a medical school there.

Yersin studied the diseases that affected the Vietnamese people, including plague. The cause and transmission of this serious disease were completely unknown at the time. In 1894, Yersin went to study an outbreak of plague in Hong Kong. Unfortunately for Yersin, Shibasaburo Kitasato, the famous colleague of Robert Koch, had arrived 3 days earlier with a large team of Japanese scientists. British authorities had given Kitasato and his colleagues access to patients and laboratory facilities. Yersin did not speak English, so it was difficult for him to communicate with the authorities. He set up a laboratory in a bamboo shack and was forced to do his research on samples from dead plague victims that he got from British soldiers, whose job it was to bury them. ⏮ Koch, ⏮ Kitasato

A week after he arrived in Hong Kong, Yersin reported his discovery of a bacillus that was always present in the swollen lymph nodes of plague victims. The bacterium could be cultivated, and it caused plague-like disease when injected into rats. The disease was transmitted from one rat to another. Kitasato, working with blood from the hearts of plague victims, also announced soon after his arrival in Hong Kong that he had isolated the bacterium that caused plague. It was later proved to be only a laboratory contaminant, but Kitasato was named co-discoverer of the plague bacillus because of his great prestige.

Yersin's plague bacillus was named *Yersinia pestis.* It was used to make a vaccine, and later, antiserum prepared against the organism was used to cure a patient with plague—the first successful treatment of a plague victim.

The cardiovascular system circulates blood and lymph fluids to tissues, providing cells with nutrients and O_2 and removing their waste products. The flow of blood also heats or cools the body, maintaining optimum temperature. Infection of this system can be serious because infectious agents become **systemic,** which means that they are carried throughout the body. Once disseminated (spread), the agents can produce disease in one or more organs. When a substance is circulating in the bloodstream, the condition is given a name that specifies the nature of the substance as in **bacteremia, viremia,** and **fungemia.** These terms do not imply a disease state—in many cases, the circulating substance does not cause any symptoms. For example, a person is often briefly bacteremic after toothbrushing, when mouth bacteria enter small abrasions caused by the brush. Normally, the immune system quickly removes microbes that enter the bloodstream.

25.1 ■ Anatomy, Physiology, and Ecology of the Blood and Lymphatic Systems

Learning Outcome

1. Describe the characteristics and functions of the heart, blood vessels, lymphatics, and spleen.

The cardiovascular system is composed of the heart, blood vessels, and blood. The lymphatic system consists of lymph, lymph vessels, lymph nodes, and lymphoid organs, including the tonsils, appendix, and spleen (**figure 25.1**). Both systems are normally sterile. As blood flows around the body, it passes alternately through the lungs and tissue capillaries. During each circuit, some fluid passes through the lymphoid organs, which contain phagocytic cells that remove infectious agents and other foreign material. ◀◀ lymphatic system

The Heart

The heart—a muscular pump enclosed in a fibrous sac called the pericardium—pumps the blood around the body.

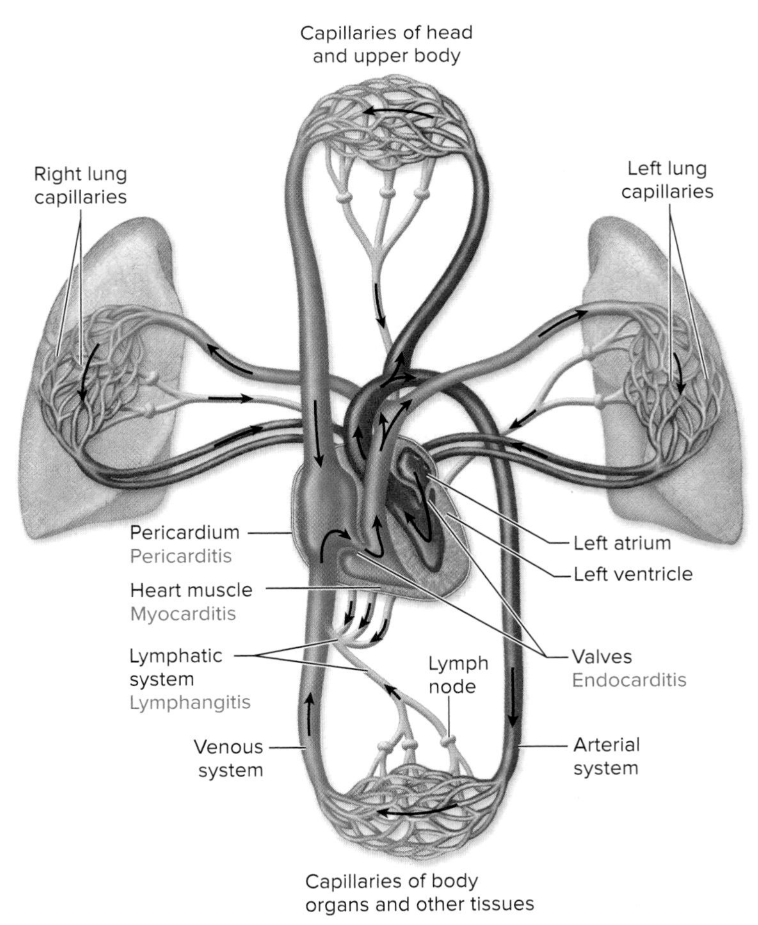

FIGURE 25.1 The Blood and Lymphatic Systems Disease conditions are indicated in red type. For simplicity, the spleen is not shown.

? What is the function of lymph nodes?

A wall of tissue (septum) separates the right and left sides of the heart, each of which is divided into two chambers—the atrium that receives blood, and the ventricle that discharges it (see figure 25.1). Blood from the right ventricle flows through the lungs and then back to the heart and into the left atrium. From there, the blood passes into the left ventricle and is then pumped through the aorta to the arteries and capillaries that supply the tissues of the body. The blood is then returned to the heart and into the right atrium by the veins. Valves at the entrance and exit of each ventricle prevent the blood flow from reversing. Although not common, infections of the heart valves, heart muscle, and pericardium can be serious because they affect blood circulation.

Blood Vessels

Blood vessels include arteries, veins, and capillaries. Arteries carry blood away from the heart and have thick muscular walls to withstand the high pressure of blood pumped by the ventricles. Exchange of materials between the blood and the tissues occurs across the very thin walls of microscopic capillaries. Veins carry blood back to the heart. Valves in veins play an important role in keeping the blood flowing in one direction, because the venous pressure is very low. Veins are easily compressed, so the action of skeletal muscles also aids the flow of venous blood.

Lymphatics (Lymphatic Vessels)

Lymphatic capillaries arise in the tissues and carry an almost colorless fluid called lymph, which comes from plasma (the fluid portion of the blood). Fluid seeps through the walls of the blood capillaries and joins interstitial fluid that bathes and feeds the tissue cells (see figure 15.4). A portion of this fluid enters the lymphatic capillaries, becoming lymph. Lymphatic capillaries are permeable to large molecules and take up foreign material such as invading microbes and their products, including toxins and other antigens. The lymphatic capillaries converge to form lymphatic vessels (lymphatics) that carry lymph to blood veins entering the heart (see figure 15.3). Many one-way valves in the lymphatic vessels keep the flow of lymph moving away from the tissues. As with the veins, lymph is propelled by skeletal muscle contractions.

As lymph is carried through the lymphatics, it passes through lymph nodes where it is exposed to phagocytes and lymphocytes that trap and destroy foreign materials such as microbial cells. When a hand or a foot is infected, a visible red streak may spread up the limb from the infection site toward the nearest lymph node. This is due to inflammation of one or more lymphatic vessels, a condition called **lymphangitis** (**figure 25.2**). ◀◀ lymph nodes

Blood and lymph both carry infection-fighting leukocytes and antimicrobial proteins, including antibodies, complement,

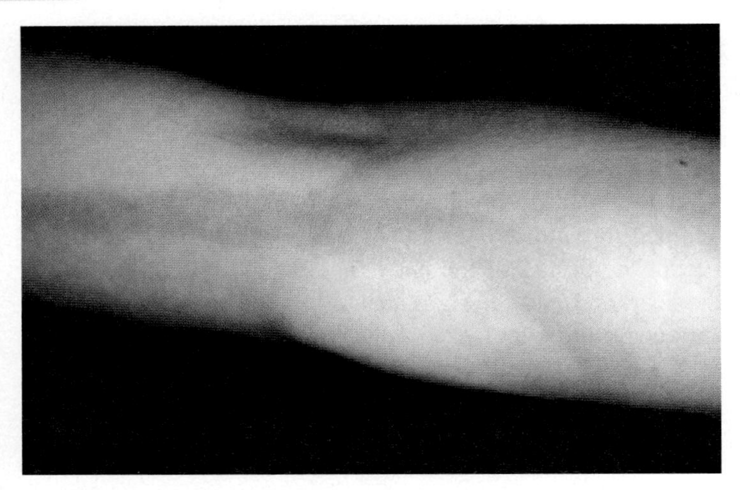

FIGURE 25.2 Lymphangitis Notice the red streak extending up the arm. ©SPL/Science Source

What causes lymphangitis?

lysozyme, and interferon. An inflammatory response may cause lymph and blood to clot in vessels close to areas of infection, preventing microbes from spreading.

Spleen

The spleen is a fist-sized organ found on the left side of the abdominal cavity, behind the stomach. It contains two kinds of tissue: red pulp (multiple blood-filled passageways) and white pulp (lymphoid tissue). The spleen has several functions. The red pulp cleans the blood by filtration, in the same way that the lymph nodes clean the lymph. Large numbers of phagocytes in red pulp remove aging or damaged red blood cells, bacteria, and other foreign materials from the blood. The red pulp also has a reserve of monocytes and produces new blood cells in rare situations where the bone marrow is unable to make enough. The white pulp, which contains both B and T lymphocytes, provides an active immune response to microbial invaders.

MicroAssessment 25.1

The heart pumps blood through arteries and veins, potentially carrying infectious agents and their toxins throughout the body. Systemic infections may disrupt the transport of O_2 and nutrients to body tissues and removal of waste products from them. The lymphatic system carries lymph through lymph nodes containing immune cells before returning it to the bloodstream. Blood is exposed to immune cells as it passes through the spleen.

1. What are the functions of the spleen?
2. What is the function of valves in the heart?
3. Why do you sometimes see swelling in a limb that is immobile?

25.2 ■ Bacterial Diseases of the Blood and Lymphatic Systems

Learning Outcomes

2. Outline the events that lead to infective endocarditis.
3. Explain the role of the inflammatory response in sepsis and septic shock.
4. Compare and contrast plague, Lyme disease, *Vibrio vulnificus* infection, tularemia, and brucellosis.

Bacteria that cause vascular infections usually originate from an infected area in the tissues. From the tissues, they enter lymph and can then be carried to the bloodstream. Some then multiply in the bloodstream, and they may colonize and form biofilms on structures such as the heart valves. Others multiply in cells of the mononuclear phagocyte system such as macrophages. In addition to the diseases discussed in this chapter, other diseases that affect the vascular systems include Rocky Mountain spotted fever and typhoid and paratyphoid fevers, which were covered in earlier chapters. ◄◄ biofilm, ◄◄ RMSF, ◄◄ typhoid and paratyphoid fevers

Infective Endocarditis

Infective endocarditis (IE) is an infection of an inner surface of the heart, often a heart valve. Predisposing factors include use of indwelling catheters, underlying heart defects such as deformed or damaged valves (possibly caused by rheumatic fever—a post-streptococcal illness), or injected drug abuse. Some types of IE progress gradually; these were previously referred to as subacute bacterial endocarditis (SBE). Other types have a more rapid progression; these were referred to as acute bacterial endocarditis. ◄◄ rheumatic fever

Signs and Symptoms

People with endocarditis usually have noticeable fatigue and slight fever. They typically become ill gradually and slowly lose energy over a period of weeks or months. Small hemorrhagic lesions may develop in the conjunctiva, in the skin, or under the nails. Strokes can be a life-threatening complication.

Causative Agent

A number of different pathogens cause endocarditis, including *Staphylococcus aureus, Staphylococcus epidermidis,* enterococci, and a variety of *Streptococcus* species, including viridans streptococci normally found in the mouth. The infecting organisms are usually shed from the infected heart valve into the blood and can be identified by cultivating samples taken from an arm vein. In cases where the causative agent cannot be cultured, PCR and nucleic acid probes may be used for identification. ◄◄ PCR, ◄◄ nucleic acid probes, ◄◄ viridans streptococci

Pathogenesis

The bacteria that cause infective endocarditis may enter the bloodstream from another infected body site or during dental procedures, tooth brushing, or trauma. These microbes adhere to heart tissue, particularly if they get trapped in the thin blood clots that often form around deformed heart valves or other areas with disturbed blood flow. They multiply, creating a biofilm that protects them from phagocytosis and antimicrobial medications. The clot grows larger around the multiplying organisms, gradually building up a fragile mass. Bacteria continually wash off the mass into the circulation, and pieces of infected clot (septic emboli) can break off. Emboli can block blood vessels, leading to death of the tissue supplied by the vessel. They can also cause a vessel to weaken and balloon out, forming an aneurysm.

People with endocarditis often have high levels of antibodies against the bacteria. These may be harmful because they lead to the formation of immune complexes, which may lodge in the skin, the eyes, and other body structures, triggering an inflammatory response. In the kidney, the complexes cause glomerulonephritis. ◀◀ immune complexes, ◀◀ glomerulonephritis

Even though the bacteria that cause endocarditis usually have little invasive ability, large numbers of them growing in the heart are sometimes able to penetrate into heart tissue, producing abscesses or damaging valve tissue, and resulting in a leaky valve.

Epidemiology

In recent years, the number of cases of endocarditis due to prosthetic heart valves or intravenous catheters has risen as the devices have become more common. Increased abuse of injected drugs also results in higher rates of the disease. In either case, people are often infected with *Staphylococcus aureus, Staphylococcus epidermidis,* or enterococci. Viridans streptococci have caused fewer cases of endocarditis, but they are often more serious because the causative organisms are becoming increasingly antibiotic-resistant. The American Heart Association once recommended prophylactic antibiotic treatment before dental procedures or certain surgeries for all patients with serious heart murmurs (abnormal sounds, often indicative of a deformed valve), but those recommendations have been relaxed.

Treatment and Prevention

Infective endocarditis is treated with combinations of intravenous antibacterial medications chosen according to the susceptibility of the causative organism. Prolonged treatment over 1 or more months is usually required. Devices covered with resistant biofilms must be replaced. Sometimes, surgery is needed to remove an infected clot or to drain abscesses.

There are no proven methods for preventing infective endocarditis. However, people with certain heart valve abnormalities are commonly given an antibacterial medication shortly before dental or other bacteremia-causing procedures. Healthcare-associated endocarditis can be prevented by strict use of sterile technique when inserting plastic intravenous catheters, moving the catheters to a new site every few days, and removing them as soon as possible. Such precautions help prevent bacterial colonization of the catheters and consequent bacteremia that could lead to heart valve infection. The main characteristics of infective endocarditis are presented in **table 25.1.** ◀◀ healthcare-associated infections

Sepsis and Septic Shock

Sepsis, an infection-induced systemic inflammatory response, is a common healthcare-associated illness. The disease is caused by the release of bacterial products that change the normally beneficial inflammatory response into an excessive damaging response that can lead to life-threatening illness. If uncontrolled, sepsis can progress to **septic shock,** a dramatic drop in blood pressure that may be fatal. The discovery that sepsis involves pro-inflammatory cytokines and not only invading microorganisms has led to advances in understanding the pathogenesis of this disease. ◀◀ endotoxic shock, ◀◀ pro-inflammatory cytokines

Signs and Symptoms

The signs and symptoms of severe sepsis include violent shaking, chills, and fever, often with rapid breathing and feelings of anxiety and confusion. If septic shock develops, urine output drops, respiration and pulse speed up, and the arms and legs become cool and dusky-colored. Bruising and bleeding can also occur.

TABLE 25.1	Infective Endocarditis
Signs and symptoms	Fever, loss of energy over a period of weeks or months; sometimes, a stroke
Incubation period	Poorly defined, usually weeks
Causative agents	*Staphylococcus aureus* or *S. epidermidis;* enterococci; streptococci including oral viridans streptococci
Pathogenesis	Normal microbiota or organisms from an infected body site enter bloodstream; turbulent blood flow causes formation of a thin clot that traps circulating organisms; a biofilm forms, protecting the organisms from phagocytes; pieces of clot break off, blocking blood vessels, leading to tissue death.
Epidemiology	People at risk are mainly those with heart abnormalities, including artificial valves or damage from rheumatic fever; may develop after dental procedures or other situations that cause bacteremia; injected drug abuse.
Treatment and prevention	Treatment: a combination of antibiotics. Prevention: in certain cases, administration of an antibiotic immediately before anticipated bacteremia, such as before dental work.

Causative Agents

Systemic infection by any microorganism can cause sepsis. Although more than half of cases are caused by Gram-positive bacteria, most fatal cases involve Gram-negative bacteria. Recall that the outer membrane of Gram-negative bacteria contains lipopolysaccharide (LPS), also referred to as endotoxin (see figure 3.34). Examples of common causes of fatal sepsis include members of the normal microbiota of the large intestine—particularly facultative anaerobes such as *Escherichia coli* and other members of the *Enterobacteriaceae* as well as anaerobes such as *Bacteroides* species. Environmental bacteria such as *Pseudomonas aeruginosa* are also common causes. **◀◀ lipopolysaccharide, ◀◀ *Enterobacteriaceae*, ◀◀ *Pseudomonas***

Pathogenesis

Sepsis almost always starts from an infection somewhere in the body other than the bloodstream. These include lung and urinary tract infections, meningitis, and infections originating from surgical wounds. When normal body defenses are compromised by medical treatments such as surgery, catheters, and some medications, even microorganisms that normally have little invasive ability are able to infect the blood.

Sepsis progresses in stages and is initially due to an overstimulation of the inflammatory response. When pattern recognition receptors (PRRs) of macrophages and neutrophils detect endotoxin or other microbe-associated molecular patterns (MAMPs), the phagocytes respond by releasing pro-inflammatory cytokines. When this occurs systemically, an uncontrolled release of cytokines—a cytokine storm—results. The complement pathway is also activated, further amplifying the inflammatory response. **◀◀ PRRs, ◀◀ MAMPs, ◀◀ complement system**

The systemic release of pro-inflammatory cytokines, along with activation of complement, results in a widespread, self-stimulating inflammatory response. Various pro-inflammatory compounds recruit additional phagocytes from the bone marrow and cause the phagocytic cells to produce more PRRs. This increases the system's sensitivity to MAMPs, leading to an even greater release of pro-inflammatory mediators. The phagocytic cells and dying host cells then release compounds that function as damage-associated molecular patterns (DAMPs). The host responses to MAMPs and DAMPs leads to multiple devastating outcomes, including the (1) inhibition of systems that normally control inflammation, (2) suppression of the adaptive immune response, and (3) activation of blood clotting mechanisms (**figure 25.3**). **◀◀ DAMPs**

Activation of blood clotting mechanisms causes small clots to form in the capillaries. This may block the blood supply to tissues, causing hypoxia (low levels of O_2) and subsequent cell death that may lead to multi-organ failure. Widespread clotting called **disseminated intravascular coagulation (DIC)** depletes the blood of clotting proteins and platelets, resulting in hemorrhage. At the same time, phagocytes release tissue-damaging lysosomal enzymes. This is particularly important in the lungs, which are seriously and irreversibly damaged by these enzymes. Lung damage often results in death even if the individual's infection has been cured. The harmful effects of the inflammatory response are often worsened by low blood pressure due to (1) decreased muscular tone of the heart and blood vessel walls and (2) the low blood volume that results from fluid leakage out of the blood vessels. Current evidence indicates that the pro-inflammatory cytokines lead to septic cardiomyopathy (disorder of the heart muscles) and heart failure. Shock results when the blood pressure falls so low that vital organs are no longer supplied with adequate amounts of blood to maintain their function.

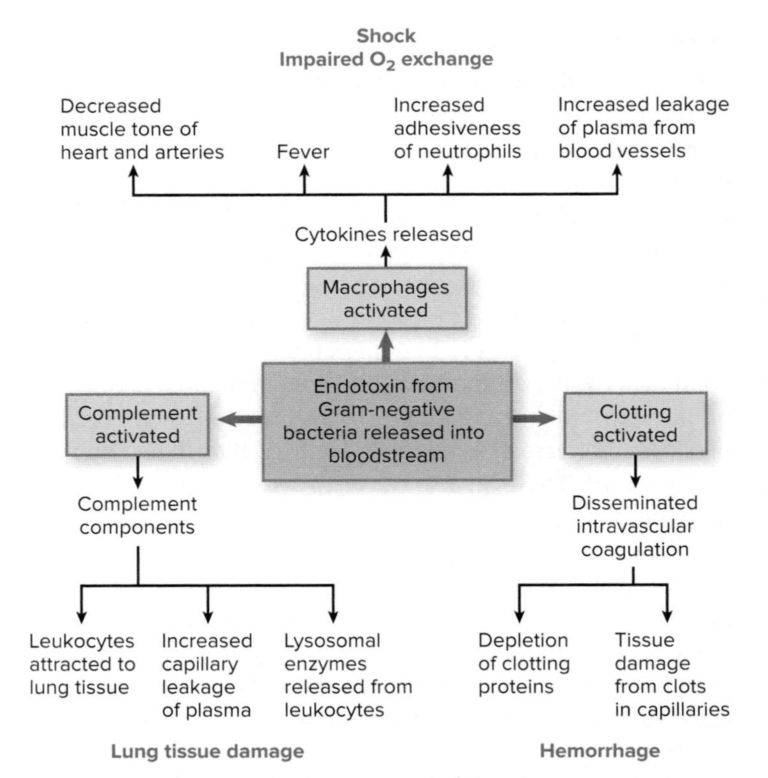

FIGURE 25.3 Events in Gram-Negative Sepsis

? What causes organ failure in sepsis?

Even if the patient survives the initial stages of sepsis, the later effects can be fatal. Neutrophils stop working properly, so bacteria are not cleared from the blood effectively. In addition, lymphocytes—particularly helper T cells—and dendritic cells undergo apoptosis, affecting the adaptive immune response and leading to immunosuppression that results in increased susceptibility to secondary infection.

Epidemiology

Sepsis is mainly a healthcare-associated disease, reflecting the high incidence of bloodstream infections in hospitalized patients with impaired host defenses. Rates of sepsis have been increasing, likely due to longer life spans, antibiotic suppression of normal microbiota, and immunosuppressive medications. Use of medical equipment where biofilms readily develop—such as ventilators, and catheters placed in blood vessels and the urinary system—also increases the risk of sepsis.

Treatment and Prevention

Treatment of sepsis is difficult, because it is a complex syndrome involving several host responses. It is currently treated with antimicrobial medications effective against the causative organism. Bactericidal antibiotics often lyse bacterial cells, however, increasing the release of endotoxin from Gram-negative organisms, and worsening the effects of the sepsis. The person is also treated for shock and tissue hypoxia. In addition, fluid replacement and support for organ dysfunction are given.

Sepsis can be prevented by quickly identifying and effectively treating localized infections, particularly in people with impaired immune defenses. Also, predisposing conditions such as bedsores and pyelonephritis (inflammation of the kidneys)—which commonly lead to sepsis in patients with cancer and diabetes—can usually be prevented. **Table 25.2** gives the main features of sepsis. ⏭ pyelonephritis

Plague ("Black Death")

Plague, once known as the "black death," caused the death of approximately a quarter of the population of Europe between 1346 and 1350. Crowded conditions in the cities and a large rat population played major roles in the spread of the disease. Plague is a potential bioterrorism disease, listed as Category A. ⏮ bioterrorism

TABLE 25.2 Sepsis

Signs and symptoms	Chills, fever, shaking, rapid breathing and pulse, confusion, and anxiety; if shock develops, drop in urine output, bruising, bleeding, and organ failure.
Incubation period	Variable
Causative agents	Usually bacteria (both Gram-positive and Gram-negative) but other microbes as well; common bacterial causes of fatal cases include *E. coli, Enterobacter* species, and *P. aeruginosa*.
Pathogenesis	Sepsis starts from infection in the body tissues; macrophages and neutrophils release pro-inflammatory cytokines in response to bacterial products; complement is activated; dysregulated inflammatory response results, leading to disseminated intravascular coagulation (DIC) and organ failure.
Epidemiology	Mainly a healthcare-associated disease.
Treatment and prevention	Treatment: multiple antibiotics as well as fluid and supportive therapy. Prevention: prompt identification and treatment of localized infections.

Signs and Symptoms

The signs and symptoms of plague depend on how the disease was acquired. Most commonly, a person is bitten by an infected flea, and symptoms develop 2 to 6 days later. The person characteristically develops significantly enlarged and tender lymph nodes called **buboes**—hence the name **bubonic plague**—in the region that receives lymph drainage from the area of the flea bite. High fever, shock, delirium, and patchy bleeding under the skin quickly develop.

If a person inhales respiratory droplets from an infected patient or animal, he or she may develop **pneumonic plague,** characterized by headache, fever, cough, and pneumonia accompanied by shortness of breath, chest pain, and possibly bloody sputum. Pneumonic symptoms arise more quickly than bubonic symptoms, typically in 1 to 3 days. ⏮ pneumonia

When the causative organism spreads via the bloodstream, the person may develop **septicemic plague.** Endotoxin released by the bacteria results in shock and disseminated intravascular coagulation (DIC), which causes bleeding into the skin and the organs, leading to a red or black patchy rash. ⏮ DIC

Causative Agent

Plague is caused by *Yersinia pestis,* which, like other members of the *Enterobacteriaceae,* is a facultatively anaerobic, Gram-negative rod. It is non-motile and grows best at 28°C. The organism resembles a safety pin when stained with certain dyes because the ends of the bacterium stain more intensely than the middle (**figure 25.4**).

Pathogenesis

Yersinia pestis forms biofilms in the digestive tract of infected fleas, often blocking the tract (**figure 25.5**). This both starves the flea—increasing the likelihood that it will try to feed again—and causes bacteria to be regurgitated into the bite

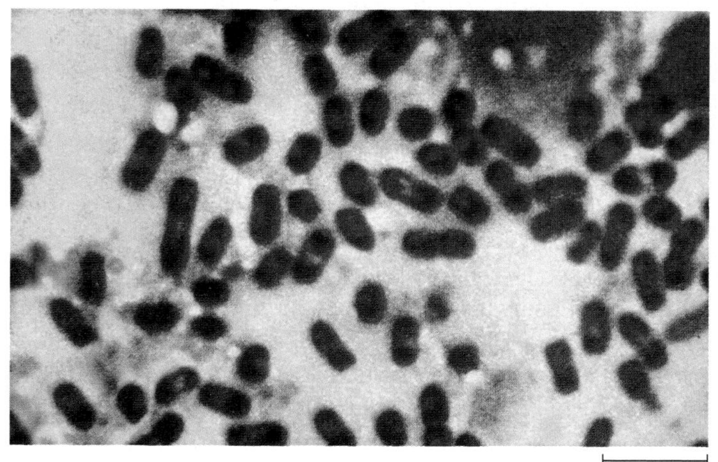

FIGURE 25.4 ***Yersinia pestis*** Notice that the cells stain more darkly at the ends. ©James Cavallini/Science Source

? **What is the function of the Yops (proteins) made by this organism?**

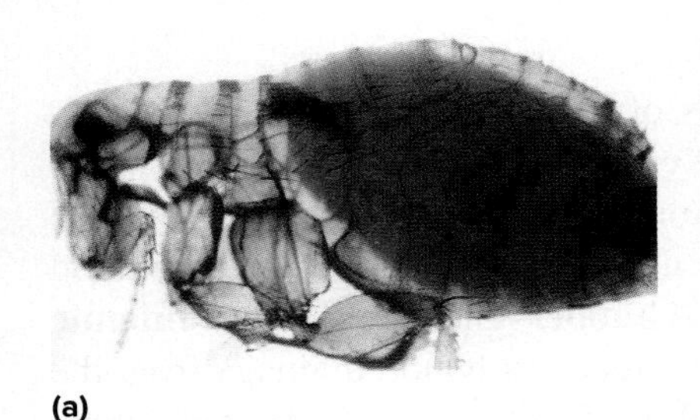
(a)

(b)

FIGURE 25.5 Fleas Following a Blood Meal **(a)** Healthy flea with fresh blood in gut. **(b)** Flea with obstruction due to *Yersinia pestis* infection. a: ©Daniel Cooper/E+/Getty Images; b: ©Science Source

? **Why is a flea with a digestive system obstruction more likely to transmit infection?**

wound as the flea attempts to feed. Even fleas that do not have an obstructed digestive tract can transmit *Y. pestis* because they release bacterial cells in their feces, and these can then be introduced into human tissue when a person scratches the flea bite.

Y. pestis has multiple virulence factors, making it extremely well equipped for avoiding mammalian host defenses. Once *Y. pestis* enters via a flea bite, it releases a protein-degrading protease (Pla) that clears the lymphatics and capillaries of clots and inactivates certain complement system components, allowing the organisms to spread. The lymphatic vessels carry the bacterial cells to the regional lymph nodes, where they are taken up by macrophages. The bacteria sense the intracellular conditions of the macrophages and respond by turning on various genes required for resisting the killing effects of the macrophages and for multiplying within them. Genes for other products that allow the organism to survive in the host are also activated. After several days, an acute inflammatory reaction develops in the lymph nodes, producing the enlargement and marked tenderness that characterizes bubonic plague.

The infected macrophages eventually die and release the bacteria, which are now "armed" to withstand host defenses. For example, the capsule allows *Y. pestis* to avoid phagocytosis, and an iron acquisition system allows the organism to trap and take up iron from the host. A variety of *Yersinia* outer proteins (Yops) are also made by the bacteria. These have multiple effects, including disrupting the host cell cytoskeleton, inhibiting phagocytosis, and blocking the production of pro-inflammatory cytokines, thereby reducing the inflammatory response. Yops (proteins) are delivered directly into host cells by a type III secretion system (TTSS). The swollen nodes may become necrotic, allowing large numbers of the bacterial cells to spread into the bloodstream. Endotoxin released by *Y. pestis* in the bloodstream causes the signs and symptoms of septicemic plague, including shock and DIC. The dark hemorrhages and dusky color of skin and mucous membranes from DIC probably inspired the common name "black death." ◀◀ type III secretion systems

In 10% to 20% of the cases, the infection spreads to the lungs, resulting in pneumonic plague. A person with pneumonic plague can transmit the disease to others in respiratory droplets. Organisms acquired this way are already fully virulent and, therefore, are especially dangerous. The virulence factors of *Y. pestis* are summarized in **table 25.3.**

Epidemiology

Plague is a zoonotic disease that is endemic in rodent populations worldwide except Australia. In the United States, the disease mostly occurs in the western states. The main reservoirs are wild rodents such as rock squirrels and prairie dogs, as well as their fleas. However, rats, rabbits, dogs, and cats can also be hosts. Hundreds of species of fleas are able to transmit plague, and the fleas can remain infectious for a year or more. Epidemics of pneumonic plague in humans are caused by respiratory droplets produced by coughing pneumonic plague patients. ◀◀ endemic disease, ◀◀ reservoir

Treatment and Prevention

Plague can be effectively treated with antimicrobial medications including gentamicin, ciprofloxacin, or doxycycline, especially if given within 24 hours of the onset of signs and symptoms. Between 50% and 80% of people with bubonic plague die if not treated. The case-fatality rate for untreated pneumonic plague is almost 100%.

Plague epidemics can be prevented by rodent control measures such as proper garbage disposal, construction of rodent-proof buildings, and rat extermination programs. Extermination programs must be combined with the use of insecticides to prevent the escape of infected fleas from dead animals. There is no current vaccine for plague, but efforts are underway to develop one that is effective against both bubonic and pneumonic

TABLE 25.3 Virulence Factors of *Yersinia pestis*

Factor	Coded by	Action
Pla (protease)	pPCP1 plasmid	Activates plasminogen; destroys C3b, C5a, and clots
Yops (proteins)	pCD1 plasmid	Interfere with phagocytosis and the immune response
V antigen	pCD1 plasmid	Controls type III secretion system that delivers Yops (proteins)
F1	pMT1 plasmid	Forms antiphagocytic capsule at 37°C
PsaA (adhesin)	Chromosome	Role in attachment to host cells
Complement resistance	Chromosome	Protects against lysis by activated complement
Iron acquisition	Chromosome	Traps iron-containing substances; stores iron compounds intracellularly

TABLE 25.4 Plague ("Black Death")

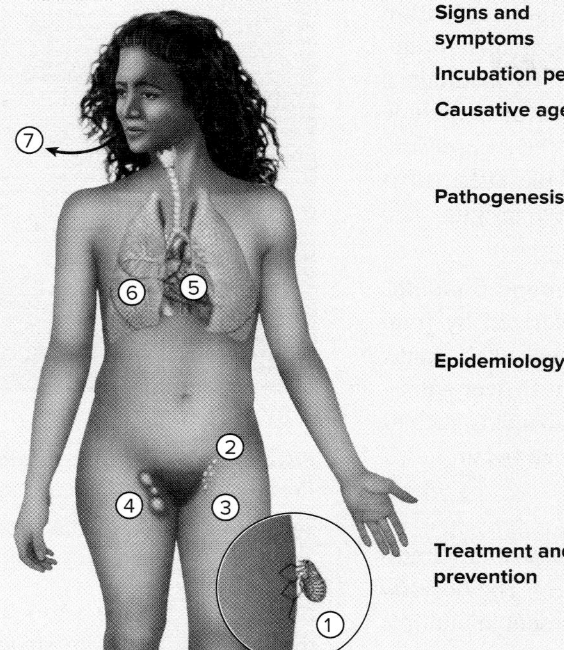

Signs and symptoms	High fever, large lymph nodes called buboes, skin hemorrhages; sometimes bloody sputum
Incubation period	Usually 1 to 6 days
Causative agent	*Yersinia pestis,* a Gram-negative rod; a member of the *Enterobacteriaceae* with multiple virulence factors
Pathogenesis	Enters the body with bite of infected flea or inhalation; bacteria taken up by macrophages. Intracellular environment causes the bacterial cells to produce multiple virulence factors that allow attachment to host cells, and provide defense against the immune system.
Epidemiology	Endemic in rodents and their fleas, particularly in the western United States. Bubonic plague is transmitted by fleas; pneumonic plague can be transmitted person to person in respiratory droplets. Pneumonic plague is the most dangerous because *Y. pestis* is fully virulent at the time of transmission.
Treatment and prevention	Treatment: prompt diagnosis and antibacterial treatment necessary to prevent high mortality. Prevention: currently no vaccine available; new vaccines are under development. Avoiding contact with wild rodents and their burrows. Insecticides and rat control.

plague. The antibiotic doxycycline can be given as a preventive for someone exposed to plague and is useful in controlling epidemics because of its immediate effect. The main features of plague are presented in **table 25.4.**

Lyme Disease

Over 30,000 new confirmed cases of Lyme disease are reported each year, making it the most common vector-borne disease in the United States. Surprisingly, Lyme disease was not recognized until the 1970s, when a cluster of cases occurred in Lyme, Connecticut. In 1982 the causative agent was first identified in ticks by Dr. Willy Burgdorfer at the Rocky Mountain Laboratories in Montana. We now know that Lyme disease was widespread long before its recognition at Lyme.

Signs and Symptoms

Signs and symptoms of Lyme disease can be divided roughly into three stages, although individual patients may be asymptomatic in one or more of these:

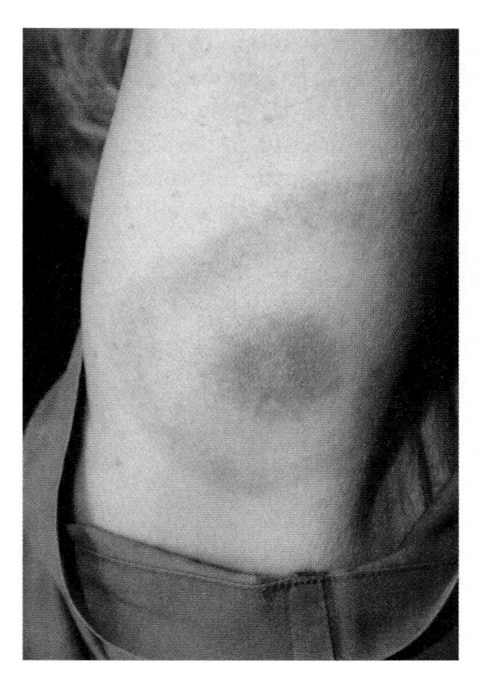

FIGURE 25.6 Erythema Migrans, the Characteristic Rash of Lyme Disease The rash usually has a target-like or bull's-eye appearance. Although the rash is highly suggestive of Lyme disease, many people with the disease do not develop the rash. Source: James Gathany/CDC

? What causes the bull's-eye pattern of this rash?

- **Early localized infection.** This stage typically begins a few days to several weeks after a bite by an infected tick. In many people (about 80%), early infection is characterized by enlargement of lymph nodes near the bite and a circular skin rash called erythema migrans (**figure 25.6**). The rash begins as a red spot or bump at the site of the tick bite and slowly enlarges to eventually reach a diameter of about 15 cm (6 inches). The advancing edge is bright red, leaving behind an area of fading redness as the lesion enlarges. This characteristic bull's-eye rash is the hallmark of Lyme disease but is not present in all cases. Moreover, the circular rash may appear after the bite of a type of tick that does not transmit Lyme disease. Most of the other signs and symptoms that occur during this stage are flu-like—malaise, chills, fever, headache, fatigue, joint and muscle pains, and backache.

- **Early disseminated infection.** This generally begins 2 to 8 weeks after the first signs and symptoms appear, and affects the nervous system. Electrical conduction within the heart is impaired, leading to dizzy spells or fainting. A temporary pacemaker may be required to maintain a normal heartbeat. Nervous system effects also include one or more of the following: paralysis of the face, severe headache, stiff neck, pain when moving the eyes, difficulty concentrating, emotional instability, fatigue, and numbness or pain in the legs or arms.
- **Late persistent infection.** This begins around 6 months after the skin rash. This stage is characterized by joint pain and swelling, usually of large joints such as the knee. These signs and symptoms slowly disappear over subsequent years. Chronic nervous system impairments such as localized pain, paralysis, and depression can occur.

Causative Agent

Lyme disease is caused by *Borrelia burgdorferi,* a Gram-negative, microaerophilic spirochete (**figure 25.7**). The *Borrelia* chromosome is unusual in that it is linear and present in multiple copies. *B. burgdorferi* also contains many plasmids that contain genes usually found on bacterial chromosomes. ◀◀ spirochetes

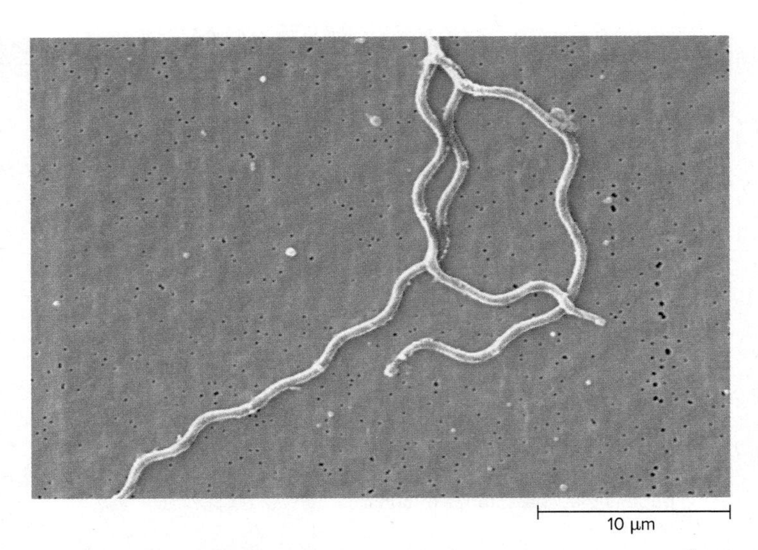

FIGURE 25.7 ***Borrelia burgdorferi*, the Cause of Lyme Disease** Color-enhanced SEM. Source: Claudia Molins/CDC

? How would you describe the shape of this organism?

Pathogenesis

The spirochetes are introduced into the skin by the bite of an infected tick. They multiply and migrate outward in a circular fashion. The lipopolysaccharide layer of these Gram-negative bacteria causes an inflammatory reaction in the skin, producing the expanding rash. The host's immune response is initially suppressed by components of the tick's saliva, allowing continued multiplication of the spirochete at the site. The bacterial cells then enter the bloodstream and spread to all parts of the body. This dissemination accounts for the flu-like signs and symptoms of the first stage. At this point, an intense immune response occurs, but after the first few weeks, it becomes very difficult to recover *B. burgdorferi* from blood or body tissues. The immune response against the bacterial antigens is probably responsible for the signs and symptoms of the second and third stages. The third stage of Lyme disease is characterized by arthritis, and the affected joints have high concentrations of reactive immune cells and immune complexes. Evidence suggests a role for autoimmune response against host tissues in some cases. ◀◀ autoimmunity, ◀◀ lipopolysaccharide

Epidemiology

Lyme disease is a zoonosis with humans being an accidental host. It is widespread in the United States, and its incidence depends on complex ecological factors (**figure 25.8**). Several

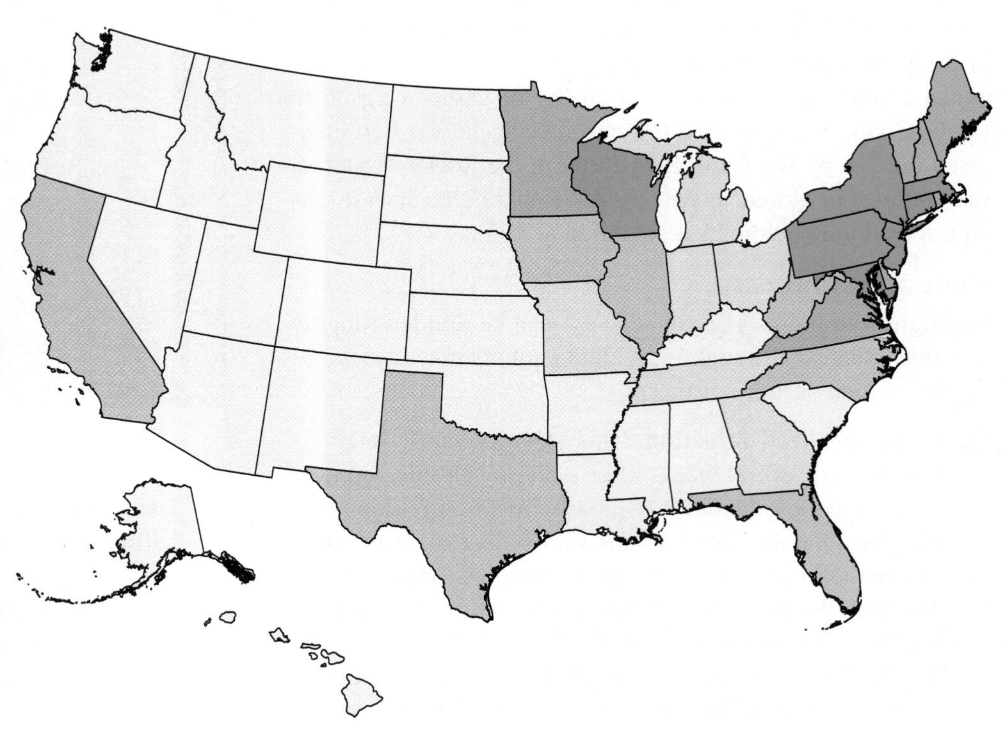

FIGURE 25.8 Lyme Disease Reported cases in the United States, 1990–2014; darker color indicates higher incidence of disease. Source: Data Compiled from CDC pub. data © 2015 Lyme Disease Association, Inc. www.lymediseaseassociation.org/about-lyme/cases-stats-maps-a-graphs/612-now-available-lyme-case-maps-1990-2014

? Why is the distribution of Lyme disease cases not uniform across the United States?

species of ticks have been implicated as vectors, but the most important in the eastern United States is the black-legged (deer) tick, *Ixodes scapularis* (**figure 25.9**). In some areas of the East Coast, 80% of these ticks are infected with *B. burgdorferi.* Because of the tick's small size (about the size of a sesame seed), they often feed and drop off their host without being detected—two-thirds of Lyme disease patients are unable to recall a tick bite.

The ticks mature during a 2-year cycle (**figure 25.10**). A larval form emerges from the egg. After growing, it molts, shedding its outer covering to become a nymph. The nymph actively seeks blood meals and is therefore mainly responsible for transmitting Lyme disease. *B. burgdorferi* is easily transferred from the tick to its preferred host, the white-footed mouse. Once infected, the mouse develops a sustained bacteremia and becomes a source of infection for other ticks. After another molt, the nymph becomes the sexually mature adult form. The spirochete is rarely passed from adult tick to its offspring. ◀◀ **bacteremia**

Infected ticks and mice are the main reservoirs of *B. burgdorferi,* but deer play an important role as well (see figure 25.10). They are the preferred host of the adult ticks and the site where tick mating occurs. Moreover, deer can quickly spread the disease over a wide area. Tick nymphs are most active from May to September, corresponding to the peak incidence of Lyme disease. Adult ticks sometimes bite humans late in the season and transmit the disease. Infectious ticks can be found in lawns as well as in wooded areas. ◀◀ **reservoirs**

Treatment and Prevention

Several different antibiotics can be used to treat patients with early stages of Lyme disease, including doxycycline,

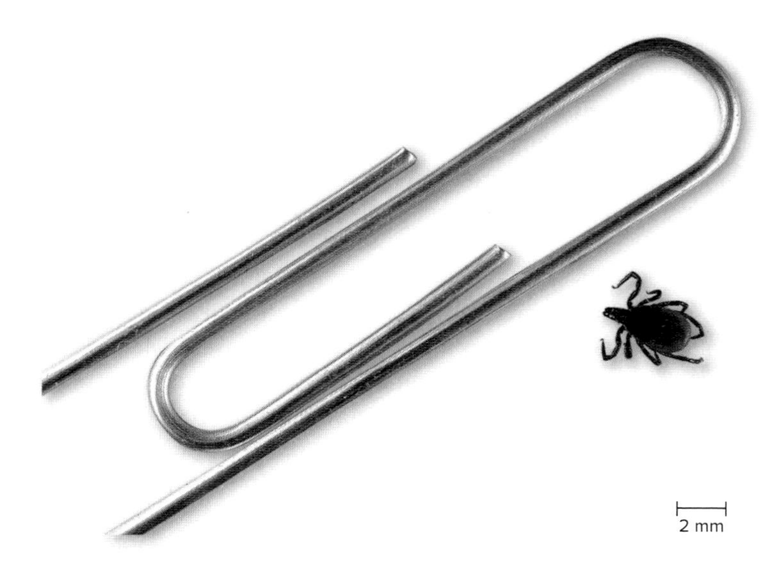

FIGURE 25.9 The Black-Legged (Deer) Tick, *Ixodes scapularis* This tick is the most important vector of Lyme disease in the eastern and north-central United States.

❓ **Why are people not always aware that they have been bitten by this tick?**

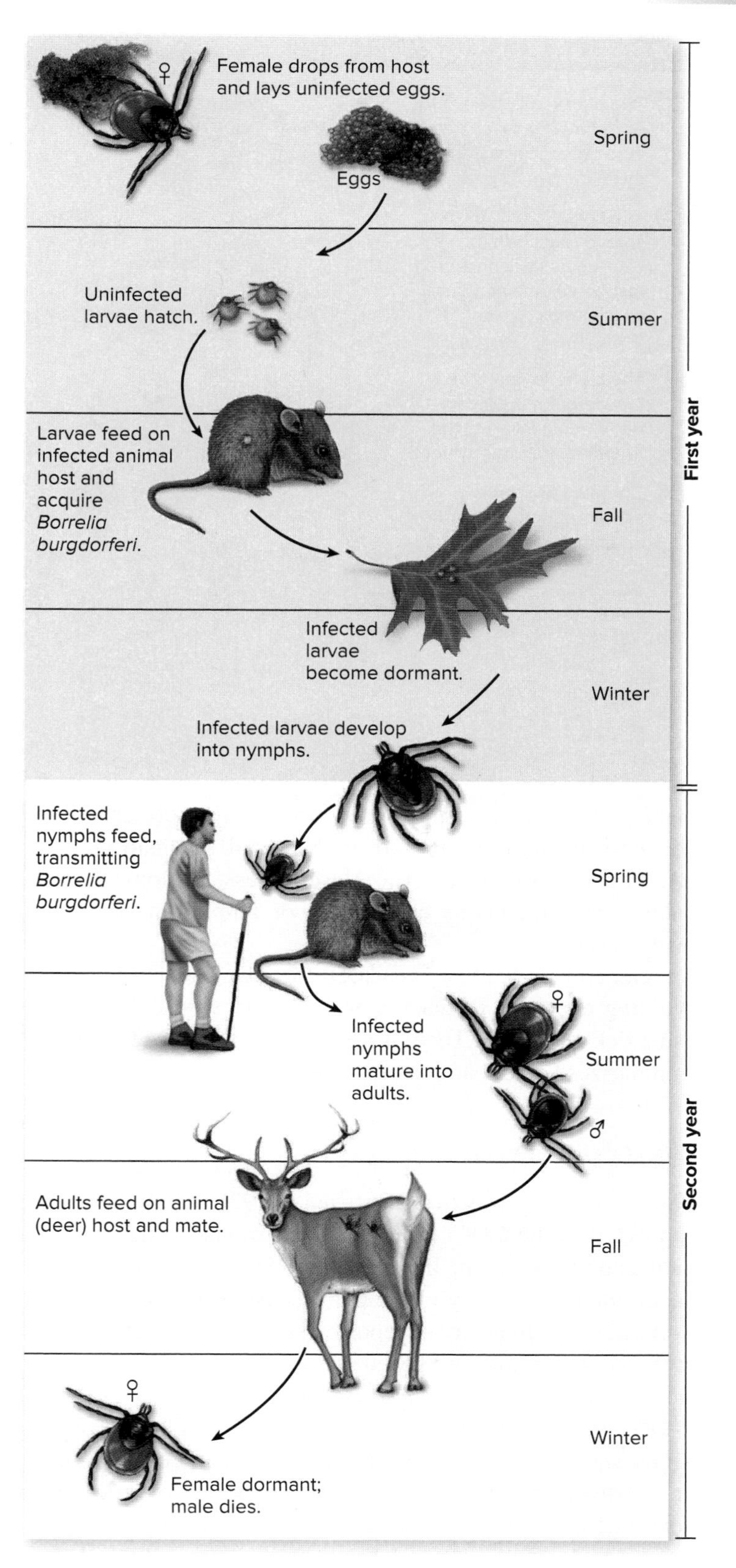

FIGURE 25.10 Life Cycle of the Black-Legged (Deer) Tick, *Ixodes scapularis* The life cycle covers 2 years. Variations In the life cycle occur, probably dependent on climate and food availability for the natural hosts.

❓ **What time of the year has the highest number of new cases of Lyme disease?**

TABLE 25.5 Lyme Disease

1. Bite of tick infected with *Borrelia burgdorferi* introduces the bacteria into the skin.
2. The bacteria multiply and spread radially in the skin, causing an expanding red rash (erythema migrans), which tends to clear centrally.
3. The bacteria enter the bloodstream; they cause fever, acute injury to the heart and nervous system.
4. Chronic symptoms develop, such as arthritis and paralysis due to persisting bacteria and the immune response to them.

Signs and symptoms	*Early localized infection:* enlarging rash that resembles a bull's eye develops at the site of the bite; lymph node enlargement near bite, flu-like symptoms. *Early disseminated infection:* heart and nervous system involvement. *Late persistent infection:* chronic arthritis and nervous system impairment.
Incubation period	Several days to several weeks
Causative agent	*Borrelia burgdorferi,* a spirochete
Pathogenesis	Spirochetes injected into the skin by an infected tick multiply and spread radially; spirochetes enter the bloodstream and are carried throughout the body; immune reaction to bacterial antigens causes tissue damage.
Epidemiology	Spread by the bite of ticks, *Ixodes* species, usually found in association with animals such as white-footed mice and white-tailed deer living in wooded areas.
Treatment and prevention	Treatment: early treatment with appropriate antibiotic; prolonged antibiotic therapy in chronic cases. Prevention: avoiding ticks, protective clothing, tick repellents. No vaccine.

amoxicillin, and cefuroxime. In late disease, antibiotic treatment is less effective, most likely because the spirochetes are not actively multiplying and antibacterial medications usually target dividing bacteria. Nevertheless, prolonged treatment with intravenous ceftriaxone or ampicillin has cured many cases.

General preventive measures for Lyme disease include avoiding exposure to ticks, wearing protective clothing, and using tick repellents. There is no available vaccine. **Table 25.5** summarizes some features of Lyme disease.

Vibrio vulnificus Infection

Vibrio vulnificus was first identified in 1976 in blood samples submitted to the CDC. Since then, the organism has become increasingly prevalent in the United States—multiple cases occur annually, mostly in the Gulf Coast region. It causes an overwhelming majority of reported seafood-associated deaths in the United States every year.

Signs and Symptoms

Signs and symptoms of *Vibrio vulnificus* infection (a type of vibriosis) vary, depending on the portal of entry and host factors. When the organism enters the bloodstream, however, the disease typically progresses rapidly and dramatically. There are two general routes of initial entry: the mouth and open wounds. When the organism is ingested, an otherwise healthy person may develop only a self-limiting case of gastroenteritis, with signs and symptoms including fever, abdominal pain, vomiting, and diarrhea. In a person who has liver disease or other predisposing conditions, the bacterium may invade the bloodstream, leading to fevers, chills, severe skin blistering, and shock due to sepsis, which is often fatal. If the organism enters the body through a cut or other open wound, blood-filled skin blisters called bullae develop; these rapidly progress to cellulitis and sometimes necrotizing fasciitis. As with infection acquired by ingesting the organism, wound infections sometimes spread to the bloodstream. ⏮ necrotizing fasciitis

Causative Agent

Vibrio vulnificus is a Gram-negative, motile, curved, rod-shaped bacterium. It is related to *Vibrio cholerae,* the causative agent of cholera, and like this organism, it is a halophile, found in warmer water in marine environments such as estuaries. Virulent strains have a polysaccharide capsule. ⏮ *Vibrio cholerae*

Pathogenesis

Vibrio vulnificus produces a wide range of virulence factors that allow it to evade the host immune responses. The capsule appears to be essential for pathogenicity—it protects the cells from the effects of complement proteins. To obtain iron from the host, *V. vulnificus* produces siderophores (that scavenge iron), as well as a hemolysin (that lyses blood cells to release iron-containing hemoglobin). Iron appears to be very important for the growth of the organism, because people with underlying diseases that elevate serum iron levels are much more susceptible to severe infections. ⏮ siderophores

V. vulnificus directly damages tissues by producing a wide range of degradative enzymes as well as toxins. These appear to be released in a coordinated manner, controlled by a quorum-sensing mechanism. One protease in particular—metalloproteinase—damages the basement membrane of blood vessels and destroys fibrin, leading to the blood-filled bullae seen in some people. A cytotoxin called RTX kills epithelial cells by forming pores in their membranes. The cytotoxicity of RTX depends on cell-to-cell contact, which involves pili and also the flagellum. As a Gram-negative organism, *V. vulnificus* produces endotoxin, which causes an uncontrolled inflammatory response that leads to endotoxic shock. ⏮ quorum sensing

Epidemiology

Vibrio vulnificus is a naturally occurring bacterium in coastal regions worldwide. It can be isolated from water, sediments, and a variety of aquatic organisms in these environments. Most cases of *V. vulnificus* infection in the United States are associated with eating contaminated oysters, although other seafood may carry the bacterium. It can also enter the body through cuts, abrasions, or other open wounds. Anyone can be at risk of *V. vulnificus* infection but those with underlying conditions affecting the immune system, particularly people with liver disease, are most likely to develop severe disease. Most infections occur during the warmer months (May through October) because the organism thrives in warmer water.

TABLE 25.6 *Vibrio vulnificus* **Infection**

Signs and symptoms	If the causative organism enters the bloodstream, it causes sepsis and septic shock, particularly in those with immunodeficiency. In most cases, the organism is ingested, causing fever, vomiting, diarrhea, and abdominal pain; it may also infect wounds, causing blood-filled skin blisters (bullae) and cellulitis.
Causative agent	*Vibrio vulnificus, a* Gram-negative motile, halophilic, curved rod
Incubation period	1 to 4 days
Pathogenesis	Encapsulated organism avoids phagocytosis and multiplies using siderophore-acquired iron from host cells; produces a variety of degradative enzymes as well as toxins that cause host cell damage and death.
Epidemiology	Found worldwide in warmer marine habitats; typically acquired by eating raw or undercooked seafood, especially oysters; can also enter via wounds; immunocompromised people, especially those with liver disease or certain underlying conditions, are at risk of developing sepsis.
Treatment and prevention	Treatment: combination of antibiotics; in wound infections, removal of infected tissue and sometimes amputation is essential; supportive care also given, including fluid replacement. Prevention: avoiding raw or undercooked seafood and wearing protective clothing when handling seafood in high-risk areas.

Treatment and Prevention

Prompt and aggressive treatment of *V. vulnificus* infections is essential because any delay significantly increases the risk of death. Patients are usually given a combination of synergistic antibiotics (such as doxycycline and a third-generation cephalosporin), along with supportive therapy including fluid replacement. Infected wounds require debridement and cleansing; sometimes amputation is necessary to stop the spread of the disease. ⏮ debridement

V. vulnificus infection can be prevented by avoiding raw seafood—especially oysters and particularly during warmer months—or by thoroughly cooking seafood. People with open wounds should avoid swimming in warm salt or brackish water. Wounds should be promptly and effectively treated. Protective clothing should be worn when handling seafood. People with liver disease should be educated on the risks of this disease. The main features of *V. vulnificus* infection are given in **table 25.6.**

Tularemia ("Rabbit Fever" or "Deer Fly Fever")

Tularemia is widespread in the United States and is found in wild animals such as rabbits, muskrats, and bobcats. The disease was first identified in Tulare County, California, following an outbreak among ground squirrels in 1911. Although tularemia affects humans only occasionally in the United States today, it is potentially a bioterrorism threat (see Focus on the Future 19.1). It is listed as Category A, the most serious threat, because it has a low infectious dose and can be transmitted through the air.

Signs and Symptoms

Signs and symptoms of tularemia depend on how the causative organism enters the body—through mucous membranes or minor cuts and scratches in the skin, or by inhalation. Typically, they appear 2 to 5 days after a person is bitten by an infected tick or insect, or handles an infected wild animal. A characteristic ulcer develops at the site where the organism enters the body (**figure 25.11**). Regional lymph nodes enlarge, and fever, chills, and headaches and muscle aches occur. Signs and symptoms usually clear in 1 to 4 weeks, but sometimes last for months.

If the organism is inhaled or infects the lung from the bloodstream, pneumonia may occur. Signs and symptoms then include a dry cough and pain beneath the sternum (breastbone), a result of enlarged lymph nodes. Tularemic pneumonia, although rare, is more serious and has a case-fatality rate as high as 30% if untreated.

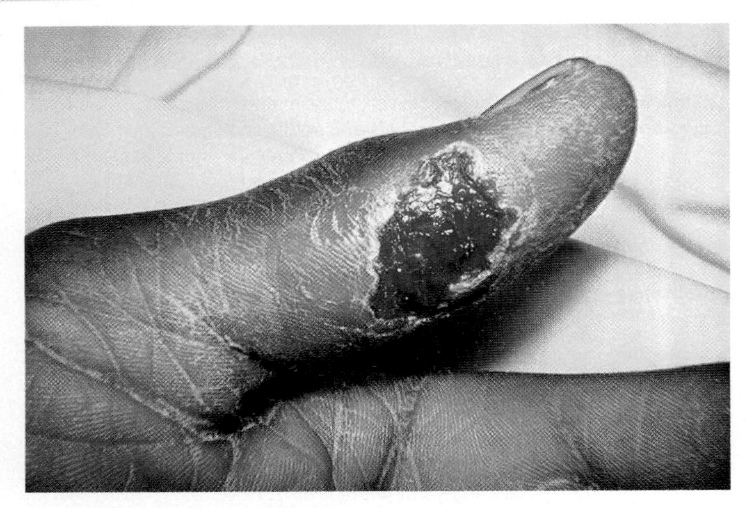

FIGURE 25.11 Tularemic Ulcer of the Thumb This kind of lesion develops when *F. tularensis* is acquired from a tick or insect bite or through skinning wild animals. ©Science Source

In what other ways can tularemia be transmitted?

TABLE 25.7 Tularemia ("Rabbit Fever" or "Deer Fly Fever")

Signs and symptoms	Ulcer at site of entry, enlarged lymph nodes in area, fever, chills, aches; pneumonia if causative agent is inhaled
Incubation period	1 to 10 days; usually 2 to 5 days
Causative agent	*Francisella tularensis;* an aerobic, Gram-negative rod.
Pathogenesis	Organisms are ingested by phagocytic cells, grow within these cells, and then spread throughout body.
Epidemiology	Present among wildlife in Northern Hemisphere; risk mainly to hunters and others who handle wildlife; acquired when the organism penetrates a mucous membrane or enters broken skin, as may occur when skinning wild animals; bite of infected insect or tick.
Treatment and prevention	Treatment: antibiotics. Prevention: avoiding bites of insects and ticks; wearing rubber gloves, goggles, when skinning animals.

Causative Agent

Tularemia is caused by *Francisella tularensis,* a non-motile, aerobic, Gram-negative rod. The organism requires a special medium enriched with the amino acid cysteine in order to grow.

Pathogenesis

Francisella tularensis enters through breaks in the skin or mucous membranes. The invading bacterial cells manage to avoid causing a strong inflammatory response, partly by having altered surface components that limit detection by pattern recognition receptors of the innate immune system. When ingested by macrophages, the bacteria escape from the phagosome and grow within the phagocyte's cytosol, protected from antibodies. Infected macrophages carry the bacteria to the regional lymph nodes, which become large, tender, and often filled with pus. The organism spreads to other parts of the body via the lymphatics and the blood vessels. Cell-mediated immunity effectively rids the host of this infection. Even without treatment, most infected people survive. ◀◀ **pattern recognition receptors,** ◀◀ **cell-mediated immunity**

Epidemiology

Tularemia is a zoonotic disease that occurs among wild animals in many areas of the Northern Hemisphere, including all the states of the United States except Hawaii. In the eastern United States, summertime human infections typically result from tick bites. However, cases do occur in the winter months in people skinning rabbits (explaining the common name, "rabbit fever"). People also contract the disease from muskrats, beavers, squirrels, deer, and other wild animals. The animals are generally free of illness but carry the causative organism. In the western United States, infections mostly result from the bites of infected ticks and deer flies (explaining the other common name, "deer fly fever") and usually occur during summer. Generally, fewer than 200 cases are reported each year across the United States. Epidemics of inhalation tularemia have occurred from dust arising from mowing or from rodent-infested buildings. Tularemia can also be contracted by eating contaminated meat.

Treatment and Prevention

Most cases of tularemia are effectively treated with streptomycin. Ciprofloxacin or gentamicin can also be used.

Tularemia can be prevented by using rubber gloves and goggles or face shields when skinning or handling wild animals. Insect repellents and protective clothing help protect against insect and tick vectors. The main features of tularemia are summarized in **table 25.7.**

Brucellosis ("Undulant Fever" or "Bang's Disease")

Brucellosis is often called "undulant fever," or "Bang's disease," after Bernhard Frederik Bang (1848–1932), a Danish veterinary professor who discovered the cause of cattle brucellosis. Relatively few cases of human brucellosis are reported each year in the United States. Like tularemia, brucellosis is a zoonotic disease and a potential bioterrorism threat. It is listed as a Category B threat (see Focus on the Future 19.1).

Signs and Symptoms

Brucellosis usually starts gradually, and the signs and symptoms are vague. Typically, patients complain of mild fever, sweating, weakness, aches and pains, enlarged lymph nodes, fatigue, depression, and weight loss. The recurrence of fevers over weeks or months in some cases gave rise to the alternative name, "undulant fever." Most people recover within 2 months even without treatment, but some people develop chronic illness.

Causative Agent

Brucella sp. are small, aerobic, non-motile, Gram-negative rods with complex nutritional requirements. The causative agent was discovered by Dr. David Bruce, the Scottish pathologist after whom it was named. Four varieties of the genus *Brucella* cause brucellosis in humans. DNA studies show that all of these fall into a single species, *Brucella melitensis,* but traditionally, the different varieties were assigned species names depending largely on their preferred host: *B. abortus* infects cattle; *B. canis,* dogs; *B. melitensis,* goats; and *B. suis,* pigs. The distinctions among the different varieties are mainly useful epidemiologically.

Pathogenesis

Brucella cells enter the body through mucous membranes (following ingestion or inhalation) or through wounds. The inflammatory response to the invasion is relatively weak, probably because some of the bacterial cells' surface features have structural modifications that limit detection by pattern recognition receptors of the innate immune system. When the bacterial cells are taken up by macrophages, they prevent phagosome-lysosome fusion and multiply within the phagocytic cell, protected from antibodies. Infected macrophages carry the bacteria to other parts of the body, including lymph nodes, spleen, liver, bone marrow, heart, and kidneys, where they infect other cells of the mononuclear phagocyte system (see figure 14.6). The mortality rate of brucellosis is low although some deaths do occur, generally due to endocarditis or meningitis (an infection of the membranes covering the brain and spinal cord). A serious but rare complication is bone infection (osteomyelitis).

Epidemiology

Brucellosis is typically a chronic infection of domestic animals. It contaminates their milk and can cause abortions in the affected animals. Worldwide, brucellosis is a major problem in animals used for food, causing yearly losses of millions of dollars.

Most brucellosis in humans is associated with occupational exposure—workers in the meat packing industry, veterinarians, and slaughterhouse workers can be exposed to the causative agents. People drinking unpasteurized milk or eating soft cheeses made from contaminated milk can also contract the disease. In the United States, hunters have acquired infections by eating meat from infected elk, moose, bison, caribou, and reindeer.

Treatment and Prevention

Brucellosis can usually be treated using doxycycline with rifampin or streptomycin for 6 weeks. In some chronic cases, extended treatment (sometimes up to 6 or more months) may be needed.

The most important control measures for brucellosis are pasteurization of dairy products and inspection of domestic animals for evidence of the disease. Veterinarians, butchers, and slaughterhouse workers should use protective gear such as goggles or face shields and rubber gloves. An attenuated vaccine effectively controls the disease in domestic animals. **Table 25.8** gives the main features of brucellosis.

TABLE 25.8 Brucellosis ("Undulant Fever" or "Bang's Disease")

1. *Brucella melitensis* enters the body through mucous membranes following ingestion or inhalation, or through skin abrasions.
2. The bacteria are taken up by phagocytes but resist digestion and grow within cells.
3. The bacteria enter the lymphatics and bloodstream and are carried throughout the body.
4. Infection is established in other body tissues, such as the heart valves, meninges, and bones.
5. Osteomyelitis is a rare but serious complication. Most deaths are due to endocarditis or meningitis.

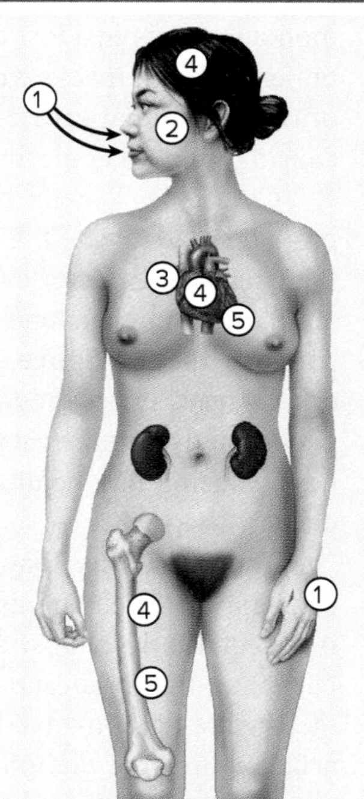

Signs and symptoms	Recurrent fever, body aches, weight loss, enlargement of lymph nodes; symptoms usually lessen even without treatment but may become chronic.
Incubation period	Usually 5 to 21 days
Causative agent	*Brucella melitensis,* a small, aerobic, Gram-negative rod
Pathogenesis	Organisms penetrate mucous membranes and are carried to heart, kidneys, meninges, and other parts of the body via the blood and lymphatic system; they resist killing by macrophages and grow within these cells.
Epidemiology	Main sources of human infections: domestic animals. Disease also occurs in wild animals such as moose, caribou, bison; butchers, farmers, veterinarians, those who drink unpasteurized milk are at risk.
Treatment and prevention	Treatment: appropriate antibiotics. Prevention: vaccination of domestic animals; pasteurization of milk and milk products; gloves and face shields.

MicroAssessment 25.2

People with indwelling catheters or heart abnormalities, or who abuse intravenous drugs, are at increased risk of developing infective endocarditis as a result of bacteremia. A systemic infection may lead to sepsis, a life-threatening inflammatory response. Plague is endemic in rodents and is transmitted to humans by flea bites. Lyme disease is the most common vector-borne disease in the United States. *Vibrio vulnificus* infection may cause severe disease, particularly in the immunocompromised. Tularemia is a disease of wild animals transmitted to humans by exposure to their blood or tissues, or by biting ticks and insects. Brucellosis is most commonly a disease of domestic animals, transmitted to humans when individuals handle the animals' flesh or drink their unpasteurized milk.

4. What is the difference between bacteremia and sepsis?
5. What is the most common way to contract *Vibrio vulnificus* infection?
6. Why might clots on the heart valves make microorganisms there inaccessible to phagocytic killing?

25.3 ■ Viral Diseases of the Blood and Lymphatic Systems

Learning Outcomes

5. Describe the characteristics of infectious mononucleosis.
6. Compare and contrast Ebola and Marburg virus diseases with yellow fever.
7. Compare and contrast dengue fever, chikungunya, and Zika virus disease.

A number of viral diseases affect the blood and lymphatic systems, causing widespread effects. Some have been studied for decades, but others are newly recognized.

Infectious Mononucleosis ("Mono" or "Kissing Disease")

Infectious mononucleosis ("mono") is a disease familiar to many students because of its high incidence among young people. The term *mononucleosis* refers to the fact that proliferating infected B cells accumulate in large numbers; these cells are mononuclear. Their abundance and appearance sometimes lead to the incorrect diagnosis of leukemia—this is disproved when the person spontaneously recovers. ⏮ leukocytes

Signs and Symptoms

Typically, signs and symptoms of infectious mononucleosis appear after a long incubation period, usually 30 to 60 days. They include fatigue, fever, a sore throat covered with pus, and enlargement of the lymph nodes. The spleen sometimes also becomes enlarged. In most cases, the fever and sore throat are gone in about 2 weeks and the swollen lymph nodes in 3 weeks. Generally, people recover fully within 4 weeks, although some suffer severe exhaustion and difficulty concentrating that affect them for months.

Causative Agent

Infectious mononucleosis is caused by the Epstein-Barr virus (EBV), named after its discoverers, Michael Anthony Epstein and Yvonne Barr. It is an enveloped double-stranded DNA virus of the family *Herpesviridae,* and although identical in appearance to the other known herpesviruses that cause human disease, it is not closely related to any of them.

MicroByte

EBV was discovered in the 1960s when it was isolated from Burkitt's lymphoma, a malignant tumor derived from B lymphocytes.

Pathogenesis

EBV initially infects the mouth and throat, where it causes the symptoms of pharyngitis (**figure 25.12**). After replicating in the mouth epithelium, salivary glands, and throat, the virus is carried to the lymph nodes. There, it infects B lymphocytes. The B-cell infection can be (1) productive, in which the virus replicates and kills the B cell; or (2) non-productive, in which the virus establishes a latent infection, maintained as either plasmid or provirus (integrated into the host cell chromosome). For most B cells, the infection is non-productive. In this infection, EBV activates the B cells, causing them to proliferate and produce immunoglobulins, including a type called a heterophile antibody. The heterophile antibody has no pathological significance, but can be used for diagnosis of mononucleosis in a test called the monospot test—the heterophile antibody reacts with an antigen on the red blood cells of certain animal species, including sheep, horses, and oxen, causing them to agglutinate.

Cytotoxic T cells respond to the infection, destroying productively infected B cells that display viral antigens on their surface. Effector cytotoxic T cells become enlarged with an atypical nucleus and can easily be seen in smears of the patient's blood (**figure 25.13**). Proliferation of the lymphocytes causes lymph node and spleen enlargement. The spleen occasionally ruptures within 3 to 4 weeks of the start of illness, causing rare deaths associated with infectious mononucleosis. ⏮ cytotoxic T cell

The possibility that EBV plays a role in causing certain cancers has been intensely investigated. The EBV genome is detectable in almost all cases of Burkitt's lymphoma and nasopharyngeal carcinoma, but evidence suggests that other factors are also involved in these cancers. EBV may also be a factor in some malignancies in patients with immunodeficiency from AIDS or organ transplantation.

FIGURE 25.12 Pathogenesis of Infectious Mononucleosis

? Why is infectious mononucleosis also called the "kissing disease"?

Epidemiology

EBV is distributed worldwide. It infects individuals in crowded, economically disadvantaged groups at an early age without causing significant illness or symptoms of infectious mononucleosis. In more affluent populations, fewer people get infected with EBV, so most lack immunity to it. Adolescents and adults who lack antibody to the virus may get infectious mononucleosis when they are exposed to it later in life, such as when they enter college. Even then, in the age group of 15 to 24 years, only about half of the EBV infections are symptomatic.

EBV is present in the saliva for up to 18 months after infectious mononucleosis and then intermittently for life. Continuous shedding of virus in saliva is common in people with AIDS or other immunodeficiencies. Mouth-to-mouth

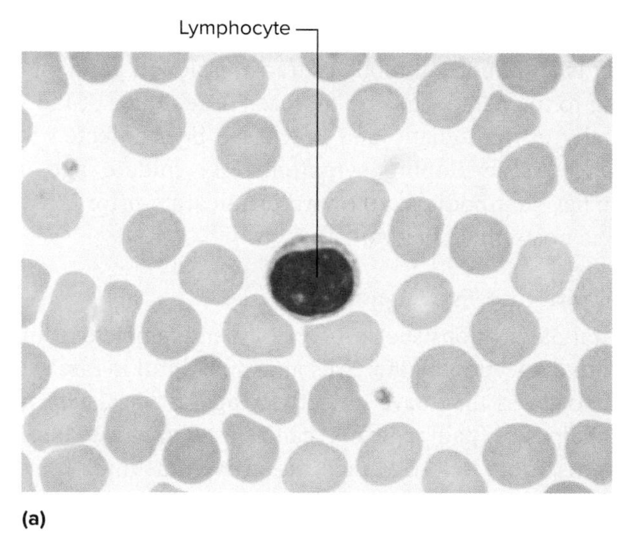

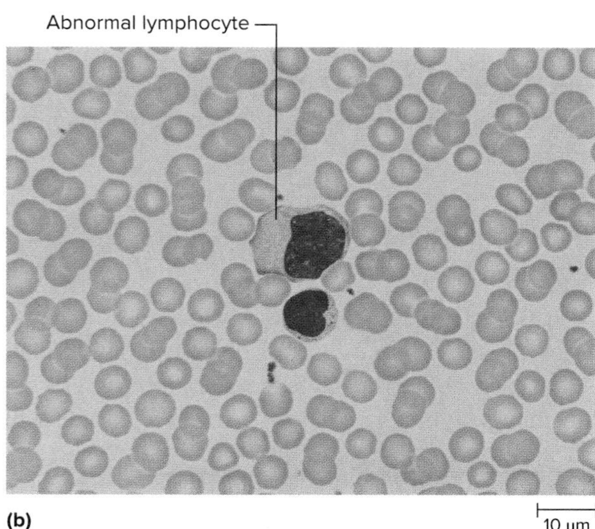

FIGURE 25.13 Normal and Atypical Lymphocytes (a) A normal lymphocyte and **(b)** an atypical lymphocyte characteristic of infectious mononucleosis. a: ©Michael Abbey/Science Source; b: ©Jacquin Carrillo-Farga/Science Source

? Why is this disease sometimes misdiagnosed as leukemia?

TABLE 25.9	Infectious Mononucleosis ("Mono" or "Kissing Disease")
Signs and symptoms	Fatigue, fever, sore throat, and enlargement of lymph nodes
Incubation period	Usually 1 to 2 months
Causative agent	Epstein-Barr virus (EBV), a DNA virus
Pathogenesis	Productive infection of epithelial cells of throat and salivary ducts; latent infection of B lymphocytes; hemorrhage from a ruptured spleen is a rare but serious complication.
Epidemiology	Spread by saliva; lifelong recurrent shedding of virus into saliva of asymptomatic, latently infected individuals.
Treatment and prevention	Treatment: analgesics and sometimes medications that reduce inflammation. Prevention: avoiding saliva of infected people.

kissing is an important mode of transmission in young adults, leading to the name "kissing disease." The donor of the virus is usually asymptomatic and may have been infected in the past without developing symptoms. By middle age, most people have antibodies to the virus, indicating prior infection. There is no animal reservoir.

Treatment and Prevention

Antiviral medications such as acyclovir and famciclovir inhibit productive infection by the virus and are helpful in rare serious cases; however, they have no activity against the latent infection. Analgesics (pain relievers) can be used to reduce localized symptoms. Cortisone-like medication is sometimes useful in relieving airway obstruction from swollen tissue.

Infectious mononucleosis can be prevented by avoiding the saliva of another person. This includes not sharing toothbrushes, drinking glasses, or any other object that may be contaminated with saliva. There is no vaccine for this disease. **Table 25.9** gives the main features of infectious mononucleosis.

Ebola Virus Disease (EVD) and Marburg Virus Disease (MVD)

Ebola virus disease (EVD; formerly called Ebola hemorrhagic fever) and Marburg virus disease (MVD; formerly called Marburg hemorrhagic fever) are both emerging diseases. Although typically found in Africa, they are a global concern because of their high case-fatality rate and person-to-person transmission. Both are considered Category A bioterrorism threats.

Signs and Symptoms

Signs and symptoms of EVD and MVD both appear about 5 to 10 days after infection (range 2 to 21 days). Patients initially have flu-like symptoms, including fever, headache, muscle pain, and sometimes a sore throat. The patient then develops abdominal pain, diarrhea, and a diffuse macular rash (small flat red spots). Hemorrhaging (bleeding) in the internal organs can also occur; not all patients experience this, however, which is why the names of the diseases were changed. The patient may bleed from mucous membranes, puncture sites, and body orifices, particularly the eyes. The case-fatality rates of the diseases vary considerably, ranging from 25% to 90%, depending on the infecting strain and the supportive treatment available. Patients with fatal disease usually die from multi-organ failure or shock within 2 weeks of disease onset.

Causative Agent

Ebola virus and Marburg virus are both members of the *Filoviridae* family, a group of filamentous, enveloped, single-stranded RNA viruses. Ebola virus was named after the Ebola River Valley in the Democratic Republic of the Congo (in Africa). Marburg virus was named for the German town of Marburg when several people developed the disease after handling infected African green monkeys.

Pathogenesis

Relatively little is known about the pathogenesis of Ebola and Marburg viruses. They primarily infect cells of the mononuclear phagocyte system, including dendritic cells and macrophages, but other cell types can be infected as well. Infection interferes with the innate and adaptive immune responses, allowing the viruses to replicate to high levels. Blood clotting is disrupted and capillaries may become leaky, leading to hemorrhaging.

Epidemiology

Ebola and and Marburg viruses are both endemic in Africa, and some evidence indicates that fruit bats are the reservoir. The diseases have been found in non-human primates but the high mortality rate in these infected animals makes it unlikely that they are natural reservoirs. People can become infected by handling an infected animal, but most cases are acquired from an infected person. Human-to-human transmission occurs through direct contact with body fluids of a symptomatic or deceased patient, a situation that puts health-care professionals and family members of infected individuals at greatest risk. Social and economic conditions in endemic areas often favor the development of outbreaks. The viruses can also be spread to other countries by infected travelers as well as in imported monkeys.

The most severe EVD outbreak to date started in 2014 in Guinea and spread to multiple countries in West Africa, including Liberia, Sierra Leone, Nigeria, and Senegal. Before the outbreak was controlled, over 11,000 deaths were reported in that region. Infected travelers carried the disease to other continents.

TABLE 25.10 Ebola Virus Disease and Marburg Virus Disease

Signs and symptoms	Fever, headache, muscle pain, possible sore throat; then abdominal pain, diarrhea, a rash, and possibly bleeding from internal organs and mucous membranes
Incubation period	Usually 5 to 10 days
Causative agents	Ebola virus and Marburg virus, filamentous, enveloped, single-stranded RNA viruses
Pathogenesis	Infect immune cells and interfere with immune responses and blood clotting.
Epidemiology	Both viruses endemic to Africa; fruit bats are possible reservoir; direct human-to-human transmission.
Treatment and prevention	Treatment: fluids and supportive therapy. Prevention: strict infection control measures; vaccines in clinical trials.

Treatment and Prevention

There is currently no standard treatment for EVD or MVD. Patients receive fluids and are given supportive therapy to maintain blood and O_2 levels. Medications may be given to control bleeding. An experimental monoclonal antibody therapy has shown promise. ⏮ monoclonal antibodies

Preventing EVD and MVD is difficult because the natural reservoirs have not been definitively established. Once EVD or MVD occurs in a community, quick diagnosis and rapid response by healthcare workers is essential to prevent spread to other individuals. Healthcare workers treating infected people must use strict infection-control measures. Vaccines against EVD and MVD are in clinical trials. The main features of these diseases are shown in **table 25.10.**

Yellow Fever

Yellow fever was first recognized with an epidemic in the Yucatan of Mexico in 1648, probably introduced there from Africa. There have been no outbreaks in the United States since 1905.

Signs and Symptoms

The signs and symptoms of yellow fever appear after 3 to 6 days, and can range from very mild to severe. Symptoms of mild disease, the most common form, are fever and a slight headache lasting a day or two. Patients suffering severe disease, however, may experience a high fever, nausea, bleeding, "black vomit" (from gastrointestinal bleeding), and jaundice (hence, the name *yellow fever*). The mortality rate of severe yellow fever cases can reach 50% or more. The reasons for the wide variation in severity of symptoms are unknown but the variation is probably due to the size of the infecting dose and the status of the host's defenses.

Causative Agent

Yellow fever is caused by an enveloped, single-stranded, RNA arbovirus (a virus transmitted by an arthropod) of the family *Flaviviridae.* The virus is transmitted by the bite of infected *Aedes* or *Haemagogus* mosquitoes, its biological vectors. ⏮ arbovirus

Pathogenesis

Once injected by a mosquito, the yellow fever virus multiplies in local host cells, enters the bloodstream, and is carried to the liver and other parts of the body. Liver damage results in jaundice and decreased production of clotting proteins. Injury to small blood vessels produces **petechiae**—tiny hemorrhages—throughout the body. Viral replication and the immune response to the infection damage the circulatory system, leading to bleeding in various tissues, and causing disseminated intravascular coagulation (DIC). Kidney failure is a common consequence of loss of circulating blood and low blood pressure. ⏮ biological vector, ⏮ DIC

Epidemiology

The reservoir of the yellow fever virus is mainly infected primates living in the tropical jungles of Central and South America and in Africa. Periodically, the disease spreads from the jungle reservoir to urban areas, where it is transmitted to humans by *Aedes* or *Haemagogus* mosquitoes.

Treatment and Prevention

There is no proven antiviral treatment for yellow fever, but supportive care addressing symptoms is important.

Yellow fever can be prevented in urban areas by spraying insecticides and eliminating the breeding sites of its main vector, *Aedes aegypti.* In the jungle, yellow fever control is almost impossible because the mosquito vectors live in the forest canopy and transmit the disease among monkeys. A highly effective attenuated vaccine is available to immunize people who might become exposed, including travelers to endemic areas. The main features of yellow fever are presented in **table 25.11.** ⏮ attenuated vaccine

Dengue Fever

Dengue fever (pronounced DENG-ee) is a mosquito-borne viral disease similar to, but milder than, yellow fever. The disease is also known as "break-bone fever" because it causes severe joint and muscle pain. The disease is found in many tropical areas and is endemic in more than 100 countries worldwide. It is the most common vector-borne viral disease in the world. Even though many infections are asymptomatic, the World Health Organization estimates that there are 50 to 100 million new cases every year and that one-half of the world's population is at risk of infection.

TABLE 25.11 Yellow Fever

Signs and symptoms	Often only headache and fever; severe cases characterized by high fever, jaundice, black vomit, and hemorrhages into the skin
Incubation period	Usually 3 to 6 days
Causative agent	Yellow fever virus, an enveloped, single-stranded RNA virus of *Flaviviridae*
Pathogenesis	Virus multiplies locally at site of introduction by an infected mosquito; spreads to the liver and throughout the body by the bloodstream; virus destroys liver cells, causing jaundice and decreased production of blood-clotting proteins; DIC and kidney failure may occur.
Epidemiology	Virus persists in forest primates and the mosquitoes that feed on them, in Africa and Central and South America; human epidemics occur when the virus infects mosquitoes that feed on humans.
Treatment and prevention	Treatment: no proven antiviral therapy is available. Prevention: highly effective attenuated vaccine.

MicroByte

The dengue outbreak in Hawaii in late 2015 resulted in over 250 confirmed cases; about 10% of those were in tourists.

Signs and Symptoms

Many dengue infections are asymptomatic or subclinical. If signs and symptoms occur, they begin with a sudden onset of fever that lasts for 2 to 7 days but may recur 12 to 24 hours later (biphasic fever). Other signs and symptoms include headache, joint and muscle pain, pain behind the eyes, and a rash that may become hemorrhagic, with development of petechiae. In mild cases, the rash does not develop, often resulting in a misdiagnosis of influenza or other viral disease. Mild mucosal bleeding may occur, typically involving the nose or the gums. Patients may also complain of sore throat. In some cases, abdominal pain, vomiting, and diarrhea occur.

Dengue fever is generally self-limiting and is rarely fatal. However, a severe form of the disease—dengue hemorrhagic fever (DHF)—can occur in patients who experience a second dengue infection. DHF may lead to dengue shock syndrome (DSS), which causes a weak, rapid pulse, moist skin, restlessness, and low blood pressure. Petechiae, bruising, and sometimes massive gastrointestinal and vaginal bleeding occur. Later, widespread blood clotting and disseminated intravascular coagulation (DIC) may develop. DHF is mostly seen in children under 15 years old and can be fatal in this population. ◀◀ DIC

Causative Agent

Dengue fever is caused by a single-stranded RNA virus of the *Flaviviridae*. There are four widely spread serotypes of the virus (DENV1, DENV2, DENV3, and DENV4). These are all transmitted by the bite of an infected mosquito of the *Aedes* family.

Pathogenesis

When a mosquito carrying dengue virus feeds on a human host, viral particles are injected into the bite wound, where they infect epidermal dendritic cells and keratinocytes (epidermal cells that produce keratin, a waterproofing protein). Some of the dendritic cells migrate to the lymph nodes, where monocytes and macrophages are recruited and also become infected. Dengue virus multiplies in those cells and the resulting viral particles are disseminated systemically within them. Humoral and cell-mediated immune responses clear the virus. The details of dengue fever pathogenesis are not well understood because there are no animal models for the disease.

The pathogenesis of dengue hemorrhagic fever (DHF) and dengue shock syndrome (DSS) is better understood. DHF and DSS generally occur in people who experience a second dengue infection with a strain of the virus different from that of their first infection. **Antibody-dependent enhancement (ADE)** of infection has been suggested to explain severe dengue disease. In this model, pre-existing dengue antibodies from a primary dengue infection recognize the virus in the second infection and bind to it, forming immune complexes. However, because the preformed antibodies are against a different serotype of the virus, they do not neutralize the second virus. Instead, they facilitate viral entry into cells that express Fc receptors, especially monocytes and macrophages, where the viruses then freely replicate. ADE thus leads to increased numbers of infected cells and a high viral load.

During a second dengue infection, the infected monocytes and macrophages produce pro-inflammatory cytokines, resulting in a cytokine storm. This, along with activated complement and cytokines released by memory T cells, causes changes in vascular permeability, leading to plasma leakage that may result in respiratory distress and dehydration. Capillaries become fragile, resulting in the observed hemorrhages (petechiae, easy bruising, and gastrointestinal and vaginal bleeding). Infected monocytes and macrophages die and release toxic products that cause blood clotting and DIC. Leukopenia (decrease in white blood cell number) occurs. Blood pressure drops, sometimes to life-threatening levels. ◀◀ cytokine storm

Epidemiology

Dengue is an emerging disease and the fastest-spreading mosquito-borne viral disease in the world. It is found in mostly tropical and subtropical regions that have high rainfall and plenty of fresh water, conditions that aid in disease development because the mosquito vector breeds in standing water. The most common vector is *Aedes aegypti,* the same species that carries yellow fever (**figure 25.14**). Another important vector of dengue virus is *Aedes albopictus,* the Asian tiger mosquito, now spreading worldwide. Geographic spread of the disease is linked to international travel, breakdown of vector control measures, and increase in mosquito-breeding habitats—such as used tires and containers in which water has accumulated.

FOCUS ON A CASE 25.1

The patients were two American boys, 9 and 11 years old, living in Thailand, who developed irritated eyes and slightly runny noses, progressing to fever, headache, and severe muscle pain. It hurt them to move their eyes, and they refused to walk because of pain in their legs. The symptoms subsided after a couple of days, only to recur at lesser intensity. No treatment was given, and they were completely normal within a week. Their illness was diagnosed as dengue, also known as "breakbone fever."

1. Why was it significant that the boys were living in Thailand rather than the United States?
2. What would be the risk if the boys became infected with a different strain of dengue virus?
3. What can the boys do to prevent reinfection with dengue?

Discussion

1. Dengue, which is spread by mosquitoes, is endemic in large areas of the tropics and subtropics around the world. It is currently found in Southeast Asia, the western Pacific, the eastern Mediterranean, the Americas (South American countries and along the Gulf Coast of the United States), and rural areas in parts of Africa and the Caribbean. Dengue is not common in the United States—most reported cases in this country have occurred in travelers returning from countries where dengue is endemic. The boys, being in Thailand, were living in a country in which dengue is endemic and were thus at risk of contracting the disease.
2. The boys had a mild form of the dengue, characterized by fever, headache, muscle aches, rash, nausea, and vomiting. If they were to become infected again, and with a different strain of the virus, they would be at risk of developing dengue hemorrhagic fever (DHF) or dengue shock syndrome (DSS). These forms of the disease are much more serious than regular dengue. People who have DHF or DSS suffer from capillary leakage, bleeding, and bruising. Blood pressure drops and disseminated intravascular coagulation (DIC) may occur—these can lead to shock and death.
3. The boys can protect themselves from getting re-infected with dengue by taking precautions to avoid mosquito bites. These precautions include sleeping under a bed net (preferably impregnated with insecticide), wearing long sleeves and pants, and avoiding areas where mosquitoes breed, such as standing water. Also, a vaccine has recently become available in some countries, including Thailand.

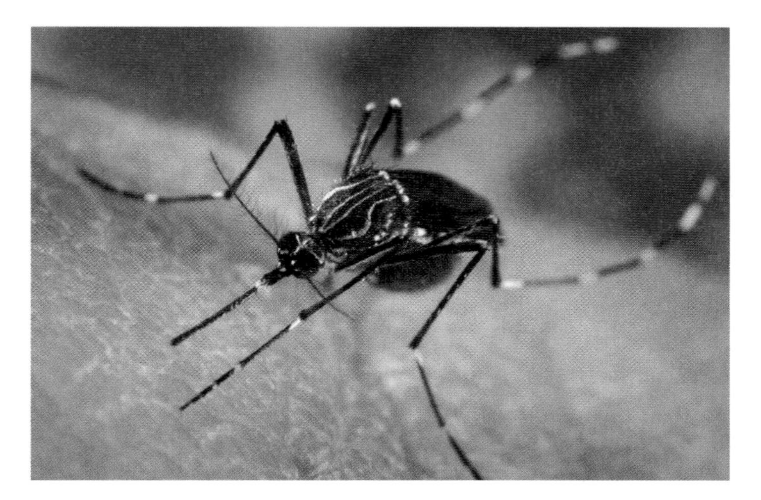

FIGURE 25.14 ***Aedes aegypti*** **mosquito taking a blood meal** Female mosquitoes require blood for development of their eggs. During a bite, she pumps saliva through her hollow mouthpart (proboscis), which increases blood flow and prevents clotting. Infectious agents picked up from the host's capillaries multiply within her body and can be transferred to a new host in a later bite.
Source: Paul I. Howell, MPH; Prof. Frank Hadley Collins/CDC

? Why is it important to avoid being bitten by a mosquito when you have a blood-borne infection?

Treatment and Prevention

There is no specific treatment for dengue fever. Patients are given analgesics (pain relievers). They should avoid aspirin and nonsteroidal anti-inflammatory drugs such as ibuprofen because these medications can worsen bleeding associated with dengue. Aspirin has been linked to Reye's syndrome in children who have the viral infection. Oral rehydration therapy is given to replace fluid lost from fever and vomiting. Platelet or blood transfusions are given for dengue hemorrhagic fever (DHF) if bleeding and plasma leakage have occurred. Efforts are underway to develop antiviral medications. ⏮ Reye's syndrome, ⏮ oral rehydration therapy

Prevention and control of dengue and DHF have become more important with their increasing geographic distribution and incidence. Measures include controlling or eradicating the mosquito population by destroying their breeding sites (standing water). Environmental insecticide use is also important, although increasing insecticide resistance is becoming a problem. Biological control of the vectors (for example, using fish that eat the mosquito larvae) has been successful in some countries. Other prevention measures include educating people in the use of insecticide-treated mosquito bed nets and insect repellents.

Significant effort has been made in developing a dengue vaccine—the first was approved for use in several countries in 2016. It is a recombinant attenuated virus that includes components from all four dengue virus strains. WHO recommends use of the vaccine in areas of highest prevalence of the disease in persons who are at least 9 years of age. Early trials indicate that it is about 60% effective in preventing disease, but may actually increase the incidence of serious disease in children 2–5 years of age. The main characteristics of dengue fever are presented in **table 25.12.**

TABLE 25.12 Dengue Fever, Chikungunya, and Zika Virus Disease Compared

	Dengue Fever	Chikungunya	Zika Virus Disease
Signs and symptoms	Often asymptomatic; fever, headache, rash, and severe joint pain; in dengue hemorrhagic fever (DHF), bleeding and shock can occur, as well as disseminated intravascular coagulation (DIC).	Similar to dengue fever, but followed by severe joint pain that may become chronic	Usually asymptomatic; mild disease with fever, rash, joint pain, red eyes; rare nervous system involvement; congenital Zika syndrome
Incubation period	Usually 4 to 7 days	Usually 3 to 7 days	2 to 14 days
Causative agents	Dengue virus serotypes DENV1, DENV2, DENV3, and DENV4; RNA virus of *Flaviviridae*	Chikungunya virus; RNA virus of *Togaviridae*	Zika virus; RNA virus of *Flaviviridae*
Pathogenesis	Pro-inflammatory cytokines cause leaky blood vessels, dehydration, and hemorrhaging. In DHF, DIC and shock may be fatal.	Release of cytokines that affect immune cells; bone destruction.	Virus binds to receptors on a variety of cells; enters fluid around fetus and brain; affects neural stem cells.
Epidemiology	Mosquito-borne; found predominantly in tropical and subtropical regions but range is increasing; DHF usually occurs in children under 15 years old.	Mosquito-borne; mainly in Africa and Asia, but now in Europe and the Americas.	Mosquito-borne and sexually transmitted; females should avoid pregnancy for 8 weeks after exposure; males should use condoms for 6 months.
Treatment and prevention	Treatment: analgesics for pain; oral rehydration therapy and blood or platelet transfusions if bleeding occurs. Prevention: vector control; vaccine in limited areas.	Treatment: analgesics for pain and oral rehydration. Prevention: vector control.	Treatment: no specific treatment. Prevention: vector control.

Chikungunya

Chikungunya (pronounced chik-en-gun-ye), commonly known as CHIK, got its name from an African word meaning "that which bends up" because people with the disease show bent posture due to severe joint pain. Although CHIK is not a new disease, there have been several recent outbreaks, making it an emerging disease.

Signs and Symptoms

The signs and symptoms of CHIK are similar to those of dengue and include fever that typically lasts 2 to 5 days, followed by severe joint pain (especially in the extremities—fingers, toes, wrists, and ankles) that can persist for weeks or months. Sometimes patients develop chronic joint pain that differentiates CHIK from dengue fever. Patients often develop a rash. Nonspecific symptoms include headache, conjunctivitis and photophobia, back pain, nausea, and general malaise. The disease is seldom fatal.

Causative Agent

CHIK is caused by chikungunya virus, an enveloped, single-stranded RNA virus of the *Togaviridae* family. It is transmitted by *Aedes* mosquito species, mostly *A. aegypti* and *A. albopictus.*

Pathogenesis

The disease mechanism of chikungunya is largely unknown, but it appears, like dengue fever, to cause its damaging effects by affecting the immune system. Infection stimulates significant release of cytokines and eventual decrease of helper T cells. Severe joint disease may be related to loss of bone cells as a result of viral infection.

Epidemiology

CHIK occurs mainly in Africa, Asia, and Southeast Asia, where the mosquito vectors are common. Recent epidemics have occurred in several other areas, however, associated with the inadvertent introduction and spread of the mosquito *A. albopictus.* This vector is now present in Brazil, Central America, the United States, and many European countries, where it has adapted to favorable environmental conditions. In 2013, Chikungunya virus was reported for the first time in Caribbean nations. Since then, it has spread throughout the Americas.

Treatment and Prevention

There is no specific treatment for CHIK. Analgesics and nonsteroidal anti-inflammatory drugs such as ibuprofen are used to reduce the joint pain. Fluids are given to reduce dehydration from fever. As with most mosquito-borne diseases, CHIK can be prevented by effective vector control—destroying the vector and protecting the population by use of insect repellents and insecticide-impregnated mosquito netting. A vaccine for CHIK is currently in clinical trials. The main characteristics of chikungunya are presented in table 25.12.

Zika Virus Disease

Zika virus (ZIKV) dominated media reports in May 2015 when it was associated with a severe birth defect called microcephaly (small brain) in Brazil as travelers from around the world were preparing to visit that country for the 2016 Summer Olympics. Zika virus was first isolated in 1947 in monkeys of the Zika Forest of Uganda. It was discovered in humans in 1952, but documented cases were rare before 2007, when an outbreak in Micronesia signaled its emerging status. Since then, the virus has rapidly appeared across the Americas. In August 2016 the CDC issued the first Zika transmission warning in the United States for an area in south Florida.

Signs and Symptoms

Zika virus infection is asymptomatic in 75–80% of cases. Mild disease may appear 2 to 14 days after the virus enters the body, characterized by fever, rash, joint pain, and red eyes. These fade in 2–7 days. Complications of Zika virus disease include a possible connection to Guillain-Barré syndrome (a rare nervous disorder), and congenital Zika syndrome in the fetus of an infected pregnant woman. ⏮ Guillain-Barré syndrome

Causative Agent

Zika virus (ZIKV) is an enveloped, single-stranded RNA virus in the family *Flaviviridae,* and is transmitted by *Aedes* mosquitoes.

Pathogenesis

Studies indicate that when Zika virus enters the host, it binds to a receptor found on a number of different human tissues, which helps to explain the potential involvement of the skin, joints, nerves, and eyes. Unlike other flaviviruses, ZIKV has been detected in the fluid surrounding a fetus as well as in its brain, regions that are typically immunologically privileged, meaning that they are isolated from destructive immune mechanisms (see Focus Your Perspective 17.1).

Microcephaly has only recently been recognized as a consequence of Zika virus infection, but since the 2015 outbreak in Brazil, researchers have recognized that the pathology in the brains of newborns from infected mothers is more extensive—a condition called congenital Zika syndrome. ZIKV preferentially infects neural cells, and in particular, neural progenitor (stem) cells from which the brain develops. Neural progenitor cells are present throughout fetal development, so infection during any trimester of pregnancy can damage the brain. Even newborns with normal head size can rapidly exhibit developmental delays and neurological abnormalities.

Epidemiology

ZIKV is transmitted by the bite of infected *Aedes* mosquitoes. Most cases involve *A. aegypti,* a species that feeds primarily on people and survives best in warm climates. *A. albopictus* probably transmits the disease less often because it feeds on various animals and therefore is less likely to bite people. It is a concern, however, because it tolerates cooler climates, and thereby has a wider geographic range. In fact, its distribution has expanded as the mosquito has inadvertently been introduced to countries around the globe.

ZIKV is also sexually transmitted. ZIKV RNA has been detected in blood, semen, saliva, and secretions of the female genital tract, as well as in other body fluids. Females should avoid getting pregnant for at least 8 weeks after possible exposure. Males should avoid unprotected sex for 6 months after exposure, as the virus can be found in the semen for that long after infection. In 2016, the CDC established the U.S. Zika Pregnancy Registry to monitor infections and to provide recommendations and services for women who are concerned about infection during pregnancy.

Treatment and Prevention

No specific treatment is used for Zika virus infection. Aspirin and non-steroidal pain relievers should be avoided until the possibility of infection with dengue fever virus has been eliminated because it could worsen the hemorrhaging associated with that disease.

There is no vaccine against ZIKV, so the best preventive measure is control of the mosquito vector. Sources of standing water where mosquitoes can breed should be eliminated, both inside and outside. Long sleeves and pants along with use of mosquito nets will help people to avoid bites. In August 2016 the FDA approved release of modified male *Aedes* mosquitoes whose offspring would die before they could reproduce. The goal was to reduce the effective vector population.

Infection with ZIKV results in immunity, but because of the devastating effect on a developing fetus, there is a great effort to develop a vaccine to to prevent the disease. Potential vaccines are under development and some have been successful at eliciting immunity in mice. The main characteristics of Zika virus disease are presented in table 25.12.

MicroAssessment 25.3

Infectious mononucleosis occurs worldwide and is transmitted from person to person by saliva. Ebola virus disease (EVD) and Marburg virus disease (MVD) are both severe emerging diseases of global concern because they can result in massive bleeding and multi-organ failure. Yellow fever virus is transmitted to humans by mosquitoes. Dengue fever (break-bone fever) is the most common vector-borne disease in the world and is similar to, but milder than, yellow fever. A severe form of the disease, dengue hemorrhagic fever, can be fatal. Chikungunya is a newly emerging disease that, like dengue, causes severe joint pain. Zika virus disease is usually mild, but can produce severe birth defects.

7. What characteristic changes occur in the blood of patients with infectious mononucleosis?
8. Compare and contrast the global spread of dengue fever, chikungunya, and Zika virus disease.
9. Why must a vaccine against dengue fever be effective against all strains of the virus?

25.4 ■ Protozoan Diseases of the Blood and Lymphatic Systems

Learning Outcomes

8. Outline the *Plasmodium vivax* life cycle.
9. Discuss why malaria has been so difficult to control.

Protozoa infect the blood and lymphatic systems of millions of people globally. Malaria is a widespread protozoan disease that is a leading cause of morbidity (illness) and mortality (death) worldwide. Other protozoa that cause blood and lymphatic diseases include *Trypanosoma brucei,* a bloodstream parasite that causes African sleeping sickness in humans, and *T. cruzi,* which causes Chagas' disease (American trypanosomiasis) that often manifests as a chronic heart infection. The most severe symptoms of *T. brucei* infection involve the central nervous system, so this parasite is covered in nervous system diseases (chapter 26).

Malaria

Malaria is an ancient disease and one of the most serious infectious diseases worldwide. The name means "bad air," but the disease is not related to air quality—rather, the disease is spread by mosquitoes that fly through the air. In 1902, Ronald Ross, an Indian-born British physician, received a Nobel Prize for determining the life cycle of the protozoan cause of malaria.

In 1955, the World Health Organization (WHO) began a program for the worldwide elimination of malaria, using insecticides such as DDT to kill the mosquito vector, and diagnosing and treating infected patients. The program was initially successful—52 nations participated, and by 1960, 10 of them had eradicated malaria. Unfortunately, the mosquito vectors began to develop resistance to insecticides, and malaria began to reappear. In 1976, the WHO acknowledged that the eradication program was a failure. A CDC surveillance report for 2015 estimated 214 million cases of malaria and 438,000 deaths globally.

MicroByte

Most of the people who die from malaria each year are children under the age of 5 who live in sub-Saharan Africa.

Signs and Symptoms

The first signs and symptoms of malaria are flu-like, with fever, headache, and pain in the joints and muscles. These generally begin about 2 weeks after being bitten by an infected mosquito, but in some cases they start many weeks later. After 2 or 3 weeks of these vague symptoms, the pattern changes and symptoms fall into three phases: (1) cold phase—the patient feels cold and develops shaking chills that can last for as much as an hour; (2) hot phase—the temperature begins to rise sharply, often reaching 40°C (104°F) or more; and (3) wet phase—the patient's temperature falls, and drenching sweating occurs. This cycle of intense symptoms—chills, fevers, and sweats—is called a **paroxysm.** Except for fatigue, the patient feels well until 24 or 48 hours later, when the next paroxysm occurs. In some rare cases, malaria affects the brain, causing life-threatening illness.

Causative Agent

Human malaria is caused by protozoa of the genus *Plasmodium,* which are transmitted by infected female mosquitoes of the genus *Anopheles.* Five species of *Plasmodium* are involved—*P. vivax, P. falciparum, P. malariae, P. ovale,* and *P. knowlesi* (this last is a simian species that is more frequently being identified as a cause of human malaria now that molecular testing is available). These species differ in microscopic appearance and sometimes in life cycle. Moreover, malaria caused by different species differs in severity and treatment.

The *Plasmodium* life cycle is complex, consisting of a liver stage (exoerythrocytic stage; *exo-* means "outside of," and *-erythrocytic* refers to red blood cells) and a red blood cell (erythrocytic) stage. The exoerythrocytic stage begins when the infectious form of the protozoan is injected into the human host (**figure 25.15** ①). This form—called a **sporozoite**—is carried by the bloodstream to the liver, where it infects hepatocytes (liver cells). ② There, each parasite enlarges and divides asexually, producing thousands of **merozoites,** which are then released into the bloodstream. Some species of *Plasmodium* are able to form hypnozoites, which can live in the hepatocytes for years before reproducing to form merozoites. The merozoites released from liver cells infect red blood cells (RBCs), beginning the red blood cell stage of the protozoan's life cycle.

③ In an RBC, the malarial parasite develops first into a ring form, then into a larger motile **trophozoite** (feeding stage) and then a **schizont** (reproducing stage). The schizont gives rise to merozoites that are released when the RBC ruptures. The merozoites then enter new RBCs and multiply, repeating the cycle.

Some merozoites that enter RBCs develop into **gametocytes** (specialized sexual forms), rather than becoming schizonts. These sexual forms cannot develop further in the human host, do not rupture RBCs, and are not important in causing the symptoms of malaria. They are, however, infectious for certain species of *Anopheles* mosquitoes and are thus ultimately responsible for transmitting malaria from one person to another.

④ When a mosquito feeds on an infected person's blood, it ingests the infected RBCs. ⑤ It digests the RBCs, releasing the gametocytes. Shortly after entering the intestine of the mosquito, and stimulated by the drop in temperature, the male and female gametocytes change in form to become gametes.

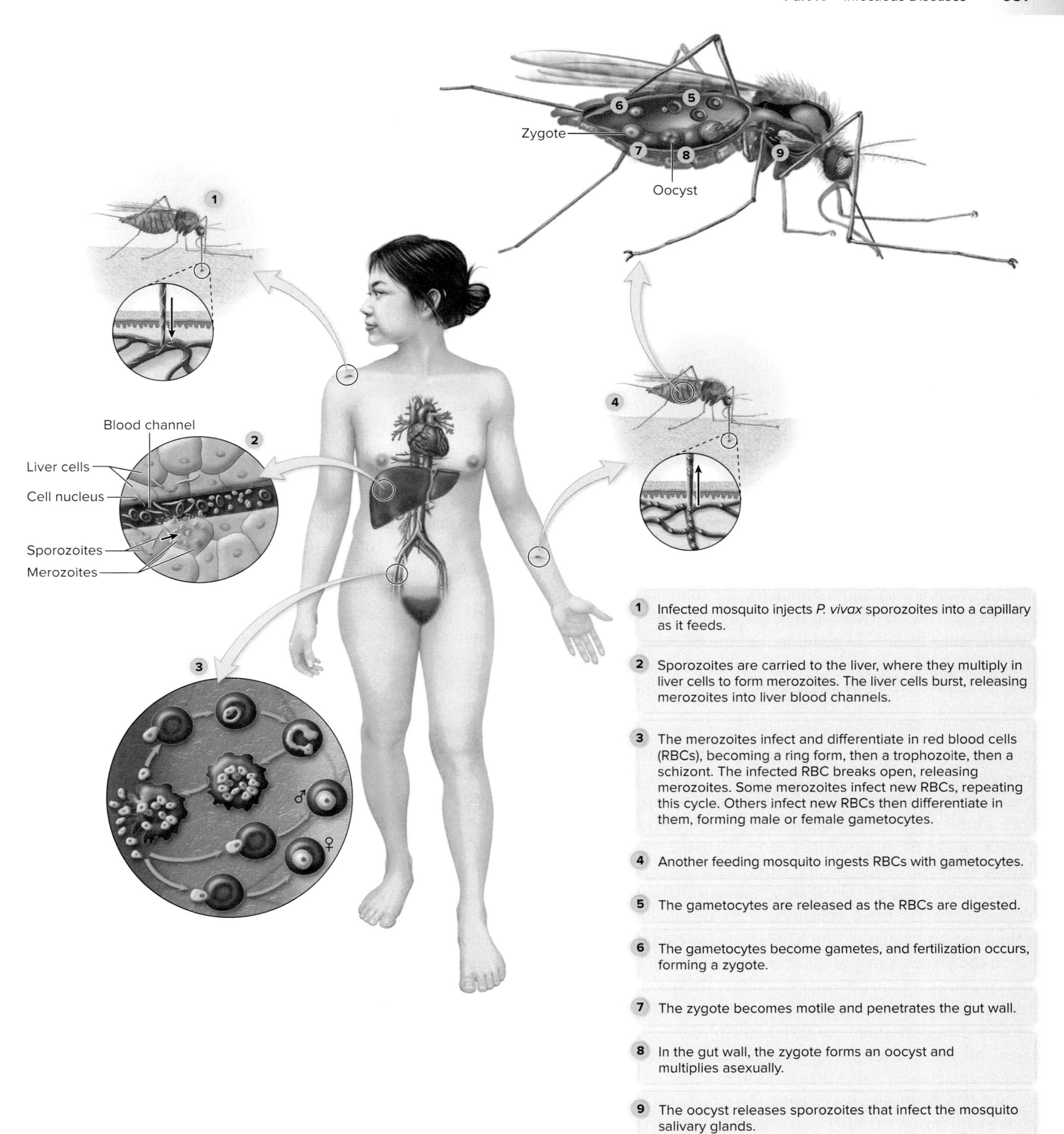

FIGURE 25.15 Life Cycle of *Plasmodium vivax*

Why do only female *Anopheles* mosquitoes transmit malaria?

⑥ The male gametocyte transforms into about six tiny, whip-like gametes that swim about until they unite with the female gamete (in much the way that sperm and ovum unite in animals), forming a zygote.

⑦ The zygote transforms into a motile form that burrows into the wall of the midgut of the mosquito and forms an oocyst. ⑧ The oocyst enlarges as the diploid nucleus undergoes meiosis, dividing asexually into numerous offspring. ⑨ The oocyst ruptures into the body cavity of the mosquito, releasing sporozoites that then find their way to the mosquito's salivary glands and saliva, from which they may be injected into a new human host.

Pathogenesis

The characteristic feature of malaria—recurrent paroxysms followed by feeling healthy again—results from the cycle of growth and release of malarial parasites from RBCs. Interestingly, the infections in all the millions of RBCs become nearly synchronous. Thus, all infected RBCs rupture to release merozoites at about the same time, causing a rapid increase of antigens in the bloodstream that triggers a fever.

Malarial parasites cause anemia by destroying RBCs and converting the iron in hemoglobin to a form not readily recycled by the body. The large amount of foreign material in the bloodstream strongly stimulates the immune system. In some cases, the overworked immune system fails and immunodeficiency results. In addition, the spleen enlarges to cope with the high levels of foreign material and abnormal RBCs that it removes from the circulation. As a result, the spleen may rupture.

P. falciparum infections are typically more severe than other types of malaria. One reason for this is that *P. falciparum* can infect all RBCs (leading to very high levels of parasitemia), whereas the other common *Plasmodium* species infect only either young or old RBCs. In addition, *P. falciparum* causes the infected RBCs to stick to the walls of capillaries. When this happens, the RBCs block the blood flow in the capillaries and thereby deprive the nearby tissues of O_2. If this occurs in the brain, it leads to a life-threatening outcome called cerebral malaria. The RBCs can also clog the capillaries in the placenta of a pregnant woman, leading to devastating outcomes for the developing fetus. *P. falciparum*–infected RBCs stick to the capillaries because the parasite produces a protein that inserts into the membrane of the infected RBCs. That protein, called *Plasmodium falciparum* erythrocyte membrane protein (pfEMP1) is an important virulence factor for *P. falciparum*. By causing the infected RBCs to stick to the capillaries, pfEMP1 prevents those cells from circulating and being cleared by the spleen. *P. falciparum* has over 60 different genes that encode pfEMP1, and this antigenic variation allows the parasite to evade the adaptive immune system. A given infected RBC will have only one variety of pfEMP1 inserted into the membrane, but different infected RBCs could have different varieties. Recent studies suggest that only some of the varieties of pfEMP1 cause cerebral malaria, and others cause the complications in pregnancy. ⏮ antigenic variation

Generally, people who live in areas where malaria is endemic develop immunity from repeated infections, giving them some protection from the disease. Newborns have partial protection from the disease because of maternal antibodies. That passive immunity decreases over time, so the greatest risk of death from malaria is to children over 6 months of age.

P. vivax and *P. ovale* malaria often relapse because the parasites form hypnozoites (dormant forms) that persist in the liver. Months or even years later, hypnozoites can begin growing in the liver, starting new erythrocytic cycles of infection after the earlier bloodstream infection has been cured.

Epidemiology

Malaria was once common in both temperate and tropical areas of the world; endemic malaria was eliminated from the continental United States only in the late 1940s. Today, malaria is mostly a disease of warm climates, but even so, almost half of the world's population lives in endemic areas (**figure 25.16**). Human-biting mosquito species of the genus *Anopheles* are biological vectors of malaria, but only the females transmit the disease because males do not feed on blood. Malaria can also be transmitted by blood transfusions or among intravenous drug users who share syringes. Malaria contracted in this manner is easier to treat because it involves only red blood cells and not the liver—only sporozoites from mosquitoes can infect the liver. Some people of black African heritage are genetically resistant to *P. vivax* malaria because their RBCs lack an antigen called the Duffy antigen that the parasite uses to enter the RBC. Also, people with certain genetically determined blood diseases such as sickle cell anemia are partially protected against the disease. ⏮ biological vector

Treatment and Prevention

Treatment of malaria is complicated by the fact that different stages in the life cycle of the parasite respond to different medications. Most medications used for treatment are active against the erythrocytic stages of the parasite. These include chloroquine, quinine, and mefloquine. However, chloroquine and mefloquine will not kill the hypnozoites of *P. vivax* or *P. ovale* or the gametocytes of *P. falciparum*. Other medications such as primaquine are generally effective against them. The same medications can be used to prevent malaria.

The development of drug-resistant malaria further complicates treatment choice. If the patient has a history of travel to a region where resistance is common, fewer treatment options are available. In an effort to prevent the development of resistance, combination therapies are now used for treating malaria. Derivatives of artemisinin, the active ingredient of an ancient Chinese herbal medicine, are now regularly given

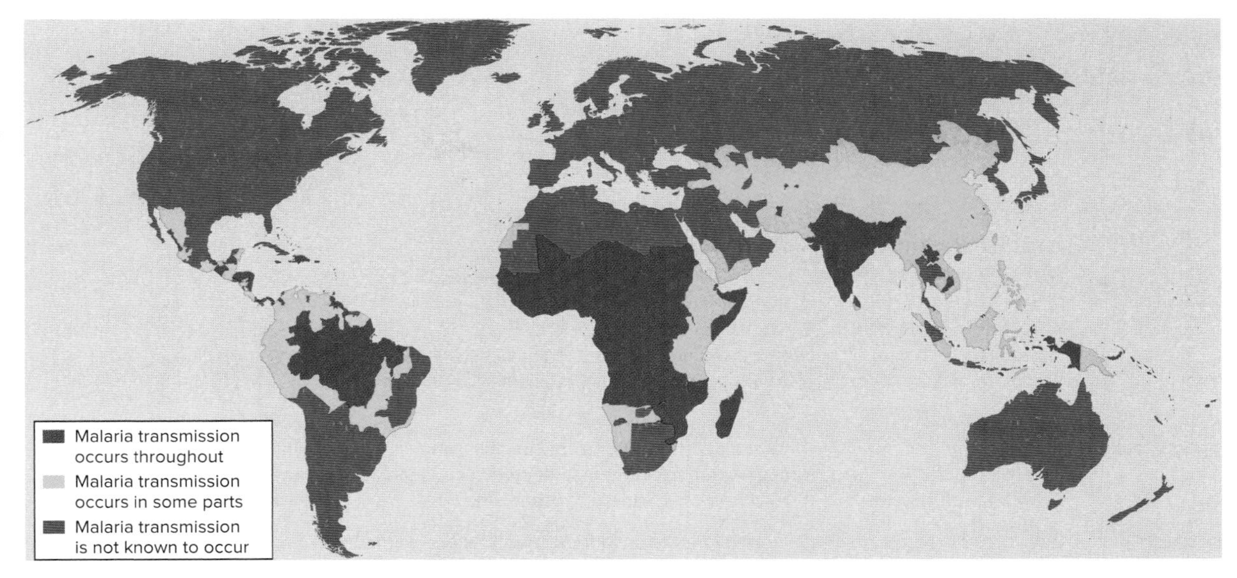

FIGURE 25.16 Distribution of Malaria Source: www.cdc.gov/malaria/malaria_worldwide/impact.html

Why might someone in the United States be diagnosed with malaria?

in combination with traditional antimalarial medication. These combinations, called ACTs (artemisinin-based combination therapies), are widely available and are helping to treat malaria.

Malaria prevention has become a global focus. In 1998, a new disease-fighting initiative called Roll Back Malaria was started (involving the WHO, the United Nations Children's Fund [UNICEF], the United Nations Development Program, and the World Bank). The goal was to reduce malaria deaths by 75% by 2015. The program included organizing and funding a sustained effort to improve access to medical care, strengthening local health facilities, and supplying medications and long-lasting insecticide-treated mosquito nets. In recent years, private foundations such as the Bill and Melinda Gates Foundation have donated significant amounts of money for work on diagnosing and preventing malaria and for vaccine development. Although the 1998 Roll Back Malaria goal was not reached, these combined efforts resulted in a 60% decline in global malaria death rates between 2000 and 2015.

For malaria to be controlled, new medications are needed. ACTs are helping reduce the incidence of malaria, but recently, artemisinin-resistant strains of *P. falciparum* have been reported. The best way to prevent malaria will be through vaccination. A number of potential vaccines are being tested, the most promising of which so far gives only 30% protection to children. Indoor spraying of the insecticide DDT, the use of insecticide-impregnated bed nets, and elimination of mosquito breeding areas are also being emphasized. Adequate funding of malaria control is a continuing problem. **Table 25.13** gives the main features of malaria.

The key features of the diseases covered in this chapter are highlighted in the **Diseases in Review 25.1** table that follows.

TABLE 25.13 Malaria

Signs and symptoms	Recurrent cycles of intense chills and fever alternating with feeling healthy
Incubation period	Varies with species; 6 to 37 days
Causative agents	Five species of protozoa of the genus *Plasmodium*.
Pathogenesis	Protozoan enters hepatocytes and multiplies in them; cells burst and release protozoa causing fever; organisms infect red blood cells, differentiating in them; spleen enlarges in response to removing large amount of foreign material and many abnormal blood cells from the circulation; with *P. falciparum* infection, red blood cells stick together and to walls of capillaries, blocking vessels and depriving tissue of O_2.
Epidemiology	Transmitted from person to person by bite of infected *Anopheles* mosquito.
Treatment and prevention	Treatment: usually ACTs; other medicines if sensitivity known. Prevention: the same medications used for treatment; eradication of mosquito vectors; mosquito netting impregnated with insecticide; vaccines under development.

MicroAssessment 25.4

There are a number of different stages in the life cycles of the protozoa that cause malaria, and they have differing microscopic appearance, antigenicity, and susceptibility to antimalarial medications. Malaria is a leading serious, infectious disease worldwide, and sustained, well-funded measures are needed to control it.

10. Why does malaria sometimes cause the spleen to rupture?
11. What causes the recurrent paroxysms that characterize malaria?
12. Why is malaria contracted from a blood transfusion easier to treat than malaria contracted from a mosquito?

Diseases in Review 25.1

Blood and Lymphatic Infections

Disease	Causative Agent	Comment	Summary Table
BACTERIAL INFECTIONS			
Infective endocarditis	Pathogens from another infected site, or normal microbiota of the skin or mouth	Results from organisms infecting a surface of the heart; people with indwelling catheters, artificial heart valves, or heart abnormalities, or who abuse intravenous drugs are most at risk.	Table 25.1
Sepsis and septic shock	Often *E. coli*, *Bacteroides*, or *Pseudomonas* species	Characterized by fever, shaking chills, and rapid pulse and respiration; due to widespread release of cytokines in response to systemic infection, involving a wide variety of organisms.	Table 25.2
Plague ("black death")	*Yersinia pestis*	Spread by flea bites (bubonic plague) and aerosols (pneumonic plague); endemic in rodent populations; Category A bioterrorism agent	Table 25.4
Lyme disease	*Borrelia burgdorferi*	Spread by ticks; often leads to "bull's-eye rash"; later symptoms Include injury to heart and nervous system, and arthritis.	Table 25.5
***Vibrio vulnificus* infection**	*Vibrio vulnificus*	If in the bloodstream, organism causes sepsis and shock, particularly in those with compromised immunity; acquired by eating contaminated seafood or through wounds.	Table 25.6
Tularemia ("rabbit fever")	*Francisella tularensis*	Can be acquired in multiple ways, often by direct contact with an infected animal; characterized by an ulcer at entry site, fever, and enlarged lymph nodes; Category A bioterrorism agent.	Table 25.7
Brucellosis ("undulant fever")	*Brucella melitensis*	Characterized by recurrent fever, weakness, enlarged lymph nodes, and weight loss; acquired via contact with an infected animal or foodborne; causative agent enters via mucous membranes or breaks in the skin.	Table 25.8
VIRAL INFECTIONS			
Infectious mononucleosis ("mono")	Epstein-Barr virus (EBV)	Long incubation period, followed by fatigue that can last for months, fever, sore throat, and enlarged lymph nodes; transmitted in saliva; infects B cells, causing both productive and latent infections.	Table 25.9
Ebola and Marburg virus diseases	Ebola viruses and Marburg virus	Severe, often fatal diseases characterized by fever and internal and external bleeding; spread person to person by body fluids; Category A bioterrorism agents.	Table 25.10
Yellow fever	Yellow fever virus	Transmitted by mosquitoes; disease can be mild or severe; severe form characterized by fever, bleeding, black vomit, and jaundice, and often fatal. Vaccine available for prevention.	Table 25.11
Dengue fever	Dengue virus	Transmitted by mosquitoes; generally mild, although symptoms include severe joint pain; later infection can result in dengue hemorrhagic fever (usually in children), which may be fatal.	Table 25.12
Chikungunya (CHIK)	Chikungunya virus	Transmitted by mosquitoes; characterized by fever and severe joint pain, which may become chronic.	Table 25.12
Zika virus disease	Zika virus	Transmitted by mosquitoes or through sexual contact; mild disease, but can cause congenital Zika syndrome in fetus.	Table 25.12
PROTOZOAN INFECTIONS			
Malaria	*Plasmodium* species	Transmitted by mosquitoes; characterized by paroxysms of chills, fevers, and sweats; *P. falciparum* infections are life-threatening, even in otherwise healthy people.	Table 25.13

Summary

25.1 ■ Anatomy, Physiology, and Ecology of the Blood and Lymphatic Systems (figure 25.1)

The Heart
The left side of the heart receives oxygenated blood from the lungs and pumps it into the thick-walled arteries for delivery to the rest of the body; the right side of the heart receives blood from the veins and pumps it to the lungs, where it is replenished with O_2.

Blood Vessels
Blood vessels include arteries, veins, and capillaries. Arteries have thick, muscular walls and carry blood away from the heart. Exchange of materials occurs across the walls of capillaries. Veins carry blood toward the heart.

Lymphatics (Lymphatic Vessels) (figure 25.2)
Lymphatics—lymph vessels—take up fluid that leaks from capillaries. They also take up bacteria, which are normally trapped by lymph nodes distributed along the course of the lymphatics.

Spleen
The spleen filters unwanted material such as bacteria and damaged RBCs from the arterial blood. It becomes enlarged in diseases such as infectious mononucleosis and malaria.

25.2 ■ Bacterial Diseases of the Blood and Lymphatic Systems

Infective Endocarditis (table 25.1)
Infective endocarditis is often caused by members of the normal microbiota in people with indwelling catheters or heart abnormalities, or who abuse intravenous drugs. It also may be caused by pathogens that have spread from another infected site.

Sepsis and Septic Shock (table 25.2; figure 25.3)
Sepsis, an infection-induced systemic inflammatory response, is commonly a healthcare-associated illness; many afflicted individuals have serious underlying illnesses such as cancer and diabetes. If uncontrolled, it can progress to **septic shock.** Most fatal cases involve Gram-negative bacteria.

Plague ("Black Death") (table 25.4)
Plague is caused by *Yersinia pestis,* a member of the *Enterobacteriaceae* with many virulence factors that interfere with phagocytosis and immunity (table 25.3, figure 25.4). **Bubonic plague** is transmitted to humans by fleas (figure 25.5). **Pneumonic plague** is transmitted from person to person by aerosols. If the bacteria enter the bloodstream, **septicemic plague** may occur.

Lyme Disease (table 25.5; figure 25.8)
Lyme disease is characterized by stages of disease and is caused by a spirochete, *Borrelia burgdorferi* (figure 25.7). A bull's-eye rash is characteristic of the early stage of the disease (figure 25.6). *B. burgdorferi* is transmitted by certain ticks (figures 25.9, 25.10). Untreated Lyme disease may lead to serious complications.

Vibrio vulnificus infection (table 25.6)
The causative agent is typically ingested in raw or undercooked contaminated seafood or enters through wounds. If it enters the bloodstream, septic shock may result, particularly in those with compromised immunity.

Tularemia ("Rabbit Fever" or "Deer Fly Fever") (table 25.7, figure 25.11)
Tularemia is usually transmitted from wild animals to humans by exposure to the animal's blood or via insects and ticks. The cause is *Francisella tularensis,* which is found throughout the United States except Hawaii.

Brucellosis ("Undulant Fever" or "Bang's Disease") (table 25.8)
Brucellosis, caused by *Brucella melitensis,* is usually acquired from cattle or other domestic animals, sometimes from wild animals. Hunters, butchers, and those who drink unpasteurized milk or milk products are at increased risk for the disease. The organisms can infect via mucous membranes and minor skin injuries.

25.3 ■ Viral Diseases of the Blood and Lymphatic Systems

Infectious Mononucleosis ("Mono" or "Kissing Disease") (table 25.9)
Infectious mononucleosis is caused by the Epstein-Barr virus (EBV), which establishes a lifelong latent infection of B lymphocytes. The disease has a high incidence in young people and can cause exhaustion that lasts for months (figures 25.12, 25.13).

Ebola Virus Disease (EVD) and Marburg Virus Disease (MVD) (table 25.10)
Ebola virus disease (EVD) and Marburg virus disease (MVD) are both severe emerging diseases that are a global concern. They start with flu-like symptoms but can result in massive bleeding and multi-organ failure.

Yellow Fever (table 25.11)
Yellow fever is a zoonosis of mosquitoes and monkeys that exists mainly in tropical jungles; it can become epidemic in humans where a suitable *Aedes* mosquito vector is present. The disease involves the heart and the blood vessels throughout the body and is characterized by fever, jaundice, and hemorrhaging. A highly effective attenuated vaccine is available for preventing the disease.

Dengue Fever (table 25.12)
Dengue fever is caused by viruses that are transmitted by *Aedes* mosquito species (figure 25.14). The disease is similar to but milder than yellow fever and is characterized by joint pain, giving the common name "breakbone fever." A more severe form of the disease, dengue hemorrhagic fever (DHF), may occur in people reinfected with dengue and can be fatal.

Chikungunya (table 25.12)
Chikungunya (CHIK) is a newly emerging *Aedes* mosquito-borne disease that has spread globally. Like dengue, it causes severe joint pain, which may become chronic.

Zika Virus Disease (table 25.12)
Zika virus disease is an emerging disease spread by *Aedes* mosquitoes. Symptoms are typically mild, but complications involve severe nervous system effects, including microcephaly in fetuses.

25.4 ■ Protozoan Diseases of the Blood and Lymphatic Systems
Malaria is the most widespread of the protozoan blood and lymphatic diseases.

Malaria (table 25.13)
Malaria is caused by five species of *Plasmodium* and is transmitted from person to person by the bite of the females of certain species of *Anopheles* mosquitoes, its biological vector. Now found mainly in warm regions of the world, malaria still occurs despite massive eradication programs. The *Plasmodium* life cycle is complex (figure 25.15); different forms of the organism invade different body cells and have different susceptibility to antimalarial medication (figure 25.16).

Review Questions

Short Answer

1. What is the significance of immune complex formation in infective endocarditis?
2. Describe disseminated intravascular coagulation (DIC).
3. What activities of humans are likely to expose them to tularemia?
4. Why is brucellosis a threat to big-game hunters?
5. Why might the *Yersinia pestis* from a patient with pneumonic plague be more dangerous than the same organism from fleas?
6. Why might rodent burrows be a source of plague months after they are abandoned?
7. What type of leukocytes does EBV infect?
8. Travelers to and from which areas of the world should be vaccinated against yellow fever?
9. Why is a second infection with dengue virus more serious than the first?
10. Which *Plasmodium* species causes the most dangerous form of malaria?

Multiple Choice

1. Which of the following infection fighters are found in lymph?
   a) Leukocytes
   b) Antibodies
   c) Complement
   d) Interferon
   e) All of the above
2. Which of the following statements about the spleen is *false*?
   a) It routinely produces new blood cells.
   b) It cleanses the blood of foreign material and damaged cells.
   c) It contains B cells.
   d) It consists of red pulp and white pulp.
   e) It enlarges in a number of infectious diseases.
3. Which of these infections is *not* transmitted by mosquitoes?
   a) Malaria
   b) Yellow fever
   c) Ebola virus disease
   d) Chikungunya
   e) All of these infections are transmitted by mosquitoes.
4. Choose the one *true* statement about sepsis.
   a) It is a rare healthcare-associated disease.
   b) The output of urine increases if shock develops.
   c) It can be caused only by anaerobic bacteria.
   d) An antibiotic that kills the causative organism can be depended on to cure the disease.
   e) Lung damage is an important cause of death.
5. Which of these statements about tularemia is *false*?
   a) It can be contracted from muskrats and bobcats.
   b) Biting insects and ticks can transmit the disease.
   c) The causative organism has growth requirements similar to those of *E. coli.*
   d) An ulcer at the site of entry of the bacteria and enlargement of nearby lymph nodes are characteristic.
   e) It can be treated effectively with antibiotics.
6. Which of the following statements about brucellosis is *false*?
   a) Fevers that come and go over a long period of time gave it the name "undulant fever."
   b) The causative agent can infect via mucous membranes.
   c) The causative agent is readily killed by phagocytes.
   d) The disease in cattle is characterized by chronic infection.
   e) Butchers are advised to wear goggles or a face shield to help protect against the disease.
7. Which statement about *Yersinia pestis* is *false*?
   a) Conditions inside human phagocytes activate virulence genes.
   b) The bacterium can form biofilms in the flea digestive system.
   c) Yops (proteins) increase phagocytosis.
   d) The organism resembles a safety pin in certain stained preparations.
   e) It was responsible for the "black death" in Europe during the 1300s.
8. Which of the following statements about yellow fever is *false*?
   a) There is no animal reservoir.
   b) The name "yellow" comes from the fact that many victims have jaundice.
   c) Certain mosquitoes are biological hosts for the causative agent.
   d) Outbreaks of the disease could occur in the United States because a suitable vector is present.
   e) An attenuated vaccine is widely used to prevent the disease.
9. The malarial form infectious for mosquitoes is called a
   a) gametocyte.
   b) trophozoite.
   c) sporozoite.
   d) schizont.
   e) merozoite.
10. Which of the following statements about malaria is *true*?
    a) Transmission cannot occur in temperate climates.
    b) Transmission usually occurs with the bite of a male *Anopheles* mosquito.
    c) The disease is currently well controlled in tropical Africa.
    d) *P. falciparum* infects only old RBCs and therefore causes milder disease than other *Plasmodium* species.
    e) The characteristic recurrent fevers are associated with release of merozoites from RBCs.

Applications

1. Some years ago, dentists and doctors began noticing an association between infective endocarditis and prior dental work, and they began advising that an antibiotic be administered at the time of dental procedures to those with known or suspected heart defects. What was the rationale for this advice and why has the practice been discontinued for many people?
2. A healthcare worker in Honduras is concerned about a potential outbreak of yellow fever in his town. A laborer from a jungle area known to be endemic for the disease had come to the town 2 weeks earlier to work and subsequently developed yellow fever. Several coworkers reported getting mosquito bites while working with him. Why is it important that the healthcare worker determine how long it has been since the workers were bitten by the mosquitoes?

Critical Thinking

1. At least four emerging mosquito-borne diseases have recently spread to the Western Hemisphere and now have global distribution. If you were assigned to control the spread of these diseases, would you focus on the mosquito population or on the causative agent that the mosquitoes carry? Explain your rationale and how you would implement your plan.
2. Even though genetically engineered mosquitoes might be developed that do not allow the reproduction of malaria protozoa, these mosquitoes would have little, if any, immediate effect on the spread of the disease. Why should this be so? What would have to happen for these mosquitoes to significantly affect the spread of malaria?

26 Nervous System Infections

Structure of West Nile virus particles. ©Science Picture Co/Getty Images

KEY TERMS

Arbovirus Arthropod-borne RNA virus, carried by vectors such as mosquitoes.

Blood-Brain Barrier Cells that work together to restrict exchange between the bloodstream and the brain.

Central Nervous System (CNS) Brain and spinal cord.

Cerebrospinal Fluid (CSF) Fluid produced in the brain that flows within and around the CNS.

Encephalitis Inflammation of the brain.

Meninges Membranes covering the brain and spinal cord.

Meningitis Inflammation of the meninges.

Peripheral Nervous System (PNS) Division of the nervous system that carries information to and from the CNS.

Transmissible Spongiform Encephalopathy (TSE) Chronic degenerative brain disease caused by prions; characterized by spongy appearance of brain tissue.

A Glimpse of History

Today it is hard to appreciate the fear and loathing once attached to leprosy (*lepros,* meaning "scaly"). The Bible refers to several disfiguring skin diseases, including leprosy, and people suffering from the diseases are portrayed as filthy, outcast, or condemned by God for sin. Moses called lepers "unclean" and proclaimed they must live away from others. In the Middle Ages, lepers attended their own symbolic burial before being sent away.

Gerhard Henrik Armauer Hansen (1841–1912) was a Norwegian physician with many interests, ranging from science to religion to polar exploration. When he was 32 years old, Hansen went into medical research, and was named assistant to Dr. Daniel C. Danielson, a leading authority on leprosy. Danielson believed that leprosy was a hereditary disease of the blood and considered the idea that the disease was contagious as a "peasant superstition." Hansen, however, disproved Danielson's hypothesis in careful studies conducted over a number of years. He found a unique bacterium associated with the disease in every leprosy patient he studied. His 1873 report of the findings marked the first time that a specific bacterium was linked to a disease—almost a decade before Koch's proof of the cause of tuberculosis.

In the United States, even during the first half of the twentieth century, people diagnosed with leprosy risked having their houses burned to destroy the source of infection. Their names were changed to avoid embarrassing their family, and they were sent to a leprosarium such as the one at Carville, Louisiana, surrounded by a 12-foot fence topped with barbed wire. Sufferers were separated from spouses and children and denied the right to marry or vote. Those who attempted to escape were captured and brought back in handcuffs. The Carville leprosarium was finally closed and converted to a military-style academy for high school dropouts in 1999.

Because the word *leprosy* carries centuries of dark overtones, many people prefer to use the term *Hansen's disease,* a name that honors the discoverer of the causative bacterium.

Nervous system infections are frightening. They threaten a person's ability to move, feel, or even think. Consider poliomyelitis, which can result in a paralyzed limb or the inability to breathe without mechanical assistance. Hansen's disease (leprosy) can result in loss of fingers or toes or deformity of the face. Infections of the brain or its covering membranes can render a child deaf or intellectually disabled. Before the discovery of antibiotics, bacterial infections of the nervous system were often fatal. Fortunately, these infections are uncommon.

26.1 ■ Anatomy, Physiology, and Ecology of the Nervous System

Learning Outcomes

1. Describe how information flows through and between neurons.
2. Differentiate between the central nervous system and the peripheral nervous system.
3. Explain how bone, cerebrospinal fluid, meninges, and the blood-brain barrier protect the central nervous system.

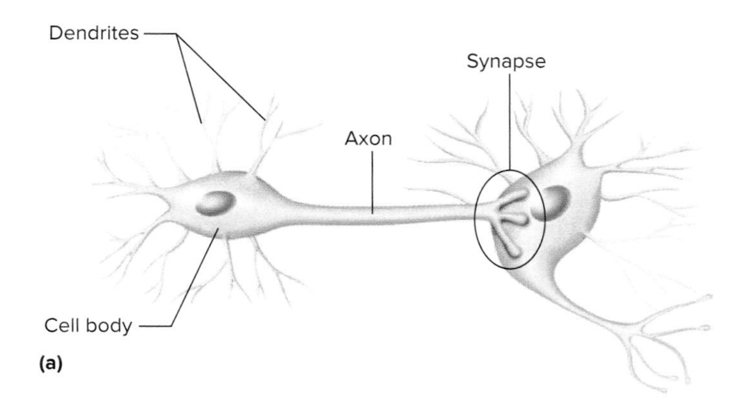

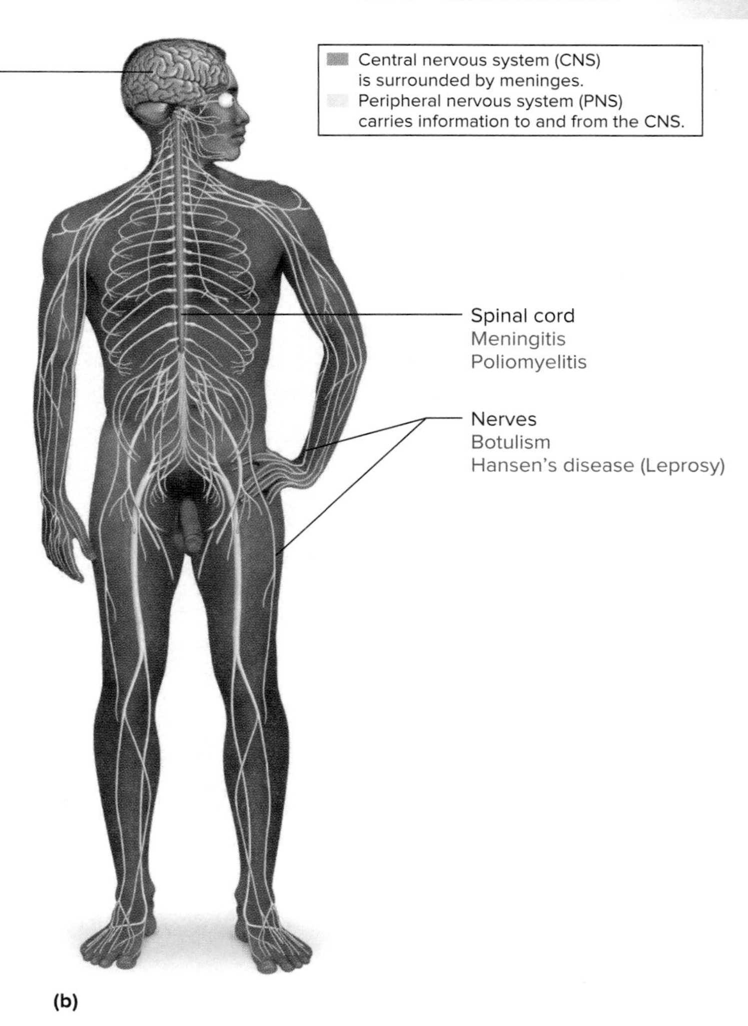

FIGURE 26.1 The Nervous System (a) Neurons receive information via dendrites and transmit impulses over axons. **(b)** The central nervous system is composed of the brain and spinal cord. Peripheral nerves carry information to and from the CNS. Nervous system infections are shown in red.

? **What generally protects the CNS from infection?**

Nerve cells work together, transmitting electrical impulses throughout the body like a highly sophisticated circuit board. Each nerve cell, or **neuron,** has different regions with distinct functions (**figure 26.1a**). Branching projections called dendrites receive information. They convey that information to the cell body, the command center that contains the cell nucleus. A long, thin extension called an axon then transmits the information from the cell body to another cell. In many cases, axons are surrounded by a fatty sheath called myelin that aids in transmission of impulses.

Most neurons communicate using **neurotransmitters,** chemicals stored in vesicles at the end of the axon. When a neurotransmitter is released, the molecules diffuse across the short distance to a neighboring cell. That cell has receptors for the neurotransmitter, allowing it to receive and respond to the signal. The region where transmission of information takes place between the two cells is called a synapse.

The brain and spinal cord (figure 26.1b) make up the **central nervous system (CNS).** The brain is a very complex structure; distinct parts have different functions. In fact, physicians can sometimes determine the location of a brain abscess by evaluating the resulting loss of function. Generalized inflammation or infection of the brain is called **encephalitis.** The spinal cord extends from the brain down the vertebral column. It conveys sensory information from receptors to the brain, and motor commands from the brain to the muscles and internal organs. It also contains nervous circuits for simple reflexes. Generalized inflammation of the spinal cord is called **myelitis.**

The **peripheral nervous system (PNS)** is made up of nerves composed of bundles of axons that carry information to and from the CNS. Axons of motor neurons carry messages from the CNS to different parts of the body and cause them to respond; axons of sensory neurons transmit sensations like touch, heat, light, and sound to the central nervous system.

The central nervous system is much better protected than the peripheral nervous system. Both the brain and spinal cord are enclosed by bone—the brain by the skull and the spinal cord by the vertebral column. Nerves can be damaged if the bones are infected at sites where the PNS penetrates this protective covering. Only rarely do infections in the bone surrounding the CNS spread to the brain or spinal cord. Sometimes, however, they extend through bone from the sinuses, mastoid air cells, or middle ear. Skull fractures commonly predispose a person to recurring infection of the CNS.

⏮ mastoid air cells, ⏮ middle ear

Deep inside the brain are four fluid-filled cavities called ventricles. The fluid within them, **cerebrospinal fluid (CSF),**

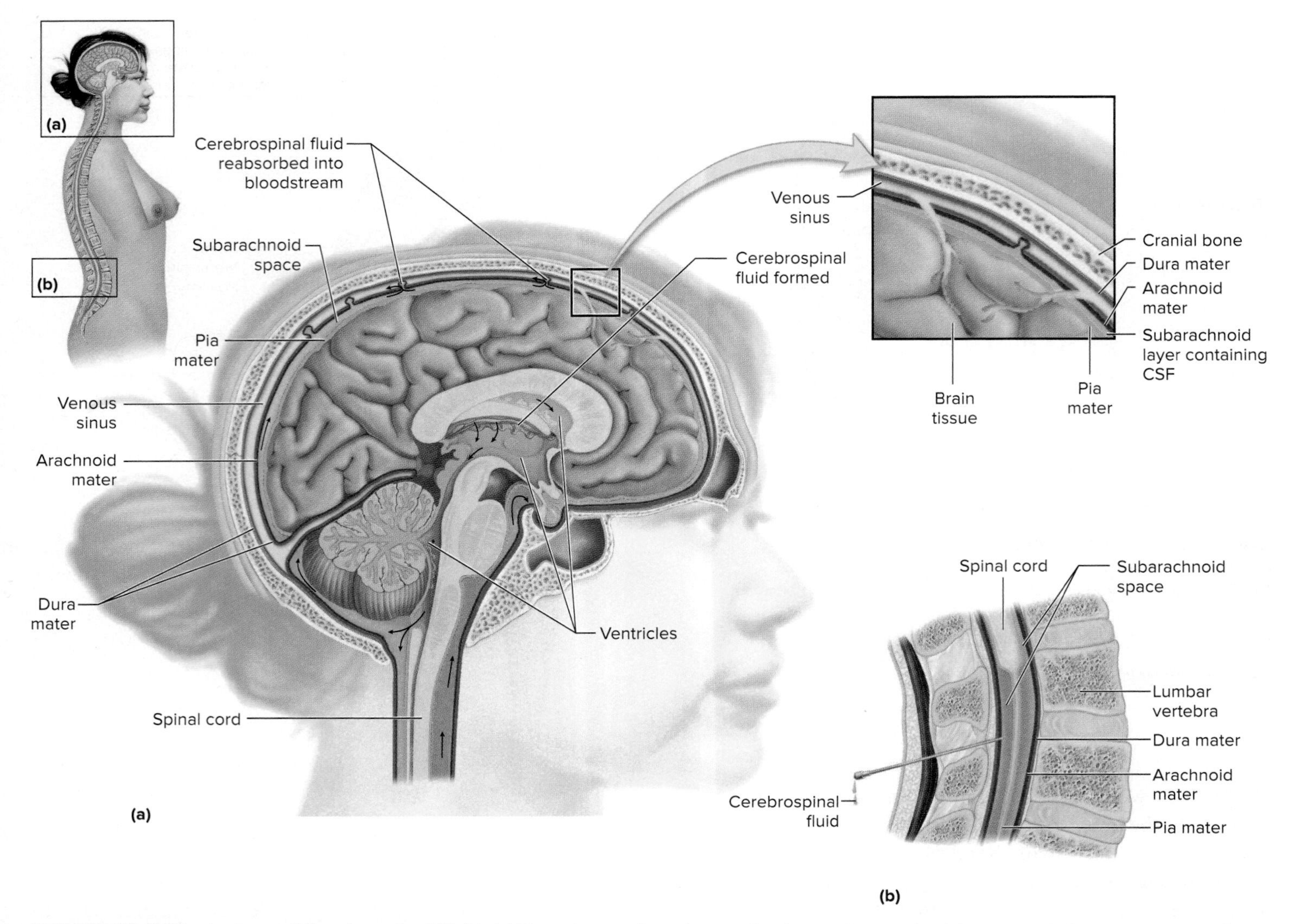

FIGURE 26.2 Meninges and Cerebrospinal Fluid **(a)** Three layers of meninges—the dura mater, arachnoid mater, and pia mater—surround the CNS. CSF is formed in the ventricles, flows within the subarachnoid space, and is reabsorbed into the bloodstream. **(b)** A sample of cerebrospinal fluid can be withdrawn and examined.

? What is the purpose of a lumbar puncture?

is continuously produced in the ventricles and circulates over the surface of the brain and spinal cord (**figure 26.2a**). CSF is reabsorbed into the bloodstream at specialized sites within a venous sinus (channel for blood flow from the brain). CSF provides cushion and support for the brain and also transports nutrients and other materials throughout the CNS.

Three layers of membranes called **meninges** cover the surface of the brain and spinal cord. The outer dura mater is tough and provides a barrier to the spread of infection from bones surrounding the central nervous system. It adheres closely to the skull and the vertebrae, and in some parts of the brain encloses a blood-filled venous sinus. The two inner membranes, the arachnoid mater and the pia mater, are separated by the subarachnoid space within which the cerebrospinal fluid flows (figure 26.2a). The innermost pia mater adheres to the brain and spinal cord. Inflammation or infection of these membranes is called **meningitis.** When both the meninges and the brain are infected, the condition is meningoencephalitis.

The nervous system lies entirely within body tissues and has no normal microbiota. CSF is generally sterile, so the presence of microbes indicates an infection. To diagnose meningitis, a sample of CSF is obtained using a procedure called a lumbar puncture (or spinal tap). A needle is inserted into the subarachnoid space between lumbar vertebrae where the spinal cord has tapered to a thread-like structure, and a sample of CSF is withdrawn (figure 26.2b).

The bloodstream is the primary source of CNS infections. However, it is difficult for infectious agents to cross from the bloodstream to the brain because of the **blood-brain barrier.** This barrier depends on special cells lining capillaries in the CNS. The blood-brain barrier generally prevents pathogens from entering nervous tissue except when a rare high concentration of the infectious agent circulates for a long time in the bloodstream. Unfortunately, the barrier also prevents many medications, including penicillin, from crossing into the CNS unless their concentrations in the blood are very high.

MicroByte

Infections on the face can move through bone to veins on the brain surface. A good rule is not to pop pimples on the upper part of the face.

MicroAssessment 26.1

Information within a neuron typically flows from dendrites to cell body to axon. Neurotransmitter molecules are released from the axon at a synapse, where they bind to receptors on a neighboring cell, initiating a response. The brain and spinal cord make up the central nervous system (CNS). Nerves in the peripheral nervous system (PNS) convey sensory and motor information to and from the CNS. The CNS is protected by bone, by circulating cerebrospinal fluid (CSF), and by membranes called meninges. The blood-brain barrier prevents passage of most pathogens from the bloodstream into the CNS, but it also can prevent access to the CNS by antimicrobial medications.

1. What molecules are released at the synapse?
2. Distinguish among the terms *encephalitis, myelitis,* and *meningitis.*
3. Why can encephalitis usually be detected in CSF obtained from the lower back?

CENTRAL NERVOUS SYSTEM INFECTIONS

26.2 ■ Bacterial Diseases of the Central Nervous System

Learning Outcomes

4. Compare and contrast the various types of bacterial meningitis.
5. Explain how newborns most often develop meningitis.

Bacteria can infect neurons, but they most often infect the meninges surrounding the central nervous system—causing meningitis. The most common species that cause meningitis are often part of the normal microbiota of the upper respiratory tract. Newborns are exposed to causative agents for meningitis from the mother while in the uterus or during passage through the birth canal.

FOCUS ON MENINGITIS

Meningitis is the inflammation of the meninges (the membranes surrounding the brain and spinal cord), and most often results from microbial infection.

Signs and Symptoms

Signs and symptoms of meningitis typically begin a few days after exposure to the causative agent, and resemble those of a mild cold. This is followed by the sudden onset of characteristic severe, throbbing headache and stiffness of the neck and back, accompanied by fever, sensitivity to strong light (photophobia), nausea, and vomiting. Deafness, confusion, loss of consciousness, and coma may develop. The case-fatality rate of untreated bacterial meningitis can approach 100% and death may occur within hours. People who recover sometimes suffer permanent disabilities, including deafness, blindness, paralysis, or mental impairment. Viral meningitis occurs more often than bacterial meningitis, but it is typically less severe and causes little permanent damage in those with normal immunity.

To diagnose meningitis, the **cerebrospinal fluid (CSF)** that bathes the central nervous system tissue is examined. A sample of CSF can be obtained by lumbar puncture (see figure 26.2b). Bacterial meningitis is a medical emergency, so specimens are examined immediately. The CSF of a person with bacterial meningitis is cloudy and opaque, rather than a clear and pale yellow. The fluid has elevated pressure and typically contains neutrophils, high levels of protein, and low levels of glucose compared to the blood. The sample can also be concentrated and examined microscopically and then inoculated onto appropriate laboratory media as part of the process to identify the bacterium causing the disease. CSF in people with viral meningitis is clear, although it contains numerous lymphocytes. Because no bacteria are present, viral meningitis is sometimes called **aseptic meningitis.**

Pathogenesis

Normal microbiota of the upper respiratory tract can occasionally enter the bloodstream. Once in the bloodstream (causing bacteremia), they may penetrate the blood-brain barrier and pass into the cerebrospinal fluid (CSF). Trauma such as skull fractures that allow contact between the CSF and the mucous membranes of the throat can also allow infection.

Much of the damage associated with bacterial meningitis is due to the inflammatory response to pathogens that have invaded the CSF. The vasodilation that characterizes inflammation allows fluids to accumulate in the area, causing brain swelling and nerve damage.

(Continued)

FOCUS ON MENINGITIS (continued)

The vasodilation also disrupts the blood-brain barrier, allowing proteins and neutrophils to enter the CSF. Although neutrophils can help clear an invader, they also cause life-threatening tissue damage in sensitive areas such as the brain and the spinal cord. In addition, the clots that form in the capillaries can block the blood supply, leading to cell death. Inflammation can also obstruct the normal outflow of cerebrospinal fluid, causing the brain to be squeezed against the skull by the buildup of internal pressure.

Viral meningitis also stimulates inflammation of the meninges, but it is much less severe and typically causes little neurological damage. The disease usually resolves itself in a few days.

Epidemiology

Meningitis is a relatively rare disease, probably because it is difficult for the causative agents to cross the blood-brain barrier protecting the CNS. As with many diseases, people with lowered immunity such as newborns and the elderly are at most risk.

Many of the bacteria that cause meningitis are transmitted through respiratory droplets from close contacts. Young adults living in crowded, confined environments such as military barracks or college dormitories are also at increased risk, particularly from highly infectious *Neisseria meningitidis,* which can cause epidemics in close quarters.

Some of the agents that cause viral meningitis are transmitted by the fecal-oral route rather than by respiratory droplets. These viruses (enteroviruses) are relatively stable in the environment, and feces of infected individuals may contain infectious viruses for weeks.

Treatment and Prevention

Due to the typically rapid progression and life-threatening nature of bacterial meningitis, treatment with a broad-spectrum antimicrobial medication is begun immediately if the disease is suspected. Once the causative agent has been identified, then the treatment may be tailored to that specific bacterium. There is no effective therapy for viral meningitis. ◀◀ broad-spectrum antimicrobial medication

Vaccines are available to prevent what were once the most significant types of bacterial meningitis, resulting in a dramatic drop in the incidence of the disease in recent years. Unvaccinated people who have been in close contact with an infected individual may be given prophylactic doses of an antimicrobial medication to help prevent the disease. If viral meningitis is reported, prevention may be enhanced with increased handwashing and avoidance of crowded areas. There are no vaccines for viral meningitis.

Pneumococcal Meningitis

Pneumococcal meningitis is caused by *Streptococcus pneumoniae,* part of the normal microbiota in the nose and throat of many healthy individuals. Although best known as a cause of pneumonia, it is also the leading cause of meningitis in adults. ◀◀ pneumococcal pneumonia

Signs and Symptoms

Pneumococcal meningitis usually begins like a mild cold. This is followed by the sudden onset of a severe, throbbing headache; fever; pain and stiffness of the neck and back; nausea; and vomiting. Deafness, confusion, loss of consciousness, and coma may develop. Death may occur rapidly due to inflammation and shock.

Causative Agent

Streptococcus pneumoniae, often called pneumococcus, is a Gram-positive diplococcus—the lancet-shaped cells are typically in pairs (see figure 21.11). Many strains are protected from phagocytosis by a polysaccharide capsule. Antigenic differences in the capsules produce over 90 recognized serotypes. When the pathogen is part of the normal respiratory microbiota, the mucociliary escalator usually prevents it from entering the lungs, where it could then enter the bloodstream. ◀◀ mucociliary escalator

Pathogenesis

Streptococcus pneumoniae is a common cause of otitis media (inflammation of the inner ear), sinusitis, and pneumonia, any of which can precede pneumococcal meningitis. Encapsulated strains resist phagocytosis and may enter the bloodstream of an infected individual, and then pass into the cerebrospinal fluid, causing meningitis. The damage associated with meningitis is largely due to the inflammatory response. Fluid accumulation (cerebral edema) and clots that form in the capillaries can lead to tissue death. Obstruction of the normal outflow of cerebrospinal fluid can cause buildup of internal pressure that squeezes the brain against the skull. Mortality and neurological damage are more common with this type of meningitis than with others, especially among the elderly.

Epidemiology

S. pneumoniae causes over half of the cases of bacterial meningitis in the United States, most of them in adults. The case-fatality rate is less than 10% in young adults, but may be twice that in the elderly. The pathogen is spread by respiratory droplets, but very few who are infected will develop meningitis.

Treatment and Prevention

Penicillin remains the medication of choice for susceptible strains of *Streptococcus pneumoniae,* but some strains are now resistant to that antibiotic. To treat meningitis caused by those strains, a third-generation cephalosporin such as ceftriaxone is administered, with the possible addition of vancomycin.

Vaccines against *S. pneumoniae* offer protection from several pneumococcal diseases, including meningitis, pneumonia, and otitis media, but are not effective against all of the different serotypes of the bacterium. The newest pneumococcal conjugate vaccine, PCV13 (for **p**neumococcal conjugate vaccine, **13** serotypes), protects against 13 serotypes and is administered to children under the age of 2, adults over age 65, and other children and adults with certain health problems. A vaccine called PPSV23 (for **p**neumococcal **p**olysaccharide **v**accine, **23** serotypes) consisting of purified capsular polysaccharides from the 23 most common pneumococcal serotypes is also available, but is not effective in children younger than age 2 because polysaccharides are T-independent antigens. ⏮ vaccines against pneumococcus

Meningococcal Meningitis

Neisseria meningitidis, often called meningococcus, is frequently responsible for epidemics of meningitis. It is commonly part of the normal respiratory microbiota of healthy individuals.

Signs and Symptoms

Symptoms of meningococcal meningitis appear after 1 to 7 days and are similar to those of pneumococcal meningitis, but often include purplish spots on the skin called **petechiae** (**figure 26.3**). These indicate capillary damage, in this case a result of meningococcal bloodstream infection that can lead to rapidly fatal septic shock. ⏮ septic shock, ⏮ petechiae

Causative Agent

Meningococcal meningitis is caused by *Neisseria meningitidis,* a Gram-negative encapsulated diplococcus. The most serious infections are due to serotypes A, B, C, W, and Y. Groups B, C, and Y are most common in the United States and Europe; type A is most common in Africa and Asia. Like *Neisseria gonorrhoeae, N. meningitidis* can vary some of its antigens and easily acquires DNA through horizontal gene transfer. ⏮ serotype, ⏮ antigenic variation, ⏮ horizontal gene transfer

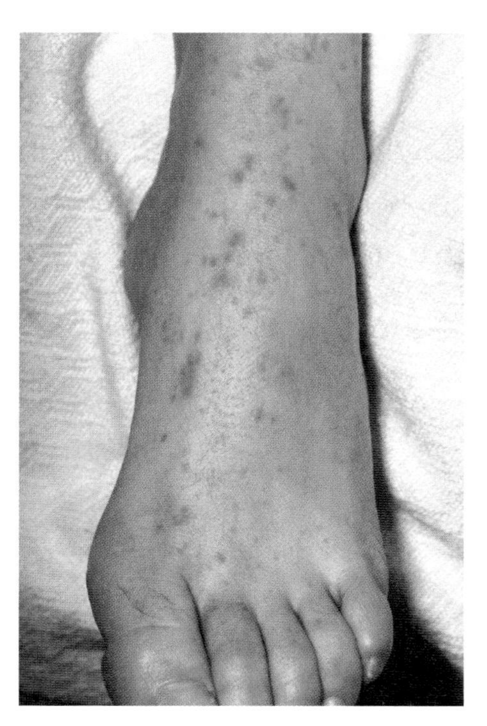

FIGURE 26.3 Petechiae of Meningococcal Disease These purple spots are caused by bleeding under the skin, a result of capillary damage caused by *Neisseria meningitidis.* ©Medical-on-Line/Alamy Stock Photo

? Why is it important for public health officials to distinguish between cases of pneumococcal and meningococcal meningitis?

Pathogenesis

Meningococci inhaled in airborne droplets attach by pili to mucous membranes and then multiply. Specific proteins in their outer membranes allow the bacterial cells to pass through epithelial cells lining the respiratory tract and into the bloodstream to the meninges, although this is rare. Circulating *Neisseria meningitidis* release blebs (vesicles) of their outer membranes, and the inflammatory response to the endotoxin results in vasodilation and capillary leakage, leading to a drop in blood pressure. Endotoxic or septic shock results when blood pressure becomes so low that circulation cannot adequately supply O_2 to vital body tissues. Capillary damage causes the petechiae associated with meningococcal meningitis. ⏮ pili, ⏮ endotoxin

The capsule helps the cells avoid phagocytosis and protects them from the lethal actions of the complement system. Large numbers of neutrophils enter the CSF in response to the infection, but the bacteria multiply faster than they can be destroyed. As in pneumococcal meningitis, tissue damage and cerebral edema often result from inflammation. ⏮ complement system

Epidemiology

Transmission of *Neisseria meningitidis* can occur when someone is exposed to a person with the disease or to an asymptomatic carrier. The organism can cause meningitis in any age group, but infants, adolescents, and young adults are at particular risk. Although most cases are sporadic, meningococcal meningitis outbreaks arise if the bacteria spread via respiratory droplets through crowded and stressed populations.

In the United States, the disease is relatively rare (fewer than 1 case per 100,000 people annually). The highest incidence worldwide occurs in the "meningitis belt" extending between Senegal and Ethiopia in sub-Saharan Africa, where the population is about 300 million (**figure 26.4**). During the dry season between December and June, incidence has reached 500 cases per 100,000 people. A campaign to administer a new meningococcal A conjugate vaccine in the African meningitis belt was initiated in 2010, resulting in immunization of over 235 million people from 16 countries. The results have been dramatic, with some areas showing a 98% reduction in the number of meningitis cases. Continued monitoring of the vaccine's impact on disease caused by this strain is important, however, as is keeping track of the incidence of meningitis caused by other strains of the bacterium.

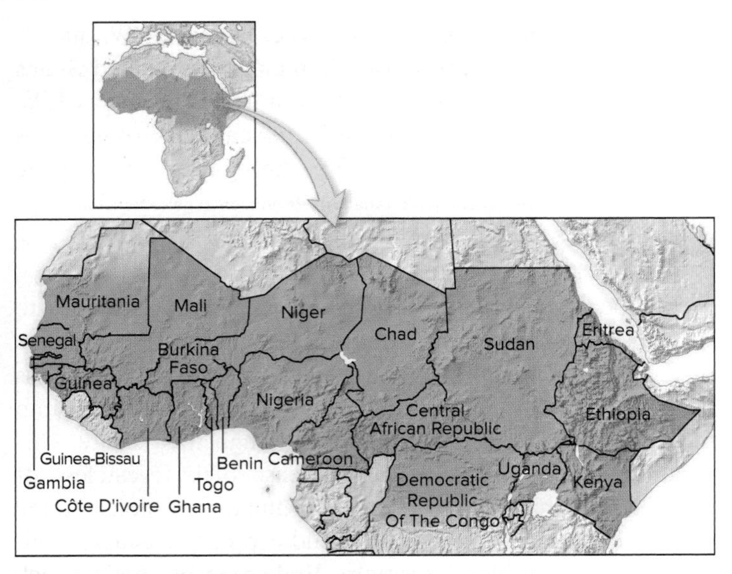

FIGURE 26.4 The Meningitis Belt Meningitis is most prevalent in Mali, Burkina Faso, Niger, Chad, Sudan, Ethiopia, and northern Nigeria.

? **Why would the incidence of meningitis be seasonal in this region?**

Treatment and Prevention

If treated in time, meningococcal meningitis can usually be cured with ceftriaxone or high doses of penicillin. Most patients recover without permanent nervous system damage. The case-fatality rate is less than 10% in treated cases.

Two general types of vaccines (conjugate and polysaccharide) are used to prevent meningococcal disease caused by serotypes A, C, W, and Y. The conjugate version, referred to as MenACWY, is recommended for all 11 to 12 year olds with a booster at age 16, and for people 2 months or older who are at increased risk due to certain medical conditions or potential exposure. The polysaccharide version, MPSV4 (for meningococcal polysaccharide vaccine, 4 serotypes), is the preferred meningococcal vaccine for some people over age 55. ⏮ conjugate vaccine, ⏮ polysaccharide vaccine

Two outbreaks of serotype B meningitis occurred on university campuses in 2013. By early 2015, a recombinant subunit vaccine against serotype B was approved in the United States to protect people who are at least 10 years of age and at increased risk for serotype B meningococcal disease, particularly in the event of an outbreak. ⏮ subunit vaccine

Antibacterial medications can be a useful preventive measure in controlling meningococcal meningitis outbreaks in confined groups such as people in schools, nursing homes, and jails. People intimately exposed to cases of meningococcal disease are routinely given prophylactic treatment with an antibiotic, typically ceftriaxone or rifampin. **Table 26.1** gives the main features of this disease.

Haemophilus influenzae Meningitis

Haemophilus influenzae once caused meningitis in about 1 out of 200 children under 5 years of age and was the leading cause of meningitis in that age group. Hib vaccine, introduced in the late 1980s, is a great success story as it has decreased incidence of this type of meningitis in children by over 99%.

TABLE 26.1 Meningococcal Meningitis

1. *Neisseria meningitidis* inhaled, infects upper airways.
2. Bacteria enter the bloodstream and are circulated throughout the body.
3. The bacteria damage skin capillaries and cause petechiae.
4. Bacteria infect the meninges causing meningitis.
5. Lysing bacteria in the circulation release endotoxin, producing shock.
6. Inflammatory response in meninges can damage nerves of hearing, causing deafness, and obstruct the flow of cerebrospinal fluid, causing increased pressure inside the brain.
7. Bacteria exit with respiratory secretions.

Signs and symptoms	Mild cold followed by headache, fever, pain, stiff neck and back, vomiting, petechiae
Incubation period	1 to 7 days
Causative agent	*Neisseria meningitidis*, the meningococcus; a Gram-negative diplococcus
Pathogenesis	Meningococci adhere by pili, colonize upper respiratory tract, enter bloodstream; carried to meninges and spinal fluid; inflammatory response obstructs normal outflow of fluid; increased pressure caused by obstructed flow impairs brain function; damage to motor nerves produces paralysis; endotoxin release leads to shock.
Epidemiology	Close contact with a case or carrier; inhalation of infectious droplets; crowding and fatigue predispose to the disease.
Treatment and prevention	Treatment: antibiotics. Prevention: age-specific vaccines; antibiotics given to people exposed.

Signs and Symptoms

As with other types of meningitis, symptoms usually begin with a mild cold and progress to severe headache, fever, and vomiting. Older children may report a stiff neck. Infants may show a bulging "soft spot" (the gap between the bones of an infant's skull). The disease may progress rapidly to coma and death.

Causative Agent

Haemophilus influenzae, a Gram-negative, non-motile coccobacillus (very short rod), was named because it was thought to have caused an influenza epidemic in the late 1890s. Although it did not cause the epidemic, it was and still is often isolated from influenza patients. Encapsulated strains labeled a through f cause most disease in children; bacteria with capsular antigens of type b, abbreviated Hib, cause the most serious disease. Unencapsulated strains are often part of the normal microbiota of the nose and throat in healthy individuals, but a small percentage of people may carry Hib. In 1991, *H. influenzae* earned the distinction of being the first free-living organism to have its small genome completely sequenced.

Pathogenesis

Encapsulated *Haemophilus influenzae* in the upper respiratory tract can bind to and penetrate the epithelium with the aid of pili and enter capillaries. Bacteremia can result in meningitis or other infections such as epiglottitis or cellulitis. Unencapsulated strains generally do not infect the bloodstream, but can cause local infections such as sinusitis and otitis media (ear infection).

Epidemiology

Unlike meningococcal meningitis, *Haemophilus influenzae* usually does not cause epidemics. Healthy adult carriers are often the source of sporadic infection by *H. influenzae* type b. Children rarely carry Hib due to the high rate of immunization. The pathogen is spread by respiratory droplets, but it rarely causes disease except in unimmunized infants. If untreated, the case-fatality rate of the disease is about 90%. Even with treatment, about 5% die and 10–30% of children have lasting neurological damage.

Treatment and Prevention

Treatment with a third-generation cephalosporin such as ceftriaxone is usually successful against *Haemophilus influenzae.* People who have been in close contact with an infected individual may be given rifampin to help prevent the disease.

The conjugate Hib vaccine contains the type b polysaccharide antigen attached to a protein such as diphtheria toxoid. The protein component makes the vaccine effective in children under age 2 years, who otherwise respond poorly to polysaccharide antigens. Infants in the United States are routinely vaccinated against Hib beginning at 2 months of age (**figure 26.5**). This vaccine is relatively expensive and must be administered more than once, making it less practical in developing countries, where Hib infections are still responsible for most deaths due to bacterial meningitis. ⏮ conjugate vaccines

Neonatal Meningitis

Most cases of neonatal meningitis (during the first month of life) are caused by bacteria that colonize the mother's birth canal. The common causes of meningitis in children and adults (*Neisseria meningitidis,* *Streptococcus pneumoniae,* and *Haemophilus influenzae*) seldom cause meningitis in newborns because most mothers have antibodies (IgG) against them. These antibodies cross the placenta and protect the baby until it is about 6 months old. ⏮ IgG

Signs and Symptoms

Meningitis symptoms in an infant may be vague, as it is difficult to tell if a baby has a headache or a stiff neck. Signs of meningitis in a newborn may include fever and vomiting, irritability, poor feeding, and lethargy. As with *Haemophilus influenzae* infection, newborns may display bulging soft spots due to increased pressure within the cranial cavity.

Causative Agents

The most common cause of meningitis in newborn infants is *Streptococcus agalactiae.* This bacterium, frequently referred to as group B streptococcus (GBS), colonizes the vagina of

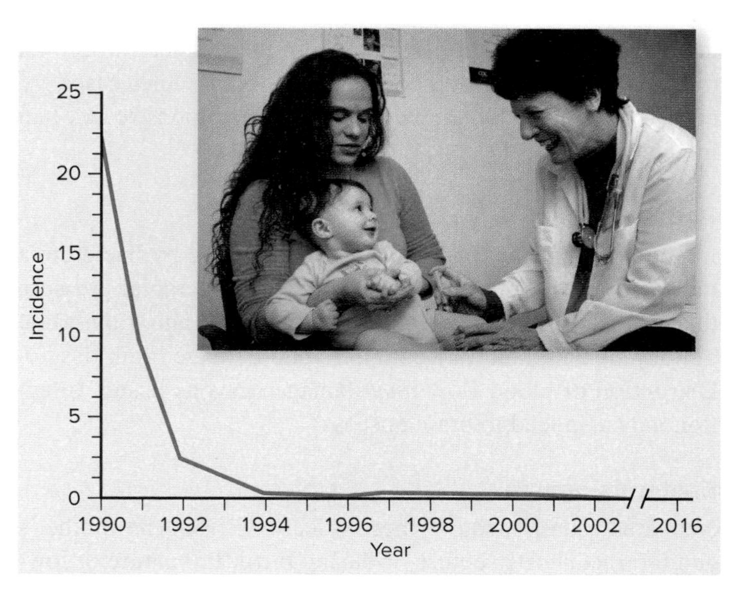

FIGURE 26.5 Rate of Serious *Haemophilus influenzae* Disease per 100,000 Children Less than Age 5, United States, 1990–2016 Before the availability of conjugate vaccines in late 1987, *H. influenzae* type b was the most common cause of bacterial meningitis in preschool children. Source: James Gathany/CDC

? What are the components of the conjugate Hib vaccine?

FOCUS ON A CASE 26.1

The patient was a 31-month-old girl admitted to the hospital because of fever, headache, drowsiness, and vomiting. She had been well until 12 hours before admission, when she developed a runny nose, fever, and loss of appetite.

Her birth and development were normal, but her routine immunizations had been neglected. On examination, her temperature was 40°C (104°F), her neck was stiff, and she did not respond to verbal commands. There was no history of head trauma.

The child's white blood cell count was elevated and showed a marked increase in the percentage of neutrophils. Her blood sugar was in the normal range. A spinal tap revealed cloudy cerebrospinal fluid containing 18,000 white blood cells per microliter (normally, there are few or none), a markedly elevated protein level, and a markedly low glucose level. Gram stain of the fluid showed tiny, Gram-negative coccobacilli.

1. What was the diagnosis, and what was the causative agent?
2. What was the prognosis in this case?
3. What age group is most susceptible to this illness?
4. Compare the incidence of this disease in children to that in adults.

Discussion

1. The patient had bacterial meningitis caused by *Haemophilus influenzae* serotype b. The fact that she had been healthy prior to her illness makes it unlikely that other pathogens could be responsible.
2. With treatment, the fatality rate is approximately 5%. Treatment is with newer cephalosporin-type antibiotics such as ceftriaxone. Unfortunately, about one-third of patients who are treated and recover from the infection are left with permanent damage to the nervous system, such as deafness or paralysis of facial nerves. Prompt diagnosis and correct choice of antibacterial treatment minimize the chance of permanent damage.
3. The peak incidence of this disease is in the age range of 6 to 18 months, corresponding to the time when maternal antibodies have diminished and adaptive immunity has not yet fully developed. Since 1987, vaccines consisting of type b capsular antigen conjugated with a protein have been used to immunize infants and thus eliminate the immunity gap. As a result, meningitis caused by *H. influenzae* type b is now rare.
4. *Haemophilus influenzae* type b strains are referred to as "invasive" strains because they establish infection of the upper respiratory tract, pass the epithelium, and gain access to the general circulation that carries them to the CNS. The Hib vaccine has largely eliminated this type of meningitis in children. Most adults are immune to type b strains, but some adults develop meningitis from non-invasive *H. influenzae* strains that gain access to the nervous system because of a skull fracture or from an infected sinus or middle ear.

many women. Certain Gram-negative rods, such as encapsulated strains of *E. coli,* originate from the mother's intestinal tract and also can cause neonatal meningitis. Other cases are caused by *Listeria monocytogenes* from the bloodstream of an infected mother.

Pathogenesis

Infection of the meninges is usually preceded by bacteremia in the newborn. Inflammation increases intracranial pressure that can block CSF flow, causing hydrocephalus (abnormal buildup of CSF that may compress and damage brain tissue). Disruption of blood flow may damage nervous tissue. Infection may also lead to brain abscess.

Epidemiology

Neonates usually acquire these infections from the mother's genital tract shortly before or during birth. Premature or low-birth-weight babies are at most risk for meningitis. Neonatal meningitis causes death in 5–20% of affected newborns. Those who survive often have long-lasting consequences such as hearing loss or mental disability.

Treatment and Prevention

Treatment of neonatal meningitis includes intravenous dosage with a mixture of antibacterial medications such as ampicillin and gentamicin effective against both group B streptococci and *E. coli.* Other medications may be used after the causative agent is identified.

The Centers for Disease Control and Prevention (CDC) recommends that the vagina and rectum of pregnant women be tested for group B streptococci late in pregnancy. Women with positive cultures can then be treated with an appropriate antibacterial medication shortly before or during labor. Screening and subsequent treatment can decrease the incidence of serious group B streptococcal disease by more than 75%.

Listeriosis

Meningitis is the most common result of listeriosis, a foodborne disease caused by *Listeria monocytogenes.* This organism generally causes only a small percentage of meningitis cases in the United States, but epidemics sometimes occur.

Signs and Symptoms

Listeria monocytogenes infections are generally asymptomatic or mild in most healthy people. Symptomatic listeriosis is usually characterized by fever and muscle aches, and sometimes nausea or diarrhea. Most of the people requiring medical attention have meningitis with fever, headache, stiff neck, and vomiting. Pregnant women who become infected often miscarry or deliver terminally ill infants. Babies infected at birth usually develop meningitis after an incubation period of 1 to 4 weeks. Babies who contract the disease in the uterus show a shorter incubation period.

Causative Agent

Listeria monocytogenes is a motile, non-spore-forming, facultatively anaerobic, Gram-positive rod that can grow at 4°C. The organism can grow in refrigerated foods even if they are vacuum-packaged.

Pathogenesis

The mode of entry of *L. monocytogenes* in sporadic cases of listeriosis is usually unclear, but during epidemics it is generally via the gastrointestinal tract. The organism can induce its own phagocytosis and then survive and reproduce within many different host cell types where it is protected from the immune system. Moreover, it can cause host cell actin to polymerize, forming a "tail" that propels the organism with enough force to push it into an adjoining cell (see Focus Your Perspective 3.1). When *L. monocytogenes* enters the bloodstream, the resulting bacteremia is the source of meningeal infection. In pregnant women, *L. monocytogenes* crosses the placenta and produces widespread abscesses in tissues of the fetus.

Epidemiology

Listeria monocytogenes is widespread in natural waters and vegetation and can be carried in the intestines of asymptomatic humans and other animals. Pregnant women, the elderly, and those with underlying illnesses such as immunodeficiency, diabetes, cancer, or liver disease are especially susceptible to listeriosis. Outbreaks have resulted when *L. monocytogenes* has contaminated foods including coleslaw, unpasteurized milk, pork tongue in jelly, some soft cheeses, hot dogs, and canteloupe. Because the organisms can grow in commercially prepared food stored at refrigeration temperatures, thousands of infections can come from a single food-processing plant.

Treatment and Prevention

Most strains of *L. monocytogenes* remain susceptible to antibacterial medications such as penicillin. Even though the disease is often mild in pregnant women, prompt diagnosis and treatment are important to protect the fetus.

Listeria monocytogenes can be killed by thoroughly cooking foods. To reduce the risk of cross-contamination, uncooked meats should not be kept with other foods; countertops and utensils should be cleaned after food preparation; and raw fruits and vegetables should be thoroughly washed before eating. Pregnant women and others at high risk are advised to avoid soft cheeses, refrigerated meat spreads, and smoked seafood. They should also heat cold cuts and hot dogs before eating them and avoid the fluids that may be in the packaging. The U.S. Food and Drug Administration has approved a food additive consisting of a mixture of bacteriophage strains that lyse *L. monocytogenes* (**figure 26.6**). The additive can be sprayed on a variety of meats during the production process, and its presence is recorded on the food label. Use of this bacteriophage additive has increased the safety of these foods by 80–99%. Some of the main features of listeriosis are presented in **table 26.2.** The main causes of acute bacterial meningitis are compared in **table 26.3.**

⏮ bacteriophage

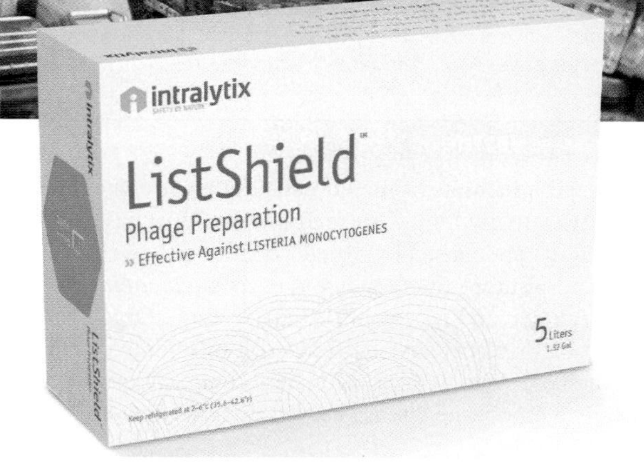

FIGURE 26.6 Bacteriophage in Food Safety When sprayed onto ready-to-eat foods, bacteriophage preparations significantly reduce the risk of listeriosis. (top): ©Tomas Del Amo/Photo Library/Getty Images; (bottom): ©Intralytix, Inc.

? What is the source of most listeriosis infections?

TABLE 26.2 Listeriosis

1. *Listeria monocytogenes* is ingested with food such as soft cheeses, unpasteurized milk, raw fruits and vegetables, hot dogs, or smoked fish.
2. The bacteria rapidly penetrate the intestinal epithelium and establish bacteremia, especially in pregnant women, the elderly, and the immunodeficient.
3. In pregnant women, circulating *L. monocytogenes* cells cross the placenta and fatally infect the fetus; bacteria transmitted to the baby at birth cause meningitis in 1 to 4 weeks. The mother usually does not have a serious illness.
4. In older people and those with underlying diseases, *L. monocytogenes* attacks the brain and meninges.

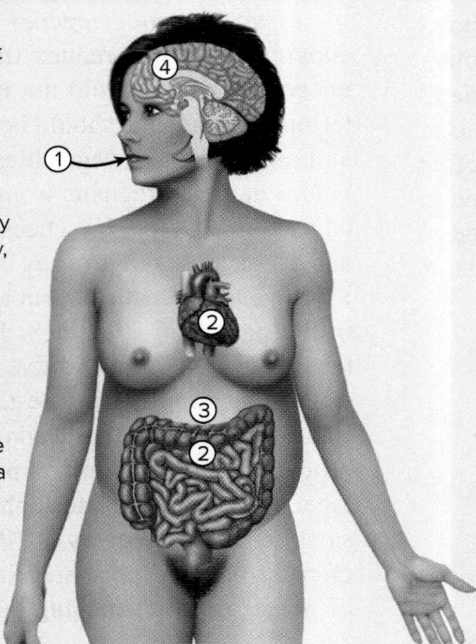

Signs and symptoms	Fever and muscle aches, with or without gastrointestinal symptoms; headache and stiff neck mark the onset of meningitis
Incubation period	A few days to 2 to 3 months; in newborn babies, 1 to 4 weeks
Causative agent	*Listeria monocytogenes,* a non-spore-forming Gram-positive rod able to grow at 4°C
Pathogenesis	Ingested *L. monocytogenes* cells penetrate the intestinal epithelium and enter the bloodstream; the resulting bacteremia spreads to the meninges, causing meningitis.
Epidemiology	Epidemics from contaminated foods. Pregnant women develop bacteremia that may result in fetal infection or miscarriage. Infants contract infection in the uterus or at birth. Elderly and those with immunodeficiency, cancer, and diabetes also at high risk.
Treatment and prevention	Treatment: antibacterial medications. Prevention: care in handling, cooking of raw meats; thorough washing of vegetables; reheating of cold cuts, hot dogs, and refrigerated leftovers.

TABLE 26.3 Main Causes of Acute Bacterial Meningitis Compared

Agent	Characteristics	Source	Vaccine?	Age Group Usually Affected
Streptococcus pneumoniae	Gram-positive diplococci	Respiratory system	Yes, against multiple serotypes	Mostly late teens and adults
Neisseria meningitidis	Gram-negative diplococci	Respiratory system	Yes, against multiple serotypes	Mostly infants, adolescents, and young adults; can cause epidemics
Haemophilus influenzae	Gram-negative coccobacilli	Respiratory system	Yes, against type b	Infants and young children
Streptococcus agalactiae	Gram-positive cocci in chains	Colon, vagina	No	Mostly neonates
Escherichia coli	Gram-negative rods	Colon	No	Mostly neonates
Listeria monocytogenes	Gram-positive rods; multiply at refrigerator temperatures	Environment; contaminated cheeses, cold cuts, other foods	No	Pregnant women, neonates; elderly

MicroAssessment 26.2

Bacterial infections of the nervous system are uncommon, but have a relatively high case-fatality rate. Most cases of bacterial meningitis are caused by species commonly carried in the upper respiratory tract (*Streptococcus pneumoniae, Neisseria meningitidis,* and *Haemophilus influenzae*). Organisms from the mother's birth canal are frequent causes in infants less than 1 month old. Meningitis is a common complication of listeriosis, usually a foodborne disease caused by *Listeria monocytogenes.*

4. Why are most newborn babies unlikely to contract pneumococcal, meningococcal, or *Haemophilus* meningitis?
5. How would a Gram stain distinguish *Streptococcus pneumoniae* from *Neisseria meningitidis*?
6. Why would you expect to find relatively high levels of protein and low levels of glucose in the CSF of someone with bacterial meningitis?

26.3 ■ Viral Diseases of the Central Nervous System

Learning Outcomes

6. Compare and contrast viral meningitis with viral encephalitis.
7. Explain the history of poliomyelitis and why it has been targeted for global elimination.
8. Outline the steps in the pathogenesis of rabies in wild animals and explain how it can be prevented in humans.

A wide variety of viruses can infect the nervous system causing meningitis or encephalitis. Poliomyelitis, now targeted for global elimination, and rabies, a disease normally found in animals, result from viral infection of neurons. Many viruses that typically affect other body systems can occasionally infect the CNS. These include the Epstein-Barr virus; the mumps, rubeola, varicella zoster, and herpes simplex viruses; and, most commonly, human enteroviruses.

Viral Meningitis

Viral meningitis is more common and much milder than bacterial meningitis. It usually does not require specific treatment, and individuals generally recover in 7 to 10 days.

Signs and Symptoms

The onset of viral meningitis is typically abrupt within 1 to 4 weeks after infection, with fever, severe headache, sensitivity to light, a stiff neck, and often vomiting. Depending on the causative agent, a sore throat, chest pain, swollen parotid salivary glands, or a skin rash may also be present. Viruses are responsible for most cases of aseptic meningitis. ⏮ aseptic meningitis

Causative Agents

Non-enveloped RNA viruses—members of the enterovirus subgroup of the family *Picornaviridae*—are responsible for at least half of the cases of viral meningitis. Of these, the most common offenders are coxsackie viruses, which can cause throat or chest pain, and echoviruses, which can cause a rash.

Pathogenesis

Enteroviruses characteristically infect the throat, intestinal epithelium, and lymphoid tissue, and then spread through the bloodstream to other sites. This can result in meningeal infection. The inflammatory response differs from bacterial meningitis in that fewer leukocytes usually enter the cerebrospinal fluid, and more are monocytes rather than neutrophils. Viral meningitis is typically less severe than bacterial meningitis and causes little lasting neurological damage.

Epidemiology

Enteroviruses are relatively stable in the environment, sometimes even persisting in chlorinated swimming pools. Enteroviral meningitis is transmitted by the fecal-oral route, unlike bacterial meningitis, which is usually transmitted by respiratory droplets. The feces of infected individuals often contain viruses for weeks.

Treatment and Prevention

No specific treatment for viral meningitis is available; most people recover on their own within 7–10 days. Handwashing and avoiding crowded swimming pools are reasonable preventive measures during outbreaks. There are no vaccines against most causes of viral meningitis. The main features of the disease are presented in **table 26.4.**

TABLE 26.4	Viral Meningitis
Signs and symptoms	Abrupt onset, fever, severe headache, stiff neck, often vomiting; sometimes sore throat, rash, or chest pain
Incubation period	Usually 1 to 4 weeks
Causative agents	Most cases: small non-enveloped RNA enteroviruses of *Picornaviridae*, usually coxsackie or echoviruses
Pathogenesis	Viruses from primary infection spread through the bloodstream to the meninges. Fewer leukocytes enter cerebrospinal fluid than with bacterial infections, and many are monocytes rather than neutrophils.
Epidemiology	Enteroviruses are transmitted by the fecal-oral route.
Treatment and prevention	Treatment: no specific treatment. Prevention: handwashing, avoiding crowded swimming pools during enterovirus outbreaks.

Viral Encephalitis

Although viral meningitis is usually mild, viral encephalitis is more likely to cause death or permanent disability. Viral encephalitis can be sporadic, resulting in a few widely scattered cases, or it can be epidemic. Sporadic encephalitis is usually due to activation of latent herpes simplex viruses. Most people recover from the disease but are left with permanent damage such as epilepsy, paralysis, deafness, or mental impairment. ⏮ herpes simplex viruses

Signs and Symptoms

Indications of viral encephalitis occur abruptly within 1 week of infection. Signs and symptoms include fever, headache, and vomiting, as well as possible disorientation, localized paralysis, deafness, seizures, or coma.

Causative Agents

Epidemic viral encephalitis is usually caused by **arboviruses** (**ar**thropod-**bo**rne viruses), a group of enveloped, single-stranded RNA viruses transmitted by insects, mites, or ticks. The leading causes of epidemic encephalitis in the United States are all transmitted by mosquitoes. These include LaCrosse encephalitis virus, St. Louis and West Nile encephalitis viruses, and eastern and western equine encephalitis viruses (**table 26.5**). ⏮ arthropods

Pathogenesis

Viruses multiply at the site of a mosquito bite and in local lymph nodes, producing mild and brief viremia. Relatively few infected individuals develop encephalitis. Some develop viral meningitis, or mild fever and headache. If viruses infect cells of the blood-brain barrier, they can enter the brain and replicate in neurons, causing destruction of brain tissue. Neutralizing

TABLE 26.5 Arboviruses That Cause Encephalitis in the United States

Virus	Family
LaCrosse encephalitis virus	*Bunyaviridae*
St. Louis encephalitis virus	*Flaviviridae*
West Nile encephalitis virus	*Flaviviridae*
Eastern equine encephalitis virus	*Togaviridae*
Western equine encephalitis virus	*Togaviridae*

antibodies stop disease progression. Case-fatality rates range from about 2% with LaCrosse encephalitis to 35–50% with eastern equine encephalitis. Emotional instability, epilepsy, blindness, or paralysis may occur in 5–50% of those who recover. The likelihood of disability depends largely on the kind of virus and the age of the patient. As with many diseases, the very young and the elderly are most affected.

Epidemiology

The common types of epidemic viral encephalitis are all zoonoses with a natural reservoir in birds or other small animals. Natural hosts can develop such high levels of virus in the bloodstream that mosquitoes can transmit the virus from one host to another. Humans are an accidental or **dead-end host**—they can acquire the virus from a mosquito bite but do not develop sufficient viremia to transmit it back to the arthropod vector. ◀◀ reservoir, ◀◀ dead-end host

Some arboviruses have a broad host range and cause disease in horses as well as humans. This gives rise to the names eastern and western equine encephalitis. The LaCrosse virus infects mosquitoes, which pass it directly from one mosquito to another in semen. They feed on and infect squirrels and chipmunks, where the virus multiplies and then spreads to uninfected mosquitoes feeding on the rodents' blood. The St. Louis virus and the West Nile virus have bird reservoirs. When the West Nile virus was introduced into New York from the Middle East in the summer of 1999, migrating birds quickly spread the virus across the country (**figure 26.7**).

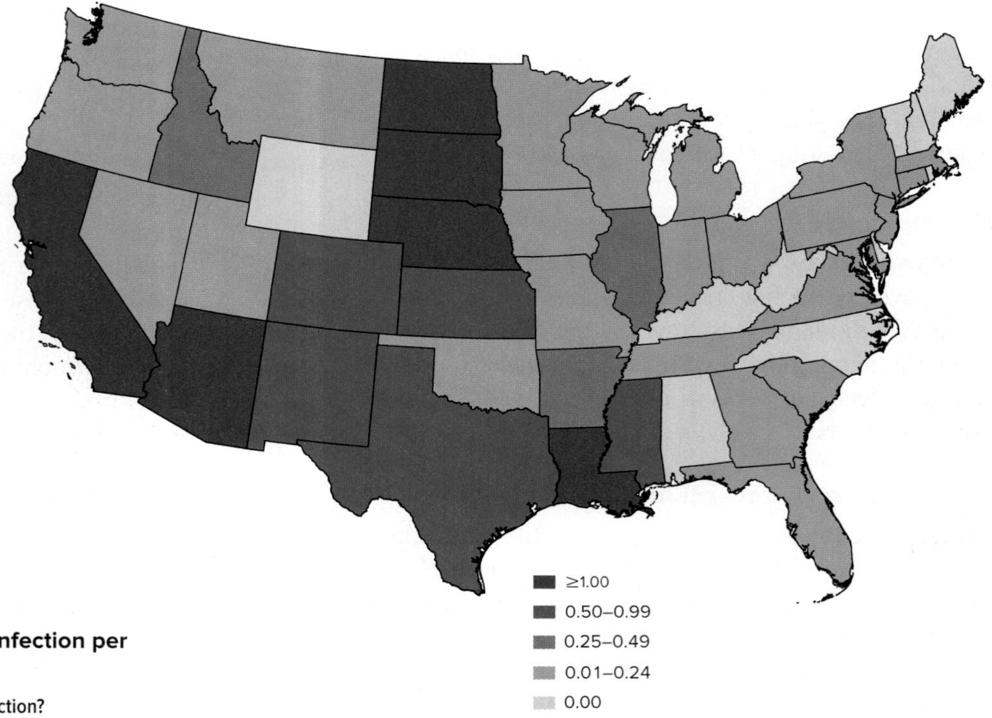

FIGURE 26.7 Incidence of West Nile Virus Infection per 100,000, United States, 2014

? What is the best preventive against West Nile virus infection?

Treatment and Prevention

There is no proven antiviral therapy for epidemic viral encephalitis. The blood of sentinel chickens in cages with free access to mosquitoes is tested periodically for evidence of arbovirus infection. A positive test would trigger an encephalitis alert, like this St. Louis encephalitis alert issued in Florida: "Avoid outdoor activity during evening and night, the peak hours of biting for the *Culex* mosquito vector. If outdoors, wear long sleeves and pants. Make sure windows and porches are properly screened. Use insect repellents and insecticides." Equine encephalitis viruses generally infect horses 1 or 2 weeks before the first human cases appear, so these cases provide a warning to increase protection against mosquitoes. Vaccines against equine encephalitis have been approved for use in horses and the emu—a large, domesticated, meat-producing bird susceptible to this virus. Clinical trials for a human vaccine against West Nile virus are underway. Currently, the best preventive is to avoid mosquito bites. The main features of epidemic viral encephalitis are presented in **table 26.6.**

Poliomyelitis

The characteristic feature of poliomyelitis (also called polio) is destruction of motor neurons, resulting in paralysis of a group of muscles, such as those of an arm or a leg. Two individuals—Albert Sabin and Jonas Salk—each developed an effective vaccine to control this terrifying disease. They were bitter rivals who both expected, but did not receive, a Nobel Prize for the work. Soon after the vaccines were introduced, the number of polio cases in the United States and other

TABLE 26.6 Viral Encephalitis

1. Infected mosquito introduces encephalitis virus.
2. Virus multiplies locally, establishes brief low-level viremia.
3. Virus crosses blood-brain barrier and preferentially attacks the brain.
4. Destruction of brain tissue causes death or permanent disabilities such as emotional instability, mental disability, paralysis of face, arm, leg.
5. There is no exit for the virus; thus, humans are dead-end hosts.

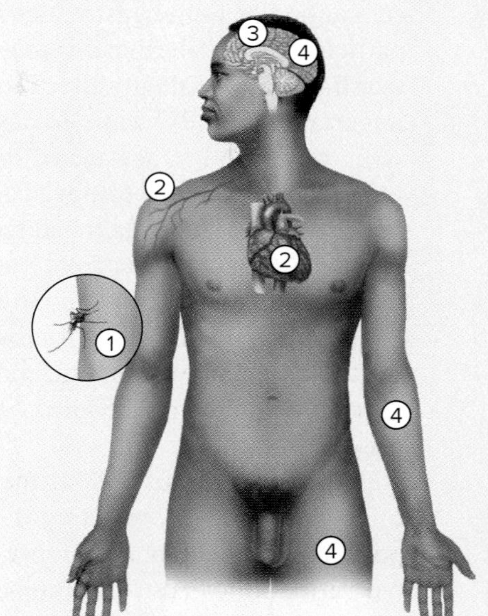

Signs and symptoms	Abrupt onset, fever, headache, vomiting, disorientation, paralysis, seizures, deafness, coma
Incubation period	First symptoms within a few days; encephalitic symptoms often within the first week
Causative agent	One of five arboviruses: LaCrosse, St. Louis, West Nile, western equine, or eastern equine
Pathogenesis	Replication of virus at the site of the mosquito bite, replication in lymph nodes, then viremia and invasion of brain tissue. Nerve cells in the brain destroyed.
Epidemiology	Viruses transmitted to humans from birds or rodents by mosquitoes.
Treatment and prevention	Treatment: no accepted treatment for arboviral encephalitis. Prevention: chicken sentinels to warn of arbovirus epidemics; insecticides and other anti-mosquito measures.

developed countries dropped dramatically. Following that success, a program to eliminate the disease from the Western Hemisphere began. Then in 1988, a public-private partnership called the Global Polio Eradication Initiative was launched, with the aim of eliminating polio worldwide. At that time, the disease was still endemic in 125 countries, with an estimated 350,000 cases of paralysis occurring each year. As of 2016, the disease is endemic in only three countries (Afghanistan, Nigeria, and Pakistan), with fewer than 40 cases worldwide. Although that reduction is impressive, eradication of polio has proven to be a very difficult challenge.

Signs and Symptoms

The outcome of infection by the virus that causes poliomyelitis is highly variable, with up to 95% of cases being asymptomatic. If signs and symptoms do develop, they typically begin 6 to 20 days after infection. About 5% of those infected experience a mild illness, characterized by non-specific signs and symptoms such as fever, headache, nausea, and malaise, which last less than a week. In a small percentage of cases (about 1%), infection spreads to the central nervous system, resulting in back pain and muscle spasms that resolve within 10 days. In a few people, these are followed by paralysis. When paralysis occurs, muscles shrink and bones do not develop normally in the affected area over a period of weeks to months. In severe cases, the respiratory muscles are paralyzed, and air must be pumped in and out of the lungs by a ventilator (**figure 26.8**). Those who survive this acute stage of the illness recover some function. Sensory neurons are not affected.

People who survive acute paralytic poliomyelitis sometimes develop post-polio syndrome. This condition is characterized by muscle pain, increased weakness, and muscle degeneration 15 to 50 years after recovering from poliomyelitis. Post-polio syndrome is not due to a reappearance of polioviruses and sometimes involves muscles not obviously affected by the original illness. Instead it is thought to be a secondary effect of the initial damage. During recovery from

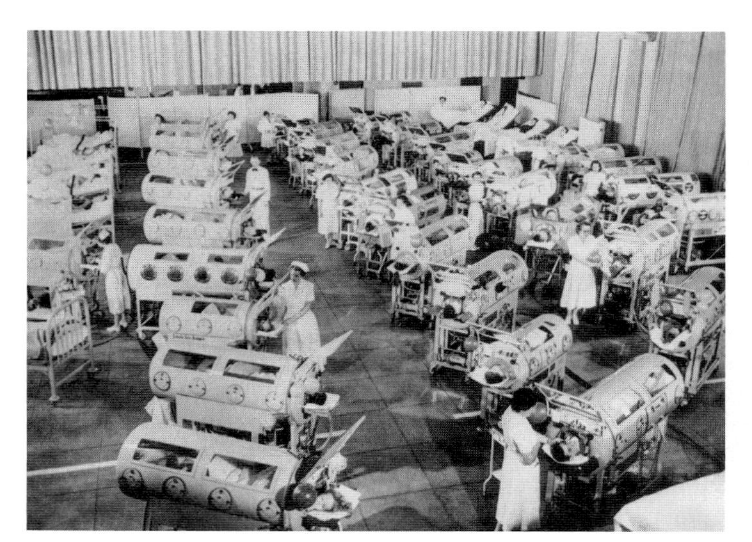

FIGURE 26.8 The Horror of Poliomyelitis These tank-like respirators ("iron lungs") were used during the 1950s epidemics of poliomyelitis to keep alive people whose respiratory muscles were paralyzed by the disease.

? How are polioviruses transmitted?

acute paralytic poliomyelitis, surviving nerve cells branch out to take over the functions of the killed nerve cells. Post-polio syndrome is probably due to the death of these nerve cells after doing double duty for many years.

Causative Agent

Poliomyelitis is caused by three serotypes of poliovirus (designated 1, 2, and 3), but type 1 might be the only one still circulating today. Type 2 was declared eradicated in 2015, after not being detected anywhere in the world since 1999. Type 3 has not been detected since 2012. The non-enveloped, single-stranded RNA virus is a member of the enterovirus subgroup of the family *Picornaviridae.* Like other enteroviruses, poliovirus is quite stable under natural conditions.

Pathogenesis

Poliovirus enters the body orally, infects cells that line the throat and intestinal tract, and then invades the bloodstream. Usually, symptoms are mild or absent, and recovery is complete. Only rarely does the virus enter the central nervous system (CNS) and bind to specific receptors on motor neurons, replicate in the cell bodies, and destroy the cells when the mature virus is released. Although the infection is in the CNS, the effects are peripheral. As the motor neurons degenerate due to infection, the muscles innervated by them are paralyzed and atrophy. Because most people infected with poliovirus do not develop nervous system involvement, a single case of poliomyelitis means that the virus is widespread.

Epidemiology

Poliovirus is transmitted by the fecal-oral route and is quite stable under natural conditions, so it spreads easily. When polio is endemic, children usually become infected within the first few months of life, a time when antibodies from the mother still protect the infant. Maternal anti-poliovirus antibodies can neutralize the virus in an infant's bloodstream, thereby preventing it from infecting the motor neurons to cause paralysis. In contrast, when a person is first infected with poliovirus later in life, maternal antibodies are no longer present so paralysis is more likely to result.

Treatment and Prevention

Two types of vaccines are used to prevent polio. Oral polio vaccine (OPV) is attenuated poliovirus and can be administered as drops on the tongue; inactivated polio vaccine (IPV) is given by injection. The OPV strains replicate in the recipient's intestinal tract and are excreted in feces, and can thereby spread to others, immunizing them as well. OPV is also less expensive to produce than IPV and is much easier to administer because it does not require an injection.

As described in chapter 18, eradication of polio requires that OPV be used in regions where the virus is endemic or likely to be reintroduced. This is because OPV prevents the spread of poliovirus in a population. If an OPV-vaccinated individual ingests poliovirus-contaminated material, mucosal antibodies in the gastrointestinal tract neutralize the viral particles before they can infect cells and replicate. IPV is unreliable at eliciting production of mucosal antibodies. The importance of OPV was demonstrated in 2013, when poliovirus was found in sewage in Israel. That country had been free of poliovirus and stopped using OPV in 2004, so the only polio vaccine the children had received was IPV. The vaccine protected the recipients from paralytic poliomyelitis, but did not prevent reintroduced poliovirus from spreading within the population. After the virus was detected in sewage, Israel started an intensive immunization program using OPV, successfully eliminating the virus once again. ⏮ **campaign to eliminate poliomyelitis**

The drawback of OPV is that a vaccine strain can mutate to become virulent, resulting in vaccine-associated paralytic polio (approximately 1 case per 2.4 million doses given). In fact, after poliovirus was eliminated in the United States, the rare cases of polio that occurred were caused by vaccine-derived poliovirus (now called VDPV) rather than wild poliovirus (WPV) (**figure 26.9**). Once WPV was eliminated from the Western Hemisphere, the small risk of developing poliomyelitis due to VDPV led the United States to change to the routine use of IPV only.

In the past, OPV was trivalent, meaning that it contained all three poliovirus serotypes. Now that type 2 has been eradicated, the trivalent vaccine has been replaced by a bivalent OPV

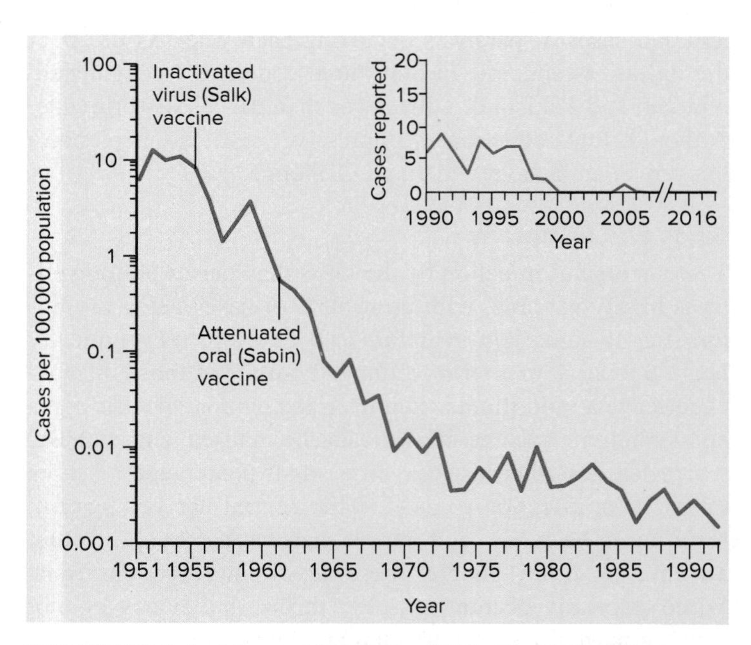

FIGURE 26.9 Incidence of Poliomyelitis in the United States, 1951–2016 Wild poliovirus (WPV) was eliminated from the United States by 1980, and from the entire Western Hemisphere by 1991. Polio cases acquired in the United States after 1980 were caused by vaccine-derived poliovirus (VDPV).

What was the cause of the polio case reported in 2005?

TABLE 26.7 Poliomyelitis

1. Poliovirus enters the body orally.
2. The virus replicates in cells lining the throat and the intestines.
3. In some cases, the virus spreads via the bloodstream, causing mild symptoms of fever, headache, and nausea.
4. In a small percentage of cases, the virus enters the CNS and may then cause the paralysis that characterizes poliomyelitis.
5. Poliovirus exits the body with feces.

Signs and symptoms	Headache, fever, stiff neck, nausea, pain, muscle spasm, followed by paralysis
Incubation period	7 to 14 days
Causative agent	Poliovirus serotypes 1, 2, 3; member of *Picornaviridae*
Pathogenesis	Virus infects the throat and intestine, circulates via the bloodstream, and enters some motor nerve cells of the brain or spinal cord (CNS); infected nerve cells lyse upon release of mature virus.
Epidemiology	Spreads by the fecal-oral route; asymptomatic and non-paralytic cases common
Treatment and prevention	Treatment: artificial ventilation for respiratory paralysis; physical therapy and rehabilitation. Prevention: vaccination

(bOPV) that contains only types 1 and 3; monovalent options against those types are also available. The removal of poliovirus type 2 from OPV is significant because most VDPV was due to that type. IPV is still trivalent, and is important because it protects against any type 2 VDPV that might still be circulating.

Now that the geographic distribution of poliovirus is limited and DNA sequencing methods are so powerful, the source of infecting strains can be traced more easily. In turn, this makes it easier for public health officials to target effective vaccination campaigns. Testing of sewage and other environments also helps officials monitor the spread of the virus. However, the fact that less than 1% of infections are symptomatic is particularly problematic with respect to eradication. A seemingly low number of 100 polio cases actually indicates over 10,000 infections. In fact, after Nigeria reported no polio cases for 2 years and was taken off the list of endemic countries, several children there were diagnosed with the disease. The infecting strains were genetically related to WPV that had circulated in the region previously, indicating that the virus had never been eliminated. As a result, Nigeria was added back to the list of endemic countries.

In addition to controlling poliovirus through vaccination, the viral particles can be inactivated by pasteurization and by chlorinating drinking water. The main features of poliomyelitis are summarized in **table 26.7.**

Rabies

Rabies is a classic zoonotic disease—one normally found in animals but that can be transmitted to humans. In the United States, immunization of dogs and cats against rabies has practically eliminated them as a source of human disease. Still, rabies has multiple wild animal hosts. These remain a constant threat to non-immunized domestic animals and humans. ⏮ zoonotic disease

Signs and Symptoms

Rabies begins with fever, head and muscle aches, sore throat, fatigue, and nausea. The characteristic symptom is a tingling or twitching sensation at the site of viral entry, usually an animal bite. These symptoms generally do not appear until 1 to 2 months after infection, but they progress rapidly to agitation, confusion, hallucinations, seizures, increased sensitivity, and encephalitis. A few days later, the individual typically goes into a coma and dies of respiratory failure or cardiac arrest.

The later stages of rabies are often characterized by increased salivation and difficulty in swallowing. This causes the "frothing at the mouth" classically associated with rabid animals and "mad dogs." Swallowing, or even the sight of fluids, often leads to severe spasms of the throat and respiratory muscles. The common name for rabies is "hydrophobia" or fear of water.

Causative Agent

The cause of rabies is the rabies virus (**figure 26.10a**), a member of the family *Rhabdoviridae*. This virus has a striking bullet shape, is enveloped, and has a single-stranded, negative-sense RNA genome.

Pathogenesis

After the rabies virus is introduced into the body, it typically multiplies in cells at the site of infection for some time before entering a sensory neuron. It then travels by retrograde transport up the axon to the spinal cord and eventually to

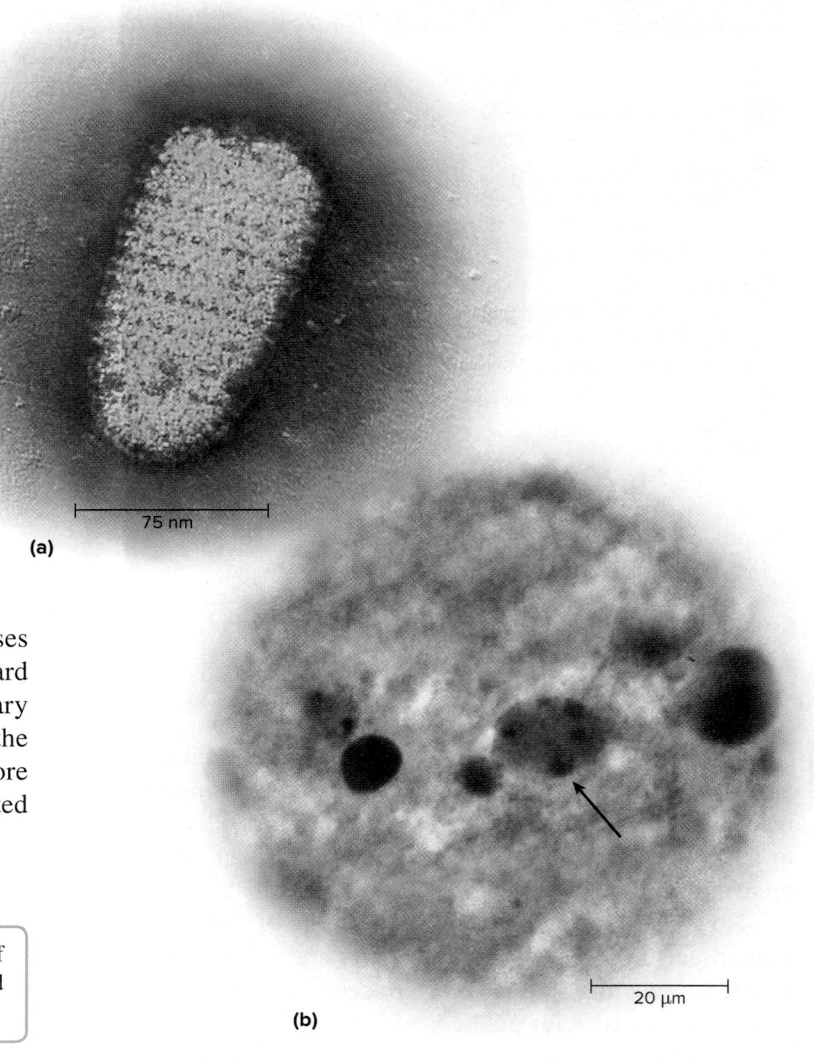

FIGURE 26.10 Rabies Virus (a) Color-enhanced transmission electron micrograph of rabies virus. Notice the bullet shape. **(b)** Stained smear of brain tissue from a rabid dog. A Negri body (see arrow) represents a site of rabies virus replication.

? **Why can an animal that appears well still transmit the rabies virus?**

the brain. The length of time before symptoms occur depends on the location of the bite, the amount of virus introduced, and the condition of the host. Individuals with head wounds, for example, tend to show symptoms sooner than those with leg wounds.

Once in brain tissue, the virus multiplies extensively, causing the symptoms of encephalitis. Characteristic inclusion bodies called **Negri bodies,** made up of viral nucleocapsids, are found in most rabies cases (see figure 26.10b). From the brain, the virus spreads outward via the nerves to various body tissues, notably the salivary glands, eyes, and fatty tissue under the skin, as well as to the heart and other vital organs. Rabies can be diagnosed before death by identifying the virus in stained smears collected from the surface of the eyes.

MicroByte

The only documented cases of human-to-human transmission of rabies have been in patients receiving transplants from infected donors.

Epidemiology

The primary mode of transmission of rabies to humans is via the saliva of a rabid animal introduced into bite wounds of the skin. It has been reported, but not documented, that individuals can also contract rabies by inhaling aerosols containing the virus, such as from bat feces. Because of the extensive rabies vaccination program for dogs in the United States, the main reservoir for rabies is wild animals such as raccoons, bats, skunks, and foxes (**figure 26.11**). Over 5,000 wild animal cases are reported in the United States each year, representing an enormous reservoir from which infection can be transmitted to domestic animals and humans. Raccoons lead the list of wildlife cases, but almost all human cases are due to contact with infected bats. Human rabies is rare in the United States, with only one to three cases reported per year.

Although rabies is rare in the United States, about 40,000–70,000 people worldwide die of rabies. Most of these are due to dog bites in areas where dogs are not routinely vaccinated.

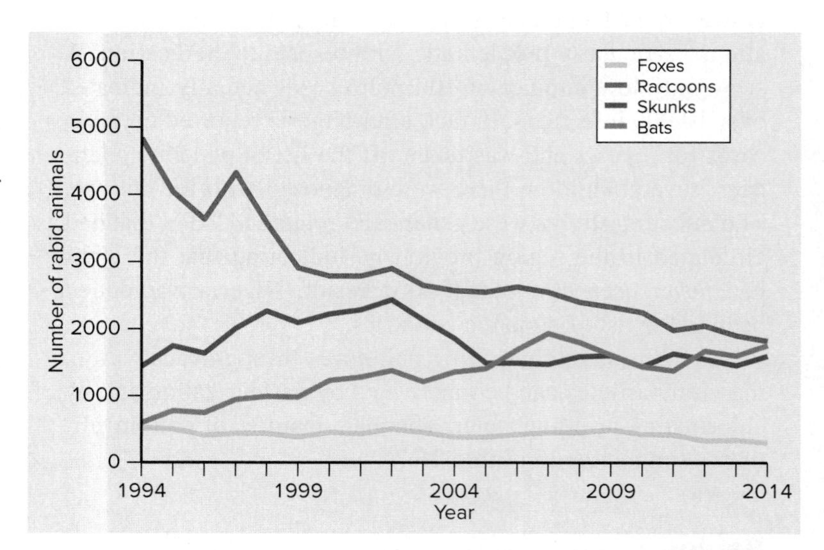

FIGURE 26.11 Cases of Rabies Among Wildlife in the United States

? **What animal is responsible for most human cases of rabies in the United States?**

FOCUS YOUR PERSPECTIVE 26.1

Rabies Survivors!

In October 2004, a 15-year-old girl was admitted to the hospital with typical signs and symptoms of rabies. A month earlier, while in church, she had picked up a bat that had fallen to the floor. The bat bit her left index finger, and she carried it outside and released it. The small bite wound was cleaned with hydrogen peroxide and healed uneventfully. Later, when she was in the hospital, a tube was placed in her trachea and hooked to a respirator because most rabies deaths result from respiratory failure. She was given medication to put her into a coma to rest her nervous system, and she was given the antiviral medication ribavirin, according to an experimental protocol. After 7 days, the coma-inducing medication was reduced and the girl was allowed to wake up. A little more than a month after her illness began, her breathing tube was removed; she regained speech, solved mathematics problems, and walked with assistance. Slow, steady improvement continued after she returned home.

In February 2009, a 17-year-old girl reported to the emergency room with a severe headache that had lasted for 2 weeks, vomiting, neck pain, and other symptoms of nervous system infection. On March 6, she was admitted to the hospital with suspected infectious encephalitis but did not respond to treatment. Two months earlier she had entered a cave in Texas and several flying bats had hit her body, although she did not notice any scratches or bite wounds at the time. On March 11, she tested positive for rabies and was given rabies immune globulin (RIG) and one dose of anti-rabies vaccine. She remained in the hospital with supportive care, but never entered intensive care and was discharged on March 22. After an emergency room visit on April 3 for headache and vomiting, she did not return for follow-up.

Survival after the onset of rabies symptoms has rarely been reported globally. Most survivors had received anti-rabies vaccine either before they were exposed to rabies or before they showed signs or symptoms of the disease. Most also suffer from persisting neurological damage, unlike these patients, who made complete recoveries. Over the years, many treatments have been tried without benefit, including rabies immune globulin, anti-rabies vaccines, and interferon. What is to be learned from the survival of these patients? Perhaps their survival had nothing to do with their treatment—but was due to their own immune systems fighting a small infectious dose or a rabies viral strain of low virulence. Is there a clue in their survival that will lead to the first effective rabies treatment in the 3,000-year history of the disease?

A person bitten by a dog with rabies virus in its saliva has about a 30% risk of developing rabies. Most rabies cases are in Asia, followed by Africa.

Most rabid dogs excrete the virus in their saliva, sometimes even a few days before they get sick. Therefore, if an unvaccinated dog bites a person, the animal should be confined for 10 days to see if it develops symptoms of rabies. Some dogs become irritable and hyperactive with the onset of a type of rabies called "furious" rabies. Animals produce excessive saliva, and attack people, other animals, and inanimate objects. Perhaps more common is the "paralytic" form of rabies, in which an infected dog simply stops eating, becomes inactive, and suffers paralysis of throat and leg muscles. Obviously, one should not try to remove a foreign body from the throat of a sick, choking, unvaccinated dog!

Treatment and Prevention

A person who has been bitten by an animal should immediately wash the wound thoroughly with soap and water, and then apply an antiseptic to avoid any kind of infection. If contact with a rabid animal is suspected, the person should receive post-exposure prophylaxis (PEP). The aim of this is to provide immediate passive immunity while also inducing active immunity, a strategy that is effective because of the long incubation period before the virus enters neurons. As part of PEP, human rabies immune globulin (HRIG) (anti-rabies antibody) is injected at the wound site and intramuscularly to neutralize free virus. In addition, four injections of rabies vaccine are given to provoke an effective immune response. The first dose of the inactivated vaccine is administered as soon as possible after exposure, and the others at 3, 7, and 14 days after the first. ⏮ passive immunity

There is no effective treatment for rabies, and very few people are known to have recovered from the disease once symptoms appear (**Focus Your Perspective 26.1**). In the United States, about 30,000 people annually receive inactivated rabies vaccine after a bite from a suspected rabid animal. Individuals at high risk for rabies, such as veterinarians or animal control personnel, should be immunized before exposure.

It is impossible to eradicate a zoonotic disease without clearing the animal hosts. Routine vaccination of domestic dogs and cats in the United States has dramatically reduced the number of cases in domestic pets. Vaccination of wild animals is more difficult since it is impossible to catch, immunize, and release all of them. Programs are underway to administer an oral vaccine to widespread wild populations by placing it into a bait food, such as peanut butter or fish meal. Many eastern states have participated in air-drop programs to vaccinate raccoons; Texas has a similar program targeting foxes and coyotes. Some features of rabies are summarized in **table 26.8.**

TABLE 26.8 Rabies

Signs and symptoms	Fever, headache, nausea, vomiting, sore throat, cough at onset; later, spasms of the muscles of mouth and throat, coma, and death
Incubation period	Usually 30 to 60 days
Causative agent	Rabies virus, single-stranded RNA; bullet shape
Pathogenesis	During incubation period, virus multiplies at site of bite, then travels via nerves to the central nervous system; it multiplies and spreads outward via multiple nerves to infect heart and other organs.
Epidemiology	Bite of rabid animal, usually a bat.
Treatment and prevention	Treatment: immediately wash wound with soap and water and apply antiseptic; inject rabies vaccine and human rabies immunoglobulin as soon as possible. No effective treatment once symptoms begin. Prevention: avoid suspect animals; immunize pets.

MicroAssessment 26.3

Many different kinds of viruses can attack the nervous system, but they generally do so in only a small percentage of infected people. At least half of viral meningitis cases are caused by the enterovirus subgroup of picornaviruses, which are generally spread by the fecal-oral route. Viral meningitis is usually mild, but viral encephalitis often causes permanent disability. Poliomyelitis, characterized by paralysis of one or more muscle groups, is caused by poliovirus. Rabies is a widespread zoonosis, almost uniformly fatal for humans, usually transmitted by animal bites.

7. Explain why epidemic viral encephalitis is called a zoonosis.
8. What are the advantages and disadvantages of OPV relative to IPV?
9. Why is rabies now rare in humans when it is still so common in wildlife?

26.4 ■ Fungal Diseases of the Central Nervous System

Learning Outcome

9. Compare and contrast the development of cryptococcal meningoencephalitis in AIDS patients and in healthy individuals.

Inhalation of fungal cells found in soil and bird droppings seldom causes serious lung disease, but phagocytic cells that ingest these cells may carry them from the lungs via the bloodstream to the brain. This can lead to inflammation of the brain and meninges.

Cryptococcal Meningoencephalitis

Cryptococcal meningoencephalitis was uncommon until the onset of the AIDS epidemic. Now the disease is among the most important HIV-related opportunistic infections. Because it is difficult to clear without T-cell involvement, AIDS patients with this infection are sometimes maintained indefinitely on antifungal medications. More recently, the incidence of cryptococcal meningoencephalitis has increased among healthy individuals in the western United States and Canada due to an emerging pathogen, *Cryptococcus gattii*.

◀◀ opportunistic infection

Signs and Symptoms

In apparently healthy people, symptoms of cryptococcal meningoencephalitis develop gradually and generally consist of difficulty in thinking, dizziness, intermittent headache, and possibly slight fever. After weeks or months, vomiting, weight loss, paralysis, seizures, and coma may appear. In people with immunodeficiency, the disease generally progresses much faster; without treatment, death can occur in as little as 2 weeks.

Causative Agent

Cryptococcal meningoencephalitis is an infection of the meninges and brain by either *Cryptococcus neoformans* or *Cryptococcus gattii*. The organisms are spherical yeasts generally 3 to 20 µm in diameter surrounded by a thick capsule that resists the immune response (**figure 26.12**). *Cryptococcus neoformans* is an opportunistic pathogen that usually causes disease only in the immunocompromised. *C. gattii,* on the other hand, causes disease even in healthy individuals.

Pathogenesis

Organisms enter the body by inhalation, and infection is first established in the lung, usually producing mild or no symptoms. Immune defenses of healthy people usually eliminate

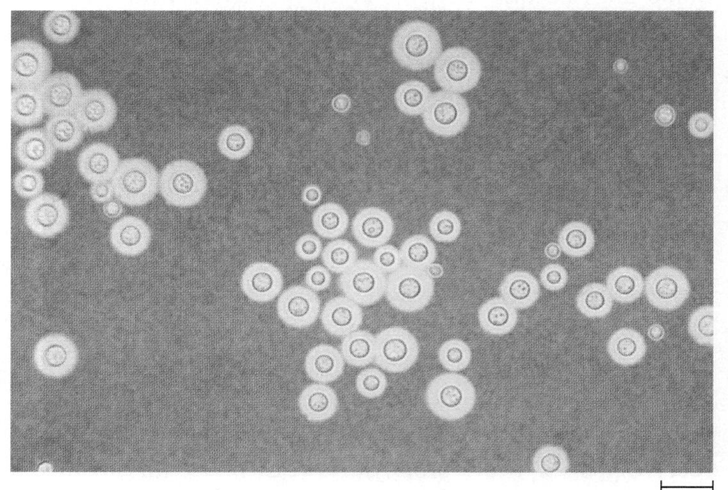

FIGURE 26.12 ***Cryptococcus neoformans*** This India ink stain reveals the capsule surrounding the yeast cells. Source: Dr. Leanor Haley/CDC

How does a capsule allow an organism to establish an infection?

the infection, but phagocytic killing is slow and inefficient due to the presence of thick capsules around the fungal cells. In some cases, particularly in immunocompromised individuals, the organisms multiply, enter the bloodstream, and are distributed throughout the body. Meningoencephalitis is the most common infection outside of the lung, but organisms in the bloodstream can also infect skin, bones, and other body tissues. Capsular material can be detected in spinal fluid and urine, aiding diagnosis. In meningoencephalitis, the organisms typically cause thickening of the meninges, sometimes hindering the flow of cerebrospinal fluid and increasing pressure within the brain. They also invade brain tissue, producing multiple abscesses.

Epidemiology

Cryptococcus neoformans is distributed worldwide in soil and vegetation contaminated with bird droppings. Humans, as well as cats, dogs, and other animals, may inhale the organism, but subsequent onset of disease is rare. Symptomatic infection is sometimes the first indication of AIDS. Person-to-person transmission of the disease does not occur.

Cryptococcus gattii was once associated with eucalyptus trees in tropical or subtropical regions of the world, but emerged in 1999 in British Columbia, Canada. Pathogenic strains have since spread in the U.S. Pacific Northwest and Canada (**figure 26.13**). Appearance of more virulent strains has increased the case-fatality rate from 5% to about 25%. These emerging strains also cause more cases of fungal pneumonia.

MicroByte

Several Dall's porpoises have been found washed up on the shores in British Columbia since 2000; all were positive for *C. gattii* infection.

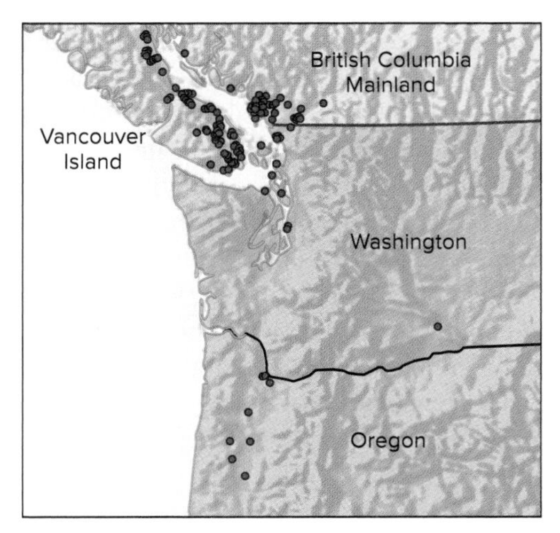

FIGURE 26.13 Spread of *Cryptococcus gattii* in the Pacific Northwest in 2008

? How does *Cryptococcus gattii* enter the body?

TABLE 26.9	Cryptococcal Meningoencephalitis
Signs and symptoms	Headache, vomiting, confusion, and weight loss; slight or no fever; symptoms may progress to seizures, paralysis, coma, and death
Incubation period	Widely variable, few to many weeks
Causative agent	*Cryptococcus neoformans, Cryptococcus gattii*—encapsulated yeasts
Pathogenesis	Infection starts in lung; encapsulated organisms multiply, enter bloodstream, and are carried to various parts of the body; phagocytosis inhibited; meninges and adjacent brain tissue become infected.
Epidemiology	Inhalation of material contaminated with the fungus; other sources; most people resistant to the disease.
Treatment and prevention	Treatment: antifungal medications. Prevention: none.

Treatment and Prevention

Treatment with the antifungal medication amphotericin B is often effective, particularly if given with flucytosine (5-fluorocytosine) followed by the oral medicine fluconazole. Amphotericin B must be given intravenously and the dose carefully regulated to minimize its toxic effects. Because amphotericin B does not reliably cross the blood-brain barrier, the medication may be administered through a plastic tube inserted through the skull into a ventricle of the brain. Treatment is successful in about 70% of cases except in AIDS patients, who respond poorly, most likely because they lack T-cell-dependent killing that normally assists the action of the antifungal medications. Unless their T-cell function can be restored by treatment, AIDS patients are rarely cured of their infection. ◀◀ antifungal medications

There is no vaccine or other preventive measure available. The main features of cryptococcal meningoencephalitis are summarized in **table 26.9.**

MicroAssessment 26.4

Cryptococcal meningoencephalitis occurs opportunistically in immunocompromised individuals such as AIDS patients. The causative organism is an encapsulated yeast, often found in soil contaminated with pigeon droppings. The disease has recently emerged among healthy individuals in the Pacific Northwest.

10. Why is it difficult to cure AIDS patients of a fungal infection?
11. Why are AIDS patients so vulnerable to cryptococcal meningoencephalitis?
12. What might cause the spread of new strains of *C. gattii* in the northwest United States?

26.5 ■ Protozoan Diseases of the Central Nervous System

Learning Outcome

10. Compare and contrast African trypanosomiasis, toxoplasmosis, and primary amebic meningoencephalitis.

Infection of the nervous system by protozoa is quite rare, but usually results in fatal disease when it occurs. Protozoan diseases are difficult to treat because medications that affect eukaryotic cells may also affect fragile neurons.

African Trypanosomiasis ("African Sleeping Sickness")

African trypanosomiasis—also known as African sleeping sickness—is transmitted by the tsetse fly, its biological vector. Sickness may persist for years, producing the symptoms for which it is named. The disease can be contracted by residents and visitors in a wide area across the middle of the African continent. ◀◀ biological vector

Signs and Symptoms

African trypanosomiasis can be chronic or acute. A tender nodule develops at the site of the bite within a week after a person is bitten by an infected tsetse fly. Regional lymph nodes might enlarge, but symptoms can disappear spontaneously. In the chronic form of the disease, recurrent fevers can continue for months or years. Involvement of the CNS is marked by gradual withdrawal, decreased activity, and indifference to food. The eyelids droop, the individual falls asleep even while eating or standing, and speech is slurred. Eventually the person becomes comatose and dies. In acute cases, neurological symptoms develop much more rapidly.

Causative Agent

African sleeping sickness is caused by the flagellated protozoan *Trypanosoma brucei.* These organisms (referred to as trypanosomes) are slender and have a wavy, undulating membrane and an anteriorly protruding flagellum (**figure 26.14**). Two subspecies are morphologically identical—*T. brucei rhodesiense* and *T. brucei gambiense.* The *rhodesiense* subspecies causes acute disease and occurs mainly in the cattle-raising areas of East Africa; the *gambiense* subspecies causes chronic disease and occurs mainly in forested areas of Central and West Africa. Both are transmitted by tsetse flies, biting insects of the genus *Glossina.*

Pathogenesis

The protozoan enters the bite wound in the saliva of an infected tsetse fly. The parasite multiplies at the site and eventually enters the circulation. Fever and antibody production lessen early symptoms. Within about a week and at roughly weekly intervals thereafter, however, the number of parasites in the blood increases. Each burst coincides with the appearance of a new glycoprotein on the surface of the trypanosomes. More than a thousand different genes code for variations of that glycoprotein, but only one is expressed at a time. Each time a new gene is expressed, a new immune response is needed to produce appropriate antibodies. Recurrent cycles of parasitemia and antibody production continue until the patient is treated or dies.

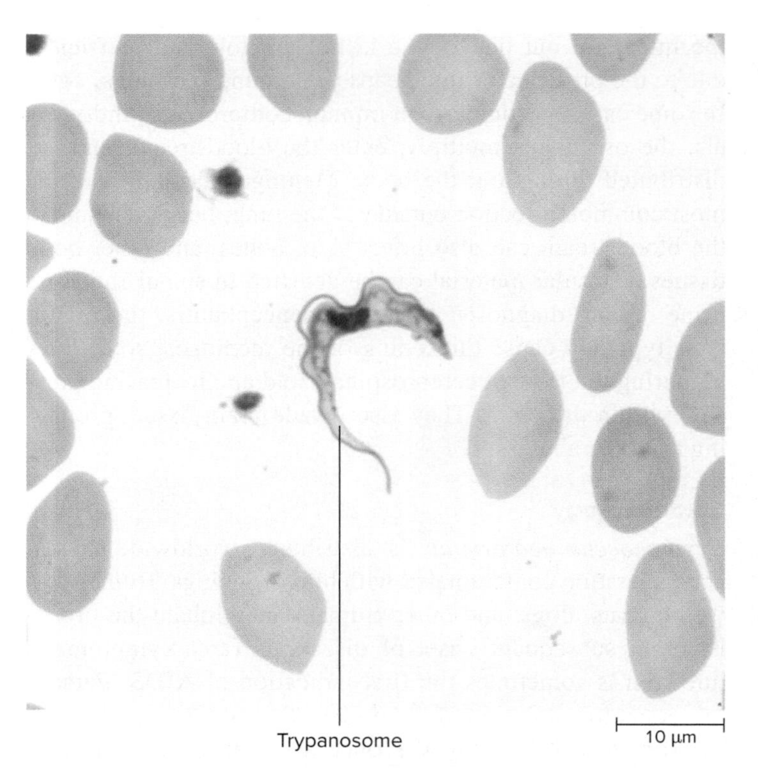

FIGURE 26.14 ***Trypanosoma brucei*** **in a Blood Smear** Notice the slender protozoan among the red blood cells of an individual with African trypanosomiasis. Source: Blaine Mathison/CDC

? Name the biological vector of *Trypanosoma brucei.*

In *T. brucei rhodesiense* infections, disease can progress rapidly. The parasite enters the heart and the brain within 6 weeks of infection. Irritability, personality changes, and mental dullness result from brain involvement, but the patient usually dies from heart failure within 6 months. With *T. brucei gambiense,* progression of the chronic infection is much slower. Years may pass before death occurs, often from secondary infection. Much of the damage to the host is due to immune complexes formed when high levels of protozoan antigen are bound by antibodies that then react with complement system proteins. ◀◀ immune complex

Epidemiology

The WHO reported a record low of 3,000 cases in 2015 for cases of African trypanosomiasis, including cases among tourists. The acute Rhodesian form of the disease is a zoonosis; the main reservoirs are wild animals. Humans are the main reservoir for the chronic Gambian form, and human-to-human transmission is more common. The presence of animal hosts makes the Rhodesian form more difficult to control.

Treatment and Prevention

As with other eukaryotic pathogens, treatment is problematic because of toxic side effects of the available medications. Nevertheless, treatment of infected people helps reduce the protozoan's reservoir. Pentamidine is used to treat early stages of the Gambian (chronic) form of the illness; suramin is used to treat early stages of the Rhodesian (acute) form. If the disease progresses to involve the central nervous system, other drugs that can cross the blood-brain barrier must be used. Early diagnosis and treatment of the Rhodesian (acute) form is important to prevent rapid damage to the nervous system. ⏮ antiprotozoan medications

Actions directed against tsetse fly vectors include use of insect repellents and protective clothing to prevent bites, use of traps containing bait and insecticides to reduce the vector population, and clearing of brush to reduce breeding habitats for the flies. The main features of African trypanosomiasis are presented in **table 26.10.**

Toxoplasmosis

The protozoan parasite that causes toxoplasmosis can be a serious problem for those who are immunosuppressed, as well as for fetuses. Even in healthy people, the immune system may not clear the organism, resulting in a latent infection that can reactivate when immunity declines.

Signs and Symptoms

Toxoplasmosis in otherwise healthy people is generally asymptomatic. Some people, however, develop signs and symptoms, including sore throat, fever, and enlarged lymph nodes and spleen, similar to those of other mild systemic infections. These subside over weeks or months and do not require treatment.

Toxoplasmosis in immunodeficient people commonly occurs due to reactivation of a latent infection. In these patients, the systemic infection can progress to life-threatening encephalitis, manifested by confusion, weakness, impaired coordination, seizures, stiff neck, paralysis, and coma. The patient may also develop brain masses much like tumors, which cause headaches and other neurological signs and symptoms such as poor coordination, seizures, sensory loss, and weakness. Infection of the retina (the light-sensitive part of the eye) is also common.

Fetal toxoplasmosis can be acquired across the placenta if a woman is infected with the protozoan for the first time during her pregnancy. Congenital toxoplasmosis does not typically occur if the woman was infected previously, most likely because of protection by maternal IgG. Infection of the fetus during the first trimester is most severe, often resulting in miscarriage or stillbirth. Babies born live may have serious birth defects. Infections during the last trimester usually have less severe effects on the fetus. Most of these infants appear normal at birth, but retinitis may occur later in life. This results in recurrent episodes of pain, sensitivity to light, and blurred vision. Intellectual disability and epilepsy may develop.

TABLE 26.10	African Trypanosomiasis ("African Sleeping Sickness")
Signs and symptoms	Tender nodule at site of tsetse fly bite; fever, enlargement of lymph nodes; later, involvement of the central nervous system, withdrawal, uncontrollable sleepiness, coma, death
Incubation period	Weeks (acute) to several years (chronic)
Causative agent	*Trypanosoma brucei,* a flagellated protozoan
Pathogenesis	The protozoa multiply at site of a tsetse fly bite, then enter the circulation; the parasites can change their surface protein, making previously formed antibodies ineffective.
Epidemiology	Bites of infected tsetse flies transmit the trypanosomes through saliva; wild animal reservoir for *T. brucei rhodesiense.*
Treatment and prevention	Treatment: antiprotozoal medications. Prevention: protective clothing, insecticides, clearing of brush where flies breed.

Causative Agent

Toxoplasmosis is caused by *Toxoplasma gondii,* an obligate intracellular protozoan that infects a wide variety of animals. The organism is an apicomplexan with a complex life cycle, as shown in **figure 26.15.** ⏮ apicomplexa

The definitive host of *T. gondii,* in which the organism reproduces sexually, is domestic cats and other felines. When an infected cat defecates, *T. gondii* oocysts are released into soil or cat litter. Millions of the oocysts are released daily, generally continuing for only a few weeks while the cat recovers from acute infection. Once an oocyst is released, it matures over a period of 1 to 5 days, becoming infectious to cats or other animals that eat it, including humans. The mature oocysts can remain viable for up to a year, contaminating soil and water and, secondarily, hands and food.

Each mature *T. gondii* oocyst that is ingested by a human or other animal contains two sporocysts. A form of the organism called a sporozoite emerges from each of these; they invade the cells of the small intestine, developing into a rapidly multiplying form called a tachyzoite (*tachy* means fast). The intestinal infection spreads by the lymphatics and blood vessels throughout the tissues of the host, infecting cells of the heart, brain, and muscles. As host immunity develops, multiplication of the parasite slows. A tough, fibrous capsule forms around infected host cells, forming a tissue cyst. Each cyst contains large numbers of a smaller form of *T. gondii* called a bradyzoite (*brady* means slow), which survive for months or years, resulting in a latent infection. Tissue cysts, like oocysts, are infectious to cats and other animals when eaten.

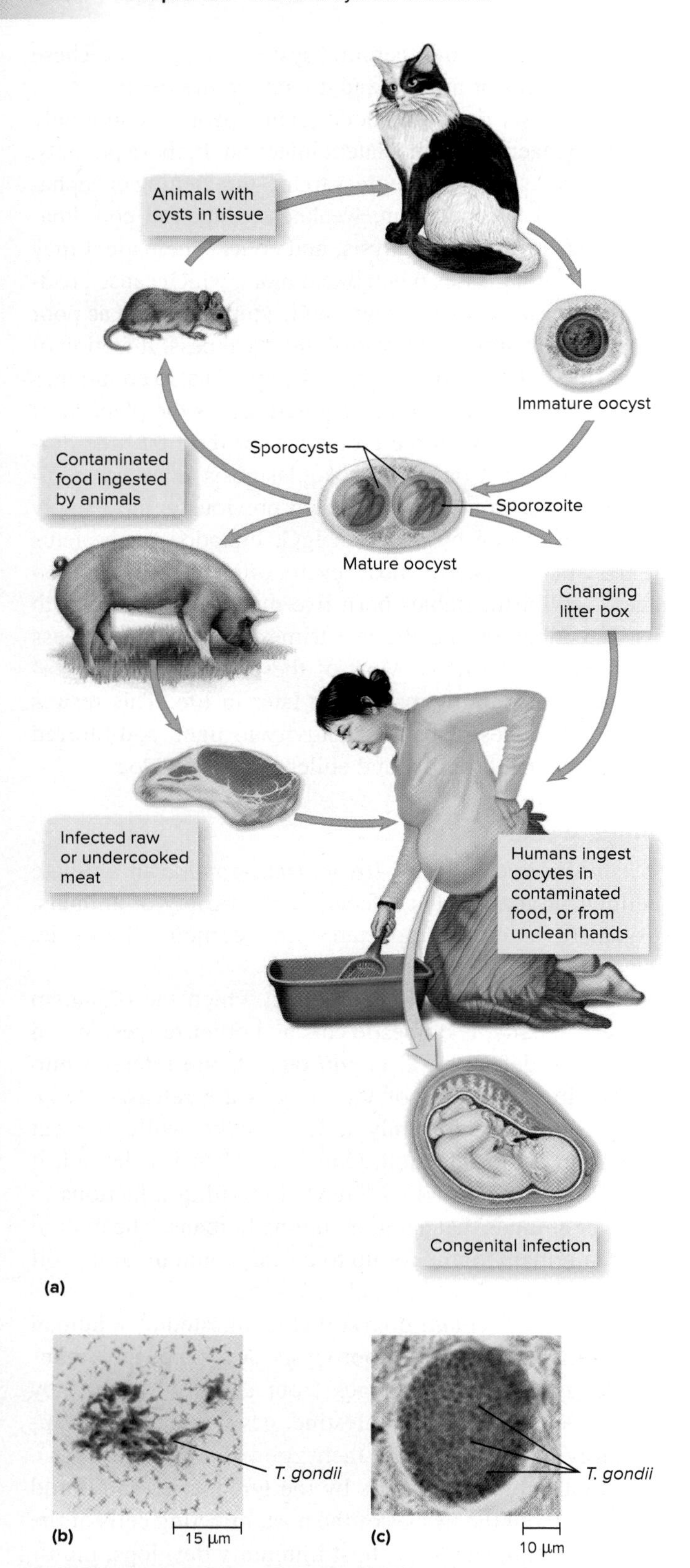

FIGURE 26.15 ***Toxoplasma gondii*** **(a)** Life cycle. Oocysts from cat feces and cysts from raw or inadequately cooked meat can infect humans and many other animals. **(b)** Tachyzoites; **(c)** Tissue cyst containing bradyzoites. b: Source: Dr. L.L. Moore, Jr./CDC; c: ©Dr. Thomas Fritsche

? **Why should a pregnant woman prevent her cat from hunting birds and small rodents?**

Pathogenesis

Toxoplasma gondii enters the mouth when either mature oocysts or undercooked meat containing tissue cysts are consumed. The organism can infect any kind of nucleated cell. Within a host cell, a tachyzoite develops and then proliferates, eventually destroying the infected cell. This process is normally brought under control by the immune response, but any tissue cysts that have formed persist. In immunodeficient patients, newly acquired infections as well as activation of latent infections can result in widespread and uncontrolled multiplication of the parasite, producing many areas of tissue necrosis. Severe toxoplasmosis can damage the brain, the eyes, or other organs.

Epidemiology

Toxoplasma gondii is distributed worldwide, and many people have latent infections, particularly in areas where eating raw meat is common. Infections are typically acquired by eating undercooked meat that contains tissue cysts or by consuming something contaminated with cat feces containing oocysts. Gardening in areas where stray cats hunt birds or rodents and then defecate in the soil can result in *T. gondii* contamination of hands or vegetables.

Treatment and Prevention

Toxoplasmosis is not treated in otherwise healthy people. In immunocompromised patients, the disease is typically treated with medications such as pyrimethamine (interferes with folate metabolism) in combination with sulfadiazine (a sulfa drug). An additional medication called folinic acid is also given to prevent bone marrow toxicity associated with pyrimethamine. Immunosuppressed patients who test positive for antibody to *T. gondii* are presumed to have a latent infection and are given prophylactic medication. Treatment does not destroy the tissue cysts. If PCR tests of amniotic fluid indicate that a fetus is infected, various medications may be used, depending on the stage of pregnancy.

Measures to prevent *T. gondii* infection are especially important for pregnant women and immunodeficient people. The organisms can be avoided by washing hands after touching raw meat, soil, or cat litter. Meat, especially lamb, pork, and venison (deer meat), should be cooked thoroughly. Fruits and vegetables should be washed before eating. Litter boxes should be cleaned regularly, before oocysts have a chance to mature. Cats should not be allowed to hunt birds and rodents, nor should they be fed undercooked or raw meat. The main features of toxoplasmosis are presented in **table 26.11.**

Primary Amebic Meningoencephalitis (PAM)

Naegleria fowleri, the cause of primary amebic meningoencephalitis (PAM), is commonly found in warm fresh water and soils. Nevertheless, fewer than 200 cases of PAM have been reported in the United States. Typically, fewer than five people per year become infected after swimming or diving in natural waters in the United States, making it a rare event, but one that is usually fatal.

TABLE 26.11	**Toxoplasmosis**
Signs and symptoms	Usually asymptomatic in healthy individuals, but sometimes sore throat, fever, enlarged lymph nodes. Encephalitis may develop in immunodeficient individuals, resulting in confusion, poor coordination, weakness, paralysis, seizures, and coma. Fetal infections can result in stillbirth, birth defects, seizures, intellectual disability, retinitis
Incubation period	Usually unknown
Causative agent	*Toxoplasma gondii,* an apicomplexan protozoan infectious for most warm-blooded animals
Pathogenesis	When mature oocysts are ingested, a form of *T. gondii* penetrates host cells, develops into a rapidly multiplying form, and spreads throughout the body. With development of immunity the infection is brought under control. Tissue cysts persist, resulting in a latent infection. Organisms are released from tissue cysts if immunity is impaired.
Epidemiology	Worldwide distribution. Infected cats discharge oocysts with their feces. Infection acquired by eating inadequately cooked meat containing tissue cysts or by ingesting mature oocysts.
Treatment and prevention	Treatment: not usually treated in healthy adults; antimicrobial medications if the patient is pregnant or immunocompromised. Prevention: avoid inadequately cooked meat and contact with material that may contain oocysts from cat feces. Antimicrobial medications are given to immunodeficient individuals with latent infections.

Signs and Symptoms

Signs and symptoms of PAM appear within a week and are similar to those of bacterial meningitis. Early indications include headache, fever, stiff neck, and vomiting. Once neurological symptoms appear—such as confusion or seizures—death quickly follows in over 95% of cases.

Causative Agent

Naegleria fowleri is one of only a few free-living protozoa pathogenic for humans. This organism can use phagocytosis to literally "eat your brain" (**figure 26.16**). The ameboid trophozoite gives rise to flagellated forms and spherical cysts (see figure 12.13). It prefers warm temperatures.

Pathogenesis

Naegleria fowleri penetrates the skull along the olfactory nerves serving the nasal mucosa. It multiplies and migrates to the brain, where it destroys nervous tissue, especially in the frontal lobes. Hemorrhage, coma, and death occur within a week.

Epidemiology

Naegleria fowleri may be carried in the normal microbiota, but it rarely causes disease. For every case of *Naegleria* meningoencephalitis, many millions of people are exposed to the organism

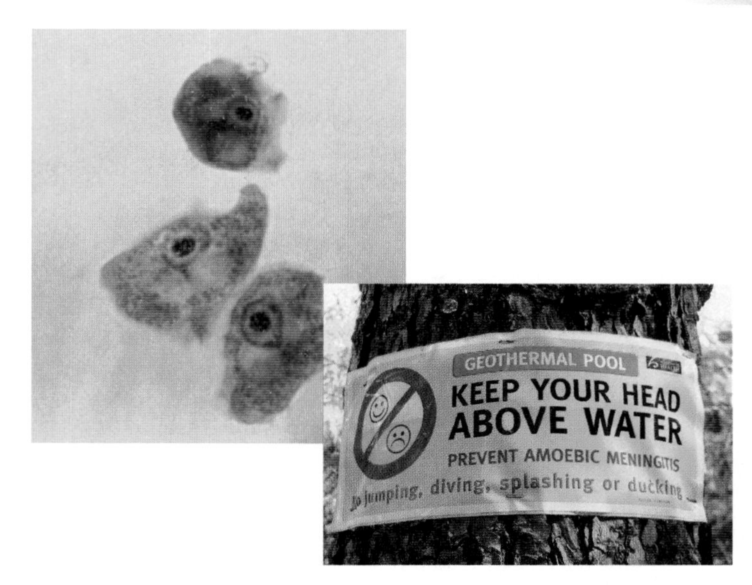

FIGURE 26.16 ***Naegleria fowleri*** (top): Source: Dr. George Healy/CDC; (bottom): ©Dr. Jay Stahl-Herz

? By what route does *Naegleria fowleri* enter the brain?

without harm. PAM is usually acquired when individuals swim or dive in warm natural fresh water. The geographic range of the organism is expanding northward and has recently been found in piped water. It is not transmitted from person to person.

MicroByte

A person is over 10,000 times more likely to die from drowning in natural waters than to die from PAM.

Treatment and Prevention

Several medications have been used to treat PAM, but with little success. In 2013, however, the CDC authorized investigational use of a drug called miltefosine in combination with existing treatments. Using this approach, two children with the disease survived—the first U.S. survivors of PAM in 35 years. There is no vaccine. The main characteristics of primary amebic meningoencephalitis are presented in **table 26.12.**

TABLE 26.12	**Primary Amebic Meningoencephalitis (PAM)**
Signs and symptoms	Headache, fever, stiff neck, vomiting, confusion, seizures
Incubation period	Less than a week
Causative agent	*Naegleria fowleri*
Pathogenesis	Destroys brain tissue
Epidemiology	Rare incidence after swimming or diving in warm, natural waters
Treatment and prevention	Treatment: investigational use of miltefosine combined with other medications. Prevention: none.

MicroAssessment 26.5

Residents and visitors to a wide region of tropical Africa are at risk of contracting African trypanosomiasis, caused by the protozoan parasite *Trypanosoma brucei,* and transmitted by the tsetse fly. The parasites may penetrate the CNS, causing sleepiness, coma, and death. Toxoplasmosis is often asymptomatic in healthy individuals, but can cause encephalitis in immunocompromised individuals. Fetal infections may result in stillbirth or a variety of nervous system disorders. Only on rare occasions can free-living amebas such as *Naegleria fowleri* cause meningoencephalitis.

13. How likely is it that a person who swims in warm fresh water will contract primary amebic meningoencephalitis?
14. Explain two ways in which a person might become infected with *Toxoplasma gondii.*
15. How can one explain repeated, abrupt increases in *T. brucei* in the blood of African trypanosomiasis victims?

26.6 ■ Diseases Caused by Prions

Learning Outcomes

11. Explain how prions differ from other infectious agents.
12. Describe how prions can cause disease.

A rare and mysterious group of chronic, degenerative brain diseases caused by prions has been seen in wild animals (mink, elk, and deer), domestic animals (sheep, goats, and cattle), and humans. Affected brain tissue has a spongy appearance, which is why these diseases are called spongiform encephalopathies. ◀◀ prions

Scrapie, a disease of sheep and goats, was named because affected animals had difficulty standing and therefore "scraped" along fences for support. Cattle, presumably fed meat and bone meal from infected sheep, developed bovine spongiform encephalopathy, or "mad cow disease." These diseases assumed new prominence in the 1990s when an outbreak of a spongiform encephalopathy in humans in the United Kingdom was related to an earlier outbreak of mad cow disease. ◀◀ chronic wasting disease

Transmissible Spongiform Encephalopathies in Humans

Transmissible spongiform encephalopathy (TSE) is rare in humans, occurring in only 0.5 to 1 case per million people. Most cases occur as Creutzfeldt-Jakob disease (CJD), which affects individuals over 45 years of age and sometimes runs in families. The disease acquired by eating affected animals is distinctively different—but has the same result and is identified as a variant of CJD, termed vCJD. Cases of CJD and vCJD are invariably fatal.

MicroByte

Another human TSE, kuru, is associated with cannibalism in New Guinea, where natives ate deceased relatives as a sign of respect.

Signs and Symptoms

Early symptoms of TSE include vague behavioral changes, anxiety, insomnia, and fatigue. Symptoms progress to characteristic muscle jerks, lack of coordination, memory loss, and dementia. Although the incubation period may last for years, once symptoms appear, death generally occurs within a year.

Causative Agent

The causative agents of spongiform encephalopathies are called proteinaceous infectious particles, or **prions.** Much smaller than a virus, a prion (PrP^{SC}) appears to be a misfolded form of a normal cellular protein (PrP^{c}). PrP^{SC} is encoded by a normal human gene and modified after transcription. The misfolding results in a protein that is protease-resistant, whereas the normal protein is protease-sensitive.

Pathogenesis

Prions increase in quantity during the incubation period of the disease as the misfolded protein (PrP^{SC}) acts as a template that promotes misfolding of the normal cellular protein (PrP^{c}) on the surface of neurons (see figure 13.24). Prions aggregate in insoluble masses in the brain, causing tissue damage. They may be taken up by neurons or phagocytic cells, but cannot be degraded by cellular proteases. Death of neurons produces the spongy appearance of affected brain tissue. Not all prions, however, act the same. They differ in host range, incubation period, and their nervous system targets. Transmissible spongiform encephalopathies can be differentiated from encephalitis because they typically do not evoke an immune response.

Epidemiology

Creutzfeldt-Jakob disease (CJD) generally occurs in individuals older than 45 years. It has been transmitted from human to human through corneal transplants, contaminated surgical instruments, and injections of human hormone replacements. It can be transmitted experimentally to chimpanzees.

Spongiform encephalopathy of sheep (scrapie) has been known for more than two centuries, without any evidence that it is directly transmissible to humans. Current evidence indicates that cattle prions can be transmitted to humans through consumption of contaminated beef and cause a variant Creutzfeldt-Jakob disease (vCJD) marked by differences in symptoms, brain pathology, and age of onset. The median age of individuals with vCJD is only 28 years. When mad cow disease is suspected, radical measures may minimize its transmission to humans. In 2003, for example, the United States banned importation of cattle from Canada when the disease was identified in several animals there.

Treatment and Prevention

There is no treatment for the spongiform encephalopathies, and they are always fatal. It is important to avoid eating the meat of any animals that show neurological symptoms. Prions are highly resistant to disinfectants, including formaldehyde. They also are resistant to heat and to ultraviolet and ionizing radiation. They can be inactivated by extended autoclaving in 1M (molar) sodium hydroxide. The main features of transmissible spongiform encephalopathies are presented in **table 26.13.**

MicroAssessment 26.6

Transmissible spongiform encephalopathies are degenerative nervous system diseases that occur in a variety of wild and domestic animals. They are rare in humans, but disease can be transmitted from animals via ingested tissues. These diseases appear to be caused by prions—infectious agents consisting only of protein and highly resistant to inactivation by heat, radiation, and disinfectants. The diseases are always fatal.

16. What is a prion?
17. What is the best way to prevent transmissible spongiform encephalopathies?
18. If you were an eye surgeon, would you rather the donor for a cornea transplant be under 35 or over 45 years of age?

TABLE 26.13 Transmissible Spongiform Encephalopathies

Signs and symptoms	Behavioral changes, anxiety, insomnia, fatigue, progressing to muscle jerks, lack of coordination, dementia
Incubation period	Usually many years
Causative agents	Proteinaceous infectious particles known as prions; lack nucleic acids; identical amino acid sequence to a normal protein, but folded differently, and relatively resistant to proteases; resistant to heat, radiation, and disinfectants
Pathogenesis	Prions increase in quantity by converting normal protein to more prions; transmission to the brain; aggregation into masses outside the nerve cells; cell malfunction and death.
Epidemiology	Human-to-human transmission by corneal transplantation and by contaminated surgical instruments; probable transmission of cattle prions to humans through consumption of contaminated beef; sporadic Creutzfeldt-Jakob disease in those over age 45 years; median age of variant Creutzfeldt-Jakob cases only 28 years.
Treatment and prevention	Treatment: none, invariably fatal. Prevention: inactivated by autoclaving in concentrated sodium hydroxide.

PERIPHERAL NERVOUS SYSTEM INFECTIONS

26.7 ■ Bacterial Diseases of the Peripheral Nervous System

Learning Outcomes

13. Explain how Hansen's disease can cause peripheral nerve damage.
14. Explain how botulism affects the nervous system.

Hansen's Disease (Leprosy)

Hansen's disease, or leprosy, is an ancient disease that has been widely feared throughout history (see **A Glimpse of History**). In reality, it appears that most of the human population is not susceptible to the disease. If infection does occur, it can be easily treated. The prevalence of Hansen's disease has dropped by 99% in the last three decades, largely due to multidrug treatment programs established by the World Health Organization. Still, over 6,000 people in the United States continue to be treated for the long-lasting effects of this disease and over 150 new cases are reported annually by the National Hansen's Disease Program.

Signs and Symptoms

Hansen's disease begins gradually, months or sometimes years after exposure. Onset is characterized by pigmentation changes and increased or decreased sensation in certain areas of skin. These areas may thicken—losing hair, sweat glands, and all sensation. The nerves of the arms and legs may become visibly enlarged with accompanying pain, later changing to numbness. Loss of nerve activity can lead to muscle wasting, ulceration, and finally loss of fingers or toes due to unnoticed or untreated injury (**figure 26.17**). In more severe cases, changes are most obvious in the face, with thickening of the nose and ears and deep wrinkling of the facial skin. Collapse of the supporting structure of the nose leads to congestion and bleeding.

Causative Agent

Mycobacterium leprae is aerobic, rod-shaped, and acid-fast (**figure 26.18**). It grows very slowly with a generation time of 12–13 days, and prefers the slightly cooler temperatures of the body's extremities. Hansen's disease is usually diagnosed based on clinical findings, but skin biopsy specimens that show acid-fast rods can provide an early indication of nerve invasion. Despite many attempts, *M. leprae* has not been grown in the absence of living cells. It can, however, grow in the footpads of mice, in armadillos, and in mangabey monkeys. Moreover, a DNA library of *M. leprae* has been made and expressed in *E. coli,* providing large quantities of the organism's antigens for study. ◀◀ **acid-fast stain,** ◀◀ ***Mycobacterium*****,** ◀◀ **DNA library**

FIGURE 26.17 Effects of Leprosy **(a)** An early symptom of leprosy is a change in skin pigmentation. **(b)** Notice the absence of fingers and sunken nose as a result of severe disease. a: ©Biophoto Associates/Science Source; b: ©Palo Koch/Science Source

Why are the fingers, toes, and nose often affected by leprosy?

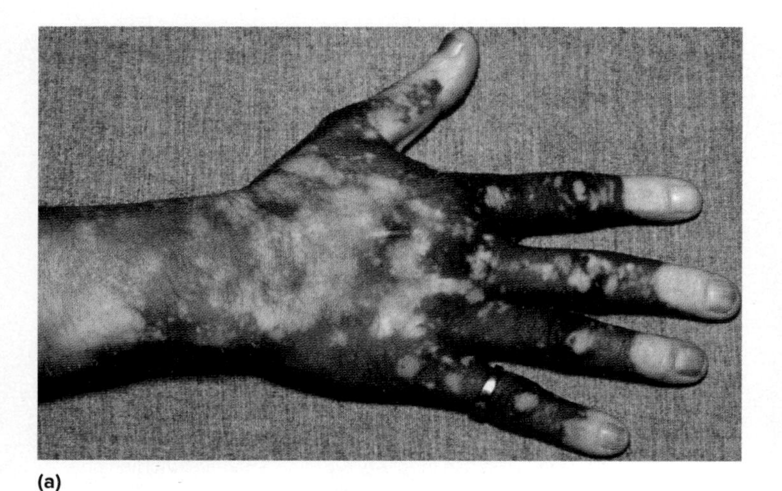

(a)

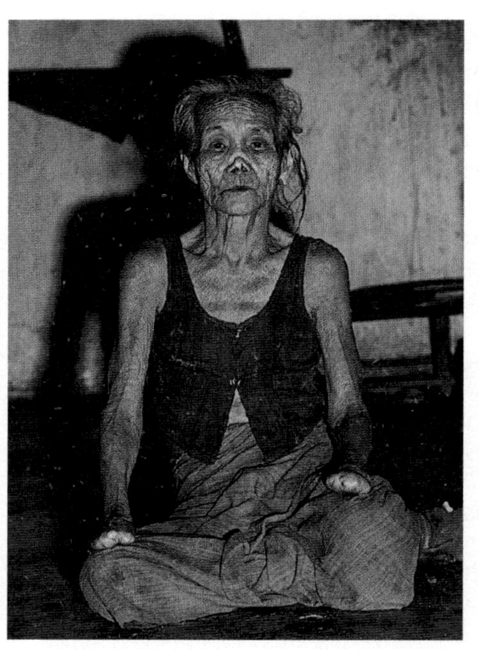

(b)

Pathogenesis

Mycobacterium leprae is the only known human pathogen that preferentially infects peripheral nerves. From there, the course of the infection depends on the immune response of the host. In most cases, cell-mediated immunity develops against the invading bacteria, and activated macrophages limit their spread. The chronically infected cells, however, are progressively damaged by attacking immune cells, leading to disabling deformities, resorption of bone, and skin ulcerations that characterize the disease. The disease often spontaneously stops progressing, and the nerve damage, although permanent, does not worsen. This limited type of Hansen's disease, in which cell-mediated immunity successfully stops the proliferating bacteria, is called **tuberculoid leprosy** (also called paucibacillary Hansen's disease). People with tuberculoid leprosy rarely, if ever, transmit the disease to others.
◀◀ macrophage, ◀◀ cell-mediated immunity

When cell-mediated immunity to *M. leprae* fails to develop or is suppressed, unrestricted growth of *M. leprae* occurs, leading to a form of Hansen's disease called **lepromatous leprosy** (also called multibacillary Hansen's disease). The bacteria first multiply in the cooler tissues of the body, notably in skin macrophages and peripheral nerves, and later throughout the body. The tissues and mucous membranes contain billions of *M. leprae,* but there is almost no inflammatory response to them. Mucus of the nose and throat contains high numbers of the pathogen, which can easily be transmitted to others.

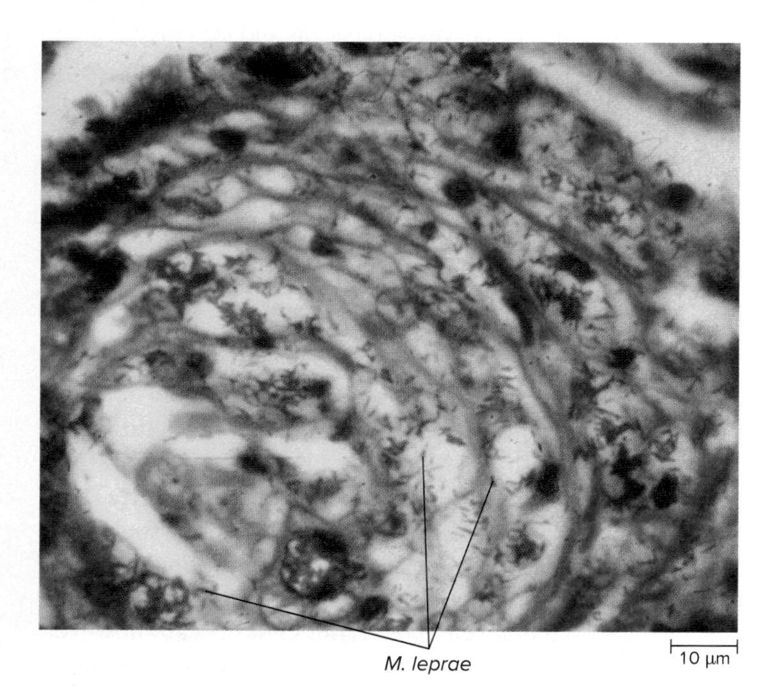

FIGURE 26.18 ***Mycobacterium leprae*** **in a Biopsy Specimen** Acid-fast stain reveals red dense masses of the bacterial cells, a typical finding in lepromatous leprosy. Source: Arthur E. Kaye/CDC

In what human cells would you expect to find these bacteria?

Epidemiology

Transmission of *Mycobacterium leprae* is by direct human-to-human contact. The source of the organisms is mainly nasal secretions of lepromatous cases, which can transport *M. leprae* to mucous membranes or skin abrasions of other individuals. Even then, the disease develops in only a tiny minority, being controlled by immune defenses in the rest. Natural infections with *M. leprae* also occur in wild nine-banded armadillos, possibly due to their relatively low body temperature suitable for replication of the bacterium. Gene-sequencing studies of animals in the southern United States confirm that the same strain of *M. leprae* is found in both humans and armadillos. Interaction with armadillos in these regions may explain cases of leprosy seen in individuals who have neither traveled to a part of the world where the disease is considered endemic nor had contact with those who have.

Elimination of Hansen's disease as a public health problem, defined as less than 1 case per 10,000 individuals, has been achieved in most countries. It is difficult to eradicate, however, due to the long generation time of *M. leprae*. This results in an incubation period of about 3 years (range, 3 months to 20 years), during which time the disease can remain undetected. Goals of the WHO's Global Leprosy Strategy include eliminating leprosy-related disabilities in children by 2020 and repeal of laws that still allow discrimination against leprosy patients because of the stigma associated with their disease. ⏮ generation time

MicroByte

During much of history, Hansen's disease was so feared that "lepers" were forced to carry a bell or horn to warn others of their presence.

Treatment and Prevention

Early treatment can keep Hansen's disease from progressing. Since 1995, the World Health Organization has provided free multidrug therapy to patients, radically reducing the number of cases. Tuberculoid leprosy can be successfully treated by a combination of dapsone and rifampin administered for 6 months. Lepromatous leprosy is generally treated for a minimum of 2 years, with addition of a third drug, clofazimine, to the treatment process. This combination therapy is required to prevent drug-resistant strains from developing. No proven vaccine to control Hansen's disease is yet available. Some features of Hansen's disease are shown in **table 26.14.**

TABLE 26.14 Hansen's Disease (Leprosy)

Signs and symptoms	Skin lesions that lack sensation, deformed face, loss of fingers or toes
Incubation period	3 months to 20 years; usually 3 years
Causative agent	*Mycobacterium leprae*, an acid-fast rod that has not been grown in the absence of living cells
Pathogenesis	Invasion of small nerves of skin; multiplication in macrophages; course of disease depends on immune response of host; in tuberculoid leprosy, activated macrophages limit the growth of the bacterium but the damage to infected nerve cells leads to disabling deformities; in lepromatous leprosy, cell-mediated immunity fails, allowing unrestrained growth of *M. leprae*.
Epidemiology	Direct contact with *M. leprae* from mucous membrane secretions.
Treatment and prevention	Treatment: antimicrobial medication. Prevention: no vaccine.

Botulism

Botulism is not a nervous system infection, but is considered in this chapter because its key symptom is paralysis. The name "botulism" comes from *botulus*, meaning "sausages," chosen because some of the earliest recognized cases occurred in people who ate contaminated sausages. Like other clostridia, *Clostridium botulinum* produces heat-resistant endospores. The endospores themselves do not cause disease, but they germinate in nutrient-rich anaerobic environments and the resulting vegetative cells produce a powerful exotoxin called botulinum toxin, which causes the characteristic disease signs and symptoms.

There are several different forms of botulism. **Foodborne botulism** results from ingesting foods containing botulinum toxin and is the most common form of the disease worldwide. It typically occurs when inadequate canning practices fail to destroy all endospores, allowing those that survive to germinate and produce vegetative cells that grow in the food. **Intestinal botulism** occurs when endospores are ingested and then germinate, producing vegetative cells that colonize the intestine. This is quite rare in adults, but sometimes happens in children, resulting in what is referred to as **infant botulism.** In **wound botulism,** endospores contaminate a wound and germinate; vegetative cells then multiply. ⏭ foodborne botulism

Signs and Symptoms

Symptoms of foodborne botulism usually begin 12 to 36 hours after an individual ingests toxin-contaminated food, but the incubation time for other forms is variable or unknown. Most cases of botulism begin with dizziness, dry mouth, and blurred or double vision, indicating eye muscle weakness. Abdominal symptoms, including pain, nausea, vomiting, and diarrhea or constipation can also occur. Infants with botulism may also drool, have trouble sucking, and have drooping eyelids. Progressive paralysis generally involves all voluntary muscles; respiratory paralysis is the most common cause of death.

Causative Agent

Botulism is caused by the strictly anaerobic, Gram-positive, endospore-forming, rod-shaped bacterium *Clostridium botulinum* (**figure 26.19**). The endospores generally survive boiling and can then germinate to form vegetative cells if the environment is favorable (nutrient-rich, anaerobic conditions, a pH above 4.5, and a temperature above 4°C). Several types of botulinum toxin, which is a neurotoxin, are produced by different strains of *C. botulinum,* and all cause paralysis. Types A, B, and E are responsible for most human cases. ⏮ neurotoxin

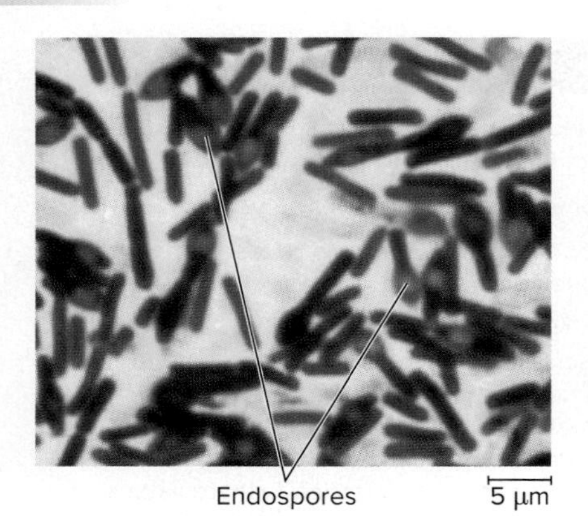

FIGURE 26.19 ***Clostridium botulinum*** Notice the lighter-colored endospores in this Gram stain. These are not reliably killed by the temperature of boiling water. ©Michael Abbey/Science Source

? **What is an endospore?**

Pathogenesis

In the case of foodborne botulism, vegetative cells of *Clostridium botulinum* growing in food release botulinum toxin, one of the most powerful poisons known. A few milligrams could kill the entire population of a large city. Indeed, cases of botulism have resulted from eating a single, contaminated string bean, and from licking a finger when tasting a food contaminated with botulinum toxin. When a person eats the contaminated food, the toxin passes through the stomach and into the small intestine, where it is absorbed into the bloodstream.

Intestinal botulism is the result of ingestion of *C. botulinum* endospores that germinate in the intestinal tract to become vegetative cells, which then multiply there and produce botulinum toxin; the toxin is then absorbed. Intestinal botulism generally occurs only in infants, particularly those between 6 weeks and 6 months of age, presumably because members of their normal intestinal microbiota are unable to inhibit growth of the germinating *C. botulinum* cells. Adult intestinal botulism can occur in immunodeficient patients whose normal microbiota has been suppressed by antibiotic treatment.

Clostridium botulinum can also colonize dirty wounds, especially those containing necrotic (dead) tissue. The bacteria do not invade, but they can multiply in necrotic tissue. Botulinum toxin then diffuses into the bloodstream.

Circulating botulinum toxin attaches to motor neurons and then stops transmission of signals to the muscles, producing paralysis. Botulinum toxin is an A-B toxin. The B portion binds to specific receptors on motor nerve endings, and the A portion enters the nerve cell, where it inactivates proteins that regulate the release of the neurotransmitter. Unlike tetanus toxin, which blocks the action of inhibitory neurons in the spinal cord and results in spastic muscle contraction, botulinum toxin stops muscle contraction, resulting in **flaccid paralysis.** Because it prevents nerve transmission to muscles, botulinum toxin has important applications. A commercial version (Botox) is used to prevent headaches and provide relief from a variety of conditions involving muscle contractions; it is also used cosmetically to reduce wrinkles (see Focus Your Perspective 30.1). ⏮ A-B toxin, ⏮ neurotransmitter, ⏮ tetanus

Epidemiology

Clostridium botulinum endospores are widely distributed in soils and aquatic sediments worldwide. Still, fewer than 30 cases of foodborne botulism per year are typically seen in the United States. Most of these are from eating preserved fish or improperly home-canned foods. Strict controls on commercial canners have drastically reduced the number of botulism outbreaks. Infant botulism is more common than foodborne botulism in the United States, but still there are generally fewer than 150 cases per year. Ingestion of honey has been implicated in infant botulism cases, leading to the recommendation that honey—a source of *C. botulinum* spores—not be given to children under 1 year old. Most cases of wound botulism in the United States are due to wounds caused by abuse of injected drugs (**figure 26.20**).

Treatment and Prevention

Foodborne botulism is treated by administering antitoxin intravenously as soon as possible after the diagnosis. The antitoxin, however, only neutralizes toxin circulating in the bloodstream. The nerves already affected recover slowly, over weeks or months. Artificial respiration with a mechanical ventilator may be required for prolonged periods until affected nerve endings can regenerate. Most intestinal botulism patients recover without receiving antitoxin treatment, although respiratory support and tube feeding may be needed until the pathogens are replaced by normal intestinal microbiota. An antitoxin specifically for babies shortens the recovery time from infant botulism. For wound botulism, surgical

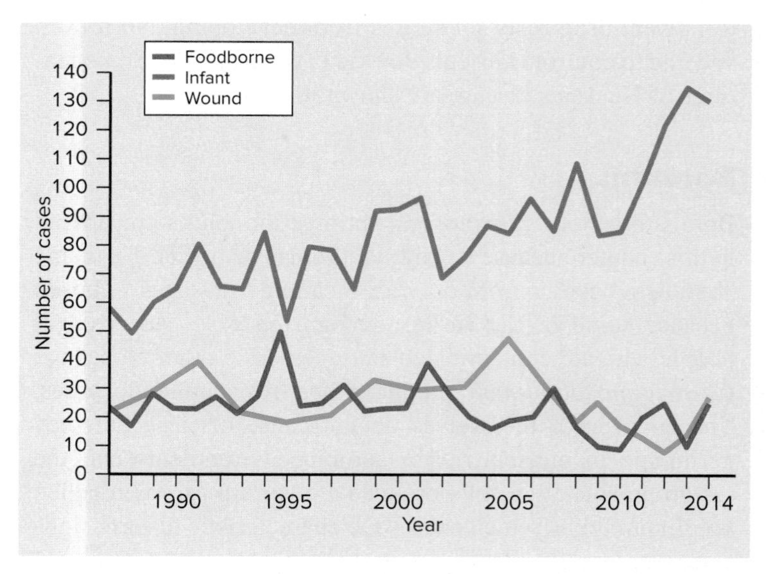

FIGURE 26.20 Botulism in the United States, 1987–2014

? **What is the most common form of botulism in the United States?**

removal of dirt and dead tissue helps to eliminate any organisms and unabsorbed toxin.

Prevention of foodborne botulism depends on proper sterilization and sealing of food at the time of canning. Contamination does not always result in a spoiled smell, taste, or appearance, but fortunately, the toxin is heat-labile (destroyed by heat). Heating home-canned, low-acid foods (pH over 4.5) to 100°C for 10 minutes just prior to serving should ensure that they are safe. Intestinal botulism is difficult to prevent, but because honey can contain *C. botulinum* endospores, it should not be fed to infants younger than 1 year of age. Wound botulism is prevented through prompt and appropriate wound care, and by discouraging injection-drug abuse. Immunity does not develop in response to the low levels of toxin, so a person may get botulism more than once. The main features of botulism are presented in **table 26.15.**

The key features of the diseases covered in this chapter are highlighted in the **Diseases in Review 26.1** table that follows.

MicroAssessment 26.7

Hansen's disease (leprosy) is caused by *Mycobacterium leprae,* which preferentially invades peripheral neurons. Botulism, a toxin-mediated disease characterized by flaccid paralysis, is caused by *Clostridium botulinum.*

19. Why is blurred vision an early sign of botulism?

20. Differentiate between the two types of Hansen's disease.

21. Why is it difficult to determine an incubation period for *Clostridium botulinum* in cases of wound botulism seen in chronic drug abusers?

TABLE 26.15 Botulism

Signs and symptoms	Blurred or double vision, weakness, nausea, vomiting, diarrhea or constipation; generalized paralysis and respiratory insufficiency
Incubation period	Usually 12 to 36 hours for foodborne; variable or unknown for other types
Causative agent	*Clostridium botulinum,* an anaerobic, Gram-positive, endospore-forming, rod-shaped bacterium
Pathogenesis	*Clostridium botulinum* endospores germinate and the resulting vegetative cells multiply in favorable environments, releasing botulinum toxin. The toxin may be consumed in food (foodborne botulism), produced by cells growing in the intestinal tract (intestinal botulism; in babies this is called infant botulism), or made by cells infecting a wound (wound botulism). Circulating toxin binds to motor nerves and stops the transmission of signals to the muscles, producing flaccid paralysis.
Epidemiology	Endospores widespread in soils and aquatic sediments. Foodborne botulism is often due to consuming an improperly processed home-canned, low-acid food. Intestinal botulism typically seen in infants (infant botulism), and may involve endospore-containing honey. Wound botulism often involves injection-drug abuse.
Treatment and prevention	Treatment: depends on the type of botulism, but may involve intravenous administration of antitoxin and artificial ventilation for respiratory paralysis. Prevention: specific for the type of botulism. Foodborne botulism is prevented through proper food handling and home-canning methods; also, boiling home-canned foods for 10 minutes just prior to serving. Intestinal botulism is difficult to prevent, but honey should not be fed to infants under 1 year of age. Wound botulism is prevented through prompt wound care.

Diseases in Review 26.1

Nervous System Diseases

Disease	Causative Agent	Comment	Summary Table
BACTERIAL DISEASES OF THE CENTRAL NERVOUS SYSTEM			
Pneumococcal meningitis	*Streptococcus pneumoniae* (pneumococcus)	Leading cause of meningitis in adults; 90 serotypes; conjugate vaccine against 13 serotypes and polysaccharide vaccine against 23 of the most common.	Table 26.3
Meningococcal meningitis	*Neisseria meningitidis* (meningococcus)	May lead to endotoxic shock; formation of petechiae due to capillary damage; associated with meningitis epidemics; vaccines available.	Table 26.1
***Haemophilus influenzae* meningitis**	*Haemophilus influenzae*	Once the most common cause of infant meningitis, now largely controlled by a conjugate vaccine.	Table 26.3
Neonatal meningitis	*Streptococcus agalactiae* *Escherichia coli*	Newborns acquire causative agent from the mother's birth canal; U.S. pregnant women are routinely screened for *S. agalactiae* before delivery and if positive are treated with antibiotics.	Table 26.3
Listeriosis	*Listeria monocytogenes*	Contaminated foods have caused epidemics; multiplies at refrigeration temperatures; infected pregnant women may miscarry; may cause abscesses in the fetus.	Table 26.2
VIRAL DISEASES OF THE CENTRAL NERVOUS SYSTEM			
Viral meningitis	Usually enteroviruses	Aseptic meningitis; more common and much milder than bacterial meningitis; often transmitted by fecal-oral route.	Table 26.4
Viral encephalitis	Arboviruses	Transmitted in the United States by mosquitoes; more likely than viral meningitis to cause death or disability; case-fatality rate varies, depending upon type of virus.	Table 26.6
Poliomyelitis	Poliovirus	Destruction of motor neurons leads to paralysis; fecal-oral transmission, but infection rarely leads to disease; vaccination has eliminated the disease in most parts of the world, and the disease is now targeted for eradication.	Table 26.7
Rabies	Rabies virus	Zoonotic disease; virus travels from bite wound through sensory neurons to CNS, causing a generally fatal infection; vaccination along with passive immunization shortly after exposure prevents disease.	Table 26.8
FUNGAL DISEASES OF THE CENTRAL NERVOUS SYSTEM			
Cryptococcal meningoencephalitis	*Cryptococcus neoformans* and *C. gattii*	Caused by inhalation of fungal cells that are carried by phagocytic cells to the brain; can be an early sign of AIDS, but also seen in healthy individuals where *C. gattii* is found.	Table 26.9
PROTOZOAN DISEASES OF THE CENTRAL NERVOUS SYSTEM			
African trypanosomiasis	*Trypanosoma brucei*	Chronic (Gambian) or acute (Rhodesian); two subspecies of protozoa transmitted by tsetse fly; loss of interest, sleepiness, coma, death.	Table 26.10
Toxoplasmosis	*Toxoplasma gondii*	Infection due to ingestion of oocysts (in cat feces) or tissue cysts (in raw meat); usually asymptomatic in healthy individuals, but fetal infections may cause stillbirth, and encephalitis may develop in the immunodeficient.	Table 26.11
Primary amebic meningoencephalitis (PAM)	*Naegleria fowleri*	Swimming or diving in water containing the organism transmits this rare but deadly disease; travels to brain via olfactory neurons where it damages tissue; no vaccine or effective treatment.	Table 26.12
PRION DISEASES			
Transmissible spongiform encephalopathies	Prions	Caused by an accumulation of destruction-resistant misfolded proteins, resulting in death of neurons; may come from contaminated animal tissues; no treatment; always fatal.	Table 26.13
BACTERIAL DISEASES OF THE PERIPHERAL NERVOUS SYSTEM			
Hansen's disease (leprosy)	*Mycobacterium leprae*	Infects peripheral nerves, causing immune system to attack them; course of disease determined by cell-mediated immune response; long incubation period; loss of limbs, blindness.	Table 26.14
Botulism	*Clostridium botulinum*	Botulinum toxin causes flaccid paralysis; foodborne botulism is the most common form worldwide; infant botulism is the most common form in the United States; wound botulism can also occur.	Table 26.15

Summary

26.1 Anatomy, Physiology, and Ecology of the Nervous System (figure 26.1)

The brain and spinal cord make up the **central nervous system (CNS);** the **peripheral nervous system (PNS)** includes nerves that carry information to and from the CNS. **Cerebrospinal fluid (CSF)** is produced in ventricles in the brain and flows out over the brain and spinal cord (figure 26.2). **Meninges** are the membranes that cover the brain and spinal cord.

CENTRAL NERVOUS SYSTEM INFECTIONS

Focus On Meningitis

Infection of the membranes surrounding the brain and spinal cord causes **meningitis.** Bacteria that cause meningitis are often part of the normal microbiota, and are transmitted through respiratory droplets. Enteroviruses that cause viral meningitis are transmitted by the fecal-oral route. Symptoms are similar to cold symptoms, then followed by abrupt onset of fever, severe headache, pain and stiffness of the neck and back, nausea, and vomiting. The case-fatality rate for untreated bacterial meningitis can approach 100%. People who recover sometimes suffer permanent disabilities. Immunization has greatly decreased the incidence of bacterial meningitis, particularly among children. Viral meningitis occurs more often, but it is typically less severe than bacterial meningitis.

26.2 Bacterial Diseases of the Central Nervous System

Pneumococcal Meningitis (table 26.3)

Streptococcus pneumoniae, or pneumococcus, is the most common cause of meningitis in adults.

Meningococcal Meningitis (figure 26.4; tables 26.1, 26.3)

Meningococcal meningitis caused by *Neisseria meningitidis* is associated with meningitis epidemics. Small hemorrhages in the skin (figure 26.3), deafness, and coma can occur. Shock results from the release of endotoxin into the bloodstream.

***Haemophilus influenzae* Meningitis** (table 26.3)

Haemophilus influenzae, once the leading cause of childhood bacterial meningitis, is largely sporadic and mostly controlled by a vaccine (figure 26.5).

Neonatal Meningitis (table 26.3)

Newborns most often acquire meningitis-causing bacteria from the mother's genital tract shortly before or during birth. Infants who survive often face long-lasting consequences of their infection.

Listeriosis (tables 26.2, 26.3)

Listeriosis is caused by *Listeria monocytogenes,* a non-spore-forming, Gram-positive rod usually associated with foodborne illness. The bacterium is widespread, commonly contaminates foods such as nonpasteurized milk, cold cuts, and soft cheeses, and can grow in refrigerated foods (figure 26.6). The bacteria penetrate the gastrointestinal mucous membranes, enter the bloodstream, and infect the meninges.

26.3 Viral Diseases of the Central Nervous System

Viral Meningitis (table 26.4)

Viral meningitis is much more common than bacterial meningitis. It is generally a mild disease for which there is no specific treatment.

Viral Encephalitis (table 26.6)

Viral encephalitis has a high case-fatality rate and often leaves survivors with permanent disabilities. Herpes simplex virus is the most important cause of sporadic encephalitis; epidemic encephalitis is usually caused by **arboviruses** (figure 26.7; table 26.5).

Poliomyelitis (figures 26.8, 26.9; table 26.7)

Infection of cell bodies of motor neurons in the CNS (brain and spinal cord) leads to degeneration of motor nerves leading to paralysis and muscle atrophy. Post-polio syndrome occurs years after poliomyelitis, and it is probably caused by the death of nerve cells that had taken over for those killed by poliovirus.

Rabies (figure 26.10; table 26.8)

Rabies is a widespread zoonosis transmitted to humans mainly through the bite of an infected animal (figure 26.11). Once symptoms appear, the disease is almost always fatal. Because of the long incubation period, prompt immunization with inactivated vaccine after a rabid animal bite is effective in preventing the disease. Passive immunization given at the same time increases protection.

26.4 Fungal Diseases of the Central Nervous System

Cryptococcal Meningoencephalitis (table 26.9)

Infection begins in the lung after a person inhales spores of *Cryptococcus neoformans* or *Cryptococcus gattii,* encapsulated yeasts that resist phagocytosis (figure 26.12). *C. neoformans* infects immunocompromised individuals, but *C. gattii* can cause disease in healthy people (figure 26.13). Treatment of these diseases is usually difficult.

26.5 Protozoan Diseases of the Central Nervous System

African Trypanosomiasis ("African Sleeping Sickness") (table 26.10)

African trypanosomiasis is a major health problem in a wide area across equatorial Africa. In its late stages, it is marked by indifference, sleepiness, coma, and death. The disease is caused by *Trypanosoma brucei* (figure 26.14) transmitted by its biological vector, the tsetse fly. During infection, the organism shows bursts of growth, each appearing with different surface proteins.

Toxoplasmosis (figure 26.15; table 26.11)

Toxoplasmosis is caused by *Toxoplasma gondii.* People contract toxoplasmosis by ingesting oocytes discharged in the feces of acutely infected cats or inadequately cooked meat containing tissue cysts. Toxoplasmosis is usually asymptomatic among otherwise healthy people, but may cause encephalitis, brain masses, and other nervous system problems among the immunodeficient. Infection of the fetus, especially during the first trimester, results in miscarriage, stillbirth, or birth defects.

Primary Amebic Meningoencephalitis (PAM) (table 26.12)

PAM is rare, but fatal. Infection with *Naegleria fowleri* (figure 26.16) can occur after swimming in warm fresh water.

26.6 Diseases Caused by Prions

Prions—abnormal proteins that are resistant to heat, radiation, and disinfectants—cause the spongiform encephalopathies, rare diseases characterized by a sponge-like appearance of brain tissue caused by loss of nerve cells.

Transmissible Spongiform Encephalopathies in Humans (table 26.13)
Examples of **transmissible spongiform encephalopathies (TSEs)** include "mad cow disease," Creutzfeldt-Jakob disease (CJD), and variant Creutzfeldt-Jakob disease (vCJD), which can be transmitted to humans from infected meat. There is no treatment for these diseases, and they are invariably fatal.

PERIPHERAL NERVOUS SYSTEM INFECTIONS

26.7 Bacterial Diseases of the Peripheral Nervous System

Hansen's Disease (Leprosy) (figure 26.17; table 26.14)
Hansen's disease is characterized by invasion of peripheral nerves by the acid-fast rod *Mycobacterium leprae,* which has not been cultivated in vitro (figure 26.18). The disease occurs in two main forms—tuberculoid and lepromatous—depending on the immune status of the individual.

Botulism (table 26.15)
Botulism is caused by a toxin produced by *Clostridium botulinum,* an anaerobic, Gram-positive rod that forms heat-resistant endospores (figure 26.19). Globally, the most common form of the disease is **foodborne botulism,** which results from ingesting botulinum toxin. It typically occurs when inadequate canning practices fail to destroy all endospores, and those surviving germinate to form vegetative cells that then grow in the food. **Intestinal botulism** occurs when endospores are ingested and then germinate to form vegetative cells that colonize the intestine. This is quite rare in adults, but sometimes results in **infant botulism,** which is the most common type of botulism in the United States (figure 26.20). In **wound botulism,** endospores contaminate a wound, often the result of injection-drug abuse, and then germinate.

Review Questions

Short Answer

1. What sign would differentiate meningococcal meningitis from pneumococcal meningitis?
2. Name and describe the organism that is the leading cause of bacterial meningitis in adults.
3. What measures can be undertaken to prevent neonatal meningitis?
4. Why is listeriosis so important to pregnant women even though it usually causes them few symptoms?
5. Can botulism be spread from person to person?
6. Give two ways in which viral meningitis usually differs from bacterial meningitis.
7. What is the difference between sporadic encephalitis and epidemic encephalitis? Name one cause of each.
8. What is an arbovirus?
9. Why is it possible to prevent rabies with vaccine given after exposure?
10. If you contract African trypanosomiasis on a visit to central Africa, what type do you most likely have?

Multiple Choice

1. Which is the best way to prevent meningococcal meningitis in individuals intimately exposed to the disease?
   a) Vaccinate them against *Neisseria meningitidis.*
   b) Give them the antibiotic rifampin.
   c) Culture their throat and hospitalize them for observation.
   d) Withdraw a sample of spinal fluid and begin antibacterial treatment if the cell count is high and the glucose level is low.
   e) Have them return to their usual activities, but seek medical evaluation if symptoms of meningitis occur.
2. Which of these statements concerning the causative agent of listeriosis is *false*?
   a) It can cause meningitis during the first month of life.
   b) It is a Gram-positive rod that can grow in refrigerated food.
   c) It is usually transmitted by the respiratory route.
   d) Infection can result in bacteremia.
   e) It is widespread in natural waters and vegetation.
3. Which of these statements concerning Hansen's disease is *false*?
   a) It is not common in the United States.
   b) An early symptom is loss of sensation, sweating, and hair in a localized patch of skin.
   c) The incubation period is usually less than 1 month.
   d) Treatment should include more than one antimicrobial medication given at the same time.
   e) The form the disease takes depends on the individual's immune status.
4. Which of these statements concerning foodborne botulism is *false*?
   a) It is not a central nervous system infection.
   b) Blurred vision is an early symptom.
   c) Food can taste normal but still cause botulism.
   d) Treatment is based on choosing the correct antibiotic.
   e) Control of the disease depends largely on proper food-canning techniques.
5. Which of the following statements about viral meningitis is *true*?
   a) Vaccines are generally available to protect against the disease.
   b) The main symptom is muscle paralysis.
   c) Transmission is often by the fecal-oral route.
   d) The causative agents do not survive well in the environment.
   e) Recovery is rarely complete.
6. Which of these statements concerning arboviral encephalitis is *false*?
   a) It is likely to occur in epidemics.
   b) Mosquitoes can be an important vector.
   c) Epilepsy, paralysis, and thinking difficulties are among the possible sequels to the disease.
   d) Use of sentinel chickens helps warn about the disease.
   e) In the United States, the disease is primarily a zoonosis involving cattle.
7. Which of these statements concerning poliomyelitis is *false*?
   a) The sensory nerves are usually involved.
   b) It can be caused by more than one type of poliovirus.
   c) Only a small fraction of those infected will develop the disease.
   d) The disease is transmitted via the fecal-oral route.
   e) A post-polio syndrome can develop years after recovery from the original illness.

8. Which of these statements concerning cryptococcal meningoencephalitis is *true*?
   a) It is caused by a yeast with a large capsule.
   b) It is a disease of trees transmissible to humans.
   c) It typically attacks the meninges but spares the brain.
   d) Person-to-person transmission commonly occurs.
   e) It is seen only in persons who are immunocompromised.
9. Which of these statements concerning African trypanosomiasis is *true*?
   a) It is transmitted by a species of biting mosquito.
   b) It is a threat to visitors to tropical Africa.
   c) The onset of sleepiness is usually within 2 weeks of contracting the disease.
   d) It is caused by free-living protozoa.
   e) Distribution of the disease is determined mainly by the distribution of standing water.
10. Which of these statements concerning Creutzfeldt-Jakob disease (CJD) and vCJD is *true*?
    a) CJD occurs in children; vCJD occurs in adults over 45.
    b) CJD and vCJD are fatal in under 50% of cases.
    c) CJD is caused by prions; vCJD is a viral infection.
    d) Only humans suffer from diseases like CJD and vCJD.
    e) Both CJD and vCJD produce a spongy appearance in affected brain tissue.

Applications

1. An outbreak of viral meningitis in a small eastern city was linked epidemiologically to a group who swam in a nonchlorinated pool in an abandoned quarry outside of town. What might public health officials surmise about the probable cause of the outbreak?
2. Two microbiologists are writing a textbook, but they cannot agree on where to place the discussion of botulism. One favored the chapter on nervous system infections, whereas the other insisted on the chapter covering digestive system infections. Where do you think the discussion should be placed, and why?

Critical Thinking

1. A pathologist stated that it was much easier to determine the causative agent of meningitis than that of an infection of the skin or of the intestine. Is her statement valid? Why or why not?
2. Why is it important to learn about rabies when only a few cases occur in the entire United States each year?

27 Genitourinary Tract Infections

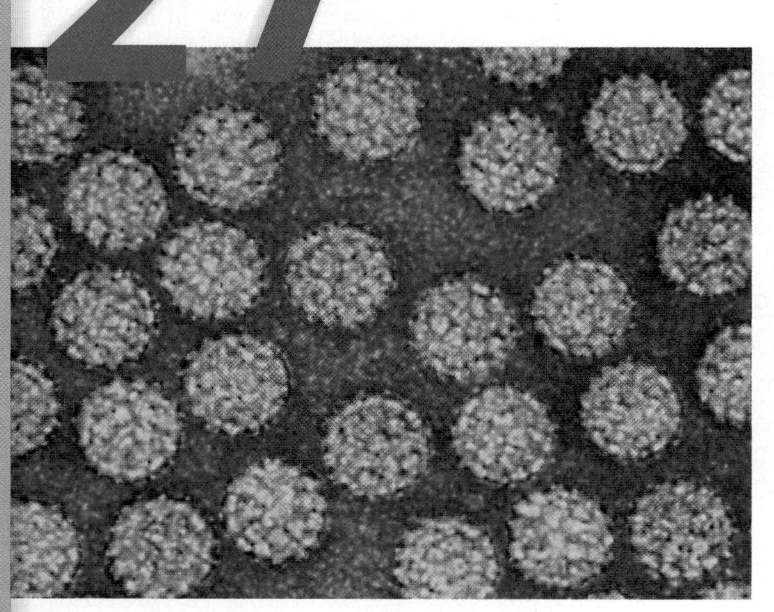

Color-enhanced TEM of human papillomavirus particles. ©*Kwangshin Kim/ Science Source*

KEY TERMS

Catheter A flexible plastic or rubber tube inserted into the bladder or other body space, in order to drain it or deliver medication.

Chancre Ulcerating sore that develops during the initial stage of syphilis.

Cystitis Inflammation of the bladder.

Papilloma A kind of tumor characterized by rough projections of tissue; warts are an example.

Pelvic Inflammatory Disease (PID) Infection of the fallopian tubes, uterus, or ovaries.

Pyelonephritis Infection of the kidneys.

Toxic Shock Syndrome Potentially life-threatening drop in blood pressure due to a circulating bacterial toxin.

A Glimpse of History

Syphilis is an ancient disease that was once very common. History and literature describe many cases of famous people who developed mental illness or were seriously disabled because of this disease. A few examples are Henry VIII (King of England, 1491–1547), Merriwether Lewis (explorer, 1774–1809), Oscar Wilde (writer, 1854–1900), and Al Capone (gangster, 1899–1947).

Syphilis was first named the "French pox" or the "Neapolitan disease" because people thought it came from France or Italy. In 1530, the Italian physician Girolamo Fracastoro wrote a poem about a shepherd named Syphilis, who had ulcerating sores all over his body. The description matched the signs and symptoms of syphilis, and from then on, the disease was known by the shepherd's name.

At first it was not known how syphilis was contracted, but gradually people realized the disease was sexually transmitted. In 1905, the German biologist Fritz Schaudinn examined fluid from a syphilitic sore. He observed some motile organisms that were very pale, slender, and difficult to see. Using dark-field microscopy, he saw that the organisms looked like a corkscrew without a handle—bacteria that are now known as spirochetes. The spirochetes were found in specimens from other cases of syphilis, and Schaudinn felt certain he had discovered the cause of syphilis, even though the organism could not be cultured on laboratory media. Schaudinn named the organisms *Spirochaeta pallida,* "the pale spirochete." This organism is now called *Treponema pallidum,* from the Greek words *trep* and *nema,* which mean "turning thread."

Infections of the reproductive and urinary tracts are very common. They often create great discomfort and can have serious consequences, sometimes without advance symptoms. Urinary tract infections (UTIs)—bacterial infections of any part of the urinary system—are the most frequent healthcare-associated infections and are the main origin of fatal, bacterial bloodstream invasions. Healthcare-associated uterine infections such as puerperal fever, once a common cause of maternal deaths from childbirth, still require strict medical attention to be prevented (see A Glimpse of History, chapter 19). Sexually transmitted infections (STIs) are widespread and some can be life-threatening. Although most STIs are transmitted by sexual contact, some can also be transmitted by drug needles, or through childbirth or breast feeding.

⏮ puerperal fever

27.1 ■ Anatomy, Physiology, and Ecology of the Genitourinary System

Learning Outcome

1. Describe the anatomy, physiology, and ecology of the urinary and genital systems.

The genital and urinary systems are referred to as the *genitourinary system* because they are positioned close to each other. Both systems are often affected by the same pathogens.

The Urinary System

The urinary system (or tract) consists of the kidneys, ureters, bladder, and urethra (**figure 27.1**). The kidneys act as a filtering system, removing waste materials from the blood and selectively reabsorbing substances that can be reused. Each kidney is connected to the urinary bladder by a ureter. The bladder acts as a holding tank for urine. Once full, it empties through the urethra. Waste materials are excreted in the urine.

Based on culture results, scientists have assumed that the urine and the urinary tract above the urethra are generally free of microorganisms in both men and women. However, recent studies using nucleic acid amplification tests to detect bacterial DNA have challenged that assumption. Regardless, the lower urethra has a normal resident microbiota that includes species of *Lactobacillus, Staphylococcus, Corynebacterium, Haemophilus, Streptococcus,* and *Bacteroides.*

The urinary system is protected from infection by a number of mechanisms. Sphincter muscles near the urethra keep the system closed most of the time and help prevent infections by stopping bacteria from ascending to the bladder. The downward flow of urine also helps clean the system by washing away microorganisms before they have a chance to multiply and cause infection. In addition, normal urine contains organic acids that may make it acidic, as well as antimicrobial substances such as antibodies. The length and position of the urethra also play a role in preventing infection. Women have a relatively short urethra (4 cm) emptying close to the anus, and because of this they get urinary tract infections (UTIs) far more frequently than men, who have a longer urethra (20 cm) more distant from the anus.

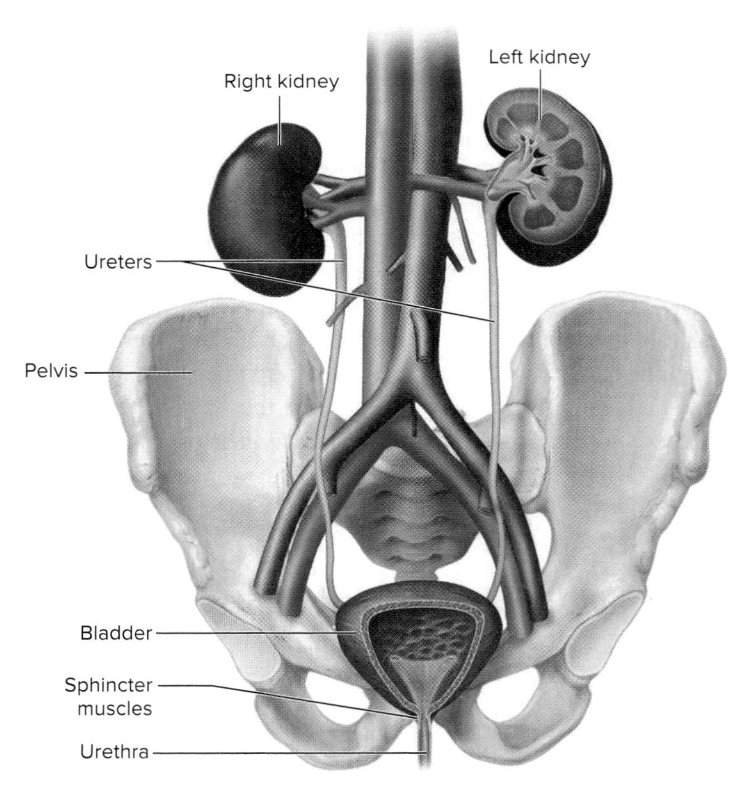

FIGURE 27.1 Anatomy of the Urinary System Urine flows from the kidneys, down the ureters, and into the bladder, which empties through the urethra.

? **How do the length and position of the urethra affect the tendency to get UTIs?**

MicroByte

If urine is alkaline, the person probably has an infection with a urease-producing bacterium that converts the urea in urine to ammonia.

The Genital System

The female genital system is composed of two ovaries, two fallopian tubes, a uterus, the vagina, and the external genitalia (vulva), including the labia and the clitoris (**figure 27.2a**). In women of childbearing age, an egg (ovum) is released from one of the two ovaries each month during ovulation and is swept into the adjacent fallopian tube. When fertilization occurs, it is normally in the fallopian tube. Ciliated epithelium of the tube then moves the fertilized ovum to the uterus, where it implants in the epithelial lining, developing into an embryo and then a fetus. If fertilization does not occur, the epithelial lining of the uterus sloughs off, producing a menstrual period. The lower end of the uterus is the cervix, which is filled with antimicrobial mucus, except during menstruation. The cervix opens to the vagina, which leads to the external genitalia.

The vagina is a portal of entry for a number of pathogens, particularly those that are sexually transmitted. Infection of the cervix frequently results in inflammation of the cervix (cervicitis). If the infection spreads to the fallopian tubes, scarring and destruction of the ciliated epithelium may occur there. That damage may interfere with efficient movement of ova to the uterus, leading to infertility. The fallopian tubes are open at both ends, so microorganisms that infect the tubes may also move into the abdominal cavity, where they can infect the liver and other organs.

In men, the reproductive organs include the testes (testicles), a variety of tubes, ducts and glands, and the penis (figure 27.2b). The testes are found outside the abdominal cavity in the scrotum. Sperm cells from each testis collect in a tightly coiled tubule called the epididymis and are carried to the urethra by a long tube, the vas deferens. At the junction of these tubes are glands called the seminal vesicles. Surrounding the urethra just below the bladder is another gland, the prostate. During ejaculation, sperm is released from the body via the urethra as semen—fluid containing many sperm, as well as secretions from the prostate gland and seminal vesicles. Sexually transmitted pathogens commonly infect the urethra, causing urethritis (inflammation of the urethra). Although prostate secretions also have antimicrobial properties, the prostate gland can be infected by urinary or sexually transmitted pathogens. In older men, the prostate often

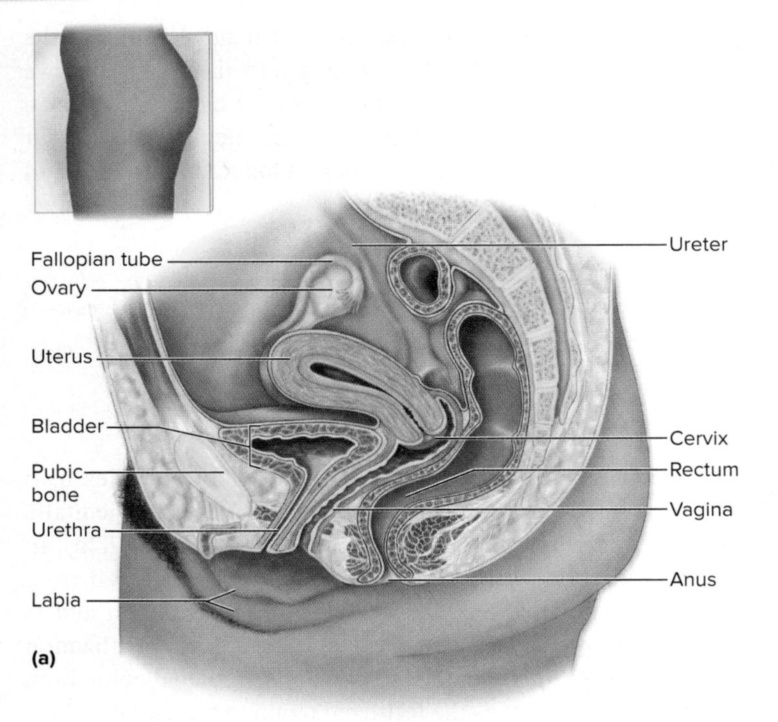

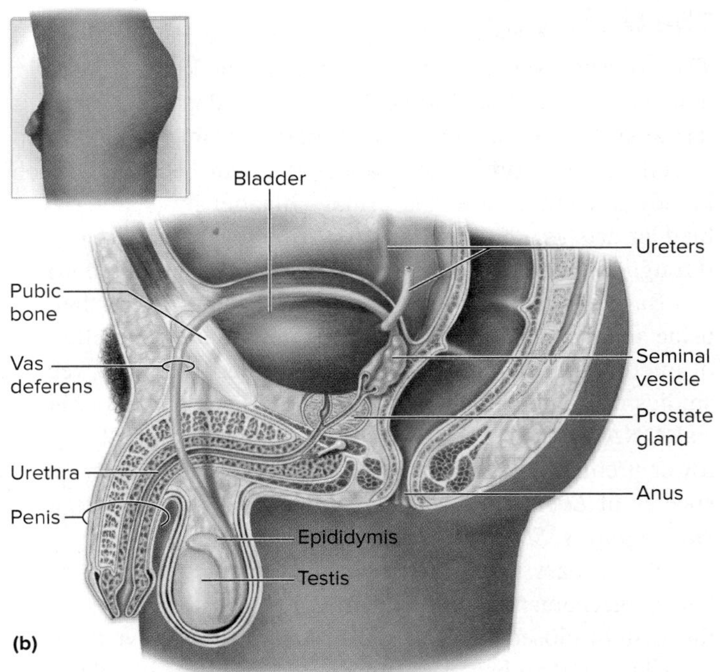

FIGURE 27.2 Anatomy of the Genital System **(a)** Female. **(b)** Male.

? **Why can infection of the fallopian tubes result in infertility?**

enlarges, compressing the urethra and slowing the flow of urine, making it more likely that a UTI will develop.

The normal microbiota of the genital tract of women is affected by the action of estrogen hormones on the epithelial cells of the vaginal mucosa. When estrogens are present, glycogen is deposited in these cells. The glycogen is converted to lactic acid by lactobacilli, resulting in an acidic pH that inhibits the growth of many potential pathogens. Lactobacilli may also release hydrogen peroxide, an inhibitor of some anaerobic bacteria. Thus, the normal microbiota and resistance to infection of the female genital tract vary considerably with the person's hormonal status. ◀◀ lactobacilli

MicroAssessment 27.1

The genitourinary system is one of the portals of entry for pathogens. The urinary system includes the kidneys, ureters, bladder, and urethra. The male reproductive system includes the testes, various glands, and the penis. The female reproductive system includes the vagina, cervix, uterus, fallopian tubes, and ovaries. In females, the cervix is a common site of STIs; inflammation of the cervix is cervicitis. In males, the urethra is a site of of STIs, resulting in urethritis.

1. Describe how the bladder is protected from pathogens.
2. Why do women get urinary tract infections more frequently than men?
3. What changes might occur in the vagina if lactobacilli were eliminated?

27.2 ■ Urinary Tract Infections

Learning Outcomes

2. Describe the features of bacterial cystitis.
3. Outline the characteristics of leptospirosis.

Urinary tract infections (UTIs) can involve the urethra, bladder, or kidneys, alone or in combination. They account for several million visits to the doctor's office each year in the United States. Any situation that causes inhibition of urination increases the risk of developing UTIs. Anesthesia and major surgery, for example, temporarily stop the reflex ability to urinate, and urine accumulates in the bladder. Even being too busy to empty the bladder may predispose a person to infection. Most cases of UTI actually occur in otherwise healthy young women with normal urine flow.

Bacterial Cystitis ("Bladder Infection")

Cystitis (inflammation of the bladder) is the most common type of UTI. Bacterial cystitis is common among otherwise healthy women, and is also a frequent healthcare-associated infection.

Signs and Symptoms

Bacterial cystitis is sometimes asymptomatic, especially among children and the elderly. When symptoms do occur,

FOCUS ON A CASE 27.1

A 32-year-old married woman complained of 1 week of burning pain on urination, and frequent release of small amounts of bloody urine. About 8 days earlier, she had completed 3 days of co-trimoxazole (TMP-SMX) treatment for similar symptoms; urine culture results before that treatment showed *Escherichia coli,* resistant only to amoxicillin. When her symptoms returned, she began drinking 12 ounces of cranberry juice three times daily but felt only slightly better. She did not have other symptoms such as chills, fever, back pain, nausea, or vomiting, and the results from the rest of her physical examination were normal.

Her medical history showed that she had suffered two or three similar episodes of urinary symptoms every year for a number of years. Sometimes the symptoms would go away when she forced herself to drink more, but at other times the symptoms would persist and she would obtain medical evaluation and treatment with an antibacterial medication. On one occasion several years before the present illness, she had chills, fever, back pain, nausea, and vomiting with her other urinary symptoms, and she was hospitalized for a "kidney infection." There was no history suggesting any underlying disease such as diabetes, cancer, or immunodeficiency.

The results of her laboratory tests included normal leukocyte count and kidney function tests. Microscopic examination of her urine showed many red and white blood cells, including neutrophils. A stained smear of the urine showed numerous rod-shaped, Gram-negative bacteria, identified as *Escherichia coli.* Culture of the urine revealed more than 100,000 colony-forming units (CFUs) per milliliter. The bacterium was resistant to amoxicillin, but sensitive to the other antibacterials useful for treating urinary tract infections. ◀◀ **neutrophils**

1. What is the diagnosis?
2. What is the treatment?
3. What is the prognosis?
4. What future preventive measures would be advisable?

Discussion

1. This woman's signs and symptoms clearly led to a diagnosis of bacterial cystitis, but clues in the presentation of this case suggest that it was not as simple as first thought. First, her signs and symptoms recurred only 1 day after completing her medication. Most patients with uncomplicated bacterial cystitis are cured by 3 days of an antibacterial medication to which the causative bacterium is susceptible. Second, she gave a past history of being hospitalized for pyelonephritis. In addition, she had signs and symptoms for a full week before she went for medical evaluation, potentially increasing the possibility of developing serious complications.

   These clues make it highly likely that she has a condition called subclinical pyelonephritis, in which her bladder infection has spread to her kidneys but has not yet produced the symptoms of pyelonephritis. As many as 30% of patients with cystitis have subclinical pyelonephritis, depending on risk factors such as the clues mentioned in this case. The diagnosis of a kidney infection can be confirmed using imaging studies, which allow the physician to view the kidney and see areas of inflammation or structural abnormalities.
2. The patient can be treated with co-trimoxazole, the same medication given before. Because infection in the kidneys takes much longer to cure than bladder infections, however, the treatment must be continued for 2 weeks or longer. A urine culture must be done 1 week after treatment is completed, to be sure that the infection is truly gone.
3. The outlook is good for a full recovery without any permanent damage to the kidneys. The patient should be advised, however, that repeated future infections of the kidneys could lead to kidney failure if not treated quickly.
4. For the future, this patient should use all the usual methods for preventing UTIs, including drinking enough to ensure urinating at least four or five times daily, avoiding delays in emptying the bladder, and urinating promptly after intercourse. Because she has recurrent infections, a preventive antimicrobial medication may be appropriate. No vaccine for this infection has been approved for use in the United States.

they typically start suddenly and include a burning pain during urination, an urgent need to urinate, and frequent release of small amounts of urine. The urine is cloudy due to accumulation of leukocytes, and may be a pale red color due to blood. It also often has a bad odor. The area above the pubic bone may be painful because of the underlying inflamed bladder.

Sometimes a more serious condition called **pyelonephritis** (infection of the kidneys) develops. Signs and symptoms of this include fever, chills, vomiting, back pain, and tenderness overlying the kidneys. Repeated episodes of pyelonephritis lead to scarring and shrinkage of the kidneys and can cause kidney failure.

Causative Agents

Bladder infections usually originate from the normal intestinal microbiota. More than 80% of cases are caused by specific uropathogenic strains of *Escherichia coli.* The remaining infections in young women are caused by other *Enterobacteriaceae* members such as *Klebsiella* and *Proteus* species (Gram-negative), or by *Staphylococcus saprophyticus* (Gram-positive). Hospitalized patients, and people with long-standing bladder **catheters** (tubes inserted into the bladder), are often chronically infected with multiple species of bacteria, such as *Serratia marcescens* and *Pseudomonas aeruginosa* (Gram-negative) and

Enterococcus faecalis and *E. faecium* (Gram-positive). Many of these species are resistant to antibiotics and are difficult to treat.

Pathogenesis

The causative agents of cystitis generally reach the bladder by moving up the urethra. Uropathogenic *E. coli* (UPEC) strains have pili that allow them to attach specifically to receptors on bladder epithelial cells. Bacterial attachment is followed by the death and sloughing of this layer of cells. The bacteria then enter the underlying epithelium by endocytosis and multiply rapidly to create biofilm-like intracellular bacterial communities (IBCs). Bacteria later detach from the outer surface of the IBC and move into the bladder lumen to attach to surrounding naive epithelium, creating new IBCs. The UPEC may eventually establish a dormant intracellular reservoir that resists antibiotics and is undetected by the immune system, often leading to chronic or recurrent infections. Filamentous forms of UPEC are observed during IBC development. It appears that these forms resist ingestion by phagocytes, thereby helping the organism avoid the innate immune response. Bacteria that move up the ureters to the kidneys can cause pyelonephritis. ⏮ fimbriae, ⏮ endocytosis

Epidemiology

About 30% of women develop cystitis at some time during their life. Factors that predispose women to UTIs include:

- **Short urethra.** The length and position of the urethra put it at risk for fecal contamination and colonization with potentially pathogenic intestinal bacteria. From the urethra, the distance to the bladder is only a few centimeters.
- **Sexual intercourse.** About one-third of UTIs in sexually active women are associated with sexual intercourse. The massaging effect of sexual intercourse on the urethra moves bacteria from the urethra into the bladder. Many women develop bladder infections after having sex for the first time.
- **Birth control devices.** Diaphragms compress the urethra and slow the flow of urine, increasing the risk of UTI.

Other factors involved in development of UTIs include:

- **Enlarged prostate.** UTIs are unusual in men until about age 50, when enlargement of the prostate gland compresses the urethra and makes it difficult for them to completely empty the bladder.
- **Catheterization.** Medical conditions may require a urinary catheter to drain the bladder, and this unfortunately allows bacteria to reach the bladder and establish UTIs. Pathogens may create a biofilm either within or on the catheter, making it difficult or impossible to kill them with antibacterial medications. Paraplegics (individuals with paralysis of the lower half of the body) cannot urinate normally due to lack of bladder control and require a catheter indefinitely to transfer their urine to a container. Because of this, they often have recurrent UTIs.

MicroByte

In the United States, nearly 100,000 hospitalized patients develop bladder infections each year, mostly after catheterization.

Treatment and Prevention

Cystitis is usually easily treated with a few days of an antimicrobial medication effective against the causative bacterium. Pyelonephritis, a more serious condition, usually requires hospitalization and intravenous antibiotic treatment.

General ways to prevent UTIs include drinking enough to ensure urinating at least four or five times daily, urinating immediately after sexual intercourse, and wiping from front to back after defecation to minimize fecal contamination of the urethra. Many antimicrobial medications accumulate in the urine, reaching concentrations higher than in the blood, so taking a low dose of an antibiotic daily can be used to prevent recurrent infections. Cranberry juice and green tea are often used to prevent bacterial cystitis, but the scientific evidence on their effectiveness is inconclusive. The main features of bacterial cystitis are presented in **table 27.1.**

Leptospirosis

Leptospirosis is mainly a disease of animals, but it can be passed to humans. The causative bacterium enters the body through a mucous membrane or wound and is then carried to the urinary system by the bloodstream. The disease is probably one of the most common of all the zoonoses, but many cases are mild, resolve without treatment, and remain undiagnosed. More severe cases require treatment and can sometimes be fatal. ⏮ zoonoses

Signs and Symptoms

Leptospirosis infections are often asymptomatic. When signs and symptoms do occur, they begin after an incubation period averaging 10 days, and often occur in two phases (biphasic). In mild cases, which are most common, symptoms are flu-like and include the sudden development of a headache, spiking fever, chills, and muscle pain. A dry cough may occur. A characteristic feature of this first phase (septicemic) is photophobia (sensitivity to light) and red eyes, caused by dilation of small blood vessels. These symptoms usually resolve within a week, and all signs and symptoms of illness are gone in a month or less. In severe cases of the disease, symptoms recur after 1 to 3 days of feeling well. Symptoms of this

TABLE 27.1 Bacterial Cystitis

Signs and symptoms	Sudden onset, burning pain on urination, frequent need to urinate; cloudy, red-colored urine that smells bad; fever, chills, back pain, and vomiting suggest pyelonephritis
Incubation period	Usually 1 to 3 days
Causative agents	Most commonly *Escherichia coli,* but also other *Enterobacteriaceae* as well as *Staphylococcus saprophyticus;* in the case of healthcare-associated infections, *Pseudomonas, Serratia,* and *Enterococcus* species
Pathogenesis	Bacteria ascend the urethra, enter the bladder, and attach by pili to receptors on epithelium; sloughing of cells and an inflammatory response follow; bacterial spread to the kidneys can occur via the ureters, causing pyelonephritis and possible kidney failure.
Epidemiology	Bacterial cystitis is common in women, promoted by a relatively short urethra, sexual intercourse, and use of a diaphragm. Older men are at risk of infection because enlargement of the prostate gland partially obstructs their urethra; catheterization often results in infection.
Treatment and prevention	Treatment: short-term antimicrobial medication usually sufficient; longer treatment for pyelonephritis. Prevention: drinking enough liquid to urinate at least four to five times daily; wiping from front to back; low daily dose of antibiotic may help prevent recurrent bacterial cystitis in women.

second (immune) phase include those of the septicemic phase as well as bleeding from various sites, vomiting, rash, and confusion.

In rare cases, people with leptospirosis develop Weil's disease, which affects the liver and kidneys. Signs and symptoms of Weil's disease include jaundice (yellowing of skin and eyes), liver and kidney failure, hemorrhage in many organs, and meningitis. People who suffer biphasic leptospirosis or Weil's disease need medical attention, since these can be fatal.

Causative Agent

Leptospirosis is caused by *Leptospira interrogans,* a slender aerobic Gram-negative spirochete with hooked ends (**figure 27.3**). There are more than 250 antigenic types of this species, some of which were given different names in the past. ⏮ **spirochetes**

Pathogenesis

Leptospira interrogans enters the body through mucous membranes, eyes, and breaks in the skin. No lesion develops at the site of entry, but the organisms multiply and spread throughout the body by way of the bloodstream. Severe pain is characteristic of this first (septicemic) phase, and patients sometimes undergo unnecessary surgery because the pain is

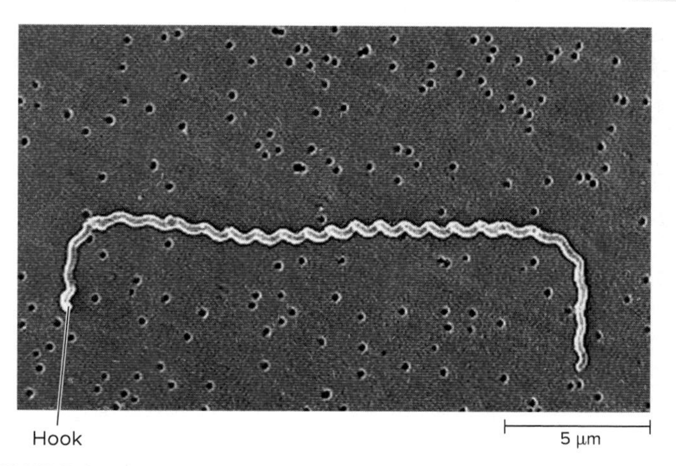

FIGURE 27.3 ***Leptospira interrogans*** This spirochete is the cause of leptospirosis, a zoonosis (SEM). Source: Rob Weyant/CDC

? **What is the main mode of transmission of this organism?**

mistaken for a symptom of appendicitis or gallbladder infection. There are no inflammatory changes or tissue damage at this stage. Within a week, the immune response destroys the organisms present in most tissues, although they continue to multiply in the kidneys. Many people recover at this stage, with no further development of symptoms.

If the disease progresses, the infected person then usually seems healthy for a few days before the second phase of illness occurs. The new symptoms are thought to be due to the immune response, so the stage is called the "immune phase." The immune phase is characterized by injury to the cells that line the lumen of tiny blood vessels, causing clotting and impaired blood flow in tissues throughout the body. This leads to most of the serious effects of the illness, including kidney failure, which is the main cause of fatalities with leptospirosis.

Epidemiology

Leptospirosis occurs around the world in all types of climates, although it is more common in the tropics. *Leptospira interrogans* infects many species of wild and domestic animals, usually causing little or no apparent illness, but sometimes causing fatal, epidemic disease. The bacteria are excreted in the animal's urine, and urine spots on the ground remain infectious for as long as they are still moist. Contaminated urine is the main mode of transmission to other hosts. *L. interrogans* can also survive in mud or water for several weeks, resulting in a correlation between the amount of rainfall and leptospirosis incidence. Humans contract leptospirosis from water, soil, or food contaminated with infected animal urine. Many cases are caused by swimming in urine-contaminated fresh water. Ingested organisms enter through the mucosa of the upper digestive system. Infected humans usually excrete the organisms for a few weeks to several months, but person-to-person transmission does not seem to occur.

TABLE 27.2 Leptospirosis

1. Water or animal urine contaminated with *Leptospira interrogans* splashes onto mucous membrane or damaged skin.
2. The bacteria enter the bloodstream and are carried throughout all body tissue, sometimes causing fever and intense pain.
3. Signs and symptoms resolve and bacteria disappear from blood and tissues, except kidneys.
4. Second phase of the biphasic illness is associated with severe damage to liver and kidneys.
5. Complete recovery occurs if kidney failure can be effectively treated.
6. Excretion of *L. interrogans* continues in urine.

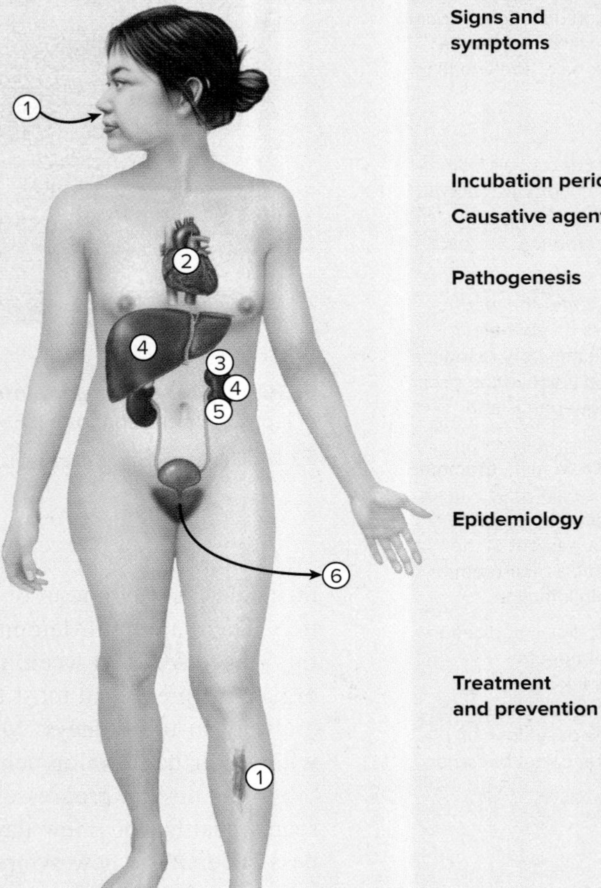

Signs and symptoms	Many mild and asymptomatic cases; others have a biphasic illness: spiking fever, headache, muscle pain, bloodshot eyes in the first (septicemic) phase, then 1 to 3 days of improvement; heart, brain, liver, and kidney damage in the second (immune) phase.
Incubation period	Usually about 10 days (range, 2 to 30 days)
Causative agent	*Leptospira interrogans,* a spirochete with many serotypes
Pathogenesis	The bacteria penetrate mucous membranes or breaks in the skin, multiply in the bloodstream, and are carried to all parts of the body. Septicemic phase: severe pain with penetration of body tissues, but little or no tissue damage. Immune phase: damage to cells that line small blood vessels and clotting of blood; causes severe damage to the liver, kidneys, heart, brain, and other organs.
Epidemiology	Worldwide distribution; wide range of animal hosts chronically excrete the bacteria in their urine, causing contamination of natural waters and soils; organisms remain infectious under warm, moist, neutral or alkaline conditions for long periods of time.
Treatment and prevention	Treatment: various antibacterial medications useful but only if given early in the disease. Prevention: avoiding contact with animal urine; vaccines prevent disease in domestic animals, but may not prevent urinary carriage; antibiotics preventive in epidemics.

Treatment and Prevention

Mild cases of leptospirosis resolve without treatment, although antibiotics can be given. A number of different antibiotics are effective for treating the disease, but only if they are started during the first 4 days of the initial illness. Once treatment begins, the person often shows a temporary worsening of symptoms due to massive release of antigens from bacterial cells lysed by the antimicrobial medication.

There are few effective measures for preventing leptospirosis other than avoiding animal urine. Multivalent vaccines (that protect against a number of different serotypes of *L. interrogans*) are available for preventing the disease in domestic animals, but they do not always prevent the animal from becoming a carrier. Small doses of a tetracycline antibiotic can prevent the disease in high-risk individuals, such as veterinarians, slaughterhouse workers, farmers, sewer workers, and water sport enthusiasts. The main features of leptospirosis are shown in **table 27.2.**

MicroAssessment 27.2

Situations that affect normal urine flow predispose a person to UTIs. Bladder infections are common, especially in women, and are usually caused by normal intestinal bacteria ascending from the urethra. Pyelonephritis is a serious complication that may cause death due to kidney failure. In leptospirosis, a widespread zoonosis spread by urine, the causative organisms enter the urinary system from the bloodstream. Although usually mild, the disease can be a severe biphasic illness characterized by tissue invasion and pain in the first phase, and tissue destruction and kidney damage in the second.

4. What organism causes the majority of bladder infections in otherwise healthy women? From where does this organism come?
5. How do people become infected with *Leptospira interrogans*?
6. How can catheterization lead to bladder infection?

27.3 ■ Genital System Diseases

Learning Outcome

4. Compare and contrast bacterial vaginosis (BV), vulvovaginal candidiasis (VVC), and staphylococcal toxic shock syndrome.

The genital tract is the portal of entry for many infectious diseases, both non-sexually and sexually transmitted. This section discusses some genital system diseases that are not generally transmitted sexually.

Bacterial Vaginosis (BV)

In the United States, bacterial vaginosis (BV) is the most common vaginal disease of women in their childbearing years. It is termed *vaginosis* rather than *vaginitis* because there are no inflammatory changes. In the United States, BV is common in pregnant women and puts them at risk for premature deliveries.

Signs and Symptoms

About half of BV cases are asymptomatic. Symptomatic BV is characterized by a thin, grayish-white, slightly bubbly vaginal discharge that has a characteristic strong fish-like smell. The bacteria associated with BV may spread to the uterus or the fallopian tubes, causing **pelvic inflammatory disease (PID),** which can lead to infertility.

Causative Agent

The cause or causes of BV are unknown, but the disease is characterized by a shift in the vaginal flora from the dominant *Lactobacillus* to a polymicrobial flora. Because most cases show a significant decrease in vaginal lactobacilli, conditions that suppress lactobacilli or promote the growth of other microbiota are thought to play a causative role. The discharge that characterizes BV contains large numbers of bacteria, including *Gardnerella vaginalis,* and members of the genera *Mobiluncus, Prevotella,* and *Peptostreptococcus.* Although these bacteria are generally present in vaginal secretions of women with BV, women voluntarily inoculated with cultures of these organisms do not always develop the disease. Also, low numbers of each of these species can occur in vaginal secretions of healthy women. However, since discharge from women with BV can lead to the disease in healthy women, these species may play a role in developing BV, even if they do not directly cause the disease. In addition, PCR-based methods have indicated that several novel, previously uncultured species of bacteria (such as BVAB1, BVBA2, and BVBA3) may also be involved in BV.

Pathogenesis

Women with BV have characteristic changes in the vagina, including a loss of acidity of the vaginal secretions (normally pH 3.8–4.2), disruption of the normal microbiota, and substantial increase in the numbers of clue cells. Clue cells are epithelial cells that have sloughed off the vaginal wall and are covered with bacteria (**figure 27.4**). There is no inflammation unless another, concurrent vaginal infection is present. The strong fishy odor is caused by metabolic products of the anaerobic bacteria. This odor can be detected using the whiff test in which a drop of potassium hydroxide is added to a slide containing a vaginal discharge sample—in women with BV, the fishy odor is produced.

Epidemiology

The epidemiology of BV is not well understood because the causative agent is not known. The disease is most common among sexually active women and sometimes occurs in children who have been sexually abused. Pregnant women are at increased risk of BV. Women who douche, have multiple sex partners, have sex with other women, have a new sex partner, or use an intrauterine device (IUD) also have an increased risk. There is no proof that bacterial vaginosis is sexually transmitted; however, virgins seldom get BV. Women with BV are at higher risk of getting other STIs such as chlamydia, gonorrhea, or HIV.

Treatment and Prevention

Most cases of BV respond quickly to treatment with antibiotics such as metronidazole or clindamycin, which can be given orally or vaginally. The disease can recur. Treatment of BV is important in pregnant women because BV may cause premature birth. Studies on using yogurt vaginally to restore lactobacilli have given conflicting results.

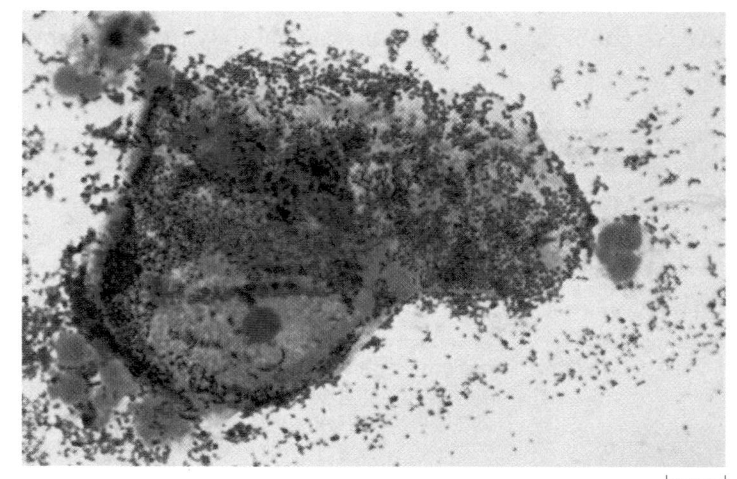

FIGURE 27.4 Clue Cell from a Woman with Bacterial Vaginosis The cell in the photograph is a stained epithelial cell that has sloughed from the vaginal wall. It is covered with rod-shaped bacteria. ©Dr. Y. Boussougan/CNRI/Science Source

? Why does the clue cell appear darker in color than other cells in a cervical smear?

BV can be prevented by abstinence, limiting the number of sex partners, and avoiding douching. Treatment of the male sex partners of patients with BV does not prevent recurrences. The main features of BV are summarized in **table 27.3.**

Vulvovaginal Candidiasis (VVC)

Vulvovaginal candidiasis (VVC), a fungal infection, is the second most common cause of vaginal symptoms after BV. Like BV, it seems to occur after a disruption of the normal microbiota. As the name indicates, VVC often involves not only the vagina, but the vulva as well.

Signs and Symptoms

The most common signs and symptoms of VVC are constant, intense itching and burning of the vagina or vulva, which appear red and swollen. Typically, there is a large amount of thick, clumpy whitish or whitish-gray vaginal discharge that may resemble cottage cheese.

Causative Agent

VVC is caused by *Candida albicans,* a yeast that is part of the normal microbiota of the vagina in about a third of all women (**figure 27.5**). Because *C. albicans* is a fungus, it has a eukaryotic cell structure. It typically grows as yeast cells but can also form hyphae or pseudohyphae. ⏮ **yeast,** ⏮ **hyphae**

Pathogenesis

Normally, vaginal colonization by *Candida albicans* causes no symptoms. The growth of the organism is usually limited by the immune system and the normal vaginal lactobacilli that occupy the same niche and compete for nutrients. When the normal microbiota balance is disturbed—as occurs during menstruation or pregnancy, or when using oral contraceptives or antibiotics— *C. albicans* can multiply freely, causing

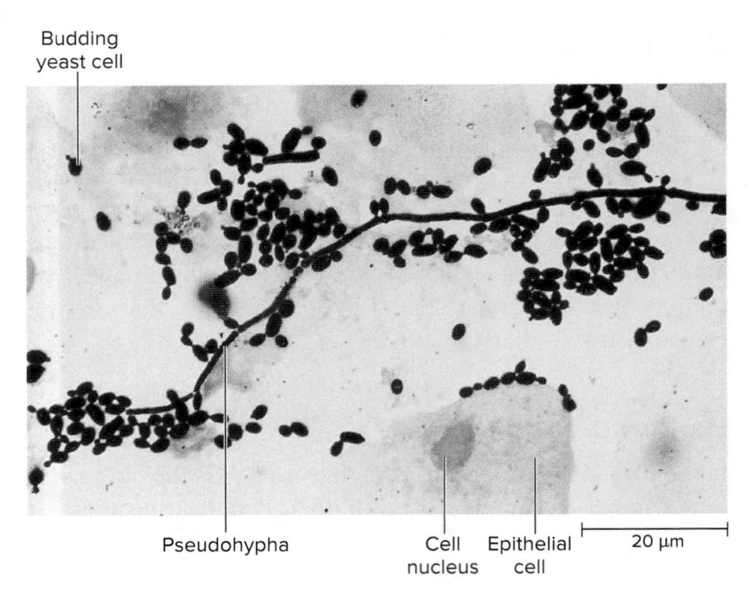

FIGURE 27.5 ***Candida albicans*** Stained smear of vaginal discharge from a woman with vulvovaginal candidiasis, showing *C. albicans*. ©John Durham/Science Source

? **What usually prevents VVC from occurring?**

an inflammatory response. The signs and symptoms of VVC occur within about 10 days.

Epidemiology

Factors that predispose a woman to *Candida* infection are late stage of pregnancy, poorly controlled diabetes, and the use of oral contraceptives or antibiotics. Hormone replacement therapy may also increase the risk of VVC. However, most patients with VVC have no known predisposing factors. The disease does not spread from person to person.

Treatment and Prevention

Vulvovaginal candidiasis can be treated with antifungal medications (such as azoles) that are applied vaginally; several formulations—including creams, gels, and suppositories—are available as over-the-counter medications, but others require a prescription. Fluconazole taken by mouth is generally safe, although it may cause side effects such as headache and nausea, and it can interact with other medications, causing rare, but serious reactions. Self-diagnosis and treatment with over-the-counter medications can lead to development of drug-resistant organisms.

Prevention of VVC depends on minimizing the use and duration of antibacterial medications and on effective treatment of underlying conditions such as diabetes. The main features of VVC are presented in **table 27.4.**

TABLE 27.3	**Bacterial Vaginosis**
Signs and symptoms	Gray-white vaginal discharge with unpleasant fishy odor
Incubation period	Unknown
Causative agent	Unknown
Pathogenesis	Uncertain; obvious change in normal microbiota composition; increased sloughing of vaginal epithelium in the absence of inflammation; odor due to metabolic products of anaerobic bacteria; may cause complications of pregnancy, including premature births.
Epidemiology	Associated with many sexual partners or a new partner, but can occur in the absence of sexual intercourse.
Treatment and prevention	Treatment: antibacterial medications. Prevention: abstinence, limiting number of sexual partners, avoiding douching.

Staphylococcal Toxic Shock Syndrome

Toxic shock syndrome was described in the late 1970s in several children with staphylococcal infections. In 1980, it

TABLE 27.4 Vulvovaginal Candidiasis

Signs and symptoms	Itching, burning, thick, white vaginal discharge, redness and swelling
Incubation period	Usually unknown; generally 3 to 10 days when associated with antibacterial medications
Causative agent	*Candida albicans,* a yeast
Pathogenesis	Inflammatory response to overgrowth of the yeast, which is often present among the normal microbiota
Epidemiology	Not contagious; associated with antibacterial therapy, use of oral contraceptives, pregnancy, and uncontrolled diabetes, but most cases have no identifiable predisposing factor.
Treatment and prevention	Treatment: intravaginal antifungal medications usually effective. Prevention: minimizing antibiotic use and treating known predisposing conditions.

became epidemic in young, healthy, menstruating women who were using a brand of high-absorbency tampon that has since been removed from the market (**figure 27.6**). The term *toxic shock syndrome* was used to describe the illness. Now that we know its cause, it is called *staphylococcal toxic shock syndrome.* This is not a new disease, but one that emerged in a new form and became much more common as a result of changes in technology and human behavior.

Signs and Symptoms

Staphylococcal toxic shock syndrome is characterized by the sudden development of high temperature, headache, muscle aches, bloodshot eyes, vomiting, diarrhea, a sunburn-like rash, and confusion. Typically, the skin peels about a week after the development of the disease. Without treatment, the blood pressure can drop, leading to multi-organ failure, coma, and sometimes death.

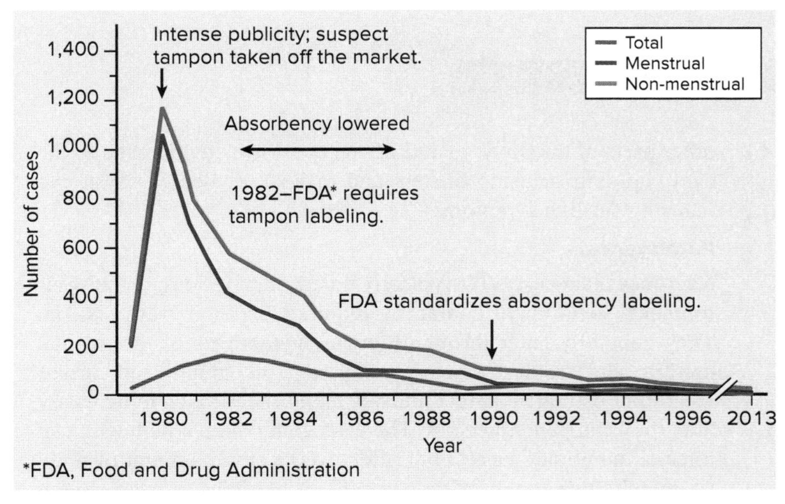

FIGURE 27.6 Staphylococcal Toxic Shock Syndrome, United States A sharp drop in cases occurred when a brand of high-absorbency tampon was taken off the market.

? How can tampon use increase the risk of toxic shock syndrome?

Causative Agent

Staphylococcal toxic shock syndrome is caused by strains of *Staphylococcus aureus* that produce toxic shock syndrome toxin-1 (TSST-1) or other related exotoxins. ◀◀ *Staphylococcus aureus,* ◀◀ exotoxins

Pathogenesis

Tampon-associated toxic shock syndrome usually begins 2 to 3 days after the start of menstruation when tampons are used. The staphylococci grow in the blood-soaked tampon. The bacteria rarely spread throughout the body, but as they multiply they produce TSST-1 or other exotoxins. Staphylococcal toxic shock syndrome results from absorption of these toxins into the bloodstream. The toxins are superantigens that cause activation of large numbers of helper T cells, leading to a massive release of cytokines (cytokine storm). This in turn causes a drop in blood pressure and multi-organ failure—the most dangerous aspect of the potentially fatal illness. ◀◀ superantigen

Epidemiology

Staphylococcal toxic shock syndrome can occur after infection with any strain of *Staphylococcus aureus* that produces one of the responsible exotoxins. However, strains that produce TSST-1 cause three-quarters of all cases. The syndrome can occur after infection of surgical wounds, infections associated with childbirth, and other types of staphylococcal infections. It does not spread from person to person. Using tampons increases the risk of staphylococcal toxic shock, and the higher-absorbency tampons may pose a greater risk. Use of intravaginal contraceptive sponges also increases risk. People who have recovered from the disease are not always immune to it. Since 1990, there has been a slow, steady decline in the incidence of staphylococcal toxic shock syndrome, now estimated to be six or fewer total cases per 100,000 people per year (see figure 27.6).

Treatment and Prevention

Staphylococcal toxic shock syndrome is a severe disease and requires hospitalization. It can be effectively treated with antibacterial medication active against the infecting *S. aureus* strain, intravenous fluid, and other measures to prevent shock and kidney damage. The source of the infection should be eliminated if possible. Although most people recover fully in 2 to 3 weeks, the disease can be fatal within a few hours.

Toxic shock syndrome associated with the use of tampons can be prevented by the appropriate use of tampons, including washing hands thoroughly before and after inserting a tampon, using tampons with the lowest practical absorbency, changing tampons at least every 6 hours, and using a pad instead of a tampon while sleeping. It is also important to avoid trauma to the vagina when inserting tampons, to recognize the signs and symptoms of staphylococcal toxic shock syndrome, and to remove any tampon immediately if symptoms occur. Women who have had staphylococcal toxic shock syndrome previously should not use tampons. **Table 27.5** describes the main features of staphylococcal toxic shock syndrome.

MicroAssessment 27.3

Bacterial vaginosis (BV), the most common vaginal disease of women in their childbearing years, is characterized by a significant change in the composition of the normal vaginal microbiota. Vulvovaginal candidiasis (VVC) often occurs as a result of antibacterial therapy suppressing normal vaginal microbiota, but many other cases arise for unknown reasons. Staphylococcal toxic shock syndrome is caused by certain strains of *Staphylococcus aureus* that produce exotoxins, which are absorbed into the bloodstream, causing the massive release of cytokines responsible for shock.

7. What is the causative agent of bacterial vaginosis (BV)?
8. Why was the incidence of staphylococcal toxic shock syndrome higher in menstruating women than in other people during the early 1980s?
9. Why would using antibiotics predispose a woman to vulvovaginal candidiasis (VVC)?

TABLE 27.5 Staphylococcal Toxic Shock Syndrome

Signs and symptoms	Fever, vomiting, diarrhea, muscle aches, a rash that peels, and low blood pressure that may lead to multi-organ failure
Incubation period	3 to 7 days
Causative agent	Certain toxin-producing strains of *Staphylococcus aureus*
Pathogenesis	Toxin (TSST-1 and others) produced by certain strains of *S. aureus;* toxins are superantigens, causing cytokine release and drop in blood pressure.
Epidemiology	Associated with certain high-absorbency tampons, leaving tampons in place for long periods of time, and abrasion of the vagina from tampon use; also as a result of infection by certain toxin-producing *S. aureus* strains in other parts of the body.
Treatment and prevention	Treatment: antimicrobial medication effective against the causative *S. aureus* strain; intravenous fluids. Prevention: awareness of symptoms; prompt treatment of *S. aureus* infections; frequent change of tampons by menstruating women.

SEXUALLY TRANSMITTED INFECTIONS

27.4 ■ Bacterial STIs

Learning Outcomes

5. Compare and contrast chlamydia, gonorrhea, and *Mycoplasma genitalium* infection.
6. Compare and contrast syphilis and chancroid.

Chlamydial Infections

Chlamydial infections are the most common bacterial STIs. These infections can cause urethritis and may cause tubal damage in both the male and the female reproductive systems, which can lead to infertility. ⏮ *Chlamydia*

FOCUS ON SEXUALLY TRANSMITTED INFECTIONS

Sexually transmitted infections (STIs) are typically spread through intimate contact with body fluids such as urethral discharge, vaginal secretions, and sometimes saliva. Because of this, they are usually transmitted through unprotected sexual contact—including vaginal, oral, and anal sex—but they can also be acquired without sexual contact, through close contact between mucous membranes.

Signs and Symptoms

Many STIs are asymptomatic, which is a problem because an infected person may unknowingly transmit the causative agent to others. When signs and symptoms do occur, they vary according to the pathogen, but may include vaginal or urethral discharge, pain with urination, genital sores, rashes, abnormal vaginal bleeding, and abdominal pain. The causative agents of some STIs can ascend the female reproductive tract to cause pelvic inflammatory disease (PID; characterized by chronic pelvic pain) or spread to other parts of the body, including the eyes (causing damage to the eye) and skin (causing blisters and rashes), or become systemic (causing flu-like symptoms).

Pathogenesis

Microbes that cause STIs typically do not survive long in the environment, so intimate contact is required for their transmission. They generally enter through mucous membranes of the vagina, urethra, and rectum, or less commonly, the mouth and throat; recall that pathogens infect mucous membranes much more easily than they can penetrate skin. The ease with which a pathogen can infect a membrane varies, but often even a small amount of body fluid contains enough infectious cells or particles to put a person at high risk of acquiring that microbe.

Because the causative agents of STIs do not survive well in the environment, their survival depends on their ability to avoid

(*Continued*)

FOCUS ON SEXUALLY TRANSMITTED INFECTIONS *(continued)*

the immune system for extended periods; mechanisms they use to do this vary according to the pathogen. Viruses that cause STIs often become latent.

The signs and symptoms associated with bacterial STIs are usually a result of damage due to inflammation, while those of viral STIs are often due to direct damage to infected cells. In some cases, the damage causes permanent problems. For example, the inflammatory response of PID can lead to scarring of the fallopian tubes, which can result in infertility or ectopic pregnancy in which the embryo develops in the fallopian tube. Ectopic pregnancy can lead to life-threatening internal hemorrhaging. Some sexually transmitted pathogens cause cancer.

Epidemiology

STIs are most commonly transmitted through body fluids, so all sexual contact should be regarded as having some risk. The risk of infection varies, depending on the causative microbe and the type of sexual activity. A person is generally more likely to transmit an STI when signs and symptoms are present, but can also transmit the infection when these are absent. Failure to recognize that STIs can be transmitted even when symptoms are not apparent has contributed to the alarming increase in STI prevalence. The more unprotected sex a person has, the greater that person's risk of acquiring an STI (**box figure 27.1**). In addition, any STI that results in open sores or inflammation increases the risk of contracting HIV; the sores allow HIV access to tissues, and the inflammation recruits cell types that HIV infects (helper T cells and macrophages).

Fetuses and newborns are at risk of acquiring certain genital pathogens through vertical transmission—via the placenta, or during the birth process. In infants, the infections may have serious consequences such as blindness or even death.

Treatment and Prevention

Common bacterial STIs are currently curable with antimicrobial medications, although some organisms are now resistant to most available options. Although viral STIs cannot be cured with medications, some antivirals can reduce the severity and duration of symptoms.

The most reliable ways to avoid contracting STIs include not having sex, or having a monogamous relationship with an uninfected person. Condom use significantly reduces the chances of acquiring an STI but does not guarantee protection. Some human papilloma virus (HPV) infections can be prevented though vaccination. Male circumcision also significantly reduces the risk of acquiring some sexually transmitted viruses such as HIV.

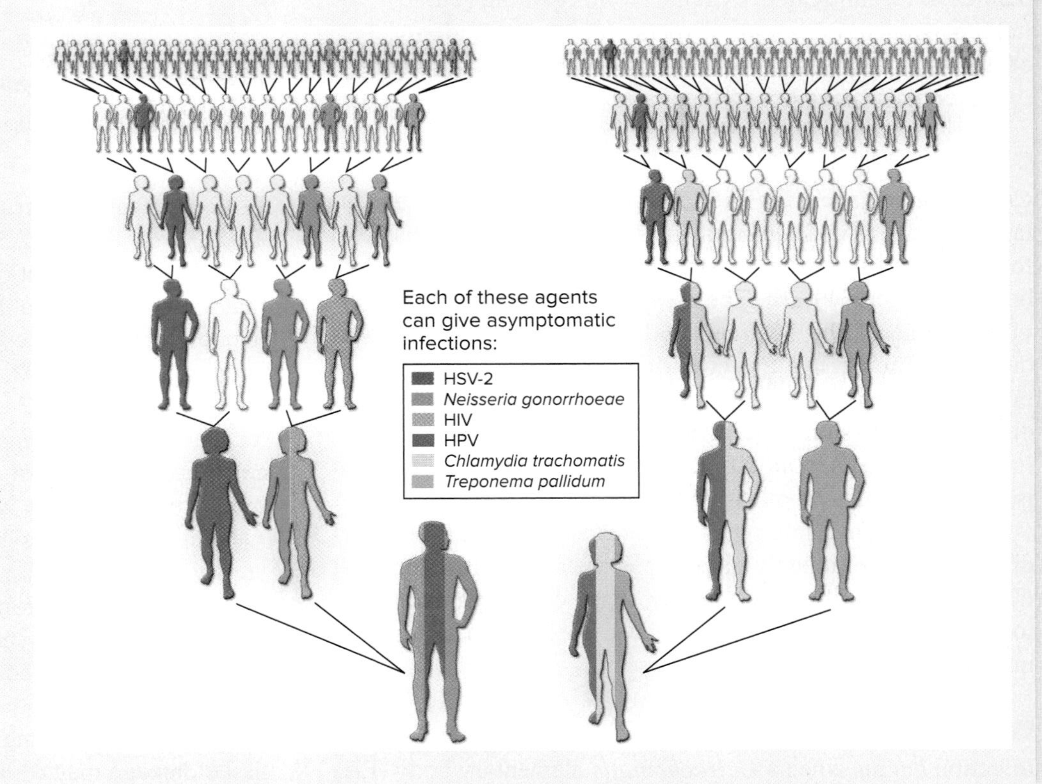

BOX FIGURE 27.1 The Possible Risk of Acquiring STIs in People Having Unprotected Sex Each partner had two previous sexual partners, and each of these partners had two previous partners, and so on. This risk of contracting an STI rises with the number of sexual partners. (HSV-2—herpes simplex virus type 2, HIV—human immunodeficiency virus, HPV—human papillomavirus)

MicroByte

Despite spending billions each year for STI control, an estimated 19 million Americans are infected annually, almost half in people 15 to 24 years old.

Signs and Symptoms

Many chlamydial infections of men and women are asymptomatic—nearly 75% of women and 50% of men do not develop symptoms and can have the disease for months or years before being diagnosed. If symptoms of *Chlamydia trachomatis* infection occur, they generally appear 7 to 14 days after exposure. In men, the main symptom is a thin, gray-white discharge from the urethra, sometimes with painful testes. Some men also develop pain on urination, and fever. Women with symptomatic genital chlamydia most commonly develop an increased vaginal discharge, sometimes with painful urination, abnormal vaginal bleeding, abdominal pain, and pain during sexual intercourse. Chlamydial infection in women can lead to pelvic inflammatory disease (PID). Occasionally the organism can move from the fallopian tubes to the abdominal cavity and infect the surface of the liver.

Babies born to mothers who have symptomatic or asymptomatic *C. trachomatis* infection can become infected during passage through the birth canal. Infection of the eyes leads to neonatal conjunctivitis, manifested by inflammation of the eye surface and a discharge from the eye that is initially watery and then becomes thick with pus. If left untreated, the infection can cause blindness. Lower respiratory infection leads to neonatal pneumonia.

Some strains of *C. trachomatis* can cause lymphogranuloma venereum, a rare disease in which the lymph nodes of the groin swell and drain pus. These enlarged nodes, called buboes, can be quite painful. If left untreated, extreme swelling of the genitalia may occur after several years, due to obstruction of the lymphatic vessels by fibrosis (thickening and scarring) that eventually develops in the nodes.

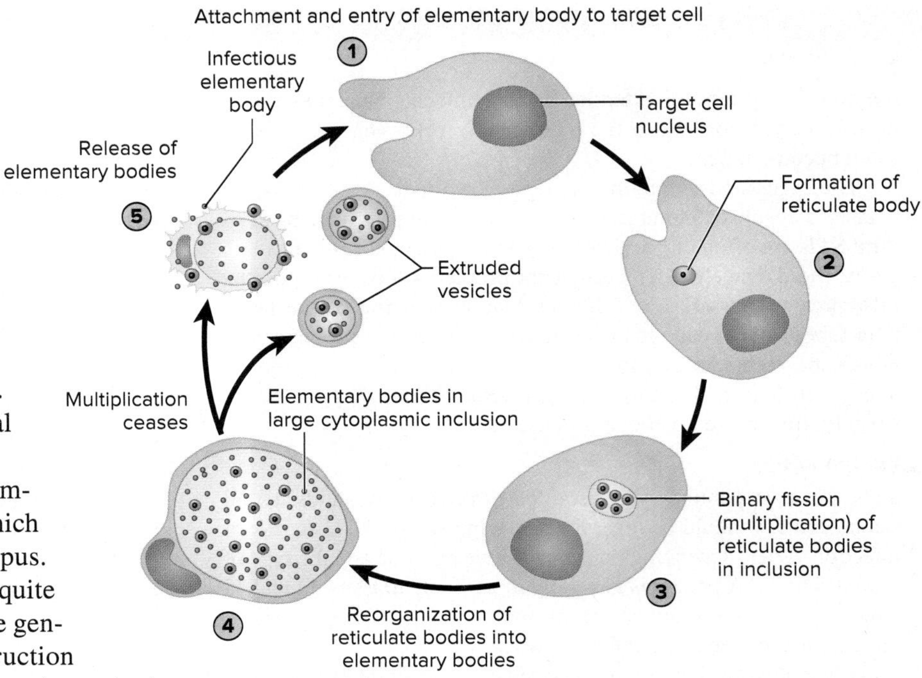

FIGURE 27.7 ***Chlamydia trachomatis*** **Life Cycle**

? How does *C. trachomatis* keep the immune system from destroying infected cells?

Causative Agent

Chlamydia trachomatis is a spherical, obligate intracellular Gram-negative bacterium. It is unusual in that it has two forms in its life cycle: an elementary body and a reticulate body (see figure 11.26). The **elementary body (EB)** can survive outside of the host cell and is the infectious form; its outer membrane has cysteine-rich proteins that cross-link, providing structural stability that enhances survival of the EB during transmission. The **reticulate body (RB)** lives within infected host cells in modified endosomes called inclusions; it is the intracellular, replicating form.

Different antigenic types of *C. trachomatis* cause distinct diseases. Approximately eight types are responsible for most *C. trachomatis* STIs. Three other types cause lymphogranuloma venereum, and four others cause trachoma, a serious eye infection that can lead to blindness.

Pathogenesis

Infection begins when a *C. trachomatis* elementary body (EB) attaches specifically to a mucosal epithelial cell, for example, a cell of the cervix, urethra, rectum, pharynx, or conjunctiva (**figure 27.7** ①). Following attachment, the EB delivers effector proteins into the host cell that induce the cell to endocytose the EB. ② Two important processes occur within an EB-containing endosome. The bacterium directs modifications that create a protective compartment called the inclusion, and the EB differentiates to become a reticulate body (RB). ③ Within the inclusion, the RB multiplies, and ④ the progeny eventually develop to become infectious EBs. ⑤ These are released from the infected cells, either when the host cell bursts, or by extrusion through the host cell membrane. The EBs can then attach to and infect other cells. ◀◀ endocytosis

Chlamydia infections can persist for months, indicating that the organism is able to avoid the host defenses. One way it does this is by preventing infected cells from presenting antigens to T cells. The bacterium synthesizes a protein that interferes with MHC synthesis and MHC-antigen complex formation. *C. trachomatis* also inhibits apoptosis in infected cells, keeping the host cells alive long enough for the RBs to replicate and develop into EBs. ◀◀ MHC

The signs and symptoms of chlamydial infections are largely due to the inflammatory response. Neutrophils recruited to the site of infection release various bacteria-killing substances that lead to tissue damage; further tissue damage results from the cell-mediated immune response. In men, the infection usually involves the urethra. From there, it may spread to the epididymis, causing acute pain and swelling. Repair of the tissue damage may result in scarring, which can lead to infertility. In women, the infection commonly involves the cervix. The organism can also spread into the upper reproductive tract, perhaps attached to sperm, leading to the problems associated with PID. Infection of the uterus results in pain and bleeding; infection of the fallopian tubes may lead to eventual scarring, increasing the risk of ectopic pregnancy or infertility.

Epidemiology

Chlamydial genital infections are the most common of all notifiable bacterial infectious diseases in the United States. The number of reported cases has been rising, probably because of increased awareness of the disease and better diagnostic tests. *Chlamydia trachomatis* can be transmitted by vaginal, oral, or anal sex.

Treatment and Prevention

Several antibiotics can be used to treat chlamydial infections. Azithromycin can be given as a single dose; tetracyclines and erythromycin are less expensive alternatives but require multiple doses. Sexual partners are also treated. Neonatal conjunctivitis and pneumonia are usually treated with oral erythromycin.

Chlamydial infections can be prevented by abstinence, monogamous relationships, and condoms used correctly. All sexually active people are advised to get tested for *Chlamydia* each year, or twice yearly if they have multiple partners or if their partner has multiple partners. The main features of chlamydial genital infections are summarized in **table 27.6.**

Gonorrhea

Gonorrhea is among the commonly reported STIs. It has become increasingly resistant to antimicrobial medications, a characteristic that the CDC considers an "urgent threat" (see table 20.2). The name of the disease originates from the Greek words *gonos* meaning "seed" and *rhoia* meaning "to flow," probably because the pus produced during infection looks like semen.

TABLE 27.6	Chlamydial Reproductive Tract Infections
Signs and symptoms	Frequently asymptomatic. Men: thin, gray-white penile discharge, painful testes. Women: vaginal discharge, vaginal bleeding, abdominal pain
Incubation period	Usually 7 to 14 days
Causative agent	*Chlamydia trachomatis,* an obligate intracellular bacterium, certain serotypes
Pathogenesis	Elementary body (EB) attaches to specific receptors on the epithelial cell, and is engulfed by that cell; becomes reticulate body (RB) in the endosome; replicates and then differentiates into EBs, which infect other cells when released; tissue damage results from inflammation and the cell-mediated immune response to infection; if scar tissue forms in the fallopian tubes, ectopic pregnancy and infertility risk increase.
Epidemiology	The leading reportable bacterial infection in the United States; large numbers of asymptomatic infections.
Treatment and prevention	Treatment: antibacterial medications. Prevention: abstinence, monogamous relationships, correct use of condoms; test sexually active people at least once yearly to rule out asymptomatic infection.

Signs and Symptoms

Gonorrhea infections are often asymptomatic in both men and women. If signs and symptoms occur, they appear after an incubation period of only 2 to 5 days. In men, these include pain during urination and a thick, pus-containing urethral discharge (**figure 27.8**). Gonorrhea follows a different course in women. If present, genital signs and symptoms include painful urination and an abnormal vaginal discharge. The causative agent grows well in the cervix, the fallopian tubes, and other areas of the genital tract, causing inflammation, pain, and possible scarring.

Gonorrhea in either men or women can lead to complications. In men, an inflammatory response to the infection can cause scar tissue formation that partially obstructs the urethra, slowing urination and creating a predisposition to urinary tract infections (UTIs). The infection may spread to the prostate gland and the testes, producing prostatic abscesses and orchitis (inflammation of the testicles). If scar tissue blocks the tubes that carry the sperm, or if testicular tissue is destroyed by the infection, infertility can result. In women, the infection can progress upward through the reproductive tract, causing pelvic inflammatory disease (PID). It is unclear how *Neisseria gonorrhoeae* (which does not have flagella) can move through the uterus to reach the fallopian tubes, but the organisms might be carried on sperm. Occasionally, the bacteria

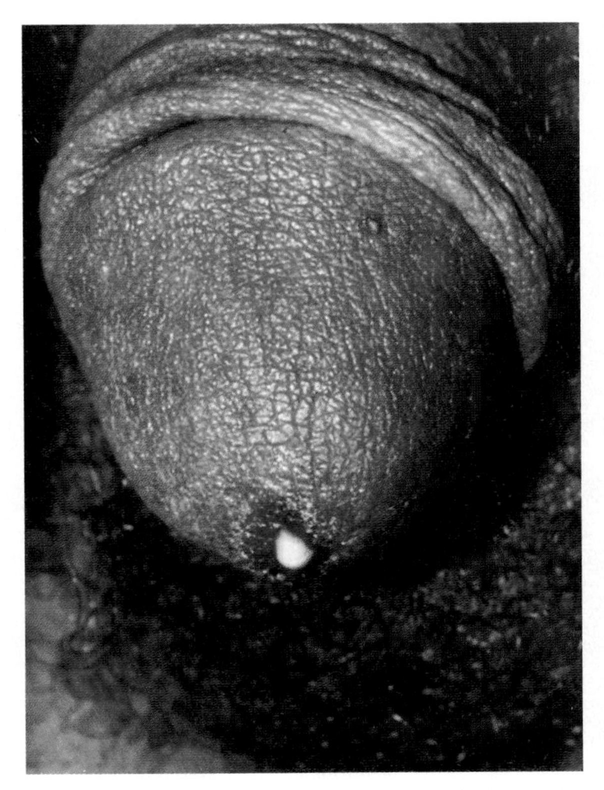

FIGURE 27.8 Urethral Discharge in a Man with Gonorrhea While symptoms of gonorrhea in men are noticeable when present, the disease is often asymptomatic. Source: CDC

? **What is the risk of asymptomatic gonorrhea in men and women?**

spread from the fallopian tubes into the abdominal cavity, where they can affect the liver or other abdominal organs.

Certain strains of *N. gonorrhoeae* can produce **disseminated gonococcal infection (DGI).** The systemic infection is characterized by fever, rash, and arthritis caused by growth of the pathogen within the joint spaces. It can also lead to infective endocarditis and meningitis. DGI is not usually preceded by urogenital symptoms. ⏮ infective endocarditis, ⏮ meningitis

Babies born to mothers who have symptomatic or asymptomatic gonorrhea can develop neonatal conjunctivitis (also called ophthalmia neonatorum). The bacteria are transmitted during the infant's passage through the infected birth canal, and the resulting signs and symptoms are similar to those of neonatal chlamydial conjunctivitis.

Causative Agent

Gonorrhea is caused by *Neisseria gonorrhoeae*—commonly called gonococcus (GC)—a fastidious Gram-negative diplococcus that requires a rich medium such as chocolate agar for cultivation (**figure 27.9**). The organism produces pili (fimbriae) for attachment to host cells (**figure 27.10**). The outer membrane of GC has the general structure of a typical Gram-negative outer membrane, but it contains lipooligosaccharide (LOS) instead of lipopolysaccharide (LPS). LOS plays a role in virulence and pathogenicity of GC. ⏮ fastidious, ⏮ chocolate agar

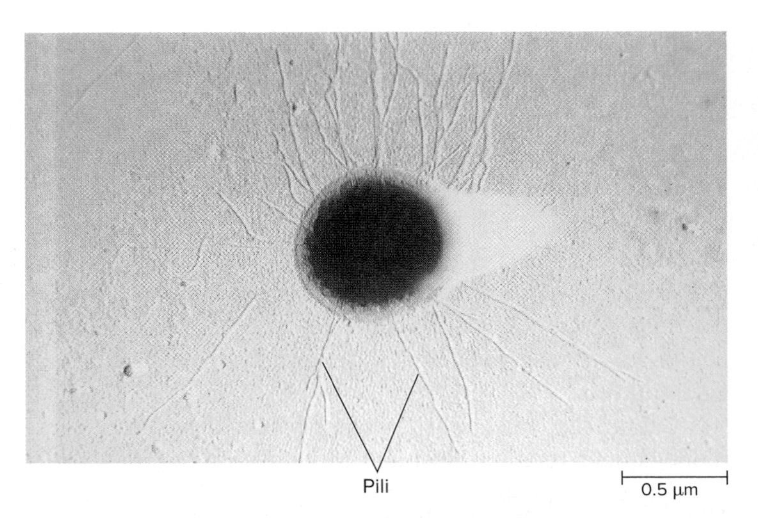

FIGURE 27.10 Pili on Single Coccus of *Neisseria gonorrhoeae* (EM) ©Stanley Falkow

? **What is the function of the pili?**

Pathogenesis

Neisseria gonorrhoeae (gonococcus, or GC) is well adapted to grow within the human host. Like other STI agents, it is a human-specific pathogen that survives poorly in the environment and therefore must be able to avoid the host defenses in order to survive.

Infection begins when GC attaches by means of its pili to receptors on non-ciliated columnar mucosal cells, including those of the urethra, cervix, mouth, pharynx, rectum, and conjunctiva. The organism cannot attach to squamous epithelium, such as that lining the adult vagina, or to ciliated cells. Once attached to mucosal epithelia, some of the bacterial cells grow on the surface, but others direct the host cells to engulf them. The intracellular bacteria can then multiply within the protected environment of the epithelial cell or be released to the other side of the mucosal barrier (the submucosa).

The symptoms of gonorrhea are primarily due to the intense inflammatory response that results from infection. As the bacterial cells grow, they release outer membrane vesicles (called blebs) that contain LOS and peptidoglycan. These compounds are recognized by the host's pattern recognition receptors (PRRs), leading to the release of pro-inflammatory cytokines. The complement system is also activated. Neutrophils are recruited to the site of infection as a result, and the microbe-destroying substances they release contribute to the tissue damage. In addition, peptidoglycan fragments released from gonococcal cells are toxic to nearby ciliated mucosal epithelial cells. ⏮ PRRs, ⏮ complement system

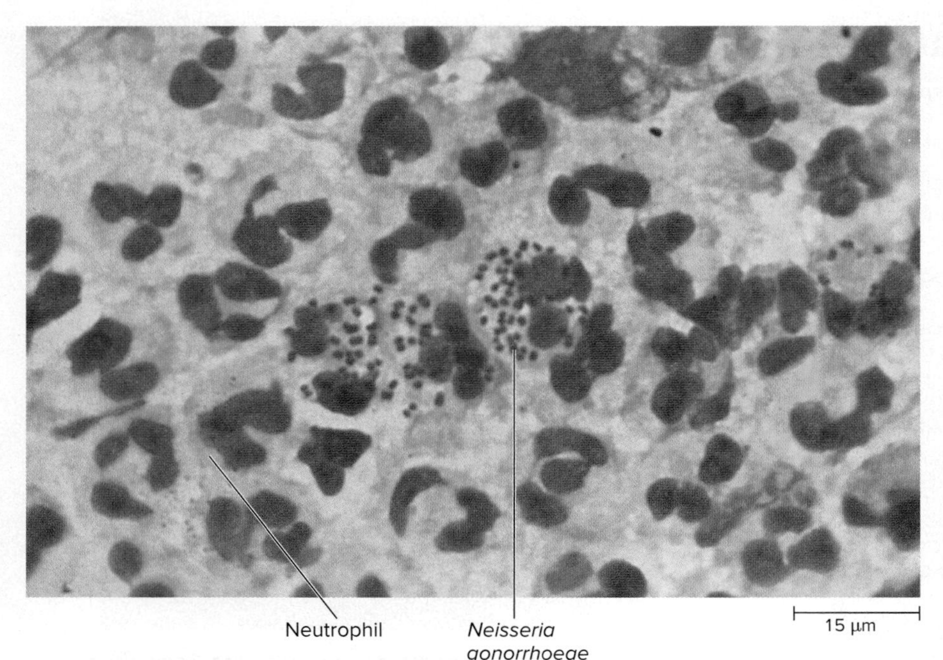

FIGURE 27.9 *Neisseria gonorrhoeae* Stained smear of pus from the urethra showing *N. gonorrhoeae*. Source: Bill Schwartz/CDC

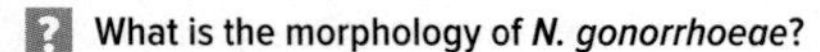

? **What is the morphology of *N. gonorrhoeae*?**

GC persists and multiplies in the host, using several mechanisms to avoid the immune response. The bacterial cells prevent efficient phagocytosis by binding to the host's complement regulatory proteins; recall that these proteins normally protect host cells by inactivating C3b that could otherwise opsonize those cells (see figure 14.13). Antigenic and phase variation of three important antigens—pili, a membrane protein called Opa (for opacity), and the carbohydrate portion of LOS—allow the bacterial cells to avoid being tagged by antibodies. By the time antibodies are produced against one version of the antigens, some gonococcal cells will have begun producing alternative versions. This genetic plasticity is not only important to GC for evading the immune response; it also allows the bacterial cell population to survive different host environments. For example, cells expressing certain versions of LOS are better able to bind sialic acid, a host cell compound that helps disguise them from the immune system, whereas others make it easier for GC to attach to host cells. ⏮ complement C3b, ⏮ opsonization, ⏮ antigenic variation

Gonococci have a variety of other significant features as well. Although the bacteria have several ways to avoid phagocytosis, they also appear to stimulate their own uptake by neutrophils. Once engulfed, many of the bacteria are killed, but some survive and even seem to multiply within neutrophils—an impressive characteristic considering that neutrophils are the most potent of the body's phagocytic cells. Several outer membrane proteins allow gonococci to attach to helper T cells, preventing the activation and proliferation of those important cells. The bacteria also produce IgA protease, which destroys secretory IgA and seems to interfere with phagosome maturation. In addition, gonococci produce membrane proteins that "steal" iron directly from the host's storage forms (transferrin and lactoferrin). Also, gonococcal cells secrete DNA and are naturally competent, which means that they continually exchange genetic information through DNA-mediated transformation. Some strains carry conjugative plasmids, providing another mechanism of horizontal gene transfer. ⏮ IgA protease, ⏮ competence, ⏮ DNA-mediated transformation, ⏮ conjugative plasmids

Epidemiology

Gonorrhea is among the most common of the STIs; in the United States, its incidence is the highest of any reportable bacterial disease other than chlamydial infection. The number of cases has declined since the emergence of HIV/AIDS, however, probably due to increased condom use.

N. gonorrhoeae infects only humans, living mainly on the mucous membranes of the host. Most strains are susceptible to UV light, cold, and desiccation and so do not survive well outside the host. For this reason, gonorrhea is transmitted almost exclusively by direct contact. Because the bacteria mainly live in the genital tract, this contact is almost always sexual— gonorrhea can be transmitted by vaginal, oral, or anal sex. It can also be passed from mother to baby during vaginal childbirth. Factors that influence the incidence of gonorrhea include:

- **Birth control pills.** Oral contraceptives without use of a condom offer no protection against STIs and may increase susceptibility to them. This is partly because the hormones in oral contraceptives cause epithelial cells to migrate from the cervical lumen onto more exposed areas of the outer cervix. Oral contraceptives also tend to increase both the pH and the moisture content of the vagina, favoring infection with GC and other agents of STIs. Besides being more susceptible to gonorrhea, women taking oral contraceptives are also more likely to develop serious complications from the disease.
- **Asymptomatic infection.** Male and females who have asymptomatic gonorrhea can unknowingly transmit the infection over months or even years. Studies show that up to 20% of men and 60% of women with gonorrhea are asymptomatic.
- **Lack of immunity.** There is little or no immunity following recovery from the disease. A person can contract gonorrhea repeatedly.

Treatment and Prevention

N. gonorrhoeae strains are now frequently resistant to many antibacterial medications, including penicillins, tetracyclines, sulfonamides, and fluoroquinolones. Because of this, the CDC listed drug-resistant *N. gonorrhoeae* as an urgent threat (see table 20.2). To prevent further development and spread of resistance, combination therapy of ceftriaxone (administered by injection) plus azithromycin is recommended.

Prevention of gonorrhea depends on abstinence, monogamous relationships, and consistent, correct use of condoms. Rapid identification and treatment of sexual contacts help prevent spread of the disease by decreasing the overall number of infected individuals. No vaccine for preventing gonorrhea has been developed due to the difficulties posed by GC's antigenic variation and the lack of a suitable animal model for studying the immune response to the disease.

Neonatal gonococcal conjunctivitis is prevented by putting an antibiotic ointment such as erythromycin directly into the eyes of all newborn infants within 1 hour of birth. Some mothers who insist that they do not have gonorrhea have challenged this practice, but because the disease is frequently asymptomatic and can cause blindness in infants, this prophylactic treatment of a baby's eyes is required by law. As a result, the disease is now unusual in the United States. **Table 27.7** describes the main features of the disease.

Mycoplasma genitalium Infections

Mycoplasma genitalium is an emerging cause of genital tract infections in men and women worldwide. The organism was discovered

TABLE 27.7 Gonorrhea

1. *Neisseria gonorrhoeae* attaches to mucous membranes of the genitalia, mouth, anus, rectum, or eyes; serious infections leading to loss of vision are a risk in newborns.
2. The cervix is the usual site of primary infection in women.
3. The outer covering of the liver is infected when gonococci enter the abdominal cavity from infected fallopian tubes.
4. Prostatic gonococcal abscesses may be difficult to eliminate.
5. Infection of the fallopian tubes results in scarring, which can cause infertility or ectopic pregnancy.
6. Organisms carried by the bloodstream infect the heart valves and joints.
7. Urethral scarring from gonococcal infection can predispose to urinary infections by other organisms.
8. Scarring of testicular tubules can cause infertility.

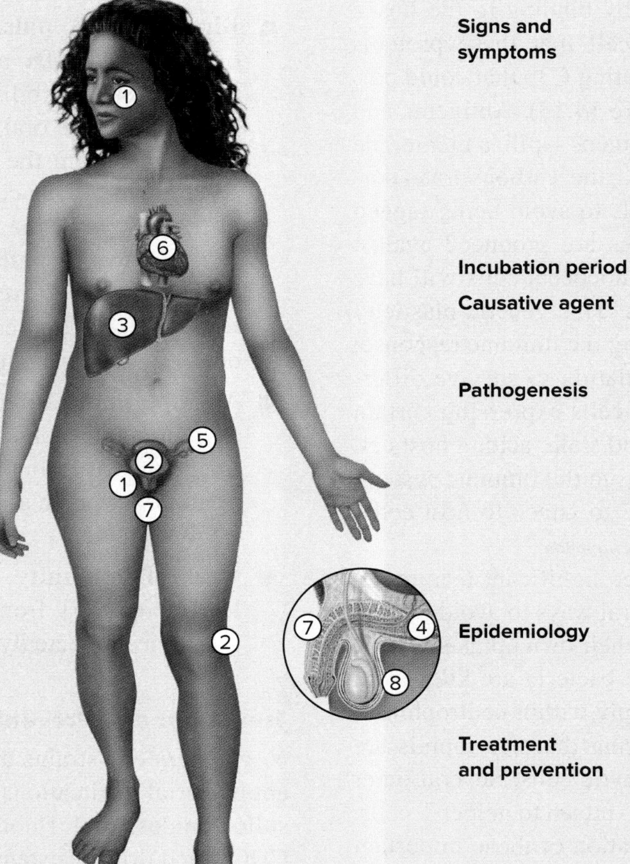

Signs and symptoms	Frequently asymptomatic. Men: pain on urination, urethral discharge; complications may include impaired urinary flow, infertility, and arthritis. Women: abnormal vaginal discharge, pain on urination; complications can include pelvic pain, infertility, ectopic pregnancy, and arthritis.
Incubation period	2 to 5 days
Causative agent	*Neisseria gonorrhoeae* (also called gonococcus or GC), a Gram-negative diplococcus
Pathogenesis	Organisms attach to certain non-ciliated epithelial cells by pili; phase and antigenic variation in surface proteins and pili allows attachment to different host cells and escape from immune mechanisms; inflammation, scarring; can spread by bloodstream.
Epidemiology	Transmitted by sexual contact; asymptomatic infections are common; no immunity following recovery.
Treatment and prevention	Treatment: combination of antibiotics; widespread resistance limits medication options. Prevention: abstinence, monogamous relationships, condoms, early treatment of sexual contacts.

and identified only in the early 1980s, partly because it cannot be detected with the Gram stain, but also because it is very difficult to culture. Little was known about its disease association until nucleic acid amplification tests (NAATs) were developed for its detection. Currently, none of these DNA- and RNA-based methods are FDA-approved for use in *M. genitalium* diagnosis, so they are primarily used for research. ⏮ NAAT

Signs and Symptoms

Like many STIs, *M. genitalium* infections are frequently asymptomatic. The incubation time is not well established, but when signs and symptoms do occur, they are similar to those of chlamydia and gonorrhea. In men, they include urethral discharge and pain during urination. In women, infection causes cervicitis (inflammation of the cervix) and vaginal discharge. In addition, the infection can result in complications, including PID in women and epididymitis in men.

Causative Agent

Mycoplasma genitalium, like all mycoplasmas, does not have a cell wall. The organism has a very small genome. As a result, it is very fastidious, and obtains many of its required nutrients from the host. It grows very slowly, with a generation time of at least 16 hours, and can take several months to culture in vitro. Even with that small genome, however, the organism produces a complex cytoplasmic projection called a terminal organelle at one end of the cell. The structure plays important roles in pathogenesis and gives the cells a characteristic bottle shape (**figure 27.11**).

Epidemiology

Transmission of *M. genitalium* is through direct mucosal contact, and occurs during unprotected sex. The infections may be more common than gonorrhea and chlamydia, but they are often undiagnosed because (1) no widely available detection methods are available; (2) infections are often asymptomatic; and (3) when signs and symptoms do occur, they can be

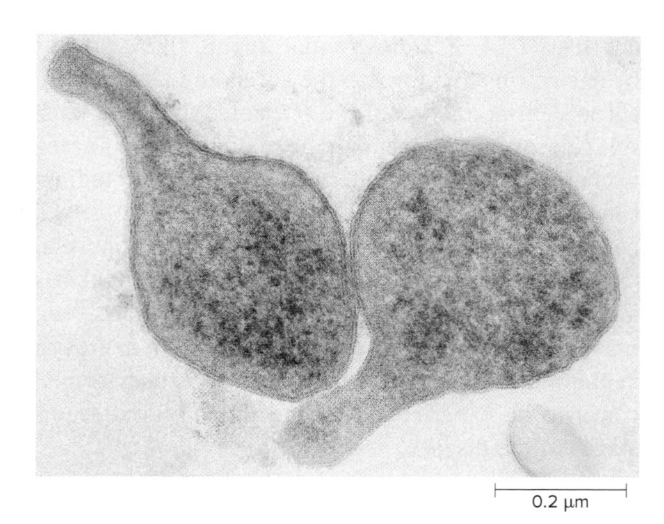

FIGURE 27.11 ***Mycoplasma genitalium*** Notice the distinctive shape caused by the terminal organelle. ©Thomas Deerinck, NCMIR/Science Source

Give two functions of the terminal organelle.

similar to those of other bacterial STIs. Unlike gonorrhea and chlamydia, *M. genitalium* infections are not nationally notifiable to the CDC.

Pathogenesis

The terminal organelle of *M. genitalium* cells is used for both a unique type of gliding movement and attachment to eukaryotic cells. Adhesins at the tips of the organelle are involved in attachment, and the genes for these can recombine with partially homologous nucleotide stretches in the chromosome, leading to antigenic variation. Many of the signs and symptoms of infection are caused by the host's inflammatory response.

Treatment and Prevention

Suspected *M. genitalium* infections are treated with antibiotics. The organism lacks a cell wall, so it has intrinsic resistance to peptidoglycan synthesis inhibitors such as penicillin. Macrolides such as azithromycin are generally the most effective for treating the infection, but resistance of *M. genitalium* to azithromycin is becoming more widespread. Moxifloxacin can also be used as an alternative if azithromycin treatment fails. Both sexual partners are treated. As with other STIs, prevention is by abstinence, monogamy, or use of condoms during intercourse. **Table 27.8** describes the main features of the disease. ⏮ intrinsic resistance

TABLE 27.8 ***Mycoplasma genitalium* Infections**

Signs and symptoms	Frequently asymptomatic. Men: urethral discharge, painful urination. Women: cervicitis, vaginal discharge.
Incubation period	Not well established.
Causative agent	*Mycoplasma genitalium,* a fastidious bacterium that grows very slowly in culture; has distinctive "flask" shape.
Pathogenesis	Causative agent attaches to host cells by adhesins in terminal organelle. Signs and symptoms result from host's inflammatory response.
Epidemiology	Transmission of *M. genitalium* is through direct mucosal contact, and occurs during unprotected sex. Infection increases a person's risk of contracting and transmitting HIV.
Treatment and prevention	Treatment: antibacterial medications. Prevention: abstinence, monogamous relationships, correct use of condoms.

Syphilis

During the first half of the twentieth century, syphilis was a major cause of mental illness and blindness, and a significant cause of heart disease and stroke. The disease was almost eradicated by the mid-1950s through a successful program aimed at locating and treating all people with syphilis and their sexual contacts. Unfortunately, the number of cases then increased in the 1990s due to factors including prostitution and drug use. This trend was reversed through renewed efforts in education, case finding, and treatment. By the end of 1998, the syphilis rate was at the lowest level since it became a notifiable disease, but rates have risen since then.

Signs and Symptoms

Syphilis causes so many different signs and symptoms that it is easily confused with other diseases and is often called "the great imitator." Its symptoms occur in defined clinical stages.

- **Primary syphilis.** The stage is characterized by a painless red ulcer called a hard **chancre** (pronounced "shanker") that appears at the site of infection about 3 weeks after exposure (**figure 27.12**). Chancres usually develop on

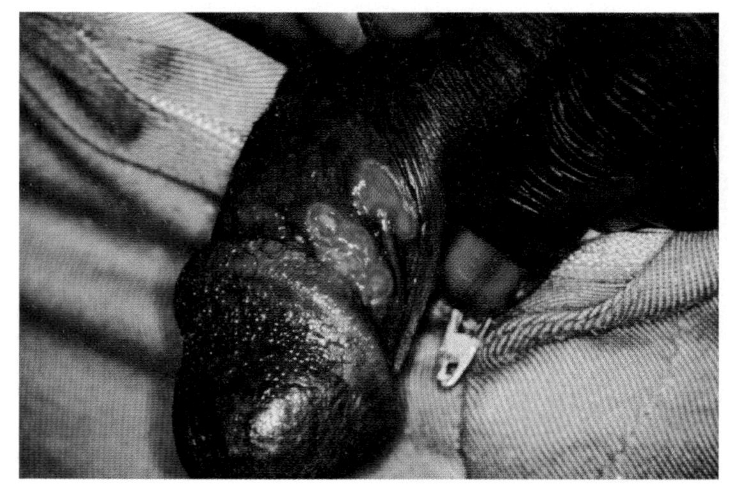

FIGURE 27.12 Syphilitic Chancre A chancre develops at the site where *Treponema pallidum* entered the body, as on the shaft of this penis. Source: M. Rein, VD/CDC

What is the risk associated with an open ulcer, as shown here?

the genitalia, but they may occur anywhere on the body. Because they are painless, the chancres often go unnoticed, particularly if they are very small or are hidden from view in the vagina or the rectum. Lymph nodes near the chancre may swell.

- **Secondary syphilis.** The symptoms of secondary syphilis usually appear 2 to 10 weeks after the primary stage. The infection has spread systemically by this time and causes many diverse signs and symptoms, the most common of which is a rash that includes the palms and soles, and white patches on the mucous membranes (**figure 27.13**). Other signs and symptoms include runny nose and watery eyes, aches and pains, sore throat, fever, malaise, weight loss, and headache. Occasionally, hepatitis (liver disease) and renal disease occur. The person is most infectious at this stage.
- **Latent syphilis.** As the term implies, there are no signs or symptoms during this stage; however, infected people have antibodies to the causative agent. A quarter of people in this stage recover; another quarter remain in this stage. The rest progress to the next stage of the disease.
- **Tertiary syphilis.** After a latent period that can last for many years, signs and symptoms of tertiary syphilis sometimes occur, and these can be grouped into three categories. In gummatous syphilis, localized areas of tissue damage develop as a result of prolonged inflammatory responses. These lesions, called **gummas,** are granulomas similar to the tubercles of tuberculosis. Gummas are chronic and may occur anywhere in the body (**figure 27.14**). Cardiovascular syphilis is characterized by aneurysms that form in the ascending aorta, caused by chronic inflammation of the arterioles supplying this vessel. Neurosyphilis occurs in a small number of cases and is characterized by a pattern of symptoms including personality change, emotional instability, delusions, hallucinations, memory loss, impaired judgment, abnormalities of the pupils of the eye, and speech defects. Other signs and symptoms of this stage include blindness, vertigo, insomnia, stroke, and meningitis. Although neurosyphilis symptoms are usually seen during the tertiary stage, they can also be experienced during the first and secondary stages of the disease. ◀◀ tubercles

In pregnant women, the bacterium that causes syphilis easily crosses the placenta and infects the fetus, causing **congenital syphilis.** Fetal infections can occur in the absence of any signs or symptoms of syphilis in the mother and at any stage of pregnancy, but damage to the fetus does not generally happen until the fourth month. Common outcomes of fetal infection include spontaneous abortion, stillbirth, and neonatal death. Almost all infected infants that survive birth develop distinctive signs and symptoms. Early on, those resemble the characteristics of severe adult secondary syphilis and include runny nose, rash, and liver and spleen enlargement. Other signs occur after 2 years. The bones, cartilage, and teeth can become deformed as the child develops, causing cleft palate, sabre shins (bent tibias), saddle nose (sunken nasal bridge), mulberry molars (many cusps), and Hutchinson teeth (notched incisors) (**figure 27.15**). Eye inflammation and deafness often occur.

Causative Agent

Syphilis is caused by *Treponema pallidum,* a very slender, highly motile spirochete that cannot be seen using the

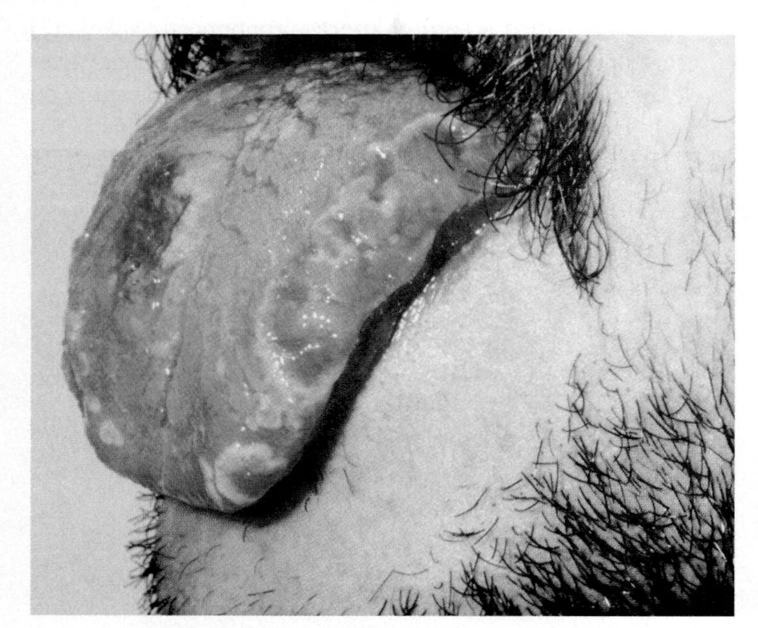

FIGURE 27.13 Secondary Syphilis Secondary syphilis lesions such as these oral mucous patches contain numerous *Treponema pallidum* cells and are highly infectious. ©Mediscan/Alamy Stock Photo

? **What causes the manifestations of secondary syphilis?**

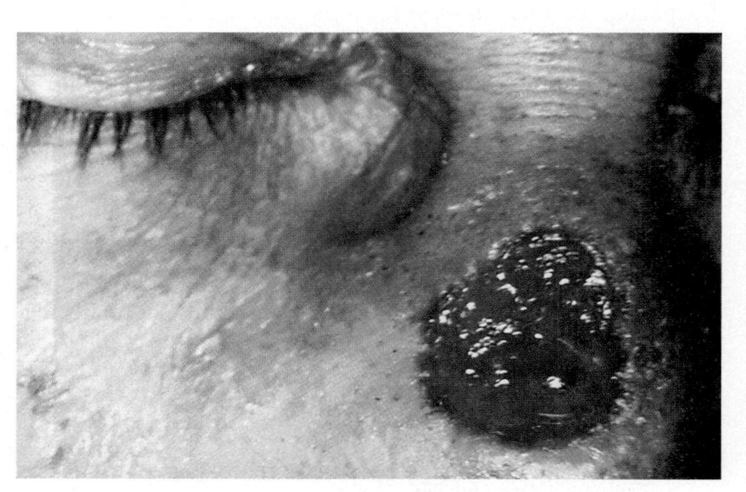

FIGURE 27.14 A Gumma of Tertiary Syphilis Gummas are formed by an inflammatory mass that can perforate tissue, as in this example. Source: J. Pledger/CDC

? **Where in the body do gummas occur?**

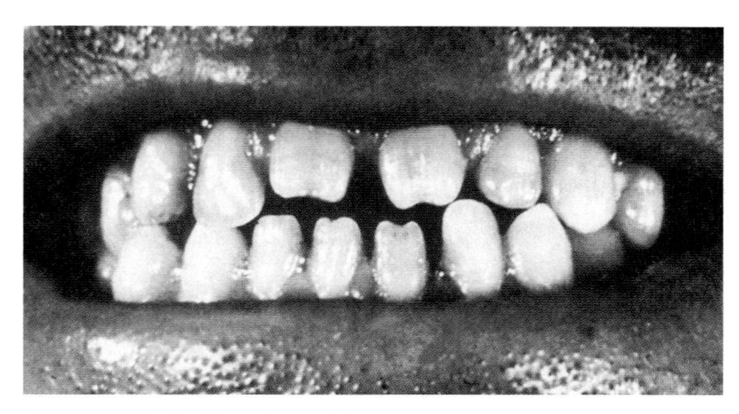

FIGURE 27.15 Hutchinson Teeth Notice the notched, deformed incisors, a late manifestation of congenital syphilis. Source: Robert Sumpter/CDC

? What are some other manifestations of congenital syphilis?

Gram stain. Instead, the organism is visualized by dark-field microscopy, immunofluorescence, or a special method used for tissue staining called a silver stain (**figure 27.16**). Like all spirochetes, *T. pallidum* cells have endoflagella that propel them in a corkscrew-like motion, and this allows them to penetrate a wide variety of tissues and organs. In nature, *T. pallidum* infects only humans. ⏮ dark-field microscopy

T. pallidum must be grown in the testicles of laboratory rabbits to study its cell division and pathogenesis, because it does not replicate in vitro. The organism lacks certain metabolic abilities and does not have enzymes needed for the TCA cycle or electron transport chain, so cannot generate much ATP. It is therefore thought to get most of its essential macromolecules from the host. ⏮ TCA cycle

Pathogenesis

Treponema pallidum easily penetrates mucous membranes and damaged skin. The infectious dose is very low—less than 100 cells. In primary syphilis, *T. pallidum* multiplies in a localized area of the genitalia, spreading from there to the lymph nodes and the bloodstream. The hard chancre is caused by an intense inflammatory response to the bacteria, which are present in high numbers in the lesion. The chancre disappears within 2 to 6 weeks, even without treatment, and patients may mistakenly believe that they have recovered from the disease. However, the organism avoids destruction by the body's defenses, and progression of the disease can continue for years. Few virulence factors have been discovered for *T. pallidum,* and the outer membrane of the cells lacks LPS, the molecule that normally triggers an immune response. There are also few proteins in the outer membrane and therefore few targets for opsonizing antibodies to bind.

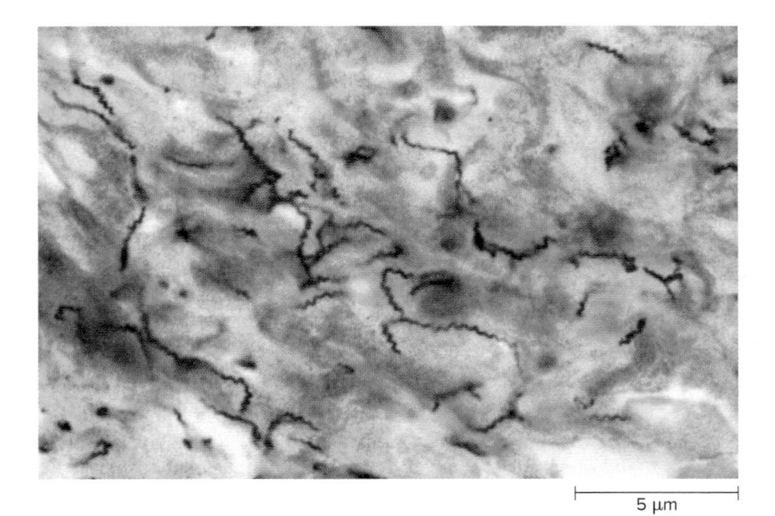

FIGURE 27.16 *Treponema pallidum* Silver-stained *T. pallidum* in rabbit testes tissue. ©Michael Abbey/Science Source

? In wet mounts, why is dark-field microscopy needed to observe *T. pallidum*?

Many of the signs and symptoms of secondary syphilis are due to immune complexes that form as specific antibodies bind to circulating *T. pallidum.* By this time, the spirochetes have become systemic, and infectious lesions occur on the skin and mucous membranes in various locations, especially in the mouth. The secondary stage lasts for weeks to months, sometimes as long as a year, and then gradually subsides. About 50% of untreated cases never progress past the secondary stage. After a latent period of from 5 to 20 years or even longer, however, some people with the disease develop tertiary syphilis. ⏮ immune complexes

Tertiary syphilis signs and symptoms occur as a result of hypersensitivity reactions to small numbers of *T. pallidum* that grow and persist in the tissues. In this stage, the patient is no longer infectious. The organisms may be present in almost any part of the body, and the outcome depends on where the hypersensitivity reactions occur. If they occur in the skin, the bones, or other areas not vital to existence, the disease is not life-threatening. If, however, they occur within the walls of a major blood vessel such as the aorta, the vessel may become weakened and even rupture, resulting in death. The main characteristics of the stages of syphilis are summarized in **table 27.9.**

TABLE 27.9 Stages of Syphilis

Stage of Disease	Main Characteristics	Infectious?
Primary	Firm, painless ulcer (hard chancre) at site of infection; lymph node enlargement	Yes
Secondary	Rash, aches, and pains; rash and mucous membrane lesions	Yes
Latent	No signs or symptoms	Early—yes; Late—no
Tertiary	Gummas; damage to large blood vessels, eyes, nervous system; mental illness	No

Epidemiology

Syphilis is usually transmitted by sexual intercourse. However, infection can occur from kissing a person with secondary syphilis, or by contact with a primary ulcer infected with *T. pallidum.*

Treatment and Prevention

Primary and secondary syphilis are easily treated with an antibiotic such as penicillin—the dose and drug type depend on the stage of the disease. Treatment is more difficult for tertiary syphilis, because most of the organisms are not actively multiplying and penicillin acts against dividing cells. As a result, large doses of intravenous penicillin are given. Treatment of tertiary syphilis limits progression of the disease but has little effect on tissue damage that has already occurred.

There is no vaccine for syphilis. Abstinence, monogamous relationships, and correct use of condoms decrease the risk of contracting the disease. Congenital syphilis can be prevented by diagnosing and treating the mother's syphilis before the fourth month of pregnancy, when *T. pallidum* starts to affect the fetus. Quick identification and treatment of sexual contacts are important in limiting spread of the disease. **Table 27.10** summarizes the main features of syphilis.

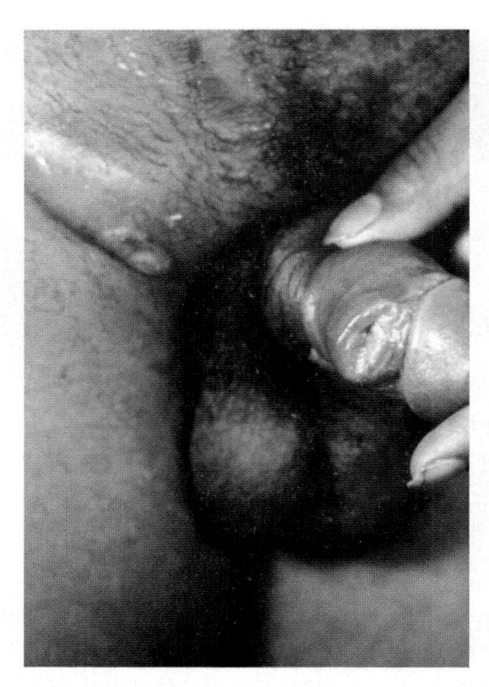

FIGURE 27.17 Chancroid A genital ulcer and a ruptured lymph node can be seen in this patient. Source: Dr. Pirozzi/CDC

? How are chancroid ulcers different from chancres?

Chancroid

Chancroid is another bacterial STI that, like syphilis, causes genital sores, thereby increasing the risk of contracting HIV. The disease is common in developing countries, usually associated with commercial sex workers. The reported incidence of chancroid in the United States is very low, but it often goes unreported or misdiagnosed and may therefore be more common than recorded.

Signs and Symptoms

Chancroid is characterized by one or more painful genital sores called soft chancres (**figure 27.17**). Typically, these begin as a small pimple at the site of bacterial entry through the skin,

TABLE 27.10 Syphilis

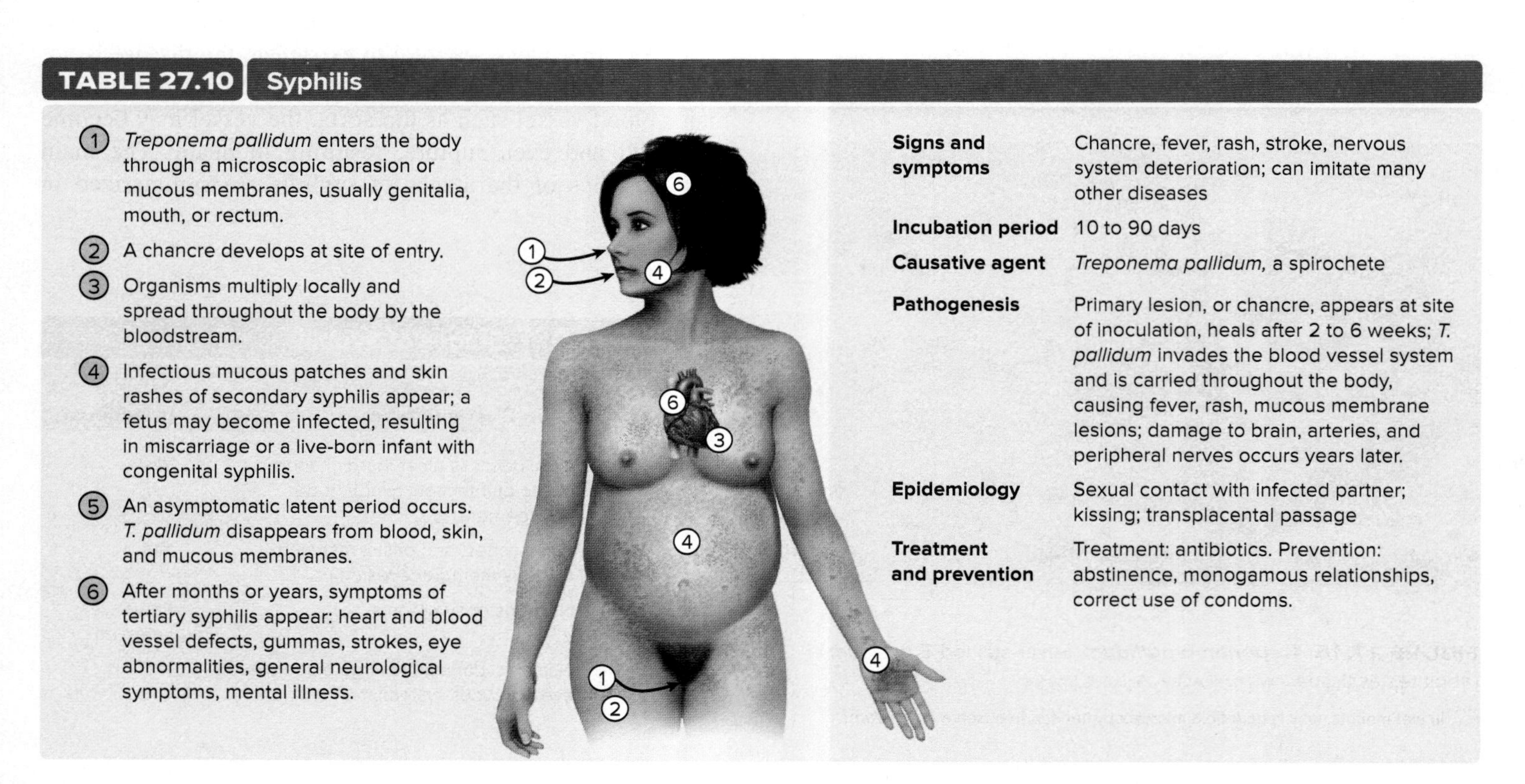

1. *Treponema pallidum* enters the body through a microscopic abrasion or mucous membranes, usually genitalia, mouth, or rectum.
2. A chancre develops at site of entry.
3. Organisms multiply locally and spread throughout the body by the bloodstream.
4. Infectious mucous patches and skin rashes of secondary syphilis appear; a fetus may become infected, resulting in miscarriage or a live-born infant with congenital syphilis.
5. An asymptomatic latent period occurs. *T. pallidum* disappears from blood, skin, and mucous membranes.
6. After months or years, symptoms of tertiary syphilis appear: heart and blood vessel defects, gummas, strokes, eye abnormalities, general neurological symptoms, mental illness.

Signs and symptoms	Chancre, fever, rash, stroke, nervous system deterioration; can imitate many other diseases
Incubation period	10 to 90 days
Causative agent	*Treponema pallidum,* a spirochete
Pathogenesis	Primary lesion, or chancre, appears at site of inoculation, heals after 2 to 6 weeks; *T. pallidum* invades the blood vessel system and is carried throughout the body, causing fever, rash, mucous membrane lesions; damage to brain, arteries, and peripheral nerves occurs years later.
Epidemiology	Sexual contact with infected partner; kissing; transplacental passage
Treatment and prevention	Treatment: antibiotics. Prevention: abstinence, monogamous relationships, correct use of condoms.

FOCUS YOUR PERSPECTIVE 27.1

The Death of Syphilis?

For 40 years (1932 to 1972), the U.S. Public Health Service conducted a study (now known as the Tuskegee Syphilis Experiment) of 399 Alabama black men with advanced syphilis. The study was done to assess the natural progression of the disease in black men. The men were not told their diagnosis nor educated about the disease and its transmission. For the first half of the study, there were no effective drugs against syphilis, and so the men remained untreated. However, even after effective therapy became available, the study was continued, and treatment was withheld from the men, in a blatant mix of bad science and racism. Twenty-eight of the men died of the disease, and another hundred died from its complications. Forty of their wives became infected, and 19 children were born with congenital syphilis. Nothing of scientific value was learned, and many people to this day remain bitter and distrustful of the U.S. Public Health Service. In 1997, President Clinton, speaking for the government, formally apologized to the surviving eight subjects.

Fortunately, in more recent years, major governmental efforts to conquer the disease have met with some success. In 2015, 23,872 cases of primary and secondary syphilis were reported, down from 50,223 in 1990. The Centers for Disease Control and Prevention (CDC) has even considered the possibility that transmission of the disease could be eliminated in the United States.

Syphilis is a good target for eradication because it has no animal reservoir, infectious cases can generally be treated with a single injection of penicillin, and the incubation period is long enough that sexual contacts of an infected person can be found and treated before they spread the disease any farther. Moreover, instead of being widely endemic, most syphilis cases are now concentrated in a relatively few areas and certain populations. Also, new techniques are now available for rapid diagnosis, treatment, and epidemiological tracing of the disease. Finally, control of syphilis is closely linked to AIDS control efforts because the risk of HIV transmission is six times as great if either sex partner has syphilis.

but they then ulcerate and quickly get bigger. The lymph nodes in the groin also enlarge and become tender (figure 27.17). Sometimes the nodes are pus-filled and may rupture, releasing the pus. In women, chancres typically occur on the labia or the thighs, causing pain during urination and sexual intercourse.

Causative Agent

Chancroid is caused by *Haemophilus ducreyi,* a fastidious, pleomorphic, Gram-negative coccobacillus that can be cultivated only on a rich medium such as chocolate agar. ◀◀ *Haemophilus*

Pathogenesis

Haemophilus ducreyi infection causes an intense inflammatory response, with accumulation of neutrophils and macrophages in the area. The bacteria survive by producing proteins that prevent phagocytes from engulfing them. In addition, proteins in the outer membrane protect the cells from activated complement proteins. The mechanism of these protective proteins is still unknown.

Epidemiology

Outbreaks in American cities associated with prostitution have sometimes occurred. The disease is now relatively rare in this country, with only a few reported cases every year.

Treatment and Prevention

Chancroid can often be treated with a single oral dose of azithromycin or a single injection of ceftriaxone, but multidose options of other medications may also be used. Effectiveness of therapy is significantly reduced if the patient has AIDS.

Chancroid can be prevented by abstinence from sexual intercourse, monogamous relationships, and correct use of condoms. **Table 27.11** gives the main features of chancroid.

TABLE 27.11 Chancroid

Signs and symptoms	One or more painful, gradually enlarging, soft chancres on or near the genitalia; large, tender regional lymph nodes
Incubation period	3 to 10 days
Causative agent	*Haemophilus ducreyi,* a pleomorphic, fastidious Gram-negative rod requiring special growth media
Pathogenesis	A small pimple appears first, which ulcerates and gradually enlarges; multiple lesions may join together; lymph nodes enlarge, liquefy, and may discharge to the skin surface.
Epidemiology	Sexual transmission; common in sex workers.
Treatment and prevention	Treatment: several antibacterial medications effective; resistance can be a problem. Prevention: abstinence, monogamous relationships; correct use of condoms.

MicroAssessment 27.4

Transmission of bacterial STIs usually requires direct person-to-person contact. Even asymptomatic infections can cause genital tract damage (including PID) and can be spread to other people. Bacterial STIs in pregnant women can be a serious risk to fetuses and newborn babies. Chlamydia, gonorrhea, and *Mycoplasma genitalium* infections are frequently asymptomatic but can cause similar signs and symptoms, including genital discharge and painful urination. Syphilis can imitate other diseases; untreated, it may lead to stroke and nervous system deterioration. Chancroid is associated mainly with sex workers.

10. Can a baby have congenital syphilis without its mother ever having had signs and symptoms of syphilis? Explain.

11. Why is it difficult to diagnose *Mycoplasma genitalium* infections?

12. Why does scarring of a fallopian tube increase the risk of an ectopic pregnancy?

27.5 ■ Viral STIs

Learning Outcomes

7. Compare the infections caused by genital herpes simplex virus and by human papillomavirus.
8. Outline the relationship between HIV infection and AIDS.

Viral STIs are probably at least as common as or more common than the bacterial diseases, and they are incurable. Their effects can be severe and long-lasting. Genital herpes can cause recurrent signs and symptoms for years, human papillomavirus infections can lead to cancer, and untreated HIV infections usually result in AIDS.

Genital Herpes

Genital herpes is one of the most common STIs. An estimated 45 million Americans are infected with the causative virus, and about 500,000 new cases are thought to occur each year.

Signs and Symptoms

Genital herpes begins 2 to 20 days (usually about a week) after exposure, with genital itching and burning, and in some cases, severe pain. Groups of small, red bumps appear on the genitalia or the anus, depending on the site of infection. These bumps become blisters surrounded by redness (**figure 27.18**). The blisters break in 3 to 5 days, leaving an ulcerated area that slowly dries and becomes crusted. This eventually heals without a scar. Less common signs and symptoms include urethral or vaginal discharge, headache, fever, muscle pain, and in the case of women, painful urination. Signs and symptoms are the most severe in the first 1 to 2 weeks and disappear within 3 weeks. They often recur, however, because the virus becomes latent and can then reactivate. The recurrences are usually not as severe as those of the first episode, and they usually happen less often with time. Some people have recurrences throughout their life, whereas others never do. ⏮ latent infections

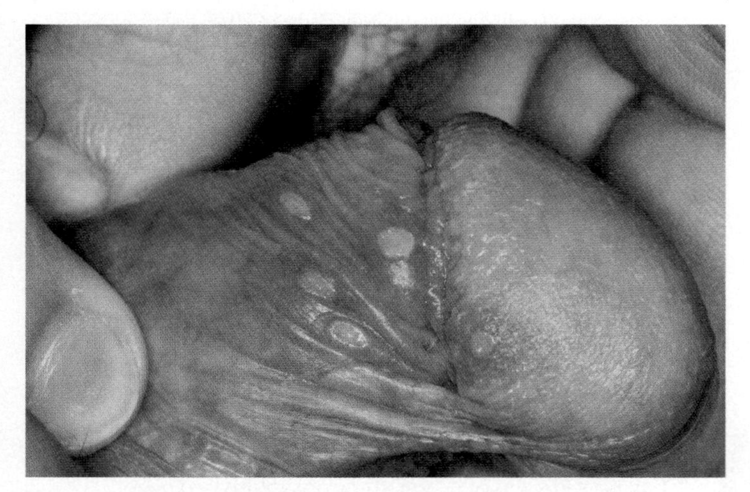

FIGURE 27.18 Genital Herpes on the Shaft of the Penis

? Can genital herpes be transmitted when there are no visible signs or symptoms of the infection?

Causative Agents

Genital herpes is usually caused by herpes simplex virus type 2 (HSV-2), an enveloped, double-stranded DNA virus of the *Herpesviridae* family. Herpes simplex virus type 1 (HSV-1), the cause of "cold sores" ("fever blisters"), can also cause genital herpes. The two virus types look the same, and their genomes are about 50% homologous. Either virus can infect the mouth and the genitalia, but HSV-2 causes more severe genital lesions with a greater frequency of recurrence. ⏮ HSV-1

Pathogenesis

Lesions begin with infection of a group of epithelial cells that lyse following viral replication, creating small, fluid-filled blisters (vesicles) containing large numbers of infectious virions. The vesicles burst, producing painful ulcers. During initial infection and for as long as 1 month after, the virus is found in genital secretions. In recurrence, the virus is usually present in large numbers for less than a week.

Latency of the disease is not completely understood but probably depends on cell-mediated immunity as well as viral products. Most of the time, the viral DNA exists within nerve cells in a circular, non-infectious form, causing no symptoms. In this state, only a single gene is expressed, and it encodes a small segment of RNA called a microRNA (miRNA) that probably allows the virus to remain latent. At times, however, the entire viral chromosome can be transcribed and complete infectious virions produced. These reinfect the area supplied by the nerve and cause symptom recurrence. ⏮ microRNA

Epidemiology

Sexual transmission of HSV is most likely to occur during the first few days of symptomatic disease, but can happen in the absence of signs or symptoms of the disease. Once infected, a person is forever at risk of transmitting the virus to another person. HSV, like other genital diseases, increases the risk of HIV infection. HSV-2 can survive for short periods on fomites or in water, but non-sexual transmission is rare. Neither HSV-1 nor HSV-2 has animal reservoirs.

Treatment and Prevention

There is no cure for genital herpes, although anti-HSV medications such as acyclovir and famciclovir can decrease the severity of the first attack and the incidence of recurrences.

Genital herpes can be prevented by abstaining from sexual intercourse or having a monogamous relationship with a genital herpes–free person. In addition, avoiding sexual intercourse during active signs or symptoms and using condoms with a spermicide that inactivates HSVs help prevent contracting the disease. However, using a condom does not always give full protection because herpes lesions can occur on parts of the genitalia that a condom does not cover.

Genital herpes can be a serious risk to newborn babies. If the mother has a primary infection near the time of delivery, the baby has about a 30% risk of acquiring the infection. The baby often dies from the infection or is permanently disabled by it. To prevent contact with the infected birth canal, these babies must be delivered by cesarean section. If the mother has recurrent disease, the risk to the baby is very low, presumably because transplacental anti-HSV antibody from the mother protects the baby, but doctors will often do a cesarean section to minimize the risk. The main features of genital herpes are presented in **table 27.12.**

Human Papillomavirus STIs: Genital Warts and Cervical Cancer

Sexually transmitted human papillomavirus (HPV) strains are among the most common of the STI agents, infecting an estimated 40 million Americans. Some HPV strains cause **papillomas**—warty growths of the external and internal genitalia. Other strains cause non-warty lesions of mucosal surfaces such as the uterine cervix, and these are a major factor in the development of cervical cancer.

TABLE 27.12 Genital Herpes

Signs and symptoms	Itching, burning pain at the site of infection, painful urination, tiny blisters with underlying redness; the blisters break, leaving a painful superficial ulcer, which heals without scarring; recurrences are common.
Incubation period	Usually 1 week (range, 2 to 20 days)
Causative agent	Usually herpes simplex virus type 2, but sometimes herpes simplex type 1 (the cold sore virus); herpesviruses are enveloped, double-stranded DNA viruses.
Pathogenesis	Lysis of infected epithelial cells results in fluid-filled blisters containing infectious virions; vesicles burst, causing a painful ulcer; the acute infection is controlled by immune defenses; viral genome persists within nerve cells in a non-infectious form; reactivation can occur and cause recurrent symptoms in the area supplied by the nerve; newborn babies can contract generalized herpes infection if their mother has a primary infection at the time of delivery.
Epidemiology	Transmission by sexual intercourse, oral-genital contact; transmission risk greatest first few days of active disease; transmission can occur in the absence of symptoms; herpes simplex increases the risk of contracting HIV.
Treatment and prevention	Treatment: no cure; anti-HSV medications help prevent recurrences, shorten duration of symptoms. Prevention: abstinence, monogamy, and correct use of condoms help prevent transmission.

MicroByte

Approximately 14 million people are newly infected with HPV in the United States every year.

Signs and Symptoms

Most people who have HPV infections clear them without ever developing signs or symptoms. In fact, they often do not know that they are infected. Some types of HPV do cause symptoms, however, and warts are the most easily recognized of these. Warts usually appear about 3 months after infection (range, 3 weeks to 8 months), developing on the head or shaft of the penis (**figure 27.19**), at the vaginal opening, or around the anus. Different kinds of lesions develop depending on the virus strain and the location of the infection. Lesions can be flat or raised, cauliflower-like, or hidden within the epithelium. Occasionally, warts become inflamed and bleed. After warts are removed, HPV persists in surrounding normal-looking epithelium and can cause additional warts. Warts sometimes partly block the urethra or, if they are very large, the birth canal. Newborn infants can become infected with HPV at birth and develop warts that block their respiratory tract, a rare but serious condition. Certain HPV strains can cause cervical cancer. Precancerous lesions are generally asymptomatic and can be detected only by examining the cervical tissues. The same strains can also cause other cancers, including oral, penile, vaginal, and anal cancers. Accumulating evidence indicates that oral sex is a risk factor for oral cancers; a CDC study implicated HPV as the cause of more than 60% of mouth and throat cancers.

FIGURE 27.19 Genital Warts The warts on the penis are the result of human papillomavirus infection. ©Biophoto Associates/Science Source

? **Would condom use always prevent the transmission of HPV?**

Causative Agents

Human papillomaviruses are non-enveloped, double-stranded DNA viruses of the *Papillomaviridae* family. There are more than 100 types of HPVs, and the different types are tissue- and species-specific. For example, HPVs that cause the common warts of the hands and feet generally do not infect the genitalia. Similarly, the HPVs that cause genital warts may infect the cervix but do not cause cancer.

At least 40 HPV types are transmitted by sexual contact. Approximately 15 of these are strongly associated with cancer of the cervix, penis, rectum, vagina, and throat. These high-risk HPVs include HPVs 16, 18, 31, and 45.

Pathogenesis

HPVs are thought to enter and infect the deeper layers of epithelium through microscopic abrasions. The genome of low-risk (wart-causing) HPV types exists in infected cells as extrachromosomal, closed DNA circles, but the mechanism by which warts arise is unknown. In contrast, the genome of high-risk (cancer-associated) HPV types can integrate into the chromosome of the host cell; they are oncogenic ("cancer causing") because they code for a protein that allows excessive cell growth. Oncogenic types of HPV tend to cause infections that persist longer than other HPV types, and to cause precancerous lesions. It takes on average 20 years for the infection to cause invasive cancer. A cancer-associated HPV is present in more than 90% of cervical cancers, but only a small percentage of infections by these viruses results in cancer. This indicates that other unknown factors must be present for cancer to develop. ⏮ oncogenes

Epidemiology

HPV strains are easily spread by sexual intercourse. Asymptomatic people infected with HPV can transfer the virus to others. HPV infection is the most common reason for an abnormal Papanicolaou (Pap) test in teenage women (see next section). A history of having multiple sex partners is the most important risk factor for acquiring genital HPV infection.

Treatment and Prevention

Warts often regress over time, but several treatments are available to remove them or hasten their clearance. Unfortunately, none of these methods cures the infection, so the warts may recur. Dermal and external genital warts can be removed by laser treatment, freezing with liquid nitrogen, surgical excision, or acid treatment using trichloroacetic. External genital warts can also be treated with creams and ointments, including (1) imiquimod (an immune modifier that induces cytokine production); (2) podofilox (blocks cell division, leading to cell death); and (3) sinecatechin-containing ointments (cause wart regression by an unknown mechanism).

HPV infections can be prevented by abstinence, monogamous relationships, or using condoms appropriately. As with HSV, condoms do not provide complete protection against HPV because the virus can be transmitted by exposure to areas not covered by the condom. Women should have a Pap smear every 12 months. This test involves removing cells from the cervix, staining them on a microscope slide, and examining them for any abnormal cells (**figure 27.20**). Cervical cancer is generally preceded by precancerous lesions—areas of abnormal cell growth that can be detected by this test. The lesion can be removed, thereby preventing development of a cancer.

Gardasil 9 (9vHPV) is a vaccine developed to protect against HPV types 16 and 18, which are responsible for about 70% of cancers, and HSV types 6 and 11, which cause more than 90% of genital warts. This vaccine also protects against HPVs 31, 33, 45, 52, and 58. 9vHPV is part of the routine vaccine schedule for adolescents at age 11 or 12, although it can be given to children as young as 9 years old. For people who have not received the vaccine previously, it is recommended for females up to age 26 and males up to age 21, and it can be given to males up to age 26. The number of doses given is age-dependent; a two-dose series is used for people under age 15, and a three-dose series is used for people 15 and older. **Table 27.13** summarizes some important features of human papillomavirus STIs.

HIV/AIDS

Acquired immunodeficiency syndrome (AIDS) is unquestionably the most important sexually transmitted infection of the past century. The AIDS epidemic was first recognized in the United States in 1981 when a number of young, previously

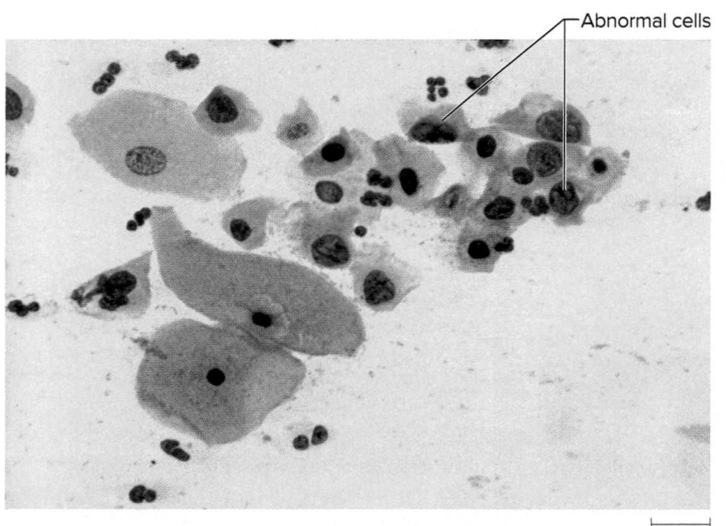

FIGURE 27.20 Abnormal Papanicolaou (Pap) Smear The pink and blue cells are squamous epithelial cells; abnormalities include double nuclei that have a clear area around them. ©Frederick C. Skvara, M.D.

? What is the most frequent cause of an abnormal Pap smear in young women?

TABLE 27.13 Human Papillomavirus STIs

Signs and symptoms	Often asymptomatic; warts of the external and internal genitalia the most common symptom; symptoms are strain-specific.
Incubation period	Usually 3 months (range, 3 weeks to 8 months)
Causative agents	Human papillomaviruses (many strains): non-enveloped DNA viruses of the papillomavirus family; different types infect different tissues and produce different lesions.
Pathogenesis	Virus enters epithelium through abrasions, infects deep layer of epithelium; establishes latency; cycles of replication occur when host cell begins maturation; cancer-associated viral types can integrate into the host cell chromosome and can cause precancerous lesions.
Epidemiology	Asymptomatic people can transmit the disease; multiple sex partners the greatest risk factor; warts can be transmitted to the mouth with oral sex, and to newborn babies at birth.
Treatment and prevention	Treatment: wart removal by multiple techniques. Prevention: latex condoms advised to minimize transmission and avoiding sexual contact with people having multiple sex partners; vaccine protects against the most common cancer- and wart-causing strains.

healthy homosexual and bisexual men developed *Pneumocystis* pneumonia, a disease caused by an organism of such low virulence that it seldom affects healthy people. As other groups of people developed unusual opportunistic infections for no apparent reason, the CDC began using the acronym AIDS to describe the unexplained increased susceptibility that indicated underlying immunosuppression. The various opportunistic infections became known as "AIDS-defining conditions"—diseases that indicate a person has AIDS. These were useful in studying the AIDS epidemic before its cause was known, and they still help physicians by alerting them to the possibility that a patient has AIDS (**table 27.14**). Fairly soon after AIDS was recognized, scientists identified the causative agent—human immunodeficiency virus (HIV)—which led to major advances in the prevention and treatment of the disease.

More than 35 million people worldwide have died of AIDS over the last three decades, and nearly 37 million are currently living with HIV (**figure 27.21**). In fact, more people have died of AIDS than died of the "Black Death" of Europe during the Middle Ages. Sub-Saharan Africa is currently the most affected region, with 1 in 20 adults living with HIV. Despite the significant advances made in understanding HIV and AIDS and the billions of dollars spent trying to control the AIDS pandemic, the disease still cannot be cured and is far from conquered.

Signs and Symptoms

The terms HIV and AIDS are not synonymous, and the distinctions are very important. AIDS is only the end stage of a complex disease that has many signs and symptoms that precede immunodeficiency (**figure 27.22**). Three separate but related terms are important to keep in mind as you read about HIV and AIDS:

- **HIV infection.** When a person first contracts HIV, he or she has an HIV infection. A person can be infected with HIV but may not be ill.
- **HIV disease.** This develops 6 days to 6 weeks after contracting HIV. Some people experience flu-like signs and symptoms, including fever, head and muscle aches, sore throat, and enlarged lymph nodes. A generalized rash may also develop. Signs and symptoms that indicate this early stage of HIV disease coincide with high levels of virus replication and are called the **acute retroviral syndrome (ARS);** they are often mild and go away by themselves in about a month. However, many people with HIV disease are asymptomatic. After the initial burst of viral replication, viral levels decrease but persist without causing symptoms for many years—a period of **clinical latency.** During this period, the virus continues to replicate, infecting and destroying helper T cells and macrophages. The concentration of virions in the bloodstream during this time remains relatively stable and is referred

TABLE 27.14 AIDS-Defining Conditions

- Cancer of the uterine cervix, invasive
- Candidiasis involving the esophagus, trachea, bronchi, or lungs
- Coccidioidomycosis, of tissues other than the lung
- Cryptococcosis, of tissues other than the lung
- Cryptosporidiosis of duration greater than 1 month
- Cytomegalovirus disease of the retina with vision loss or other involvement outside liver, spleen, or lymph nodes
- Encephalopathy (brain involvement with HIV)
- Herpes simplex virus causing ulcerations lasting a month or longer involving the esophagus, bronchi, or lungs
- Histoplasmosis of tissues other than the lung
- Isosporiasis (a protozoan disease of the intestine) of more than 1 month's duration
- Kaposi's sarcoma
- Lymphomas, such as Burkitt's lymphoma
- Mycobacterial diseases, including tuberculosis
- *Pneumocystis* (pneumonia due to *Pneumocystis jirovecii*)
- Pneumonias occurring repeatedly
- Progressive multifocal leukoencephalopathy (a brain disease caused by the JC polyomavirus)
- *Salmonella* infection of the bloodstream, recurrent
- Toxoplasmosis of the brain
- Wasting syndrome (weight loss of more than 10% due to HIV); also known as slim disease

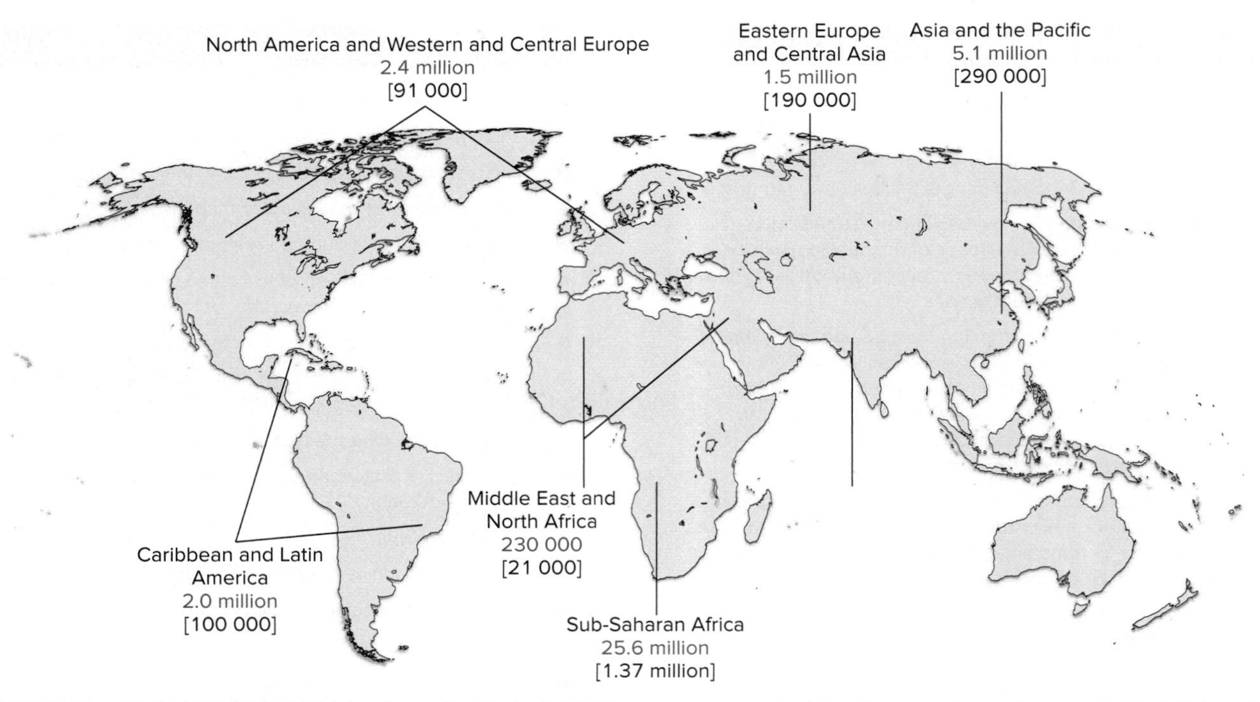

FIGURE 27.21 The Global HIV/AIDS Epidemic at the End of 2015 Altogether, nearly 37 million people are living with HIV/AIDS (red numbers). An estimated 2 million people were newly infected during the year (numbers in parentheses).

What is one factor that has led to the decrease in new HIV infection rates in many countries?

to as the **HIV (viral) set point.** This set point, which is measured as HIV RNA in plasma, is important because it serves as a predictor of disease progression—the higher the viral set point in an infected person, the more quickly he or she is likely to develop AIDS.

- **AIDS.** When helper T cells drop to low enough levels, immunodeficiency develops, a stage that characterizes AIDS. At this point, certain cancers and unusual microbial infections develop, including Kaposi's sarcoma, B-lymphocyte tumors, *Pneumocystis* pneumonia, toxoplasmosis, cytomegalovirus disease, and mycobacterial disease (see table 27.14). Most people with AIDS also suffer from weight loss, fever, fatigue, and diarrhea—a set of problems referred to as the **AIDS-related complex (ARC).** AIDS is fatal without treatment. **◀◀ *Pneumocystis* pneumonia, ◀◀ toxoplasmosis, ◀◀ mycobacterial disease**

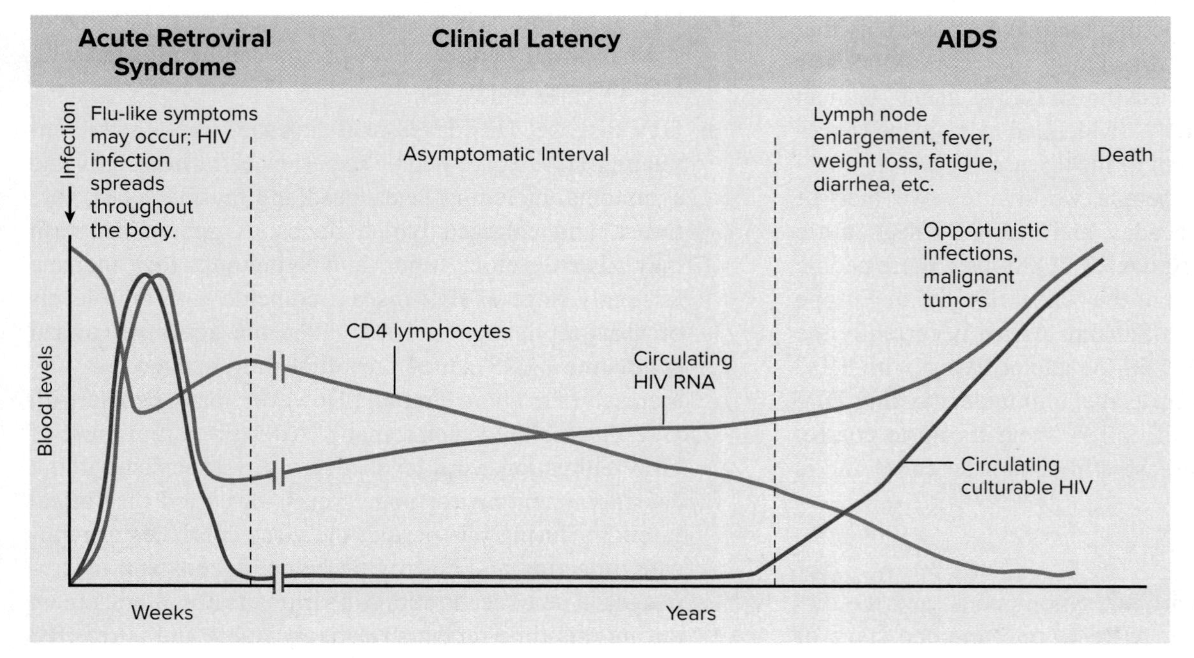

FIGURE 27.22 HIV Disease Progression During the acute retroviral syndrome, viral levels are very high; the patient is infectious, but antibody tests used for diagnosis are often negative. In the clinical latency period, the disease continues to progress, as shown by the falling CD4 cell count and rising viral RNA levels. The onset of AIDS is marked by a very low CD4 cell count.

What are some of the opportunistic infections and malignant tumors that people with AIDS contract?

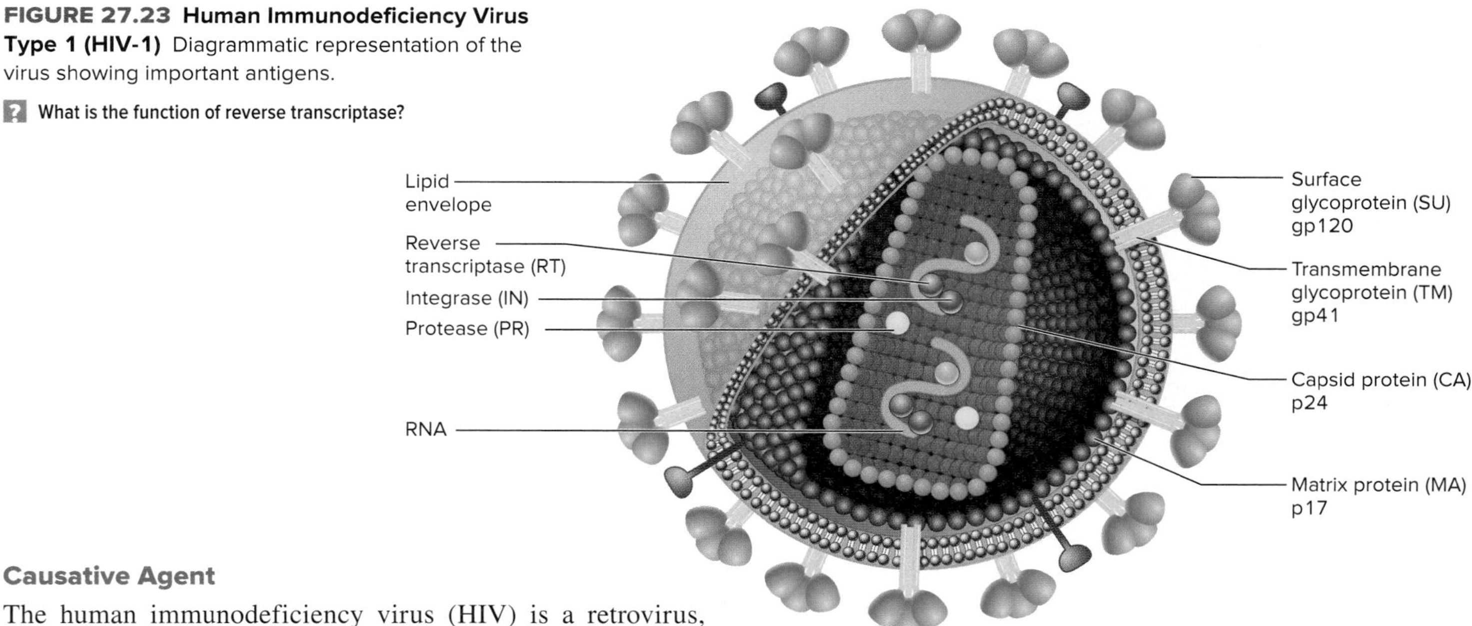

FIGURE 27.23 Human Immunodeficiency Virus Type 1 (HIV-1) Diagrammatic representation of the virus showing important antigens.

? What is the function of reverse transcriptase?

Causative Agent

The human immunodeficiency virus (HIV) is a retrovirus, the common term for members of the *Retroviridae* family. Like all members of this family, HIV is an enveloped single-stranded virus that has duplicate copies of a single-stranded RNA genome (**figure 27.23**). All retroviruses encode reverse transcriptase, an enzyme that uses an RNA genome as a template to make a complementary DNA strand. There are two distinct types of HIV (HIV-1 and HIV-2), and these have only about 50% genetic similarity. Most AIDS cases are caused by HIV-1, and more is known about the virus, so this section will focus on that type. HIV-2 is found mainly in West Africa, and it causes a less severe form of disease. ⏮ retroviruses

As shown in figure 27.23, various proteins make up the HIV structure, and these are often referred to by two alternative names (**table 27.15**). One name is an abbreviation of the functional role or location of the protein, and the other indicates the size of the protein or glycoprotein in kilodaltons (kDa). For example, the capsid protein is 24 kDa in size and is called either CA (for capsid) or p24. CA is the most abundant protein in the virus, and serum levels can be measured to detect early HIV infection.

TABLE 27.15 HIV Components

Protein	Function
Capsid protein (CA or p24)	Coat protein
Matrix protein (MA or p17)	Stabilizes virion
Transmembrane protein (TM or gp41)	Stalk-like portion of the spikes; it anchors the spikes to the viral envelope
Surface glycoprotein (SU or gp120)	Cap-like portion of the spikes, with which virus attaches to host receptors
Reverse transcriptase (RT)	Enzyme that makes copy of viral RNA into DNA
Integrase (IN)	Enzyme that inserts viral cDNA into host genome
Protease (PR)	Enzyme involved in viral protein processing

Projecting from the envelope are multiple copies of a spike made up of two types of glycoprotein subunits: (1) a transmembrane glycoprotein called TM or gp41, which anchors the spike in the viral envelope; and (2) a surface glycoprotein called SU, or gp120, which rests on the TM glycoprotein like a cap; SU allows the virus to attach to and enter host cells. Between the capsid protein (CA) and the envelope is a layer of matrix protein or MA, which functions in stabilizing the virus.

Three important HIV-encoded enzymes—reverse transcriptase (RT), protease (PR), and integrase (IN)—are significant because they are targets of anti-HIV medications. **Reverse transcriptase** is an RNA-dependent DNA polymerase, which uses the viral RNA genome to make a DNA copy. **Integrase** inserts a DNA copy of the viral genome into the host cell genome. HIV **protease** cleaves a virally encoded polyprotein into individual functional components.

Based on comparison of HIV-1 genome sequences, many different strains of the virus exist, and these can be divided into at least four general groups. By far the most common is group M (for "major"), and within this are subtypes, called clades. There are at least 13 clades, and these can be subdivided even further.

Pathogenesis

HIV can infect a variety of human cell types, but the most significant are the helper T cells. These cells, also referred to as CD4 T cells, have the CD4 surface protein to which the virus first attaches (**figure 27.24**). Monocytes, including macrophages and dendritic cells, also have CD4 protein and can

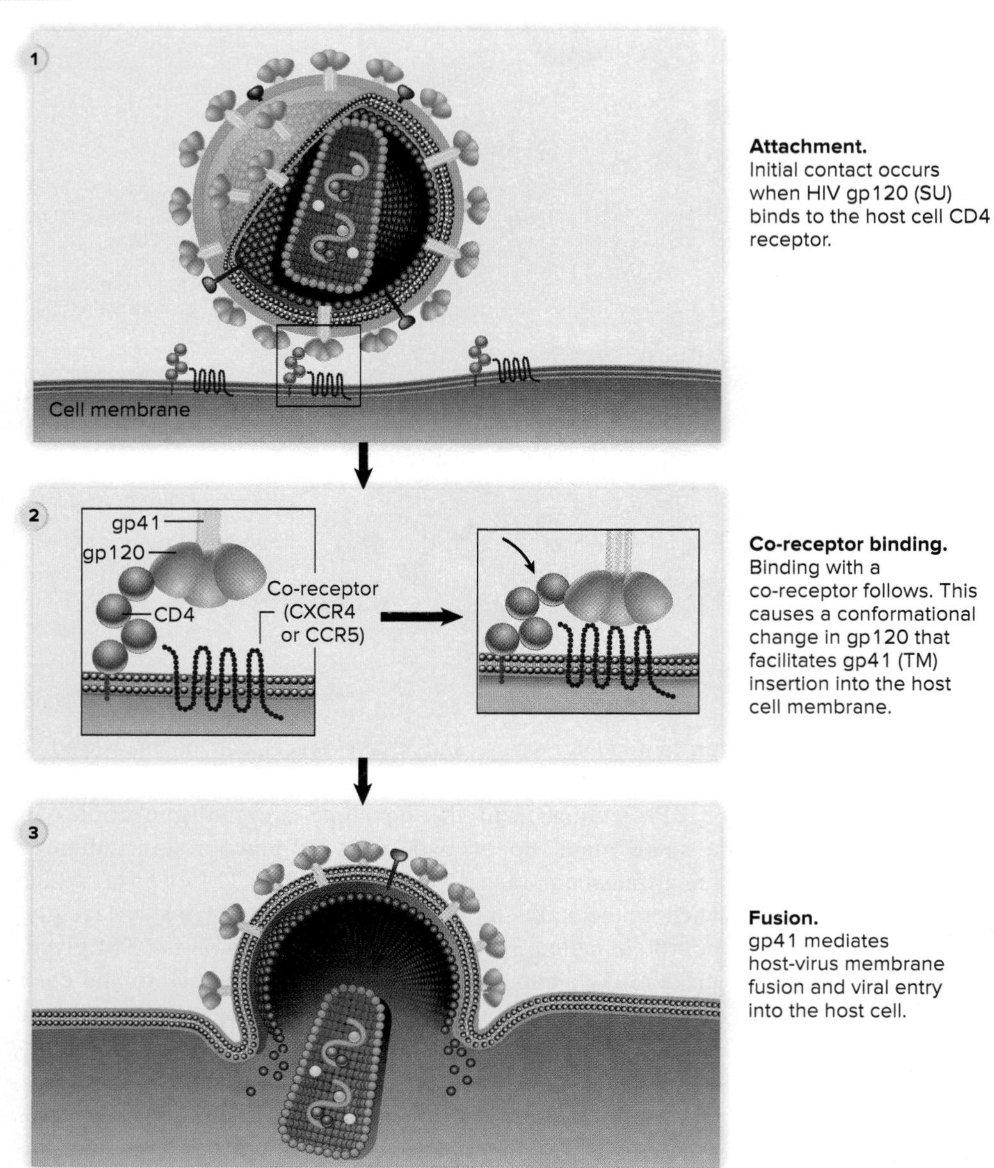

FIGURE 27.24 Attachment and Entry of HIV into a Host Cell

? How do entry inhibitor medications work?

therefore be targeted by HIV as well. After attaching to CD4, the virus must then attach to a co-receptor—either CCR5 or CXCR4, which are cytokine receptors on the host cell surface. The importance of CCR5 is illustrated by the observation that people who do not make normal amounts of CCR5 are less susceptible to HIV infection. In addition to attaching to host cells and then entering them, HIV also causes infected cells to fuse with nearby cells, forming syncytia. This allows the virus to spread from cell to cell without encountering antibodies. ⏮ syncytia

After entering the host cell, the HIV-encoded reverse transcriptase makes a DNA copy of the viral RNA genome (**figure 27.25** ① and ②). ③ The RNA template is degraded and a complementary DNA strand is made, resulting in a double-stranded DNA copy of the viral genome. The nucleotide sequence of the DNA copy is generally slightly different from that of the parent molecule because reverse transcriptase makes frequent mistakes. ④ The DNA copy enters the host nucleus, where the virally encoded enzyme integrase then inserts it randomly into a host chromosome; the integrated DNA (provirus) becomes a permanent part of the cell's genome.

⑤ After the HIV DNA copy is integrated into the host cell genome, it can be transcribed, creating RNA copies. ⑥ These serve dual roles: functioning both as RNA genomes for new viral particles and as mRNA molecules that can be translated to make the various viral proteins. ⑦ and ⑧ The developing virions bud from the host cell, gaining their envelope as they do so. Because of the mutations introduced during reverse transcription, the viral particles that bud from the cell are likely slightly different from the parent virus that entered. The high mutation rate allows HIV to evolve quickly, altering its antigens and thereby avoiding the immune response.

Helper T cells typically die as a result of HIV infection, and although the body can replace hundreds of billions of CD4 T cells, the number slowly declines over many months or years. The loss of these cells is very important because of their central role in the body's adaptive immune response. Meanwhile, infected macrophages and dendritic cells usually survive, and they release new virions over long periods of time. These cells, along with memory helper T cells, become reservoirs for replicating HIV, protecting the virus from the rest of the immune system and providing a continuing source of infectious viral particles. Although infected macrophages and dendritic cells survive, they do not function normally, further weakening the immune response. Eventually, the immune system becomes so weakened it can no longer respond to infections or cancers.

Signs and symptoms of AIDS usually appear when helper T cell counts fall below 200 cells per microliter. As the helper T cell counts decrease and the immune system stops functioning properly, the viral load rises dramatically. The patient

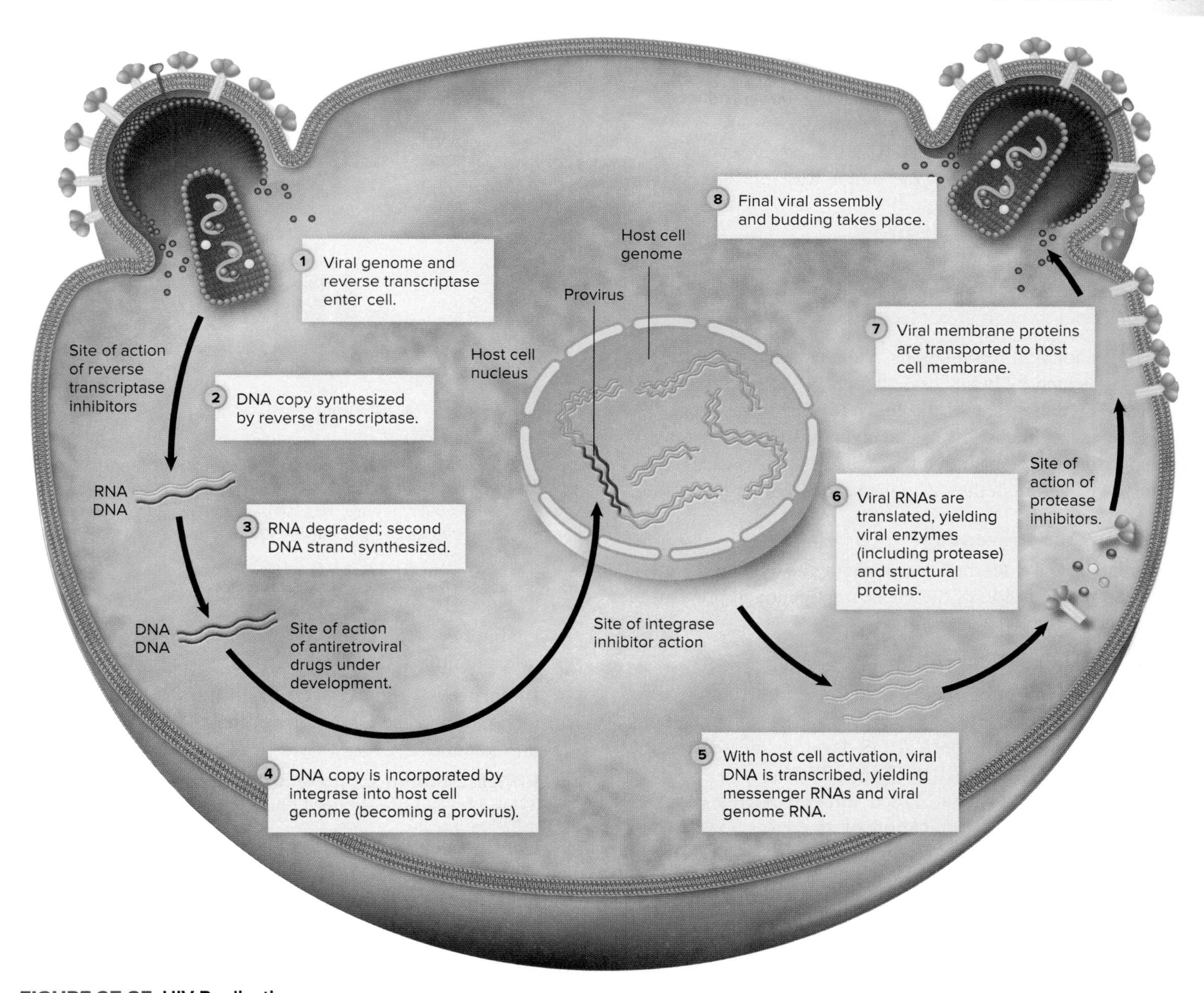

FIGURE 27.25 HIV Replication

Which cells are the main target for HIV?

now begins experiencing a variety of opportunistic infections, including malignant tumors caused by infectious agents. Half of untreated patients progress to AIDS within 9 to 10 years. Rapid progressors (about 10% of infected individuals, who have high viral set points) develop AIDS within a few years of initial infection. At the other end of the spectrum are long-term non-progressors (about 5% to 10% of infected individuals, with low viral set points). These people show no decrease in CD4 cells and maintain high levels of anti-HIV antibody and HIV-specific $CD8^+$ cytotoxic T cells over many years. At the extreme end is a group called elite controllers, who have very low viral set-point levels, below the detection limit of conventional clinical assays.

Epidemiology

HIV is present in blood, semen, and vaginal secretions in symptomatic and asymptomatic infections. Common mechanisms of person-to-person HIV spread include:

- **Sexual contact.** HIV can be transmitted by both male-to-male and male-to-female sexual contact. The risk of spread by anal sex is greater than that of vaginal sex, probably because the rectal mucosa is thin and easily damaged, allowing viral entry into the blood.
- **Blood and blood products.** HIV can be spread among IV (intravenous) drug abusers who share hypodermic needles. It can also be transmitted to healthcare workers

by a needle stick, but the risk of contracting HIV this way is extremely low (0.3%). Although HIV can be transmitted in whole blood and blood products such as plasma and clotting factors, these are now screened for antibodies against HIV; the presence of antibodies suggests that the donor was infected, so the product is not used.

- **Vertical transmission.** HIV can be transmitted vertically—from mother to fetus or newborn—during pregnancy, during passage through the birth canal, and through breast feeding. ⏮ vertical transmission

HIV is not highly contagious, and most transmissions can be prevented by behavioral changes. Risk factors for acquiring HIV are given in **table 27.16.** The virus is not spread by insect bites or by casual contact, and there is no evidence that it is transmitted through sweat, urine, tears, or saliva, even though these fluids may contain viral particles.

Treatment and Prevention

As scientists have learned more about the HIV replication cycle and transmission of the virus, treatment and prevention options have improved dramatically.

Treatment Improved anti-HIV therapy, as well as better treatment and prevention of opportunistic infections, has significantly slowed the progression of HIV disease. The treatments have also prolonged life once AIDS develops, decreasing the number of deaths due to AIDS (**figure 27.26**).

Anti-HIV medications, called **antiretrovirals (ARVs),** are designed to block replication of the virus, but do not affect viral nucleic acid already integrated in the genome of host cells. The various categories of these medications were described in chapter 20. Here are their characteristics (**table 27.17**):

- **Entry inhibitors** block access to the co-receptor CCR5 or prevent fusion of the viral envelope with the host membrane.
- **Reverse transcriptase inhibitors** prevent the virus from making a DNA copy of its RNA genome. There are two general types in this category: nucleoside reverse transcriptase inhibitors (NRTIs), and non-nucleoside reverse transcriptase inhibitors (NNRTIs).
- **Integrase strand transfer inhibitors (ISTIs)** prevent integration of the DNA copy of the HIV genome into the host genome.
- **Protease inhibitors** prevent the viral polyproteins from being cleaved into individual proteins.

Unfortunately, a significant problem with ARVs is that HIV quickly develops resistance to the medications. To help prevent this, a combination of medications—a

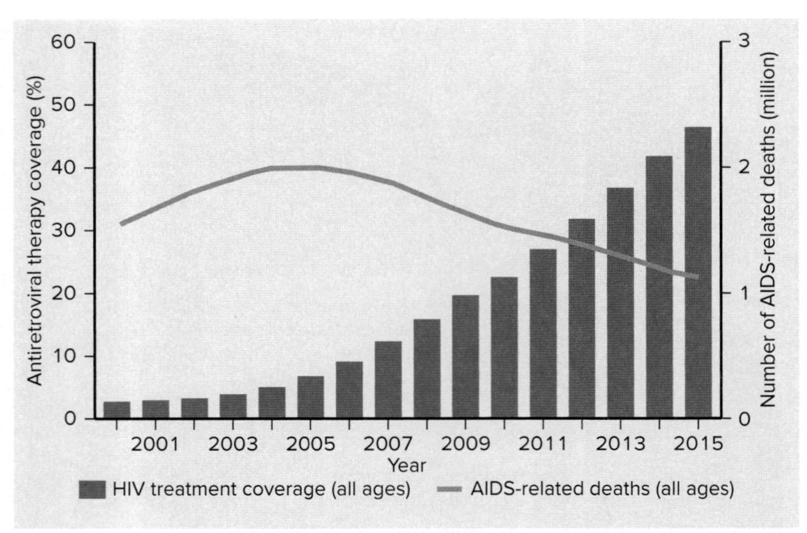

FIGURE 27.26 Global Numbers of People on ART, and Number of AIDS-Related Deaths, 2000–2015 As the number of people receiving antiretroviral therapy has increased worldwide (green bars), the number of people dying of AIDS-related conditions has decreased (blue line).

? What caused the marked drop in HIV-related deaths after 2005?

TABLE 27.16 People at Increased Risk for HIV Disease

1. Injected-drug abusers who have shared needles.
2. Sexually promiscuous men and women, especially sex workers, drug abusers, and men who have sex with men.
3. People with history of hepatitis B, syphilis, gonorrhea, or other sexually transmitted infections that may be markers for unprotected sex with multiple partners.
4. People who have had blood or sexual exposure to any of the people listed above.

TABLE 27.17 Antiretroviral Medications

Medication	Actions	Examples
Entry inhibitors	Interfere with viral entry into host cell	Enfuviritide (ENF), maraviroc (MVC)
NRTIs (nucleoside reverse transcriptase inhibitors)	Chain terminators that stop DNA replication	Abacavir (ABC), emtricitabine (FTC), lamivudine (3TC), tenofovir alafenamide (TAF), zidovudine (AZT or ZDV).
NNRTIs (non-nucleoside reverse transcriptase inhibitors)	Inhibit reverse transcriptase	Efavirenz (EFV), etravirine (ETR), nevirapine (NVP), rilpivirine (RPV)
Integrase inhibitors	Prevent integration of HIV cDNA into host genome	Dolutegravir (DTG), elvitegravir (EVG), raltegravir (RAL)
Protease inhibitors	Prevent virions from maturing	Atazanavir (ATV), darunavir (DRV), ritonavir (RTV)

"cocktail" called highly active antiretroviral therapy (HAART) or **antiretroviral therapy (ART)**— is prescribed. A person receiving ART typically takes three drugs: two nucleoside reverse transcriptase inhibitors (NRTIs) and either a protease inhibitor or a non-nucleoside reverse transcriptase inhibitor (NNRTI). In some cases, these medications are available in fixed-dose combinations—a single tablet that contains two or more different medications, for example, TDF + 3TC + EFV. In many cases, this therapy can clear the patient's blood of detectable viral nucleic acid, stop progression of the disease, and even allow partial recovery of immune function.

ART does not cure HIV disease. It can, however, improve the quality of the person's life, reduce the risk of other complicating diseases, and lower the HIV set point. The latter is particularly important because lowering the number of replicating viral particles decreases the chance that one of them will mutate to drug resistance. It is important to remember, however, that ART does not eliminate HIV provirus integrated in host cell genomes. If medications are stopped, the viral set point rebounds to pretreatment levels.

Although ART can successfully lower the viral set point, there are problems associated with this treatment. Many HIV strains are now resistant to the medications, and these strains are spreading. The medications can also have several toxic effects. For example, zidovudine (azidothymidine, or AZT) can cause anemia, low white blood cell count, vomiting, fatigue, headache, and muscle and liver damage. Other NRTIs cause painful peripheral nerve injury; inflammation of the pancreas; rash, mouth, and esophagus ulceration; and fever. Some of the protease inhibitors cause kidney stones, nausea, diarrhea, and sometimes diabetes.

Another major factor limiting the use of anti-HIV medications is their cost, which can make them unaffordable in many parts of the world. Negotiated price decreases, the approval of three generic medications by the U.S. Food and Drug Administration, and medications made available by the U.S. government to several highly affected countries have made treatment available to many more HIV patients worldwide.

Preventing Transmission via Sexual Contact Educating people about HIV and how it is transmitted is a powerful weapon against the AIDS epidemic. The virus is not highly contagious, and the risk of contracting and spreading it can be eliminated or greatly reduced by avoiding activities that might transmit it (**table 27.18**). Using latex condoms during sex significantly reduces HIV transmission. Both male and female condoms are available, although male condoms are less expensive and more readily available. People who are unsure of their HIV status, and especially those at increased risk of contracting the virus, should get tested. HIV-positive people who know their status can ensure that they do not spread the virus to others and can receive antiretroviral treatment that lowers the viral set point and helps reduce transmissibility.

TABLE 27.18 Behaviors That Help Control an AIDS Epidemic

1. Abstinence.
2. Staying in a mutually monogamous relationship.
3. Avoiding sex with people at risk for HIV infection.
4. Avoiding genital and rectal trauma. Small breaks in the skin and mucous membranes allow HIV to infect.
5. Avoiding sex when sores from herpes simplex or other causes are present. These are sites where HIV can infect.
6. Not engaging in unprotected anal intercourse. Receptive anal intercourse carries a high risk of HIV transmission.
7. Using latex condoms from beginning to end of sex. Polyurethane condoms are a reasonable alternative for those allergic to latex. Condoms made from other materials are not reliable for disease prevention, nor are those marketed in many countries outside the United States. Oil-based lubricants cannot be used with latex condoms. Condoms for women are available.
8. Being aware of HIV status before pregnancy and, if HIV-positive, taking appropriate steps during pregnancy to prevent transmission to fetus.

The World Health Organization has reported that circumcision (surgical removal of the penis foreskin) can reduce HIV acquisition by up to 60% and should also be included in prevention strategies. Circumcision changes the anatomy of the penis, reducing the number of certain cell types that are HIV targets. It also changes the normal microbiota populations under the foreskin, reducing the number of certain anaerobic species, some of which increase the risk of HIV infection by enhancing virus stability. Several countries that have high HIV infection rates are developing programs promoting circumcision as a preventive measure. Condom use is still a cheaper way of preventing spread of the disease, however, provided that reliable condoms are available. Furthermore, there is some concern that circumcised men may not realize that they are still at risk of HIV or other sexually transmitted infections if they have unprotected sex.

The most recent addition to the methods for preventing HIV infection in high-risk uninfected adults is called PrEP (preexposure prophylaxis). A single pill, taken daily, contains a fixed dose of two reverse transcriptase inhibitors (tenofovir and emtricitabine). Those medications interfere with the HIV replication cycle, so a person who has consistently taken the pills and is then exposed to HIV is much less likely to become infected. PrEP is not 100% effective, however, so it should be used in combination with other prevention methods, including proper use of condoms.

Preventing Transmission via Blood and Blood Products Many people with HIV disease do not know they are infected, and therefore all blood and blood products should be considered as potentially containing the virus. Transmission by blood and blood products has been lowered dramatically by screening potential donors for HIV risk factors and testing donated blood for antibodies to HIV. The risk is now estimated to be less than 1 in 1.5 million transfusions. Screening tests have also significantly reduced the risk of HIV transmission from artificial insemination and organ transplantation.

Hundreds of needle and syringe exchange programs help prevent the spread of HIV and other blood-borne diseases among injected-drug abusers. In these programs, people are given sterile syringes and needles in exchange for used ones. Programs also provide drug rehabilitation efforts and education about condom use and other safer sex practices.

Infectious HIV persists in samples of blood plasma for at least a week after they are taken from infected patients. HIV on objects and surfaces contaminated by body fluids is easily inactivated by high-level disinfectants and heat at 56°C or more for 30 minutes. Freshly opened household bleach (5.25% sodium hypochlorite), diluted 1:10, is a cheap and effective disinfectant for general use. ⏮ high-level disinfectant

Preventing Vertical Transmission HIV transmission from infected mother to newborn can be prevented in two-thirds of cases by giving AZT to the mother during pregnancy and to the newborn infant for 12 weeks, or by providing nevirapine (NVP) to the mother at delivery and the infant within 3 days of birth. Newer, more potent combinations of AZT and other antiretroviral (ARV) therapies are widely used and may be more effective, but their long-term safety is still being evaluated. Recent studies suggest that the use of ARVs during the breast-feeding period, in the mother or the infant, or both, can reduce transmission by this route. Elective cesarean section significantly reduces the risk of HIV transmission to the newborn baby. The risk of such transmission is significantly decreased by providing the mother anti-HIV therapy, delivering the infant by cesarean section, and advising the mother to not breastfeed the baby. In developing countries, however, stopping breast feeding is problematic because breast milk contains antibodies that protect against many other diseases common in such places. Also, clean water may not be available for the mother to make infant formula safely.

Vaccine Prospects A vaccine against HIV would prevent new infections, but creating an effective option has proven extremely difficult. One reason for this is that the virus has multiple mechanisms to avoid the adaptive immune response. In a natural infection, for example, the patient's initial immune response is strong, but is unable to clear the virus. Part of the reason for this is that critical HIV antigens change rapidly due to the high error rate of reverse transcriptase, meaning that the antibodies produced against one antigenic variant do not protect against another. In addition, HIV causes syncytia to form, allowing the virus to pass directly from cell to cell without coming into contact with antibodies. Another complicating factor in vaccine development is the lack of a good animal model for HIV/AIDS; without this model, it is difficult to test new vaccines for safety and effectiveness.

Despite the challenges, development of an effective vaccine offers the most hope for stopping the HIV/AIDS pandemic, so a great deal of research is still aimed at this effort. Recombinant vector vaccines—which are viruses that have been genetically engineered to carry HIV genes—show some promise. For example, an experimental recombinant vector vaccine appeared to protect approximately 30% of participants in a trial in Thailand. Several other HIV vaccine strategies are also being researched, including DNA vaccines and peptide vaccines. The fact that immune response of a small subset of people appears to either prevent infection or keep the viral set point extremely low encourages scientists to keep trying to develop an effective vaccine. **Table 27.19** summarizes the main features of HIV disease and AIDS.

TABLE 27.19	HIV Disease
Signs and symptoms	Often no early symptoms but over half develop flu-like symptoms, including fever, sore throat, head and muscle aches, rash, and enlarged lymph nodes, early in the infection; after a period of clinical latency, symptoms arise due to immunodeficiency and resulting infections of lung, intestine, skin, eyes, and brain, as well as certain cancers.
Incubation period	Usually 6 days to 6 weeks for flu-like symptoms; without treatment, immunodeficiency symptoms occur within 10 years in half the infections (10% within 5 years and 90% within 17 years)
Causative agent	Mainly human immunodeficiency virus type 1 (HIV-1), many subtypes and strains; HIV-2 mostly in West Africa
Pathogenesis	HIV infects various body cells, notably those vital to adaptive immunity, including CD4 T lymphocytes and antigen-presenting cells; T cells are killed—their numbers slowly decline until the immune system can no longer resist infections or tumor development.
Epidemiology	Three main routes of transmission: intimate sexual contact, transfer of blood or blood products, and from mother to child around the time of childbirth; transmission can also occur by breast milk, and possibly by oral-genital contact; other STIs such as syphilis increase transmission.
Treatment and prevention	Treatment: ART, consisting of combinations of several anti-HIV medications, slows disease progression; medications and vaccines can prevent infections that can complicate HIV disease. Prevention: no vaccine yet available; abstinence, monogamous relationships, and condom use, avoidance of sexual contact with injected-drug abusers and those with multiple partners or history of STIs; circumcision and pre-exposure prophylaxis also preventive; anti-HIV medications and cesarean section decrease mother-to-newborn transmission; other effective preventive measures include sex education for schoolchildren, and needle exchange programs for drug abusers.

MicroAssessment 27.5

Viral STIs are as common as bacterial STIs, but so far they are incurable. Genital herpes is common and, like most other STIs, can be transmitted in the absence of signs or symptoms. Human papillomaviruses cause genital warts; some play an important role in cancer of the cervix and probably cancers of the vagina, penis, anus, and throat. HIV disease generally ends in AIDS, but treatment can significantly improve longevity and quality of life.

13. What are the possible consequences of sexually transmitted human papillomavirus infections for men, women, and babies?
14. What kinds of sex partners present a high risk of transmitting HIV?
15. Why should a person be concerned about genital herpes—is it not just a cold sore on the genitals? Explain.

27.6 ■ Protozoan STIs

Learning Outcome

9. List the distinctive characteristics of the organism that causes trichomoniasis.

A number of protozoan diseases can be transmitted sexually. Most, however, are intestinal infections that can be transmitted through oral-anal contact during sexual activity. Trichomoniasis is a sexually transmitted protozoan disease that involves the genital system.

Trichomoniasis ("Trich")

Trichomoniasis is a common sexually transmitted protozoan disease that is frequently asymptomatic; however, it can cause vaginitis and preterm births in women, as well as genital discharge and painful urination in both women and men. An estimated 3.7 million Americans currently have this STI.

Signs and Symptoms

Trichomoniasis is often asymptomatic. Symptomatic infections in women are characterized by itching of the vulva and inner thighs, itching and burning of the vagina, and a frothy, sometimes smelly, yellowish-green vaginal discharge. In some cases, there is burning pain with urination. The vulva and vaginal wall are red and slightly swollen, and may develop pinpoint hemorrhages. In pregnant women, trichomoniasis may lead to preterm birth. Symptomatic men have urethral discharge, burning pain with urination, painful testes, or a tender prostate gland.

Causative Agent

Trichomoniasis is caused by *Trichomonas vaginalis,* a motile protozoan that has four anterior flagella and a posterior flagellum attached to an undulating membrane (**figure 27.27**). It also has a slender, posteriorly protruding, rigid rod called an axostyle that is used for attachment. *T. vaginalis* can be identified by its unmistakable, jerky motility when viewed using a microscope. Unlike most other pathogenic protozoa, *T. vaginalis* lacks a cyst form to aid its survival in the environment away from the host. The organism is a eukaryote, but it does not have any mitochondria. Instead, it has cytoplasmic organelles called hydrogenosomes for glucose metabolism and respiration. Enzymes within these organelles remove the carboxyl group (—COOH) from pyruvate and transfer electrons to hydrogen ions, producing hydrogen gas. ◀◀ mitochondria

Pathogenesis

The pathogenesis of trichomoniasis is not fully understood. *T. vaginalis* produces a variety of adherence factors specific for vaginal epithelium, allowing it to colonize the vulva and vagina walls. The signs and symptoms of trichomoniasis are thought to be due to mechanical trauma by the axostyle of the moving protozoa, but other factors may also be involved. The frothy discharge is probably due to the hydrogen gas produced by the organisms.

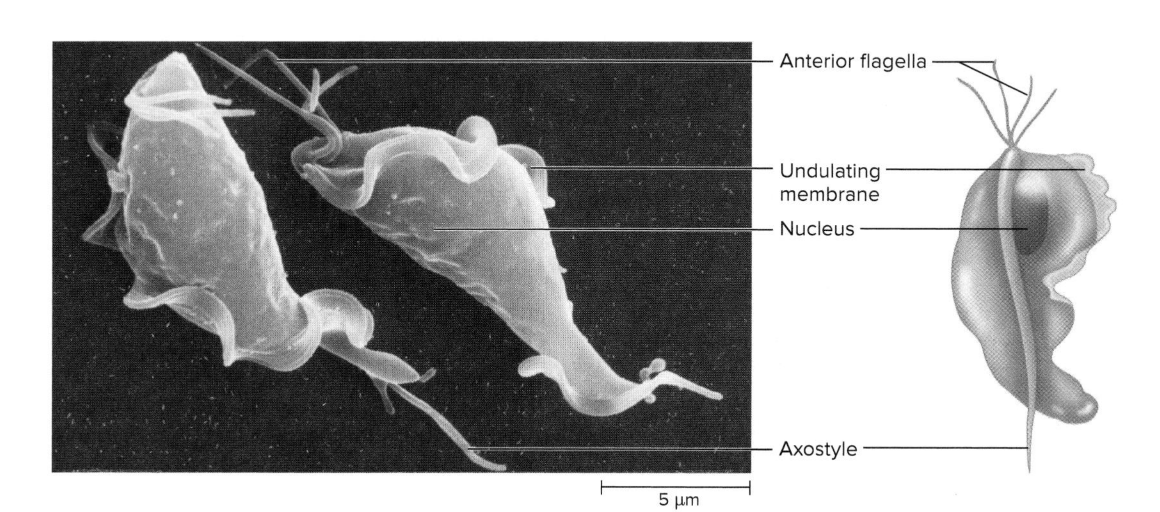

FIGURE 27.27 ***Trichomonas vaginalis*** Scanning electron micrograph and diagram of *T. vaginalis,* a common cause of vaginitis. ©David M. Phillips/Science Source

What is the possible role of the axostyle in the pathogenesis of trich?

Epidemiology

Trichomonas vaginalis is a widespread human parasite that has no other reservoirs and is usually spread by sexual contact. It is easily killed by drying, but can survive for a time on moist objects such as towels, so it can occasionally be transmitted non-sexually. Nevertheless, *T. vaginalis* infections in children should at least raise the question of sexual abuse and possible exposure to other STIs. Newborn infants can contract *T. vaginalis* from infected mothers during vaginal birth. The high percentage of asymptomatic infections, especially in men, helps transmission of the disease. Infection rates are highest in men and women with multiple sex partners. *T. vaginalis* causes tissue damage, so people with this disease are at higher risk of contracting HIV.

Treatment and Prevention

Most strains of *T. vaginalis* respond quickly to treatment with metronidazole or tinidazole, but a few are resistant. Transmission is prevented by abstinence, monogamy, and the use of condoms. **Table 27.20** gives the main features of trichomoniasis.

The key features of the diseases covered in this chapter are highlighted in the **Diseases in Review 27.1** table that follows.

TABLE 27.20 Trichomoniasis

Signs and symptoms	Many women, most men asymptomatic. Women: itching, burning, swelling, vaginal redness; frothy, sometimes bad-smelling, yellow-green discharge, and burning on urination; preterm births sometimes a complication. Men: urethral discharge, burning on urination, painful testes, tender prostate.
Incubation period	4 to 20 days
Causative agent	*Trichomonas vaginalis,* a protozoan with unmistakable jerky motility
Pathogenesis	Unexplained; inflammatory changes and pinpoint hemorrhages suggest mechanical trauma from the motile organisms.
Epidemiology	Worldwide distribution; asymptomatic infections foster spread; easily killed by drying due to lack of cyst form; transmission by intimate contact; high rate of infection with multiple sex partners; newborn infants of infected mothers can acquire the organism at birth.
Treatment and prevention	Treatment: appropriate antimicrobial agent. Prevention: abstinence, monogamy, and consistent use of condoms.

MicroAssessment 27.6

Trichomoniasis is caused by a flagellated protozoan that does not form cysts and lacks mitochondria.

16. What significance does the lack of a cyst form of *Trichomonas vaginalis* have on the epidemiology of trichomoniasis?

17. What is the relationship between being "easily killed by drying" and "transmission usually by sexual contact"?

FOCUS ON THE FUTURE 27.1

Getting Control of Sexually Transmitted Infections

Few problems are as complicated as getting control of STIs because of the psychological, cultural, religious, and economic factors involved—factors that vary from one population and culture to another. Gaining control means improving diagnosis and treatment, interrupting transmission, and increasing educational efforts. The challenge is to develop innovative approaches in each of these areas and to apply them worldwide on a sustained basis.

In the area of diagnosis, the challenge is to use molecular biological techniques to identify and produce specific antigens and antibodies that can be used to quickly identify causative agents and determine the extent of their spread. The development of nucleic acid probes to identify genes coding for resistance to therapeutic agents helps in choosing prompt treatment.

Using condoms is an effective method for preventing STI transmission, but in practice there are a number of problems with using them. When infected college students are asked why they did not use a condom, they often state that "it takes away the romance," or "it decreases pleasurable sensation." Also, many individuals are allergic to latex, and the alternatives, except for polyurethane, are unreliable for preventing disease transmission. Worldwide, as many as one out of three condoms is defective, and people in the poorest countries cannot even afford them. The challenge is to develop cheaper, more reliable, and more acceptable barrier methods, perhaps using new polymers or antimicrobial vaginal gels.

Effective vaccines could block the spread of STIs, but their development has been extremely problematic and remains a long-term goal. Sequencing the genomes of the causative agents is helping to identify appropriate antigens to include in vaccines, but using them to stimulate a protective immune response is a continuing challenge.

Educational efforts focusing on groups at high risk for STIs have had mixed success. Education can be a useful tool for gaining control of STIs, and the challenge is to better understand the reasons for its failures.

Diseases in Review 27.1

Genitourinary Infections

Disease	Causative Agent	Comment	Summary Table
UROGENITAL INFECTIONS			
Cystitis (bladder infection)	Usually uropathogenic *Escherichia coli*	Most common in women because of their relatively short urethra; characterized by painful urination; can progress to kidney infection.	Table 27.1
Leptospirosis	*Leptospira interrogans*	Spread by water contaminated with the urine of infected animals; enters via mucous membranes or breaks in skin, and then spreads to all tissues, including eyes.	Table 27.2
Bacterial vaginosis (BV)	Unknown, possibly bacterial	Characterized by thin, gray-white discharge and fishy odor; most common in sexually active women and pregnant women.	Table 27.3
Vulvovaginal candidiasis (VVC)	*Candida albicans* (a fungal disease)	Characterized by intense itching and thick, white discharge; associated with antibiotics or other conditions that disrupt the normal vaginal microbiota, allowing the overgrowth of *C. albicans.*	Table 27.4
Staphylococcal toxic shock	Toxin-producing strains of *Staphylococcus aureus*	Toxin is a superantigen and causes cytokine release, leading to drop in blood pressure.	Table 27.5
BACTERIAL STIS			
Chlamydial infections	*Chlamydia trachomatis*	Most common STI in the United States; infections are often asymptomatic; signs and symptoms include genital discharge and painful urination; can progress to PID if untreated.	Table 27.6
Gonorrhea	*Neisseria gonorrhoeae*	Infections are often asymptomatic, particularly in women; signs and symptoms include genital discharge and painful urination; can progress to PID.	Table 27.7
***Mycoplasma genitalium* infections**	*Mycoplasma genitalium*	Infections often asymptomatic; signs and symptoms resemble those of chlamydia and gonorrhea, and can cause PID in women; difficult to diagnose.	Table 27.8
Syphilis	*Treponema pallidum*	Primary syphilis is characterized by a painless chancre, secondary by a widespread rash, and tertiary by cardiovascular and nervous system damage; congenital infections also occur.	Table 27.9 Table 27.10
Chancroid	*Haemophilus ducreyi*	Characterized by painful genital sores and enlarged groin lymph nodes.	Table 27.11
VIRAL STIS			
Genital herpes	Herpes simplex virus, usually type 2 (HSV-2)	Characterized by itching, burning, painful blisters; virus becomes latent, so infections are lifelong.	Table 27.12
Human papillomavirus infections	Human papillomaviruses (HPVs)	Some strains cause genital warts, and others cause cancers. Vaccines protect against some cancer-causing strains.	Table 27.13
HIV/AIDS	Human immunodeficiency virus, usually type 1 (HIV-1)	HIV disease usually progresses to AIDS; virus infects helper T cells, resulting in immunodeficiency; ART slows disease progression but not curative.	Table 27.19
PROTOZOAN STI			
Trichomoniasis ("trich")	*Trichomonas vaginalis*	Sexually transmitted; characterized by itching, discharge, and urinary discomfort.	Table 27.20

Summary

27.1 ■ Anatomy, Physiology, and Ecology

The Urinary System (figure 27.1)
The urinary system is composed of the kidneys, ureters, bladder, and urethra. Infections occur more frequently in women than in men because of the shortness of the female urethra and its closeness to the anus.

The Genital System (figure 27.2)
The fallopian tubes provide a passageway for infections to enter the abdominal cavity. The vagina is a portal of entry for a number of infections. The normal vaginal microbiota, including lactobacilli, protect the vagina from colonization by pathogens. In men, prostate secretions have antimicrobial properties.

27.2 ■ Urinary Tract Infections

Any condition that prevents a person from urinating normally increases the risk of infection. The urinary system usually becomes infected by organisms ascending from the urethra, but it can also be infected from the bloodstream.

Bacterial Cystitis ("Bladder Infection") (table 27.1)
Most UTIs in healthy people are caused by *Escherichia coli* or other *Enterobacteriaceae* members from the person's own normal intestinal microbiota. Healthcare-associated UTIs are common and are caused by *Pseudomonas aeruginosa, Serratia marcescens,* and *Enterococcus faecalis,* which are often resistant to many antibiotics. Kidney infection (**pyelonephritis**) may complicate a bladder infection when pathogens ascend through the ureters to the kidneys.

Leptospirosis (table 27.2)
In leptospirosis, the urinary system is infected by *Leptospira interrogans* from the bloodstream (figure 27.3). This biphasic illness causes fever, bloodshot eyes, and pain in the septicemic phase. Improvement occurs, then new signs and symptoms emerge, with damage to multiple organs during the immune phase.

27.3 ■ Genital System Diseases

Bacterial Vaginosis (BV) (table 27.3, figure 27.4)
Bacterial vaginosis is the most common cause of vaginal signs and symptoms that include a gray-white discharge from the vagina and a strong fishy odor; there is no inflammation. The causative agent or agents are unknown, but BV occurs with a disruption of the normal microbiota.

Vulvovaginal Candidiasis (VVC) (table 27.4)
Vulvovaginal candidiasis signs and symptoms include itching, burning, vulvar redness and swelling, and a thick, white discharge. The causative agent is a yeast, *Candida albicans,* that is commonly part of the normal vaginal microbiota (figure 27.5). Antibacterial treatment, uncontrolled diabetes, and oral contraceptives are predisposing factors, but in most cases no such factor can be identified.

Staphylococcal Toxic Shock Syndrome (table 27.5)
Staphylococcal toxic shock syndrome became widely known with a 1980 epidemic in menstruating women who used a certain kind of tampon that has since been removed from the market (figure 27.6). Signs and symptoms include sudden fever, headache, muscle aches, bloodshot eyes, vomiting, diarrhea, a sunburn-like rash that later peels, and confusion. The blood pressure drops, and without treatment, kidney failure and death may occur.

SEXUALLY TRANSMITTED INFECTIONS

Focus On Sexually Transmitted Infections
STIs are typically spread through intimate contact with body fluids. Many are asymptomatic; signs and symptoms often include genital discharge, and pain on urination. Inflammatory reaction to the infection may cause scarring, which can partially obstruct the urethra or cause infertility in men and women. Simple measures for controlling STIs include abstinence from sexual intercourse, a monogamous relationship with an uninfected person, and consistent and correct use of latex or polyurethane condoms.

27.4 ■ Bacterial STIs

Chlamydial Infections (table 27.6, figure 27.7)
Chlamydial genital infections, caused by *Chlamydia trachomatis,* are reported more often than any other STI. Asymptomatic infections are common and easily transmitted. Signs and symptoms include genital discharge and pain.

Gonorrhea (table 27.7; figures 27.9, 27.10)
Gonorrhea, caused by *Neisseria gonorrhoeae,* is frequently asymptomatic. If present, signs and symptoms include painful urination and thick pus draining from the urethra in men (figure 27.8). Women may have similar signs and symptoms, but these tend to be milder.

***Mycoplasma genitalium* Infections** (table 27.8)
Mycoplasma genitalium (figure 27.11) infections are becoming increasingly common. Asymptomatic infections are common. Signs and symptoms, when present, are similar to those of chlamydia and gonorrhea. Diagnosis is difficult because the causative agent is fastidious and grows very slowly in vitro.

Syphilis (tables 27.9, 27.10)
Syphilis is caused by the spirochete *Treponema pallidum* (figure 27.16). Primary syphilis is characterized by a painless, firm ulceration called a hard **chancre;** the causative organism multiplies and spreads throughout the body (figure 27.12). In secondary syphilis, skin and mucous membranes show lesions that have large numbers of the organisms (figure 27.13); a latent period of months or years occurs between the secondary and tertiary phases of the disease. Tertiary syphilis is not contagious. It causes damage to the eyes and the cardiovascular and central nervous systems. An inflammatory, necrotizing mass called a **gumma** can involve any part of the body (figure 27.14). Syphilis in pregnant women can spread across the placenta to involve the fetus, resulting in congenital syphilis (figure 27.15).

Chancroid (table 27.11)
Chancroid is caused by *Haemophilus ducreyi,* and is characterized by single or multiple soft, tender genital ulcers, and enlarged, painful groin lymph nodes (figure 27.17).

27.5 ■ Viral STIs

Viral STIs are at least as common as bacterial STIs, but they are not yet curable.

Genital Herpes (table 27.12)

Genital herpes is a very common disease, caused by herpes simplex viruses (usually HSV type 1, but also HSV type 2). Signs and symptoms include blisters (figure 27.18) with itching, burning, or pain; these break, leaving ulcers. HSV establishes a latent infection in sensory nerves and cannot be cured. The virus can reactivate, causing recurrent infections; many of these have few or no signs or symptoms while others are characterized by painful blisters. Genital herpes can be transmitted in the absence of signs or symptoms, but the risk is greatest when lesions are present.

Human Papillomavirus STIs: Genital Warts and Cervical Cancer (table 27.13)

Human papillomavirus STIs, which are probably more common than any other kind of STI, are caused by human papillomaviruses (HPVs). Genital infection symptoms include warts on or near the genitalia (figure 27.19). Some HPV types cause precancerous lesions that are asymptomatic and can be detected only by medical examination. HPVs are the main cause of abnormal Pap smears in young women (figure 27.20). Several types of HPV are associated with cancer of the cervix, vagina, penis, anus, or throat. Gardasil 9 (9vHPV) vaccine protects against the most common cancer- and wart-causing strains.

HIV/AIDS (tables 27.15, 27.16, 27.17, 27.18, 27.19)

AIDS is the end stage of disease caused by human immunodeficiency virus (HIV) (figures 27.23, 27.24, 27.25). Worldwide, millions of new HIV infections and deaths from AIDS occur each year (figures 27.21, 27.26). HIV disease first manifests as a flu-like illness that develops about 6 days to 6 weeks after infection with the virus. An asymptomatic interval follows that typically lasts almost 10 years, during which the immune system is slowly and progressively destroyed. Unusual cancers and infectious diseases indicate the start of AIDS (figure 27.22). No vaccine or medical cure is yet available, but antiretroviral therapy (ART) reduces viral load and maintains health in infected people. HIV transmission can be prevented by abstinence, monogamous relationships with an uninfected partner, and consistent use of condoms, as well as by use of sterile needles by injected-drug abusers. Blood and blood products are screened before use. A significant reduction in mother-to-newborn transmission can be achieved with medication. Circumcision and pre-exposure prophylaxis with antiretroviral medication also show promise in reducing infection rates.

27.6 Protozoan STIs

Trichomoniasis ("Trich") (table 27.20)

Signs and symptoms include itching, burning, swelling, and redness of the vagina; frothy, sometimes smelly, yellow-green discharge, and burning on urination. Men may have discharge from the penis and burning on urination, sometimes accompanied by painful testes and a tender prostate gland. Many women and most men are asymptomatic. The causative agent is *Trichomonas vaginalis,* a protozoan with four anterior flagella, and a posterior flagellum attached to an undulating membrane (figure 27.27).

Review Questions

Short Answer

1. How do lactobacilli help protect the vagina from potential pathogens?
2. List four factors that increase the likelihood of developing a urinary tract infection.
3. What are the typical signs and symptoms of cystitis?
4. What danger might possibly be found in a spot on the ground where an animal urinated 1 week earlier?
5. What is a clue cell?
6. List three diseases caused by different antigenic types of *Chlamydia trachomatis.*
7. What is ophthalmia neonatorum?
8. Why is dark-field microscopy used to view wet mounts of *Treponema pallidum*?
9. Give two ways in which the chancre of chancroid differs from the chancre of syphilis.
10. What is the relationship between AIDS and HIV disease?

Multiple Choice

1. Which of the following about bacterial cystitis is *false*?
   a) About one-third of all women will have it at some time during their life.
   b) Catheterization of the bladder markedly increases the risk of contracting the disease.
   c) Individuals who have a bladder catheter in place indefinitely risk bladder infections with multiple species of intestinal bacteria at the same time.
   d) Bladder infections occur as often in men as they do in women.
   e) Bladder infections can be asymptomatic.
2. Choose the one correct statement about leptospirosis.
   a) Humans are the only reservoir.
   b) Most infections produce severe symptoms.
   c) Transmission is by the fecal-oral route.
   d) It can lead to unnecessary abdominal surgery.
   e) An effective vaccine is generally available for preventing human disease.
3. Which one of the following statements about bacterial vaginosis is *false*?
   a) It is the most common vaginal disease in women of childbearing age.
   b) In pregnant women, it is associated with premature delivery.
   c) Inflammation of the vagina is a constant feature of the disease.
   d) The vaginal microbiota shows a significant decrease in lactobacilli and a marked increase in anaerobic bacteria.
   e) The cause is unknown.
4. Pick the one *false* statement about vulvovaginal candidiasis.
   a) It often involves the external genitalia.
   b) It is readily transmitted by sexual intercourse.

c) It is caused by a yeast present among the normal vaginal microbiota in about one-third of healthy women.
d) It is associated with prolonged antibiotic use.
e) It involves increased risk late in pregnancy.

5. All of the following statements about staphylococcal toxic shock are true *except*
a) it can lead to kidney failure.
b) the causative organism usually does not enter the bloodstream.
c) it occurs only in tampon users.
d) the causative organism produces superantigens.
e) person-to-person spread does not occur.

6. Which one of these statements about chlamydial genital infections is *false*?
a) The incubation period is usually shorter than in gonorrhea.
b) Infected cells develop inclusions.
c) Pelvic inflammatory disease (PID) can be complicated by infection of the surface of the liver.
d) Tissue damage largely results from cell-mediated immunity.
e) Fallopian tube damage can occur in the absence of symptoms.

7. Which of the following statements about gonorrhea is *false*?
a) The incubation period is less than a week.
b) Disseminated gonococcal infection (DGI) is almost invariably preceded by prominent urogenital symptoms.
c) DGI can result in arthritis of the knee.
d) Phase and antigenic variation helps the causative organism evade the immune response.
e) Pelvic inflammatory disease (PID) is common in untreated women.

8. Which symptom is least likely to occur as a result of tertiary syphilis?
a) Gummas
b) White patches on mucous membranes
c) Emotional instability
d) Stroke
e) Blindness

9. Which of the following statements about HIV/AIDS is *false*?
a) HIV is transmitted by body fluids such as blood and semen.
b) Antiretroviral medications can cure AIDS.
c) Unusual cancers and infections indicate the start of AIDS.
d) HIV can be transmitted transplacentally to a fetus.
e) HIV infects white blood cells, damaging the immune system.

10. All of the following are true of "trich" (trichomoniasis) *except*
a) it can cause burning pain on urination and painful testes in men.
b) it occurs worldwide.
c) asymptomatic infections are rare.
d) transmission can be prevented by proper use of condoms.
e) individuals with multiple sex partners are at high risk of contracting the disease.

Applications

1. Religious restrictions of a small North African community are preventing a World Health Organization project from reducing the incidence of gonorrhea. The community will not allow the testing of females for the disease. They can be treated, however, if they show outward evidence of the disease. Only males are allowed to participate fully in the project, with testing for the disease and treatment. The village elders argue that eradicating the disease from males would eventually remove it from the population. What would be the impact of these restrictions on the success of the project?
2. Former President Ronald Reagan once commented at a press conference that the best way to combat the spread of AIDS in the United States was to prohibit everyone from having sexual contact for 5 years. What would be the success of such a program if it were possible to carry it out?

Critical Thinking

1. The middle curve of figure 27.6 shows the occurrence of staphylococcal toxic shock syndrome in menstruating women from 1979 to 2013. What aspect of these data argues that high-absorbency tampons were not the only cause of staphylococcal toxic shock syndrome associated with menstruation?
2. In early attempts to identify and isolate the cause of syphilis, various bacteria in the discharge from syphilitic lesions in experimental animals were isolated in pure culture. None of them, however, caused the disease when used in attempts to infect healthy animals. Why was it considered a critical step to have the cultivated bacteria reproduce the disease in the healthy animals?

28 Microbial Ecology

Farming relies on the activities of microorganisms. *Source: Tim McCabe/USDA*

KEY TERMS

Consumers Organisms that eat primary producers or other consumers.

Decomposers Organisms that digest the remains of primary producers and consumers.

Ecology Study of interactions of organisms with one another and with their environment.

Eutrophic An environment that is nutrient rich, supporting the excessive growth of algae and other organisms.

Hydrothermal Vents Undersea geysers that spew out mineral-laden hot water.

Hypoxic An environment very low in dissolved O_2.

Microbial Mat A type of microbial community characterized by distinct layers of different groups of microbes that together make up a thick, dense, highly organized structure.

Nitrogen Fixation Conversion of nitrogen gas to ammonia.

Oligotrophic A nutrient-poor environment.

Primary Producers Organisms that convert CO_2 into organic compounds, sustaining other forms of life.

Rhizosphere Zone around plant roots containing organic materials secreted by the roots.

A Glimpse of History

While many microbiologists in the late 1800s were particularly interested in studying disease-causing bacteria, the Russian microbiologist Sergei Winogradsky (1856–1953) was more intrigued by microorganisms in soil and water. Considered by some to be the founder of the field of microbial ecology, he was the first to recognize what he called chemosynthesis: the biological conversion of inorganic to organic chemicals.

Winogradsky realized that although the pure culture techniques developed by Robert Koch are useful for studying medically important bacteria, they are not adequate for studying environmental microorganisms. So instead he tried to mimic in the laboratory the natural conditions in which organisms grow. A device called a Winogradsky column illustrates his method. This in vitro ecosystem can be built using a cylindrical glass container filled with mud, carbon and sulfur sources, and water from a pond or lake. The column is covered to prevent evaporation and placed in sunlight. Eventually, layers of different colors corresponding to microbial populations with different physiological capabilities will appear in the column. The top layer consists of algae and cyanobacteria that produce O_2. Below that are the green and purple sulfur bacteria that oxidize H_2S, producing sulfur. Near the bottom of the column are the sulfate-reducers, producing H_2S and precipitating iron.

Winogradsky's work was instrumental in describing the global nutrient cycles described in this chapter. In recognition of his contributions, the bacterium *Nitrobacter winogradskyi* was named in his honor.

Microorganisms have existed on Earth for about 4 billion years. By all evidence they occupy every possible environment, from ocean depths to hot springs and from rich agricultural soil to the human gut. In fact, without microbial activities, life on Earth could not exist—we depend on them to capture energy, cycle nutrients, maintain fertile soil, and decompose dead organisms and pollutants. Without microorganisms, the earth would quickly become buried by tons of wastes, and nutrients would be depleted, halting growth and reproduction.

In view of the crucial functions microorganisms perform, it seems we should know a great deal about the diverse microbial species that inhabit our surroundings. Quite the opposite is true, however, as less than a mere 1% have been successfully grown in culture. In chapter 4, we discussed how organisms are grown as pure cultures in the laboratory under controlled conditions that promote optimal growth. In nature, however, organisms generally grow in close associations with many different species, often in biofilms. Nutrients may be in short supply, quite unlike the rich conditions in growth media.

◀◀ biofilms, ◀◀ microbial growth in laboratory conditions

When researchers work with environmental samples in the laboratory, the microorganisms that grow in culture are not necessarily the most abundant or the most important in their natural setting. They are simply the ones able to grow best under the conditions we provide. Even if all microorganisms could be cultivated in the laboratory, their behavior in culture might not accurately reflect their role in the environment. Techniques of molecular biology provide much more extensive and reliable information through the direct analysis of DNA from environmental samples. These studies provide vital clues to the diversity and roles of microorganisms in the natural world. Efforts are currently underway to sequence microbial genomes from around the globe.

Chapter 11 presented the great diversity of physiological mechanisms that prokaryotes use to survive in terrestrial and aquatic environments. This chapter will expand on some of those concepts by describing actions of microbes in those habitats. We will also describe the roles of microorganisms in cycling of nutrients and capturing energy in a habitat. Finally, we will examine some of the intimate relationships that certain microorganisms maintain with eukaryotic organisms. ⏮ ecophysiology

28.1 ■ Principles of Microbial Ecology

Learning Outcomes

1. Compare and contrast the roles of primary producers, consumers, and decomposers.
2. Explain how some microbes are able to grow in low-nutrient environments.
3. Describe the significance of microbial competition.
4. Describe how environmental changes can result in alterations in a microbial community.
5. Describe the structural organization of a microbial mat.

Ecology is the study of interactions of organisms with one another and with their physical environment. Organisms of the same type in a given location make up a **population.** All of the different organisms in the location form a **community.** An **ecosystem** consists of a community of organisms and the non-living environment with which they interact. Major ecosystems include the oceans, rivers and lakes, deserts, marshes, grasslands, forests, and tundra. Each ecosystem possesses characteristic organisms and physical conditions. All of the ecosystems on Earth make up the **biosphere.**

Within the biosphere, ecosystems vary in both **biodiversity** (number and variety of species present and their evenness of distribution) and **biomass** (the weight of all organisms present). Microorganisms play a major role in most ecosystems, and many ecosystems host microbes unique to themselves. The role an organism plays in a particular ecosystem is called its **ecological niche.**

The environment immediately surrounding an individual microbe—the **microenvironment**—is most relevant to that cell, but because microorganisms are so small, the microenvironment is difficult to identify and measure. The more easily measured gross environment—the **macroenvironment**—may be very different from the microenvironment. Consider a bacterial cell living within a biofilm (see figure 4.3); growth of aerobic organisms in the biofilm can deplete O_2, creating tiny zones where obligate anaerobes can grow. Fermenters can produce organic acids that may then be metabolized by other organisms in the biofilm. In addition, various growth factors and even genes can be transferred between organisms. Thus, microorganisms that might be unexpected in a given macroenvironment may actually thrive there within specialized microenvironments. ⏮ biofilm, ⏮ growth factor

As described in chapter 16, living organisms often interact with one another in long-lasting and intimate relationships called symbioses. In a mutualistic relationship, both organisms benefit from the relationship. For example, bacteria in the human gut benefit from a steady flow of available nutrients, but they also provide metabolic end products that our bodies can use and they compete with potential pathogens. In a commensalistic relationship, one organism benefits and the other is unaffected. We have this type of relationship with many of the microorganisms found on our skin. In a parasitic relationship, one organism benefits, but the other is harmed. Well-adapted parasites do not usually kill their host, unlike a predator. ⏮ symbiosis

Nutrient Acquisition

Organisms are categorized according to their trophic level (source of food), which is intimately related to the cycling of nutrients. There are three general trophic levels (**figure 28.1**):

- **Primary producers.** These are autotrophs; they convert CO_2 into organic materials. Producers include both photoautotrophs (use sunlight for energy) and chemolithoautotrophs (oxidize inorganic chemicals for energy). Primary producers serve as a food source for consumers and decomposers. ⏮ photoautotroph, ⏮ chemolithoautotroph
- **Consumers.** These are heterotrophs that eat primary producers or other consumers. Herbivores, which eat plants or algae, are primary consumers. Carnivores that eat herbivores are secondary consumers; carnivores that eat other carnivores are tertiary consumers. A chain of consumption is a food chain; interacting food chains form a food web.
- **Decomposers.** These are heterotrophs that digest the waste products and remains of primary producers and consumers. Decomposers specialize in digesting complex materials such as cellulose, converting them into small molecules that can more easily be used by other organisms. The complete breakdown of organic molecules into inorganic molecules such as ammonia, sulfates,

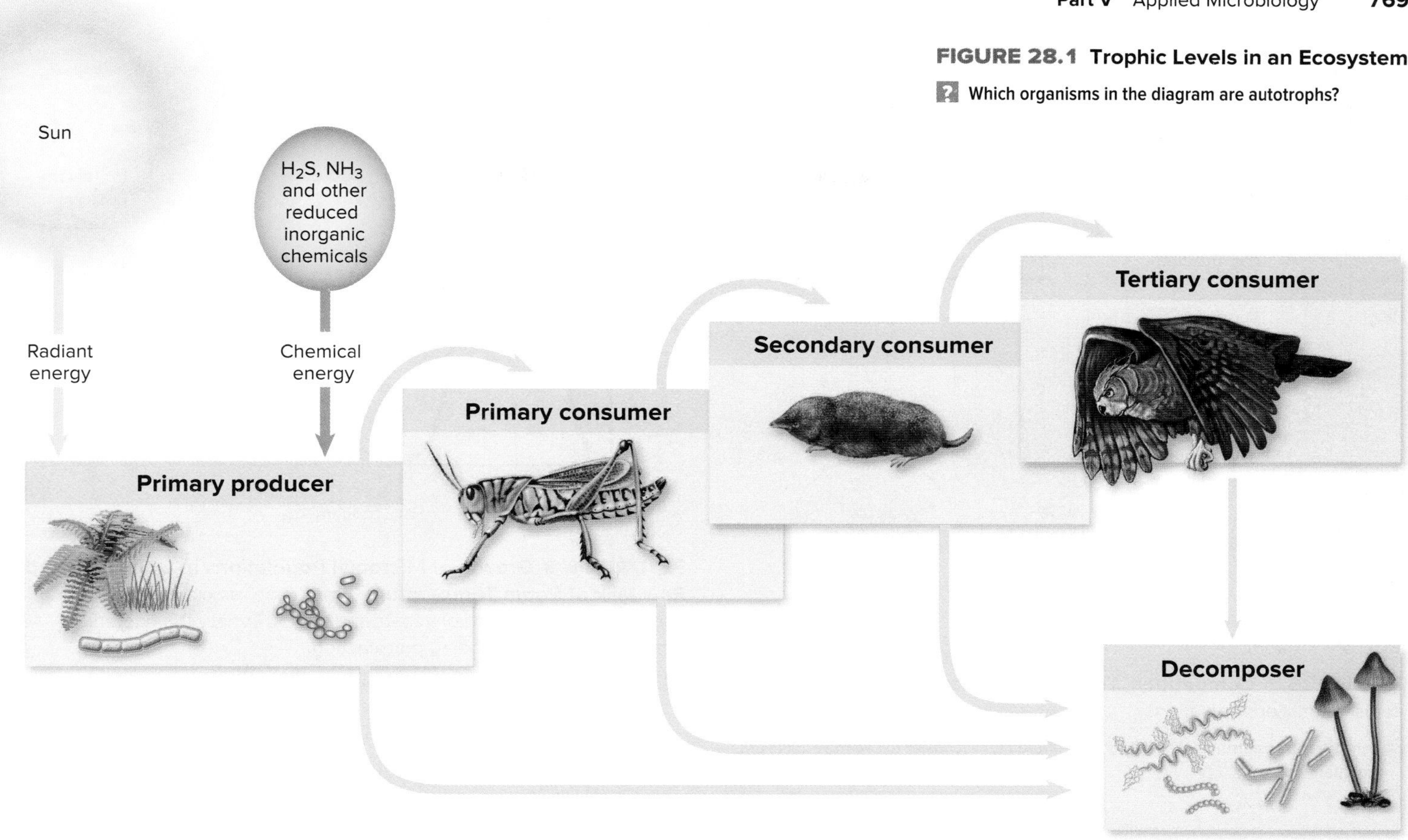

FIGURE 28.1 Trophic Levels in an Ecosystem

Which organisms in the diagram are autotrophs?

phosphates, and carbon dioxide is called mineralization. Microorganisms, particularly bacteria and fungi, play a major role in decomposition because of their unique metabolic capabilities.

Microbes in Low-Nutrient Environments

Low-nutrient environments such as lakes, rivers, and streams are common in nature, so microorganisms that can grow in dilute aqueous solutions are widespread. Most microbial growth in these settings is in biofilms, and the cells are shed from the biofilm into the aqueous solution.

Even distilled water has trace amounts of nutrients absorbed from the air, so microorganisms can grow in distilled-water reservoirs used in laboratories. Although the organisms grow slowly, they can reach concentrations as high as 10^7 per milliliter. This cell concentration is not high enough to result in a cloudy solution, so the growth usually goes unnoticed, but can have serious consequences for the success of laboratory experiments that depend on water purity.

Organisms that grow in dilute environments contain highly efficient transport systems for moving nutrients into the cell. Other mechanisms that bacteria use to thrive in dilute aquatic environments are described in chapter 11. ⏮ **transport systems,** ⏮ **thriving in aquatic environments**

Microbial Competition

Perhaps nowhere in the living world is competition more quickly evident than among microorganisms. The ability of a microorganism to compete successfully for a habitat is generally related to the rate at which it multiplies, as well as to its ability to withstand adverse environmental conditions. Because bacteria multiply logarithmically, any small differences in their generation times will result in a very large difference in the total number of cells of each species after a relatively short time (**figure 28.2**).

The outcome of microbial competition is not always simply determined by rate of replication. Some microorganisms actively compete using antagonistic strategies. In the soil, for example, some microbes resort to a type of chemical warfare, producing antimicrobial compounds such as **bacteriocins** (proteins produced by bacteria that kill closely related strains). It is tempting to speculate that the natural function of antibiotics produced by *Streptomyces* species is to inhibit competing organisms, but their role in nature is still poorly understood. Recent evidence suggests that antibiotics are important in cell-to-cell signaling. ⏮ **the genus *Streptomyces***

Microorganisms and Environmental Changes

Environmental changes often result in alterations in a community. Those organisms that have adapted to live several

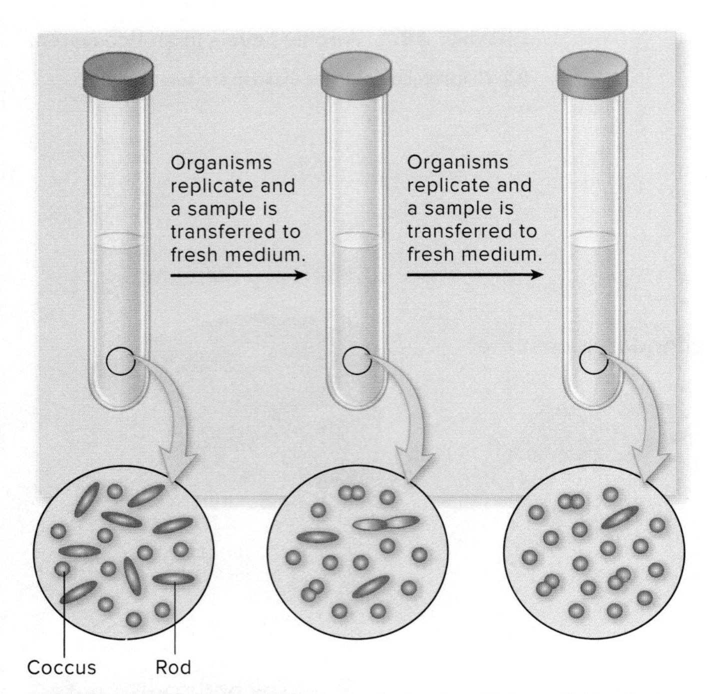

FIGURE 28.2 Competition The bacterium that multiplies faster yields the larger population.

Why does transferring only some of the culture make it easier to see the effects of competition?

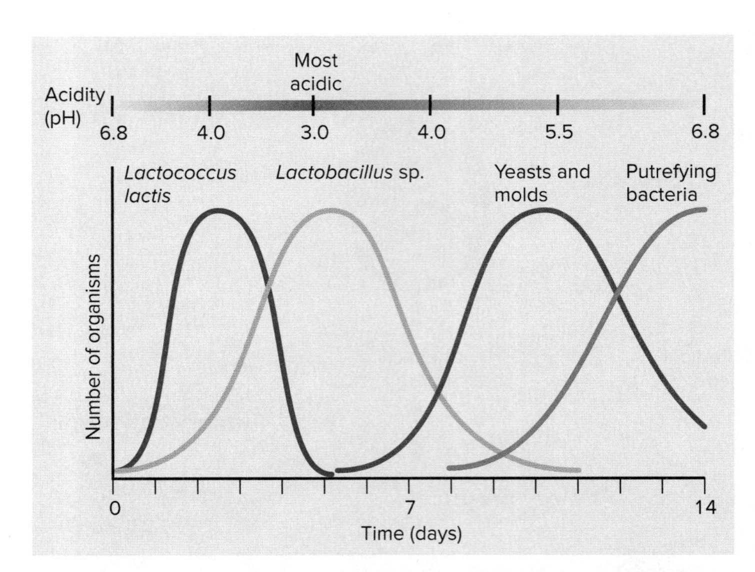

FIGURE 28.3 Growth of Microbial Populations in Unpasteurized Raw Milk at Room Temperature Production of acid causes souring and encourages growth of yeasts and molds. Eventually, bacteria digest the proteins, causing putrefaction.

Why does the pH of the milk first decrease and then increase?

inches beneath the surface of an untilled field will probably not be well suited to growth in that field if it is plowed, fertilized, and irrigated. In addition to external sources of environmental change, the growth and metabolism of organisms themselves can alter the environment dramatically. Nutrients may become depleted and toxic waste products accumulate.

In some environments, the changing conditions bring about a highly ordered and predictable succession of bacterial species. An example of this occurs in unpasteurized milk, which usually contains various species of microbes associated with cows. Initially, the dominant bacterium is *Lactococcus lactis,* which breaks down the milk sugar lactose, forming lactic acid as a fermentation end product (**figure 28.3**). This sours the milk and denatures milk proteins, causing the milk to curdle. The acid inhibits most other organisms in the milk, and eventually enough acid is produced to inhibit *L. lactis. Lactobacillus* species can multiply in this highly acidic environment, however, and these bacteria metabolize any remaining sugar, forming more acid until their growth is also inhibited. Yeasts and molds, which tolerate even more acidic conditions, then become the dominant group. They oxidize the lactic acid, which raises the pH. Most of the sugar has already been used at this point, so the lactococci and lactobacilli cannot resume multiplication. Milk protein (casein) is still available, however, and can be used by protease-producing members of the endospore-forming genus *Bacillus.* This breakdown of protein, known as putrefaction, yields a completely clear and foul-smelling product. The milk thus goes through a succession of changes with time, first souring and finally putrefying. ⏮ proteases

Microbial Communities

Microorganisms most often grow as biofilms attached to solid surfaces or at air-water interfaces. General aspects of biofilms were described in detail in chapter 4. In this section, we focus on a specific type of biofilm: a microbial mat. ⏮ biofilms

A **microbial mat** is a thick, dense, highly organized structure composed of distinct layers. Frequently they are green, reddish-pink, and black, which indicate the growth of different microbial groups (**figure 28.4**). The top green layer is typically composed of various species of cyanobacteria, and the color is due to their photosynthetic pigments. Directly below the green layer is a reddish-pink layer consisting of purple sulfur bacteria. The light-harvesting pigments of these anoxygenic phototrophs can use wavelengths of light not collected by the cyanobacteria. At the bottom is a black layer, resulting from iron molecules reacting with hydrogen sulfide produced by a group of bacteria called sulfate-reducers. These obligate anaerobes oxidize the organic compounds produced by the photosynthetic microbes growing in the mat's upper layers, using sulfate as a terminal electron acceptor. ⏮ cyanobacteria, ⏮ photosynthetic pigments, ⏮ purple sulfur bacteria, ⏮ sulfate-reducers, ⏮ terminal electron acceptor

Although microbial mats can be found in many areas, those near hot springs in Yellowstone National Park are some of the most intensively studied. The mats in these extreme areas are undisturbed by grazing eukaryotic organisms and, consequently, provide an important model for studying microbial interactions.

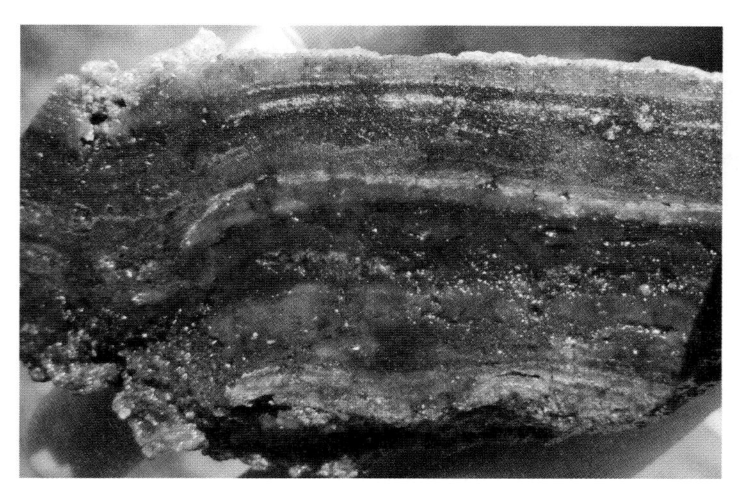

FIGURE 28.4 A Microbial Mat This dense structure is composed of distinct layers of different microbial communities. ©Reut S. Abramovich

Why are the layers of the mat different colors?

MicroAssessment 28.1

Microorganisms play a major role in most ecosystems. Organisms are categorized as primary producers, consumers, or decomposers. Many microorganisms live in low-nutrient environments where they form biofilms. Competition among microorganisms in a habitat can be intense. Microbial mats have distinct layers.

1. What are the roles of primary producers, consumers, and decomposers?
2. Describe the stages of milk spoilage.
3. What aspects of an ecosystem would increase its biodiversity?

28.2 ■ Studying Microbial Ecology

Learning Outcome

6. Explain the role of cultivation-independent methods to study microorganisms in the environment.

Estimates vary widely, but one gram of soil may contain between 10^4 and 10^7 different microbial species. Studying these is a daunting task by any method, and impossible using culture techniques (as mentioned earlier, less than 1% of microorganisms can be expected to grow under laboratory conditions). Yet scientists strive to understand the diversity of microorganisms on the planet, including their roles in nature and how they affect the environment. Researchers in microbial ecology often rely on **metagenomics,** the cultivation-independent study of communities or their members by analyzing genetic material taken directly from an environment (**figure 28.5**). Fragments of DNA are extracted from soil or another natural sample, and then one of a variety of DNA-sequencing methods is used to determine the nucleotide sequences of that DNA. The sequences are then compared with others available in public databases. One goal of current research is to expand and improve the existing databases, because the power of metagenomics depends on the quality of the databases. ◀◀ metagenomics

Metagenomic data provide two separate but related types of information about a microbial community: taxonomic diversity and genetic capabilities (see figure 28.5). Taxonomic diversity is studied by examining either the nucleotide sequence of 16S rRNA genes (18S rRNA genes for eukaryotes) or the deduced amino acid sequences of ribosomal proteins to classify and identify microorganisms. The genetic capabilities of the community as a whole are determined by analyzing the entire spectrum of sequenced genes in a sample. A significant challenge is to link the two types of information, allowing researchers to determine which uncultivated microbes have a given genetic capability. ◀◀ ribosome structure

Culture-independent methods are rapidly advancing the study of microbial ecology because sequence information learned from one species can be applied to others. For example, researchers found that variations of a gene coding for bacterial rhodopsin, a light-sensitive pigment that provides a mechanism for harvesting the energy of sunlight, are widespread in marine bacteria. This gene provides bacteria with a mechanism for phototrophy that does not require chlorophyll and might be an important mechanism for energy accumulation in ocean environments. ◀◀ bacteriorhodopsin

Additional methods to study microbial ecology are available as well. These include fluorescence in situ hybridization (FISH), which uses probe technology in combination with fluorescence microscopy to determine if a specific species or group is present in a sample (see figure 9.21), and DNA microarrays designed to detect genes with known functions (see figure 9.22). ◀◀ FISH, ◀◀ DNA microarrays

MicroAssessment 28.2

Metagenomics is the study of genetic material taken from an environment. Such cultivation-independent techniques allow researchers to better understand natural microbial communities.

4. What can researchers learn about a microbial community using metagenomics?
5. Why are molecular techniques important in studying microbial ecology?
6. How could FISH (fluorescence in situ hybridization) be used to determine the relative proportions of archaea and bacteria in a population?

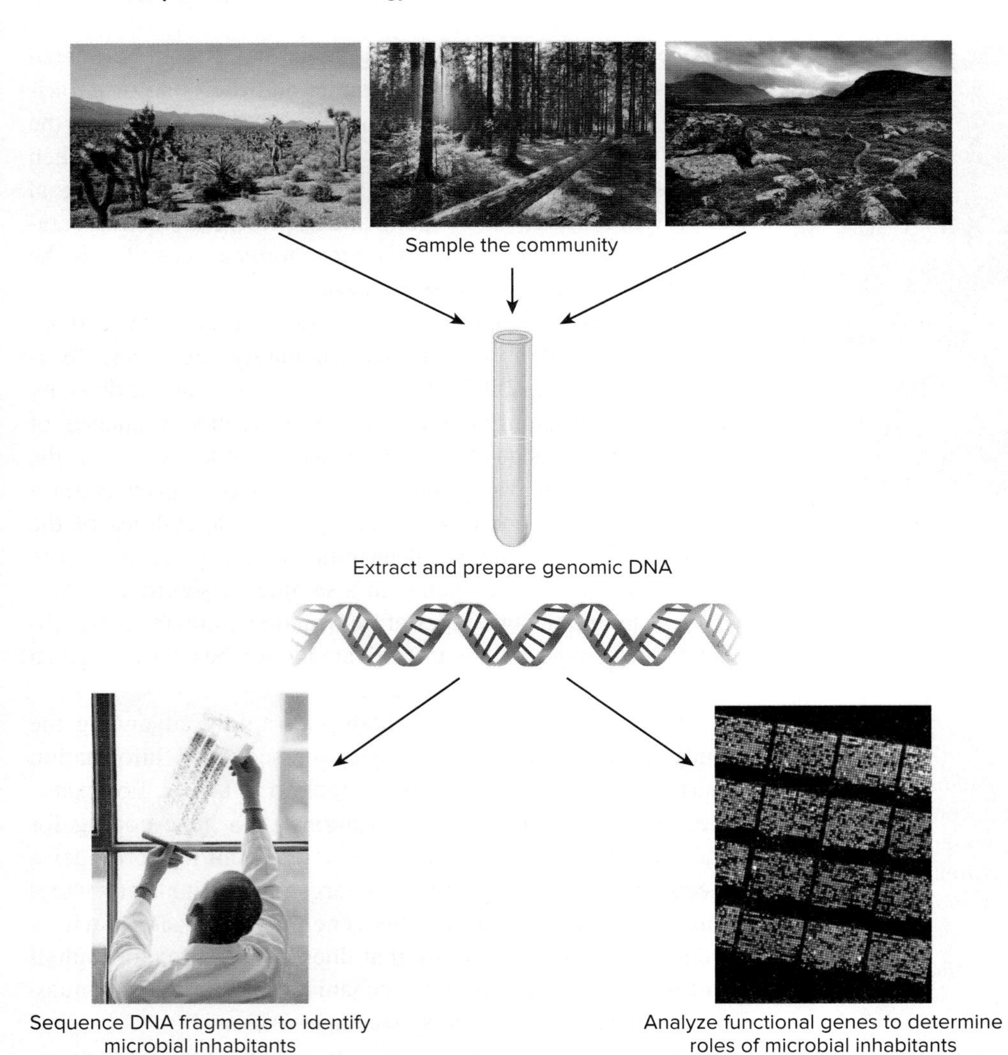

FIGURE 28.5 Metagenomics Used to Characterize Microbial Communities Gene sequences in DNA from environmental samples are compared to known gene sequences in current databases. (top left): Source: John & Karen Hollingsworth/U.S. Fish & Wildlife Service; (top middle): ©konradlew/Getty Images; (top right): ©ajliikala/Getty Images; (bottom left): ©Rob Melnychuk/Getty Images; (bottom right): ©Deco Images II/Alamy Stock Photo

What gene sequences are used to identify the organisms present in an environmental sample?

28.3 ■ Aquatic Habitats

Learning Outcome

7. Compare and contrast the habitats provided by marine, freshwater, and specialized aquatic environments.

Marine environments such as open oceans cover more than 70% of Earth's surface. They are the most abundant aquatic habitat, representing about 95% of the global water. The freshwater environments—lakes and rivers—represent only a small fraction of the total water.

Deep lakes and oceans have characteristic zones that influence the distribution of microbial populations. The uppermost layer—where sufficient light penetrates—supports the growth of photosynthetic microorganisms, including algae and cyanobacteria. The organic material synthesized by these primary producers gradually sinks and is then metabolized by heterotrophs.

The number of microorganisms in waters is influenced by the nutrient content. In **oligotrophic** waters, meaning nutrient poor, the growth of photosynthetic organisms and other autotrophs is limited by the lack of inorganic nutrients, particularly phosphate, nitrate, and iron. When waters are **eutrophic** (nutrient rich), photosynthetic organisms flourish, often forming a visible layer on the surface (**figure 28.6**). In turn, photosynthesis produces organic compounds that foster the growth of heterotrophs in lower layers. The heterotrophs consume dissolved O_2 as they metabolize the organic material. Because O_2 consumption can outpace the slow rate of diffusion of atmospheric O_2 into the waters, the environment can become **hypoxic** (very low in dissolved O_2). Insufficient O_2 leads to the death of resident fish and other aquatic animals.

Marine Environments

Marine environments range from the deep sea, where nutrients are scarce, to the shallower coastal regions, where nutrients may be abundant due to runoff from the land. Seawater contains about 3.5% salt, compared with about 0.05% for fresh water. Consequently, it supports the growth of halophilic organisms, which prefer or require high salt concentrations, and halotolerant ones. Temperatures often vary widely at the surface, but decrease with depth until reaching about 2°C in the deeper waters; an exception is the areas around hydrothermal vents. ⏮ halophile, ⏮ hydrothermal vents

Ocean waters are typically oligotrophic, limiting the growth of microorganisms. The small amounts of organic material produced by photosynthetic organisms are quickly consumed as they sink, so few nutrients reach the sediments below. Even in the deep sea, marine water is O_2-saturated due to mixing associated with tides, currents, and wind action.

The ecology of inshore areas is not as stable as the deep sea and can be dramatically affected by nutrient-rich runoff. An unfortunate example is the **dead zone**—a region lacking fish and other marine life—that forms in the Gulf of Mexico (**figure 28.7**). The Mississippi River, carrying nutrients

FIGURE 28.6 Eutrophication in a Polluted Stream Photosynthetic organisms flourish in the nutrient-rich water. The organic compounds they produce foster growth of heterotrophs in lower layers. ©Geography Photos/Universal Images Group via Getty Images

What is the likely fate of a fish living in a eutrophic stream?

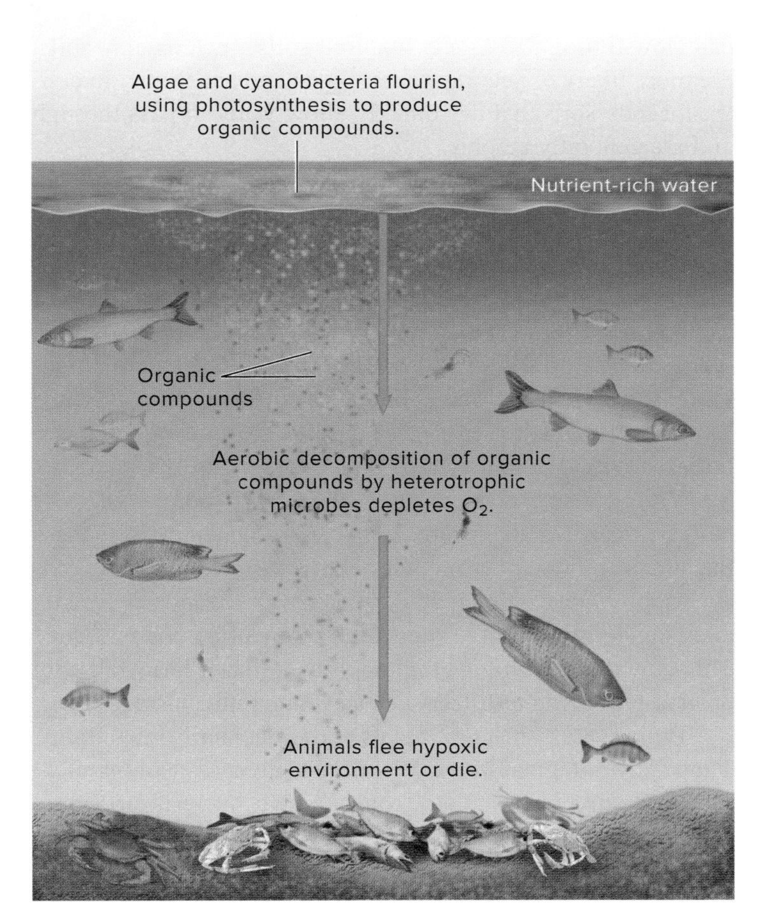

FIGURE 28.7 Dead Zone Formation

Why does decomposition by microbes deplete O_2 in water?

accumulated as it runs through agricultural, industrial, and urbanized regions, feeds into the Gulf. As a consequence of the excess nitrate and phosphate, algae and cyanobacteria flourish in the spring and summer when sunlight is also plentiful. Heterotrophic microbes then metabolize the organic compounds synthesized by these primary producers, consuming dissolved O_2 in the process. This causes a large region in the Gulf—sometimes in excess of 7,000 square miles—to become hypoxic. Animals in the area either flee or die. Dead zones are a problem in other waters as well, including fresh waters such as Lake Erie (see Focus on a Case 11.1). Nutrient enrichment of coastal waters also contributes to blooms of toxin-producing algae. ◀◀ medical importance of algae

Freshwater Environments

As with marine environments, the types and relative numbers of microbes inhabiting fresh waters depend on multiple factors, including light, concentration of dissolved O_2 and nutrients, and temperature.

Oligotrophic lakes in temperate climates may have anaerobic layers due to thermal stratification resulting from seasonal temperature changes. During the summer months, the surface water warms. This decreases the density of the water, causing it to form a distinct layer that does not mix with the cooler, denser water below. The upper layer, called the epilimnion, is generally O_2 rich due to the activities of photosynthetic organisms. In contrast, the lower layer, the hypolimnion, may be anaerobic due to the consumption of O_2 by heterotrophs. Separating these two layers is the thermocline, a zone of rapid temperature change. As the weather cools, the waters mix, providing O_2 to the deep water.

Rapidly moving waters, such as rivers and streams, are very different from lakes. They are usually shallow and turbulent, facilitating O_2 circulation, so they are generally aerobic. Light may penetrate to their bottom, making photosynthesis possible. Sheathed bacteria such as *Sphaerotilus* and *Leptothrix* species commonly adhere to rocks and other solid structures, where they then use nutrients that flow by. ◀◀ sheathed bacteria

Specialized Aquatic Environments

Specialized aquatic environments include salt lakes, such as the Great Salt Lake in Utah, which have no outlets. As water in these lakes evaporates, the salt concentrations become much higher than that in seawater. Extreme halophiles thrive in this environment. ◀◀ extreme halophiles

Other specialized habitats include iron springs that contain large quantities of ferrous ions; these springs are habitats for species of *Gallionella* and *Sphaerotilus.* Sulfur springs support the growth of both photosynthetic and non-photosynthetic sulfur bacteria. Other aquatic environments include groundwater, stagnant ponds, swimming pools, and drainage ditches, each offering its own opportunity for microbial growth.

MicroAssessment 28.3

Aquatic habitats include marine, freshwater, and specialized environments. Nutrient-rich waters can become hypoxic. Lakes often exhibit thermal stratification during summer months.

7. Compare the salt content of seawater with that of fresh water and salt lakes.
8. Explain how nutrient-rich runoff can cause waters to become hypoxic.
9. Why would the nutrient content of a body of water be more homogeneous than that of a terrestrial environment?

28.4 ■ Terrestrial Habitats

Learning Outcome

8. Describe soil as a microbial habitat.

Although microorganisms can adhere to and grow on a variety of objects on land, the focus in this section is soil—a crucial component of terrestrial ecosystems. Extreme terrestrial habitats, such as volcanic vents and fissures, and some of the extremophiles that inhabit them are described in chapter 11.

Soil is composed of finely ground rock, decaying organic material, air, and water. It is full of life, including bacteria, fungi, algae, protozoa, worms, insects, and plant roots. The top 6 inches of fertile soil may contain more than 2 tons of bacteria and fungi per acre! Scientists have known for years that the microorganisms are essential to the health and productivity of plants, but the National Microbiome Initiative now provides an exciting chance to learn much more. Just as it is difficult to unravel the relationships in the complex human microbiome, however, so it will be with the soil microbiome. ⏮ National Microbiome Initiative

An especially important part of the soil microbiome is found in the **rhizosphere** (the zone of soil that adheres to plant roots). The concentration of microbes there, particularly Gram-negative bacteria, is generally much greater than in surrounding soil. This is because the root cells secrete organic molecules that serve as a nutrient source for microbes. Specific types of bacteria appear to preferentially interact with particular plants. For example, the rhizosphere of certain grasses can have high concentrations of *Azospirillum* species, which fix nitrogen. ⏮ nitrogen fixation

Microorganisms in soil are important for reasons beyond their ecological niche. The various species of *Streptomyces* produce over 500 different antibiotic substances, at least 50 of which have useful applications in medicine, agriculture, and industry. The pharmaceutical industry has tested many thousands of soil microorganisms in search of those that produce useful antibiotics. In addition, soil microbes are being investigated for their ability to degrade toxic chemicals, an application of microbiology called bioremediation. Probably no other habitat represents such a wide range of biosynthetic and biodegradative capabilities as that of soil. ⏮ bioremediation

MicroByte

The number of microorganisms in a cup of fertile soil is greater than the number of people who have ever lived on Earth.

Characteristics of Soil

Soil forms as rock weathers. Water, temperature changes, windblown particles, and other physical forces gradually cause the rock to crack and break. Photosynthetic organisms including algae, mosses, and lichens growing on the surfaces of rocks synthesize organic compounds. Various bacteria and fungi then use these compounds as carbon and energy sources, producing acids and other chemicals that gradually decompose the rocks. As soil slowly forms, some plants begin to grow. When these die and decay, the residual organic material functions as a sponge, retaining water and thus allowing more plants to grow. Over time, more organic compounds accumulate, forming a slowly degrading complex polymeric substance called humus. ⏮ lichens

Soil represents an environment that can change abruptly and dramatically. Heavy rains, for example, can cause a soil to rapidly become waterlogged. Trees dropping their leaves can suddenly enrich the soil with organic nutrients. Farmers and gardeners rapidly change the nutrient mix by applying fertilizers.

The texture of the soil influences how much air and water can flow through it. Finely textured soils, such as clay soils, are more likely to become waterlogged and anaerobic. In contrast, sandy soils that dry quickly allow water to pass through and are generally aerobic.

Microorganisms in Soil

The density and composition of the soil microbiome are dramatically affected by environmental conditions. Wet soils, for example, are unfavorable for aerobic microbes because the spaces in the soil fill up with water, lowering the amount of air in the soil. When the water content of soil drops to a very low level, as during a drought or in a desert environment, the metabolic activity and number of soil microorganisms decrease. Many produce survival forms such as endospores and cysts that are resistant to drying. Other environmental influences that affect soil microbes include acidity, temperature, and nutrient supply. For example, acidity suppresses bacterial growth, allowing fungi to thrive with less competition for nutrients. This is why mushrooms often appear in a lawn fertilized with an acid-producing fertilizer such as ammonium chloride.

Prokaryotes are the most numerous soil inhabitants. Their physiological diversity allows them to colonize all types of soil. In general, Gram-positive bacteria are more abundant in soils than Gram-negative bacteria (except around the rhizosphere). Among the most common Gram-positive bacteria are members of the genus *Bacillus*. These form endospores, allowing them to survive long periods of adverse conditions such as drought or extreme heat. *Streptomyces* species produce conidia, which are desiccation-resistant structures. They also produce metabolites called geosmins, which give soil its characteristic musty odor. As discussed earlier, *Streptomyces* species produce many medically useful antibiotics. Other bacteria adapted to thrive in terrestrial environments—including myxobacteria and species of *Clostridium, Azotobacter, Agrobacterium,* and *Rhizobium*—were discussed in chapter 11. ⏮ endospore formers, ⏮ the genus *Streptomyces*, ⏮ thriving in terrestrial environments

Although prokaryotes are the most numerous soil microbes, the biomass of fungi is much greater. Most fungi are aerobes, so they usually grow in the top 10 cm of soil. The soil fungi degrade complex macromolecules such as lignin (the major component of cell walls of woody plants) and cellulose. Some soil fungi are free-living, and others live in symbiotic relationships. The latter include **mycorrhizas** ("fungus root"), which are fungi growing in a symbiotic relationship with certain plant roots. These important examples of symbiosis are discussed later in the chapter. ⏮ mycorrhizas

In addition to bacteria and fungi, various algae and protozoa are found in most soils. Algae depend on sunlight for energy, so they mostly live on or near the soil surface. Most protozoa require O_2, so they too are found near the surface, typically where microbes on which they feed are plentiful.

MicroAssessment 28.4

The density and composition of the soil are dramatically affected by environmental conditions. The concentration of microbes in the rhizosphere is generally much higher than that of the surrounding soil.

10. What is the significance of the rhizosphere?

11. Why are wet soils unfavorable for aerobic organisms?

12. How can the biomass of fungi in soil be greater than the biomass of bacteria, considering that bacteria are far more numerous?

28.5 ■ Biogeochemical Cycling and Energy Flow

Learning Outcomes

9. Diagram the carbon, nitrogen, sulfur, and phosphorus cycles, and describe some of the important microbial contributors.

10. Compare and contrast energy cycling in environments with sunlight versus those far removed from sunlight.

Biogeochemical cycles are the cyclical paths that elements take as they flow through living (biotic) and non-living (abiotic) components of ecosystems. These cycles are important because a fixed and limited amount of the elements that make up living cells exists on the earth and in the atmosphere. Thus, in order for an ecosystem to sustain its characteristic life-forms, elements must continually be recycled. For example, the organic carbon that animals use as an energy source is exhaled as carbon dioxide (CO_2); if this inorganic carbon were not eventually converted back to an organic form, we would run out of the organic carbon needed for growth. The carbon and nitrogen cycles are particularly important because they involve stable gaseous forms (carbon dioxide and nitrogen gas), which enter the atmosphere and thus have global impacts.

Although elements continually cycle in an ecosystem, energy does not. Instead, energy must be continually added to an ecosystem, fueling the activities required for life.

Understanding the cycling of nutrients and the flow of energy is particularly important as human activities can affect the environment. For example, industrial processes that convert nitrogen gas (N_2) into ammonia-containing fertilizers have boosted food production substantially, but they also alter the nitrogen cycle by increasing the amount of nitrogen available in organic compounds. When these nitrogen sources pollute lakes and coastal areas, eutrophication can result. As a consequence, the dissolved O_2 is depleted, leading to the death of aquatic animals. Burning of coal, oil, and other carbon-rich fossil fuels provides energy for our daily activities, but also releases CO_2 and other carbon-containing gases into the atmosphere. Fossil fuels, the ancient remains of partially decomposed plants and animals, are nutrient reservoirs that would be unavailable without human intervention, and therefore would not normally participate in biogeochemical cycles. Carbon-containing gases in the atmosphere absorb infrared radiation and reflect it back to Earth, raising global temperatures—a phenomenon called the Greenhouse Effect.

When studying biogeochemical cycles, it is helpful to consider the role of a given element in a particular organism's metabolism. Elements are used for three general purposes:

- **Biosynthesis (biomass production).** All organisms require elements for biosynthesis. As an example, nitrogen is required to produce amino acids. Plants and many prokaryotes assimilate nitrogen by incorporating ammonia (NH_3) to synthesize the amino acid glutamate (see figure 6.29a). Some prepare for this step by converting nitrate (NO_3^-) to ammonia. Once glutamate has been synthesized, the amino group can then be transferred to other carbon compounds to produce the other amino acids. Animals cannot incorporate ammonia, and instead require amino acids in their diet. Some prokaryotes can reduce atmospheric nitrogen to form ammonia—the process of nitrogen fixation. The ammonia can then be incorporated into cellular material. ⏮ amino acid synthesis
- **Energy source.** Reduced carbon compounds such as sugars, lipids, and amino acids are used as energy sources by chemoorganotrophs. Chemolithotrophs can use reduced inorganic molecules such as hydrogen sulfide (H_2S), ammonia (NH_3), and hydrogen gas (H_2) (see table 4.5). ⏮ energy source

- **Terminal electron acceptor.** In aerobic conditions, O_2 is used as a terminal electron acceptor in respiration. In anaerobic conditions, some prokaryotes can use nitrate (NO_3^-), nitrite (NO_2^-), sulfate (SO_4^-), or carbon dioxide (CO_2) as a terminal electron acceptor. ⏮ terminal electron acceptor

Carbon Cycle

All organisms are composed of organic molecules such as proteins, lipids, and carbohydrates. The carbon travels through the food chain as primary producers are eaten by primary consumers, which are then eaten by secondary consumers. Decomposers then use organic molecules in the remains of primary producers and consumers.

Carbon Fixation

A fundamental part of the carbon cycle is **carbon fixation,** the defining characteristic of primary producers (**figure 28.8**). Fixation incorporates chemical elements into molecules that can be directly used by living organisms. Primary producers convert CO_2 into an organic form, using mechanisms described in chapter 6. Without primary producers, no other organisms, including humans, could exist. We depend on them to generate the organic carbon compounds we use as an energy source and for biosynthesis. ⏮ carbon fixation

Respiration and Fermentation

When heterotrophs consume organic material, they break it down using respiration and/or fermentation to release energy, which is captured to make ATP. These processes usually also make CO_2.

The type of organic material helps dictate which species degrade it. A wide variety of organisms use sugars, amino acids, and proteins as energy sources, but rapidly multiplying bacteria often play the dominant role in decomposing these substances. In contrast, relatively few organisms (primarily fungi) can break down lignin, a major component of wood (**figure 28.9**). Aerobic conditions are required for this degradation, so wood at the bottom of marshes resists decay.

The O_2 supply has a strong influence on the carbon cycle. Not only does O_2 allow degradation of certain compounds such as lignin, it also helps determine the types of carbon-containing gases produced. When organic matter is degraded aerobically, a great deal of CO_2 is produced. When the O_2 level is low, however, as is the case in marshes, swamps, and manure piles, the degradation is incomplete, generating some CO_2 and also a variety of other products.

Methanogenesis and Methane Oxidation

In anaerobic environments, CO_2 is used by methanogens. These archaea obtain energy by oxidizing hydrogen gas, using CO_2 as a terminal electron acceptor, and generating methane (CH_4). Methane that enters the atmosphere is oxidized by ultraviolet light and chemical ions, forming carbon monoxide (CO) and CO_2. Microorganisms called methylotrophs can use methane as an energy source, oxidizing it to produce CO_2. ⏮ methanogens

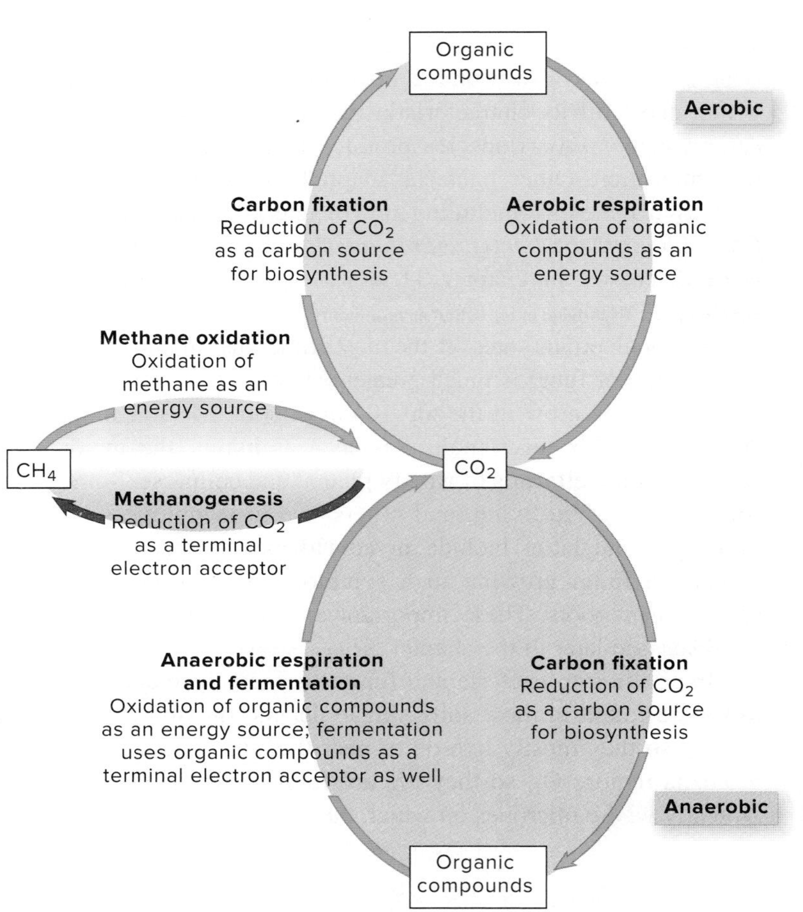

FIGURE 28.8 Carbon Cycle Blue arrows represent steps where carbon compounds are used as energy sources, red arrows represent steps where carbon compounds are used as terminal electron acceptors, and green arrows represent steps where inorganic carbon is used in biosynthesis.

? What is the role of autotrophs in the carbon cycle?

FIGURE 28.9 Wood-Degrading Fungus These fungi digest lignin, the major cell wall component of woody plants.

? Dead trees submerged in marshes resist decay. Why?

Nitrogen Cycle

Nitrogen is a component of proteins and nucleic acids. As consumers ingest plants and animals for carbon and energy, they also obtain their required nitrogen. Prokaryotes, as a group, are far more diverse in their use of nitrogen-containing compounds. Some use oxidized nitrogen compounds such as nitrate (NO_3^-) and nitrite (NO_2^-) as terminal electron acceptors; others use reduced nitrogen compounds such as ammonium (NH_4^+) as energy sources. These metabolic activities represent essential steps in the nitrogen cycle (**figure 28.10**).

Nitrogen Fixation

Nitrogen fixation is the process in which nitrogen gas (N_2) is reduced to form ammonia (NH_3), which can then be incorporated into cellular material. The process, catalyzed by the enzyme complex nitrogenase, requires a tremendous amount of energy because N_2 has a very stable triple covalent bond.

Although our atmosphere consists of approximately 80% N_2, relatively few organisms—all of which are prokaryotes—can reduce this gaseous form of the element. Thus, just as humans and other animals depend on primary producers to fix carbon, they rely on prokaryotes to convert atmospheric nitrogen to a form they can assimilate to create biomass.

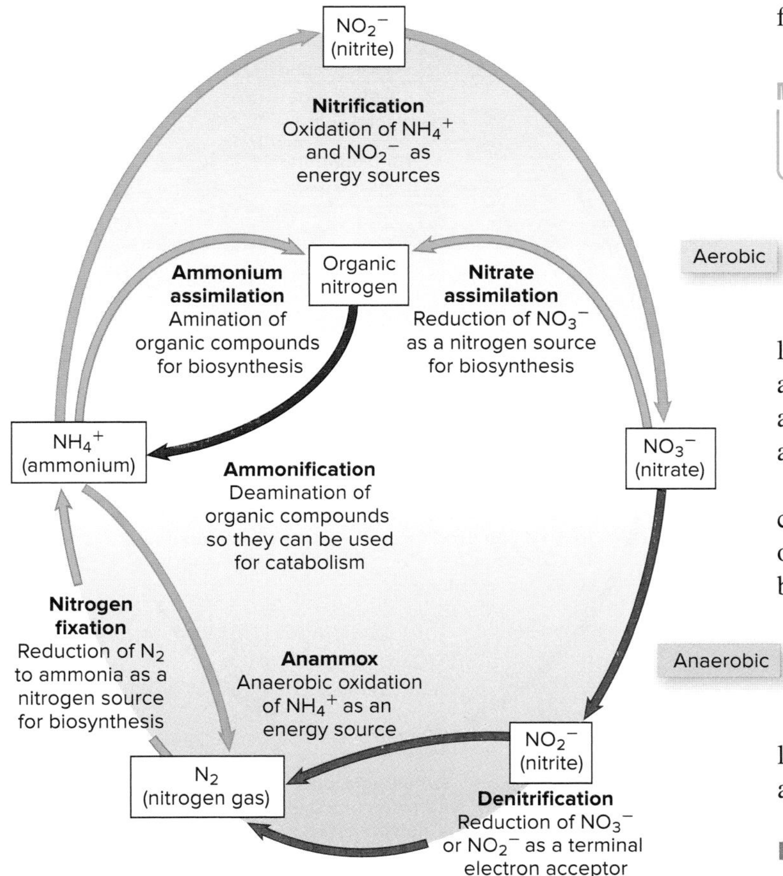

FIGURE 28.10 Nitrogen Cycle Blue arrows represent steps where nitrogen compounds are used as energy sources, red arrows represent steps where nitrogen compounds are used as terminal electron acceptors, and green arrows represent steps where inorganic nitrogen is used in biosynthesis. The black arrow represents decomposition reactions. Not all parts of the cycle are shown.

? Why are prokaryotes crucial in the nitrogen cycle?

Some nitrogen-fixing prokaryotes, or **diazotrophs,** are free-living, whereas others form symbiotic associations with higher organisms, particularly certain plants. Among the free-living examples are members of the genus *Azotobacter.* These heterotrophic, aerobic, Gram-negative rods may be the chief suppliers of fixed nitrogen in ecosystems such as grasslands that lack plants with nitrogen-fixing symbionts. The dominant free-living, anaerobic, soil diazotrophs are certain members of the genus *Clostridium.* Some cyanobacteria are diazotrophs; these photosynthetic bacteria use both nitrogen and carbon from the atmosphere. Symbiotic diazotrophs will be discussed later in the chapter. ◀◀ the genus *Azotobacter*, ◀◀ the genus *Clostridium*, ◀◀ nitrogen-fixing cyanobacteria

Energy-expensive chemical processes can fix nitrogen to make fertilizers, and these are playing an increasingly larger role in the nitrogen cycle. In fact, fixed nitrogen sources associated with human intervention, including fertilizer production and planting crops that foster the growth of symbiotic nitrogen-fixers, now surpass natural biological nitrogen fixation.

MicroByte

Nitrogenase uses approximately 16 molecules of ATP for every molecule of nitrogen fixed.

Ammonification

Ammonification is the decomposition process that converts organic nitrogen into ammonia (NH_3). In alkaline environments, such as heavily limed soil, the gaseous ammonia may enter the atmosphere. In neutral environments, ammonium (NH_4^+) is formed. This positively charged ion adheres to negatively charged particles.

Proteins, which are among the most common nitrogen-containing compounds, can be degraded by a wide variety of microbes. The microbes secrete proteolytic enzymes that break down proteins into short peptides or amino acids. These products are then transported into the cell of the decomposer, and the amino groups removed, releasing ammonium. The decomposer will assimilate much of this to create biomass and use the remaining parts for catabolism. Some ammonium will be released, however, and then be assimilated by plants and other organisms. ◀◀ deamination

Nitrification

Nitrification is the process that oxidizes ammonium (NH_4^+) to nitrate (NO_3^-). Bacteria known as **nitrifiers** do this in a cooperative two-step process, using ammonium and an intermediate nitrite (NO_2^-) as energy sources. Nitrifiers are obligate aerobes, using O_2 as a terminal electron acceptor. Consequently, nitrification does not occur in waterlogged soils or in anaerobic regions of aquatic environments.

Nitrification has some important consequences with respect to agricultural practices and pollution. Farmers often apply ammonium-containing compounds to soils as a source of nitrogen for plants. The ammonium is retained by soils, because its positive charge causes it to adhere to negatively charged soil particles. Nitrification converts the ammonium to nitrate, a form of nitrogen more readily used by plants, but rapidly leached from soil by rainwater. To slow nitrification, certain chemicals can be added to ammonium-fertilized soils.

Leaching of nitrate and nitrite from soil is a health concern if the compounds contaminate drinking water. Nitrite is toxic because it can combine with hemoglobin of the blood, reducing blood's O_2-carrying capacity. Even nitrate, which in itself is not very toxic, can be dangerous if high levels are ingested because some intestinal bacteria can use it as a terminal electron acceptor, converting it to nitrite.

Denitrification

Denitrification is the process that reduces nitrate (NO_3^-), converting it to gaseous forms such as nitrous oxide (N_2O) and molecular nitrogen (N_2). This happens when prokaryotes anaerobically respire using nitrate as a terminal electron acceptor. ⏮ anaerobic respiration

Denitrification can have negative environmental and economic consequences. Under anaerobic conditions in wet soils, for example, denitrifying bacteria will reduce the oxidized nitrogen compounds of fertilizers, releasing gaseous nitrogen to the atmosphere. In some areas, this process may represent 80% of nitrogen lost from fertilized soil, a considerable economic loss to the farmer. In addition, nitrous oxide contributes to global warming.

Denitrification is not always undesirable. The process can be actively fostered in certain steps of wastewater treatment as a means to remove nitrate. This compound could otherwise act as a fertilizer in the waters to which the wastewater is discharged, promoting algal growth. ⏭ microbiology of wastewater treatment

Anammox

Certain bacteria oxidize ammonium under anaerobic conditions, using nitrite as a terminal electron acceptor. This reaction, called **anammox** (for **a**noxic **amm**onia **ox**idation), forms N_2 and might provide an economical means of removing nitrogen compounds during wastewater treatment.

Sulfur Cycle

Sulfur is found in all living matter, primarily as a component of the amino acids methionine and cysteine. Like the nitrogen cycle, key steps of the sulfur cycle depend on the activities of prokaryotes (**figure 28.11**).

Sulfur Assimilation and Decomposition

Most plants and microorganisms assimilate sulfur as sulfate (SO_4^{2-}), reducing it and then incorporating the reduced form into biomass. Like nitrogen, organic sulfur is present chiefly as a part of the amino acids that make up proteins. Decomposition of the sulfur-containing amino acids releases hydrogen sulfide (H_2S), a gas.

Sulfur Oxidation

Hydrogen sulfide (H_2S) and elemental sulfur (S^0) can both serve as an energy source for certain chemolithotrophs. Sulfur-oxidizing prokaryotes, including species of *Beggiatoa, Thiothrix,* and *Thiobacillus,* oxidize these molecules to sulfate (SO_4^{2-}). Certain prokaryotes in anaerobic marine environments can oxidize elemental sulfur, using nitrate as a terminal electron acceptor. As discussed in chapter 11, these organisms, including *Thioploca* species and the largest known bacterium, *Thiomargarita namibiensis,* have unusual mechanisms to cope with the fact that their energy source and terminal electron acceptor are found in two different environments.
⏮ sulfur-oxidizing bacteria, ⏮ the genera *Thioploca* and *Thiomargarita*

Hydrogen sulfide and elemental sulfur are oxidized anaerobically by photosynthetic green and purple sulfur bacteria. These bacteria harvest energy from sunlight but require reduced molecules as a source of electrons to generate

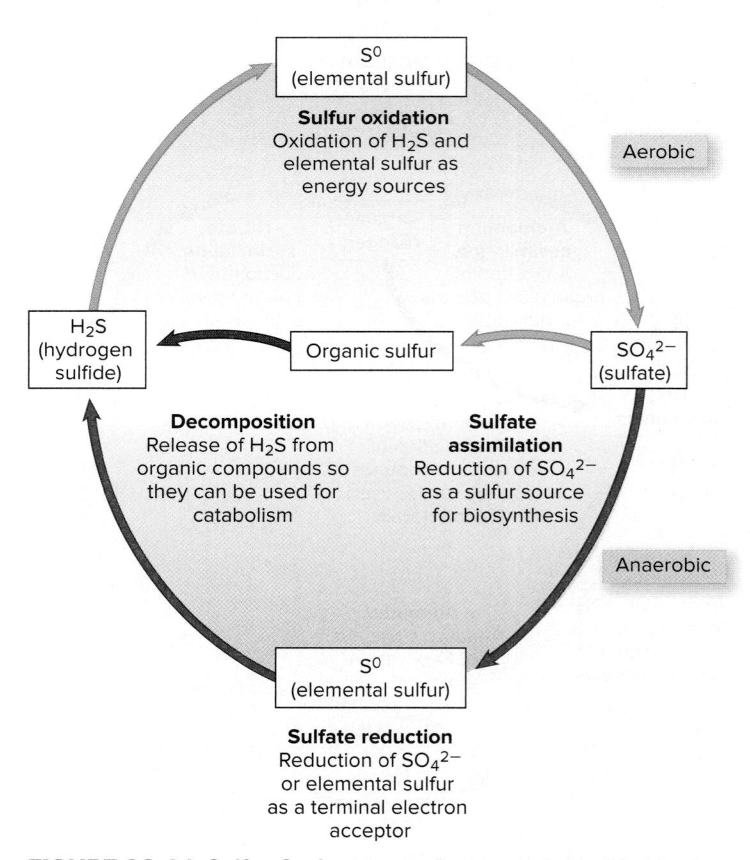

FIGURE 28.11 Sulfur Cycle Blue arrows represent steps where sulfur compounds are used as energy sources, red arrows represent steps where sulfur compounds are used as terminal electron acceptors, and green arrows represent steps where inorganic sulfur is used in biosynthesis. The black arrow represents decomposition reactions.

? Organisms that oxidize sulfur use it for what purpose in their metabolism?

reducing power. Like the chemolithotrophs that use hydrogen sulfide and elemental sulfur, the photosynthetic sulfur oxidizers produce sulfate. ◀◀ green sulfur bacteria, ◀◀ purple sulfur bacteria

Sulfur Reduction

Under anaerobic conditions, sulfate generated by the sulfur-oxidizers can then be used as a terminal electron acceptor by certain organisms. The sulfur- and sulfate-reducing bacteria and archaea use sulfate in the process of anaerobic respiration, reducing it to hydrogen sulfide (H_2S). In addition to its unpleasant odor, the H_2S is a problem because it reacts with metals, resulting in corrosion. ◀◀ sulfur- and sulfate-reducing bacteria

Phosphorus Cycle and Other Cycles

Phosphorus is a component of several critical biological compounds, including nucleic acids, phospholipids, and ATP. Most plants and microorganisms can take up phosphorus as orthophosphate (PO_4^{3-}), incorporating it into biomass. From there, the phosphorus is passed along the food web. When plants and animals die, decomposers convert organic phosphate back to the inorganic form.

In many aquatic habitats, growth of algae and cyanobacteria—the primary producers—is limited by low concentrations of phosphorus. When phosphates are added from sources such as agricultural runoff, phosphate-containing detergents, and wastewater, eutrophication can result.

Other important elements, including iron, calcium, zinc, manganese, cobalt, and mercury, are also recycled by microorganisms. Many prokaryotes contain plasmids coding for enzymes that carry out oxidation of metallic ions.

Energy Sources for Ecosystems

All chemotrophs harvest the energy trapped in chemical bonds to generate ATP. This energy cannot be totally recycled, however, because a portion is always lost as heat when bonds are broken. Thus, energy is continuously lost from biological systems. To compensate, energy must be added to ecosystems.

Photosynthesis, carried out by chlorophyll-containing plants and microorganisms, converts radiant energy (sunlight) to chemical energy in the form of organic compounds, which can then be used by chemoorganotrophs. The requirement for radiant energy has traditionally been used to explain why life is not equally abundant everywhere. However, the discovery of different types of communities far removed from sunlight has dramatically altered this idea. These communities rely on chemolithoautotrophs, which harvest the energy of reduced inorganic compounds and use it to form organic compounds. ◀◀ chemolithoautotroph

A number of **hydrothermal vents** have been discovered some thousands of meters below the ocean surface. These vents form when water seeps into cracks in the ocean floor and becomes heated by the molten rock, finally spewing out in the form of mineral-laden undersea geysers. The hydrogen sulfide discharged supports thriving deep-sea communities, oases in the otherwise desolate ocean floor (**figure 28.12**). Large numbers of sulfur-oxidizing chemolithoautotrophs are

(a)

(b)

FIGURE 28.12 Hydrothermal Vent Community **(a)** This diverse community is supported by the metabolic activities of chemolithoautotrophs. **(b)** Water escaping from the vent is rich in reduced compounds, including hydrogen sulfide, that can serve as energy sources. a: ©Kim Juniper/AFP/Newscom

? Photosynthetic organisms are normally considered the most important primary producers; which organisms are primary producers in hydrothermal vent communities?

found in and around the vents. Many of these bacteria and archaea are free-living but some live in symbiotic association with the large tube worms and clams that inhabit the areas. The chemolithoautotrophs obtain energy by oxidizing hydrogen sulfide, and they fix CO_2, providing the animals with both a carbon and an energy source.

Microbial populations have been found almost 3 km under the ground and in iron-rich volcanic rocks from nearly a thousand meters below the surface of the Columbia River. These organisms gain energy from hydrogen (H_2) produced in the subsurface. It has been estimated that if (and it is a big "if" at this point) most similar rocks contain microbes, there could be as much as 2×10^{14} tons of underground microorganisms—equivalent to a layer 1.5 meters thick over the entire land surface of the earth!

MicroAssessment 28.5

Elements are recycled as organisms incorporate them to produce biomass, oxidize the reduced forms as energy sources, and reduce the oxidized forms as terminal electron acceptors. Carbon fixation incorporates atmospheric CO_2 into organic material; the CO_2 is regenerated during respiration and some fermentations. Prokaryotes are essential for several steps of the nitrogen cycle, including nitrogen fixation, nitrification, denitrification, and anammox. Prokaryotes are also essential for several steps of the sulfur cycle, including sulfur reduction and sulfate oxidation.

13. What are the three general roles of carbon-containing compounds in metabolism?
14. Why do farmers try to prevent nitrification?
15. Although chemoautotrophs serve as the primary producers near hydrothermal vents, animals there still ultimately depend on the photosynthetic activities of plants and cyanobacteria. Why?

28.6 ■ Mutualistic Relationships Between Microorganisms and Eukaryotes

Learning Outcome

11. Describe the mutualistic relationships between fungi and plant roots, symbiotic nitrogen-fixers and plants, and microorganisms and herbivores.

As described in chapter 16, mutualism is a symbiotic association in which both partners benefit. A variety of other ecologically important symbiotic relationships exist, but mutualistic relationships highlight the vital role of microorganisms to life on this planet. ◂◂ symbiosis

Mycorrhizas

Mycorrhizas (*myco* means "fungus"; *rhizo* means "root") are fungi growing in symbiotic relationships with plant roots (**figure 28.13**). They enhance the competitiveness of plants by helping them take up phosphorus and other substances from the soil. In turn, the fungi gain nutrients for their own growth from root secretions. It is estimated that over 85% of vascular plants (plants with specialized water- and food-conducting tissues) have mycorrhizas.

There are two common types of mycorrhizal relationships:

- **Endomycorrhizas.** The fungi penetrate root cells, growing as coils or bush-like masses within the cells. These are by far the most common mycorrhizal relationships and are found in association with most herbaceous plants. Relatively few species of fungi are involved, perhaps only 100 or so, and most appear to be obligate symbionts.

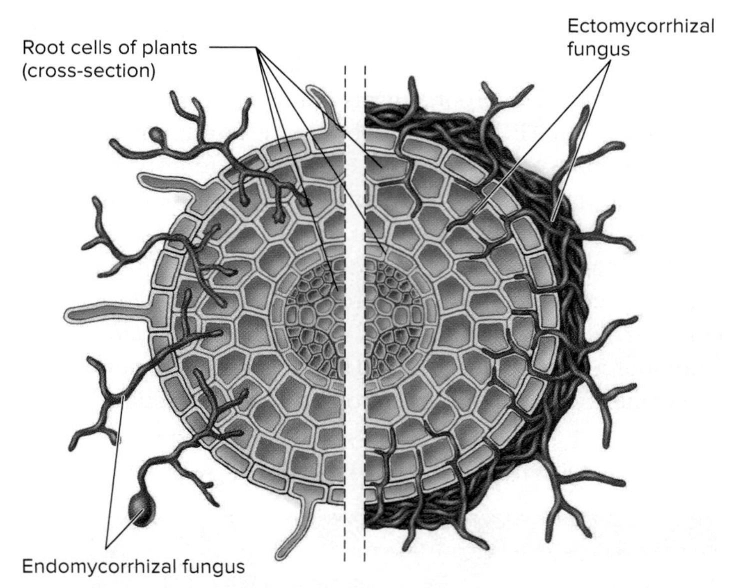

(a) Diagram of Mycorrhizas

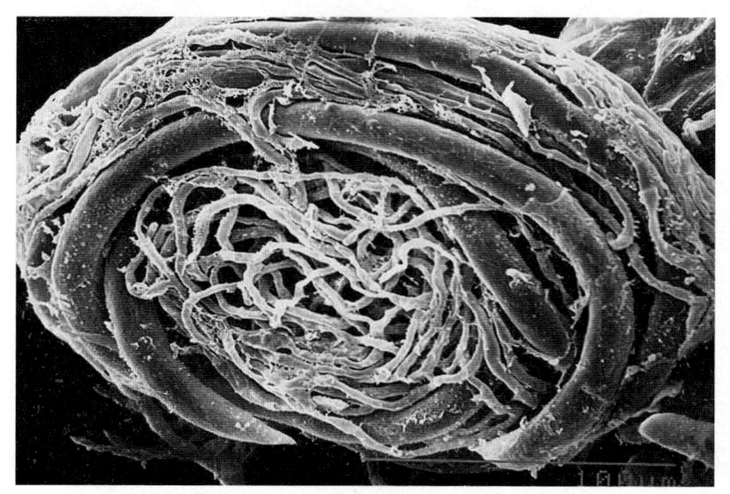

(b) Endomycorrhiza

FIGURE 28.13 Mycorrhizas (a) Diagram illustrating the two types of mycorrhizas. **(b)** In an endomycorrhizal relationship, the fungi penetrate root cells and grow within them. b: ©Stephan Imhof/Philipps-Universität

Which type of mycorrhizal relationship is common in herbaceous plants?

The relationship for some plants is also obligate; for example, most orchid seeds will not germinate without the activities of a fungal partner.

- **Ectomycorrhizas.** The fungi grow around the plant cells, forming a sheath around the root. These fungi associate mainly with certain trees, including conifers, beeches, and oaks. Over 5,000 species of fungi are involved in ectomycorrhizal relationships but many are restricted to a single type of plant.

MicroByte

Chanterelles and truffles are examples of commercially valuable ectomycorrhizal fungi.

Symbiotic Nitrogen-Fixers and Plants

Although some free-living bacteria can add fixed nitrogen to the soil, symbiotic nitrogen-fixing organisms are far more significant in benefiting plant growth and crop production. They are important in both terrestrial and aquatic habitats.

Rhizobia

Members of a diverse group of genera, including *Rhizobium, Bradyrhizobium, Sinorhizobium,* and *Azorhizobium*—collectively referred to as **rhizobia**—are the most agriculturally important symbiotic nitrogen-fixing bacteria. These grow as endosymbionts within specialized organs called **nodules** on the roots of legumes (plants that bear seeds in pods), including alfalfa, clover, peas, beans, and peanuts (**figure 28.14a**). The input of soil nitrogen from rhizobia may be about 10 times the annual rate of nitrogen fixation by non-symbiotic organisms. To foster plant growth, farmers often add the appropriate symbionts to the seeds of certain legumes. ⏮ endosymbiont

The plant cells foster the nitrogen fixation of the endosymbionts by synthesizing a protein called leghemoglobin. This binds to O_2 and regulates its concentration in the nodule, protecting the O_2-sensitive nitrogenase. The plant also provides various nutrients to the endosymbionts. Meanwhile, the endosymbionts provide fixed nitrogen to the plant.

Nodule formation involves extensive chemical communication between the rhizobia and the legume partners (figure 28.14b). First, plant root secretions attract the appropriate rhizobial species, which then colonizes the roots. In the most well-characterized nodulation systems, some of the chemicals cause the bacteria to produce Nod factors (NFs), which induce a series of events that cause the root hair to curl, trapping the bacterial cells. The NFs also cause underlying plant cells to divide. The plasma membrane of the curled root hair then invaginates, forming a tube-like structure called an infection thread. The rhizobia cells travel through this structure to underlying plant tissues, and are then engulfed by the dividing plant cells, forming membrane-bound structures called symbiosomes. Within a symbiosome, a rhizobial cell changes in shape and function to become a specialized nitrogen-fixing cell called a bacteroid.

Although the relationship between the plant and the bacterium is not obligate, it offers a competitive advantage to both partners. The rhizobia do not fix nitrogen in soils lacking legumes, and they compete poorly with other microbes, slowly disappearing from soils in which legumes are not grown. Likewise, legumes compete poorly against other plants in heavily fertilized soils.

Other Nitrogen-Fixing Symbionts

Several genera of non-leguminous trees, including alder and gingko, have nitrogen-fixing root nodules at some stages of their life cycle. The bacteria involved in the symbiosis are members of the genus *Frankia.*

In aquatic environments, the most significant nitrogen-fixers are cyanobacteria. They are especially important in flooded soils such as rice paddies. In fact, rice has been cultivated successfully for centuries without the addition of nitrogen-containing fertilizer because of the symbiotic relationship between the cyanobacterium *Anabaena azollae* and the aquatic fern *Azolla.* The bacterium grows in specialized sacs in the leaves of the fern, providing nitrogen to the fern. Before planting rice, the farmer allows the flooded rice paddy to overgrow with *Azolla* ferns. Then, as the rice grows, it eventually crowds out the ferns. As the ferns die and decompose, their nitrogen is released into the water.

Microorganisms and Herbivores

Another mutualistic relationship occurs between microbes and certain herbivores. In order to live on grass and other plant material, herbivores such as cattle and horses rely on a microbial community that inhabits a specialized digestive compartment. The microbes digest cellulose and hemicellulose, two of the major components of plant material, releasing compounds that can then serve as a nutrient source for the animal. The specialized digestive compartment in ruminants such as cattle, sheep, and deer is called a rumen, and it precedes the true stomach. Non-ruminant herbivores such as horses and rabbits have a compartment called a cecum, which serves a similar purpose; it lies between the small intestine and the large intestine. ⏮ cellulose

The rumen functions as an anaerobic fermentation vessel to which nutrients in the form of plant materials are intermittently added. Each milliliter of rumen content contains approximately 10^{10} bacteria, 10^6 protozoa, and 10^3 fungi; of the over 200 species identified, no single one accounts for more than 3% of the total number of cells. The microorganisms in the rumen degrade the ingested material, releasing sugars that are then fermented to produce various organic acids. Those acids are absorbed by the animal, providing an energy and nutrient source. Large quantities of gas—produced as a result of fermentation—are discharged when the animal belches.

After being digested in the rumen, the food mass enters another compartment (the omasum), eventually reaching the

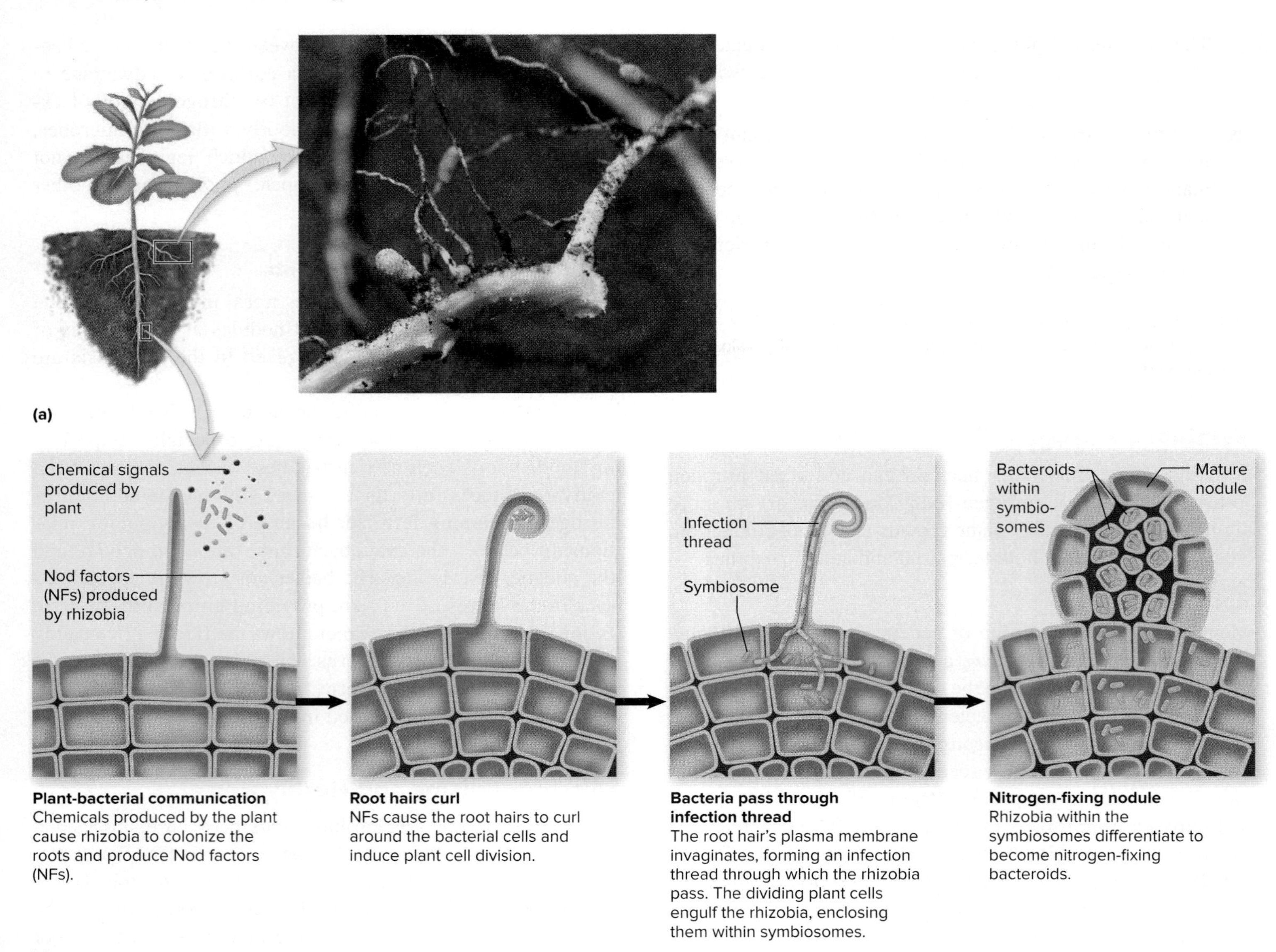

FIGURE 28.14 Root Nodules (a) Appearance. **(b)** The major steps leading to nodule formation by *Rhizobium* species. a: ©McGraw-Hill Education/Lisa Burgess

The pinkish color of the nodules is due to leghemoglobin. What is the role of this protein?

acidic true stomach (abomasum). There, organic acids are absorbed. In addition, lysozyme is secreted, allowing the animal to lyse and then digest members of the microbial population, providing even more nutrients. A crucial feature of ruminants is that the microbial population gets the first opportunity to use the ingested nutrients. The animal then uses the metabolic end products as well as the microbial cells themselves.

The cecum of non-ruminant herbivores serves a function similar to a rumen. Microbes in the cecum cannot be used as a food source, however, because of the cecum's location relative to the stomach. The benefit of the location is that the animal can digest and absorb readily available nutrients without competition from microbes.

MicroAssessment 28.6

Mycorrhizal fungi gain nutrients from plant root secretions while helping plants take up substances from soil. Symbiotic nitrogen-fixers provide plants with a source of usable nitrogen while being provided with an exclusive habitat. Microorganisms in the rumen and cecum of herbivores digest cellulose and hemicellulose, allowing the animal to subsist on plant material.

16. Describe the differences between an endomycorrhiza and an ectomycorrhiza.

17. Describe the differences between a rumen and a cecum.

18. Gardeners sometimes plant clover between productive growing seasons. Why would this practice be beneficial?

FOCUS ON A CASE 28.1

A novice farmer noticed his two steers behaving strangely; one was staggering and the other was bellowing and urinating. Both looked as if someone had blown them up like balloons. He had fed the animals a diet of hay for several weeks at the end of winter, but then moved the animals into a pasture with newly sprouted grass and lush alfalfa. The animals greedily grazed on the tender plants, but then an hour later the problems began. The farmer ran to call the veterinarian and desperately tried to remember what he had been told about something called "frothy bloat"—a type of gas buildup in the rumen.

1. Why are ruminants subject to bloat?
2. What was the likely cause of bloat in the farmer's animals?
3. Why is bloat dangerous for ruminants?
4. What can be done to prevent bloat?
5. What can be done to treat animals with bloat?

Discussion

1. When ruminants swallow freshly eaten vegetation, it enters the rumen. There, microorganisms ferment cellulose and other structural carbohydrates, producing 200 to 400 liters of gas per day (largely CO_2 and methane). This gas is expelled regularly by a type of belching called eructation that healthy cattle do about once per minute. If not released, the gas remains in the rumen, causing bloat in as little as 15 minutes.
2. Microorganisms in the rumen can digest high-protein alfalfa much faster than high-fiber grass or hay, producing greater volumes of gases, small particles, and slime associated with biofilms. Foam or froth forms, trapping the gas in the mass of small bubbles that cannot rise freely to be released by belching.
3. Trapped gas inflates the rumen, causing it to press against vital organs such as the heart and lungs. This can kill the animal, often by suffocation. In addition, rapid breakdown of the easily digested alfalfa causes buildup of organic acids, the end products of fermentation. These can diffuse into the bloodstream and cause systemic acidosis.
4. Bloat can be prevented by allowing only gradual changes in the animal's diet so that microbial communities in the rumen can adapt to new foods.
5. Walking an animal may cause it to belch and release gas. Oils that act as antifoaming agents can also be administered to break the bubbles that otherwise prevent gas from escaping. Inserting a hose down the animal's esophagus into the rumen may physically allow gas to leave under pressure through the hose. In extreme cases, a tube may be inserted directly into the rumen through the skin. This offers immediate relief for the animal, but poses a risk of infection.

Summary

28.1 Principles of Microbial Ecology

Ecosystems vary in their biodiversity and biomass. The microenvironment is most relevant to a microorganism's survival and growth.

Nutrient Acquisition

Primary producers convert CO_2 into organic material that can be used by **consumers; decomposers** digest the remains of primary producers and consumers (figure 28.1).

Microbes in Low-Nutrient Environments

Microorganisms capable of growing in dilute aqueous solutions are common in nature; often they grow in biofilms.

Microbial Competition

Microorganisms in the environment compete for the same limited pool of nutrients (figure 28.2). A species can competitively exclude others, or produce compounds that inhibit others.

Microorganisms and Environmental Changes

Environmental changes are common, and often result in the dominance of a different species (figure 28.3).

Microbial Communities

A **microbial mat** is a thick, dense, highly organized biofilm composed of distinct colored layers of different groups of microbes (figure 28.4).

28.2 Studying Microbial Ecology

Microbial ecology has been difficult to study because so few environmental prokaryotes can be successfully grown in the laboratory. **Metagenomics** now allows researchers to characterize microbial communities (figure 28.5).

28.3 Aquatic Habitats

Oligotrophic waters are nutrient poor; **eutrophic** waters are nutrient rich (figure 28.6). Excessive growth of aerobic heterotrophs may cause an aquatic environment to become **hypoxic,** resulting in the death of fish and other aquatic animals.

Marine Environments

Ocean waters are generally oligotrophic and aerobic, but inshore areas can be dramatically affected by nutrient-rich runoff (figure 28.7).

Freshwater Environments

Oligotrophic lakes may have anaerobic layers due to thermal stratification. Shallow, turbulent streams are generally aerobic.

Specialized Aquatic Environments

Salt lakes and mineral-rich springs support the growth of microbes specifically adapted to thrive in these specialized environments.

28.4 Terrestrial Habitats

The concentration of microbes in the **rhizosphere** is generally much higher than that in the surrounding soil.

Characteristics of Soil

Soil represents an environment that can change abruptly and dramatically.

Microorganisms in Soil

Environmental conditions affect the density and composition of the soil microbiome.

28.5 Biogeochemical Cycling and Energy Flow

Organisms use elements in biosynthesis to produce biomass, as sources of energy, and as terminal electron acceptors.

Carbon Cycle (figure 28.8)

One of the fundamental aspects of the carbon cycle is carbon fixation. As consumers and decomposers degrade organic material (figure 28.9), respiration and some fermentations release CO_2.

Nitrogen Cycle (figure 28.10)

The steps of the nitrogen cycle include **nitrogen fixation, ammonification, nitrification, denitrification,** and **anammox.**

Sulfur Cycle (figure 28.11)

Certain steps of the sulfur cycle—sulfur reduction and sulfate oxidation—depend on the activities of prokaryotes.

Phosphorus Cycle and Other Cycles

Most plants and microorganisms take up orthophosphate, incorporating it into biomass. Iron, calcium, zinc, manganese, cobalt, and mercury are recycled by microorganisms.

Energy Sources for Ecosystems

Photosynthetic organisms convert radiant energy to chemical bond energy in the form of organic compounds. Chemolithoautotrophs harvest energy from reduced inorganic chemicals (figure 28.12).

28.6 Mutualistic Relationships Between Microorganisms and Eukaryotes

Mycorrhizas

Mycorrhizas help plants take up phosphorus and other substances from soil; in turn the fungal partners gain nutrients for their own growth (figure 28.13). Endomycorrhizal fungi penetrate root cells; ectomycorrhizal fungi grow around root cells.

Symbiotic Nitrogen-Fixers and Plants

Rhizobia reside as nitrogen-fixing endosymbionts in **nodules** on legume roots (figure 28.14). *Frankia* species fix nitrogen in nodules of alder and gingko. A species of cyanobacteria fixes nitrogen in specialized sacs in the leaves of the *Azolla* fern.

Microorganisms and Herbivores

In order to live on a diet of grass and other plant material, herbivores rely on a community of microbes that inhabit a specialized digestive compartment, either a rumen or a cecum.

Review Questions

Short Answer

1. Describe why a microbial mat has green, reddish-pink, and black layers.
2. How do bacteriocins benefit bacteria in their natural habitat?
3. Why is there a high concentration of microbes in the rhizosphere?
4. What is the role of genetic databases in metagenomic studies of environmental samples?
5. Why does wood resting at the bottom of a bog resist decay?
6. What is the importance of nitrogen fixation?
7. Describe the relationship between ammonia oxidizers and nitrite oxidizers.
8. How do hydrothermal vents support thriving communities of microbes, clams, and tube worms?
9. Give examples of free-living and symbiotic nitrogen-fixing microorganisms. Are these prokaryotic or eukaryotic?
10. Describe the steps that lead to the formation of the symbiotic relationship between rhizobia and legumes.

Multiple Choice

1. Cyanobacteria are
   a) primary producers.
   b) consumers.
   c) herbivores.
   d) decomposers.
   e) more than one of the above.
2. Which of the following is *false*?
   a) Culture techniques are an accurate way of determining which members in a microbial community are most common.
   b) Fluorescence in situ hybridization (FISH) can be used to distinguish subsets of prokaryotes that contain a specific nucleotide sequence.
   c) Taxonomic diversity of prokaryotes can be studied by examining 16S rRNA sequences.
   d) Studying the genome of one organism can give insights into the characteristics of another.
3. Which of the following pairs that relate to aquatic environments does not match?
   a) Oligotrophic—nutrient poor
   b) Hypoxic—O_2 poor
   c) Epilimnion—O_2 poor
   d) Hypolimnion—lower layer
   e) Eutrophic—nutrient rich
4. Adding high levels of nutrients to a lake or inshore area would have all of the following effects in that environment *except*
   a) death of clams and crabs.
   b) increased growth of heterotrophic microbes.
   c) increased growth of photosynthetic organisms.
   d) increased levels of dissolved O_2.
5. Which of the following is not a matching pair?
   a) Soil—minimal biodiversity
   b) *Bacillus*—endospores
   c) *Streptomyces*—geosmin production
   d) Fungi—lignin degradation
   e) Rhizosphere—soil that adheres to plant root
6. Atmospheric nitrogen can be used
   a) directly by all living organisms.
   b) only by aerobic bacteria.
   c) only by anaerobic bacteria.
   d) in symbiotic relationships between rhizobia and plants.
   e) in photosynthesis.

7. Which process converts ammonium (NH_4^+) into nitrate (NO_3^-)?
   a) Nitrogen fixation
   b) Ammonification
   c) Nitrification
   d) Denitrification
   e) Anammox
8. Energy for ecosystems can come from
   a) sunlight via photosynthesis.
   b) oxidation of reduced inorganic chemicals by chemoautotrophs.
   c) both a and b.
9. Mycorrhizas represent associations between plant roots and microorganisms that
   a) are antagonistic.
   b) help plants take up phosphorus and other nutrients from soil.
   c) involve algae in the association with plant roots.
   d) form nodules on the plant's leaves.
   e) lead to the production of antibiotics.
10. In symbiotic nitrogen fixation by rhizobia and legumes,
    a) the amount of nitrogen fixed is much greater than by non-symbiotic organisms.
    b) neither the bacteria nor the legume can exist independently.
    c) the bacteria enter the leaves of the legume.
    d) the bacteria operate independently of the legume.

Applications

1. A farmer who was growing soybeans, a type of legume, saw an Internet site advertising an agricultural product for safely killing soil bacteria. The ad claimed that soil bacteria were responsible for most crop losses. The farmer called the agricultural extension office at a local university for advice. Explain what the extension office adviser most likely told the farmer about the usefulness of the product.
2. Recent reports suggest that human activities, such as the generous use of nitrogen fertilizers, have doubled the rate at which elemental nitrogen is fixed, raising concerns of environmental overload of nitrogen. What problems could arise from too much fixed nitrogen, and what could be done about this situation?

Critical Thinking

1. Each colony growing on an agar plate arises from a single cell (see photo). Colonies growing close together are much smaller than those that are well separated. Why would this be so?
2. An entrepreneur found an economically feasible way of collecting large amounts of sulfur from underwater hot vents in the Pacific Ocean. The sulfur will be harvested from the microorganisms found in the vent areas. A group of ecologists argued that the project would destroy the fragile ecosystem by depleting it of usable sulfur. The entrepreneur argued that the environment would not be harmed because the vents produce more than enough sulfur for the clams and tube worms in the area. Explain who is correct.

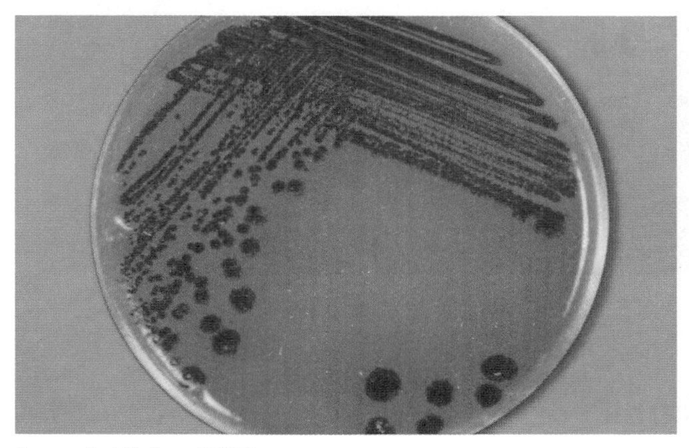

Source: Dr. V.R. Dowell/CDC

29 Environmental Microbiology: Treatment of Water, Wastes, and Polluted Habitats

A pristine mountain lake. ©*Robert Glusic/Getty Images*

KEY TERMS

Advanced Treatment In wastewater treatment, any physical, chemical, or biological purification process beyond secondary treatment.

Biochemical Oxygen Demand (BOD) The amount of O_2 required for the microbial decomposition of organic matter in a sample.

Bioremediation Process that uses microorganisms to degrade harmful chemicals.

Compost Material that results from the controlled decomposition of solid organic material.

Indicator Organisms Microbes whose presence in an environment suggests fecal contamination.

Primary Treatment A wastewater treatment process used to remove material that settles out.

Sanitary Landfill A site used for disposal of non-hazardous solid wastes in a manner that minimizes damage to human health and the environment.

Secondary Treatment A wastewater treatment process that uses microbes to convert suspended solids to inorganic compounds and removable cell mass.

Septic System An individual wastewater treatment system in which the sludge settles out and is degraded by microorganisms while the effluent percolates through a drain field.

Sludge The solid portion of wastewater that settles to the bottom of sedimentation tanks during primary and secondary treatment.

Wastewater Material that flows from household plumbing systems; municipal wastewater also includes business and industrial wastes and stormwater runoff.

A Glimpse of History

Delivering fresh water to urban areas and removing human wastes have been practiced at least since Roman times—the ruins of ancient aqueducts can still be seen today in many parts of Europe. In the 1800s, the desire for clean, clear water led to the use of sand filtration systems in London and elsewhere. Later, Robert Koch showed that this kind of filtration not only yields clear water but also removes most bacteria.

In the 1840s, Edwin Chadwick, an English activist, proposed a system to flush solid wastes away from a city. His plan was to use narrow, smooth ceramic pipes and water under pressure to move the wastes to a distant site, where they then could be used to produce fertilizer to sell to farmers. Concern about cholera in 1848 prompted the newly formed Board of Health in England to begin installing a sewage system somewhat similar to that envisioned by Chadwick. New York City did the same in 1866, again in response to a threatened cholera epidemic. By the end of the nineteenth century, most large European and U.S. cities had sewer systems to remove and treat waste materials as well as water systems to deliver safe drinking water. Cholera in the industrialized nations of Europe and North America virtually disappeared.

Most people living in developed countries take for granted that their tap water is safe to drink, that their wastes will reliably disappear into sewers or landfills for proper disposal, and that pollutants (substances that are harmful or injurious) will not accumulate in the environment. They seldom consider the role of microbes in these essential aspects of modern life.

Microorganisms are important in the treatment processes described in this chapter for two very distinct reasons. First, we benefit from the fact that microbes are the ultimate recyclers, playing an essential role in the decomposition of our wastes. Second, some microbes are pathogens and must be removed from sewage before it is discharged and from drinking water before it is safe for human consumption. Recreational waters such as swimming pools, water parks, lakes, rivers, and shorelines are also monitored to ensure they do not have harmful levels of certain pathogens.

Treatment of water, waste, and polluted habitats is a significant challenge, particularly in densely populated areas. Consider that every day the average American uses about 100 gallons of water, and generates 80 gallons of wastewater and 5 pounds of trash. This means that a city with only 1 million inhabitants is faced with the disposal of approximately 30 billion gallons of wastewater and a million tons of trash each year!

29.1 ■ Microbiology of Wastewater Treatment

Learning Outcomes

1. Describe the concept of biochemical oxygen demand (BOD).
2. Compare and contrast primary treatment, secondary treatment, advanced treatment, and anaerobic digestion.
3. Describe how septic systems function.

Wastewater, or sewage, is composed of all the material that flows from household plumbing systems, including toilet wastes as well as washing and bathing water. Municipal wastewater also includes business and industrial wastes. In many cities, stormwater runoff that flows into street drains enters the system as well.

The most obvious reason wastewater must be treated before discharge into the environment is that pathogenic microbes—including those that cause diarrheal diseases and hepatitis—can be transmitted in feces. If untreated sewage is released into a river or lake that is then used as a source of drinking water, disease can easily spread. If marine waters become contaminated in a similar manner, then eating the local shellfish can result in disease. Shellfish are filter feeders and they concentrate microbes from the waters in which they live.

The high nutrient content of wastewater can also be damaging to the receiving water. When any nutrient-rich substance is added to an aqueous environment, microorganisms quickly use the compounds as energy sources. As a result, microbes that aerobically respire consume available O_2 in the water, using it as a terminal electron acceptor (see figure 6.20). The amount of dissolved O_2 in lakes and rivers is limited and can easily be depleted during the microbial breakdown of nutrients. Fish and other aquatic animals in the environment die because they require O_2 for respiration (see figure 28.7). Thus, effective wastewater treatment must decrease the level of organic compounds substantially, in addition to eliminating pathogens and pollutants. ⏮ aerobic respiration

Biochemical Oxygen Demand (BOD)

An important goal of wastewater treatment is to decrease the environmental impact by reducing the **biochemical oxygen demand (BOD)**—the amount of O_2 required for the microbial decomposition of organic matter in a given sample. High BOD values indicate that large amounts of degradable materials are present, resulting in the consumption of correspondingly large amounts of O_2 during biological degradation of the material. The BOD of raw sewage is approximately 300 to 400 mg/liter, whereas the dissolved O_2 content of natural waters is generally 5 to 10 mg/liter. Thus, adding raw sewage to a lake could easily deplete the dissolved O_2 in that water.

To determine the BOD of a sample, the O_2 level is first measured. The sample is then incubated in a sealed container in the dark under standard conditions of time and temperature, usually 5 days at 20°C. During that time, microbes in the sample will degrade the organic material. The O_2 level is then determined again. The difference between the amount of dissolved O_2 at the beginning of the test and the amount at the end reflects the BOD. In the case of samples that have a very high BOD, they must be diluted first. Otherwise, the O_2 in the water could be depleted before the sample has been completely decomposed.

Municipal Wastewater Treatment Methods

Large-scale wastewater treatment plants in the United States use a series of two processes—primary and secondary treatment—as required by the Clean Water Act. Once treated, the **effluent** (the liquid portion) can be discharged into the receiving water. Additional steps are used to treat **sludge** (the solid portion). Industrial wastes, including organic solvents and heavy metals, can interfere with standard wastewater treatment processes, so industries must have facilities to pretreat their wastewater. The pretreatment processes are specific to the types of wastes generated and may remove organic solvents, heavy metals, and other harmful chemicals.

Primary Treatment

Primary treatment is a physical process designed to remove large objects and material that will settle out. Raw sewage is first passed through a series of screens to remove items such as sticks, rags, and trash (**figure 29.1**). Skimmers then remove scum and other floating materials, including grease and oil. After that, the sewage sits in a sedimentation tank, allowing more solids to settle out. Primary treatment removes approximately 50% of the solids and 25% of the BOD from raw wastewater. Once the settling period is complete, the material at the bottom of the tank is scraped out and forms primary sludge. The sludge is pumped to an anaerobic sludge digester (described later in this section), and the remaining liquid waste is sent for secondary treatment.

Secondary Treatment

Secondary treatment is mainly a biological process that converts most of the suspended solids to inorganic compounds and cell mass that can then be removed, eliminating as much as 95% of the BOD. Microbial growth is actively encouraged during secondary treatment, allowing aerobic organisms to oxidize the biologically degradable organic material to CO_2 and H_2O. Because secondary treatment relies on the metabolic activities of microorganisms, the processes could be devastated if too much toxic industrial waste or too many hazardous household materials were dumped into wastewater systems, killing the microbial population.

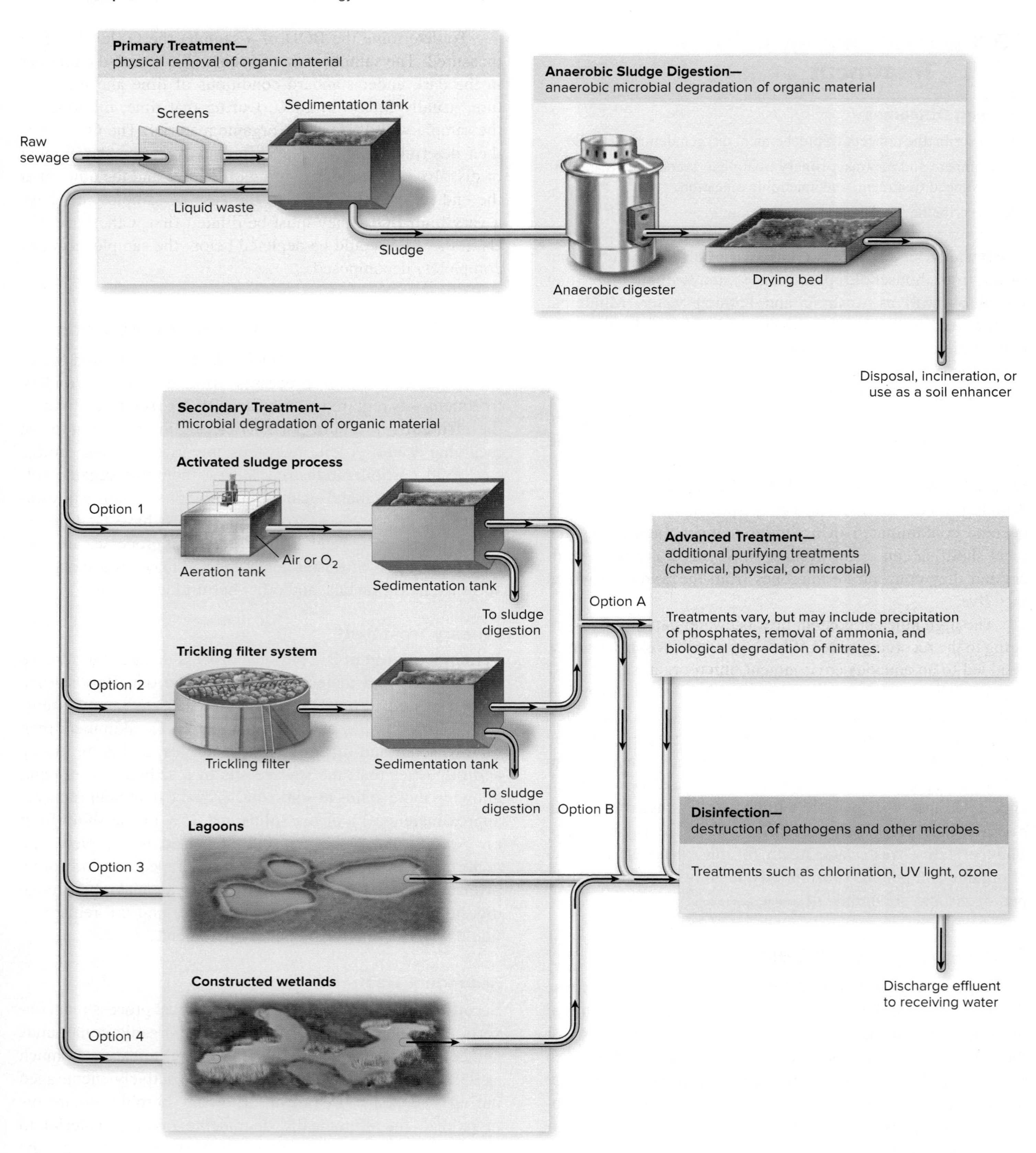

FIGURE 29.1 Municipal Wastewater Treatment The processes consist of primary treatment, secondary treatment, advanced treatment (optional), disinfection, and anaerobic sludge digestion.

Why does secondary treatment require O_2?

Methods used for secondary treatment include:

- **Activated sludge process.** This common system relies on mixed populations of aerobic microbes that grow as flocs (suspended biofilms). Although the organisms are often naturally present in wastewater, large numbers are added by introducing a small portion of leftover sludge from the previous load of treated wastes. Plenty of O_2 is supplied by mixing the wastewater in an aerator. As the microbes multiply, the organic matter is converted into both biomass and waste products such as CO_2. Following the aeration, the wastewater is again sent to a sedimentation tank. There, most of the flocs settle and the resulting secondary sludge is removed and pumped to the anaerobic sludge digester. A portion of this sludge is introduced to a new load of wastewater to act as an inoculum. ◀◀ biofilm
- **Trickling filter (TF) system.** This method is frequently used in smaller wastewater treatment plants. The TF has a rotating arm that distributes the liquid waste over a bed of plastic pieces or coarse gravel and rocks (**figure 29.2**). The surfaces of these materials become coated with a biofilm—a mix of bacteria, fungi, algae, protozoa, and nematodes—that aerobically degrades the organic material as it passes. The rate of wastewater flow can be adjusted for maximum degradation.
- **Lagoons.** The wastewater is channeled into shallow ponds, or lagoons, where it remains for several days to a month or more, depending on the design of the lagoon. Algae and cyanobacteria that grow at the surface provide O_2, allowing aerobic organisms in the ponds to degrade the organic materials.
- **Constructed wetlands.** These follow the same principles as lagoons, but their more advanced designs make them suitable habitats for birds and other wildlife (**figure 29.3**). For example, the wastewater treatment processes in Arcata, California, use a series of marshes that now attract a variety of shorebirds and serve as a wildlife sanctuary.

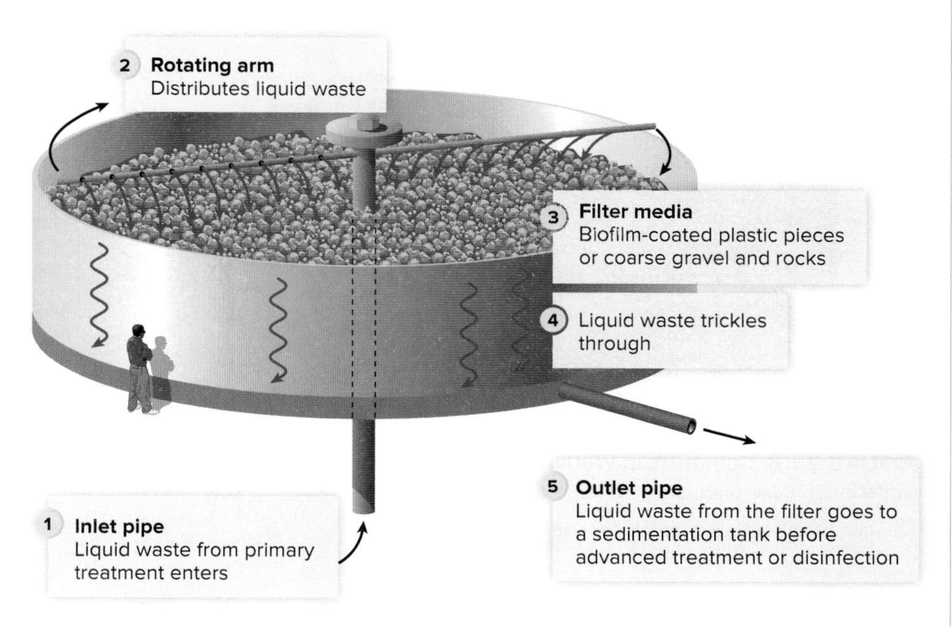

FIGURE 29.2 Trickling Filter Wastewater enters a rotating arm and then trickles out onto plastic pieces or a gravel and rock bed.

What role do biofilms play in trickling filters?

Advanced Treatment

As water supplies are becoming scarce—and in order to comply with discharge standards designed to protect the environment—many communities are finding **advanced treatment** necessary. This includes any purification process beyond secondary treatment; it may involve physical, chemical, or biological processes, or any combination of these. Advanced treatment is expensive, however, and has not been common in the past.

Advanced treatment is often designed to remove ammonia, nitrates, and phosphates—compounds that foster growth of algae and cyanobacteria in receiving waters. The concentrations of these nutrients, which are very low in unpolluted waters, normally limit the growth of the photosynthetic organisms. If the nutrients are added, photosynthetic microbes multiply to very high numbers, often leading to buoyant masses of cells that form a surface scum (see figure 11.7). Accumulations of dying photosynthetic organisms—which serve as a source of carbon for other microbes—increase the BOD and, consequently, threaten other forms of aquatic life.

Various mechanisms are used to remove the inorganic nutrients. Ammonia can be removed by a process called ammonia stripping, which frees gaseous ammonia from the water. Nitrates can be removed using denitrifying bacteria. These organisms use nitrate as a terminal electron acceptor during anaerobic respiration to form N_2—a gas that is inert, non-toxic, and easily removed. The discovery of anammox bacteria, which use ammonia as an energy source and nitrite as a terminal electron acceptor, provides an alternative means of removing inorganic nitrogen-containing compounds. Phosphates are eliminated using chemicals that combine with them to form a precipitate. ◀◀ denitrification, ◀◀ anammox

Disinfection of Effluent

Before discharge into the receiving water, the effluent is disinfected with chlorine, ozone, or UV light to decrease the numbers of microorganisms and viruses. If chlorine is used, the disinfected water can then be dechlorinated to avoid releasing

(a)

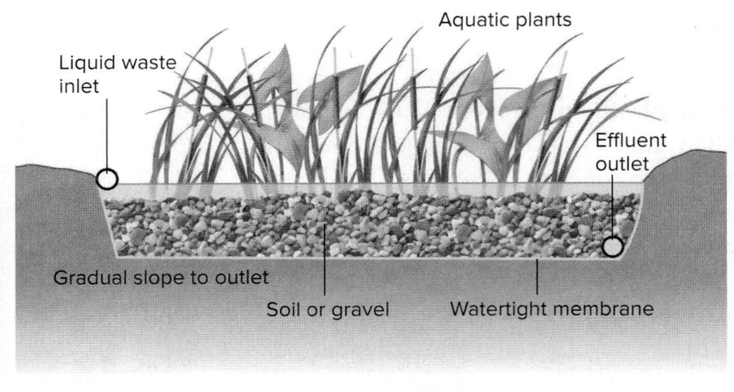

(b)

FIGURE 29.3 Constructed Wetland **(a)** Wastewater may be turned into a natural resource. **(b)** General components of a constructed wetland. a: ©Carrie Garcia/Alamy Stock Photo

What advantage does a constructed wetland have over a lagoon?

excessive amounts of the toxic chemical into the environment.

◀◀ chlorine, ◀◀ ozone, ◀◀ UV light

Anaerobic Digestion of Sludge

Sludge obtained during the sedimentation steps of primary and secondary treatment is transferred to a tank for **anaerobic digestion.** There, various anaerobic microbes convert much of the organic material to methane using the following sequence of reactions:

- Organic compounds → organic acids, CO_2, H_2
- Organic acids → acetate, CO_2, H_2
- Acetate, CO_2, H_2 → methane (CH_4)

Many wastewater treatment plants are equipped to use the methane generated, thereby avoiding the cost of other sources of energy to run their equipment.

After anaerobic digestion, water is removed from the remaining sludge, generating a nutrient-rich product called stabilized sludge. This can be incinerated or disposed of in landfills, but biosolids—a term for stabilized sludge that meets certain standards—may be used to improve soils and promote plant growth. An increasing number of wastewater treatment facilities are finding ways to recycle their treated sludge. Concerns exist, however, about heavy metals and other pollutants that can sometimes be concentrated in the product.

MicroByte

Sludge generated by the city of Milwaukee, Wisconsin, is used to produce fertilizer for lawns, gardens, golf courses, and playfields.

Individual Wastewater Treatment Systems

Rural homes usually rely on **septic systems** for wastewater treatment (**figure 29.4**). These small-scale on-site systems move household wastewater to a large tank, where much of the solid material settles and is degraded by anaerobic microorganisms. The effluent—which has a high BOD—flows to a series of perforated pipes, where it percolates (slowly passes) through a gravel-containing drain field. The drain field is designed to allow aerobic microorganisms to oxidize the organic material. Sludge that remains in the septic tank must be pumped out from time to time.

Septic systems must be properly designed and monitored to ensure they work adequately. Circumstances that prevent adequate drainage—such as a clay soil under a drain field—allow anaerobic conditions to develop, preventing the organic material from being oxidized adequately. In addition, toxic materials, including antibacterial soaps and bleach, can inhibit microbial activity in the system. Drainage from a septic tank may contain pathogens, so the tank must never be allowed to drain where it can contaminate water supplies.

MicroAssessment 29.1

Primary treatment removes material that settles out. Secondary treatment converts suspended material into inorganic compounds and microbial biomass. Advanced treatment is often designed to remove ammonia, phosphates, and nitrates. Anaerobic digestion converts much of the organic matter to methane. Septic systems are small-scale on-site wastewater treatment systems.

1. Why is it important that the BOD of wastewater is decreased before discharge?
2. What is the advantage of removing phosphates and nitrates from wastewater?
3. Why is denitrification unlikely to occur in an aeration tank used during secondary treatment?

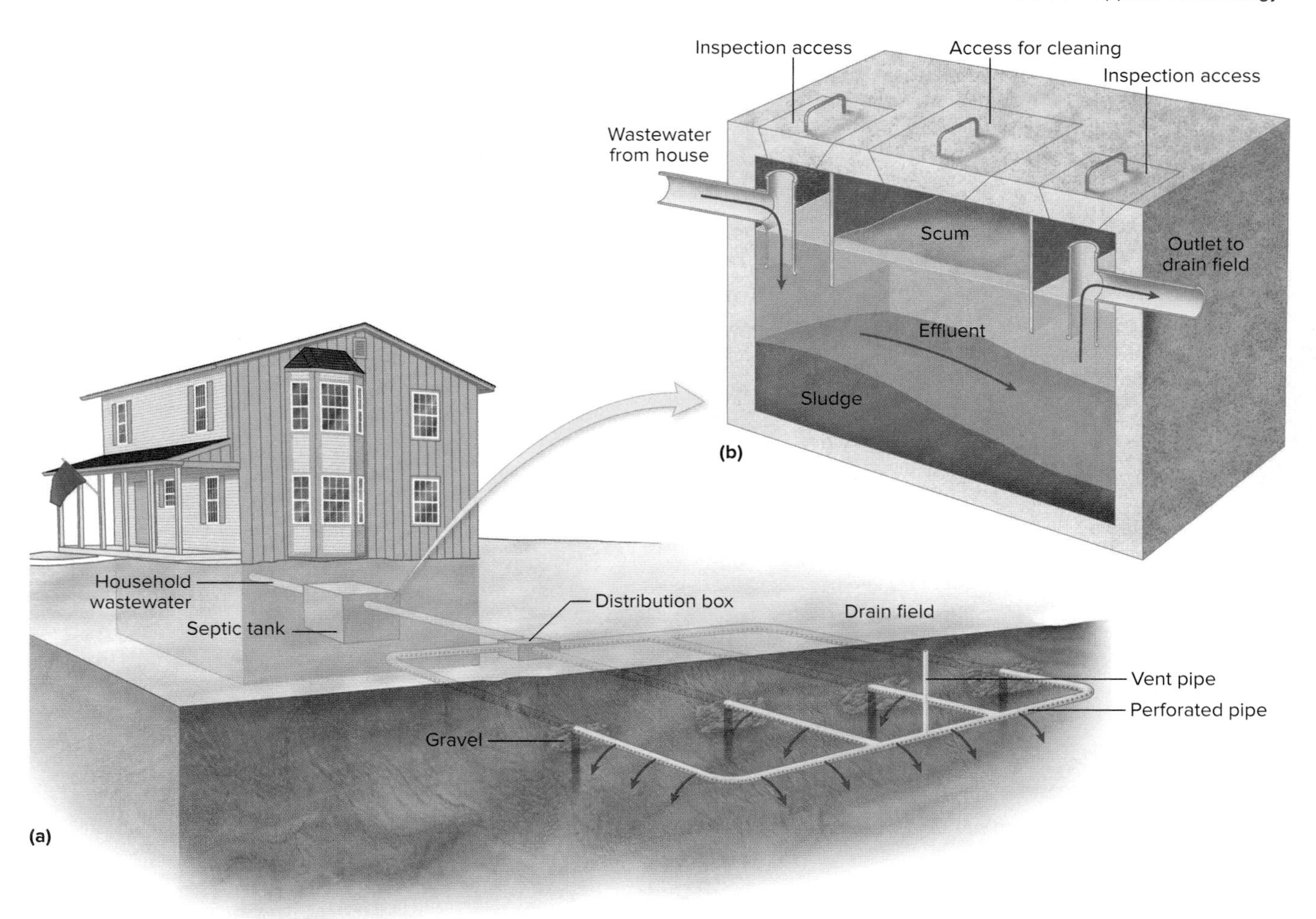

FIGURE 29.4 Septic System **(a)** The components of a septic system. **(b)** Septic tank, where anaerobic degradation takes place.

Why might a farm family avoid using bleach in their laundry?

29.2 ■ Drinking Water Treatment and Testing

Learning Outcomes

4. Describe how drinking water is typically treated.
5. Describe why and how drinking water is tested for total coliforms.

Large cities generally obtain their drinking water from surface waters such as lakes or rivers. Because surface water may also have served as the receiving water for another city's wastewater effluent, drinking water treatment is intimately connected to wastewater treatment. The quality of the surface water is also affected by the characteristics of the watershed (the land over which water flows into the river or lake). Even pristine rivers are likely contaminated with feces of animals that inhabit the watershed.

Smaller communities often use groundwater—pumped from a well—as a source of drinking water. Groundwater is in aquifers (water-containing layers of rock, sand, and gravel) and is replenished as water from rain and other sources seeps through the soil. Because aquifers are not directly exposed to animals and the atmosphere, they are somewhat protected from contamination. However, poorly located or maintained septic systems and sewer lines, as well as sludge or other fertilizers, can lead to groundwater contamination (**figure 29.5**).

Public water systems in the United States are regulated under the Safe Drinking Water Act. This gives the Environmental Protection Agency (EPA) the authority to set standards in order to control the level of contaminants in drinking water. Standards are modified in response to new concerns; for example, regulations now govern the maximum levels of *Cryptosporidium* oocysts, *Giardia* cysts, and enteric (intestinal) viruses in drinking water. ◀◀ *Cryptosporidium*, ◀◀ *Giardia*

Water Treatment Processes

Community drinking water treatment is designed to provide safe or **potable** water that lacks pathogenic microbes as well as harmful chemicals. First, water flows into a reservoir where

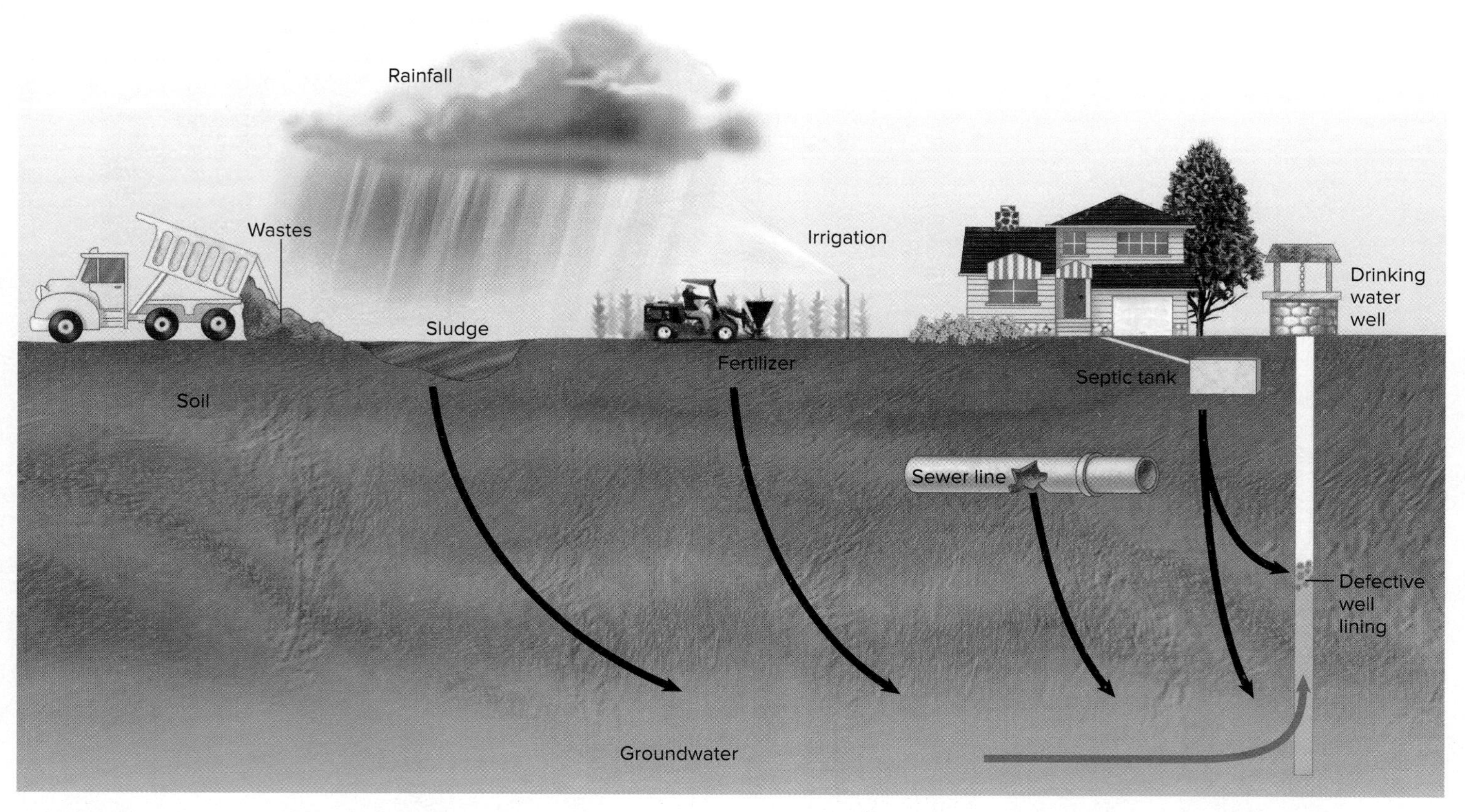

FIGURE 29.5 Groundwater Contamination Poorly located or maintained septic tanks and sewer lines, as well as sludge or other fertilizers, can lead to groundwater contamination.

How is water in the aquifer replenished?

it is allowed to stand long enough for material to settle out (**figure 29.6**). The water is then transferred to a tank where it is mixed with a chemical that causes suspended materials to coagulate (flocculate); an example of a coagulant is alum (aluminum sulfate). The mixture then flows to a sedimentation tank, where the coagulated materials are allowed to slowly sink to the bottom. As they settle, some microbes and other substances are trapped and are thereby removed as well.

Following the removal of the coagulated materials, the water is filtered—often through a thick bed of sand and gravel—to remove various microbes including bacteria and protozoan cysts and oocysts. Organic chemicals that may be harmful or give undesirable tastes and odors can be removed by additional filtration. This is done using an activated charcoal filter, which adsorbs dissolved chemicals. An added benefit of filtration is that microorganisms growing in biofilms on the filter materials use carbon from the water as it passes. This lowers the organic carbon content of the water, resulting in less microbial growth in pipes delivering the water. ◀◀ filtration

Finally, the water is treated with chlorine or other disinfectants to kill or inactivate harmful bacteria, protozoa, and viruses that might remain. A concern with using chlorine, however, is that some of the disinfection by-products (DBPs) might be carcinogenic. In response to this concern, ultraviolet radiation and ozone are increasingly being used as alternatives, but a small amount of chlorine must still be added to prevent problems associated with post-treatment contamination. Note that disinfection of waters with a high organic content requires more chlorine because organic compounds consume free chlorine. ◀◀ chlorine, ◀◀ disinfection by-products, ◀◀ ultraviolet radiation, ◀◀ ozone

Water Testing

A primary concern regarding the safety of drinking water is the possibility that it might be contaminated with any of a wide variety of intestinal pathogens, such as those discussed in chapter 24. It is not practical to test for all of the pathogens, however, so **indicator organisms** are used—these microbes are routinely found in feces, survive longer than intestinal pathogens, and are relatively easy to detect. Finding them suggests fecal contamination, and therefore a greater chance that intestinal pathogens are also present.

The most common bacterial group used as indicator organisms in the United States is **total coliforms**—lactose-fermenting members of the family *Enterobacteriaceae,* including *Escherichia coli.* The group is functionally defined as facultatively anaerobic, Gram-negative, rod-shaped, non-spore-forming bacteria that ferment lactose, forming acid and gas within 48 hours at 35°C.

Although total coliforms are routinely present in the intestinal contents of warm-blooded animals, certain species can also

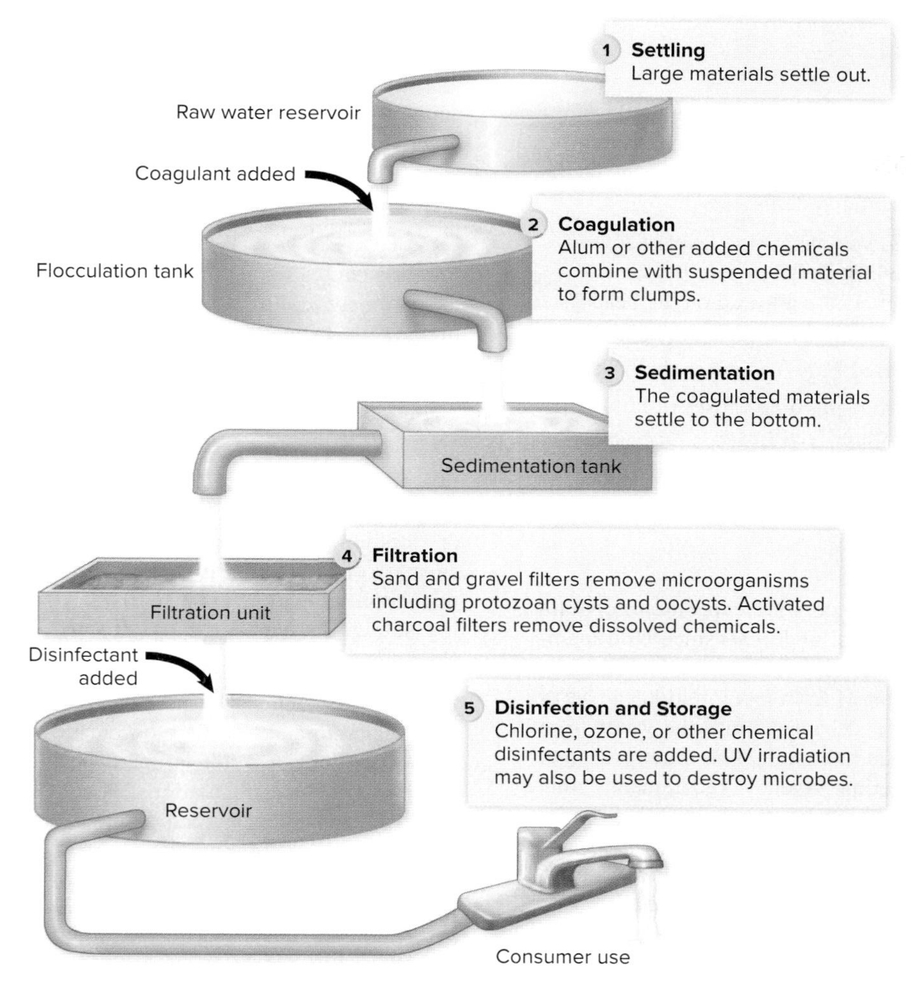

FIGURE 29.6 Treatment of Municipal Water Supplies

Considering that *Cryptosporidium* oocysts are resistant to chlorine, which step in water treatment protects us from them?

thrive in soils and on plant material. Because of this, their presence does not necessarily imply fecal pollution. To compensate for this shortcoming, **fecal coliforms,** a subset of total coliforms more likely to be of intestinal origin, are also used as indicator organisms. The most common fecal coliform is *E. coli.* Note that although some strains can cause intestinal disease, *E. coli* is used in water testing merely to indicate fecal pollution.

Methods used to detect total coliforms in a water sample include:

- **ONPG/MUG test.** A water sample is added to a medium containing ONPG (*o*-nitrophenyl-β-D-galactopyranoside) and MUG (4-methyl-umbelliferyl-β-D-glucuronide). Lactose-fermenting bacteria hydrolyze ONPG, generating a yellow compound; thus, all coliforms turn the medium yellow (**figure 29.7**). *E. coli* produces an enzyme that hydrolyzes MUG, making a fluorescent compound. Because of ONPG and MUG, a sample can be tested simultaneously for both total coliforms and *E. coli.*
- **Presence/absence test.** A 100 mL water sample is added to a lactose-containing broth that is selective for Gram-negative rods. An inverted tube is included to trap gas. If gas is produced, the broth is then tested to confirm that coliforms are present.
- **Most probable number (MPN) method.** This statistical assay of cell numbers uses successive dilutions to determine the most probable number of bacteria in a sample (see figure 4.21). To test drinking water, the broth used is similar to that in the presence/absence test, and each tube includes a small inverted tube to trap gas. MPN tubes that show gas production are further tested to confirm that they contain coliforms.
- **Membrane filtration.** A water sample is passed through a filter that retains bacteria (see figure 4.20), collecting them from a known volume of water. The filter is then placed on a lactose-containing selective and differential agar medium. ⏮ selective media, ⏮ differential media

As part of the Safe Drinking Water Act, the Revised Total Coliform Rule (RTCR) outlines the sample collection requirements for water testing. It also establishes monitoring and corrective actions if samples are positive for total coliforms or *E. coli.*

Because total coliform and fecal coliform assays cannot always predict contamination with protozoan cysts and oocysts, alternatives are being explored. Other

FIGURE 29.7 ONPG/MUG Test Coliforms hydrolyze ONPG, yielding a yellow-colored compound. *E. coli* hydrolyzes MUG, generating a blue fluorescent compound.

Which of these tested bottles would you prefer to contain a sample of your drinking water?

FOCUS ON A CASE 29.1

A large lake that serves as a popular recreation site was found to have *Escherichia coli* levels well above the U.S. Environmental Protection Agency's suggested maximum limit. In response, authorities closed a number of public beaches surrounding the lake due to health concerns. Local business owners and homeowners were angry about the closure, worrying about the effect on tourism and property values. Prospective homebuyers were hesitant to invest in the region because of the controversy.

Discussions regarding the source of the problem caused considerable confusion, as no one wanted to take responsibility. Some people felt that the results were meaningless because they followed periods of heavy rain and the resulting runoff. Others blamed the region's small wastewater treatment facilities that served local business and housing developments or the hundreds of aging septic systems that served individual homes. Yet others felt that because the lake water was clear, there was no cause for concern.

In an attempt to find the source of the high *E. coli* levels, authorities began inspecting the local wastewater treatment facilities. Over 30% of the facilities were found to be in violation of the conditions agreed to in their permits. The state then instituted a strict policy of testing and enforcement to improve the quality of the water. In addition, officials began exploring longer-term solutions, including developing a regional sewage treatment system to take the place of smaller treatment facilities and septic systems.

1. Would you expect the level of *E. coli* in the lake to fall to zero before swimmers are allowed on the beach?
2. What are the possible sources of increased *E. coli* found in the lake?
3. List three possible causes of failure of wastewater facilities.
4. How can individuals improve the quality of the lake water?

Discussion

1. Natural waters are expected to contain a low level of *E. coli* since wild mammals and birds can defecate in the lake and runoff from precipitation can carry feces and bacteria into the water from surrounding soil. The Environmental Protection Agency recommends that *E. coli* levels stay below 235 CFUs (colony-forming units) per 100 mL of water at public beaches. The water in this case contained over 2,000 CFUs per 100 mL.
2. *E. coli* contamination may result from a specific source such as a faulty septic system serving a private home or a wastewater treatment facility serving a larger population, such as in a campground or condominium development. It may also come from nonspecific sources such as surface runoff of rainfall flowing through agricultural land or storm water from city streets.
3. Possible causes for failure of wastewater treatment could include: (1) allowing filters to become clogged with sludge or vegetation; (2) failure to replace chemicals such as chlorine tablets; (3) leaks in wastewater tanks that allow wastewater to seep into the soil; (4) uncontrolled overflow of wastewater tanks or sewers due to storms or flooding; (5) failure to upgrade the capacity of the facility to keep up with population growth; or simply (6) allowing sewage to bypass the system and flow directly into the lake.
4. People with septic systems can regularly maintain and test their systems to make sure they function properly and comply with local regulations. People with lakefront property can plant vegetation at the shoreline to minimize erosion and runoff, and reduce use of pesticides, fertilizers, and other chemical pollutants. They can avoid dumping yard waste into the water and keep pets from defecating in or near the lake. Residents and tourists can pressure businesses that have been repeatedly cited for water quality violations.

microbes that can be used as indicators of fecal pollution include enterococci, some *Clostridium* species, and certain types of bacteriophages.

MicroAssessment 29.2

Drinking water may be obtained from surface water or groundwater. Treatment of drinking water is designed to eliminate pathogens and harmful chemicals. In the United States, total and fecal coliforms are the most commonly used indicator organisms.

4. What is the purpose of coagulation in drinking water treatment?

5. Describe two methods of water testing.

6. Which would be more likely to cause illness and why—a water sample that tested positive for fecal coliforms or one that tested positive for *E. coli* O157:H7? Explain your answer.

29.3 ■ Microbiology of Solid Waste Treatment

Learning Outcome

6. Compare and contrast sanitary landfills and composting programs.

In addition to ridding our environment of wastes in water, we must dispose of the solid wastes (trash and garbage) generated each day. Eliminating these has become an increasingly complex problem.

Sanitary Landfills for Solid Waste Disposal

Sanitary landfills are widely used to dispose of nonhazardous solid wastes in a manner that minimizes damage to human health and the environment. Before sanitary landfills were developed, solid wastes were often piled up on the ground in open-burning dumps, attracting insects and rodents and causing aesthetic and public health problems.

Federal standards dictate that sanitary landfills must be located away from wetlands, earthquake-prone faults, flood plains, or other sensitive areas. The excavated site is lined with plastic sheets or a special membrane on top of a thick layer of clay to prevent contaminants from seeping out into the surrounding environment. A layer of sand with drainage pipes is placed on top of this. When wastes are added to the site, they are compacted and covered with a layer of soil. Once a landfill is full, it is covered with material that minimizes the amount of storm water that will flow through the landfill. Then it is covered with soil and plants and can be used for recreation and eventually as a site for construction.

Sanitary landfills have several disadvantages with respect to waste management. For one thing, only a limited number of sites are available near urban and suburban areas. In addition, the methane gas must be removed as the organic waste material anaerobically decomposes, a process that may continue for years after the landfill is full. The methane is burned or recovered for use. If it is not removed, disastrous explosions can occur. If pollutants such as heavy metals and pesticides do leak from landfill sites, they may reach underground aquifers. It is very difficult to purify these aquifers once they have become contaminated.

Sanitary landfills have traditionally been a low-cost method of handling large quantities of solid waste. Because of increased costs and decreased availability of land, however, many cities are looking for ways to cut the amount of solid waste dumped in landfills. In some cities, the fees charged to people for garbage collection are based on the size of the container collected. The smaller the can, the lower the cost. This is intended to raise people's awareness of how much solid waste they generate as well as offer an incentive to recycle. Programs to recycle paper, plastics, glass, and metal have been implemented in many regions with great success. As a result of these programs, landfill areas will be available for a longer period of time.

(a)

(b)

FIGURE 29.8 Backyard Composting **(a)** A home compost bin. **(b)** Garden debris and many kitchen organic wastes can be composted. a: ©Wave Royalty Free/Alamy Stock Photo; b: ©Sharon Dominick/iStock Exclusive/Getty Images

? Why is it important that the compost heap be turned frequently?

Municipal and Backyard Composting—Alternative to Landfills

Composting is the process of making **compost,** the material that results from the controlled decomposition of solid organic material. Not only does it reduce the amount of organic waste added to landfills, the product can be used to improve garden soils.

Backyard composting usually starts with a supply of organic material such as leaves, grass clippings, and food scraps (**figure 29.8**). Often, some soil and water are added to facilitate the process. If composting is done correctly, the inside of the pile heats up due to the metabolic reactions of the multiplying microbes. Achieving high temperatures (55°C to 66°C) is important to kill pathogens that might be present in the solid wastes. Thermophilic organisms thrive at the high temperature, so they continue decomposing the material. Frequent aeration of the pile, accomplished by physically stirring and turning it as well as keeping it moist, speeds the process so that composting can be completed in as little as 6 weeks.

◀◀ thermophile

Composting on a large scale offers cities a way to reduce the amount of garbage sent to their landfills while also generating a commercially valuable product. In some cities, yard wastes are collected separately from the main trash and garbage. These are then processed using machinery such as grinders that help make the composting more efficient by increasing the surface area for microbial action (**figure 29.9**).

FIGURE 29.9 Municipal Composting Compost is turned for aeration at a municipal composting site. ©FLPA/Alamy Stock Photo

What is the purpose of grinding yard waste before putting it in piles?

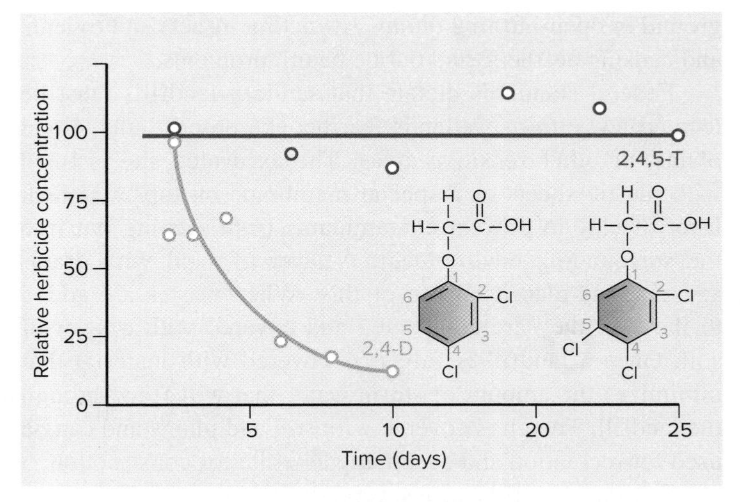

FIGURE 29.10 Comparison of the Rates of Disappearance of Two Structurally Related Herbicides, 2,4-D and 2,4,5-T

Why would 2,4,5-T be banned from use in the United States?

MicroAssessment 29.3

Sanitary landfills are used to dispose of non-hazardous solid wastes. Composting of solid organic wastes dramatically reduces the need for large landfills and generates a useful product.

7. What are the advantages and disadvantages of landfills?
8. List the steps in successful composting.
9. Why would adding soil and water facilitate composting?

29.4 ■ Microbiology of Bioremediation

Learning Outcomes

7. Describe why pollutants are a problem, and why some persist in the environment.
8. Describe how two bioremediation strategies remove pollutants.

Bioremediation is the use of microorganisms to degrade or detoxify pollutants in a given environment. It generally takes advantage of organisms already present, but in some cases specific organisms are added to the polluted environment.

Pollutants

Most naturally occurring organic compounds are biodegradable, meaning they can be degraded by one or more microbial species. As oil spills dramatically demonstrate, however, some natural materials can cause devastating effects before they are degraded.

Synthetic compounds are more likely to be biodegradable if their chemical composition is similar to that of naturally occurring substances. In contrast, xenobiotics (synthetic compounds quite different from any in nature) are more likely to persist in the environment because microorganisms are unlikely to have suitable enzymes to break them down. Such enzymes would not give a microbe a competitive advantage in a natural situation.

Relatively slight molecular changes significantly alter the biodegradability of a compound. Perhaps the best-studied example involves the herbicides 2,4-dichlorophenoxyacetic acid (2,4-D) and 2,4,5-trichlorophenoxyacetic acid (2,4,5-T). The only difference between these two compounds is the additional chlorine atom on the latter. When 2,4-D is applied to the soil, it disappears within a period of several weeks, as a result of its degradation by microbes in the soil. When 2,4,5-T is applied, however, it is often still present more than a year later (**figure 29.10**). The additional chlorine atom of 2,4,5-T blocks the enzyme that makes the initial change in 2,4-D.

Most herbicides and insecticides not only are toxic to their target, but also have damaging effects on fish, birds, and other animals. For example, the pesticide DDT accumulates in the fat of predatory birds through biological magnification (**table 29.1**). Small amounts of the pollutant in water can be

TABLE 29.1 Biological Magnification of DDT

Parts per Million DDT	Source
0.00005	Water
0.04	Plankton
0.23	Minnow
3.57	Heron
22.8	Merganser (fish-eating duck)

concentrated in the tiny plankton, which are eaten by minnows, so that even more accumulates in the fish. When large birds eat the fish, the amount of chemical in tissues is amplified. The continuing ingestion of DDT, which accumulates in fat, results in an ever-greater concentration of the pollutant as it passes upward through the food chain. DDT interferes with the reproductive process of birds, leading to the production of fragile eggs, which break before the young can hatch. Although banned in the United States, DDT is still used in other countries, particularly to control mosquitoes that transmit malaria.

Strategies of Bioremediation

Many factors influence the degradation rate of pollutants. As a general rule, any practice that favors multiplication of microorganisms will increase the rate of degradation. Thus, providing adequate nutrients, maintaining a nearly neutral pH, raising the temperature, and providing an optimal amount of moisture are all likely to promote pollutant degradation.

There are two general bioremediation strategies—biostimulation and bioaugmentation. **Biostimulation** enhances growth of resident microbes in a contaminated site by providing additional nutrients. Petroleum-degrading bacteria are naturally present in seawater, but they degrade oil at a very slow rate because the low levels of certain nutrients, including nitrogen and phosphorus, limit their growth. To enhance bioremediation of oil spills, a fertilizer containing these nutrients—and which adheres to oil—was developed. When this fertilizer is applied to an oil spill, microbial growth is stimulated, leading to at least a threefold increase in the speed of degradation (**figure 29.11**).

FIGURE 29.11 Oil Spill Bioremediation Nutrients being added to stimulate the growth of resident oil-degrading microorganisms. ©Accent Alaska.com/Alamy Stock Photo

Which kind of bioremediation is this—biostimulation or bioaugmentation?

Bioaugmentation relies on activities of microorganisms added to the contaminated material, complementing the resident population. The activated sludge process used during secondary treatment of wastewater is a form of bioaugmentation. Research is underway to develop microbial strains suited for bioaugmentation. In the laboratory, the bacterium *Burkholderia cepacia* has been used to remove 2,4,5-T from soil samples. However, microbes that thrive under laboratory conditions may not compete well in natural habitats. ⏮ activated sludge process

Successful bioremediation may also involve controlling metabolic processes by manipulating the growth conditions. For instance, aerobic conditions are important when degrading trichloroethylene (TCE), a solvent used to clean metal parts. Anaerobic degradation results in the accumulation of vinyl chloride, a compound more toxic than TCE. Some pollutants are degraded only when specific substrates are made available to the microbes. This phenomenon, called co-metabolism, occurs because the enzyme produced by the microbe to degrade the additional substrate degrades the pollutant as well. As an example, the enzymes produced by some microbes to degrade methane also degrade TCE. In this case, adding methane enhances the degradation of TCE.

Bioremediation may be done either in situ ("in place") or off-site. In situ bioremediation generally relies on biostimulation and is less disruptive. Technicians can add oxygen (O_2) to contaminated groundwater and soil either by injecting hydrogen peroxide, which rapidly decomposes to liberate O_2 and water, or by pumping air into soil. Off-site processes may be performed using a bioreactor, a large tank designed to accelerate microbial processes. Nutrients and O_2 may be added to facilitate microbial growth and metabolism, while the slurry is agitated to ensure that the microbes remain in contact with the contaminants. A slower process involves mounding the contaminated soil over a layer that traps seeping chemicals. To provide O_2, the soil can be turned occasionally or air can be forced through it.

MicroAssessment 29.4

Bioremediation uses microorganisms to degrade pollutants. Biostimulation and bioaugmentation are two methods of bioremediation.

10. Why do xenobiotics often persist in the environment?

11. How is biostimulation different from bioaugmentation?

12. In treating an oil spill, why might biostimulation be preferred over bioaugmentation?

Summary

29.1 Microbiology of Wastewater Treatment

Biochemical Oxygen Demand (BOD)

An important goal of **wastewater** treatment is the reduction of the **biochemical oxygen demand (BOD).**

Municipal Wastewater Treatment Methods (figures 29.1, 29.3)

Primary treatment is a physical process designed to remove materials that settle out. **Secondary treatment** is chiefly a biological process designed to convert most of the suspended solids to inorganic compounds and microbial biomass, removing most of the BOD. **Advanced treatment** is often designed to remove ammonia, nitrates, and phosphates. Biosolids that result from **anaerobic digestion** of **sludge** can be used to improve soils and promote plant growth.

Individual Wastewater Treatment Systems

Rural homes often use **septic systems** for wastewater treatment (figure 29.4).

29.2 Drinking Water Treatment and Testing

Water Treatment Processes

Community drinking water is treated to remove particulate and suspended matter, various microorganisms, and organic chemicals (figure 29.6). Chlorine or other disinfectants are then used to destroy harmful microbes.

Water Testing (figure 29.7)

Total coliforms and **fecal coliforms** are used as **indicator organisms;** their presence suggests the possible presence of pathogens.

29.3 Microbiology of Solid Waste Treatment

Sanitary Landfills for Solid Waste Disposal

Landfills are used to dispose of solid wastes near towns and cities. Increased recycling of solid wastes decreases the demand on landfills.

Municipal and Backyard Composting—Alternative to Landfills

Composting reduces the amount of garbage sent to landfills (figures 29.8, 29.9).

29.4 Microbiology of Bioremediation

Pollutants

Synthetic compounds are more likely to be biodegradable if they have a chemical composition similar to that of naturally occurring compounds (figure 29.10).

Strategies of Bioremediation

Biostimulation can be used to increase the effectiveness of oil degradation by naturally occurring microorganisms (figure 29.11). **Bioaugmentation** involves addition of other microbes to the contaminated area.

Review Questions

Short Answer

1. Describe how the BOD of a water sample is determined.
2. Which step of wastewater treatment removes most of the BOD?
3. Compare and contrast the activated sludge process and the trickling filter system used in secondary treatment of wastewater.
4. What is effluent?
5. How does a septic system work?
6. What is an aquifer?
7. Why do water-testing procedures look for coliforms rather than pathogens?
8. How does the ONPG/MUG test allow a sample to be assayed simultaneously for the presence of both total coliforms and *E. coli*?
9. Describe two advantages of composting.
10. Describe the use of bioremediation in the cleanup of oil spills.

Multiple Choice

1. A significant decrease in BOD during secondary treatment indicates
   a) lack of O_2 during treatment.
   b) effective aerobic decomposition during treatment.
   c) effective anaerobic decomposition during treatment.
   d) removal of all pathogenic bacteria.
   e) removal of all toxic chemicals.
2. Advanced treatment is often designed to remove
   a) BOD.
   b) nitrates and phosphates.
   c) bacteria.
   d) protozoa.
   e) methane.
3. Which of the following is not a matching pair?
   a) Potable water—presence of pathogens
   b) High BOD—high organic content
   c) Stabilized sludge—fertilizer
   d) Primary treatment—removal of material that settles
4. Which of the following is *false*?
   a) Trickling filter systems are used for primary treatment.
   b) Artificial wetlands provide a habitat for wildlife.
   c) Removal of nitrates by microorganisms requires anaerobic conditions.
   d) Methane is a by-product of anaerobic digestion.
   e) Ozone can be used to disinfect water.
5. Which of the following is not a matching pair?
   a) Surface water—watershed
   b) Groundwater—aquifer
   c) Sand and gravel filters—remove organic chemicals
   d) Alum—causes suspended material to coagulate
   e) Disinfection—chlorine or ultraviolet light

6. Septic tanks should be placed
   a) as close to the well as possible.
   b) near a lagoon.
   c) under the house.
   d) in deep clay soil.
   e) where the outflow cannot contaminate any water supply.

7. Which of the following about coliform testing methods is *true*?
   a) All determine the number of *E. coli* present in a sample.
   b) The MPN procedure precisely indicates the concentration of coliforms.
   c) The media used in these tests check for the ability to ferment lactose.
   d) A positive test indicates that pathogens are definitely present in the sample.
   e) All coliforms hydrolyze ONPG and MUG.

8. Landfills are often used to dispose of
   a) household wastewater.
   b) commercial wastewater.
   c) solid wastes.
   d) petroleum wastes.
   e) wastewater effluent.

9. Backyard composting is an excellent way to dispose of
   a) cooking fats.
   b) garden debris.
   c) spoiled meats.
   d) insecticides.
   e) cleaning supplies.

10. Synthetic compounds are most likely to be biodegradable if they
    a) are totally different from anything found in nature.
    b) have three chlorine atoms per molecule.
    c) are plastics.
    d) are present in very large amounts.
    e) are chemically similar to naturally occurring substances.

Applications

1. A developer is interested in building vacation homes on 150 acres of oceanfront property. A priority is to retain as much natural beauty of the area as possible. Safe and effective wastewater treatment must be part of the plan. What advantages and disadvantages of each of the following options must the developer consider before selecting one?
   a) Individual septic systems for each home
   b) Trickling filter system
   c) Constructed wetlands
2. A public health official is investigating waterborne diseases in Illinois. She notes that over half of the cases of waterborne diseases originating from drinking water were caused by *Giardia lamblia.* Other data showed that most cases of gastroenteritis attributed to exposure to recreational waters were caused by *Cryptosporidium parvum.* What does this suggest about controlling waterborne diseases?

Critical Thinking

1. Why is oil not degraded when in a natural habitat underground, yet susceptible to bioremediation in an oil spill?
2. The accompanying figure shows the effects of different treatments of drinking water on the incidence of typhoid fever in Philadelphia, 1890–1935. If filtration of drinking water caused such a dramatic decrease in the disease incidence, was it necessary to introduce chlorination a few years later? Why or why not?

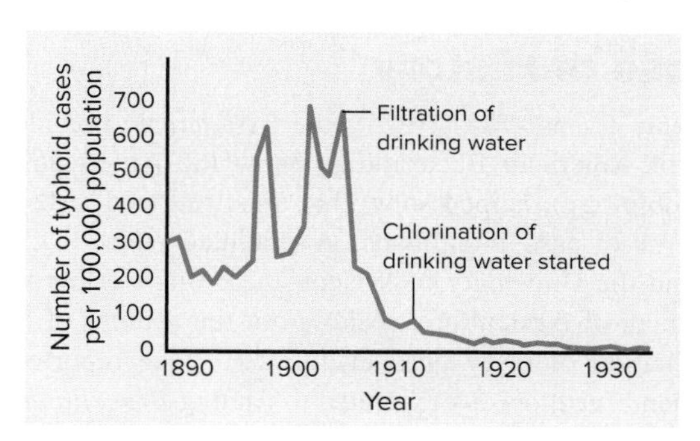

www.mcgrawhillconnect.com

Enhance your study of this chapter with study tools and practice tests. Also ask your instructor about the resources available through Connect, including the media-rich eBook, interactive learning tools, and animations.

30 Food Microbiology

Refrigeration is a method of food preservation. ©*Tang Ming Tung/DigitalVision/Getty Images*

KEY TERMS

Fermented Foods Foods intentionally altered during production by encouraging the activity of bacteria, yeasts, or molds.

Food Preservation Increasing the shelf life of foods by preventing the growth and activities of microorganisms.

Food Spoilage Undesirable biochemical changes in foods.

Foodborne Infection An illness that results from consuming a food product contaminated with microbes that colonize the host and cause disease.

Foodborne Intoxication An illness that results from consuming an exotoxin produced by a microorganism growing in a food product.

Water Activity (a_w) The relative amount of water available for microbial growth; pure water has an a_w of 1.0.

A Glimpse of History

Alice Catherine Evans (1881–1975), the first female president of the Society of American Bacteriology (now the American Society for Microbiology), helped show that unpasteurized milk could be a source of disease in humans. A graduate of both Cornell University and the University of Wisconsin, Evans worked for the U.S. Department of Agriculture, seeking out the sources of microbial contamination of dairy products. In 1917, she reported that cases of human brucellosis were related to finding *Brucella abortus* in cows' milk.

Evans's conclusion that *B. abortus* could be transmitted from cows to humans through milk conflicted with the common view of a number of prominent scientists, including Robert Koch. In 1900, Koch had declared that bovine tuberculosis and brucellosis could not be transmitted to humans. As a result, many scientists and dairy workers would not accept the increasing evidence that milk could be a source of the diseases. In the late 1930s, after a number of children of dairy workers had died of brucellosis, the problem was finally acknowledged. Today, milk is routinely pasteurized, and only very small amounts of unpasteurized milk are sold in the United States.

Everyone has experienced finding moldy bread, sour milk, or perhaps a container of long-forgotten green goo in the back of the refrigerator. Although disappointed with the fate of our food, we are seldom surprised. Just as our foods provide nutrition for our bodies, they also provide a rich nutrient source for ever-present populations of microbes.

Microorganisms on foods are not necessarily undesirable. Sometimes, their growth results in pleasant flavors or textures. Foods intentionally altered during production by carefully controlling the activity of bacteria, yeasts, or molds are called **fermented foods** (**figure 30.1**). The processes that cause food to spoil are often the same ones involved in the fermentation of foods. The souring of milk, for example, involves the same microbial processes that produce the agreeable acidic flavor of sour cream, and the mold growing on bread may be related to the one that causes the blue veining in Gorgonzola cheese. In fact, a food considered by one cultural population to be fermented may be thought of as spoiled by another. Strictly speaking, the term *fermentation* refers to only those metabolic activities that use pyruvate or another organic compound as an electron acceptor, with the result that alcohols and acids are produced. Food scientists, however, use the term more generally, to encompass any desirable change that a microorganism contributes to food. ⏮ fermentation

Foods like fresh vegetables and meats that easily support the growth of microorganisms are considered perishable. Microbes growing on these foods can lead to **food spoilage,** meaning a deterioration in the food's quality (**figure 30.2**). We can slow spoilage by changing the food's properties or by changing the conditions under which it is stored or prepared.

Slowing microbial growth maintains food quality for a longer period, and also helps prevent illnesses that result from eating contaminated food. Foodborne illness may be due to ingesting either a pathogen or a toxin produced by a pathogen growing in the food.

FIGURE 30.1 Fermented Foods ©McGraw-Hill Education/John Thoeming

How is a fermented food different from a spoiled food?

30.1 ■ Factors Influencing the Growth of Microorganisms in Foods

Learning Outcome

1. Describe five intrinsic factors and two extrinsic factors that influence the growth of microorganisms in foods.

Understanding the factors that influence microbial growth is essential to maintaining food quality. As a general rule, bacteria will predominate in fresh meats and other moist, pH-neutral, nutrient-rich foods. Yeasts and molds can also grow in these foods, but the more rapid increase of bacteria overwhelms the competitors. When conditions such as lack of moisture or high acidity restrict the growth of bacteria, fungi predominate despite their relatively slow growth.

Intrinsic Factors

Characteristics of the food itself—such as water availability, acidity, and nutrient level—are called **intrinsic factors.** They affect which microbes predominate on the product.

Water Availability

Foods vary in terms of how much water is accessible to microorganisms. Fresh meats and milk, for example, have plenty of water to support the growth of many microbes. Bread, nuts, and dried foods, on the other hand, are relatively dry. Jams, jellies, and some other sugar-rich foods are seemingly moist, but most of that water is chemically interacting with the sugar, making it unavailable for use by microbes. Highly salted foods, for similar reasons, have little available moisture.

The term **water activity (a_w)** is used to indicate the amount of water available in foods. By definition, pure water has an a_w of 1.0. Most fresh foods have an a_w above 0.98, whereas ham has an a_w of 0.91, jam has an a_w of 0.85, and some cakes have an a_w of 0.70. Most bacteria require an a_w above 0.90 for growth, which explains why fresh, moist foods spoil more quickly than dried, sugary, or salted foods. Fungi can grow at an a_w as low as 0.80, so forgotten bread, cheese, jam, and dried foods often become moldy.

Staphylococcus species, which are adapted to grow on the dry, salty surfaces of skin, can grow at an a_w of 0.86, lower than the minimum required by most common spoilage bacteria. Staphylococci normally do not compete well with other bacteria, but on salty products such as ham and other cured meats, they can multiply with little competition. Ham is a common vehicle for *S. aureus* food poisoning.

FIGURE 30.2 Examples of Spoiled Food **(a)** Rotten apples. **(b)** Moldy bread. **(c)** An ear of corn and a lemon after several weeks in the refrigerator. a: ©Melbourne Etc/Alamy Stock Photo; b: ©Westend61/Alamy Stock Photo; c: ©Martha Nester

Why was the apple spoiled by bacteria, but the bread spoiled by mold?

(a)

(b)

(c)

pH

Many bacterial species, including most pathogens, are inhibited by acidic conditions and cannot grow at a pH below 4.5. An exception is the lactic acid bacteria, which are able to grow at a pH as low as 3.5. These bacteria produce lactic acid as a result of fermentative metabolism and are used to make yogurt, sauerkraut, and some other fermented foods. Acidic end products preserve fermented food by preventing the growth of other microorganisms. Those same end products are undesirable in many foods, however, so lactic acid bacteria are also common spoilage bacteria. ⏮ lactic acid bacteria

Fungi can grow at a lower pH than most spoilage bacteria, so acidic foods may eventually become moldy. For example, the pH of lemons is approximately 2.2, which inhibits the growth of bacteria, including the lactic acid group. However, some fungi can grow at this low pH.

The pH of a food product can also determine whether toxins can be produced. *Clostridium botulinum* (the bacterium that causes botulism) does not grow or produce toxin below pH 4.5, so it is not considered a danger in highly acidic foods. This is why the canning process for acidic fruits and pickles does not call for as high a temperature as that required for canning less acidic foods. Some newer varieties of tomatoes have a higher pH than older types, so acid must be added if they are to be safely canned using the lower temperatures.

Nutrients

An organism requiring a particular vitamin cannot grow in a food lacking that vitamin. A microbe capable of synthesizing that vitamin, however, can grow if other conditions are favorable. Members of the genus *Pseudomonas* often spoil foods because they can synthesize essential nutrients and can multiply in various environments, including refrigeration.

Biological Barriers

Rinds, shells, and other coverings help protect foods from invasion by microorganisms. Eggs, for example, retain their quality much longer with intact shells. Whole lemons keep longer than slices. Even so, microorganisms will eventually break down these coverings and cause spoilage.

Antimicrobial Chemicals

Some foods contain natural antimicrobial chemicals that help prevent spoilage. Egg white, for instance, is rich in lysozyme. If lysozyme-susceptible bacteria get through the protective shell of an egg, they are destroyed by lysozyme before they can cause spoilage. Other examples of naturally occurring antimicrobial chemicals are benzoic acid in cranberries and allicin in garlic. ⏮ lysozyme

Extrinsic Factors

Environmental conditions under which a food is held—such as temperature and atmosphere—are called **extrinsic factors.** These can greatly affect the extent of microbial growth on food.

Storage Temperature

The storage temperature affects the growth rate of microorganisms. At low temperatures above freezing, many enzymatic reactions are either very slow or non-existent, with the result that microorganisms multiply slowly, if at all. Microorganisms that grow on refrigerated foods are most likely psychrophiles or psychrotrophs, such as some members of the genus *Pseudomonas*. Most pathogens prefer the warmer temperatures of the human body and do not grow well under refrigeration. ⏮ psychrophiles, ⏮ psychrotroph

Some foods can be stored for longer periods of time at freezing temperatures. Temperatures below about −23°C (about −10°F) stop the growth of most organisms because water is solid and therefore unavailable for microbial growth. Even frozen, however, foods should not be stored indefinitely.

Atmosphere

The presence or absence of O_2 affects the type of microbial population able to grow in food. Obligate aerobes cannot grow in foods stored under conditions that exclude their required O_2. Keeping O_2 out of food, however, may permit other bacteria to grow, including the obligate anaerobe *Clostridium botulinum*. People have developed botulism after eating homemade stew left at room temperature overnight. The cooking process does not destroy the endospores of *C. botulinum* and drives off the O_2, thereby creating anaerobic conditions in which the endospores can germinate and the resulting vegetative cells can multiply and produce toxin. ⏮ oxygen requirements

MicroAssessment 30.1

Intrinsic factors (such as available moisture, pH, and the presence of antimicrobial chemicals) and extrinsic factors (including storage temperature and atmosphere) influence the types of microorganisms that grow and predominate in a food product.

1. Why is *Staphylococcus aureus* more likely to be found in high numbers on ham than on fresh meat?
2. Which is important to refrigerate: fresh stew or bread? Why?
3. A camper carries dried meat, shelled pecans, and cranberries on a week-long hike. What characteristics of the foods will prevent spoilage during that time?

30.2 ■ Microorganisms in Food and Beverage Production

Learning Outcomes

2. Explain the role of the lactic acid bacteria in food production, including fermented milk products, pickled vegetables, and fermented meats.
3. Compare and contrast the production of wine, beer, distilled spirits, and vinegar.
4. Describe the production of soy sauce.

Fermented foods and beverages, such as yogurt, cheese, pickled vegetables, wine, and beer, are perceived as having a pleasant taste. In addition, various compounds produced by the fermenting organisms inhibit the growth of many spoilage organisms as well as of foodborne pathogens. Thus, fermentation historically has been, and continues to be today, an important method of food preservation.

To maintain the quality of fermentations, **starter cultures** containing one or more strains of the desired microorganisms are often used. These strains are carefully selected to produce the most pleasant flavors and textures. Precious starter cultures must be carefully maintained and protected against contamination, particularly by viruses that can damage or destroy them. The species used in starter cultures are genetically diverse, and so they are often taxonomically divided into subspecies. For example, *Streptococcus lactis* subsp. *lactis* and *S. lactis* subsp. *cremoris* are both used in cheese production.

Lactic Acid Fermentations by the Lactic Acid Bacteria

The tart taste of yogurt, pickles, sharp cheeses, and some sausages is due to the production of lactic acid by one or more members of the lactic acid bacteria (**table 30.1**). These bacteria—including members of the genera *Lactobacillus, Lactococcus, Streptococcus, Leuconostoc,* and *Pediococcus*—are obligate fermenters that characteristically produce lactic acid as an end product of their metabolism. Some also produce flavorful and aromatic compounds that contribute to the overall quality of fermented foods. ◀◀ **lactic acid bacteria,** ◀◀ **obligate fermenters**

TABLE 30.1 Foods Produced Using Lactic Acid Bacteria

Food	Characteristic
Milk Products	
Cheese (unripened)	Uses a starter culture usually containing *Lactococcus lactis* subsp. *cremoris* and *L. lactis* subsp. *lactis*
Cheese (ripened)	Uses rennin and a starter culture containing *Lactococcus lactis* subsp. *cremoris* and *L. lactis* subsp. *lactis;* ripened for weeks to years; other bacteria and/or fungi may be added to enhance flavor development
Yogurt	Uses a starter culture containing *Streptococcus thermophilus* and *Lactobacillus delbrueckii* subsp. *bulgaricus*
Vegetables	
Sauerkraut	Cabbage; succession of naturally occurring bacteria including *Leuconostoc mesenteroides, Lactobacillus brevis,* and *Lactobacillus plantarum*
Pickles	Cucumbers; naturally occurring bacteria
Poi	Taro root; naturally occurring bacteria; Hawaii
Olives	Green olives
Kimchee	Cabbage and other vegetables; Korea
Meats	
Dry and semidry sausages	Use a starter culture containing species of *Lactobacillus* and *Pediococcus;* meat is stuffed into casings, incubated, heated, and then dried

Cheese, Yogurt, and Other Fermented Milk Products

Milk from a cow's udder rapidly becomes contaminated with a variety of microorganisms during milking and handling. Various species of lactic acid bacteria are inevitably introduced because they commonly reside on the udder. If the milk is not refrigerated, these bacteria grow and ferment lactose—the main sugar in milk—producing lactic acid. The combined effect of removing the primary carbohydrate as a nutrient source and accumulating lactic acid inhibits the growth of many other microbes. Aesthetic features of the milk change as well because lactic acid lowers the pH, which in turn causes the milk proteins to coagulate or curdle, and sours the flavor. ◀◀ **milk spoilage**

Cheese Cottage cheese is one of the simplest cheeses to make. Pasteurized milk is inoculated with a starter culture, usually containing *Streptococcus lactis* subsp. *lactis* and *S. lactis* subsp. *cremoris,* and then incubated until fermentation products cause the proteins in milk to coagulate. The coagulated proteins, or **curd,** are heated and cut into small pieces to make it easier to drain the liquid portion (called whey). Unlike most cheeses that undergo further microbial processes called ripening or curing, cottage cheese is unripened.

The initial steps of ripened cheese production are the same as those of cottage cheese, except the enzyme rennin is added to the fermenting milk to speed protein coagulation (**figure 30.3**). After the proteins coagulate, the whey is removed. The curds are then salted, pressed, and shaped into the traditional forms, usually bricks or wheels. The cheese is then ripened, resulting in characteristic textural and flavor changes due to the metabolic activities of naturally occurring or starter lactic acid bacteria. Depending on the type of cheese, ripening can take from several weeks to years. Longer ripening creates more acidic, sharper cheeses.

Some cheeses are inoculated with other bacteria or fungi that give characteristics particular to the kind of cheese. For example, the bacterium *Propionibacterium freudenreichii* subsp. *shermanii* ripens Swiss cheese and gives it the characteristic holes and a nutty flavor. This bacterium ferments organic compounds to produce propionic acid and CO_2. The CO_2 gas causes the holes in the cheese, while the propionic acid gives the typical flavor. Propionic acid also inhibits spoilage organisms. Roquefort, Gorgonzola, and Stilton cheeses are ripened by the fungus *Penicillium roquefortii.* Growth of the fungus along cracks in the cheese gives these cheeses the

Coagulation— Lactic acid production and rennin activity cause the milk proteins to coagulate. The coagulated mixture is then cut so that the liquid whey will start separating from the solid curd.

Separation of curds from whey— The curd is heated and cut into small pieces. The liquid whey is removed by draining.

Aging— Curds are salted and pressed into blocks or wheels for aging.

FIGURE 30.3 Commercial Production of Cheese (top): ©Wisconsin Milk Marketing Board, Inc.; (middle): ©Wisconsin Milk Marketing Board, Inc.; (bottom): ©Wisconsin Milk Marketing Board, Inc.

? Why does a longer ripening process give rise to a sharper cheese?

distinctive bluish-green veins. Brie and Camembert are ripened by a white fungus such as *P. camemberti* inoculated on the surface of the cheese. As the fungal cells grow into the cheese, they produce enzymes that alter the texture and flavor.

Yogurt To produce yogurt, pasteurized milk is concentrated slightly by evaporation and then inoculated with a starter culture containing *Streptococcus thermophilus* and *Lactobacillus delbrueckii* subsp. *bulgaricus*. The mixture is incubated at 40°C to 45°C for several hours, during which time these thermophilic bacteria grow rapidly. They produce lactic acid and other end products that contribute to the flavor. Carefully controlled incubation conditions favoring the balanced growth of the two species ensure the proper levels of acid and flavor compounds.

MicroByte

The enzyme rennin is found in calves' stomachs, where it helps digest the mother's milk. Today, genetically engineered microbes make it.

Pickled Vegetables

The fermentation process known as pickling originated as a way to preserve vegetables such as cucumbers and cabbage. Today, pickled products such as sauerkraut (cabbage), pickles (cucumbers), and olives are valued for their flavor. Fermentation of most vegetables uses naturally occurring lactic acid bacteria from the vegetables rather than starter cultures.

One of the most well-studied natural fermentations is the production of sauerkraut. To make sauerkraut, shredded cabbage is mixed with salt and then firmly packed into a container to provide an anaerobic environment. The salt draws water and nutrients from the cabbage, creating a brine that inhibits most microbes other than the lactic acid bacteria. Under the correct conditions, natural successions of lactic acid bacteria grow. These bacteria—*Leuconostoc mesenteroides, Lactobacillus brevis,* and *Lactobacillus plantarum*—produce lactic acid, which lowers the pH, further inhibiting undesired microbes. The lactic acid and other fermentation end products that develop over a 2- to 4-week period give sauerkraut its characteristic tangy taste. When the desired flavor has developed, the sauerkraut is often canned. Similar processes are used to make some pickles, olives, and other vegetable products.

Fermented Meat Products

Fermented meat products—such as salami, pepperoni, and summer sausage—were traditionally produced by allowing the small numbers of lactic acid bacteria naturally present to multiply to the point of dominance. Relying on the natural fermentation of meat is inherently risky, however, because the incubation conditions can potentially support the growth and toxin production of pathogens such as *Staphylococcus aureus* and *Clostridium botulinum.* Starter cultures are now used because they ensure that lactic acid is rapidly produced, thereby inhibiting the growth of pathogens and improving flavor development. Starter cultures used by U.S. sausage-makers typically contain *Lactobacillus* and/or *Pediococcus* species.

To make fermented sausages, meat is ground and combined with a starter culture and other ingredients including sugar, salt, and nitrite. The sugar serves as a substrate for fermentation, because meat does not naturally contain enough fermentable carbohydrate to produce adequate lactic acid. Salt and nitrite contribute to the flavor and also inhibit the growth of spoilage microorganisms. More important, they inhibit *C. botulinum.* After thorough blending, the mixture is stuffed into a casing and incubated from one to several days. The product can then be smoked or otherwise heated to kill bacteria. Finally, it is dried.

Alcoholic Fermentations by Yeast

Some yeasts, such as members of the genus *Saccharomyces,* ferment simple sugars to produce ethanol and CO_2. They are used to make alcoholic beverages as well as vinegar and bread (**table 30.2**).

Wine

Wine is the product of the alcoholic fermentation of naturally occurring sugars in the juices of grapes or other fruits. The commercial production of wine begins by crushing grapes in a machine that removes the stems and collects the resulting solids and juices, called must (**figure 30.4**). If white wine is to be made, only the clear juices are fermented. For red wine, the entire must of red grapes is put into the fermentation vat. The color and complex flavors of these wines are derived from components of the grape skin and seeds. The solids are removed during fermentation once the desired amount of color and flavor compounds have been extracted. Rose wines get their light pink color from a short fermentation of the entire crushed red grape (about 1 day), after which the juice is removed and fermented alone.

Fermentation starts when a specially selected strain of *Saccharomyces cerevisiae* is inoculated. Sulfur dioxide (SO_2) is generally added to inhibit the growth of the natural microbial population of the grape, especially acetic acid bacteria. These bacteria convert alcohol to acetic acid (vinegar) and are a common cause of wine spoilage. The *S. cerevisiae* strains used to make wine are more resistant to the antimicrobial action of SO_2 and produce a higher alcohol content than naturally occurring yeasts.

Fermentation is carried out at a carefully controlled temperature for a period ranging from a few days to several weeks. During fermentation, most of the sugar is converted to ethanol and CO_2, generally resulting in a final alcohol content of

TABLE 30.2 Foods and Beverages Produced Using Alcoholic Fermentation by Yeast

Product	Characteristic
Alcoholic Beverages	
Wine	Sugars in grape juice are fermented by *Saccharomyces cerevisiae.*
Sake	Amylase from mold *(Aspergillus oryzae)* converts the starch in rice to sugar, which is then fermented by *S. cerevisiae.*
Beer	Enzymes in germinated barley convert starches of barley and other grains to sugar, which is then fermented by *S. cerevisiae* or *S. pastorianus.*
Distilled spirits	Sugars, or starches that are converted to sugars, are fermented by *S. cerevisiae;* distillation purifies the alcohol.
Vinegar	Alcohol produced by fermentation is oxidized to acetic acid by species of *Gluconobacter* or *Acetobacter.*
Breads	*S. cerevisiae* ferments sugar; expansion of CO_2 causes the bread to rise; alcohol evaporates during baking.

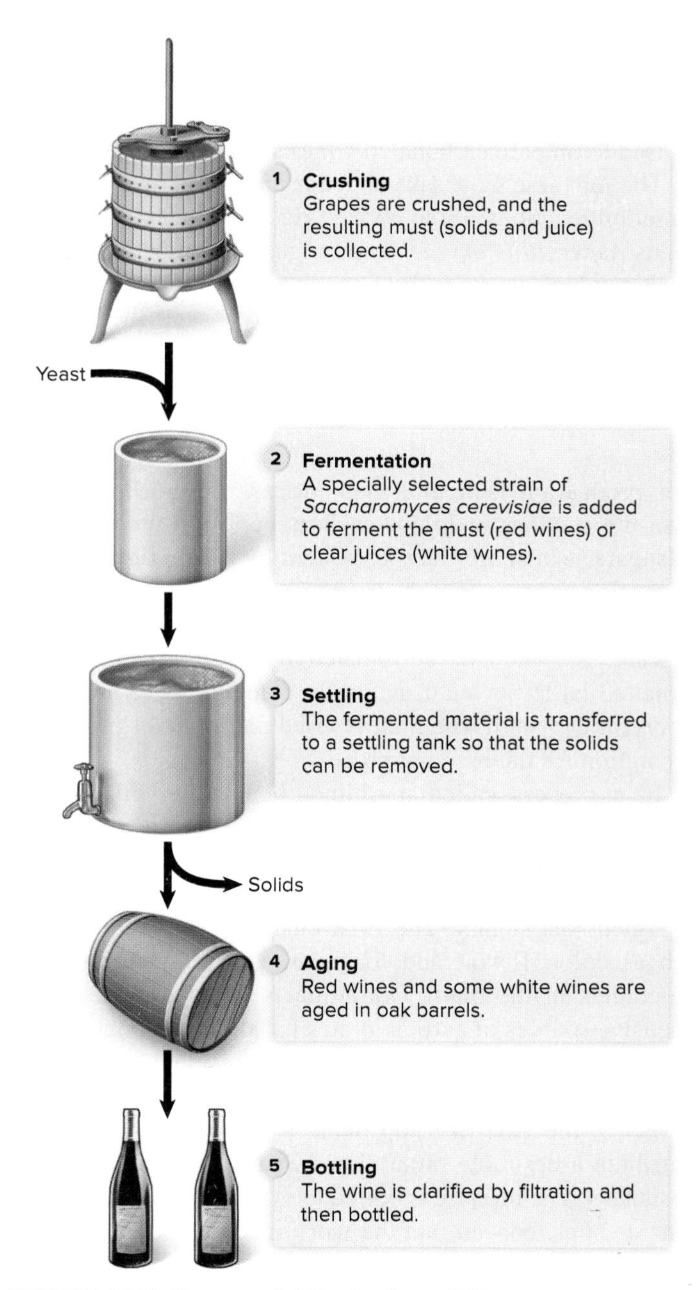

FIGURE 30.4 Commercial Production of Wine

? How is the initial step of making white wines different from that of making red wines?

less than 14%. Dry wines result from the complete fermentation of the sugar, whereas sweet wines contain residual sugar.

In addition to the alcoholic fermentation of the grape sugars, a distinctly different type of fermentation, called malolactic fermentation, can occur during wine production. Lactic acid bacteria, primarily species of *Leuconostoc,* convert malic acid to the less acidic lactic acid. Red wines made from grapes grown in cool regions tend to have high levels of malic acid, and their flavor is mellowed by this fermentation.

After fermentation, the wine is transferred to a settling tank, where fermented solids can settle out. Most red wines and some white wines are then aged in oak barrels, contributing to the complexity of the flavor. Finally, wine is clarified by filtration and then bottled. The CO_2 produced during fermentation is usually released before the wine is bottled, resulting in a "still" (non-carbonated) wine. Other processes are used to prepare carbonated wines such as champagne.

The Japanese wine sake depends on several microbial fermentation reactions. First, cooked rice is inoculated with the fungus *Aspergillus oryzae.* The fungus produces the enzyme amylase, which degrades the rice starch to sugar. Then, a strain of *Saccharomyces cerevisiae* is added to convert the sugar to alcohol and CO_2. Lactic acid bacteria add to the flavor by producing lactic acid and other fermentation end products.

Beer

Beer production is a multistep process designed to break down the starches of grains such as barley to produce simple sugars, which are then fermented by yeast (**figure 30.5**). Yeasts alone cannot convert grain to alcohol because they lack the enzymes that degrade starch, the primary carbohydrate of grain. Sprouted or germinated barley, however, known as malted barley or malt, naturally contains these and other important enzymes. The malt is dried and milled (ground) in preparation for the brewing process.

In a process called mashing, the malt is mixed with adjuncts (starches, sugars, or whole grains such as rice, corn, or sorghum), and then soaked in warm water. During mashing, enzymes of the malt act on the starches, converting them to fermentable sugars. The final characteristics of the beer, such as color, flavor, and foam, are derived entirely from compounds in the malt. The adjuncts simply serve as less expensive sources of carbohydrates for alcohol production.

After mashing, the spent grains (residual solids) are removed to yield the sugary liquid called wort. Hops, the flowers of the vine-like hop plant, are added to the wort to contribute a desirable bitter flavor and provide antibacterial substances. The mixture is boiled to extract the flavor components of hops, concentrate the wort, inactivate enzymes, kill most microbes, and precipitate proteins so that they can be removed more easily. The wort is then centrifuged to remove the solids and cooled before being transferred to the fermentation tank.

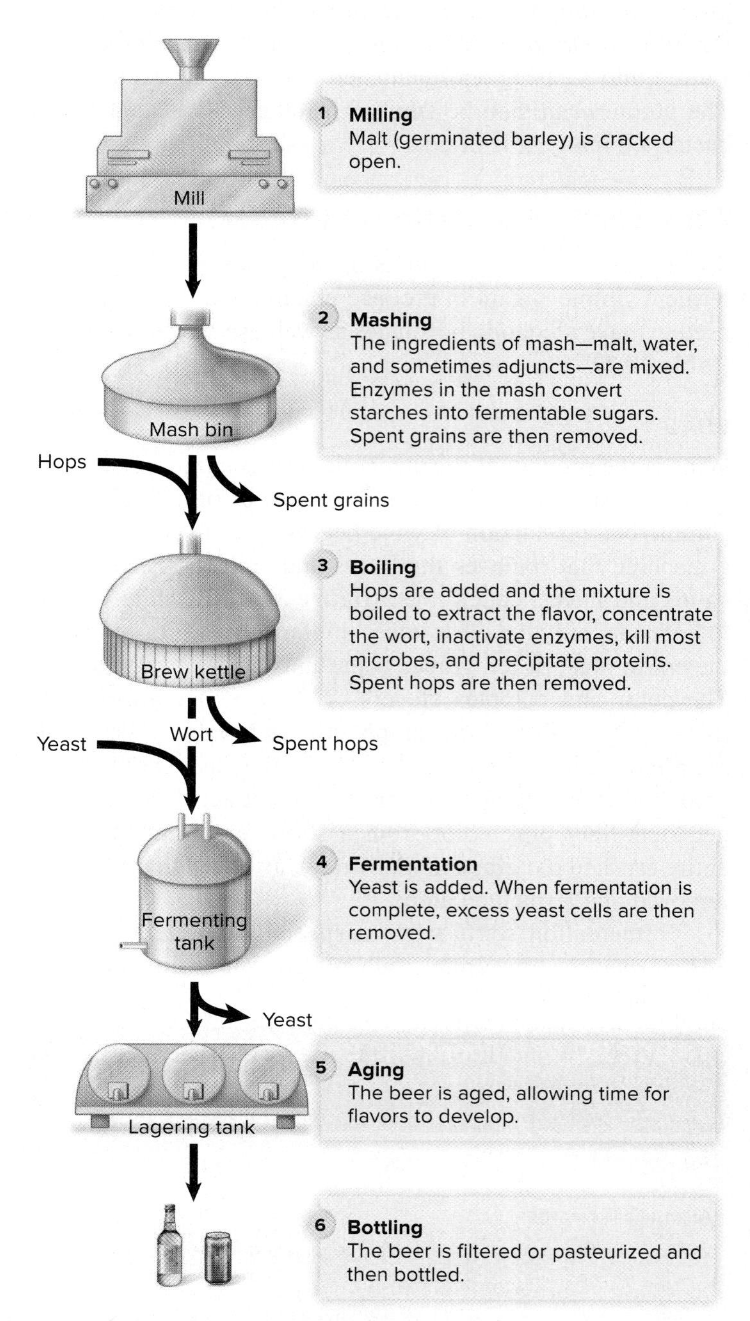

FIGURE 30.5 Commercial Production of Beer

? How are the yeasts used to make ales different from the ones used to make lagers?

For the fermentation, special strains of **brewer's yeasts** (as opposed to baker's yeasts) are added. When making lager-style beers, *Saccharomyces pastorianus* strains are used. These are "bottom yeasts," meaning that they tend to form clumps that sink to the bottom of the fermentation tank. They ferment best at relatively low temperatures (between 6°C and 12°C), and usually take about a week or two to complete the fermentation. In contrast, when making ales, porters, or stouts, *S. cerevisiae* strains are used. These are called "top-fermenting yeasts" because they are carried to the top of the tank by the rising CO_2. They ferment at higher temperatures (14°C to 23°C) and over a shorter period (5 to 7 days).

The fermentation process generates beer with an alcohol content ranging from 3.4% to 6%. Most of the yeasts settle out following fermentation and are removed. These yeasts can be sold as flavor and dietary supplements. The beer is then aged, during which time residual, unwanted flavor compounds are metabolized by remaining yeast cells or settle out. Cask-conditioned beer undergoes a second fermentation, which generates CO_2 in the cask. Other beers must be carbonated to replace the CO_2 that escapes during fermentation. After aging, microorganisms are removed or killed using membrane filtration or pasteurization, and the product is packaged.

Distilled Spirits

The manufacturing of distilled spirits such as scotch, whiskey, and gin is initially similar to that of beer, except the wort is not boiled. Consequently, enzymes in the wort continue breaking down the starches during fermentation. When fermentation is complete, the ethanol is collected by distillation.

Different types of spirits are made in different ways. For example, rum is made by fermenting sugar cane or molasses. Malt scotch whiskey is the product of the fermentation of barley that may then be aged for several years in oak sherry casks. The wood and the residual sherry contribute both flavor and color to the whiskey as it ages. Lactic acid bacteria are used to produce lactic acid in grain mash for making sour-mash whiskey. The yeast *S. cerevisiae* subsequently ferments the sour mash to form alcohol. The distilled spirit tequila is traditionally made from the fermentation of juices from the agave plant using the bacterium *Zymomonas mobilis.* This bacterium ferments sugars to ethanol and CO_2 via a pathway similar to the yeast alcoholic fermentation pathway.

Vinegar

Vinegar, an aqueous solution of at least 4% acetic acid, is the product of the oxidation of ethanol by the acetic acid bacteria—*Acetobacter* and *Gluconobacter* species. Acetic acid bacteria are strictly aerobic, Gram-negative rods, characterized by their ability to carry out a number of oxidations. They can tolerate high concentrations of acid as they oxidize alcohol to acetic acid.

Alcohol is commercially converted to vinegar using processes that provide O_2 to speed the oxidation reaction. One method uses a vinegar generator, which sprays alcohol onto loosely packed wood shavings that have a biofilm of acetic acid bacteria. As the alcohol trickles through the bacteria-coated shavings, it is oxidized to acetic acid. In principle, the vinegar generator operates much like the trickling filter used in wastewater treatment by providing a large surface area for aerobic metabolism. Another method uses a submerged culture reactor, an enclosed system that continuously pumps small air bubbles into alcohol that has been inoculated with acetic acid bacteria. ◀◀ trickling filter, ◀◀ biofilm

Bread

Yeast bread rises through the action of **baker's yeast**—strains of *Saccharomyces cerevisiae* carefully selected for the commercial baking industry. The CO_2 produced during fermentation causes the bread to rise, producing the spongy texture characteristic of yeast breads (**figure 30.6**). The alcohol evaporates during baking.

Yeast bread is made from a mixture of flour, sugar, salt, milk or water, yeast, and sometimes butter or oil. Packaged baker's yeast that can be reconstituted in warm water is readily available as pressed cakes or dried granules. An excess of yeast is added to allow adequate production of CO_2 in a time period too short to permit multiplication of spoilage bacteria.

Sourdough bread is made with a combination of yeast and lactic acid bacteria. Lactic acid is produced, as well as alcohol and CO_2, giving the bread its sour flavor.

Changes Due to Mold Growth

Molds contribute to the flavor and texture of some cheeses, as already discussed. In addition, many traditional dishes and condiments used throughout the world are produced by encouraging the growth of molds on food (**table 30.3**). Successions of naturally occurring microorganisms are often involved. The microbiological and chemical aspects of many of these foods have not been extensively studied.

Soy Sauce

Soy sauce is made by inoculating equal parts of cooked soybeans and roasted cracked wheat with a culture of either *Aspergillus oryzae* or *A. sojae.* The mixture, called koji, is allowed to stand for several days, during which time carbohydrates and proteins in the soybeans are broken down, producing a yellow-green liquid containing fermentable sugars,

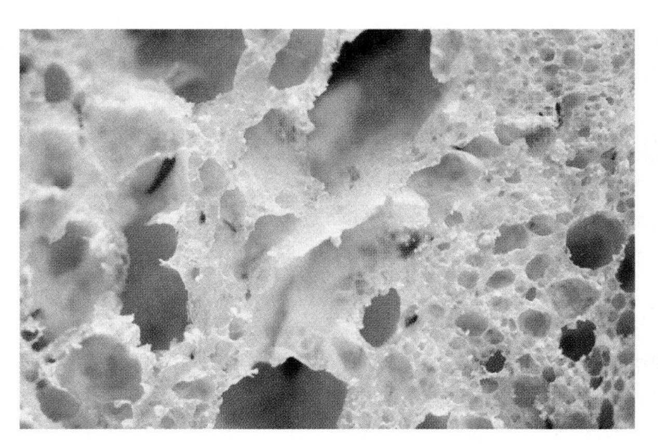

FIGURE 30.6 Bread ©Comstock Images/Getty Images

? **What creates the air pockets in bread?**

TABLE 30.3 Foods Produced Using Molds

Food	Characteristic
Soy sauce	Koji is produced by inoculating soybeans and cracked wheat with a starter culture of *Aspergillus oryzae* or *A. sojae;* this is then added to a brine and incubated for months.
Tempeh	Soybeans are fermented by lactic acid bacteria and then inoculated with a species of the mold *Rhizopus;* Indonesia.
Miso	Rice, soybeans, or barley are inoculated with *Aspergillus oryzae;* Asia.
Cheeses	
Roquefort, Gorgonzola, and Stilton	Curd is inoculated with *Penicillium roquefortii.*
Brie and Camembert	Wheels of cheese are inoculated with selected species of *Penicillium.*

peptides, and amino acids. After this, the mixture is put into a large container with an 18% NaCl solution (brine). Salt-tolerant microorganisms then grow and produce flavor changes over an extended period. Microorganisms involved in this stage of fermentation include lactobacilli, pediococci, and yeasts. After the brine mixture is allowed to ferment for 8 to 12 months, the liquid soy sauce is removed. The residual solids are used as animal feed.

MicroAssessment 30.2

Lactic acid bacteria are used to produce a variety of foods including cheese, yogurt, sauerkraut, and some sausages. Yeasts are used to produce alcoholic beverages and breads. Some cheeses and many traditional foods owe their characteristics to changes caused by molds.

4. Describe how the metabolism of lactic acid bacteria differs from that of most other microorganisms that can grow aerobically.
5. How does the use of starter cultures improve the safety of fermented meat products?
6. How could cottage cheese be produced without bacteria?

30.3 ■ Food Spoilage

Learning Outcome

5. Distinguish between fermented foods and spoiled foods.

Spoilage microorganisms produce metabolites that have undesirable tastes and odors, resulting in **food spoilage.** Although the spoilage microbes damage the quality of the food, they are generally not harmful. This is not surprising when microbial growth requirements are considered. Most human pathogens grow best near normal body temperatures of 37°C, whereas most foods are usually stored at temperatures well below that. Similarly, the nutrients available in fruits, vegetables, and other foods are generally not suitable for the optimum growth of human pathogens. As a result, the non-pathogens can easily outgrow the pathogens when competing for the same nutrients. Spoiled foods are considered unsafe to eat, however, because high numbers of spoilage organisms indicate that foodborne pathogens may be present as well.

Common Spoilage Bacteria

Numerous types of bacteria are important in food spoilage. *Pseudomonas* species can degrade a wide variety of compounds, and grow on and spoil many different kinds of foods, including meats and vegetables. Psychrotrophic species are notorious for spoiling refrigerated foods. Members of the genus *Erwinia* produce enzymes that degrade pectin, and so they commonly cause soft rot of fruits and vegetables. *Acetobacter* species transform ethanol to acetic acid, the principal acid of vinegar. Although this is very beneficial to commercial producers of vinegar, it presents a great problem to wine producers. Milk products are sometimes spoiled by *Alcaligenes* sp. that form a glycocalyx, causing strings of slime, or "ropiness," in raw milk. The lactic acid bacteria, including species of *Streptococcus, Leuconostoc,* and *Lactobacillus,* all produce lactic acid. Anyone who has unexpectedly consumed sour milk knows that this can be disagreeable. *Bacillus* and *Clostridium* species are particularly troublesome causes of food spoilage because their heat-resistant endospores survive cooking and, in some cases, canning. *B. coagulans* and *B. stearothermophilus* spoil some canned foods. ◀◀ glycocalyx

Common Spoilage Fungi

A wide variety of fungi, including species of *Rhizopus, Alternaria, Penicillium, Aspergillus,* and *Botrytis,* spoil foods. Because fungi grow in acidic as well as low-moisture environments, fruits and breads are more likely to be spoiled by fungi than by bacteria. *Aspergillus flavus* grows on peanuts and other grains, producing **aflatoxin,** a potent carcinogen monitored by the Food and Drug Administration.

MicroAssessment 30.3

The metabolites of microbes can spoil foods by generating undesirable flavors, odors, and textures.

7. What characteristics of *Pseudomonas* species allow them to spoil such a wide variety of foods?
8. Why is a grapefruit more likely to be spoiled by fungi than by bacteria?
9. How can disinfection of milking equipment by farmers postpone spoilage of milk in your refrigerator?

30.4 ■ Foodborne Illness

Learning Outcome

6. Distinguish between foodborne intoxication and foodborne infection, and give two examples of each.

Foodborne illness, commonly referred to as food poisoning, occurs when a pathogen, or a toxin it produced, is consumed in a food product. Food production is carefully regulated in the United States to prevent foodborne illness. Federal, state, and local agencies cooperate in inspections to help enforce protective laws. In spite of strict controls, millions of cases of food poisoning are estimated to occur each year from foods prepared commercially, at home, or in institutions such as hospitals and schools. The vast majority of these cases could have been prevented with proper storage, sanitation, and preparation. **Table 30.4** lists some microbes that cause foodborne illness in the United States.

To more accurately determine the burden of foodborne illness in the United States, a government program called **FoodNet** (Foodborne Disease Active Surveillance Network) now collects data on laboratory-confirmed cases of diarrheal illness in 10 states, covering approximately 15% of the population (www.cdc.gov/foodnet). By gaining a better understanding of the epidemiology of foodborne diseases, hopefully these diseases can be prevented more easily.

Foodborne Intoxication

Foodborne intoxication is an illness that results from consuming an exotoxin produced by a microorganism growing in a food product. It is the toxin that causes illness, not the living organisms (**figure 30.7**). *Staphylococcus aureus* and *Clostridium botulinum* are two examples of bacteria that cause foodborne intoxication. ◀◀ exotoxin

Staphylococcus aureus

Recall that *Staphylococcus aureus* causes a wide variety of diseases, including skin and wound infections. A very different set of circumstances leads to what is referred to as "Staph food poisoning." Many strains of *S. aureus* produce a toxin that causes nausea and vomiting when ingested. *S. aureus* does not compete well with most spoilage organisms, but it thrives in moist, rich foods in which other organisms have been killed or their growth inhibited. For example, it can grow with little competition on unrefrigerated salty products such as ham (a_w of 0.91). Creamy pastries and starchy salads stored at room temperature also offer ideal conditions for the growth of this pathogen, because most competing organisms were killed as the ingredients were cooked.

The source of *S. aureus* is usually a human carrier who has not followed adequate hygiene procedures—such as handwashing—before preparing the food. If the organism is inoculated into a food that supports its growth, and the food is left at room temperature for several hours, *S. aureus* can

TABLE 30.4 Common Foodborne Illnesses

Microbe	Signs and Symptoms	Foods Commonly Implicated
Intoxication		
Bacillus cereus	Nausea, vomiting, abdominal cramping	Rice
Clostridium botulinum	Blurred or double vision, progressive muscle weakness, nausea, vomiting, diarrhea, and inability to breathe	Low-acid canned foods such as vegetables and meats
Staphylococcus aureus	Nausea, vomiting, abdominal cramping	Cured meats, creamy salads, cream-filled pastries
Infection		
Bacillus cereus	Nausea, diarrhea, abdominal cramping	Meat, vegetables
Campylobacter species	Diarrhea, fever, abdominal pain, vomiting	Poultry, raw milk
Clostridium perfringens	Intense abdominal cramps, watery diarrhea	Meats, meat products
Cryptosporidium species	Watery diarrhea, nausea, abdominal cramping, fever, loss of appetite	Contaminated uncooked food
Cyclospora species	Watery diarrhea, loss of appetite, bloating, vomiting, fatigue, weight loss	Fresh produce
Shiga toxin–producing *Escherichia coli* (STEC)	Severe abdominal pain, bloody diarrhea; sometimes hemolytic uremic syndrome	Ground beef, raw vegetables, unpasteurized juices
Listeria monocytogenes	Fever, muscle aches; may progress to sepsis, meningitis	Raw milk, cheese, meats, raw produce
Norovirus	Nausea, vomiting, diarrhea, abdominal cramps	Shellfish, salads, raw produce
Salmonella species	Vomiting, abdominal cramps, diarrhea, fever, headache	Poultry, eggs, milk, meat
Shigella species	Abdominal cramps; diarrhea with blood, pus, or mucus; fever; vomiting, headache	Salads, raw vegetables
Vibrio vulnificus	Fever, nausea, muscle pain, abdominal cramps	Oysters
Vibrio parahaemolyticus	Diarrhea, abdominal cramps, nausea, vomiting, headache, fever, chills	Fish and shellfish

FOCUS YOUR PERSPECTIVE 30.1

Botox for Beauty and Pain Relief

The toxin produced by *Clostridium botulinum* can be very dangerous, but dilute preparations—called Botox—are used to treat various conditions associated with muscle contractions. Injections of Botox directly into the affected muscles give relief for 3 to 4 months, after which the treatment can be repeated.

One popular use of Botox is a cosmetic treatment to remove frown lines and other facial wrinkles. By paralyzing the muscles involved, the lines are effectively erased.

Botox treatment is also used to relieve a number of very painful and disabling conditions. For example, it gives relief from cervical dystonia (severe muscle cramping in the neck and shoulders), characterized by jerky movements, muscle pain, and tremors (shaking). Another example is Parkinson's disease, a condition in which certain nerve cells are lost, resulting in tremors, slow movement, stiffness, and in some cases, severe cramping. So, in spite of botulinum toxin's powerful and dangerous properties, it can also be a useful therapeutic agent.

grow and produce the toxin. Unlike most exotoxins, *S. aureus* toxin is heat-stable, so cooking the food will not destroy it.

Clostridium botulinum

Botulism is a deadly paralytic disease caused by ingesting a neurotoxin produced by the anaerobic, Gram-positive rod *Clostridium botulinum.* This soil bacterium is an endospore former found naturally on many foods. The endospores survive usual cooking methods and inadequate canning procedures. Should conditions become anaerobic in the food, toxin can be produced, and if ingested it causes botulism. Characteristics of the disease were covered with nervous system infections (chapter 26), so this section will instead focus on the events that lead to foodborne botulism. ⏮ botulism

Botulinum toxin is one of the most powerful poisons known (see **Focus Your Perspective 30.1**). The toxin blocks transmission of the neurotransmitter that triggers muscle contraction, resulting in a potentially fatal paralysis. A few milligrams of this exotoxin are enough to kill the entire population of a large city. ⏮ neurotoxin

Canning processes for low-acid foods are specifically designed to destroy the endospores of *C. botulinum.* If the endospores are not destroyed or are introduced post-processing, they can germinate in those foods. The resulting vegetative cells can then grow and produce toxin. Manufacturing errors are rare in commercially canned foods, so most cases of botulism are due to improperly processed home-canned foods. As an added safety measure, low-acid, home-canned foods should be boiled for at least 10 minutes immediately before serving. The toxin is heat-labile (sensitive), so the heat treatment will destroy any toxin that may have been produced.

Unfortunately, growth of the organism and toxin production may not result in any noticeable changes in the taste or appearance of the food. Cans of foods that are damaged should be discarded, because they might have small holes through which *C. botulinum* could enter. Likewise, bulging cans indicate gas production. Because this could be a result of microbial growth, the cans should be discarded.

Staphylococcus aureus

- Most bacteria that normally compete with *Staphylococcus aureus* are either killed by cooking or inhibited by high salt conditions.
- A food handler inadvertently transfers *S. aureus* onto food.
- *S. aureus* grows and produces toxin when food is allowed to slowly cool or is stored at room temperature.
- A person ingests the toxin-containing food. Symptoms of staph food poisoning—nausea, abdominal cramping, and vomiting—begin after 4 to 6 hours.

Clostridium botulinum

- *Clostridium botulinum* endospores, common in soil and marine sediments, contaminate many different foods.
- Endospores survive inadequate canning processes. Canned foods are anaerobic.
- Surviving *C. botulinum* endospores germinate; the resulting vegetative cells grow and produce toxin in low-acid canned foods.
- A person ingests the toxin-containing food. Symptoms of botulism, including weakness, double vision, and progressive inability to speak, swallow, and breathe, begin in 12 to 36 hours.

FIGURE 30.7 Typical Events Leading to Foodborne Intoxications by *Staphylococcus aureus* and *Clostridium botulinum* (left): ©FoodCollection; (right): ©Gaetano/Corbis

? Why is botulism primarily a problem in canned foods rather than fresh ones?

Foodborne Infection

Unlike foodborne intoxication, **foodborne infection** requires the consumption of living organisms. The symptoms of the

FOCUS ON A CASE 30.1

A 37-year-old man was admitted to the hospital with dizziness, dry mouth, blurred vision, slurred speech, and nausea. He had difficulty breathing, so he was placed on a mechanical ventilator. Based on the signs and symptoms, the physician suspected botulism, and botulinum antitoxin was immediately administered intravenously.

When asked about the foods he had consumed the previous day, the patient reported eating homemade stew and home-canned green beans. As the physician interviewed the patient, he noticed a partially healed wound on the man's elbow.

A stool sample and a specimen from the wound were obtained and cultured anaerobically for *Clostridium botulinum;* the stool was positive, but the wound was negative. In addition, leftover stew and green beans, as well as the stool sample, were tested for botulinum toxin, which was detected in the stew and in the stool sample, but not in the green beans. These results confirmed the diagnosis of foodborne botulism.

The patient was hospitalized for 6 weeks before being discharged, but required many months at home for a complete recovery.

1. How did the antitoxin help the patient, and why was it administered before the diagnosis was confirmed?
2. Why is it surprising that the implicated food was homemade stew rather than home-canned green beans, and what could have led to the problem?
3. How could this case have been prevented?
4. Why was the patient's wound cultured along with the other samples?

Discussion

1. The antitoxin is a mixture of antibodies that bind to and thereby neutralize the different serotypes of botulinum toxin. Botulism is a life-threatening disease, and the toxin is one of the most potent known, so it is essential that antitoxin be administered as soon as possible when botulism is suspected.
2. Foodborne botulism typically involves home-canned low-acid foods that have not been adequately heated to destroy the endospores. Although stews and certain other foods can also support the germination and subsequent growth of *Clostridium botulinum,* refrigeration usually prevents this from occurring. In this case, however, the stew had been cooked in a large pot, covered with a heavy lid, and then left at room temperature for 3 days. Cooking does not destroy the endospores, and the extended time at room temperature—along with an environment that provided anaerobic conditions—resulted in the opportunity for those spores to germinate and the resulting vegetative cells to multiply to high numbers and produce botulinum toxin.
3. If the stew had been transferred to smaller containers and refrigerated shortly after cooking, *Clostridium botulinum* endospores would not have germinated, nor would any vegetative cells have multiplied and produced toxin. Also, if the stew had been boiled for 10 minutes immediately before it was eaten, the heat would have destroyed any botulinum toxin that was produced. Heating the stew to destroy the toxin would have been a risky option in this case, however, because if the food also contained *Staphylococcus aureus,* the room temperature storage would have allowed its growth as well. Recall that some *S. aureus* strains produce a toxin not destroyed by heat.
4. Not all botulism is foodborne. Wound botulism also occurs, so this was another possible cause of the man's signs and symptoms.

illness, which usually do not appear for at least 1 day after eating the contaminated food, usually include diarrhea. Thorough cooking of food immediately before consuming it will kill the organisms, thereby preventing infection. *Salmonella* species, *Campylobacter* species, and Shiga toxin–producing *Escherichia coli* (STEC) are examples of organisms that cause foodborne infection (**figure 30.8**). Characteristics of these bacteria and the diseases they cause were covered with digestive system infections (chapter 24), so this section will focus on the events that lead to foodborne infection.

Salmonella and *Campylobacter*

Salmonella and *Campylobacter* are two genera commonly associated with poultry products such as chicken, turkey, and eggs. Inadequate cooking of these products can result in foodborne infection. In addition,

Salmonella, Campylobacter

- Incomplete cooking fails to kill all pathogens. Surviving *Salmonella* and/or *Campylobacter* can multiply as food is cooled slowly or stored at room temperature.
- Live organisms are ingested. They multiply in the intestinal tract and cause disease. Symptoms include diarrhea, abdominal pain, and vomiting.

Shiga toxin–producing *E. coli*

- Incomplete cooking fails to kill all pathogens. Even low numbers of surviving Shiga toxin–producing *E. coli* can cause illness.
- Live organisms are ingested. They multiply in the intestinal tract and cause disease. Symptoms include severe abdominal pain and bloody diarrhea.

FIGURE 30.8 Typical Events Leading to Foodborne Infections by *Salmonella, Campylobacter,* and Shiga Toxin–Producing *E. coli* (STEC) (left): ©D. Fischer and P. Lyons/Cole Group/Photodisc/Getty Images; (right): ©FoodCollection

Why are ground meats more common than steaks as a source of STEC infection?

cross-contamination can result in the transfer of pathogens to other foods. For example, if a cutting board on which raw chicken was cut is then immediately used to slice vegetables for a salad, the salad can become contaminated with *Salmonella* or *Campylobacter* species. ⏮ *Salmonella*, ⏮ *Campylobacter*

Shiga Toxin–Producing *Escherichia coli* (STEC)

Escherichia coli O157:H7 and other Shiga toxin–producing *E. coli* (STEC) cause bloody diarrhea. Infection sometimes results in hemolytic uremic syndrome (HUS), a life-threatening condition. ⏮ STEC, ⏮ hemolytic uremic syndrome

STEC strains can colonize the intestinal tract of healthy cattle and other livestock, and are then shed in their feces. Because of this, meats can easily become contaminated. The initial contamination normally occurs on the meat surface, and the bacterial cells are easily destroyed by searing the exterior of meats such as steaks. Grinding the meat to create beef patties distributes the pathogen throughout the product. Prevention then involves more thorough cooking so that enough heat reaches the center to kill all the *E. coli* cells, which is why hamburgers are a particularly troublesome source of infection. Outbreaks have also been linked to unpasteurized milk and various kinds of produce that were contaminated with animal manure.

MicroAssessment 30.4

Foodborne intoxication results from consuming toxins produced by microbes growing in a food. Foodborne infection results from consuming living organisms.

10. How does boiling a home-canned food immediately prior to serving it prevent botulism?
11. Which foodborne pathogen can cause hemolytic uremic syndrome?
12. Why would a large number of competing microorganisms in a food sample result in lack of sensitivity of culture methods for detecting pathogens?

30.5 ■ Food Preservation

Learning Outcome

7. Describe the methods used to preserve foods.

Preventing the growth and metabolic activities of microorganisms that cause spoilage and foodborne illness preserves the quality of food. Some methods of **food preservation,** such as drying and salting, have been known throughout the ages, whereas others have been discovered or developed more recently. The major methods of preserving foods are briefly summarized here and described in more detail in chapter 5.

- **Canning.** The canning process destroys all spoilage and pathogenic organisms capable of growth at normal storage temperatures. Low-acid foods are processed using steam under pressure (autoclaving) in order to reach temperatures high enough to destroy the endospores of *Clostridium botulinum.* Acidic foods do not require such high heat because *C. botulinum* cannot grow and produce toxin in those foods. ⏮ canning
- **Pasteurization.** Heating foods under controlled conditions at high temperatures for short periods of time destroys non-spore-forming pathogens, and reduces the numbers of spoilage organisms without significantly altering the flavor of food. ⏮ pasteurization
- **Cooking.** Cooking, like pasteurization, can destroy non-spore-forming organisms. Cooking, however, obviously alters the characteristics of food. Heat distribution may be uneven, resulting in organisms' survival in inadequately heated regions.
- **Refrigeration.** Refrigeration preserves food by slowing the growth rate of microbes. Many common organisms, including most pathogens, are unable to multiply at low temperatures. ⏮ temperature requirements
- **Freezing.** Freezing stops microbial growth because water in the form of ice is unavailable for biological reactions. Some of the microbial cells will be killed by damage caused by ice crystals, but those remaining can grow and spoil food once it is thawed.
- **Drying/reducing the a_w.** Drying foods or adding high concentrations of sugars or salts inhibits microbial growth by decreasing the available moisture. Eventually, however, molds may grow. ⏮ drying food
- **Lowering the pH.** Lowering the pH, either by adding acids or by encouraging fermentation by lactic acid bacteria, inhibits a wide range of spoilage organisms and pathogens.
- **Adding antimicrobial chemicals.** Organic acids such as propionic acid, benzoic acid, and sorbic acid are naturally occurring antimicrobial chemicals that are added to a variety of foods to inhibit fungal growth. Nitrates are added to cured meats to inhibit growth of *Clostridium botulinum* and other organisms. Wine, fruit juices, and other products are preserved by the addition of sulfur dioxide. ⏮ chemical preservatives
- **Irradiation.** Gamma radiation destroys microorganisms without significantly altering the flavor of foods such as spices and meats. ⏮ irradiation

MicroAssessment 30.5

Food spoilage can be eliminated or delayed by destroying microorganisms or altering conditions to inhibit their growth.

13. Why are the process temperatures for canning low-acid foods higher than ones for acidic foods?
14. Why are nitrates added to cured meats?
15. Microorganisms are often grouped according to their optimum growth temperatures. Which groups are most likely to spoil refrigerated foods?

Summary

30.1 Factors Influencing the Growth of Microorganisms in Foods

Intrinsic Factors

Bacteria require a high a_w. Fungi often grow when the a_w is too low to support bacterial growth. Many bacterial species, including most pathogens, are inhibited by acidic conditions. The nutritional content of a food determines the kinds of organisms that can grow in it. Rinds, shells, and other coverings aid in protecting some foods from invasion by microorganisms. Some foods contain natural antimicrobial chemicals that help prevent spoilage.

Extrinsic Factors

Low temperatures stop or inhibit the growth of most foodborne microorganisms. Psychrophiles and psychrotrophs, however, grow at refrigeration temperatures. The presence or absence of O_2 affects the type of microbial population able to grow in a food.

30.2 Microorganisms in Food and Beverage Production

Not only are fermented foods perceived as pleasant tasting, the acids inhibit the growth of many spoilage organisms and foodborne pathogens. **Starter cultures** are sometimes used.

Lactic Acid Fermentations by the Lactic Acid Bacteria (table 30.1)

The tart taste of yogurt, pickles, sharp cheese, and some sausages is due to the metabolic products of the lactic acid bacteria. Other bacteria or fungi are sometimes added to cheese to give characteristic flavors or textures (figure 30.3). Pickling relies on naturally occurring lactic acid bacteria. Commercial sausage production uses starter cultures to rapidly decrease the pH and prevent the growth of pathogens.

Alcoholic Fermentations by Yeast (table 30.2)

Wine is the product of the fermentation of sugars in fruit juices (figure 30.4). Beer production is a multistep process designed to break down the starches of grains such as barley to produce simple sugars, which can then serve as a substrate for alcoholic fermentation by yeast (figure 30.5). Distilled spirits are produced using distillation to collect the alcohol generated during fermentation. Vinegar is the product of the oxidation of alcohol by the acetic acid bacteria. In bread-making, the CO_2 produced by yeast causes bread to rise; the alcohol is lost to evaporation (figure 30.6).

Changes Due to Mold Growth (table 30.3)

Some cheeses and other foods are produced by encouraging the growth of molds on foods. Soy sauce is made by allowing an *Aspergillus* species to degrade a mixture of soybeans and wheat, which is then fermented in brine.

30.3 Food Spoilage

Food spoilage is most often due to the metabolic activities of microorganisms as they grow and use the nutrients in the food.

Common Spoilage Bacteria

Pseudomonas, Erwinia, Acetobacter, Alcaligenes, lactic acid bacteria, and bacteria that form endospores are important causes of food spoilage.

Common Spoilage Fungi

Fungi grow readily in acidic as well as low-moisture environments; therefore, fruits and breads are more likely to be spoiled by fungi than by bacteria.

30.4 Foodborne Illness (table 30.4)

Foodborne Intoxication

Foodborne intoxication results from consuming a toxin produced by a microorganism growing in a food product (figure 30.7). Many strains of *Staphylococcus aureus* produce a toxin that, when ingested, causes nausea and vomiting. Botulism is caused by ingestion of a neurotoxin produced by the anaerobic, spore-forming, Gram-positive rod *Clostridium botulinum.* As an added safety measure, low-acid home-canned foods should be boiled at least 10 minutes immediately before serving to destroy any botulinum toxin that could be present.

Foodborne Infection

Foodborne infection requires the consumption of living organisms (figure 30.8). Thorough cooking of food immediately before eating it will kill bacteria, thereby preventing infection. *Salmonella* and *Campylobacter* species are commonly associated with poultry products. Some outbreaks of Shiga toxin–producing *E. coli* have been traced to undercooked contaminated hamburger patties and produce contaminated with manure.

30.5 Food Preservation

Food spoilage can be eliminated or delayed by destroying microorganisms or altering conditions to inhibit their growth. Methods used to preserve foods include canning, pasteurization, cooking, refrigeration, freezing, reducing the a_w, lowering the pH, adding antimicrobial chemicals, and irradiation.

Review Questions

Short Answer

1. What is the purpose of rennin in cheese-making?
2. What causes bluish-green veins to form in some types of cheese?
3. What causes the holes to form in Swiss cheese?
4. Why are starter cultures used when making semidry sausages?
5. What is the purpose of the mashing step in beer-making?
6. Explain how *Alcaligenes* species cause "ropiness" in raw milk.
7. Explain the significance of *Aspergillus flavus* in grain products.
8. Explain the typical sequence of events that lead to botulism.
9. Explain the typical sequence of events that lead to staphylococcal food poisoning.
10. How does canning differ from pasteurization?

Multiple Choice

1. The a_w of a food product reflects which of the following?
   a) Acidity of the food
   b) Presence of antimicrobial constituents such as lysozyme
   c) Amount of water available
   d) Storage atmosphere
   e) Nutrient content
2. Most spoilage bacteria cannot grow below an a_w of
   a) 0.3. b) 0.5. c) 0.7. d) 0.9. e) 1.0.
3. What is a generally minimum pH for growth and toxin production by *Clostridium botulinum* and other foodborne pathogens?
   a) 8.5 b) 7.0 c) 6.5 d) 4.5 e) 2.0
4. Benzoic acid is an antimicrobial chemical naturally found in which of the following foods?
   a) Apples b) Cranberries c) Eggs
   d) Milk e) Yogurt
5. Which of the following is often added to wine to inhibit growth of the natural microbial populations on grapes?
   a) Benzoic acid b) Lactic acid c) Carbon dioxide
   d) Sulfur dioxide e) Oxygen
6. In the brewing process, the sugar and nutrient extract obtained by soaking germinated grain in warm water is called
   a) baker's yeast. b) hops. c) malt.
   d) must. e) wort.
7. Which of the following genera is used in bread, wine, and beer production?
   a) *Lactobacillus* b) *Pseudomonas* c) *Saccharomyces*
   d) *Streptococcus* e) *Staphylococcus*
8. Which group of organisms most commonly spoils breads, fruits, and dried foods?
   a) *Acetobacter* b) Fungi c) Lactic acid bacteria
   d) *Pseudomonas* e) *Saccharomyces*
9. Which of the following organisms cause foodborne intoxication?
   a) Shiga toxin–producing *E. coli*
   b) *Campylobacter* species
   c) *Lactobacillus* species
   d) *Salmonella* species
   e) *Staphylococcus aureus*
10. Canned pickles require less stringent heat processing than canned beans, because pickles
    a) contain fewer nutrients.
    b) are more acidic.
    c) have a lower a_w.
    d) contain antimicrobial chemicals.
    e) are less likely to be contaminated with endospores.

Applications

1. A small cheese-manufacturing company in Wisconsin is looking for ways to reduce the costs of disposing of whey, a cheese by-product. As a food microbiologist, what would you suggest that the company do with the thousands of liters of whey being produced per month so the company can actually profit from it?
2. A microbiologist is troubleshooting a batch of home-brewed ale that did not ferment properly. She noticed that the alcohol content was only 2%, well below the desired level. Microscopic examination showed numerous yeast cells. Chemical analysis indicated low levels of sugar, high levels of CO_2 , and large amounts of protein in the liquid. What did the microbiologist conclude was the probable cause of the problem when the beer did not come out properly?

Critical Thinking

1. It has been argued that the nature of the growth of fungi in Roquefort cheese, indicated by the appearance of bluish-green veins, is evidence that these fungi require O_2 for growth. How does this evidence lead to the conclusion?
2. In the production of sauerkraut, a natural succession of lactic acid bacteria is observed growing in the product. What causes the succession? What does this tell you about the optimal growth conditions of the different species of lactic acid bacteria?

Appendix I

Microbial Mathematics

Because prokaryotes are very tiny and can multiply to very large numbers of cells in short time periods, convenient and simple ways are used to indicate their numbers without resorting to many zeros before or after the number. This is one reason it is important to understand the metric system, which is used in scientific measurements.

The basic unit of measure is the meter, which is equal to about 39 inches. All other units are fractions of a meter:

1 decimeter is one tenth = 0.1 meter

1 centimeter is one hundredth = 0.01 meter

1 millimeter is one thousandth = 0.001 meter

Because prokaryotes are much smaller than a millimeter, even smaller units of measure are used. A millionth of a meter is a micrometer = 0.000001 meter, and is abbreviated μm. This is the most frequently used size measurement in microbiology, because most bacteria are about 1 μm wide. For comparison, a human hair is about 75 μm wide.

Because it is inconvenient to write so many zeros in front of the 1, an easier way of indicating the same number is through the use of superscript, or exponential, numbers (exponents). One hundred dollars can be written 10^2 dollars. The 10 is called the base number and the 2 is the exponent. Conversely, one hundredth of a dollar is 10^{-2} dollars; thus the exponent is negative. The base most commonly used in biology is 10 (which is designated as $\log_{10}$). The above information can be summarized as follows:

1 millimeter = 1 mm = 0.001 meter = 10^{-3} meter

1 micrometer = 1 μm = 0.000001 meter = 10^{-6} meter

1 nanometer = 1 nm = 0.000000001 meter = 10^{-9} meter

The same prefix designations can be used for weights. The basic unit of weight is the gram, abbreviated g. Approximately 450 grams are in a pound.

1 milligram = 1 mg = 0.001 gram = 10^{-3} g

1 microgram = 1 μg = 0.000001 gram = 10^{-6} g

1 nanogram = 1 ng = 0.000000001 gram = 10^{-9} g

1 picogram = 1 pg = 0.000000000001 gram = 10^{-12} g

Note that the number of zeros before the 1 is one less than the exponent.

The value of the number is obtained by multiplying the base by itself the number of times indicated by the exponent.

Thus, $10^1 = 10 \times 1 = 10$

$10^2 = 10 \times 10 = 100$

$10^3 = 10 \times 10 \times 10 = 1{,}000$

When the exponent is negative, the base and exponent are divided into 1.

For example, $10^{-2} = 1/10 \times 1/10 = 1/100 = 0.01$

When multiplying numbers having exponents to the same base, the exponents are added.

For example, $10^3 \times 10^2 = 10^5$ (not 10^6)

When dividing numbers having exponents to the same base, the exponents are subtracted.

For example, $10^5 \div 10^2 = 10^3$

In both cases, only if the bases are the same can the exponents be added or subtracted.

Appendix II
Pronunciation Key for Bacterial, Fungal, Protozoan, and Viral Names

A

Acetobacter (a-see′-toe-back-ter)
Acinetobacter (a-sin-et′-oh-back-ter)
Actinomycetes (ak-tin-oh-my′-seats)
Adenovirus (ad′-eh-no-vi-rus)
Agrobacterium tumefaciens (ag-rho-bak-teer′-ee-um too-meh-faysh′-ee-enz)
Alcaligenes (al-ka-li′-jen-ease)
Amoeba (ah-mee′-bah)
Arbovirus (are′-bow-vi-rus)
Aspergillus niger (ass-per-jill′-us nye′-jer)
Aspergillus oryzae (ass-per-jill′-us or-eye′-zee)
Azolla (aye-zol′-lah)
Azotobacter (ay-zoh′-toe-back-ter)

B

Bacillus anthracis (bah-sill′-us an-thra′-siss)
Bacillus cereus (bah-sill′-us seer′-ee-us)
Bacillus fastidiosus (bah-sill′-us fas-tid-ee-oh′-sus)
Bacillus subtilis (bah-sill′-us sut′-ill-us)
Bacillus thuringiensis (bah-sill′-us thur′-in-jee-en-sis)
Bacteroides (back′-ter-oid′-eez)
Bdellovibrio (del′-o-vib′-re-oh)
Beggiatoa (beg-gee-ah-toe′-ah)
Beijerinckia (by-yer-ink′-ee-ah)
Bordetella pertussis (bor-deh-tell′-ah per-tuss′-iss)
Borrelia burgdorferi (bor-real′-ee-ah berg-dor′-fir-ee)
Bradyrhizobium (bray-dee-rye-zoe′-bee-um)
Brucella abortus (bru-sell′-ah ah-bore′-tus)

C

Campylobacter jejuni (kam′-peh-low-back-ter je-june′-ee)
Candida albicans (kan′-did-ah al′-bi-kanz)
Caulobacter (caw′-loh-back-ter)
***Chikungunya* virus** (chik-en-gun-ye vi-rus)
Chlamydia trachomatis (klah-mid′-ee-ah trah-ko-ma′-tiss)
Chlamydophila pneumoniae (klah-mid′-o-fil-ah new-moan′-ee-ee)
Claviceps purpurea (kla′-vi-seps purr-purr′-ee-ah)
Clostridium botulinum (kloss-trid′-ee-um bot-you-line′-um)
Clostridium difficile (kloss-trid′-ee-um dif′-fi-seal)
Clostridium perfringens (kloss-trid′-ee-um per-frin′-gens)
Clostridium tetani (kloss-trid′-ee-um tet′-an-ee)
Coccidioides immitis (cock-sid-ee-oid′-eez im′-mi-tiss)
Coronavirus (kor-oh′-nah-vi-rus)
Corynebacterium diphtheriae (koh-ryne′-nee-bak-teer-ee-um dif-theer′-ee-ee)
Coxsackievirus (cock-sack-ee′-vi-rus)
Cryptococcus neoformans (krip-toe-cock′-us knee-oh-for′-manz)
Cryptosporidium parvum (krip-toe-spo-rid-ee-um par-vum)
Cutibacterium acnes (kyoo-tee-bak-teer-ee-um ak-neez)
Cyclospora (sigh-clo-spo-rah)

D

Dengue virus (deng-ee vi-rus)
Desulfovibrio (dee-sul-foh-vib′-ree-oh)

E

Entamoeba histolytica (en-ta-mee′-bah his-toh-lit′-ik-ah)
Enterobacter (en′-ter-oh-back-ter)
Enterococcus faecalis (en′-ter-oh-kock′-us fee-ka′-liss)
Enterovirus (en′-ter-oh-vi-rus)
Epidermophyton (eh-pee-der′-moh-fy-ton)
Epulopiscium (ep′-you-low-pis-se-um)
Escherichia coli (esh-er-ee′-she-ah koh′-lee)

F

Flavivirus (flay′-vih-vi-rus)
Francisella tularensis (fran-siss-sell′-ah tu-lah-ren′-siss)
Frankia (frank′-ee-ah)

G

Gallionella (gal-ee-oh-nell′-ah)
Gardnerella vaginalis (gard-nee-rel′-lah va-jin-al′-is)
Geobacillus stearothermophilus (gee′-oh-bah-sill′-us steer-oh-ther-maw′-fill-us)
Giardia lamblia (jee-are′-dee-ah lamb′-lee-ah)
Gluconobacter (glue-kon-oh-back′-ter)
Gonyaulax (gon-ee-ow′-lax)

H

Haemophilus influenzae (hee-moff′-ill-us in-flew-en′-zee)
Helicobacter pylori (he′-lih-koh-back-ter pie-lore′-ee)
Hepadnavirus (hep-ad′-nah-vi-rus)
Hepatitis virus (hep-ah-ti′-tis vi-rus)
Herpes simplex (her′-peas sim′-plex)
Herpes zoster (her′-peas zoh′-ster)
Histoplasma capsulatum (his-toh-plaz′-mah cap-su-lah′-tum)
Hyphomicrobium (high-foh-my-krow′-bee-um)

I

Influenza virus (in-flew-en′-za vi-rus)

K

Klebsiella pneumoniae (kleb-see-ell′-ah new-moan′-ee-ee)

L

Lactobacillus brevis (lack-toe-ba-sil′-lus bre′-vis)
Lactobacillus bulgaricus (lack-toe-ba-sil′-lus bull-gair′-i-kus)
Lactobacillus casei (lack-toe-ba-sil′-us kay′-see-ee)
Lactobacillus plantarum (lack-toe-ba-sil′-us plan-tar′-um)
Lactobacillus thermophilus (lack-toe-ba-sil′-us ther-mo′-fil-us)
Lactococcus lactis (lack-toe-kock′-us lak′-tiss)
Legionella pneumophila (lee-jon-ell′-ah new-moh′-fill-ah)
Leptospira interrogans (lep-toe-spire′-ah in-ter-roh′-ganz)
Leuconostoc mesenteroides (lew-kow-nos′-tok mes-en-te-roi-deze)
Listeria monocytogenes (lis-tear′-ee-ah mon′-oh-sigh-to- jen′-eze)

M

Malassezia (mal-as-seez′-e-ah)
Methanobacterium (me-than′-oh-bak-teer-ee-um)
Methanococcus (me-than-oh-ko′-kus)
Microsporum (my-kroh-spore′-um)
Mobiluncus (moh-bi-lun′-kus)
Moraxella catarrhalis (more-ax-ell′-ah kah-tah-rah′-liss)
Moraxella lacunata (more-ax-ell′-ah lak-u-nah′-tah)
Mucor (mu′-kor)
Mycobacterium leprae (my-koh-bak-teer′-ee-um lep-ree)
Mycobacterium tuberculosis (my-koh-bak-teer′-ee-um too-ber-kew-loh′-siss)
Mycoplasma pneumoniae (my-koh-plaz′-mah new-moan′-ee-ee)
Mycoplasma genitalium (my-koh-plaz′-mah gen-ih-tale′-ee-um)

N

Neisseria gonorrhoeae (nye-seer′-ee-ah gahn-oh-ree′-ee)
Neisseria meningitidis (nye-seer′-ee-ah men-in-jit′-id-iss)
Neurospora (new-rah′-spor-ah)

O

Orthomyxovirus (or-thoe-mix′-oh-vi-rus)
Oscillatoria (os-sil-la-tor′-ee-ah)

P

Papillomavirus (pap-il-oh′-ma-vi-rus)
Parainfluenza virus (par-ah-in-flew-en′-zah vi-rus)
Paramecium (pair′-ah-mee-see-um)
Paramyxovirus (par-ah-mix′-oh-vi-rus)
Parvovirus (par′-vo-vi-rus)
Pasteurella multocida (pass-ture-ell′-ah mul-toe-sid′-ah)
Pediococcus (ped-ih-oh-ko′-kus)
Penicillium camemberti (pen-eh-sill′-ee-um cam-em-bare′-tee)
Penicillium roqueforti (pen-eh-sill′-ee-um rok-e-for′-tee)
Peptostreptococcus (pep′-to-strep-to-ko-kus)
Phytophthora infestans (fy′-toe-fy-thor-ah in-fes′-tanz)
Picornavirus (pi-kor′-na-vi-rus)
Plasmodium falciparum (plaz-moh′-dee-um fall-sip′-air-um)
Plasmodium malariae (plaz-moh′-dee-um ma-lair′-ee-ee)
Plasmodium ovale (plaz-moh′-dee-um oh-vah′-lee)
Plasmodium vivax (plaz-moh′-dee-um vye′-vax)
Pneumocystis jirovecii (new-mo-sis′-tis yee′-row′-vet-ze)
Poliovirus (poe′-lee-oh-vi-rus)
Polyomavirus (po-lee-oh′-mah vi-rus)
Propionibacterium shermanii (proh-pee-ah-nee-bak-teer′-ee-um sher-man′-ee-ee)
Proteus mirabilis (proh′-tee-us mee-rab′-il-us)
Pseudomonas aeruginosa (sue-dough-moan′-ass aye-rue-gin-o′-sa)

R

Rabies virus (ray′-bees vi-rus)
Retrovirus (re′-trow-vi-rus)
Rhabdovirus (rab′-doh-vi-rus)
Rhinovirus (rye′-no-vi-rus)
Rhizobium (rye-zoh′-bee-um)
Rhizopus stolonifer (rise′-oh-pus stoh′-lon-ih-fer)
Rickettsia rickettsii (rik-kett′-see-ah rik-kett′-see-ee)
Rotavirus (row′-tah-vi-rus)
Rubella virus (rue-bell′-ah vi-rus)
Rubeola virus (rue-bee-oh′-la vi-rus)

S

Saccharomyces carlsbergensis (sack-ah-row-my′-sees karls-bergen′-siss)
Saccharomyces cerevisiae (sack-ah-row-my′-sees sara-vis′-ee-ee)
Salmonella enterica (sall-moh-nell′-ah en-ter-ih-kah)
Serratia marcescens (ser-ray′-sha mar-sess-sens)
Shigella dysenteriae (shig-ell′-ah diss-en-tair′-ee-ee)
Spirillum volutans (spy-rill′-um vol-u-tanz)
Sporothrix schenckii (spore′-oh-thrix shenk-ee-ee)
Staphylococcus aureus (staff-ill-oh-kok′-us aw′-ree-us)
Staphylococcus epidermidis (staff-ill-oh-kok′-us epi-der′-mid-iss)
Streptobacillus moniliformis (strep-tow-bah-sill′-us mon-ill-i-form′-is)
Streptococcus agalactiae (strep-toe-kock′-us a-ga-lac′-tee-ee)
Streptococcus cremoris (strep-toe-kock′-us kre-more′-iss)
Streptococcus mutans (strep-toe-kock′-us mew′-tanz)
Streptococcus pneumoniae (strep-toe-kock′-us new-moan′-ee-ee)
Streptococcus pyogenes (strep-toe-kock′-us pie-ah-gen-ease)
Streptococcus salivarius (strep-toe-kock′-us sal-ih-vair′-ee-us)
Streptococcus thermophilis (strep-toe-kock′-us ther-moh′-fill-us)
Streptomyces griseus (strep-toe-my′-seez gree′-see-us)

T

Thiobacillus (thigh-oh-bah-sill′-us)
Treponema pallidum (tre-poh-nee′-mah pal′-ih-dum)
Trichomonas vaginalis (trick-oh-moan′-as vag-in-al′-iss)
Trichophyton (trick-oh-phye′-ton)
Trypanosoma brucei (tri-pan′-oh-soh-mah bru′-see-ee)

V

Varicella-zoster virus (var-ih-sell′-ah zoh′-ster vi-rus)
Vibrio cholerae (vib′-ree-oh kahl′-er-ee)
Vibrio fischeri (vib′-ree-oh fish′-er-i)
Vibrio vulnificus (vib′-ree-oh vul-ni-fi-kus)

W

Wolbachia (wol-bach′-ee-ah)

Y

Yersinia enterocolitica (yer-sin′-ee-ah en-ter-oh-koh-lih′-tih-kah)
Yersinia pestis (yer-sin′-ee-ah pess′-tiss)

Appendix III

Metabolic Pathways

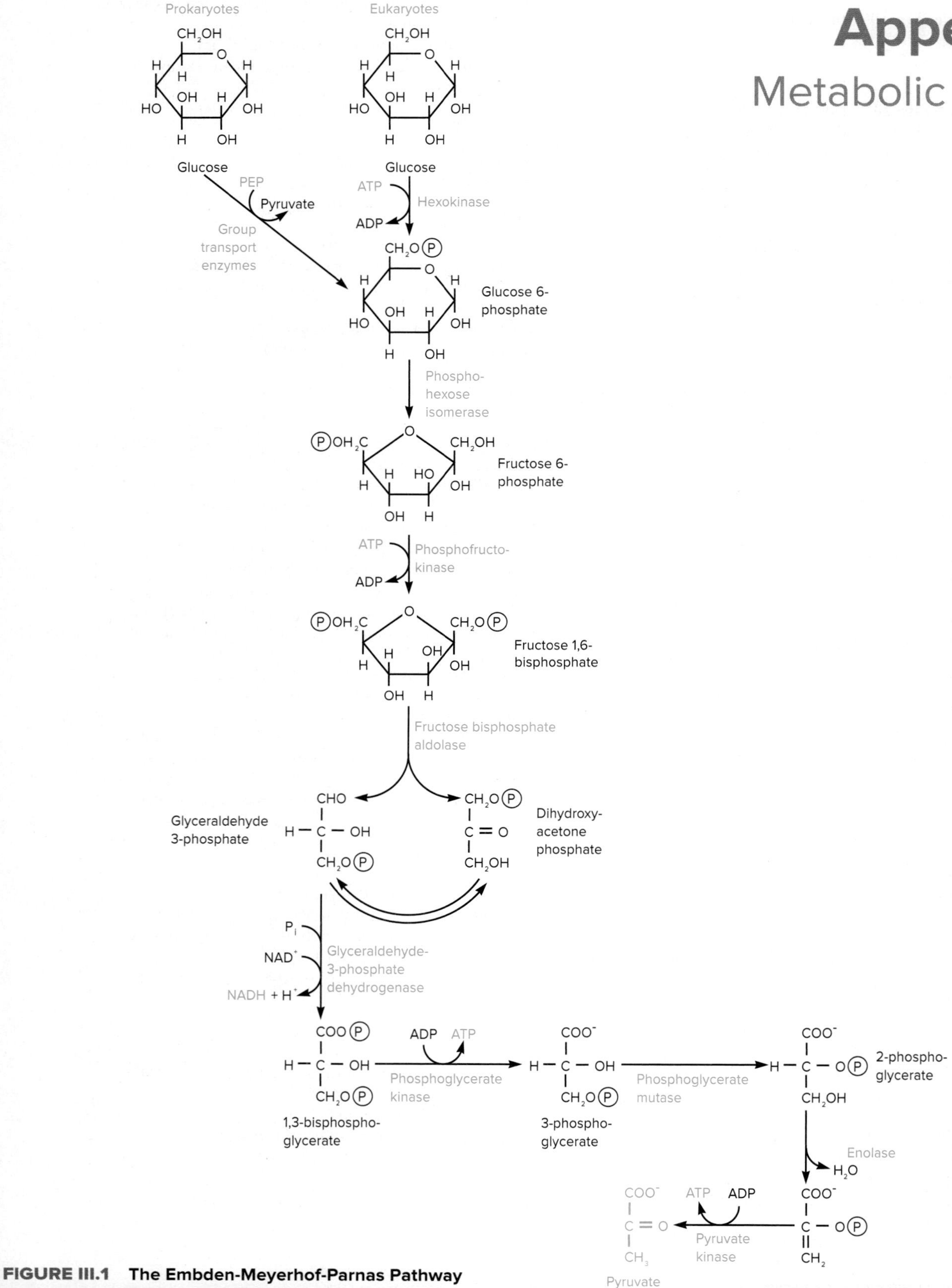

FIGURE III.1 The Embden-Meyerhof-Parnas Pathway
Commonly called glycolysis or the glycolytic pathway.

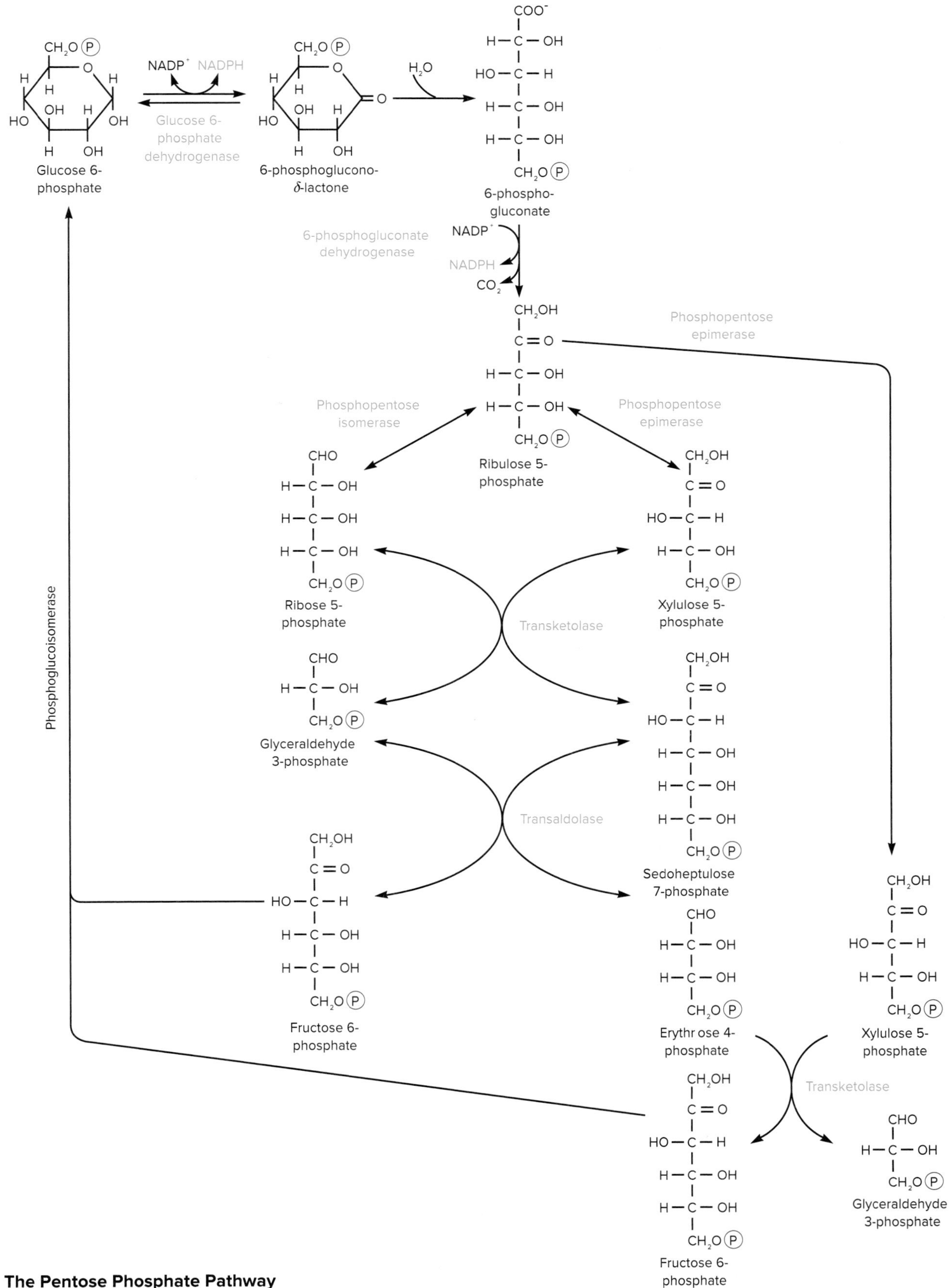

FIGURE III.2 The Pentose Phosphate Pathway

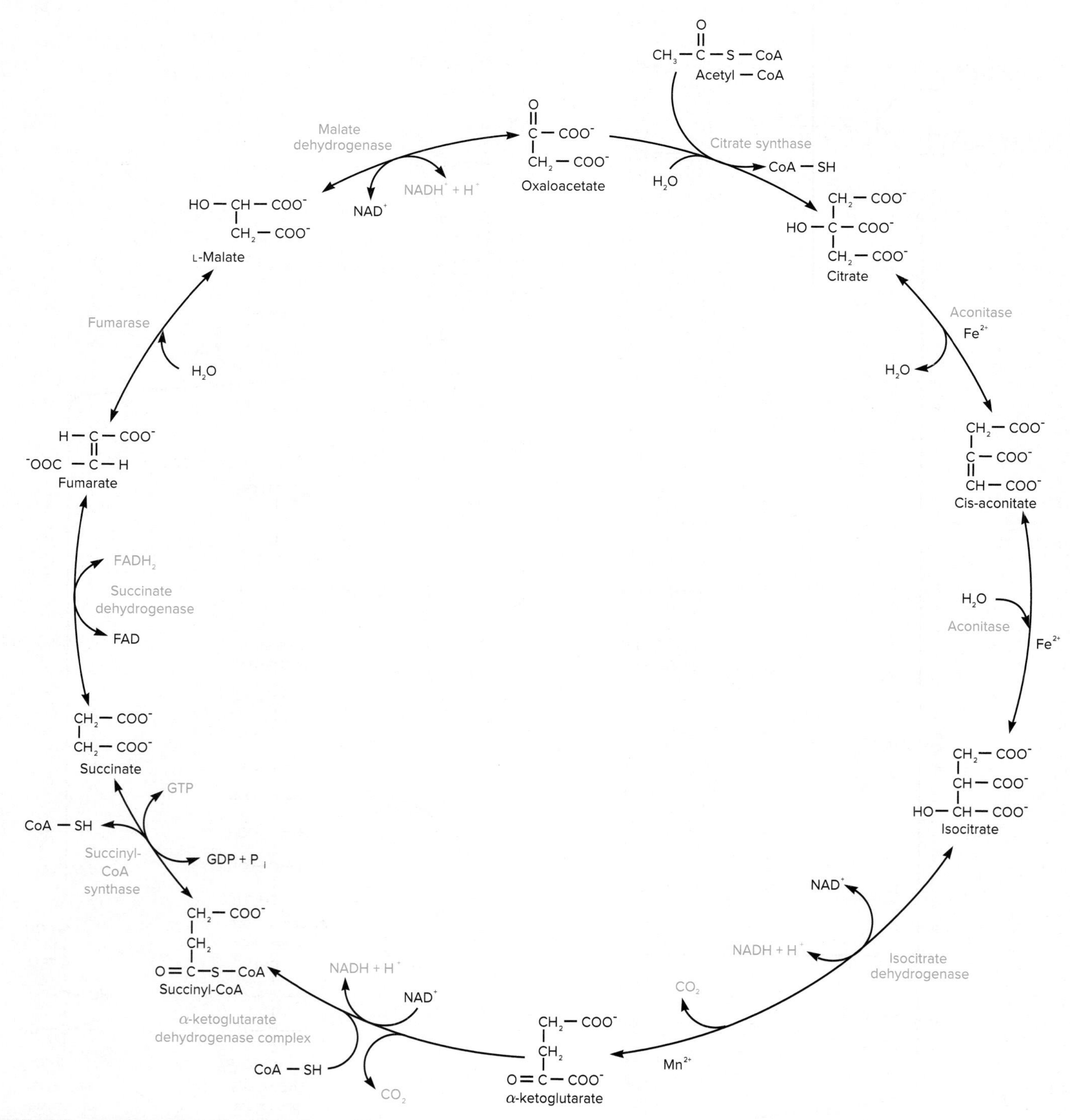

FIGURE III.3 The Tricarboxylic Acid Cycle (TCA Cycle)
Also referred to as the Krebs cycle or the citric acid cycle.

Appendix IV
Answers to Multiple Choice Questions

Chapter 1
1. C
2. C
3. C
4. A
5. C
6. C
7. C
8. C
9. A
10. D

Chapter 2
1. C
2. B
3. E
4. A
5. B
6. A
7. C
8. D
9. E
10. D

Chapter 3
1. A
2. A
3. C
4. B
5. E
6. D
7. E
8. B
9. D
10. A

Chapter 4
1. D
2. A
3. B
4. B
5. B
6. E
7. C
8. A
9. C
10. B

Chapter 5
1. C
2. D
3. A
4. B
5. B
6. A
7. E
8. D
9. C
10. A

Chapter 6
1. D
2. C
3. D
4. E
5. B
6. A
7. D
8. D
9. D
10. E

Chapter 7
1. B
2. A
3. A
4. A
5. C
6. C
7. C
8. B
9. D
10. D

Chapter 8
1. C
2. A
3. D
4. A
5. B
6. D
7. A
8. B
9. A
10. B

Chapter 9
1. C
2. C
3. E
4. B
5. B
6. C
7. B
8. B
9. B
10. D

Chapter 10
1. E
2. B
3. E
4. B
5. D
6. B
7. C
8. D
9. B
10. E

Chapter 11
1. E
2. B
3. E
4. B
5. D
6. A
7. D
8. B
9. C
10. D

Chapter 12
1. C
2. A
3. B
4. D
5. B
6. C
7. A
8. A
9. C
10. B

Chapter 13
1. C
2. A
3. C
4. D
5. E
6. C
7. D
8. A
9. A
10. D

Chapter 14
1. B
2. E
3. A
4. B
5. C
6. A
7. E
8. A
9. C
10. D

Chapter 15
1. E
2. E
3. D
4. D
5. B
6. D
7. D
8. A
9. B
10. D

Chapter 16
1. C
2. A
3. D
4. D
5. A
6. C
7. C
8. D
9. D
10. A

Chapter 17
1. A
2. E
3. A
4. B
5. D
6. E
7. C
8. C
9. A
10. C

Chapter 18
1. C
2. D
3. D
4. E
5. D
6. A
7. C
8. D
9. A
10. C

Chapter 19
1. A
2. B
3. A
4. B
5. E
6. B
7. B
8. E
9. C
10. E

Chapter 20
1. D
2. C
3. A
4. B
5. A
6. B
7. D
8. D
9. A
10. C

Chapter 21
1. E
2. C
3. E
4. A
5. B
6. A
7. A
8. C
9. D
10. D

Chapter 22
1. E
2. E
3. A
4. D
5. E
6. A
7. C
8. A
9. D
10. D

Chapter 23
1. B
2. E
3. C
4. A
5. A
6. C
7. D
8. C
9. E
10. B

Chapter 24
1. C
2. A
3. B
4. D
5. A
6. D
7. C
8. D
9. C
10. E

Chapter 25
1. E
2. A
3. C
4. E
5. C
6. C
7. C
8. A
9. A
10. E

Chapter 26
1. B
2. C
3. C
4. D
5. C
6. E
7. A
8. A
9. B
10. E

Chapter 27
1. D
2. D
3. C
4. B
5. C
6. A
7. B
8. B
9. B
10. C

Chapter 28
1. A
2. A
3. C
4. D
5. A
6. D
7. C
8. C
9. B
10. A

Chapter 29
1. B
2. B
3. A
4. A
5. C
6. E
7. C
8. C
9. B
10. E

Chapter 30
1. C
2. D
3. D
4. B
5. D
6. E
7. C
8. B
9. E
10. B

Appendix V

Microbial Terminology

Singular and Plural Forms

Singular	Plural	Singular	Plural
alg*a*	alg*ae*	fung*us*	fung*i*
amoeb*a*	amoeb*ae*	hydroly*sis*	hydroly*ses*
bacill*us*	bacill*i*	hyph*a*	hyph*ae*
bacteri*um*	bacteri*a*	inocul*um*	inocul*a*
cili*um*	cili*a*	medi*um*	medi*a*
clostridi*um*	clostridi*a*	mucos*a*	mucos*ae*
cocc*us*	cocc*i*	myceli*um*	myceli*a*
conidi*um*	conidi*a*	mycos*is*	mycos*es*
dat*um*	dat*a*	phyl*um*	phyl*a*
diagnos*is*	diagnos*es*	pil*us*	pil*i*
fimbri*a*	fimbri*ae*	nucle*us*	nucle*i*
flagell*um*	flagell*a*	sept*um*	sept*a*
foc*us*	foc*i*	synthes*is*	synthes*es*

Meanings of Prefixes and Suffixes

Prefix or Suffix	Meaning	Example
a-, an-	not, without	avirulent (lacking virulence), anaerobic (without air)
aer-	air	aerobic
anti-	against	antiseptic
-ase	enzyme	penicillinase
chlor-	green	chlorophyll
-chrom-	color	metachromatic (staining differently with the same dye)
-cide	causing death	germicide
co-, com-, con-	together	coenzyme
-cyan-	blue	pyocyanin (a blue bacterial pigment)
de-	down, from	dehydrate (remove water)
-dem-	people, district	epidemic
endo-	within	endospore (spore within a cell)
-enter-	intestine	enteritis (inflammation of the intestine)
epi-	upon	epidermis
erythro-	red	erythocyte (red blood cell)
eu-	well, normal	eukaryotic (true nucleus)
exo-	outside	exoenzyme (enzyme that acts outside the cell that produced it)
extra-	outside of	extracellular
flav-	yellow	flavoprotein (a protein containing a yellow compound)
-gen	produce, originate	antigen (a substance that induces a production of antibodies)
glyc-	sweet	glycemia (the presence of sugar in the blood)
hetero-	other	heterotroph (organism that obtains carbon from organic compounds)
homo-	common, same	homologous (similar in structure or origin)
hydr-	water	dehydrate
hyper-	excessive, above	hypersensitive

Prefix or Suffix	Meaning	Example
hypo-	under	hypotonic (having low osmotic pressure)
iso-	same, equal	isotonic (having the same osmotic pressure)
-itis	inflammation	appendicitis, meningitis
leuko-	white	leukocyte (white blood cell)
ly-, -lys, -lyt-	loosen; dissolve	bacteriolysis (dissolution of bacteria)
meso-	middle	mesophilic (preferring moderate temperatures)
meta-	after, beyond, changed	metachromatic (staining differently with the same dye)
micro-	small; one-millionth part	microscopic
milli-	one-thousandth part	millimeter (10^{-3} meter)
mito-	thread	mitochondrion (small, rod-shaped or granular organelle)
mono-	single	monotrichous (having a single flagellum)
multi-	many	multinuclear (having many nuclei)
myc-	fungus	mycotic (caused by fungus)
myx-	mucus	myxomycete (slime mold)
-oid	resembling	lymphoid (resembling lymphocytes)
-ose	a sugar	lactose (milk sugar)
-osis	disease of	coccidioidomycosis (disease caused by *Coccidioides*)
pan-	all	pandemic (widespread epidemic)
para-	beside	parasite (an organism that feeds in and at the expense of the host)
patho-	disease	pathogenic (producing disease)
peri-	around	peritrichous (having flagella on all sides)
-phag-	eat	phagocyte (a cell that ingests other cell substances)
-phil	like, having affinity for	eosinophilic (staining with the dye eosin)
-phot-	light	photosynthesis
-phyll	leaf	chlorophyll (green leaf pigment)
pleo-	more	pleomorphic (occurring in more than one form)
poly-	many	polymorphonuclear (having a many-shaped nucleus)
post-	after	postnatal (after birth)
pyo-	pus	pyogenic (producing pus)
-sta-	stop	bacteriostatic (inhibiting bacterial multiplication)
sym-, syn-	together	symbiosis (life together)
thermo-	heat	thermophilic (liking heat)
tox-	poison	toxin
trans-	through, across	transfusion
-trich-	hair	monotrichous (having a single flagellum)
-troph	nourishment	autotroph (organism that obtains carbon from CO_2)
zym-	ferment	enzyme

Glossary/Index

Page numbers followed by a *t* and *f* indicate tables and figures respectively. Bolded terms are defined here.

B

D

E

F

H

J

K

L

N

O

P

R

S

T

U